T4-AQL-645

COMPREHENSIVE ORGANIC SYNTHESIS

IN 9 VOLUMES

COMPREHENSIVE ORGANIC SYNTHESIS

Selectivity, Strategy & Efficiency in Modern Organic Chemistry

Editor-in-Chief
BARRY M. TROST
Stanford University, CA, USA

Deputy Editor-in-Chief
IAN FLEMING
University of Cambridge, UK

Volume 9

CUMULATIVE INDEXES

PERGAMON PRESS

OXFORD • NEW YORK • SEOUL • TOKYO

ELSEVIER SCIENCE Ltd
The Boulevard, Langford Lane
Kidlington, Oxford OX5 1GB, UK

First edition 1991
Second impression 1993
Third impression 1999

Library of Congress Cataloging in Publication Data

Comprehensive organic synthesis: selectivity, strategy and efficiency in modern organic chemistry/editor[s] Barry M. Trost, Ian Fleming.
p. cm.
Includes indexes.
Contents: Vol. 1.–2. Additions to C-X[pi]-Bonds — v. 3. Carbon–carbon sigma-Bond formation — v. 4. Additions to and substitutions at C-C[pi]-Bonds — v. 5. Combining C-C[pi]-Bonds — v. 6. Heteroatom manipulation — v. 7. Oxidation — v. 8. Reduction — v. 9. Cumulative indexes.
9. Organic Compounds — Synthesis I. Trost, Barry M. 1941–
II. Fleming, Ian, 1935–
QD262.C535 1991
547.2—dc20 90-26621

British Library Cataloguing in Publication Data

Comprehensive organic synthesis
9. Organic compounds. Synthesis
I. Trost, Barry M. (Barry Martin) 1941–
547.2

ISBN 0-08-040600-9 (Vol. 9)
ISBN 0-08-035929-9 (set)

∞ ™ The paper used in this publication meets the minimum requirements of American National Standard for Information Sciences — Permanence of Paper for Printed Library Materials, ANSI Z39.48-1984.

Contents

Preface vii

Contents of All Volumes ix

Cumulative Author Index 1

Cumulative Subject Index 401

Preface

The emergence of organic chemistry as a scientific discipline heralded a new era in human development. Applications of organic chemistry contributed significantly to satisfying the basic needs for food, clothing and shelter. While expanding our ability to cope with our basic needs remained an important goal, we could, for the first time, worry about the quality of life. Indeed, there appears to be an excellent correlation between investment in research and applications of organic chemistry and the standard of living. Such advances arise from the creation of compounds and materials. Continuation of these contributions requires a vigorous effort in research and development, for which information such as that provided by the *Comprehensive* series of Pergamon Press is a valuable resource.

Since the publication in 1979 of *Comprehensive Organic Chemistry*, it has become an important first source of information. However, considering the pace of advancements and the ever-shrinking timeframe in which initial discoveries are rapidly assimilated into the basic fabric of the science, it is clear that a new treatment is needed. It was tempting simply to update a series that had been so successful. However, this new series took a totally different approach. In deciding to embark upon *Comprehensive Organic Synthesis*, the Editors and Publisher recognized that synthesis stands at the heart of organic chemistry.

The construction of molecules and molecular systems transcends many fields of science. Needs in electronics, agriculture, medicine and textiles, to name but a few, provide a powerful driving force for more effective ways to make known materials and for routes to new materials. Physical and theoretical studies, extrapolations from current knowledge, and serendipity all help to identify the direction in which research should be moving. All of these forces help the synthetic chemist in translating vague notions to specific structures, in executing complex multistep sequences, and in seeking new knowledge to develop new reactions and reagents. The increasing degree of sophistication of the types of problems that need to be addressed require increasingly complex molecular architecture to target better the function of the resulting substances. The ability to make such substances available depends upon the sharpening of our sculptors' tools: the reactions and reagents of synthesis.

The Volume Editors have spent great time and effort in considering the format of the work. The intention is to focus on transformations in the way that synthetic chemists think about their problems. In terms of organic molecules, the work divides into the formation of carbon–carbon bonds, the introduction of heteroatoms, and heteroatom interconversions. Thus, Volumes 1–5 focus mainly on carbon–carbon bond formation, but also include many aspects of the introduction of heteroatoms. Volumes 6–8 focus on interconversion of heteroatoms, but also deal with exchange of carbon–carbon bonds for carbon–heteroatom bonds.

The Editors recognize that the assignment of subjects to any particular volume may be arbitrary in part. For example, reactions of enolates can be considered to be additions to C—C π-bonds. However, the vastness of the field leads it to be subdivided into components based upon the nature of the bond-forming process. Some subjects will undoubtedly appear in more than one place.

In attacking a synthetic target, the critical question about the suitability of any method involves selectivity: chemo-, regio-, diastereo- and enantio-selectivity. Both from an educational point-of-view for the reader who wants to learn about a new field, and an experimental viewpoint for the practitioner who seeks a reference source for practical information, an organization of the chapters along the theme of selectivity becomes most informative.

The Editors believe this organization will help emphasize the common threads that underlie many seemingly disparate areas of organic chemistry. The relationships among various transformations becomes clearer and the applicability of transformations across a large number of compound classes becomes apparent. Thus, it is intended that an integration of many specialized areas such as terpenoid, heterocyclic, carbohydrate, nucleic acid chemistry, *etc.* within the more general transformation class will provide an impetus to the consideration of methods to solve problems outside the traditional ones for any specialist.

In general, presentation of topics concentrates on work of the last decade. Reference to earlier work, as necessary and relevant, is made by citing key reviews. All topics in organic synthesis cannot be treated with equal depth within the constraints of any single series. Decisions as to which aspects of a topic require greater depth are guided by the topics covered in other recent *Comprehensive* series. This new treatise focuses on being comprehensive in the context of synthetically useful concepts.

The Editors and Publisher believe that *Comprehensive Organic Synthesis* will serve all those who must face the problem of preparing organic compounds. We intend it to be an essential reference work for the experienced practitioner who seeks information to solve a particular problem. At the same time, we must also serve the chemist whose major interest lies outside organic synthesis and therefore is only an occasional practitioner. In addition, the series has an educational role. We hope to instruct experienced investigators who want to learn the essential facts and concepts of an area new to them. We also hope to teach the novice student by providing an authoritative account of an area and by conveying the excitement of the field.

The need for this series was evident from the enthusiastic response from the scientific community in the most meaningful way — their willingness to devote their time to the task. I am deeply indebted to an exceptional board of editors, beginning with my deputy editor-in-chief Ian Fleming, and extending to the entire board — Clayton H. Heathcock, Ryoji Noyori, Steven V. Ley, Leo A. Paquette, Gerald Pattenden, Martin F. Semmelhack, Stuart L. Schreiber and Ekkehard Winterfeldt.

The substance of the work was created by over 250 authors from 15 countries, illustrating the truly international nature of the effort. I thank each and every one for the magnificent effort put forth. Finally, such a work is impossible without a publisher. The continuing commitment of Pergamon Press to serve the scientific community by providing this *Comprehensive* series is commendable. Specific credit goes to Colin Drayton for the critical role he played in allowing us to realize this work and also to Helen McPherson for guiding it through the publishing maze.

A work of this kind, which obviously summarizes accomplishments, may engender in some the feeling that there is little more to achieve. Quite the opposite is the case. In looking back and seeing how far we have come, it becomes only more obvious how very much more we have yet to achieve. The vastness of the problems and opportunities ensures that research in organic synthesis will be vibrant for a very long time to come.

BARRY M. TROST
Palo Alto, California

Contents of All Volumes

Volume 1 Additions to C—X π-Bonds, Part 1

Nonstabilized Carbanion Equivalents

1.1 Carbanions of Alkali and Alkaline Earth Cations: (i) Synthesis and Structural Characterization
1.2 Carbanions of Alkali and Alkaline Earth Cations: (ii) Selectivity of Carbonyl Addition Reactions
1.3 Organoaluminum Reagents
1.4 Organocopper Reagents
1.5 Organotitanium and Organozirconium Reagents
1.6 Organochromium Reagents
1.7 Organozinc, Organocadmium and Organomercury Reagents
1.8 Organocerium Reagents
1.9 Samarium and Ytterbium Reagents
1.10 Lewis Acid Carbonyl Complexation
1.11 Lewis Acid Promoted Addition Reactions of Organometallic Compounds
1.12 Nucleophilic Addition to Imines and Imine Derivatives
1.13 Nucleophilic Addition to Carboxylic Acid Derivatives

Heteroatom-stabilized Carbanion Equivalents

2.1 Nitrogen Stabilization
2.2 Boron Stabilization
2.3 Sulfur Stabilization
2.4 The Benzoin and Related Acyl Anion Equivalent Reactions
2.5 Silicon Stabilization
2.6 Selenium Stabilization

Transformation of the Carbonyl Group into Nonhydroxylic Groups

3.1 Alkene Synthesis
3.2 Epoxidation and Related Processes
3.3 Skeletal Reorganizations: Chain Extension and Ring Expansion

Author Index

Subject Index

Volume 2 Additions to C—X π-Bonds, Part 2

Uncatalyzed Additions of Nucleophilic Alkenes to C=X

1.1 Allyl Organometallics
1.2 Heteroatom-stabilized Allylic Anions
1.3 Propargyl and Allenyl Organometallics
1.4 Formation of Enolates
1.5 The Aldol Reaction: Acid and General Base Catalysis
1.6 The Aldol Reaction: Group I and Group II Enolates
1.7 The Aldol Reaction: Group III Enolates
1.8 Zinc Enolates: the Reformatsky and Blaise Reactions
1.9 The Aldol Reaction: Transition Metal Enolates
1.10 The Henry (Nitroaldol) Reaction
1.11 The Knoevenagel Reaction
1.12 The Perkin Reaction
1.13 Darzens Glycidic Ester Condensation
1.14 Metal Homoenolates
1.15 Use of Enzymatic Aldol Reactions in Synthesis
1.16 Metalloenamines

1.17 Hydrazone Anions

Catalyzed Additions of Nucleophilic Alkenes to C═X

2.1 The Prins and Carbonyl Ene Reactions
2.2 Allylsilanes, Allylstannanes and Related Systems
2.3 Formation and Addition Reactions of Enol Ethers
2.4 Asymmetric Synthesis with Enol Ethers
2.5 Reactions of Activated Dienes with Aldehydes

Addition–Elimination Reactions (Acylations)

3.1 The Aliphatic Friedel–Crafts Reaction
3.2 The Bimolecular Aromatic Friedel–Crafts Reaction
3.3 The Intramolecular Aromatic Friedel–Crafts Reaction
3.4 The Reimer–Tiemann Reaction
3.5 The Vilsmeier–Haack Reaction
3.6 Acylation of Esters, Ketones and Nitriles
3.7 The Eschenmoser Coupling Reaction

Additions of Nucleophilic Alkenes to C═NR and C═NR_2^+

4.1 The Bimolecular Aliphatic Mannich and Related Reactions
4.2 The Bimolecular Aromatic Mannich Reaction
4.3 Reactions of Allyl and Propargyl/Allenic Organometallics with Imines and Iminium Ions
4.4 The Intramolecular Mannich and Related Reactions
4.5 Additions to *N*-Acyliminium Ions
4.6 The Passerini and Ugi Reactions

Author Index

Subject Index

Volume 3 Carbon–Carbon σ-Bond Formation

Alkylation of Carbon

1.1 Alkylations of Enols and Enolates
1.2 Alkylations of Nitrogen-stabilized Carbanions
1.3 Alkylations of Sulfur- and Selenium-containing Carbanions
1.4 Alkylations of Other Heteroatom-stabilized Carbanions
1.5 Alkylations of Nonstabilized Carbanions
1.6 Alkylations of Vinyl Carbanions
1.7 Alkylations of Alkynyl Carbanions
1.8 Friedel–Crafts Alkylations
1.9 Polyene Cyclizations
1.10 Transannular Electrophilic Cyclizations

Coupling Reactions

2.1 Coupling Reactions Between sp^3 Carbon Centers
2.2 Coupling Reactions Between sp^3 and sp^2 Carbon Centers
2.3 Coupling Reactions Between sp^2 Carbon Centers
2.4 Coupling Reactions Between sp^2 and *sp* Carbon Centers
2.5 Coupling Reactions Between *sp* Carbon Centers
2.6 Pinacol Coupling Reactions
2.7 Acyloin Coupling Reactions
2.8 Kolbe Reactions
2.9 Oxidative Coupling of Phenols and Phenol Ethers

Rearrangement Reactions

3.1 Wagner–Meerwein Rearrangements

3.2 The Pinacol Rearrangement
3.3 Acid-catalyzed Rearrangements of Epoxides
3.4 The Semipinacol and Other Rearrangements
3.5 Dienone–Phenol Rearrangements and Related Reactions
3.6 Benzil–Benzilic Acid Rearrangements
3.7 The Favorskii Rearrangement
3.8 The Ramberg–Bäcklund Rearrangement
3.9 The Wolff Rearrangement
3.10 The Stevens and Related Rearrangements
3.11 The Wittig Rearrangement

Other Carbon–Carbon Bond Forming Reactions

4.1 Carbonylation and Decarbonylation Reactions
4.2 Carbon–Carbon Bond Formation by C—H Insertion

Author Index

Subject Index

Volume 4 Additions to and Substitutions at C—C π-Bonds

Polar Additions to Activated Alkenes and Alkynes

1.1 Stabilized Nucleophiles with Electron Deficient Alkenes and Alkynes
1.2 Conjugate Additions of Reactive Carbanions to Activated Alkenes and Alkynes
1.3 Conjugate Additions of Carbon Ligands to Activated Alkenes and Alkynes Mediated by Lewis Acids
1.4 Organocuprates in the Conjugate Addition Reaction
1.5 Asymmetric Nucleophilic Additions to Electron Deficient Alkenes
1.6 Nucleophilic Addition–Electrophilic Coupling with a Carbanion Intermediate
1.7 Addition of H—X Reagents to Alkenes and Alkynes
1.8 Electrophilic Addition of X—Y Reagents to Alkenes and Alkynes
1.9 Electrophilic Heteroatom Cyclizations

Nucleophilic Aromatic Substitutions

2.1 Arene Substitution *via* Nucleophilic Addition to Electron Deficient Arenes
2.2 Nucleophilic Coupling with Aryl Radicals
2.3 Nucleophilic Coupling with Arynes
2.4 Nucleophilic Addition to Arene–Metal Complexes

Polar Additions to Alkenes and Alkynes

3.1 Heteroatom Nucleophiles with Metal-activated Alkenes and Alkynes
3.2 Carbon Nucleophiles with Alkenes and Alkynes
3.3 Nucleophiles with Allyl–Metal Complexes
3.4 Nucleophiles with Cationic Pentadienyl–Metal Complexes
3.5 Carbon Electrophiles with Dienes and Polyenes Promoted by Transition Metals

Nonpolar Additions to Alkenes and Alkynes

4.1 Radical Addition Reactions
4.2 Radical Cyclizations and Sequential Radical Reactions
4.3 Vinyl Substitutions with Organopalladium Intermediates
4.4 Carbometallation of Alkenes and Alkynes
4.5 Hydroformylation and Related Additions of Carbon Monoxide to Alkenes and Alkynes
4.6 Methylene and Nonfunctionalized Alkylidene Transfer to Form Cyclopropanes
4.7 Formation and Further Transformations of 1,1-Dihalocyclopropanes
4.8 Addition of Ketocarbenes to Alkenes, Alkynes and Aromatic Systems
4.9 Intermolecular 1,3-Dipolar Cycloadditions
4.10 Intramolecular 1,3-Dipolar Cycloadditions

Author Index

Subject Index

Volume 5 Combining C—C π-Bonds

Ene Reactions

1.1 Ene Reactions with Alkenes as Enophiles
1.2 Metallo-ene Reactions

[2 + 2] Cycloadditions

2.1 Thermal Cyclobutane Ring Formation
2.2 Formation of Four-membered Heterocycles
2.3 Photochemical Cycloadditions
2.4 The Paterno–Büchi Reaction
2.5 Di-π-methane Photoisomerizations
2.6 Oxa-di-π-methane Photoisomerizations

[3 + 2] Cycloadditions

3.1 Thermal Cycloadditions
3.2 Transition Metal Mediated Cycloadditions

[4 + 2] Cycloadditions

4.1 Intermolecular Diels–Alder Reactions
4.2 Heterodienophile Additions to Dienes
4.3 Heterodiene Additions
4.4 Intramolecular Diels–Alder Reactions
4.5 Retrograde Diels–Alder Reactions

Higher-order Cycloadditions

5.1 [4 + 3] Cycloadditions
5.2 [4 + 4] and [6 + 4] Cycloadditions
5.3 [3 + 2] and [5 + 2] Arene–Alkene Photocycloadditions

Electrocyclic Processes

6.1 Cyclobutene Ring Opening Reactions
6.2 1,3-Cyclohexadiene Formation Reactions
6.3 Nazarov and Related Cationic Cyclizations

Sigmatropic Processes

7.1 Cope, Oxy-Cope and Anionic Oxy-Cope Rearrangements
7.2 Claisen Rearrangements
7.3 Consecutive Rearrangements

Small Ring Rearrangements

8.1 Rearrangements of Vinylcyclopropanes and Related Systems
8.2 Rearrangements of Divinylcyclopropanes
8.3 Charge-accelerated Rearrangements

Other Transition Metal Associated Reactions

9.1 The Pauson–Khand Reaction
9.2 Metal–Carbene Cycloadditions
9.3 Alkene Metathesis and Related Reactions
9.4 [2 + 2 + 2] Cycloadditions
9.5 Zirconium-promoted Bicyclization of Enynes
9.6 Metal-catalyzed Cycloadditions of Small Ring Compounds

Author Index

Subject Index

Volume 6 Heteroatom Manipulation

Displacement by Substitution Processes

1.1 Synthesis of Alcohols and Ethers
1.2 Synthesis of Glycosides
1.3 Synthesis of Amines and Ammonium Salts
1.4 Synthesis of Nitroso, Nitro and Related Compounds
1.5 Synthesis of Sulfides, Sulfoxides and Sulfones
1.6 Synthesis of Phosphonium Ylides
1.7 Synthesis of Halides
1.8 Synthesis of Pseudohalides, Nitriles and Related Compounds
1.9 Ritter-type Reactions

Acylation-type Reactions

2.1 Synthesis of Acid Halides, Anhydrides and Related Compounds
2.2 Synthesis of Esters, Activated Esters and Lactones
2.3 Synthesis of Amides and Related Compounds
2.4 Synthesis of Thioamides and Thiolactams
2.5 Synthesis of Thioesters and Thiolactones
2.6 Selenoesters of All Oxidation States
2.7 Synthesis of Iminium Salts, Orthoesters and Related Compounds
2.8 Inorganic Acid Derivatives

Protecting Groups

3.1 Protecting Groups

Functional Group Interconversion

4.1 Carbonyl Group Derivatization
4.2 Use of Carbonyl Derivatives for Heterocyclic Synthesis
4.3 Functional Group Transformations *via* Carbonyl Derivatives
4.4 Degradation Reactions
4.5 Functional Group Transformations *via* Allyl Rearrangement
4.6 2,3-Sigmatropic Rearrangements
4.7 Polonovski- and Pummerer-type Reactions and the Nef Reaction

Elimination Reactions

5.1 Eliminations to Form Alkenes, Allenes and Alkynes and Related Reactions
5.2 Reductive Elimination, Vicinal Deoxygenation and Vicinal Desilylation
5.3 The Cope Elimination, Sulfoxide Elimination and Related Thermal Reactions
5.4 Fragmentation Reactions

Author Index

Subject Index

Volume 7 Oxidation

Oxidation of Unactivated C—H Bonds

1.1 Oxidation by Chemical Methods
1.2 Oxidation by Nitrene Insertion
1.3 Oxidation by Remote Functionalization Methods
1.4 Oxidation by Microbial Methods

Oxidation of Activated C—H Bonds

2.1 Oxidation Adjacent to C═C Bonds
2.2 Oxidation Adjacent to C═X Bonds by Dehydrogenation
2.3 Oxidation Adjacent to C═X Bonds by Hydroxylation Methods
2.4 Oxidation Adjacent to Sulfur
2.5 Oxidation Adjacent to Nitrogen
2.6 Oxidation Adjacent to Oxygen of Ethers
2.7 Oxidation Adjacent to Oxygen of Alcohols by Chromium Reagents
2.8 Oxidation Adjacent to Oxygen of Alcohols by Activated DMSO Methods
2.9 Oxidation Adjacent to Oxygen of Alcohols by Other Methods
2.10 Vinylic and Arylic C—H Oxidation
2.11 Synthesis of Quinones

Oxidation of C═C Bonds

3.1 Addition Reactions with Formation of Carbon–Oxygen Bonds: (i) General Methods of Epoxidation
3.2 Addition Reactions with Formation of Carbon–Oxygen Bonds: (ii) Asymmetric Methods of Epoxidation
3.3 Addition Reactions with Formation of Carbon–Oxygen Bonds: (iii) Glycol Forming Reactions
3.4 Addition Reactions with Formation of Carbon–Oxygen Bonds: (iv) The Wacker Oxidation and Related Reactions
3.5 Addition Reactions with Formation of Carbon–Nitrogen Bonds
3.6 Addition Reactions with Formation of Carbon–Sulfur or Carbon–Selenium Bonds
3.7 Addition Reactions with Formation of Carbon–Halogen Bonds
3.8 Cleavage Reactions

Oxidation of C—X Bonds

4.1 Oxidation of Carbon–Boron Bonds
4.2 Oxidation of Carbon–Metal Bonds
4.3 Oxidation of Carbon–Silicon Bonds
4.4 Oxidation of Carbon–Halogen Bonds

Oxidation of C—C Bonds

5.1 The Baeyer–Villiger Reaction
5.2 The Beckmann and Related Reactions
5.3 Glycol Cleavage Reactions
5.4 The Hunsdiecker and Related Reactions

Oxidation of Heteroatoms

6.1 Oxidation of Nitrogen and Phosphorus
6.2 Oxidation of Sulfur, Selenium and Tellurium

Special Topics

7.1 Oxidation by Electrochemical Methods
7.2 Oxidative Rearrangement Reactions
7.3 Solid-supported Oxidants
7.4 Electron-transfer Oxidation

Author Index

Subject Index

Volume 8 Reduction

Reduction of C═X Bonds

1.1 Reduction of C═O to CHOH by Metal Hydrides

1.2 Reduction of C=N to CHNH by Metal Hydrides
1.3 Reduction of C=X to CHXH by Hydride Delivery from Carbon
1.4 Reduction of C=X to CHXH by Dissolving Metals and Related Methods
1.5 Reduction of C=X to CHXH Electrolytically
1.6 Reduction of C=X to CHXH by Catalytic Hydrogenation
1.7 Reduction of C=X to CHXH by Chirally Modified Hydride Reagents
1.8 Reduction of C=X to CHXH Using Enzymes and Microorganisms
1.9 Reduction of Acetals, Azaacetals and Thioacetals to Ethers
1.10 Reduction of Carboxylic Acid Derivatives to Alcohols, Ethers and Amines
1.11 Reduction of Carboxylic Acids to Aldehydes by Metal Hydrides
1.12 Reduction of Carboxylic Acids to Aldehydes by Other Methods
1.13 Reduction of C=X to CH_2 by Dissolving Metals and Related Methods
1.14 Reduction of C=X to CH_2 by Wolff–Kishner and Other Hydrazone Methods

Reduction of X=Y Bonds

2.1 Reduction of Nitro and Nitroso Compounds
2.2 Reduction of N=N, N—N, N—O and O—O Bonds
2.3 Reduction of S=O and SO_2 to S, of P=O to P, and of S—X to S—H

Reduction of C=C and C≡C Bonds

3.1 Heterogeneous Catalytic Hydrogenation of C=C and C≡C
3.2 Homogeneous Catalytic Hydrogenation of C=C and C≡C
3.3 Reduction of C=C and C≡C by Noncatalytic Chemical Methods
3.4 Partial Reduction of Aromatic Rings by Dissolving Metals and Other Methods
3.5 Partial Reduction of Enones, Styrenes and Related Systems
3.6 Partial and Complete Reduction of Pyridines and their Benzo Analogs
3.7 Partial and Complete Reduction of Pyrroles, Furans, Thiophenes and their Benzo Analogs
3.8 Partial and Complete Reduction of Heterocycles Containing More than One Heteroatom
3.9 Hydrozirconation of C=C and C≡C, and Hydrometallation by Other Metals
3.10 Hydroboration of C=C and C≡C
3.11 Hydroalumination of C=C and C≡C
3.12 Hydrosilylation of C=C and C≡C

Reduction of C—X to C—H

4.1 Reduction of Saturated Alkyl Halides to Alkanes
4.2 Reduction of Saturated Alcohols and Amines to Alkanes
4.3 Reduction of Heteroatoms Bonded to Tetrahedral Carbon
4.4 Reduction of Epoxides
4.5 Reduction of Vinyl Halides to Alkenes, and of Aryl Halides to Arenes
4.6 Reduction of Ketones to Alkenes
4.7 Hydrogenolysis of Allyl and Benzyl Halides and Related Compounds
4.8 Reduction of α-Substituted Carbonyl Compounds —CX—CO— to Carbonyl Compounds —CH—CO—

Author Index

Subject Index

Volume 9

Cumulative Author Index

Cumulative Subject Index

Cumulative Author Index

This Author Index comprises an alphabetical listing of the names of over 30 000 authors cited in the references listed in the bibliographies which appear at the end of each chapter in these volumes.

Each entry consists of the author's name, bold numbers, and other numbers which are associated with superscripts. For example

Abbott, D. E., **2**, 6[12,12c], 10[40], 573[53,54]

The bold number indicates the volume number, and the other numbers indicate the text pages on which references by the author in question are cited; the superscript numbers refer to the reference number in the chapter bibliography. Citations occurring in the text, tables and chemical schemes and equations have all been included.

Although much effort has gone into eliminating inaccuracies resulting from the use of different combinations of initials by the same author, the use by some journals of only one initial, and different spellings of the same name as a result of transliteration processes, the accuracy of some entries may have been affected by these factors.

Aalbersberg, W. G. L., **5**, 1151[132]
Aaliti, A., **1**, 331[49]
Aarts, V. M. L. J., **7**, 333[25]
Aasen, S. M., **5**, 914[110]
Abad, A., **3**, 851[64]; **6**, 644[81]
Abatjoglou, A. G., **3**, 124[273,288], 128[273,288,323], 132[273,288], 134[273,323]; **4**, 924[34]; **8**, 859[216]
Abbas, N., **7**, 473[29]
Abbaspour, A., **5**, 769[136]
Abbott, B. J., **7**, 65[71]
Abbott, D. E., **2**, 4[12,12c], 6[12,12c], 10[40], 573[53,54]
Abbott, D. J., **4**, 37[104]; **6**, 150[115]
Abbott, T. W., **6**, 964[80]
Abboud, J. L. M., **5**, 72[188]
Abboud, W., **8**, 445[53]
Abdali, A., **4**, 699[19], 700[23]
Abdallah, A. A., **1**, 567[222]
Abdallah, H., **4**, 957[24], 964[48]; **5**, 944[244]
Abdallah, M. A., **3**, 125[304], 126[304]
Abdallah, Y. M., **3**, 751[87]; **5**, 320[8]; **7**, 362[30]
Abd-el-aziz, A. S., **4**, 529[77,78], 530[78], 531[78,81]
Abdel Hady, A. F., **5**, 355[87c], 356[87c], 365[87c,96c]
Abd Elhafez, F. A., **1**, 49[3], 50[3], 182[46], 222[69], 295[50], 460[1], 678[211]; **2**, 24[96], 217[137], 666[36], 677[36]; **8**, 3[16]
Abdel-Halim, F. M., **8**, 860[221]
Abdel-Halim, H., **5**, 637[102]
Abdelkader, M., **5**, 71[138], 73[193,195,202], 461[105]
Abd Ellal, E. H. M., **4**, 413[276]
Abdel-Magid, A., **2**, 116[136], 117[136], 124[136], 436[66], 437[66b]; **5**, 516[28]; **8**, 54[160], 66[160], 74[245], 176[136], 393[110]
Abdel-Rahman, M. O., **8**, 478[38]
Abd El Samii, Z. K. M., **4**, 386[156], 387[156]
Abdel-Wahab, A. M., **3**, 325[160]
Abderhalden, E., **8**, 526[23]
Abdou, S. E., **2**, 403[37]
Abdulaev, N. F., **1**, 543[16]
Abdul-Hai, S. M., **7**, 26[54]
Abdulla, R. F., **4**, 123[210b,211], 125[210b]; **6**, 488[31], 571[31]
Abdullah, A. H., **4**, 428[73]; **8**, 586[29], 589[29]
Abdullin, K. A., **6**, 515[316]
Abdul-Majid, Q., **5**, 202[35], 221[54]
Abdul-Malik, N. F., **7**, 162[62]
Abdun-nur, A. R., **8**, 214[43]
Abdurasuleva, A. R., **3**, 303[58], 321[138]
Abe, A., **3**, 530[78], 535[78]
Abe, E., **1**, 858[60]
Abe, H., **3**, 226[193,196]; **4**, 331[11], 447[218]; **6**, 9[42]; **8**, 609[49]
Abe, J., **6**, 523[351], 524[351]
Abe, K., **1**, 422[92]
Abe, M., **4**, 30[88,88k], 121[209,209a], 261[299], 356[138], 856[100]; **7**, 102[136], 239[53]; **8**, 945[128]
Abe, R., **5**, 96[113]; **6**, 23[93]
Abe, S., **8**, 134[33], 137[33]
Abe, T., **2**, 116[140], 610[94], 611[94], 1059[78,81]; **3**, 617[15], 619[15], 621[15], 623[15], 627[15]; **5**, 92[81]; **7**, 800[34]
Abe, Y., **1**, 808[319]; **3**, 289[68], 446[77], 461[146], 541[107]; **6**, 438[47], 448[107], 491[115]; **7**, 86[16a]
Abecassia, J., **5**, 567[102,103]
Abecassis, J., **3**, 173[518]
Abed, O. H., **8**, 860[220]
Abegaz, B., **6**, 425[68], 430[68], 509[275]
Abel, D., **6**, 555[807]
Abel, E. W., **1**, 2[7,8], 125[84], 139[3], 211[2], 212[2], 214[2], 222[2], 225[2], 231[1], 428[121], 429[121], 457[121], 580[1]; **2**, 567[24], 587[24], 712[42]; **3**, 208[1,11], 210[11,11a], 219[11a], 228[214], 234[11a], 436[9,13], 524[33]; **4**, 518[2], 521[2], 735[84], 770[84], 914[1], 922[1], 925[1], 926[1], 932[1], 939[73], 941[1], 943[1]; **5**, 46[39], 56[39], 272[1], 641[131]; **6**, 690[397], 692[397], 831[11], 832[12], 848[11], 865[12]; **7**, 335[28], 594[5], 595[5], 598[5], 614[3], 629[48], 816[6a,b], 824[6], 825[6], 827[6a], 829[6a], 831[6a], 832[6a], 833[6a]; **8**, 99[110], 100[114], 443[1], 674[33], 708[42], 715[42], 717[42], 728[42]
Abele, W., **5**, 432[125,133]
Abeles, R. H., **4**, 1007[110]
Abell, A. D., **6**, 186[168]
Abell, P. I., **4**, 279[104], 284[153]
Abelman, M. M., **1**, 872[91]; **4**, 845[66], 847[66], 848[66,79]; **5**, 841[89], 843[123], 857[232]; **6**, 859[170]
Abeln, R., **1**, 749[78], 816[78]
Aben, R. W., **2**, 664[29]; **5**, 77[260,261,264], 434[142], 677[6], 684[37]; **6**, 558[847,848,854]
Abenhaim, D., **1**, 86[33,37-40]; **3**, 770[172]
Abenhaïm, D., **1**, 218[49], 220[49], 223[49,72c], 226[49a]
Aberchrombie, M. J., **4**, 310[433]

Abeysekera, B. F., **2**, 851[225]; **7**, 262[75]
Abeywickrema, A. N., **4**, 797[102], 805[140]
Abicht, H.-P., **3**, 229[224]
Abidi, S., **6**, 213[84]
Abiko, A., **1**, 161[88,89]
Abiko, S., **3**, 274[21]
Abiko, T., **3**, 956[107]
Abis, L., **5**, 504[277]
Ablenas, F. J., **5**, 441[176]; **8**, 407[60]
Abley, P., **8**, 445[44], 452[44], 459[227]
Abola, E., **4**, 5[18], 27[84,84a]
Abou-Elzahab, M. M., **4**, 45[126]
Abou-Gharbia, M. A., **6**, 498[168]
Aboujaoude, E. E., **1**, 788[257]; **2**, 482[27], 483[27]; **3**, 201[83]; **4**, 459[77], 473[77], 474[77]
Abraham, E. P., **4**, 282[139], 288[182]
Abraham, M. H., **1**, 214[24]
Abraham, N. A., **3**, 572[61]; **8**, 493[20], 497[20,38]
Abraham, T. S., **4**, 578[19,21]
Abraham, W., **8**, 595[78]
Abraham, W. D., **1**, 239[39], 864[86]; **4**, 71[20], 243[71]
Abraham, W. R., **7**, 62[50b,52b], 429[157a]
Abrahamson, E. W., **4**, 1016[205]
Abrahamsson, J., **6**, 824[123]
Abram, D. M. H., **4**, 1039[65]
Abramendo, Yu. T., **2**, 854[236]
Abramov, A. F., **2**, 933[139]
Abramov, A. I., **2**, 534[32]
Abramovitch, R. A., **3**, 681[99], 839[12], 840[12], 854[12]; **4**, 23[70], 430[89], 780[4], 790[4], 801[4], 953[8], 954[8k], 1081[72]; **5**, 64[54], 402[6], 634[75]; **6**, 140[60,61], 141[58,59], 550[677]; **7**, 21[20], 29[79], 476[64,65], 505[286], 736[1], 745[1], 749[1]
Abrams, G. D., **3**, 362[89]
Abrams, S. R., **6**, 903[137]
Abramskj, W., **1**, 838[166]
Abramson, N. L., **3**, 380[11]; **4**, 767[234]; **7**, 493[188]
Abramson, S., **5**, 223[84], 224[84], 740[150]
Abrecht, S., **4**, 764[222], 765[222], 808[155]
Abreo, M. A., **3**, 224[179]
Abril, O., **8**, 200[137]
Abrosimova, A. T., **4**, 291[208]
Absalon, M. J., **5**, 797[65]
Abskharoun, G. A., **4**, 1001[30]
Abu-El-Halawa, R., **6**, 517[327]
Abul-Hajj, Y. J., **7**, 59[41]; **8**, 475[21]
Aburaki, S., **6**, 978[24]
Aburatani, M., **4**, 249[128]
Abushanab, E., **7**, 778[415]
Abuzar, S., **7**, 78[128b]
Acampora, L. A., **4**, 476[155]
Acar, M., **5**, 560[70]
Accountius, C. E., **7**, 100[130]
Accrombessi, G. C., **8**, 882[84,85], 883[84]
Acemoglu, M., **1**, 770[184]; **5**, 311[102]; **6**, 677[313]; **7**, 410[103]
Acevedo, O. L., **6**, 554[733,738]
Acharya, A. S., **6**, 790[113-115,118]
Achatz, J., **2**, 1090[73], 1100[118], 1101[118], 1102[73,122], 1103[73,118b,122]
Achaya, K. T., **6**, 690[389]
Achenbach, H., **6**, 294[235]
Acheson, R. M., **3**, 630[55]; **4**, 55[157], 57[157b,j]; **5**, 687[66]; **8**, 589[50,53]
Achi, S., **3**, 47[256]; **4**, 746[147]
Achiba, Y., **1**, 287[18]
Achini, R., **2**, 159[127]; **5**, 71[124]; **6**, 773[45]
Achiwa, K., **1**, 385[115], 389[137]; **2**, 914[79], 915[79], 939[159], 940[161], 948[180], 994[38]; **3**, 99[184], 288[64]; **4**, 221[149], 964[46], 1089[125], 1095[152]; **5**, 100[157]; **6**, 716[96], 799[23], 811[77]; **7**, 228[95]; **8**, 146[100], 152[181], 154[190,191], 535[166], 944[125]
Achmatowicz, B., **1**, 329[39], 806[314]; **6**, 989[81]; **8**, 163[42]
Achmatowicz, O., **2**, 537[50], 538[54,64], 662[14], 664[14]
Achmatowicz, O., Jr., **2**, 535[39], 538[53,54,64], 663[28], 664[28]; **8**, 219[81]
Acholonu, K. U., **6**, 714[82]
Achrem, A. A., **7**, 160[53]
Achyutha Rao, S., **1**, 149[49b]; **4**, 898[173]; **8**, 696[121]
Aciego, R. M. D., **4**, 379[115]
Acke, M., **8**, 212[17]
Acker, K. J., **2**, 757[13], 759[13]
Acker, R.-D., **3**, 223[159], 262[162], 263[162]; **6**, 11[47]
Ackerman, J., **8**, 530[101]
Ackerman, J. H., **8**, 263[25]
Ackerman, J. J. H., **2**, 489[48], 490[48]
Ackerman, M. H., **8**, 618[120]
Ackermann, E., **6**, 247[133]
Ackermann, K., **5**, 1086[66], 1098[126], 1112[126]
Ackermann, M. N., **3**, 380[10]
Ackroyd, J., **2**, 711[30]; **3**, 1049[15], 1058[42]; **4**, 55[157], 57[157d]; **5**, 18[128]
Acott, B., **6**, 802[42], 803[42]
Acquadro, M. A., **5**, 165[86]
Acton, E. M., **6**, 1014[18]; **8**, 819[44]
Acton, N., **4**, 966[55]; **5**, 804[93]
Adachi, I., **1**, 569[248]
Adachi, J., **2**, 810[66], 851[66]
Adachi, M., **2**, 748[124]; **5**, 210[57]; **6**, 444[97]
Adachi, S., **1**, 512[40]; **3**, 147[387], 149[387], 150[387], 151[387]
Adachi, T., **6**, 76[46], 646[100b]; **8**, 384[36]
Adam, G., **2**, 1086[28], 1096[28]; **3**, 224[173]; **5**, 180[144]; **6**, 136[39]; **8**, 354[162], 537[184]
Adam, M., **8**, 370[86]
Adam, M. A., **2**, 6[25], 12[25], 13[25], 26[25], 27[25], 30[25], 31[25], 41[25], 42[25]; **6**, 7[30], 89[117]; **7**, 358[9], 400[53], 401[62]
Adam, M. J., **4**, 347[48]
Adam, W., **3**, 587[148]; **5**, 86[21], 155[35,36], 156[36], 157[36], 195[9], 198[21], 200[30], 201[31], 205[41-44], 206[45,46], 207[41,48], 209[9], 224[101], 582[179], 584[196,197], 986[37]; **6**, 121[130]; **7**, 98[97], 182[163], 185[175,177], 374[77b,d], 384[114c], 399[38], 400[38,38b], 406[38], 409[38], 415[38], 674[48], 818[16]; **8**, 398[144], 477[30,30a], 851[135]
Adamali, K. E., **1**, 544[34], 551[34], 553[34]
Adamczyk, M., **1**, 337[80], 827[67], 828[69]; **5**, 566[101]
Adames, G., **3**, 770[180]; **5**, 165[87], 936[199]
Adami, C. L., **2**, 529[20]
Adamov, A. A., **3**, 640[101]
Adamovich, S. N., **8**, 765[12]
Adamowicz, H., **5**, 108[207]
Adams, A., **7**, 49[64]
Adams, A. D., **1**, 32[157,158]; **2**, 106[47], 189[46], 209[46]; **7**, 49[63], 372[70]
Adams, B. L., **6**, 1016[28], 1036[142]
Adams, C., **3**, 642[111]
Adams, C. D., **8**, 614[88], 621[88]
Adams, C. T., **3**, 723[9], 731[9]
Adams, D. R., **2**, 527[1], 528[1], 553[1], 720[86]
Adams, E. W., **3**, 564[8], 721[3]; **7**, 884[188]
Adams, H., **1**, 310[109]
Adams, J., **1**, 822[33]; **3**, 289[66,67], 1056[36]; **4**, 1040[101], 1045[101a], 1059[154,156,157,158]; **6**, 74[30]; **7**, 408[89]

Adams, J. L., **2**, 813[75], 818[75]; **4**, 311[442], 364[1]; **8**, 853[143], 857[143]
Adams, J. M., **5**, 345[68], 346[68b]
Adams, J. T., **2**, 797[5], 829[5], 837[5], 843[5], 845[5]
Adams, K. G., **3**, 224[163]
Adams, K. H., **3**, 721[5]
Adams, M. A., **5**, 612[73]; **7**, 219[12]
Adams, R., **2**, 141[42], 145[64], 277[2], 281[2], 284[57], 294[2], 296[2], 1090[61]; **3**, 296[13], 564[8], 721[3], 825[27a]; **7**, 884[188]; **8**, 140[13,20,21,23], 142[49], 150[20], 364[22], 533[144]
Adams, R. D., **2**, 829[135]; **3**, 1012[79]; **6**, 874[12]; **8**, 847[94], 929[29]
Adams, R. M., **8**, 708[37]
Adams, T., **7**, 236[20]
Adams, T. C., Jr., **6**, 900[118]; **8**, 542[232], 928[23]
Adams, W. J., **3**, 846[47]
Adams, W. R., **7**, 96[89], 97[89]
Adamson, P. S., **4**, 2[5]
Addadi, L., **5**, 855[185]
Addess, K. J., **3**, 224[165]
Addison, C. C., **7**, 765[143], 846[85]; **8**, 842[43]
Addison, J. F., **2**, 363[196]
Addy, L. E., **2**, 530[21]
Ade, E., **3**, 45[251]
Adelakun, E., **7**, 846[83]
Adelfang, J. L., **2**, 149[86]
Adelman, R. L., **2**, 538[65], 539[65]
Adelsberger, K., **8**, 664[123]
Adembri, G., **4**, 956[18]
Ader, J. C., **6**, 245[127]
Adgar, B. M., **8**, 366[46], 368[68]
Adger, B. M., **6**, 644[89]; **7**, 743[65]
Adhikary, P., **4**, 124[214a]; **7**, 751[141]
Adickes, H. W., **2**, 492[53], 493[53]; **6**, 274[103-107]; **8**, 276[148,149,150]
Adinolfi, M., **4**, 347[95]; **7**, 438[17-19], 445[17-19,58]
Adiwidjaja, G., **5**, 115[250,251]; **6**, 419[3], 426[3,74], 448[111]
Adkins, H., **2**, 141[40], 240[8]; **3**, 823[15]; **7**, 14[126]; **8**, 140[9,24,25], 141[9,24,25], 142[25], 143[9,56,59], 148[59], 212[12], 242[42], 246[42], 452[189a], 533[139], 814[18]
Adkins, J. D., **8**, 236[6], 284[2], 285[2], 607[30]
Adlerova, E., **2**, 765[78]; **6**, 266[46]
Adlington, M. G., **8**, 318[68], 813[13], 969[97]
Adlington, R. M., **1**, 477[144,145,146], 545[46-48], 894[159]; **2**, 523[75]; **3**, 251[79], 254[79]; **4**, 111[152e], 113[170], 735[85], 744[134,136], 745[140], 822[226]; **5**, 116[269]; **6**, 96[152], 779[68], 783[86,87], 784[88-90], 961[68]; **7**, 231[153,154]; **8**, 387[56], 940[100,108]
Adolph, H. G., **4**, 426[54]; **7**, 749[121]
Adrian, F. J., **4**, 719[22]
Aebersold, D. R., **3**, 299[33]
Aebi, J. D., **3**, 40[221], 43[240], 44[240]
Aebischer, B., **6**, 1000[127]
Aerssens, M. H. P. J., **8**, 478[45], 480[45], 481[45]
Aeschimann, R., **3**, 41[227]
Affrossman, S., **8**, 286[13], 287[13]
Afghahi, F., **3**, 918[25]
Afify, A. A., **2**, 744[99], 745[99]
Afonso, A., **8**, 528[86], 895[2], 899[2]
Afonso, M. M., **5**, 434[145]
Afshari, G. M., **7**, 236[27]
Agami, C., **3**, 13[66], 21[66,130], 22[130]; **4**, 23[70]; **6**, 718[122-124]; **8**, 524[13]
Agarwal, K. L., **6**, 606[39], 625[157], 626[166]
Agarwal, R., **2**, 364[202]
Agawa, T., **1**, 332[55,56], 787[254]; **2**, 128[240], 579[95], 581[105]; **3**, 672[65]; **4**, 102[131], 247[102], 252[102], 259[102], 589[78], 630[419], 1020[239], 1023[262], 1024[263,264]; **5**, 422[82], 474[158], 1200[49,55]; **6**, 186[172]; **7**, 209[93], 453[65]; **8**, 806[106,107], 807[106], 900[31]
Agbalyan, S. G., **6**, 712[78]
Agdeppa, D. A., Jr., **3**, 747[70]
Agenas, L. B., **7**, 769[226]
Ager, D. J., **1**, 570[271], 620[65], 731[4], 786[249], 815[4]; **2**, 716[58]; **3**, 123[248,252], 125[248], 135[252,350,351,352,354,355], 136[350,351,352], 137[350,351,352,355], 140[352], 141[350,351,355], 778[4]; **4**, 113[167], 120[197], 241[55], 255[55], 682[57]; **6**, 139[52,53], 687[383]; **7**, 582[148]; **8**, 248[85], 769[24], 771[24], 782[24b]
Aggarwal, S. K., **4**, 356[135]; **6**, 527[406]; **7**, 415[113], 601[84], 602[84]
Aggarwal, V. K., **1**, 526[96]; **2**, 202[95]; **6**, 25[100], 902[126]
Agho, M. O., **1**, 420[85]
Agnello, E. J., **7**, 136[108]; **8**, 566[450]
Agnès, G., **8**, 287[22]
Agorrody, M., **4**, 753[166]
Agosta, W. C., **3**, 19[103], 901[115,116]; **4**, 611[343]; **5**, 21[144], 133[56], 136[56,69-72], 141[56], 164[75,76], 165[87], 176[75], 918[127]; **6**, 836[58], 1061[73]; **7**, 140[131]; **8**, 248[83], 948[151]
Agouridas, K., **2**, 537[49]
Agranat, I., **2**, 766[83-85]
Agster, W., **4**, 429[87]
Aguero, A., **5**, 1116[12], 1125[65]
Aguiar, A. M., **2**, 482[25,28,29], 483[25,28,29]; **4**, 119[192e], 473[145]; **8**, 461[258], 535[166]
Aguilar, D. A., **2**, 536[41]
Aguilar, E., **8**, 124[88]
Aguilar-Laurents de Guttierrez, M. I., **6**, 452[133]
Aguilo, M., **1**, 34[223]
Agwaramgbo, E. L. O., **1**, 731[3], 785[246]
Ahern, D., **4**, 342[62]
Ahern, M. F., **5**, 428[111]
Ahern, T. P., **7**, 211[97]
Ahibo-Coffy, A., **2**, 855[244]
Ahlbrecht, H., **1**, 559[153]; **2**, 60[18], 62[18d], 65[30], 510[44]; **3**, 65[6]; **4**, 80[60], 102[128a-c], 109[149], 112[160], 113[168], 116[185a], 119[195]; **6**, 706[38], 722[141,142], 724[141]
Ahlers, H., **1**, 645[124], 669[124,181,182], 670[181,182], 680[124]; **3**, 105[214]; **4**, 120[200]
Ahlfaenger, B., **8**, 755[130], 758[130]
Ahlgren, G., **5**, 788[14]
Ahlhelm, A., **7**, 247[106]
Ahmad, H. I., **7**, 262[79]
Ahmad, J., **6**, 525[381]
Ahmad, M. S., **3**, 762[145]; **7**, 92[43], 675[60]; **8**, 222[96]
Ahmad, S., **1**, 266[49]; **2**, 282[36]; **4**, 331[17]; **7**, 580[144], 586[144], 822[33]
Ahmad, S. Z., **7**, 92[43]
Ahmad, V. U., **2**, 1026[67], 1028[67]; **6**, 725[167]
Ahmed, A., **5**, 552[35]
Ahmed, F. R., **3**, 1036[83]
Ahmed, I., **4**, 485[27], 503[27]
Ahmed, M., **5**, 582[182]
Ahmed, M. G., **6**, 712[72]
Ahmed, M. T., **7**, 7[41]
Ahmed, R., **3**, 692[138], 693[142]
Ahmed, S. S., **2**, 746[112]
Ahmed, Z., **1**, 678[211]
Ahmed-Schofield, R., **2**, 1038[102,103]
Ahn, K. H., **2**, 42[148], 45[148]; **3**, 505[166]; **5**, 926[160]; **8**, 14[77,81], 16[100], 18[122], 54[152], 66[152], 241[39], 244[68],

247[39,68], 250[68], 272[113], 354[176], 536[174], 538[192], 544[278], 806[101], 880[57], 938[88], 969[93]
Ahn, S. H., **3**, 49[261]
Ahond, A., **2**, 901[31,32]; **6**, 912[22], 920[44]
Ahramjian, L., **2**, 397[10], 412[9], 413[9]
Ahrens, F., **3**, 582[111]
Ahrens, F. B., **8**, 591[63]
Ahrens, G., **6**, 519[335]
Ahrens, K. H., **5**, 410[40]; **6**, 524[355], 525[355], 532[355]
Ahuja, R. R., **4**, 231[264]
Ahuja, V. K., **8**, 375[158], 418[13], 422[13], 425[13]
Aibe, H., **8**, 190[81]
Aida, T., **2**, 605[56], 630[11], 631[11]; **4**, 404[243]; **7**, 503[275]; **8**, 856[181]
Aidhen, I. S., **4**, 810[171]
Aig, E. R., **1**, 821[29]; **3**, 289[69], 301[48]
Aigami, K., **3**, 383[45]; **6**, 270[80]; **7**, 9[70]
Aigner, H., **2**, 1089[59], 1090[72]
Aihara, S., **7**, 881[156]
Aihara, T., **1**, 558[134]
Aikawa, H., **1**, 511[32]; **4**, 113[174], 245[86], 259[86], 260[86]
Aikawa, Y., **4**, 158[78]
Aiken, J. W., **7**, 340[46]
Aime, S., **8**, 457[216]
Aimetti, J. A., **2**, 212[119]
Aimi, N., **2**, 1021[50]; **6**, 916[31]; **8**, 31[47], 66[47]
Aimino, D., **8**, 412[113]
Ainley, A. D., **7**, 595[9]
Ainslie, R. D., **8**, 697[132]
Ainsworth, C., **2**, 602[40], 606[65], 837[161a], 838[161]; **3**, 626[42]; **4**, 111[155a]
Airoldi, M., **8**, 451[172]
Aishima, I., **8**, 754[94]
Aitken, D. J., **1**, 559[146]
Aitken, R. A., **4**, 953[8], 954[8p], 961[8p]; **5**, 803[87]; **7**, 479[92]
Aiura, H., **2**, 920[95], 921[95]
Aizpurua, J. M., **5**, 94[87], 95[96]; **6**, 249[144], 250[144], 490[102], 491[117], 655[159], 810[73], 938[129], 940[129]; **7**, 275[145], 278[159,160], 283[186,187], 530[18], 531[18], 752[144], 760[24]; **8**, 19[133,134]
Ajioka, S., **5**, 297[59], 1196[38], 1197[38]
Akaba, R., **7**, 881[156]
Akabane, Y., **3**, 695[153]
Akabori, S., **1**, 543[14]; **2**, 352[88], 357[88]; **3**, 826[39]; **7**, 230[129], 660[38]; **8**, 149[117-119]
Akaboshi, S., **4**, 331[14], 344[14]
Akagi, M., **4**, 680[50]
Akai, S., **1**, 242[49-51], 243[52]; **6**, 930[87], 931[88]; **7**, 199[34], 209[89]; **8**, 837[14]
Akaishi, R., **8**, 412[117]
Akalaeva, T. V., **6**, 487[74], 489[74]
Akam, T. M., **8**, 478[44], 480[44]
Akama, T., **4**, 258[240], 261[240], 262[240]
Akamatsu, H., **4**, 30[90]
Akao, S., **4**, 1103[205]
Akasaka, K., **1**, 762[141]; **6**, 5[24]
Akasaka, T., **7**, 470[13], 498[230b], 769[222]; **8**, 392[97]
Akashi, C., **6**, 217[115,116]
Akashi, K., **7**, 309[22], 439[28,30]
Akawie, R. I., **3**, 754[107], 757[107]
Akbutina, F. A., **2**, 814[80]
Åkermark, B., **2**, 66[32]; **4**, 290[201], 302[331], 560[21,22], 572[4], 591[111], 598[184,185,188,190,209], 599[215], 600[239], 609[215], 616[111], 622[185], 623[190], 624[215], 625[239], 631[420,421], 633[111], 638[185,190,209,422], 641[215], 643[239]; **5**, 4[34]; **7**, 95[66], 474[44,45], 504[282]
Åkermark, G., **4**, 598[188]
Akgun, E., **3**, 266[194]; **5**, 692[102]
Akhmedov, K. N., **3**, 303[58]
Akhmetova, N. E., **2**, 343[12], 359[12]
Akhrem, A. A., **3**, 734[7]; **4**, 145[29b], 241[58], 379[115]; **5**, 699[2]
Akhrem, I., **3**, 297[25], 334[25]; **4**, 840[33]
Akhrem, I. S., **2**, 727[136]; **7**, 7[51]
Akhtar, J. A., **7**, 583[153]
Akhtar, M., **6**, 74[28], 943[157]; **7**, 9[68]; **8**, 561[412]
Akhtar, M. H., **3**, 927[54]; **5**, 687[59]
Akhtar, M. S., **7**, 265[103], 267[103]
Aki, L. Y., **5**, 797[65]
Aki, O., **8**, 975[133], 992[55]
Akiba, K., **1**, 120[65], 236[30], 237[30], 350[152,153], 361[35,35a,b], 362[35a,b], 436[146]; **2**, 24[93], 556[151], 615[127], 655[138], 656[138], 657[138], 905[56], 906[56], 907[59], 920[95], 921[95]; **4**, 446[212,213]; **5**, 841[97]; **6**, 821[115], 931[91]
Akiba, M., **3**, 135[344], 136[344], 137[344], 138[344]; **7**, 774[323]; **8**, 370[92], 382[8]
Akimoto, A., **8**, 452[187], 535[162]
Akimoto, H., **7**, 692[23]
Akimoto, I., **3**, 262[158]
Akimoto, K., **3**, 594[186]; **5**, 442[181]; **6**, 46[59]; **8**, 820[46]
Akimoto, M., **4**, 589[84], 590[95], 592[95], 598[84], 633[95]
Akimova, A. Y., **5**, 432[129]
Akita, H., **6**, 5[26]; **8**, 190[64], 191[95], 195[106], 197[106], 198[134], 201[139]
Akita, M., **1**, 162[101,102], 163[106]; **5**, 1172[28], 1182[28]; **6**, 16[60]; **7**, 642[8]; **8**, 787[119]
Akita, Y., **6**, 938[134]; **8**, 371[103], 902[47], 904[47], 905[47]
Akiyama, A., **2**, 456[22]; **7**, 200[42], 209[92]
Akiyama, F., **3**, 583[120]; **4**, 856[100]; **6**, 726[174]; **7**, 173[131]; **8**, 889[128]
Akiyama, M., **5**, 581[174], 862[246]; **6**, 113[71], 533[499], 843[87]; **8**, 222[97]
Akiyama, S., **2**, 152[100]; **3**, 585[137]; **8**, 14[80]
Akiyama, T., **1**, 72[68], 752[98], 753[103]; **3**, 652[221]; **5**, 1139[75]; **8**, 661[112], 797[43], 807[43]
Akiyoshi, K., **3**, 231[243]
Akiyoshi, S., **7**, 92[42], 93[42]
Akkerman, O. S., **1**, 26[132,133,134], 746[70]; **5**, 1125[57]
Akpuaka, M. U., **3**, 810[43]
Aksnes, G., **4**, 35[98a]; **5**, 76[244]; **8**, 860[221], 864[240]
Akssira, M., **5**, 578[153,154,155,157]
Aktogu, N., **2**, 125[214]; **4**, 82[62a]
Akutagawa, K., **1**, 481[160]; **3**, 71[35], 197[36]
Akutagawa, S., **4**, 609[332]; **6**, 866[208]; **8**, 154[199], 459[244], 462[267]
Akutsu, N., **8**, 187[39]
Akuzawa, K., **3**, 470[215], 473[215], 476[215]
Aladzheva, I. M., **4**, 55[156]
Alais, J., **6**, 51[107]
Alajarin, M., **4**, 440[170]
Alam, I., **2**, 801[22]
Alam, N., **4**, 459[80,86], 469[80,86]
Alam, S. K., **8**, 331[35]
Alami, M., **2**, 127[238]; **3**, 578[93]; **4**, 98[111]
Alami, N. E., **2**, 980[22], 981[22]
Alario, F., **8**, 535[166]
Alary, J., **8**, 594[70]
Al-Aseer, M., **1**, 477[133]; **3**, 67[17]
Al Ashmawy, M. I., **4**, 386[156], 387[156], 413[276]
Alauddin, M. M., **5**, 385[130,130i], 389[138], 392[138b], 682[32], 691[83,83a,89], 692[83], 693[83,106], 1031[95]; **6**, 977[11], 1007[11]
Alazard, J.-P., **3**, 681[100]
Albanbauer, J., **4**, 1081[83]

Albanesi, G. A., **5**, 1138[65]
Albano, C., **2**, 1099[115]
Albanov, A. I., **4**, 291[208]
Albarello, J. A., **7**, 30[81]
Albaugh-Robertson, P., **2**, 718[80]; **3**, 443[60]; **5**, 1053[42], 1060[42]
Al-Bayati, R., **1**, 347[130]
Albeck, M., **7**, 7[36]
Alberola, A., **5**, 478[164]; **8**, 646[50,51]
Albers, M. O., **8**, 457[214], 458[214]
Alberstron, N. F., **8**, 143[60], 148[60]
Albert, A., **6**, 543[605], 546[650]
Albert, A. H., **8**, 220[83]
Albert, P., **4**, 272[27,28]
Albert, R., **6**, 22[81]; **7**, 207[73]
Alberti, B. N., **7**, 79[135]
Al'bertinskii, G. L., **3**, 635[44]
Alberts, A. H., **3**, 124[281], 125[281]
Alberts, V., **8**, 852[141], 857[141]
Alberts-Jansen, H. J., **1**, 214[27]
Albertson, D. A., **8**, 526[28]
Albertson, N. F., **2**, 739[43a], 963[52]; **4**, 3[11]; **6**, 636[26], 637[26]
Albinati, A., **4**, 602[257,261]; **5**, 1147[112]
Albini, A., **7**, 340[48], 874[108], 882[170]
Albisetti, C. J., **5**, 3[23], 64[44]
Albizati, K. F., **6**, 1044[16b], 1048[16]
Albonico, S. M., **7**, 686[100]
Albrecht, H. A., **8**, 645[43]
Albrecht, H. P., **8**, 354[164]
Albrecht, R., **2**, 1026[69]; **5**, 402[3,3b], 403[3b,11], 422[88], 423[88]; **8**, 315[45]
Albrecht, W., **5**, 451[2]
Albrect, H., **1**, 476[114], 477[114]
Albright, J. D., **1**, 542[8], 543[8], 544[8], 547[8], 548[8], 550[8], 552[8], 553[8], 555[8,8a,109], 556[8,109], 557[8], 559[109], 560[8]; **3**, 48[258], 197[38], 198[38]; **4**, 113[168,168d]; **7**, 294[16], 295[16,19], 299[40]; **8**, 30[42], 66[42]
Albright, T. A., **4**, 538[101,102], 539[102]; **5**, 300[71]; **6**, 175[68]
Alcaide, B., **5**, 92[64]; **8**, 36[74], 38[74], 66[74]
Alcántara, M. P. D., **6**, 941[150]
Alcaraz, J. M., **8**, 865[247]
Alcock, N. W., **4**, 1032[11]; **5**, 916[119]
Aldar, K., **7**, 663[60]
Aldean, J. K., **4**, 190[107]
Alder, A., **5**, 68[98]
Alder, B., **1**, 215[33]
Alder, K., **2**, 139[31], 369[252]; **5**, 3[24], 6[24], 7[24], 29[1], 316[1], 402[2], 451[3-9], 453[59], 513[1]
Alder, R. W., **7**, 878[139]; **8**, 388[64]
Aldinucci, D., **2**, 465[104]
Aldridge, C. L., **8**, 452[190]
Alejski, K., **3**, 619[23]
Aleksandrov, A. M., **6**, 270[81]
Aleksandrowicz, P., **4**, 590[103,104]
Aleksankin, M. M., **8**, 236[7]
Alekseeva, N. V., **4**, 55[156]
Aleksejczyk, R. A., **5**, 855[189]; **7**, 365[47]
Aleksnadrov, G. G., **5**, 1174[34]
Alemagna, A., **4**, 522[54], 523[57,58]; **6**, 178[121]
Alemdaroglu, N. H., **4**, 915[11]
Alemzadeh, I., **8**, 52[137], 66[137]
Ales, D. C., **8**, 336[68]
Alessi, T. R., **5**, 514[8], 527[8]
Alewood, P. F., **7**, 31[87]
Alexakis, A., **1**, 78[5], 107[3], 112[27], 124[81], 132[107], 133[108], 343[115,117], 347[131], 348[140], 370[68], 371[68], 373[84], 374[84], 428[116-118]; **2**, 448[43], 596[3], 614[116]; **3**, 209[21], 216[73], 217[21,84], 219[21b], 223[154], 224[166,167], 225[167b], 226[203,207], 227[210], 245[30], 249[58-61], 258[125], 259[129], 263[173], 265[188], 423[72], 466[182,193], 470[214], 473[214], 476[214], 485[32-35], 486[32-35], 494[34,35,89], 516[89], 579[100]; **4**, 24[72], 152[58,59], 173[34], 183[81], 185[87], 207[59], 209[63,68,69], 210[77,78], 238[10], 249[10], 250[10,136], 254[10], 255[10], 258[232], 262[136], 866[3], 867[3], 873[3], 877[73], 893[3,159], 895[164], 896[3,168,170], 897[170], 898[3,177], 899[3], 900[3,164,179,180], 901[3], 902[190], 903[188,189,191,192,194,196,197,199], 1009[139]; **5**, 829[24], 1163[2]; **6**, 5[23], 849[122]
Alexander, C. W., **4**, 23[70]
Alexander, D. L., **7**, 254[28]
Alexander, E. R., **3**, 725[20]; **4**, 1021[244]
Alexander, J., **2**, 824[121]; **7**, 341[51,52]
Alexander, J. J., **4**, 903[200,200b], 905[200b]
Alexander, J. R., **4**, 283[144], 288[144]
Alexander, R. K., **4**, 331[9]
Alexander, R. P., **4**, 675[41], 691[75]; **6**, 142[65]
Alexandre, C., **5**, 324[23]
Alexandrou, N., **3**, 380[9]
Alexandrov, Y. A., **7**, 599[65]
Alexis, M., **7**, 395[21]
Al-Fekri, D. M., **5**, 586[204]
Alfonso, C. M., **7**, 298[36]
Alfonso, L. M., **7**, 766[176]
Alford, G., **5**, 716[85]
Alfs, H., **3**, 307[87]
Alfter, I., **7**, 748[116]
Al-Hassan, M. I., **3**, 232[260], 254[98], 257[98], 495[90,92], 503[90]
Ali, E., **6**, 487[72], 489[72]
Ali, M. B., **4**, 960[36]; **5**, 132[47]
Ali, M. R., **6**, 538[565]
Ali, M. U., **6**, 555[815]
Ali, S., **1**, 585[14]; **4**, 610[342]
Ali, S. A., **1**, 436[153]; **2**, 2[7], 100[13], 224[153], 253[42], 630[9]; **4**, 1078[50]
Ali, S. F., **6**, 934[99]; **7**, 206[66]
Ali, S. M., **4**, 510[170]; **6**, 104[7], 109[7]; **7**, 674[37]
Aliev, I. A., **6**, 462[10]
Alimardanov, R. S., **3**, 306[82]
Aliminosa, L. M., **7**, 92[48]
Alink, R. J. H., **3**, 904[129]
Al Jazzaa, A., **7**, 844[54]
Al-Kathumi, K. M., **4**, 670[15]
Alker, D., **5**, 847[135]
Al Kolla, A., **1**, 544[34], 551[34], 553[34]
Alkonyi, I., **7**, 92[41,41a,42], 93[42], 94[41]
Allain, L., **2**, 809[52], 823[52]
Allakhverdiev, M. A., **3**, 304[69]
Allamagny, Y., **4**, 30[87]
Allan, A. R., **3**, 383[48]; **4**, 1014[187]
Allan, R. D., **4**, 6[23]; **6**, 81[76], 82[76], 818[106,107]
Allavena, C., **1**, 219[61], 830[95]; **4**, 764[220]
Allen, A. A., **6**, 268[62]
Allen, A. C., **5**, 161[65], 163[65], 176[130]
Allen, A. D., **4**, 299[301]
Allen, C. F. H., **2**, 744[92]; **4**, 272[33]; **5**, 752[50]; **6**, 228[31], 960[49]
Allen, C. F. R. H., **3**, 721[4]
Allen, D. E., **2**, 536[42,48], 537[42,48]; **7**, 543[16]
Allen, D. L., **8**, 459[239]
Allen, D. S., Jr., **3**, 11[52], 17[52]; **7**, 564[93], 565[93], 568[93], 711[57]; **8**, 958[17]
Allen, F. H., **1**, 2[3], 37[3]; **6**, 436[9]
Allen, G. F., **3**, 1038[91]; **4**, 560[26,28], 561[30]

Allen, G. R., Jr., **8**, 527[41]
Allen, J. C., **7**, 13[114]
Allen, L. E., **3**, 1040[107]
Allen, M. J., **3**, 564[10], 566[32,33], 595[188], 634[17]
Allen, M. S., **7**, 58[57], 62[57], 63[57], 340[46]; **8**, 587[34]
Allen, R., **6**, 237[64], 243[64]
Allen, R. C., **4**, 439[168]
Allen, R. E., **8**, 568[471]
Allen, R. L. M., **7**, 740[40]; **8**, 382[12], 383[12]
Allen, T. G., **6**, 834[40]
Allendörfer, H., **8**, 918[120]
Allenmark, S., **7**, 196[30], 199[30]
Allenstein, E., **6**, 495[145,148]
Alleston, D. L., **4**, 1007[109]; **8**, 807[117]
Allevi, C., **5**, 1147[112]
Allevi, P., **7**, 674[42]; **8**, 565[448]
Allgeier, H., **5**, 439[166]
Alligood, D. B., **4**, 1033[27], 1049[27], 1060[27c]; **5**, 599[40], 804[94], 905[60]
Allinger, J., **1**, 141[19], 151[19]
Allinger, N. L., **1**, 2[15]; **3**, 382[36], 854[79], 905[140]; **4**, 23[70], 187[95,96]; **5**, 88[50]
Allison, N. T., **4**, 1012[167]
Allmann, R., **2**, 1053[52,53], 1055[53]; **6**, 501[197], 535[527,528]
Allport, D. C., **3**, 664[34]
Allred, E. L., **5**, 568[108]
Allwohn, J., **5**, 850[152]
Almansa, C., **5**, 1059[54], 1060[55], 1062[54c,d]
Almaski, L., **8**, 756[138]
Almassian, B., **7**, 473[27]
Almirantis, Y., **5**, 955[304]
Almond, H. R., **8**, 28[37], 66[37]
Almond, H. R., Jr., **1**, 755[116], 756[116,116b,118], 758[116], 761[116]; **8**, 568[473]
Almond, M. R., **6**, 803[47], 804[47]
Almond, S. W., **5**, 791[40]
Alnajjar, M. S., **4**, 736[89], 738[89]; **7**, 877[129], 884[181]
Al-Nuri, M., **5**, 256[57]
Alonso, C., **8**, 936[74]
Alonso, J. H., **5**, 911[91]
Alonso, M. E., **4**, 1033[26], 1035[26a], 1036[26c,44,52], 1046[26a,b], 1051[26a]; **5**, 942[231,234]
Alonso, R., **8**, 874[24]
Alonso, R. A., **4**, 453[25,34], 461[25,102], 462[25], 464[102], 465[102], 466[128], 467[102,128], 468[135], 472[34,143], 473[146], 474[25,146], 475[102], 476[157]
Alonso, T., **1**, 17[217]
Alonso-Cires, L., **7**, 93[54], 486[142], 490[176]
Alonso-López, M., **4**, 108[146h], 230[247]
Alonso-Silva, I. J., **5**, 407[27], 408[30,30b]
Alpegiani, M., **4**, 745[138]; **7**, 340[48]
Alper, H., **2**, 527[4], 528[4]; **3**, 539[102], 559[55,56], 582[116], 770[181], 1020[12], 1024[29], 1028[50], 1030[60], 1032[69], 1036[83,84], 1037[86], 1039[99]; **4**, 553[4,8], 939[79], 941[79,82,83]; **5**, 1[2], 2[2], 151[18], 948[290], 1065[1], 1066[1], 1074[1], 1083[1], 1084[1], 1093[1], 1138[59,60]; **6**, 686[367,368]; **7**, 451[17,30,31], 462[17], 482[117]; **8**, 36[82], 39[82], 66[82], 301[91], 373[125], 390[77,78], 392[105], 394[117], 449[153], 452[153], 454[201], 455[201], 683[89,91], 840[32], 847[94], 890[140]
Alpha, S. R., **8**, 86[21]
Alpoim, M. C. M. de C., **4**, 823[228]; **5**, 925[153]
Al-Razzak, L. A., **7**, 350[25], 355[25]
Alsdorf, H., **1**, 3[21]
Alsop, D. J., **8**, 901[40]
Alt, G. H., **2**, 367[230]; **6**, 211[78]
Alt, H., **8**, 513[102]
Alt, H. G., **5**, 1066[7,8], 1070[18, 11659], 1178[9]
Alt, K. O., **4**, 270[10]
Al-Talib, M., **6**, 522[346]
Altarejos, J., **7**, 634[68]
Altenbach, H.-J., **6**, 155[153], 190[195], 840[72], 841[75], 845[97,98], 903[138]; **8**, 211[6], 222[6]
Altland, H. W., **7**, 732[58]
Altman, J., **2**, 1074[146]
Altman, L. J., **3**, 99[186], 107[186]
Altnau, G., **1**, 82[24], 83[24], 97[82], 215[31]
Al'tshuler, R. A., **6**, 554[755]
Altukhov, K. V., **7**, 855[64]
Alumbaugh, R. L., **3**, 734[3]; **8**, 873[14], 875[14]
Alunni, S., **6**, 958[33]
Alvarez, C., **5**, 1076[40], 1105[159]; **7**, 693[24]
Alvarez, E., **3**, 448[95]; **4**, 373[87]; **7**, 413[118]
Alvarez, F. M., **1**, 568[227], 571[227]
Alvarez, F. S., **6**, 941[152]
Alvárez, J., **8**, 312[19]
Alvarez, M., **6**, 801[36]; **8**, 587[34]
Alvarez, N. M., **4**, 452[9]
Alvarez, R. M., **8**, 349[137], 934[53]
Alvernhe, G., **1**, 387[126,128,134], 388[126,134]; **3**, 381[28], 382[28]; **4**, 331[19], 356[137]; **6**, 94[136]; **7**, 498[222]
Alvhall, J., **8**, 678[58,65], 683[58,65], 686[58,65], 691[58]
Alwani, D. W., **2**, 149[83]
Alward, S. J., **5**, 944[239]; **8**, 114[56]
Alwell, G. J., **6**, 532[473]
Aly, M. M., **2**, 772[15]
Alyea, E. C., **8**, 459[228]
Amadou, S. T., **2**, 1090[72]
Amagasa, M., **8**, 881[82]
Amâl, H., **6**, 554[806]
Amamria, A., **3**, 495[93b]
Amann, A., **7**, 359[15]; **8**, 15[88]
Amano, A., **5**, 552[1]
Amano, E., **3**, 844[32,33]
Amano, K., **7**, 693[30], 694[30]
Amano, T., **3**, 244[25], 267[25], 494[84]; **8**, 698[138]
Amarasekara, A. S., **4**, 341[60]; **5**, 143[97,98], 252[44], 636[88]; **7**, 242[61], 496[216], 522[38], 772[287]
Amaratunga, S., **8**, 373[125]
Amarnath, V., **6**, 635[12], 643[12], 660[210], 662[210]
Amaro, J., **8**, 330[49]
Amato, J. S., **2**, 648[97], 649[97b]; **8**, 272[122]
Amatore, C., **4**, 453[29-31], 458[68], 459[29-31,80,81,85,86], 467[68], 469[68,80,81,86,136], 471[31,68,139,140,141], 472[29], 473[68,139], 475[30,150]; **7**, 850[10], 852[37,40], 854[45]
Ambach, E., **6**, 554[772,777]
Ambekar, S. Y., **2**, 381[302]
Ambler, P. W., **1**, 119[63]; **2**, 125[225], 315[46], 316[46]; **4**, 977[95]
Ambles, A., **6**, 216[109]
Ambrose, M. G., **6**, 19[69]
Ambrosetti, R., **4**, 330[3], 345[3,79]
Ambrosini, A., **6**, 487[52,54], 489[52,54]
Ambrus, G., **7**, 70[93]
Ambuehl, J., **4**, 403[239], 404[239]
Ame, P., **7**, 65[68]
Amedio, J. C., **4**, 1040[94], 1041[94]; **5**, 839[81]; **6**, 126[153]; **7**, 399[39]
Ameer, F., **3**, 421[52]
Amelotti, C. W., **3**, 572[66]; **8**, 312[22], 321[22]
Amen, K.-L., **8**, 724[175,176]
Amende, J., **3**, 888[13]
Amendola, M., **4**, 106[138e]
Amendolla, C., **8**, 528[65]

Amer, I., **8**, 535[163]
Amer, M. I., **2**, 748[125]; **6**, 291[215]
Ames, A., **1**, 95[73,74]; **2**, 737[38], 818[94], 855[240]; **6**, 466[38,39], 468[38,39], 470[39,57]
Ames, D. E., **3**, 281[43], 505[169], 530[74,75], 531[87], 534[74,75], 538[87]; **6**, 209[69]; **8**, 613[79], 926[15], 940[15]
Ames, G. R., **8**, 533[137]
Amice, P., **1**, 879[111c,d]; **4**, 972[82]; **5**, 461[103], 910[89]; **7**, 121[26]
Amici, R., **6**, 736[26]
Amick, D. R., **4**, 430[95]
Amick, T. J., **4**, 1104[212]
Amin, N. V., **7**, 220[25]
Amin, S., **7**, 350[22]
Amin, S. G., **5**, 86[13], 95[89]
Amino, Y., **2**, 215[134]; **4**, 75[44b]; **5**, 693[112]
Amit, B., **6**, 624[151], 636[22], 651[138]
Amiya, S., **5**, 787[8]
Ammanamanchi, R., **8**, 385[47]
Ammar, F., **3**, 574[77]
Ammon, H. L., **1**, 294[47]; **4**, 82[62g], 218[133], 243[76]; **5**, 480[178], 829[21]
Amon, C. M., **4**, 1018[217]; **7**, 300[57]
Amonoo-Neizer, E. H., **2**, 183[12]
Amoros, L. G., **4**, 505[147,148]
Amos, R. A., **4**, 315[523,530], 379[114], 394[189]; **7**, 163[78], 167[78]; **8**, 407[53], 412[118]
Amosova, S. V., **7**, 194[6]
Amouroux, R., **1**, 53[18], 215[43], 216[43]; **2**, 590[159]; **3**, 252[85]; **4**, 383[138,139], 390[168]
Amrani, Y., **8**, 535[166]
Amrein, W., **5**, 219[37]
Amri, H., **3**, 246[34]; **4**, 34[97], 35[97]
Amriev, R. A., **4**, 288[187]
Amrollah-Modjdabadi, A., **1**, 557[130]
Amstutz, E. D., **2**, 747[121]; **3**, 564[12]
Amstutz, R., **1**, 1[1], 3[1], 26[1], 28[142], 30[151], 37[238,240,241], 38[261], 41[151], 43[1], 299[61], 316[61]; **2**, 100[2], 280[25]; **5**, 841[104]; **6**, 112[65]
Amtmann, R., **8**, 742[42], 746[42]
Amupitan, J. A., **2**, 546[90]; **3**, 348[29], 355[53], 357[53]
Amvam-Zollo, P.-H., **6**, 51[105]
An, R., **8**, 376[161]
Anagnostopoulos, C. E., **8**, 991[49]
Ananchenko, S. N., **2**, 382[313]
Anand, N., **6**, 538[552], 550[552]; **8**, 654[84]
Anand, R. C., **5**, 944[241]
Ananda, G. D. S., **7**, 96[199], 112[198], 820[26]
Anandrut, J., **1**, 303[79]
Anani, A., **7**, 662[55]
Ananthanarayan, T. P., **3**, 816[80]; **8**, 623[148]
Anantharamaiah, G. M., **6**, 635[15], 636[15], 651[136,136c], 668[250]; **8**, 959[22]
Anantha Reday, P., **3**, 462[149]
Anastasia, M., **7**, 674[42]; **8**, 565[448]
Anastassiou, A. G., **5**, 634[77], 716[85]; **7**, 21[13], 25[46], 26[46,60], 479[95], 507[307]; **8**, 407[58], 625[161]
Anciaux, A. J., **1**, 631[47], 656[154], 658[154]; **3**, 87[92], 1047[8], 1051[8]; **4**, 318[560], 1033[16,16d,e], 1035[16e], 1051[125], 1052[16d]
Ancillotti, F., **4**, 307[396]
Andal, R. K., **8**, 446[65]
Andermann, G., **8**, 370[96], 639[19]
Anders, E., **6**, 172[11]
Anders, R. T., **6**, 498[163]
Andersen, K. K., **6**, 149[93,94,101]
Andersen, K. S., **5**, 552[14]
Andersen, L., **4**, 436[140]; **5**, 686[53]
Andersen, N. H., **1**, 506[10], 582[7]; **2**, 90[41], 381[310], 544[85], 547[85,96,110-112], 551[96,110-112], 552[85], 572[42], 710[28]; **3**, 132[334], 133[334], 136[334], 354[62], 356[56]; **4**, 373[72]; **6**, 134[29]
Andersen, P., **1**, 33[163]
Andersen, V. K., **4**, 181[73]
Anderskewitz, R., **3**, 303[53]
Anderson, A. B., **7**, 64[63]; **8**, 423[37], 458[223,223c]
Anderson, A. G., **1**, 846[16], 851[16]
Anderson, A. G., Jr., **2**, 169[164], 323[27]; **6**, 204[23]; **8**, 568[472]
Anderson, A. L., **8**, 900[32]
Anderson, B. A., **5**, 1070[29], 1074[29], 1076[44], 1080[52]
Anderson, B. C., **3**, 380[5]
Anderson, B. F., **8**, 500[51]
Anderson, C., **4**, 857[103]
Anderson, C. B., **7**, 92[50]
Anderson, C. E., **6**, 660[199]
Anderson, C. L., **7**, 597[48], 641[5]
Anderson, D., **6**, 508[286], 537[286]
Anderson, D. F., **3**, 749[81]
Anderson, D. G., **1**, 618[56]; **4**, 120[197]
Anderson, D. J., **2**, 787[53], 792[53]; **7**, 743[60]
Anderson, D. K., **2**, 763[59]; **4**, 14[46]; **5**, 1022[76]
Anderson, D. M. W., **2**, 367[225]; **3**, 826[40]
Anderson, D. R., **5**, 684[36b], 834[53]
Anderson, F. E., III, **4**, 18[59]
Anderson, G. J., **7**, 291[2], 654[9], 655[9,18]
Anderson, G. K., **4**, 915[15]; **8**, 447[116]
Anderson, G. W., **6**, 614[90], 636[26], 637[26], 664[222]
Anderson, H. J., **1**, 474[84]; **2**, 743[84,87], 780[9]; **5**, 581[176]
Anderson, J., **5**, 413[48]
Anderson, J. D., **8**, 532[132,132d]
Anderson, K., **7**, 160[49]
Anderson, K. D., **1**, 471[67]
Anderson, K. W., **3**, 736[27]
Anderson, L. G., **4**, 85[75]
Anderson, M. B., **1**, 238[37], 529[123], 735[28], 736[28]
Anderson, M. W., **1**, 366[52]; **2**, 494[56,57]; **4**, 76[47]; **6**, 516[321], 552[692]; **8**, 638[16]
Anderson, N. H., **4**, 120[201]; **7**, 228[93], 258[44]; **8**, 278[157], 374[147,148]
Anderson, O. P., **1**, 827[67]; **4**, 161[87,87b], 560[26]; **5**, 1107[168]; **7**, 399[33]; **8**, 333[57], 345[127]
Anderson, P. C., **3**, 227[208], 421[61], 422[61], 589[161,162], 610[161,162]; **4**, 18[59], 121[208], 373[82], 386[148b], 387[148], 993[161]
Anderson, P. H., **3**, 927[48]; **6**, 134[37]
Anderson, P. S., **3**, 380[11]; **5**, 382[121], 410[41]; **8**, 580[3], 584[3], 585[3]
Anderson, R., **1**, 731[4], 815[4]
Anderson, R. A., **3**, 663[26], 665[26]
Anderson, R. C., **2**, 827[126]; **5**, 350[80], 841[86]; **6**, 859[164], 978[21]; **7**, 261[67]; **8**, 566[450]
Anderson, R. J., **1**, 116[45], 128[45], 433[139], 434[139]; **2**, 159[127]; **3**, 220[121], 224[163], 226[192], 257[120], 265[187], 365[94], 419[43]; **6**, 4[20], 9[40]
Anderson, R. L., **4**, 104[135c]; **5**, 1076[35,38,46], 1079[49]
Anderson, R. S., **8**, 73[248], 74[248], 373[136]
Anderson, S. M., **8**, 802[86]
Anderson, S. W., **4**, 1007[112]
Anderson, T., **6**, 175[75]
Anderson, T. J., **8**, 851[126], 858[126]
Anderson, V. E., **6**, 455[149]
Anderson, W. K., **5**, 581[172]

Andersson, B., **7**, 331[15]
Andersson, F., **7**, 272[144], 274[144]
Andersson, K., **2**, 346[43]
Andersson, L., **6**, 20[72], 74[33]
Andersson, P. G., **4**, 371[49]
Andersson, S., **4**, 227[210]
Andisik, D., **4**, 295[262], 296[262]
Ando, A., **2**, 233[189], 455[16]; 7, 158[35]
Ando, H., **7**, 425[149a]
Ando, K., **2**, 8[37], 13[37], 20[37a], 35[37], 43[147], 44[147]; **3**, 137[376]; **4**, 21[69], 222[173,176]; **6**, 864[197]; **7**, 764[116]; **8**, 847[91]
Ando, M., **1**, 98[84], 99[84], 387[136]; **2**, 443[15], 445[24], 451[15], 576[79], 718[77], 995[45]; **3**, 1041[113]; **4**, 24[75], 25[75a], 857[104], 970[72], 972[81]; **5**, 841[88]; **6**, 542[601], 767[28], 768[28], 769[28], 984[59]; **7**, 155[25], 163[73], 641[6], 696[43], 697[43]; **8**, 43[108], 47[108], 64[220], 66[108], 67[220], 394[119]
Ando, N., **8**, 170[79]
Ando, R., **1**, 670[186]; **7**, 761[63], 771[279], 773[279]
Ando, T., **2**, 115[124]; **3**, 218[98]; **4**, 354[130], 1006[96], 1017[211], 1021[211,242]; **6**, 66[5], 498[167], 968[106]; **7**, 255[38]; **8**, 86[26], 798[54]
Ando, W., **3**, 887[6], 889[6], 893[6], 894[6], 896[6], 897[6], 900[6], 903[6,123], 919[28,32], 923[43,44], 934[44,63], 954[44], 1008[70,71]; **6**, 114[76], 440[77], 846[101]; **7**, 763[102], 769[222], 778[403], 851[24]; **8**, 979[147]
Andrac, M., **2**, 81[5]
Andrade, J. G., **3**, 499[124], 507[124], 670[58], 673[58], 693[141]; **7**, 336[34]; **8**, 459[240], 460[250]
Andre, C., **1**, 419[79], 797[292], 802[292]; **6**, 995[104], 996[104]; **7**, 400[46]
André, E., **4**, 47[133]
Andreades, S., **7**, 801[36]
Andreae, S., **7**, 746[93]
Andreani, A., **2**, 787[52]
Andree, H., **7**, 760[40]
Andree, R., **6**, 111[64]
Andreetta, A., **8**, 443[1], 446[95], 449[157], 450[157], 452[95a,b], 457[95a-c,218], 458[218]
Andreetti, G. D., **3**, 386[57]; **6**, 195[224]
Andreev, L. N., **6**, 419[6], 508[243]
Andreev, V. M., **3**, 305[72], 644[140,141]
Andreeva, L. N., **4**, 992[154]
Andreini, B. P., **3**, 525[38,40]
Andrejevic, V., **3**, 380[13]; 7, 229[112]; **8**, 872[5], 875[5,33]
Andreocci, A., **3**, 804[11]
Andreoli, P., **5**, 100[156]
Andreou, A. D., **2**, 745[102,106]
Andres, H., **5**, 13[90], 14[96]
Andres, W. W., **7**, 157[33], 158[33b,43]
Andreu, M. R., **5**, 474[158]
Andrew, R. G., **2**, 578[85]
Andrews, A., **7**, 452[49]
Andrews, D. J., **5**, 168[101], 176[131]
Andrews, D. R., **1**, 883[124]; **5**, 806[104], 1029[91]; **7**, 90[32]
Andrews, D. W., **6**, 707[42]
Andrews, G. C., **2**, 66[35], 71[55], 74[55], 442[6]; **3**, 86[16], 95[16], 96[16], 97[16], 98[16], 99[16], 104[16], 116[16], 117[16,235,236], 118[16], 155[16,235,236], 156[235,236], 168[16], 196[30], 358[67], 919[31], 934[31], 939[78], 946[87]; **5**, 890[34]; **6**, 153[142], 838[64], 873[1,5], 895[1], 901[121], 903[5], 996[109], 997[110], 1022[67]; **8**, 18[127], 537[178]
Andrews, G. D., **5**, 907[78], 908[78], 918[78], 1007[37]
Andrews, L., **5**, 704[21]
Andrews, L. H., **8**, 604[2], 605[2]
Andrews, L. J., **5**, 71[144]; **6**, 980[39]; **8**, 796[27], 888[122]
Andrews, M. A., **7**, 107[163]
Andrews, P. R., **5**, 855[182,184]
Andrews, R. C., **3**, 100[202], 168[202], 170[202], 172[202]; **5**, 173[123]; **6**, 7[34], 158[183]; **7**, 89[27]
Andrews, S. B., **1**, 213[18]
Andrews, S. L., **6**, 936[105]; **7**, 205[61]
Andrews, S. M., **4**, 439[159]
Andrews, S. W., **1**, 612[49]; **3**, 752[93]; **5**, 247[24]
Andriamialisoa, R. Z., **1**, 838[169]; **4**, 598[206]; **5**, 409[37]; **6**, 487[75], 921[49]; **7**, 164[80]; **8**, 58[174], 66[174]
Andrianome, M., **2**, 727[133]
Andrianov, K. A., **6**, 546[645]; **8**, 765[11]
Andrianova, G. M., **1**, 555[112]
Andrieux, C. P., **3**, 574[77]; **8**, 135[49]
Andriollo, A., **8**, 446[74], 452[74], 457[74]
Andrisano, R., **8**, 161[22]
Andrist, A. H., **5**, 705[23]; **8**, 743[163], 757[163]
Andrulis, P. J., **7**, 872[98]
Andrus, A., **3**, 1025[36]; **5**, 854[181]
Andrus, W. A., **4**, 107[142]; **6**, 676[307]
Andruzzi, R., **8**, 132[14]
Andrzejewski, D., **6**, 1012[7], 1013[7,11]
Aneja, R., **7**, 544[34]; **8**, 568[467]
Anet, F. A. L., **1**, 174[3], 528[113]; **2**, 811[67]; **3**, 164[475], 382[36]; **5**, 552[21]; **7**, 483[125]
Añez, M., **8**, 16[101], 537[180]
Angeletti, E., **2**, 345[20]
Angeli, A., **3**, 890[35]; **8**, 664[120]
Angelici, R. J., **8**, 608[42], 629[42], 847[96]
Angelini, G., **3**, 300[40]; **4**, 445[206]; **8**, 336[70]
Angelino, N., **7**, 749[123]
Angell, E. C., **7**, 221[31], 227[31]
Angelo, M. M., **2**, 710[23]
Angeloni, A. S., **8**, 161[22]
Angelov, C. M., **4**, 364[1,1e], 395[1e]
Angelova, O., **1**, 36[175]
Angelucci, F., **8**, 358[197]
Angély, L., **3**, 669[56]
Angermann, A., **1**, 54[20], 338[83]; **2**, 31[114,114a]; **7**, 253[18]
Angermund, K., **1**, 14[77], 37[251], 180[30], 531[132]; **5**, 497[224]; **8**, 682[83]
Angibeaud, P., **7**, 247[104]
Angibeaund, P., **8**, 479[48], 481[48], 528[56]
Angier, R. B., **7**, 30[80]
Angiolini, L., **1**, 59[35]; **2**, 948[184]
Angle, S. R., **1**, 889[143], 890[143]; **5**, 311[104], 563[89], 843[121], 934[188]
Angoh, A. G., **1**, 882[123], 885[133c]; **3**, 219[110]; **4**, 10[32], 733[79], 791[54], 824[235]
Angrick, A., **6**, 33[6], 34[6], 40[6], 46[6]
Angst, C., **1**, 343[107]; **2**, 1074[149], 1075[149]; **4**, 382[134]; **5**, 836[63]
Angus, H. J. F., **5**, 646[4]
Angus, R. H., **6**, 436[10,11]
Angus, R. O., Jr., **4**, 1010[161]; **6**, 977[18]
Angustine, R. L., **7**, 564[86]
Angyal, S. J., **1**, 2[15]; **4**, 187[96]; **6**, 648[117a]; **7**, 231[139], 666[72,73,76], 803[52]
Anh, N. T., **1**, 497[7,8], 50[7], 110[20], 153[55], 182[47], 185[47], 198[47], 222[69], 310[102]; **2**, 24[96], 217[137], 666[38], 677[38]; **4**, 70[11], 71[16a], 139[2,3]; **5**, 64[27], 260[72], 263[72], 417[63], 453[67], 683[219]; **6**, 873[2]; **8**, 3[23], 7[37], 16[104], 536[170], 541[213,214,215], 542[214], 543[213,214,215], 930[34]
Anhoury, M.-L., **8**, 237[12]
Anikina, E. V., **3**, 303[57]
Animati, F., **6**, 490[109]

Anisimov, A. V., **6**, 860[180]; **8**, 609[54]
Anisimov, K. N., **4**, 701[28,29], 702[28,29]
Anisimova, O. S., **6**, 502[208,209], 531[453], 554[763,786,791]
Anjaneyulu, B., **5**, 92[68], 95[68], 96[68]
Anjik, T., **7**, 335[32]
Anker, D., **4**, 331[19]
Anker, M., **1**, 440[171]
Anklam, E., **4**, 251[144], 257[144]
Anklekar, T. V., **1**, 477[131]; **3**, 75[43], 76[43], 77[57], 80[43]; **7**, 227[80]
Ankner, K., **8**, 994[66]
Anliker, R., **8**, 108[2], 118[2], 530[91]
An-naka, M., **3**, 461[148], 541[113], 543[113]; **6**, 46[59]
Annen, K., **7**, 773[305]
Annen, U., **5**, 1101[145]
Annenkova, V. Z., **4**, 461[100], 475[100]
Anner, G., **3**, 816[79]; **6**, 1015[23], 1059[64]; **7**, 41[20]; **8**, 974[124]
Annino, R., **8**, 988[25]
Annis, G. D., **1**, 849[28]; **2**, 546[89]; **6**, 979[26], 1004[140]
Annis, M. C., **6**, 279[136]; **7**, 699[54]
Anno, K., **8**, 269[86]
Annoura, H., **1**, 63[43,44], 64[46]; **4**, 304[356]; **5**, 736[145], 737[145], 838[74]
Annunziata, R., **1**, 519[65,66], 520[67], 523[80], 524[86-88], 765[151]; **2**, 31[108], 228[166,167], 374[276], 435[62,65], 486[42], 492[52], 515[55-57], 516[58]; **6**, 149[97,99,102], 840[71]; **7**, 442[47], 767[193], 778[411]; **8**, 72[239], 74[239], 844[67]
Anoshina, G. M., **6**, 530[421], 538[555], 550[421]
Ansari, H. R., **8**, 613[79]
Ansari, J. A., **7**, 92[43], 481[112]
Anschütz, R., **3**, 828[45]
Anschütz, W., **3**, 909[152]
Ansell, M. F., **1**, 739[37]; **3**, 326[164,165]; **4**, 313[462], 951[1], 968[1], 979[1]; **5**, 433[136a,b], 605[56], 612[77], 819[152]; **6**, 104[1,9]
Anselme, J.-P., **1**, 845[10]; **2**, 830[137]; **6**, 249[140], 809[67]; **7**, 663[59], 749[120]
Anson, F. C., **3**, 213[51]; **8**, 595[76]
Anstad, T., **6**, 242[94]
Antar, M. F., **6**, 968[113]
Antczak, K., **5**, 944[239]
Antebi, S., **4**, 710[57-59]; **8**, 847[94]
Antel, J., **2**, 345[34], 351[34], 357[34], 371[34], 373[273,274]; **5**, 14[99], 17[121], 468[125]
Anteunis, M., **5**, 1130[7]; **6**, 80[66], 237[69]; **8**, 212[17], 348[132]
Anthoine, G., **5**, 717[90d]
Anthon, N. J., **4**, 381[126b], 382[126], 383[126]
Anthonsen, T., **1**, 823[41,42], 832[41]; **8**, 528[66]
Anthony, D. R., Jr., **6**, 515[235]
Anthony, I. J., **4**, 1016[202]
Anthony, J. M., **8**, 358[202]
Anthony, N. J., **1**, 801[303]
Antipin, I. S., **5**, 76[248]
Anton, D. R., **6**, 958[34]
Antonakis, K., **2**, 555[141]; **7**, 265[97,98], 272[98]; **8**, 64[212], 66[212], 67[212], 68[212]
Antonenko, L. M., **2**, 737[31]
Antoni, G., **7**, 229[121]
Antonietta Grifagni, M., **4**, 98[114], 113[114]
Antonik, L. M., **4**, 461[100], 475[100]
Antonini, I., **5**, 92[63]
Antonioletti, R., **3**, 512[213], 515[213]; **4**, 391[176]; **5**, 771[151], 772[151]; **7**, 265[102], 266[107], 267[102,107], 530[15]
Antonjuk, D. J., **1**, 519[63]
Antonova, N. D., **2**, 534[32], 535[38]
Antonsson, T., **7**, 453[77]
Antonucci, F. R., **8**, 300[84]
Antonucci, R., **8**, 566[450]
Antony-Mayer, C., **4**, 1006[104]
Antus, S., **7**, 831[62]; **8**, 544[276]
Anwar, S., **7**, 645[23]; **8**, 9[53], 21[53]
Anwer, M. K., **8**, 904[60], 905[60]
Anzalone, L., **8**, 351[165]
Anziani, P., **8**, 143[57]
Aoai, T., **4**, 341[56]; **6**, 289[194,195], 293[194,195], 1031[110,112]; **7**, 495[207], 523[43], 771[264]; **8**, 849[114]
Aoe, K., **2**, 1066[122]; **4**, 398[217], 399[217b], 401[217b], 403[217b], 404[217b], 803[132]; **7**, 353[35], 355[35]
Aoki, H., **5**, 504[276]
Aoki, K., **5**, 1142[87,89]; **7**, 802[49]
Aoki, O., **1**, 279[86], 280[86]; **3**, 567[190], 595[190], 607[190]
Aoki, S., **1**, 212[12], 213[12], 215[12b], 217[12], 448[207]; **2**, 443[19], 447[19], 448[19], 449[19], 450[52,53], 451[56], 578[88], 587[88]; **3**, 169[509], 221[131,132], 455[125], 460[125]; **4**, 163[96], 164[96]; **5**, 74[205]; **6**, 848[110]; **8**, 846[80]
Aoki, T., **2**, 784[38], 1052[51]; **7**, 707[29], 708[29], 803[51]; **8**, 151[147], 881[82]
Aoki, Y., **1**, 373[88], 374[88]; **5**, 844[131]
Aono, M., **4**, 1095[152]
Aoshima, A., **4**, 298[288]
Aoyagi, S., **2**, 116[140], 610[94], 611[94], 1059[78]; **4**, 503[129]
Aoyagi, T., **3**, 623[39]
Aoyama, H., **3**, 1057[38]; **4**, 614[372,373], 840[37], 905[207]; **5**, 181[150]; **8**, 964[55]
Aoyama, M., **4**, 147[42]
Aoyama, T., **1**, 844[6,7], 851[40,47], 852[40], 853[47]; **3**, 900[90]; **4**, 507[151]; **6**, 121[127,-128], 127[159], 129[166], 507[240], 515[240]
Aoyama, Y., **8**, 709[45]
Aparajithan, K., **2**, 149[83]
Aparicio, C., **2**, 361[178]
Aparicio, F. J. L., **2**, 385[328]
Apchié, A., **2**, 142[44]
Apel, J., **6**, 134[19]
Apel, M., **5**, 1052[38]
Apeloig, Y., **4**, 986[132], 987[132]
Apene, I., **6**, 489[37]
Apotecher, B., **4**, 905[212]
Apoussidis, T., **8**, 806[109]
Appa Rao, J., **1**, 544[31], 548[31]
Apparao, S., **5**, 464[112], 466[112]
Apparu, M., **4**, 789[31]; **6**, 2[3], 25[3]
Appel, R., **2**, 520[66]; **4**, 21[69], 104[137], 222[168], 224[168]; **5**, 829[23,26]; **6**, 74[37], 172[12,28], 190[197,198], 196[198], 205[29,31], 525[385]; **7**, 474[49], 476[49]
Appelbaum, A., **7**, 95[80]
Appell, H. R., **3**, 309[91]; **8**, 373[128]
Appelman, E. H., **4**, 347[97]
Appelt, A., **1**, 17[213]
Appenstein, C. K., **5**, 1090[88]
Apple, D. C., **8**, 764[6]
Applebaum, M., **8**, 950[167]
Applegate, H. E., **6**, 644[93]
Appleton, R. A., **3**, 715[37], 735[19]; **4**, 6[23]
Appleyard, G. D., **4**, 55[156]
ApSimon, J. W., **2**, 323[21], 352[86], 369[86]; **3**, 35[206], 289[67]; **4**, 200[2]; **5**, 946[256], 947[256], 952[256]; **6**, 541[595], 719[126], 725[126], 739[58]; **7**, 30[82]; **8**, 159[4,9], 535[165], 541[212], 720[135]
Aquadro, R. E., **8**, 240[33]
Arad, D., **3**, 194[4]
Arad, Y., **8**, 532[130]
Aradi, A. A., **5**, 1135[49]

Arai, H., **1**, 834[124]; **4**, 298[285]; **5**, 497[225], 914[115]; **6**, 17[63], 18[63], 774[50], 1066[98]; **7**, 452[41], 628[44-46], 645[19-21], 701[64]; **8**, 788[121]
Arai, I., **2**, 96[57]; **4**, 839[30], 974[90]; **6**, 849[123], 865[200]; **8**, 223[100], 224[100]
Arai, K., **1**, 543[14]; **7**, 209[94], 414[108]
Arai, M., **2**, 184[21]; **3**, 221[132]; **4**, 903[202], 904[202]; **6**, 619[115], 848[110]
Arai, Y., **1**, 752[97], 821[28]; **2**, 823[116]; **4**, 413[274]; **5**, 185[164], 369[101], 370[101a-c], 768[131], 779[131]; **6**, 150[128]; **7**, 257[47]
Arakawa, H., **7**, 160[55]
Arakawa, K., **6**, 668[251], 669[251]
Arakelova, L. V., **3**, 304[68]
Arakelyan, S. V., **4**, 315[505,506,507]
Araki, K., **1**, 802[304]; **5**, 767[120]
Araki, M., **1**, 406[30], 407[30,32,33], 415[30], 424[100], 427[30], 430[131], 454[32]; **6**, 438[58], 439[58,67]
Araki, N., **4**, 1040[87], 1041[87]
Araki, S., **1**, 256[20,21]; **2**, 24[92,94]; **3**, 220[116]; **6**, 834[41]
Araki, T., **8**, 846[84]
Araki, Y., **1**, 83[27], 357[7], 361[7]; **2**, 476[4], 601[34]; **3**, 32[187]; **5**, 158[43,47,48], 159[49], 185[164], 337[51]; **6**, 46[66], 720[132]; **8**, 18[130]
Aramendia, M. A., **8**, 368[65]
Arancibia, L., **6**, 453[138]
Aranda, G., **8**, 111[20], 118[20]
Aranda, V. G., **7**, 486[141]
Araratyan, E. A., **3**, 318[123]
Arase, A., **2**, 112[88], 241[14]; **3**, 470[196,197], 473[196,197], 522[21]; **4**, 145[23,29a]; **7**, 16[163], 602[99], 604[130], 608[170,171]; **8**, 720[130]
Arashiba, N., **1**, 860[69]
Arata, K., **2**, 737[37]; **3**, 300[42,43]; **7**, 5[22]
Arata, Y., **3**, 1057[38]
Aratani, M., **2**, 213[124], 878[40], 1060[86]; **4**, 27[79,79a]; **6**, 266[50]; **7**, 169[107]
Aratani, T., **3**, 216[72]; **4**, 952[3], 996[3], 1011[165], 1038[56,57], 1039[56]; **8**, 807[115]
Araujo, H. C., **7**, 507[309]
Aravind, S., **4**, 350[115]; **7**, 502[258]
Arber, W., **4**, 931[56]
Arbic, W., **3**, 634[30], 644[30b]
Arbuzov, B. A., **5**, 104[183,185], 107[183], 451[40], 470[40], 485[40]
Arbuzov, Yu. A., **2**, 534[32], 535[38]
Arcadi, A., **3**, 539[96]; **4**, 411[266b]
Arcamone, F., **2**, 323[35], 762[54]; **8**, 347[141], 350[141], 358[197]
Arcelli, A., **8**, 369[73], 552[351]
Archelas, A., **7**, 59[42,43], 60[43-45,46a,47a,b], 62[47c], 64[61b], 78[126], 429[157b]
Archer, S., **2**, 758[25]; **7**, 75[113], 690[14]; **8**, 143[60], 148[60]
Archibald, T. G., **8**, 384[27]
Archie, W. C., Jr., **5**, 678[15]
Arco, M. J., **1**, 131[103]; **3**, 261[148], 264[148], 957[110], 958[113]; **5**, 611[71], 894[46]; **6**, 897[101], 898[102]
Arcoleo, J. P., **3**, 759[132]
Arct, J., **4**, 1014[192], 1021[251]
Arcus, C. L., **6**, 799[21]
Ard, J. S., **8**, 228[132]
Ardakni, M. A., **6**, 530[416]
Ardecky, R. J., **5**, 394[145a]; **6**, 977[14]
Ardid, M. I., **4**, 5[17]
Ardisson, J., **4**, 796[97]
Ardoin, N., **8**, 518[128]
Arduengo, J. E., **4**, 47[134]
Areda, A., **6**, 509[275]
Arens, A. K., **8**, 587[31]
Arens, J. F., **3**, 106[222], 113[222], 121[245], 123[245,246], 125[246], 257[115], 262[160], 263[160]; **4**, 238[2], 297[266]; **6**, 612[74], 964[82,83]
Arentzen, R., **6**, 658[189]
Arenz, T., **6**, 181[130,131], 188[183]
Arepally, A., **5**, 7[51], 514[6]
Aresta, M., **3**, 461[145]; **8**, 446[63]
Aretakis, A. J., **4**, 730[64]
Arfmann, H. A., **7**, 62[50b,52b]
Argade, N. P., **6**, 569[940]
Argade, S., **5**, 348[74a]
Argay, G., **6**, 525[382], 534[519]
Argues, A., **6**, 554[735]
Argyle, J. C., **3**, 804[5]
Argyropoulos, J. N., **3**, 824[20]; **4**, 240[50,51]
Argyropoulos, N., **6**, 175[72], 668[252], 669[252]
Arhart, R. J., **6**, 960[56]
Ariaans, G. J. A., **3**, 855[83]
Aribi-Zoviueche, L., **4**, 591[114], 617[114]
Arickx, M., **8**, 237[12]
Arief, M. M. H., **6**, 770[36]
Ariga, M., **2**, 805[43]; **6**, 535[524]
Arigoni, D., **3**, 341[2], 352[45], 360[2]; **5**, 15[109]; **7**, 86[13], 236[24]; **8**, 204[153], 528[68], 530[68]
Ari-Izumi, A., **6**, 846[100]
Arima, M., **5**, 714[72]
Arimoto, M., **4**, 155[72]; **6**, 938[132], 944[132]; **7**, 92[41,41b], 93[41b], 94[41], 340[45], 457[110]; **8**, 331[31]
Arimura, T., **3**, 329[184]
Arinich, L. V., **6**, 499[173]
Arison, B. H., **2**, 756[8]; **4**, 1072[18]
Aristoff, P. A., **1**, 739[36]; **7**, 415[111]
Arita, M., **7**, 684[93a]
Arita, Y., **5**, 963[323]
Ariyoshi, K., **7**, 91[34], 310[28], 657[22]
Arjona, O., **1**, 117[56]; **4**, 368[17]
Arkell, A., **3**, 890[32], 894[61]
Arkley, V., **3**, 807[25]
Arledge, K. W., **7**, 528[5]
Arman, C. G. V., **6**, 487[57], 489[57]
Armand, J., **7**, 218[3]
Armande, J. C. L., **7**, 225[66]
Armesto, D., **5**, 201[32], 202[33,34,36], 220[50,51], 221[52,53,56]
Armistead, D. A., **6**, 27[117]
Armistead, D. M., **1**, 103[95,96], 126[89,90], 418[72]; **2**, 578[84], 701[85]; **4**, 33[96,96d], 34[96e], 735[85]; **5**, 534[92], 574[130], 843[117], 850[163]; **7**, 237[37]
Armitage, J. B., **3**, 554[18,19]
Armstead, D. A., **2**, 523[91]
Armstrong, A., **4**, 381[126b], 382[126], 383[126]; **6**, 650[133b], 668[261]
Armstrong, C., **4**, 274[55]
Armstrong, D. R., **1**, 6[33]
Armstrong, J. C., **3**, 537[90], 538[90]
Armstrong, J. D., **5**, 842[108], 843[108]; **6**, 858[162]
Armstrong, L. J., **3**, 890[32]
Armstrong, P., **4**, 1086[116]; **5**, 257[59]
Armstrong, R., **1**, 92[65]
Armstrong, R. W., **3**, 231[253]; **6**, 632[4]; **7**, 399[33]
Armstrong, W. P., **5**, 477[159]
Arn, H., **3**, 223[155]
Arnaboldi, A. G., **3**, 871[55]
Arnaiz, D. O., **5**, 843[121]
Arnap, J., **7**, 245[75]
Arnarp, J., **6**, 23[89], 647[106]
Arnaud, P., **3**, 572[66]

Arndt, C. H., **6**, 147[85]
Arndt, D., **7**, 541[2], 851[22]
Arndt, F., **3**, 563[117], 582[117], 888[13], 891[37]; **6**, 120[121]
Arndt, H. C., **1**, 794[275]; **3**, 86[61], 88[61], 89[61], 91[61], 124[61], 918[23]; **5**, 1020[70], 1027[70]; **6**, 143[70], 991[87]; **8**, 843[62], 993[60], 994[60]
Arndt, R. R., **8**, 864[244]
Arne, K. H., **1**, 491[31], 495[31], 496[31], 497[31], 501[31]; **3**, 199[62]
Arnet, J. E., **4**, 665[7]
Arnett, C. D., **8**, 344[123]
Arnett, E. M., **1**, 41[203]
Arnett, J. F., **1**, 359[25], 364[25]; **3**, 570[54]
Arney, B. E., Jr., **4**, 1002[62]; **6**, 744[72]; **8**, 36[88], 52[149], 66[88,149]
Arnó, M., **3**, 851[64]; **6**, 780[71]
Arnold, B., **5**, 113[239]
Arnold, B. J., **5**, 682[31], 683[115], 693[113,115], 694[115], 711[57a]
Arnold, C., **6**, 228[16]
Arnold, D. R., **5**, 151[3], 152[3], 167[97], 489[199], 645[1], 650[1q,25], 651[1]; **7**, 874[107], 875[111], 878[137], 879[150]
Arnold, E. V., **2**, 88[29]; **6**, 835[46]
Arnold, L. D., **3**, 227[213]; **5**, 86[31]; **8**, 205[155]
Arnold, R. A., **5**, 891[37], 892[37]; **6**, 288[191]
Arnold, R. C., **4**, 270[17,19], 271[17,19]
Arnold, R. D., **6**, 198[236]
Arnold, R. T., **2**, 282[42], 529[19], 538[62], 539[62]; **3**, 324[152], 807[20]; **5**, 2[12], 828[5], 847[5,132], 1001[13]; **8**, 925[11], 926[11]
Arnold, S., **4**, 345[81]
Arnold, S. C., **7**, 11[83]
Arnold, W., **6**, 48[86]
Arnold, Z., **2**, 346[45], 358[156,157], 782[27], 783[34], 784[39b], 785[39c]; **3**, 890[30], 897[30]; **5**, 461[102]; **6**, 512[302]; **7**, 507[308], 824[41]; **8**, 950[168]
Arnoldi, A., **4**, 764[218]
Arnone, A., **4**, 382[131a,b], 384[131b]; **5**, 64[52]; **8**, 587[30,32], 856[171]
Arnost, M. J., **3**, 865[26]; **5**, 438[163]
Arnould, D., **3**, 168[507], 169[507], 171[507]; **6**, 157[165,166]
Arnstedt, M., **1**, 215[32]
Aronovich, P. M., **7**, 597[42,43]
Arora, A., **2**, 1094[89], 1095[89]
Arora, G. S., **3**, 405[138]
Arora, K. K., **2**, 465[103]
Arora, P., **2**, 287[68]
Arora, P. C., **3**, 649[209]
Arora, S., **5**, 95[102], 288[36]
Arora, S. K., **7**, 598[53], 600[74]; **8**, 249[95], 263[27], 269[27], 273[27], 275[27], 279
Arous-Chtara, R., **2**, 283[49], 983[28], 989[34], 990[34]
Aroyan, A. A., **6**, 507[237], 515[237]
Arpe, H. J., **3**, 1039[98]; **4**, 606[301]; **8**, 960[33]
Arques, A., **6**, 509[273]
Arreguy, B., **2**, 716[66]
Arrhenius, P., **1**, 822[32]
Arrias, E., **7**, 766[175], 768[175]
Arribas, E., **6**, 524[362]
Arrick, B. A., **5**, 494[215], 579[163]
Arrieta, A., **2**, 649[102], 1059[75]; **5**, 95[90,96,97], 96[109]; **6**, 249[144], 250[144], 251[149], 490[101-103], 491[117], 810[73], 816[101]
Arrieta, J. M., **3**, 625[41]
Arrivo, S. M., **2**, 835[159]
Arsenijevic, V., **4**, 307[393]
Arseniyadis, S., **1**, 387[134], 388[134], 542[9], 544[9], 551[9], 552[9], 553[9], 554[9], 555[9], 557[9], 558[138,139,140,141], 560[9]; **2**, 73[62], 223[151], 420[24]; **3**, 39[218], 48[218]; **4**, 292[222], 412[268a,269], 413[268a,269a]; **8**, 854[153]
Arshava, B. M., **5**, 431[122]
Artamkina, G. A., **4**, 423[8], 426[8], 444[8], 519[12]
Artaud, I., **2**, 432[57]
Arth, G. E., **6**, 219[123]; **7**, 236[22,25], 256[40]; **8**, 268[73]
Arthur, S. D., **1**, 738[39]; **2**, 814[77]
Artmann, K., **4**, 282[133], 288[133]
Artsybasheva, Yu. P., **7**, 483[127]
Arumugam, A., **4**, 350[115]
Arumugam, N., **7**, 502[258]
Arunachalam, T., **4**, 339[42]
Arundale, E., **2**, 527[2], 528[2], 553[2]
Arvanaghi, A., **6**, 564[915]
Arvanaghi, M., **6**, 237[68], 251[146], 564[915], 938[127], 944[127]; **7**, 299[47], 760[47]; **8**, 319[73], 391[89], 406[48], 988[33]
Arvanitis, G. M., **1**, 807[316]
Arvia, A. J., **3**, 636[58]
Arvidsson, L.-E., **7**, 831[64]
Arwentiew, B., **2**, 146[69]
Arya, P. S., **8**, 881[73], 882[73]
Arya, V. P., **6**, 543[618]
Arzoumanian, H., **1**, 489[17,20]; **3**, 199[57]; **7**, 92[42], 93[42], 95[71,73a], 600[73], 601[73]; **8**, 444[8], 708[35], 716[35], 717[95], 726[95]
Asaad, A. N., **5**, 76[244]
Asaad, F. M., **2**, 150[97]
Asada, A., **6**, 443[92]
Asada, H., **4**, 561[29]; **6**, 284[176]
Asada, K., **4**, 587[27]; **7**, 763[92]; **8**, 390[84], 391[84]
Asada, M., **5**, 595[12]
Asada, S., **2**, 1066[119]
Asada, T., **3**, 355[54], 357[54]
Asahara, T., **6**, 498[164]
Asahi, Y., **8**, 975[133]
Asahina, Y., **3**, 810[47]; **8**, 328[10]
Asai, H., **8**, 650[64]
Asai, K., **2**, 1066[119]
Asai, M., **3**, 564[79], 574[79]
Asaka, M., **6**, 578[981], 626[167]
Asaka, Y., **8**, 544[266]
Asakawa, M., **1**, 558[136]; **7**, 231[138]
Asaki, Y., **8**, 992[55]
Asami, K., **1**, 180[32]
Asami, M., **1**, 64[47-49], 65[51], 69[60], 336[71,72]; **4**, 93[92], 207[56]; **8**, 159[8], 168[66-68], 178[68], 179[68], 764[5]
Asami, T., **2**, 5[18], 6[18], 24[18,18a]
Asano, H., **8**, 224[104]
Asano, K., **6**, 801[28]
Asano, O., **2**, 743[82]; **3**, 261[157]; **7**, 678[70]
Asano, R., **4**, 836[4,5], 837[13,14], 841[50]
Asano, T., **3**, 891[45]; **5**, 77[267], 453[63], 454[63], 458[63]; **8**, 190[80]
Asano, Y., **4**, 258[254]
Asanuma, M., **5**, 442[185]
Asao, N., **4**, 238[14], 247[106], 257[106], 260[106]
Asao, T., **4**, 115[180e], 852[89]
Asaoka, M., **1**, 880[114]; **2**, 617[144]; **3**, 20[116]; **4**, 91[88e], 121[207], 152[55], 211[94-96], 231[94], 239[19]; **7**, 673[27]; **8**, 518[132]
Asaoka, T., **3**, 295[8]
Asato, A. E., **6**, 157[176]
Asberom, T., **1**, 409[38]
Asbóth, B., **6**, 451[128]
Ascherl, B., **5**, 422[81]; **6**, 104[5], 115[81]
Asel, S. L., **7**, 874[109]

Asensio, G., **1**, 361[32a,b]; **2**, 790[57], 980[21], 981[21]; **4**, 290[196], 291[217,220], 311[446], 315[511], 347[93], 351[93c], 354[93d], 399[224]; **6**, 494[135]; **7**, 93[54], 486[142], 490[176], 501[255], 505[285], 536[52-55]; **8**, 854[150], 856[180], 857[198]
Ash, A. B., **6**, 622[135]; **7**, 656[16]
Ash, L., **3**, 99[186], 107[186]
Ashani, Y., **8**, 977[138]
Ashbrook, C. W., **3**, 934[65], 953[65]
Ashburn, G., **8**, 904[58]
Ashburn, S. P., **2**, 113[105]
Ashby, E. C., **1**, 67[58,59], 78[14-17], 79[14-16], 117[53,54], 162[103], 223[85], 224[85], 283[2], 315[2], 325[1], 329[37], 333[1,37,59]; **2**, 798[10,11]; **3**, 213[41,42b], 214[55,58], 824[19,20,23]; **4**, 140[9], 240[41,46,50,51], 257[220], 897[172]; **7**, 884[181]; **8**, 2[11], 3[15], 14[85], 22[147], 26[27], 37[27], 66[27], 86[23], 214[32], 315[53], 483[52,53,58], 484[52,53,58], 485[52,53,58], 541[207], 543[240], 545[283], 549[283,328], 696[120], 697[130,132], 801[73], 802[79-83], 803[88,89], 806[99], 872[10], 873[10]
Ashby, J., **8**, 30[43], 66[43]
Ashcraft, A. C., **3**, 23[139]
Ashcroft, A., **5**, 41[29]
Ashcroft, A. C., **7**, 100[118]
Ashcroft, A. E., **4**, 27[83]
Ashcroft, J., **5**, 736[145], 737[145]
Ashcroft, W. R., **8**, 587[42]
Ashe, A. J., III, **1**, 490[28]; **3**, 406[145], 905[138]
Asher, V., **6**, 237[69]
Ashida, T., **1**, 555[115]
Ashkenazi, P., **3**, 628[45]
Ashley, K. R., **7**, 709[47]
Ashley, W. C., **2**, 504[5]
Ashmore, J. W., **8**, 614[91]
Ashnagar, A., **7**, 345[4]
Ashton, R., **4**, 282[138]
Ashwell, M., **1**, 569[252]
Ashwood, M. S., **4**, 85[77c]
Asindraza, P., **8**, 49[113], 66[113]
Asinger, F., **2**, 364[206]; **4**, 297[278]; **6**, 430[100]; **7**, 579[130], 760[40]; **8**, 737[25], 739[34]
Asirvatham, E., **4**, 18[61,61a], 213[114,115], 215[114], 243[68], 247[101], 249[130], 257[130], 259[259], 262[101,130]; **6**, 150[123], 934[99], 1064[90a]; **7**, 206[66], 625[41,42], 627[42,43]; **8**, 995[68]
Askani, R., **5**, 686[41]; **7**, 748[116]
Asker, W., **3**, 563[117], 582[117]
Askerov, A. K., **3**, 304[60]
Askin, D., **1**, 343[107], 402[17], 799[296]; **2**, 665[31], 666[34,35], 668[31], 674[31], 675[35], 682[31], 700[34,35]; **3**, 45[247]; **5**, 434[144], 850[160]; **7**, 416[122]; **8**, 544[277]
Aslam, M., **1**, 786[252]; **2**, 74[76]; **3**, 878[92-95], 879[92-94,96], 880[94,98], 881[94]; **4**, 348[108], 349[108c], 359[159], 771[251]; **5**, 441[179]; **6**, 161[181,182], 934[101]; **8**, 842[44,44b], 844[44b], 845[44b], 846[44b], 847[44b]
Aslanian, R., **2**, 66[32]; **4**, 598[209], 599[215], 609[215], 624[215], 638[209], 641[215]; **6**, 86[96]
Asmus, K. D., **7**, 855[63]
Aso, K., **8**, 798[55]
Aso, M., **3**, 809[41]; **4**, 159[82]
Aso, Y., **4**, 343[74], 372[58], 397[58]; **6**, 18[65], 603[11,17], 608[11]; **7**, 492[183], 497[219], 657[22], 752[151], 761[61], 774[322]; **8**, 370[93]
Asoka, M., **4**, 258[239]
Asokan, C. V., **8**, 540[201]
Asrof-Ali, S., **1**, 410[43]
Assercq, J.-M., **1**, 826[59]; **2**, 542[81]; **4**, 952[5]
Assithianakis, P., **6**, 94[145]
Aster, S. D., **8**, 50[117], 66[117]
Astik, R. R., **5**, 664[38]
Astle, M. J., **2**, 345[21]
Astles, D. J., **8**, 875[35]
Aston, J. G., **3**, 842[18,20]
Astrab, D. P., **2**, 89[34], 109[71]; **5**, 774[172,173,174], 775[172], 780[173,174]
Astruc, D., **4**, 518[8], 521[43], 985[127]
Astudillo, M. E. A., **2**, 965[65]
Asveld, E. W. H., **3**, 865[28]; **7**, 169[110]
Atabekov, T., **8**, 535[160]
Atanesyan, K. A., **2**, 720[84]
Ateeg, H. S., **5**, 633[67]
Atherton, E., **6**, 638[38], 670[38], 671[38]
Atherton, F., **6**, 624[150]
Atkins, A. R., **8**, 852[141], 857[141]
Atkins, K. E., **4**, 589[77], 590[77], 591[77], 597[181], 598[181], 638[181]
Atkins, M. P., **3**, 296[15]; **4**, 307[395]; **6**, 673[290]
Atkins, R. K., **6**, 91[121]
Atkins, T. J., **6**, 581[990]; **8**, 85[19]
Atkinson, E. F. J., **2**, 848[210]
Atkinson, E. R., **2**, 757[12]
Atkinson, J. G., **1**, 821[28]; **3**, 289[67]; **4**, 1059[155]; **5**, 646[7]; **8**, 406[43]
Atkinson, R. S., **5**, 938[205]; **7**, 474[41], 480[100,101,103], 481[100,103,106-108,110,111], 482[100,115,116], 483[41,108,128,129], 742[55], 743[55], 744[55,66,67]
Atkinson, T., **8**, 206[172]
Atkinson, T. R. S., **6**, 116[90], 245[122], 256[122]
Atland, H. W., **7**, 185[173]
Atlani, P. M., **1**, 508[20]; **3**, 96[164], 98[164], 99[164]; **8**, 597[87]
Atlay, M. T., **7**, 108[174]
Atsumi, K., **2**, 567[30]; **5**, 833[46]
Atsuta, S., **8**, 55[180], 66[180]
Atsuumi, S., **6**, 487[40], 489[40]
Attah-Poku, S. K., **4**, 24[73,73d]; **5**, 944[239]; **8**, 114[56]
Attanasi, O., **2**, 345[24], 524[82,83]; **4**, 146[36]
Attar, A., **6**, 428[86]
Atta-ur-Rahman, **4**, 483[5], 484[5], 495[5], 1058[152], 1059[152]
Attenberger, P., **3**, 531[86]
Attia, M. E. M., **5**, 938[211]
Attinà, M., **4**, 445[203]
Atton, J. G., **4**, 688[65]
Attrill, R. P., **2**, 1059[74]; **5**, 407[28]
Attwood, S. V., **1**, 797[293]; **4**, 380[121]; **6**, 996[107]; **8**, 248[82]
Attygalle, A. B., **1**, 808[320]; **6**, 188[181]
Atwell, W. H., **5**, 950[294]
Atwood, J. D., **8**, 447[96]
Atwood, J. L., **1**, 6[32], 17[206,216], 19[101], 37[178,179], 162[99], 215[33]; **8**, 447[96,137], 677[61], 679[61], 682[61], 683[93], 685[61], 687[61]
Au, A. T., **1**, 544[34], 551[34], 553[34]; **4**, 113[169]; **5**, 157[41]
Au, K. G., **2**, 904[49]
Au, M.-K., **3**, 877[89]
Aubé, J., **1**, 838[161,167,168]; **2**, 506[17], 696[80]; **4**, 1076[43]; **7**, 750[132]
Aubert, C., **2**, 546[91]; **3**, 699[161]
Aubert, F., **3**, 416[17], 417[17]
Aubouet, J., **4**, 210[70]
Aubourg, S., **4**, 505[140]
Auburn, P. R., **4**, 564[43], 599[221], 624[221], 641[221], 653[444]; **5**, 1174[37]; **6**, 450[117]
Auchus, R. J., **5**, 913[100], 1008[43]
Auderset, P. C., **3**, 583[118]

Audia, J. E., **2**, 633[33b], 640[33]; **4**, 159[82], 186[92], 243[65], 253[65]; **5**, 517[29], 519[29], 534[29], 843[116,118]; **7**, 273[135]
Audia, V. H., **3**, 226[197]; **5**, 934[187]; **6**, 10[44], 11[44], 12[44]; **7**, 416[121b]
Audin, A., **4**, 395[206]
Audin, C., **4**, 792[62]
Audin, P., **4**, 308[406,409], 311[452], 396[209], 397[209]
Audouin, M., **3**, 738[36]
Aue, D. H., **5**, 482[170], 805[100]; **7**, 477[75]
Auer, E., **6**, 914[26]; **7**, 222[40]
Auerbach, J., **6**, 647[113]
Auerbach, R. A., **2**, 128[242], 184[23,25,25b], 835[158]
Aufderhaar, E., **4**, 27[83]
Augé, C., **2**, 463[90], 464[90,95,95a,b,96], 467[90]; **6**, 662[217]
Augé, J., **1**, 831[101,102]; **2**, 6[32], 19[32b], 20[79]
Augelmann, G., **5**, 418[67]; **8**, 395[129]
Auger, J., **4**, 883[99], 884[99]
Augood, D. R., **3**, 505[162], 507[162], 512[162]
Augur, M. V., **8**, 328[13], 329[13]
August, B., **1**, 878[108]; **4**, 1005[84]; **8**, 528[55]
Augustin, M., **4**, 53[149]
Augustine, R. L., **2**, 734[5], 797[8], 808[50], 829[8], 835[8], 861[8]; **3**, 1[1], 3[1], 8[1], 11[1], 13[1], 17[1], 18[1], 21[1], 23[1], 25[1], 59[1]; **4**, 7[24]; **6**, 177[110], 181[110], 182[110], 184[110], 185[110], 200[110], 201[110]; **7**, 92[41,41a], 94[41], 96[89], 97[89], 235[10], 237[38], 542[6], 543[6], 671[3], 674[3]; **8**, 139[3], 212[8,19], 214[34], 328[6], 338[6], 339[6], 340[6], 341[6], 342[6], 343[6], 524[9], 530[9], 531[9], 533[145,151,152,154], 534[151,152], 863[238], 991[48]
Augustine, R. N., **8**, 307[3], 309[3]
Auksi, H., **6**, 926[66]; **7**, 196[11], 199[11]
Aul'chenko, I. S., **6**, 836[53]
Aumann, R., **4**, 673[31]; **5**, 1037[4], 1076[47], 1107[165,166], 1109[173,174,175,176,177,178,179,180], 1132[18]; **6**, 724[153]
Aunmiller, J. C., **2**, 166[153]; **3**, 414[5], 422[5]
Aurbach, D., **6**, 533[506]
Aurell, C.-J., **5**, 686[53], 688[68]
Auret, B. J., **7**, 78[125]
Auricchio, S., **8**, 645[42]
Aurich, H. G., **7**, 745[74]
Aurora, R., **7**, 854[57], 855[57]
Aurrecoechea, J. M., **6**, 14[57], 84[86]
Austin, E., **4**, 468[135]
Austin, G. N., **5**, 151[11]
Austin, W. B., **3**, 530[69], 533[69], 537[69]
Autze, V., **8**, 545[285]
Auvray, P., **1**, 189[75], 219[57], 220[57b]; **2**, 6[34], 20[34c], 23[34c]; **4**, 34[97], 35[97]; **6**, 164[197,198]
Auwers, K., **2**, 773[21], 957[19]; **4**, 2[3]
Au-Yeung, B.-W., **2**, 716[65]; **7**, 616[12,20]
Au-Young, Y.-K., **5**, 418[69]; **7**, 797[18]
Avaro, M., **2**, 523[79], 524[79]
Avasthi, K., **5**, 418[67]; **8**, 354[161]
Averko-Antonovich, I. G., **6**, 212[82]
Avery, M. A., **4**, 373[75], 1040[98], 1043[98]; **5**, 841[97]; **6**, 7[32]; **8**, 798[58], 856[174]
Avetisyan, A. A., **4**, 315[505]
Avila, L., **6**, 579[984]
Avila, W. B., **1**, 468[54]; **4**, 14[47,47m]
Aviron-Violet, P., **8**, 462[264]
Avnir, D., **2**, 766[84]
Avram, M., **4**, 600[234], 643[234], 963[44]
Avramenko, V. I., **2**, 787[52]; **4**, 48[138,138f]
Avramoff, M., **8**, 568[485]
Avramovíc, O., **8**, 79[1], 82[1b]
Avrutskaya, I. A., **3**, 639[80]
Awad, S. B., **7**, 162[62], 778[398]
Awad, W. I., **2**, 757[10]; **4**, 84[69]
Awano, H., **8**, 408[68], 589[51]
Awasthi, A., **1**, 835[135]
Awasthi, A. K., **1**, 835[135]; **5**, 95[100]; **7**, 145[160,161]
Awata, T., **2**, 105[42]; **4**, 439[163]
Axelrod, E. H., **3**, 201[74], 597[197]; **6**, 431[110]
Axelrod, J., **7**, 11[89]
Axén, R., **2**, 1104[132]
Axenrod, T., **8**, 271[107]
Axiotis, G., **1**, 202[99], 425[108]; **2**, 630[24], 631[24], 632[24], 930[130], 931[130]; **5**, 102[180]
Aya, T., **2**, 443[17]; **4**, 973[85]
Ayala, A. D., **4**, 770[247]
Aycard, J.-P., **3**, 892[47]
Ayer, D. E., **5**, 646[7]; **7**, 393[16], 398[16]
Ayer, W. A., **5**, 637[112]; **6**, 677[314], 1054[47], 1066[93]; **8**, 333[55], 397[141], 880[64]
Aylward, F., **8**, 472[5]
Aylward, J. B., **7**, 231[145,146]
Ayrey, G., **7**, 616[10], 620[10], 769[242], 771[242], 773[242]
Ayrey, P. M., **1**, 778[220], 782[233]
Ayyad, S. N., **4**, 45[126]
Ayyangar, N. R., **7**, 601[80]; **8**, 366[44], 367[61], 716[88]
Ayyaswami, N., **8**, 137[56-58]
Azadi-Ardakani, M., **5**, 692[94]
Azami, H., **4**, 858[110]
Azerad, R., **8**, 185[28], 187[28], 188[53], 196[114]
Azhaev, A. V., **6**, 450[121]
Azimov, V. A., **6**, 554[770,771]
Aziz, M., **7**, 406[86]
Azman, A., **1**, 837[156]
Aznar, F., **4**, 291[219], 292[224,225,228], 311[453]; **5**, 410[42], 411[42], 433[136b], 848[141], 850[162]; **6**, 555[810,811,812]; **7**, 486[141]; **8**, 856[177,178], 857[196]
Azogu, C. I., **4**, 529[75], 530[75]
Azran, J., **8**, 535[163]
Azuara, S. J., **3**, 781[12]
Azuma, H., **3**, 1041[111]
Azuma, K., **3**, 984[22], 985[22], 987[31]
Azuma, K. I., **6**, 851[129]
Azuma, K.-I., **6**, 876[31], 877[31], 882[31], 885[31], 887[61]
Azuma, S., **7**, 168[101]
Azuma, T., **6**, 240[79]; **8**, 917[116,117], 920[116,117]
Azuma, Y., **3**, 232[267], 510[185]
Azumi, N., **3**, 300[42]
Azzaro, M., **1**, 294[48]; **6**, 769[30]

B

Baake, H., **3**, 872[59]
Baan, J. L. v. d., **5**, 850[147]
Baardman, F., **3**, 302[49]; **5**, 9[70,71]; **6**, 809[65]
Baasner, B., **7**, 498[224]
Baay, Y. L., **8**, 773[66]
Baba, A., **5**, 151[18], 1200[55]; **6**, 89[120]; **8**, 20[137]
Baba, H., **3**, 95[156], 97[156]; **7**, 800[34]
Baba, N., **2**, 232[174]; **8**, 170[74], 562[423]
Baba, S., **3**, 230[239], 232[239a], 238[239a], 266[195], 274[24], 486[40,41], 495[40,91], 497[40], 498[40], 503[40], 530[54]; **4**, 587[30]; **5**, 1166[25]; **8**, 680[72], 683[72], 693[72], 755[116,120,125], 757[125], 758[125]
Baba, Y., **5**, 603[48], 608[65]; **8**, 266[57], 756[151]
Babad, H., **8**, 297[65]
Babayan, A. T., **5**, 410[40], 435[151]
Babb, R. M., **4**, 310[425]
Babcock, J. C., **8**, 528[80]
Babiak, K. A., **3**, 209[17]; **4**, 175[43]
Babiarz, J. E., **5**, 476[147]
Babichenko, L. N., **2**, 787[52]
Babin, E. P., **3**, 304[69]
Babin, P., **5**, 527[59]; **6**, 185[162]
Babine, R. E., **7**, 416[122]
Babirad, S. A., **1**, 188[95], 198[95]; **2**, 579[94]
Babler, J. H., **4**, 247[104], 259[104], 993[159a]; **6**, 835[48], 1022[68], 1054[49]; **7**, 228[98], 655[20]; **8**, 263[33,34]
Babot, O., **2**, 900[30], 901[30], 964[62]
Babston, R. E., **2**, 655[131]; **5**, 841[104,106,107], 856[208], 857[222], 872[222]
Babu, G. P., **3**, 331[195]
Babu, S., **4**, 846[75], 847[75]
Babudri, F., **2**, 87[27], 211[117], 213[127]; **3**, 461[145]
Babychev, F. S., **6**, 509[283]
Bac, N. V., **1**, 419[82]
Bacaloglu, I., **6**, 244[110]
Bacaloglu, R., **6**, 244[110]
Baccar, B., **2**, 286[65], 287[65]
Bacchetti, T., **3**, 871[55]
Baccolini, G., **4**, 86[79a], 146[36]
Baccouche, M., **7**, 95[73a]
Bach, G., **8**, 530[90]
Bach, J., **8**, 392[100], 905[61]
Bach, R. D., **1**, 2[9], 43[9], 476[126]; **3**, 66[15], 74[15], 748[79], 749[80,81]; **4**, 314[479], 315[514], 868[19]; **6**, 953[8], 1012[7], 1013[7,11]; **7**, 340[49]; **8**, 851[126], 858[126]
Bach, T. G., **7**, 132[89,93], 134[93]
Bacha, J. D., **7**, 720[12]
Bachand, B., **8**, 98[102]
Bachand, C., **1**, 123[78]; **2**, 212[120], 656[158], 1059[77]; **6**, 923[58]
Bachechi, F., **4**, 403[239], 404[239]
Bachelor, F. W., **2**, 412[10], 413[10], 414[10]; **7**, 528[8]
Bachhawat, J. M., **7**, 446[61]
Bachhuber, H., **8**, 472[8,9]
Bachi, M. D., **2**, 887[53]; **4**, 795[82,85], 798[108,109]; **5**, 92[65,71,73], 95[73]; **6**, 538[553]; **8**, 502[65], 503[65], 656[91], 796[29], 823[58]
Bachman, G. B., **2**, 744[93], 965[66]; **6**, 825[128]; **7**, 729[43]
Bachman, G. L., **8**, 459[232], 460[232], 535[166]
Bachman, P. L., **8**, 542[235]
Bachmann, G., **6**, 238[73]
Bachmann, J., **2**, 372[269], 373[269]
Bachmann, M., **5**, 497[226]
Bachmann, P., **4**, 283[150]
Bachmann, W. E., **1**, 844[3a]; **2**, 284[58], 838[175], 1090[61]; **3**, 505[161], 507[161], 564[12], 579[96], 724[13], 887[2], 888[2], 891[2], 897[2]; **4**, 41[119,119b]; **5**, 491[204], 552[13]; **8**, 308[4]
Bachrach, A., **8**, 36[49], 66[49]
Bachstadter, G., **1**, 29[145]; **2**, 508[29]; **4**, 104[137]
Baciocchi, E., **4**, 763[209]; **6**, 958[27,33]; **7**, 649[46], 765[151]
Back, M. H., **5**, 64[22], 69[103,104]
Back, R. A., **8**, 472[10]
Back, T. G., **3**, 86[53], 94[53], 114[53], 117[53]; **4**, 176[52], 339[39], 341[59], 342[39], 771[252]; **6**, 438[52], 467[47-49], 469[55,56], 745[85,86], 981[47], 1028[96]; **7**, 110[188], 519[23], 523[44], 741[51], 742[59], 771[280], 773[280], 779[422]; **8**, 410[98], 848[103], 936[71], 996[72]
Backeberg, O. G., **8**, 298[76,77]
Backer, H. J., **1**, 851[35]; **6**, 120[123]
Backinovsky, I. V., **6**, 43[52], 49[100]
Bäcklund, B., **3**, 861[1]
Backlund, S. J., **2**, 83[11]
Bäckstrom, P., **5**, 210[60]
Bäckvall, J.-E., **2**, 66[32]; **3**, 222[139]; **4**, 5[17], 290[201], 302[331], 343[72,73], 399[225,225b], 560[21,23,24], 565[44,45], 591[115], 592[115], 596[164], 597[168,169,170,171], 598[183,185,187,190,195,197,209], 599[215], 609[215], 617[115], 618[115], 621[164,168,169,170], 622[168,185,195], 623[190,197], 624[215], 633[115], 637[169], 638[183,185,190,195,209], 639[197], 641[215]; **5**, 333[46], 935[193], 936[193]; **6**, 85[89,91,93,94], 86[93,94,96], 831[9], 849[116]; **7**, 94[55,57], 438[11], 441[11], 443[11], 452[40], 474[43-45], 484[137], 490[174,175], 504[282], 519[23], 527[2], 528[2,3], 530[2]
Bacon, C. C., **7**, 778[394]
Bacon, E. R., **2**, 544[85], 547[85], 552[85]
Bacon, R. G. R., **2**, 757[14]; **3**, 499[110,112], 501[110], 505[110], 509[110], 512[110], 664[31]; **6**, 208[62], 269[73]; **7**, 228[104]
Bacon, W. E., **4**, 317[542,543]
Bacos, D., **8**, 35[65], 47[65], 66[65]
Bacquet, C., **1**, 873[92,95b]; **3**, 639[73], 651[73b]; **6**, 46[72], 47[72]
Badami, P. S., **6**, 771[43]
Badanyan, Sh. O., **2**, 720[84]; **4**, 304[362], 305[362], 315[525]
Badaoui, E., **1**, 631[54,60], 632[60], 633[60], 634[60], 635[60], 658[54,60], 659[54], 660[54], 661[54,60,167,167a,c], 662[54,60], 663[54,60], 664[54], 672[54], 705[60]; **3**, 87[72,73], 95[72]
Baddeley, G., **2**, 740[61], 745[101], 756[6], 760[6]; **5**, 766[113], 777[113]
Baddiley, J., **6**, 614[89]
Badding, V. G., **8**, 214[36], 217[36], 218[70], 219[70], 229[136]
Bade, T. R., **2**, 425[38]
Badejo, I., **1**, 522[78]; **4**, 12[37]; **5**, 1053[41], 1061[41]
Bader, A., **5**, 787[7]
Bader, A. R., **3**, 828[51]
Bader, F. E., **2**, 1022[53]; **4**, 373[68]; **5**, 341[61a]
Bader, J. M., **6**, 959[44]; **8**, 205[158,164], 561[409]
Badertscher, R., **2**, 219[144]
Badertscher, U., **2**, 194[68], 205[102,103], 206[102b,103], 219[68]
Badet, B., **3**, 159[456,461], 161[456,461], 162[456], 165[461]; **6**, 157[167]
Badger, G. M., **8**, 364[21], 568[466], 950[159]
Badger, R. A., **3**, 20[107]
Badger, R. C., **6**, 953[8]
Badorc, A., **2**, 765[74]
Badoud, R., **5**, 841[86]; **6**, 859[164], 978[21]
Badr, M. Z. A., **2**, 772[15]

Bae, S. K., **7**, 318[56], 319[56]
Baechler, R. D., **8**, 404[20]
Baeckström, P., **5**, 839[77]; **7**, 822[34]; **8**, 163[36]
Baenziger, N. C., **8**, 336[68]
Baer, D. R., **1**, 846[15], 847[15], 848[15], 850[15]; **3**, 781[11]
Baer, H. H., **2**, 321[9], 325[9], 326[9], 327[9], 328[9], 329[9], 354[109]; **4**, 598[208], 622[208]; **5**, 552[20]; **6**, 36[24]; **8**, 52[148], 66[148], 874[27], 875[35], 882[86]
Baer, J. E., **2**, 971[92]
Baer, T. A., **3**, 416[14], 417[14]
Baerends, E. J., **8**, 670[10], 671[10]
Baert, F., **1**, 34[227]
Baertschi, S. W., **5**, 781[206,208]
Baese, H. J., **6**, 244[111]; **8**, 912[91]
Baettig, K., **1**, 766[157]; **3**, 380[10]; **4**, 21[69], 107[141], 111[153]; **5**, 45[36], 532[86]
Baeyer, A., **3**, 828[42]
Baeza, J., **5**, 86[21]
Bag, A. K., **7**, 760[26]
Bagal, L. I., **6**, 795[12], 798[12], 817[12], 820[12]
Bagby, B., **4**, 956[20]
Baggaley, A. J., **7**, 794[7b]
Baggaley, K. H., **3**, 383[46]; **4**, 6[23]
Baggett, N., **6**, 2[2], 23[2]; **8**, 162[28]
Baggiolini, E. G., **1**, 822[34]; **4**, 370[29], 413[277]; **5**, 218[30], 221[30], 256[58], 257[58b], 433[137b]; **6**, 960[56]; **7**, 268[122], 564[92], 567[92]
Baghdanov, V. M., **4**, 14[47,47a]; **6**, 105[13], 944[158]
Bagheri, K., **6**, 736[30]
Bagheri, V., **1**, 95[80], 448[206]; **2**, 449[48]; **3**, 251[78], 254[78], 463[156], 920[34], 925[34a], 934[34]; **4**, 892[145], 903[198], 965[51], 1036[49,50,53]; **6**, 127[161], 873[9], 874[13], 883[13], 897[9b]; **7**, 315[43]
Bagley, E., **7**, 488[163]
Bagli, J., **6**, 941[151]
Bagnell, L., **4**, 140[9]; **8**, 316[58]
Bagotzky, V. S., **3**, 647[177], 648[177], 649[177]
Bahadori, S., **5**, 3[22]
Bahl, C. P., **6**, 625[158]
Bahr, J., **6**, 637[32,32c]
Bähr, K., **4**, 866[4]
Bahrmann, H., **3**, 1017[6]; **4**, 918[17]
Bahsas, A., **2**, 1002[57]; **5**, 408[35], 500[261], 552[40]
Bahsoun, A. A., **8**, 445[43]
Bahurel, Y., **5**, 1007[36]
Bai, D. L., **1**, 824[45]; **2**, 205[104], 206[104]; **3**, 225[187]
Baier, H., **5**, 687[58]
Baigrie, B., **4**, 487[44]
Baigrie, L. M., **1**, 418[73]; **2**, 107[59,60], 108[60], 196[76]
Bailar, J. C., Jr., **8**, 447[110-113], 453[111]
Bailey, A. R., **5**, 680[24,24a,b]
Bailey, A. S., **7**, 158[42]
Bailey, D. L., **8**, 556[376], 778[87]
Bailey, D. M., **2**, 739[43a]
Bailey, D. T., **8**, 968[86]
Bailey, E. J., **3**, 804[13], 816[79]; **7**, 159[45]
Bailey, G. C., **5**, 1116[6]
Bailey, G. M., **5**, 337[71b], 345[71b], 346[71b]
Bailey, J. R., **8**, 143[64]
Bailey, M. D., **4**, 121[208]
Bailey, N. A., **1**, 310[109]
Bailey, P. D., **6**, 737[40], 738[44,45], 739[60,61], 861[184]
Bailey, P. S., **2**, 283[51]; **4**, 1075[32], 1098[172,174]; **7**, 542[8], 543[8], 544[36], 581[141], 737[8]; **8**, 398[147], 964[50]
Bailey, R. A., **2**, 1088[47]
Bailey, R. J., **6**, 121[132]
Bailey, S. J., **5**, 514[7]
Bailey, T. R., **1**, 461[11], 473[11], 477[128], 478[11], 482[128], 584[12]; **2**, 713[47]; **3**, 69[25], 70[25], 72[25], 74[25], 77[55,58], 232[267], 514[209]; **5**, 414[54], 531[82], 539[108], 757[75,76], 778[75,76]; **6**, 780[74]; **8**, 946[140]
Bailey, W. F., **3**, 419[29,30], 422[29,30]; **4**, 871[32-34]; **8**, 224[111]
Bailie, J. C., **8**, 564[444]
Baillarge, M., **4**, 502[120]; **5**, 941[228]
Baillargeon, D. J., **1**, 227[98], 880[116]; **4**, 98[110]; **5**, 788[14], 847[133], 872[133], 1000[3,7]; **6**, 834[39]
Baillargeon, V. P., **3**, 1021[15]
Bain, J. P., **2**, 529[18]
Baine, N. H., **4**, 796[90,93]; **5**, 742[161]
Baines, D., **3**, 386[68], 399[118]
Baines, D. A., **7**, 92[44]
Baines, K. M., **4**, 760[196]
Bainton, H. P., **3**, 247[48]
Bair, K. W., **4**, 868[19]; **6**, 1013[11]
Baird, G. J., **2**, 125[219]; **3**, 47[257]; **4**, 82[62a]
Baird, M. C., **1**, 440[169], 451[216]; **8**, 674[32]
Baird, M. S., **3**, 383[48]; **4**, 284[154], 951[1], 968[1], 979[1], 1001[32], 1009[143], 1012[171,172], 1014[186,187,188,189,190], 1015[201], 1022[188,189], 1050[123]; **5**, 794[44], 947[266]; **7**, 825[43]
Baird, W. C., **7**, 530[27], 531[27]; **8**, 475[17]
Bairgrie, L. M., **1**, 418[74]
Baisheva, A. U., **7**, 767[190]
Baitz-Gács, E., **2**, 812[72], 851[223]; **7**, 746[93]
Baizer, M. M., **1**, 268[54]; **2**, 105[42]; **3**, 214[56], 564[9], 567[9], 634[13,20], 638[20], 649[20], 655[20]; **4**, 12[41], 439[163], 726[51], 809[164]; **6**, 836[58], 939[136], 942[136]; **7**, 810[88]; **8**, 129[1], 131[1], 134[1], 527[51], 532[132,132d]
Bajgrowicz, J. A., **3**, 46[254], 215[66], 251[75]
Bajorek, J. J. S., **5**, 515[14], 518[14]
Bajt, O., **6**, 514[306], 543[306]
Bajwa, J. S., **5**, 558[59]
Bak, T., **5**, 1138[65]
Bakal, Y., **4**, 710[61,62], 712[65,66,68]
Bakale, R. P., **6**, 1066[97]
Bakalova, G. V., **8**, 318[61]
Baker, A. D., **7**, 762[77], 852[43]
Baker, A. J., **8**, 500[47]
Baker, B., **4**, 845[65], 848[78]
Baker, B. J., **7**, 406[75]
Baker, B. R., **7**, 294[14]
Baker, B. W., **3**, 643[117], 644[150]
Baker, C. S. L., **6**, 213[88]
Baker, E. B., **8**, 754[82]
Baker, F. S., **4**, 1021[249]
Baker, F. W., **7**, 724[28]
Baker, G., **4**, 436[145], 437[145]
Baker, J. D., Jr., **8**, 237[13], 240[13], 244[13], 253[13], 343[118], 344[118], 356[118,187], 357[118,192,193]
Baker, J. P., **5**, 516[28]
Baker, J. W., **2**, 363[189], 530[21]; **7**, 666[70]
Baker, K. V., **1**, 808[321]
Baker, L. M., **7**, 571[120,121], 575[120,121], 576[120,121]
Baker, P., **4**, 603[266]
Baker, P. B., **7**, 58[56], 62[51,56], 63[56]
Baker, P. M., **5**, 716[86]
Baker, R., **1**, 130[94,95], 764[148], 801[301], 808[326]; **3**, 416[22], 423[76]; **4**, 578[19-21], 579[22], 586[4], 590[4], 603[265], 644[265]; **5**, 310[99], 883[16]; **6**, 5[27], 750[105], 994[98]; **7**, 400[42], 406[71]; **8**, 10[56]
Baker, R. H., **2**, 1085[22]; **8**, 425[44,45], 426[44,45]

Baker, R. J., **1**, 60[36], 75[36], 468[55]
Baker, S. R., **1**, 758[126]; **3**, 232[264]; **7**, 401[58], 712[63]
Baker, W., **3**, 382[36]
Baker, W. R., **1**, 142[28]; **5**, 133[58], 342[61b]; **6**, 1063[82]
Bakhmutov, V. I., **4**, 218[147,148]
Bakhmutskaya, S. S., **5**, 752[18,26], 757[18], 759[26], 767[26]
Bakker, B. H., **5**, 253[46,46c]; **8**, 61[189], 66[189]
Bakker, C. G., **5**, 71[160], 459[91]; **6**, 558[850,855,856]
Baklan, V. F., **3**, 383[47]
Bakos, J., **8**, 459[236], 535[166]
Bakos, V., **6**, 902[127]
Bakovetila, M., **7**, 842[21]
Bakshi, R. K., **1**, 317[143,144], 319[144]; **2**, 247[82]; **3**, 797[92,93]; **4**, 798[110], 1040[72], 1044[72]; **6**, 837[60]; **7**, 376[85], 595[127], 604[127]; **8**, 160[102], 171[102-104], 178[102], 179[102], 710[58], 722[146,148]
Bakthavachalam, V., **5**, 1151[131]; **6**, 533[477]
Bakulev, V. A., **6**, 538[551,555]
Bakunin, V. N., **4**, 610[341]
Bakuzis, M. L. F., **2**, 332[60]; **4**, 104[136b], 1058[149]; **6**, 931[92]; **7**, 120[6], 204[56]
Bakuzis, P., **2**, 332[60]; **3**, 857[91]; **4**, 104[136b]; **6**, 931[92], 1042[6]; **7**, 120[6], 204[56]
Bal, B., **2**, 193[63]
Bal, B. S., **1**, 551[69]; **5**, 942[233]; **7**, 240[57]
Bal, S. A., **1**, 415[64]; **4**, 91[88f]; **5**, 522[47]
Balaban, A. T., **2**, 708[2], 710[2], 711[2], 712[40], 727[2], 744[99], 745[99]; **3**, 331[196]
Balabane, M., **4**, 598[183], 638[183]; **6**, 85[93], 86[93], 849[115]
Balachander, N., **8**, 64[215], 394[118]
Balaji, T., **4**, 988[140]
Balakrishnan, P., **4**, 486[37], 505[37,146]; **7**, 131[84]
Balani, S. K., **7**, 78[125]
Balanikas, G., **7**, 350[22]
Balanson, R. D., **3**, 135[340], 137[340], 139[340], 141[340]; **6**, 76[42]; **8**, 384[24]
Balaram Gupta, B. G., **6**, 270[79]; **8**, 405[25,32], 406[32,51], 407[25]
Balashova, T. M., **6**, 490[106]
Balasubrahmanyam, S. N., **2**, 842[191]
Balasubramanian, K., **4**, 4[14]; **7**, 162[61], 277[156]; **8**, 52[147], 66[147], 625[165]
Balasubramanian, K. K., **2**, 782[24]; **3**, 17[88]; **5**, 834[54]
Balasubramanian, M., **2**, 842[191]
Balasubramanian, N., **6**, 816[98]
Balasubramanian, R., **3**, 380[7]
Balasubramanian, T. R., **7**, 267[117], 268[117]
Balasubranian, T. M., **6**, 546[652]
Balasuryia, A., **6**, 1056[54]
Balavoine, G., **4**, 590[91], 605[294,296,297,298], 647[297]; **5**, 289[40], 290[40]; **6**, 641[58]; **7**, 14[142], 238[40]; **8**, 21[143], 553[362]
Balbi, A., **6**, 487[47,48,51-53], 489[47,48,52,53], 543[47,48]
Balcells, J., **8**, 170[73]
Balcells, M., **5**, 72[187]
Balch, A. L., **6**, 533[510]
Balchunis, R. J., **7**, 125[55], 126[55]
Balci, M., **5**, 714[74], 736[140]; **8**, 396[139]
Balcioglu, N., **5**, 810[127]
Balczewski, P., **1**, 788[259], **4**, 113[166]
Bald, E., **5**, 850[146]; **8**, 166[63]
Baldassarre, A., **2**, 588[150]; **3**, 251[79], 254[79]; **7**, 172[128]
Baldauf, C., **2**, 360[167]
Baldauf, H.-J., **6**, 20[71], 605[35]
Balderman, D., **8**, 384[26]
Baldinger, H., **7**, 746[92], 752[92]; **8**, 829[82]
Baldoli, C., **4**, 522[54], 524[61]; **6**, 114[79], 178[121]; **8**, 185[27]
Baldwin, J., **4**, 822[226]
Baldwin, J. E., **1**, 133[109], 477[144,145,146], 512[36], 545[46-48], 621[69], 894[159]; **2**, 165[151], 523[75], 547[117], 551[117], 555[147,148], 596[1], 807[46], 1007[2], 1008[2], 1024[2], 1034[2]; **3**, 18[98], 252[81,82], 257[81,82], 382[34], 640[108], 647[108], 711[23], 871[52], 916[17,18], 932[17], 934[65], 939[77], 946[88], 953[65], 963[119], 990[34]; **4**, 27[83], 37[107,107a,b,e], 38[107e], 39[107c,e], 74[37], 83[65d], 111[152e], 113[170], 249[116], 367[10], 375[97], 735[85], 744[134,136], 745[140], 746[142], 782[10], 805[143], 1039[63], 1076[42]; **5**, 64[31], 167[94], 404[17], 419[74], 438[161], 702[11,14], 703[14], 705[23], 738[149], 740[14], 804[96], 826[159b], 829[12], 907[78], 908[78], 918[78], 973[10], 985[35], 1001[13], 1007[37]; **6**, 96[152], 198[236], 742[68], 750[68], 783[85], 915[30]; **7**, 95[70,70a], 168[105], 231[153,154], 405[68], 501[255], 545[28], 630[53,54]; **8**, 253[117], 387[56], 395[129], 542[224], 940[107]
Baldwin, J. F., **7**, 43[37]
Baldwin, J. J., **2**, 756[8]; **5**, 410[41]; **6**, 526[395]
Baldwin, M. J., **4**, 331[6]
Baldwin, S., **3**, 1058[40]
Baldwin, S. W., **1**, 131[100], 853[45]; **4**, 4[14,14a], 83[65a,b], 1076[43]; **5**, 123[1], 125[19], 126[1], 128[19,30], 129[34,36], 132[49], 134[30]; **7**, 43[35,36], 111[190,191]; **8**, 246[79], 248[87], 824[64]
Balenkova, E. S., **2**, 709[15]; **4**, 337[37,38]
Balenovic, K., **7**, 657[33], 777[366]
Balerna, M., **8**, 47[126], 66[126]
Bales, S. H., **4**, 276[69,70], 283[70]
Balestra, M., **1**, 131[98,99]; **5**, 894[42]; **6**, 859[174], 887[63]
Balf, R. J., **3**, 396[103]
Balfe, M. P., **7**, 775[341], 776[341]
Balgobin, N., **6**, 625[157], 659[191,197b]
Balicki, R., **6**, 533[478]; **8**, 390[85], 391[85]
Balk, M. A., **2**, 740[63a]
Balkeen, W. G., **4**, 520[36], 531[36]
Balko, T. W., **4**, 402[235], 404[235], 408[235]; **7**, 503[273]
Balkovec, J. M., **2**, 186[36]; **5**, 311[104], 563[89]; **6**, 888[65]
Ball, H., **2**, 740[57]
Ball, J.-A. H., **8**, 333[55]
Ball, S. S., **7**, 763[91], 769[91]
Ball, T. F., **2**, 553[131]
Ball, T. J., **7**, 384[115]
Ball, W. A., **1**, 773[203,203b]
Ballabio, M., **8**, 358[197]
Ballantine, J. A., **4**, 313[466,467]
Ballard, D. A., **3**, 825[27b]
Ballard, D. H., **4**, 306[374]
Ballenegger, M., **3**, 822[12], 831[12], 835[12b]
Ballester, M., **2**, 409[2], 410[2], 411[2,2b,8]
Ballester, P., **4**, 111[155d]
Ballesteros, A., **5**, 161[64]; **6**, 757[134]
Ballesteros, P., **2**, 353[100]; **4**, 439[160]
Balli, H., **2**, 740[56]
Ballini, R., **2**, 321[19], 323[19,38,39], 324[19], 330[19,51], 331[19], 332[19,51], 333[19,51], 598[12]; **4**, 13[44]; **6**, 104[10], 938[128], 942[128], 944[128]; **7**, 262[77]
Ballou, C. E., **2**, 456[73], 458[73]
Balls, D. M., **8**, 340[97]
Bally, C., **2**, 456[64]
Bally, I., **2**, 744[99], 745[99]
Bally, T., **5**, 704[21]
Balme, G., **2**, 106[52]; **4**, 308[408]; **5**, 797[66]; **6**, 960[59]; **7**, 453[63], 455[63]
Balog, I. M., **4**, 342[67]
Balog, M., **4**, 877[71], 905[211]
Balogh, A., **3**, 334[222]
Balogh, D., **4**, 373[67]; **6**, 1063[81]
Balogh, D. W., **3**, 380[13], 623[32], 626[32b]; **5**, 348[73b]

Balogh, M., **1**, 564[187]; **3**, 125[303]; **7**, 846[97,98]
Balogh, V., **3**, 664[28], 698[28]
Balogh-Hergovich, E., **7**, 532[30]; **8**, 447[99]
Balogh-Nair, V., **2**, 482[34], 484[34]
Bal'on, Ya. G., **5**, 425[102,103]; **6**, 577[976]
Bal Reddy, K., **2**, 762[57,58]
Balsamini, C., **2**, 904[53]
Balsamo, A., **2**, 284[53], 291[71]
Balschukat, D., **4**, 1017[215]
Balsiger, R. W., **6**, 555[816]
Balsubramaniyan, V., **6**, 569[940]
Baltazzi, E., **2**, 376[280], 396[6], 402[6,31], 403[6], 406[6b]
Baltes, H., **7**, 795[9]
Baltorowicz, M., **2**, 354[116]
Balu, M. P., **2**, 495[62], 496[62]
Balvert-Geers, I. C., **6**, 638[41]
Baltzinger, M., **8**, 54[154], 66[154]
Baltzly, R., **8**, 904[56], 907[56], 956[2]
Balzarini, J., **7**, 350[25], 355[25]; **8**, 679[66], 680[66], 681[66], 683[66], 694[66]
Bamberger, E., **3**, 806[16], 812[53]
Bamfield, P., **2**, 770[5]; **3**, 501[138]
Bamford, C. H., **2**, 735[11]; **8**, 850[121]
Bamford, W. R., **6**, 961[69]
Ban, N., **1**, 739[38]
Ban, T., **5**, 634[71,72,78], 819[156]
Ban, Y., **2**, 1022[52]; **3**, 1032[67], 1036[82], 1037[90], 1038[90,95,95b]; **4**, 30[87], 102[129], 127[220b], 680[50], 803[132], 843[53-55], 846[74], 852[53]; **5**, 687[57], 808[109]; **6**, 8[38]; **7**, 175[142], 353[35], 355[35]; **8**, 244[57], 249[97], 253[97], 620[132], 844[75]
Banach, T. E., **7**, 749[118], 854[56], 855[56], 882[167]
Banah, M., **8**, 600[103]
Banait, N., **4**, 300[309]
Banaszak, L. J., **8**, 206[172]
Banaszek, A., **6**, 77[54]
Banba, Y., **5**, 829[22]; **6**, 509[266]
Bancel, S., **5**, 851[168,169]
Bancin, A., **4**, 1099[177]
Bancìu, M., **6**, 489[85]
Band, E., **8**, 458[223,223d]
Bandara, B. M. R., **4**, 670[28], 683[59], 687[59,63,64]
Bandaranayake, W. M., **4**, 761[204]
Bando, K., **4**, 939[76]
Bando, T., **4**, 402[236]; **7**, 503[273]
Bandodakar, B. S., **3**, 380[4]
Bandurco, V. T., **8**, 365[27]
Bandy, J. A., **2**, 125[219]; **3**, 47[257]; **7**, 4[18]
Banerjee, A. K., **3**, 594[187]; **8**, 312[19]
Banerjee, D. K., **2**, 148[78]; **8**, 494[24], 526[17], 568[467]
Banerjee, P. C., **8**, 371[101]
Banerjee, S., **7**, 66[76,77], 68[76,77,83b]
Banerjee, S. N., **6**, 430[103]
Banerjee, U. K., **2**, 360[170]
Banerji, A., **2**, 801[30]
Banerji, J., **7**, 823[35]
Banert, K., **6**, 247[136]
Banes, D., **5**, 723[108a]
Banfi, L., **1**, 568[244,247]; **2**, 13[59], 35[126], 41[126], 42[126], 44[126], 221[146], 514[54], 636[53], 639[63], 640[53,63], 642[78], 643[78], 922[101], 923[101], 931[135], 933[135,136], 934[135,136], 940[135,136]; **5**, 100[153], 102[177]; **6**, 118[107], 149[100], 864[197]; **7**, 128[63]; **8**, 187[45]
Banfi, S., **2**, 435[65]; **7**, 778[411]
Banford, C. H., **6**, 950[1]
Bang, L., **7**, 247[101], 842[27,28]
Bangert, R., **7**, 157[33,33a]
Banhidai, B., **1**, 846[19b], 847[19b], 850[19b]
Bank, K. C., **3**, 134[336]
Bankaitis, D. M., **1**, 419[80]
Bankaitis-Davis, D. M., **1**, 419[81]
Bankhead, R., **2**, 757[14]
Banko, K., **2**, 430[52a], 431[52a]
Bankov, Y. I., **2**, 804[40]
Banks, B. A., **5**, 534[95]
Banks, C. H., **8**, 413[134]
Banks, C. M., **3**, 390[78,79], 392[78]
Banks, D. F., **7**, 16[162]
Banks, H., **3**, 767[164]
Banks, R., **3**, 382[36]
Banks, R. B., **3**, 439[38], 482[3]
Banks, R. E., **6**, 498[161], 836[55]; **7**, 24[23,25], 26[25]; **8**, 901[39]
Banks, R. L., **5**, 1115[1], 1116[1,6]
Bannai, K., **3**, 222[144]; **4**, 262[308]; **6**, 837[60], 942[154], 944[154]
Banner, B. L., **4**, 410[260c]
Bannister, B., **8**, 494[24], 530[101]
Banno, H., **1**, 92[60,61]; **5**, 850[150,153,154,155]
Banno, K., **2**, 613[112], 614[112], 616[131], 629[2], 630[20], 631[2,20], 632[2], 667[43]
Banno, M., **6**, 856[152]
Bannore, S. N., **7**, 375[80]
Bannou, T., **4**, 371[47]; **7**, 98[104], 770[256c], 771[256], 819[22]
Bannova, I. I., **1**, 34[232]
Banoub, J., **6**, 49[91]
Bansal, R. C., **3**, 322[141]; **8**, 354[177]
Bansal, R. K., **2**, 412[10], 413[10], 414[10]
Banthorpe, D. V., **6**, 65[2], 690[398], 692[398], 795[3,9], 797[9], 799[3,9], 806[9], 807[9], 826[9], 950[1]; **8**, 542[231]
Bantjer, A., **2**, 773[25]
Banucci, E. G., **5**, 253[46,46d]
Banville, J., **2**, 213[126], 505[12], 510[12]; **3**, 32[188]; **5**, 953[297]; **6**, 724[154], 726[174]
Banwell, M. G., **4**, 1018[217,218], 1050[122]; **5**, 714[75b], 847[136]; **7**, 300[57]
Bao, J., **8**, 864[241]
Bapat, D. S., **8**, 950[161]
Bapuji, S. A., **5**, 290[42]
Baqant, V., **8**, 274[137], 773[64]
Bär, T., **6**, 54[124]
Barabadze, Sh. Sh., **3**, 318[128]
Barabas, A., **3**, 416[15], 417[15]
Barak, G., **4**, 434[125]
Baraldi, P. G., **2**, 803[32]; **5**, 403[8]; **7**, 143[140,141]; **8**, 392[108], 394[116], 645[45]
Baran, D., **4**, 438[152]
Baran, J., **3**, 582[115]; **4**, 424[16], 426[16], 432[16,112], 1075[34]; **5**, 600[42]
Baranova, V. A., **8**, 318[65,66], 546[307,308]
Baranovskii, I. B., **7**, 108[170]
Baranowska, E., **1**, 329[39], 806[314]; **6**, 989[81]
Barany, G., **6**, 643[78], 670[270], 698[270]
Barash, L., **5**, 355[87a]; **7**, 15[147]
Baratchart, M., **3**, 1046[3]
Barbachyn, M., **2**, 697[81]
Barbachyn, M. R., **1**, 533[138], 534[139], 740[43], 741[43], 825[48]; **4**, 226[189]; **7**, 440[41], 441[41]
Barbadoro, S., **5**, 771[145], 772[145]
Barbara, C., **4**, 45[130,130a]; **5**, 18[127]
Barbara, D., **5**, 836[65]
Barbaro, G., **5**, 113[230,231], 114[241], 440[173]; **8**, 394[117]
Barbaro, S., **6**, 508[289]
Barbas, C. F., III, **8**, 187[34]

Barbeaux, P., **1**, 630[43], 631[43], 664[169,170], 665[169,170], 669[169,170], 670[169,170]; **3**, 107[228], 111[228,230,231,232]
Barber, G. N., **5**, 1008[44]
Barber, L., **5**, 730[129], 731[129]
Barber, L. L., **5**, 597[30]
Barbero, M., **8**, 277[153], 660[109]
Barbier, P., **2**, 652[127]
Barbieri, G., **7**, 557[74], 764[112], 767[112], 777[384]
Barboni, L., **6**, 115[83]
Barborak, J. C., **3**, 855[82]
Barbot, F., **1**, 219[63]; **2**, 82[8]; **4**, 84[68b], 89[84h], 95[84h], 149[52], 182[77], 183[78]
Barbry, D., **6**, 74[37], 897[100]
Barbulescu, E., **8**, 124[89]
Barbulescu, N., **8**, 124[89]
Barcelo, J., **5**, 417[65]
Barcelos, F., **5**, 256[58], 257[58b]; **7**, 268[122]
Barchi, J. J., **1**, 872[89]; **3**, 785[36,36b]
Barchielli, G., **8**, 358[197]
Barclay, L. R. C., **3**, 326[166]
Barco, A., **2**, 803[32]; **3**, 738[37]; **5**, 403[8], 451[44], 453[44], 468[44]; **7**, 143[140,141]; **8**, 392[108], 394[116], 645[45]
Bard, A. J., **3**, 634[22]; **7**, 850[3,8], 852[8]; **8**, 594[71]
Bard, R. R., **4**, 455[43], 458[67], 459[76,89], 460[89], 463[43,67], 464[76], 466[76], 473[89], 478[76]
Bardakos, V., **2**, 345[41]
Barden, T. C., **3**, 711[23]; **4**, 1039[63]; **7**, 545[28]
Bardenhagen, J., **2**, 498[75,77], 499[75]
Bardi, R., **4**, 915[15], 936[68]
Bardili, B., **3**, 752[92]
Bardon, A., **4**, 453[34], 472[34]
Bardone, F., **4**, 53[148]
Bardone-Gaudemar, F., **2**, 81[5]
Bare, T. M., **1**, 411[45]; **3**, 16[81]
Barefield, E. K., **4**, 985[126,131]
Bares, J. E., **1**, 632[66]
Baret, P., **1**, 835[136]
Baretta, A., **3**, 839[12], 840[12], 853[70], 854[12]
Barf, G., **8**, 97[98]
Barfield, M., **2**, 345[29]
Barfknecht, C. F., **8**, 146[97]
Bargar, T. M., **3**, 492[73], 497[73]; **4**, 455[44,45], 457[45], 463[44,45], 464[45], 465[45], 466[44,45], 467[45], 469[45], 472[45], 476[44,45], 532[83], 543[83], 545[83]
Bargas, L. M., **7**, 228[96]
Barger, P. T., **8**, 673[26], 676[26]
Bargiotti, A., **6**, 984[55]
Bari, S. S., **5**, 96[105,116,119]
Barieux, J.-J., **8**, 705[10], 726[10]
Barili, P., **5**, 456[87]
Barili, P. L., **2**, 291[71]; **3**, 725[19], 734[9], 743[58]
Barillier, X., **5**, 829[22]
Barinelli, L. S., **4**, 695[4]
Barker, A. J., **6**, 1045[24]
Barker, J. M., **2**, 963[54]
Barker, M. W., **5**, 113[228,232], 117[277]
Barker, P. J., **8**, 818[39]
Barker, R., **2**, 456[42], 466[42], 467[42]
Barker, W., **2**, 159[128]
Barker, W. D., **3**, 903[126]
Barkley, L. B., **1**, 446[195]
Barkley, R. M., **3**, 4[22]
Barkovich, A. J., **5**, 1151[132,136]
Barl, M., **4**, 924[33]
Barlet, R., **4**, 1000[7]; **8**, 795[21]
Barlett, P. A., **2**, 555[143]; **4**, 377[104,104a], 378[104a], 381[104a]
Barlin, G. B., **7**, 768[202]
Barlos, K., **8**, 245[73]
Barlow, A. P., **8**, 557[382]
Barlow, L., **3**, 585[137], 587[148]
Barltrop, J. A., **5**, 165[85]; **7**, 884[182]; **8**, 143[58], 493[21], 526[25]
Barluenga, J., **1**, 361[32a,b], 782[234], 830[94]; **2**, 790[57], 980[21], 981[21]; **3**, 282[48], 788[50]; **4**, 290[196,200], 291[211,212,213,214,215,216,217,218,219,220], 292[224,225,228], 295[248,249,254], 302[338], 303[341], 311[446,453], 315[511,513], 347[93], 349[110], 351[93c,124], 354[93d,110], 405[251], 735[82], 741[82], 799[111]; **5**, 161[63,64], 410[42], 411[42], 433[136b], 480[178], 484[179], 848[141], 850[162]; **6**, 184[150], 494[135], 555[810,811,812], 757[134]; **7**, 93[54], 486[141,142], 490[176], 501[255], 505[285], 533[35,36], 534[35], 536[52-55], 632[60]; **8**, 13[72], 124[87,88], 851[132], 854[150], 856[176,177,178,179,180,183], 857[187,196,198]
Barman, T. E., **2**, 456[46]
Barnah, J. N., **6**, 685[363]
Barnah, R. N., **6**, 685[363]
Barnard, D., **7**, 762[72,73], 763[87], 765[157], 766[87], 769[242], 771[242], 773[242]
Barnela, S. B., **2**, 787[52], 789[55,56]
Barner, B. A., **1**, 263[41], 359[22], 383[22], 384[22]; **4**, 77[50], 206[45,47,49], 893[155]; **6**, 25[100], 842[78]; **7**, 300[54]
Barner, R., **5**, 834[50,53], 850[146], 877[5]; **8**, 205[159], 560[405]
Barnes, D. G., **3**, 331[200a]
Barnes, G. H., **8**, 764[3], 776[3]
Barnes, J. C., **7**, 440[39,39b]
Barnes, J. H., **8**, 135[48]
Barnes, J. R., **7**, 845[77]
Barnes, K. K., **7**, 854[55], 855[55]
Barnes, R. A., **2**, 740[58]; **7**, 16[159]; **8**, 568[467]
Barnes, R. P., **8**, 286[15]
Barnett, G. H., **2**, 780[9]
Barnett, J. E. G., **7**, 239[46]
Barnett, P. G., **2**, 782[22]
Barnette, W. E., **4**, 255[192], 260[192], 368[18], 369[18b,20,24], 370[31], 371[60], 372[31,60], 374[20], 397[31], 413[271a,b,272,273]; **5**, 386[135]; **6**, 540[583], 650[132], 1031[113]; **7**, 522[41], 523[41,46], 524[54]; **8**, 343[114], 847[99], 848[99], 849[99]
Barnetzky, E., **3**, 643[124]
Barnick, J. W. F. K., **2**, 801[36]; **5**, 986[38]
Barnier, J.-P., **3**, 727[32]; **5**, 910[89], 956[305]
Barnikow, G., **6**, 422[34], 429[34], 449[113], 453[113]
Barnum, C., **4**, 245[89]; **6**, 1030[107]; **7**, 131[83-85]; **8**, 850[119]
Barnum, C. S., **1**, 642[111], 644[111], 669[111,183], 670[183], 671[111,183], 699[183]
Barnvos, D., **4**, 1040[99]
Baron, W. J., **6**, 776[55]
Barone, A. D., **6**, 619[118]
Barone, G., **4**, 347[95]; **7**, 438[17-19], 445[17-19]
Baroni, A., **7**, 774[320]
Barot, B. C., **2**, 141[39]
Barr, D., **1**, 6[33], 33[165], 38[258], 39[187]; **3**, 763[151]
Barr, P. J., **3**, 219[109], 530[72], 534[72], 547[124]
Barra, M., **4**, 426[40,41]
Barraclough, P., **5**, 71[156]
Barrage, A. K., **6**, 498[161]
Barrans, J., **8**, 663[119]
Barras, C. A., **7**, 257[49]
Barras, J.-P., **5**, 47[41], 48[41], 50[41], 362[94], 363[94b]
Barre, L., **8**, 268[72]
Barre, M., **8**, 205[165]
Barreau, M., **6**, 176[105]
Barreiro, E., **4**, 229[212]; **8**, 881[69]
Barrero, A. F., **7**, 634[68]

Barret, R., **7**, 764[114]
Barrett, A. G. M., **1**, 511[28], 569[253], 797[293]; **2**, 321[15], 325[15], 332[57], 338[75], 635[50], 640[50], 645[81], 742[77], 784[39a], 968[78], 1059[74]; **3**, 251[79], 254[79], 613[2], 615[2]; **4**, 35[98f], 380[121], 384[142]; **5**, 85[9], 116[269,270], 407[28], 1068[12,13], 1075[30]; **6**, 107[24], 473[73], 687[380], 779[68], 783[86,87], 784[88-90], 961[68], 987[72], 996[107], 1017[37]; **7**, 529[11]; **8**, 84[15], 117[74], 243[47], 248[82], 363[2], 374[151], 505[83], 816[24], 817[29], 822[52], 940[100,108]
Barrett, G. C., **1**, 820[12]; **3**, 86[21,22], 104[21], 121[21,22]; **4**, 316[535]; **6**, 423[47], 424[55], 437[32], 450[118]; **8**, 403[7]
Barrett, J., **4**, 37[107]
Barrett, J. H., **5**, 634[68,69], 687[62], 688[62]; **7**, 507[306]
Barrett, J. W., **8**, 328[9]
Barrette, E.-P., **7**, 278[165], 279[165]
Barrick, P. L., **6**, 288[184]
Barrière, F., **1**, 561[166]
Barrière, J. C., **1**, 561[166]
Barrio, J. R., **8**, 48[109], 66[109]
Barrio, M. G., **1**, 858[62]
Barrios, H., **1**, 544[35], 552[80]
Barrish, J. C., **2**, 547[100], 548[100]; **4**, 370[29], 390[171]; **5**, 433[137b]; **8**, 704[7], 713[7]
Barron, C. A., **5**, 731[130b]
Barron, H. E., **7**, 765[153]
Barros, M. T., **7**, 298[36]
Barros Papoula, M. T., **1**, 698[242], 699[242]
Barrows, R. D., **4**, 797[101]; **8**, 356[186]
Barry, C. N., **6**, 24[97]
Barry, J., **6**, 2[5], 18[5]
Barry, J. E., **2**, 971[93]
Barsoum, S., **8**, 435[71]
Barstow, D. A., **8**, 206[172]
Barstow, J. F., **4**, 370[25]; **5**, 841[91], 877[7]; **6**, 859[174]
Barta, M. A., **6**, 182[139]; **7**, 100[129], 104[129], 260[43], 779[421]
Barta, T. E., **5**, 524[49], 539[49]
Bartak, D. E., **4**, 453[27], 471[27], 872[37]
Bartel, K., **6**, 295[250]
Bartels, A. P., **8**, 114[58], 308[5,6], 309[5,6]
Bartels, H. M., **4**, 729[61], 730[61], 765[61]
Bartes, O., **7**, 235[1]
Bartetzko, R., **3**, 614[6], 623[6], 627[6]
Barth, G., **7**, 676[62]; **8**, 353[154]
Barth, J., **2**, 96[56]; **6**, 554[783]
Barth, M., **6**, 7[34]
Barth, V., **5**, 829[23]
Barthels, R., **6**, 639[53,54]
Bartholomew, D., **1**, 751[92]; **7**, 237[36]
Bartkowiak, F., **8**, 535[166]
Bartlett, N., **7**, 882[168]
Bartlett, P. A., **1**, 49[10], 109[14], 286[9], 348[142], 608[38], 749[80], 766[154], 804[313], 805[313]; **2**, 1[1], 123[200], 125[200], 193[64], 202[96], 280[24], 281[24], 555[142], 580[100], 649[106], 657[163], 904[50]; **3**, 341[5], 365[5], 366[99], 374[5]; **4**, 142[16], 311[442], 364[1,1m], 366[8], 367[15], 370[25,25a], 373[1m,78], 377[104], 378[104b,111], 380[120,120b,122,125], 381[8a,120b,122a,125a,b], 382[120b,134,134a], 383[15,134a,137], 384[15], 386[8b,125a,148a,151], 387[125a,148,148a], 390[104b,175c], 815[189]; **5**, 308[95], 827[2], 829[2], 836[65], 837[68], 841[91], 855[187,192], 859[236], 877[7,8]; **6**, 26[105-107], 855[148], 856[154], 859[174], 897[95], 938[126], 968[107], 998[120,121], 1021[50]; **8**, 7[39], 541[207], 542[226], 853[143,147], 856[147], 857[143,147]
Bartlett, P. D., **3**, 705[4], 725[18]; **4**, 5[17], 868[11], 1089[133], 1099[176]; **5**, 63[5,6], 64[32,51], 65[61], 69[5,6,105,106], 70[5,107-110], 71[5,122,143], 72[6], 74[5,208], 77[6], 211[67], 1188[15]; **7**, 98[100], 480[102]; **8**, 91[51]
Bartlett, R. A., **1**, 11[57], 23[118]
Bartlett, R. K., **6**, 961[70]
Bartlett, R. S., **8**, 60[181], 66[181], 73[248], 74[248], 373[136]
Bartlett, W. R., **3**, 362[89], 369[117], 371[114], 372[117]; **5**, 828[7], 839[7], 882[13], 888[13], 892[13], 893[13]
Bartley, W. J., **8**, 140[14], 248[89]
Bartmann, D., **4**, 386[147]
Bartmann, W., **1**, 78[12]; **2**, 240[5], 369[253], 370[253], 371[253], 372[253], 455[1]; **4**, 91[89], 792[66], 876[64]; **5**, 461[95], 464[95], 468[95], 645[1], 648[1j], 651[1]; **6**, 33[2], 34[2], 40[2], 50[2], 51[2], 53[2], 54[2], 57[2], 538[569], 655[166], 746[93], 765[17], 767[25]; **7**, 694[32], 695[35]; **8**, 5[25]
Bartmess, J. E., **1**, 632[66]; **5**, 788[15]
Bartnik, R., **1**, 387[132], 836[142]; **4**, 1086[117]
Bartocha, B., **1**, 214[25]
Bartók, M., **1**, 819[3]; **3**, 726[23]; **6**, 2[2,3], 23[2], 25[3,99]; **7**, 358[6], 372[6]; **8**, 418[5], 420[5], 423[5], 439[5], 441[5], 442[5], 883[89,90]
Bartok, W., **7**, 759[5]
Bartoletti, I., **3**, 1028[47]
Bartoli, D., **1**, 670[187], 678[187]; **4**, 437[148], 438[148]; **7**, 340[45], 770[256b], 771[256], 773[306], 779[427]
Bartoli, G., **1**, 569[262]; **4**, 85[78a], 86[78b,c,e,79a,b], 424[11], 426[37], 428[11,77-79], 429[80-82]; **6**, 115[83]; **7**, 331[16]
Bartoli, J. F., **7**, 108[176], 383[109]
Bartolini, G., **4**, 98[114], 113[114]
Bartolini, O., **7**, 762[84], 777[380]
Bartolotti, L. J., **1**, 476[127], 477[129]; **3**, 66[16], 67[16], 74[16,42], 75[42], 77[42]; **7**, 227[78], 230[78]
Barton, D. H. R., **1**, 2[6], 174[8], 175[8], 391[150], 489[18], 499[18], 542[5], 543[5], 544[5], 546[5], 561[166], 580[1], 630[10], 698[242], 699[242], 820[7,12], 822[7], 833[117]; **2**, 240[12], 321[11], 323[32], 329[11], 784[39a], 791[60], 866[4]; **3**, 86[21,22,32,33,37,38,48,62], 104[21], 121[21,22], 147[32,33], 154[32], 158[37,38], 159[37], 163[38], 164[37], 173[37,38,62], 386[66], 419[35], 422[68], 505[158,159], 613[2], 615[2], 660[8], 664[30], 681[97], 689[123], 733[2], 780[10], 831[63]; **4**, 3[8], 6[8], 24[72,72c], 27[83], 99[118d], 259[276], 316[535], 344[77], 347[104], 483[4], 484[4], 495[4], 605[293], 646[293], 674[35], 688[35], 725[43], 747[149,151,152,153], 748[155,156,157,158,159], 753[159], 765[223,226], 768[239], 790[35], 800[120], 820[221], 824[234], 953[8], 968[57], 987[133], 1002[51], 1004[76,77], 1021[76,77], 1093[149]; **5**, 223[77]; **6**, 2[1], 28[1], 70[20], 104[1,9], 111[64], 116[90], 133[5], 134[11], 171[4], 177[4], 198[4], 211[78], 225[5], 226[5], 258[5], 293[229], 419[1], 421[1], 436[15], 437[15], 442[87], 447[104], 448[15], 450[15], 451[104], 453[15], 455[15], 473[72,73], 474[72,84-86], 475[72,93], 659[195], 687[380], 690[394,399], 691[399], 692[399], 837[59], 938[130], 942[130], 954[14], 980[45], 981[47], 987[72], 1017[37,38], 1024[38]; **7**, 13[115-117,119], 14[142], 15[145], 27[62], 40[11], 41[18,22], 84[1], 85[1], 90[32], 92[41,41a,49], 94[41], 108[1,180], 110[187,188], 123[35], 129[72], 132[72,91,92,95,97-100], 133[72,92], 134[92], 144[35], 146[100], 159[45], 170[122], 171[122], 223[42], 227[87], 231[141], 244[69,70], 299[50,51], 307[16], 310[16], 318[16], 319[16], 322[16], 329[4], 343[4], 345[1], 356[48], 470[2], 529[11], 594[4], 598[4], 671[10], 673[10], 687[10], 704[12-14], 705[14], 719[6,7], 720[6], 721[7], 722[20], 723[24], 724[24], 725[7,31,33], 726[6,7,20,35-37], 727[38], 728[7,41,42], 730[45-47,49,51], 731[45,52,53,55], 732[59], 741[49], 747[94,106], 761[59], 776[59,357,362]; **8**, 26[12], 27[12], 108[3], 110[3], 116[3], 117[74], 121[77], 141[37], 243[47], 247[81], 279, 301[93], 330[45,46], 340[45], 342[46], 370[91], 392[95,109], 393[113], 394[114], 403[2,6-8], 404[2], 505[83], 525[14], 526[14], 565[448], 699[148], 704[4], 705[4], 706[4], 707[4], 710[4], 715[4], 716[4], 717[4], 722[4], 724[4], 725[4], 726[4], 728[4], 794[3], 796[28], 816[24], 818[35,36,40], 820[35,47], 821[50,51], 822[52], 823[35,55], 825[68], 830[89],

831[89-91], 836[9], 837[13a], 839[13a], 840[13a], 847[9], 848[9,105], 880[56], 935[63,69], 936[70], 937[80,84,85], 984[2]
Barton, D. L., **6**, 960[58]
Barton, F. E., **8**, 795[25]
Barton, H. D., **6**, 65[1]
Barton, J. C., **1**, 107[6], 110[6]; **3**, 223[160]
Barton, T. J., **5**, 65[73], 199[27], 587[206,208], 588[208]
Bartos, J., **2**, 354[106,107]
Bartra, M., **8**, 385[41]
Bartroli, J., **1**, 400[11]; **2**, 25[99], 250[38], 251[38,39], 260[38], 274[38], 436[67]; **7**, 300[53]; **8**, 720[136]
Bartrop, J., **5**, 153[24]
Bartsch, R. A., **6**, 220[126], 953[7], 954[19], 955[24], 959[32]; **7**, 543[13]
Bartulin, J., **5**, 478[162]
Bartz, W., **3**, 909[152]
Bartz, W. J., **6**, 535[529]
Barua, A. K., **8**, 333[56]
Barua, N. C., **3**, 380[10]; **6**, 987[70]
Baruah, J. N., **8**, 891[148]
Baruah, P. D., **6**, 539[578]
Baruah, R. N., **8**, 891[148]
Barylett, P. A., **6**, 22[84]
Baryshnikova, T. K., **7**, 596[40]
Barzoukas, M., **4**, 469[136]
Basak, A., **1**, 596[30], 601[30]; **2**, 85[15]; **4**, 158[76], 735[85], 744[136], 1023[256]; **5**, 277[14,15], 278[14,15], 279[15]; **8**, 355[182]
Basato, M., **2**, 369[245]
Basavaiah, D., **3**, 795[83,84]; **4**, 34[97], 35[97]; **7**, 596[36]; **8**, 715[85,85a], 716[85], 718[85a], 719[120]
Bascetta, E., **4**, 390[173a]; **8**, 855[158]
Basche, R. W., **4**, 170[20]
Basendregt, T. J., **8**, 144[70]
Basha, A., **1**, 92[68], 93[68], 376[94], 385[116,117,120], 386[120], 400[10]; **2**, 746[112]; **8**, 27[32], 66[32], 249[92], 301[94]
Basha, F. Z., **4**, 10[34], 113[164]; **5**, 406[23,23b]; **6**, 83[82], 814[88]; **8**, 34[62], 66[62]
Bashe, R. W., **1**, 431[134]; **3**, 248[55], 251[55], 269[55], 419[47], 494[87], 502[87]
Bashe, R. W., II, **7**, 123[33]
Bashiardes, G., **8**, 935[69], 937[84,85]
Bashir-Hashemi, A., **4**, 161[87]; **5**, 383[125]
Bashkin, J., **8**, 459[239]
Basile, T., **2**, 925[111], 926[111]; **5**, 102[174]
Baskakov, Yu. A., **2**, 854[236]
Baskar, A. J., **1**, 300[69]
Baskaran, S., **7**, 266[106], 267[106], 276[106]
Basok, S. S., **6**, 94[139]
Bass, J. D., **5**, 124[6], 125[6], 128[6]
Bass, L., **2**, 88[29]
Bass, L. S., **1**, 889[142], 890[142]; **7**, 441[42]
Bass, R. G., **4**, 48[138]
Bassani, A., **4**, 426[63]
Bassedas, M., **1**, 543[24,26]
Basselier, J.-J., **2**, 289[70], 291[70], 917[88]
Basset, J.-M., **5**, 1115[1], 1116[1,9], 1118[9]; **8**, 445[53]
Bassetti, M., **4**, 300[306]
Bassi, I. W., **1**, 303[77]
Bassignani, L., **7**, 439[25]
Bassindale, A. R., **1**, 580[2], 581[2], 582[2], 610[2a], 611[2a], 616[2a], 618[58], 621[67], 784[239,240], 815[239]; **2**, 601[35]; **3**, 419[31]
Bassler, G. C., **8**, 754[83]
Bassner, S. L., **2**, 127[236]
Basso-Bert, M., **8**, 447[122], 457[122]
Bassova, G. I., **8**, 318[63], 486[61]
Bast, H., **7**, 92[46], 154[14]
Bast, K., **4**, 1079[58]
Bast, P., **1**, 2[10]
Bastiaansen, L. A. M., **8**, 94[81], 95[82]
Bastian, J.-M., **2**, 765[71]
Bastiani, B., **1**, 34[167], 70[63], 141[22]
Bastide, J., **4**, 1073[26]; **5**, 247[26]
Bastos, C., **7**, 443[51b]; **8**, 459[228]
Bastos, H., **1**, 837[154]
Basu, A., **8**, 446[73]
Basu, B., **1**, 534[143]; **8**, 505[80]
Basu, N. K., **1**, 174[8], 175[8]; **7**, 41[22]; **8**, 796[28]
Basu, P. K., **5**, 452[55], 453[55]
Basu, S. K., **7**, 71[95]
Basyouni, M. N., **4**, 50[142]
Batchelor, M. J., **5**, 531[74]
Batcho, A. D., **2**, 530[23], 547[100], 548[100]; **4**, 390[171]; **5**, 4[39], 5[39,42,44], 8[44], 256[58], 257[58b]; **6**, 960[56]; **7**, 268[122], 564[92], 567[92]; **8**, 368[71]
Bate, N. J., **5**, 1125[54]
Bateman, L., **7**, 762[73], 763[85], 766[85]
Bates, D. J., **4**, 48[137]
Bates, G., **1**, 95[72]
Bates, G. S., **4**, 113[173]; **6**, 134[25], 438[42,57], 440[76], 463[27]; **8**, 549[327], 681[78], 683[78], 689[78], 693[78], 696[122], 801[70,71], 813[11], 938[89]
Bates, H. A., **2**, 1012[14]; **6**, 736[30]; **7**, 208[85]; **8**, 604[6], 639[18]
Bates, R. B., **4**, 495[86]
Bateson, J. H., **5**, 105[194]
Bath, S. S., **8**, 964[50]
Bather, P. A., **6**, 734[15], 923[57]
Batis, F., **7**, 43[43]
Bator, B., **6**, 668[259]
Batra, M. S., **4**, 443[194]
Batra, R., **7**, 49[65]
Batra, S., **4**, 505[132]
Batres, E., **7**, 92[42], 93[42]; **8**, 566[450]
Batroff, V., **3**, 623[35]
Bats, H., **6**, 563[895]
Bats, J. W., **2**, 547[113], 551[113]; **3**, 56[285]; **4**, 229[235,236], 1055[138]; **8**, 545[285]
Batsanov, A. S., **4**, 218[147]
Batt, D. G., **4**, 373[74]; **5**, 335[49]; **8**, 355[181], 929[32]
Batta, A. K., **3**, 644[149]
Battaglia, A., **2**, 927[120]; **5**, 100[148], 113[230,231], 114[241,242,243,244,245], 440[173]
Battegay, M., **6**, 423[43]
Batten, P. L., **7**, 231[141]
Battersby, A. R., **2**, 811[68]; **3**, 659[1], 660[1], 679[1], 681[94], 807[31]; **5**, 468[133]; **6**, 538[571], 734[9]
Bättig, K., **1**, 87[49]; **3**, 17[89]; **4**, 218[137], 501[115]; **5**, 11[80-82], 13[93], 40[26], 139[84], 414[53], 742[162], 779[198], 829[25]; **6**, 757[131], 1018[41], 1063[86]
Battigelli, L. C., **5**, 141[90]
Battioni, J.-P., **4**, 240[45]
Battioni, P., **4**, 878[75], 898[75]; **7**, 383[109], 426[148c], 477[78], 483[78,133], 484[78,133,134], 500[133]
Battista, R. A., **4**, 573[9], 614[378]
Battiste, M., **3**, 896[67]
Battiste, M. A., **1**, 789[262b], 790[262]; **4**, 587[48,49]; **5**, 531[73], 947[267], 948[267,271]
Battistini, C., **3**, 734[8,10]; **4**, 126[219]; **8**, 856[167]
Batty, J. W., **4**, 54[154]
Battye, P. J., **5**, 1003[25]
Batu, G., **2**, 332[61]; **3**, 219[112]
Batz, H.-G., **6**, 642[74]

Batzer, H., **7**, 765[167]
Bau, R., **1**, 41[266,267], 174[11], 179[11]
Baucom, K. B., **5**, 65[56]
Baudat, R., **1**, 698[249]; **4**, 339[46]
Baude, G., **8**, 548[318]
Baudin, J.-B., **6**, 162[188], 841[77], 866[77]
Baudisch, O., **3**, 890[35]
Baudouy, R., **1**, 853[50], 876[50]; **3**, 256[113]; **5**, 772[162], 774[171], 780[162]; **7**, 298[34]; **8**, 945[127]
Baudrowski, E., **4**, 289[192]
Baudry, D., **3**, 842[16]; **7**, 6[29]
Baudy-Floc'h, M., **6**, 67[11]
Bauer, B., **5**, 365[95]
Bauer, F., **8**, 568[482]
Bauer, I., **6**, 291[199]
Bauer, J., **2**, 1093[84]
Bauer, K., **3**, 828[44]
Bauer, L., **6**, 119[117], 795[15], 798[15], 821[15], 822[118]
Bauer, M., **1**, 846[12], 851[12], 852[12]
Bauer, P., **3**, 723[10]; **8**, 552[357,358]
Bauer, P. E., **8**, 333[57], 345[127]
Bauer, T., **1**, 54[19], 109[14], 183[57], 185[57], 339[84]; **2**, 31[116], 671[51], 694[77], 995[43]; **5**, 432[124], 433[139,139f], 434[139i]; **7**, 397[29], 713[73]
Bauer, V. J., **5**, 857[229]
Bauer, W., **1**, 19[103], 20[107], 23[123-125], 32[156], 34[170], 35[172]; **2**, 100[4], 508[30]; **4**, 872[40]; **6**, 419[7], 420[7], 423[7], 437[33], 440[33], 441[33], 443[33], 444[33], 445[33], 446[33], 519[338]
Bauer, W. N., Jr., **2**, 969[83,83b]
Baukov, Yu. I., **1**, 608[39]; **2**, 597[96], 607[76], 609[76,79], 610[96], 616[135], 726[122]; **8**, 547[315], 548[315]
Bauld, N. L., **3**, 423[79], 649[202], 650[202c], 652[202c]; **5**, 63[11], 453[65], 520[39], 522[43], 704[20,22], 1020[68,69], 1023[69,78]; **7**, 860[74], 879[148,149], 880[148,149,154], 882[166]; **8**, 564[446]
Baum, A. A., **5**, 220[49]
Baum, G., **1**, 18[94], 37[242]
Baum, J. S., **4**, 1033[31], 1096[159], 1097[159,167], 1098[159]
Baum, K., **7**, 747[98]; **8**, 384[27]
Baum, M., **4**, 379[116], 380[116c]
Baum, T., **5**, 196[12]
Bauman, J. G., **2**, 434[58]
Baumane, L., **8**, 595[77]
Baumann, B. C., **5**, 710[47]
Baumann, M., **5**, 768[133], 779[133]
Baumann, R., **5**, 453[66]
Baumberger, F., **2**, 324[40], 334[40]; **4**, 719[21]
Baume, E., **3**, 505[166]
Baumeister, M., **2**, 1099[110]
Baumeister, U., **6**, 190[198], 196[198]
Baumgarten, H. E., **6**, 802[43], 803[43]
Baumgarten, R. J., **3**, 567[37], 606[37], 857[91]; **6**, 1042[6]; **7**, 229[108]; **8**, 827[71]
Baumgartner, M. T., **3**, 505[156]; **4**, 460[98], 463[109], 470[137], 471[109], 477[98]
Baumgartner-Rupnik, M., **2**, 1090[73], 1102[73], 1103[73]
Baumstark, A. L., **3**, 588[157,158]; **7**, 374[76]
Baur, R., **6**, 554[802], 576[802], 581[802]
Baures, P. W., **1**, 41[195]
Baus, U., **2**, 514[50], 524[50]; **3**, 39[213]; **6**, 728[208]
Bauslaugh, P. G., **5**, 123[1], 126[1]
Bauta, W. E., **5**, 1070[20], 1071[20], 1072[20], 1074[20], 1089[84], 1098[117], 1099[117], 1100[117], 1101[117], 1110[20], 1111[20], 1112[117]
Bavley, A., **7**, 306[3]
Bawa, A., **6**, 109[38]
Baxter, A. D., **5**, 829[25]
Baxter, A. J. G., **5**, 105[194], 403[9]
Baxter, E. W., **4**, 1040[73], 1043[73]; **5**, 829[21]; **6**, 122[135]
Baxter, G. J., **2**, 355[125]; **5**, 732[133,133b]
Baxter, S. L., **8**, 838[19,19b]
Baxter, S. M., **7**, 3[8]; **8**, 696[126]
Bay, E., **4**, 446[211]; **5**, 203[39,39e-g], 204[39j], 209[39], 210[39]; **7**, 876[121]
Bayard, P., **5**, 416[56], 480[168,169], 483[169]
Baydar, A. E., **5**, 478[163]
Bayer, C., **6**, 1030[107]; **7**, 131[83]
Bayer, E., **2**, 1099[110]; **6**, 670[273]
Bayer, H. O., **4**, 1097[166]; **8**, 314[41], 315[41], 814[15]
Bayer, J., **3**, 835[80]
Bayer, M., **6**, 692[408]
Bayer, O., **6**, 233[44]
Bayer, R., **6**, 535[535], 538[535]
Bayer, R. P., **8**, 364[13]
Bayer, T., **5**, 829[26]
Bayerlein, F., **3**, 814[70]; **6**, 911[18], 912[18]
Bayet, P., **1**, 630[40], 631[40], 648[132], 649[40], 651[132], 652[132], 655[132], 656[40], 659[40], 672[40,132], 673[132], 675[132], 679[132], 684[132], 708[40], 709[40], 710[40]; **3**, 87[66,116]
Bayles, R., **5**, 461[101], 463[101]
Bayless, J., **1**, 377[97]; **6**, 727[206]; **8**, 943[120]
Bayless, P. L., **2**, 283[43], 296[43]
Bayley, H., **7**, 21[18]; **8**, 384[39]
Baylis, A. B., **4**, 34[97], 35[97]
Baylis, E. K., **3**, 329[189]
Bayne, C. K., **2**, 1099[115]
Bayner, C. M., **7**, 451[24]
Baynes, J. W., **6**, 790[116]
Baynham, M. K., **3**, 380[13]
Bayod, M., **4**, 311[453]; **5**, 848[141], 850[162]; **8**, 857[196]
Bayomi, M., **3**, 904[134]
Bayomi, S. M., **4**, 350[116,118]
Bayón, A. M., **1**, 361[32a,b]; **2**, 980[21], 981[21]; **4**, 290[196], 399[224]; **8**, 856[180]
Bays, J. P., **4**, 980[110], 982[110]
Baysdon, S. L., **5**, 1202[56]
Bažant, V., **8**, 907[72], 918[72]
Bazant, V., **8**, 544[267]
Bazbouz, A., **1**, 564[189]
Baze, M. E., **3**, 1050[18], 1060[18]
Bazhulina, V. I., **3**, 644[140,141]
Beaber, N. J., **6**, 149[92]
Beach, D. L., **8**, 774[71]
Beacham, J., **6**, 669[265]
Beacham, L. M., **6**, 669[263]
Beak, P., **1**, 23[120], 385[117], 460[5], 461[5,9,13,14], 464[9,13,14,40,41], 476[111,117,125], 477[111,117,125,133,134], 480[156,157,158,159], 483[167], 630[26], 835[132]; **2**, 528[10], 709[6]; **3**, 65[1,2], 66[12], 67[17,20], 68[2,21,22], 69[2,22], 71[2], 74[12,41], 88[130,131], 90[130,131], 105[131], 194[10,11], 580[108]; **4**, 121[205d-f], 494[83]; **5**, 2[8], 429[113b]; **6**, 65[2], 504[223], 531[462]; **7**, 225[61-64], 226[72]; **8**, 531[112]
Beal, C., **5**, 939[223], 951[223], 962[223], 964[223]
Beal, D. A., **7**, 235[3]
Beal, P. F., **3**, 888[17], 891[39], 892[39], 898[39]
Beal, R. B., **1**, 885[132]; **2**, 547[105], 550[105]; **5**, 796[56], 835[61], 1029[92]
Beale, J. M., Jr., **7**, 58[53a], 62[53,53a], 63[53a,58]; **8**, 192[99]
Beale, M. H., **3**, 222[140]; **8**, 537[183]
Beam, C. F., **2**, 190[55,56], 512[46], 522[72], 523[73,74]
Beamer, R. L., **8**, 149[120]
Beames, D. J., **1**, 430[128]; **4**, 1040[82], 1043[82]; **8**, 503[68]
Bean, F., **7**, 602[94]

Beanland, J., **5**, 180[147]
Beard, C. C., **8**, 201[145]
Beard, C. D., **7**, 747[98]
Beard, L., **4**, 746[146]
Beard, R. D., **3**, 159[467], 166[467]
Beard, R. L., **4**, 372[62]
Beard, R. M., **3**, 877[86]
Beard, W. Q., **3**, 914[12]
Bearse, A. E., **4**, 307[394], 312[456], 313[456]
Beasley, G. H., **2**, 477[12]; **3**, 431[95,96]; **5**, 856[196]
Beattie, T. R., **2**, 212[118]; **8**, 384[25]
Beatty, K. M., **4**, 315[519]
Beau, J.-M., **3**, 174[525], 196[26], 231[253]; **4**, 339[45], 792[62]; **6**, 60[147], 632[4], 978[24]; **8**, 849[111]
Beaucage, S. L., **6**, 618[108], 650[131]
Beaucaire, V. D., **2**, 348[55]
Beauchamp, J. L., **1**, 287[16]; **4**, 980[109]
Beauchamp, P. D., **5**, 805[100]
Beaucourt, J.-P., **3**, 416[17], 417[17], 462[152], 513[207]; **6**, 690[401], 692[401]; **8**, 679[66], 680[66], 681[66], 683[66], 694[66]
Beaudegnies, R., **5**, 416[56], 480[168]
Beaudoin, S., **1**, 773[202]
Beaujean, M., **1**, 700[245], 705[245], 709[245]; **4**, 758[190]
Beaulieu, F., **8**, 136[51]
Beaulieu, N., **6**, 937[117], 939[117], 940[117]
Beaulieu, P., **1**, 419[79]; **7**, 153[7]
Beaulieu, P. L., **3**, 213[44]; **4**, 183[80], 340[54], 342[63], 784[17], 789[30], 791[30], 815[190]; **7**, 520[28], 521[36]
Beaumont, A. G., **3**, 564[15]
Beaupere, D., **8**, 552[357,358]
Bebault, G. M., **6**, 43[52]
Bebb, R. L., **1**, 630[39], 631[39], 701[39], 702[39]
Bebernitz, G. E., **3**, 25[152]
Bebernitz, G. R., **2**, 189[46], 209[46]
Bebikh, G. F., **3**, 864[21]
Becalski, A., **4**, 1039[62]
Bechara, E. J. H., **3**, 588[157]
Becher, J., **2**, 150[97]; **5**, 531[74], 532[74a]
Becherer, J., **5**, 65[63]
Bechgaard, K., **7**, 801[39]
Bechstein, U., **2**, 779[5], 782[25]
Bechtolsheimer, H.-H., **6**, 659[193]
Beck, A., **5**, 744[167]
Beck, A. K., **1**, 34[167], 166[113], 314[128], 323[128], 341[98], 631[48,49], 636[49], 656[49], 658[48,49], 659[48,49], 672[48,49], 708[48,49], 710[48,49], 721[49]; **2**, 323[22,26], 332[55], 335[26], 336[26], 830[138]; **3**, 87[68,69], 89[68], 128[322], 130[322], 134[322,338], 135[338], 137[338], 144[384], 145[384], 147[384]; **5**, 79[291]; **6**, 147[84]; **7**, 774[316]
Beck, A. L., **7**, 32[91]
Beck, F., **3**, 634[25b,28], 635[25b,43], 639[25b], 640[97], 655[43], 656[43]; **7**, 248[108]
Beck, G., **6**, 538[569], 765[17]; **7**, 694[32]
Beck, H., **8**, 904[59], 907[59], 909[59]
Beck, J. F., **2**, 456[29]; **8**, 183[1], 185[1], 187[1a], 196[1a], 199[1a], 201[1a], 204[1], 207[1a], 209[1a]
Beck, J. R., **4**, 425[28], 441[172,174,175,176,178]
Beck, K. R., **2**, 5[20], 6[20], 21[20]
Beck, W., **6**, 554[772,777]
Becker, D., **4**, 45[128]; **5**, 139[85], 142[93], 143[85]
Becker, D. A., **1**, 604[33], 605[34]; **2**, 85[17]; **5**, 277[17]
Becker, D. P., **5**, 1060[57]
Becker, E. I., **3**, 262[159], 264[159]; **7**, 92[40], 94[55]; **8**, 74[250], 850[120], 902[43], 907[43], 908[43]
Becker, H., **3**, 691[135], 829[54]
Becker, H.-D., **5**, 229[117,118]; **7**, 135[106], 835[85], 877[131]
Becker, H. G. O., **2**, 896[11,12]
Becker, J., **6**, 175[77], 1058[58], 1067[101]
Becker, J. L., **7**, 801[37]
Becker, J. Y., **6**, 282[153,155,156]; **7**, 107[168], 799[25], 800[32], 801[41]
Becker, K. B., **1**, 755[115], 812[115], 813[115]; **3**, 497[102], 873[65], 874[65]; **4**, 252[159], 273[44]; **5**, 65[74]; **6**, 1036[141], 1041[1], 1042[1]; **8**, 537[185]
Becker, K. M., **7**, 724[30]
Becker, M., **3**, 348[31], 358[31]
Becker, P., **2**, 1016[25]
Becker, P. D., **1**, 835[132]; **3**, 88[131], 90[131], 105[131]
Becker, P. N., **2**, 127[233], 314[39]; **7**, 484[136]
Becker, R. S., **5**, 729[125], 741[156]
Becker, S., **4**, 155[74]; **8**, 840[38]
Becker, S. B., **8**, 754[88]
Becker, W., **6**, 476[96]
Becker, W. G., **7**, 854[61]
Becker, Y., **1**, 442[178]; **4**, 926[40], 930[50]; **8**, 395[130]
Beckert, R., **6**, 552[699]
Beckert, W. F., **3**, 319[130]
Beckett, A. H., **6**, 116[88]; **8**, 64[202], 66[202]
Beckett, B. A., **3**, 20[105]; **8**, 331[30]
Beckh, H. J., **5**, 635[86]
Beckham, L. J., **7**, 15[154]
Beckham, M. E., **5**, 676[3]
Beckhaus, H., **1**, 846[19b], 847[19b], 850[19b]
Beckhaus, H.-D., **3**, 382[36]; **4**, 717[12], 758[191]
Becking, L., **3**, 647[197,198], 649[200]; **4**, 805[142]; **7**, 806[72]
Beckley, R. S., **7**, 4[16]; **8**, 857[202]
Beckman, J. C., **2**, 356[129]
Beckmann, E., **6**, 763[1]; **7**, 689[1]
Beckmann, P., **8**, 918[120]
Beckmann, W., **3**, 628[47]
Beckwith, A. L. J., **1**, 269[59], 699[255], 894[155]; **3**, 107[227], 600[212]; **4**, 453[37], 736[88], 753[165], 780[2,3], 781[6], 784[13,14], 785[2,22], 786[6,24], 787[6,27], 790[3], 795[83], 796[91,100], 797[102], 805[139,140,141], 812[183], 815[191,197], 816[197,200], 820[219], 822[224], 827[6]; **5**, 4[32], 133[52], 918[126]; **6**, 261[13], 763[4], 802[42], 803[42]; **7**, 40[3], 96[82], 689[6], 842[29,30,32], 883[176]; **8**, 807[120], 818[39], 903[52], 906[52], 907[52], 908[52], 909[52]
Becu, C. H., **6**, 237[69]
Beddoes, R. L., **3**, 810[43]; **6**, 523[347]
Bedell, A., **5**, 639[122]
Bedenbaugh, A. O., **8**, 236[6], 249[91], 284[2], 285[2], 294[53], 607[30]
Bedenbaugh, J. H., **8**, 236[6], 249[91], 284[2], 285[2], 607[30]
Bedenbaugh, J. L., **8**, 294[53]
Bederke, K., **5**, 422[88], 423[88]
Bedeschi, A., **4**, 126[219]
Bedford, C. D., **4**, 78[52b]; **6**, 109[44], 533[475]; **7**, 746[87]
Bedi, G., **7**, 477[78], 483[78,133], 484[78,133,134], 500[133]
Bednar, R., **2**, 346[47]
Bednarski, M. D., **1**, 314[131,132,134]; **2**, 455[16], 456[22,33,39,52], 457[33], 458[33,52], 459[33,52], 460[33], 461[33], 462[33], 463[89], 464[89], 465[106], 466[33], 669[45], 670[45,46], 681[57], 682[63,64], 683[64], 686[63,64], 687[63,71], 688[69], 690[71], 692[69], 696[80]; **5**, 434[143,146], 459[92]; **7**, 175[141]
Bédos, P., **6**, 421[33]
Bedoukian, R., **5**, 17[117]
Bedoya-Zurita, M., **5**, 51[45,45a], 53[45a], 56[50a], 57[50,50a,b], 59[50b]
Beebe, T. P., Jr., **8**, 421[31]
Beebe, T. R., **7**, 706[26]
Beeck, O., **3**, 332[207]
Beedle, E. C., **2**, 445[23]; **3**, 796[89]; **6**, 77[57], 1064[88]; **7**, 824[42]

Beelitz, K., **5**, 436[157]; **6**, 472[68]
Beereboom, J. J., **3**, 707[11]; **5**, 595[15], 596[15]; **6**, 265[38]
Beermann, C., **1**, 140[8], 141[15]
Beers, S., **3**, 197[33]
Beerstecher, W., **8**, 528[71], 971[108]
Beetz, I., **3**, 1028[48]
Beetz, T., **6**, 38[37], 658[186b]
Beger, J., **4**, 294[243]; **6**, 283[168], 287[183], 288[186,187,188,189,190], 526[400,401]; **7**, 490[179]
Begland, R. W., **5**, 486[196], 687[63], 688[63]
Begley, M. J., **2**, 494[57], 534[34], 535[34]; **3**, 407[147], 586[153], 604[153], 610[153]; **4**, 76[47]; **5**, 136[67], 181[151], 432[129], 841[87]; **6**, 859[171], 1062[76], 1063[81]; **7**, 338[42]
Begley, T., **5**, 41[29]
Begley, W. J., **3**, 623[32], 677[84]; **4**, 373[67]
Bégué, J. P., **1**, 530[128]; **2**, 546[91]; **3**, 842[16]
Behaghel, O., **7**, 769[234], 770[234]
Behan, J. M., **8**, 887[115]
Behforouz, M., **2**, 121[189]; **4**, 185[86]; **7**, 355[42]
Behling, J. R., **3**, 209[17], 216[68], 224[168]; **4**, 175[43]; **6**, 831[9]
Behnam, E., **6**, 481[121]
Behner, O., **8**, 649[62]
Behnke, M., **1**, 114[35,36]; **2**, 120[177]; **4**, 98[114,114b], 113[114]
Behrens, C. H., **3**, 225[185], 264[181]; **6**, 2[3], 25[3], 88[105], 89[105,112,113], 253[157,159]; **7**, 390[10], 403[10,65], 406[77], 409[77], 414[77], 415[77], 421[77], 423[77]
Behrens, O. K., **6**, 644[90]
Behrens, U., **1**, 21[110]
Behrman, E. J., **7**, 863[85]
Behrooz, M., **2**, 109[70]; **3**, 164[478], 167[478], 168[478]; **8**, 746[62], 750[62]
Beier, B. F., **8**, 458[223,223c]
Beifuss, U., **2**, 345[33], 350[71], 354[71,103], 357[71,103], 369[255a], 372[271], 373[71,103,272,273,274]; **5**, 14[99], 17[118-121], 466[120]
Beigelman, L. N., **6**, 618[114]
Beijer, B., **6**, 604[29]
Beilefeld, M. A., **7**, 720[10]
Beiler, F., **1**, 34[228]
Beilinson, E. Yu., **8**, 623[147], 624[147]
Beirich, C., **1**, 202[102], 203[102], 234[26], 331[47]
Beiser, W., **6**, 566[922]
Beisler, J. A., **1**, 856[54]
Beissner, C., **5**, 1148[123]
Beissner, G., **3**, 482[3]
Beisswenger, H., **8**, 873[19]
Beisswenger, T., **4**, 331[7]; **6**, 142[65]
Beke, D., **1**, 370[70], 371[70]; **8**, 957[14]
Bekhazi, M., **4**, 1089[132]
Bekker, R. A., **2**, 726[122]
Belan, A., **8**, 187[32], 188[32]
Belanger, A., **2**, 169[167]; **4**, 1040[83], 1043[83]
Belanger, D., **5**, 578[156]
Bélanger, G., **3**, 635[38]
Bélanger, J., **2**, 662[20], 663[20], 664[20]; **5**, 432[130]
Bélanger, P. C., **7**, 693[26]; **8**, 315[52], 969[94]
Belaud, C., **1**, 214[21]; **2**, 980[22], 981[22]
Belavin, I. Y., **2**, 597[96], 607[76], 609[76,79], 610[96]
Beldid, B., **6**, 514[306], 543[306]
Belen'kii, L. I., **3**, 615[7]; **4**, 468[132], 469[132]; **8**, 608[40], 610[56,60], 630[56]
Beletskaya, I. P., **1**, 3[23], 276[78], 277[78], 437[157], 438[159], 441[175], 445[175]; **3**, 219[111], 436[14], 503[152], 524[29], 529[29,48], 530[73], 531[88], 534[73], 537[88]; **4**, 423[8], 426[8], 444[8], 452[7], 519[12], 610[340,341]; **8**, 2[9]
Belew, J. S., **7**, 54[3], 56[3], 66[3], 77[3], 78[3], 542[6], 543[6]
Belfoure, E. L., **7**, 738[20]
Belgaonkar, V. H., **2**, 792[66]
Belik, M. Yu., **3**, 306[77]
Belikov, V. M., **4**, 218[147,148]
Belikova, N. A., **8**, 124[92], 125[92]
Belikova, Z. V., **8**, 775[77]
Belin, B., **6**, 85[87]
Belinka, B. A., Jr., **1**, 570[264]; **8**, 384[38], 385[48]
Bell, A., **3**, 201[76], 721[4]; **5**, 71[153]; **6**, 960[49]
Bell, A. J., **6**, 819[109]
Bell, A. P., **8**, 447[119a,b], 675[36,37], 679[37], 691[36]
Bell, A. S., **3**, 510[184]
Bell, C. L., **6**, 822[118]
Bell, E. V., **7**, 763[98]
Bell, F., **2**, 367[225]; **3**, 826[40]
Bell, F. A., **5**, 63[9]; **7**, 879[146]
Bell, H. C., **3**, 505[160]; **7**, 649[45]
Bell, H. M., **8**, 214[51], 709[43], 812[3], 908[78], 909[78], 968[89]
Bell, K. H., **2**, 809[58], 811[58]; **3**, 806[17]; **8**, 60[185], 66[185], 314[27]
Bell, K. L., **1**, 593[25], 595[25]; **2**, 1030[83,84]; **8**, 537[189], 755[131], 758[131]
Bell, L. A., **4**, 85[77c]
Bell, L. T., **2**, 1049[19]; **8**, 542[221]
Bell, N. A., **1**, 2[8], 21[111]; **8**, 754[101]
Bell, R., **6**, 1024[77]
Bell, R. A., **8**, 244[66]
Bell, R. P., **7**, 709[46]
Bell, S. H., **2**, 606[69]
Bell, S. I., **5**, 423[91], 424[92]
Bell, T. W., **1**, 343[110]; **3**, 219[106], 224[165], 512[201]; **6**, 743[71]
Bell, V. L., **2**, 149[86], 355[124]
Bellamy, A., **4**, 239[35]
Bellamy, A. J., **4**, 30[87]; **8**, 114[54]
Bellamy, F., **4**, 390[174]; **5**, 948[291]
Bellamy, F. D., **8**, 371[112]
Bellamy, L. J., **6**, 724[151], 727[201]
Bellard, S., **1**, 2[3], 37[3]
Bellasoued, M., **1**, 214[22], 218[22], 220[22]; **2**, 283[49], 286[65], 287[65], 486[38], 614[117], 634[37], 640[37]; **4**, 95[102g], 880[91], 883[91,101], 884[101,106]
Belleau, B., **1**, 123[78]; **2**, 213[126], 554[133], 1056[67]; **4**, 398[216], 399[216d]; **5**, 94[85], 96[111], 418[69]; **6**, 561[871]; **7**, 797[18,19], 799[28], 800[28a]
Bellegarde, B., **2**, 607[75], 608[75]
Bellemin, A.-R., **4**, 1009[141]
Bellenbaum, M., **5**, 30[2,2f]
Belleney, J., **4**, 519[27]
Beller, H., **2**, 770[10], 771[10]
Bellesia, F., **3**, 395[98]; **4**, 337[32]; **8**, 389[74]
Bellet, P., **8**, 201[144]
Belletire, J. L., **4**, 598[205], 638[205]; **5**, 373[106,106b]
Belley, M. J., **4**, 316[540], 1040[101], 1045[101a]
Belli, A., **3**, 789[56]
Bellido, I. S., **3**, 396[115]
Bellmann, P., **2**, 850[218]
Belloli, R. C., **7**, 26[59]
Bellora, E., **6**, 550[678]
Belloto, D., **4**, 809[163]
Bellucci, C. G., **4**, 330[3], 345[3,79]
Bellucci, G., **3**, 733[1], 734[9]
Bellus, D., **1**, 830[96], 843[2b]; **3**, 848[50]; **4**, 754[174,175], 819[209]; **5**, 4[31], 68[98], 90[58], 829[14,15]; **7**, 205[65]
Bellut, H., **8**, 724[175]
Bellville, D. J., **5**, 453[65], 520[39], 704[22], 1020[68,69], 1023[69]; **7**, 879[148], 880[148,154], 882[166]
Belmonte, F. G., **8**, 478[41], 480[41], 517[123]

Belniak, K., **2**, 538[64], 663[28], 664[28]
Belokon', Y. N., **4**, 218[147,148]
Belot, G., **8**, 135[39], 137[53], 390[83]
Belotti, D., **1**, 268[55], 269[55]; **3**, 599[209,210], 602[223], 603[228]; **5**, 310[100]
Belov, P. N., **1**, 837[147]
Belov, V. N., **5**, 432[129]
Belovsky, O., **8**, 164[47]
Below, P., **2**, 163[146]
Bel'skij, I. F., **8**, 956[5]
Belsner, K., **6**, 215[102]
Belter, R. K., **7**, 453[71]
Beltrami, H., **4**, 425[24,25], 428[24], 430[24]
Belyaeva, O. Ya., **6**, 502[209], 531[453], 554[763]
Belyakova, Z. V., **8**, 763[1], 769[1b], 771[1b], 775[77], 778[88], 785[1]
Belzecki, C., **1**, 838[163,165,166]; **5**, 99[137], 100[137], 108[207], 111[224,225], 260[63a], 264[73]
Belzner, J., **4**, 1009[145]
Bemiller, J. N., **7**, 235[1]
Bemis, A., **6**, 280[143]
Ben, I., **3**, 591[171]
Benage, B., **2**, 1067[127], 1068[127]; **5**, 467[116,117], 531[80], 539[109]; **7**, 297[33]
Benaim, J., **2**, 504[2], 506[2], 509[2], 524[2]; **3**, 23[140], 30[140], 33[140], 34[140], 35[140]; **6**, 722[140], 727[192]
Benary, E., **4**, 123[210a]
Benassi, R., **5**, 439[166]
Benati, L., **4**, 336[29,30]; **7**, 493[197]
Ben-Bassat, J. M., **5**, 706[26]
Benbow, J. W., **4**, 404[244]; **7**, 410[97b]
Bencini, E., **3**, 1033[72]
Bencivengo, D. J., **8**, 850[121]
Bencze, W. L., **3**, 566[33]
Ben-David, D., **8**, 35[66], 66[66]
Bender, C. O., **5**, 207[50]
Bender, D., **4**, 852[90]
Bender, D. R., **2**, 943[171], 945[171]
Bender, H., **6**, 440[74]; **7**, 658[26]
Bender, H.-J., **4**, 13[45], 14[45b]
Bender, H. S., **5**, 949[277]
Bender, M. L., **7**, 659[36], 660[40]
Bender, S. L., **1**, 130[97], 401[12,15], 402[12]; **2**, 263[55]; **4**, 366[7], 384[7,143]; **7**, 245[79], 408[90], 418[90], 545[25]
Benderly, A., **1**, 92[70]; **2**, 111[84]; **4**, 293[235]
Benders, P. H., **4**, 45[126]; **5**, 676[4], 686[48,49], 687[48,49]
Bendich, A., **7**, 657[25]
Bendig, J., **5**, 724[110]
Bénéchie, M., **6**, 7[30]
Benecke, B., **4**, 222[181], 224[181]
Benecke, H. P., **3**, 926[46], 928[46]
Benedetti, F., **3**, 174[534,535], 176[534,535], 178[534,535]; **4**, 20[63], 21[63], 442[184]
Ben-Efraim, D. A., **8**, 935[67]
Benesova, V., **8**, 334[65]
Ben-Et, G., **2**, 1074[143]
Benetti, S., **2**, 803[32]; **3**, 738[37]; **5**, 403[8]; **7**, 143[140,141]; **8**, 392[108], 394[116], 645[45]
Benezra, C., **1**, 219[56,58]; **7**, 550[52]
Benfaremo, N., **4**, 27[83]
Benfield, F. W. S., **7**, 3[11]
Benhamou, M. C., **4**, 290[204], 291[204,210], 292[204], 311[439,440]; **7**, 632[61]; **8**, 854[154]
Benington, F., **4**, 504[131]; **8**, 146[97], 568[466]
Ben-Ishai, D., **2**, 1074[143,144,145,146,148], 1079[157]; **5**, 406[22], 407[25], 501[264]; **6**, 96[150], 636[17]; **7**, 555[66]
Benjamin, B. M., **3**, 328[179], 726[22], 782[20]
Benjamin, L., **7**, 602[102]
Benkeser, R. A., **1**, 180[28]; **2**, 5[20], 6[20], 21[20], 588[152]; **6**, 1034[134], 1035[135]; **8**, 238[24], 249[91], 251[106], 294[54], 322[110], 478[40,41], 479[40], 480[41], 516[120,121], 517[123,124], 547[314], 629[186], 764[2], 767[23], 770[2b,31], 776[79], 880[71], 881[71]
Benkó, P., **6**, 553[702,703]
Benko, Z., **4**, 753[165]
Benkovic, P. A., **2**, 956[10]
Benkovic, S. J., **2**, 956[10]; **8**, 31[46], 66[46], 93[69], 94[69], 206[168]
Benkser, R. A., **5**, 1003[20]
Benn, F. R., **5**, 2[20]
Benn, M., **7**, 31[87]
Benn, R., **1**, 2[10], 180[30]; **4**, 608[318,319]; **5**, 442[182]; **6**, 179[124]
Benn, W. R., **8**, 928[26]
Bennani, Y. L., **4**, 110[151]
Bennasar, M.-L., **2**, 765[77]; **8**, 587[34]
Benneche, T., **5**, 1012[52]; **6**, 11[46], 842[82]
Bennek, J. A., **8**, 219[77]
Benner, C. W., **5**, 821[163], 857[225]
Benner, J. P., **2**, 534[33-35], 535[34,35]; **5**, 432[129]
Benner, S., **8**, 82[6], 84[6], 197[123]
Benner, S. A., **6**, 432[120]; **7**, 672[17]
Bennet, A. J., **8**, 388[64]
Bennet, W., **1**, 205[105]
Bennetau, B., **2**, 713[44], 726[124]
Bennett, C. F., **7**, 769[209,217]
Bennett, D. A., **2**, 74[70]; **3**, 103[205], 108[205]; **7**, 207[83], 208[83], 209[83], 210[83]
Bennett, D. W., **3**, 380[7]; **7**, 544[39], 553[39], 556[39]
Bennett, F., **1**, 760[135]; **4**, 382[135,135b], 383[135a]; **5**, 736[145], 737[145]; **7**, 440[38]
Bennett, G. B., **5**, 829[26], 853[172], 877[8]; **6**, 488[33], 834[35], 855[35]
Bennett, G. M., **7**, 763[98]
Bennett, J. N., **5**, 451[30], 453[30], 464[30], 513[5], 514[5], 527[5]
Bennett, M. A., **8**, 152[164], 443[3], 453[194,195]
Bennett, M. J., **8**, 14[80]
Bennett, R. D., **1**, 293[33]
Bennett, R. H., **7**, 208[80]
Bennett, S. M., **3**, 219[110]
Bennett, W. D., **6**, 65[2]; **7**, 229[119]
Bennua, B., **6**, 49[97], 193[215]
Benny, D., **3**, 619[27]
Bennyarto, F., **3**, 909[152]
Beno, M. A., **1**, 23[119]
Benoiton, N. L., **2**, 403[36]
Bensadat, A., **7**, 498[220]
Bension, R. M., **4**, 868[17], 869[17], 877[67]; **5**, 30[3]
Benson, A. A., **8**, 143[65]
Benson, B. W., **7**, 31[86]
Benson, F. R., **4**, 296[265], 1099[180]; **6**, 262[15], 264[15], 268[15]
Benson, R. E., **4**, 1090[139]; **5**, 1141[82]
Benson, S. W., **4**, 1072[17]; **5**, 64[36], 856[193], 900[8]; **7**, 8[54,57]; **8**, 407[55]
Benson, W., **6**, 501[204]
Benson, W. R., **5**, 723[108a]
Bentham, S., **4**, 301[322], 302[322]
Bentlage, A., **2**, 385[321]
Bentley, K. W., **3**, 836[81]; **6**, 212[83]
Bentley, P. H., **6**, 669[265]; **7**, 671[6]
Bentley, R., **6**, 653[150]
Bentley, R. K., **7**, 306[4]
Bentley, T. W., **3**, 799[99]
Benton, J. L., **7**, 10[79]

Bentor, Y., **2**, 766[83]
Bentrude, W. G., **7**, 7[42]
Bentz, P., **5**, 290[41], 1189[21,22], 1190[22], 1192[22]; **8**, 773[70], 774[70]
Ben-Yakov, H., **6**, 217[111]
Benzaid, A., **4**, 313[471]
Benzing, M., **2**, 368[233]
Benzinger, E., **3**, 660[6]
Benzing-Nguyen, L., **7**, 255[32]
Benzon, M. S., **5**, 826[158], 857[224]
Beppu, K., **1**, 410[39], 568[231]; **3**, 124[280], 125[280], 126[280]; **6**, 49[96]
Beranek, I., **4**, 730[66]
Berardi, F., **6**, 787[99,100]
Bercaw, J. E., **5**, 1145[105], 1174[37], 1178[44]; **8**, 668[2], 670[11], 671[16,18], 672[2,21], 673[23-26], 675[23,46,50], 676[24,26,51-53], 682[24,46,85], 685[98], 687[85], 689[98], 690[85], 691[23,85], 696[25,126]
Berchet, G. J., **4**, 315[517]
Berchier, F., **7**, 257[49]
Berchtold, G. A., **2**, 904[49]; **4**, 5[19,19e], 7[24], 45[126,126a]; **5**, 322[12], 855[186,189]; **7**, 78[125], 365[46,47], 429[151]; **8**, 726[190]
Bercovici, T., **5**, 729[125]
Bercz, P. J., **6**, 965[90]
Berenblit, V. V., **3**, 639[82], 644[82,137]
Berenblyum, A. S., **8**, 447[106], 450[106], 963[43]
Berenbom, M., **3**, 888[17]
Berends, W., **8**, 643[37]
Berenjian, N., **5**, 126[22], 127[22], 128[22]
Berenschot, D. R., **7**, 179[153]
Béres, J., **7**, 723[25]
Beretta, M. G., **2**, 267[64], 630[21], 631[21], 632[21], 634[21], 640[21], 641[21], 642[21], 644[21], 645[21]
Berezin, I. V., **3**, 499[125]
Bereznitskii, G. K., **3**, 305[70]
Berg, H., **3**, 482[3]
Berg, H. A. J., **5**, 799[73]
Bergami, B., **5**, 73[204]
Bergan, J. J., **6**, 22[83]; **7**, 752[154]
Bergbreiter, D. E., **1**, 357[1]; **2**, 102[24], 476[4], 489[48], 490[48], 505[11], 510[11,37,38,40-42], 917[89], 919[89], 920[89], 924[89], 935[89]; **3**, 30[181], 31[185], 32[185], 34[192,193], 39[192], 418[26]; **4**, 5[17], 221[157], 738[98]; **5**, 100[147]; **6**, 531[448], 719[125], 720[125,130], 722[139], 723[146], 725[125]; **7**, 604[138]; **8**, 113[38], 720[131], 800[67], 947[142]
Bergdahl, M., **1**, 112[27]; **4**, 152[56], 255[204]
Berge, D. D., **7**, 143[147]
Berge, J. M., **5**, 689[73]
Bergen, E. J., **7**, 76[117]
Bergens, S., **7**, 416[123]
Bergenthal, M. D., **6**, 966[96]
Berger, A., **6**, 636[17]
Berger, B., **5**, 468[123]; **7**, 131[87]
Berger, C. R. A., **8**, 314[40]
Berger, D. E., **2**, 530[23]; **5**, 4[39], 5[39,44], 8[44]
Berger, G., **8**, 35[64], 52[145], 53[128], 66[64,128,145]
Berger, H., **7**, 759[14]
Berger, J., **1**, 130[96]; **7**, 54[7]
Berger, J. G., **2**, 965[68]; **8**, 59[178], 66[178], 614[88,89], 620[89,134,135], 621[88], 624[134]
Berger, K. R., **2**, 709[6]
Berger, M. H., **4**, 10[33], 23[33d], 107[144], 262[306]
Berger, S., **1**, 191[77], 272[68], 300[67], 322[67], 335[63]; **2**, 6[35], 247[35], 630[7], 631[7]; **6**, 217[114]
Bergeron, R., **6**, 83[78]
Bergeron, R. J., **7**, 668[82,83]; **8**, 62[195], 66[195], 994[63]
Bergin, W. A., **8**, 236[6], 284[2], 285[2], 607[30]
Bergius, M., **3**, 822[5], 834[5]
Bergman, A., **5**, 715[81]
Bergman, J., **3**, 511[191]; **4**, 342[64], 343[64,72], 386[155]; **5**, 721[102]; **6**, 463[24], 824[123]; **7**, 534[43]
Bergman, P. M., **8**, 364[13]
Bergman, R., **3**, 767[165]
Bergman, R. G., **2**, 125[212,213], 127[212,213,233], 310[30], 311[30], 313[36,37], 314[39], 655[143]; **5**, 588[209], 736[144], 790[21], 797[63], 820[21], 1134[38], 1145[105], 1149[38,126]; **7**, 3[12], 4[15], 8[12], 484[136]; **8**, 289[29], 290[29], 669[6], 670[6], 797[35]
Bergmann, A., **2**, 356[132]
Bergmann, E., **3**, 724[14]
Bergmann, E. D., **2**, 352[85]; **3**, 124[264], 867[35]; **4**, 3[7], 4[7], 63[7b], 65[7b], 70[2], 258[231]; **6**, 219[124]; **7**, 107[168]; **8**, 228[125]
Bergmann, H.-J., **2**, 212[121]
Bergmann, K., **8**, 319[79]
Bergmann, M., **6**, 632[3], 635[3]
Bergmann, W., **3**, 892[51]
Bergmark, W., **3**, 580[107]
Bergmeyer, H. U., **2**, 456[45,47]
Bergmeyer, J., **2**, 456[45,47]
Bergson, G., **6**, 489[96]
Bergstein, W., **4**, 874[54]; **5**, 30[2]; **8**, 460[254]
Bergstorm, F. W., **4**, 491[67]
BergStresser, G. L., **1**, 15[80]
Bergstrom, D. E., **3**, 459[135], 462[149,151], 470[218], 476[218], 513[207]
Bergstrom, F. W., **8**, 143[65]
Bergstrom, R. G., **6**, 556[26], 563[26]
Beringer, F. M., **4**, 42[122a]; **8**, 995[70]
Berk, H. C., **3**, 623[33]; **5**, 347[73a]
Berk, S. C., **1**, 212[5,8,11], 213[5,8,11], 214[5b,19], 432[138], 433[138]; **2**, 450[50]; **3**, 209[18]; **4**, 175[42]
Berka, A., **7**, 704[8]
Berkessel, A., **5**, 200[30], 201[31]
Berkowitz, L. M., **7**, 236[12], 572[114]
Berkowitz, W. F., **5**, 143[97,98], 636[88], 945[250]; **6**, 786[96]; **7**, 242[61]
Berks, A. H., **1**, 262[35], 739[41]; **2**, 441[1], 443[1]
Berkvich, E. G., **4**, 485[29]
Berlan, J., **1**, 112[27], 124[81]; **2**, 448[43]; **3**, 155[430]; **4**, 152[58], 210[70-76], 229[223], 240[45]; **5**, 557[53]; **8**, 111[21], 123[21]
Berlin, A. J., **4**, 1023[260]; **5**, 1006[35]
Berlin, A. Y., **5**, 453[59]
Berlin, K. D., **4**, 273[46], 280[46]
Berlin, Y. A., **6**, 603[21]
Berlo, R. G., **8**, 367[58]
Berman, D. A., **6**, 809[69]
Berman, E., **7**, 438[22]
Berman, E. M., **4**, 403[241]
Berman, R. J., **7**, 545[26]
Berman, Z., **7**, 79[135]
Bernady, K. F., **4**, 141[15], 142[15]; **6**, 648[124]
Bernal, I., **1**, 310[107]; **8**, 459[238]
Bernard, D., **1**, 331[50]; **3**, 482[5], 499[5], 505[5], 509[5]
Bernard, G., **1**, 631[54], 658[54], 659[54], 660[54], 661[54], 662[54], 663[54], 664[54], 672[54]; **3**, 87[72], 95[72]
Bernard-Henriet, C., **5**, 109[219]; **6**, 578[980]; **7**, 502[262]
Bernardi, A., **2**, 103[28], 221[146], 267[63], 488[43], 605[57], 614[57], 630[21], 631[21], 632[21], 634[21], 636[56], 637[56,59], 640[21,56,59], 641[21,71], 642[21,71,73,74], 643[73,74], 644[21,73], 645[21,59], 652[59], 920[98], 930[131], 931[131]; **4**, 152[58], 207[57,58]

Bernardi, F., **1**, 506[8,11]; **4**, 330[5]; **5**, 75[220,223], 114[244]; **6**, 133[4]
Bernardi, R., **1**, 55[23], 153[57], 543[28]; **8**, 187[47]
Bernardinelli, G., **1**, 307[111], 312[111], 770[187]; **2**, 631[17], 632[17], 634[17], 924[108a]; **4**, 21[69], 85[72], 111[153], 201[16], 202[16], 204[33], 218[137], 231[282], 249[120], 520[28]; **5**, 45[36], 157[38], 260[70], 263[70], 362[93], 363[93a,d], 364[93d], 365[93a,e], 543[117], 545[120]
Bernardini, A., **3**, 251[75]
Bernardon, C., **1**, 226[94], 227[99]; **4**, 98[110]
Bernardon, J.-M., **3**, 262[167]; **7**, 359[16]
Bernardou, F., **4**, 883[96,97], 884[96,97,105]
Bernasconi, C. F., **2**, 349[67-69]
Bernasconi, S., **1**, 566[216]; **2**, 833[148]; **4**, 261[285,288,290]
Bernassau, J.-M., **8**, 111[20], 118[20]
Bernath, G., **5**, 583[186], 584[194]; **6**, 525[382], 534[519]
Bernath, J., **7**, 872[98]
Bernauer, K., **2**, 965[70]
Berndl, K., **6**, 554[787,788], 572[957]
Berndt, A., **5**, 850[152]; **6**, 970[122]; **7**, 506[293]; **8**, 532[127]
Berner, D., **3**, 373[129]
Bernet, B., **2**, 303[6]; **5**, 255[50], 264[50b]; **6**, 561[878]; **7**, 493[185]
Bernett, R. G., **4**, 366[6]
Berney, D., **6**, 531[456]
Bernhard, E., **2**, 780[10]
Bernhard, W., **3**, 17[92], 18[92]; **4**, 243[64], 247[64], 255[64], 260[64]; **7**, 646[24]
Bernhardt, J. C., **1**, 450[212]; **3**, 470[204], 473[204], 484[24], 501[24]
Bernhart, C., **8**, 60[192], 66[192]
Bernini, R., **4**, 370[28]
Berno, P., **1**, 29[150]
Bernotas, R. C., **4**, 404[247], 405[247a,b]; **7**, 636[72]
Bernou, A., **7**, 477[72], 483[72]
Bernstein, M. A., **3**, 227[208]; **8**, 52[134], 66[134]
Bernstein, M. D., **6**, 724[162]; **8**, 26[26], 37[26], 47[26], 54[26], 55[26], 57[26], 60[26], 66[26], 351[151]
Bernstein, P. R., **3**, 17[86]
Bernstein, S., **6**, 959[36]; **7**, 27[71]; **8**, 566[450]
Bernstein, S. C., **2**, 278[10]; **8**, 566[450], 767[23]
Bernstein, Z., **2**, 1074[145]; **6**, 96[150]
Berrada, S., **4**, 107[143e], 259[277]
Berridge, J. C., **5**, 71[151], 637[104,105], 671[49]
Berridge, M. S., **4**, 445[205]
Berrier, C., **7**, 333[20]
Berrios, R., **5**, 844[129]
Berris, B. C., **5**, 1151[128]
Berry, D., **7**, 24[23]
Berry, D. E., **1**, 215[33]
Berry, D. J., **4**, 868[16]
Berry, F. J., **8**, 849[116]
Berry, J. M., **6**, 19[69]
Berry, K. H., **8**, 597[88]
Berry, M., **6**, 686[370]; **8**, 889[129]
Berry, R. S., **4**, 483[10]
Berryhill, S. R., **3**, 1036[81]; **4**, 401[232], 563[37]
Bersch, H. W., **5**, 342[62a]
Berse, C., **5**, 222[63]; **6**, 529[424]
Bershas, J., **4**, 125[216,216e]
Bershas, J. P., **2**, 824[119]
Berski, Z., **1**, 474[92]
Berson, J. A., **1**, 885[134]; **3**, 706[7]; **5**, 2[13], 3[13], 240[2,2a,b,3], 568[106], 603[50], 639[122,124], 714[65], 786[3], 787[3,6,9], 794[49], 805[99], 815[142], 856[209], 1009[45], 1016[61], 1025[81,81e]
Bert, L., **3**, 242[2]
Bertelli, D. J., **4**, 663[4]
Bertelo, C. A., **3**, 1027[42], 1030[42]; **8**, 681[74], 684[74], 691[74,106]
Berthelot, J., **4**, 345[81]; **8**, 133[19,20]
Berthelot, P., **8**, 660[108]
Berthet, D., **4**, 242[62], 253[62], 261[62]
Berthiaume, G., **4**, 29[85], 46[131]
Bertho, A., **8**, 568[471]
Berthold, H., **6**, 452[132]
Bertholon, G., **3**, 306[83]
Berti, G., **1**, 822[35], 824[35]; **2**, 410[6], 411[6], 415[6]; **3**, 725[19], 733[1], 741[51], 743[58], 745[65]; **5**, 456[87], 464[114,115], 466[114]; **7**, 358[13], 362[13], 363[13], 364[13], 365[13], 373[13]
Bertini, F., **1**, 206[111], 830[92], 832[92]; **7**, 16[160]
Bertini, V., **6**, 775[51]
Bertocchio, R., **4**, 4[14]
Bertolasi, V., **1**, 309[101]; **5**, 113[230]
Bertolini, G., **2**, 636[57], 637[57], 640[57], 653[57]; **4**, 159[82], 218[145]; **6**, 118[105]
Bertolo, P., **3**, 804[11]
Berton, A., **4**, 915[15]
Bertounesque, E., **1**, 202[99]; **2**, 614[117], 630[24], 631[24], 632[24], 634[37], 640[37]
Bertozzi, S., **4**, 877[68]; **5**, 1154[151,158], 1155[151]
Bertram, E., **2**, 368[244]
Bertram, J., **7**, 8[56], 793[3]
Bertran, J., **5**, 64[40], 72[183,186,187], 73[189,190]
Bertrand, H. T., **4**, 883[100], 884[100]
Bertrand, J., **2**, 281[29]; **4**, 21[65], 100[123]
Bertrand, J. L., **1**, 648[135], 653[135], 659[135], 686[135]; **3**, 88[127], 89[127], 91[127,147], 92[127], 93[127], 109[127,147], 116[127]
Bertrand, M., **2**, 547[95]; **3**, 599[207]; **4**, 55[156,157], 245[81], 249[81], 1002[56]; **5**, 515[17], 518[17], 547[17], 772[153,154,156,158,161]; **6**, 563[892]; **8**, 332[41]
Bertrand, M. P., **4**, 807[149,150]; **7**, 92[40]
Bertucci, C., **3**, 483[12]; **5**, 1152[144]
Bertz, S. H., **1**, 122[69], 377[102], 378[106], 428[121], 429[121], 432[137], 456[137], 457[121]; **2**, 381[308]; **3**, 209[12], 210[24], 211[27], 213[43,46], 215[27b], 216[46], 223[46]; **4**, 89[87], 90[87d], 107[143c], 141[13], 177[58], 180[58a], 229[230], 255[190,191]; **5**, 657[31], 903[40]; **6**, 9[40]; **8**, 940[102]
Bérubé, D., **3**, 649[190]
Bérubé, G., **3**, 380[10]; **4**, 339[44], 1011[162]; **5**, 532[86,86c]
Bérubé, N., **8**, 254[126]
Berwin, H. J., **8**, 99[107], 750[63]
Besace, Y., **1**, 112[27], 124[81]; **2**, 448[43]; **4**, 152[58], 210[71-76]; **8**, 111[21], 123[21]
Besan, J., **6**, 543[612]
Besemann, M., **7**, 846[90]
Beslin, P., **2**, 73[68], 214[130,131]; **5**, 556[52], 829[22]; **6**, 454[145]
Bespalova, G. [V., **6**, 509[272], 538[554]
Bessey, E., **6**, 443[95]
Bessho, K., **5**, 404[14]
Bessiere, J.-M., **3**, 744[61]
Bessiere, Y., **3**, 215[65]; **5**, 384[126a]
Bessiere-Chretien, Y., **1**, 377[100]; **3**, 771[183,189]
Best, W. H., **4**, 489[58]
Bestian, H., **1**, 140[8], 141[15]; **5**, 105[187], 107[187]; **8**, 755[115]
Bestmann, H. J., **1**, 368[63], 391[63], 722[272,273,274], 755[115], 808[320], 812[115], 813[115], 844[3c]; **2**, 1004[58], 1005[58]; **3**, 200[73], 286[59], 644[159], 877[88], 887[4], 888[4], 893[4,56], 897[4,76], 898[76], 900[4], 903[4]; **4**, 31[94,94d], 55[157], 115[182,182e], 259[261]; **6**, 83[81], 134[33], 171[3,5], 172[5,11,25], 174[56,58,59,61], 175[56,81], 176[85], 177[56,110,111,113], 179[81,127], 181[110,111,130,131,133], 182[56,85,110,111,113,133,134,137], 184[110,111,149], 185[110,111,154,158,159], 186[169,171], 187[158,176,177], 188[81,113,181,183], 189[186], 190[193,201,203],

191[204], 192[204], 193[85,204,207,208,209,210,211,212,213,214,217], 194[204,219,220], 195[204,212,222,223,224,225], 196[201,204], 197[169,204], 198[3,5], 199[5], 200[5,110,111], 201[5,110], 202[5], 449[115], 558[839]; **7**, 109[185], 213[101,102]; **8**, 860[223], 863[238]
Beswick, P. J., **3**, 494[88]; **4**, 524[64], 525[64], 526[64]
Betancor, C., **4**, 375[98a], 388[98,98a], 409[98a]; **7**, 41[15], 157[34]
Betancourt de Perez, R. M., **1**, 790[263]; **2**, 103[38]
Betanelli, V. I., **6**, 43[52]
Beth, A., **8**, 52[147], 66[147]
Bethel, J. R., **2**, 170[172]
Bethell, D., **3**, 770[178]; **7**, 26[50], 874[110]
Betlinetti, G. F., **7**, 882[170]
Betschart, C., **1**, 809[328]; **2**, 897[18]; **3**, 580[104]
Betterton, K., **7**, 739[35]
Bettman, B., **7**, 602[100]
Bettolo, R. M., **3**, 717[44]
Bettoni, G., **6**, 787[100]
Betts, M. J., **2**, 1059[74]; **5**, 407[28]
Betz, R., **3**, 253[93]
Betz, W., **6**, 294[243]
Betzecki, C., **6**, 520[343]
Beugelmans, R., **3**, 505[157]; **4**, 452[17], 459[71-75], 460[95-97], 464[71,73,114,123], 465[73,96], 466[17,71,73,123,125,126], 467[72,97], 469[75], 470[75,97,138], 474[95], 475[74,95,152], 476[17], 477[114], 478[71,73,96,114,125,166,167], 479[123,126,169,170,171,172], 480[74], 1089[126]; **7**, 169[108], 223[42], 878[140]
Beumel, O. F., **8**, 547[316]
Beurskens, P. T., **8**, 96[92]
Beverung, W. N., **6**, 274[105]
Bevetskaya, I. P., **4**, 594[141,144]
Bewersdorf, M., **1**, 661[167,167d], 662[167d], 677[167d], 832[108]
Bewick, A., **3**, 568[44,45], 636[59]; **7**, 494[203], 495[203,204,209]
Bey, A. E., **1**, 411[49]
Bey, P., **3**, 365[63]; **4**, 125[216,216h]; **6**, 645[96]; **7**, 324[73], 711[59]; **8**, 932[44]
Beyer, L., **6**, 526[393]
Beyermann, M., **5**, 724[110]
Beyler, R. E., **7**, 100[124], 256[40]
Bezemer, L., **5**, 707[40]
Bezmenov, A. Ya., **7**, 595[18]
Bezuidenhoudt, B. C. B., **4**, 384[142]
Bezzubov, A. A., **3**, 648[181]
Bhacca, N. S., **4**, 1090[142]; **5**, 128[27], 631[54-56]
Bhaduri, A. P., **6**, 943[157]; **7**, 265[103], 267[103]
Bhaduri, S., **7**, 766[185,186]; **8**, 446[73], 447[117]
Bhagade, S. S., **2**, 345[22]
Bhagwat, M. M., **7**, 601[82,83]; **8**, 450[170]
Bhagwat, S. S., **4**, 107[141]; **5**, 151[7], 180[7]
Bhakta, C., **8**, 938[92]
Bhakuni, D. S., **5**, 829[26]
Bhalerao, U. T., **1**, 568[237]; **4**, 391[175d]; **7**, 85[9], 86[16b], 87[17], 108[17]
Bhandal, H., **4**, 761[200]
Bhandari, K., **6**, 538[552], 550[552]
Bhanot, O. S., **6**, 624[136], 625[156]
Bhanu, S., **3**, 281[41]
Bharathi, S. N., **1**, 515[56]
Bharucha, K. R., **4**, 310[433]
Bhasin, K. K., **4**, 520[36], 531[36]
Bhat, G. A., **6**, 554[711]; **7**, 800[30]
Bhat, K. S., **1**, 192[81]; **2**, 6[23,23c], 10[23], 33[23c,123], 42[146], 43[146,146b], 44[146b,c], 45[146b,c], 52[23a]; **6**, 865[201]; **8**, 714[80], 722[145]
Bhat, N. G., **4**, 147[39]; **7**, 597[49]
Bhat, V., **4**, 239[18], 249[18], 257[18], 261[18]; **8**, 844[68]
Bhat, V. V., **7**, 803[57]
Bhatacharjee, S. S., **8**, 224[108]
Bhati, A., **2**, 760[41]; **7**, 231[144]
Bhatia, A. V., **7**, 341[51,52]
Bhatnagar, A. D., **7**, 43[37]
Bhatnagar, A. K., **6**, 915[30]
Bhatnagar, I., **7**, 231[149], 738[23]
Bhatnagar, N. Y., **3**, 505[159]
Bhatnagar, S. P., **2**, 527[1], 528[1], 553[1], 720[86]; **6**, 806[54]
Bhatt, D. G., **4**, 240[53]
Bhatt, M. V., **2**, 782[23]; **6**, 646[104], 647[104], 659[104], 660[104]; **7**, 186[182], 235[8], 800[30]; **8**, 406[44], 408[44], 568[467]
Bhattacharjee, D., **4**, 1018[225], 1019[225]; **5**, 473[153], 477[153]; **7**, 318[50]
Bhattacharjee, G., **3**, 908[146]; **4**, 1040[103]
Bhattacharjee, M. N., **7**, 267[117], 268[117]
Bhattacharjee, S. S., **6**, 456[162], 457[162]
Bhattacharya, A., **1**, 66[53,54]; **2**, 536[41,44], 538[60], 539[60]; **4**, 230[251]; **7**, 877[135]
Bhattacharya, K. K., **3**, 304[64]
Bhattacharya, S., **7**, 823[35]
Bhattacharya, S. K., **5**, 92[68], 95[68], 96[68]
Bhattacharyya, B. K., **8**, 530[101]
Bhattacharyya, N. K., **6**, 817[104]
Bhattacharyya, S., **8**, 505[80]
Bhattacharyya, S. C., **5**, 775[177], 780[177]; **7**, 239[47], 558[79], 560[79]
Bhattasharya, A., **1**, 87[50]
Bhupathy, M., **1**, 53[17], 135[116], 222[70], 239[40], 240[40], 786[248], 864[86], 887[139], 888[139]; **4**, 373[79], 374[79]; **5**, 455[75], 456[84], 1017[65], 1018[65,65a], 1020[65,65a]; **6**, 146[89]; **7**, 202[50]
Bhushan, V., **2**, 506[13], 514[51], 523[13,84], 524[84]; **3**, 728[34]; **6**, 119[110]; **7**, 187[185]
Biacchi, G. S., **4**, 510[168,169]
Bialecka-Florjańczyk, E., **2**, 537[50]
Bianchetti, G., **4**, 1099[187]; **6**, 705[27]
Bianchi, D., **7**, 429[151]
Bianchi, G., **4**, 1093[147]; **7**, 143[142]; **8**, 645[41]
Bianchi, M., **8**, 87[27], 236[3], 239[3], 552[352]
Bianchini, C., **5**, 1158[172]; **8**, 458[219]
Bianchini, J.-P., **4**, 277[82], 284[82]
Bianchini, R., **4**, 330[3], 345[3,79]
Bianco, E. J., **7**, 157[33]; **8**, 974[122]
Bianco, V. D., **8**, 446[63], 858[209]
Biasha, J., **1**, 312[112,113]
Bibby, C., **3**, 770[180]; **5**, 165[87], 936[199]
Bichelmayer, K.-P., **6**, 575[969]
Bïchi, G., **3**, 936[72]
Biciunas, K. P., **6**, 190[194]
Bickart, P., **5**, 890[34]; **6**, 152[141], 899[109]; **8**, 238[24]
Bickel, A., **2**, 1022[53]
Bickel, A. F., **4**, 1017[210]
Bickel, C. L., **4**, 41[118]
Bickel, H., **4**, 373[68]; **5**, 341[61a]
Bickelhaupt, F., **1**, 13[69], 16[86], 26[132,133,134], 218[54], 746[70]; **2**, 378[287], 801[36], 981[23]; **4**, 595[152], 869[23,24,26], 877[69], 884[108], 972[79], 1002[59], 1018[226]; **5**, 46[37], 850[147], 908[73], 986[38], 1125[57]; **6**, 86[99], 856[150]; **7**, 373[73]; **8**, 863[236]
Bicker, U., **7**, 471[23], 474[23]
Bickley, D. G., **8**, 675[48], 676[48]
Biddlecom, W. G., **3**, 953[103]
Bieber, J. B., **3**, 975[3], 980[3]
Bieber, L., **5**, 331[42,42b]
Bied, C., **1**, 253[10]
Biedenbach, B., **5**, 1199[46]

Biedermann, A., **2**, 399[21]
Bieg, T., **8**, 212[7]
Biegi, E., **5**, 381[118]
Biehl, E. R., **1**, 554[103]; **4**, 483[5], 484[5], 485[26], 486[38], 492[74,75], 493[79], 495[5], 497[5b,98], 499[5b,102]; **6**, 690[398], 692[398]
Biehl, H., **4**, 484[15]
Bielawska, A., **4**, 290[204], 291[204], 292[204], 311[440]
Bieler, A., **8**, 248[89]
Biellmann, J. F., **1**, 508[20], 512[44], 568[225], 630[30], 678[30]; **2**, 55[1], 71[1c,54]; **3**, 86[14], 94[14], 95[14,151,155], 96[151,161,164,165], 98[151,161,164], 99[151,164,165,179], 100[14,179], 101[165], 107[165], 149[404,406], 151[406], 152[404,406], 153[406], 196[29]; **4**, 1061[163]; **6**, 145[78], 832[18], 961[71]; **8**, 445[16], 597[87]
Bielski, R., **6**, 687[380], 987[72]; **8**, 822[52]
Bien, S., **3**, 896[70]; **4**, 1040[92], 1042[92]; **7**, 95[67], 107[167]
Bieniek, A., **1**, 329[38], 474[90,91]
Bienz, S., **3**, 288[65]; **5**, 372[104]
Bierbaum, V. M., **8**, 91[60]
Bierenbaum, R., **3**, 592[172]; **8**, 252[111]
Bieri, J. H., **5**, 92[78]; **6**, 539[579], 544[626]
Bierling, B., **7**, 140[130], 141[130]
Bierling, V. B., **4**, 611[345]
Bierman, M. H., **7**, 705[17]
Biermann, T. F., **7**, 729[43]
Biernat, J., **6**, 624[147]
Biesemans, M., **8**, 348[130], 349[130]
Biffin, M. E. C., **4**, 295[257]; **6**, 245[119], 246[119], 247[119], 248[119], 249[119], 251[119], 252[119], 253[119], 254[119], 255[119], 256[119], 258[119]; **8**, 501[57]
Biftu, T., **3**, 676[75]
Bigalke, J., **3**, 277[29]; **6**, 562[888]
Bigelli, C., **3**, 554[27]
Bigelow, H. E., **8**, 364[19], 365[28], 366[36]
Bigelow, L. A., **7**, 15[143]
Bigelow, S. S., **2**, 13[52], 26[100], 27[100b], 29[103], 35[128]; **6**, 46[75]
Bigelow, W. B., **3**, 565[19]
Bigg, D. Ch., **6**, 579[987]
Bigham, E., **7**, 828[51]
Bigi, F., **3**, 311[100]
Bigler, P., **5**, 664[38]
Bigley, D. B., **8**, 249[95], 263[27], 269[27,87], 273[27], 275[27], 278, 725[184]
Bigot, B., **4**, 1082[86]
Bihari, V., **7**, 71[95]
Bihlmaier, W., **4**, 1092[144], 1093[144], 1102[199]
Bihovsky, R., **3**, 216[78]
Bijev, A., **5**, 428[110]
Biletch, H., **8**, 408[63]
Bilevitch, K. A., **7**, 884[185,186]
Bilgic, S., **5**, 71[156]
Billard, C., **8**, 447[116]
Billedeau, R., **1**, 373[90], 375[90], 376[90], 477[137,138]; **2**, 1075[150]
Biller, S. A., **4**, 792[68]; **7**, 489[166]
Billhardt, U.-M., **8**, 545[285]
Billica, H. R., **8**, 140[9], 141[9], 143[9]
Billig, E., **8**, 859[216]
Billigmeier, J. E., **1**, 886[135]; **5**, 787[10]
Billimoria, J. D., **2**, 147[76]
Billingham, N. C., **7**, 721[16]
Billington, D. C., **3**, 423[73], 424[73], 426[73], 428[73], 1025[33,33a]; **5**, 1038[10], 1043[25], 1044[26], 1046[25b], 1048[25a,b,32], 1051[10,26,32,36a], 1055[45], 1056[25a], 1062[45]
Billion, A., **7**, 227[87]
Billman, J. H., **8**, 36[87], 38[86,87], 66[86,87], 568[467]
Billmers, J. M., **7**, 162[63,65], 181[65], 772[291], 779[425]
Billmers, R., **1**, 838[158]; **7**, 162[64]
Billups, W. E., **4**, 1002[55,62], 1015[193,194,196,197]; **5**, 794[46], 911[94], 948[269], 1199[47]; **7**, 16[160]
Biloski, A. J., **4**, 398[213]; **6**, 116[84]; **7**, 503[272], 745[79], 763[89]
Bilovic, D., **5**, 513[4]
Bilow, N., **3**, 530[69], 533[69], 537[69]
Biltz, H., **3**, 822[5], 826[36], 834[5]
Bimanand, A., **4**, 1076[47]
Bimanand, A. Z., **5**, 627[43], 629[49]
Bindel, T. H., **6**, 281[147]
Binder, D., **2**, 828[129]
Binder, D. A., **8**, 84[10]
Binder, H., **6**, 51[113]
Binder, J., **1**, 623[76], 788[258], 823[44c], 825[44c]; **3**, 199[54]
Binder, V., **6**, 840[71]
Bindewald, R., **6**, 668[253], 669[253]
Bindra, J. S., **5**, 560[72]; **7**, 686[99]
Bindra, R., **4**, 878[75], 898[75]; **5**, 560[72]
Bineeva, N. G., **3**, 304[68]
Binger, P., **2**, 589[158]; **5**, 63[14], 65[62], 244[15], 287[34], 288[34,36], 289[34,37,38], 290[34,39,41], 291[34,43], 293[45-47,49], 294[34,50,52], 296[34,53,54], 297[34,55,56,58], 308[34], 641[134], 901[18], 905[18], 947[18], 948[18], 1185[1], 1188[17-20], 1189[21,22], 1190[22,23,25-29], 1191[25,26,27b,28,30], 1192[22,29,31,32], 1193[30,32], 1194[33,34], 1195[34-36], 1196[37], 1197[31,40-44], 1199[46], 1204[1a]; **7**, 598[53,57]; **8**, 717[105,106], 741[37]
Bingham, E. M., **1**, 425[103]; **2**, 366[221]
Bingham, R., **2**, 840[182]
Binkley, E. S., **2**, 109[64], 159[129], 518[62], 1012[17], 1013[17]; **3**, 8[36]; **4**, 31[94], 32[94f], 241[56], 255[56], 260[56]
Binkley, R. W., **5**, 125[17]; **6**, 19[69], 34[12], 46[12], 51[12]
Binkley, S. B., **2**, 463[81], 464[81]
Binks, R., **6**, 734[9]
Binnewies, M., **5**, 577[147]
Binns, F., **8**, 584[15]
Binns, M. R., **1**, 520[69,70,72,73], 635[90], 636[90], 678[90], 681[90], 691[90]; **2**, 66[33], 75[33,82a]; **3**, 87[94]; **4**, 12[37,37a-d], 119[192a,b,193], 159[85], 226[190,191,192,194], 258[245], 260[245]; **6**, 150[114], 154[145], 864[192]
Binns, R. N., **1**, 635[83], 678[83], 681[83], 691[83]
Binns, T. D., **3**, 638[94]; **7**, 805[67]
Binsch, G., **3**, 896[71]; **5**, 714[74]; **8**, 335[67]
Biougne, J., **4**, 48[137,137c], 51[143]
Biran, C., **2**, 716[66]; **3**, 577[90]
Birbaum, J.-L., **4**, 770[248]
Birch, A. J., **2**, 170[171,173], 598[17], 619[149], 723[98], 761[51], 838[165]; **3**, 17[91], 831[63]; **4**, 6[21], 27[79,79c], 181[74], 665[8,10,11], 666[11], 667[10,11], 668[14], 669[11], 670[8,10,14-17,19,21,28], 674[8,10,17], 677[14], 683[59,60], 687[59,60,63,64], 688[65,66], 695[2], 698[22], 1018[221]; **5**, 841[100]; **6**, 690[402], 692[402]; **7**, 884[183]; **8**, 139[2], 152[2,173], 154[2], 212[18], 249[91], 293[52], 302[52], 443[1], 445[13,14], 447[1a], 452[13,184,184a], 490[4,8], 491[12], 492[4], 493[21], 499[42], 500[8,48], 501[53,55], 502[59], 507[53,84,85], 514[108-110], 515[114,115,119], 524[2], 526[19,36], 530[2,98], 531[2], 533[147], 535[147], 605[12,13], 608[35,36], 609[12], 614[12], 629[12], 910[83], 971[104]
Birch, A. M., **3**, 384[52]; **5**, 140[86], 181[151]
Birch, D. J., **4**, 745[140]
Birch, S. F., **3**, 331[198]
Birch, Z., **1**, 41[202]
Birchall, G. R., **2**, 739[43a]
Birchall, J. M., **4**, 1000[13]; **7**, 488[162,163], 750[135]
Birckelbaw, M. E., **8**, 242[41]

Bird, C. W., **2**, 771[12], 844[201]; **3**, 744[62], 807[19], 1015[2], 1027[2]; **5**, 947[264], 1130[4], 1133[27]; **6**, 276[115]
Bird, M. J., **7**, 770[254]
Bird, T. G. C., **3**, 159[457], 173[457]; **5**, 137[80], 386[135]; **6**, 217[112]; **7**, 69[92], 73[92]
Birke, A., **8**, 329[24], 336[24]
Birkhahn, M., **3**, 851[67]
Birkhofer, H., **4**, 758[191]
Birkinshaw, T. N., **5**, 50[43], 403[9]
Birkofer, L., **1**, 583[8,8b]; **2**, 725[111], 726[111], 935[147]; **5**, 99[138], 100[138]; **6**, 91[124], 653[149,150]
Birks, J. B., **5**, 650[23], 653[23]; **7**, 852[42]
Birktoft, J. J., **8**, 206[172]
Birladeanu, L., **2**, 744[99], 745[99]; **4**, 972[78]; **5**, 478[162]; **6**, 291[222], 707[42]; **7**, 236[22,25]
Birnbach, S., **6**, 639[54]
Birnbaum, G. I., **2**, 538[64]
Birnbaum, J., **7**, 429[156]
Birnbaum, S. M., **8**, 145[88]
Birr, C., **6**, 636[24]
Birr, K.-H., **8**, 725[182]
Birse, E. F., **1**, 774[209]
Birss, V. I., **8**, 848[103], 996[72]
Birum, G. H., **1**, 562[170]; **7**, 206[69]
Bisacchi, G. S., **5**, 692[91], 696[91]
Bisagni, E., **1**, 471[70], 474[109], 475[109]; **2**, 379[294]; **7**, 350[21]
Bisaha, J., **4**, 524[59]; **5**, 186[170], 361[92], 543[116,116b]
Bischofberger, K., **6**, 902[125]
Bischofberger, N., **2**, 456[33], 457[33], 458[33], 459[33], 460[33], 461[33], 462[33], 466[33]; **5**, 929[169], 930[169]
Bischoff, C., **3**, 848[52]
Bishop, J. L., **2**, 969[83,83b]
Bishop, K. C., III, **5**, 1185[1]
Bishop, P. M., **2**, 303[6]; **3**, 371[115]; **5**, 37[22a]
Bishop, R., **3**, 383[46]; **4**, 293[238,239]; **6**, 266[54], 278[131]
Biskup, M., **4**, 185[90], 186[90], 248[109]
Bisling, M., **2**, 127[239]; **8**, 889[133]
Biss, J. W., **8**, 275[141]
Bissell, R. L., **2**, 190[55,56]
Bisset, G. M. F., **2**, 780[14]
Bissinger, W. E., **6**, 204[14]
Bisson, R., **6**, 268[61]
Biswas, K. M., **6**, 737[36]; **8**, 315[55], 316[55]
Biswas, S., **3**, 783[28]
Biswas, S. G., **8**, 331[35]
Bitar, H., **7**, 95[71]
Bite, M. G., **4**, 55[157], 57[157b]
Bitenc, M., **3**, 652[223], 653[223]
Bitler, S. P., **3**, 528[46]
Bitter, I., **6**, 499[177], 514[307], 520[340]
Bitter, J., **2**, 371[261]; **5**, 76[239]
Bittler, D., **7**, 74[111], 75[111]; **8**, 881[72], 882[72]
Bittler, K., **4**, 939[74]
Bittman, R., **6**, 23[90]; **7**, 393[16], 398[16]
Bittner, A., **1**, 301[73]
Bittner, S., **2**, 856[249]; **6**, 825[126]; **7**, 692[22]
Bizzozero, N., **8**, 61[187], 66[187]
Bjelakovic, M., **8**, 875[33]
Bjeldanes, L. F., **2**, 943[171], 945[171]
Bjellquist, B., **7**, 14[138]
Bjoerkling, F., **3**, 489[61], 495[61], 504[61], 511[61], 515[61]
Bjorge, S. M., **4**, 903[202], 904[202]
Björkling, F., **6**, 811[76]; **8**, 163[36]
Bjorklund, C., **3**, 499[120]; **8**, 252[111]
Bjorkman, E. E., **4**, 560[23], 596[164], 597[171], 621[164]; **7**, 490[174,175]
Bkole, S. I., **4**, 231[265]
Black, A. Y., **4**, 18[59], 262[313]; **6**, 176[93]
Black, D. K., **5**, 797[59]
Black, D. St. C., **2**, 780[10]; **3**, 575[83], 826[34]; **4**, 12[42], 1076[37]; **5**, 474[158]; **6**, 428[81], 487[71], 489[71], 937[116], 939[116], 942[116]; **7**, 229[109]
Black, H. K., **3**, 553[12]
Black, L. A., **1**, 555[116]
Black, T. H., **2**, 835[159]; **3**, 592[172]; **4**, 261[288], 1033[34]
Blackadder, D. A., **8**, 364[18]
Blackburn, B. K., **8**, 584[16]
Blackburn, C., **5**, 597[23], 603[23], 606[23]
Blackburn, D. E., **7**, 35[103]
Blackburn, E. V., **7**, 883[175]
Blackburn, G. M., **3**, 202[85], 919[30], 933[30,62], 946[30]; **4**, 507[154]; **6**, 893[79]
Blackburn, T. F., **1**, 143[33]; **4**, 153[62c]; **7**, 453[80]; **8**, 669[4], 679[4], 687[4], 691[105]
Blackett, B. N., **3**, 742[57], 743[57,60]
Blacklock, T. J., **1**, 447[198]; **4**, 573[8], 614[376], 905[208]; **5**, 67[84,86,88,89], 795[52], 914[114], 948[270]; **6**, 789[106]
Blackman, N. A., **7**, 229[109]
Blackstock, S. C., **5**, 430[117]; **7**, 874[110]
Blackstock, W. P., **2**, 852[233]
Blackwell, J., **8**, 314[39]
Blackwood, R. K., **6**, 226[12], 265[38]
Blade, R. J., **3**, 545[122]; **6**, 902[128]; **8**, 676[69], 679[69], 681[77], 684[77], 694[77]
Blade-Font, A., **8**, 860[222]
Bladon, C. M., **2**, 555[146]; **5**, 437[159]
Bladon, P., **4**, 703[32], 704[32]; **5**, 1039[18], 1046[18], 1051[36a], 1062[59]; **7**, 582[149]; **8**, 532[128]
Blagg, J., **1**, 307[111], 312[111]; **2**, 252[41], 253[41]; **3**, 197[37]; **5**, 362[93], 365[93e]
Blagg, M., **8**, 927[21]
Blagg Cox, M., **1**, 767[180]
Blagoev, B., **2**, 211[112], 942[166], 944[166]
Blagoy, M., **2**, 211[112]
Blaha, K., **1**, 366[42]; **6**, 704[9], 801[29]
Blair, I., **5**, 918[126]
Blair, I. A., **1**, 411[46]; **4**, 27[79]; **6**, 174[60]; **7**, 200[39]; **8**, 321[99,103]
Blair, J. A., **4**, 274[55]; **8**, 70[226]
Blair, J. M., **5**, 646[2]
Blair, P. A., **1**, 476[123]; **3**, 35[204], 66[14]; **7**, 187[183]
Blais, C., **8**, 847[94]
Blais, J., **2**, 6[26]
Blaise, E. E., **2**, 297[90]; **4**, 97[107b]
Blake, A. J., **6**, 126[152]
Blake, J. F., **1**, 297[58]
Blake, K. W., **5**, 947[259]
Blakeney, A. J., **4**, 1015[196]
Blanc, A. A., **8**, 444[8]
Blanc, P.-Y., **2**, 138[23]
Blanc-Guenée, J., **3**, 147[388], 151[388]
Blanch, R., **4**, 484[15]
Blanchard, E. P., **4**, 969[63]
Blanchard, E. P., Jr., **8**, 828[76]
Blanchard, J. M., **8**, 545[296]
Blanchard, L., **3**, 629[51]
Blanchard, M., **5**, 341[60]
Blanchette, M. A., **1**, 434[140], 769[182]; **2**, 264[58]
Blanco, L., **1**, 879[111c-e]; **2**, 804[41]; **3**, 538[91]; **4**, 1005[82], 1018[82], 1019[228]; **5**, 63[13], 1151[128]; **7**, 121[26]
Blancou, H., **1**, 212[4]
Bland, J. M., **4**, 345[84]

Blandy, C., **6**, 91[125]
Blank, B., **7**, 236[14,15]
Blank, D. R., **5**, 604[54]
Blank, F., **8**, 141[33]
Blank, P. A., **8**, 141[33]
Blankenship, R. M., **3**, 623[32]; **4**, 373[67]; **6**, 1063[81]
Blankespoor, C. L., **3**, 325[161]
Blankley, C. J., **3**, 17[83]
Blanquet, S., **8**, 36[69,70], 66[69,70]
Blanshtein, I. B., **7**, 579[135]
Blanton, C. D., Jr., **3**, 629[53,54]
Blanton, J. R., **4**, 738[98]; **8**, 800[67]
Blaqevic, N., **6**, 85[87]
Blarer, S. J., **1**, 568[245]; **4**, 21[69], 224[182,183,184]; **6**, 716[101-103]
Blaschek, U., **2**, 205[101,101b]
Blaschke, G., **3**, 667[46], 687[112,113]
Blaschke, H., **7**, 24[32]
Bläser, D., **5**, 1134[41]; **8**, 447[127], 463[127]
Blaser, G., **6**, 174[57]
Blasioli, C., **3**, 159[463], 161[463], 165[463]; **4**, 314[495]
Blaskó, G., **2**, 812[72]; **3**, 682[165]
Blass, H., **8**, 70[225], 71[225]
Blass, J., **2**, 968[81]
Blaszczak, L., **8**, 531[121,121c]
Blaszczak, L. C., **3**, 218[97], 762[145]
Blaszczyk, K., **7**, 255[34]
Blatcher, P., **3**, 123[251], 124[261], 126[261], 946[87]
Blatchford, T. P., **3**, 583[119]
Blatt, A. H., **2**, 933[137]; **7**, 689[2]
Blatt, H., **3**, 575[83]
Blatt, K., **2**, 819[98]; **5**, 501[268]; **6**, 672[285]
Blatt, R. S., **8**, 643[35]
Blattner, R., **6**, 978[25]
Blazejewski, J.-C., **3**, 664[30]
Bleasdale, D. A., **2**, 809[51], 823[51]
Blechert, S., **1**, 664[169], 665[169], 669[169], 670[169]; **4**, 56[158]; **5**, 827[2], 829[2], 847[135], 867[2a], 1004[30]; **6**, 834[36], 855[36]
Blecke, R. G., **8**, 528[62]
Bleeke, J. R., **8**, 454[196,202]
Bleicher, W., **8**, 319[79]
Blenderman, W. G., **8**, 609[50,52]
Blezard, M., **1**, 524[91]; **3**, 748[77]
Blicke, F. F., **2**, 420[24], 894[1], 897[1], 953[1], 1090[61]; **3**, 781[12]; **8**, 608[38]
Blidner, B. B., **8**, 96[94]
Blizzard, T. A., **4**, 893[156]; **7**, 410[94]
Bloch, I., **6**, 435[5a]
Bloch, R., **1**, 59[34]; **2**, 194[66]; **3**, 173[518]; **4**, 38[108]; **5**, 15[105], 17[116], 21[142], 553[45,47], 555[45], 567[102,103]; **8**, 9[55]
Block, E., **1**, 630[33], 675[33], 722[33], 786[252]; **2**, 74[76], 866[5]; **3**, 86[8], 121[8], 147[8], 154[8], 173[8], 878[92-95], 879[92-94,96,97], 880[94,98], 881[94]; **4**, 331[16,17], 335[25], 345[83], 346[85], 348[108], 349[108c], 359[159], 771[251], 987[133]; **5**, 440[174], 441[174,179], 829[26]; **6**, 161[180,181,182], 686[371], 687[371], 829[3], 934[101], 982[48], 984[48]; **7**, 516[5], 517[13], 768[203]; **8**, 842[44,44b], 844[44b], 845[44b], 846[44b], 847[44b]
Block, R., **1**, 153[63]
Blöcker, H., **6**, 602[4]
Blodgett, J. K., **6**, 803[47], 804[47]
Blohm, M., **2**, 464[94]
Blohm, M. L., **8**, 374[146], 446[70]
Blok, A. P., **8**, 657[94]
Blom, J. E., **4**, 588[58]
Blom, J. H., **5**, 426[106], 451[3]
Blomberg, C., **1**, 13[69], 16[86], 219[55]
Blommerde, W. W. J. M., **6**, 249[142]
Blomquist, A. T., **1**, 630[32], 675[32], 722[32]; **2**, 529[20], 530[21]; **3**, 905[139]; **5**, 752[39]; **6**, 968[110]; **7**, 660[37]; **8**, 142[52], 568[467], 950[173]
Blondeau, D., **2**, 962[48]
Blondeau, P., **6**, 529[424]
Blondet, D., **5**, 412[47]
Bloodworth, A. J., **3**, 380[9]; **4**, 306[374,375,376,377,378,379,380,381,382,383,384,385], 307[385,390,391], 309[420,421], 310[428], 311[444], 314[480], 390[174,174a,c], 800[121]; **7**, 534[37], 632[57], 728[40]; **8**, 798[56], 854[155], 855[155,156,159], 856[155]
Bloom, A. J., **4**, 356[141,143]; **6**, 109[42]; **7**, 488[156,162], 505[287]
Bloom, B. M., **8**, 530[101]
Bloom, J. D., **1**, 129[93], 779[224]; **6**, 509[253]; **8**, 37[99], 42[99], 66[99]
Bloom, M. S., **2**, 296[87]
Bloom, S. H., **3**, 215[63], 251[80], 254[80]; **4**, 403[240], 872[38]; **6**, 783[84]; **7**, 503[270]
Bloomfield, J. J., **2**, 602[40,41], 797[6], 806[6], 808[6], 813[6], 814[6], 848[6], 849[6]; **3**, 597[200], 613[1], 614[1,5], 615[1,10], 616[1], 617[5a,14], 619[1], 620[1,14], 621[1], 622[1], 623[1], 625[1], 626[1,42], 627[1], 628[1], 629[1], 630[1]; **5**, 571[116]; **8**, 240[29], 242[41], 243[29]
Bloss, D. E., **1**, 451[217]
Blossey, E., **8**, 526[24]
Blossey, E. C., **7**, 108[179]
Blough, B. E., **1**, 418[72], 767[164], 768[167]; **3**, 226[199]; **7**, 418[127]
Blount, J. F., **2**, 384[319], 877[36]; **3**, 390[85], 392[85], 407[148], 623[32], 626[32b], 675[72], 872[57], 873[57]; **4**, 249[125], 258[125], 370[32], 371[32], 629[416]; **6**, 531[430], 1031[115]; **7**, 524[49]; **8**, 645[43], 836[10b], 847[10b], 848[10b], 849[10b], 861[225]
Blount, J. J., **4**, 1104[213]
Bloy, V., **4**, 956[17]; **8**, 166[53,54], 545[294,295]
Bludsuss, W., **7**, 483[130,131]
Blue, C. D., **8**, 724[159], 726[194]
Blues, E. T., **3**, 415[6]
Blum, D. M., **1**, 738[39]; **2**, 814[77]
Blum, J., **6**, 93[132]; **7**, 107[168], 475[50], 476[50]; **8**, 453[193], 535[163], 551[340,341,343,347], 552[340,341], 557[383]
Blum, L., **3**, 505[163]
Blum, M. S., **7**, 528[9]; **8**, 51[123], 66[123]
Blum, R. B., **2**, 600[31]; **3**, 20[120], 194[6]
Blum, S., **4**, 292[234]; **6**, 264[32,33], 268[33]
Blum, Y., **8**, 446[72]
Blumbach, J., **5**, 85[1]; **7**, 131[88]
Blumbergs, J. H., **7**, 674[43]
Blume, G., **4**, 1001[37,39], 1015[37]
Blumel, J., **1**, 476[115]; **5**, 568[110]
Blumenfeld, J., **2**, 232[175]
Blumenfeld, O. O., **8**, 52[138], 66[138]
Blumenkopf, T. A., **1**, 583[8,8a], 589[8a,19,20a,b], 591[19,20b], 592[8a,20a], 595[8a]; **2**, 555[136], 570[36], 580[96], 1030[81]; **3**, 867[34]; **4**, 155[68c]; **6**, 742[67], 752[111,113]; **7**, 256[25]
Blumn, Z., **7**, 799[25], 800[30,31], 804[59], 805[59]
Blunden, S. J., **6**, 662[216]
Blunt, J. W., **3**, 741[50,52]; **8**, 213[29]
Bluthe, N., **1**, 892[149]; **5**, 800[79], 802[82]; **8**, 857[192]
Bly, R. K., **5**, 585[200]; **8**, 103[130]
Bly, R. S., **5**, 585[200]; **8**, 103[130]
Blye, R. P., **7**, 372[71]
Blyskal, J., **5**, 406[24]
Blyston, S. L., **8**, 459[228]
Blystone, S. L., **1**, 535[144]; **4**, 689[68]
Bo, L., **1**, 227[100], 511[34]
Boag, N. M., **8**, 557[382]

Boar, R. B., **3**, 613[2], 615[2]; **7**, 170[122], 171[122]; **8**, 117[74], 243[47], 394[114], 505[83], 816[24], 837[13a], 839[13a], 840[13a], 935[63]

Boardman, L. D., **2**, 713[45]; **3**, 251[78,100], 254[78,100]; **4**, 892[145]; **5**, 1165[13]; **8**, 693[108]

Boaretto, A., **2**, 564[9]

Boate, D. R., **4**, 795[83], 820[220]

Boatman, R. J., **6**, 814[89]

Boatman, S., **2**, 837[161b], 838[161]

Boaventura, M. A., **4**, 905[210]; **5**, 21[152,153,155,156,157,158], 22[152,153,155,156,157,158]; **8**, 851[129]

Boaz, N. W., **1**, 112[27]; **2**, 120[172,173,174], 448[42]; **3**, 212[37], 213[50], 217[81], 235[81b]; **4**, 152[57], 171[29]

Bobbitt, J. M., **2**, 1018[40]; **3**, 665[40], 666[44], 681[95], 687[44,112]; **6**, 279[133], 751[106]; **7**, 709[40], 745[74], 801[41]

Bobdanov, V. S., **4**, 885[115], 886[115]

Bobe, F. W., **2**, 770[10], 771[10]

Bobek, M., **8**, 877[47], 878[47]

Boberg, F., **6**, 421[25-27], 943[156]

Bobic-Korejzl, L., **4**, 30[87]

Boccardo, D., **8**, 457[216]

Bocelli, G., **3**, 386[57]

Boch, M., **8**, 266[58]

Boche, G., **1**, 2[14], 10[47], 18[94,96], 29[145], 32[159,160,161], 33[162], 34[166], 35[14], 36[14], 37[242,243,245,246], 43[14], 44[96], 385[118], 513[50], 528[119], 531[133]; **2**, 508[29]; **3**, 277[29]; **4**, 104[137], 1007[107]; **5**, 714[76], 715[79], 716[85], 901[26]; **6**, 114[78], 116[87], 119[114,115], 562[888], 881[51]

Bocher, S., **6**, 1012[4], 1013[4]

Bochmann, G., **7**, 772[295], 773[295]

Bochu, C., **4**, 317[551]

Bocionek, P., **2**, 388[339]

Bock, B. V., **1**, 144[40]

Böck, F., **2**, 399[15]

Bock, G., **3**, 903[127]; **6**, 532[472]

Bock, H., **5**, 575[131]; **6**, 245[126]; **7**, 874[105]; **8**, 513[102], 773[62]

Bock, K., **6**, 61[150]

Bock, M. G., **2**, 482[31], 484[31], 511[45], 962[51]; **3**, 125[292], 126[292]; **6**, 647[111], 648[111], 679[329], 680[329b], 728[214]; **8**, 11[61]

Bock, W., **2**, 1088[41-43]

Bockhorn, G. H., **7**, 29[77]

Bockmain, G., **7**, 799[25,26]

Bockman, T. M., **7**, 852[35]

Bocz, A. K., **8**, 229[135]

Bodalski, R., **4**, 252[162]

Bodamer, G. W., **4**, 311[447]

Bodanszky, M., **6**, 635[11], 645[11], 665[11d], 667[11d], 668[11d], 669[11d]

Boddy, I., **1**, 770[188]

Bode, K.-D., **2**, 867[16]; **6**, 419[4], 420[4], 423[38], 424[38]

Bödeker, C., **8**, 843[55]

Bodem, G. B., **4**, 12[39]

Boden, E. P., **1**, 769[194]; **2**, 4[12,12c], 6[12,12c], 573[53,54]; **6**, 437[39]

Boden, R. M., **7**, 97[91]

Bodenbenner, K., **6**, 565[920]

Bodendorf, K., **2**, 785[45]; **4**, 6[23]

Bodennec, G., **7**, 498[220]

Bodesheim, F., **2**, 1088[38,39], 1089[39,54]; **8**, 830[85]

Bodine, J. J., **6**, 430[92], 501[202]

Bodkin, C. L., **7**, 842[29,30]

Bodnarchuk, N. D., **6**, 499[172]

Bodot, H., **3**, 892[47]

Bodrikov, I. V., **4**, 310[427], 330[5]; **7**, 494[202]

Bodurow, C., **4**, 379[116], 380[116c], 384[144]; **6**, 125[148], 127[148]

Boeckman, R. K., Jr., **1**, 127[91], 409[38], 619[63,64], 731[5], 738[39], 765[166], 791[266]; **2**, 111[78], 183[13], 370[260], 599[25], 814[77], 824[119], 828[131]; **3**, 8[45], 15[45], 16[45], 48[260], 252[82,85], 257[82], 602[220]; **4**, 6[23,23b], 8[27], 24[72,72e], 74[41], 100[41], 113[169,169c], 191[113], 192[113], 245[92], 260[92], 261[296], 309[419]; **5**, 362[93], 363[93h], 514[8], 516[20], 524[20,49], 527[8,8b,61], 531[61,72], 534[92], 539[49], 946[257], 829[25]; **7**, 164[79], 299[42], 313[36], 407[78e], 567[104], 579[131], 656[13], 673[24]; **8**, 549[326], 696[122], 854[148], 857[148]

Boeckmann, J., **3**, 247[45], 248[45]

Boeckmann, R. K., Jr., **3**, 602[220]

Boeder, C. W., **4**, 311[449]

Boehm, J. C., **2**, 1097[100]

Boehm, P., **8**, 18[123], 263[31]

Boeje, L., **6**, 464[33], 465[33]

Boekelheide, V., **3**, 124[267], 125[267], 126[267], 127[267], 877[83-85], 927[48]; **4**, 507[155,156]; **5**, 692[103]; **6**, 134[37]; **7**, 661[48], 801[37]

Boele, S., **7**, 535[47]

Boelema, E., **3**, 832[68a]

Boelens, H., **2**, 782[17,21]; **3**, 251[79], 254[79]; **4**, 18[62], 20[62h]; **6**, 714[87]; **8**, 535[161], 542[229], 946[139]

Boelhouwer, C., **3**, 260[145]; **5**, 1116[8,8b], 1118[8b]

Boens, N., **5**, 637[109]

Boente, J. M., **4**, 505[139]

Boer, V. I., **6**, 182[142]

Boerekamp, J., **6**, 984[58]

Boering, H. L., **4**, 280[129], 281[129]

Boerrigter, J. C. O., **2**, 821[110]

Boersma, J., **1**, 23[121-124], 30[153], 211[2], 212[2], 214[2,27], 222[2], 225[2]; **2**, 123[195,196], 124[204], 125[204], 280[27]; **8**, 99[110], 589[46]

Boer-Terpstra, Tj., **2**, 1062[100]

Boerwinkle, F. P., **7**, 500[237], 501[247,248,249]

Boes, M., **2**, 910[66]; **3**, 75[49], 78[60-62], 79[60,61], 81[60,61]; **7**, 224[49]

Boes, O., **7**, 762[71]

Boese, R., **1**, 36[237]; **3**, 537[90], 538[90]; **5**, 1134[41], 1143[94]; **8**, 13[75], 447[127], 463[127]

Böeseken, J., **7**, 766[175], 768[175]

Boettcher, R. J., **5**, 204[40]

Boettger, S. D., **1**, 464[37]

Boexkes, W., **3**, 307[87]

Boeyens, J. C. A., **1**, 38[259]

Bogan, R. T., **8**, 900[32]

Bogatova, N. G., **8**, 699[150]

Bogavac, M., **4**, 307[393]

Bogdan, S., **5**, 108[210], 109[210,220,221,222], 110[210,222], 111[210,222], 112[222]

Bogdanov, V. S., **7**, 597[42,44]

Bogdanovic, B., **1**, 14[77,78], 25[129]; **5**, 37[22b], 1197[39]; **6**, 184[153]; **8**, 697[135], 698[136]

Bogdanowicz, M. J., **3**, 86[61], 88[61], 89[61], 91[61], 124[61], 178[543,544], 179[543], 181[543], 761[144], 762[144], 785[32], 1040[106]; **4**, 989[143]; **5**, 910[85], 919[130], 922[130], 1020[70], 1027[70]; **6**, 143[68,70], 147[85], 1044[16b], 1048[16]

Bogdanowicz-Szwed, K., **2**, 378[292]

Boger, D. L., **2**, 495[66], 496[66], 497[66], 518[63], 540[72], 542[72], 843[194], 1056[65], 1059[65], 1070[65], 1074[65]; **3**, 356[59]; **4**, 8[30b], 76[46], 113[46], 125[217b], 183[79], 238[6], 252[165], 433[119], 740[117], 797[105], 798[107]; **5**, 65[57], 78[270,271], 266[76], 267[76,76b], 268[76], 402[1], 403[1], 404[1], 410[1], 413[1,1b], 417[1,62], 420[62], 422[1], 425[1], 426[1], 429[1], 430[1], 433[1], 434[1], 435[1], 436[1], 438[1], 440[1], 444[1], 451[15,21-23], 453[15,60,61], 454[15], 458[61], 460[61,94], 461[15,60,61], 464[15,61],

468[15], 469[15], 470[15,21-23], 473[15,155], 474[155,156,157], 476[148], 477[155], 480[15,148,176], 485[15], 486[15], 491[15,21-23], 492[22,23,238,239,240,241,242,243,244,245], 497[227], 498[231,232,238], 499[15], 501[15], 508[15], 510[15], 511[15], 531[77], 572[125,126], 573[127], 583[187,188,190,191], 594[7], 599[7], 604[7,54]; **6**, 471[65], 472[65], 559[864], 751[107], 756[123], 814[87,90]; **7**, 34[100], 260[86], 347[17], 355[17], 543[23], 544[23], 748[112]; **8**, 394[121], 618[114], 657[96]
Bogert, M. T., **3**, 898[79]; **5**, 752[46]
Bögge, H., **4**, 1017[216], 1041[105]
Boggs, N. T., III, **4**, 5[17]
Boggs, R. A., **4**, 104[135c], 170[21], 190[107], 710[52]; **5**, 1079[49]; **7**, 4[16]
Bognar, R., **5**, 438[164], 534[94]
Bogoslovskii, K. G., **3**, 648[175,183]
Bogri, T., **6**, 941[151]
Boguslavskaya, L. S., **4**, 329[1], 344[1], 347[91,93], 350[1], 351[1]
Boguth, W., **8**, 205[159,161], 560[405,406]
Böh, H., **7**, 765[147], 769[147]
Bohle, M., **4**, 434[126]
Bohlen, D. H., **8**, 341[102], 927[19]
Böhler, F., **8**, 896[18]
Bohlmann, C., **4**, 222[177]
Bohlmann, F., **1**, 564[205], 733[21]; **3**, 277[28]; **4**, 5[19], 52[147,147f]; **5**, 409[36]; **6**, 673[288], 677[313]; **7**, 95[78]; **8**, 339[91], 353[153]
Bohlmann, R., **4**, 222[177]
Böhm, I., **2**, 81[3]
Bohm, M., **7**, 169[112]
Bohme, E., **4**, 608[325]
Böhme, H., **1**, 366[43,47], 370[72], 371[72]; **2**, 365[208], 894[8], 898[8,21], 900[22-24], 961[37,39,43], 1007[1,3], 1008[4], 1050[29]; **3**, 154[421]; **5**, 410[40], 716[85]; **6**, 231[35], 238[71], 495[149], 508[279], 509[279], 519[335], 524[355], 525[355], 527[410], 532[355,471], 564[912], 570[953]; **7**, 206[68,70], 210[70], 212[68,99], 765[152]
Böhme, R., **6**, 195[225]
Bohn, B. A., **4**, 279[104]
Bohn, K. H., **1**, 34[166]
Bohonek, J., **6**, 562[884]
Böhrer, G., **3**, 587[143]
Boicelli, A. C., **4**, 429[81]
Boikess, R. S., **5**, 906[63]
Boireau, G., **1**, 86[33,37-41], 223[72c]; **3**, 770[172]
Bois-Choussy, M., **3**, 505[157]; **4**, 459[72,74,75], 460[95-97], 464[114,123], 465[96], 466[123], 467[72,97], 469[75], 470[75,97,138], 474[95], 475[74,95], 477[114], 478[96,114], 479[123,172], 480[74]
Boisden, M.-T., **8**, 663[119]
Boissier, J. R., **6**, 268[60]
Boivin, J., **7**, 13[119], 227[87]
Boivin, T. L. B., **4**, 10[32], 381[127], 733[79], 791[54]; **6**, 2[2], 23[2]
Bok, Th. R., **4**, 8[29,29g]
Bokadia, M. M., **4**, 508[158]; **8**, 545[281]
Bokanov, A. I., **6**, 487[74], 489[74]
Bokel, H. H., **3**, 124[268], 125[268], 126[268], 127[268], 131[268]
Böker, R., **3**, 809[42]
Bolan, J. L., **2**, 121[189]; **4**, 185[86]
Boland, W., **3**, 220[120]; **5**, 563[91], 973[15], 975[15]; **8**, 185[30], 798[52]
Bold, G., **2**, 309[24]
Boldeskul, I. E., **1**, 820[4]; **3**, 86[43], 179[43]; **4**, 987[133]
Boldrini, G. P., **1**, 188[73], 189[73], 192[82]; **2**, 35[130], 36[130], 566[23]; **7**, 549[42]; **8**, 36[80], 54[80], 66[80], 550[333], 551[336]
Boldt, P., **2**, 760[42]; **4**, 729[61], 730[61], 765[61]
Boles, D. L., **5**, 710[48]
Boleslawski, M., **4**, 887[125], **8**, 736[21], 739[35]
Bolesov, I. G., **4**, 310[426]
Bolesova, I. N., **4**, 521[41], 529[73]
Bolestova, G. I., **4**, 155[65]; **8**, 608[37], 610[56,58], 630[56,187]
Bolhofer, W. A., **6**, 526[395]; **8**, 148[109]
Bolikal, D., **8**, 532[133], 863[232]
Bolker, H. I., **8**, 212[13], 214[44,45], 222[44]
Bollenback, G. N., **6**, 36[16]
Bollinger, F. W., **4**, 1033[30]
Bollinger, J. M., **1**, 488[12]
Bollman, H. T., **5**, 790[35]
Bollyky, L., **3**, 154[417]; **6**, 1022[62]
Bolm, C., **1**, 614[50], 615[50]
Bolourtchian, M., **3**, 577[90]
Bol'shedvorskaya, R. L., **4**, 55[156]
Bolte, J., **2**, 464[102], 465[102]; **8**, 187[32], 188[32]
Bolte, M. L., **2**, 353[95], 365[95]
Bolton, G. L., **2**, 69[49]; **4**, 191[111]; **5**, 249[32], 517[29], 519[29], 534[29], 547[29g], 573[128], 574[129], 619[9], 624[9], 625[9]
Bolton, I. J., **5**, 839[75], 888[25]
Bolton, J. L., **8**, 93[75], 584[19], 589[48]
Bolton, R., **4**, 270[3], 329[1], 330[1a], 344[1], 350[1], 351[1], 364[2], 369[2b]
Bolton, R. E., **7**, 34[98,99]; **8**, 618[115,116]
Boltze, K. H., **6**, 264[35]
Bolze, R., **6**, 970[122]
Bombala, M. U., **4**, 331[13]
Bomhard, A., **6**, 175[81], 179[81], 188[81,182]
Bomke, U., **1**, 370[72], 371[72]; **2**, 961[39]
Bommer, R., **6**, 54[123]
Bomse, D. S., **3**, 587[141,142]
Bonacic-Koutechky, V., **5**, 72[178]
Bonadeo, M., **5**, 625[31]
Bonadies, F., **4**, 391[176]; **7**, 103[140], 240[55], 266[110,113], 267[110], 410[95]
Bonadyk, S. V., **6**, 524[372]
Bonakdar, A., **3**, 440[45]; **4**, 878[79]
Bonati, F., **6**, 295[249]
Bonavent, G., **2**, 432[55]
Bonazzi, D., **2**, 787[52]
Boncompte, F., **8**, 587[34], 621[143]
Boncza-Tomaszewski, Z., **8**, 34[60], 66[60]
Bond, A. C., Jr., **8**, 26[1], 735[9], 736[9]
Bond, F. T., **1**, 377[97]; **2**, 588[150]; **3**, 251[79], 254[79]; **4**, 273[51]; **6**, 781[78,79]; **8**, 940[103], 941[103], 946[103,141], 947[103]
Bond, G. C., **8**, 431[61], 445[18], 568[476]
Bondarenko, O. P., **4**, 288[187]
Bondavalli, F., **6**, 776[57]
Bonde, S. E., **3**, 494[85]
Bonds, W. D., Jr., **8**, 447[126], 457[126]
Bonetti, M., **6**, 1056[56]
Bonfand, A., **4**, 926[38], 928[38]
Bonfiglio, J. N., **2**, 102[22]
Bonfrèr, J. M. G., **5**, 708[41]
Bong, C.-H., **5**, 7[50]
Bongers, S. L., **6**, 836[50]
Bongini, A., **2**, 925[111], 926[111]; **4**, 344[78a], 375[94], 377[104], 386[94a,153,153a,157], 387[94a,153a,157], 388[164], 393[164c], 401[226], 407[104c,153a,157b,254]; **5**, 102[174]; **6**, 26[106], 648[124]; **7**, 493[184], 503[269]
Bongrand, J. C., **4**, 286[172], 289[172]
Bonhoeffer, K. F., **3**, 822[2], 831[2]; **8**, 87[28]
Bonhomme, M., **6**, 428[86]
Bonilavri, E., **4**, 127[220a]
Bonin, M., **1**, 557[127], 559[144]

Bonini, B. F., **5**, 440[173]
Bonini, C., **7**, 240[55], 266[113], 410[95]
Bonitz, G. H., **1**, 366[44]; **2**, 900[28], 901[28], 910[28]
Bonjoch, J., **2**, 809[56], 819[56]; **6**, 917[33]; **8**, 344[122], 621[143]
Bonjouklian, R., **2**, 662[17], 664[17]; **4**, 389[167]; **5**, 328[29], 433[134]; **6**, 807[60]
Bonk, P. J., **2**, 981[25], 982[25]; **5**, 307[91,92]
Bonneau, R., **5**, 125[13], 128[13]
Bönnemann, H., **5**, 37[22b], 1152[139,140,141], 1153[139,146,147], 1154[140,149,153]
Bonner, W. A., **8**, 433[69], 836[1,1c], 837[1], 964[60], 995[69]
Bonner, W. H., **6**, 546[651]
Bonnesen, P. V., **1**, 309[98]
Bonnet, A., **5**, 96[115]
Bonnet, G., **2**, 60[18], 62[18d]
Bonnet, J. J., **1**, 441[173]
Bonnet, M., **8**, 142[54]
Bonnet, P.-H., **4**, 299[304]
Bonnet-Delpon, D., **1**, 530[128]; **3**, 324[155]
Bonnett, R., **6**, 487[3], 488[3], 489[3], 515[3], 523[3], 524[3], 525[3], 526[3], 532[3]
Bonser, S. M., **3**, 876[78]; **7**, 208[81]
Bonsignore, S., **4**, 767[233]
Bontempelli, G., **7**, 769[215]
Bonvicini, P., **8**, 152[175]
Bonvicino, G. E., **8**, 973[120]
Boocock, J. R. B., **7**, 16[158]
Booij, M., **1**, 232[16]
Bookbinder, D. C., **5**, 707[29]
Booker, E., **8**, 580[6]
Boom, J. H., **1**, 737[30]
Booms, R. E., **8**, 410[91]
Boon, W. H., **4**, 905[211]
Boone, J. R., **8**, 2[11], 26[27], 37[27], 66[27], 541[207]
Boons, G. J. P. H., **1**, 737[30]; **6**, 17[62]
Boontanonda, P., **7**, 453[95], 831[66]
Boop, D. C., **4**, 505[137]
Boop, J. L., **3**, 392[92]
Boor, J., **5**, 1163[4]
Boord, C. E., **4**, 283[151]
Boorman, E. J., **2**, 348[65]; **4**, 282[136]
Boot, J. R., **7**, 401[58]
Booth, B. L., **2**, 748[125]; **6**, 291[212,213,214,215], 507[241,242], 517[326], 529[241,242]
Booth, P. M., **6**, 443[91]
Boothe, J. H., **8**, 973[120]
Boothe, R., **4**, 288[188], 346[86a]
Boothe, T. E., **6**, 993[94]
Boots, D. R., **1**, 608[37]
Boots, S. G., **3**, 369[119], 370[113], 372[119]; **8**, 353[156], 542[228]
Boozer, C. E., **6**, 204[15]
Bopp, H., **6**, 1013[13]
Bora, J. M., **8**, 271[110]
Borch, R. F., **2**, 427[40], 477[10]; **3**, 905[136]; **4**, 304[354]; **5**, 514[8], 527[8]; **6**, 724[162,163]; **8**, 26[26], 37[26,101], 47[26,127], 54[26], 55[26], 57[26], 60[26], 66[26,101,127], 170[92], 251[102], 253[120], 275[143], 351[151]
Borchardt, J. K., **8**, 477[35]
Borchardt, R. T., **1**, 463[32]; **7**, 333[23]
Borcherdt, G. T., **3**, 428[89]
Borchert, A. E., **5**, 1016[59]
Bordakov, V. G., **4**, 963[43]; **5**, 1198[45]
Bordas, X., **6**, 77[54]
Borden, M. R., **8**, 896[17]
Borden, W. T., **1**, 92[64], 506[10]; **3**, 380[10]; **5**, 240[2,2a], 857[227]; **7**, 737[7], 875[111]
Borders, D. B., **5**, 736[145], 737[145]
Borders, R. J., **8**, 214[48]
Bordic, S., **3**, 374[132]
Bordier, E., **7**, 92[44]
Bordignon, E., **7**, 777[386]
Bordner, J., **2**, 547[101], 548[101]; **3**, 14[77], 15[77], 363[84]; **5**, 130[42]; **8**, 542[237]
Bordwell, F. G., **1**, 531[129], 632[66]; **3**, 88[121], 756[115], 862[5], 866[5,32]; **4**, 310[434], 425[32,34], 429[34], 434[34]; **6**, 133[1,3], 161[180], 950[1]; **7**, 203[53], 206[72], 207[72], 210[72], 229[119], 765[138,148]; **8**, 407[56], 853[144]
Borel, C., **3**, 1037[87]; **5**, 1107[170], 1108[170]
Borel, D., **5**, 938[207]
Borer, A., **4**, 931[58]
Borer, C., **2**, 547[93]
Borer, M. C., **5**, 830[32]
Borer, R., **6**, 560[868]; **8**, 544[257,258]
Borer, X., **6**, 769[31]
Borg, R. M., **7**, 876[123]
Borgen, G., **2**, 808[69]
Borgen, P. C., **4**, 14[47,47k]
Borgi, A. E., **2**, 286[65], 287[65]
Borgogno, G., **8**, 409[86], 412[86]
Borgstrom, B., **2**, 456[28]
Borgulya, J., **5**, 834[50], 850[146], 877[5]
Borisov, A. E., **4**, 315[518]
Borisova, A. I., **2**, 365[214]; **4**, 317[554]
Borisova, L. N., **8**, 627[176]
Bork, K.-H., **8**, 528[71], 971[108]
Borkent, G., **4**, 278[95], 286[95], 289[95]
Borkowski, M., **4**, 272[32]
Bormann, D., **3**, 664[27]; **6**, 487[63], 489[63]; **8**, 254[125]
Bormuth, M. L., **2**, 943[169]
Bornack, W. K., **4**, 107[141]
Bornancini, E. R., **4**, 473[146], 474[146]
Borner, E., **7**, 231[140]
Borner, P., **6**, 565[920]; **8**, 758[171]
Bornmann, W. G., **4**, 373[81], 374[81]; **5**, 453[69], 455[69,79]
Bornstein, J., **3**, 760[139], 774[139]; **5**, 580[170], 582[180,181]; **8**, 896[17]
Borodin, V. S., **8**, 694[118]
Boross, F., **7**, 831[62]
Borovas, D., **6**, 644[86]
Borowitz, I. J., **2**, 610[98]; **3**, 8[42]; **6**, 716[94]; **7**, 710[51]; **8**, 990[42]
Borredon, M. E., **1**, 821[20-22]; **2**, 354[108]
Borretzen, B., **5**, 680[24], 683[24c]
Borrini, A., **5**, 1148[114]
Borrmann, D., **5**, 86[36], 87[37], 88[45]; **6**, 46[56]
Borromeo, P. S., **1**, 366[45]; **2**, 909[62], 910[62]; **3**, 246[43], 258[43]
Bors, D. A., **1**, 528[118]
Borschberg, H.-J., **2**, 1024[60], 1062[99]; **5**, 841[94]
Borsche, W., **2**, 757[20], 759[37]
Bortolini, O., **7**, 95[69], 425[147a], 762[69], 777[69b], 778[69]
Bortolussi, M., **5**, 15[105], 17[116], 21[142]
Borunova, N. V., **8**, 535[160]
Bory, S., **3**, 147[390,398], 149[390,398,408,409,411], 150[411], 151[390,408,409,411], 152[390], 155[408,409]; **7**, 777[388]
Bos, H. J. T., **2**, 85[20], 838[170]; **3**, 219[104]; **5**, 116[255], 163[72], 742[161]; **6**, 572[960]
Bos, M. E., **2**, 588[151], 589[151]; **5**, 1070[29], 1074[29]
Bos, M. G. J., **7**, 12[101]
Bos, T. J. T., **4**, 897[171], 898[171], 899[171]
Bos, W., **5**, 561[84,85]
Bosc, J.-J., **1**, 368[61], 369[61]

Bosch, E., **4**, 723[40], 747[40], 776[40], 798[108,109], 803[130]; **8**, 823[58]
Bosch, G. K., **3**, 591[166,168], 610[166,168]; **7**, 94[62], 556[73], 647[37]
Bosch, J., **2**, 765[77], 809[56], 819[56], 828[133]; **6**, 917[33]; **8**, 32[53], 66[53], 344[122], 587[34], 621[143]
Bosch, M., **5**, 179[142]
Boschelli, D., **1**, 763[143], 766[143]; **2**, 256[47], 257[47]; **7**, 162[67], 176[67]
Boschelli, D. H., **7**, 243[64], 477[71]
Boschetti, A. B., **4**, 379[116]
Boschi, T., **4**, 600[242]
Boschung, A. F., **7**, 98[97]
Bosco, M., **4**, 86[78c,e,79a], 428[77,79], 429[80-82]; **7**, 331[16]
Bose, A. J. K., **4**, 553[6]
Bose, A. K., **1**, 294[39-41]; **2**, 296[85], 919[92]; **4**, 45[126]; **5**, 86[13,14,18], 92[68,77], 95[68,89,93,95], 96[68,114,116,119,121], 98[121], 100[141]; **6**, 253[155], 744[74]; **7**, 454[99]; **8**, 817[26]
Bose, J. A., **4**, 390[175c]
Böshagen, H., **1**, 364[39]; **6**, 551[686]; **8**, 649[61]
Boshar, M., **4**, 1075[33]
Boshart, G. L., **7**, 294[12]
Boshmann, G., **7**, 772[294], 773[294]
Boska, I. M., **2**, 1079[156]
Boskin, M. J., **3**, 760[137]
Bösler, M., **3**, 825[31]
Bosman, W. P., **8**, 96[92]
Bosnich, B., **4**, 564[40,43], 567[40], 590[90], 599[221], 624[221], 641[221], 653[443,444,445], 927[44], 945[44]; **6**, 450[117], 843[88]; **7**, 416[123]; **8**, 459[234,235], 535[166]
Bosnjak, J., **7**, 831[68]
Bosold, F., **6**, 114[78], 119[114,115]
Bosone, E., **2**, 922[101], 923[101]
Boss, R., **6**, 652[146]
Bossert, F., **2**, 377[281], 384[281]
Bosshard, C., **7**, 3[5]
Bosshard, H., **3**, 815[74]
Bosshard, H. H., **2**, 748[127]; **6**, 204[20]
Bosshard, P., **2**, 964[57]; **5**, 947[288]; **8**, 606[21]
Bosshardt, H., **7**, 99[107]
Bost, H. W., **2**, 283[51]
Bostmembrun-Desrut, M., **8**, 203[148], 205[148], 559[401]
Bostock, S. B., **2**, 379[293]
Boston, M. C., **2**, 286[64]
Bos Vanderzalm, C. H., **3**, 575[83]
Boswell, G. A., Jr., **7**, 751[140]
Bosworth, N., **7**, 833[72]
Bothner-By, A. A., **2**, 145[64]; **5**, 687[65]; **8**, 159[12], 493[19], 526[33]
Bothwell, T. C., **4**, 443[189]
Botros, S., **6**, 551[687]
Bott, K., **6**, 523[348]
Bott, R. W., **3**, 564[15]; **8**, 763[1], 785[1]
Botta, M., **4**, 113[171,171g]; **6**, 490[109]; **7**, 713[71]
Bottaccio, G., **5**, 1133[23]
Bottari, F., **3**, 741[51]
Bottaro, D., **4**, 252[160]; **6**, 134[34]
Bottaro, J. C., **1**, 477[144,145], 545[46-48]; **2**, 523[75], 588[152]; **6**, 109[44], 783[85]; **7**, 231[153,154], 471[19], 746[87]; **8**, 940[107]
Bottcher, B., **1**, 6[32]
Botteghi, C., **3**, 124[286], 125[286], 127[286], 228[222], 487[45]; **4**, 915[9,15], 919[18,19], 920[20], 930[50]; **5**, 1152[143], 1153[145]; **8**, 236[3], 239[3], 552[352]
Botteron, D. G., **3**, 564[8], 727[28]
Böttger, R. C., **3**, 208[6]
Bottin, J., **8**, 536[170], 541[215], 543[215]
Bottino, F. A., **3**, 587[148]
Bottin-Strzalko, T., **8**, 865[245]
Bottom, F. H., **3**, 23[141]
Bottomley, W. E., **6**, 493[127]
Bottorff, K. J., **7**, 874[104]
Bottrill, M., **1**, 139[3]; **5**, 1136[54]
Bou, A., **5**, 1133[33]
Bouche, C., **4**, 318[561]
Boucher, R. J., **2**, 635[40], 640[40]; **3**, 26[161]; **7**, 294[15]
Bouchoule, C., **1**, 219[63]; **2**, 988[32], 989[32]
Bouda, H., **1**, 821[20]
Boudart, M., **8**, 419[19-21], 420[19-21], 424[19,21], 429[21], 430[20], 436[21], 454[198]
Boudet, B., **4**, 459[72,74], 466[125], 467[72], 475[74], 478[125], 480[74]
Boudet, R., **8**, 658[102]
Boudjouk, P., **1**, 308[96]; **2**, 279[19]; **3**, 466[189], 565[16], 578[16]; **5**, 386[132], 387[132a], 638[118], 639[119]; **6**, 977[15]; **8**, 764[4c]
Boudreaux, G. J., **3**, 174[538,539], 177[538,539], 868[42,44], 869[44], 876[44]
Boudrow, C., **4**, 557[13]
Bouet, G., **4**, 878[81]; **5**, 475[139]
Bouffard, F. A., **2**, 213[122]; **5**, 107[199]; **6**, 125[148], 127[148]
Bougeard, P., **8**, 675[48], 676[48]
Boughton, N. A., **2**, 23[89]
Bouglel, K., **8**, 884[96]
Bouhy, P., **1**, 636[94], 639[94], 672[94], 691[94], 692[94], 697[94], 723[94]
Boujlel, K., **4**, 478[167]
Boukou-Poba, J. P., **2**, 780[9]; **4**, 350[117]
Boukouvalas, J., **1**, 52[16], 134[114], 135[114]; **2**, 625[162], 631[17], 632[17], 634[17]; **5**, 157[38,39]
Boulaajaj, S., **4**, 38[109b]
Boulanger, W., **6**, 1036[144]
Boulanger, Y., **8**, 54[154], 66[154]
Boulet, C. A., **8**, 205[156]
Boulette, B., **4**, 1020[231]; **7**, 764[128]
Boulos, A. L., **8**, 566[450]
Boulton, A. J., **4**, 1099[183]; **6**, 261[10], 273[10], 280[10]
Bouma, R. J., **3**, 174[523], 175[523]
Bouman, T. D., **7**, 262[81]
Bounds, D. G., **3**, 642[112], 644[112,148,153]
Bounkhala, Z., **2**, 710[21]
Bourdois, J., **6**, 501[200]
Bourelle-Wargnier, F., **5**, 829[25], 930[175], 931[175], 932[175]
Bourgain, M., **1**, 428[116]; **3**, 274[19], 473[217], 476[217]; **4**, 896[167], 897[172]; **6**, 849[120]
Bourgain-Commercon, M., **1**, 107[3], 113[34], 428[119], 436[145]; **2**, 584[125]; **3**, 246[35], 470[223,226], 473[226], 482[4], 485[4]; **4**, 183[81], 898[178]
Bourgasser, P., **1**, 563[184]
Bourgeois, O. P., **3**, 730[44]
Bourgeois, P., **2**, 725[120]
Bourgery, G., **6**, 271[86]
Bourgoin-Lagay, D., **8**, 658[102]
Bourgois, J., **6**, 555[813]
Bourhis, M., **1**, 368[61], 369[61]
Bourhis, R., **2**, 603[43]; **8**, 556[377]
Bourne, E. J., **2**, 736[22]; **7**, 760[28]
Bourquelot, E., **3**, 660[7]
Bourrie, D. B., **8**, 781[98]
Bousquet, E. W., **7**, 138[126]
Boussinesq, J., **3**, 851[65]
Boussu, M., **1**, 427[112]
Boutagy, J., **1**, 755[115], 812[115], 813[115]; **3**, 201[75]
Boutan, P. J., **7**, 765[138]

Boutelje, J., **6**, 811[76]
Boutin, R. H., **1**, 406[28]; **6**, 803[47], 804[47]
Boutonnet, J. C., **4**, 523[56], 541[116], 543[122]
Bouveault, L., **6**, 672[283]
Bouwman, R. E., **8**, 84[11]
Bouxom, B., **2**, 464[96]
Bovara, R., **8**, 194[105]
Bovicelli, P., **1**, 754[107]; **7**, 832[69]
Bovill, M. J., **8**, 724[170]
Bowd, A., **5**, 727[120]
Bowden, E., **3**, 309[92b]
Bowden, K., **4**, 51[143], 278[97,98], 285[98], 286[97], 289[97,98]
Bowe, M. D., **1**, 661[167,167b]
Bowen, R., **7**, 145[167]
Bowen, R. D., **7**, 508[310]
Bower, J. D., **5**, 162[68]
Bowers, A., **7**, 86[16a], 136[116], 137[116], 253[17]; **8**, 530[106]
Bowers, K. G., **5**, 605[55], 612[76]
Bowers, K. W., **3**, 577[87]; **4**, 537[99,100], 538[99,100]; **8**, 524[12], 527[49], 532[12c]
Bowers, M. M., **2**, 1097[101]
Bowers, V. A., **4**, 719[22]
Bowles, S., **7**, 373[72b]
Bowles, T., **4**, 1063[168]; **8**, 393[113]
Bowlin, H. B., **5**, 856[210], 1003[22]
Bowlus, S. B., **2**, 761[48]; **7**, 87[18]
Bowmaker, G. A., **6**, 196[234]
Bowman, D. H., **7**, 605[144]
Bowman, E. R., **3**, 158[437], 159[437], 160[437], 166[437]
Bowman, E. S., **5**, 552[36], 568[107], 847[136]
Bowman, R. E., **6**, 209[69]; **8**, 926[15], 940[15]
Bowman, R. G., **8**, 447[132]
Bowman, R. M., **5**, 128[29], 650[25]
Bowman, W. R., **4**, 477[164]
Bowman Mertes, K., **6**, 71[23]
Bowne, A. T., **5**, 681[27]
Bowyer, W. J., **8**, 220[87]
Box, S. J., **2**, 758[21]
Boxer, M., **1**, 506[12]
Boxler, D., **3**, 882[105], 894[63], 943[82]; **6**, 841[76], 1021[53]
Boyajian, C. G., **3**, 747[70], 770[176]
Boyce, C. B., **8**, 584[24]
Boyce, R., **3**, 131[332]
Boyd, D. B., **5**, 96[103], 98[103], 99[103]
Boyd, D. R., **1**, 837[152]; **4**, 356[135]; **7**, 6[34], 78[125], 750[131]; **8**, 187[36], 541[212]
Boyd, G. S., **3**, 741[50]
Boyd, G. V., **5**, 478[163], 491[206]; **6**, 438[60], 493[127], 504[220]; **7**, 772[296]; **8**, 664[122]
Boyd, S. A., **7**, 845[68]
Boyd, S. D., **6**, 1000[125]
Boyer, B., **8**, 5[27], 14[27], 536[172]
Boyer, J., **4**, 100[126]; **8**, 246[79], 262[18]
Boyer, J. H., **4**, 1099[181]; **6**, 76[43], 104[1], 110[47], 525[380], 799[26]; **7**, 736[4], 737[4], 747[100], 748[100], 749[119], 750[4], 752[4]
Boyer, P. D., **2**, 456[43,48], 464[99], 466[43,117], 467[43], 468[117]
Boyer, R., **6**, 208[63]
Boyer, S., **5**, 766[115]
Boyer, S. K., **3**, 897[91], 900[91]; **8**, 392[100], 905[61]
Boyes, R. H. O., **1**, 130[95]
Boykin, D. W., **2**, 401[28]
Boykin, D. W., Jr., **5**, 220[48]
Boyle, E. A., **8**, 626[175], 629[175]
Boyle, L. W., **5**, 255[50]
Boyle, P. H., **4**, 951[1], 968[1], 979[1]; **7**, 102[134]
Boys, M. L., **1**, 741[45]
Bozell, J. J., **4**, 560[28], 844[62]; **5**, 1094[98,115], 1096[98], 1098[98], 1112[98,115]
Bozhanova, N. Ya., **2**, 787[52]
Bozsár, G., **2**, 851[223]
Bozzato, G., **5**, 223[66]
Bozzini, S., **3**, 503[149], 512[149]
Braatz, J., **7**, 452[59]
Braatz, T. A., **4**, 587[42,46,47]
Brabender, W., **5**, 154[33]
Braca, G., **5**, 1133[27]
Bracci, S., **4**, 958[28]
Brachel, H. v., **6**, 570[948,949]
Brachiand, J., **7**, 800[30]
Bracht, J., **8**, 382[10], 392[10], 394[10]
Bracken, C., **2**, 1037[100]
Brackenridge, I., **6**, 650[133b], 668[261]
Bradbury, R. B., **3**, 681[94], 807[31]
Bradbury, R. H., **4**, 382[134,134b]; **5**, 707[34], 725[34], 803[92], 979[26]
Bradbury, S., **7**, 743[65]
Brade, H., **6**, 33[7], 40[7], 57[7]
Brade, L., **6**, 33[7], 40[7], 57[7]
Braden, M. L., **1**, 476[126]; **3**, 66[15], 74[15]
Bradford, V. S., **3**, 969[134]
Bradley, C. W., **8**, 626[173]
Bradley, D. G., **8**, 201[145]
Bradley, J. N., **3**, 783[21]
Bradley, J. S., **1**, 443[179]
Bradley, W., **3**, 888[14,15]
Bradlow, H. L., **3**, 564[8]
Bradshaw, J. B., **6**, 436[28], 437[28], 447[28], 448[28], 449[28], 450[28], 452[28]
Bradshaw, J. S., **4**, 489[59,60]; **5**, 163[70]; **6**, 71[21], 283[158], 449[116]; **7**, 415[113]; **8**, 828[78], 838[19,19b]
Bradshaw, R. W., **6**, 79[65], 81[65]
Bradsher, C. K., **1**, 412[54]; **3**, 251[77], 254[77], 324[149]; **5**, 499[253,255], 501[253]; **8**, 323[114], 587[38], 588[38]
Brady, D. G., **4**, 48[138,138d], 66[138d], 597[179]
Brady, K., **6**, 252[153]
Brady, K. A., **8**, 764[6], 770[40]
Brady, L. E., **6**, 277[128], 288[128]
Brady, S. F., **3**, 362[90]
Brady, T. E., **8**, 967[83]
Brady, W. T., **3**, 848[49], 905[141]; **4**, 53[151]; **5**, 86[26], 87[38,39,43,44], 90[26,55,59], 94[59], 99[133,134,135,139], 100[133,135]
Braekman, J., **1**, 100[88]
Braekman, J. C., **5**, 456[86]; **6**, 914[27]
Braendin, H. P., **5**, 513[4]
Bragina, I. O., **6**, 542[598,599]
Brahma, R., **5**, 243[13], 244[13]
Brahme, K. C., **8**, 366[44]
Braillon, B., **5**, 576[136]
Brainina, E. M., **1**, 143[31]
Braish, T. F., **4**, 78[54,54c], 251[153]; **5**, 936[194]; **8**, 958[18]
Braitsch, D. M., **3**, 483[18], 500[18]; **8**, 839[22]
Bram, G., **1**, 34[226,227]; **6**, 2[5], 18[5]
Bramanik, B. N., **6**, 253[155]
Brambilla, R., **6**, 266[51]
Brämer, R., **2**, 360[167]
Bramley, R. K., **5**, 790[32], 802[81]
Brammer, L., **6**, 436[9]
Brams, C. T., **4**, 1004[71]
Branca, J. C., **1**, 531[129]; **3**, 864[18,22]
Branca, M., **4**, 930[50]
Branca, Q., **3**, 167[485], 168[485]; **6**, 1058[59]; **7**, 57[27]

Branca, S. J., **3**, 22[133], 906[144]; **4**, 1040[102]; **7**, 7[47], 683[89]; **8**, 795[17]
Branch, G. E. K., **7**, 602[100]
Branchadell, V., **5**, 64[40], 73[190]
Branchaud, B. P., **2**, 940[163]; **4**, 761[202,205], 768[237]; **8**, 72[238], 74[238]
Brand, M., **4**, 344[77], 347[92,94]
Brand, S., **2**, 345[34], 351[34], 357[34], 371[34,263]; **5**, 17[123], 468[124,125], 531[81], 545[121]
Brand, W. W., **3**, 86[35], 164[35], 173[35]
Brandänge, S., **2**, 233[188], 827[128]; **6**, 85[87]
Brandenburg, C. F., **5**, 830[36]
Brandes, E., **5**, 349[76], 854[179,180], 855[179]; **7**, 569[107]
Brandes, R., **2**, 957[21]
Brandi, A., **5**, 922[135]; **8**, 851[127]
Brandl, M., **7**, 49[62-64]
Brändli, U., **2**, 338[74]
Brandsma, L., **1**, 17[211], 23[121-125], 471[62]; **2**, 81[1], 82[1], 96[1], 587[145]; **3**, 87[120,213], 88[120], 105[120,213,221], 106[222], 113[222], 257[115], 271[2], 272[2], 273[2d], 521[2], 522[2], 551[6], 552[6]; **4**, 309[412], 869[22]; **5**, 772[164], 949[282]; **6**, 426[76,78], 480[111], 482[125], 962[76], 963[77], 965[77]; **8**, 478[45], 480[45], 481[45]
Brandstadter, S. M., **2**, 930[129]; **5**, 96[107], 410[41]
Brändström, A., **6**, 809[66]
Brandt, A., **7**, 439[25]
Brandt, C. A., **1**, 636[101], 640[101], 666[101]; **3**, 136[374], 141[374]; **6**, 175[74]
Brandt, J., **2**, 1099[107]; **4**, 874[48]
Brandt, P., **8**, 262[20]
Brandt, S., **3**, 500[132], 505[132]; **4**, 980[110], 982[110]
Brandvold, T. A., **5**, 1089[87], 1090[87,90], 1091[90], 1094[87], 1098[87], 1099[87,90], 1100[87], 1101[87,90], 1102[147], 1112[87], 1113[87]; **7**, 487[149]
Brannan, M. D., **8**, 28[37], 66[37]
Brannen, W. T., Jr., **7**, 765[148]
Brannfors, J. M., **7**, 270[128], 271[128]
Brannigan, L. H., **1**, 532[136]
Brannock, K. C., **3**, 21[122]; **4**, 45[126,126b,e]; **5**, 71[153,158], 686[50], 847[132], 850[151], 862[247]; **6**, 426[70]; **8**, 140[19]
Brannon, M. J., **4**, 16[51,51a]
Branton, G. R., **5**, 911[92]
Branz, S. E., **5**, 382[119b]
Bras, J.-P., **1**, 377[100]; **3**, 771[189]
Brasca, M. G., **4**, 27[81], 381[126b], 382[126], 383[126]; **5**, 524[55]
Brasch, J., **2**, 146[66]
Brasen, W. R., **3**, 918[21]
Brash, A. R., **5**, 781[206,207,208]
Braslavsky, S., **5**, 931[179], 937[179], 938[179], 947[179]
Brassard, P., **5**, 330[36]; **6**, 939[143], 941[143]
Brassier, L., **5**, 128[27]
Bratby, D. M., **5**, 818[151], 819[153]
Brattesani, D. N., **1**, 562[168]; **3**, 273[3]; **4**, 91[88a]; **6**, 1027[91]
Bratz, E., **7**, 450[15]
Bratz, M., **1**, 589[20,20a], 592[20a]; **2**, 1026[70,71]; **5**, 17[123], 468[126], 531[81]
Braude, E. A., **5**, 752[48], 754[48], 756[48], 757[48,77], 758[48,83], 759[77], 769[48]; **7**, 135[102]; **8**, 561[413]
Brauman, J. I., **3**, 315[112], 317[112,119]; **5**, 678[15]; **7**, 12[94], 282[179]; **8**, 550[334]
Braun, A. G., **7**, 365[47]
Braun, D., **7**, 822[32]
Braun, F., **2**, 943[168,169], 970[88]
Braun, H., **5**, 418[71]; **6**, 104[5]; **7**, 505[283,284]; **8**, 395[131]
Braun, L. M., **8**, 708[37]
Braun, M., **1**, 341[97]; **2**, 134[3], 190[57], 226[159], 231[170,171], 240[7], 305[10], 455[2], 761[50]; **3**, 124[255,256], 125[255], 128[255,256], 129[255]; **4**, 1007[121,123,131], 1008[123]; **5**, 151[6]; **7**, 384[114c], 399[38], 400[38], 406[38], 409[38], 415[38]
Braun, R., **5**, 344[65], 571[115], 854[175]; **8**, 91[63]
Braun, R. A., **8**, 708[37]
Braun, S., **1**, 37[247]
Bräuner, H.-J., **6**, 512[120,303], 543[120], 553[796,798], 572[796,958], 573[798,963], 576[973,974]
Bräuniger, H., **6**, 509[244]
Bräunling, H., **8**, 271[104]
Braunstein, D. M., **8**, 64[205], 66[205], 67[205], 370[87], 376[164]
Bräutigam, I., **7**, 751[141]
Braverman, S., **2**, 91[42]; **3**, 862[3]; **6**, 834[34], 837[62,63]
Bravetti, D., **3**, 717[44]
Bravo, P., **1**, 514[51]; **3**, 147[393]; **4**, 382[131a,b], 384[131b]; **8**, 836[2], 843[2f], 856[170,171]
Bravo-Borja, S., **8**, 827[75]
Brawn, N., **8**, 950[167]
Bray, T. L., **5**, 15[108], 850[144]
Braye, E. H., **5**, 1139[73]
Breant, P., **1**, 474[99]
Breau, L., **1**, 825[50]; **8**, 242[40]
Brechbiel, M., **2**, 528[12]; **5**, 2[10]; **6**, 1013[9]
Brecht, A., **4**, 55[156]; **8**, 563[434]
Brede, O., **4**, 307[389]
Bredenberg, J. B., **8**, 528[86]
Bredenkamp, M. W., **6**, 189[185]
Bredereck, H., **6**, 425[67], 515[315], 519[336], 540[590], 545[634], 546[634], 553[797], 554[799], 556[829], 567[634,829], 568[634]; **7**, 657[30], 768[200]
Bredon, L. D., **5**, 143[101], 144[101]
Bredt, J., **6**, 955[22]
Bregant, N., **7**, 777[366]
Bregeault, J. M., **7**, 452[51,53], 453[51]
Bregovec, I., **7**, 777[366]
Brehm, A., **6**, 233[43]
Brehm, W. J., **4**, 120[202]
Brehme, R., **2**, 791[61]; **6**, 727[202]
Breil, H., **1**, 139[6]
Breiman, R., **2**, 887[53]; **5**, 92[71]
Breindel, A., **8**, 754[95]
Breining, T., **8**, 15[89]
Breitenbach, J. W., **5**, 63[19]
Breitgoff, D., **7**, 397[30]
Breitholle, E. G., **1**, 856[56]; **4**, 611[358]
Breitmaier, E., **2**, 779[4], 780[4]; **3**, 509[181]; **5**, 412[47,47b], 416[58], 432[131,133]
Breitner, E., **8**, 141[29], 533[141]
Breliere, C., **8**, 262[18]
Bremer, B. W., **6**, 685[354], 959[46]
Bremholt, T., **7**, 835[85]
Bremmer, M. L., **5**, 414[52], 415[55], 539[108]
Bremner, J., **8**, 847[88]
Bremner, J. B., **2**, 421[26]
Breña, L. J., **2**, 514[51]
Brena-Valle, L. J., **6**, 119[110]
Brendel, J., **4**, 152[55]
Breneman, C. M., **1**, 428[121], 429[121], 457[121]; **3**, 209[15], 211[34a], 213[34a]; **4**, 184[82]
Brener, L., **8**, 713[75,78], 714[75]
Brennan, J., **1**, 755[115], 812[115], 813[115]; **6**, 193[215]
Brennan, M. P. J., **3**, 635[36]
Brennan, M. R., **7**, 154[16], 174[137]
Brenner, A., **7**, 154[15]
Brenner, D. G., **1**, 389[141]; **6**, 278[130]

Brenner, W., **3**, 390[84], 392[84]; **5**, 800[76], 809[113]; **6**, 685[354], 959[46]
Breque, A., **3**, 201[82]; **8**, 865[247]
Bresadola, S., **7**, 500[239]
Bresinsky, E., **4**, 52[147,147f]
Breslauer, H., **7**, 92[46]
Breslow, D. S., **3**, 1029[54]; **7**, 21[14], 24[29,30,36,38], 25[36,43], 26[43,47], 219[14]; **8**, 447[118], 454[118], 455[118], 568[481]
Breslow, R., **1**, 543[18]; **3**, 896[67], 1058[40]; **4**, 1005[92]; **5**, 344[66], 345[66], 346[66], 453[66], 493[212], 789[18], 854[175], 1133[28]; **6**, 564[908]; **7**, 24[39], 25[39], 40[8,13], 42[32,34], 43[8,35,36,38,46,47], 46[48-50], 47[50-52,54], 48[58-61], 49[62-68], 50[74], 805[66], 854[49], 855[49]
Bressan, M., **7**, 238[43]
Bresse, M., **1**, 462[16]
Bressel, U., **5**, 75[217], 78[217]
Brestensky, D. M., **4**, 254[184]; **8**, 550[330]
Breswich, M., **1**, 310[107]
Breton, P., **6**, 690[401], 692[401]
Bretsch, W., **4**, 1046[115]
Bretschneider, H., **2**, 740[62]; **7**, 678[69]; **8**, 496[32]
Bretschneider, T., **1**, 122[72]
Brett, D., **6**, 228[18]
Bretting, C., **4**, 149[52], 181[72,73]
Brettle, R., **3**, 583[126], 635[36], 638[94], 643[122]; **7**, 794[7b,d], 805[67]; **8**, 563[435]
Breu, V. A., **6**, 543[616]
Breuckmann, R. B., **5**, 64[29]
Breuer, A., **8**, 352[148]
Breuer, E., **1**, 836[141]; **5**, 959[319]; **7**, 606[153]; **8**, 315[54], 316[54], 707[25]
Breuer, S. W., **3**, 418[27]; **7**, 595[28,31]
Breuer, W., **2**, 1099[110]
Breuilles, P., **5**, 305[84,86]
Breulet, J., **5**, 703[15]
Brevilles, P., **4**, 593[136], 763[212]
Brew, S. A., **8**, 70[228]
Brewer, D., **5**, 167[94]
Brewer, E. N., **2**, 838[176]
Brewer, P. D., **3**, 251[77], 254[77]
Brewster, A. G., **7**, 110[188]
Brewster, J. H., **3**, 927[51,57]; **4**, 3[9]; **8**, 314[41,42], 315[41,42], 541[207], 814[15]
Brewster, P., **6**, 3[13], 4[13]
Brewster, R. Q., **8**, 916[113]
Brewster, W. D., **8**, 409[80]
Brey, W. S., Jr., **3**, 898[80]
Brezhnev, L. Yu., **8**, 606[24], 607[24]
Brice, M. D., **1**, 2[3], 37[3]
Brich, Z., **4**, 257[219]; **6**, 71[22], 726[176]
Brickner, S. J., **3**, 423[79], 1031[65,65b]
Brickner, W., **3**, 705[1]
Bridges, A. J., **2**, 90[40], 616[138]; **3**, 139[378], 154[378], 155[378]; **4**, 336[28], 342[28], 346[28], 347[28], 395[205]; **5**, 333[45], 683[36a]; **6**, 146[89], 1016[33], 1022[33]; **7**, 316[47], 317[47], 521[35]; **8**, 844[64,64a]
Bridon, D., **4**, 747[153], 765[223]; **7**, 726[35,37], 727[38], 728[41], 730[49]
Bridson, J. N., **2**, 111[83], 242[15], 909[64], 911[64]; **4**, 145[31,32]
Brieger, G., **5**, 451[30], 453[30], 464[30], 513[5], 514[5], 527[5]; **8**, 320[88-90], 440[82], 551[339]
Brien, D. J., **5**, 1154[159]
Briggs, L. H., **3**, 759[126]; **7**, 92[40]; **8**, 228[131]
Briggs, P. C., **3**, 347[25]; **8**, 530[109]
Brijoux, W., **5**, 1152[140,141], 1154[140,149]
Brikhahn, M., **4**, 1041[105]
Brill, G., **2**, 356[132]
Brimacombe, J. S., **1**, 760[135]; **4**, 297[272,273,274]; **7**, 262[79], 440[38,39a,b]
Brimble, M. A., **6**, 750[105]; **7**, 350[19]; **8**, 844[66]
Brindell, G. D., **8**, 755[133]
Brindle, I. D., **4**, 391[178c]
Brindle, J. R., **8**, 64[206], 66[206], 250[100], 253[100], 254[126]
Brindley, P. B., **7**, 599[66]
Brine, G. A., **3**, 677[85]
Briner, K., **6**, 54[128], 128[162]
Briner, P. H., **2**, 69[48]; **3**, 196[32]
Bringhen, A. O., **5**, 635[87]
Bringmann, G., **2**, 173[181]; **3**, 505[169], 677[82]; **6**, 489[100], 734[8], 736[8,29], 738[52-56], 751[53]; **8**, 830[89], 831[89-91]
Brini, M., **8**, 568[469]
Brinich, J. M., **1**, 488[12]
Brink, M., **7**, 306[5]
Brinker, U. H., **3**, 466[183]; **4**, 1009[146], 1012[173,176], 1022[252]; **7**, 800[30]
Brinkman, G. A., **7**, 535[47]
Brinkman, K., **2**, 125[223]
Brinkman, M. R., **7**, 26[50]
Brinkmann, A., **5**, 293[46], 1188[19], 1190[25], 1191[25,30], 1193[30], 1197[42,43]
Brinkmann, R., **5**, 37[22b], 1152[141], 1153[146,147], 1154[149]
Brinkmeyer, R. S., **1**, 608[41]; **3**, 125[291], 126[291], 223[152,153], 363[88], 373[127]; **4**, 402[235], 404[235], 408[235]; **6**, 488[31], 571[31]; **7**, 503[273]; **8**, 165[50], 545[292,293]
Brintzinger, A., **6**, 226[9]
Brintzinger, H. H., **8**, 690[102]
Brion, F., **4**, 707[43]
Brion, J.-D., **2**, 141[39]; **6**, 436[8], 515[314], 522[314]
Brisdon, B. J., **4**, 837[19]
Brisse, F., **2**, 605[61], 625[61]
Bristol, D., **7**, 41[16]
Bristol, J. A., **6**, 266[51]
Britcher, S. F., **3**, 380[11]
Britikova, N. E., **6**, 543[609]
Brito, A., **3**, 327[171]
Britt, R. W., **2**, 74[73]
Brittain, R. T., **2**, 323[33]
Brittain, W. J., **6**, 498[163]
Brittelli, P. R., **6**, 172[10]
Britten-Kelly, M. R., **4**, 259[276]; **6**, 981[47]; **8**, 836[9], 847[9], 848[9]
Britton, D., **1**, 33[164]
Britton, R. W., **2**, 843[196]; **3**, 20[108]; **8**, 121[79], 986[14]
Britton, T. C., **1**, 400[9]
Britton, Th. C., **6**, 118[103], 248[137], 256[171]
Britton, W. E., **8**, 389[72]
Brizzolara, A., **3**, 28[170], 30[170]; **4**, 6[21,21a], 7[21a]; **6**, 703[3], 714[3]
Brlik, J., **6**, 917[36]
Broach, V., **8**, 357[195], 358[195]
Broadbent, H. S., **3**, 124[282], 125[282], 128[282], 129[282]; **8**, 140[14], 248[89]
Broadbent, T. A., **4**, 24[74]
Broadhurst, M. D., **2**, 477[12]; **3**, 431[95,96]
Broadhurst, M. J., **3**, 380[10]; **4**, 709[45], 710[45,52]; **5**, 395[148]; **7**, 351[28], 355[28]
Broadley, K., **2**, 910[67], 933[140]
Brobst, S. W., **4**, 497[97]
Brocard, J., **3**, 990[34,34a]
Brock, C. P., **5**, 1068[12]
Brock, D. J. H., **3**, 699[162]

Brocker, U., **8**, 397[143]
Brockman, H. L., **2**, 456[28]
Brockmann, C. J., **3**, 634[18]
Brockmann, H., **3**, 699[160]
Brockmann, M., **8**, 764[9]
Brocksom, T. J., **1**, 641[107], 672[107], 677[107]; **3**, 222[141]; **5**, 828[7], 839[7], 882[13], 888[13], 891[37], 892[13,37], 893[13]; **7**, 239[54], 355[39]
Brocksom, U., **7**, 355[39]
Brockway, C., **3**, 224[163,163a]
Brode, W. R., **2**, 960[34]
Brodie, B. B., **7**, 11[89]
Brodie, T. D., **5**, 718[95]
Brodowski, W., **2**, 598[13]; **5**, 388[137]
Brodskaya, T. G., **8**, 99[107]
Brodt, W., **6**, 650[128]
Broeckx, W., **7**, 475[52]
Broekema, R. J., **6**, 559[860]
Broekhof, N. L. J. M., **1**, 563[172,174]; **6**, 134[32], 705[32-34]; **7**, 235[1]
Broekhuis, A. A., **2**, 662[22], 663[22], 664[22]; **5**, 76[241], 431[121], 434[121b]
Broess, A. I. A., **3**, 367[100]
Broger, E. A., **3**, 431[101]; **5**, 835[59]
Brogli, F., **7**, 867[90]
Brohi, M. I., **3**, 530[74], 534[74]
Brois, S. L., **7**, 474[37]
Broka, C. A., **1**, 104[98]; **2**, 622[155]; **3**, 125[294], 126[294], 167[294], 168[294]; **4**, 378[110], 384[110], 812[181]; **6**, 851[128], 879[43]; **7**, 246[88], 519[21], 633[66]
Brokaw, F. C., **1**, 742[46]; **3**, 766[158]
Brokke, M. E., **3**, 790[59]
Broline, B. M., **5**, 229[121]; **7**, 616[11]
Bromidge, S. M., **4**, 1089[128], 1092[128], 1093[128]
Bromley, D., **7**, 718[3], 724[3]
Broms, M., **4**, 313[461]
Bronberger, F., **5**, 430[116,116b]
Bronneke, A., **1**, 162[91]
Bronson, J. J., **5**, 847[136], 913[105], 1012[49]
Brook, A. G., **1**, 436[147], 608[39], 618[56]; **2**, 601[35]; **3**, 125[314]; **4**, 98[112], 115[182], 120[197]; **5**, 843[126]; **6**, 140[54]; **7**, 163[77], 164[77]
Brook, M. A., **1**, 180[50], 182[50], 366[49], 391[49,149]; **2**, 336[72], 616[132], 619[132], 620[151], 621[132], 630[5], 913[74], 978[10]; **4**, 161[89c]; **6**, 677[309]
Brook, P. R., **3**, 767[162], 901[109]; **7**, 676[63]
Brooke, G. M., **8**, 901[37], 904[53]
Brooke, P. K., **4**, 435[136]
Brooker, A. C., **5**, 949[278]
Brooker, L. G. S., **6**, 477[100]
Brookes, A., **3**, 380[10]
Brookes, C. J., **3**, 648[187]
Brookhart, M., **1**, 116[43]; **4**, 115[180c], 688[67], 696[5,7], 703[32-35], 704[32-34], 705[34b], 712[34,35], 976[100], 980[100j], 983[117,119], 984[120,123,124], 985[130]; **5**, 715[78], 904[49], 905[49], 1084[58], 1086[70,74], 1089[80]
Brookhart, T., **5**, 5[43,45], 8[43,45]
Brooks, D. J., **8**, 584[17]
Brooks, D. W., **1**, 385[116,120], 386[120]; **2**, 112[98], 195[71], 244[30], 246[34], 247[34], 801[31]; **6**, 446[102]; **7**, 579[132]; **8**, 190[75], 198[132], 201[140,142]
Brooks, G., **6**, 471[64]
Brooks, H. G., **1**, 619[60]; **4**, 120[197]
Brooks, J. J., **1**, 12[59]
Brooks, L. A., **2**, 838[174], 839[174]
Brooks, P., **8**, 859[211]
Brookshire, K. W., **5**, 736[145], 737[145]
Broom, A. D., **6**, 635[12], 643[12], 660[210], 662[210]; **8**, 533[154]
Broom, D. M. P., **1**, 130[95]
Broom, N. J. P., **4**, 14[47], 111[154f]
Broome, J., **8**, 315[44], 545[281]
Brophy, B. V., **3**, 901[109]; **7**, 676[63]
Brophy, G. C., **5**, 468[135]
Broschard, R. B., **8**, 973[120]
Broser, E., **5**, 394[146], 395[146]
Bross, H., **5**, 734[137]
Brossi, A., **2**, 1048[7], 1049[7]; **4**, 505[149]; **6**, 736[28], 738[53], 751[53]; **8**, 968[86]
Brossmer, R., **2**, 464[97]; **7**, 306[6]
Broster, F. A., **3**, 418[27]; **7**, 595[28]
Brot, F. E., **3**, 358[66]
Brothers, D., **7**, 130[77]
Brotherton, C. E., **4**, 433[119]; **5**, 65[57], 78[270], 266[76], 267[76], 268[76], 573[127], 594[7], 599[7], 604[7,54]; **6**, 559[864], 751[107]
Brotherton-Pleiss, C. E., **5**, 266[76], 267[76], 268[76]
Brougham, P., **7**, 194[4], 374[78], 674[41]
Broughton, H. B., **4**, 27[81], 35[98f], 380[121]; **5**, 524[55]
Brouin, J., **4**, 905[210]
Brounts, R. H. A. M., **8**, 95[82]
Brouwer, A. C., **3**, 750[82]; **5**, 116[255]
Brouwer, A. M., **5**, 707[40]
Brouwer, D. M., **2**, 727[135]; **3**, 330[191]
Brouwer, W. M., **7**, 759[10]
Brovarets, V. S., **6**, 524[365]
Brovet, D., **8**, 303[98]
Brovko, V. S., **8**, 765[13]
Brower, K. R., **4**, 433[122]
Brown, A., **5**, 857[228]
Brown, A. C., **3**, 633[4]; **6**, 999[124]; **8**, 843[56]
Brown, A. L., **2**, 771[12]
Brown, B. B., **5**, 524[50,52], 539[50], 548[50c]
Brown, B. R., **8**, 314[38], 315[44], 545[281]
Brown, B. S., **3**, 438[35]
Brown, C. A., **1**, 511[29], 566[213]; **2**, 709[12], 710[17], 799[13]; **4**, 10[34], 113[163]; **8**, 14[82,84], 18[120], 240[35], 375[158], 418[13], 422[13], 425[13], 715[85,85b], 716[85]
Brown, C. H., **8**, 756[159], 758[159]
Brown, C. J., **3**, 688[115]
Brown, C. K., **8**, 445[28]
Brown, D., **6**, 487[70], 836[58]; **7**, 131[86], 273[134]
Brown, D. J., **3**, 568[45], 636[59]; **6**, 533[489], 554[732]; **8**, 906[66], 909[66]
Brown, D. L., **4**, 404[244]; **7**, 410[97b]
Brown, D. M., **6**, 614[85]; **8**, 879[50]
Brown, D. S., **2**, 657[162]; **5**, 21[161], 23[161]
Brown, D. W., **2**, 198[85], 632[27], 640[27]; **4**, 1011[166]
Brown, E., **3**, 501[139], 509[139], 512[139]; **4**, 258[255]
Brown, E. A., **6**, 680[331]
Brown, E. D., **3**, 390[71,72], 396[111,114], 398[111,114]; **8**, 231[146]
Brown, E. G., **4**, 380[123], 386[148a], 387[148,148a]; **6**, 26[107]
Brown, E. J., **5**, 829[20]
Brown, E. V., **3**, 564[8]; **7**, 829[57]
Brown, F., **3**, 898[74]; **4**, 976[100]
Brown, F., Jr., **8**, 389[71]
Brown, F. J., **5**, 1065[2], 1066[2], 1074[2], 1083[2], 1084[2], 1093[2]
Brown, F. K., **5**, 79[292], 257[61], 347[72,72c], 516[19,22], 517[22], 518[22]
Brown, F. R., Jr., **2**, 868[20], 869[20], 871[20], 872[20], 876[20], 890[20]
Brown, G. G., Jr., **4**, 665[10], 667[10], 670[10], 674[10]

Brown, G. M., **2**, 466[125]

Brown, G. R., **8**, 563[433], 614[84]

Brown, G. W., **6**, 204[8], 209[8], 213[8]

Brown, H. C., **1**, 192[81], 223[84], 225[84c], 292[23], 343[104], 489[14,16], 563[183]; **2**, 5[22], 6[23,23c], 10[22,23], 13[56], 33[23c,123], 42[146], 43[146,146b], 44[146b,c], 45[146b,c], 52[23a], 111[83,85,86], 112[88], 241[14], 242[14b,16], 247[82], 735[11], 977[3]; **3**, 231[245], 242[6], 249[66], 257[6], 259[6], 260[143], 274[21], 299[33], 300[35,36], 421[53], 483[10,11], 522[20], 523[24], 554[24], 706[6], 779[8], 793[8,69,71-75], 794[76-78], 795[8,75,82-84], 797[75,91-95], 798[95], 799[98]; **4**, 145[23,24,29a], 146[37a,b], 147[38a,39,40], 164[99,99a], 272[32,34], 273[34,47,49,52], 294[244], 300[312,313], 301[312,313,314,315,318,328,329,330], 302[328], 303[339], 309[416,417], 314[478], 756[181,183]; **6**, 78[59,60], 283[167], 284[167,169], 724[160], 865[201], 954[16], 955[25], 958[30]; **7**, 14[134], 16[166], 253[13], 264[90], 474[40,48], 476[48], 594[3], 595[3,11-13,17,19,24,30,127], 596[35-37], 597[49], 598[3,19], 599[62,63,69-71], 600[77], 601[3,77,80,89,90], 602[92,93,103], 603[123,126], 604[127,132,135,137], 605[141,142,144,145], 606[146,147,149,152,153,154,159], 607[162,163,164,165,169], 608[169]; **8**, 1[1], 2[4,14], 14[78,79,84], 15[90], 16[105], 17[109], 18[119,120], 26[2,8,11,13,18,23,25], 27[8,11,13,18], 36[8,11,18], 37[8,11], 43[23], 64[201], 66[23,25,201], 67[25], 74[25,251], 101[123], 159[10], 170[89], 180[10], 214[31,51], 217[68], 237[8,9,15,16], 238[9], 240[8,9,15,34,35], 241[9], 244[8,9,15,16,48,53,60-62], 245[9], 247[8,9,34,48,53], 249[8,15,48,60,93,95], 250[34], 251[104], 253[8,16,48,60,104,116], 260[3,4], 261[7,11-13], 263[21-24,27], 265[21-23], 267[64,65], 269[27,87], 271[105,106,111,112], 273[27], 274[132,133], 275[27], 278, 279, 373[135], 386[51], 403[9], 408[65], 412[111], 524[1], 537[177], 541[1,206,207,211,218], 542[218], 543[218,241], 544[206,269,270,271], 703[1], 704[1-3,5,6,8], 705[1,3,5,6,9,11-13], 706[1,5,6,15,16], 707[1,9,19,21,23-30], 708[1,3,30,33,36,37,40,42], 709[1,15,15a,43,43a,47-51], 710[49,52-55,57-60], 711[62,63,65,67,68], 712[1], 713[1,70-75,77,78], 714[74,75,79-83], 715[1,42,77,84,85a], 716[1,33,85-88,91,92], 717[1,27,28,42,57,98-100,102-104], 718[65,85a,107-115], 719[62,117,118,120-123], 720[1,128,129,134], 721[1,53,54,60,134,139,140,141,142,143], 722[1,134,142,144,145,146,147,148,149], 724[1,33,152,155-167,174,177,178], 725[8a,174,178], 726[11,33,174,177,178,178b,187], 727[112,178], 728[42], 802[78], 803[94], 804[97,98], 805[98], 853[145], 854[149], 856[164,166], 875[34,36,38,39], 876[34,38,42], 877[42], 880[58], 901[36], 906[36], 907[36], 908[36], 909[36], 910[36], 934[55], 939[98], 968[89]

Brown, H. L., **4**, 30[87]; **6**, 1050[38]

Brown, I., **7**, 27[74]

Brown, J. B., **2**, 156[114]

Brown, J. D., **1**, 359[20], 383[20], 463[30,31], 466[44]; **3**, 255[109], 512[198]; **4**, 76[49], 243[66], 252[66]; **8**, 95[90], 584[18]

Brown, J. F., **7**, 488[161]

Brown, J. H., **7**, 613[1]

Brown, J. H., Jr., **8**, 212[16]

Brown, J. L., **8**, 332[40]

Brown, J. M., **1**, 808[321]; **3**, 1022[19]; **5**, 10[78], 791[42], 794[45], 803[42], 916[119], 971[2], 972[2], 973[2], 984[34]; **6**, 51[106], 1056[53]; **7**, 671[13]; **8**, 447[138,139], 448[145,145b,150], 449[152], 452[145], 460[253], 462[266], 463[268], 509[92], 510[92]

Brown, K., **5**, 196[12]

Brown, K. A., **7**, 862[81], 880[155], 888[81]

Brown, K. C., **7**, 564[83], 584[158]; **8**, 383[14], 917[115], 918[115]

Brown, K. H., **5**, 586[205]

Brown, K. L., **6**, 708[50]

Brown, L., **1**, 824[45]; **2**, 205[104], 206[104]; **3**, 225[187]; **5**, 835[59]; **6**, 174[54]

Brown, L. A., **5**, 8[56]

Brown, L. E., Jr., **2**, 175[184]

Brown, L. R., **6**, 801[35]

Brown, M., **4**, 83[65a]; **6**, 960[61]; **7**, 111[190]; **8**, 444[11]

Brown, M. J., **1**, 891[147]; **2**, 554[135]; **3**, 792[68]

Brown, N. M. D., **1**, 488[6,10]

Brown, P., **3**, 1056[34], 1062[34]; **5**, 76[243], 385[130]; **6**, 930[85]; **8**, 220[83]

Brown, P. B., **5**, 125[15], 153[25]

Brown, P. E., **4**, 1019[227]

Brown, P. J., **4**, 102[129], 317[557]

Brown, P. R., **6**, 449[116]

Brown, P. S., **6**, 546[652]

Brown, R. A., **1**, 461[13], 464[13]; **7**, 220[24]

Brown, R. D., **4**, 484[13]; **8**, 495[28]

Brown, R. E., **7**, 664[65]

Brown, R. F. C., **2**, 353[98], 355[125], 356[133,134], 838[176]; **5**, 732[133,133a,b]; **7**, 30[84], 827[48]

Brown, R. J., **6**, 89[114], 843[86]; **7**, 371[69], 418[129a,b]

Brown, R. K., **4**, 331[6]; **5**, 455[78]; **8**, 214[38], 218[71,72,74], 219[72], 221[89-92], 222[91,92,94,95], 227[118], 229[138], 230[138], 612[71]

Brown, R. S., **4**, 330[3], 345[3]; **6**, 570[950], 711[63]

Brown, R. T., **2**, 822[111], 852[233]; **3**, 380[7]; **4**, 1004[77], 1021[77]

Brown, R. W., **7**, 316[44]

Brown, S. B., **7**, 662[54,55]

Brown, S. F., **7**, 771[258]

Brown, S. H., **3**, 1047[4], 1062[4]; **7**, 5[27], 15[27]

Brown, S. L., **2**, 125[222]; **3**, 325[161]

Brown, S. M., **2**, 757[18]

Brown, S. P., **2**, 757[9]; **5**, 467[116], 531[80], 942[233]

Brown, S. S., **3**, 326[165]

Brown, T. H., **4**, 33[95,95a]

Brown, T. L., **1**, 9[43], 41[199]

Brown, W., **2**, 142[46]

Brown, W. G., **6**, 959[42]; **8**, 26[3], 27[3], 36[3], 237[20], 238[20], 241[20], 245[20], 247[20], 251[20], 253[20], 568[483], 872[2]

Brown, W. L., **4**, 368[16]; **6**, 839[68], 902[133]

Brown, W. V., **7**, 768[202]

Brownawell, M. L., **8**, 850[121]

Brownbridge, P., **1**, 879[110]; **2**, 73[65], 74[65], 281[32], 599[21], 606[64,68], 607[70], 616[132], 617[145], 619[64,68,132,146], 620[152], 621[132,154], 623[158], 630[4]; **3**, 3[12], 8[12], 25[12]; **4**, 158[78], 371[52], 378[52b]; **6**, 544[627], 932[95]; **7**, 163[74], 164[74], 167[74], 177[74], 493[200], 494[200]

Browne, A. R., **3**, 627[43]; **4**, 1013[182]; **5**, 324[21]

Browne, A. T., **4**, 484[21]

Browne, C. E., **8**, 354[177]

Browne, E. J., **6**, 533[501]; **7**, 262[74]

Browne, L. J., **4**, 173[32]; **5**, 803[90], 979[24]; **6**, 746[95]

Browne, L. M., **6**, 1066[93]; **8**, 397[141], 880[64]

Brownell, R., **7**, 236[19]

Brownfain, D. S., **7**, 300[56]

Browning, K. C., **4**, 47[133]

Brownlee, P. P., **6**, 277[124]

Brownlee, R. G., **3**, 126[318]

Brownstein, S., **7**, 856[65]

Brownstein, S. K., **4**, 892[144]

Brown-Wensley, K. A., **1**, 748[72], 749[78], 816[78]; **4**, 979[101]; **5**, 1115[3], 1122[3,30], 1123[3], 1124[3]

Brubaker, A. N., **2**, 760[43]

Brubaker, C. H., Jr., **8**, 447[126], 457[126], 676[54]

Brubraker, G. R., **7**, 225[62]

Bruce, D. W., **1**, 252[4]

Bruce, J. M., **3**, 810[43]; **7**, 345[4]; **8**, 330[27]

Bruce, M. I., **5**, 1068[14]

Bruce, M. R., **6**, 456[158]

Bruch, M., **3**, 587[141]

Bruche, L., **6**, 252[152]
Brück, B., **4**, 124[214c]; **5**, 829[23]
Bruck, M. A., **1**, 36[235]
Bruckner, C., **4**, 1035[39], 1046[39]; **6**, 456[161]
Brückner, R., **3**, 1002[58], 1008[66]; **6**, 27[116], 833[25], 834[25], 852[134,137], 873[7], 875[7], 879[45], 882[49], 883[53], 885[49,53], 886[59], 888[7], 889[68,71,72], 890[49,53,68,72,73]
Bruder, W. A., **6**, 84[85]; **7**, 229[119]
Bruderer, H., **8**, 528[68], 530[68]
Bruderlein, F., **2**, 284[54]
Bruderlein, H., **2**, 284[54]
Brüggemann, K., **2**, 356[132]
Brüggemann-Rotgans, I. E. M., **8**, 556[375]
Bruggink, A., **2**, 855[237]
Brugidou, J., **2**, 902[44]
Brugman, A., **4**, 52[146]
Bruhn, F.-R., **6**, 478[104]
Bruhn, J., **5**, 812[131], 894[43]
Bruhn, M. M. S., **8**, 544[264], 756[159], 758[159]
Bruice, T. C., **2**, 279[13], 902[41]; **7**, 763[90,91], 769[91]; **8**, 93[69], 94[69]
Bruix, M., **4**, 425[26]
Brumby, T., **2**, 351[81], 364[81], 371[263,264,265], 375[81]; **5**, 17[123,124], 468[127], 531[81,81d]
Brumfield, M. A., **7**, 876[124]
Brummel, R. N., **4**, 315[514]
Brun, L., **1**, 302[75]
Brun, P., **3**, 380[13]; **5**, 641[133]
Bruncks, N., **1**, 253[12]
Brundle, C. R., **7**, 852[43]
Brune, H.-A., **5**, 680[24], 683[24c]
Brunelet, T., **7**, 280[173], 281[173], 283[173,184], 285[173], 840[4], 844[4,63], 845[4,63]
Brunelle, D. J., **1**, 408[35], 430[35]; **2**, 121[188]; **3**, 8[41], 9[41], 15[41], 20[41], 250[70], 288[63]; **4**, 178[61], 189[105], 192[119], 197[119b], 245[87], 256[206], 372[64a], 439[162], 443[162], 444[196]; **6**, 438[66], 660[202]; **7**, 678[73]; **8**, 881[68]
Bruner, L. B., **8**, 385[46]
Brunet, J.-C., **4**, 519[20], 520[20]
Brunet, J. J., **4**, 452[19]; **8**, 16[103], 140[10], 141[10], 142[10], 418[14,15], 483[57], 485[57], 533[138], 558[391,392,393], 802[76], 909[79], 910[79]
Brunet, M., **8**, 285[6]
Brunetti, P., **2**, 463[80], 464[80]
Brungs, P., **2**, 1051[39], 1075[39]
Bruni, E., **4**, 408[259c]
Brunings, K. J., **8**, 566[450]
Bruniquel, F., **8**, 343[112]
Brunk, H. J., **6**, 561[881,882]
Brunke, E.-J., **3**, 343[18], 353[51], 354[51], 373[130]
Brunner, E., **1**, 70[63], 141[22]; **2**, 120[179]; **4**, 229[238]; **6**, 501[203], 531[203]
Brunner, G. L., **3**, 380[7]
Brunner, H., **1**, 310[107]; **4**, 230[245], 983[119]; **5**, 1062[59], 1086[74]; **7**, 401[61a]; **8**, 84[16], 91[57], 152[170], 174[121-125], 178[125], 179[125], 459[228,233], 461[260]
Brunner, R. K., **4**, 350[119]; **5**, 841[94]
Bruno, F., **3**, 636[60]
Bruno, G., **4**, 452[20]
Bruno, J. W., **7**, 881[157]
Bruno, M., **2**, 657[162]
Brunold, A., **3**, 640[97]
Brunovlenskaya, I. I., **8**, 318[67]
Bruns, L., **5**, 66[77]
Bruns, W., **6**, 677[323]
Brunskill, J. S. A., **2**, 361[177]
Brunsvold, W. R., **4**, 104[135a], 191[114]; **5**, 1076[36,37,39,42], 1082[55,56], 1083[36]
Brüntrup, G., **2**, 205[101,101b], 211[113], 597[8]; **4**, 111[155c]
Bruson, H. A., **2**, 956[11]; **4**, 3[7], 4[7,14], 6[14f], 31[7f], 59[7f]; **6**, 271[84]
Bruss, D. R., **3**, 571[58]
Brussani, G., **4**, 379[115], 380[115g], 391[115g]; **7**, 523[48]
Brussee, J., **1**, 546[52]
Brutcher, F. V., **7**, 438[14], 444[14]
Brutcher, F. V., Jr., **3**, 892[51]
Brutts, D., **3**, 197[33]
Bruza, K. J., **3**, 252[82,85], 257[82]; **4**, 191[113], 192[113]
Bruzik, K. S., **6**, 604[27], 606[40]
Bryan, C. A., **5**, 323[14]; **7**, 764[131]
Bryan, D. B., **3**, 273[15], 274[15]; **7**, 219[11]
Bryan, H., **4**, 31[93]
Bryan, H. G., **7**, 543[12], 551[12]
Bryan, R. F., **3**, 670[57,57a], 816[82]
Bryant, D. R., **4**, 924[34]; **8**, 859[216]
Bryant, J. A., **3**, 503[149], 512[149]; **4**, 427[66]
Bryant, K. E., **5**, 839[77]
Bryant, T., **5**, 909[99], 958[99]
Bryce, M. R., **4**, 509[162]; **5**, 379[112], 383[112], 384[112], 422[85]
Bryce-Smith, D., **3**, 415[6], 807[18]; **5**, 71[151], 637[104], 645[1], 646[2,4,9,10], 647[1a,e,f,11,12], 648[1e,11,12,20], 649[1f], 651[1], 654[27], 671[1a]
Bryker, W. J., **7**, 740[44]
Bryndza, H. E., **8**, 671[16]
Bryson, I., **3**, 404[136], 405[139]
Bryson, T. A., **1**, 366[44], 749[80]; **2**, 388[342], 900[28], 901[28], 910[28]; **3**, 24[146], 86[60], 88[60], 89[60], 364[93], 380[13]; **4**, 116[185d]; **5**, 850[149], 1123[37]; **6**, 501[205], 980[30]; **7**, 771[271]; **8**, 214[48]
Bryusova, L. Y., **5**, 850[146]
Brzezinski, J. Z., **1**, 474[90,92,93]
Bsata, M., **1**, 9[45]
Bub, O., **4**, 140[7b]
Buback, M., **5**, 458[72,73], 459[72]
Bube, T., **2**, 39[137], 40[137c], 44[137c]; **3**, 797[90]
Bubnov, N. N., **7**, 884[185,186]
Bubnov, Yu. N., **4**, 145[29b], 885[113-115], 886[115]; **5**, 33[8], 34[9,10]; **7**, 594[7], 595[7,20], 597[44], 598[7], 599[7], 601[7], 603[117]; **8**, 102[129], 705[14], 707[14], 725[14], 726[14], 727[196], 728[14]
Buc, S. R., **4**, 31[92,92g]; **6**, 261[7], 268[7]; **7**, 766[173]
Bucciarelli, M., **1**, 837[155], 838[160]; **7**, 747[96], 778[402]; **8**, 187[37]
Buch, A., **5**, 595[8]
Büch, H. M., **5**, 287[34], 288[34], 289[34], 290[34], 291[34], 294[34], 296[34], 297[34,55], 308[34], 641[134], 901[18], 905[18], 947[18], 948[18], 1185[1], 1192[31], 1197[31], 1204[1a]
Buch, M., **5**, 299[67]
Buchan, G. M., **7**, 158[39]
Buchanan, A. C., **3**, 328[177]
Buchanan, C. M., **1**, 66[55,56]; **2**, 536[44,46]
Buchanan, D., **1**, 174[7]
Buchanan, G. L., **2**, 897[14]; **4**, 30[87]; **5**, 794[50]; **6**, 1050[38]
Buchanan, J. G., **3**, 733[1]; **6**, 614[89]; **8**, 309[10], 311[10], 312[10], 313[10]
Buchanan, J. L., **5**, 516[25]
Buchanan, R. A., **3**, 960[116]; **5**, 829[16], 894[48]; **6**, 898[107b]
Buchart, O., **4**, 1089[131]
Buchbauer, G., **3**, 380[7]
Bucheck, D. J., **5**, 127[23]
Bucheister, A., **5**, 273[6], 277[6], 279[6]
Büchel, K. H., **5**, 457[89]; **8**, 229[135]

Büchel, T., **6**, 539[579]
Bucher, J. E., **4**, 47[133]
Bucher, W., **7**, 85[10]
Buchert, M., **4**, 981[111]; **5**, 1086[76]
Buchholz, B., **6**, 94[145], 95[146]; **7**, 470[9], 487[9]
Buchholz, M., **6**, 667[243]
Buchholz, R. F., **3**, 761[142]
Büchi, G., **2**, 157[120], 477[8,9]; **3**, 350[37], 499[121], 871[53], 882[102], 883[102], 953[101]; **4**, 24[75], 25[75a,b], 30[91,91a], 115[183], 242[62], 253[62], 261[62]; **5**, 20[138], 123[3], 129[33], 151[2], 163[71], 646[7], 741[154], 790[31], 829[18], 830[30,37], 832[40], 847[135], 857[231], 859[37b], 1004[29]; **6**, 834[33], 853[33], 875[24], 879[24]; **6**, 1035[137], 1060[71]; **7**, 85[11], 160[48], 163[73]; **8**, 526[37], 531[37], 544[275]
Buchkremer, J., **5**, 596[27], 597[27]
Buchkremer, K., **5**, 151[14], 154[14,32], 158[46], 160[14,46]; **6**, 558[853]
Buchman, O., **3**, 1021[14]; **8**, 535[163], 900[29]
Buchmeier, W., **1**, 9[45]
Buchner, E., **7**, 709[43], 710[43]
Büchner, W., **8**, 830[86]
Buchs, P., **1**, 857[58]
Buchschacher, P., **2**, 167[161]; **4**, 7[25]
Buchwald, S. L., **1**, 743[67], 746[67], 748[67,72], 749[78,81], 816[78]; **4**, 485[30], 979[101]; **5**, 851[169], 1115[3], 1122[3], 1123[3,42], 1124[3,48], 1175[39,40], 1178[39,40]; **8**, 675[41], 677[56], 678[56], 679[41,56], 680[56], 681[56], 683[95], 684[41], 685[56], 686[56,95], 688[56], 692[56], 694[56]
Buck, H. M., **1**, 661[165,166], 663[165,166], 672[166], 700[166], 704[166]; **3**, 367[101,102]; **8**, 94[81], 95[82-84], 96[93]
Buck, J. S., **1**, 542[4], 543[4], 547[4]; **2**, 143[56]; **8**, 956[2]
Buck, P., **4**, 86[78d]
Buckle, J., **1**, 305[86]
Buckler, S. A., **8**, 366[37]
Buckles, R. E., **6**, 959[44]; **7**, 720[8]
Buckleton, J. R., **4**, 1018[218]
Buckley, D. A., **6**, 277[124]
Buckley, D. G., **2**, 811[68]
Buckley, G. D., **4**, 85[76,77a]; **7**, 833[74]
Buckley, J. F., **5**, 94[82]
Buckley, J. S., Jr., **3**, 807[20]
Buckley, M. I., **2**, 156[115]
Buckley, T. F., **1**, 413[58]; **7**, 219[11], 228[99]
Buckley, T. F., III, **2**, 741[65]
Buckner, C., **3**, 693[148], 694[148]
Buckus, P., **3**, 328[178]
Buckwalter, B., **2**, 66[35]; **3**, 196[30]; **4**, 1033[26], 1035[26a], 1036[44], 1046[26a], 1051[26a], 1058[149]; **5**, 942[231]
Buco, S. N., **3**, 889[23]
Bucourt, R., **3**, 21[131]
Budanova, L. I., **6**, 554[755]
Budashevskaya, T. Y., **4**, 885[114]
Budd, D. E., **8**, 445[30]
Buddhasukh, D., **6**, 1047[33]
Buddrus, J., **2**, 598[13]; **4**, 1020[229,230]; **7**, 235[2]
Budesiinsky, M., **7**, 73[102]
Buding, B. H., **1**, 752[99]
Budylin, V. A., **8**, 612[76], 623[147], 624[147]
Budzelaar, P. H. M., **1**, 30[153], 214[27]; **2**, 123[196], 124[204], 125[204], 280[27]; **8**, 589[46]
Budziarek, R., **6**, 685[357]
Budzikiewicz, H., **2**, 355[126]; **3**, 812[58,59]; **6**, 441[84]; **7**, 222[36]; **8**, 346[125]
Bueckmann, A. F., **8**, 204[154]
Buehl, H., **3**, 909[155]
Buehler, C. A., **2**, 737[34], 766[86]; **6**, 795[4]
Buehler, N. E., **5**, 204[40]
Buell, B. G., **4**, 31[92,92e]
Buell, G. R., **1**, 619[59]; **4**, 120[197]
Buemi, G., **3**, 386[57]
Buendia, J., **8**, 227[117]
Buess, C. M., **7**, 516[9]
Buevich, V. A., **6**, 556[825,826]
Buffet, H., **1**, 835[136]
Bugden, G., **1**, 495[46], 496[48], 497[48]
Bugel, J.-P., **8**, 543[246]
Bugge, G., **6**, 435[5a]
Buggle, K., **7**, 470[12]
Bugrova, L. V., **3**, 320[134]
Buhler, R., **5**, 404[19], 405[19]
Buhlmayer, P., **2**, 186[36]; **6**, 888[65]
Buhr, G., **3**, 887[12]; **7**, 157[33,33a]
Buhro, W. E., **4**, 1033[17], 1035[37,41], 1037[17,37]
Buijle, R., **6**, 966[98]
Buisson, D., **8**, 185[28], 187[28], 188[53], 196[114]
Buist, G. J., **7**, 709[42], 710[42]
Bujnicki, B., **6**, 149[98,101]
Bujoli, B., **8**, 135[47], 136[47]
Bukhari, A., **5**, 241[4]
Bukhmutskaya, S. S., **5**, 752[9], 767[9]
Buki, K. G., **7**, 70[93]
Bukovits, G. J., **8**, 346[125]
Bukownik, R. R., **8**, 244[70]
Bula, O. A., **2**, 303[4]
Bulbulian, R. V., **2**, 745[102,106]
Bulecheva, E. V., **4**, 1051[127]
Bulenkova, L. F., **8**, 606[24], 607[24]
Bulgarevich, S. B., **6**, 226[13]
Bulina, V. M., **8**, 756[143]
Bulka, E., **8**, 413[129]
Bull, D., **3**, 530[75], 531[87], 534[75], 538[87]
Bull, J. R., **3**, 741[53]; **6**, 902[125]
Bullee, R. J., **4**, 905[209]
Bullen, N. P., **7**, 333[26], 606[155]
Bullnheimer, M., **2**, 943[169]
Bullock, W. H., **3**, 253[89], 262[89]; **5**, 252[43,44], 257[43]
Bullpitt, M. L., **8**, 856[162]
Bulls, A. R., **8**, 671[18], 696[126]
Bulman Page, P. C., **5**, 432[130], 433[130b]; **7**, 352[34], 418[129c]; **8**, 708[39]
Bu'Lock, J. D., **3**, 664[34]
Bulot, J. J., **4**, 459[77], 473[77], 474[77]
Buloup, A., **3**, 1026[38]
Bulovyatova, A. B., **6**, 560[866]
Bulusheva, E. V., **5**, 65[55]
Bulychev, A. G., **4**, 218[147,148]
Bulychev, Yu. N., **6**, 533[476], 554[745]
Bumagin, I. G., **4**, 610[341]
Bumagin, N. A., **1**, 437[157], 438[159]; **3**, 219[111], 503[152], 529[48], 530[73], 531[88], 534[73], 537[88]; **4**, 594[141,144], 610[340]
Bumagina, I. G., **3**, 503[152], 529[48]
Bumgardner, C. L., **3**, 88[139], 89[139], 91[139], 918[25]; **4**, 52[146]; **6**, 1014[20], 1016[25]; **8**, 36[76], 66[76], 829[83]
Bunce, N. J., **7**, 7[44], 15[150]
Bunce, R. A., **2**, 611[101]; **4**, 17[53], 24[72,72a], 161[86b]
Bunce, R. J., **4**, 306[374,378,379,380], 309[420,421]
Buncel, E., **1**, 595[28]; **2**, 81[1], 82[1,1a], 96[1], 100[16], 101[16], 111[16], 134[3], 182[2], 190[57], 240[3], 455[9], 510[39]; **3**, 147[399], 839[10]; **4**, 84[68d], 100[120], 426[43], 444[43]; **7**, 641[1]
Bunch, J. E., **4**, 52[146]
Bund, J., **1**, 773[203,203b]
Bunda, J., **7**, 723[23], 724[28]

Bundel, Yu. G., **6**, 555[817]
Bundle, D. R., **6**, 23[92], 41[43], 533[500], 550[500], 651[135], 652[142]
Bundy, G. L., **1**, 739[34]; **3**, 11[53]; **4**, 115[184b], 258[251], 989[142]; **7**, 152[5], 174[135]
Bundy, J. M., **4**, 345[81]
Bunel, E., **8**, 673[25], 696[25]
Bunes, L. A., **3**, 359[70]
Bunge, K., **4**, 1082[91], 1103[204]
Bunke, H., **6**, 501[203], 531[203]
Bunnell, C. A., **8**, 940[105,110], 947[144]
Bunnelle, W. H., **1**, 248[71], 735[27], 736[27]; **2**, 553[131], 584[120]; **4**, 5[17]; **5**, 552[38], 560[68], 1123[39]
Bunnenberg, E., **6**, 494[134]
Bunnett, J. F., **3**, 512[201]; **4**, 423[1], 425[1], 426[36], 452[4,21], 453[23,26,32,33], 454[38], 455[43], 456[21,26,46,48], 457[33,38,46,56-61], 458[32,58,66,67], 459[4,69,70,76,88-92], 460[89-94], 461[38], 462[107], 463[43,56,58,66,67], 464[58,69,70,76,88,94,112,113,121], 465[70,112,121], 466[76,124], 467[112], 468[38], 469[38,124], 471[124], 472[46,56,60,124], 473[21,23,48,89,121,147], 474[23,58,90,92,147], 475[61,91,149], 477[94], 478[76], 479[70], 484[22], 487[49], 488[53], 489[62], 490[65,66], 493[81], 499[66], 500[104], 765[228]; **5**, 390[140]; **6**, 240[82], 950[1]; **7**, 437[1]
Buntain, G. A., **6**, 116[85]; **8**, 28[35], 66[35]
Bunting, J. W., **8**, 93[75], 584[19], 589[48]
Bunting, S., **5**, 151[7], 180[7]
Bunton, C. A., **3**, 723[11], 724[15,16], 725[11], 726[11], 761[143]; **4**, 426[39]; **7**, 445[59], 703[2], 709[2,42], 710[2,42], 712[2], 851[19]
Bunya, M., **6**, 605[37]; **7**, 245[78]
Bunyan, P. J., **3**, 639[88]
Buono, G., **1**, 223[78], 224[78]; **4**, 55[155,157], 57[157k,l], 930[49]
Bur, D., **8**, 189[60]
Burbaum, B. W., **2**, 555[139]; **5**, 527[60], 1021[71]
Burch, M. T., **5**, 839[77]
Burch, R. R., **8**, 431[66], 459[226]
Burchardt, B., **5**, 929[168], 930[168]
Burchill, M. T., **4**, 811[175], 812[175]
Burchuk, I., **5**, 552[22]
Burckhalter, J. H., **2**, 956[13], 958[13,22-24]
Burckhardt, C. A., **3**, 566[33]
Burden, E. J., **6**, 487[42], 489[42], 543[42], 550[42], 554[42]
Burdett, J. E., Jr., **7**, 372[71]
Burdi, D. F., **5**, 534[95]
Burdisso, M., **5**, 78[281], 257[60]
Burdon, J., **3**, 522[11]; **8**, 901[40], 904[53]
Burdsall, D. C., **2**, 363[201]
Burfield, D. R., **6**, 89[107]
Burford, C., **1**, 622[73], 737[31], 828[79]; **2**, 426[39]; **3**, 198[51]
Burford, S. C., **1**, 132[106]; **4**, 896[169]
Burg, A. B., **8**, 1[1]
Burg, D. A., **4**, 121[205f]
Burgar, U., **3**, 572[64]
Burge, G. L., **3**, 762[147]
Burge, R. E., **6**, 968[110]
Burge, R. E., Jr., **8**, 950[173]
Burger, A., **8**, 376[159]
Burger, B. J., **8**, 672[21], 673[25], 696[25,126]
Burger, D. H., **1**, 420[85]
Burger, G., **4**, 588[52]
Burger, H., **1**, 149[47]
Burger, J. J., **3**, 875[72], 881[99], 882[100,101]
Burger, K., **4**, 1081[76,83]; **7**, 475[55]
Burger, M., **5**, 712[60]
Burger, T. F., **1**, 175[16]
Burger, U., **4**, 1002[54]; **5**, 560[70], 635[87], 913[108]; **8**, 109[10], 110[10], 112[10], 116[10], 120[10]
Burger, W., **3**, 593[178]
Bürger, W., **2**, 900[26], 960[35], 962[35]
Burgers, P. M. J., **6**, 602[3], 658[187]
Burgess, E., **4**, 1097[169]
Burgess, E. M., **3**, 909[154]; **6**, 960[57], 1011[3]
Burgess, G., **4**, 293[238]; **6**, 266[54]
Burgess, H., **7**, 776[360]
Burgess, J., **4**, 710[50]
Burgess, K., **4**, 1035[43]; **5**, 926[157,158]
Burghard, H., **2**, 1094[89], 1095[89]; **3**, 584[131]
Burgi, H. B., **1**, 49[8], 297[59], 307[59], 621[68]; **5**, 468[127]; **8**, 3[21], 89[43]
Burgos, C. da G., **4**, 16[51]
Burgos, C. E., **7**, 393[16], 398[16]
Burgot, J. L., **8**, 660[108]
Burgoyne, W., **3**, 753[102]; **8**, 803[93], 804[93], 826[69], 876[45], 877[45]
Burgstahler, A. W., **3**, 341[2], 360[2]; **4**, 313[463]; **5**, 706[27], 803[88], 830[29], 850[151]; **8**, 228[127], 287[20], 288[20]
Burinsky, D. J., **2**, 807[49]
Burk, M. J., **7**, 6[30]
Burk, P., **8**, 947[143]
Burk, R. M., **1**, 361[33], 592[23], 889[142], 890[142]; **2**, 941[165], 1032[87], 1043[115]; **3**, 365[96]; **6**, 742[69]
Burke, D. C., **8**, 526[34]
Burke, L. A., **5**, 72[171], 741[153]
Burke, L. D., **5**, 20[136], 70[116], 203[39], 204[39k,l], 209[39], 210[39,56], 514[9], 527[9]
Burke, P. L., **8**, 707[26], 717[99,100]
Burke, S., **8**, 112[24], 120[24]
Burke, S. D., **1**, 126[87,89,90], 418[72], 584[9], 642[110], 643[110], 757[122]; **2**, 369[249], 714[50,53]; **3**, 363[86], 1051[21]; **4**, 24[72,72f], 75[42b], 735[85], 952[7], 1049[121b], 1103[208]; **5**, 516[25,28], 531[73], 534[92], 537[98], 574[130], 841[87], 843[117], 850[163], 864[260], 905[55], 994[53], 997[53]; **6**, 859[174]; **8**, 927[20]
Burke, S. S., **5**, 949[280]
Burke, W. J., **2**, 957[16], 968[82], 969[83,83a,b]
Burkert, U., **3**, 382[36]; **6**, 524[359]; **7**, 358[14]
Burkey, J. D., **8**, 356[188]
Burkhardt, E. R., **1**, 214[20], 227[20]; **2**, 125[213], 127[213]
Burkhardt, G. N., **4**, 280[127]
Burkhardt, K., **2**, 943[168]
Burkhardt, T. J., **4**, 980[106], 981[106]
Burkhardt, U., **6**, 574[965]
Burkhart, G., **4**, 905[212]; **5**, 1138[71], 1156[166], 1157[71]
Burkhart, J. P., **4**, 602[253,254], 625[253]; **5**, 35[14], 841[95]; **7**, 324[73]
Burkholder, C. R., **1**, 837[154]
Burkholder, E. G., **8**, 492[18], 502[61], 504[72], 505[72]
Burkholder, T. P., **5**, 537[100]; **8**, 395[131]
Burkoth, T. L., **3**, 893[57]; **5**, 716[88,89]
Burks, J., **3**, 176[540]; **6**, 161[179]
Burks, J. E., Jr., **5**, 886[21,22]; **6**, 994[95], 998[95]
Burks, S. R., **4**, 295[250,252], 404[242,242a-c]
Burlachenko, G. S., **2**, 609[79], 616[135], 804[40]
Burlant, W. J., **7**, 771[259]
Burley, J. W., **2**, 443[16]
Burlinson, N. E., **4**, 887[125]; **8**, 739[35]
Burlitch, J. M., **4**, 508[159], 1000[15], 1002[52]
Burman, D. L., **4**, 74[39b], 243[72], 257[72], 260[72]; **6**, 185[161], 190[161]
Burmeister, M. S., **5**, 977[22]
Burmistrov, S. I., **6**, 428[83]

Burmistrova, M. S., **5**, 752[8,11,22,33], 757[11,22], 767[8,11,22], 768[11]
Burn, D., **7**, 136[109]
Burneleit, W., **6**, 120[120]
Burnell, D. J., **3**, 590[164]; **5**, 225[116], 227[116], 233[116], 347[72,72c]
Burnell, R. H., **3**, 325[162]
Burness, D. M., **7**, 720[8]
Burnett, D. A., **1**, 389[140]; **2**, 926[115], 937[115], 940[162]; **4**, 794[78], 823[232]; **5**, 100[152,158]; **7**, 647[31]
Burnett, E., **2**, 901[35], 908[35], 909[35], 910[35]
Burnett, M. G., **8**, 446[86-88]
Burnett, R. E., **8**, 461[258], 535[166]
Burnette, L. W., **8**, 140[27]
Burnham, J. W., **2**, 738[41], 760[40]
Burnier, J. S., **5**, 64[26], 452[56], 706[28]
Burns, C., **8**, 568[471]
Burns, C. J., **7**, 395[20b]
Burns, R. C., **8**, 675[48], 676[48]
Burns, R. H., **6**, 515[235]
Burns, S. A., **1**, 461[12]
Burns, T. P., **1**, 212[6], 213[6]; **4**, 969[64]
Burns, W., **5**, 65[68]
Burpitt, R. D., **4**, 45[126,126b,e]; **5**, 71[153,158], 686[50]
Burrage, M. E., **5**, 428[108]
Burrell, J. W. K., **7**, 306[9]
Burri, K., **3**, 414[3]
Burri, K. F., **8**, 647[57]
Burrous, M. L., **2**, 588[152]; **6**, 1034[134]; **8**, 764[2], 770[2b]
Burrow, M. J., **7**, 92[43]
Burrow, P. D., **5**, 452[57]; **7**, 861[77]
Burrows, C. J., **1**, 747[63]; **5**, 856[212,213]; **6**, 70[18]
Burrows, E. P., **3**, 414[1]; **7**, 736[5], 737[5], 745[5], 746[5], 749[5]
Burrows, W. D., **2**, 1024[57,58]; **6**, 734[4]
Bursian, N. R., **3**, 328[181]
Burstall, F. H., **7**, 775[339,344]
Burstein, K. Ya., **5**, 125[17]
Burstein, S. H., **7**, 253[20]
Bürstinghaus, R., **1**, 511[33]; **4**, 11[35], 113[172], 259[260]
Burt, E. A., **3**, 297[17], 306[86]
Burton, A., **1**, 723[281]; **3**, 87[86], 142[86], 144[86]
Burton, C. I., **2**, 142[46]
Burton, D. J., **3**, 202[86]; **5**, 680[24,24a,b]; **6**, 172[10,13,15]; **8**, 860[223], 861[227], 895[6,7], 896[11], 897[6,7], 898[7], 899[7], 900[32,33], 904[11]
Burton, H., **2**, 738[42]
Burton, R., **4**, 663[2]
Burtscher, P., **3**, 862[11], 863[11]
Burwell, R. L., **7**, 5[21]
Burwell, R. L., Jr., **8**, 419[18], 447[132]
Bury, A., **4**, 50[142,142k]
Burzlaff, H., **6**, 195[225]
Busacca, C., **5**, 372[104]
Busby, R. E., **4**, 1021[249,250]
Busch, A., **5**, 612[75]
Busch, F. R., **2**, 904[49]
Busch, M., **8**, 904[57,57a], 910[57]
Buschmann, E., **2**, 866[4]; **5**, 480[176]
Buschmann, H., **5**, 186[171]
Buse, C., **3**, 762[148], 769[148]
Buse, C. T., **1**, 141[18], 179[26], 182[26], 357[2]; **2**, 3[9], 6[9], 19[9], 29[9a], 94[54], 182[9,9c], 184[9c], 190[9c], 191[9c], 192[9c], 193[9c,63], 197[9c], 198[9c], 200[9c], 211[9c], 217[9c], 223[151], 235[9c], 236[9c], 289[69], 634[36], 640[36]; **6**, 814[95]
Buser, H. R., **3**, 223[155]
Buser, K. R., **3**, 158[440]
Bushaw, B. A., **6**, 955[24]
Bushby, R. J., **1**, 733[10]
Bushey, D. F., **3**, 19[103]
Bushkov, A. Ya., **7**, 774[335]
Bushnell, G. W., **7**, 771[281]
Bushweller, C. H., **5**, 741[153]; **7**, 94[55]
Buskkov, A. Ya., **6**, 543[616]
Buss, A. D., **1**, 774[205,206], 776[206,214], 777[217], 778[221], 780[227], 814[205b,217]; **8**, 13[67]
Buss, D., **1**, 499[52,53]
Buss, D. H., **7**, 294[14]
Buss, V., **8**, 69[221], 70[221], 647[53,54]
Bussas, R., **5**, 422[86]
Busse, U., **3**, 53[274]; **4**, 111[152d], 222[179]; **6**, 531[436]
Bussolotti, D. L., **1**, 739[35]; **6**, 998[118]
Buswell, R., **3**, 392[92]
Buswell, R. L., **1**, 377[97]
Buszek, K. R., **4**, 33[95,95b], 48[136], 49[136,141]
Butcher, M., **2**, 838[176]
Butenschön, H., **5**, 1144[98]
Buter, J., **3**, 229[224], 865[28]; **4**, 1093[150]; **6**, 70[18]; **8**, 84[12]
Büthe, H., **8**, 445[22], 459[229]
Büthe, I., **8**, 815[23]
Bütikofer, P. A., **6**, 782[81]; **8**, 946[134]
Butler, D. N., **4**, 1012[177]
Butler, F. R., **4**, 5[18]
Butler, G. B., **5**, 65[56]
Butler, K., **7**, 157[33]; **8**, 974[122]
Butler, M. E., **3**, 790[59]
Butler, P. E., **7**, 498[230a], 516[7], 517[10]
Butler, R. N., **6**, 252[151]; **7**, 696[42], 719[4], 722[4], 724[4], 727[4], 740[42]; **8**, 853[144]
Butler, R. W., **7**, 481[109]
Butler, S., **4**, 128[221]
Butler, W., **4**, 36[100], 49[100a]
Butler, W. M., **1**, 212[11], 213[11]; **4**, 664[6], 667[6b]
Butler-Ransohoff, I., **8**, 349[136]
Butsugan, Y., **1**, 256[20,21]; **2**, 24[92,94]; **3**, 220[116]; **6**, 834[41]; **8**, 190[80]
Butt, S., **8**, 198[130]
Butt, Y., **5**, 717[92]
Buttafava, A., **5**, 454[70]
Butter, S. A., **3**, 305[75b]
Butterfield, R. O., **8**, 451[180]
Buttero, P. O., **6**, 178[121]
Buttery, C. D., **3**, 261[153]
Buttery, R. G., **7**, 8[58]; **8**, 604[7]
Büttner, H., **8**, 263[26]
Buttrus, N. H., **1**, 17[91], 39[191]
Butz, L. W., **5**, 513[2], 518[2]
Buu-Hoi, N. P., **2**, 149[85]; **7**, 16[157]; **8**, 328[4], 340[4]
Buxton, S. R., **4**, 1009[143], 1014[190]
Buynak, J. D., **4**, 350[119]; **5**, 107[202,203]; **7**, 378[97]; **8**, 846[83]
Buza, M., **1**, 858[61]; **7**, 530[27], 531[27]
Buzas, A., **2**, 816[84], 828[84]
Buzilova, S. R., **7**, 774[325]
Buzykin, B. I., **1**, 378[104]
By, A. W., **4**, 357[149]; **6**, 108[33]; **7**, 502[264], 534[39]
Bychikhina, N. N., **6**, 554[771]
Byers, J. H., **3**, 785[35]; **4**, 745[141], 746[144], 819[211]; **5**, 923[138]
Byers, J. R., **6**, 213[87]
Bykova, L. U., **8**, 621[142]
Byram, S. K., **4**, 38[108,108a]
Byramova, N. F., **6**, 49[100]
Byrd, J. E., **4**, 311[451]; **7**, 462[122]
Byrd, L. R., **6**, 282[156]

Byrn, S., **4**, 79[55a], 251[152]
Byrne, K. J., **6**, 675[299]; **8**, 213[30], 214[33], 217[67], 218[69]
Byrne, L. T., **1**, 17[219], 36[234]
Byrne, N. E., **2**, 1066[122]; **5**, 459[92]
Byrne, W. L., **2**, 456[70], 457[70]
Byrom, N. T., **4**, 374[88]; **7**, 451[21]
Bystrenina, V. I., **8**, 451[177]
Bystrom, S. E., **4**, 565[44], 598[197], 623[197], 639[197]; **6**, 85[94], 86[94,96]; **7**, 94[57], 490[175]
Bystrov, V. R., **7**, 773[304]
Byun, H.-S., **7**, 393[16], 398[16]
Bzowej, E. I., **1**, 746[71]

C

Caamoño, C., **5**, 416[57]
Cabal, M. P., **2**, 656[150]; **4**, 159[84]; **5**, 410[42], 411[42], 736[145], 737[145]; **8**, 856[178]
Cabal, P., **5**, 433[136b]
Caballero, M. C., **3**, 396[115]
Cabanac, M., **8**, 212[14]
Cabaret, D., **8**, 37[91], 40[91], 44[91], 66[91]
Cabello, J. A., **2**, 345[18], 359[18], 360[18]
Cabiddu, S., **8**, 847[89]
Cabrera Escribano, F., **6**, 941[150]
Cabri, W., **7**, 429[151]
Cabrini, D., **8**, 61[187], 66[187]
Cacace, F., **4**, 445[203]; **5**, 1148[119]; **6**, 280[144]
Cacchi, S., **1**, 195[90]; **3**, 539[96], 1035[79]; **4**, 189[105], 190[105a], 411[266b], 860[112], 861[112]; **8**, 84[13], 910[82], 911[87], 933[49,51]
Caccia, G., **5**, 1153[145]
Cachia, P., **3**, 711[22]
Caciagli, V., **7**, 439[25]
Cadamuro, S., **2**, 737[39]; **8**, 277[153]
Caddy, P., **5**, 1146[107]
Cadena, R., **7**, 229[107]
Cadilla, R., **5**, 429[114]
Cadiot, P., **2**, 6[26], 583[110], 587[110]; **3**, 219[103], 551[3], 552[3], 553[11], 964[125]; **8**, 714[83]
Cadman, E., **2**, 456[42], 466[42], 467[42]
Cadmuro, S., **8**, 660[109]
Cadoff, B. C., **3**, 737[31,32]
Cadogan, J. I. G., **1**, 755[115], 812[115], 813[115]; **3**, 505[161], 507[161]; **4**, 487[44]; **5**, 803[87], 1123[35]; **6**, 20[76], 74[35], 79[35], 80[35], 173[36], 174[36], 177[112], 181[112], 199[36], 200[112], 205[28,31], 208[55], 212[55]; **7**, 13[114], 396[22], 479[92]; **8**, 916[99], 917[99], 918[99]
Cadot, P., **4**, 1003[64]
Cady, M. A., **3**, 226[198]; **6**, 10[44], 11[44], 12[44]
Cady, S. S., **7**, 845[74]
Caesar, P. D., **8**, 408[62]
Cafferata, L. F. R., **6**, 278[132]
Caflisch, E. G., Jr., **8**, 297[67]
Caglioti, L., **4**, 146[36]; **5**, 487[188]; **6**, 727[190]; **8**, 343[115,116], 344[115,120], 345[115,116,120], 346[115], 348[120], 349[116], 942[117]
Cagniant, D., **2**, 759[30], 765[81,82]; **8**, 847[97], 848[97c], 849[97c], 867[97c]
Cagniant, P., **2**, 758[23b], 759[30], 765[68,69,81,82]
Cahen, D., **8**, 418[7]
Cahiez, G., **1**, 331[50,51], 428[116]; **2**, 127[238], 596[3]; **3**, 227[210], 245[30], 247[46], 249[58,59], 263[173], 265[188], 466[182,193], 470[214], 473[214], 476[214], 482[5], 499[5], 505[5], 509[5], 578[93]; **4**, 98[111], 173[34], 250[136], 262[136], 877[73], 895[166], 896[168], 897[172], 898[177], 900[180], 902[190], 903[191,194]
Cahours, M. A., **1**, 139[1], 156[1]
Cai, D., **1**, 421[86]
Cai, K., **7**, 283[185]
Caillaux, B., **6**, 429[89], 495[144], 499[144], 515[317], 517[325], 544[325], 546[317], 552[325]
Caille, J.-C., **2**, 838[169]; **5**, 21[154], 22[154]; **6**, 781[80]; **8**, 946[134]
Cain, A. M., **7**, 779[421]
Cain, B. F., **6**, 532[473]; **7**, 92[40]; **8**, 331[36]
Cain, E. N., **7**, 722[21]
Cain, M. E., **7**, 762[73]; **8**, 542[222]
Cain, P., **2**, 163[149], 167[163]; **4**, 10[31,31c]; **6**, 718[121]
Cain, P. A., **3**, 672[66]
Cain, W. T., **6**, 14[58], 16[58]
Caine, D., **1**, 3[20], 779[226]; **2**, 442[8], 447[8], 797[8], 829[8], 835[8], 861[8]; **3**, 1[1], 3[1,10], 8[1], 11[1], 13[1], 14[71], 15[71,79], 17[1,84], 18[1], 21[1], 23[1], 25[1], 59[1], 75[52]; **6**, 1042[11], 1043[11]; **7**, 136[115], 137[115]; **8**, 337[78], 524[10], 525[10], 526[10,29], 527[29], 542[236]
Cainelli, G., **1**, 206[111], 391[148], 489[19], 498[19,50], 830[92], 832[92]; **2**, 66[32], 187[43], 613[114], 635[48], 640[48], 656[157], 807[48], 925[111], 926[111], 927[120], 935[151], 936[151], 937[156,157]; **4**, 344[78a]; **5**, 100[148,155,156], 102[165,173,174]; **6**, 21[80], 22[80], 759[140], 795[5]; **7**, 137[124], 236[24], 252[1], 280[177], 816[8], 817[8], 821[8], 824[8], 825[8]; **8**, 289[27], 550[331]
Cairncross, A., **3**, 499[122]; **5**, 486[196]
Cairns, P. M., **5**, 581[175], 803[89], 976[20]; **6**, 754[119]
Cairns, S. M., **8**, 861[226]
Cairns, T. L., **1**, 880[112]; **4**, 968[57]; **5**, 3[22], 70[114], 74[212], 905[57], 1146[110]; **6**, 288[184]
Cais, M., **8**, 451[180]
Calabrese, J., **1**, 301[74], 316[74]
Calabrese, J. C., **5**, 1171[26], 1178[26]
Calabretta, P. J., **3**, 768[167]
Calabro, M. A., **8**, 295[60]
Calas, R., **1**, 328[17,18], 580[1]; **2**, 564[1], 575[65], 576[65], 582[108], 584[116], 713[44], 716[57,61,66], 717[68-70], 718[72,75], 721[89], 725[120], 726[124], 728[140,141]; **3**, 577[90]; **8**, 409[82], 518[128,131], 777[82a], 785[115]
Calcagno, M. A., **5**, 938[215]
Calcagno, M. P., **4**, 497[96]
Calcaterra, M., **1**, 303[77]
Calder, G. V., **4**, 483[6]
Calder, M. R., **6**, 74[28]
Calderari, G., **2**, 207[106], 325[44]; **3**, 41[227]; **4**, 21[69], 109[150a], 218[144], 340[53], 342[53]
Calderazzo, J. M., **2**, 956[14], 962[14]
Calderon, J. S., **7**, 355[37]
Calderon, N., **5**, 1115[1], 1116[1,8], 1120[23]
Calderon, O., **2**, 748[120], 758[26]
Caldwell, C., **4**, 1033[36]; **6**, 124[144], 662[215]
Caldwell, C. G., **2**, 482[34], 484[34]
Caldwell, H. M. E., **2**, 145[64]
Caldwell, J., **5**, 802[81]
Caldwell, R. A., **5**, 152[22], 154[28], 164[76]; **7**, 851[28], 879[150]
Caldwell, W. B., **8**, 187[42]
Calet, S., **5**, 151[18]; **7**, 482[117]
Calhoun, A. D., **8**, 764[7], 770[7], 773[7]
Calianno, A., **3**, 246[40], 257[40]
Calienni, J., **4**, 72[26], 249[119], 257[119]
Calihan, L. E., **3**, 334[214]
Calingaert, G., **3**, 415[7]
Callahan, F. M., **6**, 664[222]
Callahan, J. F., **2**, 553[126]; **3**, 390[85], 392[85]; **5**, 131[45]; **7**, 543[12], 551[12]
Callant, P., **3**, 900[93]; **7**, 301[58]
Callen, G. R., **1**, 177[22]; **2**, 13[56], 14[56a], 19[78]
Callens, R., **6**, 80[66], 237[69]
Callet, G., **8**, 383[17]
Calligaris, M., **4**, 20[63], 21[63]; **5**, 272[2], 275[2]
Calligaro, L., **4**, 710[48]
Callighan, R. H., **7**, 544[37]
Callipolitis, A., **3**, 159[454], 161[454]
Callot, H. J., **3**, 1048[9]; **4**, 1033[18,19], 1037[18]
Calo, V., **3**, 222[145], 246[40], 257[40]; **4**, 34[97], 35[97]; **7**, 120[14], 760[49], 764[49], 854[57], 855[57]; **8**, 887[116]

Calogero, S., **1**, 305[84,85], 323[84]
Calogeropoulou, T., **2**, 103[33,34]
Calonna, F., **5**, 456[87]
Calson, G. R., **7**, 691[19]
Calundann, G. W., **7**, 738[31]
Calverley, M. J., **1**, 656[148], 699[148]; **3**, 89[144], 107[144], 109[144], 116[144]
Calvert, B. J., **8**, 90[46]
Calvin, M., **8**, 152[159], 285[5]
Calvo, C., **5**, 128[29]
Calzada, J. C., **2**, 111[83]
Cama, L., **1**, 434[141]
Cama, L. D., **3**, 48[259]; **8**, 384[25]
Camacho, C., **8**, 237[14], 240[14], 244[14], 708[41]
Camacho, F., **3**, 327[171]
Camaioni, D. M., **4**, 797[101]; **7**, 877[129]
Cambie, R. C., **1**, 753[102]; **3**, 675[74], 753[105], 759[126]; **4**, 347[96], 350[121], 351[126], 354[126-128], 369[21,22], 370[21,22], 371[21], 377[21,22,101], 545[126]; **6**, 668[254], 669[254]; **7**, 92[40], 121[24], 331[14], 438[15,16], 445[15,16], 447[16], 502[261], 530[20], 531[20], 706[25]; **8**, 944[123]
Cambillau, C., **1**, 34[226,227]
Camerino, B., **6**, 795[5]
Camerman, P., **7**, 13[107]
Cameron, A. F., **2**, 1048[11]; **4**, 30[87]
Cameron, A. F. B., **6**, 675[299]
Cameron, A. G., **1**, 570[263]; **2**, 800[17]; **5**, 841[87,94], 863[255]; **6**, 176[95,97,98], 859[171]
Cameron, D. D., **2**, 148[78]
Cameron, D. J., **6**, 650[129]
Cameron, D. W., **2**, 170[173], 606[69]
Cameron, I. R., **2**, 1048[11]
Cameron, S., **1**, 822[38]; **7**, 177[147], 550[48]; **8**, 888[125]
Cameron, T. S., **1**, 528[122], 804[308], 805[308]; **5**, 645[1], 650[1q,25], 651[1]; **7**, 258[55]
Camici, L., **7**, 330[7]
Cammarata, A., **7**, 95[80]
Camp, N. C., III, **8**, 532[130]
Camp, R. N., **5**, 72[172]
Campaigne, E., **2**, 356[129], 359[158], 361[158], 376[158], 388[158], 852[228], 853[228]; **3**, 582[111,113]; **4**, 257[226], 259[226]; **5**, 947[289]; **7**, 666[74]
Campbell, A., **3**, 914[6], 927[6]; **6**, 799[22]
Campbell, A. L., **3**, 209[17], 216[68], 224[168]; **4**, 175[43]; **8**, 497[37]
Campbell, B. K., **3**, 273[6]
Campbell, B. S., **7**, 753[159]
Campbell, C., **5**, 515[12], 516[12b]
Campbell, C. B., **4**, 375[93], 386[93]
Campbell, C. D., **4**, 488[56], 509[162]; **5**, 380[113e], 478[162]; **7**, 482[113], 743[62]
Campbell, C. L., **6**, 722[138]; **7**, 228[91]
Campbell, D. H., **7**, 485[140]
Campbell, E. A., **8**, 114[54]
Campbell, G. C., **8**, 140[14]
Campbell, H. F., **4**, 399[221]
Campbell, J. A., **8**, 528[80]
Campbell, J. B., **7**, 597[49]
Campbell, J. B., Jr., **3**, 249[66], 483[10,11]; **6**, 901[123], 1066[94]; **8**, 718[111]
Campbell, J. R., **6**, 207[47]; **7**, 571[115]
Campbell, K. N., **3**, 273[6]; **8**, 478[39], 479[39]
Campbell, M. M., **2**, 632[27], 640[27], 1048[11]; **7**, 373[72b], 473[29]
Campbell, M. W., **2**, 198[85]
Campbell, P., **6**, 423[49]; **7**, 49[66,67]
Campbell, P. G. C., **4**, 314[481,482], 603[274]; **7**, 92[47], 94[56]
Campbell, P. H., **5**, 754[67]
Campbell, S., **1**, 694[238], 697[238], 698[238], 801[302]; **6**, 1006[145]
Campbell, S. A., **4**, 38[108,108c]
Campbell, S. F., **6**, 533[490], 550[490], 994[98]
Campbell, S. J., **3**, 969[136]
Campbell, T. W., **7**, 84[1], 85[1], 108[1]; **8**, 568[470]
Campelo, J. M., **2**, 345[18], 359[18], 360[18]
Campo, M., **8**, 445[55]
Campopiano, O., **5**, 242[8], 243[13], 244[13,14]
Campos, M. P. A., **4**, 347[89]
Campos, O., **2**, 1017[34]
Campos, P. J., **3**, 282[48]; **4**, 347[93], 351[93c], 354[93d]; **6**, 494[135]; **7**, 536[52-55]; **8**, 854[150]
Campos-Neves, A. da S., **4**, 24[72,72c]
Camps, F., **5**, 36[18], 57[54]; **6**, 172[18]; **7**, 87[20], 359[18]; **8**, 500[50], 515[119], 563[425]
Camps, M., **4**, 312[457], 313[471], 315[516]
Camps, P., **6**, 563[901], 984[54]
Campus, P. J., **2**, 790[57]; **7**, 501[255]
Camus, A., **4**, 170[11]; **8**, 450[167], 552[359]
Can, N.-T. H., **8**, 540[194]
Canal, G., **3**, 282[48]
Canal, P., **7**, 500[239]
Cañas-Rodriguez, P., **8**, 568[470]
Canceill, J., **2**, 289[70], 291[70], 292[78]
Candlin, J. P., **8**, 445[61], 457[212]
Cane, D. E., **2**, 839[172]; **3**, 249[64], 1059[43], 1060[43]; **8**, 885[103]
Canepa, C., **2**, 345[20]
Canfield, J. H., **6**, 206[46]
Cann, K., **8**, 372[123], 373[123]
Cannan, R. K., **3**, 822[5], 834[5]
Cannic, G., **3**, 896[69]
Canning, L. R., **8**, 460[253]
Cannizzo, L., **1**, 743[51,68], 745[68], 746[51], 748[68,72], 749[78], 811[51], 816[78]; **2**, 1062[98]; **4**, 979[101]; **5**, 1115[3], 1121[27], 1122[3], 1123[3], 1124[3,44,49]
Cannon, J. G., **2**, 760[43]; **8**, 53[130], 66[130]
Cannon, R. D., **7**, 852[33]
Cano, A. C., **7**, 693[24]
Cano, F. H., **5**, 410[42], 411[42]
Canonica, L., **3**, 752[97]; **7**, 153[9]
Canonico, D. M., **3**, 328[179]
Canonne, P., **5**, 578[153,154,155,156,157]
Cánovas, A., **5**, 232[135,136]
Cantacužene, J., **3**, 748[75], 759[128]
Cantacuzene, D., **2**, 209[108]
Cantello, B. C. C., **5**, 985[36]
Cantin, D., **8**, 594[70]
Cantor, S. E., **7**, 567[103]
Cantow, H.-J., **3**, 557[48]
Cantrall, E. W., **7**, 27[71]; **8**, 319[74]
Cantrell, G. L., **5**, 909[97]
Cantrell, T. S., **3**, 381[32]; **5**, 161[61,65,66], 163[65,66], 176[130], 637[106]
Cantuniari, I. P., **3**, 295[10]; **7**, 5[25]
Cantwell, S. G., **8**, 332[40]
Canziani, F., **5**, 1147[112]
Cao, W., **6**, 184[150]
Capdevila, J., **7**, 378[97], 713[72]
Capek, K., **2**, 736[27]
Caperos, J., **1**, 155[65]
Capet, M., **8**, 842[42d]
Capetola, R. J., **4**, 932[63]
Capitaine, J., **3**, 846[44]; **8**, 530[90]

Capka, M., **8**, 274[137], 544[267,268], 770[38], 771[42], 773[64], 907[72], 918[72]
Caplar, V., **2**, 406[47]; **8**, 459[228], 535[165]
Caple, G., **5**, 702[9,9b], 710[51], 713[51]
Caple, R., **2**, 725[108,109]; **4**, 273[50], 282[141,142], 868[17], 869[17]; **5**, 345[70], 346[70], 453[66], 775[176], 1055[46], 1056[48]
Capler, V., **8**, 152[169]
Capmau, M.-L., **2**, 91[43]
Capon, B., **4**, 365[3], 367[3]; **6**, 35[15]
Capon, R. J., **2**, 315[45], 316[45]; **7**, 453[68]
Caporusso, A. M., **3**, 217[80,80b], 246[41], 247[41], 257[41], 439[40,41], 483[12], 491[70]; **8**, 99[108], 100[118], 348[131], 558[389], 564[440]
Capozzi, F., **4**, 331[12]
Capozzi, G., **4**, 330[5], 331[12], 335[24], 387[160], 401[231], 410[231]; **6**, 575[968]; **7**, 758[1], 759[1], 760[1]
Capparella, G., **8**, 449[157], 450[157]
Capperucci, A., **4**, 247[97], 256[97]; **5**, 438[162]
Capps, N. K., **1**, 511[28], 569[253]
Capshew, C. E., **4**, 980[107]
Capson, T. L., **6**, 814[91]
Capuano, L., **1**, 377[97]; **2**, 1087[35]; **6**, 189[187], 526[396], 575[967]
Caputo, A., **4**, 189[105], 190[105a]
Caputo, J. A., **4**, 7[24]
Caputo, R., **1**, 168[116b], 563[179]; **6**, 980[34]; **8**, 406[41], 891[147]
Carabateas, C. D., **8**, 263[25]
Caram, J., **6**, 278[132]
Caramella, P., **3**, 18[96]; **4**, 47[134,134d], 49[134d], 1078[55], 1081[81], 1082[85], 1083[55,81], 1084[55], 1085[85]; **5**, 257[61,61a,c], 258[61b], 630[53], 631[53]
Carballeira, N., **3**, 587[148]; **5**, 205[41], 207[41], 582[179]
Carbonaro, A., **5**, 1142[86]
Carbonelle, A.-C., **3**, 681[100]
Carboni, B., **5**, 336[50]; **8**, 386[54]
Carboni, R. A., **5**, 491[201]
Carcano, M., **4**, 379[118]
Carceller, E., **3**, 380[7]; **5**, 1059[54], 1060[55], 1062[54c]
Card, P. J., **4**, 279[111]; **6**, 121[132]
Card, R. J., **4**, 524[62]
Carda, M., **5**, 474[158]; **6**, 780[71]
Cardani, S., **1**, 524[88]; **2**, 103[28], 266[61], 267[63], 515[57], 516[58], 605[57], 614[57], 630[21], 631[21], 632[21], 634[21], 640[21], 641[21], 642[21,73,74,78], 643[73,74,78], 644[21,73], 645[21], 920[98]; **4**, 152[58], 207[57,58]
Carde, A. M., **1**, 514[52]
Cardé, R. T., **1**, 514[52]
Cardellach, J., **6**, 984[54]
Cardellicchio, C., **1**, 416[67], 452[220]; **3**, 463[155,166]; **4**, 93[93d]; **6**, 446[101]
Cardellina, J. H. I., **3**, 871[52]
Cardellini, M., **5**, 92[63]; **8**, 585[26]
Cardenas, C. G., **8**, 806[110], 807[110]
Cardillo, G., **2**, 66[32], 187[43]; **3**, 45[250]; **4**, 375[94], 377[104], 386[94a,153,153a,157], 387[94a,153a,157], 388[164], 389[166,166a], 393[164a,c], 401[226], 407[104c,153a,157b,254], 408[254a,b,259c]; **6**, 26[106], 533[497], 648[124]; **7**, 137[124], 252[1], 280[177], 493[184], 503[269], 530[13], 663[62], 664[63], 816[8], 817[8], 821[8], 824[8], 825[8]
Cardillo, R., **8**, 187[47]
Cardin, C. J., **2**, 348[64]
Cardin, D. B., **1**, 405[25]
Cardin, D. J., **1**, 139[3,4]
Cardone, R. A., **8**, 647[57]
Cardwell, K., **7**, 199[37]
Carefull, J. F., **7**, 418[129c]
Carel, A. B., **3**, 322[141]
Carelli, V., **5**, 64[52]; **8**, 585[26], 587[30]
Carethers, M. E., **2**, 828[130]
Carey, F. A., **1**, 622[71], 732[17], 786[251]; **3**, 135[368], 136[368], 137[368]; **4**, 349[109], 717[8]; **6**, 686[372], 982[49]; **8**, 216[65], 486[63], 487[63-66], 813[14], 814[14]
Carey, F. H., **6**, 134[12]
Carey, J. T., **2**, 448[36]; **7**, 673[30]
Carey, P. R., **6**, 436[10-12,16-18], 451[129], 455[154]
Carey, S. C., **2**, 878[40]
Cargill, R. L., **3**, 19[103], 23[134]; **5**, 130[42], 133[57], 676[3], 900[11], 901[11], 906[11], 907[11], 910[11]; **6**, 689[387]
Caringi, J. J., **3**, 590[163]
Carini, D. J., **1**, 595[29], 596[29,30], 601[30]; **2**, 85[13,15], 575[60], 579[91], 587[91]; **4**, 158[76]; **5**, 277[14,15], 278[14,15], 279[15]; **7**, 545[24]; **8**, 355[182]
Caristi, C., **4**, 387[160]
Carithers, R., **3**, 909[154]
Carl, C., **7**, 586[167]
Carlberg, D., **5**, 75[220]
Carless, H. A. J., **2**, 599[19]; **5**, 151[5], 153[24], 159[53], 164[77], 165[84,85,88], 167[5], 168[5], 178[5], 180[145,146,147,148]
Carletti, C., **2**, 524[82,83]
Carlier, P. R., **7**, 395[20a], 412[104], 413[104]
Carling, R. W., **3**, 281[46]
Carlò, V., **3**, 283[49]
Carlock, J. T., **3**, 904[133]; **6**, 283[158]
Carlon, M., **3**, 197[33]
Carlough, K. H., **5**, 133[51], 195[6]
Carls, R. R., **4**, 298[290]
Carlsen, D., **6**, 546[643]
Carlsen, L., **6**, 448[112]
Carlsen, P. H. J., **4**, 1082[87], 1083[87], 1103[87]; **6**, 789[106]; **7**, 238[42], 239[42], 240[42,58], 571[113], 572[113], 587[113]
Carlsmith, L. A., **4**, 483[1], 484[1], 486[32], 487[1], 488[1], 489[1], 491[1], 492[1], 493[1], 495[1], 506[1], 508[1]
Carlson, A. A., **6**, 236[57], 709[54]
Carlson, B. A., **3**, 794[76]; **5**, 430[116]
Carlson, C. G., **3**, 327[168]
Carlson, C. L., **2**, 744[93]
Carlson, E. H., **2**, 141[37]
Carlson, G. R., **2**, 711[35]; **5**, 768[125], 779[125]
Carlson, J. A., **5**, 129[33]
Carlson, J. G., **3**, 717[45], 752[94]; **8**, 946[137]
Carlson, J. L., **3**, 369[120]
Carlson, P. G., **7**, 167[100]
Carlson, P. H. J., **7**, 710[52]
Carlson, R., **2**, 1099[114]; **6**, 705[24]
Carlson, R. D., **6**, 208[59]
Carlson, R. G., **3**, 103[207], 874[71]; **5**, 218[35], 219[40], 221[35,59], 451[32], 453[32], 464[32], 513[5], 514[5], 527[5]; **8**, 528[62], 880[66]
Carlson, R. K., **4**, 904[204]
Carlson, R. M., **2**, 481[20]; **3**, 106[223,224], 113[223,224], 125[299], 135[367], 137[377], 278[34]; **4**, 120[201], 262[307], 308[405]; **6**, 677[316,316a]; **7**, 228[93]; **8**, 374[147]
Carlsson, A., **7**, 331[15], 831[64]
Carlsson, R., **5**, 721[102]
Carlsson, S., **8**, 541[202]
Carlton, F. E., **7**, 160[47]
Carlton, L., **4**, 161[91]
Carman, R. M., **6**, 204[24]; **7**, 352[30], 356[30]
Carmely, S., **3**, 380[13], 407[146]
Carmichael, C. S., **5**, 435[149], 524[53]
Carmody, M. A., **5**, 165[86]
Carmona, D., **8**, 445[55]

Carmosin, R., **5**, 404[18]; **8**, 28[37], 66[37]
Carnahan, J. C., Jr., **6**, 687[379]
Carnargo, W., **7**, 253[22]
Carnduff, J., **7**, 365[49]
Carnevale, G., **8**, 856[163]
Carney, P. L., **8**, 51[121], 66[121]
Carney, R. L., **3**, 370[109], 416[14], 417[14], 570[52], 572[52], 610[52]
Carniato, D., **8**, 798[51]
Caro, B., **8**, 5[27], 14[27]
Caron, H., **5**, 578[155]
Caron, M., **2**, 303[6]; **6**, 89[115], 91[115], 237[62]; **7**, 405[69]
Caronna, S., **6**, 551[682]
Caroon, J. M., **4**, 500[107]
Carothers, W. H., **4**, 315[517]; **8**, 140[13]
Carpanelli, C., **5**, 423[89]
Carpenter, A. J., **1**, 471[66], 472[73,74,76], 474[74]
Carpenter, B. K., **1**, 747[63]; **4**, 537[95], 538[102], 539[102], 1036[48]; **5**, 552[27], 595[8], 596[8b], 604[8b], 608[8b], 710[54], 788[16], 854[176], 855[176], 856[176,212,213], 905[62], 908[62], 1000[9,10], 1001[9], 1002[10], 1009[10], 1031[9]
Carpenter, C. W., **5**, 199[25]
Carpenter, F. H., **6**, 635[18], 636[18]
Carpenter, G. B., **1**, 19[105], 26[135,136], 43[136]; **2**, 100[6,7]
Carpenter, J. F., **5**, 424[96]; **6**, 838[65], 900[113]
Carpenter, N. E., **5**, 1075[30]
Carpenter, P. D., **2**, 764[64]
Carpenter, R. A., **2**, 740[59]
Carpenter, T. C., **6**, 213[84]
Carpino, L. A., **3**, 883[109]; **6**, 636[26], 637[26], 638[37,39], 639[47], 671[37], 999[124]; **7**, 480[105], 482[105], 662[50], 763[100], 766[100], 767[196]; **8**, 843[56]
Carpino, P., **2**, 373[273]
Carpino, P. A., **1**, 635[88,89], 806[315]; **3**, 104[208,209], 117[208,209]; **6**, 846[103], 905[145]
Carpio, H., **8**, 526[24]
Carpita, A., **3**, 217[91], 221[128], 439[37], 489[58], 495[58], 511[58], 515[58], 525[38-40], 527[45], 539[45,103,104], 541[45,105], 554[27]; **7**, 453[81]; **8**, 743[49]
Carr, C. S., **7**, 169[116], 171[116]
Carr, D., **8**, 65[211], 66[211]
Carr, D. B., **4**, 143[19]; **8**, 675[38], 677[38], 679[38], 681[38], 692[38], 693[38]
Carr, K., **7**, 630[52]; **8**, 344[123]
Carr, M. D., **3**, 724[15,16], 761[143]; **7**, 445[59]
Carr, P. W., **8**, 52[140], 66[140]
Carr, R. A. E., **1**, 797[293]; **6**, 996[107]
Carr, R. C., **5**, 514[9], 527[9]
Carr, R. V. C., **4**, 390[174]; **5**, 324[18a]; **6**, 1003[134]; **7**, 96[87]; **8**, 844[63]
Carra, S., **7**, 8[59]
Carrahar, P., **8**, 901[41]
Carrasco, M. C., **8**, 312[19]
Carraway, K. L., **6**, 36[30]
Carré, F., **8**, 766[20]
Carre, M. C., **1**, 563[184]; **4**, 250[138], 496[95]; **5**, 692[100]; **6**, 91[121]
Carrea, G., **8**, 194[105]
Carreira, E. M., **5**, 763[108]; **8**, 9[54], 619[130]
Carreira, L. A., **5**, 901[24]
Carrell, C. J., **1**, 8[38]
Carrell, H. L., **1**, 8[38]
Carreno, M. C., **6**, 149[105]; **8**, 15[91]
Carrera, P., **1**, 514[51]
Carreras, M., **1**, 543[24]
Carretero, J. C., **2**, 651[115,115a]; **3**, 162[487], 168[487]; **6**, 164[195]; **8**, 844[68]
Carrick, W. L., **3**, 664[33], 723[9], 731[9]; **7**, 571[120,121], 575[120,121], 576[120,121]
Carrie, R., **4**, 38[109b], 955[11], 964[48], 990[146]; **5**, 254[48], 944[244]; **6**, 76[45], 690[401], 692[401]; **7**, 476[63]; **8**, 385[42], 386[54]
Carrol, J. T., **4**, 386[152a]
Carroll, G. L., **1**, 328[21,23], 846[17]; **3**, 781[13]; **5**, 241[4], 243[10]; **6**, 682[338,340]
Carroll, M. F., **5**, 827[3], 834[3]
Carroll, P., **2**, 1097[101]
Carroll, P. J., **1**, 838[162]; **3**, 618[21]; **6**, 448[106,107]; **7**, 778[399]
Carroll, S., **7**, 723[23], 724[28]
Carrupt, P.-A., **7**, 257[49]
Carruthers, R. A., **7**, 879[147]
Carruthers, W., **1**, 428[121], 429[121], 457[121]; **3**, 1[6], 53[6], 54[6]; **4**, 405[248]; **5**, 451[45]; **7**, 543[11]; **8**, 269[78], 856[184]
Carruthers, W. H., **5**, 63[21]
Carsen, R. D., **1**, 309[99,100]
Carson, F. W., **5**, 890[34]; **6**, 152[141], 899[109]
Carson, J. F., **6**, 667[242]
Carson, J. R., **2**, 745[104]; **8**, 28[37], 66[37]
Carson, K. G., **8**, 54[160], 66[160]
Cartaya-Marin, C. P., **2**, 538[51], 547[120], 551[120], 552[120]; **5**, 4[37]; **8**, 61[188], 66[188]
Carter, C., **5**, 219[39], 230[39]
Carter, C. G., **4**, 350[119], 1039[63]; **6**, 531[441]
Carter, D., **7**, 58[56], 62[56], 63[56]
Carter, H. E., **2**, 396[6], 402[6], 403[6]; **4**, 279[108], 310[430]
Carter, I. M., **7**, 779[429]
Carter, J. P., **4**, 373[75]; **8**, 856[174]
Carter, J. S., **5**, 1123[37]
Carter, L. G., **3**, 194[10]
Carter, M. J., **1**, 623[78]; **2**, 587[137]; **5**, 335[47]
Carter, M. L. C., **8**, 70[232], 72[232]
Carter, P., **2**, 739[44]; **8**, 623[149]
Carter, P. A., **3**, 27[166]; **5**, 736[145], 737[145]
Carter, S. P., **3**, 168[496]; **4**, 1089[137], 1090[137], 1091[137]; **6**, 128[163]
Cartoon, M. E. K., **1**, 474[85]
Carturan, G., **5**, 272[2], 275[2]
Cartwright, B. A., **1**, 2[3], 37[3]
Cartwright, N. J., **3**, 693[140]
Caruso, A. J., **2**, 388[343]; **6**, 1020[48]; **8**, 852[139]
Caruso, T., **4**, 1084[97], 1104[213]
Caruso, T. C., **3**, 3[17], 7[17]
Caruthers, M. H., **6**, 554[731], 618[108], 619[118], 650[131]
Carvajal, S., **6**, 487[42], 489[42], 543[42], 550[42], 554[42]
Carver, D. R., **4**, 252[155], 426[59], 452[16], 462[104], 465[104,116,120], 466[104], 468[104], 469[104,116]
Carver, J. G., **8**, 36[76], 66[76]
Carver, J. R., **7**, 84[4], 85[4,6]
Casabo, J., **1**, 34[223]
Casadei, M. A., **3**, 55[284]; **6**, 538[572]; **8**, 135[41]
Casadevall, A., **2**, 709[10]
Casadevall, E., **2**, 851[220]
Casado, M. M., **8**, 349[137]
Casagrande, F., **3**, 789[56]
Casagrande, P., **7**, 26[48]
Casal, B., **3**, 770[179]
Casals, P.-F., **8**, 527[53]
Casamitjana, N., **1**, 564[206]; **6**, 917[33]; **8**, 621[143]
Casanova, J., **3**, 380[8]; **4**, 954[10]; **5**, 595[14], 597[29]; **6**, 776[54,56]; **7**, 601[87]; **8**, 794[12], 807[129]

Casanova, J., Jr., **2**, 141[37]; **3**, 649[202], 650[202c], 652[202c]; **6**, 294[239]
Casanova, R., **3**, 898[85]
Casara, P., **3**, 740[44]; **6**, 80[70]
Casares, A. M., **1**, 114[35,36]; **2**, 120[177]; **4**, 12[38], 98[114,114b,116], 113[114,169,169c]
Casati, P., **8**, 190[73,79], 191[73], 195[111], 203[111,149]
Casati, R., **7**, 283[181], 284[181]
Cascarano, G., **2**, 213[127]
Casella, L., **7**, 194[9], 777[382], 778[411]
Caserio, F. F., Jr., **4**, 604[284]; **7**, 564[90], 565[90]
Caserio, M. C., **1**, 586[18]; **4**, 53[151], 315[515], 332[20], 337[20,33]; **6**, 204[17]
Casey, C. P., **1**, 143[32], 426[109]; **3**, 418[24], 482[3,6]; **4**, 104[135a,c], 170[21,24], 189[104], 190[107], 191[114], 976[100], 980[105,106], 981[105,106], 982[113,114], 984[121]; **5**, 1065[1], 1066[1], 1074[1], 1076[32,35-39], 1079[49], 1082[55,56], 1083[1,36], 1084[1], 1085[65], 1086[67,69,71], 1093[1], 1094[1d]; **6**, 291[211]; **8**, 815[21]
Casey, M., **4**, 116[187]
Cashaw, J. L., **3**, 665[37]
Casini, A., **5**, 64[52]; **8**, 587[30,32]
Casiraghi, G., **3**, 311[100]
Casnati, G., **2**, 137[17]; **3**, 311[100]; **6**, 936[109]; **7**, 197[21]; **8**, 406[50]
Cason, J., **4**, 97[107c]; **7**, 92[41,41a], 94[41]; **8**, 286[14]
Cason, L. F., **1**, 619[60]; **4**, 120[197]
Caspari, I., **3**, 903[128]
Casper, E. W. R., **2**, 610[98]; **3**, 8[42]
Caspi, E., **4**, 339[42]; **7**, 154[13], 673[21], 675[21]; **8**, 237[10], 243[10]
Cass, Q. B., **2**, 350[77]; **7**, 355[39]
Cassady, J. M., **1**, 768[173]
Cassady, T. J., **5**, 639[120]
Cassani, G., **3**, 489[60], 495[60], 504[60], 511[60], 515[60]
Cassar, L., **3**, 271[1], 530[59], 539[101], 559[54], 1018[11], 1026[37], 1027[43]; **5**, 36[17], 1037[3], 1132[22], 1133[24], 1138[62], 1188[13], 1192[13]; **6**, 431[108]; **7**, 4[16], 462[122]
Casserly, E. W., **4**, 1002[62], 1015[197]
Cassidei, L., **7**, 167[186]
Cassidy, H. G., **3**, 262[166]
Cassidy, J. M., **7**, 409[91]
Cassidy, K. C., **3**, 46[255], 47[255]; **7**, 49[62], 229[122]
Cassinelli, G., **8**, 347[141], 350[141]
Cassio, C., **6**, 713[80a]
Cassis, R., **7**, 355[43]
Cast, J., **6**, 283[161]
Castagnino, E., **2**, 363[195]; **4**, 308[404]; **5**, 771[152]
Castagnoli, N., Jr., **7**, 232[157]; **8**, 618[109]
Castaing, M., **4**, 971[75]
Castaldi, G., **3**, 778[6], 788[6,54], 789[6,55-57]; **7**, 828[52], 829[55]; **8**, 111[17,18], 113[18], 117[17,18]
Castañeda, A., **1**, 589[19,20a], 591[19], 592[20a]; **2**, 555[136], 580[96]; **6**, 752[111,113]
Castanedo, N., **2**, 345[42]
Castanet, Y., **4**, 596[158], 620[158], 621[158], 636[158]
Castedo, L., **3**, 232[262], 545[121], 585[133], 586[156], 591[171], 610[156], 983[19,21], 984[21]; **4**, 483[5], 484[5], 495[5], 505[139,140], 513[179,180], 1004[78], 1020[238], 1023[238]; **5**, 384[127]; **6**, 487[76], 489[76], 533[481], 550[481]; **7**, 547[33], 746[85]; **8**, 874[24]
Casteel, D. A., **7**, 584[159]
Castelhano, A. L., **1**, 373[90], 375[90], 376[90]; **2**, 1063[106], 1075[150]; **4**, 717[13]
Castellanos, M. L., **4**, 425[26]
Castellino, S., **1**, 295[54], 296[54,55], 297[56], 306[54]; **2**, 630[8], 644[8a], 646[8a], 671[49,50]; **5**, 431[121], 434[121d]; **6**, 558[845]
Castello, G., **7**, 15[151]
Castells, J., **1**, 543[15,20,21,24,26], 547[61]; **6**, 510[299]
Castellucci, N. T., **4**, 1011[163]; **6**, 970[126]
Castiello, F. A., **5**, 718[95]
Castiglioni, M., **8**, 449[158], 457[216]
Castle, P. L., **3**, 244[24], 464[170]
Castle, R. B., **1**, 490[24]
Castle, R. N., **8**, 390[76], 905[62]
Castognino, E., **7**, 732[59]
Castonguay, L. A., **3**, 529[49]
Castrillon, J. R. A., **7**, 764[115]
Castro, B., **3**, 194[14]; **6**, 20[76], 74[35,36], 79[35], 80[35], 205[27]; **8**, 13[71]
Castro, C. E., **3**, 217[90], 219[90], 521[4], 522[4]; **6**, 544[628]; **8**, 481[50,51], 482[50,51], 483[51], 531[120], 908[75]
Castro, J., **5**, 1062[59]
Castro, J. L., **1**, 764[148], 808[326]; **4**, 1004[78], 1020[238], 1023[238]
Castro, P. P., **8**, 214[49]
Casucci, D., **1**, 480[154]
Casy, G., **3**, 875[76]; **4**, 245[91], 254[91]
Catala, A., **2**, 343[8], 363[8]
Catalano, M. M., **4**, 437[146]
Catch, J. R., **7**, 120[15]
Catelani, G., **5**, 456[87], 464[114,115], 466[114]; **7**, 245[76]
Catellani, M., **5**, 36[16]
Catino, J. J., **5**, 736[145], 737[145]
Cativiela, C., **2**, 406[45]
Catlin, W. E., **3**, 415[9]
Caton, M. P. L., **1**, 739[37]; **2**, 1064[107]; **4**, 371[52], 378[52a]; **5**, 605[56], 612[77]
Catskis, B. D., **7**, 15[144]
Catsoulacos, P., **8**, 64[219]
Cattalini, L., **7**, 777[386]
Caubère, P., **1**, 563[184]; **3**, 509[179]; **4**, 250[138], 452[19], 486[34], 496[89,93-95], 503[127]; **5**, 692[99,100], 1022[75]; **6**, 91[121]; **8**, 14[86], 16[102,103], 140[10], 141[10], 142[10], 418[14,15], 483[57], 485[57], 533[138], 558[390,391,392,393], 794[10], 802[10,76], 840[37,38], 878[48], 901[42], 908[42], 938[90]
Caulfield, T., **6**, 77[56]
Caulton, K. G., **8**, 550[329]
Caunt, P., **7**, 58[56], 62[56], 63[56]
Cauquis, G., **3**, 640[103]
Causse, M., **2**, 432[55]
Cauwberghs, S. G., **5**, 539[106]; **6**, 1055[52b]
Cava, M. P., **1**, 328[24], 544[31], 548[31]; **3**, 135[344], 136[344], 137[344], 138[344], 261[154], 890[31,34], 891[42], 901[111-113], 903[113,124]; **4**, 27[83], 45[126], 370[35], 385[35], 433[116]; **5**, 385[129a,c,d], 394[145a], 404[14], 436[155], 693[105], 724[111]; **6**, 420[13], 428[13], 478[107], 481[107], 977[13,14]; **7**, 330[11], 769[233], 774[312,313,317,323,334], 775[349], 776[349,361], 777[312,313]; **8**, 314[30], 370[92], 382[8], 410[96], 568[471], 618[111-113], 623[113,146], 628[113], 990[43], 995[43]
Cavalla, D., **1**, 774[211]
Cavallin, B., **2**, 547[95]
Cavallini, D., **8**, 192[96]
Cavanaugh, D. J., **7**, 778[417]
Cavanaugh, R., **4**, 5[19]; **5**, 322[13], 925[150]
Cavaye, B., **2**, 464[95,95b,96]
Cavazza, M., **6**, 176[103]
Cavé, A., **2**, 901[31,32]; **6**, 911[19], 912[22], 920[44]
Cave, R. J., **3**, 290[70]; **5**, 859[233], 888[25]
Cavicchio, G., **8**, 856[170]
Cavicchioli, S., **3**, 789[57]
Cavicchioni, G., **6**, 67[10], 575[966]
Cavill, G. W. K., **2**, 424[36]; **7**, 152[3], 153[10], 765[153]
Cavinato, G., **4**, 915[15]

Cavoli, P., **4**, 915[15]
Cawell, G. W., **1**, 832[109]
Cayen, C., **2**, 840[182]
Cazaux, L., **6**, 441[79], 494[137]
Cazes, B., **1**, 191[79], 192[83], 551[75]; **2**, 35[129]; **3**, 222[144], 223[183], 225[183], 991[36], 998[48,49], 999[50]; **4**, 102[128d]; **6**, 852[132,133]
Cazianis, C. T., **3**, 824[20,21], 825[21]
Ceccherelli, P., **1**, 656[151], 658[151], 846[18b], 847[18]; **2**, 823[112]; **3**, 857[92], 908[146]; **4**, 1040[77]; **8**, 880[63]
Ceccon, A., **4**, 527[67]
Cedar, F. J., **8**, 14[80]
Ceder, O., **3**, 645[169]; **7**, 574[128], 581[129]; **8**, 221[88]
Cederbaum, F. E., **1**, 162[96]; **5**, 1037[5], 1165[14,15], 1166[15-17,23], 1167[15-17], 1170[15], 1171[14,15], 1172[14], 1174[14], 1175[14,15], 1178[14-16], 1179[15]
Cedergren, R. J., **8**, 568[467]
Cedheim, L., **2**, 1051[35], 1052[35]
Cefelín, P., **8**, 806[124]
Cekovic, Q., **7**, 738[27]; **7**, 815[2], 816[2c], 824[2c], 827[2c], 831[68], 851[18]
Cekovic, Z., **4**, 820[222], 824[233]; **7**, 41[23], 703[5], 710[5]
Celebuski, J. E., **3**, 493[80]; **5**, 941[226]
Célérier, J. P., **2**, 356[131]; **6**, 501[206]; **8**, 35[65], 47[65], 66[65]
Cella, J. A., **2**, 567[25]; **7**, 574[127], 580[127]
Cellura, R. P., **8**, 625[161]
Cen, W., **6**, 185[160]
Cenini, S., **7**, 108[173]
Cense, J.-M., **1**, 632[63]
Centellas, V., **5**, 1059[54]
Cerar, D., **7**, 657[33]
Cerati, A., **4**, 763[208]
Ceré, V., **1**, 516[59,60], 517[61,62]; **3**, 147[396], 149[413], 151[413], 152[413], 153[396,413], 155[396], 865[27], 944[90,91], 946[92], 958[90,112]; **6**, 898[103]; **8**, 664[121]
Cereda, E., **6**, 550[678]
Cerefice, S. A., **5**, 1187[11,12]
Cereghetti, M., **7**, 222[36]
Cérésiat, M., **5**, 108[210], 109[210,222], 110[210,222], 111[210,222], 112[222]
Cerfontain, H., **5**, 165[79], 196[14], 212[69], 217[21], 221[21], 755[74], 760[74]
Cermier, R. A., **5**, 165[87]
Cerniglia, C. E., **7**, 75[116]
CernÓy, J., **2**, 382[313]
Cerny, M., **8**, 269[79], 272[79], 274[137], 279, 314[32-34,36], 541[207], 544[267], 907[72], 918[72], 967[80,81]
Cerny, V., **4**, 391[179]
Cerriani, A., **7**, 770[253]
Cerrini, S., **3**, 396[110], 397[110]
Ceruti, M., **8**, 660[109]
Cervantes, L., **8**, 526[24]
Cerveau, G., **2**, 572[43]
Cervenka, J., **8**, 545[286]
Cerveny, L., **8**, 424[42], 425[42]
Cervini, L. A., **4**, 872[38]
Cervinka, O., **1**, 366[42]; **6**, 704[9]; **7**, 471[20]; **8**, 161[17], 164[47], 176[127,-128,129], 545[286]
Cervino, R. M., **3**, 636[58]
Cesa, M. C., **4**, 170[24]
Cesario, M., **7**, 64[61b]
Cesarotti, E., **8**, 460[248], 683[92], 689[92]
Cessac, J., **5**, 704[20]
Cesti, P., **7**, 429[151]
Cetinkaya, M., **5**, 1185[1]
Cevasco, G., **4**, 426[52]
Cevelín, P., **6**, 685[353]
Cha, D. D., **5**, 689[71]
Cha, D. Y., **7**, 439[26,27]
Cha, J. K., **2**, 578[81]; **4**, 379[113,115], 384[113b]; **5**, 707[35], 773[168], 774[168], 864[258]; **7**, 439[31,32,34], 440[34,40]
Cha, J. S., **8**, 14[83,84], 18[120], 237[16], 240[35], 244[16], 253[16], 261[12-14], 406[42], 412[111]
Cha, Y. C., **4**, 1081[79]
Chaabouni, R., **1**, 387[129,134], 388[129,134]; **6**, 69[16], 98[154]
Chabala, J. C., **1**, 418[76]; **8**, 389[71]
Chabardes, P., **6**, 157[168], 836[54]
Chabaud, B., **6**, 134[17]; **7**, 88[24], 91[35]
Chabert, P., **3**, 203[102]; **5**, 829[26]
Chaboteaux, G., **1**, 677[222], 683[222], 700[222], 705[222], 708[222], 712[222], 722[222], 723[222]
Chabra, B. R., **7**, 271[129]
Chabrier, P., **6**, 536[548], 538[548]
Chackalamannil, S., **2**, 911[71]; **4**, 18[60,60a], 121[209], 799[112]
Chacko, E., **5**, 580[170]
Chadenson, M., **3**, 831[60]
Chadha, M. S., **7**, 453[64,82], 454[64,82]
Chadha, N. K., **2**, 547[100], 548[100]; **4**, 390[171]; **5**, 129[33]; **8**, 269[82], 722[150]
Chadha, R., **3**, 46[255], 47[255]
Chadha, V. K., **5**, 687[62], 688[62]; **7**, 220[23]
Chadrasekaran, S., **7**, 103[143]
Chadwick, D. J., **1**, 471[65,66], 472[73,74,76], 474[74,83]; **3**, 264[184]
Chadwick, J. C., **5**, 818[151]
Chae, Y. B., **1**, 798[286], 804[309]; **4**, 1040[68]
Chaffin, T. L., **8**, 531[112]
Chafin, A. P., **4**, 425[27]
Chafin, T. C., **3**, 736[29], 771[190]
Chai, O. L., **4**, 226[191]; **6**, 154[145]
Chaigne, F., **5**, 301[77]
Chaika, E. A., **6**, 510[295], 556[824]
Chaimovich, H., **3**, 897[94]
Chakhmakhcheva, O. G., **6**, 603[21,25], 604[27]
Chaki, H., **6**, 603[14]
Chakrabarti, D. K., **8**, 113[30], 117[30]
Chakrabarti, P., **1**, 8[37], 38[260,262], 306[91]; **8**, 333[56]
Chakraborti, A. K., **8**, 331[35]
Chakraborti, P. C., **8**, 331[35]
Chakraborti, R., **1**, 857[59]
Chakraborty, P. K., **6**, 487[72], 489[72]
Chakraborty, S., **8**, 113[30], 117[30], 412[108a]
Chakraborty, T., **7**, 318[49]
Chakraborty, T. K., **1**, 772[199]; **4**, 377[105b], 381[105]; **6**, 8[39], 51[104], 253[158,159]; **7**, 266[115], 267[115], 396[25], 574[122], 575[122], 576[122]
Chakraborty, U. R., **5**, 817[145]
Chakraborty, V., **4**, 854[95]
Chakravarti, R. N., **2**, 811[70], 813[70], 814[70]
Chakravarty, J., **8**, 331[35], 651[70]
Chakravarty, P. K., **4**, 394[193]
Chalais, S., **2**, 345[19], 359[19]; **3**, 322[144], 809[37]; **7**, 846[100]
Chalchat, J. C., **4**, 55[156]
Chalk, A., **4**, 849[81]
Chalk, A. J., **5**, 1142[90], 1146[106]; **8**, 699[148], 763[1], 765[17], 773[65], 785[1]
Challa, G., **3**, 552[9,10], 557[9], 559[57]
Challand, B. D., **5**, 127[24]
Challand, S. R., **4**, 435[136]; **7**, 476[64,65]
Challener, C. A., **2**, 588[151], 589[151]; **5**, 1089[87], 1090[87], 1094[87], 1098[87], 1099[87], 1100[87], 1101[87], 1102[147], 1112[87], 1113[87]

Challenger, F., **7**, 595[9], 770[254]
Challenger, S., **3**, 224[178]
Challis, B. C., **6**, 487[2], 488[2], 489[2], 515[2], 523[2], 524[2]; **7**, 746[84]; **8**, 301[93]
Challis, J. A., **6**, 487[2], 488[2], 489[2], 515[2], 523[2], 524[2]; **7**, 746[84]; **8**, 301[93]
Chaloner, P. A., **1**, 223[77], 224[77], 317[150]; **4**, 140[8]; **7**, 107[166]; **8**, 460[253]
Chaloupka, S., **4**, 1081[78]
Chaly, T., **2**, 621[153]; **6**, 677[309]
Chamberlain, G., **5**, 132[48]
Chamberlain, K. B., **4**, 665[10], 667[10], 668[14], 670[10,14], 674[10], 677[14]
Chamberlain, P., **4**, 301[321,322], 302[322]
Chamberlain, R. E., **4**, 1055[137]
Chamberlain, S. D., **4**, 465[122], 466[122,127]
Chamberlain, T., **1**, 243[53]
Chamberlain, T. R., **5**, 650[25]
Chamberlin, A. R., **1**, 523[83], 610[45]; **2**, 106[53], 228[163], 588[150], 1048[9], 1049[9,24], 1050[9], 1064[9]; **3**, 215[63], 251[79,80], 254[79,80]; **4**, 379[113,114a], 380[113a,114a], 872[38], 968[62]; **6**, 8[36], 26[106], 781[78,79], 783[84]; **7**, 358[11], 415[115c], 418[115c]; **8**, 537[190], 538[190], 844[72,72c], 940[101,103], 941[103], 946[103], 947[103], 948[101]
Chamberlin, E. M., **6**, 685[359]; **7**, 92[48]
Chamberlin, J. W., **8**, 528[63]
Chamberlin, K. S., **6**, 189[189]
Chambers, D., **7**, 706[25]
Chambers, J., **3**, 264[184]
Chambers, J. R., **6**, 806[55]
Chambers, L., **3**, 168[491], 169[491], 171[491]
Chambers, M. S., **6**, 842[79]
Chambers, R. D., **3**, 825[32]; **4**, 496[92]; **8**, 643[36], 794[3], 901[35,37], 903[35], 905[35]
Chambers, R. J., **6**, 904[142]
Chambers, R. W., **6**, 605[36], 614[79]
Chambers, V. E. M., **7**, 71[99]
Chambers, V. M. A., **8**, 852[140], 856[162]
Chamchaang, W., **1**, 787[255]; **3**, 1036[85]
Chamness, J. T., **8**, 364[24]
Chamot, E., **3**, 872[56]; **4**, 1013[182]
Champagne, P. J., **3**, 644[138], 648[138,182,184,189], 649[189]; **4**, 759[193]
Champion, E., **6**, 26[109]
Champion, J., **3**, 765[153]
Chan, A., **8**, 269[90]
Chan, A. C., **8**, 563[437]
Chan, C., **3**, 232[266], 488[54], 495[54]
Chan, C. B., **5**, 125[13], 128[13]
Chan, C. C., **2**, 771[12]; **7**, 283[183], 284[183], 760[22]
Chan, C. S., **7**, 308[20]
Chan, D., **3**, 849[60]
Chan, D. M. T., **4**, 18[59], 121[208], 247[103], 315[532], 870[28], 991[151]; **5**, 244[15,16,19], 245[20], 246[22], 298[60-63], 299[62,65,69], 300[62,63,69,72], 301[62], 302[79], 304[82], 307[88], 309[79], 311[69], 596[37], 598[37]; **6**, 176[92]
Chan, J. A., **8**, 237[10], 243[10]
Chan, K. K., **3**, 953[101]; **5**, 893[41]; **6**, 853[139], 875[23], 887[23], 888[23]
Chan, K. S., **5**, 1070[15,25], 1072[25], 1074[25], 1076[41], 1089[84,86,87], 1090[86,87,90], 1091[90], 1094[87], 1096[106,108,108c], 1098[15,87,106,108c], 1099[15,87,90,106,108c], 1100[87], 1101[87,90], 1112[15,87,106,108c], 1113[87], 1183[57]
Chan, L., **4**, 1039[65], 1053[131], 1063[131]
Chan, L.-H., **1**, 390[147]
Chan, M.-C., **8**, 840[33], 966[76]
Chan, M. F., **4**, 386[149], 387[149], 1076[42]
Chan, N., **5**, 490[192]; **8**, 389[68]
Chan, P. C.-M., **2**, 39[136]; **3**, 196[21]; **8**, 164[44]
Chan, P. S., **6**, 554[710,737]
Chan, S., **2**, 622[155]
Chan, T. H., **1**, 580[1], 623[80,82,84], 630[31], 722[31], 731[4], 732[6], 815[4]; **2**, 6[27], 57[7], 58[12,13], 68[44], 553[132], 554[132], 567[26], 588[150], 590[159], 599[24], 605[56,61], 606[64,67,68], 607[70,72,73], 617[145], 619[64,67,68,73,146,147], 620[72,151,152], 621[153,154], 622[156,157], 623[73,147,158,160], 625[61,165], 630[5,11], 631[11], 712[41], 713[49], 728[49], 853[232]; **3**, 25[157], 200[71,72], 251[79], 254[79], 257[119], 564[14], 566[29], 607[14], 929[58]; **4**, 120[197], 155[65], 347[88]; **5**, 596[31], 597[31], 608[31], 780[205]; **6**, 3[11], 652[147], 653[147], 654[147], 655[147], 677[309], 681[147], 1005[141,142,143,144]; **7**, 423[144]; **8**, 408[67], 777[82b]
Chan, T.-L., **3**, 877[89]
Chan, T. M., **7**, 172[128]
Chan, T. W., **7**, 763[90]
Chan, W., **1**, 95[72]
Chan, W. H., **5**, 131[46]; **6**, 438[44], 937[117], 939[117], 940[117]; **7**, 763[101]
Chan, Y.-L., **6**, 624[142]
Chan, Y. M., **8**, 176[131]
Chance, J. M., **8**, 764[6]
Chander, M. C., **7**, 246[95]
Chandhuri, J., **4**, 868[16]
Chandler, A., **6**, 547[658], 552[658]
Chandler, J. H., **8**, 356[183,187], 357[203], 359[203]
Chandler, L. B., **6**, 265[45]
Chandler, M., **4**, 675[38]
Chandra, V., **2**, 357[147]
Chandrachood, P. S., **1**, 463[29]
Chandragekaran, R. Y., **8**, 846[83]
Chandrakumar, N. S., **1**, 474[95], 564[202]; **5**, 376[109]; **6**, 534[518]
Chandraratna, R. A. S., **5**, 741[155]
Chandrasekaran, R. Y., **5**, 107[202]
Chandrasekaran, S., **1**, 272[66], 273[66d]; **2**, 156[118], 894[9], 912[9]; **3**, 570[53], 572[53,69], 573[53,69], 575[53], 583[53], 596[53], 602[69], 607[69], 610[53,69], 728[34], 744[63]; **4**, 373[70]; **6**, 938[123], 939[123], 942[123]; **7**, 103[144], 220[17], 266[106,108,112,115], 267[106,108,112,115], 276[106], 277[152], 318[49], 559[82], 560[82], 561[82], 562[82], 563[82], 566[100], 574[122], 575[122], 576[122], 601[91], 711[60]; **8**, 375[156], 531[125]
Chandrasekhar, J., **1**, 287[15,17], 476[125], 477[125], 487[1], 488[1], 580[2], 581[2], 582[2]; **3**, 66[12], 74[12], 194[11]
Chandrasekhar, S., **3**, 380[7]; **5**, 230[127], 232[127]; **8**, 116[69], 121[69]
Chandrasekharan, J., **7**, 595[13], 603[126]; **8**, 2[14], 708[36], 718[112], 720[128], 724[152,161-164,166,167], 727[112]
Chandross, E. A., **5**, 817[146]
Chandwadkar, A. J., **3**, 331[195]
Chandy, M. J., **3**, 396[104]
Chaney, G. A., **8**, 593[67]
Chaney, J., **4**, 1007[113], 1014[183]
Chang, B.-H., **8**, 676[54]
Chang, C.-A., **5**, 1141[85], 1143[91]
Chang, C.-C., **4**, 483[7], 1101[193]
Chang, C.-J., **2**, 384[317]
Chang, C. K., **3**, 729[38]
Chang, C.-T., **4**, 738[96], 754[177], 785[19], 787[19], 801[19], 802[19,129], 803[19], 824[237]
Chang, C. W. J., **7**, 586[165]
Chang, D. W. L., **7**, 488[155], 490[155]
Chang, E., **1**, 732[6]; **4**, 120[197]
Chang, F. C., **7**, 730[50]; **8**, 350[142,143], 352[142]

Chang, H. K., **7**, 43[39]
Chang, H. T., **1**, 557[126]; **4**, 80[60], 790[39]
Chang, J. H., **6**, 1059[70], 1066[70]; **7**, 376[83]
Chang, K. T., **4**, 1103[203]
Chang, L. H., **6**, 960[61]
Chang, L.-J., **3**, 172[513], 173[513]
Chang, L. L., **4**, 390[168]
Chang, L. T., **7**, 55[12], 56[12]
Chang, L. W., **8**, 319[72]
Chang, M. J., **6**, 575[971]
Chang, P. T. W., **4**, 152[55]
Chang, R., **7**, 74[109]
Chang, R.-C., **4**, 1101[193]
Chang, R. S. L., **2**, 962[51]
Chang, S.-C., **4**, 1012[175]; **6**, 498[162]; **8**, 743[163], 757[163]
Chang, T. C., **4**, 562[36], 576[15,16]
Chang, V. H.-T., **4**, 785[21], 786[26], 791[21]
Chang, V. S., **7**, 578[154], 584[154,157,158], 585[161]
Chang, Y.-H., **2**, 882[47]; **3**, 88[138], 89[138], 159[138], 161[138], 164[138], 165[138]; **6**, 538[568]
Chang, Y. K., **7**, 318[56], 319[56]
Chang, Y. L., **2**, 153[109]
Chang, Y. M., **4**, 1076[47]; **5**, 626[36], 629[36]
Chang, Y.-T., **8**, 148[111], 149[111,112]
Chang, Y. W., **7**, 236[19], 673[21], 675[21]
Chang, Z.-Y., **1**, 359[15], 391[15]
Chang Kuo, M. C., **5**, 451[16], 470[16]
Channon, J. A., **1**, 248[70]
Chano, K., **3**, 311[97]
Chanon, M., **5**, 913[102]; **6**, 510[294]; **7**, 851[12], 860[73]
Chanson, E., **1**, 438[158], 457[158]; **8**, 392[98]
Chansri, A., **2**, 711[38]
Chantegrel, B., **3**, 318[127]; **5**, 766[116]; **6**, 807[56]
Chantrapromma, K., **3**, 916[20], 926[47], 928[47], 930[59]
Chantrenne, M., **2**, 60[18,18b]; **4**, 117[191]
Chan-Yu-King, R., **1**, 522[79]; **4**, 119[194], 226[200,201], 227[201]; **6**, 864[193], 900[119]
Chao, B. Y.-H., **8**, 407[58]
Chao, H. S.-I., **2**, 953[3a]
Chao, K.-H., **2**, 675[52]; **4**, 695[4]
Chao, O., **3**, 901[112]
Chao, S., **5**, 829[21]
Chao, S. T., **3**, 15[79]
Chao, T. H., **7**, 14[134]
Chapa, O., **4**, 113[163]
Chapat, J.-P., **4**, 301[323], 302[323]
Chapdelaine, M. J., **1**, 698[250]; **2**, 614[116], 762[52]; **3**, 17[90], 530[77], 535[77]; **4**, 24[72], 238[7], 254[7], 256[209], 258[232]; **7**, 841[11]
Chapelle, F., **7**, 79[131]
Chapleo, C. B., **5**, 859[233], 888[25]; **8**, 615[94], 618[94]
Chapleur, Y., **4**, 386[154], 792[61]; **6**, 74[36]
Chaplin, C. A., **7**, 775[341], 776[341]
Chapman, D., **1**, 114[35,36]; **2**, 120[177]; **4**, 98[114,114b], 113[114]
Chapman, K. T., **1**, 312[112-114]; **5**, 186[170], 361[92], 543[116,116b,c]; **7**, 318[61]; **8**, 9[54], 619[130]
Chapman, N. B., **1**, 580[4]
Chapman, O. L., **3**, 698[157a], 815[73], 891[43], 892[43]; **4**, 285[157], 483[6,7]; **5**, 468[135], 597[30], 687[65], 730[129], 731[129]; **8**, 501[56], 502[56]
Chapman, R. D., **7**, 746[88]
Chapman, R. L., **7**, 62[53,53b], 63[58]
Chapman, S. L., **4**, 39[111]; **6**, 152[138], 153[138]
Chapman, T. M., **4**, 987[147]
Chapovskii, Y. A., **8**, 214[50]
Chappell, I., **5**, 2[20]
Chappell, R. L., **2**, 842[189]
Chappuis, J. L., **5**, 65[74]
Chapuis, C., **1**, 314[127]; **2**, 924[108a]; **4**, 377[104], 378[104b], 390[104b]; **5**, 356[90], 358[90b], 362[93], 363[93a], 365[93a], 432[124], 543[117]
Chaquin, P., **5**, 178[136], 455[76]
Char, H., **3**, 864[19]
Charalambides, A. A., **5**, 433[136b]
Charalampous, F. C., **2**, 456[74]
Charaux, C., **2**, 765[68]
Charbonnier, C., **6**, 437[39]
Charbonnier, F., **1**, 849[29]; **6**, 849[115]; **8**, 932[42]
Charcosset, H., **8**, 436[73]
Charette, A. B., **1**, 127[91], 409[38]; **5**, 534[92]; **7**, 579[131]
Charles, A. D., **4**, 706[38]
Charles, G., **2**, 367[227]; **8**, 27[31], 66[31], 478[44], 480[44]
Charles, J. R., **8**, 36[71], 66[71]
Charles, J. T., **3**, 564[7]; **8**, 116[70], 120[70], 121[70]
Charles, K. A., **5**, 229[121]
Charles, R., **2**, 1094[88]; **7**, 505[284]
Charleson, D. A., **7**, 177[145]
Charllampous, G. C., **2**, 466[114], 467[114]
Charlton, J. L., **5**, 385[130,130i], 389[138], 392[138b], 682[32], 691[83,83a,89], 692[83], 693[83,106], 738[148], 1031[95]; **6**, 977[11], 1007[11]
Charney, W., **7**, 55[10], 66[10], 68[10], 70[10], 71[10], 77[10]
Charpentier, R., **7**, 107[162], 452[45]
Charpentier-Morize, M., **1**, 530[128]; **3**, 324[155], 842[16]
Charpin, P., **5**, 410[41]
Charpiot, B., **3**, 664[30]
Charton, M., **1**, 580[4]
Charumilind, P., **5**, 89[51]
Charushin, V. N., **4**, 423[7]
Chasar, D. W., **7**, 763[95]; **8**, 405[30], 406[49]
Chase, B. H., **2**, 742[71]
Chase, J. A., **6**, 1045[29b]
Chasey, K. L., **3**, 874[70]
Chass, D. A., **6**, 1047[33]
Chassaing, G., **1**, 513[45,46], 531[130]; **3**, 149[407], 150[407], 151[407], 152[407], 153[407]
Chastanet, J., **4**, 479[169,170,171], 1089[126]
Chastrette, F., **4**, 383[138]
Chastrette, M., **1**, 53[18], 215[43], 216[43]; **4**, 383[138,139]
Chatani, N., **6**, 233[41]
Chatgilialoglu, C., **4**, 723[39], 739[106,107]; **8**, 801[75]
Chatt, J., **1**, 451[216]; **7**, 92[42], 93[42]
Chatterje, R. M., **7**, 137[122], 139[122]
Chatterjee, A., **3**, 856[88]; **6**, 507[234]; **7**, 823[35]; **8**, 510[97]
Chatterjee, S., **1**, 89[56], 95[80], 448[206]; **2**, 449[48]; **3**, 8[38], 12[55], 260[140], 385[56], 463[156], 469[199], 470[199], 472[199], 473[199]; **4**, 591[109], 633[109], 903[198]; **6**, 848[111]; **7**, 64[64]; **8**, 568[467], 755[122], 758[122]
Chatterjee, S. K., **8**, 510[97]
Chatterjee, S. S., **8**, 654[84]
Chattha, M. S., **2**, 482[25,28,29], 483[25,28,29]
Chattopadhyay, G., **6**, 510[297]
Chattopadhyay, J. K., **2**, 598[14]
Chattopadhyay, S., **1**, 470[61], 834[128], 838[159]; **6**, 150[114]; **7**, 778[401,401a]
Chattopadhyaya, H. B. J., **8**, 874[23]
Chattopadhyaya, J. B., **6**, 625[157], 659[191,192,197b]
Chatziiosifidis, I., **3**, 25[158,160]; **6**, 227[23], 229[23,27], 230[23]
Chau, F., **4**, 24[73,73d]
Chau, L. V., **4**, 1001[38]
Chau, W., **4**, 36[100]
Chaudhary, C., **2**, 364[202]

Chaudhary, S. K., **6**, 23[94], 650[128]
Chaudhary, S. S., **7**, 10[72]
Chaudhuri, A. P., **2**, 801[22]
Chaudhuri, M. K., **7**, 267[117], 268[117]
Chaudhuri, N. C., **8**, 874[21], 881[21]
Chaudhuri, N. K., **7**, 384[115]
Chaudhuri, S. A., **7**, 855[63]
Chaudhuri, S. R. R., **3**, 856[88]
Chauduri, N., **7**, 8[58]
Chauffaille, J., **1**, 555[111], 557[111]
Chauhan, P. M. S., **5**, 639[123]
Chaumette, P., **7**, 422[139]
ChauMont, P., **5**, 1117[15]
Chauncy, B., **3**, 507[171]
Chauqui-Offermanns, N., **1**, 357[8]
Chaussard, J., **4**, 458[68], 467[68], 469[68], 471[68], 473[68]
Chaussin, R., **5**, 829[22]
Chauvelier, J., **4**, 46[132], 53[132,148]
Chauvet, F., **7**, 452[48]
Chauvière, G., **8**, 37[91], 40[91], 44[91], 66[91]
Chauvin, J., **3**, 253[87]
Chauvin, M., **3**, 21[130], 22[130]
Chaux, R., **4**, 274[54]
Chavan, S. P., **5**, 341[58], 520[37,38]
Chavdarian, C. G., **1**, 544[40]; **3**, 252[81], 257[81], 280[40]; **4**, 390[168]; **8**, 618[109]
Chaves Das Neves, H. J., **8**, 62[196], 66[196]
Chavez, A. L., **1**, 837[154]
Chavira, R. S., **7**, 462[119]
Chawanya, H., **8**, 431[60]
Chawla, H. M., **6**, 111[58]; **7**, 843[46,47]
Chawla, H. P. S., **2**, 384[317]; **3**, 737[33]; **8**, 48[110], 66[110]
Chawla, R., **4**, 503[126]
Chaykovsky, M., **1**, 722[279], 820[5]; **3**, 147[391], 158[391], 778[2], 909[150], 918[21]; **4**, 487[41], 987[134], 988[134]; **5**, 847[139]; **6**, 812[79,80]; **7**, 194[10]; **8**, 844[70]
Che, C. M., **7**, 236[29]
Chebolu, V., **1**, 255[18]
Cheburkov, Yu. A., **2**, 739[46]
Chechina, O. N., **3**, 639[83]
Chedekel, M. R., **4**, 12[39]
Chee, O. G., **8**, 693[111]
Cheema, Z. K., **3**, 726[22]
Cheeseman, G. W. H., **1**, 474[85]; **4**, 485[27], 503[27]; **5**, 947[264]
Cheetham, I., **2**, 770[5]
Chefczynska, A., **6**, 150[117]
Cheikh, R. B., **6**, 69[16], 98[154]
Cheik-Rouhou, F., **6**, 173[46], 175[46,73]
Chekesova, M. P., **3**, 305[72]
Cheklovskaya, I. M., **8**, 236[7]
Chekulaeva, I. A., **4**, 3[10], 41[10], 47[10], 66[10]
Chelsky, R., **5**, 453[65]
Chelucci, G., **5**, 1153[145]; **8**, 91[54]
Chemburkar, S. R., **3**, 76[53], 77[57]; **7**, 227[80]
Chemerda, J. M., **6**, 685[358,359]; **7**, 92[48], 429[156]
Chen, B., **2**, 1070[137]; **4**, 126[218c]; **5**, 429[115]; **6**, 117[97]
Chen, B. C., **2**, 353[97]
Chen, C., **2**, 24[92]; **8**, 18[129]
Chen, C.-C., **8**, 332[38], 339[38]
Chen, C. H., **2**, 363[200]; **6**, 1025[79]
Chen, C. K., **1**, 555[120]; **7**, 749[117]
Chen, C.-L., **3**, 665[35]
Chen, C.-P., **1**, 317[144], 319[144]; **8**, 171[103]
Chen, C. S., **7**, 80[139]; **8**, 185[19], 190[19,85]
Chen, C.-W., **1**, 461[14], 464[14,40,41]
Chen, D., **1**, 123[74], 373[92], 375[92], 376[92]; **2**, 649[104], 1052[48], 1075[48], 1076[48]; **8**, 655[86]
Chen, E. Y., **4**, 1040[91], 1043[91]
Chen, F., **5**, 904[54]
Chen, F.-C., **3**, 500[127], 501[127]
Chen, F. M. F., **2**, 403[36]; **3**, 618[19]
Chen, G. M. S., **4**, 868[17], 869[17]
Chen, H., **6**, 220[126]
Chen, H. G., **1**, 213[17,17b], 214[19,19b]; **4**, 881[94]; **7**, 738[19]
Chen, H.-H., **1**, 66[55]; **2**, 536[43,44,46,47]; **8**, 774[76]
Chen, H.-J. C., **5**, 86[15,17], 98[127,-128]
Chen, H.-L., **1**, 771[193]; **8**, 163[41]
Chen, H.-W., **3**, 883[109]; **7**, 763[100], 766[100]
Chen, J., **3**, 50[266], 669[55]; **6**, 746[95]
Chen, J. C., **3**, 261[151]; **8**, 713[74], 714[74]
Chen, J. S., **1**, 838[158]; **7**, 162[64]
Chen, J.-T., **5**, 225[89]
Chen, K., **2**, 657[165]; **5**, 362[93], 363[93i]
Chen, K. K., **5**, 798[4]
Chen, K.-M., **6**, 717[114]; **8**, 9[48]
Chen, K. S., **5**, 901[32]
Chen, L.-C., **1**, 474[100]
Chen, L. M., **7**, 423[144]
Chen, L. S., **5**, 277[13]
Chen, M.-H., **3**, 221[129]; **4**, 754[179], 803[133], 820[133,213], 824[236,237]
Chen, M.-J., **5**, 637[103]
Chen, M. Y., **1**, 569[256,257]; **2**, 78[93,94]
Chen, N. Y., **3**, 305[75c]
Chen, P., **3**, 877[83]; **5**, 637[103]
Chen, P.-C., **8**, 542[236]
Chen, Q., **4**, 298[286], 452[11], 842[52], 858[111]
Chen, Q.-Y., **3**, 455[124], 525[42], 530[67], 531[67], 533[67]; **8**, 912[89]
Chen, S., **1**, 765[150]; **2**, 693[73], 694[74]; **4**, 593[130], 598[193], 638[193]; **5**, 394[146], 395[146]; **7**, 841[13]
Chen, S.-C., **3**, 730[44]; **8**, 124[91], 125[91]
Chen, S. F., **5**, 480[178], 596[37], 598[37], 1183[51]
Chen, S.-H., **7**, 265[95], 279[95], 280[95]
Chen, S.-M., **3**, 738[34]
Chen, S.-Y., **2**, 1090[70], 1100[70]
Chen, T., **7**, 564[96], 565[96], 568[96], 569[96], 570[96]
Chen, T. B. R. A., **3**, 875[72], 881[99], 882[100,101]
Chen, T. K., **8**, 336[68]
Chen, T. L., **3**, 566[29]
Chen, T.-M., **2**, 813[75], 818[75]
Chen, T.-S., **4**, 398[217], 399[217b], 401[217b], 403[217b], 404[217b]
Chen, W., **3**, 261[154]
Chen, W. Y., **8**, 647[57]
Chen, X., **2**, 357[143]; **7**, 761[65]
Chen, Y., **5**, 260[64], 265[64], 829[21], 950[285]; **6**, 535[543], 538[543]
Chen, Y.-Q., **3**, 369[107]
Chen, Y. S., **4**, 682[55]; **7**, 107[153,155]
Chen, Y. Y., **4**, 1076[44], 1086[118], 1088[118,121], 1089[118]; **7**, 763[99], 766[99]
Chen, Z., **4**, 452[12]
Chen, Z.-X., **5**, 75[229]
Chenard, B. L., **2**, 727[134]; **3**, 254[102]; **4**, 14[46,47f,g], 429[88]; **5**, 762[96]
Chenault, H. K., **8**, 185[15]
Chenault, J., **6**, 177[118]
Chenault, K. H., **2**, 456[32,39-41]
Chenchaiah, P. C., **7**, 76[117]
Chenets, V. V., **3**, 328[172,173,174]
Chênevert, R., **8**, 185[17], 193[100]
Cheney, L. C., **8**, 645[40]

Cheng, C., **7**, 109[181]
Cheng, C.-C., **3**, 587[148]
Cheng, C.-W., **1**, 463[33]; **7**, 107[163]
Cheng, H. N., **5**, 3[21]
Cheng, J. C.-P., **5**, 20[135]
Cheng, K.-F., **3**, 365[63]; **4**, 176[52]; **7**, 711[59]; **8**, 932[44]
Cheng, K.-M., **8**, 840[33], 966[76]
Cheng, K. P., **7**, 90[32]
Cheng, L., **4**, 429[83], 438[83], 441[83]; **6**, 1000[126]
Cheng, L. K., **3**, 643[116]
Cheng, M.-C., **2**, 119[160], 208[107]
Cheng, S., **3**, 369[108]
Cheng, T.-C., **8**, 724[153]
Cheng, W., **3**, 504[155], 511[155], 515[155]
Cheng, X.-M., **5**, 249[35]; **6**, 830[5]
Cheng, Y.-S., **5**, 473[150], 478[150], 480[178], 531[79]; **7**, 265[95], 279[95], 280[95], 841[13]
Cheng-fan, J., **5**, 949[280]
Chenier, P. J., **3**, 839[11], 853[69], 854[11]; **8**, 941[114]
Cheon, S. H., **1**, 198[91]
Cherbuliez, E., **6**, 601[1]
Cherchi, F., **7**, 774[330]
Cherest, M., **1**, 49[6], 50[6], 80[22], 109[13], 110[13], 153[54], 182[47], 185[47], 198[47], 222[69], 310[102], 678[213,214]; **2**, 24[96], 217[137], 666[37], 677[37]; **4**, 877[66]; **6**, 915[29]; **8**, 3[20,22], 5[33]
Cheriyan, U. O., **7**, 528[8]
Cherkasov, R. A., **6**, 432[119], 538[570]
Cherkasov, V. M., **6**, 552[698]
Chermomordik, Y. A., **5**, 1148[115]
Chern, C.-I., **6**, 2[8], 22[8]; **7**, 279[172], 744[68], 845[65]
Chernov, V. A., **6**, 543[611]
Chernova, T. N., **3**, 342[7], 345[21]
Chernyshev, E. A., **8**, 556[376], 769[27]
Chernyshev, V. O., **8**, 557[385]
Chernysheva, T. I., **8**, 778[86]
Chernyshov, A. I., **6**, 546[645], 554[745]
Cherpeck, R. E., **2**, 1049[26], 1064[26]
Cherry, W. R., **5**, 75[223], 207[49]
Chertkov, V. A., **2**, 710[16], 723[101]
Cheshire, D. R., **1**, 672[203], 678[203], 700[203]; **3**, 107[229]; **4**, 785[20], 791[49]
Chessa, G., **8**, 91[54]
Chetia, J. P., **6**, 552[700]
Chetty, G. L., **8**, 340[89], 526[17]
Cheung, C. K., **8**, 5[31], 597[94], 606[26]
Cheung, H.-C., **3**, 168[493], 169[493], 171[493]; **4**, 305[363]; **7**, 347[15]
Cheung, J. J., **5**, 804[93]
Cheung, L. D., **6**, 962[74]
Cheung, Y.-F., **3**, 370[112]
Chevallier, Y., **8**, 445[25,26], 452[25,26]
Chevolot, L., **2**, 1021[48]; **6**, 734[16]; **8**, 714[83]
Chexal, K. K., **2**, 541[79]
Chhabra, B. R., **2**, 353[95], 365[95]; **3**, 402[130]; **7**, 91[36]
Chi, K.-W., **5**, 886[22]
Chia, H.-A., **5**, 9[65]
Chia, H. L., **5**, 742[158]
Chia, W. N., **8**, 206[172]
Chiacchio, U., **3**, 168[496]; **5**, 252[45], 260[64], 265[64]; **6**, 178[121]
Chianelli, D., **1**, 670[187], 678[187]; **3**, 457[133], 503[146], 509[178], 513[146]; **4**, 426[51], 437[51], 441[181,182], 447[216,217]; **6**, 462[20,21]; **7**, 338[41], 770[256b], 771[256], 773[306]
Chiang, C. C., **6**, 914[28]
Chiang, C.-S., **6**, 675[301]; **7**, 845[80], 846[80]
Chiang, H. C., **3**, 380[4]
Chiang, J., **5**, 86[18], 96[121], 98[121]
Chiang, Y., **4**, 298[283], 300[307]; **6**, 569[937]
Chiang, Y.-C., **2**, 498[78], 501[78]; **6**, 531[435]
Chiang, Y. H., **7**, 656[17]
Chiao, W.-B., **6**, 1013[8], 1017[8]
Chiappe, C., **4**, 330[3], 345[3]
Chiarello, R. H., **5**, 1001[17]
Chiaroni, A., **4**, 221[165]; **8**, 58[174], 66[174]
Chiba, K., **3**, 1032[67], 1036[82], 1037[90], 1038[90]; **4**, 846[74]
Chiba, M., **1**, 846[20], 847[20], 853[49], 876[49]; **3**, 784[31]; **8**, 154[190]
Chiba, T., **1**, 110[17], 131[17], 134[17], 339[87], 564[195]; **2**, 648[96], 649[96], 789[56], 937[152,153,154,155]; **3**, 639[85]; **5**, 102[167,168]; **8**, 135[42], 253[114]
Chiba, Y., **3**, 231[248]
Chibante, F., **4**, 1040[101], 1045[101a]
Chibata, I., **2**, 456[37]
Chiboton, I., **2**, 1104[132]
Chiche, B., **2**, 736[21]
Chiche, L., **8**, 862[230]
Chicheportiche, R., **8**, 47[126], 66[126]
Chickos, J., **5**, 571[116]; **8**, 411[103]
Chida, Y., **1**, 359[14], 363[14], 384[14]
Chidambaram, N., **7**, 103[144], 277[152]
Chidambaram, R., **5**, 1068[14], 1203[58]
Chidgey, R., **5**, 595[11,15], 596[15], 603[49], 609[49]
Chieffi, G., **5**, 151[1], 152[1]
Chiem, P. V., **5**, 151[19]
Chien, C.-S., **7**, 382[108]
Chiesa, A., **8**, 460[248], 683[92], 689[92]
Chiesi-Villa, A., **1**, 29[150]; **2**, 127[229]; **8**, 683[90]
Chignell, C. F., **8**, 566[452]
Chigr, M., **5**, 473[153], 477[153]
Chiheru, K. S., **7**, 606[155]
Chihiro, M., **2**, 222[147]; **5**, 522[45], 536[97]
Chikamatsu, H., **1**, 860[68,70]; **7**, 425[149a]; **8**, 201[143]
Chikamatsu, K., **8**, 413[131]
Chikashita, H., **2**, 495[63], 496[63]; **4**, 254[181]; **6**, 541[592], 937[121]; **8**, 98[99], 291[32], 563[427]
Chikugo, T., **1**, 569[261]
Child, R. G., **3**, 767[163]
Childers, W. E., Jr., **1**, 551[69]; **5**, 456[82], 830[37]
Childress, S. J., **7**, 695[37]; **8**, 64[200], 66[200]
Childs, R. F., **1**, 294[49]; **2**, 6[35]; **5**, 597[23], 603[23], 606[23], 687[62], 688[62]
Chilikin, V. G., **7**, 493[196]
Chilot, J.-J., **4**, 395[205,205c,d]
Chimiak, A., **6**, 493[126], 494[126], 668[259]
Chimichi, S., **4**, 115[182]
Chimizu, H., **4**, 358[154]
Chin, A. W., **5**, 787[10]
Chin, C. G., **5**, 716[89]
Chin, C. S., **8**, 445[51], 452[186]
Chin, T. H., **2**, 466[113], 467[113]
Ching, K. N., **7**, 801[41]
Chini, M., **3**, 734[8]; **6**, 253[159]
Chini, P., **8**, 372[121]
Chinn, H. R., **3**, 224[163]
Chinn, L. J., **2**, 148[78]
Chinn, M. S., **3**, 920[34], 923[34b], 934[34], 1008[69]; **4**, 1036[49]; **6**, 873[9], 874[9c]
Chinn, R. L., **1**, 791[266]; **2**, 111[78]; **4**, 113[169,169c]
Chinnasamy, P., **2**, 894[9], 912[9]
Chino, K., **4**, 738[98], 792[67]
Chintani, M., **7**, 406[78b]
Chiong, K. G., **5**, 394[147]; **7**, 355[46]
Chiong, K. N., **3**, 681[95]

Chiou, B. L., **3**, 795[81]; **7**, 596[38]
Chiou, H.-S., **5**, 704[22], 1020[69], 1023[69]
Chipman, D. M., **8**, 94[79], 473[12], 977[138]
Chiraleu, F., **4**, 600[234], 643[234]
Chiriac, C. I., **6**, 490[107,112,113]
Chirkov, Yu. G., **8**, 373[132]
Chiron, R., **2**, 744[100]
Chisholm, M. H., **3**, 583[119]
Chistovalova, N., **4**, 840[33]
Chitrakorn, S., **7**, 103[142], 264[94], 265[94], 266[111], 267[111]
Chittattu, G., **1**, 699[247,254]; **3**, 107[225,226], 109[226], 386[60]; **4**, 365[4], 370[4], 372[55], 380[4], 381[4]; **6**, 470[58], 1028[94], 1031[94]; **8**, 847[97,97d], 849[97d,107,110,112]
Chittenden, G. J. F., **6**, 984[58]
Chitty, A. W., **4**, 1033[26], 1036[26c]
Chiu, A. A., **4**, 4[16]
Chiu, C.-W., **3**, 124[276], 126[276]; **6**, 676[304], 677[304]
Chiu, F.-T., **2**, 1038[101]
Chiu, I.-C., **5**, 515[12], 516[12b]
Chiu, J. J., **7**, 205[63]
Chiu, K.-W., **8**, 724[179]
Chiu, N. W. K., **4**, 955[12]; **5**, 754[67,68]
Chiu, S. K., **3**, 281[44]
Chiusoli, G. P., **3**, 423[77], 1027[43], 1030[57]; **4**, 932[65]; **5**, 36[16,17,19], 925[155], 1037[3], 1132[21,22], 1133[24], 1135[50], 1137[55], 1138[62], 1149[124], 1154[157]; **8**, 287[22]
Chiusoli, P., **5**, 1133[23]
Chivers, G. E., **7**, 750[125]
Chizhov, O., **2**, 358[156]; **5**, 817[147]
Chizhov, O. S., **6**, 271[85]
Chkir, M., **3**, 639[73], 648[180], 651[73b]
Chládek, S., **6**, 649[126]
Chliwner, I., **3**, 219[114], 499[140], 501[140], 502[140]
Chlopin, W., **7**, 775[351]
Chmielewsky, M., **2**, 663[25,26,28], 664[25,26,28], 665[25]; **4**, 36[102]; **5**, 108[207,208], 433[138,139,139a], 434[138]
Chmilenko, T. S., **4**, 48[138,138f]
Ch'ng, H. S., **5**, 408[30]
Cho, B. P., **7**, 220[23]
Cho, B.-R., **5**, 1135[52]; **7**, 543[13]
Cho, B. T., **8**, 60[184], 66[184], 159[10], 160[90], 170[89,90], 178[90], 179[90], 180[10], 237[19]
Cho, H., **1**, 523[83]; **2**, 228[163]; **4**, 111[154g]; **6**, 531[429]; **7**, 358[11]; **8**, 844[72,72c]
Cho, I., **5**, 926[160], 931[178]
Cho, I. H., **7**, 309[21]
Cho, J.-H., **5**, 664[38]
Cho, S. Y., **8**, 368[69,70], 375[70]
Cho, Y., **6**, 108[34]
Cho, Y. J., **6**, 790[118]
Cho, Y.-S., **3**, 41[224]; **4**, 357[150]
Chobanyan, Z. A., **4**, 315[525]
Chodkiewicz, W., **2**, 91[43]; **3**, 551[3], 552[3], 553[11]
Chodosh, D. F., **8**, 446[91], 455[91], 456[91]
Choi, H.-D., **2**, 556[152]; **6**, 931[89]; **7**, 200[41]; **8**, 964[56]
Choi, H. S., **4**, 689[72]
Choi, J.-K., **2**, 1049[22], 1050[22]; **4**, 794[78], 795[80]
Choi, K. N., **8**, 384[28,35]
Choi, L. S. L., **8**, 205[156]
Choi, O.-S., **6**, 446[103]
Choi, P., **6**, 245[128]
Choi, S., **1**, 259[28], 265[28], 894[154,156]
Choi, S.-C., **1**, 894[154]; **2**, 448[39]; **4**, 822[225]; **6**, 915[30]
Choi, S. K., **7**, 692[21]
Choi, S. S.-M., **2**, 555[149]
Choi, S. U., **3**, 299[33]
Choi, V. M. F., **1**, 347[137,138], 348[138], 546[59]; **6**, 237[68]
Choi, Y. B., **1**, 886[136]
Choi, Y. L., **4**, 761[202], 768[237]
Choi, Y. M., **8**, 244[48,53], 247[48,53], 249[48,93], 253[48,116], 806[128]
Chojnowski, J., **7**, 752[150]
Chokotho, N. C. J., **2**, 965[65]
Chollet, A., **5**, 791[25], 798[25], 847[140]
Chome, C., **5**, 456[86]
Chong, A., **4**, 476[163], 502[124], 766[229]
Chong, A. O., **7**, 254[31], 485[138]
Chong, B. P., **4**, 476[163], 502[124], 766[229]; **5**, 634[73]
Chong, J. A., **6**, 982[51]
Chong, J. M., **1**, 343[122], 345[122]; **2**, 39[136], 589[153]; **3**, 196[21], 225[186,190,191]; **4**, 249[122], 256[210], 901[186,186a]; **6**, 11[50]; **7**, 405[70]; **8**, 164[44]
Chong, W. K. M., **2**, 26[102], 27[102]; **3**, 683[103]; **5**, 841[97]
Chon-Yu-King, R., **2**, 66[34], 67[34]
Choo, K. Y., **8**, 765[16]
Chopard, P. A., **4**, 35[98b], 74[38b]
Choplin, F., **5**, 772[162], 780[162]; 7, 298[34]; **8**, 945[127]
Chopra, A. K., **2**, 821[108]
Chopra, C. L., **7**, 68[85], 71[85]
Chopra, S. K., **4**, 710[55], 712[67]
Chorev, M., **6**, 636[21]
Chorn, T. A., **2**, 746[109]
Chortyk, O. T., **6**, 268[67]
Chorvat, R. J., **4**, 14[49]; **6**, 554[769]
Chou, C.-H., **5**, 639[121]
Chou, C. S., **3**, 1048[13]; **7**, 809[82], 875[118]
Chou, C.-Y., **2**, 542[81]; **4**, 952[5]
Chou, D. T.-W., **5**, 942[235]
Chou, K. J., **4**, 1036[44]; **5**, 942[231]
Chou, S.-S. P., **3**, 172[514]; **5**, 333[45]; **8**, 70[232], 72[232]
Chou, T., **4**, 314[497]; **8**, 859[211]
Chou, T.-C., **5**, 768[125], 779[125]; **8**, 795[18]
Chou, T.-S., **1**, 213[17,17b], 743[65]; **2**, 105[40], 738[41], 760[40]; **3**, 172[513], 173[513,517]; **5**, 1124[43,49]; **6**, 705[30]; 7, 261[68]; **8**, 859[210,212]
Choudary, B. M., **8**, 372[117]
Choudhry, S. C., **6**, 7[32]; **8**, 798[58]
Choudhuri, L. N., **2**, 725[119]
Choudhuri, M., **8**, 626[168]
Chouinard, P. M., **2**, 904[50]
Choukroun, R., **6**, 91[125]; **8**, 447[122], 457[122]
Chovin, P., **8**, 285[6]
Chow, D., **4**, 294[246]; **6**, 283[165]
Chow, F., **1**, 631[56], 632[56], 633[56], 634[56], 635[56], 636[56,97], 638[56,97], 640[56,97], 641[56], 642[56], 645[56], 648[97], 649[97], 650[97], 656[56], 657[56], 664[56], 666[56], 672[56,97,192], 675[192], 68; **3**, 87[74,113], 95[74], 97[74], 109[74], 114[74], 116[74], 136[74,113], 141[74], 145[113], 157[113], 787[46]; **6**, 1026[88], 1027[88], 1031[116]; **7**, 765[158], 771[275]
Chow, H.-F., **1**, 586[16]; **2**, 114[114], 244[24], 245[24], 267[24], 576[78], 579[78], 645[82]; **6**, 765[18]
Chow, J., **7**, 600[74]
Chow, K., **2**, 656[150]; **4**, 159[84]
Chow, M., **5**, 240[2]
Chow, M.-S., **2**, 85[22,23]
Chow, S., **1**, 293[37]
Chow, S. W., **3**, 427[88]
Chow, W. Y., **4**, 1015[193,196]; **5**, 948[269], 1199[47]
Chow, Y., **1**, 293[32]
Chow, Y. L., **3**, 380[11], 689[123]; **7**, 40[14], 488[155], 490[155], 500[241], 736[1], 745[1], 749[1]; **8**, 248[82]
Chowdhury, R. L., **4**, 529[77,78], 530[78,79], 531[78], 541[113,114]

Chowdhury, S., **7**, 854[46]
Choy, W., **1**, 410[43], 769[182]; **2**, 2[5], 25[5], 33[5], 40[5], 111[81], 134[3], 190[57], 192[61], 221[61], 232[180,183], 240[5], 248[5b], 249[36], 260[5e], 308[20], 455[8], 652[124], 686[68], 979[17]; **4**, 1079[61,62]; **5**, 147[110], 359[91], 373[91], 374[91], 543[118], 545[118], 682[33], 691[33], 1032[99]; **6**, 8[39]; **7**, 390[8], 399[40a], 442[48]; **8**, 535[165]
Chretien, F., **4**, 386[154]
Chrisope, D. R., **2**, 528[10]; **5**, 2[8]
Christ, H., **4**, 587[18-21,37], 603[267], 604[281], 645[267], 646[281]; **7**, 107[160], 452[58]
Christ, J., **2**, 809[53]; **3**, 640[95]
Christ, R. E., **2**, 152[104]
Christ, W. J., **1**, 193[86], 198[91]; **7**, 439[31,32,34], 440[34,40]
Christen, M., **8**, 190[82]
Christense, S. B., **4**, 497[97]
Christensen, A. T., **4**, 38[108,108a]
Christensen, B. G., **1**, 434[141]; **2**, 212[118], 213[122], 1102[123]; **3**, 890[32]; **5**, 92[76], 107[199]; **6**, 125[148], 127[148]; **7**, 257[46]; **8**, 384[25]
Christensen, B. W., **7**, 777[385]
Christensen, C. G., **5**, 249[33]
Christensen, D., **7**, 752[150]
Christensen, J. J., **6**, 449[116]
Christensen, J. R., **3**, 595[191]
Christensen, L., **3**, 750[84], 762[84]
Christensen, L. W., **3**, 181[554]; **7**, 760[46]
Christensen, S. B., **4**, 250[139]; **7**, 355[44]
Christenson, B., **3**, 212[35]; **4**, 240[44]
Christenson, P. A., **6**, 1067[107]
Christenson, R. M., **6**, 264[29], 268[29], 286[29]
Christiaens, L., **1**, 644[123], 646[123], 668[123], 669[123], 695[123]; **2**, 817[95]; **6**, 462[13,14]
Christiaens, L. E. E., **4**, 50[142]
Christian, S. L., **4**, 425[27]
Christian, W., **8**, 589[52]
Christiansen, R. G., **2**, 839[173]; **8**, 566[451]
Christie, B. D., **2**, 1018[38,42]; **6**, 734[18]
Christl, M., **4**, 1010[160], 1017[214], 1079[58,59]; **5**, 489[200]; **8**, 472[6]
Christmann, K. F., **1**, 758[123]
Christner, D. F., **5**, 736[142i]
Christol, H., **2**, 902[44]; **3**, 727[29], 734[11], 740[47,48], 744[61], 813[64]; **4**, 55[156], 252[160], 315[509]; **6**, 134[17,34], 262[17], 264[28,34], 267[57], 268[28]; **8**, 36[71], 66[71], 862[228,230], 863[231,233]
Christopfel, W. C., **8**, 459[228]
Christoph, G. G., **3**, 872[56]
Christophersen, M. J. N., **1**, 28[140]
Christy, K. J., **5**, 850[161]
Christy, M. E., **3**, 380[11]; **5**, 382[121]
Christy, M. R., **1**, 827[67]
Chrovat, R. J., **6**, 554[768]
Chrusciel, R. A., **5**, 429[112]
Chrystal, E. J. T., **8**, 652[78]
Chrzanowska, M., **1**, 552[79], 564[203]
Chu, C., **3**, 305[75b]
Chu, C. C. C., **3**, 30[174]; **6**, 714[81]
Chu, C.-K., **1**, 327[13]; **2**, 727[128]; **4**, 155[66]; **8**, 621[144]
Chu, C.-Y., **1**, 116[46,50], 118[46,50]; **3**, 249[63]; **4**, 91[88b]
Chu, D.-L., **4**, 438[149]
Chu, D. T. W., **3**, 216[76]; **7**, 209[91]; **8**, 384[40]
Chu, G.-N., **2**, 877[38,39]
Chu, H.-K., **7**, 98[100]
Chu, J. Y., **7**, 774[311]
Chu, K.-H., **3**, 215[60]; **4**, 395[204]
Chu, S. C., **8**, 846[83]
Chu, S.-D., **4**, 795[84]
Chu, S. Y., **5**, 75[230]
Chuang, C.-P., **4**, 374[91], 790[41], 791[41,48]; **8**, 514[107]
Chuang, K.-S., **2**, 378[284]; **4**, 1101[193]
Chuang, Y.-H., **3**, 224[175], 752[95]
Chuaqui-Offermanns, N., **2**, 476[4]; **3**, 31[186]; **6**, 724[150,152]
Chuche, J., **3**, 578[92], 610[92]; **4**, 309[414], 393[191], 394[191]; **5**, 797[64], 829[25], 930[174,175], 931[175], 932[175,180], 933[182], 938[180,220]; **8**, 532[130]
Chucholowski, A., **2**, 205[101,101a,b], 597[8]; **4**, 21[65], 108[146e]; **6**, 46[63], 61[150]
Chudgar, R. J., **2**, 401[29]
Chui, J.-J., **6**, 934[98]
Chuilon, S., **6**, 66[3]
Chuit, C., **1**, 107[3], 113[33], 428[116]; **2**, 572[43]; **3**, 246[38], 470[221], 471[221]; **4**, 100[126], 183[81], 897[172]; **8**, 246[79]
Chujo, Y., **4**, 591[112], 611[359], 633[112]
Chukovskaya, E. C., **8**, 91[58]
Chukovskaya, E. Ts., **3**, 422[69]; **4**, 952[4]; **8**, 765[11]
Chulkov, I., **7**, 763[93]
Chum, P. W., **4**, 887[128]; **8**, 483[54], 485[54]
Chun, D. T. W., **6**, 646[100a]
Chun, M. W., **3**, 49[261]; **4**, 350[116,118]
Chung, B. C., **6**, 205[25], 210[25]
Chung, B. Y., **1**, 435[143]; **2**, 1103[130]
Chung, C.-J., **5**, 154[31]
Chung, J. Y. L., **2**, 1048[9], 1049[9], 1050[9], 1064[9]; **5**, 1183[50]; **7**, 246[93]
Chung, K.-H., **3**, 592[172]
Chung, K.-S., **8**, 795[18]
Chung, L.-L., **3**, 870[46]
Chung, M. W. L., **7**, 138[127]
Chung, S.-K., **3**, 824[19,23]; **7**, 97[96], 528[7]; **8**, 86[24], 404[17], 483[56], 485[56], 802[84]
Chung, W.-K., **3**, 49[261]
Chung, W. S., **5**, 436[155]
Chung, Y. K., **4**, 520[38], 542[38], 689[72,73], 691[74]
Chung, Y.-S., **5**, 552[15,32]
Chung, Y. W., **8**, 14[81]
Chung-heng, He., **4**, 581[28]
Chupakhin, O. N., **4**, 423[6,7], 441[6]; **8**, 580[1]
Chupp, J. P., **2**, 787[51]; **6**, 540[587]
Church, D. F., **7**, 488[158], 761[54]
Church, J. P., **5**, 835[59], 862[250]
Church, K. M., **5**, 266[75], 268[75]
Church, L. A., **4**, 1038[55]
Church, R. F., **5**, 830[29]
Churchill, M. R., **4**, 485[31]
Churi, R. H., **3**, 748[76]
Chuvatkin, N. N., **4**, 347[93]
Chvalovsky, V., **8**, 544[267,268], 770[38], 771[42], 773[64], 776[80], 907[72], 918[72]
Chvalovsky, Y., **8**, 274[137]
Chwang, W. K., **4**, 299[296]
Chylinska, B., **4**, 432[105]
Chys, J., **6**, 547[663]
Ciabattoni, J., **3**, 380[8]; **4**, 5[19,19e]; **5**, 322[12]
Ciaccio, J. A., **1**, 343[110]; **3**, 224[165]
Ciamician, G., **2**, 773[27]; **5**, 123[2]
Cianciosi, S. J., **5**, 64[31]; **7**, 545[28]
Ciani, G., **8**, 460[248]
Ciattini, P. G., **3**, 220[118], 222[118], 1035[79]; **7**, 143[148], 144[148]; **8**, 84[13], 911[87], 933[51]
Cicala, G., **1**, 834[126]
Ciccio, J. F., **7**, 820[25]

Cichowicz, M. B., **6**, 651[137]; **7**, 241[59]
Cichra, D. A., **7**, 749[121]
Ciegler, A., **7**, 62[49]
Cieplak, A. S., **1**, 67[59], 99[86], 298[60]; **7**, 363[39]; **8**, 5[29]
Ciganek, E., **1**, 709[262]; **4**, 48[137], 1080[69]; **5**, 37[21], 71[147], 78[147], 198[22], 451[34], 453[34], 464[34], 513[5], 514[5], 527[5], 714[65,66]; **6**, 173[30], 1015[21]; **8**, 251[105]
Cihova, M., **7**, 451[36]
Cilento, G., **5**, 198[21]
Cimarusti, C. M., **4**, 30[89]; **6**, 644[93]
Ciminale, F., **4**, 452[20]; **7**, 760[49], 764[49]
Cinnamon, M., **7**, 656[17]
Cinquini, M., **1**, 519[65,66], 520[67], 523[80], 524[86,87], 765[151]; **2**, 31[108], 228[166,167], 229[168], 374[276], 486[42], 492[52], 515[55,56], 516[58]; **6**, 149[97,99,102], 425[66], 840[71]; **7**, 442[47], 764[112], 767[112,193], 771[261], 772[286]; **8**, 72[239], 74[239], 844[67]
Cioni, P., **5**, 1154[152]; **8**, 690[103]
Cioranescu, E., **5**, 478[162]; **6**, 489[85]
Cipris, D., **7**, 769[216]; **8**, 408[77], 409[77]
Cipullo, M. J., **4**, 126[218b]
Ciranni, G., **6**, 280[144]
Cirillo, P. F., **4**, 161[91]
Ciskowski, J. M., **3**, 316[118], 317[118]
Cisneros, A., **7**, 706[22]
Cistone, F., **8**, 26[19,28], 27[19], 30[28], 36[19,28], 37[28], 40[28], 43[28], 44[28], 46[28], 55[28], 66[28], 357[204], 803[93], 804[93], 826[69]
Citerio, L., **6**, 555[814]
Citron, J. D., **8**, 265[52], 906[67], 907[67], 908[67]
Citterio, A., **2**, 735[15]; **4**, 719[18], 723[18], 739[112], 763[208], 764[112,217-219], 765[227], 767[219,233], 768[235,236,241], 810[168], 820[168],
823[168]; **7**, 828[52]
Ciuffreda, P., **7**, 674[42]; **8**, 565[448]
Ciufolini, M. A., **2**, 1049[16], 1066[122]; **4**, 403[241]; **5**, 335[48], 432[133], 459[92]; **7**, 365[48]; **8**, 187[34]
Ciula, R. P., **5**, 1185[2]
Civier, A., **2**, 379[294]
Clader, J. W., **1**, 544[36], 547[36], 548[36]; **2**, 61[20], 69[46], 711[37]; **5**, 766[118]; **6**, 531[430], 648[121]
Claes, P. J., **5**, 92[72]
Claeson, G., **8**, 608[45]
Claesson, A., **2**, 465[107]; **3**, 217[82], 445[72], 492[77]; **4**, 55[157], 56[157a], 308[407], 395[202,203], 411[264], 872[42]; **6**, 866[203]
Claesson, S. M., **2**, 465[107]
Claeys, M., **5**, 127[25], 689[71]
Claeyssens, M., **6**, 48[88]
Clagett, M., **4**, 987[147]
Clague, A. R., **1**, 420[85]
Claisen, L., **2**, 134[7], 150[7b], 796[1,3]; **4**, 283[148]; **5**, 827[1], 850[1]; **8**, 644[39]
Clamot, B., **6**, 543[624]
Claparède, A., **2**, 134[7]
Clapp, L. B., **7**, 751[140]
Clapper, G. L., **3**, 568[47]
Clar, E., **8**, 972[115]
Claramunt, R. M., **4**, 439[160]
Clardy, J., **1**, 28[143], 29[144], 34[169], 359[22], 383[22], 384[22], 401[14], 889[142], 890[142]; **2**, 88[29], 507[26,27], 508[27], 824[119]; **3**, 872[56]; **4**, 42[121], 206[43], 213[106,115], 215[106], 532[83], 543[83], 545[83], 581[28], 695[4], 709[45], 710[45]; **6**, 727[195,197], 835[46], 1023[73], 1054[47]; **7**, 401[61c]; **8**, 347[141], 350[141], 880[61]
Clardy, J. C., **3**, 33[189], 34[198], 39[198], 232[266], 284[54], 380[10], 395[102], 396[102], 592[174,175], 594[174], 698[157a], 872[56]; **5**, 179[141], 203[39,39a], 209[39], 210[39], 324[22], 468[135], 736[143,145], 737[145], 1090[90], 1091[90], 1099[90], 1101[90]; **7**, 441[42]
Clare, M., **7**, 255[35]
Clarembeau, M., **1**, 571[274,276], 630[44,45], 631[44-46,53,54,58,60], 632[60], 633[60,72], 634[60,72], 635[60,72], 636[45,72], 637[45], 640[45,72], 641[72], 642[72], 648[135], 653[135], 656[53], 657[53], 658[53,54,60], 659; **2**, 76[86]; **3**, 87[72,73,79,80,84,101], 88[127], 89[127], 91[127], 92[127], 93[127], 95[72,79,80], 104[101], 109[79,80,84,127], 111[79,80,230], 116[127]; **4**, 71[12]
Claremon, D. A., **1**, 122[67], 359[11], 380[11], 382[11], 409[36,37], 876[99]; **3**, 39[216]; **4**, 370[31], 372[31], 397[31]; **5**, 534[92]; **6**, 158[184], 466[44,45], 469[44,45]; **7**, 522[41], 523[41]; **8**, 388[61]
Claret, J., **5**, 1133[33]
Clark, A. C., **4**, 869[27], 870[27], 871[27]
Clark, B., **1**, 310[103]
Clark, B. C., Jr., **3**, 736[29], 771[190]
Clark, C. T., **6**, 81[74]; **7**, 11[89]
Clark, C. W., **7**, 84[1], 85[1], 108[1]
Clark, D. A., **6**, 1056[55]
Clark, D. B., **6**, 282[152]
Clark, D. E., **3**, 688[115]
Clark, D. G., **8**, 905[62]
Clark, D. M., **5**, 599[40], 804[94], 905[60]; **6**, 126[149]
Clark, D. R., **3**, 1024[27]
Clark, F. R. S., **3**, 501[135]
Clark, G., **1**, 412[51], 733[13]; **5**, 519[30]; **7**, 549[46]
Clark, G. M., **4**, 887[130]; **8**, 708[34], 716[34], 717[94], 726[34], 727[34], 756[141]
Clark, G. R., **1**, 892[149]; **2**, 186[35]; **3**, 675[74]; **4**, 517[5], 518[5], 532[83,89-91], 534[90,91], 537[91], 543[83], 545[83,126], 1018[218]; **7**, 453[94]
Clark, G. W., **3**, 1050[17]; **7**, 341[51]
Clark, H. C., **4**, 915[8,15]; **8**, 447[116]
Clark, J., **8**, 839[25]
Clark, J. D., **5**, 854[180]
Clark, J. H., **2**, 343[11], 359[11]; **3**, 54[278]; **4**, 354[130], 425[35], 439[164,166], 445[35,208]; **6**, 66[6], 939[145,146], 941[146], 942[146]; **7**, 844[54,55]; **8**, 86[26]
Clark, J. S., **1**, 744[55]; **2**, 318[50]; **3**, 39[214]; **5**, 1123[36]; **7**, 679[75]
Clark, K. J., **4**, 283[145]
Clark, L. C., Jr., **8**, 568[466]
Clark, M., **3**, 825[32]
Clark, M. C., **1**, 635[85], 636[85], 640[85], 642[85], 643[85], 672[85], 682[85], 700[85], 705[85]; **2**, 76[87], 601[36]; **3**, 87[100], 104[100], 110[100], 117[100]
Clark, M. T., **3**, 825[28]
Clark, P. D., **1**, 632[65], 633[65], 636[65], 638[65], 644[65], 645[65], 646[65], 647[65], 648[65], 669[65], 672[65], 695[65], 700[65], 705[65]; **3**, 87[107], 105[107], 106[107], 114[107], 120[107], 157[107]
Clark, P. W., **2**, 749[132]
Clark, R., **1**, 834[121b]; **2**, 116[139]
Clark, R. A., **5**, 714[65]
Clark, R. D., **1**, 366[49], 391[49], 608[35]; **2**, 913[76], 928[123,124], 929[125]; **3**, 21[129], 158[441], 159[441], 280[40], 344[19]; **4**, 45[126], 500[107]; **6**, 1033[122]; **7**, 166[86a], 673[32]; **8**, 36[98], 42[98], 66[98], 938[91]
Clark, R. H., **4**, 274[58]
Clark, R. L., **6**, 487[57], 489[57]; **7**, 863[85]
Clark, S., **2**, 456[23]; **4**, 31[92,92k]
Clark, T., **1**, 29[147], 41[196], 487[1,2], 488[1,2]; **3**, 194[4], 422[68]; **4**, 872[41]; **6**, 172[11], 500[179]; **8**, 724[168,169,169b]
Clark, T. A., **7**, 74[109]
Clark, T. J., **4**, 1038[55]
Clark, V. M., **6**, 609[57,58], 614[78,87,88]
Clark, W., **4**, 850[85]

Clark, W. M., **7**, 135[104]
Clarke, A. R., **8**, 206[172]
Clarke, C., **2**, 1072[140]; **7**, 318[60]
Clarke, D. M., **4**, 1033[27], 1049[27], 1060[27b,c]
Clarke, F. H., Jr., **8**, 531[114], 986[13]
Clarke, G. M., **8**, 656[89]
Clarke, H. T., **2**, 396[8]; **7**, 122[28]
Clarke, R. L., **6**, 566[923,924]
Clarke, S. J., **8**, 384[35]
Clarke, T., **4**, 381[126b], 382[126], 383[126]; **6**, 267[56]
Clarke, U., **6**, 67[12]
Clase, A., **2**, 547[106], 550[106]
Clasper, P., **8**, 227[118]
Class, M., **8**, 332[42]
Classen, A., **7**, 429[151]
Classon, B., **7**, 237[32]
Claudi, F., **5**, 92[63]
Claus, C. J., **8**, 213[28], 267[66]
Claus, P. K., **4**, 430[98]
Clausen, H., **5**, 531[74], 532[74a]
Clausen, K., **2**, 867[13]; **6**, 420[23], 423[23], 436[19], 437[19], 451[130], 456[130]
Clausen, M., **5**, 497[224]
Clauson, L., **5**, 1115[3], 1122[3], 1123[3], 1124[3]
Clauson-Kaas, N., **7**, 808[77]
Clauss, K., **1**, 141[13,15], 142[13], 143[13,30], 372[79]; **2**, 1049[14]; **5**, 69[99]; **8**, 755[115]
Clawson, L., **1**, 743[67], 746[67], 748[67,72], 749[78], 816[78]; **4**, 979[101]; **5**, 1123[42]
Claxton, E. E., **6**, 127[161]
Claxton, G. P., **2**, 943[173], 945[173]
Clayson, D. B., **8**, 963[48]
Clayton, F. J., **5**, 762[101]
Clayton, J. F., **3**, 254[101]
Clayton, J. O., **7**, 10[76]
Clayton, R. B., **6**, 1042[10]
Cleary, D. G., **1**, 767[175]; **3**, 226[199]; **4**, 593[134]; **5**, 301[78], 931[185], 934[185]; **6**, 10[44], 11[44], 12[44], 174[60], 187[175]
Cleary, M., **5**, 713[61]
Cleary, T. P., **2**, 553[126]
Clegg, J. M., **4**, 1101[191]
Clegg, W., **1**, 6[33], 33[165], 38[258], 39[187]; **2**, 371[262]; **5**, 468[121]; **8**, 335[67]
Cleghorn, H. P., **3**, 568[44]
Cleland, G. H., **3**, 889[24]
Clemens, A. H., **3**, 88[121]
Clemens, K. E., **5**, 595[10]; **6**, 569[935]
Clemens, R. J., **5**, 451[43], 485[43]
Clement, A., **3**, 892[47]
Clement, B., **3**, 154[421]
Clément, G., **8**, 396[137]
Clement, K. L., **6**, 450[118]
Clement, K. S., **2**, 710[18], 728[137]; **5**, 758[86], 759[86]; **6**, 1066[95]
Clement, R. A., **2**, 162[142]; **8**, 530[101]
Clement, W. H., **4**, 920[24]; **7**, 449[3], 450[3], 453[3]
Clementi, S., **2**, 735[12], 964[58]
Clements, G., **3**, 261[152], 514[211]
Clemo, N. G., **4**, 126[217c]
Clennan, E. L., **5**, 71[126], 72[177]
Cleophax, J., **1**, 561[166]; **4**, 36[102,102c]
Clerici, A., **1**, 272[66,66a-c]; **3**, 564[6,30,31], 566[30,31], 595[189], 606[189]; **8**, 113[43-45], 116[43]
Clerici, F., **8**, 72[241], 74[241]
Cleve, A., **1**, 733[21]
Cleve, C., **6**, 509[265]
Cleve, G., **7**, 773[305]
Cleveland, E. A., **3**, 499[111]
Cleveland, M. J., **6**, 245[124]
Clezy, P. S., **6**, 737[32]
Clibbens, D. A., **3**, 888[16]
Cliff, I. S., **2**, 757[12]
Clift, S. M., **1**, 807[316]; **5**, 1125[64]; **8**, 682[82]
Climent, M. S., **8**, 368[65]
Cline, J. K., **3**, 582[113]
Cline, R. E., **4**, 280[124]
Clineschmidt, B. V., **5**, 410[41]
Clinet, J.-C., **2**, 88[32]; **3**, 255[106], 257[116], 264[179], 491[72]; **4**, 76[48]; **5**, 1144[97]
Clinton, N. A., **4**, 697[11]
Clissold, D. W., **1**, 758[126]; **7**, 712[63]
Clive, D. L. J., **1**, 571[273], 630[4], 631[4], 656[146,147,149], 658[4], 672[4,203], 678[203], 699[247,254], 700[203,256,257,260], 825[57], 828[57], 882[123], 885[133c]; **2**, 76[85]; **3**, 86[45], 87[45], 94[45], 107[225,226,229], 109[226], 114[45], 117[45], 213[44], 219[110], 386[60], 421[61], 422[61], 589[161,162], 610[161,162]; **4**, 10[32], 183[80], 340[54], 364[1,1j], 365[4], 370[4], 372[55], 373[82], 380[4], 381[4], 386[148b], 387[148], 398[216], 399[216a], 401[216a], 405[216a], 410[216a], 733[79], 784[17], 785[20], 791[49,53,54,56], 815[190], 824[235]; **5**, 810[128], 812[128], 841[94]; **6**, 470[58], 980[31], 1026[81,86], 1027[81], 1028[94,97], 1029[81], 1030[81], 1031[81,86,94], 1033[81]; **7**, 119[2], 124[37], 128[37], 129[2,37], 146[37], 495[211], 522[40], 524[53], 772[289], 819[20], 826[20]; **8**, 847[97,97d], 848[97b], 849[97a,b,d,107,110,112,115], 887[114,117]
Clizbe, L. A., **3**, 816[83]; **6**, 533[496]
Cloez, C., **3**, 843[26]
Cloez, S., **6**, 243[98]
Cloke, F. G. N., **7**, 4[18]
Cloke, J. B., **5**, 945[252]
Clopton, J. C., **4**, 952[5]
Clos, N., **7**, 443[51b]; **8**, 459[228]
Closier, M. D., **7**, 479[96]
Closs, G. L., **4**, 959[32], 962[40], 966[52], 977[32]; **7**, 474[37], 883[177]
Closs, L. E., **4**, 962[40], 966[52]
Closse, A., **3**, 131[327]; **6**, 112[65]
Closson, R. D., **4**, 520[37]
Closson, W. D., **3**, 393[93], 394[95]; **6**, 687[379]
Clouet, F. L., **6**, 456[158]
Clough, J. M., **3**, 514[210]
Clough, R., **8**, 423[38], 428[38]
Clough, S., **4**, 1082[89]
Cloux, R., **2**, 630[8]
Clover, J. S., **4**, 1084[94]
Coad, J. R., **8**, 850[122]
Coard, L. C., **7**, 846[86]
Coates, C. E., **8**, 99[107]
Coates, G. A., **8**, 99[111]
Coates, G. E., **1**, 3[21], 21[111], 139[4]; **8**, 754[101]
Coates, I. H., **7**, 728[42]
Coates, R. M., **1**, 3[23], 359[15], 391[15], 506[14], 826[60], 861[72]; **2**, 105[45]; **3**, 8[44], 15[44], 714[34], 786[39]; **4**, 18[62], 21[62d], 189[106], 244[77], 255[77], 260[77]; **5**, 133[58], 636[89], 712[59], 854[178], 856[178], 872[178]; **6**, 133[2], 139[47,50], 707[43], 849[121], 860[177], 1061[74], 1063[82]; **7**, 124[50], 127[50], 186[181]; **8**, 527[44], 528[44,73,74], 993[61]
Coates, W. J., **7**, 32[91]
Coates, W. M., **7**, 272[133]
Cobb, R. L., **5**, 566[100]; **8**, 295[56,59]
Cobern, D., **8**, 545[281]
Coblens, K. E., **8**, 798[62], 800[62]

Cobler, H., **6**, 566[922]
Coburn, C. E., **2**, 763[59]; **7**, 160[49]
Coburn, J. I., **3**, 739[43]; **7**, 167[97]
Coccia, R., **8**, 880[63]
Cocevar, C., **8**, 450[167]
Cochand, C., **2**, 143[54]
Cochran, D., **4**, 957[22]
Cochran, D. W., **2**, 384[317]
Cochran, E. L., **4**, 719[22]
Cochran, J. C., **2**, 772[20]
Cockcroft, R. D., **5**, 829[25], 929[167]
Cocker, W., **4**, 280[127]; **7**, 92[44], 102[134]; **8**, 530[94]
Cockerill, A. F., **6**, 950[1], 951[4], 959[47], 1011[2]
Cockerill, G. S., **2**, 651[118,119]; **3**, 26[163]
Cockerille, F. O., **8**, 243[46]
Cocuzza, A. J., **6**, 147[83]; **7**, 684[94]
Coda, A. C., **2**, 364[204]; **5**, 454[70]
Coda, L., **6**, 255[168]
Codding, P. W., **7**, 340[46]
Coe, D. E., **7**, 494[203], 495[203,209]
Coe, J. W., **1**, 770[186]; **5**, 526[58], 539[58], 540[58], 541[58]
Coe, P. L., **3**, 522[11], 585[135], 639[84], 648[187]; **8**, 904[54]
Coelho, F., **3**, 222[141]
Coen, R., **2**, 838[179]
Coenen, J. W. E., **8**, 447[98]
Coerver, J. M., **3**, 352[43b]
Coffen, D. L., **3**, 134[336], 1038[92]; **7**, 516[6]; **8**, 337[81], 695[119]
Coffield, T. H., **4**, 520[37]
Coffin, R. L., **5**, 218[35], 219[40], 221[35,59]
Coffman, H., **3**, 790[61]
Coffman, K. J., **7**, 744[69]
Coghlan, M. J., **1**, 744[60], 745[60]; **4**, 380[122]; **5**, 210[56], 831[38], 1123[37]; **8**, 942[116]
Cogolli, P., **4**, 441[173,177]
Cohen, D., **4**, 710[53,57,59,63]; **5**, 736[145]; **6**, 612[73]; **7**, 875[119]
Cohen, G. M., **5**, 64[51], 70[107]
Cohen, H., **3**, 566[32]; **5**, 473[149]
Cohen, J., **2**, 223[148]
Cohen, J. F., **3**, 48[259]
Cohen, J. I., **3**, 567[37], 606[37]
Cohen, K. F., **8**, 544[261]
Cohen, L. A., **6**, 220[127], 734[7]
Cohen, M., **3**, 564[10]; **8**, 453[193], 551[343]
Cohen, M. J., **5**, 64[41]
Cohen, M. L., **1**, 636[96], 642[96]
Cohen, M. P., **6**, 182[138], 1021[51]; **8**, 384[27]
Cohen, N., **4**, 410[260b,c]; **5**, 893[41]; **6**, 718[119]; **7**, 346[11]; **8**, 237[11], 545[291]
Cohen, S. A., **5**, 1174[37]
Cohen, S. G., **3**, 564[11], 567[11,37], 606[37]; **6**, 3[14]
Cohen, T., **1**, 53[17], 135[116], 222[70], 239[39,40], 240[40], 786[248,249], 864[85,86], 887[139], 888[139]; **2**, 74[70], 107[57], 547[100], 548[100]; **3**, 88[128,129], 89[129], 103[204,205], 105[129,217], 108[204,205], 112[129], 124[128,253,258], 126[253], 196[28], 482[1,2], 483[1,2], 491[1,2], 499[1,2,114,117], 660[8], 753[103], 771[184], 785[36,36a], 937[74,75], 1051[20]; **4**, 71[20], 113[171], 116[185b,c], 243[71], 258[252], 259[258], 373[79], 374[79], 987[147], 992[157]; **5**, 333[44a], 455[75], 456[84], 1017[65], 1018[65,65a], 1020[65,65a,70,70c], 1023[77], 1027[70]; **6**, 140[57], 145[81], 147[86], 445[99], 893[75,80,89], 895[75]; **7**, 144[156], 202[50], 207[83], 208[83], 209[83], 210[83,95], 662[51], 720[8]; **8**, 842[47], 935[59], 970[100]
Cohen, T. C., **1**, 239[38,39]
Cohen, V. I., **6**, 473[74], 474[74,78-81], 475[94], 477[97], 479[74], 481[122]
Cohen, Z., **7**, 14[127,-128], 40[2,5,10], 842[24-26]
Cohn, H., **4**, 287[180]
Coillard, J., **3**, 831[60]
Coke, J. L., **4**, 189[103]; **6**, 954[13]
Coker, W. P., **8**, 898[21]
Colapret, J. A., **1**, 383[110], 883[124]; **2**, 479[17]; **5**, 806[104], 1029[91]; **6**, 722[137]
Colburn, C. B., **7**, 498[228]
Colby, T. H., **3**, 890[34]
Colclough, E., **1**, 499[53]
Colclough, T., **7**, 762[73]
Coldham, I., **1**, 570[269]; **6**, 25[100]
Cole, A. R., **4**, 299[300]
Cole, C. A., **7**, 85[12], 87[12]
Cole, E. R., **7**, 13[110], 338[38], 765[153]
Cole, L. L., **6**, 204[12]
Cole, P., **2**, 911[71]; **4**, 121[209]
Cole, T. E., **3**, 797[93]; **6**, 78[59]; **7**, 595[13], 606[159]; **8**, 289[26], 290[26], 371[115], 372[123], 373[123], 710[57,58], 717[57]
Cole, T. W., **3**, 855[80]
Cole, T. W., Jr., **5**, 123[5]
Cole, W., **2**, 838[175]; **6**, 959[43]
Coleman, D., **3**, 214[55]
Coleman, D. T., III, **3**, 824[19]; **8**, 86[23]
Coleman, G. H., **8**, 300[88]
Coleman, H. A., **5**, 797[60]
Coleman, J. P., **3**, 634[24], 639[24], 649[24]; **7**, 8[56]; **8**, 133[23,25], 236[5], 240[5], 242[5], 248[5], 249[5], 292[43]
Coleman, R. A., **7**, 596[37]; **8**, 715[85,85b], 716[85]
Coleman, R. S., **4**, 797[105]; **5**, 492[238,242,243,244], 498[238], 736[145], 737[145]; **7**, 34[100], 543[23], 544[23]; **8**, 618[114]
Coleman, W. E., **5**, 385[129b], 386[129b]
Colens, A., **6**, 493[129]
Coles, J. A., **5**, 752[48], 754[48], 756[48], 757[48], 758[48], 769[48]
Coleson, K. M., **8**, 458[224]
Coletti-Previero, M. A., **2**, 1104[133]; **6**, 423[48]; **8**, 52[141], 66[141]
Colgan, D., **1**, 17[210]
Colin, G., **2**, 609[83]; **8**, 547[316], 548[319]
Coll, G., **7**, 334[27], 346[8]
Coll, J., **5**, 36[18], 57[54]; **6**, 172[18]; **7**, 123[35], 144[35]; **8**, 500[50], 515[119], 563[425]
Coll, J. C., **1**, 858[62]; **7**, 87[20]
Collard, J. N., **1**, 219[58]
Collard-Charon, C., **6**, 472[69], 478[103]
Colleluori, J. R., **1**, 425[105]
Collerette, J., **2**, 213[126]
Colleuille, Y., **8**, 462[264], 535[166]
Collie, J. N., **2**, 170[170]
Collienne, R., **6**, 472[70]
Collier, T. L., **3**, 35[206]; **4**, 200[2]; **6**, 719[126], 725[126]; **8**, 159[9], 720[135]
Collignon, N., **1**, 788[257]; **2**, 482[27], 483[27]; **3**, 201[83]; **4**, 459[77], 473[77], 474[77]
Collin, J., **1**, 253[10,11], 255[19], 258[19,26,26b], 259[31], 271[19], 273[70], 274[75]; **8**, 552[360]
Collington, E. W., **1**, 366[51], 570[269], 741[45], 779[225]; **2**, 765[75]; **5**, 432[130], 433[130b]; **6**, 205[36]
Collingwood, S. P., **6**, 20[72]
Collins, A., **5**, 647[12], 648[12]
Collins, C. H., **4**, 272[41], 280[41]; **6**, 1033[129]
Collins, C. J., **3**, 722[6], 723[6,7], 724[12], 726[22], 731[6,7], 782[20], 822[8,9], 823[9], 825[29], 830[58], 831[58], 834[9], 836[8]; **6**, 3[12]
Collins, D. J., **3**, 762[147]; **4**, 609[327], 614[327], 615[327,391], 629[327,391]
Collins, J. C., **2**, 758[25]; **7**, 100[131], 256[42]

Collins, J. J., **5**, 735[139]; **6**, 866[205]
Collins, M. A., **6**, 736[28]
Collins, P., **4**, 113[176]
Collins, P. M., **2**, 385[327]; **3**, 597[199]; **8**, 817[33]
Collins, P. W., **4**, 143[20]; **8**, 544[264], 756[159], 758[159]
Collins, S., **1**, 180[42], 181[42]; **4**, 176[52], 341[59]; **6**, 467[47,48], 1028[96]; **7**, 523[44], 771[280], 773[280]; **8**, 410[98]
Collins, T. J., **3**, 213[51]
Collins Thompson, S., **7**, 347[17], 355[17]
Collman, J. P., **1**, 439[163,164], 440[167,171], 457[163]; **2**, 148[81]; **3**, 208[3], 213[3b], 1024[27], 1028[51]; **4**, 115[177], 518[1], 545[127], 547[1], 895[160]; **5**, 46[39], 56[39], 1065[1], 1066[1], 1074[1], 1083[1], 1084[1], 1093[1], 1112[1g], 1144[102], 1146[102], 1163[3], 1183[3]; **7**, 12[94], 107[167]; **8**, 289[25], 421[28], 422[28], 432[28], 435[28], 436[28], 550[334]
Collonges, F., **4**, 898[175]
Collum, D. B., **1**, 28[143], 29[144], 34[169], 336[70], 427[113]; **2**, 9[39], 31[39], 194[67], 507[20,26,27], 508[27,32], 509[20]; **3**, 33[189], 34[198,199], 39[198,199]; **4**, 380[120]; **6**, 724[156], 727[195,196,197,198], 879[44]; **8**, 851[135], 856[135b], 949[155]
Colobert, F., **4**, 651[428]; **5**, 974[16]
Colombo, L., **1**, 72[72], 524[86,87], 526[100], 527[101,102,107], 528[108]; **2**, 103[28], 221[146], 266[61,62], 267[62], 488[43], 514[54], 515[55,56], 605[57], 614[57], 630[19], 631[19], 636[19,56], 637[19,56], 640[56], 642[73], 643[73], 644[73], 930[131], 931[131]; **4**, 113[166], 159[82], 218[145], 226[187,188]; **6**, 118[105], 149[96,108]; **7**, 128[63], 441[45]
Colomer, E., **3**, 381[25], 382[25]; **8**, 766[20], 797[38]
Colomer Gasquez, E., **4**, 315[512]
Colon, C., **1**, 367[54]; **4**, 436[145], 437[145], 438[150-152]; **6**, 545[635]; **7**, 488[155], 490[155]
Colon, I., **3**, 500[130], 509[130]; **8**, 795[22], 906[64]
Colonge, J., **2**, 147[74], 286[61], 534[32]; **3**, 315[110]; **4**, 191[114]; **8**, 227[117]
Colonna, F. P., **4**, 20[63], 21[63]; **5**, 464[115]; **6**, 709[55], 710[57-59], 711[62]
Colonna, S., **1**, 828[70]; **2**, 435[60,62,65]; **4**, 37[104], 230[249], 231[249,268]; **5**, 99[131]; **6**, 105[11], 149[97], 150[115]; **7**, 194[9], 429[150a], 764[112], 767[112,193], 771[261], 772[286], 777[382], 778[411]; **8**, 170[73,75,76], 409[86], 412[86]
Colquhoun, H. M., **3**, 1017[8]
Colstee, J. H., **7**, 763[96]; **8**, 405[31]
Colter, A. K., **8**, 93[71]
Colter, M. A., **7**, 608[172]
Colthup, E. C., **5**, 1146[106]
Colthup, E. S., **5**, 1148[121]
Colton, C. D., **3**, 380[11]
Colton, F. B., **8**, 530[105], 561[408]
Colvin, E., **7**, 177[147], 816[12]
Colvin, E. W., **1**, 34[167], 544[30], 569[259], 580[1], 722[276], 731[4], 815[4,4f], 822[38]; **2**, 321[12], 323[22], 324[12], 326[12], 329[12], 335[71], 564[5], 629[3], 630[3], 633[3], 636[52], 640[52], 935[145], 937[145,158], 938[145,158], 940[145]; **3**, 200[67]; **4**, 78[52a], 155[65], 191[112], 681[53], 1088[123]; **5**, 102[179], 777[189]; **6**, 107[24], 676[306], 682[338], 760[143], 911[16]; **7**, 550[48], 671[11]; **8**, 363[1], 374[1,145], 785[114], 888[125]
Colwell, B. L., **2**, 282[34], 286[34], 287[34]
Comar, D., **4**, 445[205]; **8**, 35[64], 52[145], 53[128], 66[64,128,145]
Comasseto, J. V., **1**, 630[11], 636[101,102], 640[101], 641[102,106,107], 666[101], 669[11], 672[102,106,107], 677[102,106,107], 724[11,102,106]; **3**, 120[239,241], 136[374], 141[374]; **4**, 120[196], 370[34], 372[56], 443[192], 447[192]; **5**, 268[78]; **6**, 175[74]; **7**, 775[352a,b]; **8**, 411[101], 412[101,116], 849[116]
Comb, D. G., **2**, 466[115], 467[115]
Combe, M. G., **8**, 533[154]
Combellas, C., **4**, 453[31], 459[31,80,81,86], 469[80,81,86,136], 471[31,141], 475[150]
Comber, R. N., **5**, 1032[98]
Combes, D., **8**, 52[137], 66[137]
Combret, J.-C., **2**, 414[16], 415[16]; **3**, 242[9], 759[129]; **6**, 705[26]; **8**, 267[71]
Combrisson, S., **6**, 713[80b]
Combs, D. W., **4**, 111[154e]
Combs, L. L., **5**, 117[277]
Comer, F., **6**, 1017[38], 1024[38]
Comer, F. W., **2**, 170[174]
Comes, R. A., **5**, 960[321]
Cometti, G., **3**, 1030[57]
Comfort, D. R., **2**, 956[10]
Comi, R., **6**, 705[25]; **7**, 248[111], 801[44]
Comins, D. L., **1**, 212[16], 213[16], 418[71], 422[95], 463[30,31], 466[44]; **3**, 255[109], 512[198]; **4**, 428[73,74]; **7**, 360[22]; **8**, 586[29], 589[29]
Comisso, G., **2**, 406[47]; **8**, 152[169], 459[228], 535[165]
Comita, P. B., **5**, 736[144]
Commercon, A., **3**, 246[34], 247[46], 440[42], 441[42], 442[42], 473[217], 476[217], 485[31], 486[31], 522[18,19]; **4**, 900[179], 903[195]; **6**, 849[120]
Commeyras, A., **1**, 212[4]; **2**, 709[10]
Compagnini, A., **6**, 178[121], 508[289,290]
Compagnon, P. L., **4**, 459[79], 476[79]; **8**, 137[53]
Compernolle, F., **8**, 528[76,77]
Compos, P. J., **7**, 93[54]
Comte, M., **1**, 862[75b]
Comte, M.-T., **3**, 767[160]
Conacher, H. B. S., **3**, 752[96]
Conant, J. B., **2**, 140[35]; **3**, 565[18]; **8**, 531[124]
Conant, R., **7**, 246[92]
Conaway, R., **4**, 288[188]
Concalves, D. C. R. G., **7**, 207[74]
Concannon, P. W., **3**, 380[8]
Concellón, J. M., **1**, 830[94]; **3**, 788[50]
Concepcion, J. I., **7**, 41[15], 722[19], 723[19], 725[19]
Condon, B. D., **1**, 507[19]
Condon, S., **7**, 822[33]
Condulis, N., **8**, 639[18]
Coneely, J. A., **5**, 916[119]
Conesa, J., **5**, 1062[59]
Confalone, D. L., **4**, 31[92,92j]
Confalone, P. N., **1**, 568[235]; **4**, 31[92,92j], 413[277], 1076[49], 1078[49], 1080[71], 1086[116]; **5**, 249[30,31], 257[59,59a]; **6**, 764[12]; **7**, 347[14], 691[16], 701[66]; **8**, 608[46], 839[27], 8 82[87]
Công-Danh, N., **3**, 462[152], 513[207]
Conger, J. L., **5**, 718[95]
Congson, L. N., **3**, 714[33]
Conia, J. M., **1**, 879[111c-e]; **2**, 152[102]; **3**, 4[24], 20[104], 727[32], 765[153,154], 767[162], 832[68b], 842[17], 848[17,48]; **4**, 240[42], 905[210], 972[82], 1005[82], 1010[156], 1018[82]; **5**, 2[141], 15[105], 17[116], 20[141], 21[141,142,143,146,147,151,152,153], 22[151,152,153], 65[58], 461[103], 776[182], 789[29], 790[35], 796[55], 904[51,52], 910[89], 920[131,132]; **6**, 677[314]; **7**, 121[26], 168[103,103b], 825[44], 833[77]; **8**, 851[129]
Conley, D. L., **2**, 466[110]
Conley, R. A., **7**, 829[59]
Conley, R. T., **6**, 268[68], 279[134,135,136], 818[105], 1066[91]; **7**, 699[54,55]
Conlin, R. T., **5**, 65[71]
Conlon, D. A., **5**, 913[101], 1014[58]
Conlon, D. M. A., **2**, 348[64]
Conlon, L. E., **5**, 830[29]

Conn, E. E., **8**, 79[1]
Conn, R. E., **1**, 400[11]; **2**, 113[107], 254[43]
Conn, R. S. E., **3**, 244[26], 485[28]; **4**, 230[250], 879[84]; **5**, 7[54], 8[54], 519[35]
Connell, R., **4**, 1033[30]
Connelly, N. G., **4**, 689[72], 691[74]
Conner, D. T., **8**, 50[118], 66[118]
Connet, P. H., **7**, 839[2]
Connon, H., **8**, 224[111]
Connor, D. S., **3**, 422[70]; **5**, 1185[2]
Connor, D. T., **7**, 198[27]
Connor, R., **4**, 3[7,7a], 4[7]; **8**, 140[25], 141[25], 142[25], 212[12], 533[139], 814[18]
Connors, W. J., **3**, 665[35]
Conover, L. H., **7**, 157[33]; **8**, 314[25], 974[122]
Conrad, F., **2**, 323[24]; **7**, 500[240]
Conrad, M., **2**, 352[87], 357[87]
Conrad, N. D., **5**, 856[201,203]
Conrad, P., **3**, 960[116]; **5**, 894[48]; **6**, 898[107b]
Conrad, P. C., **4**, 251[150]
Conrad, R. C., **7**, 759[9]
Conrad, T. T., **5**, 154[33]
Conrad, W. E., **6**, 253[158], 667[238]
Conrow, K., **8**, 91[63]
Conrow, R. B., **7**, 27[71]; **8**, 527[41]
Conrow, R. E., **2**, 578[85]
Conroy, H., **8**, 895[4], 899[4]
Consiglio, G., **3**, 228[222], 229[230,230a], 246[36], 438[29], 452[110], 487[45], 1023[22]; **4**, 403[239], 404[239], 919[18,19], 926[37], 927[42], 930[42,45,48,53], 931[48,56-58], 932[64], 936[69], 939[76], 945[42,90]; **6**, 831[10], 832[10], 848[10]
Consonni, A., **2**, 783[35]
Constable, A. G., **7**, 630[51]
Constable, E. C., **4**, 837[18]
Constantinides, D., **4**, 16[51]; **6**, 1022[61]
Constantino, M., **3**, 592[172]
Constantino, M. G., **4**, 300[310], 304[361], 305[361]; **6**, 188[181]
Conte, J. S., **4**, 280[123]
Conte, V., **7**, 777[376]
Contelles, J. L. M., **1**, 766[156]
Contento, M., **2**, 187[43], 635[48], 640[48], 937[156]; **4**, 344[78a]; **5**, 100[156], 102[173]; **8**, 124[90]
Conti, F., **4**, 608[323], 839[26]; **7**, 452[61]; **8**, 443[1]
Conti, P. G. M., **8**, 33[58], 66[58]
Contreras, L., **6**, 561[876]
Contreras, R., **8**, 16[101], 54[156], 66[156], 237[14], 240[14], 244[14], 537[180], 708[41]
Conway, B. E., **3**, 634[6], 636[6,56,57], 637[6,61], 639[56b,77], 655[6]
Conway, D. C., **3**, 299[33]
Conway, P., **3**, 933[61]; **8**, 93[76]
Conway, R., **4**, 346[86a]
Conway, T. T., **4**, 398[216], 399[216d]
Conway, W. P., **6**, 1021[50]
Cook, A. G., **1**, 366[43]; **2**, 865[2]; **4**, 6[21], 1004[73]; **5**, 676[4]; **6**, 704[11]; **8**, 40[89], 66[89]
Cook, A. H., **8**, 328[8], 639[21]
Cook, B. R., **7**, 50[71,72]
Cook, C., **6**, 108[34]
Cook, C.-H., **4**, 357[150]
Cook, F., **1**, 377[97]; **4**, 115[184b]; **6**, 727[206]; **8**, 943[120]
Cook, J. M., **1**, 262[40]; **2**, 381[307], 1016[27], 1017[31,34,35], 1018[35]; **3**, 380[7]; **5**, 864[262,263]; **6**, 737[31,37,41,42], 738[42,43], 746[89]; **7**, 340[46], 544[39], 553[39], 556[39]
Cook, L. C., **3**, 407[150]
Cook, L. S., **6**, 546[654]
Cook, M. J., **7**, 228[102], 662[53,55]; **8**, 296[61]
Cook, N. C., **4**, 272[38,39], 273[38,39], 287[38,39]
Cook, P. L., **8**, 320[85], 533[148]
Cook, P. M., **5**, 137[77]; **6**, 1016[27]
Cook, R. M., **2**, 381[308]
Cook, S. J., **3**, 1022[19]; **6**, 51[106]
Cook, W. J., **7**, 841[18]
Cooke, B., **8**, 872[10], 873[10]
Cooke, B. J. A., **5**, 753[53], 754[53,70], 759[70]
Cooke, D. W., **8**, 533[146]
Cooke, F., **1**, 622[73,74], 737[31], 828[79]; **2**, 426[39], 713[46]; **3**, 198[51]; **4**, 258[251], 989[142]; **5**, 777[190,191]
Cooke, M. D., **4**, 509[162], 980[102]
Cooke, M. P., Jr., **1**, 264[43], 411[48], 492[41], 493[41], 495[41]; **2**, 107[56]; **3**, 877[87], 1021[17], 1024[28]; **4**, 35[98h], 72[28], 73[34], 74[38a,39a-c,40a-c], 75[42a], 115[178], 239[22], 243[70,72], 247[70,99], 253[174], 257[70,72], 260[72,99]; **6**, 185[161], 189[191], 190[161,192,194], 954[13]; **8**, 863[239]
Cooke, R., **3**, 419[33]
Cooke, R. G., **3**, 828[52,53]
Cooks, R. G., **2**, 807[49]; **8**, 629[186]
Cookson, C. M., **3**, 616[12]
Cookson, R. C., **2**, 530[22], 547[116], 551[116], 578[88], 587[88], 720[86]; **3**, 135[350], 136[350], 137[350], 141[350], 349[34], 623[37], 625[37]; **4**, 35[98j], 44[125], 395[205], 893[157]; **5**, 76[243], 123[4], 178[138], 221[61], 428[108], 429[112], 594[3], 595[3], 596[3], 597[29], 603[3], 618[1], 791[27], 799[27], 802[86], 882[15], 883[17]; **6**, 139[52], 836[56], 841[74], 1022[60]; **7**, 219[15]; **8**, 844[68]
Coolbaugh, T. S., **4**, 576[16]
Cooley, J. H., **4**, 12[42]; **6**, 70[20]
Cooley, N. A., **1**, 808[321]
Coolidge, M. B., **6**, 581[991]
Coombes, R. G., **2**, 321[11], 329[11]; **6**, 104[1,9]
Coombs, R. V., **2**, 184[26]; **3**, 2[8], 11[8], 16[8], 17[8], 26[8], 653[226]; **4**, 240[39], 254[39]; **8**, 527[40]
Coombs, W., **6**, 927[70]
Coon, M. J., **7**, 80[138]
Cooney, E., **4**, 394[193]
Cooney, J. V., **1**, 559[152]; **6**, 176[100]
Cooney, K. E., **5**, 835[59]
Cooper, C. F., **1**, 832[110]; **4**, 953[8,8i], 954[8i], 1031[5], 1032[5], 1035[5], 1102[198]
Cooper, C. M., **6**, 1017[38], 1024[38]
Cooper, C. S., **6**, 517[328,329]; **8**, 190[75]
Cooper, D., **5**, 847[135]; **8**, 775[77]
Cooper, D. K., **5**, 514[9], 527[9]
Cooper, E. L., **6**, 653[151]
Cooper, G. D., **3**, 866[32]; **8**, 751[64,65]
Cooper, G. F., **5**, 768[125], 779[125]
Cooper, G. M., **8**, 532[130]
Cooper, G. R., **4**, 1002[61]
Cooper, J., **5**, 843[124]; **6**, 859[173]; **8**, 196[122]
Cooper, J. L., **3**, 343[17], 348[27], 353[48], 355[17,48], 357[17,48], 358[68]
Cooper, J. W., **4**, 719[27]
Cooper, K., **1**, 391[154], 392[154]; **3**, 218[100,100b], 229[234], 444[66]
Cooper, L., **2**, 917[89], 919[89], 920[89], 924[89], 935[89]; **5**, 100[147]
Cooper, M. M., **1**, 473[80]
Cooper, M. S., **2**, 962[45], 964[45]; **7**, 194[4], 374[78], 674[41]
Cooper, M. W., **4**, 55[157], 57[157b]
Cooper, N. J., **4**, 696[7], 712[7b]
Cooper, P. J., **3**, 352[43b]; **8**, 724[170]
Cooper, P. N., **7**, 534[37]
Cooper, R., **3**, 693[144]
Cooper, R. D. G., **5**, 96[103], 98[103,129], 99[103]
Cooper, W. D., **2**, 739[43a]

Coote, S. J., **2**, 315[45], 316[45]

Cope, A. C., **1**, 709[262]; **2**, 343[9], 366[219,220], 1024[57,58]; **3**, 379[1,3], 380[1,5,13,16], 381[16a,17], 382[3b], 393[93], 414[1], 735[17], 760[139], 774[139], 856[87]; **5**, 552[24], 789[28], 791[37], 827[3], 834[3], 1146[110]; **6**, 734[4], 960[61], 961[65], 1012[4], 1013[4,15], 1014[18,20], 1015[21,24], 1016[25]; **7**, 92[42], 93[42]; **8**, 228[123], 251[105], 726[190], 957[9]

Copenhaver, J. E., **6**, 204[10]

Copinschi, G., **7**, 704[11]

Copland, D., **5**, 626[38], 629[38]

Copley, D. J., **8**, 112[25], 119[25]

Copley, S. D., **5**, 854[177], 855[185,188]

Copp, F. C., **8**, 964[57]

Copp, L. J., **3**, 217[92]

Coppi, L., **1**, 612[48]; **2**, 566[22], 567[26]

Coppinger, G. M., **7**, 84[1], 85[1], 108[1]

Coppola, B. P., **2**, 567[30]

Coppola, G. M., **2**, 81[1], 82[1], 96[1]; **3**, 223[146]; **4**, 53[151], 1010[150]; **6**, 830[4], 873[10]

Coppolino, A., **2**, 162[144]; **4**, 7[26]

Coppolino, A. P., **7**, 506[300]

Corain, B., **2**, 369[245]

Corbani, F., **7**, 170[121]

Corbeil, J., **2**, 656[158,159], 1059[77]

Corbel, B., **3**, 167[486], 168[486], 199[55], 257[117], 1052[27]; **6**, 126[150]

Corbelin, S., **1**, 34[224], 36[174]

Corbett, D. F., **2**, 758[21]; **4**, 50[142]

Corbett, J. F., **3**, 829[55]

Corbett, W. L., **5**, 474[156,157]

Corbett, W. M., **7**, 760[28]

Corbier, B., **7**, 64[60]

Corbin, U., **8**, 645[40]

Corbin, V. L., **3**, 419[43]

Corby, N. S., **6**, 607[46]

Corcoran, D. E., **2**, 442[9], 447[9,33]

Corcoran, J. W., **6**, 440[76]

Corcoran, R. J., **7**, 43[47], 46[48-50], 47[50,52,56]

Cordell, G. A., **3**, 682[165], 691[133]; **6**, 737[39]

Cordes, A. W., **8**, 435[71]

Cordes, E. H., **6**, 556[23], 563[23]

Cordiner, B. G., **4**, 6[23]

Cordova, R., **2**, 527[9], 528[9], 531[24,28], 533[24,30], 537[24]; **4**, 904[206]; **5**, 2[7], 4[7], 433[137c], 435[137c]; **6**, 894[90]

Core, M. T., **5**, 960[321]

Coreil, M., **3**, 509[179]

Corey, E., **6**, 438[64]

Corey, E. J., **1**, 87[48], 112[25,27], 223[82], 224[82], 227[97], 237[33], 268[56], 272[66], 273[66d], 313[119,120], 314[119,120], 317[143,144,151,152], 319[144,151,152], 357[6], 359[25], 364[25], 378[6], 386[122], 408[3]; **2**, 47[154], 72[57], 76[57], 91[45], 120[172,173,174,182], 156[118], 158[126], 161[136], 182[11], 228[163], 249[84], 259[83], 264[83], 355[120], 356[120], 373[273], 448[42], 482[31], 484[31], 495[66], 496[66], 497[66], 504[3], 506[14], 507[21], 509[3,35], 511[14,21,45], 512[21], 517[21], 518[35,63], 524[80], 540[72], 542[72], 564[9], 588[150,152], 599[18], 600[33], 605[33], 638[61], 640[61], 800[16], 809[54], 824[54], 839[172], 843[194], 866[5], 1100[117]; **3**, 3[9], 12[64], 17[85], 30[180], 34[180], 35[180,200], 86[26,59], 88[59,125], 89[59], 103[203,206], 108[203], 112[59], 118[125], 120[240], 121[26], 123[59,125], 124[265,266,274,275,278], 125[59,266,274,275,278,292], 126[292], 127[265,266,274], 128[265,266,278], 129[274,278], 131[326], 132[278], 134[338], 135[326,338], 137[338], 147[391], 158[391], 178[541,545], 181[545], 194[8], 201[80], 212[37,38], 213[50], 215[61,67], 216[70], 217[81,84,93], 220[126], 235[81b], 247[49,50], 248[55], 249[64], 250[49,69,70], 251[55], 257[121], 269[55], 274[20], 288[63], 356[59], 361[79], 365[96], 419[44], 420[44], 421[56], 427[85,88], 430[92-94], 431[99-102], 546[123], 570[52-54], 572[52,53,69], 573[53,69,72], 575[53], 583[53], 596[53,72], 599[72], 602[69,72], 607[69,72], 610[52,53,69], 638[90], 649[202], 650[202c], 652[202c], 709[16], 712[25], 713[29], 729[40], 744[63], 762[146], 778[2], 909[150], 918[21], 936[73]; **4**, 5[17], 12[41], 14[46,46a], 21[69], 76[46], 85[77b], 91[88c], 108[146g], 113[46], 115[179a], 129[223a], 143[18], 152[57], 171[29], 172[30], 173[36], 174[39], 176[45,47], 183[79], 185[89], 218[139,140], 229[231], 239[21], 248[110], 252[165], 256[21], 261[21], 262[110], 315[501], 331[16], 349[109], 370[26,39,43], 372[64a], 373[69,70], 377[100], 378[109], 384[143], 390[172], 487[41], 731[69], 738[93], 763[213,214], 798[110], 807[151], 809[160], 964[46], 971[73,74], 976[98], 987[134], 988[134,138], 989[141], 1033[35], 1040[70-72,85,96], 1041[96], 1043[71], 1044[70,72], 1045[85], 1048[96]; **5**, 90[53], 124[6,10], 125[6], 128[6], 130[10], 320[9], 330[38], 339[56], 347[56], 353[86], 356[88], 365[96b], 377[110,110b], 378[110b], 404[18], 429[113a], 439[169], 514[7], 522[46], 543[115], 560[77], 569[111], 738[147], 755[71], 780[71,204], 829[25], 830[37], 837[69], 843[125], 853[125a], 890[35], 910[88], 957[308], 1014[56]; **6**, 2[8], 8[36], 22[8], 76[42], 107[28], 108[37], 134[9,14], 135[9,23], 138[44], 139[45,46], 172[17], 174[60], 175[67], 438[59,66], 452[136], 561[880], 609[55], 632[6], 647[111], 648[111,118], 652[144], 655[156], 656[170], 657[173], 659[144], 660[202], 662[215], 666[247], 673[291,292], 674[292,293,294], 677[316,319], 678[324], 679[326,329], 680[329b], 682[341], 686[372], 727[193], 728[214,215], 784[91], 830[5], 837[60], 848[109], 859[175], 864[198], 865[200], 866[206], 903[134], 927[72], 960[53], 982[49,50], 983[52], 998[118], 1043[12], 1047[34], 1059[67]; **7**, 101[133], 103[139], 104[145], 120[11], 127[62], 180[155], 182[161], 194[10], 197[18], 218[6], 228[93,95], 260[63,65,85,86], 263[83], 272[131], 273[131], 278[165], 279[165], 292[10], 297[29], 298[38], 318[59], 358[4,11], 363[35], 373[75], 376[85,89], 378[95], 400[50], 418[128], 419[134a], 420[137], 438[20], 445[20], 501[250], 516[5], 534[38], 566[100], 620[27], 633[63], 634[67], 677[66], 678[73], 680[77], 686[100], 711[60], 737[16], 752[146], 768[203], 821[30], 823[39], 824[41], 831[67]; **8**, 111[23], 117[23,72], 160[102], 163[40], 171[102-106], 178[102], 179[102], 269[80,81], 319[74], 374[147,148], 384[24], 388[60], 393[111], 395[128], 448[148], 472[2], 473[2], 531[125], 537[185], 543[250], 615[95], 657[96], 725[180], 727[198], 754[75,76], 800[66], 824[61], 837[13b], 844[70,72,72c], 881[68], 885[103], 891[149], 986[11]

Corey, G. C., **7**, 706[21]

Corey, M. D., **7**, 235[3]

Corey, P., **5**, 1096[110], 1098[110]

Corey, P. F., **4**, 391[185], 393[185]

Corfield, A. P., **2**, 463[86]

Coria, J. M., **4**, 189[103]

Cork, D. G., **4**, 354[130]; **6**, 939[145,146], 941[146], 942[146]; **7**, 844[55]; **8**, 86[26]

Corkins, H. G., **2**, 418[22]; **3**, 871[51]; **4**, 115[184d]; **6**, 787[97]

Corley, E., **4**, 767[234]

Corley, E. G., **3**, 380[11]; **7**, 493[188]

Cormons, A., **4**, 395[205], 396[208], 397[208], 411[267a]

Corn, J. E., **4**, 270[13]

Cornejo, J., **7**, 228[101], 845[77]

Cornelis, A., **1**, 564[187,188]; **3**, 125[303], 322[144]; **6**, 111[60,63]; **7**, 760[25], 846[87-90,92,96,98,100]

Cornelisse, J., **5**, 647[14,18,19], 649[14,18,22], 650[19,22], 652[18,19], 653[19], 656[18,19], 707[40]

Cornell, S. C., **6**, 291[223]

Cornet, D., **8**, 142[48]

Cornforth, J., **3**, 502[141]

Cornforth, J. W., **1**, 49[4], 50[4], 153[62]; **2**, 24[96], 396[8]; **6**, 980[32]; **7**, 272[132], 477[70]; **8**, 3[17], 93[76], 532[128], 957[16]

Cornforth, R. H., **1**, 49[4], 50[4], 153[62]; **2**, 24[96]; **6**, 980[32]; **7**, 272[132]; **8**, 3[17]

Cornil, A., **7**, 704[11]
Cornish, A. J., **8**, 556[372], 779[89], 780[93,94]
Cornubert, R., **2**, 148[82]; **8**, 143[57]
Cornwall, P., **1**, 477[139]
Correa, A. G., **7**, 355[39]
Correa, I. D., **4**, 331[18]
Correa, P. E., **7**, 748[115], 765[146], 877[130]
Correia, C. R. D., **5**, 145[108], 639[125], 641[130], 1026[85]
Correia, V. R., **8**, 412[116]
Corrie, J. E. T., **5**, 419[74], 576[146]; **6**, 664[219]
Corrigan, J. R., **7**, 272[133]
Corriol, C., **6**, 563[892]
Corriu, R. J. P., **1**, 113[31], 461[12], 619[59], 623[81]; **2**, 58[11], 572[43]; **3**, 228[215], 436[4], 484[26], 492[26], 494[26], 495[26], 503[26], 513[26]; **4**, 100[126], 120[197], 248[112], 630[418]; **8**, 246[79], 262[18,19], 265[19], 546[311], 766[20], 797[38]
Corsano, S., **1**, 754[107]; **2**, 363[195]; **4**, 308[404]; **7**, 112[196], 732[59]; **8**, 244[58], 248[58]
Corsaro, A., **6**, 178[121], 508[289,290], 509[271]
Corse, J., **4**, 288[186]
Corset, J., **1**, 34[226,227], 41[200]
Cortes, D., **4**, 537[95]
Cortes, D. A., **7**, 308[20], 809[82]
Cortés, M., **7**, 90[33]
Cortese, F., **6**, 209[65]
Cortese, N., **4**, 845[68], 846[73], 847[68,73], 848[73], 855[97]
Cortese, N. A., **7**, 691[20]; **8**, 368[64], 557[386,387], 902[46], 904[46], 907[46]
Cortez, C., **1**, 466[47]; **7**, 296[26], 346[7]; **8**, 19[132]
Cortez, H. V., **8**, 807[119]
Corwin, A. H., **8**, 604[1]
Cory, R. M., **4**, 18[59], 121[208], 262[312], 991[151], 992[155], 993[161]; **5**, 249[33]; **6**, 176[92]
Coscia, C. J., **8**, 527[41]
Cose, R. W. C., **3**, 949[95]; **6**, 897[95]
Cossais, F., **2**, 816[84], 828[84]
Cossar, B. C., **2**, 914[80]
Cossec, B., **3**, 46[254]
Cossentini, M., **1**, 683[219]; **4**, 71[16a], 139[3]
Cossey, A. L., **6**, 489[98]; **8**, 503[70]
Cossío, F. P., **2**, 649[102], 1059[75]; **5**, 94[87], 95[90,98,99], 96[109], 100[146]; **7**, 275[145], 277[153,154], 554[64,65]
Cossu-Jouve, M., **2**, 138[20]
Cossy, J., **1**, 268[55], 269[55]; **3**, 176[540], 281[42], 599[209,210], 602[223], 603[228]; **4**, 48[139], 809[163]; **5**, 310[100]; **6**, 86[99], 161[179]
Costa, A., **4**, 111[155d]; **6**, 89[110]; **7**, 334[27], 346[8]
Costa, M., **5**, 925[155], 1135[50], 1137[55]
Costa, P. R. R., **2**, 744[96]
Costa, T., **6**, 190[199], 196[199,228,232,233]
Costall, B., **2**, 760[43]
Costantino, P., **3**, 741[49]
Costanzo, S. J., **3**, 260[144]
Coste, J., **3**, 740[47,48]
Costello, F., **4**, 272[40]
Costello, G., **3**, 444[67]; **4**, 878[83]
Costero, A. M., **4**, 492[73]; **5**, 474[158]; **6**, 172[19]
Costisella, B., **2**, 60[18]; **3**, 953[105]; **6**, 134[36]; **7**, 197[22]
Cota, D. J., **4**, 292[229]; **6**, 261[12], 263[12], 264[12], 267[12]; **7**, 9[71]
Côté, J., **3**, 846[43]
Cote, R., **7**, 503[281]
Cotelle, P., **6**, 897[100]
Cotsaris, E., **8**, 496[30]
Cott, W. J., **6**, 228[32]
Cottam, P. D., **6**, 1024[77]
Cottens, S., **4**, 1020[237], 1035[42]
Cotter, R. J., **3**, 393[93]
Cottier, L., **5**, 1007[36]
Cottineau, F., **6**, 70[20]
Cotton, F. A., **1**, 193[88], 416[66], 422[94]; **4**, 104[135b]; **5**, 715[77]; **7**, 844[59]
Cotton, W. D., **1**, 116[48], 118[48]
Cottrell, C. E., **1**, 885[134]; **3**, 163[472]; **5**, 210[56], 815[143], 817[150]
Cottrell, D. M., **5**, 203[39,39a,b,d], 209[39], 210[39]
Cottrell, P. T., **7**, 769[218]
Coudane, H., **3**, 754[109]; **4**, 98[108d]
Coudert, G., **5**, 1022[75]; **6**, 648[120]
Couffignal, R., **1**, 218[51], 273[71]; **2**, 286[66], 799[20]
Coughlan, M. J., **7**, 655[20]
Coughlin, D. J., **8**, 513[104]
Couladouros, E. A., **6**, 448[107]
Coulentianos, C., **4**, 900[179]
Coulombeau, A., **3**, 564[7]; **8**, 527[42]
Coulson, C. A., **5**, 900[3]
Coulson, D. R., **3**, 48[259]
Coulston, K. J., **7**, 827[48]
Coulter, A. W., **3**, 901[112]
Coulter, J. M., **8**, 813[9], 915[96], 939[95]
Coulter, M. J., **1**, 520[76], 521[76], 522[76,78]; **2**, 75[81]; **4**, 12[37,37f], 119[194], 226[198,199]; **5**, 1053[41], 1061[41]; **6**, 154[146]
Coulter, P. B., **7**, 750[131]
Coults, R. T., **4**, 293[235]
Counter, F. T., **8**, 47[124], 66[124]
Couret, C., **5**, 444[189]
Court, A. S., **1**, 622[71]; **6**, 134[12]
Court, J. J., **1**, 400[8]
Courtheyn, D., **2**, 343[15], 353[102], 357[102], 380[102]
Courtneidge, J. L., **4**, 306[375,383,384], 314[480]; **8**, 854[155], 855[155], 856[155]
Courtney, J. L., **7**, 237[38], 851[18]
Courtney, P. M., **1**, 420[85]
Courtois, G., **1**, 368[57], 369[57]; **2**, 2[6], 3[6], 6[6d], 21[6d], 23[6d], 49[6d], 980[19], 1000[53,54], 1004[60,63], 1005[60,63]; **3**, 420[51], 421[51]; **4**, 871[30], 877[30], 880[90], 883[90,96,98-100], 884[90,96,99,100]; **5**, 39[25]
Courtot, P., **5**, 708[42], 709[45], 710[52], 739[45b]
Coury, J., **1**, 367[54]
Cousin, H., **3**, 693[145]
Cousse, H., **8**, 343[112]
Cousseau, J., **4**, 272[27,28], 278[99], 286[173]; **8**, 851[125]
Coussemant, F., **6**, 263[20,24], 264[24], 267[24], 269[20]
Coutrot, P., **2**, 61[21], 429[47]; **3**, 759[129,130]; **6**, 437[39], 533[507]; **8**, 267[71]
Coutts, I. G. C., **3**, 689[118]
Coutts, R. S. P., **1**, 139[4]; **8**, 754[114]
Coutts, R. T., **7**, 79[128b]; **8**, 373[133], 376[133]
Couture, A., **4**, 317[551], 318[561]; **6**, 474[82]; **7**, 143[151], 144[151]
Couture, C., **6**, 232[39]
Couture, R., **7**, 797[19]
Couturier, D., **1**, 368[58], 369[58]; **6**, 74[37], 897[100]
Couturier, J.-C., **8**, 228[129]
Couturier, S., **8**, 447[120], 688[99], 690[101], 691[99]
Couvillon, J. L., **7**, 15[149]
Covell, A. N., **3**, 281[43]
Covell, J., **5**, 208[52]
Coveney, D. J., **1**, 846[13]; **4**, 761[206]
Covert, L. W., **8**, 212[12]
Covitz, F. H., **4**, 129[225]
Cowan, D. A., **7**, 675[55]
Cowan, D. O., **5**, 220[49], 637[102]
Cowan, J. C., **8**, 450[161], 451[180]

Cowan, P. J., **2**, 801[21]
Coward, J. K., **2**, 902[41]
Cowell, A., **3**, 1033[62]
Cowell, G. W., **1**, 844[8]; **3**, 783[21]; **4**, 953[8,8d], 954[8d], 1101[197]; **5**, 904[46], 905[46]
Cowen, K. A., **8**, 54[153], 66[153]
Cowherd, F. G., **5**, 911[93]
Cowie, M., **4**, 964[49]
Cowitz, F. H., **7**, 800[30,30a]
Cowley, A. H., **1**, 41[270], 432[137], 456[137]; **3**, 211[28], 215[28]
Cowling, A. P., **5**, 608[64], 609[64]; **6**, 764[8]
Cowling, M. P., **1**, 391[154], 392[154]
Cox, A., **3**, 689[123]; **5**, 212[70]
Cox, D. G., **8**, 860[223], 861[227]
Cox, D. J., **6**, 172[13]
Cox, D. P., **6**, 219[119]
Cox, G. B., **3**, 638[94]
Cox, J. D., **5**, 900[6]
Cox, J. H., **2**, 388[341]
Cox, J. M., **8**, 530[103]
Cox, M. T., **4**, 405[248]; **8**, 856[184]
Cox, N. J. G., **4**, 791[59]
Cox, P., **2**, 655[140]
Cox, P. B., **3**, 232[266], 488[54], 495[54]
Cox, R. A., **8**, 52[146], 66[146]
Cox, R. H., **3**, 629[53]; **6**, 23[94]
Cox, S. D., **8**, 446[93], 452[93], 534[157]
Cox, W. M., Jr., **8**, 243[46]
Cox, W. W., **5**, 221[59]; **8**, 880[66]
Coxon, J. M., **1**, 181[35]; **2**, 19[77,77b], 31[77b], 573[52]; **3**, 741[53], 742[55-57], 743[57,60], 746[66], 751[90], 752[91]; **7**, 88[23], 90[23]; **8**, 941[112]
Coyle, J. D., **5**, 123[1], 126[1], 212[70], 645[1], 648[1i], 651[1]; **7**, 877[133]
Coyle, T. D., **1**, 292[24]
Coyne, L. M., **7**, 840[8]
Coyner, E. C., **5**, 63[16]
Cozort, J. R., **8**, 437[77,78]
Cozzi, F., **1**, 519[65,66], 520[67], 523[80], 524[86,87], 765[151]; **2**, 31[108], 228[166,167], 374[276], 486[42], 515[55,56], 516[58]; **6**, 149[97,99,102], 927[73]; **7**, 442[47]; **8**, 72[239], 74[239], 844[67]
Cozzi, P. G., **2**, 605[59], 638[60], 640[60,80], 644[80], 645[60], 652[60], 653[60]
Crabb, J. N., **4**, 1075[31], 1104[210]; **7**, 508[310]
Crabb, T. A., **7**, 72[101], 75[115]
Crabbé, P., **1**, 848[25]; **3**, 257[120]; **4**, 229[212,220]; **6**, 175[69]; **8**, 526[24], 881[69]
Crabtree, R. H., **1**, 307[93,94], 310[93], 320[161]; **3**, 1047[4], 1062[4]; **6**, 958[34]; **7**, 1[1], 3[1,9,10], 4[1], 5[27], 6[30,33], 15[27]; **8**, 446[89-93], 447[90], 448[146], 452[93], 455[91], 456[91], 457[213], 534[157]
Crackett, P. H., **2**, 578[85]
Crafts, J. M., **3**, 299[30], 317[121]
Cragoe, E. J., Jr., **2**, 971[92]; **5**, 780[201]
Craig, D., **1**, 787[253]; **4**, 27[81]; **5**, 37[21], 513[5], 514[5,5i], 522[44], 524[55], 527[5]
Craig, D. C., **5**, 474[158]
Craig, J., **4**, 587[47]
Craig, J. C., **2**, 828[132]; **6**, 966[94]; **7**, 693[29]; **8**, 277[152]
Craig, J. T., **2**, 849[215]
Craig, T. A., **5**, 494[215], 579[163]
Crain, D. L., **3**, 319[131]
Cram, D. J., **1**, 49[3], 50[3], 72[71], 109[12], 110[12], 134[12], 141[19,20,22], 151[19,20], 153[20], 182[46], 222[69], 295[50], 460[1,2], 528[113,115], 678[211,212]; **2**, 24[96], 217[137], 666[36], 677[36]; **3**, 114[233], 124[281], 125[281], 154[233], 164[475], 503[149], 512[149], 727[27], 822[8], 836[8]; **4**, 230[246], 426[42], 427[42,66]; **5**, 170[111], 513[4]; **6**, 154[149], 799[24], 968[113], 1012[5,6], 1013[6], 1017[36], 1034[133]; **7**, 483[125], 777[383]; **8**, 3[16,18], 228[128], 329[23], 335[23], 410[91], 828[78]
Cramer, C. J., **5**, 539[104]
Cramer, F., **6**, 20[71], 602[6], 605[35], 607[43], 610[61], 611[68], 612[70,72,77], 614[91], 625[155]
Cramer, H. I., **8**, 140[24], 141[24]
Cramer, P., **4**, 313[465]
Cramer, R. D., **5**, 1138[65]; **8**, 447[108]
Cramer, R. E., **1**, 36[235]
Cramer, R. J., **5**, 73[197]
Crammer, B., **4**, 1043[108], 1048[108]; **5**, 1006[34,34b]
Cramp, M. C., **6**, 836[56]
Crampton, M. R., **4**, 426[43], 444[43]
Crandall, J. K., **1**, 822[30]; **3**, 380[7], 605[232], 741[54], 892[52]; **4**, 308[403], 789[31], 808[156], 809[160], 869[27], 870[27], 871[27], 878[75], 898[75,175]; **5**, 15[100], 597[30]; **6**, 2[3], 25[3], 960[61], 994[95], 998[95]; **8**, 114[53]
Crane, R. I., **2**, 355[119], 382[314]; **3**, 386[68]
Crank, G., **6**, 563[898], 570[947], 984[53]; **7**, 338[38], 738[32], 761[54]
Crans, D. C., **2**, 456[76], 461[76], 465[106]
Crass, G., **1**, 70[63], 141[22]; **2**, 120[179]; **4**, 229[238]
Cravador, A., **1**, 571[274], 631[53], 633[71], 634[71], 636[71], 637[71], 641[71], 642[71], 656[53,71,143,144], 657[53,71,144], 658[53,71], 659[53], 664[71], 672[71,199], 675[71], 702[199], 705[199], 712[199], 716[199]; **3**, 86[50], 87[76,84,86], 109[84], 136[76], 142[86], 144[76,86], 145[76]; **8**, 847[97], 848[97e], 849[97e]
Craven, R. L., **4**, 274[57], 282[57]
Craw, P. A., **3**, 675[74]; **8**, 944[123]
Crawford, H. T., **8**, 497[39]
Crawford, J. A., **4**, 745[140]
Crawford, M., **2**, 400[25], 403[33]
Crawford, R. J., **3**, 903[121]; **4**, 83[65a]; **5**, 876[4], 929[167]; **7**, 111[190]
Crawford, T. C., **3**, 799[102]; **6**, 163[193], 838[66]; **8**, 537[178]
Crawley, L. C., **6**, 960[61]
Crea, R., **6**, 620[130]
Creary, X., **1**, 425[104]; **3**, 614[5], 788[52]; **4**, 457[61], 458[67], 459[90,91], 460[90,91], 462[107], 463[67], 474[90], 475[61,91,149]; **5**, 856[216]; **6**, 451[125]
Creaser, E. H., **8**, 206[171]
Creasey, S. E., **8**, 814[16]
Creasy, W. S., **6**, 209[70]
Creck, C., **6**, 155[154]
Cree, G. M., **7**, 293[11]
Creed, D., **7**, 851[28], 879[150]
Creeke, P. I., **4**, 408[259e], 413[259e]
Creese, M. W., **3**, 750[85]
Cregg, C., **1**, 248[70]
Cregge, R. J., **4**, 10[32,32a,c], 109[148]; **5**, 839[77,84]
Creighton, D. J., **8**, 589[49]
Creighton, E. M., **3**, 913[1]
Crellin, R. A., **5**, 63[9]; **7**, 879[146,147]
Cremer, D., **2**, 267[64]
Cremer, G. A., **8**, 600[104]
Cremer, S. E., **8**, 861[224]
Cremins, P. J., **7**, 96[199]
Cremlyn, R. J. W., **3**, 781[19]
Cremonesi, P., **4**, 523[58]
Crenshaw, L. C., **4**, 486[38], 499[102]
Cresp, T. M., **6**, 1056[53,54]; **8**, 509[92], 510[92]
Cressman, E. N. K., **2**, 1064[109]
Cresson, E. L., **7**, 778[414]
Cresson, M. P., **6**, 856[156]

Cresson, P., **3**, 963[122], 991[35]; **4**, 210[71,74,75], 229[223]; **5**, 851[168,169]; **8**, 111[21], 123[21]
Crews, A. D., **4**, 436[145], 437[145], 438[150]
Crews, C. D., **8**, 583[12]
Crich, D., **3**, 380[9]; **4**, 747[149,150,151,152], 748[159], 753[159], 798[106], 799[119], 800[121]; **6**, 442[87], 472[66]; **7**, 110[187], 719[6,7], 720[6], 721[7,15], 722[20], 725[7], 726[6,7,20], 728[7,40], 730[45,48,51], 731[45]; **8**, 818[40], 825[68]
Criegee, R., **4**, 1098[171], 1099[177]; **5**, 680[24], 683[24c]; **7**, 41[17], 92[40,41a], 94[41], 111[193], 235[6], 437[4], 438[4], 543[10], 548[10], 558[10], 708[34], 709[43-45], 710[43], 851[19]
Crimmins, M. T., **1**, 419[80,81], 791[267]; **2**, 223[151]; **4**, 255[198], 372[63], 794[75], 804[137]; **5**, 123[1], 126[1], 129[34,36], 133[59], 140[87,88], 141[90,91], 143[100-102], 144[101,102]; **6**, 888[66], 905[143], 1063[83,84]
Cripe, K., **3**, 419[33]
Cripe, T. A., **4**, 425[32]; **7**, 229[119]
Crisp, G. T., **1**, 193[89]; **2**, 110[72]; **3**, 250[71], 487[49], 495[49], 529[50], 530[76], 534[76]; **5**, 763[107], 779[107]; **8**, 933[45]
Crissman, H. R., **8**, 708[37]
Crist, D. R., **1**, 836[143], 837[153], 858[62]
Cristau, H. J., **1**, 564[189]; **4**, 55[156], 252[160]; **6**, 134[17,34]; **8**, 862[228,230], 863[231,233]
Cristea, I., **3**, 482[2], 483[2], 491[2], 499[2]
Cristol, S. J., **2**, 968[79]; **3**, 164[476], 422[67]; **4**, 273[50], 1016[207]; **6**, 281[146,147,148,149], 954[15]; **7**, 718[2], 724[2,29]; **8**, 857[202]
Criswell, T. R., **8**, 950[162]
Crittenden, N. J., **7**, 340[46]
Crivello, J. V., **6**, 110[52]; **7**, 13[113]
Croce, P. D., **6**, 705[27]
Crociani, B., **4**, 600[242]
Crocker, M., **6**, 291[211]
Croft, K. D., **5**, 144[104]
Croisy, A., **6**, 462[13]
Croisy-Delcy, M., **7**, 350[21]
Crombie, L., **2**, 742[69]; **3**, 242[6], 257[6], 259[6], 431[97,98], 494[86], 545[122], 558[53], 976[9], 977[9,9a], 989[9], 990[9]; **5**, 803[89], 812[132], 976[20]; **6**, 911[14]; **7**, 156[32], 157[32e], 158[32e], 306[8]; **8**, 477[33], 676[69], 679[69], 681[77], 684[77], 694[77]
Cromwell, N. H., **2**, 149[86], 413[11], 740[61]; **7**, 471[21]
Cronin, J. P., **7**, 208[79], 211[79]
Cronin, T. H., **4**, 5[19]; **5**, 513[4], 514[4e], 527[4e]
Cronnier, A., **1**, 474[94]
Cronyn, M. W., **3**, 868[40]; **8**, 303[99]
Crook, E. M., **3**, 898[82]
Crooks, P. A., **7**, 675[55]
Crooks, S. L., **1**, 188[68]; **2**, 18[71]; **5**, 517[29], 519[29], 534[29], 538[29e], 539[29e]
Crooks, W. J., III, **8**, 815[22]
Croom, E. M., **5**, 155[37]
Crooy, P., **8**, 237[12]
Crosby, A., **3**, 88[124], 118[124]
Crosby, D. G., **8**, 568[466]
Crosby, G. A., **3**, 279[39], 758[125], 770[172]; **4**, 738[98]; **6**, 21[78], 247[131]; **8**, 800[68]
Cross, A. D., **6**, 217[114]; **7**, 86[16a], 137[121], 139[121]; **8**, 321[102]
Cross, B., **7**, 95[79]
Cross, B. E., **3**, 715[42]; **8**, 242[41]
Cross, G. A., **3**, 229[223]
Cross, P. E., **4**, 665[8], 670[8], 674[8]
Crossland, I., **6**, 545[633]
Crossley, F. S., **8**, 956[3]
Crossley, J., **7**, 306[8]
Crossley, M. J., **4**, 437[146]
Croteau, A. A., **2**, 482[35], 484[35]
Crotti, P., **2**, 284[53], 291[71]; **3**, 734[8,10], 741[49]; **6**, 253[159]; **8**, 856[167]
Crouch, R. K., **2**, 610[98]; **3**, 8[42]
Croudace, M. C., **2**, 597[7]; **5**, 1046[30], 1053[39]
Crouse, D. J., **2**, 746[114]; **8**, 843[59c]
Crouse, G. D., **1**, 700[258], 712[258], 722[258]; **3**, 445[70]; **5**, 324[18b,19], 715[83], 806[108], 1003[24]
Crout, D. H. G., **8**, 190[82]
Crouzel, C., **4**, 445[205]
Crow, E. L., **8**, 300[85]
Crow, R., **6**, 831[9]
Crow, W. D., **2**, 353[95], 365[95]
Crowder, D. M., **3**, 97[170], 117[170]
Crowe, D. F., **6**, 1043[15], 1059[15,63]; **8**, 948[149]
Crowe, W. E., **1**, 8[39]; **5**, 1055[47], 1062[59]
Crowell, J. D., **2**, 541[78]
Crowley, J. I., **2**, 815[82]
Crowley, K. J., **5**, 707[30,31], 708[41,41a]; **8**, 336[69]
Crowley, S., **4**, 111[154g]
Crozier, R. F., **4**, 1076[37]
Cruickshank, K. A., **3**, 530[79], 535[79]
Cruickshank, P. A., **7**, 294[12]
Crul, M. J. F. M., **5**, 562[87]
Crumbie, R., **2**, 323[25], 333[25]
Crump, D. R., **3**, 427[87]; **8**, 971[105]
Crumrine, A. L., **2**, 187[41]; **3**, 50[267]
Crumrine, D. S., **2**, 124[201], 184[25,25b], 235[190], 268[66], 280[23], 289[23], 311[34]
Crundwell, E., **5**, 123[4]
Cruse, W. B., **1**, 774[206,211], 776[206]
Crute, T. D., **1**, 767[164], 768[167]; **3**, 226[199]; **7**, 418[127]
Cruz, A., **8**, 881[69]
Cruz, R., **7**, 662[55]
Cruz, S. G., **1**, 743[54], 746[54], 748[54]; **5**, 1115[2], 1116[2], 1122[2b], 1123[2b], 1124[2b]
Cruz, W. O., **7**, 586[162], 844[56]
Cruz-Sanchez, J. S., **1**, 511[27], 564[194]
Csacsko, B., **5**, 991[47], 992[47]
Csapilla, J., **4**, 282[142]
Csendes, I. G., **7**, 230[131]
Csicsery, S. M., **3**, 305[75b]
Csizmadia, I. G., **1**, 506[8], 512[42,43], 528[116]; **3**, 891[41b]; **4**, 330[5]; **6**, 133[4]
Csontos, G., **8**, 445[29], 446[80], 453[29]
Csuk, R., **1**, 212[6], 213[6], 271[62,62b]; **2**, 280[21]; **3**, 570[55], 582[55], 583[55], 630[57], 631[57]; **6**, 978[22]
Csuros, Z., **8**, 140[17]
Cuadrado, P., **2**, 583[114]; **4**, 895[165], 900[165]
Cuadriello, D., **6**, 217[114]
Cuberes, M. R., **1**, 477[141]
Cubero, I. I., **7**, 296[23]
Cubero, J. J., **2**, 385[328]
Cudd, M. A., **4**, 306[387,388]; **8**, 855[157]
Cuéllar, L., **8**, 54[156], 66[156]
Cuer, A., **6**, 436[22]
Cueto, O., **7**, 185[175]
Cuevas, J. C., **4**, 797[104]
Cuffe, J., **4**, 1032[10], 1063[10]
Cuiban, F., **6**, 642[69]
Cuiz-Sánchez, J. S., **8**, 837[16]
Culbertson, T. P., **3**, 846[40]
Cullen, E., **8**, 364[26], 365[26]
Cullen, E. R., **6**, 981[47]
Cullen, W. R., **4**, 1039[62]; **8**, 783[107]
Cullin, D., **3**, 197[33]

Cullis, C. F., **7**, 759[7,8]
Cullison, D. A., **5**, 768[125], 779[125]
Culshaw, D., **4**, 381[126b], 382[126], 383[126]
Culshaw, S., **4**, 298[293], 300[293]
Cumbo, C. C., **8**, 937[79]
Cumins, C. H., **7**, 186[181]
Cummerson, D. A., **7**, 194[4], 374[78], 674[41]
Cummings, T. F., **2**, 954[8]
Cummings, W. J., **7**, 400[42]
Cummins, C. C., **7**, 3[8]
Cummins, R. W., **7**, 763[94]
Cundasawmy, N. E., **5**, 581[176]
Cun-Heng, H., **4**, 213[115], 695[4]
Cunico, R. F., **3**, 254[101]; **4**, 357[151], 487[47]; **6**, 107[30], 108[30]; **8**, 770[31], 776[79]
Cunio, R. F., **5**, 762[101]
Cunkle, G. T., **1**, 480[153,154]
Cunneen, J. I., **7**, 762[73]
Cunningham, A. F., Jr., **1**, 404[24], 568[241,242]; **4**, 532[85], 536[85]; **5**, 44[34]; **7**, 239[48]; **8**, 843[56]
Cunningham, D., **1**, 305[87]
Cunningham, M. P., **5**, 94[85]
Cunningham, R., **4**, 5[19]
Cunningham, R. H., **3**, 760[140]
Cunningham, W. C., **8**, 957[11]
Cunnington, A. V., **4**, 47[133]
Cupas, C., **5**, 794[45], 984[33]
Cupas, C. A., **3**, 334[220]
Cupery, M. E., **5**, 63[21]
Cuppen, Th. J. H. M., **5**, 723[108b], 724[110], 725[115]
Cupps, T. L., **1**, 406[28]; **2**, 555[140]
Curci, R., **1**, 834[126]; **7**, 13[125], 167[186], 374[77a], 763[88], 766[88,182], 777[376]
Curé, J., **2**, 279[15], 292[76]
Curi, C. A., **3**, 828[47], 854[74]
Curini, M., **1**, 656[151], 658[151], 846[18c], 847[18]; **2**, 823[112]; **3**, 857[92], 908[146]; **4**, 1040[77]; **8**, 880[63]
Curphey, M., **8**, 336[84], 339[84]
Curphey, T. J., **2**, 1011[10]; **3**, 30[174,177], 572[66]; **6**, 714[81]; **8**, 312[22], 321[22]
Curragh, E. F., **7**, 221[33]
Curran, A. C. W., **2**, 897[14]; **8**, 336[85]
Curran, D. P., **1**, 248[67], 270[61]; **2**, 908[61]; **3**, 55[281], 219[114], 221[129,130], 499[140], 501[140], 502[140], 603[229], 649[201], 672[64], 824[23]; **4**, 716[5], 719[19], 722[19], 723[40], 733[78], 735[5], 738[96], 741[5], 743[5], 744[133], 747[5,40], 751[5,161], 752[5], 754[177,178,179], 755[178], 776[40], 790[34], 791[34], 799[117], 802[34,126,129], 803[130,136], 808[159], 818[34], 820[213,218,223], 824[236,237,238], 829[34], 830[34], 1076[45], 1078[56], 1079[56,67], 1086[110], 1087[110], 1101[192]; **5**, 247[25], 249[34], 250[40], 255[53], 260[25,67,70], 261[67], 262[67], 263[70], 264[53], 841[89,100], 854[178], 856[178,211], 872[89b,178,211], 888[27], 1012[53]; **6**, 860[177]; **7**, 137[120], 648[40], 676[65], 769[212]; **8**, 392[106], 647[58], 802[85], 823[56], 846[80,87], 849[87], 856[173]
Curran, T. T., **2**, 121[189]; **4**, 185[86]; **5**, 474[156,157]
Curran, W. V., **7**, 30[80]
Current, S., **4**, 377[104], 390[104h]; **5**, 64[38]; **6**, 1031[114]
Currie, J. K., **3**, 1020[12], 1024[29]; **5**, 1138[59]
Currie, R. B., **1**, 425[103]
Curry, D. C., **8**, 478[37]
Curry, M. J., **5**, 679[20]
Curry, T. H., **5**, 10[75]
Curtin, D. Y., **3**, 755[110]; **5**, 876[4]; **6**, 1033[128]
Curtis, A. J., **8**, 542[231]
Curtis, G. G., **3**, 901[112]
Curtis, N. J., **1**, 699[247]; **3**, 107[226], 109[226]; **4**, 372[55]; **6**, 470[58], 570[950]; **8**, 847[97,97d], 849[97d,107]
Curtis, R. F., **3**, 553[14], 690[124]
Curtis, R. J., **4**, 390[174,174a,c]
Curtis, V. A., **3**, 714[33]; **7**, 229[108]; **8**, 827[71]
Curtius, T., **3**, 890[32]; **6**, 245[116], 968[109]
Curulli, A., **2**, 965[69]
Cusack, N. J., **8**, 312[20], 472[7]
Cusak, P. A., **6**, 662[216]
Cuscurida, M., **8**, 966[73]
Cushley, R. J., **6**, 425[62]
Cushman, M., **3**, 953[101]; **6**, 834[33], 853[33], 875[24], 879[24]; **8**, 36[97], 42[97], 66[97]
Cusic, M. E., Jr., **2**, 456[59], 457[59], 458[59]
Cusmano, G., **8**, 663[117]
Cussans, N. J., **6**, 474[84,85]; **7**, 132[97]
Cutler, A., **5**, 272[4,5], 273[4], 275[4,10], 277[10], 281[20]
Cutler, A. B., **8**, 91[50]
Cutler, F. A., **6**, 685[358]
Cutler, H. G., **8**, 621[144]
Cutler, R. A., **2**, 741[64]
Cutrone, L., **5**, 421[80]; **8**, 476[28]
Cutrufello, P. F., **4**, 348[107]
Cutter, H. B., **3**, 565[18]; **8**, 531[124]
Cutting, J., **4**, 37[107,107b]
Cutting, J. D., **1**, 564[202]
Cuvigny, T., **2**, 76[83b], 507[25]; **3**, 31[184], 45[243], 123[247], 174[536], 176[536], 178[536], 448[95]; **4**, 84[68a], 89[68a]; **6**, 720[131]; **8**, 113[33], 249[91], 842[42c,d], 844[42c], 847[42c]
Cuza, O., **8**, 124[89]
Cvejanovich, G. J., **3**, 889[24]
Cvengrosova, Z., **7**, 154[21]
Cvetanovic, R. J., **7**, 5[26]
Cvetkovic, M., **7**, 831[68]
Cvetovich, R. J., **5**, 426[104]
Cybulski, J., **6**, 745[83]
Cymerman-Craig, J., **3**, 824[18]; **6**, 966[96]; **7**, 748[113]; **8**, 249[91], 293[52], 302[52], 507[85]
Cyr, D. R., **3**, 225[186]
Cyr, T. D., **8**, 540[198]
Czarkie, D., **3**, 380[13], 407[146]; **8**, 446[72]
Czarnik, A. W., **5**, 493[212], 552[15,32]; **7**, 778[408]
Czarnocki, Z., **6**, 738[57], 739[57]; **7**, 712[67]
Czarny, M., **4**, 524[59]
Czarny, M. R., **7**, 31[86], 228[100], 229[115]
Czarny, R. J., **3**, 365[63]; **7**, 711[59]; **8**, 932[44]
Czech, A., **7**, 346[9], 365[43]
Czegeny, I., **8**, 226[113]
Czernecki, S., **4**, 297[271]; **7**, 272[142,143], 276[143,148]; **8**, 753[73]
Czeskis, B. A., **4**, 874[52]
Czochralska, B., **8**, 974[132]
Czombos, J., **8**, 418[5], 420[5], 423[5], 430[59], 437[75], 439[5], 441[5], 442[5]
Czuba, L. J., **2**, 184[24], 599[23]
Czuba, W., **7**, 675[56]
Czugler, M., **6**, 499[177]

D

Daalman, L., **5**, 1043[24], 1046[24], 1049[24], 1051[24,36b]
Dabard, R., **2**, 758[22a]; **4**, 521[43]; **8**, 322[111], 445[38], 451[180]
Dabbagh, G., **1**, 428[121], 429[121], 432[137], 456[137], 457[121]; **3**, 209[12], 210[24], 211[27], 213[43], 215[27b]; **4**, 89[87], 90[87d], 107[143c], 141[13], 177[58], 180[58a], 229[230], 255[190,191]; **6**, 9[40]; **8**, 940[102]
Dabby, R. E., **8**, 971[110]
Dabkowski, W., **6**, 602[6]
Daboun, H. A., **2**, 403[37]
Dabrowiak, J. C., **5**, 736[145], 737[145]
Dabrowski, Z., **6**, 745[83]
Daccord, G., **5**, 178[135]
Da Costa, R., **7**, 122[30], 144[30]
Dadson, W. M., **3**, 712[24]; **5**, 647[12], 648[12]
Daesslé, C., **3**, 358[65]
Dagani, D., **4**, 1000[18]
Daggett, J. U., **5**, 249[32]; **6**, 1063[87]
Dagleish, D. T., **3**, 663[26], 665[26]
Dagli, D. J., **2**, 419[23]; **6**, 439[72]
Daglish, A. F., **8**, 528[81], 529[81]
Dagnino, L., **8**, 92[68]
Daham, R., **4**, 1020[237]
Dahan, N., **8**, 21[144]
Dahl, H., **6**, 193[215]
Dahl, K., **7**, 231[148]
Dahl, R., **4**, 258[255]
Dahler, P., **3**, 623[38]; **4**, 670[16]
Dahlhoff, W. V., **6**, 662[213]; **8**, 724[176]
Dahlig, W., **8**, 756[146]
Dahma, A. S., **2**, 466[118], 469[118]
Dahmen, A., **4**, 1090[143]; **5**, 714[76], 743[163]
Dahmen, F. J. M., **6**, 558[850]
Dahmer, J., **2**, 456[32]
Dahn, H., **2**, 413[12]; **3**, 822[3,12], 831[3,12], 835[12b]; **8**, 957[15]
Dai, G.-Y., **2**, 655[147], 907[58], 908[58]
Dai, L., **8**, 879[55], 880[55]
Dai, W.-M., **2**, 610[95], 611[95], 1059[82,83]
Daignault, R. A., **8**, 214[37], 218[70], 219[70], 228[130], 230[139], 246[77]
Daigneault, S., **4**, 370[42]
Dai-Ho, G., **2**, 1035[93], 1040[105]
Daikaku, H., **6**, 528[411,412]
Dailey, O. D., Jr., **3**, 135[368], 136[368], 137[368]
Dailey, W. P., **5**, 419[73], 802[84]
Daines, R. A., **1**, 772[199]; **4**, 377[105b], 381[105]; **6**, 8[39], 51[104]; **7**, 396[25]
Dainter, R. S., **4**, 426[53]; **6**, 557[834]
Dainton, F. S., **7**, 14[137]
Daire, E., **7**, 11[87]
Daitch, C. E., **1**, 248[68]
Dajani, E. Z., **8**, 756[159], 758[159]
Dakin, H. D., **3**, 889[24]
Dalacker, V., **5**, 948[268]
d'Alarcao, M., **3**, 217[93]; **5**, 780[204], 1151[131]
Dal Bello, G., **1**, 489[19], 498[19,50]
Dale, J., **2**, 808[69]; **5**, 1148[122]; **8**, 369[84]
Dale, J. A., **7**, 40[8], 43[8,47]
Dale, W. J., **2**, 662[13], 664[13]
Dalessandro, J., **8**, 803[93], 804[93], 826[69]
Daley, R. F., **3**, 579[94], 640[110]; **4**, 140[10]
Daley, S. K., **8**, 404[20]
Daljeet, A., **3**, 503[149], 512[149]
Dallacker, F., **2**, 139[33], 782[26]
Dallaire, C., **3**, 380[10]; **5**, 532[86]
Dall'Asta, G., **5**, 1142[86]
Dallatomasina, F., **5**, 36[19]
Dall' Occo, T., **1**, 471[68]
Dalmases, P., **5**, 232[134]
DalMonte, D., **8**, 664[121]
D'Aloisio, R., **7**, 381[107]
Daloze, D., **1**, 100[88]; **5**, 456[86]; **6**, 914[27]
Dalpozzo, R., **1**, 569[262]; **4**, 86[78c,e], 424[11], 428[11,77,79], 429[80,82]; **6**, 115[83]; **7**, 331[16]
Dalsin, P. D., **6**, 556[820]
Dalton, D. R., **4**, 272[40]; **5**, 404[14]; **6**, 736[24]
Dalton, J. C., **5**, 165[78,82], 166[91], 176[78]
Dalton, J. R., **5**, 133[57]
Dalton, L. K., **4**, 51[144a]
Daltrozzo, E., **4**, 429[87]
Daluge, S., **8**, 87[30]
Daly, B., **8**, 404[20]
Daly, J., **3**, 167[485], 168[485]
Daly, J. J., **1**, 174[14], 179[14]; **2**, 385[322]; **6**, 531[451]
Daly, J. W., **2**, 876[32]; **7**, 6[34]
Daly, P. J., **7**, 490[177]
Damani, L. A., **7**, 675[55], 736[3]
Damas, C. E., **5**, 107[200]
Damasevitz, G. A., **2**, 589[155]; **8**, 738[28], 755[28]
D'Amato, C., **7**, 15[151]
Dämbkes, G., **6**, 681[335]
D'Ambrosio, M., **7**, 579[137]
Damen, H., **4**, 874[50]
Damerius, A., **4**, 1084[96]
Damiani, D., **3**, 734[8]
D'Amico, A., **3**, 45[250]
Damin, B., **7**, 498[225], 537[56,57]
Damm, L., **6**, 708[50], 831[7]
Dammann, R., **4**, 1007[121]
Dammel, R., **6**, 245[126]
Damodaran, K. M., **8**, 494[24]
Damodaran, N. P., **3**, 402[127]; **5**, 802[83], 810[83]
Damon, D. B., **5**, 841[95]
Damon, R. E., **2**, 805[45]; **3**, 380[9], 675[72]; **4**, 10[34], 113[171,171e], 249[125], 258[125]; **7**, 519[20]
D'Amore, M. B., **5**, 797[63]
Damour, D., **2**, 579[93]
Dampawan, P., **6**, 105[17], 107[20,21]; **7**, 218[7]; **8**, 925[10]
Damrauer, R., **6**, 531[428]
Dams, R., **3**, 583[122], 584[130], 587[148]; **6**, 985[64]
Dan, P., **8**, 97[96]
Dana, G., **3**, 572[60], 578[92], 610[92], 728[36]; **8**, 135[38], 532[130]
Danaher, E. B., **1**, 262[38]
Danan, A., **2**, 354[118], 355[118]
Dance, I. G., **7**, 759[9]
D'Andrea, S., **3**, 1037[87]; **4**, 561[29]; **5**, 1080[53], 1084[53], 1107[171,172], 1108[171,172]; **6**, 760[144]
Dane, E., **6**, 644[94]; **7**, 92[41,41a], 94[41]
Danen, W. C., **4**, 452[3], 453[27], 471[27], 726[53]; **7**, 40[14], 736[2], 745[2], 882[173]
Dang, H. P., **3**, 440[43], 485[30], 491[30], 527[44]
Dang, Q., **5**, 460[94], 492[245], 583[190]
Dang, T.-P., **8**, 173[118], 180[137], 459[231], 460[231], 462[264], 535[166]
D'Angeli, F., **6**, 67[10], 575[966]

d'Angelo, J., **1**, 553[87]; **2**, 101[17], 105[17], 106[17], 108[17], 182[2], 185[27], 227[161], 304[7], 709[11]; **3**, 1[3], 197[42], 198[44]; **4**, 7[25], 37[106a], 221[161,162,163,164,165], 231[276], 239[38], 240[38], 254[38], 259[38]; **5**, 327[27], 341[60], 557[53], 676[5], 732[134]; **6**, 738[51]; **7**, 96[84]; **8**, 188[53], 925[8]
Dangles, O., **6**, 641[58]
Dangyan, M. T., **4**, 315[505,506,507]
Danheiser, R. C., **7**, 598[54]
Danheiser, R. L., **1**, 272[66], 273[66d], 404[20], 428[120], 595[29], 596[29,30], 601[30,31], 602[31,32], 603[32], 604[33], 605[34], 770[191], 887[138], 888[138]; **2**, 85[13-17], 156[118], 187[38], 575[60,61], 579[91], 587[91], 1061[95]; **3**, 21[126], 22[132], 570[53], 572[53,69], 573[53,69], 575[53], 583[53], 596[53], 602[69], 607[69], 610[53,69], 744[63]; **4**, 155[71b], 158[76], 373[70], 1008[133], 1023[256,258]; **5**, 277[14-17], 278[14,15], 279[15,16], 683[82], 689[71,73,78,78a,b], 690[82], 732[135,135a], 733[135b], 806[106], 847[136], 856[210], 913[100,105], 1007[38], 1008[43], 1012[49], 1017[64], 1018[64], 1020[64], 1021[64], 1025[82], 1026[82]; **6**, 648[119]; **7**, 545[24], 566[100], 711[60]; **8**, 355[182], 531[125], 756[144]
Danheux, C., **6**, 707[45]
Daniel, B., **8**, 563[435]
Daniel, H., **1**, 759[130]
Daniel, J. R., **8**, 413[134]
Danieli, B., **7**, 153[9], 346[12]
Danieli, N., **7**, 86[16a]
Danieli, R., **2**, 807[48]
Daniels, K., **7**, 378[92]
Daniels, P. J. L., **7**, 96[87]
Daniels, R., **7**, 746[83]
Daniels, R. G., **1**, 480[154]; **2**, 82[6]; **4**, 667[13], 669[13], 677[13]; **8**, 946[134]
Daniels, S. B., **4**, 394[193]
Danielson, S. J., **8**, 52[139], 66[139]
Daniewski, A. R., **1**, 329[39], 806[314]; **4**, 5[17]; **6**, 937[117], 939[117], 940[117], 989[81]
Daniewski, W. M., **3**, 88[128], 124[128,258]; **5**, 432[132], 1023[77]; **8**, 842[47]
Daniher, F. A., **7**, 498[230a]
Danikiewicz, W., **4**, 432[107], 446[214]
Danilov, L. L., **6**, 533[498]
Danilov, S. N., **4**, 304[358]
Danilova, N. A., **8**, 680[73], 683[73]
Danion-Bougot, R., **4**, 955[11]
Danishefsky, S. J., **1**, 92[67], 103[95,96], 314[131,132,133,134], 329[35], 343[107], 425[106], 529[124], 732[16], 765[150], 787[16], 799[297]; **2**, 105[44], 163[149,150], 167[163], 455[16], 465[108], 570[38], 578[82,84], 613[113], 617[143], 633[33b], 640[33], 652[126], 656[150,151], 662[1,2,11], 664[30], 665[30-33], 666[34,35], 667[40-42], 668[31,32], 669[45], 670[45,47], 671[48], 673[32,33,40], 674[31,32,40-42], 675[33,35,40-42,52], 681[57], 682[30-33,63,64], 683[64], 686[63,64], 687[63,71], 688[69], 689[30,33], 690[1,2,71], 692[69,72], 693[73], 694[74-76], 696[78-80], 697[81], 700[34,35], 701[85,86], 702[86], 703[87], 704[88], 905[55], 907[55,57], 908[55], 910[55], 911[55,71], 1054[63]; **3**, 49[264], 503[149], 512[149], 816[83], 890[33]; **4**, 5[18,19g], 10[31,31c], 18[60,60a,b], 27[84,84a], 29[84b,c], 30[89], 33[96,96d], 34[96e], 54[152,152a,b], 121[209], 159[80,82,84], 246[96], 258[96,233], 260[96], 262[314], 295[251], 372[64b], 373[80], 374[80], 398[216], 399[216b], 403[241], 404[216b], 405[252], 561[31], 741[125], 799[112,115], 1040[95], 1041[95a,b], 1045[95a,b,109]; **5**, 268[77], 320[6,7], 322[13], 324[22], 329[33], 330[34,35], 410[40,41a,b,f], 411[41f], 434[140,141,143,144,145,146], 459[92], 683[35], 736[143,145], 737[145], 841[87], 843[116,118], 921[141,142], 925[150]; **6**, 8[35], 27[117], 48[89], 91[123], 93[123], 718[121], 859[169], 919[42], 960[57], 989[80], 995[80], 1023[72,73]; **7**, 175[141], 237[37], 245[74], 246[89], 374[77c], 438[22], 439[37], 440[37], 737[12]; **8**, 5[32], 448[143], 540[195], 542[238], 544[277], 856[171,186]
Dankleff, M. A. P., **7**, 763[88], 766[88]
Dan'kov, Y. V., **6**, 2[9], 3[9]
Dankowski, M., **8**, 858[206]
Danks, L. J., **4**, 347[104]
Danks, T. N., **4**, 115[180d]
Dankwardt, J. W., **1**, 470[61]; **5**, 1021[72]
Dann, O., **3**, 864[19]
Dannecker, R., **4**, 809[164]
Dannecker-Doerig, I., **5**, 241[5]
Danneel, E., **2**, 169[164]
Dannenberg, H., **8**, 957[10]
Dannenberg, W., **5**, 730[128]
Dannhardt, G., **5**, 410[38]
Danno, S., **4**, 836[2,3]
Dannoue, Y., **5**, 736[142g]
Danon, L., **1**, 258[26]; **8**, 115[64], 124[64], 125[64]
Dansette, P., **7**, 95[72]
Dansted, E., **7**, 88[23], 90[23]
Danzer, B., **2**, 1090[73], 1100[118], 1101[118], 1102[73], 1103[73,118b]
Danzin, C., **6**, 80[70]
Dao, G. M., **5**, 356[90]
Dao, L. H., **3**, 822[3,12], 831[3,12]; **5**, 485[184]
Dao, T. V., **8**, 240[33]
Daoust, V., **6**, 939[147], 940[147]
Dappen, M. S., **5**, 539[104], 856[220]
Dar, F. H., **7**, 584[158]
Darack, F., **8**, 365[27]
Daran, J.-C., **4**, 982[112]; **5**, 1076[40], 1086[68], 1105[161,162,163]
Darapsky, A., **8**, 144[68]
Darbre, T., **2**, 1024[60], 1062[99]
Darby, N., **3**, 711[22]; **7**, 58[57], 62[57], 63[57]
Darby, P. S., **6**, 1066[91]
Darchen, A., **8**, 660[108]
Darcy, P. J., **5**, 722[104]
Dardis, R. E., **1**, 366[44]; **2**, 900[28], 901[28], 910[28]
Dardoize, F., **2**, 294[84], 296[84], 486[38]; **5**, 100[145]
Darensbourg, M. Y., **8**, 22[146], 289[28]
Darias, J., **5**, 830[32]
Daris, J.-P., **1**, 123[78]; **2**, 212[120], 213[126], 656[158,159], 1059[77]
Darley, P. A., **6**, 660[198]
Darling, G., **7**, 663[58]
Darling, P., **7**, 281[175], 282[175]
Darling, S. D., **2**, 106[51]; **4**, 15[50], 102[133a-c]; **5**, 71[165], 130[40]; **6**, 782[82]; **8**, 478[47], 479[47,47b], 481[47], 525[15], 526[15], 626[172]
Darling, T. R., **5**, 596[28], 597[28]
Darlington, W. H., **4**, 571[2]
Darnault, G., **2**, 86[24]
Darnbrough, G., **3**, 977[9a]
Darnell, K. R., **7**, 7[42]
Darst, K. P., **1**, 757[120]; **6**, 174[60]
Dart, M. C., **8**, 429[54]
Dartmann, M., **5**, 444[188]
Daruwala, K. P., **1**, 886[137]; **5**, 1000[6]
Darwish, D., **3**, 969[136]
Das, A., **6**, 487[55], 489[55], 543[55]
Das, A. K., **7**, 823[35]
Das, B., **2**, 725[119]; **7**, 823[35]
Das, B. C., **1**, 766[152]; **8**, 333[56]
Das, G., **3**, 362[82]
Das, J., **3**, 361[79], 762[146]; **7**, 438[20], 445[20], 493[188], 633[63]
Das, K., **8**, 113[29], 116[29], 117[29], 119[29], 880[65]
Das, K. C., **5**, 803[88]

Das, K. G., **3**, 386[61], 393[61]; **8**, 515[116]
Das, P. K., **4**, 1089[131]
Das, S., **8**, 113[29,31], 116[29], 117[29], 119[29], 816[24], 880[65]
Das, T. K., **2**, 747[117]
Das, V. G. K., **2**, 727[128]
D'Ascoli, R., **7**, 265[100], 267[100], 530[14]
Da Settimo, F., **3**, 439[40,41]
Das Gupta, A. K., **7**, 137[122], 139[122]
Dasgupta, H. S., **7**, 267[117], 268[117]
Dasgupta, R., **8**, 331[35]
Dasgupta, S., **8**, 724[169,169a]
Dasgupta, S. K., **3**, 427[87]; **8**, 971[105]
Das Gupta, T. K., **5**, 501[268]; **6**, 672[285]
Dashan, L., **1**, 464[39]
Dasher, L. W., **2**, 523[74]; **4**, 615[392], 629[392]
Dashkovskaya, E. V., **6**, 524[374]
daSilva, E., **4**, 824[234]
da Silva, G. V. J., **3**, 589[162], 610[162]; **4**, 373[82]
Da Silva, R. R., **2**, 823[112]; **3**, 1051[22], 1052[22]
Da Silva Jardine, P., **1**, 421[88]
D'Astous, L., **8**, 185[17]
Dastur, K. P., **8**, 515[119], 529[88]
Date, T., **7**, 353[35], 355[35]
Date, V., **4**, 391[180,181a]
Daterman, G. E., **3**, 124[277]
Dather, S., **5**, 402[5]
Datt, D. B., **2**, 747[117]
Datta, A., **2**, 286[63]
Datta, M. K., **8**, 724[169,169a]
Datta, R., **8**, 724[169,169a]
Datta, S. K., **2**, 736[28]; **6**, 817[104]
Daub, G. W., **2**, 1035[91], 1050[32], 1072[32]; **3**, 369[121], 372[121]; **5**, 839[77], 1123[37]; **6**, 624[143]
Daub, J., **2**, 375[278]; **6**, 518[331], 519[338], 859[168]
Daub, J. P., **5**, 859[234]
Dauben, H. J., **1**, 846[16], 851[16]; **4**, 663[4]; **7**, 722[21]
Dauben, H. J., Jr., **2**, 323[27]; **3**, 651[218]
Dauben, W. G., **1**, 377[97]; **2**, 101[21], 194[69], 277[7], 281[7], 287[7], 370[259], 477[12], 510[43], 611[101], 855[243]; **3**, 23[139], 99[189], 107[189], 110[189], 407[150], 431[95,96], 843[22], 855[86], 903[120]; **4**, 17[53,53a], 45[128], 115[183], 161[86b], 241[57], 259[263]; **5**, 5[43,45], 8[43,45], 143[103], 145[106,107], 215[1], 216[1], 217[26,27], 218[1], 220[1,47], 223[1], 224[1], 226[110], 227[110], 228[110], 341[60], 342[61b,62b], 345[71c], 346[71c], 453[64], 458[71], 459[93], 686[45], 707[37,39], 708[43], 709[39,44], 712[59], 715[82], 716[84], 717[91], 737[37,44a], 739[43,44a,82], 791[25], 798[25], 830[29], 842[110], 847[140]; **7**, 100[118], 101[132], 123[34], 239[48], 258[57], 263[87], 845[64]; **8**, 476[27], 531[116], 735[17], 737[17], 746[17], 753[17], 761[17], 940[109], 947[109], 952[109]
Daudon, M., **7**, 764[114]
Daugherty, B. W., **5**, 96[103], 98[103], 99[103]; **6**, 84[85]
Daugherty, J., **4**, 1079[67]; **5**, 260[70], 263[70]
Daulton, A. L., **7**, 500[240]
Daum, H., **1**, 70[63], 141[22]; **3**, 144[384], 145[384], 147[384], 568[40], 606[40]; **8**, 166[64], 178[64], 179[64]
Daum, S. J., **3**, 854[78]
Daunis, J., **8**, 662[115]
Dauphin, G., **1**, 410[40]; **2**, 782[16,18]; **4**, 374[90]; **6**, 436[22]; **7**, 60[46b]; **8**, 203[148], 205[148], 558[399], 559[401]
Dauplaise, D. L., **5**, 557[56]
D'Auria, M., **3**, 512[192,213], 515[192,213]; **5**, 771[148,150,152], 772[148,150], 780[148]; **7**, 103[137], 260[64], 265[99-102,104], 266[105,107], 267[99-102,104,105,107], 530[14,15,17], 531[17]; **8**, 563[430]
Daussin, R. D., **6**, 281[146]
Dauter, Z., **6**, 739[60]
Dauzonne, D., **2**, 742[77], 968[78]
Dave, K. G., **2**, 1015[20], 1018[44]
Dave, M. P., **2**, 867[14]
Dave, P. R., **6**, 744[72]; **8**, 36[88], 52[149], 66[88,149]
Dave, V., **1**, 853[48]; **3**, 784[30]; **4**, 1031[6], 1043[6], 1052[6], 1063[6]; **5**, 905[56]; **7**, 673[25]
Davenport, K. G., **2**, 510[41,42]
Davenport, R. J., **2**, 657[162]
Davenport, T. W., **6**, 1066[94]
Daves, G., Jr., **4**, 839[29,30]
Daves, G. D., Jr., **3**, 124[277]; **6**, 666[232,232a], 667[232]
Davey, A. E., **1**, 564[197]
Davey, W., **8**, 533[137]
Daviaud, G., **4**, 84[68c], 95[102b-d]; **5**, 777[185]
David, G., **5**, 771[148], 772[148], 780[148]
David, J., **6**, 543[618]
David, M., **5**, 936[195]
David, S., **1**, 831[101]; **2**, 6[32], 19[32b], 463[90], 464[90,95,95b,96], 467[90], 663[23], 664[23], 681[62]; **5**, 432[126,127]; **6**, 18[64], 23[95], 662[216,217]; **7**, 88[26]
Davidovics, G., **3**, 892[47]
Davidovid, Yu. A., **6**, 533[491]
Davidowitz, B., **7**, 483[124]
Davidsen, S. K., **2**, 162[141], 479[18], 480[18], 630[10], 631[10], 632[10], 640[10], 641[10], 642[10], 646[10], 1064[111]; **7**, 228[90]
Davidsohn, W., **8**, 548[319]
Davidson, A. H., **1**, 836[145]; **3**, 201[76]; **5**, 536[96], 862[251]; **6**, 766[22]
Davidson, E. R., **1**, 506[10]; **5**, 202[38]
Davidson, F., **1**, 488[6,7,10]; **8**, 620[135]
Davidson, J. A., **4**, 871[36]
Davidson, J. G., **4**, 985[131]
Davidson, J. L., **5**, 1134[45], 1136[54]
Davidson, P. J., **1**, 140[7]; **4**, 914[4], 924[4]
Davidson, R. S., **3**, 567[37], 606[37]; **5**, 185[166]; **7**, 850[11]
Davidson, T. A., **2**, 291[73]; **3**, 679[91]; **4**, 290[197]
Davies, A. G., **5**, 901[30]; **7**, 594[2], 598[56], 599[67], 602[105,107], 604[133], 607[133], 641[1]; **8**, 726[186], 753[72]
Davies, A. M., **3**, 949[95]; **6**, 822[116], 897[95]
Davies, A. P., **6**, 20[72]
Davies, D. I., **4**, 364[1,1a], 368[1a], 373[1a], 717[8]; **7**, 732[56]
Davies, D. T., **4**, 27[83]
Davies, G. M., **3**, 511[189]; **7**, 373[72b]
Davies, H. G., **7**, 59[37]; **8**, 198[130]
Davies, H. M. L., **4**, 1033[27,31], 1038[55], 1040[79], 1049[27,79], 1051[128], 1056[139], 1060[27b,c,79b]; **5**, 599[40], 804[94], 905[60], 986[39,40]; **6**, 126[149]
Davies, I. W., **1**, 526[96]
Davies, J., **6**, 1024[77]; **8**, 196[116]
Davies, J. A., **4**, 915[8,15]; **7**, 723[23]; **8**, 447[116]
Davies, J. E., **1**, 38[253]; **3**, 638[89]; **7**, 709[36], 747[101], 765[136], 843[49]
Davies, J. M., **4**, 6[23]
Davies, J. S., **3**, 818[94]
Davies, J. W., **7**, 730[48]
Davies, L. B., **5**, 492[246]
Davies, M., **3**, 381[32]
Davies, P. S., **3**, 511[189]
Davies, R. B., **4**, 424[13]
Davies, R. V., **2**, 765[76]
Davies, S. G., **1**, 119[63], 343[114]; **2**, 125[214,215,216,217,219,220,221,222,224,225], 127[228,232], 271[77], 272[77,79,80], 315[42,44-46], 316[42,44-46], 317[44], 910[67], 933[140]; **3**, 47[257], 197[37], 1029[53]; **4**, 82[62a,b,e,f], 217[129,130,131,132], 231[130,132], 243[73-75], 250[139], 257[73-75], 260[75], 497[97], 520[32], 977[95]; **5**, 367[100]; **6**, 685[346],

686[370], 690[346], 692[346]; **7**, 355[44]; **8**, 95[90], 505[74], 797[44], 889[129]
Davies, T. M., **7**, 882[171]
Davies, W., **3**, 898[82]; **8**, 950[158]
Davies, W. L., **6**, 270[77]
Davini, E., **4**, 370[28]; **8**, 856[163]
Davis, A. G., **4**, 305[364,366,367], 306[366], 735[84], 770[84]
Davis, A. P., **3**, 380[4]; **7**, 645[23]; **8**, 9[53], 21[53]
Davis, B. R., **3**, 325[157], 353[49]; **7**, 92[40]; **8**, 312[20]
Davis, C. C., **7**, 295[21]
Davis, C. S., **2**, 1077[155]
Davis, D. D., **1**, 174[6], 175[6]; **3**, 565[19]
Davis, F. A., **1**, 86[42-44], 389[138], 390[142,143], 834[128], 837[148,150], 838[148,157,158,159,162]; **2**, 603[48], 994[40], 999[40]; **5**, 422[82]; **6**, 150[114], 1024[78]; **7**, 162[59-68], 163[69], 176[67], 181[65], 184[169,170], 330[10], 425[147b], 741[50], 746[93], 747[50], 765[156], 772[291], 778[398,399,400,401,401a,b], 779[401b,425,426]; **8**, 395[134]
Davis, G. A., **5**, 167[100]
Davis, G. E., **8**, 475[19]
Davis, G. T., **7**, 222[38]; **8**, 568[477]
Davis, H. A., **8**, 214[38]
Davis, H. B., **7**, 267[121], 269[121], 270[128], 271[121,128], 278[121]
Davis, H. R., **8**, 364[12,13]
Davis, H. S., **6**, 951[2]
Davis, J. A., **7**, 724[28]
Davis, J. E., **1**, 72[75]
Davis, J. H., **5**, 790[21], 820[21]
Davis, J. T., **1**, 350[150], 769[182]; **2**, 946[177], 947[177]; **4**, 1079[61]; **5**, 359[91], 373[91], 374[91], 543[118], 545[118]
Davis, J. W., Jr., **8**, 295[60]
Davis, K. E., **6**, 1024[74]
Davis, L., **4**, 439[168]; **5**, 404[19], 405[19], 431[121]
Davis, L. H., **7**, 15[144]
Davis, L. L., **2**, 823[112]; **3**, 1051[22], 1052[22]
Davis, M., **4**, 313[467]; **8**, 476[23], 478[23]
Davis, M. W., **8**, 448[146]
Davis, N. M., **2**, 745[104]
Davis, N. R., **3**, 125[314]
Davis, P. D., **2**, 727[132]; **5**, 70[116], 514[9], 527[9], 757[78], 762[102]
Davis, P. J., **7**, 65[65,70]; **8**, 56[168], 66[168]
Davis, R., **2**, 385[320]; **4**, 10[33,33b], 159[85], 187[100], 216[122], 261[295], 262[295]; **6**, 234[51], 235[51], 236[51]
Davis, R. C., **7**, 227[85]; **8**, 47[125], 66[125]
Davis, R. D., **8**, 938[93]
Davis, R. E., **3**, 380[10]; **4**, 980[107]
Davis, R. H., **7**, 712[65]
Davis, S. B., **8**, 428[52]
Davis, S. J., **6**, 275[111,112]
Davis, T. C., **5**, 355[87a]
Davis, V. C., **4**, 1076[37]
Davis, V. J., **7**, 800[34]
Davison, J., **4**, 844[61]
Davison, S. F., **7**, 452[56], 851[18]
Davisson, M. E., **3**, 325[161]
Davoli, V., **7**, 777[384]
Davoust, S. G., **2**, 530[23]; **5**, 4[39], 5[39]
Davrinche, C., **2**, 141[39]
Davtyan, S. Z., **4**, 315[525]
Davy, H., **6**, 437[31]
Davydova, G. V., **8**, 609[51]
Dawber, J. G., **8**, 860[221]
Dawe, R. D., **6**, 889[67]
Dawes, K., **5**, 166[91]
Dawson, A. D., **7**, 295[22]
Dawson, C. R., **8**, 564[443]
Dawson, D. J., **4**, 83[65a]; **7**, 111[190]
Dawson, I. M., **6**, 1002[133]
Dawson, J. H., **7**, 80[141]
Dawson, J. R., **4**, 70[7], 260[282]
Dawson, M. I., **2**, 547[101], 548[101]; **3**, 363[84]; **4**, 83[65a]; **7**, 111[190]; **8**, 542[237]
Dawson, M. J., **7**, 59[37]; **8**, 198[130]
Dawson, R. L., **1**, 422[90]
Dawydoff, W., **8**, 452[190]
Dax, K., **6**, 22[81]
Dax, S. L., **1**, 243[56]; **5**, 468[137]; **6**, 1006[148]
Daxner, R., **2**, 736[23]
Day, A. C., **3**, 665[36], 689[119]; **8**, 526[25]
Day, A. R., **2**, 138[24]
Day, M. J., **7**, 41[22]
Day, R. A., Jr., **3**, 577[85], 579[94], 595[192], 640[110]; **8**, 527[52]
Day, V. W., **8**, 458[223,223c]
Dayagi, S., **6**, 726[185]
Dayrit, F. M., **1**, 156[69]; **4**, 143[19], 155[63b]; **8**, 693[114]
D'Costa, R., **4**, 36[100], 49[100a]
De, A., **2**, 361[177]
De, B., **3**, 274[20]; **6**, 673[292], 674[292], 1047[34]; **7**, 376[89]
De, R. L., **1**, 41[194]
Deacon, G. B., **1**, 276[79], 277[79b,d,81,82]; **8**, 851[131]
Deakin, M. R., **7**, 854[45]
De Amici, M., **7**, 143[142]; **8**, 645[41]
Dean, D. C., **5**, 252[45]
Dean, F. M., **3**, 807[25], 900[89]; **7**, 564[111], 572[111]; **8**, 606[23]
Dean, J. A., **7**, 854[44]; **8**, 113[50], 114[50]
Dean, R. T., **2**, 1012[16]; **7**, 230[130]
Dean, W. D., **6**, 546[648]
Dean, W. P., **1**, 180[42], 181[42]
Deana, A. A., **2**, 971[92]; **5**, 385[129c,d], 693[105]
de Ancos, B., **6**, 67[13]
DeAngelis, F., **4**, 1021[246]; **6**, 490[109]
de Araujo, H. C., **8**, 351[168], 353[155]
Deardorff, D. R., **3**, 736[28]; **4**, 176[45], 597[176], 598[176], 622[176], 637[176]; **6**, 656[170]
Deardurff, L. A., **7**, 877[129]
de Armas, P., **4**, 814[187]
Deaton, D. N., **1**, 418[72]
Deb, K. K., **6**, 526[390]
DeBacker, M. G., **8**, 524[13,13c]
Debaerdemaeker, T., **6**, 551[689]
Debaert, M., **8**, 660[108]
Debal, A., **1**, 789[261], 791[265]; **3**, 263[173]
DeBardeleben, J. F., Jr., **7**, 136[115], 137[115]
de Belder, A. N., **6**, 660[200]
Debeljak-Šuštar, M., **6**, 554[782]
de Benneville, P. L., **4**, 6[20,20c]; **6**, 431[113]
Deberly, A., **1**, 86[37-40], 223[72c]; **6**, 501[200]
De Bernardi, M., **2**, 547[114], 551[114]
DeBernardis, A. R., **3**, 727[31]; **5**, 455[74]
DeBernardis, J., **3**, 946[88], 990[34]
DeBernardis, J. F., **4**, 10[34], 113[164]; **6**, 83[82]
DeBesse, J. J., **4**, 240[42]
De Beys, V., **1**, 661[167,167a]
Deblandre, C., **7**, 614[6]
DeBoer, C. D., **4**, 7[24]; **5**, 636[91], 639[91], 805[98], 1025[81]
De Boer, H. J. R., **4**, 1018[226]
DeBoer, J. A., **5**, 797[58]
de Boer, J. L., **8**, 96[92]
de Boer, T. J., **1**, 528[114], 851[35]; **4**, 423[3]; **5**, 3[26]; **6**, 120[123]; **7**, 748[110]
Debono, M., **8**, 47[124], 66[124]

DeBons, F. E., **6**, 816[102], 822[117]
de Bont, J. A. M., **7**, 429[150b]
de Brouwer, R. J., **3**, 367[102]
De Bruyn, D. J., **8**, 216[64]
DeBruyne, C. K., **6**, 48[88]
De Buyck, L., **2**, 343[15], 353[102], 357[102], 380[102], 423[34], 424[35]; **6**, 500[182], 547[663]; **8**, 36[73], 38[73], 66[73]
DeCamp, M. R., **5**, 681[25]; **6**, 776[55]
de Carvalho, H., **2**, 855[246]
Decedue, C. J., **7**, 350[25], 355[25]
Decesare, J. M., **2**, 430[53]; **3**, 167[486], 168[486], 748[77]
Dechend, F. v., **6**, 476[95]
DeCian, A., **1**, 365[41]
Decicco, C., **5**, 132[48]
DeCicco, G. J., **7**, 8[60]
Deck, J. C., **4**, 1014[183]
Decker, H., **2**, 1016[25]
Decker, O. H. W., **5**, 688[70], 689[70,77], 690[80], 733[136,136c,g], 734[136g], 829[20], 864[261]; **6**, 861[185], 862[185]
De Clercq, E., **7**, 350[25], 355[25]; **8**, 679[66], 680[66], 681[66], 683[66], 694[66]
De Clercq, J. P., **2**, 201[92], 423[33], 424[35], 838[179]; **3**, 713[28], 857[90]; **4**, 239[15], 1040[75]; **5**, 109[217],539[106], 924[146]; **6**, 690[395], 1055[52b]; **7**, 105[147], 363[33]; **8**, 122[80]
De Cock, C. J., **5**, 70[113]
De Cock, C. J. C., **2**, 353[99]; **5**, 553[41]
Decodts, G., **6**, 2[5], 18[5]
Decor, J. P., **6**, 157[168]
Decorzant, R., **2**, 166[155], 185[28]; **3**, 882[104]; **4**, 91[90], 92[90b], 242[62], 253[62], 261[62,287,293]; **6**, 161[185], 1067[106]; **8**, 358[199]
Decouzon, M., **6**, 769[30]; **8**, 536[172]
De Crescenzo, G., **4**, 438[151,152]; **6**, 545[638]
Decroix, B., **6**, 515[236], 552[695]
Dederer, B., **2**, 1093[83]
Dedier, J., **2**, 600[30]
Dedolph, D. F., **8**, 624[154], 628[154]
Dedov, A. G., **8**, 600[106], 606[25], 625[25,159]
Dedrini, P., **1**, 471[68]
Deeb, T. M., **4**, 811[175], 812[175,176]
Deeg, M. A., **2**, 481[20]
Deem, M. F., **8**, 476[23], 478[23]
Deem, M. L., **5**, 64[53]
Deeming, A. J., **1**, 451[216]
Deeter, J. B., **4**, 128[221]
Defauw, J., **4**, 98[116], 155[73]; **5**, 7[51], 20[140], 514[6]; **7**, 565[97]
Defaye, J., **7**, 247[104]
DeFeo, R. J., **4**, 93[93a,b]
Defoin, A., **5**, 417[65], 419[74], 420[75]; **8**, 652[79]
De Frees, D. J., **1**, 487[4], 488[4]
De Frees, S. A., **6**, 448[106,107]
Defusco, A., **6**, 552[695]
Deganello, G., **4**, 710[48]; **5**, 715[77]; **8**, 451[172]
Degani, I., **2**, 737[39]; **3**, 125[307]; **6**, 134[13]; **8**, 277[153,154], 660[109]
Degani, Y., **6**, 726[185]
Degen, P., **3**, 248[56], 251[56]; **4**, 170[22]
Degenhardt, C. R., **2**, 363[201], 588[150]; **3**, 254[96]; **7**, 172[127]; **8**, 946[134]
Degering, E. F., **8**, 366[34]
Deghenghi, R., **8**, 492[15], 498[15], 527[43,45], 528[45], 529[45], 530[45,96]
DeGiovani, W. F., **7**, 158[40]
Degner, D., **3**, 634[28], 640[100]; **6**, 561[874]
DeGournay, A. H., **4**, 653[434]
de Graaf, C., **4**, 900[180]
De Graaf, S. A. G., **4**, 1004[71,72]; **6**, 712[71]
de Graaf, W. L., **4**, 869[23], 972[79]; **5**, 908[73]
de Graaff, G. B. R., **4**, 493[77]
Degrand, C., **4**, 459[79,82-84], 476[79,82,83]; **7**, 497[219]; **8**, 137[53], 390[83], 594[72]
de Groot, A., **1**, 570[266,267,268]; **2**, 198[83], 817[91], 835[157], 838[170]; **6**, 1023[70]; **7**, 363[38], 376[87]
de Groot, J. A., **2**, 780[9]; **6**, 494[131]
Degtyarev, L. S., **6**, 547[664]
Deguchi, R., **4**, 600[237], 643[237], 650[424]
Degueil-Castaing, M., **4**, 754[176]; **8**, 21[144]
Degurko, T. A., **4**, 386[153], 400[228b], 413[278c]
de Haan, A., **7**, 429[150b]
DeHaan, F. P., **3**, 299[33]
de Haan, J. W., **3**, 367[101]
Dehasse-De Lombaert, C. G., **1**, 683[227], 714[227], 715[227], 717[227], 718[227]; **3**, 786[42]
de Heij, N., **7**, 851[24]
De Hemptinne, X., **8**, 321[94,95]
Dehmlow, E. V., **3**, 509[178], 851[67]; **4**, 1001[24,25,28,35,36,40], 1002[50], 1003[65], 1005[93,95], 1017[215,216], 1041[105]; **5**, 689[72,76], 770[140]; **6**, 66[7], 579[983]; **8**, 798[53]
Dehmlow, S. S., **4**, 1001[28], 1005[95]; **5**, 770[140]
Dehn, R. L., **6**, 1043[15], 1059[15,63]; **8**, 948[149]
Dehn, W. M., **3**, 825[25,27b], 826[25]
Dehnicke, K., **4**, 349[112]
DeHoff, B. S., **1**, 188[68], 767[175,178], 768[169], 772[200]; **2**, 18[71]; **3**, 1008[73], 1009[74], 1010[74]; **6**, 174[60], 187[175]
Deiko, S. A., **3**, 499[125], 669[53]
Deimer, K.-H., **6**, 665[226], 667[226], 668[226], 669[226]
Deisenroth, T. W., **4**, 401[228c,d]
Deitch, J., **2**, 799[18]
Deitrich, W., **3**, 668[50]
de Jeso, B., **4**, 21[69], 221[158,159], 754[176]; **6**, 722[143], 726[180]
de Jong, J. C., **5**, 371[103]
De Jong, K. P., **3**, 552[10]
de Jong, R., **6**, 426[76], 480[111]
de Jong, R. L. P., **1**, 471[62]
de Jonge, C. R. H., **4**, 763[210], 806[147]; **7**, 99[111], 252[2], 437[7], 438[21], 439[7], 527[1], 703[5], 710[5], 737[18], 754[18], 755[18], 815[2], 816[2b,c], 824[2b,c], 827[2c], 851[18]
de Jonge, J., **3**, 904[129]
DeJongh, D. C., **3**, 891[42]
de Jongh, H. A. P., **4**, 26[77], 27[77b]
deJongh, R. O., **8**, 447[124], 450[124]
de Kanter, F. J. J., **1**, 10[55], 11[55b]; **4**, 1018[226]
De Keijser, M. S., **2**, 114[121]
Dekerk, J. P., **5**, 117[278]; **6**, 540[584]
De Keukeleire, D., **5**, 127[25], 131[44]
De Keyser, J.-L., **2**, 353[99]; **5**, 553[41]
De Kimpe, N., **2**, 343[15], 353[102], 357[102], 380[102], 423[32-34], 424[32,35]; **3**, 857[90]; **4**, 1031[8], 1043[8]; **5**, 904[45], 905[45], 925[45], 926[45], 943[45]; **6**, 500[182], 547[663]; **8**, 36[73], 38[73], 66[73]
Dekker, H., **8**, 144[78]
Dekker, J., **1**, 30[153]; **2**, 123[195,196], 124[204], 125[204], 280[27]
DeKlein, W. J., **4**, 763[210]; **7**, 851[18]
de Kock, R. J., **5**, 700[8], 737[8]
de Kok, P. M. T., **8**, 95[82,84]
Dekoker, A., **5**, 109[214]
de Koning, A. J., **8**, 589[46]
de Koning, C. B., **7**, 355[41]
de Koning, H., **2**, 823[118], 1063[105]; **5**, 453[66]; **6**, 746[97]
de Koning, J. H., **2**, 599[20]; **6**, 652[145]
de la Fuente Blanco, J. A., **4**, 161[89d]
Delair, T., **4**, 397[211]

DeLaitech, D. M., **6**, 147[84]
de la Mare, P. B. D., **4**, 270[3], 329[1], 330[1a], 344[1], 350[1], 351[1], 364[2], 369[2b]
De La Mater, G., **6**, 488[11], 508[11], 545[11]
DeLaMater, M. R., **7**, 330[9]
Delaney, M. S., **8**, 445[32]
de Lange, B., **4**, 36[101]; **6**, 26[111]
Delannay, J., **8**, 134[30]
DeLano, J., **6**, 122[134], 128[134]
De la Paz, R., **4**, 932[63]
de la Pradilla, F., **8**, 36[74], 38[74], 66[74]
de la Pradilla, R. F., **3**, 265[191]; **4**, 1046[117]; **7**, 376[82]
de Lasalle, P., **1**, 185[53]; **2**, 29[104], 205[101,101b]
de Laszlo, S. E., **2**, 1096[98]; **7**, 362[28]
Delaude, L., **6**, 111[60]
Delaunay, J., **8**, 134[31]
de Lauzon, G., **8**, 859[213]
Delavarenne, S., **7**, 8[53]
DeLay, A., **5**, 560[70]
Delay, F., **4**, 1002[54], 1007[108]; **5**, 560[71]; **8**, 807[112]
Delbecq, F., **1**, 191[78]; **3**, 256[113]; **5**, 774[170,171]
Del Bene, J. E., **1**, 286[12,13], 322[13]
Delbianco, A., **8**, 560[404]
Delbord, A., **7**, 473[35]
Del Buttero, P., **4**, 522[54], 523[57,58], 524[61]; **6**, 114[79]
del Carmen, D., **4**, 394[196]
Del Cima, F., **1**, 857[57]
Delduc, P., **4**, 748[160]
Delektorsky, N., **3**, 725[17]
de Lera, A. R., **3**, 223[148], 586[156], 610[156]; **4**, 505[139]; **5**, 742[159b,c]
Deleris, G., **1**, 328[17,18]; **2**, 564[1], 575[65], 576[65], 717[68], 718[72]; **8**, 409[82]
del Fierro, J., **7**, 182[163]
Delgado, A., **8**, 125[94]
Delgado, P., **2**, 517[61]
Del Giacco, T., **7**, 649[46]
Del Gobbo, V. C., **4**, 871[33]
Delhon, A., **8**, 343[112]
de Liefde Meijer, H. J., **5**, 1148[117]
Del'Innocenti, A., **4**, 18[61], 115[182,182e]; 247[97], 249[130], 256[97], 257[130], 262[130]; **5**, 438[162]; **6**, 179[127], 1064[90a]; **7**, 627[43]
Delion, A., **4**, 298[289]
Dell, C. P., **1**, 477[139]; **2**, 810[61], 824[61]; **5**, 841[91], 853[91d]
Della, E. W., **1**, 855[51]; **8**, 229[137], 230[137], 231[142], 798[50,63]
Dellaria, J. F., Jr., **6**, 118[103], 208[56]; **7**, 230[124,126]; **8**, 50[120], 66[120]
Della Vecchia, L., **3**, 105[219], 113[219]
Dell'Erba, C., **4**, 426[47,52,58], 457[62-64], 460[64], 471[62], 475[148], 476[64,158,159,160,161]; **6**, 240[80]
DelMar, E. G., **2**, 148[79], 170[169]
Delmas, M., **1**, 821[20-22]; **2**, 354[108], 772[17], 775[29]; **6**, 173[46], 175[46,71,73]
Del Mazo, J. M., **6**, 579[984]
Del Mazza, D., **6**, 452[135]
Delmond, B., **2**, 727[133]; **3**, 741[50]
Delmonte, D. W., **8**, 872[6]
DeLoach, J. A., **4**, 255[198]; **5**, 133[59], 140[87]; **6**, 1063[83]
Delogu, G., **8**, 91[54]
De Lombaert, S., **2**, 651[115,115a]; **4**, 117[191]; **6**, 164[195], 509[257]; **8**, 844[68]
de Lopez-Cepero, I. M., **1**, 436[150]
Delorme, D., **1**, 773[202]; **3**, 277[27]; **6**, 210[74], 214[74]
Delpech, B., **4**, 296[263]; **6**, 284[172,174]; **7**, 381[105]
Delphey, C., **5**, 710[51], 713[51]
Delpierre, G. R., **1**, 391[150]
Del Pra, A., **4**, 915[15], 936[68]
Delpuech, J.-J., **8**, 384[27]
Del'tsova, D. P., **6**, 527[409]
DeLuca, H. F., **6**, 219[121], 989[79]; **7**, 675[54]
DeLuca, O. D., **8**, 240[33]
DeLucca, G., **6**, 210[77], 960[57]
De Lucchi, O., **4**, 50[142], 102[132]; **5**, 205[41-44], 206[45,46], 207[41], 211[61], 224[101], 324[17], 370[102], 371[102], 582[179], 584[196,197], 986[37]; **6**, 150[114], 936[108], 999[122,123]; **7**, 205[64], 777[376]; **8**, 836[2], 842[2e], 843[2e], 844[2e]
De Lue, N. R., **2**, 13[56]; **7**, 604[132], 606[146,149]
del Valle, L., **3**, 418[23]
Delyagina, N. I., **6**, 495[146]
Demachi, Y., **6**, 624[144]
De Mahieu, A. F., **1**, 648[131], 650[131]; **4**, 992[158]
Demailly, G., **1**, 513[49], 833[118]; **2**, 1024[64], 1025[65], 1026[64,72]; **3**, 343[16]; **6**, 149[109], 155[154,155], 156[155,156,157,158,159]; **8**, 12[66], 837[17]
Demain, A. L., **7**, 429[156]
de Maldonado, V. C., **2**, 604[50]
de March, P., **1**, 832[113]; **2**, 359[164]; **4**, 1089[135], 1090[135]; **6**, 173[42]
De Maria, P., **4**, 426[55]
de Mayo, P., **1**, 844[5d]; **3**, 379[3], 386[66], 706[7], 707[9,10], 815[73], 836[81], 922[37], 924[37]; **4**, 780[2], 785[2]; **5**, 124[8,9], 125[16], 126[22], 127[22,24], 128[22,26], 194[2], 196[2], 197[2], 198[2], 202[2], 209[2], 210[2], 215[4], 219[42], 223[72], 224[4], 637[112], 638[116], 738[148], 760[92], 904[53]; **6**, 795[2], 799[2]; **7**, 671[4], 689[8]; **8**, 17[114], 21[114], 115[66], 335[66]
Dembech, P., **4**, 115[182,182e], 247[97], 256[97]; **6**, 179[127]
de Meester, P., **8**, 846[83]
de Meijere, A., **1**, 664[169], 665[169], 669[169], 670[169]; **3**, 905[138]; **4**, 18[59], 545[125], 546[125], 874[53]; **5**, 78[283], 226[111], 925[149], 1039[16], 1041[16], 1043[16], 1044[16], 1046[16], 1048[16], 1049[33], 1052[33,38], 1188[15]; **7**, 842[33,34]
DeMember, J. R., **3**, 299[34b], 333[34,208,209]; **6**, 749[101]
de Mendoza, J., **4**, 425[26]
Demerac, S., **4**, 51[144a]
Demers, J. P., **6**, 112[68], 118[100]
Demerseman, B., **8**, 456[211], 458[211]
Demerseman, P., **2**, 332[54]; **7**, 16[157], 333[21]
De Mesmaeker, A., **1**, 850[31,32]; **4**, 792[63], 795[82]
DeMicheli, C., **5**, 625[31], 626[33]
De Mico, A., **3**, 512[192], 515[192]; **5**, 771[151,152], 772[151]; **7**, 265[100,102], 266[107], 267[100,102,107], 530[15]
Deming, S., **4**, 1099[185]
Demir, A. S., **2**, 520[68,69], 521[70]; **4**, 161[87], 222[172]
Demko, D. M., **1**, 127[91]; **5**, 527[61], 531[61], 534[92]; **7**, 579[131]
Demmin, T. R., **3**, 347[25]; **6**, 774[48]; **7**, 700[60]; **8**, 530[109]
Demo, N. C., **7**, 12[96]
Demole, E., **2**, 547[93]; **4**, 378[106]; **5**, 757[80], 761[80], 830[32]
Demonceau, A., **3**, 1047[7], 1051[7]; **4**, 1031[4], 1033[16,16d], 1052[16d]; **6**, 25[101]; **7**, 8[61]
DeMore, W. B., **7**, 8[57]
De Mos, J. C., **8**, 499[43], 500[43]
DeMott, D. N., **7**, 16[156]
Demoulin, A., **5**, 482[171]
Demoulin, B., **3**, 88[123], 91[123], 93[123], 109[123]
Dempster, C. J., **5**, 817[146]
De Munno, A., **6**, 775[51]
Demuth, F., **3**, 124[283], 125[283], 127[283], 132[283]
Demuth, M., **1**, 436[149], 892[149]; **3**, 216[75], 815[73]; **5**, 200[29], 215[5-11], 219[5-11,39], 221[57], 223[72], 225[5,8,93,94,114], 226[5-11,107,108], 227[114,115], 228[5,8,114,115], 229[5,7,8], 230[5-11,39,57,114,115,127,-128,129,130,131], 231[5],

232[57,93,127,129,130,131,134,135,136], 233[93,94,114,115], 234[9,10,140], 760[92], 944[240]; **7**, 650[50]
Demuynck, C., **2**, 464[102], 465[102]
Demuynck, J., **1**, 631[59], 632[59]
DeMuynck, M., **4**, 1040[75]; **7**, 363[33]
Dem'yanova, E. A., **8**, 628[177]
Den Besten, I. E., **7**, 548[76], 558[76]
Denemark, D., **4**, 798[109]
Dener, J. M., **4**, 795[81]
Denerley, P. M., **6**, 501[184]
DeNeys, R., **8**, 237[12]
Deng, C., **3**, 53[274]; **6**, 531[433]
Deng, D., **8**, 447[135], 680[70], 691[70]
Deng, Y.-X., **5**, 1148[114]
den Hertog, H. J., **4**, 493[77]
den Hertog, H. J., Jr., **2**, 821[110]
Deniau, J., **4**, 98[109d]
DeNicola, A., **5**, 1062[59]
De Nie-Sarink, M. J., **3**, 302[49]; **8**, 93[70], 94[78], 561[419]
DeNinno, M. P., **1**, 329[35], 765[150]; **2**, 465[108], 570[38], 652[126], 662[1], 690[1], 693[73], 694[74], 704[88]; **6**, 989[80], 995[80]; **7**, 246[89]
Denis, G., **2**, 772[17], 775[29]
Denis, J. M., **5**, 444[187], 576[136,214], 589[214]
Denis, J.-N., **1**, 630[41], 636[41], 641[108], 642[108], 645[41,125], 646[41,125], 647[41], 648[129], 649[129], 650[129,139], 665[108,174], 666[176], 667[108], 668[176], 669[41,125,174], 670[41], 672[41,108,129,176,199], 675; **3**, 86[50], 87[83,89-91,110,118], 89[110], 105[110], 106[83,110], 114[110], 120[83,90,110], 121[83], 155[427], 253[91]; **4**, 120[200], 349[114]; **6**, 213[90], 976[8], 980[44]; **7**, 406[76], 496[215], 773[307]; **8**, 405[27,28], 411[27,28], 806[125], 847[97], 848[97e], 849[97e,108], 886[107], 990[41]
Denis, J.-P., **2**, 961[39]; **6**, 495[149]
Denis, J. S., **1**, 215[33]
Denis, Y. St., **4**, 370[42]
Denise, B., **5**, 1105[162,163]
Denis-Garez, C., **6**, 424[57]
Denisov, E. T., **7**, 10[77]
Denivelle, L., **6**, 244[108]
Denkewalter, R. G., **6**, 664[223]
Denmark, S. E., **1**, 36[173], 158[74], 180[34], 237[31], 239[31], 294[38], 304[38], 340[93], 359[18], 380[18], 381[18], 585[13], 615[51,52], 616[52], 768[170]; **2**, 3[11], 4[15], 6[35], 564[9], 568[32], 573[49], 630[8], 655[145], 710[19], 978[12]; **3**, 21[127]; **4**, 253[168], 255[199]; **5**, 539[104], 760[91], 762[95,97-100,104,106], 763[95,98,104,108,109], 764[104,109], 765[97,110,111], 829[17,18], 847[17,137,138], 856[220], 1004[26,27a,c]; **6**, 856[158,159,160], 857[159], 858[160]; **7**, 397[27]; **8**, 388[60], 946[138]
Denne, I., **5**, 336[50]
Denney, D. B., **3**, 760[137]; **7**, 95[80]
Denney, D. Z., **7**, 95[80]
Denney, R. C., **5**, 947[259]
Denniff, P., **2**, 183[20]
Dennis, E. A., **6**, 677[311]
Dennis, N., **4**, 1093[145]; **5**, 630[52]
Dennis, W. H., Jr., **8**, 568[477]
Denniston, A. D., **1**, 506[10]; **2**, 90[41]; **3**, 132[334], 133[334], 136[334], 354[62]; **4**, 120[201]; **6**, 134[29]
Denny, R. W., **7**, 96[88], 97[88], 98[88], 110[88], 111[88], 165[82], 178[82]
Denny, W. A., **2**, 759[34a]; **4**, 435[137]; **7**, 68[84], 71[99], 72[84]; **8**, 331[36]
Deno, N. C., **4**, 1015[193]; **5**, 491[204], 754[62]; **6**, 262[18], 263[21]; **7**, 13[120], 16[160], 17[171], 235[1], 851[25]
DeNoble, J. P., **5**, 893[41]
Denoon, C. E., Jr, **2**, 832[154]
Denot, E., **8**, 530[106]
Dent, B. R., **4**, 1005[90]
Dent, S. P., **8**, 265[53]
Dent, W., **4**, 588[53], 600[229], 795[86], 1086[118], 1088[118], 1089[118]; **5**, 260[64], 265[64]
Denton, D. A., **2**, 736[28]
Denyer, C. V., **8**, 887[114]
Denzer, H., **2**, 372[266]; **5**, 468[128]
Denzer, W., **1**, 185[56]; **2**, 29[106]; **7**, 549[45]
Deol, B. S., **8**, 196[117], 197[117]
de Oliveira, A. B., **2**, 745[97]
de Oliveira-Neto, J., **1**, 757[119]
Deorha, D. S., **2**, 283[46]
Deota, P. T., **5**, 233[139]
de Pascual Teresa, J., **3**, 396[115]
De Pasquale, R. J., **6**, 496[153]
De Paulet, A. C., **3**, 854[77]
Depaye, N., **7**, 760[25], 846[92]
de Perez, C., **5**, 113[233], 116[261]
Depezay, J.-C., **1**, 553[87], 770[185]; **3**, 198[44], 258[127]; **5**, 689[71]; **6**, 93[132]; **7**, 297[32], 487[146], 495[146]
DePinto, J. T., **6**, 291[199]
Depner, M., **6**, 565[919]
De Poortere, M., **5**, 108[209], 109[209]; **7**, 476[68]
DePorter, B., **4**, 610[336]
Depp, M. R., **3**, 380[7]; **7**, 544[39], 553[39], 556[39]
Depraétère, P., **6**, 456[157]
Deprès, J. P., **1**, 848[25,27]; **3**, 222[141], 783[26]; **6**, 175[69]
DePriest, R., **3**, 214[58]; **7**, 884[181]
DePriest, R. N., **8**, 3[15], 315[53], 802[79,80]
DePue, R. T., **1**, 28[143], 29[144], 34[169]; **2**, 507[26,27], 508[27]; **3**, 33[189], 34[198], 39[198]; **6**, 727[195,197]
DePuy, C. H., **1**, 878[105]; **4**, 1016[206,207]; **6**, 692[405], 954[17], 1012[4], 1013[4], 1033[126,129]; **8**, 91[60], 856[167]
de Raditsky, P., **7**, 13[107]
Derancourt, J., **6**, 423[48]
Derdar, F., **7**, 452[51], 453[51]
Derdzinski, K., **4**, 357[148]
de Reinach-Hirtzbach, F., **2**, 430[53]; **3**, 748[77]; **7**, 764[129]
Derelanko, P., **7**, 56[20,21]
Derenberg, M., **2**, 737[33]
Derenne, S., **8**, 674[34]
Dereu, N., **7**, 774[333]
Derfer, J. M., **4**, 283[151]
Derguini, F., **2**, 482[34], 484[34]; **3**, 215[65]
Derguini-Boumechal, F., **3**, 224[161], 243[13], 262[161], 263[161], 416[20], 417[20], 464[176], 466[176,188]
Derieg, M. E., **3**, 414[1]
Derkach, N. Y., **6**, 463[26]
Dermanovic, M., **3**, 313[103]
Dermanovic, V., **3**, 313[103]
Dermer, O. C., **2**, 735[13]; **3**, 505[161], 507[161]; **7**, 470[7], 472[7], 474[7], 476[7]
Dermer, V. H., **2**, 735[13]
Dern, D., **7**, 880[153]
Dern, M., **5**, 71[135]; **7**, 880[152]
Derocque, J.-L., **3**, 396[103]
Derome, A. E., **5**, 167[94]; **8**, 354[174]
Deronzier, A., **7**, 809[83]
de Rooy, J. F. M., **6**, 602[3], 662[214]
de Rossi, R. H., **4**, 426[40,41,45,56], 452[18], 453[18], 454[18,39], 457[18], 459[39], 461[39], 463[111], 464[111], 468[18,133], 469[39], 471[18], 475[111], 502[123], 765[228]
de Rostolan, J., **2**, 901[31]; **6**, 912[22], 920[43]
Dersch, F., **4**, 868[13]
DeRussy, D. T., **1**, 240[43]; **5**, 796[53], 812[135], 815[135], 817[150]

Dervan, P. B., **5**, 64[33-35], 639[124], 805[99], 1025[81,81e]; **7**, 742[56,57]; **8**, 473[11], 886[111]
Deryabin, V. V., **4**, 529[76]
Des, R., **1**, 4[28]
des Abbayes, H., **3**, 1020[12], 1026[38,39]; **5**, 1138[59]
Desai, B. N., **6**, 554[769]
Desai, K. R., **6**, 441[80]
Desai, M. C., **2**, 638[61], 640[61]; **3**, 546[123], 797[91]; **4**, 239[21], 256[21], 261[21]; **6**, 174[60], 674[294]; **7**, 182[161], 595[12,127], 604[127], 680[77]; **8**, 48[110], 66[110], 721[142], 722[142,147]
Desai, R. C., **6**, 21[79]; **7**, 177[144]; **8**, 112[26], 118[26], 949[154]
Desai, S. R., **4**, 15[50]; **6**, 690[396]
de Salas, E., **3**, 706[5]
De Sarlo, F., **8**, 851[127]
Desauvage, S., **1**, 641[108], 642[108], 665[108], 667[108], 672[108], 700[108], 705[108], 724[108]; **3**, 87[90], 120[90]
de Savignac, A., **6**, 245[127]
Desbois, M., **5**, 412[46]
Deschamps, B., **1**, 683[219,220]; **2**, 430[51]; **4**, 71[16a,17b], 139[3]; **8**, 859[213], 865[247]
Deschler, K., **6**, 196[231]
Deschler, U., **6**, 175[67], 178[123]
DeSchryver, F. C., **5**, 637[109]; **7**, 476[68]
Descoins, C., **3**, 246[34], 416[21]; **4**, 1007[111]; **6**, 692[408]
Descotes, G., **5**, 1007[36]; **8**, 551[344], 552[355,356]
DeSelms, R. C., **1**, 878[108]; **4**, 1005[85]; **5**, 560[71]
de Sennyey, G., **7**, 88[26]
Desert, S., **4**, 259[277]
Deshayes, H., **3**, 613[3]; **5**, 848[141]; **8**, 817[25,33]
Deshmukh, A. A., **7**, 283[182], 284[182]
Deshmukh, A. R. A. S., **6**, 1022[64]
Deshmukh, M., **6**, 136[40], 150[112]
Deshmukh, M. N., **7**, 425[146], 777[377], 778[377]
Deshmukh, M. W., **1**, 568[236]
DeShong, P., **4**, 905[213], 1076[41]; **5**, 255[52,53], 260[52,63b], 264[52,53,63b]
Deshpande, M. N., **1**, 420[85]; **3**, 380[7]; **7**, 544[39], 553[39], 556[39]
Deshpande, V. H., **2**, 762[57]
Desilets, D., **4**, 110[151]
de Silva, A. P., **5**, 724[112]
DeSilva, N., **7**, 822[34]
de Silva, S. O., **1**, 466[48,49]; **7**, 355[47]
Desilvestro, H., **6**, 711[66]
Desimoni, G., **2**, 351[79], 364[79,204]; **5**, 451[24,44], 453[24,44], 454[70], 468[24,44], 485[24]
Desio, P. J., **1**, 225[88], 226[88], 326[3]; **4**, 98[108a]
Desjardins, S., **8**, 135[39]
Deskin, W. A., **7**, 16[164]
Deslongchamps, P., **1**, 766[157], 894[158]; **2**, 36[132], 169[167], 303[6], 1008[6]; **3**, 56[286], 380[10]; **4**, 23[70], 29[85], 46[131], 245[84], 258[84], 260[84], 390[170], 1011[162], 1040[83,84], 1043[83,84]; **5**, 532[86,86c]; **6**, 2[3], 25[3], 35[13], 40[13], 1043[13]; **7**, 673[31]; **8**, 211[3], 531[115], 925[5]
Desmaele, D., **4**, 221[164]; **5**, 18[127], 152[20], 169[109], 174[124], 176[125], 185[124]
DesMarteau, D. D., **4**, 347[102]; **6**, 497[157], 498[162]; **7**, 500[246], 747[95]
Desmond, R., **1**, 402[17], 791[296a], 799[296]; **2**, 482[37], 483[37], 485[37]; **4**, 98[116]; **7**, 228[92]
Desobry, V., **4**, 527[68], 532[68,84], 534[68,84], 536[68,84], 545[84], 546[84]
deSolms, S. J., **2**, 971[92]; **5**, 137[81], 780[201]
de Souza, J. P., **7**, 124[49], 127[49]
de Souza, N. J., **7**, 64[64]
de Souza Barbosa, J. C., **4**, 95[99b]
Despeyroux, B., **3**, 1030[60]; **4**, 939[79], 941[79]
de Spinoza, G. R., **6**, 204[19]
D'Esposito, L. C., **4**, 924[34]
Despreaux, C. W., **6**, 913[24]; **7**, 70[94]
Des Roches, D., **3**, 770[181]; **6**, 686[367,368]
Desrut, M., **8**, 559[401], 560[402]
Dess, D. B., **7**, 311[32], 324[32]
Dessau, R. M., **4**, 763[211]; **7**, 154[24], 870[96]
Dessauges, G., **7**, 3[5]
Dessy, R. E., **1**, 361[30]; **3**, 581[110]; **7**, 805[66]; **8**, 136[52], 725[185], 950[169]
Destro, R., **2**, 833[148]; **4**, 261[285]
de Suray, H., **4**, 953[9]
DeTar, D. F., **6**, 69[17]; **8**, 917[119]
DeTar, M. B., **4**, 1093[151], 1095[151]
Detilleux, E., **7**, 13[109]
Detre, G., **6**, 1059[63]; **8**, 934[54], 938[54], 948[149], 993[57]
Detty, M. R., **4**, 50[142,142e,l]; **6**, 11[45], 463[23], 960[62]; **7**, 771[262], 773[262], 774[315], 775[262], 777[365]; **8**, 406[36], 411[100]
Detzel, A., **3**, 829[54]
Deubelly, B., **1**, 17[213,214,215]
Deuchert, K., **3**, 197[40]; **6**, 682[340]
Deugau, K., **5**, 754[67]
Deuschel, G., **8**, 271[102], 636[2]
Deuter, J., **3**, 872[61,62]
Deutsch, E. A., **2**, 527[9], 528[9], 547[98], 548[98], 550[98]; **5**, 2[7], 4[7], 5[46], 6[46], 527[59]
Deutschman, A., Jr., **5**, 79[287]
Dev, S., **3**, 713[27], 715[37], 737[33]; **5**, 759[88], 776[88,178,179], 802[83], 810[83]; **6**, 1052[45]; **7**, 95[73a], 279[170], 375[80], 544[40], 551[40], 556[40], 676[64], 844[61], 845[61]; **8**, 48[110], 65[210], 66[110,210], 330[48], 333[52], 526[17], 530[104]
deVaal, P., **5**, 647[18], 649[18], 652[18], 656[18]
Devadas, B., **4**, 497[96]; **6**, 533[493]
Devant, R., **2**, 226[159], 231[170]
Devaprabhakara, D., **3**, 386[58,61], 393[61]; **4**, 284[156], 303[343], 311[450]; **7**, 601[82-84,91], 602[84]; **8**, 450[170], 477[31], 708[31,32], 726[193]
Devaquet, A., **4**, 1082[86]
de Vargas, E. B., **4**, 426[40,41]
Devaud, M., **8**, 412[107], 857[199]
Deveze, L., **7**, 59[42]
DeVicaris, G., **5**, 404[18]
Deville, C. G., **4**, 1082[85], 1085[85]
Devine, J., **6**, 267[56]
Devine, K. G., **6**, 604[30]
De Vita, C., **8**, 856[170]
Devitt, F. H., **4**, 74[38b]
Devlin, J. A., **1**, 130[95]
Devlin, J. P., **4**, 273[46], 280[46], 753[173]
De Voe, R. J., **5**, 71[157], 650[25]
de Voghel, G. J., **6**, 543[624]
Devos, A., **6**, 493[129]
Devos, L., **1**, 34[227]
Devos, M. J., **4**, 259[266,269], 992[153], 993[160]
de Voss, G., **3**, 829[54]
DeVoss, J. J., **4**, 303[342]; **5**, 736[145], 737[145]; **7**, 635[70]; **8**, 854[152], 856[152]
Devreese, A. A., **7**, 363[33]
de Vries, B., **8**, 453[191]
de Vries, G., **7**, 706[23]
DeVries, G. H., **2**, 463[81], 464[81]
de Vries, J. G., **2**, 855[242]; **8**, 95[91]
DeVries, K., **1**, 402[16]
de Vries, L., **3**, 406[142]
DeVries, V. G., **8**, 30[42], 66[42]

de Waal, W., **3**, 20[108]
de Waard, E. R., **2**, 812[73]; **3**, 153[414], 155[414], 875[72,77], 881[99], 882[100,101]; **6**, 753[116], 755[116]; **8**, 843[55]
Dewan, J. C., **1**, 314[129]; **5**, 1175[40], 1178[40]; **8**, 683[95], 686[95]
Dewar, J., **2**, 152[103]
Dewar, M. J. S., **1**, 580[4]; **2**, 1054[56]; **3**, 664[32], 665[32]; **4**, 270[1], 273[45], 280[45,119,120], 484[11,14], 1070[15]; **5**, 67[94], 72[179], 491[206], 516[19], 703[19], 705[19], 829[24], 856[197,198,205,214], 857[228], 900[5]; **7**, 872[98]; **8**, 724[169,169c]
de Weck, G., **5**, 221[62], 1080[53], 1084[53], 1107[171], 1108[171]
Dewey, R. S., **8**, 472[4], 476[25]
DeWilde, H., **4**, 259[268], 262[268]; **5**, 924[146]
DeWinter, A. J., **1**, 648[128], 649[128], 650[128], 672[128], 675[128], 679[128], 708[128], 710[128], 715[128], 716[128], 862[78]
de Wit, P. P., **2**, 482[33], 484[33]
DeWitt Blanton, C., Jr, **2**, 828[134]
De Witte, M., **6**, 80[66]
De Wolf, W. H., **4**, 1002[59], 1018[226]
DeWolfe, R. H., **2**, 368[239]; **6**, 488[24], 489[24], 556[24,27], 561[24,27], 562[24,27], 563[24,27], 566[24], 567[24], 571[24], 572[27]
de Woude, G. V., **8**, 348[130], 349[130]
Dexheimer, E. M., **4**, 487[47]
Dey, A. N., **8**, 312[21]
Dey, K., **2**, 743[78]
Deya, P. M., **7**, 346[8]
Deycard, S., **4**, 723[38], 738[38], 747[38]
Deyo, D., **1**, 66[54], 87[50]; **2**, 536[44], 538[60], 539[60]
Deyo, R. A., **6**, 822[116]
Deyrup, J. A., **1**, 834[131], 835[131,137], 836[131], 837[131]; **2**, 428[45]; **5**, 583[184]; **7**, 470[5], 471[5], 472[5], 473[5], 474[5], 476[5], 481[5], 483[5]; **8**, 386[50]
Déziel, R., **2**, 624[161], 1059[80], 1102[123]
Dezube, M., **4**, 379[114,114a], 380[114a]; **6**, 26[106]; **7**, 415[115c], 418[115c]
D'Haenens, L., **3**, 900[93]
Dhaliwal, G., **4**, 37[107]
Dhanak, D., **4**, 1018[224], 1019[224]
Dhani, S., **4**, 564[39]
Dhanoa, D. S., **1**, 433[226]; **3**, 212[40], 250[70]; **4**, 177[57], 789[30], 791[30]
Dhar, D. N., **5**, 71[146]; **7**, 760[26]
Dhar, R., **5**, 634[79]
Dhar, R. K., **2**, 247[82]; **7**, 267[119,120]
Dharan, M., **4**, 1081[77], 1082[88,89]
Dhararatne, H. R. W., **1**, 804[308], 805[308]
Dhareshwar, G. P., **6**, 825[129]
Dhavale, D. D., **2**, 657[161b]
Dhawan, K. L., **3**, 32[188], 265[193], 266[194]; **5**, 692[102]; **6**, 724[154]
Dheer, S. K., **6**, 625[156]
Dhillon, R. S., **3**, 231[244]
Dhimane, H., **3**, 579[101]
Dhingra, O. P., **3**, 509[176], 660[19], 670[57], 673[68], 679[19], 681[68], 683[57b], 686[68], 807[32,33]; **6**, 7[32], 419[12]; **8**, 798[58]
Dia, G., **8**, 451[172]
Diab, J., **5**, 773[167], 774[167]
Diab, Y., **1**, 387[130,131]
Diakur, J., **6**, 438[42], 463[27]
Dial, C., **4**, 288[188], 346[86a]
Dialer, K., **7**, 450[15]
Diamanti, J., **2**, 840[182]
Diamond, S. E., **7**, 452[46]
Dianin, H., **3**, 660[4]
Diaper, D. G. M., **2**, 277[3]
Dias, A. R., **8**, 671[17]
Dias, H. V. R., **1**, 11[57], 23[118]
Dias, J. R., **7**, 680[78]; **8**, 248[84]
Diatta, L., **8**, 58[174], 66[174]
Diaz, F., **3**, 327[171]
Diaz, G. E., **7**, 883[175]
Diaz De Villegas, M. D., **2**, 406[45]
Dibi Ammar, **6**, 530[422]
Di Braccio, M., **6**, 487[49-51,53], 489[53]
Dich, T. C., **3**, 124[286], 125[286], 127[286]; **5**, 1152[143]
Dick, K. F., **7**, 254[29]
Dickason, W. C., **8**, 707[30], 708[30]
Dicke, R., **3**, 495[93b]; **4**, 760[196]
Dicken, C. M., **5**, 255[52,53], 260[52], 264[52,53]
Dickenson, H. W., **6**, 818[106]
Dickenson, W. A., **6**, 281[147,148]
Dicker, D. W., **6**, 267[56]
Dickerhof, K., **8**, 113[40], 115[40], 863[237]
Dickers, H. M., **8**, 770[41]
Dickerson, D. R., **6**, 221[134]
Dickerson, J., **4**, 846[72], 849[82]
Dickerson, J. R., **7**, 264[93]
Dickerson, R. E., **2**, 547[101], 548[101]; **3**, 14[77], 15[77], 363[84]
Dickey, E. E., **3**, 693[143]
Dickey, J. B., **6**, 213[87]
Dickinson, A. D., **5**, 223[81]
Dickinson, C., **1**, 34[231]
Dickinson, R. A., **2**, 809[52], 823[52]
Dickinson, T., **3**, 634[25a]
Dickman, D. A., **1**, 477[136]; **3**, 75[50], 77[55], 78[60-62], 79[60,61], 81[60,61,66]; **7**, 224[49]
Dickson, J. K., Jr., **4**, 815[194], 817[194]; **5**, 837[70]
Dickson, L., **8**, 507[86]
Dickson, R. S., **5**, 1038[7], 1133[32], 1134[34,36,43], 1146[32]
Dickstein, J. I., **4**, 3[10], 41[10], 47[10], 66[10]
DiCosmio, R., **2**, 465[106]
Diddams, P. A., **4**, 313[466]
Diebold, J. L., **3**, 845[36]
Dieck, H., **3**, 271[1], 530[60]; **4**, 841[48], 844[56], 845[67], 850[85], 852[48]
Dieck-Abularach, T., **2**, 844[198]
Dieckmann, W., **2**, 796[2]
Dieden, R., **1**, 661[167,167a,c]; **3**, 87[87]
Diederich, F., **3**, 557[42], 927[49]
Diehl, J. W., **8**, 322[108,109]
Diehl, K., **6**, 185[164], 187[164]
Diehr, H. J., **4**, 51[145a]; **8**, 990[40]
Diels, O., **3**, 893[54]; **4**, 44[125]; **5**, 316[1], 426[106], 451[3,4], 513[1], 552[5]; **8**, 526[23]
Dien, C.-K., **4**, 282[137]
Diepers, W., **2**, 1054[60]; **5**, 501[263]
Diercks, P., **7**, 95[73a]
Diercks, R., **3**, 537[90], 538[90,92]; **5**, 1151[130]
Dieter, J. W., **8**, 540[201]
Dieter, L. H., **1**, 543[19]
Dieter, R. K., **2**, 120[181], 517[60], 838[167]; **3**, 22[133], 24[150]; **4**, 189[102], 190[108], 191[109], 192[115], 229[232]; **5**, 178[139]; **8**, 540[201], 836[2,8], 839[2d], 842[2d,8]
Dietl, H., **3**, 21[122]; **4**, 587[20,22,37]
Dietliker, K., **6**, 543[625]
Dietrich, H., **1**, 9[41,44], 12[61], 13[65], 18[93,94], 19[100], 23[123], 29[147]; **3**, 693[143]; **6**, 723[145]
Dietrich, H. W., **7**, 506[298]
Dietrich, M. W., **2**, 363[189]
Dietrich, R., **3**, 890[34], 904[131,132]; **4**, 1104[209]
Dietrich, W., **1**, 162[92]

Dietsche, T. J., **4**, 588[71], 614[380], 615[380], 628[380], 652[432]; **5**, 830[29]; **6**, 154[151], 1021[50]; **8**, 843[51], 844[51]
Dietz, F., **6**, 489[81]
Dietz, G., **2**, 361[176]
Dietz, K.-P., **7**, 471[23], 474[23]
Dietz, M., **4**, 109[149], 119[195]
Dietz, R., **7**, 810[88], 872[98]
Dietz, R. E., **5**, 1089[83], 1090[88], 1092[92], 1098[120,121], 1101[120], 1103[83,153], 1112[120,121]
Dietz, S. E., **8**, 36[48], 66[48], 347[140], 616[102], 617[102], 618[102]
Di Fabio, R., **7**, 240[55], 266[110], 267[110]
Differding, E., **1**, 683[227], 714[227], 715[227], 717[227], 718[227]; **3**, 786[42]; **5**, 113[238]
Di Furia, F., **7**, 95[69], 425[147a], 762[69,84], 777[69b,376,380], 778[69]
DiGiacomo, P. M., **8**, 798[55]
Di Giamberardino, T., **1**, 694[237]
DiGiorgio, J. B., **7**, 96[87]; **8**, 338[88]
DiGiovanni, J., **7**, 346[7]
Di Gregorio, F., **8**, 171[110]
Dijkink, J., **2**, 1054[57], 1062[57,100], 1072[57]; **3**, 361[77], 364[77], 368[104]; **8**, 273[129]
Dijkstra, D., **2**, 902[46]
Dijkstra, G., **6**, 2[7], 21[79]
Dijkstra, P. J., **2**, 821[110]
Dijkstra, R., **3**, 904[129]
Dikareva, L. M., **7**, 108[170]
Dike, M., **4**, 130[226b]
Dike, M. S., **1**, 584[9]; **2**, 714[50]; **4**, 24[72,72f], 1049[121b]; **5**, 994[53], 997[53]; **8**, 927[20]
Dike, S., **4**, 130[226b]
Dikic, B., **1**, 515[58]
Dillender, S. C., Jr., **4**, 464[118], 465[118]
Dilling, W. L., **2**, 151[98], 152[98]; **5**, 123[1], 126[1], 636[90]; **8**, 543[244,245]
Dillinger, H. J., **4**, 52[147,147e]
Dillon, P. W., **5**, 736[140]
Dillon, R. T., **4**, 288[184]
Dilworth, B. M., **3**, 862[8]; **6**, 142[66]; **7**, 208[79], 211[79], 214[105]
Dim, N. ud., **7**, 801[44]
Di Mare, A., **8**, 72[241], 74[241]
DiMare, M., **1**, 430[127], 798[288]; **2**, 846[206]; **6**, 995[101]
Di Martino, A., **6**, 69[17]
DiMatteo, F., **8**, 347[141], 350[141]
Dimcock, S. H. D., **6**, 831[9]
Dime, D. S., **1**, 584[12]; **2**, 713[47]; **4**, 973[83]; **5**, 757[75,76], 778[75,76]
DiMechele, L. M., **7**, 877[135]
Dimitriadis, E., **8**, 212[23]
Dimmel, D. R., **4**, 869[27], 870[27], 871[27]
Dimock, S. H., **1**, 114[38]
Dimroth, K., **7**, 747[103]; **8**, 568[466]
Dimroth, O., **7**, 92[41,41a], 94[41], 152[1]
Dimsdale, M. J., **2**, 421[26]
Din, Z. U., **5**, 819[152]
Dinculescu, A., **2**, 712[40]
Dinda, J. F., **4**, 1084[93]
DiNello, R. K., **5**, 71[137]
Diner, U. E., **8**, 214[38], 218[71], 221[89,90]
Dinerstein, R. J., **7**, 43[43]
Dines, M., **4**, 356[142]
Dines, M. B., **6**, 287[181,182]
Ding, Q.-J., **7**, 283[185]
Ding, W., **4**, 991[152]; **6**, 184[150]
Ding, X., **2**, 772[18]
Dingerdissen, U., **1**, 37[251], 531[132]
Dingwall, J., **3**, 848[50]
Dinh-Nguyen, N., **3**, 634[29]
DiNinno, F., **2**, 212[118], 1102[123]; **6**, 511[300]
DiNinno, F., Jr., **4**, 25[76], 46[76], 128[222]
Dinizo, S. E., **7**, 229[110]
Dinjus, E., **1**, 320[160]; **5**, 1157[168]
Dinkel, R., **5**, 710[49]
Dinkeldein, U., **6**, 545[634], 546[634], 556[829], 567[634,829], 568[634]
Dinné, E., **5**, 702[10], 716[10], 794[47], 806[47], 824[47]
Dinner, A., **2**, 828[132]
Dinnocenzo, J. P., **5**, 913[101], 1014[58]; **7**, 749[118], 854[56], 855[56], 882[167]
Dinulescu, I. G., **4**, 600[234], 643[234], 963[44]
Di Nummo, L., **2**, 211[117]
Di Nunno, L., **7**, 737[15]
Dion, R. P., **5**, 945[253]; **7**, 3[10]
Dionne, G., **4**, 125[216,216g]
DiPardo, R. M., **2**, 962[51]; **7**, 602[98]; **8**, 11[61], 949[154]
DiPasquale, F., **3**, 7[32], 8[32]
DiPasquo, V. J., **3**, 855[81]
DiPierro, M., **2**, 1017[31,34,35], 1018[35]; **6**, 737[31,41], 746[89]
DiPietro, R. A., **6**, 173[47]; **8**, 946[141]
Di Rienzo, B., **6**, 538[572]; **8**, 587[32]
Dirks, G. W., **5**, 455[77]
Dirksen, H. W., **3**, 499[113]
Dirkx, I. P., **4**, 423[3]
Dirlam, J. P., **5**, 595[14], 597[29]; **6**, 530[417]; **8**, 391[87]
Dirlan, J. P., **7**, 799[27]
Dirnens, V. V., **7**, 477[79,81]
Dirstine, P. H., **4**, 491[67]
Disanayaka, B. W., **3**, 590[163]
Discordia, R. P., **7**, 413[115e]
Disnar, J. R., **3**, 202[95]
Disselnkötter, H., **4**, 54[153a], 56[153a]
Distefano, G., **5**, 257[60]; **6**, 711[62]
Distler, H., **3**, 872[59]
Distler, J. J., **6**, 614[79]
Ditrich, K., **1**, 832[108]; **2**, 39[137], 40[137c], 44[137c], 266[60], 267[64], 571[39]; **3**, 797[90]; **6**, 864[195]
Ditson, S. L., **8**, 47[125], 66[125]
Dittami, J. P., **3**, 23[135]; **4**, 797[105], 1101[192]; **5**, 225[97], 582[177], 938[217]; **6**, 784[91]; **7**, 120[5], 378[95]; **8**, 505[82], 507[82]
Dittel, W., **5**, 418[70]
Dittman, W. R., Jr., **8**, 318[68]
Dittmann, W., **3**, 396[107], 397[107]; **7**, 832[70]
Dittmar, W., **5**, 498[228]
Dittmer, D. C., **3**, 725[20]; **5**, 476[147]; **7**, 413[115e]; **8**, 96[94]
Dittmer, K., **2**, 968[79]
Dittus, G., **3**, 284[54]
Di'Tullio, D., **8**, 191[94]
Divakar, K. J., **7**, 828[50,50b]
Divakaruni, R., **3**, 1030[58], 1031[61]; **4**, 948[96]
Divanford, H. R., **2**, 1090[70], 1100[70]; **8**, 607[31]
Diversi, P., **4**, 706[38]; **5**, 1131[12], 1148[114], 1154[152], 1155[162], 1156[162]
Diwatkar, A. B., **6**, 660[202]
Dixit, D. M., **2**, 911[70]
Dixneuf, P. H., **8**, 456[211], 458[211]
Dixon, A. J., **4**, 260[281]
Dixon, B. R., **3**, 493[80]
Dixon, D. A., **6**, 708[48]
Dixon, J., **5**, 403[9]
Dixon, J. A., **4**, 871[31]; **8**, 314[26]
Dixon, J. R., **3**, 735[19]

Dixon, N. J., **3**, 444[68]
Djahanbini, D., **3**, 223[183], 225[183]
Djega-Mariadassou, G., **8**, 419[20], 420[20], 430[20]
Djerassi, C., **1**, 698[251]; **3**, 804[2], 805[15], 810[2,51], 812[2]; **4**, 23[70]; **5**, 595[15], 596[15]; **6**, 175[69], 471[63], 494[134], 680[331]; **7**, 92[42], 93[42], 222[36], 236[13], 253[19], 254[19], 400[45], 676[62], 820[23]; **8**, 88[38], 108[1], 118[1], 333[58], 344[119], 345[119], 353[154], 355[180], 445[16], 490[2], 492[2], 493[2], 524[3], 530[3], 533[147], 535[147], 566[450], 930[33], 964[54], 991[50], 992[50]
Djokic, S., **6**, 766[23]; **7**, 698[51]
Djuric, S., **1**, 580[1]; **2**, 712[42], 920[96]; **3**, 255[105]; **6**, 646[103], 832[12], 865[12], 935[102], 1007[150]; **7**, 255[35], 816[6b], 824[6], 825[6]
Dlubala, A., **2**, 73[68]; **6**, 454[145]
Dlugonski, J., **7**, 80[142]
Dmitrenko, A. V., **2**, 387[334]
Dmitrichenko, M. Yu., **6**, 495[150]
Dmitrienko, G. I., **5**, 326[24]; **8**, 609[55]
Dmitriev, B. A., **2**, 385[325]
Dmitriev, V. I., **8**, 771[42]
Dmowski, W., **6**, 496[155]; **8**, 897[19]
Doa, M. J., **5**, 627[43], 628[44]
Doad, G. J. S., **7**, 271[130]
do Amaral, A. T., **4**, 315[503]
Do Amaral, C. F., **2**, 464[93,94]
do Amaral, L., **4**, 315[503]
Doan-Huynh, D., **4**, 405[249], 406[249]
Doat, E. G., **1**, 472[75]
Dobbin, C. J. B., **5**, 485[184]
Dobbs, H. E., **4**, 1021[248]
Dobbs, K. D., **4**, 87[82], 213[120], 240[54]
Dobeneck, H. V., **6**, 501[203], 531[203]
Döbler, Chr., **3**, 229[224]
Dobler, W., **4**, 384[144,144b]
Dobrenko, T. T., **6**, 509[283]
Dobretsova, E. K., **5**, 431[122]
Dobrev, A., **6**, 269[71,72]; **8**, 974[123]
Dobrynin, V. N., **3**, 734[7]
Dobson, A., **8**, 552[350]
Dochwat, D. M., **5**, 909[97]
Dockal, E. R., **7**, 355[39]
Dockner, T., **6**, 644[94]
Dockx, J., **4**, 1001[23]
Dodd, D., **1**, 206[110]; **8**, 850[122]
Dodd, G. H., **6**, 620[124]
Dodd, J. H., **4**, 14[47,47o]
Dodd, T. N., Jr., **8**, 606[19]
Doddrell, D., **6**, 1042[6]
Dodds, D. R., **8**, 188[54], 201[54]
Dodonov, V. A., **8**, 753[68]
Dodson, P. A., **3**, 689[121], 813[60]
Dodson, R. M., **1**, 878[107]; **3**, 807[20], 818[95]; **5**, 208[52], 828[5], 847[5]; **8**, 928[26]
Dodson, V. H., **7**, 599[63]
Dodsworth, D. J., **2**, 928[121,122], 946[121,122]; **4**, 497[96]
Doecke, C. W., **3**, 627[43]; **5**, 324[21], 817[146]
Doedens, R. J., **1**, 488[13], 889[141], 890[141], 898[141b]; **2**, 1041[108,111]
Doehner, R., **4**, 1040[95], 1041[95b], 1045[95b]; **5**, 921[142]
Doerge, R. F., **2**, 962[46]
Doering, E., **8**, 92[66]
Doering, W. von E., **1**, 632[67]; **2**, 368[243]; **3**, 822[13], 823[16], 825[16], 829[13], 894[66], 896[66,68]; **4**, 10[31], 959[30], 999[5], 1000[5], 1006[101], 1009[147]; **5**, 64[23,29,30,37,41], 67[83,94], 709[46], 714[68], 721[100], 752[47], 820[160], 856[196], 857[224,228,230], 971[1]; **6**, 707[42], 970[124]; **7**, 8[58], 159[44], 296[24]; **8**, 88[42], 428[52]
Doerjer, G., **8**, 214[41]
Doerler, G., **3**, 904[133]
Dogan, B. M. J., **5**, 64[41]
Doherty, A. M., **4**, 390[175a]; **5**, 229[126], 239[1], 270[1c]; **7**, 404[67]; **8**, 847[100b]
Doherty, J. B., **2**, 542[83]
Doherty, N. M., **8**, 668[2], 672[2]
Doherty, R. F., **6**, 953[11], 969[119]
Dohmori, R., **8**, 244[54]
Do-hyun Nam, **7**, 490[178]
Doi, J. T., **4**, 366[6], 394[196]
Doi, K., **1**, 512[39]; **8**, 185[26], 190[26]
Doi, M., **6**, 1046[31]; **7**, 544[35], 556[35], 566[35], 821[29]
Doi, T., **1**, 553[98]; **3**, 380[10]; **4**, 192[117], 227[205], 255[196]; **5**, 532[87]
Dojo, H., **4**, 286[174], 287[174], 290[174], 358[158]
Dokuchaeva, T. G., **3**, 329[183]
Dolak, L. A., **2**, 158[125]; **3**, 353[47]; **7**, 77[119]
Dolak, T. M., **3**, 86[60], 88[60], 89[60]; **4**, 116[185d]
Dolan, S. C., **7**, 90[30], 301[61]; **8**, 824[60]
Dolata, D. P., **1**, 377[103]
Dolbier, W. R., Jr., **1**, 837[154]; **5**, 65[59,60], 586[204], 680[24,24a,b], 911[91,95], 912[95]
Dolby, L., **3**, 31[183]
Dolby, L. D., **8**, 408[61]
Dolby, L. J., **8**, 615[96], 616[100], 624[96]
Dolce, D. L., **3**, 621[30]
Dolcetti, G., **8**, 445[19]
Doldouras, G. A., **8**, 828[79]
Dolence, E. K., **1**, 337[80], 827[67], 828[69]
Doleschall, G., **6**, 112[66], 775[52]; **8**, 276[151], 662[114]
Dolfini, J. E., **4**, 48[140], 83[65a], 680[49]; **6**, 644[93]; **7**, 111[190]
Dolgii, I. E., **4**, 963[43]; **5**, 65[55], 1198[45]
Dolgov, B. N., **3**, 318[122]
Dolgova, S. P., **8**, 610[59]
Dolinski, R. J., **2**, 343[13]; **5**, 776[181]
Doll, L., **8**, 366[37]
Doll, R. J., **7**, 46[50], 47[50]; **8**, 248[87], 270[99]
Dolle, R. E., **3**, 39[216], 278[31], 289[31], 1029[55]; **5**, 477[159], 534[92]; **6**, 46[59,63]; **7**, 245[72], 401[56], 554[67]; **8**, 933[50]
Dolling, U.-H., **4**, 230[251]; **8**, 54[159], 66[159]
Dolling, V.-H., **7**, 877[135]
Dollinger, G. D., **5**, 702[13]
Dollinger, H., **1**, 476[114], 477[114], 559[153]; **3**, 65[6]
Dolman, D., **5**, 207[50]
Dolph, T., **3**, 596[193]
Dolphin, D., **2**, 743[86]; **5**, 71[137]; **7**, 12[95], 13[95]; **8**, 79[1], 82[1b], 605[9]
Dolphin, J. M., **1**, 378[105]; **3**, 960[115]; **5**, 438[163]
Dolson, M. G., **4**, 14[47,47f,g]; **5**, 438[162]
Dolzine, T. W., **1**, 215[33]
Domagala, J. M., **3**, 748[79], 749[80]
Domaille, P. J., **8**, 671[16]
Domalski, M. S., **4**, 877[67]
Domány, G., **4**, 110[151]
Domb, S., **5**, 217[24], 222[24], 223[24,66]
Dombi, G., **2**, 838[171]
Dombo, B., **6**, 641[60], 671[60,278]
Dombroski, M. A., **1**, 564[196]; **2**, 218[140]; **4**, 820[217]; **8**, 839[27,27a]
Dombrovskii, V. A., **4**, 951[1], 968[1], 979[1]
Dombrovskii, V. S., **4**, 310[426]
Dombrowski, A., **2**, 957[19]
Dombrowskii, A. V., **6**, 182[142]

Domek, J. M., **4**, 968[60], 969[60]
Domelsmith, L. N., **4**, 484[21], 1076[47]; **6**, 711[65]
Domiano, P., **2**, 284[53]
Domingo, A., **2**, 828[133]; **8**, 32[53], 66[53]
Domingo, L., **1**, 543[20]; **6**, 510[299]
Domingos, A. M., **1**, 37[180]
Dominguez, D., **6**, 487[76], 489[76], 977[14]; **8**, 874[24]
Domínguez, E., **8**, 15[91]
Dominguez, G., **5**, 92[64]
Dominguez, M. J. F., **8**, 2[10]
Dominguez Aciego, R. M., **1**, 759[132]
DoMinh, T., **4**, 1090[141]
Domnin, I. N., **4**, 1051[126,127]
Domnin, N. A., **8**, 950[164]
Domsch, D., **6**, 502[217], 560[870]; **7**, 650[51]
Domsch, P., **2**, 681[58], 683[58]
Don, J. A., **8**, 418[11], 437[11]
Donald, D. S., **5**, 486[196]
Donaldson, R. E., **4**, 79[55a], 251[152]; **6**, 163[192]; **8**, 494[26]
Donaldson, W. A., **3**, 274[17]; **4**, 604[283]
Donaruma, L. G., **6**, 727[203], 763[2]; **7**, 689[5], 691[5]
Donate, P. M., **4**, 300[310], 304[361], 305[361]
Donatelli, R. A., **6**, 1025[79]
Donati, D., **4**, 956[18]
Donati, M., **4**, 608[323], 839[26]; **7**, 452[61]
Donaubauer, J., **1**, 429[126], 798[289]; **6**, 995[101]
Donda, A. F., **5**, 1148[114]
Dondio, G., **1**, 765[151]
Dondoni, A., **1**, 471[68]; **3**, 232[267], 511[186,187], 514[186]; **5**, 99[131], 113[229,230], 114[242,243,244,245], 451[42], 485[42]; **8**, 394[117]
Donegan, G., **5**, 257[59]
Donek, T., **1**, 305[87]
Donelson, D. M., **2**, 388[342]
Doner, H. E., **7**, 845[69]
Donetti, A., **6**, 550[678]
Doney, J. J., **2**, 125[212,213], 127[212,213], 313[36,37]
Dongala, E. B., **2**, 225[157], 232[157]
Dönges, R., **5**, 65[64]
Donike, M., **6**, 653[149]
Donkersloot, M. C. A., **8**, 95[82,83], 96[93]
Donne, C. D., **5**, 1138[67]
Donnelly, B., **8**, 530[94]
Donnelly, D. M. X., **3**, 691[136]; **5**, 947[287]
Donnelly, J. A., **3**, 736[29]
Donnelly, K. D., **7**, 532[31]
Donnelly, S. J., **8**, 318[64], 319[70], 487[67], 546[312]
D'Onofrio, F., **3**, 512[192,213], 515[192,213]; **5**, 771[150,152], 772[150]; **7**, 530[17], 531[17]
Donoghue, E., **4**, 45[126,126c]; **5**, 686[51], 687[51], 688[51]
Donohue, B. E., **6**, 534[523]
Donovan, S. F., **4**, 35[98g]
Donovan, V., **4**, 305[369]
Donskikh, V., **4**, 314[487]; **6**, 495[150]
Dontheau, A., **4**, 308[408,409]
Dontsova, N. E., **7**, 766[177]
Dooley, J. F., **4**, 1023[255]
Dooley, T., **2**, 364[207]
Doolittle, R. E., **6**, 279[133]
Doomes, E., **3**, 877[86]; **6**, 67[12]
Doorenbos, N. J., **3**, 781[12]
Doornbos, T., **6**, 612[74], 964[83]
Döpp, D., **5**, 194[3], 196[3], 197[3], 198[3], 202[3], 209[3]; **6**, 249[138]
Döpp, H., **6**, 249[138]
Doran, M. A., **2**, 124[206]; **4**, 96[103b], 868[14]
Dorder, I. M., **1**, 469[56], 474[56]
Dordick, J. S., **7**, 79[134]
Dordor-Hedgecock, I. M., **2**, 125[217,224], 127[228,232], 271[77], 272[77,79,80], 315[42,44,45], 316[42,44,45], 317[44]; **4**, 82[62e], 217[130,132], 231[130,132], 243[74,75], 257[74,75], 260[75]
Doré, G., **6**, 428[86]
Dorf, U., **1**, 162[97], 163[105]; **5**, 1131[12]; **8**, 675[47]
Dorfman, L., **4**, 45[126,126c]; **5**, 686[51], 687[51], 688[51]
Dorfman, R. I., **7**, 673[21], 675[21]
Doria, G., **4**, 106[138e]; **8**, 568[467]
Dorier, P. Ch., **3**, 242[2]
Dorigo, A., **4**, 954[29]; **5**, 266[75], 268[75]; **6**, 490[109]
Döring, I., **1**, 214[23]; **4**, 880[88,90], 883[90], 884[90]
Dorland, L., **6**, 34[8]
Dorling, S., **3**, 807[21]
Dormagen, W., **5**, 412[47,47b]
Dorman, L. A., **1**, 543[22]
Dorman, L. C., **6**, 709[53], 711[69]
Dormond, A., **1**, 331[49], 749[78], 816[78]; **5**, 1126[68]
Dormoy, J. R., **2**, 61[21]
Dorn, C. R., **8**, 561[408]
Dorn, J., **5**, 676[3]
Dornacher, I., **2**, 388[338]
Dornow, A., **1**, 391[153]; **2**, 360[173], 801[29]; **4**, 5[17]
Dornow, R., **4**, 729[61], 730[61], 765[61]
Dorofeenko, G. N., **2**, 712[40], 737[31]; **6**, 556[25], 563[25]
Dorofeev, I. A., **6**, 509[282]
Dorofeeva, O. V., **8**, 610[57]
Dorokhova, O. M., **4**, 387[158]
Dorokova, E. M., **5**, 425[101]
Doronzo, S., **8**, 446[63], 858[209]
Dorow, R. L., **1**, 36[173]; **2**, 113[108]; **4**, 965[51], 1033[17], 1035[37,41], 1036[53], 1037[17,37], 1038[59]; **6**, 77[56], 118[103], 248[137]; **7**, 162[68], 184[171]
Dorr, H., **1**, 70[63], 141[22]; **3**, 873[64]
Dörr, M., **5**, 205[43]
Dors, B., **7**, 42[33]
Dorsch, H. L., **6**, 968[115]
Dorsey, E. D., **5**, 99[133,134,135], 100[133,135]
Dosi, I., **2**, 535[37]
Doskotch, R. W., **3**, 390[77,79], 392[77], 396[109], 397[109]
Dossena, A., **6**, 936[109]; **7**, 197[21]; **8**, 406[50]
Dost, J., **6**, 487[66], 489[66]
Dostalek, R., **6**, 188[181]
Dostovalova, V. I., **7**, 500[236]
Do Thi, N. P., **4**, 545[125], 546[125]
Do-Trong, M., **8**, 798[57]
Doty, J. C., **8**, 496[34]
Dötz, K. H., **4**, 976[100], 980[102-104], 981[102a,103,104]; **5**, 689[73], 1065[1], 1066[1,1a,7], 1070[18,26,28], 1072[26], 1074[1,28], 1083[1], 1084[1], 1085[61,62,64], 1089[83], 1090[88], 1092[92], 1093[1,95,96], 1094[100,100a], 1095[104,105], 1096[109,109d,e,124,127], 1098[96b,100a,105,109,109c-e,120-123,125,126], 1099[109c-e,125,141], 1101[120,145], 1103[83,152,153], 1112[96b,100b,104,105,109a-c,120-123,125,126], 1113[125], 1122[32]
Dotzauer, E., **4**, 519[21]
Dötzer, R., **8**, 754[108], 755[108]
Dou, H. J.-M., **3**, 505[162], 507[162], 512[162]
Doubleday, A., **1**, 2[3], 37[3]
Doubleday, C. J., Jr., **5**, 72[172], 728[121]
Doubleday, W., **1**, 239[39]
Douch, J., **8**, 133[22]
Doucoure, A., **1**, 831[103]
Dougal, P. G., **6**, 11[49]
Dougherty, C. M., **2**, 182[10]; **4**, 486[39], 966[53]

Dougherty, E. F., **7**, 11[84]
Dougherty, J. T., **6**, 426[70]
Dougherty, T. K., **3**, 541[116], 543[116]
Doughty, A., **4**, 876[61]
Doughty, D. H., **3**, 1041[110]
Doughty, M., **7**, 772[296]
Douglas, A. J., **6**, 455[155]
Douglas, A. W., **7**, 877[135]; **8**, 54[159], 66[159]
Douglas, B., **8**, 568[471]
Douglas, E. C., **3**, 328[179]
Douglas, J. L., **2**, 170[174]
Douglass, C. H., Jr., **6**, 708[48]
Douglass, J. E., **3**, 327[170]
Douglass, J. G., III, **4**, 1089[129]
Douglass, M. L., **4**, 310[434]; **8**, 853[144]
Doukas, H. M., **8**, 220[86]
Doukas, P. H., **6**, 498[168]
Doumaux, A. R., Jr., **5**, 513[4]; **7**, 230[135,136], 766[174]; **8**, 253[112]
Doutheau, A., **4**, 308[406], 311[452], 395[205,205c,d,206], 396[209], 397[209,211]; **5**, 773[166,167], 774[167,169], 797[66]
Dove, M. F. A., **3**, 299[33]
Dovinola, V., **7**, 445[58]
Dow, R. L., **1**, 400[11]; **4**, 384[145b]; **6**, 116[94]; **7**, 315[43]; **8**, 448[142]
Dowbenko, R., **3**, 380[9]
Dowd, P., **1**, 259[28], 265[28], 894[154,156]; **2**, 448[39]; **4**, 54[153b], 822[225], 1007[110], 1041[104]; **5**, 165[79], 240[2], 244[2c]; **6**, 117[95], 677[318,318a], 686[366]; **8**, 413[121], 890[139]
Dowd, S. R., **1**, 361[34]; **3**, 30[179]; **4**, 1000[15]; **6**, 703[4], 719[129]
Dowdall, J. F., **2**, 529[19]
Dowell, D. S., **8**, 584[17]
Doweyko, A. M., **3**, 158[447], 159[447]; **8**, 87[33]
Dowle, M. D., **2**, 963[55]; **4**, 364[1,1a], 368[1a], 373[1a]
Down, J. L., **8**, 524[13]
Downe, I. M., **6**, 228[18]
Downs, A. J., **8**, 460[253]
Doxsee, K. M., **3**, 503[149], 512[149]; **4**, 427[66]
Doyama, K., **5**, 1137[56]
Doyle, I. R., **5**, 561[86]; **8**, 806[104], 807[104]
Doyle, K. M., **2**, 867[17]; **6**, 424[52]
Doyle, M. J., **5**, 1185[1]
Doyle, M. P., **2**, 710[27]; **3**, 919[29], 920[34], 923[34b], 925[34a], 934[34], 1008[69], 1056[35], 1062[35]; **4**, 953[8], 954[8n], 964[45], 965[51], 996[8n], 1031[2,3], 1032[2], 1033[17,20,22,22a], 1034[2], 1035[2,37,41], 1036[47,49,50,53], 1037[17,37], 1038[59], 1049[2], 1057[22a-c]; **5**, 904[47,48], 905[47,48], 917[123], 1084[60]; **6**, 126[154], 127[161], 208[56], 212[80], 873[9], 874[9c,13], 883[13], 897[9b]; **7**, 315[43], 740[44]; **8**, 88[41], 89[41], 105[41], 216[64], 319[70], 383[14], 486[60], 487[60,67], 546[312], 801[74], 917[115], 918[115]
Doyle, P., **3**, 386[68]
Doyle, T. D., **5**, 723[108a]
Doyle, T. W., **5**, 92[70], 94[83-85], 736[145], 737[145]; **6**, 538[573], 569[937]; **8**, 231[148], 232[148], 303[97], 843[57]
Dozorova, E. N., **6**, 554[773]
Drabowicz, J., **6**, 149[98,101]; **7**, 760[45], 762[69,74,81], 764[108], 777[69a], 778[69,407]; **8**, 403[1,3], 404[1,3,15], 406[46,52], 408[46], 410[52]
Drach, B. S., **6**, 524[360,364,365], 528[413], 532[364], 539[577]
Dradi, E., **4**, 106[138e]; **5**, 36[16]
Draggett, P. T., **8**, 445[52]
Dragisich, V., **5**, 1076[45], 1101[143], 1102[147]
Drago, R. S., **4**, 306[371]
Dragovich, P. S., **5**, 736[142d]
Dragu, E., **6**, 489[85]
Draguet, C., **6**, 462[6]
Drahnak, T. J., **5**, 199[28]
Drake, B. V., **5**, 618[1]
Drake, C. A., **6**, 284[177]
Drake, J., **3**, 380[8]
Drake, J. E., **3**, 46[255], 47[255]
Drake, R., **1**, 493[42,42b], 495[42]
Drake, S. D., **7**, 347[17], 355[17]
Drakesmith, F. G., **8**, 901[35,41], 903[35], 905[35]
Dralle, G., **3**, 647[199]
Dran, R., **2**, 816[85]
Draney, D., **2**, 345[32]
Dranga, D., **4**, 298[284]
Draper, A. L., **7**, 167[98]
Draper, R. W., **7**, 488[154], 504[154], 508[154]
Drašar, P., **2**, 382[313]
Drauz, K., **8**, 459[228]
Drebowicz, J., **7**, 778[407]
Drechsel-Grau, E., **4**, 729[61], 730[61], 765[61]
Drechsler, H. J., **6**, 524[355], 525[355], 527[410], 532[355]
Drechsler, K., **6**, 834[40]
Drees, F., **6**, 644[87]
Drefahl, G., **4**, 280[130], 281[130], 282[130], 286[130]
Dreger, L. H., **5**, 857[229]
Dreiding, A. S., **1**, 876[102]; **2**, 158[123], 711[30], 725[115]; **3**, 804[3], 807[22], 810[48], 1049[14,15], 1058[42]; **4**, 148[50]; **5**, 18[128], 689[73], 710[47], 770[137,138,139], 806[106], 829[25], 1025[83], 1026[83]; **6**, 778[63], 1058[60]; **7**, 483[120], 487[147], 493[147], 495[147]
Dreikorn, B. A., **6**, 526[392]
Drenth, W., **4**, 278[95], 286[95], 289[95]; **6**, 489[89,94], 533[508]; **7**, 95[70,70a]
Dresely, K., **1**, 149[50]
Dresely, S., **2**, 39[137,137b,139], 44[151,151a], 45[151]; **7**, 597[52]
Dress, F., **6**, 644[94]
Dressaire, G., **1**, 115[40]; **2**, 784[39a]
Dressel, J., **3**, 874[70]
Dressler, H., **2**, 362[186], 363[186]
Dressler, V., **4**, 6[23]
Dreux, J., **4**, 4[14], 191[114]; **7**, 124[41]
Dreux, M., **3**, 257[117]
Drevs, H., **4**, 841[46]
Drew, H. D. K., **7**, 774[328,329]
Drew, J., **7**, 821[31]
Drew, M. G. B., **2**, 823[115]; **5**, 648[20]
Drew, R., **4**, 303[342], 390[175b]
Drew, R. A. I., **7**, 635[70]; **8**, 854[152], 856[152]
Drewer, R. J., **4**, 953[8]
Drewes, M. W., **1**, 56[29], 460[3]; **2**, 646[84]; **6**, 644[88]
Drewes, S. E., **1**, 188[72], 189[72]; **2**, 20[81], 51[81]; **3**, 421[52]; **4**, 34[97], 35[97]; **5**, 834[56]
Drewniak, M., **6**, 84[86]
Drexler, S. A., **4**, 213[106], 215[106]
Dreyer, G. B., **5**, 656[30], 660[30], 667[43], 924[148]
Driessen, P. B. J., **6**, 294[245]
Driessen-Engels, J. M. G., **3**, 367[101]
Driggs, R. J., **3**, 470[204], 473[204]
Driguez, H., **6**, 648[123]; **7**, 499[231]
Drizina, I. A., **6**, 554[730]
Droghini, R., **2**, 648[97,97c], 649[97c]
Dromzee, Y., **8**, 847[88,88d]
Dronkina, M. I., **6**, 496[154], 543[622,623], 552[622]
Dronov, V. I., **7**, 767[190]
Drosten, G., **8**, 303[102]

Drouin, J., **4**, 189[103], 240[42], 905[210]; **5**, 21[146,147,152,153,155,156,157,158], 22[152,153,155,156,157,158]; **8**, 851[129]
Drover, J. C. G., **3**, 227[213]; **5**, 86[31]
Drozd, V. N., **4**, 50[142], 452[7]; **7**, 606[161]
Drozda, S. E., **2**, 583[111]; **5**, 524[52]
Drtina, G. J., **6**, 883[60], 884[60]; **7**, 100[116], 552[57]
Drucker, G. E., **1**, 632[66]
Druckrey, E., **5**, 708[41]
Drueckhammer, D. G., **2**, 456[49], 460[49], 461[49]; **8**, 185[12], 187[34]
Druelinger, M., **3**, 36[209]; **6**, 725[171,173], 728[171]
Drues, R. W., **5**, 229[121]
Druey, J., **3**, 927[55]; **5**, 732[132,132a]
Drummond, A. V., **7**, 154[23], 157[33], 158[33b]
Drumright, R. E., **4**, 24[72,72a]
Drusiani, A., **2**, 635[48], 640[48]
Druzhkova, G. V., **8**, 765[11]
Dryanska, V., **2**, 495[61]
Dryden, H. L., **3**, 816[79]
Dryden, H. L., Jr., **8**, 321[104], 490[5], 492[5], 493[5], 524[4], 526[4], 530[4]
Drysdale, J. J., **6**, 967[102]
D'Silva, T. D. J., **5**, 455[81]
Du, C.-J. F., **4**, 494[84], 878[76]
Du, P. C., **7**, 155[29], 875[111]
Dua, S., **8**, 746[62], 750[62]
Dua, S. K., **2**, 109[70]; **3**, 164[478], 167[478], 168[478]
Dua, S. S., **3**, 457[128]
Duax, W. L., **3**, 386[61], 393[61]; **4**, 83[65c]
Dubac, J., **5**, 2[19], 3[19]; **6**, 832[15]
Dube, D., **1**, 419[79], 797[292], 802[292]; **6**, 995[104], 996[104]; **7**, 153[7], 400[46]
Dube, S., **1**, 508[20]; **3**, 96[164], 98[164], 99[164]; **8**, 531[115]
Dubeck, M., **7**, 587[168]
Düber, E. O., **2**, 510[44]; **6**, 722[141], 724[141]
Dubey, S. K., **5**, 420[75]; **6**, 958[35]; **7**, 489[170]
Dubina, V. L., **6**, 428[83]
Dubini, R., **3**, 996[45]; **5**, 851[166]
Dubinskaya, T. P., **6**, 543[606]
DuBois, G. E., **3**, 14[76], 15[76], 347[25], 373[126]; **6**, 247[131]; **8**, 530[109]
Dubois, J. C., **2**, 286[62]; **4**, 106[138d]
Dubois, J.-E., **1**, 202[99], 427[112]; **2**, 110[75,76], 116[75], 117[75], 143[49], 144[58,62], 153[110,111], 154[112], 186[34], 190[59], 193[62], 199[59], 235[34], 245[33], 268[69], 281[30], 614[117], 630[24], 631[24], 632[24], 634[37], 640[37], 930[130], 931[130]; **3**, 257[121], 636[60], 723[10]; **5**, 102[180]
DuBois, K. P., **1**, 252[4]
Dubois, M., **2**, 153[110,111], 268[69]
Duboudin, F., **2**, 900[30], 901[30], 964[62]; **6**, 185[162]
Duboudin, J., **2**, 159[127]
Duboudin, J. G., **3**, 440[45]; **4**, 876[59], 877[72], 878[59,76,78,79]; **5**, 1166[22]
Dubovenko, Z. V., **3**, 386[68]
Dubs, P., **2**, 368[241], 866[7], 867[7], 870[7], 871[7], 872[7], 875[7], 876[7]; **6**, 937[115], 941[115]; **7**, 124[45]
Duburs, G., **8**, 595[77]
Duc, C. L., **7**, 79[131]
Duc, L., **3**, 822[12], 831[12], 835[12b]; **5**, 128[27]
Ducep, J. B., **1**, 568[225], 630[30], 678[30]; **2**, 55[1], 71[1c,54]; **3**, 86[14], 94[14], 95[14,151], 96[151,161,165], 98[151,161], 99[151,165,179], 100[14,179], 101[165], 107[165], 196[29]; **6**, 145[78], 832[18]
Duchêne, A., **3**, 196[24]
Ducker, J. W., **2**, 367[226], 378[289]; **6**, 279[138,139]
Duckworth, C. A., **6**, 546[652]
Duclos, R. I., Jr., **3**, 507[172]
Ducos, P., **5**, 561[82]; **8**, 543[246]
Ducrocq, C., **2**, 379[294]
Duddeck, H., **4**, 1055[136], 1056[136]
Dudek, V., **8**, 161[17], 176[129]
Dudfield, P., **2**, 924[108b]; **3**, 46[252]; **4**, 152[54], 201[15], 202[15]; **6**, 77[55]; **7**, 182[165]
Dudgeon, C. D., **6**, 556[832]
Dudley, C., **7**, 107[168]
Dudley, C. W., **8**, 444[10]
Dudman, C. C., **6**, 121[131], 778[64]; **8**, 297[66]
Dudzik, Z., **8**, 606[22]
Duerr, B. F., **5**, 552[15,32]
Dufaux, R., **8**, 830[86]
du Feu, E. C., **4**, 2[5]
Duff, J. M., **1**, 618[56]; **3**, 125[314]; **4**, 120[197]
Duff, S. R., **5**, 492[240]
Duffaut, N., **3**, 577[90]; **8**, 518[131]
Duffin, D., **3**, 386[62,63]
Duffy, J. P., **7**, 418[127]
Duffy, P. F., **4**, 120[201]; **6**, 134[29]
Dufour, J.-M., **3**, 325[162]
Dufour, M.-N., **8**, 13[71]
Dufour, R., **1**, 644[123], 646[123], 668[123], 669[123], 695[123]
Dufresne, C., **8**, 315[52], 316[57], 969[94]
Dufresne, Y., **6**, 210[74], 214[74]
Dugar, S., **3**, 504[154], 511[154], 515[154]; **7**, 36[108]
Dugat, D., **4**, 404[243]; **5**, 94[86]; **7**, 503[275]; **8**, 856[181]
Duggan, A. J., **5**, 459[91], 612[73]; **8**, 795[16]
Duggan, M. E., **3**, 224[174], 618[20,21]; **5**, 687[67]; **7**, 407[84b], 408[88c]
Dugger, R. W., **2**, 188[44]; **3**, 220[117]
Duggin, A. J., **7**, 219[12]
Duggleby, P. McC., **3**, 888[15]; **6**, 129[165]
Duguay, G., **6**, 554[751,752]; **8**, 658[100]
Dugue, B., **2**, 1004[60], 1005[60]
Dugundji, J., **2**, 1090[75]
Duh, H.-Y., **5**, 260[68], 261[68], 262[68]
Duhamel, L., **1**, 56[30], 366[47], 566[210]; **2**, 899[25], 900[25]; **3**, 253[87]; **6**, 710[61], 713[80b], 717[109]
Duhamel, P., **1**, 56[30], 366[47], 566[210]; **2**, 899[25], 900[25]; **4**, 159[81]; **6**, 509[256], 713[80b], 717[109]
Duhl-Emswiler, B. A., **1**, 514[52], 755[116], 756[116], 758[116,116a], 761[116]
Duisenberg, A. J. M., **1**, 10[54], 23[125], 214[27]; **2**, 124[204], 125[204]
Dujardin, R., **8**, 857[195], 858[195]
Duke, A. J., **3**, 767[162]
Duke, R. E., **4**, 1098[170]
Duke, R. E., Jr., **5**, 216[14], 219[14], 626[34]
Duke, R. K., **7**, 373[74], 375[74]
Dukesherer, D., **4**, 438[152]
Dukin, I. R., **4**, 483[8]
Dulcere, J. P., **4**, 793[72]; **5**, 772[156,158,161,163]
Dulenko, V. I., **2**, 737[31]
Dull, D. L., **2**, 225[157], 232[157]
Dulova, V. G., **6**, 836[53]
du Manoir, J., **3**, 180[550]
Dumas, D. J., **3**, 373[129]
Dumas, F., **2**, 227[161]
Dumas, P., **7**, 282[178]; **8**, 113[37]
Dummer, G., **2**, 943[168]
Dumont, C., **5**, 677[9,10]
Dumont, P., **2**, 353[99]; **5**, 553[41]
Dumont, W., **1**, 571[274], 618[57], 630[40], 631[40,47,50,53-55,60], 632[60], 633[60], 634[60], 635[60], 636[100], 639[100], 641[100],

647[55], 648[129,130,131,132], 649[40,129], 650[129,130,131,139], 651[132], 652[132], 655; **3**, 86[50], 87[66,67,72,73,77,84,86,91-93,116-119], 89[142,143], 92[143], 95[72], 109[84], 111[231], 116[142], 118[143], 120[242], 123[143], 136[77], 137[77], 141[77], 142[86], 144[77,86], 145[77]; **4**, 10[34], 318[560], 990[145], 991[150], 992[158], 1007[125]; **6**, 26[112]; **7**, 771[267], 772[267], 773[307]; **8**, 173[118], 847[97], 848[97e], 849[97e], 850[118]
Dunach, E., **1**, 543[15], 733[14], 786[14]; **3**, 255[106]; **5**, 1144[97]; **6**, 150[112]; **7**, 425[146], 777[377,378], 778[377,378]
Dunach, F., **1**, 543[21]
Dunathan, H. C., **3**, 361[75]
Duncan, D. M., **8**, 140[16], 568[470]
Duncan, D. P., **4**, 868[16]; **6**, 180[129]
Duncan, J. A., **5**, 797[65]
Duncan, J. H., **1**, 377[97]; **4**, 952[5]
Duncan, J. L., **2**, 367[225]; **3**, 826[40]
Duncan, M. P., **7**, 166[91], 222[37], 227[37,81], 833[76]
Duncan, M. W., **6**, 737[32]
Duncan, S. M., **5**, 15[108], 850[144]
Duncan, W. G., **3**, 126[318]
Duncia, J. V., **5**, 4[35,36,38], 5[35,36], 10[78]
Dundulis, E. A., **3**, 39[219], 40[219]
Dung, J.-S., **4**, 255[201]; **7**, 399[33]
Dunham, D., **6**, 715[92]
Dunham, D. J., **4**, 8[29]
Dunitz, J. D., **1**, 1[1], 3[1], 8[37], 26[1,137], 28[142], 30[151], 31[155], 34[167,170], 37[238,240,241], 38[261], 41[151], 43[1], 49[8], 297[59], 299[61], 306[91], 307[59], 316[61], 621[68]; **2**, 100[2,3], 280[25]; **4**, 202[19]; **5**, 468[127], 841[104]; **6**, 708[50]; **8**, 3[21], 89[43]
Dunkelblum, E., **8**, 551[347], 704[8], 936[73]
Dunkerton, L. V., **2**, 538[63]; **4**, 572[3]
Dunkin, I. R., **5**, 704[21]
Dunlap, N. K., **2**, 690[70], 725[106]; **4**, 161[87,87b], 212[97], 793[69]; **7**, 174[140]; **8**, 333[57]
Dunlap, R. B., **2**, 388[342]; **6**, 462[17]; **8**, 170[74]
Dunlap, R. P., **8**, 86[21]
Dunlop, A. P., **8**, 606[20]
Dunlop, M. G., **3**, 675[74]
Dunlop, R., **4**, 356[135]
Dunn, D. A., **5**, 125[13], 128[13]
Dunn, G. L., **3**, 855[81]
Dunn, J. L., **3**, 914[3,5]
Dunn, L. C., **5**, 78[273], 626[36,37], 629[36,46,47]
Dunn, W. I., III, **2**, 1099[115]
Dunnavant, W. R., **2**, 182[1]; **3**, 826[37]
Dunne, K., **2**, 709[9]; **4**, 587[25], 604[278], 608[322]
Dunne, T. J., **3**, 4[23]
Dunnigan, D. A., **4**, 312[459]
Dunning, R. W., **8**, 445[61]
Dunny, S., **8**, 341[106], 770[31], 926[16]
Dunoguès, J., **1**, 328[17,18]; **2**, 564[1,3], 575[65], 576[65], 582[108], 584[116], 712[43], 713[44], 716[57,61,66], 717[68-70], 718[74,75], 721[89], 726[124], 728[140,141], 900[30], 901[30], 964[62], 1030[80]; **3**, 577[90]; **5**, 527[59]; **6**, 185[162], 832[12], 865[12]; **8**, 409[82], 518[128,131], 785[115]
Dunstan, A. E., **3**, 331[198]
Duong, T., **7**, 40[3]
duPenhoat, C. H., **2**, 76[83b]; **8**, 842[42c,d], 844[42c], 847[42c]
Dupin, C., **3**, 734[8]
Dupin, J. F., **3**, 734[8]; **6**, 177[118]
Duplantier, A. J., **2**, 249[84]; **6**, 452[132]
Dupont, A., **1**, 100[88]
Dupont, W., **4**, 37[107,107b]
Dupont, W. A., **4**, 74[37], 249[116]
Dupre, B., **3**, 77[56]
Dupre, M., **4**, 55[155]
DuPree, L. E., Jr., **5**, 960[320]
DuPreez, N., **1**, 70[63], 141[22]; **4**, 1007[123], 1008[123]
Dupuis, D., **1**, 770[187]; **5**, 356[90], 358[90b], 362[93], 363[93d], 364[93c,d], 543[117,117b], 545[120]
Dupuis, J., **2**, 334[67]; **4**, 729[58], 735[85], 738[90], 740[116], 754[58]; **5**, 159[51], 189[51]; **7**, 883[175]
Dupuis, P., **6**, 711[68]
Dupuy, C., **1**, 219[61]; **4**, 95[99a,b], 764[220]
Dupy-Blanc, J., **8**, 560[402]
Duquette, L. G., **6**, 709[53], 711[69]
Duraisamy, M., **3**, 587[146], 610[146]; **6**, 727[194]
Duran, E., **6**, 494[137]
Duran, F., **5**, 92[61], 94[61]
Durán, M., **5**, 72[183,186,187], 73[189,190]
Durand, J., **8**, 16[104], 541[214], 542[214,225], 543[214]
Durand-Dran, R., **7**, 666[75]
Durandetta, J. L., **2**, 494[55]; **6**, 541[593,594]
Durant, E., **2**, 904[53]
Durant, F., **1**, 675[210], 677[210], 706[210], 721[210]
Dürckheimer, W., **3**, 890[34]; **5**, 85[1]
Duréault, A., **1**, 770[185]; **3**, 258[127]; **6**, 93[132]; **7**, 477[78], 483[78], 484[78], 487[146], 495[146]
Durham, D., **2**, 345[32]
Duri, Z. J., **8**, 537[181], 798[61]
Durland, J. R., **7**, 14[126]
Durman, J., **6**, 932[94,95]
Dürner, G., **2**, 547[113], 551[113]; **3**, 56[285]; **4**, 229[235,236], 1055[138]; **8**, 545[285]
Durr, H., **4**, 1104[211]
Dürr, H., **5**, 157[42], 380[114]
Dürr, M., **6**, 554[794]
Durrant, G., **8**, 865[245]
Durrwachter, J. R., **2**, 456[34,49], 460[49], 461[49], 463[34]; **7**, 312[33]
Durst, F., **6**, 26[110]
Durst, H. D., **6**, 724[162,163]; **7**, 765[156]; **8**, 26[26], 37[26], 47[26], 54[26], 55[26], 57[26], 60[26], 66[26], 351[151]
Durst, T., **1**, 123[78], 512[37,38], 524[89,90], 595[28], 821[25], 825[50], 833[117]; **2**, 81[1], 82[1,1a], 96[1], 100[16], 101[16], 111[16], 134[3], 182[2], 190[57], 240[3], 363[198], 417[18], 430[53], 455[9], 510[39]; **3**, 86[32,33,37,38], 147[32,33,392,397], 149[392,397,400,403], 150[397,403], 151[397,403], 152[392], 154[32], 158[37,38], 159[37,460], 161[460], 163[38,460], 164[37], 167[460,486], 168[486], 173[37,38], 180[550], 265[193], 266[194], 748[77], 770[181], 862[2], 979[13]; **4**, 84[68d], 100[120], 404[243], 501[112], 1057[143,144], 1058[143,144]; **5**, 105[193], 389[138], 390[140], 567[105], 692[97,98,102]; **6**, 134[11], 686[368]; **7**, 196[13], 292[4], 503[275], 653[3], 764[129], 766[181], 767[191]; **8**, 403[2,6], 404[2], 856[181]
Dusold, L. R., **3**, 891[42]; **6**, 1012[7], 1013[7]
Dussault, P., **4**, 379[114,114a], 380[114a]; **6**, 26[106]
Dussault, P. H., **1**, 744[56]; **5**, 1123[36]; **7**, 549[44], 583[44], 586[44]
Dust, J. M., **5**, 485[184]
Dutcher, J. S., **7**, 111[192]
Duteil, M., **4**, 844[58]
Duthaler, R. O., **2**, 47[153], 308[22,23], 309[23,24], 318[51], 857[250]; **4**, 382[134]; **5**, 165[83], 836[63], 841[91]; **6**, 865[201]
Dutky, S. R., **7**, 673[29]
Dutler, H., **3**, 815[74]
Dutra, G. A., **3**, 224[176], 248[57], 249[57], 251[57], 263[57]; **4**, 170[18]
Dutt, M., **4**, 486[37], 505[37]
Dutta, J., **2**, 162[142]
Dutta, N., **2**, 811[70], 813[70], 814[70]
Dutta, P. K., **7**, 502[260]

Dutta, S. P., **8**, 333[56]
Dutton, F. E., **5**, 71[164]
Dutton, G. G. S., **6**, 43[52], 647[107]
Dutton, H. J., **8**, 450[161], 453[191]
Duus, F., **2**, 812[74]; **6**, 419[1], 421[1], 436[15], 437[15], 448[15], 450[15], 453[15], 455[15]
Duval, D., **4**, 71[19]
du Vigneaud, V., **6**, 636[16], 644[90], 664[16]
Dux, F., III, **8**, 803[93], 804[93], 826[69]
Dvolaitzky, M., **3**, 807[22]
Dvorák, D., **2**, 358[157]; **4**, 5[17]; **5**, 461[102]
Dvorko, G. F., **4**, 289[190,193]
Dvortsák, P., **6**, 543[610]
Dwivedi, C. P. D., **5**, 72[184]
Dworkin, A. S., **3**, 328[177]
Dwyer, F. G., **3**, 305[75c]
Dwyer, J., **5**, 2[20]
D'Yachenko, A. I., **4**, 489[63]
D'yaknov, I. A., **5**, 948[273]
Dyall, L. K., **7**, 92[50]
Dyankonov, I. A., **4**, 1003[63]
Dye, J. L., **8**, 524[11,13,13c]
Dyer, U. C., **1**, 188[94], 198[94], 199[94]
Dyke, H., **7**, 543[22]
Dyke, H. J., **2**, 749[132]
Dyke, S. F., **2**, 749[132]; **3**, 670[60]; **4**, 811[172], 837[19], 839[28]; **5**, 723[106]; **6**, 502[211]; **7**, 231[137]
Dykes, W., **7**, 264[93]
Dykstra, C. E., **4**, 47[134]
Dykstra, S. J., **7**, 168[104]
Dyllick-Brenzinger, R., **5**, 636[101]
Dymov, V. N., **6**, 543[608]
Dymova, S. F., **4**, 992[154]; **8**, 535[160]
Dyong, I., **6**, 900[114]; **7**, 489[171]
Dyrbusch, M., **3**, 303[53]; **4**, 111[152d], 222[179], 1038[60]
Dyrkacz, G. R., **2**, 348[55]
Dyszlewski, A. D., **3**, 253[89], 262[89]
Dzhafarova, N. A., **3**, 304[69]
Dzhemilev, U. M., **4**, 589[79], 591[79], 598[199], 599[213], 638[199], 640[199,213], 875[56]; **5**, 1154[159]
Dzhemilev, V. M., **8**, 697[133,134], 698[133,140]
Dziadulewicz, E., **2**, 73[67]
Dzidic, I., **1**, 287[16]
Dzieciuch, M., **3**, 636[56,57], 639[56b,77]
Dziewonska-Baran, D., **4**, 432[112]

E

Eaborn, C., **1**, 16[90], 17[91,208], 39[191], 41[264], 214[28]; **2**, 743[78]; **3**, 419[31], 564[15]; **4**, 170[15]; **8**, 265[53], 513[101,103], 763[1], 766[20], 785[1]
Eades, R. A., **6**, 708[48]; **7**, 830[61]
Eakin, M. A., **3**, 380[13], 600[213]
Eapen, K. C., **3**, 450[104], 457[128]
Earhart, C. E., **5**, 729[125]
Earl, G. W., **7**, 660[42], 882[171,172]
Earl, R. A., **1**, 568[235]; **2**, 725[118], 757[13], 759[13], 962[50]; **4** , 1086[116]; **5**, 249[30], 1156[165]; **6**, 807[62]; **8**, 839[27]
Earl, S., **4**, 505[148]
Earley, J. V., **2**, 748[123]
Earley, W. G., **2**, 1069[136]
Earnshaw, C., **1**, 774[209]; **2**, 596[4]; **3**, 201[76]
East, M. B., **4**, 113[167]; **7**, 582[148]
Eastham, J. F., **3**, 722[6], 723[6], 731[6], 822[6,8], 823[6], 824[6], 825[29], 834[6,73], 836[6,8]; **8**, 526[30]
Eastman, R. H., **2**, 149[92]; **4**, 120[202]; **7**, 12[103]
Eastmond, R., **3**, 555[31]
Easton, C. J., **2**, 1052[44]; **4**, 744[134], 787[27]; **7**, 206[71]
Easton, N. R., Jr., **3**, 382[36]
Easton, R. J. C., **2**, 125[216]; **4**, 217[131], 243[74], 257[74]
Eastwood, F. W., **2**, 353[98], 356[133,134]; **3**, 71[28]; **6**, 563[898], 570[947], 687[373], 984[53,56], **7**, 827[48]
Eaton, D. F., **5**, 211[61], 552[30]; **7**, 874[103]
Eaton, D. R., **5**, 901[31]
Eaton, J. K., **3**, 888[14]
Eaton, J. T., **8**, 36[48], 66[48], 347[140], 616[102], 617[102], 618[102]
Eaton, P. E., **1**, 284[4], 480[152,153,154]; **2**, 711[35,36]; **3**, 573[70], 855[80]; **5**, 123[5], 124[7], 127[23], 339[55], 758[82], 768[124,125], 779[124,125], 791[27], 799[27], 817[145], 1188[13], 1192[13]; **6**, 779[66], 964[78]; **7**, 4[16], 462[122], 683[89], 691[19], 752[147]; **8**, 795[17]
Eaton, P. J., **8**, 333[54]
Eaton, T. A., **5**, 627[43]
Eavenson, C. W., **5**, 1134[44]
Ebata, T., **1**, 790[263]; **3**, 224[177]; **4**, 364[1,1h]; **6**, 862[186]; **7**, 399[37], 406[78c]; **8**, 957[13]
Eber, J., **8**, 333[53]
Eberbach, W., **5**, 929[168,170,171,172], 930[168,171], 932[170,172], 933[181]
Eberhardt, G., **8**, 270[96,97]
Eberhardt, H.-D., **3**, 904[131]
Eberle, D., **3**, 933[62]
Eberle, G., **2**, 1094[89], 1095[89], 1098[103], 1099[103,106]; **8**, 384[35]
Eberle, M., **1**, 415[62]; **2**, 325[45], 327[45], 352[90], 353[90]; **5**, 386[133]; **8**, 190[72]
Eberlein, T. H., **2**, 556[150]; **5**, 436[158,158a,g], 438[158d], 442[158], 532[85]
Ebermann, R., **3**, 687[112]
Ebersen, K., **2**, 1099[115]
Eberson, L., **2**, 1051[35], 1052[35]; **3**, 634[7,7c,19,20,22], 636[7,7a-c], 637[7,63,64], 638[19,20], 639[68-72], 643[132], 649[19,20], 655[19,20]; **4**, 452[14], 726[50]; **5**, 595[14], 597[29]; **6**, 281[150]; **7**, 796[14], 799[24,27], 800[33], 801[36,43], 852[37], 878[139,142]; **8**, 794[12]
Ebert, G. W., **1**, 426[111]; **2**, 121[190]; **3**, 209[19], 522[10], 553[17]
Ebert, L. B., **7**, 282[179]
Ebert, R. W., **7**, 482[119]
Ebetino, F. F., **3**, 564[12]; **7**, 21[15]
Ebetino, F. H., **3**, 681[100]; **7**, 347[16], 355[16]
Ebihara, K., **1**, 86[35], 223[79], 224[79]; **4**, 230[254]
Ebina, M., **6**, 491[115]
Ebine, S., **3**, 818[96]; **5**, 626[39]; **7**, 356[52]
Eble, K. S., **7**, 80[141]
Ebner, C. B., **6**, 189[190]
Ebner, T., **8**, 40[92], 66[92]
Ebnöther, A., **2**, 765[71]
Eby, L. T., **8**, 478[39], 479[39]
Eby, R., **6**, 47[78], 49[90]
Eby, S., **6**, 531[428]
Echavarren, A. M., **1**, 474[95]; **3**, 232[257,263], 233[272], 455[123], 470[211], 473[211], 475[211], 476[211], 492[94], 495[94], 504[94], 514[94], 1024[25]; **4**, 594[143], 619[143], 633[143]; **5**, 480[167]; **7**, 355[42]
Echigo, Y., **3**, 730[46]; **6**, 206[38]
Echsler, K.-J., **1**, 669[182], 670[182]
Eck, C. R., **3**, 386[68], 711[22], 712[26]; **8**, 237[10], 243[10]
Eck, S. L., **3**, 380[10]; **4**, 192[116], 983[116]
Ecke, G. G., **3**, 296[12]; **4**, 272[38,39], 273[38,39], 287[38,39], 697[8]; **8**, 873[12]
Eckell, A., **4**, 1084[94]; **8**, 336[69]
Eckersley, T. J., **5**, 217[22,23], 226[23,106]
Eckert, C. A., **5**, 77[266]
Eckert, H., **2**, 1084[19], 1094[89], 1095[89]; **6**, 638[45], 639[46], 667[243]
Eckert, K., **1**, 162[104]
Eckert, Ph., **6**, 256[170]
Eckhardt, H. H., **5**, 731[131]
Eckle, E., **5**, 595[20], 596[20]
Eckman, R. R., **1**, 314[129]
Ecknig, W., **2**, 896[11,12]
Eckrich, T. M., **3**, 194[8], 217[84], 247[50]; **4**, 971[74]
Eckroth, D. R., **5**, 478[162]
Eckstein, B., **8**, 561[412]
Eckstein, F., **6**, 625[159]
Eckstein, Z., **2**, 365[215]; **6**, 824[120-122]
Eda, S., **3**, 585[137]
Edamura, F. Y., **8**, 345[128]
Eddy, N. B., **8**, 566[452]
Edelson, S. S., **5**, 596[28], 597[28,29]
Edenhofer, A., **2**, 850[219], 854[219]
Eder, K., **7**, 92[41,41a], 94[41]
Eder, R., **5**, 1062[59]
Eder, U., **2**, 167[160], 360[171], 902[40]; **4**, 10[34]; **6**, 718[117]; **7**, 65[69]; **8**, 847[92]
Edgar, J. F., **3**, 896[67]
Edgar, K. J., **3**, 251[77], 254[77]
Edgar, M., **7**, 676[62]; **8**, 353[154]
Edge, D. J., **5**, 901[32]
Edington, C., **2**, 650[110], 651[110]
Edison, O. H., **6**, 957[26]
Edlund, U., **2**, 1099[115]
Edman, J. R., **5**, 197[20], 198[20]; **7**, 341[50]
Edmison, M. T., **3**, 505[161], 507[161]
Edmonds, A. C. F., **6**, 1019[44,45]
Edmonds, C. G., **8**, 880[59]
Edmonds, M. D., **5**, 960[321]
Edo, K., **3**, 461[147]
Edstrom, E. D., **3**, 362[81]
Edward, J. T., **6**, 923[56]; **8**, 247[80], 390[78], 394[117]
Edwards, A. G., **7**, 601[79]

Edwards, B., **3**, 897[91], 900[91]
Edwards, E. I., **7**, 24[29,30], 25[43], 26[43,47]
Edwards, J. A., **5**, 92[69]; **7**, 86[16a], 136[116], 137[116]; **8**, 490[5], 492[5], 493[5], 524[4], 526[4,24], 530[4], 541[207]
Edwards, J. D., Jr., **2**, 291[72]; **3**, 665[37]
Edwards, J. M., **2**, 381[306]
Edwards, J. O., **1**, 512[41]; **7**, 763[88], 766[88]
Edwards, J. P., **8**, 388[60]
Edwards, K., **7**, 444[55]
Edwards, M., **4**, 37[104], 231[278]; **5**, 181[155]; **7**, 154[20], 480[104]; **8**, 848[103], 996[72]
Edwards, M. P., **1**, 800[299]; **5**, 534[92]; **6**, 995[103]
Edwards, O. E., **2**, 1057[71], 1058[71]; **4**, 399[221]; **5**, 407[29]; **6**, 745[85], 807[59]; **7**, 21[10], 27[74], 30[10,82], 31[89]; **8**, 354[169]
Edwards, O. K., **7**, 771[257]
Edwards, P. D., **1**, 461[11], 473[11], 477[128], 478[11], 482[128], 832[114]; **3**, 69[25,27], 70[25], 72[25], 74[25]
Edwards, P. G., **1**, 41[266,267]
Edwards, R., **4**, 673[31]
Edwards, S., **6**, 264[36]
Edwards, T., **6**, 262[18]
Edwards, W. B., **5**, 960[321]
Eenoo, M. V., **8**, 886[107]
Effenberger, F., **1**, 546[53]; **2**, 456[66], 457[66], 460[66], 461[66], 463[78], 612[103], 739[49-51], 740[53]; **4**, 331[7], 429[87]; **5**, 108[205]; **6**, 67[9], 72[9], 76[40], 77[40], 142[65], 443[95], 488[6], 489[6], 650[128]
Effenberger, G., **6**, 54[126]
Effio, A., **4**, 785[22]
Effland, R. C., **4**, 439[168]
Efimov, V. A., **6**, 603[21,25], 604[27]
Efraty, A., **4**, 701[27], 712[69]; **6**, 495[141]
Efremova, G. G., **7**, 194[6]
Efremova, L. A., **8**, 778[88]
Egan, S. C., **4**, 6[23]
Egashira, N., **4**, 487[45]; **8**, 131[6], 132[7]
Egberink, R. J. M., **4**, 45[126]; **5**, 676[4], 686[48,49], 687[48,49]; **7**, 333[25]
Egbertson, M., **1**, 425[106]; **5**, 491[208]
Ege, G., **3**, 807[26]
Ege, S. N., **4**, 295[260], 296[260]; **7**, 372[70]; **8**, 70[232], 72[232]
Egert, E., **2**, 94[51]; **3**, 303[53]; **4**, 111[152d], 222[179,181], 224[181], 1038[60]
Eggelte, H. J., **8**, 477[30,30a], 798[56]
Eggelte, T. A., **5**, 453[66]
Eggensperger, H., **3**, 661[22]; **4**, 271[21], 272[21]
Egger, B., **3**, 871[53]
Egger, H., **2**, 365[215]
Egger, K. W., **4**, 1072[17]; **5**, 900[8]
Egger, N., **7**, 483[120]
Eggerichs, T. L., **6**, 543[624]
Eggersdorfer, M., **3**, 312[101]
Eggersichs, T., **5**, 829[22]
Eggler, J., **4**, 5[19,19g], 54[152,152a,b]
Eggleston, D. S., **2**, 116[136], 117[136], 124[136], 436[66], 437[66b]; **4**, 1018[225], 1019[225]
Egli, M., **2**, 325[45], 327[45]; **7**, 86[15], 487[147], 493[147], 495[147]
Egli, R. A., **8**, 253[118]
Eglington, G., **3**, 551[1]
Eglinton, G., **4**, 55[156]
Egloff, G., **7**, 7[37], 15[148]
Egolf, T., **5**, 1154[149]
Egorov, M. P., **4**, 423[8], 426[8], 444[8], 519[12]
Egron, M.-J., **7**, 265[98], 272[98]; **8**, 64[212], 66[212], 67[212], 68[212]
Egsgaard, H., **6**, 448[112]
Eguchi, K., **4**, 298[285]
Eguchi, M., **6**, 615[102]; **7**, 415[113]
Eguchi, S., **2**, 619[148]; **5**, 417[65], 973[14], 981[14]; **6**, 252[154], 264[37], 265[37,41,42], 273[99]; **7**, 153[6]
Eguchi, T., **8**, 204[152]
Együd, L. G., **8**, 87[30]
Ehler, D. F., **8**, 238[24]
Ehlers, J., **5**, 115[249]; **6**, 426[73], 509[277]
Ehlinger, E., **1**, 622[73], 623[77,79]; **2**, 57[6], 426[39]; **3**, 491[72]
Ehmann, W. J., **5**, 1145[104]; **8**, 478[43], 480[43]
Ehntholt, D., **4**, 709[46], 710[46]; **5**, 272[4,5], 273[4], 275[4,10], 277[10]
Eholzer, U., **2**, 1084[8], 1086[25]; **6**, 242[85,86], 243[85,86], 489[88]
Ehrenkaufer, R. E., **3**, 1040[109]; **6**, 76[47]; **8**, 84[14], 368[66]
Ehret, C., **7**, 64[60]
Ehrig, V., **1**, 70[63], 141[22]; **2**, 223[149], 334[69], 846[209]; **4**, 12[43], 13[43a,b], 102[127], 113[164]
Ehrler, R., **2**, 338[78]; **5**, 260[66], 261[66]; **8**, 70[223], 647[54]
Ehrlich, P. P., **1**, 404[23]
Ehrlich-Rogozinski, S., **8**, 794[4]
Ehrmann, E. U., **4**, 497[96]
Ehrmann, U. E., **2**, 928[121,122], 946[121,122]
Eiband, G. R., **4**, 1033[27], 1049[27], 1060[27c]; **5**, 599[40], 804[94], 905[60]
Eibl, R., **8**, 316[58]
Eibler, E., **7**, 24[27,28], 25[28]
Eich, E., **8**, 568[466]
Eichel, W., **4**, 729[61], 730[61], 765[61]; **5**, 893[41]
Eichenauer, H., **2**, 504[4], 510[42,44], 514[50,52,53], 515[53], 524[50,53,81]; **3**, 39[213]; **6**, 722[141], 724[141], 728[208]; **7**, 98[100]
Eichenberger, H., **5**, 221[62]
Eichen Conn, R. S., **5**, 8[56]
Eicher, T., **2**, 342[7]
Eichler, D., **2**, 961[43]
Eichler, E., **7**, 750[126]
Eichler, G., **2**, 15[62,62a], 32[62a], 981[26], 982[26], 994[26], 995[26]
Eickhoff, D. J., **7**, 111[194], 378[92], 819[18]
Eidels, L., **8**, 36[52], 66[52]
Eiden, F., **2**, 902[45]; **6**, 554[787,788,794], 572[957]
Eiduka, S., **2**, 345[23]
Eierdanz, H., **6**, 970[122]
Eigen, I., **3**, 898[87]; **8**, 270[97]
Eigner, D., **6**, 438[41]
Eiho, J., **4**, 125[216,216c]
Eijsurga, H., **4**, 895[163]
Eiki, T., **7**, 764[111]
Eilbracht, P., **3**, 623[38]; **7**, 3[11]
Eilhauer, H. D., **6**, 431[105]
Eilingsfeld, H., **6**, 428[84], 488[8], 495[8], 499[8], 543[8], 566[8]; **8**, 636[3]
Einhellig, S., **4**, 1081[83]
Einhorn, C., **1**, 219[60], 830[95]
Einhorn, G. L., **7**, 852[36]
Einhorn, J., **1**, 464[38], 563[185]; **2**, 23[89], 565[13], 572[13]; **7**, 333[21]
Eirín, A., **5**, 416[57]
Eis, M. J., **1**, 343[105,118]; **3**, 224[169], 262[169]; **6**, 5[23], 95[147]
Eisch, J. J., **1**, 329[34], 746[69], 748[69]; **2**, 81[4], 109[70], 589[155], 976[1]; **3**, 164[478], 165[479], 167[478], 168[478], 173[479,519], 198[50], 579[97], 759[134]; **4**, 79[57], 89[83], 192[120], 877[65], 878[77], 879[65,77], 887[125,127], 889[134,137], 892[144]; **5**, 1125[56], 1135[49]; **8**, 734[1,3-5], 736[20,21], 737[24,27], 738[28], 739[35], 740[4,39,40], 741[39,40], 742[41-43,45], 743[45,47], 744[51-53], 745[4,54], 746[4,42,52-54,62], 747[3,57,58], 748[59,60], 749[53], 750[4,53,62], 751[52,57], 752[58,66], 753[3,20,52,54,67,71],

754[109], 755[28,109], 756[39], 757[39,161], 758[43], 759[3], 838[21,21a,c], 840[21a,c,34]
Eisele, G., **1**, 368[60], 369[60]
Eisele, W., **7**, 700[61]
Eisen, N., **5**, 812[130]
Eisenberg, M., **8**, 70[233]
Eisenberg, R., **8**, 456[209]
Eisenbraun, E. J., **2**, 141[39], 738[41], 760[40]; **3**, 322[141]; **8**, 354[177]
Eisenhart, E. K., **2**, 88[31]; **5**, 519[31]
Eisenhuth, W., **4**, 14[46]
Eisenstadt, A., **4**, 926[40], 930[50]; **5**, 1130[9]; **8**, 395[130]
Eisenstein, O., **1**, 49[7,8], 50[7], 182[47], 185[47], 198[47], 222[69], 310[102], 820[14]; **2**, 24[96], 217[137]; **4**, 664[6], 667[6b]; **5**, 417[63], 453[67]; **8**, 2[7], 536[170], 541[213,215], 543[213,215]
Eiserle, R. J., **2**, 152[99]
Eisert, M. A., **4**, 966[54], 968[92], 977[92]
Eisler, K., **6**, 668[262]
Eisner, T., **7**, 86[16a], 109[182]
Eisner, U., **2**, 954[4]; **8**, 92[67], 580[6], 584[21]
Eissenstat, M. A., **1**, 383[108]; **2**, 479[19], 481[19]; **3**, 31[182]; **5**, 982[30], 983[30]; **6**, 721[134]
Eistert, B., **1**, 217[48], 844[3b,8]; **3**, 816[85], 887[3,8], 888[3,8,13], 890[35], 891[37], 893[8], 897[3,8], 900[8], 903[8,126,-128]; **4**, 953[8,8b], 954[8b]; **8**, 88[36], 382[11], 383[11]
Eitelman, S. J., **6**, 978[23]
Eiter, K., **3**, 286[57]; **6**, 958[29]
Eizember, R. F., **5**, 915[111,112]
Eizenber, R. F., **5**, 225[102]
Eizenberg, L., **5**, 65[66]
Ejiri, E., **3**, 593[181]; **6**, 531[459,460], 764[9]
Ejiri, S., **5**, 266[75], 268[75]
Ejjiyar, S., **1**, 53[18]
Ek, M., **8**, 224[107]
Ekhato, I. V., **7**, 366[51], 414[120]
Ekogha, C. B. B., **3**, 252[86]
Ekström, B., **2**, 465[107]
Ekwuribe, N. N., **8**, 277[152]
El Abed, D., **2**, 85[18]; **4**, 155[71a]
El Alami, N., **1**, 214[21]
El Ali, B., **7**, 452[53]
Elam, E. U., **6**, 426[70], 543[617]; **8**, 141[30]
ElAmin, B., **6**, 635[15], 636[15], 668[250]; **8**, 440[84], 959[22]
El-Amouri, H., **8**, 445[43]
Elander, R. P., **7**, 55[12], 56[12]
El-Batouti, M., **8**, 860[221]
El-Berins, R., **6**, 573[964]
Elberling, J. A., **3**, 849[58]
El Boouadili, A., **5**, 1126[68]
El Bouadili, A., **1**, 749[78], 816[78]
El-Bouz, M., **4**, 10[34], 73[32], 112[159], 113[164], 240[48]; **8**, 850[118]
Elder, J. W., **4**, 399[221]
Elderfield, R. C., **2**, 283[52]; **4**, 1099[181]; **7**, 477[70]; **8**, 606[19], 625[157], 626[157]
Eldik, R. V., **5**, 453[63], 454[63], 458[63]
Eldred, C. D., **5**, 584[192]
Eldridge, J. M., **7**, 583[155], 584[155]
El-Durini, N. M. K., **8**, 774[74]
Elebring, T., **7**, 331[15]
El-Enien, M. N., **8**, 478[38]
Eleuterio, H. S., **5**, 1116[7]
Eleveld, M. B., **1**, 72[73]
El-Fadl, A. A., **8**, 98[105]
El-Faghi, M. S., **2**, 362[185]
Elfahham, H. A., **2**, 379[296]
Elfehail, F., **6**, 105[17], 106[18]
El-Feraly, F. S., **3**, 390[77,79], 392[77]
Elferink, V. H. M., **5**, 742[161]; **6**, 572[960]
Elfert, K., **4**, 8[29]
Elgamal, S., **2**, 379[296]
El-Garby Younes, M., **3**, 834[78,79]
Elgemeie, G. E. H., **2**, 379[296]
Elgendy, S., **1**, 499[54], 501[54]
El Ghandour, N., **5**, 247[26]
El Gharbi, R., **6**, 173[46], 175[46,73]
ElGomati, T., **2**, 1087[34]
Elguero, J., **4**, 55[157], 439[160]; **5**, 741[153]; **6**, 579[984]; **8**, 636[7]
Elhafez, F. A. A., **1**, 141[19], 151[19]
Elhalim, M. S. A., **5**, 488[197]
El Hallaoui, A., **3**, 215[66], 251[75]
El-Hashash, M. A., **2**, 744[99], 745[99]
El-Helow, E. R., **7**, 71[97]
Eliaers, J., **8**, 237[12]
Elias, H., **3**, 890[35], 903[127]
Eliason, J., **8**, 94[79]
Eliel, E. L., **1**, 2[15], 49[11], 50[11], 57[11], 61[37], 62[38-41], 63[42], 65[11,52], 66[57], 69[57b], 70[52], 150[52], 153[57], 182[48], 285[6], 460[2]; **2**, 630[8], 964[56,56b,c]; **3**, 76[54], 124[273,287,288], 128[273,287,288,323], 132[273,288], 134[273,323]; **4**, 3[9], 187[96], 202[26]; **5**, 88[50], 754[60]; **6**, 685[353]; **7**, 549[41]; **8**, 14[87], 66[104], 141[38,45], 214[36,37], 217[36], 218[70], 219[70], 224[111], 228[130], 229[136,137], 230[137,139], 231[142,144,147,148], 232[148], 246[77], 502[63], 843[57], 872[4,6], 966[71]
Eliel, L., **2**, 455[5]
Eliev, S., **8**, 242[40]
Eliseeva, L. V., **4**, 314[486]
Elisseou, E. M., **3**, 220[119]
Elissondo, B., **1**, 479[149,150], 480[149,150]; **2**, 71[53]; **3**, 196[25], 453[115]
Elizarova, A. N., **4**, 30[87]
Eljanov, B. S., **2**, 663[24], 664[24]
El-Jazouli, M., **4**, 85[74]; **6**, 455[153]
El-Kady, I. A., **7**, 71[96]
El-Kady, M. Y., **5**, 488[197]
El-Kady, S. S., **2**, 744[99], 745[99]
El Khadem, H., **8**, 70[237]
El-Khawaga, A. M., **2**, 745[103]; **3**, 325[160]
El-Kheli, M. N. A., **1**, 39[191]
Elkik, E., **2**, 411[7]; **4**, 123[210b], 125[210b]
Elkind, V. T., **6**, 431[113]
Elks, J., **3**, 804[13]; **7**, 159[45], 582[149]; **8**, 639[20]
Ellegast, K., **2**, 368[244]
Ellenberger, M. R., **6**, 708[48]
Ellenberger, S. R., **5**, 105[197]; **7**, 441[42]
Ellenberger, W. P., **6**, 900[118]; **8**, 542[232], 928[23]
Ellermeyer, E. F., **1**, 360[28], 361[28]
Ellery, E., **4**, 85[77a]
Ellestad, G. A., **5**, 736[145], 737[145]
Elliger, C. A., **6**, 843[90]; **8**, 314[31]
Ellingboe, J. W., **2**, 192[61], 221[61], 256[47], 257[47]
Elliot, J. D., **1**, 347[136,137,138], 348[138,139,142], 546[59]; **3**, 281[45]
Elliot, M., **1**, 546[50]; **5**, 560[74]
Elliot, R., **1**, 347[136]
Elliott, A. J., **8**, 621[140,141]
Elliott, C. S., **5**, 918[125]
Elliott, D. F., **8**, 639[20]
Elliott, I. W., **3**, 686[109], 699[109]; **8**, 568[471]
Elliott, J., **1**, 776[216], 777[216b], 814[216]; **8**, 13[68]
Elliott, J. D., **2**, 555[143], 556[154], 578[85], 580[97], 650[110,111], 651[110,111a,b], 791[60]; **3**, 226[206], 360[72]; **6**, 81[74], 237[68]
Elliott, R., **2**, 580[97]

Elliott, R. C., **1**, 742[47,48]
Elliott, R. D., **3**, 790[58]
Elliott, R. L., **1**, 270[61]; **3**, 221[129,130], 603[229]; **4**, 820[218]; **5**, 841[89], 872[89b]; **7**, 137[120]; **8**, 856[173]
Elliott, S. P., **5**, 70[107,108]
Elliott, W. H., **7**, 564[112], 572[112], 587[112]
Ellis, A. F., **7**, 488[160]
Ellis, B., **8**, 986[17], 987[17]
Ellis, B. S., **4**, 54[154]
Ellis, D., **7**, 479[96]
Ellis, G. P., **6**, 231[36], 232[36], 233[36]
Ellis, G. W. L., **4**, 37[107]
Ellis, J. E., **2**, 162[144]; **4**, 7[26]
Ellis, J. W., **4**, 262[309]; **7**, 154[16], 163[70], 174[136,137,138]
Ellis, K. L., **2**, 655[134,134a,b]; **3**, 26[161]; **5**, 282[24]
Ellis, K. O., **6**, 515[235]
Ellis, M. C., **5**, 959[313,314]
Ellis, P. D., **2**, 388[342]; **3**, 19[103]
Ellis, R. J., **1**, 618[58], 621[67], 784[239,240], 815[239]; **5**, 906[65], 908[65], 910[65,82], 989[44], 1016[62]
Ellis-Davies, G. C. R., **5**, 649[22], 650[22]
Ellison, D. L., **2**, 184[25,25b]
Ellison, G. B., **3**, 4[22]
Ellison, R. A., **1**, 566[212]; **3**, 124[272,276], 125[272], 126[272,276], 145[386]; **6**, 135[24], 676[304], 677[304]; **8**, 625[160]
Ellison, R. H., **8**, 929[28]
Ellman, J. A., **1**, 400[9]; **2**, 113[108]; **6**, 77[56], 248[137]
Ellsworth, E. L., **1**, 110[23], 112[24,26], 114[38], 335[66]; **3**, 212[34b,c], 213[46], 216[46], 223[46]; **4**, 244[78], 245[78], 255[78]; **6**, 831[9]
Elltsov, A. V., **8**, 98[100]
Elmes, B. C., **2**, 835[156]; **4**, 51[144a]
Elming, N., **7**, 618[23], 802[46,47]
Elmoghayar, M. R. H., **2**, 650[108]
Elmore, D. T., **6**, 420[24], 451[127], 455[155]
El Mouhtadi, M., **5**, 72[188]
El-Mowafy, A. M., **8**, 827[75]
Elnagar, H. Y., **5**, 707[35], 735[35a], 740[35a]
Elnagdi, M. H., **2**, 379[296]
El'natanov, Yu. I., **4**, 48[136], 49[136]
Elofson, R. M., **3**, 507[170]
El-Omrani, Y. S., **8**, 17[113], 22[113], 551[338]
Elphimoff-Felkin, I., **4**, 976[93], 977[93], 978[93], 994[93]; **8**, 310[15], 311[18]
El Raie, M. H., **7**, 41[27]
El-Refai, A.-M. H., **7**, 69[87], 71[97]
Elsasser, A. F., **6**, 526[392]
El Sekeili, M., **8**, 70[237]
El Seoud, O. A., **4**, 315[503]
Elsevier, C. J., **2**, 587[148]; **3**, 217[81], 491[68,69], 531[84]; **4**, 905[209]
El-Shafei, Z. M., **8**, 70[237]
El-Sharkaway, S. H., **7**, 59[41]
El-Shenawy, H. A., **6**, 36[20]; **8**, 91[50]
Elslager, E. F., **6**, 533[505], 554[709]
Elsner, B. B., **8**, 963[45]
ElSohly, H. N., **5**, 155[37]
Elsom, L. F., **3**, 418[28], 499[123]
Elston, C. T., **6**, 825[127]
El-Telbany, F., **6**, 533[479], 550[479]; **8**, 244[67], 250[67]
El'tsov, A. V., **8**, 637[13], 638[17], 661[113]
Elvidge, J. A., **2**, 365[213]; **6**, 275[111,112]
Elving, P. J., **7**, 603[112,113]
Elvinga, P. J., **8**, 592[65], 642[30]
Elwahab, L. M. A., **5**, 488[197]
El-Wassini, M. T., **2**, 760[38]
Elzen, V. D., **3**, 979[13]
Elzey, T. K., **4**, 389[167]
Elzinga, J., **4**, 1036[54]
Elzohry, M. F., **3**, 325[160]
Eman, A., **1**, 631[47], 656[154], 658[154]; **3**, 87[92]; **4**, 318[560]; **5**, 676[5]
Emanuel, N. M., **7**, 10[77]
Emde, H., **2**, 605[63], 681[58], 683[58]; **5**, 843[120]; **6**, 502[217], 560[870]; **7**, 650[51]
Emel'yanov, M. M., **3**, 359[71]
Emerson, D., **6**, 134[16]
Emerson, D. W., **3**, 154[420]; **7**, 124[44]
Emerson, G. F., **4**, 701[26]; **6**, 690[397], 692[397]
Emerson, K., **1**, 309[99,100]
Emerson, T. R., **8**, 391[86]
Emerson, W. S., **8**, 84[9], 139[1], 144[1], 963[46]
Emery, S. E., **8**, 315[50]
Emke, A., **7**, 833[72]
Emling, B. L., **1**, 360[28], 361[28]
Emma, J. E., **6**, 554[737]
Emmer, G., **7**, 491[182]
Emmert, B., **8**, 916[97]
Emmons, W. D., **1**, 761[139]; **7**, 502[265], 673[19]
Emolaev, M. V., **8**, 963[43]
Emoto, M., **4**, 615[384]
Emoto, S., **4**, 201[8]; **8**, 144[76]
Emptoz, G., **4**, 98[109d]
Emrani, J., **5**, 841[99], 856[99,219]
Emslie, N. D., **4**, 34[97], 35[97]; **5**, 834[56]
Emziane, M., **6**, 91[125,127], 237[63]; **7**, 493[190]
Encinas, M., **5**, 153[27]
Enda, J., **2**, 68[44], 89[38], 579[95]; **3**, 197[33]; **4**, 630[419]
Ender, U., **8**, 331[32]
Enders, D., **1**, 29[145], 357[6], 359[17], 378[6], 380[17], 381[17], 476[118], 477[118], 555[123], 630[27]; **2**, 60[18], 62[18d], 455[14], 482[31], 484[31], 504[3,4], 506[13,15], 507[21,24], 508[29], 509[3], 510[42,44], 511[21,45], 512[21], 514[50-53], 515[53], 516[59], 517[21], 519[64,65], 520[66-69], 521[70], 523[13,84], 524[50,53,80,81,84], 830[139], 1077[153]; **3**, 30[180], 34[180,191], 35[180,191], 37[191], 39[191,213], 65[7]; **4**, 21[69], 104[137], 173[36], 212[99], 222[167,168,169,170,171,172], 224[167,168,170]; **5**, 485[182]; **6**, 119[110,111], 419[8], 425[8], 684[342], 716[104], 722[141], 724[141], 727[193], 728[208,209,210,211,212,213,214,215]; **7**, 98[100], 187[185], 224[55], 225[56]; **8**, 388[60]
Enders, E., **1**, 563[173]; **8**, 382[12], 383[12]
Endesfelder, A., **2**, 8[38], 15[38,62,62a], 26[101], 27[101], 32[62a], 41[101], 42[101], 977[9], 981[26], 982[26], 994[9,26,41], 995[26,42], 996[9], 997[9]
Endo, A., **7**, 77[123]
Endo, H., **3**, 421[54]
Endo, J., **2**, 1021[50]; **6**, 916[31]; **8**, 31[47], 66[47]
Endo, K., **4**, 903[202], 904[202]; **8**, 881[83]
Endo, M., **2**, 576[72], 578[87], 624[161], 648[97,97c], 649[97a,c]; **4**, 155[68b]
Endo, T., **1**, 407[31]; **4**, 792[64,65], 823[231]; **5**, 185[164]; **6**, 625[152], 807[57]; **7**, 760[31]; **8**, 332[43], 369[81-83], 382[9], 395[132], 645[47], 846[84]
Endo, Y., **4**, 522[53], 523[53]
Endres, R., **8**, 445[20]
Eneikina, T. A., **6**, 489[78]
Enescu, L. N., **4**, 963[44]
Eng, K., **1**, 104[98]
Eng, K. K., **4**, 957[21], 1101[192]
Engberts, J. B. F. N., **2**, 332[56]; **8**, 837[11], 839[11]
Engebrecht, J. R., **4**, 331[10], 345[10]; **7**, 400[43]

Engel, C. R., **3**, 846[41-44]; **4**, 125[216,216g]; **8**, 34[60], 66[60], 530[90]
Engel, K., **1**, 162[92,97,99]; **8**, 675[47]
Engel, K.-H., **8**, 190[84]
Engel, M. R., **3**, 698[157a]; **5**, 468[135]
Engel, N., **6**, 523[352], 524[352]
Engel, P., **3**, 623[32], 626[32b]
Engel, P. S., **4**, 725[44]; **5**, 178[137], 205[44], 217[18], 220[46], 221[46], 223[70], 224[99], 632[59]; **6**, 960[56]; **7**, 874[110]; **8**, 390[79]
Engel, R., **8**, 412[108a]
Engelbrecht, F., **8**, 754[108], 755[108]
Engelen, B., **1**, 9[45]
Engelhard, M., **6**, 670[274]
Engelhardt, E. L., **3**, 380[11]; **5**, 382[121]
Engelhardt, G., **1**, 215[32]
Engelhardt, L. M., **1**, 13[73], 17[207,212,219], 36[233,234], 37[177,178]
Engelhardt, V. A., **5**, 1138[65]
Engelmann, H., **4**, 274[65], 275[65], 280[65], 281[65]
Engelmann, H. M., **2**, 144[57]
Engels, J., **6**, 625[154]
Engels, R., **7**, 796[15]
Engemyr, L. B., **6**, 242[94]
Engerer, S., **7**, 155[29]; **8**, 458[224]
Enggist, P., **2**, 547[93]; **4**, 378[106]; **5**, 830[32]
England, D. C., **2**, 538[67], 539[67]; **5**, 64[44]
England, W. P., **1**, 878[109]; **4**, 1005[86]; **5**, 20[136]
Engle, R. R., **7**, 236[13]
Engler, A., **4**, 18[59]
Engler, D. A., **3**, 918[24]; **7**, 124[47], 160[52], 161[52], 176[52], 180[52], 183[52], 187[52]
Engler, R., **6**, 134[33], 187[177]; **8**, 860[223]
Engler, T. A., **1**, 766[153]; **3**, 844[35]; **5**, 24[165]; **6**, 866[206]; **8**, 448[148]
Englert, H., **6**, 487[63], 489[63]
Englert, L. F., Jr., **8**, 376[161]
Englhardt, H. F., **5**, 1066[8]
English, J., Jr., **3**, 262[166]; **6**, 678[324]
Engman, A. M., **8**, 308[6], 309[6]
Engman, L., **4**, 340[51], 342[64,70], 343[64,72,73], 386[155]; **6**, 463[24], 685[355], 976[2]; **7**, 135[101], 534[43,44], 772[297], 774[317], 776[361]; **8**, 410[96], 413[132], 806[126], 990[43], 995[43]
Engster, C. H., **5**, 947[288]
Enholm, E. H., **4**, 744[130]
Enholm, E. J., **1**, 127[91], 794[281], 795[281]; **2**, 4[12,12c], 6[12,12c], 15[65], 573[53], 581[104], 981[24], 989[24], 990[24], 1064[109]; **4**, 744[129], 799[114]; **5**, 534[92]; **6**, 990[82], 991[82]; **7**, 579[131]
Enhsen, A., **3**, 253[93]
Enjo, H., **7**, 761[55], 764[55]
Enk, M., **1**, 749[78], 816[78]
Ennen, B., **5**, 1126[66]
Ennis, M. D., **2**, 240[4], 436[67], 438[70], 846[203]; **3**, 45[249]; **8**, 11[61]
Enokiya, M., **1**, 448[205]; **2**, 749[135]
Enokiya, R., **3**, 566[26]; **8**, 404[18]
Enomiya, T., **7**, 451[20], 452[20], 454[20]
Enomoto, K., **5**, 809[120]
Enomoto, M., **3**, 45[248]
Enomoto, S., **4**, 444[198]
Enomura, K., **8**, 407[54]
Enos, H. I., Jr., **3**, 304[63]
Ens, L., **7**, 763[96]
Ensley, H. E., **1**, 87[48]; **4**, 91[88c]; **5**, 353[86], 419[72], 543[115]; **6**, 657[173]; **7**, 96[87], 131[84], 180[155], 260[85]; **8**, 111[23], 117[23]
Enslin, P. R., **7**, 156[32], 157[32a]
Ent, H., **2**, 1063[105]; **6**, 746[97]
Entenmann, G., **6**, 242[89], 243[89,104]
Entreken, E. E., **8**, 237[18], 244[18], 249[18], 250[18]
Entwhistle, E., **3**, 554[19]
Entwistle, I. D., **6**, 119[109], 642[71], 1022[62]; **7**, 772[288]; **8**, 18[123], 81[4], 91[4], 104[4], 263[31], 366[43], 367[57,62], 368[63], 440,[83], 551[339], 958[19], 959[27]
Enzell, C., **7**, 100[117]; **8**, 310[17]
Eordway, D., **4**, 305[365]
Eoyama, I., **4**, 560[27]
Epa, W. R., **7**, 145[160,161]
Epe, B., **8**, 342[110], 398[144]
Epe, M., **8**, 342[110]
Ephritikhine, M., **7**, 6[29], 533[32]; **8**, 737[26]
Epifani, E., **1**, 113[29]; **2**, 87[26], 495[64,65], 496[64,65]
Epifanio, R de A., **7**, 253[22]
Epiotis, N. D., **1**, 506[10,11]; **2**, 662[5]; **5**, 75[218,219,220,221,222,223], 686[52], 703[19], 705[19,24], 1009[46]
Epling, G. A., **5**, 196[12]; **7**, 248[111], 801[44]; **8**, 517[127]
Epple, G., **2**, 739[49-51]
Eppley, R. L., **4**, 871[31]
Epprecht, A., **7**, 657[32]
Epstein, J. W., **8**, 502[65], 503[65], 796[29]
Epstein, M., **8**, 366[37]
Epstein, W. W., **2**, 285[59]; **7**, 292[3,9], 653[2], 656[15]; **8**, 93[76], 531[116]
Epsztajn, J., **1**, 329[38], 474[90-93]; **7**, 745[73]
Epsztein, R., **1**, 595[28]; **2**, 81[1], 82[1,1a], 96[1]; **4**, 84[68d]
Erb, R., **5**, 687[65]
Erben, H. G., **5**, 1096[127]
Ercoli, R., **8**, 452[189b]
Erdelmeier, I., **1**, 37[252], 531[131], 535[145,147], 536[145,147], 773[203,203b], 788[256]; **3**, 230[237]
Erden, I., **2**, 744[91]; **4**, 969[67]; **5**, 78[283], 206[45]; **8**, 477[30]
Erdik, E., **1**, 107[3], 212[6], 213[6], 427[112]; **2**, 279[20], 282[20]; **3**, 208[10], 210[10], 215[10c], 243[12], 250[12], 262[12], 436[18]; **4**, 89[87], 141[13], 172[31], 173[31]
Erdtman, H., **3**, 660[9], 667[47]
Erfort, U., **8**, 852[139]
Erhardt, F., **3**, 665[41]
Erhardt, J. M., **3**, 226[195], 264[183]; **6**, 9[43], 581[989]; **8**, 85[19]
Erhardt, P. W., **6**, 809[63]
Erickson, A. S., **7**, 219[16]
Erickson, B. W., **2**, 72[57], 76[57]; **3**, 103[203], 108[203], 124[275], 125[275,313]; **6**, 134[14], 138[44]
Erickson, D., **6**, 204[23]
Erickson, F. B., **2**, 329[49]
Erickson, G. W., **3**, 36[209], 37[212]; **6**, 723[147], 725[171], 728[171]
Erickson, J. G., **6**, 564[911]
Erickson, K. L., **8**, 336[69]
Erickson, R. E., **7**, 247[100], 544[36]; **8**, 974[131], 988[25]
Erickson, R. L., **7**, 92[48]
Erickson, T. J., **7**, 410[98]
Erickson, W. F., **4**, 871[29], 877[67], 878[80], 884[80]
Ericsson, E., **2**, 365[210]
Eriksen, J., **7**, 881[158,159]
Erkamp, C. J. M., **3**, 210[25]
Erker, G., **1**, 162[92,97,99], 163[105]; **5**, 442[185], 589[213], 1131[12], 1174[35], 1178[35,45,46]; **8**, 675[47], 677[61], 679[61], 682[61,83], 683[88,93,95], 685[61], 686[95], 687[61]
Erlebacher, J., **1**, 8[38]
Erlenmeyer, E., **6**, 236[55]
Erlenmeyer, E., Jr., **2**, 396[5], 409[1]
Erlenmeyer, H., **8**, 907[70]
Erman, M. B., **6**, 836[53]
Erman, W. F., **3**, 706[8], 815[74]

Ermann, P., **6**, 174[59], 186[169], 197[169]; **7**, 109[185]
Ermel, J., **2**, 105[41]
Ermer, O., **3**, 382[36], 593[179]; **5**, 64[41]
Ermert, P., **1**, 373[91], 375[91], 376[91]; **2**, 996[47], 1077[154]
Ermili, A., **6**, 487[47-50,52,53], 489[47,48,52,53], 543[47,48]
Ermokhina, V. A., **3**, 321[138]
Ermolaeva, V. G., **8**, 656[92]
Ermolenko, M. S., **1**, 95[75,76]; **6**, 466[40-42], 469[40,54]; **8**, 694[118]
Ernert, P., **7**, 230[133]
Ernest, I., **2**, 765[78]; **3**, 781[14], 896[72]
Ernst, A. B., **4**, 764[215], 807[152]
Ernst, B., **1**, 843[2b], 850[31,32]; **4**, 792[63]; **5**, 596[26], 597[26], 829[15], 859[234]; **6**, 750[104], 859[165]
Ernst, F. D., **1**, 231[1]
Ernst, L., **2**, 385[321], 651[121]
Ernst, P., **3**, 660[3]
Eros, D., **5**, 10[75]
Errede, L. A., **8**, 364[12,13]
Ershov, L. V., **6**, 554[741,767,779]
Ershov, V. V., **3**, 814[69]
Ershova, I. I., **4**, 408[259b], 409[259b], 413[278c]
Ershova, Yu. A., **6**, 543[611]
Erskine, G. J., **1**, 162[103]
Erskine, R. W., **8**, 89[43]
Ertas, M., **2**, 107[55], 114[120], 193[65], 269[70]; **4**, 72[25]
Ertel, W., **2**, 359[159]
Erwin, D. K., **8**, 675[46], 682[46]
Erwin, R. W., **3**, 25[151]; **6**, 160[177]; **8**, 843[54], 847[54]
Esch, P. M., **2**, 1069[133], 1070[138,139], 1071[138], 1072[139], 1075[138], 1077[139], 1078[139], 1079[156]; **5**, 501[267]
Eschenmoser, A., **2**, 368[241], 587[138], 817[89], 866[7-9], 867[7,18], 870[7], 871[7], 872[7], 875[7], 876[7], 899[27], 901[27]; **3**, 13[68], 341[2], 345[20], 346[23], 351[39], 360[2], 916[19]; **5**, 501[268,269], 714[75a], 828[6], 836[6,64], 888[26], 893[26]; **6**, 419[10], 538[571], 672[285], 708[50], 831[7], 1042[3], 1043[14], 1056[56,57], 1058[14], 1059[62,64-66]; **7**, 482[118]; **8**, 948[148]
Escher, S. D., **3**, 126[320], 882[104]; **6**, 161[185]
Esclamadon, C., **8**, 216[58]
Escobar, G., **5**, 92[64]
Escot, M. T., **8**, 133[18]
Escudie, J., **5**, 444[189]
Escudie, N., **4**, 630[418]
Eshima, K., **2**, 387[333]
Esipov, S. E., **6**, 554[745]
Eskenazi, C., **5**, 289[40], 290[40], 464[114], 466[114]; **7**, 238[40]
Eskew, N. L., **6**, 74[39]
Eskin, N. T., **3**, 639[80]
Eskins, K., **3**, 691[129], 693[129]
Eskola, P., **7**, 93[53]
Esperling, P., **8**, 548[318]
Espidel, J., **8**, 446[74], 452[74], 457[74]
Espy, H. H., **3**, 393[93]
Ess, R. J., **8**, 978[142]
Essawy, S. A., **6**, 770[36]
Essefar, M., **5**, 72[188]
Essenfeld, A. P., **1**, 769[182]; **5**, 516[24], 517[24], 518[24], 520[36], 545[119]
Esser, F., **4**, 410[260a]
Essig, M. G., **8**, 89[45]
Essiz, M., **5**, 1022[75]
Esslinger, M. A., **6**, 452[132]
Esswein, A., **6**, 51[114], 57[142]
Estel, L., **4**, 465[115], 474[115], 478[115]
Estep, R. E., **1**, 524[91]; **2**, 417[21]; **3**, 748[77]
Ester, W., **7**, 10[80]
Estermann, H., **3**, 43[237]
Esteruelas, M. A., **8**, 764[6], 773[6b]
Estes, J. M., **4**, 629[415]
Estopà, C., **2**, 381[300]
Estreicher, H., **2**, 588[150]; **4**, 5[17], 85[77b]; **5**, 320[9]; **6**, 107[28], 108[37]; **7**, 218[6], 501[250], 534[38]
Eswarakrishnan, S., **2**, 103[32], 209[108], 211[116]; **5**, 828[8]; **6**, 858[161], 861[182]
Eswarakrishnan, V., **3**, 878[93,94], 879[93,94,97], 880[94], 881[94]; **4**, 345[83], 348[108], 349[108c], 359[159], 771[251]; **6**, 161[182]
Etemad-Moghadam, G., **1**, 774[208]; **2**, 369[251]; **4**, 290[204], 291[204,210], 292[204], 311[439,440]; **8**, 854[154]
Etheredge, S. J., **3**, 19[101]; **4**, 54[152,152a], 258[233], 403[241]; **5**, 330[34,35]
Étienne, A., **6**, 535[525,526]
Etienne, Y., **5**, 86[20]
Etievant, P., **8**, 683[87,94]
Etinger, M. Y., **4**, 885[115], 886[115]
Etkin, N., **4**, 1057[143], 1058[143]
Etlis, V. S., **6**, 560[866]
Eto, H., **3**, 691[128], 693[128,139], 697[139,155,156], 698[128]
Etter, J. B., **1**, 262[34,36], 263[42], 264[34,42], 265[42], 266[42], 267[50], 278[36]; **2**, 127[237]; **4**, 971[76], 972[76]
Etter, M. C., **1**, 41[195]
Ettlinger, M., **6**, 176[85], 182[85], 193[85]
Etzbach, K. H., **2**, 387[332]
Etzrodt, H., **1**, 18[94]
Eudy, N. H., **6**, 538[557]
Eugster, C. H., **2**, 964[57]; **5**, 342[62c], 564[97], 595[19], 596[19]; **6**, 677[313], 782[81]; **7**, 92[42], 93[42], 410[103]; **8**, 606[21], 946[134]
Eul, W., **6**, 552[694]
Eustache, J., **2**, 681[62]; **3**, 262[167]; **5**, 432[126], 438[163]; **7**, 359[16]
Evain, E. J., **6**, 70[20]
Evanega, G. R., **5**, 168[103]
Evans, A., **8**, 670[12], 671[12]
Evans, A. J., **2**, 477[10]; **5**, 514[8], 527[8]
Evans, B. E., **1**, 823[44b]; **2**, 962[51]; **3**, 380[11]; **8**, 315[49], 968[87]
Evans, B. R., **5**, 603[51], 612[74]
Evans, C. F., **3**, 589[161], 610[161]
Evans, D., **8**, 445[27,56,58], 452[58], 568[462]
Evans, D. A., **1**, 130[97], 223[73], 224[73], 227[98], 312[112-114], 314[135], 315[135], 317[135], 328[21-23], 358[10], 398[3], 399[3], 400[9,11], 401[12,14,15], 402[12,16], 424[99], 430[127], 743[52], 744[52], 747[5]; **2**, 2[4], 25[99], 30[110], 31[110], 64[27], 66[35], 71[55], 74[55], 91[49], 99[1], 100[1,15], 101[1,15], 103[1], 111[15], 112[15,99,100], 113[15,106-108], 116[135], 119[156,158], 134[3], 182[2], 190[57], 192[2c], 197[79], 211[114], 214[2c], 223[150], 232[181], 239[2], 240[2,4], 242[20], 245[20f,31,32], 246[20f,31], 247[31], 250[38], 251[38,39], 254[43], 255[44], 256[45], 257[4a], 260[38], 263[55], 274[38], 304[8], 305[8], 318[50], 436[67], 438[70], 442[6], 455[7], 475[1], 509[33], 597[9], 846[203,206], 894[10], 917[10], 918[10], 919[10], 930[10], 1024[59], 1049[26], 1064[26]; **3**, 1[2], 2[2], 3[2], 13[2], 15[2], 23[2], 25[2], 41[2], 44[2], 45[2,244,246,249], 55[2], 86[16], 95[16], 96[16], 97[16], 98[16], 99[16], 104[16], 116[16], 117[16,235,236], 118[16], 135[363], 136[363], 139[363], 142[363], 155[16,235,236], 156[235,236,363], 168[16], 196[30], 199[53], 672[66], 781[13], 799[102], 919[31], 934[31], 939[78], 946[87]; **4**, 98[110], 145[35], 366[7], 384[7,143,145b]; **5**, 186[170], 323[14], 361[92], 543[116,116b,c], 788[11,13,14], 790[34], 798[11], 814[136,137,138], 821[13], 847[133], 849[142], 872[133], 888[30], 890[34], 903[35], 1000[2-4,7,8], 1115[2], 1116[2], 1122[2a], 1123[2a]; **6**, 77[56], 118[103], 147[87], 153[142], 163[193], 248[137], 256[171], 682[338,340], 759[137,138], 834[39], 838[64,66], 854[145], 873[1,5], 895[1], 901[121], 903[5], 995[101], 996[109], 997[110], 1022[67]; **7**, 162[68], 184[171], 245[79], 300[53], 401[61b], 407[61b], 408[90],

418[90], 545[25], 602[96], 764[131]; **8**, 9[54], 11[61], 36[72], 37[72], 38[72], 44[72], 66[72], 269[91], 386[53], 448[140,141,142], 619[130], 676[80], 698[143], 720[136], 814[17], 945[129], 948[145,152]
Evans, D. E., **3**, 854[77]; **5**, 96[122], 98[122,126]
Evans, D. F., **1**, 254[13], 276[13], 278[13]
Evans, D. H., **3**, 568[47]; **7**, 805[66]
Evans, D. J., **4**, 688[65]
Evans, D. L., **7**, 143[146]
Evans, D. W., **4**, 462[105]
Evans, E. A., **5**, 758[83]
Evans, E. L., **2**, 329[47]
Evans, G., **4**, 665[9], 688[9]; **6**, 690[399], 691[399], 692[399]
Evans, G. E., **4**, 14[47]
Evans, G. G., **8**, 605[14]
Evans, G. W., **7**, 96[82]
Evans, J., **4**, 691[76]
Evans, J. B., **3**, 160[465], 164[465], 166[465]
Evans, J. C., **7**, 340[49]
Evans, J. J., **8**, 316[58]
Evans, J. M., **3**, 735[19]; **7**, 71[99]
Evans, L. T., **2**, 141[39]
Evans, M. E., **6**, 660[201,204]
Evans, M. G., **5**, 856[204]
Evans, P. L., **8**, 460[253]
Evans, R. D., **4**, 347[93], 367[14], 368[14], 369[14]; **6**, 26[104]; **7**, 535[49], 536[50]
Evans, R. J. D., **6**, 835[43]
Evans, R. M., **7**, 582[149]; **8**, 987[23]
Evans, S., **7**, 763[96]
Evans, S. A., Jr., **6**, 22[85], 24[97], 74[39]
Evans, S. V., **5**, 211[62]
Evans, T. L., **4**, 441[183], 443[193]; **7**, 765[155]
Evans, T. W., **3**, 825[25], 826[25]; **7**, 68[81]
Evans, W. H., **3**, 383[46]
Evans, W. J., **1**, 231[10,11], 251[2], 252[2b]; **8**, 447[130,137], 458[224]
Everby, M. R., **8**, 839[26a], 840[26]
Everhardus, R. H., **3**, 87[213], 105[213]
Evering, B. L., **7**, 7[38]
Evers, M., **6**, 462[13]
Evitt, E. R., **5**, 1134[38], 1149[38]
Evnin, A. B., **5**, 406[24], 489[199], 604[54]
Evrard, G., **1**, 650[139], 661[167,167a], 664[200], 672[200], 675[210], 677[210], 706[210], 718[200], 719[200], 720[200], 721[210], 722[200], 870[84]; **7**, 773[307]
Evstigneev, V. V., **4**, 426[64]
Evstigneeva, R. P., **6**, 271[88]
Evstratova, M. I., **8**, 599[101]
Ewin, G., **6**, 538[550]
Ewing, D. F., **2**, 361[177]
Ewing, J. H., **8**, 149[120]
Ewing, S. P., **1**, 506[6], 510[6]
Ewins, R. C., **7**, 390[1]
Exner, H. D., **5**, 589[213]
Exner, O., **6**, 795[15], 798[15], 821[15]
Exon, C., **5**, 1053[40,42], 1060[40a,42]
Exon, C. M., **4**, 578[19-21]
Eyer, M., **2**, 332[58], 333[58], 338[74]
Eyken, C. P., **5**, 649[22], 650[22]
Eyley, S. C., **2**, 742[73,74], 748[126], 948[183], 965[63], 966[71], 967[63,71]; **4**, 820[216]
Eyman, D. P., **8**, 238[23], 261[6]
Eyring, H., **5**, 72[180]; **7**, 852[36]
Eyring, L., **1**, 231[4], 251[1,2], 252[1]
Eyring, M. W., **1**, 308[96]
Ezaki, Y., **6**, 175[76]
Ezmirly, S. T., **5**, 788[12], 1003[21]
Ezquerra, J., **5**, 833[48]
Ezzel, M. F., **6**, 959[41]

F

Faber, K., **7**, 493[191]; **8**, 383[22]
Fabian, J. M., **7**, 762[72]
Fabian, W., **3**, 975[4], 979[4]
Fabiano, E., **6**, 79[62]
Fabio, P. F., **3**, 767[163]; **6**, 487[42], 489[42], 543[42], 550[42], 554[42]
Fabio, R. D., **7**, 103[140]
Fabra, F., **8**, 636[4]
Fabre, J.-L., **3**, 447[93], 448[94], 493[81]; **8**, 842[42a,b]
Fabriás, G., **6**, 172[18]
Fabricius, D. M., **5**, 20[134], 790[35]
Fabrissin, F., **4**, 20[63], 21[63]
Fabrissin, S., **4**, 20[63], 21[63]
Fachinetti, G., **5**, 1174[36]; **8**, 447[125], 683[90]
Factor, A., **4**, 294[247], 302[332], 314[483], 315[483]
Fadel, A., **3**, 41[227]; **7**, 843[39,40]
Fadlallah, M., **4**, 23[70]
Faehl, L. G., **7**, 765[159]
Fagan, G. P., **8**, 615[94], 618[94]
Fagan, P. J., **5**, 1139[76], 1166[20], 1167[20], 1169[20], 1170[20], 1178[20]; **6**, 291[211]; **8**, 447[131,132], 697[129]
Faggi, C., **4**, 247[97], 256[97]
Faggiani, R., **7**, 876[122]
Fagundo, R., **3**, 327[171]
Fahey, D. R., **7**, 449[5], 450[5], 452[5]; **8**, 451[176]
Fahey, R. C., **4**, 270[1,2], 273[42,43,45], 274[60], 277[84,91-93], 279[117], 280[45,119,120], 329[1], 344[1], 350[1], 351[1]
Fahidy, T. Z., **3**, 636[50]
Fahmy, A. M., **2**, 772[15]
Fahmy, S. M., **2**, 362[179]
Fahr, E., **5**, 426[105], 428[105], 429[105]
Fahrbach, G., **7**, 775[342]
Fahrenholtz, S. R., **6**, 955[23]; **7**, 96[86]; **8**, 543[247]
Fahrni, P., **5**, 850[146], 876[3]
Faillard, V. H., **2**, 464[94]
Faillebin, M., **8**, 142[47]
Failli, A., **2**, 1088[40], 1095[92], 1097[40]
Fainzilberg, A. A., **4**, 347[103]
Fairbrother, F., **3**, 299[33]
Fairfull, A. E. S., **6**, 430[97]
Fairhurst, R. A., **2**, 948[183], 959[31], 960[31], 962[45], 964[45], 965[63], 967[63]
Fairlie, D. P., **1**, 310[108]
Fairman, J., **8**, 435[71]
Faita, G., **2**, 364[204]
Faith, W. C., **7**, 503[278]; **8**, 354[172]
Fajgelj, S., **6**, 554[742]
Falbe, J., **2**, 282[35], 554[133]; **3**, 1017[6]; **4**, 914[2], 921[28], 923[2], 924[2], 925[2], 926[2], 928[2], 932[2], 939[2], 941[2], 943[2]; **6**, 37[35]
Falci, K. J., **8**, 939[97]
Falck, J. R., **1**, 408[35], 430[35,132], 864[85]; **2**, 1085[21]; **3**, 159[464], 161[464], 166[464], 288[63], 685[106,107], 771[184], 785[36,36a]; **4**, 372[64a]; **5**, 499[250,251], 500[250,251]; **6**, 11[45], 206[42], 210[42], 218[42], 536[546], 538[546], 660[202], 682[341]; **7**, 87[18,18a], 260[84], 378[97], 678[73], 713[72], 801[41]; **8**, 881[68], 935[59]
Falcone, S. J., **7**, 774[334]
Faler, G. R., **7**, 96[90], 98[90]
Fales, H. M., **7**, 528[9]; **8**, 51[123], 66[123]
Falk, H., **2**, 817[89]
Falk, J. C., **8**, 449[156]
Falk, K. G., **2**, 399[17]
Falkenburg, H. R., **6**, 635[13], 636[13]
Falkow, L. H., **4**, 18[62], 21[62e]
Falkowski, D. R., **4**, 697[10]
Falkowski, L., **6**, 554[749], 789[110]
Fall, R. R., **8**, 332[40]
Faller, A., **1**, 694[238], 697[238], 698[238]; **6**, 1006[145]
Faller, J. W., **2**, 6[33], 35[33]; **4**, 604[279,280], 695[4]; **5**, 434[148]; **8**, 443[1]
Faller, P., **6**, 463[25]
Fallert, M., **2**, 379[297]; **6**, 450[123]
Fallis, A. G., **1**, 227[100], 511[34], 856[56]; **2**, 839[181], 840[181]; **3**, 906[142]; **4**, 24[73,73d], 368[16,16b], 370[16b], 378[16b], 790[37], 795[37]; **5**, 37[21], 347[72,72f], 451[33], 453[33], 464[33], 513[5], 514[5], 527[5], 944[239]; **6**, 839[68], 902[133]; **8**, 114[56]
Fallon, B., **7**, 470[12]
Fallouh, F., **8**, 862[230]
Falls, J. W., **4**, 915[13]
Falmagne, J.-B., **1**, 123[73], 370[66]; **7**, 122[30], 144[30]
Falou, S., **4**, 173[33]; **5**, 732[134]
Falshaw, C. P., **3**, 665[36]
Falsone, G., **2**, 381[301], 382[312]
Falter, W., **1**, 766[153]; **3**, 844[35]
Fama, F., **7**, 429[151]
Famulok, M., **6**, 114[78], 119[115]
Fan, C., **3**, 222[144]
Fan, H., **2**, 355[127]
Fan, W.-Q., **3**, 197[36]; **6**, 677[320]
Fan, X., **1**, 165[112b]
Fañanás, F. J., **5**, 1138[64]
Fancelli, D., **4**, 767[233]
Fanelli, J. M., **5**, 780[201]
Fang, G., **5**, 484[179]
Fang, H. W., **1**, 758[127]
Fang, J.-M., **1**, 512[35], 557[126], 565[207], 567[220], 569[256,257]; **2**, 6[34], 21[34a,b], 23[34a], 29[34a], 78[93,94]; **4**, 80[60], 113[164], 245[88], 255[88], 259[256], 790[39]; **5**, 885[19]
Fanghänel, E., **2**, 896[11,12]
Fang-Ting Chin, **7**, 478[86]
Fankhauser, J. E., **1**, 635[88]; **3**, 104[209], 117[209]; **6**, 846[103], 905[145]
Fanshawe, W. J., **8**, 618[127], 623[127]
Fanta, P. E., **3**, 209[22], 499[110], 501[110], 505[110], 509[110], 512[110]; **4**, 52[147,147b], 386[152b]; **5**, 949[276]; **7**, 470[6], 472[6], 473[6], 474[6], 476[6]
Fanta, W. I., **3**, 11[53]
Fantin, G., **1**, 471[68]; **3**, 232[267], 511[186], 514[186]
Farachi, C., **7**, 709[37], 765[134]; **8**, 240[31]
Faraday, M., **3**, 633[1]
Faragher, R., **5**, 422[81], 717[93], 742[159a]
Farah, D., **8**, 806[109]
Farahi, J., **1**, 309[98]
Farall, M. J., **7**, 281[174], 282[174]
Faraone, F., **4**, 588[66]
Farcasiu, D., **1**, 859[66]; **3**, 330[193]; **5**, 65[71]; **6**, 819[110]
Farcasiu, M., **1**, 859[66]; **3**, 328[179]; **5**, 65[71]
Fargher, J. M., **7**, 23[24], 24[24], 26[24]
Farid, S., **5**, 154[34], 913[102]; **7**, 851[31], 854[53], 855[53], 862[81], 879[150], 880[155], 888[81]
Fariña, F., **6**, 67[13]
Farina, J. S., **3**, 265[191]; **4**, 1089[127], 1092[127]
Farina, M., **7**, 17[177]

Farina, R., **4**, 532[83,91], 534[91], 537[91,95], 543[83], 545[83]
Farina, V., **1**, 699[247], 836[145]; **3**, 107[226], 109[226], 213[44], 232[264]; **4**, 183[80], 398[216], 399[216a], 401[216a], 405[216a], 410[216a]; **6**, 470[58], 766[22]; **7**, 495[211], 524[53]; **8**, 847[97,97d], 849[97d,107,115]
Faris, B. F., **6**, 263[26], 264[26], 265[45], 270[26]
Farkas, A., **8**, 142[46], 422[34]
Farkas, L., **8**, 142[46], 422[34]
Farkas, M., **6**, 430[93], 452[131]
Farkhani, D., **1**, 223[71]
Farmar, J. G., **5**, 851[168]
Farmer, E. H., **4**, 5[18]
Farmer, J., **3**, 918[22]
Farmer, P. B., **4**, 231[274]
Farneth, W. E., **6**, 708[48]
Farnetti, E., **8**, 534[159]
Farney, R. F., **7**, 225[61,62]
Farnham, W. B., **3**, 572[62]; **4**, 710[51]; **7**, 4[16]
Farnia, S. M. F., **3**, 295[9], 334[221,221a]
Farnier, M., **2**, 780[9]; **6**, 781[80]; **7**, 27[65], 32[93]; **8**, 946[134]
Farnocchi, C. F., **5**, 1043[25], 1046[25b], 1048[25b]
Farnoux, C. C., **6**, 488[20], 517[20], 546[20], 548[20], 549[20]
Farnow, H., **2**, 529[20]; **5**, 809[114]; **8**, 566[457]
Farnsworth, D. W., **7**, 225[58], 280[167]
Farnum, D. G., **1**, 514[52]; **3**, 888[21], 890[35]
Farnung, W., **2**, 65[30]
Faro, H. P., **7**, 723[24], 724[24]
Faron, K. L., **5**, 1067[10,11], 1073[10], 1075[11], 1089[87], 1090[87], 1094[87], 1098[87], 1099[87], 1100[87], 1101[87], 1112[87], 1113[87]
Farona, M. F., **3**, 300[41], 381[31], 382[31]; **5**, 1037[5], 1165[10], 1178[10]
Farooq, O., **3**, 295[9], 332[204], 334[215,221,221a,b], 1046[1]
Farooq, S., **3**, 13[68], 916[19]
Farooqui, F., **6**, 489[38]
Farooqui, T. A., **2**, 746[112]
Farrall, M. J., **1**, 821[25]; **7**, 281[175], 282[175], 395[21], 663[58]
Farrar, D. H., **4**, 654[446]
Farras, J., **6**, 570[954]
Farrell, C. O., **6**, 644[89]
Farries, H., **7**, 479[92]
Farrington, G., **5**, 166[91]
Farrow, H., **8**, 965[68]
Farukawa, N., **7**, 470[13]
Farzaliev, V. M., **3**, 304[69]
Fasani, E., **7**, 874[108]
Fasiolo, F., **8**, 54[154], 66[154]
Fassakhov, R. Kh., **6**, 489[78]
Fasth, K.-J., **7**, 229[121]
Fataftah, Z. A., **2**, 842[193]
Fathy, N. M., **6**, 487[45], 489[45], 573[45]
Fatiadi, A. J., **2**, 354[111], 358[111], 359[111]; **5**, 71[148]; **6**, 225[7], 229[7], 233[7], 256[7], 258[7]; **7**, 143[143], 306[2], 307[12], 437[9], 438[9], 444[9], 703[3], 710[3], 738[22], 841[15], 843[15], 845[79], 851[18]
Fatti, G., **4**, 877[68]
Fatutta, S., **4**, 20[63], 21[63]
Faubl, H., **4**, 18[62], 20[62a], 868[17], 869[17]
Faucitano, F. M., **5**, 454[70]
Fauconet, M., **7**, 8[53]
Faul, D., **3**, 555[32]
Faulkner, D. J., **1**, 755[115], 812[115], 813[115]; **4**, 27[83]; **5**, 821[161], 828[7], 839[7], 862[249], 882[13], 888[13], 891[36,37], 892[13,37,38b], 893[13]
Faulkner, L. R., **7**, 850[8], 852[8]
Faulston, D., **3**, 334[220]
Faunce, J. A., **2**, 655[149]
Fauq, A. H., **1**, 772[200]; **2**, 89[34]; **3**, 224[171], 225[171], 264[182]; **5**, 774[173], 780[173]; **6**, 5[27]; **7**, 647[35]
Fauran, F., **8**, 343[112]
Faust, G., **2**, 361[176]
Faust, J. A., **2**, 420[24]
Faust, W., **8**, 141[36]
Faust, Y., **6**, 217[111]
Fauth, D. J., **8**, 535[164]
Fauvarque, J.-F., **3**, 443[58], 450[102], 454[117,118]
Fauve, A., **4**, 374[90]; **8**, 187[32], 188[32,51], 558[394]
Fava, A., **1**, 516[59,60], 517[61,62]; **3**, 147[396], 149[413], 151[413], 152[413], 153[396,413], 155[396], 865[27], 944[90,91], 946[92], 958[90,112]; **6**, 898[103]; **7**, 760[44], 764[126], 767[126]
Favero, J., **7**, 71[100]
Favier, R., **7**, 447[73]
Favini, G., **2**, 267[64]; **3**, 386[57]
Favorskii, A. E., **3**, 839[1-3], 843[27]
Favre, H., **2**, 284[54]; **3**, 358[65]
Favreau, D., **2**, 1059[80]; **8**, 242[40]
Fawcett, F. S., **6**, 955[22]
Fawcett, J., **7**, 481[110]
Fawcett, S. M., **8**, 412[118]
Fawzi, M. M., **3**, 896[68]
Fawzi, R., **5**, 1131[13], 1145[105]
Fayat, C., **6**, 745[84]
Fayat, G., **8**, 36[69,70], 66[69,70]
Fayos, J., **3**, 380[10]; **4**, 709[45], 710[45]
Fazakerley, G. V., **1**, 254[13], 276[13], 278[13]
Fazio, M. J., **6**, 74[34]; **7**, 487[148]
Fazio, R., **4**, 342[62]
Feast, W. J., **5**, 168[101], 176[131]
Fedde, C. L., **6**, 898[106]
Feder, H. M., **8**, 455[206]
Federici, G., **8**, 192[96]
Federici, W., **3**, 380[4], 735[19]
Federlin, P., **7**, 805[66]
Fedin, E. I., **6**, 279[137]
Fedoronko, M., **2**, 140[36]; **8**, 292[42]
Fedorov, L. A., **1**, 266[45]
Fedorov, V. V., **6**, 516[320]
Fedorova, A. V., **4**, 276[81], 284[81,155]
Fedorova, E. B., **4**, 218[148]
Fedorova, N. I., **8**, 956[7]
Fedorovich, A. D., **5**, 699[2]
Fedorynski, M., **2**, 429[49]; **4**, 1001[29,34,44]; **6**, 556[818]
Fedoseev, D. V., **7**, 7[39]
Feely, W., **7**, 661[48]
Feeney, R. E., **8**, 54[158], 66[158]
Feenstra, R. W., **6**, 114[73]
Feger, H., **2**, 681[58], 683[58]; **6**, 502[217], 560[870]; **7**, 650[51]
Feghouli, A., **8**, 14[86]
Feher, F. J., **7**, 3[14]
Fehlhammer, W. P., **6**, 295[250]
Fehlner, T. P., **8**, 673[27], 724[154]
Fehn, J., **4**, 1081[76,83]; **7**, 475[55]
Fehnel, E. A., **4**, 282[134], 288[134]
Fehr, C., **1**, 417[70], 418[74]; **5**, 456[85]; **6**, 1060[71]; **8**, 843[53], 844[53]
Feibush, B., **2**, 1094[88]
Feigel, M., **1**, 19[103]; **4**, 872[40]
Feigelson, G. B., **1**, 561[161], 732[16], 787[16]; **4**, 111[158b]; **6**, 919[42]; **8**, 836[5]
Feigenbaum, A., **5**, 176[129]
Feil, D., **8**, 98[104]
Feil, M., **3**, 735[20]

Feilich, H., **1**, 856[54]
Feinauer, R., **6**, 488[36]
Feinemann, H., **5**, 7[55]
Feinstein, I., **6**, 495[141]
Feiring, A. E., **4**, 128[221]; **7**, 24[39], 25[39], 520[32]
Feist, F., **4**, 41[117]
Feit, B.-A., **8**, 563[431]
Feizi, T., **6**, 33[3], 40[3]
Fekarurhobo, G. K., **5**, 180[145,146,148]
Fekih, A., **8**, 392[95], 880[56]
Fekih, F., **6**, 70[20]
Felber, H., **5**, 418[71]; **6**, 115[81]
Felberg, J. D., **3**, 333[210]; **7**, 17[177]
Felcht, U.-H., **7**, 752[157]
Feld, W. A., **4**, 439[159]
Felder, L., **8**, 190[76]
Feldhues, M., **3**, 642[115]
Feldkamp, J., **4**, 883[100], 884[100]
Feldkimel, M., **8**, 297[68]
Feldman, J., **2**, 958[26]; **5**, 1116[11], 1117[11], 1118[11]
Feldman, K. S., **4**, 378[112], 824[239,240], 825[242], 1089[130]; **5**, 266[75], 268[75], 927[161,162], 931[161,162]
Feldman, P. L., **6**, 127[160], 734[17]
Felfoldi, K., **6**, 25[99], 653[151]; **8**, 418[5], 420[5], 423[5], 439[5], 441[5], 442[5]
Feliu, A. L., **8**, 344[123]
Feliu, J. M., **5**, 1133[33]
Felix, A., **7**, 56[20,21], 80[137]
Felix, A. M., **6**, 635[14b], 636[14]; **8**, 959[21]
Felix, D., **5**, 501[268,269], 828[6], 836[6], 888[26], 893[26]; **6**, 672[285], 831[7], 1043[14], 1058[14], 1059[62,64,66]; **7**, 482[118]; **8**, 948[148]
Felix, M., **8**, 125[94]
Felker, D., **1**, 328[20]
Felkin, H., **1**, 49[6], 50[6], 80[22], 109[13], 110[13], 153[54], 182[47], 185[47], 198[47], 222[69], 310[102], 678[213,214]; **2**, 24[96], 125[214], 217[137], 666[37], 677[37]; **3**, 243[12], 246[38], 250[12], 262[12], 470[221], 471[221], 482[5], 499[5], 505[5], 509[5]; **4**, 82[62a], 869[27], 870[27], 871[27], 876[63], 877[27d,66]; **5**, 38[23a,b]; **7**, 6[29]; **8**, 3[20], 5[33], 446[89,91], 455[91], 456[91]
Fell, B., **4**, 918[17], 924[33]; **8**, 737[25], 739[34]
Fellenberger, K., **3**, 976[5,8]; **4**, 1016[209]; **5**, 1003[20]; **6**, 876[29]
Feller, D., **5**, 202[38]
Felletschin, G., **3**, 914[7], 924[7]
Fellmann, P., **2**, 110[75,76], 116[75], 117[75], 144[62], 186[34], 190[59], 193[62], 199[59], 235[34], 245[33], 268[69], 281[30]
Fellows, C., **3**, 1033[73]; **4**, 841[39,41]
Felman, S. W., **3**, 35[208]; **4**, 597[180], 622[180]; **6**, 705[22], 717[112], 725[112]
Felmeri, I., **8**, 756[138]
Felner, I., **2**, 889[55]
Fels, G., **2**, 867[19], 869[21], 870[19,21], 871[19], 876[21], 879[19], 880[21], 890[21], 1012[15]; **6**, 509[270]
Felt, G. R., **7**, 477[76]
Felty-Duckworth, A. M., **8**, 47[124], 66[124]
Feltz, T. P., **6**, 516[318]
Felzenstein, A., **5**, 71[149,150]
Fence, D. A., **1**, 314[137], 315[137]
Fendler, J. H., **4**, 426[38]
Fendrick, C. M., **7**, 3[7]
Feng, M., **7**, 655[20]
Fengl, R. W., **2**, 127[227,231], 315[43], 316[43], 934[142]; **4**, 82[62c], 217[128], 231[128]; **5**, 689[75]
Fengying, J., **1**, 543[15]
Fenical, W., **5**, 686[43]; **7**, 98[98]
Fenk, C. J., **5**, 249[34]; **7**, 676[65]; **8**, 647[58]
Fenn, D., **7**, 845[73]
Fennen, J., **2**, 371[263]
Fenoglio, D. J., **7**, 439[35]
Fenoglio, R. A., **5**, 675[2]
Fenselau, A. H., **7**, 292[6]; **8**, 496[33]
Fenske, R. F., **5**, 300[73], 302[73]
Fentiman, A. F., **8**, 47[125], 66[125]
Fentiman, A. F., Jr., **5**, 455[80]
Fenton, D. M., **4**, 939[75], 947[94]
Fenton, G. A., **5**, 648[20]
Fenton, H. S. H., **7**, 11[85]
Fenton, S. W., **3**, 379[1], 380[1]
Fenzl, W., **2**, 112[89,93,94], 240[11], 241[14], 244[11,21,22]; **4**, 145[23]; **8**, 724[175,176]
Feoktistov, L. G., **3**, 564[9], 567[9]
Ferao, A., **6**, 509[273]
Férézou, J. P., **4**, 796[97]
Ferguson, G., **3**, 381[32], 386[68]; **4**, 30[87], 48[137,137g]; **5**, 768[127]; **7**, 833[72]
Ferguson, I. E. G., **2**, 555[146]; **5**, 437[159]
Ferguson, L. N., **4**, 951[1], 968[1], 979[1]; **8**, 285[5]
Feringa, B. L., **1**, 125[85], 218[52], 223[84], 225[84e]; **3**, 586[151], 665[39], 689[120]; **4**, 36[101], 97[106], 229[240]; **5**, 371[103]; **6**, 26[111]; **7**, 454[98]; **8**, 99[113]
Ferland, J. M., **8**, 247[80]
Ferles, M., **8**, 583[14], 587[41], 590[54], 591[56,57], 596[54,81]
Fernanda, M., **6**, 507[241,242], 529[241,242]
Fernandes, J. B., **3**, 246[38], 446[81]
Fernandez, A. H., **8**, 934[53]
Fernández, F., **5**, 416[57]; **7**, 691[15]
Fernández, H., **8**, 54[156], 66[156]
Fernández, I., **3**, 87[99], 104[99]
Fernandez, I. F., **5**, 95[89], 96[114]
Fernandez, J. E., **2**, 956[14], 959[30], 962[14]
Fernandez, J. M., **1**, 309[99,100]
Fernandez, M., **5**, 90[57], 95[57]; **6**, 501[189]
Fernandez, M.-J., **8**, 764[6], 773[6b,70], 774[70]
Fernandez, S., **7**, 706[22]
Fernández-Alvarez, E., **2**, 780[12]
Fernandez de la Pradilla, R., **1**, 117[56]; **3**, 226[194,194a]; **6**, 9[42], 152[135]; **7**, 358[12]; **8**, 836[10f], 844[10f], 846[10f]
Fernandez Martin, J.-A., **5**, 201[32], 202[34]
Fernandez-Picot, I., **4**, 747[153]
Fernández-Simón, J. L., **1**, 830[94]; **3**, 788[50]
Fernando, S., **4**, 476[158]
Fernelius, W. C., **2**, 357[148]
Fernholz, E., **6**, 685[357]
Fernholz, H., **6**, 269[74]
Ferraboschi, P., **7**, 286[189], 331[17], 841[17], 845[66]; **8**, 240[30], 244[30,56], 263[29]
Ferrand, E. F., **1**, 3[18,20], 42[20c]
Ferrari, C. F., **5**, 94[85]
Ferrari, G. F., **8**, 443[1], 446[95], 449[157], 450[157], 452[95a,b], 457[95a-c,218], 458[218]
Ferrari, M., **2**, 833[148]; **3**, 752[97]; **4**, 261[285]
Ferraz, H. M. C., **1**, 642[115], 645[115]; **4**, 364[1,1n], 370[33], 372[56], 376[103], 380[1n]; **8**, 849[106]
Ferre, E., **8**, 205[165]
Ferre, G., **6**, 204[21]
Ferree, W. I., Jr., **5**, 647[13]
Ferreira, D., **3**, 831[61]; **8**, 836[9], 847[9], 848[9]
Ferreira, G. A. L., **7**, 507[309]
Ferreira, J. T. B., **4**, 443[192], 447[192]; **7**, 239[54], 586[162], 775[352a], 844[56]; **8**, 412[116]

Ferreira, T. W., **3**, 447[91], 448[97], 456[126], 503[146], 513[146]; **8**, 842[40,41], 935[61]
Ferreira, V. F., **3**, 229[233]
Ferrel, J. W., **7**, 604[136]
Ferrer, P., **6**, 172[19]
Ferreri, C., **1**, 168[116b], 563[179]; **8**, 406[41]
Ferretti, M., **2**, 291[71]; **3**, 741[49]
Ferrier, R., **5**, 850[160]
Ferrier, R. J., **6**, 34[11], 46[68], 48[85], 51[11], 836[51], 846[104], 977[19], 978[25]; **8**, 857[193]
Ferrieri, R. A., **5**, 1148[119,120]
Ferrino, S., **5**, 351[81]; **8**, 932[40]
Ferrino, S. A., **1**, 554[107]
Ferris, J. P., **6**, 923[54,55]; **8**, 300[84]
Ferro, M. P., **3**, 168[491], 169[491], 171[491]; **5**, 921[143], 976[21]
Ferroud, D., **4**, 629[417], 653[436]
Fersht, A. R., **8**, 206[169]
Fesik, S. W., **3**, 216[76]
Fessenden, J. M., **2**, 456[77]
Fessenden, J. S., **2**, 604[52]
Fessenden, R. J., **2**, 604[52]
Fessenden, R. W., **4**, 719[22]
Fessler, D. C., **7**, 124[40]
Fessler, W. A., **7**, 15[154]
Fessner, W. D., **2**, 456[33], 457[33], 458[33], 459[33], 460[33], 461[33], 462[33], 466[33]; **5**, 64[29]; **8**, 795[19]
Fetizon, M., **3**, 131[333], 664[28], 698[28]; **6**, 263[22]; **7**, 276[150], 312[34], 320[34], 738[26], 747[26], 841[9], 851[18]; **8**, 111[20], 118[20], 881[70], 930[34], 943[120]
Fetter, M. E., **5**, 949[283]
Fetzer, U., **2**, 1084[8,17], 1087[36], 1090[36]; **6**, 242[85,86], 243[85,86], 489[88]
Feuer, B. I., **7**, 27[69,72,73], 29[72]
Feuer, H., **2**, 321[9], 323[20], 325[9], 326[9], 327[9], 328[9], 329[9], 354[109]; **4**, 423[3]; **6**, 2[2], 23[2], 104[1,9], 105[15]; **7**, 736[6], 746[86], 747[100], 748[100]; **8**, 60[181], 64[205], 66[181,205], 67[205], 73[248], 74[248], 363[7], 370[87], 373[136], 376[164], 389[71]
Feuerherd, K.-H., **2**, 1026[67], 1028[67]
Feugeas, C., **3**, 124[257], 127[257]
Feustel, M., **3**, 284[53]
Feutrill, G. I., **2**, 606[69]
Fevig, J. M., **1**, 126[89,90]; **5**, 843[117]
Fevig, T. L., **1**, 270[61]; **3**, 221[130], 603[229]; **4**, 808[159], 820[218]; **7**, 137[120]
Fewkes, E. J., **1**, 214[29], 218[29]
Fex, T., **1**, 429[124]; **5**, 687[56]
Fey, P., **2**, 455[14]; **6**, 119[111], 728[210]
Fiaita, G., **7**, 794[7c]
Fiala, R. E., **6**, 581[988]
Fiandanese, V., **1**, 413[57], 416[67], 452[220]; **3**, 230[235,236], 441[48,49], 446[87], 449[48,101], 463[153,154,155,166], 485[29], 492[79], 493[29], 503[29,79], 513[29,79]; **4**, 93[93d]; **6**, 446[101]
Fiato, R. A., **4**, 587[48,49]
Fiaud, J.-C., **1**, 363[37]; **4**, 590[105], 591[110,114], 599[224], 615[105,386], 616[110], 617[114], 619[386], 622[386], 625[224,386], 629[105], 642[224], 653[434]; **8**, 159[3], 170[91]
Fibiger, R., **4**, 295[262], 296[262]
Fichter, F., **3**, 634[16], 636[16]
Fichter, K. C., **4**, 889[134]; **8**, 734[1], 744[51,52], 746[52], 747[58], 751[52], 752[58], 753[52], 757[161]
Ficini, J., **2**, 709[11]; **3**, 197[42], 896[68]; **4**, 45[130,130a], 173[33]; **5**, 18[127], 116[252,258,263,266], 557[53], 676[5], 689[71,72], 732[134], 836[65]; **6**, 738[51]; **7**, 96[84]; **8**, 925[8]
Fickes, G. N., **2**, 631[16], 711[32]
Fickling, C. S., **8**, 149[120]
Fiddler, S., **1**, 779[226]
Fidler, F. A., **3**, 331[198]
Fiecchi, A., **7**, 331[17], 674[42]; **8**, 187[38], 190[71,73], 191[73], 240[30], 244[30,56], 263[29], 565[448]
Fiedler, C., **3**, 194[14]
Fiedler, W., **2**, 361[176]
Fiegenbaum, P., **5**, 1126[66]
Field, J. A., **7**, 355[38]
Field, K. W., **7**, 741[48], 747[48]
Field, L., **3**, 86[2], 158[439,441], 159[441]; **5**, 167[94]; **6**, 133[6], 157[164], 1016[30], 1020[30]; **7**, 758[3], 760[3,29,30]; **8**, 408[64,73,74], 836[2], 839[2c], 842[2c], 843[2c], 844[2c]
Field, L. D., **5**, 791[37]
Field, S. J., **3**, 898[84]
Fielding, H. C., **3**, 639[84]
Fields, D. L., **2**, 969[84]; **3**, 905[136]
Fields, D. L., Jr, **4**, 753[169]
Fields, E. K., **5**, 379[112], 383[112], 384[112]; **7**, 507[305], 581[143]
Fields, K. W., **1**, 874[103]; **2**, 109[66], 611[100]
Fields, R., **6**, 104[1,9]
Fields, S. C., **4**, 492[71], 495[71]
Fields, T. L., **6**, 487[42], 489[42], 543[42], 550[42], 554[42,775]
Fieser, L. F., **2**, 763[61], 1090[61]; **3**, 828[49,51], 898[77]; **4**, 330[2], 344[2]; **5**, 595[15], 596[15]; **7**, 84[3], 86[16a], 92[41,41a], 94[41], 128[65], 571[118], 576[118], 709[35], 730[50], 820[24]; **8**, 26[10], 27[10], 36[10], 220[84], 330[25], 541[207], 916[110], 917[110]
Fieser, M., **3**, 828[49]; **4**, 330[2], 344[2]; **7**, 84[3], 128[65], 709[35]; **8**, 26[10], 27[10], 36[10], 220[84], 330[25], 541[207]
Fiesselmann, H., **6**, 964[79]
Fife, W., **5**, 856[217]
Figeys, H. P., **5**, 412[44]
Figge, K., **8**, 292[38]
Figge, L., **5**, 64[29]
Figueras, J., Jr., **8**, 321[105], 496[34]
Figures, W., **5**, 404[18]
Fijii, S., **6**, 734[6,7]
Fikui, S., **6**, 658[184]
Filatova, E. I., **8**, 556[378]
Filatovos, G. L., **8**, 556[372]
Filbey, A. H., **5**, 2[17]
Filby, W. G., **7**, 760[42]
Filer, C. N., **4**, 342[62]
Filin, V. N., **3**, 304[66]
Filipek, S., **2**, 663[25], 664[25], 665[25]; **5**, 342[62c], 433[139,139a]
Filipescu, N., **5**, 723[108a]
Filipini, L., **4**, 765[227]
Filipovic, L., **7**, 657[33]
Filipp, N., **5**, 735[139]; **6**, 866[205]
Filippini, L., **2**, 735[15]
Filippo, J. S., **6**, 2[8], 22[8]; **7**, 530[26]
Filippone, P., **2**, 345[24]
Filippova, A. K., **4**, 317[554]
Filippova, T. M., **7**, 774[325]
Fillebeen-Khan, T., **7**, 6[29]; **8**, 446[91], 455[91], 456[91]
Filler, R., **2**, 396[6], 402[6], 403[6,37], 404[6c], 407[6c]; **3**, 757[121]; **7**, 253[12], 878[141]
Filley, J., **3**, 4[22]
Filliatre, C., **3**, 1046[3]; **7**, 7[43]
Fillion, H., **5**, 473[153], 477[153]
Fillol, L., **8**, 227[116]
Filonova, L. K., **6**, 94[139]
Filosa, M. P., **5**, 803[89], 825[89a], 976[19], 979[19], 982[30], 983[30]
Filppi, J. A., **1**, 554[101]
Finan, J. M., **8**, 879[52]
Finch, A. M. T., Jr., **3**, 710[21]
Finch, H., **1**, 741[45], 865[87]; **5**, 432[130], 433[130b]
Finch, M. W., **3**, 840[15]; **5**, 595[16], 596[16]

Finch, N., **3**, 124[270], 127[270], 128[270], 129[270], 629[51]
Findeis, M. E., **1**, 798[287]
Findeisen, K., **6**, 244[113]
Finding, R., **6**, 562[885,886]
Findlay, A., **6**, 276[114]
Findlay, D. M., **1**, 544[43]
Findlay, J. A., **3**, 503[149], 512[149]
Findlay, J. W. A., **3**, 661[25]; **7**, 158[39]
Findlay, P., **4**, 300[309]
Finet, J.-P., **3**, 505[159]; **7**, 90[32], 356[48], 704[14], 705[14]
Finger, A., **3**, 55[280]
Finger, G. C., **6**, 221[134]
Finholt, A. E., **8**, 26[1,2], 274[134], 735[9], 736[9]
Finiels, A., **2**, 736[21]
Fink, D. M., **1**, 596[30], 601[30,31], 602[31]; **2**, 85[14]; **4**, 155[71b]; **5**, 277[15], 278[15], 279[15]; **8**, 355[182]
Fink, H., **3**, 816[85]
Fink, S. C., **2**, 362[185]
Finkbeiner, H. L., **2**, 841[188], 842[188]; **6**, 653[149]; **8**, 374[142,143], 751[64,65]
Finke, J., **2**, 211[113]
Finke, M., **5**, 391[143], 721[101]
Finke, R. G., **1**, 439[163,164], 440[167,171], 457[163]; **3**, 208[3], 213[3b]; **4**, 518[1], 547[1], 895[160]; **5**, 46[39], 56[39], 1065[1], 1066[1], 1074[1], 1083[1], 1084[1], 1093[1], 1112[1g], 1163[3], 1183[3]; **8**, 421[28], 422[28], 432[28], 435[28], 436[28], 550[334]
Finkelhor, R. S., **3**, 964[126]; **4**, 115[183], 253[176], 255[176], 259[264], 991[148]
Finkelstein, B. L., **2**, 195[72,72b]; **5**, 839[79], 843[79]
Finkelstein, H., **6**, 977[12]
Finkelstein, J., **5**, 95[91]
Finkelstein, M., **3**, 634[21], 635[38], 649[206,206b], 655[21]; **6**, 572[959]; **7**, 804[61], 805[64]
Finklea, H. O., **2**, 286[64]
Finlander, P., **6**, 533[482], 550[482]
Finlay, J. D., **3**, 159[459], 163[459]
Finlayson, A. J., **6**, 204[16]
Finley, K. T., **3**, 613[1], 614[1], 615[1], 616[1], 619[1], 620[1], 621[1], 622[1], 623[1], 625[1], 626[1], 627[1], 628[1], 629[1], 630[1]; **4**, 1099[184]; **5**, 468[134]
Finn, J., **3**, 220[124]; **5**, 350[77]; **6**, 899[110], 900[110]; **7**, 162[65], 181[65]
Finn, K., **5**, 107[202]; **8**, 846[83]
Finn, M. G., **7**, 390[2], 394[2,18], 395[2,18], 398[18], 399[18], 412[2], 413[2], 419[2], 420[2,135,136], 421[2,136,136b], 422[2], 424[2,18], 425[2], 430[159], 442[46b], 489[165]
Finnan, J. L., **6**, 619[117]
Finnegan, R. A., **4**, 35[99]; **8**, 542[235]
Finney, N. S., **5**, 736[142c]
Finocchiaro, P., **1**, 294[43], 488[9]; **3**, 583[118], 587[148]
Finseth, G. A., **8**, 568[477]
Finucane, B. W., **7**, 112[197]
Finzi, C., **4**, 763[208]
Fioravanti, J., **7**, 479[91]
Fioravanti, S., **6**, 717[111]
Fiorentino, M., **1**, 834[126]; **7**, 13[125], 374[77a]
Fiorenza, M., **4**, 98[114], 113[114], 115[182,182e]; **6**, 179[127], 238[74]
Fiorini, M., **8**, 460[247]
Fioshin, M. Y., **3**, 635[31], 636[55], 639[80], 648[172,173,174]
Firestone, R. A., **4**, 1070[11], 1072[11,18], 1083[11], 1102[202]; **5**, 493[210]; **8**, 52[151], 66[151]
Firl, J., **2**, 1087[34]; **5**, 65[70], 417[61,64], 418[61], 490[190,191]; **6**, 525[387]
Firouzabadi, H., **7**, 236[27], 266[109], 267[109], 286[190], 307[13], 561[85], 738[28,29], 760[23,27]
Firrell, N. F., **3**, 30[176]; **6**, 710[60]
Firth, B. E., **3**, 747[70], 770[176]; **7**, 778[405]
Firth, W. C., **7**, 718[2], 724[2]
Fisch, J. J., **7**, 603[114]
Fischer, A., **7**, 345[5], 845[67]
Fischer, C., **8**, 342[110]
Fischer, C. D., Jr., **8**, 87[31]
Fischer, C. M., **8**, 974[131]
Fischer, D. A., **5**, 522[44]
Fischer, E., **5**, 729[123,125]; **6**, 473[76], 631[1], 632[2], 642[64,73], 660[2], 671[279]; **8**, 292[40]
Fischer, E. O., **4**, 104[135d], 520[30], 588[52], 663[3], 976[99], 980[102,103], 981[102a,103]; **5**, 1070[16], 1076[33], 1081[54], 1085[61,62,64]
Fischer, E. V., **4**, 587[33,34]
Fischer, F., **2**, 428[44]
Fischer, G., **5**, 404[16]; **8**, 472[8,9]
Fischer, G. W., **2**, 712[40]
Fischer, H., **2**, 770[10], 771[10]; **4**, 485[28], 491[68], 492[72], 503[28], 719[18], 722[33,35], 723[18,42], 728[33,35], 730[66], 759[194], 783[12], 976[100], 985[128]; **5**, 442[182,185], 714[74], 1065[1], 1066[1,1a], 1074[1], 1083[1], 1084[1], 1086[66], 1093[1], 1094[100,100a], 1098[100a], 1101[144], 1103[152]; **6**, 104[5], 105[16], 480[116], 531[446], 680[332], 681[332]; **7**, 206[68], 212[68], 765[147,152], 769[147], 852[34]
Fischer, H. O. L., **2**, 456[73], 458[73]; **8**, 296[63], 925[9]
Fischer, J., **4**, 688[67]; **5**, 877[6]; **7**, 11[87], 107[162], 422[139], 452[45]; **8**, 445[43]
Fischer, J. C., **1**, 55[25]
Fischer, J. R., **4**, 275[67], 279[67], 287[67]
Fischer, J. W., **3**, 627[43]; **4**, 336[28], 342[28], 346[28], 347[28]; **5**, 324[21]; **6**, 535[529]
Fischer, K., **5**, 526[56], 539[107]; **8**, 747[56], 752[56]
Fischer, M., **6**, 247[133]; **8**, 344[119], 345[119]
Fischer, N., **5**, 86[20]
Fischer, N. H., **5**, 571[117]
Fischer, P., **4**, 429[87], 977[94]; **6**, 553[797], 637[32,32c]
Fischer, P. A., **6**, 70[18]
Fischer, R., **6**, 49[98]
Fischer, R. D., **4**, 663[3]
Fischer, R. H., **2**, 321[13]
Fischer, U., **5**, 714[71]
Fischer, W., **2**, 759[36]; **4**, 425[29,30]
Fischer, W. F., **1**, 118[60]; **4**, 170[20]
Fischer, W. F., Jr., **1**, 431[134]; **2**, 120[176]; **3**, 248[55], 251[55], 269[55], 419[47], 494[87], 502[87]
Fischetti, W., **4**, 849[83], 856[83,98]
Fischli, A., **2**, 866[8], 867[18]; **3**, 167[485], 168[485,498,502,503], 169[498,502,503]; **6**, 1058[59]; **8**, 253[115], 299[83], 562[424]
Fiscus, D., **4**, 1104[212]
Fiser-Jakic, L., **6**, 510[298]
Fish, M. J., **8**, 150[128]
Fish, R. H., **7**, 616[11]; **8**, 455[205], 456[205a], 600[104,105], 613[81], 629[81,184], 720[126,127]
Fish, R. W., **5**, 275[10], 277[10], 281[20]
Fishbein, R., **7**, 16[160]
Fishburn, B. B., **8**, 143[60], 148[60]
Fishel, D. L., **2**, 504[1]
Fisher, A., **7**, 843[45]
Fisher, A. M., **2**, 441[1], 443[1]
Fisher, C. D., **1**, 292[26]
Fisher, C. L., **4**, 337[33]
Fisher, H. F., **8**, 79[1]
Fisher, J., **8**, 859[213]
Fisher, J. F., **6**, 685[358]
Fisher, J. W., **7**, 521[35]

Fisher, K. J., **4**, 48[137,137g]; **5**, 670[46]
Fisher, L. P., **4**, 1023[260]; **5**, 1006[35]
Fisher, M. H., **4**, 356[136]; **7**, 93[53]
Fisher, M. J., **5**, 349[75]
Fisher, N. G., **5**, 3[23]
Fisher, R. P., **3**, 199[58]; **8**, 474[16]
Fisher, R. R., **2**, 388[342]
Fisher, T. E., **5**, 927[162], 931[162]
Fishli, A., **7**, 57[27]
Fishman, D., **2**, 612[106]
Fishman, J., **6**, 655[162]; **8**, 935[65]
Fishpaugh, J. R., **1**, 41[270], 432[137], 456[137]; **3**, 24[150], 211[28], 215[28]; **4**, 189[102], 191[109], 192[115]
Fishwick, B. R., **4**, 54[154]
Fishwick, C. W., **4**, 510[164]; **6**, 1007[151]; **7**, 508[310]
Fisk, M. T., **2**, 964[56,56c]
Fisk, T. E., **2**, 725[117]; **6**, 504[221]
Fitcher, F., **3**, 668[50]
Fitjer, L., **1**, 672[190,191], 674[190,191], 714[190,191], 715[190], 718[190,191], 722[190,191], 731[1], 867[80]; **4**, 784[15], 969[67]; **6**, 174[62], 183[147]; **7**, 543[17], 551[17], 554[17]; **8**, 335[67]
Fitt, J. J., **1**, 480[155]; **2**, 74[80]; **3**, 67[19]; **6**, 554[766]
Fittig, R., **2**, 401[27]; **3**, 563[2], 721[2]
Fittkau, K., **6**, 538[558]
Fitton, A. O., **5**, 151[10]
Fitton, H., **6**, 690[398,402], 692[398,402]
Fitton, P., **4**, 856[102]; **8**, 501[56], 502[56]
Fitzgerald, B. M., **8**, 542[225]
Fitzhugh, A. L., **2**, 838[176]
Fitzi, K., **3**, 353[46]
Fitzi, R., **3**, 41[229]
Fitzjohn, S., **2**, 739[44]; **3**, 810[43]; **8**, 623[149]
Fitzner, J. N., **1**, 635[88]; **3**, 104[209], 117[209]; **6**, 846[103], 905[145]
Fitzpatrick, F. A., **5**, 151[7], 180[7]
Fitzpatrick, J. D., **4**, 701[26]
Fitzpatrick, J. M., **3**, 250[70]; **5**, 947[260], 960[260]; **6**, 274[104]
Fitzsimmons, B. J., **4**, 1059[154]; **5**, 490[192], 841[86]; **6**, 859[164], 978[21]; **8**, 389[68]
Fitzwater, S., **1**, 367[54]
Fjeldberg, T., **1**, 17[206]
Flagg, E. M., **6**, 537[576]
Flaim, S. F., **8**, 28[37], 66[37]
Flamini, A., **5**, 1158[173]
Flanagan, D. M., **2**, 1097[101]
Flanagan, P. W., **2**, 738[41], 760[40]
Flanagan, P. W. K., **3**, 322[141]; **8**, 754[85,86]
Flanagan, V., **8**, 943[122]
Flanigan, I., **2**, 456[63], 458[63]
Flann, C., **2**, 1034[88], 1035[88], 1057[69], 1064[110]
Flann, C. J., **4**, 309[419], 974[89]; **5**, 531[72], 829[25]; **8**, 854[148], 857[148]
Flasch, G. W., Jr., **3**, 638[91], 644[91]
Flaskamp, E., **5**, 331[42], 333[42c]; **6**, 94[143]
Flatow, A., **3**, 878[90]
Flatt, S. J., **7**, 278[158]
Flaugh, M. E., **8**, 618[110]
Flechtner, T., **3**, 1058[40]; **7**, 43[36]
Fleck, C., **2**, 1090[73], 1094[91], 1095[91], 1099[109,109b,116], 1102[73], 1103[73]
Fleck, T. J., **1**, 755[116], 756[116,116f], 758[116], 761[116]
Fleet, G. W. J., **4**, 370[40]; **5**, 151[11], 835[59]; **6**, 74[31]; **7**, 104[145], 260[63], 278[158], 710[48], 725[32]; **8**, 264[37,38], 347[144], 384[35], 540[199]
Fleig, H., **6**, 556[833]
Fleischer, G. A., **8**, 988[28]
Fleischhauer, I., **2**, 194[69]; **3**, 99[189], 107[189], 110[189]; **4**, 1012[176]
Fleischmann, C., **3**, 939[76]
Fleischmann, F. K., **5**, 77[252]
Fleischmann, M., **4**, 356[141]; **6**, 282[152]; **7**, 8[55,56], 488[156], 793[2,3], 794[7c]
Fleischmann, R., **8**, 446[75-77], 453[75]
Fleming, A., **5**, 913[103], 918[128], 925[128,152], 930[176], 933[176], 937[128,201], 958[103,128], 964[176], 987[42], 993[42], 994[42]
Fleming, B. I., **8**, 212[13], 214[44,45], 222[44]
Fleming, F. F., **5**, 841[87]
Fleming, G. H., **3**, 415[7]
Fleming, I., **1**, 272[67], 358[9], 359[9], 362[9b], 580[1], 586[16], 610[44], 623[78]; **2**, 6[27], 186[32,33], 200[90], 201[91], 564[3,7], 569[34], 576[77,78], 579[78], 582[77], 583[112-114], 584[119,127], 586[132], 587[112,137,144,146], 589[153], 614[118], 616[134], 617[140,141,142], 619[134], 662[10,16], 663[10], 664[16], 669[10], 707[1], 710[19], 712[41,43], 714[54], 716[60,64,65], 1030[80], 1072[140]; **3**, 17[92], 18[92-94], 23[137], 25[154], 27[154,167], 28[154], 42[94], 198[43], 200[69], 345[22], 746[68]; **4**, 155[65], 186[93], 231[258,259], 241[55], 243[64], 247[64], 248[107], 253[173], 255[55,64], 257[173,217], 260[64], 486[33], 503[128], 675[37], 682[57], 727[55], 895[165], 900[165], 901[183,184,186], 1069[3], 1081[3], 1099[3]; **5**, 248[28], 316[3], 317[3], 335[47,49], 339[3], 391[142], 451[52], 516[23], 517[23], 762[105], 812[134], 854[173], 1037[5]; **6**, 16[61], 17[62], 108[32], 226[11], 687[383], 757[133], 829[1-3], 832[12,16,17], 833[20,21,25,26], 834[25,26,29], 850[29], 865[12], 966[96], 1004[139], 1011[1], 1019[43]; **7**, 137[119], 138[119], 144[119], 208[76], 318[60], 360[20], 616[12,13,20], 621[32], 641[4], 646[4,24-26,28,29], 647[30]; **8**, 99[107], 699[148], 769[24], 771[24], 782[24b], 784[112], 788[120], 836[6]
Fleming, J. A., **6**, 1003[136], 1004[137]
Fleming, M. P., **2**, 482[26], 483[26]; **3**, 579[125], 582[125], 583[124,125,127], 584[125], 585[124,125], 587[124,144], 588[125,127a], 595[125], 596[125], 610[125,127a]; **6**, 687[377], 980[37]; **8**, 889[127], 992[53]
Fleming, P., **7**, 431[163]
Fleming, S. A., **4**, 989[144]; **5**, 132[50]; **8**, 395[131]
Fleming, W. P., **6**, 835[45]
Flesh, G. D., **7**, 528[10]
Flesher, J. W., **8**, 979[149]
Fletcher, A. R., **6**, 644[92]
Fletcher, H. G., Jr., **6**, 36[26,27], 660[204]
Fletcher, M. T., **4**, 303[342]; **7**, 635[70]; **8**, 854[152], 856[152]
Fletcher, R. S., **8**, 217[68]
Fletcher, T. L., **7**, 85[12], 87[12], 655[11]
Fleury, J. P., **5**, 412[46]
Fliedner, L. J., Jr., **2**, 741[67]
Flinn, A., **7**, 405[68]
Flint, J. A., **6**, 614[85]
Flippen, J. L., **6**, 914[28]
Flippin, L. A., **1**, 286[7], 335[64], 564[196], 889[140], 890[140]; **2**, 5[16], 217[138,139], 218[140], 630[10], 631[10], 632[10], 640[10,64], 641[10], 642[10], 646[10], 649[64], 931[134]; **6**, 91[121]; **7**, 493[189]; **8**, 399[148], 839[27,27a]
Fliri, H., **1**, 372[79]
Flitsch, V. W., **4**, 252[161]
Flitsch, W., **2**, 282[37], 377[283], 780[9], 810[65]; **3**, 623[35]; **4**, 15[50], 16[50e]; **6**, 176[88,105], 185[163], 186[163]
Flock, F. H., **7**, 663[60]
Flodman, L., **2**, 827[128]
Flogaus, R., **5**, 595[18], 596[18]
Flood, L. A., **7**, 318[57], 319[57], 447[71], 674[50]
Flood, M. E., **4**, 435[136]
Flood, S. H., **3**, 305[71]
Flood, T. C., **3**, 583[119], 760[140]; **6**, 686[369], 985[65]; **8**, 888[121]

Florence, M. R., **6**, 644[83]
Florent'ev, V. L., **5**, 491[206]
Flores, M. C. L., **7**, 745[74]
Flores R, H., **3**, 901[112]
Florian, W., **2**, 368[234]
Floriani, C., **1**, 29[150]; **2**, 127[229]; **5**, 1174[36]; **8**, 447[125], 683[90]
Florio, E., **8**, 517[127]
Florio, S., **1**, 113[29]; **2**, 87[26,27], 211[117], 213[127], 495[64,65], 496[64,65]; **3**, 461[145]; **7**, 737[15]
Floris, B., **4**, 300[306]
Florvall, L., **8**, 617[108]
Flowers, H. M., **6**, 41[44]
Flowers, L. I., **4**, 931[57]
Flowers, M. C., **5**, 910[79]
Flowers, W. T., **6**, 526[404]
Floyd, A. J., **4**, 811[172]; **5**, 723[106]
Floyd, C. D., **3**, 200[69]; **5**, 536[96]
Floyd, D., **1**, 430[130]
Floyd, D. M., **1**, 112[25], 118[58], 129[58]; **3**, 217[87], 250[70,73]; **4**, 176[47], 183[79]; **5**, 553[43]; **6**, 9[42], 21[78]
Floyd, J. C., **1**, 412[52]
Floyd, M. B., **4**, 141[15], 142[15]; **6**, 648[124]
Fluckiger, E., **2**, 875[30]
Fludzinski, P., **3**, 284[54]; **4**, 42[121], 129[224]; **6**, 967[104]; **8**, 237[11]
Fluharty, A. L., **8**, 269[89]
Flynn, A. P., **5**, 461[101], 463[101]
Flynn, D. L., **1**, 134[113]; **2**, 824[121]
Flynn, G. A., **2**, 911[68]
Flynn, J. H., **8**, 447[107]
Flynn, J. R., **2**, 760[43]; **8**, 53[130], 66[130]
Flynn, K. E., **7**, 413[116], 416[121a]
Flynn, R. M., **6**, 172[10]
Flynn, R. R., **4**, 490[66], 499[66]
Flynn, S. T., **5**, 1156[164]
Foa, M., **3**, 1018[11], 1026[37], 1033[72], 1039[100]; **5**, 1138[62]
Fobare, W. F., **2**, 1002[56], 1027[76], 1034[89,90]; **4**, 735[85]; **5**, 408[34], 409[34], 411[34,43], 552[39]; **6**, 212[80], 859[174]
Fobker, R., **1**, 831[100]
Focella, A., **8**, 606[18]
Foces-Foces, C., **5**, 410[42], 411[42]
Fochi, R., **2**, 737[39]; **3**, 125[307]; **6**, 134[13]; **8**, 277[153,154], 660[109]
Fodor, C. H., **7**, 224[54]
Fodor, G., **2**, 735[17]; **6**, 213[84], 291[206,218,219,220], 525[375,378], 529[467]
Fodor, P., **6**, 430[95]
Foerst, D. L., **6**, 437[35]
Foerst, W., **2**, 477[7], 478[7]; **5**, 451[7]; **6**, 242[86], 243[86,100]; **8**, 536[169]
Fogagnolo, M., **3**, 232[267], 511[186,187], 514[186]
Foglia, T. A., **7**, 498[229]
Föhles, J., **6**, 668[253], 669[253]
Föhlisch, B., **5**, 595[18,20], 596[18,20], 605[62], 609[69]; **8**, 91[63]
Föhr, M., **7**, 770[251], 773[303]
Fok, C. C. M., **8**, 884[98], 885[98]
Fokin, A. V., **7**, 493[196]
Fokken, B., **6**, 509[244]
Fokkens, R., **6**, 494[131]
Foland, L. D., **5**, 689[76,77], 690[80,80c], 733[136,136c,f,g], 734[136f,g]
Folcher, G., **8**, 737[26]
Foley, H. C., **5**, 1090[89]
Foley, K. M., **8**, 767[23]
Foley, L., **5**, 75[214]
Folkers, E. A., **3**, 14[74,75], 15[74,75]
Folkers, K., **2**, 284[56]; **7**, 778[414]
Folli, U., **3**, 95[154], 119[154]; **5**, 439[166]; **7**, 777[371,372,373,384]
Folliard, J. T., **6**, 937[114]
Folsom, T. K., **8**, 502[62]
Folting, K., **3**, 583[119]; **8**, 550[329]
Foltz, R. L., **5**, 455[80]
Fomum, Z. T., **4**, 55[157], 57[157c], 251[145]
Foner, S. F., **8**, 472[8]
Fones, W. S., **3**, 888[17]
Fong, L. K., **8**, 671[16]
Fong, W. C., **3**, 854[75]
Fonken, G. J., **3**, 329[189]; **4**, 241[57]; **5**, 702[12]
Fonken, G. S., **2**, 169[166]; **7**, 54[3], 56[3], 58[35,36], 66[3], 77[3], 78[3], 429[152]
Font, J., **2**, 381[300]; **6**, 4[19], 173[42], 563[901], 984[54]
Fontain, E., **2**, 1093[84,85]
Fontaine, T. D., **8**, 220[86], 228[132]
Font-Altaba, M., **1**, 34[223]; **3**, 380[7]
Fontan, R., **4**, 478[168]
Fontana, F., **4**, 768[235,240,243], 770[244]; **7**, 778[411]
Fontanella, L., **8**, 641[27]
Fontanille, M., **4**, 980[115], 982[115]; **5**, 1103[150], 1104[150,158]
Fontecave, M., **7**, 95[72], 108[176], 383[109]
Fooks, A. G., **3**, 829[55]
Foos, J. S., **2**, 710[26]; **6**, 986[68]
Foote, C. S., **5**, 72[182], 428[107], 429[107]; **7**, 98[99], 765[166], 769[220], 881[158,159], 884[187]
Foote, R. S., **2**, 512[46], 522[72], 523[73]
Forbes, C. P., **7**, 299[51]
Forbes, E. J., **8**, 963[45]
Forbes, J. E., **8**, 393[113]
Forbes, W. F., **5**, 757[77], 758[83], 759[77]
Forbus, T. R., **3**, 328[179]
Forbus, T. R., Jr., **2**, 740[54]
Forcellese, M. L., **7**, 832[69]
Forche, E., **4**, 270[5], 271[5,23], 272[23]; **6**, 204[4]
Forchiassin, M., **4**, 20[63], 21[63]
Ford, G. P., **4**, 484[11]; **5**, 856[197]
Ford, J. A., Jr., **7**, 706[21]
Ford, M. E., **6**, 725[170], 728[170]; **7**, 665[67], 829[59]
Ford, P. W., **5**, 804[96]
Ford, T. A., **5**, 1138[65]; **8**, 814[20]
Ford, T. M., **1**, 859[65]; **4**, 1089[136]
Fordham, W. D., **8**, 542[231]
Ford-Moore, A. H., **7**, 764[113]
Fordyce, W. A., **8**, 455[205]
Foreman, G. M., **8**, 425[47], 476[22]
Foreman, M., **8**, 474[16]
Foreman, M. I., **5**, 1037[6], 1039[6], 1040[6], 1049[6], 1165[8], 1183[8]; **7**, 851[29]
Forenza, S., **5**, 736[145], 737[145]; **8**, 347[141], 350[141]
Foresti, E., **4**, 426[55]; **5**, 829[15]
Forgione, L., **4**, 83[65c]
Forkner, M. W., **5**, 386[132], 387[132c], 691[83], 692[83], 693[83]; **6**, 977[17]
Forlani, L., **4**, 426[55]
Fornasier, R., **1**, 828[70]; **2**, 435[60]; **5**, 211[61]; **8**, 170[73,75,76], 409[86], 412[86]
Forni, A., **1**, 837[155], 838[160]; **7**, 747[96], 778[402]; **8**, 187[37]
Forni, L., **7**, 8[59]
Forrest, A. K., **5**, 404[17], 428[109]; **8**, 389[67]
Forrest, D., **5**, 901[28]
Forrester, A. R., **3**, 689[122]
Forrester, J., **3**, 399[118], 404[133]; **5**, 671[49]
Forsberg, J. H., **1**, 231[3]; **6**, 546[652]

Forsek, J., **2**, 553[128]
Forsén, S., **1**, 292[30], 293[35,36]
Forster, A. M., **8**, 460[253]
Forster, D., **3**, 1018[9]
Forster, M. O., **3**, 890[31]
Forster, S., **1**, 546[53]
Förster, W.-R., **6**, 419[3], 426[3,79]
Forsyth, C. J., **3**, 232[266]
Forsyth, D. A., **2**, 963[53]
Fort, A. W., **5**, 594[2]
Fort, J.-F., **2**, 154[112], 268[69]
Fort, R. C., Jr., **3**, 334[218]; **8**, 798[48]
Fort, Y., **3**, 509[179]; **8**, 14[86], 840[38], 878[48]
Forte, P. A., **6**, 536[547], 538[547]
Fortes, C. C., **7**, 207[74]
Fortes, H. C., **7**, 207[74]
Fortgens, H. P., **2**, 89[36], 586[133], 587[133], 1049[15], 1050[30,31], 1061[93], 1072[30,31], 1078[15]
Forth, M. A., **8**, 366[46]
Fortt, S. M., **4**, 798[106]; **6**, 472[66]
Fortunak, J. M., **2**, 1072[140]; **4**, 596[167], 608[324], 610[333], 616[167], 621[167], 652[167], 836[6]; **6**, 108[32]; **7**, 318[60]
Fortunato, J. M., **4**, 254[183], 260[183]; **8**, 536[176]
Fos, E., **8**, 636[4]
Foscante, R. E., **8**, 533[154]
Fossel, E. T., **3**, 896[68]
Fossey, J., **7**, 727[39]
Foster, A., **2**, 743[87], 780[9]
Foster, A. B., **4**, 231[274]; **6**, 275[112]
Foster, A. M., **6**, 1061[73]; **8**, 948[151]
Foster, B. S., **5**, 690[81]
Foster, C. H., **6**, 543[617]; **8**, 331[34]
Foster, D. F., **2**, 125[222]
Foster, D. G., **7**, 769[225], 771[258], 772[292]
Foster, G., **7**, 12[97]
Foster, H., **1**, 12[60]
Foster, J. P., **3**, 242[4], 244[4]
Foster, N., **8**, 626[168,169]
Foster, R., **7**, 851[29], 863[84]
Foster, R. V., **4**, 305[366,367], 306[366]
Foster, R. W. G., **5**, 21[161], 23[161]
Fotsch, C. H., **4**, 872[38]
Fottinger, W., **7**, 576[124]
Foubister, A. J., **8**, 563[433], 614[84]
Foucaud, A., **1**, 821[26]; **2**, 330[52], 344[16,17], 345[17], 353[16,17], 359[16], 360[16], 363[16,17], 538[57], 1084[20]; **5**, 488[195]; **6**, 175[79], 540[585,586], 745[84]; **7**, 842[21]
Fouchet, B., **4**, 247[98], 257[98], 262[98], 1086[114]
Fougerousse, A., **3**, 740[44]
Foulger, B. E., **5**, 646[10], 654[27], 671[49]
Fouli, F. A., **4**, 245[83]
Foulon, J. P., **1**, 107[3], 113[33,34], 428[119], 436[145]; **2**, 584[125]; **3**, 482[4], 485[4]; **4**, 183[81], 898[178]
Fouquet, G., **2**, 138[25]; **3**, 215[62], 244[20], 466[184]
Fouquey, C., **1**, 59[35]; **8**, 162[31]
Four, P., **1**, 440[170]; **8**, 21[143], 265[48,49], 553[362]
Fouret, R., **1**, 34[227]
Fournari, P., **7**, 27[65], 32[93]
Fourneron, J.-D., **3**, 342[13], 351[42], 352[42]; **7**, 59[43], 60[43,46a,47a,b], 62[47c], 64[60], 78[126], 429[157b]
Fournet, G., **6**, 960[59]
Fournier, A., **3**, 380[5]
Fournier, F., **3**, 568[48]; **4**, 345[81]; **8**, 133[19-21]
Fournier, M., **3**, 568[48]; **4**, 345[81]; **8**, 133[21]
Fourrey, J.-L., **6**, 618[107]; **8**, 935[69], 937[84,85]
Fourtinon, M., **6**, 722[143]
Foust, D. F., **5**, 1135[48]
Fowler, F. W., **1**, 473[79]; **4**, 1061[162]; **5**, 473[150], 478[150], 949[279]; **7**, 473[28], 502[28,256]; **8**, 386[50], 586[28]
Fowler, J. S., **2**, 959[30]; **4**, 445[204]; **8**, 344[123]
Fowler, K. W., **3**, 219[114], 499[140], 501[140], 502[140]
Fowler, R., **2**, 901[35], 908[35], 909[35], 910[35]
Fox, B. A., **8**, 907[69]
Fox, B. L., **8**, 270[99]
Fox, C. M. J., **1**, 243[54]; **6**, 443[91]
Fox, D. N. A., **4**, 412[268e]
Fox, D. P., **1**, 268[54]; **4**, 809[164]; **6**, 966[93]
Fox, F., **7**, 108[173]
Fox, J. J., **4**, 38[108]; **6**, 425[62], 553[795], 554[795]; **7**, 265[96]
Fox, J. L., **2**, 363[200]
Fox, J. P., **2**, 349[68]
Fox, M. A., **4**, 452[22], 473[22]; **5**, 699[5], 704[5]; **7**, 247[98], 539[66], 851[12,32], 852[32]
Foxman, B. M., **1**, 308[95], 314[95]; **3**, 219[102]; **4**, 562[36], 576[16,18]
Foxton, M. W., **8**, 734[4], 737[24,27], 740[4], 745[4], 746[4], 750[4]
Fracheboud, M. G., **3**, 743[59]
Fraenkel, D., **8**, 451[180]
Fraenkel, G., **1**, 2[9], 43[9]
Fraga, B. M., **8**, 330[49], 798[61]
Fragalà, I. L., **1**, 231[5], 251[1], 252[1], 273[1c], 274[1c]
Frahm, A. W., **2**, 852[227], 957[21]; **8**, 54[162], 66[162]
Frainnet, E., **2**, 600[30], 603[43]; **8**, 216[58], 556[377], 777[82a]
Fraisse-Jullien, R., **2**, 432[55]
Frajerman, C., **3**, 246[38], 470[221], 471[221]; **4**, 877[66]
Francalanci, F., **3**, 1033[72]; **7**, 429[151]; **8**, 236[3], 239[3]
Francavilla, M., **8**, 560[404]
France, A. D. G., **8**, 613[79]
France, R., **7**, 64[61b]
Francesch, C., **2**, 411[7]
Francetic, D., **7**, 777[366]
Franchi, V., **4**, 768[236]
Francis, F., **5**, 752[51]
Francis, R. F., **2**, 912[72]; **8**, 583[12]
Francisco, C. G., **4**, 814[187], 817[203]; **7**, 157[34], 495[210], 722[19], 723[19], 725[19]
Francisco, M. A., **8**, 843[49]
Franck, B., **3**, 667[46], 687[112,113]; **5**, 457[90]
Franck, R. W., **2**, 720[85]; **4**, 239[18], 249[18], 257[18], 261[18]; **5**, 348[74a], 414[54], 499[252], 500[252], 531[82]; **6**, 705[25]; **7**, 258[53]; **8**, 939[97]
Francke, W., **3**, 223[155]; **4**, 390[175b]; **6**, 677[323]
Franck-Neumann, M., **1**, 115[42]; **2**, 723[97,99]; **4**, 698[17], 699[17,19-21], 700[20,23], 701[21], 707[43], 956[16,17]; **5**, 64[37], 622[24], 632[24]; **6**, 691[404], 692[404]
Franckson, J. R. M., **7**, 704[11]
Francois, D., **1**, 543[19]
Francois, H., **7**, 95[77]
Francois, J. P., **6**, 501[190]
Francolanci, F., **3**, 1039[100]
Francotte, E., **5**, 331[42,42b], 414[53], 422[81], 829[25]; **6**, 757[131]
Frangin, Y., **1**, 214[22], 218[22], 220[22]; **2**, 486[38]; **4**, 95[102b,f,g], 880[91], 883[91,101], 884[101,105,106]
Frangopol, P. T., **2**, 744[99], 745[99]
Franich, R. A., **3**, 753[105]
Frank, A. W., **8**, 862[229]
Frank, B. L., **5**, 736[142i]
Frank, D., **6**, 1042[5]
Frank, F. J., **7**, 100[131], 256[42]
Frank, G. A., **3**, 851[66]
Frank, R., **2**, 424[35]; **6**, 526[391]; **7**, 206[68], 212[68], 765[152]

Frank, R. K., **6**, 531[454]; **8**, 41[95], 66[95]
Frank, R. L., **4**, 6[20]
Frank, R. W., **7**, 172[126]
Frank, W., **1**, 40[193], 732[17]
Frank, W. C., **4**, 401[228c]
Franke, E. R., **8**, 513[102]
Franke, F. P., **2**, 456[58,63], 458[58,63]
Franke, K., **6**, 579[983]
Franke, L. A., **5**, 211[66]
Franke, U., **8**, 696[125]
Frankel, E. N., **8**, 451[179,180,183], 554[367], 567[367b]
Frankel, J. J., **3**, 757[124]; **6**, 271[86]
Frankel, R. B., **8**, 366[51]
Franken, S., **2**, 520[69]; **4**, 222[172]
Frankevich, Ye. L., **7**, 852[42]
Frank-Kamenetskaya, O. V., **1**, 34[232]
Frankland, E., **4**, 93[95]
Franklin, J. F., **1**, 41[203]
Franklin, R. C., **3**, 823[15]
Franko, J. B., **8**, 51[123], 66[123]
Franks, D., **8**, 504[71]
Frantz, A. M., Jr., **8**, 91[62]
Frantzen, V., **3**, 888[20], 889[20], 894[20]
Franz, J. A., **4**, 736[89], 738[89], 797[101]
Franz, J. E., **4**, 35[98c]; **6**, 419[12]; **7**, 506[298]
Franz, R., **2**, 338[78]; **8**, 647[54]
Franzen, G. R., **3**, 390[81], 392[81]
Franzen, V., **3**, 553[13]; **5**, 913[107]; **6**, 122[136]; **7**, 663[56,57], 833[75]; **8**, 87[31]
Franzmann, K. W., **8**, 964[57]
Franzus, B., **5**, 856[210], 1003[22]; **6**, 205[30]; **8**, 475[17]
Fraser, D. J. J., **4**, 588[62]
Fraser, P. J., **5**, 1038[7], 1133[32], 1146[32]
Fraser, R. R., **1**, 357[8], 462[16], 468[51], 476[120,121], 512[37]; **2**, 476[4], 505[12], 510[12,39]; **3**, 31[186], 32[188], 66[9,10], 149[400]; **4**, 100[120]; **6**, 724[150,152,154], 726[174]
Fraser-Reid, B., **1**, 732[18], 766[156]; **2**, 124[203], 232[182], 827[126,127]; **3**, 848[54], 849[54b]; **4**, 36[102], 375[94,95a,b], 391[180,181a,b,182,182a,183], 401[94c,226], 753[164,165], 813[184,185], 815[185,192,193,194], 817[185,193,194], 820[192]; **5**, 350[80], 837[70]; **6**, 8[38], 27[113,118], 40[40], 118[101], 889[67], 987[69]; **7**, 246[90,91], 300[55], 318[53], 319[53], 362[32], 378[93], 454[97], 567[102], 584[102]; **8**, 219[78-80,82], 347[138], 965[69]
Frasnelli, H., **1**, 846[19a,b], 847[19b], 850[19b]
Fráter, G., **2**, 163[148], 554[134], 925[109]; **3**, 41[228], 43[239]; **4**, 314[498]; **5**, 18[125], 707[35], 834[53], 857[226]; **7**, 418[130b]; **8**, 196[118]
Fratiello, A., **1**, 292[27,31], 293[32,34,37]
Frattini, P., **5**, 630[53], 631[53]
Frauenrath, H., **2**, 655[133]
Fray, G. I., **5**, 737[146], 818[151], 819[153]; **7**, 583[153]
Fray, M. J., **1**, 823[40]; **2**, 567[30]; **6**, 1035[138]
Frazee, W. J., **1**, 62[40], 63[42]; **6**, 1044[16b], 1048[16], 1049[35]
Frazer, M. J., **1**, 305[87]
Frazier, H. W., **7**, 664[64]
Frazier, J., **6**, 91[121]
Frazier, J. O., **2**, 286[64]; **4**, 1040[88], 1048[88,88b]; **5**, 938[212], 939[212,223], 951[223], 957[309], 962[212,223], 964[223]
Frazier, K., **1**, 529[125]; **2**, 332[59]; **4**, 119[192d], 372[65]; **8**, 245[74], 248[86]
Frechet, D. M., **5**, 348[74a]
Fréchet, J. M. J., **1**, 223[76,76b], 317[149], 319[149], 320[149], 821[25]; **6**, 670[270], 698[270]; **7**, 281[174,175], 282[174,175], 663[58]; **8**, 166[63], 372[118]
Frechette, R., **5**, 491[208], 494[215], 579[163]; **6**, 736[26]
Fréchou, C., **6**, 149[109], 156[157,158,159,161], 927[78]; **8**, 12[66]
Fredenhagen, H., **3**, 822[2], 831[2]; **8**, 87[28]
Fredericks, P. M., **6**, 217[112]; **7**, 68[84], 69[92], 72[84], 73[92]
Frederiksen, J. S., **2**, 722[94]
Frediani, P., **8**, 87[27], 236[3], 239[3], 552[352]
Freedman, H. H., **7**, 228[94]; **8**, 91[62]
Freekers, R. L., **6**, 533[496]
Freeman, A., **2**, 1104[132]
Freeman, F., **2**, 354[110], 358[110], 359[110]; **4**, 377[104], 379[104e], 380[104d,e]; **6**, 938[122]; **7**, 99[111], 252[2], 528[5,6], 815[2], 816[2b], 824[2b], 851[18]
Freeman, J. P., **1**, 387[125], 834[129], 837[129,149]; **4**, 115[183], 259[262]; **7**, 470[17], 750[131], 751[140]; **8**, 390[76], 829[83]
Freeman, P. K., **3**, 906[143]; **4**, 273[53]; **6**, 777[60]; **8**, 340[97]
Freeman, R. C., **3**, 843[23]
Freeman, W. A., **6**, 177[117]; **7**, 155[29]
Freenor, F. J., **6**, 204[23]
Freer, A. A., **5**, 441[176,176f], 575[133]
Freer, P., **4**, 288[183]
Freer, V. J., **5**, 225[116], 227[116], 233[116]
Freerken, R. W., **7**, 172[125]
Frehel, D., **2**, 765[74], 817[92]
Frei, B., **3**, 133[335], 136[335]; **5**, 223[83], 224[83], 806[102], 929[169], 930[169], 1028[90]; **6**, 93[131]; **7**, 77[120b]
Freiberg, J., **2**, 900[26], 960[35], 962[35]
Freiberg, L. A., **3**, 905[140]
Freidinger, R. M., **2**, 962[51]; **3**, 882[102], 883[102]; **5**, 790[31]; **6**, 1061[74]
Freidlin, A. Kh., **8**, 458[221]
Freidlin, G. N., **3**, 640[98,101]
Freidlin, L. Kh., **8**, 447[103,104], 450[103,104,163], 551[342]
Freidlina, R. Kh., **1**, 142[24], 143[31]; **3**, 422[69]; **4**, 288[187], 330[5], 952[4]; **6**, 153[143], 542[598]; **7**, 500[236]; **8**, 765[11]
Freiesleben, W., **7**, 94[55], 450[8]
Freifelder, M., **4**, 31[92,92c]; **6**, 724[157]; **8**, 139[5,6], 141[32], 143[55], 150[6], 251[108], 418[2], 420[2], 423[2], 431[2], 433[2], 533[136], 597[89,91], 794[6]
Freilich, S. C., **5**, 154[29], 156[29]; **8**, 724[169,169f]
Freimanis, Ya. F., **4**, 26[77]
Freire, R., **4**, 375[98b], 388[98,98b], 408[98b], 817[203]; **7**, 157[34], 722[19], 723[19], 725[19]
Freiser, H., **3**, 211[30]
Freissler, A., **1**, 185[53]; **2**, 29[104]
Freist, W., **6**, 625[155]
Freitag, S., **6**, 508[288]
Freitas, E. R., **5**, 1117[14]
Frejaville, C., **4**, 180[66]
Frejd, T., **2**, 785[42]; **3**, 260[141,142], 261[142], 466[180,181], 763[149], 768[149,166]; **4**, 211[93]; **7**, 410[101]
Fremdling, H., **7**, 770[251], 773[303]
Fremery, M. I., **7**, 507[305], 581[143]
French, L. G., **5**, 417[60]; **8**, 20[136]
French, N. I., **5**, 394[147]
Frenette, R., **3**, 771[185], 1056[36]; **4**, 1040[101], 1045[101a]; **7**, 360[21]
Fréon, P., **1**, 215[40], 216[40], 218[49], 220[49], 223[49], 226[49a,90-92,94,96], 326[6,7], 327[8], 349[143,144]; **4**, 95[97], 98[108d,109a,b,e]
Freppel, C., **7**, 447[73]
Frese, A., **4**, 5[17]
Freskos, J. N., **4**, 14[47,47e]
Freudenberg, K., **3**, 693[143,147]
Freudenberg, U., **5**, 151[15], 154[32]
Freudenberger, J. H., **1**, 749[78], 816[78]; **5**, 1116[12], 1125[65]
Freudenreich, B., **6**, 526[391]
Freund, A., **5**, 1194[33]
Freund, E., **2**, 138[21]

Freund, F., **7**, 840[8]
Freund, M., **2**, 91[42]
Freund, R., **4**, 222[178]
Frey, A. J., **2**, 1022[53]; **3**, 346[23]; **4**, 373[68]; **5**, 341[61a]; **6**, 1042[3]
Frey, D. W., **5**, 196[11]
Frey, H., **6**, 185[154]; **7**, 128[171]
Frey, H. H., **2**, 360[173]
Frey, H. M., **5**, 679[19], 790[19], 905[61], 906[61,64,65], 908[65], 910[65,79,82], 911[92], 918[125], 989[44], 1016[60,62]
Frey, M., **5**, 260[66], 261[66]; **8**, 70[223]
Freyer, A. J., **2**, 1026[68]; **5**, 425[99]; **6**, 900[112]
Friary, R. J., **2**, 841[186]; **3**, 55[283], 781[14]; **8**, 527[48], 528[57], 939[97]
Fribush, H. M., **1**, 357[3]
Frick, U., **6**, 502[217], 560[870]; **7**, 650[51]
Frick, W., **6**, 573[963]
Frickel, F., **2**, 819[98]; **6**, 839[70]
Fridkin, M., **3**, 302[51]; **6**, 625[156]
Fridman, A. L., **3**, 887[10], 888[10], 889[10], 890[10], 893[10], 897[10], 900[10], 903[10]
Friebe, T. L., **2**, 655[149]
Fried, F., **5**, 687[60]
Fried, J., **2**, 332[61]; **3**, 279[38], 572[61]; **8**, 477[36], 490[5], 492[5], 493[5,20], 497[20,38], 524[4], 526[4], 530[4], 541[207]
Fried, J. H., **3**, 9[46], 10[46]; **4**, 185[90], 186[90], 248[109], 932[62]; **6**, 216[106], 219[123], 647[112a]; **7**, 86[16a], 139[128]
Friedel, C., **3**, 299[30], 317[121]
Friedel, P., **4**, 257[225]; **8**, 943[122]
Friedel, R. A., **5**, 1138[67]
Friedländer, P., **3**, 828[42]
Friedli, F. E., **4**, 279[111]
Friedlin, L. K., **8**, 535[160]
Friedlina, R. K., **8**, 91[58]
Friedman, B. S., **4**, 316[538]
Friedman, H. A., **6**, 425[62]
Friedman, H. L., **4**, 315[504]
Friedman, H. M., **3**, 304[63]
Friedman, L., **1**, 377[97]; **3**, 244[21], 247[21], 464[173]; **4**, 488[51,52], 489[52], 511[177]; **5**, 380[113c], 383[124]; **6**, 228[17], 232[38], 727[206]; **8**, 269[92], 274[92], 941[111], 943[120]
Friedman, M. D., **2**, 198[86]
Friedman, N., **7**, 40[7]
Friedman, S., **8**, 455[206]
Friedrich, A., **7**, 742[54]
Friedrich, D., **5**, 850[158]
Friedrich, E., **2**, 514[52]; **7**, 98[100], 165[84]
Friedrich, E. C., **4**, 968[60], 969[60], 1018[222]; **6**, 210[77]
Friedrich, J. D., **5**, 439[168]
Friedrich, K., **2**, 359[159]; **3**, 271[2], 272[2]; **5**, 151[19], 436[157]; **6**, 225[4,8], 226[4], 228[4], 230[4], 231[4,8], 232[4,8,37], 233[4], 234[4], 235[4], 236[4], 238[4], 239[4], 240[4], 241[4], 258[4], 489[87]
Friedrich, L. E., **4**, 587[48]; **5**, 2[15], 162[68,69]
Friedrichsen, W., **6**, 551[689]; **8**, 493[23], 626[170]
Friege, H., **5**, 71[125]; **7**, 489[171]
Friend, C. M., **8**, 608[43], 629[43]
Fries, K., **3**, 809[42]
Fries, P., **7**, 655[20]
Fries, R. W., **1**, 442[178]
Fries, S., **5**, 74[207]
Friese, C., **4**, 992[156]
Friesen, J. D., **8**, 206[173]
Friesen, R. W., **3**, 488[52,53], 495[52,53]
Frigd, J. H., **4**, 608[325]
Frigerio, M., **4**, 382[131b], 384[131b]
Frigo, T. B., **5**, 430[117]
Frigo, T. M., **5**, 430[117]
Frimer, A. A., **7**, 168[103], 816[10], 818[10]
Fringuelli, F., **8**, 494[25]
Fringuelli, R., **1**, 656[151], 658[151], 846[18b,c], 847[18]; **4**, 308[404]
Frisbee, R., **3**, 460[142], 497[105]
Frisch, M. A., **5**, 857[229]
Frisell, C., **7**, 95[76], 613[1]
Frisque-Hesbain, A. M., **5**, 109[217]; **6**, 493[129]
Fristad, W. E., **2**, 713[47]; **4**, 763[213], 764[215], 807[152]; **5**, 757[75,76], 778[75,76]; **6**, 780[74]; **7**, 92[43], 97[92], 447[72], 487[149], 532[31], 720[13], 722[13]; **8**, 946[140]
Fristad, W. F., **1**, 584[12]
Fritch, J. R., **5**, 1134[36]
Frith, R. G., **8**, 321[103]
Fritsch, J. M., **7**, 798[23]
Fritsch, N., **5**, 324[22]; **6**, 1023[73]
Fritsch, W., **7**, 124[42]
Fritschi, H., **4**, 1039[64]; **6**, 531[444]
Fritz, G., **3**, 224[172], 312[101]; **8**, 771[43]
Fritz, H., **2**, 534[36]; **3**, 381[32], 621[31]; **5**, 404[16], 417[65], 418[67], 419[74], 420[75]; **6**, 561[881,882], 734[5]; **8**, 395[129], 621[145], 652[79]
Fritz, H.-G., **8**, 795[20]
Fritz, H. P., **7**, 799[25,26]
Fritz, S., **1**, 223[71]
Fritzberg, A. R., **3**, 892[52]; **6**, 94[140]
Fritzche, T. M., **8**, 52[147], 66[147]
Fritzen, E. L., **2**, 903[48]
Fritzsche, H., **7**, 213[102]
Frobese, A. S., **2**, 442[8], 447[8]; **8**, 337[78], 542[236]
Froborg, J., **1**, 429[124]; **5**, 687[56]
Froech, S., **2**, 32[118], 114[113], 267[64]
Froehler, B. C., **6**, 604[28], 618[110], 620[132]
Froemsdorf, D. H., **6**, 1033[129]
Froen, D. E., **5**, 1056[48]
Fröhlich, H., **8**, 242[40]
Frohlisch, B., **7**, 657[30]
Froitzheim-Kühlhorn, H., **5**, 66[77]
Fröling, A., **3**, 121[245], 123[245,246], 125[246]
Fröling, M., **3**, 121[245], 123[245]
Frolov, S. I., **7**, 597[44]
Frolow, F., **4**, 596[160], 604[286], 621[160], 626[286], 636[160], 795[85]
Fromageot, P., **2**, 232[176]; **8**, 460[254]
Frömberg, W., **8**, 682[83], 683[88,93]
Frommeld, H. D., **2**, 182[5], 477[6]; **6**, 719[128], 720[128]; **8**, 392[104]
Fronczek, F. P., **1**, 568[238]
Fronczek, F. R., **3**, 591[170]; **4**, 379[117], 462[105], 1079[65]; **5**, 627[43], 635[85]; **7**, 439[36]; **8**, 7[35]
Fronczek, R. R., **5**, 260[65], 261[65]
Fronza, C., **8**, 195[112]
Fronza, G., **1**, 185[55], 186[55], 221[68]; **2**, 29[105], 30[113], 31[113,113a], 32[119,119b], 547[114], 551[114], 998[51], 999[51]; **4**, 36[103,103c]; **8**, 195[110,113]
Frosch, J. V., **8**, 544[273]
Frost, J. W., **2**, 466[109,110]
Frostick, F. C., Jr., **2**, 182[3], 834[155]
Frostin-Rio, M., **4**, 746[147]
Fröstl, W., **5**, 527[66], 528[66a], 529[66a]
Fruchey, D. S., **6**, 205[33], 834[40]
Frühauf, H.-W., **5**, 498[230,233]
Frump, J. A., **6**, 488[19]
Fruscella, W. M., **1**, 878[109]; **4**, 1005[86]; **5**, 514[9], 527[9]
Fry, A., **3**, 723[8,9], 731[8,9]; **6**, 1013[10]

Fry, A. J., **3**, 870[46]; **4**, 294[245]; **7**, 656[14]; **8**, 125[93], 135[45], 389[72], 807[118], 856[183], 984[5,6], 986[9], 987[5,24], 988[5,6], 991[5], 992[5], 994[5], 995[5], 996[5], 997[5]
Fry, E. M., **8**, 566[452], 585[27]
Fry, J. L., **6**, 219[120]; **8**, 275[144], 318[68,69], 813[13], 969[97]
Fry, M. A., **7**, 720[13], 722[13]
Frydrych, C., **5**, 108[210], 109[210], 110[210], 111[210]
Frydrych-Houge, C. S. V., **3**, 594[187]
Frye, L. L., **1**, 751[90]; **2**, 124[211]; **4**, 86[81], 213[107,110-112], 215[107,112]; **6**, 150[121,122]; **8**, 844[71], 847[71]
Frye, R. B., **8**, 34[62], 66[62]
Frye, R. L., **3**, 649[207]
Frye, S. V., **1**, 62[41], 63[42]; **2**, 630[8]
Fryer, R. I., **2**, 329[47]; **6**, 490[104], 534[515]; **8**, 337[81]
Fryermuth, H. B., **7**, 766[173]
Frysinger, J. F., **8**, 584[16]
Fryxell, G. E., **4**, 452[22], 473[22], 1089[137], 1090[137], 1091[137]
Fryzuk, M. D., **4**, 1039[62]; **8**, 459[234,235], 535[166], 673[24], 676[24], 681[78], 682[24], 683[78], 689[78], 693[78]
Fu, C. C., **8**, 277[152]
Fu, G. C., **8**, 698[143]
Fu, J.-M., **3**, 231[250]
Fu, P. P., **2**, 760[47]; **7**, 136[107]
Fu, T.-H., **8**, 320[89,90]
Fu, W. Y., **1**, 506[12]
Fu, X., **1**, 243[53]
Fu, Y.-L., **8**, 877[47], 878[47]
Fuchigami, T., **2**, 105[42,43]; **4**, 439[163]; **8**, 134[28,33], 135[40], 137[33]
Fuchikami, T., **3**, 1027[45]; **7**, 144[158]; **8**, 765[11]
Fuchita, T., **6**, 685[363]; **8**, 885[106], 886[106]
Fuchs, B., **1**, 752[99]; **5**, 223[84-86], 224[84-86], 604[54], 740[150,151], 818[151]
Fuchs, C. F., **3**, 208[5]
Fuchs, D. L., **8**, 885[104]
Fuchs, P. L., **1**, 114[37], 238[37], 343[119], 529[123], 735[28], 736[28], 765[174]; **2**, 514[49]; **3**, 125[300], 128[300], 129[300], 133[300], 140[380], 154[380], 168[380], 174[380], 176[380], 209[16], 223[16]; **4**, 15[50,50c], 78[54,54c], 79[55a,b,56], 162[92], 192[118,118b], 251[150,151,152,153,154], 257[154], 260[154], 1040[85], 1045[85]; **5**, 134[60], 268[78], 537[100], 936[194], 1024[79], 1166[24]; **6**, 9[43], 162[189,190,191], 163[189,192,194], 164[196], 172[17], 176[102], 1056[55]; **7**, 362[29], 517[11]; **8**, 395[131], 494[26], 885[104], 940[105,110], 947[144], 958[18]
Fuchs, R., **6**, 204[12]
Fuchs, W., **3**, 643[124]
Fuegen-Koster, B., **5**, 1094[100], 1112[100b]
Fueno, T., **3**, 867[36]; **4**, 276[79], 299[297]; **5**, 72[166]; **6**, 445[100]; **7**, 794[7e], 801[36]; **8**, 87[33], 477[32]
Fuentes, A., **8**, 86[26]
Fuentes, L. M., **1**, 477[135]; **2**, 256[46], 604[50,54], 652[125], 1059[79]; **3**, 67[18], 75[47,48]; **4**, 5[17]; **5**, 99[130], 850[161]; **6**, 554[774], 759[139]
Fugami, K., **2**, 589[154]; **3**, 445[74]; **4**, 588[68], 637[68], 721[31], 725[31], 791[42], 824[241]; **5**, 927[164], 938[164,219]; **8**, 798[46], 807[46]
Fuganti, C., **1**, 55[23], 221[68], 389[139], 543[28]; **2**, 29[105], 30[113], 31[113,113a], 32[119,119b], 998[50,51], 999[50,51]; **4**, 36[103,103c]; **8**, 190[79], 195[110-113], 203[111,149,150], 560[407]
Fuganti, P., **1**, 185[55], 186[55]
Fügedi, P., **6**, 46[67], 47[67,81], 51[103]; **8**, 224[103,110], 225[103,110], 230[140]
Fugger, J., **3**, 891[38]
Fugiwara, M., **5**, 167[94]
Fuhlhage, D. W., **2**, 943[170], 970[89], 971[89]
Fuhr, K. H., **4**, 123[210b], 125[210b]; **8**, 880[60,61]
Fuhrer, W., **1**, 464[34]
Fuhrhop, J.-H., **2**, 354[114], 357[114]; **3**, 866[29]; **4**, 70[3]; **6**, 830[5]; **7**, 95[73a]
Fujami, H., **7**, 307[11]
Fuji, K., **2**, 816[83], 819[102]; **3**, 135[341,342,343], 136[341,342,343], 137[341,342], 1050[19]; **4**, 42[122b], 124[215]; **5**, 439[171], 440[171]; **6**, 647[110,112b]; **7**, 256[39], 588[174,175], 710[56]; **8**, 902[44], 908[44], 909[44], 989[37]
Fuji, M., **1**, 63[43]
Fuji, T., **4**, 254[179]
Fujihara, H., **1**, 825[50]; **3**, 510[182]; **4**, 427[69], 487[46]; **6**, 934[98]; **7**, 205[63], 425[149b]
Fujihara, Y., **2**, 128[240]; **4**, 298[291]; **7**, 453[65]
Fujihira, M., **7**, 50[69]
Fujii, A., **6**, 535[539], 538[539]
Fujii, E., **8**, 934[56]
Fujii, H., **5**, 832[39]; **7**, 879[146]; **8**, 391[90]
Fujii, K., **7**, 829[56]
Fujii, M., **2**, 332[63]; **4**, 790[36]; **5**, 623[25]; **6**, 606[40], 618[112]; **8**, 170[71], 561[417]
Fujii, N., **2**, 1099[112b]; **4**, 152[55]; **5**, 63[9]
Fujii, S., **1**, 880[113]; **2**, 367[222], 444[21]; **6**, 507[240], 515[240]; **7**, 606[156]
Fujii, T., **1**, 768[168]; **3**, 652[221]; **6**, 531[425]; **8**, 201[143], 244[50], 548[324], 549[324]
Fujii, Y., **4**, 1103[205]; **8**, 149[117-119]
Fujiki, M., **3**, 530[63], 532[63]
Fujikura, S., **3**, 244[29]; **4**, 120[203], 879[85]
Fujikura, Y., **4**, 921[26]; **6**, 270[80]; **7**, 9[70]; **8**, 331[33]
Fujimori, K., **7**, 761[58]; **8**, 36[85], 39[85], 66[85], 370[89], 409[81], 412[114]
Fujimori, M., **3**, 528[47]
Fujimoto, E., **8**, 989[36]
Fujimoto, G. I., **4**, 41[119,119b]
Fujimoto, H., **5**, 75[224,225], 714[73]; **8**, 766[19]
Fujimoto, K., **1**, 642[121], 646[121], 656[121,153], 658[121], 665[121], 667[121], 672[121]; **2**, 351[82], 357[82], 1060[89]; **3**, 1005[61-63]; **4**, 89[85], 206[52-54]; **6**, 60[145], 523[351], 524[351], 883[55], 890[69], 891[69,70]
Fujimoto, M., **8**, 444[6], 881[76], 882[76], 902[44], 908[44], 909[44], 989[37]
Fujimoto, N., **3**, 565[22]
Fujimoto, R., **2**, 877[35,36]
Fujimoto, T., **1**, 779[222]
Fujimoto, T. T., **3**, 117[235,236], 155[235,236], 156[235,236], 946[87]; **6**, 163[193], 838[66], 997[110]
Fujimoto, Y., **3**, 856[89]; **5**, 136[67,68], 137[74]; **6**, 1050[37]; **7**, 80[139], 153[6]; **8**, 967[78]
Fujimura, A., **1**, 85[30], 104[30]
Fujimura, H., **6**, 510[293]
Fujimura, N., **1**, 116[50], 118[50]
Fujimura, O., **1**, 809[329], 810[329b]
Fujimura, T., **2**, 450[52,53]; **3**, 455[125], 460[125]
Fujinaga, M., **6**, 266[47]
Fujinami, H., **7**, 452[54,55], 462[54,55]
Fujinami, T., **1**, 243[59,60], 254[15], 268[53,53a,c], 269[57]; **2**, 446[32]; **3**, 570[56]; **4**, 809[165]; **6**, 564[909]; **8**, 16[98], 988[32]
Fujinari, E. M., **5**, 197[18]
Fujino, M., **2**, 1099[112b]; **6**, 637[33], 644[82], 664[222]
Fujino, T., **4**, 753[164]; **8**, 836[7]
Fujioka, A., **3**, 437[25], 440[25], 448[25], 449[25], 450[25], 451[25], 452[25], 484[26], 492[26], 494[26], 495[26], 503[26], 513[26]
Fujioka, H., **1**, 63[43,44], 64[46]; **3**, 466[186], 751[88]; **4**, 304[356]; **5**, 838[74]; **6**, 1066[97]; **7**, 440[40]
Fujioka, T., **8**, 856[168]
Fujioka, Y., **3**, 503[142]

Fujisaki, S., **3**, 321[139]; **6**, 801[27,28]
Fujisawa, T., **1**, 85[30], 104[30], 122[68], 221[68], 233[20], 238[36], 336[73], 423[97], 424[98], 425[102], 427[112], 448[205], 568[246]; **2**, 30[113], 31[113], 507[22], 749[135], 780[8]; **3**, 124[259], 125[259], 153[416], 154[416], 218[100], 220[125], 227[210,211,212], 243[16], 245[30-32], 246[39], 257[39], 446[78], 463[158], 470[212,213], 476[212,213]; **4**, 120[201], 262[303], 898[177], 902[190]; **5**, 841[95]; **6**, 20[74], 493[128], 494[138], 505[225], 836[53]; **7**, 516[4]; **8**, 168[69,70], 178[69], 179[69], 187[40,44], 188[44], 190[77], 195[107,109], 196[77,120], 203[146], 205[157], 241[38], 263[32], 267[32], 545[302], 837[18], 889[126]
Fujise, Y., **1**, 553[94], 876[98]; **3**, 390[79], 1008[72], 1009[72], 1010[72]; **5**, 618[2], 620[14], 621[17], 622[22], 623[25-27], 627[42]; **6**, 875[18]
Fujishita, T., **4**, 8[28,28a]
Fujita, A., **1**, 410[39], 568[231]; **3**, 124[280], 125[280], 126[280]; **6**, 468[53]
Fujita, E., **1**, 834[123], 894[160]; **2**, 116[131,140], 610[94,95], 611[94,95], 718[78,79], 816[83], 819[102], 855[240], 1059[78,82]; **3**, 125[309], 135[341,342,343], 136[341,342,343], 137[341,342], 168[492], 169[492], 217[79], 1050[19]; **4**, 42[122b], 155[72]; **5**, 92[81], 945[249]; **6**, 134[28], 647[110,112b], 666[231], 667[231], 936[111], 938[132], 944[132], 1004[138], 1018[42], 1065[90b]; **7**, 92[41,41b], 93[41b], 94[41], 457[110], 588[174,175], 621[34], 623[35], 710[56], 765[149], 773[149,301]; **8**, 241[38], 272[119,120], 496[31], 514[111], 544[259], 902[44], 904[55], 908[44,55], 909[44], 910[55], 911[55], 914[55]
Fujita, H., **2**, 780[6]; **3**, 217[95], 579[101]; **4**, 958[27]; **5**, 736[145], 737[145], 817[146]
Fujita, I., **8**, 145[83]
Fujita, J., **7**, 778[404]; **8**, 535[166]
Fujita, K., **2**, 5[21], 13[21,21c], 14[21c]; **3**, 381[24], 382[24], 726[25]; **5**, 429[115]; **6**, 117[97]
Fujita, M., **1**, 474[100], 506[5], 526[5], 527[103], 751[109], 831[98]; **2**, 572[40], 575[40]; **3**, 1027[45]; **4**, 354[130], 1040[67]; **7**, 778[410]; **8**, 8[45,46], 11[63], 13[45,46], 20[45,46], 21[46], 145[82]
Fujita, N., **8**, 369[74]
Fujita, S., **1**, 366[48]; **3**, 593[182]; **4**, 115[180e], 394[194,195], 1017[213], 1019[213]; **5**, 693[111]; **6**, 487[43], 489[43]; **7**, 26[51], 219[10], 524[51], 698[50]
Fujita, T., **3**, 402[130], 404[132]; **4**, 313[460], 1023[257]; **6**, 566[929,930], 780[69]; **7**, 811[91]; **8**, 496[31], 544[259]
Fujita, Y., **1**, 803[306]; **2**, 527[6], 528[6]; **3**, 244[25], 267[25], 428[90], 470[209], 472[209], 475[209], 494[84], 992[37], 994[41]; **5**, 1[4], 2[4], 15[105], 787[8], 797[61], 799[74], 821[162], 882[14], 888[29], 889[31]; **6**, 157[172], 852[138], 876[26], 882[26], 885[26], 991[89]; **7**, 660[43]; **8**, 698[137,138]
Fujitaka, N., **2**, 21[85]
Fujitani, T., **1**, 543[29]
Fujitsu, H., **8**, 598[100]
Fujiu, T., **6**, 498[171]
Fujiwa, T., **3**, 445[71]
Fujiwara, A., **5**, 167[94]; **7**, 64[61a]
Fujiwara, A. N., **8**, 354[171]
Fujiwara, H., **8**, 844[75]
Fujiwara, I., **5**, 758[81]
Fujiwara, J., **1**, 88[52,54], 165[111]; **4**, 140[11], 209[65]; **6**, 291[217], 770[35]; **7**, 697[47]; **8**, 223[99], 224[99]
Fujiwara, K., **2**, 348[58], 357[58]; **5**, 736[142l,m]
Fujiwara, M., **6**, 89[120]; **7**, 64[61a]; **8**, 201[141]
Fujiwara, N., **6**, 498[160]
Fujiwara, S., **2**, 749[134]; **3**, 470[208], 471[208], 475[208]; **4**, 594[146]
Fujiwara, T., **1**, 561[159]; **2**, 649[105]; **3**, 137[376]; **4**, 1056[141,141b]; **5**, 736[142g]; **7**, 774[332]; **8**, 847[91]
Fujiwara, Y., **1**, 162[104], 254[14,14b], 277[80,82,83], 278[14a,84], 279[86], 280[86]; **3**, 295[11], 302[11], 567[36,190], 583[120], 595[190], 607[190], 610[36]; **4**, 836[2-5], 837[10,13-15], 841[50], 856[100]; **7**, 107[168]; **8**, 113[36], 889[128]
Fujiyama, F., **8**, 28[33], 36[33], 66[33]
Fujiyama, R., **2**, 577[80]
Fukagawa, T., **1**, 162[104], 254[14], 277[82,83], 278[14a,84]
Fukami, N., **5**, 350[79]; **7**, 255[38]
Fukamiya, N., **8**, 544[255]
Fukanaga, T., **4**, 787[28,29], 1103[206]
Fukase, H., **6**, 74[37]
Fukata, G., **3**, 329[187]
Fukatsu, S., **4**, 111[152b], 218[146]; **6**, 532[468]
Fukaya, C., **2**, 1041[109]
Fukazawa, M., **5**, 637[102]
Fukazawa, Y., **1**, 553[90,94,98]; **3**, 390[79]; **4**, 255[196], 477[165]; **5**, 532[87], 623[25-27], 736[142m]; **6**, 875[18], 932[97]
Fukomoto, K., **4**, 505[134]
Fuks, R., **4**, 45[130]; **5**, 116[256,257,265], 676[3], 689[74], 694[3a]; **6**, 517[323,324,325], 544[323,324,325], 551[684], 552[325,701]
Fukuda, E. K., **7**, 854[46]
Fukuda, H., **6**, 1022[60]
Fukuda, J., **2**, 771[13]
Fukuda, M., **6**, 487[39], 489[39], 543[39]
Fukuda, N., **4**, 845[69]
Fukuda, T., **6**, 637[33]
Fukuda, Y., **1**, 343[111]; **3**, 279[37]; **4**, 567[49], 969[68]; **5**, 829[26]
Fukui, K., **2**, 282[33], 291[72,74], 292[79], 662[8]; **3**, 556[36]; **4**, 510[167], 1073[20], 1076[20]; **5**, 75[224,225], 451[54], 678[14]; **7**, 96[90], 98[90], 877[128]
Fukui, M., **1**, 463[20]; **2**, 541[77]; **4**, 522[53], 523[53]; **5**, 412[45]; **8**, 7[42]
Fukui, T., **5**, 297[59], 1196[38], 1197[38]
Fukumori, S., **4**, 261[286]
Fukumoto, K., **2**, 222[147], 384[316], 765[67], 851[222], 888[54], 1024[61,62]; **3**, 164[480], 165[480], 167[482], 168[482], 660[17], 677[86], 679[17]; **4**, 30[88,88k-o], 121[209,209a,b], 181[70], 231[277], 239[17], 261[17,299,300,301], 333[21-23], 398[215], 401[215a,229], 500[110], 501[113,116,117], 505[135], 510[172,176]; **5**, 410[38], 473[154], 479[154], 522[45], 524[54], 531[78], 534[54,95], 536[97], 541[110,111], 681[28], 691[83,84,86], 692[83,83c,84], 693[83,107-109,111,114], 694[114], 712[57b], 723[107], 741[157,157c,d], 742[162], 841[88,98], 843[117], 847[136], 1032[100]; **6**, 756[126], 757[130], 780[70]; **7**, 493[199], 517[15], 564[89], 569[89]; **8**, 314[37], 945[128]
Fukumoto, T., **8**, 248[86]
Fukunaga, J. Y., **5**, 70[107]
Fukunaga, K., **6**, 685[348], 959[45]; **8**, 806[121]
Fukunaga, R., **1**, 860[68]
Fukunaga, T., **4**, 47[134], 424[12], 429[12], 430[12]; **5**, 552[30]
Fukunaga, Y., **7**, 58[55], 62[55], 63[55]
Fukunishi, K., **4**, 753[163]
Fukuoka, S., **2**, 725[121]
Fukushi, H., **2**, 810[60]; **3**, 216[72]; **4**, 1009[140]
Fukushima, A., **6**, 556[827]
Fukushima, D., **7**, 761[57]
Fukushima, D. K., **8**, 629[179]
Fukushima, H., **7**, 473[33], 501[33], 502[33]; **8**, 164[46], 178[46], 179[46]
Fukushima, M., **3**, 229[226]; **8**, 84[12], 459[245]
Fukushima, T., **6**, 453[143]; **7**, 120[16]
Fukuta, K., **5**, 282[28], 284[28], 601[45], 606[45]
Fukutani, H., **5**, 38[23c]
Fukutani, Y., **1**, 88[54]; **4**, 140[11], 209[64,65], 753[167], 968[59], 969[59]; **6**, 850[125]

Fukuto, J., **6**, 544[628]
Fukuyama, J., **5**, 162[67]; **6**, 734[10], 735[10]
Fukuyama, K., **8**, 535[166]
Fukuyama, T., **1**, 762[141]; **2**, 652[123b], 1069[135], 1096[97]; **4**, 377[145a], 384[145a]; **6**, 5[24], 266[50], 531[454]; **7**, 169[107], 246[82], 358[10], 371[10], 380[103]; **8**, 11[60], 41[95], 66[95]
Fukuyama, Y., **5**, 611[72]; **7**, 174[134]
Fukuzaki, K., **2**, 714[51,52], 720[52]
Fukuzawa, A., **5**, 611[72]
Fukuzawa, S., **1**, 243[59,60], 254[15], 268[53,53a,c], 269[57]; **2**, 446[32], 598[16]; **3**, 381[26], 382[26], 570[56]; **4**, 347[87], 809[165]; **7**, 95[64], 773[308,309], 774[326], 775[352c,354,355], 776[308,309,355,363]; **8**, 16[98], 988[32]
Fukuzawa, Y., **6**, 501[193]
Fukuzumi, K., **3**, 99[190]; **8**, 552[349], 557[384], 859[215]
Fukuzumi, S., **5**, 71[132,133]; **7**, 852[37], 883[180]; **8**, 95[86]
Fulcher, J. G., **8**, 889[136]
Fulka, C., **4**, 1074[29]
Fuller, C. J., **8**, 264[37]
Fuller, G., **8**, 321[106], 897[20]
Fuller, G. B., **7**, 495[209]
Fuller, R. W., **8**, 618[110], 623[151]
Fullerton, D. S., **7**, 101[132], 258[57], 845[64]
Fullerton, T. J., **3**, 753[105]; **4**, 587[26], 588[69], 614[380], 615[380], 628[380]; **6**, 154[151]; **8**, 843[51], 844[51]
Fulop, F., **5**, 584[194]; **6**, 525[382]
Fülöp, G., **6**, 534[519]
Fulton, B. S., **2**, 846[202]
Fulton, R. P., **4**, 604[285,290], 646[290]; **8**, 960[35]
Fultz, W., **1**, 301[74], 316[74]
Fu-Lung Lu, **7**, 500[241]
Funabashi, M., **4**, 85[77e]; **7**, 856[66]
Funabashi, Y., **4**, 79[59a,c], 216[126]; **6**, 27[115], 164[199]
Funabiki, T., **8**, 449[159,160], 453[191], 567[461]
Funaki, K., **1**, 790[263]; **4**, 351[125]
Funaki, Y., **8**, 388[63], 874[22]
Funakoshi, K., **2**, 540[71]; **8**, 191[95], 198[134]
Funakoshi, W., **8**, 971[109], 995[67]
Funakubo, E., **4**, 1006[106]
Funakura, M., **8**, 991[44]
Funasaka, W., **4**, 1017[211], 1021[211]
Funayama, M., **3**, 295[8]
Funayama, T., **8**, 61[190], 66[190]
Funcke, W., **6**, 790[112]
Fünfschilling, P., **5**, 710[49]
Funfschilling, P. C., **3**, 17[86]
Fung, A. P., **4**, 304[357]; **6**, 765[14]; **8**, 406[51]
Fung, D., **7**, 854[54], 855[54]
Fung, K. H., **2**, 294[80]
Fung, N. Y. M., **8**, 17[114], 21[114], 115[66]
Fung, S., **4**, 893[154]; **8**, 397[141], 880[64]
Fung, V. A., **3**, 362[89]
Funhoff, D. J. H., **3**, 927[50]
Funk, R., **5**, 513[5], 514[5], 524[5e], 527[5]
Funk, R. F., **5**, 857[232]
Funk, R. L., **1**, 872[91]; **2**, 69[49]; **3**, 58[292]; **4**, 191[111], 239[16]; **5**, 249[32], 385[130], 435[151], 517[29], 519[29,102], 534[29], 537[99], 538[102], 547[29g], 573[128], 574[129], 619[9], 624[9], 625[9], 691[83], 692[83,90], 693[83], 841[89], 843[116,122,123], 1151[132,134,135]; **6**, 757[129], 859[169,170,172], 1063[87]; **7**, 338[39]
Funke, B., **2**, 368[238]; **6**, 425[67]
Funke, C. W., **5**, 165[79]
Funke, E., **1**, 859[64]
Funke, P. T., **6**, 644[93]
Furber, M., **1**, 132[106], 752[95]; **4**, 896[169]; **7**, 90[31], 367[54], 375[54], 552[55]
Furin, G. G., **8**, 546[310]
Furlani, D., **2**, 564[11]
Furlenmeier, A., **7**, 86[16a]
Furneaux, R. H., **6**, 48[85]
Furniss, B. S., **7**, 555[70]
Fürst, A., **2**, 167[161]; **4**, 7[25], 30[89]; **7**, 86[16a]; **8**, 228[124], 367[58], 544[257]
Furst, G. T., **4**, 795[87]; **8**, 846[81]
Fürstner, A., **1**, 212[6], 213[6], 271[62,62b]; **2**, 280[21], 294[81]; **3**, 570[55,129], 582[55], 583[55,129], 630[57], 631[57]; **6**, 978[22]
Furstoss, R., **7**, 59[42,43], 60[43,45,46a,47a,b], 62[47c], 64[60,61b], 78[126], 429[157b], 503[280,281]
Furtek, B. L., **3**, 901[110]
Furth, B., **5**, 178[135]
Furuhashi, K., **7**, 429[155]
Furuhata, T., **6**, 440[77]
Furuichi, A., **8**, 195[106], 197[106]
Furuichi, K., **7**, 550[49]
Furukawa, I., **6**, 172[24]
Furukawa, J., **4**, 93[95], 601[251], 602[251], 643[251], 968[57,58], 970[58b,70], 973[58b]; **5**, 35[12], 56[51]; **7**, 400[51]; **8**, 292[39]
Furukawa, K., **6**, 766[21]
Furukawa, M., **2**, 922[103,104]; **3**, 832[69]; **4**, 231[275]; **5**, 102[164]; **6**, 119[116], 817[103]
Furukawa, N., **1**, 825[50]; **3**, 510[182]; **4**, 335[27], 427[69], 487[46]; **6**, 934[98]; **7**, 124[46], 205[63], 425[149b], 470[10,11], 498[230b], 762[80], 764[121], 777[80], 778[395]; **8**, 410[88,93]
Furukawa, S., **1**, 474[104], 477[137,138], 568[239]; **3**, 135[360,361], 136[360,361], 137[360,361], 139[360,361], 142[360,361], 143[360,361], 155[429], 156[360,361]; **4**, 497[99]; **6**, 893[82]
Furukawa, Y., **2**, 215[134]; **4**, 75[44b], 379[114,114b], 382[114b], 383[114b], 413[114b]; **6**, 501[185], 531[185]; **7**, 862[79]
Furumai, S., **8**, 392[97]
Furusaki, A., **3**, 400[119], 404[135]; **4**, 817[202]; **8**, 604[3]
Furusaki, F., **8**, 395[129]
Furusako, S., **8**, 375[155]
Furusato, M., **3**, 96[168], 104[168], 108[168], 117[168]; **8**, 966[74]
Furusawa, F., **3**, 168[494,504], 169[494,504], 170[494,504]
Furusawa, K., **6**, 606[42]
Furuta, K., **1**, 78[12], 161[81,82], 165[108,110], 509[23], 827[64a]; **2**, 22[87], 72[59], 91[46], 93[46], 94[50], 103[29]; **3**, 446[88]; **4**, 107[140d], 814[186], 976[96,97]; **5**, 355[87b], 377[111,111a,b]; **7**, 318[51]
Furuta, T., **5**, 225[103,104], 226[104]
Furutachi, N., **5**, 222[64], 223[64]
Furutani, H., **8**, 193[102], 195[104b]
Furuya, T., **4**, 208[62]
Furuyama, H., **2**, 547[121], 551[121], 552[121]; **5**, 847[136], 1032[100]; **8**, 534[158], 537[158]
Fusaka, T., **4**, 1056[141,141b]
Fusco, C., **7**, 13[125]
Fusco, R., **4**, 1085[98], 1099[187]
Fuse, M., **4**, 121[205c]
Fuse, Y., **2**, 587[139]; **3**, 222[143]
Fushiya, S., **8**, 61[190], 66[190]
Fusi, A., **7**, 108[173]
Fuson, R. C., **2**, 152[104]; **3**, 499[111], 503[149], 512[149], 563[1], 890[32]; **4**, 83[66]; **7**, 156[32]; **8**, 532[129]
Fustero, S., **2**, 5[19], 6[19]; **5**, 161[63,64], 480[178], 484[179]; **8**, 13[72], 124[87,88], 696[124]
Futagawa, T., **4**, 975[91]
Fuzesi, L., **2**, 555[144,145]; **3**, 978[11]
Fuzuzaki, K., **1**, 584[10]
Fytas, G., **5**, 578[154]

G

Gaasbeek, M. M. P., **5**, 1187[10]
Gabarczyk, J., **6**, 436[7]
Gabard, J., **2**, 292[78]
Gabe, E., **4**, 892[144]; **7**, 856[65]
Gabel, N. W., **1**, 387[135]
Gabel, R. A., **3**, 261[156], 512[202]; **4**, 428[70]
Gabhe, S. Y., **1**, 582[7]; **2**, 572[42]; **7**, 330[9]
Gable, R. A., **1**, 474[108]
Gäbler, M., **6**, 551[685]
Gabmeier, J., **3**, 380[7]
Gabor, G., **4**, 608[318]
Gabriel, J., **1**, 506[9], 631[57]; **3**, 87[75], 987[30]; **6**, 876[33], 882[33], 887[33]
Gabriel, S., **2**, 400[24]
Gadalla, K. Z., **2**, 790[58,59]; **6**, 487[45], 489[45], 500[178], 573[45]
Gadallah, F. F., **3**, 507[170]
Gadamasetti, K., **5**, 930[177], 931[177], 938[177], 940[224], 943[251], 947[177], 951[177], 963[251]
Gadek, T., **1**, 377[101]; **5**, 1144[99]
Gadelle, A., **7**, 247[104]
Gaden, H., **6**, 49[93]
Gadi, A. E., **2**, 429[47]
Gadient, F., **2**, 331[53], 332[53]
Gadola, M., **5**, 10[76]
Gadras, A., **8**, 409[82]
Gadre, S. R., **6**, 1022[64]
Gadru, K., **8**, 113[39], 115[39]
Gadwood, R. C., **1**, 648[127,-128], 649[127,-128], 650[127,-128], 672[127,-128], 675[127,-128], 679[127,-128], 708[127,-128], 710[127,-128], 715[127,-128], 716[127,-128], 862[78], 881[119], 883[127], 884[127]; **3**, 49[264], 87[114], 194[15], 196[15], 787[45]; **4**, 159[80], 246[96], 258[96], 260[96]; **5**, 806[101-103], 1026[86], 1027[89]; **6**, 901[123]; **7**, 673[20]
Gaertner, R., **8**, 972[116]
Gaeta, F. C. A., **6**, 109[43], 1066[92]
Gaetano, K., **4**, 288[188], 346[86a]
Gaeumann, T., **7**, 3[5]
Gaevoi, E. G., **6**, 439[70]
Gaft, Yu. L., **8**, 770[34]
Gagel, K., **5**, 596[26], 597[26]
Gaggero, N., **7**, 194[9], 429[150a]
Gaginkina, E. G., **3**, 635[31]
Gagné, M. R., **4**, 410[263]
Gagne, P., **4**, 476[155]
Gagnier, R. P., **3**, 419[29,30], 422[29,30]
Gagosian, R. B., **1**, 847[24]; **3**, 892[52]; **5**, 597[29]
Gahan, L. R., **4**, 298[292]
Gai, Y.-Z., **8**, 374[139], 377[168]
Gaiani, G., **5**, 423[89]
Gaibel, Z. L. F., **4**, 1013[181]; **6**, 971[128]
Gaiffe, A., **8**, 299[81]
Gaillot, J.-M., **3**, 866[30]
Gaines, D. F., **1**, 13[67]
Gainor, J. A., **3**, 511[191]
Gains, L. H., **3**, 736[28]
Gainsford, G. J., **1**, 786[250]
Gainullina, E. T., **1**, 520[68]
Gair, I. A., **8**, 197[126]
Gais, H.-J., **1**, 37[244,247,248,249,250,251,252], 528[120,121], 531[131,132], 535[145,147], 536[145,147], 773[203,203b], 788[256], 805[311]; **2**, 76[84]; **3**, 159[466], 166[466], 174[530], 230[237]; **5**, 116[264]; **6**, 438[54], 463[28], 464[29], 998[121]
Gaisin, R. L., **4**, 599[213], 640[213]
Gaitanopoulos, D. E., **5**, 404[20], 405[20]
Gaj, B. J., **2**, 183[15]
Gajda, T., **6**, 267[55]
Gajek, K., **5**, 803[87], 971[1]
Gajewski, J. J., **4**, 1036[48]; **5**, 515[17], 518[17], 547[17,17c], 699[3], 700[3], 821[163], 826[158,158b], 827[2], 829[2], 841[99], 854[176,179], 855[176,179], 856[99,176,201,203,207,208,215,219], 857[224,225,230], 859[239], 877[8], 972[7], 1016[60]
Gajewski, R. P., **4**, 128[221]; **5**, 152[22]
Gakh, A. A., **4**, 347[103]
Gal, J., **6**, 291[218,219]
Gala, K., **5**, 95[89], 96[114]
Galamb, V., **3**, 559[56]
Galambos, G., **2**, 381[305]
Galarini, R., **2**, 338[76]
Galat, A., **2**, 357[138]; **5**, 736[142k]
Galatsis, P., **5**, 345[67d]
Galbo, J. P., **7**, 710[54]
Galbraith, H. W., **8**, 366[34]
Galdecki, Z., **4**, 83[65c]
Gale, A., **3**, 816[79]
Gale, D. P., **7**, 231[137]
Galeazzi, E., **4**, 1061[166]
Galeeva, R. G., **4**, 41[119]
Galeffi, B., **5**, 850[152]
Galemmo, R. A., Jr., **2**, 838[169]; **5**, 21[154], 22[154]
Galesloot, W. G., **4**, 309[412]
Galiano-Roth, A. S., **2**, 507[20], 509[20]; **6**, 727[198]
Galin, F. Z., **8**, 680[73], 683[73]
Galindo, A., **3**, 391[90], 393[90], 395[97]
Galindo, J., **1**, 417[70], 418[74]; **5**, 456[85]
Galinovsky, F., **8**, 273[124,125]
Gall, M., **1**, 3[18,19]; **2**, 128[242], 183[17], 184[23,24], 599[23], 835[158]; **3**, 4[18], 7[18], 8[18], 11[18]; **7**, 130[76]; **8**, 30[44], 66[44]
Galla-Bobik, S. V., **4**, 342[66]
Gallacher, G., **4**, 1076[42]
Gallacher, I. M., **6**, 441[82]
Gallagher, D. W., **8**, 399[148]
Gallagher, J., **5**, 736[145]
Gallagher, M. J., **8**, 859[211]
Gallagher, P. T., **2**, 765[76], 1072[140]; **5**, 167[94], 843[124]; **6**, 859[173]; **7**, 21[6], 318[60]; **8**, 196[122]
Gallagher, S. R., **4**, 545[126]
Gallagher, T., **2**, 73[67], 655[140]; **3**, 673[70], 674[70b]; **4**, 308[410], 397[210], 412[268b-e,269,270,270b], 413[268c,269b], 562[34]; **5**, 385[130]; **6**, 930[85]; **7**, 199[37]; **8**, 623[148]
Gallagher, T. F., **8**, 991[51]
Gallardo, T., **1**, 248[69]
Gallazzi, M. C., **4**, 602[258]; **5**, 36[15]
Galle, J. E., **1**, 329[34]; **3**, 165[479], 173[479,519], 198[50], 759[134]; **4**, 79[57], 89[83], 192[120], 295[258], 297[268], 877[65], 878[77], 879[65,77]; **8**, 734[5], 742[41], 747[57], 751[57]
Gallego, C. H., **1**, 743[54], 746[54], 748[54]; **5**, 1115[2], 1116[2], 1122[2b], 1123[2b], 1124[2b]
Gallego, M. G., **5**, 202[36]
Gallegos, G. A., **2**, 367[230]
Gallenkamp, B., **7**, 753[158,159]
Gallezot, P., **8**, 436[73]
Galli, C., **3**, 55[284]; **4**, 453[32,33], 457[33], 458[32], 459[69,88], 464[69,88,113], 758[187], 804[138], 811[138]; **6**, 69[17], 208[54], 211[54]

Galli, G., **2**, 736[29]
Galli, R., **4**, 1008[134]; **6**, 171[6]; **7**, 488[153], 506[296]
Gallina, C., **3**, 213[45], 220[118], 222[118]
Gallinella, E., **7**, 500[239]
Gallis, D. E., **1**, 837[153]
Gallivan, R. M., Jr., **8**, 707[23], 934[55]
Gallmeier, H. J., **5**, 436[157]
Gallo, R., **6**, 510[294]
Gallois, P., **8**, 140[10], 141[10], 142[10], 418[14], 533[138]
Gallos, J. K., **3**, 282[47]
Galloway, J. G., **4**, 116[188b]
Galloy, J., **7**, 778[398]
Gallucci, J. C., **1**, 240[43]; **2**, 926[115], 937[115]; **4**, 374[91], 790[41], 791[41,48], 1010[157]; **5**, 100[152], 857[230]; **7**, 163[76], 164[76], 647[31]; **8**, 447[129], 463[129], 514[107], 861[224]
Gallucci, R. R., **3**, 897[92], 900[92]; **8**, 925[11], 926[11]
Galpern, E. G., **7**, 800[35]
Galt, R. H. B., **5**, 461[101], 463[101]
Galteri, M., **7**, 686[98]
Galton, S. A., **4**, 42[122a]; **8**, 995[70]
Galust'yan, G. G., **7**, 7[47]
Galuszko, K., **3**, 927[48]
Galvez, C., **4**, 478[168]
Galy, J. P., **3**, 124[257], 127[257]
Galyer, A. L., **1**, 215[31,31a]
Galynker, I., **3**, 41[226]
Gamasa, M. P., **3**, 824[19]; **8**, 86[23]
Gamba, A., **2**, 351[79], 364[79]; **5**, 78[281]
Gambacosta, A., **4**, 1021[246]
Gambale, R., **5**, 780[200]
Gambaro, A., **2**, 6[29,31,32], 18[31], 564[12], 566[20], 726[127]; **4**, 527[67]
Gambarotta, S., **8**, 683[90]
Gambaryan, N. P., **5**, 113[227,234]; **6**, 498[169], 527[409], 547[667], 552[667]; **7**, 800[35]
Gamboni, R., **7**, 162[56], 180[160]
Gammill, R. B., **3**, 24[146], 969[133]; **5**, 683[35]; **6**, 501[205], 893[88]; **7**, 452[57], 462[123], 571[119], 577[119]; **8**, 542[221,238]
Gamoh, K., **7**, 366[52]
Gamper, N. M., **4**, 317[548]
Gampp, H., **7**, 766[187]
Gan, S.-n., **6**, 89[107]
Ganazzoli, F., **5**, 1147[112]
Ganboa, I., **5**, 95[98]; **7**, 278[159], 695[34]
Gancarz, R. A., **7**, 769[244]
Gande, M. E., **4**, 1036[48]; **5**, 854[176], 855[176], 856[176]
Gander-Coquoz, M., **3**, 40[221], 41[227]
Gandhi, R. P., **5**, 687[62], 688[62]
Gandhi, S. S., **4**, 505[132,133]; **6**, 819[110]
Gandillon, G., **5**, 913[108]
Gandin, J., **4**, 602[262], 644[262]
Gandolfi, C., **4**, 106[138e]
Gandolfi, M., **7**, 828[52]
Gandolfi, R., **4**, 1093[147]; **5**, 78[281], 257[60], 625[31], 626[33]
Gandolfi, V., **2**, 369[245]
Gandour, R. D., **3**, 591[170]; **4**, 395[201]
Gandour, R. W., **4**, 1082[85], 1085[85]
Ganellin, C. R., **4**, 33[95,95a]
Ganem, B., **1**, 329[36], 343[105,118], 619[63]; **2**, 904[51]; **3**, 22[132], 224[169], 262[169], 1055[32]; **4**, 8[27], 240[53], 254[183], 260[183], 367[13], 369[19,23], 373[74], 374[19], 398[213], 404[247], 405[247a,b], 1036[48]; **5**, 335[49], 854[176], 855[176], 856[176]; **6**, 5[23,27], 95[147], 116[84], 685[364], 807[60]; **7**, 299[42], 367[58], 403[64], 406[72], 503[272], 545[27], 636[72], 656[13], 745[79], 763[89]; **8**, 29[40,41], 36[78], 37[78], 39[78], 44[78], 46[78], 54[78], 56[170], 64[199], 66[40,41,78,170,199], 253[119], 254[123], 355[181], 375[157], 499[44], 536[176], 794[11], 798[62], 800[62], 839[24], 890[143], 929[32]
Gange, D., **2**, 88[29]
Ganguli, A. N., **7**, 318[52], 319[52]
Ganguli, B. N., **7**, 64[63,64]
Ganguly, A. K., **3**, 253[89], 261[147], 262[89], 741[50]; **4**, 347[104]
Ganguly, R., **3**, 946[93]
Ganguly, S. N., **6**, 546[655], 552[655]
Gani, D., **7**, 673[28]
Ganis, P., **2**, 564[9], 566[20]; **4**, 744[131]
Gannett, P., **1**, 429[126], 797[283], 798[289]; **6**, 995[101]; **7**, 555[69]
Gannon, W. F., **1**, 846[11]
Gano, J. E., **5**, 65[66]
Gans, R., **3**, 927[56]
Gänshirt, K. H., **8**, 494[24]
Gansser, C., **7**, 693[25]
Ganter, C., **1**, 857[58]; **3**, 390[83], 392[83]; **4**, 373[86], 399[86,222]; **5**, 165[83], 596[26], 597[26]
Gantert, S., **8**, 853[144]
Gants, A., **8**, 756[138]
Ganyushkin, A. V., **7**, 641[2]
Ganz, C. R., **3**, 380[16]
Gao, Y., **3**, 223[156]; **7**, 390[4], 393[4], 394[4], 395[4], 396[4], 397[4], 398[4,32], 399[4], 400[4], 401[4], 406[4], 407[4], 410[4], 411[4], 413[4], 431[160,162]; **8**, 879[54]
Gaoni, Y., **5**, 771[144]; **6**, 990[86], 991[86]; **8**, 384[25]
Gapinski, D. M., **2**, 64[25], 478[14]
Gapinski, D. P., **5**, 428[111]
Gapinski, R. E., **3**, 103[204], 108[204]; **4**, 116[185c]; **6**, 147[86]
Gapski, G. R., **6**, 923[54,55]
Gara, W. B., **4**, 796[91]
Garad, M. V., **1**, 488[11], 492[37], 494[37], 501[37]
Garai, C., **5**, 96[121], 98[121]
Garanti, L., **4**, 1085[98,99,103]; **6**, 252[152]
Garapon, J., **7**, 498[225], 537[56,57]
Garbacik, T., **3**, 30[173]; **6**, 712[77]
Garbarino, G., **4**, 426[58], 457[63,64], 460[64], 475[148], 476[64,158,159,160,161]; **6**, 240[80]
Garbe, W., **1**, 808[320]
Garbesi, A., **3**, 179[547], 181[547]
Garbisch, E. W., **3**, 851[63]; **8**, 425[48], 474[14], 475[14], 476[14]
Garburg, K.-H., **6**, 943[156]
Garcea, R. L., **2**, 421[26]
Garcia, A., **2**, 345[18], 359[18], 360[18]
Garcia, B., **3**, 87[99], 104[99]; **4**, 768[239]; **7**, 732[59]
Garcia, B. A., **6**, 212[81]
Garcia, B. S., **4**, 438[149]
García, E., **5**, 1059[54], 1060[55], 1062[54c]
Garcia, G. A., **1**, 552[81,83]; **2**, 183[18]; **3**, 198[49]; **8**, 544[262]
Garciá, H., **2**, 747[116,118]
Garcia, J., **2**, 35[124a], 42[124], 244[25,26], 258[51]; **6**, 77[54]
García, J. I., **2**, 406[45]
Garcia, J. L., **4**, 517[5], 518[5], 535[93], 537[95], 538[93], 539[93]
Garcia, J. M., **2**, 649[102], 1059[75]; **8**, 125[94]
Garcia, J. N., **4**, 538[103]
Garcia, L. A., **4**, 1072[18]
Garcia, M. C., **4**, 1036[52]
García, M. L., **5**, 1059[54]
Garcia, T., **6**, 490[101,103]
Garcia-Garibay, M., **5**, 211[62,63]
Garcia-Luna, A., **7**, 752[152]
Garcia Mendosa, P., **2**, 348[60], 357[60]
Garcia Muñoz, G., **6**, 275[108], 277[127], 280[141]
Garcia-Ochoa, S., **1**, 759[132]
Garcia-Oricain, J., **4**, 527[69], 528[69]
Garcia-Raso, A., **3**, 825[24,24b], 835[24]; **7**, 346[8]

Garcia Ruano, J. L., **6**, 149[105]; **8**, 15[91]
Garcia Segura, R., **2**, 386[329]
Gard, G. L., **7**, 267[121], 269[121], 270[128], 271[121,128], 278[121]
Gardano, A., **3**, 1026[37], 1033[72]
Gardein, T., **1**, 34[224], 36[174]
Gardent, N. J., **8**, 973[119]
Gardette, M., **3**, 249[61], 485[35], 486[35], 494[35]; **4**, 898[177], 903[199]
Gardini, G. P., **7**, 16[160]
Gardlik, J. M., **1**, 429[123]; **3**, 593[177]
Gardlund, S. L., **4**, 872[39]
Gardlund, Z. G., **4**, 120[197], 868[17], 869[17], 872[39]
Gardner, D. V., **4**, 674[36]
Gardner, H. C., **7**, 874[102]
Gardner, J. D., **2**, 183[13], 764[62,63]
Gardner, J. H., **8**, 254[124]
Gardner, J. N., **6**, 268[64]; **7**, 160[47]; **8**, 407[57]
Gardner, J. O., **3**, 48[260]; **4**, 83[65c]
Gardner, P. D., **2**, 149[84], 736[28]; **4**, 276[80], 303[346], 304[346], 310[425], 1014[191]; **6**, 970[127]; **7**, 167[99]; **8**, 530[95], 708[31], 806[110], 807[110,119]
Gardner, R. J., **8**, 70[226]
Gardner, S. A., **5**, 520[39]
Gardner, T. S., **7**, 666[77]
Gardnier, B., **7**, 825[44], 833[77]
Gardocki, J. F., **8**, 36[75], 37[75], 38[75], 39[75], 45[75], 54[75], 66[75]
Gardrat, C., **4**, 753[166]
Garegg, J., **8**, 969[96]
Garegg, P. J., **6**, 34[11], 43[54], 46[67,70], 47[67,81], 51[11], 205[32], 620[133], 660[205,206]; **7**, 237[32], 259[59]; **8**, 224[105,107]
Gareil, M., **4**, 453[29], 459[29], 472[29]
Gareyan, L. S., **4**, 315[507]
Garg, C. P., **7**, 253[13], 600[77], 601[77]; **8**, 271[111], 274[133]
Gargano, M., **8**, 450[171]
Gargano, P., **6**, 69[17]
Garibay, M. E., **8**, 537[187]
Garibdzhanyan, B. T., **6**, 507[237], 515[237]
Gariboldi, P., **2**, 833[148]; **4**, 261[285,288,290]
Garigipati, R. S., **1**, 238[34], 404[21]; **2**, 542[84]; **3**, 511[191]; **4**, 14[47,47o]; **5**, 414[54], 424[98], 425[99], 426[104], 539[108]; **6**, 894[90], 900[112]; **7**, 491[181]
Garin, D. L., **3**, 767[161]
Garito, A. F., **6**, 510[291]
Gar'kin, V. P., **8**, 410[97]
Garland, R., **3**, 42[231]
Garland, R. B., **7**, 352[29]
Garlaschelli, L., **5**, 1147[112]
Garmaise, D. L., **3**, 781[19]
Garner, A. Y., **4**, 1002[49]
Garner, B. J., **7**, 299[50]
Garner, H. K., **6**, 215[101]
Garner, P., **1**, 759[133]; **2**, 699[83]; **4**, 1041[104]; **5**, 344[67a,b], 349[76], 854[175]; **7**, 407[80], 569[107]
Garner, R., **6**, 668[254], 669[254]
Garnier, B., **5**, 920[132]
Garnovskii, A. D., **6**, 226[13]
Garratt, D. G., **4**, 329[1], 330[1c,4], 339[41], 341[57], 342[62,63,65,69,71], 344[1], 350[1], 351[1]; **7**, 520[25,28,29], 521[36], 769[230]
Garratt, P. J., **2**, 376[279]
Garraway, J. L., **4**, 50[142]
Garrou, P. E., **8**, 859[214]
Garry, S. W., **8**, 33[57], 66[57]
Garsky, V., **2**, 538[62], 539[62]
Garst, J. F., **2**, 798[9]; **6**, 959[41]; **8**, 795[25]
Garst, M. E., **1**, 822[31,32]; **2**, 102[22]; **3**, 54[276], 370[112], 537[89]; **4**, 1052[129]; **5**, 618[7], 620[7], 621[21], 624[29], 933[183]; **6**, 176[86]
Gartenmann, T. C. C., **3**, 583[118]
Garti, N., **8**, 421[29], 422[29], 436[74], 437[76,77]
Gartiser, T., **8**, 384[27]
Gärtner, J., **3**, 383[44]; **6**, 267[58,59]
Gärtner, K.-G., **6**, 607[43], 612[77]
Garver, L., **4**, 311[451]
Garvey, B. S., **8**, 140[20], 150[20]
Garvey, D. S., **1**, 410[43,44], 436[153]; **2**, 2[7], 100[13], 113[102], 224[153], 240[13], 242[18], 245[18b], 246[18b], 253[42], 256[13], 303[5], 630[9], 926[117]; **5**, 832[41]; **7**, 257[52]
Garwood, R. F., **3**, 635[33], 640[107,107a], 647[33,107]; **7**, 306[9], 801[44]; **8**, 974[128]
Gary, J. A., **1**, 131[99]; **6**, 859[174]
Garza, T., **5**, 65[60]
Gasanov, R. G., **8**, 765[11]
Gasc, J. C., **3**, 21[131]
Gasc, M. B., **4**, 290[199]; **7**, 470[1], 488[1], 490[1]; **8**, 856[175]
Gasdaska, J., **1**, 821[27]; **4**, 1081[79]; **6**, 542[602]
Gase, R. A., **8**, 93[70,72], 94[78], 561[418,419]
Gaset, A., **1**, 821[20-22]; **2**, 354[108], 404[40], 772[17], 775[29]; **6**, 173[46], 175[46,71,73]
Gasic, G. P., **4**, 413[273]
Gasiecki, A. F., **4**, 384[142]
Gasiorek, M., **8**, 606[22]
Gaskell, A. J., **8**, 587[34]
Gaspar, P. P., **5**, 950[285]; **8**, 765[16]
Gasparrini, F., **6**, 727[190]
Gasperoni, S., **2**, 524[82,83]
Gassman, P. G., **1**, 118[61], 648[137], 654[137], 655[137], 846[17]; **3**, 334[218], 614[5], 649[206], 876[78-80], 890[31], 901[112], 903[99], 905[138], 969[131,132]; **4**, 285[161], 430[94,95,97]; **5**, 341[58], 520[37,38,40,41], 521[42], 522[41], 585[203], 856[216], 936[200], 1186[4]; **6**, 237[64], 243[64], 657[181], 672[181], 736[24]; **7**, 125[55], 126[55], 208[81], 476[62], 794[5], 874[104], 878[145]; **8**, 899[27], 994[62]
Gassner, T., **1**, 563[173]; **6**, 172[11]
Gastambide, B., **7**, 169[113]
Gasteiger, J., **5**, 71[123]
Gastiger, M. J., **7**, 13[115,116,119]
Gaston, L. K., **7**, 724[29]
Gaston, R. D., **5**, 527[63], 534[90], 535[90]
Gateau-Oleskar, A., **1**, 561[166]; **4**, 748[156]
Gatenbeck, S., **6**, 811[76]
Gates, B. C., **7**, 840[7]
Gates, J. W., Jr., **3**, 747[72]; **8**, 912[90]
Gates, M., **8**, 330[28,29]
Gatilov, Y. V., **3**, 386[68]
Gatterman, L., **3**, 582[112]
Gatti, A., **2**, 737[39]; **8**, 277[153]
Gatti, N., **8**, 900[30]
Gattow, G., **6**, 233[49], 552[694]
Gattuso, M., **4**, 387[160]
Gaude, D., **6**, 110[54]
Gaudemar, F., **4**, 55[156]
Gaudemar, M., **1**, 214[22,22a], 218[22,49], 220[22,49,65], 223[49]; **2**, 81[2,4,5], 122[193], 183[19], 277[5], 279[5,15], 280[28], 281[5], 283[49], 284[55], 286[65,66], 287[65], 291[75], 292[76], 294[84], 296[84], 486[38], 630[22], 631[22], 632[22], 634[22], 635[51], 640[51], 983[28], 989[34,35], 990[34], 992[36], 993[36]; **4**, 89[84f], 95[102b,e,g], 880[91], 883[91,101], 884[101,105]; **5**, 100[145]; **6**, 425[69]
Gaudemar-Bardone, F., **1**, 218[51]; **2**, 284[55], 291[75]; **4**, 53[148], 89[84f], 95[102e]

Gaudemer, A., **2**, 345[28], 346[28]; **4**, 746[147]; **7**, 451[35], 462[35]; **8**, 674[34]
Gaudemer, F., **7**, 451[35], 462[35]
Gaudenzi, M. L., **3**, 217[91], 539[103]
Gaudin, J.-M., **1**, 767[179]; **5**, 46[38], 47[41], 48[38,41], 49[38], 50[38,41,43], 51[45,45a], 53[45a], 55[49], 56[49]; **7**, 229[120]
Gaudino, J., **1**, 759[134], 832[106]; **3**, 511[188,212], 515[188,212]; **4**, 732[76]
Gaudry, M., **2**, 902[47], 903[47]
Gaudry, R., **8**, 527[43,45], 528[45], 529[45], 530[45]
Gaugler, R. W., **6**, 263[21]
Gaukhman, A. P., **8**, 771[45]
Gaul, M. D., **3**, 1061[46], 1062[46]; **4**, 501[114]; **5**, 854[175]
Gault, F. G., **8**, 142[48], 896[15]
Gault, H., **2**, 147[73]
Gault, R., **6**, 288[185,191]
Gault, Y., **7**, 6[29]
Gaumont, Y., **6**, 812[79,80]
Gauntlett, J. T., **1**, 310[109]
Gaur, J. N., **7**, 705[15]
Gaus, P. L., **8**, 22[146], 289[28], 371[114]
Gautheron, B., **6**, 464[31,32]; **8**, 447[120,129], 463[129], 683[87,94], 688[99], 690[101], 691[99]
Gautheron, C., **2**, 463[90], 464[90,95,95a,b,96], 467[90]
Gauthier, C., **2**, 736[21]
Gauthier, J., **2**, 169[167]
Gauthier, J. Y., **6**, 26[109], 927[70]
Gauthier, R., **6**, 529[424]
Gautier, A., **2**, 1083[3], 1084[3]; **8**, 457[213]
Gautier, H., **4**, 469[136]
Gautier, J.-A., **2**, 957[20]; **6**, 488[20], 517[20], 546[20], 548[20], 549[20]
Gautschi, F., **3**, 273[13]; **6**, 1059[64]
Gauvreau, A., **8**, 252[109]
Gavai, A. V., **2**, 638[61], 640[61]; **3**, 220[126]; **4**, 148[50]; **6**, 848[109]; **8**, 171[106]
Gavars, R., **8**, 595[77]
Gavens, P. D., **1**, 139[3]
Gavina, F., **4**, 492[73]; **5**, 474[158]; **6**, 172[19]
Gaviraghi, G., **7**, 747[105]
Gavrilenko, V. V., **5**, 1148[115]; **8**, 267[61], 271[108], 274[135], 698[147], 735[14], 741[36,38], 742[46], 747[14], 748[14], 754[78,80,99]
Gavrilov, L. D., **4**, 55[156]
Gawienowski, J. J., **4**, 95[98]
Gawinecki, R., **6**, 110[46]
Gawley, R. E., **1**, 476[127], 477[129,130,131]; **2**, 156[113], 162[113], 506[17], 523[85]; **3**, 35[205], 66[16], 67[16], 74[16,42], 75[42-44,51], 76[43,53], 77[42,57], 78[51], 80[43,65], 81[65], 685[108]; **4**, 3[8], 5[17], 6[8], 99[118c]; **6**, 727[204], 763[6], 771[39-41]; **7**, 227[78-80], 230[78], 689[11], 691[11], 695[11], 697[48], 698[11], 699[11], 700[11]
Gawronska, K., **4**, 231[270,271]
Gawronski, J., **4**, 231[270,271]
Gawronski, J. K., **2**, 297[89]; **7**, 262[81]
Gaydoul, K. R., **1**, 635[91], 637[91], 672[91], 678[91]
Gaylord, N. C., **8**, 278
Gaylord, N. G., **3**, 262[159], 264[159]; **8**, 26[4], 27[4], 36[4], 70[4], 213[26], 650[63], 812[1]
Gaythwaite, W. R., **7**, 771[257]
Gazarov, T. Sh., **8**, 772[57]
Gazis, E., **6**, 644[86]
Gazit, A., **2**, 357[150]
Ge, Y., **2**, 772[18]; **8**, 447[135]
Gebauer, H., **7**, 799[25]
Gebelein, C. G., **7**, 501[252,255]
Gebhard, J. S., **8**, 838[19]
Gebhart, H. J., Jr., **3**, 721[5]
Gebreyes, K., **3**, 878[94], 879[94,97], 880[94], 881[94]; **4**, 359[159], 771[251]
Gebrian, J. H., **5**, 716[85]
Geckler, R. D., **4**, 4[14]
Géczy, I., **8**, 140[17]
Gedge, S., **6**, 562[884]
Gedheim, L., **7**, 799[25], 800[31]
Gedye, R. N., **2**, 287[68]
Gedymin, V. V., **3**, 898[81]
Gee, K. R., **5**, 856[208,215]
Gee, S. K., **5**, 154[33], 683[82], 689[73,78,78a,b], 690[82], 732[135,135a], 733[135b], 806[106], 1025[82], 1026[82]
Gee, V., **4**, 1040[99]
Geenen, J. J. H., **8**, 56[169], 66[169]
Geer, S. M., **6**, 516[318]
Gees, T., **2**, 1086[28], 1096[28]
Geeseman, D., **5**, 455[74]
Geetha, K. V., **3**, 380[7]
Geffroy, G., **5**, 417[65], 419[74]; **8**, 652[79]
Gehlhaus, J., **5**, 75[217], 78[217]
Gehret, J.-C. E., **2**, 809[57]
Gehrlach, E., **5**, 605[62]
Gehrt, H., **1**, 391[153]
Geib, G. D., **1**, 743[54], 746[54], 748[54]; **5**, 1115[2], 1116[2], 1122[2b], 1123[2b], 1124[2b]
Geiger, G. A. P., **1**, 19[102]
Geiger, R., **6**, 635[11], 642[68], 645[11]
Geiger, R. E., **5**, 1154[155]
Geiger, W., **1**, 364[39]; **6**, 551[686]
Geijo, F., **1**, 543[26]
Geiman, I. I., **8**, 606[24], 607[24]
Geingold, D. S., **2**, 466[113], 467[113]
Geise, B., **4**, 735[87]
Geise, H. J., **3**, 583[122], 584[130], 587[148]
Geise, H. Y., **6**, 985[64]
Geisel, M., **7**, 724[30]
Geiser, F., **5**, 154[33]
Geiss, F., **3**, 890[35]
Geiss, K.-H., **1**, 510[25,26], 630[18], 826[62,63]; **2**, 55[1], 72[78]; **3**, 86[18], 94[18], 95[18,158,178], 96[158,178], 97[158,178], 99[158,178], 121[18,158], 144[384], 145[384], 147[384]; **6**, 833[22]
Geissler, G., **3**, 825[24,24a], 828[24a], 835[24]
Geissler, M., **1**, 19[104], 20[108]
Geissman, T. A., **3**, 754[107], 757[107]; **8**, 86[20]
Geister, B., **8**, 70[225], 71[225]
Geistlich, P., **8**, 530[89]
Geiszler, A. O., **3**, 813[64]
Geittner, J., **4**, 1102[200]
Gelan, J., **5**, 637[109], 1130[7]
Gelas, J., **6**, 660[202]; **8**, 211[5], 227[119]
Gelas-Mialhe, Y., **3**, 866[30]; **4**, 48[140]; **5**, 938[207,211]
Gelb, M. H., **7**, 545[26]
Gelbard, G., **6**, 105[11]; **7**, 280[173], 281[173], 283[173,184], 285[173], 840[4], 844[4,63], 845[4,63]; **8**, 97[97], 563[425,426]
Gelbein, A. P., **8**, 285[8], 293[8]
Gelfand, S., **8**, 336[75], 341[75,101], 926[17,18], 927[18]
Gelin, R., **3**, 318[127]
Gelin, S., **3**, 318[127]; **5**, 766[116]; **6**, 807[56]
Gell, K. I., **5**, 1173[31], 1178[31]; **8**, 447[121], 675[43-45], 677[60], 684[43]
Geller, J., **2**, 1102[124]
Gellerman, B. J., **7**, 532[31]
Gellert, E., **3**, 507[171]
Gellert, H. G., **3**, 194[5]; **4**, 867[8], 887[122], 888[133]; **8**, 734[2,8], 735[10], 736[19], 737[2], 739[19], 744[50], 753[2], 756[50]
Gellert, R. W., **1**, 41[266]

Gellibert, F., **1**, 64[45]
Gelmi, M. L., **8**, 72[241], 74[241]
Gemal, A. L., **3**, 249[65]; **5**, 176[127]; **6**, 676[302], 931[93]; **8**, 16[96], 17[115-117], 538[193], 540[193,196,197], 988[27]
Gembitskii, P. A., **5**, 938[208]
Gemenden, C. W., **3**, 124[270], 127[270], 128[270], 129[270]
Gemmer, R. V., **5**, 64[38]
Gemroth, T. C., **7**, 367[57]
Genco, N., **4**, 579[23]; **5**, 274[7], 275[7], 277[7], 279[7]; **8**, 392[105]
Gendler, P. L., **6**, 812[81]
Gendreau, Y., **3**, 470[225], 473[225]
Geneste, P., **2**, 736[21]; **8**, 142[54], 882[84,85], 883[84]
Genet, J.-P., **3**, 47[256]; **4**, 598[183,203,204], 629[417], 638[183,203,204], 651[428], 653[436]; **5**, 373[106,106a], 374[106a], 689[72], 974[16]; **6**, 85[93], 86[93,99], 118[108], 842[81], 849[115,119]; **7**, 229[120]
Genge, C. A., **7**, 24[36], 25[36]
Genin, D., **1**, 838[169]; **5**, 409[37]
Genizi, E., **4**, 710[53,54]
Gennari, C., **1**, 72[72], 524[86-88], 526[100], 527[101,102,107], 528[108], 764[147]; **2**, 221[146], 266[61,62], 267[62-64], 488[43], 514[54], 515[55-57], 605[59], 630[19,21], 631[19,21], 632[21], 634[21], 636[19,56,57], 637[19,56-58], 638[60], 639[58,62], 640[21,56-58,60,62,80], 641[21,71], 642[21,71,74], 643[74], 644[21,80], 645[21,60], 652[60], 653[57,60], 920[98], 930[131,132,133], 931[131], 932[132,133]; **4**, 113[166], 159[82], 218[145], 226[187,188]; **5**, 102[176,178]; **6**, 118[105], 149[96,108]; **7**, 128[63], 396[23], 441[45]
Gennaro, G., **8**, 451[172]
Genoni, F., **8**, 449[157], 450[157]
Gensch, K. H., **7**, 764[110], 778[390]
Gensike, R., **8**, 756[152]
Gensler, W. J., **2**, 801[22]
Genthe, W., **1**, 162[104], 251[2]
Gentile, A., **4**, 768[241]
Gentile, B., **1**, 759[131]; **4**, 35[98d]
Gentile, R. J., **4**, 290[197]
Geoffroy, G. L., **2**, 127[236]; **4**, 976[100]; **5**, 1090[89]
Geoffroy, M., **8**, 61[187], 66[187], 109[8,9], 110[9]
Geoffroy, P., **4**, 250[138]; **5**, 692[100]
Geoghegan, P. J., Jr., **4**, 300[312,313], 301[312,313,314,315], 303[339]; **6**, 284[169]; **8**, 853[145]
Geokjian, P. G., **1**, 188[95], 198[93,95], 199[93]
Georg, G. I., **2**, 648[95], 649[95], 925[110,113], 926[110,114]; **5**, 65[57], 100[151,160], 101[160], 102[171,172], 266[76], 267[76], 268[76]; **6**, 820[112]
George, A. V., **5**, 453[63], 454[63], 458[63]
George, A. V. E., **1**, 471[62]
George, C. F., **3**, 901[116]; **5**, 136[70]; **6**, 836[58]
George, G., **2**, 1093[82,83], 1094[89], 1095[89]
George, I. A., **4**, 794[77]; **7**, 649[43]
George, J., **8**, 375[156]
George, J. K., **4**, 1098[170]; **5**, 626[34]
George, M. V., **4**, 45[129], 52[147,147c]; **5**, 230[130], 232[130], 740[152]; **7**, 231[149], 738[23,25], 746[25], 851[18]
George, T., **5**, 829[22]
George, T. J., **6**, 1013[8], 1017[8]
Georges, M., **2**, 827[127]; **4**, 375[95a,b]
Georgescu, D., **6**, 489[85]
Georghiou, P. E., **8**, 549[327], 696[122], 801[71], 813[11]
Georgian, V., **3**, 834[77], 897[91], 900[91]; **5**, 143[103]; **7**, 236[14]
Georgiou, S., **8**, 103[131]
Georgoulis, C., **1**, 632[62]; **4**, 297[271]; **7**, 272[142], 276[148]; **8**, 753[73]
Gera, L., **6**, 534[519]; **8**, 418[5], 420[5], 423[5], 439[5], 441[5], 442[5]
Geraghty, M. B., **2**, 843[196]; **4**, 955[14]; **8**, 121[79]
Gérard, F., **1**, 219[63]
Gerasimenko, A. V., **7**, 606[160]
Gerasimova, E. S., **6**, 795[12], 798[12], 817[12], 820[12]
Gerba, S., **1**, 894[155]; **4**, 797[102]
Gerbella, M., **5**, 1135[50]
Gerber, H., **6**, 243[99]
Gerber, H.-D., **6**, 453[139]
Gerberding, K., **5**, 557[55]
Gerbing, U., **5**, 442[182,185]; **6**, 480[116]
Gercke, A., **2**, 169[164]
Gerdes, H., **6**, 1042[7]
Gerdes, H. M., **3**, 136[373], 137[373]
Gerdes, J. M., **4**, 17[53,53a]; **5**, 342[62b], 458[71]; **6**, 667[239]
Gerdes, P., **2**, 514[51]; **6**, 119[110]
Gerdil, R., **8**, 231[145]
Gere, J. A., **5**, 211[66]
Gerecs, A., **8**, 612[70], 613[70]
Geresh, S., **2**, 232[175]
Gergmann, E. D., **4**, 187[96]
Gerhardt, C., **5**, 100[160], 101[160]
Gerhart, F., **2**, 1084[11]
Gerhartl, J. F. J., **4**, 26[77], 27[77b]
Gerhold, J., **1**, 632[66]
Geribaldi, S., **4**, 71[19]; **8**, 536[172]
Gerin, B., **4**, 383[139]
Gerken, M., **6**, 60[146], 61[150]
Gerkin, R. M., **3**, 760[136,138,141], 761[141], 762[141], 763[138,141], 764[141]
Gerlach, H., **2**, 844[199]; **3**, 286[60], 747[71]; **6**, 438[65], 667[241], 673[289], 1062[79]
Gerlach, R., **6**, 153[143], 839[70], 902[124]
Gerliczy, G., **8**, 248[89]
Gerlits, J. F., **4**, 812[181]
Gerlt, J. A., **8**, 206[170]
Germain, A., **2**, 709[9,10], 744[90]
Germain, G., **1**, 838[166]; **3**, 625[41]; **5**, 109[217]
German, A. L., **7**, 759[10]
German, L. S., **3**, 647[196]
Germanas, J. P., **4**, 255[199]
Germer, A., **5**, 1185[1], 1190[23]
Germeraad, P., **7**, 35[106]
Germon, C., **1**, 370[68], 371[68], 373[84], 374[84]; **3**, 258[125]; **4**, 903[196,197]
Germroth, T. C., **2**, 161[137]; **4**, 255[193]
Gero, S. D., **1**, 561[166]; **2**, 791[60]; **3**, 126[315,316,317]; **4**, 36[102,102c], 748[156]; **7**, 239[46], 704[13]
Gerrans, G. C., **2**, 882[46], 885[48,49]
Gerritz, S. W., **8**, 371[114]
Gersdorf, J., **5**, 151[15], 154[32]; **6**, 558[853]
Gershanov, F. B., **7**, 750[129]
Gershbein, L. L., **4**, 31[92,92f]
Gerstenberger, M. R. C., **4**, 445[201]; **6**, 204[6]
Gerstmans, A., **3**, 322[144]; **6**, 111[60]; **7**, 760[25], 846[92]
Gerth, D. B., **4**, 738[93]
Gertler, S., **4**, 292[234]; **6**, 264[32]
Gertner, D., **7**, 495[206]
Gertsyuk, M. N., **6**, 543[622], 552[622]
Gervais, D., **6**, 91[125]; **8**, 447[122], 457[122]
Gerval, P., **2**, 600[30]
Gervay, J. E., **3**, 681[96]
Gerwe, R. D., **6**, 923[54,55]
Gerzon, K., **6**, 637[34]
Geschwend, H. W., **3**, 124[271], 125[271], 127[271], 128[271], 129[271]
Gesing, E. R., **4**, 15[50], 16[50e], 252[161]
Gesing, E. R. F., **3**, 583[118]; **5**, 1134[37,39,44]
Geske, D. H., **7**, 603[112]

Gess, E. J., **7**, 372[70]
Gess, N. G., **5**, 76[246]
Gessner, M., **8**, 191[92]
Gesson, J. P., **1**, 567[222]
Getman, D., **5**, 914[114]
Getmanskaya, Z. I., **3**, 643[131]
Getson, J. C., **8**, 231[143], 275[145]
Geueke, K. J., **6**, 839[70]
Geurink, P. J. A., **1**, 10[54]
Geuss, R., **4**, 876[58]
Geuther, A., **2**, 799[12]; **4**, 999[2]
Gevaza, Yu. I., **4**, 364[1,1a-d,g], 367[1g], 368[1a], 373[1a,b], 387[158], 391[177,178a,b], 395[1d], 397[1c], 399[1c], 408[1c,259b], 409[259b], 410[1c], 413[1d], 421[1c]
Gewald, K., **2**, 748[120], 758[26], 850[217,218]; **6**, 430[99]
Gewali, M. B., **1**, 753[105]; **2**, 725[105]
Geyer, E., **3**, 890[34]
Geyer, I., **8**, 654[85]
Geywitz, B., **5**, 605[62]
Ghaderi, E., **7**, 307[13]
Ghadiri, M. R., **3**, 274[25]; **6**, 165[201,202]
Ghalambar, M. A., **2**, 466[112,116], 467[112,116]
Ghanem, K. M., **7**, 71[97]
Ghannam, A., **1**, 133[111]; **6**, 654[155]
Gharbi-Benarous, J., **3**, 728[36]
Gharibi, H., **7**, 286[190]
Ghatak, U. R., **1**, 857[59]; **3**, 87[115], 898[86], 908[146]; **4**, 1040[80,103], 1043[80]; **8**, 248[83], 331[35]
Ghattas, A.-B. A. G., **8**, 541[202]
Ghavshou, M., **4**, 523[55], 524[55], 525[55], 526[55]
Ghawgi, A. B., **3**, 325[158]
Ghazarossian, V. E., **2**, 879[41]
Ghelfi, F., **4**, 337[32]
Ghenciulescu, A., **4**, 963[44]; **6**, 500[183]
Gheorghiu, C. V., **2**, 146[69]
Gheorghiu, M. D., **3**, 331[196]; **5**, 90[54]
Gherardini, E., **1**, 838[161]
Ghiacchio, U., **6**, 508[290]
Ghiaci, M., **2**, 811[67]
Ghigi, E., **2**, 368[242]
Ghiglione, C., **4**, 1002[56]
Ghilardi, C. A., **4**, 170[14]
Ghilezan, I., **7**, 68[84], 72[84]
Ghirardelli, R. G., **6**, 951[5]
Ghiringhelli, D., **3**, 575[84]; **8**, 187[47]
Ghirlando, R., **2**, 881[43]; **6**, 509[276]
Ghisalba, O., **7**, 77[120b]
Ghisalberti, E. L., **5**, 144[104]; **7**, 64[62]
Ghodoussi, V., **4**, 729[62]
Ghomi, S., **7**, 486[145]
Ghosal, S., **3**, 483[15], 500[15]
Ghose, B. N., **3**, 505[168]
Ghosez, A., **4**, 761[203], 805[144]
Ghosez, L., **1**, 683[227], 714[227], 715[227], 717[227], 718[227]; **2**, 60[18,18b], 651[115,115a]; **3**, 162[487], 168[487], 786[42], 896[71]; **4**, 117[191]; **5**, 87[41], 92[41,61], 94[61], 108[209,210,211,212], 109[41,209,210,211,213,214,215,216,217,218,219,220,221,222], 110[210,222,223], 111[210,222], 112[222,223], 113[233,236,238], 116[261,262], 410[40], 416[56], 473[152], 477[152], 480[165,166,168,169], 482[171], 483[165,169]; **6**, 164[195], 430[96], 493[129], 509[257], 520[341,342], 543[341], 544[342], 578[980]; **7**, 122[30], 144[30], 502[262]; **8**, 844[68]
Ghosh, A., **2**, 360[170]; **4**, 239[21], 256[21], 261[21]; **7**, 766[185]
Ghosh, A. K., **2**, 638[61], 640[61]; **3**, 546[123], 783[28]; **4**, 763[214], 1079[66]; **5**, 260[69], 261[69], 263[69]; **7**, 182[161], 680[77]
Ghosh, M., **5**, 96[105]
Ghosh, S., **3**, 87[115]; **7**, 239[50]
Ghosh, S. K., **2**, 586[131]; **5**, 14[98]
Ghosh, T., **4**, 496[90,91], 791[57], 878[76]; **5**, 381[117]; **7**, 87[19]
Ghoshal, M., **4**, 164[100]
Ghoshal, N., **7**, 823[35]
Ghozland, F., **4**, 229[220]
Ghrayeb, N., **8**, 354[163]
Ghribi, A., **1**, 347[131], 348[140]; **3**, 226[207], 259[129]; **4**, 209[63,69], 895[164], 900[164]; **6**, 5[23], 849[122]
Giacin, J. R., **7**, 13[123]
Giacobbe, T. J., **6**, 687[382]; **7**, 170[120]
Giacomelli, G., **3**, 285[55], 483[12]; **8**, 99[108,112], 100[118,118e], 348[131], 558[389], 564[440]
Giacomello, P., **6**, 280[144]
Giacomini, D., **1**, 391[148]; **2**, 613[114], 656[157], 925[111], 926[111], 927[120], 935[151], 936[151], 937[156,157]; **5**, 100[148,155,156], 102[173,174]; **6**, 759[140]
Giagante, N., **4**, 770[247]
Giam, C.-S., **4**, 428[71]; **5**, 418[67]
Giamalva, D. H., **3**, 1049[16], 1050[18], 1053[16], 1060[18]
Giamalva, O. H., **7**, 488[158]
Giammarino, A. S., **4**, 257[225]
Giancola, D., **8**, 566[450]
Giandinoto, S., **8**, 332[44]
Giangiordano, M. A., **1**, 390[143]; **2**, 994[40], 999[40]
Gianni, F., **2**, 716[64]
Gianni, F. L., **5**, 391[142]
Giannis, A., **3**, 25[160]
Giannoccaro, P., **8**, 450[171]
Gianotti, M. P., **2**, 853[231]
Gianturco, M. A., **4**, 257[225]; **8**, 943[122]
Giardi, M. T., **2**, 965[69]
Gibb, A. R. M., **8**, 950[159]
Gibboni, D. J., **7**, 3[10]
Gibbons, C., **2**, 110[74]; **8**, 447[126], 457[126]
Gibbons, E. G., **4**, 30[88,88e], 121[207], 243[63]; **6**, 1052[43]
Gibbs, D. E., **4**, 273[46], 280[46]
Gibbs, R. A., **6**, 903[140]
Gibby, M. G., **3**, 299[33]
Gibs, G. J., **8**, 564[443], 614[86]
Gibson, C. P., **1**, 428[121], 429[121], 457[121]; **3**, 209[12]; **4**, 89[87], 90[87d], 141[13], 255[190,191]
Gibson, D. H., **1**, 878[105]; **8**, 17[113], 22[113], 551[338]
Gibson, D. M., **5**, 947[267], 948[267]
Gibson, F., **5**, 855[182,183]
Gibson, M. S., **4**, 391[178c], 496[87]; **6**, 65[1], 79[65], 81[65], 104[1,9], 525[381], 819[110]; **8**, 367[54]
Gibson, T., **2**, 159[128]; **4**, 153[62b]; **8**, 673[29], 675[29,40], 676[40], 677[29,40], 684[29], 685[29], 688[29], 691[29], 694[40]
Gibson, T. W., **3**, 850[61]
Gibson, V. C., **8**, 445[54,54d], 459[239]
Gidaspov, B. V., **6**, 110[50], 795[13], 798[13], 817[13]; **7**, 690[13], 750[133]
Giddings, P. J., **1**, 656[152], 658[152]; **3**, 934[64], 953[64]; **8**, 849[109]
Gidley, G. C., **7**, 17[170]
Gielen, M., **3**, 587[148]; **7**, 614[6]
Gieren, A., **2**, 1093[83]; **7**, 475[55]
Giering, W. P., **5**, 272[4], 273[4], 275[4,10], 277[10], 281[20]; **6**, 686[365], 690[392]; **8**, 890[138]
Gierisch, S., **2**, 375[278]
Giersch, W., **3**, 736[29]; **7**, 306[10], 708[32]
Giersig, M., **8**, 335[67]
Gierstae, R., **8**, 864[240]
Giesbertz, K., **2**, 364[206]

Giesbrecht, E., **7**, 770[248], 772[293], 773[293], 774[336]
Giese, B., **1**, 269[60]; **2**, 334[67], 448[37,38]; **3**, 598[203], 649[201]; **4**, 48[140], 311[445], 386[147], 716[3], 717[3,10], 725[10], 727[3,56], 728[56,57], 729[56,58,60], 730[60,67], 735[3,81,85,86], 738[90,93], 739[107,109], 740[60,115,116], 741[3,81,121,123,124,126], 743[3], 744[133], 747[3], 752[3], 754[58], 761[203], 767[232], 774[3], 777[3], 780[5], 790[5], 791[5,52], 805[144], 876[62]; **5**, 159[51], 186[172], 189[51]; **6**, 442[87]; **7**, 399[36], 860[71], 883[175]; **8**, 846[87], 849[87], 852[139], 853[144,146]
Giese, R. W., **3**, 577[87]; **8**, 524[12], 527[49], 532[12c]
Giesecke, H., **6**, 579[986]
Giesemann, G., **2**, 1096[94]
Giffard, M., **8**, 851[125]
Gifkins, K. B., **5**, 203[39,39a], 209[39], 210[39]
Giga, A., **7**, 668[82]
Gigg, J., **3**, 273[7]; **7**, 246[92]
Gigg, R., **3**, 273[7]; **6**, 652[143]; **7**, 246[92]
Gigian, M. J., **7**, 761[62]
Giguere, R. J., **5**, 7[51], 15[108], 514[6], 595[15], 596[15,36], 598[36], 608[66], 850[144]; **7**, 262[78], 362[25]
Gil, G., **2**, 1086[31]; **4**, 245[81], 249[81]; **5**, 772[156,161]; **8**, 205[165]
Gil, J. B., **7**, 832[69]
Gilabert, D. M., **7**, 227[82]
Gilani, S. S. H., **5**, 429[112]
Gilardi, A., **2**, 492[52], 516[58]
Gil-Av, E., **2**, 1094[88]
Gilberg, J. C., **4**, 682[56]
Gilbert, A., **3**, 807[18]; **5**, 71[151], 585[199], 637[104,105], 645[1], 646[9,10], 647[1a-d,f,12], 648[12,20], 649[1f,22], 650[22], 651[1], 654[27], 655[29], 671[1a,49]
Gilbert, E. E., **6**, 431[107]
Gilbert, F. L., **7**, 776[359]
Gilbert, G., **4**, 84[67]
Gilbert, J. C., **2**, 477[11], 597[5], 728[137]; **3**, 1049[16], 1050[18], 1053[16], 1060[18]; **5**, 836[62], 841[100], 906[70], 908[70]; **6**, 67[15], 69[15], 705[35,36]
Gilbert, K. E., **5**, 856[207], 985[35]; **7**, 737[7]
Gilbert, L., **1**, 59[34], 153[63]; **2**, 194[66]; **5**, 553[47]; **8**, 9[55]
Gilbert, R., **3**, 903[126]
Gilbertson, S. R., **2**, 588[151], 589[151]; **5**, 1070[25], 1072[25], 1074[25], 1077[48], 1079[48], 1089[82,86], 1090[86], 1092[82,94], 1094[82,94], 1096[108,108c], 1098[82,108c], 1099[82,108c], 1111[82], 1112[48,82,108c], 1113[82,94], 1183[57]
Gilbreath, S. G., **2**, 173[182], 175[184]
Gilchrist, T. L., **3**, 486[39], 491[39], 495[39], 498[39], 503[39], 908[148]; **4**, 483[4], 484[4], 495[4], 1099[183]; **5**, 379[112], 383[112], 384[112], 422[81], 485[183], 486[185], 487[185,187], 500[260], 707[34], 717[93], 725[34,117], 742[159a], 803[92], 979[26], 1070[27], 1073[27]; **7**, 480[104], 743[60,61], 744[70]; **8**, 337[76], 364[10], 510[93], 640[25], 652[78]
Gilchrist, T. T., **6**, 104[3]
Gilday, J. P., **4**, 524[63,65], 525[63,65]; **6**, 881[50]
Gilde, H. G., **3**, 647[191,192]
Giles, R. G. F., **2**, 746[109], 748[126]; **5**, 572[122]; **7**, 355[41]
Gilge, U., **2**, 743[80]
Gilgen, P., **4**, 1081[73,78]
Gilgert, F. L., **7**, 776[359]
Gilges, S., **4**, 744[133]
Gilham, P. T., **6**, 611[64]; **7**, 765[153]
Gilje, J. W., **1**, 36[235]
Gilkerson, T., **8**, 366[43]
Gill, A., **8**, 28[37], 66[37]
Gill, D. S., **8**, 445[34,54,54c]
Gill, G. B., **1**, 844[4]; **2**, 534[33-35], 535[34,35], 538[61], 539[61]; **4**, 707[42]; **5**, 432[129]; **7**, 338[42]
Gill, H. S., **2**, 925[110,113], 926[110]; **5**, 100[160], 101[160], 102[171]
Gill, J. T., **5**, 117[277]
Gill, M., **1**, 551[74]; **3**, 247[48]
Gill, T. P., **4**, 521[45]
Gill, U. S., **4**, 521[42], 529[72,74,75,77], 530[75], 531[72]; **7**, 155[29]
Gillard, F., **7**, 564[88], 568[88]
Gillard, M., **5**, 461[105]; **7**, 122[30], 144[30]
Gillard, R. D., **4**, 611[351]; **8**, 445[33]
Gillaspey, W. D., **6**, 121[130]
Gillen, M. F., **6**, 624[146], 625[146]
Giller, S. A., **6**, 554[730]
Gilles, J.-M., **5**, 717[90d]
Gillespie, J. P., **3**, 683[101]; **4**, 505[147,148]; **5**, 68[96]
Gillespie, R. J., **1**, 292[29]; **3**, 297[23]; **4**, 1032[10], 1063[10,10a,168,171]
Gillet, J. P., **3**, 498[107]
Gillette, J. R., **7**, 778[418]
Gillhouley, J. G., **7**, 712[66]
Gilli, G., **1**, 309[101]
Gilliatt, V., **4**, 738[100]
Gillick, J. G., **8**, 526[27]
Gillie, A., **1**, 444[188], 457[188]
Gillies, I., **5**, 947[259]
Gilligan, J. M., **8**, 939[97]
Gilligan, P. J., **4**, 206[50]
Gilliom, L. R., **5**, 1121[26]
Gillis, D. J., **8**, 674[32]
Gillis, H. F., **5**, 515[11], 517[11a], 519[11a], 526[11], 543[11a]
Gillis, H. R., **5**, 515[11,16], 518[16a], 520[36], 524[16], 526[11]
Gillissen, H. M. J., **1**, 661[165,166], 663[165,166], 672[166], 700[166], 704[166]
Gillois, J., **8**, 185[27,28], 187[28]
Gillon, A., **4**, 1040[92], 1042[92]
Gillon, I., **5**, 407[25]
Gilman, B. L., **2**, 963[53]
Gilman, H., **1**, 107[7], 226[89], 383[111], 506[1,2], 622[72], 630[38,39], 631[38,39], 636[38], 701[39], 702[39]; **2**, 183[15], 294[82], 917[87], 976[1]; **3**, 208[8,9], 244[19], 415[9], 511[190], 565[17]; **4**, 70[6], 72[23], 95[23], 98[6], 106[138c], 140[7a], 148[47b]; **5**, 100[140]; **6**, 149[92]; **8**, 322[108,109], 564[444], 568[484], 626[173]
Gilman, J. W., **5**, 70[116], 514[9], 527[9]
Gilman, N. W., **2**, 748[123]; **5**, 791[24], 822[24]; **6**, 843[89]
Gilman, S., **6**, 656[171], 1029[103], 1032[119]; **7**, 674[40]
Gilmore, J., **8**, 964[57]
Gilmore, J. R., **7**, 92[43], 705[18]
Gilmore, W. F., **2**, 424[36]
Gilow, H. M., **5**, 176[132]
Gilpinand, M. L., **7**, 158[42]
Gilsdorf, R. T., **8**, 373[134], 376[134,162]
Ginak, A. I., **4**, 329[1], 344[1], 350[1], 351[1]
Gindraux, L., **7**, 252[7]
Ginebreda, A., **4**, 1001[41]; **6**, 80[68]
Giner-Sorolla, A., **7**, 657[25]
Gingras, M., **6**, 3[11]
Gingrich, H. L., **4**, 1096[159], 1097[159,169], 1098[159]
Ginn, D., **1**, 402[16]
Ginocchio, S. D., **8**, 269[90]
Ginos, J. Z., **8**, 598[97]
Ginsberg, S., **2**, 466[126]
Ginsburg, D., **2**, 352[85]; **3**, 628[45]; **4**, 3[7], 4[7], 63[7b], 65[7b], 70[2], 187[96], 258[231], 1017[212], 1021[212]; **5**, 451[54], 595[15], 596[15], 706[26]
Ginsburg, H., **4**, 459[71], 464[71], 466[71,126], 475[152], 478[71], 479[126,171,172]; **7**, 878[140]
Ginzel, K.-D., **2**, 1051[39], 1075[39]
Gioeli, C., **6**, 659[192,197b]
Giolando, D. M., **1**, 41[270], 432[137], 456[137]; **3**, 211[28], 215[28]

Giomi, D., **4**, 958[28]
Giongo, G. M., **8**, 171[110], 460[247]
Giongo, M., **4**, 753[170]
Giordano, C., **3**, 778[6], 788[6,54], 789[6,55-57]; **4**, 768[235,240]; **5**, 504[277], 758[82]; **6**, 431[108]; **7**, 828[52], 829[55]; **8**, 111[17,18], 113[18], 117[17,18]
Giordano, G., **8**, 457[213]
Giordano, R., **8**, 449[158]
Giordau, J., **7**, 874[105]
Giorgianni, P., **5**, 113[230,231], 114[241,242,243,244,245], 440[173]
Giovannini, F., **8**, 185[29], 190[69]
Giovini, R., **7**, 772[286]
Gipe, A., **4**, 345[81]
Gipson, R. M., **3**, 391[89], 393[89]
Giraldi, P., **7**, 100[121]
Girard, C., **5**, 856[217], 910[89]; **8**, 9[55]
Girard, J.-P., **3**, 851[65]; **4**, 301[323], 302[323,334]
Girard, P., **1**, 179[25], 255[16,16b,c], 256[16c], 258[16c], 259[16c,27], 261[16c,27], 265[27], 266[16c,27], 273[71], 278[16b,c], 751[112]; **6**, 980[40]; **7**, 846[91]; **8**, 113[48,49], 115[48,49], 797[31], 889[134]
Girard, Y., **1**, 821[28]; **3**, 277[27]; **4**, 1059[155]; **6**, 489[86]
Girardin, A., **6**, 149[91], 985[67]
Girault, Y., **6**, 769[30]
Girdaukas, G., **8**, 190[85]
Girgenti, S. J., **3**, 14[75], 15[75]; **5**, 634[77]
Girgis, N. S., **7**, 137[125], 138[125]
Giri, B. P., **3**, 579[99], 1048[12]
Giri, V. S., **8**, 249[96]
Girina, G. P., **3**, 636[55]
Girodeau, J. M., **2**, 537[49]
Girotra, N. N., **8**, 945[132]
Giroud, A. M., **8**, 118[75], 122[75]
Girrard, P., **8**, 552[360]
Girshovich, M. Z., **8**, 637[13]
Gisby, G. P., **5**, 571[120]
Gisin, B. F., **6**, 667[249], 670[249]
Gisler, M., **4**, 443[191]
Gislon, G., **2**, 639[62], 640[62], 930[132], 932[132]; **5**, 102[176]
Gist, R. P., **2**, 1079[158]; **5**, 528[68], 531[68]
Gittelman, M. C., **6**, 812[79]
Gittos, M. W., **2**, 765[76]; **6**, 501[190]
Giua, M., **7**, 774[330]
Giudici, T. A., **8**, 269[89]
Giuliani, A. M., **5**, 1158[173]
Giuliano, R. M., **4**, 401[228c,d]; **8**, 52[150], 66[150]
Giumanini, A. G., **1**, 832[115]
Giusti, I., **3**, 216[78]
Givens, R. S., **5**, 218[29,35], 219[40], 221[35,59], 229[120], 706[27]
Gizycki, U. V., **3**, 664[27]
Gjøystdal, A. K., **6**, 291[210]
Glacet, C., **8**, 228[129], 532[130]
Gladfelter, E. J., **7**, 17[171]
Gladfelter, W. L., **8**, 446[70,71]
Gladiali, S., **2**, 435[59]; **8**, 91[54]
Gladkowski, D. E., **4**, 155[63b]
Gladstone, M., **4**, 279[106,107]
Gladstone, W. A. F., **7**, 231[145]
Gladych, J. M., **2**, 323[32]; **6**, 65[1]
Gladysz, J. A., **1**, 307[92], 309[99,100]; **2**, 127[234], 538[59], 539[59]; **8**, 103[131], 323[117], 889[136]
Gladysz-Dmochowska, J., **7**, 77[124c]
Glahsl, G., **2**, 1090[72,73], 1102[73], 1103[73]
Glamkowski, E. J., **3**, 890[31], 903[124]
Glans, J. H., **3**, 493[80]
Glänzer, B. I., **1**, 212[6], 213[6]; **6**, 978[22]
Glasbrenner, J., **6**, 188[181]
Glaser, R., **1**, 41[197]; **2**, 232[175]; **6**, 727[199,200]
Glasgow, L. R., **8**, 526[28]
Glass, C., **3**, 691[129], 693[129]
Glass, C. A., **8**, 451[180]
Glass, D. S., **5**, 702[10], 716[10], 906[63]
Glass, R. S., **3**, 736[28]; **7**, 765[161]; **8**, 966[72]
Glasscock, K. G., **6**, 134[18]; **8**, 846[85]
Glassman, S. D., **2**, 838[176]
Glasstone, S., **3**, 636[46]
Glatt, H., **6**, 244[110]
Glattfeld, J. W. E., **8**, 292[45]
Glatz, B., **7**, 160[50]
Glaz, A. Sh., **4**, 426[64]
Glaze, W. H., **4**, 868[16]
Glazier, E. R., **7**, 120[8]
Gleason, J. G., **3**, 273[15], 274[15]; **5**, 94[82]; **7**, 122[29], 219[11]; **8**, 413[123]
Gleason, M. M., **8**, 28[37], 66[37]
Gleason, R. W., **3**, 380[5]
Glebova, Z. I., **7**, 294[18]
Gledhill, A. P., **4**, 444[200]; **8**, 916[101], 917[101], 918[101], 919[101], 920[101]
Gledinning, R. A., **3**, 334[220]
Gleich, P., **6**, 430[101]
Gleicher, G. J., **4**, 729[62], 730[64]
Gleim, R. D., **2**, 746[108], 762[56], 824[120]
Gleiter, R., **3**, 382[36], 587[148], 592[175], 614[6], 623[6], 627[6]; **4**, 355[133], 667[13], 669[13], 677[13], 1010[157]; **5**, 72[173], 812[129]; **6**, 960[52], 1044[18]; **8**, 349[136], 946[134]
Gleize, P. A., **8**, 553[361]
Glemser, O., **7**, 483[131]
Glenar, D. A., **5**, 64[31]
Glenn, A. G., **4**, 723[37]
Glenneberg, J., **8**, 545[285]
Glennie, E. L. M., **2**, 969[83,83a,b]
Glens, K., **7**, 802[46], 808[77]
Gless, R. D., **1**, 836[146]; **6**, 501[199]
Glick, A. H., **5**, 167[97]
Glick, M. D., **8**, 851[126], 858[126]
Glikmans, G., **6**, 263[20,24], 264[24], 267[24], 269[20]
Glinka, J., **4**, 1104[212]
Glinka, T., **3**, 790[61]; **4**, 432[108]; **7**, 551[53]
Glinski, M. B., **3**, 266[194]; **5**, 692[102]
Glinski, R. P., **6**, 622[135]
Glocking, F., **4**, 605[295]; **8**, 446[85], 754[111], 755[111]
Glockner, P., **5**, 1141[83]
Gloede, J., **2**, 900[26], 960[35], 962[35], 1088[44]; **6**, 488[30], 533[513,514], 566[30]
Gloggler, K. G., **8**, 52[147], 66[147]
Glogowski, M. E., **5**, 73[195]
Gloor, B. F., **4**, 454[38], 457[38,60], 461[38], 464[121], 465[121], 468[38], 469[38], 472[60], 473[121]
Glos, M., **3**, 904[131]
Glotter, E., **7**, 253[21], 445[60], 707[28]
Glover, D., **8**, 269[91]
Glover, G. M., **5**, 798[4]
Glover, S. A., **4**, 784[13]
Glowinski, R., **3**, 500[134], 509[134]
Gluchowski, C., **2**, 917[89], 919[89], 920[89], 924[89], 935[89]; **5**, 100[147]; **6**, 531[448]
Glück, C., **4**, 1006[97]
Glue, S. E. J., **1**, 560[154]
Glukhova, O. F., **8**, 214[40]
Glüsenkamp, K.-H., **2**, 358[154], 371[154]; **5**, 129[34], 461[96-98], 462[97,98], 468[133]

Glushkov, R. G., **6**, 488[17,34], 502[208,209], 507[228], 529[17], 531[453], 554[718,755,763,764,770,776,780,784,786,789,790,791,792,793]
Glusker, J. P., **1**, 8[38]
Gmelin, G., **6**, 112[65]
Gnecco Medina, D. H., **2**, 1014[19]
Gnin, D., **6**, 487[75]
Gnoj, O., **1**, 174[8], 175[8]; **7**, 160[47]
Goasdoue, C., **2**, 630[22], 631[22], 632[22], 634[22], 635[51], 640[51]; **6**, 425[69]
Goasdoue, N., **2**, 630[22], 631[22], 632[22], 634[22]
Gobao, R. A., **5**, 414[52]
Gobbi, C., **6**, 927[73]
Göbel, T., **4**, 761[203], 805[144]
Gobillon, Y., **1**, 38[259]
Gobinsingh, H., **8**, 530[94]
Gocmen, M., **4**, 89[86a], 98[109b,c]
Goddard, J., **4**, 1070[13]
Goddard, W. A., **1**, 880[116]
Goddard, W. A., III, **5**, 72[181], 788[14], 849[142], 1000[8]
Godefroi, E. F., **1**, 364[40], 371[73]; **2**, 759[33]; **3**, 158[445], 159[445], 367[102]; **6**, 984[58]; **8**, 56[169], 66[169], 638[15], 967[82]
Godel, T., **2**, 924[108b]; **3**, 46[252]; **4**, 152[54], 184[85], 201[13,15], 202[13,15]; **5**, 137[73], 356[90]; **8**, 720[136]
Godet, J.-Y., **4**, 971[75]
Godfrey, A., **2**, 120[186], 651[115]; **3**, 196[27]; **4**, 153[61a,d]
Godfrey, C. R. A., **3**, 613[2], 615[2]; **7**, 132[95]; **8**, 117[74], 243[47]
Godfrey, J. D., **4**, 8[28], 159[80]; **5**, 841[87], 859[87c]
Godfrey, J. D., Jr., **6**, 835[46]
Godfrey, K. L., **8**, 490[7]
Godfrey, P. D., **4**, 484[13]
Godhart, J. B., **6**, 660[203]
Godin, P. J., **7**, 156[32], 157[32e], 158[32e]
Godinger, N., **8**, 554[365]
Godleski, S. A., **4**, 596[160], 597[180], 598[204,205,207,210], 604[286], 621[160], 622[180], 626[286], 629[406,407,415], 636[160], 638[204,205]; **5**, 373[106,106a], 374[106a]; **6**, 86[99], 842[81]; **8**, 331[33], 334[60], 342[109], 992[54]
Godot, J. M., **8**, 53[128], 66[128]
Godovikova, T. I., **7**, 740[43]
Godoy, J., **7**, 309[23], 767[194], 773[194]
Godschalx, J. P., **2**, 727[131]; **3**, 232[271], 469[201], 470[201], 471[201], 473[201], 475[201]; **4**, 594[138,145,147], 619[138,145], 633[145]
Godtfredsen, S., **3**, 352[45]
Goe, G. L., **5**, 64[43]
Goebel, M., **2**, 1099[109,109b,112a]
Goebel, P., **6**, 707[41]; **7**, 7[48]; **8**, 726[191]
Goedecke, E., **5**, 65[64]
Goedken, V., **2**, 271[78], 933[141], 934[141]
Goehring, R. R., **1**, 564[202]; **4**, 457[49,50], 477[49,50], 503[125]
Goel, A. B., **3**, 824[20]; **4**, 240[41], 897[172], 915[15]; **8**, 3[15], 14[85], 315[53], 483[58], 484[58], 485[58], 549[328], 696[120], 801[73], 802[80]
Goel, O. P., **1**, 399[5,7]
Goel, S., **4**, 915[15]
Goeldner, M. P., **4**, 1061[163]
Goerdeler, J., **6**, 423[42], 424[61], 644[86]
Goering, H. L., **3**, 220[121], 222[121c,138,138b,142], 393[93,94], 394[95]; **5**, 856[217]; **6**, 835[42], 848[108]; **7**, 95[76]; **8**, 275[142]
Goerner, R. N., **8**, 819[44]
Goethals, E. J., **6**, 443[94]
Goethel, G., **8**, 743[166], 758[166,170]
Goetz, H., **5**, 552[23]
Goetz, R. W., **8**, 551[337]
Gogan, N. J., **2**, 743[84]
Gogerty, J. H., **6**, 523[353]
Gogins, K. A. Z., **7**, 264[89], 275[89], 843[44]
Gogoll, A., **4**, 597[170], 621[170]
Gogte, V. N., **4**, 231[264,265]; **5**, 802[86]
Goh, S. H., **4**, 962[40]; **7**, 296[25,26], 883[177]; **8**, 19[132], 807[120], 903[52], 906[52], 907[52], 908[52], 909[52]
Gohda, M., **4**, 261[286]
Gohdes, J. W., **3**, 328[179]
Goheen, D. W., **3**, 848[51]; **7**, 769[209,217]
Gohke, K., **8**, 423[40], 429[40]
Göhrt, A., **2**, 372[271]
Goicoechea-Pappas, M., **1**, 477[131]; **3**, 75[43,51], 76[43], 78[51], 80[43]; **7**, 227[79]
Going, R., **8**, 925[11], 926[11]
Goins, D. E., **3**, 297[17], 306[86]
Goji, H., **7**, 774[318]
Gokel, G., **2**, 1094[89], 1095[89], 1098[105]; **3**, 136[373], 137[373], 505[163]; **4**, 1001[26]; **6**, 242[88], 243[88]; **8**, 830[84]
Gokhale, U., **7**, 771[280], 773[280]
Gokou, C. T., **5**, 575[134]
Gokturk, A. K., **7**, 13[117]
Gokyu, K., **1**, 238[36], 568[246]
Gold, A., **5**, 165[79]
Gold, A. M., **3**, 812[55]
Gold, E. H., **4**, 426[65], 441[65]
Gold, H., **4**, 54[153a], 56[153a]
Gold, M., **8**, 206[173]
Gold, M. H., **4**, 4[14]
Gold, P. M., **1**, 636[93], 638[93], 640[93], 646[93], 647[93]; **3**, 87[103], 104[103], 106[103], 111[103], 117[103]; **7**, 826[47], 827[47]; **8**, 542[221]
Gold, V., **4**, 444[199]
Goldbach, M., **5**, 973[13]
Goldberg, A. A., **6**, 429[91]; **7**, 657[24]
Goldberg, I., **2**, 91[42]; **4**, 710[59]
Goldberg, I. H., **5**, 736[142i-k]
Goldberg, M., **2**, 456[77]
Goldberg, M. W., **1**, 130[96]; **8**, 328[15]
Goldberg, O., **5**, 92[65]
Goldberg, S. I., **3**, 154[419]; **6**, 1020[46]; **8**, 170[74]
Goldberg, Yu. Sh., **7**, 477[79,81]; **8**, 764[4a]
Goldblum, N., **1**, 300[64]
Golden, D. M., **4**, 1072[17]; **5**, 900[8]
Gol'dfarb, Ya. L., **3**, 615[7]; **4**, 468[132], 469[132]; **8**, 608[40], 609[51]
Goldhamer, D., **4**, 50[142,142a]; **5**, 63[4]
Goldhill, J., **2**, 617[141]; **4**, 675[37]; **6**, 1004[139], 1019[43]
Golding, B. T., **2**, 866[9]; **5**, 791[42], 803[42], 971[2], 972[2], 973[2]; **6**, 20[72], 79[62], 620[124], 673[290]
Gol'ding, I. R., **3**, 209[20]
Goldish, D. M., **6**, 245[121], 248[121], 249[121], 251[121]
Goldmacher, J. E., **6**, 825[128]
Goldman, A., **4**, 1041[104]; **6**, 495[141]; **7**, 707[28]
Goldman, B. E., **2**, 538[52], 542[52], 547[52,99], 548[99]; **5**, 5[47], 6[47]
Goldman, I. M., **5**, 123[3]
Goldman, L., **6**, 614[90]; **7**, 294[16], 295[16,19]
Goldman, N., **4**, 240[39], 254[39]
Goldman, N. L., **2**, 106[50], 184[26]; **3**, 2[8], 11[8], 16[8], 17[8], 26[8]; **8**, 527[39,40]
Goldman, P., **4**, 426[46]
Goldmann, S., **6**, 153[143], 839[70], 902[124]
Gol'dovskii, A. E., **3**, 305[72]
Goldschmidt, S., **3**, 647[176], 648[176]
Goldschmidt, Z., **4**, 710[53,54,56-63], 712[65,66,68], 1043[108], 1048[108]; **5**, 1006[34,34b]

Goldshleger, N. F., **7**, 17[173]
Goldsmith, B., **8**, 803[93], 804[93], 826[69]
Goldsmith, D., **4**, 376[102], 377[102,124b], 380[124,124b]
Goldsmith, D. J., **1**, 117[57], 118[57], 571[279], 822[39]; **3**, 752[95]
Goldstein, A., **8**, 142[52]
Goldstein, E., **5**, 406[22]
Goldstein, L., **2**, 1104[132]
Goldstein, M. J., **5**, 686[44], 714[68], 826[158], 857[224]
Goldstein, R. F., **7**, 7[38]
Goldstein, S. W., **2**, 1015[23]; **4**, 411[267b]; **7**, 567[104]
Golduras, G. A., **7**, 15[147]
Golebiowski, A., **2**, 671[51]; **5**, 430[118], 431[121], 433[138,139,139f], 434[138]
Golec, F. A., Jr., **2**, 381[310], 547[112], 551[112]; **4**, 373[72]
Golembeski, N., **2**, 829[135]
Golfier, M., **3**, 664[28], 698[28]; **7**, 312[34], 320[34], 738[26], 747[26], 841[9], 851[18]
Goliaszewski, A., **4**, 620[396,397], 636[396,397]
Golic, L., **1**, 320[160]
Golik, J., **5**, 736[145], 737[145]; **6**, 789[110]
Golikov, A. V., **4**, 426[64]
Golinski, J., **3**, 174[528]; **4**, 20[64], 21[64], 424[16], 426[16], 432[16,104,112]; **5**, 75[215], 79[291]
Gölitz, P., **7**, 742[58]
Goliński, J., **2**, 12[51], 431[52b], 537[50]
Gollaszewski, A., **4**, 594[139], 619[139], 634[139]
Goller, E. J., **1**, 221[67], 223[72a,b]; **4**, 72[23], 95[23]
Gollnick, K., **5**, 74[207]; **7**, 96[87], 97[94], 816[10], 818[10]
Golob, A. M., **1**, 880[116]; **5**, 788[11], 798[11], 814[136], 847[133], 872[133], 888[30], 1000[2]; **8**, 36[72], 37[72], 38[72], 44[72], 66[72], 948[152]
Golobov, Yu. G., **6**, 525[388]
Golod, E. L., **6**, 110[50]
Gololobov, Yu. G., **1**, 820[4]; **3**, 86[43], 179[43]; **4**, 987[133]
Golse, R., **1**, 368[61], 369[61]
Golubeva, G. A., **8**, 636[5]
Golubtsov, S. A., **8**, 775[77], 778[88]
Gomann, K., **4**, 1022[252]
Gombatz, K., **7**, 438[22]
Gomberg, M., **3**, 505[161], 507[161], 564[12]
Gómez, G., **5**, 416[57]
Gómez-Aranda, V., **4**, 291[216,217,219], 303[341], 315[511,513]
Gómez-Parra, V., **6**, 273[93]
Gómez-Sánchez, A., **6**, 941[150]
Gomez-Solivellas, A., **4**, 111[155d]
Gompper, R., **4**, 54[152]; **5**, 74[211], 451[28], 468[134], 482[173], 483[175], 501[28]; **6**, 226[10], 256[10], 257[10], 506[226,227], 518[332,333], 519[333,337], 531[437], 575[969,970], 705[28], 832[19]
Goncalves, J. M., **7**, 852[40]
Gonda, E., **2**, 558[161]
Gondos, G., **8**, 418[5], 420[5], 423[5], 439[5], 441[5], 442[5]
Gong, W., **4**, 848[78]
Goñi, T., **2**, 780[12]
Gonis, G., **7**, 710[51]
Gonnela, N. C., **6**, 746[95]
Gonnermann, J., **2**, 337[73]; **4**, 104[136a]
Gonschorrek, C., **2**, 94[51,52]
Gontarz, J. A., **4**, 301[319,320,324], 302[319,320], 314[484]
Gonzáles, A., **6**, 116[92]
González, A., **7**, 277[154,155]
Gonzalez, A. A., **8**, 669[8]
Gonzalez, A. G., **5**, 830[32]; **8**, 330[49], 798[61]
González, A. G., **3**, 391[90], 393[90], 395[97]; **7**, 820[25]
Gonzalez, A. M., **4**, 895[165], 900[165]
González, A. M., **5**, 478[164]; **8**, 646[50,51]
González, B., **5**, 478[164]
Gonzalez, D., **5**, 8[57,62]; **7**, 738[23]
González, E., **1**, 377[99]; **8**, 446[74], 452[74], 457[74]
Gonzalez, F. B., **2**, 902[40]; **4**, 380[122], 381[122a]
Gonzalez, F. J., **5**, 161[64], 484[179]
Gonzalez, J. M., **7**, 501[255], 505[285], 536[52-55]
González, J. M., **4**, 347[93], 351[93c], 354[93d]
Gonzalez, M. A., **3**, 71[36,37]; **7**, 224[52]
González, M. S., **3**, 396[115]
Gonzalez, M. S. P., **4**, 379[115]
González-Nogal, A. M., **2**, 583[114]
Gonzalez-Nuñez, E., **2**, 790[57]; **6**, 494[135]
Gonzenbach, H.-U., **5**, 216[16], 218[34], 219[16], 221[16]
Gonzo, E. E., **8**, 419[21], 420[21], 424[21], 429[21], 436[21]
Gooch, A., **4**, 987[147]
Gooch, E. E., **7**, 606[150,151]; **8**, 377[168]
Good, M., **5**, 2[15]
Goodacre, J., **6**, 669[264]
Goodbrand, H. B., **7**, 316[46], 317[46]; **8**, 188[55], 196[55], 199[55], 201[55]
Goodburn, T. G., **3**, 281[43]
Goodchild, J., **6**, 624[136], 625[156]
Goodfellow, C. L., **3**, 197[37]
Goodhue, C. T., **7**, 760[43]
Goodhue, T., **7**, 57[28], 58[28], 63[28]
Gooding, D., **5**, 3[25]
Goodlett, V. W., **4**, 45[126,126b,e]; **5**, 686[50]; **6**, 426[70]
Goodman, I., **8**, 568[482]
Goodman, J. M., **2**, 249[84]
Goodman, L., **4**, 367[11], 386[153]
Goodman, M., **2**, 403[32]; **6**, 804[49]
Goodman, M. M., **7**, 775[348]; **8**, 92[68]
Goodrich, J. E., **8**, 303[99]
Goodridge, R. J., **4**, 12[37,37d], 226[192]; **6**, 150[114]
Goodwin, D., **5**, 185[166]
Goodwin, H. A., **4**, 520[30]
Goodwin, R. C., **8**, 143[64]
Goodwin, S., **3**, 804[8], 810[46]
Goodwin, T. E., **3**, 97[170], 117[170]; **4**, 16[51]; **6**, 675[300]
Goossens, D., **5**, 113[236]
Goossens, H. J. M., **6**, 561[872]
Gootz, R., **6**, 49[90]
Gopal, D., **5**, 878[9]
Gopal, H., **7**, 236[20]; **8**, 745[54], 746[54], 753[54]
Gopal, M., **3**, 559[56]; **8**, 390[77]
Gopal, R., **4**, 505[141]
Gopalakrishnan, S., **7**, 374[77e]
Gopalan, A., **5**, 809[118]
Gopalan, A. S., **3**, 133[335], 136[335]; **8**, 190[83,85]
Gopalan, B., **2**, 782[24]; **4**, 373[70]
Gopalan, R., **6**, 841[74], 1022[60]
Gopichand, Y., **5**, 89[51]
Gopinath, K. W., **5**, 513[4]
Gopinathan, M. S., **5**, 73[192]
Goralski, C. T., **3**, 164[474], 180[474], 181[474]; **8**, 705[11], 726[11], 939[98]
Göran, P., **1**, 472[77]
Gorbachevskaya, V. V., **5**, 1174[34]
Gorbunkova, V. P., **3**, 644[141]
Gorbunov, A. V., **7**, 602[106]
Gorby, R. R., **4**, 4[16]
Gordeev, A. D., **6**, 524[360], 528[413]
Gordeev, M. F., **4**, 342[68]
Gordillo, B., **1**, 774[213]
Gordon, A. B., **3**, 913[1]
Gordon, D., **8**, 566[457], 568[468]
Gordon, D. A., **8**, 828[76]

Gordon, D. J., **5**, 300[73], 302[73]
Gordon, H. B., **1**, 141[14]
Gordon, J. E., **7**, 852[34]
Gordon, K. M., **1**, 712[264], 714[264]; **3**, 87[63], 114[63], 117[63]; **6**, 1026[81], 1027[81], 1029[81], 1030[81], 1031[81], 1033[81]; **7**, 87[22], 124[38], 128[38], 129[38], 775[353]
Gordon, L., **2**, 740[58]
Gordon, M. E., **4**, 1002[52]
Gordon, M. H., **8**, 707[20]
Gordon, P. N., **6**, 265[38]
Gore, J., **1**, 191[78-80], 192[83], 853[50], 876[50], 892[149]; **2**, 35[129]; **3**, 223[183], 225[183], 256[113]; **4**, 292[222], 308[406,408,409], 311[452], 395[205,205c,d,206], 396[209], 397[209,211], 412[268a], 413[268a]; **5**, 772[157,159,163], 773[166,167], 774[167,169,170,171], 797[62,66], 800[79], 802[82], 821[62], 936[196]; **6**, 210[76], 214[98], 960[59]; **7**, 684[91] **8**, 705[10], 726[10], 854[153], 857[192]
Gore, M. P., **7**, 184[172]
Gore, P. H., **2**, 745[102,106], 753[1]; **3**, 499[115]; **5**, 724[114]
Gore, W. E., **5**, 453[68]
Gorelik, M. V., **6**, 499[173]
Gorgues, A., **1**, 188[71]; **5**, 386[132,132d], 387[132d]; **6**, 977[16]; **8**, 28[36], 66[36]
Gorham, W. F., **3**, 782[16]
Gorin, P. A. J., **2**, 456[60,61,75], 457[60,61]; **6**, 36[23], 561[879]; **8**, 224[108]
Gorina, N. V., **8**, 150[132]
Gorini, C., **4**, 523[57]
Gorisch, H., **5**, 855[183]
Gorissen, H., **5**, 482[171]
Gorissen, J., **6**, 499[174]
Görlach, Y., **2**, 809[53]
Görlich, K.-J., **6**, 943[156]
Gorlier, J. P., **1**, 107[4]; **3**, 419[45], 420[45]; **4**, 176[48]
Gorman, E., **4**, 1023[260]
Gorman, M., **2**, 913[78], 915[78], 925[78]; **5**, 85[2]; **6**, 759[136], 1025[80]; **8**, 964[54]
Gorman, R. R., **7**, 340[46]
Görner, H., **5**, 154[32]
Gorrichon, E., **6**, 494[137]
Gorrichon, J. P., **2**, 404[40]
Gorrichon, L., **2**, 281[29,31]; **4**, 21[65], 100[123]
Gorrichon-Guigon, L., **2**, 478[13]
Gorrod, J. W., **7**, 736[3]
Gorski, R. A., **6**, 439[72]
Gorteli, J., **6**, 671[277]
Gorter-La Roy, G. M., **2**, 780[9]
Gorthy, L. A., **7**, 400[45]
Gorvin, J. H., **4**, 434[129,130]
Goryaev, M. I., **7**, 93[52]
Gorys, V., **2**, 605[56], 630[11], 631[11]
Gorzynski, J. D., **1**, 193[87]; **3**, 483[17]
Gorzynski Smith, J., **6**, 88[105], 89[105]; **8**, 543[250]
Gosden, A., **8**, 974[129,130]
Gosden, A. F., **5**, 819[152]
Gosh, S., **2**, 747[117]
Gosney, I., **1**, 755[115], 812[115], 813[115], 825[52]; **5**, 803[87]; **6**, 173[36], 174[36], 199[36]; **7**, 479[92]
Gossauer, A., **2**, 874[27,28], 875[28,29]
Gosselck, J., **1**, 655[142], 684[142]; **6**, 134[19]
Gosselin, P., **3**, 124[260]; **6**, 455[151]
Gosselin, R. E., **6**, 216[105]
Gosser, L. W., **8**, 451[174]
Gosteli, J., **2**, 358[153]; **3**, 781[14]; **4**, 50[142]; **6**, 667[242]
Gostevsky, B. A., **8**, 546[310]
Goswami, A., **7**, 58[53a], 62[53,53a-c], 63[53a,58]; **8**, 192[97]
Goswami, R., **2**, 442[9], 446[30], 447[9,33]; **4**, 74[38a], 253[174]
Goswani, P. P., **7**, 71[95]
Gotah, J., **4**, 609[329]
Gothe, S. A., **5**, 260[67], 261[67], 262[67]
Goti, A., **8**, 851[127]
Goto, E., **6**, 746[91]
Goto, F., **3**, 635[44]
Goto, H., **3**, 446[77]
Goto, J., **2**, 1022[52]
Goto, K., **4**, 611[362], 612[362]
Goto, M., **6**, 453[142], 454[146]
Goto, S., **8**, 885[105]
Goto, T., **1**, 101[90], 453[222]; **2**, 124[207], 232[179], 381[311], 578[83], 743[82]; **3**, 261[157], 426[82], 429[82]; **4**, 79[57,58a,b,59a-c], 96[104], 216[126], 251[149], 257[149], 260[149], 331[11]; **5**, 350[79], 829[26], 924[147]; **6**, 18[65], 27[115], 164[199], 266[50], 900[115], 1023[69]; **7**, 169[107], 370[65], 380[65], 678[70]; **8**, 412[117], 991[47]
Goto, Y., **1**, 349[149]; **2**, 924[106]; **4**, 839[27]; **8**, 168[69], 178[69], 179[69]
Gotoh, T., **5**, 73[200]
Gotoh, Y., **3**, 446[78]; **8**, 168[70], 545[302]
Gotor, V., **5**, 161[63,64], 480[178], 484[179]; **6**, 757[134]
Götschi, E., **2**, 368[241], 866[7], 867[7], 870[7], 871[7], 872[7], 875[7], 876[7]
Gott, P. G., **6**, 448[110]
Gottardi, W., **7**, 229[116]
Gottarelli, G., **5**, 99[131]
Gotteland, J.-P., **5**, 301[77], 311[106]
Gottfried, N., **8**, 950[170]
Gotthardi, F., **4**, 527[67]
Gotthardt, H., **4**, 1097[161,162,163,165,166]; **5**, 167[98,99], 185[159,160]; **6**, 454[147], 509[269]
Gottikh, B. P., **6**, 450[121]
Gottlieb, H. E., **4**, 710[53,56,57,60,63]
Gottschalk, A., **2**, 463[88]
Gottsegen, A., **8**, 544[276]
Götz, A., **2**, 605[63]; **6**, 502[217], 560[870]; **7**, 650[51]
Gotz, H., **2**, 681[58], 683[58]
Götz, J., **2**, 1090[72]
Götz, M., **2**, 1088[40], 1095[92], 1097[40]
Goubeau, J., **8**, 724[171]
Goubitz, K., **2**, 922[100], 936[100]
Goudgaon, N. M., **8**, 377[168], 386[52]
Goudie, A. C., **8**, 500[47]
Gouedard, M., **7**, 451[35], 462[35]
Gough, M. J., **4**, 370[40]
Gouin, L., **4**, 878[80-82], 884[80]
Goulaouic, P., **7**, 276[150]
Gould, E. S., **7**, 750[134], 769[237,238], 770[238], 771[259,274]
Gould, I. R., **4**, 1081[79]
Gould, K. J., **4**, 670[22]
Gould, L. D., **5**, 143[102], 144[102]; **6**, 888[66], 1063[84]
Gould, S. J., **8**, 526[37], 531[37], 648[59]
Gould, T. J., **1**, 131[99]; **2**, 102[26]; **6**, 859[174,175]
Goulet, M. T., **2**, 42[148], 45[148]; **3**, 979[14]; **6**, 8[39]; **7**, 416[124], 549[47]
Gourcy, J.-G., **1**, 410[40]; **8**, 187[32], 188[32]
Gourdol, A., **6**, 423[48]
Goure, W. F., **2**, 727[132]; **5**, 757[78], 762[102]
Gouriou, Y., **6**, 745[84]
Gourlay, N., **4**, 707[42]
Gourley, R. N., **3**, 574[78]; **8**, 532[131a]
Goutarel, R., **6**, 266[49]
Gouverneur, V., **7**, 502[262]

Gouzoules, F. H., **4**, 301[316], 303[316], 310[316]; **8**, 852[142], 853[142b], 857[142b]
Govindachari, T. R., **2**, 894[9], 912[9], 1016[26]; **3**, 396[115], 741[50]; **6**, 736[25]; **7**, 221[32]
Govindan, C. K., **7**, 761[54]
Govindan, S. V., **4**, 1040[88], 1048[88,88b,g]; **5**, 907[76]; **7**, 362[29]
Gowal, H., **3**, 822[3,12], 831[3,12], 835[12b]
Gowda, D. S. S., **3**, 386[61], 393[61]
Gowda, G., **7**, 821[31]
Gowenlock, B. G., **8**, 364[9]
Gower, M., **4**, 670[20]
Gowland, B. D., **3**, 159[460], 161[460], 163[460], 167[460], 265[193]
Gowland, F. W., **8**, 2[13]
Gowriswari, V. V. L., **4**, 34[97], 35[97]
Gozlan, A., **3**, 640[101]
Gozzo, F., **6**, 171[6]
Grab, L. A., **8**, 543[242], 940[106]
Graber, D. R., **7**, 633[64,65]
Grabhöfer, H., **2**, 801[29]
Grabiak, R. C., **8**, 390[76]
Grabley, F.-F., **6**, 426[75,79], 480[115]
Graboski, G. G., **2**, 321[15], 325[15], 332[57]; **6**, 107[24]; **8**, 363[2]
Grabow, H., **7**, 95[73a]
Grabowich, P., **7**, 139[128]
Grabowski, E. J. J., **2**, 635[49], 640[49], 648[91], 649[91], 1059[74]; **4**, 230[251]; **6**, 22[83]; **7**, 752[154], 877[135]; **8**, 54[159], 66[159]
Gracián, D., **6**, 273[93]
Grade, M. M., **7**, 765[155]
Graden, D. W., **1**, 756[118]
Gradner, M., **8**, 384[31]
Graebe, C., **8**, 949[156]
Graefe, J., **3**, 380[6], 386[59]
Graf, B., **3**, 144[384], 145[384], 147[384]
Graf, G., **4**, 55[157]; **6**, 182[137]
Graf, H., **5**, 77[249,250]
Graf, R., **5**, 105[186,192], 107[192]; **6**, 261[6]; **7**, 10[76]
Graf, R. E., **2**, 721[90], 722[93,94]; **4**, 698[15,16,18], 699[18], 701[18,24]
Graf, W., **5**, 842[112]; **6**, 470[59,60], 471[59,60]; **7**, 721[17]; **8**, 825[67]
Graff, Y., **2**, 744[100]
Gräfing, R., **3**, 87[120,213], 88[120], 105[120,213,221]
Graham, B. W., **6**, 822[116]
Graham, C. R., **4**, 688[67]; **5**, 640[127], 715[78]
Graham, D. W., **8**, 530[103]
Graham, E. S., **3**, 856[87]
Graham, G. D., **8**, 724[169,169d,f]
Graham, J. E., **2**, 362[186], 363[186]
Graham, P., **4**, 987[147]
Graham, P. J., **6**, 288[184]
Graham, R. S., **1**, 405[25]; **2**, 679[56], 699[56]
Graham, S., **6**, 897[95]
Graham, S. H., **3**, 383[46,50], 735[19]; **4**, 6[23]
Graham, S. L., **2**, 148[79]
Graham, W. A. G., **7**, 3[13]
Grahn, W., **6**, 970[122]
Graillot, Y., **5**, 527[64], 530[64]
Grakauskas, J., **7**, 723[26]
Gralak, J., **1**, 56[30]
Gramain, J.-C., **2**, 1065[116]; **4**, 434[127]; **8**, 205[163], 558[399]
Gramatica, P., **2**, 853[231]; **7**, 109[183]; **8**, 560[404]
Grampoloff, A. V., **4**, 283[150]
Gramstad, T., **3**, 297[19]; **6**, 504[222]
Gran, H. J., **7**, 206[70], 210[70], 212[99]
Granados, R., **4**, 438[154]; **6**, 801[36]; **8**, 587[34]
Granata, A., **6**, 603[13]
Granberg, K. L., **6**, 849[116]
Granberg, L., **4**, 278[94], 285[94]
Grand, P. S., **3**, 14[77], 15[77]
Grandberg, A. I., **8**, 102[129]
Grandberg, I. I., **2**, 787[52]
Grandbois, E. R., **8**, 159[7], 166[65], 170[85], 178[65], 179[65]
Grandclaudon, P., **4**, 317[551], 318[561]; **6**, 474[82]
Grandi, R., **8**, 349[145,146]
Grandjean, D., **1**, 303[79]
Grandjean, J., **5**, 1130[6]
Grandmaison, J.-L., **5**, 330[36]
Granger, R., **3**, 851[65]; **4**, 301[323], 302[323,334]
Granik, V. G., **6**, 488[17,21,34], 502[208,209], 507[228,229], 529[17], 531[453], 546[21], 553[728], 554[728,729,734,741,755,763-765,767,770,773,776,779,780,784-786,789-793]
Granja, J. R., **3**, 983[19]
Granoth, I., **8**, 864[242]
Gränse, S., **3**, 634[7], 636[7,7b], 637[7]
Grant, B., **6**, 831[9]
Grant, D., **5**, 65[67]
Grant, D. P., **4**, 629[409]
Grant, H. G., **5**, 776[180]
Grant, J., **2**, 134[8]
Grant, R. D., **6**, 108[31]
Grant, R. W., **6**, 1016[27]
Grantham, R. K., **8**, 839[25]
Gras, J.-L., **2**, 156[118], 897[16]; **3**, 572[69], 573[69], 602[69], 607[69], 610[69]; **4**, 373[70]; **5**, 515[17], 518[17], 547[17]; **6**, 648[118]; **7**, 566[100], 711[60]
Graselli, G., **1**, 185[55], 186[55]
Grashey, R., **4**, 953[8,8a], 954[8a], 1069[5], 1070[5], 1095[153], 1097[162,165]; **8**, 391[92]
Grashey, R. K., **8**, 664[123]
Graske, K.-D., **5**, 1060[58]
Grasmuk, H., **2**, 464[97]
Grassberger, M. A., **6**, 173[48], 174[48]
Grasselli, P., **1**, 55[23], 206[111], 221[68], 389[139], 543[28], 830[92], 832[92]; **2**, 29[105], 30[113], 31[113,113a], 32[119,119b], 998[50,51], 999[50,51]; **4**, 36[103,103c]; **8**, 190[79], 195[110,111,113], 203[111,149,150], 343[116], 345[116], 349[116], 560[407], 942[117]
Grassl, M., **2**, 456[45,47]
Grattan, T. J., **5**, 832[40]; **7**, 463[127]
Gratz, J. P., **7**, 113[200]
Grauert, M., **2**, 498[72,73], 499[72,73]
Gravatt, G. L., **2**, 315[45], 316[45]; **4**, 1018[217]; **7**, 300[57]
Gravel, D., **5**, 229[122]; **6**, 529[424], 709[56], 715[93]
Graves, D. M., **6**, 124[146]
Graves, J. M. H., **4**, 1018[221]
Gravestock, M. B., **1**, 601[40], 608[40]; **3**, 364[93], 369[118], 372[118]; **8**, 244[66]
Gravier-Pelletier, C., **1**, 770[185]
Gray, A. P., **8**, 254[125]
Gray, B. D., **2**, 631[15]; **8**, 798[58]
Gray, D., **4**, 288[188,189], 290[189], 346[86a]
Gray, G. A., **4**, 601[268], 603[268]; **7**, 107[161]; **8**, 852[140]
Gray, G. R., **8**, 52[139], 66[139], 219[75-77]
Gray, M. D. M., **6**, 624[146], 625[146]
Gray, P., **7**, 8[63]
Gray, R. W., **6**, 1058[60]
Graysham, R., **6**, 134[22]
Grayshan, R., **3**, 125[295,296,301], 126[295,296,301], 128[296,301], 129[301], 133[301]
Grayson, D. H., **2**, 360[165a]; **3**, 710[19]; **7**, 102[134]
Grayson, E. J., **4**, 354[129]

Grayson, J. I., **1**, 774[210]; **3**, 123[251]; **4**, 1076[39]; **5**, 329[30], 333[30], 434[140]; **6**, 932[94]
Grayson, M., **3**, 300[35]; **8**, 413[126]
Grayston, M. W., **6**, 980[37]; **8**, 889[127], 992[53]
Graythwaite, W. R., **7**, 779[423]
Graziani, M., **4**, 753[170]; **8**, 534[159]
Graziano, M. L., **4**, 1036[45]; **6**, 558[857,858]
Grdina, M. J., **7**, 816[10], 818[10]
Gream, G. E., **4**, 784[14]
Greaves, A. M., **2**, 1057[71], 1058[71]; **5**, 407[29]
Greaves, E. O., **2**, 722[91,92]; **4**, 697[12], 698[14]
Greaves, P. M., **4**, 55[157]; **6**, 235[53], 265[40]
Greaves, W. S., **3**, 643[125]
Grebenik, P. D., **8**, 459[239]
Greber, G., **8**, 535[166], 769[28]
Greci, L., **2**, 787[52]
Greck, C., **1**, 513[49], 833[118]; **3**, 174[527], 175[527c], 258[127]; **4**, 381[126b], 382[126], 383[126]; **6**, 93[132], 155[155], 156[155,156,157], 994[98]; **7**, 487[146], 495[146]; **8**, 12[66], 837[17]
Greco, A., **5**, 1142[86]; **8**, 772[61]
Grée, D., **5**, 254[48]
Grée, R., **1**, 821[23]; **2**, 97[60]; **4**, 38[109b], 957[23,24], 964[48], 990[146], 1040[66], 1041[66]; **5**, 254[48], 944[244]; **6**, 690[401], 692[401]; **2**, 97[60]; **7**, 713[72]
Greeley, A. C., **5**, 70[116], 514[9], 527[9], 805[100]
Greeley, R. H., **5**, 716[88]
Green, A. G., **7**, 558[77]
Green, B., **2**, 355[119], 382[314]; **8**, 248[83]
Green, B. R., **5**, 732[132,132b]
Green, B. S., **3**, 382[36]; **5**, 183[157]; **8**, 977[138]
Green, D. L. C., **1**, 350[155]
Green, E. E., **3**, 649[207]
Green, F. R., III, **1**, 804[313], 805[313]; **4**, 142[16]; **5**, 308[95]; **6**, 938[126], 968[107], 998[120,121]
Green, G., **5**, 453[65]
Green, G. E., **3**, 735[18]
Green, I. R., **2**, 746[109]
Green, J., **6**, 644[83]; **8**, 528[81], 529[81]
Green, J. A., II, **6**, 294[231]
Green, K. E., **4**, 1010[157]
Green, M., **1**, 310[103]; **4**, 709[44,47], 710[44,47,49], 712[49,70]; **5**, 1134[45], 1136[54], 1146[107]; **7**, 94[55]; **8**, 445[39,52], 766[18]
Green, M. J., **6**, 563[903], 1019[44]
Green, M. L. H., **1**, 139[4]; **4**, 980[108]; **6**, 686[370]; **7**, 3[11], 4[17,18]; **8**, 99[107], 459[439], 889[129]
Green, M. M., **7**, 777[369]
Green, R., **7**, 444[57]
Green, R. L., **8**, 170[74]
Green, R. M. E., **7**, 78[125]
Green, S., **8**, 101[122]
Green, S. I. E., **8**, 725[185]
Green, T. W., **1**, 548[64], 563[177], 564[177]; **4**, 308[400]; **6**, 632[6]
Greenber, D. M., **2**, 466[123,124], 469[123,124]
Greenberg, H. J., **4**, 473[145]
Greenberg, R. S., **4**, 1033[26], 1046[26b,116], 1048[120]
Greenberg, S. G., **5**, 492[246]
Greenburg, R. B., **3**, 842[18]
Greene, A. E., **1**, 848[25-27], 849[26,29], 850[26]; **2**, 160[132]; **3**, 222[141], 783[26,29]; **4**, 95[99c]; **5**, 1062[59]; **6**, 129[167], 175[69]; **7**, 121[20,21], 123[20], 145[20], 163[71], 406[76]; **8**, 357[201], 358[200,201], 881[69], 925[11], 926[11], 932[42]
Greene, F. D., **5**, 637[108]; **6**, 962[74]; **7**, 750[136]; **8**, 35[63], 66[63], 389[72]
Greene, G. L., **8**, 385[43]
Greene, J. C., **4**, 467[130]
Greene, J. L., **3**, 723[10]
Greene, J. L., Jr., **6**, 993[94]
Greene, J. M., **8**, 231[143], 275[145,146]
Greene, R. M. E., **5**, 1116[5]
Greene, R. N., **5**, 908[72]
Greene, T. W., **6**, 632[5], 635[5], 659[5], 664[5], 665[5], 675[5], 678[5], 680[5], 682[5], 749[99], 866[207]; **8**, 74[246], 211[1], 212[1], 320[87], 956[8]
Greenfield, H., **8**, 452[189c], 456[208], 459[228], 608[44]
Greenfield, S., **7**, 253[21]
Greengrass, C. W., **2**, 907[60]
Greenhalgh, P. F., **5**, 7[52]
Greenhalgh, R., **8**, 884[101]
Greenhouse, R., **8**, 587[33]
Greenland, H., **7**, 352[31], 356[31]
Greenlee, K. W., **3**, 273[5]; **4**, 283[151]; **8**, 479[49]
Greenlee, T. W., **8**, 433[69]
Greenlee, W. J., **1**, 555[114]; **5**, 351[82]
Greenslade, D., **6**, 920[46]
Greenspan, P. D., **5**, 311[105]
Greenspoon, N., **4**, 604[287], 605[287], 606[300], 626[287], 647[287]; **8**, 553[363,364], 554[364], 555[364,369], 961[38,40], 984[4], 991[4]
Greenstein, J. P., **8**, 145[88]
Greenwald, R. B., **1**, 835[137]
Greenwell, P., **8**, 52[146], 66[146]
Greenwood, G., **5**, 596[24,32], 597[24,32], 603[32]
Greenwood, J. M., **3**, 399[117], 402[117], 600[215]
Greenwood, N. N., **1**, 292[28]
Greenwood, T. D., **2**, 189[54]; **4**, 462[104], 464[118], 465[104,118], 466[104], 468[104], 469[104]
Greer, S., **7**, 603[120,121]
Greeves, N., **1**, 778[221]; **8**, 13[67]
Gregersen, N., **4**, 36[102,102e]
Gregoire, B., **4**, 496[95]; **5**, 692[100]
Gregorcic, A., **4**, 356[139]
Gregoriou, G. A., **8**, 986[11]
Gregory, B. J., **6**, 291[216]
Gregory, C. D., **3**, 211[31], 213[31]; **4**, 170[16]
Gregory, J. A., **6**, 1002[133]
Gregory, P., **1**, 19[100]
Gregory, W. A., **6**, 76[45]
Gregson, M., **3**, 357[64]; **6**, 206[44]
Grehn, L., **6**, 81[75]; **8**, 336[71]
Greiber, D., **3**, 903[128]
Greibrokk, T., **4**, 1001[30], 1003[66]
Greig, D. G. T., **6**, 1017[38], 1024[38]
Greijdanus, B., **4**, 230[242], 231[242]
Grein, F., **4**, 46[131]
Greiser, T., **1**, 16[85], 38[253,255,257], 39[188]
Gremban, R. S., **6**, 237[64], 243[64]
Grenier, L., **6**, 74[30]
Gresham, D. G., **2**, 722[94]
Gresham, T. L., **2**, 662[12]
Gressier, J.-C., **6**, 455[150], 456[156], 545[632]
Grethe, G., **5**, 499[255]; **7**, 678[72]; **8**, 968[86]
Greuter, H., **1**, 830[96]; **3**, 848[50]; **5**, 4[31], 829[13]
Grevels, F.-W., **5**, 1130[9], 1131[15]
Grevil, F. S., **6**, 545[633]
Grewal, R. S., **6**, 543[618]
Grewe, R., **2**, 1023[54,55]; **8**, 263[26], 493[23]
Grey, R., **5**, 692[103]
Grey, R. A., **1**, 448[208], 449[208]; **8**, 242[44], 252[44], 291[33], 455[205]
Grgurina, I., **6**, 490[109]
Grhosh, A. K., **6**, 674[294]

Griasnow, G., **8**, 717[106]
Gribble, G. W., **1**, 473[81], 474[86,87,98]; **2**, 744[94]; **3**, 261[155], 262[170], 311[98], 644[154]; **5**, 382[119b], 384[128], 385[128b]; **6**, 487[61], 489[61], 724[161]; **8**, 16[108], 17[108], 26[14], 27[14], 36[14,48], 55[14], 60[14,186], 66[48,186], 215[52], 315[49,50], 347[140], 581[8], 582[8], 615[96], 616[99,100,102-105], 617[102,106,107], 618[102,107,119], 619[103,105], 624[96,107], 968[87]
Grice, P., **1**, 801[303]; **3**, 174[527], 175[527c]; **4**, 381[126b], 382[126], 383[126], 1018[223], 1019[223]; **6**, 994[98]; **8**, 297[66]
Grider, R. O., **5**, 220[43-45]
Gridnev, I. D., **2**, 709[15]
Grieco, P. A., **1**, 134[113], 410[41], 473[78], 562[167], 642[110,114,117], 643[110], 645[117], 686[117]; **2**, 160[133,134], 369[249], 737[32], 1002[56,57], 1027[76], 1034[89,90]; **3**, 17[87], 20[113], 101[506], 155[431], 159[462], 161[462], 168[506], 169[506], 170[506], 212[39], 220[124], 253[92], 882[105], 894[63], 943[82], 964[126], 1051[21]; **4**, 75[42b], 112[158c], 115[183], 253[176], 255[176], 259[264], 307[398], 308[398], 673[30], 952[7], 991[148], 1103[208]; **5**, 172[118], 344[67a-c], 345[67d,e], 349[76], 350[77], 351[81], 408[34,35], 409[34], 411[34,43], 413[51], 415[51b-d], 500[261], 532[83], 534[95,95h], 539[95g], 552[39,40], 854[175,179,180], 855[179], 905[55]; **6**, 157[171], 206[44], 439[71], 466[43,44], 469[44], 648[125], 656[171], 756[127], 836[50], 841[76], 899[110], 900[110], 1021[49,53,55], 1029[103], 1030[105,106], 1032[119], 1033[123]; **7**, 105[148], 125[53], 126[53], 129[73,74], 130[73-75], 377[90], 569[107], 674[40,44,45], 682[83,84], 701[65]; **8**, 112[24], 120[24], 527[47], 540[195], 544[255], 843[50], 932[40], 945[131]
Grieder, A., **4**, 242[62], 253[62], 261[62]
Grier, D. L., **8**, 94[79]
Grierson, D. S., **1**, 367[55], 557[127,-128], 559[143,144,149], 564[206]; **2**, 63[22b], 1014[19], 1018[41,43,45]; **5**, 829[22]; **6**, 910[6], 917[34], 919[40]
Grierson, J. R., **5**, 979[28], 992[49]
Gries, J., **6**, 501[201]
Griesbaum, K., **4**, 276[77], 277[88], 284[77], 288[77], 289[77], 317[553], 770[246]; **7**, 574[140], 579[136], 581[140], 582[140,147]
Griesbeck, A., **7**, 384[114c], 399[38], 400[38,38b], 406[38], 409[38], 415[38], 818[16]
Griese, A., **8**, 589[52]
Griesinger, A., **2**, 1088[51]
Griess, P., **6**, 245[115]; **8**, 916[109], 918[109]
Griesser, H., **6**, 637[32,32c]; **8**, 535[166]
Griessmayer, V., **3**, 660[2]
Griessmeier, H., **3**, 864[19]
Griffin, C. E., **3**, 748[76]
Griffin, D. A., **4**, 675[40], 679[40]
Griffin, G. W., **3**, 572[67]; **4**, 1089[131], 1090[140,141,142]; **5**, 128[27], 199[26], 200[26], 208[52], 947[267], 948[267]; **7**, 372[70]
Griffin, I. M., **4**, 306[376,381,382], 310[428]
Griffin, J. H., **3**, 229[224,224b], 920[34], 923[34b], 934[34], 1008[69]; **4**, 965[51], 1035[37], 1036[49,53], 1037[37]; **6**, 873[9], 874[9c]; **8**, 84[12]
Griffin, T. S., **8**, 366[50]
Griffini, A., **4**, 764[218]
Griffith, D. L., **6**, 958[30]
Griffith, E. J., **8**, 413[126]
Griffith, J. R., **6**, 244[107]
Griffith, R., **4**, 37[104]
Griffith, R. C., **2**, 291[73]; **4**, 231[278], 290[197]
Griffith, R. K., **6**, 173[47]
Griffith, W. P., **7**, 311[30], 312[30], 439[24], 489[172]
Griffiths, D., **6**, 487[70]
Griffiths, G., **4**, 54[154]
Griffiths, J., **5**, 223[78,79], 730[126]
Griffiths, R., **4**, 33[95,95a]
Grigat, E., **6**, 243[100,101]
Grigat, H., **8**, 63[197], 64[197], 66[197]
Grigg, R., **2**, 1017[32]; **3**, 484[21], 770[180]; **4**, 374[88], 753[173], 848[78], 854[93], 1086[114,116]; **5**, 165[87], 257[59], 790[32], 802[81], 936[199], 1149[125]; **7**, 451[21], 453[95], 831[66]
Grignard, V., **2**, 136[14]; **3**, 243[15]
Grignon, J., **4**, 743[128]
Grignon-Dubois, M., **2**, 717[69], 718[74], 728[140,141]
Grigor'ev, A. B., **6**, 502[208], 554[792]
Grigor'eva, N. V., **1**, 34[228,232]
Grigoryan, D. V., **5**, 410[40]
Grigoryan, G. V., **6**, 712[78]
Grigoryan, M. S., **4**, 885[115], 886[115]; **5**, 34[10]; **8**, 727[196]
Grigos, V. I., **5**, 480[177]
Grill, W., **6**, 291[205]
Griller, D., **1**, 274[72]; **3**, 69[26]; **4**, 717[13,14], 719[27], 722[34], 728[34], 739[106], 785[22], 1081[80]; **5**, 901[29]; **8**, 801[75]
Grillot, G. F., **1**, 370[67]; **2**, 1004[61]; **3**, 258[124]
Grimaldi, J., **4**, 394[189], 395[205], 396[189d,208], 397[208], 411[267a]; **5**, 772[153,154,158]
Grimaldo Moron, J. T., **8**, 860[223]
Grimm, D., **1**, 372[79]; **2**, 1049[14]
Grimm, E. L., **4**, 27[79], 1048[119]; **5**, 539[105]
Grimm, K., **3**, 595[191]
Grimm, K. G., **6**, 147[87]
Grimm, R. A., **8**, 836[1,1c], 837[1], 995[69]
Grimme, W., **5**, 10[78], 552[29,37], 702[10], 716[10], 794[47], 806[47], 824[47], 847[136]
Grimmett, M. R., **7**, 750[124]
Grimova, J., **6**, 524[368]
Grimshaw, C. E., **5**, 855[186]
Grimshaw, J., **3**, 568[41], 574[78], 577[87], 677[83,84]; **5**, 724[112]; **6**, 639[46]; **8**, 524[12], 527[49], 532[12c,131a]
Grimshire, M. J., **7**, 481[106]
Grimwade, M. J., **2**, 782[19]
Grinberg, S., **6**, 825[126]; **7**, 692[22]
Grinberg, V. A., **3**, 647[177,196], 648[177,178], 649[177]
Grindley, T. B., **1**, 476[121]; **2**, 167[158]; **3**, 66[10]
Grinev, A. N., **2**, 785[42]
Gringauz, A., **7**, 674[35]
Gringore, O., **4**, 91[90]; **5**, 561[81]; **6**, 689[385]; **8**, 543[246]
Grinvald, A., **1**, 292[26]
Griot, R. G., **6**, 523[353]
Grippi, M., **5**, 404[18]
Grisar, J. M., **2**, 943[173], 945[173]; **3**, 380[16], 735[17]
Grisdale, P. J., **1**, 580[4]; **8**, 496[34]
Grisebach, H., **6**, 294[235]
Grisenti, P., **8**, 244[56]
Grishin, Y. K., **8**, 99[107]
Grison, C., **6**, 437[39]
Grisoni, S., **6**, 849[119]
Grisso, B. A., **2**, 655[149]
Grissom, J. W., **4**, 1086[111]; **8**, 651[68]
Griswold, P. H., Jr., **6**, 678[324]
Grivas, J. C., **6**, 546[649]
Grivet, C., **4**, 532[88], 534[88], 537[88], 538[88], 539[88]
Grizik, S. I., **6**, 554[773]
Grob, C. A., **1**, 894[158]; **2**, 331[53], 332[53], 1023[56], 1026[56], 1047[4]; **4**, 273[44], 285[162]; **5**, 809[111]; **6**, 912[21], 1041[1,2], 1042[1,2,8], 1043[8]; **7**, 694[31], 700[61], 724[30]
Grob, J., **6**, 1015[23]
Grobe, J., **5**, 442[183], 444[188], 577[147]
Grobe, K. H., **7**, 10[80]
Gröbel, B. I., **4**, 120[201]
Gröbel, B. T., **1**, 542[3], 544[3], 563[3], 564[3], 566[3], 568[3], 569[3], 636[95], 642[95], 665[95], 672[95]; **3**, 86[20], 87[82,88],

104[20], 105[82], 112[82], 120[20,82], 121[20], 125[20], 126[20], 128[322], 130[322], 131[324,331], 134[322], 135[324]; **4**, 113[164], 120[201]; **6**, 134[19,35], 678[321], 679[321]
Grochulski, P., **4**, 83[65c]
Grodkowski, J., **7**, 850[10]
Grodski, A., **7**, 686[99]
Groen, M. B., **3**, 361[76], 367[76,100,103]
Groen, S. H., **4**, 1005[89]
Groenen, L. C., **8**, 98[104]
Groenewegen, P., **2**, 101[19]; **3**, 20[121], 25[121]; **7**, 125[57]
Grogg, P., **6**, 671[277]
Groginsky, C. M., **5**, 803[88]
Groh, B. L., **3**, 232[265]
Grol, C. J., **2**, 902[46]
Groliere, C. A., **8**, 560[402]
Grollman, A. P., **4**, 48[139]
Gromek, J. M., **1**, 35[171], 341[96], 477[132], 482[132]
Gromov, S. P., **4**, 424[19]
Gromova, G. A., **3**, 639[80]
Grondey, H., **1**, 180[30]
Grondin, J., **1**, 212[4]
Gronert, S., **1**, 41[197]
Gröniger, K. S., **4**, 791[52]
Gröning, C., **4**, 1017[216]
Gröninger, K. S., **5**, 159[51], 189[51]
Gronowitz, S., **2**, 765[70]; **3**, 232[259], 511[188], 515[188]; **8**, 384[37], 608[39], 678[58,64,65], 683[58,64,65], 686[58,64,65], 691[58], 908[74]
Gronwall, S., **7**, 163[72]
Gronyu, M. W., **6**, 496[152]
Groody, E. P., **6**, 625[162]
Gropen, O., **1**, 488[8]
Gros, E. G., **6**, 278[132]
Grosclaude, J.-P., **5**, 216[16], 219[16], 221[16]
Grosheintz, J. M., **8**, 296[63]
Grosjean, F., **3**, 579[100]
Gross, A. W., **2**, 169[164], 182[11], 600[33], 605[33]; **3**, 3[9], 891[40], 900[40]; **4**, 370[43], 763[213]; **5**, 843[125], 853[125a]; **7**, 737[16]; **8**, 395[128]
Gross, B., **1**, 219[62]; **2**, 977[4]; **6**, 74[36], 790[111]
Gross, E., **2**, 1094[86]; **6**, 635[11], 645[11], 664[224], 665[227], 668[227], 669[227], 670[270], 698[270]
Gross, G., **4**, 484[15]; **6**, 440[74]
Gross, G. W., **4**, 1000[13]
Gross, H., **2**, 60[18], 900[26], 960[35], 962[35], 1088[44]; **6**, 134[36], 488[30], 533[513], 566[30]; **7**, 235[1]
Gross, N., **2**, 933[137]
Gross, R. S., **4**, 161[87], 793[69]
Gross, S., **2**, 547[113], 551[113]; **3**, 56[285]; **4**, 229[236]
Grossa, M., **3**, 817[90,92]
Grosse, A. V., **3**, 331[197]; **4**, 270[7,8], 271[24], 272[24]
Grosse Brinkhaus, K.-H., **6**, 561[874]
Grossel, M. C., **8**, 504[71]
Grosser, J., **1**, 873[92,95a]
Grossert, J. S., **1**, 528[122], 804[308], 805[308]; **3**, 865[28]; **8**, 403[4], 404[4], 407[4]
Grossi, L., **1**, 569[262]; **4**, 424[11], 428[11,77]
Grossman, J., **4**, 1061[162]
Grossman, N., **3**, 804[4]
Grostic, M. F., **4**, 273[53]
Grote, J., **5**, 517[29], 519[29], 534[29]; **7**, 273[135]
Grotemeier, G., **3**, 45[251]
Grotewold, J., **7**, 605[140]
Groth, P., **1**, 34[230]
Groth, U., **2**, 498[71,74,76,78], 499[71,74,76], 501[78]; **3**, 53[274]; **6**, 531[431,433,435,443]
Grothaus, P. G., **7**, 579[132]; **8**, 201[140,142]
Grotjahn, D. B., **1**, 582[7]; **2**, 572[42]; **4**, 120[201]; **5**, 1143[93], 1144[93]; **6**, 134[29]
Groutas, W. C., **1**, 328[20]
Grove, J. F., **3**, 736[25]
Grovenstein, E., Jr., **5**, 646[6]; **8**, 828[76]
Grover, E. R., **6**, 581[989]; **7**, 800[30]; **8**, 85[19]
Grover, S. K., **7**, 143[150], 144[150]
Groves, J. K., **2**, 708[3], 710[3]; **4**, 238[4]; **5**, 581[176], 777[183]
Groves, J. T., **4**, 1006[100], 1021[240]; **7**, 11[92], 12[93], 50[73], 95[73b], 426[148b]; **8**, 807[114]
Growe, W. E., **6**, 692[409]
Groweiss, A., **3**, 380[13], 395[102], 396[102], 407[146]
Grözinger, K., **2**, 1103[130]
Grubb, P. W., **5**, 714[65]
Grubb, S. D., **8**, 880[59]
Grubbs, E. J., **1**, 846[11]
Grubbs, R. H., **1**, 743[51-53,66-68], 744[52], 745[68], 746[51,62,67,71], 747[52], 748[67,68,72,74], 749[52,78,81], 811[51,52], 816[78], 850[31]; **2**, 309[26], 597[9], 1062[98]; **4**, 979[101]; **5**, 948[271], 1115[1-3], 1116[1,2], 1118[18], 1120[21], 1121[18,26-28], 1122[2a,3,30,31], 1123[2a,3,42], 1124[3,28,43,44,48,49], 1126[1d], 1131[16]; **8**, 447[126], 457[126], 676[54,80]
Grübel, B. T., **1**, 666[173], 695[173]
Grubenmann, W., **8**, 907[70]
Gruber, J. M., **2**, 542[81,82]; **4**, 952[5]; **7**, 167[93], 177[145,146], 178[149], 182[164], 186[179], 673[24]
Gruber, L., **2**, 529[17]; **7**, 723[25]
Gruber, P., **4**, 12[40], 42[120]
Grubmüller, B., **5**, 598[33]
Grudoski, D. A., **2**, 102[22]
Grudzinski, Z., **7**, 674[33]
Gruenanger, P., **5**, 626[33], 630[53], 631[53]
Grueter, H.-W., **5**, 636[98]
Gruetzmacher, G., **3**, 969[131,132]; **4**, 430[94]
Gruetzmacher, H. F., **4**, 484[18]
Grugel, C., **3**, 571[59]
Gruhl, A., **4**, 663[1]
Grummitt, O., **2**, 142[48]; **6**, 204[18]
Grumüller, P., **8**, 320[80]
Grunanger, P., **4**, 1078[54,55], 1083[55], 1084[55]
Grünbaum, W. T., **5**, 209[55]
Grund, H., **2**, 338[78]; **8**, 647[54]
Gründemann, E., **6**, 727[202]
Grundke, G., **6**, 116[89]
Grundkötter, M., **3**, 904[131]
Grundman, L., **4**, 483[4], 484[4], 495[4]
Grundmann, C., **4**, 1078[54]; **5**, 379[112], 383[112], 384[112]; **6**, 225[3], 226[3], 227[3], 228[3], 229[3], 231[3], 232[3], 233[3], 234[3], 235[3], 236[3], 237[3], 238[3], 240[3], 241[3], 242[92], 243[92], 245[118], 246[118], 247[118], 248[118], 249[118], 252[118], 253[118], 254[118], 255[118], 258[3], 259[3], 817[104]; **8**, 392[104]
Grundon, M. F., **4**, 393[187]; **7**, 227[83]; **8**, 170[93], 336[73], 337[73], 338[73], 339[73], 341[73]
Grunenwald, G. L., **7**, 474[36]
Gruner, I., **8**, 856[167]
Grunert, R. R., **6**, 270[77]
Gruning, R., **1**, 6[32], 37[179]
Grunwald, F. A., **8**, 408[64]
Grunwald, J., **8**, 185[10]
Grunwell, J. R., **2**, 745[107]; **6**, 437[35]
Gruse, W. A., **7**, 7[38]
Gruska, R., **5**, 95[89]
Grüssner, A., **6**, 37[32]
Grütter, H., **3**, 345[20]
Grutter, P., **4**, 1055[137]

Grüttner, S., **3**, 303[53]
Grützmacher, H.-Fr., **3**, 587[147]
Grutzner, J. B., **1**, 42[204]; **2**, 5[20], 6[20], 21[20], 524[77]; **5**, 552[36], 568[107], 847[136]; **8**, 767[23]
Grzegorzek, M., **2**, 105[41]
Grzegorzewska, U., **6**, 276[117]
Grzejszczak, S., **3**, 953[105]; **6**, 134[36], 150[117]; **7**, 197[22]
Grzeskowaik, N. E., **1**, 464[36], 478[36]
Gschneidner, K. A., Jr., **1**, 231[4], 251[1,2], 252[1]
Gschwend, H. W., **1**, 23[120], 460[4], 463[4], 464[34], 471[4], 472[4], 473[4], 480[155]; **2**, 74[80]; **3**, 67[19], 193[2], 194[2], 261[146], 264[146]; **5**, 527[64,65], 530[64,65]; **6**, 554[766]
Gschwend-Steen, K., **5**, 828[6], 836[6], 888[26], 893[26]
Gu, C. I., **7**, 769[220], 884[187]
Gu, J.-M., **5**, 829[21]
Gu, K., **6**, 71[23]
Gu, X., **2**, 109[71]
Gu, Y., **8**, 447[135]
Gu, Z., **4**, 991[152]
Gua, M., **4**, 1060[160]
Guaciaro, M. A., **2**, 388[340], 651[122]; **3**, 26[165]
Guademar, M., **1**, 179[23]
Guadino, J. J., **4**, 791[44]
Guajardo, R., **1**, 767[177]
Guan, X., **6**, 820[112]
Guanti, G., **1**, 527[101,102], 568[244,247]; **2**, 636[53], 639[63], 640[53,63], 642[78], 643[78], 922[101], 923[101], 931[135], 933[135,136], 934[135,136], 940[135,136]; **4**, 426[52]; **5**, 100[153], 102[177]; **6**, 118[107], 149[100]; **8**, 187[45]
Guaragna, A., **1**, 568[244]
Guardado, P., **4**, 710[50]
Guare, J. P., **5**, 410[41]; **7**, 673[30]
Guarna, A., **8**, 851[127]
Guarneri, M., **7**, 143[140]
Guarnieri, A., **2**, 787[52]
Guastini, C., **1**, 29[150]; **2**, 127[229]; **5**, 1079[51]; **6**, 178[121]; **8**, 683[90]
Gubaidullin, L. Yu., **8**, 697[133], 698[133]
Gubenko, N. T., **4**, 115[180b]
Gubernantorov, V. K., **7**, 505[289]
Gubernator, K., **5**, 812[129]
Gubernick, S., **1**, 390[142]; **7**, 330[10]
Gubler, B., **2**, 157[120]; **8**, 544[275]
Guedin-Vuong, C., **4**, 261[292]
Guedin-Vuong, D., **7**, 679[74]
Guenard, D., **7**, 169[108]
Guenot, P., **5**, 444[187]
Guenther, C., **8**, 191[92]
Guenther, T., **2**, 353[96], 388[96]
Guenther, W. H. H., **6**, 462[19]
Guenzet, J., **4**, 312[457]
Guenzi, A., **7**, 764[126], 767[126]
Guerchais, V., **4**, 985[127]
Guerin, C., **1**, 113[31], 619[59], 623[81]; **2**, 58[11]; **4**, 120[197], 630[418]
Guerin, P., **7**, 426[148c]
Guerin, P. M., **3**, 223[155]
Guéritte, F., **6**, 921[47], 1067[100]
Guerrero, A., **6**, 172[18]; **8**, 563[425]
Guerrero, A. F., **7**, 530[16]
Guerrier, L., **1**, 559[143]
Guerrieri, F., **3**, 1027[43]
Guerriero, A., **7**, 579[137]
Guerry, P., **8**, 597[86]
Guessous, A., **6**, 150[127,130]
Guest, I. G., **3**, 741[53]
Guest, M. F., **1**, 9[43]; **4**, 484[16]
Guette, J. P., **2**, 286[62]; **4**, 106[138d]; **5**, 527[64], 530[64]; **6**, 763[7]; **7**, 876[125]
Guével, A., **6**, 554[751,752]
Gugelchuk, M., **2**, 368[240], 873[25]; **5**, 1070[21], 1072[21]; **8**, 447[129], 463[129]
Guggenberger, L. J., **1**, 13[71], 749[76]; **5**, 1124[45]; **8**, 614[88], 621[88]
Guggenheim, T. L., **6**, 237[64], 243[64]
Guggisberg, A., **6**, 1058[61]
Guggisberg, D., **4**, 1009[135]
Guglielmetti, R., **6**, 568[932,933]
Guha, P. C., **3**, 894[64]
Guhl, D., **5**, 1140[77]
Guiard, B., **5**, 178[135]
Guibe, F., **1**, 440[170]; **4**, 590[91], 605[294,296,297,298,299], 647[297]; **6**, 641[58], 659[196]; **8**, 21[143], 265[48,49], 553[362]
Guibé-Jampel, E., **6**, 639[48]
Guida, A. R., **8**, 237[18], 244[18], 249[18], 250[18]
Guida, W. C., **8**, 237[18], 240[33], 244[18], 249[18], 250[18]
Guida-Pietra Santa, F., **8**, 862[230]
Guigina, N. I., **3**, 648[175]
Guignard, H., **8**, 227[117]
Guilard, R., **2**, 780[9]; **6**, 781[80]; **8**, 946[134]
Guiles, J., **3**, 79[64]
Guilford, W. J., **5**, 855[188]
Guilhem, J., **4**, 748[158], 800[120]; **6**, 718[122]; **7**, 731[53]
Guillaumet, G., **4**, 496[94]; **5**, 1022[75]; **6**, 648[120]
Guillemin, J. C., **5**, 444[187], 576[214], 589[214]
Guillemonat, A., **4**, 55[155], 277[82], 284[82]; **7**, 84[2], 85[2], 500[242]
Guillemot, M., **5**, 289[40], 290[40]
Guillerm, D., **3**, 217[94]
Guillerm, G., **2**, 575[63]
Guillet, J. E., **5**, 161[59,60]
Guilmet, E., **7**, 12[94]
Guimbal, G., **4**, 95[102f]
Guin, H. W., **3**, 391[89], 393[89]
Guindi, L. H. M., **8**, 363[4], 375[4]
Guindon, Y., **1**, 821[28]; **3**, 227[208], 289[67], 771[185]; **4**, 370[42], 379[115,115e], 380[115e], 381[129], 382[129], 1059[155]; **6**, 48[84]; **7**, 360[21]; **8**, 406[43], 662[115]
Guingant, A., **2**, 1054[63]; **4**, 7[25], 221[161,165]; **5**, 327[27]; **6**, 738[51]
Guinguene, A., **1**, 474[99]
Guir, C., **3**, 226[207]; **4**, 209[69], 895[164], 900[164]; **6**, 849[122]
Guise, G. B., **3**, 831[59]; **6**, 440[73]
Guisnet, M., **6**, 563[897]
Guitard, M., **2**, 432[55]
Guitart, J., **8**, 563[425]
Guither, W. D., **8**, 905[62]
Guitian, E., **4**, 483[5], 484[5], 495[5], 513[179,180]; **5**, 384[127]
Guittet, E., **1**, 554[106], 664[170,200], 665[170], 669[170], 670[170], 672[200], 718[200], 719[200], 720[200], 722[200], 870[84]; **3**, 147[395], 155[395]; **4**, 113[169,169c], 820[220]; **6**, 838[65]
Guivisdalsky, P. N., **6**, 23[90]; **7**, 393[16], 398[16]
Guixer, J., **2**, 435[62]; **6**, 80[68]
Guk, Y. V., **6**, 110[50]
Gülaçar, F. O., **8**, 134[36]
Gulácsi, E., **7**, 723[25]
Guladi, P., **2**, 1099[115]
Gulevich, Yu. V., **1**, 438[159]
Guli'nski, J., **8**, 764[8], 774[72,73]
Gull, M.-R., **2**, 498[71], 499[71]
Gulles, J., **7**, 224[50]
Gullotti, M., **7**, 194[9], 777[382]

Gulta, V. S., **7**, 695[36]
Gul'tyai, V. P., **8**, 611[64]
Gum, C. R., **5**, 1117[14]
Gumbel, H., **4**, 1075[33]; **6**, 523[349]
Gumby, W. L., **1**, 3[18,21]
Gumulka, M., **8**, 34[60], 66[60]
Gunar, V. I., **6**, 489[93]
Gunatilaka, A. A. L., **4**, 380[121], 674[35], 688[35]; **6**, 690[399], 691[399], 692[399]
Gunawardana, D., **3**, 71[28]
Gundermann, K. D., **5**, 64[46-49]; **7**, 758[4]
Gundiah, S., **5**, 1116[4]
Gundlach, K. B., **4**, 596[160], 604[286], 621[160], 626[286], 636[160]
Gung, W. Y., **1**, 188[68]; **2**, 39[136], 574[57]; **8**, 164[45], 537[186], 546[186]
Güngör, T., **1**, 474[97,102]
Gunn, B. G., **5**, 515[12,18], 516[12b], 547[18]
Gunn, D. M., **2**, 242[15]; **4**, 145[30]
Gunn, P. A., **3**, 715[37]
Gunn, V. E., **7**, 663[59]
Gunning, H. E., **3**, 892[48]
Gunsalus, I. C., **7**, 80[140]
Gunsher, J., **8**, 965[67]
Gunstone, F. D., **3**, 752[96]; **4**, 310[423,424], 390[173a]; **7**, 437[2], 438[2]; **8**, 855[158]
Gunter, M. J., **2**, 367[226], 378[289]; **6**, 279[139]
Günter, W., **2**, 163[148]
Günthard, H. H., **4**, 277[85], 285[85], 288[85]; **6**, 968[111]
Günther, H., **1**, 2[10], 37[247]; **5**, 714[74], 929[165]; **6**, 448[111]; **8**, 205[158,164], 561[409]
Gunther, H. J., **3**, 390[82], 392[82]; **4**, 380[122]
Günther, K., **6**, 288[186]; **7**, 760[42]
Günther, P., **3**, 872[61]
Günther, W., **3**, 41[228], 43[239]
Günther, W. H. H., **1**, 630[3]; **6**, 461[1,2], 462[11,12]; **7**, 774[311]; **8**, 370[88]
Guntrum, E., **4**, 91[89], 380[122]
Gunzinger, J., **3**, 813[62]
Guo, B.-S., **1**, 239[39]; **2**, 547[100], 548[100]
Guo, D., **4**, 741[120], 768[238]; **8**, 852[136]
Guo, G., **8**, 879[55], 880[55]
Guo, M., **2**, 146[70]; **4**, 107[141]; **5**, 356[90], 358[90b]
Guo, Q., **8**, 677[59], 685[59]
Guo, T., **5**, 344[66], 345[66], 346[66]; **7**, 48[61], 49[64]
Guo, W., **6**, 575[971]
Guo-giang, L., **7**, 844[58]
Gupta, A. K., **3**, 797[94]; **7**, 595[127], 604[127]; **8**, 722[146,149]
Gupta, A. R., **3**, 304[64]
Gupta, A. S., **3**, 402[127]
Gupta, B., **7**, 17[177]
Gupta, B. G. B., **2**, 728[138]; **4**, 356[145]; **6**, 210[75], 214[93], 215[93], 647[108,109], 654[152], 940[148]; **7**, 752[152], 765[141], 769[210]
Gupta, D., **7**, 544[40], 551[40], 556[40]
Gupta, D. N., **5**, 124[11], 129[11]; **7**, 709[36], 747[101], 765[136], 843[49]
Gupta, I., **4**, 1012[177]; **6**, 557[837]
Gupta, K., **2**, 296[85]; **5**, 100[141]
Gupta, P., **2**, 283[46]; **3**, 661[25]
Gupta, P. K., **5**, 92[79], 95[101]
Gupta, R. B., **5**, 499[252], 500[252]
Gupta, R. C., **4**, 510[164]; **5**, 374[107,107a], 376[107b]; **7**, 352[33]
Gupta, R. K., **4**, 45[129]
Gupta, S. C., **7**, 155[26], 179[153]
Gupta, S. K., **2**, 757[16], 855[238]; **8**, 533[146], 710[55], 717[102-104], 720[129]
Gupta, S. N., **4**, 721[30], 725[30]
Gupta, S. P., **8**, 654[84]
Gupta, V. K., **3**, 509[179]
Gupta, Y. N., **5**, 627[43], 628[44,45], 629[48,49]
Gupta, Y. P., **4**, 486[37], 505[37,144,145]
Gupte, S. S., **5**, 428[108]
Gupton, J. T., **1**, 367[54]; **3**, 246[44]; **4**, 436[145], 437[145], 438[149,150,151,152]; **6**, 545[635,638]; **8**, 542[236]
Gurfinkel, E., **6**, 965[86]
Gurien, H., **8**, 287[17]
Guritz, D. M., **6**, 955[24]
Gurjar, M. K., **8**, 823[54]
Gurjar, V. G., **3**, 635[42]
Gurowitz, W. D., **6**, 708[49]
Gurskii, M. E., **8**, 102[129]
Gürsoy, A., **6**, 554[805,806]
Gurudas, S. Z., **4**, 119[194]
Gurudutt, K. N., **8**, 875[31], 889[135]
Gurusamy, N., **6**, 172[13]; **8**, 860[223]
Gurusiddappa, S., **8**, 959[22]
Gurvich, I. A., **8**, 526[35], 530[35,107]
Gurvich, L. V., **7**, 852[42]
Gurzoni, F., **3**, 788[54]
Gusar, N. I., **6**, 550[679]
Gusarova, N. K., **7**, 194[6]
Gusev, A. I., **8**, 769[27]
Guseva, V. V., **5**, 768[122,135]
Gus'kova, T. A., **6**, 554[791]
Gust, D., **1**, 488[9]
Gustafsson, B., **4**, 201[9], 229[213,214,216,221]
Gustafsson, H., **6**, 802[41]
Gustavson, G., **4**, 287[179], 999[1]
Gustavson, L. M., **3**, 1000[54]; **5**, 1001[16]
Gusten, H., **7**, 867[91]
Gustowski, W., **8**, 205[156]
Gut, M., **3**, 427[87]; **7**, 145[168], 401[55]; **8**, 971[105]
Gut, R., **2**, 587[138]; **5**, 836[64]
Gut, S. A., **1**, 29[144]; **2**, 507[26]; **3**, 34[198], 39[198]; **6**, 727[195]; **7**, 399[39]
Guthrie, D. J. S., **5**, 1138[68]; **6**, 420[24], 451[127]
Guthrie, J. P., **2**, 134[10,11], 140[11]
Guthrie, R. D., **4**, 423[10], 437[10]; **6**, 27[114]; **7**, 709[39]; **8**, 814[16]
Gutierrez, A. J., **2**, 1050[28]; **3**, 391[90], 393[90]; **5**, 1133[30]
Gutierrez, C. G., **6**, 134[18]; **8**, 214[49], 845[79], 846[79,85], 970[99]
Gutman, A. L., **3**, 628[45]
Gutman, G., **8**, 476[26]
Gutmann, H. R., **7**, 737[14]
Gutmann, V., **3**, 299[33]
Gutowski, F. D., **3**, 422[71], 423[71]
Gutowski, P., **6**, 454[147]
Guts, S. S., **6**, 266[53]
Gutsche, C. D., **1**, 722[277], 832[111], 843[1], 844[9], 845[9], 846[9], 847[1,9], 850[9], 851[36], 861[71], 896[1]; **2**, 148[78]; **3**, 892[51], 896[68], 900[96], 903[96]
Guttieri, M. J., **8**, 289[23], 612[74]
Gutzwiller, J., **3**, 736[25]; **4**, 7[25], 384[143]; **6**, 266[48], 531[452]; **8**, 445[16], 533[147], 535[147]
Guy, A., **5**, 527[64], 530[64]; **6**, 763[7]; **7**, 484[135], 876[125]
Guy, J. K., **8**, 917[115], 918[115]
Guy, J. T., Jr., **3**, 380[7]; **7**, 544[39], 553[39], 556[39]
Guy, R. G., **7**, 516[8]; **8**, 413[128]
Guyer, A., **8**, 248[89]
Guyon, R., **3**, 771[192]; **8**, 872[7]
Guyot, J., **2**, 348[54]
Guyton, C. A., **5**, 67[83]

Guziec, F. S., Jr., **1**, 635[76]; **3**, 862[3]; **6**, 461[3], 980[45], 981[47]; **7**, 252[3], 258[54], 260[61], 267[61], 269[127], 270[127], 288[3]
Guzik, H., **8**, 621[140,141], 935[65]
Guzman, A., **4**, 1061[166]
Gvaliya, T. Sh., **8**, 772[59]
Gverdtsiteli, D. D., **3**, 318[128]
Gverdtsiteli, I. M., **8**, 772[60]
Gvinter, L. I., **8**, 535[160]
Gvozdeva, H. A., **1**, 632[61]
Gwynn, D., **5**, 856[217]
Gybin, A. S., **5**, 345[70], 346[70], 453[66], 1055[46], 1056[48], 1057[51], 1062[51]
Gygax, P., **5**, 501[268,269]; **6**, 672[285]
Gymer, G. E., **4**, 1099[183]; **7**, 744[70]
Gyoung, Y. S., **8**, 16[99,107], 17[107], 238[22], 241[22], 242[22], 244[22], 247[22], 251[22], 253[22], 272[114]
Gysel, U., **4**, 207[61], 208[61]

H

Ha, D., **4**, 378[109]
Ha, D.-C., **1**, 237[33]; **2**, 648[93], 649[93], 925[112], 926[112], 935[148], 936[148], 937[112,148], 948[179], 1049[22], 1050[22]; **4**, 795[80]; **5**, 100[150,159], 101[150], 102[169,170]; **7**, 419[134a]
Ha, D. S., **7**, 530[16]
Ha, T. K., **1**, 286[11]; **3**, 41[230]; **4**, 207[61], 208[61]
Haack, A., **2**, 779[3]
Haack, J. L., **4**, 31[94]; **7**, 682[81]
Haack, R. A., **6**, 1022[68]
Haaf, A., **7**, 153[11], 154[11b]
Haaf, W., **6**, 270[78], 291[202,203]
Haag, T., **7**, 548[61], 553[61]
Haag, W., **6**, 685[362]; **8**, 885[102]
Haag, W. O., **3**, 330[194]
Haage, K., **8**, 756[153]
Haag-Zeino, B., **5**, 464[113], 466[113]
Haaima, G., **4**, 803[134]
Haak, P., **6**, 966[93]
Haake, M., **2**, 894[8], 898[8], 1008[4]
Haas, A., **4**, 445[201]; **5**, 442[185]; **6**, 204[6], 456[159]
Haas, C. K., **4**, 1000[16,18]
Haas, D. J., **1**, 38[260]
Haas, G., **6**, 509[246]; **7**, 160[50]
Haas, M. A., **4**, 665[10], 667[10], 670[10], 674[10]
Haas, W. L., **6**, 637[34]
Haase, K., **6**, 233[47]
Haase, L., **6**, 488[30], 560[867], 566[30]
Haase, M., **2**, 785[42]
Habaschi, F., **8**, 61[187], 66[187]
Habashi, A., **2**, 378[291]
Habaue, S., **4**, 239[28], 883[95]
Habeck, D., **5**, 409[36]; **6**, 523[353]
Habeeb, J. J., **1**, 215[35]
Haber, R. G., **6**, 685[353]
Haberfield, P., **5**, 76[247]
Habermahl, G., **7**, 153[11], 154[11b]
Haberman, L. M., **5**, 936[200]; **6**, 237[64], 243[64]
Habermas, K., **1**, 585[13]
Habermas, K. L., **3**, 21[127]; **5**, 762[95,97,104], 763[95,104], 764[104], 765[97]
Habib, M. J. A., **5**, 1038[8,9], 1047[9], 1049[8], 1051[8]
Habib, M. M., **8**, 535[164]
Habib, M. S., **2**, 960[36]
Habich, A., **5**, 834[53], 857[226]
Häbich, D., **2**, 649[99]; **4**, 372[53], 1006[105]; **5**, 859[235]
Habig, K., **3**, 904[135]
Habus, I., **8**, 460[246]
Haces, A., **3**, 223[147,148]; **6**, 150[131], 151[131]
Hacini, S., **2**, 710[20,21]; **5**, 777[184], 779[184]
Hacker, R., **1**, 12[61], 17[211]
Hackett, S., **1**, 622[75], 787[255]
Hackler, L., **2**, 838[171]; **6**, 276[122], 534[519]
Hackler, R. E., **1**, 512[36]; **3**, 916[17,18], 932[17], 963[119]
Hackley, B. E., Jr., **7**, 656[16]
Hackney, M. A., **3**, 509[179]
Hacksell, U., **4**, 839[29]; **7**, 831[64]
Haddad, N., **5**, 142[93]
Haddadin, M. J., **1**, 834[129], 837[129,149]; **7**, 470[17], 750[131]
Haddock, N. F., **6**, 220[126]
Haddon, R. C., **5**, 855[182]
Haddon, V. R., **3**, 347[25]
Hadel, L., **4**, 1081[80]
Hadfield, J. R., **3**, 380[7]
Hädicke, E., **6**, 561[881]
Hadjiarapoglou, L., **6**, 172[8]; **7**, 374[77b,d]
Hadler, E., **6**, 824[123]
Hadler, H. I., **8**, 530[101]
Hadley, C. R., **1**, 772[198]; **2**, 195[72,72b]; **5**, 839[79], 843[79]; **8**, 925[6]
Hadley, S. B., **8**, 20[136]
Hadley, S. W., **2**, 544[85], 547[85], 552[85]
Hädrich, J., **4**, 758[191]
Hadwick, T., **3**, 723[11], 725[11], 726[11]
Haeckl, F. W., **4**, 270[14]
Haede, W., **7**, 124[42]
Haelg, P., **4**, 930[45]
Haelters, J.-P., **3**, 1052[27]; **6**, 126[150]
Haemmerlé, B., **3**, 640[103]
Haenel, M. W., **3**, 878[90]
Haeseler, P. R., **4**, 31[92]
Haesslein, J.-L., **1**, 365[41]
Hafele, B., **5**, 260[66], 261[66]; **7**, 416[122]; **8**, 70[223]
Hafez, M. S., **2**, 757[10]
Haff, R. F., **6**, 270[77]
Haffer, G., **2**, 167[160], 360[171]; **8**, 331[32], 615[92], 618[123]
Haffmanns, G., **4**, 45[130,130c]; **6**, 542[602], 572[961]
Hafiz, M., **7**, 365[49]
Hafner, A., **2**, 722[96]
Hafner, K., **2**, 366[218], 782[30]; **5**, 65[64], 116[264], 229[126]; **6**, 499[175]; **7**, 29[77]; **8**, 318[59], 322[59], 568[482]
Häfner, L., **2**, 900[23]
Hafner, W., **4**, 588[64]; **6**, 193[218], 558[843]; **7**, 449[1,2], 450[1,2]
Hafner-Schneider, G., **8**, 568[482]
Haga, N., **8**, 836[4], 842[4], 931[39], 993[59]
Haga, T., **2**, 116[130], 424[37], 425[37]
Haga, Y., **4**, 333[23], 398[215]; **7**, 493[199]
Hagadone, M. R., **6**, 294[236]
Hagaman, E., **2**, 384[317]; **4**, 876[63]; **5**, 38[23a]
Hagan, C. P., **6**, 835[45]
Hagedorn, I., **2**, 1084[7], 1086[25]
Hagedorn, L., **8**, 596[83]
Hagedorn, M., **6**, 247[136]
Hagel, I., **2**, 943[169]
Hagelee, L. A., **2**, 589[157]
Hageman, H. A., **6**, 208[58], 212[58]
Hageman, W. E., **6**, 531[445], 538[445]
Hagemann, H., **5**, 708[41]; **7**, 498[224]
Hagen, G., **2**, 566[21]
Hagen, H., **6**, 492[125], 522[125], 554[802], 556[833], 576[802], 581[802]
Hagen, H. E., **2**, 957[21]
Hagen, J. P., **2**, 194[68], 205[102,103], 206[102b,103], 219[68,144], 221[145], 223[151]; **3**, 865[26], 957[109,110]; **5**, 894[44,46]
Hagen, S., **1**, 823[41,42], 832[41]
Hagen, T. J., **7**, 340[46]
Hagen, W., **7**, 330[13]
Hagenah, J. A., **4**, 51[143,143c]
Hagenbach, A., **8**, 542[237]
Hager, D. C., **5**, 635[85]
Hager, G. F., **2**, 144[57]
Hager, R. B., **3**, 572[67]
Haghani, A., **4**, 1022[252]

Hagihara, N., **1**, 447[204], 458[204]; **2**, 725[110]; **3**, 217[89], 219[89], 271[1], 521[7], 530[7,61], 531[61], 532[7,61], 537[61], 552[8], 554[26], 557[48,49]; **5**, 1133[31], 1174[33]; **8**, 456[210], 457[210]
Hagihara, T., **4**, 595[153], 614[382], 615[153], 619[153], 653[440,442]; **7**, 564[91], 565[91], 582[138], 616[18]; **8**, 568[467]
Hagihara, Y., **4**, 315[502]
Haginiwa, J., **2**, 1021[50]; **6**, 916[31] ; **8**, 31[47], 66[47]
Hagio, K., **1**, 367[53]; **2**, 780[10]
Hagishita, S., **6**, 787[98]
Hagiwara, A., **2**, 913[73]; **4**, 802[128]
Hagiwara, D., **6**, 637[29]
Hagiwara, H., **2**, 641[72]; **4**, 18[59], 30[88], 121[205c], 258[240,241,250], 261[240], 262[240] ; **6**, 144[77]
Hagiwara, I., **3**, 12[62]
Hagiwara, K., **2**, 370[256]
Hagiwara, Y., **2**, 859[252]
Hagler, A. T., **5**, 183[157]
Haglid, F., **2**, 1018[44]
Hagmann, W. K., **8**, 34[62], 66[62]
Hahl, R. W., **4**, 971[73]; **7**, 634[67]
Hahn, C. S., **7**, 309[21], 318[54-56], 319[54-56]
Hahn, E., **1**, 162[104]
Hahn, G., **1**, 220[66]; **2**, 82[7], 91[48] ; **5**, 937[202]; **7**, 400[49]; **8**, 115[67], 847[93], 883[94], 884[94], 987[22], 992[22a], 994[22]
Hahn, H.-D., **6**, 430[100]
Hahn, J. E., **4**, 976[100]
Hahn, T., **5**, 185[167]
Hahn, V., **5**, 513[4]
Hahn, W., **6**, 204[4]
Hahne, W. F., **5**, 836[65], 837[68]; **6**, 856[154]
Hähnle, R., **4**, 270[11], 271[21], 272[21]
Hahnvajanawong, V., **3**, 154[424]
Haiduc, I., **5**, 1134[44]
Haidukewych, D., **2**, 343[13]; **3**, 883[108] ; **6**, 674[295]; **8**, 651[69]
Haile, L., **1**, 360[28], 361[28]
Hain, U., **2**, 748[120], 758[26]; **6**, 430[99]
Haines, A. H., **2**, 167[158]; **6**, 2[2], 23[2,87], 27[87]; **7**, 235[11], 305[1], 437[5], 438[5], 439[5], 541[1], 543[1], 564[1], 815[1], 816[1]
Haines, R. M., **7**, 159[44]
Haines, S. R., **4**, 213[112], 215[112]; **6**, 46[60,66], 150[122]; **8**, 857[193]
Haire, M. J., **8**, 73[247], 74[247]
Hajdu, J., **6**, 534[518]; **8**, 589[49]
Hájídek, J., **1**, 893[153]; **4**, 364[2]
Hajjaji, N., **6**, 175[71]
Hajós, A., **6**, 724[159]; **8**, 1[2], 26[7], 27[7], 36[7], 37[7], 64[7], 70[7], 278, 307[2], 382[2], 383[2], 384[2], 541[210], 547[315], 548[315], 550[315b], 580[2]
Hajos, Z. G., **2**, 167[159]; **3**, 23[142] ; **4**, 7[25]; **6**, 718[118]; **8**, 534[156], 544[256]
Hakam, K., **1**, 753[101]
Hakamada, I., **6**, 150[126]
Hake, H., **5**, 257[60]
Hakiki, A., **5**, 589[210]
Hakimelahi, G. H., **2**, 74[72]; **5**, 92[75]; **6**, 655[161]
Hakomori, S., **2**, 323[34]; **6**, 33[3], 40[3]
Hakoshima, T., **3**, 751[88]
Hakotani, K., **7**, 764[124], 844[52]
Hakozaki, S., **7**, 796[12]
Halaska, V., **1**, 10[49]
Halazy, S., **1**, 635[77], 636[77], 638[77,104], 640[77,104], 648[133,136], 652[77,133,141], 653[77,133,136,141], 654[141], 656[77,104], 657[77], 659[104,136,159,160,161,162,163], 672[77,104,136,159,160,161,162,163,199,205,206], 673[159,206]; **3**, 86[50], 87[70,71,78,117], 88[122], 89[70,71,122], 90[70,71], 91[71,122], 92[70,71,122], 93[122], 116[70,71], 119[70,71], 136[78,122], 144[78], 145[78,122], 766[156], 786[44]; **4**, 342[63], 1007[125]; **5**, 677[7], 923[139]; **8**, 623[149], 847[97], 848[97e], 849[97e]
Halberstadt, I. K., **5**, 598[33]
Halberstadt, J., **8**, 144[70]
Halcomb, R. L., **7**, 374[77c], 737[12]
Halczenko, W., **1**, 389[141]; **2**, 1027[75] ; **3**, 258[126], 513[204]
Hale, K. J., **5**, 362[93], 363[93i]; **7**, 712[62]
Hale, R. L., **4**, 18[62], 21[62e]
Hales, N. J., **5**, 383[123]
Hales, R. H., **4**, 489[59,60]
Haley, B. E., **4**, 1033[23]
Haley, G. J., **3**, 592[174,175], 594[174]
Haley, R. C., **3**, 246[35]
Haley, T. J., **1**, 252[4]
Halgren, T. A., **5**, 741[153]
Halim, H., **3**, 623[36]
Hall, C. D., **2**, 424[36]
Hall, D., **1**, 776[216], 777[216b], 814[216] ; **5**, 501[269]; **8**, 13[68]
Hall, G. E., **5**, 714[65]
Hall, H. K., Jr., **5**, 71[138,154], 73[193,194,195,196,197,198,199,200,201,202,203,204], 78[273], 79[287], 405[21], 461[105]
Hall, H. T., Jr., **4**, 522[51,52], 523[51], 526[51,52], 532[52,83], 543[83], 545[83]
Hall, I. R., **8**, 114[54]
Hall, J. A., **4**, 1082[85], 1085[85]
Hall, J. H., **4**, 1101[191]; **6**, 558[859] ; **7**, 23[24], 24[24], 25[40], 26[24]
Hall, L. D., **4**, 347[48]; **6**, 19[69]; **7**, 550[50]; **8**, 52[134], 66[134]
Hall, M. L., **8**, 99[107]
Hall, P. L., **7**, 845[74]; **8**, 241[38]
Hall, R. D., **5**, 15[104]
Hall, R. H., **4**, 272[35], 273[35]; **6**, 620[131], 625[131], 978[23]
Hall, S. A., **8**, 448[145,145b,150], 452[145]
Hall, S. E., **5**, 515[16], 518[16b], 519[16b], 524[16], 526[16b]
Hall, S. S., **1**, 124[83]; **4**, 600[239], 625[239], 643[239], 1013[181]; **5**, 459[91], 659[32]; **6**, 971[128]; **8**, 87[33], 114[58], 308[5,6], 309[5-7], 605[17], 795[16], 971[102]
Hall, T. C., **4**, 611[350]; **7**, 228[105]
Hall, T. K., **7**, 824[40]
Hall, T. W., **2**, 651[122]; **3**, 20[106], 26[165]; **7**, 760[48]
Hallam, B. F., **4**, 663[2]
Hallam, H. E., **6**, 478[106]
Hallas, R., **8**, 27[30], 66[30]
Hallberg, A., **2**, 783[33]; **3**, 495[95]; **4**, 846[70], 857[103]; **8**, 678[58,65], 683[58,65], 686[58,65], 691[58]
Hallcher, R. C., **3**, 666[44], 687[44]
Halleday, J. A., **8**, 503[68]
Halleday, J. E., **4**, 1040[82], 1043[82]
Hallenbeck, L. E., **8**, 743[47], 838[21,21c], 840[21c,34]
Hallenga, K., **3**, 587[148]
Hallensleben, M. L., **8**, 769[28]
Haller, A., **2**, 148[82]
Haller, K. J., **2**, 547[92]; **4**, 982[114]
Haller, R., **8**, 161[19], 541[212]
Hallett, A., **6**, 637[28]
Hallett, P., **1**, 570[269]; **5**, 859[233], 888[25]
Halleux, A., **6**, 966[98]
Hallinan, N., **4**, 710[50]
Hallman, P. S., **8**, 375[153], 445[57,58], 452[57,58], 568[462]
Hallmark, R. K., **5**, 394[147]
Hallnemo, G., **2**, 120[170,171]; **4**, 171[26,27], 178[60], 229[221]; **7**, 331[15]
Hallock, J. S., **8**, 851[135], 856[135b]
Halloran, L. J., **7**, 845[75]
Halls, A. L., **6**, 526[404]
Halls, T. D. J., **5**, 942[229]

Halmann, M., **6**, 602[9]
Halpern, B., **6**, 644[94], 645[94b]
Halpern, J., **5**, 1188[13], 1192[13]; **7**, 4[16], 462[122]; **8**, 429[53], 455[204,205,206], 460[252], 672[20], 674[31]
Halpern, Y., **3**, 299[34b], 322[142a], 332[206], 333[34]; **6**, 749[101]
Halsall, T. G., **7**, 253[17]; **8**, 530[97]
Halstenberg, M., **6**, 205[31]; **7**, 474[49], 476[49]
Halteren, B. W. V., **7**, 535[47]
Halterman, R. L., **1**, 192[81], 733[14], 786[14] ; **2**, 8[37], 13[37], 20[37a], 35[37], 42[144,145] ; **4**, 345[82], 367[12]; **5**, 1134[40,41], 1144[97], 1151[133]; **6**, 864[197]; **8**, 447[127], 463[127]
Haltiwanger, R. C., **5**, 441[176,177]; **7**, 553[60]
Halton, B., **1**, 786[250]; **2**, 849[215]; **4**, 1005[90], 1050[122]; **5**, 947[267], 948[267,271]
Haluska, R. J., **7**, 507[306]
Halweg, K. M., **5**, 8[60], 515[14], 518[14], 522[14a], 524[54,54e], 534[54]
Ham, B. M., **6**, 452[132]
Ham, G. E., **6**, 94[138]; **7**, 470[7], 472[7], 474[7], 476[7]; **8**, 898[21]
Ham, P., **4**, 675[39], 679[45,46]
Ham, W. H., **1**, 413[56]; **2**, 710[25]; **4**, 973[83]; **6**, 837[60]
Hamachi, I., **1**, 158[74], 180[39], 181[39] ; **2**, 4[13], 567[31], 977[7]; **3**, 438[30]
Hamada, A., **7**, 144[157]
Hamada, H., **8**, 95[88]
Hamada, K., **3**, 168[495], 169[495]
Hamada, M., **4**, 21[66]
Hamada, T., **1**, 183[51], 763[145]; **2**, 10[40] ; **6**, 652[140]; **7**, 246[83,84,86]
Hamada, Y., **1**, 386[122], 555[113], 560[155] ; **2**, 801[34]; **4**, 1024[264]; **6**, 235[52], 637[32,32b], 811[78], 816[99]; **7**, 172[124], 506[302]; **8**, 49[115], 66[115]
Hamaguchi, F., **1**, 559[150,151]; **2**, 1049[18] ; **7**, 227[76]
Hamaguchi, H., **1**, 366[48]; **2**, 1051[36], 1052[36]; **7**, 227[74], 707[29], 708[29], 797[18], 798[18b], 802[49], 803[51,53], 804[59], 805[59]
Hamaguchi, M., **4**, 1089[138], 1091[138]; **6**, 568[934]
Hamaguchi, S., **7**, 56[17,18], 57[18]
Hamajima, R., **7**, 109[186]
Hamamoto, I., **2**, 333[65], 334[68], 360[168] ; **4**, 13[44,44b], 591[108], 599[219], 633[108], 641[219], 790[36]; **6**, 1000[128]; **8**, 962[41]
Hamamura, K., **2**, 350[73], 363[73]; **3**, 153[415]
Hamamura, T., **2**, 152[100]
Hamana, H., **2**, 244[23], 245[23], 708[5] ; **7**, 545[27]
Hamana, M., **2**, 348[63]; **4**, 429[84,85] ; **7**, 598[61]
Hamanaka, E., **3**, 431[99,100]
Hamanaka, N., **1**, 850[34]; **4**, 370[27]
Hamanaka, S., **5**, 812[132]; **8**, 323[112]
Hamann, P. R., **4**, 79[56], 251[151,154], 257[154], 260[154]; **5**, 151[7], 180[7], 736[145], 737[145]
Hamann, R., **2**, 1023[54]
Hamano, K., **7**, 77[122]
Hamano, S., **1**, 751[93]
Hamano, S.-I., **8**, 244[71], 247[71], 251[71], 253[71]
Hamaoka, T., **7**, 605[145]
Hamashi, R., **4**, 607[312]
Hamatsu, T., **6**, 46[59]
Hamberg, M., **5**, 780[203]
Hamberger, H., **4**, 1090[143]
Hambley, T. W., **1**, 779[223]; **2**, 66[33], 75[33]; **4**, 119[193], 159[85], 226[195] ; **6**, 154[145]
Hambly, G. F., **2**, 625[165]
Hamböck, H., **2**, 367[224]; **3**, 826[41]
Hambrecht, J., **5**, 1139[72], 1140[72]
Hamdan, A., **8**, 376[163]
Hamdouchi, C., **6**, 156[162]
Hamel, E., **4**, 1018[218]
Hamel, N., **3**, 1028[50]
Hamel, P., **2**, 743[84]
Hamelin, J., **4**, 247[98], 257[98], 262[98], 1086[114], 1096[157]; **5**, 254[48]; **7**, 471[18]
Hamer, J., **1**, 391[150], 834[130]; **5**, 86[25], 90[25], 402[1,1d], 403[1], 404[1], 410[1], 413[1], 416[1d], 417[1], 422[1], 425[1], 426[1], 429[1], 430[1], 433[1], 434[1], 435[1], 436[1], 438[1], 440[1], 444[1], 451[14], 468[14], 470[14], 594[1], 601[1], 604[1]
Hamer, M., **8**, 754[91]
Hamer, N. K., **5**, 154[34]; **6**, 614[85]
Hamersak, Z., **3**, 746[67]
Hamersma, J. A. M., **2**, 1062[101]
Hames, R. A., **4**, 386[153,153b]
Hamill, B. J., **5**, 803[87]
Hamill, T. G., **4**, 213[115]
Hamilton, C. L., **4**, 876[58]
Hamilton, D. E., **4**, 306[371]
Hamilton, G., **8**, 584[25]
Hamilton, G. A., **3**, 660[20], 661[20], 699[20] ; **7**, 11[90], 12[99], 13[123], 851[20]
Hamilton, G. S., **8**, 618[117]
Hamilton, H., **3**, 595[192]
Hamilton, J. G., **5**, 1120[22]
Hamilton, L., **4**, 1085[107]
Hamilton, R., **3**, 334[220], 677[83]; **5**, 801[80]
Hamilton, R. J., **8**, 505[77,78]
Hamilton, W., **7**, 832[71]
Hamlet, A. B., **2**, 363[198]
Hamlet, J. C., **5**, 767[119]; **8**, 987[23]
Hamlet, Z., **7**, 696[41]
Hamlin, J. E., **8**, 445[54,54d]
Hammann, W. C., **3**, 415[8]; **4**, 93[93c]
Hammar, W. J., **4**, 301[318,329]; **8**, 856[164,166]
Hammell, M., **2**, 182[4]
Hammen, R. F., **3**, 953[102]; **6**, 893[83]
Hammer, B., **4**, 230[245], 983[119]; **5**, 1086[74]
Hammer, C. F., **8**, 612[72]
Hammer, G. N., **2**, 968[82]
Hammer, H., **5**, 945[247]
Hammer, J., **7**, 530[24], 531[24]
Hammerer, S., **2**, 478[13]
Hammerich, O., **6**, 282[154]; **7**, 42[31], 801[39], 854[47], 855[47], 856[67]
Hammerschmidt, F., **7**, 57[30], 58[30], 63[30]
Hammerschmidt, F.-J., **3**, 343[18], 353[51], 354[51]
Hammerum, S., **7**, 42[31]
Hammes, W., **1**, 373[85], 374[85]
Hamming, M. C., **2**, 738[41], 760[40]
Hammond, D. A., **7**, 131[88]
Hammond, G., **1**, 767[180]
Hammond, G. B., **2**, 103[33-35]
Hammond, G. S., **4**, 272[41], 279[118], 280[41,118], 284[152], 725[46], 959[31]; **5**, 133[51], 165[80], 167[98,99], 636[91], 639[91], 805[98], 856[200], 1025[81]; **6**, 1033[129]
Hammond, M., **1**, 838[161,167]
Hammond, S. M., **2**, 465[107]
Hammond, W. B., **3**, 892[52]; **5**, 596[28], 597[28]
Hammoud, A., **3**, 487[47], 530[68], 533[68]
Hamoaka, T., **7**, 606[152]
Hamon, D. P. G., **4**, 738[94]; **7**, 410[100]
Hamon, L., **1**, 107[4]; **3**, 216[69], 419[45], 420[45]; **4**, 176[48]
Hamor, T. A., **5**, 151[11]; **7**, 725[32]
Hampel, K., **6**, 176[105]
Hamper, B. C., **6**, 185[165]

Hampson, N. A., **7**, 228[106]
Hampton, J., **7**, 252[9]
Hampton, K. G., **1**, 3[20]; **2**, 507[18]; **7**, 187[184]
Hamsen, A., **1**, 202[102], 203[102], 234[26], 331[47], 669[181], 670[181]
Hamuro, J., **7**, 16[167]
Han, B.-H., **2**, 279[19]; **4**, 426[50]; **5**, 386[132], 387[132a], 638[118], 639[119]; **6**, 977[15]; **8**, 368[69,70], 375[70], 764[4c]
Han, C.-Q., **8**, 191[94]
Han, C. Y., **5**, 92[66], 94[66]
Han, G., **8**, 404[17]
Han, G. R., **7**, 107[153,154]
Han, G. Y., **7**, 480[105], 482[105]
Han, K. I., **5**, 1135[49]; **8**, 743[47], 838[21,21c], 840[21c]
Han, L. P.-B., **3**, 829[56]
Han, N. F., **8**, 783[107]
Han, O., **8**, 344[121,121b]
Han, W. T., **6**, 899[111]
Han, X.-M., **8**, 769[26]
Han, Y. K., **2**, 546[88], 548[88]
Han, Y. X., **4**, 492[75]
Hanack, M., **1**, 416[68]; **3**, 381[18], 396[103,104]; **4**, 270[11,12], 271[21,22], 272[21]; **6**, 19[68], 72[24], 172[20], 500[183], 1012[4], 1013[4], 1033[125]; **8**, 349[137], 873[19], 933[47]
Hanafusa, M., **7**, 426[148a]
Hanafusa, T., **1**, 116[50], 118[50], 555[115]; **4**, 972[78]; **5**, 79[286]; **6**, 233[41]
Hanagan, M. A., **1**, 60[36], 75[36], 468[55]
Hanaki, A., **5**, 98[125]
Hanaki, K., **8**, 535[166]
Hanamoto, T., **2**, 119[162], 846[204]; **3**, 225[187], 1000[55], 1004[60]; **6**, 8[39], 852[135], 877[39], 878[39], 883[39], 885[56], 887[39], 890[56]; **7**, 379[99], 382[99]; **8**, 9[49], 860[223]
Hanaoka, M., **7**, 155[29,29c]
Hanaya, K., **8**, 248[82], 369[74]
Hanayama, K., **8**, 205[157]
Hanazaki, Y., **3**, 566[26]; **8**, 371[110], 404[18]
Hancock, E. M., **2**, 366[219]; **8**, 228[123]
Hancock, J. E. H., **2**, 742[71]; **4**, 4[16]
Hancock, K. G., **2**, 6[24,26]; **5**, 220[43-45]; **8**, 717[94]
Hancock, W. S., **6**, 668[251], 669[251]
Handa, S., **2**, 124[203], 232[182]; **3**, 219[113], 503[144], 505[167]; **5**, 528[67]
Handa, Y., **1**, 271[63]; **3**, 566[28], 571[28], 578[28]; **8**, 412[108b]
Handel, H., **8**, 2[8]
Handel, T. M., **7**, 410[97a]
Handley, F. W., **4**, 426[62]
Handley, J. R., **7**, 684[92]
Handoo, K. L., **8**, 113[39], 115[39]
Handy, C. T., **5**, 1146[110]
Hane, J. T., **1**, 770[189]; **7**, 503[278]
Haneda, A., **8**, 795[24]
Haneda, T., **5**, 356[89]
Hanefeld, W., **6**, 104[1]
Häner, R., **1**, 418[73]; **2**, 107[58], 108[58], 196[76], 197[77]; **4**, 20[64], 21[64], 72[31], 100[124]
Hanes, R. M., **1**, 452[218]
Hanessian, S., **1**, 55[26], 153[57], 419[79], 773[202], 797[292], 798[285], 802[292]; **2**, 323[30], 330[30], 332[30], 555[140], 911[70]; **3**, 512[200], 734[6]; **4**, 27[80], 110[151], 113[171,171f-i], 126[219], 384[143], 745[138], 789[30], 791[30]; **6**, 18[64], 35[14], 46[72], 47[72], 48[84], 49[91], 642[67], 650[128], 651[136], 656[171], 660[208], 662[216], 984[55], 995[104], 996[104]; **7**, 153[7], 162[57], 261[66], 295[20], 299[39], 400[46], 713[71], 722[18]; **8**, 245[73]
Haney, W. A., **7**, 877[132], 878[136], 881[162]
Hangauer, D. G., Jr., **7**, 313[38]
Hangeland, J. J., **5**, 736[145], 737[145]
Hanicak, J. E., **4**, 868[16]
Hanifin, J. W., **6**, 554[724]
Hanisch, G., **3**, 753[103]
Hankinson, B., **2**, 744[95]
Hanko, R., **1**, 180[31]; **2**, 10[43], 21[43], 22[43]
Hanlon, T. L., **4**, 602[258]; **5**, 36[15]
Hann, A. C. O., **2**, 347[49]
Hann, R. A., **2**, 345[26], 357[26]
Hanna, I., **7**, 276[150]; **8**, 111[20], 118[20], 881[70]
Hanna, J., **2**, 613[110]; **3**, 565[23], 570[23], 583[23]; **8**, 168[66]
Hanna, J. M., Jr., **3**, 513[208]; **8**, 935[68]
Hanna, M. T., **8**, 860[221]
Hanna, R., **1**, 760[135]; **7**, 440[38,39a]
Hanna, Z. S., **4**, 598[208], 622[208]; **8**, 882[86]
Hannaby, M., **8**, 15[92]
Hannack, M., **7**, 825[45]
Hannaford, A. J., **7**, 555[70]
Hannah, D. J., **6**, 134[20]
Hannah, J., **8**, 561[413]
Hannart, J., **8**, 618[125]
Hannebaum, H., **3**, 634[28]; **8**, 624[153]
Hannessian, S., **6**, 210[74], 214[74]
Hannick, S. M., **2**, 297[92], 298[92]
Hannon, F. J., **1**, 223[82], 224[82], 313[119,120], 314[119,120], 317[151,152], 319[151,152]; **2**, 120[182]; **3**, 212[37,38]; **4**, 172[30], 229[231]; **5**, 134[66], 137[66]
Hannon, S. J., **8**, 99[107], 643[35]
Hannum, C. W., **4**, 279[107], 280[125], 287[176], 288[125,176]
Hanocq, M., **6**, 517[325], 544[325], 552[325]
Hanotier, J., **7**, 13[107]
Hanotier-Bridoux, M., **7**, 13[107]
Hanquet, C., **8**, 395[135]
Hänsel, R., **3**, 691[134]
Hanselaer, R., **6**, 25[103]
Hansen, B., **6**, 546[640]
Hansen, D. W., Jr., **6**, 71[23]; **7**, 35[105], 352[29]
Hansen, E. B., Jr., **7**, 75[116]
Hansen, F., **2**, 150[97]
Hansen, G., **2**, 387[332]
Hansen, H., **6**, 668[254], 669[254]
Hansen, H.-J., **3**, 809[39,40], 874[69]; **4**, 1033[21], 1037[21], 1040[21], 1081[73,74,78], 1084[95]; **5**, 681[27], 707[32], 709[45], 712[45d], 713[32], 799[72], 822[164], 829[20], 834[50,52], 837[67], 850[146], 856[67,199], 857[67,199,226,227], 858[199], 877[5], 1130[8]; **6**, 185[166], 535[542], 538[542]
Hansen, H. V., **7**, 664[65]
Hansen, J., **1**, 34[170], 35[171], 341[96], 345[124], 477[132], 482[132]; **2**, 150[97]
Hansen, J. F., **3**, 507[174], 677[85]
Hansen, J. H., **4**, 30[91,91a]
Hansen, M. M., **2**, 123[200], 125[200], 280[24], 281[24], 1024[63]; **4**, 239[25], 247[25], 258[25], 260[25]; **5**, 249[32]; **6**, 1063[87]
Hansen, M. R., **7**, 608[172]
Hansen, P.-E., **6**, 173[40], 473[72], 474[72], 475[72,87]
Hansen, R., **8**, 285[8], 293[8]
Hansen, R. T., **4**, 143[19]; **7**, 247[100]
Hansen, S. C., **8**, 140[16]
Hansen, S. W., **5**, 680[24,24a]
Hanson, A. W., **5**, 8[64]
Hanson, G. J., **1**, 56[28]; **2**, 555[143], 556[154]; **3**, 281[45], 360[72]
Hanson, J. R., **1**, 174[4], 177[4]; **3**, 380[13], 715[38,41,42], 738[35]; **4**, 12[42]; **8**, 371[102], 531[117,122], 537[181], 798[61], 987[19]
Hanson, P., **4**, 509[162]; **5**, 424[94]

Hanson, R. M., **3**, 223[156]; **7**, 390[4], 393[4,14], 394[4,14], 395[4], 396[4,14], 397[4], 398[4], 399[4], 400[4], 401[4], 406[4], 407[4], 410[4], 411[4], 413[4]
Hanson, R. N., **2**, 588[152]
Hanson, S. W., **5**, 680[24,24b]
Hansske, F., **7**, 259[60]; **8**, 819[43], 820[43]
Hansson, A.-T., **3**, 257[122]; **4**, 201[9], 229[217,219]
Hansson, B., **7**, 581[129]
Hansson, S., **4**, 631[420,421]; **7**, 453[77]
Hanstein, W., **8**, 750[63]
Hantawong, K., **5**, 944[245]
Hantelmann, O., **2**, 1088[43]
Hantke, K., **4**, 116[189]
Hantzsch, A., **6**, 291[204]
Hanus, J., **8**, 472[1]
Hanyu, Y., **7**, 763[102]
Hanzawa, Y., **1**, 429[123]; **3**, 593[177] ; **6**, 217[117], 221[117]
Hanzel, R. S., **8**, 978[142]
Hao, N., **8**, 675[48], 676[48]
Hao Ku, **7**, 483[121]
Happe, W., **3**, 975[1]
Happel, J., **7**, 7[38]
Happer, D. A. R., **2**, 350[72], 363[72]
Haque, F., **4**, 541[111], 689[69]
Haque, M. S., **1**, 86[42-44], 874[103,104]; **2**, 109[66-68]; **5**, 524[54,54f], 526[57], 534[54] ; **7**, 163[69]
Haque, S. M., **8**, 545[282]
Hara, D., **1**, 215[36]
Hara, H., **3**, 672[65]; **4**, 1017[213], 1019[213] ; **7**, 339[43]
Hara, K., **3**, 135[345], 174[345]; **4**, 964[46]
Hara, M., **8**, 773[68], 778[68]
Hara, N., **6**, 989[78], 993[78]
Hara, R., **4**, 159[82]
Hara, S., **1**, 851[39], 852[39]; **2**, 819[97], 823[97]; **3**, 231[248], 251[78], 254[78], 262[158], 443[56], 490[66], 511[66], 515[66], 523[25]; **4**, 147[38b,41,42], 148[45b], 250[137], 286[174], 287[174], 290[174], 358[153,154,155,156,157,158], 886[118]; **7**, 169[114], 248[112]
Hara, T., **3**, 501[136]; **7**, 473[33], 501[33], 502[33], 750[127]
Harada, A., **7**, 451[32]
Harada, F., **7**, 761[56]; **8**, 450[164]
Harada, K., **1**, 360[26], 364[26]; **2**, 894[7], 916[7], 933[7]; **4**, 494[84], 878[76]; **6**, 704[5]; **7**, 124[46], 474[46,47]; **8**, 26[5,22], 27[5], 36[5], 37[22], 66[22], 123[84], 124[84], 144[78,79], 145[80-82,85], 146[85,91-96,101], 147[91,102], 148[91,92,95,96,103,104,106,110], 149[115], 535[166], 642[34]
Harada, M., **3**, 421[53]
Harada, N., **4**, 227[204]; **7**, 761[56]
Harada, S., **3**, 469[216], 470[216], 476[216] ; **6**, 516[319]
Harada, T., **1**, 215[36]; **2**, 18[72], 31[112], 82[10], 211[115], 215[115], 216[135], 572[46,48], 610[89], 651[113,114], 657[164], 867[15]; **3**, 1047[6]; **4**, 76[44c], 100[122], 249[117], 257[117], 958[26], 960[26], 1014[185]; **5**, 791[26], 889[32]; **6**, 425[65], 509[248,250] ; **8**, 150[121-125,129,133,135], 151[121,129,133,135,145,146,148,149,150,151,152,153,154]
Haraguchi, K., **1**, 471[71]
Harakal, M. E., **6**, 150[114]; **7**, 162[62], 778[400,401,401a]
Haraldsson, M., **7**, 245[75]
Harama, M., **2**, 1000[53,54], 1004[63], 1005[63]
Haran, N. P., **2**, 345[22]
Harangi, J., **6**, 660[207]; **8**, 226[112-114]
Harano, K., **5**, 634[71,72,80,81], 819[156]; **6**, 997[115]
Harano, Y., **4**, 1024[264]
Haraoubia, R., **6**, 456[156]
Harasuwa, S., **6**, 227[20], 236[20]
Harata, J., **6**, 936[112]
Harayama, T., **2**, 157[119], 810[60]; **3**, 216[72]; **4**, 1009[140]; **5**, 850[146]; **6**, 1023[72]; **7**, 438[22], 569[108]
Harbert, C. A., **2**, 1026[73]; **3**, 363[85], 369[117], 372[117]
Hardacre, J. M., **4**, 467[129]
Harde, C., **6**, 677[313]
Hardee, J. R., **6**, 110[56]
Hardegger, E., **8**, 141[33], 794[14]
Hardenbergh, E., **2**, 366[220]
Harder, R. J., **6**, 967[102]
Harder, S., **1**, 23[121-125]
Harder, U., **6**, 277[129], 288[129]
Harding, C. E., **3**, 396[103]
Harding, D. R. K., **6**, 668[251], 669[251]
Harding, K., **8**, 374[148]
Harding, K. E., **2**, 710[18], 728[137], 1077[155] ; **3**, 341[4], 343[17], 348[27], 349[40], 351[40], 353[46,48,50], 354[4c,50], 355[17,48], 357[17,48], 358[68], 369[40], 788[53]; **4**, 295[250,252], 404[242,242a,b,246], 405[246], 407[256a-d], 408[257b,258]; **5**, 758[86], 759[86]; **6**, 1066[95]; **7**, 254[29], 490[178]; **8**, 854[151]
Harding, L. B., **5**, 72[181]
Harding, M., **6**, 126[152]
Harding, P. J. C., **8**, 264[37,38], 347[144], 540[199]
Hardinger, S. A., **4**, 192[118,118b], 590[102] ; **5**, 1024[79]; **6**, 162[190]
Hardstone, J. D., **6**, 533[490], 550[490]
Hardtmann, G., **6**, 509[246]; **7**, 160[50]; **8**, 9[48]
Hardy, A. D. U., **4**, 697[13]
Hardy, A. T., **5**, 223[81]
Hardy, F. E., **7**, 762[82]
Hardy, J.-C., **5**, 723[109]
Hardy, P. M., **2**, 1096[96]; **6**, 639[51], 666[51], 667[51]
Hardy, R., **3**, 242[6], 257[6], 259[6]
Haremsa, S., **3**, 874[68]
Harger, M. J. P., **6**, 114[77]; **7**, 746[92], 752[92]; **8**, 829[82], 864[243]
Hargrave, K. R., **7**, 763[85], 766[85]
Hargrove, R. J., **3**, 380[4]; **6**, 966[91]
Harirchian, B., **3**, 105[215]; **5**, 63[11], 522[43], 704[22], 1020[69], 1023[69,78]; **6**, 1023[71]; **7**, 771[266], 772[266]
Hark, R. R., **6**, 448[108]
Harkema, S., **4**, 45[126]; **5**, 676[4], 686[46-49], 687[46,48,49]; **7**, 333[25]; **8**, 98[103,105]
Harkin, S. A., **4**, 159[82]; **6**, 1031[109]
Harkins, J., **7**, 247[100]
Harland, P. A., **5**, 374[107,107a]; **6**, 83[79]
Harley-Mason, J., **3**, 807[21]; **6**, 966[96]
Harling, J., **5**, 736[145], 737[145]
Harling, J. D., **4**, 823[227]
Harlow, R. L., **5**, 1171[27], 1172[27], 1178[27]
Harman, D., **4**, 317[549]
Harman, R. E., **2**, 757[16]
Harman, W. D., **1**, 310[108]; **8**, 519[133]
Harmata, M. A., **5**, 736[142f], 829[17], 847[17,137], 1004[26,27a]; **6**, 856[158,159], 857[159]
Harmon, C., **5**, 65[60]
Harmon, C. A., **3**, 380[10]; **8**, 390[77]
Harmon, J., **7**, 138[126]
Harmon, R. E., **8**, 533[146]
Harmony, J. A. K., **7**, 883[178,179]
Harms, K., **1**, 10[47], 18[96], 29[145], 32[159,160,161], 33[162], 37[242,243,245,246], 44[96], 306[90], 460[3], 528[119]; **2**, 345[34], 351[34], 357[34], 371[34,262,264], 508[29]; **4**, 104[137] ; **5**, 461[96], 468[121,125], 1096[127], 1101[145] ; **6**, 881[51], 889[72], 890[72]
Harms, R., **7**, 232[155]
Harmuth, C. M., **8**, 329[19]
Harn, N. K., **4**, 1036[50]; **6**, 874[13], 883[13]

Harnfeinst, M., **7**, 306[3]
Harnirattisai, P., **2**, 760[43]
Harnisch, J., **3**, 625[41]
Harnsberger, H. F., **7**, 766[180]
Harp, J. J., **5**, 277[12]
Harper, R. J., Jr., **6**, 677[315]
Harper, R. W., **7**, 43[37]
Harpold, M. A., **7**, 765[133]
Harpp, D. N., **2**, 605[56], 630[11], 631[11] ; **5**, 439[167]; **7**, 122[29]; **8**, 408[67], 413[123]
Harre, M., **3**, 42[231]; **4**, 211[88]
Harrell, W. B., **2**, 962[46,47,49]
Harreus, A., **6**, 657[179]
Harries, C., **2**, 146[68]
Harriman, A., **7**, 877[133]
Harring, L., **1**, 268[52]
Harrington, C. K., **4**, 285[161]; **5**, 856[216]
Harrington, J. K., **3**, 164[476]
Harrington, K. J., **2**, 356[133,134]; **6**, 687[373], 984[56]
Harrington, P. J., **3**, 498[108]
Harrington, P. M., **5**, 492[249]
Harriott, P., **4**, 605[295]
Harris, A. S., **8**, 329[22], 338[22]
Harris, B. D., **3**, 278[31], 289[31]
Harris, C. E., **2**, 523[74]
Harris, C. J., **7**, 480[104]
Harris, C. M., **2**, 171[176], 173[180,182], 175[184], 189[52], 381[303], 619[150], 832[153], 841[187] ; **3**, 58[289]
Harris, D. J., **2**, 6[25], 12[25], 13[25], 26[25,102], 27[25,102], 30[25], 31[25], 41[25], 42[25]; **7**, 401[62]
Harris, E. E., **5**, 493[210]; **8**, 50[117], 66[117]
Harris, F., **1**, 92[65]
Harris, F. L., **4**, 799[116]
Harris, H. A., **1**, 13[67]
Harris, J., **2**, 601[35]
Harris, J. F., **7**, 14[132]
Harris, J. F., Jr., **4**, 279[103], 770[245]; **5**, 797[67]
Harris, M., **1**, 557[128]; **2**, 63[22b], 1018[43] ; **6**, 917[34]
Harris, P. L., **4**, 285[166]
Harris, R. L. N., **6**, 487[46,62], 489[46,62,98]
Harris, S. A., **7**, 306[7]
Harris, S. H., **3**, 1040[105]
Harris, S. J., **6**, 426[77]
Harris, T. D., **1**, 463[27]; **3**, 255[107]
Harris, T. M., **1**, 3[20]; **2**, 171[176,177,178,179], 173[180,182], 175[184], 189[52,53], 381[303], 619[150], 821[107], 832[153], 837[161b], 838[161], 841[187]; **3**, 58[288,289]; **4**, 48[139], 74[36], 500[105]; **5**, 781[206,208]; **7**, 374[77e]
Harrison, A. W., **1**, 739[36]; **7**, 415[111]
Harrison, B., **5**, 410[39]
Harrison, C. H., **7**, 709[38]
Harrison, C. R., **2**, 310[29], 630[6]; **3**, 274[23], 799[99]; **6**, 489[99], 525[99], 545[635], 767[24]; **7**, 763[103]
Harrison, I. T., **2**, 856[249]; **4**, 608[325], 932[62], 969[63]; **7**, 239[45]; **8**, 278, 544[273]
Harrison, J. J., **4**, 517[5], 518[5], 543[121]
Harrison, L. W., **4**, 395[200], 411[200], 597[175,177], 621[175], 623[175]; **8**, 851[130]
Harrison, M. J., **3**, 753[100]; **6**, 861[184]
Harrison, P., **2**, 911[71]; **4**, 18[60,60b], 121[209], 262[314]
Harrison, P. G., **1**, 305[86]
Harrison, P. J., **4**, 405[252]; **6**, 960[57]
Harrison, R. G., **5**, 839[75], 888[25]; **6**, 950[1], 951[4]
Harrison, S., **7**, 239[45]; **8**, 278
Harrison, W., **3**, 217[88]; **4**, 213[113] ; **6**, 150[131], 151[131,133]
Harrison, W. F., **5**, 714[69]
Harrit, N., **6**, 480[110]
Harrod, J. F., **8**, 699[148], 763[1], 765[17], 785[1]
Harrold, S. J., **8**, 132[10], 134[10]
Harrowfield, J. M., **4**, 298[292]
Harruff, L. G., **5**, 692[103]
Harsanyi, M. C., **2**, 743[83]
Hart, D. A., **8**, 36[52], 66[52]
Hart, D. J., **1**, 389[140], 390[144]; **2**, 368[240], 648[93,94], 649[93,94], 872[24], 873[25], 877[37], 882[24,44,45], 924[107], 925[112], 926[112,115], 935[146,148], 936[146,148], 937[112,115,148], 940[162], 948[179], 999[39], 1048[8], 1049[8,21,22], 1050[21,22], 1060[84], 1063[103,104]; **3**, 598[203]; **4**, 45[128], 115[183], 259[263], 374[91], 398[216], 716[2], 739[110], 760[195,197], 765[197], 780[1], 790[41], 791[41,48], 794[78], 795[80,81]; **5**, 100[149,150,152,158,159], 101[150], 102[166,169,170] ; **6**, 746[96]; **7**, 204[59], 350[26], 647[31], 677[68], 731[54]; **8**, 57[171], 66[171], 514[107]
Hart, D. W., **1**, 143[33]; **4**, 153[62c]; **8**, 669[4], 673[28], 675[28], 676[28], 677[28], 679[4], 685[97], 687[4], 688[100], 691[100], 692[28,100]
Hart, F. A., **8**, 445[41]
Hart, G. C., **1**, 477[129,131]; **3**, 74[42], 75[42,43,51], 76[43], 77[42], 78[51], 80[43] ; **7**, 227[78,79], 230[78]
Hart, H., **2**, 715[55]; **3**, 738[34], 753[101], 839[8]; **4**, 493[78], 494[84], 496[90,91], 791[57], 878[76]; **5**, 210[57], 223[75,76,78,79,81], 229[119], 381[116,117], 382[120], 383[125], 552[28], 558[61], 730[126], 915[113]; **7**, 87[19], 743[63]
Hart, K. W., **8**, 206[172]
Hart, P. A., **7**, 253[19], 254[19]
Hart, R. B., **7**, 219[13]
Hart, R. J., **5**, 722[104]
Hart, T. W., **1**, 862[75b]; **3**, 767[160] ; **6**, 656[169]
Hartenstein, J. H., **5**, 708[41]
Hartfiel, U., **2**, 356[132]
Hartke, K., **2**, 367[223], 379[297], 898[21], 1050[29]; **5**, 475[145,146]; **6**, 436[20,21], 450[20,21,122-124], 453[139], 454[21], 455[20,21,124], 501[192,194], 509[284], 548[671]
Hartlev, D., **6**, 65[1]
Hartley, D., **2**, 323[32]
Hartley, F. R., **1**, 215[39], 218[39], 225[39], 326[4], 327[13]; **2**, 727[128]; **3**, 228[217], 436[7,10]; **4**, 70[9], 72[24], 83[63], 93[95], 139[1], 140[8], 144[22], 155[65,66], 552[2], 553[2], 588[60], 867[7], 879[86], 883[86], 903[200,200b], 905[200b]; **5**, 1037[3], 1132[21], 1144[101]; **8**, 459[228], 670[13]
Hartley, S. G., **2**, 811[68]
Hartley, W. M., **7**, 439[27]
Hartman, B. C., **3**, 263[177]; **6**, 8[35] ; **8**, 873[15], 874[15]
Hartman, G. D., **2**, 1027[75]; **3**, 258[126], 513[204], 863[14], 865[14], 884[14]; **5**, 71[163]
Hartman, J. S., **1**, 292[29], 293[35]
Hartman, M. E., **1**, 846[21,22], 851[21], 853[21], 856[22], 859[21], 861[21], 896[21]; **3**, 783[105]
Hartman, P. H., **5**, 682[34a], 683[34a]
Hartman, R. D., **3**, 863[14], 865[14], 884[14]
Hartman, W. W., **6**, 213[87]
Hartmann, A., **6**, 668[262]
Hartmann, A. A., **3**, 124[273], 128[273,323], 132[273], 134[273,323]
Hartmann, G. D., **6**, 452[132]
Hartmann, H., **2**, 785[43]; **5**, 355[87c], 356[87c], 365[87c,96a]; **6**, 480[112]; **8**, 298[72], 725[182]
Hartmann, J., **2**, 5[20,20a,21], 6[20], 13[21], 21[20], 66[37]; **3**, 99[182], 101[182]
Hartmann, M., **8**, 638[14]
Hartmann, R., **3**, 17[85]
Hartmann, W., **5**, 160[56], 646[5,6], 1188[15] ; **6**, 119[117], 254[164]
Hartner, F. M., Jr., **8**, 682[81,82]

Hartner, F. W., **1**, 749[75], 807[316]
Hartner, F. W., Jr., **5**, 1124[46], 1125[62,64]
Hartog, F. A., **1**, 219[55]
Hartshorn, M. P., **2**, 835[156]; **3**, 741[50,52,53], 742[55,57], 743[57,60], 746[66], 751[90], 752[91] ; **7**, 88[23], 90[23]; **8**, 213[29], 941[112]
Hartter, D. R., **5**, 486[196], 714[65]
Hartter, P., **7**, 53[2], 66[2], 67[2], 68[2], 70[2], 75[2], 77[2], 80[2]
Hartung, H., **4**, 55[157]; **6**, 134[33], 182[137], 187[177]; **8**, 860[223]
Hartung, J., **2**, 334[67]; **7**, 883[175]
Hartung, W. H., **6**, 651[136,136a]; **8**, 148[111], 149[111,112], 956[1,3], 957[1]
Hartwell, J. L., **3**, 828[49]
Hartwig, J. F., **4**, 1002[48]
Hartwig, W., **2**, 649[99]; **3**, 53[274]; **6**, 447[104], 451[104], 531[431,432,443], 987[71] ; **8**, 817[27], 818[27], 820[27], 821[51]
Hartz, G., **4**, 111[155c]
Hartzel, L. W., **4**, 296[265]; **6**, 268[70], 271[70]
Hartzell, S. L., **1**, 789[260]
Harui, N., **6**, 176[90]
Harukawa, T., **7**, 86[16a]
Haruki, E., **8**, 371[113]
Haruna, M., **7**, 88[25]
Harusawa, S., **1**, 544[32,37], 548[66], 555[113], 560[37,155], 561[37,156]; **6**, 1053[46]; **7**, 172[124]
Haruta, J., **4**, 155[75], 160[86a], 350[116]
Haruta, J.-I., **7**, 829[56,56c]
Haruta, R., **1**, 165[108]; **2**, 91[46], 93[46], 96[58]
Harutunian, V., **1**, 544[34], 551[34], 553[34]
Harvey, A. B., **7**, 774[321]
Harvey, D. F., **2**, 667[40-42], 671[48], 673[40], 674[40-42], 675[40-42,54], 701[86], 702[86], 703[87] ; **4**, 373[80], 374[80]; **5**, 434[145]; **8**, 856[171]
Harvey, G. R., **4**, 55[156]; **7**, 471[22]
Harvey, R. G., **1**, 466[46,47]; **2**, 760[47] ; **7**, 136[107], 296[25,26], 329[3], 346[7], 358[3], 365[44], 833[73], 884[183]; **8**, 19[132], 490[3], 492[3], 496[3], 497[3,36], 530[99], 910[85], 911[85], 949[157]
Harvey, S., **1**, 17[217]
Harvey, S. M., **3**, 325[161]
Harville, R., **6**, 134[15], 151[15]; **7**, 764[118]
Harwood, L. M., **1**, 243[55]; **4**, 359[160] ; **8**, 337[77]
Hasagawa, T., **1**, 739[38]
Hasan, I., **1**, 436[148], 473[79]; **5**, 327[28]
Hasan, N. M., **5**, 493[213]
Hasan, S. K., **3**, 154[418], 155[418]; **7**, 571[115]; **8**, 404[19], 410[19], 411[19]
Hase, T. A., **1**, 542[13], 544[13], 563[13]; **2**, 55[1], 442[7]; **3**, 86[30], 121[30]; **4**, 113[162], 247[105], 253[105], 257[105], 262[105], 377[100], 731[72]; **5**, 516[25], 517[25c]; **6**, 679[327], 833[24], 862[24]; **7**, 453[75], 686[96]; **8**, 18[125]
Hasebe, K., **2**, 176[185], 832[152]; **4**, 21[66,66c], 62[66c], 107[146c], 218[143]
Hasebe, M., **7**, 719[5], 720[14], 732[5,57]
Hasegawa, A., **6**, 36[28]
Hasegawa, H., **7**, 196[29]; **8**, 191[91,93], 650[67]
Hasegawa, J., **7**, 56[17,18], 57[18,23,29], 58[23,29], 63[23,29]
Hasegawa, K., **7**, 125[60]
Hasegawa, M., **1**, 88[54]; **4**, 140[11], 209[65] ; **6**, 14[52]
Hasegawa, S., **5**, 830[31]
Hasegawa, T., **7**, 242[62]; **8**, 244[50]
Hasegawa, Y., **2**, 859[252]; **8**, 405[23]
Hasek, R. H., **8**, 141[30]
Haselbach, E., **5**, 704[21]
Haseltine, J. N., **5**, 736[145], 737[145]
Haseltine, R. P., **5**, 597[30]
Hasenhündl, A., **6**, 518[331], 519[338]
Hashem, M. A., **4**, 1002[58], 1023[259]; **7**, 359[19]
Hashiba, N., **3**, 404[135]
Hashigaki, K., **2**, 167[157]
Hashiguchi, S., **2**, 810[64]; **5**, 96[118], 1118[19]; **8**, 836[10a], 837[10a]
Hashiguchi, Y., **4**, 435[138]
Hashill, T. H., **8**, 843[48], 846[48]
Hashimato, S., **2**, 282[40]
Hashimoto, A., **2**, 656[154]; **5**, 100[142]
Hashimoto, C., **8**, 936[70]
Hashimoto, H., **1**, 733[20]; **2**, 68[44], 567[28], 718[76], 901[36,37], 908[36,37], 909[36,37] ; **3**, 124[263], 126[263], 219[113], 243[18], 244[18], 247[18], 400[121], 464[169], 500[133], 503[144], 505[167]; **4**, 590[94], 598[196] ; **5**, 297[57,59], 1157[169], 1196[37,38], 1197[38] ; **6**, 74[34], 86[98], 560[869]; **7**, 550[49] ; **8**, 170[72], 185[24], 535[166], 795[23], 857[191], 906[65], 907[65], 908[65], 909[65], 910[65]
Hashimoto, I., **3**, 304[61], 483[8]; **5**, 1138[63]
Hashimoto, K., **1**, 766[161]; **2**, 576[69], 600[32]; **3**, 168[497], 170[497]; **6**, 995[99] ; **8**, 349[135], 354[175], 554[366]
Hashimoto, M., **1**, 101[90], 860[69,70]; **2**, 213[124], 1060[86]; **3**, 100[199]; **5**, 96[106,117] ; **6**, 93[133], 615[98,99], 780[70]; **7**, 255[38], 493[198]; **8**, 34[62], 66[62]
Hashimoto, N., **1**, 851[40], 852[40]; **2**, 899[27], 901[27]; **7**, 692[23]
Hashimoto, S., **1**, 266[48], 314[121], 359[19], 382[19,19a-e], 386[122]; **2**, 114[122], 269[71] ; **3**, 36[211], 246[35], 349[33], 354[60], 358[33]; **4**, 85[75], 210[80-86], 211[82,85], 433[124]; **5**, 516[28], 609[68], 925[154] ; **6**, 46[61], 172[24], 721[136], 723[148], 724[148], 726[177,178,183,184]; **8**, 146[99], 356[185], 986[16]
Hashimoto, S.-I., **5**, 376[108a,b]
Hashimoto, T., **2**, 649[103], 1059[76]; **7**, 707[29], 708[29], 803[51]
Hashimoto, Y., **2**, 810[60]; **4**, 161[91,91c,d] ; **6**, 54[131]
Hashizume, A., **1**, 563[171]; **2**, 65[29] ; **3**, 197[41], 199[41]
Hashizume, K., **2**, 384[319]; **3**, 136[370], 137[370], 138[370], 139[370], 140[370]; **8**, 563[425]
Hashmet Ali, M., **5**, 24[165]
Hasiak, B., **6**, 897[100]
Haslam, E., **6**, 2[4], 657[174], 665[174], 667[174]
Haslanger, M., **8**, 346[124]
Haslanger, M. F., **3**, 288[63]; **8**, 881[68]
Haslego, M. L., **2**, 356[128]; **4**, 89[84i]
Haslegrave, J. A., **8**, 196[116]
Haslinger, E., **2**, 346[47], 1088[48], 1089[48] ; **3**, 380[7]; **7**, 498[223]
Haslouin, J., **5**, 555[50], 577[148]
Hass, H. B., **2**, 321[5,6], 326[5,6]; **7**, 659[36], 660[40]
Hassall, C. H., **3**, 660[15], 690[124], 818[94] ; **5**, 395[148]; **7**, 351[28], 355[28], 671[1], 672[1], 674[1], 684[1]
Hassan, A., **8**, 117[73]
Hassan, F., **4**, 1002[61]
Hassanali, A., **6**, 802[42], 803[42]
Hassdenteufel, J. R., **8**, 933[47]
Hasse, K., **5**, 677[12]
Hassel, P., **6**, 706[37]; **8**, 285[10]
Hassel, T., **7**, 226[69]
Hasselmann, D., **5**, 64[37]
Hassen, W. D., **3**, 497[101], 505[101]
Hassid, A. I., **8**, 47[127], 66[127]
Hassig, R., **1**, 631[57]
Hässig, R., **1**, 20[107], 506[9]; **3**, 87[75]; **4**, 1008[132]
Hasskerl, T., **4**, 729[58], 754[58]
Hassner, A., **1**, 544[41,42], 570[264], 819[1,3], 834[1,129,131], 835[1,131], 836[131], 837[129,131,149], 838[171]; **2**, 148[80], 149[86], 486[39] ; **3**, 901[112]; **4**, 36[100], 49[100a], 295[256,258,262], 296[262], 297[268], 341[60], 349[111], 350[111,120], 357[147]; **5**, 86[11], 105[188], 107[188], 252[43,44], 257[43], 413[48], 949[279,280];

6, 94[141], 98[141], 104[8], 246[130], 247[134], 288[185,191], 289[192], 658[188]; **7**, 21[4], 186[178], 473[28,31,32], 475[56], 496[216], 500[237,238], 501[31,32,247,248,249,251,253,254], 502[28,253,254,256,257,259,263], 506[294], 522[38], 598[55], 750[131], 772[287] ; **8**, 64[219], 384[38], 385[48], 386[50], 536[168], 713[76], 718[116], 925[11], 926[11]
Hasso-Henderson, S. E., **5**, 1156[164]
Hasty, N. M., **7**, 96[90], 98[90]
Hasukichi, H., **1**, 750[88]
Haszeldine, R. N., **3**, 297[19]; **4**, 271[25], 272[25], 274[63,64], 275[63,64], 278[25], 280[63,131], 281[63,131], 285[25], 286[25], 288[25,131], 304[353], 1000[13]; **6**, 220[130]; **7**, 94[55], 488[162,163], 750[135], 800[34]; **8**, 770[41], 775[77], 900[34], 901[39]
Hata, G., **1**, 881[118]; **4**, 597[178], 598[194] ; **5**, 810[126], 812[126]; **8**, 758[168]
Hata, H., **6**, 606[41]; **8**, 190[66,67], 994[64]
Hata, K., **5**, 729[123]; **7**, 660[38]; **8**, 554[366]
Hata, N., **5**, 808[109]; **8**, 844[75]
Hata, T., **1**, 563[171]; **2**, 65[29], 830[145] ; **3**, 197[41], 199[41]; **6**, 563[899], 566[927], 604[32,33], 606[38,40,42], 607[45], 608[50], 609[51,56], 612[75,76], 614[97], 615[45,100], 618[112], 624[137,139], 626[164,168]
Hata, Y., **3**, 902[103]; **5**, 422[83], 585[201] ; **6**, 96[151]
Hatada, K., **6**, 928[82]; **7**, 202[45]
Hatajima, T., **1**, 248[62], 735[25]
Hatakeyama, S., **1**, 343[106], 436[148], 569[260], 763[144]; **2**, 814[81], 824[81]; **3**, 278[30] ; **4**, 376[99], 381[130], 387[99]; **5**, 327[28] ; **7**, 416[122], 441[44]
Hatamura, M., **8**, 405[24]
Hatanaka, H., **1**, 101[90]
Hatanaka, K., **4**, 247[100], 257[100], 260[100]
Hatanaka, M., **2**, 826[122,123,125], 1102[121a,b], 1103[121]; **8**, 405[23,24]
Hatanaka, N., **5**, 96[107,112,113], 914[114]; **6**, 9[42]; **7**, 475[57]
Hatanaka, Y., **1**, 232[13], 233[13], 234[13], 253[9], 276[9], 278[9]; **2**, 312[35], 720[83] ; **3**, 233[273,274], 538[95], 539[97], 567[34], 570[34]; **5**, 181[154]; **7**, 308[19], 877[133] ; **8**, 786[116]
Hatanaku, Y., **8**, 113[47]
Hatano, M., **1**, 256[20,21]
Hatayama, Y., **7**, 137[118], 138[118]
Hatch, L. F., **3**, 304[59]
Hatch, M. J., **3**, 918[21]; **8**, 228[128]
Hatch, R. L., **7**, 603[122]; **8**, 102[124], 537[189]
Hatch, R. P., **6**, 563[903]
Hatch, W. E., **4**, 5[18], 27[84,84a]
Hatcher, A. S., **3**, 106[224], 113[224]; **4**, 308[405]
Hatekeyama, S., **1**, 766[155]
Hatem, J., **8**, 802[87]
Hatenaka, Y., **7**, 843[48]
Hatfield, G. L., **3**, 602[226]; **5**, 1017[66] ; **6**, 210[73]
Hathaway, B. J., **3**, 499[118]
Hathaway, S., **8**, 459[228]
Hathaway, S. J., **5**, 618[5]
Hathway, D. E., **7**, 582[149]
Hatjiarapoglou, L., **4**, 1032[11]
Hatky, G. G., **1**, 307[93], 310[93]
Hatsui, T., **1**, 187[61,64]; **5**, 619[12], 620[12], 621[19], 622[23]
Hatsuki, T., **5**, 308[96]
Hatsuya, S., **2**, 625[164]; **3**, 443[62]
Hatta, A., **8**, 149[113-115]
Hattingh, W. C., **5**, 501[270]
Hatton, J., **7**, 5[23]
Hattori, I., **1**, 390[145], 391[145]
Hattori, K., **1**, 98[84], 99[84], 215[36], 387[136]; **2**, 995[45]; **3**, 789[70], 1047[6] ; **6**, 509[263], 542[601], 767[28], 768[28], 769[28,32], 770[33]; **7**, 696[43,44], 697[43,46], 773[309], 776[309]; **8**, 43[108], 47[108], 64[220], 66[108], 67[220], 394[119], 837[15a]
Hattori, M., **4**, 1054[133]; **8**, 154[190], 698[141]
Hattori, R., **6**, 454[146]
Hattori, T., **8**, 858[205]
Hatzigrigoriou, E., **1**, 561[162]; **4**, 112[158d,159], 259[270,271,272]
Haubenstock, H., **8**, 159[6]
Hauber, M., **6**, 576[973]
Haubrich, D. R., **8**, 623[150]
Haubrich, G., **6**, 172[28]
Hauck, A. E., **4**, 428[71]
Hauck, F. P., Jr., **7**, 221[26]
Hauck, M., **6**, 960[52]
Hauck, P. R., **7**, 691[20]
Haufe, G., **3**, 379[3], 380[6,12,13], 381[28], 382[28,38], 386[59], 849[56]; **4**, 331[19], 356[137]
Haufe, J., **3**, 634[25b], 635[25b], 639[25b], 640[97]
Haug, E., **2**, 368[238]; **6**, 229[25], 430[93], 452[131], 512[120,303], 543[120], 553[796,798], 554[800,802], 572[796,958], 573[798,963], 576[802,973,974], 581[802]
Haugen, R. D., **3**, 589[162], 610[162]; **4**, 373[82]
Haumaier, L., **6**, 450[120]
Haung, W., **3**, 596[194]
Hauptmann, H., **1**, 746[61]; **5**, 809[121] ; **8**, 836[1,1b], 837[1], 847[1b], 964[53]
Hauptmann, S., **3**, 890[33], 894[59], 900[59] ; **6**, 122[138]
Hauschild, K., **5**, 30[2,2f]
Hause, N. L., **6**, 954[15]
Hauser, A., **1**, 218[49], 220[49], 223[49] ; **4**, 242[62], 253[62], 261[62]
Hauser, C. R., **1**, 3[20], 463[23]; **2**, 182[1,3], 189[53], 190[55,56], 268[69], 280[22], 283[43], 296[43,87], 507[18], 512[46], 522[72], 523[73,74], 735[16], 797[4,5], 802[27], 829[5], 834[155], 837[5,161b], 838[161,162], 843[5], 845[5]; **3**, 58[288], 158[438], 159[438,467], 160[438], 166[438,467], 914[8,12], 915[14,15], 918[21], 957[108], 967[14], 969[108], 975[2], 976[2], 979[2], 980[2] ; **4**, 73[33], 500[105]; **8**, 329[21], 564[443]
Hauser, F. M., **2**, 547[104], 549[104]; **4**, 14[47,47a-d], 111[154c-e], 258[247,249], 373[77] ; **5**, 105[197]; **6**, 105[13], 137[41], 900[118], 944[158]; **7**, 441[42]; **8**, 928[23]
Hauser, G., **2**, 855[242]
Hauser, G. R., **7**, 187[184]
Häuser, H., **5**, 75[217], 78[217]
Hauser, J. W., **4**, 1016[206]
Haushalter, R., **7**, 12[93]
Hauske, J. R., **8**, 146[98]
Haüsler, J., **2**, 866[4]
Hausmann, J., **2**, 141[43]
Häussermann, M., **2**, 137[16], 138[16]
Haussinger, P., **5**, 417[64]
Haustveit, G., **2**, 456[57], 458[57]
Haut, S. A., **8**, 246[79], 248[87], 824[64]
Hautala, R., **5**, 166[91]
Hauth, H., **2**, 1015[22]
Haveaux, B., **5**, 109[213,214]; **8**, 374[144]
Havel, M., **2**, 382[313]; **6**, 959[39]
Haven, A. C., **5**, 552[24]
Havens, J. L., **6**, 182[139]; **7**, 100[129], 104[129], 260[43]
Havinga, E., **5**, 700[8], 708[41], 737[8]
Havinga, E. E., **3**, 552[9], 557[9]; **4**, 12[39]
Havlin, R., **3**, 217[90], 219[90]
Hawari, J. A., **4**, 739[106]
Hawkes, G. E., **3**, 635[40], 637[65]
Hawkins, A., **8**, 437[78]
Hawkins, C. M., **5**, 821[163], 857[225]
Hawkins, D. W., **8**, 837[13a], 839[13a], 840[13a], 935[63]
Hawkins, E. G. E., **3**, 770[177]

Hawkins, J., **1**, 461[14], 464[14]
Hawkins, J. A., **8**, 423[39], 431[67]
Hawkins, L. D., **1**, 198[91]
Hawkins, R. T., **7**, 596[33b]
Hawkins, S. C., **4**, 293[239]; **6**, 278[131]
Hawks, G. H., III, **3**, 54[275]
Hawley, D. M., **3**, 386[68]
Hawley, M. D., **4**, 453[27], 471[27]; **8**, 135[48]
Hawley, R. C., **2**, 434[58]; **5**, 152[20], 736[145], 737[145]; **8**, 933[48]
Haworth, R. D., **3**, 693[140], 818[98]
Hawrelak, S. D., **6**, 657[180]
Hawthorne, J. R., **4**, 2[5]
Hawthorne, M. F., **4**, 145[28]; **7**, 330[12], 599[72], 673[19]; **8**, 445[32], 710[56,61]
Hay, A. S., **3**, 552[7], 557[44,45]
Hay, B. P., **4**, 812[183], 815[197], 816[197]
Hay, D. R., **1**, 477[133,134]; **3**, 67[17,20] ; **7**, 225[63]
Hay, G. W., **5**, 439[168]
Hay, J. M., **4**, 717[8]
Hay, J. N., **5**, 79[291]
Hay, J. V., **4**, 465[119], 466[119], 467[119,131]
Hay, R. W., **6**, 46[68]
Haya, K., **6**, 116[88]
Hayakawa, A., **6**, 554[708]
Hayakawa, H., **1**, 471[71]
Hayakawa, I., **2**, 368[235]; **6**, 554[727]
Hayakawa, K., **3**, 809[41]; **4**, 159[82]; **5**, 537[99], 1130[8]; **6**, 552[690]
Hayakawa, N., **5**, 282[33], 286[33], 605[58]
Hayakawa, S., **2**, 780[6]; **6**, 516[319], 682[337]
Hayakawa, T., **7**, 462[119-121]
Hayakawa, Y., **4**, 27[79,79a]; **5**, 282[25,27-33], 283[25,27,30], 284[28-30], 285[25,27,31,32], 286[30,33], 594[5], 595[11], 596[11a], 601[44-47], 603[5,47,48e], 605[5,46,57,58,60,60b,61,63], 606[45], 608[5,65], 609[60,60b,c,61,68], 611[57]; **6**, 18[65], 603[11,15,17], 608[11], 614[82,94], 619[120], 620[129], 624[15,120,129,148,149] ; **7**, 682[86], 750[131]; **8**, 356[185], 987[18], 991[18,44]
Hayama, N., **1**, 450[211]; **3**, 463[163], 483[19], 484[20,24], 500[19,131], 501[24], 509[19]
Hayama, T., **1**, 619[62]; **3**, 45[248]; **6**, 109[39-41]; **7**, 496[217], 497[218]
Hayami, H., **4**, 892[143]; **7**, 378[96]
Hayami, J., **5**, 600[41]; **6**, 1000[129]
Hayano, K., **3**, 382[39], 402[126], 404[135], 405[138]; **7**, 91[36]
Hayano, M., **7**, 145[168]
Hayasaka, E., **5**, 356[89]
Hayasaka, K., **3**, 404[137]
Hayasaka, T., **1**, 86[35], 223[79], 224[79] ; **4**, 230[254,255], 510[173], 558[17]
Hayase, Y., **1**, 95[72]; **2**, 482[24], 483[24] ; **8**, 493[22]
Hayashi, G., **6**, 289[196], 293[196]; **7**, 495[208]
Hayashi, H., **1**, 546[54], 559[151], 568[234] ; **2**, 68[42]; **5**, 153[26]; **7**, 227[76] ; **8**, 643[38], 875[30]
Hayashi, I., **7**, 153[11]; **8**, 546[309]
Hayashi, J., **3**, 23[143], 24[143]; **5**, 222[65], 223[65]; **7**, 796[12], 806[75], 808[80], 809[81,85]
Hayashi, K., **2**, 547[108], 550[108]; **3**, 623[39]; **4**, 159[81], 589[87,88], 598[191], 599[218], 638[191], 640[218], 1006[96], 1020[239] ; **5**, 369[101], 370[101d], 835[59]; **6**, 715[89,90], 980[43]; **7**, 453[65]; **8**, 900[31]
Hayashi, M., **1**, 850[34]; **2**, 823[116]; **3**, 96[169], 104[169], 108[169], 117[169] ; **4**, 370[27], 413[274]; **6**, 46[61], 457[163], 744[76], 746[76]; **8**, 100[117], 309[12], 311[12], 545[284], 840[35], 899[26], 906[26], 907[26], 913[26], 914[26], 966[75]
Hayashi, N., **3**, 891[45]; **5**, 1188[16]; **8**, 191[95]
Hayashi, R., **2**, 18[73]; **8**, 391[90]
Hayashi, S., **3**, 832[69]; **4**, 508[157]; **5**, 355[87b]; **6**, 172[15]; **7**, 100[115] ; **8**, 353[152]
Hayashi, T., **1**, 158[74], 180[39], 181[39], 319[159], 320[158,159], 610[46], 611[46], 617[53] ; **2**, 4[13], 17[67,68], 21[85], 38[67], 73[66], 114[115,117], 213[125], 233[186], 317[48,49], 318[48,49], 345[39], 346[39], 352[39], 455[19], 567[31], 568[33], 572[45], 584[118,122,124], 716[62,63], 977[7]; **3**, 17[95], 95[156,157], 97[156,172,173,177], 108[177], 109[177], 114[157], 116[157,172,173], 136[369], 138[369], 228[219,220,221], 229[225,226,227,228,229,231], 246[37], 436[19], 438[30,33], 441[47], 442[50], 445[69,71], 450[33], 452[33], 455[122], 457[129], 470[219], 471[219], 472[219], 492[76,78], 495[78], 1016[3], 1039[101]; **4**, 231[260], 571[1], 572[1], 588[67], 595[153], 596[159], 610[339], 614[382], 615[153,387], 619[153], 620[387], 626[339], 635[159,387], 649[339], 653[437,439,440,441,442], 682[57], 841[47], 915[10], 918[10], 930[47], 931[55], 945[91]; **5**, 305[87], 850[146,147], 890[33] ; **6**, 88[104], 842[83,84]; **7**, 564[91], 565[91], 582[138], 616[18], 642[12]; **8**, 84[12], 152[178,179,180], 173[112,113], 459[245], 461[262], 549[325], 556[373], 568[467], 783[106-109], 784[110], 820[45], 822[52]
Hayashi, Y., **1**, 243[56], 554[108]; **2**, 804[42]; **3**, 361[78], 365[97], 469[202], 470[202], 473[202], 922[37], 924[37]; **4**, 154[64b], 257[222], 261[286], 262[222], 520[33-35], 546[129], 595[156], 620[156], 635[156]; **5**, 24[164,168], 369[101], 370[101b], 1098[114], 1112[114] ; **6**, 54[131]; **7**, 24[35], 219[10], 257[47] ; **8**, 693[113,117], 857[189], 889[131]
Hayashimatsu, M., **6**, 531[429]
Hayashiya, T., **2**, 651[114]
Hayashizaki, K., **3**, 229[228,229]
Hayatsu, H., **6**, 611[65]
Hayatsu, R., **5**, 637[112]
Haydn, J., **2**, 139[31]
Hayes, B. R., **5**, 1130[7]
Hayes, D. M., **4**, 510[166]
Hayes, J. E., **6**, 173[43]
Hayes, J. F., **7**, 26[50], 400[42]
Hayes, M. L., **2**, 456[42], 466[42], 467[42]
Hayes, N. F., **8**, 533[140]
Hayes, R., **2**, 787[51], 963[55]
Hayes, R. A., **3**, 891[43], 892[43]
Hayes, R. J., **8**, 52[142], 66[142]
Hayes, T. K., **4**, 802[127]
Hayez, E., **4**, 1033[16]
Haymi, J. I., **6**, 42[46]
Hay Motherwell, R. S., **6**, 837[59]; **7**, 13[119], 40[11]; **8**, 821[51], 823[55], 830[89], 831[89]
Haynes, L., **3**, 380[13]
Haynes, N. B., **7**, 100[126]
Haynes, R. K., **1**, 37[239], 520[69,70,72-74], 635[83,90], 636[90], 678[83,90], 681[83,90], 691[83,90], 779[223]; **2**, 66[33], 75[33,82a]; **3**, 87[94] ; **4**, 12[37,37a-d], 119[192a,b,193], 159[85], 226[190,191,192,193,194,195], 249[129], 258[129,244,245,246], 260[245]; **6**, 150[114], 154[145], 439[68], 864[192]
Haynes, S. K., **4**, 259[265]
Hays, H. R., **1**, 476[112]; **3**, 65[4]; **8**, 860[219], 862[219]
Hays, J. T., **2**, 144[57]
Hays, S. L., **7**, 474[36]
Hayward, R. C., **4**, 377[101]; **7**, 121[24], 331[14], 438[15], 445[15], 502[261], 530[20], 531[20]
Hayward, R. J., **5**, 720[97,98]
Haywood, B. D., **8**, 532[130]
Haywood, D. J., **2**, 599[19]; **5**, 134[62], 159[53], 180[148]
Haywood, R. C., **8**, 504[71]
Hazato, A., **3**, 222[144]; **4**, 258[248], 261[248], 262[308]; **6**, 837[60], 942[154], 944[154]
Hazato, H., **4**, 253[169]

Hazdra, J. J., **6**, 1034[134], 1035[135]
Hazen, G. G., **4**, 1033[30]
Hazra, B. G., **3**, 676[75]; **8**, 515[116]
Hazum, E., **4**, 674[33]
Hazuto, A., **4**, 13[44,44c]
He, D.-W., **5**, 37[22b]
He, J., **4**, 730[67]
He, J.-F., **5**, 531[74], 532[74c]
He, W., **5**, 833[48]
He, X.-C., **1**, 62[39]
He, Y., **8**, 300[86]
He, Y.-B., **3**, 455[124], 525[42]; **8**, 912[89]
He, Z.-M., **5**, 344[67a]
Heacock, D. J., **4**, 598[207]; **7**, 760[46]
Heacock, R. A., **8**, 315[48]
Head, D. B., **3**, 220[122], 224[175], 416[12,13], 417[13], 752[95]
Head, J. C., **6**, 5[24]; **7**, 406[71]
Heah, P. C., **2**, 127[234]
Heald, P. W., **8**, 581[8], 582[8]
Healey, A. T., **4**, 5[18]
Healy, E. F., **5**, 72[179], 856[205,214]
Heaney, F., **5**, 257[59]
Heaney, H., **2**, 707[1], 742[73,74], 744[95], 748[126], 901[39], 948[183], 959[31], 960[31], 962[45], 964[45,60,61], 965[63], 966[61,71], 967[61,63,71]; **4**, 477[164], 489[64], 510[165,174]; **5**, 383[123]; **7**, 194[4], 374[78], 674[41]
Heap, N., **3**, 735[18]
Heard, N., **5**, 913[103], 955[302], 958[103]
Heard, N. E., **8**, 764[6]
Hearn, M. J., **8**, 374[144]
Hearst, J. E., **8**, 626[167]
Heasley, G. E., **4**, 345[81], 347[90,95]; **7**, 500[243], 530[28], 531[28]
Heasley, L. E., **4**, 347[90]
Heasley, V. L., **4**, 345[81], 347[90,95]; **7**, 530[28], 531[28]
Heath, E. C., **2**, 466[112,116,126], 467[112,116]
Heath, M. J., **1**, 377[97,97a]; **6**, 704[7], 727[205]; **8**, 943[120]
Heath, P., **5**, 655[29]
Heath, R. R., **3**, 45[245]
Heath, T. D., **8**, 36[50], 66[50]
Heath-Brown, B., **8**, 564[443], 614[87]
Heathcock, C. H., **1**, 2[16], 3[24], 49[8], 141[18], 179[26], 182[26], 286[7-9], 335[64], 357[2], 399[6], 405[26], 460[1], 544[40], 562[168], 608[35], 771[196], 772[198], 824[45]; **2**, 2[4], 3[9], 5[16], 6[4a,9], 19[9], 25[4a], 29[9a], 49[4a], 94[54], 100[14], 101[14], 109[64], 111[14], 123[199,200], 125[200,212,213], 127[212,213], 134[3], 148[79], 159[129,130], 161[137], 162[141,144], 182[9,9c], 184[9c], 188[44], 190[9c,57], 191[9c], 192[9c], 193[9c,63,64], 194[68], 195[70,72,72b,73-75], 197[9c], 198[9c], 200[9c], 201[93], 205[102-104], 206[102b,103,104], 211[9c], 217[9c,138], 218[141], 219[68,144], 221[70,145], 223[57d,151], 226[158], 235[9c], 236[9c,57d], 237[57d], 238[57d], 240[3], 248[3a], 277[8], 280[24], 281[24], 282[39], 284[39], 289[8,69], 301[1], 310[30], 311[30], 313[36,37], 455[9], 475[1], 518[62], 570[36], 577[80], 580[100], 611[101], 630[4,4a,10], 631[10], 632[10], 633[4a], 635[4a], 637[4a], 639[4a], 640[10,64,67], 641[10], 642[4a,10,75,77], 643[75,77], 646[10], 649[64,106], 652[4a], 655[143,148], 677[55], 678[55], 915[81], 931[134], 946[81], 1012[17], 1013[17], 1024[63], 1064[111]; **3**, 8[36], 20[107], 21[129], 39[220], 178[542], 179[542], 225[187], 252[81], 257[81], 273[3], 280[40], 344[19]; **4**, 7[26], 10[34], 21[67,67a,b], 31[94], 32[94f], 72[31], 91[88a], 100[123,123b], 103[134], 106[139], 107[139], 115[182], 155[68c], 158[79], 161[86b], 194[122], 239[25], 241[56], 247[25], 255[56,193], 258[25], 260[25,56], 297[270]; **5**, 11[84], 170[113], 808[110], 839[79], 843[79,114], 914[109]; **6**, 814[95], 1027[91], 1033[122]; **7**, 111[192], 158[37], 166[86a], 256[25], 367[57], 473[31,32], 500[238], 501[31,32,251,253,254], 502[253,254], 574[125], 673[32]; **8**, 4[24], 536[168], 925[6], 938[91], 946[136], 948[150]
Heathcock, S. M., **4**, 709[47], 710[47,49], 712[49]
Heather, J. B., **8**, 194[103], 544[264,265], 546[304], 561[304]
Heaton, P. C., **7**, 705[18]
Heavner, G. A., **6**, 619[117]; **8**, 385[43]
Hebeisen, P., **5**, 21[159], 23[159]
Hebel, D., **4**, 347[98]
Hebert, E., **1**, 555[111], 557[111]; **3**, 216[77]
Hecht, S. M., **7**, 143[146]; **8**, 34[62], 66[62]
Hecht, S. S., **3**, 380[8]; **4**, 313[475]; **7**, 350[22]
Hechtl, W., **5**, 714[76]
Heck, F. R., **3**, 271[1], 530[60]
Heck, G., **1**, 844[8]; **3**, 887[8], 888[8], 893[8], 897[8], 900[8], 903[8,128]; **4**, 953[8,8b], 954[8b]; **8**, 382[11], 383[11]
Heck, H. E., **2**, 495[58]
Heck, R. F., **3**, 228[214], 436[15], 470[220], 471[220], 521[8], 524[8], 539[98], 1021[13], 1028[47], 1029[54], 1032[68], 1034[78], 1037[89]; **4**, 115[181a], 552[1], 564[41], 598[186], 603[273], 638[186], 834[1a,b], 837[1a], 838[21-23], 839[21,23-25], 841[21,48], 842[51], 844[56,57,64], 845[1b,64,67,68], 846[71-73], 847[51,68,71,73], 848[73,80], 849[80,82,83], 850[84], 852[48,57,90], 853[91], 854[94], 855[96,97], 856[83,98], 903[201,201b], 904[205]; **5**, 1037[3], 1132[20], 1138[61], 1163[5]; **7**, 450[12]; **8**, 368[64], 557[386,387], 902[46], 904[46], 907[46], 959[28]
Heckendorn, D. K., **4**, 113[169,169c]
Heckendorn, R., **5**, 161[62]
Hecker, E., **3**, 810[49], 812[52]
Hecker, S. J., **1**, 405[26], 772[198]; **4**, 194[122]; **8**, 948[150]
Heckl, B., **5**, 1085[64]
Hedaya, E., **3**, 623[33]
Hedayatullah, M., **6**, 243[105], 244[108]; **7**, 484[135]
Hederich, V., **6**, 565[921]
Hedge, J. A., **2**, 355[121], 356[130]
Hedge, P., **7**, 763[103]
Hedgecock, H. C., **7**, 597[47], 606[149]
Hedges, S. H., **6**, 502[211]
Hedrick, S. T., **5**, 73[198]
Hedstrand, D. M., **8**, 93[77], 94[77]
Heerding, D. A., **2**, 1035[91], 1050[32], 1072[32]
Heeren, J. K., **3**, 760[140]
Heeres, H. J., **1**, 232[16]
Heerze, L. D., **3**, 589[162], 610[162]; **4**, 373[82]
Heeschen, J. P., **2**, 151[98], 152[98]; **4**, 337[34]
Heffe, W., **6**, 555[816]
Heffner, T. A., **4**, 1079[67]; **5**, 260[70], 263[70]
Heffron, P. J., **7**, 606[158]
Heflich, R. H., **7**, 75[116]
Hegarty, A. F., **2**, 107[61]; **6**, 252[153], 291[208], 293[224,225], 547[658], 552[658]; **7**, 671[15]
Hégazi, E., **6**, 423[43]
Hegde, M. S., **5**, 452[55], 453[55]
Hegde, V. R., **5**, 86[14], 95[95], 96[105,116,119,121], 98[121]
Hegedus, L. S., **1**, 189[74], 439[163,164], 440[167,171], 451[217], 457[163]; **2**, 596[2]; **3**, 208[3], 213[3b], 423[75], 426[82], 429[82], 485[38], 486[38], 491[38], 492[38], 495[38], 498[108], 503[38], 1024[30], 1037[87], 1038[91]; **4**, 12[41], 115[179a,181b], 174[39], 401[234a], 518[1], 547[1], 558[18], 560[21,22,25,26,28], 561[29,30,32], 566[46], 571[1,2], 572[1,4,5], 589[81], 598[81,188], 600[226], 638[81], 695[3], 834[1c], 841[43,47], 895[160], 903[200]; **5**, 46[39], 56[39], 85[7], 1065[1], 1066[1,5], 1074[1], 1076[34], 1080[53], 1083[1], 1084[1,53], 1087[77], 1093[1], 1105[167], 1107[167,168,169,170,171,172], 1108[170,171,172], 1111[34], 1112[1g], 1163[3], 1183[3]; **6**, 117[98], 284[176], 535[543], 538[543], 760[144], 849[114]; **8**, 278, 421[28], 422[28], 432[28], 435[28], 436[28]
Heggie, W., **3**, 386[64,65]
Heggs, R. P., **7**, 763[89]

Hehre, W. J., **1**, 297[57], 487[4], 488[4], 610[45]; **4**, 87[82], 213[119,120], 240[54], 379[113], 380[113a], 484[19], 968[62]; **5**, 72[167], 349[75], 856[208], 901[33]
Heiba, E. I., **4**, 763[211]; **7**, 154[24], 870[96]
Heibel, G. E., **5**, 125[15], 153[25]
Heibl, C., **6**, 426[72]
Heicklen, J., **5**, 931[179], 937[179], 938[179], 947[179]
Heid, H., **6**, 184[149], 509[247]
Heide, F. R., **4**, 1033[22], 1057[22c]
Heidelberger, C., **8**, 643[35]
Heider, K., **3**, 324[151]
Heidlas, J., **8**, 190[84]
Heigl, U. W., **4**, 1002[60]
Heijman, H. J., **8**, 285[9], 292[9], 293[9]
Heil, B., **4**, 925[36], 927[41], 930[41], 939[41]; **8**, 91[55], 152[177], 535[166]
Heilbron, I., **2**, 143[51,56]; **7**, 254[30]; **8**, 639[21]
Heilbronner, E., **2**, 144[59]; **4**, 537[98], 538[98]; **7**, 867[90]
Heiligenmann, G., **3**, 45[251]
Heilman, W. J., **7**, 167[98]
Heilmann, S. M., **1**, 548[65], 551[72,74]; **2**, 283[44], 298[44], 323[31]; **6**, 229[24]
Heilweil, E., **6**, 533[488], 550[488]; **8**, 895[5], 898[5]
Heim, N., **8**, 545[285]
Heim, P., **8**, 244[51], 249[51], 261[7,8], 263[8], 273[8], 875[39]
Heimann, M. R., **6**, 651[137]; **7**, 241[59]
Heimbach, H., **3**, 25[158,160]; **4**, 162[94c]
Heimbach, K., **2**, 139[31]
Heimbach, P., **3**, 390[84], 392[84]; **4**, 887[126]; **5**, 641[132], 800[76], 809[113]
Heimer, E. P., **6**, 635[14b], 636[14]; **8**, 959[21]
Heimgartner, H., **1**, 843[2e]; **4**, 1081[73,74], 1084[95]; **5**, 92[78], 438[165], 482[172], 707[32], 709[45], 712[45d], 713[32], 812[131], 894[43]; **6**, 539[579], 540[582], 543[625], 544[626]; **7**, 831[63]
Hein, R. W., **2**, 345[21]
Hein, S. J., **8**, 648[59]
Heindel, N. D., **8**, 626[168,169]
Heine, H.-G., **5**, 160[56], 1188[15]; **6**, 254[164]
Heine, H. W., **1**, 838[171]; **4**, 1085[106]; **5**, 474[158], 938[209], 949[277,278,283]; **6**, 96[152], 182[140]
Heine, W. H.-G., **5**, 646[5,6]
Heinekey, D. M., **1**, 215[33]
Heinemann, F., **6**, 7[34]
Heinemann, H., **8**, 600[105]
Heinemann, U., **5**, 482[173], 483[175]
Heinen, H., **5**, 1076[47], 1109[174,175,176,177,178,179,180]
Heinhold, G., **2**, 850[217]
Heinicke, G. W., **2**, 809[55]; **6**, 186[168]
Heinicke, R., **2**, 782[29]
Heinis, T., **7**, 854[46]
Heinke, B., **6**, 438[40]
Heinsohn, G., **4**, 140[9], 257[220]; **6**, 639[46], 659[46c]
Heinz, E., **6**, 51[112,113]
Heinze, J., **7**, 852[39]; **8**, 131[3]
Heinze, P. L., **5**, 680[24,24b]
Heinzer, J., **3**, 75[46]
Heinzman, S. W., **8**, 253[119]
Heisey, L. V., **2**, 965[66]
Heisler, R. Y., **6**, 253[158]
Heiss, H., **3**, 904[132]
Heissler, D., **2**, 816[86]; **7**, 517[14], 564[88], 568[88]
Heiszwolf, G. J., **2**, 101[18], 597[11]
Heitkämpfer, P., **5**, 157[42]
Heitler, C., **2**, 365[212]
Heitmann, P., **1**, 546[58]; **2**, 249[84], 654[129]
Heitmeier, D. E., **8**, 254[125]
Heitz, M.-P., **1**, 64[45], 115[42], 594[27], 595[27]; **2**, 1064[110]
Heitz, W., **4**, 461[99], 475[99]
Heitzmann, M., **5**, 386[133,133a]
Helbert, M., **6**, 489[82]
Helbig, R., **6**, 625[155]
Helbig, W., **2**, 33[121]; **8**, 196[121]
Helbling, A. M., **2**, 102[25]; **7**, 674[38]
Helder, R., **4**, 230[241]
Heldeweg, R. F., **5**, 79[284]
Heldrich, F. J., **3**, 464[174]
Heldt, W. Z., **6**, 727[203], 763[2]; **7**, 689[5], 691[5]
Helfer, A. P., **5**, 950[285]
Helfer, D. L., II, **6**, 553[723], 554[723]
Helferich, B., **6**, 39[38], 42[45], 49[90]
Helfrich, O. B., **7**, 768[197]
Helg, R., **8**, 141[44]
Helgée, B., **2**, 1051[35], 1052[35]; **4**, 455[42]; **7**, 801[36]
Hell, W., **6**, 526[396]
Helle, M. A., **1**, 768[171]; **4**, 408[257a]
Heller, H. G., **5**, 721[103], 722[104,105], 729[124]
Heller, M., **6**, 959[36]
Hellier, M., **8**, 568[476]
Hellin, M., **4**, 298[289]; **6**, 263[20,24], 264[24], 267[24], 269[20]
Helling, J. F., **4**, 521[44]
Hellman, H. M., **4**, 20[63], 21[63]
Hellman, T. M., **7**, 13[123]
Hellmann, G., **1**, 37[249], 773[203,203b]
Hellmann, H., **1**, 367[56], 368[56], 370[56], 371[75]; **2**, 894[3], 897[3], 933[3], 953[1,2], 966[2], 1088[46], 1090[46]; **3**, 933[62]; **7**, 804[63]
Hellou, J., **2**, 839[181], 840[181]
Hellring, S., **1**, 482[164,165]; **3**, 71[34], 72[38]; **6**, 545[629], 551[681]; **7**, 224[48]
Hellrung, B., **2**, 740[56]
Hellström, H., **6**, 602[2]
Hellwinkel, D., **7**, 775[342]
Helmchen, G., **1**, 168[116a], 303[78], 307[78]; **2**, 35[127], 455[12], 636[54], 637[54], 640[54], 681[59], 1108[78]; **3**, 45[251]; **4**, 152[54], 202[17,23]; **5**, 353[85], 355[87c], 356[87c], 365[87c,95,95b,96a,c], 543[113], 555[49]; **6**, 679[328], 863[191]; **7**, 160[50]
Helmchen-Zeier, R. E., **8**, 196[115]
Helmers, R., **2**, 849[216], 850[216], 857[251]
Helmkamp, G. K., **6**, 277[125,126]; **8**, 614[91]
Helmlinger, D., **3**, 383[44]
Helmut, H., **1**, 551[71]
Helmy, E., **6**, 1019[44]
Helps, I. M., **5**, 1043[25], 1048[25a], 1051[36a], 1056[25a]
Helquist, P., **1**, 193[87], 415[64], 428[115,116]; **2**, 66[32], 125[223], 448[36]; **3**, 135[367], 137[377], 243[17], 249[17], 251[77], 254[77], 263[17,172], 423[80], 430[93], 483[17], 499[128], 500[128,132], 505[132], 509[128]; **4**, 91[88f], 107[141], 120[201], 212[99], 250[135], 251[150], 255[135], 598[209], 599[215], 609[215], 624[215], 638[209], 641[215], 898[174], 899[174], 976[100], 980[100k,110], 982[110], 983[118], 984[121,122], 985[125]; **5**, 522[47], 1086[72,73]
Helson, H. E., **3**, 538[91,93]; **5**, 1151[128,129]
Hem, S. L., **7**, 845[77]
Hembre, R. T., **3**, 1026[39]
Hemetsberger, H., **3**, 804[10]; **5**, 198[23], 209[54], 210[57], 413[49]; **7**, 32[92]
Hemling, T. C., **6**, 845[96]
Hemo, J. H., **8**, 27[31], 66[31]
Hems, B. A., **8**, 639[20]
Henbest, H. B., **2**, 156[114]; **3**, 734[4], 741[53]; **4**, 301[325], 302[336], 315[508]; **5**, 767[119]; **6**, 1042[10]; **7**, 152[4], 153[4], 221[33-35], 236[17], 390[1], 582[149], 768[205], 769[205,211]; **8**, 336[73], 337[73], 338[73], 339[73], 341[73], 429[54], 533[154]

Hencken, G., **1**, 9[40]
Henderson, G. N., **7**, 345[5], 843[45], 845[67] ; **8**, 343[113]
Henderson, M. A., **4**, 21[67,67a], 106[139], 107[139]
Henderson, M. J., **1**, 17[218]
Henderson, R., **8**, 883[91]
Henderson, W. A., Jr., **5**, 712[60]
Henderson, W. W., **6**, 244[106]
Hendi, M. S., **4**, 464[118], 465[118]
Hendi, S. B., **7**, 29[79]
Hendley, E. C., **3**, 822[10], 825[28], 829[10]
Hendric, R., **7**, 365[49]
Hendrick, G. W., **3**, 771[187]
Hendrick, M. E., **4**, 1011[166]; **6**, 776[55]
Hendricks, R. T., **5**, 797[65]
Hendrickson, J. B., **1**, 894[158]; **3**, 158[436], 159[436], 174[538,539], 177[538,539], 831[63], 868[42-44], 869[43,44], 876[44], 1048[10]; **4**, 6[23,23b], 52[147,147a], 1033[33], 1089[127], 1092[127]; **5**, 700[7]; **6**, 83[78], 125[147] ; **7**, 299[41], 668[82,84]; **8**, 916[102], 917[102], 918[102], 919[102], 920[102], 992[52], 994[63]
Hendrickson, W. A., **4**, 521[44]
Hendriks, K. B., **6**, 43[47], 633[9]
Hendrikse, J. L., **8**, 447[98]
Hendrikx, G., **4**, 45[130,130d]
Hendrix, J., **1**, 404[23]
Hendry, D., **6**, 73[27], 76[51]
Henecka, H., **6**, 507[231]
Henegar, K. E., **5**, 137[83]
Henegeveld, J. E., **7**, 209[91]
Henery, J., **5**, 677[11], 695[11]
Heng, K. K., **2**, 185[31]; **4**, 243[67], 244[67], 260[67], 261[67]
Hengartner, U., **8**, 606[18], 817[28], 930[36]
Henggeler, B., **7**, 219[16]
Hengrasmee, S., **8**, 783[107]
Henin, F., **4**, 857[105]
Henkal, J., **7**, 613[1]
Henke, B. R., **1**, 294[38], 304[38]; **2**, 6[35], 568[32]
Henke, K., **1**, 66[53]; **2**, 536[41]
Henkel, J. G., **7**, 503[278]; **8**, 354[172]
Henkler, H., **7**, 506[303]
Henly, T. J., **7**, 8[65]
Henn, L., **7**, 27[64]
Henne, A. L., **3**, 273[5]; **4**, 270[4,14,16-19], 271[4,16,17,19,26], 272[4,26], 274[62], 275[62], 280[62], 285[163]; **5**, 216[17], 217[17,19,20], 219[17]; **8**, 479[49]
Henneberg, D., **4**, 868[12], 874[50,51,54,55], 887[12]; **5**, 30[2]
Henneke, K.-W., **7**, 232[155]
Hennen, W. J., **2**, 463[91], 464[91]; **6**, 474[83], 814[96]
Hennequin, L., **4**, 159[81]
Hennessy, B. M., **1**, 822[34]; **5**, 256[58], 257[58b]; **6**, 960[56]; **7**, 268[122], 564[92], 567[92]
Hennessy, M. J., **3**, 1055[31], 1062[31,47]; **4**, 1033[32]; **6**, 123[141], 125[141]
Hennige, H., **6**, 502[215], 531[215]
Henning, R., **2**, 63[23], 337[73]; **3**, 851[68] ; **4**, 12[43], 13[43b], 104[136a,c]; **5**, 596[36], 598[36], 609[70]; **6**, 16[61], 284[171] ; **7**, 360[22], 646[26]; **8**, 788[120]
Henninger, M., **2**, 523[90]; **8**, 384[30]
Hennion, G. F., **3**, 273[9]; **4**, 276[78], 277[86], 303[345], 315[522]
Henoch, F. E., **2**, 507[18]; **7**, 187[184] ; **8**, 329[21]
Henretta, J. P., **4**, 1040[79], 1049[79]; **5**, 599[40], 804[94], 986[40]
Henrichson, C., **6**, 46[70], 620[133]
Henrici-Olivé, G., **8**, 447[105]
Henrick, C. A., **1**, 116[45], 128[45], 433[139], 434[139]; **3**, 220[121], 224[163], 257[120], 416[21], 419[43]; **7**, 661[47]; **8**, 430[58]
Henrie, R. N., II, **8**, 31[46], 66[46]
Henriksen, L., **6**, 464[33], 465[33], 480[113], 509[249]
Henriksen, U., **5**, 741[153]; **6**, 481[117]
Henriksson, U., **1**, 292[30]
Henrion, A., **8**, 595[78]
Henri-Rousseau, O., **4**, 1073[26]; **5**, 247[26]
Henrot, S., **3**, 41[225], 263[175]; **6**, 3[17] ; **8**, 196[114]
Henry, D. W., **8**, 254[125], 354[171], 819[44]
Henry, J. A., **8**, 964[54]
Henry, J. P., **5**, 1025[81]
Henry, L., **2**, 321[1]
Henry, M. C., **8**, 548[319]
Henry, P. M., **3**, 381[32]; **4**, 552[1], 596[165], 834[1d]; **7**, 94[55], 450[10], 451[33], 541[4], 564[4]
Henry-Basch, E., **1**, 86[33,41], 215[40], 216[40], 218[49], 220[49], 223[49], 226[49a,90-92,94,96], 326[6,7], 327[8], 349[143,144]; **4**, 95[97], 98[108d,109d-f]
Henseke, G., **3**, 753[103]
Hensel, M. J., **1**, 765[174]
Henshall, A., **7**, 506[298]
Henssen, G., **5**, 475[145]
Hentges, S. G., **7**, 438[12], 441[12], 443[12], 489[164]
Hentschel, P., **2**, 598[13]
Henz, K. J., **2**, 1056[64], 1070[64]; **5**, 404[18]; **7**, 673[30]
Henzel, R. P., **5**, 225[102], 915[112]
Henzen, A. V., **1**, 774[212]
Hepp, E., **6**, 261[3,4], 270[3,4]
Heppke, G., **6**, 462[15]
Herald, C. L., **7**, 153[11]
Herb, G., **4**, 1039[62]
Herberhold, M., **6**, 450[120]; **7**, 774[319]
Herbert, D. J., **2**, 851[225]; **7**, 262[75]
Herbert, K. A., **4**, 1018[218]
Herbert, R., **8**, 472[6]
Herbert, R. B., **3**, 670[61], 681[94], 807[31] ; **4**, 435[136]; **6**, 1002[133]
Herbert, W., **8**, 297[65]
Herbig, K., **4**, 48[138,138a], 66[138a,b]
Herbst, P., **5**, 157[42]
Herbst, R. M., **8**, 145[87]
Herczegh, P., **5**, 438[164]
Herdewijn, P., **5**, 92[72]
Hergenrother, W. L., **6**, 960[54]
Herges, R., **2**, 1090[72]; **5**, 77[251]
Hergott, H. H., **6**, 502[217], 560[870]; **7**, 650[51]
Hergrueter, C. A., **3**, 251[77], 254[77]
Hering, G., **2**, 1090[72]
Hering, H., **8**, 310[16]
Herissey, H., **3**, 693[145]
Heritage, G. L., **4**, 84[67], 106[138a,b]
Herkes, F. E., **4**, 49[141]
Herktorn, N., **1**, 476[115]
Herlem, D., **8**, 618[126]
Herlihy, K. P., **3**, 725[21], 726[21]
Herlinger, H., **1**, 542[6], 546[6]; **2**, 1089[57], 1090[68,69], 1091[69], 1094[89], 1095[89]
Herlt, A. J., **4**, 298[292]
Herman, B., **1**, 661[167,167c]
Herman, D. F., **1**, 139[2], 140[8]
Herman, F., **7**, 24[39], 25[39]
Herman, G., **4**, 987[147]; **8**, 935[59]
Herman, J. J., **4**, 609[330]; **5**, 86[35]
Herman, L. W., **2**, 323[21]
Hermanek, S., **1**, 489[15]
Hermann, C. K. F., **4**, 463[110], 468[110], 469[110]
Hermann, C. W., **2**, 1066[122]
Hermann, E. C., **6**, 270[77]
Hermann, H., **4**, 522[49], 523[49]

Hermann, K., **4**, 230[243]
Hermann, R. B., **3**, 905[140]
Hermecz, I., **2**, 789[56]; **6**, 499[177], 520[340]; **7**, 846[97]
Hermeling, D., **3**, 644[142], 651[217]
Hermes, M. E., **7**, 474[39], 480[98]
Hermolin, J., **8**, 594[73], 595[73,74]
Hermosin, M. C., **7**, 845[77]
Hernandez, A., **1**, 790[262]; **8**, 707[22], 936[75,76], 937[76]
Hernandez, D., **2**, 111[79]; **8**, 707[22], 936[75]
Hernandez, E., **5**, 1140[77]; **8**, 936[74]
Hernandez, H., **6**, 554[735]
Hernandez, J. E., **7**, 706[22]
Hernandez, M., **4**, 1033[26], 1046[26b]
Hernandez, O., **3**, 135[368], 136[368], 137[368] ; **6**, 23[94], 650[128]
Hernandez, R., **2**, 1049[16]; **4**, 342[61], 814[187]; **7**, 41[15], 722[19], 723[19], 725[19]
Herndon, J. H., **5**, 277[12]
Herndon, J. W., **1**, 248[68]; **2**, 587[140] ; **4**, 82[62g], 218[133], 243[76], 580[24], 581[26,27], 607[309], 626[309], 647[309], 695[4] ; **5**, 1105[164]
Herndon, W. C., **2**, 662[7]; **4**, 1073[24] ; **5**, 552[18]
Hernot, D., **3**, 1052[27]; **6**, 126[150]
Herocheid, J. D. M., **7**, 535[47]
Herold, L. L., **4**, 744[136]
Herold, P., **2**, 308[23], 309[23], 318[51] ; **4**, 382[134]; **5**, 836[63]
Herold, T., **1**, 192[81]; **2**, 33[121]; **4**, 130[226b]
Herout, V., **5**, 809[114]; **8**, 334[65]
Herr, C. H., **8**, 328[14], 329[14]
Herr, D., **8**, 986[9]
Herr, H.-J., **5**, 744[165]
Herr, M. E., **2**, 169[166]
Herr, M. F., **3**, 739[43]; **7**, 167[97], 168[101]
Herr, R., **6**, 4[19,20], 11[47], 176[91], 196[227,234]
Herr, R. R., **3**, 242[6], 257[6], 259[6]
Herr, R. W., **3**, 223[158], 226[192], 263[177], 265[187]; **4**, 170[19]
Herradon, B., **1**, 759[132]
Herranz, E., **7**, 489[167,168,169]
Herrera, A., **5**, 1130[2]
Herrera, F. J. L., **4**, 379[115]
Herrick, A. B., **4**, 1021[244]
Herrick, J. J., **1**, 418[71]
Herrin, T. R., **3**, 369[117], 372[117]
Herrinton, P. M., **1**, 826[61], 891[147]; **2**, 554[135]; **3**, 224[172,172b], 325[162], 416[19], 792[68]
Herrmann, G., **6**, 583[995,996]
Herrmann, J., **7**, 763[96]
Herrmann, J. H., **4**, 113[166]
Herrmann, J. L., **1**, 527[105,106], 564[191], 567[224] ; **2**, 187[41]; **3**, 43[238], 135[358], 136[358], 137[358], 139[358], 142[358], 143[358]; **4**, 10[32,32a-g,33], 11[32e], 23[33d], 107[144], 109[148], 125[216,216f], 262[306]; **6**, 134[26]
Herrmann, R., **2**, 1084[10], 1090[72,73,76], 1098[104], 1099[109,109b], 1102[73], 1103[73]; **6**, 489[97] ; **7**, 778[397]
Herrmann, S., **8**, 85[17]
Herron, D. K., **8**, 727[198]
Herscheid, J. D. M., **6**, 113[70]; **7**, 230[134] ; **8**, 60[183], 61[183], 62[183], 66[183]
Herschenson, F. M., **1**, 359[21], 383[21], 384[21]
Herscovici, J., **2**, 555[141]; **7**, 265[97,98], 272[98]; **8**, 64[212], 66[212], 67[212], 68[212]
Hersh, W. H., **1**, 309[98]; **5**, 588[209]
Hershberg, E. B., **3**, 242[3], 244[3]; **8**, 286[14]
Hershberger, J. W., **4**, 738[100], 739[104], 746[146], 771[254]; **8**, 398[145]
Hershberger, P. M., **3**, 328[179]; **5**, 134[64]
Hershberger, S. A., **8**, 398[145]
Hershberger, S. S., **3**, 497[100], 498[100]; **4**, 602[254], 739[104], 840[32], 904[203]
Hershenson, F. M., **3**, 750[84], 762[84]; **4**, 1097[168]; **5**, 320[6]
Hershkowitz, R. L., **6**, 205[25,26], 210[25]
Hertel, G., **4**, 438[152]
Hertel, M., **7**, 740[45]
Hertenstein, U., **1**, 547[63], 548[63]; **2**, 69[45]; **3**, 197[40]; **6**, 681[333], 682[340]
Herter, R., **5**, 595[18,20], 596[18,20], 609[69]
Hertler, W. R., **2**, 619[148]
Hertz, E., **6**, 425[63]
Hertzberg, R. P., **8**, 643[35]
Herunsalee, K., **3**, 154[425], 155[425]
Hervé, Y., **4**, 748[158], 800[120]; **7**, 722[20], 725[31], 726[20,37], 731[53]
Hervé du Penhoat, C., **3**, 174[536,537], 176[536,537], 178[536,537], 448[95]
Herweh, J. E., **3**, 765[152]
Herwig, W., **1**, 174[10]
Herynk, J., **7**, 723[23], 724[28]
Herz, C., **6**, 37[34]
Herz, J. E., **1**, 377[99]
Herz, W., **2**, 146[66], 965[67], 968[67,79] ; **7**, 259[58], 821[27], 834[81]
Herzberg-Minzly, Y., **8**, 502[65], 503[65], 796[29]
Herzé, P. Y., **7**, 846[88]
Herzig, K., **4**, 587[21]
Herzig, P. T., **8**, 228[124]
Herzig, U., **2**, 346[47]
Herzog, C., **4**, 1017[214]
Herzog, H., **7**, 235[2]
Herzog, H. L., **7**, 55[10], 66[10], 68[10], 70[10], 71[10], 77[10]
Herzog, L., **4**, 4[14]
Hes, J., **6**, 624[140]
Hesbain-Frisque, A.-M., **5**, 473[152], 477[152], 480[165], 482[171], 483[165]
Heschel, M., **6**, 487[66], 489[66]
Heslop, J. A., **1**, 3[21]
Hess, A., **2**, 805[44], 815[44]
Hess, B. A., Jr., **3**, 901[105]; **5**, 702[14], 703[14], 740[14]
Hess, H. M., **8**, 541[218], 542[218], 543[218]
Hess, K., **2**, 141[41]
Hess, L. D., **5**, 226[109]
Hess, P., **1**, 473[82]
Hess, T. C., **3**, 891[43], 892[43]
Hess, W. W., **7**, 100[131], 256[42]
Hesse, G., **3**, 851[62]; **7**, 601[86,88]; **8**, 274[136], 989[39]
Hesse, K.-D., **5**, 500[258]
Hesse, M., **1**, 145[45], 843[2a,g], 893[152,153], 898[2a]; **4**, 629[414]; **6**, 1058[61]
Hesse, R. H., **1**, 174[8], 175[8]; **4**, 344[77], 347[104,105]; **7**, 15[145], 41[21,22], 90[32], 741[49], 747[94]; **8**, 796[28]
Hessell, E. T., **4**, 443[188]
Hessing, A., **8**, 440[85]
Hessler, E. J., **3**, 342[6]; **8**, 558[396]
Hester, J. B., Jr., **7**, 691[17]; **8**, 53[129], 66[129]
Hetflejš, J., **8**, 770[38]
Hettrick, C. M., **5**, 712[57d]
Heuberge, C., **5**, 842[112]
Heuberger, C., **6**, 470[59], 471[59]; **8**, 825[67]
Heuberger, G., **7**, 721[17]
Heublein, A., **2**, 424[35]
Heuck, K., **4**, 741[123]; **8**, 852[139], 853[146]
Heuck, U., **5**, 15[101]
Heuckeroth, R. O., **2**, 1039[104]; **7**, 876[123]
Heude, J. P. M., **7**, 485[139]

Heuer, W., **2**, 1088[43]; **6**, 547[661]
Heumann, A., **2**, 711[34]; **7**, 95[65,66], 452[47,48]
Heumann, H., **1**, 631[52]
Heuschmann, M., **3**, 870[45]; **5**, 64[42]
Heusler, K., **2**, 156[117], 358[153]; **4**, 410[262]; **6**, 667[242], 685[361]; **7**, 41[19,20] ; **8**, 974[124]
Heusser, H., **8**, 228[124], 530[89,91]
Heusser, H. L., **8**, 108[2], 118[2]
Heveling, J., **8**, 847[94]
Hevesi, L., **1**, 571[274], 630[13], 631[53], 635[87], 641[108], 642[108], 656[53,143], 657[53], 658[13,53], 659[53], 661[167,167c], 664[87], 665[87,108,175], 667[108], 672[13,87,108,195,199], 679[87], 682[87], 698[195,241]; **3**, 86[50,55], 87[81,84,86,87,90], 89[142,143], 92[143], 94[55], 104[81], 109[84], 116[142], 118[143,238], 120[90,244], 123[143,238], 142[86,244], 144[86]; **4**, 50[142], 342[63] ; **6**, 1027[90], 1031[112]; **7**, 515[1], 523[1] ; **8**, 847[97], 848[97e], 849[97e]
Hewawasam, P., **2**, 547[104], 549[104]
Hewawassam, P., **4**, 258[249]
Hewetson, D. W., **2**, 481[20]
Hewgill, F. R., **3**, 669[51]
Hewitt, B., **7**, 199[37]
Hewitt, C. D., **4**, 445[202]
Hewitt, G., **6**, 1017[38], 1024[38]
Hewitt, G. M., **8**, 205[156]
Hewitt, J. M., **8**, 192[99]
Hewson, A. T., **1**, 570[263]; **2**, 800[17] ; **4**, 15[50]; **6**, 176[95-99]
Hextall, P., **8**, 501[53], 507[53]
Hey, D. H., **2**, 742[71]; **3**, 505[162], 507[162], 512[162], 639[88], 807[27], 813[61]; **7**, 13[114], 120[15]
Hey, H., **4**, 606[301]; **5**, 641[132]; **8**, 960[33]
Hey, J. P., **3**, 58[291]; **5**, 134[66], 137[66,82]
Heydkamp, W., **6**, 911[18], 912[18]; **7**, 606[153]
Heydt, H., **3**, 909[152]; **4**, 1075[33], 1101[196], 1102[196]; **7**, 752[149]
Heyer, D., **7**, 43[38], 47[51], 48[60], 50[74]
Heyer, E. W., **5**, 908[72]
Heyes, J., **8**, 901[37]
Heyl, D., **5**, 789[28]
Heyland, D., **7**, 306[7]
Heyman, M. L., **7**, 747[105], 751[137]
Heymann, H., **8**, 916[110], 917[110]
Heymanns, P., **1**, 191[77], 272[68], 300[67], 322[67], 335[63]; **2**, 6[35], 247[35], 630[7], 631[7]
Heyn, A. S., **4**, 871[29], 877[67], 878[80], 884[80]
Heyn, M., **3**, 822[5], 834[5]
Heyns, K., **6**, 120[122]
He Youn, C., **5**, 347[72,72d]
Hibbert, H., **4**, 311[454]
Hiberty, P. C., **4**, 1070[14]
Hibi, T., **5**, 297[57], 1196[37]
Hibino, J., **2**, 588[151], 589[151,154]; **4**, 901[186]; **5**, 1124[51], 1125[58]
Hibino, J.-i., **1**, 749[87], 750[87,88], 812[87]
Hibino, K., **7**, 707[29], 708[29], 803[51]
Hibino, S., **2**, 363[193]; **5**, 402[1], 403[1], 404[1], 406[23,23b], 410[1], 413[1,1c], 417[1], 422[1], 425[1], 426[1], 429[1], 430[1], 433[1], 434[1], 435[1], 436[1], 438[1], 440[1], 444[1], 531[77], 725[118]; **6**, 780[72], 814[88] ; **8**, 64[207a], 65[207a], 66[207], 237[17], 240[17], 249[17], 369[77], 389[75], 934[57]
Hibino, T., **6**, 921[48]
Hickey, D. M. B., **7**, 27[64]
Hickinbottom, W. J., **4**, 3[12]; **7**, 16[158] ; **8**, 314[39]
Hickling, A., **3**, 634[9], 636[46], 657[9]
Hickmott, P. W., **2**, 1008[5]; **3**, 28[171], 30[171,176,178], 35[171], 39[171]; **4**, 45[126] ; **5**, 331[39]; **6**, 704[13-16], 710[60], 711[67], 712[72], 714[84], 719[15,16], 725[15,166]
Hicks, D. R., **6**, 27[113], 987[69]
Hida, T., **1**, 738[40]; **4**, 391[179]; **6**, 998[117]; **7**, 162[58], 243[66]
Hidai, M., **5**, 1158[173]; **8**, 446[94], 452[94], 460[254], 554[366]
Hidaka, A., **2**, 603[46]
Hidaka, T., **3**, 528[47]
Hidber, A., **1**, 40[193]; **2**, 197[80]; **4**, 77[51]
Hideaki, S., **4**, 507[150]
Hideg, K., **7**, 566[99]
Hiebert, J. D., **7**, 135[105], 136[105], 137[105], 145[105]
Hiegel, G. A., **3**, 653[226]; **6**, 1042[9], 1044[9]; **8**, 947[143]
Hiemstra, H., **1**, 371[75], 372[75e], 617[54], 771[192]; **2**, 89[36,37], 558[162], 586[133], 587[133], 652[123a], 971[91], 1048[6,7], 1049[6,7,15,22], 1050[22,30,31], 1053[6], 1061[93], 1062[6], 1064[108], 1065[113-115,117], 1069[133,134], 1070[138], 1071[138], 1072[30,31], 1075[138,151], 1078[15], 1079[151,156] ; **3**, 223[183], 225[183], 832[67]; **4**, 231[261] ; **5**, 501[267]; **6**, 118[99], 744[73]
Hienuki, Y., **3**, 380[10]
Hietbrink, B. E., **1**, 252[4]
Hifgt, H., **2**, 466[119], 469[119]
Higaki, M., **1**, 512[40]; **3**, 147[387], 149[387], 150[387], 151[387]; **8**, 185[18], 190[18]
Higashi, F., **8**, 150[144]
Higashi, H., **8**, 191[89]
Higashimura, H., **3**, 1032[70]; **4**, 408[259d], 558[19], 841[45]
Higashimura, T., **5**, 1148[114]
Higashiyama, K., **1**, 166[114], 369[64a]; **8**, 652[73]
Higby, R. G., **5**, 243[9]
Higgins, J., **6**, 210[72]
Higgins, R. H., **5**, 581[171]
Higgs, H., **1**, 2[3], 37[3]
Higgs, L. A., **5**, 955[303]
High, J., **3**, 197[33]
Highes, A. N., **8**, 445[37]
Hightower, L. E., **8**, 526[28]
Hignett, G. J., **1**, 473[80]
Hignett, R. R., **4**, 914[4], 924[4]
Higuchi, H., **1**, 480[152]; **5**, 841[92]
Higuchi, K., **1**, 424[98]; **6**, 493[128]
Higuchi, M., **6**, 533[486], 554[706,707,708]
Higuchi, N., **6**, 17[63], 18[63]; **7**, 645[20,21] ; **8**, 788[121]
Higuchi, T., **1**, 119[64]; **2**, 1060[90]; **3**, 228[220], 441[47]; **7**, 764[110], 778[390] ; **8**, 857[189]
Higurashi, K., **8**, 534[158], 537[158]
Hihira, T., **7**, 73[104]
Hii, G. S. C., **1**, 181[35]
Hii, P., **7**, 706[26]
Hiiragi, M., **4**, 495[85], 505[136], 510[173] ; **7**, 453[76]
Hiiro, T., **8**, 881[83]
Hijfte, L. V., **4**, 809[164]
Hikima, H., **8**, 170[88]
Hikino, H., **3**, 396[115], 748[73]
Hikita, T., **8**, 446[94], 452[94]
Hikita, Y., **5**, 412[45]
Hikota, M., **7**, 246[85]
Hilaire, L., **8**, 896[15]
Hilbert, G. E., **8**, 973[117]
Hilbert, S. D., **2**, 387[337]
Hild, W., **1**, 341[97]
Hildebrand, R., **3**, 621[31]
Hildebrand, U., **6**, 441[84]
Hildebrandt, A., **6**, 518[334]
Hildebrandt, B., **2**, 44[151], 45[151]
Hildenbrand, D. L., **1**, 252[7]
Hildenbrand, K., **5**, 200[30]
Hilinski, E. F., **5**, 71[134]; **7**, 851[14], 855[63], 856[66], 865[87]
Hill, A. E., **5**, 596[24], 597[24], 608[24a]

Hill, A. S., **8**, 828[80]
Hill, C. L., **3**, 332[204], 1047[5]; **7**, 8[65,66], 9[67], 632[58], 637[58]; **8**, 852[138]
Hill, C. M., **8**, 937[77]
Hill, D. G., **3**, 748[74]
Hill, D. R., **6**, 1024[75], 1025[79]
Hill, E. A., **3**, 906[145]; **4**, 871[36], 876[57,61], 878[57]
Hill, H. A. O., **3**, 499[110], 501[110], 505[110], 509[110], 512[110]; **6**, 208[62]
Hill, H. S., **4**, 311[454]
Hill, J., **4**, 27[78]; **5**, 151[10]; **7**, 760[18]
Hill, J. B., **6**, 971[129]
Hill, J. E., **2**, 821[107]
Hill, J. G., **7**, 375[79], 394[19]
Hill, J. H., **3**, 284[54]; **4**, 42[121]
Hill, J. H. M., **2**, 186[32]; **3**, 18[94], 42[94]; **6**, 16[61]
Hill, J. O., **6**, 538[550]
Hill, J. S., **4**, 27[79,79c]
Hill, J. W., **7**, 23[24], 24[24], 25[40], 26[24]
Hill, K., **5**, 205[41], 206[45,46], 207[41,48], 224[101]
Hill, K. A., **3**, 11[51]; **7**, 120[17], 123[17]
Hill, M. E., **8**, 267[69], 937[77]
Hill, M. P., **7**, 827[48]
Hill, R. J., **1**, 563[186]
Hill, R. K., **1**, 880[115], 898[115]; **2**, 538[58], 539[58]; **3**, 743[59], 929[58], 943[83], 953[83,100], 994[42]; **4**, 4[15] ; **5**, 2[14,16], 451[13], 470[13], 513[2], 518[2], 786[2], 791[24], 819[2], 822[24], 827[2], 829[2,20], 835[59], 877[8]; **6**, 268[67], 279[135], 834[27], 843[89], 855[27], 861[183], 873[4], 1066[91]; **7**, 601[79]
Hill, W. E., **7**, 498[228]
Hille, A., **6**, 734[8], 736[8,29], 738[56]
Hillemann, C. L., **5**, 984[32]
Hillenbrand, G. F., **7**, 177[144]
Hiller, W., **5**, 1131[13], 1145[105]
Hillgärtner, H., **4**, 760[196]
Hillhouse, G. L., **8**, 383[15], 673[24], 676[24,51,52], 682[24]
Hillhouse, J. H., **3**, 867[36]
Hillhouse, M. C., **4**, 985[126,131]
Hilliard, R. L., III, **5**, 692[90], 1150[127], 1151[132]
Hillier, I. H., **1**, 9[43]; **4**, 484[16]
Hilliger, E., **3**, 324[148]
Hillis, L. R., **1**, 477[143]
Hillman, M. E., **4**, 34[97], 35[97]
Hillman, W. S., **5**, 63[16]
Hillyard, R. A., **8**, 445[18]
Hiltbrünner, K., **4**, 1007[116]
Hilton, C., **5**, 710[55], 713[61]
Hilty, T. K., **8**, 786[118]
Hilvert, D., **1**, 70[63], 141[22]; **2**, 120[179] ; **4**, 229[238]; **5**, 855[191]
Himbert, G., **3**, 284[53], 555[32]; **5**, 116[259], 441[176,176f]; **6**, 185[164], 187[164], 549[672], 550[675,676], 552[675,691]
Himelstein, N., **8**, 140[18]
Himeno, Y., **5**, 176[128]
Himeshima, Y., **4**, 487[48]
Himics, R. J., **2**, 530[21]
Himmele, W., **3**, 640[97], 1022[18]; **4**, 920[22,23], 921[22], 922[23b], 923[22], 924[22], 925[22]
Himmelsbach, R. J., **4**, 427[68]
Himrichs, R., **5**, 1120[23]
Hinckley, J. A., Jr., **4**, 279[105,106]
Hinde, A. L., **8**, 491[12]
Hinder, M., **7**, 543[19], 546[19]
Hindersinn, R. R., **3**, 11[52], 17[52]
Hindley, K. B., **2**, 171[176], 381[303]
Hindriksen, B., **3**, 361[76], 367[76]
Hindsgaul, O., **6**, 51[110]
Hine, J., **2**, 773[23], 774[23]; **3**, 505[166], 829[56]; **4**, 999[3,4]; **6**, 556[820], 707[44], 951[5]; **8**, 87[31]
Hiner, R. N., **1**, 127[92], 243[54], 427[114] ; **2**, 225[155]
Hiner, S., **5**, 320[7]
Hines, J. N., **6**, 687[374]
Hinman, C. W., **2**, 738[41], 760[40]
Hinman, R. L., **7**, 505[291]; **8**, 70[236], 637[12]
Hino, K., **5**, 15[105], 799[74]
Hino, M., **2**, 737[37]; **7**, 5[22]
Hino, T., **2**, 6[28], 17[28,28d], 116[125], 323[23], 331[23], 332[23]; **4**, 285[164], 289[164]; **6**, 914[28]; **7**, 96[87], 335[32]
Hinrichs, E. v., **2**, 1094[89], 1095[89]
Hinsching, S., **2**, 495[60]
Hinshaw, B. C., **6**, 474[83]
Hinshaw, J. C., **5**, 568[108]
Hinshelwood, C., **8**, 364[18], 371[108]
Hinsken, W., **5**, 225[92-94], 232[92,93], 233[92-94]
Hintz, G., **7**, 603[116]
Hintz, H. L., **7**, 705[16]
Hintze, H., **8**, 440[85]
Hintze, M. J., **1**, 27[138,139], 30[152]; **2**, 100[8]
Hinz, A., **2**, 143[53]
Hinz, J., **5**, 1188[15]
Hinz, W., **5**, 95[94]; **6**, 175[75]
Hinze, R.-P., **2**, 874[27]
Hioki, H., **4**, 231[280]
Hioki, T., **2**, 185[30], 211[115], 215[115,133], 216[133], 217[30,136]; **3**, 229[226]; **4**, 85[73], 249[118], 257[118]; **6**, 425[65] ; **8**, 84[12]
Hippeli, C., **2**, 486[40]; **8**, 652[76]
Hippich, S., **6**, 554[802], 576[802], 581[802]
Hirabayashi, T., **5**, 575[131]
Hirabayashi, Y., **6**, 464[34-36], 465[34-36], 469[35] ; **7**, 774[318,332]
Hirabe, T., **8**, 863[235], 864[235], 965[66]
Hirai, H., **2**, 213[124], 771[13], 1060[86] ; **4**, 1040[93], 1041[93]; **5**, 260[71]; **6**, 152[136]; **8**, 431[60]
Hirai, K., **1**, 642[121], 646[121], 656[121,153], 658[121], 665[121], 667[121], 672[121]; **2**, 1060[89]; **3**, 88[132,133], 90[132,133], 95[132,132a,133], 99[132a,133], 101[132a,133], 107[132a,133], 118[132], 894[63], 1027[45]; **4**, 927[43], 930[43], 945[43]; **5**, 410[41]; **8**, 173[111], 445[54,54d], 784[111]
Hirai, S., **7**, 476[61]; **8**, 530[100]
Hirai, Y., **2**, 913[73]; **4**, 27[82], 222[166], 501[116], 802[128]; **5**, 693[108], 841[87], 843[115]; **7**, 406[78b], 776[356]; **8**, 600[102]
Hiraki, N., **4**, 973[84]
Hiraki, Y., **8**, 150[123], 151[146,154]
Hirako, Y., **4**, 218[134]
Hirakura, M., **3**, 810[45]; **4**, 23[70]
Hirama, M., **1**, 436[153], 529[126], 797[294] ; **2**, 75[82b], 112[92], 113[102], 197[81], 198[82], 242[18,20], 245[18b,20d], 246[18b,20d], 253[42], 926[117]; **4**, 38[109a], 119[192c], 231[280], 386[150], 387[150], 406[253], 408[253] ; **5**, 324[22], 539[107], 736[142l,m]; **6**, 89[118], 101[118], 509[254], 994[97], 1023[73] ; **7**, 247[102,103], 257[51], 438[22], 489[165], 503[271]; **8**, 191[88]
Hiramatsu, H., **2**, 348[58], 357[58]
Hirano, M., **7**, 92[44]
Hirano, S., **1**, 174[9], 177[9], 179[9], 234[25] ; **2**, 3[9], 6[9], 19[9]; **4**, 1017[213], 1019[213]; **5**, 760[89,90]; **7**, 219[10], 299[44]
Hirao, A., **1**, 317[138,139,140,141,142], 390[145], 391[145]; **8**, 18[126], 160[100], 170[82-84,96-101], 176[135], 178[100]
Hirao, I., **1**, 343[100-103,112,113], 419[78]; **2**, 174[183], 363[199]; **3**, 277[33], 279[35,36] ; **4**, 10[32], 21[66,66a,b], 107[146a,b], 108[146d], 988[139]; **5**, 151[17], 166[17]; **6**, 7[33], 176[89]; **8**, 860[223]
Hirao, J., **4**, 247[102], 252[102], 259[102]
Hirao, K., **1**, 858[60]; **6**, 777[61]

Hirao, K.-I., **7**, 686[97]
Hirao, T., **1**, 332[55,56]; **2**, 128[240], 579[95], 581[105]; **4**, 589[78], 600[241], 611[354], 614[373], 630[419], 840[35,37], 905[207], 1020[239], 1023[262], 1024[263,264]; **7**, 141[133], 144[133], 453[65]; **8**, 806[106,107], 807[106], 900[31]
Hiraoka, H., **1**, 165[110]; **2**, 94[50]; **7**, 318[51]
Hiraoka, T., **1**, 123[77], 372[80,81]; **2**, 649[101]; **3**, 24[147]; **7**, 741[52]
Hirashima, T., **3**, 1034[77]; **4**, 444[197], 878[83]; **8**, 366[45], 367[60]
Hirata, F., **5**, 817[146]
Hirata, K., **5**, 308[96]; **6**, 668[260]; **8**, 552[354]
Hirata, M., **4**, 377[104], 378[104f], 383[104f], 557[10]; **6**, 787[102]
Hirata, N., **1**, 872[89]; **3**, 785[36,36b]
Hirata, T., **4**, 33[96]; **6**, 554[719]; **8**, 353[157], 383[21]
Hirata, Y., **2**, 360[166]; **4**, 27[79,79a], 958[25]; **6**, 8[37], 121[133]; **7**, 440[40]; **8**, 309[11-13], 310[11,13], 311[12], 803[95], 885[105]
Hiratake, J., **4**, 230[244], 231[244]
Hiratsuka, H., **8**, 874[28], 875[28]
Hirayama, H., **2**, 427[43]
Hirayama, M., **4**, 111[152b], 218[146]; **6**, 563[900]; **7**, 366[52]
Hiremath, S. P., **6**, 771[43]
Hiremath, S. V., **6**, 687[381]
Hiriart, J. M., **5**, 1012[50]
Hiriyakkanavar, J. G., **4**, 52[147,147c]
Hiro, E., **5**, 945[246]
Hirobe, M., **1**, 119[64]; **2**, 1060[90]; **6**, 438[47]; **7**, 759[11]; **8**, 366[52], 392[94]
Hiroi, K., **3**, 943[82]; **4**, 221[149,155], 654[447,448,449]; **6**, 149[103], 154[150], 716[96-100], 718[120], 725[169], 774[47], 855[146], 893[76,84], 1020[47], 1033[123]; **7**, 124[43], 125[43], 126[43], 127[61], 701[65]; **8**, 934[56]
Hiroi, M., **8**, 369[80]
Hiroi, T., **7**, 771[283]
Hiroi, Y., **8**, 546[309]
Hirokawa, S., **5**, 136[67]
Hirokawa, T., **3**, 667[49]
Hiroki, O., **6**, 252[154]
Hiron, F., **6**, 3[13], 4[13]
Hironaka, K., **7**, 883[180]
Hirooka, S., **6**, 764[10]
Hirose, H., **4**, 34[97], 35[97,97i]
Hirose, M., **6**, 624[149]
Hirose, T., **4**, 206[54]; **8**, 774[75]
Hirose, Y., **3**, 347[26], 391[87]; **8**, 552[349]
Hiroshima, T., **3**, 168[490], 169[490]
Hirota, E., **8**, 422[35]
Hirota, H., **4**, 405[250a,b]; **5**, 809[119,120]; **7**, 239[49], 543[21]
Hirota, K., **2**, 790[57]; **3**, 219[107]; **4**, 436[142,143]; **7**, 877[135]; **8**, 908[76]
Hirota, S., **8**, 535[166]
Hirota, T., **2**, 780[6,13]
Hirotsu, K., **1**, 188[66], 189[66]; **2**, 23[91]; **3**, 228[220], 423[79], 441[47], 872[56]; **4**, 520[35], 532[83], 543[83], 545[83]; **7**, 301[60]; **8**, 857[189]
Hirotsu, Y., **6**, 566[925]
Hiroya, K., **2**, 1024[61,62]; **5**, 681[28], 712[57b], 741[157,157c,d]; **6**, 757[130]; **7**, 564[89], 569[89]
Hirsch, A. F., **6**, 660[199]
Hirsch, J. A., **5**, 537[101]; **8**, 351[165,166]
Hirsch, L. D., **4**, 181[75]
Hirsch, S., **5**, 139[85], 143[85], 501[264]
Hirschbein, B. L., **8**, 189[61]
Hirschberg, K., **3**, 894[59], 900[59]; **6**, 122[138]
Hirschhorn, A., **7**, 17[175]
Hirschmann, R., **3**, 890[32]; **6**, 219[122], 635[23], 636[23], 664[223]
Hirsekorn, F. J., **8**, 450[168], 454[168,202]
Hirsh, S., **2**, 1074[148]; **7**, 555[66]
Hirshfield, J., **1**, 823[44b]; **4**, 957[22]; **8**, 50[117], 66[117]
Hirst, G. C., **1**, 892[148]; **3**, 779[7], 792[7]
Hirthammer, M., **3**, 538[91]; **5**, 1151[128]
Hirukawa, T., **1**, 100[89]
Hiruta, K., **5**, 406[22]
Hirwe, S. N., **6**, 173[43]
Hisakawa, H., **1**, 544[32]
Hisano, T., **6**, 997[115]
Hiscock, S. M., **6**, 462[18]
Hiskey, C. F., **3**, 843[22]
Hiskey, R. G., **4**, 5[17]; **6**, 664[220,224], 669[263]; **7**, 765[133]; **8**, 146[89], 148[105]
Hissom, B. R., Jr., **3**, 325[161]
Hitch, E. F., **8**, 366[34]
Hitchcock, P. B., **1**, 16[90], 17[91,208], 39[191], 41[264]; **3**, 666[42]; **4**, 170[15]
Hite, G., **3**, 788[51], 845[37]
Hite, G. A., **1**, 585[13]; **3**, 21[127]; **5**, 760[91], 762[95,104], 763[95,104], 764[104]; **6**, 677[311]
Hiti, J., **8**, 798[48]
Hitomi, K., **2**, 575[66]
Hitzel, E., **8**, 444[9]
Hixon, R. M., **2**, 964[56]; **8**, 140[27]
Hixson, S. S., **5**, 194[1], 196[1], 197[1], 198[1], 210[1], 211[66]; **7**, 875[117]
Hiyama, T., **1**, 174[9], 177[9,19], 179[9,19,27], 180[27], 181[27], 182[27], 193[84], 198[84], 202[100], 234[25], 366[46], 751[109], 825[51], 831[98], 876[100]; **2**, 3[9], 5[9b,c], 6[9], 19[9,9b,c], 20[9b,c], 24[95], 29[9c], 298[93], 572[40], 575[40], 718[71], 849[214]; **3**, 202[96,97], 216[70], 229[230], 233[273,274], 246[33], 257[33], 421[63], 538[95], 539[97], 565[17]; **4**, 1007[118,119,122,124,126], 1008[126], 1009[118,136,137], 1017[213], 1018[219], 1019[213], 1040[67]; **5**, 760[89,90], 768[128,130,134], 769[128], 770[141,142], 771[142,143], 779[128], 780[141], 943[237], 1007[41]; **6**, 563[905], 677[312]; **7**, 219[10]; **8**, 8[45,46], 9[49], 11[63], 13[45,46], 20[45,46], 21[46], 773[63], 786[116], 797[45]
Hiyoshi, T., **8**, 591[59], 614[83]
Hjeds, H., **8**, 604[8], 605[8]
Hjorth, S., **7**, 831[64]
Hlasta, D. J., **1**, 400[8]; **8**, 623[150]
Hlubucek, J. R., **6**, 664[219]
Hnevsová, V., **3**, 896[72]
Ho, B. T., **8**, 376[161]
Ho, C. D., **2**, 84[12]
Ho, C.-K., **7**, 413[107b,c]
Ho, C. Y., **6**, 531[445], 538[445]
Ho, D., **1**, 768[173]; **7**, 409[91]
Ho, E., **5**, 480[178], 531[79]
Ho, H. C., **6**, 134[17]
Ho, H. Y., **4**, 596[160], 604[286], 621[160], 626[286], 636[160]
Ho, I., **3**, 579[98]; **8**, 123[86]
Ho, I. H., **7**, 462[119,120]
Ho, K. M., **8**, 840[33]
Ho, L.-K., **1**, 118[62]; **3**, 221[127]
Ho, L. L., **3**, 864[16,18,20], 866[16], 883[16]
Ho, P.-T., **2**, 139[30]; **6**, 807[59]
Ho, S., **1**, 748[72], 749[78], 816[78]; **4**, 979[101]; **5**, 1115[3], 1122[3], 1123[3], 1124[3]
Ho, T.-I., **3**, 500[127], 501[127]; **5**, 71[157]
Ho, T.-L., **1**, 174[4], 177[4], 563[186], 566[212]; **2**, 161[138], 523[86,88], 728[138]; **3**, 86[13], 159[13], 421[62], 563[1]; **4**, 12[42], 70[11], 139[2]; **5**, 553[46], 835[59]; **6**, 4[21], 134[17], 226[13], 658[190], 665[229], 667[229]; **7**, 231[150,152], 235[4], 581[139], 760[48], 761[51], 765[150], 851[18]; **8**, 15[93], 113[46], 116[46],

371[98,107], 383[18], 384[30], 388[18], 404[13,14,16], 531[119], 794[7], 797[42], 987[20,21], 988[31]
Hoagland, P. D., **6**, 57[135]
Hoagland, S., **1**, 824[45]; **2**, 205[104], 206[104], 219[144,144b]; **3**, 225[187]
Hoán, N., **8**, 328[4], 340[4]
Hoang, H., **5**, 1123[40]
Hoard, D. E., **6**, 614[92]
Hoare, D. G., **6**, 824[124]
Hoare, J. H., **2**, 904[49]; **5**, 855[186]
Hoashi, K., **3**, 246[44]; **4**, 589[80], 591[80]
Hobart, K., **7**, 47[53]
Hobbs, A. J. W., **3**, 558[53]; **8**, 681[77], 684[77], 694[77]
Hobbs, C. C., **7**, 11[84]
Hobbs, C. F., **3**, 415[8], 1023[22]; **4**, 930[51]; **8**, 459[228]
Hobbs, S. J., **5**, 854[178], 856[178], 872[178] ; **6**, 849[121], 860[177]
Hobe, M., **1**, 663[168], 664[168], 666[168] ; **3**, 95[160], 97[160], 120[160]
Hoberg, H., **4**, 905[212], 962[39]; **5**, 1130[2], 1138[64,71], 1140[77], 1141[78,79], 1156[163,166], 1157[71,167], 1158[163a]; **8**, 756[135,136,137,139]
Hobi, R., **6**, 708[50]
Hobkirk, J., **6**, 441[82]
Hoblitt, R. P., **7**, 501[251]
Höbold, W., **6**, 288[190]
Hobson, J. D., **8**, 90[46]
Hoch, H., **6**, 704[10]
Hoch, M., **4**, 740[115]; **6**, 51[112,113]
Hochstein, F. A., **8**, 568[483]
Hochstetler, A. R., **8**, 278[157]
Höchstetter, H., **3**, 872[60]
Hochuli, E., **8**, 197[128]
Hock, H., **7**, 111[193]
Hock, R., **5**, 442[185]
Hocker, J., **4**, 1033[29]; **6**, 124[143], 125[143], 579[986], 582[994]
Hockerman, G. H., **3**, 325[161,161a]
Hocking, M. B., **6**, 799[25]
Hocks, P., **7**, 86[16a]
Hodder, O. J. R., **1**, 38[263]
Hodge, C. N., **3**, 594[176], 610[176]
Hodge, H. C., **6**, 216[105]
Hodge, P., **2**, 737[33]; **4**, 231[269]; **6**, 83[79], 489[99], 525[99], 767[24]; **7**, 333[26], 709[36,38], 747[101], 765[136], 843[49]; **8**, 924[4]
Hodges, J. C., **1**, 753[100]
Hodges, M. L., **7**, 739[35]
Hodges, P. J., **1**, 419[79], 797[292], 802[292] ; **6**, 995[104], 996[104]; **7**, 400[46]
Hodges, R., **7**, 350[19]
Hodgins, T., **3**, 319[131]; **7**, 738[31]
Hodgson, D. M., **7**, 555[71], 564[71]
Hodgson, G. L., **3**, 349[32], 427[86], 712[26]
Hodgson, H. H., **4**, 426[62]; **6**, 208[53], 239[78]; **8**, 370[95], 916[98,100,106,107], 917[107], 918[100,106,107], 920[106]
Hodgson, J. C., **7**, 599[66]
Hodgson, S. M., **1**, 6[33]
Hodgson, S. T., **3**, 1036[80]
Hodjat, H., **4**, 290[207]
Hodjat-Kachani, H., **4**, 290[198], 292[221]
Hodosan, F., **3**, 416[15], 417[15]
Hodson, D., **3**, 888[18]
Hoechstetter, C., **1**, 214[19,19b]
Hoederath, W., **6**, 453[139]
Hoeger, C. A., **8**, 448[149]
Hoek, A., **8**, 150[139]
Hoekstra, A., **7**, 535[47]
Hoekstra, M. S., **2**, 334[70], 830[138]
Hoekstra, W., **1**, 571[279]; **6**, 836[58] ; **7**, 273[134], 822[32]; **8**, 850[119]
Hoelzel, C. B., **7**, 760[30]
Höenl, H., **8**, 242[45]
Hoerger, F. D., **3**, 317[120]
Hoermer, R. S., **6**, 121[129]
Hoesch, L., **7**, 483[120], 487[147], 493[147], 495[147]
Hoet, P., **6**, 578[980]
Hoeve, W. T., **8**, 349[134]
Hoey, J. G., **3**, 736[29]
Hofer, H., **3**, 636[47]
Hofer, P., **7**, 124[40]
Hofer, R. M., **3**, 846[40]
Hofer, W., **7**, 753[158,159]
Hoff, C. D., **8**, 459[236], 669[6,8], 670[6]
Hoff, D. J., **1**, 563[174]
Hoff, D. R., **6**, 219[123]
Hoff, S., **3**, 257[115]; **8**, 657[94]
Hoffee, P., **2**, 466[122], 469[122]
Hoffer, R. K., **1**, 218[52]
Hoffman, A., **2**, 140[35]
Hoffman, C., **2**, 407[49]; **4**, 374[91], 790[41], 791[41]; **8**, 64[207b], 66[207], 514[107]
Hoffman, C. H., **2**, 284[56]
Hoffman, F., **7**, 100[124]
Hoffman, H., **4**, 51[145a]
Hoffman, J. H., **8**, 617[107], 618[107], 624[107]
Hoffman, J. M., **6**, 526[395]
Hoffman, L., Jr., **1**, 822[33]
Hoffman, L. K., **5**, 857[230]
Hoffman, N., **5**, 10[77], 186[171]
Hoffman, N. E., **7**, 5[20]
Hoffman, P. G., **7**, 668[83]
Hoffman, R. A., **3**, 505[161], 507[161]
Hoffman, R. V., **6**, 116[85]; **7**, 169[115,116], 171[115,116], 229[107], 738[20]; **8**, 28[35], 66[35]
Hoffman, R. W., **2**, 571[39]; **7**, 597[52]
Hoffman, W. D., **3**, 574[76]
Hoffmann, A., **3**, 890[35]
Hoffmann, A. K., **1**, 632[67]; **4**, 999[5], 1000[5]; **7**, 854[47], 855[47]
Hoffmann, C., **5**, 847[132], 1001[13]
Hoffmann, C. E., **6**, 270[77]
Hoffmann, E. G., **8**, 100[116]
Hoffmann, F. W., **6**, 221[133]; **8**, 978[142]
Hoffmann, H., **3**, 201[77], 904[131]; **8**, 652[71], 990[40]
Hoffmann, H. M. R., **1**, 761[138], 773[204]; **2**, 527[3], 528[3], 708[4]; **4**, 34[97], 35[97], 797[104]; **5**, 1[1], 2[1], 6[1], 7[55], 9[73], 15[1], 341[58], 594[4], 595[8-11,15], 596[8b,9,15,24,25,32,34,36], 597[24,25,32], 598[34,36], 602[4], 603[9,32,49], 604[8b], 608[4,8b,24a,66], 609[49,70], 612[75]; **6**, 204[22], 215[102], 233[47], 569[935]; **7**, 262[78], 362[25], 429[157a]
Hoffmann, J. M., **5**, 789[18], 790[34]
Hoffmann, K., **4**, 41[119,119a], 980[106], 981[106] ; **6**, 502[217]
Hoffmann, K.-L., **6**, 779[66]
Hoffmann, P., **2**, 1094[89], 1095[89], 1104[134] ; **6**, 242[87,88], 243[87,88]; **8**, 830[84]
Hoffmann, P. T., **6**, 294[231]
Hoffmann, R., **3**, 914[9]; **4**, 510[166], 538[101], 1016[204], 1070[10], 1075[10], 1093[10] ; **5**, 64[25], 65[63], 66[25], 72[167,173], 75[216,217], 78[217], 318[4], 336[50], 379[112], 380[113d], 383[112], 384[112], 451[50,51], 618[3,4], 619[3], 635[3], 678[13], 699[1], 714[67], 743[1], 754[61], 760[61], 830[27], 838[73], 857[223], 1002[19], 1009[19], 1131[11], 1186[6] ; **6**, 436[20,21], 450[20,21,124], 454[21], 455[20,21,124], 501[192,194], 509[284],

1011[1], 1044[18]; **7**, 422[140], 438[10], 441[10]; **8**, 80[3], 669[5], 670[5], 671[5,19]
Hoffmann, R. W., **1**, 78[8], 79[8], 141[17], 158[17], 159[17], 180[33,37], 181[37,49], 192[81], 440[190], 445[190], 661[167], 832[108]; **2**, 2[3,6], 3[6,10], 5[6b], 6[10b], 8[38], 12[3,6b,10], 13[10,53,55,60], 14[10,53,55], 15[38,61,62a], 25[97], 26[97,97a,100,100a,101], 27[97,100a,101], 31[97], 32[62a,118], 33[3,97b,121], 36[61], 39[137,137a,b,139], 40[137c,141], 41[97b,101,121a], 42[101,137a], 43[97b], 44[137c,151,151a], 45[151], 68[39], 114[113], 240[6], 266[60], 267[64], 819[98], 977[9], 981[26], 982[26], 994[9,26,41], 995[26,42], 996[9], 997[9]; **3**, 438[32], 797[90], 953[100], 976[7]; **4**, 483[3], 484[3], 488[55], 495[3], 878[76], 1089[134]; **6**, 153[143,144], 540[591], 834[28], 839[70], 864[194,195], 873[3], 899[3], 902[124]; **8**, 196[121]
Hoffmann, S. R., **4**, 1033[20]
Hoffmann, W., **5**, 15[106], 768[133], 779[133], 835[59], 911[96], 912[96]; **6**, 37[33]
Hoffmann, W. R., **1**, 661[167,167d], 662[167d], 677[167d]
Hofheinz, W., **2**, 1101[119]
Höfle, G., **2**, 596[1]; **3**, 252[81], 257[81]; **4**, 42[120]; **6**, 242[95-97], 657[176]
Hofman, H., **7**, 232[158]
Hofmann, A. W., **2**, 1083[4], 1084[4]; **3**, 660[5]; **6**, 294[232], 955[21]
Hofmann, C. M., **2**, 366[220]
Hofmann, D., **6**, 238[73]
Hofmann, H., **6**, 531[446]
Hofmann, J., **5**, 442[185]
Hofmann, K., **2**, 605[63], 681[58], 683[58]; **6**, 560[870]; **7**, 650[51]
Hofmann, P., **4**, 538[101], 792[63], 976[100]; **5**, 1065[1], 1066[1,1a], 1074[1], 1083[1], 1084[1], 1093[1]
Hofmeister, H., **7**, 74[111], 75[111], 773[305]; **8**, 881[72], 882[72]
Hofstraat, R. G., **6**, 558[846]
Hogan, J. C., **3**, 244[27]; **8**, 771[50]
Hogan, K. T., **2**, 657[169]
Hoganson, E. D., **5**, 687[65]
Högberg, H.-E., **8**, 163[36]
Hogen-Esch, T. E., **6**, 723[149]
Hoger, E., **7**, 709[45]
Hogeveen, H., **1**, 72[73]; **4**, 1036[54]; **5**, 78[275], 79[284,285], 1187[10]; **6**, 294[245]
Hogg, D. R., **3**, 86[6,10,11]; **6**, 437[32], 453[140]
Hogsed, M. G., **5**, 3[23]
Hogsed, M. J., **5**, 64[44]
Hohenlohe-Oehringen, K., **2**, 740[62]; **6**, 276[116]; **7**, 678[69]
Hohmann, S., **8**, 446[83]
Höhn, F., **6**, 435[5a]
Hohne, G., **5**, 436[157]
Ho Hyon, H., **2**, 648[94], 649[94]
Hoiness, C. M., **1**, 880[112]; **4**, 968[57]; **5**, 905[57]
Hojatti, M., **4**, 300[309]
Hojjat, M., **3**, 943[83], 953[83]
Hojo, K., **5**, 716[86,89], 804[93]; **6**, 214[95]; **7**, 299[46]
Hojo, M., **1**, 543[29], 563[178]; **2**, 199[87]; **3**, 1032[70], 1033[71], 1040[103]; **4**, 379[115,115b], 380[115b], 402[236], 403[237], 404[237], 408[259d], 435[135], 557[14], 558[19], 562[33], 841[45], 948[95]; **6**, 510[292]; **7**, 503[273], 764[124], 843[50], 844[51,52]; **8**, 18[128], 245[75], 856[185]
Hokari, H., **2**, 901[36,37], 908[36,37], 909[36,37]
Hoke, D., **8**, 812[4], 967[84]
Hol, C. M., **7**, 12[101]
Holah, D. G., **3**, 499[118]; **8**, 445[30,31,37]
Holan, G., **2**, 741[66]; **8**, 316[58]
Holbert, G. W., **4**, 369[19,23], 374[19]; **8**, 499[44]
Holbrook, J. J., **8**, 206[172]
Holcomb, W. F., **2**, 958[23]
Holden, J. R., **1**, 34[231]
Holden, K. G., **5**, 94[82], 95[91]; **6**, 759[136]; **7**, 219[11]
Holden, R. W., **1**, 3[23]
Holder, D. A., **5**, 161[60]
Holder, N. L., **6**, 987[69]; **7**, 318[53], 319[53]
Holder, R. W., **3**, 905[141]; **8**, 813[8]
Holdgrün, X., **2**, 372[266]; **5**, 468[128]
Holdren, R. F., **2**, 964[56]; **8**, 140[27]
Holík, M., **8**, 583[14]
Holker, J. S. E., **2**, 541[79]
Holl, P., **6**, 180[129]
Holl, R., **3**, 581[109]
Holla, W., **2**, 358[154], 371[154]; **5**, 461[98], 462[98]
Holland, G. W., **4**, 164[99,99a]; **7**, 728[42]
Holland, H. J., **5**, 605[62]
Holland, H. L., **1**, 366[49], 391[49], 447[200]; **2**, 913[74,75]; **7**, 65[67], 68[83a], 69[89], 72[83a], 76[117], 779[429]; **8**, 254[127]
Holland, M. J., **8**, 344[123]
Hollander, J., **8**, 364[25]
Holle, S., **4**, 596[161]
Hollenberg, D. H., **7**, 265[96]
Holler, H. V., **3**, 898[75]
Holley, A. D., **6**, 642[70]
Holley, R. W., **6**, 642[70]
Holliman, F. G., **3**, 699[162]; **4**, 435[136]
Hollinger, W. M., **2**, 523[74]
Hollingshead, D. M., **3**, 1036[80]
Hollingsworth, D. R., **4**, 407[256a,b], 408[257b]
Hollins, R. A., **4**, 347[89]
Hollinshead, D. M., **3**, 613[2], 615[2]; **7**, 53[1], 63[1]; **8**, 117[74], 243[47]
Hollinshead, J. H., **5**, 383[123]
Hollinshead, S. P., **5**, 864[262]; **6**, 737[40], 739[60,61]
Hollis, W. G., Jr., **1**, 419[81]; **4**, 372[63]; **6**, 905[143]
Hollister, K. R., **3**, 88[137], 95[137], 165[137], 167[137]
Hollowood, F. S., **3**, 334[220]
Hollstein, W., **1**, 37[245,246]
Holly, F. W., **2**, 284[56]; **3**, 644[167]
Holm, A., **6**, 243[102], 244[102,114], 294[236], 480[110]
Holm, K. H., **4**, 382[134,134a], 383[134a], 1012[170]; **5**, 949[275]
Holm, R. H., **8**, 366[51]
Holm, R. R., **1**, 292[23]
Holman, R. J., **3**, 647[195]; **8**, 133[23,24]
Holmberg, B., **6**, 424[50]
Holmberg, G. A., **4**, 89[84c]
Holmberg, K., **5**, 2[18]
Holme, G., **1**, 821[28]; **4**, 1059[155]
Holme, K. B., **7**, 550[50]
Holmes, A., **6**, 541[595]
Holmes, A. B., **1**, 744[55]; **2**, 355[124], 370[258], 725[114]; **3**, 39[214], 281[46], 284[52], 555[30]; **5**, 403[9], 1123[36]; **7**, 679[75], 683[90]
Holmes, C. P., **2**, 202[96], 657[163]; **4**, 380[125], 381[125a], 386[125a], 387[125a]
Holmes, H. L., **5**, 513[2], 518[2]
Holmes, R. H., **8**, 861[224]
Holmes, R. R., **8**, 364[13]
Holmes, S. J., **5**, 1037[5], 1165[11,15], 1166[11,15], 1167[11,15], 1170[15], 1171[15], 1175[15], 1178[11,15], 1179[15]
Holmes, S. W., **3**, 297[17], 306[86]
Holmes-Smith, R., **7**, 6[29]
Holmlund, C. E., **7**, 157[33], 158[33b,43]
Holmquist, C. R., **6**, 129[168]
Holmquist, H. E., **5**, 1138[65]
Holst, A., **6**, 644[86]
Holstein, W., **5**, 198[23]

Holsten, J. R., **3**, 158[441], 159[441]
Holt, D. A., **1**, 885[133b]; **3**, 242[7], 257[7]; **8**, 566[454]
Holt, E. M., **6**, 1066[91]; **7**, 3[10]
Holt, G., **3**, 888[15,18], 893[55]; **6**, 104[1,9], 129[165]
Holtan, R. C., **5**, 1014[55]
Holtkamp, H. C., **1**, 16[86]
Holton, R. A., **1**, 447[201]; **2**, 749[131] ; **3**, 8[39,40], 232[256], 365[95], 600[216], 679[88,90], 683[90,104], 744[64]; **4**, 17[54], 63[54], 215[121], 573[10,11], 837[20]; **6**, 144[76], 152[139], 153[139], 1044[17], 1052[44], 1053[17,46]; **8**, 843[59b,c]
Holtz, H. D., **7**, 724[28]
Holtzkamp, E., **1**, 139[6]
Holum, J. R., **7**, 256[41]
Holweger, W., **7**, 825[45]
Holy, A., **6**, 507[238], 515[238], 615[103]
Holy, N., **3**, 619[24]; **4**, 987[147]
Holy, N. L., **1**, 366[44]; **2**, 901[35], 908[35], 909[35,63], 910[35]; **4**, 262[310]; **7**, 453[79], 861[76], 882[171]; **8**, 454[199]
Holzapfel, C. W., **2**, 866[4]; **5**, 501[270] ; **6**, 108[36]
Holzer, L., **3**, 818[93]
Holzinger, H., **4**, 1035[39], 1046[39]
Holzkamp, E., **8**, 744[50], 756[50]
Homann, W. K., **2**, 345[25]
Hömberger, G., **4**, 954[29]
Homer, J., **4**, 274[55]
Homeyer, A. H., **2**, 800[15]
Hommes, H., **2**, 587[145]
Homnick, C. F., **1**, 823[44b]
Homoto, Y., **4**, 1014[185]
Hon, J. F., **4**, 347[100]
Hon, M.-Y., **5**, 221[58], 226[58,112], 900[12], 901[12], 903[12], 905[12], 907[12], 913[12], 921[12], 926[12], 943[12], 1006[33]; **7**, 815[3], 824[3], 833[3]
Hon, Y.-S., **4**, 31[94,94e]
Honan, M. C., **2**, 1072[140]; **6**, 1056[54] ; **7**, 318[60]
Honda, H., **5**, 196[15]
Honda, K., **3**, 903[125]
Honda, M., **5**, 308[96]; **6**, 527, 217[117], 221[117]; **7**, 400[44], 408[44], 415[115a]
Honda, S., **4**, 309[418], 314[494]; **6**, 604[33]
Honda, T., **2**, 547[121], 551[121], 552[121], 1049[13]; **4**, 500[110], 510[176], 795[84] ; **5**, 92[74], 410[38], 524[54], 534[54], 536[97], 541[110], 693[109,110], 694[110]; **7**, 243[68], 423[142], 476[59]; **8**, 340[100], 534[158], 537[158], 821[48]
Honda, Y., **1**, 89[58], 90[57,58], 151[53,53b], 152[53], 158[53], 168[53b], 566[208]
Hondo, M., **7**, 297[30]
Hondrogiannis, G., **4**, 347[86b]
Hondu, T., **4**, 1076[38]
Honeychuck, R. V., **1**, 309[98]
Honeyman, J., **6**, 36[25]
Honeywood, R. I. W., **4**, 27[78]
Hong, B.-C., **1**, 512[35], 565[207]; **2**, 6[34], 21[34a,b], 23[34a], 29[34a]
Hong, B. C., **1**, 565[207]
Hong, P., **5**, 1135[52], 1137[57,58], 1155[160,161], 1156[160]
Hong, P.-K., **3**, 392[92]
Hong, R., **4**, 537[95]
Hong, R.-Y., **5**, 75[229]
Hong, W.-P., **2**, 872[24], 882[24,44], 1063[103] ; **8**, 57[171], 66[171]
Hong, Y. H., **4**, 703[35], 712[35]
Hongo, A., **4**, 1046[111]
Honig, E. D., **4**, 518[3], 689[70], 691[74]
Honig, H., **7**, 493[191]; **8**, 383[22]
Honig, M. L., **3**, 620[29]
Honigberg, J., **2**, 81[2]
Honkanen, E., **2**, 354[117], 357[117]
Honma, A., **1**, 555[110]; **2**, 61[19]
Honma, H., **5**, 623[28]
Honma, S., **4**, 145[29a]
Honma, T., **6**, 42[45]
Honma, Y., **8**, 338[82], 339[82]
Honnick, W. D., **4**, 925[35]
Honti, K., **4**, 33[96,96b]
Honty, K., **2**, 817[90]
Honwad, V. K., **3**, 217[90], 219[90]; **8**, 445[12]
Honzl, J., **7**, 884[184]
Hoobler, M. A., **2**, 489[48], 490[48]
Hoodless, I. M., **8**, 445[31]
Hoogenboom, B. E., **2**, 362[185]
Hoogzand, C., **5**, 1146[110,111], 1147[111]
Hook, J. M., **8**, 212[20], 490[6], 492[6], 493[6], 502[58], 505[58,81], 513[6], 520[6]
Hook, S. C. W., **7**, 604[133], 607[133]
Hoole, R., **3**, 421[52]
Hoole, R. F. A., **1**, 188[72], 189[72]; **2**, 20[81], 51[81]
Hoong, L. K., **2**, 13[59], 35[125], 41[125], 42[149], 45[149]
Hooper, J. W., **2**, 352[86], 369[86]
Hooper, M., **5**, 408[30]
Hoornaert, C., **4**, 795[85]
Hoornaert, G., **2**, 723[100], 817[89]; **3**, 332[203]
Hootele, C., **4**, 91[89]; **6**, 914[27]
Hooton, S., **8**, 367[58]
Hoover, D. J., **1**, 98[83]; **5**, 841[95], 890[35]; **6**, 903[134], 927[72]; **7**, 197[18] ; **8**, 637[9]
Hoover, J. R. E., **3**, 855[81]
Hoover, T. E., **8**, 502[63]
Hooz, J., **2**, 111[82-84], 242[15], 830[143], 909[64], 911[64]; **3**, 54[277], 363[83], 794[80]; **4**, 141[14,14b], 145[30-34], 180[68]; **6**, 2[2], 23[2]; **8**, 14[80], 756[158], 758[158]
Hope, H., **1**, 22[113,117], 23[119], 41[265,268] ; **2**, 770[10], 771[10]; **4**, 170[13]; **5**, 517[27], 538[27], 829[20], 1039[11,17], 1050[17], 1052[17], 1133[26], 1146[26]
Hope, M. A., **6**, 129[165]
Hopf, H., **2**, 81[3]; **5**, 344[65], 734[137], 736[141,142], 948[268]; **6**, 830[4]
Hopff, H., **4**, 314[490,491]
Hopkins, M. H., **1**, 891[146,147]; **2**, 554[135] ; **3**, 792[67,68]
Hopkins, P. B., **1**, 430[132], 635[88,89], 733[9], 806[315]; **3**, 104[208,209], 117[208,209]; **5**, 849[143], 1001[16]; **6**, 662[215], 682[341], 846[103], 905[145], 983[52]; **7**, 517[11]
Hopkins, R. B., **7**, 452[40]
Hopkinson, A. C., **1**, 487[3], 488[3]; **5**, 720[98]
Hopla, R. E., **3**, 365[95]
Hoppe, D., **1**, 161[85], 162[91], 180[31], 630[28]; **2**, 10[42,43], 21[42,43b], 22[42b,43], 38[134a], 39[134b], 42[150], 55[1], 61[1e], 62[1e], 68[40,41,43], 94[51-53], 361[174], 445[26] ; **3**, 88[134], 196[29]; **6**, 863[187,188,189]
Hoppe, H., **6**, 450[122]
Hoppe, I., **1**, 55[26]; **5**, 116[267,268], 117[272], 187[174]
Hoppmann, A., **3**, 124[268,285], 125[268,285], 126[268,285], 127[268], 131[268]
Hopps, H. B., **8**, 814[15]
Hopton, J. D., **7**, 759[7,8]
Horaguchi, T., **8**, 625[166], 627[166]
Horák, M., **5**, 809[114]
Horak, V., **6**, 120[126]; **7**, 228[96]
Horcher, L. H. M., **5**, 249[32]; **6**, 1063[87]
Horder, J. R., **2**, 242[18]; **8**, 756[154]
Horeau, A., **2**, 232[173]; **5**, 186[169]; **6**, 725[168]; **8**, 161[18]
Horecker, B. L., **2**, 456[43,44,65,71], 458[65,71], 466[43,122,126], 467[43], 469[122]

Horgan, A. G., **5**, 829[22]
Horgan, S. W., **5**, 727[119]; **7**, 143[145], 346[6]
Hörhold, H.-H., **6**, 564[907]
Hori, F., **4**, 487[45]; **8**, 131[6], 132[7]
Hori, H., **6**, 647[110]
Hori, I., **1**, 834[121a,122]; **3**, 97[173], 116[173], 136[369], 138[369]; **4**, 127[220b]; **5**, 890[33]; **8**, 12[64,65]
Hori, K., **1**, 188[70], 553[88], 554[104]; **3**, 1041[113]; **4**, 382[132,132b], 553[7,9], 857[104]; **7**, 452[43], 462[43], 465[130]
Hori, M., **3**, 969[135]; **5**, 504[276]; **6**, 510[293], 893[87], 927[80], 936[106]; **8**, 996[71]
Hori, T., **1**, 359[14], 363[14], 384[14]; **2**, 370[257]; **4**, 340[52]; **6**, 1026[87], 1027[87], 1031[87]; **7**, 91[37], 110[37]
Hori, Y., **4**, 313[470]
Horibe, I., **3**, 386[57]; **5**, 809[115]
Horie, K., **4**, 430[96]
Horie, S., **3**, 638[94]
Horie, T., **4**, 435[134]
Horiguchi, Y., **1**, 112[27]; **2**, 90[39], 117[149], 310[27], 448[44,45,47], 452[47], 651[112]; **3**, 257[118], 464[175]; **4**, 152[55,57]; **5**, 1022[74]
Horihata, M., **8**, 174[126], 178[126], 179[126]
Horii, S., **6**, 74[37]
Horii, Y., **6**, 88[103]
Horii, Z., **3**, 677[81], 686[81]; **4**, 91[89]; **8**, 568[466]
Horiie, T., **1**, 347[133,134]; **4**, 23[71], 162[92]; **6**, 237[67], 564[914]
Horiike, T., **6**, 684[344]
Horikawa, H., **2**, 1051[41]; **3**, 650[210c,212], 651[210c,216]; **7**, 806[74]
Horikawa, M., **7**, 761[56]
Horike, H., **2**, 655[135]
Horiki, K., **6**, 437[38], 438[38]
Horikoshi, K., **8**, 195[106], 197[106]
Horino, H., **1**, 176[18]; **4**, 852[89], 903[202], 904[202]
Horita, K., **6**, 23[93], 652[140]; **7**, 245[73,80], 246[81]; **8**, 963[49]
Horito, S., **6**, 560[869]
Horiuchi, C. A., **4**, 603[270]; **7**, 95[65], 107[165], 530[22], 531[22]
Horiuchi, S., **2**, 603[47]; **8**, 173[117], 555[370]
Horiuti, I., **8**, 420[24], 422[24]
Horler, H., **2**, 448[37,38]; **4**, 729[60], 730[60], 740[60], 741[126]
Horn, D. H. S., **1**, 337[80], 828[69]; **3**, 553[12]
Horn, F., **6**, 432[122]
Horn, H., **6**, 538[561]
Horn, K. A., **5**, 856[210], 910[81], 912[81]; **6**, 146[88]
Horn, R. K., **3**, 634[28]
Horn, U., **6**, 1059[66]
Hornback, J. M., **3**, 735[16]; **8**, 356[186]
Hornberger, P., **8**, 273[123]
Horne, D. A., **7**, 503[279]; **8**, 216[54], 224[54]
Horne, K., **3**, 196[27]; **4**, 153[61a]
Horne, S., **1**, 373[90], 375[90], 376[90]; **2**, 1075[150]; **4**, 73[35]
Horner, J. H., **1**, 42[204]; **2**, 524[77]; **5**, 552[33]
Horner, L., **1**, 761[138], 773[204]; **3**, 201[77], 580[105,106], 596[195], 890[31,34], 891[40], 900[40,95], 901[111], 904[135], 909[155]; **4**, 317[555]; **6**, 644[81], 840[71]; **7**, 763[86], 765[142]; **8**, 113[40], 115[40], 135[46], 242[45], 249[90], 278[158], 388[65], 397[142], 445[22,24], 459[229], 532[131b], 863[234,237], 898[22], 904[59], 907[59], 909[59]
Hörnfeldt, A.-B., **3**, 232[259]; **7**, 596[33a]; **8**, 384[37]
Horng, A., **3**, 199[58]; **8**, 214[46], 717[96,97]
Horng, J. S., **8**, 623[151]
Horning, D. E., **6**, 249[143]
Horning, E. C., **2**, 354[105]; **7**, 166[92]
Horning, M. G., **2**, 354[105]
Hornischer, B., **2**, 367[224]; **3**, 826[41]
Hornish, R. E., **3**, 99[192], 103[192], 107[192]; **8**, 837[12], 842[12]
Hornke, G., **7**, 689[10]
Hornung, N. L., **4**, 982[113]; **5**, 1086[69]
Horowitz, A., **7**, 856[66]
Horowitz, H. H., **7**, 451[37]
Horri, Z., **8**, 274[130]
Horsewood, P., **5**, 421[79]
Horsham, M. A., **3**, 545[122], 558[53]; **6**, 911[14]; **8**, 676[69], 679[69], 681[77], 684[77], 694[77]
Horspool, M., **5**, 645[1], 647[1b], 651[1]
Horspool, W. M., **2**, 1037[96]; **5**, 123[1], 126[1], 201[32], 202[33,34,36], 220[50,51], 221[52,53,56], 819[152]
Horstmann, H., **6**, 424[61]
Horstschäfer, H. J., **7**, 598[57]
Hortmann, A. G., **1**, 555[120]; **5**, 738[147]; **7**, 749[117]; **8**, 390[80]
Horton, D., **1**, 55[25], 564[193]; **6**, 48[83], 660[202], 789[108], 977[19]; **7**, 703[1], 709[1], 710[1]; **8**, 568[469]
Horton, I. B., III, **5**, 534[90], 535[90]
Horton, M., **4**, 27[84], 29[84d], 102[127,127b]; **5**, 779[199]
Horton, W. J., **2**, 764[62,63]; **8**, 568[467]
Horvat, J., **6**, 54[132]
Horvat, Š., **6**, 54[132]
Horváth, A., **2**, 789[56]
Horvath, B., **7**, 85[7]
Horváth, I. T., **5**, 1138[65,69]
Horvath, K., **8**, 827[73]
Horváth, K., **2**, 381[305]; **6**, 543[610]
Horvath, M., **1**, 379[107], 385[107]
Horvath, R. F., **3**, 200[72]
Horvath, R. J., **1**, 360[28], 361[28]
Horwell, D. C., **5**, 583[184]; **8**, 394[114]
Horwitz, J. P., **3**, 757[121]
Hosaka, H., **4**, 1017[211], 1021[211]
Hosaka, K., **2**, 370[256]
Hosaka, M., **7**, 339[43]
Hosaka, S., **4**, 600[232,233,238], 643[238], 945[88]
Hosakawa, T., **7**, 178[150]
Hosaki, T., **6**, 976[9]
Hosangadi, B. D., **6**, 825[129]
Hoshi, K., **1**, 561[160]
Hoshi, M., **3**, 470[196,197], 473[196,197], 522[21]; **7**, 16[163], 604[130]; **8**, 720[130]
Hoshi, N., **2**, 363[191]; **3**, 946[87]; **4**, 106[140b]
Hoshino, K., **8**, 410[88]
Hoshino, M., **2**, 368[236], 386[331]; **3**, 571[74,75], 574[74,75], 586[140], 594[186], 883[110]; **4**, 507[153], 509[161]; **5**, 442[181]; **8**, 806[122], 836[3]
Hoshino, O., **3**, 672[65]; **7**, 339[43]; **8**, 29[38], 49[115], 50[119], 66[38,115,119]
Hoshino, S., **1**, 658[158], 659[158], 664[158], 665[158], 672[158]
Hoshino, T., **4**, 1078[57], 1080[57]
Hoshino, Y., **2**, 556[155]; **8**, 770[39]
Hoshito, T., **4**, 218[136]
Hoskin, D. H., **6**, 501[186], 502[186]
Hosking, J. W., **7**, 107[167]
Hosmane, R. S., **6**, 533[477,478,483,487], 553[722,723], 554[721,722,723]
Hosoda, H., **6**, 655[162]
Hosoda, Y., **2**, 291[74]
Hosogai, T., **4**, 378[108]
Hosoi, A., **7**, 184[168]
Hosojima, S., **6**, 508[285]
Hosokawa, T., **4**, 310[435], 377[104], 378[104f], 383[104f], 393[188], 557[10,11], 611[353]; **7**, 94[58], 107[164], 419[134b], 451[18], 452[50], 454[18]; **8**, 856[170]

Hosomi, A., **1**, 180[43], 181[43], 327[14], 346[14], 357[7], 361[7]; **2**, 6[28], 17[28], 68[44], 476[4], 565[14,16], 566[17], 567[27,28], 572[43,45], 576[27,72,79], 578[81,87], 582[27,109], 601[34], 718[76,77], 719[82], 721[87,88], 901[38], 908[38]; **3**, 32[187], 246[44], 437[28], 485[27]; **4**, 98[113,113b], 155[67,68a,b], 589[80], 591[80], 1088[124]; **5**, 337[51], 431[120], 596[22], 597[22], 598[35], 603[22], 1166[18]; **6**, 83[82], 720[132], 832[12], 865[12]; **7**, 458[113], 641[6]; **8**, 774[75]

Hosoya, K., **3**, 153[415], 224[170]

Hossain, A. M. M., **8**, 103[130], 881[67]

Hostapon, W., **7**, 14[138]

Hosten, N., **8**, 348[132]

Hotelling, E. B., **3**, 781[12]

Hotoda, H., **6**, 606[38,40,41]

Hotta, H., **5**, 196[16], 197[16]

Hotta, Y., **1**, 749[83,86], 750[83,86], 812[83]; **5**, 1124[50]

Hou, C. T., **7**, 56[20,21], 80[137]

Hou, K. C., **6**, 254[163]; **7**, 155[30]

Hou, W., **7**, 446[64]

Hou, Z., **1**, 277[83], 279[86], 280[86]; **3**, 567[36,190], 595[190], 607[190], 610[36]; **8**, 113[36]

Houben, J., **6**, 435[5b]

Houge, C., **5**, 109[217]

Hough, L., **2**, 456[54,56,61,69,75], 457[61,69], 460[69]; **6**, 73[27], 76[51], 662[212]; **7**, 712[62]; **8**, 247[81]

Houghton, D. S., **7**, 765[162]

Houghton, L. E., **8**, 315[55], 316[55]

Houghton, P. G., **4**, 85[77c]

Houghton, R. P., **4**, 524[60]

Houk, K. N., **1**, 41[198], 49[8], 80[23], 92[64], 109[13], 110[13], 191[77], 287[20], 288[20], 289[20], 357[8], 462[17], 463[17], 476[125], 477[125], 610[45]; **2**, 24[96], 258[50], 476[4], 662[9]; **3**, 4[21], 12[21], 18[96], 31[186], 66[12], 74[12], 194[4,11], 587[142], 985[26b]; **4**, 47[134,134d], 49[134d], 202[20], 379[117], 484[21], 729[59], 781[7], 782[9], 787[7], 827[7], 872[41], 954[29], 1070[8], 1072[18], 1073[19,23], 1075[30], 1076[19,47], 1079[65], 1081[81], 1082[85], 1083[81], 1085[85], 1097[164], 1098[170]; **5**, 71[127,-128,129], 79[292], 203[39,39c], 204[39h-j], 209[39], 210[39], 215[3], 216[14], 218[33], 219[14], 224[3], 241[6], 247[26], 248[26a,29], 249[26a], 257[61,61a,c], 258[61b], 260[65,68,70], 261[65,68], 262[68], 263[70], 347[72,72c], 436[158,158b], 442[158], 451[53], 452[58], 454[58], 515[13,13a,15], 516[13a,19,22,26], 517[13a,b,22,26], 518[13a,b,15,22,26], 519[15], 620[15], 621[15,21], 622[15], 625[30,32], 626[32,34-37,40], 627[43], 628[44,45], 629[36,46-49], 630[40,50], 631[54-56], 632[62,63], 647[16], 649[16], 653[16], 678[18], 679[18], 680[18,22], 681[18], 682[34b], 683[22,34b], 685[18], 703[15], 733[15b], 819[155], 857[227], 1031[96,97]; **6**, 711[65], 724[150]; **7**, 439[36]; **8**, 5[25], 6[34], 7[35], 89[43], 171[109], 723[151], 724[151,169,169g]

Houlden, S. A., **3**, 891[41b]

Houlihan, W. J., **2**, 133[1], 134[1], 136[15], 147[77], 149[87], 150[93]; **6**, 523[353]

Houmounou, J. P., **6**, 91[121]

Hountondji, C., **8**, 36[69,70], 66[69,70]

Houpis, I. N., **1**, 264[43]; **2**, 638[61], 640[61]; **3**, 546[123]; **4**, 14[46,46a], 239[21], 256[21], 261[21], 976[98]; **6**, 674[294]; **7**, 182[161], 680[77]

House, H. E., **7**, 437[3]

House, H. O., **1**, 3[18,19,23], 116[46,47,49,50], 118[46,47,49,50,60], 123[47], 124[47], 411[45], 431[134], 433[225], 683[218], 820[6], 822[6], 846[11]; **2**, 109[62], 120[169,176], 124[201], 128[242], 183[14,16,17], 184[16,22,25b], 235[190], 268[66], 311[34], 342[4], 396[7], 410[3], 424[36], 599[23], 756[7,7a,b], 797[7], 829[7], 835[7,158], 837[7], 865[1], 897[13,15], 902[13]; **3**, 1[4], 2[4,7], 3[16], 4[18], 7[18,34], 8[18,43], 11[18,54], 13[4], 14[70,73], 15[73], 16[54,81], 17[54,83], 18[99], 19[100], 20[118], 23[4], 26[54], 31[118], 39[4], 54[4], 55[4], 248[55], 249[63], 250[72], 251[55], 264[72], 265[72], 269[55], 419[47], 494[87], 502[87], 563[1], 564[7], 577[87], 606[1b], 746[69], 748[69], 754[108], 755[108,111], 851[66]; **4**, 3[7,7c], 4[7], 5[19], 6[20,20a,21], 31[94,94a], 35[99,99a], 59[7c], 70[2,5], 71[17a], 91[88b], 139[3], 148[47a,48], 164[48], 169[2], 170[9,20,23], 171[9], 178[23,63], 187[94], 259[278], 272[30], 277[30], 279[30]; **5**, 513[4], 514[4e], 527[4e]; **6**, 684[343], 786[96], 959[48], 1027[89]; **7**, 120[10], 123[33,36], 130[76], 145[159], 154[14], 168[102], 170[119], 178[151], 179[151,152], 186[180], 252[4], 671[5], 682[81]; **8**, 108[4], 109[4], 110[4], 111[4], 112[4], 113[4], 114[4], 116[4], 117[71], 120[4], 123[83], 124[83], 309[14], 478[42], 501[54], 502[54], 524[5,12], 526[31], 527[49], 530[5], 531[5,123], 532[5,12c], 533[5], 544[5], 573[5], 794[2], 795[2], 812[1], 986[8], 988[8]

Houser, F. M., **8**, 542[232]

Houser, R. W., **3**, 866[30]; **5**, 687[58]

Houston, A. H. J., **3**, 914[6], 927[6]

Houston, B., **7**, 12[100]

Houston, T. L., **1**, 520[69,70], 635[83], 678[83], 681[83], 691[83]; **4**, 12[37,37b,c]

Houwen-Claassen, A. A. M., **5**, 562[87]

Houwing, H. A., **6**, 538[549]

Houze, J. B., **6**, 123[139], 124[139]

Hovakeemian, G. H., **5**, 1151[128]

Hovey, M. M., **7**, 605[143]

Hoveyda, A. H., **5**, 168[105], 170[112], 171[116], 174[105], 176[105,112,116], 180[105], 181[105], 461[99], 462[99]; **8**, 698[143]

Hovnanian, N., **8**, 859[218]

Howard, A. S., **2**, 830[140], 881[43], 882[46], 885[48,49,51]; 6, 509[276]

Howard, C., **6**, 655[165]; **7**, 674[33]

Howard, C. C., **3**, 290[70]

Howard, E., Jr., **7**, 769[213]

Howard, J., **3**, 380[10]

Howard, J. A. K., **8**, 766[18]

Howard, P. N., **1**, 892[148]; **3**, 779[7], 792[7]

Howard, R. W., **4**, 16[51]

Howard, S. I., **8**, 159[7], 166[65], 170[85], 178[65], 179[65]

Howard, T. R., **1**, 743[53]

Howard, W. L., **8**, 212[16]

Howarth, T. T., **2**, 171[177]

Howatson, J., **1**, 21[111]

Howbert, J. J., **5**, 647[15], 653[15], 656[15], 657[15], 665[41], 666[42], 916[116,117], 956[117]; **8**, 123[82]

Howe, G. P., **1**, 339[89]

Howe, J. P., **8**, 770[40]

Howe, R., **8**, 526[17], 528[69], 530[69]

Howe, R. K., **5**, 438[162]

Howell, A. R., **4**, 384[142]

Howell, F. H., **3**, 306[85]

Howell, J. A. S., **4**, 115[180a], 665[9], 673[31], 687[64], 688[9]

Howell, J. O., **7**, 852[40], 854[45]

Howell, S. C., **7**, 53[1], 63[1], 307[14]

Howell, W. C., **6**, 228[32]

Howes, D. A., **6**, 673[290]

Howes, P. D., **4**, 54[154]

Howk, B. W., **5**, 1138[65]

Howles, F. H., **4**, 4[13]

Howsam, B. W., **4**, 587[28]

Howsam, R. W., **8**, 533[154]

Howton, D. R., **7**, 712[65]

Hoy, R. C., **8**, 956[7]

Hoyano, J. K., **7**, 3[13]

Hoye, T. R., **2**, 350[76], 388[343]; **3**, 342[9]; **4**, 187[97], 980[115], 982[115]; **5**, 166[92], 1104[160]; **6**, 1020[48]; **7**, 404[66]; **8**, 852[139]

Hoyer, D., **1**, 359[22], 383[22], 384[22]; **4**, 77[50], 206[46,49]; **6**, 501[188]

Hoyer, E., **6**, 441[86]

Hoyer, G. A., **8**, 187[42]
Hoyle, J., **1**, 528[122]; **7**, 766[169]
Hoyle, K. E., **5**, 789[28]
Hoyng, C. F., **2**, 1094[90], 1095[90]; **6**, 639[50]
Hoz, S., **4**, 452[21], 456[21], 473[21], 1013[179] ; **6**, 533[506]; **7**, 875[119]
Hoz, T., **6**, 980[37]; **8**, 889[127], 992[53]
Hozumi, T., **6**, 603[23]
Hrib, N. J., **2**, 555[145], 1011[11]; **3**, 978[11]; **5**, 8[59], 342[63]
Hrnjez, B., **4**, 846[73], 847[73], 848[73]
Hropot, M., **6**, 554[748]
Hrovat, D. A., **7**, 875[111]
Hrusovsky, M., **7**, 154[21], 451[25,27,36]
Hrutford, B. F., **4**, 490[65]
Hrytsak, M., **4**, 1057[143,144], 1058[143,144]
Hseu, T. H., **7**, 367[56]
Hsi, R. S. P., **8**, 497[40]
Hsiao, C.-N., **2**, 113[105], 586[130]; **3**, 161[470], 167[470]; **6**, 1003[135]; **8**, 844[69]
Hsiao, Y., **8**, 460[255], 461[257]
Hsieh, D.-Y., **4**, 1101[193]
Hsieh, H. H., **3**, 390[83], 392[83]; **4**, 297[277]
Hso, E. T., **7**, 798[23]
Hsu, C. K., **7**, 180[157], 182[157]
Hsu, C.-L. W., **8**, 487[66]
Hsu, C. T., **5**, 1096[110], 1098[110]; **8**, 188[52]
Hsu, C.-Y., **4**, 915[6]
Hsu, E. T., **8**, 991[49]
Hsu, F., **8**, 53[130], 66[130]
Hsu, G. J.-H., **1**, 438[160]; **3**, 252[83]
Hsu, H., **1**, 2[9], 43[9]
Hsu, H. B., **3**, 918[25]
Hsu, H. C., **7**, 605[139]
Hsu, J. N. C., **8**, 475[18]
Hsu, K. C., **4**, 492[74]
Hsu, L.-Y., **2**, 882[44]; **8**, 57[171], 66[171]
Hsu, M. H., **4**, 597[175], 621[175], 623[175]
Hsu, S. Y., **4**, 698[22]; **7**, 107[153,155], 377[91]
Hsu, Y. F. L., **8**, 898[25], 899[25]
Hsu, Y. L., **5**, 1070[21], 1072[21]
Hsu Lee, L. F., **8**, 194[103], 544[264,265], 546[304], 561[304]
Hu, C. K., **8**, 986[12]
Hu, C.-Y., **8**, 769[26]
Hu, H., **7**, 155[26], 179[153,154]
Hu, J., **7**, 446[64]
Hu, K.-C., **7**, 155[27]
Hu, L., **6**, 879[43]; **7**, 633[66]
Hu, L.-Y., **3**, 219[106], 512[201]; **6**, 743[71]
Hu, N. X., **4**, 343[74], 372[58], 397[58] ; **7**, 492[183], 752[151], 761[61]
Hu, Y., **7**, 451[32]
Hua, D. A., **4**, 226[201], 227[201]
Hua, D. H., **1**, 123[76], 425[107], 515[56], 520[76], 521[76], 522[76-79], 523[83]; **2**, 66[34], 67[34], 75[81], 228[163]; **3**, 866[31], 868[39]; **4**, 12[37,37e,f], 119[194], 226[196,197,198,199,200] ; **5**, 1053[41], 1061[41]; **6**, 154[146], 656[170], 864[193], 900[119]; **7**, 358[11], 552[58], 554[58]; **8**, 813[9], 844[72,72c]
Huang, C.-W., **5**, 650[25]
Huang, E. C. Y., **3**, 790[59]
Huang, F., **3**, 1040[108]
Huang, F.-C., **3**, 693[148], 694[148]; **8**, 237[10], 243[10]
Huang, H., **2**, 116[133]
Huang, H.-C., **4**, 374[91], 739[110]; **5**, 365[96b]; **7**, 350[26]
Huang, H.-N., **7**, 747[95]
Huang, J.-T., **5**, 225[89]
Huang, N. Z., **6**, 114[72]
Huang, P. C., **2**, 456[68], 460[68]
Huang, P. Q., **1**, 558[138,139,140,141]
Huang, R. L., **3**, 242[6], 257[6], 259[6]
Huang, S., **4**, 869[27], 870[27], 871[27]
Huang, S.-B., **1**, 743[65]; **5**, 1124[43,49]
Huang, S. D., **4**, 1076[38]
Huang, S. J., **3**, 681[95]; **7**, 798[23], 801[41]; **8**, 995[70]
Huang, S.-L., **7**, 297[28], 396[24]
Huang, S.-P., **8**, 314[37], 647[56]
Huang, T.-N., **8**, 453[194,195]
Huang, T. T.-S., **6**, 205[30]
Huang, W., **3**, 596[194]
Huang, W. H., **7**, 873[99]
Huang, X., **2**, 353[97]; **4**, 126[218c]; **7**, 283[183], 284[183], 760[22]; **8**, 300[86]
Huang, Y., **4**, 991[152]; **6**, 185[160]; **8**, 678[63], 685[63], 686[63]
Huang, Y. C. J., **5**, 515[17], 518[17], 547[17,17c]
Huang, Y.-H., **1**, 307[92]
Huang, Y.-Z., **2**, 24[92]; **8**, 18[129]
Huang, Z., **2**, 772[16]
Huang-Minlon, **8**, 328[5,16,17], 330[16,17], 338[17], 339[17], 340[98], 568[486]
Huba, F., **7**, 800[34]
Hubbard, J. L., **8**, 14[78]
Hubbard, J. S., **4**, 426[59], 462[104], 465[104,116,120], 466[104], 468[104], 469[104,116]
Hubbard, W. N., **5**, 857[229]
Hubbell, J. P., **3**, 407[150]
Hubbs, J. C., **2**, 553[125]; **3**, 380[10]
Hubbuch, A., **6**, 668[253], 669[253]
Hübel, E., **5**, 1139[73]
Hübel, W., **5**, 1133[27], 1139[73], 1146[110,111], 1147[111]
Hubele, A., **8**, 141[40]
Hübener, G., **2**, 1090[72], 1098[104]; **7**, 778[397]
Hübenett, F., **7**, 765[142]
Huber, B., **5**, 1070[18]; **6**, 480[116]
Huber, C. P., **6**, 436[12]
Huber, D., **8**, 370[96], 639[19]
Huber, F. X., **5**, 492[238], 498[238]
Huber, G., **3**, 818[99]; **6**, 48[86]; **7**, 709[45]
Huber, H., **4**, 48[138,138a,140], 66[138a,b], 1085[108,109] ; **5**, 743[163]
Huber, I., **5**, 108[210], 109[210], 110[210], 111[210]; **6**, 430[96]
Huber, I. M. P., **1**, 482[162]; **3**, 79[63]
Huber, L. E., **1**, 683[218]
Huber, U., **3**, 99[191], 103[191a], 107[191]
Huber, W., **2**, 410[5]; **5**, 339[56], 347[56] ; **8**, 163[40], 269[80]
Huber, W. J., **4**, 370[26]
Huber-Patz, U., **3**, 872[61,63]
Hubert, A. J., **3**, 290[71], 1047[7,8], 1051[7,8] ; **4**, 886[116], 1031[4], 1033[16,16d,e], 1035[16e], 1051[125], 1052[16d]; **5**, 1148[116,122]; **6**, 25[101]; **7**, 8[61]; **8**, 727[199]
Hubert, E., **8**, 52[144], 66[144]
Hubert, I., **5**, 110[223], 112[223]
Hubert, J. C., **2**, 1049[17], 1050[17]; **6**, 745[80]; **8**, 273[128]
Hubert, M., **3**, 290[71]
Hubert, T. D., **8**, 238[23], 261[6]
Hubert-Pfalzgraf, L. G., **8**, 859[218]
Hubertus, G., **3**, 872[61]
Hubino, J., **1**, 205[105]
Hubner, F., **3**, 25[158]
Hübner, H. H., **3**, 693[147]
Hübner, J., **6**, 441[84]
Hübsch, T., **5**, 458[73]
Hubschwerlen, C., **5**, 96[104,120], 97[104], 98[104]
Huche, M., **3**, 991[35]; **4**, 210[70,72], 229[223]
Huchting, R., **5**, 64[48]

Huckaby, J. L., **4**, 937[72]
Hückel, W., **2**, 169[164]; **8**, 141[40,41,43], 374[146], 496[32], 596[83], 629[185]
Huckin, S. N., **2**, 189[49], 832[151]; **3**, 58[290]
Huckstep, M. R., **4**, 371[52], 378[52a]
Hudac, L. D., **1**, 360[28], 361[28]
Huddleston, P. R., **2**, 963[54]
Hudec, J., **3**, 744[62]; **4**, 44[125]; **5**, 123[4], 618[1], 791[27], 799[27], 802[86]
Hudec, T. M., **7**, 336[33]
Hudlicky, M., **6**, 204[2,3]; **8**, 26[9], 27[9], 36[9], 37[9], 64[9], 70[9], 139[8], 278, 307[1], 367[56], 595[79], 597[79], 896[8,10,16], 904[10]
Hudlicky, T., **2**, 282[34], 286[34,64], 287[34]; **4**, 18[59], 467[130,131], 952[6], 970[71], 1040[88,99], 1043[107], 1048[88,88a-c,g,107]; **5**, 11[81], 211[65], 221[58], 226[58,112], 239[1], 900[9,12], 901[9,12], 903[12,41], 904[42], 905[9,12,41], 906[9,66], 907[9,12,41,66,76], 908[66], 909[9,41,97-99], 913[9,12,103], 916[41,122], 918[41,128,129], 921[12], 925[122,128,152], 926[12], 930[176,177], 931[177], 933[176], 937[41,128,201], 938[177,212], 939[41,212,221,222,223], 940[41,222,224,225], 942[232], 943[12,251], 947[9,42,177], 951[41,177,223], 954[298], 955[302,303], 957[309,310,311], 958[99,103,128], 962[212,221,222,223], 963[222,225,251], 964[176,223,324], 993[52], 994[52], 1006[33,34]; **6**, 204[2]; **7**, 324[72], 557[74,75], 815[3], 824[3], 833[3]
Hudnall, P. M., **1**, 212[6], 213[6]; **8**, 907[73]
Hudrlik, A. M., **1**, 731[3], 785[246], 828[81]; **2**, 524[78]; **3**, 34[196], 349[40], 351[40], 369[40], 759[132], 760[135]; **4**, 292[226], 299[303], 303[350], 313[472], 315[521]; **6**, 1066[97]; **7**, 701[64]
Hudrlik, P. F., **1**, 436[150], 580[1], 620[66], 731[3], 783[238], 784[242,243,244], 785[246], 828[81]; **2**, 109[63], 110[63], 184[24], 524[78], 599[22], 602[39]; **3**, 8[35], 34[196], 224[182], 244[27], 759[131,132], 760[135]; **4**, 83[65a], 120[197], 258[235], 292[226], 299[303], 303[350], 313[472], 315[521]; **5**, 942[235]; **6**, 655[160], 1066[97]; **7**, 111[190], 701[64]; **8**, 771[50], 937[87]
Hudson, B. E., Jr., **2**, 797[4]
Hudson, C. B., **2**, 756[7,7a]; **8**, 605[15], 624[152]
Hudson, C. S., **6**, 36[22,29]
Hudson, H. R., **6**, 204[19]
Hudson, R. F., **4**, 35[98b]; **5**, 417[63], 453[67]
Hudson, R. L., **8**, 472[8]
Hudspath, J. P., **7**, 57[25], 407[81]
Hudspeth, J. P., **1**, 783[235]; **5**, 791[41]
Huebner, C. F., **4**, 45[126,126c]; **5**, 686[51], 687[51], 688[51]
Hueck, K., **4**, 311[445]
Huehnermann, W., **5**, 634[79]
Huel, C., **7**, 350[21]
Huesmann, P. L., **3**, 369[123]; **5**, 829[21]
Hueso-Rodriguez, J., **5**, 51[45,45a], 53[45a]
Huestis, L., **7**, 738[30]
Huestis, L. D., **8**, 269[75]
Huet, F., **1**, 226[90-92], 326[6,7], 825[46]; **2**, 616[137], 785[44]; **3**, 765[154]; **4**, 98[108d,109d,e]; **6**, 677[314]; **7**, 168[103,103b]
Huet, J., **1**, 361[31]; **2**, 980[20], 981[20], 992[20], 993[20]; **5**, 938[216], 948[216]; **8**, 7[37], 16[104], 541[214], 542[214,225], 543[214]
Huff, B., **1**, 3[20], 122[70], 376[93]; **7**, 330[13]
Huff, B. J. L., **3**, 14[71], 15[71]; **8**, 526[29], 527[29], 528[54]
Huff, J. R., **5**, 410[41]
Hufferd, R. W., **2**, 141[42]
Huffman, C. W., **8**, 754[91]
Huffman, J. C., **1**, 305[89], 310[105], 311[89], 514[53,54]; **3**, 217[95,95a], 583[119]; **4**, 398[218,218c], 399[218c]; **5**, 523[48], 605[59], 736[145], 737[145], 778[192], 1076[45], 1101[143], 1175[39,42], 1178[39,42]; **6**, 930[85]; **7**, 199[37], 865[87]; **8**, 550[329], 679[67]
Huffman, J. W., **1**, 272[65]; **2**, 818[96]; **3**, 564[5,7], 572[64], 607[5]; **4**, 159[81]; **6**, 21[79], 714[85]; **7**, 111[195], 112[195], 177[144]; **8**, 108[5], 109[5,7,13], 110[5,7], 111[5], 112[5,7,25-27], 113[5,27,35], 114[5], 116[5,70], 118[26], 119[5,25], 120[5,70], 121[70], 949[154]
Huffman, K. R., **5**, 712[60]
Huffman, W. F., **3**, 369[119], 372[119], 394[96]; **4**, 31[93]; **5**, 94[82]; **6**, 1055[51]; **7**, 543[12], 551[12]
Hufford, C. D., **3**, 390[79], 396[109], 397[109]
Hufnal, J. M., **3**, 136[373], 137[373]
Hug, D. H., **4**, 176[45]
Hug, K. T., **2**, 630[10], 631[10], 632[10], 640[10], 641[10], 642[10], 646[10], 931[134]
Hug, R., **5**, 681[27]
Hug, R. P., **7**, 154[20]
Hugel, G., **6**, 755[120]
Hugel, H., **1**, 149[50]
Huggins, R. A., **7**, 282[179]
Hughes, A. N., **5**, 444[187]; **8**, 445[30,31]
Hughes, D., **8**, 640[25]
Hughes, D. A., **4**, 425[32]
Hughes, D. L., **4**, 425[34], 429[34], 434[34]; **6**, 22[83]; **7**, 752[154]
Hughes, E. D., **4**, 287[180]; **6**, 3[13], 4[13], 951[3]
Hughes, G., **5**, 90[57], 95[57]
Hughes, G. B., **5**, 552[36], 568[107], 847[136]
Hughes, G. J., **3**, 380[4]
Hughes, G. K., **3**, 499[115]
Hughes, J. M., **3**, 131[330]
Hughes, J. W., **5**, 534[95]
Hughes, L., **4**, 148[45a]; **7**, 595[23], 600[23]
Hughes, L. R., **3**, 369[120,122,124], 372[122,124]; **4**, 45[128]
Hughes, M. T., **5**, 829[25]
Hughes, N., **4**, 426[53]; **6**, 557[834]
Hughes, N. W., **5**, 883[17]
Hughes, P., **5**, 179[141]; **7**, 401[61c]
Hughes, R., **3**, 554[23]
Hughes, R. E., **4**, 1024[265]
Hughes, R. J., **8**, 944[123]
Hughes, R. P., **3**, 380[10]; **4**, 587[44], 601[249,250], 602[255,256,259], 643[249]; **5**, 35[12,12b,13], 46[13], 56[13]; **7**, 36[107]
Hughman, J. A., **2**, 907[60]
Hugl, H., **7**, 508[311]
Hugo, V. I., **2**, 746[109]
Huguenin, R., **3**, 131[327]
Huguerre, E., **6**, 474[82]
Huh, T.-S., **7**, 574[140], 581[140], 582[140]
Huheey, J. E., **1**, 251[3], 252[3]
Huhn, G. F., **7**, 220[25]
Huhtasaari, M., **3**, 640[109], 647[109,197]
Hui, B. C., **8**, 445[30,31,37]
Hui, R. A. H. F., **7**, 110[188], 132[92,99], 133[92], 134[92]; **8**, 185[16]
Hui, R. C., **1**, 273[69], 544[45]; **3**, 1024[31]; **4**, 115[177], 174[37,38], 184[37,38], 192[37,38]; **6**, 548[670]
Huie, E. M., **1**, 506[13]; **4**, 1076[49], 1078[49]; **5**, 333[44b]; **6**, 764[12]; **7**, 691[16]
Huisgen, R., **1**, 832[113]; **2**, 753[3], 1102[120]; **3**, 893[53a,b], 896[71]; **4**, 48[138,138a,140], 66[138a,b], 490[66], 493[80], 499[66], 500[106], 872[43], 953[8,8a], 954[8a], 1069[4,5,7], 1070[4,5,7,9], 1072[9], 1073[27], 1074[27,29], 1078[4], 1079[58,59], 1081[75], 1082[4,91], 1083[4,7,92], 1084[94], 1085[100,101,108,109], 1086[113], 1089[135], 1090[135,142,143], 1092[144], 1093[144], 1097[161,162,165,166], 1098[7], 1099[186], 1100[189,190], 1102[199,200,201], 1103[4,204]; **5**, 70[115], 71[123], 75[213], 76[234,235,236,237,238,242], 77[245,249,250,255,259], 78[276,277,278,279], 79[115], 248[29], 250[39], 254[29b], 391[143],

430[116,116b], 451[46], 714[76], 715[79], 743[163], 947[262] ; **6**, 249[139], 911[18], 912[18]; **7**, 24[32], 475[51], 477[73]; **8**, 391[92]
Huisman, H. O., **2**, 812[73], 823[118]; **3**, 153[414], 155[414], 875[72,77], 882[100,101]; **4**, 52[147,147d]; **5**, 453[66]; **6**, 753[116], 755[116]; **8**, 843[55]
Huizinga, W. B., **5**, 560[78]
Hula, R. B., **8**, 875[32], 876[32]
Hulburt, H. M., **8**, 447[107]
Hulce, H., **8**, 844[71], 847[71]
Hulce, M., **3**, 564[6]; **4**, 86[81], 213[106-110], 215[106,107,109], 238[7], 254[7]; **6**, 150[121] ; **8**, 112[28], 113[28,41], 116[28,41], 844[71], 847[71]
Hulin, B., **4**, 817[206]; **6**, 1067[103]; **7**, 361[23]
Hull, C., **2**, 573[51], 575[51]
Hull, K., **4**, 98[116], 155[73]; **5**, 806[105], 809[105], 913[104], 1030[93]; **7**, 822[33]
Hull, R., **5**, 99[132]
Hull, V. J., **7**, 881[164]
Hull, W. E., **2**, 1099[110]
Hullen, A., **6**, 184[153]
Hüllmann, M., **1**, 52[15], 142[27], 149[27], 150[27], 152[27], 153[61,64], 191[77], 272[68], 295[53], 296[53], 300[67], 322[67], 331[46], 334[46], 335[63], 336[75]; **2**, 6[35], 247[35], 307[18], 630[7,8], 631[7]
Hullot, P., **8**, 862[230]
Hulme, A. N., **5**, 843[113]
Hulshof, L. A., **6**, 1013[14]
Hülsmeyer, K., **8**, 303[100]
Hulstkamp, J., **3**, 615[8]
Hult, K., **6**, 811[76]
Hultberg, H., **6**, 660[206]; **8**, 224[105,107], 969[96]
Humber, L. G., **8**, 654[83]
Humbert, H., **5**, 589[213]
Humer, K., **6**, 543[614]
Humffray, A. A., **7**, 765[162], 767[195]
Hümke, K., **7**, 758[4]
Hummel, C., **5**, 791[39]
Hummel, H.-U., **1**, 34[228]
Hummel, K., **5**, 70[107,108]
Hummel, W., **8**, 189[56,57]
Hummelen, J. C., **2**, 435[61]
Hummelink, T., **1**, 2[3], 37[3]
Hummelink-Peters, B. G., **1**, 2[3], 37[3]
Humphrey, M. B., **4**, 703[32], 704[32], 984[124] ; **5**, 1089[80]
Humphreys, D. J., **2**, 764[64]; **3**, 689[118] ; **8**, 314[35]
Hundley, H. K., **8**, 618[109]
Hunds, A., **6**, 567[931]
Hundt, B., **2**, 382[312]
Huneck, S., **3**, 834[75], 890[31], 901[114], 903[114]
Hünerbein, J., **5**, 829[23]
Hung, D. T., **1**, 312[114]; **5**, 543[116,116c]
Hung, H. K., **8**, 333[53]
Hung, J. C., **3**, 30[174,177]; **6**, 714[81]
Hung, M.-H., **5**, 435[149], 524[53], 1083[57]
Hung, N. C., **1**, 471[70]
Hung, N. M., **4**, 303[348]
Hung, S. C., **7**, 261[68]; **8**, 859[211,212]
Hung, T., **4**, 601[245]
Hung, T. V., **8**, 53[131], 66[131]
Hung, W. M., **8**, 918[121], 919[121]
Hungate, R., **1**, 103[95]; **2**, 578[84], 696[78], 701[85]; **4**, 33[96], 34[96e]; **7**, 237[37], 245[74]
Hunger, J., **7**, 49[62-64]
Hunger, K., **6**, 635[19], 636[19], 668[252], 669[252]
Hungerbühler, E., **4**, 200[5]; **6**, 25[102]
Hünig, S., **1**, 547[63], 548[63,67]; **2**, 69[45]; **3**, 197[40], 586[139]; **4**, 113[169]; **5**, 526[56], 539[107] ; **6**, 229[24], 233[45], 234[45], 681[333], 682[340], 704[10], 961[67]; **7**, 762[71] ; **8**, 472[3], 474[15], 476[29], 657[98]
Hunkler, D., **5**, 744[167]
Hunma, R., **3**, 822[12], 831[12]
Hunold, R., **5**, 850[152]
Hunsberger, J. M., **7**, 655[19]
Hunt, A. H., **4**, 389[167]
Hunt, D. A., **3**, 251[77], 254[77]; **4**, 615[392], 629[392]
Hunt, D. F., **4**, 697[9,10]; **7**, 451[28], 637[74,75]
Hunt, E., **3**, 936[71]; **6**, 471[64], 854[142]
Hunt, G. E., **7**, 167[100]
Hunt, J. D., **7**, 154[18], 828[51]
Hunt, J. S., **6**, 675[299]; **8**, 987[23]
Hunt, P. A., **4**, 408[259a]
Hunt, P. G., **2**, 73[65], 74[65]; **6**, 932[94,95]
Hunt, R. L., **7**, 872[98]
Hunter, B. H., **1**, 162[103]
Hunter, G. L. K., **3**, 736[29], 771[188,190]
Hunter, J. E., **1**, 480[157,159]; **2**, 1067[127], 1068[127]; **5**, 467[117], 539[109]
Hunter, N. R., **5**, 125[21], 128[21]; **7**, 174[139]
Hunter, R., **4**, 161[91]; **6**, 929[83]; **7**, 199[32,33], 202[33]
Hunter, W. E., **1**, 162[99]; **8**, 447[137], 677[61], 679[61], 682[61], 683[93], 685[61], 687[61]
Huntington, R. D., **3**, 380[7]
Hunton, D. E., **4**, 587[44]
Huntress, E. H., **2**, 757[12]
Huntsman, W. D., **5**, 10[75], 15[104], 17[114], 797[58,63]
Hunziker, H., **4**, 277[85,87], 285[85], 288[85]
Hunziker, P., **2**, 1099[110]
Huong, K. C., **8**, 332[41]
Hüper, F., **5**, 457[90]
Hupfeld, B., **4**, 1038[60]
Huppatz, J. L., **6**, 487[62], 489[62,98]
Huppes, N., **4**, 921[28]
Huq, E., **3**, 355[53], 357[53], 769[170], 771[170]
Hur, C.-U., **4**, 744[137]
Hurd, C. D., **2**, 323[37]; **4**, 31[92,92f]
Hurd, C. T., **5**, 790[35]
Hurd, R. N., **2**, 849[212]; **6**, 488[11], 508[11], 545[11]
Hurlbut, S. L., **2**, 746[114]
Hurnaus, R., **3**, 319[131]
Hurst, G. D., **2**, 39[138]
Hurst, J. R., **5**, 699[5], 704[5]
Hurst, K. M., **2**, 64[27]; **3**, 199[53]
Hursthouse, M. B., **5**, 608[66]
Hurt, W. S., **5**, 127[23]
Hurwitz, J., **2**, 466[120,126], 469[120]
Husain, A., **6**, 647[109]; **7**, 17[177]; **8**, 406[39], 989[34]
Husband, S., **3**, 69[26]
Husebye, S., **6**, 504[222]
Husemann, E., **8**, 769[29]
Husemann, W., **3**, 587[147]
Huser, D. L., **8**, 491[13,14], 496[35], 504[72], 505[72]
Husinec, S., **2**, 369[254]; **8**, 393[113]
Husk, G. R., **4**, 877[65], 879[65], 887[127], 983[117,119], 984[120,123,124], 985[130]; **5**, 1086[70,74]; **8**, 742[43,45], 743[45], 753[67], 758[43]
Huss, O. M., **6**, 509[269]
Hussain, H. H., **4**, 1015[201]
Hussain, N., **3**, 503[149], 512[149]; **7**, 350[22], 363[36]
Hussain, S., **5**, 164[75], 176[75]
Hussey, A. S., **2**, 279[14]; **3**, 242[6], 257[6], 259[6]; **8**, 425[44,45], 426[44,45]

Hussey, B. J., **8**, 18[123]
Hussmann, G., **5**, 587[208], 588[208]
Hussmann, G. P., **7**, 763[99], 766[99]
Husson, A., **6**, 734[16]
Husson, H. P., **1**, 367[55], 555[119], 557[127,-128], 558[137,138,139,140,141], 559[142,143,144,146,147,148], 564[206]; **2**, 63[22b], 901[31], 1014[19], 1018[43,45], 1021[48]; **5**, 829[22]; **6**, 734[16], 912[22], 917[34], 919[40], 920[43] ; **8**, 587[35]
Husstedt, U., **4**, 345[80]; **6**, 685[347,356]
Husted, C. A., **8**, 115[61], 510[96]
Hustedt, E. J., **4**, 4[16]
Huston, R., **5**, 806[106], 1025[83], 1026[83]
Hutchings, M. G., **3**, 798[96,97]; **4**, 311[444] ; **7**, 479[97], 595[22], 600[75]
Hutchins, J. E. C., **8**, 84[10]
Hutchins, L., **2**, 1017[31]; **6**, 737[31], 746[89]
Hutchins, M. G., **6**, 959[37]
Hutchins, M. G. K., **8**, 383[14]
Hutchins, R. O., **3**, 753[102], 850[61]; **4**, 604[285,290,291], 646[290,291], 647[291]; **6**, 533[479], 550[479], 959[37]; **7**, 841[18] ; **8**, 14[76], 26[16,19,20,24,28], 27[16,19,20], 30[28], 36[19,20,24,28], 37[16,24,28,103], 39[24], 40[28], 43[28], 44[24,28], 46[24,28], 47[16], 54[20,155], 55[16,20,28], 60[16,20], 66[24,28,103,155], 67[24], 70[16,20], 72[243], 74[243,245], 176[136], 244[67], 250[67], 264[39], 343[117], 344[117], 347[139], 349[139], 350[139,149,150], 351[117,149], 352[139], 354[117,149], 355[117,149,179], 357[195,204], 358[195], 383[14], 393[110], 538[191], 803[93], 804[93], 806[100], 812[4,6], 826[69], 840[31], 848[31], 876[45], 877[45], 929[31], 960[35,36], 967[84]
Hutchins, R. R., **1**, 786[249]; **3**, 103[204], 108[204]; **8**, 842[47]
Hutchinson, C. R., **1**, 268[56]; **2**, 25[99], 157[121], 240[5], 455[6]; **3**, 599[211]; **4**, 790[35]; **5**, 109[215], 129[34], 350[78]
Hutchinson, D. B., **7**, 78[127]
Hutchinson, D. K., **1**, 114[37]; **3**, 209[16], 223[16]; **4**, 79[55b], 192[118,118b]; **5**, 1166[24]; **6**, 162[189,190,191], 163[189]
Hutchinson, D. W., **6**, 609[57,58]
Hutchinson, E. G., **4**, 6[21]; **8**, 499[42]
Hutchinson, J., **3**, 878[94], 879[94], 880[94,98], 881[94]; **4**, 359[159], 771[251]
Hutchinson, J. H., **2**, 651[117]; **3**, 50[265] ; **5**, 812[133]; **6**, 1045[26-28,29a]
Hutchison, D. A., **2**, 5[20], 6[20], 21[20]
Hutchison, J., **5**, 212[70]
Hutchison, J. D., **6**, 203[1]
Huth, A., **3**, 1028[48]
Huth, H.-U., **8**, 621[145]
Hüther, E., **8**, 754[104], 755[104]
Hutley, B. G., **8**, 89[43]
Hutmacher, H.-M., **8**, 795[20]
Huton, J., **8**, 545[290]
Hutson, A. C., **8**, 132[11]
Hutson, D., **6**, 48[83]
Hutt, J., **1**, 564[200]; **6**, 141[62], 149[91], 156[159,161], 927[78]; **8**, 563[432]
Hüttel, R., **4**, 587[18-23,37], 603[267], 604[281,282], 645[267], 646[281], 939[80], 941[80]; **7**, 107[160], 452[58]
Hüttenhain, S., **2**, 614[119]
Huttner, G., **1**, 221[68]; **2**, 205[101,101b] ; **4**, 21[65], 36[103,103b], 108[146e]; **5**, 428[109]; **6**, 517[327], 522[346], 524[359] ; **8**, 389[67], 690[102]
Hutton, J., **7**, 824[41]
Hutton, R. E., **2**, 443[16]
Hutton, T. W., **3**, 892[51]
Hutzinger, O., **8**, 315[48]
Huu, F. K., **8**, 618[126]
Huurdeman, W. F. J., **6**, 134[16]
Huxol, R. F., **3**, 88[136], 91[136], 179[136], 181[136]
Huy, N. H. T., **5**, 1098[131], 1103[151]
Huybrechts, G., **5**, 571[114]
Huyer, E. S., **4**, 305[365]
Huynh, C., **3**, 224[161], 244[27], 262[161], 263[161], 264[180], 438[34], 464[34], 466[34,188], 901[112], 934[68], 943[84], 969[130]; **4**, 102[128d]; **6**, 853[140]
Huynh, V., **1**, 461[12]
Huys, F., **6**, 495[142,143], 496[143,156], 497[143], 514[156]
Huyser, E., **7**, 883[178]
Huyser, E. L. S., **4**, 726[53]
Huyser, E. S., **4**, 717[8], 752[162], 765[162] ; **7**, 15[155], 16[156,162], 851[25], 883[179]
Huys-Francotte, M., **5**, 386[133], 681[26]
Hvidt, T., **2**, 642[78], 643[78]; **7**, 580[146]
Hwang, C.-K., **3**, 618[20,21], 751[90]; **5**, 687[67], 736[145], 737[145]; **6**, 206[45], 448[106,107,109]; **7**, 401[61d], 407[84b], 408[88c]
Hwang, D.-R., **4**, 445[207]; **8**, 344[123]
Hwang, K.-J., **1**, 532[136]; **3**, 174[522] ; **5**, 839[77], 886[21,22]; **8**, 341[103], 928[24]
Hwang, Y. C., **5**, 473[150], 478[150]
Hwu, J. R., **2**, 583[115]; **3**, 160[469], 161[469], 373[131]; **6**, 116[93], 938[125], 940[125]
Hyatt, J. A., **2**, 735[14]; **5**, 558[60]; **6**, 787[104]; **7**, 738[24]
Hyer, P. K., **3**, 14[74], 15[74]
Hylarides, M. D., **7**, 605[139]
Hyldahl, C., **5**, 1089[87], 1090[87], 1094[87], 1098[87], 1099[87], 1100[87], 1101[87], 1112[87], 1113[87]
Hylton, T. A., **3**, 124[267], 125[267], 126[267], 127[267], 927[48]; **6**, 134[37]
Hyne, R. V., **2**, 809[58], 811[58]
Hynes, M. J., **4**, 710[55], 712[67]
Hyodo, C., **4**, 1095[152]
Hyodo, N., **6**, 602[7], 603[7]
Hyon Yuh, Y., **3**, 647[179], 648[179]
Hyuga, S., **3**, 231[248], 251[78], 254[78], 443[56], 490[66], 511[66], 515[66]; **4**, 147[42], 358[158]

I

Iacobelli, J. A., **1**, 822[34]; **5**, 256[58]; **6**, 960[56]; **7**, 564[92], 567[92]
Iacobucci, G. A., **8**, 314[28,29]
Iacona, R. N., **7**, 654[7,8]
Iarossi, D., **5**, 439[166]; **7**, 777[371,372,373,374]
Iavarone, C., **4**, 370[28]; **8**, 856[163]
Ibana, I. C., **4**, 293[239]; **6**, 278[131]
Ibarra, C. A., **8**, 2[10]
Ibata, T., **1**, 853[46]; **4**, 433[123], 1089[138], 1091[138]; **5**, 79[288]; **6**, 66[8], 568[934]
Ibberson, P. N., **4**, 435[136]
Ibe, M., **4**, 112[160], 116[185a]
Ibers, J. A., **1**, 18[95], 441[173]; **4**, 964[49]; **8**, 366[51], 418[7], 458[225]
Iborra, S., **2**, 747[118]
Ibragimov, A. G., **4**, 589[79], 591[79], 598[199], 638[199], 640[199]
Ibragimov, I. I., **5**, 1056[48], 1057[50]
Ibragimov, M. A., **6**, 141[64]
Ibraheim, N. S., **2**, 378[291]
Ibrahim, B., **5**, 630[52]
Ibrahim, B. E., **4**, 48[137,137g], 1093[145]
Ibrahim, I. H., **8**, 34[60], 66[60]
Ibrahim, M. H., **8**, 587[37]
Ibrahim, M. K., **6**, 556[822]
Ibrahim, N., **6**, 745[86]; **7**, 779[422]
Ibuka, E., **3**, 503[142]
Ibuka, T., **2**, 876[33,34], 877[38,39]; **3**, 222[136,137], 226[200], 623[39]; **4**, 148[51], 149[51], 150[53], 179[65], 180[67], 188[101], 189[101], 739[111]; **6**, 4[19], 764[11], 848[108]; **7**, 417[130c]
Ichiba, M., **1**, 34[168]; **2**, 792[64]; **7**, 342[54]
Ichibori, K., **3**, 919[32], 923[44], 934[44], 954[44], 1008[70]
Ichida, H., **1**, 243[56]
Ichihara, A., **3**, 693[149], 694[149], 747[72]; **5**, 451[38], 516[25], 524[54], 534[54,95], 553[42], 563[92,93], 564[94,96,97], 571[121], 578[152]; **6**, 689[384,386], 690[384]
Ichihara, J., **1**, 555[115]
Ichihara, S., **5**, 578[150]
Ichikawa, K., **1**, 512[39]; **3**, 311[97], 313[106], 315[113], 769[169]; **4**, 302[337]; **7**, 154[17], 451[29]; **8**, 476[24]
Ichikawa, M., **5**, 196[17]
Ichikawa, Y., **2**, 578[83], 650[109]; **3**, 829[57]; **4**, 79[59a]; **6**, 27[115], 164[199]; **7**, 370[65], 380[65]
Ichimoto, I., **8**, 979[148]
Ichimura, K., **6**, 535[544], 538[544]
Ichino, K., **7**, 356[50]
Ichino, T., **8**, 452[187], 535[162]
Ichinohe, Y., **3**, 826[39]
Ichinose, I., **4**, 378[108]
Ichinose, Y., **4**, 721[31], 725[31], 770[248], 771[253], 789[32], 791[42]; **8**, 699[149], 798[46], 807[46]
Ichioba, M., **6**, 533[486]
Ida, H., **4**, 410[261]; **7**, 474[42]
Idacavage, M. J., **3**, 7[32], 8[32]; **4**, 889[135]; **6**, 14[51]
Iddon, B., **1**, 471[64]; **2**, 765[76]; **7**, 21[6]; **8**, 65[211], 66[211], 629[180]
Ide, H., **8**, 291[32]
Ide, J., **2**, 547[115], 551[115], 823[117]; **3**, 939[79]
Ide, W. S., **1**, 542[4], 543[4], 547[4]
Ideses, R., **1**, 758[125]
Idogaki, Y., **5**, 839[76]; **7**, 713[70]
Idoux, J. P., **4**, 436[145], 437[145], 438[149,150,151,152]; **6**, 545[638]
Idrissi, M. E., **7**, 554[63]
Iemura, S., **3**, 246[35], 354[60]
Ienaga, K., **6**, 533[489]
Iesaki, K., **6**, 113[71]
Iffert, R., **5**, 202[37]
Iffland, D. C., **1**, 72[75]; **6**, 226[12]; **7**, 231[143]; **8**, 916[111], 918[111]
Iflah, S., **8**, 453[193], 557[383]
Ifzal, S. M., **7**, 294[17]
Igaki, M., **3**, 900[97]
Igami, M., **8**, 291[36,37], 292[36], 293[46]
Igarashi, K., **4**, 201[10]; **6**, 42[45]
Igarashi, M., **4**, 239[26], 251[26], 257[26]; **5**, 583[183]
Igarashi, S., **4**, 359[161]
Igarashi, Y., **4**, 405[250a]
Igeta, H., **3**, 461[146], 541[107]; **5**, 497[225], 914[115]; **6**, 664[220]; **8**, 641[26]
Iglauer, N., **8**, 453[192]
Ignatova, E., **3**, 45[243]
Igolen, J., **4**, 502[120]
Iguchi, H., **2**, 578[87], 582[109]; **4**, 155[68b]; **5**, 337[51]
Iguchi, K., **2**, 370[256]; **3**, 342[12], 347[12], 381[30], 382[30], 421[59], 422[59]; **7**, 828[54]
Iguchi, M., **2**, 553[124]; **3**, 390[74,86], 392[74], 395[100], 396[113,115], 397[116], 398[113], 665[38], 691[128,131], 693[128,139], 697[38,131,139,155,156], 698[128], 769[171]; **8**, 152[161,162]
Iguchi, S., **8**, 100[117], 545[284]
Iguchi, Y., **2**, 823[116]
Iguertsira, L. B., **4**, 1089[126]
Igumnov, S. M., **6**, 495[146]
Ihama, M., **8**, 562[421]
Ihara, J., **7**, 92[42], 93[42]
Ihara, M., **2**, 222[147], 819[103], 888[54]; **4**, 30[88,88k-o], 121[209,209a,b], 231[277], 239[17,27], 257[27], 261[17,27,299,300,301], 333[21-23], 398[215], 401[215a,229]; **5**, 473[154], 479[154], 522[45], 531[78], 534[95], 536[97], 693[114], 694[114]; **6**, 74[37]; **7**, 229[113], 493[199], 517[15]; **8**, 314[37], 647[56], 945[128]
Ihara, R., **7**, 476[66]
Ihde, A. J., **2**, 770[9]
Ihle, N. C., **5**, 639[126], 640[128], 641[130]
Ihn, W., **7**, 480[99]
Ihrig, K., **5**, 555[49]
Ihrig, P. J., **2**, 362[185]
Ii, M., **4**, 439[165]
Iida, H., **1**, 55[24], 506[5], 526[5], 527[103], 555[110], 557[129,131], 558[129,132,135]; **2**, 61[19], 74[75]; **3**, 136[370,371], 137[370], 138[370,371a], 139[370,371], 140[370,371,371a,b], 143[371,371a,b], 144[371a], 507[175]; **4**, 18[58], 259[257], 434[132], 502[121,122], 597[173,174], 637[173,174], 847[77]; **5**, 74[206], 256[55,56], 421[78]; **6**, 81[72]; **7**, 95[65], 297[31], 778[410]; **8**, 277[155], 395[126], 652[80]
Iida, K., **1**, 143[37], 158[37], 159[37,79], 180[41], 181[41], 340[90]; **2**, 5[17], 6[17], 22[17,17a], 23[88]; **8**, 410[92], 797[43], 807[43]
Iida, M., **1**, 269[57]; **7**, 63[59]
Iida, S., **1**, 425[102]; **3**, 246[39], 257[39]; **5**, 634[74,76]
Iida, T., **4**, 85[77e]; **8**, 350[142,143], 352[142]
Iihama, T., **2**, 74[75]
Iijima, M., **4**, 227[206,207], 255[200]
Iijima, S., **1**, 143[37,38], 158[37], 159[37,38], 180[41], 181[41], 340[90]; **2**, 5[17,19], 6[17,19], 22[17,17a], 23[19a], 901[38], 908[38]
Iimori, T., **2**, 113[104], 926[116,118,119]; **7**, 399[40b], 410[96]
Iimura, Y., **2**, 386[331]; **6**, 677[323]
Iino, K., **2**, 649[103], 1059[76]
Iinuma, M., **7**, 136[111], 137[111]

Iio, H., **1**, 767[176]; **2**, 381[311], 509[34], 585[128,129]; **3**, 35[201], 39[201], 355[54], 357[54]; **5**, 924[147]; **6**, 5[24], 836[58], 900[115], 1023[69]

Iitaka, Y., **1**, 512[39]; **2**, 656[154]; **4**, 378[107]; **5**, 100[142]; **7**, 255[36], 362[31], 377[31], 438[13], 443[13]; **8**, 49[115], 66[115], 201[139]

Iitaki, Y., **3**, 380[9], 675[73]

Iizuka, K., **5**, 621[20], 632[64]; **6**, 814[92], 820[111]

Iizuka, M., **8**, 616[97]

Iizumi, K., **6**, 22[86]

Ijima, I., **4**, 680[50]

Ikada, M., **7**, 606[156]

Ikan, R., **7**, 121[27], 123[27]

Ikariya, T., **7**, 314[41], 315[41,42]; **8**, 239[28]

Ikawa, T., **4**, 610[338], 649[338]

Ike, K., **3**, 1040[104]

Ikeda, A., **7**, 795[8,10], 796[12]

Ikeda, H., **1**, 553[96]; **2**, 157[119], 709[7]; **3**, 49[263], 904[134], 1033[76]; **4**, 430[96], 921[26]; **5**, 137[74]; **7**, 453[86], 455[86]

Ikeda, I., **7**, 471[24]

Ikeda, K., **1**, 385[115]; **2**, 914[79], 915[79], 939[159], 940[161], 948[180], 1096[99]; **5**, 100[157]; **6**, 625[152]

Ikeda, M., **2**, 556[153], 580[99]; **3**, 904[134]; **4**, 350[116,118], 826[244], 837[11]; **5**, 133[53,54], 439[170]; **6**, 744[76], 746[76], 764[13], 910[3], 930[87], 931[88,90]; **7**, 199[31,35], 200[42], 208[87], 209[89,92], 391[13], 411[13], 412[13], 413[13], 746[90]; **8**, 829[81], 964[59]

Ikeda, N., **1**, 161[80-82,86,87,90], 165[108-110], 509[23], 790[264], 827[64a]; **2**, 22[87], 65[31], 72[59], 74[71], 84[47], 88[33], 91[44,46,47], 93[46,47], 94[50], 96[57,58], 269[72], 615[126], 631[18]; **3**, 446[85,86,88]; **4**, 971[77], 976[96]; **6**, 865[200]; **7**, 318[51], 537[59]

Ikeda, O., **7**, 423[145], 424[145b]

Ikeda, S., **1**, 174[13], 202[13], 346[127]; **6**, 453[143], 644[85], 711[64]; **8**, 445[62], 460[254]

Ikeda, T., **1**, 268[56], 463[20]; **3**, 599[211]; **7**, 247[105], 674[46]

Ikeda, Y., **1**, 161[80-83,86,87,90], 509[23], 827[64a]; **2**, 22[87], 65[31], 72[59], 74[71]; **3**, 446[85,86,88]; **4**, 347[95], 930[46,47]; **7**, 751[139]

Ikefuji, Y., **8**, 598[98]

Ikegami, S., **1**, 568[240]; **2**, 495[63], 496[63]; **3**, 45[248], 135[364], 139[364], 142[364], 143[364], 1000[55]; **4**, 301[330], 331[14,15], 344[14]; **5**, 516[28], 925[154]; **6**, 5[25], 21[79], 778[62], 780[62], 877[39], 878[39], 883[39], 887[39]; **7**, 246[94], 419[133], 455[105], 617[21], 620[26], 621[30]; **8**, 580[7], 880[58]

Ikegawa, A., **4**, 231[262]

Ikegawa, S., **8**, 883[92]

Ikeguchi, M., **8**, 93[72], 94[80]

Ikehara, M., **6**, 604[26,34], 614[96], 626[167]

Ikehira, H., **3**, 95[152]; **7**, 747[97]; **8**, 817[34], 847[90]

Ikehira, T., **8**, 860[223]

Ikekawa, N., **2**, 187[42]; **5**, 151[9]; **6**, 219[121], 989[78], 993[78], 996[105]; **7**, 366[52], 675[54], 680[76]; **8**, 967[78]

Ikemoto, Y., **2**, 976[2], 981[2], 982[2]; **3**, 579[101]

Ikemura, Y., **2**, 651[113]

Ikenaga, K., **3**, 495[96], 497[103]

Ikenaga, S., **1**, 407[31]

Ikeno, M., **8**, 979[147]

Ikeshima, H., **7**, 153[11]; **8**, 557[380]

Ikezaki, M., **8**, 64[216], 67[216]

Ikezawa, K., **2**, 953[3b]

Ikhlobystin, O. Y., **1**, 212[4]

Ikizler, A., **7**, 228[102]

Ikka, S. J., **5**, 935[192]

Ikota, N., **2**, 904[51]; **3**, 1055[32]; **5**, 98[125], 376[108b]; **7**, 545[27]; **8**, 146[100]

Ikunaka, M., **7**, 239[52]; **8**, 190[70]

Ikuta, S., **1**, 642[120], 645[120], 672[120], 708[120]; **3**, 301[47]

Ila, H., **2**, 286[63], 495[62], 496[62]; **6**, 456[162], 457[162]; **7**, 154[12]; **8**, 540[201], 839[26b], 840[26]

Ilijev, D., **4**, 820[222]

Il'in, M. M., **6**, 546[645]

Il'in, V. F., **3**, 305[75a]

Ilingin, O. V., **6**, 494[133]

Il'inskaya, L. V., **8**, 765[11]

Illig, C. R., **4**, 1078[50]

Illuminati, G., **4**, 786[25]; **6**, 24[96]

Ilyakhina, T. V., **8**, 530[107]

Ilyushin, M. A., **6**, 110[50]

Im, K. R., **8**, 838[21,21a], 840[21a]

Im, M.-N., **7**, 230[125]

Imada, M., **6**, 89[120]

Imada, T., **7**, 768[199]

Imada, Y., **4**, 393[188], 598[198], 640[423]; **6**, 76[53], 86[97], 113[69], 253[156]; **7**, 74[110]; **8**, 395[123], 600[102]

Imafuku, K., **5**, 1022[74]

Imagawa, T., **1**, 752[98]; **3**, 652[221]; **5**, 168[106]; **6**, 561[875]

Imai, E., **6**, 927[80]

Imai, H., **1**, 544[44], 551[77]; **4**, 162[93,94a]; **6**, 238[72]; **8**, 552[349], 557[384]

Imai, I., **3**, 919[32], 923[43,44], 934[44], 954[44], 1008[70]

Imai, J., **6**, 614[81,82], 625[161]

Imai, K., **1**, 387[133]; **3**, 159[455], 161[455]; **4**, 300[305]

Imai, N., **4**, 1089[125], 1095[152]

Imai, R., **6**, 554[719]

Imai, S., **8**, 150[140,143], 151[155,156,157], 625[164]

Imai, T., **1**, 563[183]; **2**, 258[49,50], 348[59], 362[59], 719[82]; **3**, 231[245], 797[91]; **5**, 86[29]; **7**, 57[22]; **8**, 159[108], 171[108], 178[108], 179[108], 720[138], 721[138], 722[138]

Imai, Y., **4**, 439[165]; **6**, 510[293], 546[641,642], 936[107]; **8**, 144[74], 190[80]

Imai, Z., **8**, 249[98], 253[98], 369[75]

Imaida, M., **8**, 150[133], 151[133,145]

Imaizumi, S., **1**, 447[214], 450[214]; **4**, 302[333]; **8**, 418[12], 422[12], 856[165]

Imaizumi, T., **8**, 144[71]

Imamoto, T., **1**, 232[12-14], 233[13,14,18,19], 234[12-14,21], 235[27,28], 243[57], 248[61,62], 253[9], 255[17], 259[29], 260[32,32a], 261[32a,c], 276[9], 278[9], 332[52-54], 561[160], 735[25], 829[87], 831[104]; **2**, 311[33], 312[33,35]; **3**, 567[34], 570[34]; **4**, 229[224], 973[86]; **6**, 214[94], 428[82], 438[49], 457[163], 799[19,20]; **7**, 308[19], 843[48]; **8**, 113[47], 115[65], 405[23], 551[335], 803[92]

Imamura, A., **5**, 72[167]

Imamura, J., **7**, 155[25]

Imamura, S., **4**, 588[65], 600[231]; **7**, 463[125]

Imamura, Y., **7**, 463[129]

Imanaka, T., **4**, 941[81]; **7**, 107[168]; **8**, 142[50], 150[137,138], 419[17], 430[17]

Imanishi, T., **4**, 1040[89,90], 1045[89,90]; **6**, 1046[31]; **7**, 178[148], 455[104], 544[35], 550[51], 556[35], 566[35], 803[54], 821[29]; **8**, 837[15a]

Imao, T., **4**, 1056[141]

Imaoka, A., **8**, 52[143], 66[143]

Imaoka, M., **6**, 979[29]

Imazawa, A., **2**, 150[95]

Imbeaux-Oudotte, M., **4**, 123[210b], 125[210b]

Imberger, H. E., **7**, 767[195]

Imbert, D., **7**, 876[125]

Imhof, R., **6**, 531[451]

Imi, K., **2**, 584[126]; **4**, 120[203], 300[305], 879[85]; **6**, 237[60], 243[60]

Immer, H., **2**, 1088[40], 1095[92], 1097[40]

Imoto, E., **6**, 425[64]; **7**, 761[55], 764[55]; **8**, 994[64]

Imoto, H., **8**, 411[105]
Imoto, M., **6**, 53[118], 57[141]
Imoto, T., **7**, 693[27]
Imperiali, B., **1**, 410[43]; **2**, 232[183], 249[36], 303[5], 308[20]
Imre, J., **6**, 660[207]; **8**, 226[112], 227[115]
Imrich, J., **6**, 195[224]
Imuta, M., **3**, 734[8]; **8**, 187[41], 203[147]
Imwinkelried, R., **1**, 314[128], 323[128]; **2**, 259[83], 264[83], 578[85,86]; **3**, 580[103]; **4**, 209[67]; **5**, 377[110,110b], 378[110b], 1066[5]
Inaba, M., **3**, 833[70]; **6**, 77[54], 83[82], 938[134]; **8**, 244[50], 371[103]
Inaba, S., **1**, 349[146,148], 453[223], 555[117]; **3**, 421[55]; **5**, 386[132], 387[132c], 691[83], 692[83], 693[83]; **6**, 49[95]
Inaba, S.-i., **2**, 635[45], 640[45], 646[86], 647[87], 929[126], 930[126,-128], 931[126]; **3**, 499[116]; **5**, 102[175,181]; **6**, 977[17]
Inaba, T., **1**, 527[103]; **5**, 850[146]
Inada, Y., **4**, 557[11]
Inaga, J., **6**, 668[260]
Inagaki, H., **1**, 359[14,16], 363[14,14a], 379[16c], 384[14]
Inagaki, M., **4**, 230[244], 231[244]; **7**, 24[37], 25[37,41,42,45], 26[41,52,58]
Inagaki, S., **4**, 510[167]; **5**, 75[224,225]; **7**, 96[90], 98[90]
Inagaki, Y., **6**, 487[43], 489[43]; **7**, 698[50]
Inage, M., **6**, 603[14]; **8**, 150[122], 151[148]
Inaishi, M., **7**, 765[165]
Inaki, H., **8**, 244[57], 249[97], 253[97], 620[132]
Inamdar, P. K., **7**, 64[64]
Inami, K., **5**, 101[163]; **6**, 566[925]
Inamoto, N., **1**, 781[230]; **4**, 446[213]; **5**, 829[23]; **6**, 475[88-90]
Inamoto, Y., **3**, 383[45], 1041[113]; **4**, 857[104], 921[26]; **6**, 270[80]; **7**, 9[70]; **8**, 331[33]
Inamura, Y., **4**, 393[186]
Inanaga, J., **1**, 256[22], 257[23], 258[24,25], 259[24], 260[32,32b], 261[33], 266[47], 268[53,53b], 270[25], 271[63], 275[25,76], 421[89], 751[110,111], 831[105]; **3**, 566[28], 571[28], 578[28]; **4**, 606[305], 607[305,311,313], 626[311], 647[305], 648[311]; **6**, 7[31], 175[76], 648[116], 980[41]; **7**, 662[52]; **8**, 412[108b], 540[195], 797[33], 883[95], 884[95], 960[32], 987[22], 992[22b], 994[22]
Inaoka, T., **6**, 445[100]
Inayama, S., **4**, 23[70]; **8**, 201[139], 935[64]
Inazu, T., **7**, 356[49]
Inbal, Z., **5**, 407[25]
Inbasekaran, M. N., **4**, 430[89]
Inch, T. D., **1**, 55[26]; **4**, 93[91b]; **6**, 36[17]; **8**, 224[102], 541[212]
Indorato, C., **6**, 439[68]
Inenaga, M., **7**, 227[81]
Ineyama, T., **2**, 1066[119]
Ingall, A. H., **5**, 469[138]
Ingemann, S., **8**, 89[43]
Ingham, H., **2**, 848[210]
Ingham, K. C., **8**, 70[228]
Ingham, R. K., **7**, 718[1], 731[1]; **8**, 568[484]
Ingham, S., **5**, 683[39], 684[39]
Inglis, R. P., **4**, 310[423,424]
Ingold, C. F., **7**, 236[28], 237[28], 768[204], 844[57]
Ingold, C. K., **2**, 134[9]; **3**, 822[7], 965[127]; **6**, 3[13], 4[13]
Ingold, K. U., **1**, 274[72], 699[252]; **3**, 69[26]; **4**, 717[14], 719[27], 722[34], 723[38,39], 728[34], 736[88], 738[38,101], 747[38], 780[2], 785[2,22], 812[178]; **5**, 901[28,29]; **6**, 960[57]; **8**, 264[47], 857[200]
Ingraham, J. N., **3**, 309[92c]
Ingram, C. D., **5**, 781[206]
Ingram, D. D., **6**, 959[32]
Ingrosso, G., **1**, 113[29]; **2**, 87[26], 495[64,65], 496[64,65]; **3**, 733[1]; **5**, 1131[12], 1148[114], 1154[152], 1155[162], 1156[162]
Ingwalson, P., **8**, 340[89]
Inman, C. G., **5**, 151[2]
Inman, K. C., **4**, 797[103]
Inners, R. R., **1**, 755[116], 756[116,116b], 758[116], 761[116]; **4**, 38[108,108c]
Inokawa, S., **3**, 500[131]; **6**, 134[27], 186[172]; **8**, 411[105]
Inokuchi, T., **1**, 551[78]; **2**, 187[40], 655[132,135]; **4**, 159[82], 383[140]; **5**, 42[31]; **8**, 216[61]
Inomata, K., **2**, 112[90,91], 240[10], 242[10], 614[115]; **3**, 159[451], 160[451], 161[451]; **4**, 359[161], 595[151], 599[212], 604[289], 640[212], 641[212], 646[289], 753[164], 756[182]; **6**, 641[62]; **7**, 262[82], 299[45], 564[95], 568[95], 709[37]; **8**, 840[30,30b], 960[37]
Inone, N., **4**, 852[89]
Inone, T., **8**, 245[75]
Inoue, A., **8**, 902[47], 904[47], 905[47]
Inoue, H., **1**, 223[81], 224[81]; **2**, 889[57]; **3**, 565[22]; **5**, 693[114], 694[114]; **6**, 530[415], 626[167]; **7**, 453[76]; **8**, 338[82], 339[82], 371[113], 994[64]
Inoue, I., **2**, 1051[41]; **4**, 680[50]; **6**, 76[46], 646[100b]; **7**, 806[74]; **8**, 384[36]
Inoue, J., **1**, 561[164]
Inoue, K., **1**, 346[128], 546[56]; **2**, 348[59], 362[59]; **5**, 500[259]; **7**, 95[71], 314[39], 804[58], 808[78-80]; **8**, 190[68], 191[87], 332[43]
Inoue, M., **1**, 314[124-126], 347[132,134], 415[61], 474[100]; **2**, 599[26]; **3**, 244[29], 311[97], 313[106], 463[167], 529[52], 555[28]; **4**, 23[71], 162[92], 444[198], 879[85]; **5**, 377[110], 378[110a]; **6**, 237[67], 564[913,914]; **7**, 340[45]; **8**, 658[99]
Inoue, N., **4**, 903[202], 904[202]; **7**, 606[154]
Inoue, S., **1**, 546[54]; **2**, 384[319], 610[97], 1059[78]; **3**, 168[494,501,504], 169[494,501,504], 170[494,501,504], 426[82-84], 428[90,91], 429[82,91]; **4**, 20[63], 21[63], 213[100], 430[96]; **6**, 266[50], 510[296]; **7**, 109[186]; **8**, 154[199], 371[109], 462[267]
Inoue, T., **2**, 112[95-97], 242[20], 650[108]; **6**, 960[63]; **7**, 764[109]; **8**, 18[128], 245[75]
Inoue, Y., **1**, 349[147], 447[214], 450[214], 750[88]; **2**, 116[140], 610[94], 611[94], 1059[78,81]; **3**, 500[133], 1024[30]; **4**, 590[94,95], 592[95], 598[196], 633[95]; **5**, 92[81], 297[57,59], 618[2], 1157[169], 1196[37,38], 1197[38]; **6**, 74[34], 86[98]; **7**, 800[34]; **8**, 369[78], 795[23], 906[65], 907[65], 908[65], 909[65], 910[65]
Inouye, H., **5**, 468[132]
Inouye, K., **4**, 201[10]; **7**, 80[138]; **8**, 191[86]
Inouye, M., **4**, 447[219]; **6**, 976[3]; **8**, 370[94], 806[127]
Inouye, Y., **2**, 232[174]; **4**, 200[7]; **5**, 137[75,76], 143[75,76]; **6**, 843[91], 1016[26]; **7**, 242[60]; **8**, 170[77-79], 562[423], 853[144]
Insalaco, M. A., **6**, 249[141]
Intille, G. M., **8**, 320[83]
Intrito, R., **1**, 303[77]
In't Veld, P. J. A., **7**, 763[96]; **8**, 405[31]
Inubushi, Y., **2**, 157[119], 810[60], 876[33,34]; **3**, 623[39]; **6**, 551[680], 764[11]; **7**, 569[108]
Inui, S., **7**, 107[164], 178[150]
Inukai, N., **4**, 208[62]
Inukai, T., **5**, 4[29], 339[57a], 345[57]
Inuzuka, N., **8**, 609[49]
Invergo, B. J., **7**, 228[98]; **8**, 263[33], 618[114]
Ioanid, N., **4**, 84[67]
Ioannou, P. V., **6**, 620[123,124]
Ioffe, B. V., **3**, 305[73], 310[93], 311[93]; **7**, 483[126,127], 742[55], 743[55], 744[55]
Ioffe, D. V., **8**, 500[49]
Ioffe, S. L., **4**, 145[25]
Iogagnolo, M., **1**, 471[68]
Iokubaitite, S. P., **6**, 423[40], 424[40], 428[40], 432[40]
Ioramashvili, D. Sh., **8**, 772[55,58,59]
Iorio, E. J., **5**, 787[5], 798[4]
Iorio, L. C., **6**, 523[353]
Iovel, I. G., **8**, 764[4a]

Ip, H. S., **8**, 242[40]
Ip, W. M., **7**, 17[177]
Ipach, I., **8**, 910[84], 916[84]
Ipaktschi, J., **4**, 115[183], 259[263]; **5**, 209[53], 215[1], 216[1], 218[1,31,32], 220[1], 223[1], 224[1,31,32], 345[69], 346[69]; **6**, 189[188]; **8**, 64[209], 65[209], 66[209], 67[209], 68[209]
Ipatieff, V. N., **3**, 309[91], 329[188], 331[197]; **4**, 316[538]; **5**, 10[77]; **8**, 814[19]
Ippen, J., **8**, 992[54]
Ippolito, R. M., **1**, 582[7]; **2**, 572[42]
Iqbal, A. F. M., **8**, 372[120]
Iqbal, J., **2**, 616[134], 619[134]; **3**, 26[162], 27[167], 198[43]
Iqbal, M., **4**, 518[7], 521[42], 529[7,75], 530[75], 1021[249,250]
Iqbal, M. N., **5**, 595[9], 596[9], 603[9]
Iqbal, R., **7**, 561[84]
Iqbal, S. M., **8**, 231[146]
Iqbal, T., **5**, 524[54,54g], 534[54], 535[54g]
Iranpoor, N., **7**, 266[109], 267[109], 760[23]
Irany, E. P., **2**, 144[60]
Irei, H., **3**, 135[346]
Ireland, R. E., **1**, 126[88], 131[101,102], 447[199], 744[56], 747[64], 829[88]; **2**, 101[20], 102[20,27], 182[9], 200[88], 547[101], 548[101], 604[55], 837[164], 838[164], 868[20], 869[20], 871[20], 872[20], 876[20], 890[20], 935[150]; **3**, 14[77], 15[77], 21[123], 363[84,87], 365[63,87]; **4**, 83[65a,b], 372[53,54], 665[10], 667[10], 670[10], 674[10], 1006[105]; **5**, 828[9], 830[29], 840[9], 841[9,9c,86,87,90,93,97], 842[108,111], 843[108,125], 847[9], 853[125a], 856[9], 857[9a], 859[9,87c,90,93,234,235,238,241], 863[238,254], 886[20], 893[20], 1001[12], 1123[36,37,39,41]; **6**, 464[30], 471[30,61], 750[103,104], 858[162], 859[163,164,165,166,167,168], 860[178], 978[21], 984[57]; **7**, 111[190,191], 301[62], 549[44], 565[98], 567[98], 583[44], 586[44], 711[59]; **8**, 244[69], 2 69[91], 528[58,75], 542[233,237], 544[260], 566[451], 817[28], 930[35,36], 932[44]
Irgolic, K., **7**, 774[314], 785[314]
Iriarte, A., **8**, 52[144], 66[144]
Iriarte, J., **7**, 820[23]
Irie, A., **2**, 176[185]
Irie, H., **2**, 1099[112b]; **4**, 27[79,79d,e]; **6**, 820[111]; **7**, 156[32], 175[143]
Irie, K., **2**, 150[94,95], 1066[122]; **6**, 509[258]
Irie, S., **7**, 618[22]
Irie, T., **5**, 210[56]
Iriguchi, J., **4**, 115[177]
Irikawa, H., **8**, 803[95]
Irirmetura, R. S., **7**, 93[52]
Irisawa, J., **6**, 42[45]
Irismetov, M. P., **7**, 93[52]
Iritani, K., **4**, 394[192], 567[51]
Iritani, N., **4**, 315[502]
Iriuchijima, S., **1**, 513[47], 515[57]; **3**, 147[394]; **6**, 150[110], 926[67], 927[75]
Iriye, R., **2**, 166[156]
Irmscher, K., **8**, 528[71], 971[108]
Irngartinger, H., **3**, 872[61-63]; **4**, 1041[104]
Irvine, J. L., **5**, 834[53]
Irvine, R. W., **1**, 554[102]; **7**, 380[102]
Irving, E. M., **8**, 384[35]
Irving, J. R., **8**, 409[83]
Irving, K. C., **6**, 96[152], 182[140]
Irwin, A. J., **8**, 200[136]
Irwin, W. J., **7**, 739[34]
Irwin, W. L., **8**, 201[142]
Isaac, K., **2**, 651[116]
Isaacs, L. D., **5**, 1076[44]
Isaacs, N. S., **1**, 243[55]; **3**, 733[1]; **5**, 77[253], 85[4], 379[112], 383[112], 384[112], 453[63], 454[63], 458[63]; **6**, 207[48]; **8**, 860[220], 872[2]
Isaacs, S. T., **8**, 626[167]
Isaacson, P. J., **7**, 845[76]
Isacescu, D. A., **3**, 321[137]
Isagawa, K., **4**, 1001[43]; **8**, 698[141,142,145], 709[45,45a]
Isak, H., **4**, 331[7]; **6**, 142[65], 554[802], 576[802], 581[802]
Isaka, M., **4**, 895[162]; **5**, 977[23]
Isaksson, R., **3**, 499[120]; **5**, 686[53]
Isbell, A. F., **6**, 806[55]
Ische, F., **1**, 391[153]
Ise, T., **8**, 153[187]
Iseitlin, G. M., **6**, 494[133]
Iseki, K., **7**, 455[105]
Iseki, T., **3**, 380[9]
Iselin, B., **6**, 636[25]
Iselin, M., **4**, 125[216,216a]
Isemura, S., **4**, 379[114]
Isenring, H. P., **2**, 817[89], 1101[119]
Ishag, C. Y., **4**, 48[137,137g]
Ishaq, M., **4**, 980[108]
Ishibashi, H., **2**, 556[152,153], 580[99]; **5**, 133[53,54], 439[170], 504[274]; **6**, 744[76], 746[76], 930[87], 931[88-90]; **7**, 199[31,34,35], 200[41,42], 208[86,87], 209[89,92], 211[86]; **8**, 837[14], 964[56,59]
Ishibashi, K., **2**, 617[144]; **4**, 239[19], 258[239]
Ishibashi, M., **1**, 513[47]; **3**, 147[394]; **6**, 150[110]; **8**, 432[68]
Ishibashi, Y., **5**, 597[23], 603[23], 606[23]
Ishida, A., **2**, 616[139]; **3**, 585[136]; **6**, 509[258]
Ishida, H., **2**, 967[74,75]; **3**, 224[169]; **6**, 94[144]; **7**, 209[93]
Ishida, K., **8**, 902[47], 904[47], 905[47]
Ishida, M., **1**, 766[161]; **6**, 450[119], 453[143], 454[119,146], 461[5]
Ishida, N., **1**, 123[77], 372[80]; **3**, 24[147], 200[68], 438[31], 453[31], 459[31]; **4**, 116[188a], 682[57]; **7**, 643[15], 647[34,36,38]; **8**, 788[120]
Ishida, S., **8**, 385[45]
Ishida, T., **4**, 592[127], 633[127]; **5**, 281[19], 935[191], 936[191]; **6**, 88[103]; **7**, 340[45]; **8**, 658[99]
Ishida, Y., **2**, 113[104], 926[118]; **4**, 30[88,88k,n], 121[209,209a], 239[17], 261[17,299]; **6**, 528[414], 767[26], 771[37]; **7**, 696[38], 697[49]; **8**, 945[128]
Ishido, Y., **5**, 158[43,47,48], 159[49], 185[164]; **6**, 23[91], 46[66], 49[95], 54[132]
Ishidoya, M., **7**, 607[168]
Ishige, M., **8**, 418[9]
Ishige, O., **1**, 544[39]; **8**, 302[96]
Ishiguro, H., **3**, 523[25]; **4**, 358[157]
Ishiguro, M., **1**, 165[108,109], 708[253], 790[264]; **2**, 84[47], 88[33], 91[44,46,47], 93[46,47], 96[58], 161[136]; **3**, 244[25], 267[25], 454[120]; **6**, 991[89]; **7**, 318[59], 680[76]
Ishiguro, T., **3**, 406[141], 675[73]; **4**, 249[126], 258[126]
Ishihara, H., **2**, 614[115]; **6**, 453[143], 464[34-37], 465[34-37], 467[50,51], 469[35], 473[77]; **7**, 774[318,332]
Ishihara, J., **5**, 839[76]
Ishihara, K., **2**, 655[148]; **6**, 849[123]; **8**, 223[100], 224[100], 227[120], 659[106]
Ishihara, M., **2**, 368[236]; **7**, 356[50]
Ishihara, S., **3**, 919[33], 954[33]; **6**, 897[97]
Ishihara, T., **1**, 751[109]; **2**, 115[124]; **3**, 218[98]; **4**, 1006[96], 1021[242]; **8**, 798[54]
Ishihara, Y., **1**, 117[55], 124[82], 329[32]; **2**, 4[12], 6[12], 10[12b], 573[55]; **4**, 148[50], 149[50b], 179[64], 182[64c], 184[64c]; **7**, 579[134]; **8**, 95[87,89]
Ishii, A., **3**, 571[75], 574[75], 883[110]; **6**, 475[88-90]
Ishii, H., **2**, 780[10]; **5**, 850[146,148]
Ishii, M., **2**, 585[129]; **7**, 878[144]
Ishii, T., **6**, 273[99]; **7**, 806[71]

Ishii, Y., **1**, 441[173]; **2**, 369[250]; **3**, 446[76]; **4**, 566[46], 588[56], 596[162,163], 601[247], 614[374], 621[163], 637[163]; **5**, 35[12], 487[193], 1185[1], 1186[3,3a]; **6**, 117[98]; **7**, 309[26], 314[41], 315[41,42], 708[31]; **8**, 678[63], 685[63], 686[63], 698[146]
Ishikawa, H., **1**, 131[105], 346[125-127]; **3**, 598[202]; **4**, 501[113]; **7**, 86[16a]; **8**, 967[77,79]
Ishikawa, K., **4**, 1090[142]; **6**, 66[4]; **7**, 136[111], 137[111]
Ishikawa, M., **1**, 258[25], 270[25], 275[25], 787[254]; **3**, 231[252], 844[33]; **4**, 251[146]; **5**, 152[21], 210[57], 809[117]; **7**, 173[132]; **8**, 95[86], 797[33]
Ishikawa, N., **2**, 635[40], 640[40]; **3**, 244[23], 420[50], 444[64], 452[108], 503[147]; **4**, 102[130], 128[221], 216[123], 508[157], 595[150], 1015[199]; **5**, 839[85], 841[95]; **6**, 172[15], 498[160], 556[822], 967[100]; **8**, 560[403]
Ishikawa, R., **3**, 583[120]; **8**, 889[128]
Ishikawa, S., **5**, 406[22]; **6**, 432[115]
Ishikawa, T., **5**, 850[146]; **7**, 462[119,120]
Ishikura, K., **5**, 152[21]
Ishikura, M., **3**, 498[109], 1038[95,95b]
Ishimaru, T., **2**, 826[122,123,125], 1102[121a,b], 1103[121]; **8**, 143[61], 148[108], 405[24]
Ishimoto, S., **2**, 833[147]; **4**, 159[85], 256[208,212], 261[208,284], 262[212]; **7**, 54[8]; **8**, 843[59a], 993[58]
Ishino, Y., **4**, 878[83]; **8**, 366[45]
Ishitsuka, M., **2**, 553[129]; **3**, 380[13]
Ishiwari, H., **1**, 752[98]; **3**, 652[221]
Ishiyama, J., **8**, 418[12], 422[12]
Ishiyama, K., **5**, 439[170], 504[274]; **6**, 930[87]; **7**, 199[34], 209[89]
Ishiyama, M., **2**, 152[101]
Ishiyama, T., **3**, 231[249,252], 446[89], 489[64], 490[64], 495[64], 496[64], 498[64], 511[64], 515[64]
Ishizaka, S., **6**, 531[460]; **7**, 77[120a]
Ishizaki, K., **2**, 138[19]
Ishizaki, M., **8**, 50[119], 66[119]
Ishizu, J., **6**, 86[100]
Ishizu, T., **5**, 603[53], 604[53]
Ishizuka, N., **4**, 249[128]
Ishizuka, S., **5**, 927[163]
Ishizuka, T., **1**, 119[64]; **2**, 1060[90]
Ishizumi, K., **4**, 369[19], 374[19]; **8**, 241[37]
Isida, T., **6**, 261[11]; **8**, 661[112]
Isidor, J. L., **3**, 106[223], 113[223], 125[299]; **6**, 677[316,316a]
Isied, S. S., **6**, 671[280,281]
Isihara, M., **3**, 125[297,298], 126[298], 128[297], 129[297], 130[297,298], 133[297], 137[298]
Isiyama, S., **4**, 885[111]
Iskander, G. M., **4**, 48[137,137g]; **7**, 695[36]
Iskikian, J. A., **4**, 347[95]
Islam, A. M., **5**, 753[55], 757[55], 769[55]
Islam, I., **7**, 266[106], 267[106], 276[106]
Islam, M. M., **5**, 223[80]
Islam, Q., **4**, 1019[227]
Isler, O., **2**, 410[5], 612[104,105]; **3**, 698[159]; **6**, 836[55], 965[85]
Ismagilova, G. S., **3**, 887[10], 888[10], 889[10], 890[10], 893[10], 897[10], 900[10], 903[10]
Ismail, M. F., **4**, 84[69]
Ismail, Z. M., **5**, 9[73]; **6**, 233[47]
Ismailov, R. G., **3**, 306[82,84]
Ismailzade, I. G., **3**, 304[60]
Iso, T., **7**, 760[50]
Isobe, K., **2**, 69[47]; **3**, 224[169]; **4**, 121[205c], 258[250], 520[33,34], 535[94]; **5**, 323[15]; **6**, 94[144], 525[379]; **8**, 245[72]
Isobe, M., **2**, 124[207], 381[311], 578[83]; **4**, 79[57,58a,b,59a-c], 96[104], 216[126], 251[149], 257[149], 260[149]; **5**, 350[79], 829[26], 924[147]; **6**, 27[115], 164[199], 647[114], 900[115], 1023[69]; **7**, 370[65], 380[65]
Isobe, S., **6**, 676[303]
Isobe, T., **4**, 387[161], 388[161]
Isobe, Y., **3**, 219[107]
Isoe, S., **1**, 822[36], 894[160]; **2**, 749[134]; **3**, 198[52], 470[208], 471[208], 475[208], 628[49]; **4**, 35[98i], 594[146], 759[192], 763[192]; **7**, 624[37], 625[39], 650[47-49]
Isogai, K., **8**, 881[83]
Isogami, Y., **4**, 433[123]; **5**, 79[288]; **6**, 66[8]
Isoya, T., **3**, 640[99]
Isoyama, T., **4**, 931[55]
Israel, M., **8**, 989[35]
Israel, R. J., **1**, 872[90]
Israeli, Y., **2**, 348[51], 365[211]
Isser, S. J., **3**, 20[111]
Issidorides, C. H., **6**, 209[66]; **7**, 84[3], 760[20]; **8**, 214[43]
Issleib, K., **8**, 858[208]
Istomina, Z. I., **7**, 479[89]
Itabashi, K., **6**, 443[89], 493[139,140], 508[285], 509[263], 753[115], 754[115]; **8**, 650[67]
Itabayashi, K., **8**, 323[113]
Itagaki, K., **2**, 584[121]
Itahara, T., **4**, 837[7,8,11]
Itai, A., **1**, 512[39]
Itai, J., **8**, 837[18]
Itai, T., **8**, 390[76]
Itakura, K., **6**, 603[22], 618[113], 625[158]
Itani, H., **6**, 3[10], 30[10], 835[47]
Itatani, H., **8**, 447[110,112,113]
Itaya, N., **3**, 715[39]
Itazaki, H., **8**, 332[50], 340[50]
Itkin, E. M., **4**, 347[106]
Itnizda, V., **3**, 415[7]
Ito, A., **5**, 439[171], 440[171]; **8**, 266[57]
Ito, E., **3**, 202[85]
Ito, F., **1**, 593[25], 595[25]; **2**, 1030[84]; **8**, 537[189]
Ito, H., **1**, 256[20,21]; **2**, 17[68], 24[92,94], 584[124], 716[62]; **8**, 406[40]
Ito, K., **1**, 242[45], 243[58], 317[138,139,140,141,142], 385[114]; **3**, 565[21]; **6**, 264[37], 265[37], 533[510]; **7**, 88[25], 314[40], 315[40], 356[50]; **8**, 18[126], 64[208], 65[208], 66[208], 67[208], 151[149,150,151,154], 160[100], 170[98-101], 176[135], 178[100], 394[119], 406[38]
Ito, M., **2**, 357[145], 358[145], 821[109]; **3**, 168[490], 169[490]; **5**, 720[96], 741[157]; **6**, 765[15]; **8**, 422[35]
Ito, M. M., **5**, 483[174]
Ito, R., **5**, 605[57], 611[57]; **8**, 782[105], 783[107]
Ito, S., **1**, 797[294], 876[98]; **2**, 197[81], 198[82]; **3**, 100[193,194,195,196,197,200,201], 103[193,194,195,200,201], 107[194,195,201], 390[69,70], 395[101]; **4**, 38[109a], 231[280], 406[253], 408[253]; **5**, 618[2], 620[14], 621[17,19], 622[22], 623[25-28], 829[20]; **6**, 89[118], 101[118], 145[80], 501[193], 900[116,117], 1045[21,22]; **7**, 95[71], 299[48], 311[29], 489[165], 503[271]; **8**, 191[88], 856[171]
Ito, T., **1**, 176[17,18]; **4**, 601[244], 643[244]; **7**, 168[101]; **8**, 459[244], 460[254], 535[166]
Ito, W., **1**, 121[66]; **2**, 11[49], 32[120,120a], 95[55], 979[15], 983[15], 984[15], 985[15], 986[15,31], 987[15,31], 992[15,37], 993[37], 995[46]; **4**, 388[162]
Ito, Y., **1**, 113[30], 223[75], 319[159], 320[158,159], 544[44], 546[49], 551[77], 624[85], 880[113]; **2**, 17[67], 23[90,90b], 29[90b], 38[67], 59[16], 123[197,198], 233[186], 292[77], 317[48,49], 318[48,49], 444[21,22], 455[19], 572[45], 716[63], 846[205], 1059[76]; **3**, 17[95], 45[248], 225[184], 227[210], 229[228,229], 245[30], 262[165], 450[105], 463[167], 529[52], 555[28], 623[39]; **4**, 162[93,94a,b], 221[160], 231[260], 241[60], 254[60], 260[60], 600[241], 610[339], 611[354], 613[370], 614[372,373], 626[339], 649[339], 653[439,440,441], 682[57], 840[35-37], 905[207]; **5**, 98[123,124], 282[21,22], 305[87], 386[134],

391[134], 392[134], 473[151], 479[151], 595[12], 693[112], 841[92], 850[147], 1132[19], 1183[55]; **6**, 17[63], 18[63], 46[65], 51[108], 53[108,120], 60[148], 88[104], 238[72], 295[251], 471[62], 533[509,512], 540[509], 546[509], 551[680], 717[108], 757[132], 842[83,84], 1007[149]; **7**, 141[133], 144[133], 237[34], 402[63], 452[44], 474[42], 530[29], 643[14], 645[19-21], 684[93a]; **8**, 11[61,62], 461[262], 783[107], 788[121], 830[88]

Ito, Z. I., **6**, 489[93]

Itoh, A., **1**, 101[93,94], 103[93,94]; **2**, 269[73], 600[27]; **3**, 349[33], 356[58], 358[33]; **4**, 33[96,96c], 257[227], 261[227], 795[84]; **7**, 615[8]; **8**, 986[16]

Itoh, F., **1**, 391[152]; **2**, 611[102], 643[79], 644[79a], 647[88a,b]; **4**, 161[91]; **6**, 935[104]

Itoh, H., **6**, 554[720]

Itoh, I., **3**, 572[62]

Itoh, K., **1**, 714[267], 717[267]; **2**, 495[63], 496[63], 576[76], 580[101], 584[121]; **3**, 262[164], 446[76], 466[191]; **4**, 93[94], 107[141], 238[12], 254[181], 255[194], 794[74]; **6**, 46[66], 541[592], 774[49,50], 787[101,102], 937[121], 1066[96,98]; **7**, 262[80], 628[44-46], 649[42], 701[64]; **8**, 98[99], 189[63], 291[32], 563[427]

Itoh, M., **2**, 112[88], 241[14], 417[19]; **3**, 221[134], 274[21], 421[53]; **4**, 145[23,26,29a], 148[44], 164[99,99a], 784[16]; **5**, 926[159]; **6**, 453[143], 637[29]; **7**, 603[108-111], 607[162]; **8**, 101[120], 386[51], 881[74]

Itoh, N., **2**, 953[3b]; **7**, 539[67]; **8**, 64[216], 67[216], 170[94], 250[101], 253[117], 826[70]

Itoh, O., **4**, 302[337]; **6**, 924[61,63], 925[64], 926[65]; **7**, 197[20,25]

Itoh, S., **1**, 101[94], 103[94]

Itoh, T., **1**, 78[18], 79[21], 80[21], 81[21], 82[21], 100[18], 221[68], 283[3], 316[3], 333[60,61], 335[60,61], 336[73], 418[75]; **2**, 30[113], 31[113], 685[67]; **3**, 124[259], 125[259], 153[416], 154[416], 463[158], 470[212], 476[212]; **5**, 434[147], 497[225]; **6**, 711[64]; **7**, 743[64], 808[80], 823[36]; **8**, 187[40,44], 188[44], 190[77], 195[109], 196[77], 203[146]

Itoh, Y., **5**, 1157[169]; **6**, 923[60]

Itokawa, M., **4**, 115[178]

Iton, T., **3**, 286[56a]

Itooka, T., **3**, 853[72]

Itsuno, S., **1**, 223[76,76b], 317[138,139,140,141,142,149], 319[149], 320[149], 385[114]; **8**, 18[126], 64[208], 65[208], 66[208], 67[208], 160[100], 170[82,84,96-101], 176[135], 178[100], 394[119]

Ittah, Y., **6**, 93[132]; **7**, 475[50], 476[50]

Ittel, S., **1**, 301[74], 316[74]; **4**, 691[74]

Ittyerah, P. I., **2**, 149[88]

Iuchi, Y., **3**, 168[508], 169[508]

Ivanenko, T. L., **8**, 518[130]

Ivanics, J., **2**, 381[305]

Ivanov, C., **2**, 495[61]; **6**, 269[71,72]; **8**, 492[17], 974[123]

Ivanov, D., **2**, 210[110]

Ivanov, J., **6**, 519[336]

Ivanov, K. I., **7**, 10[81]

Ivanov, L. L., **8**, 735[14], 747[14], 748[14], 756[143]

Ivanov, P. Yu., **6**, 487[74], 489[74]

Ivanov, S. K., **7**, 95[73a]

Ivanova, Zh. M., **6**, 550[679]

Ivanyk, G. D., **6**, 216[107]

Ivein, K. J., **5**, 1116[5]

Iversen, P. E., **8**, 285[4], 293[4], 294[4]

Iversen, T., **6**, 23[92], 533[500], 550[500], 651[135], 652[142]

Ives, D. A. J., **8**, 330[45,46], 340[45], 342[46]

Ives, J. L., **7**, 96[88], 97[88], 98[88], 110[88], 111[88], 816[10], 818[10]

Ivin, F. J., **5**, 1115[1], 1116[1]

Ivin, K. J., **5**, 1116[5], 1120[22]; **7**, 14[137]

Iwaake, N., **2**, 651[114]

Iwabuchi, R., **5**, 621[19]

Iwabuchi, Y., **4**, 387[158,158b]

Iwahara, T., **6**, 16[60]; **7**, 642[8], 645[18]; **8**, 787[119]

Iwai, I., **1**, 368[59], 369[59]

Iwai, K., **4**, 231[266,267]; **6**, 1022[58]

Iwai, S., **6**, 604[34], 626[167]

Iwaki, S., **4**, 817[207]; **6**, 1065[90b]; **7**, 615[9], 624[36]

Iwakiri, H., **2**, 615[123], 633[30], 635[30], 640[30]; **6**, 498[160]

Iwakuma, T., **2**, 953[3b]; **8**, 64[216], 67[216], 170[94], 176[132,133,134], 250[101], 253[117], 826[70]

Iwakura, C., **3**, 635[44]

Iwakura, H., **2**, 1048[12]

Iwakura, Y., **5**, 949[281]

Iwama, K., **2**, 367[222]

Iwamatsu, K., **6**, 524[361]

Iwami, F., **2**, 790[57]

Iwami, H., **6**, 619[115]

Iwamoto, H., **4**, 76[44c], 249[117], 257[117]

Iwamoto, M., **4**, 1014[185]

Iwamoto, N., **4**, 588[55]; **8**, 865[249]

Iwamura, H., **5**, 209[54], 210[58]; **7**, 771[260], 772[260], 779[260]

Iwamura, M., **5**, 210[58]

Iwanaga, K., **2**, 103[29]; **4**, 976[97]; **5**, 355[87b], 377[111,111a]

Iwano, Y., **1**, 642[121], 646[121], 656[121,153], 658[121], 665[121], 667[121], 672[121]; **2**, 1060[89]

Iwanowicz, E. J., **1**, 3[24], 32[157,158]; **2**, 189[47]; **4**, 48[139,139f]; **6**, 717[113]

Iwao, A., **1**, 520[71]

Iwao, J., **7**, 760[50]

Iwao, M., **1**, 466[48,49], 474[105]; **7**, 333[22]

Iwaoka, T., **8**, 820[45], 822[52]

Iwasa, A., **6**, 685[363]; **8**, 885[106], 886[106]

Iwasaki, F., **7**, 808[79,80]

Iwasaki, G., **2**, 113[104], 271[76], 348[63], 920[93], 921[93], 922[93], 923[93]; **4**, 429[84,85]; **5**, 101[162]

Iwasaki, H., **2**, 907[59]

Iwasaki, M., **1**, 860[69,70]

Iwasaki, S., **7**, 385[118], 400[51]

Iwasaki, T., **2**, 1051[41]; **3**, 650[210c,212], 651[210c,216]; **7**, 806[74]

Iwasawa, H., **4**, 401[233]; **6**, 751[108]

Iwasawa, N., **1**, 834[121b]; **2**, 116[127,-128,129,130,132,133,134,139], 233[185], 424[37], 425[37], 436[68], 437[68], 610[90,92,93], 611[92], 633[32], 640[32]; **4**, 85[70], 202[25], 230[256,257]; **5**, 15[102], 377[110], 378[110a], 543[114]; **6**, 26[108]; **8**, 216[60]

Iwasawa, Y., **8**, 150[129], 151[129], 418[16], 421[16]

Iwase, R., **6**, 614[97]

Iwase, S., **5**, 812[132]

Iwashashi, H., **7**, 477[75]

Iwashita, M., **4**, 406[253], 408[253]; **7**, 503[271]

Iwashita, T., **3**, 135[341], 136[341], 137[341]

Iwata, C., **1**, 753[103]; **3**, 677[81], 686[81]; **4**, 1040[89,90], 1045[89,90], 1056[141,141b]; **6**, 1046[31]; **7**, 178[148], 455[104], 544[35], 550[51], 556[35], 566[35], 821[29]; **8**, 274[130], 837[15a]

Iwata, K., **6**, 614[86]

Iwata, N., **3**, 483[9]; **6**, 927[80]

Iwata, S., **1**, 287[18]

Iwata, T., **4**, 610[339], 626[339], 649[339]

Iwatsubo, H., **4**, 102[130], 216[123]

Iwayama, A., **7**, 451[19], 452[19], 454[19]

Iwayanagi, T., **8**, 850[123]

Iwema Bakker, W. I., **2**, 1079[156]

Iyengar, R., **5**, 3[27], 320[5], 347[5]

Iyer, K. N., **7**, 586[165]

Iyer, P. S., **3**, 298[28], 329[185]; **5**, 552[34]; **6**, 251[146]; **7**, 674[39]; **8**, 216[57]

Iyer, R. S., **3**, 878[94], 879[94], 880[94,98], 881[94]; **4**, 359[159], 771[251], 898[174], 899[174], 985[125]; **5**, 1086[73]

Iyer, S., **4**, 602[263], 644[263]; **5**, 47[40], 689[77], 690[77a], 733[136,136b], 734[136b], 1202[56]
Iyobe, A., **6**, 150[129]
Iyoda, J., **3**, 381[21]
Iyoda, M., **3**, 421[60]
Izatt, R. M., **6**, 71[21], 449[116]
Izawa, K., **2**, 1066[119]; **4**, 276[79]
Izawa, M., **7**, 168[101]
Izawa, T., **2**, 213[123], 612[108], 613[109], 681[57]; **4**, 382[133], 388[133]; **8**, 272[117,118]
Izawa, Y., **3**, 891[45]; **4**, 960[34]; **6**, 66[4], 240[79]; **7**, 9[69]; **8**, 917[116,117], 920[116,117]
Izdebski, J., **5**, 433[139]
Izmailov, B. A., **8**, 765[11]
Izukawa, H., **7**, 618[22]
Izumi, M., **1**, 512[39]
Izumi, T., **3**, 530[81-83], 536[81-83], 594[185]; **4**, 557[12], 558[17], 611[352], 837[9], 839[27], 845[69], 847[76], 858[110], 903[202], 904[202]; **8**, 18[128], 245[75]
Izumi, Y., **2**, 232[178], 310[31], 311[31], 576[75], 587[136], 615[124,125,130], 630[23], 631[12,14,23], 635[14,44], 640[44], 655[139]; **4**, 161[88]; **6**, 89[108], 93[130], 237[66], 254[160]; **7**, 539[67]; **8**, 149[117-119], 150[121,122,126,130,131,133,135,144], 151[121,133,135,145,148,151], 533[150], 786[118], 789[123]
Izumisawa, Y., **7**, 92[42], 93[42]
Izumiya, N., **2**, 1094[88]; **6**, 636[20]; **8**, 145[88]
Izzat, A. R., **3**, 664[31]

J

Jablonski, C. R., **6**, 692[405]
Jabri, M., **4**, 183[81]
Jabri, N., **1**, 107[3], 428[118]; **3**, 217[84], 226[203], 485[32-35], 486[32-35], 494[34,35,89], 516[89]; **4**, 903[188,189,199]
Jachiet, D., **1**, 343[115,117]; **3**, 224[166,167], 225[167b]; **4**, 903[192]; **6**, 5[23]
Jack, D., **2**, 323[33]
Jack, T., **4**, 601[250]; **5**, 35[12]
Jackisch, J., **7**, 746[82]
Jackman, D., **8**, 581[9]
Jackman, L. M., **1**, 2[4], 3[26], 4[27], 22[115], 41[201], 43[4,26]; **2**, 100[10-12], 345[38]; **3**, 3[15], 4[23], 54[15]; **5**, 855[182]; **6**, 226[10], 256[10], 257[10]; **7**, 135[103], 306[9]
Jacknow, B. B., **7**, 17[169]
Jackson, A. C., **2**, 533[29], 538[51,52], 542[52], 547[52], 709[8]
Jackson, A. E., **6**, 635[14a], 636[14]; **8**, 368[63], 958[20]
Jackson, A. H., **3**, 680[93], 807[28,30]; **4**, 1004[76], 1021[76]; **6**, 737[33-36]; **7**, 846[83]; **8**, 315[51,55], 316[55]
Jackson, B. G., **6**, 936[105]; **7**, 205[61]; **8**, 531[113]
Jackson, B. L. J., **3**, 743[60]
Jackson, C. B., **3**, 407[147], 591[169]
Jackson, D. A., **5**, 689[73], 770[137,138,139]; **7**, 352[33]
Jackson, D. K., **5**, 1022[76]
Jackson, D. Y., **5**, 855[192]
Jackson, E. L., **6**, 36[29]
Jackson, H. L., **6**, 227[15], 242[15]
Jackson, J. L., **4**, 128[221]
Jackson, L. M., **7**, 135[102]
Jackson, P. F., **5**, 946[257]; **7**, 567[104]
Jackson, R. A., **3**, 564[15]; **7**, 721[16]; **8**, 513[101,103], 774[74], 825[66]
Jackson, R. F. W., **1**, 449[209]; **2**, 645[82]; **3**, 226[204]; **4**, 370[36]; **6**, 650[133b], 668[261]
Jackson, R. W., **5**, 157[41]; **7**, 86[16a]
Jackson, T. E., **3**, 23[134]
Jackson, W., **1**, 821[24], 825[24]; **3**, 168[491], 169[491], 171[491], 757[122]
Jackson, W. P., **1**, 436[152]; **2**, 240[13], 256[13], 257[13b]; **3**, 342[11], 351[44], 352[44]; **4**, 254[188], 261[188], 391[176]; **5**, 21[150], 22[150]; **8**, 849[113]
Jackson, W. R., **1**, 520[69,70], 635[83], 678[83], 681[83], 691[83]; **4**, 12[37,37b,c], 23[70], 538[103], 588[59], 601[268], 603[268], 609[327,328], 614[327], 615[327,391], 629[327,391,403,404]; **7**, 107[161]; **8**, 533[154], 536[173], 538[173], 542[173], 852[140], 856[162]
Jacob, G. S., **6**, 74[31]
Jacob, L., **6**, 624[145]
Jacob, P., III, **8**, 101[120], 716[90]
Jacob, P. W., **5**, 829[19]
Jacob, T. A., **7**, 92[48]
Jacob, T. M., **6**, 612[71]
Jacober, W. J., **3**, 299[33]
Jacobi, E., **2**, 943[169], 970[88]
Jacobi, P. A., **1**, 406[29]; **5**, 491[208], 494[215,216,217], 495[218], 579[162,163,164,165]; **8**, 540[195]
Jacobs, H. J. C., **5**, 707[40], 708[41]; **7**, 12[101]
Jacobs, I., **3**, 629[51]
Jacobs, J. W., **5**, 855[192]; **8**, 206[167]
Jacobs, P., **4**, 146[37c], 147[38a]
Jacobs, P. B., **5**, 841[89], 872[89b]; **8**, 392[106], 856[173]
Jacobs, S. A., **1**, 466[46,47]
Jacobs, T. L., **3**, 273[11]; **4**, 276[73,76], 277[73,76,83], 284[76], 285[76,83], 299[299], 303[344], 308[402], 395[202,205]; **7**, 506[295]
Jacobsen, E. J., **1**, 889[140,141], 890[140,141,144], 898[141b,144]; **2**, 1041[107,110-112]; **4**, 202[22], 1088[122]; **5**, 41[30], 800[77], 822[77]; **6**, 735[19], 740[19], 741[65,66]; **8**, 34[61], 53[133], 66[61,133]
Jacobsen, E. N., **7**, 428[148g], 429[158], 430[158,159], 442[46a,b], 489[165]
Jacobsen, G., **2**, 1023[54]
Jacobsen, G. E., **1**, 17[219], 36[233,234]
Jacobsen, J. P., **5**, 531[74], 532[74a]
Jacobsen, N., **3**, 642[111]
Jacobsen, P., **6**, 570[945]; **8**, 604[8], 605[8]
Jacobsen, R. R., **5**, 856[217]
Jacobsen, W. N., **7**, 229[119]
Jacobsen-Bauer, A., **2**, 655[142]
Jacobson, B. M., **5**, 3[22], 70[110]
Jacobson, E. C., **8**, 274[134]
Jacobson, H. W., **8**, 814[20]
Jacobson, J. L., **5**, 226[109]
Jacobson, R. A., **4**, 315[517]
Jacobson, R. M., **1**, 542[8], 543[8], 544[8,36], 547[8,36], 548[8,36], 550[8], 552[8], 553[8], 555[8,8b], 556[8], 557[8], 560[8]; **2**, 61[20], 69[46], 711[37]; **4**, 12[42]; **5**, 766[118], 769[136]; **6**, 648[121]
Jacobson, S. E., **7**, 674[51]
Jacobsson, U., **5**, 433[139]
Jacobus, J., **4**, 738[95]; **5**, 890[34]; **6**, 152[141], 899[109]; **8**, 726[186]
Jacot-Guillarmod, A., **1**, 155[65]
Jacquasy, J. C., **1**, 567[222]
Jacquemin, W., **3**, 825[24,24a], 828[24a], 835[24]
Jacques, J., **1**, 59[35]; **2**, 289[70], 291[70], 292[78]; **8**, 162[31]
Jacquesy, J.-C., **3**, 810[50]; **7**, 333[20]
Jacquesy, R., **3**, 810[50]; **6**, 216[109]
Jacquet, I., **8**, 166[51,52], 178[52], 179[52]
Jacquet, J.-P., **2**, 816[84], 828[84]
Jacquier, R., **3**, 46[254], 215[66], 251[75], 813[64], 839[4]; **4**, 55[157]; **5**, 766[115]; **6**, 262[17], 267[57]; **8**, 270[95], 662[115]
Jacquignon, P., **2**, 149[85]; **6**, 462[13]
Jacquin, D., **8**, 137[53]
Jacquinet, J.-C., **6**, 41[44], 43[50]
Jacquot, R., **3**, 324[155]
Jacyno, J., **6**, 804[48]
Jadach, T., **6**, 824[120]
Jadhav, K. P., **3**, 104[211], 107[211], 111[211]
Jadhav, K. S., **5**, 775[177], 780[177]
Jadhav, P. K., **2**, 6[23,23c], 10[23], 33[23c,123], 52[23a]; **3**, 600[214]; **7**, 595[12,127], 604[127]; **8**, 334[61], 338[94], 541[207], 710[59], 720[134], 721[134,139,141,142,143], 722[134,142,144,145,147]
Jadodzimski, J. J., **2**, 578[85]
Jadot, J., **7**, 13[109]
Jaeger, D. A., **8**, 532[133], 863[232]
Jaeger, E., **4**, 430[98]
Jaeger, R. H., **8**, 532[128]
Jaen, J. C., **3**, 226[194], 264[183], 265[190]; **7**, 564[94], 566[94]
Jaenicke, L., **5**, 563[90,91], 973[15], 975[15]; **8**, 798[52]
Jaffe, E. K., **5**, 855[185]
Jaffer, H. J., **5**, 1043[23], 1051[23]
Jagadale, M. H., **8**, 537[179]
Jagdmann, E., Jr., **8**, 392[100], 905[61]
Jagdmann, G. E., Jr., **1**, 461[10,11], 473[11], 478[10,11]; **3**, 693[141]; **7**, 336[34]
Jäger, E., **6**, 637[31]
Jager, H. J., **7**, 12[104]

Jäger, K. F., **4**, 791[52]
Jäger, V., **2**, 338[78]; **3**, 271[2], 272[2], 390[82], 392[82]; **4**, 299[302], 303[349], 379[117], 380[122], 1076[46], 1079[64,65]; **5**, 260[65,66], 261[65,66], 451[12]; **6**, 962[75], 964[84]; **7**, 374[77d], 416[122], 439[36]; **8**, 69[221], 70[221,222,223], 647[53-55]
Jaggi, D., **1**, 52[16], 134[114], 135[114]; **2**, 625[162], 631[17], 632[17], 634[17]; **5**, 157[38]
Jagner, S., **4**, 227[210], 532[88], 534[88], 537[88], 538[88], 539[88]
Jagt, J. C., **5**, 416[56]
Jagusztyn-Grochowska, M., **4**, 429[86]
Jahangir, **1**, 366[49], 391[49,149]; **2**, 913[74-76], 928[124], 929[125]; **8**, 36[98], 42[98], 66[98]
Jahn, E. P., **6**, 822[116]
Jahngen, E., **6**, 921[48]
Jähnisch, K., **7**, 470[15]
Jahnke, D., **8**, 756[156,157]
Jahreis, G., **4**, 53[149]
Jaime, C., **6**, 4[19]
Jain, A., **3**, 640[106]
Jain, A. U., **2**, 523[75]; **4**, 113[170]
Jain, A. V., **1**, 545[47]
Jain, P. C., **6**, 538[552], 550[552]; **8**, 654[84]
Jain, T. C., **3**, 390[75,76,78,79], 392[75,76,78]
Jain, V., **4**, 505[144]
Jaisli, F., **6**, 1056[56,57]
Jakiela, D. J., **4**, 212[99]
Jakob, P., **2**, 1090[73], 1102[73], 1103[73]
Jakob, R., **4**, 1022[253]
Jakobsen, H. J., **6**, 462[16]; **7**, 330[8]
Jakobsen, P., **6**, 547[657], 570[943]; **7**, 95[80]
Jakopdid, K., **6**, 510[298]
Jakovac, I. J., **7**, 316[46,48], 317[46,48], 318[48]; **8**, 188[55], 196[55], 199[55], 201[55]
Jakovljevic, M., **3**, 380[13]
Jakubke, H.-D., **6**, 635[11], 645[11], 665[11c], 667[11c], 668[11c,254], 669[11c,254]
Jakubowski, A. A., **7**, 258[54]
Jakubrova, J., **6**, 524[368]
Jakupovic, J., **8**, 339[91]
Jalali-Araghi, K., **8**, 399[148]
Jalander, L., **4**, 313[461]
Jallageas, J.-C., **2**, 851[220]
Jamali, F., **7**, 79[128b]
Jamart-Gregoire, B., **4**, 250[138]; **5**, 692[100]
James, A. P., **8**, 449[152]
James, B. D., **8**, 541[207]
James, B. G., **3**, 587[148]
James, B. R., **4**, 1039[62]; **7**, 108[174]; **8**, 91[55], 152[171], 443[1], 446[69], 533[134], 597[93], 600[93]
James, D., **2**, 536[44]; **5**, 424[96]
James, D. E., **4**, 946[92,93], 947[93]
James, D. R., **8**, 884[100], 926[13]
James, D. S., **4**, 52[147,147b]
James, F. G., **8**, 383[13]
James, F. L., **3**, 826[37]
James, K., **1**, 449[209]; **6**, 43[47], 633[9]
James, L. B., **6**, 644[94], 645[94b]
James, N., **5**, 560[74]
James, R., **3**, 681[97]
James, R. B., **6**, 530[417]
James, S. P., **5**, 152[22]
James, S. R., **8**, 794[3]
Jamison, J. D., **7**, 664[66]; **8**, 319[77]
Jamison, M. M., **8**, 144[77]
Jamison, W. C. L., **8**, 63[198], 64[198], 66[198], 67[198], 69[198]
Janairo, G., **6**, 91[129]
Janda, K. D., **4**, 495[86]
Janda, L., **6**, 524[368]
Janda, M., **8**, 274[138]
Jander, G., **8**, 842[43]
Jander, J., **7**, 741[47]
Jane, D. E., **5**, 151[10]
Janedková, E., **8**, 583[13]
Janes, J. M., **7**, 500[243]
Janes, N. F., **1**, 546[50]; **3**, 831[59]
Jang, Y. M., **3**, 49[261]
Janiga, E. R., **3**, 88[136], 91[136], 179[136,548], 181[136]
Janik, T. S., **8**, 447[96]
Jankovic, J., **8**, 872[5], 875[5]
Jankowska, J., **6**, 614[95]
Jankowski, K., **1**, 314[123]; **2**, 662[20], 663[20], 664[20], 685[66]; **4**, 339[44]; **5**, 432[128,130]
Janks, C. M., **1**, 294[44,45]
Jano, P., **4**, 1033[26], 1046[26b]
Jánossy, L., **8**, 227[115]
Janot, M.-M., **6**, 920[45]; **7**, 222[36]
Janousek, Z., **3**, 890[30], 897[30]; **4**, 758[189,190,191]; **5**, 70[111-113]; **6**, 429[87], 495[142,143], 496[143,156], 497[143], 506[226], 514[156], 521[344]
Janout, V., **4**, 304[352]; **6**, 685[353]; **8**, 806[124]
Janowicz, A. H., **7**, 3[12], 8[12]
Jans, A. W. H., **5**, 649[22], 650[22]
Jänsch, H.-J., **2**, 850[218]
Jansen, A. B. A., **6**, 667[236]
Jansen, B. J. M., **1**, 570[266,267,268]; **2**, 835[157], 838[170]; **6**, 1023[70]
Jansen, E. F., **4**, 288[186]
Jansen, G., **6**, 268[69], 271[69]
Jansen, J. F. G. A., **1**, 125[85], 218[52]; **4**, 97[106], 229[240]
Jansen, J. R., **3**, 505[169], 677[82]; **6**, 738[52,54,56]
Jansen, R. H. A. M., **6**, 489[94]
Jansen, U., **5**, 151[19]
Jansons, E., **6**, 421[28], 424[28], 436[29], 453[29], 455[29]
Jansse, P. L., **6**, 662[214]
Janssen, C. G. M., **3**, 158[445], 159[445]; **8**, 967[82]
Janssen, E., **4**, 874[52,54]; **5**, 30[2,2b]
Janssen, H. H., **6**, 161[180]
Janssen, J., **3**, 587[142]; **7**, 742[58]
Janssen, J. W. A. M., **6**, 1025[79]
Janssen, M. J., **3**, 862[5], 866[5]; **6**, 420[15], 436[24], 437[24]
Janssen, P. A., **4**, 932[63]
Janssen, R., **8**, 737[25]
Janssens, F., **2**, 723[100]
Jansson, A. M., **2**, 465[107]
Januszkiewicz, K., **4**, 553[4,8]; **7**, 451[17,30,31], 462[17]; **8**, 449[153], 452[153], 454[201], 455[201]
Janzen, E. G., **7**, 884[189]
Jaouen, G., **4**, 519[19], 520[19,31], 522[19]; **6**, 286[178,179], 287[179,180]; **8**, 5[27], 14[27], 185[27,28], 187[28], 451[180]
Jaouhari, R., **3**, 1046[3]; **7**, 7[43]
Japp, F. R., **2**, 142[46], 146[69]; **3**, 828[45]; **5**, 753[54]; **6**, 261[5], 275[5], 276[114]
Jaques, B., **4**, 485[27], 502[119], 503[27]
Jaquier, R., **8**, 636[7]
Jaquinet, J.-C., **6**, 54[129]
Jardim-Barreto, V. M., **6**, 436[18]
Jardine, F. H., **8**, 152[165,166,167], 443[1,2,5], 444[5b], 445[1g,5,56], 449[5a], 452[5b,184], 456[5a], 568[478]
Jardine, I., **8**, 445[44,45,59], 452[44,185], 456[207], 458[220], 533[154]
Jardine, P. D. S., **7**, 821[30]; **8**, 171[105]
Jarecki, C., **1**, 733[10]
Jarman, M., **4**, 231[274], 439[158]

Jarosz, S., **2**, 694[77]; **5**, 169[107,108], 170[110], 433[139], 434[139i]
Jarrar, A., **7**, 760[20]
Jarret, R. M., **4**, 871[33]
Jarrett, A. D., **6**, 707[41]
Jarupan, P., **6**, 1022[59]
Jarvi, E. T., **2**, 195[72,72b,73], 205[102], 206[102b]; **3**, 557[40,41]; **5**, 839[79], 841[95], 843[79]
Jarvie, A. W. P., **5**, 761[94]
Jarvinen, G., **7**, 15[144]
Jarvis, B., **3**, 736[25]
Jarvis, B. B., **3**, 422[67]
Jarvis, J. A. J., **5**, 514[7]
Jarvis, T. C., **2**, 965[65]
Jason, M. E., **5**, 66[75,76]
Jasor, Y., **2**, 902[47], 903[47]
Jasperse, C., **4**, 733[78], 799[117], 819[210]
Jasperse, C. P., **1**, 248[67]; **5**, 841[100]
Jasserand, D., **4**, 301[323], 302[323,334]
Jastrzebski, J. T. B. H., **1**, 25[126], 28[140], 30[154], 41[269]; **2**, 296[86], 922[99,100], 923[99], 936[99,100]; **3**, 210[25,26], 211[32]; **4**, 170[12]; **5**, 101[161]
Jaudon, P., **8**, 943[120]
Jaun, B., **7**, 805[66]
Jaunin, R., **3**, 581[109]
Jaurand, G., **6**, 60[147]; **8**, 849[111]
Jautelat, M., **1**, 213[17], 722[280]; **3**, 88[125], 118[125], 123[125], 178[541,545], 181[545]; **4**, 989[141]; **6**, 139[46]
Javeri, S., **6**, 1042[7]
Jaw, B.-R., **1**, 72[74]
Jaw, J. Y., **4**, 74[40c]; **6**, 190[192], 466[44], 469[44]; **8**, 932[40]
Jawanda, G. S., **7**, 271[129]
Jawdosiuk, M., **2**, 1017[31]; **3**, 127[321], 380[7]; **4**, 429[86]; **7**, 544[39], 553[39], 556[39]
Jaworska-Sobiesiak, A., **4**, 73[35]
Jaworski, A., **7**, 80[142]
Jaworski, K., **2**, 579[92]
Jaxa-Chamiec, A. A., **2**, 350[77]
Jay, E., **6**, 603[24], 625[156]
Jay, J., **5**, 164[76]
Jayaram, C., **1**, 464[36], 477[140], 478[36]
Jayasinghe, L. R., **7**, 399[39]
Jayathirtha Rao, V., **4**, 391[175d]
Jayne, H. W., **2**, 401[27]
Jaynes, B. H., **4**, 40[113], 53[113]; **5**, 531[75], 549[75]; **7**, 257[48], 376[81]
Jean, E., **8**, 61[187], 66[187]
Jean, M., **3**, 443[57]; **5**, 990[46], 991[46]
Jeanloz, R. W., **6**, 41[44], 646[102]; **8**, 244[52]
Jeannin, Y., **4**, 528[71], 982[112]; **5**, 1086[68], 1103[151]; **8**, 847[88,88d]
Jebaratnam, D. J., **5**, 16[113]
Jecko, G., **2**, 765[81]
Jedham, S., **1**, 766[152]
Jedlinski, Z., **2**, 105[41]
Jefferies, P. R., **3**, 818[98]; **7**, 64[62], 254[27]
Jeffery, B. A., **3**, 699[162]
Jeffery, E., **1**, 77[1]
Jeffery, E. A., **1**, 95[77]; **2**, 114[118,119], 268[67,68], 531[26], 545[26]; **4**, 140[9], 257[221], 887[120]
Jeffery, T., **3**, 544[119]; **4**, 852[88]
Jeffery-Luong, T., **3**, 491[71,72], 531[85]
Jefford, C. W., **1**, 52[16], 134[114], 135[114]; **2**, 625[162], 631[17], 632[17], 634[17]; **3**, 1057[37]; **4**, 1002[53,54], 1007[108], 1061[165,167]; **5**, 157[38,39], 560[70], 680[21], 1031[97]; **7**, 98[97], 165[85], 169[111], 313[35]; **8**, 540[194], 807[112], 965[67]
Jeffrey, D., **5**, 736[142f]
Jeffrey, D. A., **3**, 25[151]; **8**, 843[54], 847[54]
Jeffrey, E. A., **8**, 671[15]
Jeffrey, P. D., **2**, 1102[123]; **6**, 641[59], 659[59], 670[59]
Jeffrey-Luong, T., **3**, 218[96]
Jeffreys, E., **6**, 801[32]
Jeffries, P. M., **8**, 371[114]
Jeffries, P. R., **5**, 144[104]; **7**, 154[14]
Jeffs, P. W., **3**, 380[11], 507[174], 677[85]; **6**, 1042[6]; **7**, 691[20]
Jefson, M., **6**, 1030[108]; **7**, 131[81]
Jeganathan, A., **3**, 247[47], 253[93]; **4**, 1033[23]
Jeganathan, S., **1**, 757[119]
Jeger, O., **3**, 341[2], 360[2], 815[74]; **5**, 218[30], 221[30], 222[63], 229[122,123], 929[169], 930[169]; **7**, 236[24]; **8**, 108[2], 118[2], 293[48], 336[74], 340[74], 528[68], 530[68,89]
Jegou, E., **4**, 36[102,102c]
Jeker, N., **6**, 1062[80]
Jelenick, M. S., **3**, 380[13]
Jelich, K., **6**, 960[57]
Jelinski, L. W., **4**, 107[143c]
Jellal, A., **2**, 85[18]; **4**, 155[71a], 394[189], 396[189d]
Jellinek, F., **1**, 159[76,77]; **2**, 5[19], 6[19]; **8**, 696[123]
Jemilev, U. M., **7**, 750[129]
Jeminet, G., **1**, 410[40]
Jemison, R. W., **3**, 916[16], 949[94]; **6**, 897[96]
Jempty, T. C., **3**, 669[52], 683[52]; **7**, 801[37], 843[43,44]
Jen, K. Y., **7**, 769[233]
Jenck, J., **8**, 535[166]
Jencks, W. P., **2**, 955[9]; **5**, 856[202]; **6**, 438[53], 726[187]
Jendralla, H., **6**, 960[57]
Jendrzejewski, S., **1**, 373[83]; **6**, 745[79]
Jenevein, R. M., **3**, 564[10], 568[43]
Jenker, H., **8**, 754[110]
Jenkins, C. L., **7**, 725[34]
Jenkins, G., **5**, 639[123]
Jenkins, I. D., **4**, 670[21]; **6**, 27[114]
Jenkins, I. H., **3**, 514[211]
Jenkins, J. A., **5**, 639[124], 805[99], 1025[81,81e]; **7**, 144[156]
Jenkins, J. W., **8**, 265[51]
Jenkins, K. F., **8**, 398[145]
Jenkins, P. A., **8**, 477[33]
Jenkins, P. R., **1**, 391[154], 392[154]; **2**, 587[138]; **5**, 531[75], 549[75], 836[64]; **6**, 831[7]
Jenkins, R., Jr., **1**, 838[158]
Jenkins, R. H., Jr., **1**, 837[148], 838[148]; **6**, 1024[78]; **7**, 162[60,64,66,67], 176[67], 778[398]
Jenkins, R. W., Jr., **5**, 960[321]
Jenkins, W. L., **6**, 723[149]
Jenkitkasemwong, Y., **5**, 578[149]; **6**, 690[388]
Jenkner, H., **8**, 754[105]
Jenner, E. L., **8**, 447[108]
Jenner, G., **5**, 552[19]
Jenneskens, L. W., **4**, 1002[59], 1018[226]
Jenni, J., **6**, 430[102]
Jennings, K. F., **3**, 846[41]
Jennings, M. N., **6**, 554[775]
Jennings, W. B., **1**, 837[151,152]; **4**, 356[135]
Jennings-White, C., **5**, 841[97]
Jenny, C., **5**, 92[78]
Jenny, E. F., **3**, 927[55]; **5**, 732[132,132a]
Jenny, T., **6**, 175[67]
Jenny, W., **3**, 414[3]; **7**, 770[247]
Jensen, B. L., **7**, 268[123]; **8**, 20[136]
Jensen, C. B., **3**, 901[102]
Jensen, F., **1**, 92[64]; **5**, 385[129b], 386[129b], 428[107], 429[107], 857[227]
Jensen, F. R., **4**, 315[510]; **5**, 30[3]; **8**, 850[121], 857[194,202]

Jensen, H., **1**, 141[15]
Jensen, H. P., **7**, 86[13]
Jensen, J. L., **4**, 298[295]
Jensen, K. A., **6**, 243[102], 244[102], 424[56], 461[1], 464[33], 465[33], 477[101], 478[101], 481[117]
Jensen, K. M., **1**, 548[65]; **2**, 323[31]
Jensen, L. H., **8**, 89[45]
Jensen, N. P., **3**, 363[83]
Jensen, U., **3**, 644[134b,158]; **7**, 806[70]
Jensen, W. L., **7**, 10[76]
Jensen-Korte, U., **3**, 644[160]; **6**, 116[90]
Jenson, T. M., **1**, 768[169]; **3**, 994[39], 1008[73], 1009[74], 1010[74]; **6**, 837[61]; **7**, 410[99], 421[99]
Jentzsch, W., **6**, 555[809]
Jeong, J., **1**, 97[81]
Jeong, K.-S., **5**, 260[70], 263[70]
Jeong, N., **1**, 212[5], 213[5,18], 214[5b]
Jephcote, V. J., **1**, 832[114]; **2**, 39[135], 566[19], 574[59], 575[19], 587[142]; **6**, 863[190]
Jepson, J. B., **6**, 420[21], 424[21], 429[21]
Jereczek, E., **6**, 789[110]
Jeremic, D., **3**, 380[13]
Jerina, D. M., **7**, 6[34], 362[26]
Jerkunica, J., **8**, 99[107]
Jernberg, K. M., **5**, 788[12], 1003[21]
Jernow, J. L., **7**, 728[42]
Jernstedt, K. K., **4**, 366[8], 386[8b,148a,151], 387[148,148a]; **6**, 26[105,107]
Jerris, P. J., **2**, 189[45]; **3**, 24[149], 25[149]; **6**, 174[63]; **8**, 940[104]
Jershel, D., **2**, 495[58]
Jerussi, R. A., **4**, 20[63], 21[63]; **5**, 1146[106]; **7**, 84[1], 85[1], 108[1]
Jerzak, B., **6**, 556[818]
Jerzowska-Trzebiatowska, B., **3**, 563[1]
Jesce, M. R., **6**, 558[858]
Jeschke, R., **5**, 480[176], 829[15]
Jeschkeit, H., **6**, 635[11], 645[11], 665[11c], 667[11c], 668[11c], 669[11c]
Jeske, G., **8**, 447[133,134,136], 696[127,-128]
Jesser, F., **4**, 229[225,226,228]
Jessop, J. A., **3**, 965[127]
Jessup, D. W., **8**, 514[106]
Jessup, P. J., **5**, 331[43], 333[43b]; **6**, 811[75]
Jesthi, P. K., **1**, 489[23], 494[43]
Jeuenge, E. C., **7**, 235[3]
Jew, S., **4**, 18[61], 116[188c], 249[130], 257[130], 262[130], 357[150], 391[184], 393[184a,b], 397[184b]; **6**, 108[34], 1064[90a]; **7**, 625[42], 627[42,43]
Jewell, C. F., Jr., **5**, 689[77], 690[77a], 733[136,136b], 734[136b], 1202[56]; **8**, 41[95], 66[95]
Jewers, K., **6**, 719[127]
Jewett-Bronson, J., **8**, 20[136]
Jeyaraman, R., **1**, 834[125,127]; **3**, 736[24]; **7**, 13[124], 750[129]
Jhingan, A. K., **5**, 1148[118]
Jhoer, A., **2**, 775[29]
Ji, G.-J., **2**, 765[73]
Ji, S., **3**, 298[26]
Jiang, B., **7**, 166[86b]
Jiang, J. B., **4**, 435[139]
Jiang, J.-L., **6**, 726[181,182]
Jiang, S., **3**, 298[27]
Jiang, X., **4**, 590[89], 613[89]
Jiang, Y.-Y., **8**, 769[26]
Jianshe, K., **1**, 543[17]
Jibil, I., **6**, 517[327]
Jibodu, K. O., **2**, 748[125]; **6**, 291[212,214], 507[241,242], 529[241,242]
Jibril, I., **1**, 221[68]; **2**, 205[101,101b]; **4**, 21[65], 36[103,103b]; **5**, 428[109]; **6**, 522[346]; **8**, 389[67]
Jida, S., **6**, 505[225]
Jie, C., **5**, 856[205]
Jigajinni, V. B., **7**, 604[134], 607[167]; **8**, 910[81]
Jikihara, T., **7**, 423[145], 424[145b]
Jilek, J. O., **2**, 765[78]
Jiménez, C., **4**, 291[214,218], 295[248,249,254], 405[251]; **8**, 368[65], 856[183], 857[187]
Jimenez, J. L., **5**, 859[239]
Jimenez, M. S., **8**, 764[6], 773[6b]
Jimenez-Barbero, J., **4**, 108[146h]
Jin, H., **1**, 193[86], 198[91]; **6**, 46[69]
Jinbo, T., **8**, 246[78], 284[1]
Jinbo, Y., **4**, 650[425]
Jingren, T., **1**, 543[15]
Jintoku, T., **1**, 277[83]
Jinuai, K., **6**, 1016[35]
Jira, R., **4**, 588[57]; **7**, 94[55], 449[1,2], 450[1,2,8], 451[34]
Jirousek, M. R., **6**, 441[85]
Jisheng, L., **1**, 248[69]
Jit, P., **4**, 505[142]
Jitsuhiro, K., **4**, 1089[138], 1091[138]
Jitsukawa, K., **4**, 603[275], 626[275], 645[275]; **7**, 321[66], 587[171], 823[36]
Jizba, J., **8**, 590[55]
Jo, L. F., **2**, 354[108]
Jo, S., **6**, 134[31]
Joag, S. D., **4**, 50[142,142k]
Jobe, P. G., **3**, 573[70]
Jobling, W. H., **4**, 311[447]
Jochims, J. C., **6**, 517[327], 522[346], 524[359]
Jochum, C., **2**, 1099[107]
Jochum, P., **2**, 1099[107]
Jodal, I., **8**, 224[109]
Jodál, I., **6**, 660[207]
Jodoi, Y., **7**, 73[105]
Joe, D., **5**, 850[160]
Joerg, J., **3**, 813[65]
Joern, W. A., **7**, 219[13]
Joffe, M. L., **5**, 850[146]
Joglar, J., **5**, 161[63,64], 480[178], 484[179]
Jogun, K. H., **4**, 429[87]
Joh, T., **5**, 1137[56]
Johannes, J., **1**, 551[71]
Johannesen, R. B., **8**, 217[68]
Johansen, J. E., **2**, 1018[38,42]; **6**, 734[18]
Johansen, N. L., **6**, 637[31]
Johansen, Ø. H., **2**, 365[209]
Johansen, T., **2**, 150[97]
Johanson, E., **2**, 1099[115]
Johansson, J.-A., **5**, 2[18]
Johansson, R., **6**, 205[32], 652[139], 660[139]; **7**, 237[33]; **8**, 224[106], 969[95]
John, A. M., **4**, 295[261], 296[261]
John, C., **5**, 1117[15]
John, D. I., **1**, 656[152], 658[152], 832[114]; **3**, 934[64], 953[64]; **8**, 831[92], 849[109]
John, G. R., **4**, 670[15,20]
John, J. A., **2**, 6[33], 35[33]
John, J. P., **4**, 4[14]; **8**, 530[108]
John, L. S., **7**, 155[29]
John, R. A., **3**, 12[55,58]; **4**, 591[109], 633[109]; **6**, 848[111]
John, T. K., **1**, 822[39]
John, T. V., **7**, 172[126], 258[53]
Johncock, W., **4**, 1061[165]

Johne, S., **6**, 574[965]
Johns, A., **4**, 820[222]; **5**, 830[35]
Johns, I. B., **8**, 140[27]
Johns, N., **6**, 441[83]
Johns, R. B., **6**, 618[109]
Johns, W. F., **2**, 158[124]; **8**, 495[27], 530[92,102]
Johnson, A. L., **2**, 357[135]; **5**, 634[70]; **7**, 138[126]
Johnson, A. P., **2**, 1050[27]; **7**, 654[10]
Johnson, A. T., **1**, 822[31]
Johnson, A. W., **1**, 630[25], 675[25], 722[25], 820[8], 825[8]; **3**, 86[40], 665[36], 921[36]; **4**, 707[42]; **5**, 687[62], 688[62]; **6**, 171[2], 198[2], 200[2]
Johnson, B. F., **3**, 125[294], 126[294], 167[294], 168[294]; **7**, 246[88]
Johnson, B. F. G., **4**, 115[180a], 665[9], 673[31], 688[9], 691[76], 706[36-40], 707[40,41]; **6**, 690[399,403], 691[399], 692[399,403]
Johnson, C. A., **7**, 673[30]
Johnson, C. D., **2**, 965[65]; **4**, 37[107]
Johnson, C. K., **3**, 749[81]
Johnson, C. R., **1**, 112[27], 233[17], 238[17], 248[65], 433[226], 531[129], 532[134,135,137], 533[138], 534[139,140,141,142], 535[146], 621[70], 722[278], 734[24], 735[24], 736[24], 737[32,33], 738[33], 739[42], 740[43]; **2**, 418[22], 425[38], 448[46], 726[126]; **3**, 9[48], 11[48], 86[62], 88[136], 91[136], 173[62,521], 179[136,548], 181[136], 212[40], 223[158], 224[176], 226[192], 248[57], 249[57], 250[70], 251[57], 263[57,177], 265[187], 786[40], 871[51], 942[81b], 1008[67]; **4**, 115[184d], 152[56], 170[18,19], 177[57], 226[189], 245[85], 987[133,135,136], 989[136]; **5**, 439[166], 765[112], 855[187], 905[58]; **6**, 4[19,20], 9[41], 11[47], 139[50], 874[15], 998[116], 1066[97]; **7**, 194[5,7], 204[7], 205[7], 292[7], 363[39], 440[41], 441[41], 621[33], 764[119,127,130], 767[119], 778[394]
Johnson, D., **5**, 687[64]; **7**, 723[23], 724[28]; **8**, 425[47], 476[22]
Johnson, D. A., **1**, 474[86,87]; **3**, 262[170]
Johnson, D. C., **7**, 705[16]
Johnson, D. K., **7**, 829[59]; **8**, 51[121], 66[121]
Johnson, D. M., **6**, 237[65], 243[65]
Johnson, D. W., **5**, 794[48], 809[48]; **6**, 1055[52a]
Johnson, E. F., **6**, 707[43]
Johnson, F., **2**, 1011[12], 1031[12]; **3**, 13[69], 14[69]; **4**, 48[139], 292[231]; **5**, 1096[111], 1098[111]; **6**, 236[56,57], 261[10], 273[10], 280[10], 685[357], 709[52-54], 711[69], 751[109], 876[25]; **7**, 160[49]; **8**, 43[106], 66[106], 524[6], 528[82], 529[82], 530[6], 563[428], 566[457], 568[468]
Johnson, F. H., **3**, 743[59]
Johnson, F. M., **2**, 735[13]
Johnson, G., **2**, 1048[11]; **7**, 90[32]
Johnson, G. B., **8**, 254[127]
Johnson, G. M., **1**, 733[11]
Johnson, H. E., **1**, 851[36]; **8**, 568[466]
Johnson, J. A., **7**, 602[94]
Johnson, J. E., **6**, 291[223]
Johnson, J. H., **8**, 140[14]
Johnson, J. L., **5**, 478[161]; **6**, 533[505], 554[709]; **7**, 548[62], 553[62]; **8**, 36[48], 66[48], 347[140], 616[102], 617[102], 618[102,119]
Johnson, J. R., **2**, 142[48], 395[1], 396[1,8], 399[1,20], 400[1], 406[1], 1090[61]; **3**, 273[11]; **4**, 311[447]; **7**, 595[8], 599[8a]; **8**, 364[22], 724[172]
Johnson, J. W., **8**, 454[196]
Johnson, L. F., **5**, 128[27]
Johnson, M., **4**, 856[102]; **5**, 553[44]
Johnson, M. A., **5**, 854[181]
Johnson, M. D., **1**, 206[110]; **4**, 746[147]; **8**, 674[34], 850[122]
Johnson, M. G., **3**, 807[18]
Johnson, M. I., **5**, 1021[71]
Johnson, M. R., **4**, 301[326,327], 314[485]; **6**, 5[24]; **7**, 57[26], 369[64]; **8**, 267[68], 536[171]
Johnson, M. W., **7**, 166[87]
Johnson, N. A., **5**, 829[22]; **7**, 750[134]
Johnson, P. C., **2**, 547[110], 551[110]
Johnson, P. D., **7**, 415[111]
Johnson, P. R., **2**, 287[67]
Johnson, P. Y., **2**, 279[18]; **3**, 380[9], 625[40], 629[51,52]
Johnson, R. A., **4**, 231[273], 384[143], 398[219]; **6**, 2[8], 22[8]; **7**, 54[3,4], 56[3,4], 57[34], 64[4], 66[3,4], 71[4], 72[4], 75[4], 77[3,4], 78[3,4], 80[4], 340[46], 393[16], 394[18], 395[18], 398[16,18], 399[18], 424[18], 429[152], 633[65]
Johnson, R. E., **2**, 739[43a]; **7**, 14[131]
Johnson, R. G., **7**, 718[1], 731[1]
Johnson, R. L., **8**, 895[6,7], 897[6,7], 898[7], 899[7]
Johnson, R. N., **4**, 276[73], 277[73], 303[344]
Johnson, R. P., **4**, 1010[159,161]; **5**, 212[69]; **6**, 961[70], 977[18]
Johnson, R. T., **4**, 922[25]
Johnson, R. W., **7**, 800[30]
Johnson, S., **1**, 753[100]
Johnson, S. E., **8**, 474[16]
Johnson, S. J., **3**, 325[157], 353[49]
Johnson, T. A., **5**, 834[53]; **8**, 366[47]
Johnson, T. B., **2**, 916[82]; **3**, 890[35]
Johnson, T. H., **8**, 161[21]
Johnson, T. R., **3**, 555[31]
Johnson, W. L., **6**, 205[33], 834[40]
Johnson, W. M. P., **2**, 741[66]
Johnson, W. S., **1**, 347[136,137,138], 348[138,139,142], 546[59], 601[40], 608[36,37,40]; **2**, 148[78], 158[125], 162[142], 555[142,143], 556[154], 578[85], 580[97], 650[110,111], 651[110,111a,b], 753[1], 838[166], 1026[73]; **3**, 11[52], 17[52], 126[320], 133[335], 136[335], 226[206], 281[45], 341[3], 349[40], 351[40], 353[3,46,47], 358[66,67], 360[72], 361[75], 362[89,90], 363[83,85], 364[91,93], 366[99], 369[40,107,108,117-125], 370[109-113], 371[114,116], 372[117-119,121,122,124,125], 373[126,-128,129]; **4**, 4[15], 24[72,72b], 31[92,92e], 83[65a], 312[459], 386[149], 387[149]; **5**, 828[7], 839[7], 857[229], 882[13], 888[13], 891[37], 892[13,37,38a,40], 893[13]; **6**, 237[68]; **7**, 111[190], 167[100], 169[113], 564[93], 565[93], 568[93], 711[57]; **8**, 58[175], 66[175], 353[156], 494[24], 495[27], 530[101,103], 542[226,228], 545[293], 564[442], 566[451]
Johnston, A. D., **8**, 514[113]
Johnston, B. D., **7**, 238[41], 401[54]
Johnston, B. H., **1**, 409[38]
Johnston, D. B. R., **5**, 107[199]; **6**, 219[123], 659[197a]
Johnston, D. E., **3**, 334[220]
Johnston, F., **4**, 272[36], 273[36]
Johnston, G. A. R., **6**, 81[76], 82[76], 818[106,107]
Johnston, J. D., **7**, 157[33]; **8**, 974[122]
Johnston, L. J., **5**, 164[76]; **6**, 960[57]
Johnston, M. D., Jr., **1**, 294[44]
Johnston, M. I., **2**, 547[105], 550[105]; **4**, 377[104]
Johnstone, R. A. W., **3**, 229[232]; **6**, 119[109], 635[14a], 636[14], 834[32], 1022[62]; **7**, 772[288]; **8**, 18[123], 81[4], 91[4], 104[4], 263[30,31], 366[43], 367[57], 368[63], 440[83], 551[339], 887[115], 958[19,20], 959[29]
Johri, K. K., **7**, 723[27]
Jokic, A., **7**, 92[41,41a], 94[41]
Jokubaityte, S. P., **6**, 570[944]
Jolidon, S., **5**, 829[20]
Joliveau, C., **6**, 509[274]
Jolly, B. S., **1**, 13[73], 37[177]
Jolly, P. W., **3**, 228[214], 436[9]; **4**, 596[161], 601[245], 608[318,319], 939[73], 980[107]; **5**, 641[131], 1142[86]
Jolly, R. S., **4**, 803[131]
Jolly, W. L., **8**, 244[64], 253[64], 526[32]
Jommi, G., **1**, 566[216]; **2**, 833[148]; **4**, 261[285,288,290]
Jonas, D. A., **3**, 383[46,50]
Jonas, J., **5**, 453[63], 454[63], 458[63]; **7**, 235[1]; **8**, 214[35]

Jonas, K., **8**, 747[56], 752[56]
Jonassen, H. B., **5**, 800[75]; **8**, 452[190]
Jonczyk, A., **2**, 76[83a], 429[49,50], 430[52a], 431[52a], 432[56]; **3**, 158[448], 159[448], 168[448], 174[529]
Jondahl, T. P., **5**, 710[50]
Jones, A. B., **1**, 529[124]; **4**, 381[126b], 382[126], 383[126]; **6**, 994[98]
Jones, A. J., **5**, 536[96]; **7**, 31[87]
Jones, A. S., **1**, 569[252]
Jones, B., **7**, 689[3]
Jones, B. A., **6**, 436[28], 437[28], 447[28], 448[28], 449[28,116], 450[28], 452[28]; **8**, 838[19]
Jones, C. M., **6**, 478[106]
Jones, C. R., **4**, 190[107], 980[105], 981[105]; **5**, 1085[65]
Jones, D. H., **3**, 383[46]; **8**, 340[97]
Jones, D. M., **2**, 800[15]
Jones, D. N., **3**, 114[234], 354[61]; **4**, 102[129], 258[243], 317[556,557], 587[29], 603[271], 605[292], 626[292], 646[292]; **5**, 7[52]; **6**, 1017[38], 1019[44,45], 1024[38,75-77], 1025[79], 1026[82], 1029[82]; **7**, 152[4], 153[4], 302[63], 629[49], 766[178], 771[284], 772[284]
Jones, D. S., **6**, 94[140]
Jones, D. W., **2**, 809[51], 823[51]; **3**, 690[124]; **5**, 618[6], 1003[25]; **6**, 1007[151]
Jones, E. M., **2**, 958[23]
Jones, E. R. H., **2**, 143[51], 156[114]; **3**, 554[18,19]; **4**, 55[156]; **5**, 767[119]; **6**, 217[112]; **7**, 68[82,84], 69[92], 71[82,99], 72[84], 73[92], 120[15], 158[42], 253[17], 254[30], 306[4], 582[149]
Jones, F. N., **1**, 463[23]; **3**, 484[22]
Jones, G., **1**, 243[55]; **2**, 342[3], 343[3], 352[3], 354[3], 356[3], 357[3], 358[3], 359[3], 360[3], 361[3], 362[3], 765[75]; **6**, 759[135]; **7**, 877[132]; **8**, 587[43]
Jones, G., Jr., **5**, 151[4], 158[4], 161[4], 165[86], 166[90], 168[4], 176[132], 178[4], 190[4], 687[64], 815[142]; **7**, 120[12], 123[12], 851[12], 854[61],
Jones, G. A., **7**, 247[96]
Jones, G. B., **7**, 32[97], 33[97]
Jones, G. C., **3**, 914[12]; **7**, 800[30,30b]
Jones, G. E., **2**, 725[114]; **3**, 284[52], 555[30]
Jones, G. H., **2**, 139[29]; **4**, 38[108,108a]
Jones, G. R., **6**, 116[88]
Jones, G. R. N., **2**, 385[323]
Jones, G. W., **4**, 1103[203]
Jones, J. B., **2**, 455[21], 456[21,25,29]; **3**, 125[295,296], 126[295,296], 128[296], 783[24]; **6**, 134[22], 560[868]; **7**, 79[133], 145[166], 158[41], 316[46-48], 317[46-48], 318[48]; **8**, 183[1], 185[1,11,16], 187[1a], 188[54,55], 189[60], 196[1a,55,116], 197[126], 199[1a,55], 200[136], 201[1a,54,55], 204[1], 206[173], 207[1a], 209[1a]
Jones, J. H., **2**, 756[8]; **6**, 81[74], 644[92]
Jones, J. K. N., **2**, 456[53-56,60-62,69,75], 457[55,60,61,69], 458[53,62], 460[69]; **6**, 207[47]; **8**, 794[15]
Jones, J. R., **2**, 365[213]
Jones, K., **5**, 528[67]; **6**, 538[571]
Jones, L. A., **8**, 568[472]
Jones, L. B., **3**, 242[4], 244[4]
Jones, L. D., **3**, 499[128], 500[128,132], 505[132], 509[128]; **4**, 476[163], 502[124], 766[229]; **5**, 692[101]
Jones, M., **1**, 885[134]; **4**, 33[95,95a], 976[100], 1089[136], 1101[195]
Jones, M., Jr., **2**, 838[176]; **3**, 897[92], 900[92], 903[123]; **4**, 483[4], 484[4], 495[4], 953[8,8f], 954[8f,m], 961[8f,m], 1001[45], 1002[45,48], 1011[166], 1012[174], 1014[184]; **5**, 677[8], 681[25], 716[87], 786[3], 787[3], 1065[1], 1066[1], 1074[1], 1083[1], 1084[1], 1093[1], 1094[1d]; **6**, 776[55], 1036[143]
Jones, M. D., **5**, 300[74], 307[93]
Jones, M. E., **3**, 4[22]
Jones, M. F., **2**, 822[111]; **4**, 791[46]
Jones, M. H., **4**, 287[180]
Jones, M. J., Jr., **5**, 65[71]
Jones, M. M., **8**, 152[158]
Jones, M. W., **4**, 407[256d]
Jones, N., **5**, 766[114]
Jones, N. D., **4**, 128[221], 389[167]
Jones, N. R., **1**, 568[226]; **3**, 124[278], 125[278], 128[278], 129[278], 132[278]
Jones, P. A., **2**, 958[23]
Jones, P. F., **3**, 124[289], 125[289,314]; **6**, 134[12]
Jones, P. G., **2**, 68[40]; **8**, 446[73], 987[23]
Jones, P. R., **1**, 221[67], 223[72a,b], 225[88], 226[88], 326[3]; **3**, 135[359], 136[359], 137[359], 139[359], 142[359], 260[144]; **4**, 72[23], 95[23], 98[108a,110], 434[131]; **7**, 762[79]; **8**, 707[22], 770[31]
Jones, R., **6**, 213[86]
Jones, R. A., **1**, 41[270], 432[137], 456[137]; **3**, 211[28], 215[28]; **6**, 175[75]
Jones, R. B., **2**, 801[30]
Jones, R. C. F., **1**, 300[70], 366[52]; **2**, 494[56,57], 742[69]; **4**, 76[47]; **6**, 516[321], 552[692]; **8**, 638[16]
Jones, R. E., **6**, 219[122]
Jones, R. G., **1**, 107[7]; **3**, 208[9], 244[19]; **4**, 72[23], 95[23], 148[47b]
Jones, R. H., **2**, 125[217], 127[232], 315[42], 316[42]; **4**, 217[132], 231[132]; **5**, 260[62]; **6**, 1031[109]; **7**, 630[54]
Jones, R. J., **8**, 618[113], 623[113,146], 628[113]
Jones, R. L., **4**, 1021[245]
Jones, R. R., **5**, 736[144]
Jones, R. S., Jr., **3**, 927[57]
Jones, R. V. H., **3**, 390[70]; **5**, 809[116]
Jones, R. W., **3**, 106[224], 113[224]; **4**, 308[405]
Jones, S. B., **8**, 675[49], 676[49]
Jones, S. O., **4**, 316[541]
Jones, S. R., **4**, 588[60]; **7**, 14[136]
Jones, S. S., **6**, 656[168]
Jones, T. H., **7**, 528[9]; **8**, 51[123], 66[123]
Jones, T. K., **1**, 401[14], 402[17], 585[13], 799[296]; **5**, 762[95,98-100,106], 763[95,98,109], 764[109]; **7**, 397[27]; **8**, 879[49], 946[138]
Jones, T. R., **6**, 26[109]
Jones, W., **4**, 313[466]
Jones, W. D., **7**, 3[14]; **8**, 289[29], 290[29], 797[35]
Jones, W. H., **8**, 451[180]
Jones, W. J., **7**, 291[2], 654[9], 655[9,18]
Jones, W. M., **4**, 967[56], 1012[173]; **5**, 736[140]; **6**, 276[118]; **7**, 800[30]
Jong, M. E., **6**, 214[91]
Jong, T. T., **4**, 313[473]
Jonkers, F. L., **6**, 705[32,33]
Jonsson, H. G., **8**, 608[45]
Jonsson, L., **7**, 878[139,142]
Jonsson, N. Å., **6**, 802[40]
Joo, Y. J., **4**, 462[105]
Joos, R., **6**, 1059[66]; **7**, 482[118]
Jordaan, A., **6**, 978[23]
Jordaan, J. H., **6**, 677[317]
Jordan, A., **8**, 216[54], 224[54]
Jordan, A. D., Jr., **4**, 38[108]; **6**, 174[55]
Jordan, E., **8**, 141[43]
Jordan, F., **8**, 87[33]
Jordan, K. D., **5**, 452[57]; **7**, 861[77]
Jordan, M., **4**, 1086[116]; **8**, 388[65]
Jordis, U., **7**, 143[146]; **8**, 661[110]
Jordy, J. D., **2**, 169[168]
Jorgensen, K. A., **7**, 358[8b], 422[140], 438[10], 441[10], 752[150]; **8**, 541[202]
Jorgensen, P. M., **4**, 149[52]

Jorgensen, W. L., **1**, 287[15], 297[58], 580[2], 581[2], 582[2]; **4**, 425[31]; **5**, 64[26], 257[61], 452[56], 706[28]; **7**, 816[7]
Jorgenson, M. J., **1**, 398[2], 410[42]; **5**, 164[74], 914[109]
Josan, J. S., **6**, 687[373], 984[56]
Joseph, M. A., **3**, 760[139], 774[139]; **6**, 708[49]
Joseph-Nathan, P., **8**, 537[187]
Josephson, K. O., **5**, 451[3]
Josephson, S., **2**, 233[188]; **6**, 41[43], 625[157], 659[191,197b]
Josey, A. D., **5**, 39[24], 71[121]
Joshi, B. C., **8**, 566[452]
Joshi, B. S., **3**, 396[115]; **8**, 339[96]
Joshi, B. V., **7**, 246[95]
Joshi, G. S., **4**, 505[143]
Joshi, K. K., **3**, 383[49]
Joshi, N. N., **1**, 223[84], 225[84c]
Joshi, V. S., **6**, 1022[64]
Joshua, A. V., **4**, 350[122,123]; **8**, 821[48]
Joshua, C. P., **8**, 390[79]
Joshua, H., **3**, 927[56]
Josse, D., **4**, 469[136]
Jössung-Yanagida, A., **7**, 693[25]
Jost, P., **5**, 178[136]
Josty, P. L., **4**, 115[180a], 665[9], 688[9]
Jouannetaud, M. P., **7**, 333[20]
Joucla, M., **4**, 247[98], 257[98], 262[98], 1086[114]; **5**, 254[48], 422[82]
Jouin, P., **8**, 13[71], 95[91]
Jouitteau, C., **7**, 280[173], 281[173], 283[173,184], 285[173], 844[63], 845[63]
Joukhadar, L., **3**, 613[2], 615[2]; **8**, 117[74], 243[47], 816[24]
Joule, J. A., **1**, 473[80]; **4**, 1004[77], 1021[77]; **8**, 587[34,42], 618[118]
Joullie, M. M., **2**, 1090[70], 1096[95], 1097[101], 1100[70]; **4**, 364[1], 371[60], 372[60]; **6**, 1031[113]; **7**, 523[46]; **8**, 607[31], 609[50,52], 847[99], 848[99], 849[99]
Jourdan, G. P., **6**, 526[392]
Jourdian, G. W., **2**, 463[80], 464[80]
Jousseaume, B., **1**, 438[158], 457[158], 479[149], 480[149]; **3**, 440[45]; **4**, 876[59], 877[72], 878[59,76,78,79]; **5**, 1166[22]; **8**, 392[98]
Joussot-Dubieu, J., **5**, 125[13], 128[13]
Jovanovic, M. V., **4**, 492[75]
Joy, D. R., **5**, 596[32], 597[32], 603[32]
Joyce, R. M., **5**, 3[23], 64[44]
Joyce, R. P., **3**, 511[191]
Joyeux, M., **3**, 1017[4]
Józwiak, A., **1**, 474[90,92,93]
Jrundmann, C., **2**, 342[6]
Juaristi, E., **1**, 511[27], 774[213]; **6**, 73[26]; **8**, 837[16]
Juaristi, M., **7**, 278[160], 283[187], 530[18], 531[18], 752[144], 760[24]
Jubault, M., **8**, 134[32], 135[47], 136[47], 137[32,59,60]
Jucker, E., **2**, 765[71]; **7**, 446[68], 704[10]
Juday, R. E., **3**, 568[42]
Judd, D. B., **2**, 963[55]
Judge, J. M., **7**, 123[32]
Judkins, B. D., **7**, 474[41], 483[41,128], 744[67]
Judy, W., **5**, 1116[8]
Jug, K., **5**, 72[184,185], 73[191,192], 74[209,210], 75[227], 202[37]
Juge, S., **3**, 47[256]; **4**, 653[436]; **6**, 118[108]; **7**, 229[120]
Jugelt, W., **3**, 891[36], 909[153]; **8**, 900[30]
Jui, H., **4**, 1104[212]
Jukes, A. E., **3**, 208[10], 210[10], 499[110], 501[110], 505[110], 509[110], 512[110], 522[14]
Julia, M., **1**, 792[271], 793[272], 804[308], 805[308]; **2**, 76[83b], 967[72]; **3**, 159[454,456,461,463], 160[471], 161[454,456,461,463,471], 162[456], 163[471], 165[461,463], 167[481], 168[481,507], 169[507], 170[481], 171[507], 172[481], 174[536,537], 176[536,537], 178[536,537], 342[13], 351[42], 352[42], 381[25], 382[25], 447[93], 448[94,95], 493[81], 882[103]; **4**, 314[495], 315[512], 359[160], 502[120], 599[211,217], 640[211], 785[18], 796[97], 801[122,123], 844[58], 1007[111]; **5**, 847[132], 941[228], 1001[13]; **6**, 157[165,166,167,169,170], 161[186], 162[187,188], 502[212], 624[145], 672[287], 987[73]; **8**, 337[77], 505[73], 839[23], 842[42a-d], 844[42c], 847[42c,88,88b,d]
Juliá, S., **1**, 551[75], 554[106]; **2**, 435[62]; **3**, 147[395], 155[395], 252[86], 757[124], 896[69], 934[68], 943[84], 963[123], 964[123,125], 969[130], 991[36], 998[48,49], 999[50]; **4**, 102[128d], 113[169,169c], 231[268], 293[237], 1001[41], 1054[134]; **5**, 847[132], 1001[13]; **6**, 80[68], 271[86,87], 579[984], 838[65], 841[77], 852[132,133], 853[140], 854[143], 866[77], 893[78]
Julian, P. L., **2**, 406[48]; **6**, 959[43]; **8**, 612[67]
Juliano, B., **8**, 604[7]
Julius, M., **1**, 661[167]
Julliard, M., **5**, 913[102]; **7**, 860[73]
Jullien, J., **5**, 579[160,161]
Jullien, R., **4**, 180[66]
Jun, J.-G., **8**, 219[76]
Jun, Y. M., **1**, 480[157,159]; **3**, 587[141,143]
Junek, H., **2**, 367[224], 789[54]; **3**, 824[22], 826[22,41]; **4**, 4[16,16b], 440[171]; **6**, 553[761], 554[726,761,762]
Jung, A., **1**, 153[60], 157[60], 336[76,77,79], 340[77,79], 614[50], 615[50]; **2**, 31[117], 32[117], 117[148], 307[16], 310[16], 570[37], 640[66], 641[70], 646[66,70,83,85]; **8**, 568[469]
Jung, C. J., **8**, 544[264]
Jung, D. M., **4**, 299[300]
Jung, F., **5**, 567[105]; **7**, 121[23]
Jung, G. L., **4**, 1040[78], 1049[78a]; **5**, 804[95], 987[41], 988[41], 989[43], 993[50], 994[50]
Jung, K.-H., **6**, 51[113]
Jung, M., **2**, 555[148]; **5**, 155[37]; **6**, 80[70]
Jung, M. E., **1**, 476[123], 783[235]; **2**, 156[113], 162[113], 381[309], 505[12], 510[12,36], 600[31], 657[169]; **3**, 11[49,50], 20[120], 23[49], 35[204], 66[14], 194[6], 504[154], 511[154], 515[154], 602[226]; **4**, 3[8], 6[8], 7[8a], 8[8a], 30[89], 33[95,95b], 40[116], 48[136], 49[136,141], 51[143,143c], 65[8a], 99[118b], 158[78], 486[36], 489[36], 501[111]; **5**, 8[60], 329[32], 404[19], 405[19], 431[121], 515[14], 518[14], 522[14a], 524[54,54e], 534[54], 692[95], 791[41], 1017[66]; **6**, 210[73], 647[108], 665[229], 667[229], 676[307], 714[86], 752[110], 767[27], 775[27]; **7**, 144[152], 187[183], 316[44], 696[39]
Jung, S.-H., **1**, 786[248]; **4**, 355[134]; **5**, 524[54], 534[54]; **7**, 566[101]
Jung, Y. H., **3**, 504[154], 511[154], 515[154]
Jung, Y.-W., **2**, 15[63,64], 996[49], 999[52]
Junga, M., **2**, 900[24]
Junge, B., **1**, 364[39]
Jungers, J. C., **8**, 419[22], 420[22], 430[22], 436[22]
Junghans, K., **8**, 568[474]
Jungheim, L. N., **1**, 753[100]; **3**, 785[34]; **6**, 117[96], 783[83]; **8**, 545[279]
Jungk, H., **3**, 300[36]
Junius, M., **6**, 575[970]
Junjappa, H., **2**, 286[63], 495[62], 496[62]; **6**, 456[162], 457[162]; **7**, 154[12]; **8**, 540[201], 839[26b], 840[26]
Junk, P. C., **1**, 13[73], 17[217]
Juntunen, S. K., **4**, 5[17]; **5**, 935[193], 936[193]
Juo, R. R., **6**, 1053[46]
Juraristi, E., **1**, 564[194]
Jurayj, J., **4**, 1036[48]; **5**, 854[176], 855[176], 856[176,208,215]
Jurczak, J., **1**, 54[19], 109[14], 183[57], 185[57], 314[127], 339[84,86]; **2**, 31[116], 662[18], 663[25-28], 664[18,25-28], 665[25], 671[51], 694[77], 995[43]; **5**, 108[207], 342[62c], 430[118], 431[121],

432[124,132], 433[138,139,139a,f], 434[138,139i], 453[63], 454[63], 458[63]; **6**, 70[19], 532[474]; **7**, 397[29], 568[105], 713[73]
Jurek, J., **6**, 773[44]; **8**, 928[25]
Jurgeleit, W., **8**, 397[142]
Jurgens, E., **7**, 763[86]
Juri, P. N., **6**, 220[126]
Juric, P., **6**, 554[725]
Jurion, M., **3**, 131[333]; **8**, 930[34]
Jurjev, V. P., **7**, 750[129]
Jurlina, J. L., **4**, 350[121]; **7**, 121[24], 530[20], 531[20]
Jursic, B., **6**, 227[22], 228[22], 229[22]
Jurss, C. D., **4**, 436[145], 437[145], 438[150]
Just, G., **2**, 74[72], 1103[130]; **3**, 259[132], 380[10], 541[115], 846[41]; **4**, 262[302], 740[118,119], 903[199]; **5**, 94[86,88], 95[88,94], 96[110], 421[80]; **6**, 176[101], 642[72]; **7**, 231[148], 272[141], 713[68]; **8**, 476[28]
Jutand, A., **3**, 443[58], 450[102], 454[117,118]; **7**, 854[45]
Jutland, A., **4**, 591[111], 616[111], 633[111]
Jütten, P., **5**, 187[173]
Jutz, C., **2**, 777[1], 779[1], 780[1], 781[1], 782[15,29,31], 783[1], 786[1], 787[1], 789[1], 791[1], 792[1]; **5**, 710[56,56c], 719[56], 742[160], 744[56]; **6**, 487[4], 488[4], 489[4], 522[4]
Jutzi, P., **2**, 743[80]
Juve, H. D., Jr., **7**, 177[145], 182[164]

K

Kaas, N. C., **7**, 802[46,47]
Kaathawala, F. G., **6**, 94[142]
Kaba, T., **1**, 86[34,36], 223[80], 224[80], 317[146,147], 319[147]
Kabachnik, M. I., **4**, 317[548]
Kabalka, G. W., **2**, 111[85], 141[39], 241[14], 242[14b], 321[16,18], 324[18], 325[16,18]; **4**, 140[10], 145[23,24,27,28,29a], 164[99,99a], 288[188,189], 290[189], 346[86a], 347[86b]; **6**, 107[24,27], 938[135], 939[135,142]; **7**, 597[47,50], 599[69], 602[104,104b], 604[136], 605[139], 606[149,150,151,157]; **8**, 70[229], 237[13], 240[13], 244[13], 253[13], 343[118], 344[118], 356[118,183,184,187,189], 357[118,184,189,192,193,194,195,203], 358[195], 359[203], 363[3,4], 373[137,138], 374[139,140], 375[4], 376[140,165,166], 377[137,167,168], 386[52], 720[132], 726[186]
Kabasakalian, P., **3**, 649[202]
Kabasawa, Y., **5**, 225[103,104], 226[104]
Kabass, G., **2**, 152[103]
Kabat, M. M., **2**, 382[315]
Kabay, M. C., **2**, 102[23]
Kabbe, H. J., **2**, 1103[131]
Kabeta, K., **1**, 158[74], 180[39], 181[39]; **2**, 4[13], 567[31], 584[118], 977[7]; **3**, 229[227], 438[30]; **7**, 616[18]; **8**, 568[467], 783[108,109]
Kabir, A. K. M. S., **1**, 760[135]; **7**, 440[38,39a]
Kabiraj, A., **8**, 331[35]
Kabir-ud-Din, **8**, 91[52]
Kabli, R. A., **3**, 325[158]
Kabo, A., **4**, 330[4]; **7**, 520[25]
Kabore, I., **4**, 296[263]; **6**, 264[31], 266[49], 278[31]
Kabuto, C., **2**, 198[82]; **3**, 395[101]; **4**, 8[28], 30[88,88o], 231[280], 261[301]
Kabuto, H., **6**, 467[50]
Kabuto, K., **8**, 161[23], 170[72]
Kachensky, D. F., **1**, 794[281], 795[281]; **2**, 581[104]; **5**, 527[59]; **6**, 990[82], 991[82]
Kacher, M., **8**, 355[179]
Kacher, M. L., **7**, 769[220]
Kachinski, J. L. C., **5**, 1001[16]; **6**, 851[131]
Kachkovskii, A. D., **6**, 509[255]
Kacprowicz, A., **4**, 1001[44], 1004[70]
Kaczmarek, C. S. R., **1**, 406[29]; **5**, 494[217], 579[164]; **8**, 540[195]
Kaczmarek, L., **8**, 390[85], 391[85], 392[96]
Kada, R., **2**, 362[180], 363[187]
Kadaba, P. K., **7**, 475[53], 476[53]
Kadam, S. R., **1**, 268[56], 269[56a]; **3**, 602[225]; **8**, 114[51,52]
Kadano, S., **2**, 187[40]
Kaddah, A. M., **2**, 744[99], 745[99]; **3**, 325[158]
Kader, A. T., **6**, 638[40]
Kadib-Elban, A., **4**, 84[68b], 149[52], 182[77], 183[78]
Kadin, S. B., **8**, 536[175]
Kadish, V., **8**, 595[77]
Kadkhodayan, M., **3**, 464[174]
Kadokawa, Y., **8**, 97[95]
Kadokura, M., **4**, 602[264], 609[264], 644[264]
Kadonaga, J. T., **5**, 913[100], 1008[43]
Kadono, Y., **5**, 461[107], 464[107], 466[107]
Kadota, I., **1**, 766[162]; **8**, 227[121]
Kadow, J. F., **1**, 248[65]; **5**, 736[143]; **7**, 621[33]; **8**, 341[103], 928[24]
Kadowaki, T., **3**, 321[139]
Kadunce, W. M., **1**, 411[45]
Kadushkin, A. V., **6**, 554[776]
Kadyrov, Ch. Sh., **1**, 543[16]; **3**, 306[79], 315[109,111], 316[116], 317[116]; **7**, 7[47]
Kadzyauskas, P. P., **4**, 357[146]
Kaeding, W. W., **3**, 305[75b,c]
Kaempchen, T., **5**, 634[79]
Kaenel, H. R., **3**, 298[29]
Kaeseberg, C., **4**, 877[66]
Kaesler, R. W., **5**, 1070[20], 1071[20], 1072[20], 1074[20], 1094[102], 1102[102], 1104[156], 1110[20], 1111[20], 1113[156]
Kaesz, H. D., **1**, 214[25]; **8**, 600[103]
Kaftory, M., **8**, 451[180]
Kaga, H., **7**, 680[79]
Kaga, K.-i., **6**, 450[119], 454[119]
Kagabu, S., **8**, 408[79]
Kagami, H., **8**, 392[99]
Kagami, M., **8**, 93[72,73], 561[416]
Kagan, B. S., **4**, 344[75]
Kagan, H., **4**, 302[335]; **7**, 846[91]
Kagan, H. B., **1**, 179[25], 231[4-6], 251[1], 252[1], 253[10,11], 255[16,16b,c,19], 256[16c], 258[16c,19,26,26b], 259[16c,27,31], 261[16c,27], 265[27], 266[16c,27], 271[19,62], 273[1c,70,71], 274[1c,73,75], 278[16b,c], 328[27]; **2**, 232[173,176], 286[62], 294[83], 297[91], 345[28], 346[28], 917[88], 918[90,91], 935[91,149]; **3**, 567[35], 570[35]; **4**, 106[138d], 653[434]; **5**, 100[143,144], 102[143], 186[169]; **6**, 150[112,113], 237[61], 980[40]; **7**, 282[180], 381[106], 425[146], 777[377,378,379,381], 778[377,378,379]; **8**, 113[48,49], 115[48,49,64], 124[64], 125[64], 159[3], 161[18], 170[91], 173[118], 180[137], 459[228,231], 460[231,254], 535[166], 552[360], 797[31,32], 889[134]
Kagan, J., **3**, 739[38], 747[70], 770[176]
Kagan, M. B., **1**, 546[57]
Kaganowitch, M., **5**, 729[123]
Kagaruki, S. R. F., **6**, 556[822]
Kagechika, K., **2**, 1051[43]; **3**, 650[213]; **7**, 804[60]
Kagei, K., **7**, 136[111], 137[111]
Kageyama, H., **1**, 14[74-76], 569[255]
Kageyama, M., **2**, 224[154], 225[154], 249[84]; **4**, 370[44,45]; **5**, 524[50], 539[50], 548[50c]
Kageyama, T., **4**, 823[231]; **7**, 322[68], 533[33], 765[137]
Kagi, A., **7**, 667[79]
Kagotani, M., **2**, 215[134]; **3**, 512[200]; **4**, 75[44a,b], 599[214], 606[308], 607[308]; **6**, 846[105]; **8**, 459[245]
Kahan, G. J., **8**, 364[15]
Kahle, G. G., **1**, 415[62]
Kahle, G. R., **2**, 283[50]
Kahn, B. E., **3**, 563[1], 570[1h]
Kahn, M., **1**, 884[128], 898[128]; **2**, 613[113], 911[71], 1070[137]; **4**, 794[73]; **5**, 429[112,115], 806[107], 1027[87], 1028[87]; **6**, 117[97]; **7**, 439[33], 648[41]
Kahn, S. D., **1**, 297[57]; **4**, 87[82], 213[119,120], 240[54], 379[113], 380[113a], 968[62]; **5**, 349[75], 856[208]
Kahne, D., **1**, 29[144]; **2**, 507[26]; **3**, 34[198], 39[198]; **6**, 727[195]
Kahne, D. E., **8**, 448[147], 534[157], 814[17]
Kai, M., **6**, 247[131]
Kai, Y., **1**, 14[75,76], 19[98], 34[167], 162[93,102]; **3**, 672[65]; **4**, 589[87], 598[191], 638[191]; **5**, 1200[49]; **6**, 976[9]
Kaiho, T., **1**, 410[44], 436[151,152]; **2**, 240[13], 256[13], 257[13b]; **7**, 257[52], 743[64]
Kaihoh, T., **5**, 497[225]
Kaimanakova, E. F., **6**, 554[765]
Kaiser, A., **2**, 740[62]; **7**, 678[69]
Kaiser, C., **6**, 811[74]

Kaiser, E., Sr., **6**, 637[32,32d]
Kaiser, E. M., **2**, 268[69]; **3**, 158[438], 159[438,467], 160[438], 166[438,467]; **8**, 329[21], 516[121,122], 517[124], 880[59], 971[101]
Kaiser, E. T., **3**, 86[44], 649[202], 650[202c], 652[202c]; **4**, 537[99], 538[99]; **7**, 850[1]
Kaiser, G. V., **3**, 934[65], 953[65]
Kaiser, J., **1**, 320[160]
Kaiser, J. K., **4**, 1093[150]
Kaiser, R., **5**, 15[107], 65[64]
Kaiser, S., **8**, 407[57]
Kaiser, W., **2**, 857[251]; **3**, 381[18]; **4**, 271[22]; **6**, 91[124]
Kaito, M., **2**, 166[154], 270[74]; **3**, 639[75]; **4**, 609[331]; **7**, 453[85,89,90], 455[89,90]; **8**, 961[39]
Kajansky, B. A., **7**, 595[20]
Kajfeq, F., **6**, 85[87]
Kajfez, F., **7**, 232[158]
Kaji, A., **2**, 73[64], 74[64], 321[17], 323[17], 330[17], 332[63,64], 333[64-66], 334[68], 348[56], 350[73,74], 360[168], 362[56], 363[56,73]; **3**, 99[180], 100[180], 153[415], 159[455], 161[455], 174[526], 224[170]; **4**, 13[44,44b], 31[92,92i], 40[114], 64[92i], 78[54], 86[54b], 251[147], 591[108], 599[219], 633[108], 641[219], 790[36]; **5**, 320[10,11], 600[41]; **6**, 87[102], 161[178], 839[69], 926[68,69], 927[68], 1000[128,129,130], 1022[63]; **7**, 197[17,19], 883[174]; **8**, 962[41], 969[98]
Kaji, K., **3**, 155[428], 168[497], 170[497], 220[115]; **6**, 548[669], 995[99]; **8**, 349[135], 354[175]
Kaji, S., **1**, 820[16]
Kajigaeshi, S., **3**, 321[139]; **4**, 1005[89], 1015[199]; **6**, 801[27,28]
Kajihara, Y., **1**, 162[95,100]
Kajikawa, A., **2**, 187[42]
Kajikawa, Y., **1**, 858[60]
Kajimoto, T., **4**, 600[240], 643[240]; **6**, 968[108]; **7**, 739[33], 746[81]; **8**, 287[21]
Kajitani, M., **5**, 1139[75]
Kajiura, T., **8**, 212[9], 222[9]
Kajiwara, A., **3**, 751[88]
Kajiwara, K., **3**, 135[341], 136[341], 137[341]
Kajiwara, M., **2**, 384[316], 851[222]; **4**, 501[117]; **5**, 410[38]; **8**, 190[74], 204[152]
Kajtár-Peredy, M., **6**, 917[36]
Kakehi, K., **4**, 309[418]
Kakhniashvili, A. I., **8**, 772[55,58,59]
Kakihana, M., **1**, 759[128,129]; **4**, 589[88], 599[219], 641[219], 823[231]
Kakihana, T., **5**, 552[28]
Kakimoto, M., **2**, 348[59], 362[59], 1040[106]; **6**, 247[131], 546[641,642], 936[107]; **8**, 935[60]
Kakinami, T., **6**, 801[28]
Kakinuma, A., **7**, 59[38]
Kakinuma, K., **3**, 999[51]
Kakis, F. J., **8**, 490[2], 492[2], 493[2]
Kakisawa, H., **2**, 553[129]; **3**, 380[13]; **5**, 137[75,76], 143[75,76]; **7**, 242[60]; **8**, 645[44]
Kakiuchi, K., **3**, 380[9]; **6**, 976[9], 1036[145]
Kakodkar, S., **4**, 1104[212]
Kakoi, H., **2**, 384[319]; **6**, 266[50]
Kakudo, M., **4**, 310[435]; **8**, 856[170]
Kakui, T., **3**, 469[203], 470[203], 473[203], 483[16]; **4**, 840[34]; **6**, 16[60]; **7**, 642[8,9]; **8**, 787[119]
Kakushima, M., **2**, 743[84], 809[52], 823[52]; **7**, 360[21]
Kalabina, A. V., **6**, 495[150]; **8**, 771[42]
Kalantar, T. H., **7**, 442[46c]; **8**, 205[155]
Kalaus, G., **6**, 917[36]
Kalb, L., **3**, 835[80]
Kalbacher, H., **6**, 637[35]
Kalbfus, H. J., **5**, 1081[54]
Kaldor, S., **1**, 401[13,14]
Kale, A. V., **7**, 143[147]
Kale, N., **2**, 760[41]
Kale, V. N., **1**, 700[260]
Kalechits, I. V., **8**, 454[198], 963[43]
Kalesse, M., **4**, 121[207]
Kaleya, R., **7**, 46[50], 47[50]
Kaliba, C., **1**, 372[77]; **5**, 598[33]; **6**, 501[195]
Kalicky, P., **3**, 1058[40]; **7**, 40[8], 43[8,36,47]
Kalikhman, I. D., **2**, 365[214]; **6**, 550[673,674]
Kalinichenko, N. V., **8**, 373[132]
Kalinin, V. N., **3**, 898[81]; **6**, 524[360,373], 528[413]; **8**, 765[11]
Kalinina, G. S., **7**, 641[2]
Kalinkin, M. I., **8**, 610[60], 611[66]
Kalinovski, I. O., **1**, 437[157]; **3**, 530[73], 534[73]
Kalinowski, H.-O., **1**, 70[63], 141[22]; **3**, 66[11], 74[11]; **6**, 419[8], 425[8]
Kalir, A., **7**, 657[31]; **8**, 384[26]
Kalish, J., **6**, 261[2], 262[2], 266[2]
Kaliska, V., **8**, 86[26]
Kalkote, U. R., **8**, 366[44]
Kallen, R. G., **2**, 955[9]
Kallenberg, H., **2**, 101[19]; **3**, 20[121], 25[121]; **7**, 125[57]
Kallman, N., **6**, 724[156]
Kallmerten, J., **1**, 131[98,99], 777[219]; **2**, 102[26], 763[60]; **3**, 1007[64]; **4**, 14[48]; **5**, 827[2], 829[2], 867[2b], 894[42]; **6**, 859[174,175], 878[40], 883[40], 887[62,63]; **7**, 546[30], 580[30]
Kallmunzer, A., **5**, 444[186]
Kallo, D., **4**, 298[287]
Kálmán, A., **6**, 525[382], 534[519]
Kalman, J. R., **3**, 505[160]; **7**, 649[45]
Kalnins, M. A., **4**, 120[199]
Kalos, A. N., **7**, 155[31a]
Kaloustian, M. K., **6**, 430[92], 452[133,134], 501[202], 503[219]
Kalra, B. L., **5**, 64[31]
Kalsi, P. S., **7**, 271[129]; **8**, 338[92], 339[92]
Kalsuki, T., **7**, 710[52]
Kaltia, S. A. A., **5**, 516[25], 517[25c]
Kaluza, Z., **5**, 108[207,208]
Kalvin, D. M., **7**, 574[126]
Kalvin'sh, I. Ya., **4**, 48[140]
Kalvoda, J., **5**, 179[143]; **6**, 1015[23], 1059[64]; **7**, 41[19,20,24]; **8**, 873[17], 974[124]
Kalwinsch, I., **4**, 1074[29]
Kalyan, Y. B., **5**, 850[148]
Kalyanam, N., **6**, 283[159]; **7**, 276[149]
Kalyanasundaram, S. K., **6**, 685[355]
Kamachi, H., **8**, 645[44]
Kamada, M., **3**, 498[109]
Kamada, S., **2**, 1020[46]; **5**, 864[257]
Kamada, T., **7**, 811[91]
Kamaike, K., **6**, 604[31]
Kamaishi, T., **8**, 144[71,72]
Kamal, A., **5**, 105[191]
Kamamoto, T., **8**, 549[325]
Kamasheva, G. I., **5**, 516[19]
Kamat, V. N., **3**, 396[115]; **8**, 339[96]
Kamata, A., **7**, 74[110]
Kamata, K., **6**, 725[170], 728[170]
Kamata, M., **4**, 1103[205]; **7**, 875[116]
Kamata, S., **5**, 611[72]; **6**, 438[42], 440[75]; **8**, 493[22], 836[4], 842[4], 931[39], 993[59]
Kamatani, T., **7**, 423[142]
Kambach, C., **2**, 1090[73], 1102[73], 1103[73]
Kambe, N., **2**, 450[54]; **3**, 1034[77]; **4**, 566[46]; **6**, 117[98], 479[108]; **7**, 131[80]; **8**, 323[112], 370[90], 382[7], 412[119], 413[119]

Kambe, S., **2**, 916[83]
Kamber, B., **6**, 668[262]; **7**, 236[24]
Kamber, M., **7**, 268[125]
Kambouris, J. G., **2**, 814[79]
Kamdar, B. V., **8**, 30[44], 66[44]
Kameda, T., **6**, 641[62]
Kamemura, I., **4**, 969[65]
Kamenar, B., **6**, 766[23]; **7**, 698[51]
Kamernitskii, A. V., **4**, 241[58]; **7**, 479[89]
Kameswaran, V., **3**, 509[176], 670[57,57a], 673[68], 681[68], 686[68], 807[33], 816[82]
Kametani, T., **1**, 372[78]; **2**, 222[147], 384[316], 547[121], 551[121], 552[121], 765[67], 819[100,103], 824[100], 851[222], 888[54], 1016[28], 1020[46], 1024[61,62], 1049[13]; **3**, 164[480], 165[480], 167[482], 168[482], 660[17], 677[86], 679[17]; **4**, 30[88,88k-o], 121[209,209a,b], 181[70], 231[277], 239[17,27], 257[27], 261[17,27,299,300,301], 333[21-23], 398[215], 401[215a,229], 487[42], 495[85], 500[110], 501[113,116-118], 504[130], 505[134,135,136], 510[172,173,176], 795[84], 1076[38]; **5**, 92[74], 385[130], 402[1], 403[1], 404[1], 410[1,38], 413[1,1c], 417[1], 422[1], 425[1], 4 26[1], 429[1], 430[1], 433[1], 434[1], 435[1], 436[1], 438[1], 440[1], 444[1], 473[154], 479[154], 522[45], 524[54], 531[77,78], 534[54,95], 536[97], 541[110,111], 681[28], 691[83,84,86], 692[83,83c,84,96], 693[83,107-111,114], 694[110,114], 712[57b], 723[107], 741[157,157c,d], 742[162], 839[82], 843[117], 847[136], 864[257], 1031[95], 1032[100]; **6**, 74[37], 739[58], 756[126], 757[130], 780[70], 896[93]; **7**, 21[15], 229[113], 243[68], 453[76], 476[59], 493[199], 564[89], 569[89]; **8**, 314[37], 534[158], 537[158], 647[56], 945[128]
Kameyama, M., **4**, 753[172]; **7**, 518[19]
Kameyama, Y., **2**, 649[103], 1059[76]
Kamigata, N., **4**, 753[172]; **7**, 518[18,19], 779[428]
Kamikawa, T., **4**, 258[254]; **7**, 355[40]; **8**, 544[266]
Kamimura, A., **2**, 332[64], 333[64,65], 334[68], 360[168]; **4**, 13[44,44b], 31[92,92i], 37[105], 64[92i], 790[36]; **5**, 320[10,11]; **6**, 161[178], 1000[130]; **8**, 962[41]
Kamimura, J., **8**, 314[43], 968[88], 988[29], 989[38]
Kamimura, Y., **3**, 295[8]
Kaminski, F. E., **5**, 390[140]
Kaminski, M., **6**, 496[155]
Kaminski, Z. J., **6**, 276[117]
Kamishiro, J., **6**, 619[115]
Kamitori, Y., **1**, 543[29], 563[178]; **8**, 18[128], 245[75]
Kamiya, T., **5**, 96[106,117]; **6**, 637[29]; **7**, 493[198]
Kamiya, Y., **1**, 248[62], 735[25]; **7**, 235[5]
Kamiyama, N., **7**, 875[112]
Kamiyama, S., **4**, 302[333]; **8**, 856[165]
Kamiyama, T., **4**, 444[198]
Kamm, J. J., **7**, 778[418]
Kamm, O., **6**, 209[64]
Kammerer, H., **3**, 660[6]; **8**, 898[22], 904[59], 907[59], 909[59]
Kamneva, A. I., **3**, 648[172,173]
Kamonah, F. S., **5**, 724[114]
Kämpchen, T., **5**, 475[146]; **6**, 502[213]
Kampe, W., **6**, 610[59,60]
Kampmeier, J. A., **3**, 1040[105]; **4**, 719[23,25]; **5**, 2[15]
Kamysheva, T. P., **3**, 304[60]
Kamyshova, A. A., **3**, 422[69]; **4**, 952[4]
Kan, K., **5**, 108[206]
Kan, R. O., **2**, 759[28]; **3**, 823[14], 835[14]; **5**, 637[111,114]
Kan, T., **3**, 100[199]
Kan, T. Y., **6**, 680[331]
Kanabus-Kaminska, J. M., **4**, 739[106]
Kanagasabapathy, V. M., **5**, 1123[34]
Kanai, A., **8**, 146[90]
Kanai, H., **4**, 973[84]
Kanai, K., **1**, 390[144]; **2**, 877[37], 882[45], 935[146], 936[146], 999[39]; **5**, 100[149]
Kanai, M., **8**, 144[73,75]
Kanai, T., **6**, 657[180]
Kanai, Y., **7**, 628[44]
Kanakarajan, K., **5**, 475[144]
Kanakura, A., **3**, 202[97]; **4**, 1007[126], 1008[126]
Kanamaru, H., **3**, 259[130]
Kanamori, Y., **2**, 792[63]
Kanaoka, Y., **5**, 181[153,154,156]; **6**, 18[65]; **7**, 42[28,29], 877[133,134]
Kanata, S., **5**, 611[72]
Kanatani, R., **1**, 619[61]; **4**, 120[202]; **6**, 16[60]; **7**, 642[8], 643[13], 645[17,18]; **8**, 787[119]
Kanavarioti, A., **2**, 349[68,69]
Kanaya, I., **3**, 1021[16]
Kanayama, S., **7**, 451[18], 454[18]
Kanazawa, H., **1**, 34[168]; **6**, 533[486]
Kanazawa, R., **8**, 267[63], 268[63]
Kanazawa, S., **5**, 839[76]
Kanazawa, T., **2**, 784[38], 1052[51]; **7**, 804[62]
Kanazawa, Y., **6**, 119[116]
Kanbara, H., **1**, 243[58]; **7**, 371[68], 379[100]
Kanbe, T., **7**, 794[7e]
Kanda, N., **1**, 317[142]; **4**, 803[132], 843[53,54], 852[53]; **8**, 170[101]
Kanda, Y., **1**, 553[90,94]; **3**, 390[79], 1008[72], 1009[72], 1010[72,75]; **6**, 875[18]; **8**, 588[45]
Kandall, C., **8**, 992[52]
Kandasamy, D., **8**, 538[191], 803[93], 804[93], 806[100], 826[69]
Kandror, I. I., **6**, 542[598,599]
Kane, P. D., **4**, 380[119]
Kane, R., **2**, 134[5]
Kane, V. V., **1**, 511[30]; **4**, 10[34], 71[13]; **7**, 136[113], 137[113]
Kaneda, K., **3**, 1041[111]; **4**, 603[275], 626[275], 645[275], 856[100], 941[81]; **6**, 711[64]; **7**, 309[25], 321[66], 587[171], 823[36]; **8**, 419[17], 430[17]
Kaneda, M., **8**, 334[64]
Kaneda, S., **4**, 433[121]
Kaneda, T., **3**, 556[35]
Kanefusa, T., **4**, 413[275]; **7**, 519[22]
Kanehira, K., **2**, 527[6], 528[6]; **3**, 229[226]; **4**, 653[437,442]; **5**, 1[4], 2[4]; **7**, 564[91], 565[91]; **8**, 84[12], 698[137]
Kanehisa, N., **1**, 162[93,102]
Kaneko, C., **2**, 967[77]; **4**, 208[62]; **5**, 134[65], 356[89]; **6**, 559[863]; **7**, 335[29]; **8**, 391[90], 397[140], 528[86], 641[26]
Kaneko, H., **6**, 534[516]; **8**, 531[110]
Kaneko, K., **6**, 753[118], 936[111]; **7**, 765[149], 773[149]
Kaneko, T., **1**, 223[84], 225[84a]; **2**, 102[24], 879[42]; **4**, 125[216,216d]; **5**, 803[92], 979[25]; **6**, 538[573], 637[36], 928[81,82]; **7**, 201[44], 202[45]; **8**, 99[113], 303[97]
Kaneko, Y., **1**, 561[159], 828[68]; **2**, 649[105]; **6**, 91[122], 240[79]
Kanellis, P., **7**, 274[139]
Kane-Maguire, L. A. P., **2**, 964[59]; **4**, 518[3], 542[117], 670[15,20,22], 688[65]
Kanemasa, S., **2**, 482[21]; **3**, 201[84]; **4**, 16[52b,c], 75[43a], 100[43], 111[152c], 120[197], 1086[114,115]; **5**, 758[81]; **6**, 542[603]
Kanemase, S., **1**, 770[190]
Kanematsu, K., **3**, 383[41,44], 809[41]; **4**, 159[82], 297[276]; **5**, 537[99], 603[53], 604[53], 621[20], 627[41], 632[64], 634[71,72,78,80,81], 819[156]; **6**, 814[92]; **7**, 462[124]
Kanemoto, S., **3**, 484[24], 501[24], 730[43]; **7**, 267[118], 268[118], 275[146,147], 276[147], 281[176], 282[176], 283[118], 284[118], 308[17], 378[96], 379[101]; **8**, 886[113]
Kanemoto, Y., **3**, 404[135]
Kanevskii, L. S., **3**, 635[35], 647[177,196], 648[181], 649[177]
Kan-Fan, C., **2**, 901[31,32]; **6**, 734[16], 912[22], 920[44]

Kanfer, S., **5**, 154[31]
Kanfer, S. J., **3**, 219[114], 499[140], 501[140], 502[140]
Kang, D. B., **8**, 423[37]
Kang, G. J., **2**, 616[132], 619[132], 621[132], 622[156]
Kang, H.-J., **5**, 833[44]; **6**, 859[169]; **8**, 14[77], 17[111]
Kang, H.-Y., **1**, 198[93], 199[93]
Kang, J., **1**, 97[81]; **3**, 217[84], 250[69]; **6**, 175[67]; **8**, 18[121], 517[123]
Kang, J. W., **5**, 1134[42], 1144[102], 1146[102]; **8**, 154[197], 155[197], 445[34], 453[191]
Kang, K., **6**, 686[366]; **8**, 890[139]
Kang, M., **4**, 239[21], 256[21], 261[21], 807[151]; **6**, 7[32]; **7**, 182[161]; **8**, 798[58]
Kang, M.-c., **2**, 638[61], 640[61]; **3**, 546[123]; **6**, 674[294]; **7**, 680[77]
Kang, S. H., **1**, 798[286], 804[309]; **3**, 225[187]; **4**, 1040[68]; **6**, 5[28]
Kang, S. I., **6**, 462[22]
Kang, S.-Z., **4**, 279[116]
Kang, Y. H., **7**, 770[246]
Kang, Y. S., **6**, 524[371]
Kanghae, W., **2**, 417[20]; **6**, 1022[65]
Kanh, P., **4**, 173[33]
Kanischev, M. I., **2**, 725[108]
Kanjilal, P. R., **3**, 87[115]; **5**, 232[134]
Kannan, P. S. M., **8**, 137[55]
Kannan, R., **1**, 468[52]; **5**, 382[119a]
Kanner, C. B., **4**, 48[139]
Kanno, S., **4**, 820[215]; **8**, 369[77]
Kano, S., **1**, 790[263]; **2**, 363[193], 1060[88,91], 1073[141,142]; **4**, 364[1,1h], 401[233], 487[42], 505[138]; **5**, 725[118]; **6**, 749[98], 751[108], 780[72], 879[43]; **8**, 64[207a], 65[207a], 66[207], 237[17], 240[17], 249[17], 389[75], 934[57], 957[13]
Kano, Y., **1**, 174[13], 202[13]
Kanoaka, Y., **7**, 877[134]
Kanoh, M., **4**, 104[137], 227[208]
Kanoh, S., **8**, 170[71,95]
Kant, J., **2**, 648[95], 649[95], 925[110], 926[110]; **5**, 102[171,172]; **6**, 820[112]
Kanters, J. A., **1**, 23[121-125]; **8**, 95[82]
Kantlehner, W., **2**, 368[238]; **6**, 229[25], 430[93], 452[131], 487[5], 488[5,9,10,13,14,28], 489[5,28], 490[5], 491[5], 492[5,125], 493[5], 494[5], 495[5,9], 496[5,9], 497[9], 498[9], 499[9], 500[9], 501[10], 502[10], 503[10], 504[10], 505[10], 506[10], 507[10], 508[13], 510[13], 511[13], 512[14,120,303], 513[14], 514[14], 515[10,13,14,315], 517[14], 518[14], 519[14,336], 521[14], 522[14,125], 524[5,9], 526[5,9], 529[10], 536[13], 543[5,14,120], 544[5], 545[9,10,13,14,634], 546[634,653], 553[796,797,798], 554[799,800,802], 556[821,829,830], 562[887,891], 566[9,28], 567[10,28,634,829], 568[14,28,634], 570[28], 571[28], 572[28,796,958], 573[28,798,963], 574[28], 575[28], 576[28,802,973,974], 577[28], 578[28], 579[28], 580[28], 581[28,802], 672[282]; **8**, 85[18]
Kantner, S. S., **3**, 222[138,138b,142]; **6**, 848[108]
Kantor, E. A., **8**, 214[40]
Kantor, S. W., **3**, 914[8], 915[14], 918[21], 967[14], 975[2], 976[2], 979[2], 980[2]
Kanuma, N., **6**, 559[861], 996[105]
Kao, J. C., **3**, 853[69]
Kao, L., **4**, 852[90], 854[94], 855[96,97]
Kao, S. C., **8**, 22[146], 289[28]
Kapassakalidis, J. J., **6**, 546[653], 556[821,830], 562[887,891]
Kapecki, J. A., **5**, 68[95]
Kapeisky, M. Y., **6**, 618[114]
Kapil, R. S., **7**, 261[71]
Kapkan, L. M., **6**, 516[320]
Kaplan, F., **3**, 891[44]
Kaplan, J., **3**, 824[23]; **4**, 723[40,41], 747[40], 757[185], 776[40], 803[130,135], 811[173]
Kaplan, L., **5**, 646[3,8], 662[35]
Kaplan, M., **5**, 947[260], 960[260]
Kaplan, M. S., **5**, 826[159b]
Kaplan, R. B., **7**, 500[240]
Kapnang, H., **8**, 27[31], 66[31]
Kapon, M., **4**, 1040[92], 1042[92]
Kapoor, S. K., **7**, 220[19]
Kapoor, V. M., **3**, 373[127]; **8**, 165[50], 510[98], 545[292,293]
Kapovits, I., **7**, 764[120]
Kappan, L. S., **5**, 736[142i,j]
Kappeler, H., **3**, 345[20]; **4**, 43[123]; **6**, 637[27]
Kappert, M., **1**, 221[68]; **4**, 36[103,103b], 991[149]
Kappey, C.-H., **3**, 373[130]
Kaptein, B., **8**, 97[98]
Kapur, J. C., **5**, 95[89]
Kapuscinski, M., **2**, 456[58,63], 458[58,63]
Karabatsos, G. J., **1**, 49[5]; **2**, 24[96], 217[137]; **3**, 839[8]; **6**, 1034[132]; **7**, 439[35]; **8**, 3[19]
Karabelas, K., **4**, 846[70]
Karabinos, J. V., **7**, 760[33]; **8**, 293[47]
Karady, S., **2**, 648[97], 649[97b]; **3**, 30[175], 380[11]; **4**, 230[250,251], 767[234]; **7**, 493[188]; **8**, 272[122]
Karafiath, E., **5**, 802[85]
Karaghiosoff, K., **6**, 509[265]
Karahanov, R. A., **6**, 530[418]
Karakasa, T., **5**, 475[140,141]
Karakhanov, E. A., **8**, 600[106], 606[25], 625[25,159], 628[177]
Karakhanov, R. A., **5**, 480[177]; **6**, 439[70]; **8**, 606[24,27], 607[24]
Karalis, P., **7**, 155[28]
Karaman, B., **6**, 510[298]
Karaman, H., **7**, 564[83]
Karapetyan, K. G., **3**, 648[181]
Karapinka, G. L., **3**, 664[33]
Karas, G. A., **7**, 96[85]
Karasaki, Y., **6**, 477[98]
Karasawa, T., **8**, 613[78]
Karasawa, Y., **2**, 967[77]
Karashima, D., **8**, 618[109]
Karaulova, E. N., **4**, 364[1,1f], 413[1f]
Karbach, S., **6**, 42[45]
Kardos, J., **2**, 812[72]
Karge, R., **2**, 681[59]; **5**, 353[85], 365[96c], 543[113]
Karger, M. H., **2**, 739[47]
Karickhoff, S., **2**, 765[80]
Karila, M., **2**, 91[43]
Karim, A., **8**, 460[248]
Karimov, K. G., **8**, 450[163]
Karimov, K. K., **8**, 458[221]
Kariv, E., **8**, 313[23]
Kariv-Miller, E., **1**, 268[54], 269[54c]; **3**, 568[57], 570[57], 599[205,222], 602[222]; **4**, 809[162]; **8**, 132[10-14], 134[10,12,13], 517[125], 624[154], 628[154], 630[188]
Kariyone, T., **8**, 212[9], 222[9]
Karkour, B., **3**, 620[28]; **5**, 1021[72]
Karl, E., **8**, 690[102]
Karl, H., **6**, 60[146]
Karl, R., **2**, 1090[73], 1099[109,109b], 1102[73], 1103[73]
Karle, I. L., **1**, 11[56], 300[65,66]
Karlin, K. D., **4**, 706[38]
Karlsson, F., **4**, 278[94], 285[94]
Karlsson, J. O., **2**, 785[42]; **5**, 690[80,80c], 733[136,136f], 734[136f]
Karlsson, M., **4**, 55[157], 56[157a]
Karlsson, S., **6**, 423[44]
Karni, M., **4**, 986[132], 987[132]

Karns, T. K. B., **5**, 404[15]
Karo, W., **6**, 294[233], 685[346], 690[346], 692[346], 726[186]; **7**, 741[46,50], 746[46], 747[50,99,100], 748[99,100]; **8**, 364[11,23], 365[30], 382[5]
Karodia, N., **5**, 834[56]
Karoglan, J. E., **3**, 595[191]; **8**, 528[60]
Kárpáti-Adam, E., **6**, 514[307]
Karpeiskii, M. Ya., **5**, 491[206]
Karpeles, R., **8**, 408[74]
Karpenko, T. F., **4**, 289[190]
Karpf, M., **2**, 711[30], 725[115]; **3**, 1049[14,15], 1058[42]; **4**, 148[50]; **5**, 18[128]
Karputschka, E. M., **7**, 746[92], 752[92]; **8**, 829[82]
Karra, S. R., **5**, 714[74]
Karras, M., **2**, 527[9], 528[9], 541[76], 544[76], 546[76], 547[76,118], 551[118]; **3**, 244[26], 485[28]; **4**, 879[84], 892[144]; **5**, 2[7], 4[7]
Karrer, P., **2**, 143[54]; **7**, 92[42], 93[42], 657[32]; **8**, 531[121], 812[2], 813[2]
Karrick, G. L., **3**, 325[161,161a]
Karsch, H. H., **1**, 17[213,214,215], 36[236,237]; **6**, 172[27]
Kartashov, A. V., **4**, 347[93]
Kartashov, V. R., **4**, 310[427]
Karten, M. J., **1**, 411[45]
Karten, M. T., **7**, 372[71]
Kärtner, A., **2**, 657[160], 1052[50], 1053[50], 1067[50]
Kartoon, I., **6**, 825[126]
Karube, I., **7**, 429[155]
Karunaratne, V., **3**, 20[114,115]; **4**, 255[203]
Kasafirek, E., **6**, 639[46]
Kasahara, A., **3**, 530[81-83], 536[81-83], 594[185]; **4**, 557[12], 558[17], 611[352], 837[9], 839[27], 845[69], 847[76], 858[110], 903[202], 904[202]; **6**, 726[179]
Kasahara, C., **2**, 167[162]; **4**, 387[159], 393[159]
Kasahara, I., **5**, 221[55]; **8**, 462[267], 463[269]
Kasahara, T., **2**, 13[56]
Kasai, N., **1**, 14[74-76], 19[98], 162[93,102]; **3**, 672[65]; **4**, 607[317], 615[384]; **5**, 275[11], 422[82], 1200[49]; **6**, 976[9]
Kasai, R., **7**, 43[40,41]
Kasal, A., **7**, 71[99]
Kasamatsu, Y., **7**, 25[44]
Kasatani, R., **8**, 817[34]
Kasatkin, A. N., **4**, 594[144], 610[340]
Kasatkin, P. N., **4**, 594[141]
Kascheres, A., **3**, 735[20]; **6**, 94[141], 98[141]
Kaschube, W., **1**, 749[78], 816[78]
Kase, K., **3**, 649[205]
Kasel, W., **3**, 872[61,63]
Kasha, M., **7**, 98[102]
Kashihara, H., **2**, 209[109]
Kashima, C., **1**, 373[88], 374[88]; **2**, 153[105]; **7**, 229[123]; **8**, 540[200], 642[32,34]
Kashimura, M., **8**, 614[82]
Kashimura, N., **7**, 299[44]
Kashimura, S., **1**, 346[128], 544[39], 804[310]; **2**, 138[19]; **3**, 54[279]; **4**, 130[226c], 247[100], 257[100], 260[100]; **5**, 500[259]; **6**, 801[37]; **7**, 796[13], 798[22], 804[58]; **8**, 302[96]
Kashimura, T., **3**, 1026[40]
Kashin, A. N., **1**, 437[157]; **3**, 503[152]; **4**, 610[341]
Kashiwaba, N., **5**, 323[16]
Kashiwabara, K., **8**, 535[166]
Kashiwabara, M., **5**, 86[34]
Kashiwaga, M., **4**, 145[26]
Kashiwagi, H., **5**, 151[18]
Kashiwagi, K., **2**, 655[141], 948[182]
Kashiwagi, M., **8**, 95[87]
Kashiwagi, T., **5**, 1158[173]
Kashman, K., **5**, 603[52], 604[52]
Kashman, Y., **3**, 380[13], 395[102], 396[102], 407[146]
Kashti-Kaplan, S., **8**, 594[73], 595[73]
Kashu, M., **8**, 375[155]
Kasina, S., **6**, 94[140]
Kasiwagi, I., **3**, 824[17], 825[17]
Kaska, D. D., **3**, 579[97]
Kaska, W. C., **4**, 889[137]; **8**, 740[39,40], 741[39,40], 756[39], 757[39]
Kaskimura, S., **6**, 991[88]
Kaslow, C. E., **8**, 905[63]
Kasonyi, A., **7**, 154[21]
Kaspar, F., **5**, 402[5]
Kašpar, J., **8**, 534[159]
Kasper, A. M., **5**, 473[155], 474[155], 477[155]
Kasper, D., **2**, 464[97]
Kasperowicz, S., **8**, 52[150], 66[150]
Kass, N. C., **7**, 618[23]
Kass, S. R., **3**, 4[22]
Kasten, R., **7**, 880[152]
Kastner, B., **3**, 890[32]
Kastner, M. R., **4**, 191[109]
Kasturi, T. R., **2**, 378[288]; **8**, 214[42], 220[42], 246[76], 248[76,83]
Kasuga, K., **2**, 710[24]; **3**, 95[153], 107[153], 114[153], 115[153]; **6**, 137[43], 604[31], 1022[57]; **7**, 453[73,84], 454[73,84]; **8**, 173[113], 836[10a], 837[10a]
Kasuga, O., **6**, 489[92]
Kasuga, T., **2**, 119[161]; **3**, 1004[59], 1005[62,63]; **6**, 883[54], 890[69], 891[54,69,70]
Kaszonyi, A., **7**, 451[27]
Kaszynski, P., **4**, 765[224]
Katada, T., **6**, 252[154], 453[142,144]
Kataev, E. G., **4**, 318[562]; **7**, 521[37]
Katagiri, K., **6**, 441[85]
Katagiri, M., **8**, 418[10]
Katagiri, N., **2**, 757[15], 761[49], 762[49]; **5**, 90[56], 356[89], 451[43], 485[43]; **6**, 487[40], 489[40], 603[22], 618[113], 625[158]
Katagiri, T., **7**, 407[83]
Katahiro, D. A., **8**, 847[94]
Katakawa, J., **4**, 27[79,79d]; **7**, 156[32], 175[143]
Katakis, D., **3**, 565[20]
Kataoka, F., **3**, 583[127], 587[143]; **5**, 76[240]
Kataoka, H., **1**, 187[63]; **3**, 575[82], 883[107]; **4**, 611[361], 629[410,411], 638[410]; **5**, 736[145], 737[145], 790[22], 820[22], 935[190]; **6**, 11[46], 77[56]
Kataoka, M., **2**, 152[100]; **3**, 585[137]; **6**, 968[106]; **7**, 700[59]
Kataoka, N., **4**, 611[362], 612[362]
Kataoka, S., **3**, 557[49]
Kataoka, T., **5**, 504[276], 627[41]; **6**, 510[293], 927[80], 936[106]; **8**, 996[71]
Kataoka, Y., **1**, 205[108], 807[322], 808[322], 809[329], 810[329a,b]; **2**, 603[42]; **8**, 146[95,96], 148[95,96]
Katayama, E., **1**, 184[52], 185[54], 248[64]; **2**, 10[40], 29[106]; **3**, 730[42,42b]; **6**, 14[53]; **7**, 298[35]; **8**, 10[57]
Katayama, H., **4**, 230[252], 231[252]; **5**, 461[107], 464[107], 466[107]; **6**, 753[118]
Katayama, S., **2**, 388[344]
Katayama, T., **3**, 843[25]
Katekar, G. F., **3**, 88[136], 91[136], 179[136], 181[136]
Katerinopoulos, H., **3**, 217[94]
Kates, M. R., **2**, 977[5]
Kates, S. A., **4**, 764[216], 807[148,153]
Kathawala, F., **5**, 514[8], 527[8,8b]; **6**, 509[246]; **7**, 160[50]; **8**, 9[50]
Katjar-Peredy, M., **7**, 831[62]
Kätker, H., **5**, 1119[20]
Kato, E., **7**, 760[50]

Kato, H., **2**, 357[136]; **4**, 162[93]; **6**, 533[512], 619[120], 624[120,148]; **7**, 693[27]
Kato, J., **8**, 216[60]
Kato, K., **6**, 559[861,862]; **7**, 489[173]; **8**, 371[109]
Kato, M., **1**, 759[128]; **2**, 68[44], 113[111], 244[27], 245[27], 601[36], 605[58], 1066[119]; **3**, 88[124], 118[124], 197[33], 224[162]; **4**, 243[69], 244[69], 245[69], 258[69], 370[44,45]; **5**, 755[73], 761[73], 1188[16]; **6**, 147[85,86], 1046[32a], 1047[32b]; **8**, 836[10c,d], 837[10d], 916[108], 917[108], 918[108], 920[108]
Kato, N., **1**, 187[61-64], 188[62]; **2**, 666[34,35], 675[35], 700[34,35]; **3**, 213[42], 224[162], 264[185], 573[71], 575[81,82], 610[71]; **4**, 176[54]; **5**, 24[169], 434[144], 790[22], 820[22]; **6**, 531[426]; **7**, 506[302]; **8**, 544[277]
Kato, R., **1**, 238[36], 568[246]
Kato, S., **4**, 430[89]; **6**, 450[119], 453[142,143,144], 454[119,146], 461[5], 464[37], 465[37], 467[50], 473[77]; **8**, 253[122]
Kato, T., **1**, 58[33], 100[89], 135[117], 881[118]; **2**, 204[97], 587[136], 711[29], 757[15], 761[49], 762[49], 789[56], 819[99]; **3**, 446[77], 512[199], 730[45]; **4**, 358[154], 373[73], 378[108], 490[66], 499[66], 510[172]; **5**, 90[56], 225[103,104], 226[104], 451[43], 485[43], 637[115], 704[21], 810[126], 812[126]; **6**, 487[40], 489[40], 533[492], 559[861,862], 569[938], 602[5,10]
Kato, Y., **5**, 693[109,114], 694[114]; **8**, 188[49], 193[49], 563[429]
Katoaka, Y., **5**, 1125[60]
Katogi, M., **4**, 30[88,88m], 261[300]
Katoh, A., **6**, 113[71]; **7**, 533[33]; **8**, 642[32,34]
Katoh, E., **1**, 561[164]
Katoh, S., **7**, 811[91]
Katoh, T., **1**, 860[68]; **4**, 606[307], 607[307,315], 647[307]; **7**, 243[68]; **8**, 976[135]
Katoh, Y., **5**, 410[38]
Katonak, D. A., **7**, 516[6]; **8**, 337[81]
Katou, T., **3**, 530[81,82], 536[81,82]
Katritzky, A. R., **1**, 357[4], 463[28], 464[36], 469[59], 471[69], 474[89], 477[140], 478[36], 481[160]; **2**, 786[47,48]; **3**, 71[35], 197[36], 282[47]; **4**, 48[137,137g], 113[175], 425[24,25], 428[24], 430[24,90-93], 436[144], 440[170], 1093[145], 1099[183]; **5**, 469[138], 491[206,207], 630[52], 947[263,264,265,274,286,287,289]; **6**, 84[86], 255[169], 261[10], 273[10], 280[10], 419[11,12], 532[469], 677[320]; **7**, 138[127], 226[73], 228[102], 305[1], 358[1], 384[1], 470[4], 472[4], 473[4], 474[4], 476[4], 662[53-55], 739[37-39], 745[73], 750[130]; **8**, 296[61], 392[103], 587[37], 653[82], 827[72-75], 843[49]
Katsaros, N., **3**, 565[20]
Katsifis, A. A., **4**, 226[190,191]; **6**, 154[145]
Katsifis, A. G., **1**, 520[72-74]; **2**, 66[33], 75[33]; **4**, 119[192b,193], 159[85], 226[194,195], 258[246]; **6**, 154[145], 864[192]
Katsobashvili, V. Y., **5**, 64[24]
Katsuhara, J., **3**, 771[191]
Katsuhara, Y., **4**, 347[102]
Katsuki, M., **3**, 303[55]
Katsuki, T., **2**, 119[159,162,163], 304[9], 305[9], 725[121], 846[204,205]; **3**, 45[248], 223[156], 225[185,187], 264[181], 1000[55], 1001[56], 1004[60]; **5**, 185[163], 366[98]; **6**, 2[3], 5[25,27], 8[39], 25[3], 88[105], 89[105,111], 175[76], 668[260], 852[135], 877[38,39], 878[39], 880[46], 883[38,39], 885[56], 887[38,39], 890[56], 927[76]; **7**, 198[26], 238[42], 239[42], 240[42], 246[94], 297[30], 379[99], 382[99], 390[3,12], 391[13], 397[3], 399[3], 400[3,41,44], 401[59], 403[59], 406[3,59,77], 407[3,41], 408[44], 409[3,77], 410[3], 411[13], 412[13], 413[13,107a], 414[77], 415[77], 417[131], 419[133], 421[77], 422[141], 423[77,141], 571[113], 572[113], 587[113]; **8**, 11[62], 145[83]
Katsumata, N., **7**, 356[52]
Katsumata, S., **1**, 287[18]
Katsumi, K., **7**, 851[27]
Katsumura, A., **8**, 152[180]
Katsumura, S., **2**, 749[134]; **3**, 470[208], 471[208], 475[208]; **4**, 594[146]
Katsuno, H., **8**, 698[139]
Katsurada, M., **7**, 423[145], 424[145b]
Katsuro, Y., **2**, 584[122]; **3**, 445[69,71], 455[122], 492[76,78], 495[78]; **7**, 642[12]; **8**, 783[106]
Katsuura, K., **1**, 567[221], 698[248]; **4**, 398[220]
Kattenberg, J., **3**, 875[77]
Katterman, L. C., **4**, 121[205b]
Katti, S. B., **6**, 606[39]
Katz, A., **7**, 100[127]
Katz, A. H., **4**, 411[266a], 567[48]
Katz, J.-J., **3**, 794[76], 795[82]; **8**, 706[15], 709[15,47], 710[52]
Katz, R. B., **2**, 881[43], 885[51]; **6**, 509[276]
Katz, S., **2**, 710[26]; **6**, 986[68]
Katz, T. J., **4**, 966[55]; **5**, 794[51], 804[93], 1094[103], 1102[146a], 1104[157], 1113[146], 1115[1], 1116[1], 1187[11,12]; **7**, 884[191]
Katzenellenbogen, J. A., **2**, 187[41], 718[80]; **3**, 50[267], 288[64], 443[60]; **4**, 185[89], 248[110], 262[110], 315[523,529,530,533], 379[114], 394[189,189b,193]; **5**, 850[161]; **6**, 219[118]; **7**, 87[18], 163[78], 167[78]; **8**, 754[75,76], 872[8]
Kaubisch, S., **2**, 792[67]
Kauder, O., **8**, 957[16]
Kauer, J. C., **6**, 244[106]
Kaufer, J. N., **2**, 323[34]
Kauffman, G. B., **3**, 209[13]
Kauffman, W. J., **1**, 221[67], 223[72a,b]; **3**, 765[152]; **4**, 72[23], 95[23]
Kauffmann, H. F., **5**, 63[19]
Kauffmann, T., **1**, 202[101,102], 203[102], 234[22,26], 253[8], 331[47,48], 630[20,21], 635[91], 637[91], 645[124], 669[124,181,182], 670[181,182], 672[91], 678[91], 680[124], 718[20], 734[22], 749[78], 755[113], 816[78]; **2**, 127[239]; **3**, 105[214], 203[100], 253[91], 482[3,5], 499[5], 505[5], 509[5,196], 512[196], 522[16]; **4**, 120[200], 485[28], 491[68], 492[72], 503[28]; **5**, 1126[66]; **7**, 506[303], 746[82]; **8**, 568[486], 889[133]
Kaufhold, G., **2**, 1090[64], 1091[64], 1108[64]
Kaufman, C., **4**, 1007[110]; **6**, 677[318,318a]
Kaufman, D. C., **4**, 1076[47]
Kaufman, G., **1**, 377[97]; **6**, 727[206]; **8**, 943[120]
Kaufman, M. J., **1**, 41[197]
Kaufmann, D., **5**, 1188[15]
Kaufmann, E., **4**, 872[41]
Kaufmann, H., **6**, 1059[64,65]
Kaufmann, K., **3**, 735[20]
Kaufmann, S., **3**, 805[15]; **6**, 685[357]
Kaul, C. L., **6**, 543[618]
Kaulen, J., **6**, 20[72]
Kaupmann, W., **8**, 57[172], 66[172]
Kaupp, G., **5**, 636[98,101], 707[38], 723[106], 726[106d]
Kaur, A. J., **6**, 563[896]
Kaura, A. C., **4**, 1014[188], 1022[188]
Kautzner, B., **4**, 153[62a]
Kavaliunas, A. V., **4**, 588[62,63]
Kavapetyan, A. V., **4**, 304[362], 305[362]
Kavarnos, G. J., **7**, 851[15]
Kavka, M., **5**, 955[303]
Kawabata, A., **1**, 166[114]; **5**, 176[128]
Kawabata, H., **5**, 736[142g]
Kawabata, K., **8**, 241[38], 272[119,120]
Kawabata, N., **4**, 93[95], 968[57,58], 969[65], 970[58b,70], 971[77], 972[80], 973[58b], 1032[9], 1051[9]; **7**, 765[163]
Kawabata, T., **5**, 98[123,124], 439[171], 440[171]; **7**, 588[174,175]; **8**, 902[44], 908[44], 909[44], 989[37]
Kawabata, Y., **4**, 915[10], 918[10], 931[55,57]; **8**, 463[270]

Kawada, M., **1**, 802[304], 803[307]; **2**, 74[74]; **3**, 135[345,346,347,348], 136[347,348], 137[347], 139[347,348], 141[348], 144[347,348], 168[508], 169[508], 174[345], 586[138], 650[211]; **4**, 609[329], 964[46]; **5**, 767[120], 830[31], 833[49]; **6**, 927[77], 1016[35], 1022[60]
Kawada, N., **7**, 641[6]
Kawafuji, Y., **1**, 116[50], 118[50]
Kawagishi, H., **3**, 693[149], 694[149]; **5**, 534[95]
Kawagishi, T., **4**, 13[44,44c], 177[59], 238[13], 253[169], 257[230], 261[13,230], 262[308]; **8**, 544[254]
Kawaguchi, A., **4**, 239[27], 257[27], 261[27]; **5**, 473[154], 479[154], 522[45], 531[78]
Kawaguchi, A. T., **1**, 312[114]; **5**, 543[116,116c]
Kawaguchi, K., **1**, 415[63]; **4**, 278[100], 286[100]
Kawaguchi, S., **4**, 587[30]
Kawaguchi, T., **7**, 248[112]
Kawaguti, T., **7**, 219[10]
Kawahara, I., **3**, 167[484], 168[484], 361[74]
Kawahara, K., **8**, 185[24]
Kawahara, T., **4**, 36[102,102a]
Kawaharada, H., **7**, 56[17]
Kawahata, Y., **5**, 151[9]
Kawai, A., **7**, 692[23]
Kawai, K., **1**, 78[20], 223[76], 224[76a], 317[145,155], 319[145], 320[155]; **8**, 652[73]
Kawai, K.-I., **1**, 34[168]; **7**, 43[40]
Kawai, M., **2**, 576[75], 615[130], 631[14], 635[14], 655[139]; **4**, 161[88]; **6**, 66[4], 89[108]; **8**, 142[50], 190[80]
Kawai, N., **7**, 506[301]
Kawai, T., **2**, 381[311]; **4**, 599[219], 641[219]; **5**, 829[26], 924[147]; **6**, 900[115]; **7**, 762[80], 777[80]
Kawakami, J. H., **4**, 301[328,329,330], 302[328]; **8**, 880[58]
Kawakami, N., **6**, 619[115], 734[11]
Kawakami, S., **3**, 597[201]
Kawakami, Y., **3**, 244[22], 465[179], 494[89], 516[89]; **4**, 95[96,100]; **5**, 168[106]
Kawakita, T., **1**, 223[84], 225[84b]
Kawamata, T., **8**, 935[64]
Kawamato, T., **2**, 303[6]
Kawami, Y., **4**, 179[65]
Kawamori, M., **8**, 338[82], 339[82]
Kawamoto, F., **1**, 880[113]
Kawamoto, I., **1**, 553[86]; **3**, 198[46]; **4**, 113[164]
Kawamoto, K., **3**, 987[32,33], 1000[52]; **6**, 883[58], 887[58,64]; **8**, 773[70], 774[70,71]
Kawamoto, S., **6**, 438[48]
Kawamoto, T., **1**, 85[28]; **2**, 114[115,116], 584[126]; **3**, 259[137]; **4**, 254[180,182]
Kawamura, C., **3**, 640[99]
Kawamura, F., **4**, 147[41]
Kawamura, J., **7**, 197[14]
Kawamura, K., **1**, 258[24], 259[24], 266[47]
Kawamura, M., **2**, 823[116]; **6**, 554[707]
Kawamura, N., **6**, 927[80]; **8**, 461[262]
Kawamura, S., **2**, 185[30], 217[30]; **3**, 1033[71]; **4**, 85[73], 249[118], 257[118], 377[104], 379[104j,114,114b,115,115b], 380[104j,115b], 382[114b], 383[114b], 402[236], 413[114b], 557[14]; **7**, 197[15], 503[273]
Kawamura, T., **7**, 642[11]
Kawana, M., **4**, 201[8]; **8**, 144[76]
Kawanami, Y., **1**, 421[89]; **3**, 45[248]; **5**, 366[98]; **6**, 7[31]; **8**, 145[83]
Kawanishi, G., **8**, 33[56], 66[56]
Kawanishi, Y., **7**, 309[25], 321[66]
Kawanisi, M., **1**, 752[98]; **2**, 86[25]; **3**, 652[221], 903[125]; **5**, 168[106]; **6**, 561[875]
Kawano, Y., **2**, 157[122]
Kawara, T., **3**, 227[210], 245[30], 446[78]; **4**, 262[303], 898[177], 902[190]; **6**, 836[53]
Kawas, E. E., **8**, 964[50]
Kawasaki, A., **4**, 589[76], 598[189]; **6**, 85[92]
Kawasaki, H., **1**, 566[215]; **2**, 846[208]; **3**, 41[224], 43[235]; **6**, 137[74]
Kawasaki, K., **4**, 254[179]; **6**, 438[55]; **8**, 548[324], 549[324]
Kawasaki, M., **1**, 241[44]; **8**, 166[61]
Kawasaki, S., **4**, 21[66]
Kawasaki, T., **5**, 725[118]; **7**, 335[31], 382[108]
Kawasaki, Y., **1**, 192[82]; **2**, 566[23]
Kawase, M., **7**, 333[23]; **8**, 60[182], 61[182], 62[182], 64[182], 66[182], 70[230], 71[230], 357[191]
Kawashima, H., **7**, 628[44]
Kawashima, K., **1**, 188[70]; **2**, 540[70]
Kawashima, M., **3**, 218[100], 220[125], 227[210], 245[30-32], 446[78], 470[212,213], 476[212,213]
Kawashima, S., **3**, 380[13]
Kawashima, T., **1**, 781[230]; **2**, 805[43]; **5**, 829[23]; **6**, 535[524]
Kawashima, Y., **5**, 297[57,59], 1196[37,38], 1197[38]
Kawata, K., **7**, 476[61]
Kawauchi, H., **5**, 1186[3], 1187[8]
Kawazoe, K., **6**, 538[562]
Kawazoe, T., **3**, 470[215], 473[215], 476[215]
Kawazoe, Y., **1**, 387[133]; **8**, 405[23]
Kawazura, H., **1**, 441[173]
Kay, D. G., **7**, 211[97]
Kay, D. J., **8**, 650[63]
Kay, G., **6**, 420[24], 451[127]
Kay, I. T., **1**, 560[154], 566[218], 751[92]; **2**, 553[130], 554[130]; **4**, 55[157], 57[157f], 249[127], 258[127]; **7**, 237[35,36]
Kayama, Y., **7**, 473[33], 501[33], 502[33]
Kaydos, J. A., **7**, 208[84]
Kaye, A. D., **3**, 985[27]; **6**, 876[27], 880[27]
Kaye, H., **7**, 586[166]
Kaye, I. A., **3**, 724[12]
Kaye, P., **5**, 403[9]
Kaye, P. T., **3**, 421[52]
Kaye, R. L., **3**, 896[68]; **5**, 64[37]
Kaye, S., **4**, 274[62], 275[62], 280[62]
Kayser, M. M., **8**, 2[7], 239[27], 240[27], 242[27,40]
Kayser, R. H., **6**, 954[19]
Kazakova, L. I., **3**, 639[80]
Kazakova, V. V., **6**, 546[645]
Kazama, H., **5**, 1157[169]
Kazanskii, B. A., **3**, 381[32]; **4**, 885[114]; **5**, 33[8]
Kazarinov, V. E., **3**, 636[55]
Kazda, S., **8**, 592[64]
Kazi, A. B., **3**, 353[52]
Kazimirchik, I. V., **3**, 864[21]
Kazimirova, V. F., **3**, 723[10]
Kazlauskas, R., **4**, 816[200]; **6**, 81[76], 82[76], 818[106]; **8**, 205[166], 206[166], 544[261]
Kazmaier, P. M., **6**, 924[62], 926[66]; **7**, 196[11], 199[11]
Kazubski, A., **1**, 405[25]; **7**, 603[115,125]; **8**, 102[126,128]
Kazuta, Y., **2**, 1089[57]
Ke, Y. Y., **1**, 391[152]; **2**, 611[102]
Keal, C. A., **1**, 569[251]; **3**, 125[301], 126[301], 128[301], 129[301], 133[301]
Keana, J. F. W., **1**, 377[103], 392[157], 393[157]; **7**, 43[43]
Kearley, F. J., Jr., **4**, 489[61]
Kearns, D. R., **5**, 216[15], 219[15]; **7**, 96[90], 98[90,98]
Keating, M., **4**, 488[57]; **7**, 743[65]
Keats, G. H., **8**, 321[97]
Keavy, D. J., **5**, 382[119b]

Keay, B. A., **3**, 488[52], 495[52]; **4**, 249[122]
Keay, J. G., **4**, 425[25], 430[91]; **8**, 580[3], 584[3], 585[3], 646[48]
Kebarle, P., **1**, 287[16]; **3**, 301[47]; **7**, 854[46]
Keberle, W., **4**, 270[12]
Keblys, K. A., **7**, 587[168]
Keck, G. E., **1**, 295[54], 296[54,55], 297[56,57], 306[54], 769[194], 794[281], 795[281]; **2**, 4[12,12c], 6[12,12c], 10[40], 15[65], 156[118], 573[53,54], 581[104], 630[8], 644[8a], 646[8a], 981[24], 989[24], 990[24], 1064[109]; **3**, 572[69], 573[69], 602[69], 607[69], 610[69]; **4**, 373[70], 744[129,130], 745[139,141], 746[144], 799[114], 819[211], 823[232]; **5**, 204[40], 420[77], 527[59], 576[138,139,140,141,142,143,144]; **6**, 115[80], 437[39], 990[82], 991[82]; **7**, 566[100], 711[60]; **8**, 395[131]
Keckeis, H., **2**, 278[9], 285[9]
Keda, T., **6**, 563[900]
Kedrinskii, M. A., **8**, 373[132]
Keefer, L. K., **6**, 119[109]; **7**, 224[54]; **8**, 373[129], 383[19,20], 387[20], 389[19,20], 392[19], 597[95]
Keefer, R. M., **5**, 71[144]
Keehn, M., **7**, 318[62]
Keehn, P. M., **3**, 167[483]; **7**, 143[139], 247[107], 706[19,20]
Keek, J., **6**, 111[64]
Keeley, D. E., **3**, 86[61], 88[61], 89[61], 91[61], 124[61], 785[32,33]; **5**, 1020[70], 1027[70]; **6**, 143[68-71], 147[83,85]; **7**, 684[94]
Keely, S. L., **8**, 28[37], 66[37]
Keely, S. L., Jr., **3**, 396[109], 397[109]; **5**, 959[318]
Keen, R. B., **4**, 579[22]; **5**, 310[99]
Keenan, R. M., **4**, 1009[142]; **5**, 249[36], 431[119]
Keene, B. R. T., **7**, 750[124]
Kees, K., **1**, 824[45]; **2**, 205[104], 206[104]; **3**, 225[187], 579[125], 582[125], 583[125], 584[125], 585[125], 588[125,159], 591[165], 595[125], 596[125], 610[125,165]; **4**, 595[157]
Keese, R., **2**, 810[62], 829[62,136], 866[9]; **3**, 380[4]; **5**, 65[72], 664[38]; **8**, 941[113]
Keese, W., **6**, 116[89]
Keewe, B., **1**, 33[163]
Kegelman, M. R., **3**, 564[8]
Kegley, S. E., **4**, 984[120], 985[130]
Kehne, H., **6**, 531[432]
Keichiro, F., **4**, 239[27], 257[27], 261[27]
Keiko, N. A., **6**, 577[979]
Keiko, V. V., **4**, 291[208]
Keil, G., **2**, 773[21]
Keinan, E., **2**, 608[78]; **3**, 12[59], 229[231]; **4**, 591[106], 594[137], 596[160], 598[200,202], 604[286,287], 605[287], 606[300], 615[385], 616[106,385], 619[137,385], 620[385], 621[160,385], 623[200,398], 626[286,287,385], 628[399], 633[106,399], 634[137,399], 636[160], 637[385], 638[200,202], 647[287]; **6**, 85[90], 86[95], 848[111], 937[120], 939[120], 942[120]; **7**, 14[127,-128], 40[2,5], 218[5], 465[131], 737[9], 842[24,25,31,35-37], 843[41,42]; **8**, 187[33], 553[361,363,364], 554[364,368], 555[364,368,369], 782[101], 961[38,40], 984[4], 991[4]
Keiser, J. E., **5**, 439[166]; **7**, 764[130]
Keitel, I., **2**, 1088[44]
Keith, D. D., **6**, 543[621]; **8**, 460[249]
Keitzer, G., **2**, 900[22], 961[37]
Kelarev, V. I., **6**, 530[418,422], 534[517]
Kelbar, G. R., **6**, 687[381]
Keldsen, G. L., **8**, 861[226]
Kell, D. R., **3**, 381[22]
Kelland, J. G., **8**, 205[155]
Kelland, J. W., **1**, 139[3]
Kelleghan, W. J., **3**, 530[69], 533[69], 537[69]
Kelleher, R. G., **2**, 146[71]
Keller, J., **7**, 236[24]
Keller, K., **4**, 443[190], 500[109]; **5**, 10[79], 19[79], 26[79], 27[79], 264[74], 385[131], 527[66], 529[66b], 681[29], 691[88]
Keller, L., **3**, 500[132], 505[132]; **4**, 82[61], 115[179b], 257[224], 519[17], 520[17], 540[17], 541[17], 546[128]; **5**, 1094[98], 1096[98], 1098[98], 1112[98]
Keller, L. S., **5**, 692[92]
Keller, O., **6**, 637[28]
Keller, P. C., **8**, 650[66]
Keller, R. T., **7**, 92[42], 93[42]
Keller, T. H., **5**, 51[45,45b], 57[50,50b], 59[50b]
Keller, U., **5**, 686[41]
Keller, W., **4**, 1039[64]
Kellert, C. A., **7**, 482[119]
Keller-Wojtkiewicz, B., **3**, 390[83], 392[83]
Kellett, R. E., **7**, 774[327]
Kelley, D. F., **5**, 240[3]
Kelley, D. R., **8**, 979[149]
Kelley, E. A., **5**, 1135[47], 1147[112]
Kelley, J. A., **8**, 798[60]
Kelley, M. D., **6**, 751[107]
Kelling, H., **2**, 345[42]
Kellner, J., **3**, 531[86]
Kellog, M. S., **3**, 369[107]
Kellogg, M. S., **2**, 212[119]; **5**, 220[47], 707[39], 709[39,44], 737[44a], 739[44,44a]
Kellogg, R. M., **1**, 299[62]; **3**, 229[223,224,224a,b], 512[199], 865[28]; **4**, 1074[28], 1093[28,150]; **6**, 2[7], 21[79], 70[18], 980[46]; **7**, 169[110]; **8**, 82[5,6], 84[6,12], 93[74,77], 94[77], 95[91], 96[92], 97[98], 98[101]
Kellogg, R. P., **2**, 195[71]; **8**, 190[75]
Kellom, D. B., **6**, 825[127], 1033[128]
Kelly, B. J., **7**, 480[100,103], 481[100,103,106,107], 482[100,116], 744[66]
Kelly, C. A., **5**, 71[153]
Kelly, C. C., **7**, 16[165]
Kelly, D. P., **3**, 330[192], 916[17], 932[17], 963[119]
Kelly, D. R., **3**, 290[70]; **4**, 375[97], 746[142]
Kelly, F. W., **5**, 552[11]; **6**, 1033[124], 1034[124], 1035[124]
Kelly, J. A., **4**, 682[56]
Kelly, J. D., **2**, 544[85], 547[85], 552[85]
Kelly, J. F., **5**, 107[201]
Kelly, J. T., **3**, 330[190]
Kelly, J. W., **6**, 24[97], 74[39]
Kelly, K. P., **7**, 452[49]
Kelly, L. F., **4**, 670[16,17,26], 674[17], 683[59,60], 687[59,60,64], 688[65]
Kelly, M. J., **2**, 88[30,31]; **5**, 1014[55]
Kelly, P. B., **2**, 958[27]
Kelly, R. B., **2**, 456[62], 458[62]; **3**, 20[105]; **8**, 331[30], 333[53]
Kelly, R. C., **3**, 651[219]; **6**, 116[94]; **7**, 439[26]
Kelly, R. E., **4**, 41[118]
Kelly, S. E., **1**, 420[83], 568[230]; **4**, 53[150]; **5**, 167[95]
Kelly, T. A., **5**, 836[62], 841[100]
Kelly, T. R., **1**, 474[95], 564[202]; **2**, 521[71]; **4**, 164[100], 261[291]; **5**, 12[87], 376[109]; **7**, 355[38,42]
Kelly, W., **6**, 429[91]
Kelly, W. J., **4**, 766[230]; **5**, 382[119b]; **7**, 219[16]; **8**, 315[50]
Kelly, W. L., **1**, 3[18,21]
Kelm, H., **5**, 77[252,255,259], 453[63], 454[63], 458[63]
Kelman, R. D., **3**, 597[200]
Kelner, M. J., **2**, 725[108]
Kelsey, D. R., **3**, 500[130], 509[130]; **8**, 839[22]
Kelsey, R., **7**, 710[51]
Kelso, P. A., **7**, 41[26]
Kelso, R. G., **4**, 283[151]
Kelson, A. B., **2**, 650[111], 651[111b]

Kemal, C., **7**, 763[90]
Kemal, O., **5**, 522[44]
Kemball, C., **4**, 914[4], 924[4]; **8**, 431[64]
Kemin, M. D., **6**, 526[402]
Kemmitt, R. D. W., **5**, 300[74], 307[93]; **8**, 674[33]
Kemp, D. S., **6**, 531[450], 639[49,50]; **8**, 389[69,71]
Kemp, J. E. G., **7**, 479[96], 750[135]
Kemp, K. C., **2**, 711[32]
Kemp, T. J., **5**, 212[70]
Kempe, T., **3**, 872[56]
Kempe, U. M., **5**, 501[268]; **6**, 672[285]
Kemper, B., **2**, 13[53,55], 14[53,55]
Kempf, D. J., **1**, 477[133], 480[156,158]; **3**, 67[17]
Kemppainen, A. E., **7**, 41[26]
Kendall, E. C., **8**, 988[28]
Kendall, M. C. R., **4**, 611[349]
Kendall, P. E., **3**, 418[26]; **4**, 170[20]
Kendall, P. M., **1**, 451[217]
Kende, A. S., **1**, 424[101], 447[198], 464[37], 753[100]; **2**, 106[49], 762[53]; **3**, 50[266], 51[269], 217[95], 219[114], 284[54], 483[18], 499[140], 500[18], 501[140], 502[140], 672[64], 674[71], 681[100], 807[23], 839[6]; **4**, 14[48], 16[51], 42[121], 129[224], 573[6-9], 614[375,376,377,378,379], 810[166], 841[38], 905[208], 1015[195]; **5**, 21[159,160], 23[159,160], 736[143]; **6**, 783[83], 814[89], 967[104], 1022[61]; **7**, 347[16], 355[16], 409[102], 410[102], 551[54]; **8**, 237[11], 839[22]
Kendrick, D. A., **3**, 555[30]
Kenion, G. B., **7**, 488[158]
Kennard, O., **1**, 2[3], 37[3], 774[206], 776[206]; **6**, 436[9]
Kenne, L., **6**, 23[89]
Kennedy, J. F., **6**, 34[11], 51[11]
Kennedy, J. H., **3**, 762[148], 769[148]
Kennedy, J. P., **3**, 331[199]
Kennedy, M., **4**, 1055[135,136], 1056[136]; **7**, 194[3], 200[40], 208[88]
Kennedy, P., **8**, 413[121]
Kennedy, R. A., **5**, 597[23], 603[23], 606[23]
Kennedy, R. M., **1**, 191[77]; **2**, 258[50]; **4**, 17[54], 63[54], 215[121]; **6**, 144[76], 1044[17], 1053[17]; **8**, 159[108], 171[108,109], 178[108], 179[108], 843[59b,c]
Kenner, G. W., **3**, 770[178]; **6**, 607[46], 644[84], 668[257], 669[265]; **8**, 514[112], 932[43]
Kennerly, G. W., **5**, 1146[106]
Kennewell, P. D., **1**, 469[56], 474[56]; **5**, 13[92]
Kenney, P. M., **6**, 284[175]
Kenny, C., **1**, 264[44], 269[44,58], 270[58], 271[44,64], 273[44]; **3**, 574[73], 575[73], 599[73,208], 610[73]; **4**, 809[161]
Kenny, M. J., **4**, 27[80]
Kent, G. J., **7**, 502[263]; **8**, 342[109]
Kent, S. B. H., **6**, 670[274]
Kentgen, G., **5**, 348[73b]
Kenyon, J., **3**, 914[6], 927[6]; **6**, 799[21,22]; **7**, 771[257], 772[296], 779[423]; **8**, 971[110]
Keogh, J., **4**, 350[120]; **7**, 506[294]; **8**, 812[4], 967[84]
Keough, A. H., **3**, 380[5]
Keramat, A., **3**, 759[128]
Kerb, U., **4**, 182[76]; **7**, 47[55], 86[16a]
Kerber, R. C., **4**, 429[83], 438[83], 441[83], 452[2], 709[46], 710[46], 984[121,122]; **7**, 882[171,173]
Kerdesky, F. A. J., **1**, 410[43]; **2**, 232[183], 249[36], 308[20]; **5**, 394[145a]
Kerekes, I., **4**, 271[20]; **6**, 216[108], 219[108]
Kergomard, A., **2**, 348[54]; **7**, 60[46b], 92[48,51]; **8**, 203[148], 205[148,162,163], 558[394,399], 559[401], 560[402], 881[75]
Kerkman, D. J., **3**, 629[51,52]
Kern, D. H., **3**, 568[38,39]
Kern, J. M., **7**, 805[66]
Kern, J. R., **2**, 385[320]; **4**, 216[122]
Kern, J. W., **8**, 142[49], 533[144]
Kern, R., **6**, 238[73]
Kern, R. J., **3**, 741[52]
Kernaghan, G. F. P., **5**, 596[32], 597[32], 603[32]
Kernebeck, K., **6**, 664[221]
Kerr, C. A., **3**, 316[115]
Kerr, J. B., **7**, 801[37]
Kerr, K. M., **7**, 65[70]
Kerr, R. G., **6**, 467[48], 469[55,56]; **7**, 742[59]; **8**, 410[98]
Kerr, W. J., **5**, 1043[25], 1046[25b], 1048[25b,32], 1051[32]
Kerschen, J. A., **3**, 750[86]
Kertesz, D. J., **8**, 198[131]
Kerton, N. A., **3**, 494[86]
Kerwin, J. F., Jr., **2**, 578[82], 664[30], 665[30], 666[34], 682[30], 689[30], 700[34]; **5**, 410[41,41a], 434[141], 843[116]; **6**, 859[169]; **7**, 236[14,15]; **8**, 542[238], 544[277]
Keskin, H., **2**, 139[32]
Kesling, H. S., **6**, 172[15]
Kessabi, J., **6**, 690[401], 692[401]
Kessar, S. V., **3**, 1057[39]; **4**, 483[5], 484[5,5a], 486[37], 488[5d], 495[5], 497[100], 503[126], 505[5a,37,132,133,141,142,143,144,145,146]; **8**, 341[107]
Kessel, C. R., **5**, 342[62b]
Kessel, S., **2**, 1051[40]
Kesseler, K., **1**, 153[59], 154[59], 295[51,52], 336[74,76,78,79], 338[81], 340[74,79], 612[47]; **2**, 117[148], 307[15,16], 310[15,16], 507[23], 512[23], 570[37], 640[65a,66,68], 641[65a,68,70], 642[65a], 644[65a], 646[65a,66,70,85]
Kessler, H., **2**, 547[113], 551[113]; **3**, 56[285]; **4**, 229[236]; **6**, 667[246]
Kesten, S., **1**, 399[7]
Kestner, M. M., **4**, 429[83], 438[83], 441[83], 766[230]
Keszler, D. A., **8**, 648[59]
Ketcha, D. M., **2**, 744[94]
Ketcham, R., **4**, 317[546]; **5**, 64[28]
Ketelaar, P. E. F., **5**, 86[30], 88[30]
Ketley, A. D., **4**, 587[42,46,47], 1023[260]; **5**, 1006[35]; **7**, 452[59]
Ketlinskii, V. A., **8**, 98[100]
Kettenring, J., **3**, 628[45]
Keul, H., **4**, 1099[177]; **7**, 579[136]
Keulks, G. W., **8**, 425[44,45], 426[44,45]
Keumi, T., **2**, 736[24,25]; **8**, 626[174]
Keung, E. C., **2**, 527[4], 528[4]; **5**, 1[2], 2[2]
Kevan, L., **4**, 537[99], 538[99]; **7**, 850[1]
Kevelam, H. J., **3**, 552[10]
Kexel, H., **7**, 778[419]
Keyaniyan, S., **5**, 1052[38]
Keyes, G. H., **6**, 477[100]
Keyes, M., **4**, 1007[112]; **8**, 802[86]
Keys, B., **5**, 707[29]
Keys, D. E., **6**, 960[56]; **7**, 874[110]
Kezar, H. S., III, **6**, 1030[107]; **7**, 131[83]
Keziere, R. J., **2**, 843[196]; **8**, 121[79], 944[124]
Khafizov, U. R., **5**, 1154[159]
Khai, B. T., **8**, 369[73], 552[351]
Khaidem, I. S., **2**, 355[119], 382[314]
Khajavi, M. S., **2**, 919[92]; **5**, 95[89,93]
Khalaf, A. A., **3**, 294[5], 299[5], 300[5], 303[5], 304[5], 306[80], 323[145], 324[147], 325[156,158,160], 327[170]
Khalid, M., **5**, 556[51]
Khalikov, S. S., **3**, 306[79]
Khalil, A. A. M., **6**, 770[36]
Khalil, A. H., **2**, 287[68]
Khalil, F. Y., **8**, 860[221]

Khalil, M. H., **2**, 422[30], 423[30], 432[54]
Khalil, Z. H., **5**, 166[90]
Khalilov, L. M., **4**, 589[79], 591[79]; **8**, 699[150]
Khalilova, S. F., **6**, 515[316]
Khambata, B. S., **5**, 552[25]
Khambay, B. P. S., **1**, 546[50]
Khan, A. U., **7**, 98[102]
Khan, A. W., **7**, 71[95]
Khan, H. A., **7**, 208[77,82]
Khan, I. A., **7**, 675[60]
Khan, J. A., **4**, 306[385], 307[385]; **8**, 855[156,159]
Khan, K. M., **5**, 841[91], 853[91d]
Khan, L. D., **8**, 497[40]
Khan, M. A., **2**, 655[144], 780[9]; **4**, 102[129], 317[557], 695[4]
Khan, M. N., **8**, 924[4]
Khan, M. Y., **2**, 359[163]
Khan, N., **5**, 731[130b]; **7**, 483[128]
Khan, N. H., **8**, 387[57]
Khan, R., **3**, 1022[19]; **6**, 51[106]
Khan, R. H., **8**, 938[92]
Khan, S. A., **7**, 768[205], 769[205]
Khan, S. D., **1**, 610[45]
Khan, S. H., **7**, 76[117]
Khan, S. I., **1**, 41[267], 174[11], 179[11]
Khan, T., **8**, 457[213]
Khan, W. A., **1**, 476[113]; **3**, 65[5]
Khan, Z. U., **6**, 535[540], 538[540]
Khanapure, S. P., **4**, 483[5], 484[5], 485[26], 486[38], 495[5], 497[5b,98], 499[5b,102]; **5**, 439[171], 440[171]; **8**, 515[116]
Khanapure, S. R., **1**, 554[103]
Khand, I. U., **3**, 1024[32]; **4**, 521[39], 541[109,110]; **5**, 1037[6], 1038[8,9], 1039[6,12-14,18], 1040[6], 1041[20], 1043[13], 1044[14], 1045[20,28,29], 1046[18,20], 1047[9,12], 1048[14], 1049[6,8,12,20,28], 1050[12,14], 1051[8,20], 1138[68]
Khand, M. N. I., **4**, 581[28,29]
Khandekar, G., **1**, 568[228]
Khandelwal, Y., **7**, 64[64], 384[116]
Khanna, P. L., **7**, 46[50], 47[50]
Khanna, P. N., **1**, 546[51]
Khanna, R. K., **4**, 747[148], 768[238]; **8**, 851[134], 858[204]
Khanna, R. N., **2**, 747[115]
Kharana, H. G., **6**, 625[156]
Kharasch, M. S., **3**, 208[4,5], 210[4]; **4**, 70[4], 89[4], 148[46], 274[61,65], 275[61,65], 279[61,105-107,109], 280[61,65,121,125,126], 281[61,65], 282[61,109], 283[147], 285[160], 287[176,177], 288[125,176,177]; **7**, 14[133,134], 16[166], 95[74], 483[132]; **8**, 505[75]
Kharasch, N., **3**, 505[164], 507[164,173], 512[164], 515[164]; **7**, 516[9], 760[41]
Kharchenko, V. G., **6**, 538[554]
Khar'kovskaya, V. A., **2**, 387[334]
Kharrat, A., **4**, 753[166]; **8**, 584[20]
Khathing, D. T., **7**, 267[117], 268[117]
Khatri, H. N., **5**, 829[20]; **6**, 861[183]
Khatri, N. A., **5**, 406[23,23b], 415[55], 539[108]; **6**, 814[88]
Kheifits, L. A., **6**, 836[53]
Khemani, K. C., **3**, 867[36]
Khetan, S. K., **4**, 45[129], 52[147,147c]
Khettskhaim, A., **6**, 554[730]
Khidekel, M. L., **8**, 451[177], 963[43]
Khitrov, A. P., **6**, 970[120]
Khitrov, P. A., **4**, 1010[155]
Khlebnikov, A. F., **6**, 498[166]
Khmelnitskii, L. I., **7**, 740[43]
Khodabocus, A., **5**, 731[130c]
Khoilova, E. M., **8**, 771[46]
Khomutov, R. M., **4**, 314[486]
Khoo, L. E., **8**, 824[59]
Khor, T. C., **4**, 670[28], 687[64]
Khorana, H. G., **6**, 603[19], 605[36], 607[44], 611[63-65], 612[71], 614[79,80], 622[134], 625[153], 626[166], 643[75], 650[130]
Khorlin, A. Ya., **6**, 271[85]
Khorlina, I. M., **8**, 214[39], 260[2], 266[54], 267[54,61,67], 272[115], 274[139], 275[140], 746[55], 753[55]
Khoshdel, E., **4**, 231[269]; **6**, 498[161]
Khosrowshahi, J. S., **7**, 488[150], 828[52], 829[52a]
Khotimskaya, G. A., **8**, 778[84]
Khouri, F., **6**, 452[134]
Khoury, G., **3**, 753[100]
Khramova, I. V., **6**, 501[207]
Khrimyan, A. P., **4**, 304[362], 305[362], **7**, 415[115d]
Khripach, V. A., **4**, 145[29b]
Khripak, S. M., **4**, 342[66]
Khromov, S. I., **3**, 381[32]
Khrustalev, V. A., **6**, 487[41], 489[41], 515[310,311,312]
Khrustova, Z. S., **3**, 644[140]
Khuang-Huu, F., **6**, 7[30]
Khudyakov, I. V., **7**, 850[5]
Khuong-Huu, Q., **4**, 296[263,264], 405[249,250a,b], 406[249]; **6**, 264[31], 266[49], 278[31], 284[172,174]; **7**, 27[63]; **8**, 856[182]
Khurana, A. L., **8**, 646[49]
Khurshudyan, S. A., **2**, 723[101]
Khusainova, N. G., **4**, 41[119], 55[156]
Khusid, A. K., **6**, 676[305]
Khwaja, H., **7**, 766[185,186]; **8**, 446[73]
Khwaja, T. A., **6**, 624[141]
Kianpour, A., **8**, 895[1], 898[1]
Kibayashi, C., **1**, 55[24]; **3**, 507[175]; **4**, 410[261], 502[121,122], 503[129], 847[77]; **5**, 256[55,56], 421[78]; **6**, 81[71,72]; **7**, 297[31]; **8**, 395[126], 652[80]
Kice, J. L., **7**, 765[159], 769[244], 770[246,250]; **8**, 409[85]
Kida, S., **6**, 546[647]; **7**, 415[114]
Kidd, D. A. A., **6**, 643[76]
Kidd, J. M., **3**, 297[19]
Kido, F., **3**, 956[107]; **4**, 8[28,28a]; **7**, 564[87], 565[87]; **8**, 121[79]
Kidokoro, K., **6**, 533[503]
Kidwai, A. R., **8**, 387[57]
Kidwell, R. L., **8**, 691[107]
Kieboom, A. P. G., **8**, 418[1], 419[1], 420[1], 423[1], 427[50], 431[1], 432[1], 433[1], 437[1], 438[1], 439[1]
Kieczykowski, G. R., **2**, 187[41]; **3**, 33[190], 34[197]; **4**, 10[32,32f,g], 109[148]; **6**, 647[113]
Kiedel, J., **5**, 340[57c], 345[57]
Kiedrowski, G. V., **5**, 17[122]; **7**, 131[87]
Kiefel, M. J., **1**, 551[74]
Kiefer, B., **8**, 299[82]
Kiefer, E. F., **4**, 311[448]
Kiefer, G., **5**, 108[205]
Kiefer, G. E., **3**, 509[179]
Kiefer, H., **5**, 714[69], 715[80]
Kiefer, H. B., **7**, 601[87]
Kiehlmann, E., **2**, 144[63]
Kiehs, K., **3**, 890[34], 903[126]
Kiel, W. A., **1**, 699[247]; **3**, 107[226], 109[226]; **4**, 372[55], 398[216], 399[216a], 401[216a], 405[216a], 410[216a]; **6**, 470[58]; **7**, 495[211]; **8**, 847[97,97d], 849[97d,107,115]
Kielbasinski, P., **7**, 762[69], 777[69a], 778[69]
Kiely, D. E., **7**, 255[32]
Kiely, J. S., **3**, 466[189]
Kienhuis, H., **4**, 51[145b]
Kienitz, L., **2**, 368[238]

Kienlen, J.-C., **6**, 1035[136]
Kienzle, F., **3**, 505[165]; **4**, 428[76]; **5**, 329[31]; **7**, 728[42], 732[58], 828[51]
Kieper, G., **5**, 1126[66]
Kierstead, R. W., **2**, 1022[53]; **3**, 643[117]; **4**, 373[68]; **5**, 341[61a]; **7**, 54[7]
Kiesel, R. J., **8**, 264[42]
Kiesewetter, D. O., **6**, 219[118]
Kieslich, K., **4**, 182[76]; **7**, 54[5], 55[5], 58[5], 59[5], 62[5,50b,52b], 63[5], 69[91], 70[91], 78[5], 429[157a]; **8**, 185[6]
Kiessling, L. L., **3**, 217[95], 545[120]; **5**, 514[9], 527[9], 736[143,145], 737[145]
Kiffen, A. A., **2**, 727[135]
Kigasawa, K., **4**, 495[85], 505[136], 510[173]; **7**, 453[76]
Kigawa, Y., **2**, 888[54]; **6**, 74[37]
Kigma, A. J., **4**, 249[120]
Kigoshi, H., **1**, 738[40]; **4**, 391[179]; **6**, 998[117]; **7**, 162[58], 243[65,66]
Kiguchi, T., **8**, 936[70]
Kihara, M., **3**, 677[87]; **7**, 423[145]
Kihira, K., **3**, 644[149]
Kihlberg, J., **5**, 433[139]
Kii, N., **7**, 587[171]
Kiji, J., **3**, 555[29], 1027[46]; **4**, 600[230,232,238], 601[251], 602[251], 606[303], 643[238,251], 646[303]; **5**, 35[12], 56[51]
Kijima, M., **8**, 369[81-83], 382[9], 395[132], 645[47]
Kijima, S., **8**, 241[37]
Kijima, Y., **1**, 277[80]
Kikawa, S., **8**, 548[319]
Kikkawa, I., **1**, 417[69]
Kikuchi, H., **4**, 313[460]
Kikuchi, K., **8**, 244[65], 250[99]
Kikuchi, M., **6**, 438[48]
Kikuchi, O., **4**, 810[170], 823[230]; **5**, 740[152]
Kikuchi, R., **5**, 71[120], 1139[75]
Kikuchi, T., **1**, 755[113]; **8**, 331[31]
Kikugawa, Y., **6**, 548[669]; **8**, 55[179], 60[182], 61[182], 62[182], 64[182], 66[179,182], 70[230], 71[230], 244[67], 249[96], 250[67], 357[191], 580[4,5,7], 581[5], 587[5], 614[82], 619[131], 620[136,137,138]
Kikui, J., **6**, 498[167]
Kikui, T., **6**, 71[21],
Kikukawa, K., **3**, 495[96], 497[103], 530[78], 535[78], 1026[41]; **4**, 587[27], 841[49], 856[99,101]; **7**, 764[116]
Kilaas, L., **1**, 823[41,42], 832[41]
Kiladze, T. K., **8**, 214[40]
Kilbourn, M. R., **6**, 219[118]
Kilburn, J. D., **2**, 186[33], 200[90], 201[91]; **7**, 646[29], 647[30]
Kilger, R., **3**, 53[274]; **5**, 225[96]; **6**, 531[436]
Kilian, R. J., **3**, 391[88]; **6**, 1054[48]
Kiliani, H., **7**, 252[6]
Kill, R. J., **8**, 977[139]
Killian, D. B., **4**, 315[522]
Killinger, T. A., **2**, 23[89]; **5**, 8[56], 12[88], 16[111], 18[111], 461[104]
Killough, J. M., **8**, 113[38]
Kilmer, A. E. H., **2**, 145[64]
Kilmer, G. W., **3**, 898[77]
Kilpatrick, M., **3**, 297[22]
Kilpert, C., **5**, 877[6]
Kilroy, M., **8**, 140[18]
Kiltz, H. H., **2**, 961[40]; **7**, 223[45]
Kilwing, W., **8**, 70[225], 71[225]
Kim, B., **5**, 636[96]
Kim, B. H., **4**, 1079[67]; **5**, 260[70], 263[70], 841[89], 872[89b]; **6**, 448[108]; **8**, 856[173]
Kim, B. M., **2**, 35[124a], 42[124], 112[101], 244[25,26], 258[48,49,51], 261[48]; **6**, 19[70], 79[63]; **7**, 431[161], 442[46c]; **8**, 171[107], 720[138], 721[138], 722[138]
Kim, C., **1**, 411[47], 894[160]; **7**, 625[38]
Kim, C.-K., **3**, 673[68], 681[68,98], 686[68,110], 807[32,33], 815[77], 816[81]
Kim, C.-M., **3**, 509[176]
Kim, C. R., **4**, 384[144,144b]
Kim, C. S., **4**, 1000[11], **6**, 754[119]
Kim, C. U., **2**, 648[98], 649[98], 1059[76]; **7**, 204[57], 292[10], 297[29], 298[38]
Kim, C.-W., **2**, 445[23]; **6**, 1064[88]; **7**, 182[164], 824[42]; **8**, 514[105]
Kim, D., **3**, 49[261]; **4**, 731[68], 754[178,179], 755[178], 803[133,136], 820[133,223]; **5**, 406[23,23b], 1056[49]; **6**, 814[88]
Kim, E. K., **7**, 856[65], 862[80]
Kim, G., **2**, 597[9]; **5**, 1124[49]
Kim, H., **3**, 380[13]; **6**, 108[34]
Kim, H.-B., **4**, 215[121]; **6**, 152[139], 153[139], 1053[46]; **8**, 843[59b]
Kim, H.-D., **4**, 357[150]
Kim, H.-J., **7**, 530[16], 587[170], 823[37]
Kim, H. K., **3**, 727[33]; **7**, 372[71]
Kim, H. L., **2**, 1011[10]
Kim, H.-O., **6**, 446[103]
Kim, H. S., **3**, 49[261]; **4**, 1048[120]
Kim, I. B., **8**, 452[186]
Kim, I. O., **3**, 49[261]
Kim, J., **4**, 852[90], 856[98]
Kim, J. C., **7**, 692[21]
Kim, J. D., **8**, 406[42]
Kim, J. E., **8**, 237[16], 244[16], 253[16], 261[14], 406[42]
Kim, J. H., **4**, 505[137]; **5**, 699[6]
Kim, J. K., **1**, 586[18]; **4**, 332[20], 337[20,33], 452[1], 453[1], 457[1], 470[1], 471[1], 472[1], 493[76], 494[76]
Kim, K., **1**, 97[81]; **6**, 727[191]; **7**, 744[71], 752[143], 874[106], 881[164]
Kim, K. E., **4**, 254[178]; **8**, 18[121], 412[112], 806[103]
Kim, K. H., **4**, 510[163]; **5**, 413[49]
Kim, K. S., **7**, 274[140], 309[21], 318[54-56], 319[54-56]
Kim, K.-W., **6**, 78[59]; **7**, 606[159]; **8**, 710[57], 717[57]
Kim, M., **2**, 406[46]; **4**, 793[69]
Kim, M.-J., **2**, 456[22,30,33], 457[33], 458[33], 459[33], 460[33], 461[33], 462[33], 463[79,91], 464[91], 466[33]; **8**, 189[58]
Kim, M. Y., **1**, 377[95]; **2**, 1049[23], 1050[23]; **7**, 183[166]
Kim, N.-J., **4**, 24[75]
Kim, S., **1**, 408[35], 430[35,132], 435[142,143], 563[182], 846[20], 847[20], 853[49], 865[88], 876[49]; **2**, 406[46]; **3**, 288[63]; **4**, 163[98], 372[64a]; **6**, 660[202], 682[341]; **7**, 278[157], 678[73]; **8**, 14[77,81], 16[100], 17[111], 18[122], 54[152], 66[152], 241[39], 244[68], 247[39,68], 250[68], 261[15-17], 272[113], 354[176], 536[174], 538[192], 544[278], 806[101], 880[57], 881[68], 938[88], 969[93]
Kim, S.-C., **8**, 18[119,120], 26[25], 66[25], 67[25], 74[25], 240[35], 244[68], 247[68], 250[68], 541[207], 718[107]
Kim, S.-G., **6**, 799[20]; **8**, 95[85]
Kim, S. J., **5**, 773[168], 774[168]; **7**, 318[54], 319[54]
Kim, S. O., **8**, 541[207]
Kim, S. S., **1**, 563[182]; **2**, 47[154], 249[84]; **6**, 864[198]; **8**, 261[15,16]
Kim, S. W., **3**, 784[31]
Kim, T. H., **1**, 822[36]; **4**, 35[98i]; **7**, 73[102]
Kim, T.-S., **6**, 834[39]
Kim, W. J., **1**, 798[286], 804[309]; **4**, 1040[68]
Kim, Y., **3**, 571[58], 596[193], 728[37]; **5**, 455[77]; **8**, 423[37]
Kim, Y. G., **4**, 379[113,115], 384[113b]
Kim, Y. H., **6**, 208[57], 212[57], 727[191]; **7**, 744[71], 752[143], 759[15], 761[57,58], 765[160], 766[189], 769[214]

Kim, Y. J., **8**, 18[122], 54[152], 66[152], 354[176], 536[174], 806[101], 969[93]
Kim, Y. K., **8**, 781[98]
Kim, Y. S., **8**, 14[83]
Kimball, J. P., **7**, 661[45]
Kimball, S. D., **5**, 1096[111], 1098[111]; **7**, 160[49]
Kimbrough, D. R., **4**, 1036[48]; **5**, 854[176], 855[176], 856[176]
Kimel, W., **5**, 827[3], 834[3]
Kimling, H., **6**, 968[114]
Kimmel, T., **2**, 651[115,115a]; **6**, 164[195]
Kimmich, R., **6**, 91[129]
Kimoto, H., **6**, 734[6,7]
Kimoto, S., **2**, 759[27]
Kimura, A., **2**, 152[101], 464[98]
Kimura, B. Y., **1**, 9[43]
Kimura, G., **4**, 608[325]
Kimura, I., **5**, 98[124]
Kimura, J., **6**, 22[86]
Kimura, K., **1**, 177[19], 179[19,27], 180[27], 181[27], 182[27], 193[84], 198[84], 287[18]; **2**, 3[9], 5[9b,c], 6[9], 19[9,9b,c], 20[9b,c], 29[9c], 641[72]; **3**, 638[94]; **4**, 969[68]; **5**, 817[146]; **8**, 33[56], 66[56], 87[33]
Kimura, M., **1**, 109[14], 341[94], 838[158]; **2**, 1[1]; **3**, 1038[95]; **5**, 211[67], 223[82], 480[177], 530[71], 636[100], 637[102]; **6**, 453[143], 531[461]; **7**, 162[64], 230[128], 384[114a], 390[9], 765[165]
Kimura, R., **1**, 336[72]; **5**, 524[54], 534[54], 553[42], 564[94]
Kimura, S., **4**, 120[201]
Kimura, T., **1**, 563[178]; **4**, 354[130], 611[367], 636[367]; **7**, 761[65]; **8**, 86[26], 93[72], 94[80], 95[85]
Kimura, Y., **1**, 242[47,48]; **3**, 977[10], 985[26a], 992[38], 993[38], 999[51]; **4**, 1001[43]; **5**, 851[165], 889[31]; **6**, 851[130], 877[36], 879[36], 882[47], 883[36], 885[47], 886[36], 958[28]
Kinas, R., **4**, 231[274]
Kinast, G., **2**, 900[29], 901[29], 902[29]
Kinastowski, S., **2**, 348[61,62], 354[62], 399[13]
Kinberger, K., **8**, 711[66], 718[66]
Kindaichi, Y., **4**, 601[244], 643[244]
Kindler, K., **2**, 735[17]; **6**, 431[106], 477[99]
Kindon, N. D., **2**, 569[34]; **4**, 231[258]; **7**, 646[28]; **8**, 784[112]
King, A. O., **1**, 214[27]; **3**, 266[195], 450[103], 453[103], 486[44], 495[44], 503[150,151], 524[35-37], 529[53]; **4**, 767[234], 892[141], 893[153]; **5**, 1166[19]; **6**, 966[95]; **8**, 693[112], 755[117], 756[147], 758[117], 950[171]
King, F. D., **2**, 1010[9]; **3**, 500[126], 512[126]
King, F. E., **2**, 837[163b], 838[163]; **3**, 831[59]; **6**, 643[76]; **8**, 143[58]
King, H. F., **5**, 72[172]
King, J. A., **8**, 143[60], 148[60]
King, J. A., Jr., **5**, 1143[91]
King, J. C., **2**, 384[317]; **6**, 716[95]
King, J. F., **3**, 707[10], 867[36]; **5**, 571[118], 904[53]
King, J. L., **6**, 687[375]
King, J. R., **3**, 330[190]
King, K., **5**, 451[38], 552[8,10]
King, L. G., **4**, 437[146]
King, M. V., **8**, 366[39]
King, R. B., **2**, 81[4]; **3**, 208[3], 380[10]; **5**, 1134[44]; **8**, 390[77], 459[236], 736[20], 753[20]
King, R. R., **7**, 204[60]; **8**, 884[101]
King, R. W., **6**, 1012[4], 1013[4], 1033[126]
King, S. A., **4**, 593[131,133]; **5**, 300[76], 307[90]
King, S. M., **8**, 675[41], 679[41], 684[41]
King, S. W., **4**, 957[22]
King, T. J., **1**, 305[86]; **2**, 837[163b], 838[163]; **3**, 665[36]
Kingma, A. J., **4**, 85[72], 204[33,35]
Kingma, R. F., **5**, 78[275]
Kingsbury, C. A., **1**, 506[18]; **2**, 345[32]; **3**, 114[233], 154[233]; **6**, 154[149], 1017[36]; **7**, 341[50]; **8**, 5[26]
Kingsbury, W. D., **2**, 1097[100]; **7**, 764[119], 767[119], 778[394]
Kingston, J. F., **2**, 839[181], 840[181]
Kinishi, R., **8**, 170[77,78]
Kinloch, E. F., **8**, 478[42], 531[123]
Kinloch, S. A., **1**, 554[102]
Kinnard, R. D., **4**, 441[183]
Kinney, C. R., **1**, 383[111]
Kinney, R. J., **8**, 289[29], 290[29], 797[35]
Kinney, W. A., **1**, 744[60], 745[60]; **3**, 168[505]; **5**, 324[18b], 831[38], 1123[37]; **8**, 942[116]
Kinnick, M. D., **4**, 13[45], 14[45a]; **6**, 1022[63]
Kino, T., **1**, 101[90]
Kinomura, K., **6**, 658[184]
Kinoshita, H., **3**, 159[451], 160[451], 161[451]; **4**, 127[220b], 359[161], 595[151], 604[288,289], 646[289], 647[288], 753[164]; **6**, 641[62]; **8**, 840[30,30b], 960[37]
Kinoshita, K., **4**, 148[51], 149[51], 179[65]; **7**, 26[54]
Kinoshita, M., **1**, 513[48], 564[201], 569[254]; **2**, 263[54], 350[75]; **6**, 27[119], 46[59], 60[145], 1022[65]; **7**, 350[23], 778[391,392,393]; **8**, 820[46]
Kinoshita, S., **7**, 743[64]
Kinoshita, T., **8**, 607[33,34], 620[139]
Kinowski, A. K., **4**, 432[107,108]
Kinsel, E., **2**, 187[43]
Kinsley, S. A., **6**, 546[652]
Kinsman, D. V., **3**, 735[14]
Kinson, P. L., **3**, 891[42], 903[122]
Kinstle, T. H., **4**, 273[48], 279[48], 280[48]; **7**, 548[76], 558[76]
Kinter, C. M., **5**, 839[83]; **6**, 150[131], 151[131]
Kinter, M. R., **7**, 92[42], 93[42]
Kinugawa, N., **1**, 215[37]; **2**, 23[90,90a]; **4**, 607[314]
Kinzebach, W., **8**, 568[466]
Kinzel, E., **8**, 459[240], 460[250,251], 535[166]
Kinzer, G. W., **5**, 455[80]
Kinzig, C. M., **3**, 273[15], 274[15]
Kinzy, W., **6**, 53[117]
Kiovsky, T. E., **6**, 291[209]
Kiplinger, J. P., **5**, 788[15]
Kipnis, F., **4**, 317[547]
Kipphardt, H., **2**, 455[14], 514[51]; **6**, 119[110,111], 728[210]
Kira, M., **1**, 180[43], 181[43]; **2**, 6[28], 17[28,28b,d], 572[44], 790[58,59], 792[65]; **6**, 487[45], 489[45], 500[178], 573[45], 832[12], 865[12]; **7**, 641[7]; **8**, 20[141], 478[38], 547[313]
Kirby, A. J., **8**, 211[2]
Kirby, G. W., **2**, 555[146,149]; **3**, 681[97], 689[123]; **5**, 419[72,74], 420[76], 421[79], 437[159], 441[176,176f], 442[184], 575[133], 576[146]; **6**, 104[4], 212[83], 609[57], 614[78]; **7**, 748[111]
Kirby, J. E., **1**, 383[111]
Kirby, K. C., Jr., **8**, 990[42]
Kirby, R., **4**, 70[6], 98[6], 140[7a]
Kirby, S., **5**, 461[101], 463[101]
Kirchen, R. P., **3**, 374[133]; **8**, 83[7]
Kirchhof, W., **7**, 832[70]
Kirchhoff, R., **6**, 938[133]
Kirchhoff, R. A., **1**, 737[32,33], 738[33], 742[33]; **3**, 786[40]
Kirchlechner, R., **2**, 817[88], 824[88]
Kirchmeier, R. L., **6**, 569[936]; **8**, 864[241]
Kirchmeyer, S., **6**, 237[68], 254[162], 564[915]
Kirchner, J. J., **5**, 849[143], 1001[16]
Kireev, S. L., **5**, 1056[49]
Kirihara, M., **1**, 242[51], 243[52]
Kirihara, T., **4**, 239[27], 257[27], 261[27]; **5**, 473[154], 479[154], 531[78]
Kirilov, M., **1**, 34[224], 36[174,175]; **3**, 202[89]

Kirino, K., **6**, 233[40]
Kirisawa, M., **5**, 95[92]
Kirk, B. E., **5**, 9[65,66]
Kirk, D. N., **2**, 835[156]; **3**, 741[50,52], 791[62,63]; **6**, 996[106]; **7**, 136[109], 168[101], 332[18,19]; **8**, 89[44], 881[67], 941[112]
Kirk, J. M., **6**, 650[133b], 668[261]
Kirk, K. L., **6**, 220[127]
Kirk, L. G., **2**, 423[31]
Kirk, T. C., **2**, 527[9], 528[9], 531[24], 533[24], 537[24], 547[119], 551[119]; **4**, 181[69]; **5**, 2[7], 4[7], 8[57]
Kirkiacharian, B. S., **2**, 354[118], 355[118]
Kirkley, R. K., **7**, 480[105], 482[105]
Kirkpatrick, D., **3**, 274[23]; **4**, 1007[108]; **6**, 207[48]; **8**, 807[112]
Kirkpatrick, E. C., **5**, 752[45]
Kirmse, R., **6**, 441[86]
Kirmse, W., **1**, 528[111], 844[5c], 878[106]; **3**, 706[6], 887[9], 890[31], 893[9], 894[9], 896[9], 897[9], 900[9], 901[111], 903[9], 905[9,136], 909[155]; **4**, 953[8,8e], 954[8e,29], 961[8e], 968[8e], 1000[8], 1101[195]; **5**, 151[19], 225[124], 228[124], 229[124], 682[34b], 683[34b,38], 684[38], 856[210], 911[96], 912[96]; **6**, 961[72], 1044[19]; **7**, 835[82]
Kirn, B., **3**, 904[133]
Kirollos, K. S., **2**, 538[61], 539[61]
Kirosawa, H., **4**, 620[394]
Kirowa-Eisner, E., **8**, 594[73], 595[73,74]
Kiroya, K., **5**, 681[28]
Kirpichenko, S. V., **4**, 291[208]
Kirrmann, A., **3**, 739[42]
Kirrstetter, R. G. H., **3**, 619[25]
Kirsanov, A. V., **6**, 550[679]
Kirsch, G., **6**, 463[25]; **7**, 775[348]; **8**, 847[97], 848[97c], 849[97c], 867[97c]
Kirsch, H. P., **5**, 1134[34]
Kirsch, P., **8**, 451[182]
Kirsch, S., **6**, 980[36]; **8**, 880[62]
Kirschbaum, E., **7**, 775[340]
Kirschke, K., **2**, 364[203]; **4**, 611[345,348]; **7**, 140[130], 141[130]
Kirschleger, B., **1**, 830[91]; **3**, 202[91]
Kirschnick, B., **1**, 144[40]
Kirschning, A., **2**, 58[14]
Kirson, I., **7**, 707[28]
Kirst, H. A., **3**, 431[102]; **8**, 47[124], 64[207c], 66[124,207], 754[76]
Kirst, W., **8**, 321[98]
Kirste, B., **5**, 1060[58]
Kirtane, J. G., **1**, 819[2]; **3**, 223[157], 262[159], 264[159]; **6**, 2[3], 25[3], 88[105], 89[105]; **7**, 358[2], 366[2], 378[2], 384[2]
Kirusu, Y., **4**, 607[314]
Kis, Z., **8**, 197[129]
Kisaki, T., **7**, 58[55], 62[55], 63[55]
Kisan, W., **7**, 493[192]
Kischa, K., **6**, 1061[72]; **7**, 588[172]
Kise, H., **7**, 10[75], 24[34], 477[81]
Kise, M. A., **7**, 15[154]
Kise, N., **1**, 831[99]; **2**, 492[51]; **4**, 613[370], 810[169], 840[36]; **8**, 134[35]
Kiselev, S. S., **6**, 554[767]
Kiselev, V. D., **5**, 71[145], 76[246], 552[16]
Kiselev, V. G., **7**, 595[18], 597[43,44]
Kiseter, V. G., **7**, 603[117]
Kisfaludy, L., **6**, 637[31]
Kishi, M., **3**, 380[12]
Kishi, N., **1**, 584[11]; **2**, 569[35], 715[56]; **3**, 985[26a], 992[38], 993[38]; **5**, 888[29]; **6**, 860[179], 879[41], 882[47], 885[47]
Kishi, Y., **1**, 153[57], 182[44,45], 183[45], 188[94,95], 193[86], 197[92], 198[91,93-95], 199[93,94], 200[97], 436[148], 762[141,142]; **2**, 29[106], 30[106a,b], 31[106b], 219[143], 297[92], 298[92], 496[68], 497[68], 572[46], 578[81], 579[94], 877[35,36], 878[40], 879[42]; **3**, 125[294], 126[294], 167[294], 168[294], 225[187], 231[253], 466[186]; **4**, 377[145a], 384[145a]; **5**, 327[28], 844[127]; **6**, 5[24,28], 8[37], 89[116], 147[83], 266[50], 563[902], 632[4]; **7**, 57[26], 169[107], 246[88], 358[10], 369[64], 371[10], 376[84], 380[103], 401[62,62a], 406[62a,78a], 439[31,32,34], 440[34,40], 684[94]; **8**, 11[60], 879[52], 991[47]
Kishida, Y., **3**, 88[132,133], 90[132,133], 95[132,132a,133], 97[171], 99[132a,133], 101[132a,133], 107[132a,133], 118[132,171], 939[79], 963[120]
Kishimota, S., **7**, 59[38]
Kishimoto, H., **1**, 242[49-51], 243[52]
Kishimoto, K., **7**, 120[16]
Kishimoto, S., **5**, 92[67], 96[118]
Kishimura, K., **4**, 148[45b]
Kishimura, T., **6**, 1036[145]
Kishino, H., **2**, 611[102], 643[79], 644[79a], 647[88a]
Kishner, N., **8**, 328[1,7], 341[1], 926[14]
Kisielowski, L., **6**, 185[158,159], 187[158]
Kisilenko, A. A., **6**, 524[360], 528[413]
Kisin, A. V., **8**, 769[27]
Kisliuk, R. L., **6**, 812[79,80]
Kiso, M., **6**, 36[28]
Kiso, Y., **3**, 228[218,222], 437[21,25], 440[25], 448[25], 449[25], 450[25], 451[25,106], 452[25,106,107,109,111], 460[107], 484[26], 487[45], 492[26], 494[26], 495[26], 503[26], 513[26]; **8**, 764[10], 772[52], 773[10,67], 782[105], 783[107]
Kiss, G., **8**, 458[222]
Kiss, J., **2**, 735[17]; **6**, 48[86]
Kiss, M., **6**, 917[36]
Kissel, C. L., **6**, 960[61]
Kissel, T., **5**, 475[145]
Kissick, T. P., **7**, 256[24]
Kistenbrügger, L., **8**, 303[101], 304[101]
Kita, Y., **1**, 64[46], 242[49-51], 243[52], 391[152], 474[100]; **2**, 611[102], 643[79], 644[79a], 647[88a,b]; **3**, 939[79]; **4**, 14[46], 55[157], 57[157h], 155[75], 160[86a], 161[91], 249[114], 257[114], 261[286], 304[356]; **5**, 133[53,54], 451[18], 470[18], 473[153], 477[153], 838[74]; **6**, 935[104], 1066[97]; **7**, 202[46], 382[108]
Kitada, C., **6**, 644[82], 664[222]
Kitadani, M., **7**, 500[241]
Kitade, Y., **2**, 790[57]; **3**, 219[107]
Kitaev, Y. P., **1**, 378[104]
Kitagawa, A., **2**, 748[124]
Kitagawa, I., **1**, 188[70]; **3**, 751[88]
Kitagawa, O., **2**, 604[53], 643[79], 644[79b], 656[154]; **3**, 421[54]; **5**, 100[142]
Kitagawa, T., **3**, 45[248]; **6**, 251[147], 438[55]
Kitagawa, Y., **1**, 266[48]; **2**, 114[122], 269[71], 282[40]; **3**, 246[35], 349[33], 354[60], 358[33]; **4**, 389[166,166b]; **8**, 986[16]
Kitaguchi, H., **7**, 16[168]
Kitahara, E., **3**, 985[25]; **6**, 876[34], 885[34]
Kitahara, H., **4**, 115[180e]; **7**, 564[87], 565[87]
Kitahara, S.-I., **8**, 244[54]
Kitahara, T., **1**, 561[163], 733[12]; **2**, 905[55], 907[55], 908[55], 910[55], 911[55]; **3**, 49[262], 287[62], 871[54]; **4**, 30[89], 262[305]; **5**, 330[34], 410[40], 683[35]; **6**, 8[35], 74[29]; **8**, 196[119]
Kitahara, Y., **2**, 711[29], 819[99]; **4**, 373[73]
Kitahonoki, K., **8**, 66[214]
Kitai, M., **2**, 589[157], 726[123]; **3**, 799[100]; **8**, 785[113]
Kitajima, H., **2**, 736[24,25]; **8**, 626[174]
Kitajima, N., **7**, 17[179]
Kitajima, T., **1**, 714[267], 717[267]; **4**, 794[74]; **7**, 649[42]
Kitajima, Y., **1**, 569[261]
Kitami, M., **8**, 96[94]
Kitamoto, M., **6**, 716[97], 725[169]

Kitamura, A., **6**, 960[56]; **7**, 874[110]
Kitamura, K., **1**, 553[90,92,93]; **2**, 482[32], 484[32]; **3**, 51[271], 198[48], 751[88]; **4**, 650[425]
Kitamura, M., **1**, 78[20], 223[76], 224[76a], 317[145,155], 319[145], 320[155]; **4**, 79[57,58a,b,59b], 251[149], 257[149], 260[149]; **6**, 1023[69]; **7**, 370[65], 380[65]; **8**, 154[198,199,200,202], 459[228], 460[255], 461[257,261], 463[269], 817[30]
Kitamura, T., **3**, 219[105], 522[22]; **4**, 250[134], 255[134], 260[134], 903[195]; **5**, 168[106]; **6**, 283[163,164]
Kitani, A., **7**, 761[64]
Kitano, Y., **1**, 131[104], 185[54]; **3**, 223[183], 225[183]; **5**, 439[170]; **6**, 6[29]; **7**, 400[52], 412[105], 414[105,105b,c,108,109], 418[105b], 423[142,143], 712[64]
Kitao, O., **5**, 438[164]
Kitao, T., **6**, 443[92]; **7**, 197[14,15]
Kitaoka, M., **6**, 150[118,119], 151[119], 902[129]
Kitaoka, Y., **7**, 197[14]
Kitatani, K., **3**, 202[96], 216[70]; **4**, 1007[118,119,122], 1009[118,136,137]; **6**, 677[312]
Kitaura, K., **5**, 1145[103], 1153[103]
Kitayama, R., **4**, 654[448,449]
Kitayama, T., **1**, 14[74]
Kitazume, T., **1**, 551[73]; **2**, 656[152]; **3**, 420[50], 444[64]; **4**, 102[130], 216[123], 595[150]; **6**, 967[100]; **8**, 560[403]
Kitchin, J., **2**, 555[148]
Kitchin, J. P., **7**, 307[16], 310[16], 318[16], 319[16], 322[16], 704[12], 851[18]
Kitching, W., **1**, 610[42]; **2**, 587[140]; **4**, 303[342], 390[175b], 744[131], 817[208]; **7**, 92[40], 94[56], 616[17], 625[40], 635[70]; **8**, 850[120], 852[141], 854[152], 856[152,162], 857[141]
Kitélko, A., **8**, 271[102]
Kitiani, K., **5**, 1007[41]
Kitihara, Y., **4**, 378[108]
Kitos, P. A., **7**, 347[17], 355[17]
Kitzen, J. M., **4**, 439[168]
Kitzing, R., **5**, 708[41]
Kiuchi, K., **8**, 588[45]
Kiwi, J., **8**, 97[96]
Kiyoi, T., **2**, 17[67], 38[67], 572[45]; **3**, 229[228]; **5**, 1157[171]
Kiyomoto, A., **2**, 953[3b]
Kiyooka, S., **1**, 415[63]; **2**, 570[36], 577[80], 640[67]; **4**, 124[212], 155[68c]; **8**, 9[51], 388[60]
Kjaer, A., **6**, 424[53,54]; **7**, 777[385]
Kjeldsen, G., **2**, 713[48]; **5**, 778[195]
Kjell, D. P., **7**, 862[81], 888[81]
Kjellgren, J., **7**, 13[111]
Kjonaas, R. A., **1**, 218[52]; **2**, 124[209,210]; **4**, 96[105], 573[10,11], 841[40]
Klabunde, K. J., **1**, 212[7], 213[7]; **6**, 456[160]; **8**, 890[141]
Klabunovskii, E. I., **8**, 150[127,132]
Klaeren, S. A., **4**, 582[30]
Kläger, R., **6**, 53[119]
Klages, C.-P., **6**, 475[91,92], 482[91]
Klages, F., **2**, 737[40]; **6**, 291[205]
Klages, U., **8**, 335[67]
Klahre, G., **1**, 761[138], 773[204]
Klaman, D., **5**, 115[246]
Klamann, J.-D., **6**, 646[99,99b]
Klamar, D., **2**, 598[13]
Klandermann, B. H., **7**, 660[41]; **8**, 950[162]
Klang, J. A., **7**, 720[13], 722[13]
Klärner, F. G., **5**, 64[41]
Klasinc, L., **7**, 852[40], 867[91]
Klaubert, D. H., **3**, 499[121]; **6**, 118[100]
Klauenberg, G., **8**, 278[156]
Klauke, E., **7**, 498[224]
Klausener, A., **7**, 262[76]
Klausner, Y. S., **6**, 636[21]
Klaver, W. J., **1**, 617[54], 771[192]; **2**, 89[37], 922[100], 936[100], 1065[113-115], 1069[133,134]; **3**, 223[183], 225[183]
Klavins, M., **2**, 345[23]
Klayman, D. L., **1**, 630[3]; **5**, 155[37]; **6**, 461[1,2]; **7**, 769[235,236], 770[235]; **8**, 366[50]
Klebach, T. C., **6**, 1036[143]
Klebe, G., **1**, 34[166]
Klebe, J. F., **6**, 653[149]
Kleemann, A., **7**, 397[31]; **8**, 459[228], 460[254]
Kleev, B. V., **5**, 432[129]
Klehr, M., **4**, 1023[261]
Kleier, D. A., **5**, 714[74]; **8**, 89[43]
Kleiger, S. C., **4**, 274[61], 275[61], 279[61], 280[61], 281[61], 282[61]
Kleijn, H., **1**, 428[116]; **2**, 85[20], 584[125], 587[147]; **3**, 217[82], 491[68], 531[84]; **4**, 895[163], 897[171], 898[171,178], 899[171], 900[182], 905[209]; **8**, 743[48]
Kleiman, R., **3**, 691[129], 693[129]
Kleimann, H., **2**, 1090[68], 1094[89], 1095[89]
Klein, C. F., **3**, 790[59]
Klein, F. M., **8**, 707[17]
Klein, G., **3**, 383[43]; **5**, 817[146]
Klein, H., **7**, 842[38]
Klein, H. A., **2**, 70[50], 77[89]; **3**, 202[99]; **5**, 381[118]
Klein, H. P., **7**, 85[7]
Klein, J., **1**, 165[107]; **2**, 81[1], 82[1], 96[1]; **3**, 16[82], 872[61,62]; **4**, 186[91], 245[82], 248[82], 262[82]; **5**, 92[73], 95[73]; **6**, 965[86]; **8**, 704[8], 707[18], 936[73]
Klein, J. L., **6**, 705[26]
Klein, K. C., **8**, 161[21]
Klein, K. P., **6**, 774[48]; **7**, 700[60]
Klein, L., **1**, 217[48]
Klein, L. L., **6**, 8[37]; **8**, 837[15b]
Klein, M., **2**, 1090[73], 1102[73], 1103[73]
Klein, P., **2**, 801[23], 859[253], 867[10]; **5**, 839[77]; **6**, 509[281]
Klein, R. F. X., **7**, 228[96]
Klein, R. S., **6**, 553[795], 554[759,795]; **7**, 265[96]
Klein, S. I., **2**, 547[103], 549[103]; **8**, 341[103], 928[24]
Klein, U., **5**, 877[6]
Klein, W., **4**, 806[147]
Klein, W.-R., **6**, 448[112]
Kleinberg, J., **3**, 564[12], 574[76]; **7**, 16[162]
Kleineberg, G., **6**, 275[109,110]
Kleingeld, J. C., **8**, 89[43]
Kleinman, E. F., **1**, 358[9], 359[9], 362[9b]; **2**, 159[129,130], 518[62], 1012[17], 1013[17]; **4**, 31[94], 32[94f]
Kleinschroth, J., **2**, 81[3]
Kleinstück, R., **6**, 74[37]
Kleinwächter, I., **8**, 764[9]
Klemarczyk, P., **1**, 308[95], 314[95]; **5**, 273[6], 277[6], 279[6]
Klemm, E., **6**, 560[867], 564[907]
Klemm, L. H., **5**, 513[4]; **6**, 450[118]
Klenk, H., **2**, 740[53]
Klenke, K., **4**, 722[32]
Kleschick, W. A., **1**, 357[2]; **2**, 182[9,9c], 184[9c], 190[9c], 191[9c], 192[9c], 193[9c], 197[9c], 198[9c], 200[9c], 211[9c], 217[9c], 235[9c], 236[9c], 289[69], 509[33], 634[36], 640[36]; **4**, 31[94,94a], 103[134]; **6**, 814[95], 1027[89]
Klesney, S. P., **5**, 194[5], 196[5], 197[5], 198[5]
Klessing, K., **3**, 872[59]
Kleveland, K., **4**, 1010[154]
Klibanov, A. M., **7**, 79[134,135]; **8**, 185[10,13], 206[13]
Klich, M., **2**, 920[97], 921[97]
Kliegel, W., **1**, 386[124]; **2**, 1088[41,43]; **7**, 598[58,60]
Klieger, E., **3**, 898[87]

Kliem, U., **5**, 155[35,36], 156[36], 157[36]
Kliemann, H., **6**, 242[87], 243[87]
Klier, K., **5**, 418[71]; **8**, 395[133]
Klima, W. L., **3**, 486[42], 498[42]; **5**, 1166[19]; **6**, 966[95]; **8**, 756[147], 950[171]
Klimova, E. I., **2**, 534[32], 535[38]
Klimova, L. Z., **6**, 439[70]
Kline, D. N., **3**, 168[496]; **5**, 250[37], 252[37,42], 255[37]
Kline, E., **5**, 587[206]
Kline, M. L., **1**, 586[18]
Kline, M. W., **3**, 927[51]
Kline, S. J., **8**, 51[121], 66[121]
Klinedinst, K. A., **8**, 454[198]
Klinedinst, P. E., Jr., **8**, 561[414]
Klinge, M., **4**, 222[181], 224[181]
Klingebeil, U., **1**, 38[183]
Klinger, F. D., **1**, 109[15]
Klinger, T. C., **3**, 86[35], 164[35], 173[35]
Klingler, F. D., **7**, 544[38], 551[38]
Klingler, L., **5**, 181[155]
Klingler, R. J., **8**, 19[135]
Klingler, T. C., **4**, 337[34]
Klingsberg, A., **2**, 734[6]
Klingsberg, E., **3**, 840[13]; **6**, 207[50]
Klingstedt, T., **3**, 260[141,142], 261[142], 466[180,181]
Klinke, P., **3**, 898[87]
Klinkert, G., **3**, 290[70]
Klinot, J., **3**, 757[124]
Kliseki, Y., **4**, 856[101]
Klix, R. C., **3**, 749[81]; **5**, 765[110]; **7**, 340[49]
Klobucar, L., **7**, 667[80]
Klobucar, W. D., **3**, 872[57], 873[57]
Kloek, L. A., **3**, 369[122], 372[122]
Kloeters, W., **6**, 190[203]
Kloetzel, M. C., **3**, 898[83]; **5**, 513[2], 518[2], 552[2,13]
Kloetzer, W., **3**, 904[133]
Kloosterman, D. A., **5**, 851[170]
Kloosterman, M., **6**, 57[139], 658[183]
Kloosterziel, H., **2**, 101[18], 597[11]; **5**, 709[45]
Kloosterziel, J., **4**, 1000[12]
Klopfenstein, C. E., **5**, 513[4]
Klopmann, G., **3**, 4[20]; **4**, 537[96]; **7**, 6[28]; **8**, 724[173]
Klopotova, I. A., **8**, 449[155]
Klosa, J., **3**, 826[33,35,38]
Klose, G., **5**, 208[52]
Klose, T. R., **7**, 741[49]
Klose, W., **7**, 95[78]
Kloss, J., **2**, 323[30], 330[30], 332[30]; **4**, 384[143]
Kloss, P., **6**, 684[343]
Klotzenburg, R., **5**, 654[27]
Klötzer, W., **6**, 119[113]; **7**, 746[92], 752[92]; **8**, 829[82]
Kluender, H. C., **6**, 1055[52a]
Klug, J. T., **2**, 612[106]; **7**, 765[161], 803[57]
Kluge, A. F., **5**, 92[69], 741[155]; **6**, 647[112a]; **8**, 198[131], 477[36]
Kluge, H., **6**, 551[685]
Kluge, M., **3**, 890[33]
Kluger, E. W., **1**, 838[158]; **7**, 162[59,64]
Klumpp, G. W., **1**, 10[54,55], 11[55b], 218[54]; **2**, 981[23]; **4**, 52[146], 595[152], 869[23-26], 877[69], 884[108], 972[79]; **5**, 46[37], 906[70], 908[70,73,74], 986[38]; **6**, 86[99]; **7**, 373[73]
Klun, R. T., **5**, 209[55]
Klun, T. P., **4**, 651[430,431]; **6**, 848[112]
Klunder, A. J. H., **1**, 858[61]; **3**, 855[83]; **4**, 317[556]; **5**, 560[78], 561[80,84,85], 562[87], 568[109]; **8**, 836[2], 843[2f]
Klunder, J. M., **3**, 223[156], 224[180]; **7**, 390[4], 393[4,15,16], 394[4], 395[4], 396[4], 397[4,15,15a], 398[4,16,16a], 399[4], 400[4], 401[4], 406[4], 407[4], 410[4], 411[4], 413[4]
Klünenberg, H., **3**, 644[157]
Klusacek, H., **2**, 1094[89], 1095[89]
Klusener, P. A. A., **1**, 23[125]
Klutchko, S., **6**, 1021[51]
Klyne, W., **3**, 781[17]; **7**, 100[119]
Klynev, N. A., **6**, 494[132]
Kmiecik, J. E., **8**, 372[119]
Knabe, J., **8**, 588[44]
Knap, F. F., Jr., **7**, 775[348]
Knap, J. E., **8**, 754[81]
Knaper, A. M., **5**, 474[157]
Knapp, D. R., **2**, 943[171], 945[171]
Knapp, G. C., **2**, 970[90]
Knapp, K. K., **8**, 650[66]
Knapp, S., **1**, 872[89]; **3**, 785[36,36b]; **4**, 72[26], 155[69], 249[119], 257[119], 375[96a-c], 376[96b], 398[217], 399[217a,223], 401[96b,227], 403[223], 406[227a]; **5**, 326[26], 432[133], 552[31], 564[31]; **7**, 127[62], 493[186], 503[268]; **8**, 844[73]
Knapp, S. K., **7**, 536[51]
Knaus, E. E., **5**, 418[67], 420[75]; **7**, 489[170]; **8**, 92[68], 587[36]
Knaus, G., **2**, 489[46], 490[46]; **6**, 725[170], 728[170]
Knaus, G. N., **5**, 634[75]
Knauss, E., **5**, 379[113b], 380[113b]
Knecht, D. A., **5**, 64[22]
Knecht, E., **8**, 371[100], 531[121]
Kneen, G., **3**, 431[97,98]; **5**, 618[6]
Kneip, M., **6**, 666[244], 667[244]
Kneisley, A., **1**, 744[57]; **4**, 380[123]; **5**, 1123[36]; **6**, 14[58], 16[58]
Knesel, G. A., **3**, 390[81], 392[81]
Knieqo, L., **6**, 195[224]
Knifton, J. F., **3**, 1029[56], 1030[59], 1037[88]; **4**, 915[7], 936[67], 939[77,78], 943[86], 945[87]; **8**, 375[152,154]
Knight, A. R., **7**, 761[67]
Knight, D. J., **6**, 899[108], 901[120]
Knight, D. W., **1**, 477[139]; **2**, 810[61], 824[61]; **3**, 242[6], 257[6], 259[6], 261[153], 278[32]; **4**, 382[135,135b], 383[135a]; **5**, 841[87,91,94], 843[124], 853[91d], 863[255]; **6**, 859[171,173]; **8**, 196[122]
Knight, H. M., **3**, 330[190]
Knight, J. C., **7**, 124[40], 564[111], 572[111]; **8**, 269[83]
Knight, J. G., **1**, 779[225]
Knight, L. S., **7**, 355[41]
Knight, N. C., **2**, 329[49]
Knights, E. F., **8**, 705[9], 707[9,29]
Knipe, A. C., **4**, 521[47], 522[47], 530[47], 953[8], 954[8p], 961[8p]
Knittel, D., **5**, 413[49]; **7**, 32[92]
Knittel, P., **4**, 298[282], 299[296]
Knizhnik, A. G., **8**, 447[106], 450[106]
Knobeloch, J. M., **2**, 835[159]
Knobler, C., **4**, 527[69], 528[69]; **5**, 1105[161,162]; **8**, 445[32]
Knoch, F., **2**, 520[66]; **4**, 21[69], 104[137], 222[168], 224[168]; **5**, 829[23,26]; **6**, 172[28], 190[198], 196[198]
Knochel, P., **1**, 115[41], 189[75], 205[106], 206[106], 212[5,5a,8,10,11,13], 213[5,8,10,11,13,14,17,17b,18], 214[5b,c,19,19b], 215[10,42], 216[10], 217[10,13,46], 219[42,56,56c,57,59], 220[57b], 221[10], 432[138], 433[138]; **2**, 6[34], 20[34c], 23[34c], 325[43,44], 326[43], 442[14], 449[14], 450[14,50]; **3**, 209[18], 257[122]; **4**, 14[49,49c], 34[97], 35[97], 78[53], 102[127], 175[42], 192[120], 340[53], 342[53], 880[92,93], 881[93,94], 882[92,93], 903[193]; **5**, 32[7], 829[24]; **6**, 164[197,198]; **7**, 453[67], 738[19]
Knoess, H. P., **2**, 278[11], 280[11]
Knoevenagel, E., **2**, 341[1], 342[2], 347[50]; **4**, 2[3,4], 239[32]
Knoflach, J., **7**, 746[92], 752[92]; **8**, 829[82]

Knol, D., **6**, 533[508]
Knol, K. E., **3**, 552[9], 557[9]
Knölker, H.-J., **5**, 1143[94]
Knoll, A., **6**, 425[68], 430[68], 552[693], 553[760], 554[760]
Knoll, F. M., **4**, 596[166]
Knoll, K., **5**, 1116[11], 1117[11], 1118[11]
Knolle, J., **3**, 640[105,107], 641[107b], 644[162,163,165], 647[107], 648[107b]; **4**, 370[39]; **5**, 837[69]; **6**, 538[569], 678[324], 765[17]; **7**, 694[32]
Knoop, F., **8**, 144[67,69], 145[86]
Knopp, J. E., **7**, 502[263]
Knoppová, V., **2**, 362[180], 363[187]
Knops, G. H. J. N., **7**, 535[47]
Knorr, A., **8**, 592[64]
Knorr, E., **6**, 37[36]
Knorr, H., **6**, 509[264]
Knorr, R., **3**, 587[143]; **6**, 294[244], 502[210], 706[37], 721[133], 723[133,145], 724[155]; **8**, 285[10]
Knors, C., **2**, 448[36]; **4**, 212[99]
Knothe, L., **5**, 744[166,167]
Knott, E. B., **2**, 496[67], 497[67], 866[6], 868[6]; **4**, 1093[146]
Knotter, D. M., **3**, 210[25]
Knotter, M., **2**, 1049[22], 1050[22]
Knouzi, N., **6**, 76[45]; **8**, 385[42]
Knowles, J. R., **2**, 466[109]; **5**, 854[177], 855[185,186,188]; **7**, 43[45]; **8**, 384[39]
Knowles, W. S., **1**, 446[195]; **2**, 233[184]; **3**, 1023[22]; **4**, 930[51]; **7**, 556[72]; **8**, 459[228,230,232], 460[230,232], 535[166]
Knox, G. R., **2**, 722[91,92,95]; **4**, 82[61], 519[22], 697[12], 698[14], 701[30]; **5**, 513[4], 1037[6], 1039[6], 1040[6], 1049[6], 1138[68], 1165[8], 1183[8]; **6**, 690[400], 692[400], 1012[6], 1013[6]; **8**, 329[23], 335[23]
Knox, I., **8**, 743[163], 757[163]
Knox, L. H., **6**, 217[114]; **8**, 526[24]
Knox, S. A. R., **3**, 380[10]
Knox, S. D., **4**, 587[29], 603[271], 605[292], 626[292], 646[292]; **6**, 1019[45]; **7**, 629[49]
Knözinger, H., **7**, 840[5-7]
Knudsen, C. G., **1**, 413[58,59]
Knudsen, J. S., **2**, 713[48]; **5**, 778[195]
Knudsen, M. J., **3**, 1025[34]; **5**, 1039[11,17], 1041[19], 1046[19], 1050[17], 1052[17,19], 1057[51,53], 1062[51,53], 1133[26], 1146[26]
Knudsen, R. D., **4**, 438[153]; **5**, 163[70]
Knunyants, I. L., **2**, 739[46]; **6**, 104[2], 437[34], 440[34], 443[34], 495[146]; **8**, 896[12-14], 898[13]
Knupfer, H., **2**, 1084[8]; **4**, 1085[100,101]; **6**, 242[85,86], 243[85,86], 489[88]
Knupp, G., **8**, 54[162], 66[162]
Knuth, K., **7**, 229[119]
Knutsen, L. J. S., **8**, 504[71]
Knutsen, R. L., **5**, 167[100]
Knutson, D., **4**, 30[91,91a]
Knutson, K. K., **8**, 528[79]
Knutsson, L., **3**, 843[24], 892[52]
Ko, A. I., **5**, 515[11], 517[11a], 519[11a], 526[11], 543[11a]
Ko, J., **3**, 848[49]
Ko, J. S., **8**, 18[122], 54[152], 66[152], 354[176], 536[174]
Ko, K.-Y., **1**, 63[42]; **7**, 549[41]
Ko, O.-H., **1**, 768[173]; **7**, 409[91]
Ko, S. S., **2**, 370[260]; **5**, 249[31], 516[20], 524[20]; **6**, 8[37], 89[116]; **7**, 401[62,62a], 406[62a]
Ko, S. Y., **3**, 223[156]; **6**, 89[112], 927[71]; **7**, 198[26], 390[4], 393[4,15,17], 394[4], 395[4], 396[4], 397[4,15,15a], 398[4], 399[4], 400[4], 401[4], 402[63], 403[65], 406[4], 407[4], 410[4], 411[4], 413[4]
Ko, T., **1**, 63[43]
Kobal, V. M., **3**, 135[340], 137[340], 139[340], 141[340]
Kobashi, H., **7**, 856[66]
Kobata, H., **6**, 1022[60]
Kobayakawa, S., **8**, 426[49]
Kobayashi, H., **2**, 967[76], 1061[92], 1069[92], 1071[92]; **4**, 120[197], 487[48], 893[151]; **7**, 458[113], 805[65]; **8**, 86[25]
Kobayashi, J., **6**, 611[67]
Kobayashi, K., **2**, 68[44], 110[73], 114[73], 298[93], 849[214]; **3**, 197[33]; **4**, 30[88,88j], 121[207], 253[175], 258[175], 817[202]; **5**, 1132[19], 1183[55]; **6**, 447[105], 450[105]
Kobayashi, M., **1**, 214[23]; **2**, 6[28], 17[28], 572[44], 833[147]; **3**, 231[241], 266[196], 443[51,53,54], 453[53], 628[49], 747[72], 875[73-75]; **4**, 159[85], 256[208,212], 261[208,284], 262[212], 314[496], 893[150]; **5**, 563[93]; **6**, 689[386]; **7**, 518[18,19], 771[273], 773[299], 779[299,424,428,430,431]; **8**, 389[70], 556[375], 756[148], 843[59a], 993[58]
Kobayashi, N., **4**, 231[266,267]; **6**, 46[66]
Kobayashi, R., **5**, 601[44]; **8**, 991[44]
Kobayashi, S., **1**, 57[32], 70[64], 347[135]; **2**, 615[122], 632[29a,b], 633[31,33a], 635[41], 640[29,31,33,41], 655[136,137], 656[137], 657[161a,167,168], 664[30], 665[30], 682[30], 689[30], 917[85,86], 920[86]; **3**, 295[7], 302[52], 638[92], 677[87], 1017[7]; **4**, 30[90], 50[142,142b], 158[79], 159[82,83], 161[90], 189[106], 190[106b], 244[80], 258[80,237,238], 261[238], 285[164], 289[164], 382[133], 387[161], 388[133,161]; **5**, 434[141]; **6**, 214[96], 233[48], 283[163,164], 820[111]; **7**, 5[22], 125[59], 423[145], 680[79]; **8**, 168[66,67], 542[238], 544[277], 830[88]
Kobayashi, T., **1**, 123[77], 372[80], 551[78], 563[180]; **2**, 357[149], 711[29]; **3**, 380[4], 529[51], 1016[3], 1033[71], 1039[101,102]; **4**, 359[161], 373[73], 379[115], 520[33,34], 557[14], 629[410], 638[410]; **5**, 221[55]; **6**, 283[162], 524[354], 527[405]; **7**, 324[71], 741[52]; **8**, 252[111], 301[92], 394[116], 971[106]
Kobayashi, Y., **1**, 58[33], 110[17], 131[17,104], 134[17], 135[117], 159[78,79], 160[78], 161[78], 185[54], 339[87], 546[54], 564[195], 784[244]; **2**, 23[88], 204[97], 538[66,68], 539[66,68], 604[53], 643[79], 644[79b], 656[154]; **3**, 100[201], 103[201], 107[201], 224[162], 421[54], 489[62], 495[62], 504[62], 511[62], 515[62], 639[85]; **4**, 45[130], 510[175], 629[405], 1005[88], 1020[234,235]; **5**, 100[142], 847[136], 935[190]; **6**, 6[29], 11[46], 217[115-117], 221[117], 493[139,140], 900[116], 989[78], 993[78], 996[105]; **7**, 255[36], 371[68], 379[100], 400[52], 414[108,109], 423[142,143], 461[118], 712[64], 750[127]; **8**, 557[381], 698[137], 782[102]
Kober, B. J., **7**, 372[70]
Kober, R., **1**, 373[85,86], 374[85,86]; **2**, 1077[153]; **5**, 485[182]; **6**, 716[104]
Kober, W., **2**, 600[29], 681[58], 683[58]; **5**, 843[120]; **6**, 502[216,217], 560[870]; **7**, 650[51]
Kobler, H., **6**, 227[21], 228[21], 229[21], 230[21], 231[21], 234[21]
Koblik, A. V., **2**, 712[40]
Kobori, T., **7**, 516[4]
Kobori, Y., **3**, 228[220], 441[47]
Kobrakov, K. I., **8**, 778[86]
Kobrehel, G., **6**, 766[23]; **7**, 698[51]
Köbrich, G., **1**, 214[30], 830[89], 873[92,95a]; **3**, 553[15]; **4**, 86[78d], 1007[115,128]; **6**, 960[60], 968[105]
Kobs, H.-D., **1**, 313[115]; **8**, 756[155]
Kobs, U., **4**, 760[196]
Kobsa, H., **2**, 746[113]
Kobylecki, R. J., **8**, 133[25]
Kocevar, M., **6**, 554[725]
Koch, B., **6**, 508[288]
Koch, D., **7**, 797[17]
Koch, G. L., **5**, 855[183]
Koch, H., **5**, 185[161,165]; **6**, 291[202], 558[851]
Koch, H. P., **7**, 762[72]
Koch, K., **3**, 681[100], 807[23]; **4**, 810[166]

Koch, M., **5**, 185[167]
Koch, P., **3**, 158[443], 159[443], 160[443], 161[443], 167[443], 168[443]
Koch, R. W., **3**, 581[110]; **8**, 136[52]
Koch, T. H., **5**, 581[171], 684[36b]; **6**, 529[463]
Koch, V. R., **6**, 281[151]; **7**, 794[4], 801[44], 810[87]
Koch, W., **4**, 461[99], 475[99]
Köcher, M., **3**, 593[179]
Kocheskov, K. A., **7**, 596[32], 632[57]; **8**, 851[132]
Kochetkov, N. K., **1**, 95[75,76]; **2**, 385[325]; **6**, 43[52], 49[100], 271[85], 466[40-42], 469[40,54], 533[498]; **8**, 694[118]
Kochetov, G. A., **2**, 464[100,101]
Kochhar, K. S., **4**, 492[71], 495[71]; **7**, 240[57]
Kochi, J., **3**, 415[10], 418[10]
Kochi, J. K., **1**, 174[5-7], 175[6], 193[87], 310[105]; **3**, 210[23], 228[53], 243[13], 244[19], 418[25], 436[1,2,6], 437[20,24], 438[2,20], 439[20], 441[20], 442[20], 464[24], 482[7], 494[84], 499[7]; **4**, 717[11], 719[20], 761[198], 763[207], 770[11], 773[11], 777[11], 805[145], 808[157], 962[41], 1032[14], 1033[14]; **5**, 71[132,133,134,136], 77[269], 901[31,32], 1012[51]; **6**, 280[143,145], 980[39], 993[93]; **7**, 13[108], 95[80], 383[110], 437[8], 527[1], 628[47], 719[4], 720[12], 722[4], 724[4], 725[34], 727[4], 816[5], 850[6,10], 851[14,17], 852[35,37,38,40], 854[46,50,59], 855[50,59,62-64], 856[65], 860[69,72], 862[80], 863[82], 864[86], 865[87], 867[92], 868[93], 869[94,95], 872[98], 874[102,108,110], 875[114], 877[135], 878[136], 881[162,163], 882[165], 886[69b], 887[62]; **8**, 19[135], 796[26,27], 850[121], 888[122]
Köchling, J., **6**, 269[73]
Kochloefl, K., **8**, 274[137], 544[267,268], 907[72], 918[72]
Kochmann, E. L., **2**, 757[11]
Kochs, P., **4**, 604[282]
Kochta, J., **5**, 829[26]
Kocián, O., **8**, 587[41], 591[56,57]
Kocieński, P. J., **1**, 329[33], 570[270], 694[238], 697[238], 698[238], 793[272,273], 794[273c,d,278], 797[273d], 801[302], 804[273], 809[330]; **2**, 651[116,118,119]; **3**, 26[163], 135[349], 136[349], 137[349], 140[349], 141[349], 159[452], 162[452], 163[452], 218[100,100b], 224[163,163a], 229[234], 264[178], 444[66-68], 934[69]; **4**, 390[168,169], 878[83]; **5**, 798[4], 1014[54]; **6**, 935[103], 987[74,75], 989[76,77], 990[74,77,85], 992[90], 993[77,91,92], 994[92,98], 996[108], 997[111], 998[90], 1002[77,132], 1006[145], 1059[69]; **8**, 843[61]
Kock, U., **4**, 44[125]
Köckritz, P., **2**, 359[161]
Kocovsky, P., **4**, 5[17]; **7**, 73[102], 94[59,61], 367[55]
Kocsi, W. P., **3**, 333[211b]
Koda, S., **7**, 255[38]
Kodadek, T., **7**, 427[148f]
Kodaira, K., **7**, 800[34]
Kodakek, T., **7**, 12[94]
Kodama, A., **3**, 168[490], 169[490]; **5**, 720[96]
Kodama, H., **1**, 87[47]; **3**, 463[165]; **8**, 192[96], 698[144], 755[123]
Kodama, M., **3**, 99[190], 100[193,194,195,196,197,200,201], 103[193,194,195,200,201], 107[194,195,201], 390[69,70], 395[101]; **5**, 620[16]; **6**, 145[80], 900[116,117], 1045[21,22]; **7**, 743[64]; **8**, 856[171]
Kodama, S., **3**, 437[25], 440[25], 448[25], 449[25], 450[25], 451[25], 452[25,107], 459[137], 460[107,137], 461[137], 484[26], 492[26], 494[26], 495[26], 503[26], 510[206], 513[26,206]
Kodama, T., **4**, 18[59]; **7**, 57[22]
Kodama, Y., **1**, 512[39]
Kodera, M., **6**, 438[49]
Kodera, Y., **6**, 113[69]; **8**, 395[122,123]
Kodevar, M., **6**, 554[714,746]
Kodovsky, P., **3**, 591[167], 610[167], 709[15]; **4**, 364[2], 367[9], 374[92], 376[9], 391[179], 397[9]
Kodpinid, M., **2**, 807[47]; **5**, 560[67]
Koebernick, W., **6**, 980[33]
Koegel, R. J., **8**, 145[88]
Koehl, W. J., Jr., **3**, 649[203]
Koehler, K. A., **4**, 5[17]
Koehler, K. F., **5**, 249[36], 250[37], 252[37], 255[37]
Koehn, W., **5**, 66[79]
Koelichen, K., **2**, 140[34]
Koelliker, U., **7**, 686[100]
Koelsch, C. F., **6**, 280[142]; **8**, 568[466]
Koenig, G., **3**, 915[13], 965[13]; **5**, 69[99]
Koenig, K. E., **8**, 152[172], 459[228], 460[228b]
Koenig, T., **7**, 852[34]
Koenigkramer, R. E., **1**, 544[38], 562[38,169]
Koenigs, W., **6**, 37[36]
Koepp, E., **2**, 402[30]
Koermer, G. S., **8**, 99[107]
Koerner, M., **3**, 209[17], 213[48], 251[76], 261[76], 264[76]; **4**, 175[43], 177[55]; **5**, 564[95]; **6**, 831[9]
Koesling, V., **3**, 866[29]
Kofler, M., **2**, 410[5]
Kofron, J. T., **5**, 163[71]
Kofron, W. G., **1**, 476[122]; **3**, 35[203], 66[13]
Koft, E. R., **4**, 24[74]; **5**, 145[105]; **8**, 353[158]
Koga, K., **1**, 72[69,70], 314[121], 342[99], 359[19], 382[19,19a-e], 566[214,215], 823[44a]; **2**, 105[39], 163[147], 558[160], 846[208], 1018[37]; **3**, 36[211], 41[224], 43[235], 217[95], 675[73]; **4**, 10[34], 21[69], 76[49], 85[75], 111[157], 113[164], 200[1], 203[31], 210[79-86], 211[82,85,87], 222[173,174,175,176], 229[239], 249[126], 252[163,164], 258[126], 391[184], 393[184b], 397[184b]; **5**, 134[61], 376[108a,b], 736[145], 737[145]; **6**, 4[18], 137[74], 721[136], 723[148], 724[148], 726[177,178,183,184], 738[47]; **7**, 142[138], 438[13], 442[50], 443[13]; **8**, 166[55-57], 170[80], 178[56], 179[56], 241[37], 541[207], 545[288,289], 546[288]
Koga, N., **7**, 108[176]
Koga, T., **3**, 380[13]
Koga, Y., **8**, 248[82]
Kogai, B. E., **7**, 505[289]
Kogami, K., **2**, 547[108], 550[108]; **4**, 159[81]; **6**, 715[89,90], 980[43]
Kogan, L. M., **7**, 474[38]
Kogan, T. P., **4**, 213[109,112], 215[109,112]; **6**, 109[43], 150[122], 1024[76]; **7**, 302[63], 766[178]; **8**, 844[71], 847[71]
Kogatani, M., **4**, 564[41]
Kogawa, K., **7**, 719[5], 732[5]
Kogen, H., **1**, 359[19], 382[19,19c-e], 766[152]; **4**, 85[75], 210[82,85,86], 211[82,85]; **5**, 249[36]; **6**, 723[148], 724[148], 726[183]; **7**, 370[67]
Kögl, F., **3**, 829[54]; **8**, 144[70]
Kögler, K., **7**, 92[46]
Kogotani, M., **6**, 538[567]
Koguchi, T., **2**, 82[9], 575[64]
Kogure, I., **8**, 173[120]
Kogure, T., **1**, 63[42]; **2**, 232[177], 603[47]; **4**, 202[26], 254[187]; **8**, 20[139], 36[84], 39[84], 66[84], 152[181,182,183], 173[114,116,117,119], 460[254], 535[166], 555[370], 556[374], 763[1], 779[1d], 782[101], 785[1]
Koh, H. S., **7**, 307[11]
Kohada, H., **3**, 919[32]
Kohama, H., **5**, 841[95]
Kohama, M., **5**, 833[49]
Kohama, T., **3**, 396[115]
Kohara, N., **8**, 185[26], 190[26]
Koharski, D., **8**, 812[4], 967[84]
Kohashi, Y., **8**, 609[53]
Kohda, A., **7**, 538[64], 539[65]
Kohen, S., **4**, 1040[92], 1042[92]

Kohler, E. P., **3**, 756[113]; **4**, 5[18], 84[67], 85[76], 106[138a-c], 239[29-31]; **6**, 157[163], 280[140]; **7**, 156[32]
Köhler, F. H., **1**, 476[115]; **4**, 869[20]; **5**, 568[110]
Köhler, H.-J., **1**, 287[14]
Köhler, J., **1**, 215[32]
Kohler, R.-D., **6**, 556[833]
Kohlheim, K., **3**, 322[143]
Kohli, V., **6**, 602[4]
Kohll, C. F., **4**, 600[228], 601[246]
Kohlmaier, G., **6**, 294[238]
Kohl-Mines, E., **6**, 185[166]
Kohmoto, S., **5**, 157[38]; **7**, 618[22]
Kohmoto, Y., **7**, 537[61]
Köhn, A., **6**, 846[102]
Kohn, D. H., **8**, 451[180]
Kohn, H., **2**, 1060[87], 1062[97]; **4**, 355[134]; **5**, 89[51]; **7**, 49[68], 479[93,94]
Kohn, P., **8**, 269[88,90]
Kohne, B., **6**, 462[16]
Kohno, M., **2**, 647[88a]; **7**, 245[78]
Kohno, S., **2**, 579[95], 581[105]; **8**, 806[107]
Kohoda, H., **3**, 923[44], 934[44], 954[44], 1008[70]
Kohra, S., **1**, 180[43], 181[43]; **2**, 6[28], 17[28], 572[43,45]; **4**, 589[80], 591[80]; **6**, 83[82]
Kohzuki, K., **5**, 524[54], 534[54]
Koike, N., **8**, 432[68]
Koike, T., **6**, 820[111]
Koikov, L. N., **6**, 555[817]
Koita, N. K., **7**, 769[232]
Koitz, G., **4**, 440[171]
Koizumi, N., **3**, 125[293], 126[293], 127[293], 128[293]; **7**, 675[54]
Koizumi, S., **5**, 839[82]
Koizumi, T., **1**, 752[97]; **5**, 260[71], 369[101], 370[101a-d], 768[131], 779[131]; **6**, 150[126,128,129], 152[136]; **7**, 153[11], 257[47]; **8**, 241[37], 546[309], 557[381], 782[102]
Kojer, H., **7**, 449[1], 450[1]
Kojima, E., **1**, 221[68]; **2**, 30[113], 31[113]; **8**, 187[44], 188[44], 203[146]
Kojima, K., **2**, 852[234]
Kojima, M., **4**, 356[138]; **7**, 778[404], 881[161]
Kojima, S., **8**, 652[74]
Kojima, T., **3**, 100[200], 103[200]; **5**, 137[76], 143[76], 339[57a], 345[57]; **6**, 533[502]; **7**, 242[60]
Kojima, Y., **3**, 213[42], 264[185]; **4**, 176[54], 436[142,143]; **7**, 655[12]
Kojo, S., **8**, 34[62], 66[62]
Kok, D. M., **5**, 78[275]
Kok, J. J., **2**, 1049[22], 1050[22]
Kok, P., **4**, 1040[75]; **7**, 105[147]
Kokel, B., **6**, 499[176]
Kokil, P. B., **4**, 370[46]
Kokil, P. D., **7**, 155[31c]
Kokko, B., **1**, 385[117]
Kokorin, A., **8**, 458[223,223d]
Köksal, Y., **1**, 474[95]; **4**, 211[91]
Kokubo, T., **7**, 443[51a], 778[409]
Kol, M., **4**, 347[98]
Kolár, C., **6**, 41[44], 43[49]
Kolasa, T., **1**, 386[123]; **6**, 22[83], 114[74], 493[126], 494[126]
Kolb, M., **2**, 96[56]; **3**, 131[324,331], 135[324]; **4**, 113[161], 120[201]; **6**, 554[783]; **7**, 450[15]
Kolb, N., **6**, 531[458]
Kolb, R., **3**, 872[59]
Kolb, V. M., **3**, 864[16,23], 866[16], 883[16]
Kolbah, D., **6**, 85[87]; **7**, 232[158]
Kolbasenko, S. I., **4**, 347[106]
Kolbe, H., **2**, 321[3]; **3**, 633[2]
Kolbon, H., **4**, 231[271]
Kolc, J., **4**, 483[7]
Koldobskii, G. I., **6**, 795[12,13], 798[12,13], 817[12,13], 820[12]; **7**, 690[13]
Kole, P. L., **7**, 333[24]
Kole, S. L., **4**, 371[51], 670[23], 673[32], 687[23]
Kolesar, T. F., **2**, 711[33]
Kolesov, V. S., **5**, 1148[115]; **8**, 698[147]
Kolewa, S., **4**, 55[157]
Kolhe, J. N., **1**, 268[56], 269[56a], 477[145], 545[47,48]; **2**, 523[75]; **3**, 602[225]; **4**, 113[170]; **7**, 231[154]; **8**, 114[51,52]
Kolinsky, M., **8**, 652[75]
Kolka, A. J., **3**, 296[12]
Kolka, S., **6**, 789[109]
Köll, P., **6**, 1035[139,140]
Koll, W., **5**, 426[106], 451[3]
Kollar, L., **4**, 926[37], 932[64]
Kollenz, G., **6**, 524[356]
Koller, E., **3**, 350[37]; **4**, 30[91,91a]; **5**, 20[138], 163[71]
Koller, J., **1**, 837[156]
Koller, M., **2**, 725[115]
Koller, W., **1**, 123[79], 372[80]
Kollman, P. A., **2**, 1007[3]
Kollmannsberger-von Nell, G., **4**, 429[87]
Kollmeier, J., **5**, 1138[68]
Kollonitsch, J., **1**, 326[5]; **4**, 98[108b,c]; **7**, 15[147]; **8**, 828[79]
Kolm, H., **3**, 893[56]
Kolobova, N. E., **4**, 701[29], 702[29]
Kolodiazhnyi, O. I., **6**, 172[14]
Kolodka, T. V., **6**, 524[366]
Kolodny, N. H., **3**, 577[87]; **8**, 524[12], 527[49], 532[12c]
Kolodyazhnyi, O. I., **6**, 525[389]
Kolodziejski, W., **3**, 322[144]
Kolomnikov, I. S., **5**, 1174[34]; **8**, 457[215], 557[385]
Kolonits, P., **8**, 266[55]
Kolonko, K. J., **1**, 377[97]; **8**, 949[153]
Kolosov, M. N., **6**, 603[21]
Kolozyan, K. R., **6**, 270[83]
Kolsaker, P., **4**, 1035[40]
Kolt, R. J., **4**, 399[221]
Kolthammer, B. W. S., **8**, 797[39]
Koltzenburg, G., **5**, 66[81], 636[92,93]
Koltzenburg, R., **5**, 646[10]
Komadina, K. H., **8**, 22[145], 683[86]
Komalenkova, N. G., **8**, 769[27]
Komar, D. A., **4**, 892[144]; **8**, 753[71]
Komarewsky, V. I., **3**, 331[197]
Komarov, N. V., **7**, 762[75]
Komarovskaya, I. A., **4**, 314[487]
Komatsu, H., **2**, 556[153]; **5**, 681[28], 741[157,157d]; **6**, 931[90]; **7**, 199[31]
Komatsu, M., **1**, 787[254]; **4**, 152[57]; **5**, 474[158], 1200[49]; **6**, 530[423]
Komatsu, T., **1**, 121[66], 185[60], 221[68], 359[13], 361[13], 362[13]; **2**, 11[49], 15[66], 30[107], 31[107], 32[120,120a,b], 978[11], 979[11,15,16], 983[15,16], 984[15,16,29,30], 985[15,16,29,30], 986[15,16], 987[15,30], 988[30], 989[11], 990[11], 991[11], 992[11,15], 993[11]; **4**, 388[162], 401[162a]; **6**, 864[194]
Komendantov, M. I., **4**, 1051[127]; **5**, 948[273]
Komeno, T., **3**, 380[12]
Komeshima, M., **1**, 314[121,122]
Komeshima, N., **4**, 210[81,84]; **5**, 376[108a,b]
Komin, A. P., **4**, 458[65], 462[65,104], 465[65,104,116], 466[65,104], 468[104], 469[104,116,134], 472[134], 473[134], 475[134]
Kominami, K., **4**, 221[160]; **6**, 717[108]
Kominek, L. A., **7**, 67[78], 68[81]

Komiya, S., **8**, 16[97], 445[62]
Komiya, Z., **3**, 901[108]; **6**, 976[10]
Komiyama, M., **2**, 771[13]
Komiyama, W. E., **4**, 505[138]
Komnenos, T., **4**, 1[1], 3[1]
Komori, T., **7**, 415[115a], 778[406]
Komornicki, A., **1**, 92[64]; **4**, 1070[13]; **5**, 703[17], 710[17], 857[227]
Komoto, H., **3**, 639[81]
Komoto, R. G., **3**, 1028[51]; **8**, 550[334]
Kompis, I., **2**, 385[322]
Kompter, H.-M., **4**, 113[168]
Komzak, A., **2**, 943[169], 970[88]
Kon, G. A. R., **2**, 141[38]
Konakahara, T., **1**, 624[86], 836[139]
Konda, M., **6**, 738[48]
Kondo, A., **1**, 767[176]; **3**, 383[41,44]; **4**, 297[276]; **7**, 462[124]
Kondo, H., **1**, 63[43,44], 64[46], 881[118]; **4**, 304[356]; **5**, 810[126], 812[126]; **8**, 451[180], 568[465]
Kondo, I., **3**, 483[9]
Kondo, K., **1**, 751[109]; **3**, 99[187], 107[187], 110[187], 159[458], 161[458], 162[458], 167[458], 202[85,88], 437[22], 438[22,36], 440[36], 771[186], 923[42], 949[97]; **4**, 30[89], 1040[67,86,87], 1041[87], 1045[86]; **5**, 461[107], 464[107], 466[107], 924[144], 945[246]; **6**, 145[79], 159[174], 247[131], 845[95], 865[95]; **7**, 97[93]; **8**, 370[90], 382[7], 844[74]
Kondo, M., **3**, 891[45]; **6**, 569[938]; **8**, 174[126], 178[126], 179[126]
Kondo, N., **4**, 1056[141]; **7**, 774[318]
Kondo, S., **2**, 124[207]; **3**, 919[32], 923[44], 934[44], 954[44], 1008[70,71]; **4**, 96[104], 462[106], 475[106]
Kondo, T., **1**, 750[88]; **3**, 500[133], 1028[49]; **4**, 331[11], 941[84]; **6**, 23[89]; **7**, 856[66]; **8**, 131[6], 795[23], 906[65], 907[65], 908[65], 909[65], 910[65]
Kondo, Y., **1**, 561[164], 860[69]; **2**, 598[15]; **3**, 406[141], 460[143], 461[143,148], 530[64], 533[64], 541[108-111,113,114], 543[108-111,113,117,118]; **6**, 801[30]; **7**, 230[128]; **8**, 902[45]
Kondratenko, N. V., **6**, 496[154]
Kondrat'eva, G. Ya., **5**, 491[202,203]
Kondrat'eva, L. V., **4**, 3[10], 41[10], 47[10], 66[10]
Kondrat'yev, V. N., **7**, 852[42]
Kondrikov, N. B., **3**, 648[183]
Konen, D. A., **7**, 185[175]
Konepelski, J. P., **7**, 239[44]
Kong, F., **8**, 413[122]
Kong, N. P., **4**, 719[23]
Kong, S., **5**, 531[73]
Kongkathip, B., **4**, 374[88,89]; **7**, 451[21,23]
Kongkathip, N., **4**, 374[89]; **7**, 451[23]
Konieczny, M., **2**, 760[47]; **8**, 910[85], 911[85], 949[157]
Konieczny, S., **5**, 950[285]
Konig, B., **4**, 1093[148]
König, C., **2**, 366[218], 782[30]
König, D., **4**, 124[214e]; **6**, 937[118,119], 939[118,119], 942[118], 943[118]
König, G., **2**, 740[53]; **5**, 933[181]
König, H., **3**, 893[54], 896[71]; **4**, 500[106]; **6**, 609[55]
König, J., **3**, 509[196], 512[196]; **5**, 908[71]
König, K., **1**, 328[19]; **2**, 564[2], 572[2], 576[2]; **6**, 244[113]
König, L., **3**, 466[183]
König, R., **1**, 202[101], 331[48], 734[22], 828[80]
König, S., **2**, 1104[134]
König, W., **2**, 1099[109]; **4**, 42[120]; **6**, 635[11], 645[11], 664[221]
König, W. A., **4**, 390[175b]
Königstein, F.-J., **7**, 657[34]; **8**, 271[101,103], 965[70]
Königstein, M., **8**, 273[127]
Konijn, M., **1**, 25[126]
Konings, M. S., **6**, 291[211]
Konishi, A., **4**, 1033[21], 1037[21], 1040[21]
Konishi, H., **3**, 555[29], 1027[46]; **4**, 606[303], 646[303]
Konishi, J., **8**, 188[49], 193[49], 561[411]
Konishi, K., **4**, 444[197]
Konishi, M., **1**, 610[46], 611[46]; **2**, 17[68], 568[33], 584[124], 716[62]; **3**, 228[219,220], 229[226,227,231], 246[37], 438[33], 441[47], 442[50], 450[33], 452[33], 470[219], 471[219], 472[219]; **4**, 588[67], 596[159], 614[382], 615[387], 620[387], 635[159,387]; **5**, 736[145], 737[145]; **8**, 84[12], 152[180], 437[79], 459[245]
Konishi, S., **8**, 205[160]
Konishi, Y., **1**, 386[122]; **2**, 823[116]
Konno, A., **6**, 531[449]; **7**, 876[120]
Konno, C., **3**, 396[115]
Konno, S., **2**, 364[205]; **3**, 461[147]; **6**, 530[423]
Kono, T., **7**, 419[134b]
Konomoto, K., **6**, 439[69]
Kononov, N. F., **3**, 305[72]
Kononova, V. V., **6**, 525[377]
Konopelski, J. P., **1**, 698[251]; **3**, 709[17,18]; **8**, 931[38]
Konovalov, A. I., **5**, 76[246,248], 516[19], 552[16]
Konovalov, E. V., **4**, 391[178a]
Konowal, A., **2**, 662[18], 663[27,28], 664[18,27,28]
Konrad, P., **6**, 644[94]
Konradsson, P., **4**, 391[181b,182,182a]; **6**, 40[40]
Könst, W. M. B., **6**, 714[87]
Konstantatos, J., **3**, 565[20]
Konstantinou, C., **4**, 439[159]
Konstantinovic, S., **6**, 184[153]
Konta, H., **1**, 833[119]; **6**, 155[155], 156[155]; **8**, 12[66]
Konuma, K., **8**, 650[67]
Konwenhoven, A. P., **4**, 587[38,39]
Konya, N., **2**, 350[73,74], 363[73]
Konz, E., **6**, 767[25]; **7**, 695[35]
Konz, W. E., **3**, 365[95]
Konzelmann, F. M., **8**, 645[43]
Koo, J., **8**, 390[80]
Kooistra, D. A., **8**, 216[64], 487[67]
Kool, E., **1**, 543[18]
Koola, J. D., **7**, 383[110]
Koolpe, G. A., **1**, 630[42], 633[73], 636[73], 637[73], 639[73], 645[42], 646[73], 647[73], 648[73], 656[73], 657[73], 658[73], 669[42], 670[42], 672[73], 686[73], 688[73], 690[73], 692[73], 694[42], 708[42], 7; **3**, 87[108], 89[145], 90[145], 105[145], 106[108], 114[108], 120[108], 136[108], 144[108]; **4**, 120[200]
Koomen, G.-J., **2**, 718[81]; **6**, 181[132], 182[132]; **7**, 684[95]
Koosha, K., **3**, 155[430]; **4**, 240[45]
Kooy, M. G., **5**, 562[87]
Kopchik, R. M., **4**, 719[23]
Kopecky, K. R., **1**, 49[3], 50[3], 109[12], 110[12], 134[12], 141[20], 151[20], 153[20], 182[46], 460[2]; **2**, 24[96]; **4**, 959[31]; **8**, 3[18]
Kopecky, W. J., Jr., **1**, 294[47]
Kopf, J., **1**, 16[85], 20[108], 22[116], 34[224], 36[174], 38[253], 39[188]
Kopka, I., **2**, 616[133]; **6**, 531[439]
Koplick, A. J., **1**, 276[79], 277[79b,d]
Köpnick, H., **2**, 1023[55]
Kopola, N., **1**, 472[77]
Kopp, G., **5**, 71[135]; **7**, 880[153]
Kopp, R., **2**, 1090[75]
Koppang, M. D., **4**, 872[37]
Koppang, R., **6**, 543[619], 553[705]
Koppel, G. A., **2**, 422[29]; **4**, 5[19,19g], 13[45], 14[45a], 27[84], 29[84b]; **6**, 1022[63]
Köppelmann, E., **3**, 482[3]
Koppes, M. J. C. M., **5**, 217[21], 221[21]
Koppes, W. M., **2**, 348[55]; **4**, 426[54]

Koppetsch, G., **1**, 14[77,78]
Kopping, B., **4**, 739[107]
Kopp-Mayer, M., **3**, 842[19]
Kopylova, B. V., **6**, 542[598]
Kopylova, L. I., **8**, 770[33,38], 771[42]
Kopyttsev, Yu. A., **8**, 447[103,104], 450[103,104]
Korbukh, I. A., **6**, 533[476], 554[745]
Korchevin, N. A., **6**, 509[282]
Kordish, R. J., **8**, 819[42], 820[42]
Koreeda, M., **2**, 4[12], 6[12], 10[41], 573[50], 575[50]; **3**, 24[148], 125[293], 126[293], 127[293], 128[293], 996[46], 997[46a], 1000[46,53]; **4**, 794[77]; **5**, 335[48], 432[133], 519[34], 542[112], 549[34], 822[165], 829[18], 835[59], 848[18c], 849[18c], 851[18c], 1001[15]; **6**, 174[54], 883[57], 887[57]; **7**, 365[48], 649[43]
Koren, B., **6**, 514[306], 543[306], 554[712]
Korenova, A., **1**, 86[38]
Korenstein, R., **5**, 729[125]
Korepanov, A. N., **8**, 630[187]
Koreshkov, Y. D., **4**, 1005[91]; **8**, 557[385]
Korf, J., **8**, 53[132], 66[132]
Korff, R., **6**, 845[97,98]
Korfmacher, W. A., **7**, 75[116]
Korhummel, C., **6**, 172[20]
Kori, M., **8**, 189[63]
Korker, O., **5**, 804[94], 986[39]
Korkor, O., **4**, 1033[27], 1049[27]
Korkowski, P. F., **4**, 980[115], 982[115]; **5**, 1104[160]
Kormer, M. V., **4**, 300[308]
Kormer, V. A., **8**, 727[197]
Kornberg, H. A., **2**, 933[138]
Kornblum, N., **4**, 12[42], 429[83], 438[83], 441[83], 452[2,6], 467[6], 765[228], 766[230]; **6**, 226[12], 668[256], 669[256], 1000[125,126]; **7**, 218[4], 219[16], 220[24], 291[2], 654[9], 655[9,18], 660[42], 664[64], 665[68,69], 882[171,172,173]; **8**, 363[6], 916[111,114], 918[111,114]
Korner, H., **1**, 487[2], 488[2]
Körner, W., **5**, 225[96]
Kornet, M. J., **8**, 637[8]
Kornienko, A. G., **3**, 635[44], 648[172,173,174]
Kornienko, G. V., **8**, 373[132]
Kornienko, V. L., **8**, 373[132]
Korniski, T. J., **3**, 154[420]; **7**, 124[44]
Korobeinicheva, I. K., **3**, 386[68]
Korobitsyna, I. K., **3**, 887[7], 890[7], 892[7], 893[7], 896[7], 897[7,73], 900[7], 902[7], 903[7], 905[7]
Korobko, V. G., **6**, 603[21]
Korol'chenko, G. A., **4**, 314[492,493]
Koroleva, E. V., **7**, 483[126]
Koroleva, T. I., **4**, 992[154]
Koroniak, H., **5**, 586[204], 680[24,24a,b]
Korostova, S. E., **3**, 259[128]
Korp, J., **8**, 459[238]
Korpiun, O., **8**, 411[103]
Korshak, V. V., **2**, 387[334]; **3**, 319[132]; **5**, 1148[115]; **6**, 533[491]; **8**, 769[30]
Korsmeyer, R. W., **3**, 1052[25], 1059[25]; **6**, 127[156]
Korst, J. J., **2**, 162[142]; **7**, 157[33]; **8**, 974[122]
Kort, W., **5**, 421[80]
Korte, D. E., **4**, 558[18]
Korte, F., **2**, 282[35], 554[133]; **5**, 453[66], 457[89]; **6**, 116[90], 428[86]; **7**, 3[4]; **8**, 229[135], 649[62]
Korte, W. D., **3**, 419[32,33]
Korth, H.-G., **1**, 266[46]; **2**, 342[6]; **5**, 71[135], 159[51], 189[51]; **6**, 524[357], 526[357], 527[357], 534[522]; **7**, 880[153], 883[175]
Korth, T., **8**, 266[58]
Korver, G. L., **5**, 834[51]
Koryak, E. B., **4**, 391[177,178b]
Korytnyk, W., **7**, 749[123]
Korzeniowski, S. H., **3**, 505[163]
Kos, A. J., **1**, 19[102], 41[196]; **3**, 194[4]; **6**, 172[11]
Kosack, S., **5**, 441[176,176f]; **6**, 550[676]
Kosaka, S., **6**, 510[292]
Kosaka, T., **7**, 794[7a]
Kosar, W. P., **4**, 982[113]; **5**, 1086[69]
Kosarych, Z., **3**, 937[74,75]; **5**, 96[105], 333[44a]; **6**, 445[99], 893[75,80,89], 895[75]
Kosbahn, W., **5**, 417[64]; **6**, 525[387]
Kosch, E., **3**, 890[35], 903[127]
Kosch, W., **6**, 671[278]
Koschatzky, K. H., **6**, 172[25]
Koscielak, J., **6**, 33[3], 40[3]
Koseki, K., **6**, 74[29]
Kosel, C., **7**, 506[303]; **8**, 568[486]
Koser, G. F., **3**, 829[56]; **4**, 370[46]; **6**, 805[52,53], 806[52]; **7**, 155[31a-c], 179[31b]
Kosfeld, H., **2**, 112[94], 244[22]
Koshchii, V. A., **3**, 306[77]
Koshelev, V. I., **6**, 543[606]
Kösher, R., **6**, 662[213]
Koshihara, A., **6**, 487[40], 489[40]
Koshiji, H., **8**, 195[106], 197[106]
Koshino, H., **4**, 358[155]
Koshiro, A., **8**, 97[95]
Koshland, D. E., Jr., **6**, 824[124]
Koshy, K. M., **4**, 299[296]
Koskimies, J. K., **1**, 62[40], 63[42], 153[57], 542[13], 544[13], 563[13]; **3**, 86[30], 121[30]; **6**, 679[327]
Koskinen, A., **2**, 1020[47]; **6**, 910[4], 913[25], 939[4]
Kosley, R. W., Jr., **5**, 851[169]
Kosolapoff, G. M., **6**, 171[3], 198[3], 601[1]; **8**, 858[207], 860[219], 862[219]
Kosower, E. M., **8**, 338[95], 357[95], 828[77]
Kossa, W. C., Jr., **4**, 876[60]
Kossanyi, J., **5**, 123[1], 126[1], 178[135,136], 455[76], 723[107]
Kossmehl, G., **2**, 388[338,339]; **6**, 175[78]
Kost, A. M., **7**, 75[114]
Kost, A. N., **4**, 424[19]; **8**, 612[76], 616[98], 623[147], 624[147], 636[5]
Koster, A. S., **8**, 95[83]
Koster, D. F., **2**, 538[62], 539[62]
Köster, F.-H., **3**, 348[28]
Köster, G., **6**, 864[195]
Köster, H., **1**, 10[52], 12[64], 18[92]; **6**, 602[4], 624[147], 625[160], 675[299]
Köster, R., **2**, 112[89,93,94], 240[11], 241[14], 244[11,21,22], 589[157,158]; **4**, 145[23], 885[109,110]; **5**, 288[36]; **6**, 173[48], 174[48]; **7**, 596[41], 597[46], 598[53,57]; **8**, 717[105,106], 724[175,176], 754[97,107]
Koster, W. H., **6**, 644[93]
Kostikov, R. R., **5**, 904[50]; **6**, 498[166]
Kostitsyn, A. B., **4**, 964[47]
Kostova, K., **1**, 145[45]
Kostyanovskii, R. G., **1**, 837[147]; **4**, 48[136], 49[136]
Kosuge, T., **8**, 917[119]
Kosuge, Y., **8**, 374[150]
Kosugi, H., **1**, 155[68], 156[68]; **2**, 282[38], 363[191]; **3**, 946[87]; **4**, 18[59], 106[140b], 121[205c], 154[64a], 257[223], 258[250], 743[128]; **6**, 150[118,119], 151[119], 155[155], 156[155], 902[129], 1022[58]; **7**, 205[64]; **8**, 12[66], 679[68], 680[68], 683[68], 693[68], 695[68]
Kosugi, K., **3**, 446[80], 456[80], 470[80], 471[80], 493[82]; **6**, 564[909]; **8**, 609[54]

Kosugi, M., **1**, 436[154], 438[160,161], 452[219], 833[119], 834[124]; **3**, 12[62], 453[113,114], 454[120], 463[160,164], 469[216], 470[215,216], 473[215], 476[215,216]
Kosychenko, L. I., **8**, 611[63]
Koszalka, G. W., **4**, 115[180c]
Koszinowski, J., **2**, 780[9]; **4**, 1102[201]
Koszyk, F. J., **4**, 1040[88], 1048[88,88g]; **5**, 906[66], 907[66,76], 908[66], 909[97], 916[122], 925[122]
Kotai, A., **6**, 668[254], 669[254]
Kotake, H., **3**, 159[451], 160[451], 161[451]; **4**, 359[161], 595[151], 599[212], 604[288,289], 640[212], 641[212], 646[289], 647[288], 753[164]; **6**, 641[62]; **7**, 262[82], 564[95], 568[95], 709[37]; **8**, 840[30,30b], 960[37]
Kotaki, H., **4**, 1089[125]
Kotani, E., **3**, 679[89], 683[102,105], 693[89], 695[153], 696[154], 807[35]; **7**, 801[39]
Kotani, S., **3**, 541[106], 558[50]
Koteel, C., **6**, 939[144], 942[144]; **7**, 844[53]
Kotelko, A., **8**, 636[2]
Kotera, K., **8**, 49[115], 66[115,214]
Kotera, M., **6**, 509[256]; **7**, 50[75]
Koteswara Rao, M. V. R., **8**, 530[104]
Kothiwal, A. S., **1**, 305[83]
Kotian, K. D., **8**, 216[62]
Kotick, M. P., **4**, 191[110]
Kotkowska-Machnik, Z., **6**, 271[89]
Kotlan, J., **3**, 817[91]
Kotlarek, W., **8**, 490[9]
Kotlyarevskii, I. L., **5**, 752[15-17,23-25,30], 757[17,30], 767[15,23-25], 768[25]
Kotnis, A. S., **4**, 24[74]
Koto, H., **1**, 260[32,32a], 261[32a], 831[104]
Koto, M., **3**, 918[27], 968[128]
Koto, S., **6**, 40[39], 47[77]
Kotok, S. D., **3**, 897[73]
KotÓynek, O., **8**, 176[129]
Kotsonis, F. N., **8**, 625[160]
Kotsuji, K., **7**, 127[61]
Kotsuki, H., **1**, 107[6], 110[6], 568[229], 766[162]; **8**, 215[53], 217[53], 227[121], 240[32], 244[70], 620[133], 624[133]
Kottenhahn, A., **7**, 768[200]
Kötter, H., **4**, 874[55]; **5**, 30[2]
Köttner, J., **8**, 535[166]
Kotusov, V. V., **6**, 450[121]
Kotynek, O., **8**, 161[17]
Kouba, J. K., **4**, 970[69]
Koukoua, G., **3**, 380[8]
Koul, A. K., **8**, 271[110]
Koul, V. K., **1**, 390[142]; **6**, 97[153]
Koulkes, M., **4**, 304[359]
Koumaglo, K., **1**, 623[80]; **2**, 58[12,13]; **3**, 200[71]; **4**, 347[88]
Koussini, R., **5**, 727[119]
Koutechky, J., **5**, 72[178]
Kouwenhoven, A. P., **5**, 9[70,71], 212[69]
Kouwenhoven, C. G., **4**, 45[127,127b]; **5**, 584[195], 687[63], 688[63]
Kovac, B., **7**, 867[91]
Kovac, F., **6**, 554[712]
Kovác, J., **2**, 362[180], 363[187]
Kovacic, P., **6**, 1016[28], 1036[142,143]; **7**, 10[72], 741[48], 747[48]
Kovacik, V., **2**, 140[36]
Kovacs, C. A., **2**, 968[80]
Kovács, G., **2**, 381[305], 529[17]
Kovacs, K., **6**, 653[151]
Kovacs, M., **6**, 543[612]
Kovad, I., **3**, 904[133]
Kovalev, V. A., **6**, 524[360,364], 528[413], 532[364]
Kovaleva, L. F., **4**, 291[209]
Koval'ova, L. I., **4**, 310[427]
Kovar, R., **8**, 545[283], 549[283]
Kovats, E. Sz., **5**, 10[76]
Kovelesky, A. C., **2**, 492[53], 493[53]; **4**, 76[45]; **6**, 274[107]; **8**, 276[149]
Kövesdi, J., **7**, 777[389]
Kovsman, E. P., **3**, 640[98,101], 647[177], 648[177], 649[177]
Kovtunenko, V. A., **6**, 509[283]
Kow, R., **1**, 491[29], 492[38], 498[29], 501[29], 502[38]; **2**, 56[3], 57[3]; **3**, 199[60,66]
Kowalczuk, M., **2**, 105[41]
Kowalczyk-Przewloka, T., **2**, 547[113], 551[113]; **3**, 56[285]; **4**, 229[235,236], 1055[138]
Kowalski, B. R., **2**, 1099[115]
Kowalski, C. J., **1**, 874[103,104]; **2**, 109[66-69], 547[101], 548[101], 611[100]; **4**, 255[201]
Kowalski, J., **7**, 752[150]
Kowollik, A., **6**, 91[129]
Koya, K., **7**, 354[36], 355[36]
Koyabu, Y., **4**, 960[34]
Koyama, H., **4**, 970[70]
Koyama, J., **5**, 492[247,248]
Koyama, K., **6**, 94[135], 487[43], 489[43]; **7**, 25[44], 26[54-56], 42[28], 476[67], 698[50]
Koyama, T., **2**, 780[13]; **5**, 442[185]; **8**, 556[375]
Koyanagi, T., **5**, 600[41]
Koyano, H., **1**, 133[110], 753[104]; **4**, 96[105], 159[85]; **5**, 1124[53]; **7**, 274[137]
Koyano, K., **7**, 473[33], 501[33], 502[33]; **8**, 459[244]
Koyoma, K., **2**, 427[43]
Koź, T., **5**, 171[115]
Kozak, J., **5**, 432[124]
Kozar, L. G., **3**, 344[19]
Kozarich, J. W., **5**, 736[142i,j]
Kozawa, A., **8**, 190[80]
Kozawa, K., **5**, 1175[38], 1178[38]
Kozerski, L., **3**, 382[36]
Koziara, A., **6**, 76[45], 79[64], 267[55]; **8**, 385[44], 857[188]
Kozik, T. A., **6**, 543[615]
Kozikowski, A. P., **1**, 95[73,74], 506[13], 772[201], 794[279], 795[280]; **2**, 69[47], 370[259], 578[81], 737[38], 818[94], 855[240], 967[74,75], 1068[128]; **3**, 131[326], 135[326], 224[169]; **4**, 38[108], 115[183], 259[263], 339[44], 535[94], 1076[44], 1079[66], 1080[70]; **5**, 249[35], 252[41], 260[69,69b], 261[69], 263[69], 330[38], 333[44b], 341[60], 453[64], 493[213], 524[54], 534[54], 814[136]; **6**, 94[144], 284[170], 466[38,39], 468[38,39], 470[39,57]; **7**, 246[87], 346[13], 520[30], 566[101]; **8**, 44[105], 66[105], 190[76], 392[107], 459[228], 540[195], 647[52]
Kozima, S., **2**, 86[25]; **8**, 661[112]
Koziski, K. A., **7**, 330[6]
Kozlikovskii, Ya. B., **3**, 306[77]
Kozlov, N. G., **6**, 270[82]
Kozlov, N. S., **8**, 366[38]
Kozlov, V. V., **7**, 770[252]
Kozlova, L. M., **3**, 310[94], 311[94]
Kozlowska-Gramsz, E., **7**, 479[87]
Kozlowski, J. A., **1**, 107[5], 110[5], 124[80], 131[5], 343[116], 428[121], 429[121], 457[121]; **2**, 119[166], 120[183,184]; **3**, 209[15], 211[34a], 213[34a,46,46c,54], 216[46], 223[46], 224[167], 250[72], 251[74], 264[72,186], 265[72], 491[70]; **4**, 148[49], 170[3], 176[3,50], 177[50,56], 178[3,62], 180[62], 184[82], 196[3], 197[3], 256[214,215], 386[148c], 387[148], 903[192]; **5**, 931[186]; **6**, 4[22], 5[23], 8[39], 9[22], 10[22]
Koźluk, T., **5**, 168[104], 169[104], 342[62c]

Kozma, E. C., **5**, 389[138]
Koz'min, A. C., **6**, 2[9], 3[9]
Koz'min, A. S., **4**, 342[67], 347[101,103], 356[144]
Kozuka, S., **6**, 536[545], 538[545]; **7**, 764[107]; **8**, 392[97]
Kozyrod, R. P., **4**, 735[85]
Kpegba, K., **4**, 107[143a]
Kpoton, A., **8**, 262[18]
Kraatz, A., **5**, 710[56,56c], 719[56], 744[56]
Kraatz, U., **6**, 423[37], 424[59], 432[114], 541[596]
Krabbendam, H., **1**, 10[55], 11[55b]
Krabbenhoft, H. O., **5**, 87[40], 345[71c], 346[71c], 459[93]
Krack, W., **6**, 508[279], 509[279]
Kraevskii, A. A., **6**, 450[121]
Krafft, G. A., **3**, 86[12], 862[2]; **4**, 315[533], 394[189,189b]; **5**, 437[160], 442[180,180d]; **6**, 81[73]; **8**, 846[81]
Krafft, M., **4**, 215[121]
Krafft, M. E., **3**, 8[39,40], 232[256], 1025[33,33c]; **5**, 1041[21], 1044[21,27], 1045[27], 1049[27]; **6**, 144[76]; **8**, 815[22], 843[59b]
Kraft, B., **8**, 444[9]
Kraft, C., **7**, 709[44]
Kraft, K., **5**, 66[81], 636[92,93], 646[10], 654[27]
Krageloh, H., **2**, 681[58], 683[58]
Krägeloh, K., **6**, 502[217], 560[870]; **7**, 650[51]
Krahe, F., **7**, 766[180]
Krajniak, E. R., **2**, 773[26]
Krakenberger, B., **4**, 631[420,421]
Krakowiak, K. E., **6**, 71[21]
Král, V., **2**, 346[45]
Kramar, J., **4**, 869[22]
Kramar, V., **2**, 183[14]; **3**, 2[7]
Kramarova, E. N., **2**, 616[135]
Kramer, A., **4**, 572[4]; **6**, 535[543], 538[543]
Kramer, A. V., **1**, 443[179,180]
Kramer, G. M., **8**, 91[65]
Kramer, G. W., **2**, 5[22], 10[22], 977[3]; **7**, 595[30]; **8**, 704[6], 705[6], 706[6], 714[82,83]
Kramer, J. B., **5**, 151[8]
Kramer, J. D., **2**, 6[24,26]
Kramer, L., **7**, 7[38]
Kramer, M. P., **6**, 456[160]
Kramer, M. S., **6**, 507[237], 515[237]
Kramer, R., **1**, 428[116]; **3**, 263[172]
Kramer, R. A., **4**, 1002[50]
Krämer, T., **1**, 162[91]; **2**, 38[134a], 42[150], 68[41]; **6**, 863[188]
Kramer, W., **4**, 1006[97]
Krammer, R., **3**, 277[28]
Krampitz, L. O., **7**, 153[8]
Kranch, C. H., **7**, 769[219]
Krantz, A., **1**, 373[90], 375[90], 376[90]; **2**, 1063[106], 1075[150]; **3**, 217[92]; **4**, 394[189,189c]
Kranzlein, G., **3**, 294[3]
Krapchatov, V. P., **8**, 773[69]
Krapcho, A. P., **1**, 843[2f]; **2**, 840[182]; **3**, 39[219], 40[219], 727[29]; **7**, 583[155], 584[155]; **8**, 493[19], 526[33]
Krasavtsev, I. I., **3**, 643[118]; **6**, 559[865], 563[865]
Krasnaya, Zh. A., **6**, 503[218], 576[972]
Krasnitskaya, T. A., **4**, 408[259f]
Krasnobajew, V., **7**, 77[118]
Krasnosel'skii, V. N., **8**, 624[155]
Krasovskii, A. N., **6**, 494[132]
Krasuskaya, M. P., **8**, 896[12-14], 898[13]
Kratchanov, C. G., **2**, 398[12]
Kratky, C., **4**, 1039[64]; **6**, 553[761], 554[761], 708[50]
Kratochvil, M., **7**, 235[1]
Kratzer, H. J., **5**, 1089[80]
Krauch, H., **7**, 689[10]
Kraus, A., **6**, 789[107]
Kraus, C. A., **3**, 299[33]
Kraus, G. A., **2**, 183[18], 187[39], 332[59], 830[141], 846[202], 1060[85]; **4**, 14[47,47j], 30[91], 31[94,94e], 111[154a,b,g], 119[192d], 125[217a], 372[65], 682[56], 754[176], 812[182]; **5**, 16[110], 834[55]; **6**, 122[134], 128[134], 680[330], 682[339]; **8**, 245[74], 248[86], 544[262,263], 657[93]
Kraus, H. J., **6**, 177[119], 178[119], 179[125]
Kraus, J. L., **5**, 689[76]
Kraus, M., **8**, 274[137], 424[41], 436[41], 544[267,268], 907[72], 918[72]
Kraus, W., **2**, 711[34]; **5**, 423[90]; **6**, 1033[125]
Krause, J. F., **8**, 807[119]
Krause, N., **5**, 936[197]
Krauss, G. A., **1**, 529[125]
Krauss, P., **3**, 909[155]
Krauss, S. R., **4**, 171[28]
Kravchenko, M. I., **4**, 1058[150]
Kravetz, M., **4**, 729[62]
Kravtsova, V. N., **8**, 451[177]
Krawczyk, S. H., **6**, 554[733,738]
Krawczyk, Z., **5**, 99[137], 100[137]
Krawiecka, B., **6**, 184[153]
Kray, L. R., **2**, 912[72]; **7**, 221[28]
Krayushkin, M. M., **7**, 493[195]
Kreager, A., **5**, 1056[48]
Krebs, A., **3**, 556[33]; **6**, 968[114]; **7**, 358[14]; **8**, 951[174]
Krebs, B., **5**, 444[188]
Krebs, E.-P., **2**, 106[48], 616[134], 619[134]; **3**, 198[43]; **5**, 65[72], 768[125], 779[125]
Krebs, J., **3**, 1037[87]; **4**, 1009[135]; **5**, 1107[170], 1108[170]
Krebs, L. S., **8**, 341[106], 926[16]
Kreder, J., **5**, 95[89]
Kreevoy, M. M., **8**, 84[10], 237[17], 240[17], 249[17]
Kreft, A., **3**, 220[121]
Kreft, A. F., III, **8**, 564[441]
Kreher, R., **2**, 739[43b]; **6**, 502[215], 531[215], 919[41]; **7**, 29[77]
Kreis, W., **4**, 283[149]
Kreiser, W., **2**, 163[146]; **8**, 798[57]
Kreisley, A., **4**, 650[426]
Kreissel, F. R., **5**, 1065[1], 1066[1,1a], 1070[16], 1074[1], 1083[1], 1084[1], 1093[1]
Kreissl, F. R., **4**, 976[100]
Kreisz, S., **4**, 55[156]; **8**, 563[434]
Kreiter, C. G., **4**, 520[30]; **5**, 633[65,66], 634[82-84], 635[84], 1070[16], 1076[31]; **6**, 724[153]
Kremer, K. A. M., **1**, 29[145]; **2**, 508[29], 514[50], 524[50]; **3**, 39[213]; **4**, 104[137], 983[118], 984[121]; **5**, 1086[72]; **6**, 728[208]
Krenkler, K. P., **6**, 518[331]
Krepski, L. R., **1**, 548[65], 551[72,74]; **2**, 283[44], 298[44], 323[31]; **3**, 179[546], 579[125], 582[125], 583[125], 584[125], 585[125], 586[152], 588[125,152], 595[125], 596[125], 610[125,152], 918[21]
Kresge, A. J., **4**, 298[283], 300[307,309]; **5**, 186[168]
Krespan, C. G., **6**, 967[102]
Kress, A. O., **7**, 29[79]
Kress, J., **5**, 1116[5,10,12], 1125[65]
Kress, T. J., **7**, 364[41b]
Krestanova, V., **6**, 707[40]
Krestel, M., **2**, 1053[52]; **6**, 535[527]
Kresze, G., **2**, 1026[69]; **5**, 402[3,3b,4], 403[3b,4,11], 417[61,64], 418[61,70,71], 421[80], 422[81,86,88], 423[88,90], 424[95], 425[103], 428[110], 552[23]; **6**, 104[5], 115[81], 512[301], 517[301], 547[668], 846[102]; **7**, 505[283,284], 762[70]; **8**, 395[131,133]
Kresze, I. G., **6**, 1016[31], 1020[31]

Kretchmer, R. A., **2**, 120[178]; **3**, 391[88], 500[134], 509[134]; **4**, 91[90], 92[90c,d], 229[211], 261[294]; **5**, 809[112]; **6**, 1054[48]; **7**, 313[37], 490[177]

Kretzschmar, G., **4**, 741[121], 747[151]; **6**, 442[87]; **7**, 730[45], 731[45]

Kreuder, M., **2**, 782[30]

Kreuger, J. E., **8**, 956[6]

Kreutzberger, A., **6**, 554[803,804,805,806], 555[807,808]

Kreutzberger, E., **6**, 554[804,806]

Kreutzer, K. A., **5**, 277[12]

Kreutzkamp, N., **1**, 370[69], 371[69]; **2**, 1086[27], 1088[27]

Kreutzmann, A., **6**, 436[13]

Kreuzer, M., **6**, 46[62]

Kreuzfeld, H. J., **3**, 229[224]

Kreysig, D., **5**, 724[110]

Krezschar, R., **6**, 501[201]

Križ, O., **8**, 545[286]

Kricheldorf, H. R., **6**, 249[145], 809[71]

Kricka, L. J., **5**, 63[10]

Kricks, R. J., **4**, 1001[33]

Krief, A., **1**, 461[6], 542[2], 571[2,272,274,275,276,278], 618[57], 630[7,12,13,16,40,41,43,45], 631[7,12,16,40,43,45-47,50,53-55,58,60], 632[7,60], 633[7,60,71,72], 634[7,12,16,60,71,72,74], 635[7,60,72], 636[41,45,71,72,100], 637[45,71], 638[12,104], 639[100], 640[45,72,104], 641[7,12,16,71,72,100,108], 642[71,72,74,108,112,113], 644[74], 645[41], 646[41], 647[41,55], 648[129,130,131,132,133,134,135,136], 649[40,129], 650[129,130,131,139], 651[132], 652[132,133,141], 653[7,133,134,135,136,141], 654[141], 655[132], 656[7,12,16,40,53,55,71,100,104,143,144,145,154], 657[12,53,55,71,144], 658[7,12,13,16,50,53-55,60,71,154], 659[40,50,53-55,72,104,135,136,159,160,161,162,163,164], 660[54], 661[7,16,54,60,167,167a,c], 662[54,60], 663[54,60], 664[54,71,100,169,170,171,200], 665[108,169,170,174], 666[171,176,177], 667[108,177], 668[176,177], 669[7,16,41,169,170,174,178], 670[41,169,170], 672[7,12,13,16,40,41,45,50,54,55,71,72,74,100,104,108,129,130,132,136,159,160,161,162,163,164,171,176,177,188,193,194,195,196,197,198,199,200,204,205,206,208], 673[7,12,16,55,132,159,193,194,196,206], 674[12,160,188,206], 675[12,55,71,100,132,171,188,193,194,206,209,210], 676[12,136,159,160], 677[12,188,206,210,221,222,225], 678[45,188,206,215], 679[45,132], 680[12,162,206a], 681[206,215], 682[7], 683[7,12,221,222,224,225], 684[12,132,221,224], 685[221], 686[7,16,112,113,135,228,230,231,232,233,234], 687[7,16,112,113,230,231,233], 688[7,230,231,232,233,234], 689[230], 690[231,233], 691[112,113,231], 692[113,231], 693[113,230,234], 694[12,112,113,162,230,232,235,236,237], 695[112,176], 696[113], 697[12,162], 698[7,12,159,160,161,163,188,194,195,198,206,240,241,244,246], 699[7,12,188,193,195,240,243,246], 700[7,12,16,74,108,112,136,159,162,171,176,188,193,194,196,206,221,222,236,241], 701[136,159,188,198,204,234,244], 702[12,74,130,159,161,162,176,199,206], 703[159,188], 704[12,55,159,160,176,188,193,194], 705[7,12,60,74,108,162,171,193,194,199,206,221,222,224,225,236], 706[176,194,210], 707[74], 708[7,40,136,188,195,222,243], 709[7,40,136,195,243], 710[40,188,195], 711[136,195,206a,b], 712[7,12,16,55,130,136,163,188,196,199,206,215,221,222,236,263,265,266], 713[55,188,196,206,221,266], 714[12,16,130,196,197,204,206,206b,225,244,266], 715[160,188,197,198,204,244], 716[199,244], 717[12,159,160,197,198,199,204,225,244], 718[160,197,200,204,206,225,244,266], 719[200,206,266], 720[200], 721[7,210], 722[7,16,159,188,194,196,200,206,206a,221,222], 723[222,233,235,281], 724[7,108,171], 825[56], 826[58], 828[56,58,73,74,76-78], 862[74,75a,76], 866[79], 867[79,81], 868[81], 869[83], 870[84]; **2**, 76[86]; **3**, 86[23,25,49,50,52,55], 87[23,25,66,67,70-73,76-80,83,84,86,89-93,101,110,116-119], 88[25,127], 89[23,25,52,70,71,110,127,140,142,143], 90[25,70,71,140], 91[23,25,52,71,127,140,148,149,150], 92[23,70,71,127,140,143,148], 93[25,127], 94[23,25,52,55], 95[23,52,72,79,80], 104[101], 105[110], 106[83,110], 107[140,228], 109[79,80,84,127,140], 111[79,80,228,230,231], 114[23,52,110], 116[52,70,71,127,140,142], 117[52], 118[143,237,238], 119[52,70,71], 120[83,90,110,149,150,242,243], 121[83], 123[143,238], 124[25], 136[76-78], 137[77], 141[77], 142[86,243,379], 144[76-78,86], 145[25,76-78], 155[427], 193[1], 253[91], 766[156], 778[3], 785[3], 786[43,44];**4**, 10[34], 71[12,21], 106[140c], 113[165], 115[21], 120[200], 259[266,269], 318[560], 349[113,114], 350[113], 990[145], 991[150], 992[153,158], 993[160], 1007[125];**5**, 116[266], 151[10], 677[7], 901[17], 905[17], 923[17,139]; **6**, 26[112], 213[90], 976[8], 980[44], 1027[90], 1031[112]; **7**, 110[189], 473[30], 496[215], 515[1], 523[1], 771[267], 772[267], 773[307], 846[99]; **8**, 405[27,28], 411[27,28], 806[125], 847[97], 848[97c,e], 849[97c,e,108], 850[118], 867[97c], 886[107], 888[120], 990[41]

Krieg, C.-P., **5**, 1145[105]

Krieger, C., **3**, 877[82]; **6**, 979[27]

Krieger, J. K., **1**, 143[32]; **3**, 248[55], 251[55], 269[55], 482[6]

Kriegesmann, R., **1**, 755[113]

Kriessmann, I., **4**, 439[157]

Krijnen, E. S., **5**, 649[22], 650[22]

Krimen, L. I., **4**, 292[229]; **6**, 261[12], 263[12], 264[12], 267[12]; **7**, 9[71]

Krimer, M. Z., **3**, 342[7]; **5**, 850[148]

Krimm, S., **5**, 201[31]

Krimmer, H.-P., **3**, 891[43], 892[43]; **6**, 499[175]

Krings, P., **7**, 760[40]

Krishan, K., **2**, 404[41]

Krishna, A., **7**, 248[114]

Krishna, M. V., **4**, 771[252]; **7**, 519[23], 771[280], 773[280]; **8**, 848[103], 996[72]

Krishna, R. R., **3**, 737[33]

Krishnakumar, V. K., **6**, 603[18]

KrishnaMurthy, M. S. R., **7**, 62[52a]

Krishnamurthy, N., **3**, 604[230]; **5**, 759[87]; **7**, 573[116], 710[53]

Krishnamurthy, R., **4**, 374[91]

Krishnamurthy, S., **8**, 2[4], 14[79,82], 16[105], 17[112], 18[119,120], 26[18,23,25], 27[18], 36[18], 43[23], 66[23,25], 67[25], 74[25], 237[9,15], 238[9], 240[9,15,35], 241[9], 244[9,15,68], 245[9], 247[9,68], 249[15,94], 250[68], 261[10], 278, 403[9], 412[113], 524[1], 537[177], 541[1,207], 802[77,78], 803[94], 804[97,98], 805[98], 813[8], 875[34], 876[34], 901[36], 906[36], 907[36], 908[36], 909[36], 910[36]

Krishnamurthy, V. V., **3**, 334[215]

Krishnamurty, H. G., **8**, 242[40]

Krishnamurty, V. V., **6**, 110[51]

Krishnan, K., **5**, 839[77]

Krishnan, L., **4**, 553[6]; **5**, 96[105,116]; **7**, 454[99]

Krishnan, V., **8**, 137[56-58]

Krishna Rao, G. S., **2**, 764[66]; **8**, 332[39], 526[17], 530[104]

Kristen, H., **6**, 509[244]

Kristensen, E. W., **7**, 854[45]

Kristian, P., **6**, 195[224]

Kristinsson, H., **8**, 661[111]

Krivoruchko, R. M., **4**, 314[493]

Kriwetz, G., **6**, 524[356]

Kriz, O., **8**, 544[272]

Krizan, T. D., **1**, 468[50]

Krnjevic, H., **2**, 362[181]

Kroeger, C. F., **7**, 7[50]

Kroger, C.-F., **2**, 747[119], 749[119]

Kröger, C.-F., **3**, 322[142b]

Krogh-Jespersen, K., **7**, 49[62]

Krohn, J., **6**, 541[597]

Krohn, K., **5**, 393[144], 394[146], 395[146]; **6**, 34[10], 51[10]; **7**, 345[2]
Krohn, W., **4**, 48[137]
Kröhnke, F., **2**, 368[244]; **7**, 231[140], 657[23]; **8**, 391[88]
Krolikiewicz, K., **2**, 889[56,58]; **4**, 433[120]; **6**, 20[73], 22[73], 49[97], 637[32,32a], 669[32a]; **8**, 392[102]
Kroll, L. C., **8**, 447[126], 457[126]
Kroll, W. R., **8**, 100[116], 454[198], 739[33], 744[50], 756[50]
Krollpfeiffer, F., **2**, 759[32]
Krolls, U., **1**, 399[5,7]
Krolski, M. E., **3**, 232[269]
Kron, J., **5**, 205[43]
Kronberger, K., **3**, 577[87]; **8**, 524[12], 527[49], 532[12c]
Kronenthal, D., **5**, 92[66], 94[66]
Kröner, M., **5**, 1197[39]
Kronis, J. D., **5**, 855[189]
Kronja, O., **3**, 374[132]
Kronzer, F. J., **6**, 47[78]
Kropf, H., **3**, 563[1]; **7**, 95[73a], 111[193], 437[4], 438[4]; **8**, 236[2], 237[2], 238[2], 240[2], 241[2], 242[2], 243[2], 244[2], 245[2], 247[2], 249[2], 253[2]
Kropp, F., **8**, 738[31], 753[31]
Kropp, J. E., **7**, 501[251]
Kropp, K., **5**, 1178[45]; **8**, 677[61], 679[61], 682[61], 685[61], 687[61]
Kropp, P. J., **3**, 815[73,74]; **4**, 968[61], 969[61]; **8**, 938[93]
Krouse, S. A., **5**, 1116[11], 1117[11], 1118[11]
Krow, G. R., **1**, 843[2c], 853[2c], 896[2c]; **2**, 1056[64], 1070[64]; **5**, 113[226], 403[7], 404[7,18], 1003[23]; **6**, 744[77]; **7**, 671[9], 672[9], 673[9,30], 695[33], 831[63]; **8**, 354[174]
Krsek, G., **8**, 452[189a]
Krubiner, A. M., **8**, 407[57], 950[170]
Kruck, P., **7**, 709[45]
Krueger, D. S., **4**, 30[89]; **8**, 986[15]
Krueger, S. M., **5**, 738[149]
Kruerke, U., **1**, 231[3]
Krüger, B.-W., **7**, 753[158,159]
Krüger, C., **1**, 14[77,78], 25[129], 37[250,251], 162[92], 180[30], 531[132]; **2**, 182[7], 183[7]; **4**, 905[212]; **5**, 232[135,136], 480[178], 497[224], 1109[174,180], 1138[71], 1157[71]; **6**, 177[119], 178[119]; **8**, 682[83], 683[95], 686[95]
Kruger, D., **8**, 5[26]
Krüger, G., **5**, 161[63]; **6**, 535[533]
Krüger, H.-W., **5**, 72[185]
Krüger, K., **1**, 310[106]
Krüger, M., **1**, 832[108]
Kruglik, L. I., **6**, 525[388]
Kruizinga, W. H., **6**, 2[7], 21[79]; **8**, 93[77], 94[77]
Kruk, C., **5**, 3[26], 196[14], 649[22], 650[22]
Krukle, T. I., **6**, 554[730]
Krull, I. S., **6**, 687[382]
Krumel, K. L., **6**, 1034[132]
Krumkalns, E. V., **6**, 637[34]
Krumpolc, M., **3**, 757[124]
Kruper, W. J., **7**, 12[93], 95[73b]
Krupp, F., **8**, 735[11], 738[29], 739[11], 754[29]
Kruse, C. G., **3**, 242[10]; **6**, 134[32]; **7**, 235[1]
Kruse, C. W., **2**, 355[121], 356[130]
Kruse, L. I., **2**, 807[46]; **3**, 18[98], 1029[55]; **4**, 37[107,107b,e], 38[107e], 39[107e]; **7**, 554[67]; **8**, 542[224], 933[50]
Krushch, A. P., **7**, 17[173]
Krusic, P. J., **5**, 901[31]
Krustalev, V. A., **6**, 515[313]
Krutii, V. N., **8**, 551[342]
Krutov, S. M., **8**, 334[65]
Kryczka, B., **2**, 283[48]
Krymowski, J., **5**, 692[103]
Krynitsky, J. A., **8**, 301[89]
Krysan, D. J., **2**, 629[1], 635[1]; **6**, 850[126]
Kryuchkova, V. E., **3**, 305[74]
Krzeminski, J., **6**, 824[121]
Krzyzanowska, B., **8**, 72[242], 74[242,244], 393[110]
Ksander, G. M., **5**, 553[44], 854[181]
Ku, A., **3**, 255[104]; **4**, 1082[87], 1083[87], 1103[87]
Ku, A. Y., **5**, 203[39], 204[39h-j], 209[39], 210[39]
Ku, H., **3**, 255[104]; **4**, 1103[206]
Ku, T., **7**, 295[22]
Ku, V., **8**, 423[38], 428[38]
Ku, Y. Y., **3**, 512[203]
Kuan, F.-H., **6**, 46[66]
Kuang, S.-W., **2**, 962[49]
Kubak, E., **5**, 432[132]
Kubas, R., **1**, 350[155]
Kubasskaya, L. A., **4**, 314[492,493]
Kubayashi, H., **4**, 1057[142]
Kubiak, C. P., **8**, 456[209]
Kubiak, G., **2**, 381[307]
Kubiak, T. M., **6**, 637[32,32d]
Kubler, D. G., **2**, 662[21], 664[21]; **5**, 433[137a]
Kübler, W., **2**, 651[121]
Kubo, A., **6**, 734[11]; **7**, 350[27], 355[27]
Kubo, I., **7**, 355[40]
Kubo, M., **1**, 477[138]; **6**, 20[75]
Kubo, N., **4**, 843[53], 852[53]
Kubo, R., **1**, 293[37]
Kubo, Y., **4**, 843[55]; **5**, 181[154]
Kubota, F., **6**, 564[918]
Kubota, H., **2**, 650[109]; **5**, 714[68]; **8**, 315[46], 978[146]
Kubota, M., **7**, 107[167]
Kubota, N., **4**, 1046[111]
Kubota, S., **2**, 826[124]
Kubota, T., **4**, 1061[167]; **5**, 165[81]; **7**, 806[71]; **8**, 544[266]
Kubota, Y., **3**, 75[48]; **6**, 554[774]
Kubrak, D., **7**, 673[30]
Kuc, T. A., **8**, 445[39]
Kucera, D. J., **1**, 589[20]
Kucerovy, A., **3**, 196[23], 257[115]
Kucherov, V. F., **1**, 555[112]; **2**, 723[101], 725[107,109]; **3**, 342[7,14], 345[21], 346[24], 351[41], 361[73,80]; **5**, 775[175,176]; **6**, 503[218]; **8**, 530[107]
Kucherova, N. F., **8**, 618[121], 619[121], 627[176]
Kuchert, E., **5**, 1109[177]
Kuchynka, D. J., **7**, 854[50], 855[50]
Kucinski, P., **7**, 108[179]
Kuck, J. A., **2**, 283[52]; **7**, 595[8]
Kucsman, Á., **7**, 764[120,122], 777[389]
Kuczkowski, R. L., **4**, 1073[22], 1098[173]; **7**, 543[9], 548[9], 558[9]
Kudaka, T., **4**, 1021[242]
Kudav, N. A., **2**, 736[30]; **6**, 110[57]
Kudelska, W., **8**, 887[118]
Kudera, J., **3**, 890[30], 897[30]; **7**, 507[308]
Kudo, H., **8**, 248[82], 369[74]
Kudo, K., **3**, 1026[40]
Kudo, M., **4**, 1033[21], 1037[21], 1040[21]
Kudo, S., **2**, 198[84]
Kudo, T., **6**, 27[114], 724[165]; **7**, 598[61]; **8**, 408[66]
Kudo, Y., **2**, 582[109]
Kudon, N., **4**, 858[110]
Kuduk, J. A., **1**, 18[95]
Kuehne, M. E., **1**, 838[170]; **2**, 865[2]; **3**, 30[173], 55[282], 380[9]; **4**, 611[350]; **5**, 75[214], 116[260]; **6**, 704[12], 705[19], 712[12,77],

716[95]; **7**, 125[58], 170[120], 228[105], 503[279], 519[20]; **8**, 36[96], 41[96], 66[96], 251[102], 501[53], 507[53], 526[38]
Kuétegan, M., **3**, 219[103]
Kufner, U., **7**, 418[130a]
Kugatova, G. P., **2**, 143[50]
Kugel, W., **6**, 553[797]
Kuhara, M., **7**, 451[24]; **8**, 9[52]
Kühl, U., **4**, 111[155c]
Kuhla, D. E., **5**, 634[68]; **7**, 507[306]
Kühle, E., **4**, 330[5]
Kühlein, K., **1**, 143[30]; **6**, 419[7], 420[7], 423[7], 437[33], 440[33], 441[33], 443[33], 444[33], 445[33], 446[33]
Kuhler, T., **6**, 20[72], 74[33]
Kuhlmann, D., **3**, 482[3]
Kuhlmann, H., **4**, 12[39], 18[57]
Kuhn, D., **1**, 864[85]; **3**, 771[184], 785[36,36a]; **4**, 987[147]
Kuhn, H., **2**, 736[23]
Kuhn, H. J., **7**, 816[10], 818[10]
Kuhn, L. P., **3**, 574[80]
Kuhn, R., **6**, 36[24], 651[134]
Kuhn, S. J., **2**, 728[142], 749[136]; **3**, 305[71], 320[135], 331[200a]
Kuhn, W., **5**, 1070[18,26], 1072[26], 1095[105], 1098[105,123], 1112[105,123]
Kühne, H., **6**, 430[102]
Kühne, R. O., **1**, 286[11]
Kuhnen, F., **7**, 724[30]
Kuhnen, L., **7**, 766[183]
Kuhnke, J., **1**, 564[205]
Kuhnle, J. A., **3**, 666[45]
Kuhr, H., **1**, 3[21]
Kuilman, T., **8**, 53[132], 66[132]
Kuipers, J. A. M., **8**, 405[22]
Kuivila, H. G., **1**, 345[123]; **2**, 6[30], 240[8]; **4**, 738[92]; **7**, 595[15,16], 602[102], 884[181]; **8**, 264[41,43-45], 547[315,316], 548[315], 781[97], 798[55], 806[109,128], 825[65], 845[77], 991[45]
Kukenhohner, T., **1**, 167[115]
Kukhanova, M. K., **6**, 450[121]
Kukhar, V. P., **6**, 270[81], 499[172], 500[180], 543[623], 550[679]; **7**, 470[3]
Kukkola, P., **4**, 247[105], 253[105], 257[105], 262[105]
Kukolev, V. P., **8**, 557[385]
Kuksis, A., **2**, 277[3]
Kula, M. R., **8**, 189[56,57], 204[154]
Kulagowski, J. J., **4**, 434[128]
Kulcsar, L., **6**, 543[612]
Kulenovic, S. T., **2**, 282[42]; **8**, 925[11], 926[11]
Kuleshova, E. F., **6**, 554[765]
Kuleshova, N. D., **6**, 437[34], 440[34], 443[34], 499[173]
Kulicki, W., **3**, 849[59]
Kulig, M., **5**, 947[267], 948[267]
Kulig, M. J., **7**, 96[87]
Kulinkovich, O. G., **2**, 709[14], 740[61]; **4**, 1023[254]; **6**, 557[835,836]
Kulka, K., **2**, 152[99]
Kulka, M., **2**, 740[60]
Kulkarni, A. K., **1**, 828[81]; **2**, 602[39]; **3**, 760[135]; **8**, 937[87]
Kulkarni, B. K., **1**, 463[24,26]
Kulkarni, B. S., **2**, 544[86]
Kulkarni, G. H., **1**, 546[51]; **6**, 687[381]; **7**, 84[3]
Kulkarni, K. S., **8**, 528[70], 530[70]
Kulkarni, S. B., **2**, 740[61]; **3**, 328[175], 331[195]; **5**, 776[179]
Kulkarni, S. M., **2**, 381[304]
Kulkarni, S. U., **3**, 795[84]; **6**, 646[104], 647[104], 659[104], 660[104]; **7**, 235[8], 264[90], 596[36], 601[89,90], 602[92,93]; **8**, 17[109], 261[11], 707[24], 708[37], 711[67,68], 715[85,85a], 716[85], 718[85a,109], 719[117,120,122,123]
Kulkarni, Y. S., **5**, 558[63], 1021[71]
Kulla, H., **7**, 79[130]
Kullmer, H., **4**, 124[214d]
Kullnig, R. K., **4**, 957[21]; **7**, 261[70]
Kulomzina-Pletneva, S. D., **5**, 128[31]
Kulp, T., **4**, 1040[88], 1048[88]
Kulpe, S., **6**, 436[13]
Kulprecha, S., **7**, 73[104]
Kulsa, P., **7**, 160[48]
Kumabayashi, H., **8**, 154[199], 459[244]
Kumada, M., **1**, 113[32], 158[74], 180[39], 181[39], 610[46], 611[46], 617[53], 619[61]; **2**, 4[13], 17[68], 567[31], 568[33], 584[118,122,124], 716[62], 977[7]; **3**, 200[68], 228[216,217,218,219,220,221,222], 229[225,226,227,231], 246[37], 436[3,8,10,19], 437[21,25,26], 438[30,31,33], 440[25], 441[47], 442[50], 445[69,71], 448[25], 449[25,99,100], 450[25,26,33], 451[25,106], 452[25,33,106,107,111], 453[31], 455[122], 457[129,130], 459[31,137,139], 460[107,130,137], 461[137], 464[171], 469[203], 470[203,219], 471[219], 472[219], 473[203], 483[16], 487[45], 492[74,76,78], 495[78], 497[104], 503[145], 510[183,206], 513[205,206], 524[27,32]; **4**, 120[202], 588[67], 596[159], 615[387], 620[387], 635[159,387], 653[437,442], 840[34]; **6**, 16[60]; **7**, 453[93], 455[93], 564[91], 565[91], 616[18], 641[3], 642[8-12], 643[13,15], 644[16], 645[17], 647[34]; **8**, 84[12], 152[179,180], 173[112,113], 459[245], 556[373], 568[467], 764[10], 770[32], 772[52,53], 773[10,67], 778[85], 780[53,95], 782[105], 783[106,107,109], 784[110], 787[119], 788[120]
Kumadaki, I., **2**, 538[66,68], 539[66,68]; **4**, 377[105c], 381[105]; **6**, 217[117], 221[117]; **7**, 203[52], 750[127]
Kumadaki, S., **8**, 588[45]
Kumagai, M., **2**, 603[47], 717[67]; **8**, 173[117,119], 555[370], 770[37], 780[92], 782[103]
Kumagai, T., **2**, 116[131,140], 610[94], 611[94], 1059[78,81]; **5**, 92[81], 196[15-17], 197[16]; **6**, 531[449], 570[955]
Kumagai, Y., **5**, 293[44], 1186[3,7], 1190[24]
Kumagawa, T., **6**, 156[162]
Kumai, S., **3**, 726[25]
Kumamoto, T., **7**, 125[59], 196[29]; **8**, 609[49,54], 951[177]
Kumamoto, Y., **2**, 114[116]; **4**, 254[180]; **7**, 763[102]
Kumanotani, J., **4**, 31[92]
Kumar, A., **1**, 820[15]; **7**, 400[42]
Kumar, C. V., **5**, 125[15], 153[25]
Kumar, G., **7**, 749[119]
Kumar, K., **6**, 185[159]
Kumar, N., **2**, 780[10]; **5**, 474[158]; **6**, 487[71], 489[71]
Kumar, P., **2**, 396[6], 402[6], 403[6], 404[42], 405[42]; **4**, 503[126]; **6**, 524[358]
Kumar, R., **5**, 515[18], 547[18]; **6**, 215[103]; **7**, 884[190]
Kumar, S., **6**, 958[35]; **7**, 333[24], 346[9], 365[43]
Kumar, Y., **8**, 617[108]
Kumara, M., **4**, 614[382]
Kumarasingh, L. T., **4**, 350[115]; **7**, 502[258]
Kumaraswamy, G., **7**, 223[43], 227[43]
Kumar Das, V. G., **1**, 327[13]; **4**, 155[66]; **8**, 693[111]
Kumazawa, T., **3**, 883[106]; **4**, 12[43], 18[59], 161[89a]; **5**, 736[145], 737[145]; **6**, 77[56]; **7**, 220[21]
Kumazawa, Z., **2**, 166[156]
Kume, A., **1**, 563[171]; **2**, 65[29], 830[145]; **3**, 197[41], 199[41]; **6**, 563[899]
Kume, M., **7**, 710[56]
Kume, T., **2**, 907[59]
Kumemura, M., **7**, 693[27]
Kümin, A., **6**, 831[7]
Kumler, P. L., **7**, 774[310]
Kumli, K. F., **8**, 860[222]
Kumobayashi, H., **4**, 609[332]; **6**, 866[208]; **8**, 462[267]
Kump, W. G., **7**, 221[32]

Kunada, M., **4**, 682[57]
Kunakova, R. V., **4**, 599[213], 640[213]
Kunath, D., **2**, 1088[44]
Kuncl, K., **1**, 878[107]
Kuncová, G., **8**, 776[80]
Kundig, E. P., **1**, 568[241,242]; **4**, 253[171,172], 520[28], 527[68], 532[68,84,85,88], 534[68,84,88,92], 536[68,84,85,92], 537[88], 538[88], 539[88], 545[84,124,125], 546[84,125], 548[92], 696[6]; **5**, 37[22a]; **8**, 843[56]
Kundu, N., **3**, 834[77]
Kundu, N. G., **2**, 725[119]; **8**, 643[35]
Kunert, F., **6**, 565[920]; **8**, 918[120]
Kunerth, D. C., **4**, 437[147]
Kunesch, G., **3**, 263[173]; **6**, 66[3]; **8**, 384[24]
Kung, F. E., **6**, 204[14]
Kunieda, N., **1**, 513[48]; **2**, 350[75], 374[277]; **6**, 149[104], 902[130], 1022[65]; **7**, 778[391,392,393]
Kunieda, T., **1**, 119[64]; **2**, 1060[90]; **6**, 438[47], 614[84]; **8**, 797[34]
Kunimoto, M., **7**, 335[29]
Kunioka, E., **8**, 450[169], 568[463]
Kunisch, F., **1**, 546[58]; **2**, 249[84], 654[129]
Kunishige, M., **5**, 1175[38], 1178[38]
Kunishima, M., **3**, 286[56a], 1050[19]; **4**, 42[122b]
Kunnen, K. B., **1**, 478[148]
Kunng, F. A., **5**, 829[21], 1089[82], 1092[82], 1094[82,99], 1096[99], 1098[82,99,130], 1099[82,99], 1111[82], 1112[82,99,130], 1113[82]; **7**, 350[20]; **8**, 911[88], 933[52]
Kuntz, I., **8**, 364[14]
Kunwar, A. C., **8**, 823[54]
Kunyants, I. L., **5**, 113[234]
Kunz, D., **6**, 462[8], 472[71]
Kunz, F. J., **2**, 346[48], 347[48], 353[48], 355[48], 356[48], 358[48], 365[48], 367[48], 369[48], 374[48]
Kunz, G., **7**, 748[108]
Kunz, H., **2**, 656[155,156], 1099[111], 1102[123]; **4**, 140[12]; **5**, 366[99], 410[41], 411[41i]; **6**, 3[16], 33[6], 34[6], 40[6], 46[6,63], 633[8], 634[8,10], 638[42,43], 639[52-54], 640[55,57], 641[8,10,55,57,60,63], 646[8,10], 652[8], 657[178,179], 659[43,193], 666[244,245,248], 667[243,244,245], 670[10,63,266,267], 671[55,57,60,266,278]; **8**, 863[234]
Kunz, R. A., **6**, 1016[34]
Kunz, T., **1**, 110[21]
Kunz, W., **7**, 689[10]
Kunze, K., **2**, 147[76]
Kunze, M., **3**, 382[36]
Kunze, O., **6**, 453[139]
Künzer, H., **6**, 227[23], 229[23], 230[23]
Kuo, C. H., **2**, 746[110]; **3**, 689[119]; **8**, 357[198], 358[198]
Kuo, D. L., **2**, 651[121]; **3**, 26[164]; **5**, 51[45,45b], 52[46], 812[133]
Kuo, E., **8**, 431[62]
Kuo, E. Y., **5**, 736[142c]
Kuo, G.-H., **4**, 984[121,122], 985[125]; **5**, 1086[73]; **6**, 11[46], 736[27], 842[82]
Kuo, P. C., **3**, 694[150]
Kuo, S., **8**, 798[58]
Kuo, S.-C., **4**, 820[213]
Kuo, S. J., **3**, 219[114], 499[140], 501[140], 502[140]
Kuo, Y., **1**, 844[7]
Kuo, Y.-C., **3**, 900[90]
Kuo, Y. H., **3**, 693[146], 694[150]
Kuo, Y. N., **4**, 111[155a]
Kupchan, S. M., **3**, 507[173], 509[176], 670[57,57a,59], 673[68], 681[68,98], 683[57b], 686[68,110], 807[32,33], 815[77], 816[81,82]; **6**, 687[382]; **8**, 888[123], 895[2], 899[2]
Kupchik, E. J., **8**, 264[42]
Kupchik, I. P., **4**, 391[177]
Kuper, D. G., **3**, 906[143]
Kuper, S., **4**, 1038[60]
Kupfer, R., **1**, 372[77]; **2**, 1052[49], 1053[49,52,53], 1055[53]; **6**, 501[195,196,197,198], 535[527,528]
Kupin, B. S., **4**, 304[360]
Küpper, W., **5**, 1118[17]
Küppers, H., **1**, 641[105]; **3**, 194[13]
Kura, H., **5**, 637[102]
Kuraishi, T., **1**, 474[105]
Kurakawa, Y., **7**, 882[169]
Kuramitsu, T., **5**, 1139[74], 1140[74], 1142[74], 1146[74]
Kuramoto, M., **8**, 580[4,5], 581[5], 587[5]
Kuramoto, Y., **2**, 116[140], 610[94], 611[94], 1059[78]
Kuran, W., **1**, 440[171]; **4**, 598[182]
Kurasawa, Y., **8**, 620[139]
Kurata, H., **1**, 733[12]; **8**, 196[119]
Kurata, K., **5**, 180[149]
Kurata, Y., **1**, 506[17]
Kurath, P., **3**, 834[72]; **7**, 100[123]
Kurauchi, M., **1**, 561[160]
Kurbanov, M., **3**, 342[14], 361[80]
Kurcherov, V. F., **3**, 342[14]
Kurcok, P., **2**, 105[41]
Kurek, A., **2**, 382[315]
Kurek, J. T., **4**, 294[244], 301[314,315], 303[339], 309[416,417]; **6**, 283[167], 284[167]
Kurek-Tyrlik, A., **7**, 649[44]
Kuretani, M., **2**, 826[124]
Kurfürst, A., **8**, 584[22], 589[22]
Kuribayashi, H., **1**, 174[13], 202[13]
Kurihara, H., **2**, 1053[55]; **4**, 1046[111]; **6**, 1046[32a]
Kurihara, K., **3**, 528[47]; **4**, 405[250a]
Kurihara, T., **1**, 544[32,37], 548[66], 560[37,155], 561[37,156]; **2**, 357[149], 801[25]; **3**, 530[63], 532[63]; **6**, 227[20], 236[20], 609[51], 626[168], 1045[21,22]
Kurihara, Y., **3**, 848[54]; **7**, 801[45]
Kuriki, N., **7**, 765[165]
Kurino, K., **4**, 743[128]
Kurisaki, A., **5**, 736[142l]
Kurita, A., **7**, 642[9]; **8**, 787[119]
Kurita, H., **1**, 557[129], 558[129,135]
Kurita, K., **5**, 755[72], 760[72]
Kurita, M., **2**, 709[7]
Kurita, Y., **1**, 122[68]; **3**, 218[100]
Kuritiara, T., **7**, 172[124]
Kuriyama, K., **3**, 386[57]; **5**, 369[101], 370[101a]; **6**, 787[98]
Kurland, D. B., **5**, 758[85]
Kurmangalieva, R. G., **6**, 515[316]
Kurobashi, M., **2**, 115[124]
Kurobe, H., **3**, 164[480], 165[480], 167[482], 168[482]; **4**, 181[70]; **6**, 780[70]
Kuroboshi, M., **1**, 751[109]
Kuroda, A., **1**, 101[90]
Kuroda, C., **4**, 331[10], 345[10]; **5**, 809[120]; **7**, 400[43]
Kuroda, H., **2**, 655[135]; **6**, 438[55]; **8**, 9[51]
Kuroda, K., **2**, 403[36]; **3**, 698[157b]; **5**, 468[135]; **6**, 173[38], 174[38]; **8**, 392[101], 562[422]
Kuroda, R., **4**, 1018[224], 1019[224]
Kuroda, S., **3**, 593[181,182], 1000[55], 1001[56]; **6**, 764[10], 877[39], 878[39], 880[46], 883[39], 887[39]
Kuroda, T., **1**, 193[84,85], 195[85], 198[84,85], 201[98]; **6**, 172[24]
Kurokawa, H., **2**, 651[113]
Kurokawa, M., **8**, 531[110]
Kurokawa, N., **4**, 379[114], 382[132]
Kurokawa, Y., **2**, 655[132]; **4**, 159[82]
Kuroki, Y., **7**, 125[60], 202[48]

Kuromizu, K., **8**, 959[23]
Kurono, M., **3**, 380[4]; **6**, 927[80]
Kurosaki, A., **4**, 497[99]
Kurosawa, E., **5**, 180[149]
Kurosawa, H., **4**, 595[148,149], 607[316,317], 615[384]; **5**, 275[11]; **8**, 858[205]
Kurosawa, T., **8**, 883[92]
Kurosky, J. M., **2**, 39[138]
Kurotani, A., **5**, 864[257]
Kuroyama, Y., **3**, 530[61], 531[61], 532[61], 537[61]
Kurozumi, S., **2**, 833[147]; **3**, 222[144]; **4**, 13[44,44c], 159[85], 253[169], 256[208,212], 258[248], 261[208,248,284], 262[212,308]; **5**, 953[295]; **6**, 837[60], 942[154], 944[154]; **7**, 54[8]; **8**, 843[59a], 934[56], 993[58]
Kurr, B. G., **1**, 506[12]
Kurras, E., **1**, 174[12], 202[12]
Kurs, A., **8**, 843[49]
Kursanov, D. N., **4**, 1005[91]; **5**, 752[32,35,36], 754[32,35,36], 756[35,36]; **8**, 216[55,56], 318[60-63,65-67], 486[59,61,62], 487[59], 546[306,307,308], 608[37], 610[56,58-60], 611[66], 630[56,187], 778[84], 813[12]
Kurtev, B. J., **2**, 211[112], 398[12]
Kurth, M., **5**, 543[115]
Kurth, M. J., **2**, 117[147], 350[76]; **3**, 6[30], 224[179], 342[9]; **4**, 372[62], 380[123], 383[141], 384[141a], 390[141b], 397[141b], 403[240]; **5**, 517[27], 538[27], 829[20], 864[261]; **6**, 859[174], 861[185], 862[185]; **7**, 503[270]; **8**, 843[52], 852[139]
Kurts, A. L., **6**, 226[10], 256[10], 257[10]
Kurtz, D. W., **5**, 210[60]
Kurtz, J. L., **2**, 963[53]
Kurtz, P., **3**, 274[18]; **4**, 54[153a], 56[153a]; **6**, 225[2], 226[2], 228[2], 229[2], 230[2], 231[2], 233[2], 234[2], 235[2], 236[2], 238[2], 239[2], 240[2], 241[2], 242[84], 243[84], 258[2], 507[231]
Kurumaya, K., **4**, 446[213]; **8**, 204[152]
Kurumi, M., **6**, 507[240], 515[240]
Kurusu, Y., **1**, 215[37]; **2**, 18[73-75], 23[90,90a]; **4**, 607[312]; **6**, 837[59]; **7**, 299[43], 321[65]
Kuryatov, N. S., **6**, 554[790]
Kurys, B. E., **5**, 96[121], 98[121]
Kurz, A. L., **1**, 3[23]
Kurz, H., **5**, 633[65,66]
Kurz, J., **5**, 99[136], 100[136]; **8**, 141[41]
Kurz, K. G., **8**, 91[50]
Kurz, L. J., **2**, 385[320]; **4**, 216[122]
Kurz, W., **6**, 838[67], 902[132]
Kurzer, F., **2**, 867[17]; **6**, 268[62], 424[51,52], 454[148]
Kürzinger, A., **8**, 174[125], 178[125], 179[125]
Kurzmann, H., **8**, 568[471]
Kusabayashi, S., **7**, 543[14], 766[188]; **8**, 863[235], 864[235], 965[66]
Kusakabe, M., **1**, 58[33], 110[17-19], 131[17], 134[17], 339[87]; **2**, 204[97]; **7**, 423[142,143]
Kusama, O., **4**, 495[85], 505[136], 510[173]
Kusamran, K., **6**, 159[175], 1022[65]
Kusano, Y., **8**, 87[34]
Kusashio, K., **6**, 616[105]
Kusayanagi, Y., **8**, 134[28,29]
Kuse, T., **8**, 446[94], 452[94]
Kushida, K., **5**, 158[48]
Kushner, A. S., **8**, 541[207]
Kushner, S., **2**, 284[58]
Kustanovich, I. M., **8**, 535[160]
Küsters, W., **5**, 224[100], 1062[59]
Kusuda, K., **8**, 987[22], 992[22b], 994[22]
Kusume, K., **1**, 561[164]
Kusumi, T., **2**, 553[129]; **3**, 380[13]
Kusumoto, M., **4**, 159[82]
Kusumoto, S., **6**, 33[7], 40[7], 53[118], 57[7,141], 603[14]; **8**, 150[122], 151[148]
Kusumoto, T., **1**, 232[13,14], 233[13,14,19], 234[13,14,21], 243[57], 253[9], 276[9], 278[9], 332[52]; **2**, 311[33], 312[33,35]; **3**, 567[34], 570[34]; **6**, 214[94]; **8**, 113[47], 773[63], 803[92]
Kusumoto, Y., **6**, 1016[35]
Kusunose, N., **6**, 57[141]
Kusurkar, S. S., **2**, 782[23]
Kutateladze, A. G., **4**, 335[26], 347[106]
Kutchan, T. M., **4**, 952[6], 1043[107], 1048[107]; **5**, 211[65], 900[9], 901[9], 905[9], 906[9], 907[9], 909[9], 913[9], 916[122], 925[122], 947[9], 954[298], 1006[34]
Kuthan, J., **6**, 533[480], 550[480]; **8**, 92[67], 583[13], 584[21,22], 589[22]
Kutney, G. W., **7**, 213[104]
Kutney, J. P., **2**, 1021[51]; **3**, 380[7]; **4**, 1039[62]; **6**, 777[59], 921[48]; **8**, 205[156], 587[33], 821[48], 943[121], 945[130]
Kuwahara, K., **4**, 370[44]
Kuwahara, S., **3**, 396[115]
Kuwajima, I., **1**, 112[27], 212[12], 213[12], 215[12b,c], 217[12], 327[11], 448[207], 506[17], 584[10], 658[158], 659[158], 664[158], 665[158], 670[184,185,186], 671[185], 672[158]; **2**, 68[44], 89[38], 90[39], 103[37], 109[65], 113[111], 117[149,150], 184[21], 244[27], 245[27], 310[27,28], 441[4], 442[10,12], 443[19], 445[10,27,28], 446[28], 447[12,19], 448[19,40,41,44,45,47], 449[19,40], 450[52,53], 451[56], 452[47,57,58], 567[30], 576[69], 600[32], 601[36], 605[58], 609[85-87], 615[128], 624[86], 630[6], 633[6a,34a,b], 634[34b,35], 640[34,35], 650[108], 651[112], 714[51,52], 720[52,83], 728[139], 830[144]; **3**, 6[28], 8[28], 12[63], 14[28], 197[33], 221[131], 257[118], 454[119], 455[125], 460[125], 617[15], 619[15], 621[15], 623[15,34], 627[15], 727[33]; **4**, 116[186], 152[55,57], 163[96], 164[96]; **5**, 833[46], 844[131], 1200[51,52]; **6**, 147[86], 615[101], 624[101]; **7**, 307[15], 310[15], 323[15,69], 523[42], 771[279], 773[279]
Kuwano, H., **7**, 77[121,122]
Kuwata, F., **1**, 544[39]; **4**, 130[226c]; **8**, 302[96]
Kuwata, S., **2**, 1089[57]
Kuwayama, S., **6**, 150[128]
Kuyama, M., **3**, 167[484], 168[484], 361[74]
Kuyper, L. F., **3**, 124[288], 128[288], 132[288]
Kuz'micheva, L. K., **6**, 507[237], 515[237]
Kuz'min, V. A., **7**, 850[5]
Kuz'mina, N. A., **8**, 91[58], 765[11]
Kuzna, P. C., **2**, 175[184], 619[150]
Kuznetsov, M. A., **7**, 742[55], 743[55], 744[55]
Kuznetsov, S. G., **6**, 268[63]
Kuznetsova, A. I., **5**, 752[3]
Kuznetsova, T. A., **6**, 577[979]
Kuznicki, R. E., **2**, 348[55]
Kuzovkin, V. A., **6**, 554[755]
Kuzuhara, H., **8**, 48[112], 66[112]
Kuzuya, M., **5**, 210[57]
Kvarsnes, A., **4**, 1035[40]
Kvasyuk, E. I., **4**, 379[115]
Kvintovics, P., **8**, 91[55]
Kvita, K., **2**, 759[36]
Kvita, V., **4**, 425[29,30]
Kwa, T. L., **3**, 260[145]
Kwan, T., **8**, 453[191]
Kwart, H., **2**, 6[27], 423[31], 528[12], 565[15], 575[15], 582[15]; **3**, 822[9], 823[9], 834[9]; **5**, 2[10], 451[38], 552[8,10,22], 829[22,26]; **6**, 1013[8,9], 1017[8], 1025[79], 1036[143]; **7**, 706[21], 764[105]
Kwart, H. W., **4**, 279[115], 280[115]
Kwart, L. D., **2**, 282[34], 286[34], 287[34]; **4**, 1040[88], 1048[88]; **5**, 38[23b], 829[22], 938[212], 939[212,223], 942[229], 951[223], 957[309], 962[212,223], 964[223]; **7**, 557[74]

Kwasigroch, C. A., **1**, 595[29], 596[29], 602[32], 603[32]; **2**, 85[13,16], 575[60,61], 1061[95]; **5**, 277[16], 279[16]; **7**, 545[24]
Kwasnik, H. R., **8**, 338[88]
Kwass, J. A., **2**, 547[105], 550[105]
Kwast, A., **2**, 429[50]
Kwast, E., **3**, 790[61]; **7**, 226[68], 551[53]
Kwiatek, J., **8**, 154[196], 449[159], 453[191], 568[480]
Kwiatkowski, G. T., **3**, 664[33]; **5**, 164[74]
Kwoh, S., **7**, 728[42]
Kwon, H. B., **2**, 272[81], 315[47], 316[47]
Kwon, S., **4**, 350[116,118], 1001[43]
Kwon, T., **8**, 445[51]
Kwon, Y. C., **3**, 983[16-18]; **6**, 876[21], 887[20-22]; **8**, 720[137]
Kwong, C. D., **1**, 822[39]
Kwong, K. S., **5**, 797[65]
Kwun, O. C., **3**, 299[33]
Kyba, E. P., **4**, 295[256,261], 296[261]; **5**, 477[160]; **7**, 21[2,20]; **8**, 57[173], 66[173], 865[246]
Kyburz, E., **6**, 531[451]
Kyle, D., **7**, 40[4]
Kyler, K., **6**, 175[67], 648[117b]
Kyler, K. S., **1**, 542[9], 544[9], 551[9], 552[9], 553[9], 554[9], 555[9], 557[9], 560[9], 767[177]; **2**, 73[62], 420[24]; **3**, 39[218], 48[218], 215[67], 483[15], 500[15]; **5**, 429[114]; **7**, 406[85], 409[85]
Kyotani, Y., **4**, 27[79]; **8**, 885[105]
Kyowa-Hakko, **7**, 79[128c]
Kyriakakou, G., **2**, 414[15]
Kyriakides, L. P., **2**, 136[13], 139[13], 140[13]
Kyrides, L. P., **2**, 144[61]
Kyung, S. H., **1**, 142[27], 149[27], 150[27], 152[27], 162[103], 169[120,121], 331[46], 334[46]; **2**, 35[131]
Kyz'mina, L. G., **5**, 1055[46]

L

Laabassi, M., **4**, 1040[66], 1041[66]
Laakso, L. M., **5**, 362[93], 363[93i]
Laali, K., **2**, 745[105]; **3**, 334[215]
Laane, C., **8**, 185[14], 206[14]
Laarhoven, W. H., **5**, 211[68], 707[36], 723[106,108b,109], 724[110], 725[36,115], 729[106b]
Laasch, P., **6**, 263[23], 273[23,92], 277[92], 291[23]
Laatikainen, R., **2**, 346[45]
Labadie, J. W., **1**, 442[176], 443[182,184], 444[185,186,187,189], 446[196,197], 457[182,184,186]; **2**, 727[130], 749[133]; **4**, 175[44]
Labadie, S. S., **2**, 611[99], 727[132]; **5**, 757[78], 762[102]
LaBahn, V. A., **7**, 26[59]
Laban, G., **5**, 439[166]
Labana, L. L., **3**, 736[23]
Labana, S. S., **3**, 736[23]
Labar, D., **1**, 672[188,189,195,199,204,206], 673[189,206], 674[188,189,206], 675[188,189,206,210], 677[188,189,206,210,222], 678[188,206], 680[206a], 681[206], 683[222], 698[188,189,195,206], 699[188,189,195,243], 700[188,189,206,222], 701[188]; **3**, 86[50], 87[117], 89[142], 116[142]; **6**, 1027[90], 1031[112]; **8**, 847[97], 848[97e], 849[97e]
Labarca, C. V., **7**, 35[101]
Labaree, D., **5**, 829[21]
Labaudiniere, L., **8**, 863[233]
Labaudinière, R., **6**, 134[17]
Labaziewicz, H., **5**, 417[61], 418[61,69]
L'abbé, G., **4**, 1099[182]; **5**, 113[237], 117[278]; **6**, 247[134], 540[584]; **7**, 21[3,19], 475[52]
Labeish, N. N., **7**, 506[292]
LaBelle, B. E., **5**, 1039[11,17], 1050[17], 1052[17], 1133[26], 1146[26]; **8**, 883[94], 884[94]
Labelle, M., **4**, 379[115,115e], 380[115e], 381[129], 382[129]; **6**, 709[56], 715[93]
Laber, D., **7**, 110[189]
Labhart, M. P., **3**, 873[65], 874[65]
Labia, R., **4**, 31[93]; **5**, 85[8]
Labinger, J. A., **1**, 443[179]; **8**, 22[145], 668[1], 669[1,3], 670[11], 673[1], 676[1], 683[86], 684[1], 685[1,98], 686[1], 687[1], 688[1,100], 689[98], 691[1,100,105], 692[1,100], 697[3]
Lablanche-Combier, A., **7**, 143[151], 144[151]; **8**, 628[178]
Laborde, E., **3**, 226[194,194a], 265[191]; **4**, 120[196], 795[89], 1035[38], 1046[117], 1048[38]; **5**, 943[238]; **6**, 9[42]; **7**, 358[12], 376[82]
Laboue, B., **4**, 98[111]
Laboureur, J.-L., **1**, 648[130], 650[130], 664[200], 672[130,197,198,200,204,206], 673[206,223], 674[206], 675[206], 677[206,222,223,225], 678[206], 680[206a], 681[206], 683[222,223,225], 698[198,206,223,244], 700[206,222], 701[198,204], 825[56], 828[56,78], 862[74], 866[79], 867[79,81], 868[81], 869[83], 870[84]; **3**, 87[117,119], 778[3], 785[3], 786[43]; **6**, 26[112]
Labouta, I. M., **8**, 604[8], 605[8]
Labovitz, J., **7**, 97[95], 112[95]; **8**, 527[47]
Labovitz, J. N., **3**, 364[92], 419[43]; **6**, 655[164b]
Labunskaya, V. I., **6**, 509[272]
Labuschagne, A. J. H., **6**, 184[152]
Lacey, R. N., **5**, 86[23], 90[23]; **6**, 266[52]
Lach, D., **6**, 506[226]
Lachance, P., **3**, 846[44]
Lacher, B., **4**, 748[156]; **7**, 722[22], 725[33], 726[36], 731[52]
Lacher, J. R., **4**, 280[132]; **8**, 895[1], 898[1]
Lachhein, S., **4**, 728[57]
Lachkova, V., **3**, 202[89]
Lachmann, B., **7**, 749[123]
Lack, R. E., **5**, 752[52]; **8**, 119[76]
Lackey, P. A., **4**, 282[134], 288[134]
Lacko, R., **5**, 347[72,72d]
Lacombe, S., **6**, 94[136]; **7**, 498[222]
LaCount, R. B., **4**, 44[125]
Ladduwahetty, T., **3**, 220[119], 224[174]; **6**, 61[150]
Ladenburg, A., **8**, 595[80]
Ladika, M., **2**, 725[113]; **6**, 966[92]
Ladjama, D., **2**, 600[28]; **7**, 121[22,23]
Ladlow, M., **4**, 791[58]
Ladner, D. W., **2**, 544[85], 547[85], 552[85]
Ladner, W., **2**, 25[97], 26[97], 27[97], 31[97], 33[97b,121], 41[97b], 43[97b], 206[105], 571[39]; **8**, 196[121]
Ladner, W. E., **7**, 429[151]
Ladouceur, G., **5**, 833[47]
Ladsic, B., **6**, 658[182]
Ladurée, D., **6**, 487[64], 489[64]
Laemmle, J. T., **1**, 67[58], 78[14-17], 79[14-16], 325[1], 333[1]
Laerum, T., **6**, 509[262]
Laffan, D. O. P., **1**, 772[197]; **2**, 649[107]
Laffey, K. J., **4**, 604[279,280]
Laffitte, J.-A., **3**, 878[94], 879[94], 880[94], 881[94]; **4**, 359[159], 771[251]
Laffosse, M. D., Jr., **1**, 507[19]
LaFlamme, P. M., **4**, 959[30], 1006[101], 1009[147]; **6**, 970[124]
Lafon, C., **5**, 575[132]
Lafontaine, J., **4**, 1040[83,84], 1043[83,84]
Lagain, D., **5**, 556[52]
Lagally, R. W., **2**, 916[84]
Lage, N., **4**, 85[74]
Lagerlund, I., **2**, 1098[103], 1099[103]; **8**, 384[35]
Lagerlund, J., **6**, 667[243]
Lago, J. M., **6**, 251[149], 490[103]
Lagow, R. J., **7**, 15[144]
Lagrenée, M., **6**, 941[149]
Laguerre, M., **2**, 717[70], 718[74,75]; **8**, 518[131]
Laguna, M. A., **5**, 478[164]; **8**, 646[50,51]
Lahav, M., **3**, 382[36]; **7**, 40[7]
Lahm, G. P., **1**, 544[36], 547[36], 548[36]; **2**, 69[46]; **5**, 766[118]
Lahousse, F., **3**, 870[49]; **5**, 70[113]
Lahrmann, E., **3**, 812[56]
Lahti, P. M., **7**, 742[56]
Lai, C. K., **7**, 401[55]
Lai, C. Y., **2**, 456[43], 466[43], 467[43]
Lai, H. K., **1**, 118[62]
Lai, J. T., **5**, 947[258], 960[258]
Lai, Y.-H., **3**, 563[1], 606[1g], 607[1g], 877[83]; **4**, 83[64]; **5**, 1151[128]; **8**, 320[86]
Lai, Y.-S., **3**, 220[123]; **4**, 686[61]
Laidlaw, G. M., **2**, 758[25]
Laidler, D. A., **6**, 2[2], 23[2]
Laidler, K. J., **5**, 72[180]
Laikos, G. D., **2**, 740[63a]
Lain, M. J., **8**, 135[43]
Laine, R., **4**, 844[56]
Laine, R. M., **8**, 289[23]
Laing, J. W., **4**, 483[10]
Laipanov, D. Z., **3**, 316[116], 317[116]
Laird, A. A., **6**, 531[454]
Laird, A. E., **5**, 576[146]
Laird, B. B., **3**, 380[10]

Laird, T., **1**, 542[5], 543[5], 544[5], 546[5]; **3**, 916[20], 949[94,96]; **6**, 897[96]; **7**, 345[1], 671[16]
Laishes, B. A., **2**, 352[86], 369[86]
Lajis, N. H., **6**, 647[115], 655[157]
LaJohn, L. A., **1**, 528[117]
Lakhan, R., **5**, 491[206]
Lakhvich, F. A., **3**, 24[145]
Lakomov, F. F., **4**, 426[64]
Lakomova, N. A., **4**, 426[64]
Laksham, M., **6**, 254[160]
Lakshmikantham, M. V., **5**, 394[145a], 436[155]; **7**, 774[313], 777[313]
Lal, A. R., **4**, 113[171]
Lal, B., **6**, 253[155]
Lal, D., **3**, 69[26]
Lal, G. S., **1**, 874[104]; **2**, 109[69], 603[48]; **4**, 401[227]
Lal, K., **8**, 656[90]
Lal, S., **1**, 834[128], 838[159,162]
Lal, S. G., **7**, 765[156], 778[399]
Lalancette, J.-M., **7**, 282[178]; **8**, 64[206], 66[206], 113[37], 250[100], 253[100]
Lalancette, R. A., **2**, 280[26]
Lalande, R., **2**, 159[127]; **7**, 95[77]
Lalezari, I., **6**, 969[118]
Laliberte, B. R., **8**, 548[319]
Lalima, N. J., Jr., **3**, 250[68]
Lalko, O. R., **5**, 609[70]
Lallemand, J. Y., **1**, 791[265]; **2**, 967[72]; **3**, 224[174,174c], 263[173]
Lalloz, L., **4**, 503[127]
Lally, D. A., **1**, 551[74]
Lalonde, J. J., **4**, 5[17]
Lalonde, M., **5**, 1154[155]; **6**, 652[147], 653[147], 654[147], 655[147], 681[147], 1005[144]
La Londe, R. T., **3**, 649[202], 650[202c], 652[202c]; **6**, 914[26]; **7**, 222[40]
Lam, A., **5**, 406[24]
Lam, C. H., **4**, 310[422]
Lam, H.-L., **2**, 1094[89], 1095[89]
Lam, J. N., **1**, 471[69]
Lam, L. K. P., **8**, 185[16], 197[126], 205[155]
Lam, P. W. K., **6**, 502[214], 562[214]
Lam, P. Y.-S., **5**, 162[69], 692[95]
LaMaire, S. J., **8**, 675[41], 677[56], 678[56], 679[41,56], 680[56], 681[56], 684[41], 685[56], 686[56], 688[56], 692[56], 694[56]
Lamanna, W., **4**, 703[32], 704[32]
Lamare, V., **7**, 64[60,61b]
Lamarre, C., **3**, 635[38]
Lamatsch, B., **8**, 190[69]
LaMattina, J. L., **6**, 432[121], 524[369], 530[420], 533[369], 554[717], 566[926], 787[103]
Lamaty, G., **7**, 473[35]; **8**, 5[27], 14[27]
Lamazouère, A.-M., **6**, 421[33]
Lamb, H. H., **7**, 840[7]
Lamb, S. E., **7**, 578[154], 584[154,158]
Lamba, D., **3**, 396[110], 397[110]
Lambert, A., **2**, 365[216]
Lambert, B. F., **8**, 501[53], 507[53]
Lambert, C., **1**, 343[111]; **3**, 279[37]; **4**, 393[190], 394[190,192], 567[50,51]; **6**, 429[89], 495[144], 499[144], 515[317], 546[317]
Lambert, D. E., **2**, 75[82a]; **4**, 258[244]
Lambert, J., **7**, 485[139]
Lambert, J. B., **2**, 965[64]; **5**, 20[133,134], 761[93], 790[35]; **8**, 347[133], 348[133]
Lambert, J. L., **2**, 441[1], 443[1]
Lambert, R. F., **8**, 517[124]
Lambert, R. L., Jr., **3**, 200[69]; **4**, 1007[127]
Lamberth, C., **4**, 740[115]
Lamberts, J. J. M., **5**, 211[68]
Lambie, A. L., **6**, 607[47]
Lambrecht, J., **6**, 524[359]
Lambros, T. J., **6**, 635[14b], 636[14]; **8**, 959[21]
Lamchen, M., **1**, 391[150]
Lamed, R., **7**, 465[131]; **8**, 187[33]
Lamendola, J., Jr., **1**, 838[158]; **7**, 162[64]
Lamiot, J., **1**, 34[227]
Lamke, W. E., **8**, 873[13]
Lamm, B., **5**, 686[53], 688[68]; **6**, 809[66]; **8**, 994[66]
Lammer, O., **3**, 224[164], 225[164]; **4**, 21[65], 108[146e]; **6**, 11[48]
Lämmerhirt, K., **2**, 1086[27], 1088[27]
Lammerink, B. H. M., **8**, 405[22]
Lammers, R., **2**, 612[107], 629[1], 635[1]
Lammertsma, K., **3**, 333[210]
Lamothe, S., **3**, 56[286], 380[10]; **5**, 532[86]
Lamotte, G., **8**, 830[89], 831[89]
Lampariello, L. R., **4**, 956[18]
Lampe, J., **1**, 357[2]; **2**, 94[54], 182[9,9c], 184[9c], 190[9c], 191[9c], 192[9c], 193[9c], 195[75], 197[9c], 198[9c], 200[9c], 201[93], 211[9c], 217[9c], 223[151], 235[9c], 236[9c], 289[69], 634[36], 640[36]; **4**, 72[31]; **5**, 170[113]; **6**, 814[95]
Lampert, M. B., **7**, 372[70]
Lampilas, M., **3**, 224[174,174c]
Lampman, G. M., **3**, 414[5], 422[5]; **5**, 1185[2]
Lampos, P. J., **8**, 857[198]
Lamy-Schelkens, H., **5**, 480[166]
Lan, A. J. Y., **2**, 1039[104], 1040[105]; **7**, 876[123]
Lancaster, J., **4**, 20[63], 21[63]
Lancelin, J.-M., **1**, 419[77]; **4**, 379[116], 792[62]
Lancelot, J. C., **6**, 487[64], 489[64]
Landais, Y., **3**, 509[177]
Landau, E. F., **2**, 144[60]
Lande, S., **8**, 144[78]
Landen, H., **5**, 257[60]
Lander, N., **6**, 625[162]; **8**, 352[148], 542[234]
Lander, S. W., Jr., **5**, 255[53], 264[53]
Lander-Schouwey, M., **5**, 18[126]
Landes, M. J., **6**, 554[724]
Landesberg, J. M., **6**, 690[390]; **7**, 300[52]
Landesman, H., **3**, 28[170], 30[170]; **4**, 6[21,21a,23,23a], 7[21a]; **6**, 703[2,3], 714[3]
Landgraf, B., **2**, 1090[73], 1102[73], 1103[73]
Landgrebe, K., **4**, 754[176]
Landini, D., **4**, 275[68], 276[68], 279[68], 288[68], 444[195], 523[57]; **6**, 19[67], 204[11], 221[131], 236[54]; **7**, 253[16], 663[61], 771[261]; **8**, 806[123]
Landis, C. R., **8**, 429[53], 455[204], 460[252]
Landmann, B., **2**, 39[137,137a], 40[141], 42[137a]; **3**, 438[32]
Landmesser, N. G., **1**, 853[45]
Landolf, M., **1**, 283[1], 289[1]
Landolt, R. G., **5**, 908[72]
Landor, P. D., **3**, 522[12]; **4**, 55[157], 57[157c], 251[145], 1010[151]; **6**, 213[88], 265[40]
Landor, S. R., **2**, 81[1], 82[1], 96[1]; **3**, 223[146], 522[12]; **4**, 53[151], 55[157], 57[157c], 251[145], 276[76], 277[76,83], 284[76], 285[76,83], 299[299], 308[402], 395[202], 1010[151]; **5**, 600[43], 797[59]; **6**, 213[88], 235[53], 265[40], 830[4], 835[43]; **7**, 506[295]; **8**, 64[204], 66[204], 67[204], 161[14,15,24], 162[25-27], 176[130,131], 545[298,299,300]
Landquist, J. K., **3**, 786[38]; **6**, 453[140]
Landry, N. L., **2**, 662[20], 663[20], 664[20]; **5**, 432[130]
Lane, A. C., **8**, 615[94], 618[94]
Lane, C. A., **7**, 606[157]

Lane, C. F., **6**, 724[164]; **7**, 604[129,135,137,148], 605[141,142], 606[148]; **8**, 26[15,17], 27[15,17], 47[15], 55[15], 60[15], 237[8], 240[8], 244[8], 247[8], 249[8], 253[8], 541[207], 705[12], 706[15], 708[37], 709[15], 713[71], 720[132], 877[37]
Lane, J. F., **3**, 892[50]; **6**, 795[6], 796[6], 801[6]
Lane, N. T., **2**, 558[161]
Lane, R. W., **8**, 366[51]
Lane-Bell, P. M., **8**, 205[155]
Lang, D., **5**, 71[125], 498[237]
Lang, E. S., **8**, 411[101], 412[101,116]
Lang, J., **7**, 346[10], 356[10]
Lang, K. L., **1**, 819[3]; **6**, 2[3], 25[3]; **7**, 358[6], 372[6]
Lang, L., **4**, 445[207]
Lang, P., **6**, 636[24]
Lang, P. C., **5**, 17[114]
Lang, R., **4**, 1017[214]
Lang, R. W., **5**, 829[13]; **6**, 185[166]
Lang, S., **7**, 111[193]
Lang, S. A., Jr., **3**, 767[163]; **6**, 487[42], 489[42], 538[559], 543[42], 550[42], 554[42,710,724,756,757,758,775]; **7**, 739[39]
Lang, T. J., **6**, 953[8]
Langa, F., **5**, 202[33,34], 221[52,53,56]
Langan, J. R., **5**, 722[105]
Langbehaim, M., **5**, 71[152]
Langbein, G., **8**, 856[167]
Langdale-Smith, R. A., **5**, 797[60]
Lange, B., **6**, 226[10], 242[95,96], 256[10], 257[10]
Lange, B. C., **1**, 4[27]; **2**, 100[11,12]; **3**, 3[15], 54[15]
Lange, B. M., **1**, 2[4], 43[4]
Lange, G., **2**, 124[205]; **4**, 96[103a], 98[103a]; **8**, 609[55]
Lange, G. G. A., **7**, 483[131]
Lange, G. L., **2**, 553[127]; **3**, 380[10]; **5**, 21[143], 66[80], 129[35], 132[48], 179[142]; **7**, 541[4], 564[4]
Lange, J. H. M., **5**, 568[109]
Lange, R. J., **6**, 279[134]
Langemann, A., **7**, 86[16a]
Langendoen, A., **7**, 684[95]
Langer, E., **5**, 1148[123]
Langer, F., **3**, 817[86,88]
Langer, M. E., **2**, 670[47]; **5**, 410[41,41b]; **8**, 5[32]
Langer, V., **3**, 709[15]
Langer, W., **1**, 70[63], 141[22], 218[52], 223[52a]; **2**, 124[208]; **4**, 229[237]
Langford, G. E., **8**, 620[135]
Langhals, E., **4**, 1073[27], 1074[27]
Langhals, H., **7**, 720[11]
Langlais, M., **4**, 98[109a]
Langler, R. F., **3**, 587[141]; **7**, 211[97]
Langley, D. R., **5**, 736[143]
Langley, J. A., **2**, 564[7]
Langlois, M., **8**, 383[17]
Langlois, N., **1**, 838[169]; **4**, 598[206]; **5**, 409[37]; **6**, 487[75], 916[32], 921[47,49], 1067[100]; **7**, 164[80]; **8**, 58[174], 66[174], 180[137]
Langlois, Y., **1**, 115[40], 419[82], 838[169]; **4**, 189[104], 598[206]; **5**, 409[37]; **6**, 487[75], 916[32], 920[44], 921[47,49], 922[50], 1067[100]; **7**, 164[80]
Langner, M., **8**, 70[233]
Langry, K. C., **2**, 780[14]; **8**, 851[128]
Langsjoen, A. N., **2**, 362[185]
Langston, J. W., **4**, 270[9]
Langstrom, B., **6**, 489[96]; **7**, 229[121]
Lankford, P. J., **5**, 1070[20], 1071[20], 1072[20], 1074[20], 1110[20], 1111[20]
Lankin, D. C., **7**, 143[145], 372[70]
Lanneau, G. F., **8**, 262[19], 265[19]
Lannoye, G., **1**, 262[40]; **2**, 381[307]
Lanoiselee, M., **1**, 218[51]
Lansard, J.-P., **4**, 95[99c]
Lansbury, P. T., **1**, 564[204]; **2**, 74[73], 904[52]; **3**, 14[76], 15[76], 25[151,152], 341[4], 347[25], 358[69], 975[3], 980[3]; **5**, 4[36], 5[36], 780[204], 1130[10]; **6**, 160[177]; **7**, 164[81], 576[123], 601[81], 691[18], 710[54]; **8**, 43[107], 66[107], 530[109], 583[11], 843[54], 847[54]
Lansbury, R. T., **7**, 313[38]
Lantos, I., **2**, 116[136], 117[136], 124[136], 436[66], 437[66b]; **3**, 500[127], 501[127]; **4**, 1018[225], 1019[225]; **7**, 401[57]
Lantseva, L. T., **6**, 498[170], 500[170]
Lantsova, O. I., **6**, 543[616]
Lantzsch, R., **6**, 243[103], 244[103], 258[103]
Lanyiova, Z., **5**, 704[21]
Lanz, J. W., **2**, 26[100,100a], 27[100a], 39[137,137b], 94[53]
Lanza, T., **3**, 541[112], 543[112]; **6**, 487[57], 489[57]
Lanzetta, R., **4**, 347[95]
Lanzilotta, R. P., **8**, 201[145]
Lanzilotti, A. E., **8**, 618[127], 623[127]
Laohathai, V., **7**, 657[35]
Laonigro, G., **4**, 347[95]
Laos, I., **7**, 100[120]
Lapalme, R., **5**, 959[317], 961[317]; **6**, 1059[70], 1066[70]; **7**, 376[83]
Lapid, A., **8**, 451[180]
Lapierre, J. C., **3**, 327[168]
Lapierre, R., **2**, 284[54]
LaPierre, R. B., **3**, 328[179]
Lapin, Y. A., **4**, 335[26]
Lapinte, C., **4**, 985[127]; **7**, 450[16]
Lapointe, J.-P., **3**, 903[126]
Lapointe, P., **2**, 213[126]
Laporterie, A., **5**, 2[19], 3[19]; **6**, 832[15]
Lapouyade, R., **5**, 727[119]
Lappert, M. F., **1**, 16[89], 17[206,207,212,218], 37[178], 139[3,4], 140[7,10], 292[25], 543[25], 547[25]; **2**, 242[18]; **3**, 124[289], 125[289]; **4**, 588[61], 953[8,8c], 954[8c]; **6**, 134[12]; **8**, 556[372], 756[154], 771[44,51], 779[89], 780[93,94]
Lapporte, S. J., **8**, 454[197,198], 455[197]
LaPrade, J. E., **8**, 112[26], 118[26]
Lapshin, S. A., **6**, 516[320]
Lapworth, A., **2**, 347[49]; **6**, 681[334]
Larbig, W., **8**, 100[116], 717[106], 738[29], 739[33], 754[29]
Larcheveque, M., **1**, 789[261], 791[265]; **2**, 507[25]; **3**, 31[184], 41[225], 45[243], 215[66], 225[189,190], 263[174,175]; **4**, 89[84j], 653[434]; **6**, 3[15,17], 11[47], 720[131]; **8**, 113[33], 196[114], 249[91], 479[48], 481[48], 524[12]
Lardelli, G., **8**, 336[74], 340[74]
Lardicci, L., **3**, 217[80,80b], 246[41], 247[41], 257[41], 285[55], 439[40,41], 483[12], 491[70]; **5**, 942[230]; **8**, 99[108,112], 100[118,118e], 348[131], 558[389], 564[440]
Lardici, L., **1**, 214[26]
Lardy, H. A., **2**, 456[70], 457[70], 464[99]
Laredo, G. C., **1**, 554[105]
Large, R., **3**, 770[177]
Largis, E. E., **8**, 30[42], 66[42]
Laricchiuta, O., **1**, 834[126]
Larin, N. A., **3**, 318[122]
Lariviere, H. S., **4**, 180[66]
Larkin, D. C., **7**, 346[6]
Larkin, D. R., **8**, 526[30]
La Rochelle, R. W., **3**, 179[546], 918[21]
Larock, R., **6**, 829[3]
Larock, R. C., **1**, 225[86], 450[212]; **2**, 85[22,23]; **3**, 470[204], 473[204], 484[23-25], 489[56], 497[23,56,100], 498[100], 500[23],

501[23,24], 509[23], 1033[73]; **4**, 290[194], 292[194], 294[242], 295[242], 297[267], 300[311], 303[311], 306[372], 309[415], 310[415], 311[415], 314[477], 315[477,519], 395[200], 411[200], 559[20], 588[50,51], 597[175,177], 602[253,254], 621[175], 623[175], 625[253], 839[31], 840[31,32], 841[39,41,42], 845[65], 846[75], 847[75], 848[78], 851[87], 903[201,202], 904[202,203,204,205]; **5**, 35[14], 935[192]; **7**, 92[42], 93[42], 533[34], 604[131], 631[56], 632[56], 637[56]; **8**, 851[130,132]
Larock, R. L., **7**, 613[2], 631[2], 632[2]
Laronze, J.-Y., **4**, 40[115]; **5**, 528[69]
Larossi, D., **3**, 95[154], 119[154]
Larpent, C., **8**, 445[38]
Larscheid, M. E., **7**, 801[44]
Larsen, A. A., **8**, 263[25]
Larsen, B. D., **6**, 480[110]
Larsen, C., **5**, 439[167]
Larsen, D. S., **4**, 351[126], 354[126]
Larsen, E. H., **7**, 330[8]
Larsen, E. R., **3**, 898[75]
Larsen, J. W., **8**, 319[72]
Larsen, L., **5**, 731[130d]
Larsen, P. K., **8**, 604[8], 605[8]
Larsen, R. D., **1**, 425[105]; **3**, 571[58]
Larsen, R. D., Jr., **7**, 227[86]
Larsen, S., **2**, 150[97]; **4**, 1086[119], 1087[119]
Larsen, S. D., **1**, 875[97]; **2**, 1002[56], 1027[76]; **4**, 373[70], 673[30]; **5**, 408[34], 409[34], 411[34,43], 413[51], 415[51b], 436[156], 532[83], 913[106]; **6**, 756[124], 1006[146]; **7**, 580[144], 586[144]
Larson, E., **5**, 434[144]; **6**, 91[123], 93[123]; **7**, 439[37], 440[37]
Larson, E. R., **2**, 665[31,32], 666[35], 668[31,32], 673[32], 674[31,32], 675[35], 682[31,32], 694[75], 700[35]; **5**, 687[54]; **8**, 540[195]
Larson, G., **5**, 850[161]
Larson, G. L., **1**, 329[41], 436[150], 731[4], 784[243], 785[247], 790[262,263], 815[4]; **2**, 103[38], 111[79], 604[50,54]; **3**, 200[70], 418[23]; **5**, 844[129]; **8**, 113[34], 408[78], 478[46], 481[46], 707[22], 936[74-76], 937[76]
Larson, H. O., **6**, 104[9]; **7**, 291[2], 655[18], 736[6]
Larson, J. R., **7**, 583[155], 584[155]
Larson, K. A., **4**, 505[148]
Larson, K. D., **7**, 228[101]
Larson, M. L., **2**, 737[36]
Larson, N. R., **6**, 959[40]
Larson, S., **4**, 288[188], 346[86a], 347[86b]
Larsson, I.-M., **5**, 219[37]
Larsson, L.-G., **2**, 465[107]
Larsson, P. O., **7**, 145[162]
Lartey, P. A., **2**, 570[38]; **6**, 57[136]
Larue, M., **1**, 821[28]; **4**, 1059[155]; **6**, 984[55]
Lasch, M., **2**, 242[18]
Laschi, F., **8**, 458[219]
Laskin, A. I., **7**, 56[20,21], 80[137]
Laskos, S. J., Jr., **6**, 441[85]
Lasne, M.-C., **5**, 451[38], 552[9], 560[65,66], 565[99], 576[136,137,214], 583[185], 589[214]; **6**, 689[384], 690[384]
Lasouris, U. I., **6**, 569[936]
Lassila, J. D., **5**, 597[30], 730[129], 731[129]
Lasswell, L. D., **5**, 63[18]
Laswell, W. L., **1**, 275[77]; **4**, 795[79]; **5**, 99[130]; **6**, 755[121], 759[139]
Laszlo, P., **1**, 564[187,188], 571[277]; **2**, 345[19], 359[19], 523[87]; **3**, 125[303], 142[379], 300[44], 310[44], 311[44], 322[144], 582[115], 809[37]; **5**, 345[68,68a], 346[68a], 453[62], 1130[4,6,7]; **6**, 111[59-63]; **7**, 744[72], 760[25], 839[1], 840[3], 844[62], 846[3,84,87-90,92-100]
Latham, D. S., **8**, 509[89]
Latham, D. W., **3**, 643[122]
Latham, H. G., Jr., **8**, 228[133,134]
Latham, J. V., **8**, 901[39]
Lathbury, D., **4**, 412[268e,269,270,270b], 413[269b], 562[34], 820[223]
Lathbury, D. C., **7**, 546[32]
Latif, N., **2**, 359[160], 368[232]; **7**, 137[125], 138[125]; **8**, 542[225]
Latimer, L. H., **1**, 822[37], 862[75b]; **2**, 70[51]; **3**, 766[157]; **7**, 124[51], 127[51]
Latourette, H. K., **2**, 753[2]
Latrofa, A., **8**, 657[97]
Lattes, A., **1**, 838[161,166]; **4**, 290[195,198,199,202,203,204,207], 291[204,210], 292[204,221], 311[439,440], 383[136], 401[230]; **6**, 175[71], 245[127]; **7**, 470[1], 488[1], 490[1], 632[61,62]; **8**, 252[109], 854[154], 856[175]
Lattimer, N., **8**, 28[34], 66[34]
Lattrell, R., **5**, 85[1]
Lattuada, L., **5**, 1079[51]
Latypova, F. M., **8**, 610[61], 611[66]
Lau, C. K., **5**, 979[28]; **7**, 360[21]; **8**, 315[52], 316[57], 969[94]
Lau, H., **4**, 841[41,42]
Lau, J. C. Y., **1**, 621[67], 784[239], 815[239]
Lau, K.-L., **5**, 63[15], 123[1], 126[1]; **7**, 815[4], 824[4], 833[4]
Lau, K. S. Y., **1**, 439[166], 442[178]; **3**, 530[69], 533[69], 537[69], 541[116], 543[116]; **4**, 614[381]
Lau, M. P., **8**, 329[24], 336[24]
Lau, P. W. K., **1**, 623[82]; **2**, 57[7], 68[44], 605[56], 630[11], 631[11], 713[49], 728[49]
Lau, W., **7**, 868[93], 869[94,95]
Laub, R. J., **3**, 799[99]
Laubach, G. D., **7**, 136[108]; **8**, 566[450], 956[12], 957[12]
Laube, R., **2**, 871[23]; **5**, 841[104]
Laube, T., **1**, 26[137], 28[142], 30[151], 31[155], 32[156], 34[170], 37[238], 38[261], 41[151], 299[61], 316[61], 418[73]; **2**, 100[2-4], 107[58], 108[58], 196[76], 280[25]; **4**, 20[64], 21[64], 72[31], 100[124]; **5**, 79[291]
Laucher, D., **1**, 359[20,22], 383[20,22], 384[22]; **2**, 120[180]; **4**, 76[49], 77[50], 206[48,49], 229[226,227,228,229], 243[66], 252[66,167], 257[167]
Laude, B., **8**, 337[80]
Lauer, M., **5**, 493[212]
Lauer, R. F., **1**, 634[75], 642[75], 644[75], 652[140], 708[261], 712[264], 714[264]; **3**, 87[63], 114[63], 117[63]; **6**, 1026[81,83,85], 1027[81,83,92], 1028[83,85], 1029[81,101], 1030[81], 1031[81,85], 1033[81]; **7**, 86[13], 87[22], 124[38,39], 128[38,39], 129[38], 130[39], 131[39], 146[170], 522[40], 769[227], 771[227,269,270,272], 772[272], 775[353], 779[269], 819[21]
Lauer, W. M., **4**, 279[113], 282[113]; **5**, 834[53], 876[2]
Laufer, S., **8**, 957[10]
Laugal, J. A., **8**, 629[186]
Laughlin, R. G., **6**, 1016[29]; **7**, 8[58]
Laugraud, S., **5**, 327[27]
Lauher, J. L., **4**, 212[99]
Lauher, J. W., **8**, 669[5], 670[5], 671[5]
Lauke, H., **1**, 253[12]; **8**, 447[133,136], 696[127,-128]
Laumen, K., **7**, 397[30]
Laumenskas, G. A., **2**, 143[50]
Launay, J.-C., **6**, 717[109]
Launay, M., **1**, 804[308], 805[308]; **8**, 847[88]
Laur, D., **7**, 777[369]
Lauren, D. R., **8**, 333[54]
Laurence, G., **7**, 765[139]
Laurent, A., **1**, 387[126,128,129,130,131,132,134], 388[126,129,134]; **2**, 283[48], 942[167], 943[167], 944[167]; **3**, 381[28], 382[28], 635[41]; **4**, 331[19], 356[137]; **6**, 69[16], 94[136,137], 98[154], 264[34]; **7**, 470[8], 498[222]
Laurent, D. R. S., **1**, 567[223]; **4**, 10[34]
Laurent, E., **3**, 635[41]; **7**, 498[220], 538[63]

Laurent, H., **4**, 155[73]; **7**, 74[111,112], 75[111,112], 773[305]; **8**, 548[318], 881[72], 882[72]
Laurenzo, K. S., **4**, 436[144]
Laurino, J. P., **8**, 618[117]
Lauritzen, S. E., **3**, 864[24], 865[24], 872[24]
Lauro, A. M., **1**, 507[19]
Lauron, H., **8**, 847[88,88d]
Lautens, M., **4**, 599[225], 607[225], 625[225], 642[225]; **5**, 16[113], 435[149], 524[53], 1183[49]
Lauteri, S., **6**, 776[57]
Lauterschlaeger, F., **7**, 516[3]
Lautzenheiser, A. M., **5**, 3[22]
Lauwers, M., **6**, 213[90]
Lavagnino, E. R., **6**, 936[105]; **7**, 205[61]
Laval, J. P., **4**, 290[207], 401[230]
Lavallée, C., **5**, 86[28], 88[49]
Lavallée, J.-F., **4**, 29[85], 46[131], 245[84], 258[84], 260[84]
Lavallee, P., **7**, 299[39]
Lavallee, P. L., **6**, 656[171]
Lavanish, J., **5**, 1185[2]
Lavaud, C., **2**, 1017[33]; **6**, 735[20], 739[20]
LaVaute, T., **3**, 564[6]; **8**, 112[28], 113[28], 116[28]
Lave, D., **4**, 599[217]
Lavé, D., **3**, 882[103]; **6**, 161[186]
Lavie, D., **3**, 693[144]; **7**, 253[21]; **8**, 228[125]
Lavielle, G., **6**, 845[94]
Lavielle, S., **3**, 149[408,409], 151[408,409], 155[408,409]; **6**, 641[58], 644[91]; **7**, 777[388]
Lavieri, F. P., **3**, 815[78]; **5**, 225[97,98]; **7**, 276[151]; **8**, 505[82], 507[82]
Lavilla, R., **8**, 587[34]
Lavoie, A. C., **3**, 446[84]; **6**, 905[146]; **7**, 172[130], 173[130]
LaVoie, E. J., **3**, 503[149], 512[149]; **7**, 346[9], 365[43]
Lavrenyuk, T. Y., **5**, 425[101]
Lavrinovich, E. S., **8**, 587[31]
Lavrinovich, L. I., **4**, 885[115], 886[115]
Lavrushko, V. V., **7**, 4[19], 8[19], 17[173]
Law, C. K., **5**, 955[301]
Law, K. W., **4**, 176[52]; **7**, 771[280], 773[280]
Lawesson, S.-O., **2**, 867[12,13]; **5**, 71[155]; **6**, 420[23], 423[23], 436[19], 437[19], 451[130], 456[130], 509[252], 545[636]; **7**, 95[76,80], 163[72], 330[8,12], 613[1]; **8**, 541[202], 924[2]
Lawin, P. B., **8**, 132[11]
Lawitz, K., **3**, 511[188], 515[188]
Lawler, D. M., **4**, 260[280]
Lawrence, D. S., **6**, 1066[92]
Lawrence, G. C., **7**, 59[37]; **8**, 198[130]
Lawrence, G. W., **4**, 426[54]
Lawrence, R. F., **5**, 839[77]
Lawrence, R. M., **2**, 536[43,47]
Lawrence, T., **4**, 787[27]
Lawrie, W., **7**, 832[71]
Lawrynowicz, W., **6**, 219[119]
Laws, D. R. J., **6**, 235[53]
Lawson, A., **6**, 420[21], 424[21], 429[21], 454[148]
Lawson, J. E., **6**, 157[164]; **7**, 760[29]
Lawson, J. K., Jr., **5**, 557[54]
Lawson, J. P., **1**, 791[269]; **2**, 486[30], 488[30]; **4**, 252[155]
Lawson, K., **2**, 645[82]
Lawson, K. R., **5**, 475[143]
Lawson, T., **1**, 797[283]; **7**, 555[69]
Lawston, I. W., **4**, 93[91b]
Lawton, E. L., **8**, 36[76], 66[76]
Lawton, G., **3**, 902[119]; **4**, 6[21], 680[50]; **8**, 389[71]
Lawton, R. G., **2**, 352[90], 353[90]; **4**, 8[29,29b]; **5**, 943[236]; **6**, 715[91,92]; **8**, 346[124]
Lawton, V. D., **6**, 420[21], 424[21], 429[21]
Laxma Reddy, N., **2**, 762[57]
Lay, W. P., **5**, 582[178], 737[146]
Laycock, D. E., **8**, 683[89,91]
Layer, R. W., **1**, 360[26], 364[26], 390[26a]; **2**, 894[6], 916[6], 933[6]; **3**, 324[153]; **8**, 26[21], 66[21]
Layloff, T. P., **3**, 572[66]; **8**, 312[22], 321[22]
Layman, W. J., **3**, 246[44]
Layton, R. B., **4**, 141[14,14b], 180[68]; **8**, 756[158], 758[158]
Lazar, J., **1**, 709[262]
Lazarevski, G., **6**, 766[23]; **7**, 698[51]
Lazarow, A., **3**, 822[5], 834[5]
Lazarus, H. M., **6**, 622[135]
Lazarus, R. A., **8**, 31[46], 66[46]
Lazbin, I. M., **6**, 805[52], 806[52]
Lazdunski, M., **8**, 47[126], 66[126]
Lazennec, I., **4**, 47[133]
Lazhko, E. I., **4**, 337[38]
Lazier, W. A., **7**, 306[3]
Lazrak, T., **1**, 223[71]
Lazukina, L. A., **6**, 499[172], 543[623]; **7**, 470[3]
Lazurin, E. A., **3**, 306[81]
Lazzaroni, R., **4**, 877[68]; **5**, 1154[151,158], 1155[151]
Le, P. H., **7**, 73[103]
Le, T., **2**, 846[203]; **8**, 11[61]
Lê, T. Q., **1**, 650[138], 651[138], 672[138,206], 673[206], 674[206], 675[206], 677[206], 678[206], 680[206a], 681[206], 686[138], 698[206], 700[206], 702[206], 705[206], 711[206a], 712[206], 713[206], 714[206], 71; **3**, 87[117]
Lea, R. E., **7**, 632[59]; **8**, 851[133], 852[133a]
Leach, D. R., **3**, 489[56], 497[56]; **4**, 559[20], 903[202], 904[202]
Leach, W. A., **8**, 364[15]
Leadbetter, M. R., **6**, 93[133], 531[439]
Leake, P. H., **3**, 507[170]
Leanna, M. R., **7**, 364[41b]
Leanza, W. J., **8**, 384[25]
Leapheart, T., **7**, 105[149]
Leardini, R., **4**, 86[78b,79b]
Learn, K., **4**, 604[285,290,291], 646[290,291], 647[291]; **8**, 26[20], 27[20], 36[20], 54[20], 55[20], 60[20], 70[20], 244[67], 250[67], 840[31], 848[31], 960[35,36]
Learn, K. S., **1**, 240[42]; **5**, 798[69], 817[148], 1029[91]
Lease, M. F., **8**, 476[25]
Leatham, M. J., **7**, 595[31]
Leatherbarrow, R. J., **8**, 206[169]
Leavell, K. H., **5**, 911[94], 948[269]
Leaver, D., **5**, 626[38], 629[38]
Leaver, J., **8**, 198[130]
Leavitt, R. K., **3**, 46[255], 47[255]
Leban, I., **6**, 554[716]
Lebaud, J., **5**, 829[22]
LeBel, N. A., **4**, 1077[53]; **5**, 253[46,46d]; **6**, 1015[24]; **8**, 873[12]
Le Belle, M. J., **3**, 979[13]
Leber, J. D., **6**, 1029[102]
Leber, P. A., **8**, 356[188]
Le Berre, A., **6**, 535[525,526]
Le Berre, N., **2**, 873[26]
Le Berre, V., **3**, 669[56]
Le Bigot, Y., **6**, 173[46], 175[46,71,73]
Le Bihan, J.-Y., **2**, 758[22a]
Lebioda, L., **3**, 380[8], 627[44]; **4**, 582[30]; **5**, 173[123]
LeBlanc, B. F., **6**, 123[140]
Leblanc, E., **1**, 174[3]
LeBlanc, J. C., **8**, 290[30]
Leblanc, Y., **1**, 773[202]; **4**, 1059[154]; **5**, 490[192]; **8**, 389[68]
Leboff, A., **3**, 681[100]

Leborgne, A., **5**, 86[28], 88[49]
Le Borgne, J. F., **2**, 507[25]
Lebouc, A., **8**, 134[31], 137[59]
Leboul, J., **4**, 36[102,102c]
Le Breton, G. C., **7**, 220[23]
Lebreton, J., **1**, 768[169]; **3**, 380[10], 1009[74], 1010[74,76,77], 1011[78], 1012[80]; **6**, 874[11]; **7**, 89[28]
Lebrun, A., **3**, 509[177]
Lebrun, H., **5**, 558[62], 560[62]; **6**, 706[39]
LeBrun, J., **4**, 247[98], 257[98], 262[98]
Lebuhn, R., **6**, 43[55]
Lecas, A., **7**, 878[140]
Lecat, J.-L., **8**, 412[107]
Lecavalier, P., **8**, 166[63]
Lecea, B., **2**, 649[102], 1059[75]; **5**, 94[87], 95[97], 96[109]; **7**, 278[160], 283[187], 530[18], 531[18], 554[65], 752[144], 760[24]; **8**, 19[133,134]
Lechevallier, A., **3**, 765[154]; **6**, 677[314]; **7**, 168[103,103b]
Lechleiter, J. C., **4**, 611[356]; **5**, 125[18], 128[18]
Lecker, S. H., **5**, 1151[136]
Leckta, T. C., **5**, 829[19]
Leclaire, J., **7**, 95[72]
LeClef, B., **6**, 645[97]; **7**, 229[119]
Leclerc, G., **8**, 370[96], 639[19]
Lecocq, M., **7**, 666[75]
Lecomte, P., **1**, 664[169], 665[169], 669[169], 670[169]; **3**, 111[231]
Leconte, M., **5**, 1115[1], 1116[1,9], 1118[9]
Le Coq, A., **1**, 188[71]; **5**, 386[132,132d], 387[132d]; **6**, 977[16]; **8**, 28[36], 66[36]
Le Corre, G., **6**, 639[48]
Le Corre, M., **1**, 759[130]
Lecoupanec, P., **7**, 14[142]; **8**, 456[211], 458[211]
Lecour, L., **5**, 851[168]
LeCoz, L., **5**, 410[41]
L'Ecuyer, Ph., **8**, 364[26], 365[26]
Leder, J., **3**, 466[186]; **7**, 440[40]
Lederer, E., **6**, 52[116]
Lederer, K., **7**, 774[331], 775[338,343]
Ledford, N. D., **4**, 4[15]
Ledger, R., **6**, 645[98], 1014[19]
Leditschke, H., **4**, 373[85]; **7**, 632[57]; **8**, 851[132]
Lednicer, D., **2**, 182[1]; **3**, 957[108], 969[108]; **8**, 540[195]
Ledon, H., **3**, 889[27]; **4**, 1054[134]
Ledoussal, B., **1**, 188[71]
Ledoux, I., **4**, 469[136]
Leduc, P., **7**, 383[109]
Ledwith, A., **1**, 832[109], 844[8]; **3**, 783[21]; **4**, 953[8,8d], 954[8d]; **5**, 63[9,10], 904[46], 905[46]; **7**, 850[2,3], 854[51], 855[51], 879[146,147], 888[2]
Ledwith, Q., **4**, 1101[197]
Lee, A. O., **3**, 105[219], 113[219]; **5**, 527[64], 530[64]
Lee, A. W. M., **3**, 225[185], 264[181]; **6**, 2[3], 8[39], 25[3], 88[105], 89[105], 927[71,76]; **7**, 198[26], 401[59], 402[63], 403[59], 406[59,77], 409[77], 414[77], 415[77], 421[77], 423[77], 763[101]
Lee, B., **3**, 68[21]
Lee, C., **4**, 376[102], 377[102]; **5**, 153[24]; **7**, 737[10]
Lee, C. B., **5**, 9[68]
Lee, C. C., **3**, 825[30]; **4**, 521[42,46], 529[74,75,77,78], 530[75,78,79], 531[46,78,81], 541[113,114]; **6**, 204[16]
Lee, C. M., **4**, 430[90,92]
Lee, C.-S., **2**, 648[94], 649[94], 924[107], 1049[22], 1050[22]; **4**, 795[80]; **5**, 102[166]
Lee, D. C., **1**, 825[47]; **3**, 792[66]; **5**, 921[140]
Lee, D. G., **7**, 92[41,41a], 94[41], 235[10], 236[18], 237[38], 444[52], 541[3], 542[5], 558[3,5], 559[5], 564[83,86,96,109], 565[96], 568[96], 569[96], 570[96], 571[3], 578[154], 583[156], 584[154,157,158], 585[161], 586[163], 587[169], 768[207,208], 773[208], 841[19], 845[78], 851[20,21]
Lee, D. H., **5**, 513[4]; **6**, 859[175], 865[200]
Lee, D.-J., **4**, 277[84,91-93]
Lee, D. Y., **1**, 471[63]
Lee, E., **4**, 744[137]; **6**, 834[39]; **7**, 43[39]
Lee, E. K., **2**, 406[46], 456[38]
Lee, F., **7**, 856[65]
Lee, F. K. C., **5**, 171[117]
Lee, F. L., **4**, 892[144]
Lee, G. A., **4**, 1090[142]; **7**, 228[94]
Lee, G. C. M., **3**, 630[55]
Lee, G. E., **5**, 256[57]
Lee, G. R., **5**, 1066[6]
Lee, H., **6**, 436[10,11,17], 455[154]; **7**, 43[39], 365[44], 775[349], 776[349]
Lee, H. B., **8**, 445[34]
Lee, H. D., **8**, 719[117,122,123]
Lee, H. H., **4**, 354[127,-128]; **6**, 960[61]; **8**, 824[59]
Lee, H.-J., **6**, 619[119]
Lee, H. K., **7**, 766[189]
Lee, H. L., **4**, 370[29]; **5**, 433[137b]
Lee, H. S., **8**, 698[142], 709[45,45a]
Lee, H. T. M., **4**, 580[25]
Lee, H. W., **1**, 568[238]; **6**, 563[902]
Lee, H. Y., **4**, 1009[142]; **5**, 249[36], 431[119]
Lee, J., **1**, 97[81]; **5**, 343[64]; **6**, 504[223], 531[462]; **7**, 346[10], 356[10], 666[77]
Lee, J. B., **1**, 743[53]; **6**, 228[18]; **7**, 228[106], 671[7]
Lee, J. G., **7**, 530[16]
Lee, J. I., **1**, 435[142,143]
Lee, J. T., **2**, 711[35]; **7**, 691[19]
Lee, J. Y., **4**, 372[64b]; **5**, 931[178]; **6**, 723[146]
Lee, J. Y.-C., **2**, 175[184], 619[150]
Lee, K., **1**, 162[95], 163[106]
Lee, K. C., **2**, 610[98]
Lee, K.-J., **6**, 173[43], 182[143], 185[143]; **8**, 636[1]
Lee, K. K., **6**, 799[23]
Lee, K. S., **1**, 10[55]
Lee, K. W., **8**, 14[83], 237[16], 244[16], 253[16], 261[14]
Lee, K. Y., **7**, 854[50], 855[50]
Lee, L. C., **4**, 347[86b]
Lee, L. F., **1**, 116[49], 118[49]; **5**, 438[162]
Lee, L. F. H., **8**, 237[10], 243[10]
Lee, L. G., **8**, 187[31], 195[31], 196[31], 204[31]
Lee, L. T. C., **2**, 111[87], 242[17]
Lee, M., **7**, 483[129]
Lee, M. K., **8**, 452[186]
Lee, M. L., **7**, 415[113]
Lee, M. W., **5**, 552[18]
Lee, N. H., **7**, 318[55], 319[55]
Lee, P., **7**, 30[81]
Lee, P. E., **8**, 102[125]
Lee, P. H., **4**, 163[98]
Lee, P. L., **3**, 736[29], 771[190]
Lee, R. A., **3**, 21[125]; **4**, 30[87,87g]
Lee, R.-S., **4**, 38[110]
Lee, S., **3**, 167[483], 738[34]
Lee, S. D., **2**, 623[159]
Lee, S. F., **7**, 676[62]; **8**, 353[154]
Lee, S.-H., **4**, 24[75]
Lee, S.-J., **3**, 172[514], 173[517]; **4**, 16[51]; **5**, 1094[103], 1115[1], 1116[1]; **6**, 1022[61]; **8**, 14[77]
Lee, S. L., **8**, 242[41]
Lee, S. P., **6**, 438[44], 1050[39]
Lee, S. S., **8**, 544[264,265], 546[304], 561[304]

Lee, S. Y., **5**, 1021[71]; **7**, 552[59]
Lee, T., **5**, 282[23,24]; **8**, 53[130], 66[130]
Lee, T. D., **1**, 392[157], 393[157]
Lee, T. G., **3**, 643[121]
Lee, T. S., **5**, 75[230]; **7**, 821[27]
Lee, T. V., **1**, 248[70]; **2**, 617[142], 635[40], 640[40], 655[134,134a,b]; **3**, 26[161]; **4**, 31[94], 178[63]; **5**, 299[68]; **7**, 166[90], 294[15], 674[37]; **8**, 206[172]
Lee, V., **1**, 844[5e]; **2**, 597[9]; **3**, 898[78]; **5**, 1124[49]
Lee, W. J., **6**, 879[43]; **7**, 633[66]
Lee, W.-K., **7**, 226[72]
Lee, Y. B., **4**, 825[242]
Lee, Y. C., **6**, 635[23], 636[23]
Lee, Y.-S., **3**, 418[23]
Lee, Y. Y., **7**, 230[131]
Leech, R. E., **8**, 754[81]
Leedham, K., **4**, 304[353]
Leeds, J. P., **5**, 1202[56]; **8**, 64[207c], 66[207]
Leeds, M. W., **8**, 373[131]
Leemans, W., **5**, 571[114]
Leeming, S. A., **1**, 243[55]
Leenay, T. L., **7**, 298[37]
Leeper, F. J., **4**, 14[47]
Leermakers, P. A., **4**, 959[31]; **5**, 127[23]
Lees, W., **2**, 456[33], 457[33], 458[33], 459[33], 460[33], 461[33], 462[33], 466[33]
Leese, R. M., **8**, 315[49], 616[105], 619[105], 968[87]
Leeson, P., **4**, 305[369]
Leete, E., **4**, 12[39]
Lefebvre, Y., **2**, 284[54]
Le Fevre, G., **4**, 1096[157]
Lefferts, J. L., **3**, 200[69]
Leffew, R. L. B., **7**, 268[123]
Leffler, J. E., **7**, 671[8]
Lefker, B. A., **2**, 1049[24]; **3**, 42[234]
Lefloch, P., **5**, 444[186], 1098[131], 1103[151]
Le Floc'h, Y., **2**, 1074[147]
Lefort, D., **7**, 727[39]
Lefour, J.-M., **1**, 49[8]; **4**, 70[11], 139[2]; **5**, 417[63], 453[67]; **8**, 541[213], 543[213]
Lefrancier, P., **6**, 52[116]
Leftin, M. H., **2**, 587[146]; **3**, 447[92], 513[208]; **8**, 935[68]
Le Gall, T., **4**, 38[109b]
Legatt, T., **8**, 530[93]
Legault, R., **3**, 770[181]; **4**, 404[243]; **6**, 686[368]; **7**, 503[275]; **8**, 856[181]
Leger, S., **6**, 26[109]
Legford, T. G., **8**, 564[443]
Leggetter, B. E., **8**, 221[89,91,92], 222[91,92], 229[138], 230[138]
Leginus, J. M., **4**, 1076[41]; **5**, 255[52,53], 260[52,63b], 264[52,53,63b]
Legler, G., **7**, 573[117]
Legler, J., **3**, 482[3]; **7**, 746[82]
Le Goaller, R., **3**, 572[66]; **6**, 80[69], 110[54]
Le Goff, E., **4**, 44[125]; **5**, 531[76]; **6**, 189[189]
LeGoff, M. T., **7**, 878[140]
Le Goff-Hays, O., **5**, 710[52]
Le Goffic, F., **4**, 48[137], 502[120]; **8**, 798[51]
Legrand, L., **6**, 420[20], 424[57,58]
Le Gras, J., **4**, 55[156,157]
Legrel, P., **6**, 67[11]
Legris, C., **3**, 759[130]
Legters, J., **7**, 473[26]
LeGuennec, M., **5**, 444[187]
Legueut, C., **1**, 791[265]
Le Guillanton, G., **8**, 133[27]
Legzdins, P., **8**, 797[39]
Lehky, P., **7**, 79[130]
Lehman, G. K., **8**, 132[14]
Lehmann, B., **5**, 496[221]
Lehmann, C. H., **3**, 829[56]
Lehmann, J., **8**, 382[4], 383[4], 384[4], 388[4], 396[4]
Lehmann, R., **3**, 194[7]; **4**, 869[22]
Lehmeier, T., **2**, 13[53], 14[53]
Lehmkuhl, H., **1**, 180[30], 214[23], 313[115]; **2**, 5[19], 6[19]; **4**, 868[12], 874[46-55], 875[47], 880[87-90], 883[90], 884[90], 887[12,122,124], 888[133]; **5**, 30[2,2b,c,f], 31[4]; **8**, 696[124], 734[2], 735[10], 737[2], 753[2], 754[96-98], 756[155]
Lehn, J.-M., **1**, 49[8], 631[59], 632[59]; **3**, 557[39]; **5**, 468[127]; **6**, 133[4]; **8**, 3[21], 89[43], 318[59], 322[59]
Lehn, W. L., **7**, 661[48]
Lehnert, W., **2**, 343[14], 351[78], 354[112,113], 357[14], 359[162], 363[78], 364[162]
Lehninger, A. L., **2**, 456[72], 457[72], 459[72]
Lehong, N., **6**, 46[72], 47[72]
Lehr, F., **2**, 63[23], 321[12], 324[12], 326[12], 329[12,48], 335[48], 336[48], 337[73]; **4**, 78[52a], 104[136c]; **6**, 107[24], 911[16]; **8**, 363[1], 374[1]
Lehr, P., **3**, 53[274]; **6**, 531[436]
Lehr, R. E., **4**, 1073[23]; **5**, 71[129], 87[41], 92[41], 109[41], 703[15,19], 705[19,23], 754[59], 758[84], 760[84]; **6**, 254[160]
Lehrer, M., **3**, 692[138]
Lehrfeld, J., **6**, 650[129]
Lei, X., **8**, 677[59], 685[59]
Leibfritz, D., **6**, 531[455]
Leibner, J. E., **4**, 738[95]
Leiby, R. W., **6**, 533[484]; **8**, 60[186], 66[186]
Leicht, C. L., **3**, 857[91]; **6**, 1042[6]
Leigh, P. H., **6**, 450[118]
Leight, R. S., **5**, 686[44]
Leighton, J. P., **3**, 522[12]
Leighton, R. S., **2**, 355[119]
Leighton, V. L., **6**, 964[81]
Leikauf, U., **2**, 636[54], 637[54], 640[54]
Leimgruber, W., **5**, 714[75a]; **8**, 368[71]
Leimner, J., **3**, 584[132]
Lein, B. I., **5**, 1146[109]
Leininger, R., **7**, 810[89]
Leipprand, H., **6**, 275[108,113]
Leiserowitz, L., **7**, 40[7]
Leising, M., **2**, 334[67]; **4**, 729[60], 730[60], 740[60,116]
Leismann, H., **5**, 185[165]; **6**, 558[851]
Leistner, S., **6**, 424[60]
Leisung, M., **7**, 883[175]
Leitch, L. C., **6**, 912[20]
Leitereg, T. J., **1**, 109[12], 110[12], 134[12], 678[212]
Leiterig, T., **5**, 170[111]
Leitich, H., **5**, 76[239]
Leitich, J., **2**, 352[83], 371[261]; **3**, 814[67]; **5**, 66[78]
Leitner, P., **7**, 774[319]
Leitner, W., **8**, 84[16], 461[260]
Leitz, C., **5**, 404[18]; **6**, 744[77]
Leitz, H. F., **4**, 12[43], 13[43a,b], 113[164]
Leland, D. L., **4**, 191[110]
Lelandais, D., **3**, 639[73], 648[180], 651[73b]
Lelandais, P., **7**, 14[142]
Leliveld, C. G., **4**, 45[127]; **5**, 687[58,58b], 688[58b]
Lelli, M., **4**, 86[78b]
Lellouche, J. P., **3**, 416[17], 417[17]; **6**, 690[401], 692[401]
LeMahieu, R., **5**, 124[6], 125[6], 128[6]; **7**, 54[7]
Lemaire, M., **3**, 229[224]; **5**, 527[64], 530[64]; **7**, 876[125]; **8**, 84[12]

Lemal, D. M., **3**, 158[435], 173[435]; **5**, 741[153]; **7**, 656[14]; **8**, 605[16]
Le Marechal, J. F., **8**, 737[26]
Le Maux, P., **8**, 451[180], 462[265], 535[166]
Lemay, G., **5**, 86[28], 578[153,156,157]
Lemberton, J.-L., **6**, 563[897]
Le Men, J., **6**, 920[43]; **7**, 222[36]; **8**, 618[125]
Le Men-Oliver, L., **8**, 618[125]
Le Merrer, Y., **1**, 770[185]
Lemieux, R. U., **6**, 23[89], 40[39], 41[42], 42[46], 43[47], 48[86], 54[132], 60[144], 633[9], 643[76], 646[99], 648[123]; **7**, 535[48], 564[93], 565[93], 568[93], 586[164], 710[55], 711[57]
Lemin, A. J., **7**, 253[17]
Lemke, R., **2**, 345[27], 387[27]
Lemmen, P., **2**, 1090[72,73], 1102[73], 1103[73]; **3**, 587[149]
Lemmen, T. H., **8**, 550[329]
Lemmens, B., **6**, 535[525,526]
Lemmens, J., **7**, 742[58]
Lemmens, J. M., **6**, 249[142]
Lemmer, D., **5**, 730[128]
Lempers, E. L. M., **5**, 649[22], 650[22]
Lenain, V., **3**, 509[177]
LeNard, G., **1**, 474[101]
Lendel, V. G., **4**, 342[67]
Lenfant, M., **3**, 30[175]
Lenfers, J. B., **1**, 566[211]
Leng, J. L., **4**, 442[184]
Lenhard, R. H., **6**, 959[36]
Leniewski, A., **8**, 346[126]
Lenkinski, R. E., **1**, 294[42]
Lennartz, H.-W., **5**, 64[29], 911[96], 912[96]
Lennon, M., **8**, 336[84], 339[84]
Lennon, P., **4**, 562[36], 576[13,14]; **5**, 272[4,5], 273[4], 275[4]
Lennox, J., **7**, 778[405]
le Noble, W. J., **1**, 3[23]; **5**, 77[265,267], 436[155], 453[63], 454[63], 458[63], 619[10], 620[10]; **8**, 5[31], 7[36], 597[94], 606[26]
Lenoir, D., **2**, 107[59]; **3**, 583[128], 584[131], 587[142,149]; **5**, 65[69,70]
LeNouen, D., **5**, 417[65]
Lenox, R. S., **8**, 872[8]
Lentsch, S. E., **7**, 802[50]
Lentz, C. M., **2**, 762[52]; **3**, 6[27], 8[41], 9[41], 15[41], 17[27,90], 20[41]; **4**, 178[61], 241[59], 245[87], 255[59], 256[209], 260[59], 262[59]
Lenz, B. G., **5**, 441[176,176d]; **6**, 538[574,575]
Lenz, G., **5**, 123[1], 126[1], 687[65]; **7**, 831[65]
Lenz, G. R., **5**, 723[107]
Lenz, R. W., **5**, 86[27]
Lenz, W., **1**, 144[40]; **5**, 185[159,160]
Lenznoff, C. C., **7**, 453[66]
Leo, A., **7**, 236[19], 252[9]
Leobardo, C. R., **7**, 462[120]
León, E. I., **4**, 342[61], 375[98a,b], 388[98,98a,b], 408[98b], 409[98a]; **7**, 495[210]
Leon, P., **2**, 655[137], 656[137]
León, V., **8**, 446[74], 452[74], 457[74]
Leonard, D., **3**, 1032[69]; **4**, 941[82,83]
Leonard, J., **7**, 363[36]
Leonard, J. A., **3**, 807[27], 813[61]
Leonard, N. J., **1**, 836[143], 858[62]; **3**, 380[11]; **5**, 186[168], 1151[131]; **6**, 36[30], 255[167], 276[121], 277[121,128], 288[128], 533[477,478,493], 553[722,723], 554[721,722,723]; **7**, 194[5], 221[26,27,29], 764[130]; **8**, 48[109], 66[109], 321[104-107], 336[75,86], 341[75,101], 926[17,18], 927[18]
Leonard, R., **6**, 545[638]
Leonard, W. R., **4**, 257[229], 791[55]
Leonard-Coppens, A. M., **1**, 694[236], 700[236], 705[236], 712[236]
Leonczynski, J., **4**, 1014[192]
Leone, A., **7**, 876[121]
Leone, R., **7**, 26[47]
Leone-Bay, A., **2**, 1037[100]; **4**, 446[211]; **5**, 768[129], 779[129]
Leong, A. Y. W., **4**, 111[152a], 113[168]; **7**, 229[118]
Leong, V. S., **4**, 696[7], 712[7b]
Leong, W., **1**, 438[160]; **4**, 901[186]; **6**, 829[3]
Leonhardt, D., **6**, 441[85]
Leoni, M. A., **7**, 882[170]
Leoni, P., **4**, 170[14]
Leont'eva, L. M., **4**, 145[25]
Leopold, E. J., **3**, 349[40], 351[40], 363[83], 369[40]; **7**, 379[98]
LePage, T. J., **1**, 290[21], 321[21], 322[21]
Le Perchec, P., **5**, 2[141], 20[141], 21[141], 65[58], 790[35]
Lepeska, B., **8**, 724[153]
LePetit, J., **8**, 205[165]
Leplawy, M. T., **6**, 276[117]
Lepley, A. R., **1**, 476[113]; **3**, 65[5]; **4**, 500[106], 1015[194]; **5**, 794[46]; **6**, 1016[27]
Leppard, D. G., **4**, 82[61], 519[22]
Lepper, H., **7**, 706[24]
Leppert, E., **6**, 809[71]
Leppkes, R., **5**, 493[212]
Lequan, M., **2**, 575[63]
LeQuesne, P. L., **5**, 3[22]
Le Quesne, P. W., **2**, 1090[73], 1102[73], 1103[73]; **4**, 1058[152], 1059[152]; **8**, 242[41]
Lera, A. R., **4**, 505[140]
Lerche, H., **4**, 124[214e]; **6**, 937[119], 939[119]
Lerchen, H.-G., **6**, 3[16], 638[43], 659[43]
Leriverend, M.-L., **3**, 765[155]; **5**, 776[182]
Leriverend, P., **3**, 765[155]; **5**, 796[55]
Lerman, O., **4**, 347[98]
Lermontov, S. A., **7**, 4[19], 8[19]
Lerner, L. M., **8**, 269[88,90]
Lerner, R. A., **8**, 206[168]
Leroi, G. E., **4**, 483[9], 484[20], 485[25]
Lerouge, P., **1**, 656[156], 658[156], 676[156]
Leroux, J., **6**, 49[99]
Leroux, Y., **3**, 104[210], 111[210], 242[9], 257[115]
Leroy, G., **4**, 953[9]; **5**, 72[171], 741[153]
Lerstrup, K. A., **6**, 509[249]
Lesage, M., **4**, 739[106]; **8**, 801[75]
Lesch, J. S., **6**, 189[185]
Leschinsky, K. L., **6**, 540[587]
Lescosky, L. J., **4**, 486[35], 496[35]
Lescure, P., **1**, 825[46]
Lesimple, P., **3**, 196[26]
Lesko, P. M., **5**, 959[317], 961[317]
Leskovac, V., **8**, 373[127]
Leslie, V. J., **5**, 433[136a], 819[152]
Lesma, G., **2**, 535[37]; **7**, 65[68], 346[12]; **8**, 563[435]
Lespagnol, C., **6**, 499[176]
L'esperance, R. P., **4**, 1089[136]; **5**, 72[177]
Lessard, J., **7**, 499[231,232,234], 503[281]; **8**, 135[39]
Lessi, A., **3**, 541[105]
Lester, D., **7**, 209[91]
Lester, D. J., **1**, 698[242], 699[242]; **3**, 505[158]; **7**, 110[188], 129[72], 132[72,91,98,99], 133[72], 307[16], 310[16], 318[16], 319[16], 322[16], 704[12], 747[106]
Lester, E. W., **4**, 967[56]
Lester, M. G., **2**, 782[19]
Lester, W., **8**, 354[174]
Lestina, G., **4**, 310[436]

LeSuer, W. M., **4**, 317[542,543]; **7**, 666[74]
Lesuisse, D., **2**, 547[109], 550[109]; **5**, 109[218]; **6**, 520[342], 544[342]
Lesur, B., **4**, 117[191]; **6**, 509[257]
Lesur, B. M., **2**, 26[102], 27[102], 60[18,18b]
Leszczweski, D., **3**, 221[129]
Leszczynska, E., **6**, 824[122]
Letendre, L. J., **2**, 553[125]; **3**, 226[195], 264[183], 380[10]; **5**, 20[132]; **6**, 9[43]
Lethbridge, A., **4**, 315[499,500]; **7**, 828[50]
Le Thuillier, G., **6**, 162[187]
Letourneau, F., **6**, 291[218]
Letsinger, R. L., **6**, 603[20], 619[117], 625[162], 658[187]; **7**, 664[66]; **8**, 319[77], 385[43]
Lett, R., **1**, 408[35], 430[35], 513[45,46], 881[119], 883[127], 884[127]; **3**, 147[398], 149[398,407,411,412], 150[407,411], 151[407,411], 152[407], 153[407], 288[63]; **4**, 372[64a]; **6**, 660[202], 734[14]; **7**, 202[48], 381[105], 673[20], 678[73]; **8**, 881[68]
Lett, R. M., **5**, 806[102,103], 1026[86], 1027[89]
Letters, R., **6**, 614[89]
Leu, L.-J., **3**, 328[180]
Leuck, D. J., **4**, 597[177]; **8**, 54[153], 66[153]
Leuenberger, H. G. W., **8**, 205[159,161], 560[405,406]
Leung, C. W. F., **4**, 425[25], 430[91,93]
Leung, K. K., **6**, 1018[40]
Leung, T., **2**, 83[11]
Leung, W. H., **7**, 236[29]
Leung, W. S., **8**, 847[95]
Leung, W. Y., **6**, 765[18]
Leung-Toung, R., **1**, 418[74]
Leuros, J. Y., **4**, 615[386], 619[386], 622[386], 625[386]
Leusink, A. J., **8**, 547[316], 548[319,320]
Leutenegger, U., **4**, 231[284], 1039[64]; **6**, 531[444]; **8**, 462[263]
Leutert, T., **6**, 487[73], 489[73]
Leuthardt, F., **2**, 456[64]
Leutzow, A., **7**, 341[51]
Lev, I. J., **4**, 1090[142]
Le Van, D., **5**, 577[147]
Levand, O., **7**, 291[2], 655[18]
Levanova, E. P., **4**, 50[142,142g]
Levas, E., **1**, 561[158]; **6**, 238[75]
Levason, W., **3**, 213[51]
Levchenko, E. S., **4**, 387[158]; **5**, 425[101-103]
Levenberg, P. A., **3**, 24[149], 25[149]
Levene, P. A., **2**, 800[14]; **6**, 36[31]
Levene, R., **4**, 186[91], 245[82], 248[82], 262[82]; **8**, 936[73]
Levenson, T., **4**, 120[202]
Lever, J. G., **4**, 372[63]; **6**, 905[143]
Lever, J. R., **3**, 88[139], 89[139], 91[139], 211[33]; **4**, 170[17]
Lever, O. W., Jr., **1**, 542[12], 544[12], 547[12], 563[12], 567[12], 630[24]; **2**, 65[28], 596[1]; **3**, 86[29], 121[29], 124[29], 125[29], 127[29], 252[81,82], 257[81,82]; **4**, 11[35], 12[35c], 113[162]; **7**, 168[105]
Leveson, L. L., **3**, 891[41a]
Levesque, G., **4**, 50[142,142p]; **6**, 454[145], 455[150], 456[156,157], 545[632]
Levey, S., **3**, 822[5], 834[5]
Levi, A., **8**, 152[175,176], 155[176]
Levi, B. A., **1**, 487[4], 488[4]
Levi, E. J., **5**, 724[110]
Levif, G., **3**, 47[256]
Levin, A. I., **3**, 639[83]
Levin, D., **1**, 781[231]; **3**, 201[76]
Levin, D. Z., **3**, 305[72,75a]
Levin, J., **1**, 92[69], 103[97], 400[10], 890[144], 898[144]; **2**, 1041[107]; **6**, 735[19], 740[19]; **8**, 53[133], 66[133]
Levin, J. I., **5**, 451[20], 494[214]; **6**, 745[78], 756[78]
Levin, L. N., **7**, 763[93], 766[179]
Levin, R. H., **4**, 483[4], 484[4,21], 495[4]; **5**, 681[25,27]; **8**, 242[43], 293[49-51]
Levina, I. S., **4**, 145[29b]
Levinan, R. H., **6**, 776[55]
Levine, A. M., **3**, 735[20]
Levine, B. H., **5**, 663[37], 666[37]; **6**, 1045[30]
Levine, I. E., **6**, 690[393]
Levine, L., **2**, 403[32]
Levine, P., **8**, 428[52]
Levine, R., **1**, 411[45]; **2**, 182[4]; **3**, 261[147,150]; **4**, 10[31,31b], 27[84], 29[84b], 31[92,92a], 493[79]; **8**, 452[190]
Levine, S., **6**, 60[144]
Levine, S. G., **1**, 884[129]; **5**, 806[107], 808[107b], 809[107b], 1027[88], 1028[88]; **6**, 685[360]
Levine-Pinto, H., **8**, 460[254]
Levins, P. L., **8**, 587[40]
Levinson, A. S., **7**, 30[83]
Levinson, M., **5**, 436[155]
Levinson, M. I., **6**, 420[13], 428[13]
Levisalles, J., **1**, 107[4]; **2**, 523[79], 524[79]; **3**, 13[66], 21[66,130], 22[130], 216[69], 419[45], 420[45], 738[36], 822[11], 834[11,74,75]; **4**, 23[70], 176[48], 543[122]; **7**, 533[32]
Levitan, S. R., **8**, 170[92]
Levitin, G., **4**, 841[46]
Levitt, L. S., **7**, 769[213]
Levitz, M., **5**, 752[46]
Levitz, R., **1**, 542[8], 543[8], 544[8], 547[8], 548[8], 550[8], 552[8], 553[8], 555[8,8b], 556[8], 557[8], 560[8]
Levorse, A. T., **4**, 398[217], 399[217a]; **5**, 432[133]; **7**, 503[268]
Levy, A. B., **1**, 473[79]; **2**, 111[83]; **3**, 250[68], 262[158]; **6**, 78[60]; **7**, 474[48], 476[48], 501[248], 607[163,164,165]; **8**, 386[51], 704[6], 705[6], 706[6]
Levy, A. L., **8**, 639[21]
Levy, E. C., **3**, 693[144]
Levy, J., **2**, 1017[33]; **4**, 40[115]; **5**, 528[69]; **6**, 735[20], 739[20], 755[120]; **8**, 618[125]
Levy, L. A., **6**, 288[185]; **7**, 502[256]
Levy, M., **8**, 532[130]
Levy, P. F., **3**, 643[116]
Levy, W. J., **7**, 833[74]
Levy-Appert-Colin, M. C., **8**, 618[125]
Lew, G., **3**, 489[57], 495[57], 523[23], 799[101]; **4**, 164[99,99c]
Lew, W., **1**, 568[229]
Lewandowski, S., **4**, 313[468]
Lewars, E. G., **3**, 892[48]; **5**, 571[118]; **7**, 358[1], 384[1]
Lewellyn, M., **3**, 627[44]
Lewellyn, M. E., **6**, 687[376], 985[60]
Lewin, A. H., **3**, 499[114]
Lewin, N., **2**, 842[189]
Lewis, A., **6**, 129[165]
Lewis, A. J., **3**, 742[55]
Lewis, B., **4**, 932[62]
Lewis, B. B., **7**, 236[14,15]
Lewis, C. C., **2**, 965[65]
Lewis, C. P., **4**, 497[97]
Lewis, D. E., **4**, 195[125]
Lewis, D. K., **5**, 64[31]
Lewis, D. O., **4**, 6[23]
Lewis, D. P., **8**, 532[130]
Lewis, D. W., **2**, 684[65]
Lewis, E. S., **5**, 911[94], 948[269]; **6**, 204[15]
Lewis, F. D., **1**, 305[89], 311[89]; **5**, 71[157], 650[25]; **7**, 21[1], 877[130], 881[157,161]

Lewis, G. E., **8**, 364[21]
Lewis, G. J., **6**, 36[17]
Lewis, H. B., **3**, 507[173]
Lewis, H. H., **3**, 888[16]
Lewis, J., **4**, 115[180a], 665[8,9], 670[8], 673[31], 674[8], 688[9], 691[76], 706[36-40], 707[40,41]; **6**, 690[398,399,403], 691[399], 692[398,399,403]; **8**, 524[13], 827[74]
Lewis, J. A., **4**, 303[342], 390[175b]; **8**, 854[152], 856[152]
Lewis, J. B., **3**, 771[187]
Lewis, J. J., **3**, 18[93]
Lewis, J. R., **3**, 660[11], 661[25], 688[114], 690[114,124], 818[95]
Lewis, J. W., **8**, 705[10], 726[10], 915[96], 939[94,95]
Lewis, K. E., **5**, 703[16a], 709[16a]
Lewis, K. K., **5**, 71[126]
Lewis, L. N., **7**, 645[22]; **8**, 446[64]
Lewis, M. C., **3**, 635[34], 638[34]
Lewis, M. D., **1**, 182[45], 183[45]; **2**, 29[106], 30[106a], 578[81]; **7**, 369[62], 418[127]
Lewis, N., **7**, 645[22]
Lewis, N. G., **8**, 821[48]
Lewis, N. J., **6**, 644[89]; **7**, 330[9]
Lewis, P. H., **7**, 686[98]
Lewis, R. A., **5**, 441[176,176f], 575[133]; **8**, 411[103]
Lewis, R. L., **6**, 913[24]
Lewis, R. N., **6**, 227[29]
Lewis, R. N. A. H., **8**, 353[160]
Lewis, R. T., **3**, 217[95,95a]; **5**, 290[42], 736[145], 737[145], 1011[48]
Lewis, S. C., **5**, 864[258]
Lewis, S. N., **7**, 671[3], 674[3]
Lewis, T. W., **6**, 569[939]
Lewis, W., **8**, 754[79], 757[79]
Lewite, A., **8**, 187[35]
Lewton, D. A., **6**, 1017[38], 1024[38,75], 1025[79]
Lex, J., **3**, 593[179]; **4**, 1007[129]; **7**, 725[33]
Lex, L., **7**, 566[99]
Ley, S. V., **1**, 752[96], 780[228], 787[253], 800[299], 801[303]; **2**, 185[29], 657[162]; **3**, 174[527,527a,b], 175[527a-c], 342[11], 351[44], 352[44], 380[10], 1036[80]; **4**, 27[81], 254[188], 261[188], 331[13], 379[115], 380[115g], 381[126b], 382[126], 383[126], 390[168,175a], 391[115g,176], 510[174], 709[45], 710[45,51,52]; **5**, 21[150], 22[150], 524[55], 534[92]; **6**, 443[91,93], 474[84-86], 994[98], 995[103]; **7**, 53[1], 63[1], 110[188], 129[72], 132[72,91,92,94,95,97-99], 133[72,92], 134[92], 307[14], 311[30,31], 312[30], 352[34], 363[34], 404[67], 523[47,48], 524[55], 747[106], 761[59], 776[59,362]; **8**, 410[95], 847[98,100a,b,101], 849[98,113]
Leyba, C., **2**, 760[47]
Leyendecker, F., **2**, 120[180]; **4**, 189[103], 229[225,226,227,228,229], 240[42]; **5**, 21[146,147]
Leygue, N., **6**, 441[79]
Leznoff, C. C., **5**, 720[97,97b,98]; **6**, 726[181,182]
Lezzi, A., **3**, 527[45], 539[45], 541[45]
Lhim, D. C., **7**, 278[157]
Lhomme, J., **5**, 585[202]
Lhommet, G., **2**, 356[131]; **6**, 501[206]; **8**, 35[65], 47[65], 66[65]
L'Honore, A., **2**, 6[26]
Lhoste, J. M., **1**, 471[70]; **2**, 379[294]
Lhoste, P., **6**, 91[127], 237[63]
Li, C.-S., **4**, 805[143]; **5**, 252[41]; **8**, 190[76]
Li, G. S., **8**, 238[24], 251[106]
Li, H., **5**, 436[155]
Li, H.-t., **6**, 471[63]
Li, H.-Y., **3**, 999[51]
Li, J.-S., **1**, 623[84]; **2**, 553[132], 554[132], 567[26]
Li, L., **5**, 839[77]
Li, L. Q., **5**, 909[99], 958[99]
Li, M. K., **8**, 840[33], 966[76]
Li, M. Y., **8**, 850[121]
Li, N.-H., **8**, 372[118]
Li, P. K., **2**, 363[196]
Li, P. T. J., **1**, 212[6], 213[6]; **4**, 969[64]
Li, Q., **7**, 355[38]
Li, Q.-L., **6**, 677[320]
Li, T., **3**, 369[117], 372[117]; **7**, 397[28]; **8**, 430[57]
Li, T.-T., **4**, 14[47,47h,i], 24[74]; **5**, 342[61c], 828[7], 839[7], 882[13], 888[13], 891[37], 892[13,37], 893[13]; **7**, 362[27]
Li, W., **5**, 529[70]
Li, W. S., **6**, 8[39], 206[45]; **7**, 396[25]
Li, X., **4**, 1074[29]
Li, Y., **1**, 41[198], 462[17], 463[17]; **3**, 596[194]; **5**, 79[292]
Li, Y.-C., **4**, 38[110]
Liak, T. J., **2**, 911[70]; **5**, 96[110]; **6**, 651[136]
Liang, C. D., **1**, 359[21], 383[21], 384[21]
Liang, D. W. M., **5**, 841[87]
Liang, G., **3**, 587[145]
Liang, W. C., **3**, 20[118], 31[118]
Liang, Y., **8**, 386[52]
Liang Chen, Y. P., **3**, 24[148]
Lian-niang, L., **8**, 336[79], 337[79]
Liao, C.-C., **5**, 225[89], 231[133]; **7**, 97[92], 367[56]
Liao, L.-F., **1**, 512[35], 565[207]; **2**, 6[34], 21[34b]
Liao, Q., **1**, 825[55]
Liao, W.-P., **1**, 272[65]; **8**, 109[7], 110[7], 112[7,27], 113[27]
Liao, Z.-K., **2**, 1060[87], 1062[97]
Liard, J.-L., **8**, 254[126]
Liaw, B. R., **2**, 583[115]
Libbey, W. J., **4**, 1099[185]
Liberatore, F., **5**, 64[52]; **8**, 31[45], 36[45], 66[45], 585[26], 587[30,32]
Liberek, B., **6**, 665[230], 667[230]
Libertini, E., **3**, 583[118], 587[148]
Libit, L., **3**, 249[64]
Libman, J., **3**, 383[44]; **5**, 636[94-96], 646[10], 647[10c], 654[27], 817[146]; **6**, 441[79]; **7**, 873[100]
Libman, N. M., **6**, 268[63]
Licandro, E., **4**, 522[54], 523[57,58]; **5**, 1079[51]; **6**, 114[79], 178[121]
Licchelli, M., **8**, 772[61]
Lichtenberg, D., **8**, 707[18]
Lichtenberg, D. W., **5**, 277[13]
Lichtenberg, F., **2**, 10[42], 21[42], 22[42b]
Lichtenthaler, F. W., **2**, 321[7], 323[7], 326[7], 327[7], 328[7], 329[7]; **6**, 20[71], 611[62]; **7**, 712[61]
Lichtenwalter, G. D., **3**, 565[17]
Licini, G., **4**, 252[166], 260[166]; **7**, 425[147a], 762[69], 777[69b,376,380], 778[69]
Lico, I. M., **6**, 646[100a]; **8**, 384[40]
Lida, H., **4**, 503[129]
Liddell, R., **2**, 435[64]
Lidert, Z., **8**, 510[93]
Lidor, R., **2**, 486[41]
Lidy, W., **1**, 328[25], 343[120], 345[120], 390[146]; **6**, 237[58]
Lieb, R. I., **2**, 958[24]
Liebe, J., **6**, 979[27]
Lieben, A., **2**, 139[26]
Lieber, E., **6**, 251[150]
Lieberknecht, A., **6**, 637[32,32c]; **8**, 535[166]
Liebermann, C., **8**, 949[156]
Liebermann, S. V., **2**, 954[7]
Liebeskind, L. S., **1**, 404[22]; **2**, 125[218], 127[226,227,230,231], 271[78], 315[40,41,43], 316[41,43], 933[141], 934[141,142], 1061[94,96]; **3**, 47[257], 483[18], 500[18], 672[64], 674[71]; **4**, 16[51], 82[62c,d], 217[127,-128], 231[127,-128], 411[265a,b], 486[35], 496[35], 976[100], 980[100k]; **5**, 689[75,77], 690[77a,81], 692[93],

693[93], 696[93], 733[136,136b], 734[136b], 1065[3], 1066[3], 1068[14], 1093[3], 1094[3], 1096[3], 1099[3], 1101[3], 1102[3], 1112[3], 1135[51], 1202[56], 1203[57,58]; **6**, 1022[61]; **8**, 839[22]
Liebig, J., **6**, 233[42]
Liebscher, J., **2**, 359[161], 779[5], 782[25], 785[43]; **6**, 425[68], 430[68], 509[275], 517[322], 543[613,620], 552[693], 553[760], 554[760]
Lied, T., **1**, 805[311]; **6**, 464[29], 998[121]
Liedhegener, A., **3**, 889[26], 890[29], 909[152]; **4**, 1033[29]; **6**, 124[143], 125[143]
Lieke, W., **2**, 1083[1]; **6**, 242[93]
Lie Ken Jie, M. S. F., **4**, 310[422]
Lien, A. P., **3**, 297[20], 298[20]
Lien, M. H., **1**, 487[3], 488[3]
Lienig, D., **7**, 770[256a], 771[256]
Liepa, A. J., **3**, 575[83], 670[57,57a,59], 671[62], 816[82]; **4**, 670[15,21]
Liepins, E., **7**, 477[81]; **8**, 587[31]
Liepin'sh, E. E., **4**, 48[140]
Liepmann, H., **6**, 525[376]
Lier, E. F., **7**, 92[41,41a], 94[41]
Liesching, D., **6**, 722[142]
Liese, T., **4**, 18[59]; **5**, 1049[33], 1052[33]
Liesenfelt, H., **8**, 445[16]
Lietje, S., **1**, 788[257]; **3**, 201[83]
Lietz, H., **8**, 354[164]
Lietz, H. F., **4**, 109[149]
Liew, K. Y., **7**, 14[131]
Liew, W.-F., **4**, 817[206]; **6**, 14[56], 16[56], 1067[102]; **7**, 676[61]
Liggero, S. H., **1**, 859[64]
Light, J. P., **4**, 746[146]
Light, J. R. C., **4**, 877[70]; **8**, 754[111], 755[111]
Light, K. K., **6**, 209[70]; **8**, 604[5]
Light, L., **5**, 404[19], 405[19], 431[121]
Lightner, D. A., **7**, 262[81]
Ligon, R. C., **3**, 353[50], 354[50]
Liguori, A., **8**, 965[62]
Li Hsu, Y.-F., **5**, 347[72,72d,e]
Likholobov, V. A., **8**, 608[47]
Li-Kwon-Ken, M. C., **8**, 92[68]
Lilie, W., **5**, 720[97,97b]
Lilienfeld, W. M., **4**, 89[84a], 98[109g]
Lilje, K. C., **4**, 432[111]
Liljefors, S., **2**, 783[33]
Lilley, G. L., **3**, 396[106], 397[106]
Lilley, K. J., **3**, 396[106], 397[106]
Lillie, T. S., **7**, 414[119], 834[78]
Lilly, M. D., **8**, 185[14], 206[14]
Lillya, C. P., **2**, 721[90], 722[93,94]; **4**, 697[9-11], 698[15,16,18], 699[18], 701[18,24]; **5**, 741[155]
Lim, B. B., **6**, 533[483,487]
Lim, C.-E., **3**, 742[56]
Lim, D., **1**, 10[48], 41[194]
Lim, D. Y., **7**, 43[39]
Lim, G. C., **6**, 953[9]
Lim, H., **3**, 588[160], 610[160]; **5**, 886[23,24]
Lim, M.-I., **6**, 553[795], 554[759,795]
Lim, S. T., **1**, 563[182]; **8**, 261[16]
Lim, T. F. O., **2**, 584[127]
Lim, Y. Y., **6**, 295[246]
Liman, U., **3**, 866[29]
Limasset, J.-C., **4**, 972[82]
Limborg, F., **7**, 802[46], 808[77]
Limburg, W. W., **1**, 436[147]; **5**, 843[126]
Limdell, S. D., **2**, 578[85]
Limosin, D., **7**, 809[83]
Lin, A., **6**, 624[142]
Lin, C., **5**, 791[38]
Lin, C. C., **1**, 557[126]; **4**, 790[39]
Lin, C. H., **7**, 254[28]
Lin, C. J., **6**, 543[605]
Lin, C. M., **4**, 1018[218]
Lin, C. T., **8**, 795[18]
Lin, D. C. T., **4**, 285[158]
Lin, D. K., **6**, 938[122]
Lin, H. C., **2**, 709[9], 744[90]; **7**, 10[74]
Lin, H.-N., **5**, 571[117]
Lin, H.-S., **1**, 240[42]; **3**, 56[286]; **5**, 798[69], 817[148,149]; **7**, 163[76], 164[76], 367[56]
Lin, I., **2**, 138[24]
Lin, J., **1**, 733[13]; **2**, 186[35]; **7**, 549[46]
Lin, J. J., **1**, 117[54], 333[59]; **3**, 213[42,42b]; **8**, 483[52,53,58], 484[52,53,58], 485[52,53,58], 543[240], 545[283], 549[283,328], 696[120], 801[73], 803[88,89]
Lin, J. K., **4**, 581[27]
Lin, K., **1**, 883[125]; **3**, 178[545], 181[545]
Lin, K.-C., **5**, 810[125]
Lin, L. C., **1**, 473[79]; **2**, 583[115]
Lin, L.-H., **4**, 315[514]
Lin, L.-J., **3**, 691[133]; **4**, 1002[55,62]
Lin, L. P., **4**, 1015[197]
Lin, L.-S., **7**, 35[103]
Lin, M.-H., **8**, 5[31], 7[36]
Lin, M.-S., **4**, 45[127,127a]
Lin, M.-T., **3**, 196[28]; **8**, 970[100]
Lin, N.-H., **2**, 1030[85]; **5**, 862[252]; **6**, 742[70]; **7**, 415[112]; **8**, 540[195]
Lin, P., **1**, 791[268]; **2**, 189[46], 209[46], 482[36], 484[36], 811[71], 824[71]; **6**, 186[170], 901[120]
Lin, P. N., **6**, 124[145]
Lin, R., **8**, 382[6], 406[45]
Lin, S. T., **3**, 693[146], 694[150]; **8**, 435[70,71], 437[76]
Lin, W. S., **3**, 639[76]
Lin, Y., **3**, 767[163]; **6**, 487[42], 489[42], 538[559], 543[42], 550[42], 554[42,756,757,758,775]; **7**, 6[32]; **8**, 554[366]
Lin, Y.-i., **7**, 739[39]
Lin, Y. J., **8**, 540[201], 836[8], 842[8]
Lin, Y.-T., **4**, 378[110], 384[110]; **5**, 515[13,13a], 516[13a], 517[13a,b], 518[13a,b]; **7**, 633[66]
Linares, A., **2**, 828[133]; **8**, 32[53], 66[53]
Lincoln, F. H., **7**, 86[16a]
Lind, F. K., **8**, 366[37]
Lind, H., **5**, 426[105], 428[105], 429[105]; **7**, 765[167]; **8**, 661[111]
Lind, J., **3**, 1038[92]
Linday, L. B., **3**, 88[137], 95[137], 165[137], 167[137]
Lindbeck, A., **6**, 237[64], 243[64]
Lindbeck, A. C., **1**, 480[151]
Lindberg, A. A., **6**, 34[11], 51[11]
Lindberg, B., **6**, 23[89]
Lindberg, J. G., **1**, 3[21], 45[21e]
Lindberg, P., **7**, 831[64]
Lindberg, T., **1**, 56[28]; **5**, 743[164], 744[164]; **6**, 648[119]
Lindberg, W., **2**, 1099[115]
Linde, R. G., II, **1**, 529[124]
Lindell, S. D., **1**, 348[139]; **3**, 226[206], 369[125], 372[125], 373[128]
Linden, H., **8**, 270[97]
Lindenmann, A., **7**, 446[68], 704[10]
Linder, H. J., **4**, 740[116]
Linder, J., **2**, 780[10]
Linder, S.-M., **6**, 707[44], 1067[108,109]
Linderman, R. J., **1**, 133[111], 431[133], 791[269]; **2**, 120[186], 486[30], 488[30], 651[115], 833[150]; **3**, 196[27], 217[87], 263[176]; **4**, 153[61a-d], 194[123], 195[124], 248[108,113], 252[156,157],

256[113], 260[108], 261[113]; **5**, 757[79], 766[79]; **6**, 124[146], 654[155]
Lindert, A., **1**, 642[118,119]; **2**, 182[8], 282[41]; **3**, 7[31], 8[31]; **5**, 828[10], 841[102]; **7**, 120[18], 121[18]
Lindfors, K. R., **5**, 417[61], 418[61]
Lindh, I., **6**, 620[133]
Lindig, C., **7**, 198[28]
Lindlar, H., **2**, 612[105]; **6**, 836[55]
Lindley, J., **3**, 522[15]; **7**, 94[55]
Lindley, P. F., **5**, 478[163]
Lindner, D. L., **2**, 542[83]
Lindner, E., **5**, 1130[1], 1131[13], 1145[105]; **8**, 754[98]
Lindner, H. J., **1**, 37[244,247,248,249,252], 528[120,121], 531[131], 535[147], 536[147]
Lindner, U., **7**, 772[294], 773[294]
Lindow, D. F., **7**, 833[73]
Lindoy, L. F., **4**, 298[292]
Lindsay, D. A., **4**, 723[38], 738[38,101], 747[38]
Lindsay, J. K., **2**, 182[1]
Lindsay, K. L., **4**, 887[123], 888[123]; **8**, 100[114]
Lindsay, R. J., **6**, 65[1]
Lindsay Smith, J. R., **6**, 734[15]
Lindsell, W. E., **1**, 2[8]
Lindsey, R. V., Jr., **3**, 648[171]; **5**, 491[201], 1141[82]; **8**, 447[108,109]
Lindstaedt, J., **6**, 426[79]
Lindstedt, E.-L., **1**, 112[27]; **3**, 213[47]; **4**, 152[56], 255[204]
Lindsten, G., **3**, 499[120]
Lindstom, M. J., **4**, 743[127], 744[127]
Lindström, J.-O., **6**, 824[123]
Lindy, L. B., **3**, 158[444], 161[444], 164[444], 167[444]
Linebarrier, D. L., **2**, 6[33], 35[33]
Linek, E. V., **6**, 511[300]
Lines, R., **3**, 634[24], 639[24], 649[24]; **7**, 801[44]
Ling, L. C., **8**, 604[7]
Lingham, I. N., **2**, 1096[96]
Lingnau, J., **5**, 64[45]
Linhartová, Z., **3**, 896[72]
Link, H., **5**, 809[111]
Link, J. C., **4**, 497[97]
Link, R. W., **6**, 22[81]
Linke, S., **2**, 111[82], 242[15]; **3**, 794[80]; **7**, 24[26], 25[26], 27[75], 477[77]
Linker, T., **4**, 739[109]
Linkies, A., **1**, 123[79], 372[80]
Lin'kova, M. G., **6**, 437[34], 440[34], 443[34]
Linn, C. B., **4**, 270[7,8], 271[24], 272[24]
Linn, C. J., **2**, 362[185]
Linn, D. E., Jr., **8**, 455[205]
Linn, W. J., **4**, 1090[139]; **5**, 71[147], 78[147]
Linn, W. S., **4**, 315[515]
Linnell, S. M., **7**, 883[178]
Linsay, E. C., **7**, 25[43], 26[43]
Linstead, R., **8**, 561[413]
Linstead, R. P., **2**, 348[65], 742[71]; **3**, 351[38], 642[112], 643[117,125,130], 644[112,147,148,148a,150,151,153,166]; **4**, 282[136]; **5**, 752[47]; **7**, 135[102]; **8**, 312[21], 328[8,9,11], 428[52]
Linstrumelle, G., **1**, 832[110]; **2**, 88[32]; **3**, 215[65], 217[91,94], 218[96], 224[161], 243[13], 244[27], 247[45,46], 248[45,55], 251[55], 256[113], 257[114,116], 262[161], 263[161], 264[179,180], 269[55], 416[20], 417[20], 437[27], 438[34], 440[43], 449[98], 464[34,176], 466[34,176,188,192], 485[30], 487[45-47], 491[30,71,72], 527[44], 530[68], 531[85], 533[68], 539[99], 545[99], 896[69], 963[123], 964[123]; **4**, 953[8,8i], 954[8i], 1031[5], 1032[5], 1035[5], 1054[134], 1102[198]; **5**, 847[132], 1001[13]; **6**, 893[78]
Linz, G., **5**, 365[96c]
Lion, C., **1**, 427[112]; **3**, 124[262], 257[121]; **4**, 350[117], 877[66]
Liotard, D., **5**, 72[188]
Liotta, C., **4**, 1001[27]
Liotta, C. L., **3**, 3[17], 7[17]; **5**, 333[45]
Liotta, D., **1**, 571[279], 630[5,17,34], 631[17], 632[17], 633[17], 634[17], 635[76], 636[17], 641[17], 642[111], 644[17,111], 656[17], 658[17], 669[111,183], 670[17,183], 671[111,183], 672[5,17], 699[183], 828[72]; **3**, 86[54], 87[54], 94[54], 95[54], 117[54]; **4**, 38[108,108c], 245[89], 317[558], 339[39], 340[49], 342[39], 376[102], 377[102,124b], 379[115], 380[115h,124,124b], 383[115h], 438[152]; **5**, 1203[58]; **6**, 836[58], 1030[107]; **7**, 97[92], 130[77], 131[83-86], 132[89,94], 273[134], 515[1], 520[24,26], 521[34], 523[1,45], 741[51], 819[20], 822[32], 826[20]; **8**, 849[117], 850[119]
Liotta, R., **8**, 713[72,75], 714[75,79,82,83], 715[84]
Lipczynska-Kochany, E., **6**, 824[120-122]
Lipinski, C. A., **2**, 547[101], 548[101]
Lipinsky, E. S., **5**, 151[2]
Lipisko, B., **2**, 617[143], 907[57]
Lipkowitz, K. B., **5**, 453[69], 455[69,74,77]
Lipman, A. L., Jr., **4**, 519[26]
Lipovich, T. V., **7**, 774[325]
Lipovich, V. G., **3**, 328[172,173,174]; **8**, 454[198]
Lipp, M., **2**, 139[33]
Lipp, P., **5**, 596[27], 597[27]
Lippard, S. J., **1**, 314[129]; **4**, 230[253]; **7**, 421[138], 424[138], 766[187]
Lippert, J. L., **2**, 363[200]
Lippmaa, E., **2**, 346[43]
Lippsmeyer, B. C., **4**, 394[196]
Lipscomb, W. N., **8**, 89[43], 724[169,169d,f]
Lipshutz, B. H., **1**, 107[5,6], 110[5,6,23], 112[24-26], 114[38], 115[39], 122[70,71], 124[80], 131[5], 138[39], 335[66], 343[116], 376[93], 428[121], 429[121], 430[130], 433[227], 457[121], 568[229]; **2**, 119[166], 120[183,184]; **3**, 209[15,17], 211[34a], 212[34b,c], 213[34a,46,46c,48,54], 214[56,57], 216[46,68], 223[46,160], 224[167,168], 250[70,72,73], 251[74,76], 261[76], 264[72,76,186], 265[72], 491[70]; **4**, 148[49], 152[60], 170[3,4], 175[43], 176[3,4,47,50], 177[50,55,56], 178[3,62], 180[62], 184[82,84], 185[84], 196[3], 197[3,4], 244[78], 245[78], 255[78], 256[213,214,215], 386[148c], 387[148], 903[192]; **5**, 330[37], 931[186]; **6**, 4[22], 5[23], 8[36,39], 9[22], 10[22], 648[121], 831[9], 1006[147]; **7**, 180[156], 183[156], 330[13]; **8**, 395[124]
Lipsky, S. D., **8**, 114[58], 308[5], 309[5]
Lipták, A., **6**, 41[43], 51[103], 660[207]; **8**, 224[103,109,110], 225[103,110], 226[112-114], 227[115], 230[140]
Lipton, M., **1**, 92[68], 93[68], 400[10]; **6**, 784[92]
Lipton, M. F., **2**, 506[16], 510[16], 513[16]
Lipton, M. S., **5**, 686[44]
Liptrot, R. E., **1**, 308[95], 314[95]
Liras, S., **5**, 834[55]
Lis, H., **6**, 33[4], 34[4], 40[4]
Lis, L. G., **3**, 24[145]
Lis, R., **3**, 220[124]; **5**, 326[26], 350[77]; **6**, 899[110], 900[110]; **8**, 187[42], 844[73], 932[40]
Lis, T., **1**, 303[81]
Lishchiner, I. I., **3**, 305[72,75a]
Liso, G., **6**, 255[169]; **8**, 31[45], 36[45], 66[45], 657[97]
Lissel, M., **3**, 284[51]; **6**, 834[40]; **8**, 798[53]
Lissi, E. A., **7**, 605[140]
Lister, M. A., **1**, 403[18]; **2**, 113[109], 249[84], 263[56]
Lister, S., **2**, 655[140]
Lister, S. G., **1**, 800[299]; **5**, 534[92]; **6**, 995[103]
Lister-James, J., **4**, 344[77]
Listl, M., **6**, 638[45]
Litchmann, W. M., **4**, 1084[93]
Litinas, K. E., **6**, 173[45], 175[72]

Litle, R. L., **3**, 890[31,34], 901[111]
Litt, M. H., **5**, 114[240]
Littell, R., **8**, 566[450], 618[127], 623[127]
Litterer, W. E., **4**, 171[25]
Little, F. L., **8**, 451[180]
Little, R. D., **1**, 268[54]; **3**, 160[468], 214[56]; **4**, 12[41], 70[7], 153[61e], 194[121], 260[282], 809[164]; **5**, 241[4,5], 242[7,8], 243[4b,7,9-13], 244[13]; **6**, 836[58], 939[136], 942[136]
Little, R. Q., Jr., **4**, 274[57], 282[57]
Little, W. F., **5**, 1144[102], 1146[102]
Little, W. T., **2**, 400[25], 403[33]
Littler, J. S., **7**, 154[22], 530[12], 707[27], 851[19]
Littman, J. B., **2**, 960[34]
Litvin, E. F., **8**, 450[163], 458[221]
Litvinenko, L. M., **6**, 516[320]
Liu, C., **4**, 298[286], 841[42]
Liu, C. F., **4**, 817[202]
Liu, C.-H., **2**, 536[44]
Liu, C.-L., **2**, 536[46]; **3**, 172[514]
Liu, C.-M., **1**, 66[55]
Liu, D., **6**, 84[85]
Liu, G., **7**, 668[81]
Liu, G. K., **6**, 546[652]
Liu, H., **2**, 521[71]; **4**, 261[291]; **8**, 344[121,121b]
Liu, H. F., **7**, 14[131]
Liu, H.-J., **1**, 118[62], 564[190], 849[30], 850[33], 853[33]; **2**, 844[198]; **3**, 221[127], 783[27]; **4**, 337[31]; **5**, 125[20,21], 128[21], 131[46]; **6**, 438[43,44], 1050[39]; **8**, 244[70], 555[371], 928[22]
Liu, H. T., **4**, 820[223]
Liu, J., **1**, 389[140]; **2**, 940[162]; **5**, 100[158]
Liu, J.-C., **5**, 86[21]; **7**, 185[177]
Liu, K.-T., **4**, 273[47,49,52], 314[478]; **6**, 954[16]; **7**, 253[13], 759[12], 765[135], 778[135,396]
Liu, L., **8**, 344[121]
Liu, L. C., **4**, 301[318]; **8**, 856[164]
Liu, M., **4**, 1089[138], 1091[138]
Liu, M. S., **4**, 719[25]
Liu, R. S., **7**, 14[131]
Liu, R. S. H., **1**, 765[165]; **5**, 70[117], 133[51], 514[6], 709[45], 717[92]; **6**, 157[176]
Liu, S., **3**, 1058[40]; **7**, 43[36,47]
Liu, S.-H., **4**, 92[91a]
Liu, S. P., **7**, 40[8], 43[8]
Liu, S.-T., **8**, 865[246]
Liu, T., **2**, 803[33]
Liu, T. L., **2**, 481[20]
Liu, W.-C., **5**, 86[32]
Liu, W. G., **7**, 62[53,53b,c]; **8**, 192[97]
Liu, W.-L., **7**, 265[95], 279[95], 280[95], 841[13]
Liu, W. T., **7**, 842[22]
Liu, Y., **8**, 677[59], 685[59]
Liu, Z.-Y., **7**, 105[150]
Livak, J. E., **2**, 916[82]
Livantsova, L. I., **2**, 726[122]
Liverton, N. J., **4**, 35[98j], 893[157]; **5**, 342[63]
Livinghouse, T., **1**, 622[75], 787[255], 846[17]; **3**, 263[177], 287[61], 362[81]; **4**, 116[189], 257[229], 791[55], 803[131], 1088[120]; **5**, 1183[58]; **6**, 8[35], 229[26], 295[252,253,254], 546[656]
Livingston, D. A., **4**, 27[80]
Livingston, J. R., **6**, 960[54]
Livingston, J. R., Jr., **7**, 764[123]
Livingston, R., **5**, 1070[27], 1073[27]
Livneh, M., **7**, 875[119]; **8**, 856[172]
Liz, R., **4**, 292[224,228]; **5**, 848[141], 850[162]; **6**, 555[810,811,812]; **8**, 856[177,178]
Lizuka, Y., **7**, 171[123]
Lizzi, M. J., **6**, 545[635]
Ljunggren, S. O., **4**, 302[331]
Ljungqvist, A., **5**, 4[34]
Lledós, A., **5**, 72[187]
Llera, J. M., **3**, 848[54], 849[54b]; **5**, 837[70]
Llewellyn, D. R., **3**, 723[11], 725[11], 726[11]
Llinas, J. R., **4**, 55[157], 57[1571]
Llitjos, H., **1**, 547[61]
Llobera, A., **7**, 346[8]; **8**, 393[113]
Llobet, A., **1**, 34[223]
Lloyd, D., **1**, 825[52]; **2**, 784[40]; **6**, 522[345]
Lloyd, D. H., **8**, 368[72], 375[72]
Lloyd, W. G., **7**, 449[4], 451[4], 453[78]
Lloyd-Williams, P., **7**, 345[4]
Lo, K. M., **1**, 506[12]
Lo, S. M., **1**, 451[217]
Loacher, J. R., **3**, 898[75]
Loader, C. E., **1**, 474[84]; **2**, 743[84,87], 780[9]
Loar, M. K., **1**, 440[190], 445[190]
Loban, S. V., **2**, 787[52]
Löbberding, A., **8**, 818[40]
Lobeeva, T. S., **5**, 1174[34]
Löbering, H.-G., **5**, 710[56,56c], 719[56], 742[160], 744[56]
Lobkina, V. V., **3**, 304[68]
Lobo, A. M., **6**, 114[78]
Lochead, A. W., **2**, 555[146], 1064[107]; **5**, 437[159]
Locher, R., **2**, 107[55], 193[65], 385[322]; **4**, 72[25,25b]
Lochert, P., **7**, 805[66]
Lochinger, W., **6**, 636[24]
Lochman, R., **1**, 287[14]
Lochmann, L., **1**, 10[47-49], 41[194]
Lo Cicero, B., **3**, 13[66], 21[66]
LoCierco, J. C., **4**, 922[25]
Lock, C. J. L., **7**, 876[122]
Lock, G., **2**, 399[15]; **8**, 330[26]
Lock, G. A., **2**, 553[131]
Lock, R. L., **3**, 906[144]
Locke, J. M., **7**, 760[30]
Locke, M. J., **5**, 407[26]
Locker, R. H., **8**, 228[131]
Lockhart, L. B., **2**, 770[8]
Lockhart, R. L., **7**, 488[155], 490[155]
Lockhart, T. P., **5**, 736[144]
Lockhoff, O., **6**, 43[48,53,55]
Lodaya, J. S., **4**, 370[46]; **7**, 155[31b,c], 179[31b]
Lodder, G., **5**, 215[1], 216[1], 217[26,27], 218[1], 220[1], 223[1], 224[1], 647[18], 649[18], 652[18], 656[18]
Lodge, E. P., **1**, 49[8], 286[8], 460[1]; **2**, 218[141], 677[55], 678[55]; **8**, 4[24]
Lodi, L., **1**, 192[82]; **2**, 35[130], 36[130], 566[23]; **7**, 549[42]
Loeb, W. E., **4**, 4[15]
Loebach, J. L., **7**, 428[148g]
Loeliger, P., **2**, 875[30]
Loesch, C., **1**, 286[11]
Loeschorn, C. A., **1**, 845[10]
Loev, B., **3**, 297[16], 500[127], 501[127]; **8**, 92[68], 564[443]
Loevenich, J., **7**, 770[251], 773[303]
Loew, L., **8**, 271[107]
Loew, P., **5**, 891[37], 892[37,38a]; **7**, 747[102]
Loewe, L., **2**, 413[12]
Loewe, M. F., **1**, 482[166], 672[207], 700[207]; **3**, 72[39], 75[49], 81[67,68]
Loewe, S., **8**, 140[23]
Loewen, P. C., **8**, 218[72,74], 219[72], 222[95]
Loewenstein, P. L., **5**, 960[320]

Loewenthal, H. J. E., **2**, 844[200]; **6**, 675[298]; **8**, 502[65], 503[65], 505[76], 533[154], 796[29]
Lofgren, C. S., **1**, 568[227], 571[227]
Loftfield, R. B., **3**, 840[14], 845[38]
Logan, C. J., **6**, 638[38], 670[38], 671[38]
Logan, R. T., **3**, 846[45]; **6**, 967[101]
Logani, S. C., **8**, 222[96]
Logemann, E., **8**, 214[41]
Logerman, W., **7**, 100[121]
Loginova, N. A., **6**, 526[402,403]
Logothetis, A. L., **5**, 938[210]
Logue, M. W., **1**, 447[203], 458[203]; **2**, 799[19]; **4**, 426[50]
Logullo, F. M., **4**, 488[51,52], 489[52]; **5**, 380[113c], 383[124]
Logusch, E. W., **1**, 532[136]; **3**, 24[144]
Loh, J.-P., **1**, 506[16]
Loh, K.-L., **4**, 964[45]
Loh, T.-P., **2**, 556[157], 558[158]
Lohberger, S., **2**, 1086[29], 1088[29], 1093[85], 1099[109,109b]
Löher, H. J., **1**, 313[117]; **4**, 152[54], 184[85], 189[85a], 201[12,13], 202[13]
Lohmann, J.-J., **1**, 481[161], 483[168]; **3**, 71[30,31,33]; **4**, 484[18]; **7**, 225[65]
Löhmann, L., **3**, 975[1]
Lohner, W., **6**, 462[16]
Lohray, B. B., **2**, 507[24]; **7**, 431[162], 442[46c]
Lohri, B., **1**, 62[40], 63[42], 153[57]
Lohrmann, R., **6**, 603[19], 611[65]
Lohwasser, H., **3**, 902[117]
Loibner, H., **6**, 206[40], 210[40]
Loim, L. M., **8**, 216[55]
Loim, N. M., **8**, 36[83], 66[83], 216[56], 318[60-63,65-67], 486[59,61], 487[59], 546[306,307,308], 778[84], 813[12]
Loim, N. N., **5**, 938[208]
Loizou, G., **5**, 834[56]
Lok, K. P., **6**, 560[868]; **7**, 316[46,48], 317[46,48], 318[48]; **8**, 188[55], 196[55], 199[55], 201[55]
Löken, B., **6**, 685[357]
Lokensgard, J. P., **3**, 906[145]; **6**, 535[529]
Lökös, M., **2**, 372[271]
Loktev, A. S., **8**, 600[106], 606[25], 625[25,159]
Löliger, P., **2**, 866[9]
Lolkema, L. D. M., **2**, 558[162]
Lolla, E. D., **4**, 1080[71]
Lollar, D., **4**, 413[277]
Lollar Confalone, D., **7**, 701[66]
Lomas, D., **5**, 65[59,60]
Lombard, R., **7**, 446[62]
Lombardo, L., **1**, 749[87], 750[87], 812[87]; **4**, 121[209,209c]; **5**, 162[67], 1124[50,52], 1125[50a,52], 1130[6]; **6**, 124[142], 734[10], 735[10]; **8**, 503[70]
Lombet, A., **8**, 47[126], 66[126]
Lömker, F., **3**, 825[24,24a], 828[24a], 835[24]
Lommes, P., **1**, 266[46]
Lomölder, R., **3**, 642[115]
Lonchambon, G., **6**, 535[525,526]
Loncharich, R. J., **1**, 287[20], 288[20], 289[20]; **4**, 202[20]; **5**, 79[292], 260[70], 263[70], 516[19]
Long, A., **8**, 863[231]
Long, A. G., **8**, 987[23]
Long, A. K., **6**, 632[6]
Long, B. H., **5**, 736[145], 737[145]
Long, C., Jr., **6**, 660[201]
Long, D. J., **8**, 618[122]
Long, G. W., **1**, 744[58]; **5**, 1123[38]
Long, J. K., **4**, 27[79], 120[196]
Long, J. P., **2**, 760[43]; **8**, 53[130], 66[130]
Long, J. R., **1**, 231[8], 251[1], 252[1]
Long, N. R., **2**, 606[66]
Long, R., **4**, 588[53], 600[229]
Long, R. F., **8**, 273[127]
Longeray, R., **7**, 124[41]
Longford, C. P. D., **2**, 558[161]
Long-Mei, Z., **6**, 767[27], 775[27]; **7**, 696[39]
Longone, D. T., **5**, 1007[42]
Longstaff, P. A., **8**, 152[164], 443[3]
Longuet-Higgins, H. C., **4**, 1016[205]; **5**, 647[11], 648[11]
Lonikar, M. S., **4**, 194[123], 195[124]
Lonitz, M., **8**, 353[153]
Lönn, H., **6**, 40[41], 42[41], 46[67], 47[67,79,80]
Lönngren, J., **6**, 23[89], 40[41], 42[41], 647[106]; **7**, 245[75]
Loo, P.-W., **2**, 144[63]
Look, G. C., **1**, 589[20,20a,b], 591[20b], 592[20a]; **5**, 458[71]
Looker, J. H., **7**, 341[50]
Loomis, G., **8**, 842[42d]
Loomis, G. L., **2**, 182[9]
Loontjes, J. A., **3**, 868[41]
Loosli, H. R., **1**, 41[202]; **4**, 257[219]; **5**, 68[97]
Loots, M. J., **1**, 155[67,68], 156[68]; **3**, 483[14]; **4**, 153[63a], 154[64a], 257[223]; **8**, 679[68], 680[68], 683[68], 693[68,110,116], 695[68]
Lopatina, K. I., **6**, 279[137]
Lopatinskaya, Kh. Ya., **6**, 494[132]
Lopatinskii, V. P., **4**, 291[209]
Loper, J. T., **4**, 816[201]; **6**, 1067[105]
Lopes, C. C., **2**, 744[96]
Lopes, R. S. C., **2**, 744[96]
Lopez, A. F., **4**, 454[39], 459[39], 461[39], 463[111], 464[111], 468[133], 469[39], 475[111]
Lopez, B. O., **5**, 7[51], 514[6]
López, C., **7**, 277[154]
Lopez, F., **1**, 782[234]; **6**, 184[150]; **8**, 125[94]
Lopez, G., **7**, 760[49], 764[49]
López, J., **7**, 90[33]
Lopez, J. C., **7**, 630[55]
Lopez, L., **3**, 222[145], 246[40], 257[40], 283[49]; **4**, 34[97], 35[97]; **7**, 120[14], 167[186], 854[57], 855[57]; **8**, 887[116]
López, M. C., **7**, 277[153]
Lopez, R. C. G., **2**, 555[147]; **5**, 438[161]
Lopez Aparicio, F. J., **2**, 348[60], 357[60,139]
Lopez-Calahorra, F., **1**, 543[20,24,26]; **6**, 510[299], 801[36]
Lopez de la Vega, R., **8**, 669[8]
Lopez-Espinosa, M. T. P., **7**, 296[23]
Lopez Herrera, F. J., **1**, 759[132]; **2**, 357[139], 386[329]
Lopez Herrera, H., **2**, 386[330]
López-Mardomingo, C., **8**, 36[74], 38[74], 66[74]
Lopez Nieves, M. I., **7**, 165[83]
López-Prado, J., **8**, 854[150]
Lopez-Rodriguez, M. L., **8**, 827[75]
Lopotar, N., **6**, 766[23]; **7**, 698[51]
Lopresti, R. J., **5**, 893[41]; **7**, 346[11]; **8**, 237[11], 545[291]
Lorah, D. P., **6**, 814[89]
Lorand, J. P., **6**, 1016[27]
Lora-Tamayo, M., **6**, 273[93,94], 275[108,113], 277[127]
Lorber, M., **1**, 377[97]; **2**, 278[10]; **7**, 101[132], 258[57], 501[251,253], 502[253], 845[64]
Lorberth, J., **1**, 215[38]
Lord, E., **8**, 990[42]
Lord, P. D., **8**, 36[48], 66[48], 347[140], 616[102], 617[102], 618[102]
Lorenc, L., **2**, 553[128]; **5**, 179[143]; **7**, 92[41,41a], 94[41], 703[5], 710[5], 738[27], 815[2], 816[2c], 824[2c], 827[2c], 851[18]; **8**, 873[17]
Lorente, A., **4**, 5[17]
Lorenz, B., **4**, 485[29]

Lorenz, D. H., **8**, 74[250]
Lorenz, G., **4**, 12[39]
Lorenz, K. T., **5**, 704[22], 1020[69], 1023[69]; **7**, 879[149], 880[149]
Lorenz, R., **2**, 138[21], 901[35], 908[35], 909[35], 910[35]
Lorenzen, N. P., **1**, 40[192], 41[272]
Lorenzi, G. P., **8**, 937[78]
Lorenzi-Riatsch, A., **1**, 145[45]
Loreto, M. A., **6**, 717[111]; **7**, 479[91]
Lorey, H., **2**, 901[33,34], 911[33]; **5**, 1200[53]
Lorig, W., **2**, 852[226]
Loritsch, J. A., **4**, 292[223]
Lorke, M., **2**, 1088[43]
Lorne, R., **3**, 247[46], 416[20], 417[20]
Losey, E. N., **2**, 655[149]
Loskutov, M. P., **5**, 516[19]
Lösler, A., **5**, 64[49]
Losse, G., **6**, 642[65,66]
Lothrop, W. C., **3**, 505[166]
Lotspeich, F. J., **2**, 765[79,80]
Lott, J., **8**, 872[10], 873[10]
Lott, R. S., **4**, 611[358]
Lottenbach, W., **2**, 308[23], 309[23]
Lotti, V. J., **2**, 962[51]; **5**, 410[41]
Lotvin, B. M., **3**, 648[178]
Lotz, W. W., **1**, 655[142], 684[142]
Lou, B., **8**, 879[55], 880[55]
Lou, J.-D., **7**, 253[14], 279[168,169], 280[168,169]
Loubinoux, B., **4**, 496[93]; **6**, 648[120]; **8**, 483[57], 485[57], 558[391]
Loudon, G. M., **4**, 344[76]; **6**, 803[46,47], 804[47,50], 816[100,102], 822[117]
Loudon, J. D., **2**, 765[78], 866[3]
Lough, W. J., **4**, 1039[65]
Loughhead, D. G., **8**, 873[16]
Louis, J.-M., **7**, 312[34], 320[34], 738[26], 747[26], 851[18]
Louis-Andre, O., **8**, 563[425,426]
Lounasmaa, M., **2**, 399[14], 1020[47]; **6**, 910[4], 913[25], 939[4]
Loupy, A., **4**, 70[11], 139[2]; **6**, 2[5], 18[5]; **8**, 15[94], 541[219], 542[219]
Lourak, M., **3**, 509[179]
Loutfy, R. O., **5**, 124[9], 128[26]
Louw, J. v. d., **4**, 877[69]
Louw, R., **6**, 679[325], 1013[8], 1017[8]; **7**, 765[140]
Love, B. E., **4**, 40[116]
Love, C. J., **8**, 876[43]
Love, R. F., **2**, 765[78]
Love, S. G., **2**, 1052[44]
Loveitt, M. E., **4**, 306[377], 307[390,391]; **8**, 855[159]
Lovelace, T. C., **5**, 239[1], 903[41], 905[41], 907[41], 909[41], 916[41], 918[41], 930[176], 933[176], 937[41], 939[41,222], 940[41,222], 951[41], 962[222], 963[222], 964[176]; **7**, 324[72], 557[75]
Lovell, A. V., **1**, 425[103]; **4**, 230[250]
Lovell, B. J., **7**, 582[149]
Lovell, F. M., **8**, 618[127], 623[127]
Lovell, M. F., **6**, 554[758]
Loven, R. P., **5**, 402[5]
Lovesey, A. C., **8**, 613[79]
Lovich, S. F., **3**, 525[43]
Loving, B. A., **8**, 530[95]
Lovy, J., **7**, 884[184]
Löw, I., **6**, 651[134]
Low, J., **5**, 176[132]
Löw, M., **6**, 637[31]
Löw, P., **6**, 706[37], 721[133], 723[133], 724[155]; **8**, 285[10]
Löw, W., **2**, 399[19]
Lowe, A., **1**, 476[123]; **2**, 365[216]
Lowe, C., **4**, 744[134]
Lowe, G., **3**, 902[118,118a]; **6**, 664[219]
Lowe, J. A., **3**, 35[204], 66[14]; **7**, 187[183]
Lowe, J. L., **6**, 430[97]
Lowe, J. P., **6**, 955[23]
Lowe, J. U., Jr., **3**, 319[130]
Löwe, U., **2**, 614[119]; **3**, 25[160]
Lowell, J. R., Jr., **6**, 277[125,126]
Lowen, G. T., **2**, 381[309]; **4**, 486[36], 489[36], 501[111]
Lowenberg, A., **4**, 598[185], 622[185], 638[185]; **7**, 504[282]
Lowenthal, R. E., **6**, 1053[46]
Lowery, M. K., **7**, 741[48], 747[48]
Lowman, O., **2**, 796[1]
Lown, E. M., **3**, 892[48]
Lown, J. W., **3**, 927[54]; **4**, 350[122,123], 1085[105], 1086[105,112]; **5**, 687[59], 937[204]; **7**, 231[147], 341[53], 769[235], 770[235]
Lowndes, P. R., **3**, 380[4]
Lowrey, C. H., **6**, 790[117]
Lowrie, G. B., III, **6**, 96[152], 182[140]
Lowrie, S. F. W., **8**, 445[52]
Lowry, B. R., **3**, 890[34]
Lowry, C. D., **7**, 15[148]
Lowry, T. H., **1**, 528[109,110]; **4**, 240[49], 717[8]; **5**, 703[18]
Lowry, T. M., **7**, 776[359]
Loy, M., **5**, 712[60]
Loy, R. S., **1**, 373[87], 374[87]
Loza, R., **5**, 948[270]
Lozac'h, N., **6**, 420[20], 424[57,58]
Lozanova, A. V., **5**, 345[70], 346[70], 453[66]; **8**, 611[64]
Lozar, L. D., **1**, 608[35]
Lozinova, N. A., **6**, 547[664]
Lozinskii, M. O., **6**, 524[373,374]
Lozzi, L. L., **2**, 607[71]
Lu, K., **8**, 678[63], 685[63], 686[63]
Lu, L., **4**, 589[85], 590[93], 592[93]
Lu, L. D.-L., **1**, 436[152]; **2**, 113[102], 240[13], 242[18], 245[18b], 246[18b], 256[13], 257[13b], 801[31], 926[117]; **6**, 446[102]; **7**, 394[18], 395[18], 398[18], 399[18], 422[141], 423[141,141b,c], 424[18], 748[114]
Lü, Q.-H., **5**, 296[54], 1194[34], 1195[34]
Lu, S., **8**, 860[223]
Lu, S.-B., **4**, 18[61,61a,b], 243[68], 247[101], 262[101]; **6**, 176[94]
Lu, S.-L., **4**, 1021[241]
Lu, X., **4**, 589[85], 590[89,93], 592[93], 599[222], 613[89], 629[408], 641[222], 753[168]; **6**, 845[99]; **7**, 6[32]; **8**, 554[366]
Lu, Y.-F., **5**, 703[15]
Lübben, S., **4**, 722[32]
Lubell, W. D., **3**, 44[242]
Luberoff, B. J., **7**, 449[4], 451[4]
Lubineau, A., **2**, 632[28b], 640[28], 655[146], 663[23], 664[23], 681[62]; **5**, 432[126,127]
Lubinskaya, O. V., **2**, 710[16]; **3**, 342[7]
Luborsky, F. E., **3**, 297[22]
Lubosch, W., **1**, 474[106]; **3**, 66[11], 74[11], 194[9]; **6**, 176[88], 419[8,9], 425[8,9], 509[278]; **7**, 225[67]
Luca, C., **4**, 298[284]
Lucarelli, M. A., **8**, 840[34]
Lucarini, L., **1**, 214[26]
Lucas, H., **8**, 407[57]
Lucas, H. J., **4**, 279[112,114], 288[114,184]; **6**, 215[101], 951[2]
Lucas, R. A., **8**, 32[54], 66[54]
Lucas, T. J., **6**, 47[78]
Lucchesini, F., **6**, 775[51]
Lucchetti, C., **1**, 571[278]

Lucchetti, J., **1**, 571[274,275], 631[53], 634[74], 638[104], 640[104], 642[74,112,113], 644[74], 656[53,104], 657[53,157], 658[53], 659[53,104], 660[157], 661[157], 662[157], 672[74,104,199], 686[112,113,157,228,230,231,232,233,234]; **3**, 86[50], 87[78,84], 91[148], 92[148], 109[84], 120[243], 136[78], 142[243], 144[78], 145[78]; **4**, 10[34], 113[165]; **5**, 345[68,68a], 346[68a], 453[62]; **8**, 847[97], 848[97e], 849[97e]
Lucchini, V., **4**, 50[142], 298[280], 330[5]; **5**, 155[35], 370[102], 371[102], 408[33]; **6**, 150[114], 999[123]; **7**, 384[114c], 399[38], 400[38], 406[38], 409[38], 415[38]
Lucci, R. D., **1**, 506[12]
Luce, E., **4**, 339[44]
Lucente, G., **6**, 1017[38], 1024[38]
Luche, J. L., **1**, 219[60,61], 464[38], 563[185], 830[95]; **2**, 23[89], 294[83], 565[13], 572[13], 917[88], 918[90,91], 935[91,149]; **3**, 249[65]; **4**, 95[99a-c], 229[212,220], 764[220]; **5**, 100[143,144], 102[143], 176[127]; **6**, 676[302]; **7**, 333[21]; **8**, 16[96], 17[115-117], 538[193], 540[193,196,197], 988[27]
Luche, M., **1**, 848[26], 849[26], 850[26]
Luche, M.-J., **2**, 902[47], 903[47]; **3**, 149[408,409], 151[408,409], 155[408,409], 783[29]; **6**, 129[167]; **7**, 406[76], 777[388]
Lucherini, A., **5**, 1131[12], 1148[114], 1154[152], 1155[162], 1156[162]
Luchetti, J., **4**, 106[140c]
Lucius, G., **7**, 92[46]
Lucken, E. A. C., **8**, 231[145]
Luckenbach, R., **7**, 752[155]; **8**, 861[225]
Lückenhaus, W., **5**, 99[138], 100[138]
Luckner, M., **6**, 746[92]
Ludden, C. T., **2**, 971[92]
Luders, L., **5**, 145[106]
Ludman, C. J., **7**, 800[35]
Ludovici, D. W., **8**, 28[37], 66[37]
Ludwiczak, S., **4**, 432[102,110]
Ludwig, J. W., **2**, 510[37]; **3**, 34[193]
Ludwig, U., **3**, 927[52]
Ludwikow, M., **4**, 429[86]
Luedtke, A. E., **3**, 587[141]
Luehr, G. W., **4**, 394[196]
Luengo, J. I., **3**, 996[46], 997[46a], 1000[46]; **4**, 602[252], 643[252], 644[252]; **5**, 53[47], 519[34], 542[112], 549[34], 829[18], 848[18c], 849[18c], 851[18c], 1001[15]
Lueoend, R., **4**, 253[168]
Luetolf, J., **7**, 12[104]
Luft, R., **2**, 144[58]
Luftmann, H., **3**, 640[109], 647[109]; **7**, 42[33]
Lugade, A. G., **8**, 367[61]
Luger, P., **5**, 176[133]
Lugovoi, Yu. M., **8**, 773[69]
Lugtenburg, J., **2**, 780[9]; **5**, 708[41]; **6**, 494[131]
Luh, B., **2**, 648[98], 649[98], 1059[76]
Luh, B.-Y., **5**, 92[70], 94[84,85]
Luh, T.-Y., **6**, 765[18]; **8**, 798[49,64], 840[33], 842[41], 847[94,95], 966[76]
Luh, Y., **5**, 947[261], 961[261,322], 962[322]
Lui, B., **4**, 484[17]
Lui, H. S., **7**, 299[49]
Luibrand, R. T., **5**, 197[18], 229[121]
Luidhardt, T., **4**, 288[188], 346[86a], 347[86b]
Luijten, J. G. A., **8**, 264[40], 547[316]
Luis, A., **1**, 749[80]; **5**, 855[187]
Luis, J. G., **8**, 330[49]
Luis, S. V., **4**, 492[73]; **5**, 474[158]; **6**, 172[19]
Luk, K., **1**, 832[114]
Lukacs, A., **1**, 116[43]; **4**, 703[32], 704[32]
Lukacs, G., **3**, 126[316,317]; **7**, 630[55]
Lukanich, J. M., **4**, 467[129]
Lukanov, L. K., **6**, 744[75], 746[75,88]
Lukas, J. H., **4**, 587[38,39], 588[58]; **5**, 9[70,71]
Lukas, K. L., **2**, 76[84]; **3**, 159[466], 166[466], 174[530]
Lukaschewicz, C., **6**, 435[2]
Lukasczyk, G., **2**, 737[40]
Luke, G. P., **3**, 483[15], 500[15]
Lüke, H.-W., **6**, 722[144]
Luke, R. W. A., **2**, 1050[27]
Lukehart, C. M., **1**, 300[69]; **4**, 104[135b]
Lukenbach, E. R., **3**, 124[276], 126[276]; **6**, 676[304], 677[304]
Lukes, R., **8**, 590[55]
Lukevits, E. Ya., **4**, 48[140]; **7**, 477[79,81]; **8**, 556[376], 763[1], 764[4a], 769[1b], 771[1b,45,48], 785[1]
Lukin, K. A., **3**, 864[21]; **4**, 969[66]
Lukyanenko, N. G., **6**, 94[139]
Luly, J. R., **2**, 1015[21]
Lum, R. T., **4**, 485[30]; **5**, 1175[40], 1178[40]
Lumb, A. K., **4**, 505[142]
Lumin, S., **7**, 260[84], 713[72]
Lumma, P. K., **1**, 122[67], 359[11], 380[11], 382[11]; **8**, 388[61]
Lumma, W. C., **6**, 526[395]
Lumma, W. C., Jr., **2**, 1086[30]
Luna, D., **2**, 345[18], 359[18], 360[18]
Luna, H., **4**, 18[59]; **5**, 940[225], 963[225]; **7**, 557[74]
Lund, E. C., **3**, 1025[34]; **5**, 1041[19], 1046[19], 1050[35], 1052[19]
Lund, E. D., **8**, 161[20]
Lund, H., **3**, 564[9], 567[9], 634[22]; **4**, 240[52], 726[51]; **7**, 810[88]; **8**, 129[1], 131[1], 134[1], 285[4], 293[4], 294[4], 532[128], 591[60], 592[66], 594[72], 640[23], 641[28], 642[31,33], 644[33], 645[46], 900[30], 974[127], 975[134], 984[1]
Lund, T., **4**, 240[52]
Lundeen, A. J., **6**, 960[50]
Lundell, G. F., **5**, 382[121]
Lundt, B. F., **6**, 637[31]
Lung, K. R., **8**, 764[7], 770[7,40], 773[7]
Lunin, A. F., **6**, 530[418,422]
Lüning, U., **4**, 311[445]; **8**, 852[139]
Lunn, G., **6**, 119[109]; **8**, 373[129], 383[19], 389[19], 392[19], 399, 597[95], 605[11], 613[11]
Lunn, W. H., **3**, 353[46]
Lunsford, C. D., **8**, 652[72]
Lunsford, J. H., **7**, 14[131]
Lunsford, W. B., **6**, 619[117]
Lunt, E., **6**, 255[169]; **7**, 745[73]
Lunt, J. C., **3**, 643[130], 644[151,166]
Lunt, L. C., **3**, 644[147]
Luo, F.-T., **1**, 95[80], 448[206]; **2**, 120[185], 449[48], 584[123], 586[123]; **3**, 12[56,57], 440[44], 460[142], 463[156], 485[37], 486[37], 491[37], 495[37], 497[105], 503[37]; **4**, 256[216], 903[198]
Luo, J., **4**, 348[108], 349[108c]; **5**, 383[125]
Luo, M., **2**, 355[127]
Luo, T., **2**, 805[45]
Luo, T.-T., **3**, 525[43]
Luong-Thi, N., **4**, 841[44]
Luong-Thi, N. T., **1**, 17[206]
Lupi, A., **3**, 717[44]
Lupin, M. S., **4**, 587[41]
Lupo, A. T., Jr., **2**, 657[165]; **5**, 473[150], 478[150]
Luppold, E., **7**, 777[364]
Lurquin, F., **5**, 109[222], 110[222], 111[222], 112[222]
Lusby, W. R., **7**, 673[29]
Lusch, M. J., **2**, 165[151], 547[117], 551[117], 807[46]
Lusinchi, X., **6**, 70[20], 915[29]; **8**, 392[95], 395[135], 565[448], 848[105], 880[56], 936[70]
Lusis, V., **8**, 595[77]

Luskus, L. J., **4**, 1075[30], 1097[164]; **5**, 247[26], 248[26a], 249[26a], 626[35,40], 630[40,50], 631[54-56]
Lüssi, H., **4**, 314[488]
Lustgarten, D. M., **7**, 664[65]
Lustgarten, R. K., **5**, 585[202]
Lusztyk, E., **8**, 264[47]
Lusztyk, J., **4**, 723[38], 736[88], 738[38,101], 747[38]; **8**, 264[47]
Luteijn, J. M., **7**, 363[38], 376[87]
Lutener, S. B., **5**, 929[167]
Luthardt, H., **4**, 1089[134]
Luthardt, P., **6**, 760[141]
Luthe, H., **4**, 729[61], 730[61], 765[61]
Luthman, K., **2**, 465[107]; **4**, 411[264]
Luthra, N. P., **6**, 69[17], 462[17]
Lüthy, J., **8**, 204[153]
Lütke, H., **6**, 193[212,213], 195[212]
Lutomski, K. A., **1**, 60[36], 75[36]; **3**, 53[273], 503[149], 512[149]; **4**, 205[41], 206[48], 252[167], 257[167], 427[67]; **6**, 501[187]; **8**, 162[32]
Lutsenko, I. F., **2**, 597[96], 607[76], 609[76,79], 610[96], 616[135], 726[122], 804[40]; **4**, 314[486]; **8**, 547[315,316], 548[315]
Lutsenko, L. F., **1**, 608[39]
Luttinger, D., **8**, 623[150]
Lüttke, W., **3**, 587[142]; **8**, 364[9], 390[82]
Luttmann, C., **2**, 228[165]; **8**, 844[72]
Lüttringhaus, A., **3**, 499[113]; **7**, 741[47]
Lutz, E. F., **5**, 337[71b], 345[71b], 346[71b], 1148[114]
Lutz, R. E., **4**, 282[137]; **5**, 220[48]; **6**, 211[79]; **8**, 904[58]
Lutz, R. P., **5**, 798[70], 799[70a,73], 800[70], 827[2], 829[2], 850[2f], 867[2f], 1000[5]; **6**, 834[37], 856[37]
Lutz, W., **2**, 514[52]; **7**, 98[100], 165[84]
Luu, B., **3**, 158[443], 159[443], 160[443], 161[443], 167[443], 168[443], 406[140]; **7**, 359[15], 548[61], 553[61]; **8**, 15[88]
Luvisi, J. P., **4**, 276[71], 283[71], 313[464]
Lux, R., **5**, 425[103]; **6**, 547[668]
Luxen, A., **6**, 462[14]
Luxen, A. J., **4**, 50[142]
Luyten, M., **2**, 829[136]; **3**, 380[4]
Luyten, M. A., **8**, 189[60], 206[173], 941[113]
Luyten, W. C., **6**, 620[130]
Luzzio, F. A., **7**, 252[3], 260[61], 267[61], 269[127], 270[127], 288[3], 752[146]
L'Vova, S. D., **6**, 489[93]
Lwande, W., **1**, 823[42]
Lwowski, W., **4**, 1099[178], 1100[178], 1103[207]; **5**, 938[206]; **6**, 245[120], 249[120], 251[120], 252[120]; **7**, 21[1,5,8-14], 24[26,31,33], 25[26,33,42], 26[61], 27[8,75], 30[10], 35[5], 477[72,76,77,80], 478[80,83,84], 479[95], 483[72]
Ly, N. D., **3**, 261[149]
Ly, U. H., **3**, 810[50]
Lyakhovetsky, Yu. I., **8**, 610[59-61], 611[66], 778[84]
Lyapina, N. K., **8**, 608[48], 610[61]
Lydiate, J., **8**, 839[25]
Lyerla, R. O., **4**, 273[46], 280[46]
Lyga, J. W., **6**, 815[97]
Lygo, B., **1**, 780[228]; **3**, 174[527,527a,b], 175[527a,b]; **4**, 381[126b], 382[126], 383[126], 390[168,175a], 391[176]; **7**, 523[47]; **8**, 847[98,100a,b], 849[98]
Lyle, G. G., **1**, 357[3]
Lyle, R. E., **1**, 357[3]; **8**, 580[3], 584[3,24], 585[3]
Lyle, S. B., **3**, 325[161,161a]
Lyle, T. A., **3**, 369[121], 372[121]; **5**, 223[83], 224[83], 806[102], 1028[90]
Lynch, B., **8**, 354[174]
Lynch, G. J., **4**, 301[315,318], 303[339]; **8**, 675[39], 691[107], 854[149], 856[164]
Lynch, J., **1**, 805[312]; **5**, 245[21], 308[94,97]; **6**, 990[84]
Lynch, J. E., **1**, 62[38,40], 63[42]; **2**, 227[160], 821[104], 852[104]; **5**, 99[130]; **6**, 759[139]
Lynch, L. E., **1**, 551[72], 665[172], 668[172]; **2**, 283[44], 298[44]
Lynch, P. P., **8**, 915[96], 939[94,95]
Lynch, T. J., **6**, 723[146]; **8**, 86[21], 600[103]
Lynch, V., **2**, 655[131]; **5**, 841[104]
Lynd, R., **4**, 887[130]
Lynd, R. A., **3**, 259[134]; **8**, 755[118,129], 756[150]
Lyness, S. J., **6**, 790[117]
Lyness, W. I., **3**, 147[389]
Lynn, J. T., **3**, 681[98], 816[81]
Lynn, J. W., **6**, 273[101]
Lyon, D. R., **3**, 382[36]
Lyon, J., **6**, 671[280]
Lyon, J. T., **1**, 826[59]
Lyons, J. E., **5**, 1141[84]; **7**, 95[70]; **8**, 239[27], 240[27], 242[27], 445[60], 446[78,79], 766[19], 906[67], 907[67], 908[67]
Lyons, J. R., **4**, 611[351]
Lysenko, V. P., **1**, 820[4]; **3**, 86[43], 179[43]; **4**, 987[133]
Lysenko, Z., **2**, 1090[70], 1100[70]; **4**, 371[60], 372[57,60]; **6**, 1031[113,115]; **7**, 517[16], 523[46]; **8**, 847[99], 848[99], 849[99]
Lyster, M. A., **6**, 647[108], 665[229], 667[229], 752[110]
Lythgoe, B., **1**, 329[33], 780[229], 793[273], 794[273c,276], 804[273]; **3**, 936[71]; **5**, 839[75], 859[233], 888[25]; **6**, 854[142], 860[176], 989[77], 990[77], 992[90], 993[77,91,92], 994[92], 997[112], 998[90,119], 1002[77]; **8**, 544[273], 823[53]

M

Ma, D., **7**, 6[32]
Ma, E. C.-L., **5**, 950[285]
Ma, K., **2**, 772[16]
Ma, K. W., **4**, 1006[100], 1021[240]; **8**, 807[114]
Ma, P., **4**, 255[192], 260[192]; **5**, 386[135]; **6**, 8[39], 927[76]; **7**, 198[26], 401[59,60], 403[59], 406[59]; **8**, 879[51], 880[51]
Ma, R. M., **8**, 228[132]
Ma, S., **4**, 753[168]; **7**, 446[64]
Ma, W., **2**, 772[19]
Maag, H., **2**, 385[322], 899[27], 901[27]
Maak, N., **6**, 153[144], 839[70]
Maartin, W. B., **7**, 804[63]
Maas, D. D., **8**, 927[21]
Maas, G., **1**, 844[5a,8]; **4**, 953[8], 954[8o], 1009[146], 1031[1], 1032[1], 1033[28], 1034[1], 1035[1], 1036[1], 1040[1], 1049[1], 1050[1], 1063[1], 1075[33]; **5**, 791[39]; **6**, 120[118], 126[154], 504[221], 523[349], 550[676], 552[691]
Maas, M., **4**, 768[235]
Maasböl, A., **4**, 976[99]
Maassen, J. A., **7**, 748[110]
Maatta, E. A., **8**, 447[131], 697[129]
Mabey, W. R., **6**, 277[126]
Mabon, G., **2**, 725[102-104]; **5**, 777[186,187,188]
Mabrouk, A. F., **8**, 450[161], 453[191]
Mabury, S. A., **6**, 437[39]
McAdams, L. V., III, **3**, 883[109]
McAfee, F., **7**, 770[250]; **8**, 409[85]
McAfee, M. J., **4**, 1040[79], 1049[79], 1060[79b]; **5**, 599[40], 804[94], 986[40]
Macaione, D. P., **3**, 557[47]
McAlister, D. R., **5**, 1145[105]; **8**, 673[23], 675[23,46], 682[46], 691[23]
McAlpine, G. A., **4**, 259[267]; **5**, 125[21], 128[21], 768[126], 779[126]; **6**, 745[85], 1059[68]
Macaluso, A., **1**, 391[150]
McAndrew, B. A., **2**, 710[17]; **5**, 835[59]
McAndrews, C., **3**, 21[125]
McAninch, T. W., **5**, 907[75], 908[75], 911[93], 945[75]
McArdle, P., **4**, 707[41], 710[50,55], 712[64,67]
McArthur, C. R., **6**, 726[181,182]; **7**, 453[66]
Macaulay, J. B., **5**, 347[72,72f]
McAuley, A., **7**, 760[18,19]
McAuliffe, C. A., **7**, 632[57]
McBee, E. T., **3**, 319[131]; **4**, 280[124]; **5**, 513[4]; **7**, 206[69], 738[31]
Macbeth, A. K., **7**, 84[3], 154[14]; **8**, 141[42]
McBride, B. J., **1**, 822[31]; **3**, 54[276], 537[89]; **4**, 1089[129]
McBride, L. M., **6**, 554[731]
McCabe, J. R., **5**, 77[266]
McCabe, P. H., **8**, 528[66]
McCabe, R. W., **5**, 345[68], 346[68b]
McCabe, T., **8**, 347[141], 350[141]
Maccagnani, G., **5**, 440[173]
McCague, R., **3**, 232[260]; **4**, 439[158]; **6**, 781[77]; **8**, 948[147]
McCall, J. M., **8**, 392[96]
McCall, M. A., **4**, 120[202]
McCall, R. B., **8**, 626[171], 627[171]
McCallum, J. S., **5**, 1070[25], 1072[25], 1074[25], 1089[82,86], 1090[86], 1092[82], 1093[96], 1094[82,99], 1096[99,108,108c], 1098[82,96a,99,108c,130], 1099[82,99,108c], 1111[82], 1112[82,96a,99,108c,130], 1113[82], 1183[57]; **7**, 350[20]; **8**, 911[88], 933[52]
McCallum, K. S., **7**, 673[19]
McCann, P. J., **4**, 119[194], 226[199]; **8**, 813[9]
McCann, S., **5**, 854[180]
McCann, S. F., **2**, 1027[74], 1036[95], 1047[3]
McCants, D., Jr., **7**, 194[7], 204[7], 205[7], 764[127]
McCapra, F., **3**, 681[96], 689[121], 813[60]; **4**, 305[369]
McCarron, E. M., **7**, 800[35]
McCarry, B. E., **3**, 369[118], 370[113], 372[118]; **8**, 542[228]
McCarten, P., **4**, 985[126,131]
McCarthy, D. G., **6**, 293[225]
McCarthy, F. C., **2**, 553[127]
McCarthy, J. R., **3**, 492[73], 497[73]
McCarthy, K., **5**, 162[67]; **6**, 734[10], 735[10]
McCarthy, K. E., **1**, 107[6], 110[6]; **3**, 264[186]; **4**, 176[50], 177[50]; **6**, 675[297]
McCarthy, M., **8**, 587[40]
McCarthy, P. A., **2**, 1[2], 224[154], 225[154], 240[5]; **7**, 378[94]
McCartney, M., **4**, 839[28]
McCartney, R. L., **3**, 572[66]; **8**, 312[22], 321[22]
McCarty, C. G., **2**, 734[3]; **6**, 704[6], 763[5], 786[94]; **7**, 689[7]
McCarty, C. T., **4**, 510[165]
McCarty, J. E., **6**, 1012[5]
McCarty, L. P., **4**, 337[34]
McCaskie, J. E., **5**, 476[147]
McCaulay, D. A., **3**, 297[20], 298[20]
McCauley, J. P., Jr., **6**, 150[114]; **7**, 778[400,401,401a], 779[426]
McCaully, R. J., **8**, 388[60], 615[95]
McChesney, J. D., **8**, 49[114], 66[114], 497[37]
Macchia, B., **2**, 284[53], 291[71]; **3**, 725[19], 734[9], 743[58]; **8**, 856[167]
Macchia, F., **2**, 284[53], 291[71]; **3**, 725[19], 734[8-10], 741[49], 743[58]; **6**, 253[159]; **8**, 856[167]
Maccioni, A., **7**, 777[368,370]
McClanahan, R. J., **7**, 66[72]
McClard, R. W., **3**, 201[81]
McCleery, D. G., **8**, 170[93]
McCleland, C. W., **4**, 347[95]
McClellan, W. R., **4**, 3[7,7a], 4[7]
McClelland, B. J., **7**, 861[77]
McClelland, R. A., **5**, 1123[34]; **6**, 502[214], 504[224], 562[214,884]
McClory, M. R., **1**, 512[37,38]; **3**, 147[397], 149[397,400], 150[397], 151[397]
McCloskey, A. L., **4**, 312[459]; **7**, 602[105]
McCloskey, C. J., **3**, 17[84], 588[158]
McCloskey, C. M., **6**, 651[134]
McCloskey, J. E., **3**, 390[75,76,78,79], 392[75,76,78]
McCloskey, P. J., **1**, 769[195], 845[10]; **8**, 449[151]
McClure, C. K., **2**, 113[109,110], 249[84], 263[56], 264[57]; **5**, 436[158,158g], 442[158]
McClure, D. E., **2**, 756[8]
McClure, N. L., **2**, 655[147], 907[58], 908[58]
McClusky, J. V., **4**, 443[189], 1009[144]
Macco, A. A., **3**, 367[101,102]
McCollum, G. J., **1**, 632[66]; **3**, 864[18,22]
McCollum, G. W., **7**, 606[157]
McCollum, J. D., **8**, 91[51]
McColm, E. M., **6**, 228[30]
McCombie, H., **4**, 283[144], 288[144]
McCombie, S. W., **2**, 1102[123]; **3**, 253[89], 261[147], 262[89]; **6**, 641[59], 659[59], 670[59]; **8**, 370[91], 818[35], 820[35], 822[52], 823[35]
McCombs, C. A., **4**, 30[89]; **5**, 329[32]

McComsey, D. F., **8**, 36[75], 37[75], 38[75], 39[75], 45[75], 54[75], 55[164,165], 59[164], 66[75,164,165], 618[128,129], 619[129], 620[128,129], 624[129]
McConaghy, J. S., Jr., **7**, 477[80], 478[80,84]
MacConaill, R. J., **7**, 696[40]
McConnell, W. B., **7**, 768[208], 773[208]
McConnell, W. V., **3**, 735[15]
McCormack, M. T., **6**, 293[224]
McCormick, A. S., **1**, 554[102]
McCormick, J. E., **6**, 927[79]; **7**, 197[24]
McCormick, J. P., **6**, 960[58]; **7**, 166[87]
McCormick, M., **4**, 377[124b], 380[124,124b]
MacCorquodale, F., **3**, 380[9]; **4**, 791[51]
MacCoss, M., **6**, 650[129]
McCoubrey, A., **8**, 964[51]
McCowan, J. D., **1**, 162[103]
McCowan, J. R., **3**, 492[73], 497[73]
McCoy, K., **7**, 453[75]
McCoy, L. L., **4**, 251[143]
McCoy, M. A., **5**, 1123[37]
McCoy, P. A., **3**, 396[106], 397[106]
McCrae, D. A., **1**, 582[7]; **2**, 90[41], 572[42]; **3**, 31[183], 132[334], 133[334], 136[334]
McCrae, D. M., **7**, 163[77], 164[77]
McCrae, W., **3**, 551[1]
McCready, R., **7**, 842[20]
McCready, R. J., **4**, 389[165]; **8**, 8[44]
McCreary, M. D., **2**, 684[65]
McCrindle, R., **3**, 715[37]; **7**, 64[63]; **8**, 528[66]
McCullagh, L., **3**, 909[154]
McCulloch, A. W., **4**, 49[141]; **5**, 8[63,64]
McCullough, D. W., **1**, 786[248]
McCullough, J. D., **7**, 769[237,238], 770[238], 771[274]
McCullough, J. J., **5**, 125[17], 128[29], 385[130], 650[25]; **7**, 854[54], 855[54], 876[122]
McCullough, K. J., **1**, 429[123]; **3**, 593[177]
McCurry, C., **4**, 438[150]
McCurry, P., **3**, 99[191], 103[191a], 107[191]
McCurry, P. M., Jr., **2**, 161[139], 547[97]; **4**, 7[24,24c]; **5**, 129[37]
McCutcheon, J. W., **6**, 976[5]
McDade, C., **5**, 1178[44]; **8**, 682[85], 687[85], 690[85], 691[85]
McDaniel, K., **8**, 395[129]
McDaniel, R. L., **1**, 884[129]
McDaniel, R. L., Jr., **5**, 806[107], 808[107b], 809[107b], 1027[88], 1028[88]
McDaniel, R. S., **7**, 483[122]
McDaniel, R. S., Jr., **4**, 1036[51]
McDaniel, W. C., **2**, 756[7,7b]; **3**, 564[7]; **7**, 682[81]
McDermott, J. X., **5**, 1131[14], 1173[32]
MacDiarmid, A. G., **8**, 763[1], 773[66], 785[1]
McDonagh, A. F., **2**, 969[85]
MacDonald, A. A., **8**, 568[472]
McDonald, C. E., **3**, 220[117]
Macdonald, D., **4**, 85[75]
Macdonald, D. I., **4**, 501[112]; **5**, 390[140], 692[97,98]
MacDonald, D. L., **2**, 456[73], 458[73]
McDonald, E., **7**, 104[146]; **8**, 374[149]
McDonald, F. E., **5**, 249[36]
McDonald, F. J., **7**, 830[61]
McDonald, G., **7**, 440[39,39b]
MacDonald, I., **2**, 753[2,2d]
McDonald, I. A., **1**, 441[174]
Macdonald, J. E., **1**, 462[18]; **4**, 483[8]; **5**, 839[77]
MacDonald, J. G., **5**, 439[167]; **6**, 108[35]
McDonald, J. H., **6**, 879[44]
McDonald, J. H., III, **1**, 51[13], 52[13], 108[10], 153[57], 336[68,70], 427[113], 460[2]; **2**, 9[39], 31[39], 194[67]; **4**, 380[120]
MacDonald, K. I., **7**, 228[106]
McDonald, P. D., **3**, 660[20], 661[20], 699[20]; **7**, 851[20]
McDonald, R., **2**, 743[84]
McDonald, R. N., **3**, 739[39-41], 748[74], 828[47]; **7**, 834[80]; **8**, 475[19], 568[470]
McDonald, R. T., Jr., **8**, 409[80]
MacDonald, S. F., **2**, 376[280]
MacDonald, S. J. F., **3**, 497[101], 505[101]
MacDonald, T., **1**, 55[22]
Macdonald, T. L., **1**, 108[11], 109[11], 117[51,52], 134[11,51], 283[2], 315[2], 328[28], 333[58], 338[82], 339[82], 879[111a,b]; **2**, 55[2], 62[22a], 66[2,36], 68[36], 78[92], 601[37]; **3**, 196[23,30], 223[152,153]; **4**, 45[128], 158[77], 816[201], 1005[81,83], 1018[81,83]; **5**, 687[55], 798[68]; **6**, 1067[105]; **7**, 226[71]; **8**, 557[379]
McDonald, W. S., **7**, 630[51]; **8**, 445[35]
McDonnell, P. D., **2**, 965[65]
McDonnell Bushnell, L. P., **5**, 1134[38], 1149[38]
McDougal, P. G., **1**, 507[19]; **2**, 547[92]; **3**, 252[84]; **4**, 355[132], 681[52], 682[52]; **5**, 348[74b], 685[40]; **6**, 21[79], 143[73]; **7**, 518[17]
McDougall, D. C., **2**, 555[146]; **5**, 437[159]
McDowell, D. C., **7**, 603[122]; **8**, 102[124], 537[189]
Macdowell, D. W. H., **2**, 758[22b], 759[31]; **5**, 790[36], 791[37]
McDowell, J. W., **8**, 36[87], 38[86,87], 66[86,87]
McDowell, R. S., **6**, 175[67]
McDowell, S. T., **4**, 55[156]
McElhaney, R. N., **8**, 353[160]
McElhinney, R. S., **2**, 960[36]; **4**, 302[336]; **6**, 927[79]; **7**, 197[24]
McElroy, A. B., **1**, 781[230,230b]; **7**, 369[61]
McElvain, S. M., **2**, 149[89]; **3**, 619[22]; **4**, 25[76], 31[92,92b], 46[76], 270[9], 1005[87], 1020[87]; **6**, 566[923,924]; **8**, 604[2], 605[2]
McElvain, S. S., **5**, 953[296]
McEnroe, F. J., **5**, 659[32]; **8**, 114[58], 308[5], 309[5]
McEuen, J. M., **4**, 8[29,29b]; **6**, 715[91]
McEvoy, F. J., **1**, 555[109], 556[109], 559[109]
McEwen, A. B., **8**, 289[23]
MacEwen, G., **1**, 38[184]
McEwen, J., **4**, 298[294], 300[294]
McEwen, W. E., **1**, 559[152]; **2**, 368[243], 753[1,1c]; **3**, 564[12], 574[76]; **6**, 176[100], 253[158]; **8**, 295[56,59,60], 568[471], 860[222], 861[226]
McEwen, W. L., **6**, 209[67]
McFadyen, J. S., **8**, 297[64]
McFarland, J. W., **2**, 1086[14], 1088[14], 1090[14], 1093[14], 1106[14]; **3**, 158[439]; **4**, 1083[92]; **6**, 157[164]; **8**, 391[87]
McFarland, P. E., **1**, 108[8], 116[8], 429[122]; **3**, 226[201]
McFarlin, R. F., **8**, 263[21,22], 265[21,22]
MacFerrin, K. O., **4**, 597[176], 598[176], 622[176], 637[176]
McGahen, J. W., **6**, 270[77]
McGahey, L. F., **8**, 850[121]
McGahren, W. J., **5**, 736[145], 737[145]
McGarraugh, G. V., **6**, 247[131]
McGarrity, J. F., **1**, 10[50], 41[202]; **4**, 257[219]; **7**, 123[35], 144[35]
McGarry, D., **6**, 760[143]
McGarry, D. G., **2**, 636[52], 640[52], 935[145], 937[145,158], 938[145,158], 940[145]; **3**, 618[20]; **4**, 795[87]; **5**, 102[179]; **6**, 448[108]; **8**, 846[81]
McGarry, L., **4**, 589[82]
McGarvey, B. R., **8**, 375[153], 445[57], 452[57]
McGarvey, D. J., **5**, 829[19]
McGarvey, G., **3**, 254[98], 257[98]; **4**, 344[78b]
McGarvey, G. J., **1**, 109[14], 127[92], 329[41], 341[94], 427[114]; **2**, 1[1], 225[155]; **3**, 18[97], 42[97], 43[97], 196[23], 254[98], 257[98]; **5**, 558[59], 841[86]; **6**, 859[164], 978[21]; **7**, 390[9]

McGee, J., **8**, 457[217]
McGee, L. R., **2**, 119[156,158], 211[114], 240[4], 302[2], 303[2], 304[8], 305[8], 436[67]; **7**, 347[14]
McGhie, J. F., **3**, 613[2], 615[2]; **7**, 92[41,41a], 94[41], 170[122], 171[122], 231[141]; **8**, 117[74], 243[47], 394[114], 505[83], 816[24], 837[13a], 839[13a], 840[13a], 935[63]
McGill, C. K., **8**, 596[82]
McGill, J. M., **4**, 403[238], 404[238], 405[238], 406[238]; **7**, 503[276]
McGillivray, G., **2**, 780[11]; **6**, 487[58,59], 489[59]; **7**, 185[173]; **8**, 41[93], 66[93]
McGlinchey, M. J., **3**, 380[10]; **6**, 287[180]; **7**, 24[23]; **8**, 675[48], 676[48]
McGlynn, K., **8**, 404[20]
McGowan, D. A., **2**, 904[49]; **7**, 365[46]
McGown, W. T., **8**, 431[64]
McGrath, D. V., **7**, 3[10]
McGreer, D. E., **4**, 955[12]
McGreer, J. F., **4**, 541[112]
McGregor, A. C., **6**, 636[26], 637[26]
McGregor, D. N., **7**, 204[57]; **8**, 645[40]
MacGregor, I. R., **3**, 324[153]
MacGregor, R. R., **3**, 1040[109]; **8**, 344[123]
McGregor, S. D., **3**, 158[435], 173[435]; **6**, 276[118]; **8**, 605[16]
McGrew, F. C., **8**, 814[20]
McGuinness, J., **4**, 521[47], 522[47], 530[47]
McGuire, E. J., **2**, 463[87]
McGuire, J. S., Jr., **8**, 561[415]
McGuire, M. A., **4**, 571[1], 572[1,5], 841[43]; **5**, 1076[34], 1105[167], 1107[167,168], 1111[34]
McGuirk, P. R., **1**, 428[115,116]; **3**, 243[17], 249[17], 263[17,172]; **4**, 250[135], 255[135], 898[174], 899[174]
Machacek, J., **8**, 544[272]
Machajewski, T., **6**, 838[65]
Machat, R., **5**, 501[266]
Machell, G., **3**, 831[66]
McHenry, B. M., **1**, 786[249]; **8**, 842[47]
McHenry, W. E., **5**, 113[228]
Macher, B. A., **8**, 36[50], 66[50]
Machida, H., **3**, 594[186]; **4**, 85[71], 203[28,30]; **8**, 806[122]
Machida, K., **2**, 1059[81]
Machida, M., **5**, 181[156]; **7**, 877[133]
Machida, Y., **6**, 2[8], 22[8], 76[42]; **8**, 384[24]
Machii, D., **2**, 633[34b], 634[34b], 640[34]
Machii, Y., **7**, 471[24]
Machin, P. J., **5**, 497[222]
Machinek, R., **5**, 468[126], 531[81]
Machleder, W. H., **3**, 741[54], 892[52]; **4**, 308[403]
McHugh, C. R., **3**, 791[62]
McHugh, M., **8**, 205[156]
Macias, A., **1**, 3[23]
Macicek, J., **1**, 36[175]
Maciejewski, L., **3**, 990[34,34a]
Maciejewski, S., **4**, 36[102]
Macielag, M., **3**, 51[270], 815[78]; **7**, 276[151]; **8**, 490[10], 508[87]
McInnes, A. G., **4**, 49[141]; **5**, 8[63,64]
McInnis, E. L., **5**, 707[37], 737[37]
MacInnis, W. K., **7**, 876[122]
McIntosh, A. V., Jr., **8**, 242[43], 293[49-51]
McIntosh, C. L., **4**, 483[6]; **5**, 904[53]
McIntosh, J. M., **1**, 551[75], 559[145]; **2**, 363[197], 432[54]; **3**, 46[255], 47[255], 382[40], 735[18], 998[47]; **6**, 176[87]; **7**, 229[122]; **8**, 342[111]
McIntyre, D., **2**, 753[2,2b]
McIntyre, D. K., **7**, 742[57]
McIntyre, S., **6**, 25[100]
McIsaac, W. M., **8**, 376[161]
McIver, J. M., **1**, 131[100]
McIver, J. W., Jr., **5**, 72[172], 703[17], 710[17]
McIver, R. T., Jr., **7**, 854[46]
Mack, W., **4**, 1079[58]
McKay, A. F., **6**, 120[124]
McKay, B., **8**, 988[25]
Mackay, D., **4**, 5[17], 375[95a]; **5**, 485[184]
McKay, F. C., **6**, 636[26], 637[26]
Mackay, K. M., **1**, 252[5]
Mackay, R. A., **1**, 252[5]
McKay, W. R., **5**, 408[32]; **8**, 72[240], 74[240]
McKean, D. R., **3**, 232[268], 495[93a]; **5**, 176[126], 925[151], 944[243]; **6**, 212[81], 932[96]
McKearin, J. M., **4**, 401[234a], 561[32]
McKee, D., **4**, 345[81]
McKee, M. L., **5**, 589[212], 856[197]; **8**, 724[169,169c]
McKee, R., **2**, 905[55], 907[55], 908[55], 910[55], 911[55]; **4**, 1040[95], 1041[95a], 1045[95a]; **5**, 410[40]
Mackellar, F. A., **5**, 637[110]
McKellin, W. H., **8**, 407[56]
McKelvey, R. D., **5**, 210[59], 715[82], 739[82]
McKelvey, T., **5**, 552[15]
McKenna, E. G., **1**, 758[127], 761[137]
McKenna, J., **6**, 719[127]; **8**, 392[100], 905[61]
McKenna, J. C., **4**, 368[18]
McKenna, J. F., **3**, 316[117], 317[117]
McKenna, J. M., **8**, 528[78]
McKenna, R. S., **8**, 445[61]
McKenna, T., **6**, 1014[19]
McKennis, H., Jr., **3**, 158[437], 159[437], 160[437], 166[437]
McKenzie, A., **1**, 49[1], 774[209]; **3**, 721[5]
MacKenzie, A. R., **7**, 32[95], 35[101], 349[18], 355[18]
McKenzie, A. T., **4**, 79[55a], 251[152]
MacKenzie, B. D., **2**, 710[23]
McKenzie, J. R., **4**, 153[61b,c]
Mackenzie, K., **4**, 953[8,8g]; **5**, 582[178], 737[146]; **7**, 739[36]
Mackenzie, P. B., **1**, 746[71]; **2**, 629[1], 635[1], 655[149]; **4**, 653[443,444,445]; **6**, 850[126]
MacKenzie, R. K., **3**, 382[36]
McKenzie, S., Jr., **8**, 140[23]
McKenzie, T. C., **3**, 363[87], 365[87], 497[101], 505[101]; **4**, 308[399], 665[10], 667[10], 670[10], 674[10]; **8**, 534[155], 618[127], 623[127]
McKeon, J. E., **4**, 856[102]; **7**, 230[135], 766[174]
McKeown, E., **7**, 98[101]
McKervey, M. A., **2**, 146[71]; **3**, 334[220], 379[3], 380[13], 382[3b], 402[128], 422[68], 862[8]; **4**, 1033[24], 1053[130], 1055[24,135,136], 1056[136]; **5**, 65[67,68], 768[127]; **6**, 142[66]; **7**, 208[79], 211[79], 213[103], 214[105], 390[1]; **8**, 320[80], 541[212], 957[9]
Mackie, D. M., **7**, 293[11]
McKie, J. A., **2**, 66[34], 67[34]; **4**, 119[194], 226[201], 227[201]; **6**, 864[193]
Mackie, R. K., **6**, 205[28]
Mackiewicz, P., **7**, 503[280,281]
McKillop, A., **2**, 855[237]; **3**, 54[275], 418[28], 499[123,124], 505[165], 507[124], 670[58], 673[58,69], 693[141]; **5**, 794[50]; **7**, 154[18-20], 279[166], 335[28], 336[34], 614[3], 665[67], 674[49], 712[63], 718[3], 724[3], 732[58], 737[13], 816[6a], 824[6], 825[6], 827[6a], 828[51], 829[6a,59], 831[6a], 832[6a], 833[6a], 845[80-82], 846[80], 851[18], 872[97], 888[97]; **8**, 365[33]
McKillop, T. F. W., **3**, 386[67,68]
McKinley, S. V., **2**, 5[20], 6[20], 21[20]
McKinney, J. A., **4**, 398[216]; **5**, 736[142e]; **8**, 447[128], 463[128]
McKinney, M. A., **1**, 855[53], 856[53]; **3**, 783[22,25]; **4**, 1007[112]; **8**, 802[86]

McKinnie, B. G., **3**, 88[130], 90[130]; **6**, 531[462], 1066[91]; **7**, 225[64]
Mackinnon, D. J., **7**, 850[4]
Mackinnon, J. W. M., **5**, 419[74]
Mackinnon, L. W., **5**, 576[146]
Mackinnon, P. I., **1**, 277[81]
McKittrick, B. A., **7**, 367[58]
McKnight, M. V., **1**, 474[83]
Macko, E., **8**, 92[68], 568[471]
Mäcková, N., **2**, 362[180]
McKown, W. D., **4**, 1015[199]
Mac'kowska, E., **8**, 774[73]
McKusick, B. C., **5**, 3[22], 70[114], 76[232]; **6**, 227[15], 242[15]; **7**, 156[32]
McLain, S. J., **4**, 485[31]
McLamore, W. H., **2**, 156[117]
McLane, R., **1**, 214[23], 385[112]; **4**, 880[89,90], 883[90], 884[90]
McLaren, F. R., **4**, 18[59], 121[208], 993[161]
McLaren, K. L., **4**, 815[189]
McLaughlin, D. E., **8**, 218[73], 221[73]
McLaughlin, L. A., **8**, 87[30]
McLaughlin, L. M., **5**, 803[87]
McLaughlin, M. L., **8**, 447[128], 463[128]
McLaughlin, S., **8**, 70[233]
McLay, N. R., **6**, 737[40]
MacLean, D. B., **1**, 366[49], 391[49]; **2**, 913[74-76], 946[174,175]; **6**, 561[876], 738[57], 739[57]; **7**, 580[146], 712[67]; **8**, 346[126]
McLean, D. C., **6**, 1013[15]
McLean, E. A., **2**, 757[9]
McLean, J., **7**, 832[71]
McLean, R. L., **2**, 463[92], 464[92]
McLean, S., **4**, 152[55]
McLean, W. N., **3**, 229[232]
MacLeay, R. E., **8**, 43[107], 66[107]
McLennan, D. J., **6**, 953[9]
McLeod, A. L., **4**, 106[138b]
Macleod, I., **2**, 183[20]
McLeod, J., **6**, 669[265]
MacLeod, J. K., **2**, 315[45], 316[45], 456[58,63], 458[58,63]; **7**, 453[68]; **8**, 625[162]
MacLeod, T. K., **2**, 465[103]
Mcllugh, M., **5**, 688[69]
McLoughlin, J. I., **5**, 243[13], 244[13], 410[41], 411[41i]; **7**, 603[124]; **8**, 102[127]
McMahon, G. W., **4**, 1101[192]
McMahon, R. E., **2**, 149[89]
McMahon, R. J., **3**, 891[43], 892[43]
McMahon, W. G., **7**, 261[70]
McManis, J. S., **8**, 56[170], 66[170]
McManus, J., **6**, 624[147]
McManus, N. T., **7**, 489[172]
McManus, S. P., **4**, 365[3], 367[3], 386[152a,b,153,153b]; **5**, 379[112], 383[112], 384[112]; **6**, 35[15]; **7**, 505[286]
McMeeking, J., **1**, 139[3]; **5**, 1188[19], 1197[40,41]
McMichael, K. D., **5**, 834[51]
McMillan, F. L., **7**, 883[179]
MacMillan, J., **3**, 715[43]; **7**, 90[30], 301[61]; **8**, 537[182,183], 824[60]
Macmillan, J. G., **4**, 372[62]; **7**, 111[192]
McMills, M. C., **1**, 733[11], 826[61]; **4**, 379[114,114a], 380[114a]; **5**, 851[170]; **6**, 26[106]
McMorris, T. C., **3**, 226[193]; **7**, 73[103]
McMullen, C. H., **4**, 48[138,138c], 66[138c]
McMullen, C. W., **2**, 956[11]
McMullen, G., **3**, 621[31]
McMullen, G. L., **2**, 353[98], 356[134]; **5**, 732[133,133a,b]
McMullen, L. A., **1**, 248[68]
McMurry, J. E., **1**, 836[145]; **2**, 162[144]; **3**, 15[80], 20[110-112], 218[99], 219[99], 232[256], 239[99], 436[17], 487[48], 489[48], 492[48], 495[48], 579[125], 582[125], 583[118,124-127], 584[125], 585[124,125], 586[152], 587[124,144], 588[125,127a,152,159], 591[165,166,167,168], 592[174,175], 594[174,176], 595[125], 596[125], 610[125,127a,152,165,166,167,168,176], 630[56], 631[56,58], 762[145], 1025[36]; **4**, 7[26], 12[41], 35[98g], 107[142], 258[242], 595[157]; **5**, 147[110], 553[44], 854[181]; **6**, 108[32], 139[49], 665[230], 667[230], 687[377], 766[22], 938[131], 939[139], 942[139,153], 944[131], 980[37], 985[62,63]; **7**, 94[59,61,62], 220[18,22], 502[266,267], 506[300], 556[73], 647[37]; **8**, 371[105,106], 387[58], 527[46], 531[118,121,121c], 889[127], 992[53]
McMurry, T. B. H., **8**, 530[94]
McNab, H., **2**, 352[89], 378[285], 784[40]; **6**, 522[345]
McNab, J. G., **4**, 285[160]
McNab, M. C., **4**, 280[126], 285[160]
McNaghten, E., **2**, 1017[32]
McNally, H. M., **1**, 753[102]
McNamara, J. M., **1**, 436[148]; **5**, 327[28]
McNamara, P. E., **6**, 531[450]
McNeely, K. H., **4**, 311[443]
McNeely, S. A., **8**, 938[93]
McNeil, D., **3**, 623[33]; **8**, 329[22], 338[22]
MacNicol, D. D., **3**, 382[36]
MacNicol, M., **3**, 913[1]
McNicolas, C., **7**, 482[116]
Macoll, A., **6**, 1034[131]
Macomber, D. W., **5**, 1083[57], 1103[149], 1142[86]
Macomber, R., **4**, 395[205]
Macomber, R. S., **5**, 1134[39]; **6**, 845[96]
McOmie, J. F. W., **6**, 632[5], 635[5], 659[5], 664[5], 665[5], 675[5,298], 678[5], 680[5], 682[5], 684[345], 685[345], 687[345]
Macor, J. E., **6**, 554[778]
McOsker, C. C., **8**, 319[70], 486[60], 487[60], 546[312], 801[74]
McPartlin, M., **1**, 305[87]
McPhail, A. T., **2**, 124[203], 232[182]; **3**, 407[149], 572[64]; **4**, 36[100], 49[100a], 306[388]; **5**, 155[37]; **8**, 109[12], 614[89], 620[89], 621[141], 838[20]
McPhail, D. R., **5**, 155[37]
McPhee, D. J., **2**, 357[140]; **6**, 745[86]; **7**, 779[422]
MacPhee, J. A., **1**, 427[112]
McPhee, W. D., **3**, 840[13]; **6**, 207[50]
McPherson, C. A., **1**, 846[18a], 847[18]; **4**, 273[43]
MacPherson, D. T., **4**, 15[50], 593[134]; **5**, 596[37], 598[37]; **6**, 176[96,99]
McPherson, E., **8**, 448[144]
MacPherson, L. J., **7**, 800[30]
McQuaid, L. A., **4**, 373[78]
Macquarrie, D. J., **4**, 445[208]
McQuillin, F. J., **3**, 381[22]; **4**, 2[5], 587[25,28], 604[278], 608[322], 951[1], 968[1], 979[1]; **7**, 449[6], 451[6], 452[6], 453[6]; **8**, 269[76], 445[44,45,59], 452[44,185], 456[207], 458[220], 459[227], 524[7], 526[17,18], 528[69], 530[7,69], 533[153,154], 876[43]
McQuillin, J. F., **3**, 23[141]
Macquitty, J. J., **8**, 779[89]
MacRae, D. M., **1**, 436[147]; **5**, 843[126]
Macrae, R., **8**, 974[130]
McSwain, C. M., Jr., **4**, 486[35], 496[35]
MacSweeney, D. F., **3**, 349[32], 427[86], 712[26]; **8**, 495[29]
McVeigh, P. A., **4**, 12[42]
McVey, J. K., **5**, 636[97]
McVey, S., **5**, 1134[42]
McWhinnie, W. R., **8**, 849[116]
MacWhorter, S. E., **5**, 1043[22], 1049[22]

McWhorter, W. W., Jr., **3**, 225[187]; **6**, 5[28], 8[37]
Madan, P. B., **6**, 490[104], 534[515]
Madawinata, K., **7**, 232[155]
Maddaluno, J., **2**, 227[161]; **4**, 37[106a], 231[276]
Maddison, J. A., **8**, 663[116]
Maddock, J., **1**, 526[96]
Maddocks, P. J., **8**, 717[101]
Maddox, I. S., **7**, 74[109]
Maddox, M. L., **2**, 58[13]; **4**, 38[108,108a]
Maddox, V. H., **1**, 371[73]
Madeja, R., **5**, 850[146]
Madeleyn, E., **8**, 544[252]
Madenwald, M. L., **4**, 438[149]
Mader, H., **4**, 1085[109]
Madesclaire, M., **7**, 194[8], 762[68], 777[68], 778[68]; **8**, 403[5], 404[5]
Madge, N. C., **7**, 683[90]
Madhava, K., **7**, 71[98]
Madhavarao, M., **2**, 934[143]; **4**, 562[36], 576[13]; **5**, 272[5]; **8**, 890[138]
Madhava Reddy, S., **3**, 587[146], 610[146]
Madhusudhana Rao, J., **7**, 595[23], 600[23]
Madin, A., **4**, 231[284], 381[126b], 382[126], 383[126]; **8**, 462[263]
Madison, N. L., **5**, 17[114]
Madja, W. S., **4**, 299[298]
Madjdabadi, A., **8**, 802[83]
Madon, R. J., **8**, 419[19], 420[19], 424[19]
Mador, I. L., **8**, 373[126], 568[480]
Mador, K. L., **8**, 449[159]
Madov, I. L., **8**, 453[191]
Madrigal, D., **6**, 73[26]
Madronero, R., **4**, 292[231]; **6**, 261[10], 273[10,93,94], 275[108,113], 277[127], 280[10], 524[367]
Madsen, J. Ø., **5**, 71[155]
Madumelu, C. B., **8**, 874[27], 882[86]
Madura, J. D., **5**, 257[61]
Madyastha, K. M., **7**, 62[52a], 71[98]
Maeda, A., **6**, 936[111]; **7**, 765[149], 773[149]
Maeda, H., **2**, 386[331], 556[152], 819[99]; **5**, 439[170], 504[274]; **6**, 71[21], 439[69], 814[94], 930[87], 931[88,89]; **7**, 59[39,40], 199[34], 209[89]; **8**, 189[62], 837[14]
Maeda, I., **7**, 657[35]
Maeda, K., **1**, 113[32]; **3**, 496[99], 498[99], 511[99], 515[99]; **4**, 611[353]; **6**, 753[118]; **7**, 172[129], 644[16], 816[13]; **8**, 61[190], 66[190], 773[67], 788[120]
Maeda, M., **3**, 592[173]; **4**, 356[138]; **8**, 838[20]
Maeda, N., **2**, 4[12], 6[12], 10[12b,47], 11[47], 573[55]; **7**, 835[83]
Maeda, S., **4**, 795[84]
Maeda, T., **1**, 834[121a]; **2**, 569[35]; **3**, 984[23], 985[23]; **4**, 251[146]; **5**, 851[164], 1001[16]; **6**, 860[179], 877[37]; **8**, 12[64]
Maeda, Y., **5**, 96[118]; **7**, 314[40], 315[40]; **8**, 49[116], 66[116]
Maehara, M., **8**, 408[79]
Maehara, N., **5**, 323[16]
Maehling, K. L., **2**, 362[185]
Maehr, H., **7**, 243[63]
Maekawa, E., **4**, 394[194,195], 413[275], 744[135]; **7**, 519[22], 524[51]
Maekawa, H., **7**, 796[13]
Maekawa, K., **3**, 96[168], 104[168], 108[168], 117[168]; **8**, 966[74]
Maekawa, T., **1**, 559[150]
Maemura, K., **4**, 856[101]
Maeno, H., **8**, 196[120]
Maercker, A., **1**, 9[45], 722[269], 755[115], 812[115], 813[115]; **3**, 778[5]; **4**, 868[12], 871[35], 876[35b,58], 887[12]; **5**, 1122[29], 1123[29]; **6**, 171[1], 198[1]; **8**, 842[43,43b], 847[43b], 857[195], 858[195]
Maerkl, G., **5**, 635[86]
Maerten, G., **6**, 432[123]
Maeshima, T., **6**, 538[556]; **8**, 407[54]
Maestro, M. C., **6**, 67[13]
Maeta, M., **3**, 303[56]
Maetzke, T., **1**, 37[239]; **3**, 41[227]
Maffei, S., **6**, 255[168]
Maffi, S., **8**, 683[92], 689[92]
Maffrand, J.-P., **2**, 765[74], 817[92]
Mafunda, B. G., **3**, 216[73], 423[72]; **4**, 1009[139]
Magaha, H. S., **1**, 822[30]
Magari, H., **5**, 809[119]; **7**, 543[21]
Magarramov, A. M., **7**, 494[202]
Magat, E. E., **6**, 263[26], 264[26], 265[43,45], 268[66], 269[75], 270[26]
Magdesiava, N. N., **1**, 632[68], 644[68]
Magdesieva, N. M., **4**, 342[68]
Magdzinski, L., **6**, 8[38]; **7**, 362[32]; **8**, 565[448]
Magee, A. S., **2**, 388[343]; **4**, 187[97]; **6**, 1020[48]
Magee, J. W., **4**, 367[14], 368[14], 369[14]; **6**, 26[104]
Magee, W. L., **4**, 1104[212]
Magelli, O. L., **4**, 305[368]
Magennis, S., **4**, 849[81]
Mager, H. I. X., **8**, 643[37]
Magerlein, B. J., **2**, 409[2], 410[2], 411[2]
Mageswaran, S., **3**, 923[45], 949[98], 954[98], 963[121]; **5**, 847[135]; **6**, 897[98]
Maggio, J. E., **3**, 751[89]; **4**, 970[69]
Maggiora, G. M., **1**, 314[137], 315[137]; **8**, 89[43]
Maghin, G., **6**, 48[86]
Magi, M., **8**, 343[115], 344[115], 345[115], 346[115]
Magi, S., **5**, 464[114], 466[114]
Magid, R. M., **3**, 257[121], 467[194]; **6**, 205[33,34], 830[6], 834[40]; **8**, 965[65]
Magnane, R., **8**, 886[107]
Magnani, A., **6**, 959[43]; **7**, 236[14]
Magnenat, J.-P., **3**, 581[109]
Magnin, D. R., **4**, 785[21], 790[40], 791[21,40]; **5**, 516[28]
Magno, F., **7**, 769[215]
Magnol, E., **4**, 794[76]
Magnus, G., **4**, 10[31,31b]
Magnus, P. D., **1**, 580[1], 622[73,74], 623[77,79], 630[10], 731[4], 737[31], 815[4], 828[79], 829[86]; **2**, 57[6], 88[29], 426[39], 597[6], 712[42], 713[46], 716[59], 739[44], 920[96]; **3**, 27[166], 86[36,48], 105[215], 158[36], 163[36], 169[36], 173[36], 174[36], 198[51], 217[95,95a], 255[105], 491[72], 673[70], 674[70b], 816[80], 862[2], 1025[33], 1027[44]; **4**, 78[54], 115[184b], 258[251], 989[142]; **5**, 385[130], 531[75], 549[75], 581[175], 736[145], 737[145], 777[190,191], 778[192,193,194], 779[194], 809[118], 1053[40,42], 1060[40a,42,56,57]; **6**, 239[77], 646[103], 659[195], 754[119], 832[12], 865[12], 930[85], 1007[150], 1023[71], 1047[33]; **7**, 105[149], 128[64], 146[64], 199[37], 244[69,70], 456[107], 771[266], 772[266], 816[6b], 824[6], 825[6]; **8**, 403[8], 836[2], 842[2a], 843[2a], 844[2a]
Magnusson, G., **1**, 429[124], 566[219]; **3**, 763[149], 767[165], 768[149,166]; **4**, 211[92,93]; **5**, 687[56], 942[229]; **7**, 410[101]
Magolda, R. L., **1**, 409[37], 876[99]; **4**, 371[60], 372[60], 413[271a,b,272,273]; **5**, 534[92]; **6**, 158[184], 1031[113]; **7**, 131[78], 523[46], 524[54]; **8**, 847[99], 848[99], 849[99]
Magomedov, G. K., **4**, 701[28,29], 702[28,29]
Magomedov, G. K. I., **8**, 765[11]
Magoon, E., **3**, 813[63]
Magoon, E. F., **8**, 735[16]
Magriotis, P. A., **1**, 446[193]; **4**, 218[140]; **5**, 522[46]; **7**, 278[165], 279[165]
Mague, J. T., **1**, 440[169], 451[216]
Maguet, M., **6**, 568[932,933]

Mah, H., **6**, 573[964]
Mah, R., **7**, 648[41]
Mah, T., **4**, 486[33]; **5**, 391[142]
Mahachi, T. J., **1**, 268[54], 269[54c]; **3**, 568[57], 570[57], 599[205,222], 602[222]; **4**, 809[162]; **8**, 132[12,13], 134[12,13], 624[154], 628[154]
Mahadevan, C., **1**, 305[83]
Mahadevan, S., **5**, 419[72]
Mahaffey, R. L., **7**, 507[307]
Mahaffy, C. A. L., **4**, 519[18], 520[18], 703[32], 704[32]; **5**, 1045[29]
Mahain, C., **7**, 158[37]
Mahajan, J. R., **2**, 855[245-248], 856[248]; **7**, 507[309]; **8**, 351[168], 353[155]
Mahajan, M. P., **5**, 441[179]; **6**, 539[578], 552[700]
Mahajan, S. N., **8**, 965[67]
Mahalanabis, K. K., **1**, 466[49]; **2**, 103[30]; **3**, 44[241], 728[35]; **4**, 72[27], 74[27], 249[115], 257[115], 260[115]; **5**, 11[80]; **6**, 1018[41]; **7**, 174[135]
Mahalingam, S., **1**, 328[28]; **2**, 175[184], 619[150]; **4**, 158[77]
Mahan, J. E., **5**, 566[100]
Mahapatro, S. N., **4**, 1033[22,22a], 1057[22a]; **8**, 917[115], 918[115]
Mahara, R., **8**, 883[92]
Mahatantila, C. P., **3**, 1036[83]
Mahato, S. B., **2**, 740[61]; **5**, 176[133]; **7**, 66[76,77], 68[76,77,83b]
Mahdavi-Damghani, Z., **4**, 72[27], 74[27], 249[115], 257[115], 260[115]
Mahdi, W., **1**, 12[61], 13[65], 18[93,94], 19[100], 23[123], 29[147]; **6**, 723[145]
Mahe, C., **4**, 603[272], 626[272], 645[272]
Mahendran, M., **4**, 707[42]
Mahgoub, S. A., **2**, 772[15]
Mahidol, C., **1**, 526[94]
Mahiou, B., **4**, 730[64]
Mahjoub, A., **4**, 50[142,142p]
Mahler, H. R., **2**, 466[119], 469[119]
Mahler, J. E., **4**, 664[5]
Mahler, U., **2**, 226[159]
Mahmood, T., **8**, 864[241]
Mahmoudi, M., **3**, 990[34,34a]
Mahon, M., **7**, 53[1], 63[1], 307[14]
Mahon, M. F., **1**, 526[96]; **2**, 655[140]; **4**, 412[268e]
Mahoney, L. R., **3**, 661[24]
Mahoney, W. S., **4**, 254[184]; **8**, 550[330]
Mahoungou, J. R., **3**, 420[51], 421[51]
Mahrwald, R., **6**, 654[154]
Mahy, J.-P., **7**, 477[78], 483[78,133], 484[78,133,134], 500[133]
Mahy, M., **6**, 540[584]
Mai, J., **8**, 916[112], 918[112]
Mai, K., **1**, 555[114,118]; **4**, 278[96]
Maia, A., **6**, 19[67], 236[54]
Maienfisch, P., **2**, 857[250]; **4**, 372[54]; **5**, 863[254]; **7**, 77[120b], 565[98], 567[98]; **8**, 542[233]
Maienthal, M., **8**, 64[203], 66[203]
Maier, D. P., **2**, 380[298]
Maier, G., **1**, 303[80]; **4**, 869[20]; **5**, 680[23], 686[42], 713[63], 714[63]; **8**, 13[74,75]
Maier, L., **6**, 171[3], 198[3], 601[1]; **8**, 754[89,90,92], 858[207], 860[219], 862[219]
Maier, M., **6**, 37[34]; **8**, 141[43]
Maier, M. E., **5**, 461[110], 464[110-113], 466[110-113]
Maier, R. D., **2**, 782[26]
Maier, T., **2**, 368[238]; **6**, 546[653], 556[821,830], 562[887]
Maier, W. F., **3**, 25[160]; **5**, 1148[118]; **7**, 144[154]; **8**, 289[23], 319[79], 320[80], 431[62], 612[74]
Maierhofer, A., **1**, 373[87], 374[87]
Maignan, C., **6**, 150[125,127,130]
Maigrot, N., **1**, 555[122]; **6**, 70[20]
Maikov, S. I., **6**, 535[534]
Mailhe, A., **8**, 285[6]
Maillard, B., **4**, 723[38], 738[38,101], 747[38], 753[166], 754[176]; **5**, 901[28]; **7**, 7[43]; **8**, 264[47]
Mains, B., **4**, 987[147]
Maio, G. D., **8**, 875[29]
Maiolo, F., **4**, 441[177]
Maione, A. M., **7**, 237[31], 310[27]
Maiorana, S., **4**, 522[54], 523[57,58], 524[61], 710[52]; **5**, 1079[51]; **6**, 114[79], 178[121]; **8**, 185[27]
Maiorova, V. E., **4**, 1010[155]
Mairanovskii, S. G., **8**, 611[63]
Mairanovsky, V. G., **6**, 659[194]
Mais, R. H. B., **3**, 383[49]
Maiti, S. B., **3**, 856[88]
Maiti, S. N., **3**, 741[51]; **6**, 76[50]; **8**, 384[29], 412[115]
Maitland, P., **2**, 170[172]; **6**, 970[123]
Maitland, W., **2**, 146[69]; **5**, 753[54]
Maitlis, P. M., **4**, 586[14], 590[14], 834[1e]; **5**, 1134[42], 1135[47], 1136[53], 1146[53], 1147[112,113]; **7**, 94[55], 450[9], 452[56], 851[18]; **8**, 445[34,54,54c,d], 454[200], 773[70], 774[70]
Maitra, A. K., **5**, 165[84]
Maitra, S. K., **2**, 740[61]
Maitra, U., **5**, 344[66], 345[66], 346[66], 453[66]; **7**, 43[38], 47[54], 48[58]
Maitte, P., **3**, 324[154]; **5**, 464[114], 466[114]; **6**, 675[301]
Majchrzak, M. W., **2**, 742[76], 743[85], 965[64]
Majerski, Z., **1**, 859[63]; **3**, 746[67]; **4**, 955[15]; **8**, 356[190], 357[190]
Majert, H., **7**, 603[116]
Majerus, G., **3**, 564[7]
Majeski, E. I., **7**, 801[42]
Majeste, R., **6**, 962[74]
Majetich, G., **1**, 114[35,36], 580[1]; **2**, 120[177], 160[133]; **3**, 20[113], 212[39], 253[92]; **4**, 98[114,114b,116], 113[114], 155[73]; **5**, 7[51], 15[108], 20[140], 514[6], 806[105], 809[105], 850[144], 913[104], 1030[93]; **7**, 565[97], 821[28], 822[33]
Majetich, G. F., **7**, 377[90]
Majewski, M., **3**, 50[268], 589[162], 610[162]; **4**, 373[82], 386[148b], 387[148]; **8**, 88[39]
Majewski, P., **8**, 411[106]
Majewski, R. W., **7**, 229[108]
Majid, T. N., **1**, 212[5,5a], 213[5,14]
Majima, J., **7**, 878[138]
Majima, T., **5**, 154[28]
Majori, L., **4**, 36[103,103c]
Majumdar, A., **2**, 725[119]
Majumdar, D., **3**, 199[63]
Majumdar, D. J., **1**, 491[32-34], 498[33], 501[33], 502[33]
Majumdar, M. P., **2**, 736[30]
Majumdar, S. P., **1**, 850[33], 853[33]; **3**, 783[27]
Mak, A. L. C., **4**, 293[235]
Mak, C.-P., **1**, 372[79]
Mak, K., **4**, 853[91], 854[94], 856[98]
Mak, K. T., **6**, 687[375]
Mak, T. C. W., **3**, 557[38], 877[89]
Makarova, L. G., **7**, 632[57]; **8**, 850[120], 851[132]
Makerov, P. V., **6**, 550[673,674]
Makhno, L. P., **8**, 616[98]
Makhon'kova, G. V., **6**, 2[9], 3[9]
Makhubu, L. P., **8**, 218[72], 219[72]
Maki, T., **8**, 285[7]
Maki, Y., **3**, 219[107]; **4**, 436[142,143]; **6**, 734[6,7]; **7**, 877[135]; **8**, 244[65], 250[99], 908[76]

Makik, N., **5**, 515[18], 547[18]
Makin, G. I., **7**, 602[106]
Makin, H. L. J., **6**, 996[106]
Makin, M. I. H., **7**, 738[32]
Makin, S. M., **2**, 662[15,19], 663[24], 664[15,19,24]; **5**, 431[122]; **6**, 489[80]; **7**, 660[39]
Makino, K., **6**, 221[132]
Makino, N., **2**, 541[75]
Makino, S., **2**, 603[46]; **5**, 282[33], 286[33], 595[11], 596[11a], 601[47], 603[47,48a,e], 605[58,60,60a,61], 608[65], 609[60,60c,61]
Makino, T., **1**, 349[147]; **8**, 562[423]
Makisumi, Y., **3**, 976[6]
Makita, Y., **7**, 87[18]
Makosza, M., **2**, 362[183], 429[50], 430[52a], 431[52a,b], 432[56]; **3**, 127[321], 158[446], 159[446], 174[528]; **4**, 424[14-16], 426[15,16], 429[86], 431[15,101], 432[15,16,102-110,112-115], 433[117,118], 446[214], 1001[20-22,29,44], 1004[70]; **6**, 533[485], 556[818]
Makovetskii, K. L., **5**, 1146[109]
Makovetskii, Yu. P., **6**, 500[180]
Maksimova, N. G., **8**, 780[91]
Maksimovic, Z., **7**, 92[41,41a], 94[41]
Mal, D., **2**, 547[104], 549[104]; **4**, 14[47,47c], 258[247,249], 373[77]
Malacria, M., **1**, 892[149]; **3**, 440[46]; **4**, 308[408], 794[76]; **5**, 301[77], 311[106], 772[157,158,159,160], 774[169], 797[66], 800[79], 802[82], 936[196], 1144[100]; **8**, 857[192]
Malamas, M. S., **1**, 434[140]; **2**, 264[58]
Malamidou-Xenikaki, E., **3**, 380[9]
Malanco, F. L., **1**, 552[84]
Malanga, C., **5**, 942[230]
Malaprade, L., **7**, 708[33]
Malarek, D. H., **2**, 808[50]
Malassa, I., **5**, 404[13]
Malassiné, B., **8**, 505[73]
Malatesta, L., **6**, 295[249]
Malatesta, M. C., **5**, 1147[112]
Malatesta, V., **7**, 488[153]
Malavaud, C., **8**, 663[119]
Malchenko, S., **4**, 681[53]
Malcherek, R., **2**, 365[208]
Maldonado, L., **3**, 197[39]
Maldonado, L. A., **1**, 547[62], 552[62,84], 554[105,107]; **4**, 12[38], 71[16b], 113[16b,169], 139[3], 259[274]
Malecot, Y.-M., **5**, 559[64]
Malek, F., **7**, 721[16]; **8**, 825[66]
Målek, J., **8**, 403[10]
Málek, J., **2**, 268[65]; **8**, 2[5], 19[5], 238[21], 241[21], 242[21], 245[21], 247[21], 251[21], 253[21], 254[21], 269[79], 272[79], 274[137], 279, 314[32-34,36], 541[205,207], 542[205], 544[267], 907[72], 918[72], 967[80,81]
Maleki, M., **2**, 65[28]
Malek-Yazdi, F., **6**, 473[75], 480[109]
Malen, C., **6**, 268[60]
Malewski, G., **7**, 768[201]
Malewski, M., **4**, 313[468]
Malherbe, J. S., **6**, 189[185]
Malherbe, R., **5**, 639[124], 805[99], 829[14], 1025[81,81e]; **7**, 205[65]
Malhotra, N., **6**, 527[406]
Malhotra, R., **2**, 736[20]; **6**, 104[9], 109[44], 210[75], 214[93], 215[93], 647[108], 654[152]; **8**, 406[47]
Malhotra, S., **3**, 602[218], 607[218]
Malhotra, S. K., **3**, 23[138]; **8**, 563[428]
Mali, R. S., **1**, 461[7], 463[24-26]
Malicky, J. L., **8**, 373[133], 376[133]
Malik, A., **6**, 91[129]
Malik, A. A., **8**, 384[27]
Malik, F., **7**, 561[84]
Malinowski, M., **3**, 583[122], 584[130]; **6**, 985[64]; **8**, 390[85], 391[85], 392[96]
Malkin, L. I., **2**, 466[123], 469[123]
Malkin, L. S., **8**, 770[41]
Mal'kina, A. G., **4**, 55[157], 57[157o]
Mall, T., **7**, 470[9], 487[9], 495[205]
Mallaiah, B. V., **7**, 136[112], 137[112]
Mallamo, J. P., **2**, 74[80], 363[192], 388[192]; **3**, 155[433], 157[433], 230[238]; **4**, 18[59], 86[81], 213[105-107], 215[105-107], 262[313]; **6**, 150[121], 176[93]; **8**, 844[71], 847[71]
Mallart, S., **3**, 47[256]; **6**, 118[108]
Malleron, A., **2**, 464[95,95b]
Malleron, J. L., **4**, 590[105], 591[110], 615[105], 616[110], 629[105]
Mallet, A. I., **7**, 707[27]
Mallet, M., **1**, 474[96,103]
Mallick, I. M., **1**, 648[128], 649[128], 650[128], 672[128], 675[128], 679[128], 708[128], 710[128], 715[128], 716[128], 862[78]
Mallon, B. J., **5**, 709[46]; **6**, 707[41]; **8**, 726[192]
Mallory, C. W., **5**, 699[4], 700[4], 723[4], 724[4,113], 726[4], 729[4]
Mallory, F. B., **5**, 699[4], 700[4], 723[4], 724[4,113], 726[4], 729[4]
Mallory, H. E., **8**, 366[35]
Malloy, T. B., Jr., **5**, 901[24]
Malmberg, H., **3**, 219[114], 512[195]; **4**, 176[51], 227[209], 229[218,222]
Malmberg, M., **2**, 1052[38], 1053[54]
Malmberg, W.-D., **5**, 436[157]; **6**, 475[92]
Malon, P., **6**, 801[29]
Malone, G. R., **2**, 492[53,54], 493[53,54]; **6**, 274[104,107]; **8**, 276[149]
Malone, J. F., **4**, 1086[116]
Malone, T. C., **1**, 592[22,24]; **2**, 1031[86], 1034[88], 1035[88], 1057[69]
Maloney, K. M., **6**, 294[240]
Malova, O. V., **6**, 530[418]
Malova, T. N., **3**, 304[65]
Malpass, D. B., **8**, 736[21]
Malpass, J. M., **6**, 65[1]
Malpass, J. R., **5**, 105[198]; **7**, 480[101], 483[129]
Malquori, S., **5**, 1155[162], 1156[162]
Malrieu, J.-P., **6**, 120[119], 172[9]
Malsumoto, M., **1**, 546[56]
Malte, A. M., **3**, 217[90], 219[90], 863[15], 864[15]
Maltenieks, O. J., **6**, 965[90]
Mal'tsev, V. V., **2**, 616[135], 804[40]
Malunowicz, I., **8**, 533[154]
Malusà, N., **4**, 20[63], 21[63]
Malwitz, D., **3**, 587[142]
Maly, N. A., **3**, 381[31], 382[31]
Malzieu, R., **4**, 95[102a]
Mamada, A., **8**, 847[91]
Mamdapur, V. R., **2**, 381[304]; **3**, 416[16], 417[16]; **7**, 453[64,82], 454[64,82]
Mamedaliev, G. M., **3**, 306[84]
Mamedov, I. M., **3**, 304[60]
Mami, I. S., **3**, 680[92]
Mammarella, R. E., **2**, 70[50], 77[90], 587[143]; **6**, 175[66], 182[66]
Mamoli, L., **8**, 558[395]
Mamyan, S. S., **5**, 1057[50]
Man, E. H., **2**, 834[155]
Man, T. O., **5**, 834[55]
Manabe, H., **3**, 566[26]; **8**, 370[94], 404[18]
Manabe, K., **3**, 222[144]; **4**, 13[44,44c], 253[169], 262[308]; **6**, 837[60], 942[154], 944[154]; **8**, 463[269]
Manabe, O., **8**, 87[34], 95[88], 364[20], 367[60]
Manabe, S., **4**, 27[79]; **7**, 366[53]

Manadhar, M. D., **4**, 222[177]
Manage, A. C., **4**, 116[187]
Manami, H., **7**, 774[332]
Manas, A.-R. B., **1**, 511[31]; **4**, 11[36], 113[171]; **6**, 140[56]
Mancelle, N., **1**, 366[47], 566[210]; **2**, 899[25], 900[25]
Manchand, D. S., **4**, 629[416]
Manchand, P. S., **3**, 168[491], 169[491], 171[491], 407[148]; **8**, 695[119]
Mancinelli, P. A., **1**, 389[138]; **7**, 162[61]
Mancini, V., **8**, 494[25]
Mancuso, A. J., **7**, 292[5], 297[28], 299[5], 300[5,56], 396[24]
Mancuso, N. R., **7**, 691[18]
Mandai, T., **1**, 802[304], 803[307]; **2**, 74[74], 166[154], 270[74], 374[277]; **3**, 135[345,346,347,348], 136[347,348], 137[347], 139[347,348], 141[348], 144[347,348], 168[508], 169[508], 174[345], 586[138], 639[75], 652[220]; **4**, 590[101], 609[329,331], 964[46]; **5**, 767[120], 830[31], 833[49]; **6**, 149[104], 902[130], 927[77], 1022[60]; **7**, 453[83,88-90], 454[83,96], 455[88-90]; **8**, 961[39]
Mandal, A. K., **1**, 820[15]; **7**, 595[24]; **8**, 541[207], 708[37,40], 709[50,51], 710[54], 720[134], 721[54,134,143], 722[134,144]
Mandal, A. N., **6**, 507[234]
Mandal, S. B., **6**, 620[125-127], 625[163]; **8**, 249[96], 943[119]
Mandava, N. B., **7**, 673[29]
Mandel, G., **2**, 846[203]
Mandel, G. S., **5**, 814[136], 841[86]; **6**, 464[30], 471[30], 859[166]; **8**, 11[61], 36[72], 37[72], 38[72], 44[72], 66[72], 948[152]
Mandel, N., **2**, 846[203]
Mandel, N. G., **3**, 135[363], 136[363], 139[363], 142[363], 156[363]
Mandel, N. S., **5**, 814[136], 841[86]; **6**, 464[30], 471[30], 859[166]; **8**, 11[61], 36[72], 37[72], 38[72], 44[72], 66[72], 948[152]
Mandelbaum, A., **3**, 741[51], 745[65]
Mandell, G. S., **1**, 880[116]
Mandell, L., **3**, 577[85], 579[94], 595[192], 640[110]; **6**, 685[358]; **8**, 527[52]
Mandell, N. S., **1**, 880[116]
Mander, L. N., **1**, 411[46], 752[95], 861[73]; **2**, 839[180]; **3**, 21[123], 715[40], 934[70]; **4**, 27[79,80], 121[209,209c], 373[71], 1040[81,82], 1043[81,82]; **5**, 1125[54]; **6**, 124[142], 893[85], 895[91], 896[91,94], 1056[53]; **7**, 90[31], 199[36], 200[39], 367[54], 375[54], 552[55]; **8**, 212[20], 490[6], 492[6], 493[6], 500[51], 502[58], 503[68,70], 505[58,77,81], 509[90,92], 510[92], 513[6], 520[6]
Mandeville, W. H., **4**, 176[46], 256[205]
Mandler, D., **8**, 97[96]
Mandolini, L., **3**, 55[284]; **4**, 786[25]; **6**, 24[96], 69[17]
Mandon, M., **1**, 567[222]
Mandre, G., **8**, 451[178]
Mandville, G., **5**, 21[151], 22[151]
Mane, R. B., **2**, 365[213], 764[66]; **8**, 537[179]
Manecke, G., **2**, 388[338,339]
Manek, M. B., **7**, 770[246]
Manescalchi, F., **4**, 344[78a]; **6**, 21[80], 22[80]; **8**, 289[27]
Manfre, R. J., **4**, 892[144]
Manfredi, A., **2**, 229[168]; **6**, 425[66]; **7**, 194[9], 429[150a], 777[382]
Manfredini, S., **8**, 394[116]
Mangan, F. R., **8**, 626[175], 629[175]
Mangeney, P., **1**, 107[3]; **3**, 223[154], 226[207], 579[100]; **4**, 152[59], 183[81], 207[59], 209[68,69], 210[77,78], 895[164], 900[164,179]; **6**, 849[122], 921[49], 922[50]
Mangiaracina, P., **8**, 817[26]
Mangini, A., **1**, 506[8]; **4**, 330[5]; **6**, 133[4]
Mangold, D., **7**, 674[36]
Mangold, R., **3**, 914[7], 924[7]
Mangoni, L., **4**, 347[95]; **6**, 980[34]; **7**, 438[17-19], 445[17-19,58]; **8**, 891[147]
Mangravite, J. A., **2**, 6[30]; **7**, 616[16]
Mangum, M. G., **6**, 966[93]
Manhas, M. S., **2**, 296[85], 919[92]; **4**, 45[126], 553[6]; **5**, 86[13,14,18], 92[68,77], 95[68,89,93,95], 96[68,105,114,116,119,121], 98[121], 100[141]; **6**, 253[155], 744[74]; **7**, 454[99]
Mani, J., **5**, 664[38]
Mani, R. S., **4**, 1033[23]
Mani, S. R., **5**, 629[48]
Mania, D., **8**, 254[125]
Manion, M. L., **3**, 891[46]
Manisse, N., **5**, 797[64], 930[174], 932[180], 938[180]
Manitto, P., **2**, 853[231]; **7**, 109[183], 153[9]; **8**, 560[404]
Maniwa, K., **1**, 515[57]; **6**, 926[67], 927[75]
Manjula, B. N., **6**, 790[118]
Man Lee, C., **4**, 425[25]
Manmade, A., **3**, 741[50]
Mann, A., **2**, 1072[140]; **3**, 356[57]; **5**, 335[49]; **7**, 318[60]
Mann, B., **2**, 142[46]
Mann, B. E., **1**, 440[171]; **7**, 452[56]
Mann, C. K., **7**, 769[218], 803[56,57], 854[55], 855[55]
Mann, C. M., **7**, 283[188], 285[188]
Mann, F. G., **2**, 149[88]; **3**, 382[36]
Mann, G., **3**, 379[3]
Mann, I. S., **3**, 514[210]
Mann, J., **2**, 523[91], 819[101], 823[115]; **3**, 840[15]; **4**, 380[119], 956[19]; **5**, 594[6], 595[16,17,20], 596[16,17,20], 605[55,62], 608[6,64], 609[64], 612[76]; **6**, 764[8]
Mann, K. R., **4**, 521[45]
Mann, M. E., **8**, 624[156]
Manna, S., **2**, 1085[21]; **5**, 499[250,251], 500[250,251]; **6**, 11[45], 206[42], 210[42], 218[42], 536[546], 538[546]; **7**, 87[18,18a], 378[97]
Mannafov, T. G., **7**, 521[37]
Mannich, C., **2**, 1090[60]
Männig, D., **1**, 301[73]; **8**, 720[130]
Mannin, G. I., **7**, 599[64,65]
Manning, B., **3**, 889[24]
Manning, C., **5**, 720[97,97b]
Manning, D. T., **3**, 647[193]; **5**, 797[60]
Manning, J. M., **6**, 790[113,114]
Manning, M. J., **3**, 254[102]
Manning, R. E., **7**, 661[44]
Mano, E., **5**, 210[57]
Mano, T., **7**, 381[104]
Manoharan, P. T., **8**, 446[65]
Manoli, F., **4**, 527[67]
Manor, S., **6**, 219[124]
Manov-Yuvenskii, V. I., **3**, 1039[97]
Manring, L. E., **5**, 72[182]; **7**, 881[159]
Manriquez, J. M., **8**, 447[131], 671[18], 673[23], 675[23], 691[23], 697[129]
Mansfield, C. A., **2**, 964[59]; **4**, 670[15]
Mansfield, G. H., **4**, 55[156]
Mansilla, H., **3**, 391[90], 393[90], 395[97]
Manske, R., **6**, 134[19]
Manske, R. H., **8**, 243[46]
Mansour, T. S., **1**, 462[16]
Mansouri, A., **4**, 1019[228]
Mansouri, L. M., **5**, 692[95]
Mansuri, M. M., **2**, 1[2]
Mansuy, D., **7**, 95[72], 108[176], 297[32], 383[109], 426[148c], 477[78], 483[78,133], 484[78,133,134], 500[133]
Manta, E., **4**, 373[87]; **7**, 413[118]
Manteuffel, E., **7**, 359[19]
Manthey, M. K., **7**, 355[37]
Mantione, R., **3**, 104[210], 111[210]
Mantlo, N., **1**, 92[67], 463[30]
Mantlo, N. B., **3**, 255[109], 512[198]; **5**, 736[143,145], 737[145]

Mantz, I. B., **5**, 904[54]
Mantzaris, J., **5**, 1130[6]
Manuel, G., **8**, 873[18], 874[20]
Manwaring, R., **6**, 860[176]
Manyik, R. M., **2**, 834[155]; **4**, 589[77], 590[77], 591[77], 597[181], 598[181], 638[181]
Manz, F., **4**, 1081[83]
Manzano, C., **1**, 117[56]
Manzini, G., **4**, 20[63], 21[63]
Manzocchi, A., **6**, 227[19], 228[19]; **7**, 279[171], 709[37], 765[134], 844[60]; **8**, 187[38], 190[71,73], 191[73], 240[30,31], 244[30,56], 263[29]
Mao, C.-L., **2**, 834[155]
Mao, D. T., **5**, 523[48], 526[57], 788[12], 954[298], 1003[21], 1016[63]; **6**, 723[146]
Mao, M. K. T., **2**, 186[36]; **5**, 922[136]; **6**, 888[65]
Mapelli, C., **4**, 988[137]
Maples, P. K., **8**, 447[101,102], 450[101,102]
Maquestiau, A., **2**, 351[80], 364[80]
Mar, E. K., **3**, 225[186]
Marais, C. F., **6**, 108[36]
Marais, D., **4**, 48[137]
Marakowski, J., **5**, 403[7], 404[7,18]; **6**, 744[77]
Maran, F., **6**, 575[966]
Maraschin, N. J., **7**, 15[144]
Maravigna, P., **1**, 294[43]
Marazza, F., **5**, 11[83]
Marbet, R., **5**, 828[4], 830[29], 862[29d]; **6**, 836[55]
Marburg, S., **3**, 262[166]
Marbury, G. D., **6**, 904[142]
Marcaccioli, S., **7**, 778[402]
Marcantoni, E., **4**, 86[78e]; **6**, 115[83], 938[128], 942[128], 944[128]
Marcantonio, A. F., **7**, 24[36], 25[36]
Marcelli, M., **3**, 1046[1]
March, J., **3**, 777[1], 908[147], 909[147]; **5**, 721[99]; **6**, 226[10,14], 240[81], 256[10], 257[10]; **7**, 119[1]; **8**, 410[94]
Marchalin, S., **6**, 533[480], 550[480]
Marchand, A. P., **4**, 238[3], 1073[23]; **5**, 71[129], 87[41], 92[41], 109[41], 703[15,19], 705[19,23], 754[59], 1130[7]; **6**, 107[23], 744[72]; **8**, 36[88], 52[149], 66[88,149]
Marchand, E., **5**, 488[195]; **6**, 540[585,586]
Marchand-Brynaert, J., **5**, 108[209,210,211], 109[209,210,211,216,218,222], 110[210,222,223], 111[210,222], 112[222,223], 113[236], 410[40]; **6**, 430[96], 520[341,342], 543[341], 544[342]
Marchart, G., **8**, 212[11]
Marchelli, R., **6**, 936[109]; **7**, 197[21]; **8**, 406[50]
Marchenko, A. P., **6**, 500[180]
Marchenko, N. B., **6**, 507[228,229], 554[755,764,784,786,792]
Marchese, G., **1**, 413[57], 416[67], 452[220]; **3**, 208[2], 217[2], 230[235,236], 283[49], 436[11], 441[48], 446[87], 449[48,101], 463[153,154,155,166], 485[29], 492[79], 493[29], 503[29,79], 513[29,79]; **4**, 93[93d]; **6**, 446[101]
Marchese, J. S., **6**, 705[18]
Marchesini, A., **8**, 349[145]
Marchetti, F., **5**, 1174[36]
Marchetti, M., **4**, 939[76]; **5**, 1152[144]
Marchi, M., **4**, 600[243], 601[243]
Marchington, A. P., **4**, 159[82]
Marchini, P., **8**, 31[45], 36[45], 66[45]
Marchiori, M. L. P. F. C., **1**, 411[45]
Marchioro, C., **4**, 50[142]; **5**, 370[102], 371[102]
Marchioro, G., **1**, 480[153]; **6**, 150[114], 936[108]; **7**, 205[64]
Marchon, J.-C., **7**, 384[114b]
Marciniec, B., **8**, 764[8], 765[13], 774[72,73]
Marcinow, Z., **8**, 114[59], 509[91], 510[94]
Marco, J. A., **6**, 172[19], 780[71]
Marco, J. L., **1**, 555[119], 559[147,148]
Marco-Contelles, J., **2**, 636[55], 637[55], 640[55]
Marconi, W., **8**, 171[110]
Marçot, B., **8**, 587[39]
Marcotullio, M. C., **4**, 1040[77]
Marcus, R. A., **7**, 852[36]
Marcuzzi, F., **4**, 277[89,90], 285[90], 298[280]
Marczak, S., **8**, 163[42]
Marder, T. B., **4**, 315[532]
Mardis, W. S., **3**, 103[207]
Marecek, J. F., **6**, 70[18], 620[121-127], 625[121,163]
Maréchal, E., **3**, 331[199]
Mareda, J., **5**, 257[61,61c], 560[70], 628[45], 913[108]
Marei, A., **6**, 421[25]
Marek, I., **3**, 223[154], 226[207]; **4**, 209[69], 895[164], 900[164,179]; **6**, 849[122]
Mares, F., **7**, 452[46], 674[51]
Maresca, L. M., **7**, 43[46]
Marfat, A., **1**, 415[64], 428[115,116]; **3**, 243[17], 249[17], 263[17,172], 500[132], 505[132]; **4**, 91[88f], 250[135], 255[135], 898[174], 899[174]
Marfisi, C., **3**, 892[47]
Margaretha, P., **2**, 346[48], 347[48], 349[66], 353[48,94], 355[48,126], 356[48,66], 357[141,142], 358[48], 365[48], 367[48], 369[48], 374[48]; **3**, 579[95]; **4**, 251[144], 257[144]; **5**, 123[1], 126[1], 128[28], 164[75], 176[75], 225[96]; **7**, 876[126]; **8**, 134[36]
Margaryan, A. Kh., **5**, 128[31]
Margerum, D. W., **3**, 213[51]
Margolis, E. T., **4**, 283[147]
Margolis, N. V., **1**, 34[228,232]
Margot, C., **4**, 869[22]; **6**, 174[57]
Margraf, B., **5**, 257[60]
Margrave, J. L., **5**, 857[229]
Margulies, H., **3**, 804[7], 809[38], 815[75]
Margulis, M. A., **4**, 969[66]
Marhold, H., **2**, 362[184]
Mariano, P. S., **2**, 1035[93], 1037[96-100], 1038[101-103], 1039[104], 1040[105]; **4**, 753[165]; **5**, 194[1], 196[1], 197[1], 198[1], 210[1], 480[178], 531[79], 829[21]; **6**, 756[122], 760[122]; **7**, 854[52], 855[52], 876[121,123,124], 887[52]
Maricich, T. J., **7**, 24[33], 25[33]
Marikawa, T., **5**, 847[136]
Marinas, J. M., **2**, 345[18], 359[18], 360[18]; **3**, 825[24,24b], 835[24]; **8**, 368[65]
Marinelli, E. R., **1**, 473[79]; **3**, 262[158]
Marinelli, F., **3**, 539[96]; **4**, 411[266b]
Maring, C., **1**, 314[131]; **8**, 540[195]
Maring, C. J., **2**, 455[16], 665[33], 667[40], 673[33,40], 674[40], 675[33,40], 681[57], 682[33,64], 683[64], 686[64], 689[33], 694[76]; **4**, 373[80], 374[80]; **5**, 434[145]
Maringgele, W., **6**, 526[394]
Marini-Bettòlo, G., **5**, 130[39]
Marinier, A., **1**, 766[157]; **3**, 380[10]; **5**, 532[86]
Marino, G., **2**, 735[12], 964[58]
Marino, J. P., **1**, 118[58], 129[58], 431[133]; **2**, 833[150]; **3**, 217[87], 226[193,194,194a,196], 263[176], 264[183], 265[190,191], 667[48], 687[48], 807[34]; **4**, 27[79], 120[196], 121[205b], 125[216,216d], 148[48], 164[48], 173[32], 183[79], 248[108,113], 256[113], 260[108], 261[113], 795[89], 1035[38], 1046[117], 1048[38]; **5**, 268[78], 468[137], 553[43], 757[79], 766[79], 803[90,92], 921[143], 943[238], 976[21], 979[24,25]; **6**, 9[42], 152[134,135], 910[11], 924[11], 934[100], 1006[148]; **7**, 205[65], 227[86], 358[12], 376[82], 564[94], 566[94]; **8**, 836[10e,f], 844[10f], 846[10f]
Marinoni, G., **1**, 543[28]
Marinovic, N., **1**, 642[110], 643[110]; **8**, 509[92], 510[92]
Marioka, S., **6**, 626[167]

Marioni, F., **4**, 330[3], 345[3]
Marita, M., **5**, 581[174]
Mark, H. B., Jr., **3**, 459[138]
Mark, V., **5**, 64[39]; **6**, 844[92]
Markarov-Zemlyanski, Ya. Ya., **2**, 737[35]
Markby, R., **5**, 1138[67]
Marker, R. E., **8**, 220[85]
Markezich, R. L., **3**, 370[113]; **8**, 542[228]
Märki, H.-P., **1**, 824[45]; **2**, 205[102,104], 206[102b,104]; **3**, 225[187]
Markides, K. E., **7**, 415[113]
Markies, P. R., **1**, 26[132,133,134]
Markiewicz, W. T., **6**, 662[214]
Markl, G., **5**, 444[186], 604[54], 687[58]
Märkl, G., **2**, 369[248]; **3**, 531[86], 593[178]; **6**, 178[121]; **8**, 865[248]
Märkl, R., **5**, 1094[100,100a], 1098[100a]
Markó, I., **3**, 979[12]; **7**, 429[158], 430[158,159], 442[46a,b], 489[165]
Markó, L., **5**, 1138[66]; **8**, 152[177], 447[99], 452[190], 459[236], 551[345], 554[367]
Markov, P., **8**, 492[17]
Markova, V. V., **8**, 727[197]
Markovac, A., **7**, 656[16]
Markovac-Prpic, A., **3**, 898[88]
Markowitz, M., **4**, 604[285]; **8**, 54[155], 66[155], 264[39]
Markowski, V., **4**, 1090[142,143]
Marks, J., **2**, 102[22]
Marks, M. J., **6**, 790[119]
Marks, M. W., **1**, 41[266]
Marks, T. J., **1**, 231[1,2,5], 251[1], 252[1], 273[1c], 274[1c]; **4**, 410[263]; **7**, 3[7], 881[157]; **8**, 447[131,132,133,134,136], 670[9], 671[9], 696[127,-128], 697[129]
Marktscheffel, F., **7**, 709[45]
Markus, G. A., **8**, 616[98]
Markussen, J., **6**, 637[31]
Markwalder, J. A., **8**, 844[65]
Markwell, R. E., **7**, 15[145]; **8**, 626[175], 629[175]
Marky, M., **4**, 1084[95]
Marletta, M. A., **7**, 79[134]
Marlewski, T. A., **2**, 348[55]
Marlin, J. E., **5**, 829[18], 847[138], 1004[27,27c]; **6**, 856[160], 858[160]
Marlowe, C. K., **6**, 533[496]
Marman, T. H., **4**, 404[246], 405[246], 408[258]; **7**, 490[178]; **8**, 854[151]
Marmor, S., **4**, 35[99]
Marning, L. E., **7**, 884[187]
Marnung, T., **2**, 365[210]
Maroldo, S. G., **1**, 41[203]
Maroni, P., **2**, 281[29,31]; **4**, 21[65], 100[123]
Maroni, S., **3**, 752[97]; **7**, 153[9]
Maroni, Y., **8**, 873[18], 874[20]
Maroni-Barnaud, Y., **2**, 428[44]; **8**, 7[37]
Marotta, E., **2**, 338[76]
Maroulis, A. J., **5**, 645[1], 650[1q,25], 651[1]; **7**, 874[107], 878[137]
Marples, B. A., **3**, 741[53]; **5**, 21[161,162,163], 23[161,162,163]; **7**, 62[50a], 429[151]
Marquarding, D., **2**, 1090[69,71,75], 1091[69], 1092[71], 1093[71,83], 1094[71,87,89], 1095[89,93], 1096[71], 1098[71,105], 1099[106,108], 1100[71]; **6**, 242[87,88], 243[87,88]; **8**, 830[84]
Marquardt, D. J., **4**, 811[173]
Marquardt, F.-H., **6**, 264[36]
Marques, M. M., **6**, 114[78]
Marquet, A., **1**, 513[45,46], 531[130]; **2**, 902[47], 903[47]; **3**, 147[390,398], 149[390,398,407,408,409,410,411,412], 150[407,410,411], 151[390,407,408,409,410,411], 152[390,407], 153[407], 155[408,409,410]; **6**, 641[58], 644[91]; **7**, 777[388]
Marquet, B., **2**, 283[48]; **7**, 538[63]
Marquez, C., **7**, 693[24]
Marquez, E., **1**, 552[80]
Marquez, V. E., **8**, 798[60]
Marquis, E. T., **6**, 970[127]
Marr, D. H., **7**, 801[36]
Marr, G., **3**, 160[465], 164[465], 166[465]
Marra, A., **7**, 245[76]
Marra, J. M., **3**, 69[24]
Marren, T. J., **1**, 112[27]; **2**, 448[46]; **4**, 152[56]
Marrero, J. J., **2**, 1049[16]; **4**, 817[203]
Marrero, R., **7**, 182[162,164], 185[176], 186[179]
Marron, B. E., **3**, 618[20]; **6**, 448[107]
Marrs, P. S., **3**, 443[57]; **5**, 990[45,46], 991[46]
Marsaioli, A. J., **1**, 748[73], 812[73]
Marsais, F., **1**, 474[88,94,96,97,99,101,102]; **3**, 261[146], 264[146]; **4**, 465[115], 474[115], 478[115]
Marsch, M., **1**, 10[47], 18[94,96], 29[145], 32[159,160,161], 33[162], 34[166], 37[242,243,245,246], 44[96], 528[119]; **2**, 508[29]; **4**, 104[137]; **6**, 881[51]
Marschall, H., **6**, 1042[5,7]
Marschall-Weyerstahl, H., **3**, 752[92]; **4**, 1002[58]
Marschner, F., **4**, 1005[95]
Marschoff, C. M., **6**, 278[132]
Marsden, E., **8**, 916[100], 918[100]
Marsden, R., **2**, 946[174,175]
Marsh, B. K., **5**, 704[22], 1020[69], 1023[69]
Marsh, C. R., **3**, 522[11]
Marsh, D. G., **7**, 774[311]
Marsh, F. D., **7**, 21[13], 474[39], 479[95], 480[98]
Marsh, G., **5**, 216[15], 219[15]
Marsh, W. C., **7**, 833[72]
Marshalkin, M. F., **8**, 388[62]
Marshall, C. W., **7**, 100[120]
Marshall, D. J., **8**, 492[15], 498[15], 530[96]
Marshall, D. R., **2**, 547[101], 548[101]; **4**, 442[184]; **7**, 123[31]
Marshall, D. W., **8**, 755[133]
Marshall, G. R., **6**, 644[83], 671[277]
Marshall, J. A., **1**, 188[68], 767[164,175,178], 768[167,169], 772[200], 851[37]; **2**, 18[71], 39[136], 160[132], 162[143], 541[80], 547[110,111], 551[110,111], 574[57], 710[28], 837[164], 838[164], 911[68]; **3**, 11[53], 100[202], 135[365,366], 136[365,366], 139[365,366], 142[365,366], 168[202], 170[202], 172[202], 223[149], 226[197,199], 356[55], 380[8,10], 394[96], 592[172], 627[44], 750[86], 783[23], 943[89], 983[21], 984[21,21a], 985[26b], 994[39], 999[51], 1000[51b], 1008[73], 1009[74], 1010[74,76,77], 1011[78], 1012[79,80]; **4**, 18[62], 20[62a-c], 181[75], 186[92], 243[65], 253[65], 307[397], 380[122], 868[17], 869[17]; **5**, 20[137], 173[123], 517[29], 519[29], 534[29], 538[29e], 539[29e], 830[29], 931[185], 934[185,187]; **6**, 10[44], 11[44], 12[44], 158[183], 174[60], 187[175], 687[376], 831[8], 834[29], 837[61], 848[8], 850[29], 874[11,12], 985[60,61], 1055[50,51]; **7**, 89[27,28], 152[5], 174[135], 273[135], 364[41a], 410[99], 413[116], 416[121a,b], 421[99]; **8**, 58[175], 66[175], 164[45], 252[111], 278[157], 528[75,83], 537[186], 542[220], 546[186], 844[65], 929[28,29], 971[107]
Marshall, J. L., **2**, 345[30]; **3**, 19[102]; **8**, 502[62]
Marshall, J. M., **6**, 7[34]
Marshall, J. P., **7**, 86[16a]
Marshall, P. A., **8**, 875[40]
Marshall, R., **8**, 398[145]
Marsham, P., **3**, 753[99]
Marsheck, W. J., **8**, 561[408]
Marsheck, W. J., Jr., **7**, 66[75a], 69[90], 74[90]
Marsi, K. L., **8**, 411[104]
Marsi, M., **3**, 218[101]; **4**, 576[17]
Marsich, N., **4**, 170[11]

Marsico, J. W., **6**, 554[710], 614[90]
Marsili, A., **3**, 741[51], 745[65]
Marson, C. M., **2**, 786[47,48]; **8**, 827[72]
Marson, S., **3**, 99[186], 107[186]
Marson, S. A., **8**, 389[71]
Marston, C. R., **6**, 507[232]
Martel, A., **1**, 123[78]; **2**, 212[120], 213[126], 656[158,159], 1059[77]; **5**, 92[70], 94[84]
Martel, B., **5**, 1012[50]
Martell, A. E., **7**, 851[23]
Martelli, G., **1**, 391[148]; **2**, 613[114], 656[157], 807[48], 925[111], 926[111], 927[120], 935[151], 936[151], 937[157]; **4**, 452[20]; **5**, 100[148,155,156], 102[174]; **6**, 21[80], 22[80], 759[140]
Martelli, J., **4**, 957[23], 990[146]
Martelli, P., **7**, 65[68]
Marten, D. F., **2**, 934[143]; **4**, 189[104], 190[107], 579[23]; **5**, 272[5], 274[7], 275[7], 277[7], 279[7]
Marten, K., **4**, 14[46], 18[57]
Martens, D., **5**, 716[85]
Martens, F. M., **8**, 96[93]
Martens, H., **2**, 723[100]; **3**, 332[203]; **5**, 637[109]
Martens, J., **6**, 462[9,15]; **8**, 459[228], 460[254]
Marth, C. F., **1**, 755[116], 756[116,116d,e], 758[116,124], 761[116]
Marti, F., **4**, 31[92,92k]
Marti, M. J., **6**, 245[127]
Martigny, P., **7**, 797[19], 808[76]; **8**, 642[31]
Martin, A. A., **6**, 422[34], 429[34], 449[113], 453[113]
Martin, A. R., **2**, 902[46]; **4**, 317[546]; **7**, 202[47]
Martin, B. D., **7**, 746[83]
Martin, C., **7**, 452[51], 453[51]
Martin, C. A., **7**, 395[20b]
Martin, D., **6**, 243[105], 244[110], 612[69]; **8**, 445[31]
Martin, D. F., **2**, 357[148]
Martin, D. J., **4**, 1011[163]; **5**, 797[67]; **6**, 970[126]
Martin, D. T., **8**, 797[39]
Martin, E. L., **8**, 309[8], 310[8]
Martin, G., **4**, 125[216,216b]
Martin, G. J., **2**, 725[102-104]; **5**, 777[185,186,187,188]
Martin, H., **2**, 1090[73], 1102[73], 1103[73]; **8**, 735[11], 736[19], 739[11,19]
Martin, H. A., **1**, 139[6], 159[76,77]; **2**, 5[19], 6[19]; **8**, 447[124], 450[124], 696[123]
Martin, H.-D., **3**, 382[36], 872[60]; **5**, 257[60], 571[115]; **6**, 1015[22]
Martin, J., **3**, 380[13], 386[67,68], 595[191], 600[213]; **7**, 452[51,53], 453[51]; **8**, 269[77]
Martin, J. A., **3**, 680[93], 807[28,30]
Martin, J. A. F., **5**, 201[32], 220[50,51], 221[53]
Martin, J. C., **1**, 468[50], 471[63]; **2**, 740[54]; **6**, 448[110], 814[86], 960[56]; **7**, 311[32], 324[32]; **8**, 141[30]
Martin, J. D., **4**, 373[87]; **5**, 830[32]; **6**, 959[38]; **7**, 413[118], 820[25]
Martin, J. G., **5**, 451[13], 470[13], 513[2], 518[2]
Martin, J. R., **8**, 27[30], 66[30]
Martin, K., **5**, 345[68], 346[68b]
Martin, K. J., **8**, 829[83]
Martin, L. D., **3**, 466[190], 1032[66]
Martin, M., **2**, 280[28]; **6**, 67[13]
Martin, M. G., **6**, 685[364]; **8**, 254[123], 890[143]
Martin, M. L., **6**, 489[82]
Martin, M. M., **3**, 379[3], 382[3b]
Martin, M. R., **4**, 425[26]; **6**, 67[13]
Martín, M. V., **6**, 67[13]
Martin, N., **2**, 380[299]
Martin, O. R., **2**, 642[78], 643[78]; **7**, 258[55]; **8**, 91[50]
Martin, P., **3**, 848[50]; **4**, 754[175]
Martin, R. A., **2**, 602[41]; **3**, 614[5], 617[5a]
Martin, R. H., **6**, 707[45]
Martin, R. M., **1**, 480[153]
Martin, R. S., **7**, 96[87]
Martin, R. T., **8**, 374[149]
Martin, S. F., **1**, 41[270], 275[77], 383[110], 432[137], 456[137], 542[1], 884[130]; **2**, 55[1], 410[4], 475[3], 479[17,18], 480[18], 496[69], 498[69], 1067[127], 1068[127], 1079[158]; **3**, 211[28], 215[28]; **4**, 15[50], 314[497], 795[79]; **5**, 467[116,117], 528[68], 529[70], 530[71], 531[68,80], 539[109], 796[57], 815[57], 841[101]; **6**, 690[396], 705[28-30], 722[137,138], 755[121]; **7**, 228[90,91], 297[33]
Martin, S. J., **3**, 283[50]; **4**, 337[36]; **6**, 239[76]
Martin, S. R. W., **4**, 5[18]
Martin, T., **2**, 821[105]; **7**, 32[96]
Martin, T. R., **1**, 16[89]
Martin, V. S., **3**, 225[185], 264[181]; **6**, 2[3], 8[39], 25[3], 88[105], 89[105], 927[76]; **7**, 198[26], 238[42], 239[42], 240[42], 390[12], 391[13], 401[59,60], 403[59], 406[59,77], 409[77], 411[13], 412[13], 413[13], 414[77], 415[77], 421[77], 423[77], 571[113], 572[113], 587[113], 710[52]; **8**, 879[51], 880[51]
Martin, W., **5**, 717[90b,c]
Martin, W. B., **1**, 371[75]; **2**, 971[91]
Martina, D., **1**, 115[42]; **4**, 707[43]; **5**, 622[24], 632[24]; **6**, 691[404], 692[404]
Martina, V., **1**, 416[67]; **3**, 463[154]
Martinelli, J. E., **6**, 812[79,80]
Martinelli, L. C., **3**, 629[53,54]
Martinelli, M. J., **6**, 516[318]; **7**, 364[41b]
Martinengo, S., **8**, 372[121]
Martinengo, T., **3**, 460[141]
Martinet, P., **8**, 133[17,18]
Martinetti, G., **2**, 345[20]
Martínez, A., **2**, 780[12]
Martinez, A. G., **8**, 349[137], 886[109], 934[53]
Martínez, A. G., **6**, 835[44]
Martinez, F., **6**, 432[122]
Martinez, G. C., **7**, 462[120]
Martinez, G. R., **1**, 410[41], 473[78]; **2**, 737[32]; **3**, 918[26]; **4**, 1086[111,119], 1087[119]; **6**, 175[82], 893[81]
Martinez, M., **2**, 849[213]
Martinez, R. A., **3**, 727[33]
Martinez, V. C., **7**, 462[119]
Martinez-Carrion, M., **8**, 52[144], 66[144]
Martinez-Davila, C., **1**, 887[138], 888[138]; **5**, 856[210], 913[100], 1007[38], 1008[43], 1017[64], 1018[64], 1020[64], 1021[64]
Martinez-Gallo, J. M., **4**, 302[338], 349[110], 351[124], 354[110]; **7**, 533[35,36], 534[35]
Martinho Simões, J. A., **8**, 671[17]
Martin-Lomas, M., **1**, 759[132]; **4**, 108[146h], 230[247]; **8**, 227[116]
Martins, M. E., **6**, 278[132]
Martinsen, A., **6**, 242[94]
Martirosyan, G. T., **5**, 410[40]
Martirosyan, V. O., **6**, 270[83]
Martius, C., **8**, 145[86]
Marton, D., **2**, 6[29,32], 564[9,11,12], 566[20], 572[47], 726[127]
Marton, M. T., **5**, 165[86]
Martos-Bartsai, M., **1**, 370[70], 371[70]
Martynov, A. V., **6**, 550[673,674]
Martynov, V. F., **2**, 411[8], 420[24]
Marui, S., **7**, 209[90]
Marumoto, R., **6**, 501[185], 531[185]
Maruoka, H., **7**, 672[18]
Maruoka, K., **1**, 78[10,12,13,18], 79[21], 80[21], 81[21], 82[21], 83[27], 88[51,52,54,55], 92[60,61], 98[84], 99[84,85], 100[18], 165[111], 266[48], 283[3], 316[3], 333[60,61], 335[60,61], 348[141], 387[136]; **2**,

114[122], 269[71], 282[40], 541[74], 556[155], 685[67], 995[45]; **3**, 483[13], 750[86]; **4**, 140[8,11], 143[21], 209[64-66], 254[177], 753[167], 968[59], 969[59]; **5**, 434[147], 609[68], 850[150,153,154,155]; **6**, 5[23], 14[52], 65[2], 91[126], 254[161], 291[217], 528[414], 542[601], 767[26,28], 768[28], 769[28,29,32], 770[33-35], 771[37,38], 850[125], 856[152]; **7**, 696[38,43,44], 697[43,45-47,49]; **8**, 18[130], 43[108], 47[108], 64[213,220], 66[108,213], 67[220], 100[117], 223[99], 224[99], 356[185], 394[119], 545[284]

Marusawa, H., **1**, 101[90]

Maruta, M., **4**, 128[221]

Maruta, R., **4**, 8[28]

Maruthamuthu, M., **8**, 698[136]

Maruyama, F., **7**, 168[101]

Maruyama, H., **2**, 649[101]

Maruyama, K., **1**, 78[9], 110[22], 113[28], 115[22], 117[55], 121[66], 124[82], 143[36], 158[36,73,74], 159[36,75], 176[17], 179[24], 180[33,38,40], 181[38,40], 185[60], 221[68], 329[32], 335[65,67], 340[92], 359[13]; **2**, 2[6], 3[6], 4[12,12a,14], 5[18], 6[12,18,30], 10[12b,44,45b,46-48], 11[44,47,49,50], 13[56], 15[66], 18[30b,69], 22[45b], 24[18,18b], 30[18b,107], 31[107], 32[120,120a,b], 57[5], 58[8,10], 61[8], 67[8,38], 71[8], 72[8], 76[8], 95[55], 117[146], 119[146,157], 128[241], 302[3], 303[3], 313[38], 314[38], 564[10], 566[18], 573[18,55,56], 574[18,58], 576[70], 611[101], 632[28a], 640[28], 977[8], 978[11], 979[11,15,16], 983[15,16], 984[15,16,29,30], 985[15,16,29,30], 986[15,16,31], 987[15,30,31], 988[30], 989[11], 990[11], 991[11], 992[11,15,37], 993[11,37]; **3**, 43[236], 87[97,98], 99[98,183], 100[98,183], 105[183], 157[98], 196[31], 483[10]; **4**, 27[83], 145[35], 148[50], 149[50b], 155[70], 179[64], 182[64a,c], 184[64a-c], 185[88], 186[88], 201[11], 388[162], 401[162a]; **5**, 181[154], 936[198], 963[323]; **6**, 848[107], 864[194]; **7**, 226[70], 408[88b], 427[148e], 453[70], 579[134]; **8**, 353[159], 676[79], 725[181]

Maruyama, L. K., **6**, 960[52]

Maruyama, M., **4**, 413[278a,b]; **6**, 509[268]; **8**, 888[123]

Maruyama, O., **4**, 837[10]

Maruyama, T., **4**, 8[28]; **7**, 229[123]; **8**, 145[82], 170[101], 989[38]

Marvel, C. S., **2**, 139[28], 143[55]; **3**, 825[27a]; **4**, 317[550]; **5**, 752[39-45]; **6**, 209[64], 228[30]; **8**, 568[471], 965[64]

Marvell, E. N., **1**, 880[117]; **3**, 380[4], 735[19], 813[63,64]; **5**, 20[135], 675[1], 678[1], 683[1], 695[1], 699[3], 700[3,3a], 702[9,9b], 710[3a,51,55], 713[51,61], 714[3a], 740[3a], 741[3a], 743[3a,163], 791[38,40], 796[54,55], 830[34], 834[53], 1030[94]; **7**, 397[28]; **8**, 430[57]

Marx, B., **1**, 215[40], 216[40], 226[90], 326[6], 327[8,9]; **4**, 95[97], 98[108d,109d,e]

Marx, E., **8**, 35[65], 47[65], 66[65]

Marx, J. N., **2**, 388[341]; **3**, 804[5,9]; **4**, 162[92]; **7**, 128[66]

Marx, M., **4**, 24[72,72d], 31[92,92k], 63[72d]; **7**, 302[65]

Marx, P., **6**, 48[87]

Marx, R., **1**, 38[254]

Marxer, A., **1**, 379[107], 385[107]

Marxmeier, H., **7**, 506[304]

Maryanoff, B. E., **1**, 755[114,116], 756[116,116b,118], 757[114], 758[116,116a], 759[114], 760[114], 761[114,116], 790[114], 812[114], 813[114]; **4**, 38[108,108c], 379[115], 380[115h], 383[115h], 1032[13], 1061[13,164]; **6**, 174[55]; **7**, 523[45]; **8**, 36[75], 37[75,103], 38[75], 39[75], 45[75], 54[75], 55[164,165], 59[164], 66[75,103,164,165], 343[117], 344[117], 350[150], 351[117], 354[117], 355[117], 568[473], 618[128,129], 619[129], 620[128,129], 624[129], 806[100], 812[6], 929[31]

Maryanoff, C. A., **6**, 552[697]; **8**, 54[160], 66[160], 803[93], 804[93], 806[100], 826[69]

März, J., **6**, 640[57], 641[57], 671[57]

Marzabadi, M. R., **1**, 555[120]; **7**, 749[117]

Marzocchi, S., **8**, 161[22]

Mas, J. M., **5**, 936[196]

Masada, G. M., **8**, 568[472]

Masada, H., **3**, 1021[16]

Masagutov, R. M., **8**, 608[48]

Masai, H., **5**, 1174[33]

Masaki, N., **2**, 876[34]

Masaki, Y., **3**, 101[506], 155[428], 159[462], 161[462], 168[497,506], 169[506], 170[497,506], 220[115], 286[56a]; **6**, 157[171], 995[99]; **8**, 349[135], 354[175], 843[50]

Masalov, N. V., **2**, 740[61]; **4**, 1023[254]; **6**, 557[835]

Masamune, H., **3**, 223[156]; **7**, 390[4], 393[4,17], 394[4], 395[4], 396[4], 397[4], 398[4], 399[4], 400[4], 401[4], 406[4], 407[4], 410[4], 411[4], 413[4]

Masamune, S., **1**, 95[72], 191[77], 410[43,44], 434[140], 436[151,152,153], 763[143], 766[143], 769[182]; **2**, 1[2], 2[5,7], 25[5,99], 33[5], 35[124a,b], 40[5], 42[124], 100[13], 111[81], 112[92,98,101], 113[102,103], 134[3], 190[57], 192[61], 221[61], 224[153], 232[180,183], 240[5,13], 242[18,20], 244[25,26,30], 245[18b,20d,e], 246[18b,20d,e,34], 247[20e,34], 248[5b], 249[36,84], 253[42], 256[13,47], 257[13b,47], 258[48-51], 259[52], 260[5e], 261[48,52], 264[58], 265[59], 303[5], 308[20], 455[8], 630[9], 652[124], 686[68], 801[31], 926[117], 979[17]; **3**, 894[65]; **4**, 145[35], 1079[61,62]; **5**, 359[91], 373[91], 374[91], 543[118], 545[118], 611[72], 716[86,89], 804[93]; **6**, 8[39], 438[42,57], 440[75,76], 446[102], 463[27], 667[236], 927[71,76]; **7**, 31[85], 198[26], 257[52], 390[8], 399[40a], 401[59,60], 402[63], 403[59], 406[59], 442[48], 722[21]; **8**, 16[99], 159[108], 171[107-109], 178[108], 179[108], 535[165], 542[230], 543[230], 549[327], 696[122], 720[138], 721[138], 722[138], 801[70,71], 813[11], 879[51], 880[51], 938[89]

Masamune, T., **1**, 161[88,89], 566[209], 823[43]; **2**, 159[131]; **3**, 125[306], 126[306], 735[22]; **4**, 238[11], 245[11], 255[11], 260[11]; **6**, 1049[36]; **7**, 253[23], 680[80]; **8**, 334[59], 528[67], 607[32]

Masamura, M., **4**, 558[17]

Masana, J., **2**, 435[62]; **6**, 80[68]

Masaoka, M., **7**, 453[91]

Masaracchia, J., **5**, 66[79]

Mas Cabré, F. R., **8**, 561[419]

Mascarella, S. W., **4**, 255[198], 804[137]; **5**, 143[100,101], 144[101]

Mascareñas, J. L., **3**, 983[21], 984[21]

Maschke, A., **5**, 422[88], 423[88]

Masci, B., **6**, 110[53]

Mascolo, G., **3**, 230[235], 446[87]

Mase, M., **6**, 821[115]

Mase, T., **2**, 547[94]

Mash, E. A., **1**, 237[32]; **4**, 974[87-89]

Mashima, K., **1**, 162[93,95,100], 163[106], 180[32]; **2**, 5[18], 6[18], 24[18,18a]; **5**, 1172[28], 1182[28]; **8**, 459[244], 678[62], 683[62], 686[62]

Mashimo, K., **7**, 137[123], 139[123]

Mashkina, A. V., **8**, 608[47,48], 629[182,183]

Mashkovskii, M. D., **6**, 553[728], 554[728,741,776,780,793]

Mashraqui, S., **7**, 143[139]; **8**, 98[101]

Masilamani, D., **5**, 78[280]; **8**, 806[100]

Masjedizadeh, M. R., **5**, 241[5]

Maskell, R. K., **1**, 543[25], 547[25]; **8**, 771[51], 779[89]

Maskens, K., **4**, 301[317], 303[317], 310[431]

Maslak, P., **7**, 874[109]

Maslennikov, V. P., **7**, 599[64,65], 602[106]

Masler, W. F., **8**, 459[243], 535[166]

Maslin, D. N., **8**, 267[61], 271[108], 274[135]

Masnovi, J. M., **5**, 71[134], 636[99]; **7**, 851[14], 854[59], 855[59,63,64], 865[87], 867[92], 874[108], 881[163], 882[165]

Mason, J. R., **7**, 80[136]

Mason, J. S., **1**, 739[37]; **5**, 605[56], 612[77]

Mason, K. G., **5**, 21[161], 23[161]

Mason, N. R., **8**, 618[110], 623[151]

Mason, R., **1**, 777[217], 778[221], 814[217]; **4**, 664[6]; **8**, 13[67]

Mason, R. F., **8**, 971[110]
Mason, R. W., **6**, 860[177]
Masquelier, M., **6**, 495[143], 496[143], 497[143]
Massa, W., **1**, 18[94,96], 37[242], 44[96], 191[77], 272[68], 300[67], 322[67], 335[63]; **2**, 6[35], 247[35], 630[7], 631[7]; **5**, 850[152]; **6**, 881[51], 970[122]
Massardier, J., **8**, 436[73]
Massardo, P., **3**, 489[60], 495[60], 504[60], 511[60], 515[60]
Masse, G., **6**, 116[86]; **7**, 745[80]
Masse, J. P., **3**, 228[215], 436[4], 484[26], 492[26], 494[26], 495[26], 503[26], 513[26]
Massengale, J. T., **3**, 297[16]
Massey-Westropp, R. A., **5**, 561[86]
Massicotte, M. P., **8**, 331[37], 340[37]
Massiot, G., **2**, 765[77], 1017[33,36]; **6**, 735[20], 738[49,50], 739[20,50]
Massoli, A., **4**, 437[148], 438[148]; **7**, 340[45]
Masson, A., **4**, 883[98]; **5**, 39[25]
Masson, P., **1**, 821[28]; **4**, 1059[155]
Masson, S., **3**, 124[260]; **4**, 85[74]; **5**, 575[135]; **6**, 455[151,152,153]
Massoneau, V., **8**, 462[265], 535[166]
Massoudi, M., **2**, 209[108]; **7**, 727[39]
Massoussa, B., **2**, 765[77]
Massuda, D., **2**, 588[150]; **3**, 251[79], 254[79]; **6**, 1005[141]; **7**, 172[128]
Massy, M., **4**, 95[102c]
Massy-Bardot, M., **4**, 95[102d]
Massy-Westropp, R. A., **6**, 186[168]; **8**, 212[23-25]
Mastafanova, L. I., **8**, 599[101]
Mastalerz, H., **3**, 960[115]; **6**, 670[268]
Mastalerz, P., **6**, 801[38]
Mastatomo, I., **1**, 466[42], 473[42]
Masters, C., **8**, 445[35,36]
Masters, N. F., **1**, 463[21]
Mastrocola, A. R., **5**, 64[30]
Mastrorilli, E., **3**, 733[1]; **5**, 456[87]
Mastryukova, T. A., **4**, 317[548]
Masua, K., **4**, 1056[141]
Masubuchi, K., **7**, 350[27], 355[27]
Masuda, C., **1**, 359[14], 363[14], 384[14]
Masuda, H., **1**, 317[141]; **6**, 614[86]; **8**, 160[100], 170[100], 176[135], 178[100]
Masuda, K., **2**, 810[66], 851[66]; **5**, 768[131], 779[131]
Masuda, R., **1**, 563[178]; **4**, 435[135]; **6**, 510[292]; **7**, 764[124], 843[50], 844[51,52]; **8**, 18[128], 245[75], 315[46]
Masuda, S., **3**, 300[46], 302[46], 314[108], 318[129]; **5**, 167[94], 473[153], 477[153]
Masuda, T., **1**, 187[64]; **5**, 1148[114]; **6**, 867[209]
Masuda, Y., **3**, 470[196,197], 473[196,197], 522[21]; **5**, 829[24]; **7**, 602[99], 604[130], 608[170,171]; **8**, 720[130]
Masui, K., **5**, 56[51]
Masui, M., **6**, 439[69]; **7**, 158[35], 248[112], 752[153], 809[84]; **8**, 375[155]
Masui, Y., **6**, 637[36]; **7**, 745[78]
Masumi, F., **2**, 558[160]; **4**, 211[87], 252[163]
Masumori, H., **7**, 384[114a]
Masumoto, H., **6**, 453[143]
Masumoto, M., **6**, 764[10]
Masunaga, T., **4**, 1020[239]; **8**, 806[106], 807[106], 900[31]
Masunaga, Y., **6**, 765[19]
Masure, D., **5**, 848[141]
Masuyama, Y., **1**, 215[37], 642[116], 645[116]; **2**, 18[73-75], 23[90,90a]; **4**, 607[312,314]; **6**, 837[59]; **7**, 299[43], 320[64], 321[65], 771[265], 772[265], 773[265]
Mataga, N., **7**, 856[66]
Matar, A., **6**, 209[66]
Matar, S., **3**, 304[59]
Matarasso-Tchiroukhine, E., **2**, 583[110], 587[110]
Matassa, L. C., **1**, 559[145]
Matassa, V. G., **6**, 831[7]
Matasubara, Y., **2**, 225[155]
Matecka, D., **1**, 564[203]
Mateer, R. A., **5**, 128[27]
Mateescu, G. D., **3**, 330[192]
Mateos, A. F., **4**, 161[89d]
Mateos, J. L., **3**, 901[112]
Matern, A. I., **8**, 580[1]
Math, S. K., **4**, 974[89]
Mathai, I. M., **6**, 959[48]
Matharu, S. S., **5**, 13[92]
Matheny, N. P., **8**, 340[99]
Mather, A. N., **7**, 412[106]
Mather, A. P., **8**, 770[41]
Matheson, N. K., **2**, 456[55], 457[55]
Mathew, C. P., **8**, 240[34], 247[34], 250[34]
Mathew, C. T., **8**, 338[88]
Mathew, K. K., **1**, 49[4], 50[4], 153[62]; **2**, 24[96]; **6**, 980[32]; **8**, 3[17]
Mathew, L., **5**, 901[27]
Mathey, F., **3**, 201[82]; **4**, 688[67]; **5**, 444[186]; **8**, 859[213], 865[247]
Mathian, B., **7**, 764[114]
Mathias, L. J., **6**, 20[72], 74[32]
Mathias, R., **4**, 1001[31,39]
Mathies, R. A., **5**, 702[13]
Mathieson, A. McL., **4**, 202[18]
Mathieu, J., **1**, 367[56], 368[56], 370[56]; **3**, 521[3], 901[112]; **7**, 804[63]; **8**, 201[144], 541[212]
Mathis, J. B., **2**, 466[125]
Mathre, D. J., **1**, 425[105]; **2**, 240[4], 436[67], 438[70]; **3**, 45[249]
Mathur, H. H., **7**, 558[79], 560[79]
Mathur, N. K., **7**, 446[61]; **8**, 271[110]
Mathvink, R. J., **4**, 740[117], 798[107]; **6**, 471[65], 472[65]
Mathy, A., **2**, 345[19], 359[19]; **3**, 300[44], 310[44], 311[44], 322[144], 809[37]; **5**, 412[44]; **7**, 846[100]
Matier, W. L., **1**, 555[116]
Matikainen, J. K. T., **5**, 516[25], 517[25c]
Matl, V. G., **6**, 669[263]
Matlack, A. S., **7**, 219[14]; **8**, 447[118], 454[118], 455[118], 568[481]
Matlack, E. S., **4**, 288[185], 298[185]
Matlin, A. R., **5**, 829[19]; **6**, 836[58]
Matlin, S. A., **3**, 894[60]; **4**, 1039[65], 1053[131], 1063[131]
Matlock, P. L., **8**, 550[334]
Matloubi, F., **1**, 523[84]; **4**, 226[185]
Matloubi-Moghadam, F., **2**, 228[164,165]; **8**, 844[72]
Matoba, K., **3**, 853[72]; **6**, 529[464]
Matos, J. R., **7**, 316[45]
Matser, H. J., **5**, 163[72]
Matsnaga, K., **6**, 564[909]
Matsubara, H., **6**, 548[669], 918[37]
Matsubara, S., **2**, 19[76], 584[126], 588[151], 589[151,154]; **3**, 484[24], 501[24]; **4**, 607[310], 626[310], 647[310], 901[186]; **6**, 237[61], 563[905]; **7**, 169[117], 275[146,147], 276[147], 308[17], 674[47]; **8**, 886[113]
Matsubara, Y., **1**, 127[92], 427[114]; **6**, 538[556]; **8**, 407[54]
Matsud, M., **4**, 837[15]
Matsuda, A., **1**, 792[270]; **4**, 945[89]; **6**, 530[415], 563[900]
Matsuda, F., **1**, 241[44]; **2**, 1050[28]; **8**, 532[130]
Matsuda, H., **3**, 88[133], 90[133], 95[133], 99[133], 101[133], 107[133]; **5**, 151[18]; **6**, 89[120]; **7**, 98[105]; **8**, 20[137]

Matsuda, I., **2**, 310[31], 311[31], 369[250], 587[136], 615[124,125], 630[23], 631[12,23], 635[44], 640[44]; **3**, 262[163]; **5**, 487[193]; **8**, 786[118], 789[123]
Matsuda, K., **1**, 836[140]; **4**, 16[52c]; **6**, 542[603]
Matsuda, O., **6**, 533[510]
Matsuda, S., **3**, 483[9]; **4**, 155[75]; **7**, 774[332]; **8**, 548[319]
Matsuda, S. P. T., **5**, 780[204]; **7**, 358[4]
Matsuda, T., **2**, 374[275]; **3**, 484[26], 492[26], 494[26], 495[26,96], 497[103], 503[26], 513[26], 530[78], 535[78], 757[123], 1026[41]; **4**, 587[27], 841[49], 856[99,101]; **6**, 801[33]; **7**, 92[42], 93[42]
Matsuda, Y., **1**, 223[84], 225[84a]; **2**, 73[66]; **5**, 839[82]; **6**, 104[7], 109[7]; **7**, 16[163]; **8**, 99[113], 837[13c]
Matsue, H., **1**, 823[43]; **3**, 125[306], 126[306]; **8**, 607[32]
Matsue, T., **7**, 50[69]
Matsuhashi, Y., **3**, 168[494,504], 169[494,504], 170[494,504]
Matsui, K., **3**, 202[85], 539[97]; **7**, 451[19], 452[19], 454[19]
Matsui, M., **3**, 99[191], 107[191], 287[62], 644[161], 715[39]; **4**, 18[62], 20[62i], 33[96], 262[305], 1040[89,90,93], 1041[93], 1045[89,90]; **6**, 18[66], 435[4], 657[177]; **7**, 455[104], 550[51]
Matsui, S., **2**, 111[77], 117[77], 121[77], 124[77], 186[37], 655[136,141], 948[182]; **3**, 174[526]
Matsui, T., **3**, 714[32]; **4**, 30[88], 121[207], 253[175], 258[175]; **8**, 931[39]
Matsui, Y., **7**, 441[44]
Matsukawa, M., **1**, 261[33], 275[76], 751[110]; **6**, 980[41]
Matsukawa, T., **7**, 768[199]
Matsuki, M., **1**, 836[139]
Matsuki, Y., **3**, 100[193,194], 103[193,194], 107[194], 390[69,70]; **6**, 145[80]; **8**, 856[171]
Matsukura, H., **6**, 5[26]
Matsumiya, K., **2**, 505[8]; **3**, 34[195], 35[202]
Matsumoto, E., **8**, 152[160]
Matsumoto, H., **2**, 112[88], 241[14]; **3**, 229[225], 483[16], 499[116]; **4**, 501[113], 610[337]; **5**, 92[74], 524[54], 534[54], 691[84], 692[84]; **6**, 447[105], 450[105], 546[647], 579[982]; **7**, 415[114], 642[9]; **8**, 770[39], 787[119]
Matsumoto, K., **2**, 18[69], 329[50], 443[15], 451[15], 564[10], 611[101], 632[28a], 640[28], 830[144], 1051[41], 1089[57]; **3**, 650[210c,212], 651[210c,216], 721[5]; **4**, 17[53], 24[75], 25[75b], 161[86c], 230[252], 231[252], 433[124]; **5**, 77[262], 341[59]; **6**, 14[55], 489[92], 547[659], 814[94]; **7**, 806[74]; **8**, 144[78], 146[91,92], 147[91,102], 148[91,92,106], 963[42]
Matsumoto, K. E., **7**, 163[73]
Matsumoto, M., **2**, 967[76]; **3**, 99[187], 107[187], 110[187], 698[157b]; **4**, 145[23], 794[74], 810[167], 923[31], 924[31], 925[31], 1057[142]; **5**, 468[135]; **6**, 145[79], 173[38], 174[38], 510[296], 774[50], 1036[145], 1066[98]; **7**, 95[71], 97[93], 308[18], 311[29], 628[46], 649[42], 701[64]; **8**, 185[24], 292[44], 453[191], 567[461]
Matsumoto, S., **3**, 303[54]; **5**, 833[49]; **6**, 614[86]
Matsumoto, T., **1**, 131[104], 184[52], 248[64], 339[88], 784[243,244]; **2**, 10[40], 29[106], 282[33], 291[72,74], 292[79], 547[122], 553[122], 1050[28]; **3**, 100[199], 223[183], 225[183], 303[55,56], 382[39], 386[57], 400[119-124], 402[125,126,130,131], 404[132,135,137], 405[138], 558[50], 714[35]; **4**, 314[496]; **5**, 564[97]; **6**, 14[55], 214[94], 536[545], 538[545], 780[69]; **7**, 91[36], 109[184], 168[101], 298[35], 406[87], 412[105], 414[105,105b,108,109], 418[105b]; **8**, 350[143], 625[164], 857[191]
Matsumoto, Y., **2**, 17[67], 38[67], 572[45], 716[63]; **3**, 17[95]; **4**, 231[260], 682[57]; **6**, 91[128]; **8**, 783[107]
Matsumura, C., **3**, 555[29]
Matsumura, H., **1**, 123[75], 373[82]; **6**, 444[97]
Matsumura, M., **2**, 765[72]
Matsumura, N., **6**, 425[64]
Matsumura, Y., **1**, 98[84], 99[84,85], 346[128], 387[136], 804[310]; **2**, 613[111], 784[38], 971[94], 995[45], 1051[33,36], 1052[36,51], 1061[92], 1066[118,119], 1067[123], 1069[92,132], 1070[118], 1071[92]; **4**, 247[100], 257[100], 260[100], 587[43]; **5**, 500[259]; **6**, 291[217], 542[601], 767[28], 768[28], 769[28,32], 770[34,35], 771[37], 801[37], 991[87,88]; **7**, 227[74,75,77], 248[109], 696[43], 697[43,45-47,49], 707[29], 708[29], 794[6], 797[16,18], 798[18b], 801[45], 802[47-49], 803[51,53-55], 804[58,59,62], 805[59,65], 806[75], 808[78-80], 809[81,85], 811[91]; **8**, 43[108], 47[108], 64[220], 66[108], 67[220], 170[81], 394[119], 533[150], 817[32]
Matsunaga, H., **4**, 36[103,103a]
Matsunaga, I., **3**, 172[516], 173[516]
Matsunaga, S., **3**, 198[52]; **7**, 650[47,48]
Matsunaga, T., **8**, 432[68]
Matsunami, N., **6**, 467[51]
Matsuno, A., **6**, 49[95]
Matsuo, M., **6**, 795[7], 796[7], 801[7,33]; **8**, 850[123]
Matsuo, N., **7**, 551[54]
Matsuo, T., **5**, 92[67]
Matsuo, Y., **6**, 765[19]
Matsuoka, H., **4**, 446[212]
Matsuoka, R., **2**, 90[39]; **3**, 257[118]
Matsuoka, Y., **1**, 453[221]; **3**, 640[99]
Matsura, T., **7**, 881[160]
Matsushima, H., **3**, 402[126]
Matsushima, Y., **6**, 266[47]
Matsushita, H., **3**, 12[55], 259[138], 260[139,140], 443[52,53], 453[53], 460[142], 469[198,199,200], 470[198,199,200,210], 472[199], 473[198,199,200], 475[200,210], 497[105]; **4**, 591[109], 595[154], 606[304], 619[154], 620[154], 633[109], 635[154], 893[152]; **6**, 534[516], 717[110], 848[111]; **8**, 755[121,122], 758[121,122], 960[34]
Matsushita, K., **1**, 243[58]
Matsushita, Y., **2**, 157[122]; **6**, 489[79]; **7**, 100[115]
Matsuura, A., **7**, 452[44]
Matsuura, F., **8**, 52[143], 66[143]
Matsuura, K., **5**, 158[47,48], 159[49]
Matsuura, S., **7**, 136[111], 137[111]; **8**, 155[203]
Matsuura, T., **1**, 544[44], 546[49]; **4**, 162[94a,b]; **6**, 564[918]; **7**, 84[3], 227[89], 228[97], 381[104], 452[44], 474[42]; **8**, 817[34]
Matsuura, Y., **4**, 310[435]; **6**, 57[141]; **7**, 615[9]; **8**, 856[170]
Matsuyama, H., **3**, 875[73-75]; **8**, 32[55], 66[55], 389[70]
Matsuyama, N., **6**, 149[103]
Matsuyama, Y., **6**, 734[11]
Matsuzaki, E., **3**, 934[63]
Matsuzaki, J., **6**, 604[32], 606[38]
Matsuzaki, K., **1**, 779[222]; **8**, 899[28]
Matsuzaki, Y., **1**, 834[121a,122]; **8**, 12[64,65]
Matsuzawa, M., **4**, 152[55]
Matsuzawa, S., **1**, 112[27]; **2**, 448[44,45]; **4**, 895[162]; **5**, 977[23]
Mattay, J., **5**, 151[14,15], 154[14,32], 158[46], 160[14,46], 161[62], 645[1], 647[1m-p], 648[1m-p], 649[1m-p], 650[1m-p], 651[1,1m], 676[3], 686[41]; **6**, 558[852,853]; **7**, 851[26]
Matteazzi, J., **2**, 742[75]
Mattenberger, A., **6**, 7[30]
Matteoli, U., **8**, 87[27], 236[3], 239[3], 552[352]
Matter, Y. M., **4**, 729[61], 730[61], 765[61]
Mattes, H., **1**, 219[56]; **7**, 550[52]
Mattes, K., **4**, 483[6]
Mattes, S. L., **5**, 913[102]; **7**, 851[31], 854[53], 855[53], 879[150], 880[155]
Matteson, D. J., **2**, 368[237]
Matteson, D. S., **1**, 489[22,23], 490[24,25,27], 491[30-35], 494[22,43], 495[22,31,45,47], 496[31], 497[30,31], 498[33], 501[22,31,33], 502[33], 623[83], 830[93]; **2**, 13[56], 14[54], 39[138], 970[87], 996[48], 1088[47]; **3**, 196[22], 199[59,61-65], 780[9], 795[86,87], 796[9,87-89], 797[9,86,87]; **4**, 144[22]; **6**, 77[57,58], 98[58], 864[196]; **7**, 439[29], 597[45], 602[101,104,104a], 604[101]; **8**, 101[119], 850[120,121]
Matteucci, M. D., **6**, 604[28], 618[110], 620[132]
Matthei, J., **5**, 596[25,34], 597[25], 598[34]

Matthes, H. W. D., **3**, 406[140]
Matthews, D. P., **6**, 487[69], 489[69], 512[304]
Matthews, F. J., **6**, 714[85]
Matthews, G. J., **7**, 473[28], 502[28]; **8**, 386[50]
Matthews, J. D., **1**, 305[87]
Matthews, R. S., **3**, 14[74,75], 15[74,75], 382[37], 384[51], 393[37]; **7**, 674[38]
Matthews, W. S., **1**, 632[66]; **3**, 863[15], 864[15,16,22], 866[16], 883[16]
Matthies, D., **5**, 404[13]
Mattice, J. D., **6**, 822[116]
Mattingly, P. G., **8**, 395[125]
Mattingly, T. W., **7**, 24[31]
Mattox, J. R., **3**, 322[141]
Mattson, M. N., **4**, 980[110], 982[110]
Mattson, R. J., **8**, 54[153], 66[153]
Matturro, M. G., **5**, 66[76]; **8**, 813[8]
Matulic-Adamic, J., **8**, 794[13]
Maturova, E., **6**, 524[368]
Matusch, R., **2**, 379[297]
Matuszak, C. A., **8**, 507[86]
Matuszewski, B., **5**, 706[27]
Matuyama, Y., **3**, 125[297], 128[297], 129[297], 130[297], 133[297]
Matveeva, E. D., **8**, 612[76]
Matveeva, Z. M., **6**, 515[313]
Matyas, B. T., **4**, 443[185]
Matyushecheva, G. I., **6**, 510[295]
Matz, J. R., **1**, 786[249], 887[139], 888[139]; **3**, 88[129], 89[129], 105[129], 112[129], 196[28], 591[165], 592[174,175], 594[174], 610[165]; **4**, 595[157]; **5**, 456[84], 1017[65], 1018[65,65a], 1020[65,65a,70,70c], 1027[70]; **6**, 145[81], 146[89], 985[63]; **7**, 210[95]; **8**, 842[47]
Matzinger, M., **5**, 250[37], 252[37], 255[37]
Matzinger, P., **5**, 595[19], 596[19]
Matzita, T., **6**, 653[150]
Matzke, M., **6**, 76[52]
Maue, M., **3**, 310[95], 311[95]
Mauer, W., **5**, 10[78]
Mauermann, H., **8**, 447[133,136], 696[127,-128]
Maugé, R., **6**, 268[60]
Mauger, J., **8**, 244[58], 248[58]
Maugh, T. H., **8**, 459[228]
Maughan, W., **6**, 83[79]
Maul, A., **4**, 212[99]
Mauldin, C. H., **6**, 690[398], 692[398]
Mauleon, D., **4**, 438[154]; **8**, 125[94], 587[34]
Mauli, R. M., **3**, 736[25]
Maume, G. M., **7**, 166[92]
Maumy, M., **4**, 801[123]
Maurel, R., **8**, 424[43]
Maurer, B., **1**, 218[49], 220[49], 223[49]
Maurer, F., **7**, 753[158,159]
Maurer, P. J., **1**, 413[59,60], 733[15]; **6**, 112[67]
Mauri, M. M., **4**, 307[396]
Maurin, R., **5**, 109[219]
Maury, G., **7**, 60[45]
Maury, L. C., **7**, 5[21]
Maus, S., **1**, 83[26], 145[42], 146[42], 148[42], 149[42,51], 155[42], 170[42]; **2**, 22[86]
Mautner, H. G., **6**, 462[19]
Mauzé, B., **1**, 831[103], 835[134]; **2**, 77[88,91], 980[18], 982[27], 988[33], 989[33], 1004[62], 1005[62]; **3**, 202[98]; **4**, 871[30], 875[30b], 877[30], 878[80], 883[96], 884[80,96], 987[162], 993[162]
Mavrodiev, V. K., **8**, 699[150]
Mavrov, M. V., **6**, 556[823]
Mavunkel, B., **6**, 550[677]
Mawby, A., **4**, 689[71]
Mawby, R. J., **4**, 518[9], 542[9], 689[71]
Maxa, E., **7**, 498[223]
Maxim, N., **4**, 84[67]
Maxwell, A. R., **5**, 906[69]
Maxwell, J. R., **1**, 808[325]
Maxwell, R. J., **4**, 348[108], 349[108b]
May, A. S., **1**, 37[178]
May, C., **4**, 408[259a]; **5**, 384[128,128a]
May, E. L., **8**, 566[452]
May, G. L., **3**, 505[160]
May, H. J., **6**, 501[201]
May, K. D., **3**, 874[71]
May, L. M., **7**, 254[29]
May, P. D., **5**, 832[41]
May, S. W., **7**, 99[108-110], 429[153], 778[420]
Mayall, J., **7**, 59[37]
Maycock, C. D., **1**, 144[39], 145[39], 146[39,44], 148[44], 149[39], 152[39], 165[44]; **7**, 298[36], 704[13]
Mayeda, E. A., **2**, 971[93]; **7**, 248[110], 801[44], 852[41], 853[41]
Mayer, B., **3**, 872[60]; **5**, 257[60]; **6**, 1015[22]
Mayer, C. F., **5**, 15[100]
Mayer, F., **3**, 324[150]
Mayer, H., **1**, 392[156], 393[156]; **2**, 143[52]; **3**, 168[498,502,503], 169[498,502,503], 698[159]; **5**, 598[33]
Mayer, J., **6**, 269[76]
Mayer, J. M., **5**, 1065[4]
Mayer, K., **8**, 568[466]
Mayer, K. K., **5**, 635[86]
Mayer, R., **2**, 147[75], 785[45]; **3**, 666[43]; **4**, 436[141]; **5**, 439[166]; **6**, 420[16], 421[29], 422[36], 423[41,45], 436[14,25,26], 437[25,26], 448[14,25], 449[14,25], 450[14,25], 452[25,132], 453[14,25,26], 454[14], 455[14,26], 456[14,26], 462[8], 472[71], 552[699]
Mayer, R. P., **3**, 726[24]
Mayer, U., **7**, 95[76]
Mayer, W. J., **2**, 956[13], 958[13]
Mayers, D. A., **3**, 747[70]
Maynard, G. D., **1**, 240[43]; **5**, 857[230]
Maynard, S. C., **2**, 648[89], 649[89]; **3**, 25[152]; **5**, 407[28]
Mayne, P. M., **5**, 403[9]
Mayo, B. C., **1**, 294[47]
Mayo, F. R., **4**, 274[61,65], 275[61,65], 279[61,102,109], 280[61,65,121,126], 281[61,65], 282[61,109], 283[147], 287[177], 288[177], 316[537], 716[1], 751[1]; **5**, 63[17]; **8**, 505[75]
Mayoral, J. A., **2**, 406[45]
Mayr, A., **7**, 777[367]
Mayr, H., **2**, 566[21], 612[107], 629[1], 635[1]; **3**, 331[200b,202]; **4**, 238[5], 1002[60], 1075[34]; **5**, 600[42], 732[132,132c]
Mayrhofer, R., **2**, 212[121]
Mays, R. P., **8**, 652[72]
Mazaki, Y., **5**, 623[25,27]
Mazaleyrat, J. P., **1**, 72[71], 141[22], 555[122]; **6**, 70[20]
Mazdiyasni, H., **7**, 579[132]; **8**, 201[140]
Maze, C., **6**, 441[85]
Mazenod, F. P., **2**, 456[51], 460[51], 462[51]
Mazharuddin, M., **2**, 757[16]
Maziere, M., **8**, 52[145], 53[128], 66[128,145]
Mazius, Z. Z., **7**, 10[77]
Mazloumi, A., **6**, 481[120]
Mazo, G. Y., **7**, 108[170]
Mazouz, A., **6**, 464[31,32]
Mazumder, S. N., **5**, 441[179]; **6**, 552[700]
Mazur, D. J., **1**, 564[204]; **5**, 436[158,158g], 442[158]
Mazur, M. R., **5**, 240[3]
Mazur, P., **4**, 83[65c]
Mazur, R. H., **6**, 680[331]

Mazur, U., **4**, 315[514]
Mazur, Y., **2**, 739[47], 838[177], 839[177]; **3**, 669[52], 683[52], 791[64], 805[14]; **6**, 937[120], 939[120], 942[120]; **7**, 14[127,-128,130], 40[2,5,9,10], 86[16a], 218[5], 737[9], 842[23-26,31,36,37], 843[41-44]; **8**, 528[84], 529[84]
Mazza, D. D., **4**, 488[54]
Mazzanti, G., **5**, 440[173]
Mazzei, M., **6**, 487[49-51,54], 489[54]
Mazzenga, G. C., **8**, 35[66], 66[66]
Mazzieri, M. R., **2**, 6[33], 35[33]
Mazzocchi, P. H., **1**, 294[47]; **3**, 224[172]; **5**, 181[155]; **6**, 810[72]
Mazzocchin, G. A., **7**, 769[215]
Mazzu, A., **4**, 1082[87], 1083[87], 1103[87]; **5**, 728[121]
Mazzu, A. L., Jr., **7**, 143[146]
Mazzuckelli, T. J., **5**, 132[49]
Mbiya, K., **5**, 480[166]
M'Boula, J., **1**, 113[31], 623[81]; **2**, 58[11]
Mead, E. J., **8**, 263[24]
Mead, K., **1**, 55[22], 108[11], 109[11,16], 134[11], 338[82], 339[82]
Mead, T. C., **2**, 148[80]
Meade, E. M., **8**, 328[11]
Meader, A. L., Jr., **3**, 888[19], 891[19], 900[19]
Meador, M.-A., **5**, 383[125]
Meadows, J. D., **4**, 386[148a], 387[148,148a]; **6**, 22[84], 26[107]
Meadows, J. H., **8**, 447[137]
Méa-Jacheet, D., **6**, 725[168]
Meakins, G., **3**, 264[184]
Meakins, G. D., **1**, 632[64]; **6**, 217[112]; **7**, 68[84], 69[92], 71[99], 72[84], 73[92]
Meakins, S. E., **2**, 710[17]
Meana, M. C., **6**, 175[69]
Means, G. E., **6**, 635[15], 636[15], 668[250]; **8**, 440[84], 959[22]
Meanwell, N. A., **1**, 532[137], 739[42], 740[43], 741[43], 742[48]; **4**, 102[129], 317[557]
Mechizuki, M., **7**, 667[78]
Mechkov, T. D., **4**, 85[77d]
Mechoulam, R., **4**, 370[30], 372[30]; **6**, 776[58]; **7**, 535[45]; **8**, 352[148], 542[234]
Meckler, H., **4**, 1078[50]; **5**, 253[47]; **6**, 807[58]; **8**, 395[127]
Medard, J.-M., **6**, 436[8]
Medema, D., **4**, 600[228], 601[246,248], 643[248]
Medici, A., **1**, 471[68]; **3**, 232[267], 511[186,187], 514[186]; **4**, 86[79b]; **5**, 99[131]; **6**, 490[114]; **8**, 70[234], 354[173,178], 357[178]
Medina, J. C., **1**, 767[177]; **5**, 429[114]; **6**, 648[117b]; **7**, 406[85], 409[85]
Mednikov, E. V., **6**, 535[534]
Medvedev, B. A., **6**, 554[790]
Medvedev, V. A., **7**, 852[42]
Medvedeva, A. S., **2**, 365[214]
Medvedeva, V. G., **7**, 606[160]
Medwid, J. B., **7**, 673[24]
Mee, A., **7**, 8[64]
Meegan, M. J., **5**, 947[287]
Meehan, G. V., **5**, 915[111,112]
Meek, D. W., **8**, 459[228]
Meek, J. S., **6**, 954[15]
Meen, R. H., **6**, 448[110]
Meer, R. K. V., **1**, 568[227], 571[227]
Meerholz, C. A., **2**, 830[140], 885[49]; **6**, 474[86]; **7**, 761[59], 776[59,362]; **8**, 410[95]
Meerwein, H., **1**, 144[40], 215[44], 216[44]; **2**, 368[234]; **3**, 705[2], 725[17]; **6**, 120[120], 263[23], 273[23,92], 277[92], 291[23], 529[22], 556[22], 561[22], 562[22], 563[22], 565[920,921]; **7**, 603[116]; **8**, 918[120]
Meese, C. O., **6**, 482[124], 509[251]; **8**, 40[92], 66[92]
Meetsma, A., **1**, 232[16]; **8**, 96[92]
Meeuwissen, H. J., **8**, 863[236]
Mégard, P., **5**, 1062[59]
Meges, D. L., **6**, 958[30]
Meghani, P., **8**, 618[118]
Meguri, H., **8**, 366[48,49]
Meguriya, N., **1**, 161[81,82], 509[23], 827[64a]; **2**, 22[87], 72[59]; **3**, 446[88]
Meguro-Ku, O., **4**, 189[106], 190[106b]
Mehendale, A. R., **2**, 762[57,58]
Mehl, A. F., **6**, 245[124]
Mehl, W., **4**, 730[67]
Mehler, A. H., **2**, 456[59], 457[59], 458[59]
Mehler, K., **4**, 874[50,52,55]; **5**, 30[2,2c]
Mehra, R. K., **4**, 257[226], 259[226]
Mehra, U., **5**, 92[79,80], 95[102]
Mehrotra, A. K., **2**, 745[105]; **4**, 12[42]; **5**, 3[27], 320[5], 347[5]; **6**, 289[193]; **8**, 406[39], 989[34]
Mehrotra, I., **7**, 601[82-84], 602[84]; **8**, 708[32]
Mehrotra, K. N., **3**, 579[99]
Mehrotra, M. M., **5**, 560[77]
Mehrsheikh-Mohammadi, M. E., **5**, 856[216]; **6**, 451[125]
Mehta, A. M., **8**, 510[98]
Mehta, G., **3**, 21[124], 384[54], 385[55], 604[230]; **4**, 297[275], 303[347], 304[347]; **5**, 3[28], 5[28], 7[28], 225[91,95], 233[95], 667[44], 714[74], 759[87]; **6**, 836[58]; **7**, 220[19], 453[74], 455[74,103], 502[260], 573[116], 710[53]; **8**, 123[81], 566[455,456]
Mehta, R. R., **6**, 176[83]
Mehta, S., **2**, 747[117]
Mehta, S. M., **7**, 738[30]
Mehta, T. N., **4**, 5[18]
Mehta, Y. P., **7**, 700[58]
Mei, A., **2**, 345[24]
Meidar, D., **3**, 300[45]
Meienhofer, J., **2**, 1094[86,88]; **6**, 635[11,14b], 636[14], 645[11], 664[224], 665[227], 668[227], 669[227], 670[270], 698[270]; **8**, 959[21,23]
Meier, G. P., **1**, 592[22]; **2**, 1040[106], 1041[108]; **3**, 960[116]; **5**, 894[48]; **6**, 898[107b]
Meier, H., **1**, 844[5b]; **2**, 360[165b]; **3**, 666[42], 887[11], 891[42], 892[11], 893[11], 897[11], 898[11], 900[11], 903[11], 905[11], 909[155,156]; **4**, 1006[104], 1032[12], 1084[95]; **5**, 123[1], 126[1], 475[144]; **6**, 104[1]; **7**, 747[100], 748[100]; **8**, 950[172], 951[175]
Meier, H.-P., **5**, 527[64,65], 530[64,65]
Meier, I. K., **1**, 807[316]
Meier, J., **2**, 345[36]
Meier, M., **4**, 384[144,144b]; **6**, 294[241,242]
Meier, M. M., **5**, 498[232]
Meier, M. S., **4**, 761[202,205]
Meier, R., **8**, 896[18]
Meijer, E. M., **6**, 658[183]
Meijer, E. W., **2**, 769[1], 770[1], 771[1], 773[1]; **4**, 12[39]
Meijer, J., **2**, 85[19,21], 587[148]; **3**, 217[82], 254[96], 491[68], 531[84]; **4**, 895[163], 897[171], 898[171], 899[171], 900[180,182], 905[209]; **5**, 772[164], 949[282]; **6**, 426[78]
Meijer, K., **2**, 589[153]
Meijer, L. H. P., **8**, 562[420]
Meijer-Veldman, M. E. E., **5**, 1148[117]
Meijs, G. F., **4**, 796[91], 805[139,141]; **7**, 883[176]; **8**, 806[104,105], 807[104,105]
Meijs, G. G., **4**, 1007[114]
Meilahn, M. K., **6**, 498[163]
Meiler, W., **1**, 287[14]
Meili, J. E., **5**, 791[37]
Meinders, H. C., **3**, 552[10], 559[57]
Meindl, H., **6**, 275[109,110]

Meinhardt, D., **1**, 748[72], 749[78], 816[78]; **4**, 979[101]; **5**, 1115[3], 1122[3], 1123[3], 1124[3]
Meinhardt, N. A., **4**, 317[543]
Meinhart, J. D., **4**, 598[207,210]; **6**, 86[99]
Meinke, P. T., **5**, 437[160], 442[180,180d]; **8**, 846[81]
Meinwald, J., **1**, 377[98]; **3**, 395[102], 396[102], 736[23], 737[31,32], 890[31], 900[98], 901[106,112], 903[99], 906[143]; **4**, 298[281], 1024[265]; **5**, 68[95], 195[7], 612[73]; **6**, 535[529], 1030[108]; **7**, 86[16a], 109[182], 131[81], 219[12], 660[37]; **8**, 336[69]
Meinwald, Y. C., **3**, 900[98]
Meinzer, E. M., **8**, 242[43], 293[50,51]
Meise, W., **7**, 221[30]
Meisels, A., **3**, 638[89]
Meisenheimer, J., **8**, 364[16]
Meisinger, R. H., **5**, 609[67]
Meislich, E. K., **3**, 755[110]
Meissner, B., **7**, 595[26]
Meister, C., **7**, 399[34]
Meister, M., **3**, 309[90]
Meisters, A., **2**, 114[118,119], 268[67,68]; **4**, 140[9], 257[221]
Meixner, J., **4**, 729[58], 754[58]
Mejer, S., **8**, 115[60,62], 508[88], 510[94]
Mekelburger, H. B., **3**, 3[13]
Mekki, M. S. T., **3**, 325[158]
Meklati, B., **3**, 771[183]
Melamed, U., **8**, 563[431]
Melandri, A., **4**, 429[81]
Melany, M. L., **2**, 553[131]
Melchers, H. D., **5**, 1107[166]
Melching, K. H., **1**, 771[192]; **2**, 1069[134]
Meldahl, H. F., **3**, 781[18]
Melega, W. P., **2**, 588[150]; **3**, 254[96]; **7**, 172[127]; **8**, 946[134]
Meléndez, E., **2**, 406[45]
Melger, W. C., **4**, 493[77]
Melhado, L. L., **6**, 255[167]
Meli, A., **5**, 1158[172]; **8**, 458[219]
Melián, D., **2**, 1049[16]
Melian, M. A., **5**, 830[32]
Melika, Yu. V., **4**, 400[228b], 407[255]
Melikian, G., **2**, 720[84]; **6**, 188[181]
Melillo, D. G., **1**, 425[105]; **2**, 803[33]
Melillo, J. T., **7**, 777[369]
Melis, S., **8**, 847[89]
Mellea, M. F., **1**, 307[94]; **7**, 3[9]
Meller, A., **6**, 526[394]
Mellier, D., **6**, 644[81]
Mellini, M., **5**, 1174[36]
Mello, R., **7**, 13[125], 167[186], 374[77a]
Melloni, G., **4**, 277[89,90], 285[90]
Mellor, J. M., **1**, 469[56], 474[56]; **4**, 356[141,143], 386[156], 387[156], 408[259e], 413[259e,276]; **5**, 345[71a], 346[71a], 531[74], 985[36]; **6**, 109[42]; **7**, 14[136], 92[43], 488[156], 494[203], 495[203,204,209], 505[287]; **8**, 388[66], 637[11]
Mellor, M., **3**, 355[53], 357[53], 769[170], 771[170]; **5**, 136[67]
Mellows, G., **2**, 801[30]
Melnick, M. J., **2**, 1026[68], 1079[159]; **5**, 485[181], 531[79]
Mel'nikov, V. V., **1**, 34[228]
Mel'nikova, V. I., **1**, 520[68]
Melnitskii, I. A., **8**, 214[40]
Melot, J.-M., **2**, 330[52]
Meloy, G. K., **3**, 891[44]
Melpolder, J., **4**, 848[80], 849[80]
Mels, S. J., **8**, 249[91], 294[54]
Melser, W. F., **8**, 459[228]
Melstein, D., **4**, 315[532]
Melton, J., **4**, 12[41]; **6**, 939[139], 942[139,153]; **7**, 220[22]; **8**, 371[105]
Meltz, C. N., **1**, 350[150,151], 359[24], 364[24]; **2**, 946[176,177], 947[177,178], 948[178]
Meltzer, P. C., **7**, 384[116]
Meltzer, R. I., **7**, 664[65]
Melumad, D., **5**, 959[319]
Melville, M. G., **2**, 772[20]
Melvin, L. S., Jr., **1**, 408[35], 430[35], 630[32], 675[32], 722[32], 820[9], 822[9]; **3**, 86[41], 87[41], 178[41], 179[41], 288[63], 921[36]; **4**, 115[184a], 372[64a], 507[152], 987[133]; **5**, 905[59]; **6**, 438[64], 660[202], 854[141]; **7**, 678[73], 823[39]; **8**, 881[68]
Melvin, T., **3**, 380[9]; **4**, 800[121]; **7**, 728[40]
Melzer, H., **5**, 418[70]
Mena, P. L., **6**, 175[69]
Menachem, Y., **5**, 1130[9]
Menachery, M., **4**, 45[126]; **5**, 436[155]
Menapace, H., **3**, 381[31], 382[31]
Menapace, L. W., **4**, 738[92]; **8**, 991[45]
Ménard, G., **6**, 515[314], 522[314]
Menard, M., **1**, 123[78]; **2**, 212[120], 213[126], 656[158], 1059[77]
Mencel, J. J., **4**, 113[164]; **5**, 11[85]
Menchen, S. M., **1**, 656[146,147,149], 699[247]; **3**, 107[226], 109[226]; **4**, 398[216], 399[216a], 401[216a], 405[216a], 410[216a]; **6**, 470[58]; **7**, 495[211]; **8**, 847[97,97d], 849[97d,107,115], 887[117]
Menchi, G., **4**, 930[50]; **8**, 87[27], 236[3], 239[3], 552[352]
Mende, U., **4**, 964[50], 965[50]
Mendelsohn, J., **2**, 607[75], 608[75]
Mendelson, L., **3**, 500[132], 505[132]
Mendelson, L. T., **1**, 889[140], 890[140]; **2**, 1041[110]; **6**, 741[65]; **8**, 34[61], 66[61]
Mendelson, S. A., **1**, 786[248]; **4**, 116[185b]; **8**, 842[47]
Mendenhall, G. D., **7**, 228[101]
Mendive, J. J., **6**, 669[265]
Mendoza, A., **1**, 491[35], 495[47]; **2**, 368[237]; **3**, 199[64]; **7**, 602[101], 604[101]
Mendoza, L., **8**, 16[101], 54[156], 66[156], 537[180]
Menear, K. A., **5**, 531[75], 549[75]
Menes, R., **7**, 369[62]
Meney, J., **7**, 832[71]
Menge, W., **1**, 223[84], 225[84e]
Mengech, A. S., **7**, 262[79]
Mengel, R., **6**, 657[180]
Menger, F. M., **7**, 737[10]
Menicagli, R., **2**, 325[41,42]; **4**, 142[17a-c], 143[17b]; **5**, 942[230]; **8**, 100[118]
Menichetti, S., **4**, 331[12]
Menini, E., **7**, 100[128]
Menn, J. J., **8**, 754[93]
Menon, B. C., **2**, 144[63]
Men'shikova, N. G., **4**, 291[209]
Mensi, N., **6**, 671[281]
Menta, E., **6**, 178[121]
Mentzer, C., **3**, 831[60]
Menz, F., **3**, 890[29]
Menzel, I., **3**, 909[156]; **8**, 950[172], 951[175]
Menzies, W. B., **5**, 626[38], 629[38]
Meou, A., **2**, 1086[31]
Mérand, Y., **3**, 846[43,44]
Merault, G., **2**, 725[120]
Mercantoni, E., **7**, 331[16]
Merchant, J. R., **2**, 851[224]
Mercier, D., **7**, 239[46]
Mercier, J., **7**, 452[51], 453[51]
Mercier, R., **4**, 519[27]
Merczegh, P., **5**, 534[94]

Merényi, R., **3**, 870[49]; **4**, 758[189,190,191]; **5**, 70[111-113], 116[257], 422[81]; **6**, 495[142,143], 496[143,156], 497[143], 514[156], 515[317], 546[317]
Meresz, O., **3**, 891[41b]
Mereyala, H. B., **5**, 223[83], 224[83], 806[102], 1028[90]; **6**, 93[131]
Mergard, H., **7**, 3[4]
Mergelsberg, I., **3**, 1040[105]; **5**, 329[31]
Mergen, W. W., **6**, 229[25], 519[336]
Merger, F., **2**, 138[25]
Mergler, M., **6**, 671[277]
Merienne, C., **4**, 112[159]
Meriwether, L. S., **5**, 1146[106], 1148[121]
Merk, B., **7**, 252[6]
Merkel, C., **3**, 123[250], 124[250], 125[250]
Merkel, P. B., **7**, 777[365]
Merkel, W., **8**, 254[125]
Merkle, H. R., **1**, 214[30]
Merkley, J. H., **4**, 877[65], 878[77], 879[65,77]
Merle, G., **2**, 765[69]
Merlic, C. A., **6**, 165[200]
Merling, G., **4**, 2[4]; **8**, 302[95]
Merlini, L., **3**, 691[132,134]
Merour, A., **2**, 816[84], 828[84]
Merrifield, J. H., **4**, 594[147]
Merrifield, R. B., **6**, 633[7], 637[32,32d], 643[78], 666[233], 667[233], 670[269,270,274], 698[270]; **8**, 166[62]
Merrill, R. E., **8**, 459[228]
Merritt, A., **4**, 16[51,51a]
Merritt, J. E., **4**, 807[148]
Merritt, L., **8**, 623[151]
Merritt, R. F., **4**, 1013[178]
Merritt, V. Y., **3**, 621[30]; **5**, 647[14], 649[14,21], 658[21]
Merritt, W. D., **4**, 426[36]
Mersch, R., **6**, 263[23], 273[23,92], 277[92], 291[23]
Mersereau, J. M., **3**, 640[101]
Merslavid, M., **6**, 554[739]
Merten, R., **2**, 1054[62]; **5**, 404[12]; **6**, 570[948], 582[994]
Mertens, A., **6**, 237[68], 254[162]
Mertes, K., **3**, 220[120]
Mertes, M. P., **6**, 71[23], 624[140]; **7**, 350[25], 355[25]
Merz, A., **6**, 647[105]
Merz, K. M., **5**, 829[24]
Merz, W., **2**, 1088[51], 1090[13], 1106[13]
Meschke, R. W., **5**, 1130[10]
Meshgini, M., **5**, 1001[17]
Meshram, H. M., **6**, 555[815]
Meshram, N. R., **3**, 328[175]
Meshulam, H., **2**, 887[53]
Meskens, F. A. J., **6**, 677[308]
Meske-Schüller, I., **6**, 968[116]
Meslem, J. M., **8**, 802[87]
Meslin, J.-C., **6**, 554[753,754,781]
Mesnard, D., **2**, 1004[60], 1005[60]; **3**, 420[51], 421[51]; **4**, 871[30], 877[30]
Messeguer, A., **7**, 359[18]
Messerotti, W., **3**, 386[57]; **8**, 349[146]
Messier, A., **3**, 253[87]
Messing, A. W., **8**, 681[76], 689[76]
Messing, C. R., **8**, 387[59]
Messwarb, G., **2**, 360[173]
Mestdagh, H., **3**, 538[91]; **5**, 1151[128]; **6**, 502[212], 672[287]
Mestre, F., **1**, 770[185]
Mestres, R., **4**, 111[155d]
Mestroni, G., **8**, 91[56], 450[167], 552[359]
Mészaros, Z., **2**, 789[56]; **6**, 499[177], 520[340]; **7**, 846[97]
Metcalf, B. W., **2**, 813[75], 818[75]; **3**, 126[320]; **4**, 127[220a]; **5**, 841[95]; **6**, 80[70]
Metcalfe, A. R., **4**, 453[36], 455[36], 472[36]
Metcalfe, D. A., **5**, 859[233], 888[25]; **6**, 656[169]
Metelitza, D. I., **7**, 160[53]
Metelko, B., **8**, 356[190], 357[190]
Meth-Cohn, O., **2**, 777[2], 780[2], 784[36], 787[50,51], 792[62]; **4**, 1063[170], 1064[174]; **5**, 422[84]; **6**, 489[77,84], 543[607], 922[51]; **7**, 21[6,16], 305[1]; **8**, 98[106]
Metler, J., **4**, 41[118]
Metler, T., **4**, 46[132], 53[132,132a]
Metlesics, W., **3**, 812[54,58,59]
Metlin, S., **8**, 455[206], 456[208], 608[44]
Metlushenko, V. F., **6**, 552[696]
Métra, P., **3**, 322[144]; **7**, 471[18]
Metten, K.-H., **6**, 7[34]
Metter, J. O., **1**, 303[78], 307[78]; **5**, 365[95,95b]
Metternich, R., **2**, 13[53,60], 14[53], 15[61], 26[100,100a], 27[100a], 36[61], 318[50]
Mettile, F. J., **8**, 896[11], 900[33], 904[11]
Metwali, R. M., **7**, 71[96]
Metysova, J., **2**, 765[78]
Metz, F., **3**, 1048[9]; **4**, 1033[18,19], 1037[18]
Metz, G., **2**, 759[35]
Metz, H., **8**, 528[71], 971[108]
Metz, H. J., **6**, 515[309]
Metz, J. T., **5**, 79[292]; **8**, 723[151], 724[151]
Metz, P., **1**, 770[189]; **5**, 837[71], 838[71]
Metz, S., **2**, 787[51]
Metz, W., **8**, 112[24], 120[24]
Metzger, C., **5**, 99[136], 100[136]
Metzger, H., **6**, 104[1], 111[64]
Metzger, J., **3**, 505[162], 507[162], 512[162], 698[159]; **7**, 92[42], 93[42], 95[71], 637[73]; **8**, 444[8]
Metzger, J. O., **4**, 722[32]
Metzger, J. V., **4**, 113[175]; **6**, 419[11], 1035[139,140]
Metzher, P. J., **4**, 587[31]
Metzler, D. E., **8**, 185[20]
Metzner, G., **6**, 501[203], 531[203]
Metzner, P., **2**, 214[130]; **4**, 107[143a,b,d,e], 259[277]; **6**, 437[31]
Metzner, P. J., **7**, 94[60]
Meuche, D., **2**, 144[59]
Meudt, W. J., **7**, 673[29]
Meulendijks, G. H. W. M., **8**, 95[82]
Meunier, A., **4**, 152[54], 184[85], 201[13], 202[13]
Meunier, B., **7**, 12[94]; **8**, 908[77], 909[77], 910[77]
Meunier, F., **7**, 238[40]
Meunier, P., **6**, 464[31,32]; **8**, 447[129], 463[129], 688[99], 691[99]
Meurers, W., **5**, 1152[141], 1154[149]
Meusel, W., **8**, 533[143]
Meusinger, R., **4**, 387[163c]
Meuwly, R., **6**, 939[140], 1000[127]
Mews, R., **6**, 497[158]; **7**, 483[130]
Meyer, A., **4**, 520[31]
Meyer, B., **3**, 587[142]; **6**, 1035[139,140]
Meyer, D., **2**, 232[176]; **8**, 460[254]
Meyer, E., **2**, 655[146]; **3**, 812[52]; **7**, 876[121]
Meyer, E. W., **6**, 959[43]; **8**, 612[67]
Meyer, F. J., **2**, 124[205]; **4**, 96[103a], 98[103a]
Meyer, G., **2**, 143[53]
Meyer, G. M., **2**, 800[14]
Meyer, G. R., **4**, 279[116]; **7**, 482[119]; **8**, 15[95], 16[95]
Meyer, H., **2**, 189[51], 377[281], 384[281]; **6**, 515[308], 535[531]
Meyer, I., **2**, 1077[154]

Meyer, J., **1**, 373[91], 375[91], 376[91]; **2**, 996[47]; **6**, 476[96], 646[101]
Meyer, J. D., **8**, 243[46]
Meyer, L. A., **4**, 5[17]
Meyer, M. W., **7**, 12[104]
Meyer, N., **1**, 631[49], 636[49], 656[49], 658[49], 659[49], 672[49], 708[49], 710[49], 721[49]; **3**, 87[69], 195[18], 255[108]; **6**, 147[84]
Meyer, R., **1**, 846[19b], 847[19b], 850[19b]; **2**, 281[29,31]; **4**, 89[84g], 277[85], 285[85], 288[85]
Meyer, R. T., **7**, 429[156]
Meyer, S., **8**, 511[100]
Meyer, T. J., **7**, 158[40], 851[23]
Meyer, V., **2**, 321[2]; **5**, 664[38]; **6**, 937[113]
Meyer, V. B., **8**, 625[157], 626[157]
Meyer, W. C., **5**, 107[204]
Meyer, W. E., **6**, 554[737]
Meyer, W. L., **2**, 148[78], 166[152]; **4**, 16[51,51a]; **7**, 30[83]
Meyerhoff, G., **5**, 64[45]
Meyers, A. G., **5**, 957[308]
Meyers, A. I., **1**, 23[120], 60[36], 66[57], 69[57b], 75[36], 359[20-22], 366[50,51], 367[55], 383[20-22], 384[21,22], 422[95], 460[5], 461[5,8,10,11], 468[54,55], 473[11], 474[108], 477[128,135,136], 478[10,11,148], 482; **2**, 200[89], 214[132], 252[40], 257[40], 377[282], 455[10], 482[26], 483[26], 486[30], 488[30], 489[44-50], 490[46-49], 491[50], 492[53,54], 493[53,54], 494[55], 648[90], 821[106], 1026[66], 1049[24], 1058[72,73]; **3**, 31[185], 32[185], 36[209], 37[212], 42[231,232,233,234], 53[272,273], 67[18], 68[23], 69[24,25,27], 70[25], 71[34,36,37], 72[25,38-40], 74[25,41], 75[47-50], 77[55,56,58], 78[60-62], 79[60,61,64], 81[40,60,61,66-68], 125[291], 126[291], 181[552], 255[103], 261[103,156], 288[65], 503[149], 512[149,197,202]; **4**, 14[47,47m], 76[45,49], 77[50], 205[37-41], 206[42-49], 243[66], 250[140,141], 252[66,155,156,157,166,167], 257[167], 260[166], 294[240,241], 305[363], 425[21], 426[21], 427[21,67,68], 428[21,70,72], 494[82,83], 498[101], 872[44], 989[144]; **5**, 132[50], 372[104], 407[28,28b]; **6**, 71[22], 134[30], 205[36], 261[14], 263[14], 272[14], 273[95-97], 274[102-107], 275[14], 276[95], 277[123], 280[14,141], 501[187,188], 530[419], 534[521], 541[593,594], 545[629,630,631], 551[681], 554[774], 667[237], 674[295,296], 723[147], 725[170,171,173], 726[175,176], 728[170,171], 740[64]; **7**, 57[25], 143[146], 224[47-52], 360[22], 407[81], 580[145]; **8**, 92[68], 95[90], 162[32], 231[143], 275[145,146], 276[147,148,149,150], 584[18,23], 651[69], 653[81], 654[81]
Meyers, C. Y., **3**, 863[15], 864[15-18,20,22,23], 866[16,17,31], 868[39], 883[16,17]; **7**, 235[1]
Meyers, H. V., **1**, 200[96], 329[40], 798[290]; **5**, 466[119], 467[118,119], 545[122]; **6**, 717[115,116]
Meyers, M., **3**, 757[119], 964[126]; **4**, 243[71], 258[252]
Meyers, M. B., **3**, 689[121], 813[60]
Meyers, P. L., **3**, 50[265]
Meyers, R. F., **3**, 370[110]
Meynier, F., **6**, 718[122]
Meyr, R., **2**, 1085[23], 1086[32], 1087[36], 1090[36]
Meyring, W., **5**, 444[188]
Meystre, C., **7**, 41[20], 128[171]; **8**, 974[124]
Mezey-Vándor, G., **7**, 829[60]
Mezheritskii, V. V., **2**, 712[40]; **6**, 487[44], 489[44], 556[25], 563[25]
Mezzetti, A., **5**, 331[41]
Miah, M. A. J., **1**, 466[43]; **3**, 242[5]
Miao, C. K., **5**, 491[208]
Micas-Languin, D., **1**, 770[185]
Miccoli, G., **3**, 441[49], 485[29], 493[29], 503[29], 513[29]
Micera, G., **4**, 930[50]
Micetich, R. G., **6**, 76[50]; **8**, 384[29], 412[115]
Michael, A., **2**, 400[24]; **4**, 1[1], 2[1a,2], 3[1], 41[2], 47[133], 70[1], 272[31], 279[31], 280[31], 286[170], 287[31], 288[183], 1099[179]; **6**, 953[12]
Michael, G., **8**, 595[78]
Michael, J. P., **2**, 830[140], 881[43], 885[48,51]; **4**, 161[91], 378[111]; **6**, 509[276]
Michael, K. W., **1**, 619[60]
Michaelides, M. R., **2**, 26[102], 27[102]
Michaelis, A., **7**, 602[95]
Michaelis, K., **3**, 644[159]
Michaelis, R., **3**, 644[159], 653[227], 654[227]
Michaelis, W., **4**, 729[61], 730[61], 765[61]
Michaelson, R. C., **5**, 773[165]; **7**, 167[95]
Michaely, W. J., **4**, 808[156]
Michalak, R., **8**, 549[326], 696[122]
Michalik, M., **2**, 345[42]
Michalovic, J., **8**, 988[25]
Michalska, M., **8**, 887[118]
Michalska, Z. M., **8**, 765[13]
Michalski, J., **6**, 602[6]
Michalski, T., **4**, 252[162]
Micha-Screttas, M., **3**, 88[126], 95[126], 96[126], 107[126], 109[126], 123[126], 125[126], 824[21], 825[21]; **4**, 316[539]; **8**, 842[45]
Michaud, D. P., **7**, 362[26]
Micheel, F., **2**, 385[326]; **6**, 46[56]
Michejda, C. J., **6**, 245[124]; **7**, 485[140]
Michel, E., **8**, 753[73]
Michel, J., **1**, 226[90,96]; **4**, 98[108d,109d-f]; **6**, 50[101,102], 51[101,102], 54[125], 59[143]
Michel, M. A., **7**, 797[19]
Michel, R. E., **4**, 452[2]; **7**, 882[173]
Michel, S. T., **3**, 194[15], 196[15]
Micheli, R. A., **3**, 23[142]
Michelin, R. A., **7**, 426[148d]
Michelot, D., **3**, 256[113], 257[114], 485[30], 491[30,72], 934[68], 963[123], 964[123], 969[130]; **6**, 893[78]
Michelot, R., **6**, 911[17,19]
Michelotti, E. L., **2**, 587[146]; **3**, 229[233], 246[38], 444[65], 446[81], 447[91,92], 456[126], 492[75], 503[75], 513[208]; **8**, 842[41], 935[68]
Michels, D. G., **5**, 133[57]
Michels, E., **5**, 634[82-84], 635[84]
Michelson, A. M., **6**, 616[104]
Michl, J., **4**, 765[224]; **5**, 72[178], 199[28], 741[156]
Michler, W., **6**, 435[3]
Michman, M., **4**, 877[71], 905[211]
Michna, P., **5**, 326[26]; **8**, 844[73]
Michniewicz, J., **6**, 625[158]
Michno, D. M., **5**, 686[45], 707[37], 716[84], 717[91], 737[37]; **7**, 263[87]
Mickel, S., **5**, 105[196]
Micovic, V. M., **8**, 270[98]
Middlesworth, F. L., **4**, 682[56]
Middleton, D. J., **8**, 413[124]
Middleton, D. S., **4**, 355[131]
Middleton, S., **7**, 686[98]
Middleton, W. J., **1**, 425[103]; **2**, 366[221], 555[144]; **5**, 436[154], 439[154], 441[179,179b]; **6**, 217[110]
Midgley, I., **8**, 930[33]
Midgley, J. M., **7**, 833[72]
Midková, R., **6**, 495[147]
Midland, M. M., **1**, 405[25]; **2**, 6[24], 38[133], 111[83], 679[56], 699[56]; **3**, 274[21], 421[53], 799[98], 983[16-18], 987[29,30], 990[29], 993[29]; **4**, 345[82], 367[12], 756[181]; **5**, 410[41], 411[41i], 434[145]; **6**, 78[60], 875[17], 876[17,21,33,35], 877[17], 882[17,33], 885[17], 887[20-22,33,35]; **7**, 599[62,69-71],

602[97], 603[115,118-122,124,125], 607[162,163,164,169], 608[169]; **8**, 17[110], 101[119-122], 102[124,-128], 386[51], 537[189], 704[6], 705[6], 706[6], 720[137], 726[187]

Midorikawa, H., **2**, 146[67]; **3**, 97[177], 108[177], 109[177], 136[369], 138[369]

Midura, W., **3**, 953[105]; **6**, 150[117]; **7**, 197[22], 764[108]

Mieczkowski, J., **6**, 1013[17]

Miehling, W., **8**, 174[121]

Mielert, A., **5**, 498[237]

Mielke, D., **6**, 508[287,288]

Mierop, A. J. C., **4**, 52[146]

Miesch, M., **4**, 956[16]

Miethchen, R., **2**, 747[119], 749[119]; **3**, 322[142b,143]; **7**, 7[50]

Mietzsch, F., **5**, 715[79]

Miftakhov, M. S., **2**, 814[80]; **8**, 676[55], 677[57], 680[73], 682[55], 683[73], 689[55,57], 691[55]

Migachev, G. I., **6**, 438[50]

Migaj, B., **4**, 1014[192]

Migalina, Y. V., **4**, 342[66,67]

Migdal, S., **3**, 628[45]

Migdalof, B. H., **4**, 27[84], 29[84c]

Migge, A., **1**, 144[40]

Miginiac, L., **1**, 215[39], 218[39,51], 225[39], 326[4], 385[113], 831[103]; **2**, 2[6], 3[6], 6[6d], 21[6d], 23[6d], 49[6d], 77[91], 81[5], 579[92,93], 980[18,19], 988[33], 989[33], 995[44], 1004[60,62,63], 1005[60,62,63]; **3**, 420[51], 421[51]; **4**, 93[95], 868[15], 871[30], 877[30], 878[80], 879[86], 880[90], 883[86,90,96-100], 884[15,80,90,96,97,99,100,105], 987[162], 993[162]

Miginiac, P., **1**, 218[50], 219[50,63], 368[57], 369[57]; **2**, 82[8], 988[32], 989[32], 1000[53,54]; **4**, 84[68b,c], 89[84h], 95[84h,102b-d], 149[52], 182[77], 183[78]

Miginiac-Groizeleau, L., **1**, 218[50], 219[50]

Migita, T., **1**, 436[154], 438[160,161], 452[219], 834[124]; **2**, 446[31]; **3**, 12[62], 453[113,114], 454[120], 463[160,164], 469[216], 470[215,216], 473[215], 476[215,216], 923[43,44], 934[44,63], 954[44], 1008[70,71]; **4**, 743[128]; **8**, 824[63], 825[63]

Migita, Y., **7**, 42[28,29], 877[134]

Migliara, O., **6**, 551[682]

Migliorese, K. G., **4**, 347[97]

Migliorini, D. C., **8**, 533[154]

Mignani, S. M., **1**, 770[184]; **4**, 593[129,132,134], 758[190]; **5**, 245[21], 299[66], 303[66], 304[66]

Mignard, M., **7**, 422[139]

Migniac, L., **5**, 39[25]

Mignon, L., **4**, 527[69], 528[69]

Migrdichian, V., **2**, 770[6]

Mihaila, G., **4**, 298[284]

Mihailovic, M. L., **2**, 553[128]; **3**, 380[13]; **5**, 179[143]; **7**, 92[41,41a], 94[41], 229[112], 231[142], 236[24], 703[5], 710[5], 738[27], 815[2], 816[2c], 824[2c], 827[2c], 851[18]; **8**, 270[98], 872[5], 873[17], 875[5,33]

Mihailovic, M. M., **8**, 875[33]

Mihara, M., **8**, 661[112]

Mihelcic, J. M., **1**, 307[94]; **7**, 3[9]

Mihelich, E. D., **1**, 359[21], 383[21], 384[21]; **2**, 489[45]; **4**, 91[90], 92[90c], 261[294]; **6**, 534[521], 674[295,296], 677[311]; **7**, 111[194], 378[92], 413[117], 819[18]

Mihoubi, M. N., **4**, 793[72]

Mijngheer, R., **5**, 539[106], 924[146]

Mijs, W. J., **4**, 763[210], 806[147]; **5**, 580[166]; **7**, 252[2], 437[7], 438[21], 439[7], 527[1], 703[5], 710[5], 737[18], 754[18], 755[18], 815[2], 816[2b,c], 824[2b,c], 827[2c]

Mikaelian, G. S., **5**, 1055[46], 1056[48]

Mikajiri, T., **1**, 558[135]

Mikami, A., **7**, 63[59]

Mikami, K., **1**, 584[11]; **2**, 119[161], 455[18], 556[156,157], 558[158,159], 569[35], 715[56]; **3**, 215[64], 942[80], 976[7], 984[22,23], 985[22,23,26a], 987[31-33], 992[37,38], 993[38], 994[41], 999[51], 1000[52], 1002[57], 1004[59], 1005[61-63], 1008[65]; **5**, 11[86], 16[112], 24[166,167], 55[48], 821[162], 833[49], 850[147,159], 851[164,165], 888[28,29], 889[31], 1001[16]; **6**, 14[51], 834[30], 850[30], 851[129,130], 852[30,136,138], 853[30], 854[144], 856[151], 860[179], 873[7], 874[14], 875[7], 876[26,31], 877[14,31,36,37], 879[36,41], 882[26,31,47], 883[14,36,52,54,55,58], 885[26,31,47], 886[36], 887[58,61,64], 888[7], 890[52,69], 891[54,69,70], 892[74], 896[74]

Mikami, T., **8**, 224[104]

Mikami, Y., **7**, 58[55], 62[55], 63[55]

Mikawa, H., **5**, 71[131]; **7**, 851[28]

Mikaya, A. I., **6**, 530[422]

Mikesa, L. A., **2**, 527[2], 528[2], 553[2]

Mikhail, G., **1**, 865[87]; **3**, 785[37]; **5**, 215[11], 219[11], 221[57], 226[11], 230[11,57,130,131], 232[57,130,131,134,137], 944[240]; **7**, 650[50]

Mikhailenko, F. A., **6**, 509[255]

Mikhailopulo, I. A., **4**, 379[115]

Mikhailov, B. M., **2**, 5[22], 10[22]; **4**, 145[28,29b], 885[113-115], 886[115]; **5**, 33[8], 34[9,10], 480[177]; **7**, 594[7], 595[7,18,20], 596[40], 597[42-44], 598[7], 599[7], 601[7], 603[117]; **8**, 705[14], 707[14], 725[14,183], 726[14], 727[196], 728[14]

Mikhailov, I. E., **6**, 552[696]

Mikhailov, S. N., **6**, 618[114]

Mikhailova, L. N., **8**, 449[155]

Miki, D., **8**, 244[71], 247[71], 251[71], 253[71]

Miki, K., **1**, 14[74-76], 162[93,102]; **4**, 607[317], 615[384]; **5**, 275[11]

Miki, M., **5**, 516[25]; **8**, 881[74]

Miki, T., **2**, 538[68], 539[68], 647[88a]; **4**, 161[91]; **6**, 1022[60]; **7**, 86[16a], 202[46], 203[52]

Miki, Y., **5**, 829[23]

Mikol, G. J., **6**, 799[26]; **7**, 196[12], 215[12], 222[39]; **8**, 843[58]

Mikolajczyk, M., **1**, 788[259]; **3**, 953[105]; **4**, 113[166]; **6**, 134[36], 149[98], 150[117]; **7**, 197[22], 760[45], 762[69,74], 764[108], 765[132], 777[69a], 778[69,407]; **8**, 403[3], 404[3,15], 406[52], 410[52]

Mikolajczyk, N., **6**, 149[101]

Mikstais, U., **6**, 489[37]

Mikulec, R. A., **1**, 226[95]

Milan, **8**, 443[1]

Milani, F., **7**, 283[181], 284[181]

Milart, P., **2**, 369[247]

Milas, N. A., **4**, 305[368]

Milat, M.-L., **6**, 54[129], 66[3]

Milchereit, A., **5**, 1140[77]

Milczanowski, S. E., **8**, 815[22]

Milenkov, B., **6**, 1058[61]

Miles, D. E., **3**, 136[373], 137[373]

Miles, D. H., **3**, 364[93]

Miles, G. J., **6**, 714[84]

Miles, J. H., **7**, 709[42], 710[42]

Miles, M. L., **2**, 189[53]

Miles, W. H., **3**, 498[108]; **4**, 984[121]; **5**, 829[22], 1086[71]

Milesi, L., **8**, 806[123]

Milewska, M., **6**, 493[126], 494[126], 807[61]

Milewski, C. A., **6**, 959[37]; **8**, 37[103], 66[103], 343[117], 344[117], 350[150], 351[117], 354[117], 355[117], 812[6], 929[31]

Milewski-Mahrla, B., **6**, 175[67], 179[125], 196[232,233]

Milionis, J. P., **2**, 765[79]

Miljkovic, D., **2**, 866[9]

Milkova, T., **7**, 47[53]

Milkowski, J., **6**, 664[223]

Milkowski, W., **6**, 525[376]

Millan, A., **8**, 773[70], 774[70]

Millar, D. J., **6**, 86[99]
Millar, J. G., **3**, 223[155], 224[174]
Millar, R., **5**, 151[10]
Millard, A. A., **3**, 443[59]
Millard, B. J., **3**, 927[53]
Miller, A., **5**, 420[75]
Miller, A. C., **6**, 690[396]
Miller, A. E. G., **8**, 275[141]
Miller, A. J., **6**, 245[126]
Miller, A. R., **5**, 1125[54]
Miller, A. S., **5**, 737[146]
Miller, A. W., **6**, 667[240]
Miller, B., **3**, 803[1], 804[7], 809[1b,36,38], 815[72,75,76], 817[1b]; **5**, 790[33], 799[71]
Miller, B. J., **8**, 161[14,15,24], 162[25-27], 545[298,299,300]
Miller, C. B., **8**, 895[3], 898[3]
Miller, C. H., **2**, 843[195]; **7**, 87[18]
Miller, D., **3**, 577[85]; **8**, 527[52]
Miller, D. B., **2**, 1026[66]; **3**, 72[40], 81[40]; **6**, 740[64]; **7**, 224[51]
Miller, D. D., **3**, 41[227], 583[126], 630[56], 631[56,58]; **7**, 144[157]
Miller, D. J., **4**, 598[210], 629[415]
Miller, D. W., **7**, 63[58], 75[116]
Miller, E., **7**, 14[141]
Miller, E. G., **3**, 890[31]; **5**, 890[34]; **6**, 152[141], 899[109]
Miller, F. D., **8**, 682[84]
Miller, I. J., **5**, 754[67]
Miller, J., **1**, 78[6], 95[6,80]; **4**, 295[257], 423[2], 425[2], 541[111], 689[69], 857[107]; **6**, 245[119], 246[119], 247[119], 248[119], 249[119], 251[119], 252[119], 253[119], 254[119], 255[119], 256[119], 258[119]; **8**, 86[22]
Miller, J. A., **1**, 448[206]; **2**, 65[28], 449[48]; **3**, 13[67], 246[35], 259[133], 463[156], 486[40], 495[40], 497[40], 498[40], 503[40], 553[16]; **4**, 141[14], 297[272,273,274], 884[107,107b], 891[107b,139], 892[146], 893[107b,148], 903[198]; **5**, 32[6,6a], 1037[5], 1165[11,12,15], 1166[11,15,21], 1167[11,15], 1170[15], 1171[15], 1175[15], 1178[11,15], 1179[15]; **6**, 836[55]; **8**, 675[42], 677[42], 678[42], 681[42], 685[42], 697[42], 735[17], 737[17], 742[44], 743[164], 746[17], 753[17], 757[164], 758[164], 761[17]
Miller, J. B., **2**, 969[84]
Miller, J. G., **5**, 71[145]; **6**, 838[67], 902[132]
Miller, J. J., **4**, 6[20,20c], 315[510]; **8**, 857[202]
Miller, J. L., **6**, 546[652]
Miller, J. M., **3**, 54[278]; **6**, 66[6]
Miller, J. W., **6**, 632[6]
Miller, L. A., **7**, 221[29]
Miller, L. L., **3**, 669[52], 683[52,101], 685[106,107]; **5**, 195[8], 197[8]; **6**, 281[151], 282[153,155,156,157]; **7**, 42[30], 248[110], 264[89], 275[89], 778[405], 794[4], 799[25], 800[25a,29,32], 801[37,41,44], 810[87,90], 843[43,44], 852[41], 853[41]; **8**, 584[20]
Miller, M. A., **3**, 854[79], 905[140]
Miller, M. J., **1**, 386[123]; **2**, 74[77], 113[105]; **4**, 398[214], 589[86]; **6**, 22[83], 112[67], 114[72,74], 822[117]; **7**, 503[274]; **8**, 64[207b], 66[207], 395[125]
Miller, M. L., **5**, 299[67], 300[75]
Miller, M. W., **3**, 753[103]
Miller, O. N., **2**, 456[68], 460[68]
Miller, P., **2**, 916[84]; **6**, 971[128]
Miller, P. G., **8**, 532[131a]
Miller, P. S., **6**, 658[187]
Miller, R., **4**, 876[61]; **6**, 219[122]; **8**, 568[467]
Miller, R. A., **5**, 1070[20], 1071[20], 1072[20], 1074[20], 1110[20], 1111[20]
Miller, R. B., **2**, 897[17]; **3**, 232[260], 254[98], 257[98], 489[61], 495[61,90], 503[90], 504[61,154], 511[61,154], 515[61,154], 602[219]; **4**, 344[78b], 878[77], 879[77]; **7**, 36[108]
Miller, R. C., **7**, 854[54], 855[54]
Miller, R. D., **2**, 631[16]; **3**, 251[74], 621[30]; **5**, 680[21], 925[151], 944[243], 1031[97]; **6**, 932[96]; **7**, 742[58]; **8**, 406[40]
Miller, R. E., **2**, 139[32]; **7**, 247[99]
Miller, R. F., **4**, 824[240]; **5**, 927[161], 931[161]
Miller, R. G., **4**, 487[50]
Miller, R. K., **7**, 764[105]
Miller, R. L., **3**, 483[13]; **8**, 755[124], 757[124], 758[124]
Miller, R. W., **3**, 407[149]; **4**, 306[388]
Miller, S. A., **2**, 1049[24]; **5**, 806[101]
Miller, S. I., **4**, 3[10], 41[10,118], 46[132], 47[10,135], 53[132,132a], 65[10a], 66[10,10a]; **6**, 959[48]
Miller, S. R., **5**, 1182[48]
Miller, T., **5**, 4[36], 5[36]
Miller, T. G., **3**, 804[10]
Miller, T. L., **8**, 558[396]
Miller, V. P., **8**, 344[121,121b]
Miller, W. H., **1**, 188[95], 198[95]
Miller, W. T., Jr., **4**, 1000[11]
Millet, G. H., **7**, 228[102], 662[53,55]
Millidge, A. F., **3**, 351[38]
Milliet, P., **8**, 395[135], 848[105], 936[70]
Milligan, B., **7**, 84[3], 154[14]
Millikan, R., **1**, 480[152]
Millon, J., **3**, 247[46], 466[192]
Millor, J. M., **7**, 760[36], 761[36]
Mills, H. H., **3**, 382[36]
Mills, J. A., **8**, 141[42]
Mills, J. E., **3**, 672[64]
Mills, L. S., **7**, 406[74]
Mills, O. S., **3**, 810[43]; **5**, 1138[67], 1146[111], 1147[111]; **6**, 523[347]
Mills, R. J., **1**, 462[19]; **4**, 204[34]; **7**, 646[27]; **8**, 784[112]
Mills, R. W., **3**, 427[86], 712[26]
Mills, S., **1**, 101[91], 402[17], 461[14], 464[14], 477[133], 791[296a], 799[296]; **3**, 67[17]; **8**, 54[157], 66[157]
Mills, S. D., **4**, 791[59]
Mills, S. G., **2**, 482[37], 483[37], 485[37], 1064[111]; **7**, 228[92]
Mills, W. H., **6**, 970[123]
Millward, B. B., **8**, 499[41], 566[453]
Mil'man, I. A., **7**, 57[33]
Milne, G. M., **2**, 159[127]; **3**, 201[74]
Milner, J. R., **6**, 860[176]
Milolajczyk, M., **7**, 762[81]
Milone, L., **8**, 457[216]
Milovanovic, J., **8**, 872[5], 875[5]
Milowsky, A. S., **4**, 83[65c]; **5**, 255[49], 581[172]
Mils, W. J., **7**, 99[111]
Milstein, D., **1**, 437[155], 440[155], 442[155,177], 443[181], 445[155], 446[155,194], 447[202], 457[155]; **2**, 749[133]; **3**, 453[112], 463[161,162], 504[153]; **4**, 600[235], 643[235], 738[95], 936[66], 937[66,72]
Milstein, N., **3**, 313[104], 316[114]
Milton, K. M., **4**, 282[143], 283[143]
Milton, S. V., **7**, 856[66]
Miltz, W., **1**, 122[72]; **2**, 1077[153]; **5**, 485[182]; **6**, 716[104]
Mil'vitskaya, E. M., **5**, 900[10], 901[10], 906[10], 907[10], 972[6], 982[6], 984[6], 989[6]
Mimoun, H., **7**, 11[87,91], 95[68], 107[162], 160[53], 358[8a], 381[106], 422[139], 450[11], 452[11,42,45]
Mimura, T., **3**, 999[51]; **5**, 851[165]; **6**, 780[73,75], 851[130], 877[36], 879[36], 883[36], 886[36]; **8**, 352[147], 934[58], 949[58]
Mina, G., **4**, 45[126]
Minachev, Kh. M., **3**, 305[72]
Minagawa, M., **3**, 466[187]
Minailova, O. N., **8**, 518[130]
Minakata, H., **3**, 222[137]; **4**, 148[51], 149[51], 179[65]; **6**, 764[11]

Minakata, M., **3**, 623[39]
Minami, H., **3**, 99[190]
Minami, I., **1**, 553[97]; **4**, 591[117,118], 592[119-125], 611[361,363,364,365,366,367], 612[368], 613[117,369], 636[366,367]; **6**, 20[77], 641[61]; **7**, 141[134], 142[134,135,136,137], 453[87], 455[87]; **8**, 960[31]
Minami, K., **6**, 535[530]
Minami, M., **5**, 864[259]; **6**, 89[116]
Minami, N., **2**, 184[21]; **3**, 617[15], 619[15], 621[15], 623[15], 627[15]; **7**, 401[62,62a], 406[62a]; **8**, 241[37]
Minami, T., **1**, 554[108], 569[261]; **2**, 176[185], 363[199], 832[152]; **3**, 672[65]; **4**, 21[66,66c], 62[66c], 107[146c], 218[143], 247[102], 252[102], 259[102], 520[33-35], 546[129], 988[139]; **5**, 422[82], 1098[114], 1112[114]; **6**, 176[89,90], 186[172]; **7**, 209[93]; **8**, 131[6], 535[166], 860[223]
Minamide, H., **6**, 577[978]
Minamikawa, H., **1**, 546[55]; **2**, 635[42], 640[42]; **7**, 806[71]
Minamikawa, J., **6**, 764[13]; **7**, 606[156], 746[90]; **8**, 829[81]
Minamoto, K., **6**, 554[720]
Minaskanian, G., **2**, 651[120]; **4**, 162[92]; **7**, 128[66]
Minasz, R. J., **4**, 1000[15]
Minato, A., **3**, 437[25], 440[25], 448[25,96], 449[25], 450[25], 451[25], 452[25], 457[129,130,131], 459[137,139], 460[130,137], 461[137], 484[26], 492[26], 494[26], 495[26], 497[104], 503[26], 510[183,206], 513[26,205,206]
Minato, H., **5**, 809[115,117]; **7**, 595[14], 597[14]
Minato, I., **3**, 638[94]
Minato, M., **3**, 386[57]; **7**, 452[39], 462[39], 809[86]
Minatodd, M., **4**, 553[5]
Mincione, E., **1**, 754[107]; **4**, 611[344]; **7**, 832[69]; **8**, 263[28]
Mincuzzi, A., **8**, 887[116]
Mine, N., **1**, 254[14,14b], 277[83]; **3**, 295[11], 302[11]
Minematsu, Y., **2**, 1094[88]
Minemura, K., **5**, 847[136], 1032[100]
Mineo, I. C., **8**, 295[60]
Miner, T. G., **8**, 30[42], 66[42]
Miners, J. O., **7**, 68[84], 72[84]
Ming, Y., **2**, 401[28]
Mingos, D. M. P., **4**, 710[49], 712[49]
Minguillon, C., **4**, 438[154]; **8**, 125[94]
Minh, H. T. H., **7**, 338[38]
Minh, T. Q., **2**, 817[95]; **7**, 143[151], 144[151]
Minieri, P. P., **4**, 292[233]; **6**, 261[1], 262[1], 265[1], 266[1]
Minisci, F., **4**, 719[18], 723[18], 730[66], 739[112], 748[160], 751[161], 753[170], 758[191], 763[207,209], 764[112,219], 767[219], 768[235,236,240,241,242,243], 770[244], 802[126], 810[168], 812[179], 820[168,220], 823[168]; **7**, 16[160], 488[151], 498[151], 499[151], 506[296]
Minkiewicz, J., **4**, 855[97]
Minkin, V. I., **6**, 543[616], 552[696]; **7**, 774[335]; **8**, 410[97]
Minnella, A. E., **8**, 971[102]
Minnetian, O. M., **2**, 770[10], 771[10]
Minnis, R. L., Jr., **6**, 251[150]
Minns, R. A., **5**, 70[107,108]
Minobe, M., **4**, 379[114,114b], 382[114b], 383[114b], 413[114b]
Minoda, Y., **7**, 57[22]
Minoguchi, M., **4**, 18[58], 259[257]
Minoli, G., **7**, 882[170]
Minore, J., **8**, 597[94], 606[26]
Minot, C., **5**, 64[27], 453[67], 725[115]; **8**, 536[170], 541[213,215], 543[213,215]
Minoura, Y., **7**, 473[25]; **8**, 126[95], 161[16]
Minowa, N., **1**, 192[82]; **2**, 35[130], 36[130]; **3**, 168[501], 169[501], 170[501]; **4**, 111[152b], 161[88], 218[146]
Minsker, D. L., **4**, 589[79], 591[79], 598[199], 638[199], 640[199]
Minster, D. K., **7**, 143[146]
Mintas, M., **5**, 221[60]
Minter, D. E., **2**, 198[86], 482[22]; **8**, 583[10], 584[16,17], 587[10]
Minto, L. A., **7**, 12[98]
Minton, M. A., **1**, 66[55]; **2**, 536[44-46], 537[45]; **5**, 725[116]; **6**, 712[75], 717[114]
Mintz, E. A., **6**, 954[19]
Minutillo, A., **5**, 1154[152]
Mio, S., **4**, 79[59a]; **6**, 27[115], 164[199]
Miocque, M., **2**, 957[20], 961[38,41,42]; **3**, 147[388], 151[388]; **4**, 303[348]; **6**, 488[20], 517[20], 546[20], 548[20], 549[20]
Mioskowski, C., **1**, 64[45], 523[81,82], 821[28], 825[53]; **2**, 225[157], 227[162], 228[165], 232[157]; **3**, 203[102]; **5**, 499[250,251], 500[250,251]; **6**, 26[110], 149[95,106,107], 536[546], 538[546]; **8**, 844[72,72a]
Miquel, M., **2**, 817[92]
Miranda, E. I., **1**, 563[181]
Miranda, M. A., **2**, 747[116,118]
Mirbach, M. F., **5**, 207[49], 650[24]
Mirbach, M. J., **5**, 207[49], 216[17], 217[17], 219[17], 650[24]; **8**, 451[180]
Mirek, J., **2**, 369[247]
Miri, A. Y., **4**, 444[199]
Mirkind, L. A., **3**, 635[44], 648[172,173,174,175,183]
Mir-Mohamad-Sadeghy, B., **6**, 561[877]
Mironov, V. A., **5**, 699[2]
Mironov, V. F., **8**, 556[376], 780[91]
Mironova, D. F., **6**, 526[402,403], 547[664]
Miropo'skaya, M. A., **8**, 956[7]
Mirrington, R. N., **2**, 294[80]; **3**, 20[109]
Mirskova, A. N., **6**, 550[673,674]
Miryan, N. I., **8**, 771[47]
Mirza, N. A., **2**, 530[22], 578[88], 587[88]; **5**, 802[86]
Mirza, S., **5**, 13[95]
Mirza, S. M., **4**, 102[129]
Mirzaeva, A. K., **3**, 309[89]
Mirzai, H., **6**, 482[124]
Mirzoyan, R. G., **6**, 507[237], 515[237]
Misawa, H., **1**, 803[305,306]; **6**, 157[172,173]
Misbach, P., **8**, 747[56], 752[56]
Mischke, P., **8**, 303[102]
Mischler, S., **7**, 79[130]
Misco, P. F., **7**, 204[57]
Mise, T., **5**, 1137[57,58]; **8**, 152[179], 459[245]
Mishchenko, A. I., **1**, 837[147]
Mishima, H., **2**, 1053[55]; **8**, 541[207]
Mishima, T., **1**, 825[51]
Mishina, T., **6**, 685[363]; **8**, 885[106], 886[106]
Mishra, A., **5**, 938[206]; **7**, 478[83]
Mishra, H. D., **2**, 404[42], 405[42]
Mishra, P., **1**, 891[147]; **2**, 554[135]; **3**, 46[255], 47[255], 792[68]
Mishra, R. K., **6**, 526[390]
Mishriki, N., **2**, 359[160], 368[232]; **7**, 137[125], 138[125]
Misitri, D., **4**, 189[105], 190[105a]
Mislankar, D. G., **6**, 782[82]
Mislankar, S. G., **3**, 24[148]
Mislin, R., **8**, 197[125], 201[125]
Mislow, K., **1**, 488[9]; **2**, 640[170]; **3**, 643[127], 927[56]; **5**, 860[245], 890[34], 1133[30]; **6**, 152[141], 899[109]; **7**, 777[369]; **8**, 411[102,103]
Misner, R. E., **8**, 967[83]
Mison, P., **1**, 387[130,131]; **2**, 942[167], 943[167], 944[167]; **6**, 69[16], 98[154]; **7**, 810[87]
Misono, A., **5**, 1158[173]; **8**, 446[94], 447[97], 450[162], 451[162], 452[94]
Misono, T., **6**, 819[108]
Misra, P., **4**, 564[39]

Misra, R. A., **3**, 640[106]
Misra, R. N., **3**, 759[132], 760[135]; **5**, 1002[18]; **8**, 20[138]
Misra, S. C., **3**, 613[2], 615[2]; **8**, 117[74], 243[47], 816[24]
Misra, V. S., **2**, 364[202]
Misrock, S. L., **5**, 637[108]
Missakian, M. G., **4**, 317[546]
Missianen, P., **5**, 539[106]
Misterkiewicz, B., **2**, 1099[106]; **7**, 726[36]
Mistry, K. M., **5**, 812[132]
Mistry, N., **4**, 390[174,174c]
Mistysyn, J., **5**, 975[17]
Misu, D., **1**, 332[55,56]
Misumi, A., **2**, 103[29]; **4**, 976[97]
Misumi, F., **6**, 276[119]
Misumi, S., **3**, 400[119-122], 402[126], 404[135], 556[35]; **7**, 109[184], 605[144]; **8**, 857[191]
Mita, N., **3**, 437[22], 438[22], 485[36], 491[36], 494[36], 497[36]
Mita, T., **1**, 232[13], 233[13], 234[13], 253[9], 276[9], 278[9]; **2**, 312[35]; **8**, 551[335]
Mitamura, S., **4**, 20[63], 21[63], 37[104], 213[100,101]
Mitani, M., **1**, 268[54], 269[54a]; **2**, 427[43]; **3**, 599[204], 600[217]; **4**, 809[162]; **7**, 25[44], 26[54,55], 476[67], 778[405]; **8**, 133[26]
Mitch, C. H., **1**, 836[144]; **2**, 1024[59]
Mitchel, E., **4**, 459[69], 464[69]
Mitchell, A. R., **6**, 670[274]
Mitchell, D., **5**, 690[81]
Mitchell, D. T., **5**, 752[40]
Mitchell, J., **3**, 592[174], 594[174]; **5**, 1090[90], 1091[90], 1099[90], 1101[90]
Mitchell, J. R., **4**, 487[44]
Mitchell, J. W., **7**, 745[73]
Mitchell, M., **4**, 839[31], 840[31,32], 851[87]
Mitchell, M. A., **4**, 588[50,51], 904[203]
Mitchell, M. B., **6**, 644[89]
Mitchell, M. L., **3**, 891[44]
Mitchell, P. R. K., **2**, 746[109]
Mitchell, R. C., **4**, 33[95,95a]
Mitchell, R. H., **1**, 635[78,79], 636[78,79], 640[78,79]; **3**, 87[104,105], 95[104,105], 97[104], 109[105], 114[104], 116[104,105], 505[168], 870[47,48], 877[83,84], 927[48]; **5**, 69[101,102]; **7**, 771[281]; **8**, 320[86]
Mitchell, R. S., **6**, 602[8], 608[8]
Mitchell, R. W., **8**, 445[21], 446[68], 456[21]
Mitchell, T. D., **8**, 338[83]
Mitchell, T. N., **3**, 495[93b]
Mitchell, T. R. B., **5**, 801[80]
Mitgau, R., **8**, 270[100]
Mitoh, H., **3**, 232[267], 510[185]
Mitra, A., **3**, 982[15]; **4**, 71[14]; **5**, 740[152]; **6**, 851[127], 875[19], 876[19], 879[19,44], 888[19], 894[19]
Mitra, G., **8**, 881[67]
Mitra, R. B., **1**, 546[51]; **3**, 729[40], 878[91]; **5**, 124[6,10], 125[6], 128[6], 130[10]; **6**, 1022[64], 1043[12]
Mitrofanova, E. V., **8**, 753[69,70,74]
Mitscher, L. A., **2**, 824[121]; **7**, 341[51,52], 347[17], 355[17], 548[68], 555[68], 557[68]
Mitscherlich, E., **4**, 426[61]
Mitschler, A., **4**, 688[67]; **7**, 107[162], 452[45]; **8**, 859[213]
Mitsudo, T., **4**, 313[470], 602[264], 609[264], 644[264], 849[83], 856[83], 930[52]; **8**, 36[81], 54[81,161], 55[180], 66[81,161,180], 289[24], 291[35-37], 292[36], 293[46]
Mitsudo, T. A., **2**, 357[146], 358[146,151]
Mitsue, Y., **3**, 1040[104]
Mitsuhashi, K., **4**, 398[220]; **6**, 528[411,412]
Mitsuhashi, S., **4**, 609[332]
Mitsuhira, Y., **2**, 74[80]
Mitsui, H., **1**, 391[151], 392[155]; **7**, 745[76]
Mitsui, S., **8**, 144[71-75], 146[90], 423[40], 429[40], 445[15], 881[76,77], 882[76]
Mitsui, Y., **4**, 148[51], 149[51], 179[65]; **6**, 764[11]
Mitsunobu, O., **6**, 22[82,86], 27[114], 79[61], 80[61], 206[39,41], 210[41], 607[45], 614[86], 615[45,102], 619[115], 620[128], 825[125]; **7**, 752[154]; **8**, 224[104]
Mitsuo, N., **8**, 315[47], 369[78]
Mitt, T., **4**, 370[29]; **5**, 433[137b]
Mittal, R. S., **7**, 843[46,47]
Mittal, R. S. D., **6**, 111[58]; **8**, 237[10], 243[10]
Mittelbach, M., **2**, 789[54]; **6**, 553[761], 554[761,762]
Mittendorf, J., **2**, 498[79]
Mitzinger, L., **7**, 768[201]
Mitzlaff, M., **2**, 1051[37]
Mitzner, E., **6**, 543[613]
Miura, H., **3**, 856[89]; **4**, 1046[111]; **6**, 842[84]; **7**, 127[61], 153[6]
Miura, I., **1**, 553[91,92]; **3**, 135[341], 136[341], 137[341]; **7**, 355[40]
Miura, K., **2**, 363[192], 388[192]; **3**, 871[54]; **4**, 86[81], 213[105], 215[105], 721[31], 725[31], 791[42], 824[241]; **5**, 927[164], 938[164]; **6**, 614[97]; **8**, 798[46], 807[46]
Miura, M., **2**, 74[80]; **3**, 446[79,80,82], 456[79,80], 459[136], 460[136], 461[136], 470[79,80], 471[80], 473[79]; **7**, 766[188]; **8**, 842[41]
Miura, N., **3**, 543[118]
Miura, S., **2**, 833[147]; **4**, 159[85], 256[208,212], 261[208,284], 262[212]
Miura, T., **1**, 648[137], 654[137], 655[137]; **4**, 313[460]; **7**, 384[114a], 771[273]
Miura, Y., **4**, 601[251], 602[251], 643[251]; **5**, 35[12], 90[56]
Miwa, H., **4**, 411[266d], 567[47]
Miwa, T., **2**, 68[42], 360[166]; **6**, 734[11]; **7**, 550[49]; **8**, 607[33,34], 916[108], 917[108], 918[108], 920[108]
Miwa, Y., **5**, 355[87b], 377[111,111a,b]
Mix, G. R., **2**, 481[20]
Mix, K., **8**, 989[39]
Miyachi, N., **5**, 808[109]
Miyachi, Y., **7**, 407[82]
Miyadera, T., **5**, 105[195]
Miyagi, J., **7**, 94[58]
Miyagi, S., **4**, 310[435]; **8**, 856[170]
Miyahara, Y., **7**, 356[49]
Miyai, T., **8**, 185[25], 190[65], 195[108]
Miyaji, K., **1**, 546[54]; **7**, 239[49], 414[108]
Miyaji, Y., **8**, 249[98], 253[98], 369[75]
Miyajima, K., **8**, 944[125]
Miyake, A., **1**, 881[118]; **4**, 597[178], 598[194]; **5**, 810[126], 812[126]; **6**, 787[102]; **8**, 451[180], 568[465], 758[168]
Miyake, F., **1**, 635[89], 806[315]; **3**, 104[208], 117[208]; **6**, 846[103]
Miyake, H., **2**, 332[63], 333[65,66]; **3**, 769[168]; **4**, 13[44,44b], 790[36]; **5**, 320[10]; **6**, 1000[128]; **7**, 883[174]; **8**, 969[98]
Miyake, J., **6**, 477[98], 479[108], 481[123]; **8**, 887[119]
Miyake, K., **6**, 578[981]
Miyake, M., **3**, 380[10]; **5**, 95[92]; **7**, 239[51]
Miyake, N., **3**, 228[222], 484[26], 492[26], 494[26], 495[26], 503[26], 513[26]; **8**, 772[52]
Miyake, T., **7**, 774[318]
Miyake, Y., **6**, 914[28]
Miyakoshi, T., **4**, 31[92]; **6**, 939[138], 941[138]; **8**, 698[137]
Miyama, A., **5**, 108[206]
Miyamoto, C., **5**, 167[94]
Miyamoto, I., **7**, 693[27]
Miyamoto, K., **3**, 172[516], 173[516]
Miyamoto, N., **4**, 885[111]
Miyamoto, O., **3**, 168[494,504], 169[494,504], 170[494,504]; **4**, 430[96]
Miyamoto, S., **7**, 454[96]

Miyamoto, T., **3**, 939[79]; **4**, 125[216,216c]; **5**, 163[73]
Miyamoto, Y., **3**, 99[190]
Miyanaga, S., **2**, 60[17]
Miyane, T., **5**, 1157[171]
Miyano, K., **1**, 387[127]; **3**, 686[110], 815[77]; **6**, 918[37]; **8**, 607[33]
Miyano, M., **6**, 935[102]; **7**, 255[35]; **8**, 561[408]
Miyano, S., **2**, 901[36,37], 908[36,37], 909[36,37]; **3**, 219[113], 503[144], 505[167]; **5**, 297[59], 1196[38], 1197[38]; **7**, 422[141], 423[141,141b,c], 748[114]; **8**, 170[72], 185[24], 535[166]
Miyao, A., **3**, 751[88]
Miyaoka, T., **6**, 625[152]
Miyasaka, T., **1**, 422[93], 471[71]; **6**, 563[900]
Miyashi, T., **3**, 901[107]; **4**, 1103[205]; **5**, 206[47], 552[35], 815[142], 826[159a]; **7**, 875[113,115,116], 876[120]
Miyashita, A., **5**, 1131[16]; **6**, 866[208]; **8**, 459[244], 535[166]
Miyashita, K., **1**, 753[103]; **4**, 1056[141]
Miyashita, M., **1**, 642[114], 851[39], 852[39], 855[52]; **2**, 321[14], 325[14], 541[75]; **4**, 12[43], 18[59], 111[155b], 158[78], 161[89a,b], 307[398], 308[398]; **6**, 107[26], 648[125], 1021[55], 1030[105,106]; **7**, 129[74], 130[74], 218[9], 220[21], 458[114], 618[22]; **8**, 544[255]
Miyata, K., **3**, 844[32]; **5**, 766[117]
Miyata, N., **8**, 33[56], 66[56], 392[94]
Miyata, S., **5**, 473[151], 479[151]; **6**, 757[132]
Miyata, T., **8**, 366[45]
Miyauchi, K., **1**, 853[46]
Miyauchi, Y., **8**, 562[421]
Miyaura, N., **3**, 221[134], 231[246,247,249,252], 249[66], 274[21], 446[89], 465[178], 469[206], 470[178,205,206,207], 473[178,205,206,207], 489[58,59,61,63,64], 490[64,65], 495[58,59,61,63-65,97], 496[64,65,98,99], 498[59,64,65,98,99], 504[61,63,97,154], 511[58,59,61,63-65,97-99,154], 515[58,59,61,63-65,97-99,154], 530[57]; **4**, 145[26], 148[44], 250[137], 358[156], 886[118]; **5**, 117[273], 926[159]; **7**, 816[5]; **8**, 101[120], 786[117]
Miyawaki, T., **5**, 92[67]
Miyazaki, A., **6**, 1022[58]
Miyazaki, F., **3**, 683[102], 807[35]
Miyazaki, H., **4**, 856[100]; **6**, 1049[36]; **7**, 253[23], 765[168]
Miyazaki, J., **7**, 778[409]
Miyazaki, K., **1**, 317[141], 385[114]; **8**, 160[100], 170[100], 176[135], 178[100], 394[119]
Miyazaki, M., **6**, 1066[97]; **8**, 98[99], 563[427]
Miyazaki, S., **3**, 644[139]
Miyazaki, T., **1**, 98[84], 99[84], 387[136]; **2**, 995[45]; **6**, 542[601], 767[28], 768[28], 769[28,29,32]; **7**, 696[43], 697[43,46]; **8**, 43[108], 47[108], 64[213,220], 66[108,213], 67[220], 394[119]
Miyazawa, M., **5**, 55[48]; **6**, 14[54], 849[117]; **7**, 406[73], 458[112]; **8**, 7[43]
Miyazawa, T., **2**, 1089[57]
Miyazawa, Y., **2**, 717[67]; **3**, 875[73-75]; **8**, 780[92]
Miyoshi, H., **2**, 582[109]; **4**, 315[520]; **7**, 534[41]; **8**, 851[124]
Miyoshi, K., **2**, 603[44]; **8**, 550[332]
Miyoshi, M., **2**, 1051[41]; **3**, 650[210c,212], 651[210c,216]; **7**, 806[74]
Miyoshi, N., **2**, 578[86]; **3**, 771[186]; **7**, 131[80]
Mizobuchi, Y., **5**, 195[8], 197[8]
Mizoguchi, M., **7**, 809[81,85]
Mizoguchi, T., **6**, 664[220]; **7**, 42[29], 877[134]
Mizokami, N., **6**, 74[37]
Mizono, K., **7**, 851[24]
Mizugaki, M., **2**, 353[101]; **6**, 530[423]
Mizugami, M., **3**, 461[147]
Mizuguchi, Y., **8**, 252[111]
Mizuhara, Y., **8**, 190[64]
Mizukami, F., **7**, 155[25]
Mizuki, Y., **5**, 439[170], 504[274]; **7**, 752[153]
Mizuno, A., **6**, 235[52]
Mizuno, K., **4**, 826[244]; **7**, 875[112], 878[140,144]
Mizuno, M., **2**, 384[319]
Mizuno, S., **8**, 190[80]
Mizuno, T., **3**, 1034[77]
Mizuno, Y., **2**, 1096[99]; **4**, 27[79,79d,e]; **6**, 611[67], 625[152]; **7**, 156[32], 175[143]
Mizusawa, Y., **1**, 564[201]; **6**, 27[119]
Mizuta, M., **6**, 453[142,144], 454[146]
Mizuta, N., **3**, 554[20]
Mizuta, Y., **8**, 554[366]
Mizutaki, S., **7**, 773[309], 776[309]
Mizutani, M., **2**, 211[115], 215[115]; **4**, 379[114,114b], 382[114b], 383[114b], 413[114b], 564[42]; **6**, 425[65]
Mizutani, Y., **8**, 533[150]
Mjalli, A. M. M., **1**, 865[87]
Mkrtchyan, R. S., **5**, 435[151]
Mladenov, I., **4**, 85[77d]
Mladenova, M., **1**, 218[51]; **2**, 211[112], 284[55]; **4**, 89[84f], 95[102e]
Mlakar, B., **2**, 1026[69]; **5**, 403[11]
Mlinaric-Majerski, K., **3**, 876[79,80]
Mlochowski, J., **7**, 657[21]
Mloston, G., **4**, 1074[29], 1086[117]
Mlostón, G., **1**, 836[142]
Mloston, R., **4**, 1073[27], 1074[27]
Mlotkiewicz, J. A., **3**, 404[134,136]
Mlotkowska, B., **6**, 134[36]
Mo, S.-H., **5**, 950[285]
Mo, Y. K., **3**, 330[192], 332[206], 333[209]; **7**, 17[178]; **8**, 724[173]
Moad, G., **4**, 786[24]; **5**, 4[32]
Moakley, D. F., **8**, 563[428]
Moberg, C., **4**, 596[164], 597[171], 621[164]; **7**, 453[77]
Mobilio, D., **1**, 187[65], 188[65]; **2**, 20[83]; **4**, 155[69]
Mobius, L., **4**, 1099[186], 1100[189]; **7**, 475[51], 477[73]
Mocadlo, P. E., **1**, 174[5,7]; **8**, 796[26]
Mocali, A., **2**, 465[104]
Mocanu, M., **2**, 744[99], 745[99]
Mochalin, V. B., **5**, 418[67]; **7**, 660[39]
Mochida, D., **8**, 241[36]
Mochida, I., **8**, 598[99,100]
Mochida, K., **8**, 19[135]
Mochizuki, A., **4**, 614[373], 840[37], 905[207]; **6**, 546[641,642]
Mochizuki, D., **8**, 170[82]
Mochizuki, F., **8**, 426[49]
Mochizuki, H., **8**, 170[83,84]
Mochizuki, K., **7**, 438[23], 444[54], 559[81], 560[81], 562[81]
Mock, G. A., **2**, 370[258]
Mock, W. L., **1**, 846[21,22], 851[21], 853[21], 856[22], 859[21], 861[21], 896[21]; **3**, 783[105]; **5**, 424[93,98]; **8**, 472[2], 473[2]
Mockel, A., **5**, 109[217]
Möckel, G., **6**, 551[689]
Modelli, A., **5**, 257[60]
Modena, G., **4**, 50[142], 102[132], 277[89], 298[280], 330[5], 425[33]; **5**, 370[102], 371[102]; **6**, 150[114], 936[108], 999[122,123]; **7**, 95[69], 205[64], 425[147a], 758[1], 759[1], 760[1], 762[69,84], 766[182], 777[69b,376,380], 778[69]; **8**, 152[176], 155[176]
Moderhack, D., **1**, 386[124]; **2**, 1088[43]; **6**, 547[661,662]; **7**, 657[28,29]
Modi, M. N., **8**, 271[110]
Modi, S. P., **4**, 83[65c]
Modro, T. A., **7**, 483[124]
Mody, P. N., **2**, 867[14]
Moe, O. A., **2**, 156[115]; **4**, 239[36]
Moebus, M., **1**, 367[54]
Moell, N., **4**, 14[46]

Moeller, K. D., **4**, 194[121]; **5**, 243[9,11], 311[103]
Moeller, P. D. R., **1**, 748[73], 812[73]
Moeller, T., **1**, 231[3]
Moenius, T., **6**, 184[149]
Moëns, L., **2**, 423[34], 424[35]; **3**, 857[90]; **4**, 809[164]; **6**, 836[58]; **8**, 136[51]
Moerck, R., **2**, 713[46]; **5**, 347[73a], 777[191]
Moerck, R. E., **3**, 623[33]
Moerck, R. K., **6**, 1023[71]
Moerikofer, A. W., **8**, 716[91]
Moering, U., **6**, 56[134], 57[134,137,138]
Moerlein, S. M., **4**, 445[207]
Moersch, G. W., **3**, 790[59]; **7**, 185[174]
Moest, M., **3**, 636[47]
Moffat, J., **5**, 1147[113]
Moffatt, F., **5**, 543[115]
Moffatt, J. G., **2**, 139[29]; **4**, 38[108,108a]; **6**, 603[16], 605[36], 614[79,80], 622[134], 662[217]; **7**, 291[1], 292[6], 293[1]
Moffatt, M. E., **3**, 305[71]
Moffett, R. B., **4**, 276[74]
Moffitt, W. E., **5**, 900[3]
Mogelli, N., **4**, 126[219]
Moghadam, G. E., **7**, 632[61]
Mohacsi, E., **6**, 571[956]
Mohamad, S., **3**, 904[134]
Mohamadi, F., **1**, 29[144]; **2**, 507[26]; **3**, 34[198], 39[198]; **4**, 1003[67]; **6**, 727[195]; **8**, 851[135], 856[135b], 949[155]
Mohammad, T., **4**, 486[37], 505[37]
Mohammadi, N. A., **8**, 449[154]
Mohammed, A. Y., **4**, 791[56]; **5**, 841[94]
Mohan, L., **7**, 13[124], 737[11]
Mohan, R., **3**, 26[162]
Mohandas, J., **5**, 468[135]
Mohanty, S., **7**, 143[150], 144[150]
Mohar, A. F., **3**, 635[37], 639[87]
Mohareb, R. M., **2**, 362[179], 378[291]
Mohler, D. L., **2**, 553[131]
Mohmand, S., **5**, 575[131]
Mohr, P., **7**, 429[151]
Mohr, R., **5**, 412[45]
Mohri, K., **8**, 245[72]
Mohri, M., **4**, 604[289], 646[289]; **8**, 449[160], 840[30,30b], 960[37]
Mohri, S., **4**, 55[157], 57[157h], 249[114], 257[114]
Möhring, E., **6**, 553[797]
Mohrle, H., **2**, 905[54], 960[33]; **8**, 332[42]
Mohsen, A., **6**, 423[46]
Mohsen, K. A., **2**, 359[160]
Moine, G., **6**, 152[137]
Moini, M., **4**, 484[20]
Moir, M., **3**, 404[133]; **6**, 441[82]
Moir, R. Y., **3**, 147[399]
Moisak, I. E., **6**, 212[82]
Moise, C., **1**, 331[49], 749[78], 816[78]; **5**, 1126[68]; **8**, 290[30]
Moiseenkov, A. M., **3**, 181[553], 734[7]; **4**, 874[52]; **5**, 345[70], 346[70], 453[66]; **6**, 174[57]; **8**, 611[64], 971[111]
Moiseev, I. I., **7**, 451[38]; **8**, 447[106], 450[106]
Moiseeva, L. V., **8**, 318[66], 546[308]
Moison, H., **1**, 821[26]; **2**, 344[16,17], 345[17], 353[16,17], 359[16], 360[16], 363[16,17]; **6**, 175[79]
Moizumi, M., **5**, 524[54], 534[54]
Mojé, S., **3**, 217[90], 219[90]
Mojé, S. J., **8**, 481[51], 482[51], 483[51], 531[120]
Mojé, S. W., **5**, 429[113b]
Mojica, C. A., **2**, 904[52]
Mokhi, M., **2**, 723[97,99]; **4**, 698[17], 699[17,20,21], 700[20], 701[21]
Mokhtar, A., **3**, 834[79]
Mokrosz, M., **5**, 566[101]
Mokruschin, V. S., **6**, 530[421], 538[551,555], 550[421]
Mokry, P., **2**, 1017[31,34]; **6**, 737[31], 746[89]
Molander, G. A., **1**, 218[53], 262[34,36], 263[42], 264[34,42], 265[42], 266[42], 267[50], 269[58], 270[58], 271[64], 278[36], 612[49]; **2**, 30[111], 31[111], 127[237]; **3**, 494[85], 522[20], 523[24], 574[73], 575[73], 599[73,208], 610[73], 752[93]; **4**, 147[40], 809[161], 884[107], 971[76], 972[76]; **5**, 32[6], 246[23], 247[23,23a,24], 599[39], 935[189], 937[202]; **6**, 11[46], 86[101]; **7**, 400[49]; **8**, 115[67], 724[178], 725[178], 726[178], 727[178], 847[93], 883[94], 884[94], 987[22], 992[22a], 994[22]
Moldonado, L., **6**, 681[333]
Moldowan, J. M., **8**, 333[57], 345[127]
Mole, T., **1**, 77[1], 78[19], 95[77], 325[2]; **2**, 114[118,119], 268[67,68], 531[26], 545[26]; **3**, 894[66], 896[66]; **4**, 140[9], 257[221], 887[120]; **8**, 316[58], 671[15]
Molho, D., **3**, 831[60]
Molin, M., **3**, 147[392], 149[392], 152[392]; **5**, 567[105], 800[76]
Molina, G., **7**, 691[20]
Molina, G. A., **8**, 916[99], 917[99], 918[99]
Molina, M. T., **5**, 834[55]
Molina, P., **4**, 440[170]; **6**, 509[273], 554[735]
Molinari, H., **2**, 229[168], 435[62]; **6**, 425[66]
Molines, H., **2**, 209[108]; **4**, 391[176], 1020[236]; **6**, 527[408]; **8**, 847[98], 849[98]
Molino, B., **6**, 27[118]
Molino, B. F., **7**, 246[90], 362[32]
Moll, H., **3**, 822[12], 831[12], 835[12b]
Moll, N., **6**, 790[111]
Mollema, K., **1**, 214[27]
Moller, F., **7**, 689[4]
Möller, F., **6**, 261[8], 795[1], 958[29]
Möller, K., **3**, 904[131,132]
Moller, K. E., **7**, 8[52]
Möller, M., **5**, 115[251]
Möller, W., **2**, 385[326]
Mollere, P. D., **3**, 587[142]
Moller Jorgensen, P., **4**, 181[72,73]
Mollov, N., **2**, 971[95]
Mollov, N. M., **6**, 744[75], 746[75,88]
Molloy, K. C., **1**, 526[96]; **2**, 655[140]; **4**, 412[268e]
Molnar, A., **3**, 726[23]; **6**, 25[99]; **8**, 418[5], 420[5], 422[32], 423[5], 425[32], 426[32], 428[32], 429[32], 430[32], 433[32], 434[32], 435[32], 436[32], 439[5,32], 441[5], 442[5]
Molnar, E. M., **6**, 189[185]
Moloney, M. G., **3**, 286[56b]; **7**, 620[28,29]
Moloy, K. G., **8**, 889[132]
Möltgen, E., **8**, 298[75], 299[75]
Moltzen, E. K., **6**, 456[160]
Molyneux, R. J., **6**, 74[31]
Momongan, M., **3**, 34[192], 39[192]
Momose, D., **3**, 421[59], 422[59]
Momose, T., **3**, 810[47], 853[72]; **5**, 832[39], 841[87], 843[115]; **7**, 406[78b]
Momot, V. V., **6**, 499[172]
Monaco, S., **5**, 404[17]
Monaghan, F., **7**, 365[49]
Monagle, J. J., **7**, 654[6]
Monahan, M. W., **4**, 273[42]
Monahan, R., III, **4**, 377[124b], 379[115], 380[115h,124,124b], 383[115h]; **6**, 836[58]; **7**, 131[86], 273[134], 523[45], 822[32]
Monakov, Yu. B., **8**, 699[150]
Mondelli, R., **5**, 64[52]; **8**, 587[30]
Monden, M., **2**, 509[34]
Mondena, G., **8**, 152[175]
Mondon, A., **2**, 1056[68]; **8**, 342[110]

Mondon, M., **3**, 35[201], 39[201], 257[121]; **6**, 836[58]; **7**, 499[234]
Money, T., **2**, 170[174,175], 547[106], 550[106], 651[121]; **3**, 26[164], 349[32], 427[86], 681[96], 710[20], 711[22], 712[24,26]; **5**, 812[133]; **6**, 1045[25a,26-28,29a,b]; **7**, 58[57], 62[57], 63[57]
Mong, G. M., **7**, 167[93]
Monge, A., **2**, 780[12]
Mongrain, M., **4**, 1040[83,84], 1043[83,84]; **8**, 925[5]
Moniot, J. L., **3**, 216[76]; **7**, 256[24]
Mönius, T., **6**, 193[207,213,214]
Monk, P., **4**, 1076[42]
Monkiewicz, J., **1**, 760[136]; **3**, 201[78]; **4**, 252[162]
Monkovic, I., **4**, 398[216], 399[216d]; **6**, 923[58]; **7**, 777[366]
Monn, J. A., **3**, 71[29]; **7**, 224[53]
Monneret, C., **4**, 405[249], 406[249]; **6**, 266[49]
Monnier, C., **5**, 742[158]
Monobe, H., **1**, 422[93]
Monot, M. R., **3**, 578[92], 610[92]; **8**, 532[130]
Monpert, A., **4**, 990[146]
Monro, M. H. G., **3**, 681[94]
Monsan, P., **8**, 52[137], 66[137]
Montana, A. F., **8**, 568[472]
Montaña, A. M., **2**, 655[144]; **5**, 1047[31], 1052[31], 1054[43]
Montana, J. G., **1**, 865[87]
Montanari, F., **1**, 523[80]; **2**, 228[166,167]; **4**, 438[155], 444[195]; **6**, 19[67], 149[99], 204[11], 221[131], 236[54]; **7**, 253[16], 764[112], 767[112], 777[367,368,371,372,384]; **8**, 844[67]
Montanari, S., **1**, 566[216]
Montanucci, M., **3**, 509[178]; **4**, 426[51], 437[51], 441[181], 447[216,217]; **7**, 338[41]
Montaudo, G., **1**, 294[43]
Montaudon, E., **4**, 753[166]
Montavon, M., **2**, 612[105]; **6**, 965[85]
Monte, W. T., **4**, 12[41], 70[7]; **6**, 939[136], 942[136]
Montecalvo, D., **3**, 382[36]
Monteil, R. L., **6**, 504[220]
Monteils, Y., **3**, 321[136]
Monteiro, H. J., **3**, 1052[26]; **4**, 258[234]; **6**, 126[151], 931[93]; **7**, 124[49], 127[49]
Monteiro, M. B., **2**, 855[247]
Montelatici, S., **8**, 445[23]
Monteleone, M. G., **5**, 96[121], 98[121]
Montero, J. L. G., **6**, 554[711]
Montes, J. R., **4**, 653[436]; **7**, 229[120]
Montes de Lopez-Cepero, I., **1**, 784[243]; **2**, 111[79]
Montevecchi, P. C., **4**, 336[29,30]; **7**, 493[197]
Montforts, F. P., **6**, 538[571]
Montgomery, A. M., **6**, 237[65], 243[65]
Montgomery, C. R., **4**, 1081[80]
Montgomery, J., **4**, 572[4]
Montgomery, L. K., **5**, 65[65], 69[105,106]
Montgomery, S. H., **2**, 195[70], 201[93], 205[102], 206[102b], 221[70], 642[75], 643[75]; **4**, 72[31]; **5**, 170[113]
Montgrain, F., **8**, 98[102]
Montheard, J.-P., **4**, 313[471], 315[516]
Monthony, J. F., **5**, 715[77]
Monti, D., **7**, 109[183]
Monti, H., **6**, 563[892]; **8**, 332[41]
Monti, L., **3**, 725[19], 743[58]; **5**, 464[114,115], 466[114]
Monti, S. A., **2**, 711[31]; **3**, 730[44]; **5**, 907[75], 908[75], 911[93], 945[75]; **8**, 616[101], 624[101]
Montillier, J. P., **8**, 408[67]
Montoya, R., **1**, 837[154]
Montrasi, G., **8**, 457[218], 458[218]
Montury, M., **7**, 684[91]
Monzycki, J., **7**, 13[119]
Mooberry, J. B., **5**, 853[171]; **6**, 509[246]; **7**, 160[50]
Moodie, R. B., **6**, 291[216]; **7**, 602[107]
Moody, C. J., **2**, 821[105]; **3**, 902[119]; **4**, 408[259a]; **5**, 384[128,128a], 426[105], 428[105], 429[105], 486[189], 827[2], 829[2], 867[2e]; **6**, 127[160], 781[77]; **7**, 27[64,66], 32[91,94-97], 33[97], 34[98,99], 35[101,102], 194[3], 200[40], 208[88], 349[18], 355[18], 748[107]; **8**, 337[76], 389[71], 618[115,116], 948[147]
Moody, G. W., **7**, 62[51]
Moody, R. J., **1**, 489[23], 491[30], 495[45], 497[30]; **3**, 199[61]; **7**, 602[104,104a]
Mooiweer, H. H., **2**, 558[162], 1049[15], 1075[151], 1078[15], 1079[151,156]; **3**, 217[81]
Mook, R., Jr., **4**, 792[68], 796[95,98,99], 820[212]
Moolenaar, M. J., **1**, 617[54]; **2**, 89[37], 1065[113]; **3**, 153[414], 155[414], 223[183], 225[183]
Moolweer, H. H., **6**, 118[99]
Moon, M. P., **4**, 455[43], 456[47], 463[43], 469[134], 472[134], 473[134], 475[134]
Moon, M. W., **6**, 625[153]
Moon, S., **2**, 711[33]; **3**, 380[13,16], 414[1]; **7**, 169[112]
Moon, S.-H., **6**, 619[119]
Moon, S.-S., **3**, 226[193]
Moon, Y. C., **8**, 14[77], 244[68], 247[68], 250[68], 538[192]
Mooney, B. A., **8**, 53[131], 66[131]
Moore, B., **8**, 524[13]
Moore, C. J., **4**, 303[342], 390[175b]; **7**, 24[23], 635[70], 833[72]; **8**, 854[152], 856[152]
Moore, C. W., **2**, 848[211]
Moore, D. R., **3**, 17[87]
Moore, D. S., **8**, 552[350]
Moore, D. W., **8**, 597[88]
Moore, G. A., **6**, 644[84]
Moore, G. G., **6**, 57[135]
Moore, H. W., **2**, 1087[35]; **3**, 828[48], 829[48]; **5**, 90[54,57], 95[57], 407[26], 688[70], 689[70,72,76,77,79], 690[80,80c], 733[136,136c-g], 734[136f,g]; **6**, 245[121], 247[135], 248[121], 249[121], 251[121]; **7**, 35[106]
Moore, J. A., **3**, 844[30], 889[22]; **8**, 338[83]
Moore, J. L., **1**, 795[282]; **6**, 997[113]; **8**, 846[86]
Moore, J. W., **4**, 725[49]
Moore, L., **5**, 3[25]
Moore, L. D., **8**, 965[63]
Moore, L. L., **6**, 91[121]
Moore, M. A., **8**, 70[227], 71[227]
Moore, M. L., **4**, 31[93], 868[16]; **6**, 734[3]; **7**, 543[12], 551[12]; **8**, 84[8]
Moore, M. W., **3**, 443[55]; **4**, 892[140]; **6**, 965[90]
Moore, R. E., **3**, 438[35]; **5**, 563[90], 803[88], 975[17], 976[18]
Moore, R. H., **6**, 939[141], 942[141]; **8**, 807[118]
Moore, R. N., **5**, 86[22]
Moore, T. L., **5**, 618[8], 619[8], 624[8], 625[8]; **6**, 927[74]
Moore, W. H., **3**, 735[15]
Moore, W. R., **4**, 1010[148], 1013[178,181]; **5**, 736[140]; **6**, 970[125], 971[128,129]
Moorhoff, C. M., **1**, 760[136]
Moorhouse, S., **1**, 214[23]
Moormann, A. E., **6**, 1054[49]
Moorthy, K. B., **7**, 144[157]
Moorthy, S. N., **3**, 386[61], 393[61]; **4**, 284[156]; **8**, 477[31]
Moos, W. H., **1**, 836[146]
Moosavipour, H., **7**, 236[27]
Mooser, G., **8**, 589[49]
Moosmayer, A., **3**, 661[23]
Mootoo, D. R., **4**, 391[180,181a,182,182a,183]; **6**, 27[118], 40[40]; **7**, 246[91], 362[32], 378[93]; **8**, 347[138]
Mootz, D., **1**, 6[32]
Mope, N. S., **7**, 801[39]

Moracci, F. M., **5**, 92[63]; **6**, 538[572]; **8**, 31[45], 36[45], 66[45], 587[32]
Morales, A., **4**, 1033[26], 1036[26c]; **5**, 942[234]
Morales, H. R., **8**, 54[156], 66[156]
Morales, O., **3**, 883[108]
Moran, D. B., **1**, 555[109], 556[109], 559[109]
Moran, G., **4**, 712[67]
Moran, J. R., **4**, 1074[29]
Moran, M. D., **7**, 138[126]
Moran, T. A., **1**, 780[229]; **6**, 860[176]
Morand, P., **1**, 564[189]; **7**, 821[31]; **8**, 239[27], 240[27], 242[27,40]
Morandini, F., **3**, 229[230,230a], 246[36], 438[29], 452[110], 1023[22]; **4**, 930[48], 931[48]
Moravskiy, A., **1**, 440[190], 445[190], 457[190c]
Mörch, L., **2**, 233[188]
Morcinek, R., **2**, 782[26]
Mordenti, L., **8**, 16[103], 483[57], 485[57], 558[391,392,393]
Mordini, A., **1**, 612[48]; **2**, 566[22], 586[135]
More, K. M., **7**, 764[117]
Moreau, B., **3**, 147[398], 149[398,408,409,411], 150[411], 151[408,409,411], 155[408,409]; **6**, 644[91]; **7**, 777[388]
Moreau, C., **8**, 536[172]
Moreau, J. J. E., **1**, 461[12]; **4**, 248[112]
Moreau, J. L., **1**, 220[64,65d]; **2**, 81[1], 82[1], 96[1], 294[84], 296[84], 799[20], 983[28], 989[34,35], 990[34], 992[36], 993[36]; **4**, 95[102b]; **5**, 100[145]
Moreau, N., **6**, 263[22]
Moreau, R.-C., **6**, 420[14], 430[95]
Moreau-Hochu, M. F., **4**, 496[89]
Morehead, B. A., **6**, 546[651]
Morehouse, F. S., **1**, 174[8], 175[8]; **8**, 796[28]
Morel, D., **8**, 535[166]
Morel, G., **5**, 488[195]; **6**, 540[585,586]
Morel, J., **1**, 644[122], 646[122], 668[122], 669[122], 695[122]
Moreland, D. W., **2**, 101[21], 510[43]; **5**, 842[110]
Morella, A. M., **2**, 809[55]; **4**, 340[50]; **7**, 534[42], 772[298]
Morelli, I., **3**, 741[51], 745[65]
Moreno-Manas, M., **1**, 477[141], 547[61]; **2**, 359[164], 381[300]; **4**, 590[92], 616[393], 629[393]; **8**, 964[58]
Morera, E., **1**, 195[90]; **3**, 1035[79]; **4**, 860[112], 861[112]; **7**, 143[148], 144[148]; **8**, 84[13], 910[82], 911[87], 933[49,51]
Moret, E., **2**, 13[58], 14[58]; **4**, 869[22]
Moreto, J. M., **5**, 36[18], 57[54]
Moretti, G., **5**, 1148[114]; **8**, 754[103]
Moretti, I., **1**, 837[155], 838[160]; **7**, 747[96], 777[384], 778[402]; **8**, 187[37]
Moretti, R., **3**, 209[17]; **4**, 152[54], 175[43], 184[85], 201[13,16], 202[13,16]; **5**, 362[94]; **6**, 77[55], 118[106], 248[137]
Morey, J., **7**, 334[27], 346[8]
Morey, M. C., **1**, 122[71]; **8**, 395[124]
Morgan, A. R., **6**, 40[39], 60[144]; **7**, 535[48]
Morgan, B., **2**, 743[86]
Morgan, B. P., **6**, 677[311]
Morgan, C. R., **8**, 86[21]
Morgan, D. D., **5**, 727[119]
Morgan, E. D., **6**, 188[181]
Morgan, G. T., **7**, 774[327,328], 775[339,344], 776[360]
Morgan, J. W., **5**, 2[16]; **6**, 843[89]
Morgan, J. W. W., **3**, 831[59]
Morgan, K., **7**, 347[15]
Morgan, K. D., **3**, 168[493], 169[493], 171[493]
Morgan, L. R., **3**, 699[161]; **7**, 27[62]
Morgan, P. H., **6**, 116[88]; **8**, 64[202], 66[202]
Morgan, S. E., **7**, 401[58]
Morgan, T., **3**, 224[172]; **7**, 801[37]
Morganroth, W., **7**, 723[23]
Morgans, D., Jr., **1**, 561[161]; **3**, 770[174]; **4**, 111[158b]; **8**, 836[5]
Morganti, G., **6**, 176[103]
Morgat, J.-L., **2**, 232[176]; **7**, 805[68]; **8**, 460[254]
Morge, R. A., **5**, 157[40]
Morgenthau, J. L., Jr., **8**, 213[28], 267[66]
Mori, A., **1**, 88[52], 165[111], 348[141]; **2**, 68[44], 113[111], 244[27], 245[27], 601[36], 901[36], 908[36], 909[36]; **3**, 197[33], 573[71], 610[71]; **4**, 566[46], 974[90]; **5**, 297[59], 620[16], 622[23], 627[42], 1196[38], 1197[38]; **6**, 117[98], 849[123]; **8**, 223[99,100], 224[99,100], 227[120], 659[106]
Mori, H., **4**, 126[218a]; **7**, 243[67]; **8**, 535[162]
Mori, I., **1**, 92[62,63], 286[9]; **2**, 193[64], 580[100], 649[106], 718[71]; **3**, 565[17]; **4**, 390[175c]; **5**, 850[149]; **6**, 856[149]
Mori, K., **1**, 561[163], 733[12]; **2**, 291[74], 619[148]; **3**, 49[262], 99[191], 107[191], 124[263], 126[263], 224[162,177], 287[62], 396[115], 557[37], 639[86], 644[161], 715[39], 871[54]; **4**, 126[218a], 373[83], 893[154]; **5**, 417[65]; **6**, 74[29], 115[82], 657[177], 677[318,318b], 862[186]; **7**, 57[32], 239[51,52], 243[67], 399[37], 406[78c,d], 407[84a], 410[93], 418[125,126], 451[22], 634[69]; **8**, 49[115,116], 66[115,116], 188[50], 190[70], 196[119], 201[141], 429[55]
Mori, M., **2**, 357[149], 1051[43]; **3**, 650[213], 1032[67], 1036[82], 1037[90], 1038[90,95,95b]; **4**, 803[132], 843[53-55], 846[74], 852[53]; **5**, 603[53], 604[53]; **6**, 46[65], 74[29]; **7**, 804[60]
Mori, S., **1**, 436[153]; **2**, 112[98], 244[30], 246[34], 247[34], 253[42], 1048[12]; **3**, 1026[40]; **5**, 714[70]; **6**, 121[127]
Mori, T., **1**, 423[97], 424[98]; **5**, 86[34]; **6**, 493[128], 494[138]; **7**, 242[62]; **8**, 241[38], 263[32], 267[32]
Mori, Y., **1**, 87[46], 408[34], 422[91], 569[255]; **2**, 805[43]; **6**, 186[172], 535[524]; **7**, 451[24]; **8**, 9[52]
Moriarty, K. J., **5**, 524[50], 539[50], 548[50c]
Moriarty, R. M., **3**, 512[203]; **4**, 529[72], 531[72]; **6**, 118[102], 177[117], 254[163]; **7**, 92[40], 145[160,161], 155[26-30], 166[91], 179[153,154], 222[37], 227[37,81], 236[20], 488[150], 748[109], 827[49], 828[52], 829[52a], 833[76]
Moriconi, E. J., **3**, 574[80]; **5**, 107[201,204]; **7**, 698[52]; **8**, 967[83]
Morii, S., **3**, 426[82], 428[91], 429[82,91]
Morikawa, A., **7**, 318[58], 319[58], 320[58]
Morikawa, I., **8**, 783[107]
Morikawa, K., **4**, 91[89]
Morikawa, M., **2**, 805[43]; **4**, 590[97,98]; **6**, 535[524]
Morikawa, S., **5**, 1157[170], 1183[56]; **8**, 607[29]
Morikawa, T., **3**, 421[54]; **4**, 377[105c], 381[105], 1005[88], 1020[234,235]; **7**, 255[36]
Morimoto, A., **4**, 382[134,134a], 383[134a], 386[148a], 387[148,148a]; **8**, 975[133], 992[55]
Morimoto, H., **3**, 557[49]
Morimoto, K., **2**, 816[87]
Morimoto, M., **6**, 26[107]; **7**, 180[159]
Morimoto, T., **1**, 368[62], 389[137], 391[62]; **2**, 913[77], 914[77], 994[38], 1004[59]; **3**, 304[61], 844[34]; **4**, 810[169]; **7**, 92[44]; **8**, 134[35], 154[190,191]
Morimoto, Y., **6**, 811[77]; **7**, 255[38], 406[87]
Morimura, S., **6**, 492[121-123], 566[927]
Morin, C., **4**, 31[93]; **5**, 85[8], 412[47], 1062[59]; **7**, 60[44]
Morin, J. G., **8**, 536[167]
Morin, J. M., Jr., **1**, 404[20], 428[120]; **4**, 1023[256,258]; **5**, 1007[38]
Morin, L., **5**, 829[22]; **6**, 712[74]
Morin, R. B., **2**, 913[78], 915[78], 925[78]; **5**, 85[2]; **6**, 759[136], 936[105], 1025[80]; **7**, 205[61]
Morin, R. D., **4**, 307[394], 312[456], 313[456], 504[131]; **8**, 146[97], 568[466]
Morinaga, K., **6**, 217[116]
Morini, G., **4**, 768[240]
Morioka, M., **1**, 564[201]; **6**, 27[119]
Morisaka, K., **8**, 976[135]

Morisaki, K., **3**, 501[137], 509[137]; **4**, 606[307], 607[307,315], 647[307]
Morisaki, M., **2**, 187[42]; **7**, 675[54], 680[76]
Morisaki, Y., **1**, 802[304]; **5**, 767[120]
Morishima, A., **5**, 1138[70]
Morishima, H., **2**, 917[85]
Morishima, T., **5**, 623[26]
Morishita, T., **4**, 335[27]; **8**, 410[88]
Morisset, V. M., **7**, 521[36]
Morisson, J. D., **4**, 252[164]
Morita, A., **3**, 380[9]
Morita, E., **1**, 90[57], 566[208]
Morita, K., **4**, 34[97], 35[97,97i]
Morita, K.-I., **7**, 698[53]
Morita, N., **4**, 115[180e]
Morita, S., **8**, 370[90]
Morita, T., **6**, 214[92], 654[153]; **7**, 856[66]; **8**, 392[101]
Morita, Y., **1**, 824[45]; **2**, 205[104], 206[104]; **3**, 4[26], 5[26], 10[26], 225[187]; **4**, 96[105], 97[105b], 159[85]; **5**, 637[115]; **6**, 535[530], 937[121]; **7**, 406[75], 597[46], 774[332]; **8**, 163[39]
Moritake, M., **7**, 92[42], 93[42]
Moritani, I., **3**, 436[5], 437[5]; **4**, 590[99], 613[371], 836[2-5], 837[13-15], 841[50], 959[33], 1006[106]; **6**, 74[36], 86[99], 955[25]
Moritani, T., **6**, 431[112]
Moritz, A. G., **8**, 501[57]
Moriuchi, F., **8**, 190[81]
Moriuti, S., **4**, 963[42], 1038[61]
Moriwake, T., **1**, 751[93]; **5**, 833[49]; **6**, 77[54]; **8**, 244[50,71], 247[71], 251[71], 253[71]
Moriwaki, H., **6**, 516[319]
Moriwaki, M., **6**, 1016[26]
Moriya, H., **1**, 143[37], 158[37], 159[37], 180[41], 181[41], 340[90]; **2**, 5[17], 6[17], 22[17,17a]
Moriya, O., **4**, 738[98], 792[67], 823[231]; **6**, 577[978]
Moriya, T., **1**, 367[53]; **2**, 780[10]; **6**, 547[659,660]
Moriya, Y., **8**, 149[114]
Moriyama, K., **4**, 413[278a,b]; **6**, 509[268]
Moriyama, M., **7**, 778[395]
Moriyama, T., **1**, 803[307]; **2**, 74[74]; **3**, 135[348], 136[348], 139[348], 141[348], 144[348]
Moriyama, Y., **3**, 741[50]; **4**, 405[249,250a,b], 406[249], 606[303], 646[303]; **8**, 856[182]
Moriyasu, K., **5**, 564[94]
Moriyasu, M., **2**, 86[25]
Morizawa, Y., **2**, 19[76], 575[67], 588[151], 589[151]; **3**, 730[43]; **4**, 607[310], 626[310], 647[310], 901[185,186], 1007[126], 1008[126]; **5**, 917[124], 926[124], 938[219], 943[237]; **7**, 180[158], 378[96]
Morizur, J.-P., **5**, 455[76]
Mørkved, E. H., **6**, 496[152], 524[363]
Morley, C., **4**, 675[41], 691[75]
Morley, J. O., **2**, 744[89]; **7**, 356[51]
Mornet, R., **4**, 878[80-82], 884[80]
Moro, G., **2**, 630[21], 631[21], 632[21], 634[21], 640[21], 641[21], 642[21], 644[21], 645[21]
Moroder, F., **5**, 403[8]
Moroder, L., **6**, 637[28]
Moroe, M., **8**, 881[80], 882[80]
Morokuma, K., **5**, 1145[103], 1153[103]; **8**, 724[169,169e]
Moron, J., **8**, 597[87]
Moro-Oka, Y., **4**, 610[338], 649[338]; **7**, 160[55], 851[24]
Morosawa, S., **5**, 223[82], 636[100], 637[102]; **6**, 531[461]
Morosin, B., **1**, 21[111]
Moroz, E., **3**, 901[112]
Morper, M., **5**, 428[110]
Morris, A. D., **4**, 823[228]; **5**, 925[153]
Morris, D. F. C., **2**, 745[106]
Morris, D. S., **6**, 690[395]
Morris, G. A., **8**, 460[253]
Morris, G. E., **8**, 446[89,91,92], 455[91], 456[91]
Morris, G. F., **2**, 189[54]; **4**, 469[134], 472[134], 473[134], 475[134]; **6**, 954[17]
Morris, H. F., **1**, 3[23]
Morris, J., **1**, 130[97], 343[108], 401[12], 402[12]; **2**, 263[55]; **3**, 816[83]; **4**, 366[7], 384[7,143]; **7**, 245[79], 408[90], 418[90], 545[25]; **8**, 542[238]
Morris, J. I., **6**, 959[41]
Morris, M. D., **4**, 579[22]
Morris, M. R., **5**, 223[74,80]
Morris, P. J., **7**, 666[76]
Morris, T. H., **5**, 160[55]
Morris-Natschke, S., **1**, 61[37], 62[40]
Morrison, A., **5**, 618[1]
Morrison, A. L., **8**, 273[127]
Morrison, D. C., **8**, 323[115,116]
Morrison, D. J., **1**, 608[38]
Morrison, D. R., **2**, 152[103]
Morrison, E. D., **2**, 127[236]
Morrison, G. A., **1**, 2[15]; **3**, 735[21]; **4**, 187[96]
Morrison, H., **3**, 890[33]; **5**, 125[14], 645[1], 647[1h,13], 649[1h], 651[1], 654[1h], 661[1h]
Morrison, J., **5**, 543[113]
Morrison, J. D., **1**, 2[16], 49[9,11], 50[11], 56[9], 57[11], 58[9], 60[36], 65[11], 67[9], 70[62], 75[36], 86[31,32,45], 285[6], 334[62], 359[19], 382[19], 460[2], 825[48], 827[65], 833[116], 837[148], 838[1]; **2**, 2[4], 6[4a], 25[4a], 49[4a], 99[1], 100[1,14], 101[1,14], 103[1], 111[14], 134[3], 182[2], 190[57], 192[2c], 214[2c], 223[57d], 236[57d], 237[57d], 238[57d], 240[3], 248[3a], 277[8], 289[8], 301[1], 338[77], 455[4], 456[25], 475[1], 506[15], 510[38], 555[142], 630[4,4a], 633[4a], 635[4a], 637[4a], 639[4a], 642[4a], 652[4a], 681[61], 915[81], 946[81], 979[13], 1090[74]; **3**, 1[2], 2[2], 3[2], 13[2], 15[2], 23[2], 25[2], 30[181], 34[191], 35[191], 37[191], 39[191], 41[2,224], 44[2], 45[2], 53[273], 55[2], 76[54], 228[221], 341[5], 365[5], 374[5], 436[19]; **4**, 145[35], 200[1,6], 205[41], 213[102], 221[157], 226[189], 245[90], 257[90], 260[90], 364[1,1m], 373[1m], 927[43], 930[43], 945[43], 1038[57], 1079[60]; **5**, 96[108], 356[84], 451[37], 827[2], 829[2], 877[8]; **6**, 149[109], 684[342], 719[125], 720[125], 725[125], 728[212], 834[27], 855[27], 873[4]; **7**, 390[2,5], 394[2], 395[2], 412[2], 413[2], 419[2], 420[2], 421[2], 422[2], 424[2], 425[2]; **8**, 87[32], 88[40], 101[119], 145[84,85], 146[84,85], 152[172], 159[1,7], 166[65], 173[111], 178[65], 179[65], 459[228,243], 460[228b,252], 461[258], 535[166], 541[212], 721[139]
Morrison, J. H., **1**, 182[48]
Morrison, J. J., **2**, 226[158]
Morrison, P. A., **2**, 542[82]
Morrison, R. J., **8**, 446[86-88]
Morrison, W. H., III, **6**, 295[248]
Morrissey, M. M., **7**, 162[68], 184[171]; **8**, 448[140,141,142], 814[17]
Morrocchi, S., **4**, 1085[102]; **8**, 645[42]
Morrow, B. A., **4**, 313[474]
Morrow, C. J., **2**, 482[28], 483[28]; **8**, 461[258], 535[166]
Morrow, D. F., **3**, 790[59], 846[40]
Morrow, G. W., **1**, 554[101]; **3**, 695[152]; **4**, 14[47,47e]
Morrow, S. D., **4**, 590[102]
Morschel, H., **6**, 565[921]; **8**, 918[120]
Morse, D. F., **2**, 456[44]
Mörte, A., **6**, 173[50], 175[67]
Mortelli, J., **6**, 690[401], 692[401]
Mortiani, I., **4**, 598[201], 638[201]
Mortier, J., **5**, 1068[13]
Mortikov, E. S., **3**, 305[72,75a]
Mortimer, C. T., **8**, 670[12], 671[12]

Mortimore, M., **1**, 809[330]; **4**, 390[169]
Mortland, M. M., **7**, 845[68-71,73-75]
Mortlock, S. V., **2**, 40[142], 573[51], 575[51]; **6**, 1028[98]
Morton, C. J., **7**, 738[31]
Morton, D. R., **3**, 370[110]; **5**, 157[40], 166[91]
Morton, D. R., Jr., **1**, 608[36,37], 742[46]; **3**, 766[158]
Morton, G. H., **5**, 130[42]
Morton, G. O., **3**, 767[163]; **5**, 736[145], 737[145]
Morton, H. E., **3**, 248[52,53]; **4**, 173[35], 174[40], 189[103], 370[42], 381[129], 382[129]; **5**, 803[91], 979[27], 980[27], 981[27], 982[27]; **8**, 406[43]
Morton, J., **1**, 383[111]
Morton, J. A., **3**, 351[44], 352[44]; **4**, 391[176]; **5**, 414[54], 424[98], 539[108]; **8**, 847[98], 849[98]
Morton, J. B., **7**, 31[88]
Morton, M., **4**, 868[10]
Morton, T. H., **3**, 587[141,142]; **8**, 197[123]
Mortreux, A., **3**, 583[121]; **4**, 930[49]; **8**, 460[248]
Morvillo, A., **7**, 238[43]
Mory, R., **6**, 204[20]
Morzycki, J. W., **6**, 773[44], 989[79]; **7**, 13[117], 132[95], 236[21,23]; **8**, 928[25]
Mosaku, D., **1**, 2[10]
Mosandl, A., **8**, 191[92]
Mosandl, T., **5**, 155[36], 156[36], 157[36]
Mosbach, E. H., **3**, 644[149]
Mosbach, K., **2**, 456[35,36]; **7**, 145[162]; **8**, 185[9]
Moseley, K., **5**, 1136[53], 1146[53]
Moseley, R. H., **6**, 529[463]
Moser, G. A., **4**, 519[25,26]
Moser, J.-F., **3**, 365[63]; **7**, 711[59]; **8**, 932[44]
Moser, W. R., **4**, 915[14], 1032[15], 1039[15]; **5**, 736[140]
Moses, L. M., **4**, 877[67]
Moses, P., **6**, 802[40]
Moses, S. R., **4**, 379[117], 1079[65]; **5**, 260[65,68], 261[65,68], 262[68]; **7**, 439[36]
Mosettig, E., **8**, 278, 286[12], 291[34], 293[34], 295[34], 297[34], 298[34], 300[34], 301[34]
Moshenberg, R., **2**, 1074[146]
Mosher, H. S., **1**, 49[9], 56[9], 58[9], 67[9], 86[45], 182[48], 425[103], 833[116]; **2**, 323[25,29], 333[25], 655[147], 907[58], 908[58], 979[13], 1090[74]; **4**, 200[6]; **6**, 206[43], 210[43], 214[43], 495[151]; **8**, 87[32], 88[40], 145[84], 146[84], 159[1], 165[48,49], 178[49], 179[49], 459[228], 541[212]
Mosher, W. A., **7**, 576[123]
Mosin, V. A., **8**, 557[385]
Moskal, J., **1**, 571[281]; **4**, 16[52d]; **5**, 713[62], 728[62], 729[62]
Moskau, D., **1**, 37[247]
Moskowitz, H., **3**, 147[388], 151[388]
Mosmuller, E. W. J., **6**, 658[183]
Moss, G. I., **6**, 563[898]
Moss, G. P., **2**, 821[108]; **7**, 699[56]
Moss, N., **3**, 421[61], 422[61]; **4**, 1040[78], 1049[78a,121a]; **5**, 804[95], 993[50,51], 994[50,51], 995[51], 996[51], 997[54]
Moss, R. A., **4**, 483[4], 484[4], 495[4], 952[2], 953[8,8f], 954[8f,m], 959[32], 960[35], 961[8f,m], 976[100], 977[32], 1001[45], 1002[45,48], 1012[174], 1101[195]; **5**, 1065[1], 1066[1], 1074[1], 1083[1], 1084[1], 1093[1], 1094[1d]; **6**, 575[971], 776[55]
Moss, R. E., **8**, 388[64]
Moss, R. J., **5**, 691[85]
Mosset, P., **1**, 821[23]; **2**, 97[60]; **6**, 11[45], 690[401], 692[401]; **7**, 713[72]
Mossman, A., **1**, 648[137], 654[137], 655[137]
Mossman, A. B., **6**, 752[110]
Mossman, C. J., **5**, 537[99]
Mostafavipoor, Z., **7**, 738[28]
Mostecky, J., **8**, 200[138]
Mosterd, A., **5**, 163[72]
Mostowicz, D., **1**, 838[163,165,166]
Motegi, M., **2**, 810[63], 824[63]
Motegi, T., **4**, 610[337]
Motherwell, R. S. H., **6**, 447[104], 451[104]
Motherwell, W. B., **1**, 698[242], 699[242]; **3**, 505[158,159], 594[187], 664[30]; **4**, 747[149], 790[35], 823[227,228]; **5**, 290[42], 925[153], 1011[48]; **6**, 442[87], 447[104], 451[104], 837[59], 938[130], 942[130]; **7**, 13[115-117,119], 40[11], 132[95,100], 146[100], 307[16], 310[16], 318[16], 319[16], 322[16], 704[12,14], 705[14], 719[7], 721[7], 725[7], 726[7], 728[7]; **8**, 392[109], 818[36], 821[50,51], 823[55], 830[89], 831[89-91], 924[3]
Motherwell, W. D. S., **1**, 2[3], 37[3]
Mothes, K., **6**, 746[92]
Mothes, V., **6**, 455[152]
Motohashi, S., **6**, 156[162]; **7**, 530[21], 531[21]
Motoi, M., **8**, 170[71,95]
Motoki, S., **3**, 639[74], 643[123], 644[146]; **5**, 441[178], 475[140,141]; **8**, 392[99]
Motoyama, N., **2**, 1089[57]
Motoyama, T., **4**, 1089[138], 1091[138]
Mott, R. C., **2**, 445[23]; **6**, 1064[88]; **7**, 121[25], 530[19], 531[19], 824[42]; **8**, 986[15]
Motter, R. F., **3**, 105[212]; **4**, 120[199]
Motto, M. G., **2**, 482[34], 484[34]
Mottus, E. H., **8**, 321[107]
Moubacher, R., **7**, 230[127]
Moufid, N., **4**, 792[61]
Moukimou, A., **2**, 61[21]
Moulik, A., **1**, 139[5]
Moulineau, C., **3**, 257[121]
Moulines, F., **8**, 556[377]
Moulines, J., **2**, 159[127]; **4**, 290[207]
Mountain, A. E., **8**, 89[43]
Moura Campos, M., **4**, 315[503]
Mourad, M. S., **6**, 939[142]; **8**, 373[138], 376[165], 377[167]
Moural, J., **6**, 495[147]
Moureau, H., **8**, 285[6]
Moureu, C., **4**, 47[133], 274[54], 286[172], 289[172]
Mourgues, P., **7**, 738[26], 747[26], 851[18]
Mouriño, A., **3**, 232[262], 545[121], 983[19,21], 984[21]; **7**, 547[33]; **8**, 514[113]
Moursounidis, J., **5**, 581[175], 584[193]
Mouseron, M., **8**, 270[95]
Mouslouhouddine, M., **6**, 705[26]
Mousseron, M., **3**, 813[64], 828[47]; **5**, 766[115]; **6**, 264[34], 267[57]
Mousseron-Canet, M., **8**, 270[95]
Mousset, G., **8**, 133[17,22]
Moustrou, C., **4**, 807[150]
Moutet, J.-C., **7**, 809[83]
Mouzin, G., **8**, 343[112]
Mowat, E. L. R., **4**, 282[139]
Mowat, R., **6**, 212[80]
Mowry, D. T., **6**, 225[1], 226[1], 231[1], 265[44], 268[65], 271[44]; **7**, 764[104]
Moya-Gheorghe, S., **8**, 124[89]
Moyano, A., **3**, 380[7]; **5**, 1047[31], 1052[31], 1054[43], 1059[54], 1062[54d,59]; **8**, 932[42]
Moya-Portuguez, M., **5**, 108[210], 109[210,218,222], 110[210,222,223], 111[210,222], 112[222,223], 113[236]; **6**, 430[96], 520[342], 544[342]
Moyer, B. A., **7**, 158[40]
Moyer, M. P., **6**, 127[160]
Moyle, M., **3**, 824[18]; **6**, 966[94]

Moynehan, T. M., **3**, 807[27]
Moyse, H. W., **4**, 279[112]
Mozdzen, E. C., **8**, 251[106]
Mozingo, R., **4**, 282[135]; **5**, 752[45]; **8**, 286[12]
Mozolis, V. V., **6**, 423[40], 424[40], 428[40], 432[40]
Mozumi, M., **4**, 398[217], 399[217b], 401[217b], 403[217b], 404[217b], 413[278b]; **5**, 829[22]; **6**, 509[266]
Mpango, G. B., **3**, 50[268]; **4**, 55[157], 57[157c], 72[27,30], 74[27], 249[115], 257[115], 258[253], 260[115]
M'Pati, J., **6**, 922[50]
Mroczyk, W., **2**, 348[61]
Mrotzeck, U., **1**, 528[111]; **7**, 835[82]
Mrotzek, H., **6**, 426[73], 448[112], 482[124]
Mrozack, S. R., **7**, 229[119]
Mrozik, H., **4**, 356[136]; **7**, 93[53]
Mualla, M., **3**, 605[232]; **4**, 809[160]; **8**, 114[53]
Mubarak, A. M., **8**, 879[50]
Mubarak, M. S., **8**, 857[199]
Muccino, R. R., **8**, 333[58]
Muchmore, C. R., **2**, 1049[19]; **5**, 1101[133], 1112[133]
Muchmore, D. C., **8**, 817[28], 930[36]
Muchmore, S., **3**, 216[71]
Muchow, G., **1**, 223[78], 224[78]
Muchowski, J. M., **1**, 469[57], 473[82]; **2**, 58[13], 739[45]; **4**, 1061[166]; **6**, 176[104], 249[143], 546[644]
Muck, D. L., **3**, 635[39]
Muckensturm, B., **6**, 1035[136]
Mudd, A., **3**, 791[63]; **8**, 89[44]
Mudryk, B., **1**, 239[38]; **4**, 432[105,107,109]
Mudumbai, V. A., **8**, 384[40]
Mueller, H., **8**, 881[79]
Mueller, H. R., **8**, 624[153]
Mueller, L. G., **5**, 943[236]
Mueller, P. H., **4**, 1072[18]; **5**, 257[61,61c]
Mueller, R. A., **6**, 936[105], 1042[6]; **7**, 205[61]
Mueller, R. H., **1**, 126[88]; **2**, 101[20], 102[20], 182[9], 200[88], 604[55], 711[36], 935[150]; **5**, 828[9], 840[9], 841[9,9c], 847[9], 856[9], 857[9a], 859[9], 886[20], 893[20], 1001[12]; **6**, 858[162], 860[178]; **7**, 256[24], 602[98], 607[166]; **8**, 526[27], 949[154]
Mueller, W. H., **4**, 317[553]; **7**, 516[7], 517[10]
Mues, C., **5**, 837[71], 838[71]
Muetterties, E. L., **8**, 431[65,66], 450[168], 454[168,196,202], 458[223,223b-d], 459[226]
Mugdan, M., **7**, 446[65]
Mugrage, B., **6**, 782[82]; **8**, 190[76]
Muha, G. M., **8**, 597[88]
Mühlbauer, E., **2**, 737[40]
Muhlbauer, G., **1**, 746[61]; **5**, 809[121]
Muhlemeier, J., **5**, 1093[96], 1094[100,100a], 1095[104], 1098[96b,100a], 1112[96b,104]
Mühlenbein, H., **6**, 264[35]
Mühlstädt, M., **2**, 360[167], 902[43]; **3**, 379[3], 380[6,12], 386[59]; **4**, 387[163a-c]
Muhm, H., **5**, 1148[123]
Muhn, R., **4**, 739[109]
Muhs, M. A., **3**, 843[22]
Mui, J. Y.-P., **4**, 1000[15], 1002[52]
Muir, C. N., **3**, 741[53], 743[60]
Muira, H., **7**, 686[97]
Mukai, C., **3**, 904[134]; **5**, 736[142e,f]
Mukai, K., **4**, 599[220], 642[220]
Mukai, T., **3**, 901[107]; **4**, 1103[205]; **5**, 196[15-17], 197[16], 206[47], 552[35], 634[74,76], 714[68,70], 819[154], 826[159a]; **6**, 531[449]; **7**, 875[113,115,116]
Mukaiyama, S., **7**, 662[52]
Mukaiyama, T., **1**, 54[21], 64[47-49], 65[50,52], 69[60], 70[52,64], 71[65], 72[66-68], 141[22], 192[82], 327[16], 336[71], 339[85], 346[16,125-127], 347[135], 349[149], 406[30], 407[30-33], 415[30], 424[100], 427[30]; **2**, 2[4], 5[22], 10[22,22c,45,45a], 18[72], 30[112a], 31[109,112], 35[130], 36[130], 68[42], 70[52], 82[10], 111[80], 112[90,91,95-97], 116[126,-128,129,130,132,133,134,137,138,139,142,143], 117[144,145], 133[2], 233[185], 240[10], 242[10,19,20], 244[28], 351[82], 357[82], 424[37], 425[37], 436[68], 437[68], 455[15], 572[46,48], 576[71,74], 578[86], 605[62], 610[88-93], 611[92], 612[108], 613[109,110,112], 614[112,115], 615[122,123], 616[131,139], 629[1,2], 630[4,20], 631[2,20], 632[2,29a,b], 633[30-32,33a], 635[1,30,41,42], 640[29-33,41,42], 655[136,137,141], 656[137], 657[161a,167,168], 667[43], 744[88], 802[38], 804[42], 816[87], 920[94], 921[94], 922[94,102,105], 923[102], 924[105,106], 948[182]; **3**, 25[153], 86[15], 96[15,168,169], 104[168,169], 108[15,168,169], 117[15,168,169], 125[308], 226[205], 227[205], 286[58], 426[82], 429[82], 563[1], 565[23], 570[23], 583[23], 585[136], 598[202], 730[46]; **4**, 30[90], 50[142,142b], 85[70], 89[85], 93[92], 100[125], 158[78,79], 159[82,83], 161[88,90,91c,d], 189[106], 190[106b], 202[25], 206[51-55], 207[56], 218[134,135,136], 229[224], 230[256,257], 231[262,263], 244[80], 258[80,237,238], 261[238], 377[105a], 381[105], 756[182], 1002[46], 1078[57], 1080[57]; **5**, 15[102], 543[114], 850[146]; **6**, 18[65], 20[75], 26[108], 46[58,73], 54[131], 83[83,84], 139[48], 206[37,38], 214[95,96], 237[70], 438[56,58,62,63], 439[58,67], 558[838], 607[45], 608[50], 612[75,76], 615[45,98,99,101], 624[101,137], 715[88], 960[63], 966[99], 979[29]; **7**, 125[59], 141[132], 209[90], 299[46], 318[58], 319[58], 320[58], 760[31]; **8**, 159[8], 168[66-68], 178[68], 179[68], 216[60], 238[23], 260[5], 272[5,116-118], 413[126], 840[35], 899[26], 906[26], 907[26], 913[26], 914[26], 951[177], 966[74,75], 967[77], 991[46]
Mukaiyama, Y., **1**, 65[51]
Mukamal, H., **7**, 689[9]
Mukerjee, A. K., **2**, 396[6], 402[6], 403[6], 404[42], 405[42-44]; **5**, 85[5,6], 86[12]
Mukerjee, S. K., **7**, 544[34]; **8**, 568[467]
Mukerjee, S. L., **8**, 669[8]
Mukerji, I., **6**, 9[40]
Mukherjee, D., **3**, 846[44]; **4**, 1076[47]; **5**, 625[32], 626[32], 629[46]; **8**, 7[35], 505[80]
Mukherjee, D. K., **8**, 333[56]
Mukherjee, P. C., **7**, 318[52], 319[52]
Mukherjee, P. N., **3**, 329[182]
Mukherji, S. M., **3**, 325[159]
Mukhin, O. N., **3**, 329[183]
Mukhina, N. A., **6**, 554[734]
Mukhopadhyay, T., **2**, 323[26], 335[26], 336[26], 782[23]; **5**, 842[109]
Mukhtar, R., **5**, 77[265]
Mukhtarov, I. A., **8**, 896[14]
Mukkanti, K., **8**, 372[117]
Mukkavilli, L., **5**, 95[89]
Mukoyama, M., **6**, 441[78], 443[78]
Mulamba, T., **2**, 1017[36]; **3**, 466[182]; **6**, 738[49,50], 739[50]
Mularski, C. J., **6**, 432[121], 566[926]
Mulder, J. J. C., **5**, 647[19], 650[19], 652[19], 653[19], 656[19]
Mulder, R. J., **3**, 909[151]
Mulder, T., **8**, 93[74]
Muleka, K., **2**, 555[141]
Mulhaupt, R., **1**, 301[74], 316[74]
Mulhauser, M., **3**, 882[103]; **4**, 599[217]; **6**, 161[186]
Mulhern, L. J., **7**, 40[4]
Mulhern, T. A., **4**, 561[29]; **6**, 284[176]
Mulholland, D. L., **1**, 294[49]; **2**, 6[35]
Mulholland, R. L., Jr., **4**, 379[113], 380[113a]
Mulholland, T. P. C., **4**, 27[78]; **8**, 140[26]
Muljiani, Z., **1**, 546[51]; **6**, 1022[64]

Mullally, D., **5**, 72[172]
Mullane, M., **6**, 252[153]
Mullen, D. L., **8**, 618[110]
Mullen, G., **8**, 542[238]
Mullen, G. P., **6**, 462[17]
Müllen, K., **3**, 594[184]; **4**, 1007[116]
Mullen, P. W., **2**, 710[27]
Müller, A., **2**, 152[104]; **3**, 322[143], 816[85]; **4**, 1017[216]
Muller, A. J., **8**, 52[140], 66[140]
Muller, B., **2**, 477[12]; **3**, 431[95,96]; **5**, 366[99]; **6**, 294[242]
Muller, B. L., **6**, 1067[108]
Müller, C., **4**, 1001[37], 1015[37]; **5**, 1133[33]; **8**, 856[167]
Muller, C. L., **2**, 1010[8]
Muller, D. G., **5**, 563[90]
Müller, E., **1**, 846[12], 851[12], 852[12]; **2**, 277[6], 609[82], 782[31], 1086[26]; **3**, 324[151], 414[4], 554[25], 563[1], 661[22,23], 666[43], 891[42], 909[155]; **4**, 1081[83]; **5**, 451[11,12], 1135[46], 1136[46], 1139[46,72], 1140[46,72], 1148[46,123]; **6**, 116[90], 204[5]; **7**, 777[364]; **8**, 248[88], 249[88], 251[88], 253[88], 254[88], 300[87], 734[2], 737[2], 753[2]
Müller, E. P., **7**, 473[34], 501[34]
Muller, F., **1**, 41[269]; **3**, 211[32]; **4**, 170[12]; **8**, 367[53]
Müller, G., **1**, 17[211,213,214,215], 19[103], 35[172], 36[236,237]; **2**, 1054[62]; **3**, 901[112]; **4**, 355[133], 872[40]; **5**, 404[12], 568[110], 850[152], 1070[18,28], 1074[28], 1096[109,127], 1098[109,109c,126], 1099[109c], 1112[109c,126]; **6**, 175[67], 179[125], 480[116]; **7**, 477[73]; **8**, 859[217]
Müller, G. H., **1**, 146[68]
Muller, G. W., **5**, 241[4], 242[7], 243[7]
Müller, H., **2**, 943[168]; **4**, 611[348], 888[132], 889[132,137]; **8**, 735[12,13], 738[30], 740[12,13,30], 741[13], 753[30], 756[13], 757[13]
Muller, H. R., **8**, 474[15]
Müller, H.-R., **2**, 784[37]
Müller, I., **5**, 260[66], 261[66]; **8**, 70[223], 647[54,55]
Muller, J., **7**, 418[130b]
Müller, J., **1**, 253[12]; **5**, 1085[64]
Muller, J.-C., **7**, 121[20,21], 123[20], 145[20], 163[71]; **8**, 925[11], 926[11]
Müller, K., **2**, 866[9]; **6**, 711[66]
Muller, K. A., **7**, 10[81]
Müller, K. H., **8**, 758[171]
Müller, L. L., **1**, 834[130]; **5**, 86[25], 90[25]
Müller, N., **3**, 373[130], 648[188]; **5**, 768[133], 779[133]; **6**, 116[90]
Müller, O., **6**, 532[471]
Müller, P., **4**, 1013[181], 1015[198]; **5**, 972[8], 989[8]; **7**, 227[82], 235[7], 236[7], 247[7], 309[23], 767[194], 773[194]
Müller, P. L., **5**, 73[191]
Müller, R., **8**, 472[3]
Müller, R. H., **5**, 768[125], 779[125]
Muller, R. K., **7**, 482[118]
Müller, R. K., **6**, 1059[66]
Muller, R. N., **2**, 351[80], 364[80]
Müller, S., **2**, 521[70]; **3**, 41[227]
Müller, T., **4**, 229[234]
Müller, U., **2**, 163[148]; **3**, 41[228], 43[239], 652[222], 653[227], 654[227]
Müller, W., **2**, 782[15,31], 844[199]; **3**, 747[71]; **7**, 13[121,122], 247[106]
Müller, W. M., **6**, 42[45]
Müller-Remmers, P. L., **5**, 202[37]
Müller-Starke, H., **6**, 227[23], 229[23], 230[23], 231[34], 238[34]
Mullican, M. D., **4**, 8[30b], 125[217b]; **5**, 497[227], 572[125,126]
Mulligan, P. J., **2**, 828[132]
Mulliken, R. S., **7**, 863[83], 865[83], 866[89], 868[83]
Mullins, M., **6**, 140[55], 898[107a,b]
Mullins, M. J., **2**, 387[336]; **3**, 918[24], 957[111], 960[116]; **5**, 894[47,48]
Mullis, J. C., **1**, 343[121], 345[121]; **6**, 237[59], 257[59]
Mulvaney, J. E., **4**, 120[197], 868[17], 869[17], 872[39,39b]; **5**, 73[197]
Mulvaney, M., **6**, 692[408]
Mulvey, R. E., **1**, 6[33], 33[165], 38[258], 39[187]; **3**, 763[151]
Mulzer, J., **1**, 54[20], 185[53,56], 221[68], 338[83]; **2**, 29[104,106], 31[114,114a], 205[101,101a,b], 211[113], 597[8]; **3**, 224[164], 225[164]; **4**, 21[65], 36[103,103b], 108[146e], 111[155c], 991[149]; **5**, 75[231], 1060[58]; **6**, 11[48], 752[112]; **7**, 253[18], 549[45]
Mumtaz, M., **2**, 103[30]; **3**, 44[241]
Munakata, K., **6**, 490[111]
Munasinghe, V. R. N., **3**, 597[199]
Munasinghe, V. R. Z., **8**, 817[33]
Munavu, R., **2**, 494[55]; **6**, 541[594]
Munchausen, L. L., **4**, 1072[18]; **5**, 71[128]
Munch-Petersen, J., **4**, 73[33], 149[52], 181[72,73], 184[83]
Mund, S. L., **8**, 447[106], 450[106]
Munderloch, K., **2**, 141[41]
Mundill, P. H. C., **1**, 411[46]; **4**, 27[79]; **7**, 199[36], 200[39]
Mundlos, E., **2**, 900[22], 961[37]
Mundy, B. P., **3**, 571[58], 596[193], 727[30,31], 728[37], 849[55]; **4**, 373[81], 374[81]; **5**, 453[69], 455[69,74,77,79]
Mundy, D., **3**, 114[234]; **6**, 1026[82], 1029[82]; **7**, 771[284], 772[284]
Munegumi, T., **8**, 144[79], 145[80-82]
Munekata, E., **6**, 637[36]
Munemori, M., **8**, 151[147]
Muneyuki, R., **6**, 421[31]
Mungall, W. S., **7**, 412[104], 413[104], 429[158], 430[158], 442[46a]; **8**, 385[43]
Munger, J. D., **6**, 859[170]
Munger, J. D., Jr., **5**, 249[36], 841[89], 843[123], 857[232]; **6**, 859[170,172]
Münger, K., **4**, 719[18], 723[18,42]
Munger, P., **1**, 570[264]; **8**, 385[48]
Mungiovino, G., **2**, 787[52]
Munoz, B., **7**, 76[117]
Munoz, H., **1**, 552[83]
Muñoz, H., **3**, 198[49]
Munoz-Madrid, F., **3**, 564[8]
Munro, G. A. M., **4**, 542[119,120], 703[32], 704[32]
Munro, M. H. G., **3**, 741[50], 807[31]; **8**, 213[29]
Munroe, J. E., **5**, 829[25]
Munslow, W. D., **5**, 712[58]
Münster, P., **1**, 122[72], 373[89], 375[89]; **2**, 1052[46]
Munsterer, H., **5**, 422[86]
Muntyan, G. E., **3**, 361[80]
Mura, A. J., Jr., **2**, 74[70]; **3**, 103[205], 108[205]; **7**, 207[83], 208[83], 209[83], 210[83]; **8**, 935[59]
Mura, L. A., **3**, 794[79]
Murabayashi, S., **3**, 586[140]
Murad, E., **1**, 252[7]
Murada, T., **6**, 676[303]
Murae, T., **3**, 395[99], 402[129]; **8**, 330[47], 340[100], 925[7]
Murago, G., **2**, 758[23b]
Murahashi, E., **5**, 736[142h]
Murahashi, S., **3**, 437[23], 438[36], 440[36]; **4**, 310[435], 377[104], 378[104f], 383[104f], 393[188], 557[10,11], 589[76], 590[99], 598[189,198,201], 611[353], 613[371], 638[201], 640[423], 959[33], 1006[106]; **7**, 451[18], 452[50], 454[18]; **8**, 856[170], 876[44], 877[44]
Murahashi, S.-I., **1**, 391[151], 392[155], 551[70]; **2**, 13[56], 587[139,146], 1052[45]; **3**, 222[143], 223[150], 249[67], 259[130], 436[5], 437[5,22], 438[22], 483[10], 485[36], 491[36], 494[36],

497[36], 1040[104], 1041[112]; **6**, 74[36], 76[53], 85[92], 86[97,99], 113[69], 253[156]; **7**, 94[58], 107[164], 178[150], 227[88], 314[40], 315[40], 419[134b], 745[76-78]; **8**, 395[122,123], 600[102], 806[102]
Murahayashi, A., **7**, 415[114]
Murai, A., **1**, 161[88,89], 566[209], 823[43]; **2**, 159[131]; **3**, 125[306], 126[306], 735[22]; **4**, 373[84]; **5**, 158[48]; **6**, 1049[36]; **7**, 253[23], 680[80]; **8**, 334[59], 528[67]
Murai, F., **3**, 571[75], 574[75]; **8**, 514[111]
Murai, H., **3**, 891[43], 892[43]
Murai, K., **5**, 55[48]; **6**, 265[38]
Murai, S., **2**, 442[11], 443[15,17], 445[24], 451[15,55], 603[46]; **3**, 771[186]; **4**, 115[177], 444[197], 973[85]; **5**, 442[185,185a], 461[107], 464[107], 466[107], 532[84], 601[44]; **6**, 477[98], 479[108], 481[123], 684[344]; **7**, 125[60], 131[80], 137[118], 138[118]; **8**, 370[90], 412[119], 413[119], 773[70], 774[70,71], 789[122], 887[119], 991[44]
Murai, T., **6**, 453[143], 461[5], 467[50]; **8**, 253[122]
Murai, Y., **6**, 46[58]
Murakami, K., **5**, 196[16], 197[16], 571[121]
Murakami, M., **1**, 489[21], 497[21], 546[49]; **2**, 116[138], 117[144,145], 242[19], 244[28], 576[71,74], 615[122], 629[1], 632[29a,b], 635[1,41,42], 640[29,41,42], 744[88], 922[102], 923[102]; **3**, 226[205], 227[205], 463[167], 529[52], 555[28]; **4**, 162[94b]; **5**, 841[92], 916[121], 917[121], 1175[38], 1177[43], 1178[38,43]; **6**, 558[838]; **8**, 154[197], 155[197], 453[191]
Murakami, N., **1**, 188[70]; **7**, 78[128a]
Murakami, S., **4**, 847[76]; **5**, 687[57]; **8**, 150[123,124], 151[146,153]
Murakami, T., **2**, 792[69]; **6**, 524[354]
Murakami, Y., **2**, 736[26], 743[81], 780[10]
Murakata, C., **4**, 393[186]
Muraki, M., **2**, 112[90,91], 240[10], 242[10]; **4**, 756[182]; **8**, 238[23], 260[5], 266[60], 272[5,60,116]
Murali, C., **8**, 206[171]
Muralidharan, F. N., **4**, 102[133a-c]
Muralidharan, K. R., **4**, 771[252]; **7**, 519[23]
Muralidharan, S., **3**, 211[30]
Muralidharan, V. B., **2**, 904[51]; **4**, 102[133a-c]; **8**, 798[62], 800[62]
Muralidharan, V. P., **6**, 914[26]; **7**, 222[40]
Muralimohan, K., **4**, 472[144]
Muramatsu, S., **4**, 113[164]
Murano, K., **5**, 838[74]
Murao, K., **6**, 626[167]
Muraoka, O., **3**, 853[72]
Murase, H., **1**, 823[43]; **3**, 125[306], 126[306]
Murata, C., **8**, 626[174]
Murata, E., **3**, 552[8], 557[49]
Murata, I., **6**, 531[438], 932[97]; **7**, 743[64]; **8**, 609[53]
Murata, K., **2**, 464[98]
Murata, M., **5**, 151[9]; **6**, 609[56], 816[99]; **7**, 642[9,10]; **8**, 787[119]
Murata, N., **7**, 26[56]
Murata, R., **7**, 877[133]
Murata, S., **1**, 328[26], 882[121]; **2**, 369[250], 615[121], 635[39,39c], 640[39], 650[39c,d]; **3**, 25[159], 402[130], 404[135]; **4**, 379[115], 381[126a], 382[126], 383[126]; **5**, 809[122]; **7**, 524[50], 650[51]; **8**, 155[203]
Murata, Y., **3**, 159[451], 160[451], 161[451], 402[125], 405[138], 714[35]
Murato, K., **6**, 813[84,85]
Murayama, E., **1**, 328[29], 586[17], 587[17]; **2**, 582[106], 614[120]; **7**, 208[78], 539[65]; **8**, 99[107]
Murayama, H., **8**, 418[12], 422[12]
Murayama, K., **5**, 717[94]
Murdock, K. C., **6**, 487[42], 489[42], 543[42], 550[42], 554[42]
Murdock, T. O., **1**, 212[7], 213[7]
Murdzek, J. S., **5**, 1116[11], 1117[11,16], 1118[11], 1121[16]
Muria, A., **4**, 238[11], 245[11], 255[11], 260[11]
Murikawa, M., **4**, 600[230]
Muriyama, E., **1**, 755[113]
Muroi, H., **8**, 190[81]
Muroi, M., **2**, 232[174]
Murphy, C., **7**, 602[102]
Murphy, C. J., **6**, 981[47]
Murphy, D. J., **4**, 653[438]
Murphy, D. K., **3**, 734[3,5]; **8**, 873[14], 875[14]
Murphy, E., **5**, 1045[28], 1049[28]; **8**, 336[84], 339[84]
Murphy, F. G., **3**, 691[136]
Murphy, F. X., **6**, 272[91]
Murphy, G. J., **2**, 77[88]; **3**, 202[98]
Murphy, G. K., **5**, 207[50]
Murphy, G. W., **5**, 856[194]
Murphy, J. A., **4**, 14[47], 820[222]; **5**, 830[35]
Murphy, J. G., **8**, 566[452]
Murphy, P. J., **1**, 755[115], 812[115], 813[115]; **2**, 117[155], 309[25]; **5**, 841[97]; **6**, 193[215]; **7**, 412[106]
Murphy, R., **3**, 799[104]
Murphy, R. A., Jr., **4**, 27[83], 433[116]
Murphy, R. F., **6**, 455[155]
Murphy, R. S., **8**, 652[72]
Murphy, W. S., **2**, 73[69], 150[96], 151[96]; **3**, 131[328,329,332], 132[329], 135[328,329], 683[104]; **5**, 944[245]; **7**, 606[153]; **8**, 109[11], 112[11], 113[11], 114[55], 973[121]
Murphy, Z. L., **6**, 547[666]
Murrall, N. W., **4**, 629[409]
Murray, A. W., **1**, 774[209]; **2**, 422[28]; **5**, 827[2], 829[2], 867[2d]; **7**, 372[72a]
Murray, B. J., **4**, 50[142,142l]; **7**, 774[315]
Murray, C. K., **4**, 981[111]; **5**, 1070[20,22,23], 1071[20], 1072[20,22,23], 1074[20], 1085[63], 1086[22], 1110[20], 1111[20]
Murray, G. J., **8**, 52[136], 66[136]
Murray, H. C., **7**, 57[34]; **8**, 558[397,398]
Murray, H. H., **4**, 695[4]
Murray, K. J., **8**, 724[174], 725[174], 726[174]
Murray, L. T., **8**, 724[165]
Murray, M., **4**, 1010[152]
Murray, M. J., Jr., **3**, 643[121]
Murray, P. J., **1**, 752[96], 798[285]; **3**, 503[149], 512[149]; **4**, 113[171,171f,h,i]; **7**, 162[57], 524[55]; **8**, 245[73], 847[101]
Murray, R. D. H., **8**, 528[66]
Murray, R. E., **6**, 244[109]
Murray, R. K., **1**, 859[65,67], 872[90]
Murray, R. K., Jr., **1**, 262[38,39]; **5**, 229[119], 558[61]
Murray, R. W., **1**, 834[125,127]; **3**, 736[24]; **4**, 1098[175]; **7**, 13[124], 374[77f], 737[11], 745[75], 750[129], 778[405]; **8**, 398[146], 726[188]
Murray, T. F., **3**, 1031[64]; **4**, 937[70], 938[70], 941[85]
Murray, T. P., **2**, 173[180], 832[153]; **4**, 74[36]
Murray, T. S., **6**, 261[5], 275[5]
Murray, W. V., **8**, 566[457], 568[468]
Murray-Rust, J., **3**, 404[134]; **6**, 1024[76]
Murray-Rust, P., **3**, 404[134]; **6**, 1024[76]
Mursakulov, I. G., **3**, 349[35]
Murtas, S., **5**, 1131[12]
Murthy, A. N., **6**, 836[58]; **8**, 123[81]
Murthy, A. R. K., **8**, 503[66,69]
Murthy, K. S. K., **3**, 589[162], 610[162]; **4**, 373[82]; **5**, 105[190], 252[43], 257[43]
Murthy, P. S. N., **8**, 494[24]
Murti, V. A., **6**, 538[552], 550[552]
Murtiashaw, C. W., **1**, 584[9]; **2**, 714[50,53]; **3**, 363[86]; **4**, 24[72,72f], 1049[121b]; **5**, 841[87], 994[53], 997[53]; **8**, 927[20]
Murugesan, N., **5**, 162[67]; **6**, 734[10], 735[10]

Musada, R., **1**, 543[29]
Musallam, H. A., **7**, 155[28]
Muschaweck, R., **6**, 554[748]
Musco, A., **1**, 440[171]; **4**, 598[182]
Musgrave, O. C., **7**, 235[9]
Musgrave, W. K. R., **3**, 898[74]; **8**, 643[36], 901[35,37], 903[35], 905[35]
Mushak, P., **4**, 587[49]
Mushika, Y., **6**, 608[50], 609[52,53], 624[137]; **7**, 806[74]
Musial, S. T., **5**, 736[145], 737[145]
Musil, V., **5**, 143[96]
Musker, W. K., **4**, 366[6]; **7**, 221[27], 765[161]
Muskopf, J. W., **2**, 105[45]; **5**, 636[89]
Musliner, W. J., **3**, 747[72]; **8**, 336[86], 912[90]
Muslukov, R. R., **4**, 598[199], 638[199], 640[199]
Musoiu, M., **8**, 150[136]
Musolf, M. C., **8**, 765[15], 773[15]
Musser, A. K., **5**, 134[60], 154[33]
Musser, J. H., **2**, 541[78]; **3**, 762[145]; **4**, 107[142]; **5**, 854[181]; **6**, 108[32]; **7**, 502[266]
Musser, M. T., **7**, 882[171]
Musso, H., **1**, 752[99]; **3**, 660[12], 664[27]; **5**, 736[141]; **8**, 292[38], 795[20]
Mustafa, A., **3**, 887[5], 897[5], 900[5], 903[5]; **8**, 625[158], 626[158]
Mustafaev, E. Kh., **3**, 306[82]
Mustafaeva, M. T., **3**, 342[14]
Muszkat, K. A., **5**, 705[25], 723[106], 725[106a,115], 729[125]
Mutak, S., **2**, 362[181]
Muth, H., **8**, 383[16]
Muth, K., **3**, 890[31], 901[111]; **5**, 385[129d]
Muthanna, M. S., **3**, 894[64]
Muthard, D. A., **2**, 1102[123]
Muthard, J. L., **3**, 614[6], 623[6,32], 627[6]; **4**, 373[67]; **6**, 1063[81]
Muthukrishnan, R., **2**, 66[37]; **3**, 99[182], 101[182]
Mutin, R., **8**, 445[53]
Muto, S., **6**, 464[37], 465[37]
Muto, T., **7**, 384[114a]
Mutschler, E., **2**, 381[308]
Mutter, M., **2**, 1099[110]; **6**, 174[55], 670[273]
Mutter, M. S., **1**, 755[116], 756[116,116b], 758[116], 761[116]
Muxfeldt, H., **5**, 853[171]; **6**, 509[246], 679[328]; **7**, 157[33,33a], 160[50]; **8**, 278[156]
Muzart, J., **2**, 141[39]; **4**, 603[276,277], 610[334,335,336], 645[276,277]; **7**, 92[39], 94[39], 95[39], 96[39], 106[152], 107[39,157,158,159], 278[161,162,163,164]
Muzychenko, V. O., **8**, 657[95]
Mwesigye-Kibende, S., **5**, 180[147]
Mwinkelried, R. I., **3**, 40[223], 41[223], 42[223]
Mychajlowskij, W., **2**, 590[159], 713[49], 728[49]; **3**, 257[119]
Myer, L., **2**, 66[34], 67[34]; **4**, 119[194], 226[201], 227[201]; **6**, 864[193]
Myers, A. G., **4**, 971[73], 1033[35], 1040[96], 1041[96], 1048[96]; **5**, 736[142b,d], 1014[56]; **7**, 363[35], 410[97a]
Myers, H. K., **5**, 1141[84]
Myers, J. K., **8**, 707[22]
Myers, M., **1**, 786[248]; **4**, 71[20], 116[185b], 992[157]; **8**, 615[94], 618[94], 842[47]
Myers, M. J., **2**, 741[67]
Myers, P. L., **5**, 536[96]
Myers, R. F., **1**, 608[36,37]; **3**, 364[93], 393[94]
Myers, R. L., **2**, 139[28]
Myers, R. S., **2**, 106[46]; **7**, 426[148b]
Myerson, J., **4**, 366[8], 380[120,120b,125], 381[8a,120b,125b], 382[120b]; **6**, 26[106]; **8**, 853[147], 856[147], 857[147]
Myhre, P. C., **1**, 292[26]
Mylari, B. L., **2**, 823[112]; **3**, 1051[22], 1052[22]
Myles, D. C., **2**, 703[87]; **4**, 597[176], 598[176], 622[176], 637[176]
Mynott, R., **1**, 14[77,78]; **4**, 596[161], 608[318,319]; **5**, 297[55], 641[134], 1130[2], 1154[149], 1192[31], 1197[31]; **6**, 179[124], 184[153]
Myong, S. O., **3**, 160[468]; **4**, 1101[192]; **5**, 582[177], 938[217,218]
Myoung, Y. C., **2**, 406[46]
Myrbach, K., **2**, 464[99]
Myrboh, B., **7**, 154[12]; **8**, 540[201], 839[26b], 840[26]
Myshkin, V. E., **8**, 773[69]
Mysorekar, S. V., **7**, 90[29]
Mysov, E. I., **2**, 727[136]; **4**, 840[33]; **7**, 7[51]; **8**, 896[12-14], 898[13]

N

Naab, P., **5**, 1130[8]
Naae, D. G., **6**, 172[10,15]
Naaktgeboren, A. J., **6**, 489[89]
Nabeya, A., **2**, 492[53], 493[53]; **5**, 949[281]; **6**, 274[102-104,107], 807[57]; **8**, 276[147,148,149]
Nabeyama, K., **3**, 380[12]
Nabi, Y., **4**, 258[255]
Nabih, I., **8**, 629[185]
Nabney, J., **3**, 689[119]
Nacco, R., **3**, 124[286], 125[286], 127[286]; **5**, 1152[143]
Nace, H. R., **2**, 814[78], 818[78], 823[78,114]; **3**, 758[125], 833[70,71], 854[76]; **6**, 960[55], 1033[127], 1035[127]; **7**, 654[6-8], 657[27]
Nachtwey, P., **3**, 563[117], 582[117]
Nadamuni, G., **8**, 514[109]
Naddaka, V. I., **8**, 410[97]
Nadebaum, P. R., **3**, 636[50]
Nader, B., **2**, 720[85]; **4**, 259[273], 261[273]; **5**, 414[54], 531[82]
Nader, F. W., **4**, 55[156]; **8**, 563[434]
Nader, R. B., **6**, 452[133], 503[219]
Naderi, M., **7**, 561[85], 738[29], 760[27]
Nadi, A.-I., **5**, 766[116]
Nadir, U. K., **1**, 390[142], 838[158]; **4**, 505[132]; **6**, 97[153]; **7**, 162[59,61,64], 741[50], 747[50]
Nadjo, L., **8**, 552[357,358]
Nadkarni, D. V., **6**, 254[160]
Nadolski, D., **6**, 642[66]
Nadolski, K., **6**, 244[110]
Nadzan, A. M., **5**, 832[41]
Naef, R., **1**, 313[119], 314[119]; **2**, 58[13], 120[182], 207[106], 910[66]; **3**, 40[221,222], 41[227], 212[38]; **4**, 172[30], 229[231]
Naegele, W., **4**, 276[77], 284[77], 288[77], 289[77]
Naegeli, P., **2**, 358[153]; **5**, 15[107], 20[139]; **6**, 667[242]
Naemura, K., **1**, 860[68-70]; **8**, 201[143]
Naengchomnong, W., **1**, 558[133]
Näf, F., **2**, 166[155], 185[28]; **3**, 248[56], 251[56], 882[104]; **4**, 91[90], 92[90b], 170[22], 261[287,293]; **5**, 972[8], 989[8]; **6**, 161[185], 1067[106]; **8**, 358[199]; 526[37], 531[37]
Näf, R., **2**, 553[123]
Nafti, A., **2**, 942[167], 943[167], 944[167]; **6**, 69[16], 98[154]
Naga, T., **7**, 203[52]
Nagabhushan, T. L., **4**, 1040[76]
Nagabhushana-Reddy, G., **1**, 403[19]
Nagae, H., **3**, 848[53]
Nagahara, H., **8**, 437[79]
Nagahara, T., **8**, 647[56]
Nagahara, Y., **8**, 535[166]
Nagahisa, Y., **8**, 881[77]
Nagai, H., **6**, 618[112]; **8**, 253[114]
Nagai, K., **8**, 461[261]
Nagai, M., **5**, 524[54], 534[54], 693[111]
Nagai, N., **5**, 963[323]
Nagai, S., **6**, 531[426]
Nagai, T., **2**, 538[66,68], 539[66,68]; **7**, 25[41], 26[41,53,58]
Nagai, W., **2**, 360[166]; **4**, 958[25]; **6**, 121[133], 220[127]
Nagai, Y., **1**, 349[148]; **2**, 603[45,47]; **3**, 729[39]; **4**, 254[189], 257[189], 610[337]; **6**, 801[33,34]; **8**, 36[84], 39[84], 66[84], 173[114-116], 546[305], 555[370], 613[78], 764[5], 770[37,39], 782[103]
Nagakura, A., **8**, 881[80], 882[80]
Nagakura, I., **3**, 212[39]; **4**, 173[35], 176[52]; **5**, 803[91], 955[301], 979[27,28], 980[27], 981[27], 982[27], 992[48]
Nagal, A., **6**, 766[23]
Nagamatsu, T., **4**, 435[138]; **6**, 614[84]
Nagami, K., **1**, 366[48]
Naganathan, S., **4**, 346[85]
Nagano, H., **1**, 848[25]; **6**, 765[19]
Nagano, T., **7**, 759[11]; **8**, 366[52]
Nagao, K., **1**, 846[20], 847[20]; **3**, 784[31]
Nagao, S., **2**, 846[207]; **7**, 537[58]
Nagao, Y., **1**, 894[160]; **2**, 116[131,140], 610[94,95], 611[94,95], 816[83], 819[102], 855[240], 859[252], 1059[78,81-83]; **3**, 125[309], 217[79], 286[56a], 1050[19]; **4**, 42[122b], 817[207]; **5**, 92[81]; **6**, 134[28], 819[108], 936[111], 1065[90b]; **7**, 227[81], 615[9], 621[34], 623[35], 624[36], 710[56], 765[149], 773[149]; **8**, 241[38], 272[119,120], 544[259]
Nagaoka, H., **1**, 182[44], 762[142]; **2**, 29[106], 30[106b], 31[106b], 572[46], 576[74], 744[88]; **4**, 30[88,88h,j], 36[103,103a], 121[207], 253[175], 258[175], 373[70]; **6**, 5[24]; **7**, 406[78a]
Nagaoka, T., **8**, 320[82]
Nagarajan, K., **2**, 894[9], 912[9]; **6**, 712[76]; **7**, 221[32]
Nagarajan, M., **4**, 739[108], 791[47]
Nagarajan, S. C., **3**, 194[15], 196[15]
Nagarajarao, G. K., **8**, 875[32], 876[32]
Nagarkatti, J. P., **7**, 709[47]
Nagasaka, T., **1**, 559[150,151]; **2**, 1049[18]; **7**, 227[76]
Nagasaki, F., **6**, 533[502]
Nagasawa, H., **4**, 124[215]
Nagasawa, H. T., **3**, 849[58]
Nagasawa, J., **5**, 158[43]
Nagasawa, K., **1**, 242[45], 243[58]; **8**, 406[38]
Nagasawa, N., **2**, 124[207]; **4**, 96[104]
Nagase, H., **1**, 738[40]; **4**, 27[79,79a], 391[179]; **6**, 998[117]; **7**, 162[58], 243[66]; **8**, 885[105]
Nagase, K., **8**, 716[92]
Nagase, S., **4**, 47[134], 729[59]; **6**, 510[292]; **7**, 800[34]; **8**, 724[169,169e]
Nagase, T., **1**, 781[230]; **4**, 1038[56,57], 1039[56]
Nagase, Y., **2**, 116[140], 610[94], 611[94], 1059[78]
Nagashima, E., **7**, 173[132]
Nagashima, H., **1**, 642[109], 643[109]; **4**, 93[94], 553[7,9], 597[172], 637[172], 753[171], 837[17]; **6**, 548[669], 1032[118]; **7**, 95[63], 452[43,62], 462[43], 463[126], 465[130]
Nagashima, S., **4**, 610[337]
Nagashima, T., **1**, 553[89,96]; **3**, 49[263], 51[271], 198[47], 396[115], 1033[75]; **6**, 137[42]
Nagasuna, K., **1**, 162[93,100], 163[106]; **5**, 1172[28], 1182[28]
Nagata, J., **5**, 71[131]
Nagata, R., **1**, 758[126]; **5**, 736[142h]; **7**, 381[104]
Nagata, S., **4**, 1023[262], 1024[263,264]
Nagata, T., **4**, 814[186]; **8**, 410[89], 412[114], 994[64]
Nagata, W., **1**, 123[75], 373[82]; **2**, 482[24], 483[24]; **4**, 23[71,71a], 162[92]; **6**, 3[10], 30[10], 745[87], 835[47]; **7**, 476[61]; **8**, 332[50], 340[50], 493[22], 530[100], 836[4], 842[4], 931[39], 993[59]
Nagato, S., **7**, 245[80]
Nagatsuma, M., **2**, 648[96], 649[96], 937[152,154]; **5**, 102[167]
Nagayasu, E., **5**, 637[102]
Nagel, A., **2**, 163[149]; **4**, 10[31,31c]
Nagel, D. L., **1**, 797[283]; **7**, 471[21], 555[69]
Nagel, K., **8**, 736[19], 739[19]
Nagel, M., **2**, 1053[53], 1055[53]; **6**, 501[197], 535[528]
Nagel, U., **6**, 554[772]; **8**, 459[243], 460[250,251], 535[166]
Nagendrappa, G., **2**, 524[78]; **3**, 34[196], 380[4], 386[58]; **4**, 303[343]; **7**, 582[147]; **8**, 477[31]
Nager, M., **4**, 285[163]

Nageshwar, G. D., **2**, 345[22]
Nagibina, T. D., **5**, 752[20,34], 757[20], 768[34]
Nagira, K., **3**, 1026[41]; **4**, 856[99]
Nagl, A., **6**, 744[72]; **7**, 698[51]; **8**, 36[88], 66[88]
Nagler, M., **5**, 139[85], 143[85]
Nagoa, K., **1**, 853[49], 876[49]
Nagubandi, S., **6**, 291[206,220,221], 525[375,378], 529[467], 1017[37]
Naguib, Y. M. A., **4**, 18[59], 121[208], 991[151]; **6**, 176[92]; **7**, 500[241]
Nagumo, S., **6**, 1052[42b]
Nagy, F., **8**, 453[191]
Nagy, J. O., **6**, 122[134], 128[134]; **8**, 657[93]
Nagy, T., **3**, 35[205]
Nagy-Magos, Z., **8**, 554[367]
Nah, H.-S., **2**, 1094[89], 1095[89]
Nahm, S., **1**, 399[4], 405[4]; **4**, 1084[97], 1085[104], 1103[104]; **5**, 3[21]; **8**, 272[121]
Nahum, R., **8**, 532[130]
Naidenova, N. M., **7**, 760[38]
Naidoo, B., **6**, 737[35]; **8**, 315[51]
Naidoo, K., **2**, 844[201]
Naidu, M. V., **2**, 782[24]; **6**, 96[150]
Nai-Jue, Z., **5**, 484[179]
Naik, A. R., **7**, 693[29]
Naik, H. A., **2**, 789[55]
Naik, R. G., **3**, 797[92]; **8**, 447[139], 714[81]
Naiman, A., **5**, 1154[154,159]
Naipawer, R. E., **8**, 430[56], 814[17]
Nair, M., **4**, 1104[212]; **8**, 297[69]
Nair, M. G., **7**, 834[81]
Nair, M. S., **1**, 126[87], 757[122]; **8**, 566[456]
Nair, P., **4**, 837[19]
Nair, V., **4**, 465[122], 466[122,127], 510[163], 1005[80], 1018[80]; **5**, 413[49]; **6**, 517[328,329]; **7**, 506[297]; **8**, 70[231], 351[167], 355[167], 942[118]
Naispuri, D., **8**, 596[84]
Naithani, V. K., **6**, 668[253], 669[253]
Naito, A., **7**, 77[124a]
Naito, H., **8**, 49[116], 66[116]
Naito, S., **3**, 137[376]; **8**, 422[36]
Naito, T., **5**, 134[65]; **8**, 244[54]
Naito, Y., **4**, 8[27], 227[202], 261[297]; **7**, 460[116], 461[117]
Najafi, A., **3**, 700[163]
Najai, C., **3**, 647[194]
Najdi, S. D., **5**, 1062[59]
Najera, C., **3**, 253[90]; **4**, 291[211,212,213,214,215,218], 295[248,249,254], 302[338], 349[110], 351[124], 354[110], 405[251]; **5**, 755[71], 780[71]; **7**, 519[23], 533[35,36], 534[35], 630[53,548]; 8, 856[179,183], 857[187]
Naka, H., **3**, 469[216], 470[216], 476[216]
Naka, K., **3**, 1032[70]; **4**, 558[19]; **8**, 778[85]
Naka, M., **4**, 969[65]
Nakabayashi, S., **6**, 49[96]
Nakada, M., **1**, 57[32]; **5**, 72[166]
Nakada, S., **6**, 531[460]
Nakada, T., **6**, 531[459], 764[9]
Nakada, Y., **5**, 717[94]; **6**, 615[100]
Nakadaira, Y., **2**, 603[44]; **3**, 23[143], 24[143], 380[4]; **5**, 222[65], 223[65]
Nakadate, M., **8**, 390[82]
Nakae, I., **7**, 642[12]; **8**, 783[106]
Nakagawa, A., **6**, 918[37]
Nakagawa, E., **8**, 353[152]
Nakagawa, H., **1**, 561[159]
Nakagawa, I., **6**, 615[100]
Nakagawa, K., **4**, 14[46], 102[131]; **7**, 77[121,122], 229[111], 774[322]
Nakagawa, M., **2**, 152[100], 323[23], 331[23], 332[23], 455[17]; **3**, 245[31], 556[34,36], 585[137]; **4**, 285[164], 289[164], 379[114]; **6**, 914[28], 968[106]; **7**, 96[87], 335[32]
Nakagawa, N., **3**, 201[84]
Nakagawa, S., **7**, 57[22]
Nakagawa, T., **4**, 972[80]; **6**, 801[27]; **7**, 751[138]; **8**, 853[144]
Nakagawa, Y., **1**, 749[80]; **2**, 105[43]; **5**, 855[187]; **6**, 88[103]; **7**, 26[56], 645[21], 797[16]; **8**, 56[166], 66[166]
Nakaguchi, O., **5**, 96[106,117]; **7**, 493[198]
Nakahama, S., **1**, 317[138,139,140,141,142], 390[145], 391[145]; **8**, 18[126], 160[100], 170[82-84,96-101], 176[135], 178[100]
Nakahara, M., **5**, 71[139,140,141,142], 76[233], 77[257]
Nakahara, S., **1**, 860[69]; **5**, 474[156]
Nakahara, Y., **1**, 410[39], 568[231]; **3**, 124[280], 125[280], 126[280]; **4**, 18[62], 20[62i]; **5**, 351[82]; **6**, 51[108], 53[108], 468[53]
Nakahashi, K., **4**, 249[128]; **5**, 86[15,17], 96[107], 98[127]; **8**, 251[104], 253[104]
Nakahata, M., **8**, 150[122], 151[148]
Nakahira, H., **6**, 605[37]; **7**, 245[78]
Nakahira, T., **5**, 219[38]
Nakai, E., **3**, 985[24,25a], 987[24,33], 988[33a], 989[25a], 993[25a]; **6**, 876[32,34], 882[48], 885[34,48], 887[32], 890[48]
Nakai, H., **4**, 373[76]; **7**, 42[29]; **8**, 100[117], 135[42], 545[284]
Nakai, M., **7**, 451[20], 452[20], 454[20]; **8**, 187[40]
Nakai, S., **1**, 88[55]; **4**, 140[11], 209[66]; **6**, 850[125]
Nakai, T., **1**, 584[11]; **2**, 116[141], 119[161], 455[18], 538[68], 539[68], 556[156,157], 558[158,159], 569[35], 635[40], 640[40], 648[96], 649[96], 653[128], 656[153], 657[166], 715[56], 937[152,153,154,155], 1059[76]; **3**, 97[174,175,176], 103[175,176], 108[174], 109[175,176], 117[174], 136[372], 942[80], 976[7], 977[10], 984[22,23], 985[22,25a,26a], 986[28], 987[24,31-33], 988[33a], 989[25a], 992[37,38], 993[25a,38], 994[41], 999[51], 1000[52], 1002[57], 1004[59], 1005[61-63], 1008[65]; **4**, 128[221]; **5**, 16[112], 24[166,167], 102[168], 821[162], 833[49], 839[85], 841[95], 850[147,159], 851[164,165], 888[28,29], 889[31], 1001[16]; **6**, 14[51], 172[15], 780[73,75], 834[30], 846[100], 850[30], 851[129,130], 852[30,136,138], 853[30], 854[144], 856[151], 860[179], 873[7], 874[14], 875[7], 876[26,31,32,34], 877[14,31,36,37], 879[36,41], 882[26,31,47,48], 883[14,36,52,54,55,58], 885[26,31,34,47,48], 886[36], 887[32,58,61,64], 888[7], 890[48,52,69], 891[54,69,70], 892[74], 896[74]; **7**, 263[88]; **8**, 352[147], 934[58], 949[58]
Nakaido, S., **3**, 919[32], 923[43,44], 934[44], 954[44], 1008[70,71]
Nakaji, T., **8**, 364[20]
Nakajima, H., **6**, 77[54]
Nakajima, I., **3**, 437[25], 440[25], 448[25], 449[25], 450[25], 451[25], 452[25], 459[137], 460[137], 461[137], 484[26], 492[26], 494[26], 495[26], 503[26], 510[206], 513[26,206]
Nakajima, K., **6**, 96[149]; **7**, 62[51], 778[404]
Nakajima, M., **1**, 72[69,70], 342[99], 563[171], 845[10]; **2**, 65[29], 830[145]; **3**, 197[41], 199[41]; **6**, 989[78], 993[78]; **7**, 307[11], 438[13], 442[50], 443[13], 749[120]; **8**, 285[7]
Nakajima, N., **1**, 551[70]; **2**, 10[40]; **3**, 1041[112]; **6**, 23[93]; **7**, 246[86]
Nakajima, R., **3**, 501[136]
Nakajima, S., **7**, 340[45], 353[35], 355[35]
Nakajima, T., **3**, 135[345], 174[345], 300[46], 302[46], 313[105], 314[108], 315[113], 318[129], 769[169]; **5**, 438[161], 442[185,185a], 532[84]; **6**, 17[63], 18[63,65]; **7**, 645[19,20]; **8**, 788[121]
Nakajima, Y., **1**, 349[147]; **5**, 473[153], 477[153]; **8**, 170[77]
Nakajo, E., **1**, 113[30], 624[85]; **2**, 23[90,90b], 29[90b], 59[16]; **3**, 225[184], 262[165]; **5**, 473[151], 479[151]; **7**, 643[14]
Nakajo, T., **3**, 901[107]; **8**, 170[95]
Nakakita, M., **4**, 120[201]
Nakakyama, K., **2**, 323[23], 331[23], 332[23]
Nakama, Y., **6**, 989[78], 993[78]
Nakamaye, K. L., **5**, 30[3]

Nakaminami, G., **4**, 379[114]
Nakamine, T., **1**, 797[294]
Nakamizo, N., **6**, 554[719]; **8**, 353[157]
Nakamo, M., **4**, 431[100]
Nakamoto, H., **4**, 249[128]
Nakamoto, Y., **3**, 313[105]
Nakamura, A., **1**, 19[98], 162[93-95,98,100-102], 163[94,106], 164[94], 180[32], 223[74]; **2**, 5[18], 6[18], 24[18,18a,95], 60[17]; **4**, 615[384], 964[49], 1033[21], 1037[21], 1040[21]; **5**, 1148[114], 1172[28-30], 1182[28-30]; **7**, 178[148]; **8**, 450[165], 460[254], 535[166], 971[109], 995[67]
Nakamura, C. Y., **6**, 134[21]
Nakamura, E., **1**, 112[27], 212[12], 213[12], 215[12b,c], 217[12], 327[11], 448[207], 584[10]; **2**, 109[65], 117[149,150], 310[27,28], 441[4], 442[10,12], 443[19], 445[10,27,28], 446[28], 447[12,19], 448[19,40,41,44,45,47], 449[19,40], 450[52,53], 451[56], 452[47,57,58], 576[69], 600[32], 615[128], 630[6], 633[6a,34a,b], 634[34b,35], 640[34,35], 651[112], 714[51,52], 720[52]; **3**, 6[28], 8[28], 14[28], 221[131,132], 455[125], 460[125], 623[34], 727[33]; **4**, 152[55], 163[96], 164[96], 895[162]; **5**, 266[75], 268[75], 310[101], 977[23], 1200[51,52]; **6**, 847[106], 848[110]
Nakamura, F., **8**, 817[30,31]
Nakamura, H., **2**, 73[63]; **4**, 27[79,79a]; **6**, 8[38]; **7**, 761[56]; **8**, 903[51], 906[51], 907[51]
Nakamura, I., **7**, 474[46,47]; **8**, 198[133]
Nakamura, K., **1**, 248[61,62], 512[40], 735[25]; **3**, 147[387], 149[387], 150[387], 151[387]; **4**, 606[306], 607[306], 647[306]; **7**, 168[101]; **8**, 185[18,25], 190[18,65,68], 191[86,87], 195[108], 531[110], 561[417], 909[80], 917[118], 918[118], 919[118], 977[140,141]
Nakamura, K. H., **7**, 802[49]
Nakamura, M., **3**, 136[371], 138[371a], 139[371], 140[371,371a], 143[371,371a], 144[371a]; **6**, 531[459], 764[9]; **7**, 350[27], 355[27], 368[59]
Nakamura, N., **1**, 422[92]; **2**, 547[115], 551[115]; **7**, 543[14]
Nakamura, R., **8**, 447[132]
Nakamura, S., **3**, 573[71], 610[71]; **8**, 496[31]
Nakamura, T., **1**, 212[9], 213[9,15], 215[41], 216[41], 448[207]; **2**, 442[13,14], 449[13,14], 450[13,14,51], 816[83], 819[102]; **3**, 231[242], 420[48], 443[61], 445[61], 463[159]; **4**, 810[167]; **5**, 809[120]; **6**, 509[258], 538[556]; **7**, 239[49], 710[56]; **8**, 174[126], 178[126], 179[126]
Nakamura, T. Y., **5**, 4[29]
Nakamura, W., **1**, 636[101], 640[101], 666[101]; **3**, 136[374], 141[374]
Nakamura, Y., **5**, 637[115], 638[117]; **8**, 149[116], 838[20]
Nakane, M., **2**, 157[121]; **5**, 129[34]
Nakane, R., **3**, 300[38]
Nakanishi, A., **1**, 268[53,53a,c], 269[57]; **4**, 809[165]; **8**, 389[73]
Nakanishi, K., **1**, 187[61]; **2**, 370[257], 482[34], 484[34]; **3**, 23[143], 24[143], 380[4], 575[81]; **4**, 1033[36]; **5**, 222[64,65], 223[64,65]; **6**, 124[144], 711[70]; **7**, 238[43]
Nakanishi, M., **4**, 462[106], 475[106]
Nakanishi, S., **3**, 565[21]; **8**, 394[117]
Nakanishi, T., **4**, 674[35], 688[35]; **6**, 690[399], 691[399], 692[399]; **7**, 667[78]
Nakano, J., **2**, 757[15], 761[49], 762[49]; **8**, 205[156]
Nakano, K., **3**, 35[202]; **5**, 219[39], 230[39]
Nakano, M., **1**, 317[141,142]; **3**, 968[128]; **4**, 431[99]; **6**, 893[86]; **8**, 160[100], 170[100], 176[135], 178[100]
Nakano, N., **4**, 560[27]
Nakano, S., **5**, 79[288]; **6**, 647[112b]
Nakano, T., **4**, 505[134,135]; **6**, 53[120]; **7**, 154[17], 309[26]
Nakano, Y., **4**, 239[26], 251[26], 257[26]; **5**, 71[131], 583[183]
Nakanobo, T., **4**, 115[178]
Nakao, A., **5**, 40[27]
Nakao, H., **2**, 455[17]
Nakao, K., **3**, 789[70]; **7**, 829[56]
Nakao, R., **8**, 248[86]
Nakao, T., **3**, 222[136]; **4**, 972[80]; **6**, 848[108]
Nakashima, H., **4**, 21[66]
Nakashima, K., **1**, 766[160]
Nakashima, M., **5**, 377[110], 378[110a]
Nakashima, Y., **6**, 561[875]
Nakashita, Y., **1**, 145[45], 893[152]; **3**, 677[81], 686[81]
Nakasone, A., **7**, 878[138,143], 888[138a]
Nakata, M., **5**, 812[132]; **7**, 350[23]
Nakata, S., **1**, 853[46]
Nakata, T., **1**, 766[160], 834[121a,122]; **2**, 846[207], 879[42]; **3**, 834[76]; **6**, 5[24,26], 531[460], 979[29], 995[102]; **7**, 57[26]; **8**, 7[38,41,42], 11[38,59], 12[64,65], 500[52]
Nakatana, H., **7**, 208[87]
Nakatani, H., **2**, 580[99]; **5**, 439[170]; **8**, 964[59]
Nakatani, K., **1**, 894[160]; **3**, 1053[29]; **4**, 1056[140]
Nakatani, M., **1**, 436[146]
Nakatani, Y., **2**, 540[70]; **3**, 158[443], 159[443], 160[443], 161[443], 167[443], 168[443]
Nakati, T., **1**, 153[57]
Nakatini, K., **7**, 625[39]
Nakato, E., **4**, 599[214]
Nakatsubo, F., **6**, 147[83], 266[50]; **7**, 169[107], 684[94]
Nakatsugawa, K., **1**, 349[146]; **2**, 603[47]; **8**, 555[370]
Nakatsuji, S., **2**, 152[100]
Nakatsuji, S.-i., **3**, 585[137]
Nakatsuji, Y., **6**, 71[21]
Nakatsuka, M., **1**, 880[113]; **2**, 444[22]; **3**, 450[105]; **4**, 241[60], 254[60], 260[60], 613[370], 840[36]; **5**, 386[134], 391[134], 392[134], 473[151], 479[151], 693[112]; **6**, 757[132], 1007[149]; **7**, 530[29]
Nakatsuka, S., **3**, 261[157]
Nakatsuka, S.-i., **2**, 232[179]; **7**, 678[70]
Nakatsuka, S.-L., **2**, 743[82]
Nakatsuka, T., **3**, 452[107], 460[107]; **6**, 46[73], 658[184], 1000[129]; **7**, 141[132], 209[90]
Nakatsukasa, S., **1**, 188[72], 189[72], 201[98], 202[103], 203[103]; **2**, 20[80], 21[80], 589[154]
Nakawa, H., **5**, 79[288]
Nakaya, T., **3**, 664[32], 665[32]
Nakayama, A., **4**, 1076[38]; **5**, 92[74]; **8**, 934[56]
Nakayama, E., **3**, 919[33], 954[33]; **6**, 897[97]
Nakayama, J., **2**, 368[236]; **3**, 571[74,75], 574[74,75], 586[140], 594[186], 883[110]; **4**, 507[153], 509[161]; **5**, 211[67], 442[181]; **8**, 806[122], 836[3], 916[103], 917[103], 918[103], 919[103], 920[103]
Nakayama, K., **3**, 919[32], 923[44], 934[44], 954[44], 1008[70,71]; **4**, 18[59], 30[88]; **5**, 282[21]; **7**, 362[31], 377[31], 761[56]; **8**, 903[51], 906[51], 907[51]
Nakayama, M., **7**, 100[115]
Nakayama, S., **8**, 33[56], 66[56]
Nakayama, T., **6**, 507[240], 515[240]
Nakayama, Y., **2**, 74[74]; **3**, 135[348], 136[348], 139[348], 141[348], 144[348]; **7**, 564[95], 568[95], 709[37]
Nakazaki, M., **1**, 860[68-70]; **3**, 592[173], 628[49]; **8**, 164[46], 178[46], 179[46]
Nakazawa, K., **6**, 855[146], 893[84]; **8**, 934[56]
Nakazawa, M., **6**, 989[78], 993[78]
Nakazawa, T., **3**, 483[9]; **6**, 753[115], 754[115]; **7**, 743[64]
Nakazumi, H., **6**, 443[92]
Nakomori, S., **7**, 196[29]
Nakonieczna, L., **6**, 493[126], 494[126]
Nakova, N. Zh., **6**, 556[825,826]
Nalesnik, T. E., **7**, 453[79]
Nallaiah, C., **3**, 617[18]
Nally, J., **7**, 729[44]

Nam, D., **4**, 407[256c], 408[258]
Nam, H. H., **4**, 483[9], 485[25]
Nam, N. H., **8**, 726[189]
Namata, H., **1**, 766[155]
Namba, R., **6**, 989[78], 993[78]
Namba, T., **2**, 780[6]; **7**, 10[75], 24[34], 477[81]
Nambiar, K. P., **1**, 430[132]; **5**, 326[24]; **6**, 682[341]
Nambu, H., **3**, 242[6], 257[6], 259[6], 260[143], 794[77,78]; **8**, 951[177]
Nambu, Y., **8**, 369[81-83], 382[9], 395[132], 645[47]
Nambudiry, M. E. N., **1**, 780[229]; **6**, 860[176]
Namen, A. M., **5**, 7[51], 514[6]
Nametkin, N. S., **8**, 778[86]
Nametkin, S., **3**, 725[17]
Namikawa, M., **3**, 402[129]
Namiwa, K., **5**, 623[26]
Namy, J.-L., **1**, 179[25], 231[4,6], 251[1], 252[1], 253[10], 255[16,16b,c,19], 256[16c], 258[16c,19,26,26b], 259[16c,27], 261[16c,27], 265[27], 266[16c,27], 271[19,62], 274[73], 278[16b,c], 751[112]; **3**, 567[35], 570[35], 770[172]; **6**, 980[40]; **8**, 113[48,49], 115[48,49,64], 124[64], 125[64], 552[360], 797[31,32], 889[134]
Nánási, P., **6**, 41[43], 51[103], 660[207]; **8**, 224[103,109], 225[103], 226[112-114], 227[115]
Nanba, K., **8**, 315[47]
Nanbu, A., **8**, 423[40], 429[40]
Nanbu, H., **2**, 19[77,77a]
Nandi, K., **6**, 423[42]; **7**, 775[341], 776[341]
Nangia, A., **5**, 832[42]
Nanimoto, H., **5**, 1200[49]
Naniwa, Y., **4**, 124[215]
Nanjappan, P., **5**, 552[15,32]; **7**, 574[126]
Nanjo, K., **4**, 1000[14]; **8**, 564[439]
Nann, B., **5**, 229[122,123]
Nanninga, T. N., **4**, 593[129,133,134], 870[28]; **5**, 299[65,66], 300[73,76], 302[73], 303[66,80,81], 304[66,80], 307[88], 310[98], 596[37], 598[37]; **8**, 945[126]
Nannini, G., **2**, 284[53]
Nanoshita, K., **5**, 850[154]
Nanri, H., **2**, 725[121]
Nantz, M. H., **1**, 434[140]; **2**, 249[84], 264[58]; **3**, 140[380], 154[380], 168[380], 174[380], 176[380], 209[16], 223[16]
Nan Xing Hu, **7**, 497[219]
Naoki, H., **7**, 73[105]
Naora, M., **8**, 330[47]
Naoshima, Y., **8**, 191[91,93], 353[152]
Naota, T., **1**, 551[70]; **2**, 1052[45]; **3**, 437[23], 1041[112]; **7**, 227[88], 314[40], 315[40]
Nap, I., **3**, 217[82]
Napier, D. R., **3**, 890[31,34], 901[111]; **5**, 385[129a]; **6**, 977[13]
Napier, J. J., **1**, 765[166]
Napier, R., **7**, 95[80]
Napoli, J. J., **5**, 20[133,134], 790[35]
Napolitano, J. P., **3**, 296[12]
Napper, A., **8**, 206[168]
Naqvi, S. M., **3**, 757[121]; **4**, 952[6], 970[71], 1043[107], 1048[107]; **5**, 211[65], 900[9], 901[9], 905[9], 906[9], 907[9], 909[9], 913[9], 947[9], 1006[34]
Nar, H., **1**, 535[144]; **4**, 689[68]
Nara, M., **7**, 27[76]
Narain, R. P., **6**, 563[896]
Naraine, H. K., **4**, 443[187]
Narang, C. K., **8**, 271[110]
Narang, S. A., **6**, 603[22,23], 618[113], 624[136], 625[156,158]
Narang, S. C., **2**, 736[20]; **6**, 104[9], 110[51], 210[75], 214[93], 215[93], 492[124], 647[108,109], 654[152]; **7**, 752[152], 765[141]; **8**, 403[11], 405[25,32], 406[32,47,51], 407[25], 408[73,74], 959[25]
Narasaka, K., **1**, 192[82], 243[56], 314[124-126], 546[55]; **2**, 35[130], 36[130], 68[42], 225[156], 613[112], 614[112], 629[2], 631[2], 632[2], 650[109], 667[43], 816[87]; **3**, 96[168,169], 104[168,169], 108[168,169], 117[168,169], 125[308], 286[58]; **4**, 261[292]; **5**, 24[164,168], 377[110], 378[110a], 850[146]; **7**, 166[89], 442[49], 679[74]; **8**, 9[47], 64[218], 67[218], 899[26], 906[26], 907[26], 913[26], 914[26], 966[74,75]
Narasaka, N., **4**, 158[78], 230[257]
Narasimhan, L., **5**, 1022[76]
Narasimhan, N. S., **1**, 461[7], 463[24,26,29]; **2**, 782[23]; **4**, 810[171]; **8**, 385[47]
Narasimhan, S., **8**, 244[48,53,61,62], 247[48,53], 249[48,93], 253[48,116], 709[43,43a], 875[36]
Narasimhan, V., **7**, 266[108], 267[108]
Narayana, C., **3**, 1017[5]; **8**, 709[44]
Narayana, V. L., **5**, 95[100]
Narayanan, B. A., **1**, 248[71], 735[27], 736[27]; **2**, 78[92], 584[120]; **7**, 226[71]
Narayanan, K., **3**, 497[100], 498[100]; **4**, 904[204]
Narayanan, K. V., **5**, 768[121]; **6**, 836[52]
Narayanan, N., **7**, 267[117], 268[117]
Narayana Rao, M., **5**, 107[202,203]
Narbona, K., **2**, 538[60], 539[60]
Nardin, G., **4**, 170[11]; **5**, 272[2], 275[2]
Nardini, M., **8**, 192[96]
Nared, K. D., **5**, 855[191]
Narimatsu, S., **2**, 580[101]; **3**, 262[164], 466[191]
Narine, B., **2**, 787[50]; **6**, 543[607]
Narisada, M., **1**, 123[75], 373[82]; **6**, 745[87]; **8**, 492[16], 493[22], 508[16], 509[16]
Narisano, E., **1**, 526[100], 527[101,102], 568[244,247]; **2**, 636[53], 639[63], 640[53,63], 642[78], 643[78], 922[101], 923[101], 931[135], 933[135,136], 934[135,136], 940[135,136]; **5**, 100[153], 102[177]; **6**, 118[107], 149[96,100]; **8**, 187[45]
Narita, K., **5**, 504[276]
Narita, M., **2**, 917[85]
Narita, S., **5**, 623[28]
Naritomi, T., **2**, 482[21]
Narr, B., **3**, 666[43]
Naruchi, K., **4**, 315[526]
Naruchi, T., **3**, 829[57]
Narula, A. S., **1**, 99[87]; **4**, 75[43b], 100[43], 670[16,17], 674[17], 688[65]; **6**, 774[46]; **7**, 247[101], 368[60]
Narula, C., **4**, 853[91], 854[94]
Naruse, K., **1**, 448[205]; **2**, 749[135]; **3**, 227[211], 243[16], 245[30,32], 470[213], 476[213]; **4**, 902[190]
Naruse, M., **2**, 589[157]; **3**, 274[22], 799[100]; **7**, 601[85]
Naruse, N., **6**, 658[186a]; **7**, 700[62]
Naruse, Y., **2**, 269[72], 615[126], 631[18]; **8**, 660[107]
Naruta, A. S., **7**, 842[28]
Naruta, Y., **1**, 158[74], 179[24], 180[38], 181[38]; **2**, 4[12,12a], 6[12], 566[18], 573[18], 574[18], 977[8]; **4**, 27[83], 155[70]; **5**, 936[198], 963[323]; **7**, 408[88b], 427[148e]
Naruto, M., **6**, 658[186a]
Narwid, T. A., **5**, 835[59]
Nasada, H., **8**, 568[475]
Násái, P., **6**, 660[207]
Naser-ud-Din, **3**, 635[33], 640[107,107a], 643[120], 644[120,134c], 647[33,107]; **5**, 113[235]; **8**, 974[128]
Nash, S. A., **7**, 452[57], 462[123], 571[119], 577[119]; **8**, 542[221]
Nashed, N. T., **6**, 423[49]; **7**, 362[26]
Nasiak, L. D., **8**, 780[90]
Nasielski, J., **7**, 614[6]; **8**, 451[182]
Nasipuri, D., **3**, 362[82]
Nasman, J. H., **1**, 472[77]

Naso, F., **1**, 452[220]; **3**, 208[2], 217[2], 230[235,236], 436[11], 441[48,49], 446[87], 449[48,101], 485[29], 492[79], 493[29], 503[29,79], 513[29,79]; **4**, 93[93d]
Nassr, M., **2**, 760[38]; **4**, 311[441]; **6**, 54[129]; **7**, 635[71]; **8**, 854[150]
Nasuno, I., **6**, 566[929]
Natale, N. R., **1**, 231[7], 251[1], 252[1], 359[22], 383[22], 384[22]; **4**, 206[43,44]; **7**, 841[18]; **8**, 26[16], 27[16], 37[16], 47[16], 55[16], 60[16], 70[16], 347[139], 349[139], 350[139,149], 351[149], 352[139], 354[149], 355[149], 357[195], 358[195], 394[115]
Natalie, K. J., Jr., **1**, 447[201]; **2**, 749[131]; **5**, 798[68]
Natarajan, G. S., **5**, 1154[153]
Natarajan, R. K., **3**, 380[7]
Natarajan, S., **2**, 578[85], 894[9], 912[9]; **4**, 189[103]
Natatini, K., **7**, 624[37]
Natchev, I. A., **2**, 1097[102]
Natchus, M. G., **2**, 282[34], 286[34], 287[34]; **4**, 1040[88], 1048[88,88c]; **5**, 909[98], 925[152], 955[302], 957[310,311], 987[42], 993[42,52], 994[42,52]
Natekar, M. V., **3**, 878[91]
Nath, B., **8**, 532[133]
Nathan, W. S., **7**, 666[70]
Nath Dhar, D., **5**, 105[190]
Natile, G., **7**, 777[386]
Nations, R. G., **3**, 825[29]; **8**, 141[30]
Nativi, C., **3**, 356[57]
Natori, Y., **8**, 154[194]
Natsugari, H., **5**, 426[104]; **6**, 906[148]; **7**, 486[143]
Natsume, M., **1**, 412[53]; **2**, 1068[129,130,131]; **3**, 512[198]; **5**, 439[168]; **8**, 588[45]
Natu, A. A., **4**, 231[264]
Natu, A. D., **7**, 384[112]
Nau, P. F., **3**, 851[65]
Näumann, F., **2**, 486[39]
Naumann, K., **8**, 411[102]
Navasaka, K., **8**, 840[35]
Naves, Y.-R., **3**, 349[36]; **4**, 283[150]
Nawa, M., **8**, 535[166]
Nawata, Y., **3**, 172[516], 173[516]; **7**, 362[31], 377[31]
Nayak, U., **5**, 798[4]
Nayak, U. R., **2**, 142[47]; **3**, 406[144], 600[214]; **8**, 330[48], 333[52], 334[61], 338[94], 339[90,93], 851[135], 943[120]
Naylor, C. A., Jr., **8**, 254[124]
Naylor, R. J., **2**, 760[43]
Nazarov, D. V., **7**, 660[39]
Nazarov, I. N., **2**, 143[50]; **4**, 30[87]; **5**, 752[1-38], 754[32,35,36], 756[35,36], 757[7,11,17,18,20,22,27,30], 759[26], 767[4,6,8-11,13-15,19,22-26,31], 768[11,25,31,34]; **8**, 526[35], 530[35]
Nazarov, J. N., **7**, 660[39]
Nazarova, E. B., **2**, 854[236]
Nazarova, N. M., **8**, 447[103,104], 450[103,104]
Nazeer, M., **5**, 420[76]
Nazer, B., **8**, 14[84], 18[120], 26[20], 27[20], 36[20], 54[20], 55[20], 60[20], 70[20], 237[16], 240[35], 244[16], 253[16], 261[12,13], 412[111]
Ncube, S., **1**, 569[258]; **3**, 124[261], 126[261]
Ndebeka, G., **1**, 563[184]
Ndibwami, A., **3**, 380[10]; **5**, 532[86]
Neal, G. W., **8**, 237[13], 240[13], 244[13], 253[13], 357[192]
Neale, R. S., **7**, 505[291]
Nealy, D. L., **3**, 381[17]
Neber, M., **6**, 680[332], 681[332]
Nebout, B., **4**, 21[69]
Nechiporenko, V. P., **3**, 648[175]
Nechvatal, A., **3**, 346[23]
Nechvatal, G., **1**, 463[22]
Neckers, D. C., **4**, 721[30], 725[30]
Necula, A., **3**, 331[196]
Nedelec, J.-Y., **7**, 727[39]
Nedelec, L., **3**, 21[131]
Nederlof, P. J. R., **3**, 153[414], 155[414], 373[129]
Née, G., **2**, 818[93], 855[93]
Needham, L. L., **4**, 369[20], 374[20]; **6**, 650[132]; **8**, 343[114]
Needleman, S. B., **5**, 451[16], 470[16]
Neef, G., **2**, 167[160], 360[171], 902[40]; **4**, 10[34]; **7**, 65[69], 383[111]; **8**, 331[32], 847[92]
Neeland, E. G., **4**, 5[17]; **5**, 485[184]
Neeman, M., **3**, 734[13]
Neenan, T. X., **3**, 557[46]
Neese, H. J., **1**, 149[47]
Neeson, S. J., **3**, 955[106]; **6**, 897[99]
Nef, J. U., **2**, 321[4], 324[4], 1083[5]; **6**, 911[12,13]; **7**, 218[2]
Nefedov, B. K., **3**, 1039[97]
Nefedov, O. M., **4**, 489[63], 963[43], 964[47], 1058[150], 1059[153], 1063[172]; **5**, 65[55], 1056[49], 1198[45]
Nefkens, G. H. L., **6**, 643[77], 667[235]
Negashi, E., **3**, 104[211], 107[211], 111[211]
Negishi, A., **3**, 202[88]; **6**, 845[95], 865[95]
Negishi, E., **1**, 77[2,4], 89[56], 95[80], 143[29], 162[96], 212[3], 214[3,23,27], 222[3], 448[206], 749[81]; **2**, 120[185], 449[48], 584[123], 586[123], 713[45], 726[125]; **3**, 7[32], 8[32,38], 12[55-58], 21[128], 208[1], 230[239], 231[241,243], 232[239a], 233[1b], 238[239a], 251[78,99,100], 254[78,99,100], 259[138], 260[139,140], 266[195,196], 274[24], 436[12], 440[44], 443[51-55], 450[103], 453[53,103], 460[142], 463[156], 469[198,199,200], 470[198,199,200,210], 472[199], 473[198,199,200], 475[200,210], 485[37], 486[37,40-43], 489[57], 491[37], 495[37,40,43,57,91], 497[40,105], 498[40,42], 503[37,40,150,151], 521[9], 523[23], 524[9,35,36], 530[9,56], 793[71,73,74], 795[81,82,85], 799[101], 1025[35], 1030[35]; **4**, 72[22], 139[5], 146[37a,b], 164[99,99b,c], 249[132,133], 250[133], 256[216], 591[109], 595[154], 602[263], 606[304], 619[154], 620[154], 633[109], 635[154], 644[263], 756[183], 854[92], 857[107], 866[1,2], 867[1,2,5], 883[5,102], 884[102-104,107,107b], 889[135,136], 890[136,138], 891[107b,139], 892[104,136,140,141,145,146], 893[107b,136,147,149,150,151,152], 895[161], 903[198]; **5**, 32[6,6a,c], 47[40], 1037[5], 1124[47], 1125[63], 1163[1,6], 1165[1,11-15], 1166[11,15-17,19,21,23,25], 1167[11,15-17], 1170[15], 1171[14,15], 1172[14], 1174[14], 1175[14,15,38,41], 1177[43], 1178[1,11,14-16,38,41,43], 1179[15], 1180[47], 1181[47], 1182[48]; **6**, 14[51], 848[111], 966[95]; **7**, 594[5], 595[5], 596[38], 598[5]; **8**, 261[9], 263[9], 269[9], 275[9], 278, 675[42], 677[42], 678[42], 680[72], 681[42], 683[72], 685[42], 690[104], 693[72,104,108,112], 697[42], 706[15], 707[24,26-28,30], 708[30], 709[15,15a,47,48], 710[52], 717[27,28,98-100], 724[179], 755[116,117,120-122,125,128], 756[147,148,160], 757[125], 758[117,121,122,125], 801[72], 950[171], 960[34]
Negishi, Y., **3**, 454[120]; **4**, 446[213]
Negoro, K., **4**, 8[30a], 102[131]; **5**, 464[108,109], 466[109]; **6**, 1022[56]; **7**, 773[300]
Negoro, T., **4**, 347[95]
Negre, M., **5**, 527[64], 530[64]
Negrebetskii, V. V., **4**, 992[154]
Negri, D. P., **2**, 219[143], 496[68], 497[68]; **7**, 380[103]; **8**, 11[60]
Negrini, E., **3**, 511[187]
Negron, G., **4**, 1089[126]
Nehl, H., **1**, 180[30], 214[23]; **4**, 880[88-90], 883[90], 884[90]; **5**, 31[4]
Nehrings, A., **5**, 160[57,58], 185[162,167]
Neidert, E., **5**, 66[80], 129[35]
Neidlein, R., **1**, 466[45]; **2**, 780[10], 1088[52]; **3**, 873[64]; **4**, 1006[97]; **6**, 238[71], 509[247], 531[458], 564[912], 570[953]
Neier, R., **8**, 597[86]
Neill, A. B., **2**, 747[121]
Neill, D. C., **8**, 187[36]

Neilson, D. G., **6**, 488[15,16], 507[16,230], 529[15,16], 532[16], 533[15,16], 534[16], 536[16], 537[16], 545[15], 562[15,16]; **8**, 33[57], 66[57]
Neilson, T., **8**, 369[76]
Neimeyer, C., **2**, 240[8], 249[84]
Neisser, M., **7**, 205[65]
Neizel, J. J., **6**, 67[12]
Nel, M., **4**, 599[211], 640[211]
Nelan, D. R., **3**, 698[158]
Nelander, D. H., **3**, 833[71]; **7**, 657[27]
Nelke, J. M., **2**, 602[41]; **3**, 613[1], 614[1,5], 615[1], 616[1], 617[5a,14], 619[1], 620[1,14], 621[1], 622[1], 623[1], 625[1], 626[1], 627[1], 628[1], 629[1], 630[1]; **4**, 55[156,156n]; **8**, 240[29], 243[29], 266[58]
Nellans, H. N., **7**, 230[124]; **8**, 50[120], 66[120]
Nelsen, S. F., **2**, 138[18], 240[9]; **5**, 68[96], 430[117], 576[145]; **7**, 40[14], 851[24], 860[75]; **8**, 70[235], 71[235], 388[62]
Nelson, A. J., **4**, 426[53]; **6**, 557[834]
Nelson, A. L., **3**, 304[63]
Nelson, C. H., **7**, 85[6]
Nelson, D. A., **8**, 345[127]
Nelson, D. J., **1**, 287[19]; **3**, 352[43b]; **7**, 498[226], 503[226]; **8**, 713[73], 724[158,166,170], 726[194]
Nelson, G. L., **4**, 91[90]; **5**, 561[81], 1016[61]; **6**, 689[385]
Nelson, G. O., **4**, 115[180c], 688[67]; **5**, 1089[80]
Nelson, H. H., **8**, 964[50]
Nelson, J. A., **2**, 169[164]; **3**, 55[282]; **8**, 884[97]
Nelson, J. D., **4**, 868[17], 869[17]
Nelson, J. E., **8**, 685[98], 689[98]
Nelson, J. F., **1**, 226[89]
Nelson, J. V., **1**, 227[98], 358[10], 398[3], 399[3], 424[99], 880[116]; **2**, 2[4], 91[49], 100[15], 101[15], 111[15], 112[15,99,100], 113[15], 134[3], 190[57], 197[79], 239[2], 240[2], 242[20], 245[20f,31,32], 246[20f,31], 247[31], 436[67], 455[7], 475[1], 894[10], 917[10], 918[10], 919[10], 930[10]; **4**, 98[110], 145[35]; **5**, 788[13], 821[13], 847[133], 872[133], 1000[3,4]; **6**, 834[39]; **8**, 948[145]
Nelson, J. V. J., **7**, 602[96]
Nelson, K. A., **4**, 974[88]
Nelson, L. E., **2**, 588[152]; **8**, 565[447], 764[2], 770[2b]
Nelson, N. A., **5**, 157[41]; **6**, 1013[15]; **8**, 497[40]
Nelson, N. R., **7**, 516[6]
Nelson, P., **4**, 932[62]
Nelson, P. H., **6**, 176[104]
Nelson, R. E., **3**, 242[2]
Nelson, R. F., **8**, 591[62]
Nelson, R. P., **4**, 8[29,29b]; **6**, 715[91]
Nelson, R. V., **2**, 353[96], 388[96]; **5**, 789[30]
Nelson, S. J., **8**, 934[54], 938[54], 993[57]
Nelson, T. R., **8**, 797[37]
Nelson, V., **2**, 1088[40], 1097[40]; **4**, 1015[199]
Nelson, W. K., **1**, 139[2], 140[8]
Nemery, I., **2**, 651[115,115a]; **6**, 164[195]
Nemo, T. E., **7**, 11[92]
Nemoto, H., **1**, 553[90-92,94,95]; **2**, 888[54], 1073[142]; **3**, 51[271], 164[480], 165[480], 167[482], 168[482], 198[48], 226[200], 390[79], 1008[72], 1009[72], 1010[72,75]; **4**, 181[70], 501[113]; **5**, 385[130], 524[54], 534[54], 691[83,84], 692[83,83c,84,96], 693[83,107,111,114], 694[114], 841[88,98], 843[117], 847[136], 1031[95], 1032[100]; **6**, 4[19], 780[70], 875[18], 879[43]; **7**, 452[62]
Nemwcek, C., **7**, 777[378], 778[378]
Nenitzescu, C. D., **2**, 708[2], 710[2], 711[2], 727[2]; **3**, 295[10], 321[137]; **5**, 478[162]; **7**, 5[20,25]; **8**, 91[53,61]
Nentwig, J., **4**, 293[236]; **6**, 273[92], 277[92]
Nenz, A., **7**, 500[239]
Nepomnia, V. V., **8**, 780[91]
Nerdel, F., **2**, 598[13]; **4**, 1020[229]; **5**, 388[137]; **6**, 1042[5]
Neri, O., **6**, 980[34]; **8**, 891[147]
Nerlekar, P. G., **8**, 770[31]
Nerz-Stormes, M., **2**, 117[153], 224[152], 232[152], 308[21]
Nes, W. R., **3**, 406[143]
Nesbitt, S. L., **6**, 147[87]
Nesi, R., **4**, 958[28]
Nesmeyanov, A. N., **1**, 142[24], 143[31], 820[4]; **3**, 86[43], 179[43]; **4**, 115[180b], 315[518], 521[40,41], 529[73,76], 701[28], 702[28], 987[133]; **7**, 596[32], 606[160,161], 632[57]; **8**, 851[132]
Nesmeyanova, O. A., **4**, 885[114]; **5**, 33[8]; **7**, 595[20]
Nesovic, H., **2**, 553[128]
Nesser, J. R., **4**, 347[48]
Nesterov, G. A., **8**, 753[68]
Nesterova, T. N., **3**, 304[65]
Nestle, M. O., **6**, 692[406]
Nestler, G., **3**, 781[14]; **6**, 1061[72]; **7**, 588[172,173]
Nestrick, T. J., **8**, 320[88,90], 440[82], 551[339]
Nesty, G. A., **5**, 752[41-43]
Neszmály, A., **6**, 41[43]
Neszmélyi, A., **6**, 660[207]
Neta, P., **7**, 850[10]
Nethercott, W., **4**, 1015[201]
Netscher, T., **8**, 813[10]
Netzel, M. A., **2**, 73[62]
Neubauer, D., **4**, 939[74]
Neubauer, H.-J., **3**, 53[274]; **6**, 531[434]
Neuberg, C., **8**, 187[35]
Neuberg, M. K., **8**, 459[243]
Neubert, M. E., **6**, 441[85]
Neuenschwander, K., **2**, 1060[85]; **3**, 257[115]; **8**, 248[86]
Neuenschwander, M., **4**, 1007[120], 1008[134], 1009[135]; **5**, 632[59]; **6**, 562[889,890]
Neufang, K., **2**, 139[31]
Neugebauer, D., **5**, 633[66], 1090[88], 1092[92]; **6**, 190[202], 196[202]
Neugebauer, F. A., **7**, 736[2], 745[2]
Neugebauer, W., **1**, 19[102], 20[108], 25[128]
Neuhaus, D., **2**, 185[29]
Neukam, W., **5**, 552[37], 847[136]
Neuklis, W. A., **3**, 790[59]
Neukom, C., **3**, 168[493], 169[493], 171[493], 289[69]; **5**, 893[41]; **8**, 545[291]
Neuman, W. P., **8**, 36[77], 66[77]
Neumann, B., **2**, 785[43]
Neumann, F. W., **8**, 905[63]
Neumann, H., **3**, 247[47]; **6**, 644[81]; **8**, 249[90]
Neumann, H. C., **2**, 839[173]
Neumann, H.-J., **8**, 824[62]
Neumann, H. M., **1**, 78[17]
Neumann, M., **2**, 372[266]; **5**, 468[128]
Neumann, P., **3**, 414[2]
Neumann, R., **2**, 772[14]; **6**, 526[400,401]; **7**, 50[73]
Neumann, S. M., **3**, 494[84]; **5**, 1076[35]
Neumann, W., **6**, 501[201]
Neumann, W. L., **1**, 416[65]; **3**, 792[65]; **4**, 89[86b]
Neumann, W. P., **1**, 328[19]; **2**, 564[2], 572[2], 576[2], 587[141], 609[82]; **3**, 571[59]; **4**, 735[80], 738[80], 760[196], 770[80,247], 791[43]; **8**, 21[142], 264[46], 265[50], 278, 548[319,320], 754[112], 755[112], 794[9], 798[9], 818[38], 845[77], 849[77c]
Neumayer, E. M., **6**, 270[77]
Neumeister, J., **7**, 574[140], 579[136], 581[140], 582[140]
Neumeyer, J. L., **8**, 587[40]
Neumüller, O. A., **7**, 96[87]
Neunhoeffer, H., **4**, 1095[154]; **5**, 491[207], 496[219,220,221], 497[223,224,226], 498[230,233], 583[189]; **6**, 515[309], 612[72], 614[91]

Neunteufel, R. A., **7**, 879[150]
Neupokoev, V. I., **8**, 150[132]
Neureiter, N. P., **5**, 907[77]
Neürrenbach, A., **4**, 920[22], 921[22], 923[22], 924[22], 925[22]
Neuse, E. W., **5**, 732[132,132b]
Neuss, N., **2**, 143[54]
Neustern, F.-U., **5**, 209[54]
Neuth, J. F., **6**, 111[64]
Neuwirth, Z., **2**, 844[200]
Nevalainen, V., **2**, 346[45]
Nevedov, O. M., **4**, 1058[151]
Nevell, T. P., **3**, 706[5]
Neville, D. M., Jr., **8**, 52[136], 66[136]
Neville, O. K., **3**, 822[10,13], 825[28], 829[10,13], 830[58], 831[58]
Nevitt, T. D., **4**, 279[118], 280[118]; **6**, 835[42]
New, J. S., **8**, 253[121]
Newallis, P. E., **6**, 431[107]
Newaz, S. S., **2**, 355[119], 382[314]
Newberg, J. H., **8**, 125[93]
Newbold, B. T., **8**, 364[17], 365[17], 382[1], 383[1], 390[1]
Newbold, G. T., **8**, 528[82], 529[82]
Newbold, R. C., **5**, 21[159,160], 23[159,160]
Newbould, J., **4**, 83[65a]; **7**, 111[190]
Newburg, N. R., **7**, 24[38]
Newcomb, M., **1**, 357[1]; **2**, 102[24], 476[4], 489[48], 490[48], 505[11], 510[11,37,38,40-42], 917[89], 919[89], 920[89], 924[89], 935[89]; **3**, 30[181], 31[185], 32[185], 34[193], 824[23]; **4**, 221[157], 719[19], 722[19], 723[37,40,41], 747[40], 757[185], 776[40], 785[23], 803[130,135], 811[173,175], 812[175,176]; **5**, 100[147], 790[35]; **6**, 442[87], 531[448], 719[125], 720[125,130], 722[139], 723[146], 725[125]; **8**, 802[84,85], 947[142]
Newcombe, P. J., **4**, 424[20]
Newington, I. M., **1**, 477[146], 545[48]; **8**, 387[56]
Newitt, D. M., **2**, 348[65]
Newkirk, J. D., **3**, 842[20]
Newkom, C., **1**, 821[29]
Newkome, G. R., **1**, 568[238]; **2**, 504[1]; **3**, 509[179], 587[143]; **4**, 462[105]; **5**, 635[85]; **6**, 507[232]; **8**, 113[42]
Newland, M. J., **8**, 775[77]
Newlander, K. A., **7**, 543[12], 551[12]
Newman, B. C., **8**, 231[147], 843[57]
Newman, H., **2**, 725[116]; **8**, 756[149]
Newman, L. W. J., **4**, 283[146]
Newman, M. S., **1**, 423[96], 468[52,53], 844[5e]; **2**, 279[14], 283[50], 409[2], 410[2], 411[2], 749[130]; **3**, 814[70,71], 888[17], 890[32], 891[39], 892[39], 894[61], 898[39,78]; **4**, 7[24], 89[84b]; **5**, 382[119a]; **6**, 677[315], 968[112]; **7**, 295[21]; **8**, 297[67], 749[61], 918[121], 919[121], 950[166], 972[116]
Newman-Evans, D. D., **8**, 819[42], 820[42]
Newport, G. L., **7**, 877[133]
Newton, B. N., **4**, 429[83], 438[83], 441[83]
Newton, C., **3**, 369[125], 372[125]
Newton, C. G., **5**, 528[67]
Newton, D. J., **4**, 872[39,39b]
Newton, M. D., **4**, 4[14,14a], 484[12]
Newton, M. G., **7**, 753[159]
Newton, R. F., **3**, 290[70]; **4**, 260[281], 385[146], 413[146]; **5**, 829[25], 1043[24], 1046[24], 1049[24], 1051[24,36b]; **6**, 655[165], 1024[76]; **7**, 302[63], 674[33], 682[85], 766[178]; **8**, 198[135]
Newton, R. J., Jr., **8**, 357[203], 359[203], 726[186]
Newton, S. A., Jr., **8**, 371[108]
Newton, T. W., **2**, 589[153]; **4**, 257[217], 901[183]; **5**, 762[105]; **8**, 769[24], 771[24]
Neyer, G., **2**, 1090[73], 1100[118], 1101[118], 1102[73], 1103[73,118b]
Neyer, J., **7**, 230[133]
Nezhat, L., **4**, 116[187]
Nezu, Y., **1**, 389[137]; **2**, 994[38]
Ng, C. T., **8**, 840[33]
Ng, D. K. P., **8**, 847[95]
Ng, G. S. Y., **7**, 316[47], 317[47]
Ng, H. C., **5**, 161[59]
Ng, J. S., **3**, 209[17], 216[68], 224[168]; **4**, 175[43]
Ng, K.-K. D., **4**, 878[76]
Ng, K. S., **4**, 369[22], 370[22], 377[22]; **8**, 683[91]
Ng, L. K., **1**, 476[120]; **3**, 66[9]
Ng, S. Y.-W., **6**, 247[131]
Ngochindo, R. I., **1**, 471[65,66], 474[83]
Ngoviwatchai, P., **4**, 744[132], 746[143], 747[148], 771[254]; **6**, 832[14]
Nguyen, C. H., **7**, 766[181]
Nguyen, D., **3**, 216[68], 224[168]
Nguyen, D. H., **2**, 609[79], 1048[9], 1049[9], 1050[9], 1064[9]
Nguyen, N. H., **3**, 503[149], 512[149]; **5**, 1151[133,136]; **8**, 102[125]
Nguyen, N. V., **5**, 90[57], 95[57], 690[80], 733[136]
Nguyen, S., **5**, 805[100]
Nguyen, S. L., **1**, 107[6], 110[6], 343[116]; **3**, 224[167], 264[186]; **4**, 176[50], 177[50], 903[192]; **6**, 5[23]
Nguyen, T., **4**, 1020[236]
Nguyen Thi, K. H., **6**, 540[585,586]
Nguyen-van-Duong, K., **4**, 746[147]
Nhu Phu, T., **8**, 451[180]
Ni, J., **5**, 636[99]
Ni, Z., **4**, 599[222], 641[222]
Ni, Z.-J., **8**, 842[41]
Nibbering, M. N., **8**, 89[43]
Nicaise, O., **8**, 388[60]
Niccolai, G. P., **6**, 291[211]
Nicely, V. A., **8**, 524[13,13c]
Nichikova, P. R., **3**, 643[131]
Nicholas, D. L., **8**, 364[13]
Nicholas, K., **5**, 272[5]
Nicholas, K. M., **2**, 655[144]; **3**, 216[71]; **4**, 304[355], 695[4], 956[20]; **5**, 1055[44]; **6**, 690[391], 692[407,408]
Nicholas, P. P., **7**, 479[88]
Nicholls, B., **4**, 301[325], 315[508], 519[15], 522[15]; **7**, 236[17]
Nichols, D. E., **8**, 146[97], 368[72], 375[72]
Nichols, M. A., **1**, 41[203]
Nicholson, A. A., **5**, 125[16]
Nicholson, E. M., **5**, 949[283]
Nicholson, J. K., **4**, 588[54]
Nickel, S., **3**, 582[112]
Nickel, W.-U., **7**, 358[14]
Nickell, D., **8**, 395[131]
Nickell, D. G., **5**, 420[77], 576[143]
Nickelson, S. A., **4**, 276[69,70], 283[70]
Nickisch, K., **4**, 155[73]; **7**, 74[111], 75[111], 95[78]; **8**, 881[72], 882[72]
Nickolson, R., **7**, 74[111], 75[111]; **8**, 881[72], 882[72]
Nickon, A., **1**, 856[55]; **2**, 441[1], 443[1]; **3**, 164[477], 386[66], 709[14], 946[93]; **4**, 2[6]; **6**, 779[65], 961[73]; **7**, 96[87,88], 97[88], 98[88], 110[88], 111[88], 165[82], 178[82]; **8**, 335[51], 338[88], 345[128], 828[80]
Nickson, T. E., **8**, 370[97]
Niclas, H. J., **4**, 434[126]
Nicodem, D. E., **1**, 411[45]
Nicolaides, D. N., **6**, 173[45], 175[72]
Nicolaidis, S. A., **7**, 699[56]
Nicolaou, K. C., **1**, 227[97], 408[35], 409[36,37], 430[35], 630[1], 672[1], 772[199], 779[226], 808[319], 876[99]; **2**, 42[148], 45[148], 388[345]; **3**, 39[216], 86[56], 94[56], 117[56], 217[94], 220[119], 224[174], 278[31], 288[63], 289[31,68], 558[51,52], 618[20,21],

751[90], 883[106,107]; **4**, 255[192], 260[192], 317[559], 339[40], 370[31,32], 371[32,60], 372[31,57,60,64a], 377[105b], 381[105], 397[31], 413[271a,b,272,273], 795[87]; **5**, 386[135], 534[92], 687[67], 736[145], 737[145], 743[164], 744[164]; **6**, 2[8], 8[39], 22[8], 46[59,63,76], 47[76], 48[76], 51[104], 61[150], 76[42], 77[56], 158[184], 206[45], 438[51,59,64], 448[106-109], 466[44,45], 469[44,45], 660[202], 918[38], 1031[113,115]; **7**, 131[78], 245[72], 254[26], 396[25], 401[56,61d], 407[84b], 408[88c], 515[1], 517[16], 522[41], 523[1,41,46], 524[49,54], 678[73]; **8**, 384[24], 836[10b], 846[81], 847[10b,99,102], 848[10b,99], 849[10b,99,102], 881[68]
Nicoletti, R., **4**, 1021[246]; **6**, 490[109]
Nicoll-Griffith, D., **1**, 751[89]
Nicoloff, N., **2**, 210[110]
Nicotra, F., **4**, 38[108], 379[115,116,118], 380[115j], 382[115k]; **7**, 274[138]
Nidy, E. G., **4**, 384[143]; **6**, 2[8], 22[8]; **7**, 340[46], 393[16], 398[16], 633[65]
Nie, P.-L., **8**, 827[74]
Niedermeyer, U., **8**, 185[30]
Niedernhuber, A., **5**, 1062[59]
Niederprüm, H., **2**, 183[13]
Nieh, E., **4**, 492[74]
Nih, M. T., **3**, 583[119]; **6**, 686[369], 985[65]; **8**, 888[121]
Niel, G., **6**, 864[195]
Nielsen, A. T., **2**, 110[74], 133[1], 134[1], 136[15], 138[23], 147[77], 149[87], 150[93], 167[23a,b], 323[20]; **4**, 78[52b], 425[27]; **6**, 104[9], 533[475]; **8**, 597[88]
Nielsen, F. E., **4**, 436[140]
Nielsen, H. C., **5**, 531[74], 532[74a]
Nielsen, P. H., **6**, 477[101], 478[101]
Nielsen, R. B., **4**, 485[30]; **8**, 675[41], 679[41], 683[95], 684[41], 686[95]
Nielsen, S. D., **7**, 85[8], 100[8]
Nielsen, S. W., **7**, 92[47]
Nielsen, T., **1**, 292[26]
Niem, T., **3**, 369[122], 372[122]
Niemann, C., **3**, 889[24]
Niemczyk, M., **5**, 166[91]
Nieminen, T. E. A., **8**, 18[125]
Nienhouse, E. J., **7**, 601[81]
Nierenstein, M., **3**, 888[16]
Nierlich, M., **5**, 410[41]
Niermann, H., **8**, 548[319]
Nierth, A., **6**, 1012[4], 1013[4]
Nieto Sampedro, M., **1**, 759[132]
Nietzki, R., **3**, 828[43]
Nieuwenhuizen, M. S., **1**, 294[46]
Nieuwenhuyse, H., **6**, 679[325]
Nieuwland, J. A., **3**, 273[9]; **4**, 285[159], 303[351], 315[522]
Nieves, I., **8**, 936[74]
Niewind, H., **8**, 528[85]
Nigh, W. G., **3**, 551[4], 552[4]; **7**, 120[7], 851[20]
Nightingale, D. V., **2**, 329[49], 740[59]
Nigita, T., **3**, 919[32]
Nigrey, P. J., **6**, 510[291]
Nihira, T., **7**, 73[105]
Nihonyanagi, M., **2**, 603[47]; **8**, 555[370]
Nii, Y., **6**, 233[48]
Niibo, Y., **1**, 511[32]; **4**, 113[174], 245[86], 259[86], 260[86]
Niijima, J., **5**, 442[181]
Niimi, K., **8**, 190[80]
Niimura, K., **1**, 386[122]
Niiyama, K., **1**, 738[40]; **4**, 391[179]; **6**, 998[117]; **7**, 162[58], 243[65,66]
Nijhuis, W. H. N., **2**, 379[295]; **8**, 98[103-105]
Nikado, N., **4**, 598[209], 638[209]
Nikaido, M., **2**, 66[32], 186[35]
Nikaido, M. M., **1**, 733[13]; **7**, 549[46]
Nikaido, T., **8**, 205[156]
Nikam, S. S., **2**, 84[12], 88[28]; **4**, 395[204]; **8**, 384[40]
Nikawa, J.-I., **8**, 244[70]
Nikolaev, V. A., **3**, 897[73]
Nikolaeva, I. N., **2**, 785[42]
Nikolaeva, N. A., **4**, 145[28]
Nikolajewski, H. E., **2**, 791[61]; **6**, 727[202]
Nikolic, N. A., **2**, 558[161]
Nikolova, M., **2**, 971[95]
Nikrad, P. V., **8**, 367[61]
Nikulin, A. V., **4**, 356[144]
Nilakantan, R., **5**, 736[145], 737[145]
Nile, T. A., **8**, 556[372], 764[6,7], 770[7,40], 771[44], 773[7], 780[93,94]
Nilsen, N. O., **4**, 1009[143]
Nilson, M. E., **2**, 323[37]
Nilsson, A., **3**, 671[63], 676[80], 685[63], **6**, 705[24]; **7**, 800[29], 801[40]
Nilsson, H. G., **3**, 645[169]; **7**, 574[128]
Nilsson, J. L. G., **7**, 831[64]
Nilsson, L., **8**, 58[177], 66[177]
Nilsson, M., **1**, 112[27]; **3**, 213[47], 219[114], 257[122], 499[119,120], 512[193,195]; **4**, 152[56], 176[51], 227[209,210], 229[213,218,219,222], 255[204]; **6**, 20[72], 74[33]
Nilsson, N. H., **6**, 423[44]
Nilsson, S., **3**, 639[71]
Nilubol, N., **7**, 73[104]
Nimal Gunaratne, N. Q., **2**, 1017[32]
Nimgirawath, S., **1**, 555[121]; **4**, 113[168]; **6**, 1022[65]
Nimitz, J. S., **1**, 425[103]
Nimkar, S., **5**, 1203[58]
Nimmesgern, H., **4**, 792[60], 795[86], 1086[118], 1088[118], 1089[118,137], 1090[137], 1091[137]; **6**, 128[163]
Nimrod, A., **8**, 561[412]
Ning, R. Y., **6**, 490[104], 534[515]
Ninniss, R. W., **7**, 76[117]
Ninokogu, G. C., **4**, 598[192]
Ninomiya, I., **4**, 91[89]; **8**, 936[70]
Ninomiya, K., **6**, 251[148], 797[17,18], 811[17,18], 812[18,82,83], 813[83], 816[18]
Ninomiya, T., **8**, 150[144]
Ninomiya, Y., **4**, 75[43a], 100[43]
Nio, N., **5**, 578[150,152]
Nisai, S., **4**, 611[359]
Nisar, M., **4**, 591[118], 611[363], 612[368]
Nisbet, H. B., **2**, 902[42]
Nisbet, M. A., **8**, 530[94]
Nishi, H., **5**, 406[22]
Nishi, K., **4**, 155[75]
Nishi, M., **1**, 223[75]; **8**, 99[107]
Nishi, S., **2**, 211[115], 215[115], 216[135], 1066[119]; **4**, 591[112], 592[126], 617[126], 618[126], 633[112,126]
Nishi, S.-i., **6**, 425[65]
Nishi, T., **1**, 766[152]; **2**, 60[17]; **7**, 246[85], 261[73], 370[66,67]; **8**, 154[202]
Nishida, A., **1**, 894[156]; **3**, 321[139]; **6**, 801[27]; **7**, 620[26]
Nishida, I., **1**, 158[74]; **2**, 135[12], 615[129], 634[38], 640[38]; **3**, 6[29]
Nishida, K., **5**, 817[146]
Nishida, M., **6**, 27[114]; **8**, 934[56]
Nishida, R., **8**, 244[50]
Nishida, S., **3**, 380[10], 583[127], 587[143], 676[77], 901[108]; **4**, 951[1], 953[1h], 954[1h], 961[1h], 968[1], 979[1], 1001[19], 1006[19], 1007[19]; **5**, 71[120], 76[240], 77[268], 78[272], 86[29],

152[21], 585[198], 721[99], 904[44], 905[44], 916[121], 917[121], 927[163]; **6**, 127[157], 976[10]; **8**, 856[168]
Nishida, T., **5**, 15[105], 787[8], 797[61], 799[74], 882[14]; **6**, 991[89]
Nishide, H., **2**, 387[333]; **3**, 361[78]; **8**, 857[189]
Nishide, K., **6**, 647[110], 666[231], 667[231]; **8**, 904[55], 908[55], 910[55], 911[55], 914[55]
Nishigaichi, Y., **7**, 408[88b]
Nishigaki, S., **2**, 792[63,64]; **6**, 533[486]; **7**, 342[54]
Nishiguchi, I., **1**, 268[54], 269[54a]; **2**, 709[7]; **3**, 599[204], 602[221], 726[25], 1034[77]; **4**, 95[101], 251[142], 260[142], 444[197], 809[162]; **5**, 901[34]; **7**, 170[118], 795[11], 797[20], 798[21]
Nishiguchi, T., **4**, 1040[73], 1043[73]; **5**, 949[281]; **8**, 552[349], 557[384], 859[215]
Nishihara, T., **1**, 217[47], 327[12]
Nishihara, Y., **2**, 905[56], 906[56]
Nishihata, K., **1**, 512[39]; **3**, 149[401,402,405], 150[402], 151[402,405]; **6**, 532[468]; **7**, 777[387]
Nishii, S., **1**, 121[66], 335[67]; **2**, 32[120,120a], 576[70], 579[89], 979[15], 983[15], 984[15], 985[15], 986[15], 987[15], 992[15]; **3**, 222[136]; **4**, 89[84d], 150[53], 155[68e], 188[101], 189[101], 388[162], 739[111]; **6**, 848[108]
Nishijima, M., **6**, 533[504]
Nishikawa, K., **3**, 586[138]
Nishikawa, M., **4**, 615[383]
Nishikawa, T., **5**, 350[79], 829[26]
Nishiki, M., **8**, 315[47], 369[78]
Nishikimi, Y., **6**, 8[38]; **7**, 399[40b]
Nishimura, A., **2**, 374[277]; **6**, 149[104], 902[130,131]
Nishimura, H., **4**, 24[75], 25[75b]; **7**, 163[73]
Nishimura, J., **3**, 311[99], 464[175]; **4**, 968[58], 970[58b,70], 973[58b]
Nishimura, K., **1**, 174[13], 202[13], 820[17]; **3**, 391[87]; **4**, 247[102], 252[102], 259[102], 589[76], 598[189]; **6**, 85[92], 682[337]
Nishimura, M., **2**, 819[100], 824[100]; **6**, 896[93]
Nishimura, O., **6**, 637[33], 664[222]
Nishimura, S., **3**, 460[143], 461[143]; **4**, 433[121]; **8**, 141[28], 319[76], 391[91], 418[6,8-10], 422[33], 426[49], 427[33], 429[55], 430[6], 432[68], 438[6], 439[6], 450[163], 452[187], 533[154], 535[162]
Nishimura, T., **2**, 1018[39]; **3**, 939[79]; **4**, 125[216,216c]; **8**, 334[59]
Nishimura, Y., **4**, 611[355]; **5**, 953[296]
Nishinaga, A., **7**, 227[89], 228[97]
Nishino, C., **7**, 366[53]
Nishino, K., **8**, 609[53]
Nishino, M., **1**, 881[118]; **5**, 810[126], 812[126]
Nishino, Y., **2**, 902[40]
Nishio, H., **7**, 208[78]
Nishio, K., **1**, 366[46]; **4**, 1007[124]
Nishio, M., **1**, 512[39]; **3**, 149[401,402,405], 150[402], 151[402,405]; **7**, 777[387]
Nishio, T., **6**, 67[14], 69[14]; **8**, 541[203], 840[29]
Nishioka, E., **6**, 842[84]
Nishioka, T., **7**, 774[322]
Nishitani, K., **1**, 205[105]; **4**, 893[155]; **6**, 842[78], 1030[108]; **7**, 63[59], 131[79], 300[54], 773[302]
Nishitani, T., **2**, 1051[41]; **7**, 806[74]
Nishitani, Y., **1**, 763[143], 766[143]
Nishiumi, W., **3**, 1027[46]
Nishiura, K., **2**, 555[138]
Nishiura, P. Y., **8**, 528[63]
Nishiwaki, K.-i., **6**, 454[146]
Nishiwaki, T., **3**, 421[54]; **7**, 255[36]; **8**, 28[33], 36[33], 66[33]
Nishiyama, A., **3**, 390[74], 392[74], 665[38], 691[128,131], 693[128,139], 697[38,131,139,155,156], 698[128]
Nishiyama, E., **2**, 838[178]; **8**, 193[102], 194[104a]
Nishiyama, H., **2**, 576[76], 580[101], 584[121]; **3**, 262[164], 466[191]; **4**, 107[141], 238[12], 255[194], 794[74]; **6**, 774[49,50], 1066[96,98]; **7**, 262[80], 628[44-46], 649[42], 701[64]; **8**, 174[126], 178[126], 179[126]
Nishiyama, K., **1**, 714[267], 717[267]; **3**, 380[9]
Nishiyama, S., **3**, 676[79], 690[125], 695[79]; **6**, 563[899]; **7**, 337[35,36]
Nishiyama, T., **7**, 26[55]
Nishiyama, Y., **8**, 323[112], 412[119], 413[119]
Nishizaki, I., **6**, 89[118], 101[118]
Nishizawa, E. E., **5**, 157[41]
Nishizawa, M., **1**, 568[234], 642[110], 643[110], 833[120]; **2**, 135[12]; **3**, 361[78], 365[97]; **5**, 609[68]; **6**, 626[168], 1032[119]; **7**, 129[73], 130[73]; **8**, 112[24], 120[24], 159[5,34], 162[34], 163[35,37,38,43], 164[34], 178[34], 179[34], 356[185], 545[287,297,301], 546[303], 857[189], 945[131]
Nishizawa, R., **7**, 489[173]
Nishizawa, Y., **4**, 1103[205]
Nitasaka, T., **6**, 134[18]; **8**, 846[85]
Nitsch, H., **5**, 418[71]
Nitta, H., **2**, 826[122,123], 1102[121a,b], 1103[121]; **8**, 405[24]
Nitta, K., **1**, 185[76], 190[76], 191[76], 205[107], 807[318], 808[323]; **2**, 20[82]; **6**, 979[28]; **8**, 937[86]
Nitta, M., **5**, 221[55], 826[159a]; **7**, 798[21]; **8**, 394[116]
Nitta, Y., **7**, 693[28], 761[52]; **8**, 150[137,138]
Nittala, S. S., **8**, 396[139]
Nitti, P., **5**, 331[41]
Nittono, H., **8**, 883[92]
Nivard, R. J. F., **2**, 662[22], 663[22], 664[22], 867[11]; **5**, 71[159,160,161], 76[241], 431[121], 434[121b], 459[91], 724[110], 725[115]; **6**, 114[73], 518[330], 558[846,849,855,856], 643[77], 667[235]; **7**, 230[134]
Niven, M. L., **7**, 355[41], 483[124]
Nivert, C., **4**, 878[80], 884[80]
Niwa, H., **1**, 738[40], 739[38]; **4**, 391[179]; **6**, 998[117]; **7**, 162[58], 242[62], 243[65,66]
Niwa, M., **2**, 553[124]; **3**, 369[121], 372[121], 390[74,86], 392[74], 395[100], 396[113,115], 397[116], 398[113], 769[171]; **5**, 569[113], 1018[67], 1022[67]; **7**, 552[59]; **8**, 331[31]
Niwa, N., **7**, 407[82]
Niwa, R., **5**, 90[56]
Niwa, S., **1**, 223[81,83], 224[81,83]; **8**, 150[140,143], 151[155,156,157]
Nix, G., Jr., **8**, 460[249]
Nix, M., **4**, 738[90], 740[116]
Nixon, A., **1**, 294[49]; **2**, 6[35]
Nixon, J. R., **4**, 306[386,387]; **8**, 855[157]
Nixon, N. S., **4**, 55[157]
Nizamuddin, S., **6**, 524[370], 532[370]
Niznik, G. E., **6**, 295[247,248]; **8**, 74[249], 830[87]
Njoroge, F. G., **2**, 740[63a,b]
Nkengfack, A. E., **4**, 251[145]
Nkunya, M. H. H., **1**, 828[71]; **2**, 435[63a,b]
No, B. I., **6**, 535[534]
Noack, K., **7**, 268[125]
Noack, R., **5**, 102[182], 1070[28], 1074[28]
Noad, T., **2**, 198[82]
Noall, W. I., **4**, 347[96]
Noar, J. B., **7**, 763[91], 769[91]
Nobayashi, Y., **1**, 343[101,102]; **3**, 279[36]; **5**, 151[17], 166[17]
Nobbe, M., **5**, 210[57]
Nobbs, M. S., **4**, 603[265,266], 644[265]; **7**, 456[107]
Noble, D., **7**, 59[37]
Noble, M. C., **8**, 435[71]
Nobles, E. L., **2**, 954[6], 958[6b]
Nobori, T., **6**, 603[15], 614[82,94], 620[129], 624[15,129]
Nobs, F., **5**, 630[51]
Noda, A., **4**, 262[303]; **6**, 836[53]
Noda, I., **5**, 1172[28], 1182[28]; **6**, 453[142]

Noda, K., **6**, 636[20]
Noda, T., **2**, 197[81]; **6**, 8[38]; **8**, 191[88]
Noda, Y., **4**, 8[28]
Node, M., **4**, 124[215]; **5**, 439[171], 440[171]; **6**, 647[110], 666[231], 667[231]; **7**, 256[39], 588[174,175]; **8**, 902[44], 904[55], 908[44,55], 909[44], 910[55], 911[55], 914[55], 989[37]
Noding, S. A., **1**, 67[59], 117[54], 283[2], 315[2], 329[37], 333[37]; **8**, 14[85], 22[147], 697[130]
Noe, E. A., **2**, 438[69a,b]
Noel, M., **3**, 242[9]
Noel, M. B., **8**, 376[161]
Noell, J. O., **4**, 484[12]
Noels, A. F., **3**, 1047[7,8], 1051[7,8]; **4**, 609[330], 1031[4], 1033[16,16d,e], 1035[16e], 1051[125], 1052[16d]; **5**, 86[35]; **6**, 25[101]; **7**, 8[61]
Nofal, Z. M., **2**, 792[65]
Nogami, H., **8**, 328[10]
Nogami, T., **5**, 71[131]
Nogi, T., **5**, 1137[55]
Nogina, O. V., **1**, 142[24]
Nogradi, M., **1**, 49[12], 56[12], 58[12], 67[12], 78[11]; **3**, 35[207]; **4**, 200[3]; **7**, 829[60], 831[62]; **8**, 7[40], 159[11], 544[276]
Noguchi, A., **5**, 74[206], 210[57]
Noguchi, H., **1**, 359[16], 379[16b]; **2**, 922[103]
Noguchi, I., **3**, 665[40], 681[95]; **8**, 568[471]
Noguchi, J., **7**, 801[41]
Noguchi, K., **2**, 1012[13]; **6**, 447[105], 450[105]; **8**, 338[82], 339[82]
Noguchi, M., **3**, 321[139]; **6**, 534[516], 717[110], 801[27]
Noguchi, S., **1**, 422[93]; **7**, 59[38]
Noguchi, T., **8**, 49[116], 66[116]
Noguchi, Y., **2**, 922[104]; **5**, 102[164]; **6**, 817[103]
Nogues, P., **5**, 461[105]
Noguez, J. A., **4**, 12[38], 259[274]; **6**, 1033[123]
Noh, S. K., **4**, 696[5,7], 703[34,35], 704[34], 705[34b], 712[34,35]
Nohara, Y., **4**, 203[27]
Nohe, H., **3**, 634[28]; **8**, 624[153]
Nohira, H., **2**, 810[63], 824[63]; **6**, 726[179]
Nohira, N., **4**, 1046[111]
Nohria, V., **2**, 760[43]
Noire, J., **3**, 197[42]; **8**, 925[8]
Nojima, M., **4**, 271[20], 356[145]; **6**, 216[108], 219[108]; **7**, 543[14], 766[188]; **8**, 863[235], 864[235], 965[66]
Nojima, S., **8**, 332[43]
Nokai, H., **7**, 877[134]
Nokami, J., **1**, 513[48], 520[71,75]; **2**, 19[77,77c], 82[9], 350[75], 374[275,277], 572[46], 575[64]; **3**, 650[211]; **4**, 258[248], 261[248], 964[46], 1040[98], 1043[98]; **6**, 149[104], 902[130,131], 939[137], 1016[35], 1022[64,65]; **7**, 454[96], 537[60]
Nolan, M. C., **8**, 696[126]
Nolan, R. L., **8**, 276[149]
Nolan, S. M., **4**, 113[171]; **6**, 140[57]
Nolan, S. P., **8**, 669[6,8], 670[6]
Noland, W. E., **4**, 12[42], 253[315]; **7**, 219[13]; **8**, 612[72]
Nolen, E. G., Jr., **2**, 843[195]
Nolen, R. L., **2**, 489[45], 492[53], 493[53]; **6**, 274[107], 674[296]
Noll, W., **2**, 759[37]
Noller, C. R., **8**, 79[2]
Nolley, J., Jr., **4**, 842[51], 847[51]
Nolte, E., **2**, 1023[54]
Nolte, R. J. M., **6**, 489[89,94]
Noltemeyer, M., **4**, 222[181], 224[181]
Noltes, J. G., **1**, 428[121], 429[121], 457[121]; **3**, 208[11], 210[11,11a], 219[11a], 234[11a]; **8**, 264[40], 547[316,316e], 548[319,320]
Noma, Y., **8**, 559[400]
Nominé, G., **3**, 12[65]; **7**, 66[73]; **8**, 201[144], 533[154]
Nomizu, S., **8**, 244[50]
Nomoto, K., **8**, 48[111], 66[111]
Nomoto, S., **8**, 144[79]
Nomoto, T., **1**, 26[133,134]; **3**, 169[510], 172[510], 173[510], 380[11], 556[36]; **5**, 413[51]; **6**, 240[79], 609[56], 913[23]; **8**, 917[116,117], 920[116,117]
Nomura, H., **7**, 692[23]
Nomura, K., **1**, 101[92], 554[100,104]; **2**, 810[66], 851[66]; **7**, 200[42], 209[92], 700[63]
Nomura, M., **4**, 298[291]
Nomura, O., **5**, 1141[81], 1145[103], 1153[103]
Nomura, R., **8**, 535[166]
Nomura, Y., **1**, 619[62]; **2**, 635[46], 640[46]; **4**, 341[55]; **5**, 480[177]; **6**, 91[128], 109[39-41], 466[46], 564[916]; **7**, 475[57], 496[217], 497[218], 522[39]
Nonaka, T., **2**, 105[42]; **3**, 564[79], 574[79]; **4**, 439[163], 886[117]; **6**, 821[113]; **7**, 267[118], 268[118], 283[118], 284[118], 379[101], 778[406]; **8**, 134[28,29,33], 135[40], 137[33], 886[113]
Nonaka, Y., **5**, 725[118]
Nondek, L., **2**, 268[65]
Nonet, S., **3**, 95[159]
Nonhebel, D. C., **3**, 663[26], 665[26]; **4**, 717[8]; **7**, 860[71]
Nonomura, S., **8**, 559[400]
Nonoshita, K., **1**, 79[21], 80[21], 81[21], 82[21], 92[60,61], 333[61], 335[61]; **4**, 143[21], 254[177]; **5**, 434[147], 850[150,153]; **6**, 856[152]
Nooi, J. R., **6**, 964[82]
Noori, G. F. M., **2**, 748[125]; **6**, 291[213,215]
Norbeck, D. W., **4**, 372[53], 1006[105]; **5**, 151[8], 841[86], 843[125], 853[125a], 859[235]; **6**, 464[30], 471[30], 859[166]; **7**, 301[62]
Norberg, A., **2**, 827[128]
Norberg, B., **1**, 650[139], 664[200], 672[200], 675[210], 677[210], 706[210], 718[200], 719[200], 720[200], 721[210], 722[200], 870[84]; **7**, 773[307]
Norberg, T., **6**, 46[67,70], 47[67]
Norbury, A., **1**, 493[42,42a], 495[42]
Norcross, B. E., **8**, 561[414]
Norcross, R. D., **2**, 249[84]
Nord, F. F., **2**, 138[24], 139[32]; **8**, 187[35], 373[134], 376[134,162]
Nordahl, J. G., **4**, 1040[79], 1049[79]; **5**, 599[40], 804[94], 986[40]
Nordberg, R. E., **4**, 565[44], 591[115], 592[115], 596[164], 597[168,169,171], 598[187,190,195], 617[115], 618[115], 621[164,168,169], 622[168,195], 623[190], 633[115], 637[169], 638[190,195]; **6**, 85[89,91]; **7**, 94[57]
Nordblom, G. D., **7**, 852[41], 853[41]
Nordenson, S., **4**, 1063[173]
Nordin, I. C., **5**, 830[29]
Nordlander, J. E., **2**, 5[20], 6[20], 21[20], 740[63a,b]; **4**, 876[58]; **5**, 43[33]
Nore, P., **2**, 354[117], 357[117]
Norell, J. R., **2**, 746[111]; **3**, 381[20]; **5**, 476[147]; **6**, 263[25], 264[27], 295[25,27], 526[399]
Norin, T., **3**, 390[73], 489[61], 495[61], 504[61], 511[61], 515[61]; **5**, 839[77]; **6**, 811[76]; **7**, 822[34]; **8**, 163[36], 857[190]
Norinder, U., **3**, 594[184]
Noritake, Y., **4**, 745[141]
Norman, A. W., **3**, 816[80]; **8**, 681[76], 689[76]
Norman, B., **1**, 367[54]
Norman, B. H., **5**, 252[42], 410[39]
Norman, J. A., **7**, 92[43]
Norman, J. F., **6**, 233[46]
Norman, L. R., **2**, 388[341]; **3**, 804[5]
Norman, M. H., **4**, 158[79]
Norman, R. O. C., **3**, 501[135]; **4**, 315[499,500], 453[37]; **6**, 734[15], 923[57]; **7**, 94[55], 231[145], 828[50]; **8**, 336[85]
Norman, T. C., **3**, 223[148]

Normant, H., **1**, 3[23]; **2**, 507[25]; **3**, 123[247], 194[14], 242[9]; **4**, 84[68a], 89[68a], 107[145]; **6**, 720[131]; **8**, 479[48], 481[48], 524[12]
Normant, J. F., **1**, 78[5], 107[3], 113[33,34], 132[107], 133[108], 189[75], 205[106], 206[106], 217[46], 219[56,56c,57,59], 220[57b], 331[50,51], 343[115], 347[131], 348[140], 370[68], 371[68], 373[84], 374[84], 427; **2**, 6[34], 20[34c], 23[34c], 119[168], 427[42], 584[125], 596[3]; **3**, 202[87,92,93,95], 208[10], 209[21], 210[10], 215[21a], 216[73], 217[21,84], 219[21a,b], 223[154], 224[166], 226[203,207], 227[210], 242[1], 243[1,14], 245[30], 246[35], 247[46], 249[14,58-61], 250[72], 258[125], 259[129], 263[14,173], 264[72], 265[72,188], 274[19], 419[39-41], 423[72], 440[42], 441[42], 442[42], 464[172], 466[182,193], 470[214,223,225,226], 473[214,217,225,226], 476[214,217], 482[4,5], 485[4,31-35], 486[31-35], 487[45], 494[34,35,89], 498[107], 499[5], 505[5], 509[5], 516[89], 522[18,19], 525[41], 579[100], 788[49]; **4**, 34[97], 35[97], 148[48], 152[59], 164[48], 170[7,8], 173[34], 183[81], 185[87], 192[120], 207[59], 209[63,68,69], 210[77,78], 238[10], 249[10], 250[10,136], 254[10], 255[10], 262[136], 866[3], 867[3], 873[3], 877[73], 880[92], 882[92], 893[3,158,159], 895[164,166], 896[3,167,168,170], 897[170,172], 898[3,177,178], 899[3], 900[3,164,179,180], 901[3], 902[190], 903[188,189,191-197,199], 1009[139]; **5**, 32[7], 829[24], 848[141], 1163[2]; **6**, 5[23], 164[197,198], 849[120,122], 965[87,89]; **7**, 453[67]
Normant, J. M., **4**, 543[122]
Normura, Y., **5**, 483[174]
Noro, T., **8**, 196[120]
Norris, A. F., **6**, 690[395]
Norris, J. F., **3**, 309[92c]; **6**, 204[9]
Norris, R. K., **2**, 743[83]; **4**, 424[20], 452[10], 467[10]
Norrish, H. K., **3**, 201[76]
Nortey, S. O., **4**, 38[108,108c], 379[115], 380[115h], 383[115h]; **6**, 174[55]; **7**, 523[45]; **8**, 55[165], 66[165], 618[129], 619[129], 620[129], 624[129]
North, P. C., **7**, 406[74]
Northington, D. J., **5**, 216[14], 218[33], 219[14]
Northington, D. J., Jr., **3**, 390[81], 392[81]
Northrop, R. C., Jr., **8**, 146[89], 148[105], 228[126], 249[94]
Norton, J., **4**, 857[106]
Norton, J. A., **4**, 287[177], 288[177]
Norton, J. R., **1**, 439[163,164], 440[167,171], 457[163]; **3**, 208[3], 213[3b], 1031[63,64]; **4**, 518[1], 547[1], 895[160], 915[12], 937[70,71], 938[70,71], 941[85]; **5**, 46[39], 56[39], 1065[1], 1066[1], 1074[1], 1083[1], 1084[1], 1093[1], 1112[1g], 1163[3], 1183[3]; **8**, 421[28], 422[28], 432[28], 435[28], 436[28]
Norton, N. H., **7**, 759[6]
Norton, S. J., **3**, 848[49]
Norton, T. R., **8**, 143[65]
Norymberski, J., **7**, 100[128]; **8**, 293[48]
Nosaka, Y., **5**, 704[21]
Nose, A., **6**, 724[165]; **7**, 598[61]; **8**, 408[66]
Nosova, V. V., **6**, 607[48]
Nossin, P. M. M., **2**, 1062[101], 1063[102]; **3**, 367[102]
Noteboom, M., **4**, 1093[150]
Nöth, H., **1**, 301[73]; **4**, 886[119]; **8**, 720[130]
Notheisz, F., **8**, 418[5], 420[5], 423[5], 439[5], 441[5], 442[5], 883[89,90]
Nott, A. P., **3**, 261[153]
Notté, P., **5**, 109[219]
Notzumoto, S., **3**, 976[6]
Nouls, J. C., **6**, 707[45]
Nour, M., **7**, 497[219]
Noureldin, N. A., **7**, 768[208], 773[208], 845[78]
Nouri-Bimorghi, R., **3**, 739[42], 759[127]
Nov, E., **8**, 60[193], 62[193], 66[193]
Novack, V. J., **1**, 825[54]; **2**, 318[50]; **3**, 203[101]; **8**, 540[195]
Novak, B. M., **5**, 1118[18], 1121[18]
Novak, J., **3**, 464[168]; **5**, 418[68]; **8**, 541[204]
Novak, P. M., **1**, 872[91]
Novakovskii, E. M., **3**, 643[131]
Novelli, R., **5**, 477[159]
Novi, M., **4**, 426[47,58], 457[62-64], 460[64], 461[101], 471[62], 475[101,148], 476[64,158,159,160,161]; **6**, 240[80]
Novick, S., **4**, 438[152]
Novick, W. J., Jr., **4**, 439[168]
Novikov, E. G., **8**, 593[68]
Novikov, N. A., **3**, 648[175]
Novikov, S. S., **3**, 887[10], 888[10], 889[10], 890[10], 893[10], 897[10], 900[10], 903[10], 1039[97]; **7**, 493[195]
Novikov, Y. D., **8**, 486[62]
Novikov, Yu. N., **6**, 836[53]
Novikova, N. N., **8**, 618[121], 619[121]
Novitskii, K. Yu., **6**, 543[609]
Novitt, B., **8**, 613[79]
Novkova, S., **2**, 942[166], 944[166]
Novokhatka, D. A., **4**, 426[64]
Nowack, E., **5**, 185[167]
Nowack, G. P., **4**, 1099[185]; **8**, 425[44], 426[44]
Nowacki, A., **2**, 399[13]
Nowak, B. E., **7**, 699[55]; **8**, 218[70], 219[70], 230[139]
Nowak, K., **2**, 403[34,35]
Nowak, M., **1**, 769[183]
Nowell, I. W., **1**, 21[111]
Nowick, J. S., **1**, 770[191]
Nowlan, V. J., **4**, 297[279]
Nowoswiat, E. F., **6**, 533[494]
Noyce, D. S., **4**, 241[57]; **6**, 206[46]
Noyori, R., **1**, 78[20], 133[110], 158[74], 223[76], 224[76a], 317[145,155], 319[145], 320[155], 328[26], 347[129], 753[104], 833[120]; **2**, 72[57], 76[57], 135[12], 576[73], 577[73], 609[84], 615[121,128,129], 634[35,38], 635[39,39c], 640[35,38,39], 650[39c,d]; **3**, 4[26], 5[26], 6[29], 9[47], 10[26,47], 25[159], 103[203,206], 108[203], 402[130], 404[135], 771[182]; **4**, 13[44,44c], 96[105], 97[105b], 159[85], 177[59], 211[90], 238[9,13], 239[9], 245[9], 253[169], 254[9], 255[195,197], 257[230], 259[197], 260[197], 261[13,230], 262[308], 393[197,197b,c], 394[197b,c], 587[45], 609[326], 963[42], 1011[165], 1038[61]; **5**, 282[25-33], 283[25-27,30], 284[28-30], 285[25,27,31,32], 286[30,33], 293[44,48], 294[51], 594[5], 595[11], 596[11a,b], 601[44-47], 603[5,47,48a,e], 605[5,46,57,58,60,60a,b,61,63], 606[45], 608[5,65], 609[60,60b,c,61,68], 611[5b,57], 755[72,73], 760[72], 761[73], 1124[53], 1186[3,3a,5,7], 1187[8], 1188[16], 1190[24,24a], 1200[54]; **6**, 11[45], 18[65], 46[61], 138[44], 139[45], 603[11,15,17], 608[11], 614[82,94], 619[120], 620[129], 624[15,120,129,148,149], 866[208], 942[154], 944[154]; **7**, 26[51], 220[20], 274[137], 406[75], 650[51], 682[86], 750[131]; **8**, 154[198,199,200,201,202], 159[5,34], 162[29,33,34], 163[35,37-39,43], 164[34], 178[34], 179[34], 216[59], 217[59], 356[185], 459[228], 460[255,256], 461[257,261], 462[267], 463[269], 535[166], 537[186], 544[254], 545[287,297,301], 546[186,303], 807[115], 987[18], 991[18,44]
Nozaki, H., **1**, 77[3], 78[3], 92[62,63,66], 95[78,79], 101[93,94], 103[93,94], 174[9], 177[9,19], 179[9,19,27], 180[27], 181[27], 182[27], 188[72], 189[72], 193[84,85], 195[85], 198[84,85], 201[98], 202[100], 205[10]; **2**, 3[9], 5[9b,c], 6[9], 19[9,9b,c,76], 20[9b,c,80], 21[80], 29[9c], 72[56], 114[122,123], 269[71,73], 271[75], 282[40], 575[67], 581[103], 584[126], 588[151], 589[151,154,155,156,157], 599[26], 600[27], 718[71], 726[123]; **3**, 96[162,166,167], 99[162,163,166,167], 103[166,167], 104[162,163], 105[166,216], 112[216], 120[162,163,166], 121[162,163], 131[162,163], 155[166], 202[96,97], 216[70], 243[11], 244[29], 246[35], 254[96,97], 259[135], 274[22], 279[37], 349[33], 354[60], 356[58], 358[33], 421[63], 445[73], 449[73], 484[24], 501[24], 565[17], 730[43], 759[133], 787[47,48], 799[100,103]; **4**, 23[71], 30[89], 96[103c], 113[174], 120[203], 162[92], 257[227], 261[227], 393[190], 394[190,192], 411[266d], 431[101], 567[47,49-51], 588[68], 607[310],

626[310], 637[68], 647[310], 792[66], 879[85], 885[111], 892[143], 900[182], 901[185,186], 963[42], 1007[118,119,122,124,126], 1008[126], 1009[118,136,137], 1011[165], 1017[213], 1018[219], 1019[213], 1038[61]; **5**, 176[128], 609[68], 755[72], 760[72,89,90], 768[128,130,134], 769[128], 770[141,142], 771[142,143], 779[128], 780[141], 850[149], 917[124], 926[124], 938[219], 943[237], 1007[41], 1124[50,51], 1125[58]; **6**, 7[30], 143[73], 237[67], 563[905], 564[913,914], 677[312], 837[60], 856[149], 861[181], 960[64]; **7**, 26[51], 169[117], 219[10], 267[118], 268[118], 275[146,147], 276[147], 281[176], 282[176], 283[118], 284[118], 308[17], 309[24], 322[67], 324[70], 369[63], 378[63,96], 379[101], 601[85], 615[8], 674[47]; **8**, 356[185], 705[13], 734[6], 755[119,126,127], 757[6], 758[126,127], 785[113], 797[45], 807[115,116], 886[113], 929[30], 986[16]

Nozaki, K., **2**, 456[49], 460[49], 461[49]; **4**, 721[31], 725[31], 756[184], 770[248,249], 791[42], 796[96], 886[117]; **7**, 259[62]; **8**, 187[34], 195[108], 407[54], 798[46], 807[46], 818[41], 820[41], 823[58]

Nozaki, N., **4**, 33[96,96c]

Nozaki, Y., **2**, 348[56], 362[56], 363[56]; **3**, 1047[6]

Nozawa, H., **2**, 810[63], 824[63]

Nozawa, S., **3**, 421[53]; **4**, 164[99,99a]

Nozawa, Y., **3**, 593[182]; **6**, 764[10]

Nozoe, S., **2**, 158[126]; **3**, 713[29], 816[79,79a]; **7**, 184[168]; **8**, 61[190], 66[190]

Nozoe, T., **7**, 796[13]

Nozomi, M., **8**, 413[135]

Nozulak, J., **2**, 498[71,73,74,76], 499[71,73,74,76]; **3**, 53[274]

Nsunda, K. M., **1**, 635[87], 664[87], 665[87,175], 672[87], 679[87], 682[87]; **3**, 87[81], 104[81]

Nübling, C., **1**, 359[17], 380[17], 381[17]; **2**, 1072[140]; **6**, 108[32]; **7**, 318[60]; **8**, 388[60]

Nucciarelli, L., **7**, 530[14]

Nuck, R., **6**, 175[78]

Nudelman, A., **8**, 410[91]

Nudelman, N. S., **4**, 426[49]

Nuenke, N. F., **6**, 263[19]

Nugent, M. J., **2**, 636[52], 640[52], 935[145], 937[145], 938[145], 940[145]; **6**, 760[143]

Nugent, R. A., **7**, 105[151]

Nugent, R. M., **5**, 424[93,98]

Nugent, S., **4**, 70[7]

Nugent, S. T., **3**, 214[56]

Nugent, W. A., **4**, 808[158]; **5**, 1065[4], 1139[76], 1166[20], 1167[20], 1169[20], 1170[20], 1171[26,27], 1172[27], 1178[20,26,27]; **8**, 850[121]

Nugiel, D. A., **6**, 448[109]

Nukada, T., **6**, 51[108], 53[108], 862[186]

Nukina, S., **8**, 218[73], 221[73]

Numakunai, M., **8**, 93[72]

Numakunai, T., **8**, 561[416]

Numata, H., **1**, 763[144]; **2**, 814[81], 824[81], 1035[92]

Numata, S., **4**, 595[148,149]; **8**, 528[67]

Numata, T., **3**, 953[104]; **6**, 753[117], 910[8,9], 924[8,9,61,63], 925[64], 926[65]; **7**, 196[13], 197[20,23,25], 762[83], 778[395]; **8**, 403[1], 404[1], 405[34], 408[68], 409[81], 410[34]

Numazawa, M., **8**, 64[217]

Nunami, K., **6**, 489[90,92], 547[659]

Nunes, F., **8**, 896[17]

Nunes, J. J., **2**, 652[123b], 1069[135]; **5**, 854[175]

Nunez, A., **4**, 426[45]

Nuñez, O. S., **2**, 928[122], 946[122]

Nunn, C. M., **1**, 41[270], 432[137], 456[137]; **3**, 211[28], 215[28]

Nunn, K., **2**, 97[60]

Nunn, M. J., **6**, 836[55]

Nunokawa, O., **2**, 728[143]

Nunokawa, Y., **3**, 470[197], 473[197]; **8**, 720[130]

Nunomoto, S., **3**, 244[22], 465[179], 494[89], 516[89]

Nunomura, S., **6**, 53[120], 507[240], 515[240]

Nurenberg, R., **4**, 491[68], 492[72]

Nurimoto, S., **2**, 1012[13]

Nurmi, T. T., **4**, 871[34]

Nurrenbach, A., **5**, 835[58]; **6**, 37[33]

Nuss, J. M., **5**, 195[10], 645[1], 648[1r], 651[1], 653[1r]

Nussbaum, A. L., **6**, 626[165]

Nussbaumer, C., **2**, 554[134], 1024[60], 1062[99]; **5**, 18[125]

Nussbutel, U., **4**, 760[196]

Nussim, M., **1**, 878[108]; **3**, 791[64]; **4**, 1005[84]; **8**, 528[55,84], 529[84]

Nussler, C., **1**, 70[63], 141[22]

Nutaitis, C. F., **3**, 311[98]; **6**, 724[161]; **8**, 16[108], 17[108], 26[14], 27[14], 36[14], 55[14], 60[14], 215[52], 616[103,105], 617[106], 619[103,105]

Nutland, J. H., **2**, 141[38]

Nutt, H., **2**, 360[165b]

Nutt, R. F., **2**, 1097[101]; **3**, 644[167]

Nutter, D. E., **4**, 423[10], 437[10]

Nützel, K., **1**, 211[1], 212[1], 214[1], 215[1], 220[1], 222[1], 225[87], 226[87]; **2**, 277[6]

Nuvole, A., **8**, 589[50,53]

Nuzzo, R. G., **8**, 404[12]

Nvak, J., **4**, 1058[147]

Nwaji, M. N., **4**, 1060[159]

Nwaukwa, S. O., **3**, 167[483]; **7**, 247[107], 318[62], 706[19,20]

Nwokogu, G. C., **3**, 530[70,71], 534[70,71]

Nyangulu, J. M., **4**, 337[31]

Nyarguhi, M., **7**, 299[49]

Nyathi, J. Z., **5**, 1136[54]

Nyberg, K., **2**, 1051[35], 1052[35,38], 1053[54]; **3**, 634[22], 637[64]; **6**, 281[150], 572[959]; **7**, 799[24,25], 800[30,31], 801[42,43], 804[59], 805[59]

Nyburg, S. C., **8**, 858[206]

Nyce, J. L., **4**, 279[115], 280[115]

Nye, M. J., **5**, 594[3], 595[3], 596[3], 597[29], 603[3]

Nye, S. A., **3**, 538[94]

Nyfelder, R., **6**, 671[277]

Nyi, K., **3**, 573[70]

Nyikos, S. J., **8**, 504[72], 505[72]

Nyman, F., **3**, 383[49]

Nyns, C., **3**, 870[49]

Nysted, L. N., **8**, 530[105]

Nyström, J. E., **4**, 598[183,187,195], 599[215], 609[215], 622[195], 624[215], 638[183,195], 641[215]; **6**, 85[91,93], 86[93]; **7**, 94[57]

Nystrom, R. F., **8**, 314[40], 726[189]

Nyu, K., **7**, 21[15]

Nyuyen, S. L., **4**, 177[56]

O

Oae, K., **7**, 764[121]
Oae, S., **3**, 86[7,19,31,35,42,44], 104[31], 121[7,19], 147[31], 154[31], 164[35], 173[35], 862[3], 953[104]; **4**, 335[27], 358[152]; **6**, 208[57], 212[57], 753[117], 910[8-10], 924[8-10,61,63], 925[64], 926[65]; **7**, 124[46], 196[13], 197[14,15,20,23,25], 470[10,11,13], 498[230b], 758[3], 759[15], 760[3], 761[57,58], 762[80,83], 763[92], 764[107,116,121], 769[214], 777[80], 778[395]; **8**, 36[85], 39[85], 66[85], 370[89], 389[73], 390[84], 391[84], 392[97], 403[1,3], 404[1,3], 405[34], 406[46], 408[46,68-71,76], 409[81], 410[34,88,92,93], 411[99], 412[114]
Oakes, V., **2**, 443[16]
Oare, D. A., **1**, 3[24]; **4**, 21[67,67a,b], 100[123,123b], 106[139], 107[139]
Obana, M., **8**, 239[28]
O'Bannon, P. E., **5**, 419[73]
Obara, Y., **1**, 438[160]
Obayashi, M., **1**, 347[134]; **2**, 24[95], 572[40], 575[40], 599[26], 718[71]; **3**, 202[85], 254[96], 565[17], 759[133]; **4**, 23[71], 162[92], 900[182]; **6**, 237[67], 564[914]
Obaza-Nutaitis, J. A., **5**, 384[128], 385[128b]
Oberender, H., **4**, 611[345]; **7**, 140[130], 141[130]
Oberhansli, P., **3**, 407[150]
Oberhänsli, W. E., **2**, 850[219], 854[219]; **4**, 30[89]
Oberrauch, H., **7**, 100[122]
Oberster, A. E., **7**, 100[124]
Oberti, R., **5**, 78[281]
Obeyama, J., **4**, 79[59c], 216[126]
Obeyesekere, N. U., **2**, 603[49]
Obha, N., **7**, 158[36a,b], 175[36b]
O'Boyle, J. E., **6**, 692[408]
Obrecht, D. M., **1**, 447[199]
Obrecht, J.-P., **1**, 373[91], 375[91], 376[91]; **2**, 996[47], 1077[154]; **3**, 352[45]; **7**, 230[133]
Obrecht, R., **2**, 1084[10], 1090[72,73], 1102[73,125,126], 1103[73,125]; **6**, 489[97]
O'Brien, C., **6**, 786[93]
O'Brien, E., **5**, 1146[107]
O'Brien, J. B., **6**, 149[101]
O'Brien, M. J., **2**, 117[147]; **3**, 6[30]; **5**, 517[27], 538[27]; **8**, 843[52]
O'Brien, M. K., **4**, 670[24,25], 682[58]
O'Brien, R. E., **8**, 937[80]
O'Brien, S., **8**, 614[85]
Obrzut, M. L., **3**, 1033[74]; **4**, 923[32]
O'Callaghan, C. N., **2**, 348[64], 362[182]
Ochi, M., **1**, 766[162]; **6**, 801[34]; **8**, 215[53], 217[53], 227[121], 240[32], 244[70], 620[133], 624[133]
Ochi, T., **1**, 767[176]
Ochiai, E., **7**, 749[122], 750[122]; **8**, 390[85], 391[85]
Ochiai, H., **1**, 212[9], 213[9,15], 215[41], 216[41], 217[47], 327[12], 448[207], 449[210]; **2**, 442[13,14], 449[13,14,49], 450[13,14,51]; **3**, 221[131], 231[242], 420[48,49], 443[61], 445[61], 463[157,159], 1023[23], 1033[71]; **4**, 379[115], 557[14], 599[214]
Ochiai, M., **1**, 834[123], 894[160]; **2**, 116[131,140], 157[122], 610[94,95], 611[94,95], 718[78,79], 816[83], 819[102], 859[252], 1059[78,81-83]; **3**, 168[492], 169[492], 217[79], 286[56a], 1050[19]; **4**, 42[122b], 155[72], 817[207]; **5**, 92[67,81], 96[118], 945[249]; **6**, 936[111], 938[132], 944[132], 1004[138], 1018[42], 1065[90b]; **7**, 92[41,41b], 93[41b], 94[41], 227[81], 457[110], 518[17], 615[9], 621[34], 623[35], 624[36], 710[56], 765[149], 773[149,301]; **8**, 975[133], 992[55]
Ochiai, R., **5**, 1139[75]
Ochrymowycz, L. A., **6**, 679[325]
O'Connell, E. M., **6**, 1016[27]
O'Connell, J. P., **8**, 419[19], 420[19], 424[19]
O'Conner, A. W., **8**, 333[54]
O'Conner, B., **4**, 262[302], 854[92], 903[199]
O'Connor, B., **3**, 21[128], 259[132]; **6**, 176[101]; **7**, 272[141], 713[68]
O'Connor, C., **8**, 445[27], 568[479]
O'Connor, D. E., **3**, 147[389]
O'Connor, D. T., **7**, 202[51]
O'Connor, E. J., **4**, 980[110], 982[110], 984[121]
O'Connor, J., **4**, 850[85]
O'Connor, J. M., **2**, 127[235]
O'Connor, J. P., **8**, 457[217]
O'Connor, M. J., **8**, 851[131]
O'Connor, S., **1**, 212[16], 213[16]; **3**, 512[198]; **5**, 133[57]
O'Connor, U., **4**, 7[24], 155[69]
O'Connor, W. F., **3**, 574[80]
Oda, D., **1**, 759[128,129]; **4**, 589[87,88], 598[191], 599[218], 638[191], 640[218]
Oda, H., **2**, 575[67]; **4**, 901[185]; **8**, 699[149]
Oda, I., **4**, 843[53,54], 852[53]
Oda, J., **1**, 349[147]; **2**, 232[174]; **4**, 201[10], 230[244], 231[244]; **6**, 843[91], 1016[26]; **8**, 170[77-79], 562[423], 853[144]
Oda, K., **3**, 747[72]; **5**, 181[156], 563[92,93], 564[94]; **6**, 689[386], 1052[41]; **7**, 406[87]
Oda, M., **2**, 744[91]; **3**, 421[60]; **4**, 1054[133]; **5**, 394[145b]; **7**, 172[124], 874[101]; **8**, 5[30], 341[104]
Oda, N., **6**, 531[426]
Oda, O., **2**, 540[71], 547[115], 551[115]
Oda, R., **3**, 381[24], 382[24], 922[37], 924[37]; **4**, 511[178], 587[43]; **6**, 276[119]; **7**, 16[167], 797[19]
Oda, T., **3**, 391[88]; **6**, 1054[48]; **7**, 745[78]
Oda, Y., **4**, 609[326], 958[27]; **6**, 11[45]
Odagi, T., **5**, 294[51], 1190[24,24a]
Odaira, Y., **3**, 380[9], 638[94]; **4**, 969[68]; **5**, 163[73], 817[146]; **6**, 976[9], 1036[145]; **7**, 877[128]
Odaka, T., **3**, 639[74], 643[123]
Odashima, K., **8**, 170[80]
Ode, R. H., **7**, 223[44]
O'Dea, J., **3**, 906[145]
Odeh, I. M. A., **5**, 494[216], 579[162]
Odell, B. G., **5**, 72[173]
O'Dell, C., **2**, 962[49]
O'Dell, D. E., **2**, 78[92]; **4**, 816[201]; **6**, 1067[105]
Odenigbo, G., **5**, 1148[123]
Odham, G., **3**, 643[129]
Odiaka, T. I., **4**, 670[20]
Odic, Y., **2**, 608[77]; **3**, 7[33], 20[119]
Odinokov, V. N., **7**, 543[18], 579[18], 581[18]; **8**, 396[138], 398[138]
Odintsova, T. I., **8**, 606[27]
Odom, H. C., Jr., **4**, 18[62], 20[62f]
Odom, J. D., **6**, 462[17]
O'Donnell, J., **6**, 1050[38]
O'Donnell, M. J., **1**, 123[73], 370[66]; **5**, 87[41], 92[41], 109[41]; **6**, 65[2], 84[85], 645[97]; **7**, 229[119]
O'Donoghue, D. A., **7**, 696[42]
Odriozola, J. M., **5**, 100[146]; **7**, 554[64]
Odyek, O., **6**, 265[40]
Oechsner, H., **6**, 186[169], 190[201], 193[208], 196[201], 197[169]
Oeckl, S., **6**, 232[37]
Oediger, H., **2**, 153[106]; **6**, 958[29]
Oehlschlager, A. C., **1**, 135[115]; **3**, 223[155], 259[136]; **4**, 564[39], 901[186]; **7**, 238[41], 401[54], 478[85], 483[122]
Oehme, H., **6**, 512[305]

Oehrlein, R., **5**, 829[15]

Oei, H.-A., **1**, 70[63], 141[22]; **3**, 568[40], 569[49], 606[40], 607[49]

Oei, H.-Y., **1**, 370[69], 371[69]

Oelberg, D. G., **6**, 836[55]

Oelschläger, H., **2**, 735[17]; **8**, 141[31], 903[50]

Oertle, K., **2**, 308[23], 309[23]; **4**, 315[519], 602[253], 625[253]; **5**, 35[14]; **6**, 656[172]; **8**, 776[81a,b]

Oesterle, T., **2**, 642[78], 643[78], 681[58], 683[58]; **6**, 502[217], 560[870]; **7**, 650[51]

Offenhauer, R. D., **2**, 138[18], 240[9]

Oesterlin, H., **8**, 144[67,69]

Oestreich, T. M., **4**, 426[57], 457[53]

Oettle, W. F., **5**, 218[29], 219[40], 229[120]

Oettmeier, W., **1**, 373[87], 374[87]

Oexle, J., **5**, 595[18], 596[18]

Oez, H., **7**, 3[5]

Oezkar, S., **5**, 633[66]

O'Fee, R., **4**, 377[104], 389[104g]; **6**, 980[36]; **8**, 880[62]

Ofee, R. P., **4**, 562[35]

Ofele, K., **4**, 519[21,23], 546[23]

Ofenberg, H., **8**, 628[178]

Offermann, K., **2**, 1084[8], 1088[45], 1089[53,57], 1090[67-69], 1091[69], 1094[89], 1095[89], 1098[45]; **6**, 242[85,86], 243[85,86], 294[243], 489[88]

Offermanns, H., **5**, 453[59]; **6**, 430[100]; **7**, 760[40]

Officer, D. L., **2**, 849[215]; **8**, 844[66]

Ofner, S., **6**, 538[571]

Ofstead, E., **5**, 1116[8]

Oftedahl, E. N., **3**, 747[72]

Oftedahl, M. L., **2**, 363[189]

Ofusu-Asante, K., **8**, 319[78]

Ogaki, M., **7**, 227[75], 804[62]

Ogard, A. E., **8**, 274[134]

Ogasawara, H., **6**, 559[861,862]

Ogasawara, K., **1**, 90[59]; **2**, 167[162], 372[268,270,271], 373[268,270], 505[10], 1018[39], 1035[92,94], 1040[94]; **3**, 34[194], 224[181], 934[67]; **4**, 387[158,158b,159], 393[159], 501[118], 504[130]; **5**, 410[38], 468[129,130,131], 742[162], 839[85], 862[246], 864[257]; **6**, 509[254], 533[499], 746[91], 843[87], 893[77]; **7**, 160[48], 180[159], 299[45], 463[129], 682[82], 713[69]; **8**, 222[97]

Ogata, I., **4**, 915[10], 918[10], 930[46,47], 931[55], 945[91]; **8**, 152[178], 450[162], 451[162], 463[270]

Ogata, M., **1**, 477[142]; **6**, 546[647], 579[982]; **7**, 408[88a], 415[114]; **8**, 824[63], 825[63]

Ogata, S., **6**, 546[641,642]

Ogata, T., **3**, 453[113]; **6**, 134[27], 186[172]

Ogata, Y., **2**, 348[53]; **7**, 9[69], 230[128], 247[105], 384[113], 385[113], 438[21], 493[193], 674[46], 748[113], 766[172], 769[221], 851[18]; **8**, 371[111]

Ogawa, A., **2**, 445[24,25], 450[54]; **3**, 1034[77]; **5**, 442[185,185a]; **6**, 477[98], 479[108], 481[123]; **8**, 244[57], 249[97], 253[97], 323[112], 412[119], 413[119], 620[132], 887[119]

Ogawa, H., **2**, 187[40], 1099[112b]; **7**, 452[54,55], 454[96], 462[54,55], 693[27]; **8**, 427[51]

Ogawa, K., **1**, 86[34,36], 223[80], 224[80], 317[146,147], 319[147]; **2**, 655[132]; **3**, 216[72]; **4**, 1009[140]

Ogawa, M., **1**, 412[53]; **2**, 1068[130,131]; **3**, 672[65]; **5**, 439[168], 812[132]; **7**, 309[26], 708[31]

Ogawa, S., **2**, 364[205]; **5**, 839[76], 864[259]; **6**, 530[423], 726[179], 959[38]; **7**, 365[45], 713[70]; **8**, 388[63], 874[22]

Ogawa, T., **1**, 410[39], 561[164], 568[231]; **2**, 555[140]; **3**, 124[280], 125[280], 126[280]; **4**, 505[138]; **6**, 46[65], 49[96], 51[108], 53[108,120], 60[148], 468[53], 471[62]; **7**, 237[34]; **8**, 364[20]

Ogawa, Y., **1**, 772[199]; **2**, 547[94]; **3**, 883[106,107]; **4**, 377[105b], 381[105]; **5**, 736[145], 737[145]; **6**, 51[104], 655[163], 914[28]; **8**, 567[458]

Ogg, J. E., **4**, 505[148]

Ogi, Y., **7**, 774[318]

Ogiku, T., **6**, 53[118]

Ogilvie, K. K., **6**, 603[20], 624[146], 625[146], 655[161,167], 656[170]

Ogilvie, W. W., **1**, 825[50]; **6**, 448[107]

Ogimura, Y., **4**, 227[206,207], 255[200], 650[424]

Ogino, K., **6**, 536[545], 538[545]

Ogino, T., **1**, 849[30]; **5**, 125[20]; **7**, 438[23], 444[54], 559[81], 560[81], 562[81]; **8**, 881[83]

Ogiso, A., **8**, 789[123]

Ogiwara, H., **7**, 42[29], 877[134]

Ogle, C. A., **1**, 10[50], 41[202]; **4**, 257[219]

Ogle, J., **8**, 595[77]

Ogliaruso, M. A., **6**, 2[4], 438[61]

Ogloblin, K. A., **6**, 498[166]; **7**, 477[74]

Ognyanov, I., **3**, 390[73]; **5**, 809[114]; **8**, 857[190]

Ognyanov, V. I., **4**, 629[414]

Ogosawara, T., **7**, 245[77]

Ogoshi, H., **8**, 709[45]

O'Grodnick, J. S., **3**, 734[13]

Ogumi, N., **1**, 223[74]

Ogumi, Z., **3**, 647[194]

Oguni, N., **1**, 223[74,84], 225[84a]; **5**, 101[163]; **8**, 99[113]

Ogura, F., **3**, 135[353], 136[353], 137[353], 141[353], 142[353]; **4**, 343[74], 372[58], 397[58]; **7**, 91[34], 310[28], 492[183], 497[219], 657[22], 752[151], 761[60,61], 765[60], 774[322]; **8**, 370[93], 413[131]

Ogura, H., **1**, 418[75]

Ogura, I., **2**, 780[7]

Ogura, K., **1**, 415[61], 524[92], 526[92,97-99], 527[103,104], 555[110], 557[129,131], 558[129,132,134,135,136], 561[165], 568[239], 865[87]; **2**, 61[19], 74[75], 363[194]; **3**, 135[356,357,360,361,362], 136[356,357,360,361,362,370,371], 137[357,360,361,370], 138[357,370,371a], 139[356,360,361,362,370,371], 140[370,371,371a,b], 142[357,360,361], 143[357,360,361,371,371a,b], 144[371a], 155[429], 156[360,361,362]; **4**, 10[33], 18[58], 20[63], 21[63], 37[104], 213[100,101], 259[257], 597[173,174], 637[173,174], 753[164]; **6**, 893[82]; **7**, 95[65], 231[138], 762[78], 778[410]; **8**, 277[155], 556[375], 836[7]

Ogura, M., **7**, 56[17,18], 57[18]

Ogura, T., **4**, 587[30]

Oguri, T., **1**, 642[110], 643[110]; **3**, 46[253]; **5**, 172[118]; **6**, 645[95]; **7**, 674[40,45]

Oh, C. H., **8**, 18[122], 54[152], 66[152], 354[176], 536[174]

Oh, T., **1**, 109[14], 127[92], 427[114]; **2**, 1[1], 1069[136]; **4**, 845[66], 847[66], 848[66]; **7**, 390[9]

Oh, Y.-I., **4**, 355[132]

Ohanessian, G., **4**, 1070[14]

O'Hanion, P. J., **1**, 791[267]

Ohannesian, L., **8**, 319[73]

Ohara, M., **2**, 771[13]

Ohara, S., **1**, 543[29]; **8**, 500[52]

Ohara, T., **2**, 417[19]

O'Hare, D., **7**, 4[18]

Ohashi, K., **1**, 188[70]; **3**, 470[215], 473[215], 476[215]

Ohashi, M., **7**, 862[78,79], 877[127], 882[169]; **8**, 645[44]

Ohashi, S., **3**, 466[187]

Ohashi, T., **1**, 546[56]; **5**, 108[206]

Ohashi, Y., **4**, 590[96], 591[116], 592[120,124,125], 593[135], 594[96], 633[96]; **5**, 281[18], 304[83], 1200[48], 1201[48]; **6**, 20[77]

Ohba, M., **1**, 768[168]; **2**, 657[165]

Ohba, N., **7**, 132[96]

Ohba, S., **7**, 350[23]

Ohbayashi, A., **3**, 464[175]

Ohbuchi, S., **1**, 187[63]; **3**, 575[82]; **5**, 790[22], 820[22]

Ohdoi, K., **1**, 436[146]

Ohe, H., **1**, 243[52]

Ohe, K., **1**, 664[201], 672[201], 712[201], 714[201], 828[75], 862[77]; **6**, 467[52]; **7**, 775[352c,354,355], 776[355,356,358]
Ohe, M., **4**, 1086[115]
Ohfune, Y., **2**, 160[133,134]; **3**, 20[113], 382[39], 400[119-121], 402[126]; **4**, 379[114], 382[132,132b]; **6**, 124[145], 1029[103]; **7**, 377[90]; **8**, 48[111], 66[111], 857[191]
Ohgo, Y., **1**, 371[74]; **8**, 116[68], 154[192,193,194,195]
Ohhara, H., **6**, 984[59]
Oh-hashi, N., **3**, 99[184]
Ohi, S., **2**, 492[51]
Ohira, M., **8**, 598[99,100]
Ohira, N., **3**, 135[353], 136[353], 137[353], 141[353], 142[353]; **7**, 774[322]; **8**, 370[93]
Ohira, S., **6**, 655[164a]; **7**, 100[115]
Ohishi, J., **4**, 331[14], 344[14]
Oh-ishi, T., **6**, 509[258]
Ohizumi, N., **6**, 997[115]
Ohkata, K., **6**, 821[115]
Ohkatsu, Y., **7**, 108[175]
Ohkawa, S., **2**, 656[154]; **5**, 100[142]
Ohkawa, T., **1**, 90[59]; **2**, 372[270], 373[270]; **5**, 468[130]
Ohki, E., **2**, 213[125]; **3**, 816[79,79a]; **8**, 820[45], 822[52]
Ohki, H., **2**, 24[93], 615[127], 655[138], 656[138], 657[138]
Ohki, M., **3**, 99[191], 107[191]
Ohki, T., **7**, 628[45]
Ohkubo, H., **7**, 773[299], 779[299]
Ohkubo, K., **8**, 552[353,354]
Ohkuma, H., **5**, 736[145], 737[145]
Ohkuma, T., **1**, 238[35], 240[41], 420[84]; **8**, 154[198,199,200,202]
Ohkura, T., **8**, 201[139], 935[64]
Ohl, H., **5**, 497[224]
Ohlendorf, H. W., **8**, 57[172], 66[172]
Ohler, E., **1**, 188[73], 189[73]
Öhler, E., **1**, 219[56]; **8**, 860[223]
Ohloff, G., **1**, 754[106], **2**, 529[20], 540[69,73]; **3**, 736[29]; **5**, 10[76], 456[83,85], 757[80], 761[80], 803[88], 809[114], 853[171], 881[12], 906[67], 972[8], 973[11], 989[8]; **6**, 1043[14], 1058[14,58], 1059[62,64], 1060[71], 1067[101]; **7**, 84[3], 97[94], 306[10], 708[32]; **8**, 566[457], 929[27], 948[148], 965[68]
Ohlson, S., **7**, 145[162]
Ohlsson, B., **4**, 532[86-88], 534[87,88], 535[87], 537[88], 538[87,88], 539[87,88]
Ohmasa, N., **4**, 447[218]
Ohme, R., **6**, 494[130], 562[883]
Ohmizu, H., **1**, 268[54], 269[54a], 346[128]; **3**, 597[201], 599[204]; **4**, 809[162]; **5**, 500[259]; **6**, 11[49]; **7**, 407[79], 804[58]
Ohmori, H., **6**, 439[69]; **7**, 752[153]; **8**, 375[155]
Ohmori, K., **6**, 834[41]
Ohmori, M., **7**, 410[92]
Ohmori, S., **2**, 780[13]
Ohnaka, Y., **6**, 531[429]
Ohnesorge, W. E., **7**, 801[37,42]
Ohnishi, H., **4**, 615[384]; **5**, 497[225]
Ohnishi, K., **7**, 423[145]
Ohnishi, S., **7**, 174[135]
Ohnishi, Y., **5**, 436[153], 755[73], 761[73]; **8**, 93[72,73], 95[85], 96[94], 561[416]
Ohno, A., **1**, 450[211], 512[40]; **3**, 147[387], 149[387], 150[387], 151[387], 463[163]; **4**, 606[306], 607[306], 647[306]; **5**, 436[153]; **8**, 93[72,73], 94[80], 95[85,87,89], 185[18,25], 190[18,65,68], 191[86,87], 195[108], 561[416,417], 589[48], 591[58], 909[80], 917[118], 918[118], 919[118], 977[140,141]
Ohno, H., **8**, 168[67]
Ohno, K., **3**, 1040[105]; **6**, 658[186a]; **8**, 287[21], 773[68], 778[68]
Ohno, M., **1**, 57[32], 223[84], 225[84b]; **2**, 157[119], 213[123], 619[148], 917[85,86], 920[86]; **3**, 729[40]; **4**, 201[10], 373[76], 382[133], 387[161], 388[133,161], 1004[69]; **5**, 417[65], 451[17], 469[17], 470[17], 597[23], 603[23], 606[23], 973[14], 981[14]; **6**, 115[82], 233[48], 614[93]; **7**, 680[79], 684[93a], 700[59,62]
Ohno, R., **2**, 631[14], 635[14]
Ohno, S., **5**, 504[276]
Ohnuki, T., **4**, 719[26]
Ohnuma, T., **4**, 24[75], 25[75a], 30[87], 102[129]; **5**, 808[109]; **7**, 175[142], 353[35], 355[35]; **8**, 844[75]
Ohowa, M., **8**, 170[82]
Ohrr, J., **7**, 841[18]
Ohrt, J. M., **5**, 186[169]
Ohrui, H., **4**, 38[108,108a]
Ohsaka, T., **3**, 667[49]
Ohsawa, A., **3**, 461[146], 541[107]; **5**, 497[225]; **6**, 217[117], 221[117]; **7**, 743[64]
Ohsawa, H., **3**, 172[511,512], 173[512]
Ohsawa, K., **4**, 30[88,88h]
Ohsawa, T., **2**, 819[103]; **4**, 333[21-23], 378[107], 398[215], 401[229]; **7**, 229[113], 493[199]; **8**, 252[111], 795[24]
Ohshima, E., **3**, 883[110]; **6**, 219[121]
Ohshima, M., **2**, 116[138], 576[74], 578[86], 744[88]; **6**, 558[838]; **7**, 141[132], 761[64]
Ohshima, T., **5**, 323[16], 1022[73,73c]
Ohshiro, K., **1**, 566[208]
Ohshiro, Y., **1**, 787[254]; **2**, 128[240], 579[95], 581[105]; **3**, 672[65]; **4**, 589[78], 630[419], 1020[239]; **5**, 474[158], 1200[49,55]; **7**, 209[93], 453[65]; **8**, 806[106,107], 807[106], 900[31]
Ohsugi, M., **6**, 270[80]; **7**, 9[70]
Ohta, A., **4**, 1102[201]; **6**, 237[66], 938[134]; **7**, 667[78], 750[128]; **8**, 371[103], 902[47], 904[47], 905[47]
Ohta, B., **7**, 768[199]
Ohta, H., **1**, 820[16,17]; **3**, 222[143]; **7**, 429[154], 778[416]; **8**, 188[49], 193[49,101], 561[411], 889[126]
Ohta, K., **1**, 779[222]; **8**, 899[28], 902[44], 904[55], 908[44,55], 909[44], 910[55], 911[55], 914[55]
Ohta, M., **2**, 728[144]; **6**, 535[544], 538[544]; **7**, 750[128]; **8**, 460[255]
Ohta, N., **4**, 600[241], 611[354]
Ohta, S., **2**, 759[27]; **6**, 516[319], 682[337]
Ohta, T., **3**, 306[78], 498[109], 681[100]; **7**, 184[168], 335[31], 451[18], 452[50], 454[18]; **8**, 154[199,201], 460[255,256], 461[261], 462[267], 591[59], 614[83]
Ohta, Y., **2**, 464[98]
Ohtani, E., **8**, 798[54]
Ohtani, H., **5**, 623[28]
Ohtani, M., **2**, 157[119]
Ohtani, T., **4**, 359[161]
Ohtani, Y., **8**, 444[6]
Ohteki, H., **1**, 561[160]
Ohtomi, M., **1**, 543[14]; **3**, 826[39]
Ohtsu, M., **7**, 470[13]
Ohtsuka, E., **6**, 604[26,34], 611[65], 625[153], 626[167]
Ohtsuka, H., **8**, 7[42], 176[134]
Ohtsuka, S., **4**, 609[332]
Ohtsuka, T., **2**, 547[122], 553[122]; **3**, 400[120-124], 402[125], 404[132,135,137], 405[138], 714[35]; **6**, 780[69]; **7**, 91[36], 109[184]; **8**, 244[70], 857[191]
Ohtsuka, Y., **3**, 834[76]; **6**, 533[502]; **7**, 678[71]
Ohtsuki, K., **3**, 136[371], 138[371a], 139[371], 140[371,371a], 143[371,371a], 144[371a]; **4**, 18[58], 259[257]
Ohtsuru, M., **5**, 809[115]
Ohuchi, K., **1**, 447[214], 450[214]
Ohuchi, Y., **3**, 244[25], 267[25], 494[84]; **6**, 777[61]; **8**, 698[138]
Ohuchida, S., **1**, 850[34]; **4**, 370[27]
Ohue, Y., **7**, 834[79]
Ohuma, T., **7**, 163[73]
Ohwa, M., **1**, 63[42]

Ohwada, T., **3**, 306[78]
Ohya, T., **4**, 820[215]
Ohyama, T., **3**, 300[39]
Ohyoshi, A., **4**, 615[383]
Oi, R., **1**, 755[115], 812[115], 813[115]
Oiarbide, M., **6**, 938[129], 940[129]
Oida, H., **7**, 774[318]
Oida, S., **2**, 213[125], 649[100], 1059[74]
Oida, T., **3**, 95[152]; **8**, 847[90]
Oikawa, A., **5**, 569[112]
Oikawa, H., **5**, 564[97]
Oikawa, T., **8**, 967[79]
Oikawa, Y., **1**, 183[51], 763[145]; **2**, 10[40], 801[37]; **5**, 834[57]; **6**, 652[140], 660[209], 930[84]; **7**, 244[71], 245[73,80], 246[81,83-86], 370[66]; **8**, 963[49]
Oishi, T., **1**, 463[20], 766[160], 834[121a,122]; **2**, 21[85], 73[66], 846[207], 1022[52]; **3**, 95[157], 97[172,173], 114[157], 116[157,172,173]; **4**, 30[87], 102[129], 127[220b], 522[53], 523[53], 680[50]; **5**, 687[57], 890[33]; **6**, 5[26], 979[29], 995[102]; **7**, 175[142], 489[165]; **8**, 7[38,41,42], 11[38,59], 12[64,65], 190[64], 191[95], 195[106], 197[106], 198[134], 201[139], 252[111], 795[24]
Ojima, I., **1**, 349[146,148], 555[117]; **2**, 232[177], 272[81], 315[47], 316[47], 603[45,47], 646[86], 647[87], 717[67], 929[126], 930[126,-128,129], 931[126]; **3**, 923[42], 949[97], 1027[45]; **4**, 254[187], 927[43], 930[43], 945[43]; **5**, 86[15,17], 96[107,112,113], 98[127,-128], 102[175,181], 410[41]; **7**, 443[51b]; **8**, 20[139], 36[84], 39[84], 66[84], 152[181,182,183], 173[111,114-117,119,120], 251[104], 253[104], 459[228,242], 460[254], 461[259], 535[166], 555[370], 556[374], 763[1], 765[11], 770[37], 779[1d], 780[92], 782[101,103], 784[111], 785[1]
Ojima, J., **2**, 152[101]; **3**, 593[181,182]; **6**, 531[459,460], 764[9,10]
Ojosipe, B. A., **5**, 619[10], 620[10]
Ok, D., **5**, 382[120]; **7**, 743[63]
Ok, H., **4**, 1033[36]; **6**, 124[144]
Oka, K., **2**, 819[97], 823[97]; **3**, 870[50]; **7**, 169[114]
Oka, S., **1**, 450[211], 512[40]; **3**, 147[387], 149[387], 150[387], 151[387], 463[163], 484[24], 501[24]; **4**, 606[306], 607[306], 647[306]; **8**, 93[72], 94[80], 95[85,87,89], 185[18,25], 190[18,65,68], 191[86,87], 195[108], 561[417], 591[58], 600[107], 977[140,141]
Oka, Y., **6**, 787[101,102]
Okabe, H., **1**, 568[240]; **3**, 135[364], 139[364], 142[364], 143[364]; **6**, 21[79]
Okabe, K., **3**, 306[78]
Okabe, M., **7**, 102[136], 239[53]
Okada, E., **4**, 435[135]
Okada, H., **3**, 843[25]; **5**, 38[23c]; **8**, 858[205]
Okada, J., **3**, 557[37,38]
Okada, K., **1**, 123[75], 373[82]; **2**, 384[319], 509[34]; **3**, 35[201], 39[201], 644[134a]; **4**, 893[154], 1054[133]; **6**, 836[58]; **7**, 410[93], 874[101]; **8**, 5[30], 563[425]
Okada, M., **3**, 503[142]; **4**, 8[30a], 592[126], 617[126], 618[126], 633[126]; **5**, 439[170], 464[108], 497[225]; **6**, 1022[56]; **7**, 200[42], 209[92]; **8**, 552[354]
Okada, N., **3**, 227[212]; **5**, 377[110], 378[110a]
Okada, S., **1**, 78[20], 317[155], 320[155]
Okada, S.-i., **3**, 503[144]
Okada, T., **2**, 789[56]; **4**, 350[118]; **7**, 856[66]; **8**, 975[133], 992[55]
Okada, Y., **2**, 115[124]; **5**, 850[146]; **8**, 496[31], 535[166], 550[332]
Okahara, M., **6**, 71[21]; **7**, 471[24]
Okai, H., **6**, 658[184]
Okajima, H., **7**, 96[87]
Okajima, T., **1**, 553[94]; **3**, 390[79]; **6**, 875[18]; **8**, 289[24]
Okamoto, H., **5**, 637[102]
Okamoto, K., **5**, 714[72,73]; **7**, 874[101]
Okamoto, M., **5**, 77[254,256]; **6**, 516[319], 682[337]
Okamoto, S., **1**, 784[244]; **7**, 412[105], 414[105,105c,108,109], 712[64]
Okamoto, T., **1**, 450[211]; **2**, 187[40]; **3**, 463[163], 484[24], 501[24]; **4**, 433[124], 610[342]; **6**, 524[361], 801[28]; **8**, 600[107], 672[20]
Okamoto, Y., **1**, 14[74-76], 610[46], 611[46], 617[53]; **3**, 229[227], 445[71], 455[122], 492[78], 495[78]; **6**, 214[92], 487[39], 489[39], 543[39], 654[153], 928[82], 955[25]; **7**, 202[45], 616[18], 778[416]; **8**, 392[101], 568[467], 620[139], 865[249]
Okamura, H., **2**, 72[58], 74[80]; **3**, 446[79,80,82], 456[79,80], 459[136], 460[136], 461[136], 470[79,80,224], 471[80], 472[224], 473[79]; **4**, 589[83], 596[83]; **8**, 842[41]
Okamura, N., **3**, 222[144]; **4**, 13[44,44c], 253[169], 262[308]; **6**, 837[60], 942[154], 944[154]
Okamura, T., **4**, 30[88,88j]
Okamura, W. H., **1**, 780[229]; **3**, 223[147,148]; **5**, 707[35], 715[77], 735[35a,138a,b], 740[35a,138], 741[155], 742[159b,c], 1005[32]; **6**, 155[152], 903[139,140]; **8**, 448[149], 514[113], 681[76], 689[76], 913[93], 914[93], 936[72]
Okano, A., **4**, 258[240,241], 261[240], 262[240]
Okano, K., **2**, 213[123], 913[77], 914[77]; **5**, 847[136], 913[105], 1012[49]; **6**, 233[48]
Okano, M., **1**, 648[126]; **2**, 555[138], 598[16]; **3**, 87[111], 95[152], 106[111], 114[111], 381[26,27,29], 382[26,27,29]; **4**, 315[520], 340[47], 341[56,58], 347[87], 349[58]; **6**, 276[119], 289[194,195,197], 291[201], 293[194,195,197], 1030[104], 1031[110,112], 1032[121]; **7**, 95[64], 128[68], 129[70], 443[51a], 451[29], 495[207], 496[214], 505[288], 520[27], 521[33], 523[43], 530[23,25], 534[40,41], 760[32], 771[264], 778[409]; **8**, 170[81], 476[24], 849[114], 851[124]
Okano, T., **3**, 555[29], 1027[46]; **4**, 606[303], 646[303]; **8**, 153[184], 252[110]
Okarma, P. J., **3**, 590[163]; **4**, 871[33]; **5**, 66[76], 802[84]
Okatani, T., **5**, 492[248]
Okauchi, T., **4**, 161[91,91c]
Okawa, K., **6**, 96[149], 523[351], 524[351]
Okawa, M., **7**, 170[118], 795[11], 797[20]
Okawara, M., **1**, 642[116], 645[116]; **2**, 578[88], 587[88,149]; **3**, 97[174,175,176], 103[175,176], 108[174], 109[175,176], 117[174], 136[372], 169[509]; **4**, 738[98], 792[65,67]; **7**, 263[88], 318[58], 319[58], 320[58], 322[68], 533[33], 616[19], 764[109], 765[137], 771[265], 772[265], 773[265]; **8**, 369[81,82], 382[9], 413[135], 846[80], 935[60]
Okawara, R., **2**, 19[77,77c], 82[9], 572[46], 575[64]; **3**, 650[211]; **4**, 595[148]; **6**, 1022[64]
Okawara, T., **2**, 922[103,104]; **4**, 231[275]; **5**, 102[164]; **6**, 119[116], 817[103]
O'Kay, G., **1**, 766[158]
Okazaki, A., **5**, 637[102]
Okazaki, H., **1**, 570[265]; **6**, 620[122]; **8**, 476[24], 598[98-100]
Okazaki, K., **1**, 101[92], 554[100,104]
Okazaki, M. E., **1**, 889[141], 890[141]; **2**, 1040[106], 1041[112]; **6**, 741[66]
Okazaki, R., **1**, 370[65]; **6**, 475[88-90], 814[93], 923[59,60]; **7**, 222[41]; **8**, 392[93]
Okazaki, T., **7**, 77[124a]; **8**, 49[116], 66[116]
Okazoe, T., **1**, 205[105,108,109], 206[109], 749[79,87], 750[87,88], 807[322,324], 808[322,324], 809[327,329], 810[327,329a], 812[87]; **2**, 597[10], 603[42]; **5**, 1124[51], 1125[58-60]
Okecha, S., **7**, 822[34]
Okhlobystin, O. Y., **1**, 215[32]; **8**, 754[87,106,113], 755[106,113]
Okhlobystin, O. Yu., **7**, 884[185,186]
Ōki, M., **5**, 186[168]
Oki, M., **2**, 157[119]; **7**, 209[94], 771[260], 772[260], 779[260]; **8**, 971[109], 995[67]
Oki, T., **5**, 736[145], 737[145]
Okida, Y., **3**, 644[133]
Okimoto, M., **8**, 135[42], 253[114]
Okinaga, N., **2**, 635[47], 640[47]
Okinoshima, H., **8**, 772[53], 780[53]

Okisaki, K., **7**, 407[83]
Okita, H., **8**, 781[96]
Okita, M., **3**, 1036[82]
Okita, T., **5**, 603[48]
Okomoto, H., **4**, 600[236]
Okorie, A., **1**, 601[40], 608[40]
O'Krongly, D., **5**, 373[106,106b]
Oku, A., **1**, 215[36]; **2**, 651[113,114], 657[164]; **3**, 464[175], 1047[6]; **4**, 958[26], 960[26], 1014[185]; **5**, 552[28]
Oku, M., **5**, 324[20]
Oku, T., **4**, 350[119]; **5**, 96[106,117]; **7**, 493[198]
Okubo, T., **3**, 676[77]
Okuda, C., **4**, 557[11]
Okuda, F., **6**, 233[40]
Okuda, K., **8**, 150[131]
Okuda, S., **7**, 385[118], 400[51]
Okuda, T., **5**, 167[94], 210[57], 618[2]
Okuda, Y., **1**, 95[78], 188[72], 189[72], 450[213]; **2**, 3[9], 5[9c], 6[9], 19[9,9c], 20[9c,80], 21[80], 29[9c]; **8**, 568[475], 886[113]
Okude, Y., **1**, 174[9], 177[9,19], 179[9,19], 202[100], 234[25]; **3**, 421[63]; **8**, 797[45]
Okuhara, K., **4**, 128[221]
O'Kuhn, S., **3**, 218[97]
Okukado, N., **1**, 214[27]; **3**, 230[239], 232[239a], 238[239a], 260[139], 443[52], 450[103], 453[103], 486[42,44], 495[44], 498[42], 503[150], 524[35], 525[43], 529[53], 530[54]; **4**, 892[141], 893[149]; **5**, 1166[25]; **8**, 680[72], 683[72], 693[72,112], 755[117,128], 758[117]
Okuma, K., **1**, 820[16,17]; **6**, 237[64], 243[64]
Okumoto, H., **4**, 227[203,204], 261[298], 629[405]; **8**, 216[61], 994[65]
Okumura, F. S., **6**, 431[112]
Okumura, K., **3**, 100[201], 103[201], 107[201]; **6**, 900[116]
Okumura, T., **2**, 495[63], 496[63]; **7**, 476[61]
Okumura, W. H., **6**, 14[57]
Okumura, Y., **8**, 837[18]
Okuno, H., **3**, 891[45]
Okura, M., **5**, 439[170]
Okura, S., **2**, 581[103]
Okutani, T., **6**, 921[48]
Okutome, T., **6**, 507[240], 515[240]
Okuyama, N., **2**, 780[10]
Okuyama, S., **4**, 261[286]
Okuyama, T., **4**, 276[79], 299[297]; **5**, 72[166]; **6**, 445[100]; **8**, 87[33], 477[32]
Olah, G. A., **1**, 488[12]; **2**, 523[86,88-90], 527[2], 528[2], 553[2], 708[2], 709[9], 710[2], 711[2], 727[2], 728[138,142], 733[1], 735[10], 738[10], 740[1], 744[90], 745[105], 749[136], 753[1]; **3**, 294[1,2], 295[2,6,7,9], 297[18,24], 298[2,28], 299[34a,b], 300[37,39,45], 302[52], 303[58,58b], 305[71,76], 311[99], 320[135], 321[2], 322[142a], 326[166], 327[168], 329[185], 330[192], 331[200a], 332[204,205,206], 333[34,208,209,210,212], 334[213,215,216,221,221a,b], 335[1,6], 339[6], 421[62,64], 587[142,145], 706[6], 1017[7], 1046[1]; **4**, 271[20], 304[357], 356[145]; **6**, 104[9], 107[22], 110[48,51], 207[51], 210[75], 214[93], 215[93], 216[108], 219[108], 237[60,68], 243[60], 251[146], 254[162], 270[79], 291[209], 456[158], 492[124], 564[915], 647[108,109], 654[152], 665[229], 667[229], 685[351], 726[188], 749[101,102], 765[14], 938[127], 940[148], 944[127]; **7**, 2[3], 5[20], 6[28], 7[40,49], 10[73,74], 14[129], 17[174,177,178], 231[150,151,152], 235[4], 299[47], 674[39], 752[152], 760[47], 765[141], 769[210], 800[34]; **8**, 91[53,61,64], 113[46], 116[46], 216[57,62,63], 217[63], 319[73], 383[18], 384[30], 388[18], 391[89], 403[11], 404[14], 405[25,32], 406[32,39,47,48,51], 407[25], 408[73,74], 568[467], 724[173], 797[41,42], 850[121], 959[25], 987[21], 988[33], 989[34]
Olah, J. A., **2**, 736[20]; **3**, 295[9], 300[37,39], 303[58,58b], 322[142a], 333[209], 1017[7]; **6**, 216[108], 219[108]
Oláh, V. A., **8**, 226[114]
Olaj, O. F., **5**, 63[19]
Olano, B., **8**, 13[72], 124[87,88]
Olbrich, G., **5**, 216[12], 219[12], 221[12]
Olbrysch, O., **4**, 868[12], 874[54], 880[87], 887[12]; **5**, 30[2]
Ol'decap, Y. A., **7**, 15[153]
Oldenburg, C. E. M., **4**, 1040[79], 1049[79], 1060[79b]; **5**, 599[40], 804[94], 986[40]
Oldenziel, O.H., **2**, 1084[11]
Oldham, A. R., **8**, 445[61], 457[212]
Oldham, M. A., **6**, 36[25]
Olea, M. D. P., **2**, 385[328]
O'Leary, V., **2**, 152[102]
Olechowski, J. R., **5**, 800[75]
Oleinikova, E. B., **4**, 55[157], 57[157o]
Olejniczak, B., **8**, 857[188]
Olekhnovich, E. P., **6**, 556[25], 563[25]
Olekhnovich, L. P., **6**, 552[696]
Oleneva, G. I., **2**, 726[122]
Oleownik, A., **8**, 771[43]
Olesker, A., **7**, 239[46], 630[55]
Olin, G. R., **5**, 850[156]; **7**, 854[48], 855[48]
Olin, S. S., **5**, 568[106]
Olinski, R., **4**, 764[221]
Oliva, A., **2**, 517[61]; **5**, 64[40], 73[190]; **6**, 453[138]
Olivé, J.-L., **8**, 882[84,85], 883[84]
Olive, S., **8**, 447[105]
Olivella, S., **4**, 1070[15]
Oliver, J. D., **8**, 459[237]
Oliver, J. E., **6**, 207[49], 976[6]; **7**, 673[22]
Oliver, J. P., **1**, 17[205], 215[33]; **3**, 208[3]; **4**, 70[9], 139[1]
Oliver, M. A., **3**, 591[170]; **4**, 395[201]
Oliver, R., **3**, 219[104], 522[17]
Olivero, A. G., **1**, 146[44], 148[44], 165[44,112a]; **5**, 670[47]
Oliveros, E., **1**, 838[161,166]
Oliveto, E. P., **1**, 174[8], 175[8]; **3**, 23[142]; **8**, 530[93], 544[256], 950[170]
Olk, B., **4**, 387[163a-c]; **6**, 441[86]
Olk, R.-M., **6**, 441[86]
Oller, M., **1**, 547[63], 548[63]; **4**, 113[169]; **6**, 681[333], 961[67]
Ollerenshaw, J., **1**, 377[103]
Olli, L. K., **2**, 829[135]; **3**, 901[110]
Ollinger, J., **1**, 826[60]; **6**, 133[2], 139[50]; **7**, 656[15]; **8**, 993[61]
Ollis, D. F., **8**, 150[128]
Ollis, W. D., **1**, 2[6], 391[150], 489[18], 499[18], 542[5], 543[5], 544[5], 546[5], 580[1], 630[10], 820[7,12], 822[7], 833[117]; **2**, 240[12], 321[11], 323[32], 329[11]; **3**, 86[21,22,32,33,37,38,48,62], 104[21], 121[21,22], 147[32,33], 154[32], 158[37,38], 159[37], 163[38], 164[37], 173[37,38,62], 419[35], 422[68], 660[15], 688[115], 780[10], 862[2], 914[10], 916[16,19,20], 919[30], 923[45], 926[47], 928[47], 930[10,59], 931[10], 932[60], 933[30,62], 946[30], 949[94-96,98,99], 954[98], 963[121]; **4**, 3[8], 6[8], 99[118d], 316[535], 483[4], 484[4], 495[4], 507[154], 725[43], 953[8], 968[57], 987[133], 1002[51], 1004[76,77], 1021[76,77], 1096[160], 1097[160], 1098[160]; **5**, 847[135]; **6**, 2[1], 28[1], 104[1,9], 111[64], 116[90], 133[5], 134[11], 171[4], 177[4], 198[4], 225[5], 226[5], 258[5], 293[229], 419[1], 421[1], 436[15], 437[15], 440[73], 448[15], 450[15], 453[15], 455[15], 893[79], 897[95,96,98]; **7**, 41[18], 84[1], 85[1], 108[1,180], 329[4], 343[4], 345[1], 470[2], 594[4], 598[4], 671[10], 673[10], 687[10]; **8**, 26[12], 27[12], 247[81], 279, 301[93], 403[2,6-8], 404[2], 699[148], 704[4], 705[4], 706[4], 707[4], 710[4], 715[4], 716[4], 717[4], 722[4], 724[4], 725[4], 726[4], 728[4], 794[3], 984[2]
Ollivier, J., **4**, 1046[118]; **5**, 911[90], 923[137], 954[90], 956[307]; **7**, 14[135]
Ollmann, G., **2**, 900[26], 960[35], 962[35]

Olmstead, H. D., **1**, 3[18,19]; **2**, 124[201], 183[17], 184[24,25], 235[190], 268[66], 280[23], 289[23], 311[34], 599[23]; **3**, 4[18], 7[18], 8[18], 11[18], 14[70]; **7**, 130[76]

Olmstead, M. M., **1**, 11[58], 23[119], 34[220,221,222], 41[265,271], 488[13]; **3**, 213[43], 1025[34]; **4**, 372[62]; **5**, 1039[17], 1041[19], 1043[22], 1046[19], 1049[22], 1050[17], 1052[17,19]

Olmstead, T. A., **3**, 58[292]; **5**, 843[122]

Olmsted, A. W., **6**, 204[9]

Olofson, R. A., **1**, 755[116], 756[116], 758[116], 761[116]; **2**, 182[10]; **4**, 486[39], 966[53]; **5**, 1008[44]

Olofson, R. S., **6**, 501[186], 502[186]

Olofsson, B., **3**, 634[7], 636[7,7b], 637[7]

Olsen, B. A., **3**, 854[76]

Olsen, D. J., **3**, 1038[91]; **4**, 561[30]

Olsen, D. K., **6**, 498[163]

Olsen, D. O., **6**, 835[48]

Olsen, E. G., **5**, 709[44], 716[84], 739[44]

Olsen, H., **5**, 78[282]; **6**, 691[404], 692[404]; **8**, 365[27], 390[81]

Olsen, R. J., **1**, 418[72]

Olsen, R. K., **4**, 190[107]; **6**, 446[103], 814[96]

Olsen, R. S., **2**, 842[192,193]

Olson, A., **6**, 824[124]

Olson, D. H., **3**, 296[14], 330[194]

Olson, E. S., **7**, 673[23]

Olson, G. L., **3**, 168[493], 169[493], 171[493], 362[90]; **4**, 305[363]; **7**, 347[15]

Olson, K. D., **7**, 878[145]

Olson, R. E., **1**, 636[93], 638[93], 640[93], 646[93], 647[93]; **2**, 88[31], 601[36], 602[38]; **3**, 87[103], 104[103], 106[103], 111[103], 117[103]; **5**, 1014[55]; **7**, 826[47], 827[47]

Olsson, L., **4**, 395[203]

Olsson, L. F., **4**, 560[22]

Olsson, L. I., **3**, 217[82]; **4**, 308[407], 395[202], 872[42]; **6**, 652[147], 653[147], 654[147], 655[147], 681[147], 866[203]

Olsson, T., **1**, 112[27]; **2**, 120[170]; **4**, 18[56], 152[56], 171[26,27], 202[24], 255[204], 256[207], 262[207]; **6**, 632[6]

Olszanowski, A., **3**, 619[23]

Oltay, E., **4**, 915[11]

Oltmann, K., **1**, 661[167]

Oltvoort, J. J., **2**, 599[20]; **6**, 57[139], 652[145]

Omae, I., **8**, 548[319]

O'Mahoney, R., **1**, 419[80]

O'Mahony, M. J., **1**, 130[94], 801[301]; **6**, 994[98]

O'Mahony, R., **4**, 794[75]

O'Malley, G. J., **4**, 27[83], 370[35], 385[35]

O'Malley, R. F., **7**, 800[35]

O'Malley, S., **7**, 427[148f]

Omar, M. E., **6**, 423[46]

Omar, M. T., **4**, 50[142]

Omata, T., **7**, 470[11]

O'Meara, D., **3**, 822[2], 831[2]

Omi, T., **1**, 223[74]

Omietanski, G. M., **2**, 524[76]

Omizu, H., **3**, 602[221]; **8**, 173[113]

Omkaram, N., **5**, 211[62]

Omori, K., **2**, 97[59]

Omori, Y., **6**, 227[20], 236[20]

Omote, Y., **1**, 373[88], 374[88]; **3**, 1057[38]; **5**, 181[150]; **6**, 67[14], 69[14]; **8**, 541[203], 642[32,34], 840[29], 964[55]

Omoto, H., **7**, 806[75]

Omoto, S., **7**, 341[51,52]

Omura, H., **2**, 451[55]; **7**, 25[43], 26[43]

Omura, K., **6**, 501[185], 531[185]; **7**, 297[27], 298[27], 302[64]

Omura, S., **2**, 1[2], 240[5]; **5**, 151[9]; **6**, 918[37]; **7**, 406[87]

Omura, T., **7**, 11[88]

Onak, T., **7**, 594[6], 598[6]

Onak, T. P., **1**, 292[27]

Onaka, M., **1**, 453[221,222]; **2**, 576[75], 615[130], 631[14], 635[14], 655[139]; **3**, 426[82], 429[82]; **4**, 161[88], 1002[46]; **6**, 66[4], 89[108], 93[130], 237[66], 254[160]

Onaka, S., **8**, 150[133], 151[133,145]

Onami, T., **3**, 224[180]; **7**, 393[16], 398[16,16a]

Onan, K. D., **2**, 217[139]; **5**, 348[74a], 499[252], 500[252]; **8**, 614[89], 620[89]

Onda, H., **8**, 709[45]

Ondik, H., **1**, 8[35]

O'Neal, H. E., **5**, 856[193]

O'Neil, I. A., **2**, 742[77], 968[78]; **4**, 744[134]

O'Neill, B. T., **6**, 960[57]

O'Neill, H. J., **7**, 95[74]

O'Neill, M. E., **1**, 2[8]; **3**, 208[1]

O'Neill, P., **2**, 107[61]

Ong, B. S., **5**, 596[31], 597[31], 608[31], 780[205]; **6**, 1005[142,143]

Ong, C. C., **2**, 710[27]

Ong, C. W., **4**, 667[12], 668[14], 670[14], 675[39], 677[14], 678[42], 682[12]

Ong, J., **5**, 834[53]

Ongania, K. H., **5**, 92[62]; **6**, 543[614]

Ongoka, P., **2**, 77[91]; **4**, 987[162], 993[162]

O'Niel, B., **4**, 405[252]

Onishchenko, A. S., **5**, 451[10], 513[2], 518[2]

Onishi, A., **3**, 168[501], 169[501], 170[501]

Onishi, H., **6**, 237[61]

Onishi, T., **1**, 803[306]; **5**, 15[105], 787[8], 797[61], 799[74], 882[14]; **6**, 157[172], 991[89]; **7**, 660[43]

Onishi, Y., **7**, 764[107]

Onisko, B. C., **4**, 390[168]

Onistschenko, A., **6**, 95[146]; **7**, 470[9], 487[9]; **8**, 563[436]

Onitsuka, K., **5**, 1137[56]

Onken, D. W., **4**, 48[140]

Ono, A., **8**, 314[43], 369[79,80], 968[88], 988[29], 989[36,38]

Ono, H., **8**, 528[67]

Ono, M., **1**, 255[17], 566[209]; **3**, 541[106], 735[22]; **4**, 238[1], 371[47]; **5**, 753[58]; **6**, 552[690]; **7**, 98[104], 537[61], 770[256c], 771[256], 819[22]; **8**, 97[95], 115[65], 607[32]

Ono, N., **2**, 321[17], 323[17], 330[17], 332[63,64], 333[64-66], 334[68], 360[168]; **4**, 13[44,44b], 31[92,92i], 37[105], 64[92i], 591[108], 599[219], 633[108], 641[219], 790[36]; **5**, 320[10,11]; **6**, 161[178], 926[69], 1000[128,129,130]; **7**, 197[19], 883[174]; **8**, 962[41], 969[98]

Ono, S., **5**, 79[286]

Ono, T., **1**, 176[17], 520[71,75]; **2**, 363[193], 374[275]; **4**, 258[248], 261[248]; **6**, 780[72]; **8**, 934[57]

Onoda, Y., **1**, 328[29]

Onodera, A., **8**, 29[38], 66[38]

Onodera, K., **7**, 299[44]

Onoe, A., **3**, 381[29], 382[29]; **7**, 530[25]; **8**, 476[24]

Onomura, O., **2**, 1067[123]; **7**, 227[75], 248[109], 803[55], 804[62]

Onopchenko, A., **8**, 764[4b], 774[71]

Onoue, H., **1**, 123[75], 373[82]; **4**, 590[99]

Onozuka, J., **1**, 561[165]

Onrust, R., **5**, 714[75b], 847[136]

Onuma, K., **3**, 243[18], 244[18], 247[18], 464[169]; **8**, 460[254]

Onuma, K.-i., **8**, 535[166]

Onyestyak, G., **4**, 298[287]

Onyido, I., **4**, 425[23]

Onyiriuka, O. S., **4**, 1060[159]

Oohara, T., **1**, 524[93], 526[93,95]; **2**, 429[48]; **6**, 93[134], 156[152]; **7**, 425[149c]

Ooi, H. C., **6**, 570[951]

Ooi, N. S., **7**, 846[83]

Ooi, T., **3**, 750[86]

Ooi, Y., **8**, 369[78]

Ookawa, A., **1**, 86[34,36], 223[80], 224[80], 317[146,147], 319[147]; **4**, 85[71], 203[27-29]; **6**, 76[44]; **8**, 13[70], 244[55,59], 248[55], 250[59], 384[34], 412[110], 874[25,28], 875[28,30]
Ookita, M., **4**, 903[202], 904[202]
Oommen, P. K., **8**, 502[64], 503[67], 568[467]
Ooms, P. H. J., **2**, 867[11]; **5**, 71[159,160]; **6**, 558[849]
Ooms, P. H. M., **6**, 558[856]
Oon, S.-M., **3**, 1056[35], 1062[35]; **4**, 1033[22], 1057[22c]
Ooshima, M., **4**, 557[12]
Oosterwijk, R., **1**, 571[280]
Oostveen, E. A., **4**, 465[117]
Oostveen, J. M., **3**, 217[85]
Oota, O., **2**, 348[58], 357[58]
Ootake, A., **1**, 552[82]
Ootake, K., **3**, 159[455], 161[455]
Opalko, A., **3**, 505[169]
Oparaeche, N. N., **3**, 597[199]
Oparina, L. A., **6**, 509[282]
op den Brouw, P. M., **6**, 556[831]
Openshaw, H. T., **6**, 624[150]
Opheim, K., **7**, 86[16a], 109[182]
op het Veld, P. H. G., **5**, 723[109]
Opitz, G., **1**, 367[56], 368[56], 370[56]; **2**, 894[3], 897[3], 933[3], 953[1,2], 966[2], 1088[46,51], 1090[13,46], 1106[13]
Opitz, K., **5**, 391[143], 721[101]
Opitz, R. J., **6**, 281[147,149]
Oplinger, J. A., **2**, 714[53]; **3**, 363[86]; **4**, 1049[121b]; **5**, 516[28], 814[140], 994[53], 997[53]; **8**, 927[20]
Oppenauer, R., **6**, 37[32]
Oppenauer, R. V., **7**, 100[122]
Oppenheim, E., **6**, 441[85]
Oppenländer, T., **5**, 195[9], 209[9]; **6**, 530[419]; **8**, 95[90]
Opperman, M., **8**, 708[37]
Oppolzer, W., **1**, 87[49], 214[25], 223[25b], 225[25b], 289[110], 307[111], 311[110], 312[111], 313[117,118], 317[154], 320[154], 404[24], 767[179], 770[187]; **2**, 69[48], 106[54], 231[172], 252[41], 253[41], 358[153], 455[11], 527[5], 528[5], 540[5], 544[5], 636[55], 637[55], 640[55], 681[60], 924[108a,b], 1015[22]; **3**, 17[89], 46[252], 159[457], 173[457], 178[541], 196[32], 728[35]; **4**, 21[69], 85[72], 107[141], 111[153], 152[54], 184[85], 189[85a], 201[12-16], 202[13-16,21,22], 204[21,32-36], 218[137,138], 231[282,283], 249[120,121], 257[121], 500[108,109], 501[115], 602[262], 644[262], 876[64], 1076[39], 1077[52], 1078[52], 1079[68], 1095[155], 1096[156]; **5**, 1[3], 2[3], 9[3], 10[79], 11[80-83], 13[90,91,93-95], 14[96,97], 15[3], 19[3,79], 26[79], 27[3,79], 37[20,21,22a], 38[22c], 40[26,27], 41[28-30], 42[31], 43[32], 44[34], 45[35,36], 46[38], 47[41], 48[38,41], 49[38], 50[38,38c,41-44], 51[42,45,45a,b], 52[46], 53[45a], 55[42,49], 56[42,49,50a], 57[50,50a,b], 59[50b], 71[124], 123[1], 126[1], 137[73,80], 139[84], 141[89], 186[170], 247[25], 253[46,46c], 260[25,70], 263[70], 264[74], 331[42,42b], 333[42c], 351[83], 354[84a], 356[84,90], 358[90b], 362[93,94], 363[93a,b,d,g,94b], 364[93c,d], 365[93a,e], 376[84a], 385[130,131], 386[130c,135], 390[130a], 391[141], 414[53], 435[151], 451[31,35], 453[31], 464[31], 513[5], 514[5], 515[17], 518[17], 524[5c], 527[5,62,66], 528[66a], 529[66a-c], 543[113,115,117,117b], 545[120], 547[17], 681[29], 682[30], 691[83,83b,84,87,88], 692[83,84], 693[83], 742[162], 779[198], 829[25], 1031[95]; **6**, 77[55], 94[143], 118[106], 248[137], 667[242], 756[125], 757[131], 879[42], 1018[41], 1050[40], 1063[85,86]; **7**, 174[135], 182[165], 646[27]; **8**, 61[189], 66[189], 537[188], 784[112]
Oprean, I., **3**, 416[15], 417[15]
Or, Y. S., **7**, 683[89]; **8**, 623[148], 795[17]
Orahovats, A. S., **3**, 1038[93]; **6**, 74[37]
Oram, D., **1**, 41[268]; **4**, 170[13]
Orama, O., **5**, 1093[96], 1098[96b], 1112[96b]
Orange, C., **6**, 2[5], 18[5]
Orata, H., **4**, 610[338], 649[338]
Orban, J., **2**, 809[55]
Orbaugh, B., **3**, 616[13]
Orbe, M., **2**, 465[107]
Orbovic, N., **7**, 231[142]
Orchin, M., **4**, 915[6], 920[24]; **5**, 724[110], 727[119], 948[272]; **8**, 452[190], 455[206], 456[208], 551[337], 608[44]
Ord, W. O., **8**, 533[153]
Ordsmith, N. H. R., **7**, 729[44]
Ordubadi, M. D., **8**, 124[92], 125[92]
O'Reilly, J. E., **8**, 642[30]
Oren, J., **5**, 223[85,86], 224[85,86], 740[151]
Orena, M., **3**, 45[250]; **4**, 375[94], 377[104], 386[94a,153,153a,157], 387[94a,153a,157], 388[164], 389[166,166a], 393[164a,c], 401[226], 407[104c,153a,157b,254], 408[254a,b,259c]; **6**, 26[106], 533[497], 648[124]; **7**, 280[177], 493[184], 503[269], 663[62], 664[63]
Orfanopoulos, M., **2**, 528[11]; **5**, 2[9]; **7**, 816[10], 818[10]; **8**, 216[66], 318[68], 813[13], 969[97]
Orfanos, V., **5**, 716[85]
Org, H. H., **8**, 565[449]
Organ, H. M., **1**, 780[228]
Orgel, L. E., **7**, 866[89]
Orger, B. H., **5**, 646[9], 647[12], 648[12]
Orgis, J., **6**, 420[16]
Orgura, K., **1**, 506[5], 526[5]
Oribe, T., **8**, 557[381], 782[102]
O'Riordan, E. A., **3**, 131[332]
Orita, H., **7**, 462[119-121]
Oritani, T., **2**, 198[84]
Orito, K., **2**, 384[317]; **6**, 766[21]; **7**, 834[79]; **8**, 528[67]
Orito, Y., **8**, 150[140,143], 151[155,156,157]
Oriyama, T., **1**, 192[82], 489[21], 497[21]; **2**, 35[130], 36[130], 244[28]; **8**, 991[46]
Orlando, C. H., **5**, 158[16], 159[16], 170[16]
Orlando, J., **8**, 27[32], 66[32]
Orlek, B. S., **2**, 882[46]
Orlemans, E. O. M., **8**, 33[58], 66[58]
Orliac-Le Moing, A., **8**, 134[30,31]
Orlinkov, A., **3**, 297[25], 334[25]
Orlinkov, A. V., **2**, 727[136]; **7**, 7[51]
Orlinski, R., **4**, 12[41], 764[222], 765[222], 808[155]; **6**, 551[683]
Ormiston, R. A., **1**, 825[52]
Ornaf, R. M., **5**, 552[31], 564[31]
Ornfelt, J., **4**, 317[547]
Ornstein, P. L., **3**, 446[83]; **6**, 214[91], 676[307]; **8**, 840[39], 935[62]
Oro, L. A., **8**, 445[55], 764[6], 773[6b]
Oroshnik, E. W., **7**, 656[17]
Orozio, A. A., **5**, 945[250]
Orpen, A. G., **6**, 436[9]
Orr, D. E., **7**, 486[145]
Orr, G., **4**, 483[6]
Orr, J. C., **7**, 136[116], 137[116]
Orr, R., **4**, 345[81]
Orrom, W. J., **5**, 129[35]
Ors, J. A., **5**, 196[12,13], 647[14], 649[14,21], 658[21], 670[48]
Orsini, F., **2**, 279[16,17], 280[16], 283[17,45]; **3**, 421[65]
Ort, B., **3**, 586[139]
Ort, M. R., **8**, 532[132]
Ortaggi, G., **4**, 611[344]
Ortar, G., **1**, 195[90]; **3**, 1035[79]; **4**, 860[112], 861[112]; **7**, 92[41,41b], 93[41b], 94[41], 143[148], 144[148]; **8**, 17[118], 84[13], 910[82], 911[87], 933[49,51]
Ortega, J. P., **5**, 70[115], 79[115]
Ortega, M., **5**, 64[40]
Ortez, B., **1**, 552[80]
Ortiz, B., **1**, 544[35]
Ortiz, C. V., **1**, 377[99]

Ortiz, E., **1**, 785[247]
Ortiz, M., **7**, 505[286]; **8**, 408[78]
Ortiz, M. J., **5**, 202[36]
Ortiz de Montellano, B. R., **8**, 530[95]
Ortiz de Montellano, P. R., **4**, 988[138]; **7**, 180[157], 182[157]
Orttung, F. W., **8**, 408[63]
Ortuño, R. M., **6**, 4[19], 984[54]
Ortwine, D. F., **8**, 70[232], 72[232]
Orvis, R. L., **8**, 958[17]
Osa, T., **7**, 50[69]
Osada, M., **6**, 535[538], 538[538]
Osaka, M., **4**, 89[85]
Osaka, N., **6**, 774[50], 1066[96,98]; **7**, 701[64]
Osakada, K., **7**, 314[41], 315[41]; **8**, 239[28], 838[20], 963[42]
Osaki, K., **4**, 27[79,79d,e]
Osaki, M., **2**, 605[62]; **4**, 206[51]
Osammor, M. I., **2**, 800[17]
Osanai, K., **1**, 763[144], 766[155]
Osawa, E., **3**, 383[45], 386[57], 402[131]; **8**, 331[33], 334[60], 342[109]
Osawa, K., **3**, 644[134a]
Osawa, M., **8**, 149[115]
Osawa, T., **2**, 941[164], 942[164]; **5**, 100[154], 102[154]; **8**, 150[125], 151[152], 620[139]
Osawa, Y., **7**, 174[135], 862[79], 877[127]; **8**, 64[217]
Osborn, C. L., **2**, 149[84]; **5**, 66[79]; **8**, 806[110], 807[110,119]
Osborn, J. A., **1**, 440[169], 443[179,180], 451[216]; **5**, 1116[5,10,12], 1125[65]; **8**, 152[163,165,166,167,174], 443[2,5], 444[5b], 445[5,23,33,43,46-50,56,58], 446[66], 449[5a], 450[48], 452[5b,58], 456[5a], 458[48,50], 568[462,478]
Osborn, M. E., **8**, 548[323]
Osborn, R. B. L., **1**, 310[103]
Osborne, D. J., **7**, 401[58]
Osborne, J. E., **4**, 280[131], 281[131], 288[131]
Osby, J. O., **8**, 253[119], 254[123], 375[157], 794[11], 839[24]
Oscarson, S., **8**, 224[107]
Oseda, H., **6**, 531[438]
Osek, J., **6**, 576[975]
Osella, D., **8**, 457[216]
Osgood, E., **6**, 787[97]
O'Shea, D. M., **1**, 894[155]; **4**, 796[100], 822[224], 823[228]; **5**, 925[153]
O'Shea, K. E., **7**, 98[99]
O'Shea, M. G., **4**, 817[208]; **7**, 625[40]
Oshida, J.-V., **6**, 996[105]
Oshikawa, T., **6**, 760[142]; **8**, 411[105]
Oshiki, T., **1**, 233[19]
Oshima, H., **3**, 303[54]
Oshima, K., **1**, 92[62,63,66], 95[78,79], 101[93,94], 103[93,94], 188[72], 189[72], 193[85], 195[85], 198[85], 201[98], 218[52], 450[213], 508[22], 749[79,83,86], 750[83,86], 789[260], 809[327], 810[327], 812[83]; **2**, 19[76], 20[80], 21[80], 59[15], 72[56], 114[123], 269[73], 271[75], 575[67], 588[151], 589[151,154], 597[10], 600[27]; **3**, 96[162,166,167], 99[162,163,166,167], 103[166,167], 104[162,163], 105[166,216], 112[216], 120[162,163,166], 121[162,163], 131[162,163], 155[166], 356[58], 445[73,74], 449[73], 484[24], 501[24], 583[123], 730[43]; **4**, 33[96,96c], 96[103c], 257[227], 261[227], 588[68], 607[310], 626[310], 637[68], 647[310], 721[31], 725[31], 756[184], 770[248,249], 771[253], 789[32], 791[42], 796[96], 824[241], 886[117], 892[143], 901[185,186]; **5**, 850[149], 917[124], 926[124], 927[164], 938[164,219], 943[237], 1124[50], 1125[59,60]; **6**, 7[30], 563[905], 856[149], 861[181], 980[38]; **7**, 254[31], 259[62], 267[118], 268[118], 275[146,147], 276[147], 281[176], 282[176], 283[118], 284[118], 308[17], 309[22,24], 322[67], 324[70], 369[63], 378[63,96], 379[101], 485[138], 615[8]; **8**, 699[149], 755[126,127], 758[126,127], 798[46], 807[46,116], 818[41], 820[41], 823[58], 886[113]
Oshima, M., **4**, 611[352]; **7**, 761[64]
Oshima, T., **8**, 332[43]
Oshino, H., **1**, 212[12], 213[12], 215[12b], 217[12], 327[11], 448[207]; **2**, 68[44], 443[19], 445[28], 446[28], 447[19], 448[19], 449[19]; **3**, 197[33], 221[131]; **4**, 163[96], 164[96]; **5**, 1200[52]
Oshino, N., **7**, 791[1]
Oshiro, Y., **5**, 422[82]; **6**, 186[172]
Oshshiro, Y., **4**, 102[131]
Oshuki, S., **5**, 537[99]
Osibov, O. A., **6**, 226[13]
Oskay, E., **7**, 306[9]
Oslapas, R., **7**, 720[10]
Osman, A., **1**, 215[35]
Osman, A. M., **3**, 834[78,79]
Osman, A. N., **6**, 551[687]
Osman, M. A., **4**, 314[490,491]
Osman, S. F., **8**, 814[15]
Osowska, K., **7**, 483[123]; **8**, 857[188]
Osowska-Pacewicka, K., **6**, 76[45], 79[64], 116[91]; **8**, 385[44]
Osowska-Pacewicza, A., **7**, 500[244]
Osselton, E. M., **5**, 649[22], 650[22]
Ossowski, P., **6**, 43[54]
Ostarek, R., **1**, 83[26], 145[42], 146[42], 148[42], 149[42], 155[42], 170[42]; **2**, 22[86]
Ostaszewski, B., **8**, 765[13]
Ostaszewski, R., **6**, 70[19]
Oster, B. W., **5**, 1140[77], 1156[163], 1158[163a]
Oster, T. A., **2**, 175[184]
Osterberg, A. C., **6**, 554[724]
Osterbury, G., **4**, 298[290]
Osterhout, M. H., **4**, 403[238], 404[238], 405[238], 406[238]; **7**, 503[276]
Ostermann, G., **3**, 921[35], 922[35a,38], 924[35a], 927[52]
Osteryoung, J., **8**, 595[74,75]
Osteryoung, R. A., **2**, 757[13], 759[13]
Ostrander, R. A., **4**, 119[194], 226[199]; **8**, 813[9]
Ostrovskii, V. A., **6**, 795[13], 798[13], 817[13]; **7**, 690[13]
Ostrow, R. W., **6**, 1059[69]
Ostrowicki, A., **6**, 70[18]
Ostrowski, P. C., **1**, 511[30]; **4**, 10[34], 71[13]
Osuch, C., **4**, 35[98c]; **7**, 506[298]
Osuga, D. T., **8**, 54[158], 66[158]
Osugi, J., **5**, 71[139,140,141,142], 76[233], 77[254,257,258]
Osuka, A., **4**, 447[218]; **8**, 315[46], 370[91], 405[29], 408[75], 806[108], 807[108], 978[146]
Osuka, M., **7**, 774[322]
O'Sullivan, A., **5**, 929[169], 930[169]
O'Sullivan, A. C., **7**, 77[120b]
O'Sullivan, M. J., **5**, 105[193]
O'Sullivan, R. D., **3**, 244[23]
O'Sullivan, W. I., **7**, 205[62], 764[125]
Oszczapowicz, J., **6**, 551[683]
Ota, S., **3**, 428[90]
Ota, T., **3**, 494[84]; **4**, 430[96]; **5**, 473[153], 477[153]; **8**, 698[138]
Otaka, K., **3**, 246[44], 485[27]; **4**, 589[80], 591[80], 607[312]; **5**, 431[120], 596[22], 597[22], 603[22]
Otaka, M., **7**, 171[123]
Otake, K., **2**, 18[73,74]
Otani, G., **4**, 221[150,151,152,153,154]; **6**, 717[106,107]
Otani, S., **2**, 443[17]; **4**, 433[124], 973[85]
Otera, J., **1**, 192[82], 511[32], 563[180], 570[265], 802[304], 803[305,306,307]; **2**, 197[77,77c], 74[74], 566[23], 572[46], 581[103]; **3**, 135[345,346,347,348], 136[347,348], 137[347], 139[347,348], 141[348], 144[347,348], 168[508], 169[508], 174[345], 586[138]; **4**, 113[174], 245[86], 259[86], 260[86]; **5**, 176[128], 767[120]; **6**, 157[172,173], 927[77], 1022[60]; **7**, 660[43]

Oterson, R., **2**, 943[172], 945[172]; **7**, 227[84]
Otey, M. C., **8**, 145[88]
Oth, J. F. M., **3**, 75[46]; **5**, 717[90a-d]
Otomasu, H., **1**, 34[168], 359[16], 379[16a]
Otonnaa, D., **7**, 444[55]
Ototani, N., **2**, 819[99]
Otsubo, K., **1**, 258[24], 259[24], 268[53,53b]; **8**, 883[95], 884[95]
Otsubo, T., **3**, 135[353], 136[353], 137[353], 141[353], 142[353], 877[83,85]; **4**, 343[74], 372[58], 397[58], 507[155,156]; **7**, 91[34], 310[28], 492[183], 497[219], 657[22], 752[151], 761[60,61], 765[60], 774[322]; **8**, 370[93], 413[131]
Otsuji, A., **2**, 1067[126]
Otsuji, Y., **3**, 565[21]; **4**, 826[244]; **6**, 425[64]; **7**, 851[24], 875[112], 878[144]; **8**, 394[117], 698[141,142,145], 709[45,45a]
Otsuka, H., **3**, 421[60]
Otsuka, M., **2**, 917[85,86], 920[86]; **5**, 419[74]; **6**, 774[47]; **8**, 395[129]
Otsuka, S., **1**, 62[40], 65[52], 66[57], 69[57b], 70[52]; **2**, 455[5]; **4**, 964[49], 1033[21], 1037[21], 1040[21]; **6**, 866[208]; **7**, 426[148a]; **8**, 153[184,185,186], 154[189], 252[110], 450[165], 458[225]
Otsuka, T., **7**, 418[126]
Otsuki, M., **8**, 253[117]
Ott, D. G., **3**, 828[46]; **6**, 614[92]
Ott, E., **3**, 284[54]
Ott, J., **6**, 450[120]
Ott, J. L., **8**, 47[124], 66[124]
Ott, K. C., **1**, 746[71]
Ott, K.-H., **5**, 803[87], 971[1]
Ott, R. A., **8**, 275[144]
Ott, W., **6**, 524[356]
Ottana, R., **6**, 575[968]
Otten, J., **4**, 869[21]
Ottenheijm, H. C. J., **5**, 829[22]; **6**, 113[70], 114[73]; **7**, 230[134], 763[96]; **8**, 60[183], 61[183], 62[183], 66[183], 405[31]
Otter, B. A., **6**, 554[759]
Otto, C. A., **7**, 355[46]
Otto, E., **6**, 632[2], 642[73], 660[2]
Otto, H.-H., **2**, 212[121]; **4**, 30[88,88q], 253[170], 261[170]
Otto, S., **7**, 663[56]
Ottow, E., **6**, 22[84]
Oturan, M. A., **4**, 453[29,30], 459[29,30], 472[29], 475[30,150]
Ötvös, I., **5**, 1138[66]
Otvos, J. W., **3**, 332[207]
Otvös, L., **7**, 723[25]
Otzenberger, R. D., **3**, 727[31]; **5**, 455[74]
Ou, K., **8**, 371[112]
Ouannes, C., **1**, 115[40]; **4**, 189[104]
Ouchi, A., **8**, 836[10d], 837[10d]
Ouchi, S., **6**, 602[10]
Oudenes, J., **2**, 111[84], 830[143]; **4**, 145[33,34]
Ouellette, D., **3**, 1051[20]
Ouellette, R. J., **7**, 851[20]
Ouerfelli, O., **5**, 18[127]
Oughton, J. F., **3**, 804[13]; **6**, 675[299]; **7**, 582[149]; **8**, 987[23]
Oumar-Mahamat, H., **4**, 807[150]
Oumar-Mahamet, H., **7**, 92[40]
Ourisson, G., **3**, 158[443], 159[443], 160[443], 161[443], 167[443], 168[443], 383[44], 406[140], 564[7]; **7**, 84[3], 121[20,21], 123[20], 145[20], 163[71], 247[101], 359[15], 842[27,28]; **8**, 15[88], 925[11], 926[11]
Ours, C. W., **2**, 956[18], 957[18]
Ouseto, F., **4**, 837[8]
Ousset, J. B., **1**, 825[53]; **2**, 89[34]; **3**, 203[102]; **5**, 774[173], 780[173]
Out, G. J. J., **1**, 218[54]; **2**, 981[23]; **4**, 877[69]; **6**, 86[99]
Outcalt, R., **4**, 31[94]; **5**, 154[33]
Outlaw, J. F., Jr., **8**, 421[29], 422[29], 436[74], 437[77,78]
Outurquin, F., **2**, 787[50]
Ouyang, S.-L., **4**, 1101[193]
Ovadia, D., **4**, 1040[92], 1042[92]
Ovchinnikow, M. V., **6**, 43[52]
Ovenall, D. W., **2**, 619[148]
Overberger, C. G., **3**, 649[202]; **5**, 1016[59]; **7**, 586[166], 763[94]; **8**, 408[63], 532[130], 568[471]
Overberger, C. J., **5**, 557[55]
Overbergh, N., **7**, 475[52]
Overchuck, N. A., **3**, 305[71]
Overend, W. G., **2**, 385[327]
Overheu, W., **5**, 412[45], 422[82], 634[79]
Overman, J. D., **8**, 413[125]
Overman, L. E., **1**, 242[46], 361[33], 583[8,8a], 589[8a,19,20a,b], 591[19,20b,21], 592[8a,20a,22-24], 593[25], 594[27], 595[8a,25-27], 767[163], 889[140,141,142,143], 890[140,141,142,143,144,145], 891[146,147], 892[148], 898[141b,144]; **2**, 547[109], 550[109], 554[135], 555[136,137], 580[96], 941[164,165], 942[164], 1009[7], 1015[23], 1018[7], 1027[74], 1028[77-79], 1029[77], 1030[81-85], 1031[7,86], 1032[87], 1034[82,88], 1035[88,91], 1036[95], 1040[106], 1041[107-112], 1042[113,114], 1043[115], 1047[2,3], 1050[32], 1057[69], 1069[136], 1072[32]; **3**, 779[7], 792[7,67,68]; **4**, 375[93], 386[93], 389[165], 411[267b], 563[38], 564[38], 576[12], 596[166], 845[66], 847[66], 848[66,79], 1088[122]; **5**, 100[154], 102[154], 331[43], 333[43b], 349[75], 798[70], 800[70,70b,77,78], 822[77], 847[134], 850[145], 862[252]; **6**, 91[121], 533[496], 734[12,14], 735[19], 739[59], 740[19,62], 741[12,65,66], 742[67,69,70], 743[59], 752[111,113,114], 811[75], 835[49], 836[49], 843[85]; **7**, 415[112], 493[189]; **8**, 8[44], 34[61], 53[133], 66[61,133], 96[94], 413[125], 537[189], 755[131], 758[131], 857[192]
Overton, B. M., **6**, 67[12]
Owa, M., **1**, 317[142]
Owada, H., **1**, 648[126]; **3**, 87[111], 106[111], 114[111]; **6**, 289[195], 293[195], 1030[104], 1031[110,112], 1032[121]; **7**, 128[68], 129[70], 495[207], 523[43], 771[264]; **8**, 849[114]
Owczarczyk, Z., **4**, 432[115]
Owen, D. M., **3**, 648[187]
Owen, D. W., **8**, 445[41]
Owen, G. R., **2**, 709[13]; **6**, 602[3], 661[211]
Owen, L. N., **8**, 231[146]
Owen, N. D. S., **4**, 439[166]
Owen, N. E. T., **6**, 204[23]
Owens, G. D., **3**, 213[51]
Owers, A. J., **7**, 884[182]
Owings, F. F., **7**, 236[14,15]
Owsia, S., **7**, 352[30], 356[30]
Owsley, D. C., **2**, 602[40]; **3**, 217[90], 219[90], 597[200], 613[1], 614[1], 615[1], 616[1], 619[1], 620[1], 621[1], 622[1], 623[1], 625[1], 626[1,42], 627[1], 628[1], 629[1], 630[1]; **8**, 240[29], 243[29]
Owston, P. G., **3**, 383[49]
Owton, W. M., **7**, 494[203], 495[203,204]
Oxenrider, B. C., **6**, 774[48]; **7**, 700[60]
Oxford, A. W., **6**, 273[98]
Oxley, P. W., **8**, 366[46]
Oxman, J. D., **1**, 305[89], 311[89]
Oya, E., **7**, 761[64]
Oya, M., **7**, 760[50]
Oyama, N., **3**, 667[49]; **8**, 595[76]
Oyamada, H., **8**, 170[86-88], 178[86], 179[86], 244[55], 248[55], 874[25]
Oyamada, T., **3**, 831[60]
Oyler, A. R., **3**, 278[34]; **4**, 262[307]
Ozainne, M., **5**, 879[11]; **6**, 1064[89]
Ozaki, A., **7**, 160[55]

Ozaki, H., **8**, 150[130,133,135,142], 151[133,135,145]
Ozaki, K., **7**, 153[11]; **8**, 193[101], 557[380]
Ozaki, N., **4**, 93[94]
Ozaki, S., **6**, 244[112], 602[7], 603[7], 605[37]; **7**, 245[77,78], 248[112], 809[84]
Ozaki, Y., **4**, 960[34]; **6**, 436[12,16-18], 451[129], 455[154]
Ozasa, S., **3**, 457[131], 503[142]
Ozawa, F., **3**, 528[47]; **4**, 560[27]
Ozawa, S., **1**, 101[93,94], 103[93,94]; **2**, 114[123], 269[73], 271[75]; **4**, 33[96,96c], 257[227], 261[227]
Ozawa, T., **4**, 1032[9], 1051[9]; **8**, 422[33], 427[33]
Özbal, H., **6**, 106[19]
Ozbalik, N., **4**, 765[226]; **7**, 13[119], 731[55], 776[357]
Ozeki, H., **7**, 199[35]
Ozols, A. M., **6**, 450[121]
Ozorio, A. A., **4**, 113[168]; **5**, 910[86]

P

Paal, B., **6**, 546[650]
Paal, M., **6**, 49[96], 646[99,99b]
Paalzow, L., **7**, 831[64]
Pabon, H. J. J., **3**, 249[62]; **8**, 431[63]
Pabon, P., **3**, 1048[12]
Pabon, R., **7**, 879[149], 880[149]
Pabon, R. A., **7**, 880[154], 882[166]
Pabon, R. A., Jr., **5**, 453[65], 520[39], 704[22], 1020[68,69], 1023[69]
Pac, C., **5**, 154[28], 650[25]; **7**, 878[138,140,143], 888[138a]; **8**, 517[126], 562[421]
Pacansky, J., **4**, 483[6], 1072[16]
Pacaud, R. A., **4**, 272[33]
Pace, S. J., **8**, 642[30]
Pachaly, B., **4**, 738[98]; **8**, 800[69]
Pachinger, W., **4**, 204[34]; **5**, 50[44], 51[45,45b]; **7**, 646[27]; **8**, 784[112]
Pachter, I. J., **2**, 741[67], 965[68]; **4**, 398[216], 399[216d]
Paciello, R. A., **8**, 671[16]
Pacifici, J. A., **6**, 959[41]
Pacifici, J. G., **8**, 389[72]
Packard, A. B., **4**, 560[26]
Pacofsky, G. J., **5**, 864[260]; **6**, 859[174]
Pacreau, A., **5**, 1076[40]
Pacut, R., **8**, 115[60,62], 510[94], 630[188]
Paddon-Row, M. N., **1**, 37[239], 41[198], 49[8], 610[45]; **3**, 4[21], 12[21], 18[96]; **4**, 729[59]; **5**, 79[292], 257[61,61a,c], 258[61b], 632[61]; **7**, 821[27]; **8**, 496[30], 723[151], 724[151,169,169g]
Padeken, H. D., **7**, 747[99], 748[99]
Padeken, H. G., **2**, 342[6]; **6**, 104[9]; **7**, 752[142]
Paderes, G. D., **7**, 816[7]
Padgett, H., **2**, 1012[16]; **4**, 12[41]; **6**, 939[139], 942[139]; **7**, 220[22], 230[130,131]
Padias, A. B., **5**, 73[198,200,201,202,203,204], 78[273], 79[287]
Padmanabhan, S., **2**, 789[56]; **4**, 304[355]
Padwa, A., **1**, 357[4], 821[27]; **3**, 168[496], 253[89], 255[104], 262[89], 580[107], 1048[13]; **4**, 16[52a], 730[65], 792[60], 795[86], 1060[161], 1063[161a], 1069[6,7], 1070[6-8], 1073[22], 1075[31], 1076[35], 1077[51], 1078[51,55], 1081[72,74,77,79,82], 1082[87-90], 1083[7,55,87], 1084[55,97], 1085[51,104,105,107], 1086[105,118], 1088[118,121], 1089[6,118,131,137], 1090[137], 1091[137], 1093[147], 1095[153], 1096[158], 1097[158,169], 1098[7,158,173], 1099[178], 1100[178], 1101[196], 1102[196], 1103[87,104,206], 1104[210,213]; **5**, 10[78], 66[79], 67[84-92], 159[50], 247[25], 248[29], 250[37], 252[37,42-45], 254[29b], 255[37,51], 257[43], 260[25,64], 265[64], 410[39], 630[51], 707[36], 725[36], 728[121], 795[52], 914[114], 937[204], 947[262,274], 948[270,292], 949[284], 950[284]; **6**, 108[35], 128[163], 542[602], 570[955], 572[961], 734[1], 759[135], 789[106]; **7**, 470[4], 472[4], 473[4], 474[4], 476[4], 483[121], 690[12], 854[52,53], 855[52,53], 875[118], 887[52]; **8**, 394[120]
Padykula, R. E., **5**, 855[189]; **7**, 480[105], 482[105]
Padyukova, N. S., **6**, 618[114]
Paerels, G. B., **7**, 40[6]
Paetzold, P., **2**, 242[18]
Paetzold, R., **7**, 769[241], 770[256a], 771[256], 772[294,295], 773[294,295]
Pagani, A., **4**, 767[233]
Pagano, A. H., **4**, 12[42]; **6**, 943[155]
Paganou, A., **6**, 666[232], 667[232]
Page, A. D., **8**, 296[61]
Page, B. M., **6**, 637[28]
Page, G., **5**, 1130[6]
Page, G. A., **8**, 568[466]
Page, M., **5**, 72[172]
Page, M. I., **5**, 109[220,221]; **6**, 49[92]
Page, P. C. B., **1**, 568[243]; **3**, 225[188]; **4**, 159[82]; **7**, 261[72], 451[24]
Page, T. F., Jr., **5**, 455[80]
Pagès, O., **2**, 227[161]
Paget, W. E., **3**, 511[189]; **7**, 604[134]; **8**, 711[64]
Paglia, P., **4**, 532[85], 536[85], 545[125], 546[125]
Paglietti, G., **5**, 687[66]; **8**, 589[50,53]
Pagni, R. M., **2**, 141[39]; **4**, 288[188,189], 290[189], 346[86a], 347[86b]
Pagnoni, U. M., **3**, 386[57], 395[98], 752[97]; **4**, 337[32]; **8**, 349[145,146], 389[74]
Pagnotta, M., **6**, 707[42]
Paguer, D., **5**, 829[22]
Pahde, C., **1**, 202[101], 234[22], 253[8], 331[48], 734[22]
Pahl, A., **8**, 354[170]
Pahwa, P. S., **4**, 505[144,145]
Pai, B. R., **2**, 894[9], 912[9]; **7**, 221[32]
Pai, F.-C., **8**, 9[47]
Pai, G. C., **7**, 603[123]
Pai, G. G., **8**, 101[123], 714[81]
Paik, Y. H., **4**, 54[153b]; **6**, 117[95], 677[318,318a]
Paik Hahn, Y. S., **3**, 804[9]
Paikin, D. M., **4**, 317[548]
Pailler, J., **6**, 244[108]
Paine, A., **6**, 232[39]
Paine, J. B., III, **2**, 743[86]; **8**, 605[9]
Painter, G. R., III, **1**, 822[39]
Pairaudeau, G., **4**, 390[168], 395[168e]
Paisley, S. D., **3**, 220[121], 222[121c]; **6**, 848[108]
Pajanhesh, H., **8**, 846[83]
Pajerski, A. D., **1**, 15[80,82], 16[88]
Pajouhesh, H., **5**, 107[202]
Pak, C. S., **7**, 676[62]; **8**, 353[154], 563[435], 615[93]
Pak, H., **4**, 815[194], 817[194]
Pakhomov, A. S., **8**, 610[59]
Pakhomov, V. P., **6**, 554[790]
Pakhomova, I. E., **3**, 305[72]
Pakkanen, T. A., **2**, 346[45]
Paknikar, S. K., **1**, 819[2]; **3**, 223[157], 262[159], 264[159]; **6**, 2[3], 25[3], 88[105], 89[105]; **7**, 358[2], 366[2], 378[2], 384[2]
Pakrashi, S. C., **6**, 487[72], 489[72]; **8**, 249[96], 943[119]
Pakusch, J., **4**, 758[191]
Pal, D., **4**, 497[100]
Pal, R. S., **4**, 508[158]
Palacios, F., **1**, 782[234]; **6**, 184[150]
Palacios, S. M., **4**, 453[24], 461[103], 462[103], 474[24], 475[103], 476[157]
Paladini, J.-C., **5**, 933[182]
Palágyi, J., **5**, 1138[65]
Palazzi, C., **2**, 630[19], 631[19], 636[19], 637[19]
Palcic, M. M., **8**, 205[155]
Pale, P., **4**, 309[414], 393[191], 394[191], 603[276,277], 610[335], 645[276,277]; **7**, 107[157]
Palecek, M., **6**, 707[40]
Paledek, J., **8**, 200[138]
Palei, B. A., **8**, 741[38], 754[78]
Palermo, R. E., **7**, 439[28]
Paleveda, W. J., Jr., **6**, 635[23], 636[23]
Paley, R. S., **4**, 795[89]

Palinko, I., **8**, 418[5], 420[5], 423[5], 439[5], 441[5], 442[5]
Palio, G., **2**, 586[135]
Palitzsch, P., **2**, 428[44]
Palkowitz, A. D., **1**, 192[81], 413[59], 764[149]; **2**, 8[37], 13[37], 20[37a], 25[98], 30[98], 31[98,115], 35[37], 42[98,149], 43[98,147], 44[98,147], 45[115,149], 46[152]; **6**, 864[197]
Palla, F., **8**, 100[118,118e]
Palla, O., **6**, 171[6]
Palladino, N., **8**, 171[110]
Pallai, P., **6**, 804[49]
Pallaud, R., **7**, 805[68]
Palleroni, N. J., **7**, 70[94]
Palleros, D., **4**, 426[49]
Pallini, L., **5**, 925[155], 1149[124], 1154[157]
Pallini, U., **2**, 783[35]
Pallos, L., **4**, 110[151]; **6**, 553[702,703]
Pally, M., **3**, 380[13]
Palmer, A., **8**, 366[36]
Palmer, B. D., **1**, 752[96], 800[299]; **5**, 534[92]; **6**, 995[103]; **7**, 331[14]
Palmer, C. J., **2**, 742[69]
Palmer, D. C., **6**, 552[695]
Palmer, G. E., **4**, 314[481]
Palmer, J., **8**, 423[38], 428[38]
Palmer, J. R., **8**, 756[159], 758[159]
Palmer, M. A. J., **2**, 25[98], 30[98], 31[98], 35[125], 41[125], 42[98,149], 43[98], 44[98], 45[149]
Palmer, M. H., **4**, 313[462]; **7**, 479[92]
Palmer, M. J., **6**, 533[490], 550[490]
Palmer, R., **3**, 714[33]
Palmere, R. M., **6**, 818[105]
Palmertz, I., **6**, 809[66]
Palmieri, G., **4**, 86[78c]
Palmisano, G., **1**, 118[59]; **2**, 535[37]; **7**, 65[68], 346[12]; **8**, 563[435]
Palmquist, U., **3**, 671[63], 676[80], 685[63]; **7**, 800[29], 801[40]
Palominos, M. A., **2**, 840[184]; **6**, 453[137]
Palomo, A. L., **6**, 204[21]
Palomo, C., **2**, 649[102], 1059[75]; **5**, 94[87], 95[90,96-99], 96[109], 100[146]; **6**, 249[144], 250[144], 251[149], 490[101-103], 491[117], 655[159], 810[73], 816[101], 938[129], 940[129]; **7**, 275[145], 277[153,154,155], 278[159,160], 283[186,187], 530[18], 531[18], 554[64,65], 695[34], 752[144], 760[24]; **8**, 19[133,134]
Palomo-Coll, A., **6**, 491[116]
Palomo-Coll, A. L., **6**, 491[116]
Palomo-Nicolau, C., **6**, 491[116]
Palop, J. A., **2**, 780[12]
Pals, M. A., **5**, 382[119b]
Palumbo, G., **1**, 168[116b], 563[179]; **6**, 980[34]; **8**, 406[41], 891[147]
Palumbo, P. S., **3**, 174[538,539], 177[538,539], 868[42-44], 869[43,44], 876[44]
Pályi, G., **5**, 1138[65,66,69]
Palyulin, V. A., **4**, 342[67]
Pan, B.-C., **1**, 425[107]; **4**, 176[45]; **6**, 656[170]
Pan, H., **7**, 655[11]
Pan, H.-L., **7**, 85[12], 87[12]
Pan, L., **4**, 155[68d]
Pan, X., **3**, 596[194], 638[93]
Pan, X.-F., **7**, 166[86b]
Pan, Y., **3**, 209[16], 223[16]
Pan, Y.-G., **4**, 158[78]; **5**, 329[32]; **7**, 144[152]
Panasenko, A. A., **8**, 699[150]
Pancoast, T. A., **1**, 514[52]
Pancrazi, A., **4**, 296[263,264], 796[97]; **6**, 264[31], 278[31]; **7**, 27[63]
Panda, M., **2**, 349[68]
Panday, P. N., **7**, 220[19], 502[260]
Pandell, A. J., **3**, 317[119]
Pandey, B., **1**, 892[149]; **5**, 226[107]
Pandey, G., **4**, 391[175d]; **7**, 223[43], 227[43], 248[114]
Pandey, G. D., **3**, 1038[94]
Pandey, P. N., **4**, 297[275]; **8**, 907[71], 909[71], 910[71]
Pandey, P. S., **6**, 564[908]
Pandey, U. C., **3**, 365[98]
Pandian, R., **8**, 170[74]
Pandiarajan, P. K., **2**, 247[82]
Pandit, R. S., **2**, 787[52]
Pandit, U. K., **2**, 718[81]; **4**, 48[139], 52[147,147d], 1004[71,72]; **6**, 712[71]; **7**, 225[66], 684[95]; **8**, 93[70,72], 94[78], 561[418,419], 562[420]
Pandl, K., **6**, 176[105], 185[163], 186[163]
Pandya, A., **2**, 867[14]; **5**, 73[201]
Pandy-Szekeres, D., **2**, 718[72]
Panek, E. J., **1**, 426[109]; **3**, 418[24], 482[3]
Panek, J. S., **2**, 303[4]; **4**, 98[115], 433[119], 799[115]; **5**, 492[238,239,240], 497[227], 498[231,232,238]; **6**, 814[87,90]
Panella, J. P., **6**, 236[56,57]
Panesgrau, P. D., **4**, 498[101]
Panetta, J. A., **6**, 443[96], 564[906]
Panfil, I., **5**, 260[63a], 264[73]
Panichanun, S., **4**, 113[168]
Panitkova, E. S., **3**, 639[82], 644[82,137]
Panizzon, L., **8**, 638[14]
Pankiewicz, K., **4**, 231[274]
Pankova, M., **6**, 1013[9]
Pankowski, J., **1**, 329[39], 806[314]; **6**, 989[81]
Pankratova, A. F., **3**, 309[89]
Pannekock, W. J., **4**, 521[46], 531[46]
Pannell, K. H., **4**, 588[61]
Pannella, H., **5**, 404[18]
Panossian, S., **6**, 431[108]
Panouse, J. J., **7**, 100[125]
Pansard, J., **1**, 220[65]
Panse, D., **3**, 744[61]
Pansegrau, P. D., **4**, 250[141], 494[82], 872[44]; **7**, 255[37]
Pant, B. C., **7**, 774[337], 776[337]
Pant, C., **3**, 224[163,163a]
Pantaloni, A., **8**, 13[71]
Panunto, T. W., **1**, 838[158]; **7**, 162[64,66], 778[398]
Panunzio, M., **1**, 391[148]; **2**, 613[114], 635[48], 640[48], 656[157], 807[48], 925[111], 926[111], 927[120], 935[151], 936[151], 937[156,157]; **5**, 100[148,155,156], 102[165,173,174]; **6**, 21[80], 22[80], 759[140]; **8**, 36[80], 54[80], 66[80], 550[331]
Panyachotipun, C., **1**, 526[94], 835[138]; **3**, 154[424]
Panza, L., **4**, 379[115,116,118], 382[115k]; **7**, 274[138]
Panzer, H. P., **5**, 557[56]
Panzica, R. P., **6**, 554[711]
Panzone, G., **8**, 90[47]
Paoletti, F., **2**, 465[104]
Paolucci, C., **1**, 516[59,60], 517[61,62]; **3**, 147[396], 153[396], 155[396], 865[27], 944[90,91], 946[92], 958[90,112]; **6**, 898[103]
Paone, S., **6**, 66[6]
Pap, A. A., **3**, 24[145]
Pap, G., **8**, 372[124]
Papa, D., **2**, 734[6]; **8**, 320[84]
Papadakis, P. E., **8**, 212[10]
Papadopoulos, E. P., **6**, 546[648]; **7**, 760[20]
Papadopoulos, K., **2**, 519[64], 520[66], 1077[153]; **4**, 21[69], 104[137], 222[167,168], 224[167,168]; **5**, 485[182]; **6**, 716[104]
Papadopoulos, M., **5**, 552[19]
Papadopoulos, P., **5**, 909[98], 957[311], 993[52], 994[52]
Papaefthymiou, G. C., **8**, 366[51]

Papageorgiou, C., **6**, 769[31]
Papageorgiou, G., **2**, 901[39], 948[183], 959[31], 960[31], 962[45], 964[45,60,61], 965[63], 966[61,71], 967[61,63,71]
Papagni, A., **4**, 231[268], 524[61]; **5**, 1079[51]; **6**, 178[121]
Papahadjopoulos, D., **8**, 36[50], 66[50]
Papahatjis, D. P., **1**, 409[36,37]; **3**, 39[216]; **5**, 534[92]; **6**, 8[39], 46[59,76], 47[76], 48[76], 158[184]; **7**, 396[25]
Papaioannou, D., **8**, 245[73]
Papaleo, S., **4**, 958[28]
Papantoniou, C., **4**, 293[237]
Papasergio, R. I., **1**, 17[209,210,218,219], 36[234]
Pape, P. G., **8**, 899[27]
Papenmeier, J., **8**, 950[167]
Papies, O., **5**, 552[29]
Papile, C. J., **4**, 915[14]
Papillon-Jegou, D., **2**, 1074[147]
Papini, A., **6**, 238[74]
Papoula, M. T. B., **3**, 505[158]; **7**, 307[16], 310[16], 318[16], 319[16], 322[16], 704[12]
Pappalardo, P., **3**, 491[72]; **5**, 385[130]; **7**, 128[64], 146[64]
Pappaldo, S., **4**, 462[105]
Pappas, J. J., **2**, 420[24]
Pappas, S. P., **5**, 646[3]
Pappo, R., **2**, 352[85]; **3**, 11[52], 17[52]; **4**, 3[7], 4[7], 14[49], 24[72,72b], 63[7b], 65[7b], 70[2], 143[20], 258[231]; **7**, 169[113], 352[29], 564[93], 565[93], 568[93], 600[76], 711[57]; **8**, 494[24], 495[27], 544[264], 756[159], 758[159]
Paquer, D., **5**, 435[152]; **6**, 712[74]
Paquet, F., **4**, 381[128]; **7**, 245[76]; **8**, 856[161]
Paquette, L. A., **1**, 240[42,43], 413[56], 429[123], 567[223], 580[1], 584[12], 672[202], 683[226], 684[226], 685[226], 700[202,258], 701[202], 705[202,226], 712[258], 714[226], 717[226], 718[226], 719[226], 720[226],; **2**, 82[6], 546[88,89], 548[88], 572[41], 588[150], 681[61], 710[25], 713[47], 728[146], 838[169]; **3**, 158[434], 159[453], 160[453], 163[434,453,472,473], 168[505], 200[67], 254[96], 334[222], 380[10,13], 381[33], 383[43], 445[70], 572[62], 586[154,155], 593[177], 610[155], 614[6], 623[6,32,33], 626[32b], 627[6,43], 751[89], 786[41], 862[4,6], 864[21], 866[4,6,30], 867[6,33], 868[4,6], 872[56-58], 873[33,57,66,67], 874[70], 876[81], 983[20], 984[20], 1058[41], 1062[41]; **4**, 10[34], 85[75], 181[71], 373[67], 390[174], 593[134], 667[13], 669[13], 677[13], 709[45], 710[45,51,52], 951[1], 968[1], 973[83], 979[1], 1010[157], 1013[182], 1040[69]; **5**, 21[149,154], 22[149,154], 71[119], 96[108], 203[39,39a-g], 204[39h-l], 209[39], 210[39,56], 225[90,102], 229[125,126], 239[1], 270[1a-c], 301[78], 324[18a,b,19,21], 347[72,72b,73a], 348[73b], 356[84], 451[37], 543[113], 560[75], 609[67], 618[5], 634[68,69], 637[110,113], 687[58,62,63], 688[62,63], 715[83], 757[75,76], 768[129], 778[75,76], 779[129,197], 796[53], 798[69], 806[104,108], 810[125], 812[135], 814[140], 815[135,143], 816[144], 817[146,147,148,149,150], 831[38], 833[44,47,48], 850[157,158], 856[210], 857[230], 910[81], 912[81], 915[111,112], 954[299,300], 1003[24], 1029[91], 1070[21], 1072[21], 1123[37]; **6**, 146[88], 161[180], 283[162], 780[74], 837[60], 859[169], 881[50], 960[51,57], 979[26], 1003[134], 1004[140], 1023[71], 1063[81]; **7**, 4[16], 97[92], 100[116], 102[135], 163[76], 164[76], 172[127], 211[98], 212[100], 255[37], 261[69], 377[91], 378[91b], 507[306], 552[57], 667[80]; **8**, 61[191], 66[191], 447[128,129], 463[128,129], 548[323], 844[63], 880[60,61], 942[116], 946[134,140]
Paquot, C., **7**, 108[178]
Paradisi, C., **4**, 425[33], 426[60], 438[156], 518[11], 519[11]
Paradisi, M. P., **1**, 734[23]; **8**, 17[118]
Parady, T. E., **3**, 864[16], 866[16], 883[16]
Para-Hake, M., **4**, 608[320], 646[320]
Paraiso, E., **4**, 89[84h], 95[84h]
Paramasivam, K., **5**, 728[122]
Paranjape, B. V., **2**, 823[113]
Paranjpe, M. G., **6**, 555[815]
Paranyuk, V. E., **6**, 577[976]
Parcell, R. F., **1**, 371[73]; **6**, 789[105]
Pardasani, R., **1**, 490[26], 492[26], 494[26], 495[46], 498[26]
Pardini, V. L., **6**, 176[83]
Pardo, M., **5**, 407[27], 408[30,30b]
Pardo, R., **2**, 710[20,21]; **4**, 155[68f]; **5**, 777[184], 779[184]
Pardo, S. N., **2**, 538[55,56], 539[55,56]
Paré, J. R., **2**, 662[20], 663[20], 664[20]
Pare, J. R. J., **5**, 432[130]
Paredes, R., **1**, 837[154]; **3**, 898[80]
Parente, A., **7**, 87[20]
Parfitt, R. T., **8**, 65[211], 66[211]
Parham, H., **7**, 266[109], 267[109], 760[23]
Parham, M. E., **6**, 803[46], 804[50], 816[100]
Parham, W. E., **1**, 412[54], 878[105,107]; **2**, 773[28]; **3**, 105[212]; **4**, 120[199], 1000[6], 1005[6,79,89], 1006[103], 1015[199], 1016[6,203], 1023[255]; **5**, 692[101], 828[5], 847[5]; **6**, 147[84]; **8**, 269[75], 965[61], 978[144]
Parikh, J. R., **7**, 296[24]
Paris, J.-M., **1**, 792[271]; **6**, 157[169], 987[73]
Parish, E. J., **7**, 103[141,142], 264[91-94], 265[94], 266[111], 267[111,116], 277[116]; **8**, 872[9], 873[9], 881[81], 882[81]
Parish, R. V., **8**, 770[41]
Parish, W. W., **5**, 163[70]
Pariza, R. J., **4**, 15[50,50c]; **6**, 176[102]
Park, C. Y., **7**, 442[46c]
Park, G., **4**, 395[207a], 396[207a,b], 558[16]
Park, J., **5**, 1088[79], 1092[93], 1102[93]
Park, J. C., **2**, 13[59], 35[125], 41[125]
Park, J. D., **3**, 898[75]; **4**, 280[132]; **8**, 895[1], 898[1]
Park, J. H., **1**, 865[88]; **8**, 52[147], 66[147]
Park, J. M., **7**, 407[80]
Park, J. W., **8**, 440[84]
Park, K. B., **8**, 16[107], 17[107]
Park, K. P., **7**, 751[140]
Park, M. H., **7**, 238[43]
Park, M.-K., **3**, 565[16], 578[16]
Park, P., **2**, 1068[128]; **3**, 125[294], 126[294], 167[294], 168[294]; **7**, 246[88]
Park, P.-u., **8**, 540[195]
Park, S. B., **4**, 254[178]
Park, S.-U., **3**, 824[23]; **4**, 723[41], 803[135], 811[173]; **6**, 442[87]; **8**, 802[84]
Park, S. W., **3**, 49[261]
Park, W.-S., **2**, 798[10]; **8**, 159[10], 160[90], 170[89,90], 178[90], 179[90], 180[10], 806[99]
Parkanyi, C., **8**, 628[178]
Parkash, N., **4**, 505[113]
Parker, D., **2**, 186[32]; **3**, 18[94], 42[94], 213[54], 251[74]; **5**, 931[186]
Parker, D. A., **1**, 107[6], 110[6], 343[116]; **3**, 213[48], 224[167], 251[76], 261[76], 264[76,186]; **4**, 176[50], 177[50,55,56]; **6**, 5[23], 16[61], 831[9]
Parker, D. G., **6**, 690[403], 692[403]; **7**, 14[129], 449[6], 451[6], 452[6], 453[6]
Parker, D. T., **5**, 411[43], 413[51], 415[51c,d], 534[95,95h], 539[95g], 552[39]; **6**, 756[127]
Parker, G., **7**, 4[17]
Parker, J. E., **7**, 274[139]
Parker, K. A., **2**, 183[13], 763[60], 853[229]; **3**, 48[260], 370[111]; **4**, 14[48], 377[104], 389[104g], 562[35], 797[103], 820[214]; **5**, 524[54,54g], 534[54], 535[54g], 851[168,169]; **7**, 330[6], 350[24], 355[24], 584[159]
Parker, R. E., **3**, 733[1]; **8**, 872[2]
Parker, R. H., **8**, 890[142]
Parker, S. D., **5**, 217[23], 226[23,105]

Parker, V. D., **3**, 672[67], 676[80]; **4**, 455[42]; **6**, 282[154]; **7**, 799[27], 800[30], 801[38-40], 854[47], 855[47], 856[67], 874[110]
Parker, W., **3**, 380[13], 386[67,68], 399[118], 404[133,134], 600[213]; **8**, 269[77]
Parker, W. L., **5**, 86[32]
Parkhurst, C. S., **4**, 597[180], 622[180]
Parkin, C., **7**, 228[106]
Parkin, G., **8**, 673[25], 696[25]
Parkin, J. G., **7**, 805[67]
Parkins, A. W., **3**, 244[23]; **4**, 706[36]; **5**, 1156[164]
Parks, G. L., **3**, 771[188]
Parks, J. E., **6**, 533[510]
Parlar, H., **5**, 453[66]
Parlier, A., **4**, 980[115], 982[115]; **5**, 1066[9], 1076[40], 1103[150], 1104[150,158], 1105[159,161,162,163]
Parlman, R. M., **3**, 1024[28]; **4**, 115[178]; **6**, 954[19]
Parmigiani, G., **3**, 734[9]
Parnell, C. A., **4**, 91[88c]; **5**, 543[115], 1150[127], 1154[156]; **7**, 338[40]; **8**, 111[23], 117[23]
Parnell, C. P., **1**, 307[93], 310[93]; **7**, 6[33]
Parnell, D. R., **3**, 557[47]
Parnes, H., **8**, 798[65], 800[65]
Parnes, Z. N., **4**, 155[65]; **5**, 752[32,35,36], 754[32,35,36], 756[35,36]; **8**, 216[55,56], 318[60-63,65-67], 486[59,61], 487[59], 546[306,307,308], 608[37], 610[56,58-61], 611[66], 623[147], 624[147], 630[56,187], 778[84], 813[12]
Parola, A., **3**, 564[11], 567[11]
Parr, J. E., **3**, 88[137], 95[137], 165[137], 167[137]
Parra, M., **2**, 849[213]
Parra, T., **3**, 396[115]
Parra-Hake, M., **5**, 925[156]
Parratt, M. J., **3**, 202[85]
Parreno, U., **5**, 92[64]
Parrick, J., **4**, 1021[249,250]
Parrilli, M., **4**, 347[95]; **7**, 438[17-19], 445[17-19,58]
Parrinello, G., **3**, 232[268], 495[93a], 1022[21]; **4**, 931[59], 932[60]
Parris, C. L., **6**, 264[29], 268[29], 286[29]
Parrish, C. I., **7**, 7[37]
Parrish, D. R., **2**, 167[159]; **3**, 23[142]; **4**, 7[25]; **6**, 718[118]; **8**, 460[249], 534[156], 544[256], 606[18]
Parrish, F. W., **6**, 660[201]
Parrott, M. J., **4**, 717[8]
Parrott, S. J., **2**, 534[33-35], 535[34,35]
Parry, F. H., **5**, 99[134]
Parry, M. J., **1**, 739[35]; **6**, 998[118]
Parry, R. J., **1**, 601[40], 608[40]; **3**, 364[93]; **6**, 96[150]
Parry, S., **7**, 415[113]
Parsens, P. J., **4**, 820[223]
Parshall, G. W., **1**, 140[7], 174[15], 743[51], 746[51], 811[51]; **4**, 587[17]; **5**, 1115[2], 1116[2,2c], 1121[2c], 1122[2c], 1123[2c]; **8**, 447[109], 451[173,175], 551[339]
Parshin, V. A., **6**, 554[755]
Parsonage, J. R., **7**, 616[10], 620[10]
Parsons, G. H., Jr., **3**, 564[11], 567[11]
Parsons, J. L., **8**, 533[146]
Parsons, P. J., **2**, 907[60]; **4**, 390[168], 395[168e,205], 820[216]; **6**, 836[56]; **7**, 546[32], 555[71], 564[71]
Parsons, W. H., **1**, 838[170]; **4**, 30[88]; **7**, 105[151]
Partale, H., **2**, 371[261]; **5**, 76[239]
Partale, W., **3**, 888[13]
Partch, R., **2**, 753[2,2c]
Partch, R. E., **3**, 380[13]
Parthasarathy, P. C., **2**, 842[189]
Parthasarathy, R., **5**, 186[169]
Partis, R. A., **8**, 392[105]
Parton, B., **5**, 486[185], 487[185]
Partridge, J. J., **1**, 780[229], 851[37]; **3**, 783[23]; **5**, 129[33]; **6**, 913[24]; **8**, 269[82], 722[150]
Partsch, R. E., **7**, 13[112]
Partyka, R. A., **2**, 648[98], 649[98], 1059[76]
Parvez, M., **1**, 15[80,82], 16[88]; **3**, 58[292]; **4**, 48[137,137g], 824[239]; **5**, 426[104], 843[122]; **6**, 894[90]
Paryzek, Z., **7**, 31[89], 255[34]; **8**, 354[169], 886[110]
Pasau, P., **1**, 664[169], 665[169], 669[169], 670[169], 700[259], 705[259], 708[259], 722[259]; **3**, 111[231]; **4**, 991[150]
Pascal, Y.-L., **8**, 133[19,20]
Pascali, V., **8**, 840[36], 844[36], 913[94], 914[94]
Pascard, C., **6**, 718[122]; **7**, 64[61b]
Pascard-Billy, C., **1**, 34[227]
Paschal, J. W., **8**, 514[106], 623[151]
Pascher, F., **5**, 29[1]
Pascual, A., **5**, 223[83], 224[83], 234[140], 806[102], 1028[90]
Pascual, C., **2**, 345[36]
Pascual, J., **3**, 564[8]; **8**, 500[50], 515[119]
Pascual, O. S., **4**, 279[110]
Pasedach, H., **5**, 15[106], 835[59]
Paserini, N., **6**, 487[54], 489[54]
Pasha, M. A., **8**, 889[137]
Pashayan, D., **3**, 580[107]
Pasini, A., **7**, 108[173]
Pasiut, L. A., **7**, 810[89]
Pasquali, M., **4**, 170[14]
Pasqualini, R., **8**, 860[223]
Pasquato, L., **4**, 102[132]; **5**, 324[17]; **6**, 999[123]; **8**, 836[2], 842[2e], 843[2e], 844[2e]
Pasquini, M. A., **6**, 80[69]
Pass, M., **7**, 34[99]
Passannanti, S., **1**, 476[121]; **3**, 66[10]
Passarotti, C., **8**, 568[467]
Passer, M., **2**, 529[20]
Passerini, M., **2**, 1083[6], 1084[6,6a]
Passerini, N., **6**, 487[52], 489[52]
Passerini, R., **7**, 770[253]
Pasta, P., **8**, 194[105]
Pasteels, J., **1**, 100[88]; **6**, 914[27]
Pastel, M., **5**, 736[145], 737[145]
Pasternak, V. I., **6**, 500[180], 543[623]
Pasto, D. J., **2**, 242[15]; **3**, 491[67]; **4**, 145[34], 279[116], 301[319,320,324], 302[319,320], 314[484]; **6**, 830[4]; **7**, 600[74]; **8**, 367[59], 472[2], 473[2,12], 474[13], 477[34,35], 705[10], 707[17], 724[153], 726[10], 937[79]
Pastor, S. D., **1**, 320[162]; **4**, 443[187,188]
Pastour, P., **1**, 644[122], 646[122], 668[122], 669[122], 695[122]; **6**, 515[236]
Pasulto, M. F., **7**, 78[128b]
Pasynkiewicz, S., **8**, 756[146], 757[162]
Patachke, H. P., **5**, 422[88], 423[88]
Patai, P., **3**, 521[1], 551[5], 552[5], 556[34,35]
Patai, S., **1**, 165[107], 215[39], 218[39], 220[64], 225[39], 326[4], 327[13], 360[26], 364[26], 571[272], 580[2], 581[2], 582[2], 583[8,8b], 610[2a], 611[2a], 616[2a], 630[6,14,16], 631[16], 634[16], 641[16]; **2**, 81[1], 82[1], 96[1], 342[5], 348[51,52], 349[5], 352[5], 363[52], 365[211], 727[128], 734[3], 1102[120]; **3**, 86[52,53], 89[52], 91[52], 94[52,53], 95[52], 114[52,53], 116[52], 117[52,53], 119[52], 208[2], 217[2], 223[146], 271[2], 272[2], 436[7,10,11], 582[111], 634[19], 638[19], 649[19], 655[19], 722[6], 723[6], 726[23], 731[6], 822[8], 828[48], 829[48], 836[8], 839[9], 862[3,7], 867[34,37], 872[37], 883[37], 884[37], 887[6], 889[6,25], 890[25], 893[6,25], 894[6], 896[6], 897[6], 900[6], 903[6], 919[28]; **4**, 3[7,10], 4[7,16,16a], 41[10], 47[10], 53[151], 66[10], 70[9], 71[19], 78[54], 86[54b], 93[95], 139[1], 155[66], 238[3], 295[257], 299[303], 303[350], 316[534], 317[534], 329[1], 330[1c,d], 339[39], 342[39,64], 343[64], 344[1], 350[1], 351[1],

452[6,10], 467[6,10], 483[4], 484[4,23], 485[23], 495[4], 552[2], 553[2], 867[7], 879[86], 883[86], 903[200,200b], 905[200b], 946[92], 953[8,8a,g,i], 954[8a,i], 1000[10], 1002[10], 1016[10], 1031[5], 1032[5], 1033[28], 1035[5]; **5**, 86[23,26], 90[23,26], 113[228], 211[64], 468[134], 1037[3], 1132[21], 1144[101]; **6**, 2[2,4], 23[2], 65[1], 105[12,15], 116[90], 204[2,8], 208[52], 209[8], 211[52], 213[8], 225[7], 229[7], 233[7], 239[78], 243[102], 244[102], 245[119,121], 246[119], 247[119], 248[119,121], 249[119,121], 251[119,121], 252[119], 253[119], 254[119], 255[119], 256[7,119], 258[7,119], 420[15], 422[35], 436[6,24], 437[6,24], 438[60,61], 444[6], 445[6], 448[6], 449[6], 450[6], 453[6], 454[6], 455[6], 456[6], 461[3], 487[3], 488[3,16,28], 489[3,28], 507[16], 515[3], 523[3], 524[3], 525[3], 526[3], 529[16], 532[3,16], 533[16], 534[16], 536[16], 537[16], 556[23,26], 562[16], 563[23,26], 566[28], 567[28], 568[28], 570[28], 571[28], 572[28], 573[28], 574[28], 575[28], 576[28], 577[28], 578[28], 579[28], 580[28], 581[28], 704[5,6], 726[185], 753[117], 786[94], 795[3,9], 796[16], 797[9], 799[3,9], 806[9], 807[9], 826[9], 830[4], 834[34], 837[62,63], 910[9], 924[9], 950[1], 951[4], 958[27], 1007[151]; **7**, 21[11,20], 135[106], 196[13], 235[7], 236[7], 247[7], 541[4], 564[4], 689[7], 736[5], 737[5], 739[36], 740[41], 741[46], 742[53], 745[5,74], 746[5,46,84,91], 749[5], 751[53], 758[1], 759[1], 760[1], 761[66,67], 762[69], 766[169], 777[69a], 778[69], 842[35], 851[29]; **8**, 19[131], 26[5,22], 27[5], 36[5], 37[22], 66[22], 85[18], 123[84], 124[84], 236[5], 239[25], 240[5,25], 241[25], 242[5], 248[5], 249[5], 292[43], 363[7], 365[29], 367[54], 382[1], 383[1,14], 384[23], 390[1], 396[136,139], 403[4], 404[4], 407[4], 412[109], 413[127,-128,129,130], 670[13], 763[1], 785[1], 794[4,12], 827[71]

Pataki, J., **3**, 805[15]

Patapoff, T. W., **4**, 1039[63]

Patchett, A. A., **8**, 50[117], 66[117], 357[198], 358[198], 495[28]

Patchornik, A., **6**, 624[151], 636[22], 651[138], 658[188]

Pate, B. D., **4**, 347[48]

Patel, A. D., **2**, 1094[90], 1095[90]; **5**, 87[38]

Patel, A. N., **6**, 213[88]

Patel, B., **4**, 844[57], 845[68], 847[68], 849[82], 852[57,90], 855[96,97]

Patel, B. A., **3**, 539[98]; **8**, 959[28]

Patel, D. I., **6**, 535[540,541], 538[540,541]

Patel, D. J., **4**, 876[58]

Patel, D. K., **3**, 846[47]

Patel, D. V., **1**, 429[126], 798[289]; **4**, 375[96a,b], 376[96b], 401[96b,227], 406[227a]; **6**, 995[101]; **7**, 493[186], 536[51]

Patel, G., **6**, 502[214], 504[224], 562[214]

Patel, G. J., **6**, 554[724]

Patel, J. R., **6**, 525[380]

Patel, K. M., **3**, 21[125]

Patel, M., **5**, 417[62], 420[62], 451[23], 470[23], 491[23], 492[23,241], 497[227], 583[188]

Patel, P. D., **6**, 174[54]

Patel, P. P., **1**, 855[53], 856[53]; **3**, 783[22,25]

Patel, R. C., **6**, 255[169]

Patel, R. N., **7**, 56[19-21], 80[137]

Patel, S. K., **4**, 241[55], 255[55], 682[57]; **6**, 687[383]; **8**, 769[24], 771[24], 782[24b]

Patel, S. V., **3**, 219[106], 512[201]

Patel, V. F., **4**, 761[201,206]

Paterno, E., **5**, 151[1], 152[1]

Paterson, E. S., **4**, 867[7]

Paterson, I., **1**, 770[188], 772[197]; **2**, 1[2], 113[109,110], 202[94], 249[84], 263[56], 264[57], 584[127], 614[118], 617[141], 649[107], 716[60]; **3**, 25[156,157], 26[156], 28[168]; **4**, 331[9], 675[37]; **5**, 171[117], 843[113]; **6**, 142[65]; **7**, 137[119], 138[119], 144[119], 208[76,77,82]; **8**, 836[6]

Paterson, T. M., **5**, 420[75]; **8**, 386[55]

Pathak, S., **7**, 84[3]

Pathak, V. P., **2**, 747[115]

Pati, U. K., **2**, 547[103], 549[103]; **8**, 341[103], 928[24]

Patil, D. G., **3**, 737[33]

Patil, D. S., **1**, 140[10]

Patil, G., **1**, 555[114,118]; **4**, 278[96]

Patil, S., **5**, 690[80,80c], 733[136,136f], 734[136f]

Patil, S. L., **1**, 463[25]

Patil, S. R., **7**, 95[67,172], 108[172], 774[326]

Patil, V. D., **7**, 264[90]

Patin, A., **4**, 605[293], 646[293]

Patin, H., **4**, 603[272], 626[272], 645[272], 674[35], 688[35]; **6**, 690[399], 691[399], 692[399]; **8**, 322[111], 445[38]

Patnekar, S. G., **7**, 239[47]

Patney, H. K., **8**, 798[50,63]

Paton, A. C., **6**, 709[54]

Paton, J., **8**, 198[135]

Paton, J. M., **7**, 499[232]

Paton, R. M., **7**, 739[37]

Patra, A., **3**, 396[112], 398[112]

Patra, S. K., **3**, 87[115]

Patrianakou, S., **8**, 245[73]

Patricia, J. J., **3**, 419[30], 422[30]; **4**, 764[216], 807[153,154], 871[32-34]

Patrick, D. W., **3**, 87[63], 114[63], 117[63]; **6**, 1026[81], 1027[81], 1029[81], 1030[81], 1031[81], 1033[81]; **7**, 124[38], 128[38], 129[38], 489[166], 775[353]

Patrick, J. B., **7**, 546[31]

Patrick, J. E., **3**, 946[88], 990[34]

Patrick, T. B., **2**, 758[22b]; **4**, 344[75]; **7**, 723[27]; **8**, 366[41]

Patrie, W. J., **7**, 315[43]

Patronik, V. A., **6**, 439[72]

Patsaev, A. K., **4**, 50[142]

Patsch, M., **3**, 927[52]; **5**, 1007[39], 1008[39]

Patten, A. D., **3**, 503[149], 512[149]

Pattenden, G., **1**, 133[112]; **2**, 124[202], 651[117]; **3**, 50[265], 278[32], 384[52,53], 400[53], 407[147], 431[97,98], 494[86], 585[137], 586[153], 587[148], 591[169], 603[227], 604[153], 605[231], 610[153], 976[9], 977[9,9a], 985[27], 989[9], 990[9]; **4**, 27[84], 29[84d], 102[127,127b], 126[217c], 518[10], 761[200,201,204,206], 791[58,59], 805[146], 809[160]; **5**, 133[55], 136[67], 140[86], 181[151], 779[199], 803[89], 976[20]; **6**, 876[27], 880[27], 1045[24], 1062[75,76]; **7**, 338[42]

Patterman, S. P., **1**, 11[56]

Patterson, D. B., **8**, 474[14], 475[14], 476[14]

Patterson, I., **1**, 403[18]

Patterson, J. W., **7**, 502[267]

Patterson, J. W., Jr., **3**, 9[46], 10[46], 20[107], 822[5], 834[5]

Patterson, L. A., **4**, 282[135]

Patterson, M. A. K., **8**, 87[35]

Patterson, R. T., **5**, 629[49]

Patterson, W., **3**, 266[196]; **4**, 893[151]; **5**, 1166[19]; **6**, 966[95]; **8**, 756[147]

Pattison, F. L. M., **3**, 639[79]; **6**, 228[32]; **8**, 903[49]

Pattison, J. B., **6**, 213[86]

Pattison, V. A., **3**, 975[3], 980[3]

Patton, A. T., **2**, 127[234]

Patton, L., **5**, 241[4]

Pau, C. F., **1**, 610[45]; **4**, 968[62]

Paudler, W. W., **5**, 637[111]; **7**, 267[121], 269[121], 270[128], 271[121,128], 278[121]

Paugam, J.-P., **3**, 199[55], 257[117]

Paukstelis, J. V., **1**, 366[43], 522[79]; **4**, 226[200]; **6**, 900[119]

Paul, B. D., **8**, 51[122], 66[122]

Paul, D. B., **4**, 295[257]; **6**, 245[119], 246[119], 247[119], 248[119], 249[119], 251[119], 252[119], 253[119], 254[119], 255[119], 256[119], 258[119]; **8**, 501[57]

Paul, E. G., **8**, 568[467]

Paul, H., **2**, 162[140]; **4**, 722[35], 728[35]

Paul, M., **7**, 137[122], 139[122]
Paul, R., **6**, 664[222]
Paul, V., **8**, 494[24]
Paul, W., **6**, 558[840,841,844]
Paulen, W., **5**, 829[23]
Pauling, L., **1**, 8[34], 581[6], 582[6]; **4**, 70[9], 139[1]
Paulissen, R., **4**, 1033[16]
Paull, K. D., **7**, 124[40]
Paulmier, C., **1**, 630[15], 644[122], 646[122], 656[156], 658[156], 668[122], 669[122], 676[156], 695[122]; **2**, 787[50]; **3**, 86[51]; **4**, 339[40], 364[1,1k], 373[1k], 447[215]; **6**, 461[4], 873[8], 1026[81], 1027[81], 1029[81], 1030[81], 1031[81], 1033[81]; **7**, 84[1], 85[1], 108[1], 128[67], 129[67], 131[67], 515[1], 523[1], 769[224], 783[224]; **8**, 847[97], 848[97f], 849[97f]
Pauls, H. W., **4**, 375[94], 401[94c,226]
Paulsen, H., **4**, 78[52c]; **6**, 33[5], 34[5], 37[5], 39[5], 40[5], 41[5,44], 42[5], 43[5,48,49,53,55], 46[5], 48[82], 49[5,96], 51[82,109], 54[5], 57[140], 76[52], 646[99,99b], 652[141], 657[141], 679[329,329a], 789[108], 978[20], 980[33]; **8**, 821[49], 824[62]
Paulshock, M., **6**, 270[77]
Paulson, D. R., **4**, 951[1], 968[1], 979[1]; **5**, 929[173]
Paulus, E. F., **4**, 1079[64]; **8**, 647[55]
Paulus, H., **1**, 37[249]
Pauolvic, V., **5**, 179[143]
Paushkin, J. M., **8**, 214[40]
Pauson, P. L., **2**, 722[91,92], 958[27], 962[44]; **3**, 663[26], 665[26], 1024[32]; **4**, 82[61], 519[18,22], 520[18,36], 521[39], 531[36,82], 541[109-111], 542[118-120], 663[2], 689[69], 697[12], 698[14], 703[32], 704[32]; **5**, 1037[6], 1038[8,9], 1039[6,12-16,18], 1040[6], 1041[16,20], 1043[13,16,23-25], 1044[14,16,26], 1045[20,28,29], 1046[16,18,20,24,25b], 1047[9,12], 1048[14,16,25a,b,32], 1049[6,8,12,15,20,24,28], 1050[12,14,15], 1051[8,20,23,24,26,32,36a,b], 1056[25a], 1062[59], 1138[68]; **6**, 831[11], 848[11]
Paust, J., **4**, 1016[209]; **5**, 1007[39], 1008[39]; **6**, 37[33]; **7**, 228[93]; **8**, 374[147]
Pautard, A. M., **6**, 22[85]
Pautet, F., **7**, 764[114]
Pavela-Vrancic, M., **8**, 794[13]
Pavia, A. A., **8**, 882[85]
Pavia, M. R., **6**, 918[38]; **7**, 254[26]
Pavkovic, S. F., **5**, 1098[117], 1099[117], 1100[117], 1101[117], 1112[117]
Pavlish, N. Y., **8**, 98[100]
Pavlov, A. V., **5**, 418[67]
Pavlov, S., **4**, 307[393]
Pavlov, V. A., **4**, 218[148]
Pavlova, G. A., **4**, 55[156]
Pavlova, L. A., **6**, 501[207]
Pavlovic, V., **8**, 873[17]
Pawellek, F., **8**, 918[120]
Pawelzik, K., **6**, 20[71]
Pawlak, J., **6**, 789[110]
Pawlak, J. L., **5**, 855[189]; **7**, 429[151]
Pawlak, P., **8**, 765[13]
Pawlowski, U., **2**, 785[42]
Payet, C. R., **3**, 580[108]
Payling, D. W., **8**, 725[184]
Payne, C. W., **6**, 705[30]
Payne, D. A., **8**, 331[34]
Payne, G. B., **2**, 139[31]; **7**, 167[96], 446[66,70], 675[52]
Payne, J. C., **2**, 852[228], 853[228]
Payne, M. J., **2**, 740[63a]
Payne, M. T., **4**, 277[84]
Payne, N. C., **4**, 654[446]
Payne, O. A., **7**, 800[30,30b]
Payne, S., **7**, 246[92]
Payne, T. L., **6**, 677[323]
Payton, A. L., **8**, 249[91], 294[53]
Paz, V., **6**, 624[142]
Peace, B. W., **4**, 1033[25], 1036[51]
Peach, J. M., **5**, 151[11]; **7**, 725[32]
Peacock, J. A., **6**, 440[73]
Peacock, N. J., **3**, 868[39]; **7**, 854[61]
Peagram, M. J., **5**, 255[50]; **6**, 687[374]
Peak, D. A., **6**, 430[97]
Peake, S. L., **7**, 765[158]
Pearce, A., **2**, 576[77], 582[77], 714[54]; **3**, 345[22]
Pearce, A. A., **8**, 705[10], 726[10], 939[96]
Pearce, B. C., **2**, 885[52]; **6**, 531[440]
Pearce, C. J., **7**, 239[46]
Pearce, D., **5**, 1125[54]
Pearce, D. S., **5**, 407[26]
Pearce, G. T., **5**, 453[68]
Pearce, H. L., **1**, 708[253]; **2**, 509[35], 518[35]; **3**, 35[200]; **4**, 315[501], 373[69]
Pearce, R., **1**, 140[7]; **8**, 513[101,103]
Pearl, G. M., **8**, 904[54]
Pearlman, B. A., **1**, 436[148]; **5**, 143[99], 327[28]; **6**, 1003[136], 1004[137]; **7**, 677[67]
Pearlman, M. R. J., **8**, 568[467]
Pearlman, P. S., **8**, 264[36]
Pearlman, W. H., **8**, 568[467]
Pearlstein, R. M., **8**, 47[125], 66[125]
Pearson, A. J., **1**, 535[144]; **2**, 723[98], 814[76], 833[149]; **3**, 220[123]; **4**, 371[48,51], 581[28,29], 664[6], 667[6b,12], 668[14], 670[14,18,19,23-25,27], 673[32], 674[36], 675[27,38-41], 677[14], 678[42], 679[40,43-46], 680[47,48,51], 682[12,55,58], 686[61,62], 687[23], 689[68], 695[2,4], 698[22], 702[31]; **6**, 690[402], 692[402]; **7**, 107[153,154,155], 377[91], 453[72]
Pearson, C. J., **3**, 902[119]; **8**, 389[71]
Pearson, D. E., **2**, 737[34], 766[86]; **5**, 752[43]; **6**, 795[4]; **8**, 52[147], 66[147]
Pearson, D. L., **1**, 846[16], 851[16]; **2**, 323[27]
Pearson, I., **7**, 516[8]
Pearson, J. H., **8**, 895[3], 898[3]
Pearson, J. R., **3**, 371[115]
Pearson, N. R., **2**, 82[7]; **8**, 719[119]
Pearson, R. G., **1**, 94[71], 252[6], 512[41]; **3**, 4[19], 211[31], 213[31]; **4**, 170[16], 725[49]; **6**, 226[13]; **8**, 541[216], 988[26]
Pearson, W. H., **1**, 480[151]; **2**, 119[160], 208[107], 667[40-42], 673[40], 674[40-42], 675[40-42], 692[72]; **3**, 493[80]; **4**, 373[80], 374[80]; **5**, 434[145], 938[213], 939[213], 941[226], 962[213]; **7**, 355[45]; **8**, 385[49], 856[171]
Pease, J., **8**, 798[65], 800[65]
Peavy, R. E., **4**, 1085[106]
Pechacek, J. T., **1**, 243[53]
Pechal, M., **7**, 154[21]
Pechet, M. M., **1**, 174[8], 175[8]; **4**, 344[77], 347[104]; **7**, 15[145], 41[22], 90[32], 741[49], 747[94]; **8**, 796[28]
Pechine, J. M., **5**, 579[160,161]
Peck, D. R., **5**, 854[178], 856[178], 872[178]; **6**, 860[177]
Peck, D. W., **5**, 455[81]
Pecka, K., **6**, 707[40]
Peckham, P. E., **2**, 964[56,56b]
Pecora, A. J., **3**, 12[56]
Pecquet-Dumas, F., **2**, 304[7]
Pecunioso, A., **2**, 325[41,42]; **4**, 142[17a-c], 143[17b]
Peddle, G. J. D., **5**, 587[207]
Pedersen, B. S., **2**, 867[12,13]; **6**, 436[19], 437[19]
Pedersen, C., **4**, 36[102,102e]; **6**, 61[150], 424[56]
Pedersen, C. J., **8**, 524[13]

Pedersen, E. B., **4**, 436[140]; **6**, 533[482], 545[636,637], 546[639,640,643], 550[482]
Pedersen, K., **7**, 95[80]
Pedersen, K. J., **5**, 1123[34]
Pedersen, P. R., **5**, 1123[34]
Pedersen, S. F., **1**, 314[129]; **3**, 579[102], 596[196], 597[196], 610[102]; **7**, 421[138], 424[138]
Pederson, R. L., **2**, 463[79]
Pedler, A. E., **3**, 648[187]
Pedley, J. B., **1**, 140[10]; **5**, 71[156]
Pedlow, G. W., **7**, 100[114]
Pedlow, G. W., Jr., **4**, 71[15]
Pednekar, P. R., **8**, 244[70]
Pedoussaut, M., **3**, 754[109]
Pedrini, P., **3**, 232[267], 511[186], 514[186]
Pedro, J. R., **3**, 87[99], 104[99], 851[64]
Pedrocchi-Fantoni, G., **1**, 55[23], 221[68], 389[139]; **2**, 29[105], 30[113], 31[113,113a], 32[119,119b], 998[50,51], 999[50,51]; **4**, 36[103,103c]; **8**, 195[112]
Pedrosa, R., **6**, 77[55], 248[137]
Peek, M. E., **4**, 488[57]
Peek, R., **7**, 805[67]
Peel, M. R., **2**, 726[126]; **3**, 354[61]; **5**, 765[112]
Peel, R., **2**, 529[16]
Peel, T. E., **3**, 297[23]
Peeler, R. L., **4**, 305[368]
Peeling, M. G., **4**, 287[180]
Peels, M. R., **4**, 258[243]
Pees, K. J., **4**, 140[12]
Peet, J. H. J., **8**, 879[49]
Peet, N. P., **2**, 128[242], 184[23], 835[158]; **7**, 324[73]
Peet, W. G., **3**, 458[134]
Peeters, H., **2**, 810[65]
Peevey, R. M., **1**, 635[88]; **6**, 846[103], 905[145]
Peevey Pratt, D. V., **3**, 104[209], 117[209]
Pegg, N., **2**, 124[202]
Pegg, N. A., **1**, 240[43]
Pegg, N. J., **5**, 796[53]
Pegg, W. J., **8**, 989[35]
Pegram, J. J., **6**, 648[121], 1006[147]; **8**, 62[195], 66[195]
Pegues, J. F., **8**, 548[323]
Pehk, T., **2**, 346[43]
Pei, G.-K., **7**, 225[58], 280[167]
Peiffer, G., **4**, 55[155,157], 57[157k], 930[49]; **7**, 500[242]
Peiren, M., **2**, 742[75]
Peiseler, B., **6**, 852[134], 879[45]
Peishoff, C. E., **4**, 425[31]
Pekarek, R. S., **8**, 47[124], 66[124]
Pekhk, T. I., **8**, 124[92], 125[92]
Pelah, Z., **8**, 344[119], 345[119]
Pelcman, B., **3**, 511[191]
Pelegrina, D. R., **7**, 163[75]
Pélerin, G., **3**, 599[207]
Peleties, N., **1**, 630[35,36], 631[35,36], 632[36], 633[35,36], 635[36], 636[35,36], 639[35,36], 641[35,36], 642[35,36], 656[35,36], 657[35,36], 658[35,36], 659[35,36], 664[35,36], 665[35,36], 667[36], 672[35,36]; **3**, 87[64,65], 136[64,65], 144[64,65], 145[64,65], 147[64]; **6**, 171[7]
Pelinski, L., **3**, 990[34,34a]
Pelizza, F., **3**, 1017[7]
Pelizzoni, F., **2**, 279[16,17], 280[16], 283[17,45]; **3**, 421[65], 752[97]
Pellacani, L., **6**, 717[111]; **7**, 26[48,49,57], 479[90,91]
Pellegata, R., **1**, 118[59]; **2**, 535[37]
Pellerin, B., **5**, 444[187]
Pelletier, S. W., **2**, 842[189], 1048[8], 1049[8]; **7**, 586[165], 678[71]
Pellicciari, R., **1**, 656[151], 658[151], 846[18b,c], 847[18]; **2**, 823[112]; **3**, 857[92]; **4**, 308[404]; **8**, 880[63]
Pellissier, H., **2**, 1086[31]
Pellissier, N., **2**, 942[167], 943[167], 944[167]
Pelosi, M., **4**, 438[155]
Pelosi, S. S., Jr., **6**, 515[235]
Pelter, A., **1**, 347[130], 488[5], 489[14,15,18], 490[26], 492[5,26,37,39,40], 493[42,42b], 494[26,37], 495[42,44,46], 496[48], 497[48], 498[26,51], 499[18,51-54], 501[37,54], 502[39], 566[218], 569[258]; **2**, 57[4], 240[12]; **3**, 124[261], 126[261], 199[56,66], 261[152], 274[23], 511[189], 514[211], 554[22,23], 691[134], 780[10], 797[95], 798[95-97], 799[99]; **4**, 113[176], 145[35], 148[45a], 249[127], 258[127], 670[22]; **7**, 594[3,4], 595[3,22,23,25,27], 596[34], 598[3,4,25], 600[23,75], 601[3], 607[167], 654[10]; **8**, 26[11,12,20], 27[11,12,20], 36[11,20], 37[11], 54[20,157], 55[20], 60[20], 66[157], 70[20], 214[47], 237[16], 244[16], 253[16], 703[1], 704[1,4], 705[1,4], 706[1,4], 707[1,4], 708[1], 709[1,46], 710[4,54], 711[69], 712[1], 713[1], 715[1,4], 716[1,4,86], 717[1,4,101], 720[1], 721[1,54], 722[1,4], 724[1,4], 725[4], 726[4], 728[4]
Peltier, D., **8**, 134[32], 137[32,60]
Pelz, K., **2**, 765[78]
Pelzer, R., **5**, 187[173]
Pena, M., **4**, 861[113]
Peña, M. R., **3**, 487[51], 495[51]
Penadés, S., **4**, 108[146h], 230[247]
Penas, Y., **5**, 221[56]
Penco, S., **8**, 358[197]
Pendery, J. J., **7**, 355[46]
Pendleton, A. G., **2**, 745[101]
Penenory, A. B., **4**, 453[35], 454[35], 476[35,154,156]
Peng, X., **1**, 243[53]
Penman, K. G., **4**, 744[131]
Penmasta, R., **7**, 145[160,161], 748[109]
Penn, R. E., **7**, 768[203]
Pennanen, S. I., **2**, 739[48]; **3**, 39[215]; **5**, 116[254]; **8**, 993[56]
Penner, G. H., **5**, 389[138], 392[138b], 682[32], 693[106]
Pennetreau, P., **1**, 571[277]; **3**, 142[379]; **6**, 111[62,63]; **7**, 846[99]
Penney, C. L., **6**, 655[167]
Penney, M. R., **5**, 485[184]
Penning, T. D., **1**, 534[142]; **3**, 9[48], 11[48]; **4**, 245[85]
Penninger, J. L., **4**, 915[11]
Pennings, M. L. M., **3**, 367[101]; **8**, 60[194], 62[194], 64[194], 66[194]
Pennington, W. T., **1**, 272[65]; **3**, 572[64]; **8**, 109[13]
Penrose, A. B., **3**, 386[68]
Pensionerova, G. A., **6**, 495[150]
Pentegova, V. A., **3**, 386[68]
Penton, H. R., Jr., **6**, 960[57], 1011[3]
Penzhorm, R. D., **7**, 760[42]
Penzlin, G., **4**, 70[3]; **6**, 830[5]
Pepe, G., **6**, 510[294]; **7**, 876[121]
Pepermans, H., **3**, 587[148]
Peperzak, R. M., **1**, 570[266,267,268]
Pépin, Y., **6**, 920[43]
Peppard, D. J., **2**, 477[12]; **3**, 431[95,96]
Peppler, H. J., **7**, 55[12], 56[12]
Perchec, P. L., **4**, 1010[156]
Perchonock, C. D., **5**, 95[91]
Percival, A., **2**, 587[137], 662[16], 664[16]; **5**, 335[47]
Percy, R. K., **1**, 632[64]
Perdoncin, G., **8**, 111[17,18], 113[18], 117[17,18]
Perego, R., **4**, 379[115], 380[115j]
Pereillo, J.-M., **2**, 765[74]
Pereira, M., **6**, 487[76], 489[76]
Perekalin, V. V., **2**, 321[8], 325[8], 326[8]; **4**, 85[77d]; **6**, 556[826]; **7**, 855[64]
Perera, C. P., **3**, 1036[83]

Perera, S. A. R., **1**, 223[77], 224[77], 317[150]

Peresleni, E. M., **6**, 531[453]

Pereyre, M., **1**, 437[156], 445[192], 479[150], 480[150]; **2**, 3[8], 6[8,8a], 18[8a], 71[53], 564[6], 574[58], 607[74,75], 608[75,77], 609[81,83]; **3**, 3[14], 7[33], 8[14], 20[119], 195[16], 196[25], 453[115], 524[30], 529[30]; **4**, 735[83], 743[128], 770[83], 971[75]; **5**, 901[30]; **7**, 614[4], 616[15], 621[4]; **8**, 278, 547[316,316j,317], 548[319], 794[8], 798[8], 845[77]

Perez, A. D., **6**, 152[134], 934[100]; **8**, 836[10e]

Perez, C., **6**, 959[38]

Pérez, C., **7**, 691[15]

Perez, D., **8**, 554[368], 555[368,369], 782[101], 984[4], 991[4]

Perez, F., **4**, 1059[154]; **5**, 579[160,161]

Perez, J. J., **5**, 689[78,78b], 732[135], 733[135b]

Perez, M., **4**, 339[45]

Perez, M. A., **6**, 512[301], 517[301]

Perez, M. G., **6**, 273[94]

Pérez, M. S. A., **8**, 2[10]

Perez, S., **1**, 117[56]

Pérez-Blanco, D., **2**, 411[8]

Perez-Dolz, R., **1**, 543[26]

Pérez-Juárez, M., **8**, 54[156], 66[156]

Perez Machirant, M. M., **7**, 95[68]

Pérez-Ossorio, R., **5**, 201[32], 202[33,34,36], 220[50,51], 221[52,53]; **8**, 2[10], 36[74], 38[74], 66[74]

Perez-Prieto, J., **4**, 290[196]12291[220], 311[446], 399[224]; **8**, 856[180]

Perfetti, R. B., **7**, 320[63], 841[10]; **8**, 241[38]

Pergreffi, P., **5**, 1137[55]

Periasamy, M., **1**, 149[49b]; **3**, 1017[5]; **4**, 898[173]; **6**, 242[91], 243[91]; **7**, 800[30]; **8**, 696[121], 709[44,45], 720[130]

Pericàs, M. A., **4**, 1005[94]; **5**, 1047[31], 1052[31], 1054[43], 1059[54], 1062[59], 1133[33]

Perich, J. W., **6**, 618[109]

Perie, J. J., **4**, 290[198,199,202,203,207], 292[221], 401[230]; **7**, 470[1], 488[1], 490[1]; **8**, 252[109], 856[175]

Perinello, G., **5**, 176[126]

Perisamy, M., **4**, 254[186]

Perizzolo, C., **6**, 262[18]

Perkin, A. G., **7**, 768[198]

Perkin, W. H., **4**, 285[165], 289[165]; **5**, 899[2]

Perkin, W. H., Jr., **2**, 149[91], 395[2], 399[23]

Perkins, M., **2**, 523[74]

Perkins, M. J., **3**, 505[162], 507[162], 512[162], 803[1]; **4**, 37[107], 766[231]; **8**, 383[13]

Perkins, M. V., **4**, 390[175b]

Perkinson, N. A., **6**, 554[758]

Perl, C., **2**, 757[17]

Perlberger, J.-C., **5**, 972[8], 989[8]

Perlier, S., **6**, 456[157]

Perlin, A., **7**, 293[11]

Perlin, A. S., **6**, 36[23], 49[99], 603[13]; **7**, 703[1], 709[1,41], 710[1]; **8**, 959[26]

Perlman, D., **2**, 456[29]; **5**, 752[46]; **7**, 55[12], 56[12]; **8**, 185[7]

Perlman, K., **5**, 1096[110], 1098[110]

Perlmutter, H. D., **5**, 2[13], 3[13]

Perlmutter, P., **1**, 390[142]; **4**, 34[97], 35[97,97d]

Perlova, T. G., **4**, 992[154]

Pernet, A. G., **2**, 555[140]; **3**, 216[76]; **4**, 27[80], 249[128]; **7**, 299[39]

Perni, R. B., **1**, 765[166]; **2**, 828[131]; **6**, 487[61], 489[61]

Pero, F., **4**, 426[52]

Peron, U., **7**, 760[44]

Perot, G., **6**, 563[897]

Perregaard, J., **8**, 924[2]

Perret, A., **2**, 138[23]

Perret, C., **4**, 520[28]; **5**, 37[22a], 632[60]

Perret, R., **1**, 417[70]

Perreten, J., **5**, 947[267], 948[267]

Perri, S. T., **5**, 689[76,77], 690[80,80c], 733[136,136c,d,f], 734[136f]

Perrier, M., **5**, 553[48]

Perrier, S., **7**, 877[135]

Perrin, C., **8**, 383[17]

Perrin, C. L., **5**, 821[161]; **7**, 800[33]; **8**, 527[50]

Perrin, M., **8**, 133[17]

Perrin, P., **3**, 416[17], 417[17]

Perrin, R., **3**, 306[83]

Perrio, S., **5**, 829[22]

Perrior, T. R., **4**, 675[39,40], 679[40,44]

Perriot, P., **2**, 427[42]; **3**, 202[92,93,95], 788[49]; **6**, 965[87,89]

Perron, Y. G., **4**, 398[216], 399[216d]

Perrone, E., **8**, 409[83]

Perrone, M. H., **8**, 623[150]

Perrone, R., **6**, 787[99,100]

Perrot, A., **2**, 534[32]

Perrot, M., **8**, 262[19], 265[19]

Perrotta, A., **7**, 243[63]

Perry, C. W., **3**, 350[37]; **5**, 20[138]

Perry, D. A., **4**, 248[107]; **5**, 436[158,158b,g], 442[158]; **6**, 1019[43]

Perry, F. M., **8**, 839[25,25a], 914[95]

Perry, G., **2**, 749[132]; **4**, 305[369]

Perry, J. J., **7**, 56[15]

Perry, M. B., **6**, 939[147], 940[147]

Perry, M. W. D., **1**, 477[144], 545[46-48]; **2**, 523[75]; **4**, 113[170]; **7**, 231[153]

Perry, R., **3**, 289[66]; **4**, 1059[156]

Perry, R. J., **4**, 115[181b]

Perry, W. L., **4**, 315[504]

Pershin, D. G., **8**, 102[129]

Pershin, G. N., **6**, 543[609], 554[791]

Persia, F., **7**, 479[90]

Persianova, I. V., **6**, 507[228], 554[729,773]

Persico, F. J., **6**, 531[445], 538[445]

Person, W. B., **7**, 863[83], 865[83], 868[83]

Persons, P. E., **1**, 768[173]; **7**, 409[91]

Perst, H., **5**, 730[128], 731[131]; **6**, 749[102]

Perthuis, J., **8**, 858[203]

Perumal, P. T., **2**, 33[123]; **8**, 722[145]

Perumattam, J., **5**, 143[97,98], 636[88]

Perumattam, J. J., **7**, 242[61]

Peruzzo, V., **2**, 6[31,32], 18[31], 564[12], 566[20], 572[47], 726[127]

Peruzzotti, G. P., **8**, 194[103], 544[253,264,265], 546[304], 561[304]

Perveen, N., **6**, 725[167]

Perz, R., **4**, 100[126]; **8**, 246[79], 546[311]

Pesaro, M., **5**, 714[75a]

Pescarollo, E., **4**, 307[396]

Pesce, G., **3**, 222[145], 246[40], 257[40], 283[49]; **4**, 34[97], 35[97]; **7**, 120[14]; **8**, 887[116]

Pesce, M., **8**, 534[159]

Peseckis, S. M., **1**, 800[300]; **2**, 13[57]; **5**, 515[12], 516[12a], 534[91], 538[12a]; **6**, 7[30], 995[100]

Peseke, K., **2**, 345[42]; **6**, 509[244]

Pesez, M., **2**, 354[106,107]

Pesnelle, P., **2**, 477[12]; **3**, 431[95,96]

Pesotskaya, G. V., **6**, 499[172]

Pessolano, A. A., **6**, 487[57], 489[57]

Pestit, Y., **6**, 3[15]

Petasis, N. A., **1**, 630[1], 672[1], 746[71]; **2**, 388[345]; **3**, 86[56], 94[56], 117[56], 558[51,52]; **4**, 317[559], 339[40]; **5**, 743[164], 744[164]; **6**, 466[45], 469[45]; **7**, 515[1], 523[1]

Pete, B., **5**, 444[187]; **6**, 499[177], 520[340]

Pete, J. P., **1**, 268[55], 269[55]; **3**, 281[42], 602[223], 603[228], 613[3]; **4**, 48[139], 603[276,277], 610[334-336], 645[276,277], 809[163], 857[105]; **5**, 176[129], 310[100]; **6**, 644[81], 961[71]; **7**, 107[157,158,159]; **8**, 817[25,33]

Pete, J. R., **3**, 599[209]

Peter, D., **5**, 498[237]

Peter, R., **1**, 83[26], 141[21], 145[21,41,42], 146[21,42], 148[42], 149[21,42], 150[41], 151[21,41,53,53a], 152[21,53,53a], 153[41], 154[41], 155[42], 157[53a], 158[53,53a], 170[42,122], 333[57], 335[57], 337[57]; **2**, 22[86], 117[151], 306[13], 307[14], 345[35], 351[35], 363[35], 512[47]; **3**, 421[58]; **5**, 735[138b], 740[138], 1005[32]; **6**, 903[139]

Petermann, J., **6**, 428[80]

Peters, D. G., **8**, 857[199]

Peters, E. M., **1**, 168[116a]; **2**, 35[127]; **4**, 1022[253]; **5**, 155[36], 156[36], 157[36], 200[30], 206[45,46], 224[101]; **6**, 121[130], 863[191]; **8**, 354[170]

Peters, F. N., **8**, 606[20]

Peters, J. A., **1**, 294[46]; **8**, 287[18,19], 288[19]

Peters, J. A. M., **3**, 371[116]

Peters, J. W., **7**, 765[166]

Peters, K., **1**, 168[116a]; **2**, 35[127]; **4**, 121[207], 758[191], 1022[253]; **5**, 154[29,30], 155[36], 156[29,36], 157[36], 206[45,46], 224[101]; **6**, 121[130], 863[191]; **8**, 354[170]

Peters, K. S., **7**, 851[16]

Peters, M. A., **2**, 763[61]

Peters, N. K., **8**, 491[13], 496[35]

Peterse, A. J. G. M., **2**, 817[91]

Petersen, J. L., **4**, 194[121]; **5**, 242[8], 243[10,11]; **8**, 675[49], 676[49]

Petersen, J. S., **1**, 191[77]; **2**, 2[5], 25[5], 33[5], 40[5], 134[3], 190[57], 232[180], 240[5], 248[5b], 258[49,50], 455[8], 652[124], 686[68], 867[19], 870[19], 871[19], 879[19], 979[17], 1012[15]; **6**, 8[39]; **7**, 399[40a], 442[48]; **8**, 171[107,109], 535[165], 720[138], 721[138], 722[138]

Petersen, M. R., **5**, 828[7], 839[7], 862[249], 882[13], 888[13], 891[36], 892[13,38b], 893[13]

Petersen, R. C., **7**, 804[61], 805[64]

Petersen, U., **5**, 457[90]

Peterson, B., **2**, 622[155]

Peterson, C. J., **3**, 579[97]

Peterson, D., **3**, 224[182], 760[135]

Peterson, D. H., **8**, 558[397,398]

Peterson, D. J., **1**, 345[124], 436[150], 476[112,116], 479[116], 506[4], 618[55], 620[55,66], 630[19], 731[2], 732[2], 783[238], 784[242,243,244], 815[2]; **3**, 65[4], 86[57,58], 88[58]; **4**, 120[197,198]; **5**, 1126[67]; **8**, 860[219], 862[219]

Peterson, E. R., **8**, 50[117], 66[117]

Peterson, G. A., **5**, 1089[87], 1090[87], 1094[87,102], 1098[87,130], 1099[87], 1100[87], 1101[87], 1102[102,147], 1112[87,130], 1113[87]; **8**, 911[88], 933[52]

Peterson, G. E., **1**, 300[70]

Peterson, J., **5**, 692[103]

Peterson, J. C., **3**, 380[8], 627[44]

Peterson, J. O., **8**, 583[11]

Peterson, J. R., **1**, 683[226], 684[226], 685[226], 705[226], 714[226], 717[226], 718[226], 719[226], 720[226], 722[226], 868[82], 869[82]; **3**, 786[41]; **4**, 262[307], 763[213], 764[215]; **7**, 92[43], 447[72], 487[149], 532[31]

Peterson, M. L., **3**, 648[171], 796[88]; **6**, 77[58], 98[58]

Peterson, M. W., **7**, 95[71], 108[177]

Peterson, P. E., **3**, 281[44], 380[5,16], 381[16a], 735[17]; **4**, 312[458]; **5**, 1123[34]; **7**, 268[123]; **8**, 726[190], 815[21]

Peterson, R., **4**, 1005[92]

Peterson, R. T., **4**, 21[69], 108[146g], 218[139]

Peterson, W. D., **4**, 239[31]

Peterson, W. R., Jr., **6**, 809[68,69]

Peticolas, W. L., **6**, 450[118]

Petiniot, N., **4**, 1033[16,16e], 1035[16e], 1051[125]; **6**, 25[101]

Petisi, J. P., **8**, 973[120]

Petit, A., **6**, 2[5], 18[5]

Petit, F., **3**, 583[121]; **4**, 930[49]; **8**, 460[248]

Petit, J., **7**, 59[42]

Petit, M., **3**, 583[121]; **4**, 930[49]

Petit, T., **4**, 596[158], 620[158], 621[158], 636[158]

Petit, Y., **3**, 215[66], 225[189,190], 263[174]; **4**, 89[84j]; **6**, 3[17], 11[47]

Petraglia, S. P., **2**, 583[111]

Petragnani, N., **1**, 636[101,102], 640[101], 641[102,106,107], 642[115], 645[115], 656[150], 658[150], 666[101], 672[102,106,107], 677[102,106,107], 724[102,106]; **3**, 39[217], 87[85], 120[241], 136[374], 141[374]; **4**, 120[196], 300[310], 304[361], 305[361], 364[1,1n], 370[33,34], 372[56], 380[1n], 508[160]; **5**, 268[78]; **7**, 775[352b]; **8**, 849[106]

Petraitis, J. J., **2**, 853[229]; **3**, 242[8], 257[8]; **5**, 851[169]; **7**, 350[24], 355[24]

Petrakis, K. S., **2**, 332[61]

Petranyi, G., **8**, 29[39], 66[39]

Petraud, M., **6**, 185[162]

Petre, J. E., **7**, 603[115]; **8**, 102[126]

Petrenko, A. E., **6**, 270[81]

Petri, O. P., **8**, 547[316]

Petrid, A., **6**, 554[712,740]

Pétrier, C., **1**, 219[60,61]; **2**, 23[89], 565[13], 572[13]; **3**, 249[65]; **4**, 95[99a-c], 764[220]

Petrillo, E. W., Jr., **4**, 1014[184]

Petrillo, G., **4**, 457[62-64], 460[64], 461[101], 471[62], 475[101,148], 476[64,160,161]; **6**, 240[80]

Petrini, M., **2**, 323[39], 598[12]; **4**, 428[79]; **6**, 104[10], 115[83], 938[128], 942[128], 944[128]

Petrocchi-Fantoni, G., **1**, 185[55], 186[55]

Petropoulos, C. C., **2**, 380[298]

Petrosrich, J. P., **8**, 527[51]

Petrov, A. A., **4**, 284[155], 286[168], 289[168], 304[360]; **7**, 505[290], 506[292]; **8**, 727[197], 772[54]

Petrov, A. D., **8**, 556[376,378], 782[99]

Petrov, E. S., **1**, 276[78], 277[78]

Petrov, G., **1**, 34[224], 36[174,175]

Petrov, K. A., **6**, 419[6], 508[243], 564[910]

Petrov, M. L., **4**, 50[142]

Petrov, O., **2**, 651[121]

Petrov, V., **7**, 136[109]

Petrov, V. N., **4**, 318[562]

Petrov, Yu. I., **8**, 150[132]

Petrova, N. V., **6**, 530[422]

Petrova, T. D., **6**, 525[386], 527[386,407]

Petrovanu, M., **3**, 921[36]

Petrovich, J. P., **3**, 564[9], 567[9]; **8**, 532[132,132d]

Petrovskii, P. V., **4**, 115[180b]; **8**, 778[84]

Petrow, V., **3**, 846[47]; **8**, 986[17], 987[17]

Petrun'kova, T. I., **4**, 48[138,138f]

Petruso, S., **6**, 551[682]

Petrusova, M., **2**, 140[36]

Petrzilka, K., **4**, 501[115]

Petrzilka, M., **1**, 698[249]; **2**, 809[54], 824[54]; **3**, 17[89]; **4**, 339[43,46], 397[212]; **5**, 253[46,46c], 329[30], 333[30], 351[83], 434[140], 501[269], 514[7], 742[162], 830[33]; **6**, 1032[117]; **8**, 61[189], 66[189], 117[72]

Petrzilka, T., **7**, 503[277]

Petterson, R. C., **5**, 208[52]

Petterson, T., **7**, 800[29]

Pettersson, B., **3**, 643[129]

Pettersson, L., **7**, 410[101]

Pettersson, T., **3**, 671[63], 685[63]
Petti, M. A., **4**, 429[88]
Pettibone, D. J., **5**, 410[41]
Pettig, D., **4**, 111[152d], 222[179,180,181], 224[181]
Pettit, G. R., **3**, 126[310]; **7**, 124[40], 153[11], 680[78]; **8**, 214[42], 220[42,83,87], 244[63], 246[63,76], 248[63,76,84], 803[91], 836[1], 837[1], 935[66]
Pettit, R., **1**, 743[54], 746[54], 748[54]; **3**, 855[82]; **4**, 664[5], 665[7], 701[25,26], 980[107]; **5**, 677[11], 695[11], 1115[2], 1116[2], 1122[2b], 1123[2b], 1124[2b], 1130[5]; **6**, 690[397], 692[397,407]; **8**, 289[26], 290[26], 371[115], 372[123], 373[123]
Pettit, T. L., **7**, 539[66]
Pettitt, D. J., **6**, 277[126]
Pettus, J. A., Jr., **5**, 803[88], 975[17]
Petty, C. B., **5**, 331[43]; **6**, 811[75]
Petty, E. H., **3**, 1051[24], 1052[24], 1060[44], 1062[44]; **6**, 127[155]
Petty, S. R., **6**, 554[756]
Petukhov, G. G., **8**, 753[70,74]
Petukhova, N. P., **4**, 317[545]; **7**, 766[177]
Petzoldt, K., **7**, 65[69], 74[107,111,112], 75[111,112]; **8**, 187[42], 881[72], 882[72]
Pevarello, P., **5**, 257[60]
Pez, G. P., **8**, 242[44], 252[44], 455[205]
Pezzanite, J. O., **4**, 125[217a]
Pezzone, M. A., **4**, 404[247], 405[247b]
Pfaendler, H. R., **4**, 50[142], 285[162]; **5**, 619[11], 620[11], 631[57,58]
Pfaff, K., **4**, 80[60], 102[128a]; **6**, 8[37]
Pfäffli, P., **2**, 1015[22]
Pfaltz, A., **4**, 231[284], 1039[64]; **6**, 7[30], 531[444]; **8**, 462[263]
Pfander, H., **7**, 268[125]
Pfau, A. S., **2**, 169[164]; **3**, 273[12]
Pfau, M., **4**, 7[25], 221[161,162,163]
Pfeffer, M., **2**, 784[39a]
Pfeffer, P. E., **2**, 187[43]; **4**, 313[475]; **6**, 2[6], 57[135]; **7**, 185[175]
Pfeifenschneider, R., **7**, 598[59]
Pfeifer, P., **7**, 840[3], 846[3]
Pfeiffer, B., **6**, 624[145]
Pfeiffer, C., **6**, 509[259]
Pfeiffer, J. G., **3**, 381[19]
Pfeiffer, P., **6**, 953[10]
Pfeiffer, T., **2**, 345[34], 351[34], 357[34], 363[190], 371[34,265]; **5**, 468[124,125], 531[81], 545[121]
Pfeiffer, U., **1**, 828[70]; **2**, 435[60]; **8**, 170[76]
Pfeil, E., **3**, 825[24,24a], 828[24a], 835[24]; **6**, 239[78], 277[129], 288[129]; **7**, 506[304]
Pfenniger, A., **7**, 390[6]
Pfenninger, A., **2**, 810[62], 829[62]; **6**, 88[105], 89[105]
Pfenninger, E., **5**, 10[79], 19[79], 26[79], 27[79], 71[124], 222[63]
Pfenninger, J., **5**, 842[112]; **6**, 470[59,60], 471[59,60]; **7**, 721[17]; **8**, 825[67]
Pfister, G., **8**, 528[58], 930[35]
Pfister, J. R., **2**, 138[18], 385[320]; **4**, 216[122]; **6**, 74[38], 809[64]
Pfister, K., III, **8**, 143[62,63], 148[62,63,107]
Pfister, T., **7**, 753[158,159]
Pfister-Guillouzo, G., **5**, 575[132]
Pfitzner, K. E., **7**, 291[1], 293[1]
Pflaumbaum, W., **6**, 454[147]
Pfleiderer, W., **6**, 625[154]
Pflieger, D., **6**, 1035[136]
Pflieger, P., **6**, 26[110]
Pflüger, F., **7**, 854[45]
Pfluger, R. W., **6**, 1036[141]
Pflughaupt, K. W., **6**, 789[108]
Pfrengle, O., **4**, 663[1]
Pfrengle, W., **2**, 656[155], 1099[111]; **5**, 410[41], 411[41i]
Pfyffer, J., **4**, 1015[198]
Pham, C. V., **3**, 459[138]
Pham, H.-P., **7**, 830[61]
Pham, K. M., **8**, 54[153], 66[153]
Pham, P. Q., **3**, 466[185]; **4**, 91[88d]
Pham, P. T. K., **3**, 682[164]
Pham, T. N., **8**, 315[53], 802[79-83], 806[99]
Phan, X. T., **5**, 166[90]; **7**, 877[132]
Pharis, R. P., **5**, 1125[54]
Philipp, W., **2**, 529[20]; **5**, 809[114]
Philippi, K., **7**, 50[70]
Philippo, C. M. G., **5**, 850[157]
Philips, C., **4**, 314[479]; **8**, 535[166]
Philips, J. C., **3**, 883[108]; **5**, 324[20]
Philips, R. F., **1**, 254[13], 276[13], 278[13]
Phillipou, G., **5**, 918[126]; **8**, 321[99,103]
Phillippe, M., **4**, 181[75]
Phillipps, G. H., **3**, 816[79]
Phillips, A. P., **8**, 904[56], 907[56]
Phillips, B. A., **6**, 291[218,219]
Phillips, B. T., **1**, 122[67], 359[11], 380[11], 382[11]; **2**, 1027[75]; **3**, 513[204]; **8**, 388[61]
Phillips, C., **8**, 461[258]
Phillips, C. F., **3**, 752[95]
Phillips, C. J., **1**, 385[112]
Phillips, D. D., **2**, 529[20]
Phillips, G. B., **2**, 531[27,28], 538[27], 570[38], 910[65]; **5**, 5[40,41], 15[103], 63[12], 433[137c], 435[137c]
Phillips, G. W., **1**, 383[110]; **2**, 479[17]; **6**, 690[396], 722[137]; **7**, 228[90]
Phillips, H., **3**, 324[150]; **7**, 771[257], 775[341], 776[341], 779[423]
Phillips, J. G., **1**, 514[53-55]; **4**, 893[155]; **6**, 25[100], 842[78]; **7**, 300[54]
Phillips, J. N., **6**, 489[98]
Phillips, L., **2**, 801[30]
Phillips, L. V., **4**, 93[93a]
Phillips, M. L., **4**, 389[167]
Phillips, R. B., **5**, 790[20], 820[20], 822[20], 826[20], 857[228]
Phillips, R. S., **7**, 99[109], 778[420]
Phillips, S., **3**, 381[32]
Phillips, W. G., **7**, 292[7]
Phillips, W. V., **3**, 3[16], 18[99]; **5**, 217[18]
Phillipson, J. D., **3**, 77[59]
Phinney, B. O., **8**, 537[183]
Pho, H. Q., **3**, 1056[35], 1062[35]; **4**, 1033[22,22a], 1057[22a-c]
Phongpradit, T., **2**, 711[38]
Photaki, J., **6**, 664[220]
Photis, J. M., **1**, 551[76]; **3**, 159[453], 160[453], 163[453], 867[33], 873[33,66,67]
Phull, G. S., **6**, 497[159]
Phuoc Du, N., **3**, 828[47]
Pi, R., **2**, 359[164]
Pia-Caliagno, M., **2**, 928[121,122], 946[121,122]
Piacenti, F., **4**, 930[50]; **8**, 87[27], 236[3], 239[3], 552[352]
Piade, J. J., **5**, 579[160,161]
Piancatelli, G., **3**, 512[192,213], 515[192,213]; **5**, 771[145,146,147,148,149,150,151,152], 772[145,148,150,151], 780[148,149]; **7**, 103[137], 112[196], 260[64], 265[99-102,104], 266[105,107], 267[99-102,104,105,107], 530[14,15,17], 531[17]; **8**, 244[58], 248[58], 563[430]
Piantadosi, C., **6**, 660[199]
Piatak, D. M., **2**, 632[26], 640[26]; **7**, 154[13]; **8**, 244[63], 246[63], 248[63]
Piazzesi, A. M., **4**, 915[15], 936[68]
Picard, C., **6**, 441[79]

Picard, J.-P., **2**, 718[72]
Piccardi, P., **3**, 489[60], 495[60], 504[60], 511[60], 515[60]
Piccinni-Leopardi, C., **1**, 838[166]
Piccolo, O., **2**, 735[15]; **3**, 229[230,230a], 246[36], 312[102], 452[110], 460[141]
Pichat, L., **3**, 462[152], 513[207]; **8**, 679[66], 680[66], 681[66], 683[66], 694[66]
Pichl, R., **6**, 175[67]
Pichon, C., **7**, 704[14], 705[14]
Pickard, M. A., **8**, 205[155]
Pickardt, J., **1**, 162[104], 243[55], 253[12]
Picken, H. A., **5**, 829[19]
Pickenhagen, W., **5**, 10[76], 803[88], 972[8], 973[11], 989[8]
Picker, K., **6**, 473[72], 474[72], 475[72]
Pickering, M. W., **2**, 765[76]
Pickering, W. F., **7**, 760[19]
Pickles, G. M., **7**, 595[29]
Pickles, W., **5**, 766[113], 777[113]
Picon, M., **3**, 273[4]
Pictet, A., **2**, 1016[24]; **6**, 736[21]
Pictet, J., **3**, 273[12]
Pidacks, C., **8**, 527[41], 564[443], 614[86]
Pidcock, A., **8**, 265[53]
Piechocki, C., **4**, 1033[19]
Piechucki, C., **6**, 233[46]
Piehl, D. H., **8**, 924[1]
Pienemann, T., **8**, 54[163], 66[163], 136[50]
Pienkowski, J. J., **2**, 348[55]
Pienta, N. J., **4**, 968[61], 969[61]
Pieper, U., **1**, 38[183]
Piepers, O., **8**, 82[5]
Pierce, J., **2**, 456[42], 466[42], 467[42]
Pierce, J. D., **4**, 24[72,72a]
Pierce, J. K., **7**, 167[100]
Pierce, J. R., **1**, 851[43], 886[43]; **5**, 1001[17]
Pierce, O. R., **4**, 280[124]
Pierce, T. E., **7**, 96[87]
Pierch, O. R., **8**, 781[98]
Pierini, A. B., **3**, 505[156]; **4**, 453[35], 454[35,40], 457[52,54,55], 458[40], 460[98], 461[40,54], 463[109], 469[52], 470[137], 471[109], 476[35,54,55,153,154], 477[98]
Pierle, R. C., **4**, 6[20]
Piermattei, A., **7**, 765[151]
Pieronczyk, W., **8**, 459[238]
Pierre, F., **1**, 885[134]; **5**, 815[143]
Pierre, J. L., **1**, 835[136]; **3**, 572[66]; **6**, 80[69], 110[54]; **8**, 2[8]
Pierron, P., **8**, 373[130]
Piers, E., **1**, 763[146]; **2**, 588[151], 589[151,153], 843[196], 851[225]; **3**, 20[106,108,114,115], 212[39], 215[60], 248[52-54], 253[94,95], 380[7], 443[57], 488[52,53], 489[55], 495[52,53,55]; **4**, 148[50], 173[35], 174[40], 176[52], 189[103], 242[61], 248[111], 249[122,123], 255[61,202,203], 256[111,210], 260[61,111], 901[186,186a], 955[14], 1040[78], 1049[78a,121a]; **5**, 246[23], 247[23], 270[23b], 703[15], 803[91], 804[95], 841[87], 906[69], 953[297], 955[301], 977[22], 979[27,28], 980[27], 981[27], 982[27], 983[31], 987[41], 988[41], 989[43], 990[45,46], 991[46], 992[48,49], 993[50,51], 994[50,51], 995[51], 996[51], 997[54]; **7**, 262[75]; **8**, 121[79], 334[63], 944[124]
Piers, K., **5**, 904[53]
Pierson, C., **7**, 16[160]
Pierson, W. G., **3**, 564[10], 595[188]; **4**, 45[126,126c]; **5**, 686[51], 687[51], 688[51]
Piet, P., **7**, 759[10]
Pieter, R., **1**, 510[25]; **2**, 514[53], 515[53], 524[53]; **3**, 95[178], 96[178], 97[178], 99[178]; **6**, 833[22]
Pietra, F., **1**, 857[57]; **6**, 176[103]; **7**, 579[137]
Pietrasanta, F., **3**, 727[29]
Pietraszkiewcz, M., **6**, 70[19]
Piétré, S., **8**, 315[52], 969[94]
Pietropaolo, R., **4**, 588[66]
Pietrusiewicz, K. M., **1**, 760[136]; **3**, 201[78,79]; **4**, 252[162], 594[140], 634[140]
Pietruszkiewicz, A. M., **6**, 526[395]
Pietsch, H., **1**, 123[79], 372[80]
Piette, J. L., **6**, 462[7,14]; **7**, 774[324]
Piettre, S., **5**, 70[113]
Pifferi, G., **7**, 747[105]
Pigeon, D., **6**, 2[5], 18[5]
Pigiere, Ch., **3**, 46[254], 215[66], 251[75]
Pigman, W., **6**, 789[108], 977[19]; **7**, 703[1], 709[1], 710[1]
Pignataro, S., **6**, 711[62]
Pignolet, L. H., **3**, 1041[110]; **8**, 443[1]
Pigott, H. D., **6**, 133[2]; **7**, 124[50], 127[50]; **8**, 993[61]
Pigou, P. E., **1**, 699[255], 855[51]; **3**, 107[227]; **4**, 736[88]
Pihuleac, J., **6**, 119[117]
Pike, D., **8**, 266[58]
Pike, G. A., **4**, 1032[11]
Pike, J. E., **7**, 86[16a]; **8**, 564[442], 957[16]
Pike, M. T., **8**, 971[107]
Pike, P. W., **4**, 738[100], 739[104]
Pike, R. A., **6**, 1013[15]
Pike, S., **4**, 115[182]
Pikul, S., **1**, 54[19], 109[14], 183[57], 185[57], 339[84,86]; **2**, 31[116], 259[83], 264[83], 995[43]; **5**, 350[79], 377[110,110b], 378[110b]; **7**, 397[29], 568[105], 713[73]
Pilati, T., **4**, 152[58], 207[58]
Pilato, L. A., **8**, 229[136]
Pilcher, G., **5**, 900[6]; **8**, 670[13]
Pile, J. D., **6**, 66[6]
Pilet, O., **6**, 175[69]
Pilichowska, S., **6**, 83[77]
Pilichowski, J. F., **4**, 434[127]
Piliero, P. A., **8**, 458[224]
Pilipauskas, D., **6**, 71[23]
Pilipovich, D., **4**, 347[100]
Pilkington, J. W., **3**, 804[6]
Pilla, L. T., **1**, 471[67]
Pillai, C. N., **4**, 249[131], 873[45]; **5**, 552[34]; **8**, 568[467]
Pillai, K. M. R., **4**, 48[139]; **8**, 566[457], 568[468]
Pillai, P. M., **3**, 740[45]
Pillai, T. P., **7**, 749[119]
Pillai, V. N. R., **6**, 668[255], 669[255]
Pillay, K. S., **3**, 380[11]; **7**, 488[155], 490[155]
Pillay, M. K., **1**, 834[127]; **3**, 736[24]
Pilli, R. A., **2**, 194[68], 205[103], 206[103], 219[68,144], 934[144], 940[144]
Pillot, J.-P., **2**, 582[108], 713[44], 716[57,61,66], 717[68], 718[72], 721[89], 726[124]; **8**, 785[115]
Pilotte, J., **6**, 1016[27]
Pilotti, A., **6**, 672[286]
Pil'shchikov, V. A., **3**, 304[65]
Pilz, M., **5**, 850[152]
Pim, F. B., **3**, 331[198]
Pinchas, S., **6**, 602[9]; **8**, 228[125]
Pinck, L. A., **8**, 973[117]
Pincock, A. L., **7**, 247[98]
Pincock, J. A., **4**, 286[167]; **7**, 247[98]
Pinder, A. R., **4**, 18[62], 20[62f], 24[72], 1006[99]; **8**, 212[22], 515[117], 794[1], 971[103], 973[118], 984[3], 991[3]
Pine, R. D., **1**, 743[54], 746[54], 748[54]; **5**, 1115[2], 1116[2], 1122[2b], 1123[2b], 1124[2b]
Pine, S. H., **1**, 743[52,54], 744[52], 746[54], 747[52], 748[54], 749[52], 811[52]; **2**, 597[9]; **3**, 921[36]; **5**, 1115[2], 1116[2], 1122[2a,b],

1123[2a,b,33,40], 1124[2b,49]; **6**, 834[31], 854[31]; **7**, 777[383]; **8**, 676[80]
Pineau, R., **2**, 537[49]
Pinedo, A., **4**, 1061[166]
Pines, A. N., **8**, 778[87]
Pines, H., **3**, 309[91], 329[188], 331[197]; **4**, 316[538]; **5**, 10[77]; **7**, 5[20]; **8**, 814[19]
Pines, S. H., **2**, 741[64,64b]
Pinetti, A., **4**, 337[32]; **7**, 777[374]; **8**, 389[74]
Pinet-Vallier, M., **4**, 878[76]
Pingolet, L. H., **4**, 915[13]
Pinhas, A. R., **1**, 116[43]; **3**, 1036[85]; **4**, 703[32], 704[32]; **5**, 1138[63]
Pinhey, J. T., **3**, 286[56b], 505[160]; **6**, 108[31]; **7**, 352[31], 356[31], 620[28,29]; **8**, 544[261], 906[68], 907[68], 908[68]
Pini, D., **4**, 877[68]; **5**, 1152[144]
Pinkerton, A. A., **1**, 535[144]; **4**, 682[58], 689[68]
Pinkina, L. N., **5**, 752[7,13,14,27,31], 757[7,27], 767[13,14,31], 768[31]
Pinkney, P. S., **5**, 752[41,43,44]
Pinkus, A. G., **1**, 3[21], 45[21e]; **2**, 749[130]
Pinna, F., **7**, 426[148d]
Pinnavaia, T. J., **7**, 845[70,71,73,74]
Pinney, J. T., **7**, 649[45]
Pinnick, H. W., **1**, 551[69]; **2**, 757[9], 882[47]; **3**, 88[138], 89[138], 159[138], 161[138], 164[138], 165[138]; **4**, 429[83], 438[83], 441[83], 492[71], 495[71]; **5**, 456[82], 830[37], 942[233]; **6**, 538[568], 647[115], 655[157], 911[15], 937[15], 1000[125]; **7**, 186[178], 218[1], 219[1], 240[57], 660[42], 882[172]; **8**, 409[80]
Pino, P., **3**, 438[29], 1023[22]; **4**, 914[3], 919[18,19], 920[20], 926[37], 927[42], 930[42,45,48,53], 931[48,56,57], 932[64], 936[69], 945[42,90]; **5**, 1037[3], 1132[22], 1133[27], 1146[110]; **8**, 372[122], 690[103], 699[148], 763[1], 785[1], 937[78]
Pino Gonzalez, M. S., **1**, 759[132]
Pinschmidt, R. K., Jr., **5**, 705[23]
Pinson, J., **4**, 453[28-30], 458[68], 459[28-30,78,80,81,85], 467[68], 469[68,80,81], 471[68,78,139,140], 472[29], 473[68,139], 475[30,78,150]
Pinson, V. V., **6**, 515[310,311,312,313]
Pinsonnault, J., **3**, 25[157]
Pinto, A. C., **7**, 253[22]
Pinto, A. V., **2**, 744[96]
Pinto, B. M., **2**, 167[158]
Pinto, D. J. P., **5**, 255[49]
Pinto, H.-L., **2**, 232[176]
Pinto, I., **4**, 820[223]; **7**, 546[32]
Pinza, M., **7**, 747[105]
Piorko, A., **4**, 518[7], 529[7,74,77,78], 530[78,79], 531[78,81], 541[113,114]
Piotrowska, H., **4**, 590[103,104]; **6**, 523[350]
Piotrowska, K., **4**, 1039[62]
Piotrowski, A., **1**, 746[69], 748[69]; **5**, 1125[56]; **7**, 856[65]
Piotrowski, A. M., **4**, 892[144]; **8**, 736[21]
Piotrowski, D. W., **1**, 237[31], 239[31], 359[18], 380[18], 381[18]
Piotrowski, V., **4**, 426[65], 441[65]
Piper, J. U., **5**, 689[72]
Piper, S. E., **6**, 1045[26,28]
Pipereit, E., **6**, 943[156]
Pippin, W., **5**, 702[9,9b]
Piquard, J. L., **3**, 120[244], 142[244]
Piras, P. P., **8**, 847[89]
Piret, P., **1**, 38[259]
Pirie, D. K., **1**, 385[119], 386[119]; **2**, 939[160]
Piringer, O., **7**, 3[5]
Pirisi, F. M., **6**, 236[54]
Pirkle, W. H., **1**, 98[83], 838[164]; **2**, 648[94], 649[94]; **4**, 311[449]; **7**, 777[375]; **8**, 146[98], 476[25], 637[9]
Pirrung, M. C., **1**, 357[2]; **2**, 94[54], 182[9,9c], 184[9c], 190[9c], 191[9c], 192[9c], 193[9c], 197[9c], 198[9c], 200[9c], 201[93], 205[102], 206[102b], 211[9c], 217[9c], 221[145], 223[151], 235[9c], 236[9c], 289[69], 634[36], 640[36]; **3**, 713[30], 766[159], 942[81a], 1008[68]; **4**, 72[31]; **5**, 134[63], 137[78,79], 143[94,95], 170[113], 179[140]; **6**, 814[95], 874[16]; **7**, 376[88], 549[43]
Pisanenko, D. A., **3**, 305[70]
Pisano, J. M., **8**, 52[151], 66[151]
Pischel, H., **6**, 507[238], 515[238]
Pisciotti, F., **2**, 564[1], 716[66], 717[69]
Piscopio, A. D., **5**, 516[25], 864[260]
Pisipati, J. S., **4**, 457[50], 477[50], 503[125]
Pistorius, R., **3**, 637[62], 647[170], 648[170]
Piszkiewicz, L. W., **8**, 444[11]
Pitacco, G., **4**, 20[63], 21[63]; **5**, 331[41]; **6**, 709[55], 710[57-59], 711[62]
Pitchen, P., **6**, 150[112]; **7**, 425[146], 777[377,378], 778[377,378]
Pitchford, A., **1**, 499[52]; **3**, 421[52]
Pitis, P. M., **8**, 28[37], 66[37]
Pitman, G. B., **5**, 455[80]
Pitman, I. H., **7**, 778[390]
Pitombo, L. R. M., **7**, 774[336]
Pitt, B. M., **7**, 203[53], 206[72], 207[72], 210[72]
Pitt, C. G., **1**, 581[5]; **3**, 125[305], 126[305], 127[305]
Pitteloud, R., **1**, 766[157]; **3**, 380[10]; **4**, 21[69], 111[153], 218[137,138], 339[43]; **5**, 43[32], 45[36], 532[86], 830[33]; **6**, 1032[117]
Pitteroff, W., **3**, 194[13]
Pittman, C. U., Jr., **1**, 452[218]; **4**, 386[152b], 925[35], 931[57]; **5**, 754[62]; **8**, 457[217]
Pittol, C. A., **5**, 418[70]
Pitts, J. N., Jr., **5**, 226[109]
Pitts, W. J., **4**, 1040[73], 1043[73]
Pitzenberger, S. M., **2**, 1027[75]
Piveteau, D., **1**, 558[139]
Pivnitakii, K. K., **8**, 518[130]
Pivnitskii, K. K., **1**, 520[68]
Piwinski, J. J., **5**, 790[23], 791[23], 885[18]
Piyasena, H. P., **5**, 260[70], 263[70]
Pizey, J. S., **1**, 832[112], 844[8]; **4**, 347[95]; **7**, 306[2], 481[109]; **8**, 26[6], 27[6], 36[6], 237[8,20], 238[20], 240[8], 241[20], 244[8], 245[20], 247[8,20], 249[8], 251[20], 253[8,20], 872[3], 877[37]
Pizey, S. S., **6**, 204[13]; **8**, 213[27]
Pizzala, L., **3**, 892[47]
Pizzini, L. C., **4**, 424[13]
Pizzo, C., **6**, 897[95]
Pizzo, C. F., **4**, 370[25,25a]; **5**, 859[236]
Pizzo, F., **4**, 1060[160]
Pizzolato, G., **4**, 31[92,92j], 370[29], 384[143], 413[277], 1080[71]; **5**, 433[137b]; **6**, 531[452]; **7**, 701[66]; **8**, 608[46], 882[87]
Place, P., **1**, 191[78,80]; **5**, 797[62], 821[62]; **6**, 210[76], 214[98]
Plachky, M., **8**, 886[112]
Plackett, J. D., **3**, 933[62]
Pla-Dalmau, A., **4**, 747[148]
Pladziewicz, J. R., **8**, 917[115], 918[115]
Plakhotnik, V. A., **3**, 305[75a]
Plamer, M. A. J., **1**, 192[81]
Plamondon, J. E., **8**, 214[46], 717[97], 726[195]
Planas, T., **5**, 232[138]
Planat, D., **2**, 782[18]
Plante, R., **2**, 456[31]; **8**, 189[59]
Plappert, P., **1**, 366[47]
Plaquevent, J.-C., **6**, 710[61], 717[109]
Plasek, E., **8**, 391[91]
Plashkin, V. S., **3**, 644[137]
Plastun, I. A., **2**, 787[52]

Plat, M., **7**, 222[36]
Plata, D. J., **3**, 1007[64], 1008[64a]; **6**, 887[62]
Plate, A. F., **5**, 900[10], 901[10], 906[10], 907[10], 972[6], 982[6], 984[6], 989[6]; **8**, 124[92], 125[92]
Plate, R., **5**, 829[22]
Platem, M., **7**, 808[76]
Plath, M., **5**, 186[171]
Plati, J. T., **8**, 645[43]
Platonov, V. E., **6**, 525[386], 527[386,407]
Platoshkin, A. M., **2**, 739[46]
Platt, A. W. G., **5**, 300[74]
Platt, J., **5**, 710[51], 713[51]
Plattner, J. J., **7**, 86[16b]
Plattner, P., **2**, 169[164]; **3**, 273[12]
Plattner, P. A., **8**, 228[124]
Platz, M. S., **4**, 1081[80]
Platz, R., **2**, 138[25]
Platzer, N., **4**, 980[115], 982[115]; **5**, 1103[150], 1104[150,158]
Plau, B., **8**, 827[73]
Plaumann, H. P., **8**, 244[49]
Plaut, H., **6**, 16[61], 265[39]; **7**, 646[26]; **8**, 788[120]
Plavac, F., **3**, 39[220]; **7**, 574[125]
Plavec, F., **5**, 11[84]
Plazzogna, G., **2**, 6[31], 18[31], 572[47]
Pleixats, R., **8**, 964[58]
Plénat, F., **3**, 734[11], 740[47,48]; **4**, 315[509]; **8**, 862[230]
Plentl, A. A., **3**, 898[79]
Plepys, R. A., **8**, 543[244,245]
Plesch, P. H., **8**, 91[52]
Plesnicar, B., **2**, 734[4]; **7**, 358[7], 372[7], 671[2], 672[2], 673[2], 674[2], 675[2]
Plesnidar, B., **1**, 837[156]
Plessi, L., **2**, 635[48], 640[48]; **6**, 21[80], 22[80]
Plessner, T., **5**, 522[44]
Pletcher, D., **6**, 282[152]; **7**, 8[55,56], 253[15], 276[15], 793[2,3], 794[7c]; **8**, 132[8], 135[41,43,44], 321[96]
Pletcher, J., **4**, 5[18], 27[84,84a]
Plevey, R. G., **6**, 497[159]
Plevyak, J., **4**, 845[68], 846[71,72], 847[68,71]
Plieninger, H., **2**, 163[145]; **3**, 807[26]; **5**, 404[19], 405[19], 687[60]; **6**, 428[80], 573[964]; **8**, 299[80,82]
Pliml, J., **8**, 590[54], 596[54]
Pliura, D. H., **6**, 436[12], 451[129]
Plöchl, J., **2**, 395[4]
Ploner, K.-J., **3**, 390[84], 392[84]
Plonka, J. H., **8**, 890[141]
Ploss, G., **2**, 366[218], 782[30]
Plotka, M. W., **1**, 474[91]
Plouzennec-Houe, I., **6**, 563[897]
Plueddeman, E. P., **4**, 270[18], 271[26], 272[26]
Pluim, H., **4**, 14[47,47k], 231[272]
Plumet, J., **1**, 117[56]; **4**, 368[17]; **5**, 92[64]; **8**, 36[74], 38[74], 66[74]
Plummer, M., **3**, 815[78]
Plusquellec, D., **2**, 1074[147]
Poarch, J. W., **2**, 225[155]
Pobiner, H., **7**, 759[16]
Pocar, D., **4**, 1099[187]; **6**, 582[993], 705[27], 712[79], 713[80a]; **8**, 72[241], 74[241]
Poch, M., **5**, 1062[59]
Pochat, F., **1**, 561[158]; **6**, 238[75]
Pochini, A., **2**, 137[17], 960[32]
Pöchlauer, P., **7**, 473[34], 501[34]
Pöckel, I., **3**, 705[4]
Pocker, Y., **3**, 721[5], 723[11], 725[11], 726[11], 736[26,27], 761[142]; **4**, 272[37], 273[37]
Podányi, B., **2**, 789[56]
Podder, G., **2**, 740[61]
Podder, S., **7**, 68[83b]
Podestá, J. C., **4**, 770[247]
Podhorez, D. E., **8**, 508[87]
Podlaha, J., **3**, 709[15]
Podlahová, J., **3**, 709[15]
Podraza, K. F., **5**, 835[59]
Poeckel, I., **3**, 725[18]
Poel, D. E., **5**, 222[63]
Poeth, T., **3**, 482[1], 483[1], 491[1], 499[1]
Poetsch, E., **5**, 409[36]
Pogonowski, C. S., **3**, 34[197]; **6**, 1021[53]
Pogrebnoi, S. I., **5**, 850[148]
Pohjala, E., **2**, 147[72], 399[22]
Pohl, D. G., **7**, 12[96], 13[120], 17[171]
Pohland, A., **4**, 31[92,92d]; **8**, 165[48]
Pohlen, E. K., **3**, 831[64]
Pohlke, R., **6**, 967[103]
Pohmakotr, M., **1**, 510[26], 826[63]; **2**, 72[78], 711[38]; **6**, 1022[59]
Pohmer, L., **4**, 483[2], 484[2], 500[103]; **5**, 380[115], 381[115]
Poignant, S., **6**, 489[82], 554[752]
Poignee, V., **4**, 1006[97]
Poindexter, G. S., **1**, 462[18]; **6**, 71[22]
Pointner, A., **2**, 597[8]
Poirier, J. M., **4**, 159[81], 650[427]
Poirier, M., **8**, 246[79]
Poirier, M.-A., **8**, 873[18], 874[20]
Poirier, N., **4**, 159[81]
Poirier, Y., **2**, 758[23a]; **6**, 568[932]
Poisel, H., **7**, 230[132]
Poisson, P., **8**, 858[203]
Poitier, M., **8**, 262[18]
Poje, J. A., **8**, 916[113]
Pojer, P. M., **6**, 215[104], 648[117a]; **8**, 36[79], 66[79]
Pol, E. H., **6**, 611[64]
Polanc, S., **6**, 554[713]; **8**, 384[31,32]
Poland, J. S., **4**, 953[8,8c], 954[8c]
Polansky, O. E., **2**, 346[47,48], 347[48], 349[66], 352[83,92], 353[48,94], 355[48,122,126], 356[48,66], 357[141,142], 358[48], 365[48], 367[48], 369[48], 371[261], 374[48]; **5**, 76[239]
Polanyi, M., **8**, 420[24], 422[24]
Polaski, C. M., **1**, 367[54]
Poletto, J. F., **4**, 141[15], 142[15]; **6**, 648[124]; **8**, 527[41]
Polevy, J., **7**, 602[102]
Polezhaeva, A. I., **6**, 554[776,780,793]
Polgár, L., **6**, 451[128]
Polgar, N., **4**, 288[181], 301[317], 303[317], 310[431]
Poli, G., **1**, 72[72], 524[86-88], 770[187]; **2**, 103[28], 106[54], 266[61], 515[55-57], 516[58], 605[57], 614[57], 641[71], 642[71,73,74], 643[73,74], 644[73]; **4**, 85[72], 152[58], 204[33,35], 207[57,58], 231[282,283], 249[120]; **5**, 260[70], 263[70], 362[93], 363[93d], 364[93d], 545[120]; **7**, 441[45]; **8**, 537[188]
Poli, L., **8**, 560[404]
Poli, N., **5**, 1158[173]
Policastro, P. P., **2**, 904[49]
Polichnowski, S. W., **4**, 980[105], 981[105]; **5**, 1085[65]
Polievktov, M. K., **6**, 502[208], 554[792]
Poling, B., **4**, 953[8], 954[8k]
Polishchuk, N. V., **6**, 538[551,555]
Polishchuk, V. P., **3**, 647[196]
Polishchyuk, V. R., **3**, 647[177], 648[177], 649[177]
Politanskii, S. F., **4**, 969[66]; **6**, 216[107]
Polito, A. J., **6**, 291[199]
Politzer, I. R., **2**, 492[53], 493[53]; **6**, 274[103-107]; **8**, 276[148,149,150]
Polizzi, C., **3**, 217[80,80b], 246[41], 247[41], 257[41], 491[70]

Poljakova, L. A., **7**, 884[186]
Poljakowa, A. M., **3**, 892[49]
Polk, D. E., **1**, 367[54]; **6**, 545[635]
Poll, T., **1**, 303[78], 307[78]; **5**, 365[95,95b,96a,c]
Polla, E., **2**, 362[181], 523[87]; **3**, 374[132]; **6**, 283[160]; **7**, 744[72], 846[93-95]
Pollack, S. J., **8**, 206[167]
Pollack, S. K., **1**, 487[4], 488[4]; **4**, 484[19]
Pollak, I. E., **1**, 370[67]; **2**, 1004[61]; **3**, 258[124]
Pollard, A., **2**, 149[91]
Pollart, D. J., **5**, 689[72,76]
Pollart, K. A., **7**, 247[99]
Poller, R. C., **7**, 614[5], 616[10], 620[10]
Pollet, P. L., **2**, 801[26]
Polley, J. S., **1**, 859[67]
Pollicino, S., **1**, 516[59,60], 517[61,62]; **3**, 147[396], 149[413], 151[413], 152[413], 153[396,413], 155[396], 865[27], 944[90,91], 946[92], 958[90,112]; **6**, 898[103]; **7**, 764[126], 767[126]
Pollina, G., **7**, 506[296]
Pollini, G. P., **2**, 803[32]; **3**, 738[37]; **5**, 403[8], 451[44], 453[44], 468[44]; **7**, 143[140,141]; **8**, 392[108], 645[45]
Pollok, T., **6**, 196[230]
Polman, H., **5**, 163[72]
Polniaszek, R. P., **4**, 968[61], 969[61]
Polo, E., **2**, 803[32]; **8**, 392[108]
Polonovski, M., **6**, 910[1,2]
Polonski, J., **3**, 691[136]
Polonski, T., **6**, 807[61]
Polster, R., **8**, 758[169]
Polston, N. L., **3**, 483[12], 489[57], 495[57]; **5**, 710[51], 713[51]; **7**, 597[51]; **8**, 708[34], 716[34,93], 717[93], 726[34], 727[34,93]
Polt, R. L., **1**, 19[105], 462[17], 463[17,33]; **7**, 229[119]
Polunin, E. V., **3**, 181[553]; **8**, 971[111]
Polyachenko, V. M., **6**, 525[377]
Polyakova, A. M., **2**, 387[334]; **8**, 769[30]
Polyakova, I. A., **6**, 490[106]
Polyakova, S. G., **3**, 644[140,141]
Polyakova, V. P., **8**, 150[132]
Pomerantseva, M. G., **8**, 763[1], 769[1b], 771[1b], 775[77], 778[88], 785[1]
Pomerantz, M., **5**, 682[34a], 683[34a]
Pomme, G., **4**, 41[117]
Pommelet, J.-C., **3**, 870[49]; **5**, 930[174], 938[220]
Pommer, H., **1**, 722[271], 755[115], 812[115], 813[115]; **5**, 15[106], 835[58,59]
Pommerville, J., **3**, 788[53]
Pommier, J.-C., **2**, 609[80]; **4**, 21[69], 221[158,159]; **6**, 722[143], 726[180]
Pomogaev, A. I., **6**, 489[80]
Pompliano, D. L., **2**, 466[110]
Pomykacek, J., **2**, 765[78]
Pon, R. T., **6**, 618[111]
Ponaras, A. A., **2**, 514[48]; **5**, 847[135,140], 1004[28]
Poncet, J., **8**, 13[71]
Poncin, M., **5**, 736[145], 737[145]
Ponder, J. W., **4**, 384[143]
Pong, R. Y., **4**, 968[60], 969[60]
Pong, S. F., **6**, 515[235]
Pongor-Csákvári, M., **2**, 789[56]
Ponkshe, N. K., **1**, 474[89]
Ponomarenko, V. A., **8**, 556[376]
Ponomarev, S. V., **8**, 547[316]
Ponomareva, E.-V., **6**, 509[272]
Ponomareva, L. A., **8**, 624[155]
Ponomareva, T. K., **4**, 426[64]
Ponomaryov, A. B., **3**, 219[111], 531[88], 537[88]
Pons, J.-M., **1**, 749[82]; **3**, 563[1], 572[63], 577[85,86,91], 607[1i]
Ponsati, O., **6**, 563[901], 984[54]
Ponsford, R. J., **3**, 1051[23], 1056[23], 1062[23]; **6**, 669[264]
Ponsinet, G., **6**, 176[105]
Ponsold, K., **7**, 480[99]
Ponti, F., **6**, 80[68]; **7**, 279[171], 844[60]
Ponticello, G. S., **5**, 382[121]
Ponton, J. P., **4**, 107[141], 251[150]
Ponty, A., **7**, 47[53]
Poochaivatananon, P., **3**, 154[426]
Poole, V. D., **8**, 528[81], 529[81]
Poon, Y.-F., **3**, 493[80]; **5**, 941[226]
Poorker, C., **3**, 747[70]
Poos, G. I., **2**, 158[124]; **7**, 256[40]
Pop, L., **3**, 416[15], 417[15]
Popall, M., **5**, 1096[109,109d,e,124], 1098[109,109c-e,126], 1099[109c-e], 1112[109a-c,126]
Popien, D., **2**, 762[55]
Popjak, G., **7**, 272[132]; **8**, 925[6]
Popkov, K. K., **8**, 778[88]
Popkova, T. V., **4**, 337[37,38]
Poplavskaya, I. A., **6**, 515[316]
Popov, A. I., **7**, 16[164]
Popov, N., **5**, 179[143]
Popova, M. N., **4**, 314[487]
Popova, O. A., **4**, 50[142]
Popovich, T. P., **6**, 524[360], 528[413], 539[577]
Popovitcz-Biro, K., **7**, 40[7]
Popp, F. D., **2**, 343[8], 363[8], 753[1,1c]; **7**, 318[50]; **8**, 236[4], 242[4], 247[4], 248[4], 249[4], 295[57,58], 296[62], 297[58]
Poppi, A. L., **5**, 257[60]
Poppinger, D. J., **4**, 1070[12]
Poquet-Dhimane, A.-L., **5**, 176[129]
Porath, J., **2**, 1104[132]
Porcelli, J., **8**, 454[203]
Porcher, H., **7**, 160[50]
Porco, J. A., Jr., **1**, 420[83], 568[230]; **4**, 53[150]; **5**, 152[20], 158[44], 171[114], 176[125], 736[143,145], 737[145]
Porfir'eva, Yu. I., **4**, 286[168,169], 289[168,169]; **7**, 506[292]
Pornet, J., **1**, 385[113]; **2**, 567[29], 575[62], 579[92,93], 995[44]
Porri, L., **4**, 602[258]; **5**, 36[15]
Pörschke, K. R., **1**, 310[106]; **4**, 985[129]
Porskamp, P. A. T. W., **5**, 441[177]
Porta, O., **1**, 272[66,66a-c]; **3**, 564[6,30,31], 566[30,31], 595[189], 606[189]; **4**, 719[18], 723[18]; **8**, 113[43-45], 116[43], 383[13]
Porteau, P. J., **7**, 410[97a]
Portella, C., **1**, 268[55], 269[55]; **3**, 599[209]; **6**, 644[81]; **8**, 817[33]
Porter, A. E. A., **2**, 369[254]; **4**, 1032[10], 1063[10,10a,168,169,171]; **5**, 497[222]; **8**, 393[113], 830[89], 831[89]
Porter, B., **3**, 908[146]
Porter, C. R., **8**, 600[103]
Porter, J., **6**, 23[94]
Porter, J. R., **1**, 248[70]; **5**, 299[68]
Porter, N. A., **4**, 306[386,387,388], 390[173b,c], 785[21], 786[26], 790[40], 791[21,40]; **5**, 64[32]; **8**, 855[157,160]
Porter, Q. N., **8**, 950[158]
Porter, R., **2**, 456[23]
Porter, R. D., **3**, 330[192]
Porter, R. J., **2**, 958[27]
Porter, S., **8**, 880[61]
Portis, L. C., **7**, 803[57]
Portnov, Yu. N., **6**, 487[68], 489[68]
Portnova, S. L., **3**, 361[73]
Portnoy, N. A., **2**, 482[28], 483[28]
Portnoy, R. C., **2**, 492[53], 493[53]; **6**, 274[107]; **8**, 276[149]
Portoghese, P. S., **8**, 161[13]

Porwell, J. P., **4**, 313[473]
Porzi, G., **4**, 375[94], 386[94a], 387[94a], 388[164], 407[254], 408[254a]; **6**, 26[106], 533[497]
Porzio, W., **5**, 1131[12]
Pöschmann, C., **7**, 490[179]
Posey, I. Y., **1**, 3[18,21]
Posin, B., **8**, 677[60]
Posler, J., **2**, 841[186]; **3**, 55[283]; **8**, 527[48]
Posner, B. A., **6**, 687[375]
Posner, G. H., **1**, 82[25], 86[31], 107[1,2,7], 108[8], 115[1,2], 116[8], 125[1,2], 155[66], 398[1], 426[110], 428[121], 429[121,122], 432[135,136], 433[224], 457[121], 698[250]; **2**, 74[80], 119[164,165], 121[187,188], 124[211], 182[9], 363[192], 388[192], 614[116], 762[52]; **3**, 6[27], 8[41], 9[41], 15[41], 17[27,90], 20[41], 91[146], 155[433], 157[433], 208[10], 210[10], 212[39], 214[55], 216[70,74], 217[88], 226[39a,d,201], 230[238], 248[55], 250[70,72], 251[55], 257[121], 264[72], 265[72], 269[55], 419[37,38,42,44,46], 420[44,46], 421[56], 521[5,6], 522[5,6], 733[1]; **4**, 18[55,59,61,61a,b], 24[72], 70[8], 86[81], 89[87], 109[147], 116[188c], 121[204], 128[87a,b], 141[13], 148[48], 164[48], 169[1], 170[1,6,10], 172[1], 173[1], 176[49], 178[1,49,61], 181[1], 187[1], 189[105], 192[119], 197[119b], 213[102-117], 215[105-107,109,112,114,116,117], 239[23], 240[43], 241[59], 243[68], 245[87,90], 247[101], 249[130], 252[23], 254[43], 255[43,59], 256[206,209,211], 257[90,130], 258[232], 259[259,279], 260[59,90], 262[23,59,101,130,313], 1009[138]; **5**, 373[105], 595[13], 839[83], 1037[1]; **6**, 46[60,66], 89[109], 150[116,120-124,131,132], 151[116,131,133], 161[132,178], 176[93,94], 934[99], 1064[90a]; **7**, 206[66], 320[63], 625[41,42], 627[42,43], 841[10-12]; **8**, 548[321], 564[438], 754[75], 844[71], 847[71], 860[223], 995[68]
Posner, T., **2**, 399[16]
Pospelova, T. A., **6**, 530[421], 550[421]
Pospischil, K.-H., **6**, 531[432]
Pospisek, J., **6**, 801[29]
Pospisil, J., **1**, 10[48]
Pospišil, J., **3**, 664[29]
Poss, A., **2**, 189[46], 209[46]
Poss, A. J., **7**, 105[151], 453[71], 463[128]
Poss, K. M., **5**, 829[25]
Poss, M., **2**, 811[71], 824[71]
Poss, M. A., **1**, 32[158], 791[268]; **2**, 189[46], 209[46], 221[46a], 482[36], 484[36]; **5**, 541[111]; **6**, 186[170], 1013[16]
Possel, O., **3**, 158[449], 174[449,523], 175[449,523]; **7**, 232[156]
Post, H. W., **8**, 265[51], 780[90]
Postel, M., **7**, 11[87]
Posthumus, T. A. P., **3**, 371[116]
Postigo, C., **4**, 292[228]
Postlethwaite, J. D., **3**, 499[118]
Postovskii, I. Ya., **4**, 423[6], 441[6]; **8**, 580[1]
Posvic, H., **2**, 838[166]
Potenza, D., **2**, 637[59], 640[59], 642[78], 643[78], 645[59], 652[59]
Potenza, J. A., **2**, 280[26]; **5**, 432[133]
Poteruca, J. J., **6**, 546[652]
Potier, P., **2**, 901[31,32], 1021[48]; **4**, 747[152], 748[158], 800[120]; **6**, 734[16], 912[22], 916[32], 920[43-45], 921[47,49], 1067[100]; **7**, 722[20], 725[31], 726[20,37], 731[53]; **8**, 58[174], 66[174]
Potlock, S. J., **6**, 189[190]
Potnis, S. M., **4**, 159[81]
Pototskaya, A. E., **7**, 767[190]
Potter, G. J., **4**, 347[96]
Potter, H., **7**, 156[32]
Potter, N. H., **7**, 235[1]
Potter, S. E., **3**, 877[88], 927[48]
Potthoff, B., **5**, 416[58], 432[131,133]
Potti, N. D., **2**, 954[6], 958[6b]
Potti, P. G. G., **7**, 749[123]
Potts, G. O., **2**, 839[173]
Potts, H. A., **2**, 954[5]
Potts, K. T., **3**, 513[208], 538[94]; **4**, 126[218b], 1096[158], 1097[158,167], 1098[158]; **5**, 473[153], 477[153], 947[263]; **6**, 422[35]; **8**, 935[68]
Potvin, P., **2**, 74[72]; **8**, 882[86]
Potzolli, B., **6**, 637[32,32c]
Pougny, J.-R., **4**, 311[441], 379[116]; **6**, 54[129], 529[466], 978[24]; **7**, 635[71]; **8**, 854[150]
Pouilhes, A., **1**, 838[169]; **4**, 115[180f]; **5**, 409[37]; **6**, 487[75]
Pouillen, P., **8**, 133[17,18]
Poulain, M., **1**, 303[79]
Pouli, D., **7**, 769[216]; **8**, 408[77], 409[77]
Poulin, J.-C., **2**, 232[176]; **8**, 173[118], 460[254], 535[166]
Pouliquen, J., **5**, 116[258,263]
Poulos, A. T., **1**, 18[95]
Poulson, R., **8**, 79[1], 82[1b]
Poulter, C. D., **1**, 366[45]; **2**, 909[62], 910[62]; **3**, 246[43], 258[43]; **6**, 814[91]
Poulter, D. D., **3**, 131[330]
Poulton, G. A., **5**, 130[39]; **8**, 540[198]
Poupaert, J. H., **5**, 553[41]
Poupart, J., **2**, 169[167], 353[99]
Poupart, M., **1**, 885[131]
Poupart, M.-A., **1**, 740[44], 741[44]; **5**, 816[144]; **6**, 74[30], 938[124]
Pourabass, S., **6**, 474[78]
Pourcelot, G., **1**, 632[62,63]; **3**, 964[125]; **4**, 210[70-73,75], 229[223]
Pourcin, J., **3**, 892[47]
Pourreau, D. B., **4**, 976[100]
Pourzal, A. A., **4**, 299[304]
Poutsma, M., **4**, 726[53]
Pouwer, K. L., **4**, 12[39]
Pouzar, V., **2**, 382[313]
Povarov, L. S., **5**, 451[27], 453[27], 480[27,177]
Powell, A. L., **8**, 86[21]
Powell, C., **2**, 959[30]
Powell, D. W., **3**, 957[110], 960[114]; **5**, 894[46]; **6**, 897[101]; **8**, 846[82]
Powell, H. M., **1**, 38[263]
Powell, J., **4**, 587[41], 588[54], 601[249,250], 602[255,256,259], 643[249]; **5**, 35[12,12b,13], 46[13], 56[13]
Powell, J. E., Jr., **5**, 129[33], 830[37], 857[231], 859[37b]
Powell, J. W., **6**, 961[73]
Powell, L. A., **3**, 577[88]; **8**, 134[37]
Powell, L. H., **7**, 418[129c]
Powell, P., **4**, 691[76]
Powell, R. E., **7**, 846[86]
Powell, R. W., **8**, 343[114]
Powell, V. H., **7**, 143[144]
Power, J. M., **1**, 41[270]; **3**, 211[28], 215[28]
Power, J. R., **1**, 432[137], 456[137]
Power, P. P., **1**, 2[13], 11[57,58], 22[113,117], 23[118,119], 41[265,268,271], 488[13]; **3**, 213[43]; **4**, 170[13]
Powers, D. B., **2**, 8[37], 13[37], 20[37a], 35[37]; **6**, 864[197]
Powers, J. W., **4**, 808[157]; **7**, 291[2], 655[18]
Powers, P. J., **3**, 770[178]
Powers, S. K., **7**, 777[365]
Powers, W. J., **8**, 526[29], 527[29]
Powner, T. H., **5**, 531[73], 537[98]
Pozas, R., **3**, 901[112]
Pozdeeva, A. G., **8**, 593[68]
Prabhakar, S., **6**, 114[78]
Prabhu, A. V., **1**, 116[49], 118[49]; **3**, 3[16]
Pracejus, H., **8**, 460[246]
Pradella, G., **4**, 426[55]; **5**, 829[15]
Pradere, J.-P., **4**, 123[210b], 125[210b]; **5**, 475[139], 575[134]; **6**, 554[751,752]; **8**, 658[100]

Pradhan, B. P., **8**, 113[30], 117[30,73]
Pradhan, S. K., **1**, 268[56], 269[56a]; **3**, 572[64], 602[224,225]; **7**, 136[117], 137[117]; **8**, 108[6], 109[6,12], 110[6], 111[6,19], 112[6], 113[6], 114[6,51,52], 116[6], 119[6], 120[6], 122[19]
Pradilla, R. F. d. l., **4**, 368[17]
Praefcke, K., **5**, 436[157]; **6**, 462[9,15,16], 472[67,68]; **7**, 204[58]; **8**, 858[206]
Prager, B., **8**, 9[50]
Prager, R. H., **2**, 809[55]; **3**, 799[104]; **4**, 1040[81], 1043[81]; **8**, 53[131], 56[167], 66[131,167], 875[40]
Pragnell, J., **7**, 71[99]
Pragst, F., **8**, 595[78]
Prahlad, J. R., **8**, 333[52]
Praill, P. F. G., **2**, 738[42]
Prakash, C., **6**, 174[60]
Prakash, D., **7**, 834[81]
Prakash, G. K. S., **2**, 728[138]; **3**, 295[6], 297[18], 298[28], 329[185], 332[204], 334[215,221,221b], 335[6], 339[6], 421[64], 587[142,145], 706[6], 1046[1]; **6**, 207[51], 251[146], 685[351], 726[188], 938[127], 944[127]; **7**, 231[150], 674[39]; **8**, 797[41], 959[25]
Prakash, I., **6**, 118[102], 177[117]; **7**, 145[160], 155[27-29], 748[109]
Prakash, O., **6**, 177[117]; **7**, 155[28,29], 166[91], 827[49], 828[52], 829[52a]
Prakesh, G. K. S., **6**, 237[60], 243[60]
Praly, J. P., **8**, 552[356]
Pramod, K., **5**, 667[44], 814[138]
Prandi, J., **7**, 381[106]
Prange, T., **2**, 816[85]; **3**, 691[136]; **4**, 375[98a], 388[98,98a], 409[98a]
Prankprakma, V., **1**, 787[255]
Prapansiri, V., **1**, 526[94], 558[133], 835[138]
Prasad, A. S., **8**, 720[130]
Prasad, C. V. C., **2**, 599[24], 622[157], 853[232]; **3**, 751[90]; **6**, 448[108]; **7**, 401[61d], 407[84b]
Prasad, G., **5**, 798[68]
Prasad, J. S., **1**, 404[22]; **2**, 1061[94,96]; **3**, 589[162], 610[162]; **4**, 373[82], 411[265a,b]
Prasad, J. V. N., **3**, 797[94]; **4**, 249[131]
Prasad, K., **7**, 503[277]; **8**, 9[48,50]
Prasad, R., **2**, 801[22]; **3**, 304[64]
Prasad, S., **2**, 357[147]
Prasanna, S., **4**, 14[47,47b]; **6**, 137[41]
Prasanya, S., **4**, 111[154d]
Praschak, I., **2**, 1067[125]
Prashad, M., **7**, 300[55]
Prat, D., **4**, 210[73]; **7**, 381[105]
Prater, A. N., **4**, 279[114], 288[114]
Prather, J., **8**, 214[32]
Prathiba, V., **7**, 277[156]
Prati, L., **2**, 488[43]
Prato, M., **4**, 426[63]; **5**, 408[33]
Pratt, A. C., **5**, 202[35], 207[51], 221[54]
Pratt, A. J., **2**, 18[70], 39[135], 574[59]; **6**, 863[190]
Pratt, D. R., **4**, 489[59]
Pratt, D. V., **1**, 635[88], 733[9]; **5**, 849[143], 1001[16]; **6**, 846[103], 905[145]
Pratt, J. M., **8**, 382[3]
Pratt, L., **4**, 663[2], 689[71]
Pratt, R. F., **2**, 279[13]
Pratt, T. M., **8**, 405[30]
Pratt, W., **2**, 977[5]
Pray, A. R., **1**, 232[15]
Preckel, M. M., **5**, 1025[84]
Preece, M., **7**, 108[174]
Preftitsi, S., **6**, 233[47]
Pregaglia, G. F., **7**, 452[61]; **8**, 446[95], 449[157], 450[157], 452[95a,b], 457[95a-c,218], 458[218]
Pregosin, P. S., **4**, 403[239], 404[239], 915[9], 930[54]; **8**, 271[107]
Preiss, A., **6**, 508[287]
Prejzner, H., **6**, 523[350]
Prelle, A., **7**, 679[74,74b]
Prelog, V., **1**, 49[2], 180[50], 182[50]; **2**, 455[13], 897[20], 1108[78,79]; **3**, 379[2,3], 564[4], 566[4], 849[57]; **5**, 79[290]; **6**, 968[111]; **8**, 187[46], 293[48]
Premila, M. S., **2**, 894[9], 912[9]
Prempree, P., **4**, 108[146f]
Premuzic, E., **8**, 531[122], 987[19]
Prenant, C., **8**, 35[64], 52[145], 53[128], 66[64,128,145]
Prenton, G. W., **4**, 496[87]
Preobrashenski, N. A., **3**, 892[49]
Preobrashenski, W. A., **3**, 892[49]
Preobrazhenskaya, M. N., **6**, 533[476], 554[745]; **8**, 568[471]
Prescher, G., **8**, 459[240], 460[250]
Press, J. B., **6**, 538[557], 809[70]
Pressler, W., **2**, 358[152]
Prest, R., **4**, 459[79], 476[79]; **7**, 497[219]
Prestidge, R. L., **6**, 668[251], 669[251]
Preston, H. D., **4**, 27[78]
Preston, J., **6**, 661[211]
Preston, P. N., **7**, 356[51]
Preston, S. B., **2**, 6[24], 38[133]; **8**, 384[27]
Preston, S. C., **4**, 243[74], 257[74]
Prestwich, G. D., **2**, 204[100]; **3**, 41[225], 364[92], 395[102], 396[102]; **6**, 655[164b]; **8**, 542[227]
Preti, G., **8**, 609[50]
Pretor, M., **2**, 351[81], 364[81], 375[81]; **5**, 17[124], 468[127], 531[81,81d]
Pretzer, W. R., **8**, 458[223,223c,d]
Preus, M. W., **7**, 73[103]
Preuschhof, H., **5**, 15[101]
Preuss, H., **4**, 48[137,137a]
Preuss, R., **6**, 60[149]
Previc, E. P., **2**, 141[37]
Previdoli, F., **6**, 711[66]
Previero, A., **6**, 423[48]; **8**, 52[141], 66[141]
Previero, E., **2**, 1104[133]
Prévost, C., **1**, 218[50], 219[50,62]; **2**, 81[2,5]
Prewo, R., **3**, 313[103]; **5**, 92[78], 418[71]; **6**, 539[579], 544[626]; **7**, 160[50]
Prezant, D., **7**, 48[59]
Prezeli, M., **4**, 298[290]
Prhavc, M., **6**, 514[306], 543[306]
Pri-Bar, I., **1**, 443[183]; **3**, 1021[14]; **8**, 900[29]
Pribish, J. R., **7**, 255[33]
Price, A., **7**, 602[102]
Price, C. C., **3**, 299[31], 300[31], 309[90], 316[118], 317[118]; **7**, 123[32], 473[25], 760[39]
Price, D. T., **7**, 220[25]; **8**, 70[224]
Price, E. M., **3**, 178[542], 179[542]
Price, J. A., **2**, 802[28]
Price, J. D., **4**, 1010[159]; **5**, 940[225], 943[251], 963[225,251]
Price, J. R., **3**, 828[53]
Price, M. E., **5**, 1052[37]
Price, M. F., **5**, 829[18], 835[60], 847[135], 1001[14]; **6**, 865[202]
Price, M. J., **4**, 51[143], 278[97,98], 285[98], 286[97], 289[97,98]
Price, P., **3**, 783[24]; **4**, 187[98]; **8**, 194[103], 544[253,264,265], 546[304], 561[304]
Price, R., **4**, 524[60]
Price, R. T., **2**, 527[9], 528[9], 533[30], 541[76], 544[76], 546[76], 547[76]; **5**, 2[7], 4[7]
Price, S. J., **4**, 24[72]

Price, T., **3**, 1036[81]; **4**, 563[37]
Pricipe, P. A., **5**, 86[33]
Prickett, J. E., **5**, 948[290]
Pricl, S., **4**, 20[63], 21[63]
Priddy, D. B., **3**, 304[62]
Pride, E., **8**, 526[36]
Pridgen, H. S., **5**, 847[132]
Pridgen, L. N., **2**, 116[136], 117[136], 124[136], 436[66], 437[66b]; **3**, 457[132], 460[140], 503[143]; **7**, 401[57]
Priebe, H., **6**, 246[129]
Priebe, W., **1**, 564[193]
Priebs, B., **2**, 365[217]
Priepke, H., **3**, 1002[58]; **6**, 852[137], 886[59], 889[68,71,72], 890[68,72]
Pries, P., **5**, 451[4]
Priest, M. A., **3**, 380[9], 625[40]
Priest, N. A., **3**, 380[9]
Priester, C. U., **6**, 734[5]
Priestley, H. M., **7**, 483[132]
Prieto, J. A., **1**, 785[247], 790[262]; **5**, 844[129]
Prignano, A. L., **8**, 765[13]
Prijs, B., **6**, 430[102]
Prikazchikova, L. P., **6**, 552[698]
Prikota, T. I., **4**, 379[115]
Prilezhaeva, E. N., **4**, 316[536], 317[544,545,548]; **7**, 766[177]
Prill, E. J., **8**, 532[132]
Primeau, J. L., **2**, 827[126]; **5**, 350[80]
Primo, J., **2**, 747[116,118]
Prince, M., **6**, 685[354], 959[46]
Prince, T. L., **3**, 582[116]; **8**, 840[32]
Principe, L. M., **3**, 1025[33]; **5**, 1053[40], 1060[56]
Pring, B. G., **2**, 465[107]
Prins, H. J., **4**, 274[66], 275[66]
Printy, H. C., **8**, 612[67]
Prinz, E., **1**, 141[15]
Prinzbach, H., **3**, 621[31]; **4**, 565[44]; **5**, 64[29], 404[16], 708[41], 714[71], 731[130a], 744[165,166,167]; **8**, 795[19], 813[10]
Prior, M. J., **1**, 787[253]
Prisbylla, M., **2**, 617[143], 907[57], 1054[63]; **5**, 320[7]
Pristach, H. A., **3**, 325[161]
Pritchard, M. C., **1**, 566[218]; **4**, 249[127], 258[127]
Pritzkow, W., **7**, 10[80,81], 24[22], 493[192]
Privett, J. E., **8**, 314[42], 315[42]
Prizant, L., **8**, 526[32]
Probner, H., **7**, 760[36], 761[36]
Probstl, A., **4**, 1074[29]
Prochazka, M., **6**, 120[126], 707[40], 902[127]
Prochazka, M. P., **2**, 1099[114]
Prochazka, Z., **7**, 73[102]
Procházková, J., **8**, 583[13]
Procter, G., **1**, 339[89]; **2**, 117[155], 309[25]; **3**, 224[178]; **7**, 412[106]
Proctor, G., **5**, 420[75], 841[97]; **7**, 729[44]
Proctor, G. R., **2**, 753[2,2b,d], 764[64], 1064[107]; **5**, 408[32], 688[69]; **8**, 72[240], 74[240], 314[35], 336[84], 339[84]
Proenç[illegible], J. R. P., [illegible], 507[241,242], 529[241,242]
Proença, M. F., **2**, 748[125]; **6**, 291[212,214,215], 517[326]
Proffit, J. A., **8**, 565[449]
Profft, E., **7**, 666[71]
Proia, R. L., **8**, 36[52], 66[52]
Prokai, B., **8**, 807[111]
Prokai-Tatrai, K., **4**, 925[36], 927[41], 930[41], 939[41]
Prokipcak, J. M., **6**, 536[547], 538[547]
Prokof'ev, A. K., **4**, 489[63]
Prokofiev, E. P., **1**, 555[112]; **6**, 503[218]
Prokopiou, P. A., **3**, 613[2], 615[2]; **8**, 117[74], 243[47], 505[83], 816[24], 817[29]
Proksch, E., **7**, 842[33,34]
Promé, J.-C., **2**, 855[244]
Promonenkov, V. K., **4**, 992[154]
Pronian, M. S., **6**, 533[494]
Pronina, N. V., **4**, 1051[126]
Pronkshe, N. K., **6**, 532[469]
Proshkina, V. N., **2**, 787[52]
Proskow, S., **5**, 74[212]
Proskurovskaya, I. V., **8**, 611[64]
Pross, A., **1**, 487[4], 488[4]; **8**, 937[83]
Prossel, G., **1**, 372[79]; **2**, 1049[14]
Prosser, T. J., **7**, 24[36], 25[36]
Prost, M., **2**, 742[75]
Prosyanik, A. V., **1**, 837[147]; **7**, 747[96]
Prosypkina, A. P., **7**, 477[74]
Proteau, P. J., **5**, 736[142b]
Protiva, J., **7**, 67[79]
Protiva, M., **2**, 765[78]; **6**, 266[46]
Protschuk, G., **2**, 334[70]
Proud, J., **8**, 766[18]
Proulx, P., **7**, 821[31]
Prousek, J., **4**, 452[15]
Proust, M., **6**, 455[150], 545[632]
Proust, S. M., **4**, 50[142]
Prout, F. S., **2**, 343[10], 348[55], 358[10], 367[231]; **4**, 98[109h]
Prout, K., **2**, 125[217,219], 127[232], 315[42], 316[42]; **3**, 47[257]; **4**, 217[132], 231[132], 1076[42]; **5**, 151[11]; **7**, 4[18]
Provelenghiou, C., **4**, 297[271]
Provencher, L. R., **8**, 189[60]
Prowse, K. S., **6**, 675[297]
Prudent, N., **1**, 49[6], 50[6], 80[22], 109[13], 110[13], 153[54], 182[47], 185[47], 198[47], 222[69], 310[102], 678[213]; **2**, 24[96], 217[137], 666[37], 677[37]; **8**, 3[20,22]
Prudhomme, M., **1**, 410[40]
Prud'homme, R. E., **5**, 86[28], 88[48,49]
Pruess, D. L., **6**, 913[24]
Pruett, R. L., **4**, 520[29], 914[5], 921[27], 922[5], 923[5], 924[5], 925[5]
Pruett, W. P., **2**, 387[337]
Prugh, S., **4**, 111[154g]
Pruitt, J. R., **2**, 1013[18]; **7**, 407[78e]
Pruskil, I., **4**, 980[104], 981[104]; **5**, 1095[104], 1098[122], 1112[104,122]
Pruss, G. M., **6**, 955[24]
Prussin, C., **5**, 618[7], 620[7], 624[29]
Pryanishnikov, A. P., **5**, 768[122,135]
Pryce, R. J., **5**, 418[70]
Pryde, C. A., **6**, 247[132], 253[158]; **7**, 32[90]
Pryor, W. A., **4**, 717[8,9]; **5**, 63[18]; **7**, 488[158], 761[54], 860[70]
Psarras, T., **7**, 805[66]
Pscheidt, R. H., **8**, 36[51], 66[51]
Pshezhetskii, V. S., **8**, 600[106], 606[25], 625[25]
Psiorz, M., **6**, 116[84]; **7**, 738[21]; **8**, 63[197], 64[197], 66[197]
Puapoomchareon, P., **7**, 634[69]
Puar, M. S., **6**, 644[93]; **8**, 621[141]
Pucci, S., **8**, 683[90]
Puchalski, C., **6**, 182[138]
Puchot, C., **6**, 718[122,123]
Puckett, P. M., **3**, 348[27], 353[48], 355[48], 357[48], 358[68]
Puckett, W. E., **3**, 509[179]
Puckette, T. A., **1**, 383[110]; **2**, 479[17,18], 480[18]; **6**, 722[137]
Puda, J. M., **2**, 348[55]
Puddephatt, R. J., **8**, 670[12], 671[12]
Pudova, O. A., **8**, 771[45,48]

Pudovik, A. N., **4**, 41[119], 55[156]; **6**, 432[119], 538[570]
Puff, H., **2**, 520[69], 1077[153]; **4**, 222[172]; **5**, 485[182]; **6**, 716[104]
Pugh, S., **7**, 145[167]
Pugia, M. J., **7**, 543[13]
Pugin, B., **4**, 401[234b]
Puglia, G., **2**, 960[32]
Puglis, J., **5**, 256[54], **8**, 803[93], 804[93], 826[69]
Puglisi, V. J., **3**, 568[47]
Pugniere, M., **8**, 52[141], 66[141]
Puig, S., **4**, 1101[192]
Pujalte, R. S., **4**, 932[63]
Pujol, D., **4**, 746[147]
Pujol, F., **1**, 547[61]
Pukhnarevich, V. B., **8**, 765[12], 770[33,34,38], 771[42], 782[104]
Puleo, R., **6**, 1030[107]; **7**, 131[83]
Pulido, F. J., **2**, 583[114]; **4**, 895[165], 900[165]; **5**, 478[164]; **8**, 646[50,51]
Pullman, B., **1**, 300[64]
Pulman, D. A., **1**, 546[50]
Pulst, M., **2**, 785[42,46]; **6**, 489[81,83]
Pulwer, M., **6**, 789[106]
Pummer, H., **4**, 920[22], 921[22], 923[22], 924[22], 925[22]
Pummer, W. J., **3**, 804[3]
Pummerer, R., **6**, 910[7]; **7**, 194[1], 202[1]
Punja, N., **4**, 55[157], 57[157f]
Punk, P. C., **1**, 37[177]
Pura, J. L., **6**, 687[373], 984[56]
Purcell, T. A., **1**, 569[259]
Purello, G., **6**, 178[121]
Purkayastha, M. L., **8**, 907[71], 909[71], 910[71]
Purlei, I. I., **8**, 699[150]
Purmort, J. I., **2**, 5[20], 6[20], 21[20]
Purnaprajna, V., **2**, 789[55], 792[65]
Purnell, J. H., **4**, 313[466,467]
Purohit, M. G., **6**, 771[43]
Purpura, J. M., **5**, 790[36]
Purrello, G., **6**, 508[289,290], 509[271]
Purrington, S. T., **3**, 88[139], 89[139], 91[139], 918[25]; **4**, 331[18], 344[75]
Purushothaman, K. K., **7**, 748[113]
Purvis, S. R., **6**, 579[987]
Pusch, J., **3**, 563[117], 582[117]
Pushkareva, Z. V., **6**, 530[421], 538[551,555], 550[421]
Pusino, A., **4**, 1057[145]
Pusset, J., **7**, 169[108], 878[140]
Put, J., **5**, 637[109]
Puterbaugh, W. H., **2**, 182[1], 280[22]; **3**, 915[15]
Putsykin, Yu. G., **2**, 854[236]
Putt, S. R., **6**, 1003[136], 1004[137]
Puttaraja, **3**, 386[61], 393[61]
Pütter, M., **3**, 872[60]
Pütter, R., **6**, 243[100]
Puttner, R., **7**, 29[77]
Pyatnova, Y. B., **8**, 756[143]
Pye, E. K., **2**, 1104[132]
Pye, W. E., **5**, 406[23,23b]; **6**, 814[88]
Pyke, R. G., **4**, 272[35], 273[35]
Pyle, J. L., **2**, 823[113,114]
Pyle, R. E., **8**, 300[88]
Pyman, F. L., **7**, 769[240], 770[240]
Pyne, S. G., **1**, 268[56], 411[46], 513[50], 515[58]; **3**, 573[72], 596[72], 599[72], 602[72], 607[72]; **4**, 27[79], 37[104], 39[111,111c], 216[124,125], 231[278,279,281], 251[148], 809[160]; **6**, 152[138], 153[138]; **7**, 400[50]; **8**, 503[70]
Pynn, H. Y., **7**, 763[88], 766[88]
Pyryalova, P. S., **8**, 298[73]
Pyszczek, M. F., **8**, 447[96]
Pytlewski, D., **8**, 26[20], 27[20], 36[20], 54[20], 55[20], 60[20], 70[20]
Pyun, C., **1**, 97[81]; **3**, 797[92]; **4**, 488[53]; **5**, 403[7], 404[7,18]; **6**, 744[77]; **8**, 240[34], 247[34], 250[34]

Q

Qazi, A. H., **8**, 507[86]
Qian, C., **8**, 447[135], 678[63], 680[70], 685[63], 686[63], 691[70]
Qian, Y.-M., **1**, 165[112b]
Qiang, L. G., **5**, 742[161]
Qicheng, F., **8**, 336[79], 337[79]
Qiu, W., **1**, 825[55]; **6**, 185[167], 187[167]
Qiu, X., **5**, 86[15]
Qiu, Z., **4**, 842[52]
Quabeck, U., **1**, 731[1]; **6**, 174[62], 183[147]; **7**, 543[17], 551[17], 554[17]
Quader, A., **3**, 445[72], 492[77]
Quadri, M. L., **8**, 806[123]
Quagliato, D., **2**, 716[59]; **5**, 778[192,193,194], 779[194]
Quallich, G., **2**, 105[44], 671[48], 698[82]; **4**, 611[357]
Quan, D. Q., **2**, 957[20]
Quan, P. M., **3**, 501[138]; **5**, 404[15]
Quanic, M., **8**, 356[190], 357[190]
Quante, J., **5**, 475[146]
Quante, J. M., **4**, 615[392], 629[392]
Quartucci, J., **8**, 873[11]
Quast, H., **2**, 424[35], 809[53]; **3**, 640[95], 870[45]; **4**, 1022[253]; **6**, 256[170], 526[391]; **8**, 657[98]
Quayle, P., **2**, 635[50], 640[50], 1059[74]; **3**, 174[533], 177[533], 179[533]; **5**, 116[269,270], 407[28], 1012[52]; **6**, 783[87]
Queguiner, G., **1**, 472[72], 474[88,94,96,97,101-103]; **3**, 261[146], 264[146]; **4**, 465[115], 474[115], 478[115]
Quelet, R., **7**, 666[75]
Quendo, A., **4**, 164[100]
Quenemoen, K., **5**, 202[38]
Quenguiner, G., **1**, 474[99]
Quesada, A. M., **2**, 928[122], 946[122]
Quesada, M. L., **4**, 30[88]; **6**, 1020[48]
Quesnelle, C., **4**, 797[104]
Qui, N. T., **2**, 728[145]
Quick, J., **2**, 943[172], 945[172]; **4**, 54[152,152a,c]; **7**, 227[84], 384[116]
Quick, L. M., **5**, 64[22]
Quignard, F., **5**, 1116[9], 1118[9]
Quillen, S. L., **2**, 1039[104]; **7**, 876[123,124]
Quimpére, M., **1**, 314[123]; **2**, 685[66]; **5**, 432[128]
Quin, L. D., **4**, 252[162]; **5**, 404[15], 444[187]
Quiniou, H., **4**, 123[210b], 125[210b]; **5**, 475[139,142], 575[134]; **6**, 554[751,752,753,781]
Quinkert, G., **2**, 547[113], 551[113]; **3**, 56[285]; **4**, 229[233,235,236], 1055[138]; **5**, 223[73,77], 385[130], 391[143], 721[101], 731[130a]; **8**, 545[285]
Quinn, H. A., **4**, 588[59]
Quinn, J. M., **2**, 143[55]
Quinn, N. R., **2**, 583[111]
Quinn, R. H., **7**, 12[103]
Quintard, J.-P., **1**, 437[156], 438[158], 445[192], 457[158], 479[149,150], 480[149,150]; **2**, 71[53], 564[6], 607[74]; **3**, 3[14], 8[14], 195[16], 196[24,25], 453[115], 524[30], 529[30]; **4**, 735[83], 770[83]; **7**, 614[4], 616[15], 621[4]; **8**, 278, 794[8], 798[8], 845[77]
Quintela, J. M., **6**, 533[481], 550[481]
Quintero, L., **4**, 466[125], 478[125]
Quintero-Cortes, L., **4**, 479[171]
Quintily, U., **4**, 426[63], 438[156]; **5**, 408[33]
Quirici, M. G., **3**, 217[91], 489[58], 495[58], 511[58], 515[58], 539[103]
Quirion, J. C., **5**, 829[24]
Quirk, J., **1**, 307[94]
Quirk, J. M., **7**, 3[9]; **8**, 446[91,93], 452[93], 455[91], 456[91], 534[157]
Quirk, R. P., **7**, 632[59]; **8**, 851[133], 852[133a,137]
Quiroga, M. L., **8**, 2[10]
Qureshi, I. H., **3**, 600[215]
Qureshi, M. I., **8**, 269[78]

R

Ra, C. S., **5**, 225[90]; **6**, 881[50]
Raab, G., **1**, 619[60]
Raab, W., **6**, 706[38]
Raap, R., **5**, 117[271]
Raaphorst, J. S. T., **4**, 1004[71]
Raasch, M., **4**, 345[84]; **5**, 115[247], 436[155], 439[166]
Raban, M., **2**, 438[69a,b]; **7**, 777[369]
Rabasco, D., **5**, 408[30,30b]
Rabe, J., **4**, 34[97], 35[97]
Raber, D. J., **1**, 287[17], 294[44-46]; **8**, 240[33]
Raber, N. K., **1**, 287[17], 294[44]
Rabideau, P. W., **3**, 613[4], 619[4]; **8**, 114[59], 115[61], 491[13,14], 492[18], 496[35], 497[36], 502[61], 504[72], 505[72,79], 509[91], 510[96], 514[106]
Rabie, A., **2**, 782[26]
Rabiller, C., **2**, 725[102-104]; **5**, 777[186,187,188]
Rabinovitch, B. S., **6**, 294[238,240], 959[40]
Rabinovitch, R. F., **5**, 453[59]
Rabinovitz, M., **1**, 292[26]; **3**, 619[26,27]; **5**, 2[14]; **8**, 251[107], 252[107], 253[107], 298[70], 299[70], 950[163]
Rabinowitz, J., **5**, 708[43], 739[43]
Rabinsohn, Y., **5**, 183[157]
Rabjohn, N., **3**, 638[91], 644[91]; **4**, 93[93a,b]; **7**, 84[1], 85[1], 108[1], 132[90]; **8**, 300[85]
Rabo, J. A., **3**, 305[75b]
Rach, N. L., **4**, 370[46]
Racherla, U. S., **1**, 343[104]; **6**, 865[201]; **8**, 704[2], 716[87], 718[110,113]
Rachlin, A. I., **1**, 130[96]; **8**, 287[17]
Rachon, J., **2**, 1084[18a], 1097[18b]
Raciszewski, Z., **7**, 856[66]
Racker, E., **2**, 456[48,77], 464[99], 466[127]
Raczko, J., **5**, 433[139]
Radak, S. E., **6**, 554[768]
Radatus, B., **8**, 52[148], 66[148], 219[78,79,82], 965[69]
Radau, M., **6**, 548[671]
Radcliffe, M. D., **5**, 1133[30]
Raddatz, P., **4**, 211[88,91]; **8**, 192[98]
Rädecker, G., **6**, 837[59], 856[157]
Radeglia, R., **6**, 552[693]
Radel, P. A., **2**, 195[72,72b,74]; **5**, 839[79], 843[79,114]
Radell, J., **8**, 364[25]
Rademacher, P., **1**, 191[77], 272[68], 300[67], 322[67], 335[63]; **2**, 6[35], 247[35], 630[7], 631[7]; **6**, 487[67]; **8**, 637[10]
Rader, C. P., **8**, 436[72]
Radesca, L., **4**, 18[59]; **5**, 918[128], 925[128], 937[128], 958[128]
Radesca-Kwart, L., **5**, 909[99], 958[99]
Radford, H. D., **2**, 740[59]
Radha, S., **7**, 765[154]
Radhakrishna, A. I., **2**, 801[22]
Radhakrishna, A. S., **6**, 727[189], 803[46,47], 804[47], 806[54]
Radha Krishna, P., **7**, 415[115d]
Radhakrishnan, T. V., **1**, 268[56], 269[56a]; **3**, 602[224,225]; **8**, 114[51]
Radinov, R. N., **1**, 214[25], 223[25b], 225[25b], 317[154], 320[154]
Radisson, X., **3**, 140[380], 154[380], 168[380], 174[380], 176[380]
Radke, M., **6**, 421[30], 422[30], 424[30]
Radlick, P., **5**, 686[43], 716[85]; **6**, 801[35]; **7**, 98[98]; **8**, 497[39]
Radner, F., **7**, 800[33], 879[151]
Radom, L., **1**, 487[4], 488[4]; **8**, 491[12]
Radonovich, L. J., **1**, 308[96]
Radscheit, K., **7**, 124[42]
Raduchel, B., **4**, 964[50], 965[50]; **6**, 22[81]
Radunz, H. E., **8**, 192[98], 266[58]
Radviroongit, S., **4**, 108[146f]
Radwan-Pytlewski, T., **2**, 76[83a]; **3**, 158[448], 159[448], 168[448], 174[529]
Radzkowska, T. A., **4**, 430[89]
Rae, B., **6**, 725[166]
Rae, I. D., **2**, 712[39]; **3**, 316[115]; **4**, 538[103]; **6**, 478[105]
Rae, W. J., **3**, 746[66]
Raffelson, H., **1**, 446[195]
Rafii, S., **1**, 359[22], 383[22], 384[22]; **4**, 206[43]
Rafikov, S. R., **7**, 750[129]
Rafka, R. J., **7**, 258[55]
Raggio, M. L., **7**, 172[125]
Raggon, J., **3**, 325[162]; **5**, 815[141]
Raghavachari, R., **8**, 354[174]
Raghavan, M., **7**, 266[106], 267[106], 276[106]
Raghavan, P. R., **5**, 219[39], 230[39]
Raghu, S., **4**, 579[23]; **5**, 272[4,5], 273[4], 274[7], 275[4,7,10], 277[7,10], 279[7]
Ragnarsson, U., **6**, 81[75]
Ragni, G., **2**, 1083[6], 1084[6]
Ragone, K. S., **7**, 372[70]
Ragoussis, N., **2**, 352[84], 357[84]
Raguse, B., **4**, 230[248]
Rahamim, Y., **8**, 446[72]
Raheja, A., **7**, 229[108]
Rahimi, P. M., **3**, 1046[2]; **7**, 7[45]
Rahimi-Rastgoo, S., **1**, 474[89]
Rahimtula, A. D., **8**, 561[412]
Rahm, A., **1**, 437[156], 445[192]; **2**, 564[6], 607[74]; **3**, 195[16], 524[30], 529[30]; **4**, 735[83], 770[83]; **5**, 431[121]; **7**, 614[4], 621[4]; **8**, 21[144], 278, 794[8], 798[8], 845[77]
Rahman, A., **8**, 301[94]
Rahman, A. F. M. M., **1**, 17[205]
Rahman, A. U., **2**, 359[163], 757[17], 758[24]; **6**, 725[167]; **8**, 249[92]
Rahman, M. A., **1**, 263[41]
Rahman, M. O. A., **2**, 792[65]
Rahman, M. T., **4**, 18[56], 229[217], 256[207], 262[207]
Rahn, B. J., **3**, 494[85]
Rai, M., **2**, 404[41]
Rai, R., **6**, 538[564]
Raimondi, L., **1**, 519[65,66], 520[67], 765[151]; **2**, 31[108], 374[276]; **6**, 927[73]; **7**, 442[47]
Rainer, H., **5**, 736[145,145r], 737[145]
Raines, S., **2**, 968[80]
Rainey, W. T., **3**, 724[12]
Rainville, D. P., **7**, 604[138]; **8**, 720[131]
Raithby, P. R., **2**, 1072[140]; **5**, 403[9]; **6**, 690[403], 692[403]; **7**, 318[60]
Raj, K., **4**, 111[155e]; **8**, 499[46]
Rajadhyaksha, S. N., **7**, 737[11]
Rajagopalan, K., **2**, 782[24]; **3**, 380[7]; **5**, 809[124], 878[9]
Rajagopalan, R., **7**, 42[34]
Rajagopalan, S., **2**, 586[134], 587[145]; **4**, 147[39]
Rajamannar, T., **3**, 17[88]
Rajan, S., **4**, 457[51], 469[51], 472[144]
Rajan, V. P., **7**, 375[80]
RajanBabu, T. V., **1**, 744[59]; **2**, 619[148]; **4**, 159[82], 258[236], 424[12], 429[12,88], 430[12], 787[28,29], 791[45], 808[158]; **5**,

552[30], 1123[38], 1166[20], 1167[20], 1169[20], 1170[20], 1178[20]; **8**, 823[57]
Rajanikanth, B., **7**, 763[97]
Rajappa, S., **4**, 124[213]; **6**, 107[25], 712[76]; **7**, 221[32]
Rajca, I., **8**, 605[10], 606[10], 608[10]
Rajendra, G., **4**, 398[214]; **5**, 847[138]; **7**, 503[274]
Rajeswari, K., **3**, 304[67]
Rajeswari, S., **2**, 894[9], 912[9]
Rajfeld, J. E., **2**, 663[24], 664[24]
Rajsner, M., **2**, 765[78]
Raju, B., **2**, 785[41], 786[48,49]; **8**, 332[39]
Raju, B. N. S., **5**, 233[139]
Raju, M. S., **4**, 1046[116]
Raju, N., **3**, 215[67], 380[7]; **6**, 561[880], 673[291]
Rajyaguru, I. H., **8**, 87[29]
Rakhmanchik, T. M., **8**, 124[92], 125[92]
Rakhmankulov, D. L., **8**, 214[40]
Rakiewicz, D. M., **3**, 221[129]; **4**, 820[213]
Rakitin, O. A., **7**, 740[43]
Rakotomanana, F., **4**, 753[166]
Rakotonirina, R., **4**, 107[143a,b,d]
Rakowski, M. C., **8**, 450[168], 454[168]
Rakowski DuBois, M., **8**, 454[202]
Raksha, M. A., **6**, 217[113]
Rakshit, D., **4**, 115[180d]
Raksis, J. W., **6**, 801[31]
Raley, J. H., **1**, 174[6], 175[6]
Ralhan, N. K., **6**, 273[96,97]
Rall, G. J. H., **3**, 693[141]; **7**, 336[34]
Rall, K. B., **7**, 505[290]
Ralli, P., **4**, 126[218b]
Ralls, J. W., **4**, 5[18]
Ralston, A. W., **3**, 617[16]
Ram, B., **5**, 95[89,100]
Ram, S., **6**, 76[47]; **8**, 84[14], 320[91], 368[66]
Ramachandran, J., **6**, 436[27], 437[27], 453[27], 455[27], 456[27]
Ramachandran, K., **4**, 1060[160]
Ramachandran, P. V., **7**, 595[13], 603[126]; **8**, 159[10], 180[10], 713[77], 715[77]
Ramachandran, S., **3**, 380[7]; **4**, 7[24]
Ramachandran, V., **3**, 683[101]; **6**, 282[157]; **7**, 42[30], 801[41]
Ramadas, S. R., **6**, 436[27], 437[27], 453[27], 455[27], 456[27]
Ramage, E. M., **8**, 30[43], 66[43]
Ramage, R., **2**, 358[153]; **3**, 681[94], 807[31]; **5**, 707[33]; **6**, 644[83,84], 667[242]; **8**, 495[29]
Ramage, W. I., **6**, 644[92]
Ramah, M., **8**, 337[80]
Ramaiah, M., **1**, 843[2d]; **3**, 598[203]; **4**, 716[4], 735[4], 743[4], 747[4], 790[33], 791[33], 1040[76]; **5**, 266[75], 268[75], 753[57], 757[57]; **7**, 164[79], 673[24]; **8**, 818[37], 823[37], 842[44,44a], 845[44a], 849[44a]
Ramakanth, S., **1**, 759[133]
Ramakrishnan, K., **8**, 238[24]
Ramakrishnan, V. T., **8**, 530[108]
Ramakrishnasubramanian, S., **5**, 728[122]
Ramalingam, K., **7**, 574[126]
Ramamoorthy, B., **3**, 383[42]
Ramamurthy, V., **5**, 70[117], 185[158], 211[61], 514[6], 709[45]; **7**, 40[1]
Raman, H., **3**, 781[14]
Raman, K., **3**, 1054[30], 1060[44], 1061[46], 1062[30,44,46]; **4**, 501[114], 1040[94], 1041[94]; **6**, 126[153]
Raman, P. S., **7**, 316[47], 317[47]
Ramana, M. M. V., **6**, 110[57]
Ramana Rao, V. V., **7**, 601[91]
Ramanathan, E., **8**, 137[55]
Ramanathan, H., **4**, 111[155e], 375[96c], 797[105]; **8**, 499[46], 509[92], 510[92]
Ramanathan, S., **8**, 384[40]
Ramanathan, V., **3**, 261[147,150]
Ramani, B., **8**, 555[371]
Rama Rao, A. V., **1**, 568[236]; **2**, 762[57,58]; **7**, 90[29], 415[115d]; **8**, 823[54]
Ramasamy, K., **6**, 685[355]
Ramasseul, R., **7**, 384[114b]
Ramasubbu, A., **4**, 753[173]
Ramaswamy, M., **3**, 274[17]
Ramaswamy, S., **4**, 113[173]; **6**, 134[25]
Rambaldi, M., **2**, 787[52]
Rambaud, M., **1**, 830[91]; **3**, 202[91,94]
Rambaud, R., **5**, 941[227]
Rambault, D., **3**, 509[177]
Ramberg, L., **3**, 861[1]
Ramdas, P. K., **8**, 390[79]
Ramegowda, N. S., **8**, 271[110]
Ramer, S. E., **5**, 86[22]
Ramer, W. H., **1**, 471[69]
Ramesh, K., **7**, 261[71]
Ramesh, R., **7**, 776[357]
Ramesh, S., **2**, 1049[22], 1050[22]; **4**, 795[80,81]
Ramesh Babu, J., **8**, 406[44], 408[44]
Ramey, K., **5**, 403[7], 404[7]; **6**, 744[77]
Ramezani, S., **2**, 578[85]
Ramezanian, M., **5**, 405[21]
Ramiandrasoa, F., **6**, 66[3]
Ramirez, F., **6**, 620[121-127], 625[121,163]
Ramirez, F. A., **8**, 376[159]
Ramirez, J. A., **8**, 458[219]
Ramirez, J. S., **8**, 565[448]
Ramirez-Munoz, M., **3**, 882[103]; **4**, 599[217]; **6**, 161[186]
Ramos, D. E., **5**, 7[51], 514[6]
Ramos, S. M., **8**, 98[102]
Ramos-Morales, F. R., **1**, 511[27], 564[194]; **8**, 837[16]
Ramos Tombo, G. M., **7**, 77[120b]
Ramp, F. L., **5**, 552[24]
Rampal, A. L., **8**, 341[107]
Rampal, J. B., **2**, 369[248]; **5**, 417[66]; **7**, 752[145]
Rampazzo, P., **4**, 426[63]
Rampersad, M., **7**, 696[41]
Ramphal, J., **1**, 808[319]; **3**, 289[68]
Rampi, R. C., **4**, 436[145], 437[145], 438[150]
Ramsden, C., **4**, 1096[160], 1097[160], 1098[160]
Ramsden, N. G., **6**, 74[31]
Ramsey, J. S., **3**, 568[41]
Ramun, J., **5**, 139[85], 143[85]
Ramussen, G. H., **7**, 111[190]
Ramuz, H., **8**, 336[72]
Ran, R., **3**, 298[27]
Rana, K. K., **4**, 33[95,95a]
Ranaivonjatovo, H., **5**, 444[189]
Ranbom, K., **4**, 439[168]
Rance, M. J., **2**, 760[39]
Rancourt, G., **7**, 295[20]
Rand, C. L., **2**, 584[123], 586[123], 713[45]; **3**, 104[211], 107[211], 111[211], 251[78,100], 254[78,100], 440[44], 443[53,55], 453[53]; **4**, 249[133], 250[133], 883[102], 884[102], 892[140,145]; **5**, 1165[13]
Rand, L., **2**, 343[13]; **3**, 635[37], 639[87], 652[224]; **5**, 776[181]
Randaccio, L., **4**, 170[11]
Randad, R. S., **2**, 42[146], 43[146,146b], 44[146b,c], 45[146b,c]; **6**, 865[201]; **8**, 722[145]
Randall, C. J., **1**, 786[250]
Randall, E. W., **3**, 382[36]

Randall, G. L. P., **4**, 706[36,37,39,40], 707[40,41]
Randall, J. L., **6**, 46[59,63], 61[150]
Randall, W. C., **5**, 410[41]
Randhawa, R., **4**, 505[133]
Rando, R. R., **3**, 894[62]
Randolph, C. L., **8**, 767[22], 773[22]
Randrianoelina, B., **2**, 575[62], 579[92]
Raneberger, J., **6**, 119[113]
Raneburger, J., **7**, 746[92], 752[92]
Rangaishenvi, M. V., **8**, 705[11], 726[11], 939[98]
Ranganathan, D., **4**, 12[42]; **5**, 3[27], 320[5], 347[5]; **6**, 289[193], 489[38]
Ranganathan, R., **8**, 330[48]
Ranganathan, S., **2**, 358[153]; **4**, 12[42]; **5**, 3[27], 320[5], 347[5]; **6**, 289[193], 667[242]
Ranganayakulu, K., **3**, 304[67], 374[133]
Range, P., **6**, 174[61]
Rani, A., **5**, 829[26]
Ranise, A., **6**, 776[57]
Ranjan, H., **5**, 944[241]
Rank, B., **7**, 709[44]
Rannala, E., **5**, 77[253]
Ranneva, Y. I., **1**, 632[68], 644[68]
Ransley, D. L., **3**, 318[124,126]
Ranu, B. C., **1**, 564[192,199], 857[59]; **3**, 714[31]; **4**, 970[71], 1040[88], 1048[88,88a]; **5**, 909[98], 955[302,303], 957[311], 993[52], 994[52]
Ranus, W. J., Jr., **8**, 476[26]
Ranzi, B. M., **8**, 560[404]
Rao, A. D., **6**, 490[105]
Rao, A. S., **1**, 819[2]; **2**, 544[86]; **3**, 223[157], 262[159], 264[159]; **6**, 2[3], 25[3], 88[105], 89[105]; **7**, 358[2], 364[40], 365[50], 366[2], 376[50,86], 378[2], 384[2], 828[50,50b]; **8**, 528[70], 530[70]
Rao, A. S. C. P., **1**, 826[59]
Rao, A. V. R., **6**, 136[40]; **7**, 683[87]
Rao, B., **3**, 396[103]
Rao, B. C. S., **8**, 408[65]
Rao, B. N., **5**, 185[158]
Rao, C. B., **4**, 1104[212]; **5**, 3[27], 320[5], 347[5]
Rao, C. G., **6**, 727[189], 806[54]; **7**, 264[90], 601[84,89,90], 602[84,92,93]; **8**, 261[11]
Rao, C. N. R., **1**, 38[260,262]; **5**, 452[55], 453[55]; **6**, 251[150]
Rao, C. P., **4**, 230[253]
Rao, C. S., **2**, 867[14]; **3**, 652[224]
Rao, C. T., **6**, 289[198]; **7**, 201[43]; **8**, 874[21], 881[21,73], 882[73]
Rao, D. V., **5**, 646[6]
Rao, G. S. K., **2**, 782[20,23,24,28,32], 785[41], 786[48,49]
Rao, G. V., **6**, 690[389]
Rao, J. A., **1**, 328[24]
Rao, J. M., **4**, 148[45a]; **5**, 125[13], 128[13]
Rao, K. R. N., **7**, 846[83]
Rao, K. S., **3**, 21[124], 384[54], 385[55]; **7**, 453[74], 455[74,103]; **8**, 566[455]
Rao, K. V. B., **8**, 581[9], 798[60]
Rao, M. N., **8**, 846[83]
Rao, M. R. R., **6**, 487[55], 489[55], 543[55]
Rao, M. S. C., **2**, 782[32]
Rao, M. V., **2**, 782[23]
Rao, P. A., **6**, 3[13], 4[13]
Rao, P. N., **3**, 727[33], 757[124]; **7**, 100[123], 372[71]
Rao, R. L. N., **7**, 769[229]
Rao, S. A., **4**, 254[186]
Rao, S. J., **1**, 568[237]
Rao, S. N., **1**, 882[123]
Rao, S. P., **7**, 705[15]; **8**, 91[50]
Rao, V. B., **3**, 901[116]; **4**, 578[19-21]; **5**, 136[69,70], 164[76]
Rao, V. R., **7**, 875[117]
Rao, V. S., **8**, 959[26]
Rao, V. V., **4**, 311[450]
Rao, Y. K., **4**, 791[47]
Rao, Y. S., **2**, 396[6], 402[6], 403[6,37,39], 404[6c,39], 407[6c]
Raoult, E., **8**, 134[32], 137[32,60]
Raper, G., **8**, 445[35]
Raphael, R. A., **1**, 569[259]; **2**, 369[246], 370[258]; **3**, 226[204], 273[10], 382[36], 386[67], 551[1]; **4**, 45[128], 259[267], 370[36], 681[53]; **5**, 687[54], 753[55], 757[55], 768[126], 769[55], 779[126], 794[50]; **6**, 150[125], 676[306]; **7**, 36[107], 338[42], 493[196]; **8**, 269[77], 365[33]
Rapoport, H., **1**, 406[28], 413[58-60], 733[15], 836[146]; **2**, 149[90], 434[58], 741[65], 815[82], 867[19], 869[21,22], 870[19,21], 871[19], 876[21], 879[19], 880[21], 881[22], 890[21], 943[171], 945[171], 1012[14-16], 1015[21], 1018[38,42]; **3**, 44[242], 507[172]; **4**, 29[86], 46[86], 310[432]; **6**, 127[160], 443[96], 501[199], 509[270,280], 564[906], 734[17,18], 812[81]; **7**, 85[9], 86[16b], 87[17], 108[17], 228[99], 230[130,131]; **8**, 319[75], 604[6], 626[167]
Rapp, K. M., **6**, 519[338]
Rapp, R., **7**, 710[51]
Rappa, A., **8**, 596[82], 880[71], 881[71]
Rappe, C., **2**, 145[64], 346[43]; **3**, 839[9], 843[24,29], 844[31], 892[52]; **7**, 120[9]
Räpple, E., **6**, 724[155]
Rappo, R., **4**, 187[96]
Rappoldt, M. P., **5**, 700[8], 737[8]
Rappoport, Z., **1**, 580[2], 581[2], 582[2], 583[8,8b], 610[2a], 611[2a], 616[2a], 630[6,14,16], 631[16], 634[16], 641[16], 656[16], 658[16], 661[16], 669[16], 672[16], 673[16], 686[16], 687[16], 700[16], 712[16]; **2**, 357[150]; **3**, 86[52,53], 89[52], 91[52], 94[52,53], 95[52], 114[52,53], 116[52], 117[52,53], 119[52], 208[2], 217[2], 271[2], 272[2], 436[11], 521[1], 551[5], 552[5], 862[3,7], 867[34,37], 872[37], 883[37], 884[37]; **4**, 3[7], 4[7,16,16a], 71[19], 78[54], 86[54b], 128[221], 292[230], 294[240], 342[64], 343[64], 452[10], 467[10], 483[4], 484[4,23], 485[23], 495[4], 951[1], 953[1h], 954[1h], 961[1h], 968[1], 979[1], 986[132], 987[132], 1000[10], 1001[19], 1002[10], 1004[74], 1006[19], 1007[19,107,130], 1013[179], 1016[10], 1017[212], 1018[222], 1021[212], 1031[8], 1043[8], 1045[8a], 1046[8a], 1050[122]; **5**, 71[147], 78[147], 211[64], 900[13], 901[13,14,21,23,25,26], 904[21,44,45], 905[13,21,44,45,62], 907[13], 908[62], 913[14], 921[21], 925[45], 926[45], 943[21,45], 947[13], 952[21], 972[4], 1006[33]; **6**, 67[13], 225[4,7], 226[4], 228[4], 229[7], 230[4], 231[4], 232[4], 233[4,7], 234[4], 235[4], 236[4], 238[4], 239[4], 240[4], 241[4], 256[7], 258[4,7], 261[14], 263[14], 272[14], 275[14], 280[14], 283[163], 284[173], 293[228], 834[34], 837[62,63], 958[27], 1007[151]; **7**, 92[38,50], 94[56], 762[69], 766[169], 777[69a], 778[69], 856[66]; **8**, 251[107], 252[107], 253[107], 298[70], 299[70], 383[14], 403[4], 404[4], 407[4], 763[1], 785[1]
Rapson, W. S., **4**, 2[5], 99[119]
Rasala, D., **8**, 843[49]
Rasalka, D., **6**, 110[46]
Rascher, L., **7**, 656[17]
Rasetti, V., **6**, 538[571]
Rash, F. H., **6**, 426[70]
Rashidyan, L. G., **4**, 315[506]
Rasmussen, G. H., **4**, 83[65a]; **7**, 236[22,25]
Rasmussen, J. B., **6**, 509[252]
Rasmussen, J. K., **1**, 548[65], 551[72,74]; **2**, 283[44], 298[44], 323[31], 599[21], 630[4]; **4**, 357[147]; **5**, 105[188], 107[188], 158[45]; **6**, 104[8], 229[24]; **7**, 816[14]
Rasmussen, J. R., **8**, 819[42], 820[42]
Rasmussen, P. W., **5**, 128[29]
Rasmussen, R. R., **3**, 853[71]
Rasmusson, G. H., **6**, 684[343]
Raso, A. G., **4**, 111[155d]

Rasoul, H. A. A., **5**, 73[194], 461[105]
Raspel, B., **4**, 608[318,319]
Rassat, A., **3**, 564[7]; **8**, 118[75], 122[75]
Rastall, M. H., **4**, 18[59], 121[208], 991[151]; **6**, 176[92]
Rastelli, A., **5**, 257[60]
Rastetter, W. H., **7**, 408[89]
Raston, C. L., **1**, 13[73], 16[89], 17[207,209,210,217,218,219], 36[233,234], 37[177,178], 139[3,4]; **3**, 436[7]; **4**, 83[63]; **5**, 144[104]
Ratajczak, A., **1**, 528[113]; **2**, 1099[106]; **3**, 164[475]
Ratananukul, P., **1**, 555[121]; **4**, 113[168]; **7**, 213[103]
Ratcliff, D. G., **3**, 97[170], 117[170]
Ratcliff, M. A., Jr., **6**, 280[145]
Ratcliffe, A. H., **6**, 921[48]
Ratcliffe, B. E., **2**, 1026[73]; **3**, 358[66], 363[85]
Ratcliffe, C. T., **8**, 372[124], 373[126]
Ratcliffe, N. M., **7**, 53[1], 63[1], 72[101]
Ratcliffe, R., **7**, 103[138], 257[45], 258[45]
Ratcliffe, R. M., **6**, 646[99]
Ratcliffe, R. U., **6**, 41[42]
Ratcliffe, R. W., **2**, 213[122], 1102[123]; **5**, 92[76]; **6**, 125[148], 127[148]; **7**, 257[46]
Rathke, J., **7**, 604[128]
Rathke, M. W., **1**, 491[29], 492[38], 498[29], 501[29], 502[38], 642[118,119], 769[183], 789[260]; **2**, 56[3], 57[3], 111[85,86], 112[88], 122[194], 182[6,8], 187[41], 241[14], 242[14b,16], 277[7], 279[12], 281[7], 282[41], 287[7], 604[51], 605[60], 606[66], 616[133,136], 799[18], 801[21], 803[39], 830[142], 842[192,193]; **3**, 7[31], 8[31], 199[60,66], 257[121], 443[59], 793[69,72]; **4**, 97[107a], 145[23]; **5**, 828[10], 841[102]; **7**, 120[18], 121[18], 144[152], 606[147,154]
Rathman, T. L., **2**, 189[54]
Rathnum, M. L., **5**, 86[32]
Rathore, R., **6**, 938[123], 939[123], 942[123]; **7**, 103[143], 220[17], 266[108,112], 267[108,112], 559[82], 560[82], 561[82], 562[82], 563[82]
Ratier, M., **4**, 971[75]; **5**, 901[30]
Ratledge, C., **7**, 56[14]
Ratnasamy, P., **3**, 328[175], 331[195]; **7**, 840[6]
Ratovelomanana, R., **6**, 854[143]
Ratovelomanana, V., **3**, 217[91], 437[27], 449[98], 487[45-47], 530[68], 533[68], 539[99], 545[99], 943[84], 964[125]; **4**, 102[128d]
Ratovskii, G. V., **6**, 495[150]
Ratts, K. W., **3**, 918[21], 922[39], 939[39]; **4**, 55[156]
Ratz, R., **7**, 203[55]
Rau, A., **4**, 955[13]; **5**, 804[97], 972[9], 973[9,12], 974[9]
Rau, S., **5**, 552[23]
Raubenheimer, H. G., **3**, 566[27]
Rauch, E., **7**, 506[303]
Raucher, S., **1**, 630[42], 633[73], 636[73], 637[73], 639[73], 645[42], 646[73], 647[73], 648[73], 656[73], 657[73], 658[73], 669[42,179], 670[42], 672[73], 686[73], 688[73], 690[73], 692[73], 694[42], 708[42]; **2**, 801[23], 859[253], 867[10]; **3**, 87[108], 89[145], 90[145], 105[145], 106[108], 114[108], 120[108], 136[108], 144[108], 1000[54]; **4**, 120[200]; **5**, 839[77,78], 886[21,22], 1001[16]; **6**, 509[281]; **7**, 608[172]
Rauchschwalbe, G., **7**, 99[106,107]
Rauchschwalbe, R., **7**, 596[39]
Rauckman, B. S., **2**, 871[23]
Rauenbusch, C., **5**, 1119[20]
Rauhut, M. M., **6**, 240[82]
Rauk, A., **1**, 512[42,43], 528[116]; **3**, 147[399]; **5**, 754[59]; **8**, 670[14]
Raulin, F., **6**, 540[581]
Raulins, N. R., **1**, 880[115], 898[115]; **5**, 786[1], 798[1], 877[8], 972[5], 1000[1]; **6**, 834[35], 855[35]
Rault, S., **6**, 487[64], 489[64]
Raunio, E. K., **4**, 41[119,119b]
Rausch, M. D., **1**, 141[14]; **3**, 564[12]; **4**, 519[24-26], 520[29], 697[9,10], 905[211]; **5**, 1134[35], 1135[48], 1165[9], 1178[9]
Rautenstrauch, C. W., **5**, 344[65]
Rautenstrauch, V., **1**, 180[29], 181[29], 272[65,65a]; **2**, 21[84]; **3**, 100[198], 103[198], 570[51], 572[64], 606[51], 946[86], 976[9], 977[9], 989[9], 990[9], 1017[4]; **4**, 1002[47]; **5**, 768[123], 1062[59]; **6**, 836[57], 876[28]; **7**, 385[117], 818[17]; **8**, 109[8-10,14], 110[9,10,15,16], 112[10], 116[10,16], 120[10,16], 121[16]
Rautureau, M., **1**, 474[109], 475[109]; **7**, 350[21]
Rauwald, W., **6**, 48[82], 51[82]
Ravasi, M., **7**, 331[17]
Rave, P., **2**, 139[27]
Raveendranath, P. C., **8**, 949[154]
Ravelo, J. L., **6**, 959[38]
Ravenek, W., **8**, 670[10], 671[10]
Ravenscroft, P. D., **8**, 10[56]
Raverty, W. D., **1**, 276[79]; **4**, 688[66], 698[22]
Ravid, U., **3**, 248[57], 249[57], 251[57], 263[57]; **7**, 121[27], 123[27]
Ravikumar, P. R., **3**, 693[148], 694[148]; **8**, 237[10], 243[10]
Ravikumar, V. T., **5**, 809[124]
Ravindran, N., **7**, 605[145]; **8**, 711[62,65,67], 718[65,108,109,114,115], 719[62]
Ravindran, R., **7**, 606[152]
Ravindranath, B., **7**, 763[97]; **8**, 875[31], 889[135,137]
Ravindranathan, T., **6**, 661[211]; **7**, 831[67]; **8**, 725[180]
Ravn-Petersen, L. S., **2**, 713[48]; **5**, 778[195]
Rawal, V. H., **1**, 328[24], 544[31], 548[31]; **3**, 135[344], 136[344], 137[344], 138[344]; **8**, 618[111-113], 623[113], 628[113]
Rawalay, S. S., **7**, 228[103]
Rawlings, F. F., **3**, 643[121]
Rawlins, A. L., **2**, 958[23]
Rawlinson, D. J., **7**, 92[40], 95[75], 96[75], 152[2], 153[2], 154[2], 158[38], 171[38]
Rawn, J. D., **8**, 185[22]
Rawson, D. I., **5**, 595[8], 596[8b], 604[8b], 608[8b,66]
Rawson, D. J., **1**, 772[197]; **2**, 649[107]
Rawson, R. J., **4**, 969[63]
Ray, A. K., **5**, 76[247]
Ray, D. G., III, **4**, 370[46]; **7**, 155[31c]
Ray, F. E., **3**, 918[22]
Ray, J., **4**, 1050[124]
Ray, N. K., **8**, 724[169,169e]
Ray, P. H., **2**, 465[105], 466[121], 469[121]
Ray, P. S., **3**, 219[108]; **7**, 219[15]
Ray, R., **7**, 439[29]
Ray, R. E., **7**, 100[120]
Ray, S. K., **7**, 203[54]
Ray, T., **4**, 371[48,51], 673[32], 686[62]; **7**, 107[153,155], 413[107c], 453[72]
Rayanakorn, M., **4**, 313[467]
Raybush, S. A., **7**, 12[94]
Raychaudhuri, S. R., **6**, 507[234]; **7**, 239[50]; **8**, 510[97]
Rayez, J.-C., **5**, 727[119]
Rayford, R., **2**, 635[49], 640[49], 1059[74]
Raymond, A. L., **8**, 530[105]
Raymond, F. A., **4**, 273[53]
Raynaud, J. P., **1**, 698[251]
Rayner, C. M., **3**, 225[188]
Raynham, T. M., **1**, 770[187]; **2**, 645[81]; **5**, 51[45,45a], 53[45a], 362[93], 363[93d], 364[93d], 545[120]
Raynier, B., **7**, 499[233]
Raynolds, P. W., **1**, 118[61]; **2**, 735[14]; **3**, 254[102]; **6**, 448[110]
Raza, Z., **8**, 460[246]
Razaq, M., **8**, 132[8], 135[44], 321[96]
Razmilic, I., **7**, 90[33]
Razniak, S. L., **1**, 226[95]; **6**, 508[286], 537[286,576]
Raznikiewicz, T., **8**, 908[74]
Razumovskii, S. D., **7**, 542[7], 543[7]

Razuvaev, G. A., **7**, 641[2]; **8**, 753[68-70,74]
Razzell, W. E., **6**, 611[64]
Re, A., **4**, 230[249], 231[249]
Re, L., **7**, 439[25]
Re, M. A., **2**, 482[22]; **8**, 583[10], 587[10]
Read, A. T., **7**, 14[133]
Read, G., **7**, 95[70,70a], 107[168]; **8**, 444[10]
Read, J., **2**, 152[103]
Read, R. W., **6**, 473[73], 819[109]
Readio, P. D., **4**, 280[128], 281[128]
Reagan, D. R., **3**, 223[153]
Reagan, J., **1**, 883[126], 898[126]; **3**, 227[209]
Reale, A., **3**, 587[148]
Reamer, R. A., **1**, 402[17], 791[296a], 799[296]; **2**, 482[37], 483[37], 485[37], 648[97], 649[97b]; **3**, 45[247]; **5**, 410[41], 850[160]; **6**, 22[83], 278[130]; **7**, 416[122], 752[154]; **8**, 945[132]
Reames, D. C., **2**, 523[74]
Réamonn, L. S. S., **7**, 205[62], 764[125]
Reap, J. J., **6**, 1021[49], 1033[123]; **7**, 125[53], 126[53]
Rebane, E., **7**, 769[239], 770[239,255]
Rebarchak, M. C., **8**, 28[37], 66[37]
Rebeck, J., Jr., **2**, 809[57]; **5**, 260[70], 263[70]; **7**, 842[20]
Rebell, J., **5**, 971[3], 973[3]
Rebeller, M., **8**, 396[137]
Rebello, H., **6**, 542[602]
Reber, G., **1**, 17[211]; **5**, 850[152]; **8**, 859[217]
Reboul, J. P., **6**, 510[294]
Reboul, O., **5**, 848[141]
Rebovic, L., **7**, 155[31a]
Rebsdat, S., **6**, 540[590]
Reby, C., **8**, 973[119]
Recca, A., **1**, 294[43]; **3**, 583[118], 587[148]
Rechka, J. A., **1**, 808[325]; **4**, 1063[168]
Reckling, G., **6**, 431[105]
Redaelli, D., **4**, 768[243]
Reday, P. A., **3**, 513[207]
Redda, K., **8**, 587[36]
Reddy, A. V., **5**, 3[28], 5[28], 7[28]; **6**, 836[58]; **8**, 123[81], 384[29], 566[456]
Reddy, C. P., **2**, 792[68]
Reddy, D. B., **4**, 988[140], 1040[88], 1048[88]
Reddy, D. S., **6**, 836[58]; **8**, 123[81]
Reddy, E. J., **6**, 490[105]
Reddy, G. J., **5**, 95[100]
Reddy, G. N., **3**, 215[59]; **6**, 23[92]
Reddy, G. S., **1**, 743[51], 744[59], 746[51], 811[51]; **2**, 619[148]; **3**, 457[127]; **4**, 424[12], 429[12], 430[12], 787[28]; **5**, 1115[2], 1116[2,2c], 1121[2c], 1122[2c], 1123[2c,38]; **7**, 186[182]
Reddy, G. V., **6**, 253[158,159]
Reddy, K. A., **8**, 823[54]
Reddy, K. B., **3**, 618[20]; **5**, 687[67]
Reddy, K. S., **1**, 768[173]; **7**, 409[91]
Reddy, M. P., **2**, 782[20,23,28]
Reddy, N. L., **7**, 646[25]
Reddy, P. A., **3**, 459[135]
Reddy, P. S., **7**, 260[84]
Reddy, R. T., **1**, 554[103]; **4**, 486[38], 497[98]
Reddy, S. M., **4**, 39[112]
Reddy, T., **1**, 834[128], 838[159]
Reddy, V. M., **6**, 490[105]
Reddy, V. P., **5**, 702[11,14], 703[14], 740[14]
Redemann, C. E., **6**, 120[125]
Redlich, H., **1**, 566[211]; **6**, 677[323], 679[329,329a]; **8**, 821[49], 824[62]
Redmond, J. W., **6**, 802[42], 803[42]
Redmond, W., **6**, 220[128]
Redmond, W. A., **8**, 916[104,105], 917[104], 918[104], 919[104], 920[104]
Redmore, D., **1**, 843[1], 847[1], 896[1]; **3**, 900[96], 903[96]; **7**, 100[126]
Redwood, M. E., **1**, 3[21]
Reece, C. A., **3**, 126[318]
Reed, C. F., **7**, 14[139]
Reed, D., **1**, 6[33], 3[illegible][165]; **7**, 778[415]
Reed, G., **7**, 56[13], 65[13], 66[13], 67[13], 70[13]; **8**, 185[6]
Reed, H. W. B., **5**, 63[20]
Reed, J. N., **1**, 469[58]; **7**, 333[22]; **8**, 385[47], 405[21]
Reed, J. W., **5**, 239[1], 903[41], 904[42], 905[41], 907[41], 909[41], 916[41], 918[41], 937[41], 939[41], 940[41], 947[42], 951[41]; **8**, 244[64], 253[64]
Reed, L. A., **4**, 1079[61,62]; **5**, 359[91], 373[91], 374[91]; **7**, 198[26]
Reed, L. A., III, **5**, 543[118], 545[118]; **6**, 927[71]; **7**, 402[63]
Reed, L. E., **4**, 1015[197]
Reed, M. W., **5**, 689[76], 733[136,136e]
Reed, R. G., **8**, 135[45]
Reed, S. F., **6**, 134[15], 151[15]
Reed, S. F., Jr., **7**, 764[118]
Reedich, D. E., **5**, 647[17], 651[17], 656[17]
Reeg, S., **7**, 58[53a], 62[53,53a], 63[53a]
Reerink, E. H., **8**, 528[87]
Rees, A. H., **2**, 739[43a]
Rees, C. W., **1**, 357[4]; **3**, 807[27], 813[61], 908[148]; **4**, 113[175], 434[128], 488[56,57], 509[162], 1021[245,247]; **5**, 379[112], 380[113e], 383[112], 384[112], 422[82], 469[138], 478[162], 491[206,207], 707[34], 725[34], 803[92], 938[205], 947[263,264,265,274,286,287,289], 979[26]; **6**, 245[128], 419[11,12], 781[77]; **7**, 27[64], 32[95], 34[98,99], 35[101], 194[3], 200[40], 208[88], 305[1], 349[18], 355[18], 470[4], 472[4], 473[4], 474[4], 476[4], 480[104], 482[113], 743[60-62,65], 744[70]; **8**, 318[59], 322[59], 337[76], 391[86], 392[103], 510[93], 618[115,116], 653[82], 948[147]
Rees, D. C., **4**, 675[39], 679[45], 680[48,51]
Rees, D. W., **5**, 1070[27], 1073[27]
Rees, L., **2**, 753[2,2b]
Rees, L. G., **2**, 764[64]
Rees, R., **4**, 52[147,147a]
Rees, R. W., **8**, 884[100], 926[13], 972[113]
Rees, T. C., **4**, 871[36], 876[60]
Reese, C. B., **2**, 709[13]; **4**, 1012[171,172], 1018[223,224], 1019[223,224]; **5**, 794[44]; **6**, 121[131], 602[3], 604[30], 624[138,141], 635[12], 643[12], 650[127], 656[168], 658[185,189], 660[210], 661[211], 662[210], 778[64], 779[67]; **8**, 297[66], 472[7], 948[147]
Reese, P. B., **3**, 738[35]
Reetz, M. T., **1**, 52[15], 55[27], 56[29], 57[27], 58[27], 83[26], 109[14], 140[11], 141[12,16,21], 142[25,27], 143[16,34,35], 144[11], 145[16,21,41,42], 146[21,42], 148[16,42,46], 149[21,27,42,49a,50,51], 150[11,16,27,41,46],; **2**, 4[14], 5[17], 6[17,35], 22[17,17e,86], 31[117], 32[117], 33[122], 35[131], 117[148,151,154], 247[35], 249[84], 305[11,12], 306[13], 307[14-18], 310[15,16,32], 311[32], 446[29], 455[3], 505[6], 507[23], 512[23,47], 570[37], 614[119], 630[4,7,8], 631[7], 640[65a,66,68], 641[65a,68,69a,70], 642[65a], 644[65a,69a], 645[69a], 646[65a,66,69,70,83-85], 654[129,130], 667[44], 979[14]; **3**, 25[155,158,160], 27[155], 421[57,58]; **4**, 162[94c], 331[8]; **6**, 141[63], 142[67], 227[23], 229[23,27], 230[23], 231[34], 238[34], 644[88], 864[199]; **7**, 144[153,154], 517[12]; **8**, 886[112]
Reeve, E. W., **8**, 143[59], 148[59]
Reeve, W., **2**, 957[15]
Reeves, P. C., **6**, 690[398], 692[398]
Reeves, R. L., **2**, 342[5], 349[5], 352[5]
Reeves, W. P., **4**, 443[189,190]
Refn, S., **4**, 149[52], 181[72,73]
Reformatsky, S. N., **2**, 122[192], 277[1], 282[1], 297[88]

Regan, B. A., **6**, 554[724]
Regan, J., **4**, 1040[95], 1041[95b], 1045[95b]; **5**, 921[142]
Regan, M. T., **3**, 1040[108]
Regan, T. H., **8**, 496[34]
Regberg, T., **6**, 620[133]
Rege, S., **4**, 5[18], 61[18g]; **5**, 618[8], 619[8], 624[8], 625[8]
Regel, W., **6**, 249[145]
Regeling, H., **5**, 441[176,176d]
Regen, S. L., **4**, 304[352], 1001[42]; **6**, 939[144], 942[144], 958[28]; **7**, 844[53]; **8**, 551[346]
Reger, D. L., **4**, 582[30]; **8**, 535[164]
Reginato, G., **5**, 438[162]
Regis, R. R., **3**, 158[447], 159[447]
Regitz, M., **1**, 844[5a,8]; **3**, 887[8], 888[8], 889[25,26], 890[25,28,29], 893[8,25], 894[58], 897[8], 900[8], 903[8], 905[28,58], 909[152]; **4**, 953[8,8b], 954[8b], 1033[28,29], 1075[33], 1101[196], 1102[196]; **5**, 113[239], 1101[145]; **6**, 120[118], 124[143], 125[143], 129[164], 171[5], 172[5], 198[5], 199[5], 200[5], 201[5], 202[5], 245[123], 523[349]; **7**, 742[53], 751[53], 752[148,149]; **8**, 382[11], 383[11]
Reglier, M., **7**, 95[65], 452[47]
Regnier, B., **6**, 213[90]
Regondi, V., **2**, 737[39]; **3**, 125[307]; **6**, 134[13]; **8**, 277[153], 660[109]
Rehder-Stirnweiss, W., **8**, 446[75,77,84], 453[75]
Rehfuss, R., **5**, 736[145], 737[145]
Rehling, H., **1**, 123[79], 372[80]; **2**, 1051[37]
Rehm, D., **5**, 650[26]; **7**, 854[58], 855[58]
Rehm, H.-J., **7**, 56[13], 65[13], 66[13], 67[13], 70[13]; **8**, 185[6]
Rehman, Z., **4**, 277[88]
Rehn, D., **2**, 1090[72], 1094[89], 1095[89], 1099[106]
Rehn, H., **2**, 1099[106]
Rehnberg, N., **1**, 566[219]; **3**, 768[166]; **4**, 211[92,93]
Reho, A., **8**, 31[45], 36[45], 66[45], 657[97]
Rehwinkel, H., **1**, 773[203,203a]
Rei, M.-H., **4**, 272[34], 273[34], 309[416,417], 314[478]; **8**, 5[28]
Reibel, I. M., **7**, 95[73a]
Reibenspies, J., **1**, 123[74], 373[92], 375[92], 376[92], 827[67]; **4**, 408[257b]; **8**, 333[57], 345[127], 655[86]
Reich, H. J., **1**, 630[8,9,17], 631[9,17,56], 632[17,56,65,69], 633[17,56,65,69,70], 634[9,17,56], 635[56,69,70,80,84,85], 636[17,56,65,70,80,84,85,92,93,96-98,103], 637[80,84], 638[56,65,93,97], 639[70], 640[56,80,84,85,93,97], 641[9,17,56,98],; **2**, 76[87], 88[30,31], 601[36], 602[38], 1002[55]; **3**, 86[46,47,54], 87[47,54,74,95,96,100,102,103,106,107,109,112,113], 94[54], 95[46,54,74,106], 97[74], 104[95,96,100,102,103], 105[107], 106[103,107], 109[74,106], 110[95,100], 111[102,103], 114[47,74,107], 116[74], 117[46,47,54,95,100,102,103], 120[107], 136[74,106,113], 141[74,106], 145[113], 157[107,112,113], 248[51], 380[10], 787[46]; **4**, 262[304]; **5**, 439[166], 519[31], 714[66], 1014[55]; **6**, 154[148], 903[136], 904[141], 966[97], 1021[54], 1026[84,88], 1027[88,93], 1028[95,99,100], 1030[84,100], 1031[93,111,112,116]; **7**, 119[3], 129[3,71], 130[71], 131[71,82], 135[71], 146[3], 520[31], 522[40], 675[58], 765[158], 769[223,231], 770[249], 771[231,275,276,277,278,282,285], 772[231,290], 819[20], 826[20,47], 827[47]; **8**, 409[84], 849[116]
Reich, I. L., **1**, 635[84], 636[84], 637[84], 640[84], 672[84]; **2**, 1002[55]; **3**, 87[96], 104[96], 248[51], 380[10]; **4**, 262[304]; **6**, 1026[84], 1028[99,100], 1030[84,100]; **7**, 129[71], 130[71], 131[71], 135[71], 520[31], 675[58], 769[231], 771[231,285], 772[231,290]
Reich, J., **4**, 317[555]
Reich, M. F., **8**, 30[42], 66[42]
Reich, P., **7**, 772[294], 773[294]
Reich, R., **3**, 208[7]
Reich, S. H., **1**, 749[80]; **2**, 106[53]; **5**, 855[187,192]; **7**, 415[115c], 418[115c]; **8**, 537[190], 538[190]
Reichardt, C., **2**, 358[152]; **4**, 71[19]
Reichel, B., **4**, 492[69]
Reichel, C. J., **1**, 366[44]; **2**, 900[28], 901[28], 910[28]
Reichel, C. L., **8**, 765[14]
Reichel, L., **7**, 775[340]
Reichelt, H., **2**, 69[45], 387[332]
Reichelt, I., **4**, 1046[112,113]; **5**, 1200[53]
Reichenbach, G., **7**, 760[44]
Reichenbach, T., **4**, 878[77], 879[77]
Reichert, B., **2**, 894[2], 897[2], 1090[62]
Reichert, D. E. C., **7**, 519[21]
Reichert, U., **6**, 1062[77]
Reichlin, D., **5**, 356[90], 543[115]
Reichmanis, E., **5**, 634[77]
Reichrath, M., **6**, 56[133,134], 57[133,134,137,138]
Reichsfel'd, V. O., **5**, 1146[109]
Reichstein, I., **8**, 531[121]
Reichstein, T., **3**, 898[85]; **6**, 37[32]
Reid, D. E., **6**, 968[112]; **8**, 950[166]
Reid, D. H., **3**, 86[1,3,5]
Reid, E. E., **4**, 316[541]; **7**, 758[2], 760[2], 761[2], 768[197]; **8**, 243[46]
Reid, G. R., **5**, 128[32], 130[32], 864[256]
Reid, J. C., **8**, 285[5]
Reid, J. G., **3**, 960[117]; **4**, 971[73]; **5**, 894[47]
Reid, K. J., **8**, 410[90]
Reid, M. W., **5**, 689[79]
Reid, R. G., **2**, 422[28]; **7**, 372[72a]
Reid, S. T., **3**, 815[73]; **5**, 134[62], 637[112]
Reid, W., **3**, 902[117]; **4**, 1104[209]; **8**, 636[2]
Reidel, **8**, 443[1]
Reider, P. J., **2**, 200[89], 489[47], 490[47], 635[49], 640[49], 648[91], 649[91], 1059[74]; **6**, 667[237]
Reif, D. J., **3**, 746[69], 748[69]
Reif, W., **1**, 306[90], 460[3]; **5**, 15[106], 835[59]
Reiff, H., **2**, 182[5], 477[7], 478[7]
Reiff, H. E., **2**, 773[28]; **4**, 1016[203]
Reiffen, M., **6**, 540[591]
Reihlen, H., **4**, 663[1]
Reikhsfel'd, V. O., **8**, 765[13]
Reil, S., **2**, 1099[109,109b]
Reilly, J., **5**, 1003[23]
Reilly, P. J., **1**, 797[283]; **7**, 555[69]
Reim, H., **5**, 498[229,235]
Reimann, B., **5**, 216[13], 219[13], 221[13]
Reimann, E., **6**, 651[136,136b]
Reimann, W., **8**, 445[53], 806[109]
Reimer, K., **2**, 769[2], 770[2]
Reimlinger, H., **2**, 1102[120]; **4**, 1033[16]
Rein, B. M., **6**, 205[25,26], 210[25]
Rein, K., **1**, 477[131]; **3**, 75[43], 76[43,53], 80[43]
Rein, T., **4**, 599[215], 609[215], 624[215], 641[215]
Reinäcker, R., **8**, 755[132]
Reinbach, H., **2**, 352[87], 357[87]
Reinecke, E., **4**, 239[32]
Reinecke, M. G., **2**, 912[72]; **4**, 485[24], 488[54], 489[24], 495[24]; **7**, 221[28]
Reinehr, D., **4**, 868[12], 874[47-51,54,55], 875[47], 887[12]; **5**, 30[2]
Reineke, C. E., **7**, 834[80]
Reineke, L. M., **7**, 57[34]
Reiner, J., **8**, 765[11], 773[11d], 789[11d]
Reiner, L. M., **2**, 466[110]
Reiner, T. W., **4**, 4[14], 6[14f]
Reinert, K., **8**, 754[97]
Reinert, T. J., **7**, 50[71]
Reingold, I. D., **3**, 380[8]; **4**, 611[349]
Reinhardt, D. V., **5**, 255[49]

Reinheckel, H., **8**, 756[152,153,156,157]
Reinheimer, H., **5**, 1147[113]
Reinhold, T. L., **5**, 141[91]
Reinhoudt, D. N., **1**, 461[15], 464[15]; **2**, 379[295], 821[110]; **4**, 45[126,127b]; **5**, 584[195], 676[4], 686[46-49], 687[46,48,49,58,58b,61,63], 688[58b,63]; **7**, 333[25]; **8**, 33[58], 60[194], 62[194], 64[194], 66[58,194], 98[103-105]
Reininger, K., **1**, 188[73], 189[73], 219[56]
Reinking, P., **7**, 706[26]
Reinshagen, H., **4**, 36[102]
Reinstein, M., **4**, 45[130]
Reintjes, M., **4**, 145[28]
Reis, H., **4**, 939[74]
Reischl, W., **5**, 735[138a,b], 740[138], 742[159b,c], 1005[32]; **6**, 903[139]
Reisdorff, J., **4**, 31[92,92k]
Reisman, D., **2**, 91[42]
Reiss, J. A., **4**, 37[107], 39[107c]
Reissenweber, G., **7**, 674[36]; **8**, 13[73]
Reissert, A., **8**, 295[55], 296[55]
Reissig, H.-U., **1**, 110[21]; **2**, 448[34,35], 486[40], 901[33,34], 911[33]; **4**, 27[79], 980[102], 981[111], 1007[130], 1031[8], 1035[39], 1043[8,106], 1045[8a], 1046[8a,39,106,112-115], 1048[119], 1092[144], 1093[144], 1102[199]; **5**, 211[64], 901[16,21], 904[16,21], 905[16,21], 921[16,21], 925[16], 927[16], 943[16,21], 952[16,21], 972[4], 977[22], 1006[33], 1086[75,76], 1200[53]; **6**, 456[161]; **8**, 652[76]
Reitano, M., **6**, 705[25]; **7**, 248[111], 801[44]
Reiter, B., **8**, 174[123]
Reiter, F., **4**, 1097[163]
Reiter, S. E., **5**, 626[37]
Reith, J. E., **6**, 263[26], 264[26], 265[45], 270[26]
Reitman, G. A., **3**, 304[68]
Reitman, L. N., **3**, 855[86]
Reitz, A. B., **1**, 755[114,116], 756[116,116b,118], 757[114], 758[116,116a], 759[114], 760[114], 761[114,116], 790[114], 812[114], 813[114]; **4**, 38[108,108c], 379[115], 380[115h], 383[115h]; **6**, 174[55]; **7**, 523[45]
Reitz, D. B., **1**, 471[67], 476[111,117], 477[111,117], 630[26]; **3**, 65[1,2], 68[2], 69[2], 71[2], 88[130], 90[130], 194[10]; **7**, 225[64]
Reitz, R. R., **5**, 1186[4]
Reitz, T. J., **1**, 514[52]
Reitze, J. D., **3**, 762[147]
Rej, R. N., **3**, 503[149], 512[149]
Rejoan, A., **8**, 451[180]
Rejowski, J. E., **7**, 229[108]
Rejtö, M., **5**, 183[157]
Relenyi, A. G., **7**, 155[31a]
Reliquet, A., **3**, 202[87]; **6**, 554[753,754,781]
Reliquet, F., **6**, 554[753,754,781]
Rellahan, W. L., **1**, 3[18,21]
Relya, D. I., **7**, 760[37], 761[37]
Remberg, G., **2**, 351[81], 364[81], 375[81]; **5**, 17[124], 461[96], 468[127], 531[81]
Remers, W. A., **8**, 564[443], 612[71], 614[86]
Remijnse, J. D., **8**, 447[115], 453[115]
Remington, S., **8**, 460[249]
Remion, J., **1**, 672[193,194,199], 673[193,194], 675[193,194], 686[228], 698[194], 699[193], 700[193,194], 702[199], 704[193,194], 705[193,194,199], 706[194], 712[199], 716[199], 717[199], 722[194]; **3**, 86[50]; **6**, 493[129]; **8**, 847[97], 848[97e], 849[97e], 888[120]
Remiszewski, S. W., **5**, 424[97], 425[100]
Rempel, C. A., **2**, 223[148]
Rempel, G. L., **8**, 445[53], 449[154], 552[348]
Remuson, R., **2**, 1065[116]
Remy, D. C., **3**, 380[11]; **4**, 957[22]
Remy, D. E., **5**, 582[180,181]
Remy, P., **8**, 54[154], 66[154]
Ren, W. Y., **6**, 554[759]
Renaldo, A. F., **1**, 442[176]; **3**, 232[268,269], 495[93a]; **5**, 176[126], 798[70], 800[70,78]; **8**, 406[40]
Renard, G., **3**, 734[11]; **8**, 862[230]
Renard, M., **1**, 635[87], 664[87], 665[87], 672[87], 679[87], 682[87]; **3**, 87[81], 104[81]; **4**, 50[142]
Renard, M. F., **8**, 203[148], 205[148,162,163], 558[399], 559[401], 560[402]
Renard, S. H., **6**, 536[548], 538[548]
Renaud, A., **2**, 742[77], 968[78]
Renaud, J.-P., **7**, 426[148c]
Renaud, M., **8**, 54[154], 66[154]
Renaud, P., **2**, 332[55], 1051[42], 1066[120], 1067[124], 1070[124]; **3**, 645[168], 650[210b]; **8**, 190[78]
Renaud, R. N., **3**, 643[119], 644[138], 648[138,182,184,189], 649[189,190]; **4**, 759[193]; **6**, 912[20]; **8**, 978[143]
Renaut, P., **1**, 805[312]; **5**, 245[21], 299[70], 308[70,94,97]; **6**, 990[84]; **8**, 932[41]
Rendenbach, B. E. M., **2**, 519[65], 520[66-68]; **4**, 21[69], 104[137], 222[168,169,170,171], 224[168,170]; **6**, 728[213]
René, L., **6**, 495[143], 496[143], 497[143]
Renfrew, A. H., **2**, 379[293]
Renfroe, H. B., **4**, 14[46]
Renfrow, W. B., **7**, 24[38]
Reng, G., **7**, 429[157a]
Renga, J. M., **3**, 957[110,111], 958[113]; **4**, 262[304]; **5**, 894[46]; **6**, 897[101], 898[102], 1021[54], 1026[84], 1028[99,100], 1030[84,100], 1031[111]; **7**, 129[71], 130[71], 131[71,82], 135[71], 675[58], 769[231], 771[231,277,285], 772[231,290]
Renge, T., **2**, 443[17]
Renger, B., **7**, 225[56]
Renk, E., **4**, 43[123]; **7**, 700[61]
Renk, H.-A., **8**, 267[70]
Renkema, J., **1**, 232[16]
Renken, T. L., **7**, 768[203]
Renko, Z. D., **6**, 85[94], 86[94]
Renneboog, R. M., **4**, 18[59], 121[208], 262[312], 991[151], 992[155]; **6**, 176[92]
Renneke, R. F., **3**, 1047[5]; **7**, 9[67]
Renner, H., **2**, 476[5]; **4**, 872[43]
Renner, R., **7**, 17[178]
Rennhard, H. H., **6**, 265[38]
Renoll, M. W., **4**, 270[15], 271[15]
Rens, M., **5**, 109[214]
Renson, M., **1**, 644[123], 646[123], 668[123], 669[123], 695[123]; **2**, 817[95]; **4**, 50[142]; **6**, 462[6,7,13,14], 472[69,70], 478[103]; **7**, 774[324,333]
Rentzepis, P. M., **5**, 71[134], 240[3]; **7**, 851[14], 855[63], 856[66], 865[87]
Reonchet, J. M. J., **8**, 61[187], 66[187]
Repid, O., **8**, 9[48,50]
Repin, A. G., **2**, 737[35]
Repke, D. B., **1**, 366[49], 391[49]; **2**, 913[76]
Repke, K., **7**, 198[28]
Reppe, W., **4**, 313[476]; **5**, 1133[25], 1141[80], 1145[80]
Rerick, M. N., **8**, 214[34,36], 217[36], 872[4], 966[71]
Resck, I. S., **2**, 855[248], 856[248]
Reske, E., **6**, 563[893,894]
Resnati, G., **1**, 514[51], 527[107], 528[108]; **3**, 147[393]; **4**, 113[166], 226[187,188], 382[131a,b], 384[131b]; **8**, 836[2], 843[2f]
Respess, W. L., **1**, 116[46], 118[46]; **4**, 70[5], 148[47a], 169[2]
Respondek, S., **8**, 511[100]
Ressig, H.-U., **5**, 539[105]
Ressler, C., **6**, 430[103]

Restelli, A., **1**, 523[80]; **2**, 228[166,167], 229[168], 374[276], 486[42]; **6**, 149[99,102], 425[66], 840[71], 927[73]; **8**, 844[67]
Restivo, R., **3**, 381[32]
Resvukhin, A. I., **6**, 712[78]
Rétey, J., **8**, 204[153]
Retrakul, V., **3**, 154[425], 155[425]
Retta, N., **1**, 214[28]
Rettig, M. F., **4**, 608[320], 646[320]; **5**, 925[156]; **8**, 684[96]
Rettig, S. J., **1**, 300[71]; **8**, 446[69]
Retuert, P. J., **6**, 432[122]
Reuben, J., **1**, 294[42]
Reubke, K. J., **6**, 420[19], 421[19], 424[19]
Reucroft, J., **1**, 755[115], 812[115], 813[115]
Reuman, M., **1**, 461[8]; **3**, 255[103], 261[103], 503[149], 512[149]; **4**, 250[140], 425[21], 426[21], 427[21], 428[21]; **5**, 78[280]; **7**, 704[9]
Reus, H. R., **6**, 753[116], 755[116]
Reusch, R. N., **5**, 1146[106]
Reusch, W., **2**, 169[168], 734[5]; **3**, 21[125], 23[136], 24[136], 595[191], 749[81]; **4**, 30[88]; **5**, 196[11], 712[58]; **8**, 328[6], 338[6], 339[6], 340[6], 341[6], 342[6], 343[6], 528[60]
Reuscher, H., **6**, 738[55,56]
Reuschling, D., **1**, 123[79], 372[80]
Reuss, R. H., **7**, 186[178]; **8**, 925[11], 926[11]
Reuter, H., **2**, 1077[153]; **5**, 485[182]; **6**, 716[104]
Reuter, J. M., **7**, 107[169]
Reuter, W., **5**, 493[210]
Reuterhall, A., **6**, 672[286]
Reuther, W., **6**, 174[64]
Reutov, O. A., **1**, 3[23], 437[157]; **3**, 503[152]; **4**, 297[269], 306[373]; **6**, 226[10], 256[10], 257[10], 283[166]; **8**, 99[107], 850[120,120a]
Reutrakul, V., **1**, 526[94], 555[121], 558[133], 835[138]; **2**, 417[20], 711[38]; **3**, 154[422,423,424,426], 155[422,423]; **4**, 113[168]; **6**, 159[175], 1022[65,66]
Reuvers, J. T. A., **2**, 198[83]; **6**, 1023[70]
Revel, J., **4**, 315[509]
Reverberi, S., **5**, 1137[55]
Reverdatto, S. V., **6**, 603[25], 604[27]
Revesz, C., **8**, 527[43]
Revial, G., **1**, 882[121]; **2**, 227[161]; **4**, 7[25], 221[161,162,163]; **5**, 341[60], 809[122]; **8**, 188[53]
Revill, J. M., **4**, 382[134,134b]
Revis, A., **8**, 786[118]
Revol, J.-M., **4**, 1040[88], 1048[88,88a]; **5**, 955[302]
Rewcastle, G. W., **1**, 463[28], 469[59]; **2**, 759[34a]; **4**, 435[137]
Rey, M., **1**, 876[102]; **3**, 914[10], 930[10], 931[10]; **5**, 689[73], 770[137,138,139], 806[106], 829[25], 1025[83], 1026[83]; **6**, 677[313], 778[63]; **7**, 410[103]
Reychler, A., **3**, 242[9]
Reye, C., **2**, 572[43]; **4**, 100[126]; **8**, 246[79], 546[311]
Reynaud, P., **2**, 141[39]; **6**, 420[14], 430[95], 436[8], 515[314], 522[314]
Reynen, W. A. P., **2**, 757[19]
Reyniers, M.-F., **6**, 80[66]
Reynolds, B. E., **7**, 227[83]
Reynolds, C. H., **2**, 1054[56]; **4**, 484[11]
Reynolds, D. D., **2**, 914[80], 959[29], 969[29,84], 970[29]
Reynolds, D. P., **3**, 290[70]; **6**, 655[165]; **7**, 682[85]; **8**, 198[135]
Reynolds, D. W., **5**, 63[11], 704[22], 1020[69], 1023[69,78]
Reynolds, G. A., **2**, 380[298], 802[27]
Reynolds, G. F., **7**, 236[22,25]
Reynolds, L. J., **6**, 677[311]
Reynolds, M. A., **8**, 409[80]
Reynolds, M. E., **4**, 790[38]
Reynolds, R. M., **6**, 441[85]
Reynolds-Warnhoff, P., **8**, 90[48]
Rhee, B., **8**, 222[94]
Rhee, C. K., **8**, 16[106], 17[106], 18[124]
Rhee, I., **2**, 451[55]; **3**, 554[20,21]
Rhee, R. P., **4**, 14[47,47d], 111[154c]; **6**, 137[41]
Rhee, S. G., **8**, 744[53], 745[54], 746[53,54], 748[59,60], 749[53], 750[53], 753[54]
Rheinboldt, H., **1**, 630[2]; **7**, 770[248], 772[293], 773[293]
Rheingold, A. L., **2**, 127[235]; **4**, 856[98], 905[213]; **5**, 1131[12]; **6**, 173[43]; **8**, 447[128], 463[128]
Rheinheimer, J., **2**, 505[6]
Rhine, W. E., **1**, 9[42], 19[99]
Rhoads, S. J., **2**, 598[14], 814[79]; **5**, 786[1], 798[1], 829[25], 830[36], 877[8], 972[5], 1000[1]; **6**, 834[35], 855[35]
Rhodes, C. J., **7**, 854[60]
Rhodes, K. F., **8**, 28[34], 66[34]
Rhodes, S. J., **1**, 3[23], 880[115], 898[115]
Rhodes, S. P., **7**, 595[19], 598[19]
Rhouati, S., **2**, 787[50]; **7**, 477[72], 483[72]
Rhyne, L. D., **3**, 499[116]; **4**, 588[62]
Riahi, A., **7**, 107[158,159]
Riba, M., **6**, 172[18]; **8**, 563[425]
Ribas, J., **4**, 616[393], 629[393]
Ribeiro, A. A., **1**, 41[203]
Ribéreau, P., **1**, 472[72]
Riberi, B., **2**, 363[188]
Ribo, J., **3**, 892[47]
Ricard, D., **3**, 748[75]
Ricard, M., **1**, 821[26]; **2**, 344[16], 353[16], 359[16], 360[16], 363[16]; **6**, 175[79]; **7**, 841[16], 842[16]
Ricart, G., **6**, 74[37]
Ricca, A., **4**, 1085[102]; **8**, 645[42]
Ricca, D. J., **3**, 1000[53]; **5**, 519[34], 549[34]; **6**, 883[57], 887[57]
Ricca, G., **2**, 279[16,17], 280[16], 283[17]; **3**, 421[65]
Ricci, A., **1**, 612[48]; **2**, 567[26], 586[135], 607[71]; **4**, 98[114], 113[114], 115[182,182e], 247[97], 256[97]; **5**, 438[162]; **6**, 18[66], 179[127], 238[74]; **7**, 330[7]
Ricci, G., **8**, 192[96]
Ricci, M., **7**, 708[30]
Rice, E. M., **2**, 583[111]
Rice, F. A. H., **7**, 723[23]
Rice, F. O., **6**, 120[125]
Rice, J. E., **3**, 503[149], 512[149]
Rice, K. C., **3**, 71[29]; **7**, 224[53]
Rice, L. E., **2**, 286[64]
Rice, R. M., **4**, 24[72]
Rice, S. N., **5**, 938[206]; **7**, 478[83]
Rich, D. H., **1**, 766[159]; **3**, 369[117], 372[117]; **5**, 891[37], 892[37]; **6**, 6[29]; **7**, 400[48]
Rich, E. M., **3**, 888[16]
Rich, J. D., **5**, 199[28]
Richard, C., **5**, 69[103]
Richard, H., **3**, 380[11]
Richard, J. M., **8**, 594[70]
Richard-Neuville, C., **3**, 757[124]
Richards, D., **7**, 76[117]
Richards, E. E., **1**, 632[64]
Richards, G. N., **3**, 822[2], 831[2,66]
Richards, I. C., **4**, 674[36]
Richards, J. A., **1**, 305[86]
Richards, J. H., **3**, 342[15], 1048[10]
Richards, K. D., **4**, 190[107]
Richards, K. E., **3**, 741[53], 742[55,57], 743[57]
Richards, K. R., **4**, 738[94]
Richards, P., **5**, 487[187]
Richards, P. J., **4**, 102[129]
Richards, R. W., **2**, 170[173]

Richardson, A. C., **6**, 73[27], 76[51]; **7**, 712[62]; **8**, 247[81]
Richardson, D. P., **4**, 380[120,120b], 381[120b], 382[120b]; **8**, 853[147], 856[147], 857[147]
Richardson, G., **1**, 562[170], 797[293]; **6**, 996[107]
Richardson, G. M., **3**, 822[5], 834[5]
Richardson, K. A., **2**, 655[134,134b]; **3**, 26[161]; **5**, 282[23,24]
Richardson, K. S., **4**, 240[49], 717[8]; **5**, 703[18]
Richardson, L. J., **3**, 325[161]
Richardson, S., **1**, 32[158], 791[268]; **2**, 482[36], 484[36]; **6**, 186[170]
Richardson, S. K., **3**, 247[47], 253[93]; **4**, 1033[23]
Richardson, T. J., **7**, 882[168]
Richardson, W. H., **3**, 585[134]; **7**, 851[19]
Richardson, W. S., **5**, 166[92]
Richaud, M. G., **2**, 356[131]; **6**, 501[206]
Riche, C., **1**, 34[226,227]; **4**, 221[165]; **6**, 550[677]; **8**, 58[174], 66[174]
Riche, M. A., **7**, 206[67]
Richen, W., **4**, 1084[96]
Richer, J.-C., **6**, 268[61]; **7**, 447[73]; **8**, 542[220], 873[18], 874[20]
Richey, F. A., Jr., **7**, 123[36], 186[180]
Richey, H. G., **5**, 910[84,87]
Richey, H. G., Jr., **1**, 15[79,80,82], 16[88], 385[112]; **3**, 438[35], 735[14]; **4**, 868[17], 869[17], 871[29,36], 874[52], 876[60], 877[67], 878[74,77,80], 879[77], 884[80]; **5**, 30[3], 856[210], 1007[40]
Richman, J. A., Jr., **8**, 652[72]
Richman, J. E., **1**, 527[105,106], 564[191], 567[224]; **3**, 34[197], 135[358], 136[358], 137[358], 139[358], 142[358], 143[358]; **4**, 10[32,32a,b,d,e], 11[32e], 113[166], 125[216,216f]; **6**, 134[26]; **8**, 393[111]
Richman, J. F., **8**, 70[232], 72[232]
Richman, R. M., **7**, 95[71], 108[177]
Richmond, G. D., **8**, 994[62]
Richmond, J. P., **5**, 1052[38]
Richmond, M. H., **6**, 33[7], 40[7], 57[7]
Richmond, R. E., **3**, 906[144]; **4**, 1040[102]
Richmond, R. R., **8**, 568[477]
Richmond, T. G., **3**, 213[51]
Richter, F., **4**, 30[88,88q], 253[170], 261[170]
Richter, F. W., **2**, 152[99]
Richter, P., **6**, 507[233,239], 515[239]
Richter, R., **5**, 117[274,275]; **6**, 491[118], 531[427], 796[16], 823[119]
Richter, R. F., **4**, 314[479]
Richter, W., **5**, 837[66]; **6**, 179[124], 856[153]
Richter, W. J., **5**, 1191[30], 1193[30]; **8**, 222[98], 224[98], 659[105]
Rick, J.-D., **4**, 434[126]
Rickard, C. E. F., **4**, 1018[218]
Rickards, R. W., **3**, 222[135], 247[48]; **4**, 391[179]; **7**, 373[74], 375[74], 771[263]; **8**, 605[13]
Rickborn, B., **1**, 846[14]; **3**, 263[177], 734[3,5], 760[136,138,141], 761[141], 762[141], 763[138,141], 764[141]; **4**, 301[326,327], 314[485]; **5**, 382[122], 564[95], 580[169], 691[85]; **6**, 8[35], 9[41], 11[45], 561[877], 960[61]; **8**, 267[68], 536[171], 673[30], 873[11,13-15], 874[15], 875[14]
Rickert, H. F., **2**, 369[252]; **5**, 552[4]
Rico, I., **6**, 175[71], 245[127]
Rico, J. G., **3**, 252[84]; **5**, 348[74b]
Ridaura, V. E., **8**, 587[33]
Ridby, J. H., **3**, 86[61], 88[61], 89[61], 91[61], 124[61]
Riddell, F., **5**, 418[69]
Riddell, F. G., **3**, 404[134]
Riddle, J. E., **6**, 662[212]
Rideal, E. K., **8**, 422[34]
Ridella, J., **3**, 197[33]
Ridenour, M., **4**, 738[100], 746[146]
Rideout, D. C., **5**, 344[66], 345[66], 346[66], 453[66], 854[175]
Rider, M. E., **6**, 487[70]
Rider, P., **6**, 604[29]
Ridge, D. N., **6**, 554[724]
Ridley, D. D., **1**, 3[21], 508[21], 519[63,64]; **2**, 72[60]; **3**, 902[118,118a]; **4**, 12[37,37d], 50[142], 226[192], 230[248]; **6**, 150[114], 531[457]; **8**, 99[111], 196[117], 197[117]
Riebiro, A. A., **6**, 904[142]
Riebsomer, J. L., **7**, 488[157]
Rieche, A., **4**, 305[370], 307[389]; **7**, 613[1]
Rieck, J. A., **2**, 745[107]
Riecke, E. F., **4**, 1015[195]
Riecke, K., **2**, 1023[54]
Ried, W., **2**, 495[60], 744[98], 757[20]; **3**, 890[34]; **6**, 420[22], 451[126], 509[264], 1013[13]; **7**, 657[34], 658[26]; **8**, 271[101-103], 965[70]
Ried, W. B., **3**, 582[114]
Riede, J., **1**, 17[214]; **5**, 442[182], 850[152], 1096[124,127]; **6**, 176[91]
Rediker, M., **2**, 47[153], 308[22,23], 309[23,24], 318[51]; **3**, 469[202], 470[202], 473[202]; **4**, 309[413], 312[455], 393[197,197a], 394[197a], 595[156], 620[156,395], 635[156,395]; **5**, 181[152]; **6**, 865[201]; **8**, 680[71], 693[71,113,115], 694[71]
Riedmüller, S., **4**, 104[135d]; **6**, 473[76]
Riefling, B., **3**, 484[25]
Riegel, B., **4**, 89[84a], 98[109g]; **7**, 100[120]; **8**, 530[105]
Rieger, H., **5**, 394[146], 395[146]
Riegl, J., **5**, 113[235]
Riego, J., **6**, 89[110]
Riehl, J. J., **2**, 600[28], 816[86]; **3**, 740[44]; **7**, 121[22,23], 517[14], 564[88], 568[88]
Rieke, R. D., **1**, 212[6], 213[6], 214[20], 227[20], 426[111], 453[223]; **2**, 121[190,191]; **3**, 209[19], 226[202], 263[171], 421[55], 499[116], 522[10], 553[17], 563[1], 570[1h]; **4**, 83[64], 175[41], 588[62,63], 969[64]; **5**, 386[132], 387[132c], 691[83], 692[83], 693[83]; **6**, 2[3], 25[3], 977[17]; **8**, 907[73]
Rieker, A., **3**, 661[22,23], 666[42,43]; **7**, 800[30]
Rieker, W. F., **1**, 477[128,135], 482[128]; **3**, 67[18], 69[25], 70[25], 72[25], 74[25]; **4**, 250[141]; **5**, 10[78], 67[87,89-92]; **7**, 875[118]
Riemenschneider, K., **4**, 729[61], 730[61], 765[61]
Riemer, J., **4**, 14[48]
Riemer, R., **6**, 188[181]
Riemer, W., **2**, 352[83], 371[261]; **5**, 76[239]
Riemland, E., **7**, 69[89]
Riemschneider, R., **6**, 291[200]
Rienäcker, R., **4**, 887[129]; **5**, 810[127]
Riener, E., **6**, 271[84]
Riener, T., **6**, 271[84]
Riepl, G., **8**, 174[122-124]
Riéra, A., **5**, 362[93], 363[93i], 1062[59]
Rieser, J., **2**, 359[159]
Riess, G., **2**, 969[86]
Riess-Maurer, I., **6**, 41[43]
Rietschel, E. Th., **6**, 33[7], 40[7], 57[7]
Rieu, J. P., **8**, 343[112]
Riew, C. K., **4**, 187[95]
Riezebos, G., **5**, 835[59]
Rifi, M. R., **4**, 129[225]; **8**, 321[92]
Rigamonti, J., **6**, 220[128]; **8**, 916[105]
Rigaudy, J., **5**, 194[5], 196[5], 197[5], 198[5], 417[65]
Rigby, H. L., **4**, 195[125], 1040[88], 1048[88]
Rigby, J. H., **1**, 248[66], 742[49], 752[94], 851[41], 852[41]; **2**, 911[69]; **4**, 5[18], 61[18g], 1009[141]; **5**, 326[25], 435[150], 618[8], 619[8], 624[8], 625[8], 633[67], 815[141], 922[134], 1020[70], 1027[70]; **6**, 147[85], 816[98]
Rigby, J. M., **6**, 143[72,73]
Rigby, R. D. G., **8**, 906[68], 907[68], 908[68]
Rigby, W., **7**, 703[6,7], 704[7]
Rigden, O. W., **7**, 498[227]

Riggs, R. M., **6**, 803[46]
Righetti, P. P., **2**, 351[79], 364[79,204]; **5**, 454[70]
Righi, P., **2**, 338[76]
Riguera, C., **4**, 1020[238], 1023[238]
Riguera, R., **4**, 1004[78]; **6**, 533[481], 550[481]; **7**, 746[85]
Rihs, G., **4**, 382[134]; **5**, 836[63]; **8**, 795[19]
Rijsenbrij, P. P. M., **5**, 402[5]
Rilatt, J. A., **7**, 635[70]
Riley, D. A., **6**, 57[136]
Riley, D. P., **7**, 748[115], 765[144,145,146], 851[23]; **8**, 451[180], 459[237], 535[166]
Riley, E. F., **2**, 321[6], 326[6]
Riley, P. E., **4**, 980[107]
Riley, P. I., **1**, 139[3]
Riley, R. G., **1**, 411[45]
Riley, T. A., **6**, 474[83]
Rilling, H. C., **8**, 93[76]
Rilo, R. P., **4**, 288[187]
Rimbault, C. G., **5**, 680[21], 1031[97]; **7**, 98[97], 165[85], 169[111]; **8**, 540[194]
Rimmelin, J., **5**, 552[19]
Rimpler, M., **6**, 116[89]
Rinaldi, P. L., **1**, 838[164]; **7**, 777[375]
Rindone, B., **7**, 170[121]
Rinehart, K. L., Jr., **4**, 97[107c]
Ringele, P., **5**, 742[158]
Ringler, B. I., **4**, 73[33]
Ringold, C., **4**, 155[73]; **5**, 20[140]; **7**, 565[97]
Ringold, H. J., **1**, 846[16], 851[16]; **2**, 323[27]; **3**, 23[138]; **7**, 101[133], 136[110,114,117], 137[117], 145[168], 253[20]; **8**, 530[106]
Ringwald, E. L., **6**, 265[44], 271[44]
Riniker, B., **6**, 668[262]
Rink, H.-P., **3**, 505[169]; **6**, 738[52]
Rio, G., **5**, 723[109]
Riondel, A., **1**, 563[184]
Riordan, J. C., **7**, 696[40]
Ripamonti, M. C., **8**, 347[141], 350[141]
Ripka, W. C., **5**, 257[59,59a]
Ripley, S., **3**, 582[116]; **8**, 840[32]
Ripoll, J.-L., **3**, 848[48]; **5**, 451[38], 552[6,9], 556[51], 557[57], 558[62], 559[64], 560[62,65,66], 565[99], 575[132,135], 576[136,137,214], 579[158,159], 583[185], 589[210,211,214]; **6**, 706[39]
Ripoll, L.-L., **6**, 689[384], 690[384]
Riquelme, R. M., **6**, 428[86]
Risaliti, A., **3**, 503[149], 512[149]; **4**, 20[63], 21[63]; **6**, 710[58,59]
Riscado, A. M. V., **8**, 62[196], 66[196]
Risch, N., **2**, 853[230]
Rischer, M., **2**, 372[271]
Rischke, H., **6**, 740[63]
Rise, F., **5**, 1183[52]
Rising, A., **3**, 806[16]
Risius, A. C., **8**, 472[7]
Riskibaev, E. R., **3**, 303[58]
Risley, E. A., **6**, 487[57], 489[57]
Risley, H. A., **7**, 760[43]
Risse, S., **8**, 527[53]
Risse, W., **4**, 461[99], 475[99]
Rissi, E., **2**, 765[71]
Rissler, W., **2**, 345[32]
Rist, G., **5**, 829[14]
Rist, H., **4**, 872[43]
Ritchey, W. M., **5**, 581[173]
Ritchie, A. C., **2**, 323[33]
Ritchie, E., **2**, 773[26]; **3**, 831[59]; **6**, 215[104]
Ritchie, P. D., **8**, 184[4]
Ritchie, T. J., **7**, 721[15]
Ritschel, W., **6**, 547[665]
Ritscher, J. S., **5**, 646[3]
Rittel, W., **6**, 668[262]
Ritter, A., **2**, 725[111], 726[111]; **6**, 115[81], 653[150]
Ritter, A. R., **6**, 237[65], 243[65]
Ritter, F. J., **8**, 556[375]
Ritter, J. J., **4**, 292[233], 294[241]; **6**, 261[1,2], 262[1,2,15], 264[15,30], 265[1,39], 266[1,2], 267[30], 268[15,70], 271[70], 272[91], 273[95,100], 276[95]
Ritter, K., **1**, 416[68]; **5**, 755[71], 780[71], 841[96]
Ritter, R. H., **3**, 1051[20]
Ritterskamp, P., **3**, 216[75]; **5**, 225[114], 226[108], 227[114,115], 228[114,115], 230[114,115,129], 232[129,134], 233[114,115]
Ritter-Thomas, U., **2**, 352[83], 371[261]; **5**, 76[239]
Rittle, K. E., **1**, 823[44b]; **2**, 962[51]
Rittmeyer, P., **7**, 874[105]
Rittweger, K. R., **7**, 70[94]
Riva, M., **5**, 1025[81]
Riva, R., **5**, 524[50,50b], 539[50,50b], 548[50b,c]; **8**, 563[435]
Riva, S., **8**, 194[105], 563[435]
Rivadeneira, E., **2**, 249[84]
Rivalle, C., **2**, 379[294]
Rivera, A. P., **7**, 820[25]
Rivera, A. V., **4**, 706[38]
Rivera, E. G., **4**, 222[177]
Rivera, I., **8**, 717[101]
Rivera, M., **5**, 480[166]; **7**, 502[262]
Rivera, V., **7**, 693[24]
Rivers, D. S., **7**, 95[71], 108[177]
Rivers, G. T., **8**, 940[109], 947[109], 952[109]
Rivett, D. E. A., **6**, 441[81]
Rivett, J. E., **4**, 382[134,134b]
Riviere, H., **3**, 734[8]; **4**, 841[44]; **7**, 450[16]; **8**, 265[48], 528[56]
Rivière, M., **1**, 838[161,166]
Rivlin, V. G., **7**, 709[46]
Rivola, G., **8**, 347[141], 350[141]
Rivolta, A. M., **6**, 255[168]
Rizk, I., **6**, 625[159]
Rizk, M., **3**, 976[8]
Rizvi, S. H. M., **7**, 71[95]
Rizvi, S. Q. A., **6**, 1024[78]; **7**, 778[398]
Rizvi, S. Q. R., **7**, 162[66]
Rizzi, G. P., **2**, 742[68], 762[53]
Rizzi, J., **4**, 14[48]
Rizzi, J. P., **7**, 409[102], 410[102]
Rizzo, C. J., **2**, 725[106]
Ro, R. S., **6**, 959[48]; **8**, 141[38]
Roach, A. G., **8**, 615[94], 618[94]
Roach, B. L., **3**, 957[110]; **5**, 894[46]
Roark, D. N., **5**, 587[207]
Roark, W. H., **4**, 36[102,102g]
Robarge, K. D., **4**, 377[104], 379[104e], 380[104d,e]; **5**, 453[60,61], 458[61], 460[61,94], 461[60,61], 464[61]
Robas, V. I., **7**, 500[236]
Robb, E. W., **5**, 646[7]
Robba, M., **6**, 428[86], 487[64], 489[64]
Robbiani, C., **1**, 87[49]; **5**, 13[91,93], 391[141]
Robbins, C. M., **3**, 380[4]
Robbins, J. D., **5**, 210[59], 217[26,27]
Robbins, M. D., **1**, 345[124]
Robbins, W., **1**, 301[74], 316[74]
Robeerst, A. J. M. S., **7**, 759[10]
Roberge, G., **2**, 648[92], 649[92]
Robert, A., **6**, 67[11]; **8**, 244[58], 248[58]
Robert, F., **4**, 528[71]; **5**, 1103[151]
Robert, P. C., **6**, 1035[136]

Roberto, D., **3**, 1037[86]
Roberts, A., **8**, 315[44], 545[281]
Roberts, B. P., **7**, 598[56], 599[67], 604[133], 607[133]; **8**, 726[186], 753[72], 857[200]
Roberts, B. W., **2**, 353[100]; **4**, 262[311]; **7**, 710[50]
Roberts, C. W., **2**, 527[2], 528[2], 553[2]; **6**, 263[19]
Roberts, D. A., **1**, 793[273], 804[273]; **3**, 159[457], 173[457], 510[184]; **5**, 386[135]; **6**, 993[91]
Roberts, D. C., **6**, 639[49]
Roberts, D. H., **3**, 600[212]; **4**, 820[219]
Roberts, D. L., **3**, 407[150]
Roberts, D. T., Jr., **3**, 334[214]
Roberts, E. F., **1**, 420[85]
Roberts, F. E., **8**, 839[28], 968[90]
Roberts, J., **4**, 435[139]
Roberts, J. C., **1**, 434[140]; **2**, 249[84], 264[58], 760[39]
Roberts, J. D., **2**, 5[20], 6[20], 21[20]; **3**, 382[36], 782[16], 825[30]; **4**, 483[1], 484[1], 486[32], 487[1], 488[1], 489[1], 491[1], 492[1,70], 493[1], 495[1], 506[1], 508[1], 871[35], 876[58]; **5**, 43[33], 63[3], 65[65], 69[3], 714[65,66]; **6**, 204[17], 1013[12]; **7**, 564[90], 565[90], 742[56]
Roberts, J. L., **1**, 366[45]; **2**, 111[84], 909[62], 910[62]; **4**, 377[101]; **7**, 438[15], 445[15]
Roberts, J. L., Jr., **3**, 246[43], 258[43], 824[23]
Roberts, J. S., **3**, 386[67], 404[133,134,136], 405[139]; **4**, 681[53]; **6**, 111[64], 676[306]; **8**, 890[141]
Roberts, J. T., **8**, 608[43], 629[43]
Roberts, K. A., **1**, 665[172], 668[172]
Roberts, L. D., **7**, 759[6]
Roberts, L. W., **5**, 1125[54]
Roberts, M. F., **3**, 77[59]
Roberts, M. R., **4**, 30[88,88g]; **8**, 925[12]
Roberts, P. M., **5**, 105[194]
Roberts, R., **6**, 120[125]
Roberts, R. A., **4**, 181[71], 1040[69]; **6**, 883[60], 884[60]; **7**, 100[116], 552[57]
Roberts, R. E., **3**, 123[249]
Roberts, R. M., **2**, 740[52], 745[103]; **3**, 294[5], 299[5], 300[5], 303[5], 304[5], 306[80], 323[145], 324[147], 325[156], 327[170], 328[179], 329[189]; **5**, 834[53], 908[72]
Roberts, S. M., **1**, 865[87]; **3**, 232[266], 290[70], 488[54], 495[54], 985[27]; **4**, 385[146], 413[146], 791[46]; **5**, 418[70], 560[73], 829[21,25]; **6**, 655[165], 876[27], 880[27]; **7**, 59[37], 671[12], 674[33,37], 682[85]; **8**, 198[130,135], 584[15]
Roberts, T. G., **1**, 776[215]
Roberts, V. A., **4**, 1052[129]; **5**, 618[7], 620[7], 621[21], 624[29]
Roberts, W. J., **8**, 52[144], 66[144]
Roberts, W. L., **3**, 918[25]
Roberts, W. P., **6**, 781[76]; **8**, 945[133]
Robertson, A., **3**, 807[25], 900[89]
Robertson, A. K., **4**, 487[44]
Robertson, A. V., **8**, 605[15], 624[152]
Robertson, B. W., **7**, 57[31], 58[31], 63[31]
Robertson, D. N., **2**, 367[228]
Robertson, G., **2**, 1042[114]
Robertson, G. M., **3**, 586[153], 604[153], 605[231], 610[153]; **4**, 809[160]; **6**, 1062[76]
Robertson, J., **1**, 894[159]; **4**, 822[226]
Robertson, J. D., **1**, 753[102]; **4**, 350[121]
Robertson, J. M., **3**, 386[68]
Robertson, L. W., **7**, 66[72]
Robertson, M., **7**, 844[54]
Robertson, R. E., **2**, 602[40]; **3**, 626[42]
Robeson, C. D., **3**, 698[158]
Robev, S. K., **2**, 852[235], 854[235]; **6**, 554[750]
Robey, R. L., **2**, 1024[59]
Robichaud, A. J., **1**, 595[26]; **2**, 1009[7], 1018[7], 1031[7], 1042[114]; **6**, 739[59], 743[59]
Robien, W., **3**, 380[7]
Robin, J.-P., **3**, 501[139], 509[139,177], 512[139]
Robins, B. D., **2**, 1096[97]
Robins, M. J., **3**, 219[109], 530[72], 534[72], 547[124]; **6**, 657[180], 936[110]; **7**, 259[60]; **8**, 819[43], 820[43]
Robins, R. K., **6**, 474[83], 478[102]
Robinson, A., **2**, 787[50]
Robinson, B., **8**, 612[68], 616[68]
Robinson, B. L., **7**, 500[243]
Robinson, C. A., **8**, 143[62,63], 148[62,63,107]
Robinson, C. H., **1**, 174[8], 175[8], 751[90]; **7**, 366[51], 414[120]; **8**, 525[14], 526[14], 883[91]
Robinson, C. N., **2**, 363[196]
Robinson, D. B., **8**, 364[19]
Robinson, D. T., **4**, 603[265,266], 644[265]
Robinson, E. D., **3**, 983[21], 984[21,21a], 1011[78], 1012[79,80]; **6**, 874[11,12]; **8**, 929[29]
Robinson, G., **5**, 1138[67], 1146[111], 1147[111]
Robinson, G. C., **1**, 568[228]; **4**, 887[123], 888[123]; **5**, 260[62]; **8**, 100[114]
Robinson, G. M., **3**, 246[42], 258[42]
Robinson, J., **3**, 415[9]; **6**, 149[92]
Robinson, J. C., Jr., **5**, 473[149]
Robinson, J. E., **6**, 902[128]
Robinson, M., **7**, 170[122], 171[122]; **8**, 394[114]
Robinson, M. D., **2**, 765[76]
Robinson, M. J. T., **8**, 526[16], 707[20]
Robinson, M. M., **5**, 686[51], 687[51], 688[51]
Robinson, N. G., **4**, 111[152e]; **6**, 96[152]
Robinson, P. D., **1**, 515[56]
Robinson, P. H., **6**, 24[97]
Robinson, P. W., **5**, 277[13]
Robinson, R., **2**, 149[91], 396[8], 402[31], 838[165]; **3**, 246[42], 258[42], 828[53], 888[14,15]; **4**, 2[5], 99[119], 288[181]; **8**, 957[16], 973[118]
Robinson, R. E., **8**, 516[120]
Robinson, R. P., **4**, 350[119]; **6**, 531[441,442]; **8**, 389[70]
Robinson, S. D., **4**, 587[35]; **8**, 552[350]
Robinson, S. R., **4**, 444[199]
Robinson, V. J., **4**, 394[189,189c]
Robinson, W., **2**, 1088[40], 1097[40]
Robinson, W. T., **4**, 691[74]
Robison, M. M., **4**, 45[126,126c]
Roblin, J., **8**, 477[33]
Robson, J. H., **4**, 294[246]; **6**, 283[165]
Robson, J. N., **1**, 248[63]
Robson, M. J., **3**, 832[68b]; **5**, 904[52]
Robson, P., **7**, 762[82]
Robveille, S., **4**, 453[31], 459[31], 471[31,141], 475[150]
Roby, J., **8**, 136[51]
Robyr, C., **2**, 338[75]; **5**, 51[45,45a], 53[45a]
Roc, M., **2**, 817[92]
Rocas, J., **2**, 435[62]; **4**, 231[268]
Rocco, V. P., **5**, 145[107]
Roch, G., **7**, 745[73]
Roch, R., **8**, 558[395]
Rocha, E. K., **2**, 780[9]
Rochas, P., **3**, 315[110]
Roche, E. G., **3**, 286[56b]; **7**, 620[29]
Roche, M., **2**, 969[86]
Rochefort, M. P., **3**, 691[134]
Rocherla, U. S., **7**, 102[135]
Rocherolle, U., **7**, 630[55]
Roches, D. D., **8**, 890[140]

Rochester, C. H., **8**, 319[71]
Rochin, C., **2**, 900[30], 901[30], 964[62]; **6**, 107[22]
Rochlitz, J., **5**, 478[162]
Rochow, E. G., **1**, 390[147]; **2**, 182[7], 183[7]
Rocke, A. J., **2**, 770[9]
Rockell, C. J. M., **2**, 635[40], 640[40]; **3**, 26[161]; **7**, 294[15]
Rocterdink, F., **4**, 484[23], 485[23]
Roddick, D. M., **8**, 673[24], 676[24], 682[24]
Rodé, L., **3**, 353[50], 354[50]
Rode, T., **3**, 509[181]
Rodebaugh, R., **5**, 403[7], 404[7,18]
Rodeheaver, G. T., **7**, 451[28], 637[74,75]
Rodehorst, R., **7**, 103[138], 257[45], 258[45]
Roden, B. A., **1**, 535[144]; **4**, 689[68]
Roden, F. S., **1**, 248[70]; **5**, 299[68]
Rodenhouse, R. A., **4**, 1038[59]
Röder, A., **3**, 625[41]
Roder, H., **1**, 168[116a]; **2**, 35[127]; **6**, 863[191]; **8**, 278[158]
Roder, P., **2**, 1023[55]
Rodes, R., **4**, 292[224]; **6**, 555[810]; **8**, 856[177]
Rodes, T. M., **4**, 1021[244]
Rodewald, H., **3**, 872[61,63]
Rodewald, L. B., **3**, 635[34], 638[34]
Rodewald, P. G., **3**, 296[14]; **4**, 763[211]
Rodewald, W. J., **6**, 773[44]; **7**, 236[21,23]; **8**, 928[25]
Rodgers, J. D., **3**, 960[118]; **6**, 898[105], 901[123], 905[144]
Rodgers, J. R., **1**, 2[3], 37[3]
Rodgers, S. L., **4**, 345[81]
Rodgers, T. R., **5**, 223[75]
Rodgman, A., **3**, 407[150]; **4**, 310[433]
Rodier, N., **5**, 21[155], 22[155]; **6**, 436[8]
Rodin, A. P., **8**, 606[27]
Rodin, J. O., **3**, 126[318]
Rodin, W. A., **4**, 1002[62]
Rodina, L. L., **3**, 887[7], 890[7], 892[7], 893[7], 896[7], 897[7], 900[7], 902[7], 903[7], 905[7]
Rodini, D. J., **2**, 527[9], 528[9], 531[24,25], 533[24], 534[25], 537[24], 541[76], 544[76], 546[76], 547[76]; **5**, 2[7], 4[7], 6[48], 7[54], 8[54,58,62], 15[48], 20[48], 519[35]
Rodionov, A. P., **8**, 621[142]
Rodler, M., **4**, 484[13]
Rodrigo, R., **4**, 73[35]; **8**, 244[49]
Rodrigo, R. G. A., **7**, 65[67]
Rodrigues, R., **1**, 641[106,107], 672[106,107], 677[106,107], 724[106]
Rodriguez, A., **3**, 946[93]; **4**, 1103[206]; **7**, 674[48]; **8**, 477[30]
Rodriguez, B., **6**, 85[87]; **8**, 333[55]
Rodriguez, D., **1**, 589[20,20a], 592[20a]
Rodriguez, H. R., **1**, 23[120], 460[4], 463[4], 471[4], 472[4], 473[4]; **3**, 193[2], 194[2], 261[146], 264[146]
Rodriguez, I., **2**, 252[41], 253[41]; **5**, 362[93], 365[93e]
Rodriguez, J., **4**, 793[72]
Rodriguez, M., **8**, 142[54]
Rodríguez, M. A., **4**, 347[93], 354[93d]
Rodriguez, M. J., **4**, 383[141], 384[141a]
Rodriguez, M. L., **4**, 373[87]
Rodriguez, M. S., **4**, 817[203]
Rodriguez, O., **5**, 125[14]
Rodriguez, R., **2**, 840[184]; **4**, 1103[206]; **6**, 453[137]
Rodriguez-Hahn, L., **2**, 849[213]
Rodriguez Mieles, L., **1**, 784[243]
Rodrique, L., **1**, 38[259]
Rodriques, K. E., **1**, 385[120], 386[120]; **4**, 399[223], 403[223]; **5**, 552[31], 564[31]
Rodriquez, I., **1**, 307[111], 312[111]
Rodriquez, M. L., **7**, 413[118]
Rodwell, P. W., **5**, 649[22], 650[22]
Roe, D. C., **8**, 672[22]
Roe, D. K., **3**, 577[87]; **8**, 524[12], 527[49], 532[12c]
Roe, E. T., **6**, 262[16], 263[16], 264[16], 266[16]
Roe, R., **3**, 296[13]
Roe, S. P., **1**, 528[122]
Roeber, H., **2**, 367[223]
Roebke, H., **8**, 528[83]
Roedig, A., **4**, 279[101], 280[101], 282[101], 283[101], 284[101], 285[101], 287[175]; **6**, 547[665]; **8**, 267[70]
Roefke, P., **5**, 1191[30], 1193[30]
Roekens, B., **2**, 651[115,115a]; **6**, 164[195]
Roeland, S., **6**, 552[701]
Roelen, O., **3**, 1015[1]
Roelens, S., **6**, 16[66]
Roelofs, W. L., **4**, 6[20,20a]
Roelofsen, D. P., **6**, 705[21]
Roeske, R. W., **6**, 665[227], 668[227,258], 669[227]
Roesle, A., **2**, 810[62], 829[62]
Roesrath, U., **6**, 453[139]
Roessler, F., **2**, 589[153], 874[28], 875[28]; **4**, 186[93], 257[217], 901[183]; **5**, 762[105]; **8**, 769[24], 771[24]
Roffey, P., **5**, 412[44], 498[236]
Rogalska, E., **5**, 111[224,225]; **6**, 520[343]
Rogalski, W., **8**, 278[156]
Roger, G., **1**, 569[251]
Roger, R., **3**, 721[5]; **6**, 488[15], 507[230], 529[15], 533[15], 545[15], 562[15]
Rogers, A. E., **8**, 956[6]
Rogers, A. J., **5**, 71[156]
Rogers, A. O., **3**, 242[2]
Rogers, B. D., **5**, 854[178], 856[178], 872[178]; **6**, 860[177]
Rogers, D. N., **4**, 37[107], 425[25], 430[91]
Rogers, D. Z., **3**, 733[1]
Rogers, H. R., **8**, 807[129]
Rogers, J. A., Jr., **2**, 152[99]
Rogers, J. L., **2**, 965[67], 968[67]
Rogers, N. A. J., **5**, 217[22,23], 226[23,105,106]
Rogers, N. H., **1**, 791[267]
Rogers, N. R., **5**, 178[138], 221[61], 882[15]
Rogers, P. E., **8**, 51[121], 66[121]
Rogers, R. D., **1**, 6[32], 37[178], 240[43]; **5**, 347[72,72b], 850[158], 857[230], 1066[8], 1083[57], 1142[86]
Rogers, R. J., **6**, 220[129]
Rogers, T., **7**, 778[405]
Rogers, V., **5**, 600[43]; **7**, 555[70]
Rogers, W. J., **6**, 489[99], 525[99], 767[24]
Rogers-Evans, M., **7**, 429[151]
Rogerson, T. D., **4**, 476[162,163], 502[124], 766[229]
Roggo, S., **1**, 166[113], 314[128], 323[128], 341[98]
Rogic, M. M., **2**, 111[85,86], 112[88], 241[14], 242[14b,16]; **3**, 242[6], 257[6], 259[6], 260[143], 794[77]; **4**, 145[23]; **5**, 78[280]; **6**, 140[60], 774[48]; **7**, 606[147], 700[60]; **8**, 229[137], 230[137], 231[142]
Rogier, E. R., **8**, 269[93], 530[101]
Roginski, E., **8**, 141[29], 533[141]
Rogozhin, S. V., **6**, 533[491]
Rogozinski, S., **8**, 52[138], 66[138]
Rohde, C., **3**, 194[4]
Rohde, J. J., **4**, 143[21]
Rohde, R., **7**, 383[111]
Röhle, H., **6**, 34[9], 35[9], 49[9]
Rohloff, J. C., **1**, 421[88]; **7**, 821[30]; **8**, 171[105]
Rohm, A., **3**, 3[14], 8[14]
Röhr, A., **5**, 115[250]; **6**, 441[86]
Rohr, O., **8**, 530[91]

Rohrer, C., **1**, 271[62,62b]; **3**, 570[55], 582[55], 583[55], 630[57], 631[57]
Röhrig, P., **3**, 593[179]
Rohrmann, E., **8**, 220[85]
Röhse, C., **3**, 322[142b]
Rohwedder, W., **3**, 691[129], 693[129]; **8**, 453[191]
Roitburd, G. V., **2**, 725[107]; **5**, 775[175]
Rojas, A. C., **4**, 869[27], 870[27], 871[27]
Rokach, J., **1**, 821[28]; **2**, 743[84]; **3**, 273[16], 274[16], 277[27], 289[66,67]; **4**, 1059[154,155,156,157,158]; **5**, 490[192]; **6**, 93[132], 489[86]; **7**, 360[21]; **8**, 389[68], 540[195]
Rol, C., **7**, 649[46]
Roland, D. M., **7**, 360[22]
Roland-Gosselin, P., **5**, 725[115]
Rolando, C., **6**, 162[187,188], 624[145]
Roldan, F., **7**, 277[155]
Rolf, D., **8**, 219[75,77]
Roling, P. V., **1**, 78[16], 79[16]
Rolla, F., **4**, 275[68], 276[68], 279[68], 288[68], 438[155], 444[195]; **6**, 19[67], 76[44], 204[11], 221[131]; **7**, 253[16], 663[61]; **8**, 384[33], 803[96], 806[123]
Rollema, H., **7**, 331[15]
Roller, G. G., **3**, 393[93]
Rollin, G., **7**, 282[178]; **8**, 113[37]
Rollin, P., **6**, 253[155]
Rollinson, S. W., **4**, 315[530], 379[114]
Rom, P., **8**, 365[31]
Roma, G., **6**, 487[47-54], 489[47,48,52-54], 543[47,48]
Romadane, I., **3**, 310[94], 311[94]
Romaguoni, L., **4**, 307[396]
Romaine, J. L., **5**, 817[148]
Roman, C., **3**, 257[115]
Román, M. N., **8**, 329[20], 336[20]
Romanelli, A. L., **4**, 824[240]; **5**, 927[161], 931[161]
Romanet, R. F., **4**, 10[32,32a,f,g,33,33a], 109[148]
Romaniko, S. V., **4**, 347[103]
Romann, A. J., **5**, 829[13]
Romano, L. J., **7**, 530[26]
Romañuk, M., **3**, 643[128]
Romao, M. J., **4**, 905[212]; **5**, 161[63], 480[178], 1138[71], 1157[71]
Romashin, J. N., **6**, 557[836]
Romberger, M. L., **1**, 786[248]
Romeo, A., **7**, 92[41,41b], 93[41b], 94[41], 237[31], 310[27]
Romeo, G., **6**, 575[968]
Römer, R., **3**, 625[41]
Romero, A. G., **1**, 757[122]; **6**, 86[101]; **7**, 361[24]
Romero, J. R., **1**, 557[127]
Romero, M., **3**, 45[250]
Romero, R. R., **4**, 505[148]
Romey, G., **8**, 47[126], 66[126]
Romig, J. R., **2**, 933[138]
Romine, J. L., **1**, 240[42]; **4**, 989[144]; **5**, 798[69], 817[149], 1029[91]
Romines, K. R., **4**, 1040[79], 1049[79], 1051[128]; **5**, 599[40], 804[94], 986[40]
Rommel, S., **2**, 896[12]
Rommelspacher, H., **6**, 736[22]
Rømming, C., **3**, 864[24], 865[24], 872[24]; **6**, 291[210]
Romney-Alexander, T. M., **6**, 231[36], 232[36], 233[36]
Romo, J., **3**, 805[15], 846[46]; **7**, 92[42], 93[42]; **8**, 566[450]
Romo de Vivar, A., **3**, 846[46]
Ron, E., **2**, 528[13]; **5**, 2[11], 9[74], 19[74]
Rona, P., **3**, 257[120]
Rona, R. J., **1**, 620[66], 784[244]; **3**, 224[182], 759[132]
Ronald, B. P., **2**, 725[105]; **3**, 721[5], 736[26,27]
Ronald, R. C., **1**, 202[104], 477[143], 753[105]; **2**, 725[105]; **7**, 414[119], 834[78]
Ronan, B., **7**, 777[378], 778[378]
Ronayne, J., **2**, 346[44]
Ronca, P., **5**, 1154[152]
Roncetti, L., **3**, 229[230,230a]
Ronchetti, F., **4**, 38[108], 379[115], 380[115j], 382[115k]; **7**, 274[138]
Ronchi, A. U., **1**, 830[92], 832[92]; **2**, 187[43]; **7**, 137[124]
Ronco, A., **2**, 410[5]
Rondan, N. G., **1**, 49[8], 92[64], 357[8], 476[125], 477[125], 610[45]; **2**, 476[4]; **3**, 18[96], 31[186], 66[12], 74[12], 194[4,11]; **4**, 379[117], 484[21], 729[59], 872[41], 1079[65]; **5**, 79[292], 203[39], 204[39j], 209[39], 210[39], 257[61,61a,c], 258[61b], 260[65], 261[65], 436[158,158b], 442[158], 621[21], 678[18], 679[18], 680[18], 681[18], 682[34b], 683[34b], 685[18], 703[15], 733[15b], 857[227], 1031[96]; **6**, 724[150]; **7**, 439[36]; **8**, 723[151], 724[151,169,169g]
Rondarev, D. S., **3**, 639[82], 644[82,137]
Rondestvedt, C. S., Jr., **3**, 497[101], 505[101], 889[24]; **4**, 758[188]; **5**, 2[17]; **8**, 366[47]
Rondestvedt, E., **3**, 324[152]
Rone, A.-M., **3**, 274[19]
Ronlan, A., **3**, 667[47], 671[63], 672[67], 676[80], 685[63]; **7**, 800[29], 801[38-40]
Ronman, P., **3**, 255[111]
Ronneberg, H., **3**, 222[135]; **4**, 391[179]
Rönsch, E., **7**, 769[241]
Ronzeau, M., **6**, 244[108]
Ronzhin, L. K., **8**, 963[43]
Ronzini, L., **2**, 87[27]; **3**, 230[235,236], 441[48,49], 446[87], 449[48,101], 461[145], 463[153,154,155,166], 485[29], 492[79], 493[29], 503[29,79], 513[29,79]; **4**, 93[93d]; **6**, 446[101]
Ronzioni, L., **1**, 413[57], 416[67], 452[220]
Roobeek, C. F., **3**, 1022[20]; **4**, 921[21]
Rood, S. H., **5**, 1125[54]
Roof, A. A. M., **5**, 74[208]
Rooks, W., **4**, 932[62]
Rooney, C. S., **7**, 750[126]
Rooney, J. J., **3**, 334[220]; **4**, 588[59], 601[268], 603[268]; **5**, 65[67], 801[80], 1116[5], 1120[22]; **7**, 107[161]
Roos, E. C., **1**, 546[52]
Roos, G. H. P., **5**, 572[122]
Roos, J., **6**, 811[75]
Roos, M., **6**, 50[102], 51[102]
Root, R. L., **7**, 312[33]
Roozpeikar, B., **8**, 472[7]
Roper, J. M., **3**, 587[143]; **8**, 113[42]
Roques, J., **6**, 535[525,526]
Rorig, K., **4**, 31[92,92b]
Rosai, A., **2**, 284[53]
Rosamond, J. D., **5**, 113[232]
Rosan, A., **4**, 562[36], 576[13,14]; **5**, 272[4,5], 273[4], 275[4]
Rosan, M., **4**, 563[37]
Rosario, O., **3**, 200[70]
Rosati, R. L., **7**, 160[48]
Rosazza, J. P., **7**, 55[11], 56[11], 63[58], 65[66], 66[11]; **8**, 56[168], 66[168]
Rosazza, J. P. N., **7**, 58[53a], 62[53,53a], 63[53a]; **8**, 192[97,99]
Rosca, S., **4**, 522[50]
Rosca, S. I., **4**, 522[50]
Rosch, L., **1**, 82[24], 83[24], 97[82], 215[31]
Röscheisen, G., **3**, 414[4]
Rose, A. H., **1**, 543[27]; **7**, 66[74], 70[74]
Rose, B. F., **3**, 679[90], 683[90]
Rose, C. B., **3**, 265[187], 390[80], 392[80], 396[105], 397[105]
Rose, E., **4**, 523[56], 527[69,70], 528[69-71], 541[116], 543[122]
Rose, E. H., **1**, 804[313], 805[313]; **6**, 968[107], 998[120]

Rose, I. A., **2**, 466[111], 467[111]; **8**, 87[31]
Rose, M. E., **8**, 336[85]
Rose, U., **2**, 464[97]
Roselli, A., **8**, 683[90]
Roseman, S., **2**, 463[80], 464[80], 466[115], 467[115]; **6**, 614[79]
Rose-Munch, F., **4**, 523[56], 527[69,70], 528[69-71]
Rosen, B. I., **5**, 718[95]
Rosen, H., **8**, 532[130]
Rosen, M., **5**, 687[58]
Rosen, O. M., **2**, 466[122], 469[122]
Rosen, P., **2**, 106[50], 184[26]; **3**, 2[8], 11[8], 16[8], 17[8], 26[8]; **4**, 240[39], 254[39]; **7**, 728[42]; **8**, 527[39,40], 647[57]
Rosen, R. E., **4**, 187[97]
Rosen, S., **7**, 501[252]
Rosen, T., **1**, 772[198]; **3**, 216[76]; **4**, 10[34]; **6**, 646[100a]; **8**, 384[40], 925[6], 946[136]
Rosen, W., **4**, 7[24]
Rosenbaum, D. E., **7**, 473[27]
Rosenberg, A., **2**, 463[87]
Rosenberg, D. W., **7**, 92[48]
Rosenberg, G., **6**, 642[72]
Rosenberg, S. H., **4**, 29[86], 46[86]
Rosenberger, M., **1**, 821[24,29], 825[24]; **3**, 168[491], 169[491], 171[491], 289[69], 757[122]
Rosenblatt, D. H., **7**, 40[14], 222[38], 736[5], 737[5], 745[5], 746[5], 749[5]; **8**, 568[477]
Rosenblatt, E. F., **8**, 142[53]
Rosenblum, L. D., **1**, 116[45], 128[45], 433[139], 434[139]; **3**, 419[43]
Rosenblum, M., **1**, 308[95], 314[95]; **2**, 934[143]; **3**, 218[101], 219[102], 1036[81]; **4**, 48[137], 401[232], 562[36], 576[13-18], 579[23], 695[3]; **5**, 272[3-5], 273[4,6], 274[7-9], 275[4,7,10], 277[6-8,10], 279[6,7], 281[20]; **6**, 686[365]; **8**, 890[138], 950[167]
Rosenblum, S. B., **7**, 208[85]
Rosenbrook, W., Jr., **6**, 57[136]
Rosencrantz, D. R., **8**, 408[61]
Rosendahl, F. K., **2**, 1089[54,56]
Rosendahl, K., **2**, 1087[33]
Rosenfeld, D. D., **3**, 892[51]
Rosenfeld, J. C., **4**, 1016[208]
Rosenfeld, M. J., **4**, 1054[132]
Rosenfeld, R. S., **8**, 991[51]
Rosenfelder, W. J., **7**, 92[49]
Rosenfeldt, F., **1**, 162[92]
Rosenheim, O., **7**, 86[16a]
Rosenkilde, S., **6**, 480[110]
Rosenkranz, G., **3**, 805[15]; **6**, 685[357]; **7**, 92[42], 93[42]; **8**, 108[1], 118[1], 528[65], 566[450]
Rosenman, H., **7**, 107[168]
Rosenmund, K. W., **8**, 286[11]
Rosenquist, N. R., **4**, 285[157], 483[7]
Rosenthal, D., **5**, 163[71]; **7**, 113[200], 139[128]; **8**, 338[87]
Rosenthal, J. W., **8**, 92[66]
Rosenthal, R. J., **3**, 1048[13]; **5**, 10[78], 67[91,92], 205[43]; **6**, 121[130]
Rosenthal, S., **7**, 261[72]; **8**, 708[39]
Rosenthale, M. E., **4**, 932[63]
Rosentreter, U., **5**, 945[247]; **8**, 589[47]
Roser, J., **5**, 929[170,172], 932[170,172]
Rosevear, D. T., **4**, 27[78]
Rosevear, J., **3**, 329[186]
Rosewater, W. H., **3**, 499[114]
Roshchina, L. F., **6**, 554[776,780,793]
Rosi, D., **2**, 758[25], 839[173]; **7**, 75[113]
Rosich, R. S., **7**, 254[27]
Rosini, C., **5**, 1152[144]
Rosini, G., **2**, 321[19], 323[19,38,39], 324[19], 330[19,51], 331[19], 332[19,51], 333[19,51], 338[76], 598[12]; **4**, 13[44], 86[78b,79b], 146[36]; **5**, 487[188], 829[15]; **6**, 490[114], 938[128], 942[128], 944[128]; **7**, 262[77]; **8**, 70[234], 354[173,178], 357[178]
Roska, A., **2**, 345[23]
Roskamp, E. J., **3**, 579[102], 596[196], 597[196], 610[102], 942[81b], 1008[67]; **6**, 129[168], 874[15]
Rosnati, V., **4**, 1057[145]
Rösner, M., **3**, 553[15]; **5**, 945[247]
Ross, C. B., **2**, 530[21]
Ross, D. S., **6**, 110[56]
Ross, F. P., **8**, 681[76], 689[76]
Ross, G. A., **4**, 872[37]
Ross, J. A., **5**, 741[153]
Ross, K. H., **8**, 881[79]
Ross, L. O., **8**, 267[69]
Ross, M., **4**, 262[311]
Ross, N. C., **4**, 31[92,92a]
Ross, P. A., **3**, 770[173]
Ross, R. J., **1**, 683[226], 684[226], 685[226], 705[226], 714[226], 717[226], 718[226], 719[226], 720[226], 722[226], 868[82], 869[82]; **3**, 786[41]; **7**, 377[91], 378[91b]
Ross, S. D., **2**, 971[93]; **3**, 634[21], 635[38], 649[206,206b], 655[21]; **6**, 572[959]; **7**, 803[57], 804[61], 805[64]; **8**, 364[14,15]
Ross, W. C. J., **3**, 341[1]
Ross, W. J., **7**, 401[58]
Rossano, L. T., **1**, 131[99]; **3**, 1007[64], 1008[64a]; **6**, 859[174], 887[62]
Rossazza, J. P. N., **7**, 62[53,53b,c]
Rosseels, G., **2**, 742[75]
Rossel, J., **4**, 383[136]
Rosser, A., **3**, 634[30], 644[30b]
Rosser, R., **1**, 496[48], 497[48]
Rosser, R. M., **8**, 54[157], 66[157]
Rossert, M., **5**, 423[90]; **6**, 512[301], 517[301]
Rossi, A., **8**, 542[220]
Rossi, A. R., **5**, 196[12]
Rossi, G., **7**, 410[95]
Rossi, J. C., **4**, 302[334]
Rossi, L. M., **6**, 582[993]
Rossi, M., **7**, 777[380]; **8**, 450[171]
Rossi, R., **3**, 217[91], 221[128], 439[37], 489[58], 495[58], 505[156], 511[58], 515[58], 525[38-40], 527[45], 539[45,103,104], 541[45,105], 554[27]; **7**, 453[81]; **8**, 743[49]
Rossi, R. A., **4**, 426[56], 452[18], 453[18,24,25,34,35], 454[18,35,39-41], 457[18,41,52,54,55,57-59], 458[40,58], 459[39], 460[98], 461[25,39-41,54,102,103], 462[25,41,103], 463[58,109,111], 464[58,102,111], 465[102], 466[124,128], 467[102,128], 468[18,41,133,135], 469[39,52,124], 470[137], 471[18,109,124], 472[34,124,143], 473[146], 474[24,25,58,146], 475[102,103,111], 476[35,54,55,153,154,156,157], 477[98], 502[123], 765[228], 1021[243]
Rossier, J.-C., **5**, 157[39]
Rossini, G., **4**, 115[182]; **6**, 104[10]
Rossini, S., **2**, 807[48]
Rossiter, B. E., **2**, 338[77]; **5**, 185[163]; **7**, 364[42], 368[42], 375[79], 390[5], 394[19], 400[41], 407[41], 415[113], 419[132]
Rosslein, L., **7**, 429[151]
Rossman, P., **6**, 543[621]
Rossy, P., **2**, 1103[130]; **6**, 37[33]; **8**, 801[70], 813[11], 938[89]
Rost, W., **5**, 441[176]
Roswell, D. F., **6**, 291[199]
Roszkowski, A., **4**, 932[62]
Rotello, V., **5**, 162[67]; **6**, 734[10], 735[10]; **7**, 407[80]
Rotermund, G. W., **7**, 596[41]
Roth, B., **1**, 447[198]; **2**, 871[23]; **3**, 301[48]; **4**, 111[154a,g], 573[6,8], 614[375,376], 841[38], 905[208]

Roth, B. D., **4**, 36[102,102g]; **8**, 248[86]
Roth, C., **8**, 13[74]
Roth, G. P., **1**, 359[22], 383[22], 384[22], 463[27]; **3**, 255[107]; **4**, 77[50], 206[49]; **8**, 460[249]
Roth, H. D., **3**, 891[46]; **4**, 1002[57]; **5**, 704[22]; **7**, 851[28], 854[60], 875[113]
Roth, H. J., **2**, 957[21]; **3**, 737[30]; **7**, 41[27]
Roth, K., **3**, 644[159]; **6**, 134[33], 176[85], 182[85], 187[177], 189[186], 193[85,217], 195[222], 558[839]; **8**, 860[223]
Roth, M., **2**, 368[241], 866[7], 867[7], 870[7], 871[7], 872[7], 875[7], 876[7]; **8**, 530[91]
Roth, W., **8**, 904[57,57b], 910[57]
Roth, W. D., **5**, 10[76]
Roth, W. R., **5**, 64[29], 589[213], 709[46], 714[68], 715[80], 820[160], 857[224,228], 908[71], 911[96], 912[96], 971[1]
Roth, Z., **3**, 229[231]; **4**, 615[385], 616[385], 619[385], 620[385], 621[385], 626[385], 637[385]
Rothberg, I., **4**, 301[329]; **6**, 980[36]; **8**, 880[62]
Rothe, W., **6**, 294[234]
Rothenberg, A. S., **5**, 557[56]
Rothenberg, S. D., **4**, 434[131]
Rothenberger, S. D., **3**, 223[149]; **7**, 364[41a]
Rothman, A. M., **4**, 878[74]
Rothman, E. S., **4**, 313[475]; **6**, 57[135]
Rothman, L. A., **8**, 902[43], 907[43], 908[43]
Rothrock, T. S., **3**, 723[8], 731[8]
Rotman, A., **7**, 40[9]
Rotscheidt, K., **5**, 412[47,47b]
Rotstein, D., **8**, 803[93], 804[93], 826[69]
Röttele, H., **5**, 717[90a,b]
Rotunno, D., **3**, 230[236]
Roubineau, A., **2**, 609[80]
Roudier, J.-F., **2**, 538[57]
Rouessac, A., **5**, 451[38], 552[6]
Rouessac, F., **3**, 20[104], 342[10]; **4**, 91[90]; **5**, 324[23], 451[38], 552[6], 553[48], 555[50], 561[81-83], 577[148]; **6**, 150[127,130], 689[385]; **7**, 406[86]; **8**, 543[246,248,249]
Rougier, M., **3**, 572[66]
Rougny, A., **5**, 473[153], 477[153]
Roulet, D., **1**, 155[65]
Roulet, R., **7**, 257[49]
Roullier, L., **6**, 176[83]; **8**, 322[111]
Roumestant, M.-L., **2**, 547[95]; **5**, 772[159,160,161,163], 797[62], 821[62]; **6**, 210[76], 214[98]
Rouse, R. A., **5**, 571[116]
Roush, D., **4**, 1097[169]
Roush, D. M., **3**, 178[542], 179[542]; **5**, 8[57,61,62], 12[88], 13[89], 461[104]
Roush, W. R., **1**, 192[81], 764[149], 769[182], 770[186], 800[300]; **2**, 6[25], 8[37], 12[25], 13[25,37,57,59], 20[37a], 25[98], 26[25,102], 27[25,102], 30[25,98], 31[25,98,115], 35[37,125,126], 41[25,125,126], 42[25,98,126,144,145,149], 43[98,147], 44[98,126,147], 45[115,149], 46[152]; **4**, 893[156]; **5**, 513[5], 514[5], 515[11,12,16], 516[12a,24], 517[11a,24], 518[16a,b,24], 519[11a,16b], 520[36], 524[16,50,50b,52], 526[11,16b,58], 527[5], 532[88], 534[91], 538[12a,103], 539[50,50b,58], 540[58], 541[58], 543[11a], 545[119], 548[50b,c]; **6**, 7[30], 89[114,117], 832[16], 833[20], 843[86], 864[197], 995[100]; **7**, 358[9], 371[69], 400[53], 401[62], 410[94], 415[110], 418[129a,b]
Rousse, G., **3**, 246[38], 470[221], 471[221]
Rousseau, G., **1**, 879[111e]; **2**, 444[20], 804[41]; **4**, 164[100]; **5**, 63[13]
Roussel, J., **4**, 401[230]
Roussel, M., **7**, 452[42]
Rousset, C. J., **4**, 602[263], 644[263]; **5**, 47[40], 1180[47], 1181[47]
Roussi, G., **4**, 459[73], 464[73], 465[73], 466[73], 478[73,166,167], 479[169,170,171], 877[66], 1089[126]; **7**, 878[140]
Rouwette, P. H. F. M., **3**, 174[524], 175[524]; **4**, 14[47,47k]; **6**, 489[95]
Roux, A., **2**, 414[13]
Roux, D. G., **3**, 831[61]
Roux, J. P., **1**, 38[259]
Roux-Schmitt, M.-C., **1**, 235[29]; **2**, 414[13,14], 428[44]; **4**, 73[32], 111[156], 112[159], 113[159b], 259[270]
Rove, J. M., **4**, 587[36]
Rover-Kevers, M., **6**, 496[156], 514[156]
Rovnyak, G., **5**, 925[150]
Rowe, C. D., **6**, 714[85]; **7**, 35[105]; **8**, 385[46]
Rowe, J. E., **6**, 532[470], 539[470]
Rowe, K., **3**, 798[96,97]; **7**, 595[22], 600[75]; **8**, 711[69], 717[101]
Rowe, W., **2**, 364[207]
Rowe-Smith, P., **3**, 824[18]
Rowland, A. T., **7**, 654[7]
Rowland, C., **5**, 72[174]
Rowland, R. L., **3**, 407[150]; **4**, 315[504]
Rowland, S. J., **1**, 248[63]
Rowlands, M., **1**, 499[54], 501[54]; **3**, 261[152], 514[211]; **7**, 596[34]
Rowlands, R. T., **7**, 69[88]
Rowlett, R. J., Jr., **8**, 904[58]
Rowley, A. G., **1**, 755[115], 812[115], 813[115]; **6**, 173[36], 174[36], 199[36]
Rowley, E. G., **4**, 72[29]; **5**, 1057[52], 1062[52]
Rowley, M., **1**, 200[97]; **2**, 583[114], 587[144]; **4**, 253[173], 257[173]; **5**, 844[127]; **7**, 376[84]; **8**, 99[107]
Rowley, R. J., **2**, 567[24], 587[24]
Roy, B. C., **2**, 760[46]
Roy, B. L., **2**, 303[6]
Roy, D. A., **6**, 208[55], 212[55]
Roy, G., **1**, 737[31], 828[79], 829[86]; **2**, 597[6]; **3**, 198[51]; **6**, 539[578]
Roy, J., **7**, 771[268], 772[268]; **8**, 231[144]
Roy, M. A., **8**, 857[198]
Roy, N., **7**, 267[117], 268[117]
Roy, R. G., **3**, 846[45]; **6**, 967[101]
Roy, S., **1**, 534[143]; **5**, 432[133], 433[133a]
Roy, S. C., **4**, 1040[80], 1043[80]
Royall, S. E., **5**, 571[120]
Royals, E. E., **3**, 422[66]
Royer, G. P., **6**, 635[15], 636[15], 668[250]; **8**, 959[22]
Royer, J., **1**, 555[119], 558[137], 559[142,143,144,146,147,148]
Royer, R., **2**, 332[54]
Royer, Y., **8**, 663[118]
Royo, G., **8**, 262[18]
Rozé, J. C., **6**, 554[751,752]; **8**, 135[47], 136[47], 658[100]
Rozeboom, M. D., **4**, 47[134]; **5**, 203[39], 204[39h,i], 209[39], 210[39], 788[15]
Rozema, M. J., **1**, 213[17,17b,18]
Rozen, S., **3**, 124[264]; **4**, 344[77], 347[92,94,98]; **6**, 217[111]; **7**, 15[145]
Rozenberg, S. G., **8**, 618[121], 619[121], 621[142], 627[176]
Rozhinskii, Yu. I., **6**, 509[255]
Rozhkov, I. N., **7**, 800[35]
Rozhkova, T. I., **8**, 187[48]
Rozing, G., **2**, 823[118]; **4**, 1040[75]
Rozinov, V. G., **6**, 495[150]
Rozwadowska, M. D., **1**, 544[33], 551[33], 552[79], 564[203]
Rozwadowski, J., **2**, 537[50]
Rozzell, J. D., **8**, 561[410]
Rozzell, J. D., Jr., **7**, 672[17]
Rua, L., **8**, 355[179]
Ruback, W., **6**, 509[251]
Ruben, R. A., **4**, 453[24], 474[24]

Rubenstein, K. E., **4**, 473[145]
Rubin, I. B., **2**, 1099[115]
Rubin, M. B., **3**, 597[198], 628[50]; **5**, 713[64]
Rubino, M. R., **3**, 194[15], 196[15]
Rubinskaya, T. Ya., **8**, 611[64]
Rubinstein, M., **6**, 624[151], 658[188]
Rubio, A., **6**, 149[105]; **8**, 15[91]
Rubiralta, M., **1**, 564[206]
Rubottom, G. M., **1**, 411[47]; **2**, 445[23]; **4**, 30[89]; **5**, 386[132], 387[132b]; **6**, 1064[88]; **7**, 121[25], 163[75], 165[83], 167[93,94], 177[145,146], 178[149], 182[162,164], 185[176], 186[179], 476[58], 481[58], 530[19], 531[19], 673[24], 816[9], 824[9,42], 827[9], 851[21]; **8**, 269[91], 986[15]
Ruccio, M., **8**, 663[117]
Ruch, E., **2**, 1090[65], 1092[65], 1108[65]
Rüchardt, C., **4**, 487[43], 717[12], 758[191], 876[58]; **6**, 294[241,242]; **7**, 720[9,11]
Rücker, C., **2**, 91[45]; **5**, 71[125]; **6**, 556[819]
Rücker, D., **2**, 1090[68]
Rücker, E., **6**, 43[51]
Rücker, G., **4**, 883[100], 884[100]
Ruckle, R., **5**, 411[43], 552[39]
Ruckle, R. E., Jr., **3**, 1052[28], 1053[28], 1055[28], 1057[28], 1062[47]; **4**, 824[240], 1033[32], 1040[100], 1056[100]; **5**, 266[75], 268[75], 927[161], 931[161]; **6**, 123[141], 125[141], 127[158]
Rudakov, E. S., **7**, 12[98]
Rudashevskaya, T. Yu., **5**, 33[8]; **7**, 595[20]
Rudaya, L. I., **8**, 661[113]
Rudd, E. F., **3**, 634[21], 635[38], 655[21]
Rudd, E. J., **7**, 804[61]
Ruddick, J. D., **8**, 445[21], 456[21]
Ruddock, K. S., **3**, 510[184]
Rüden, E., **5**, 453[59]
Ruden, R., **2**, 160[132]
Ruden, R. A., **2**, 662[17], 664[17]; **4**, 18[62], 20[62b], 171[25], 181[75]; **5**, 328[29], 433[134]; **6**, 677[316]
Ruder, S. M., **7**, 414[119]
Rudi, A., **5**, 603[52], 604[52]
Rudis, J. A., **4**, 443[189]
Rudisill, D. E., **3**, 232[269], 529[49]; **4**, 411[266c], 696[5]
Rudler, H., **2**, 523[79], 524[79]; **4**, 980[115], 982[112,115]; **5**, 1066[9], 1076[40], 1079[50], 1086[68], 1103[150], 1104[150,158], 1105[159,161,162,163]; **8**, 252[111]
Rudler-Chauvin, M., **2**, 159[127]; **5**, 1079[50]
Rudolf, K., **5**, 680[22], 683[22]
Rudolph, W., **8**, 301[90]
Rudorf, W.-D., **4**, 53[149], 129[223b]
Rudrow, E. A., **6**, 245[124]
Rudzik, A. D., **8**, 53[129], 66[129]
Ruediger, E. H., **4**, 1040[78]; **5**, 803[91], 983[31], 987[41], 988[41]; **6**, 819[110]
Rüeger, H., **1**, 275[77]; **4**, 795[79]; **6**, 755[121]
Rueger, W., **6**, 767[25]
Ruegg, H. J., **4**, 915[9]
Rüegg, R., **2**, 612[105]; **3**, 698[159]; **6**, 965[85]
Ruel, O., **3**, 252[86]; **4**, 102[128d]
Ruest, L., **1**, 853[50], 876[50]; **2**, 303[6]; **4**, 308[406], 395[206], 1040[84], 1043[84]
Ruf, H.-R., **4**, 764[222], 765[222], 808[155]
Rüf, W., **2**, 367[229]
Ruff, F., **7**, 764[122], 777[389]
Ruffer, U., **5**, 416[58]
Rufinska, H., **1**, 2[10]
Ruge, B., **5**, 229[117,118]
Rüger, W., **7**, 358[14], 695[35]
Ruggeri, M. V., **5**, 476[147]
Ruggeri, R., **1**, 755[116], 756[116,116d], 758[116], 761[116]; **5**, 436[158,158g], 442[158]
Ruggeri, R. B., **2**, 123[199], 282[39], 284[39], 1024[63]; **4**, 239[25], 247[25], 258[25], 260[25]
Ruggli, P., **2**, 399[18]
Ruhemann, S., **4**, 41[117], 47[133]
Ruhland, B., **4**, 229[225,228]
Ruhlen, J. L., **5**, 127[23]
Rühlmann, A., **2**, 373[274]; **5**, 14[99]
Rühlmann, K., **2**, 443[18]; **3**, 616[11], 618[11], 619[11], 620[11], 623[11], 625[11]; **5**, 1200[50]
Ruhm, D., **2**, 138[22]
Ruhoff, J. R., **8**, 243[46]
Ruhr, M., **6**, 943[156]
Rui, T., **8**, 336[79], 337[79]
Ruicheng, R., **3**, 298[26]
Ruitenberg, K., **2**, 85[19,21], 587[147,148], 589[153]; **3**, 491[68], 531[84]
Ruiz, M. O., **6**, 835[44]; **8**, 886[109]
Ruiz-Hitzky, E., **3**, 770[179]
Ruiz Montes, J., **3**, 47[256]
Ruiz-Perez, C., **4**, 373[87]; **7**, 413[118]
Ruiz-Sanchez, J., **4**, 356[136]
Rukachaisirikul, V., **3**, 154[422], 155[422]
Ruland, A., **6**, 174[64]
RuLin, F., **5**, 239[1], 903[41], 905[41], 907[41], 909[41], 916[41], 918[41], 937[41], 939[41], 940[41,225], 951[41], 963[225]
Rumanowski, E. J., **6**, 431[107]
Rumbaur, G., **6**, 524[355], 525[355], 532[355]
Rumin, R., **5**, 709[45], 710[52], 739[45b]; **8**, 782[100]
Rummel, S., **4**, 485[29]
Rumpf, P., **8**, 587[39], 663[118]
Rund, J. V., **8**, 650[66]
Rundel, W., **1**, 391[150]
Rundle, R. E., **1**, 13[66,71], 25[130]
Runge, T. A., **2**, 23[89]; **4**, 651[429]; **5**, 847[134]
Runquist, A. W., **2**, 614[116]; **7**, 320[63], 841[10]; **8**, 564[438]
Runsink, J., **2**, 655[133]; **5**, 160[57], 161[62], 185[161,162,165,167], 186[171]; **6**, 558[851,852]; **7**, 262[76]
Rupaner, R., **4**, 735[86,87]; **7**, 399[36]
Rupani, P., **7**, 595[25], 598[25]
Rupe, H., **4**, 125[216,216a]
Rupert, J. P., **7**, 845[72]
Rupp, R. H., **6**, 538[569], 765[17]; **7**, 64[64], 694[32]
Ruppert, J. F., **4**, 1040[97,98], 1043[97,98]
Ruppert, R., **8**, 85[17]
Ruppin, C., **5**, 1140[77]
Ruqidka, V., **8**, 424[42], 425[42]
Rusch, G. M., **1**, 873[93]; **3**, 757[120]
Rusek, J. J., **2**, 602[38]
Rusek, P. E., **8**, 990[42]
Rush, P. K., **4**, 696[7]
Rushkes, A. M., **7**, 7[47]
Rusiecki, V., **6**, 93[133]
Rusik, C. A., **3**, 47[257]
Rusina, M. N., **6**, 490[106]
Rusling, J. F., **3**, 566[32]
Russ, M., **7**, 236[30]; **8**, 405[33], 406[37]
Russ, P. L., **8**, 228[126], 249[94]
Russegger, P., **1**, 286[10]
Russell, A., **2**, 770[8]
Russell, A. T., **2**, 117[155], 309[25]; **7**, 412[106]
Russell, C. E., **2**, 596[2]; **3**, 485[38], 486[38], 491[38], 492[38], 495[38], 503[38]

Russell, C. G., **1**, 699[247], 700[256,257], 882[123]; **3**, 107[226], 109[226]; **4**, 365[4], 370[4], 380[4], 381[4]; **5**, 810[128], 812[128]; **6**, 470[58], 980[31]; **8**, 847[97,97d], 849[97d,107,112]
Russell, D. N., **4**, 1053[130]
Russell, D. R., **7**, 481[110]; **8**, 674[33]
Russell, G. A., **3**, 127[321]; **4**, 452[3,5], 453[36], 455[36], 472[36], 726[52], 741[120], 744[132,136], 746[143], 747[148], 768[238], 771[254]; **6**, 679[325], 832[14]; **7**, 196[12], 215[12], 222[39], 882[173], 884[189]; **8**, 843[58], 852[136], 857[201]
Russell, G. B., **1**, 337[80], 828[69]
Russell, G. E., **8**, 185[15]
Russell, J. R., **4**, 348[108], 349[108b]
Russell, L. J., **4**, 691[76]
Russell, M. A., **2**, 332[57]; **4**, 384[142]
Russell, M. J., **8**, 454[200]
Russell, R. A., **1**, 554[102]; **4**, 14[47,47l]; **7**, 380[102], 821[27]
Russell, R. J., **5**, 829[21]
Russell, R. K., **3**, 717[45], 752[94]; **5**, 10[76]; **8**, 946[137]
Russell, S. T., **6**, 901[120]
Russell, T. J., **6**, 667[236]
Russell, T. W., **8**, 140[16], 568[470], 956[7]
Russkamp, P., **2**, 282[37]; **6**, 176[105]
Russmann, H., **2**, 900[24]
Russo, A. J., **8**, 726[189]
Russo, G., **4**, 38[108], 379[115,116,118], 380[115j], 382[115k]; **7**, 153[9], 274[138]
Russo, H. F., **2**, 971[92]
Russo, T. J., **1**, 3[18,20,21], 42[20c]
Russo, U., **1**, 305[84,85], 323[84]
Russowsky, D., **2**, 934[144], 940[144]
Rust, F. F., **7**, 10[82]
Rustamov, K. M., **8**, 772[56]
Rusterholz, D. B., **6**, 645[97]; **8**, 146[97]
Ruston, S., **1**, 329[33], 793[273], 804[273]; **6**, 989[77], 990[77], 993[77], 1002[77]
Rutavicius, A. I., **6**, 570[944]
Rüter, J., **3**, 890[28], 894[58], 905[28,58]; **6**, 129[164]
Ruth, J. A., **3**, 394[96]; **6**, 1055[51]; **8**, 542[220]
Ruth, J. L., **3**, 470[218], 476[218]
Ruther, F., **7**, 86[15]
Ruther, M., **2**, 354[103], 357[103], 369[255b], 373[103,274], 374[255b]; **5**, 14[99], 17[120]
Rutherford, K. G., **6**, 220[128]; **8**, 916[104,105], 917[104], 918[104], 919[104], 920[104]
Rutjes, F. P. J. T., **6**, 118[99]
Rutkowski, A. J., **8**, 754[84]
Rutledge, M. C., **8**, 72[243], 74[243], 393[110]
Rutledge, P. S., **1**, 753[102]; **3**, 672[64], 675[74], 759[126]; **4**, 347[96], 350[121], 351[126], 354[126-128], 369[21,22], 370[21,22], 371[21], 377[21,22,101], 545[126]; **7**, 92[40], 121[24], 438[15,16], 445[15,16], 447[16], 502[261], 530[20], 531[20], 706[25]; **8**, 944[123]
Rutledge, T. F., **3**, 273[8], 551[2], 552[2]; **4**, 3[10], 41[10], 47[10], 66[10,10f], 67[10f]; **6**, 962[75]
Rutsch, W., **6**, 5[24]
Rutschmann, S., **5**, 386[133,133a], 692[104]
Rüttimann, A., **2**, 143[52]
Ruttinger, R., **7**, 449[1], 450[1]
Ruyle, W. V., **7**, 92[48]
Ruzicka, L., **3**, 341[2], 360[2], 705[3], 781[18]; **4**, 239[34]; **6**, 685[357]; **8**, 328[15]
Ruzziconi, R., **4**, 763[209]; **7**, 765[151]
Ryabov, A. D., **3**, 499[125], 669[53]
Ryan, C. M., **8**, 624[154], 628[154]
Ryan, G., **8**, 137[54]
Ryan, J. D., **3**, 358[68]
Ryan, J. J., **5**, 92[60]; **6**, 291[218]
Ryan, J. W., **8**, 771[49], 776[78]
Ryan, K., **2**, 803[33]
Ryan, K. J., **8**, 819[44]
Ryan, K. M., **3**, 45[247]; **4**, 230[251]; **5**, 410[41]; **7**, 416[122]
Ryan, M. D., **2**, 430[53]; **3**, 748[77]; **4**, 342[63]; **7**, 520[28], 765[161]
Ryan, R. C., **8**, 457[217]
Ryan, R. J., **6**, 271[86,87]
Ryang, M., **2**, 451[55]; **3**, 483[8], 554[20,21]; **5**, 1138[63]
Ryback, G., **5**, 418[70]
Rybak, W. K., **7**, 95[73a]
Rybakova, L. F., **1**, 276[78], 277[78]
Rybakova, N. A., **7**, 500[236]
Rybchenko, L. I., **6**, 552[698]
Rybczynski, P. J., **4**, 905[213]
Rybin, L. V., **4**, 115[180b]
Rybinskaya, M. I., **4**, 115[180b]
Rychnovsky, S. D., **4**, 383[137], 792[68]
Ryckman, D. M., **1**, 188[95], 198[95]; **8**, 34[59], 66[59]
Rydberg, D. B., **4**, 980[115], 982[115]; **5**, 1104[160]
Ryde-Petterson, G., **3**, 637[63]
Ryder, D. J., **8**, 237[16], 244[16], 253[16], 709[46], 710[54], 721[54]
Rydjeski, D. R., **3**, 14[77], 15[77]
Rydon, H. N., **4**, 282[136], 283[146]; **6**, 213[89], 639[51], 666[51], 667[51]
Rydzewski, R. M., **7**, 255[35]
Rykowski, A., **4**, 424[14], 426[48], 432[103,104]
Rylander, P. N., **6**, 651[136,136d], 724[158]; **7**, 236[12], 564[110], 572[114]; **8**, 139[4,7], 140[15,18], 141[29], 236[1], 239[1,26], 242[1], 246[1], 248[1], 367[55], 382[4], 383[4], 384[4], 388[4], 396[4], 418[3], 420[3], 422[3], 423[3], 424[3], 430[3], 431[3], 433[3], 436[3], 438[3], 439[3], 459[228], 533[135,141], 597[90,92], 598[90], 794[5]
Rynbrandt, R. H., **2**, 1049[19]; **3**, 883[109]; **5**, 71[164]
Ryono, L. S., **3**, 126[319], 500[132], 505[132]; **6**, 134[38]
Ryrfors, L.-O., **2**, 146[65]
Rytina, A. W., **5**, 513[2], 518[2]
Rytz, G., **4**, 130[226a]
Ryu, I., **2**, 441[3], 442[11], 443[15,17], 445[24,25], 449[3], 450[54], 451[15,55]; **4**, 115[177], 973[85]; **5**, 442[185,185a], 461[107], 464[107], 466[107]; **6**, 684[344]; **7**, 137[118], 138[118]
Ryzhkina, T. E., **7**, 699[57]
Ryzhov, M. G., **4**, 218[147]
Ryzhova, G. L., **8**, 629[182,183]
Rzepa, H. S., **4**, 1063[168,169], 1070[15]; **5**, 856[197]; **8**, 86[22], 87[29]
Rzucidlo, E., **4**, 45[126]

S

Saa, C., **4**, 513[179,180]; **5**, 384[127]
Saa, J. M., **3**, 585[133], 591[171]; **4**, 45[126], 505[139], 513[179,180]; **5**, 384[127]; **6**, 89[110], 487[76], 489[76]; **7**, 334[27], 346[8]; **8**, 314[30]
Saadatmandi, A., **6**, 189[188]
Saak, W., **2**, 853[230]
Saakyan, A. G., **6**, 507[237], 515[237]
Saalbaum, H., **6**, 187[176]
Saalfrank, R. W., **2**, 97[60]; **5**, 441[176]; **6**, 193[218], 247[133], 558[840,841,842,843,844]
Saavedra, J. E., **1**, 357[3], 476[119], 477[119]; **3**, 65[8]; **7**, 225[58,60], 280[167]; **8**, 251[103]
Saba, A., **4**, 930[50], 1057[145,146]
Saba, S., **5**, 164[75], 176[75]
Sabacky, M. J., **2**, 233[184]; **8**, 459[230,232], 460[230,232], 535[166]
Sabadie, J., **8**, 551[344]
Sabari, M., **7**, 314[41], 315[41,42]
Sabat, M., **2**, 655[149]; **5**, 1068[13]; **8**, 670[9], 671[9]
Sabatier, P., **8**, 285[6]
Sabatucci, J. P., **5**, 946[257]; **7**, 567[104]
Sabbioni, G., **4**, 1008[134]; **6**, 562[889,890]
Sabeena, M. S., **8**, 249[96]
Sabel, A., **7**, 449[2], 450[2]
Sabelus, G., **5**, 552[23]
Sabesan, S. I., **6**, 438[43]
Sable, H. Z., **3**, 733[1]
Sabnis, S. D., **7**, 558[79], 560[79]
Sabo, E. F., **6**, 216[106]; **7**, 139[128]
Sabol, J. S., **5**, 839[77,84]
Sabol, M. A., **8**, 249[91], 294[54]
Sabol, M. R., **2**, 690[70]; **7**, 174[140]; **8**, 333[57]
Sabourault, B., **1**, 86[40]
Sabourin, E., **7**, 197[16]; **8**, 764[4b], 774[71]
Saboz, J. A., **5**, 223[66]
Sabry, S., **7**, 69[87]
Sabuni, M., **7**, 505[283,284]
Saburi, M., **5**, 418[71], 1175[38], 1177[43], 1178[38,43], 1180[47], 1181[47]; **6**, 534[516]; **8**, 239[28], 395[131]
Saccarello, M. L., **6**, 555[814]
Saccomano, N. A., **4**, 24[73,73b,c], 218[141,142]; **5**, 519[30]
Sacerdoti, M., **1**, 309[101]
Sacha, A., **2**, 365[215]
Sachdev, H. S., **1**, 636[99]; **3**, 136[375], 141[375]; **6**, 643[79], 658[184], 1059[67]; **8**, 388[60], 615[95]
Sachdev, K., **1**, 636[99]; **3**, 136[375], 141[375]; **5**, 165[79]
Sachdeva, Y. P., **4**, 457[50], 462[104], 463[110], 465[104], 466[104], 468[104,110], 469[104,110], 477[50], 503[125]
Sachleben, A., **2**, 1096[97]
Sachs, D. H., **3**, 828[49]
Sachs, W. H., **2**, 145[64]
Sachtler, W. M. H., **8**, 150[139]
Sacki, S., **4**, 429[84,85]
Sacks, C. E., **2**, 509[33]; **3**, 135[363], 136[363], 139[363], 142[363], 156[363]
Sackville, M. A., **1**, 569[251]; **3**, 125[301], 126[301], 128[301], 129[301], 133[301]
Sacripante, G., **4**, 740[118,119]
Sada, I., **5**, 108[206]
Sadana, K. L., **6**, 656[170]
Saddler, J. C., **1**, 343[119]; **4**, 79[55a], 251[152]; **5**, 936[194]; **6**, 163[192]
Saddler, J. S., **6**, 9[43]
Sadee, W., **7**, 232[157]
Sadeghi, M. M., **6**, 79[62]
Sadekov, I. D., **7**, 774[325,335]
Sadhu, K. M., **1**, 830[93]; **2**, 13[56], 39[138]; **3**, 796[88]; **6**, 77[58], 98[58]
Sadikh-Sadé, S. I., **8**, 556[378]
Sadikov, G. B., **7**, 599[64]
Sadle, A., **2**, 957[15]
Sadler, A. C., **6**, 784[92]
Sadler, D. E., **5**, 216[12,13], 219[12,13], 221[12,13]
Sadler, P., **4**, 1014[190]
Sado, M., **8**, 609[49]
Sadovaya, N. K., **7**, 494[202]
Sadozai, K. K., **6**, 51[108], 53[108]
Sadozai, S. K., **8**, 185[12]
Sadri, A. R., **2**, 740[52]
Sadykh-Zade, S. I., **3**, 304[60]; **8**, 556[376], 771[46], 782[99]
Saebö, J., **6**, 504[222]
Saeedi-Ghomi, M. H., **6**, 1054[47]
Saeed-ur-Reiman, **8**, 153[188], 155[188]
Saegebarth, K. A., **7**, 558[78], 562[78]
Saegusa, K., **1**, 759[128,129]
Saegusa, T., **1**, 85[28,29], 544[44], 551[77], 880[113]; **2**, 114[115-117], 444[21,22]; **3**, 217[86], 259[137], 450[105]; **4**, 162[93,94a], 221[160], 241[60], 254[60], 260[60], 591[112], 592[126,127], 600[241], 611[354,359], 613[370], 614[372,373], 617[126], 618[126], 633[112,126,127], 840[35-37], 905[207]; **5**, 282[21,22], 386[134], 391[134], 392[134], 473[151], 479[151], 595[12], 693[112], 935[191], 936[191], 1157[170,171], 1183[56]; **6**, 88[103], 238[72], 295[251], 533[509,512], 540[509], 546[509], 551[680], 717[108], 757[132], 1007[149]; **7**, 141[133], 144[133], 530[29]; **8**, 548[324], 549[324,325], 830[88]
Saeki, H., **6**, 668[260]
Saeki, S., **2**, 348[63]; **7**, 672[18]
Saeki, T., **8**, 191[91]
Saeman, M., **4**, 532[91], 534[91], 537[91]
Saengchantara, S. T., **4**, 215[118]; **6**, 152[140], 153[140]
Saeva, F. D., **5**, 405[21], 850[156]; **7**, 854[48], 855[48]
Safarik, I., **3**, 891[43], 892[43]
Saffhill, R., **6**, 602[3], 650[127]
Safronova, Z. V., **6**, 498[169]
Saga, H., **2**, 736[24]
Sagara, S., **5**, 636[100]
Sagawa, Y., **2**, 633[31], 640[31]; **4**, 30[90], 159[83], 161[90], 258[238], 261[238]
Sage, J.-M., **5**, 326[25], 815[141]
Sager, W., **2**, 1099[111]; **6**, 46[63]
Sági, G., **7**, 723[25]
Sagi, M., **2**, 353[101]
Sagitullin, R. S., **4**, 424[19]; **8**, 608[48]
Saha, B., **3**, 908[146]; **4**, 1040[103]
Saha, C. R., **6**, 539[580]
Saha, M., **3**, 216[71]; **4**, 956[20]
Sahai, D., **4**, 401[227]
Sahai, M., **4**, 628[399], 633[399], 634[399]
Sahara, M., **1**, 477[137,138]
Sahasrabudhe, A. D., **8**, 384[40]
Sahbari, J. J., **1**, 34[220,221,222]
Sahlberg, C., **3**, 445[72], 492[77]; **4**, 411[264]
Sahli, M. S., **3**, 154[419]; **6**, 1020[46]
Sahm, W., **3**, 482[3]

Sahoo, S. P., **4**, 113[171,171f,g]; **5**, 94[86], 130[39]; **7**, 162[57], 722[18]; **8**, 245[73]

Sahraoui-Taleb, S., **1**, 683[227], 714[227], 715[227], 717[227], 718[227]; **3**, 786[42]

Sahu, N. P., **5**, 176[133]

Sahyun, M. R. V., **6**, 1012[6], 1013[6]; **8**, 329[23], 335[23]

Sai, M., **6**, 666[231], 667[231]

Saicic, R., **4**, 824[233]

Said, F. F., **1**, 215[35]

Said, H., **1**, 892[149]

Saida, Y., **7**, 335[30]

Saidi, M. R., **8**, 890[138]

Saidov, O. O., **7**, 521[37]

Saigo, K., **2**, 605[62], 612[108]; **4**, 381[130], 1046[111]; **6**, 726[179]; **7**, 318[58], 319[58], 320[58]

Saijo, K., **1**, 343[106]; **3**, 278[30]

Saikachi, H., **6**, 251[147]

Saiki, M., **7**, 618[22]

Saiko, O., **3**, 814[68]

Saillard, J. Y., **8**, 451[180]

Saimoto, H., **1**, 366[46], 876[100]; **4**, 1007[124]; **5**, 768[128,134], 769[128], 779[128]; **6**, 7[30], 563[905]; **7**, 281[176], 282[176]

Saindane, M., **1**, 642[111], 644[111], 669[111], 671[111]; **4**, 40[113], 53[113], 245[89]; **5**, 531[75], 549[75]; **7**, 130[77], 131[84,85], 257[48], 376[81], 520[26]; **8**, 850[119]

Saino, T., **7**, 489[173]

Sainsbury, M., **3**, 499[110], 501[110], 505[110], 509[110], 512[110], 660[18], 676[78], 686[78], 700[163]; **7**, 373[72b], 473[29]; **8**, 653[82]

Sainte, F., **5**, 416[56], 480[165,166,168], 483[165]; **7**, 502[262]

Saintmileux, Y., **4**, 605[299]

Saint M'Leux, Y., **6**, 659[196]

Saint-Ruf, G., **2**, 149[85]

Sainz, C., **7**, 60[46b]

Saitner, H., **5**, 417[64]

Saito, G., **4**, 839[27]; **8**, 93[71]

Saito, H., **4**, 243[69], 244[69], 245[69], 258[69]; **5**, 622[22]; **8**, 383[21], 423[40], 429[40], 836[10c]

Saito, I., **5**, 736[142g,h]; **6**, 564[918]; **7**, 381[104], 881[160]; **8**, 249[96], 396[139], 580[5], 581[5], 587[5], 817[34]

Saito, K., **3**, 426[82,84], 429[82]; **5**, 634[74,76]; **6**, 995[102]; **8**, 55[179], 66[179], 408[79], 620[138]

Saito, M., **2**, 721[87,88]; **3**, 437[28]; **4**, 394[195]; **5**, 337[51]; **8**, 412[117]

Saito, N., **1**, 243[56]; **6**, 734[11]; **7**, 537[60]

Saito, O., **4**, 589[84], 590[95], 592[95], 598[84], 633[95]

Saito, R., **3**, 594[186]; **4**, 837[9], 839[27]

Saito, R. M., **7**, 473[33], 501[33], 502[33]

Saito, S., **1**, 751[93]; **2**, 917[85], 1066[122]; **4**, 31[92]; **5**, 833[49]; **6**, 77[54], 1021[52]; **8**, 244[50,71], 247[71], 251[71], 253[71], 338[82], 339[82]

Saito, T., **2**, 456[33], 457[33], 458[33], 459[33], 460[33], 461[33], 462[33], 466[33]; **5**, 441[178]; **6**, 532[468]; **7**, 615[8]; **8**, 153[187], 447[97]

Saito, Y., **1**, 159[75], 509[24], 635[81,82], 636[81,82], 637[81,82], 640[81,82], 672[81,82], 678[81,82], 679[81,82], 680[82], 681[82,217], 682[217], 700[82], 705[82]; **2**, 10[45,45b,46,48], 22[45b], 58[8,10], 61[8], 67[8,38], 71[8], 72[8], 76[8], 629[1], 635[1]; **3**, 87[97,98], 99[98], 100[98], 157[98], 196[31]; **7**, 350[23]; **8**, 850[123]

Saitoh, T., **5**, 541[111]; **8**, 252[111]

Saitoh, Y., **4**, 405[250a,b]; **8**, 856[182]

Saji, I., **2**, 876[34]

Sajus, L., **7**, 160[53]; **8**, 445[25,26], 452[25,26]

Saka, T., **5**, 1001[16]

Sakaeda, T., **8**, 458[225]

Sakagami, T., **4**, 299[297]

Sakagawa, K., **6**, 20[75]

Sakaguchi, H., **7**, 628[46]; **8**, 174[126], 178[126], 179[126]

Sakaguchi, K., **4**, 759[192], 763[192]

Sakaguchi, M., **1**, 215[41], 216[41]

Sakaguchi, R., **7**, 132[96], 158[36a,b], 175[36b]

Sakaguchi, S., **3**, 1000[55]; **6**, 877[39], 878[39], 883[39], 887[39]

Sakai, I., **6**, 121[127]

Sakai, K., **1**, 893[153]; **2**, 209[109], 540[71], 547[115], 551[115], 780[8], 823[117], 852[234]; **3**, 400[122]; **4**, 159[81], 587[27], 597[172], 637[172]; **6**, 249[140], 439[69], 809[67], 1052[41,42a,b]; **7**, 95[63], 109[184], 463[126], 672[18], 774[322], 877[133]; **8**, 190[64], 198[133,134], 253[113], 533[149]

Sakai, M., **1**, 89[58], 90[58]; **5**, 605[60,60b], 609[60,60b]; **6**, 233[40], 559[861,862]; **8**, 450[164]

Sakai, R., **5**, 563[92]

Sakai, S., **1**, 243[59,60], 254[15], 268[53,53a,c], 269[57]; **2**, 446[32], 1021[50], 1068[129]; **3**, 512[198], 570[56]; **4**, 588[56], 596[162,163], 601[247], 614[374], 621[163], 637[163], 809[165]; **5**, 35[12]; **6**, 564[909], 746[90], 916[31], 917[35]; **8**, 16[98], 31[47], 66[47], 988[32]

Sakai, T., **1**, 769[182]; **3**, 843[25], 844[32,33]; **4**, 229[215]; **5**, 766[117]; **8**, 191[95], 568[466], 899[28]

Sakai, Y., **5**, 166[89]; **6**, 1036[145]

Sakairi, N., **6**, 23[91]

Sakaitani, M., **3**, 421[60]; **4**, 382[132,132b]

Sakaki, K., **8**, 411[99]

Sakakibara, H., **3**, 918[27], 968[128]

Sakakibara, J., **4**, 588[56]; **6**, 531[426]; **8**, 658[99]

Sakakibara, M., **4**, 893[154]

Sakakibara, S., **6**, 636[17], 670[272]

Sakakibara, T., **1**, 642[120], 645[120], 672[120], 708[120]; **3**, 1039[99]; **4**, 23[70], 36[102,102a,b], 837[11]; **6**, 442[88]

Sakakibara, Y., **1**, 450[211]; **3**, 463[163], 484[24], 501[24]; **6**, 233[40]; **8**, 450[164]

Sakakiyama, T., **7**, 94[56]

Sakakura, T., **2**, 816[87]; **3**, 529[51], 1016[3], 1039[101,102]; **4**, 261[292]; **6**, 527[405]; **7**, 6[31], 324[71], 679[74]

Sakamaki, T., **4**, 36[103,103a]

Sakamoto, H., **8**, 422[33], 427[33]

Sakamoto, K., **2**, 651[113]; **4**, 249[128]

Sakamoto, M., **5**, 181[150], 725[118]; **7**, 382[108]; **8**, 964[55]

Sakamoto, N., **2**, 159[131]; **4**, 238[11], 245[11], 255[11], 260[11]; **7**, 680[80]

Sakamoto, T., **1**, 561[164]; **2**, 353[101], 598[15]; **3**, 460[143], 461[143,147,148], 530[64], 533[64], 541[108-111,113,114], 543[108-111,113,117,118]; **4**, 433[121]; **6**, 543[604], 801[30]; **8**, 902[45]

Sakamoto, Y., **2**, 357[149]; **3**, 826[39]

Sakamura, S., **3**, 693[149], 694[149], 747[72]; **5**, 516[25], 524[54], 534[54,95], 553[42], 563[92,93], 564[94,96,97], 571[121], 578[150]; **6**, 689[386]

Sakan, K., **5**, 515[15], 518[15], 519[15,33], 620[14], 621[17]

Sakan, T., **1**, 408[34], 422[91,92]

Sakane, S., **1**, 98[84], 99[84], 387[136]; **2**, 541[74], 995[45]; **6**, 542[601], 767[28], 768[28], 769[28], 771[37,38]; **7**, 696[43], 697[43,49]; **8**, 43[108], 47[108], 64[220], 66[108], 67[220], 394[119]

Sakane, T., **8**, 253[122]

Sakanishi, K., **3**, 587[148]; **8**, 598[99,100], 851[135]

Sakano, I., **1**, 117[57], 118[57]

Sakano, M., **3**, 295[8]

Sakari, M., **1**, 768[168]

Sakari, S., **4**, 615[383]

Sakashita, T., **3**, 125[308]

Sakata, G., **2**, 282[33]

Sakata, J., **1**, 158[74]; **2**, 135[12], 615[128,129], 634[35,38], 640[35,38]; **3**, 6[29]

Sakata, K., **6**, 91[122]

Sakata, S., **1**, 407[32,33], 424[100], 454[32]; **5**, 516[28]; **6**, 439[67]

Sakata, T., **7**, 307[11]
Sakata, Y., **2**, 578[81]; **4**, 1088[124]; **5**, 431[120]
Saki, K., **7**, 675[53]
Sakito, Y., **1**, 64[47,48], 65[50,51], 317[153]; **3**, 327[167]; **5**, 1014[57]
Sakiyama, F., **2**, 1095[93]
Sako, H., **8**, 860[223]
Sako, M., **4**, 436[142,143]; **7**, 877[135]; **8**, 908[76]
Sakoda, K., **3**, 829[57]
Saksena, A. K., **6**, 563[903]; **8**, 544[273]
Sakuda, Y., **7**, 84[3]
Sakuma, K., **3**, 155[428], 220[115]; **8**, 349[135], 354[175]
Sakuma, Y., **6**, 489[79]; **8**, 97[95]
Sakuraba, M., **6**, 616[105]
Sakuragi, H., **7**, 881[156]
Sakurai, A., **3**, 95[157], 114[157], 116[157]
Sakurai, H., **1**, 180[43], 181[43], 357[7], 361[7], 580[1]; **2**, 6[28], 17[28,28b,d], 68[44], 476[4], 565[14,16], 566[17], 567[27,28], 572[44], 576[27,72,79], 578[81,87], 582[27,109], 601[34], 635[43], 640[43], 718[76,77], 719[82], 721[87,88], 901[38], 908[38]; **3**, 32[187], 246[44], 437[28], 485[27]; **4**, 98[113,113b], 155[68a,b], 589[80], 591[80], 1088[124]; **5**, 165[81], 166[89], 168[102], 337[51], 338[52], 431[120], 596[22], 597[22], 603[22], 650[25], 1166[18]; **6**, 16[59], 17[59], 18[59], 83[82], 214[92], 487[39], 489[39], 543[39], 654[153], 720[132], 832[12], 865[12]; **7**, 458[113], 641[3,4,6,7], 646[4], 878[138,140,143], 888[138a]; **8**, 20[141], 392[101], 517[126], 547[313], 562[421], 774[75]
Sakurai, K., **1**, 343[106], 569[260]; **3**, 278[30]; **4**, 381[130]; **7**, 416[122], 441[44]
Sakurai, M., **1**, 79[21], 80[21], 81[21], 82[21], 88[55], 333[61], 335[61]; **4**, 140[11], 209[66]; **6**, 850[125]
Sakurai, S., **8**, 149[117,118]
Sakurai, Y., **1**, 120[65], 350[152,153], 361[35,35a,b], 362[35a,b]; **6**, 507[240], 515[240]; **8**, 64[208], 65[208], 66[208], 67[208], 287[16]
Sakuta, K., **2**, 580[101]; **4**, 107[141]; **6**, 774[49,50], 1066[96,98]; **7**, 701[64]
Sakuta, M., **4**, 238[12]
Sakya, S. M., **5**, 583[191]
Sala, R., **8**, 568[467]
Saladin, E., **6**, 214[97]
Salamond, W. G., **7**, 100[129], 104[129]
Salański, P., **6**, 70[19]
Salanski, P., **5**, 108[207]
Salas, M., **6**, 801[36]
Salaski, E. J., **1**, 404[20], 428[120]; **4**, 1023[258]
Salaun, J., **4**, 1046[118]; **7**, 825[44], 833[77], 843[39,40]
Salaün, J., **5**, 911[90], 923[137], 954[90], 955[304], 956[305,307]
Salaun, J. P., **6**, 26[110]
Salaün, J. R., **3**, 620[28], 765[153], 767[162], 842[17], 848[17]
Salaün, J. R. Y., **5**, 211[64], 708[42], 901[14,15], 904[51], 905[15], 911[15], 912[15], 913[14], 919[15], 920[15,131,132], 921[15], 1006[33], 1011[47], 1021[72]
Salazar, J. A., **4**, 342[61], 375[98a,b], 388[98,98a,b], 408[98b], 409[98a], 814[187]; **7**, 41[15], 495[210], 722[19], 723[19], 725[19]
Saldana, A., **4**, 1061[166]
Saldaña, M., **8**, 368[67], 374[67]
Saleh, S., **3**, 1052[25], 1059[25]; **4**, 13[45], 14[45b]; **5**, 534[89]; **6**, 127[156], 174[60]
Salem, G., **3**, 436[7]; **4**, 83[63]; **6**, 492[124]; **8**, 408[73]
Salem, L., **1**, 49[8]; **4**, 1082[84]; **5**, 72[170,174,175,176], 1009[45]
Salemink, C. A., **3**, 464[168]; **8**, 541[204]
Salemnik, G., **8**, 950[163]
Salen, G., **3**, 644[149]
Salerno, G., **2**, 137[17]; **3**, 423[77]; **5**, 36[16,19], 1037[3], 1132[21], 1135[50], 1137[55]
Salfeld, J. C., **3**, 505[166]
Salgado-Zamora, H., **4**, 411[266a], 567[48]
Salimgareeva, I. M., **8**, 699[150]
Salinger, R. M., **1**, 361[30]
Saliou, C., **8**, 35[65], 47[65], 66[65]
Salisbury, K., **5**, 721[103], 729[124]
Salisbury, L., **7**, 231[143]
Salisbury, L. F., **6**, 263[26], 264[26], 265[45], 268[66], 270[26]
Salisbury, P., **7**, 58[57], 62[57], 63[57]
Salituro, F. G., **1**, 441[174]
Saljoughian, M., **2**, 365[213]
Sall, D. J., **7**, 415[115c], 418[115c]; **8**, 940[101], 948[101]
Sallam, L. A. R., **7**, 69[87]
Salle-Sauerländer, **4**, 200[5]
Salmin, L. A., **3**, 648[172,173,174]
Salmond, W. G., **6**, 182[139]; **7**, 260[43], 779[421]
Salomaa, P., **8**, 222[93]
Salomon, M. F., **2**, 538[55,56], 539[55,56], 855[241]; **4**, 261[289], 603[269], 645[269]; **5**, 432[133], 433[133a]; **6**, 648[117b]; **7**, 258[56], 630[50]
Salomon, R. G., **1**, 534[143]; **2**, 538[55,56], 539[55,56], 855[241]; **4**, 187[98], 261[289], 962[41], 1032[14], 1033[14]; **5**, 147[109], 432[133], 433[133a], 1001[16]; **6**, 851[131]; **7**, 107[169], 239[50]; **8**, 354[161], 513[104], 544[253]
Salowey, C., **5**, 225[97]; **8**, 505[82], 507[82]
Saltykova, L. E., **4**, 1058[151]
Saltzman, M., **5**, 143[103]
Salunkhe, M. M., **8**, 537[179]
Saluzec, E. J., **5**, 1098[117], 1099[117], 1100[117], 1101[117], 1112[117]
Saluzzo, C., **4**, 331[19]
Salvador, J., **8**, 242[40]
Salvadori, P., **4**, 445[204]; **5**, 1152[144], 1154[151,158], 1155[151]
Salvatori, T., **3**, 752[97]; **7**, 153[9]
Salvesen, K., **6**, 438[53]
Salvino, J. M., **1**, 28[141], 29[149], 37[176], 43[141]; **2**, 100[5]
Salz, R., **4**, 596[161]
Salzberg, P. L., **6**, 230[33]
Salzer, A., **2**, 722[96]; **4**, 670[20]; **6**, 690[400], 692[400]
Salzmann, T. N., **2**, 213[122], 256[46], 652[125], 1059[79]; **6**, 125[148], 127[148], 154[150], 1018[39], 1020[39,47], 1044[16b], 1048[16]; **7**, 124[43], 125[43,52], 126[43,52], 257[46]
Sam, D. J., **7**, 585[160]
Sam, J., **2**, 149[83]
Sam, T. W., **3**, 380[9], 390[72], 396[111,114,115], 398[111,114]
Samaan, S., **8**, 861[224]
Samaddar, A. K., **1**, 564[195]
Samain, D., **3**, 246[34]; **6**, 692[408]
Samaki, H., **2**, 464[102], 465[102]
Samama, J.-P., **6**, 420[14]
Saman, E., **6**, 48[88]
Samanen, J. M., **3**, 807[34]
Samanich, D., **7**, 444[55]
Samarai, L. I., **6**, 543[622], 552[622]
Samaritano, R. H., **8**, 269[88]
Samartino, J. S., **2**, 103[31], 209[108]; **5**, 841[95]
Samartseva, I. V., **6**, 501[207]
Sambasivarao, K., **2**, 381[307]
Samek, Z., **8**, 334[65]
Samizu, K., **5**, 862[246]; **7**, 299[45]
Samkoff, D. E., **3**, 380[10]
Sammakia, T., **1**, 420[83], 568[230], 800[298]; **4**, 53[150]; **5**, 1055[47]; **6**, 692[409], 1067[103]; **7**, 361[23]
Samman, N. G., **5**, 65[67]
Sammes, M. P., **4**, 425[25], 430[90-93]; **7**, 138[127]; **8**, 587[37]
Sammes, P. G., **1**, 755[115], 812[115], 813[115]; **2**, 350[77], 928[121,122], 946[121,122]; **3**, 894[60]; **4**, 14[47], 111[154f], 376[98c,d], 388[98], 401[226,226d,228a], 497[96], 1004[77], 1021[77], 1089[128], 1092[128], 1093[128]; **5**, 13[92], 85[3], 123[1], 126[1],

388[136], 389[136], 492[246], 497[222], 571[120], 682[31], 683[115], 693[113,115], 694[115], 711[57a]; **6**, 1002[133], 1017[38], 1024[38], 1025[80]; **7**, 6[35], 493[187], 728[42]; **8**, 249[93]
Samoilova, M. Y., **4**, 1058[150]
Samokhvalov, G. I., **8**, 956[7]
Samori, B., **5**, 99[131]
Sampath, V., **3**, 1025[34]; **5**, 1041[19], 1043[22], 1046[19], 1049[22], 1052[19]
Sampedro, M. N., **4**, 379[115]
Samplavskaya, K. K., **3**, 319[132]
Sampson, P., **2**, 103[35,36]
Samsel, E. G., **3**, 1031[64]; **4**, 761[198], 805[145], 857[106], 937[71], 938[71], 941[85]; **8**, 413[124]
Samson, M., **4**, 259[268], 262[268]; **6**, 25[103]; **8**, 122[80], 460[254]
Samuel, C. J., **5**, 199[24], 210[59], 597[30]
Samuel, O., **6**, 150[113]; **7**, 425[146], 777[378,381], 778[378]
Samuel, P. A., **6**, 1016[27]
Samuels, S. B., **4**, 48[137], 562[36]
Samuelsson, B., **6**, 205[32], 652[139], 660[139]; **7**, 237[32,33], 259[59], 272[144], 274[144], 752[146]; **8**, 224[106], 969[95]
Samyn, C., **7**, 475[52]
Sana, M., **5**, 741[153]
Sancassan, F., **4**, 426[47,58]
Sanchez, D., **7**, 831[64]
Sanchez, E. L., **4**, 1033[26], 1035[26a], 1046[26a], 1051[26a]
Sánchez, F., **6**, 67[13]
Sanchez, F.-J., **7**, 359[18]
Sanchez, I. H., **4**, 119[192e]
Sanchez, J. F., **7**, 634[68]
Sanchez, J. P., **6**, 789[105]
Sanchez, M. G., **5**, 1123[37]
Sanchez, R., **8**, 390[79]
Sanchez, R. M., **3**, 824[23]; **4**, 757[185]
Sanchez Ballesteros, J., **2**, 357[140]
Sánchez-Delgado, R. A., **8**, 446[74], 452[74], 457[74]
Sánchez-Ferrando, F., **2**, 381[300]; **6**, 184[150]
Sanchez-Obregon, R., **1**, 544[35], 552[80]
Sancilio, F. D., **7**, 516[6]
Sanda, F., **1**, 449[210]; **2**, 492[51]; **3**, 463[157]
Sandberg, B., **3**, 639[69]
Sande, A. R., **8**, 537[179]
Sandefur, L. O., **3**, 8[44], 15[44]; **4**, 189[106], 244[77], 255[77], 260[77]
Sandel, V., **4**, 520[37]
Sander, J., **5**, 1126[66]
Sander, M., **7**, 762[76]
Sander, W., **6**, 960[52]; **8**, 349[136]
Sanders, A., **6**, 690[392]
Sanders, G. L., **4**, 744[134]
Sanders, H. P., **6**, 779[67]; **8**, 948[147]
Sanderson, D. R., **2**, 727[134]; **5**, 762[96]
Sanderson, J. E., **3**, 643[116]
Sanderson, J. J., **2**, 735[16]
Sanderson, P. E. J., **4**, 682[57]; **6**, 17[62]; **7**, 647[32]; **8**, 788[120]
Sandhu, M. A., **2**, 962[44]
Sandifer, R. M., **2**, 512[46], 523[74]
Sandler, S. R., **4**, 315[524], 1018[220], 1020[233]; **6**, 294[233], 685[346], 690[346], 692[346], 726[186]; **7**, 741[46,50], 746[46], 747[50,99,100], 748[99,100]; **8**, 364[11,23], 365[30], 382[5]
Sandman, D. J., **4**, 476[155]
Sandmeier, D., **6**, 193[208], 449[115]
Sandmeier, R., **4**, 31[92,92k]
Sandorfy, C., **6**, 711[68]
Sandoval, C., **1**, 544[35]
Sandoval, S. B., **4**, 573[9], 614[378]
Sandré-Le Craz, A., **5**, 789[29]
Sandri, E., **1**, 516[59,60], 517[61,62]; **3**, 147[396], 149[413], 151[413], 152[413], 153[396,413], 155[396], 865[27], 944[90,91], 946[92], 958[90,112]; **6**, 898[103]; **7**, 764[126], 767[126]; **8**, 664[121]
Sandri, S., **3**, 45[250]; **4**, 375[94], 377[104], 386[94a,153,153a,157], 387[94a,153a,157], 388[164], 389[166,166a], 393[164a], 401[226], 407[104c,153a,157b,254], 408[254a,b,259c]; **6**, 26[106], 533[497], 648[124]; **7**, 280[177], 493[184], 503[269], 663[62], 664[63]
Sandrin, F., **2**, 607[72], 620[72]
Sandrin, J., **2**, 1017[31]; **6**, 737[31], 746[89]
Sandrini, P. L., **4**, 710[48]
Sandris, C., **2**, 345[41]
Sands, R. D., **3**, 727[28]; **5**, 4[33]
Sandstrom, J., **5**, 686[53]; **6**, 423[44]
Sandstrom, W. M., **8**, 292[41]
Sandu, A. F., **7**, 95[73a]
Sanechika, K., **3**, 530[80], 535[80]
Saneii, H., **6**, 431[104]
Sanemitsu, Y., **4**, 564[42]
Saneyoshi, M., **6**, 76[46]; **8**, 384[36], 391[91]
San Filipo, J. S., **4**, 170[20]
San Filippo, J., Jr., **1**, 116[44], 426[109], 431[134]; **2**, 280[26]; **3**, 248[55], 251[55], 269[55], 418[24], 419[47], 482[3], 494[87], 502[87]; **7**, 279[172], 744[68], 845[65]; **8**, 404[12], 850[121], 851[133], 852[133b]
Sanfilippo, L. J., **3**, 862[3]
Sanfilippo, P., **1**, 447[198], 753[100]
Sanfilippo, P. J., **4**, 573[6,8], 614[375,376,379], 841[38], 905[208]
Sang, H. V., **4**, 1007[111]
Sanger, A. R., **8**, 445[42], 457[42]
Sang-Hun Jung, **7**, 479[93,94]
Sanghvi, Y. S., **7**, 364[40], 365[50], 376[50,86]
Saniere, M., **7**, 297[32]
Sanjoh, H., **3**, 342[12], 347[12], 381[30], 382[30]; **7**, 828[54]
San Juan, C., **8**, 52[141], 66[141]
Sankaraiah, B., **4**, 988[140]
Sankarappa, S. K., **3**, 380[7]
Sankararaman, S., **7**, 855[62], 874[108], 877[135], 878[136], 881[162], 882[165], 887[62]
Sanna, P., **8**, 589[50,53]
Sannai, A., **5**, 564[97]
Sanner, C., **3**, 225[190]
Sanner, M. A., **2**, 1064[111]; **4**, 21[67,67a], 106[139], 107[139]
Sanner, R. D., **8**, 673[23], 675[23], 682[84], 691[23]
Sanneskog, O., **7**, 878[142]
Sannié, C., **7**, 100[125]
Sano, A., **4**, 348[108], 349[108c]
Sano, H., **1**, 438[160]; **2**, 446[31]; **3**, 381[21], 453[113], 454[120], 469[216], 470[215,216], 473[215], 476[215,216]; **4**, 753[167]; **6**, 91[126], 254[161]; **7**, 616[19]; **8**, 245[72], 698[141], 824[63], 825[63]
Sano, S., **5**, 627[42]
Sano, T., **3**, 244[25], 267[25], 300[43], 494[84]; **5**, 323[15,16], 1022[73,73c,74]; **8**, 37[100], 42[100], 66[100], 698[138]
Sansbury, F. H., **6**, 25[100]
Sansone, E. B., **6**, 119[109]; **8**, 373[129], 383[19], 389[19], 392[19], 597[95]
Sansoulet, J., **1**, 530[128]; **6**, 2[5], 18[5]
Sanstead, J. K., **3**, 644[154]
Santa, L. E., **1**, 115[41], 212[10], 213[10], 215[10], 216[10], 217[10], 221[10], 432[138], 433[138]
Santa, T., **8**, 392[94]
Santaballa, J. A., **4**, 300[307]
Santana, M., **8**, 312[19]
Santaniello, E., **6**, 80[68], 227[19], 228[19]; **7**, 279[171], 283[181], 284[181], 286[189], 331[17], 709[37], 765[134], 841[17], 844[60,62], 845[66]; **8**, 187[38], 190[71,73], 191[73], 240[30,31], 244[30,56], 263[29]

Santarsiero, B. D., **3**, 213[51]; **7**, 230[126]; **8**, 672[21], 696[126]
Santelli, C., **4**, 55[157]
Santelli, M., **1**, 749[82]; **2**, 85[18], 710[19-22], 718[73]; **3**, 563[1], 572[63], 577[85,86,91], 607[1i]; **4**, 155[68f,71a], 394[189], 396[189d]; **5**, 753[56], 777[184], 779[184]; **7**, 554[63]
Santelli-Rouvier, C., **2**, 571[39], 710[19]; **5**, 753[56]
Santhanakrishnan, T. S., **8**, 330[48], 493[20], 497[20]
Santhanam, K. S. V., **8**, 592[65]
Santi, R., **4**, 600[243], 601[243], 763[208], 764[219], 767[219,233], 810[168], 820[168], 823[168]; **8**, 99[112]
Santiago, A. N., **4**, 1021[243]
Santiago, C., **5**, 203[39,39c], 204[39h], 209[39], 210[39]
Santiago, M. L., **3**, 88[135], 90[135]; **6**, 254[165], 542[600], 821[114]
Santillan, R. L., **8**, 537[187]
Santilli, D. S., **3**, 328[176]; **5**, 64[33-35]
Santini, C., **4**, 688[67]; **5**, 809[121,123]
Santo, K., **1**, 544[37], 548[66], 560[37], 561[37,156]
Santone, P., **2**, 651[120]
Saotome, K., **3**, 639[81]
Sapi, J., **4**, 40[115]; **5**, 528[69]
Sapino, C., **6**, 422[35]
Sapino, C., Jr., **3**, 232[264]
Sapiro, R. H., **2**, 348[65]
Sappa, E., **8**, 449[158]
Saquet, M., **2**, 86[24]; **8**, 268[72]
Saraceno, A. J., **1**, 360[28], 361[28]
Saradarian, A., **7**, 760[23]
Saraf, S.-D., **5**, 499[254]
Sarancha, V. N., **6**, 216[107]
Sarandeses, L. A., **3**, 232[262], 545[121]
Sarangan, S., **8**, 49[114], 66[114]
Sarasohn, I. M., **3**, 822[9], 823[9], 834[9]
Sard, H., **1**, 887[138], 888[138]; **5**, 683[87], 689[73], 690[82], 806[106], 856[210], 1017[64], 1018[64], 1020[64], 1021[64], 1025[82], 1026[82]; **8**, 756[144]
Sarda, P., **4**, 976[93], 977[93], 978[93], 994[93]; **8**, 310[15], 311[18]
Sardarian, A., **7**, 236[27], 266[109], 267[109], 286[190], 561[85], 760[27]
Sardella, D. J., **5**, 580[170]
Sardina, F. J., **7**, 547[33]; **8**, 514[113]
Sarel, S., **4**, 292[234]; **5**, 71[149,150,152], 916[118], 1006[35], 1133[29]; **6**, 264[32,33], 268[33]
Sarel-Imber, M., **5**, 1006[35]
Sarett, L. H., **2**, 158[124]; **6**, 219[123]; **7**, 100[124], 256[40]
Sargent, G. D., **3**, 753[100]
Sargent, L. J., **8**, 568[468]
Sargent, M. V., **5**, 723[109]
Sarges, R., **3**, 124[269]
Sargeson, A. M., **4**, 298[292]
Sariaslani, F. S., **7**, 58[53a], 62[53,53a-c], 63[53a]
Sarin, R., **6**, 734[13]
Saris, L. E., **6**, 478[107], 481[107]
Sarkanen, K. V., **3**, 691[130], 693[130]
Sarkar, A., **8**, 113[31]
Sarkar, A. K., **2**, 569[34], 586[132]; **7**, 360[20]
Sarkar, D. C., **1**, 564[192,199]; **3**, 714[31]
Sarkar, M., **3**, 87[115]
Sarkar, T., **2**, 712[42]; **3**, 728[35]; **6**, 832[12], 865[12], 1007[150]; **7**, 816[6b], 824[6], 825[6]
Sarkar, T. K., **1**, 580[1], 582[7]; **2**, 572[42], 586[131]; **3**, 255[105], 356[56]; **5**, 14[98]
Sarkisian, G. M., **5**, 223[71]
Sarma, D. N., **3**, 380[10]; **8**, 891[146], 968[91,92]
Sarma, J. C., **6**, 76[49]
Sarmah, P., **3**, 365[98]
Sarngadharan, M. G., **8**, 978[145]
Sarnowski, R., **8**, 875[41]
Saroli, A., **4**, 395[205,205d]
Sarroff, A., **8**, 859[211]
Sarrygina, O. A., **2**, 662[15], 664[15]
Sartirana, M. L., **4**, 461[101], 475[101]
Sartor, K., **5**, 355[87c], 356[87c], 365[87c]
Sartoretti, J., **4**, 412[269], 413[269a]; **5**, 772[162], 773[166], 780[162]; **7**, 298[34]; **8**, 945[127]
Sartori, G., **3**, 311[100]; **5**, 1025[81]
Saryev, G. A., **8**, 772[57]
Sas, W., **4**, 590[103,104]; **8**, 857[197]
Sasa, M., **2**, 538[63]
Sasada, Y., **7**, 473[33], 501[33], 502[33]
Sasai, H., **5**, 562[88]
Sasaki, A., **5**, 98[124]
Sasaki, H., **3**, 231[252]; **6**, 438[55]; **8**, 369[79]
Sasaki, J., **6**, 1045[22]
Sasaki, J.-I., **5**, 337[51]
Sasaki, K., **2**, 635[43], 640[43], 780[6]; **3**, 483[19], 500[19], 509[19]; **5**, 337[51]; **7**, 761[64]
Sasaki, M., **1**, 253[11], 259[31], 274[75], 366[48]; **2**, 709[7]; **4**, 95[101], 251[142], 255[194], 260[142]; **5**, 71[139], 76[233], 77[254,256,257,258]; **7**, 262[80]; **8**, 459[228], 925[7]
Sasaki, N., **5**, 926[159]
Sasaki, O., **5**, 829[20]; **7**, 530[23]
Sasaki, T., **3**, 383[41,44]; **4**, 297[276]; **5**, 451[17], 469[17], 470[17], 597[23], 603[23], 606[23], 621[20], 627[41], 632[64], 973[14], 981[14]; **6**, 115[82], 252[154], 264[37], 265[37,41,42], 273[99], 554[720], 814[92]; **7**, 153[6], 462[124]
Sasaki, Y., **8**, 626[174]
Sasakura, K., **2**, 244[23,29], 245[23], 478[15], 748[122,124]
Sasaoka, H., **2**, 577[80]
Sasaoka, M., **7**, 537[58,60]
Sasaoka, S., **4**, 359[161]
Sasatani, S., **6**, 528[414], 767[26], 769[29]; **7**, 696[38]; **8**, 64[213], 66[213]
Sasazawa, K., **3**, 453[114]
Sashida, H., **6**, 535[539], 538[539]; **8**, 49[115], 66[115]
Sasho, M., **1**, 242[49-51], 243[52]; **4**, 14[46]
Saski, K., **1**, 755[113]
Sasoka, S., **4**, 595[151]
Sass, V. P., **3**, 639[82], 644[82,137]
Sassaman, M. B., **6**, 237[60], 243[60]; **8**, 216[62,63], 217[63]
Sasse, K., **6**, 430[98], 964[79]; **7**, 752[156]
Sasse, M. J., **8**, 366[46]
Sasse, W. H. F., **3**, 509[180]; **5**, 637[107]
Sasson, I., **7**, 97[95], 112[95]
Sasson, M., **4**, 807[148]
Sasson, Y., **2**, 772[14]; **4**, 434[125]; **6**, 93[132]; **8**, 453[193], 551[340,341,343,347], 552[340,341,348], 557[383]
Sastre, J., **8**, 35[64], 52[145], 66[64,145]
Sastry, K. A. R., **7**, 605[139], 606[157]
Sastry, V. V. S. K., **6**, 436[27], 437[27], 453[27], 455[27], 456[27]
Satake, K., **5**, 172[119,120,122], 637[102]; **6**, 531[449,461], 903[135]
Satake, M., **5**, 297[57], 1196[37]
Satchell, D. P. N., **2**, 754[5]
Satge, J., **5**, 444[189]
Satish, A. V., **4**, 159[81]
Satish, S., **7**, 276[149]
Sato, A., **6**, 554[719]; **8**, 353[157]
Sato, E., **2**, 819[100], 824[100]; **5**, 839[82], 864[257]; **6**, 896[93]
Sato, F., **1**, 58[33], 87[46,47], 110[17-19], 131[17,104,105], 134[17], 135[117], 143[37,38], 158[37], 159[37,38,78,79], 160[78], 161[78,84], 180[41], 181[41], 185[54], 339[87,88], 340[90], 415[61], 564[195], 784[243,244]; **2**, 5[17,19], 6[17,19], 22[17,17a], 23[19a,88], 204[97]; **3**, 41[224], 223[183], 224[162], 225[183], 244[27,28], 463[165], 464[177],

489[62], 494[84], 495[62], 504[62], 511[62], 515[62]; **6**, 6[29]; **7**, 371[68], 379[100], 400[52], 412[105], 414[105,105b,c,108,109], 418[105b], 423[142,143], 458[111], 712[64]; **8**, 246[78], 284[1], 483[55], 484[55], 485[55], 697[131], 698[131,137,138,139,144,146], 755[123], 797[43], 807[43], 967[79]

Sato, H., **2**, 363[191], 558[159]; **3**, 23[143], 24[143], 380[4], 946[87]; **4**, 106[140b]; **5**, 222[64,65], 223[64,65]; **6**, 450[119], 454[119]; **7**, 750[127]

Sato, K., **1**, 180[43], 181[43], 836[139]; **2**, 6[28], 17[28,28b], 572[44]; **3**, 168[494,501,504], 169[494,501,504], 170[494,501,504], 426[82-84], 428[90,91], 429[82,91]; **4**, 430[96], 558[17], 600[236], 753[171]; **6**, 832[12], 865[12], 931[88]; **7**, 641[6,7], 655[12], 761[56]; **8**, 20[141], 547[313]

Sato, M., **1**, 87[46,47], 92[66], 131[105], 143[37,38], 158[37], 159[37,38,79], 161[84], 180[41], 181[41], 340[90], 415[61]; **2**, 5[17,19], 6[17,19], 22[17,17a], 23[19a,88], 575[67], 1067[123]; **3**, 244[27,28], 445[73], 446[77], 449[73], 463[165], 464[177], 583[123]; **4**, 147[42], 208[62]; **5**, 134[65], 534[93], 626[39]; **6**, 559[861,862,863], 980[38]; **7**, 356[52], 378[96], 458[111]; **8**, 246[78], 284[1], 483[55], 484[55], 485[55], 698[144], 755[123,126], 758[126], 797[43], 807[43], 967[79]

Sato, M.-A., **6**, 994[96]

Sato, N., **3**, 649[204,205]; **6**, 491[115]; **8**, 405[29], 412[120], 978[146]

Sato, O., **3**, 592[173]

Sato, R., **7**, 11[88]; **8**, 412[117]

Sato, S., **1**, 506[17]; **2**, 163[147], 310[31], 311[31], 587[136], 615[124,125], 631[12], 635[44], 640[44]; **3**, 735[22]; **4**, 239[26], 251[26], 257[26], 430[96], 654[447,448,449], 928[39], 929[39], 941[39]; **5**, 583[183], 1138[70]; **6**, 46[65], 51[108], 53[108,120], 464[36], 465[36], 538[562], 746[91], 774[47]; **7**, 127[61]; **8**, 222[97], 483[55], 484[55], 485[55], 698[144], 786[118], 789[123], 934[56]

Sato, T., **1**, 71[65], 72[66,67], 122[68], 141[22], 188[69], 221[68], 238[36], 328[29], 336[73], 371[74], 422[92], 423[97], 424[98], 425[102], 427[112], 448[205], 563[180], 568[246], 570[265], 755[113]; **2**, 30[113], 31[113], 112[101], 184[21], 246[34], 247[34], 258[48-50], 261[48], 507[22], 581[102,103], 605[58], 749[135]; **3**, 124[259], 125[259], 153[416], 154[416], 218[100], 227[210,211,212], 243[16], 245[30-32], 246[39], 257[39], 259[131], 380[9,11], 446[78], 463[158], 470[212,213], 476[212,213], 565[23], 570[23], 583[23], 617[15], 619[15], 621[15], 623[15], 627[15]; **4**, 115[179b], 120[201], 257[224], 262[303], 898[177], 902[190]; **5**, 595[11], 596[11b], 605[57], 611[57], 729[123], 841[95], 1094[98], 1096[98], 1098[98], 1112[98]; **6**, 20[74], 147[85,86], 493[128], 494[138], 505[225], 744[76], 746[76], 836[53]; **7**, 208[78], 538[64], 539[65], 660[38], 682[86], 693[30], 694[30], 801[45]; **8**, 99[107], 159[108], 168[69,70], 171[107,108], 173[117], 178[69,108], 179[69,108], 187[39,40,44], 188[44], 190[77], 195[107,109], 196[77,120], 203[146], 205[157], 241[38], 263[32], 267[32], 545[302], 720[138], 721[138], 722[138], 837[18]

Sato, W., **7**, 537[59]

Sato, Y., **2**, 575[66], 610[97]; **3**, 918[27], 968[128,129], 969[135]; **4**, 45[130], 431[99,100], 500[110], 507[150,151], 510[175]; **6**, 604[33], 893[86,87]; **7**, 42[29], 778[391,392,393], 877[134]; **8**, 228[133,134], 341[108]

Satoda, S., **2**, 801[25]

Satoh, A., **5**, 841[88]

Satoh, F., **4**, 501[116], 510[176]; **5**, 693[108,114], 694[114]

Satoh, J. Y., **4**, 603[270]; **6**, 134[21]; **7**, 95[65], 107[165], 530[22], 531[22], 700[63]

Satoh, K., **1**, 569[260]; **2**, 780[7]

Satoh, M., **3**, 231[246,247,249,252], 489[64], 490[64,65], 495[64,65], 496[64,65,99], 498[64,65,99], 511[64,65,99], 515[64,65,99]

Satoh, S., **2**, 372[268,270,271], 373[268,270], 1035[94], 1040[94]; **5**, 468[129,130,131]; **7**, 713[69]

Satoh, T., **1**, 524[93], 526[93,95], 828[68]; **2**, 417[19], 429[48]; **4**, 784[16]; **5**, 303[80], 304[80]; **6**, 91[122], 93[134], 156[162]; **7**, 132[96], 158[36a,b], 175[36b], 425[149c]; **8**, 249[98], 253[98], 315[47], 369[75,78]

Satoh, Y., **3**, 523[25], 677[86], 1037[87]; **4**, 147[38b,41], 250[137], 358[153,155,156,157], 886[118]; **5**, 1107[170], 1108[170]

Satomi, H., **2**, 114[115,117], 185[30], 217[30]; **4**, 85[73], 249[118], 257[118]; **5**, 438[164]; **8**, 549[325]

Satomi, M., **7**, 530[21], 531[21]

Satou, M., **3**, 530[81], 536[81]

Satsangi, R., **6**, 539[578]

Satsuk, É. N., **8**, 770[38], 771[42]

Sattar, A., **3**, 404[133,134], 405[139]; **5**, 707[33]

Sattari, S., **6**, 481[119]

Sattler, H.-J., **6**, 554[806]

Sattler, R., **2**, 359[161]

Sattsangi, P. D., **8**, 48[109], 66[109]

Sattur, P. B., **5**, 105[191]

Satyanarayana, N., **6**, 744[72]; **8**, 36[88], 52[149], 66[88,149], 709[45], 943[120]

Satyanarayana, P., **4**, 113[176]

Sau, A. C., **6**, 639[47]

Saucy, G., **1**, 821[24], 825[24]; **3**, 168[491,493], 169[491,493], 171[491,493], 757[122], 953[101]; **5**, 828[4], 830[29], 862[29d], 893[41]; **6**, 560[868], 836[55], 853[139], 875[23], 887[23], 888[23]; **7**, 346[11], 347[15]; **8**, 237[11], 544[257,258], 545[291]

Sauer, C. W., **6**, 1013[12]

Sauer, G., **2**, 167[160], 360[171]; **6**, 718[117]; **7**, 383[111]; **8**, 331[32], 615[92], 618[123]

Sauer, H., **5**, 66[77]

Sauer, J., **3**, 571[59]; **4**, 490[66], 499[66], 1069[2,5], 1070[5], 1083[92]; **5**, 71[125,130], 76[130], 340[57c], 344[65], 345[57], 451[47-49], 491[205,209], 492[238], 498[228,229,234,235,237,238], 513[2], 516[2i], 518[2], 552[7,12,17], 594[1], 601[1], 604[1], 714[74], 854[175]; **7**, 24[27,28], 25[28], 252[8], 482[114]

Sauer, J. C., **5**, 7[49], 1138[65], 1146[110]

Sauerbier, M., **5**, 1148[123]; **8**, 383[16]

Sauermann, G., **1**, 12[62,63]

Sauers, R. R., **1**, 856[54]; **3**, 638[90]; **8**, 986[12]

Sauerwald, M., **1**, 159[72], 340[91]; **2**, 4[14]

Saugier, R. K., **2**, 194[69]; **3**, 99[189], 107[189], 110[189]

Sauleau, A., **5**, 936[195], 938[216], 948[216]

Sauleau, J., **5**, 936[195], 938[216], 948[216]

Saulnier, M. G., **1**, 473[81], 474[98]; **2**, 742[70]; **3**, 261[155]; **5**, 311[105], 384[128], 385[128b], 736[143]; **8**, 618[119]

Saumtally, I., **4**, 350[117]

Saunders, B. C., **8**, 903[49]

Saunders, D., **4**, 33[95,95a]

Saunders, J., **1**, 366[52]; **2**, 494[57]; **4**, 76[47]; **6**, 516[321], 552[692]; **8**, 638[16]

Saunders, J. H., Jr., **2**, 139[28]

Saunders, J. K., **8**, 346[126]

Saunders, J. O., **1**, 584[9]; **2**, 714[50,53]; **3**, 363[86]; **4**, 1049[121b]; **5**, 994[53], 997[53]; **8**, 927[20]

Saunders, K. H., **7**, 740[40]; **8**, 382[12], 383[12]

Saunders, M., **2**, 977[5]; **3**, 706[6]; **4**, 1016[208]

Saunders, V. R., **1**, 9[43]

Saunders, W. H., **7**, 21[1]

Saunders, W. H., Jr., **6**, 950[1], 954[18], 955[23], 957[26], 959[47], 1011[2], 1013[8], 1017[8]

Saupe, T., **3**, 877[82]

Saurborn, E., **5**, 730[127]

Saus, A., **5**, 650[24]; **6**, 430[100]; **7**, 760[40]; **8**, 451[180]

Sausen, G. N., **3**, 11[52], 17[52]; **5**, 7[49]

Saussime, L., **4**, 599[211], 640[211]

Saussine, L., **6**, 162[187]; **7**, 11[87], 422[139]

Sauter, F., **8**, 242[40]

Sauter, H., **8**, 13[73]

Sauter, R., **7**, 543[22]
Sauve, D. M., **8**, 478[40], 479[40], 516[120]
Sauvé, G., **2**, 303[6], 873[26]
Sauvetre, R., **1**, 235[29]; **3**, 498[107], 525[41];; **4**, 71[16b], 111[156], 113[16b], 139[3]; **5**, 848[141]
Savard, M., **3**, 648[189], 649[189]
Savard, M. E., **8**, 609[55]
Savard, S., **1**, 468[51]
Savariar, S., **5**, 689[71]
Savchenko, T. I., **6**, 525[386], 527[386,407]
Savéant, J.-M., **3**, 574[77]; **4**, 452[13], 453[13,28-31], 458[68], 459[28-31,78,80,81,85], 467[68], 469[68,80,81], 471[31,68,78,139,140,141], 472[29], 473[68,139], 475[30,78,150]; **8**, 135[49]
Savel, W. L., **4**, 1099[180]
Savelli, G., **4**, 426[39]
Saver, J., **4**, 953[8,8a], 954[8a]
Savignac, P., **2**, 482[27], 483[27]; **3**, 201[82,83], 257[117]; **4**, 459[77], 473[77], 474[77]
Saville, B., **4**, 317[552]
Savinova, V. K., **7**, 10[81]
Savitskaya, L. N., **6**, 557[836]
Savoca, A. C., **4**, 1008[133]
Savoea, A. C., **7**, 598[54]
Savoia, D., **1**, 188[73], 189[73]; **2**, 507[19]; **3**, 168[488], 169[488]; **6**, 685[350], 976[4]; **7**, 841[14]; **8**, 124[90], 252[111], 797[40], 842[46], 843[46]
Savon, M.-C., **2**, 138[20]
Savost'yanova, I. A., **2**, 726[122]
Savu, P. M., **3**, 50[267]
Sawa, Y., **5**, 1138[63]
Sawada, H., **1**, 825[51]; **2**, 713[45]; **3**, 251[78,100], 254[78,100]; **4**, 884[103], 892[145]; **5**, 1165[13]; **7**, 73[106], 518[18]; **8**, 798[54]
Sawada, K., **2**, 213[124], 1060[86]; **6**, 566[925]
Sawada, M., **4**, 1089[138], 1091[138]
Sawada, S., **3**, 381[26], 382[26]; **4**, 347[87], 379[115,115b], 380[115b]
Sawada, T., **2**, 810[63], 824[63]
Sawahara, K., **8**, 408[79]
Sawai, H., **6**, 614[93]; **7**, 684[93a]
Sawaki, S., **8**, 29[38], 49[115], 50[119], 66[38,115,119]
Sawaki, Y., **7**, 384[113], 385[113], 438[21], 748[113], 765[165], 769[221]
Sawal, K. K., **4**, 486[37], 505[37]
Sawamura, H., **3**, 300[42]
Sawamura, M., **2**, 233[186], 317[48,49], 318[48,49], 455[19]; **3**, 54[279]; **4**, 221[160]; **6**, 717[108]
Sawamura, T., **1**, 319[159], 320[158,159]
Sawanishi, H., **6**, 535[537,538,539], 538[537,538,539]
Sawdaye, R., **2**, 818[96]; **8**, 949[154]
Sawhney, B. L., **7**, 845[76]
Sawicki, R. A., **3**, 380[11]; **4**, 290[206], 295[253], 398[218,218c], 399[218a-c]; **5**, 605[59]
Sawicki, Y., **7**, 851[18]
Sawistowska, M., **8**, 472[5]
Sawitzki, G., **4**, 121[207]
Sawkins, L. C., **7**, 630[51]
Sawyer, D. T., **3**, 824[23]; **7**, 766[170,171], 851[23]
Sawyer, J. A., **4**, 968[61], 969[61]
Sawyer, J. F., **5**, 225[116], 227[116], 233[116]
Sawyer, J. S., **3**, 196[23]; **5**, 798[68]
Sawyer, T. W., **7**, 346[7]
Sax, K. J., **3**, 892[51]; **8**, 566[450]
Sax, M., **4**, 5[18], 27[84,84a]; **7**, 763[95]
Saxena, M. P., **7**, 579[136]
Saxena, N., **7**, 103[143], 266[112], 267[112]
Saxena, R. K., **8**, 405[26]
Saxon, J. E., **2**, 1016[30]
Saxton, J. E., **4**, 6[21], 680[50]; **8**, 493[21]
Saxton, R. G., **5**, 818[151]
Sayama, S., **4**, 970[72], 972[81]
Sayed, G. H., **5**, 488[197]
Sayed, Y., **3**, 125[305], 126[305], 127[305]; **5**, 692[101]
Sayed, Y. A., **8**, 978[144]
Sayeed, V. A., **4**, 443[186]; **8**, 413[133]
Sayer, B. G., **8**, 675[48], 676[48]
Sayer, T. S. B., **3**, 18[99], 19[100]
Sayo, N., **3**, 977[10], 985[24,25], 986[28], 987[24,31]; **5**, 889[31]; **6**, 876[30,32,34], 885[34], 887[32,61]; **8**, 154[199], 462[267]
Sayre, L. M., **8**, 857[194]
Saytzeff, A., **6**, 435[2], 955[20]
Sazonova, V. A., **7**, 606[160,161]
Scacchi, G., **5**, 69[103,104]
Scaglioni, L., **6**, 171[6]
Scahill, T. A., **5**, 851[170], 1098[118,119], 1099[118,119], 1104[155], 1112[118,119], 1113[155]
Scaiano, J. C., **1**, 699[252]; **4**, 723[39], 736[88], 811[174], 1081[80]; **5**, 153[27], 164[76], 639[122]
Scala, A., **7**, 674[42]; **8**, 565[448]
Scalone, M., **3**, 1023[22]; **4**, 930[48], 931[48]
Scamehorn, R. G., **4**, 453[26], 456[26,46], 457[46], 467[129], 472[46]
Scandström, J., **2**, 365[210]
Scanga, S. A., **7**, 676[65]
Scanio, C. J. V., **4**, 18[62], 20[62g]
Scanlan, T. S., **4**, 599[223], 625[223], 642[223]
Scanlon, T. S., **1**, 827[64b]
Scanlon, W. B., **6**, 936[105]; **7**, 205[61]
Scarborough, R. M., Jr., **3**, 24[149], 25[149]; **5**, 944[242]; **7**, 238[39], 243[64]
Scardiglia, F., **4**, 492[70]
Scarmoutzos, L. M., **1**, 22[115], 41[201]
Scarpa, N. M., **1**, 345[123]; **2**, 6[30]
Scarpati, R., **4**, 1036[45]; **6**, 558[857,858]
Scarpone, S., **5**, 453[69], 455[69]
Scatturin, A., **7**, 777[386]
Scavo, F., **2**, 66[32]; **4**, 598[209], 638[209]
Scechter, H., **2**, 323[24]
Scettri, A., **3**, 512[213], 515[213]; **4**, 391[176]; **5**, 771[145,146,147,148,150,151], 772[145,148,150,151], 780[148]; **7**, 103[137], 260[64], 265[99-102,104], 266[105,107], 267[99-102,104,105,107], 530[14,15,17], 531[17]; **8**, 563[430]
Schaad, L., **3**, 845[38]
Schaad, L. J., **5**, 702[14], 703[14], 740[14]
Schaad, R. E., **4**, 295[259]; **7**, 15[148]
Schaaf, J. v. d., **4**, 869[26]
Schaaf, T. K., **1**, 739[35]; **3**, 364[91]; **4**, 370[26]; **5**, 339[56], 347[56]; **6**, 998[118]; **7**, 686[100]; **8**, 163[40], 269[80,81]
Schaal, V., **8**, 267[70]
Schaap, A., **3**, 262[160], 263[160]; **4**, 238[2]
Schaap, A. P., **7**, 96[90], 98[90]
Schaart, F. J., **1**, 571[282]; **6**, 489[91]
Schabacker, V., **8**, 390[82]
Schachschneider, G., **8**, 49[113], 66[113]
Schacht, U., **7**, 573[117]
Schachtschneider, J. H., **5**, 900[7]
Schach von Wittenau, M., **6**, 265[38]
Schack, C. J., **4**, 347[100]
Schade, C., **1**, 2[11], 8[11], 13[65], 19[11,100,104], 22[11]
Schade, G., **4**, 31[94,94d]; **5**, 809[114]; **6**, 193[207,208,210,211,213]; **7**, 97[94]; **8**, 566[457], 965[68]
Schade, U., **6**, 33[7], 40[7], 57[7]
Schädel, A., **5**, 595[18], 596[18]

Schaefer, D., **4**, 905[212]; **5**, 1138[71], 1157[71,167]
Schaefer, F. C., **4**, 1097[166]
Schaefer, F. M., **7**, 661[44]
Schaefer, G., **4**, 977[94]
Schaefer, H. F., **4**, 484[17], 1070[13]
Schaefer, H. F., III, **5**, 703[15]
Schaefer, J. P., **2**, 797[6], 806[6], 808[6], 813[6], 814[6], 848[6], 849[6]; **3**, 615[10]; **6**, 209[71], 210[72]; **7**, 85[7]
Schaefer, W. E., **7**, 167[98]
Schaefer, W. P., **8**, 519[133], 696[126]
Schaefer-Ridder, M., **5**, 634[77]
Schaeffer, D. J., **2**, 162[143]
Schaeffer, J. R., **7**, 57[28], 58[28], 63[28], 760[43]
Schaeffer, R., **7**, 604[128]
Schaeffer, T. L., **8**, 332[40]
Schaer, B., **6**, 490[110]
Schaer, B. H., **3**, 1038[92]
Schaer, H. P., **7**, 77[120b]
Schäfer, A., **8**, 690[102]
Schäfer, B., **5**, 294[52], 296[53], 1191[30], 1192[32], 1193[30,32]
Schäfer, F., **8**, 270[97]
Schäfer, H., **2**, 748[120], 758[26]; **3**, 975[4], 979[4]; **5**, 63[14], 1197[41]; **8**, 391[88]
Schäfer, H. J., **3**, 564[9], 567[9], 598[9c], 634[12,14,15], 640[105,109], 642[115], 643[12], 644[135,142,157,158,159,160,162,165], 646[135], 647[109,170,197,198], 648[170,186], 649[200], 651[217], 653[15,227], 654[227], 904[131], 975[4], 979[4]; **4**, 345[80], 759[192], 763[192], 805[142], 1023[261]; **6**, 685[347,356]; **7**, 42[33], 236[26], 795[9], 796[14,15], 797[17], 806[69,70,72]; **8**, 54[163], 66[163], 136[50]
Schäfer, P., **6**, 526[391]
Schafer, W., **3**, 592[175]; **4**, 1010[157]; **8**, 654[85]
Schafer, W. M., **4**, 91[90], 92[90d]
Schafer, W. R., **7**, 231[143]
Schäfer-Ridder, M., **8**, 397[143]
Schaffer, S. A., **8**, 30[42], 66[42]
Schaffhausen, J. G., **1**, 118[61]; **7**, 476[62]
Schaffhauser, T., **1**, 286[11]
Schaffner, K., **3**, 216[75], 815[73]; **5**, 200[29], 215[2,5,8], 216[12,13,15-17], 217[2,17,19,20,24,25], 218[30], 219[5,8,12,13,15-17,37,39], 221[12,13,16,30], 222[24,25,63], 223[24,66,72], 224[2], 225[5,8,114], 226[5,8,107,109], 227[114,115], 228[5,8,114,115], 229[5,8,122,123], 230[5,8,39,114,115,127,-128,129], 231[5], 232[127,129], 233[114,115], 760[92]
Schakel, M., **4**, 52[146]; **5**, 906[70], 908[70,74]
Schall, C., **3**, 636[48]; **8**, 321[98]
Schaller, H., **6**, 614[91]
Schaller, R., **2**, 69[45]; **6**, 233[45], 234[45]
Schallhorn, C. H., **8**, 476[27]
Schallner, O., **3**, 623[32]; **4**, 373[67]; **6**, 116[90], 1063[81]
Schaltenbrand, R., **2**, 905[54]
Schambach, R. A., **7**, 720[8]
Schamp, N., **2**, 343[15], 353[102], 357[102], 380[102], 423[32-34], 424[32,35]; **3**, 857[90]; **6**, 500[182], 547[663]; **8**, 36[73], 38[73], 66[73]
Schane, H. P., **2**, 839[173]
Schank, K., **2**, 852[226]; **3**, 174[526], 753[104], 862[7]; **4**, 14[46]; **7**, 740[41]; **8**, 88[36]
Schantz, E. J., **2**, 879[41]
Schanzenbach, D., **2**, 656[156]; **5**, 366[99]
Schaper, W., **3**, 877[88]; **6**, 186[171]
Schappert, R., **4**, 1041[104]
Schardt, B. C., **7**, 8[65]
Schardt, R., **6**, 421[26,27]
Scharf, H.-D., **5**, 160[57], 185[161,162,165,167], 186[171], 187[173], 224[100], 676[3], 686[41], 736[145,145r], 737[145]; **6**, 558[851]; **7**, 262[76], 399[34], 400[47], 429[151]
Scharfbillig, I. M., **4**, 744[134]
Scharfman, R., **7**, 762[77]
Scharkov, V. I., **8**, 140[12]
Scharver, J. D., **3**, 380[11]
Schat, G., **1**, 13[69], 16[86], 746[70]; **5**, 1125[57]
Schatterkerk, C., **6**, 619[116]
Schatz, B., **5**, 702[9,9b]
Schätzke, W., **6**, 172[25]
Schätzlein, P., **6**, 193[218], 558[842]
Schaub, B., **1**, 755[115], 757[119], 812[115], 813[115]; **6**, 173[39], 174[57], 175[39,67]
Schaub, F., **3**, 530[62], 532[62]
Schaub, R. E., **4**, 91[88g]; **8**, 527[41,45], 528[45], 529[45], 530[45], 531[111]
Schauble, J. H., **4**, 347[93], 367[14], 368[14], 369[14]; **7**, 535[49], 536[50]; **8**, 17[114], 21[114], 115[66], 536[167]
Schauder, J. R., **1**, 677[222], 678[215,216], 681[215,216], 683[222], 700[222], 705[222], 708[222], 712[215,216,222], 722[222], 723[222]; **6**, 980[44]
Schauer, R., **2**, 463[84,86], 464[93]
Schaufstätter, E., **8**, 649[61]
Schaumann, E., **2**, 58[14]; **3**, 105[218]; **4**, 992[156]; **5**, 64[28], 115[246,248,249,250,251]; **6**, 419[2,3], 420[17,18], 421[18,32], 424[18], 426[2,3,71,73-75,79], 448[112], 449[114], 454[114], 480[115], 509[277], 538[558], 540[589], 646[104], 647[104], 659[104], 660[104]
Schaumburg, K., **6**, 553[705]
Schaus, J. M., **1**, 383[109]; **2**, 479[19], 481[19]; **3**, 31[182]; **5**, 814[139]; **6**, 721[135]
Schauss, E., **5**, 1131[13]
Schechter, H., **1**, 377[97]; **4**, 12[42], 279[111], 1103[203]; **7**, 500[240]
Scheck, D. M., **5**, 1076[35,37]
Schecker, Von H.-G., **6**, 922[53]
Scheel, D., **4**, 124[214a]
Scheele, J. J., **4**, 459[87]
Scheer, R., **4**, 1085[109]
Scheer, W., **4**, 1085[108]
Scheeren, H. W., **5**, 71[161], 77[261], 434[142], 677[6], 684[37]; **6**, 558[846,848,849,854]
Scheeren, J. W., **2**, 662[22], 663[22], 664[22,29], 867[11]; **5**, 71[159,160,162], 76[241], 77[260,264], 151[13], 431[121], 434[121b], 459[91]; **6**, 518[330], 556[828,831], 558[847,850,855,856], 561[872], 563[895]
Schefczik, E., **2**, 387[332]
Scheffel, D. J., **4**, 299[300]
Scheffer, A., **4**, 35[99]
Scheffer, J. R., **5**, 211[62,63]; **7**, 98[103]
Scheffler, K., **3**, 661[22,23], 666[43]
Scheffold, R., **1**, 142[23]; **2**, 630[4], 681[59]; **3**, 40[223], 41[223], 42[223], 209[21], 213[49], 215[21a], 217[21], 219[21a]; **4**, 12[41], 130[226a,b], 200[5], 209[67], 764[221,222], 765[222], 808[155]; **5**, 200[29], 215[8,9], 219[8,9], 225[8], 226[8,9], 228[8], 229[8], 230[8,9], 234[9], 353[85], 543[113]; **6**, 25[102], 214[97], 652[146]; **8**, 459[228]
Schegolev, A. A., **2**, 723[101], 725[107-109]; **5**, 775[175,176]
Scheibler, H., **6**, 565[919], 566[922]
Scheiblich, S., **6**, 426[74]
Scheibye, S., **2**, 867[12,13]; **6**, 436[19], 437[19]
Scheidt, F., **5**, 683[38], 684[38], 856[210]
Scheigetz, J., **7**, 693[26]; **8**, 315[52], 316[57], 969[94]
Scheinbaum, M. L., **4**, 356[142]; **5**, 856[210], 1003[22]; **6**, 287[181,182]; **7**, 488[159]
Scheiner, A. C., **4**, 484[17]
Scheiner, P., **3**, 381[17]; **4**, 282[142], 1084[93], 1099[185]; **5**, 938[214]; **7**, 478[82]
Scheiner, S., **8**, 89[43]
Scheinmann, F., **3**, 281[41]; **4**, 55[157], 57[157d]; **5**, 560[73], 829[25], 834[50]; **6**, 656[169]

Scheit, K. H., **6**, 610[60], 612[72], 625[159]; **8**, 963[47]
Scheithauer, S., **6**, 421[29], 423[41,45], 436[14,25,26], 437[25,26], 448[14,25], 449[14,25], 450[14,25], 452[25], 453[14,25,26], 454[14], 455[14,26], 456[14,26], 462[8], 472[71]
Schelble, J., **4**, 54[152]; **6**, 519[337]
Schell, F. M., **5**, 137[77]
Schell, H. G., **5**, 689[76]
Schellenbaum, M., **4**, 6[21]
Schellenberger, H., **7**, 709[45]
Scheller, D., **6**, 526[400]
Schellhamer, D. F., **7**, 530[28], 531[28]
Schenach, T. A., **4**, 604[284]
Schenck, G. O., **2**, 169[164]; **4**, 1058[148]; **7**, 96[87], 97[94], 769[219]
Schenck, T. G., **4**, 564[40], 567[40], 590[90]; **6**, 843[88]
Schengrund, C.-L., **2**, 463[87]
Schenk, G., **8**, 354[164]
Schenk, H. P., **3**, 857[91]
Schenk, W. N., **3**, 969[132]; **6**, 657[181], 672[181]
Schenker, E., **8**, 536[169]
Schenker, G., **4**, 608[319]
Schenker, K., **2**, 889[55]; **3**, 379[2], 849[57]; **5**, 732[132,132a]; **6**, 968[111]
Schenkluhn, H., **5**, 1153[147]
Schenone, P., **6**, 776[57]
Schepp, H., **6**, 961[72]
Scherer, H., **6**, 245[123]
Scherer, K. V., **3**, 854[75]
Scherer, K. V., Jr., **8**, 898[24]
Scherer, O., **8**, 755[134]
Scherer, P., **1**, 610[43]
Scherkenbeck, J., **6**, 7[34]
Scherm, H. P., **6**, 183[148]
Scherrer, R. A., **3**, 427[88]
Schertler, P., **5**, 1185[2]
Scheuer, P. J., **6**, 294[236]; **7**, 406[75]
Scheuermann, H.-J., **1**, 672[190,191], 674[190,191], 714[190,191], 715[190], 718[190,191], 722[190,191], 867[80]
Scheunemann, K. H., **5**, 85[1]
Scheurs, P. H. M., **5**, 949[282]
Scheutzow, D., **5**, 986[37]; **8**, 267[70], 657[98]
Scheweizer, W. B., **1**, 299[61], 316[61]
Schexnayder, M. A., **5**, 178[137], 217[18], 220[46], 221[46], 223[70], 224[99]
Schiavelli, M. D., **4**, 299[300]; **6**, 836[55]
Schick, H., **3**, 55[280]; **6**, 654[154]; **7**, 586[167]
Schick, K. P., **4**, 608[319]
Schick, L. E., **5**, 787[10]
Schicke, H. G., **6**, 428[85], 432[118]
Schickedantz, P. D., **6**, 431[111]
Schieb, T., **7**, 725[33]
Schiedl, G., **1**, 773[203,203b]
Schield, J. A., **8**, 366[41]
Schiemenz, G. P., **2**, 735[18,19], 760[44]; **6**, 175[77]
Schier, A., **1**, 10[46]; **6**, 173[50]
Schierling, P., **6**, 193[218], 558[840,841,842,843,844]
Schierloh, C., **2**, 547[113], 551[113]; **3**, 56[285]; **4**, 229[236]
Schiess, M., **1**, 149[48], 170[48]; **2**, 897[18,19], 1086[28], 1096[28]
Schiess, P., **4**, 123[210b], 125[210b]; **5**, 386[133,133a,b], 388[133b], 394[146], 395[146], 681[26], 692[104], 710[49], 741[153], 742[158], 809[111]
Schiess, P. W., **1**, 894[158]; **6**, 1041[2], 1042[2]
Schiesser, C. H., **4**, 781[6], 786[6], 787[6], 820[219], 827[6]
Schiessler, R. W., **8**, 328[14], 329[14]
Schiffer, R., **4**, 1102[201]
Schiffmann, D., **8**, 398[144]
Schifman, A. L., **6**, 655[167]
Schigeko, E. J., **3**, 530[66], 533[66]
Schiketanz, A., **3**, 331[196]
Schildcrout, S. M., **8**, 425[48], 474[14], 475[14], 476[14]
Schildknecht, H., **7**, 576[124]
Schill, G., **1**, 15[81], 656[150], 658[150]; **3**, 87[85], 123[250], 124[250], 125[250], 135[359], 136[359], 137[359], 139[359], 142[359], 247[45], 248[45], 628[47]; **4**, 565[44]; **5**, 731[130a]; **6**, 734[5]; **7**, 762[79]; **8**, 214[41]
Schiller, A. M., **8**, 532[130]
Schilling, G., **3**, 872[59]
Schilling, L. M., **7**, 875[113]
Schilling, P., **3**, 322[142a]; **7**, 17[174]
Schilling, S. L., **1**, 41[203]
Schilling, W., **1**, 95[72]; **5**, 611[72]; **6**, 440[75]
Schillinger, E., **2**, 902[40]
Schillinger, W. J., **7**, 682[83]
Schilt, W., **4**, 283[149]
Schimdt, G., **1**, 551[68]
Schimperna, G., **2**, 103[28], 605[57], 614[57], 637[58], 639[58,62], 640[58,62], 642[73], 643[73], 644[73], 930[132,133], 932[132,133]; **4**, 159[82], 218[145]; **5**, 102[176,178]
Schimpf, R., **4**, 887[126]; **8**, 758[167]
Schimpff, G. W., **8**, 292[45]
Schindewolf, U., **8**, 114[57], 524[11]
Schindler, H., **5**, 555[49]
Schink, K., **6**, 525[387]
Schinski, W. L., **5**, 904[54]
Schinz, H., **3**, 345[20], 351[39], 358[65]; **5**, 833[45]; **7**, 154[15]; **8**, 141[44]
Schinzel, E., **3**, 807[24], 813[66]
Schinzer, D., **2**, 89[35]; **3**, 348[30,31], 358[31]; **4**, 98[116], 115[182], 121[207], 155[73,74]; **6**, 734[2]
Schipchandler, M. T., **2**, 332[62]
Schipper, E., **7**, 656[17]
Schipper, P., **1**, 661[165,166], 663[165,166], 672[166], 700[166], 704[166]; **8**, 967[82]
Schirch, P. F. T., **5**, 618[5]
Schirlin, D., **2**, 71[54]
Schisla, R. M., **4**, 93[93c]
Schjånberg, E., **4**, 274[56]
Schlachter, S. T., **1**, 759[134]
Schlademan, J., **2**, 753[2,2c]
Schlaf, H., **8**, 657[98]
Schläfer, L., **8**, 898[22]
Schlageter, M. G., **6**, 1059[70], 1066[70]; **7**, 376[83]
Schlamann, B., **2**, 371[261]; **5**, 76[239]
Schlapbach, A., **6**, 859[175]
Schlapkohl, K., **6**, 444[98]
Schlatmann, J. L. M. A., **5**, 700[8], 737[8]
Schlecht, M. F., **2**, 611[101]; **3**, 226[194], 265[189], 380[13]; **4**, 161[86b]; **7**, 158[37], 530[16], 587[170], 823[37]; **8**, 332[44]
Schlecker, R., **1**, 474[107]; **3**, 194[9]
Schlegel, H. B., **1**, 506[8]; **4**, 1070[14]; **6**, 133[4]; **8**, 2[7]
Schleich, K., **5**, 1154[155]
Schleifer, L., **5**, 223[85], 224[85], 740[151]
Schleitzer-Rust, E., **1**, 231[3]
Schlesener, C. J., **7**, 850[10], 852[37]
Schlesinger, A. H., **7**, 764[104]
Schlesinger, H. I., **8**, 1[1], 26[1,2], 735[9], 736[9]
Schlessinger, R. H., **1**, 3[24], 32[157,158], 527[105,106], 564[191], 567[224], 791[268]; **2**, 106[47], 187[41], 189[46,47], 209[46], 221[46a], 482[36], 484[36], 805[45], 811[71], 824[71]; **3**, 33[190], 34[197], 43[238], 135[358], 136[358], 137[358], 139[358], 142[358], 143[358], 675[72]; **4**, 10[32,32a-g,33,33a,34], 11[32e], 23[33d], 30[88,88g], 48[139,139f], 107[144], 109[148], 113[166,171,171e],

125[216,216f], 249[125], 258[125], 262[306], 611[357]; **5**, 338[54], 541[111]; **6**, 134[26], 186[170], 647[113], 717[113], 1013[16], 1020[48]; **7**, 105[151]; **8**, 925[12]
Schletter, I., **6**, 116[94]
Schleyer, P. von R., **1**, 2[11], 8[11], 12[61], 13[65], 17[211], 19[11,100,102-104], 20[108], 22[11], 23[123-125], 25[128], 29[147], 35[172], 41[196], 287[17], 476[125], 477[125], 487[1,2], 488[1,2], 859[64,66]; **3**, 66[12], 74[12], 194[4,11], 334[219,220]; **4**, 872[40,41], 1016[208,209]; **5**, 65[71], 794[45], 850[152], 984[33]; **6**, 172[11], 749[102]; **7**, 26[47]; **8**, 91[53,61,64], 318[59], 319[79], 320[80], 322[59], 331[33], 334[60], 342[109], 724[168,169,169b], 904[57,57b], 910[57]
Schlicher, J. W., **2**, 547[111], 551[111], 710[28]
Schlieper, A., **3**, 822[5], 834[5]
Schlimgen Davis, K., **5**, 212[69]
Schlingloff, G., **3**, 667[46]
Schlitt, R., **5**, 856[217]
Schloemer, G., **5**, 758[82]
Schlögl, K., **2**, 365[215]
Schlosberg, R. H., **3**, 299[34b], 330[192,193], 333[34,208,211b], 334[213]; **6**, 749[101]; **7**, 6[28]
Schlosser, M., **1**, 2[5], 10[51], 180[28], 755[115], 757[119], 758[123], 812[115], 813[115]; **2**, 5[20,20a,21], 6[20], 13[21,21c,58], 14[21c,58], 21[20], 66[37], 977[6]; **3**, 99[182], 101[182], 194[7], 215[62], 244[20], 261[149], 466[184]; **4**, 70[9], 139[1], 869[22], 1001[38], 1020[237], 1035[42]; **6**, 173[39], 174[57], 175[39,67], 685[349], 958[31], 959[31], 976[1]; **7**, 99[106,107], 596[39], 856[68]
Schlosser, W., **6**, 190[193]
Schlubach, H. H., **3**, 553[13]
Schlude, H., **2**, 760[45]
Schlund, R., **8**, 682[83], 683[95], 686[95]
Schlunke, H.-P., **3**, 822[12], 831[12], 835[12b]
Schluter, G., **7**, 221[30]
Schlüter, K., **6**, 970[122]
Schmack, L. G., **4**, 1016[206]
Schmakel, C. O., **8**, 592[65]
Schmalstieg, L., **4**, 1007[129]
Schmalz, D., **1**, 2[10]
Schmalz, H.-G., **2**, 547[113], 551[113]; **3**, 56[285]; **4**, 229[235,236], 1055[138]
Schmalzl, K. J., **2**, 294[80]; **3**, 20[109]
Schmelzer, H.-G., **3**, 890[31], 901[111], 904[135]
Schmerling, L., **3**, 320[133], 331[201]; **4**, 276[71], 283[71], 313[464]; **7**, 7[46,49], 15[152]
Schmersahl, P., **8**, 903[50]
Schmetzer, J., **6**, 518[331]
Schmickler, H., **3**, 593[179]; **7**, 725[33]
Schmid, A., **3**, 223[155]
Schmid, B., **3**, 105[220], 113[220], 155[432]
Schmid, C. R., **7**, 308[20]
Schmid, G., **5**, 96[104], 97[104], 98[104]; **6**, 5[24], 186[169], 193[207,208], 195[223,224,225], 197[169], 449[115]
Schmid, G. H., **4**, 329[1], 330[1c,d,5], 339[41], 342[62,65,69,71], 344[1], 350[1], 351[1]; **7**, 769[230]
Schmid, H., **2**, 138[22]; **3**, 809[39,40], 957[109]; **4**, 14[46], 587[23], 1081[73,78], 1084[95]; **5**, 596[26], 597[26], 608[26a], 638[117], 681[27], 707[32], 709[45], 712[45d], 713[32], 799[72], 812[131], 822[164], 834[50,53], 837[67], 850[146], 856[67,199], 857[67,199,226,227], 858[199], 876[3], 877[5], 894[43,45], 1130[8]; **6**, 898[104]; **8**, 812[2], 813[2]
Schmid, J., **4**, 985[128]
Schmid, K., **5**, 876[3]
Schmid, M., **5**, 799[72]; **6**, 204[20]; **8**, 205[159], 560[405]
Schmid, P., **1**, 699[252]
Schmid, R., **3**, 369[123,124], 372[124], 957[109]; **5**, 596[26], 597[26], 608[26a], 894[45]; **6**, 898[104]
Schmidbaur, H., **1**, 10[46]; **6**, 173[50], 175[67], 176[91], 177[119], 178[119,123], 179[125], 180[128,129], 182[141], 183[148], 188[180], 190[199,200,202], 196[199,202,227,228,230,231,232,233,234]; **8**, 859[217]
Schmidhauser, J. C., **5**, 721[100]
Schmidle, C. J., **2**, 782[22]
Schmidlin, C., **5**, 419[74], 420[75]
Schmidlin, J., **8**, 268[74]
Schmidpeter, A., **6**, 196[229], 509[265]
Schmidt, A., **4**, 1007[128]; **6**, 531[437], 566[922]
Schmidt, A. H., **8**, 405[33], 406[37]
Schmidt, B., **3**, 580[104]
Schmidt, C., **8**, 15[89]
Schmidt, D., **2**, 68[40], 94[51]; **3**, 891[36], 909[153]
Schmidt, D. L., **4**, 337[34]
Schmidt, E., **6**, 420[22]
Schmidt, E. A., **6**, 569[935]
Schmidt, F., **3**, 557[39]
Schmidt, G., **1**, 762[141]; **7**, 272[131], 273[131], 503[277]; **8**, 837[13b]
Schmidt, G. M. J., **3**, 382[36]
Schmidt, H., **4**, 45[130,130c]; **7**, 99[113], 221[32]
Schmidt, H.-J., **2**, 476[5]; **4**, 872[43]; **6**, 269[74]; **7**, 236[26]
Schmidt, H. L., **7**, 778[419]
Schmidt, H.-W., **4**, 440[171]; **6**, 554[726]
Schmidt, J., **3**, 828[44]
Schmidt, J. G., **2**, 134[6]
Schmidt, K., **2**, 399[15]; **7**, 725[33]
Schmidt, M., **1**, 70[63], 141[22]; **2**, 902[45]; **3**, 134[339], 135[339]; **6**, 177[113], 182[113], 188[113], 194[219,220]
Schmidt, O., **4**, 5[19]
Schmidt, O. Th., **6**, 36[18]
Schmidt, P., **6**, 519[338]; **8**, 659[103]
Schmidt, R., **4**, 1007[112]; **6**, 501[203], 531[203]; **8**, 802[86]
Schmidt, R. R., **2**, 662[3], 1054[58], 1070[58]; **3**, 105[220], 113[220], 155[432], 253[93]; **4**, 121[205a], 740[115]; **5**, 428[109], 430[118], 431[123,123a,b], 432[125,133], 433[135,135b], 451[29], 461[110], 464[110-113], 466[110-113], 485[180], 492[29], 501[29,262,265,266,271], 502[272,273]; **6**, 33[1,2], 34[1,2], 37[1,34], 40[1,2], 43[51], 46[1], 48[1], 49[1], 50[1,2,101,102], 51[1,2,101,102,111-114], 52[115], 53[2,117,119,121,122], 54[1,2,123-127], 56[133,134], 57[1,2,133,134,137,138,142], 59[143], 60[149], 62[127], 73[25], 533[495], 534[520], 655[166], 846[102]; **7**, 418[130a]; **8**, 389[67], 616[101], 624[101], 640[24]
Schmidt, S. J., **3**, 1029[55]; **8**, 933[50]
Schmidt, T., **1**, 405[27]; **5**, 307[90]
Schmidt, U., **1**, 188[73], 189[73], 219[56]; **2**, 219[142], 735[18,19], 760[44]; **6**, 430[101], 443[90], 562[885,886], 637[32,32c]; **8**, 535[166]
Schmidt, V., **3**, 903[126]
Schmidt, W., **4**, 682[58]; **5**, 705[25]; **7**, 248[113], 808[76]
Schmidtberger, S., **1**, 153[59], 154[59], 295[51,52], 336[74], 340[74]; **2**, 307[15], 310[15], 640[68], 641[68]
Schmiegel, K. K., **4**, 83[65a]; **7**, 111[190]; **8**, 528[61]
Schmiegel, W. W., **8**, 528[57,61]
Schmierer, R., **3**, 45[251]
Schmiesing, R. J., **1**, 794[279]; **4**, 38[108], 339[44]; **5**, 814[136]; **7**, 520[30]; **8**, 44[105], 66[105]
Schmit, C., **1**, 683[227], 714[227], 715[227], 717[227], 718[227]; **3**, 786[42]
Schmitt, E., **2**, 424[35]; **6**, 526[391]
Schmitt, G., **8**, 755[130], 758[130]
Schmitt, H., **6**, 51[113]
Schmitt, H.-G., **3**, 174[526]
Schmitt, J., **8**, 383[17]
Schmitt, J. L., **3**, 95[155]
Schmitt, M., **8**, 476[29]
Schmitt, R. J., **6**, 109[44], 110[56]; **7**, 746[87]; **8**, 91[60]

Schmitt, R. K., **8**, 13[74]
Schmitt, S. M., **5**, 107[199]
Schmitthenner, H. F., **5**, 567[104]; **6**, 756[128]
Schmitz, A., **1**, 56[29]; **2**, 646[84]; **5**, 29[1]; **6**, 561[873], 644[88]
Schmitz, E., **2**, 364[203]; **4**, 307[389]; **5**, 15[101]; **6**, 494[130], 562[883]; **7**, 470[15,16], 471[23], 474[23], 746[89,93]
Schmitz, H., **4**, 1104[211]
Schmitz, R., **8**, 370[85]
Schmolka, I. R., **8**, 231[141]
Schmook, F. P., **2**, 355[126]
Schmuff, N. R., **2**, 74[77]; **3**, 168[489], 169[489]; **4**, 589[86], 599[216], 624[216]
Schmuff, R. N., **4**, 681[52], 682[52]
Schmunes, E., **2**, 385[324]
Schmüser, W., **6**, 444[98]
Schnaithmann, M., **5**, 933[184]
Schnatter, W. F. K., **5**, 1102[148], 1103[148], 1105[164]
Schneebeli, J., **1**, 373[91], 375[91], 376[91]; **2**, 996[47], 1077[154]; **7**, 230[133]
Schneider, A., **5**, 1141[84]; **7**, 363[37]
Schneider, C. S., **4**, 54[152]; **6**, 506[227], 519[337]
Schneider, D. F., **1**, 760[136]; **3**, 623[32]; **4**, 710[52]; **6**, 184[152], 189[185]
Schneider, F., **1**, 524[85]; **2**, 229[169], 652[127]
Schneider, G., **2**, 838[171]; **6**, 276[122], 612[72]; **8**, 192[98]
Schneider, H.-J., **4**, 274[60]; **7**, 13[121,122], 50[70], 247[106]; **8**, 161[19], 541[212]
Schneider, H.-P., **3**, 666[42]
Schneider, J., **1**, 424[101]; **2**, 329[47]; **5**, 66[77]; **8**, 260[1], 736[19], 739[19], 744[50], 756[50]
Schneider, J. A., **1**, 52[14], 108[9], 109[12], 110[9,12], 126[9], 134[9,12], 153[56], 336[69], 460[2]
Schneider, K., **8**, 260[1]
Schneider, L., **6**, 980[36]; **8**, 880[62]
Schneider, M., **3**, 587[141]; **4**, 955[13]; **6**, 463[25]
Schneider, M. J., **4**, 370[37]
Schneider, M. P., **2**, 456[27]; **5**, 804[97], 933[184], 971[3], 972[9], 973[3,9,12,13], 974[9], 991[47], 992[47]; **7**, 397[30]
Schneider, M. R., **2**, 740[57]
Schneider, P., **2**, 785[46]; **4**, 204[36], 869[22]; **5**, 38[22c], 45[35]
Schneider, R., **3**, 530[62], 532[62]; **5**, 734[137]; **6**, 535[536], 538[536]; **7**, 741[47]
Schneider, R. A., **4**, 23[70]
Schneider, R. S., **2**, 157[120]; **5**, 853[171]; **8**, 544[275]
Schneider, S., **5**, 307[89]; **6**, 489[100], 736[29]
Schneider, U., **3**, 495[93b]
Schneider, W. P., **2**, 148[78]
Schneiders, G. E., **3**, 219[112], 676[76]
Schnell, H., **4**, 293[236]
Schnelle, G., **2**, 33[121]
Schneller, J., **6**, 273[96]
Schneller, S. W., **2**, 359[158], 361[158], 376[158], 388[158]
Schnering, H. G. V., **1**, 168[116a]; **4**, 758[191]
Schnieder, M., **2**, 465[106]
Schnoes, H. K., **2**, 879[41]
Schnoes, H. N., **6**, 989[79]
Schober, P. A., **1**, 520[72]; **2**, 66[33], 75[33,82a]; **4**, 119[192b,193], 159[85], 226[190,191,194], 249[129], 258[129,244]; **6**, 154[145], 864[192]
Schöberl, A., **6**, 540[588]
Schöberl, V., **2**, 369[248]
Schobert, R., **3**, 286[59]; **6**, 193[209], 194[219]; **8**, 889[130]
Schoch, W., **3**, 915[13], 965[13]
Schoch-Grübler, U., **2**, 482[23], 483[23]
Schock, L. E., **8**, 447[134], 670[9], 671[9]
Schöde, D., **6**, 288[187]
Schoeller, W., **5**, 857[228]
Schoemaker, H. E., **2**, 1049[20,25], 1050[20], 1054[57], 1062[57,100], 1064[25], 1072[57]; **6**, 658[183], 745[82]
Schoenberg, A., **3**, 1021[13], 1028[47], 1034[78]
Schoeneck, W., **8**, 568[486]
Schoenen, F. J., **1**, 126[87,90], 757[122]; **5**, 736[143,145], 737[145], 843[117]
Schoenenberger, H., **1**, 360[27]
Schoenewaldt, E. F., **3**, 822[13], 829[13]; **8**, 54[159], 66[159]
Schoenheimer, R., **6**, 685[352]
Schoening, C. E., **3**, 804[12]
Schoenthaler, A. C., **4**, 274[57], 282[57]
Schöffer, A., **3**, 625[41]
Schoffstall, A. M., **5**, 252[45]
Schofield, C. J., **4**, 744[134]
Schofield, J. A., **3**, 818[94]
Schofield, K., **3**, 689[118]; **6**, 110[45], 291[216]
Schofield, R. A., **4**, 1040[73], 1043[73]
Schohe, R., **4**, 379[117], 1079[65]; **5**, 260[65,66], 261[65,66]; **7**, 439[36]; **8**, 70[223], 647[54]
Scholes, G., **4**, 115[180c], 688[67]; **5**, 715[78], 985[36]
Scholl, B., **4**, 1033[21], 1037[21], 1040[21]
Scholl, H.-J., **4**, 1002[47]
Scholl, P. C., **7**, 43[42], 802[50]
Scholl, T., **3**, 737[30]; **5**, 15[109]
Schollenberger, C. S., **2**, 529[20]
Scholler, D., **8**, 817[33]
Schöllkopf, U., **1**, 55[26], 630[22,29], 641[105], 722[275], 846[19a,b], 847[19b], 850[19b]; **2**, 361[175], 498[70-78], 499[71-76], 501[78], 1084[11]; **3**, 53[274], 194[13], 195[17], 197[34], 303[53], 419[36], 921[35], 922[35a,38], 924[35a], 927[52], 975[4], 976[5,7,8], 979[4]; **4**, 89[84g], 111[152d], 116[189], 222[179,180,181], 224[181], 966[54], 1016[209], 1038[60]; **5**, 116[267,268], 117[272], 187[174], 1003[20], 1007[39], 1008[39]; **6**, 531[431,432,433,434,435,436,443], 535[532], 876[29]; **7**, 232[155]
Scholmann, N., **4**, 784[15]
Scholten, H. P. H., **7**, 230[134]
Scholten, J. J. F., **8**, 418[11], 437[11]
Scholz, A., **6**, 226[9]
Scholz, D., **1**, 424[101]; **3**, 862[9-11], 863[11-13]; **6**, 438[41], 508[287]; **7**, 768[206]
Scholz, K.-H., **5**, 160[56]
Scholz, M., **7**, 355[43]
Scholz, S., **2**, 648[90], 1058[72]; **5**, 407[28,28b]
Schomaker, J. H., **4**, 1099[185]
Schomaker, V., **4**, 275[67], 279[67], 287[67]
Schomburg, D., **2**, 651[121]; **6**, 509[265]
Schomburg, G., **4**, 868[12], 874[50,51,54,55], 887[12]; **5**, 30[2]
Schön, N., **2**, 368[234]
Schonber, A., **7**, 230[127]
Schönberg, A., **3**, 563[117], 564[11], 567[11], 582[111,112,117]; **4**, 1093[148]; **5**, 433[135,135a]; **6**, 519[339]; **7**, 144[155]
Schönberg, H., **5**, 1157[168]
Schönberger, N., **6**, 846[102]
Schone, R., **4**, 1079[64]
Schönfelder, M., **3**, 482[3]
Schönfelder, W., **5**, 1146[108]
Schönhammer, B., **8**, 459[238]
Schoolenberg, J., **8**, 533[146]
Schoon, D., **5**, 90[57], 95[57]
Schöpf, C., **2**, 943[168,169], 970[88]
Schor, J. M., **2**, 741[67]
Schore, N. E., **2**, 597[7]; **3**, 1025[34]; **4**, 72[29]; **5**, 153[24], 166[91], 561[79], 1037[2], 1039[11,17], 1041[19], 1043[22], 1046[19,30], 1049[22], 1050[17,34], 1051[34], 1052[17,19,37], 1053[39], 1057[51-53], 1062[51-53,59], 1133[26], 1146[26]

Schormann, N., **8**, 335[67]
Schors, A., **7**, 706[23]
Schorta, R., **5**, 229[122]
Schortt, A. B., **6**, 495[151]
Schossig, J., **3**, 921[35], 922[35a,38], 924[35a]
Schostarez, H., **5**, 21[149], 22[149], 779[197]
Schott, A., **3**, 482[3]
Schotten, T., **5**, 973[15], 975[15]
Schouten, A., **2**, 124[204], 125[204]
Schouten, H. G., **8**, 614[90]
Schow, S., **6**, 1059[70], 1066[70]; **7**, 376[83]
Schow, S. R., **1**, 129[93], 779[224]; **2**, 651[122]; **3**, 26[165]; **7**, 73[103]
Schowen, R. J., **8**, 89[43]
Schowen, R. L., **1**, 314[137], 315[137]
Schrader, G., **6**, 432[118]
Schrader, L., **5**, 646[5,6]
Schrader, T., **1**, 373[86], 374[86]
Schrall, R., **3**, 691[135]
Schramm, J., **2**, 935[147]
Schramm, S., **7**, 471[23], 474[23]
Schramm, S. B., **3**, 325[161,161a]
Schrauzer, G. N., **4**, 761[199]; **5**, 1141[83]
Schreck, J. O., **6**, 556[820]
Schreck, M., **4**, 1010[160]
Schrecken, H., **3**, 482[3]
Schreiber, E. C., **8**, 566[450]
Schreiber, F. G., **3**, 693[142]
Schreiber, J., **2**, 899[27], 901[27]; **5**, 714[75a]; **6**, 831[7], 1059[62,64,66]; **7**, 482[118]
Schreiber, R. S., **6**, 288[184]
Schreiber, S. L., **1**, 8[39], 200[96], 297[58], 329[40], 420[83], 568[230,233], 768[172], 798[290], 800[298], 883[126], 898[126]; **2**, 42[148], 45[148], 505[9]; **3**, 217[95], 227[209], 545[120], 979[14]; **4**, 53[150], 817[206]; **5**, 130[41], 152[20], 167[95], 169[109], 170[112,121], 171[114,116,121], 172[119-122], 174[124], 176[112,116,125], 184[121], 185[124], 461[99], 462[99], 466[119], 467[118,119], 514[9], 527[9], 545[122], 736[143,145], 737[145], 809[121,123], 843[119], 1055[47], 1062[59]; **6**, 8[39], 14[56], 16[56], 692[409], 717[115,116], 903[135], 1067[102,103]; **7**, 361[23], 396[26], 416[26,124], 549[47], 676[61]; **8**, 224[101], 933[48]
Schreiber, T. S., **7**, 396[26], 416[26]
Schreibmann, A. A. P., **3**, 564[13]
Schrempf, G., **6**, 495[145]
Schreuder, A. H., **8**, 33[58], 66[58]
Schreurs, P. H. M., **4**, 309[412]
Schrier, J. A., **7**, 66[72]
Schriesheim, A., **7**, 759[5,13,16], 760[36], 761[36]
Schrinner, E., **6**, 33[7], 40[7], 57[7]
Schriver, G. W., **1**, 41[203]
Schrock, R. R., **1**, 140[7], 174[15], 743[50], 749[78], 812[50], 816[78]; **4**, 485[31]; **5**, 1115[2], 1116[2,11,12], 1117[11,16], 1118[11], 1121[16], 1122[2d], 1125[2d,61,65]; **8**, 152[174], 445[46-50], 450[48], 458[48,50]
Schrödel, R., **8**, 274[136]
Schroder, C., **5**, 164[75], 176[75]
Schröder, E., **3**, 848[52]
Schröder, G., **5**, 709[46], 717[90a-c]; **7**, 412[104], 413[104], 429[158], 430[158], 442[46a], 446[62]
Schröder, J., **6**, 421[27]
Schröder, M., **7**, 166[88], 437[6], 438[6], 439[6,24], 866[88], 867[88]
Schröder, S., **6**, 119[115]
Schröder, W., **1**, 310[106]
Schroeck, C. W., **1**, 532[134], 535[146], 825[49]; **4**, 987[136], 989[136]
Schroeder, B., **4**, 760[196]
Schroeder, D. C., **6**, 423[39], 424[39], 428[39], 432[39]
Schroeder, D. R., **4**, 1033[36]; **6**, 124[144]
Schroeder, F., **3**, 174[526]
Schroeder, M. A., **8**, 451[181,182], 567[459,460], 765[11], 778[11g]
Schroeder, M. C., **1**, 635[84], 636[84], 637[84], 640[84], 672[84]; **3**, 87[96], 104[96]
Schroeder, T., **8**, 446[73]
Schroeder, W., **8**, 881[79]
Schroedter, G., **5**, 752[49]
Schroek, C. W., **1**, 722[278]
Schroepfer, G. J., Jr., **7**, 264[91]; **8**, 872[9], 873[9], 881[81], 882[81]
Schröer, U., **2**, 587[141]
Schroeter, S. H., **5**, 158[16], 159[16], 170[16]; **7**, 97[94]; **8**, 14[87], 141[45]
Schroetter, H. W., **8**, 100[116]
Schroll, G., **8**, 478[40], 479[40]
Schröppel, F., **7**, 482[114]
Schroter, D., **5**, 260[66], 261[66]; **7**, 416[122]; **8**, 70[223]
Schröter, G., **3**, 887[1b], 890[1], 891[1,1b], 894[1], 897[1b], 905[1b]
Schröter, R., **2**, 953[1]; **8**, 367[53]
Schroth, G., **4**, 608[319], 874[48,50,51,55]; **5**, 30[2], 297[55], 641[134], 1192[31], 1197[31,41]; **6**, 179[124]
Schroth, W., **2**, 712[40]; **6**, 508[287,288], 518[334], 551[685]
Schrott, U., **2**, 33[121]
Schubert, B., **1**, 20[106,108,109], 21[110], 39[190]; **7**, 253[18]
Schubert, H., **1**, 359[17], 380[17], 381[17]; **2**, 514[50], 524[50]; **3**, 39[213]; **6**, 728[208,209]; **8**, 388[60]
Schubert, H.-J., **2**, 744[98]
Schubert, H. W., **2**, 1088[51]
Schubert, K., **6**, 564[907]
Schubert, M., **3**, 687[112]
Schubert, P. R., **4**, 95[98]
Schubert, R., **2**, 662[6]; **4**, 1073[21], 1076[21], 1090[21], 1092[21], 1098[21], 1102[21]
Schubert, R. M., **8**, 875[34], 876[34]
Schubert, U., **1**, 10[46], 25[128]; **3**, 369[108]; **4**, 976[100]; **5**, 689[73], 1065[1], 1066[1,1a], 1074[1], 1083[1], 1084[1], 1090[88], 1093[1,96], 1098[96b], 1112[96b]; **6**, 196[232], 500[179]
Schubert, W., **4**, 1072[16]
Schubert, W. M., **1**, 292[26]; **2**, 753[2]; **6**, 959[40]
Schuchardt, J. L., **2**, 801[24]; **3**, 1060[45]
Schuchardt, U., **5**, 289[37,38], 290[39], 293[45], 1185[1], 1188[20], 1190[27], 1191[27b], 1195[35], 1197[40,44]
Schuchardt, W., **3**, 724[14]
Schuck, J. M., **8**, 497[40]
Schücking, G., **2**, 143[53]
Schuda, A. D., **6**, 810[72]
Schuda, A. D. C., **3**, 224[172]
Schuda, P., **5**, 330[34]; **7**, 241[59], 438[22]
Schuda, P. F., **2**, 905[55], 907[55], 908[55], 910[55], 911[55]; **5**, 410[40]; **6**, 189[190], 651[137]; **7**, 241[59]
Schudde, E. P., **4**, 1036[54]
Schudel, P., **2**, 612[104]; **3**, 698[159]; **5**, 714[75a]
Schueller, K., **5**, 69[106]
Schüep, W., **8**, 794[14]
Schuerch, C., **6**, 36[20], 47[78], 49[90]
Schuett, W. R., **8**, 454[197], 455[197]
Schuette, H., **8**, 189[56,57]
Schug, R., **5**, 76[242], 78[276,277,278]
Schuh, K., **6**, 531[456]
Schul, W., **2**, 372[269], 373[269]
Schulbach, H., **8**, 308[4]
Schule, G., **7**, 416[124]
Schulenberg, J. W., **2**, 758[25]; **7**, 690[14]

Schuler, F. W., **5**, 856[194]
Schüler, H., **6**, 558[841]
Schuler, R. H., **4**, 719[22]
Schüll, V., **4**, 181[71], 1040[69]
Schüller, M., **6**, 57[140]
Schulman, S., **4**, 24[72,72b]
Schulte, G., **2**, 675[52]; **5**, 736[143,145], 737[145]; **6**, 900[114], 1067[103]; **7**, 237[37], 245[74], 361[23]; **8**, 224[101]
Schülte, K. E., **4**, 317[555], 883[100], 884[100]
Schulte-Elte, K. H., **2**, 169[164], 540[69]; **5**, 10[76], 757[80], 761[80]; **6**, 1058[58], 1059[64], 1067[108]; **7**, 818[17]
Schulte-Frohlinde, D., **3**, 665[41]
Schulten, W., **2**, 377[283]
Schultess, A. H., **6**, 535[542], 538[542]
Schultheiss-Reimann, P., **6**, 657[178]
Schulthess, W., **4**, 70[1]
Schultz, A. G., **1**, 506[12], 769[195]; **3**, 23[135], 51[270]; **4**, 8[28], 10[34], 14[46,46b], 106[140a], 159[80], 370[38], 372[38], 390[38], 957[21], 1093[151], 1095[151], 1101[192,194]; **5**, 225[97,98], 558[63], 571[119], 572[124], 582[177], 938[217,218]; **6**, 835[46]; **7**, 120[5], 261[70], 276[151]; **8**, 449[151], 490[10], 505[82], 507[82], 508[87]
Schultz, E. M., **2**, 971[92]
Schultz, H., **7**, 248[108]
Schultz, H. P., **8**, 236[4], 242[4], 247[4], 248[4], 249[4]
Schultz, H. S., **7**, 766[173]
Schultz, J., **3**, 673[70], 674[70b]; **8**, 623[148]
Schultz, J. A., **5**, 338[54]
Schultz, M., **6**, 646[99,99b]
Schultz, P. G., **5**, 855[190,192]; **8**, 206[167]
Schultz, R. G., **4**, 587[40]
Schultze, H., **2**, 759[32]
Schultze, K. M. L., **6**, 435[5b]
Schultze, L. M., **5**, 1076[34], 1107[168,169], 1111[34]
Schulz, A., **8**, 853[144]
Schulz, B., **6**, 436[13]
Schulz, C. R., **8**, 40[89], 66[89]
Schulz, D., **5**, 1131[15]
Schulz, G., **2**, 782[30]; **7**, 498[223]
Schulz, H., **1**, 669[181,182], 670[181,182]; **6**, 182[134]
Schulz, H.-J., **3**, 509[178]
Schulz, J., **7**, 506[303]
Schulz, M., **4**, 611[345]; **7**, 140[130], 141[130]
Schulz, W., **6**, 554[802], 576[802], 581[802]
Schulz, W. H., **7**, 725[33]
Schulze, A., **2**, 153[106]
Schulze, B., **2**, 360[167]
Schulze, E., **4**, 222[181], 224[181]
Schulze, K., **3**, 382[38]; **6**, 1044[20]
Schulze, P.-E., **7**, 47[55]
Schulze, T., **1**, 185[56]; **2**, 29[106]; **7**, 549[45]
Schulze, U., **6**, 462[9], 472[67]
Schulz-Popitz, C., **4**, 1006[104]
Schumacher, D. P., **8**, 605[17]
Schumacher, J., **5**, 497[227]
Schumacher, J. N., **3**, 407[150]
Schumacher, M., **5**, 451[5,9], 513[3], 514[3]
Schumaker, R. R., **3**, 135[340], 137[340], 139[340], 141[340]
Schuman, S. C., **8**, 608[41]
Schumann, D., **4**, 5[19]; **5**, 409[36]
Schumann, H., **1**, 231[9], 251[2], 253[12]; **4**, 738[98]; **8**, 447[133,134,136], 696[127,-128], 800[69]
Schumann, I., **3**, 1028[48]
Schumann, K., **4**, 587[44]
Schumann, R. C., **2**, 249[84]
Schumann, U., **1**, 18[93], 22[116], 34[224], 36[174], 39[186]
Schumm, J. S., **2**, 835[159]
Schunck, K., **8**, 321[94,95]
Schunn, R. A., **5**, 57[53]
Schunselaar, W., **2**, 902[46]
Schupp, W., **4**, 45[130,130e]
Schurig, V., **4**, 390[175b]; **6**, 677[323]
Schuster, D. I., **3**, 815[73]; **4**, 75[43b], 100[43]; **5**, 125[13,15], 128[13], 153[25], 215[4], 221[60], 224[4]; **8**, 563[437]
Schuster, F., **5**, 344[65]
Schuster, G. B., **5**, 162[69]; **7**, 169[109], 854[61]; **8**, 890[144]
Schuster, H. F., **2**, 81[1], 82[1], 96[1]; **3**, 223[146]; **4**, 53[151], 1010[150]; **5**, 581[171]; **6**, 830[4], 873[10]
Schuster, I., **2**, 346[46]
Schuster, K. H., **6**, 227[21], 228[21], 229[21], 230[21], 231[21], 234[21]
Schuster, P., **1**, 286[10]; **2**, 346[46], 349[66], 352[91,92], 355[122], 356[66], 357[141], 358[91]
Schuster, R. E., **1**, 292[27,31], 293[33]; **6**, 294[239]
Schuster, T., **1**, 142[25]
Schut, J., **8**, 837[11], 839[11]
Schütte, H. R., **6**, 746[92]
Schutz, A., **2**, 1103[129,131]
Schütz, G., **7**, 144[155]
Schütz, J., **6**, 535[536], 538[536]
Schütz, O., **3**, 582[112]
Schwab, G. M., **8**, 451[178]
Schwab, H., **5**, 478[162]
Schwab, J. M., **4**, 285[158]; **7**, 413[107b,c]
Schwab, P. A., **3**, 739[39]
Schwab, W., **4**, 1076[46]; **8**, 69[221], 70[221,222], 647[53,54]
Schwabe, R., **5**, 690[80,80c], 733[136,136f], 734[136f]
Schwager, H., **3**, 587[150]
Schwager, I., **4**, 915[7]
Schwall, H., **1**, 844[8]; **3**, 887[8], 888[8], 893[8], 897[8], 900[8], 903[8,126]; **4**, 953[8,8b], 954[8b]; **8**, 382[11], 383[11]
Schwan, A., **6**, 829[3]
Schwartz, A., **2**, 169[164]; **3**, 667[48], 687[48]; **5**, 822[165]; **7**, 445[60]
Schwartz, C. E., **6**, 898[106]
Schwartz, E., **5**, 436[158,158g], 442[158]
Schwartz, H. M., **8**, 185[11], 192[98]
Schwartz, J., **1**, 143[33], 155[67,68], 156[68,69], 749[75], 807[316], 808[320]; **3**, 469[202], 470[202], 473[202], 483[14], 1027[42], 1030[42], 1031[63]; **4**, 10[34], 113[176], 143[19], 153[62c,63a], 154[64a,b], 155[63b], 249[124], 257[222,223], 258[124], 262[222], 309[413], 312[455], 393[197,197a], 394[197a], 594[139], 595[155,156], 619[139], 620[155,156,395,396,397], 634[139,155], 635[155,156,395], 636[396,397]; **5**, 181[152], 1124[46], 1125[62,64], 1173[31], 1178[31]; **7**, 17[179], 453[80]; **8**, 447[121], 668[1], 669[1,3,4], 673[1,28], 675[28,38,43-45], 676[1,28], 677[28,38,60], 679[4,38,68], 680[68,71], 681[38,74], 682[81,82], 683[68], 684[1,43,74], 685[1,97], 686[1], 687[1,4], 688[1,100], 691[1,74,100,105,106], 692[1,28,38,100], 693[38,68,71,109,110,113-117], 694[71], 695[68], 697[3], 889[131]
Schwartz, J. A., **2**, 204[98]
Schwartz, J. L., **5**, 829[22]
Schwartz, L. H., **3**, 735[20]
Schwartz, M., **7**, 17[175]
Schwartz, M. A., **2**, 541[78], 841[186]; **3**, 55[283], 679[88,90], 680[92], 682[164], 683[90]; **7**, 336[33]; **8**, 527[48]
Schwartz, N. N., **7**, 674[43]
Schwartz, N. V., **1**, 608[39]; **7**, 516[3]
Schwartz, R. D., **7**, 78[127], 429[153]
Schwartz, R. H., **1**, 116[44]; **2**, 602[39]; **3**, 244[27]; **8**, 771[50]
Schwartz, S. B., **5**, 165[86]
Schwartz, T., **4**, 374[91]

Schwartz, T. R., **1**, 287[20], 288[20], 289[20]; **4**, 202[20]
Schwartz, V., **7**, 67[79]
Schwartzentrinber, K. M., **7**, 13[119]
Schwartzman, L. H., **8**, 275[141]
Schwartzman, S., **4**, 305[363]
Schwartzmann, S. M., **7**, 299[41]
Schwarz, G.-U., **3**, 872[59]
Schwarz, H., **1**, 162[104]; **3**, 55[280], 623[36]; **6**, 462[9], 668[251], 669[251]
Schwarz, J., **7**, 42[34], 805[66]
Schwarz, M., **2**, 345[40], 350[40]; **6**, 269[76]
Schwarz, R., **2**, 736[27], 1090[72]; **4**, 53[149], 129[223b]; **8**, 310[16]
Schwarz, R. A., **3**, 159[467], 166[467]
Schwarz, S., **3**, 55[280]; **7**, 586[167]
Schwarz, U., **6**, 33[7], 40[7], 57[7]
Schwarz, V., **6**, 495[147]
Schwarzenbach, K., **1**, 215[34]; **4**, 962[38]
Schwarzenberg, K., **4**, 98[110]
Schwarzenbrunner, U., **6**, 543[614]
Schwechten, H. W., **7**, 850[1]
Schweckendieck, W. J., **5**, 1141[80], 1145[80]
Schweiger, E. J., **3**, 883[106]; **5**, 736[145], 737[145]
Schweiter, M. J., **7**, 399[38], 400[38], 406[38], 409[38], 415[38]
Schweitzer, H., **3**, 640[104]
Schweitzer, R., **7**, 92[41,41a], 94[41], 152[1]
Schweizer, B., **1**, 37[240]
Schweizer, E. E., **1**, 878[105]; **4**, 1000[6], 1005[6,79], 1006[103], 1016[6]; **5**, 938[215]; **6**, 173[43], 175[68], 182[143], 185[143], 209[70], 1014[20]; **8**, 604[5], 636[1]
Schweizer, W. B., **1**, 1[1], 3[1], 26[1], 28[142], 30[151], 34[170], 37[238], 38[261], 41[151], 43[1]; **2**, 100[2], 107[55], 193[65], 197[77], 280[25], 897[19], 910[66]; **4**, 21[69], 72[25], 202[19], 224[182]; **5**, 841[104]; **6**, 716[101]; **8**, 190[78]
Schwellnus, K., **4**, 162[94c]
Schwengers, D., **4**, 887[129]
Schwenk, E., **2**, 734[6]; **3**, 812[55]; **8**, 320[84]
Schwenk, R., **2**, 762[55]
Schwentner, J., **6**, 60[146]
Schwepler, D., **7**, 350[25], 355[25]
Schwetlick, K., **5**, 102[182]
Schwickardi, M., **1**, 14[77]; **8**, 697[135]
Schwickerath, W., **6**, 549[672], 550[675], 552[675,691]
Schwier, J. R., **8**, 709[49], 710[49,53], 721[53,143]
Schwille, D., **8**, 658[101]
Schwindeman, J., **2**, 713[46]; **5**, 777[190,191]
Schwindemann, J. A., **6**, 239[77]
Schwob, J. M., **2**, 969[86]
Schwobel, A., **5**, 422[86]
Schwotzer, W., **6**, 708[47]
Schwyzer, R., **6**, 637[27], 668[251], 669[251]
Sciacovelli, O., **7**, 374[77a]
Sciaky, R., **2**, 783[35]
Scialdone, M. A., **6**, 1066[97]
Sciano, J. C., **7**, 605[140]
Scilingo, A., **2**, 547[114], 551[114]
Scola, P. M., **2**, 1054[59,61], 1070[59], 1071[61]; **5**, 485[181], 531[79]
Scolastico, C., **1**, 72[72], 524[86,87], 527[101,102,107], 528[108]; **2**, 103[28], 221[146], 266[61,62], 267[62-64], 515[55,56], 516[58], 605[57], 614[57], 630[21], 631[21], 632[21], 634[21], 636[56], 637[56,59], 640[21,56,59], 641[21,71], 642[21,71,73,74,78], 643[73,74,78], 644[21,73], 645[21,59], 652[59], 920[98], 922[101], 923[101], 930[131], 931[131]; **4**, 113[166], 152[58], 207[57,58], 226[187,188]; **6**, 149[100,108]; **7**, 170[121], 441[45]
Scollar, M. P., **8**, 185[10]
Scopes, D. I. C., **5**, 92[69]; **6**, 535[541], 538[541]
Scorrano, G., **4**, 425[33], 426[60,63], 438[156]; **5**, 408[33]; **8**, 152[175,176], 155[176]
Scott, A., **6**, 668[256], 669[256]
Scott, A. D., **7**, 264[92,93]
Scott, A. E., **3**, 660[15]
Scott, A. I., **2**, 170[174]; **3**, 660[14], 679[91], 681[96], 689[119,121], 813[60]; **4**, 24[72,72c]; **6**, 920[46]; **7**, 97[96]
Scott, A. S., **1**, 786[251]
Scott, B. S., **8**, 583[12]
Scott, C., **5**, 1136[54]
Scott, C. J., **3**, 635[33], 640[107,107a], 647[33,107]
Scott, E. J. Y., **3**, 328[179]
Scott, F., **3**, 216[73], 243[14], 249[14], 263[14], 423[72], 464[172], 4 66[182]; **4**, 1009[139], 1040[75]
Scott, F. L., **2**, 291[73]; **4**, 290[197]; **7**, 696[40]
Scott, J. W., **1**, 825[48], 837[148], 838[148]; **8**, 152[168], 159[2], 459[228], 460[249], 541[212], 606[18]
Scott, K. U., **5**, 1116[8]
Scott, L. T., **1**, 116[48], 118[48]; **2**, 744[91]; **4**, 1055[137]; **5**, 716[87]; **7**, 8[60]; **8**, 341[104]
Scott, M. D., **8**, 336[73], 337[73], 338[73], 339[73], 341[73]
Scott, P. W., **2**, 856[249]
Scott, R., **5**, 1149[125]
Scott, R. B., **7**, 5[21]
Scott, R. D., **2**, 1010[8]; **5**, 141[92]
Scott, R. M., **1**, 512[36]; **4**, 753[173]
Scott, S. W., **3**, 679[88,90], 683[90]
Scott, T. W., **4**, 30[89]
Scott, W., **4**, 861[113]; **8**, 390[79]
Scott, W. E., **1**, 130[96]
Scott, W. J., **1**, 193[89], 836[145]; **2**, 110[72]; **3**, 218[99], 219[99], 232[256,258,270], 239[99], 250[71], 436[17], 446[75], 454[116], 487[48-50], 489[48], 492[48], 495[48-50]; **4**, 258[242]; **5**, 712[57d], 763[107], 779[107]; **6**, 766[22]; **8**, 933[45]
Scott, W. L., **5**, 814[137,138]; **8**, 945[129]
Scotton, M., **4**, 956[18]
Scotton, M. J., **7**, 599[66]
Scouten, C. G., **8**, 713[72,73], 715[84], 724[155,157]
Scovell, E. G., **2**, 546[90]; **3**, 348[29], 355[53], 357[53], 382[35], 769[170], 770[175], 771[170]
Scozzafava, M., **5**, 1131[17]
Screttas, C. G., **3**, 88[126], 95[126], 96[126], 107[126], 109[126], 123[126], 125[126], 194[12], 824[20,21], 825[21]; **4**, 316[539]; **8**, 842[45]
Scriabine, A., **2**, 971[92]
Scribner, R. M., **8**, 40[90], 66[90]
Scrimin, P., **6**, 67[10]; **7**, 95[69]
Scripko, J., **6**, 284[170]; **8**, 392[107]
Scrivanti, A., **4**, 915[9,15]; **5**, 272[2], 275[2]
Scriven, C. E., **3**, 585[135]
Scriven, E. F. V., **4**, 295[255,256]; **5**, 451[23], 470[23], 491[23], 492[23], 937[203]; **6**, 76[41], 98[41], 245[117], 246[117], 247[117], 248[117], 249[117], 252[117], 253[117], 254[117], 256[117], 535[540,541], 538[540,541]; **7**, 21[2,4,6,7,12,18,21], 35[7], 475[54,56], 476[60], 477[72], 481[108], 483[72,108], 487[60], 488[60], 491[60], 504[60], 742[55], 743[55], 744[55], 750[130]; **8**, 384[23]
Scrowston, R. M., **8**, 629[180,181]
Scudder, P. H., **1**, 715[268], 716[268], 717[268]; **5**, 856[210], 910[83], 911[83], 912[83], 921[133], 922[133], 1007[40]
Scully, F. E., Jr., **1**, 364[38]; **7**, 227[85]
Seaborn, C. J., **8**, 321[99,103]
Seager, J. L., **3**, 643[121]
Seagusa, T., **4**, 254[179,180,182]
Seale, P. W., **8**, 663[116]
Sealfon, S., **5**, 7[54], 8[54], 519[35]
Seamon, D. W., **2**, 553[131]

Searcy, A. M., **8**, 293[51]
Searle, R. J. G., **4**, 74[38b]
Searles, S., **3**, 314[107], 889[24]
Searles, S., Jr., **5**, 828[5], 847[5]; **8**, 218[73], 221[73]
Sears, A. B., **5**, 900[11], 901[11], 906[11], 907[11], 910[11]; **6**, 689[387]
Seaton, J. C., **6**, 441[82]
Sebastian, M. J., **4**, 375[96c]
Sebastiani, G. V., **7**, 649[46]
Sebastiano, R., **4**, 763[208]
Sebban, M., **1**, 212[4]
Sebek, O. K., **7**, 77[119]
Sebti, S., **2**, 1084[20]
Secci, M., **7**, 777[368]
Seck, M., **4**, 38[108]
Seckinger, K., **6**, 554[801], 576[801], 581[801]
Seconi, G., **4**, 98[114], 113[114], 115[182,182e], 247[97], 256[97]; **6**, 179[127]
Secrist, J. A., **3**, 288[63]; **8**, 881[68]
Sedavkina, V. A., **6**, 509[272], 538[554]
Seddon, D., **7**, 805[67]
Sedelmeier, G., **3**, 621[31]
Seden, T. P., **7**, 500[235]
Sedergran, T. C., **1**, 838[158]; **7**, 162[64]
Sedivy, Z., **2**, 765[78]
Sedlaczek, L., **7**, 66[75b], 80[142]
Sedlmeier, J., **4**, 588[57]; **7**, 449[1,2], 450[1,2]
Sedrani, R., **3**, 226[207]; **4**, 207[59], 209[69], 895[164], 900[164]; **6**, 849[122]
Sedrati, M., **2**, 723[97,99]; **4**, 698[17], 699[17,19-21], 700[20,23], 701[21], 956[17]
See, J., **5**, 412[46]
Seebach, D., **1**, 1[1], 2[12], 3[1], 20[107], 26[1,137], 28[142], 30[151], 31[155], 32[156], 34[167,170], 35[171], 37[238,239,240,241], 38[261], 40[193], 41[151], 43[1,12], 70[63], 140[9], 141[9,22], 142[9,23], 144; **2**, 5[17], 6[17], 12[51], 22[17,17c,87], 23[17b,d], 55[1], 63[23], 72[78], 100[2-4], 107[55,58], 108[58], 114[114,120], 120[179], 124[208], 189[51], 193[65], 196[76], 197[77,80], 207[106], 223[149], 239[1], 244[24], 245[24], 267[24], 269[70], 280[25], 321[12], 323[22,26], 324[12], 325[43-45], 326[12,43], 327[45], 329[12,48], 332[55,58], 333[58], 334[69], 335[26,48,71], 336[26,48,72], 337[73], 338[74], 442[5], 455[13], 507[28], 508[28,30,31], 578[85,86], 630[4], 645[82], 830[138], 846[209], 897[18-20], 910[66], 1051[42], 1052[47], 1057[70], 1066[120], 1067[124], 1070[124], 1086[28], 1096[28], 1108[79]; **3**, 40[221,222,223], 41[223,227,229,230], 42[223], 43[237,240], 44[240], 65[3,7], 66[11], 71[30-33], 74[11], 75[46], 79[63], 86[18,20,27,28,59], 87[64,65,68,69,75,82,88], 88[59], 89[59,68], 94[18], 95[18,158,178], 96[158,178], 97[158,178], 99[158,178], 104[20], 105[27,82], 112[59,82], 120[20,82], 121[18,20,27,28,158], 123[59], 124[27,28,255,256,265,266,274,278,279,283,284], 125[20,27,28,59,255,266,274,278,279,283,284,290,313], 126[20,279], 127[27,28,265,266,274,283], 128[255,256,265,266,278,322], 129[255,274,278], 130[322], 131[324,331], 132[278,283], 134[28,322,338], 135[324,338], 136[64,65], 137[338], 144[28,64,65,381,382,384], 145[64,65,381,382,384,385], 147[64,381,382,384], 194[9], 224[173], 247[47], 255[108], 257[122], 564[4], 566[4,27], 568[40], 569[49], 580[103,104], 606[40], 607[49], 645[168], 650[210b]; **4**, 11[35], 12[35b,43], 13[43a,b], 14[49,49c], 16[51,51a], 20[64], 21[64,69], 72[25,25b,31], 77[51], 78[52a,53], 99[117], 100[124], 102[127], 104[136a,c], 109[149,150a], 110[150b], 113[161,164,172], 115[177], 120[201], 161[89c], 200[5], 207[60,61], 208[61], 209[67], 218[144], 224[182,183,184], 229[237,238], 259[260], 340[53], 342[53], 731[73], 733[73], 773[73], 867[9], 1007[116,121,123,131], 1008[123,132]; **5**, 63[7], 75[215], 79[289,290,291], 680[24], 683[24c], 841[104,105], 842[109], 903[36], 936[197]; **6**, 25[102], 107[24], 134[8-10,19,35], 135[9,10,23], 136[39], 147[84], 171[7], 419[8,9], 425[8,9], 509[278], 678[321], 679[321,326], 716[101-103], 833[22], 849[124], 911[16]; **7**, 124[48], 125[48], 126[48], 224[55], 225[56,57,65,67], 226[69], 774[316]; **8**, 166[64], 178[64], 179[64], 185[29], 190[69,72,78], 354[162], 363[1], 374[1,145], 852[142]
Seeber, R., **7**, 769[215]
Seefelder, M., **2**, 784[37]; **6**, 428[84], 430[94], 431[94], 488[8], 495[8], 499[8], 512[119], 543[8,119], 555[809], 566[8]
Seeger, A., **7**, 65[69]; **8**, 847[92]
Seeger, W., **8**, 141[43]
Seeholzer, K., **2**, 1090[72], 1099[110]
Seel, F., **2**, 734[7], 738[7]
Seeles, H., **5**, 596[27], 597[27]
Seeley, F. L., **7**, 731[54]
Seeliger, A., **6**, 36[24]
Seeliger, W., **2**, 1054[60]; **5**, 501[263]
Seely, F. L., **1**, 768[171]; **4**, 760[195,197], 765[197]
Seely, J. H., **6**, 668[257]
Seeman, J. I., **2**, 125[220,222]; **3**, 75[45]; **4**, 815[195]; **5**, 218[28], 220[47], 223[70], 707[39], 709[39]
Seemuth, P. D., **6**, 705[31]
Seese, W. S., **3**, 901[112]
Sefton, M. A., **7**, 64[62]
Sega, A., **4**, 956[18]
Segal, G. A., **5**, 72[168]
Segal, J. A., **4**, 531[80,82], 542[118]
Segal, R., **3**, 396[108], 397[108]
Segal, Y., **3**, 896[70]; **7**, 95[67], 107[167]
Segall, Y., **8**, 864[242]
Segawa, J., **4**, 160[86a]
Seger, G., **5**, 725[115], 729[123]
Segi, M., **3**, 314[108], 318[129]; **5**, 438[161], 442[185,185a], 532[84]
Segmuller, B. E., **1**, 307[93], 310[93]; **2**, 692[72]; **4**, 18[60,60b], 262[314], 403[241]
Segner, J., **2**, 211[113]
Segnitz, A., **1**, 142[26]; **2**, 342[6]; **3**, 554[25]; **5**, 1148[123]; **6**, 104[9]; **7**, 752[142]
Segoe, K., **1**, 551[77]; **6**, 238[72]
Seguin, P., **7**, 446[67]
Seguin, R., **7**, 360[21]
Seguin, R. P., **8**, 159[4], 535[165], 541[212]
Sehgal, R. K., **7**, 473[27]
Sehrer, J. C., **5**, 1116[4]
Sei, T., **1**, 19[98]
Seibel, W. L., **3**, 215[61]
Seibert, G., **6**, 33[7], 40[7], 57[7]
Seibert, H., **7**, 769[234], 770[234]
Seibert, R. A., **8**, 143[65]
Seibert, W. E., III, **8**, 688[100], 691[100], 692[100]
Seibl, J., **3**, 13[68], 916[19]
Seidel, B., **5**, 69[105]; **7**, 374[77d]
Seidel, G., **8**, 724[176]
Seidel, M., **4**, 1083[92], 1085[101]
Seidel, M. C., **3**, 737[31]
Seidel, P., **2**, 1095[93]
Seidel, W., **3**, 644[135,155,160,162], 646[135]
Seiders, R. P., **5**, 741[153]
Seidig, K.-D., **3**, 890[33]
Seidl, H., **5**, 391[143], 715[79]; **8**, 391[92], 773[62]
Seidler, M. D., **8**, 406[36]
Seidler, P. F., **8**, 673[24], 676[24], 682[24]
Seidlova, V., **2**, 765[78]
Seidner, R. T., **5**, 716[89], 804[93]; **8**, 16[99], 542[230], 543[230]
Seifert, C. M., **3**, 767[163]
Seifert, P., **3**, 571[59]; **5**, 833[45]
Seifert, W. J., **5**, 730[128]

Seifert, W. K., **6**, 107[29]; **8**, 333[57], 345[127]
Seifter, S., **8**, 52[138], 66[138]
Seigle-Murandi, F., **7**, 79[131]
Seijas, J. A., **3**, 586[156], 591[171], 610[156]
Seikaly, H. R., **2**, 107[59,60], 108[60], 196[76]; **5**, 87[42]
Seiklay, H. R., **1**, 418[73]
Seiler, M. P., **3**, 99[185]
Seiler, P., **1**, 35[171], 37[240], 341[96], 477[132], 482[132]; **2**, 197[77]
Seilz, C., **7**, 253[18]
Seino, S., **2**, 765[67]
Seip, H. M., **1**, 488[8]
Seipp, U., **1**, 303[80]; **8**, 13[75]
Seitz, A. H., **4**, 488[51]; **5**, 380[113c]
Seitz, D. E., **2**, 588[152]
Seitz, E. P., **1**, 851[42], 886[42,135], 898[42]; **5**, 1001[17], 1002[17b]
Seitz, E. P., Jr., **6**, 848[108]
Seitz, F., **8**, 767[21], 773[21]
Seitz, G., **4**, 1006[98]; **5**, 412[45], 417[59], 422[82], 634[79]; **6**, 502[213]
Seitz, L. M., **1**, 9[43]
Seitz, S. P., **1**, 343[109], 425[107], 568[232]; **4**, 370[31,32], 371[32], 372[31], 397[31]; **6**, 46[76], 47[76], 48[76], 918[38], 1031[115]; **7**, 254[26], 522[41], 523[41], 524[49]; **8**, 836[10b], 847[10b], 848[10b], 849[10b]
Seitz, T., **1**, 153[61], 295[53], 296[53], 336[75]; **2**, 630[8]; **4**, 331[8]; **6**, 142[67]; **7**, 517[12]
Seitzinger, N. K., **3**, 97[170], 117[170]
Seiwell, R., **2**, 345[30]
Sek, B., **6**, 554[713]
Seki, F., **4**, 23[70]
Seki, H., **6**, 619[115], 620[128]
Seki, K., **4**, 102[129]; **7**, 175[142]
Seki, T., **5**, 1180[47], 1181[47]
Seki, Y., **2**, 603[46]; **8**, 773[70], 774[70,71]
Sekiguchi, A., **6**, 440[77]
Sekiguchi, S., **4**, 435[134]; **6**, 498[171]
Sekihachi, J., **1**, 242[49,50]
Sekimura, Y., **1**, 415[63]
Sekine, F., **8**, 150[137,138]
Sekine, M., **1**, 563[171]; **2**, 65[29], 830[145]; **3**, 197[41], 199[41]; **6**, 563[899], 604[32,33], 606[38,40-42], 614[97], 618[112], 626[164]
Sekine, T., **3**, 649[204,205]; **8**, 321[93]
Sekine, Y., **5**, 167[94]; **7**, 750[127]
Sekiya, A., **3**, 244[23], 452[108], 503[147]; **6**, 497[157]
Sekiya, K., **1**, 212[12], 213[12], 215[12b], 217[12], 448[207]; **2**, 443[19], 447[19], 448[19], 449[19], 452[57]; **3**, 221[131,132]; **4**, 163[96], 164[96]; **6**, 847[106], 848[110]
Sekiya, M., **1**, 368[62], 370[71], 371[71], 385[115], 389[137], 391[62]; **2**, 913[77], 914[77,79], 915[79], 939[159], 940[161], 948[180], 994[38], 1004[59]; **3**, 844[34]; **4**, 1000[14]; **5**, 100[157]; **6**, 500[181], 533[510]; **7**, 173[132]; **8**, 564[439]
Sekizaki, H., **1**, 436[148]; **5**, 327[28]
Seko, S., **4**, 817[202]
Selby, W. M., **3**, 617[16]
Self, C. R., **3**, 12[60]; **4**, 591[107], 633[107]
Self, D. P., **4**, 423[5], 426[5]
Selick, C., **3**, 216[78]
Seligman, A. M., **3**, 828[49]
Seligman, M. A., **8**, 568[471]
Selim, A., **3**, 45[251]
Sélim, M., **8**, 663[118]
Selim, V. R., **4**, 84[69]
Selimov, F. A., **5**, 1154[159]
Selin, T. G., **8**, 764[2], 776[78], 777[2a]
Selke, E., **8**, 451[180], 453[191]
Selke, R., **8**, 460[246]
Sell, C. S., **1**, 543[22]; **2**, 710[17]
Sellars, P. J., **6**, 673[290]
Selle, B. J., **7**, 532[31]
Selleck, M. E., **3**, 326[164]
Sellen, M., **3**, 222[139]; **4**, 399[225,225b]; **6**, 831[9]
Sellers, S. F., **3**, 639[84]; **5**, 586[204], 911[95], 912[95]
Selley, D. B., **5**, 164[74]
Selman, L. H., **8**, 445[12]
Selman, S., **3**, 822[6], 823[6], 824[6], 825[26], 834[6,73], 836[6]
Selnick, H. G., **1**, 103[95], 329[35]; **2**, 578[84], 652[126], 701[85], 704[88]; **4**, 33[96], 34[96e]; **5**, 495[218], 579[165]; **6**, 989[80], 995[80]; **7**, 237[37], 246[89]
Seltzer, S., **3**, 522[13]
Seltzman, H. H., **3**, 125[305], 126[305], 127[305]
Selva, A., **8**, 942[117]
Selve, C., **6**, 74[36]; **8**, 384[27]
Selwitz, C. M., **7**, 449[3], 450[3], 453[3]
Selwood, D. L., **5**, 883[16]
Semard, D., **1**, 644[122], 646[122], 668[122], 669[122], 695[122]
Semenov, A. A., **6**, 525[377]
Semenov, V. P., **7**, 477[74]
Semenova, E. A., **6**, 560[866]
Semenovsky, A. V., **3**, 181[553], 342[7,8,14], 345[21], 346[24], 349[35], 351[41], 361[73,80]; **4**, 874[52]; **8**, 609[51], 611[64], 971[111]
Semenow, D. A., **4**, 486[32]
Semigran, M. J., **3**, 588[157]
Semikolenov, V. A., **8**, 608[47]
Semkow, A., **5**, 575[131]
Semler, G., **2**, 967[73]
Semmeler, F. W., **2**, 400[26]
Semmelhack, M. F., **1**, 188[66,67], 189[66,67], 193[87], 214[29], 218[29], 553[99]; **2**, 23[91], 158[125], 710[26]; **3**, 124[254], 126[254,319], 137[254], 353[47], 423[74,79,80], 426[74], 427[85], 430[94], 483[17], 499[128], 500[128,132], 505[132], 509[128], 712[25], 1031[65,65b]; **4**, 82[61], 115[179b], 254[185], 257[224], 379[116], 380[116c,124], 384[143,144,144b], 455[44,45], 457[45], 463[44,45], 464[45], 465[45], 466[44,45], 467[45], 469[45], 472[45], 476[44,45,162,163], 502[124], 517[5], 518[4-6], 519[17], 520[17], 522[51,52], 523[51], 524[59], 525[66], 526[51,52], 532[52,83,89,91], 534[91], 535[93], 537[91,95], 538[93,103], 539[93,108], 540[17], 541[17], 543[83,121,123], 545[83], 546[128], 557[13,15], 580[24,25], 581[26,27], 695[1,4], 766[229]; **5**, 736[145], 1066[6], 1088[78,79], 1092[93], 1094[98,115], 1096[98], 1098[98], 1102[93,148], 1103[148], 1112[98,115]; **6**, 134[38], 639[46], 659[46c], 986[68]; **7**, 301[60], 308[20], 359[17], 809[82]; **8**, 20[138], 542[223], 544[223,251], 548[322]
Semmler, K., **4**, 1009[145]
Semple, J. E., **2**, 1090[70], 1100[70]; **4**, 364[1]
Semra, A., **4**, 527[69,70], 528[69-71]
Sen, A., **1**, 255[18]
Sen, A. K., **6**, 510[297]
Sen, M., **3**, 741[51]; **8**, 113[31], 816[24]
Sen, R. N., **2**, 760[46]
Senanayake, C., **1**, 248[66], 851[41], 852[41]; **4**, 5[18], 61[18g]; **5**, 435[150]
Senaratne, K. P. A., **5**, 253[47], 256[57]; **8**, 395[127]
Senaratne, P. A., **4**, 1078[50]
Senda, S., **2**, 790[57]
Senda, Y., **4**, 302[333]; **8**, 66[104], 418[12], 422[12], 423[40], 429[40], 445[15], 856[165]
Sendi, E., **5**, 742[158]
Seneci, P. F., **8**, 190[79]
Seng, F., **4**, 259[261]
Senga, K., **1**, 34[168]; **2**, 792[63,64]; **6**, 533[486]; **7**, 342[54]
Sengers, H. H. W. J. M., **2**, 838[170]

Sengupta, P., **3**, 741[51]; **8**, 113[29,31], 116[29], 117[29], 119[29], 816[24], 880[65]
Sengupta, S., **1**, 115[39], 138[39]; **7**, 226[73]
Sengupta, S. K., **7**, 473[27]
Senior, M., **8**, 545[290]
Senkus, M., **8**, 228[122]
Senn, J., **7**, 160[50]
Senna, K., **5**, 159[49]
Senning, A., **6**, 423[44], 456[160]
Seno, K., **3**, 125[309]; **6**, 134[28]; **8**, 241[38], 272[120]
Seno, M., **6**, 498[164]; **7**, 10[75], 24[34], 477[81]
Sen Sharma, D. K., **3**, 301[47]
Senter, P. A., **5**, 636[89]
Senter, P. D., **5**, 133[58]; **6**, 1063[82]
Sentman, R. C., **5**, 73[196], 461[105]
Seo, W., **4**, 21[69], 222[176]
Seoane, A., **5**, 939[223], 951[223], 962[223], 964[223]
Seoane, C., **2**, 361[178], 380[299]
Seoane, E., **3**, 851[64]
Seoane, G., **5**, 939[221,222,223], 940[222], 943[251], 951[223], 962[221,222,223], 963[222,251], 964[223]; **7**, 324[72], 557[75]
Seoane, P., **1**, 770[184]
Seong, C. M., **4**, 824[237]
Seong, S. Y., **8**, 60[184], 66[184]
Sepelak, D. J., **2**, 77[89], 584[127]; **3**, 202[99]
Seper, K. W., **2**, 209[108], 631[13]
Sephton, H. H., **2**, 456[53], 458[53]
Sepiol, J., **2**, 360[169], 376[169]
Sépulchre, A. M., **3**, 126[315,316,317]
Sequeira, R. M., **4**, 1016[207]
Sequin, U., **7**, 363[37]
Sera, A., **2**, 611[101]; **4**, 161[86c], 230[252], 231[252]; **5**, 77[262], 341[59]
Seraglia, R., **7**, 777[376]
Serbine, J. P., **1**, 786[249]
Sercel, A. D., **4**, 14[47,47e,f]
Serckx-Poncin, B., **5**, 473[152], 477[152], 480[165], 483[165]
Sere, V., **7**, 764[126], 767[126]
Serebrennikova, T. A., **5**, 453[59]
Serebryakov, B. R., **3**, 304[66]
Serebryakov, E. P., **5**, 125[17], 128[31]; **7**, 479[89], 723[24], 724[24]
Serebryanskaya, A. I., **1**, 632[68], 644[68]
Sérée de Roch, I., **7**, 11[91], 95[68], 160[53]
Serelis, A. K., **3**, 358[69]; **4**, 785[22], 787[27]
Sereno, J. F., **5**, 256[58], 257[58b]; **6**, 960[56]; **7**, 268[122], 564[92], 567[92], 678[72]
Seres, P., **6**, 775[52]
Serfontein, W. J., **6**, 677[317]
Sergeev, V. A., **5**, 1148[115]
Sergeeva, O. R., **3**, 328[173,174]
Sergent, M., **8**, 925[5]
Sergent-Guay, M., **4**, 1040[84], 1043[84]
Sergheraert, C., **4**, 519[20], 520[20]
Sergi, S., **4**, 588[66]
Sergutina, V. P., **6**, 487[41], 489[41], 515[311,312]
Serianni, A. S., **2**, 456[42], 466[42], 467[42]
Serimin, P., **7**, 762[84]
Serini, A., **6**, 675[299]
Serizawa, H., **4**, 147[38b], 250[137], 358[153,156], 886[118]
Serizawa, N., **7**, 77[121,122]
Serizawa, Y., **6**, 995[99]
Sermo, A. J., **4**, 572[3]
Serra, A. A., **1**, 848[26], 849[26], 850[26]; **3**, 783[29]; **6**, 129[167]; **7**, 406[76]
Serra, G., **5**, 1148[114]
Serra-Errante, G., **7**, 6[35]
Serratosa, F., **3**, 380[7]; **4**, 1005[94]; **5**, 1047[31], 1052[31], 1054[43], 1059[54], 1060[55], 1062[54c,d], 1133[33]
Serravalle, M., **4**, 768[240,241]
Serve, D., **8**, 594[70]
Servens, C., **2**, 3[8], 6[8,8a], 18[8a], 574[58]; **4**, 743[128]
Servi, S., **8**, 195[110-113], 203[111]
Servin, M., **7**, 771[267], 772[267]
Servin, R., **2**, 1051[35], 1052[35]; **3**, 637[64]; **6**, 572[959]
Servis, K. L., **7**, 710[50]
Sesana, G., **4**, 719[18], 723[18]
Sesartic, L., **6**, 658[182]
Seshadri, R., **8**, 989[35]
Seshadri, S., **2**, 777[1], 779[1], 780[1], 781[1], 783[1], 786[1], 787[1,52], 789[1,55,56], 791[1], 792[1,65]; **6**, 487[55,56], 489[55,56], 543[55,56]
Seshadri, T. R., **7**, 544[34]; **8**, 568[467], 978[145]
Seshasayee, M., **1**, 305[83]
Sessink, P. J. M., **5**, 560[78]
Sessions, R. B., **8**, 388[64]
Set, L., **1**, 672[203], 678[203], 700[203]; **3**, 107[229]; **4**, 791[49], 815[190]
Seta, A., **7**, 751[139]
Seter, J., **2**, 323[36]; **3**, 905[138]
Setescak, L. L., **4**, 439[168]
Seth, A. K., **6**, 603[24]
Seth, K. K., **7**, 465[131]; **8**, 187[33]
Seth, M., **7**, 265[103], 267[103]
Sethi, D. S., **8**, 726[193]
Sethi, S. C., **7**, 95[73a], 384[112]
Sethi, S. P., **2**, 839[180]; **4**, 27[80]; **6**, 774[46]; **8**, 505[77]
Sethna, S., **2**, 753[1]
Setiabudi, F., **2**, 760[42]
Setkina, V. N., **8**, 486[62]
Seto, H., **5**, 136[67,68], 137[74]
Seto, K., **3**, 454[121]
Seto, S., **8**, 244[65], 250[99], 445[15], 556[375]
Setoi, H., **6**, 93[133], 1050[37]
Setterquist, R. A., **2**, 413[11]
Settine, R. L., **3**, 771[188]
Setyoama, T., **8**, 285[7]
Setzer, A., **4**, 430[98]
Setzer, W. N., **1**, 2[11], 8[11], 19[11], 22[11]
Seu, Y. B., **7**, 418[125], 451[22]
Seubert, J., **3**, 380[4], 735[19]; **5**, 743[163]; **6**, 919[41]
Seufert, W., **4**, 82[61], 546[128]
Seuferwasserthal, P., **7**, 493[191]; **8**, 383[22]
Seuner, S., **5**, 1101[144]
Seuring, B., **1**, 510[25], 826[62]; **2**, 72[78]; **3**, 95[158,178], 96[158,178], 97[158,178], 99[158,178], 121[158], 124[279], 125[279], 126[279]; **6**, 833[22]
Seuron, N., **4**, 112[159], 113[159b], 259[275]
Severin, T., **4**, 124[214a-e]; **6**, 937[118,119], 939[118,119], 942[118], 943[118]; **7**, 751[141]; **8**, 370[85,86], 910[84], 916[84]
Sevestre, H., **4**, 793[69]; **6**, 718[123]
Sevin, A., **4**, 1082[86]
Sevodin, V. P., **6**, 543[611]
Sevost'yanova, V. V., **7**, 493[195]
Sevrin, M., **1**, 630[41], 636[41], 645[41], 646[41], 647[41], 666[177], 667[177], 668[177], 669[41,178,180], 670[41], 672[41,177,199], 673[180], 698[241,246], 699[246], 700[241], 702[199], 705[199], 712[199], 716[199],; **3**, 86[50], 87[110], 89[110,140,141,143], 90[140,141], 91[140,141,149,150], 92[140,143], 105[110], 106[110], 107[140], 109[140], 114[110], 116[140], 118[141,143,237,238], 120[110,149,150,242], 123[141,143,238], 253[91]; **4**, 120[200],

259[266], 993[160]; **5**, 151[10]; **6**, 976[8]; **8**, 806[125], 847[97], 848[97e], 849[97e]
Sewell, W. G., **7**, 768[198]
Sexsmith, S. R., **8**, 734[1], 747[58], 752[58]
Seyam, A. M., **8**, 447[131], 670[9], 671[9], 697[129]
Seybold, G., **6**, 426[72]; **8**, 636[3]
Seydel, U., **6**, 33[7], 40[7], 57[7]
Seyden-Penne, J., **1**, 235[29], 561[162], 683[219,220], 769[181], 774[208]; **2**, 369[251], 414[13-15], 428[44], 430[51], 432[57]; **4**, 71[16a,b,17b], 111[156], 112[159], 113[16b,159b,168,168e], 139[3], 240[48], 259[270,275]; **5**, 410[41]; **8**, 7[37], 15[94], 541[217,219], 542[219], 850[118], 865[245]
Seyfarth, H. E., **7**, 613[1]
Seyferth, D., **1**, 213[18], 273[69], 411[49], 544[45], 553[99], 619[60], 630[18]; **2**, 55[1], 70[50], 77[88-90], 567[29], 584[127], 587[143]; **3**, 86[18], 94[18], 95[18], 121[18], 200[69], 202[98,99], 208[3], 1024[31]; **4**, 115[177], 174[37,38], 184[37,38], 192[37,38], 508[159], 968[92], 977[92], 1000[15-18], 1001[45], 1002[45,52], 1007[109,127]; **5**, 604[54], 762[103]; **6**, 175[66], 180[129], 182[66], 548[670], 692[406]; **8**, 693[109], 807[111,117]
Seyler, J. K., **8**, 449[159], 453[191], 568[480]
Sgarabotto, P., **3**, 386[57]
Sgarra, R., **2**, 87[26]
Sha, C.-K., **2**, 378[284,286]; **4**, 38[110], 1101[192,193,194]; **5**, 582[177], 938[217]; **8**, 309[7]
Shaaban, A. F., **6**, 770[36]
Shaban, M., **6**, 646[102]
Shabana, R., **6**, 509[252]
Shabanova, M. P., **4**, 317[548]
Shabarova, Z. A., **6**, 607[48]
Shabica, A. C., **8**, 143[62,63], 148[62,63]
Shadbolt, R. S., **8**, 642[29]
Shadmanov, K. M., **3**, 318[125]
Shadrova, V. N., **3**, 643[131]
Shaefer, C. G., **8**, 287[20], 288[20]
Shafeeva, I. V., **4**, 461[100], 475[100]
Shafer, P. R., **4**, 4[15], 876[58]
Shafer, S. J., **4**, 460[93]
Shaffer, E. T., **6**, 209[70]
Shaffer, G. W., **5**, 226[110], 227[110], 228[110]
Shafiee, A., **6**, 481[118-121], 969[118]
Shafiullah, S., **7**, 481[112]
Shafizadeh, F., **5**, 350[79]; **8**, 89[45]
Shafran, Yu. M., **6**, 530[421], 550[421]
Shah, A., **2**, 364[207]
Shah, D. H., **2**, 849[212]
Shah, G. M., **7**, 700[58]
Shah, J. N., **3**, 125[304], 126[304]; **7**, 700[58]
Shah, K. H., **8**, 659[104]
Shah, M., **4**, 370[46]
Shah, S. K., **1**, 631[56], 632[56], 633[56,70], 634[56], 635[56,70], 636[56,70,92,93,97,98], 638[56,93,97], 639[70], 640[56,93,97], 641[56,98], 642[56,70], 643[70], 645[56], 646[93], 647[93], 648[97,98], 649[97,98], 650[97]; **3**, 87[74,102,103,106,112,113], 95[74,106], 97[74], 104[102,103], 106[103], 109[74,106], 111[102,103], 114[74], 116[74], 117[102,103], 136[74,106,113], 141[74,106], 145[113], 157[112,113]; **6**, 860[177], 1027[93], 1031[93,116]; **7**, 771[276,278], 826[47], 827[47]
Shahak, I., **2**, 840[183]; **3**, 124[264], 867[35]; **6**, 93[132], 219[124], 980[45]; **7**, 475[50], 476[50]
Shahkarami, N., **4**, 746[147]
Shahriari-Zavareh, H., **2**, 376[279]; **5**, 847[135]
Shaik, S., **5**, 75[221,223], 703[19], 705[19]
Shainoff, J. R., **8**, 52[135], 66[135]
Shakhatuni, R. K., **8**, 618[124]
Shakhgel'diev, M. A., **6**, 462[10]
Shakhidayatov, Kh., **1**, 543[16], 555[112]
Shakir, R., **1**, 17[206,216], 37[178]
Shaligram, A. W., **8**, 340[89]
Shalit, H., **8**, 608[41]
Shalon, Y., **7**, 564[112], 572[112], 587[112]
Sham, H. L., **2**, 223[148]
Shambayati, S., **1**, 8[39], 297[58], 299[63], 300[72], 303[76], 305[82], 308[97], 314[97]; **5**, 1062[59]
Shamma, M., **2**, 357[148], 916[84]; **3**, 680[93], 807[29]; **6**, 680[331]
Shamouilian, S., **5**, 383[125]
Shams, H. Z., **2**, 378[291]
Shand, W., Jr., **4**, 275[67], 279[67], 287[67]
Shangraw, W. R., **5**, 552[38], 560[68], 1123[39]
Shani, A., **1**, 758[125]; **2**, 169[165], 612[106]; **3**, 244[21], 247[21], 391[91], 464[173]
Shank, R. C., **3**, 499[121]
Shanka, C. G., **7**, 136[112], 137[112]
Shankar, B. B., **3**, 253[89], 261[147], 262[89]
Shankar, B. K. R., **4**, 1054[132]; **7**, 683[89]
Shankar, C. R., **6**, 490[105]
Shankaran, K., **3**, 265[192]; **5**, 534[92], 574[130], 850[163]; **6**, 2[3], 25[3]
Shankland, R. V., **3**, 724[13]
Shanklin, J. R., **1**, 535[146], 722[278], 737[32], 825[49], 826[60]; **3**, 786[40]; **4**, 987[136], 989[136]; **6**, 139[50]
Shanklin, M. S., **3**, 1056[35], 1062[35]; **4**, 1033[22,22a], 1057[22a-c]; **8**, 837[15b]
Shanmugam, P., **5**, 728[122]; **6**, 685[355]
Shanmugasundaram, G., **8**, 949[154]
Shannon, J. S., **7**, 84[3]
Shannon, P. J., **8**, 36[96], 41[96], 66[96], 251[102]
Shannon, P. V. R., **7**, 102[134]
Shanzer, A., **6**, 441[79]
Shapiro, B. L., **1**, 163[106]; **5**, 1172[28], 1182[28]
Shapiro, E., **8**, 530[93]
Shapiro, E. S., **4**, 317[544]
Shapiro, G., **5**, 131[43], 133[43], 145[43,106,107]
Shapiro, H., **3**, 415[7]
Shapiro, I. O., **1**, 632[68], 644[68]
Shapiro, M. J., **8**, 9[48,50]
Shapiro, R., **7**, 376[83]
Shapiro, R. H., **1**, 377[97,97a,101]; **2**, 506[16], 510[16], 513[16]; **3**, 804[2], 810[2], 812[2]; **4**, 952[5], 954[10]; **5**, 778[196]; **6**, 704[7], 727[205], 776[53], 779[53], 784[92], 961[68], 1059[70], 1066[70]; **8**, 345[129], 940[99], 943[120], 949[153], 952[99]
Shapiro, S. H., **2**, 838[174], 839[174]
Shapley, J. R., **8**, 445[49]
Shapley, P. A., **7**, 283[188], 285[188]
Sharadbala, P. D., **1**, 463[25]
Sharafi-Ozeri, S., **5**, 725[115]
Sharama, R. K., **6**, 219[120]
Sharanin, Yu. A., **2**, 854[236]
Sharapov, V. A., **8**, 769[27]
Sharf, V. Z., **8**, 551[342]
Sharipov, A. Kh., **8**, 608[48]
Sharma, A. K., **5**, 806[108], 1003[24]; **7**, 295[22], 302[64]
Sharma, B. L., **6**, 487[65]
Sharma, C. S., **7**, 95[73a]
Sharma, D. S., **5**, 95[101]
Sharma, G. M., **3**, 681[96]
Sharma, G. V. M., **1**, 568[236]; **6**, 136[40]
Sharma, G. V. R., **4**, 793[71]
Sharma, K. R., **8**, 447[117]
Sharma, K. S., **3**, 325[159]
Sharma, M., **5**, 441[179]
Sharma, M. L., **3**, 416[18], 417[18]
Sharma, M. M., **6**, 88[106], 89[106], 603[18]

Sharma, N. D., **7**, 750[131]
Sharma, N. K., **7**, 764[129]
Sharma, R. B., **1**, 534[143]; **6**, 441[85]
Sharma, R. K., **3**, 505[164], 507[164], 512[164], 515[164]; **4**, 276[80], 303[346], 304[346]
Sharma, R. L., **6**, 97[153]
Sharma, R. P., **2**, 744[95]; **3**, 365[98], 380[10]; **5**, 421[79]; **6**, 76[49], 685[363], 987[70]; **8**, 891[146,148], 968[91,92]
Sharma, S., **1**, 386[123]; **3**, 213[46], 216[46], 223[46], 259[136]; **4**, 901[186]
Sharma, S. C., **8**, 119[76]
Sharma, S. D., **5**, 92[79,80], 95[102]
Sharma, S. K., **7**, 705[15]
Sharma, V. K., **2**, 376[279], 378[288]; **8**, 367[61]
Sharma, V. L., **6**, 943[157]
Sharman, S. H., **8**, 726[190]
Sharnin, G. P., **6**, 489[78]
Sharon, N. S., **6**, 33[4], 34[4], 40[4]
Sharp, J. C., **5**, 439[166]
Sharp, J. T., **3**, 255[110]; **4**, 483[4], 484[4], 487[44], 495[4], 953[8], 1002[51]; **6**, 126[152]
Sharp, M. J., **2**, 1028[77-79], 1029[77]; **3**, 231[250], 504[155], 511[155], 515[155]; **6**, 740[62]
Sharp, P. R., **4**, 485[31]
Sharp, R. L., **8**, 704[8], 707[23], 725[8a]
Sharpe, J., **1**, 797[283]; **7**, 555[69]
Sharpe, L. A., **3**, 380[10]
Sharpe, L. R., **4**, 467[129]
Sharpe, P. E., **5**, 850[156]
Sharpless, K. B., **1**, 78[12], 314[129], 343[122], 345[122], 634[75], 642[75], 644[75], 652[140], 708[261], 712[264], 714[264]; **2**, 240[5], 455[1]; **3**, 87[63], 114[63], 117[63], 223[156], 224[180], 225[185,191], 264[181], 583[119], 770[174]; **4**, 231[273], 340[52], 377[104], 390[104h]; **5**, 185[163], 773[165]; **6**, 2[3], 8[39], 11[50], 19[70], 25[3], 33[2], 34[2], 40[2], 50[2], 51[2], 53[2], 54[2], 57[2], 79[63], 88[105], 89[105,111-113,115], 91[115], 237[62], 253[157,159], 655[166], 686[369], 906[147], 927[71,76], 980[42], 985[65], 1026[81,83,85,87], 1027[81,83,87,92], 1028[83,85], 1029[81,101], 1030[81], 1031[81,85,87,114], 1032[120], 1033[81]; **7**, 86[13], 87[21,22], 88[24], 91[35,37], 110[37], 120[19], 121[19], 123[19], 124[38,39], 128[38,39,69], 129[38], 130[39], 131[39], 146[170], 160[54], 167[95], 198[26], 238[42], 239[42], 240[42], 254[31], 309[22], 358[5], 364[5,42], 368[5,42], 375[5,79], 376[5], 378[5], 390[2-4,10,11], 391[13], 393[4,14-17], 394[2,4,14,18,19], 395[2,4,18,20a,b], 396[4,14], 397[3,4,15,15a], 398[4,16,16a,18,32], 399[3,4,18,38], 400[3,4,38,41], 401[4,59,60], 402[63], 403[10,59,65], 405[69,70], 406[3,4,38,59,77], 407[3,4,41], 409[3,38,77], 410[3,4], 411[4,13], 412[2,13,104], 413[2,4,13,104], 414[77], 415[38,77], 417[131], 419[2,132], 420[2,135,136], 421[2,77,136,136b,138], 422[2,141], 423[77,141,141b,c], 424[2,18,138], 425[2], 429[158], 430[158,159], 431[160,161,162,163], 438[11,12], 439[28,30], 441[11,12], 442[46a-c], 443[11,12], 485[138], 486[144], 489[164,165,166,167,168,169], 522[40], 527[2,4], 528[2,3], 530[2], 571[113], 572[113], 587[113], 710[52], 748[114], 769[227,228], 771[227,228,269,270,272,283], 772[272], 775[353], 779[269], 819[21], 843[44]; **8**, 5[25], 879[51,54], 880[51], 888[121,124]
Sharrah, M. L., **4**, 280[132]
Sharrard, F., **6**, 554[781]
Sharrocks, D. N., **8**, 711[69]
Sharrygina, O. A., **2**, 662[19], 664[19]
Sharts, C. M., **4**, 270[6], 271[6]; **5**, 63[3], 69[3], 70[107]; **7**, 229[117]
Shashkov, A. S., **2**, 710[16]; **5**, 1055[46], 1056[48], 1057[50,51], 1062[51]
Shashkova, E. M., **8**, 727[196]
Shastin, A. V., **2**, 709[15]; **4**, 337[37,38]
Shastri, R. K., **8**, 384[40]
Shatenshtein, A. I., **1**, 632[61,68], 644[68]
Shatzmiller, S., **2**, 486[41]; **5**, 501[269]
Shauble, J. H., **6**, 26[104]
Shavandin, Yu. A., **3**, 328[181]
Shavel, J., Jr., **6**, 182[138], 1021[51]
Shaver, A., **3**, 566[29]
Shavrygina, O. A., **5**, 431[122]
Shavva, A. G., **7**, 699[57]
Shaw, A., **3**, 530[77], 535[77]; **4**, 259[267]; **5**, 768[126], 779[126]
Shaw, A. C., **6**, 903[137]
Shaw, A. N., **5**, 477[159]
Shaw, B. L., **1**, 451[216]; **4**, 587[35,41], 588[54], 1002[61]; **7**, 630[51]; **8**, 445[35,36]
Shaw, C. J. G., **4**, 1021[249,250]
Shaw, C. K., **7**, 144[156]
Shaw, D. A., **7**, 673[30]; **8**, 354[174]
Shaw, D. C., **5**, 855[183]
Shaw, G., **8**, 649[60]
Shaw, G. S., **5**, 680[24,24a,b]
Shaw, I. M., **6**, 204[24]
Shaw, J., **7**, 95[70,70a]
Shaw, J. E., **4**, 18[62], 21[62d], 437[147]; **5**, 992[48]; **8**, 528[73,74,79], 598[96], 612[73], 624[73]
Shaw, M. J., **7**, 228[106]
Shaw, P. E., **8**, 161[20], 494[24]
Shaw, R. A., **2**, 183[12]; **7**, 203[54]
Shaw, R. E., **1**, 3[23]
Shaw, T. J., **2**, 505[12], 510[12,36]
Shawe, T. T., **5**, 689[75]
Shay, A. J., **7**, 157[33], 158[33b,43]
Shchegoleva, T. A., **8**, 727[196]
Shchekotikhina, N. A., **6**, 564[910]
Shchelkunov, A. V., **2**, 420[24]
Shchepinov, S. A., **8**, 769[27]
Shcherbakova, L. I., **6**, 543[609]
Shcherbik, P. K., **3**, 643[131]
Shchukina, L. A., **3**, 828[50]
Shchukina, M. N., **8**, 656[92]
Shea, C.-M., **6**, 766[20]
Shea, K. J., **1**, 878[109]; **2**, 1050[28]; **3**, 466[185]; **4**, 91[88d], 1005[86]; **5**, 20[136], 67[93], 70[116], 514[9,10], 527[9], 790[20,21], 805[100], 820[20,21], 822[20], 826[20], 857[228]; **6**, 531[447], 535[447]
Shea, R. G., **1**, 635[88]; **3**, 104[209], 117[209]; **6**, 846[103], 905[145]
Shealy, N. L., **2**, 523[74]
Shearer, B. G., **4**, 243[65], 253[65]; **5**, 517[29], 519[29], 534[29], 538[29e], 539[29e]
Shearer, H. M. M., **1**, 21[111], 318[157], 320[157]
Shearing, E. A., **3**, 689[117]
Sheats, J. R., **1**, 451[217]
Shebaldova, A. D., **6**, 509[272]; **8**, 451[177]
Shechter, H., **2**, 586[130]; **3**, 161[470], 167[470]; **4**, 287[178], 753[169], 1054[132], 1104[212]; **6**, 121[132], 213[85], 214[99,100], 228[17], 232[38], 727[206], 943[155], 1003[135]; **7**, 228[103]; **8**, 297[69], 844[69], 941[111], 943[120]
Shedrinksy, A., **5**, 526[57]
Sheehan, J. C., **2**, 1100[117]; **5**, 90[53], 92[60]; **6**, 666[232,232a], 667[232]; **7**, 294[12]; **8**, 956[12], 957[12]
Sheehan, J. J., **4**, 276[78], 303[345]
Sheehan, J. T., **6**, 570[946]
Sheehan, M. N., **8**, 60[186], 66[186]
Sheeran, P. J., **5**, 116[260]
Sheets, R. M., **7**, 267[121], 269[121], 270[128], 271[121,128], 278[121]
Sheffels, P., **5**, 1183[58]
Sheffold, R., **6**, 849[124]
Sheffy, F. K., **2**, 727[131]; **3**, 232[271], 469[201], 470[201], 471[201], 473[201], 475[201]; **4**, 594[142,145], 619[142,145], 633[142,145]

Shefter, E., **5**, 468[127]
Shei, J. C., **7**, 763[99], 766[99]
Sheinker, Yu. N., **6**, 531[453], 554[789]; **8**, 599[101]
Sheinkman, A. K., **6**, 494[132]; **8**, 616[98]
Shekhirev, Y. P., **4**, 291[209]
Shekoyan, B. M., **8**, 551[342]
Shelberg, W. E., **2**, 148[78]
Sheldon, B. G., **5**, 516[21], 531[21]
Sheldon, J. C., **7**, 765[143]
Sheldon, R. A., **7**, 437[8], 527[1], 628[47], 719[4], 722[4], 724[4], 727[4], 851[17,24], 860[72]; **8**, 396[136]
Sheldrake, G. N., **5**, 437[159]
Sheldrake, P. W., **1**, 408[35], 430[35]; **3**, 288[63]; **4**, 372[64a]; **6**, 660[202]; **7**, 678[73]; **8**, 881[68]
Sheldrick, G. M., **1**, 37[180,243], 528[119]; **2**, 345[34], 351[34], 357[34], 371[34,262], 372[271], 373[273,274]; **4**, 706[38]; **5**, 14[99], 17[121], 461[96], 468[121,125]; **6**, 690[403], 692[403]; **8**, 446[73]
Sheldrick, W. S., **2**, 354[114], 357[114], 385[321]; **5**, 634[84], 635[84]; **6**, 196[229]
Shellhamer, D. F., **4**, 345[81], 347[90,95]; **7**, 477[75]
Shellhammer, A. J., Jr., **4**, 24[72,72a]
Shelly, K. P., **1**, 569[250], 751[91]
Shelton, E. J., **4**, 342[62]
Shelton, G., **3**, 770[180]
Shelton, J. R., **2**, 345[21], 954[8]; **6**, 1024[74]
Shemyakin, M. M., **3**, 828[50]
Shen, B., **4**, 298[286]
Shen, C. C., **8**, 61[191], 66[191]
Shen, C.-U., **7**, 294[13]
Shen, G. J., **7**, 57[22]
Shen, G. S., **5**, 1123[40]
Shen, G.-Y., **3**, 223[148]; **6**, 155[152]
Shen, J., **3**, 298[27], 332[206], 333[209]
Shen, K. W., **5**, 66[82]
Shen, M., **5**, 571[119], 572[124]
Shen, T., **6**, 851[128], 879[43]; **7**, 633[66]
Shen, T. Y., **4**, 932[61]; **6**, 487[57], 489[57]
Shen, W., **4**, 820[223]
Shen, Y., **1**, 825[55]; **2**, 24[92]; **4**, 991[152]; **6**, 185[160,167], 187[167]; **7**, 68[80]; **8**, 18[129]
Shenk, W. J., Jr., **3**, 890[32]
Shenoy, P. K., **6**, 210[72]
Shenoy, S. J., **6**, 543[618]
Shenton, K. E., **3**, 1031[63]
Shenvi, A., **1**, 836[145]; **3**, 286[60], 369[122], 372[122]; **6**, 766[22]; **8**, 353[156]
Shepard, K. L., **1**, 389[141]; **4**, 486[40]; **5**, 379[113a]; **6**, 278[130]
Shepard, M. E., **8**, 838[20]
Shepelavy, J. N., **7**, 21[13], 26[60], 479[95]
Shepelin, A. P., **8**, 608[47]
Shephard, B. R., **3**, 643[125], 644[147,151]
Shephard, K. B., **6**, 901[122]
Shephard, K. P., **6**, 840[73]
Shepherd, J. P., **7**, 444[53]
Shepherd, R. A., **1**, 876[101]; **5**, 804[93]
Shepherd, R. G., **8**, 30[42], 66[42]
Sheppard, A. C., **1**, 390[142], 837[150]; **7**, 330[10], 425[147b], 746[93]
Sheppard, C., **6**, 1024[75]
Sheppard, G., **6**, 714[84]
Sheppard, G. S., **2**, 318[50]; **5**, 1143[92]
Sheppard, J. H., **8**, 237[16], 244[16], 253[16], 709[46], 710[54], 721[54]
Sheppard, R. C., **6**, 638[38], 669[265], 670[38], 671[38,276]
Sheppard, R. N., **4**, 381[126b], 382[126], 383[126]
Sheppard, W. A., **3**, 499[122]; **4**, 270[6], 271[6]; **5**, 430[116], 486[196]; **8**, 86[21], 408[72]
Sher, F., **3**, 125[299]
Sher, F. T., **4**, 7[24]
Sher, P. M., **1**, 771[193]; **4**, 721[29], 738[29], 753[29], 765[29], 796[98], 823[229]; **7**, 648[41], 731[54]; **8**, 163[41]
Sheradsky, T., **2**, 866[4]; **4**, 56[158]; **5**, 78[274]; **7**, 746[91]; **8**, 60[193], 62[193], 66[193], 384[23]
Sherbine, J. P., **1**, 786[248]; **4**, 116[185b]; **8**, 842[47]
Sheridan, J. B., **4**, 689[72]
Sheridan, R. S., **5**, 647[17], 651[17], 655[28], 656[17]; **6**, 123[140]; **7**, 862[81], 888[81]
Sherif, S. M., **2**, 378[291]
Sherk, K. W., **4**, 295[260], 296[260]; **8**, 328[12,13], 329[12,13]
Sherlin, S. M., **5**, 453[59]
Sherman, D. H., **4**, 162[92]
Sherman, P. D., **4**, 98[110]
Sherr, A. E., **3**, 842[21]
Sherrill, M. L., **4**, 288[185], 298[185]
Sherstyannikova, L. V., **8**, 765[12]
Sherwin, P. F., **7**, 768[203]
Sheta, A. E., **3**, 834[78]
Sheth, J. P., **4**, 1040[88,99], 1048[88]; **5**, 916[122], 925[122]
Shetty, R. V., **5**, 2[16]
Sheu, J. T., **5**, 945[254], 946[254], 947[261], 961[261]
Shevchenko, I. B., **6**, 500[180]
Shevchenko, M. V., **6**, 500[180]
Shevchuk, V. U., **6**, 216[107]
Shevelev, S. A., **6**, 226[10], 256[10], 257[10]
Shevlin, D. B., **8**, 890[142]
Shevlin, P. B., **5**, 589[212]; **6**, 993[94]
Shew, D., **6**, 685[358]
Shi, G., **2**, 772[16]
Shi, L., **4**, 854[94]; **8**, 680[70], 691[70]
Shi, Y.-J., **2**, 384[318]; **5**, 810[125]
Shiara, T., **7**, 172[124]
Shiba, S., **7**, 761[56]
Shiba, T., **4**, 958[27]; **6**, 18[65], 53[118], 57[141], 566[925], 603[14], 637[36]; **8**, 150[122], 151[148], 244[70]
Shibaev, V. N., **6**, 533[498]
Shibahashi, H., **6**, 453[142]
Shibasah, M., **7**, 57[24]
Shibasaki, M., **2**, 113[104], 271[76], 547[94], 920[93], 921[93], 922[93], 923[93], 926[116,118,119], 1051[43]; **3**, 650[213]; **4**, 331[15], 370[39]; **5**, 101[162], 562[88], 837[69]; **6**, 2[8], 22[8], 655[163], 678[324], 778[62], 780[62]; **7**, 353[35], 355[35], 399[40b], 410[96], 455[105], 617[21], 620[26], 621[30], 804[60]; **8**, 452[188], 457[188], 554[366,367], 555[367c], 567[458]
Shibata, A., **3**, 381[30], 382[30]
Shibata, I., **8**, 20[137]
Shibata, J., **7**, 209[93]
Shibata, N., **6**, 746[90]
Shibata, S., **1**, 317[143]; **4**, 378[107]; **6**, 534[516]; **8**, 160[102], 171[102-104], 178[102], 179[102]
Shibata, T., **2**, 649[103], 1059[76]; **3**, 1026[40]; **4**, 337[35,36]; **5**, 78[272], 585[200]; **6**, 239[76], 538[562], 614[93]; **7**, 442[46c], 493[201], 684[93a]
Shibata, Y., **7**, 675[53]; **8**, 152[160]
Shibato, K., **1**, 343[112]; **2**, 174[183]
Shibib, S. M., **8**, 563[435]
Shibilkina, O. K., **7**, 75[114]
Shibita, Y., **8**, 658[99]
Shibutani, M., **3**, 593[181]
Shibutani, T., **3**, 510[182]; **4**, 487[46]
Shibuya, H., **1**, 188[70]; **3**, 751[88]
Shibuya, M., **2**, 826[124]; **4**, 379[114]; **6**, 1056[56,57]; **8**, 496[31]

Shibuya, N., **4**, 597[173,174], 637[173,174]; **7**, 95[65]
Shibuya, S., **1**, 790[263]; **2**, 363[193], 765[67], 1060[88,91], 1073[141,142]; **4**, 364[1,1h], 401[233], 487[42], 505[136,138]; **5**, 725[118]; **6**, 749[98], 751[108], 780[72], 879[43]; **8**, 389[75], 934[57], 957[13]
Shi-Ching Chang, **7**, 500[246]
Shick, C. H., **5**, 99[139]
Shida, T., **5**, 704[21]
Shida, Y., **8**, 170[79]
Shieh, C. H., **5**, 90[59], 94[59]
Shieh, H.-M., **2**, 204[100]; **3**, 41[225]
Shieh, W.-R., **8**, 190[83,85]
Shields, C. J., **8**, 890[144]
Shields, J. E., **3**, 760[139], 774[139]; **5**, 582[181]; **6**, 635[18], 636[18]
Shields, T. C., **4**, 1014[191], 1015[193,194]; **5**, 794[46]; **8**, 530[95], 806[110], 807[110,119]
Shigehisa, T., **5**, 841[97]
Shigematsu, K., **8**, 364[20]
Shigematu, K., **5**, 736[142m]
Shigemitsu, Y., **5**, 163[73]
Shigemoto, T., **4**, 38[109a]; **6**, 89[118], 101[118]
Shigi, M., **7**, 801[36]
Shih, C., **3**, 256[112]; **4**, 390[172]
Shih, C. N., **5**, 857[230]
Shih, E. M., **3**, 738[34]
Shih, H. C., **3**, 503[149], 512[149]
Shih, J., **6**, 207[51]
Shih, N. Y., **3**, 770[176]
Shih, S., **7**, 870[96]
Shih, T. L., **2**, 250[38], 251[38], 260[38], 274[38], 436[67]; **4**, 356[136]
Shih, Y., **8**, 344[121]
Shih, Y.-S., **2**, 766[83-85]
Shiihara, I., **8**, 781[96]
Shiina, K., **3**, 541[106], 558[50]
Shiio, I., **7**, 56[16]
Shiiza, M., **7**, 323[69]
Shikano, E., **6**, 644[85]
Shikhiev, I. A., **8**, 771[46]
Shikhmamedbekova, A. Z., **2**, 580[98], 581[98]
Shilcrat, S. C., **7**, 401[57]
Shiley, R. H., **6**, 221[134]
Shilov, A. A., **7**, 17[173]
Shilov, A. E., **7**, 1[2], 4[19], 8[19], 13[118], 17[2,172]
Shilov, E. A., **4**, 289[190,193]
Shim, S. B., **6**, 727[191]; **7**, 744[71], 752[143]
Shim, S. C., **1**, 563[182]; **8**, 54[161], 66[161], 261[16], 384[28,35], 847[94]
Shima, K., **4**, 91[88e], 152[55], 211[94,95], 231[94]; **5**, 154[31], 165[81], 166[89], 168[102], 650[25]; **6**, 531[429]; **7**, 673[27]; **8**, 518[132]
Shimada, A., **3**, 310[95], 311[95]; **6**, 175[76]
Shimada, J., **1**, 212[12], 213[12], 215[12c], 217[12]
Shimada, J.-i., **2**, 117[149], 310[27], 442[12], 447[12], 651[112]
Shimada, K., **3**, 100[196,197], 395[101]; **5**, 442[185]; **6**, 900[117]; **7**, 877[135]
Shimada, N., **1**, 161[88]
Shimada, S., **5**, 24[168], 729[123], 741[157]
Shimada, Y., **6**, 616[106], 624[139]
Shimadzu, M., **7**, 462[119,120]
Shimagaki, M., **1**, 834[121a,122]; **2**, 830[141]; **6**, 979[29]; **8**, 12[64,65], 500[52]
Shimahara, M., **8**, 533[154]
Shimamura, N., **8**, 29[38], 66[38]
Shimamura, T., **4**, 598[201], 638[201]; **6**, 86[99]
Shimano, M., **4**, 373[70]
Shimasaki, C., **3**, 295[8]
Shimasaki, Y., **8**, 9[51]
Shimazaki, K., **8**, 369[80]
Shimazaki, M., **6**, 14[52,54]; **7**, 56[17], 667[78]; **8**, 10[58]
Shimazaki, T., **1**, 784[244]; **3**, 224[162], 489[62], 495[62], 504[62], 511[62], 515[62]; **7**, 414[108,109], 712[64]
Shimazaki, Y., **2**, 505[10]; **3**, 34[194]
Shimizu, F., **5**, 282[28,29], 284[28,29], 601[45,46], 605[46], 606[45], 609[68]; **8**, 356[185]
Shimizu, H., **1**, 72[66], 141[22]; **2**, 1059[78]; **3**, 227[210], 245[30], 816[79]; **5**, 504[276]; **6**, 510[293], 936[106]; **8**, 370[91], 996[71]
Shimizu, I., **3**, 530[81,83], 536[81,83], 594[185]; **4**, 590[96], 591[113,116,118], 592[119,121-125], 593[135], 594[96], 597[172], 611[360,361,363,364,365,366,367], 613[369], 633[96], 636[366,367], 637[172]; **5**, 281[18], 304[83], 850[148], 1200[48], 1201[48]; **6**, 20[77]; **7**, 95[63], 142[135,136,137], 450[7], 451[7], 455[7,106], 456[108], 457[108,109], 458[7], 459[108], 460[116], 461[117]; **8**, 960[31]
Shimizu, K., **1**, 553[98]; **3**, 380[10]; **4**, 192[117], 227[205], 255[196]; **5**, 196[15], 532[87], 581[174]; **6**, 113[71]; **7**, 360[22]
Shimizu, M., **1**, 72[68], 658[158], 659[158], 664[158], 665[158], 670[184,185], 671[185], 672[158]; **2**, 615[128], 633[34a], 634[35], 640[34,35]; **3**, 6[28], 8[28], 14[28], 174[531], 176[531], 177[531], 179[531]; **5**, 338[53]; **7**, 247[103], 307[15], 310[15], 323[15], 523[42], 530[13], 771[279], 773[279], 801[45]; **8**, 613[78], 642[32,34]
Shimizu, N., **4**, 30[88,88n], 239[17], 261[17], 1089[133]; **5**, 76[240], 86[15], 152[21], 596[21], 597[21], 601[21], 603[21], 606[21], 608[21], 927[163]; **6**, 927[80]; **8**, 803[90], 807[90], 856[168]
Shimizu, S., **1**, 477[142]; **5**, 377[111,111b]; **7**, 145[165], 415[114]; **8**, 190[66,67]
Shimizu, T., **1**, 564[201]; **7**, 228[97], 773[299], 779[299,424,428,430,431]; **8**, 320[81,82]
Shimizu, Y., **1**, 436[154], 438[161], 452[219], 474[104]; **3**, 453[114], 463[160,164]
Shimo, N., **4**, 611[353]
Shimobayashi, A., **6**, 233[40]
Shimoji, K., **1**, 789[260]; **4**, 390[172], 413[274]; **6**, 674[293]
Shimokawa, K., **1**, 350[154], 359[12], 360[12], 361[12], 362[12]
Shimomura, C., **7**, 73[104]
Shimomura, O., **3**, 743[59]
Shimonishi, Y., **6**, 636[17]
Shimosaka, M., **2**, 464[98]
Shimozono, K., **6**, 564[918]
Shimpuku, T., **2**, 657[161a]
Shimuk, N. P., **3**, 328[181]
Shimuzu, N., **8**, 201[139]
Shin, C., **1**, 733[20]; **8**, 374[150]
Shin, D. M., **7**, 874[106]; **8**, 368[69,70], 375[70]
Shin, D.-S., **7**, 260[84]
Shin, I.-J., **6**, 834[39]
Shin, S., **4**, 939[76]
Shinagawa, S., **2**, 1099[112b]; **6**, 637[33]
Shinaki, T., **7**, 24[37], 25[37,42,45]
Shindo, H., **7**, 200[41], 208[86], 211[86]; **8**, 964[56]
Shindo, M., **4**, 76[49]
Shine, H. J., **3**, 803[1], 804[12]; **6**, 240[83]; **7**, 167[98,100], 850[3], 881[164]; **8**, 364[24], 366[35]
Shine, R. J., **6**, 461[2]
Shiner, C. S., **1**, 412[51], 463[33], 739[41]; **2**, 441[1], 443[1]; **3**, 24[144]; **4**, 24[73,73a]; **5**, 519[30]; **6**, 2[8], 22[8]
Shing, T. K. M., **5**, 534[95], 731[130c]; **7**, 568[106], 569[106], 710[48], 711[58], 712[66], 713[74]
Shing, T. S., **2**, 385[327]
Shingaki, T., **7**, 25[41], 26[41,52,53,58]
Shinhama, K., **6**, 208[57], 212[57]; **7**, 761[57,58]
Shinjo, K., **6**, 441[85]

Shinkai, I., **1**, 402[17], 791[296a], 799[296]; **2**, 227[160], 256[46], 482[37], 483[37], 485[37], 652[125], 803[33], 821[104], 852[104], 1059[79]; **3**, 45[247]; **5**, 99[130], 410[41], 850[160]; **6**, 759[139]; **7**, 228[92], 416[122]
Shinkai, S., **7**, 761[63]; **8**, 87[34], 95[88], 364[20]
Shinkai, Y., **2**, 810[60]
Shinke, S., **7**, 100[115]
Shinker, W. M., **3**, 665[35]
Shinma, N., **6**, 917[35]
Shinmi, Y., **1**, 566[214]; **4**, 10[34], 113[164], 229[239]
Shinna, N. D., **4**, 1040[75]
Shinoda, E., **1**, 474[104]
Shinoda, M., **1**, 366[46], 876[100]; **4**, 1007[124], 1018[219]; **5**, 768[128,130], 769[128], 770[142], 771[142,143], 779[128]; **6**, 563[905]
Shinoda, N., **3**, 391[87]
Shinoda, T., **5**, 925[154]
Shinohara, S., **4**, 229[215]
Shinohara, T., **7**, 419[134b]
Shinohara, Y., **2**, 780[13]
Shinoki, H., **2**, 209[109]; **7**, 339[43]
Shinonaga, A., **6**, 684[344]
Shinozaki, A., **8**, 643[38]
Shinozaki, H., **1**, 766[161]
Shinozaki, K., **6**, 644[85]; **8**, 975[133], 992[55]
Shinozuka, K., **6**, 554[732]
Shintani, Y., **3**, 501[136]
Shinzo, K., **7**, 43[40,41]
Shiobara, Y., **3**, 99[190]; **7**, 476[67]
Shiohara, T., **5**, 1137[56]
Shioi, S., **4**, 379[114]
Shioiri, T., **1**, 555[113], 560[155], 844[6,7], 851[40,47], 852[40], 853[47]; **2**, 233[189], 455[16], 801[34]; **3**, 46[253], 900[90]; **6**, 121[127,-128], 127[159], 129[166], 235[52], 251[148], 438[45,46], 637[32,32b], 645[95], 690[394], 738[48], 797[17,18], 811[17,18,78], 812[18,82,83], 813[83-85], 816[18,99]; **7**, 158[35], 506[301,302]
Shiokawa, M., **6**, 979[29]
Shiokawa, T., **5**, 623[25-27]
Shiomi, Y., **5**, 167[94]; **7**, 64[61a]
Shiono, H., **3**, 97[174,175,176], 103[175,176], 108[174], 109[175,176], 117[174]
Shiono, M., **1**, 506[15]; **3**, 244[25], 267[25], 470[209], 472[209], 475[209], 494[84], 730[46]; **4**, 1002[46]; **6**, 139[48]; **8**, 698[137,138]
Shiono, S., **8**, 145[82], 146[101], 148[110]
Shiosaki, K., **2**, 869[21,22], 870[21], 876[21], 880[21], 881[22], 890[21]; **6**, 509[270,280]
Shiota, M., **6**, 765[19]; **8**, 418[9], 533[154]
Shiota, T., **1**, 392[155]; **7**, 745[76,77]
Shiotani, N., **7**, 245[77]
Shiozaki, M., **2**, 649[101]; **3**, 715[39]
Shiozawa, A., **6**, 543[604]
Shipchandler, M. T., **6**, 105[14]
Shipman, M., **5**, 290[42]
Shippey, M. A., **5**, 1115[1], 1116[1]; **8**, 886[111]
Shirafuji, T., **8**, 807[116]
Shiragami, H., **1**, 343[111]; **2**, 584[126]; **3**, 279[37]; **4**, 120[203], 567[49], 879[85]
Shirahama, H., **1**, 766[161]; **2**, 547[122], 553[122]; **3**, 100[199], 382[39], 386[57], 400[119-124], 402[125,126,130,131], 404[132,135,137], 405[138], 714[35]; **6**, 780[69]; **7**, 91[36], 109[184], 168[101], 406[87]; **8**, 269[81], 857[191]
Shirahata, A., **2**, 565[14], 566[17], 601[34]; **4**, 98[113,113b]; **5**, 596[22], 597[22], 603[22]
Shirai, F., **2**, 116[141], 653[128], 656[153], 657[166], 1059[76]; **3**, 986[28]; **6**, 876[30]
Shirai, H., **4**, 45[130], 507[151], 510[175]
Shirai, K., **7**, 196[29]; **8**, 609[49,54]
Shirai, M., **8**, 371[113]
Shirai, N., **3**, 968[129], 969[135]; **6**, 893[87]
Shirai, R., **2**, 105[39]; **7**, 142[138]
Shirai, Y., **8**, 394[117]
Shiraishi, H., **1**, 279[86], 280[86]; **3**, 567[190], 595[190], 607[190]
Shiraishi, M., **7**, 410[92]
Shiraishi, S., **6**, 498[164]
Shiraishi, T., **2**, 353[101]
Shiraishi, Y., **1**, 350[154], 359[12], 360[12], 361[12], 362[12]
Shiraiwa, M., **2**, 598[15]; **3**, 541[114]
Shiraiwa, Y., **6**, 531[460]
Shiraki, C., **8**, 383[21]
Shirasaka, T., **2**, 556[155], 685[67]; **5**, 434[147]
Shiratori, Y., **4**, 501[116]; **5**, 536[97], 693[114], 694[114]
Shirazi, A., **1**, 112[26]
Shirin, E., **8**, 864[242]
Shirlattl, V., **3**, 946[93]
Shirley, D. A., **1**, 398[3], 399[3]; **3**, 511[190]
Shirley, N. J., **7**, 410[100]
Shiro, M., **2**, 859[252]; **3**, 1050[19]; **4**, 42[122b], 377[105c], 381[105]; **5**, 369[101], 370[101a]; **6**, 787[98]; **7**, 415[114], 615[9]
Shirota, Y., **5**, 71[131]; **7**, 851[28]
Shirouchi, Y., **6**, 936[112]; **7**, 829[56,56c]
Shiroyama, K., **4**, 501[113]
Shiroyan, F. R., **8**, 618[124]
Shirra, A., **8**, 96[94]
Shishido, K., **2**, 1024[61,62]; **5**, 404[19], 405[19], 431[121], 541[110,111], 681[28], 712[57b], 741[157,157c,d]; **6**, 757[130]; **7**, 564[89], 569[89]
Shishiyama, Y., **1**, 347[132,134]; **2**, 599[26]; **4**, 23[71], 162[92]; **6**, 237[67], 564[913,914]
Shishkin, V. E., **6**, 535[534]
Shitara, E., **5**, 681[28], 741[157], 841[98], 843[117]
Shitov, O. P., **4**, 145[25]
Shiue, C., **7**, 751[140]
Shiue, C.-Y., **4**, 445[204,206]
Shiuey, S.-J., **1**, 780[229]
Shivanyuk, A. F., **6**, 524[373,374]
Shive, W., **3**, 391[89], 393[89]
Shizuri, Y., **3**, 694[151], 696[151]
Shkarin, E. G., **8**, 628[177]
Shkol'nik, O. V., **8**, 765[11]
Shlaefer, F. W., **3**, 905[139]
Shmatov, Y. N., **4**, 286[168,169], 289[168,169]
Shmelev, L. V., **3**, 342[14]
Shmidt, F. K., **8**, 454[198]
Shmueli, U., **5**, 223[84,85], 224[84,85]
Shoaf, C. J., **8**, 271[111], 274[132]
Shockor, J. P., **8**, 406[49]
Shoda, S., **1**, 347[135], 349[149]; **6**, 206[37]
Shoda, S.-I., **2**, 187[2], 572[46,48], 610[88], 924[106]; **6**, 46[58,73], 54[131], 438[56]
Shoham, G., **2**, 542[83]; **6**, 781[76]; **8**, 945[133]
Shoji, F., **3**, 461[147]
Shoji, K., **6**, 265[41]
Shoji, S., **4**, 230[255]
Shoji, T., **8**, 552[354]
Shoji, Y., **4**, 510[176]
Shokhor, I. N., **1**, 34[228]
Shokoka, E. A., **3**, 381[32]
Shold, D. M., **5**, 636[97]
Shome, M., **6**, 273[98]
Shomina, F. N., **6**, 560[866]
Shone, R. L., **3**, 592[172]
Shono, T., **1**, 268[54], 269[54a], 346[128], 366[48], 377[96], 544[39], 804[310], 831[99]; **2**, 138[19], 492[51], 613[111], 635[47], 640[47],

709[7], 784[38], 971[94], 1051[33,36], 1052[36,51], 1061[92], 1066[118,119], 1067[123], 1069[92,132], 1070[118], 1071[92]; **3**, 54[279], 597[201], 599[204], 600[217], 602[221], 726[25]; **4**, 95[101], 130[226c], 247[100], 251[142], 257[100], 260[100,142], 587[43], 809[162], 810[169]; **5**, 500[259], 901[34]; **6**, 801[37], 991[88]; **7**, 170[118], 227[74,75,77], 248[109], 707[29], 708[29], 761[66], 791[1], 794[6,7a], 795[8,10,11], 796[12,13], 797[16,18-20], 798[18b,21,22], 801[45], 802[47-49], 803[51,53-55], 804[58,59,62], 805[59,65], 806[75], 808[78-80], 809[81,85], 811[91]; **8**, 133[26], 134[35], 302[96], 817[32]
Shook, C. A., **1**, 239[38]
Shook, D., **5**, 181[155]
Shook, D. A., **4**, 1033[31]
Shoolery, J. N., **3**, 380[7]; **7**, 167[98], 820[23], 823[35]; **8**, 117[73]
Shoop, E. V., **6**, 530[417]
Shooshani, I., **8**, 566[457], 568[468]
Shoppee, C. W., **3**, 781[19], 854[77]; **5**, 752[52], 753[53], 754[53,70], 759[70]; **7**, 666[70]; **8**, 119[76], 343[113], 884[100], 926[13]
Shore, S. G., **8**, 707[25]
Short, E. L., **2**, 745[106], 958[25]
Short, J. H., **2**, 956[18], 957[18]
Short, R. P., **2**, 35[124b], 42[124], 249[84], 259[52], 261[52]; **4**, 1040[88], 1048[88,88a]; **5**, 918[129], 955[302]
Shorter, J., **1**, 580[4]; **3**, 867[34]
Shortridge, T. J., **6**, 915[30]; **8**, 389[69]
Shostakovskii, M. F., **4**, 317[544,545,548,554]; **7**, 762[75]
Shostakovsky, A. E., **4**, 1058[151]
Shostakovsky, V. M., **4**, 1058[150], 1059[153], 1063[172]
Shostenko, A. G., **8**, 773[69]
Shoulders, B. A., **4**, 276[80], 303[346], 304[346]; **8**, 806[110], 807[110,119]
Shoup, T. M., **7**, 602[104,104b]
Showalter, H. D. H., **5**, 942[229,232]
Showell, J. S., **2**, 143[55]; **5**, 2[12]
Showler, A. J., **6**, 660[198]
Shreeve, J. M., **6**, 569[936]; **8**, 864[241]
Shreve, A. P., **1**, 301[74], 316[74]
Shridhar, D. R., **5**, 95[100]
Shriner, R. L., **2**, 277[2], 281[2], 294[2], 296[2], 504[5]; **4**, 6[22]; **8**, 140[21], 142[49], 533[144], 905[63]
Shriver, D. A., **4**, 932[63]
Shroff, H. N., **5**, 500[256]
Shroot, B., **3**, 262[167]; **7**, 359[16]
Shrubsall, P. R., **7**, 401[58]
Shteinshneider, A. Yu., **5**, 65[55], 1198[45]
Shternberg, I. Ya., **4**, 26[77]
Shu, A., **4**, 850[85]
Shu, C.-S., **3**, 500[127], 501[127]
Shu, P., **1**, 490[28]
Shubert, D. C., **1**, 218[53]; **2**, 30[111], 31[111]; **3**, 494[85]; **5**, 246[23], 247[23,23a], 599[39]
Shudo, K., **3**, 306[78]; **6**, 524[361]
Shue, R. S., **4**, 837[16]
Shuetake, T., **7**, 425[149a]
Shuey, C. D., **4**, 25[76], 46[76], 128[222]
Shugihara, Y., **7**, 425[149b]
Shujkin, N. I., **8**, 956[5]
Shukis, W. F., **1**, 425[105]
Shukla, S. N., **3**, 636[57]
Shukys, J. G., **5**, 1138[67]
Shulgin, A. T., **8**, 376[160]
Shull, D. W., **5**, 910[87], 1007[40]
Shulman, E. M., **4**, 18[61,61a], 247[101], 262[101]
Shulman, J. I., **3**, 120[240]; **4**, 349[109]; **5**, 830[37]; **6**, 982[50]
Shulman, S., **2**, 823[112]; **3**, 1051[22], 1052[22]
Shultz, A. G., **3**, 815[78]
Shuman, R. T., **4**, 496[88]
Shumann, H., **1**, 162[104]
Shumate, R. E., **7**, 765[145]; **8**, 459[237a], 535[166]
Shunk, C. H., **2**, 284[56], 838[163]
Shuojian, J., **3**, 298[26]
Shur, V. B., **4**, 485[29]
Shuraeva, V. N., **5**, 64[24]
Shurubura, A. K., **6**, 524[366]
Shushunov, V. A., **7**, 602[106]
Shusterman, A. J., **4**, 980[105], 981[105], 982[113]; **5**, 1085[65], 1086[67]; **8**, 431[66], 459[226]
Shustov, G. V., **1**, 837[147]
Shutske, G. M., **2**, 852[228], 853[228]; **4**, 439[168]
Shutske, M., **4**, 440[169]
Shvarts, G. Ya., **6**, 553[728], 554[728,741,773]
Shvedov, V. I., **6**, 487[74], 489[74]
Shvekhgeimer, G. A., **6**, 507[237], 515[237], 534[517]
Shvekhgeimer, S. M., **6**, 530[422]
Shvetsova-Shilovskaya, K. D., **4**, 992[154]
Shvo, Y., **4**, 674[33]; **8**, 289[23], 395[130], 446[72]
Shyama Sundar, N., **8**, 115[63]
Shyamsunder Rao, Y., **7**, 253[12]
Shymanska, M. V., **8**, 764[4a]
Siahaan, T. J., **1**, 110[23], 112[24,26], 335[66]; **3**, 212[34b,c]; **4**, 244[78], 245[78], 255[78]
Sibarov, D. A., **3**, 329[183]
Sibi, M. P., **1**, 466[43], 469[60], 470[61]; **3**, 242[5], 683[104]; **5**, 384[128], 385[128b], 1021[72]
Sibtain, F., **8**, 847[94]
Sice, J., **2**, 456[72], 457[72], 459[72]
Sicheneder, A., **7**, 401[61a]
Sicher, J., **2**, 841[185]; **6**, 959[39]; **8**, 726[191,192]
Sicken, M., **3**, 593[179]
Sicking, W., **7**, 880[152]
Sidani, A. R., **5**, 109[214]
Siddall, J. B., **3**, 220[121], 257[120]; **4**, 185[90], 186[90], 248[109], 893[154]; **7**, 86[16a]
Siddall, T. L., **6**, 81[73]
Siddhanta, A. K., **7**, 378[97]
Siddiqui, M. A., **3**, 231[251]; **6**, 176[87]
Siddiqui, S., **3**, 892[48]; **4**, 463[108], 471[142], 472[142]
Siddiqui, T., **1**, 464[36], 478[36]
Siddons, P. T., **3**, 816[79]
Sidell, M. D., **8**, 389[69]
Siden, J., **4**, 342[64], 343[64], 386[155]
Siderov, N. N., **8**, 677[57], 689[57]
Sidgwick, N. V., **6**, 293[230]
Sidisunthorn, P., **3**, 807[25]
Sidler, D. R., **4**, 905[213]
Sidler, J. D., **3**, 975[3], 980[3]
Sidorova, N. G., **3**, 309[89]
Sidot, C., **4**, 452[19]
Siebeling, W., **8**, 910[86]
Siebenthall, F., **6**, 537[576]
Sieber, P., **6**, 636[25], 637[27], 667[241], 668[251,262], 669[251], 1062[78]
Sieber, R., **7**, 449[1,2], 450[1,2]
Sieber, W., **4**, 1081[78]; **5**, 709[45], 712[45d]
Siebert, A. E., **5**, 676[3]
Siebert, C., **6**, 970[121]
Siebert, W., **8**, 711[66], 718[66]
Sieburth, S. M., **1**, 885[133a]; **3**, 242[7,8], 257[7,8]; **5**, 789[17], 1005[31]
Sieczkowski, J., **6**, 690[390]; **7**, 300[52]
Siefert, J.-M., **6**, 657[180]
Siegel, A., **2**, 278[9], 285[9]; **3**, 817[87]

Siegel, B., **6**, 195[223]; **7**, 49[68]
Siegel, C., **2**, 117[152], 224[152], 232[152], 308[19]; **5**, 373[106,106c], 374[106c]
Siegel, E., **7**, 657[30]
Siegel, H., **1**, 830[90]; **3**, 312[101], 1022[18]; **4**, 920[22,23], 921[22], 922[23b], 923[22], 924[22], 925[22], 1007[116], 1008[132]; **6**, 864[194]; **8**, 445[22,24], 459[229]
Siegel, J., **2**, 640[170]; **5**, 1133[30]
Siegel, M. M., **5**, 736[145], 737[145]
Siegel, S., **4**, 89[84a], 98[109g]; **8**, 420[23], 421[25,29], 422[29], 423[23,38], 425[47], 426[23], 428[23,38], 429[23], 431[67], 433[23], 434[23], 435[23,71], 436[74], 437[23,77,78], 474[16], 476[22]
Siegel, T. M., **6**, 282[155]
Siegel, W., **2**, 219[142]
Siegfried, B., **6**, 208[56]
Siegl, A., **4**, 48[138,138a], 66[138a]
Siegl, W. O., **8**, 289[25]
Siegmann, K., **4**, 1039[64]
Siegmeier, H., **6**, 667[246]
Sieler, H.-J., **6**, 441[86]
Sieler, J., **1**, 320[160]; **5**, 1157[168]
Sieler, R. A., **2**, 363[197]
Sieloff, R. F., **3**, 564[7]
Siemann, H. J., **8**, 856[167]
Siemieniewski, H., **2**, 403[35]
Siemion, I. Z., **2**, 403[34,35]
Siemionko, R. K., **5**, 603[50]
Sierakowski, A. F., **3**, 502[141]
Sierra, J., **7**, 90[33]
Sierra, M. A., **5**, 1087[77]
Sieveking, S., **6**, 540[589]
Sievers, R. E., **2**, 1099[110]
Siew, P. Y., **5**, 217[19,20], 768[127]
Sifferd, R. H., **6**, 636[16], 664[16]
Sigalov, A. B., **1**, 276[78], 277[78]
Sigalov, M. V., **4**, 55[157], 57[157o]; **8**, 770[38]
Sigg, H. P., **3**, 736[25]
Siggel, L., **5**, 645[1], 648[1r], 651[1], 653[1r]
Siggins, J. E., **8**, 263[25]
Siglmüller, F., **2**, 1090[72], 1098[104], 1099[109,109b]
Sigman, D. S., **8**, 589[49]
Sigmund, F., **8**, 212[11,15]
Sigmund, G., **2**, 855[242]
Sigmund, S. K., **6**, 117[96]
Signorelli, A. J., **8**, 373[126]
Sigrist, R., **1**, 876[102]
Sigurdson, E. R., **7**, 58[57], 62[57], 63[57]
Sih, C., **5**, 1096[110], 1098[110]
Sih, C. J., **1**, 429[126], 798[289]; **2**, 746[108], 762[56], 824[120]; **3**, 693[148], 694[148]; **4**, 187[98]; **6**, 995[101]; **7**, 40[12], 80[139], 145[169], 778[415]; **8**, 185[19], 188[52], 190[19,83,85], 191[94], 194[103], 237[10], 243[10], 544[253,264,265], 546[304], 561[304]
Sih, C. S., **2**, 456[29]
Sih, J. C., **3**, 279[38]; **7**, 633[64,65]
Siirala-Hansen, K., **4**, 560[21,25,26]; **7**, 474[44]
Siiteri, P., **2**, 240[8]
Sik, V., **5**, 418[70]
Siklosi, M. P., **5**, 1003[20]
Sikora, D. J., **5**, 1135[48]
Sikorski, J. A., **1**, 471[67]; **8**, 237[16], 244[16], 253[16], 719[117,118,121,122]
Sikorsky, P., **8**, 697[135]
Silber, J. J., **7**, 881[164]
Silber, P., **5**, 123[2]
Silber, R. H., **6**, 219[123]
Silberman, L., **4**, 37[107,107b,e], 38[107e], 39[107e]; **8**, 542[224]
Silbernagel, M. J., **8**, 623[150]
Silbert, L. S., **2**, 187[43]; **4**, 313[475], 348[108], 349[108b]; **6**, 2[6]; **7**, 185[175]
Silenko, I. D., **8**, 618[121], 619[121]
Siler, P., **7**, 219[11]
Siles, S., **5**, 253[46,46c]; **8**, 61[189], 66[189]
Šilhánková, A., **8**, 587[41]
Silks, L. A., **4**, 189[102], 190[108], 191[109]
Sille, F., **6**, 495[148]
Sillion, B., **7**, 498[225], 537[56,57]
Silva, G. V. J., **8**, 849[106]
Silvani, A., **1**, 440[171]
Silveira, A., Jr., **3**, 7[32], 8[32], 12[56], 266[196], 524[36], 795[81]; **4**, 893[151]; **5**, 1166[19]; **6**, 966[95]; **7**, 596[38]; **8**, 756[147]
Silveira, C., **4**, 120[196]; **5**, 268[78]
Silver, J. E., **8**, 7[36]
Silver, M. S., **4**, 876[58]
Silver, S. M., **1**, 731[5]
Silverman, G. S., **8**, 103[130]
Silverman, I. R., **2**, 650[110], 651[110]
Silverman, L. S., **8**, 356[188]
Silverman, R. B., **8**, 36[67], 66[67]
Silverman, S. B., **8**, 318[68,69]
Silversmith, E. F., **4**, 2[6], 18[61,61a], 247[101], 262[101]; **6**, 835[42]
Silverstein, R. M., **1**, 411[45]; **3**, 126[318], 248[57], 249[57], 251[57], 263[57]; **5**, 453[68]; **7**, 684[92]
Silverthorn, W. E., **4**, 519[14], 520[14]
Silverton, J. V., **2**, 1017[34]; **7**, 366[51], 414[120]; **8**, 798[60]
Silvester, M. J., **4**, 445[202]
Silvestri, M., **2**, 613[113], 911[71]; **4**, 18[60,60a,b], 121[209], 262[314]
Silvestri, M. G., **6**, 980[37]; **8**, 371[106], 387[58], 889[127], 946[135], 992[53]
Silvestri, T. W., **5**, 225[90]
Sim, G. A., **2**, 722[92]; **4**, 697[13], 698[14]
Sim, K.-Y., **2**, 801[30]
Simakov, S. V., **6**, 543[608,611,615]
Simalty, M., **4**, 55[155]
Simamura, O., **4**, 719[24,26]; **8**, 916[103], 917[103], 918[103], 919[103], 920[103]
Simandi, L., **8**, 453[191]
Simandi, L. I., **7**, 558[80], 559[80], 560[80], 561[80]
Simanek, Y., **4**, 505[149]
Simanyi, L. H., **1**, 364[40]
Simard, M., **5**, 850[152]
Simay, A., **2**, 787[50]; **6**, 543[610]
Simchen, G., **1**, 368[60], 369[60]; **2**, 600[29], 605[63], 642[78], 643[78], 655[142], 681[58], 683[58], 742[76], 743[85]; **5**, 843[120]; **6**, 227[21], 228[21], 229[21], 230[21], 231[21], 234[21], 425[67], 487[7], 488[7,29,32,35], 489[7,29,32,35], 490[7], 491[7], 493[7], 495[7], 496[7], 497[7], 499[7], 500[7], 501[7], 502[7,216,217], 503[7], 504[7], 505[7], 507[7], 508[7], 511[7], 512[7], 513[7], 514[7], 515[7], 517[7], 518[7], 519[7], 524[7], 529[7], 533[7], 540[590], 554[799], 556[7], 560[870], 562[7], 563[7], 566[7,32,35], 567[7,32,35], 568[7,29,32,35], 570[7,29], 571[7,32,35], 572[7,29,35], 573[29,35,962], 574[7,29,35,962], 575[7,29,35], 576[7,29,35], 577[29,35,977], 578[29,35,977], 579[7,29,35], 580[7,29,35], 581[7,29,35], 582[7]; **7**, 650[51], 653[1]
Simeone, J. F., **8**, 528[57]
Simic, D., **6**, 173[48], 174[48]
Simig, G., **4**, 452[8]
Simmie, J., **4**, 712[67]
Simmons, A., Jr., **4**, 443[190]
Simmons, D. P., **4**, 253[172], 527[68], 532[68,84,85], 534[68,84], 536[68,84,85], 545[84,124,125], 546[84,125]; **5**, 41[28], 515[17], 518[17], 547[17]
Simmons, H. D., Jr., **4**, 520[30], 1000[15]

Simmons, H. E., **1**, 880[112]; **3**, 380[5], 751[89]; **4**, 49[141], 483[1], 484[1], 486[32], 487[1], 488[1], 489[1], 491[1], 492[1], 493[1], 495[1], 506[1], 508[1], 968[57], 969[63], 970[69]; **5**, 74[212], 634[70], 905[57]; **7**, 21[13], 25[46], 26[46], 479[95]; **8**, 756[140]
Simmons, H. F., **7**, 585[160]
Simmons, T., **7**, 777[369]
Simmons, T. C., **8**, 978[142]
Simmrock, K. H., **5**, 1152[140], 1154[140]
Simms, J. A., **4**, 47[135]
Simms, N., **4**, 576[16]
Simon, A., **4**, 121[207]
Simon, C. D., **4**, 161[91]; **6**, 929[83]; **7**, 199[32,33], 202[33]
Simon, E. J., **8**, 344[123]
Simon, E. S., **2**, 456[22,31,33,39,41], 457[33], 458[33], 459[33], 460[33], 461[33], 462[33], 463[89], 464[89], 465[106], 466[33]; **8**, 185[15], 189[59], 392[109]
Simon, H., **2**, 60[18]; **6**, 462[9,15], 789[107]; **8**, 205[158,164], 395[133], 561[409]
Simon, H. J., **8**, 404[12]
Simon, J., **4**, 469[136]
Simon, J. A., **4**, 294[245]; **8**, 856[183]
Simon, J. D., **5**, 154[30]
Simon, K., **2**, 789[56]; **6**, 499[177]
Simon, M., **7**, 8[53]
Simon, N., **4**, 844[61]
Simon, R., **1**, 524[85]; **2**, 229[169]
Simon, R. G., **7**, 219[13]
Simon, R. M., **2**, 77[89]; **3**, 202[99]
Simon, W., **2**, 345[36]; **3**, 168[503], 169[503]
Simon, Z., **2**, 744[99], 745[99]
Simòndi, L. I., **3**, 521[1], 551[5], 552[5]
Simoneau, B., **2**, 656[151]; **4**, 159[84]; **6**, 939[143], 941[143]
Simonelli, F., **4**, 443[192], 447[192]; **8**, 412[116]
Simonet, J., **3**, 669[56]; **4**, 478[167]; **7**, 797[19], 808[76]; **8**, 134[30,31], 640[23], 884[96], 975[134]
Simonet-Guéguen, N., **3**, 669[56]
Simoni, D., **2**, 803[32]; **5**, 403[8]; **8**, 392[108], 394[116], 645[45]
Simonian, S. O., **5**, 1056[48]
Simonidesz, V., **2**, 381[305], 529[17]
Simonis, M.-T., **6**, 1035[136]
Simonneaux, G., **8**, 451[180], 462[265], 535[166]
Simonoff, R., **6**, 651[136,136a]; **8**, 956[1], 957[1]
Simons, J. D., **7**, 851[16]
Simons, L. H. J. G., **8**, 967[82]
Simons, S. S., Jr., **6**, 803[44]
Simonsen, J. L., **4**, 2[5], 285[165], 289[165]
Simonsen, O., **1**, 34[225]; **5**, 531[74], 532[74a]
Simonsen, S. H., **3**, 391[89], 393[89]; **5**, 530[71]
Simonson, J., **3**, 341[1]
Simonyan, L. A., **5**, 113[234]; **6**, 498[169]
Simonyan, S. O., **2**, 720[84]; **5**, 1056[48], 1057[50]
Simonyi, M., **2**, 812[72]
Simova, S., **1**, 218[51]
Simpkins, N. S., **1**, 787[253]; **2**, 185[29]; **3**, 173[520], 253[88]; **4**, 260[280], 355[131]
Simpson, G. W., **6**, 531[457]; **8**, 196[117], 197[117]
Simpson, I., **7**, 479[92]
Simpson, J. H., **3**, 232[259,259b,261]
Simpson, J. M., **5**, 856[210], 910[84], 1007[40]
Simpson, P. L., **8**, 533[153]
Simpson, R., **8**, 613[79]
Simpson, R. E., **4**, 824[239]
Simpson, T. J., **2**, 541[79]
Sims, C. L., **3**, 939[78]; **5**, 323[14]; **6**, 854[145], 873[1], 895[1]; **7**, 764[131]
Sims, J., **4**, 1076[47], 1097[164], 1098[170]; **5**, 247[26], 248[26a], 249[26a], 626[40], 630[40]
Sims, J. J., **2**, 671[49,50]; **5**, 431[121], 434[121d]; **6**, 558[845]; **8**, 445[12]
Sims, L. B., **6**, 1013[10]
Sims, L. L., **4**, 280[129], 281[129]
Sims, V. A., **6**, 959[40]
Simson, J. M., **7**, 26[61]
Sinai-Zingde, G., **1**, 520[76], 521[76], 522[76]; **2**, 75[81]; **4**, 12[37,37e,f], 119[194], 226[196,198,199], 1040[88], 1048[88,88c]; **5**, 909[98], 925[152], 939[221], 957[310,311], 962[221], 987[42], 993[42,52], 994[42,52]; **6**, 154[146,199]; **8**, 813[9]
Sinay, P., **1**, 419[77]; **3**, 174[525], 196[26]; **4**, 311[441], 379[116], 381[128]; **6**, 43[50], 54[129,130], 60[147], 529[466], 978[24]; **7**, 245[76], 635[71]; **8**, 849[111], 854[150], 856[161]
Sinbandhit, S., **4**, 1096[157]
Sinclair, J. A., **3**, 274[21], 554[24], 799[98]; **4**, 147[40]
Sinclair, P. J., **1**, 123[74], 373[92], 375[92], 376[92]; **2**, 649[104], 1052[48], 1075[48], 1076[48,152]; **8**, 655[86]
Sindona, G., **8**, 965[62]
Singaram, B., **1**, 488[5], 492[5,37,39], 494[37], 498[51], 499[51], 501[37], 502[39]; **2**, 57[4]; **3**, 199[66], 793[75], 795[75], 797[75,91,92]; **6**, 78[59]; **7**, 595[13], 606[159]; **8**, 14[78], 705[11], 709[49], 710[49,53,58], 716[86], 721[53,143], 722[147,148], 726[11], 939[98]
Singaram, B. J., **2**, 247[82]
Singer, E., **4**, 1093[148]; **5**, 433[135,135a]; **6**, 519[339]
Singer, L. A., **4**, 719[23]; **5**, 167[100]
Singer, M. S., **3**, 164[476]
Singer, S. P., **1**, 712[264], 714[264]; **3**, 87[63], 114[63], 117[63], 865[25], 957[110,111]; **5**, 894[46]; **6**, 897[101], 906[147], 1026[81], 1027[81], 1029[81], 1030[81], 1031[81], 1033[81]; **7**, 87[22], 120[19], 121[19], 123[19], 124[38], 128[38], 129[38], 486[144], 775[353]
Singer, S. S., **6**, 140[60,61]
Singh, A., **1**, 17[212], 37[178], 699[247]; **2**, 404[41]; **3**, 421[61], 422[61]; **4**, 284[156], 365[4], 370[4], 380[4], 381[4], 398[216], 399[216a], 401[216a], 405[216a], 410[216a], 1001[42]; **6**, 470[58]; **7**, 495[211], 524[53]; **8**, 847[97,97d], 849[97d,107,112,115]
Singh, A. K., **4**, 798[110], 1040[71,72], 1043[71], 1044[72]; **5**, 85[6], 86[12]; **6**, 124[144], 777[59], 837[60]; **7**, 376[85]; **8**, 943[121], 945[130]
Singh, B. B., **6**, 727[189], 806[54]
Singh, B. P., **7**, 674[39]; **8**, 406[39], 989[34]
Singh, G., **2**, 505[7]; **3**, 125[313]; **4**, 505[142]; **6**, 456[162], 457[162]
Singh, H., **4**, 443[194]; **6**, 527[406], 579[985], 734[13]; **8**, 656[90]
Singh, H. S., **7**, 437[7], 439[7], 851[18]
Singh, J., **1**, 619[63]; **3**, 8[37], 602[220]; **4**, 8[27,27a], 24[72,72e], 74[41], 100[41]; **7**, 271[129]; **8**, 48[110], 66[110]
Singh, K., **6**, 734[13]
Singh, L. W., **8**, 839[26b], 840[26]
Singh, M., **2**, 541[75]; **4**, 497[100], 505[141,142,146]; **5**, 164[76]; **7**, 745[75], 775[346]; **8**, 752[66]
Singh, M. D., **5**, 1133[30]
Singh, M. P., **6**, 76[50]; **8**, 384[29], 412[115]
Singh, N. N., **7**, 79[128b]
Singh, N. P., **6**, 54[124]
Singh, P., **4**, 443[194], 460[94], 464[94], 477[94], 503[126], 505[144,145]; **5**, 383[123]; **6**, 579[985]
Singh, P. K., **3**, 304[67]
Singh, P. R., **6**, 215[103]; **7**, 884[190]; **8**, 851[134], 858[204]
Singh, R., **1**, 820[15]; **3**, 380[10], 541[115]; **6**, 538[565]
Singh, R. K., **2**, 161[139], 547[97]; **4**, 7[24,24c], 1040[95], 1041[95a], 1045[95a]; **5**, 683[35]; **6**, 1023[72]
Singh, R. P., **7**, 279[170], 844[61], 845[61]
Singh, S., **1**, 367[55]; **7**, 747[95]

Singh, S. B., **3**, 1048[12]; **4**, 1036[46]
Singh, S. K., **3**, 242[8], 257[8]; **5**, 664[39], 665[40]
Singh, S. M., **1**, 135[115], 343[104]
Singh, S. P., **3**, 739[38], 747[70]; **8**, 405[26]
Singh, T., **3**, 810[44], 1057[39]; **4**, 44[124]
Singh, U. P., **4**, 1097[167]
Singh, V. K., **1**, 317[144], 319[144]; **5**, 233[139]; **8**, 171[103]
Singleton, A., **5**, 475[143]
Singleton, D. A., **4**, 439[162], 443[162]; **5**, 266[75], 268[75], 341[58], 520[38,40,41], 521[42], 522[41]
Singleton, D. M., **5**, 1142[86]; **6**, 980[39], 993[93]; **8**, 796[27], 888[122]
Singleton, E., **3**, 896[67]; **8**, 457[214], 458[214]
Singleton, T., **3**, 464[174]
Singleton, V. D., **6**, 959[41]
Sinha, A., **7**, 107[169]
Sinha, A. K., **6**, 524[370], 532[370]
Sinha, A. M., **5**, 736[145], 737[145]
Sinha, G., **3**, 898[86]
Sinha, N. D., **3**, 219[114], 499[140], 501[140], 502[140]; **6**, 624[147]
Sinha, S. C., **3**, 956[107]
Sinhababa, A., **1**, 463[32]
Sinhababu, A. K., **7**, 333[23]; **8**, 70[231], 351[167], 355[167], 942[118]
Sinigalia, R., **7**, 426[148d]
Sinisterra, J. V., **1**, 821[22]; **3**, 825[24,24b], 835[24]; **8**, 86[26]
Sinitsa, A. D., **6**, 524[366,372]
Sink, C. W., **7**, 774[321]
Sinnema, Y. A., **4**, 297[266]
Sinnott, M. L., **8**, 211[4]
Sinnreich, D., **4**, 292[234]; **6**, 264[32]; **7**, 765[167]
Sinott, M. L., **6**, 49[92]
Sinou, D., **6**, 91[125,127], 237[63]; **7**, 493[190]; **8**, 535[166], 552[355,356]
Sinoway, L., **2**, 121[188]; **3**, 250[70]
Sinsky, M. S., **4**, 48[138]
Sioda, R. E., **3**, 577[89]
Sipilina, N. M., **8**, 618[121], 619[121], 621[142]
Sipio, W. J., **4**, 370[32], 371[32,60], 372[60]; **6**, 1031[113,115]; **7**, 131[78], 523[46], 524[49]; **8**, 499[45], 836[10b], 847[10b,99], 848[10b,99], 849[10b,99]
Sipos, F., **2**, 841[185]
Sipos, W., **3**, 872[61,62]; **4**, 799[113]
Siragusa, J. A., **3**, 564[10], 595[188]
Sirala-Hansen, K., **1**, 189[74]
Sirazova, M. M., **4**, 599[213], 640[213]
Sircar, J. C., **6**, 261[14], 263[14], 272[14], 275[14], 280[14]
Sirear, J. C., **4**, 294[240]
Siret, P., **2**, 156[118]; **3**, 572[69], 573[69], 602[69], 607[69], 610[69]; **4**, 373[70]; **6**, 713[80b]; **7**, 566[100], 711[60]
Siriwardane, U., **4**, 499[102]
Sirna, A., **4**, 611[344]
Siro, M., **1**, 767[176]
Siroi, T., **7**, 537[60]
Sironi, A., **8**, 460[248]
Sirorkina, E. I., **4**, 529[76]
Sirotkina, E. E., **4**, 291[209]
Sirrimanne, S. R., **7**, 99[110]
Sisak, A., **8**, 458[222]
Sisani, E., **1**, 846[18b,c], 847[18]
Sisbarro, M. J., **8**, 533[154]
Sisido, K., **6**, 261[11]; **7**, 595[21]; **8**, 661[112]
Sisk, S. A., **5**, 350[78]
Siskin, M., **3**, 333[211a,b]; **8**, 454[203]
Siskind, M., **7**, 851[25]
Sisko, J., **5**, 403[10]
Sisler, H., **7**, 100[130], 769[232]
Sisler, H. H., **2**, 524[76]; **8**, 754[82]
Sisti, A. J., **1**, 873[93]; **3**, 756[115], 757[116-120]; **8**, 956[4]
Sisti, J., **2**, 662[13], 664[13]
Sisti, M., **1**, 566[216]; **2**, 833[148]; **4**, 261[285,288,290]
Sit, S.-Y., **1**, 205[105], 820[13], 823[13]
Sita, G. E., **7**, 92[48]
Sita, L. R., **2**, 2[5], 25[5], 33[5], 40[5], 134[3], 190[57], 232[180], 240[5], 248[5b], 455[8], 652[124], 686[68], 979[17]; **6**, 8[39]; **7**, 399[40a], 442[48]; **8**, 535[165]
Sitrin, R., **3**, 781[14]
Sitzman, M. E., **4**, 426[54]
Siu, T.-W., **3**, 877[89]
Sivakumar, R., **8**, 26[28], 30[28], 36[28], 37[28], 40[28], 43[28], 44[28], 46[28], 55[28], 66[28], 357[204]
Sivanandaiah, K. M., **6**, 651[136,136c]; **8**, 959[22]
Sivaram, S., **6**, 283[159]
Sivaramakrishnan, H., **5**, 342[63]
Sivasubramanian, S., **4**, 350[115]; **7**, 502[258]
Sivavec, T. M., **5**, 1102[146a], 1104[157], 1113[146]
Siverns, M., **3**, 738[35]
Siwapinyoyos, T., **4**, 245[94], 262[94]; **5**, 560[67]
Sizov, Yu. A., **6**, 104[2]
Sizova, O. S., **6**, 543[609], 554[718]
Sjöberg, K., **2**, 1090[15], 1103[128], 1106[15]; **4**, 560[21]; **7**, 474[44]
Sjogren, E. B., **1**, 400[11]; **2**, 30[110], 31[110], 113[106,107], 254[43]; **5**, 98[126]; **6**, 759[137,138]
Sjoholm, R., **4**, 89[84c]
Sjöström, M., **2**, 1099[115]
Skalarz, B., **7**, 703[4]
Skaletz, D. H., **3**, 580[105,106], 596[195]; **8**, 135[46], 532[131b]
Skameikina, T. I., **3**, 306[84]
Skapski, A. C., **8**, 446[67]
Škare, D., **1**, 859[63]
Skaric, V., **8**, 794[13]
Skarstad, P., **4**, 1024[265]
Skattebøl, L., **3**, 864[24], 865[24], 872[24]; **4**, 1001[30], 1009[143], 1010[149,153,154,158], 1012[168,169,170], 1013[180], 1020[231,232], 1021[251]; **5**, 19[130], 113[235], 797[67], 949[275], 950[293]; **6**, 971[128]; **7**, 764[128]; **8**, 807[113]
Skeean, R. W., **3**, 351[43a]; **5**, 712[57e]
Skell, P. S., **3**, 390[82], 392[82]; **4**, 280[128], 281[128], 959[30], 1002[49], 1018[220]; **8**, 890[141]
Skelly, R. K., **5**, 679[19]
Skelton, B. W., **1**, 13[73], 16[89], 17[217,219], 36[234], 37[177]; **2**, 606[69]
Skerlj, R. T., **3**, 215[60], 253[94,95], 489[55], 495[55]; **4**, 249[123]; **8**, 95[90]
Skerrett, E. J., **8**, 963[44], 972[114]
Sket, B., **5**, 829[19]
Skibbe, V., **5**, 1178[46]
Skibo, E. R., **5**, 422[82]
Skidmore, S., **2**, 495[59]
Skinner, C. E. D., **3**, 255[110]
Skinner, C. G., **3**, 391[89], 393[89]
Skinner, H. A., **7**, 593[1]; **8**, 670[13]
Skinner, J. F., **8**, 459[239]
Skipka, G., **5**, 69[100]
Skipper, P. L., **3**, 158[436], 159[436]
Skita, A., **8**, 140[22], 141[35,36], 533[142]
Skjold, A. C., **4**, 445[210]
Skoda, J., **2**, 147[73]
Skolnick, P., **7**, 340[46]
Skorcz, J. A., **4**, 490[66], 499[66], 500[104]; **5**, 390[140]
Skorna, G., **2**, 1084[9], 1089[58]

Skorova, A. E., **4**, 330[5]; **6**, 153[143]
Skotnicki, J., **8**, 36[48], 66[48], 347[140], 395[129], 616[102], 617[102], 618[102]
Skovlin, D. O., **2**, 183[12]
Skowrońska-Ptasińska, M., **1**, 461[15], 464[15]; **7**, 333[25]
Skramstad, J., **4**, 1063[173]
Skrydstrup, T., **6**, 7[30]
Skuballa, W., **2**, 381[308]; **4**, 964[50], 965[50]
Skulnick, H. I., **2**, 801[35]
Skundin, A. M., **3**, 648[181]
Skupsch, J., **1**, 773[203,203a]
Skvortsov, N. K., **8**, 765[13]
Skvortsov, Yu. M., **4**, 55[157], 57[157o]
Sky, A. F., **5**, 856[216]
Slabaugh, M. R., **2**, 159[128]
Slack, S. C., **2**, 240[8]
Slade, C. J., **7**, 332[18,19]
Slade, J., **1**, 66[57], 359[21], 383[21], 384[21]; **8**, 35[66], 66[66]
Slade, M. J., **1**, 37[178]
Sladkov, A. M., **3**, 209[20]
Slagle, J. D., **6**, 205[30]
Slama, J. T., **4**, 310[432]
Slanian, R. A., **4**, 598[197], 623[197], 639[197]
Slater, G. P., **7**, 152[4], 153[4]
Slater, M., **3**, 1027[44]
Slater, M. J., **5**, 1060[56]
Slater, S., **8**, 458[223,223b]
Slates, H. L., **2**, 160[135], 746[110]; **3**, 689[119]; **7**, 675[59]
Slaugh, L. H., **1**, 174[6], 175[6]; **8**, 568[464], 735[15,16]
Slavinskaya, R. A., **6**, 432[116]
Slavinskaya, V. A., **8**, 606[24], 607[24]
Slawin, A. M. Z., **4**, 381[126b], 382[126], 383[126]; **5**, 374[107], 376[107b]; **7**, 112[198], 820[26]
Slayden, S. W., **7**, 597[50]
Slaytor, M., **4**, 27[83]; **5**, 468[135]; **8**, 249[91], 293[52], 302[52], 507[85]
Slebocka-Tilk, H., **4**, 330[3], 345[3]; **7**, 770[250]; **8**, 409[85]
Sleddon, G. J., **8**, 754[100]
Sleevi, M. C., **4**, 455[43], 457[49,50], 463[43], 477[49,50], 503[125]
Sleezer, P. D., **7**, 92[50], 94[56]
Slegeir, W., **8**, 372[123], 373[123]
Sleiter, G., **2**, 965[69]; **8**, 336[70]
Slepushkina, A. A., **7**, 500[236]
Slessor, K. N., **7**, 238[41]; **8**, 341[105]
Sletter, H., **7**, 7[48]
Sletzinger, M., **2**, 803[33]; **8**, 54[159], 66[159], 272[122]
Sliede, J., **3**, 831[62]
Slinckx, G., **4**, 314[489]
Slinger, C. J., **8**, 819[42], 820[42]
Sliskovic, D. R., **6**, 538[559], 554[775]
Sliwa, E., **8**, 757[162]
Sliwa, H., **2**, 962[48]; **5**, 790[35]; **7**, 661[49]
Sliwinski, W. F., **4**, 1016[209]
Sliwka, H.-R., **3**, 874[69]
Slizhov, Yu. G., **8**, 629[182,183]
Sloan, A. B. S., **2**, 765[78]
Sloan, A. D. B., **2**, 866[3]
Sloan, C. P., **4**, 797[104]
Sloan, M. F., **7**, 24[38]; **8**, 447[118], 454[118], 455[118], 568[481]
Sloan, R. B., **5**, 929[173]
Slob, C., **1**, 10[55], 11[55b]
Slobbe, J., **3**, 651[215]; **5**, 841[100]; **8**, 500[48], 501[55], 502[60], 605[12], 608[35,36], 609[12], 614[12], 629[12]
Sloboda, A. E., **6**, 554[724]
Slobodin, Ya. M., **4**, 1010[155]; **6**, 970[120]
Slomp, G., **5**, 637[110,113]
Slomp, G., Jr., **7**, 548[62], 553[62]
Sloneker, J., **3**, 691[129], 693[129]
Slonka, E., **4**, 432[106]
Slopianka, M., **2**, 875[29]; **4**, 51[144b], 1001[35]
Slosse, P., **4**, 91[89]
Slotin, L. A., **6**, 601[1]
Slough, G. A., **2**, 310[30], 311[30], 655[143]; **4**, 905[213]
Slougui, N., **1**, 879[111e]; **2**, 444[20]
Slowey, P. D., **4**, 1009[143]
Sluboski, B. C., **6**, 490[104], 534[515]
Slusarchyk, W. A., **3**, 680[93], 807[29]; **6**, 644[93]
Slusarska, B., **2**, 378[290]
Slusarska, E., **6**, 83[77]
Sluski, R. J., **6**, 529[463]
Slutsky, J., **1**, 859[66]; **2**, 6[27], 565[15], 575[15], 582[15]; **6**, 1036[143]
Smaardijk, A. A., **1**, 223[78], 224[78], 317[148]
Smadja, W., **4**, 276[75], 284[75], 308[401], 395[202]; **5**, 772[155], 774[155]
Smal, E., **2**, 780[11]; **8**, 41[93], 66[93]
Smal, M. A., **1**, 508[21], 519[63,64]; **2**, 72[60]
Smale, T. C., **3**, 1056[33], 1062[33]; **5**, 105[194]
Small, G. H., **8**, 971[102]
Small, L. F., **8**, 568[468]
Smalla, H., **5**, 422[88], 423[88]
Smalley, R. K., **4**, 439[167]; **6**, 530[416], 535[540,541], 538[540,541]; **8**, 386[55]
Smallheer, J., **7**, 243[63]
Smallridge, M. J., **2**, 494[56]; **4**, 76[47]
Smart, B. E., **5**, 441[179,179b]; **6**, 172[10]
Smart, C. J., **5**, 434[148]
Smart, J., **6**, 618[114], 649[126]
Smart, J. B., **1**, 215[33]
Smart, L. E., **5**, 1146[107]
Smart, W. D., **8**, 143[55]
Smeets, F., **2**, 740[55]
Smeets, F. L. M., **3**, 750[83]
Smeets, W. J. J., **1**, 25[127], 26[132,133,134]
Smegal, J. A., **7**, 8[65,66]
Smegel, J. A., **1**, 807[316]
Smeley, V., **3**, 896[72]
Smentowski, F. J., **7**, 761[53]
Smetanin, A. V., **3**, 1039[97]
Smetankina, N. P., **8**, 771[47]
Smets, G., **4**, 314[489]; **7**, 475[52]
Smetskaya, N. I., **6**, 554[734]
Smidt, J., **4**, 588[64]; **7**, 449[1,2], 450[1,2]
Smiles, S., **3**, 689[117]
Smiley, R. A., **6**, 226[12], 228[16]
Smillie, R. D., **2**, 843[196]
Smirnov, S. K., **8**, 253[113]
Smirnov, Y. D., **8**, 253[113]
Smirnov, Yu. I., **6**, 516[320]
Smirnova, T. S., **5**, 948[273]
Smirnova, V. V., **3**, 644[140,141]
Smirnow, V. N., **4**, 885[115], 886[115]
Smissman, E. E., **3**, 750[85], 788[51], 845[36,37]
Smit, C. J., **3**, 568[46,47], 569[50]; **4**, 459[87]; **8**, 131[4,5], 132[4], 134[4]
Smit, R., **5**, 77[264]
Smit, V. A., **3**, 342[7,8,14], 345[21], 346[24], 349[35], 351[41], 361[73,80]
Smit, W. A., **2**, 710[16], 723[101], 725[107-109]; **4**, 330[5], 356[140]; **5**, 345[70], 346[70], 453[66], 775[175,176], 850[148], 1055[46], 1056[48,49], 1057[50,51], 1062[51]
Smith, A., **8**, 568[471], 612[77]

Smith, A. B., III, **1**, 129[93], 779[224]; **2**, 189[45], 388[340], 541[77], 651[120,122], 657[165], 725[106]; **3**, 22[133], 24[149], 25[149], 26[165], 709[17,18], 906[144]; **4**, 212[97,98], 1040[102]; **5**, 145[105], 178[139], 342[63], 362[93], 363[93i], 944[242]; **6**, 174[63]; **7**, 7[47], 162[67], 176[67], 238[39], 239[44], 243[64], 298[37]; **8**, 248[83], 353[158], 931[38], 940[104]

Smith, A. E., **6**, 737[33]

Smith, A. G., **2**, 124[202]

Smith, A. H., **2**, 814[78], 818[78], 823[78]

Smith, A. J., **4**, 6[21], 680[50]; **6**, 1014[19]

Smith, A. K., **4**, 976[100]; **8**, 445[53], 453[195]

Smith, A. L., **3**, 75[51], 77[57], 78[51]; **5**, 736[145], 737[145]; **7**, 227[79,80]

Smith, A. S., **1**, 423[96]

Smith, B. C., **2**, 183[12]; **7**, 203[54]

Smith, B. F., **2**, 897[17]; **7**, 763[99], 766[99]

Smith, B. H., **3**, 628[48]; **7**, 660[37]

Smith, B. V., **8**, 383[13]

Smith, C., **3**, 919[30], 933[30,62], 946[30], 949[95]; **4**, 507[154]; **5**, 829[21]; **6**, 893[79], 897[95]

Smith, C. A., **3**, 217[95], 681[100]; **4**, 810[166]; **5**, 736[143]

Smith, C. D., **5**, 1187[9]

Smith, C. F. C., **8**, 615[94], 618[94]

Smith, C. J., **8**, 206[172]

Smith, C. V., **5**, 1199[47]

Smith, C. W., **6**, 637[30]; **7**, 446[66]

Smith, C. W., Jr., **3**, 265[187]

Smith, C. Z., **8**, 132[9]

Smith, D., **1**, 8[35]

Smith, D. A., **5**, 515[15], 518[15], 519[15,33]; **7**, 530[28], 531[28]

Smith, D. B., **1**, 41[201]; **5**, 342[62b], 843[119]; **7**, 396[26], 416[26]

Smith, D. C. C., **8**, 614[85]

Smith, D. G., **8**, 290[31]

Smith, D. J. H., **3**, 159[459], 163[459], 1028[50], 1036[84]; **4**, 553[8], 939[79], 941[79]; **5**, 904[53]; **6**, 171[4], 177[4,112], 181[112], 198[4], 200[112], 624[146], 625[146]; **7**, 451[17], 462[17]; **8**, 290[31]

Smith, D. K., **4**, 425[35], 445[35]

Smith, D. L., **4**, 50[142]; **7**, 43[44], 208[84], 774[315]

Smith, D. M., **6**, 208[55], 212[55]

Smith, D. R., **3**, 825[30]; **8**, 64[203], 66[203]

Smith, E. H., **5**, 160[55]; **6**, 245[128]

Smith, E. M., **1**, 366[50]

Smith, F. A., **7**, 666[77]

Smith, F. X., **2**, 356[128]; **4**, 89[84i]

Smith, G., **3**, 816[79]; **6**, 659[195]; **7**, 244[69]

Smith, G. A., **1**, 477[131]; **3**, 75[43], 76[43], 80[43,65], 81[65], 685[108]

Smith, G. B. L., **8**, 373[131]

Smith, G. D., **5**, 855[184]

Smith, G. F., **2**, 954[5]

Smith, G. G., **3**, 828[46]; **5**, 552[11]; **6**, 1033[124], 1034[124], 1035[124]

Smith, G. M., **2**, 743[84]

Smith, G. P., **3**, 328[177]

Smith, G. V., **3**, 864[16], 866[16], 883[16]; **8**, 150[136], 883[90]

Smith, G. W., **5**, 195[7]

Smith, H., **8**, 212[22], 490[1,8], 492[1], 500[8], 515[117], 524[2], 526[19,28,29,36], 527[29], 530[2], 531[2], 842[43], 971[103], 972[113]

Smith, H. A., **3**, 15[79]; **8**, 436[72]

Smith, H. D., **4**, 1033[27,31], 1049[27]; **5**, 804[94], 986[39]

Smith, H. E., **2**, 969[85]; **6**, 105[12]

Smith, H. L., **6**, 802[43], 803[43]

Smith, H. W., **4**, 310[432]

Smith, I. R., **4**, 33[95,95a]

Smith, J., **3**, 54[277]

Smith, J. C., **4**, 282[138,139], 285[166], 288[182]

Smith, J. D., **1**, 16[90], 17[91,208], 39[191], 41[264], 214[28]; **4**, 170[15]

Smith, J. G., **2**, 583[111]; **3**, 223[157], 262[159], 264[159], 500[132], 505[132], 579[98]; **6**, 2[3], 25[3]; **7**, 263[83], 677[66]; **8**, 123[86], 244[49], 331[37], 340[37], 626[171], 627[171], 871[1], 884[1]

Smith, J. K., **1**, 292[25], 357[1]; **2**, 476[4]; **3**, 31[185], 32[185]; **6**, 531[448], 720[130], 722[139]

Smith, J. L., **7**, 878[145]

Smith, J. N., **7**, 530[28], 531[28]

Smith, J. R. L., **6**, 923[57]; **8**, 336[85]

Smith, J. W., **8**, 880[59]

Smith, K., **1**, 489[14,18], 499[18,54], 501[54], 569[258]; **2**, 193[63], 240[12]; **3**, 124[261], 126[261], 554[22,23], 780[10], 797[95], 798[95-97]; **7**, 594[3,4], 595[3,22], 598[3,4], 600[75], 601[3], 604[134], 607[167]; **8**, 26[11,12], 27[11,12], 36[11], 37[11], 703[1], 704[1,2,4,5], 705[1,4,5], 706[1,4,5], 707[1,4], 708[1,38], 709[1], 710[4], 711[64,69], 712[1], 713[1], 715[1,4], 716[1,4], 717[1,4,101], 720[1], 721[1], 722[1,4], 724[1,4], 725[4], 726[4], 728[4]

Smith, K. D., **1**, 19[101]

Smith, K. M., **2**, 770[10], 771[10], 780[14], 1079[158]; **5**, 528[68], 531[68]; **7**, 256[25]; **8**, 851[128]

Smith, K. R., **2**, 477[11]; **5**, 906[70], 908[70]

Smith, L., **5**, 87[43,44]

Smith, L. B., **8**, 895[3], 898[3]

Smith, L. H., **6**, 209[68]

Smith, L. I., **4**, 41[118]; **8**, 269[93]

Smith, L. L., **7**, 96[87]

Smith, L. R., **3**, 901[106]; **6**, 429[88]; **8**, 119[76]

Smith, M., **3**, 734[4]; **4**, 181[74]; **6**, 603[12], 1042[10]; **8**, 212[19], 340[89], 524[9], 530[9], 531[9]

Smith, M. A., **7**, 750[136]; **8**, 35[63], 66[63]

Smith, M. B., **5**, 500[256]

Smith, M. G., **1**, 829[88]; **5**, 841[90], 859[90]; **6**, 750[103]

Smith, M. J., **4**, 887[130,130b]

Smith, M. R., **7**, 765[146], 851[23]; **8**, 383[15]

Smith, N. M., **8**, 388[66], 637[11]

Smith, N. P., **6**, 566[928]

Smith, N. R., **7**, 720[8]

Smith, P., **6**, 737[34,35]; **8**, 315[51]

Smith, P. A., **4**, 1101[191]

Smith, P. A. S., **1**, 360[26], 364[26], 846[15], 847[15], 848[15], 850[15]; **2**, 748[128], 759[28]; **3**, 781[11], 823[14], 835[14]; **6**, 116[90], 245[125], 249[125], 763[3], 795[2,8], 797[8], 798[8], 799[2], 806[8], 807[8]; **7**, 21[5,7], 35[5,7,105], 475[54], 671[4], 689[8]; **8**, 363[8], 372[116], 385[46], 387[59]

Smith, P. H. G., **4**, 477[164]

Smith, P. J., **3**, 794[79]; **4**, 1089[132]; **6**, 662[216]; **7**, 614[5]

Smith, P. W., **6**, 273[98]

Smith, P. W. G., **7**, 555[70]

Smith, R., **4**, 1088[120]; **6**, 295[253], 546[656]

Smith, R. A. J., **1**, 114[38], 511[31]; **2**, 185[31]; **4**, 11[36], 113[171], 243[67], 244[67], 260[67], 261[67], 279[117]; **6**, 134[20], 140[56], 831[9]; **8**, 526[26]

Smith, R. C., **6**, 487[70]

Smith, R. D., **4**, 969[63]; **8**, 756[140]

Smith, R. F., **6**, 516[318]; **7**, 744[69]

Smith, R. G., **3**, 124[277]; **4**, 429[83], 438[83], 441[83]; **6**, 1000[125]; **8**, 32[54], 66[54]

Smith, R. H., **6**, 530[416]

Smith, R. H., Jr., **6**, 245[124]

Smith, R. K., **1**, 359[21], 383[21], 384[21]; **2**, 489[50], 491[50]; **4**, 77[50], 205[40], 206[42], 428[74]

Smith, R. L., **2**, 971[92]; **8**, 803[91]

Smith, R. S., **1**, 223[85], 224[85]; **3**, 437[20], 438[20], 439[20], 441[20], 442[20]; **4**, 240[41], 897[172]

Smith, S., **8**, 249[93]

Smith, S. A., **2**, 547[116], 551[116]; **3**, 349[34], 623[37], 625[37]; **8**, 626[175], 629[175]

Smith, S. C., **8**, 764[7], 770[7], 773[7]
Smith, S. E., **2**, 547[96], 551[96]
Smith, S. F., **1**, 287[15]
Smith, S. G., **1**, 477[133,134]; **3**, 67[17,20]; **4**, 171[28]; **7**, 225[63]; **8**, 2[12]
Smith, S. J., **2**, 85[23]
Smith, S. W., **6**, 677[315]
Smith, T. A. K., **4**, 799[118]
Smith, T. H., **8**, 354[171]
Smith, T. K., **4**, 1033[27], 1049[27], 1060[27b]; **5**, 599[40]; **6**, 126[149]
Smith, T. L., **2**, 464[97]
Smith, W. A., **6**, 141[64]
Smith, W. B., **3**, 647[179,191,192,193], 648[179], 724[12]; **8**, 877[46]
Smith, W. E., **2**, 772[17]; **3**, 723[10]; **8**, 767[23]
Smith, W. H., **8**, 594[71]
Smithen, C. E., **4**, 1021[247]
Smithers, R., **2**, 1030[80]; **6**, 832[12], 865[12]
Smithers, R. H., **2**, 564[3], 712[43]; **5**, 595[10]; **6**, 89[107], 569[935]
Smith-Palmer, T., **7**, 502[261]
Smith-Strickland, S. M., **1**, 584[9]; **2**, 714[50]
Smithwick, E. L., Jr., **4**, 496[88]
Smiunic, J. L., **5**, 1138[63]
Smolanka, I. V., **4**, 407[255], 408[259f]
Smolanoff, J., **4**, 1081[77,82], 1082[88-90]; **5**, 255[51], 612[73], 948[292], 949[284], 950[284]; **7**, 219[12]
Smolii, O. B., **6**, 524[365]
Smolinsky, G., **6**, 247[132], 253[158]; **7**, 27[67-69,72,73], 29[72], 32[90]; **8**, 319[75]
Smonou, I., **8**, 216[66]
Smoot, J., **8**, 413[125]
Smurny, G. N., **6**, 245[124]
Smushkevich, Y. I., **5**, 432[129]
Smyrniogtis, P. Z., **2**, 456[65,71], 458[65,71]
Smyser, T. E., **2**, 355[119]
Smyth, M. S., **7**, 463[128]
Smythe, G. A., **6**, 737[32]
Smythe, J. A., **7**, 194[2], 202[2]
Snader, K. M., **8**, 92[68]
Snaith, R., **1**, 6[33], 33[165], 38[258], 39[187]; **3**, 763[151]
Snapper, M. L., **5**, 640[129]
Snatzke, G., **2**, 553[128]; **5**, 1146[108]; **7**, 252[11], 649[44]
Snead, T. E., **5**, 225[98]
Sneed, R., **5**, 95[91]
Sneeden, R. P. A., **1**, 173[1], 174[14], 175[16], 179[14]
Sneen, R. A., **3**, 12[64]; **8**, 340[99]
Sneen, R. H., **7**, 101[133]
Snell, J. M., **3**, 619[22]
Snider, B. B., **1**, 885[132]; **2**, 527[8,9], 528[8,9,13], 531[24,25,27,28], 533[24,29,30], 534[25], 535[40], 537[24], 538[27,51,52], 541[76], 542[40,52], 544[76], 546[76], 547[52,76,98,99,105,118-120], 548[98,99], 550[98,105], 551[118-120], 552[120], 555[139,144,145], 709[8], 910[65]; **3**, 244[26], 485[28], 978[11], 994[44]; **4**, 181[69], 377[104], 764[216], 807[148,153,154], 820[217], 879[84], 892[144], 904[206]; **5**, 2[6,7,11], 4[7,30,35-38], 5[30,35,36,40,41,44,46,47], 6[46-48], 7[53,54], 8[44,54,56-59,61,62], 9[72,74], 10[78], 12[88], 13[89], 15[48,103], 16[111], 18[111], 19[74], 20[48], 63[12], 429[113a], 433[137c], 435[137c], 461[100,104], 463[100], 519[35], 527[59,60], 796[56], 835[61], 1018[67], 1021[71], 1022[67], 1029[92]; **6**, 154[147]; **7**, 46[48-50], 47[50,52], 552[59]; **8**, 61[188], 66[188]
Snieckus, V., **1**, 461[9], 462[19], 464[9], 466[43,48,49], 469[58,60], 470[61], 472[75], 477[137,138], 567[221], 698[248]; **2**, 103[30], 527[5], 528[5], 540[5], 544[5], 745[97]; **3**, 44[241], 50[268], 231[250,251], 242[5], 265[192], 504[155], 511[155], 515[155]; **4**, 45[127,127a], 72[27,30], 74[27], 249[115], 257[115], 258[253], 260[115], 797[104]; **5**, 1[3], 2[3], 9[3], 15[3], 19[3], 27[3], 37[20], 1021[72]; **6**, 2[3], 25[3]; **7**, 333[22], 355[47]; **8**, 385[47]
Snitman, D. L., **1**, 410[43]; **2**, 2[7], 100[13], 224[153], 630[9]; **8**, 334[62]
Sno, M. H. A. M., **2**, 1064[108]
Snoble, K. A. J., **8**, 885[104]
Snoddy, H. D., **8**, 623[151]
Snook, M. E., **7**, 13[123]
Snow, A. I., **1**, 13[66]
Snow, A. W., **6**, 244[107]
Snow, D. H., **7**, 882[171]
Snow, J. T., **4**, 889[137]; **8**, 717[96], 726[195], 755[132]
Snow, R. A., **5**, 203[39,39a-d], 209[39], 210[39]; **7**, 667[80]
Snowden, R. L., **1**, 754[106]; **2**, 69[48], 553[123]; **3**, 196[32], 264[184], 345[22]; **5**, 253[46,46c], 331[40], 456[83], 515[17], 518[17], 547[17]; **6**, 1067[108,109]; **8**, 61[189], 66[189]
Snyder, D. C., **8**, 778[83]
Snyder, E. I., **6**, 205[35]
Snyder, E. S., **2**, 489[48], 490[48]
Snyder, H. R., **2**, 355[121], 356[130], 753[4], 838[174], 839[174], 933[138], 970[87], 1090[61]; **4**, 438[153]; **5**, 473[149]; **6**, 825[127]; **7**, 595[8]; **8**, 637[12]
Snyder, J. K., **5**, 343[64]; **7**, 166[89], 441[43], 442[43]
Snyder, J. P., **5**, 736[145], 737[145], 741[153]; **6**, 448[106,107], 691[404], 692[404]; **7**, 747[105], 751[137]; **8**, 50[117], 66[117], 365[27], 390[81]
Snyder, R., **8**, 705[10], 726[10]
Snyder, R. G., **5**, 900[7]
So, J.-H., **3**, 565[16], 578[16]
So, Y. H., **3**, 683[101]; **6**, 282[153,156]; **7**, 799[25], 800[25a,32]
Soai, K., **1**, 62[40], 63[42], 70[64], 71[65], 72[66-68], 86[34-36], 141[22], 223[75,79-81,83], 224[79-81,83], 317[146,147], 319[147]; **4**, 85[71], 158[78], 203[27-30], 230[254,255]; **6**, 76[44]; **8**, 13[70], 170[86-88], 178[86], 179[86], 241[36], 244[55,59], 248[55], 250[59], 384[34], 412[110], 643[38], 874[25,26,28], 875[28,30]
Soares, C. J., **5**, 829[20], 864[261]
Sobala, M. C., **7**, 779[421]
Sobata, T., **4**, 587[30]
Sobczak, A., **5**, 365[96a]
Sobieray, D. M., **8**, 542[221]
Sobolov, S. B., **2**, 547[107], 550[107]
Sobota, P., **1**, 303[81]; **3**, 563[1]
Sobotka, W., **4**, 310[429]
Sobotta, R., **6**, 518[333], 519[333]
Sobti, R., **3**, 715[37]; **7**, 676[64]
Sobukawa, M., **6**, 501[193]
Soccolini, F., **2**, 435[59]; **5**, 1153[145]; **8**, 91[54], 589[53]
Sochacka, R., **7**, 775[350]
Sochneva, E. O., **6**, 553[728], 554[728,729,785,792]
Sodano, C. S., **8**, 533[154]
Sodeoka, M., **2**, 547[94]; **7**, 399[40b], 410[96]; **8**, 452[188], 457[188], 554[367], 555[367c], 567[458]
Soderberg, B. C., **4**, 600[239], 625[239], 643[239]
Soderquist, J. A., **1**, 438[160], 544[41,42], 563[181]; **3**, 252[83]; **7**, 597[48], 598[55], 641[5]; **8**, 707[21], 713[70,76], 717[101], 718[116], 856[169]
Sodeyama, T., **7**, 6[31]
Sodhi, K. S., **2**, 742[73,74]
Sodola, F. H., **4**, 279[113], 282[113]
Soeda, M., **8**, 598[98,99]
Soeder, R. W., **1**, 878[107]
Soedigdo, S., **8**, 979[149]
Soejima, T., **3**, 54[279]
Soelch, R. R., **6**, 516[318]
Soeldner, J. S., **6**, 790[117]
Soerens, D., **2**, 1017[31,34]; **6**, 737[31], 746[89]

Soffer, M. B., **8**, 328[12], 329[12]
Soffer, M. D., **8**, 328[12,13], 329[12,13]
Sofia, M. A., **4**, 315[529]
Sofia, M. J., **4**, 394[193], 794[73]; **7**, 648[41]
Sofuku, H., **7**, 765[163]
Soga, T., **2**, 655[141]; **6**, 237[70]
Sogah, D. Y., **2**, 619[148]
Sogah, G. D. Y., **4**, 230[246]
Sogo, S. G., **5**, 855[186]
Sohani, S. V., **1**, 268[56], 269[56a]; **3**, 602[225]; **8**, 111[19], 114[51], 122[19]
Sohar, P., **5**, 583[186]; **6**, 276[122], 525[382], 534[519]; **8**, 54[159], 66[159]
Sohda, T., **1**, 672[207], 700[207]; **3**, 81[68]
Sohn, E., **2**, 739[50]
Sohn, J. E., **1**, 357[2]; **2**, 182[9,9c], 184[9c], 190[9c], 191[9c], 192[9c], 193[9c], 197[9c], 198[9c], 200[9c], 211[9c], 217[9c], 223[151], 235[9c], 236[9c], 289[69], 634[36], 640[36]; **6**, 814[95]
Sohn, M. B., **5**, 677[8]; **6**, 776[55]
Sohn, W. H., **4**, 368[18], 369[18b,24]
Soicke, H., **6**, 155[153], 840[72], 841[75], 903[138]
Soifer, G. B., **6**, 524[360], 528[413]
Soilleux, S. L., **7**, 75[115]
Soja, P., **1**, 656[155], 658[155]; **6**, 667[238]; **7**, 125[56], 129[56], 130[56]
Sojka, S., **4**, 308[403], 844[61]
Sokolenko, V. A., **7**, 505[289]
Sokolik, R., **7**, 596[32]
Sokoloff, S., **3**, 396[108], 397[108]
Sokolov, S. D., **7**, 760[38]
Sokolov, S. V., **3**, 639[82,83], 644[82,137]
Sokolov, V. I., **4**, 297[269], 306[373]; **6**, 283[166]
Sokolova, N. I., **6**, 607[48]
Sokolova, N. P., **8**, 150[132]
Sokolovsky, M., **2**, 1104[132]
Sokolowska, T., **6**, 668[259]
Solaja, B., **3**, 313[103]
Solans, X., **1**, 34[223]; **3**, 380[7]
Solas, D. R., **2**, 739[45]
Soled, S. L., **8**, 373[126]
Soledad Pino Gonzalez, M., **2**, 386[330]
Soleiman, M., **8**, 862[228]
Solheim, B., **2**, 824[119]; **3**, 395[102], 396[102]
Soliman, F., **6**, 193[214]
Soll, C. E., **5**, 499[252], 500[252]
Söll, D., **6**, 611[65], 625[153]
Söll, M., **8**, 744[50], 756[50]
Soll, R. M., **1**, 343[109], 568[232]
Solladié, G., **1**, 70[62], 86[32], 513[49], 523[81,82,84], 564[200], 825[53], 827[65], 833[118]; **2**, 225[157], 227[162], 228[164,165], 232[157], 1024[64], 1025[65], 1026[64,72]; **3**, 86[34], 147[34], 149[34], 150[34], 151[34], 343[16], 439[39]; **4**, 226[185,186]; **6**, 141[62], 148[90], 149[90,91,95,106,107,109], 152[137], 155[154,155], 156[155-162], 206[43], 210[43], 214[43], 264[28], 268[28], 927[78], 985[67]; **7**, 778[413]; **8**, 12[66], 36[71], 66[71], 563[432], 837[17], 844[72,72a]
Solladié-Cavallo, A., **1**, 223[71], 359[23], 364[23], 365[41]; **3**, 315[112], 317[112]; **4**, 538[104,105,107], 539[104]; **6**, 155[154]; **7**, 490[175]; **8**, 837[17]
Söllhuber, M. M., **6**, 501[189]
Solly, R. K., **5**, 790[19], 906[64]
Solms, U., **8**, 957[15]
Solo, A. J., **4**, 83[65c]
Solodar, A. J., **4**, 930[51]; **5**, 692[92]
Solodar, J., **4**, 510[171]; **8**, 111[22], 112[22], 113[22], 117[22], 445[40]
Soloducho, J., **6**, 498[165]
Solokov, V. I., **8**, 187[48]
Solomko, Z. F., **2**, 787[52]; **4**, 48[138,138f]
Solomon, D. H., **7**, 152[3], 153[10], 765[153]
Solomon, D. M., **4**, 54[152,152b]
Solomon, M., **8**, 850[119]
Solomon, M. D., **3**, 390[71], 399[117], 402[117]
Solomon, M. F., **2**, 823[112]; **3**, 1051[22], 1052[22]; **5**, 916[120]
Solomon, P. A., **3**, 382[37], 393[37]
Solomon, R. D., **8**, 358[196]
Solomon, R. G., **5**, 916[120]
Solomon, S., **4**, 1010[158]; **5**, 797[67]; **7**, 764[128]
Solomon, V. C., **5**, 10[75]
Solomonov, B. N., **5**, 76[248]
Solov'eva, N. P., **6**, 553[728], 554[728,765,785]
Solov'eva, O. N., **2**, 464[101]
Solov'yanov, A. A., **8**, 2[9]
Soloway, S., **4**, 719[25]
Solter, L. E., **3**, 159[467], 166[467]
Solyom, S., **3**, 348[30,31], 358[31]; **4**, 155[74]
Soma, N., **3**, 919[33], 954[33]; **6**, 897[97]
Somal, P., **7**, 68[85], 71[85]
Soman, R., **3**, 943[83], 953[83]; **7**, 544[40], 551[40], 556[40]
Somawardhana, C., **4**, 286[167]
Somayaji, V., **8**, 709[43,43a], 875[36]
Somei, M., **2**, 967[77]; **7**, 87[18], 335[29-31]; **8**, 371[109]
Somerfield, G. A., **8**, 545[281]
Somers, P. K., **3**, 751[90]; **4**, 795[87]; **6**, 448[108]; **7**, 401[61d]; **8**, 846[81]
Somers, T. C., **1**, 259[30]; **2**, 545[87]; **3**, 828[52,53]
Somerville, L. F., **2**, 744[92]
Somerville, R. F., **4**, 369[21], 370[21], 371[21], 377[21]
Someswara Rao, C., **7**, 283[182], 284[182]
Someya, K., **8**, 371[99]
Somfai, P., **1**, 434[140], 798[291]; **2**, 249[84], 264[58]; **4**, 202[24], 399[225,225a]; **6**, 93[132], 990[83]; **7**, 399[35], 473[26]
Sommaruga, H., **7**, 778[411]
Sommelet, M., **3**, 914[11], 965[11]
Sommer, A., **7**, 10[80]; **8**, 563[436]
Sommer, J., **3**, 295[6], 297[18,24], 335[6], 339[6]; **7**, 8[53]
Sommer, J. J., **1**, 568[233]
Sommer, L. H., **3**, 419[32], 855[84]; **6**, 653[148]; **8**, 766[19], 906[67], 907[67], 908[67]
Sommer, N. B., **8**, 905[63]
Sommer, R., **2**, 609[82]; **8**, 548[319,320]
Sommer, S., **5**, 117[276], 487[186], 490[190,191]
Sommerlade, R. H., **6**, 116[87], 119[114]
Somorjai, G. A., **8**, 421[27,30]
Somovilla, A. A., **4**, 27[81]; **5**, 524[55]
Son, J. C., **5**, 151[11]; **7**, 725[32]; **8**, 240[34], 247[34], 250[34]
Sonada, A., **7**, 94[58]
Sondengam, B. L., **8**, 27[31], 66[31], 478[44], 480[44]
Sondheimer, F., **2**, 156[117], 376[279], 838[177], 839[177]; **3**, 273[10], 805[14]; **5**, 69[101,102]; **6**, 685[357]; **7**, 86[16a], 254[30]; **8**, 108[1], 118[1], 338[87], 528[65,84], 529[84], 935[67]
Sondhi, S. M., **7**, 341[53]
Sone, K., **1**, 836[140]
Sone, T., **4**, 221[155,156]; **6**, 491[115], 716[100], 717[105]; **7**, 174[135]
Sone, Y., **7**, 747[104]
Sonegawa, M., **5**, 516[28]
Sonehara, S., **3**, 216[72]
Song, A.-T., **2**, 765[73]
Song, H., **4**, 848[78]
Song, I. H., **7**, 662[51]
Song, K. M., **8**, 447[97]
Song, Y. H., **7**, 309[21], 318[54,55], 319[54,55]
Song, Z., **1**, 477[134]; **2**, 528[10]; **3**, 67[20]; **5**, 2[8]; **7**, 225[63]

Songkai, X., **1**, 543[15]
Songstad, J., **6**, 226[13], 242[94]
Sönke, H., **7**, 603[116]
Sonn, A., **8**, 300[87]
Sonnay, P., **1**, 754[106]; **5**, 456[83]
Sonnenberg, F. M., **4**, 273[48], 279[48], 280[48]; **5**, 850[146]
Sonnenbichler, J., **6**, 501[203], 531[203]
Sonnet, P. E., **3**, 45[245]; **6**, 207[49], 976[6,7], 980[35]; **7**, 673[22]; **8**, 884[99], 885[108], 886[108], 890[145]
Sonntag, A. C., **2**, 285[59]
Sonntag, F. I., **5**, 730[127]
Sono, S., **7**, 607[162]; **8**, 386[51]
Sonobe, A., **3**, 466[187]
Sonobe, T., **1**, 752[98]
Sonoda, A., **2**, 13[56]; **3**, 222[143], 249[67], 259[130], 483[10]; **4**, 310[435], 377[104], 378[104f], 383[104f], 557[10], 611[353], 837[13]; **8**, 806[102], 856[170], 876[44], 877[44]
Sonoda, N., **2**, 441[3], 442[11], 443[15,17], 445[24,25], 449[3], 450[54], 451[15,55], 603[46]; **3**, 771[186], 1034[77]; **4**, 115[177], 444[197], 973[85]; **5**, 438[161], 442[185,185a], 461[107], 464[107], 466[107], 532[84], 601[44]; **6**, 477[98], 479[108], 481[123]; **7**, 131[80], 137[118], 138[118], 446[69], 829[58]; **8**, 323[112], 370[90], 382[7], 412[119], 413[119], 773[70], 774[70,71], 789[122], 887[119]
Sonoda, N. J., **6**, 684[344]
Sonoda, T., **4**, 487[48]; **6**, 939[137]
Sonoda, Y., **8**, 341[108]
Sonogashira, K., **2**, 725[110]; **3**, 217[89], 219[89], 271[1], 521[7], 530[7,61], 531[61], 532[7,61], 537[61], 541[106], 552[8], 554[26], 557[48,49], 558[50]; **4**, 520[30]; **5**, 1174[33]; **8**, 456[210], 457[210]
Sonogashita, K., **1**, 447[204], 458[204]
Sonola, O. O., **8**, 64[204], 66[204], 67[204], 176[130,131]
Sonveaux, E., **5**, 116[262]
Sood, H. R., **5**, 600[43]
Sood, R., **4**, 187[98]; **8**, 194[103], 544[253,264,265], 546[304], 561[304]
Sookkho, R., **7**, 451[23]
Soong, J., **3**, 639[76]
Soong, L. T., **8**, 213[29]
Sooriyakumaran, R., **3**, 466[189]; **5**, 639[119]
Soós, J., **8**, 606[28]
Sootome, N., **1**, 755[113]
Sopchik, A., **2**, 345[32]; **8**, 5[26]
Sopher, D. W., **8**, 133[15,16], 135[16]
Sorba, J., **7**, 727[39]
Sore, N. E., **4**, 33[95,95a]
Sorensen, C. M., **6**, 784[92]
Sörensen, H., **5**, 1062[59]
Sorensen, P. E., **5**, 1123[34]
Sorensen, T. S., **3**, 374[133], 708[13]; **5**, 754[59,63-68]; **8**, 83[7]
Sorenson, R. J., **7**, 202[51]
Sorgeloos, D., **5**, 113[237]
Sorgi, K. L., **1**, 794[279], 795[280]; **2**, 578[81]; **4**, 38[108], 339[44]; **7**, 520[30]
Soria, J., **4**, 377[124b], 380[124,124b]
Sorkina, T. I., **5**, 752[38]
Sorm, F., **2**, 784[39b]; **3**, 643[128]; **4**, 1058[147]; **5**, 809[114]
Sorochinskii, A. E., **6**, 270[81]
Soroka, M., **6**, 801[38]
Sorokin, V. D., **4**, 347[101]
Sorokin, V. L., **2**, 709[14]
Sorokina, L. P., **8**, 266[56]
Sorokina, N. Y., **6**, 676[305]
Soroos, H., **3**, 415[6,7]
Sorrell, T. N., **8**, 264[35,36]
Sorrenti, P., **2**, 323[39], 330[51], 332[51], 333[51]
Soshin, V. A., **8**, 366[38]
Soskova, L. M., **7**, 767[190]
Sosnovsky, G., **7**, 92[40], 95[74,75], 96[75,85], 152[2], 153[2], 154[2], 158[38], 171[38], 613[1]; **8**, 51[122], 66[122]
Sosnowski, J. J., **1**, 262[38]
Sosonkin, I. M., **4**, 426[64]
Soteropoulous, P., **5**, 3[22]
Sotgiu, F., **8**, 847[89]
Soth, S., **7**, 27[65], 32[93]
Sotheeswaran, S., **3**, 154[424]
Sotirious, C., **3**, 729[38]
Sotiropoulos, J., **6**, 421[33]
Soto, A., **8**, 296[62]
Soto, J. L., **2**, 361[178], 380[299]; **4**, 5[17]; **5**, 407[27], 408[30,30b]
Sotowicz, A. J., **4**, 311[444]
Sotoyama, T., **3**, 262[158]
Sotta, B., **7**, 88[26]
Souchi, T., **5**, 595[11]; **8**, 459[244], 535[166]
Soucy, C., **8**, 242[40]
Soucy, M., **4**, 955[14]
Soucy, P., **2**, 303[6]
Souey, D., **6**, 531[428]
Soula, G., **4**, 439[161], 443[161]
Soulie, J., **2**, 6[26]; **3**, 224[174,174c]; **8**, 714[83]
Soum, A., **4**, 980[115], 982[115]; **5**, 1103[150], 1104[150,158]
Souma, Y., **3**, 381[21]
Soundararajan, R., **7**, 498[226], 503[226]
Souppe, J., **1**, 258[26,26b], 271[62], 274[73]; **3**, 567[35], 570[35]; **8**, 115[64], 124[64], 125[64], 552[360]
Sousa, L. R., **5**, 210[59]
Soussan, G., **1**, 226[93]; **4**, 98[109b,c]; **8**, 541[213], 543[213]
Soussan, S., **4**, 89[86a]
South, M. S., **5**, 692[93], 693[93], 696[93], 1135[51], 1202[56], 1203[57]
Southam, D. A., **3**, 963[121]
Southam, R. M., **6**, 273[98]
Souther, B. L., **6**, 280[140]
Southgate, R., **3**, 1051[23], 1056[23,34], 1062[23,34]; **5**, 105[194]
Southon, I. W., **5**, 717[93], 742[159a]
Southwick, P. L., **2**, 141[37]; **8**, 542[225]
Souto-Bachiller, F., **6**, 438[57]
Soverini, M., **8**, 70[234], 354[173]
Sovocool, G. W., **5**, 152[22]
Sowa, F. J., **3**, 316[117], 317[117]
Sowa, T., **6**, 602[10]
Sowden, J. C., **3**, 831[64,65]; **7**, 218[8]; **8**, 925[9]
Sowell, G., **4**, 809[164]
Sowerby, D. B., **3**, 299[33]
Sowerby, R. L., **1**, 506[14], 861[72]; **3**, 786[39]; **6**, 139[47]; **8**, 527[44], 528[44]
Sowin, T. J., **2**, 648[90], 1049[24], 1058[72,73]; **5**, 407[28,28b]; **7**, 580[145]
Sowinski, A. F., **3**, 599[206]; **7**, 530[26]; **8**, 113[32]
Soyka, M., **8**, 568[466]
Soysa, H. S. D., **8**, 406[35]
Sozzani, P., **6**, 227[19], 228[19]
Spada, A. P., **4**, 893[156]; **7**, 415[110]
Spada, R., **4**, 850[85]
Spadoni, G., **2**, 904[53]
Spadoni, M., **7**, 429[150a], 777[382]
Spagnolo, P., **4**, 336[29,30]; **7**, 477[69], 493[197]; **8**, 385[46]
Spagnuolo, C. J., **3**, 25[152]
Spaite, D. W., **4**, 347[95]
Spalink, F., **3**, 903[127]
Spaltenstein, A., **1**, 635[88,89], 806[315]; **3**, 104[208,209], 117[208,209]; **6**, 846[103], 905[145]
Spanagel, H.-D., **3**, 661[22]

Spandau, H. S., **8**, 842[43]
Spangenberg, R., **6**, 638[44]
Spangenberger, H., **8**, 388[62]
Spangler, B., **5**, 710[50]
Spangler, C. W., **5**, 707[29], 710[48,50,53], 906[68], 982[29], 984[29], 989[29]
Spangler, D., **1**, 314[137], 315[137]
Spangler, F. W., **7**, 156[32]
Spangler, L. A., **5**, 1022[76]
Spangler, R. J., **3**, 901[111]; **4**, 505[137]
Spanka, C., **4**, 992[156]
Spanton, S., **4**, 10[34], 113[164]
Sparapani, C., **3**, 300[40]
Sparkes, G. R., **7**, 24[25], 26[25]
Sparks, M. A., **4**, 98[115]
Spassky, N., **5**, 86[28], 88[49]
Spassky-Pasteur, A., **8**, 528[72], 532[131c]
Späth, E., **2**, 138[21,22]
Spatola, A. F., **6**, 420[23], 423[23], 431[104], 451[130], 456[130]; **8**, 904[60], 905[60]
Spaulding, D. W., **7**, 673[29]
Spayd, R. W., **3**, 822[9], 823[9], 834[9]
Spaziano, V. T., **6**, 546[652]
Speake, R. N., **3**, 715[42]
Speakman, P. R. H., **7**, 762[82]
Spear, K. L., **3**, 957[110]; **5**, 894[46]; **6**, 901[123]
Spear, R. J., **2**, 345[29]
Spears, C. P., **6**, 462[22]
Spears, G. W., **3**, 14[72], 15[72], 16[72], 17[72]
Specian, A. C., Jr., **5**, 893[41]
Speckamp, W. N., **1**, 371[75], 372[75e], 617[54], 771[192]; **2**, 89[36,37], 558[162], 586[133], 587[133], 652[123a], 971[91], 1048[6,7], 1049[6,7,15,17,20,22,25], 1050[17,20,22,30,31], 1053[6], 1054[57], 1061[93], 1062[6,57,100,101], 1063[102,105], 1064[25,108], 1065[113-115,117], 1069[133,134], 1070[138], 1071[138], 1072[30,31,57], 1075[138,151], 1078[15], 1079[151,156]; **3**, 223[183], 225[183], 341[4], 361[77], 364[77], 368[104]; **4**, 8[29,29g]; **5**, 402[5], 501[267]; **6**, 118[99], 744[73], 745[80,82], 746[93,94,97]; **8**, 273[128,129]
Speelman, J., **8**, 96[92]
Speer, H., **3**, 155[432]
Speert, A., **5**, 1130[7]
Speeter, M., **2**, 294[82], 917[87]; **5**, 100[140]
Speh, P., **2**, 368[238]; **6**, 512[120,303], 543[120], 553[796,798], 572[796,958], 573[798,963], 576[973,974]
Spehar, A., **8**, 709[43], 812[3], 908[78], 909[78]
Spehr, M., **5**, 442[185]
Speier, G., **7**, 532[30]; **8**, 447[99], 551[345]
Speier, J. L., **1**, 619[60]; **8**, 699[148], 763[1], 764[3], 765[15], 771[49], 773[15], 776[1c,3,78], 785[1]
Spek, A. L., **1**, 10[54,55], 11[55b], 13[69], 16[86], 25[127], 26[132,133,134], 30[153], 214[27]; **2**, 123[196], 124[204], 125[204]
Spellmeyer, D. C., **4**, 729[59], 781[7], 787[7], 827[7]; **5**, 79[292], 241[6], 257[61], 680[21,22], 683[22], 1031[97]
Speltz, L. M., **5**, 692[95]
Spencer, A., **4**, 844[60,63], 847[63], 858[108], 922[29]; **7**, 108[171]; **8**, 446[68]
Spencer, C. B., **1**, 318[157], 320[157]
Spencer, C. F., **3**, 379[1], 380[1]; **6**, 1013[15]
Spencer, E. Y., **6**, 208[63]
Spencer, H. K., **2**, 538[58], 539[58]; **7**, 774[312,313], 777[312,313]
Spencer, J. L., **5**, 841[97]; **8**, 766[18]
Spencer, J. T., **5**, 442[180]
Spencer, R. W., **3**, 217[92]; **4**, 394[189,189c]
Spencer, T. A., **2**, 841[186]; **3**, 20[117], 55[283], 785[35]; **4**, 4[14,14a], 261[283], 611[349]; **5**, 923[138], 933[183]; **6**, 176[86]; **7**, 31[86], 43[44]; **8**, 526[26], 527[48], 528[57,61,78], 884[97], 958[17], 986[14], 988[30]
Spengler, T., **2**, 1016[24]; **6**, 736[21]
Spensley, C. R. C., **4**, 370[40]; **5**, 835[59]
Speranza, G., **2**, 853[231]; **7**, 109[183]
Speranza, M., **3**, 300[40]; **4**, 445[206]; **5**, 1148[119]
Sperber, N., **8**, 292[41]
Spero, D. M., **4**, 797[103], 820[214]
Spero, G. B., **7**, 86[16a]; **8**, 293[49]
Spessard, G. O., **1**, 403[19]; **3**, 215[59]; **6**, 23[92], 237[65], 243[65]
Speth, D. R., **7**, 86[14]
Spevak, P., **8**, 384[29], 412[115]
Speziale, A. J., **6**, 429[88]; **8**, 974[125]
Speziale, V., **4**, 290[204], 291[204,210], 292[204], 311[439,440], 383[136]; **7**, 632[61,62]; **8**, 854[154]
Spialter, L., **1**, 619[59]; **4**, 120[197]; **8**, 364[25]
Spicer, L. D., **8**, 320[91]
Spicer, M. D., **3**, 213[51]
Spichiger, S., **4**, 520[28], 545[125], 546[125]
Spidada, P. K., **6**, 23[88]
Spiegel, B. I., **3**, 486[44], 495[44], 529[53]; **8**, 693[112], 755[117], 758[117]
Spiegelberg, H., **2**, 850[219], 854[219]
Spiegelman, G., **5**, 92[77]; **6**, 744[74]
Spies, J. W., **3**, 582[113]
Spiess, A., **3**, 822[12], 831[12], 835[12b]
Spiess, E. J., **4**, 115[179b], 257[224]; **5**, 1094[98], 1096[98], 1098[98], 1112[98]
Spiess, G., **6**, 261[3], 270[3]
Spiessens, L., **6**, 80[66]
Spietschka, E., **3**, 891[40], 900[40,95]
Spiewak, J. W., **3**, 883[109]
Spillane, R. J., **8**, 264[35]
Spille, J., **6**, 263[23], 273[23], 291[23]
Spilleers, D., **5**, 417[65]
Spilling, C. D., **2**, 338[75]; **5**, 21[162,163], 23[162,163]; **8**, 84[15], 374[151]
Spina, K. P., **4**, 247[104], 259[104], 993[159a]
Spindell, D. K., **5**, 8[62], 9[72]
Spineti, G. G., **5**, 829[15]
Spinzak, Y., **8**, 568[485]
Spitz, U. A., **7**, 841[19]
Spitzer, U. A., **7**, 12[106], 587[169]
Spitzer, W. A., **5**, 220[47]
Spitznagel, G. W., **1**, 487[1], 488[1]; **3**, 194[4]
Spitzner, D., **2**, 384[318]; **4**, 18[59], 30[88], 121[207]
Spitzner, R., **6**, 508[287,288], 518[334]
Splettstösser, G., **4**, 18[59]
Spletzer, E., **4**, 824[237]
Spliethoff, B., **5**, 37[22b]
Spoerri, P. E., **8**, 231[141]
Spogliarich, R., **8**, 534[159]
Spohn, R. F., **5**, 345[67d]
Sponsler, M. B., **7**, 4[15]
Sporn, M. B., **6**, 622[135]
Sprague, J. T., **3**, 382[36]
Sprague, L. G., **3**, 202[86]
Spray, C. R., **8**, 537[183]
Spreafico, F., **3**, 312[102]; **4**, 36[103,103c]; **8**, 195[111], 203[111,149]
Sprecher, C. M., **8**, 425[48], 474[14], 475[14], 476[14]
Sprecher, M., **7**, 50[75]; **8**, 297[68]
Spreutels, S., **1**, 698[239], 700[239], 705[239], 709[239], 712[239], 722[239]
Spring, D. J., **3**, 825[32]; **4**, 496[92]
Spring, F. S., **6**, 685[357], 802[39]; **8**, 528[82], 529[82], 566[450]
Springer, D. M., **7**, 567[104]

Springer, J., **2**, 811[71], 824[71]
Springer, J. M., **2**, 738[41], 760[40]
Springer, J. P., **1**, 3[24], 32[158], 564[204], 823[44b]; **2**, 189[47], 667[40], 673[40], 674[40], 675[40], 694[75]; **3**, 698[157a], 872[56]; **4**, 18[60,60a], 48[139,139f], 246[96], 258[96], 260[96], 373[80], 374[80], 381[129], 382[129], 581[26], 695[4], 957[22], 1040[101], 1045[101a]; **5**, 134[64], 137[81], 333[44b], 434[145], 468[135], 541[111], 857[230], 1029[91], 1092[94], 1094[94], 1102[148], 1103[148], 1113[94]; **6**, 91[123], 93[123], 717[113]; **7**, 255[37], 346[13], 377[91], 378[91b], 439[37], 440[37], 566[101]; **8**, 50[117], 66[117], 940[106]
Sprinzak, Y., **8**, 568[485]
Sproat, B. S., **6**, 604[29]
Sprügel, W., **5**, 596[26], 597[26]
Spry, D. D., **6**, 490[108]
Spry, D. O., **4**, 86[80]; **7**, 470[14], 764[106]
Spunta, G., **1**, 391[148]; **2**, 613[114], 656[157], 807[48], 935[151], 936[151], 937[157]; **5**, 100[155]; **6**, 759[140]
Spur, B., **2**, 381[301]
Spurlin, H. M., **2**, 144[57]
Spurling, T. H., **5**, 637[107]
Spurlock, L. A., **7**, 823[38]
Spyroudis, S., **3**, 538[91]; **4**, 425[22]; **5**, 1151[128]
Squillacote, M. A., **7**, 742[56]
Squillacote, M. E., **5**, 715[81]
Squiller, E. P., **1**, 15[79]
Squires, T. G., **7**, 763[99], 766[99]
Srairi, D., **7**, 60[45]
Srebnik, M., **1**, 223[84], 225[84c]; **3**, 797[93,94]; **4**, 347[99], 370[30], 372[30]; **6**, 78[59], 776[58]; **7**, 535[45,46]; **8**, 352[148], 710[57,58], 717[57]
Sredojevic, S., **7**, 231[142]
Sreekumar, C., **1**, 757[120]; **3**, 195[20]; **6**, 174[60]
Sridar, V., **3**, 380[9]; **4**, 791[50], 801[125], 820[50]
Sridaran, P., **4**, 457[51], 469[51]
Sridharan, V., **4**, 848[78]
Srikrishna, A., **4**, 793[70,71]; **5**, 225[91], 768[124], 779[124], 839[77]; **8**, 566[456]
Srimannarayana, G., **7**, 136[112], 137[112]
Srinivasa, R., **3**, 596[193], 727[30,31]
Srinivasachar, K., **5**, 90[57], 95[57], 636[96]
Srinivasan, A., **4**, 190[107]; **6**, 94[140]
Srinivasan, C., **7**, 765[154]
Srinivasan, C. V., **1**, 410[41], 473[78]; **2**, 737[32]
Srinivasan, K., **5**, 678[17]
Srinivasan, K. G., **3**, 380[7]
Srinivasan, K. V., **8**, 366[44]
Srinivasan, N. S., **7**, 768[207]
Srinivasan, P. R., **1**, 294[39-41]
Srinivasan, P. S., **6**, 436[27], 437[27], 453[27], 455[27], 456[27]; **7**, 283[182], 284[182]
Srinivasan, R., **5**, 133[51], 195[6], 196[12,13], 586[205], 647[14], 649[14,21], 658[21], 661[34], 670[48], 730[127]; **8**, 475[18]
Srinivasan, V., **3**, 380[7]
Srinivastava, R. C., **7**, 775[346]
Srivastava, A. K., **6**, 524[358]
Srivastava, P. C., **6**, 478[102]
Srivastava, R. C., **3**, 826[34]; **5**, 85[5]
Srivastava, R. D., **7**, 247[96]
Srivastava, S., **5**, 436[155]; **8**, 5[31], 597[94], 606[26]
Srivastava, T. N., **7**, 775[346]
Srivastava, T. S., **1**, 292[28]
Srivastava, V., **6**, 51[110]
Srivatsas, J., **7**, 71[98]
Srnak, A., **4**, 970[71]
Srnic, T., **7**, 41[23]
Srogl, J., **8**, 274[138]
Ssebuwufu, P. J., **4**, 605[295]
Staab, E., **7**, 384[114c], 399[38], 400[38,38b], 406[38], 409[38], 415[38]
Staab, H. A., **1**, 423[96]; **3**, 877[82], 927[49,50]; **6**, 614[91]; **7**, 595[26]; **8**, 271[104]
Stabba, R., **8**, 747[56], 752[56]
Stabler, R. S., **8**, 9[50]
Stabler, S. R., **8**, 50[118], 66[118]
Stacey, F. W., **4**, 279[103], 770[245]
Stacey, M., **2**, 736[22]; **7**, 15[146], 760[28]
Stacey, N. A., **1**, 568[228]; **6**, 1028[98]
Stach, H., **1**, 843[2a], 898[2a]
Stache, U., **7**, 124[42]
Stachel, J., **1**, 286[11]
Stachulski, A. V., **6**, 644[92]
Stacino, J.-P., **8**, 847[88,88b,d]
Stackebrandt, J., **2**, 1090[72]
Stacy, G. W., **1**, 226[95]; **7**, 760[39]
Stadler, H., **6**, 778[63]
Stadler, M., **5**, 850[152]
Stadler, P. A., **3**, 126[311,312], 346[23], 351[39]; **6**, 494[136], 919[39]
Stadlwieser, J., **5**, 329[31]; **6**, 119[113]; **7**, 746[92], 752[92]
Staeglich, H., **1**, 38[256]
Stäheli, R., **5**, 386[133,133a]
Stahl, G. L., **6**, 637[30]
Stahl, I., **6**, 134[19]
Stahl, R. E., **7**, 660[37]
Stahle, M., **1**, 180[28]; **2**, 5[21], 13[21], 977[6]; **4**, 869[22]; **7**, 99[107]
Stahly, B. C., **4**, 432[111]
Stahly, G. P., **4**, 423[9], 429[9]
Stahly, J. P., **4**, 432[111]
Stahnke, G., **6**, 636[24]
Stahnke, M., **7**, 47[55]
Staib, R. R., **2**, 662[4], 1056[66], 1070[66], 1074[66]; **5**, 402[1], 403[1], 404[1], 410[1], 413[1], 417[1], 422[1], 425[1], 426[1], 429[1], 430[1], 433[1], 434[1], 435[1], 436[1], 438[1], 440[1], 444[1], 451[19]
Stajer, G., **5**, 583[186], 584[194]
Stakem, F. G., **4**, 598[186], 638[186], 839[25], 850[84], 855[96], 856[98], 904[205]
Staklis, A., **6**, 802[43], 803[43]
Staley, R. H., **1**, 287[16]
Staley, S. W., **5**, 901[22]; **6**, 953[11], 969[119]
Stalke, D., **1**, 38[183]
Stallberg-Stenhagen, S., **2**, 742[72]; **3**, 643[126]
Stallman, B., **4**, 850[85]
Stalmann, L., **8**, 530[101]
Stam, C. H., **1**, 25[126], 28[140], 41[269]; **3**, 210[25], 211[32]; **4**, 170[12]
Stamm, E., **5**, 664[38]
Stamm, H., **2**, 283[47]; **6**, 94[145], 95[146], 96[148]; **7**, 470[9], 487[9], 495[205]; **8**, 563[436]
Stamm, W., **8**, 754[95]
Stammer, C. H., **4**, 345[84], 611[358], 988[137]; **6**, 804[48]
Stamos, I. K., **6**, 930[86]; **7**, 200[38]
Stamouli, P., **4**, 764[222], 765[222], 808[155]
Stamper, J. G., **1**, 17[91]
Stämpfli, U., **4**, 1007[120], 1008[134], 1009[135]
Stanaback, R. J., **7**, 664[65]
Stanaszek, R. S., **8**, 27[30], 66[30]
Stancino, J.-P., **1**, 804[308], 805[308]
Standnichuk, M. D., **8**, 772[54]
Standring, D. N., **8**, 384[39]
Stanek, J., **3**, 896[72]
Stanetty, P., **8**, 242[40]
Stanfield, C. F., **7**, 274[139]

Stanford, J. E., **5**, 725[117]
Stanford, R. H., Jr., **4**, 665[10], 667[10], 670[10], 674[10]
Stang, A. F., **8**, 754[83]
Stang, P. J., **1**, 665[172], 668[172], 786[250]; **2**, 725[113,117]; **3**, 219[105], 380[4], 522[22]; **4**, 250[134], 255[134], 260[134], 903[195]; **6**, 19[68], 72[24], 504[221], 966[91-93]; **8**, 933[46]
Stange, A., **6**, 447[104], 451[104]; **8**, 821[50,51]
Stangeland, L. J., **6**, 242[94]
Stangl, H., **3**, 893[53b]; **7**, 451[34], 477[73]
Stangle, H., **4**, 1081[75]
Stanhop, B., **1**, 779[226]
Stanier, A., **3**, 786[38]
Staninets, V. I., **4**, 342[66,67], 364[1,1a-d,g], 367[1g], 368[1a], 373[1a,b], 386[153], 387[158], 391[177,178a,b], 395[1d], 397[1c], 399[1c], 400[228b], 407[255], 408[1c,259b], 409[259b], 410[1c], 413[1d,278c], 421[1c]
Stanko, M. K., **6**, 281[148]
Stanko, V. I., **8**, 214[50]
Stanley, D. R., **8**, 447[130]
Stanley, G., **8**, 542[227]
Stanley, K., **4**, 588[61]
Stanley, P., **6**, 836[55]
Stanley, W., **7**, 774[310]
Stannard, A.-M., **5**, 500[260]
Stanonis, D., **2**, 1085[22]
Stanovnik, B., **3**, 904[133]; **6**, 480[114], 514[306], 543[306], 554[712,713,714,715,716,725,739,740,742,743,744,746,747,782]; **7**, 475[53], 476[53]; **8**, 384[31,32]
Stansfield, F., **4**, 1018[221]; **6**, 509[260], 523[347]
Stanton, J., **3**, 933[61]; **8**, 93[76], 618[120]
Stanton, J. L., **4**, 159[85]; **8**, 35[66], 66[66]
Stanton, S. A., **4**, 597[180], 622[180]
Stapleford, K. S. J., **8**, 212[21], 515[118], 605[13]
Stapleton, A., **7**, 338[42]
Stapp, P. R., **2**, 528[14,15], 553[15]; **7**, 828[53]; **8**, 598[96], 612[73], 624[73]
Staral, J. S., **3**, 322[142a]
Staring, E. G. J., **5**, 86[30], 88[30,46,47]
Stark, C. J., Jr., **1**, 532[135], 740[43], 741[43]; **3**, 173[521]
Stark, H., **5**, 385[130]
Stark, J. C., **4**, 476[155]
Stark, S. R., **7**, 500[243]
Stárka, L., **7**, 96[82,83]
Starke, H. M., **1**, 561[157]
Starkemann, C., **4**, 85[72], 204[33], 231[282], 249[120]; **5**, 260[70], 263[70]
Starkovsky, N. A., **6**, 643[79], 658[184], 709[54]
Starks, C. M., **4**, 1001[27]
Starling, W. W., **7**, 86[16a]
Starner, W. E., **1**, 390[143]; **2**, 994[40], 999[40]
Staros, J. V., **7**, 21[18]
Staroscik, J., **6**, 9[41]
Starr, L. D., **1**, 226[95]
Starratt, A. N., **7**, 27[62], 41[22]
Starrett, J. E., **7**, 183[166]
Starrett, J. E., Jr., **1**, 377[95]; **2**, 1049[23], 1050[23]
Starrett, R. M., **4**, 18[62], 20[62g]
Stasi, F., **2**, 87[26], 213[127]
Staskun, B., **3**, 369[117], 372[117]; **8**, 298[76-78], 299[78,79]
Stass, W. H., **7**, 823[38]
Stassinopoulou, C. F., **3**, 565[20]
Staticu, S., **4**, 600[234], 643[234]
Staub, A., **2**, 399[18]; **3**, 734[6]; **6**, 650[128]
Staudacher, W., **5**, 1096[127]
Staudinger, H., **4**, 283[149]; **5**, 90[52]; **6**, 646[101]
Staudinger, J. J. P., **4**, 292[227]
Stauffacher, D., **3**, 345[20]
Stauffer, R. D., **3**, 500[132], 505[132]; **4**, 254[185], 476[162,163], 502[124], 766[229]; **8**, 542[223], 544[223,251], 548[322]
Staum, M. M., **3**, 782[20]
Staunton, J., **4**, 14[47]
Stavchansky, S., **1**, 92[70]
Stavinoha, J. L., **2**, 1037[96,99,100]; **7**, 876[121]
Stavrovskaya, V. I., **2**, 957[17]
Stawinski, J., **6**, 603[23], 614[95], 620[133]
Steadman, T. R., **2**, 662[12]
Steblina, E. I., **8**, 624[155]
Stec, W. J., **4**, 231[274]; **6**, 604[27], 606[40]; **8**, 72[242], 74[242,244], 393[110]
Steckbeck, S. R., **8**, 884[97]
Steckhan, E., **2**, 1051[39], 1075[39]; **4**, 371[61]; **6**, 561[874]; **7**, 248[113], 795[9], 796[15], 797[17], 808[76]; **8**, 85[17], 797[30]
Stedronsky, E. R., **8**, 986[10]
Steege, S., **3**, 322[142b]; **7**, 7[50]
Steel, P., **7**, 543[22]
Steel, P. G., **1**, 568[228]
Steel, P. J., **2**, 19[77,77b], 31[77b], 573[52]
Steele, B. R., **3**, 194[12]; **4**, 274[64], 275[64]
Steele, D. R., **8**, 140[15]
Steele, J., **2**, 650[111], 651[111a]; **3**, 373[128]
Steele, J. A., **3**, 406[143]
Steele, R. B., **8**, 736[22], 740[22], 754[22], 756[142]
Steele, R. W., **2**, 1050[27]
Steen, K., **5**, 828[6], 836[6], 888[26], 893[26]
Steenhoek, A., **8**, 431[63]
Steenson, B. E., **2**, 350[72], 363[72]
Stefanelli, S., **1**, 519[66], 520[67]; **2**, 31[108]; **7**, 442[47]
Stefani, A., **4**, 887[131], 919[18,19]; **5**, 31[5]
Stefanovic, M., **3**, 313[103], 905[136]; **7**, 92[41,41a], 94[41]
Stefanovsky, Y. N., **4**, 21[68]
Stefanska, B., **6**, 554[749], 789[110]
Steffek, R. P., **7**, 62[53,53b,c]; **8**, 192[97]
Steffen, J., **3**, 348[30]; **4**, 155[74]
Steffens, J. J., **7**, 58[53a], 62[53,53a,c], 63[53a]; **8**, 94[79], 192[97]
Steffgen, F. W., **7**, 576[123]
Stegelmeier, H., **2**, 371[264]
Stegenga, S., **2**, 1050[30], 1072[30]
Steglich, W., **1**, 122[72], 373[83,85-87,89], 374[85-87], 375[89]; **2**, 1052[46], 1077[153]; **4**, 12[40], 42[120]; **5**, 480[176], 485[182], 877[6]; **6**, 437[37], 523[352], 524[352], 642[74], 657[176], 716[104], 745[79]
Stegmann, H. B., **3**, 661[22]
Steigel, A., **5**, 257[60], 498[228,229,234,235], 714[74]
Steigerwald, M. L., **1**, 880[116]; **5**, 788[14], 849[142], 1000[8]
Steiman, R., **7**, 79[131]
Steimecke, G., **6**, 441[86]
Stein, A. R., **6**, 295[246], 529[465]
Stein, C. A., **7**, 762[77]
Stein, G., **5**, 451[4,6]
Stein, H., **2**, 223[148]
Stein, J., **6**, 487[66], 489[66]
Stein, J.-L., **7**, 499[233]; **8**, 798[47]
Stein, N. L., **8**, 639[18]
Stein, P. D., **2**, 465[106]
Stein, P. M., **8**, 320[80]
Stein, U., **7**, 874[105]
Steinacker, K. H., **6**, 673[288]
Steinbach, G., **6**, 839[70]
Steinbach, K., **2**, 33[121]
Steinbach, R., **1**, 83[26], 141[21], 145[21,41,42], 146[21,42], 148[42,46], 149[21,42], 150[41,46], 151[21,41,53,53a], 152[21,53,53a], 153[41,59], 154[41,59], 155[42], 157[53a], 158[53,53a], 165[46],

169[117,118], 170[42], 215; **2**, 22[86], 307[15], 310[15], 507[23], 512[23], 640[68], 641[68]; **3**, 421[57,58]
Steinberg, H., **1**, 528[114]; **3**, 727[27]; **5**, 77[263]; **7**, 602[105]; **8**, 719[124]
Steinberg, I. V., **3**, 643[127]
Steinberg, M., **8**, 530[93]
Steinberg, N. G., **3**, 890[32]
Steinborn, D., **1**, 215[33]
Steinbrückner, C., **2**, 1087[36,37], 1088[37,50], 1089[50,55], 1090[36,55], 1100[55]
Steindl, F., **6**, 443[90]
Steindl, H., **3**, 380[7]
Steiner, E., **4**, 754[175]
Steiner, G., **5**, 76[238], 77[245], 78[276,278]
Steiner, H., **5**, 703[16a], 709[16a]
Steiner, R. P., **5**, 804[93]
Steiner, U., **5**, 1130[8]
Steinlein, E., **5**, 1066[8]
Steinman, D. H., **1**, 805[312]; **5**, 245[21], 308[97]
Steinman, D. M., **6**, 990[84]
Steinmetz, A., **7**, 842[38]
Steinmetz, M. G., **5**, 204[40], 209[55]
Steinmetz, R., **4**, 1058[148]; **5**, 167[98]
Steinmüller, D., **3**, 124[283,284], 125[283,284], 127[283], 132[283]
Steinwand, P. J., **4**, 947[94]
Steinweg, E., **6**, 1035[139,140]
Stelakatos, G. C., **6**, 666[232], 667[232], 668[252], 669[252]
Steliou, K., **3**, 693[141]; **6**, 938[124]; **7**, 336[34]
Stella, L., **1**, 557[130]; **4**, 758[189], 812[180], 820[220]; **5**, 70[111,112]; **6**, 495[142,143], 496[143], 497[143]; **7**, 499[233]; **8**, 798[47]
Stelling, G. D., **3**, 358[66]
Steltenkamp, R. J., **7**, 816[11]
Stelzel, P., **6**, 665[226], 667[226], 668[226], 669[226]
Stemke, J. E., **1**, 377[97]; **2**, 588[150]; **3**, 251[79], 254[79]; **6**, 781[78]
Stemniski, M. A., **7**, 698[52]
Stenberg, J. F., **8**, 251[105]
Stener, A., **3**, 503[149], 512[149]
Stenger, R. J., **2**, 741[64]
Stenhagen, E., **3**, 643[129]
Stenkamp, R. E., **8**, 89[45]
Stensiö, K.-E., **7**, 820[24]
Stenstrøm, Y., **5**, 19[130]
Stenvall, K., **5**, 686[53]
Stenzel, T. T., **4**, 475[151]
Stenzel, W., **6**, 41[44]
Step, G., **5**, 65[67]
Stepanov, B. I., **6**, 438[50]
Stepanov, D. E., **4**, 314[487]
Stepanov, F. N., **3**, 383[47]; **6**, 266[53]
Stepanov, R. G., **8**, 636[5]
Stepanova, E. E., **6**, 546[645]
Stepanova, I. P., **8**, 600[106], 606[25], 625[25]
Stepanyan, A. A., **6**, 712[78]
Stepanyan, G. M., **6**, 507[237], 515[237]
Stephan, D., **5**, 386[132,132d], 387[132d]; **6**, 977[16]; **8**, 28[36], 66[36]
Stephan, E., **4**, 210[74]
Stephan, W., **6**, 519[339]; **7**, 144[153,154]
Stephanatou, J. S., **6**, 440[73]
Stephanian, M., **5**, 1123[34]
Stephanidou-Stephanatou, J., **8**, 636[6]
Stephen, E., **8**, 271[109]
Stephen, H., **8**, 271[109], 298[71]
Stephens, C. J., **3**, 649[190]
Stephens, C. R., **6**, 265[38]
Stephens, F. A., **8**, 446[67]
Stephens, J. A., **8**, 974[125]
Stephens, J. C., **8**, 978[143]
Stephens, J. R., **4**, 310[437,438]
Stephens, R., **4**, 407[256a]; **7**, 760[28]
Stephens, R. D., **3**, 521[4], 522[4]; **8**, 481[50,51], 482[50,51], 483[51], 531[120]
Stephens, T. B., **6**, 147[82]; **7**, 210[96], 211[96], 212[96]
Stephenson, B., **6**, 206[43], 210[43], 214[43]
Stephenson, D. S., **8**, 335[67]
Stephenson, G. R., **4**, 665[9], 670[15,16], 675[41], 688[9,65,66], 691[75]; **6**, 690[403], 692[403]
Stephenson, L., **3**, 804[13]; **8**, 987[23]
Stephenson, L. M., **2**, 528[11]; **5**, 2[9], 64[38], 640[127]; **7**, 86[14], 816[10], 818[10], 819[19]
Stephenson, M., **3**, 334[221,221a]
Stephenson, M. A., **5**, 942[235]
Stephenson, R. A., **3**, 279[39], 770[172]; **6**, 247[131]
Stephenson, R. W., **4**, 190[107]
Stepovik, L. P., **8**, 753[68]
Steppan, H., **3**, 887[12], 904[132]
Steppan, W., **2**, 681[58], 683[58]; **6**, 502[217], 560[870]; **7**, 650[51]
Steppel, R. N., **3**, 739[41]
Stepuk, E. S., **4**, 314[487]
Stercho, Y. P., **8**, 26[28], 30[28], 36[28], 37[28], 40[28], 43[28], 44[28], 46[28], 55[28], 66[28], 74[245], 176[136], 244[67], 250[67], 357[204], 393[110]
Sterk, H., **4**, 4[16,16b]
Sterling, J. J., **1**, 432[135]; **2**, 121[187]; **3**, 8[41], 9[41], 15[41], 20[41], 212[39], 226[39a], 250[70]; **4**, 176[49], 178[49,61], 245[87], 256[206,211]; **5**, 595[13]
Stermitz, F. R., **3**, 461[144], 683[101], 685[106,107]; **4**, 505[147,148]; **7**, 801[41], 873[99]
Stermitz, T. A., **4**, 505[148]
Stern, A., **2**, 770[10], 771[10]
Stern, A. G., **1**, 856[55]
Stern, A. J., **4**, 143[21]
Stern, E. W., **8**, 447[101,102], 450[101,102]
Stern, M., **3**, 302[51]; **6**, 246[130]
Stern, P., **5**, 724[111]
Stern, R., **3**, 262[166]; **8**, 445[25,26], 452[25,26]
Sternbach, D., **6**, 83[78], 1056[56]; **7**, 668[82,84]
Sternbach, D. D., **5**, 534[95]; **8**, 63[198], 64[198], 66[198], 67[198], 69[198]
Sternbach, L. H., **1**, 130[96]; **6**, 922[52]
Sternberg, E., **5**, 291[43], 1144[96,97], 1190[29], 1192[29]
Sternberg, E. D., **4**, 691[77], 898[176]
Sternberg, H. W., **5**, 1138[67]
Sternberg, J. A., **1**, 768[170]; **4**, 253[168]
Sternerup, H., **2**, 1051[35], 1052[35]
Sternfeld, E., **8**, 505[75]
Sternfeld, F., **3**, 174[527,527a], 175[527a]; **7**, 363[34]
Sternhell, S., **2**, 345[29]; **3**, 505[160]; **5**, 468[135]; **7**, 352[31], 356[31], 649[45]; **8**, 501[53], 507[53], 937[80,83]
Sterzycki, R., **4**, 310[429]
Stetin, C., **4**, 221[159]; **6**, 726[180]
Stetter, H., **1**, 542[7], 543[7], 546[7], 547[7], 548[7], 556[7]; **2**, 342[6], 354[104], 363[188]; **3**, 383[44], 890[34], 903[126]; **4**, 8[29,29e], 12[39], 13[45], 14[45b,46], 18[57]; **6**, 267[58,59], 269[76], 563[893,894,904], 567[931], 673[288], 681[335], 682[336]; **7**, 17[175,176]; **8**, 388[62]
Steudle, H., **2**, 283[47]
Steudle, O. W., **8**, 100[116], 739[33]
Steven, D. R., **4**, 307[392]
Stevenaert, N., **3**, 963[124]
Stevenart-De Mesmaeker, N., **4**, 795[82]

Stevens, A. E., **4**, 980[109]; **8**, 454[202]
Stevens, C. L., **3**, 740[45], 790[58,60], 842[21], 843[23]; **4**, 274[59], 282[59]; **6**, 622[135], 959[48]; **7**, 168[104,106], 170[106], 272[142], 276[148], 656[16]
Stevens, C. M., **6**, 640[56], 641[56]
Stevens, D. R., **7**, 7[38]
Stevens, H. C., **6**, 204[18]
Stevens, I. D. R., **5**, 428[108], 429[112], 679[20]; **6**, 566[928]
Stevens, J. A., **5**, 486[185], 487[185]
Stevens, K. D., **4**, 272[37], 273[37]
Stevens, K. E., **2**, 728[146]; **5**, 225[87,88], 231[87,88,132]
Stevens, M. F. G., **6**, 554[803]
Stevens, R. E., **7**, 760[43]
Stevens, R. V., **2**, 1011[10,11], 1013[18], 1015[20], 1016[29]; **3**, 287[61]; **4**, 510[168,169]; **5**, 249[33], 692[91], 696[91], 945[254,255], 946[254,255,256], 947[255,256,258,260,261], 952[255,256], 959[313,314,315,316,317], 960[258,260,320], 961[261,317,322], 962[322]; **6**, 284[175], 937[117], 939[117], 940[117], 1059[70], 1066[70,92]; **7**, 318[61], 376[83]; **8**, 34[59], 56[166], 66[59,166]
Stevens, R. W., **2**, 116[127,130,137], 117[144], 436[68], 437[68], 610[90,92], 611[92]; **4**, 100[125]
Stevens, T. B., **7**, 207[75], 208[75]
Stevens, T. E., **6**, 803[45]; **7**, 229[114], 502[265]
Stevens, T. S., **3**, 822[8], 836[8], 913[1,2], 914[3-5], 921[35], 927[53]; **6**, 283[161], 786[95], 961[69,70]; **8**, 297[64]
Stevenson, B. K., **8**, 340[97]
Stevenson, D., **8**, 974[129]
Stevenson, D. P., **3**, 332[207]
Stevenson, J. R., **3**, 386[67]
Stevenson, J. W. S., **1**, 749[80]; **5**, 850[149], 1123[37]
Stevenson, P., **3**, 484[21]; **4**, 753[173], 848[78], 854[93]; **5**, 1149[125]
Stevenson, P. J., **3**, 955[106]; **6**, 897[99]
Stevenson, R., **3**, 219[112], 676[75,76], 692[138], 693[142]; **8**, 566[450]
Stevenson, R. W., **8**, 828[76]
Stevenson, T., **1**, 313[118]; **2**, 924[108b]; **3**, 46[252]; **4**, 152[54], 201[14,15], 202[14,15], 204[34], 850[85]; **6**, 879[42]; **7**, 646[27]; **8**, 784[112]
Stevenson, T. T., **8**, 89[45]
Steward, K. R., **4**, 170[17]
Steward, L. J., **4**, 288[188]
Stewart, C. A., Jr., **5**, 64[50], 71[118], 1025[81]
Stewart, D., **4**, 393[187]; **7**, 228[104]
Stewart, D. G., **8**, 972[115]
Stewart, F. H. C., **6**, 645[98], 668[256], 669[256]
Stewart, J. C., **8**, 242[41]
Stewart, J. C. M., **6**, 624[141], 658[185]
Stewart, J. D., **4**, 492[71], 495[71]
Stewart, K. R., **3**, 211[33]
Stewart, K. T., **2**, 943[173], 945[173]
Stewart, L. J., **4**, 288[189], 290[189], 346[86a]
Stewart, M., **1**, 493[42,42b], 495[42]
Stewart, O. J., **3**, 499[112]
Stewart, P., **7**, 595[25], 598[25]
Stewart, R., **4**, 274[58]; **7**, 12[105,106], 578[150], 851[19]
Stewart, R. F., **3**, 683[101]; **7**, 800[29], 801[41]
Stezhko, T. V., **6**, 554[776,780,793]
Stezowski, J. J., **1**, 19[102]; **2**, 68[40]; **4**, 429[87]; **5**, 595[20], 596[20]; **7**, 160[50]
Stibbard, J. H. A., **4**, 370[36]
Stibor, I., **8**, 200[138]
Stichter, H., **1**, 218[54]; **2**, 981[23]; **4**, 877[69]; **6**, 86[99]
Stick, R. V., **6**, 43[47], 475[93], 633[9]; **8**, 394[114]
Stickler, M., **5**, 64[45]
Stieborova, I., **4**, 367[9], 376[9], 397[9]
Stieltjes, H., **4**, 877[69]; **5**, 46[37]
Stier, M., **1**, 399[7]
Stierli, F., **6**, 544[626]
Stierman, T. L., **6**, 961[70]
Stilbs, P., **1**, 293[35,36]
Stiles, A. B., **7**, 247[96]
Stiles, A. W., **8**, 297[65]
Stiles, J. I., **8**, 568[467]
Stiles, M., **2**, 153[109], 841[188], 842[188]; **3**, 726[24]; **4**, 487[50]; **8**, 374[142,143], 956[4]
Stilkerieg, B., **2**, 371[261]; **5**, 76[239]
Still, B., **4**, 1040[88], 1048[88]
Still, I. W. J., **4**, 443[186]; **5**, 441[176]; **7**, 213[104]; **8**, 404[19], 405[21,22], 407[59,60], 410[19,87], 411[19], 413[133]
Still, W. C., **1**, 51[13], 52[13,14], 108[9,10], 109[12], 110[9,12], 117[51,52], 126[9], 134[9,12,51], 153[56,57], 187[65], 188[65], 283[2], 315[2], 333[58], 336[68-70], 421[86], 427[113], 460[2], 478[148], 757[120]; **2**, 9[39], 20[83], 31[39], 55[2], 66[2,36], 68[36], 194[67], 601[37]; **3**, 41[226], 195[19,20], 196[30], 203[101], 602[220], 982[15]; **4**, 24[72,72e], 71[14], 240[47], 257[47,228], 370[37], 380[120], 681[54], 1003[67]; **5**, 151[7,12], 180[7], 809[121,122]; **6**, 11[49], 174[60], 851[127], 875[19], 876[19], 879[19,44], 888[19], 894[19], 1045[23]; **7**, 361[24], 396[23], 407[79], 614[7], 615[7], 619[24,25], 621[7], 622[7]; **8**, 540[195], 704[7], 713[7], 851[135]
Stille, C., **6**, 455[149]
Stille, J. K., **1**, 193[89], 437[155], 438[162], 439[166], 440[155,190], 442[155,176,177,178], 443[181,182,183,184], 444[185,186,187,188,189], 445[155,190,191], 446[155,194,196,197], 447[202], 457[155,182,184,186,188,190c]; **2**, 110[72], 576[68], 611[99], 727[129,130,131,132], 749[133]; **3**, 213[49], 231[254], 232[257,258,259,259b,261,263,265,268,269,271], 233[272], 250[71], 436[16], 446[75], 453[112], 455[123], 463[161,162], 466[190], 469[201], 470[201,211], 471[201], 473[201,211], 475[201,211], 476[211], 485[30], 487[49-51], 491[30], 492[94], 495[49-51,93a,94], 504[94,153], 514[94], 524[28,31], 529[28,31], 755[112], 1021[15], 1022[21], 1023[24], 1024[25], 1030[58], 1031[61], 1032[66], 1033[62], 1040[108]; **4**, 175[44], 273[48], 279[48], 280[48], 411[266c], 594[138,142,143,145,147], 614[381], 619[138,142,143,145], 633[142,143,145], 738[95], 861[113], 926[40], 930[50], 931[59], 932[60], 946[92,93], 947[93], 948[96], 979[101]; **5**, 74[205], 176[126], 480[167], 757[78], 762[102], 763[107], 779[107]; **8**, 933[45]
Stille, J. R., **1**, 743[66], 746[62], 748[72,74], 749[78], 816[78], 850[31]; **2**, 309[26]; **5**, 1115[3], 1121[28], 1122[3], 1123[3], 1124[3,28,43]
Stiller, E. T., **6**, 36[31]
Stillings, M. R., **4**, 423[5], 426[5]; **8**, 615[94], 618[94]
Štimac, A., **6**, 514[306], 543[306], 554[747]
Stingelin-Schmid, R. S., **5**, 165[83]
Stipanovic, R. D., **2**, 1026[73]; **3**, 363[85]; **5**, 945[248]
Stirchak, E. P., **6**, 14[58], 16[58]
Stirling, C., **6**, 834[34], 837[62,63]
Stirling, C. J. M., **3**, 86[42], 159[450], 164[450], 174[534,535], 176[534,535], 178[534,535], 862[3,7], 867[34,37], 872[37], 883[37], 884[37]; **4**, 37[104], 48[138,138c], 50[142,142k], 54[154], 55[156,156c], 66[138c], 78[54], 86[54b], 115[184c], 442[184]; **6**, 150[115], 638[40], 667[240], 753[117], 910[9], 924[9]; **7**, 196[13], 762[69], 766[169], 777[69a], 778[69]; **8**, 403[4], 404[4], 407[4]
Stirling, I., **6**, 669[264]
Stirrup, J. A., **8**, 28[34], 66[34]
St-Jacques, M., **3**, 382[36]
Stobart, S. R., **1**, 215[33]
Stobbe, H., **2**, 141[43]; **4**, 2[5], 99[118a]
Stober, M. R., **1**, 619[60]
Stock, L. E., **7**, 686[98]
Stock, L. M., **2**, 735[11]; **7**, 724[28]; **8**, 319[78], 798[49,64]
Stockburn, W. A., **5**, 424[94]

Stocker, F. B., **2**, 963[53]
Stocker, J. H., **3**, 564[10], 568[38,39,43]
Stockhammer, P., **3**, 817[87]
Stöckigt, J., **6**, 175[77]
Stockis, A., **5**, 1130[6], 1131[11]
Stöckl, E., **3**, 647[176], 648[176]
Stöcklein-Schneiderwind, R., **2**, 1090[72]
Stockmann, C., **4**, 562[36], 576[16]
Stockwell, D. L., **3**, 530[79], 535[79]
Stockwell, P. B., **8**, 445[33]
Stoddart, J. F., **6**, 2[2], 23[2], 440[73]
Stoepel, K., **2**, 385[321]
Stoerk, H. C., **6**, 219[123]
Stoessel, A., **4**, 861[113]
Stoffer, J. O., **5**, 513[4]
Stoffregen, A., **8**, 187[43]
Stofko, J. J., Jr., **5**, 791[42], 803[42], 971[2], 972[2], 973[2]
Stohrer, W.-D., **6**, 1044[18]
Stoilava, V., **6**, 74[37]
Stojda, R. J., **7**, 160[50]
Stojiljkovic, A., **3**, 380[13], 905[136,139]; **7**, 229[112], 231[142]
Stokes, L. L., **8**, 764[7], 770[7], 773[7]
Stokker, G. E., **2**, 971[92]
Stokkingreef, E. H. M., **6**, 114[73]
Stokoe, J., **6**, 707[45]
Stolarski, V., **4**, 980[110], 982[110]
Stolberg, U. G., **8**, 447[108,109]
Stoll, A., **1**, 95[80]; **7**, 446[68], 704[10]
Stoll, A. P., **2**, 765[71]
Stoll, A. T., **1**, 448[206]; **2**, 449[48]; **3**, 251[99], 254[99], 463[156]; **4**, 892[145], 895[161], 903[198]
Stoll, M., **3**, 615[8], 644[145]; **7**, 543[19], 546[19]
Stolle, W. T., **1**, 357[5], 378[5,105]; **8**, 948[146]
Stoller, H., **5**, 1154[155]
Stoller, H.-J., **3**, 168[503], 169[503]
Stolyarov, B. V., **3**, 305[73]
Stolz, F., **4**, 289[191]
Stolzenberg, A. M., **5**, 1131[17]
Stone, A. J., **8**, 89[43]
Stone, C., **5**, 57[50,50b], 59[50b]; **8**, 681[78], 683[78], 689[78], 693[78]
Stone, D. B., **7**, 15[149]
Stone, F. G. A., **1**, 2[7,8], 125[84], 139[3], 211[2], 212[2], 214[2,25], 222[2], 225[2], 231[1], 292[24], 310[103], 428[121], 429[121], 457[121], 580[1]; **2**, 712[42]; **3**, 208[1,11], 210[11,11a], 219[11a], 228[214], 234[11a], 380[10], 436[9,13], 524[33]; **4**, 518[2], 521[2], 735[84], 770[84], 914[1], 922[1], 925[1], 926[1], 932[1], 939[73], 941[1], 943[1]; **5**, 46[39], 56[39], 272[1], 641[131], 1134[45], 1136[54]; **6**, 831[11], 832[12], 848[11], 865[12]; **7**, 335[28], 594[5], 595[5], 598[5], 614[3], 629[48], 816[6a,b], 824[6], 825[6], 827[6a], 829[6a], 831[6a], 832[6a], 833[6a]; **8**, 99[110], 100[114], 443[1], 557[382], 674[33], 708[42], 715[42], 717[42], 728[42], 766[18]
Stone, G. B., **2**, 759[29]
Stone, G. R., **8**, 143[55]
Stone, H., **4**, 287[178]; **6**, 213[85], 214[99,100]
Stone, J. F., Jr., **4**, 85[76]
Stone, K. J., **5**, 241[4], 243[12]
Stone, K. M., **8**, 526[28]
Stone, M. J., **4**, 545[126]
Stone, M. P., **7**, 374[77e]
Stone, T. J., **3**, 661[21]
Stoneberg, R. L., **8**, 264[45]
Stones, P. A., **7**, 555[71], 564[71]
Stoodley, R. J., **2**, 1048[10]; **5**, 374[107,107a], 376[107b]; **7**, 96[199], 112[198], 352[33], 820[26]; **8**, 409[83]
Stopp, G., **2**, 368[234]
Storace, L., **6**, 787[97]
Storck, W., **1**, 23[123]
Storer, A. C., **6**, 436[10-12,16-18], 451[129], 455[154]
Storesund, H. J., **4**, 1035[40]
Storey, P. R., **7**, 824[40]
Storflor, H., **4**, 1063[173]
Stork, G., **1**, 19[105], 361[34], 412[51], 462[17], 463[17,33], 542[8], 543[8], 544[8], 547[8,62], 548[8], 550[8], 552[8,62], 553[8,85-87], 555[8,8b], 556[8], 557[8], 560[8], 619[63], 749[317], 771[193], 807[317]; **2**, 106[50,51], 109[63], 110[63], 183[13,18], 184[24,26], 185[27], 187[38,39], 420[24], 504[2], 506[2], 509[2], 514[48], 524[2], 599[22], 844[197], 1021[49], 1022[49]; **3**, 2[8], 8[35,37], 11[8,50], 14[78], 15[78,80], 16[8], 17[8], 21[126], 22[132], 23[140], 24[144], 26[8], 28[170], 30[140,170,179], 33[140], 34[140], 35[140], 48[259,260], 125[302], 134[302], 197[39], 198[44-46], 341[2], 351[39], 360[2], 602[218,220], 607[218], 638[89], 896[68], 902[118], 1053[29]; **4**, 6[21,21a,23,23a], 7[21a], 8[27,27a], 12[38], 14[46,46b], 24[72,72d,e,73,73a-c], 27[78,80], 44[78a], 45[128], 63[72d], 71[16b], 74[41], 91[90], 100[41], 111[152a], 113[16b,168], 128[222], 139[3], 162[92], 191[112], 218[141,142], 240[39,40], 254[39,40], 258[235], 304[354], 313[463], 373[68], 650[427], 680[49], 721[29], 738[29], 753[29], 765[29], 790[38], 792[66,68], 794[73], 796[90,93,95,99], 820[212], 823[229], 1004[68], 1005[80,81,84], 1018[80,81]; **5**, 16[110], 130[40], 171[117], 351[82], 519[30], 561[81], 687[55], 839[78]; **6**, 206[44], 647[114], 655[160], 681[333], 682[339], 689[385], 703[1-4], 714[3], 719[129], 722[140], 727[192], 838[67], 902[132]; **7**, 229[118], 439[33], 647[33], 648[41], 731[54]; **8**, 163[41], 448[147], 478[47], 479[47,47a,b], 481[47], 499[41], 525[15], 526[15], 527[39,40,46,47], 528[55,59], 531[114], 534[157], 544[262,263], 557[379], 564[441], 645[44], 814[17], 986[13]
Stork, L., **3**, 648[186]
Storm, D. L., **8**, 526[26]
Storme, P., **7**, 301[58]
Storr, R. C., **4**, 488[57], 510[164], 1075[31], 1104[210]; **5**, 639[123]; **7**, 508[310], 743[61,65]
Story, P. R., **7**, 96[86]; **8**, 543[247]
Stössel, D., **2**, 607[73], 619[73,147], 623[73,147,160]
Stostakovskii, M. F., **4**, 316[536]
Stothers, J. B., **3**, 639[79]; **5**, 637[112]; **7**, 673[25]
Stotter, P. L., **2**, 198[86]; **3**, 11[51], 99[192], 103[192], 107[192]; **7**, 120[17], 123[17]; **8**, 584[16], 837[12], 842[12]
Stotz, D. S., **1**, 3[18,20], 42[20c]
Stouch, T. R., **5**, 425[100]
Stoudt, G. S., **7**, 257[50], 268[124]
Stout, D. M., **2**, 377[282]; **3**, 181[552], 512[197]; **8**, 92[68], 584[23]
Stout, D. S., **1**, 555[116]
Stout, T. J., **1**, 401[14]; **5**, 736[145], 737[145]
Stoutland, P. O., **7**, 4[15]; **8**, 669[6], 670[6]
Stöve, H., **8**, 904[57,57a], 910[57]
Stover, C. S., **1**, 293[34]
Stover, L. R., **4**, 445[210]
Stowe, M. E., **4**, 967[56]
Stowell, J. C., **2**, 55[1,1d], 61[1d], 67[1d], 68[1d]; **3**, 1[5], 53[5], 54[5], 193[3], 208[3]; **4**, 3[7], 4[7], 60[7d], 71[18], 117[190], 139[4], 163[95]; **6**, 116[90], 452[132]; **7**, 661[46], 741[46], 746[46]
Stowell, M. H. B., **3**, 201[81]
Stowers, J., **5**, 402[6]; **6**, 550[677]; **8**, 170[74]
Stoyanovich, F. M., **8**, 609[51]
Strachan, C. H., **2**, 369[254]
Strachan, P., **4**, 45[126,126c]; **5**, 686[51], 687[51], 688[51]
Strachan, R. G., **3**, 644[167]
Stradi, P., **6**, 712[79], 713[80a]
Stradi, R., **6**, 582[993], 712[73]
Stradyns, J., **8**, 595[77]
Strahm, R. D., **7**, 599[72]
Strakov, A. Ya., **3**, 831[62]
Strakova, I. A., **3**, 831[62]

Straley, J. M., **3**, 208[8]
Strand, G., **6**, 658[188]
Stransky, W., **6**, 174[58,61]
Strappaveccia, G. P., **2**, 363[195]
Strasak, M., **7**, 451[25]
Strasser, B. L., **3**, 381[32]
Stratford, M. J. W., **7**, 221[35]
Strating, J., **3**, 649[208], 832[68a], 909[151]; **5**, 440[175]; **7**, 98[97]
Straub, A., **2**, 456[66], 457[66], 460[66], 461[66], 463[78]
Straub, H., **4**, 373[85]; **7**, 632[57]; **8**, 851[132]
Straub, J. A., **2**, 42[149], 45[149]; **6**, 843[86]; **7**, 371[69]
Straub, P. A., **4**, 537[98], 538[98]
Straus, D., **1**, 748[72], 749[78], 816[78]; **4**, 979[101]; **5**, 1115[3], 1122[3,31], 1123[3], 1124[3]
Straus, F., **4**, 47[133]
Strauss, H., **2**, 144[59]
Strauss, H. E., **8**, 963[45]
Strauss, H. F., **5**, 41[28], 43[32]
Strauss, J. U. G., **4**, 629[403,404]
Strauss, J. V., **4**, 609[328]
Strauss, M. J., **4**, 426[43], 444[43]; **6**, 552[695]
Strauss, R. S., **2**, 838[176]
Strausz, O. P., **3**, 891[43], 892[43,47,48]
Strautinya, A. K., **8**, 606[24], 607[24]
Strazewski, P., **2**, 100[9]
Strazzolini, P., **1**, 832[115]
Strebelle, M., **6**, 517[323], 544[323], 551[684]
Streck, R., **5**, 1117[15], 1118[17], 1120[24]
Strecker, A., **1**, 185[56]; **2**, 29[106]; **7**, 549[45]
Streckert, G., **6**, 659[195]; **7**, 244[69,70]
Street, D. L., **4**, 347[95]
Street, J. D., **8**, 618[118]
Street, L. J., **4**, 1089[128], 1092[128], 1093[128]
Street, S. D. A., **1**, 801[302]; **6**, 994[98]
Streg, P., **8**, 843[51], 844[51]
Strege, P. E., **1**, 794[275]; **4**, 588[70,72], 614[380], 615[72,380], 628[380], 631[72], 652[433]; **6**, 154[151], 991[87], 1021[50]; **8**, 843[62], 993[60], 994[60]
Streib, H., **3**, 924[41]
Streicher, W., **4**, 36[102]
Streichfuss, D., **5**, 1148[123]
Streinz, L., **3**, 643[128]
Streith, J., **4**, 565[44], 754[175]; **5**, 417[65], 418[67], 419[74], 420[75], 731[130a]; **8**, 395[129], 652[79]
Streitweiser, A., Jr, **1**, 41[197], 506[6,7], 510[6], 528[118]; **2**, 757[18]; **6**, 133[4], 175[67], 727[199,200]; **8**, 332[38], 339[38]
Strekowski, L., **1**, 789[262b], 790[262]; **5**, 531[73]
Streng, K., **8**, 459[238]
Stretch, W., **8**, 652[78]
Stretton, G. N., **8**, 851[131]
Streukens, M., **7**, 805[67]
Stribblehill, P., **8**, 162[28]
Strickland, J. B., **3**, 788[53]
Strickland, R. C., **6**, 134[30]; **8**, 501[54], 502[54]
Strickland, S. M. S., **5**, 531[73], 537[98]
Strickler, H., **5**, 10[76]
Strickler, J. R., **5**, 1145[104]
Strickler, R., **5**, 381[118]
Stridsberg, B., **6**, 489[96]; **7**, 196[30], 199[30]
Striepe, W., **3**, 331[200b]; **4**, 238[5]
Strijtveen, B., **6**, 21[79]
Strike, D. P., **6**, 1042[4]; **7**, 621[31]
Stringer, O. D., **1**, 838[157]; **7**, 162[63,67], 176[67], 772[291], 778[398], 779[425,426]; **8**, 395[134]
Stringham, R. A., **6**, 134[18]; **8**, 846[85]
Strobel, M. P., **6**, 712[74]
Stroecker, M., **7**, 5[24]
Stroezel, M., **3**, 634[28]
Stroh, R., **4**, 272[29], 274[29], 276[29], 277[29], 278[29]; **6**, 204[4], 243[99]; **7**, 741[48], 747[48]
Strohmeier, W., **4**, 519[16], 520[16]; **8**, 444[9], 445[20,29], 446[75-77,80,83,84], 450[166], 453[29,75,192]
Strohmeyer, T. W., **5**, 96[121], 98[121]; **6**, 538[559]
Strojanac, Ž. Q., **5**, 513[4]
Strologo, S., **8**, 683[90]
Stromar, M., **7**, 232[158]
Strömberg, R., **6**, 620[133]
Strömqvist, M., **6**, 705[24]
Strong, F. M., **2**, 879[41]
Strong, J. S., **6**, 431[113]
Strong, P. L., **8**, 720[125]
Stroömberg, R., **6**, 620[133]
Stropnik, C., **6**, 480[114], 554[715]
Stroud, E. D., **4**, 428[74]
Stroud, H. H., **8**, 84[11]
Strouf, O., **8**, 544[272]
Strozier, R. W., **1**, 357[8]; **2**, 476[4]; **3**, 31[186]; **4**, 47[134], 1098[170]; **5**, 452[58], 454[58]
Strub, H., **5**, 418[67]
Strube, E., **8**, 798[57]
Strubin, T., **5**, 394[146], 395[146]
Strubinger, L. M., **5**, 1090[89]
Struble, D. L., **4**, 23[70], 784[14]; **6**, 141[58,59]
Struchkov, Yu. T., **4**, 218[147]; **5**, 1055[46], 1174[34]
Strugnell, C. J., **8**, 446[88]
Strukul, G., **7**, 108[174], 426[148d]
Strunk, J., **4**, 987[147]
Strunk, R. J., **8**, 798[55]
Strunk, S., **6**, 190[200]
Strunz, G. M., **2**, 764[65]
Struve, G., **3**, 522[13]
Struve, W. S., **1**, 844[3a]; **3**, 887[2], 888[2], 891[2], 897[2]
Struwe, H., **3**, 343[18], 353[51], 354[51]
Stryer, L., **8**, 185[21]
Stryker, J. M., **4**, 254[184]; **8**, 550[330]
Strzelecka, H., **4**, 55[155]
Stuart, A. D., **5**, 144[104]
Stuart, A. P., **3**, 327[169]
Stuart, J. D., **7**, 801[37,42]
Stuart, S. J., **7**, 350[19]
Stuart-Webb, I. A., **3**, 846[47]
Stubbe, J., **5**, 736[142i,j]
Stubbe, M., **4**, 78[52c]
Stubbs, C. E., **6**, 964[78]
Stubbs, F. J., **8**, 371[108]
Stüber, O., **2**, 321[2]
Stüber, S., **2**, 1094[89], 1095[89]
Stuchal, F. W., **7**, 660[42], 882[172]
Stucki, C., **2**, 578[85], 996[47], 1057[70], 1077[154]; **7**, 230[133]
Stucky, G., **1**, 373[91], 375[91], 376[91]; **5**, 79[289]
Stucky, G. D., **1**, 9[42], 10[53], 11[56], 12[59], 13[70,72], 16[87], 19[99], 25[130]
Studabaker, W. B., **4**, 976[100], 980[100j], 984[123,124]; **5**, 904[49], 905[49], 1084[58], 1086[70]
Studebaker, J., **5**, 586[205]
Studenikov, A. N., **7**, 477[74]
Studnev, Yu. N., **7**, 493[196]
Studnicka, B. J., **4**, 868[18]
Studt, W. L., **5**, 128[32], 130[32], 864[256]
Stuhl, L. S., **8**, 454[202]
Stuhl, O., **1**, 583[8,8b]
Stühler, H., **6**, 180[128]

Stuhmiller, L. M., **3**, 226[193]
Stull, P. D., **5**, 252[43], 257[43]; **6**, 128[163], 281[146]
Stults, B. R., **5**, 438[162]
Stults, J. S., **5**, 436[158,158g], 437[158f], 442[158]
Stumpf, B., **7**, 62[50b], 429[157a]
Stumpf, W., **7**, 832[70]
Stumpp, M., **6**, 51[111], 54[127], 62[127]
Stunic, Z., **6**, 554[736]
Stupnikova, T. V., **6**, 494[132]
Sturc, A., **8**, 292[44]
Sturgeon, M. E., **5**, 126[22], 127[22], 128[22]
Sturgess, M. A., **5**, 85[9], 1068[12,13]
Sturgis, B. M., **2**, 406[48]
Sturkovich, R. Ya., **8**, 771[45,48]
Sturm, B., **6**, 233[49]
Sturm, E., **2**, 782[30]
Sturm, H. J., **3**, 893[53b], 896[71]; **4**, 1081[75]; **5**, 488[194]
Sturm, K., **6**, 554[748]
Sturm, W., **5**, 1066[7], 1096[124], 1098[125], 1099[125], 1112[125], 1113[125]
Sturmer, D., **3**, 380[4], 735[19]
Stürmer, R., **1**, 832[108]; **2**, 39[137], 40[137c], 44[137c]; **3**, 797[90]; **6**, 864[195]
Sturtz, G., **2**, 64[26]; **3**, 199[55], 1052[27]; **6**, 116[86], 126[150], 845[94]; **7**, 745[80]
Stürzenhofecker, F., **3**, 396[107], 397[107]
Stüssi, R., **6**, 937[115], 941[115]; **7**, 124[45]
Stütten, J., **4**, 1001[36]
Stütz, A., **8**, 29[39], 66[39]
Stutz, A. E., **6**, 22[81]
Stütz, P., **3**, 126[311,312]; **6**, 919[39]
Stylianides, N., **6**, 448[106]
Stytsenko, T. S., **6**, 576[972]
Su, B. M., **1**, 2[9], 43[9]
Su, S., **4**, 73[32], 111[158a]
Su, T., **5**, 429[112]
Su, T. M., **4**, 1016[208]
Su, W., **4**, 129[223a], 976[98]; **5**, 356[88]; **8**, 891[149]
Su, W.-g., **2**, 800[16]; **4**, 14[46,46a]; **7**, 373[75], 400[50]; **8**, 824[61]
Su, W.-Y., **8**, 26[24,28], 30[28], 36[24,28], 37[24,28], 39[24], 40[28], 43[28], 44[24,28], 46[24,28], 55[28], 66[24,28], 67[24], 357[204]
Suami, T., **2**, 323[28], 333[28], 386[331]; **5**, 839[76]; **6**, 677[323]; **7**, 713[70]; **8**, 388[63], 874[22]
Suarato, A., **8**, 358[197]
Suárez, E., **2**, 1049[16]; **4**, 342[61], 375[98a,b], 388[98,98a,b], 408[98b], 409[98a], 814[187], 817[203]; **7**, 41[15], 157[34], 495[210], 722[19], 723[19], 725[19]
Suau, R., **3**, 585[133], 591[171]; **4**, 505[139,140], 513[179]; **6**, 487[76], 489[76]; **8**, 335[66]
Subba Rao, B. C., **8**, 244[60], 249[60], 253[60], 263[23], 265[23], 950[161]
Subba Rao, G. S. R., **2**, 761[51]; **3**, 17[91]; **4**, 6[21], 111[155e], 796[94]; **5**, 814[138]; **8**, 115[63], 212[18], 490[4], 492[4], 499[42,46], 503[66,69], 510[95], 511[95,99], 514[110], 515[115], 524[2], 530[2], 531[2], 910[83]
Subba Rao, H. N., **5**, 802[83], 810[83]; **7**, 279[170], 375[80], 844[61], 845[61]
Subbarao, R., **6**, 690[389]
Subba Rao, Y. V., **8**, 372[117]
Subero, C., **8**, 227[116]
Subirato, J. R., **8**, 125[94]
Sublett, R. L., **2**, 740[59]
Subrahamanian, K. P., **3**, 595[191]
Subrahmanyam, C., **3**, 799[99]; **8**, 710[54], 711[69], 717[101], 721[54]
Subrahmanyam, D., **5**, 225[95], 233[95], 667[44]
Subrahmanyam, G., **5**, 597[29], 649[21], 658[21]
Subrahmanyam, S., **5**, 594[3], 595[3], 596[3], 603[3]
Subramanian, C. S., **4**, 239[18], 249[18], 257[18], 261[18]; **5**, 348[74a]; **7**, 453[64,82], 454[64,82]
Subramanian, L. R., **6**, 19[68], 72[24]; **8**, 349[137], 934[53]
Subramanian, N., **4**, 15[50]; **5**, 71[165]
Subramanian, P., **7**, 765[154]
Subramanian, R., **3**, 602[224]; **6**, 475[93]; **8**, 820[47]
Subramanian, R. S., **5**, 834[54]
Subramanyam, R., **5**, 74[208]
Suchan, V., **8**, 161[17], 176[129]
Suchanek, P., **2**, 182[5], 477[6]; **6**, 719[128], 720[128]
Suchiro, T., **2**, 163[145]
Suchkova, M. D., **8**, 769[30]
Suciu, D., **8**, 656[88]
Suciu, E. N., **7**, 751[137]
Suckling, C. J., **2**, 456[26]; **7**, 54[6], 145[164]; **8**, 96[94]
Suckling, I. D., **2**, 851[225]; **7**, 262[75]
Suckling, K. E., **7**, 54[6], 145[164]
Sucrow, W., **4**, 51[144b]; **5**, 837[66]; **6**, 116[90], 673[288], 837[59], 856[153,157]; **7**, 741[46], 746[46]; **8**, 544[274]
Sucsy, A. C., **6**, 968[110]; **8**, 950[173]
Suda, H., **8**, 170[71,95]
Suda, M., **4**, 125[216], 960[37], 962[37], 963[37]; **6**, 172[16], 556[827], 856[155]
Suda, S., **2**, 810[66], 851[66]
Suda, Y., **2**, 656[154]; **5**, 100[142]
Sudani, M., **1**, 566[214]; **4**, 10[34], 113[164], 229[239]
Sudau, W., **8**, 821[49]
Sudborough, J. J., **4**, 286[171]
Sudcliffe, F., **8**, 330[27]
Suder, B. J., **2**, 286[64]
Sudhakar, A. R., **7**, 491[180]
Sudheendranath, C. S., **2**, 60[18]
Sudo, T., **2**, 19[77,77c], 572[46]
Sudoh, K., **2**, 736[26]
Sudoh, R., **1**, 642[120], 645[120], 672[120], 708[120]; **4**, 23[70], 36[102,102a,b]; **6**, 442[88]; **7**, 751[138,139]
Sudol, J. J., **3**, 623[33]
Sudow, I., **4**, 30[88,88n], 239[17], 261[17]; **5**, 534[95]
Sudweeks, W. B., **3**, 124[282], 125[282], 128[282], 129[282]
Suemitsu, R., **1**, 569[249]; **2**, 505[8]; **3**, 34[195], 35[202], 1029[52]; **4**, 115[178]; **8**, 550[332], 563[429]
Suemune, H., **2**, 209[109]; **4**, 159[81]; **6**, 1052[41,42a,b]; **7**, 672[18]; **8**, 190[64], 191[95], 198[133,134]
Suen, R., **8**, 205[156]
Suen, Y.-H., **2**, 297[91]; **4**, 302[335]
Suenaga, T., **1**, 766[160]; **4**, 203[31]
Suess, R., **8**, 274[131]
Suëtaka, W., **8**, 149[113-115]
Sueyoshi, S., **8**, 390[82], 640[22]
Suffert, J., **1**, 223[71], 365[41]; **4**, 538[105]; **5**, 736[142f]
Suffness, M. I., **2**, 159[127]
Suga, A., **2**, 323[28], 333[28]
Suga, H., **1**, 770[190]; **3**, 201[84], 769[168]
Suga, K., **4**, 313[460], 1023[257]; **6**, 566[929]
Suga, S., **1**, 78[20], 223[76], 224[76a], 317[145,155], 319[145], 320[155]; **3**, 300[46], 302[46], 313[105], 314[108], 315[113], 318[129], 769[169]; **5**, 438[161], 442[185,185a], 532[84]
Suga, T., **7**, 84[3]
Sugahara, K., **4**, 313[460], 1023[257]
Sugahara, S., **2**, 89[38]
Sugahara, T., **5**, 837[69]; **7**, 745[78]
Sugai, K., **6**, 979[29]
Sugai, S., **1**, 890[145]; **2**, 1042[113]; **6**, 524[354], 734[12], 741[12]
Sugai, T., **4**, 510[176]; **6**, 677[318,318b]; **7**, 57[32]; **8**, 49[116], 66[116]

Sugakara, T., **6**, 678[324]
Sugano, H., **6**, 814[94]
Sugano, K., **2**, 801[37]; **5**, 834[57]
Suganuma, H., **4**, 247[102], 252[102], 259[102]
Sugasawa, K., **2**, 748[122]
Sugasawa, S., **8**, 292[44], 964[52]
Sugasawa, T., **2**, 244[23,29], 245[23], 478[15], 479[16], 708[5], 748[124]; **6**, 444[97]
Sugasawara, R., **5**, 855[192]
Sugawara, F., **8**, 48[112], 66[112]
Sugawara, R., **3**, 446[77]
Sugawara, T., **1**, 670[186]; **4**, 384[143], 837[9]; **7**, 771[279], 773[279]
Sugaya, T., **2**, 444[22]; **7**, 209[90]
Suggett, A., **6**, 441[82]
Suggs, J. W., **1**, 10[55]; **2**, 599[18]; **4**, 738[93]; **6**, 652[144], 659[144], 677[316,319]; **7**, 103[139], 260[65,85]; **8**, 446[93], 452[93], 534[157], 800[66]
Sugi, Y., **4**, 939[76]; **8**, 881[76,77], 882[76]
Sugihara, H., **4**, 251[147]; **6**, 926[68], 927[68], 1022[63]; **7**, 197[17], 667[79]; **8**, 189[63]
Sugihara, T., **2**, 372[271]; **3**, 224[181]; **5**, 468[129], 862[246]
Sugihara, Y., **1**, 825[50]; **2**, 784[38], 1052[51]; **6**, 438[42]; **8**, 817[32]
Sugimori, A., **5**, 1139[75]; **6**, 533[504]
Sugimori, J., **5**, 377[110], 378[110a]
Sugimori, T., **2**, 463[82,83], 464[82,83]
Sugimoto, A., **8**, 397[140]
Sugimoto, H., **4**, 14[47,47j], 30[91], 111[154b,g]; **7**, 766[170,171]
Sugimoto, K., **1**, 803[305]; **4**, 1023[257]; **6**, 157[173]; **8**, 889[126]
Sugimoto, M., **6**, 265[42]; **7**, 84[3]
Sugimoto, T., **4**, 405[250a]; **6**, 134[31]; **7**, 443[51a], 778[409]; **8**, 155[203], 170[81]
Sugimura, H., **2**, 564[8]; **3**, 446[82], 459[136], 460[136], 461[136], 462[150], 472[222], 513[208]
Sugimura, T., **3**, 983[20], 984[20], 1058[41], 1062[41]; **4**, 975[91]; **7**, 261[69]
Sugimura, Y., **2**, 649[103], 1059[76]; **6**, 538[562]
Sugino, E., **5**, 725[118]; **8**, 64[207a], 65[207a], 66[207], 237[17], 240[17], 249[17], 369[77], 389[75]
Sugino, H., **7**, 59[38]
Sugino, K., **2**, 74[74]; **3**, 135[348], 136[348], 139[348], 141[348], 144[348], 649[204]; **8**, 321[93]
Suginome, H., **1**, 262[37], 894[157]; **2**, 110[73], 114[73]; **3**, 469[206], 470[205,206,207], 473[205,206,207], 489[59,61,63], 495[59,61,63], 496[99], 498[59,99], 504[61,63], 511[59,61,63,99], 515[59,61,63,99], 530[58]; **4**, 290[205], 404[245], 812[177], 817[202,204,205]; **5**, 195[8], 197[8]; **6**, 766[20,21], 1067[104]; **7**, 834[79], 835[83,84]; **8**, 528[67], 604[3]
Suginome, J., **7**, 683[88]
Sugioka, H., **8**, 190[80]
Sugishita, M., **3**, 757[123]
Sugita, K., **3**, 300[38], 400[120]; **6**, 93[130], 237[66], 254[160]
Sugita, N., **1**, 664[201], 672[201], 712[201], 714[201], 828[75], 862[77]; **3**, 1026[40]; **6**, 467[52]; **7**, 776[356]
Sugita, T., **3**, 311[97], 313[106], 315[113], 769[169]; **4**, 302[337]; **5**, 492[247]
Sugiura, M., **4**, 45[130], 510[175]; **5**, 95[89]; **6**, 801[34]
Sugiura, S., **3**, 222[144]; **4**, 13[44,44c], 253[169], 262[308]; **6**, 266[50], 837[60], 942[154], 944[154]
Sugiura, T., **4**, 592[124]; **6**, 20[77]
Sugiura, Y., **1**, 232[12,13], 233[13], 234[12,13,21], 235[27,28], 253[9], 276[9], 278[9], 332[53,54]; **2**, 312[35]
Sugiuza, M., **2**, 1089[57]
Sugiyama, H., **5**, 736[142g]; **8**, 190[74], 244[65], 250[99], 445[15]
Sugiyama, K., **7**, 346[13]
Sugiyama, N., **2**, 153[105]
Sugiyama, S., **5**, 619[12,13], 620[12,13,16]; **7**, 415[115a]
Sugiyama, T., **3**, 421[59], 422[59]
Sugiyama, Y., **4**, 379[114]
Sugizaki, T., **4**, 823[231]
Sugumi, H., **4**, 161[91,91c]
Suh, E. M., **1**, 420[83], 568[230]; **4**, 53[150]
Suh, H., **1**, 744[58]; **5**, 1123[38]
Suh, J., **2**, 406[46]; **8**, 621[144]
Suh, Y.-G., **5**, 856[211], 872[211], 888[27]
Suhadolnik, J. C., **7**, 404[66]; **8**, 508[87]
Suhara, H., **4**, 753[164]
Suhara, Y., **3**, 644[139]
Sukata, K., **6**, 229[28]
Sukenik, C. N., **4**, 303[340]; **8**, 64[215], 394[118], 856[172]
Sukhai, R. S., **6**, 426[76,78], 480[111], 482[125]
Sukh Dev, **3**, 402[127]
Sukhoverkhov, V. D., **3**, 383[47]
Sukoca, H., **5**, 308[96]
Suksamrarn, A., **7**, 104[146]
Sukumaran, K. B., **2**, 760[47]; **5**, 740[152]
Sukuta, K., **6**, 234[50]
Sulbaran de Carrasco, M. C., **3**, 594[187]
Suld, G., **3**, 327[169]
Suleman, N. K., **4**, 736[89], 738[89]
Suleske, R. T., **6**, 787[103]
Sulikowski, G. A., **4**, 212[97]
Sulimov, I. G., **4**, 85[77d]
Sulkowski, T. S., **7**, 695[37]; **8**, 64[200], 66[200]
Sullins, D. W., **2**, 141[39]
Sullivan, A. C., **1**, 16[90], 17[91,208], 39[191]; **4**, 170[15]
Sullivan, D., **2**, 187[41]
Sullivan, D. E., **3**, 643[119]
Sullivan, D. F., **1**, 789[260]; **2**, 279[12], 604[51], 616[136], 803[39]; **5**, 92[69]; **7**, 144[152]; **8**, 109[11], 112[11], 113[11]
Sullivan, H. R., **4**, 31[92,92d]
Sullivan, J. A., **4**, 390[173b]
Sullivan, J. W., **3**, 644[154]
Sullivan, M. F., **5**, 1144[102], 1146[102]
Sullivan, P. S., **8**, 447[96]
Sulmon, P., **2**, 423[32,33], 424[32,35]; **3**, 857[90]
Sulsky, R., **6**, 112[68]
Sulsky, R. B., **3**, 33[190]
Sulston, J. E., **6**, 650[127]
Sultana, M., **3**, 512[203]; **6**, 725[167]
Sultana, N., **6**, 725[167]
Sultanbawa, M. U. S., **2**, 158[125]; **3**, 353[47]
Sultanov, R. A., **2**, 580[98], 581[98]; **8**, 772[56,57]
Sultanov, R. M., **8**, 697[133,134], 698[133,140]
Sum, F. W., **3**, 99[188], 107[188], 110[188]
Sumarokova, T. N., **6**, 432[116]
Sumi, K., **2**, 718[79]; **3**, 135[341,342,343], 136[341,342,343], 137[341,342], 217[79]; **5**, 945[249]; **6**, 1018[42]
Sumi, M., **7**, 86[16a]
Sumida, Y., **2**, 387[335]
Sumiki, Y., **3**, 715[39]
Sumita, M., **7**, 262[82]
Sumitani, K., **3**, 228[216,218], 436[3], 437[25,26], 440[25], 448[25], 449[25], 450[25,26], 451[25,106], 452[25,106], 503[145], 524[27]; **8**, 773[67]
Sumitomo, H., **3**, 99[190]
Sumiya, F., **3**, 969[135]; **6**, 893[87]
Sumiya, R., **5**, 1157[170], 1183[56]; **6**, 17[63], 18[63]; **7**, 645[19,20]; **8**, 788[121]
Sumiya, T., **1**, 438[160]; **3**, 12[62], 453[113]
Summerbell, R. K., **4**, 310[436,437,438]
Summerhays, L. R., **6**, 134[18]; **8**, 845[79], 846[79], 970[99]

Sümmermann, K., **5**, 1140[77], 1141[78,79]
Summers, A. J. H., **8**, 664[122]
Summers, D. R., **6**, 766[22]
Summers, J. B., **1**, 385[120], 386[120]
Summers, J. C., **7**, 138[126]
Summers, L. A., **2**, 765[78]
Summers, M. C., **8**, 52[142], 66[142]
Summers, S. T., **8**, 356[189], 357[189]
Summersell, R. J., **3**, 486[39], 491[39], 495[39], 498[39], 503[39]; **5**, 725[117]
Summons, R. E., **3**, 671[62]
Sumners, D. R., **1**, 836[145]
Sumpter, C. A., **2**, 744[91]
Sun, C., **2**, 145[64]
Sun, C.-Q., **1**, 766[159]
Sun, D.-J., **5**, 333[45]
Sun, J., **4**, 590[93], 592[93]
Sun, R. C., **5**, 560[76]
Sun, S. F., **6**, 937[114]
Sunagawa, M., **5**, 98[124]
Sunami, M., **6**, 902[131]
Sunay, U., **6**, 27[118]; **7**, 454[97]
Sund, E. H., **6**, 534[523]
Sundar, N. S., **8**, 503[66,69], 510[95], 511[95,99]
Sundar, R., **4**, 443[187]
Sundaram, P. K., **6**, 88[106], 89[106]
Sundararajan, G., **4**, 229[230]
Sundararaman, P., **3**, 51[270]; **4**, 370[38], 372[38], 390[38]; **5**, 225[97]; **7**, 259[58], 821[27]; **8**, 505[82], 507[82], 508[87]
Sundberg, J. E., **4**, 459[70], 462[107], 464[70,112], 465[70,112], 467[112], 479[70]
Sundberg, K. R., **8**, 724[169,169d]
Sundberg, R. J., **2**, 885[52]; **4**, 717[8], 1040[73], 1043[73]; **5**, 947[286]; **6**, 122[135]; **7**, 27[70], 35[103,104], 851[30]; **8**, 37[99], 42[99], 66[99], 584[25], 612[69], 618[117], 632[69]
SunderBabu, G., **4**, 793[70]
Sunderbery, R. J., **6**, 509[253]
Sundermann, R., **6**, 244[113]
Sundermeyer, W., **1**, 328[25], 343[120], 345[120], 390[146]; **5**, 440[172]; **6**, 237[58]
Sundin, A., **3**, 763[149], 768[149]
Sundin, C. E., **6**, 1055[52a]
Sundram, U. N., **6**, 1044[16b], 1048[16]
Sundt, E., **6**, 1059[64]
Suneel, Y., **2**, 851[224]
Sung, T. V., **5**, 180[144]
Sung-Nung Lee, **7**, 480[105], 482[105]
Sunita, M., **5**, 95[101]
Sunjic, V., **2**, 406[47]; **6**, 85[87]; **7**, 232[158]; **8**, 152[169], 459[228], 460[246], 535[165]
Supniewski, J. V., **6**, 230[33]
Sura, T. P., **6**, 110[57]
Surbeck, J.-P., **8**, 134[36]
Surberg, B. W., **5**, 1025[84]
Surendrakumar, S., **5**, 257[59]
Suri, S. C., **1**, 700[257], 882[123]; **5**, 810[128], 812[128]; **6**, 980[31]; **7**, 746[88]
Surkar, S., **7**, 823[35]
Surleraux, D., **1**, 664[169], 665[169], 669[169], 670[169]; **3**, 111[231]
Surov, I., **8**, 645[46]
Surridge, J. H., **7**, 530[27], 531[27]; **8**, 475[17]
Surtees, J. R., **1**, 78[19], 325[2]
Sury, E., **4**, 41[119,119a]
Surya Prakash, G. K., **2**, 523[88]; **8**, 216[57,62,63], 217[63], 404[14], 568[467]
Suryawanshi, S. N., **3**, 406[144]; **4**, 162[92]; **5**, 829[26]; **6**, 164[196]; **8**, 339[90,93], 851[135]
Surzur, J.-M., **4**, 780[4], 790[4], 801[4], 807[149,150]; **7**, 92[40], 499[233]; **8**, 798[47]
Süs, O., **3**, 904[129,130,131,132]
Suschitzky, H., **2**, 736[28], 765[76]; **4**, 426[53]; **5**, 451[23], 470[23], 491[23], 492[23]; **6**, 220[125], 535[540,541], 538[540,541], 557[834], 570[951], 704[13], 922[51]; **7**, 21[6], 750[125]; **8**, 65[211], 66[211], 98[106], 386[55], 584[15]
Suschitzky, J. L., **4**, 55[157]
Suseela, Y., **8**, 720[130]
Sushchinskaya, S. P., **8**, 769[25], 771[25]
Susi, P. V., **6**, 227[29]
Susilo, R., **6**, 736[22]
Susla, M., **8**, 987[24]
Suslick, K. S., **4**, 95[98]; **7**, 50[71,72]
Süss, D., **8**, 562[424]
Suss, H., **1**, 746[61]
Süss, J., **6**, 172[25]
Süsse, M., **1**, 893[153]
Süss-Fink, G., **6**, 583[995,996]; **8**, 765[11], 773[11d,e], 789[11d]
Sussman, L. G., **6**, 790[114,115]
Sustmann, R., **1**, 266[46]; **2**, 342[6], 662[6]; **4**, 1073[21,25], 1076[21], 1079[58], 1082[91], 1084[94], 1085[100], 1090[21], 1092[21], 1098[21], 1100[188], 1102[21], 1103[204]; **5**, 71[125,130,135], 76[130], 159[51], 189[51], 247[26], 248[27], 451[49], 513[2], 516[2i], 518[2], 552[12]; **6**, 524[357], 526[357], 527[357], 534[522]; **7**, 880[152,153], 883[175]
Susuki, T., **4**, 505[134], 600[238], 611[346], 643[238]
Susz, B., **2**, 734[8,9], 738[8,9]; **3**, 273[12]
Sutcliffe, R., **4**, 812[178]; **5**, 901[30]
Suter, A. K., **5**, 596[32], 597[32], 603[32]
Suter, C., **2**, 477[12]; **3**, 431[95,96], 781[14]
Suter, C. M., **2**, 741[64]
Sutherland, B. L. S., **3**, 741[53]
Sutherland, I. O., **3**, 225[188], 877[88], 914[10], 916[19,20], 919[30], 923[45], 926[47], 927[48], 928[47], 930[10,59], 931[10], 933[30,62], 946[30], 949[94-96,98,99], 954[98], 963[121]; **4**, 507[154]; **6**, 893[79], 897[95,96,98]; **7**, 418[129c], 451[24], 671[14]; **8**, 248[85]
Sutherland, J. K., **2**, 529[16], 546[90]; **3**, 341[4], 342[4d], 348[29], 353[52], 355[53], 357[53], 371[115], 379[3], 380[9], 382[35], 386[3c,62-65], 390[71,72], 396[111,114,115], 398[111,114], 399[117], 402[117], 600[215], 753[99], 769[170], 770[175], 771[170]; **4**, 27[83]; **5**, 515[14], 518[14], 731[130b-d]; **7**, 630[52]; **8**, 545[282], 865[245]
Sutherland, M. D., **3**, 390[70]; **5**, 809[116]
Sutherland, R. G., **4**, 518[7], 521[42,46], 529[7,74,75,77,78], 530[75,78,79], 531[46,78,81], 541[113,114]
Sutin, N., **7**, 852[36]
Sutoh, S., **7**, 761[64]
Sutowardogo, K. I., **7**, 493[190]
Sutrisno, R., **6**, 502[213]
Suttajit, M., **2**, 463[92], 464[92]
Sutter, J. K., **4**, 303[340]; **8**, 856[172]
Sutter, M. A., **2**, 645[82]
Sutton, J. R., **7**, 794[7d]
Sutton, K. H., **2**, 127[232]; **3**, 197[37]; **4**, 82[62e], 217[130,132], 231[130,132], 243[74,75], 257[74,75], 260[75]
Sutton, L. E., **3**, 382[36]
Sutuki, H., **4**, 610[338], 649[338]
Suvorov, N. N., **6**, 543[608,611,615]
Suvorova, S. E., **7**, 770[252]
Suwa, K., **8**, 153[184,186]
Suwa, S., **7**, 862[79], 877[127]
Suya, K., **4**, 654[447]
Suzukamo, G., **1**, 317[153]; **3**, 327[167]; **5**, 1014[57]

Suzuki, A., **1**, 558[132], 561[160]; **2**, 112[88], 241[14]; **3**, 221[134], 231[244,246,247,248,249,252], 249[66], 251[78], 254[78], 262[158], 274[21], 326[163], 421[53], 443[56], 446[89], 465[178], 469[206], 470[178,205,206,207], 473[178,205,206,207], 489[58,59,61,63,64], 490[64-66], 495[58,59,61,63-65,97], 496[64,65,98,99], 498[59,64,65,98,99], 504[61,63,97,154], 511[58,59,61,63-66,97-99,154], 515[58,59,61,63-66,97-99,154], 523[25,26], 530[57]; **4**, 145[23,26,29a], 147[38b,41,42], 148[43,44,45b], 164[99,99a], 250[137], 286[174], 287[174], 290[174], 358[152,153,154,155,156,157,158], 886[118]; **5**, 117[273], 926[159]; **7**, 603[108-111], 604[130], 607[162,168], 608[170]; **8**, 101[120], 386[51], 786[117], 881[74]

Suzuki, D., **2**, 765[72]

Suzuki, E., **7**, 109[186]

Suzuki, F., **2**, 746[108], 762[56], 824[120]; **7**, 778[415]

Suzuki, H., **1**, 223[84], 225[84d], 350[154], 359[12], 360[12], 361[12], 362[12], 561[164]; **2**, 116[142,143], 450[54], 920[94], 921[94], 922[94,105], 924[105]; **3**, 169[510], 172[510], 173[510], 566[26]; **4**, 447[218,219]; **6**, 76[48], 110[49], 685[363], 976[3]; **7**, 415[115b], 460[116], 617[21], 765[168]; **8**, 315[46], 370[91,94], 371[110], 385[45], 404[18], 405[29], 408[75], 412[120], 450[163], 451[163c], 806[108,127], 807[108], 885[106], 886[106], 978[146]

Suzuki, I., **2**, 948[181]; **8**, 390[82], 640[22]

Suzuki, J., **3**, 816[79], 923[43]

Suzuki, K., **1**, 14[74], 54[21], 71[65], 72[66,67], 141[22], 184[52], 185[54], 238[35], 240[41], 248[64], 339[85], 420[84]; **2**, 10[40], 29[106], 30[112a], 31[112]; **3**, 168[508], 169[508], 303[55,56], 448[96], 457[129,130], 459[137,139], 460[130,137], 461[137], 497[104], 510[183,206], 513[205,206], 730[42,42b], 740[46], 937[75]; **4**, 231[262], 377[105a], 381[105], 1000[14]; **5**, 693[107], 847[136], 1032[100]; **6**, 14[52-55], 445[99], 648[116], 893[80]; **7**, 173[132], 298[35], 315[42], 761[52]; **8**, 10[57,58], 388[60], 453[191], 564[439]

Suzuki, M., **1**, 133[110], 242[47,48], 328[26], 347[129], 438[160], 568[239], 569[255], 738[40], 753[104], 865[87]; **2**, 576[73], 577[73], 609[84], 615[121], 635[39,39c], 640[39], 650[39c], 736[26]; **3**, 4[26], 5[26], 9[47], 10[26,47], 25[159], 135[356,360,361], 136[356,360,361,371], 137[360,361], 139[356,360,361,371], 140[371,371b], 142[360,361], 143[360,361,371,371b], 156[360,361], 565[22], 771[182]; **4**, 13[44,44c], 27[79], 30[88,88o], 96[105], 97[105b], 159[85], 177[59], 211[90], 238[9,13], 239[9], 245[9], 253[169], 254[9], 255[195,197], 257[230], 259[197], 260[197], 261[13,230,301], 262[308], 391[179], 393[197,197b,c], 394[197b,c], 609[326]; **5**, 180[149], 850[146], 1124[53]; **6**, 11[45], 489[90,92], 547[659], 563[900], 814[94], 942[154], 944[154], 998[117]; **7**, 162[58], 220[20], 243[66], 274[137], 406[75], 441[44], 451[24], 650[51], 802[49]; **8**, 9[52], 163[39], 216[59], 217[59], 537[186], 544[254], 546[186]

Suzuki, N., **1**, 234[21]; **3**, 748[73]; **6**, 240[79]; **8**, 36[68], 66[68], 212[9], 222[9], 314[43], 917[116,117], 920[116,117], 968[88], 988[29]

Suzuki, R., **4**, 599[214]

Suzuki, S., **1**, 803[306]; **2**, 527[6], 528[6], 632[25a]; **3**, 470[209], 472[209], 475[209]; **4**, 358[158]; **5**, 1[4], 2[4]; **6**, 157[172]; **7**, 429[155], 655[12], 660[43]; **8**, 249[98], 253[98], 315[47], 369[75], 698[137], 881[78], 882[78]

Suzuki, T., **1**, 359[14], 363[14], 384[14], 553[86], 733[20]; **2**, 232[177], 819[99,100], 824[100], 1020[46]; **3**, 172[511,512,515,516], 173[512,515,516], 198[46], 443[63], 503[144], 667[49]; **4**, 177[59], 238[13], 261[13], 379[115], 381[126a], 382[126], 383[126], 413[278b], 435[134], 600[227]; **5**, 134[65], 180[149], 206[47], 839[82], 864[257], 1186[3,5]; **6**, 7[30], 509[261], 538[560], 896[93]; **7**, 337[36], 341[51,52], 369[63], 378[63], 524[50]; **8**, 20[137], 411[105], 453[191], 460[254], 535[166], 544[254], 625[166], 627[166]

Suzuki, Y., **1**, 161[84], 359[14,16], 363[14,14a,b,e], 379[16d], 384[14]; **2**, 23[88], 547[121], 551[121], 552[121]; **3**, 690[125]; **5**, 693[110], 694[110], 1130[3]; **6**, 498[164]; **8**, 166[61], 249[98], 253[98], 369[75], 388[60], 647[56]

Suzuki, Z., **4**, 34[97], 35[97,97i]; **7**, 698[53]

Suzumoto, T., **1**, 831[99]; **4**, 810[169]; **8**, 134[35]

Suzuta, Y., **5**, 492[247,248]

Svadkovskaya, G. E., **3**, 634[11], 635[11]

Svanholm, U., **7**, 801[39], 854[47], 855[47]

Svata, V., **6**, 902[127]

Svatek, E., **2**, 765[78]

Svec, H. J., **7**, 528[10]

Sveda, M., **7**, 14[140]

Svedberg, D. P., **5**, 886[21]

Svedi, A., **8**, 455[206]

Svendsen, J. S., **7**, 430[159], 442[46b], 489[165]

Svenson, R., **8**, 678[58,64,65], 683[58,64,65], 686[58,64,65], 691[58]

Svensson, K., **7**, 331[15]

Svensson, U., **7**, 831[64]

Svetlakov, N. V., **6**, 212[82]

Svétlik, J., **6**, 524[368]

Sviridov, A., **1**, 95[75,76]

Sviridov, A. F., **6**, 466[40-42], 469[40,54]; **8**, 694[118]

Svoboda, J. J., **2**, 1050[28]; **6**, 531[447], 535[447]

Svoboda, M., **6**, 959[39], 1013[9]; **8**, 726[191,192]

Svoboda, P., **8**, 773[64]

Swafford, R. L., **5**, 197[19]

Swahn, C. G., **6**, 660[205]

Swain, C. G., **6**, 220[129]; **8**, 86[21]

Swain, C. J., **1**, 130[94,95], 801[301]; **6**, 5[27], 994[98]; **7**, 406[71]; **8**, 10[56]

Swallow, A. J., **7**, 7[41]

Swallow, J. C., **7**, 95[70,70a]

Swallow, W. H., **3**, 742[55], 751[90], 752[91]

Swamer, F. W., **2**, 797[5], 829[5], 837[5], 838[162], 843[5], 845[5]

Swaminathan, S., **2**, 782[24]; **3**, 380[7]; **4**, 4[14]; **5**, 72[173], 768[121], 809[124]; **6**, 834[38], 836[52]; **8**, 530[108]

Swan, C. J., **7**, 759[8]

Swan, G. A., **5**, 500[257]; **8**, 273[126]

Swandborough, K. F., **7**, 237[38]

Swanezy, E. F., **8**, 148[107]

Swann, B. P., **7**, 154[19]

Swann, D. A., **5**, 727[120]; **6**, 462[18]

Swann, S., Jr., **3**, 635[32]; **8**, 321[104-107], 366[40]

Swanson, D. D., **7**, 765[161]

Swanson, D. R., **5**, 560[76], 1037[5], 1166[16], 1167[16], 1175[38,41], 1177[43], 1178[16,38,41,43], 1180[47], 1181[47]

Swanson, D. S., **5**, 1165[15], 1166[15], 1167[15], 1170[15], 1171[15], 1175[15], 1178[15], 1179[15]

Swanson, E. D., **5**, 1055[46]

Swanson, R., **2**, 1037[99,100]

Swanson, S., **1**, 741[45]

Swanson, S. B., **4**, 437[147]

Sward, K., **4**, 861[113]

Swart, E. A., **8**, 145[87]

Swarts, F., **8**, 903[48]

Swartz, J. E., **3**, 599[222], 602[222]; **4**, 453[23], 456[48], 473[23,48], 474[23], 475[151], 809[162]; **8**, 132[10,12], 134[10,12]

Swartz, T. D., **8**, 338[88]

Swartzendruber, J. K., **4**, 128[221]

Swarzenbach, G., **3**, 888[14]

Swatton, D. W., **5**, 223[76]

Swayze, J. K., **7**, 548[68], 555[68], 557[68]

Sweany, R. L., **8**, 674[31]

Sweat, F. W., **7**, 292[3,9], 653[2]

Swedlund, B. E., **7**, 502[261]

Sweeley, C. C., **6**, 653[150]

Sweeney, J. B., **4**, 744[134], 745[140]

Sweeney, K. M., **2**, 745[103]; **3**, 328[179]

Sweeney, T. J., **5**, 833[44]

Sweeney, W. A., **2**, 753[2]

Sweeny, J. G., **5**, 421[79]; **8**, 314[28,29]

Sweers, H. M., **2**, 456[49], 460[49], 461[49], 463[91], 464[91]
Sweet, C. S., **2**, 971[92]
Sweet, F., **5**, 455[78]
Sweeting, O. J., **7**, 203[55]
Sweigart, D. A., **4**, 518[3], 520[38], 542[38,117], 689[70,72,73], 691[74]
Swenson, J. S., **6**, 822[116]
Swenson, K. E., **6**, 777[60]; **8**, 132[14], 517[125]
Swenson, R. E., **1**, 767[179]; **5**, 47[41], 48[41], 50[41,44], 55[49], 56[49]
Swenson, W., **3**, 284[54]; **4**, 42[121]
Swenton, J. S., **1**, 554[101]; **2**, 763[59]; **3**, 254[102], 256[112], 695[152]; **4**, 14[46,47e-g], 143[21]; **5**, 71[122], 730[127], 1022[76], 1032[98]; **7**, 160[49]
Swepston, P. N., **8**, 447[133,134], 696[128]
Swern, D., **4**, 725[43]; **6**, 262[16], 263[16], 264[16], 266[16]; **7**, 24[35], 95[75], 96[75], 292[5], 295[22], 297[27,28], 298[27], 299[5], 300[5,56], 302[64], 396[24], 479[97], 498[229], 501[252]
Swierczewski, G., **3**, 243[12], 246[38], 250[12], 262[12], 470[221], 471[221], 482[5], 499[5], 505[5], 509[5]; **4**, 869[27], 870[27], 871[27], 877[27d,66]; **5**, 38[23b]
Swieton, G., **5**, 77[259]
Swigar, A. A., **7**, 684[92]
Swiger, R. T., **7**, 660[42], 882[172]
Swigor, J. E., **8**, 645[40]
Swincer, A. G., **5**, 1068[14]
Swindell, C. S., **3**, 229[233], 246[38], 444[65], 446[81], 492[75], 503[75]; **5**, 137[81]
Swinden, G., **8**, 382[3]
Swingle, R. B., **1**, 512[37]; **3**, 149[400]
Swisher, J. V., **2**, 588[152]; **3**, 883[108]; **8**, 764[2], 770[2b], 774[76]
Swistok, J., **6**, 709[51], 714[83]
Swithenbank, C.,
Swoboda, J. J., **3**, 361[75]
Switzer, C., **4**, 213[116], 215[116]; **6**, 150[124]
Switzer, C. Y., **5**, 56[50a], 57[50,50a]
Switzer, F. L., **4**, 988[137]
Swoboda, G., **2**, 352[91], 355[123], 358[91]
Swoboda, J., **2**, 355[123]
Sword, I. P., **2**, 357[144]
Sworin, M., **1**, 416[65], 883[125], 889[142], 890[142]; **2**, 1043[115]; **3**, 792[65]; **4**, 89[86b]; **5**, 810[125]
Sy, A. O., **6**, 801[31]
Sy, W.-W., **2**, 1057[71], 1058[71]; **4**, 357[149]; **5**, 407[29]; **6**, 108[33]; **7**, 534[39]
Syamata, M. S., **5**, 185[158]
Sydnes, L., **5**, 126[22], 127[22], 128[22]; **8**, 807[113]
Syfrig, M. A., **1**, 341[95], 481[161], 482[162,163]; **3**, 71[31,33]; **7**, 225[65]
Syhora, K., **7**, 96[83]
Sykes, P., **8**, 656[89]
Sykes, R. B., **5**, 86[33]
Syldatk, C., **8**, 187[43]
Sylvester, A. P., **7**, 742[56]; **8**, 473[11]
Sylvestre-Panthet, P., **2**, 547[95]
Symons, M. C. R., **8**, 524[11,11a]
Synder, F. F., **5**, 128[29]
Synder, H. R., **8**, 724[172]
Synder, J. K., **7**, 346[10], 356[10]
Synder, J. P., **6**, 935[102]
Synerholm, M. E., **5**, 2[16], 835[59]; **6**, 843[89]
Syner-Lyons, M. R., **4**, 816[200]
Syper, L., **7**, 657[21], 684[93b,c]
Syrkin, V. G., **8**, 765[11]
Syrkin, Y. K., **7**, 451[38]
Syrota, A., **8**, 35[64], 66[64]
Syrova, G. P., **7**, 773[304]
Sytin, V. N., **2**, 933[139]
Syubaev, R. D., **6**, 553[728], 554[728,741,773]
Szabo, A. E., **5**, 583[186], 584[194]
Szabo, A. G., **7**, 821[31]
Szabó, G., **7**, 777[389]
Szabo, K., **8**, 754[93]
Szabó, L., **2**, 817[90], 851[223]; **6**, 917[36]
Szabo, W. A., **5**, 105[189]
Szajewski, R. P., **3**, 902[118]; **8**, 87[35], 939[97]
Szántay, C., **2**, 812[72], 817[90], 851[223]; **4**, 33[96,96b]; **6**, 917[36]; **7**, 746[93]; **8**, 266[55]
Szarek, W. A., **2**, 167[158], 642[78], 643[78]; **6**, 738[57], 739[57]; **7**, 258[55], 274[140], 580[146], 712[67]; **8**, 794[15]
Szari, A. C., **3**, 124[289], 125[289]; **6**, 134[12]
Szczepanski, S. W., **2**, 1056[64], 1070[64]; **5**, 404[18]; **7**, 673[30], 695[33]; **8**, 354[174]
Sze, S. N., **4**, 930[54]
Szechner, B., **2**, 535[39]; **8**, 219[81]
Szeimies, G., **3**, 625[41]; **4**, 1009[145], 1085[108], 1099[186], 1100[189,190]; **5**, 64[54]; **7**, 475[51], 477[73]
Szeja, W., **8**, 212[7]
Szejtli, J., **8**, 224[103], 225[103]
Szekely, I., **1**, 708[253]; **2**, 529[17]
Szekerke, M., **2**, 735[17]
Szeleczky, Z., **7**, 80[139]
Szent-Gyogyi, A., **8**, 87[30]
Szeto, K. S., **3**, 927[48]
Szeverenyi, N. M., **1**, 3[26], 43[26]; **2**, 100[10]
Szeverenyi, Z., **7**, 558[80], 559[80], 560[80], 561[80]
Szezepanski, H., **4**, 399[222]
Szilagyi, L., **5**, 438[164], 534[94]
Szilagyi, S., **8**, 407[59]
Szmant, H. H., **8**, 329[18-20,24], 336[20,24]
Szmat, H. H., **7**, 764[115], 766[176,180]
Szmuszkovicz, J., **3**, 28[170], 30[170]; **4**, 6[21,21a], 7[21a]; **6**, 703[1,3], 704[8], 711[8], 714[3]; **7**, 571[118], 576[118]; **8**, 530[101]
Szöcs, G., **3**, 223[155]
Szostak-Rzepiak, B., **2**, 787[52]
Szpilfogel, S., **8**, 293[48]
Sztajnbok, P., **8**, 111[21], 123[21]
Szuchnik, A., **7**, 775[350]
Szulzewsky, K., **6**, 436[13]
Szurdoki, F., **7**, 746[93]
Szwarc, M., **7**, 851[13]
Szychowski, J., **2**, 538[64]
Szymanski, R., **8**, 436[73]
Szymoniak, J., **2**, 538[53]
Szymonifka, M. J., **5**, 984[32]
Szymula, M. B., **5**, 225[97]; **8**, 505[82], 507[82]

T

Taba, K. M., **6**, 662[213]
Tabacchi, R., **3**, 813[62]
Tabakovic, I., **6**, 554[736]; **8**, 591[60], 641[28]
Tabakovic, K., **6**, 554[736]
Tabata, A., **7**, 534[40]
Tabata, M., **3**, 554[22,23]; **7**, 607[168]
Tabba, H. D., **2**, 780[14]
Tabche, S., **2**, 25[97], 26[97], 27[97], 31[97], 33[97b], 41[97b], 43[97b]
Tabei, H., **3**, 530[63], 532[63]
Tabei, K., **8**, 365[32]
Tabei, T., **3**, 942[80], 1008[65]; **6**, 874[14], 877[14], 883[14]
Tabenkin, B., **7**, 54[7]
Taber, D. F., **2**, 527[7], 528[7], 801[24]; **3**, 602[220], 1051[24], 1052[24,25,28], 1053[28], 1054[30], 1055[28], 1057[28], 1059[25], 1060[44,45], 1061[46], 1062[30,44,46,47]; **4**, 13[45], 14[45b], 24[72,72d,e], 63[72d], 501[114], 1033[32], 1040[94,100], 1041[94], 1056[100]; **5**, 1[5], 2[5], 9[5], 513[5], 514[5], 515[12,18], 516[12b], 527[5], 534[89], 547[5h,18], 924[145], 1166[20], 1167[20], 1169[20], 1170[20], 1178[20]; **6**, 123[141], 125[141], 126[153], 127[155,156]; **8**, 358[202]
Taber, T. R., **1**, 358[10], 424[99]; **2**, 2[4], 100[15], 101[15], 111[15], 112[15,100], 113[15], 134[3], 190[57], 223[150], 239[2], 240[2], 245[31,32], 246[31], 247[31], 436[67], 455[7], 475[1], 509[33], 894[10], 917[10], 918[10], 919[10], 930[10]; **4**, 145[35]
Tabibian, S., **3**, 889[24]
Tabor, A. B., **5**, 403[9]
Tabor, T. E., **3**, 739[40]
Tabti, B., **5**, 324[23]
Tabuchi, T., **1**, 256[22], 257[23], 260[32,32b], 261[33], 266[47], 751[110,111], 831[105]; **4**, 606[305], 607[305,311,313], 626[311], 647[305], 648[311]; **6**, 980[41]; **8**, 960[32]
Tabushi, E., **4**, 148[51], 149[51], 180[67]; **6**, 764[11]
Tabushi, I., **3**, 381[24], 382[24]; **4**, 511[178], 753[163]; **7**, 16[167,168], 108[176]
Tacconi, G., **2**, 351[79], 364[79,204]; **5**, 451[24,44], 453[24,44], 454[70], 468[24,44], 485[24]
Tachdjian, C., **4**, 748[156]
Tachibana, K., **3**, 741[50]
Tachibana, Y., **2**, 282[33]
Tachimori, Y., **7**, 751[139]
Tacke, P., **6**, 267[59]
Tacke, R., **2**, 385[321]; **8**, 187[43]
Tada, H., **3**, 386[57], 390[79]; **5**, 839[82], 864[257]
Tada, M., **4**, 810[167]; **5**, 569[112]; **6**, 658[184]; **7**, 102[136], 239[53]
Tada, N., **6**, 787[102]; **7**, 537[58]
Tada, S., **2**, 718[79]
Tada, S.-I., **6**, 1018[42]
Tada, T., **7**, 255[38]
Tada, Y., **6**, 447[105], 450[105]
Tadanier, J., **8**, 27[30], 66[30]
Tadano, K., **2**, 386[331]; **5**, 839[76], 864[259]; **7**, 713[70]
Tadano, K.-I., **2**, 323[28], 333[28]
Taddei, F., **3**, 95[154], 119[154]
Taddei, M., **1**, 612[48]; **2**, 566[22], 567[26], 586[135], 607[71]; **3**, 356[57]; **4**, 901[184,186]; **5**, 335[49]; **6**, 238[74]; **7**, 330[7]
Taddia, R., **3**, 738[37]
Tadeno, K.-I., **6**, 677[323]
Tadj, F., **1**, 561[166]
Tadros, W., **8**, 566[450]
Tadzhimukhamedov, Kh. S., **3**, 303[58]
Taeger, E., **3**, 194[14]
Tafesh, A. M., **4**, 746[144]
Taffer, I. M., **3**, 753[102]; **8**, 876[45], 877[45]
Taft, R. W., Jr., **8**, 749[61]
Taga, J., **6**, 525[379]
Taga, J.-I., **8**, 245[72]
Taga, T., **2**, 116[140], 610[94], 611[94], 1059[78,81]; **4**, 27[79,79d,e]; **5**, 92[81]; **6**, 764[11]
Tagahara, K., **5**, 492[247,248]
Tagaki, W., **3**, 86[19], 121[19]; **7**, 764[107,111,116]; **8**, 589[51]
Tagami, K., **2**, 1069[132]; **6**, 150[118,119], 151[119]
Tagashira, M., **1**, 193[85], 195[85], 198[85]
Tagat, J., **3**, 251[77], 254[77]
Tagle, B., **6**, 1054[47]
Tagliavini, E., **1**, 188[73], 189[73], 192[82]; **2**, 35[130], 36[130], 507[19], 566[23], 657[161b]; **6**, 685[350], 976[4]; **7**, 549[42]; **8**, 252[111], 797[40]
Tagliavini, G., **2**, 6[29,31,32], 18[31], 564[9,11,12], 566[20], 572[47]; **4**, 744[131]
Tagmann, E., **4**, 41[119,119a]
Tagmazyan, K. T., **5**, 435[151]
Tago, H., **4**, 898[177]
Tagoshi, H., **1**, 187[61]
Tagoshi, T., **1**, 187[64]
Taguchi, H., **1**, 789[260], 830[97], 873[92], 874[96]; **3**, 787[47,48]; **7**, 73[104,105]
Taguchi, K., **6**, 705[20]
Taguchi, M., **4**, 598[196]; **6**, 74[34], 86[98]
Taguchi, T., **1**, 387[127,133]; **2**, 72[58], 73[63], 604[53], 643[79], 644[79b], 656[154]; **3**, 421[54]; **4**, 1005[88], 1020[234,235]; **5**, 100[142], 847[136]; **6**, 83[83], 989[78], 993[78], 996[105]; **7**, 603[108,109]
Taguchi, Y., **6**, 609[52,53]
Taha, A. A., **5**, 167[94]
Tahara, A., **3**, 834[76]; **8**, 500[52], 528[86]
Tahbaz, P., **8**, 614[89], 620[89]
Tahk, F. C., **5**, 959[318]
Tai, A., **2**, 232[178]; **4**, 975[91]; **8**, 150[122-124,129,130,133,135], 151[129,133,135,145,146,148,149,150,151,153,154]
Tai, D. F., **2**, 26[102], 27[102]
Tai, W. T., **3**, 854[73]
Taikov, B. F., **3**, 643[131]
Tailby, G. R., **5**, 1134[36]
Tailhan, C., **4**, 748[160]
Taillefer, M., **8**, 909[79], 910[79]
Taimr, L., **3**, 664[29]
Tainturier, G., **8**, 447[120], 683[87,94], 690[101]
Tait, B. D., **1**, 233[17], 238[17], 621[70], 734[24], 735[24], 736[24]; **7**, 363[39]
Tait, S. J. D., **7**, 253[15], 276[15]
Tait, T., **3**, 331[198]
Taits, S. Z., **3**, 615[7]; **8**, 611[63]
Tajika, M., **8**, 459[245]
Tajima, K., **6**, 535[538], 538[538]; **8**, 190[80]
Tajima, T., **8**, 568[463]
Tajima, Y., **2**, 649[100], 1059[74]; **8**, 450[169]
Tajiri, A., **4**, 115[180e]
Tajiri, T., **4**, 509[161]
Tajlor, W. C., **6**, 215[104]
Takabe, K., **7**, 223[46], 407[83]; **8**, 798[59]
Takacs, F., **3**, 812[57]
Takacs, J. M., **1**, 768[171]; **2**, 64[27], 240[4], 436[67]; **3**, 45[244], 199[53]; **4**, 408[257a]
Takacs, K., **2**, 787[50]; **6**, 543[610]

Takada, S., **3**, 834[76]
Takadoi, M., **5**, 369[101], 370[101c]
Takagaki, T., **6**, 959[38]; **7**, 365[45]; **8**, 795[24]
Takagi, K., **1**, 450[211]; **3**, 463[163], 483[19], 484[20,24], 500[19,131], 501[24], 509[19]; **4**, 230[252], 231[252], 602[254], 840[32], 904[203,205]; **6**, 233[40]; **8**, 620[139]
Takagi, M., **4**, 841[49]; **7**, 429[155]
Takagi, S., **5**, 760[90]
Takagi, T., **4**, 290[205], 812[177]; **6**, 453[142,144]; **7**, 766[184]; **8**, 453[191], 604[3]
Takagi, U., **8**, 418[6], 430[6], 438[6], 439[6]
Takagi, Y., **1**, 624[86]; **3**, 426[82], 429[82], 871[54]; **4**, 18[62], 20[62i], 262[305]
Takagishi, S., **2**, 655[132,135]; **4**, 159[82]; **8**, 216[61]
Takahara, J. P., **2**, 18[75]; **6**, 837[59]
Takahashi, A., **6**, 150[119], 151[119], 902[129]
Takahashi, E., **3**, 446[77]
Takahashi, F., **8**, 134[34]
Takahashi, H., **1**, 166[114], 359[14,16], 363[14,14a,b,e], 369[64a], 379[16a-d], 384[14], 418[75], 894[156]; **3**, 96[166], 99[166], 103[166], 105[166,216], 112[216], 120[166], 155[166], 530[56]; **4**, 102[131], 590[97,98], 600[240], 643[240]; **6**, 46[59], 467[52], 765[15], 968[108]; **7**, 739[33], 746[81], 776[358]; **8**, 134[34], 154[190,191], 388[60], 652[73], 781[96]
Takahashi, I., **1**, 314[122]; **5**, 376[108b]; **8**, 170[80]
Takahashi, J., **3**, 644[161]
Takahashi, K., **1**, 506[5], 526[5], 527[103], 555[110], 557[129,131], 558[129,132,134,135,136]; **2**, 61[19], 74[75], 558[159], 584[121]; **3**, 136[370,371], 137[370], 138[370,371a], 139[370,371], 140[370,371,371a], 143[371,371a], 144[371a], 826[39]; **4**, 18[58], 259[257], 434[132], 505[134], 592[123,124], 597[174,178], 598[194], 611[367], 636[367], 637[174], 753[164]; **5**, 16[112], 24[166,167], 55[48], 86[34], 541[110], 714[74], 850[147]; **6**, 20[77], 856[151]; **7**, 95[65], 142[137], 229[113], 231[138], 408[88a], 415[114]; **8**, 86[25], 277[155], 383[21], 394[117], 836[7]
Takahashi, M., **2**, 505[10]; **3**, 34[194], 644[134a,150,152], 934[67]; **5**, 412[45], 438[161], 442[185], 532[84], 864[257]; **6**, 137[43], 619[115], 744[76], 746[76], 756[126], 893[77]; **7**, 321[65], 453[73,93], 454[73], 455[93], 642[9], 682[82]; **8**, 135[40], 770[32], 787[119]
Takahashi, N., **2**, 960[36]
Takahashi, O., **1**, 58[33], 110[17], 131[17], 134[17], 135[117], 339[87]; **2**, 119[161]; **3**, 942[80], 1002[57], 1004[59], 1008[65]; **5**, 1001[16]; **6**, 852[136], 874[14], 877[14,37], 883[14,52,54], 890[52], 891[54]; **7**, 429[155]
Takahashi, R., **8**, 266[57]
Takahashi, S., **1**, 546[56]; **3**, 454[121], 530[61], 531[61], 532[61], 537[61], 552[8], 557[48,49], 681[99]; **5**, 1137[56]; 7, 451[32]; **8**, 771[44]
Takahashi, T., **1**, 143[29], 162[96], 368[62], 391[62], 552[82], 553[85,86,88-98], 642[109], 643[109], 749[81]; **2**, 482[32], 484[32], 710[24], 851[222], 1004[59]; **3**, 49[263], 51[271], 95[153], 100[200], 103[200], 107[153], 114[153], 115[153], 198[45-48], 230[239], 232[239a], 238[239a], 380[10], 390[79], 395[99,101], 402[129], 486[41], 503[151], 530[54,55], 741[50], 1008[72], 1009[72], 1010[72], 1033[75,76]; **4**, 8[27], 30[88,88n], 192[117], 227[202,203,204,205], 239[17], 255[196], 261[17,297,298], 405[250a,b], 501[118], 629[411], 650[425], 850[86], 867[5], 883[5], 889[136], 890[136], 892[136], 893[136,147]; **5**, 410[38], 532[87], 742[162], 809[119,120], 1037[5], 1165[14,15], 1166[15,16,25], 1167[15,16], 1170[15], 1171[14,15], 1172[14], 1174[14], 1175[14,15,38,41], 1177[43], 1178[14-16,38,41,43], 1179[15], 1180[47], 1181[47]; **6**, 134[21], 137[42,43], 849[117], 875[18], 1022[57], 1032[118], 1045[21]; **7**, 239[49], 406[73], 452[50], 453[84-87,91], 454[84], 455[86,87], 458[112], 461[118], 543[21], 700[63]; **8**, 7[43], 330[47], 340[100], 680[72], 683[72], 690[104], 693[72,104], 836[10a], 837[10a], 856[182]
Takahashi, W., **4**, 258[239]
Takahashi, Y., **3**, 326[163]; **4**, 258[250], 588[56], 596[162,163], 601[247], 614[374], 621[163], 637[163]; **5**, 35[12], 77[269], 394[145b]; **7**, 603[109-111], 855[62], 874[108], 875[113-115], 876[120], 887[62]
Takahaski, S., **4**, 850[86]
Takahata, H., **1**, 413[60]; **4**, 398[217], 399[217b], 401[217b], 403[217b], 404[217b], 413[278a,b]; **5**, 829[22]; **6**, 509[261,266,267,268], 538[560]; **7**, 503[273]
Takahatake, Y., **5**, 924[144]
Takai, H., **4**, 309[418]
Takai, K., **1**, 92[62,63,66], 95[79], 185[76], 190[76], 191[76], 193[84,85], 195[85], 198[84,85], 201[98], 202[103], 203[103], 205[105,107-109], 206[109], 749[79,83,86,87], 750[83,86-88], 807[318,322,324], 808[322,323,324], 80; **2**, 20[82], 114[123], 271[75], 587[146], 597[10], 603[42], 642[77], 643[77]; **3**, 445[73], 449[73]; **4**, 251[146]; **5**, 850[149], 1124[50,51], 1125[58-60]; **6**, 856[149], 979[28]; **7**, 169[117], 275[146,147], 276[147], 308[17], 309[24], 324[70], 674[47]; **8**, 755[126,127], 758[126,127], 937[86]
Takaido, T., **6**, 509[263]
Takaishi, N., **3**, 1041[113]; **4**, 857[104], 921[26]; **6**, 270[80]
Takajo, T., **2**, 916[83]
Takaki, K., **4**, 8[30a], 102[131]; **5**, 464[108,109], 466[109]; **6**, 1022[56]; **7**, 646[25], 773[300]
Takaku, H., **6**, 438[48], 604[31], 609[54,56], 616[106], 624[139]
Takamatsu, M., **6**, 500[181]
Takamatsu, S., **5**, 474[158]
Takamatsu, T., **4**, 398[217], 399[217b], 401[217b], 403[217b], 404[217b], 413[278b]; **7**, 503[273]
Takami, H., **1**, 564[201]; **6**, 27[119]
Takami, Y., **4**, 601[244], 643[244]
Takamine, K., **1**, 279[86], 280[86]; **3**, 567[36,190], 595[190], 607[190], 610[36]; **8**, 113[36]
Takamiya, H., **8**, 432[68]
Takamo, S., **1**, 766[155]
Takamoto, T., **7**, 751[138,139]
Takamuku, S., **8**, 865[249]
Takamura, I., **8**, 144[71]
Takamura, N., **3**, 1055[32]; **4**, 240[53], 373[74]; **8**, 355[181], 929[32]
Takano, K., **4**, 124[212]
Takano, S., **1**, 90[59], 343[106], 569[260], 763[144]; **2**, 167[162], 372[267,268,270,271], 373[267,268,270], 505[10], 814[81], 824[81], 1018[39], 1035[92,94], 1040[94]; **3**, 34[194], 224[181], 278[30], 934[67]; **4**, 376[99], 381[130], 387[99,158,158b,159], 393[159,186]; **5**, 468[129,130,131], 862[246], 864[257]; **6**, 509[254], 533[499], 746[91], 843[87], 893[77]; **7**, 180[159], 299[45], 416[122], 441[44], 463[129], 682[82], 713[69]; **8**, 222[97]
Takano, Y., **2**, 609[87]; **4**, 116[186]
Takao, K., **7**, 107[168]
Takao, S., **5**, 92[81]
Takaoka, A., **6**, 498[160], 556[822]
Takaoka, K., **6**, 76[48]; **7**, 92[42], 93[42]
Takaoka, Y., **3**, 286[56a]
Takasaki, K., **4**, 314[496]
Takasaki, M., **8**, 145[82]
Takase, I., **2**, 86[25]
Takase, K., **4**, 970[72], 972[81]; **5**, 714[74]; **6**, 984[59]
Takase, M., **8**, 244[55], 248[55], 874[25]
Takashi, Y., **8**, 754[94]
Takashima, T., **2**, 657[161a]
Takasu, M., **4**, 107[140d]; **8**, 660[107]
Takasu, Y., **2**, 556[151]; **6**, 931[91]
Takata, J., **2**, 971[94]; **7**, 802[47]
Takata, T., **4**, 30[88,88h]; **7**, 759[15], 763[102], 769[214], 778[403]; **8**, 410[92]
Takata, Y., **3**, 639[85]; **5**, 442[185]; **8**, 135[42], 253[114]

Takatani, M., **2**, 157[119]; **3**, 225[185], 264[181]; **6**, 2[3], 25[3], 88[105], 89[105]; **7**, 406[77], 409[77], 414[77], 415[77], 421[77], 423[77], 569[108]
Takatori, M., **4**, 847[76]
Takatsuto, S., **5**, 151[9]; **6**, 219[121]; **7**, 680[76]
Takaya, H., **4**, 587[45], 963[42], 1038[61]; **5**, 282[28,33], 284[28], 286[33], 293[44], 294[51], 595[11], 596[11a], 601[44,45], 603[48,48a], 605[58,60,60a,61], 606[45], 609[60,60c,61], 1186[3,3a,5,7], 1187[8], 1190[24,24a], 1200[54]; **6**, 866[208]; **7**, 26[51]; **8**, 154[199,200,201], 459[244], 460[255,256], 461[257,261], 462[267], 463[269], 535[166], 678[62], 683[62], 686[62], 991[44]
Takaya, T., **3**, 181[552]; **5**, 603[48,48e]; **6**, 93[133]; **7**, 761[55], 764[55]
Takayama, H., **3**, 169[510], 172[510,511,512,515,516], 173[510,512,515,516], 380[11], 443[63]; **4**, 487[49]; **5**, 369[101], 370[101a,b,d], 413[51]; **6**, 150[129], 913[23]; **7**, 257[47], 362[31], 377[31], 410[92]
Takayama, K., **4**, 208[62], 743[128]
Takayama, M., **8**, 331[31]
Takayanagi, H., **8**, 951[176]
Takazawa, O., **2**, 547[108], 550[108]; **4**, 159[81]; **6**, 715[88-90], 980[43]
Takeaishi, N., **7**, 9[70]
Takebayashi, M., **4**, 1089[138], 1091[138]; **7**, 24[37], 25[37,42,45], 26[52], 473[25]
Takebayashi, N., **6**, 615[100]
Takebayashi, T., **5**, 15[102], 543[114]
Takebe, M., **3**, 314[108]
Takechi, H., **7**, 877[133]
Takechi, K., **8**, 806[108], 807[108]
Takechi, S., **2**, 1048[12]; **7**, 399[37]
Takeda, A., **2**, 199[87], 838[178]; **3**, 843[25], 844[32,33], 848[53,54]; **4**, 229[215]; **5**, 766[117]; **6**, 867[209]; **8**, 185[26], 190[26], 191[89,90], 193[102], 194[104a], 195[104b], 205[160]
Takeda, H., **1**, 894[156]
Takeda, K., **1**, 243[56]; **3**, 386[57], 390[79]; **4**, 351[125]; **5**, 534[93], 768[131], 779[131], 809[115,117]; **6**, 994[96]; **8**, 530[100]
Takeda, M., **2**, 953[3b], 1012[13]; **8**, 176[132,133,134], 338[82], 339[82]
Takeda, N., **1**, 894[156]; **2**, 213[125], 649[100], 1059[74]; **7**, 747[104]; **8**, 820[45], 822[52]
Takeda, R., **2**, 103[37]; **7**, 238[43]
Takeda, S., **5**, 850[146]; **6**, 533[492]; **7**, 120[16]
Takeda, T., **1**, 561[159]; **2**, 351[82], 357[82], 374[277], 649[105]; **3**, 137[376]; **4**, 89[85], 206[51-55], 218[134,135,136], 837[9], 903[202], 904[202]; **6**, 149[104], 643[76], 902[130], 937[117], 939[117], 940[117]; **7**, 368[59]; **8**, 847[91]
Takeda, Y., **1**, 110[17], 131[17], 134[17], 339[87,88], 784[243,244]; **5**, 329[32]; **6**, 6[29]; **7**, 412[105], 414[105]
Takegami, Y., **2**, 357[146], 358[146,151]; **3**, 1021[16]; **4**, 930[52]; **8**, 36[81], 54[81,161], 55[180], 66[81,161,180], 289[24], 291[35-37], 292[36], 293[46], 568[475]
Takehara, S., **4**, 1007[122]
Takehara, Z., **3**, 647[194]
Takehashi, K., **5**, 833[49]
Takehashi, O., **5**, 851[164]
Takehashi, T., **3**, 1010[75]
Takehira, K., **7**, 462[119-121]
Takehira, Y., **3**, 491[67]; **5**, 381[116], 383[125]
Takei, H., **1**, 406[30], 407[30,32,33], 415[30], 424[100], 427[30], 430[131], 454[32], 880[114]; **2**, 72[58], 73[63], 74[80], 617[144]; **3**, 20[116], 446[79,80,82], 456[79,80], 459[136], 460[136], 461[136], 462[150], 470[79,80,224], 471[80], 472[222,224], 473[79], 513[208]; **4**, 91[88e], 121[207], 152[55], 211[94-96], 231[94], 239[19], 258[239], 589[83], 596[83]; **6**, 438[58], 439[58,67], 578[981]; **7**, 673[27]; **8**, 413[126], 518[132], 842[41]
Takei, Y., **3**, 875[73,74]
Takeishi, S., **2**, 197[81]
Takemasa, T., **1**, 763[143], 766[143]
Takematsu, A., **3**, 300[38]
Takematsu, T., **7**, 423[145], 424[145b]
Takemoto, M., **8**, 944[125]
Takemoto, T., **3**, 396[115], 406[141], 748[73]
Takemoto, Y., **6**, 1046[31]; **7**, 178[148], 544[35], 556[35], 566[35], 821[29]
Takemura, H., **1**, 314[122]; **5**, 376[108b]
Takemura, K. H., **5**, 342[62b]
Takemura, M., **5**, 742[162]
Takemura, S., **7**, 235[5]
Takenaga, J., **4**, 487[45]
Takenaka, A., **7**, 473[33], 501[33], 502[33]
Takenaka, H., **8**, 857[189]
Takenaka, S., **4**, 1086[115]
Takenaka, Y., **3**, 124[263], 126[263]
Takenishi, T., **6**, 602[5]
Takeno, H., **5**, 96[106,117]; **6**, 93[133]
Takeno, N., **7**, 120[16]
Takenoshita, H., **2**, 655[141]; **6**, 237[70]
Takenouchi, K., **1**, 880[114]; **3**, 20[116]; **4**, 152[55], 211[96]
Takeshima, K., **8**, 388[60]
Takeshima, T., **2**, 21[85]
Takeshita, H., **1**, 187[61-64], 188[62]; **3**, 573[71], 575[81,82], 610[71]; **5**, 24[169], 619[12,13], 620[12,13,16], 621[19], 622[23], 627[42], 790[22], 820[22]; **8**, 883[92]
Takeshita, K., **8**, 773[70], 774[70,71]
Takeshita, M., **3**, 135[347], 136[347], 137[347], 139[347], 144[347]; **4**, 48[139]; **8**, 187[39]
Takeshita, T., **3**, 300[43]
Takeshita, Y., **8**, 33[56], 66[56]
Taketomi, T., **6**, 866[208]; **8**, 459[244]
Takeuchi, A., **1**, 569[255]; **7**, 451[24]; **8**, 9[52]
Takeuchi, H., **2**, 427[43]; **6**, 94[135], 124[145], 658[186a]; **7**, 25[44], 26[54-56], 476[66,67]
Takeuchi, J., **6**, 87[102]
Takeuchi, K., **5**, 714[72,73], 841[87]
Takeuchi, M., **2**, 507[22], 810[66], 851[66]; **3**, 227[212], 470[212], 476[212]
Takeuchi, N., **3**, 683[105]
Takeuchi, R., **1**, 63[43,44]; **3**, 1018[10], 1021[10]
Takeuchi, S., **8**, 116[68], 154[192,193,194,195]
Takeuchi, T., **4**, 249[128]; **7**, 406[78d]
Takeuchi, Y., **1**, 619[62]; **2**, 635[46], 640[46], 836[160]; **4**, 341[55], 1093[145]; **5**, 480[177], 483[174]; **6**, 20[75], 91[128], 109[39-41], 150[128], 466[46], 564[916], 624[144]; **7**, 475[57], 496[217], 497[218], 522[39]; **8**, 395[129]
Takeue, S., **4**, 507[153]
Takeya, T., **3**, 676[77], 695[153], 696[154]
Takeyama, T., **1**, 259[29], 260[32,32a], 261[32a], 829[87], 831[104]; **8**, 803[92]
Takezawa, K., **4**, 239[26], 251[26], 257[26]; **5**, 583[183]
Taki, H., **7**, 227[88], 314[40], 315[40]
Taki, K., **3**, 295[8]
Takido, T., **6**, 443[89], 493[139], 508[285]; **8**, 323[113]
Takigawa, H., **7**, 57[22]
Takigawa, T., **3**, 17[87], 124[263], 126[263]; **4**, 1005[88], 1020[235]; **6**, 657[177], 836[50]; **7**, 682[83,84]
Takikawa, Y., **5**, 442[185]
Takinami, S., **3**, 251[78], 254[78], 490[66], 511[66], 515[66]; **4**, 286[174], 287[174], 290[174], 358[158]
Takino, T., **5**, 915[113]
Takino, Y., **4**, 231[277]
Takita, S., **7**, 132[96], 158[36a]

Takita, T., **2**, 917[85]; **7**, 489[173]
Takita, Y., **8**, 132[7]
Takiura, K., **4**, 309[418], 314[494]
Takiyama, N., **1**, 232[12], 234[12,21], 248[61,62], 260[32], 261[32c], 332[53], 735[25]; **4**, 973[86]
Takizawa, T., **5**, 1130[3]; **8**, 797[34]
Takken, H. J., **2**, 782[17,21]; **3**, 251[79], 254[79]; **8**, 535[161], 542[229], 946[139]
Takle, A., **6**, 989[76]
Takuma, K., **6**, 1066[97]
Takundwa, C., **3**, 531[87], 538[87]
Takusagawa, F., **1**, 838[161,167]; **5**, 417[62], 420[62]
Takyu, M., **6**, 87[102]
Tal, D., **7**, 842[31], 843[42]
Talakvadze, T. G., **8**, 772[60]
Talamas, F. X., **6**, 677[314]
Talanov, V. N., **6**, 546[645]
Talanov, Yu. M., **8**, 150[132]
Talapatra, B., **3**, 396[112], 398[112]
Talapatra, S. K., **3**, 396[112], 398[112]
Talaty, E. R., **1**, 420[85]
Talbert, J., **1**, 212[8], 213[8], 432[138], 433[138]; **2**, 450[50]; **4**, 175[42]
Talbiersky, J., **4**, 121[205a]
Talbot, H. P., **4**, 282[140], 288[140]
Tallec, A., **8**, 135[47], 136[47], 137[59]
Talley, J. J., **1**, 846[17]; **4**, 48[139]; **5**, 468[136], 534[95], 585[203]
Tallman, E. A., **4**, 14[48]
Talma, A. G., **8**, 95[91], 96[92]
Talwalker, P. K., **6**, 543[618]
Talwar, K. K., **8**, 338[92], 339[92]
Tam, C. C., **2**, 72[79]; **3**, 131[325]; **4**, 113[164], 259[256]; **5**, 885[19]
Tam, J. P., **6**, 637[32,32d]
Tam, K.-F., **5**, 63[15], 123[1], 126[1]; **7**, 815[4], 824[4], 833[4]
Tam, S. W., **6**, 765[18]
Tam, S. Y.-K., **6**, 553[795], 554[795], 987[69]; **7**, 318[53], 319[53]; **8**, 219[80], 965[69]
Tam, T.-F., **2**, 827[127]; **3**, 217[92]; **4**, 394[189,189c]
Tam, W., **3**, 457[127], 458[134]; **8**, 671[16], 684[96]
Tamada, Y., **7**, 828[54]
Tamagnan, G., **4**, 89[84j]
Tamai, K., **3**, 245[32], 470[213], 476[213]
Tamai, S., **2**, 116[140], 610[94], 611[94], 1059[78]
Tamaki, K., **3**, 789[70]; **7**, 773[309], 776[309]
Tamano, H., **1**, 559[150]; **2**, 1049[18]
Tamano, T., **8**, 916[108], 917[108], 918[108], 920[108]
Tamao, K., **1**, 113[30,32], 619[61], 624[85]; **2**, 23[90,90b], 29[90b], 59[16], 584[118]; **3**, 200[68], 225[184], 228[216,217,218,222], 229[225], 262[165], 436[3,10], 437[21,25,26], 438[31], 440[25], 448[25,96], 449[25,99,100], 450[25,26], 451[25,106], 452[25,106,107,111], 453[31], 457[129,130], 459[31,137,139], 460[107,130,137], 461[137], 464[171], 469[203], 470[203], 473[203], 483[16], 487[45], 492[74], 497[104], 503[145], 510[183,206], 513[205,206], 524[27]; **4**, 116[188a], 120[202], 682[57], 840[34]; **5**, 1132[19], 1183[55]; **6**, 16[59,60], 17[59,63], 18[59,63]; **7**, 172[129], 453[93], 455[93], 641[3], 642[8-12], 643[13-15], 644[16], 645[17-21], 647[34,36,38], 816[13]; **8**, 764[10], 770[32], 772[52], 773[10,67], 782[105], 783[106,107,109], 787[119], 788[120,121]
Tamaoka, T., **2**, 82[9], 374[275], 575[64]
Tamara, K., **4**, 611[359]
Tamara, Y., **4**, 948[95]; **7**, 314[39]
Tamariz, J., **1**, 552[83]; **3**, 198[49]
Tamarkin, D., **3**, 619[26,27]
Tamaru, R., **4**, 599[218], 640[218]
Tamaru, Y., **1**, 212[9], 213[9,15], 215[41], 216[41], 217[47], 327[12], 449[210]; **2**, 185[30], 211[115], 215[115,133], 216[133,135], 217[30,136], 442[13,14], 449[13,14,49], 450[13,14,51], 867[15]; **3**, 221[131], 231[242], 420[48,49], 443[61], 445[61], 463[157,159], 1023[23], 1032[70], 1033[71], 1040[103]; **4**, 75[44a], 76[44c], 85[73], 100[122], 249[117,118], 257[117,118], 377[104], 379[104j,114,114b,115,115b], 380[104j,115b], 382[114b], 383[114b], 402[236], 403[237], 404[237], 408[259d], 413[114b], 557[14], 558[19], 564[41,42], 599[214,220], 606[308], 607[308], 642[220], 841[45], 858[109]; **5**, 347[72,72a], 353[72a], 438[164], 791[26], 889[32]; **6**, 425[65], 509[248,250], 538[567]; **7**, 125[54], 503[273]; **8**, 856[185]
Tamary, T., **7**, 7[36]
Tamás, J., **2**, 817[90], 851[223]; **7**, 746[93]
Tamasi, M., **2**, 1094[89], 1095[89]
Tamasik, W., **7**, 166[87]
Tamblyn, W. B., **8**, 240[33]
Tamblyn, W. H., **3**, 920[34], 925[34a], 934[34]; **4**, 1033[17,20], 1035[37,41], 1036[47,49], 1037[17,37]; **6**, 873[9], 897[9b]
Tamborski, C., **3**, 450[104], 457[128]
Tamburasev, Z., **6**, 766[23]; **7**, 698[51]
Tamburini, H. J., **1**, 294[47]
Tambute, A., **4**, 869[27], 870[27], 871[27], 877[27d]
Tamm, C., **2**, 100[9]; **3**, 736[25]; **6**, 1062[80]; **7**, 120[13], 123[13], 162[56], 180[160], 429[151]; **8**, 315[45]
Tammer, T., **2**, 1087[35]
Tamnefors, I., **4**, 55[157], 56[157a]
Tamori, S., **2**, 372[270], 373[270]; **5**, 468[130]
Tamoto, K., **2**, 810[59]; **7**, 881[160]
Tamres, M., **8**, 218[73], 221[73]
Tamura, C., **6**, 566[927]
Tamura, H., **2**, 547[108], 550[108]; **3**, 635[44]; **4**, 159[81]; **5**, 79[288]
Tamura, M., **2**, 633[31], 640[31]; **3**, 210[23], 243[13], 244[19], 415[10], 418[10,25], 436[1,2,6], 437[24], 438[2], 464[24], 482[7], 494[84], 499[7]; **4**, 158[79], 159[83], 161[90], 923[31], 924[31], 925[31]; **5**, 1012[51]; **6**, 658[184], 804[48]; **7**, 595[21]; **8**, 146[94], 535[166]
Tamura, N., **2**, 157[122]; **7**, 132[96], 158[36a]
Tamura, O., **1**, 391[152]; **2**, 611[102], 643[79], 644[79a], 647[88a,b]; **4**, 161[91]; **6**, 935[104]; **7**, 202[46]
Tamura, R., **1**, 759[128,129]; **2**, 333[65], 334[68]; **4**, 589[81,87,88], 598[81,191], 600[226], 638[81,191], 790[36]; **5**, 1088[78], 1102[148], 1103[148]; **6**, 1000[128,129]; **7**, 883[174]; **8**, 598[98], 969[98]
Tamura, S., **6**, 552[690]
Tamura, T., **2**, 249[84], 258[50], 348[56], 350[74], 362[56], 363[56]; **8**, 159[108], 171[108], 178[108], 179[108], 350[143], 797[34]
Tamura, Y., **1**, 63[43,44], 64[46], 242[49-51], 243[52], 391[152], 448[207], 474[100]; **2**, 215[134], 556[152], 611[102], 643[79], 644[79a], 647[88a,b]; **3**, 386[57], 904[134], 939[79]; **4**, 14[46], 55[157], 57[157h], 75[44b], 125[216,216c], 155[75], 160[86a], 161[91], 163[97], 249[114], 257[114], 261[286], 304[356], 350[116,118], 562[33]; **5**, 133[53,54], 439[170], 451[18], 470[18], 473[153], 477[153], 504[274], 838[74]; **6**, 764[13], 846[105], 910[3], 930[87], 931[88,89], 935[104], 936[112]; **7**, 199[34], 200[41], 202[46], 208[86], 209[89], 211[86], 382[108], 606[156], 746[90], 829[56,56c]; **8**, 274[130], 568[466], 829[81], 837[14], 964[56]
Tamura, Z., **1**, 34[168]
Tamuru, R., **4**, 599[219], 641[219]
Tan, C., **4**, 740[118]
Tan, C. C., **4**, 487[43]
Tan, E. W., **4**, 744[134]
Tan, J. L., **8**, 455[205], 456[205a], 613[81], 629[81,184]
Tan, S.-H., **6**, 529[465]
Tan, S. L., **5**, 132[48]
Tan, S. Y., **6**, 66[6]
Tan, T. S., **7**, 412[106]
Tanabe, K., **3**, 300[43]
Tanabe, M., **2**, 505[8]; **3**, 34[195]; **5**, 835[59]; **6**, 1043[15], 1059[15,63]; **8**, 528[63], 531[116], 934[54], 938[54], 948[149], 993[57]

Tanabe, T., **8**, 150[131]
Tanabe, Y., **2**, 802[38]; **3**, 470[207], 473[207]; **8**, 287[16]
Tanác, B., **7**, 764[120]
Tanaguchi, M., **7**, 335[32]
Tanahashi, Y., **3**, 741[50]
Tanaka, A., **7**, 415[115b]
Tanaka, C., **4**, 792[65]
Tanaka, H., **1**, 101[90], 471[71]; **2**, 73[61], 976[2], 981[2], 982[2]; **3**, 99[181], 101[181], 501[137], 509[137], 579[101], 650[211], 652[220]; **4**, 254[189], 257[189], 606[307], 607[307,315], 647[307]; **5**, 73[201], 473[153], 477[153], 1200[49]; **6**, 552[690], 563[900], 656[171], 836[50]; **7**, 356[50], 537[58,60,61], 761[60], 765[60]; **8**, 976[135], 994[65]
Tanaka, I., **2**, 292[79]
Tanaka, J., **2**, 482[21]; **4**, 120[197]
Tanaka, K., **1**, 223[84], 225[84d], 546[60]; **2**, 73[64], 74[64], 357[143], 736[26], 743[81], 876[34]; **3**, 99[180], 100[180], 159[455], 161[455], 174[526]; **4**, 40[114], 78[54], 86[54b], 128[221], 251[147], 402[236]; **5**, 839[85]; **6**, 528[411,412], 1022[63]; **7**, 59[38], 503[273], 761[65], 766[172]; **8**, 162[30], 427[51], 562[422], 859[215]
Tanaka, K.-I., **8**, 421[26], 422[26]
Tanaka, M., **1**, 823[44a], 881[118]; **2**, 105[39], 348[58], 357[58,145], 358[145], 726[123]; **3**, 226[200], 507[175], 529[51], 1016[3], 1024[26], 1039[101,102]; **4**, 159[81], 930[46,47,52], 945[91], 1095[152]; **5**, 134[61], 256[56], 596[21], 597[21], 601[21], 603[21], 606[21], 608[21], 810[126], 812[126]; **6**, 4[19], 527[405], 1052[42a]; **7**, 6[31], 142[138], 324[71], 417[130c]; **8**, 36[81], 54[81], 55[180], 66[81,180], 152[178], 289[24], 291[35], 301[91,92], 463[270]
Tanaka, N., **1**, 14[74]; **3**, 159[455], 161[455]; **4**, 378[107]; **5**, 196[15]
Tanaka, O., **4**, 378[107]; **7**, 43[40,41]
Tanaka, R., **4**, 3[10], 41[10], 47[10], 65[10a], 66[10,10a], 278[100], 286[100], 370[44]
Tanaka, S., **1**, 65[51], 187[62,63], 188[62], 779[222]; **3**, 575[82], 740[46]; **4**, 21[66,66a,c], 62[66c], 107[146b,c], 218[143], 315[526]; **5**, 24[169], 790[22], 820[22]; **6**, 960[64]; **7**, 760[32]; **8**, 397[140]
Tanaka, T., **1**, 183[51], 658[158], 659[158], 664[158], 665[158], 672[158], 763[145]; **2**, 10[40], 232[177], 567[30], 649[103], 833[147], 1059[76], 1066[122]; **3**, 222[144]; **4**, 13[44,44c], 159[85], 253[169], 256[208,212], 261[208,284], 262[212,308], 278[100], 286[100], 682[57], 1089[138], 1091[138]; **5**, 105[195], 833[46]; **6**, 96[149], 614[96], 625[162], 652[140], 837[60], 942[154], 944[154]; **7**, 137[123], 139[123], 155[25], 245[73], 246[81,83,84,86], 643[15], 645[19], 693[28], 806[71], 883[180]; **8**, 7[41], 11[59], 95[86], 152[183], 460[254], 500[52], 535[166], 788[120], 963[49]
Tanaka, Y., **1**, 820[16]; **2**, 4[12], 6[12], 10[41], 368[235], 446[31], 573[50], 575[50]; **3**, 244[27,28], 464[177]; **4**, 359[161]; **5**, 822[165]; **6**, 554[727]; **7**, 458[111], 675[54]; **8**, 64[207a], 65[207a], 66[207], 237[17], 240[17], 244[54], 249[17], 369[77], 389[75]
Tancrede, J., **5**, 272[4], 273[4], 275[4]; **8**, 890[138]
Tancrede, T., **6**, 686[365]
Tandon, S. K., **6**, 487[65]
Tane, J. P., **5**, 1134[37,41]
Taneja, H. R., **3**, 125[299]
Taneja, K. L., **3**, 325[159]
Tanemura, K., **8**, 625[166], 627[166]
Tang, C.-P., **7**, 40[7], 367[56]
Tang, F. Y. N., **5**, 929[173]
Tang, H., **3**, 638[93]
Tang, J.-Y., **6**, 619[118]
Tang, P. C., **2**, 547[100], 548[100]; **4**, 390[171]; **5**, 1070[25], 1072[25], 1074[25], 1089[86], 1090[86], 1093[96], 1094[102], 1096[106,108,108c], 1098[96a,106,108a,c], 1099[106,108c], 1100[116], 1102[102], 1112[96a,106,108a,c,116], 1183[57]
Tang, P.-F. L., **2**, 632[26], 640[26]
Tang, P.-W., **2**, 74[80]; **3**, 155[433], 157[433], 230[238]; **6**, 150[116], 151[116]
Tang, Q., **3**, 505[157]; **4**, 464[114], 470[138], 477[114], 478[114]
Tang, R., **7**, 13[108]
Tang, R. T., **7**, 872[98]
Tang, W., **7**, 586[163]
Tang, X., **2**, 772[16]
Tanguy, G., **3**, 1026[39]
Tani, F., **7**, 427[148e]
Tani, H., **1**, 214[27]; **8**, 385[45], 408[75], 978[146]
Tani, K., **6**, 866[208]; **7**, 426[148a]; **8**, 153[184,185,186,187], 154[189]
Tani, M., **2**, 736[26], 743[81]
Tani, S., **3**, 286[56a]
Tanida, H., **5**, 210[56], 585[201], 619[11], 620[11], 621[18], 631[57,58]; **6**, 421[31]
Tanielian, C., **5**, 154[31]
Tanigawa, E., **8**, 153[184,185], 154[189]
Tanigawa, H., **4**, 163[97]
Tanigawa, S., **4**, 159[82]
Tanigawa, Y., **2**, 587[139,146]; **4**, 589[76], 598[189,198], 640[423]; **3**, 222[143], 223[150], 259[130], 437[23]; 6, 74[36], 76[53], 85[92], 86[97], 253[156]; **7**, 92[41,41a], 94[41], 173[133]
Taniguchi, H., **1**, 162[104], 254[14,14b], 277[80,82,83], 278[14a,84], 279[86], 280[86]; **3**, 295[11], 302[11], 567[36,190], 595[190], 607[190], 610[36]; **4**, 837[10], 856[100]; **6**, 283[163,164]; **8**, 113[36]
Taniguchi, K., **2**, 573[56]
Taniguchi, M., **1**, 198[91], 564[201]; **2**, 116[125], 139[29], 323[23], 331[23], 332[23]; **4**, 285[164], 289[164]; **6**, 8[38], 27[119]
Taniguchi, R., **2**, 736[24,25]
Taniguchi, T., **2**, 1089[57]; **6**, 86[97]
Taniguchi, Y., **2**, 357[136]; **3**, 45[248]; **4**, 598[198], 640[423]; **6**, 76[53], 113[69], 176[89,90], 253[156]; **7**, 693[27]; **8**, 145[83], 395[123]
Tanikaga, R., **2**, 348[56], 350[73,74], 362[56], 363[56,73]; **3**, 153[415], 224[170]; **4**, 13[44], 251[147]; **6**, 87[102], 839[69], 926[68,69], 927[68], 1022[63]; **7**, 197[17,19], 667[79]
Tanimoto, M., **8**, 422[36]
Tanimoto, N., **2**, 159[131]; **4**, 238[11], 245[11], 255[11], 260[11]; **7**, 680[80]
Tanimoto, S., **2**, 555[138], 792[68]; **3**, 95[152]; **4**, 310[435], 510[170], 610[342]; **6**, 104[7], 109[7], 134[31]; **7**, 443[51a], 747[97], 778[409]; **8**, 170[81], 847[90], 856[170]
Tanimoto, Y., **8**, 159[34], 162[33,34], 164[34], 178[34], 179[34]
Tanimura, H., **6**, 563[899], 626[164]
Tanino, H., **2**, 384[319], 879[42]; **6**, 266[50]; **8**, 563[425]
Tanis, S. P., **1**, 733[11], 826[61]; **3**, 220[122], 224[172,172b,175], 325[162], 416[11-13,19], 417[13], 751[87], 752[95]; **5**, 320[8], 851[170]; **7**, 362[30]
Taniyama, E., **4**, 295[251], 561[31], 741[125]; **8**, 856[186]
Tanji, K., **6**, 530[423]
Tanke, R. S., **8**, 64[207b], 66[207]
Tanko, J., **5**, 221[58], 226[58,112], 900[12], 901[12], 903[12], 905[12], 907[12], 913[12], 921[12], 926[12], 943[12], 1006[33]; **7**, 815[3], 824[3], 833[3]
Tanner, D., **1**, 779[226], 798[291]; **3**, 593[180,183], 594[184]; **4**, 202[24], 399[225,225a,b]; **6**, 93[132], 990[83]; **7**, 7[44,45], 17[170], 399[35], 473[26]
Tanner, D. D., **3**, 1046[2]; **7**, 883[175]; **8**, 20[140], 584[20]
Tannert, A., **1**, 202[101], 234[22], 253[8], 331[48], 734[22]
Tanno, N., **8**, 166[55-60], 178[56,60], 179[56,60], 545[288,289], 546[288]
Tanny, S. R., **4**, 1061[162]
Tanoguchi, M., **7**, 340[45]
Tanol, M., **1**, 838[167]
Tanoury, G. J., **1**, 750[108]; **5**, 1125[55], 1143[91]
Tanoury, G. L., **5**, 1183[53]
Tanouti, M., **7**, 595[21]
Tansey, L. W., **8**, 376[161]

Tantillo, A. W., **5**, 195[10]
Tanzella, D. J., **5**, 841[91]; **6**, 859[174]
Tao, E. V. P., **4**, 312[458]
Tao, F., **4**, 1011[164]; **7**, 446[64]
Tao, K., **4**, 695[4]
Tao, N. S., **6**, 423[45]
Tao, T., **5**, 796[55]
Tao, X., **4**, 590[89], 613[89]
Tao, Y.-T., **3**, 172[514]; **6**, 954[18]
Taoka, A., **7**, 429[155]
Tapia, R., **6**, 155[152]; **7**, 355[43]
Tapiolas, D., **4**, 1039[62]
Tapolczay, D. J., **6**, 1031[109]
Tapp, W. J., **5**, 473[149]
Taqui Khan, M. M., **8**, 446[65]
Tarakanova, A. V., **5**, 900[10], 901[10], 906[10], 907[10], 972[6], 982[6], 984[6], 989[6]
Tarama, K., **8**, 449[159,160], 453[191], 567[461]
Taran, M., **3**, 741[50]
Tarantello, M., **6**, 508[289], 509[271]
Tarara, G., **2**, 68[40], 94[53]
Tarasco, C., **1**, 192[82]; **2**, 35[130], 36[130], 566[23]; **7**, 549[42]
Tarasov, V. A., **5**, 1056[48,49], 1057[50]
Tarazi, M., **8**, 98[102]
Tarbell, D. S., **2**, 802[28]; **4**, 120[202]; **6**, 249[141], 277[124], 637[28]; **7**, 567[103], 760[41]; **8**, 314[25], 568[466]
Tarbin, J. A., **7**, 674[49], 737[13]
Tardella, P. A., **3**, 4[25], 7[25]; **6**, 717[111]; **7**, 26[48,49,57], 479[90,91]
Tardif, S., **8**, 316[57]
Tardival, R., **7**, 498[220], 538[63]
Tardivat, J.-C., **7**, 92[51]
Targos, T. S., **5**, 1090[89]
Tarhouni, R., **1**, 830[91]; **3**, 202[91]
Tarka, S. M., **7**, 764[123]
Tarkhanov, G. A., **3**, 640[101]
Tarlin, H. I., **8**, 896[17]
Tarlton, E. J., **2**, 376[280]
Tarnchompoo, B., **1**, 787[255]; **4**, 239[24], 258[24], 261[24]; **5**, 565[98]
Tarnowski, B., **2**, 777[2], 780[2], 787[50,51]; **6**, 489[77]
Tarrant, P., **5**, 65[59,60]
Tartar, A., **4**, 519[20], 520[20]; **7**, 661[49]
Tarttakovskii, V. A., **4**, 145[25]
Tarverdiev, Sh. A., **8**, 772[57]
Tarvin, R. F., **5**, 74[205]
Tarzia, G., **2**, 904[53]; **4**, 347[104]; **8**, 90[47]
Tasaka, A., **5**, 850[148]
Tasaka, K., **6**, 612[76]
Taschner, E., **6**, 665[230], 667[230], 668[259]
Taschner, M. J., **4**, 10[34], 370[46]; **7**, 817[15]; **8**, 248[86], 925[6], 946[136]
Tashika, H., **4**, 115[178]
Tashiro, J., **4**, 124[212]; **5**, 710[51], 713[51]
Tashiro, M., **3**, 295[7], 302[52], 307[88], 329[88,184,185,187]; **7**, 354[36], 355[36]; **8**, 903[51], 906[51], 907[51]
Tashtoush, H. I., **4**, 744[132], 746[143], 747[148], 771[254]; **8**, 857[201]
Tasi, Y.-M., **1**, 602[32], 603[32]
Tasker, P. A., **5**, 687[66]
Tasovac, R., **3**, 905[136,139]
Tata, J. R., **1**, 32[157,158]; **2**, 106[47], 189[46], 209[46]
Tatarsky, D., **8**, 451[180]
Tatchell, A. R., **7**, 555[70]; **8**, 64[204], 66[204], 67[204], 161[14,15,24], 162[25-27], 176[130,131], 545[298,299,300]
Tatcher Borden, W., **5**, 202[38]
Taticchi, A., **8**, 494[25]
Tatke, D. R., **6**, 487[55,56], 489[55,56], 543[55,56]
Tatlow, J. C., **2**, 736[22]; **3**, 522[11], 639[84], 648[187]; **6**, 497[159]; **7**, 15[146]; **8**, 897[20], 901[40], 904[53,54]
Tatone, D., **3**, 124[286], 125[286], 127[286]; **5**, 1152[143]
Tatsukawa, A., **4**, 111[152c]
Tatsumi, C., **8**, 559[400], 979[148]
Tatsumi, K., **1**, 162[98], 163[106], 440[190], 445[190]; **3**, 867[36]; **4**, 615[384]; **5**, 1172[28,29], 1182[28,29]; **8**, 698[145]
Tatsumi, S., **8**, 150[134]
Tatsumi, T., **8**, 554[366]
Tatsumo, T., **7**, 153[6]
Tatsuno, T., **3**, 856[89]; **5**, 136[67,68], 137[74]; **6**, 1050[37]
Tatsuno, Y., **4**, 1033[21], 1037[21], 1040[21]; **8**, 153[185,187], 154[189]
Tatsuta, K., **1**, 569[254]; **2**, 263[54]; **6**, 46[59], 60[145]; **8**, 820[46]
Tatsuzaki, Y., **7**, 423[142]
Tat'yanina, E. M., **3**, 328[181]
Taub, D., **2**, 156[117], 746[110]; **3**, 689[119]; **7**, 675[59]
Taub, W., **6**, 268[69], 271[69]
Taube, H., **1**, 310[108]; **8**, 519[133]
Taube, R., **1**, 215[33]; **4**, 841[46]
Tauber, J. D., **8**, 494[24]
Tauber, S. J., **4**, 868[11]
Taufer-Knöpfel, I., **2**, 636[54], 637[54], 640[54]
Taulov, I. G., **4**, 1093[145]
Taunton, J., **6**, 127[161]
Taurins, A., **6**, 546[649]
Tavakkoli, K., **1**, 39[191]
Tavanaiepour, I., **6**, 150[114]; **7**, 778[401,401a]
Tavares, D. F., **1**, 524[91], 530[127]; **2**, 416[17], 417[21]; **3**, 748[77]
Tavecchia, P., **4**, 261[290]
Taveras, A. G., **3**, 51[270]; **8**, 508[87]
Tavernier, D., **8**, 348[132]
Tavs, P., **5**, 422[82]
Tawara, K., **4**, 591[112], 633[112]; **7**, 415[114]
Tawara, Y., **4**, 1005[88], 1020[235]
Tawarayama, Y., **1**, 232[13], 233[13], 234[13], 253[9], 276[9], 278[9]; **2**, 312[35]
Tawney, P. O., **3**, 208[4], 210[4]; **4**, 70[4], 89[4], 148[46]
Tawrynowics, W., **4**, 1002[48]
Tay, M. K., **2**, 482[27], 483[27]
Taya, K., **7**, 452[54,55], 462[54,55]; **8**, 366[42], 427[51]
Taya, S., **3**, 992[37], 994[41], 999[51]; **5**, 821[162], 851[165], 888[29], 889[31]; **6**, 851[130], 852[138], 876[26], 877[36], 879[36], 882[26], 883[36], 885[26], 886[36]
Tayano, T., **4**, 147[41], 358[155]
Tayim, H. A., **8**, 447[111], 453[111]
Taylor, A., **5**, 167[94]
Taylor, A. P., **2**, 198[85], 632[27], 640[27]
Taylor, A. W., **4**, 259[267]; **5**, 768[126], 779[126]
Taylor, B. J., **7**, 278[158]
Taylor, B. S., **2**, 583[111]; **4**, 604[283]
Taylor, C. E., **3**, 53[272]
Taylor, D. A., **2**, 866[4]; **3**, 26[161]; **5**, 282[23]
Taylor, D. A. H., **8**, 537[183], 950[165]
Taylor, D. R., **5**, 9[65-69]
Taylor, E. A., **6**, 960[57], 1011[3]
Taylor, E. C., **2**, 496[69], 498[69]; **3**, 54[275], 219[108], 418[28], 499[123,124], 505[165], 507[124], 551[1], 670[58], 673[58,69], 693[141]; **4**, 51[144c], 411[266a], 567[48], 1056[139]; **5**, 417[60,66], 478[162], 491[207], 492[249], 637[111,114]; **6**, 432[117], 524[369], 533[369], 554[717,778], 675[301]; **7**, 154[19,20], 185[173], 292[4], 335[28], 336[34], 358[6], 372[6], 516[2], 614[3], 718[3], 724[3], 732[58], 752[145], 816[6a], 824[6], 825[6], 827[6a], 828[51], 829[6a,59], 831[6a], 832[6a], 833[6a], 845[80], 846[80], 851[18],

872[97], 888[97]; **8**, 365[33], 395[129], 490[4], 492[4], 625[158], 626[158]
Taylor, E. G., **6**, 896[92]
Taylor, E. J., **8**, 355[180]
Taylor, E. R., **7**, 122[28]
Taylor, G., **5**, 647[12], 648[12]
Taylor, G. A., **2**, 354[115]; **6**, 690[402], 692[402]
Taylor, G. F., **5**, 331[43]
Taylor, G. J., **2**, 1075[150]
Taylor, G. N., **5**, 637[105]
Taylor, H. M., **8**, 564[443]
Taylor, H. T., **5**, 766[113,114], 777[113]
Taylor, I. D., **1**, 760[135]; **7**, 440[38]
Taylor, J. A., **3**, 553[14]
Taylor, J. B., **5**, 13[92], 687[60]
Taylor, J. E., **7**, 444[55-57]
Taylor, J. S., **2**, 385[324]
Taylor, K. B., **7**, 99[108]
Taylor, K. G., **1**, 832[107], 873[94]; **4**, 1007[113,117], 1012[117], 1013[181], 1014[183]; **6**, 971[128]
Taylor, M. D., **2**, 651[120]
Taylor, M. K., **5**, 92[66], 94[66]
Taylor, M. V., **6**, 1017[38], 1024[38]
Taylor, N. J., **4**, 5[17], 315[532]; **5**, 485[184]
Taylor, P. C., **5**, 422[85]
Taylor, P. D., **8**, 455[206]
Taylor, P. G., **1**, 580[2], 581[2], 582[2], 610[2a], 611[2a], 616[2a], 618[58], 621[67], 784[239,240], 815[239]
Taylor, R., **1**, 2[3], 37[3]; **2**, 735[11]; **6**, 104[9], 436[9], 1033[130]; **8**, 850[121]
Taylor, R. G., **5**, 1043[24], 1046[24], 1049[24], 1051[24,36b]
Taylor, R. J., **6**, 127[160]
Taylor, R. J., Jr., **8**, 36[75], 37[75], 38[75], 39[75], 45[75], 54[75], 66[75]
Taylor, R. J. K., **1**, 132[106], 564[197], 865[87]; **2**, 119[167], 832[146]; **3**, 3[11], 8[11], 875[76]; **4**, 187[99], 197[99], 238[8], 245[91], 254[8,91], 259[8], 260[281], 261[8], 371[52], 378[52a], 896[169]
Taylor, R. L., **6**, 530[420]
Taylor, R. T., **2**, 588[150]; **3**, 254[96]; **4**, 116[188b]; **5**, 796[53], 857[230]; **7**, 172[127], 318[57], 319[57], 447[71], 674[50]; **8**, 946[134]
Taylor, S. H., **8**, 445[39]
Taylor, S. K., **3**, 325[161,161a], 390[80], 392[80], 396[105,106], 397[105,106]
Taylor, S. L., **1**, 471[63]
Taylor, T. I., **3**, 822[13], 829[13]
Taylor, W. C., **2**, 773[26]; **3**, 831[59]
Taylor, W. G., **4**, 293[235]
Taylor, W. I., **3**, 659[1], 660[1], 679[1]
Taymaz, K., **2**, 505[12], 510[12]
Tazaki, H., **5**, 516[25]
Tazoe, M., **5**, 167[94]; **7**, 64[61a]
Tazuma, J., **7**, 168[106], 170[106]
Tchir, M. F., **1**, 544[43]; **5**, 125[16]
Tchoubar, B., **1**, 851[38], 852[38]; **2**, 818[93], 855[93]; **3**, 781[15], 839[5], 845[39]; **8**, 479[48], 481[48], 528[56]
Tea-Gokou, C., **6**, 554[754]
Teague, S. J., **1**, 133[112]; **3**, 384[53], 400[53], 603[227]; **5**, 133[55]; **6**, 1062[75]
Tebbe, F. N., **1**, 743[51], 746[51], 749[76], 811[51]; **5**, 1115[2], 1116[2,2c], 1121[2c], 1122[2c], 1123[2c], 1124[45]
Tebby, J. C., **8**, 860[221]
Tecilla, P., **5**, 408[33]
Tecle, B., **1**, 17[205]
Tedder, J. M., **2**, 736[22]; **4**, 717[8], 727[54], 729[63]; **5**, 819[152]
Tedeschi, R., **8**, 92[68]
Tedjo, E. M., **6**, 1023[70]
Tee, O. S., **5**, 703[16b]
Teetz, T., **8**, 339[91]
Tefertiller, B. A., **3**, 14[70]
Tefertiller, N. B., **2**, 151[98], 152[98]
Tegg, D., **2**, 139[29]
Tegmo-Larsson, I.-M., **5**, 216[16], 218[36], 219[16], 221[16], 632[62]
Teichmann, K.-H., **3**, 890[34]
Teisseire, P., **3**, 599[207]; **7**, 61[48], 64[48]
Teissier, B., **3**, 661[22]
Teitei, T., **5**, 637[107]
Teitel, S., **8**, 968[86]
Teixeira, M. A., **8**, 881[69]
Teixeira-Dias, J. J. C., **6**, 436[18]
Teixidor, F., **1**, 34[223]
Tejero, T., **3**, 579[100]
Tekezawa, N., **8**, 86[25]
Teleha, C. A., **1**, 883[126], 898[126]; **5**, 857[230]
Telfer, S. J., **2**, 369[246]; **3**, 369[108]
Telford, R. P., **8**, 263[30,31], 366[43], 368[63]
Telkowski, L., **2**, 977[5]
Tellado, F. G., **8**, 798[61]
Teller, R. G., **8**, 431[66], 459[226]
Teller, S. R., **2**, 965[68]; **8**, 614[88], 621[88]
Tellier, F., **3**, 525[41]
Tellier, J., **8**, 424[43]
Telschow, J. E., **3**, 23[136], 24[136]
Telshow, J. E., **7**, 160[52], 161[52], 176[52], 180[52], 183[52], 187[52]
Tember, G. A., **3**, 643[131]
Tember-Kovaleva, T. A., **6**, 432[116]
Temnikova, T. I., **6**, 276[120]
Temple, D. L., **2**, 489[45]; **6**, 674[295,296]
Temple, J. S., **3**, 469[202], 470[202], 473[202]; **4**, 595[155,156], 620[155,156,395], 634[155], 635[155,156,395]; **8**, 693[113,115]
Temple, R., **2**, 957[16]
Temple, R. W., **8**, 528[69], 530[69]
Templeton, D. H., **3**, 178[542], 179[542]
Templeton, J. F., **4**, 52[147,147a]; **7**, 159[45]
Templeton, J. L., **3**, 47[257]
Templeton, L. K., **3**, 178[542], 179[542]
Tenaglia, A., **5**, 641[133]; **6**, 11[46], 24[98], 86[101]
TenBrink, R. E., **8**, 392[96]
ten Brocke, J., **7**, 429[156]
Tenca, C., **8**, 406[50]
Tendick, F. H., **2**, 958[23]
Tener, G. M., **6**, 611[64]
Teng, C.-Y. P., **2**, 904[51]
Teng, K., **1**, 447[203], 458[203]
Tenge, B. J., **1**, 635[88]; **3**, 104[209], 117[209]; **6**, 846[103], 905[145]
Tengi, J., **6**, 543[621]
Tenhaeff, H., **3**, 915[13], 965[13]
Ten Hoeve, W., **3**, 68[23], 71[34]; **4**, 26[77]; **6**, 551[681]
Tennant, G., **1**, 391[150]; **6**, 225[5], 226[5], 242[90], 243[90], 258[5], 293[229], 551[688]
Tenneson, S. M., **5**, 96[111]
Tenney, L. P., **5**, 220[48]
Tennigkeit, J., **6**, 612[72]
Tenor, E., **8**, 625[163]
Tenud, L., **3**, 13[68], 916[19]
Teo, C. C., **1**, 390[142]; **4**, 34[97], 35[97,97d]
Teplyakov, M. M., **2**, 387[334]
Teppa, F., **2**, 138[23]
Te Raa, J., **2**, 762[55]
Terada, A., **3**, 963[120]; **6**, 288[191]
Terada, I., **8**, 552[353,354]
Terada, K., **3**, 503[142]; **6**, 604[33]; **8**, 187[39], 676[79]

Terada, M., **2**, 455[18], 556[156]
Terada, S., **4**, 124[215]; **6**, 636[20]
Terada, T., **2**, 801[25], 913[73]; **4**, 27[82], 222[166], 802[128]
Terada, Y., **2**, 553[124]; **3**, 264[185], 665[38], 691[128,131], 693[128,139], 697[38,131,139,155], 698[128]
Teradd, T., **7**, 309[26]
Terahara, A., **7**, 77[121,122]
Teraji, T., **7**, 493[198]
Terakado, M., **5**, 55[48]
Teramura, D. H., **5**, 839[77]
Teranaka, T., **6**, 979[29]
Terando, N. H., **4**, 402[235], 404[235], 408[235]; **7**, 503[273]
Teranishi, A. Y., **1**, 634[75], 642[75], 644[75]; **2**, 124[201], 184[25], 235[190], 268[66], 280[23], 289[23], 311[34]; **6**, 1029[101]; **7**, 124[39], 128[39], 130[39], 131[39], 438[11], 441[11], 443[11], 527[2,4], 528[2], 530[2], 771[272], 772[272]
Teranishi, K., **8**, 320[81,82]
Teranishi, S., **3**, 583[120], 1041[111]; **4**, 603[275], 626[275], 645[275], 836[2,4,5], 837[13-15], 841[50], 856[100], 941[81]; **6**, 711[64]; **7**, 107[168], 309[25], 321[66], 587[171], 823[36]; **8**, 142[50], 150[137,138], 419[17], 430[17], 889[128]
Terao, K., **1**, 648[126]; **3**, 87[111], 106[111], 114[111]; **4**, 370[41], 387[158], 398[216], 399[216e], 401[41,158c,216e,232,232a], 405[41,158c,216e,232a], 406[41,158c,216e,232a]; **6**, 289[196], 293[196,226,227], 1030[104]; **7**, 129[70], 495[208,212,213], 496[212,213], 524[52]; **8**, 848[104], 849[104]
Terao, Y., **1**, 370[71], 371[71]; **4**, 1089[125], 1095[152]
Terasaki, T., **8**, 152[182]
Terasawa, M., **8**, 419[17], 430[17]
Terasawa, T., **8**, 530[100]
Terashima, M., **3**, 498[109], 1038[95,95b]
Terashima, S., **1**, 241[44], 242[47,48]; **2**, 123[197,198], 163[147], 292[77], 810[59], 1059[76]; **4**, 221[156], 391[184], 393[184a,b], 397[184b]; **5**, 98[123,124]; **6**, 716[97], 717[105], 725[169], 799[23], 960[53]; **7**, 27[76], 29[78], 174[135]; **8**, 166[55-61], 178[56,60], 179[56,60], 269[81], 332[43], 545[288,289], 546[288]
Teratani, S., **7**, 452[55], 462[55]; **8**, 427[51]
Terauchi, M., **2**, 73[64], 74[64]; **3**, 99[180], 100[180]
Terawaki, Y., **2**, 922[103,104]; **4**, 231[275]; **5**, 102[164]; **6**, 817[103]
Terayama, Y., **5**, 578[152]
ter Borg, A. P., **4**, 1017[210]; **5**, 709[45]
Terem, B., **3**, 577[89]
Teremura, K., **6**, 498[167]
Terenghi, G., **5**, 925[155], 1149[124], 1154[157]
Terent'ev, P. B., **6**, 555[817]
Teresawa, I., **7**, 700[62]
Tereshchenko, G. F., **6**, 795[12], 798[12], 817[12], 820[12]
Ter-Gabrielyan, E. G., **5**, 113[234]; **6**, 547[667], 552[667]
Termin, A., **6**, 53[117]
Termine, E. J., **2**, 506[17], 523[85]; **6**, 771[39-41]; **7**, 697[48]
Termini, J., **2**, 482[35], 484[35]
Termont, D., **5**, 127[25]
Ternai, B., **5**, 491[206]
Ternansky, R. J., **5**, 348[73b], 662[36], 789[17], 956[306], 1005[31]
Ternay, A. L., Jr., **7**, 763[95,96], 777[369]
Ternovskoi, L. A., **4**, 347[93]
Terpinski, J., **6**, 219[119]
Terpko, M., **4**, 845[68], 847[68]
Terpstra, J. W., **4**, 1038[59]
Terrell, R., **3**, 28[170], 30[170]; **4**, 6[21,21a], 7[21a]; **6**, 703[1,3], 714[3]
Terrett, N. K., **1**, 610[44]; **2**, 583[112], 587[112]; **5**, 812[134]; **7**, 616[13]
Terrier, F., **4**, 426[43,44], 444[43]
Terry, L. W., **3**, 1048[13]
Ter-Sarkisyan, G. S., **4**, 145[28]
Terss, R. H., **8**, 568[471]
Terui, K., **8**, 902[47], 904[47], 905[47]
Terunuma, D., **2**, 810[63], 824[63]
Tesarová, A., **8**, 596[81]
Tesch, U. H., **6**, 554[805]
Teschen, H., **8**, 558[395]
Teschner, M., **2**, 223[149]; **7**, 124[48], 125[48], 126[48]
Tesser, G. I., **6**, 638[41,41b], 643[77], 667[235]
Tessier, D., **3**, 574[77]
Testa, E., **8**, 641[27]
Testa, R., **6**, 487[51]
Testaferri, L., **1**, 670[187], 678[187]; **3**, 457[133], 493[83], 503[146], 509[178], 513[83,146]; **4**, 426[51], 437[51,148], 438[148], 441[173,177,179,180,181,182], 447[216,217], 452[20]; **6**, 462[20,21]; **7**, 338[41], 340[45], 770[256b], 771[256], 773[306], 779[427]
Tetaz, J. R., **7**, 666[73,76]
Tetenbaum, M. T., **6**, 774[48]; **7**, 700[60]
Teter, L. A., **3**, 209[13]
Tetsukawa, H., **7**, 429[154]
Teuben, J. H., **1**, 232[16]
Teufel, E., **5**, 636[98]
Teulade, M. P., **3**, 201[83]
Teuscher, P., **6**, 208[60]
Teutsch, G., **2**, 920[97], 921[97]; **5**, 96[115]
Tewari, R. S., **1**, 835[135]
Texier, F., **6**, 555[813]; **7**, 476[63]
Texier-Boullet, F., **1**, 821[26]; **2**, 330[52], 344[16,17], 345[17], 353[16,17], 359[16], 360[16], 363[16,17]; **6**, 175[79]
Teyssie, P., **3**, 1047[7,8], 1051[7,8]; **4**, 609[330], 1033[16,16d,e], 1035[16e], 1051[125], 1052[16d]; **5**, 86[35]; **6**, 25[101]; **7**, 8[61]
Tezuka, H., **3**, 381[26], 382[26]; **4**, 347[87]
Tezuka, T., **7**, 488[155], 490[155], 765[168]
Thach Duong, **7**, 842[29,30,32]
Thachet, C. T., **7**, 155[28]
Thacker, C. M., **7**, 14[141]
Thaisrivongs, S., **1**, 744[56]; **2**, 102[27], 645[82]; **5**, 841[86,87,93], 859[87c,93,234], 1123[36]; **6**, 859[163,164], 978[21]; **7**, 549[44], 583[44], 586[44]; **8**, 720[133], 852[142]
Thaker, V. B., **1**, 268[56], 269[56a]; **3**, 602[225]; **8**, 114[51]
Thakker, K. R., **3**, 572[64]; **8**, 109[12]
Thakor, M. R., **7**, 283[182], 284[182]
Thakore, P. V., **2**, 381[303]
Thakur, D. K., **5**, 504[275]
Thakur, S. B., **5**, 775[177], 780[177]
Thal, C., **3**, 681[100]; **5**, 725[115]
Thalhammer, F., **5**, 491[209]
Thaller, V., **5**, 167[94]; **7**, 306[4]
Thalmann, A., **6**, 438[65]
Thamm, P., **6**, 665[226], 667[226], 668[226], 669[226]
Thamnusan, P., **3**, 154[423], 155[423]; **6**, 1022[66]
Thanassi, J. W., **8**, 957[11]
Thanedar, S., **5**, 1037[5], 1165[10], 1178[10]
Thangavelu, S., **8**, 137[55]
Thaning, M., **2**, 1066[121]
Thanos, I., **8**, 561[409]
Thanupran, C., **4**, 30[88,88p], 121[206], 245[93], 262[93]; **6**, 144[75]
Tharp, G. A., **1**, 822[30]; **5**, 154[33]
Thatcher, K. S., **3**, 816[84]
Thayer, F. K., **2**, 395[3]
Thebtaranonth, C., **1**, 787[255]; **4**, 30[88,88p], 121[206], 239[24], 245[93], 258[24], 261[24], 262[93]; **6**, 144[75]
Thebtaranonth, T., **5**, 565[98]
Thebtaranonth, Y., **1**, 787[255]; **2**, 807[47]; **3**, 916[19], 949[99], 963[121]; **4**, 30[88,88p], 108[146f], 121[206], 239[24], 245[93,94], 258[24], 261[24], 262[93,94], 517[5], 518[5], 519[17], 520[17], 540[17], 541[17], 543[121]; **5**, 560[67,69], 578[149,151]; **6**, 144[75], 690[388]

Theil, F., **7**, 198[28]
Theiland, F., **6**, 654[154]
Theis, A. B., **3**, 209[14]
Theis, R. J., **8**, 950[169]
Theisen, P. D., **1**, 399[6], 772[198]
Theissen, D. R., **4**, 120[199]
Theissen, F., **8**, 739[34]
Theissen, F. J., **4**, 611[347]
Theissen, R. J., **4**, 876[61]; **7**, 140[129], 141[129]
Theobald, D. W., **8**, 530[97]
Theobald, F., **4**, 905[210]; **5**, 21[155], 22[155]
Theobald, P. G., **8**, 913[93], 914[93], 936[72]
Theobald, R. S., **6**, 662[212]
Theodore, L. J., **1**, 582[7]; **2**, 572[42]
Theodore, M. S., **1**, 872[89]; **3**, 785[36,36b]
Theodoridis, G., **4**, 126[218b]
Theodoropoulos, S., **3**, 623[33]
Theopold, K. H., **2**, 127[233], 314[39]
Thepot, J.-Y., **4**, 985[127]
Theramongkol, P., **4**, 331[10], 345[10]; **7**, 400[43]
Therien, M., **1**, 419[79]
Theron, F., **4**, 48[137,137c], 51[143], 55[156,156g]
Thesing, J., **1**, 392[156], 393[156]; **2**, 967[73]; **6**, 233[43]
Thetford, D., **4**, 376[98c,d], 388[98], 401[226,226d,228a]; **7**, 493[187]
Theunisse, A. W. G., **5**, 829[22]
Thewaldt, U., **1**, 139[5]; **6**, 583[995]
Thi, H. C. N., **4**, 1015[198]
Thiagarajan, V., **6**, 707[44]
Thianpatanagul, S., **5**, 257[59]
Thibault, J., **6**, 749[100]
Thibaut, P., **6**, 462[13]
Thibblin, A., **6**, 952[6]
Thiboutot, S., **8**, 193[100]
Thiébault, A., **4**, 453[29-31], 458[68], 459[29-31,80,81,85,86], 467[68], 469[68,80,81,86,136], 471[31,68,139,140,141], 472[29], 473[68,139], 475[30,150]
Thiebault, H., **7**, 538[63]
Thieffry, A., **2**, 681[62]; **5**, 432[126]; **6**, 23[95]
Thiel, U., **6**, 7[34]
Thiele, J., **2**, 144[59], 399[17]
Thiele, K. H., **1**, 215[32]
Thiele, W. E., **5**, 552[5]
Thielert, K., **3**, 1028[48]
Thielke, D., **8**, 58[176], 66[176]
Thielmann, M., **1**, 753[101]
Thielmann, T., **1**, 753[101]
Thiem, J., **3**, 563[1]; **6**, 46[57,62], 49[94], 60[146], 61[150], 978[20]
Thieme, H. K., **2**, 359[159]
Thieme, P. C., **1**, 755[115], 812[115], 813[115]; **8**, 13[73]
Thiensathit, S., **1**, 892[149]; **7**, 453[94]
Thier, W., **8**, 472[3]
Thierrichter, B., **3**, 824[22], 826[22]
Thierry, J., **4**, 748[158], 800[120]; **7**, 722[20], 725[31], 726[20,37], 731[53]
Thies, H., **1**, 360[27]; **3**, 623[36]
Thies, I., **8**, 320[80]
Thies, R. W., **1**, 851[42-44], 886[42,43,135,136,137], 898[42]; **3**, 392[92]; **5**, 787[10], 803[91], 979[27], 980[27], 981[27], 982[27], 1000[6], 1001[17], 1002[17b]
Thieser, R., **6**, 66[7]
Thiessen, G. W., **3**, 643[121]
Thiessen, W. E., **3**, 407[150]
Thijs, L., **3**, 750[82,83]; **4**, 721[30], 725[30]; **5**, 440[175]; **6**, 249[142]; **7**, 473[26]
Thimma Reddy, R., **7**, 778[401,401b], 779[401b]
Thind, S. S., **6**, 255[169]
Thirase, G., **1**, 12[62]
Thirring, K., **3**, 683[103]; **5**, 14[97]
Thoennes, D., **1**, 10[52], 16[85], 22[112], 25[131], 39[188,189]
Thoer, A., **2**, 772[17]
Thom, I. G., **2**, 722[95]; **4**, 701[30]; **6**, 690[400], 692[400]
Thoma, K., **6**, 555[816]
Thomalla, M., **3**, 635[41]
Thomas, A., **2**, 819[101], 823[115]; **3**, 734[4]; **4**, 956[19]; **7**, 221[33,34]
Thomas, A. F., **3**, 996[45]; **5**, 18[126], 632[60], 830[29], 851[166], 853[171], 878[10], 879[11], 881[12], 882[10]; **6**, 1064[89]; **7**, 85[10]
Thomas, A. P., **2**, 583[113], 587[146]; **7**, 360[20]
Thomas, B. R., **8**, 330[45,46], 340[45], 342[46]
Thomas, C. A., **3**, 294[4]; **7**, 172[125]
Thomas, C. B., **3**, 501[135]; **4**, 315[499,500]; **7**, 92[43], 94[55], 828[50]
Thomas, C. W., **2**, 512[46]; **3**, 891[41a]
Thomas, D., **5**, 482[170]
Thomas, D. G., **1**, 390[144]; **2**, 935[146], 936[146], 999[39]; **5**, 100[149]
Thomas, E., **2**, 323[26], 335[26], 336[26]; **4**, 394[189,189c]
Thomas, E. J., **1**, 568[228], 656[152], 658[152], 823[40]; **2**, 18[70], 39[135], 40[142], 566[19], 573[51], 574[59], 575[19,51], 587[142]; **3**, 380[4], 934[64], 953[64]; **5**, 260[62], 514[7]; **6**, 501[184], 687[374], 842[79], 863[190], 1028[98], 1031[109]; **7**, 543[22]; **8**, 831[92], 849[109]
Thomas, E. M., **1**, 447[200]; **3**, 217[92]
Thomas, E. W., **2**, 1049[19,26], 1064[26]; **7**, 313[36]
Thomas, F. L., **8**, 451[180,183]
Thomas, G. H., **7**, 582[149]
Thomas, G. J., **5**, 395[148]; **7**, 351[28], 355[28]
Thomas, H. G., **2**, 1051[40]; **4**, 8[29,29e]; **6**, 561[873]; **7**, 805[67]
Thomas, I., **5**, 7[52]; **7**, 490[179]
Thomas, J., **1**, 349[143,144,145], 361[29]
Thomas, J. A., **3**, 714[32]; **4**, 10[34]; **8**, 925[6], 946[136]
Thomas, J. B., **5**, 140[88]
Thomas, J. M., **4**, 313[467]
Thomas, J. T., **4**, 313[466]
Thomas, M., **7**, 356[48]
Thomas, M. G., **8**, 458[223,223c,d]
Thomas, M. J., **4**, 687[64]; **8**, 674[32]
Thomas, M. T., **3**, 50[268]
Thomas, O. H., **8**, 516[120]
Thomas, P. D., **7**, 221[29]
Thomas, P. J., **3**, 1059[43], 1060[43]; **4**, 682[56]; **6**, 1034[131]; **7**, 453[64,82], 454[64,82]
Thomas, R., **1**, 755[115], 812[115], 813[115]; **3**, 201[75], 854[75]; **5**, 64[47], 1148[123]; **6**, 277[124]
Thomas, R. C., **2**, 903[48], 1024[59]; **3**, 799[102]; **4**, 37[107,107b,e], 38[107e], 39[107e]; **6**, 163[193], 838[66]; **8**, 542[224]
Thomas, R. D., **2**, 90[40]; **4**, 395[205]
Thomas, R. M., **1**, 243[55]
Thomas, S. E., **4**, 115[180d,f]; **8**, 797[44]
Thomas, S. L. S., **3**, 643[125]
Thomas, T. K., **6**, 534[523]
Thomas, T. L., **2**, 291[73]
Thomasco, L. M., **1**, 759[134]
Thömel, F., **3**, 390[84], 392[84]
Thomi, S., **5**, 362[94]
Thommen, W., **2**, 166[155]; **8**, 109[10], 110[10], 112[10], 116[10], 120[10], 358[199]
Thompson, A. S., **1**, 129[93], 589[20,20a], 591[21], 592[20a], 767[163], 779[224]; **2**, 555[136,137]; **6**, 752[111,114]
Thompson, B., **5**, 847[132]
Thompson, B. B., **2**, 894[5], 895[5], 954[6]
Thompson, C. B., **8**, 637[12]

Thompson, C. M., **1**, 350[155]
Thompson, D. J., **3**, 1017[8]; **4**, 665[10], 667[10], 668[14], 670[10,14], 674[10,34], 677[14]
Thompson, D. T., **4**, 914[4], 924[4]
Thompson, D. W., **2**, 553[131], 558[161]
Thompson, E. A., **6**, 656[170]
Thompson, G., **2**, 764[62]
Thompson, H., **3**, 602[218], 607[218]
Thompson, H. W., **4**, 187[94]; **6**, 709[51], 714[83]; **7**, 154[14]; **8**, 430[56], 448[144], 568[467], 814[17]
Thompson, J., **5**, 403[9]
Thompson, J. L., **7**, 86[16a]
Thompson, J. M., **3**, 1037[89]
Thompson, K. J., **4**, 286[171]
Thompson, K. L., **4**, 309[416,417]; **8**, 261[10], 856[169]
Thompson, M., **5**, 468[133]
Thompson, M. E., **3**, 180[551]; **8**, 671[18], 696[126], 949[154]
Thompson, M. J., **6**, 110[55]; **7**, 673[29]
Thompson, M. R., **1**, 532[136]
Thompson, M. S., **7**, 158[40]
Thompson, N., **7**, 194[4], 374[78], 674[41]
Thompson, P., **8**, 274[134]
Thompson, P. A., **2**, 13[56], 14[56a]; **3**, 1033[74]; **4**, 923[32]
Thompson, Q. E., **1**, 446[195]; **7**, 556[72]; **8**, 974[125]
Thompson, R. C., **6**, 639[51], 666[51], 667[51]
Thompson, S. R., **7**, 487[149]
Thompson, W. J., **3**, 511[188,212], 515[188,212]; **7**, 313[37]; **8**, 244[69]
Thompson, W. W., **8**, 814[19]
Thoms, H., **8**, 910[86]
Thomsen, I., **8**, 924[2]
Thomsen, W. F., **8**, 964[50]
Thomson, J. B., **7**, 112[197]
Thomson, J. S., **8**, 354[177]
Thomson, M. J., **8**, 626[175], 629[175]
Thomson, R. H., **3**, 689[122]; **7**, 355[37]
Thomson, S. A., **3**, 766[159]; **5**, 134[63]; **7**, 376[88]
Thomson, S. J., **8**, 286[13], 287[13]
Thomson, T., **3**, 913[2], 914[3,5]
Thomson, W., **5**, 1043[25], 1048[25a], 1051[36a], 1056[25a]
Thon, E., **1**, 33[163]
Thoraval, D., **2**, 605[61], 625[61]
Thorbek, P., **8**, 604[8], 605[8]
Thorburn, I. S., **8**, 446[69]
Thorel, P.-J., **1**, 267[51], 268[51]
Thoren, S., **1**, 429[124]; **3**, 767[165]; **5**, 687[56]
Thormodsen, A. D., **8**, 455[205], 456[205a], 600[104], 613[81], 629[81,184]
Thorn, D. L., **5**, 1171[27], 1172[27], 1178[27]; **8**, 458[223,223c], 671[19]
Thornber, C. W., **4**, 675[39], 680[51]
Thorne, A. J., **1**, 17[206]
Thornmen, W., **3**, 572[64]
Thorns, J. F., **6**, 502[211]
Thornton, E. R., **2**, 117[152,153], 224[152], 232[152], 308[19,21]; **5**, 373[106,106c], 374[106c], 699[6], 856[206]
Thornton, R. E., **8**, 526[19]
Thornton, S. D., Jr., **4**, 97[107c]
Thornton-Pett, M., **4**, 1002[61]
Thorpe, F. G., **7**, 333[26], 595[29,31], 606[155]
Thorpe, J. F., **2**, 848[210,211]; **4**, 4[13]
Thorpe, S. R., **6**, 790[116]
Thorsen, M., **6**, 420[23], 423[23], 451[130], 456[130]
Thorsen, P. T., **8**, 237[17], 240[17], 249[17]
Thorsett, E., **3**, 461[144]; **5**, 249[33]
Thorsett, E. D., **8**, 50[117], 66[117]
Thorstenson, P. C., **3**, 1050[17]
Thottathil, J. K., **1**, 744[57]; **3**, 216[76]; **5**, 1123[36]; **6**, 14[58], 16[58]; **7**, 256[24]
Threadgill, M. D., **4**, 444[200]; **8**, 916[101], 917[101], 918[101], 919[101], 920[101]
Threlfall, T. L., **8**, 907[69]
Threlkel, R. S., **8**, 676[51]
Throckmorton, J. R., **1**, 878[107]
Throop, L. J., **8**, 321[101,102]
Thuan, S. L. T., **3**, 728[36]
Thuillier, A., **2**, 86[24]; **3**, 124[260]; **4**, 85[74]; **5**, 558[62], 559[64], 560[62,65,66], 576[137], 589[210,211]; **6**, 453[141], 455[141,151,152,153], 706[39]; **8**, 268[72]
Thuillier, G., **8**, 587[39]
Thuillier, G. L., **4**, 359[160]
Thummel, R. P., **6**, 11[45], 960[61]
Thun, K., **3**, 890[32]
Thurkauf, A., **7**, 458[115]
Thurmaier, R. J., **6**, 959[44]
Thurmes, W. N., **7**, 553[60]
Thurn, R. D., **6**, 954[17]
Thurston, J., **4**, 812[182]
Thuy, V. M., **6**, 675[301]
Thweatt, J. G., **4**, 45[126,126b,e]; **5**, 71[158], 686[50]
Thyagarajan, B. S., **1**, 528[112]; **3**, 158[442], 164[442], 507[173], 660[10], 803[1], 809[1b], 817[1b]; **4**, 3[10], 41[10], 47[10], 53[151], 65[10a], 66[10,10a]; **5**, 826[158,158b]; **6**, 834[32], 844[92]; **7**, 84[1], 85[1], 92[40], 108[1], 196[12], 215[12]; **8**, 541[207]
Thyes, M., **6**, 93[133]; **8**, 374[144]
Tiberi, R., **6**, 626[165]
Tice, C. M., **2**, 161[137]; **4**, 255[193]; **5**, 808[110]; **7**, 367[57]; **8**, 29[40,41], 66[40,41]
Tichman, P., **7**, 478[85]
Tichy, M., **2**, 841[185]; **5**, 709[46]
Ticozzi, C., **1**, 514[51]; **8**, 856[170,171]
Tidbury, R. C., **4**, 370[36]
Tidd, E., **2**, 495[59]
Tideswell, J., **1**, 780[229]; **6**, 860[176], 996[108]
Tidwell, M. Y., **2**, 871[23]
Tidwell, T. T., **1**, 418[73,74]; **2**, 107[59,60], 108[60], 196[76]; **3**, 587[141]; **4**, 297[279], 298[282], 299[296,301], 313[465], 1031[8], 1043[8]; **5**, 87[42], 901[23]; **7**, 302[65]
Tiecco, M., **1**, 670[187], 678[187]; **3**, 447[90], 456[90], 457[133], 493[83], 503[146], 509[178], 513[83,146]; **4**, 426[51], 437[51,148], 438[148], 441[173,177,179,180,181,182], 447[216,217], 452[20]; **6**, 462[20,21]; **7**, 338[41], 340[45,47], 343[47], 770[256b], 771[256], 773[306], 779[427]; **8**, 842[43,43c], 847[43c], 848[43c], 849[43c]
Tiedeman, T., **7**, 724[29]
Tiedje, M., **4**, 1040[88], 1048[88]; **5**, 939[223], 951[223], 957[309], 962[223], 964[223]
Tieghi, G., **4**, 602[257,260,261]
Tien, H.-J., **6**, 821[113]
Tien, J. M., **7**, 655[19]
Tien, R. Y., **4**, 284[153]
Tien, T. P., **4**, 484[14]
Tiensripojamarn, A., **6**, 1022[65]
Tierney, J., **4**, 272[40]
Tietjen, D., **8**, 269[94]
Tietze, E., **5**, 876[1]
Tietze, L. F., **2**, 342[7], 345[34], 348[57], 350[71], 351[34,81], 352[93], 354[71,93,103], 356[132], 357[34,57,71,103], 358[154,155], 360[165b], 363[190], 364[81], 369[253,255a,b], 370[253], 371[34,57,154,155,253,262,263,264,265], 372[253,266,269,271], 373[71,103,269,272,273,274], 374[255b], 375[81], 900[29], 901[29], 902[29], 1026[70,71]; **5**, 14[99], 17[118-124], 129[33,34], 457[88],

458[72,73], 459[72], 461[95-98,106], 462[97,98], 464[95], 466[120], 468[95,121,-128,133], 531[81,81d], 534[95]; **6**, 48[87], 49[98], 1062[77]; **7**, 131[87]
Tiffeneau, M., **3**, 754[106], 756[114]
Tigchelaar, M., **4**, 905[209]
Tigerstrom, R. V., **6**, 603[12]
Tihomirov, S., **8**, 214[49]
Tijerina, T., **1**, 743[54], 746[54], 748[54]; **5**, 1115[2], 1116[2], 1122[2b], 1123[2b], 1124[2b]
Tijhuis, M. W., **7**, 230[134]; **8**, 60[183], 61[183], 62[183], 66[183]
Tikhomirov, B. I., **8**, 449[155]
Tikhonov, A. Ya., **6**, 114[75]
Tikhonova, N. A., **6**, 564[910]
Tilak, B. D., **1**, 568[237]; **8**, 659[104]
Tilakraj, T., **2**, 381[302]
Tiley, E. P., **7**, 231[137]; **8**, 663[116]
Tilhard, H.-J., **1**, 645[124], 669[124,181,182], 670[181,182], 680[124]; **3**, 105[214]; **4**, 120[200]
Tilichenko, M. N., **7**, 578[151]
Tiller, T., **2**, 498[77]
Tilley, J. W., **2**, 547[101], 548[101]; **3**, 530[65], 533[65], 1038[92]; **6**, 531[430]
Tillmanns, E.-J., **6**, 273[100]
Tillotson, A., **7**, 12[100]
Tillyer, R. D., **4**, 248[111], 256[111], 260[111]
Tilney-Bassett, J. F., **8**, 901[38], 907[38], 908[38]
Tilton, M., **5**, 710[55]
Timant, B., **2**, 423[33]
Timberlake, J. W., **3**, 587[141,143]; **6**, 116[90]; **7**, 739[35], 741[46], 746[46]
Timko, J. M., **4**, 159[85]; **5**, 1098[132], 1112[132]
Timm, D., **4**, 611[348]; **7**, 24[22]
Timmerman, H., **4**, 110[151]
Timmers, D., **4**, 983[119]; **5**, 1086[74]
Timmers, F. J., **4**, 703[33-35], 704[33,34], 712[34,35]
Timmons, C. J., **5**, 723[109]; **7**, 100[126]
Timmons, R. J., **8**, 472[4], 475[20]
Timms, G. H., **8**, 371[104], 393[112]
Timms, R. N., **4**, 609[327], 614[327], 615[327,391], 629[327,391]
Timofeeva, G. N., **6**, 496[154]
Timokhin, B. V., **8**, 771[42]
Timokhina, L. V., **6**, 509[245]
Timony, P. E., **4**, 446[211]
Timori, T., **7**, 57[24]
Timoschtschuk, A., **7**, 160[53]
Timpe, W., **8**, 812[5]
Tin, K.-C., **1**, 524[90]; **2**, 430[53]; **3**, 748[77]; **7**, 767[191]
Tinant, B., **4**, 1040[75]
Tinapp, P., **8**, 298[74,75], 299[75]
Tinembart, O., **4**, 764[222], 765[222], 808[155]
Tiner-Harding, T., **6**, 531[448]
Ting, J.-S., **3**, 214[55], 216[74]
Ting, P. C., **1**, 766[154]; **4**, 367[15], 377[104,104a], 378[104a,111], 381[104a], 383[15], 384[15], 815[189]
Tingoli, M., **1**, 670[187], 678[187]; **3**, 229[233], 444[65], 447[90], 456[90], 457[133], 492[75], 493[83], 503[75,146], 509[178], 513[83,146]; **4**, 426[51], 437[51,148], 438[148], 441[173,177,179,180,181,182], 447[216,217]; **6**, 462[20,21]; **7**, 338[41], 340[45], 770[256b], 771[256], 773[306], 779[427]
Tinker, A., **7**, 6[35]
Tinker, H. B., **4**, 930[51]
Tinker, J. F., **5**, 752[50]
Tinley, E. J., **7**, 479[92]
Tinucci, L., **2**, 735[15]
Tipper, C. F. H., **2**, 735[11]; **8**, 850[121]
Tippett, C. F. H., **6**, 950[1]
Tippett, J. M., **8**, 53[131], 56[167], 66[131,167]
Tipping, A. E., **7**, 800[34]
Tipson, R. S., **6**, 36[21], 687[378]; **8**, 813[7]
Tipsword, G. E., **5**, 736[145], 737[145]
Tipton, J., **8**, 64[203], 66[203]
Tirado-Rives, J., **3**, 591[170]
Tiriliomis, A., **5**, 1101[145]; **6**, 480[116]
Tirouflet, J., **8**, 290[30]
Tirpak, J. G., **3**, 499[117]
Tirpak, R. E., **2**, 830[142], 842[192]
Tischenko, I. G., **4**, 1023[254]
Tischler, A., **3**, 541[112], 543[112]
Tischler, W., **8**, 205[158,164], 561[409]
Tishchenko, I. G., **2**, 709[14], 740[61], 933[139]; **6**, 557[835,836]
Tishchenko, N. A., **7**, 12[98]
Tishchenko, N. P., **6**, 463[26]
Tishler, M., **4**, 239[30]; **7**, 92[48], 156[32], 258[54]; **8**, 143[62,63], 148[62,63,107]
Tišler, M., **3**, 756[113], 904[133]; **6**, 157[163], 252[151], 480[114], 514[306], 543[306], 554[712,713,714,715,716,725,739,740,742,743,744,746,747,782]; **7**, 475[53], 476[53]; **8**, 384[31,32]
Tisler, T., **7**, 196[13]
Tisnes, P., **6**, 441[79], 494[137]
Tissler, M., **3**, 864[17], 866[17], 883[17]
Tissot, P., **3**, 579[95], 639[78], 644[156]; **7**, 876[126]; **8**, 134[36]
Tisue, G. T., **7**, 24[26], 25[26], 477[77]
Titanyan, S. O., **4**, 315[507]
Titiv, Y. A., **3**, 839[7], 848[7]
Titmas, R. C., **6**, 624[138]
Titov, A. I., **7**, 8[62]
Titov, M. I., **2**, 411[8]
Titov, Yu. A., **4**, 145[29b]; **5**, 513[2], 518[2]
Titterington, D., **5**, 830[34]
Titus, P. E., **6**, 536[547], 538[547]
Tius, M. A., **1**, 585[14,15], 772[200]; **2**, 89[34], 109[71]; **3**, 224[171], 225[171], 264[182], 361[79], 497[106], 762[146]; **5**, 774[172,173,174], 775[172], 780[173,174]; **6**, 5[27], 666[247]; **7**, 458[115], 633[63], 647[35]
Tiwari, K. K., **3**, 329[182]
Tiwari, K. P., **3**, 1038[94]
Tizané, D., **8**, 636[7]
Tjuttschew, J., **6**, 435[1]
Tkachenko, V. V., **6**, 487[44], 489[44]
Tkachev, A. V., **3**, 386[68]
Tkacz, M., **5**, 432[124], 433[139]
Tkatchenko, I., **3**, 822[11], 834[11,74,75]; **4**, 914[1], 922[1], 925[1], 926[1], 932[1], 941[1], 943[1]
Tllari, S., **7**, 65[68]
Toan, V. V., **5**, 692[104]
Tobe, M. L., **7**, 860[73]
Tobe, Y., **3**, 380[9]; **5**, 817[146]; **6**, 976[9], 1036[145]
Tobey, S. W., **2**, 345[37]; **4**, 1015[200]
Tobinaga, S., **3**, 676[77], 679[89], 683[102,105], 693[89], 695[153], 696[154], 807[35]; **7**, 801[39]
Tobita, M., **3**, 219[113], 505[167]
Tobito, Y., **7**, 533[33]
Tochio, H., **5**, 949[281]
Tochtermann, T., **4**, 70[10], 140[6]
Tochtermann, W., **5**, 381[118]; **6**, 979[27]; **8**, 354[170]
Toczek, J., **7**, 166[90]
Toda, F., **1**, 546[60]; **3**, 491[67], 557[37,38], 900[97]; **8**, 162[30]
Toda, H., **5**, 714[74]
Toda, J., **5**, 323[16], 1022[73,73c,74]; **6**, 525[379]
Toda, M., **4**, 1054[133]; **8**, 309[11,12], 310[11], 311[12]
Toda, S., **4**, 826[244]

Toda, T., **4**, 389[166,166b]; **5**, 714[68]; **7**, 791[1], 797[19]
Todaro, L., **6**, 543[621]
Todd, A. R., **6**, 607[46], 609[57,58], 611[63,66], 614[78,87], 620[131], 624[150], 625[131]
Todd, D., **6**, 728[207]; **8**, 328[3], 338[3], 339[3], 343[3]
Todd, H. R., **4**, 6[22]
Todd, J. S., **6**, 277[124]
Todd, L. J., **4**, 968[92], 977[92]
Todd, M., **6**, 935[103], 987[75], 990[85]
Toder, B. H., **2**, 106[49], 651[122]; **3**, 22[133], 26[165], 51[269], 906[144]; **4**, 1040[102]; **7**, 7[47]
Todeschini, R., **2**, 266[62], 267[62,64], 630[21], 631[21], 632[21], 634[21], 640[21], 641[21], 642[21], 644[21], 645[21]
Todesco, P. E., **1**, 569[262]; **4**, 424[11], 426[37], 428[11], 429[82]; **7**, 737[15], 760[49], 764[49]
Todesco, R., **5**, 637[109]
Toga, T., **7**, 86[16a]
Togashi, S., **8**, 889[136]
Togni, A., **1**, 320[162]
Togo, H., **4**, 748[157], 768[239], 810[170], 823[230]; **6**, 442[87]; **7**, 730[46,47], 732[59]; **8**, 403[3], 404[3], 405[34], 408[69-71,76], 409[81], 410[34]
Togo, N., **2**, 540[71]
Toh, H. T., **1**, 780[229]; **7**, 747[94]
Tohda, Y., **1**, 447[204], 458[204]; **2**, 725[110], 805[43]; **3**, 217[89], 219[89], 271[1], 521[7], 530[7], 532[7], 554[26]; **6**, 535[524]
Tohidi, M., **4**, 1103[206]
Tohjima, K., **2**, 1051[43]; **3**, 650[213]; **7**, 804[60]
Tohjo, T., **2**, 859[252]
Tohoma, M., **8**, 883[92]
Toi, H., **8**, 806[102], 876[44], 877[44]
Toi, N., **6**, 252[154]
Tojo, G., **3**, 585[133], 591[171]; **7**, 34[98,99]; **8**, 618[115,116]
Tokai, M., **4**, 592[127], 633[127]; **5**, 935[191], 936[191]
Tokarev, B. V., **6**, 494[133]
Tokarev, Yu. I., **3**, 306[84]
Tökes, L., **2**, 817[90]; **3**, 257[120]; **4**, 33[96,96b]; **6**, 514[307]; **8**, 266[55], 321[101], 957[14]
Toki, S., **5**, 165[81], 168[102]
Tokiai, T., **4**, 298[285]
Tokita, S., **5**, 406[22]
Tokitoh, N., **1**, 370[65]; **6**, 814[93], 923[59]; **7**, 222[41]; **8**, 392[93]
Tokiura, S., **6**, 276[119]
Tokizawa, M., **5**, 38[23c]
Tokles, M., **2**, 120[181]; **4**, 229[232]; **7**, 166[89], 441[43], 442[43]
Tokmakov, G. P., **2**, 787[52]
Tokoroyama, T., **1**, 767[176]; **2**, 509[34], 585[128,129]; **3**, 35[201], 39[201], 355[54], 357[54]; **4**, 155[68d], 258[254]; **6**, 836[58]; **7**, 174[134], 368[59]; **8**, 267[63], 268[63]
Tokoyama, M., **7**, 308[19]
Tokuda, M., **4**, 290[205], 404[245], 812[177]; **5**, 195[8], 197[8]; **7**, 603[110,111]; **8**, 604[3]
Tokugawa, N., **5**, 534[95]
Tokumaru, K., **7**, 851[24], 881[156]
Tokumasu, S., **2**, 363[199]; **4**, 988[139]
Tokumoto, T., **6**, 820[111]
Tokunaga, Y., **7**, 6[31]
Tokutake, N., **5**, 95[92]
Tolbert, L. M., **4**, 463[108], 471[142], 472[142], 960[36]; **5**, 132[47]; **7**, 851[30]
Tolchinskii, S. E., **4**, 300[308]
Toledano, C. A., **4**, 982[112]; **5**, 1086[68]
Toliopoulos, E., **1**, 749[78], 816[78]
Tölle, R., **5**, 187[174]
Tollens, B., **2**, 139[27,27a]
Tolley, M. S., **1**, 837[151]
Tolman, C. A., **1**, 440[172], 441[172]; **4**, 918[16]; **8**, 425[46], 426[46], 439[46]
Tolman, C. D., **4**, 915[13]
Tolman, R. L., **3**, 262[166]
Tolman, V., **8**, 896[9]
Tolson, S., **4**, 712[70]
Tolstikov, G. A., **2**, 814[80]; **4**, 875[56]; **7**, 93[52], 543[18], 579[18], 581[18], 750[129]; **8**, 396[138], 398[138], 676[55], 677[57], 680[73], 682[55], 683[73], 689[55,57], 691[55]
Toma, K., **3**, 395[99]
Toma, L., **4**, 38[108], 379[115], 380[115j], 382[115k]; **7**, 274[138]
Toma, S., **2**, 722[92]; **8**, 86[26]
Tomago, S., **5**, 101[163]
Tomaiqid, A., **6**, 990[86], 991[86]
Tomás, A., **6**, 80[68]
Tomás, M., **5**, 161[64]; **6**, 572[961], 757[134]
Tomasewski, A. J., **2**, 158[123]; **3**, 804[3]
Tomasi, R. A., **1**, 622[72]
Tomasic, J., **6**, 658[182]
Tomasic, V., **7**, 777[366]
Tomasik, P., **4**, 430[89]
Tomasini, C., **4**, 377[104], 386[153,153a,157], 387[153a,157], 388[164], 393[164a], 401[226], 407[104c,153a,157b,254], 408[259c]; **7**, 493[184], 503[269]
Tomasz, M., **4**, 27[78], 44[78a], 128[222]
Tomcufcik, A. S., **6**, 554[737]
tom Dieck, H., **4**, 545[125], 546[125], 874[53]; **5**, 1039[16], 1041[16], 1043[16], 1044[16], 1046[16], 1048[16]; **8**, 764[9]
Tomer, K., **1**, 377[97]
Tomesch, J. C., **1**, 188[66], 189[66]; **2**, 23[91]; **3**, 124[254], 126[254], 137[254], 423[79]; **7**, 301[60]
Tometzki, G. B., **4**, 608[321]; **8**, 545[282]
Tomic, S., **6**, 658[182]
Tomida, I., **1**, 808[320]
Tomii, K., **8**, 134[34], 291[37]
Tomilov, A. P., **3**, 639[80]; **8**, 253[113]
Tomilov, Yu. V., **4**, 963[43], 964[47]; **5**, 1198[45]
Tomimori, K., **3**, 222[144]; **6**, 837[60]
Tominaga, H., **8**, 554[366]
Tominaga, M., **6**, 657[177]
Tominaga, T., **7**, 773[302]
Tominaga, Y., **1**, 180[43], 181[43]; **2**, 6[28], 17[28], 572[43,45]; **3**, 246[44]; **4**, 589[80], 591[80]; **6**, 83[82]; **8**, 837[13c]
Tomino, I., **1**, 833[120]; **8**, 159[34], 162[33,34], 163[35,37], 164[34], 178[34], 179[34], 545[287], 546[303]
Tomioka, H., **3**, 891[45]; **4**, 483[7], 960[34]; **6**, 7[30], 240[79]; **7**, 9[69], 309[24], 322[67], 369[63], 378[63]; **8**, 917[117], 920[117]
Tomioka, K., **1**, 72[69,70], 342[99], 359[19], 382[19,19c-e], 566[214,215], 823[44a]; **2**, 482[26], 483[26], 558[160], 846[208], 1018[37]; **3**, 41[224], 43[235], 217[95], 675[73]; **4**, 10[34], 21[69], 76[49], 85[75], 111[157], 113[164], 200[1], 203[31], 210[82,85,86], 211[82,85,87], 222[173,174,175,176], 229[239], 249[126], 252[163,164], 258[126]; **5**, 376[108b], 736[145], 737[145]; **6**, 137[74], 723[148], 724[148], 726[183], 738[47]; **7**, 438[13], 442[50], 443[13]
Tomioka, M., **5**, 134[61]
Tomioka, N., **1**, 512[39]
Tomioka, T., **5**, 552[1]
Tomita, B., **3**, 347[26]
Tomita, K., **1**, 359[16], 379[16a,b]; **2**, 792[69]; **3**, 751[88]; **4**, 1056[141,141b]; **6**, 524[354]
Tomita, M., **7**, 156[32], 175[143]; **8**, 48[111], 66[111], 514[111]
Tomita, S., **6**, 893[77]; **7**, 299[45]; **8**, 5[30]
Tomita, S.-I., **3**, 934[67]
Tomita, T., **5**, 829[23]; **7**, 601[85]
Tomiyoshi, N., **2**, 1050[28]; **7**, 168[101]
Tomizawa, G., **2**, 538[68], 539[68]

Tomizawa, K., **7**, 247[105], 674[46]
Tomljanovic, D., **2**, 362[181]
Tomo, Y., **2**, 632[25a-c], 638[25b,c], 640[25c], 654[25c]
Tomoda, S., **1**, 619[62]; **2**, 635[46], 640[46]; **4**, 341[55]; **5**, 480[177], 483[174]; **6**, 91[128], 109[39-41], 466[46], 564[916,917]; **7**, 496[217], 497[218], 522[39]
Tomolonis, A., **4**, 932[62]
Tomooka, K., **1**, 184[52], 248[64]; **2**, 10[40], 29[106]; **3**, 730[42]; **6**, 14[55]; **7**, 298[35]
Tomorkeny, E., **7**, 70[93]
Tömösközi, I., **2**, 529[17]; **7**, 723[25]
Tomotaki, Y., **2**, 73[61]; **3**, 99[181], 101[181]
Tomozane, H., **2**, 642[76], 643[76], 836[160]
Tompkins, G. M., **8**, 561[415]
Toms, S., **4**, 698[14]
Tomuro, Y., **1**, 87[47]; **3**, 463[165]; **8**, 967[79]
Tondys, H., **4**, 434[133]
Tone, H., **5**, 1141[81]; **7**, 246[85]
Tonegawa, F., **6**, 644[85]
Toney, J., **1**, 13[70,72], 16[87]
Tong, Y. C., **4**, 443[185]; **7**, 759[12], 765[135], 778[135,396], 842[22]
Toniolo, L., **4**, 915[9,15], 936[68]
Tönjes, H., **2**, 1084[7]
Tonker, T. L., **3**, 47[257]
Tonnard, F., **5**, 254[48]
Tonnis, J., **8**, 568[467]
Ton That, T., **7**, 71[100]
Toofan, J., **7**, 266[109], 267[109], 760[23]
Toogood, J. B., **2**, 143[51]
Toogood, P. L., **4**, 27[81]; **5**, 524[55]
Toole, A. J., **5**, 1080[52]
Toomey, J. E., Jr., **8**, 591[61], 592[61], 593[67], 602[61]
Toong, Y.-C., **3**, 595[191]
Toops, D., **1**, 240[43]
Tooyama, T., **6**, 1047[32b]
Top, A. W. H., **6**, 561[872]
Top, S., **4**, 519[19], 520[19], 522[19]; **6**, 286[178,179], 287[179,180]; **8**, 185[27]
Topchii, V. A., **3**, 302[50]
Toporcer, L. H., **8**, 518[128,129], 565[447], 725[185]
Toppet, S., **5**, 113[237]
Topsom, R. K., **1**, 580[3]
Toraya, T., **4**, 1011[165]
Torck, B., **4**, 298[289]; **6**, 263[20,24], 264[24], 267[24], 269[20]
Tordeux, M., **2**, 749[129]; **6**, 526[397,398]
Toren, G. A., **2**, 765[82]
Torgov, I. V., **2**, 382[313]; **5**, 752[21]
Tori, K., **3**, 386[57], 407[149]; **5**, 809[115]
Torigoe, M., **8**, 935[65]
Torii, S., **1**, 551[78], 751[93]; **2**, 19[77,77a], 73[61], 187[40], 655[132,135], 976[2], 981[2], 982[2]; **3**, 99[181], 101[181], 125[297,298], 126[298], 128[297], 129[297], 130[297,298], 133[297], 137[298], 167[484], 168[484,495,499,500], 169[495,499,500], 361[74], 501[137], 509[137], 579[101], 634[14,23], 649[23], 650[211], 652[220]; **4**, 159[82], 254[189], 257[189], 371[47], 383[140], 606[307], 607[307,315], 647[307], 1040[98], 1043[98]; **7**, 98[104,105], 537[58,60-62], 765[164], 770[256c], 771[256], 819[22]; **8**, 216[61], 244[71], 247[71], 251[71], 253[71], 976[135], 994[65]
Torimitsu, S., **7**, 700[62]
Torimoto, N., **7**, 24[37], 25[37,45], 26[52,53]
Torisawa, Y., **1**, 568[240]; **3**, 135[364], 139[364], 142[364], 143[364]; **6**, 21[79]; **7**, 617[21], 621[30]
Toriumi, K., **8**, 459[244], 535[166]
Toriyama, M., **6**, 443[89]
Torizuka, K., **7**, 801[45]
Torkelson, S., **2**, 606[65]
Tornare, J.-M., **5**, 384[126b], 1096[108], 1098[108d], 1099[108d], 1112[108d]
Toro, J., **3**, 498[108]; **5**, 1107[169]
Torok, D. S., **4**, 974[87]
Toromanoff, E., **1**, 561[162]; **2**, 153[108]; **3**, 21[131]; **4**, 187[96]
Toros, S., **4**, 925[36], 927[41], 930[41], 939[41]; **8**, 152[177]
Torr, R. S., **1**, 774[207]; **3**, 201[76]
Torras, J., **5**, 36[18], 57[54]
Torre, A., **3**, 390[71,72], 399[117], 402[117]
Torre, D., **4**, 1061[166]
Torre, G., **1**, 837[155], 838[160]; **7**, 747[96], 777[371,372,373,384], 778[402]; **8**, 187[37]
Torre, M. d. C., **2**, 758[24]
Torregrosa, R. E., **8**, 452[189b]
Torrence, P. F., **6**, 614[81], 625[161]
Torrents, A., **5**, 1059[54], 1062[54c]
Torres, E., **1**, 329[41]
Torres, L. E., **1**, 436[150]; **2**, 111[79], 604[50]
Torres, M., **3**, 891[43], 892[43,47,48]
Torrey, J. V. P., **2**, 283[52]
Torri, J., **1**, 294[48]
Torrielles, E., **1**, 564[189]
Torrini, I., **1**, 734[23]
Torssell, K., **2**, 713[48]; **4**, 36[102], 1076[48], 1077[48], 1078[48], 1080[48]; **5**, 778[195]; **6**, 672[286]; **7**, 292[8], 654[4,5]
Tortorella, S., **8**, 587[32]
Tortorella, V., **6**, 787[99,100]
Toru, T., **1**, 227[97]; **2**, 833[147]; **4**, 13[44,44c], 159[85], 253[169], 256[208,212], 261[208], 262[212,308], 394[194,195], 413[275], 744[135]; **6**, 264[37], 265[37]; **7**, 54[8], 519[22], 524[51]; **8**, 843[59a], 993[58]
Toscano, V. G., **3**, 201[77]; **4**, 508[160]; **5**, 856[196]
Toshemitsu, A., **8**, 413[130]
Toshida, Y., **8**, 951[176]
Toshima, K., **1**, 569[254]; **2**, 263[54]
Toshima, N., **8**, 431[60]
Toshimasa, T., **4**, 507[150]
Toshimitsu, A., **1**, 648[126]; **2**, 598[16]; **3**, 87[111], 106[111], 114[111], 381[26,27], 382[26,27]; **4**, 340[47], 341[56,58], 347[87], 349[58], 370[41], 387[158], 398[216], 399[216e], 401[41,158c,216e,232,232a], 405[41,158c,216e,232a], 406[41,158c,216e,232a]; **6**, 289[194,195,196,197], 293[194,195,196,197,226,227], 1030[104], 1031[110,112], 1032[121]; **7**, 95[64], 128[68], 129[70], 495[207,208,212,213], 496[212,213,214], 505[288], 520[27], 521[33], 523[43], 524[52], 534[40,41], 771[264], 773[308], 776[308]; **8**, 848[104], 849[104,114]
Tosk, E., **7**, 674[35]
Tost, W., **5**, 458[72,73], 459[72]
Toteberg-Kaulen, S., **4**, 371[61]
Toth, B., **1**, 797[283]; **7**, 555[69]
Tóth, G., **2**, 789[56]; **6**, 499[177], 520[340]
Tóth, I., **2**, 817[90], 851[223]; **8**, 535[166]
Toth, J. E., **3**, 213[51]; **4**, 79[56], 251[151,154], 257[154], 260[154]; **6**, 163[194]; **8**, 47[124], 66[124]
Toth, K., **8**, 861[225]
Toth, L., **2**, 787[50]
Toth, M., **3**, 223[155]
Tóth, T., **8**, 612[70], 613[70]
Totleben, M., **4**, 808[159]
Totton, E. L., **8**, 532[130]
Toube, T. P., **2**, 821[109]
Toubiana, M.-J., **3**, 407[149]
Toubiana, R., **3**, 407[149]
Touboul, E., **3**, 572[60]; **8**, 135[38], 532[130]
Touchard, D., **7**, 499[234]
Touillaux, R., **6**, 578[980]

Toupance, G., **6**, 540[581]
Toupet, L., **4**, 38[109b], 985[127]; **6**, 690[401], 692[401]; **8**, 134[31]
Tour, J. M., **2**, 713[45]; **3**, 251[78,100], 254[78,100], 1025[35], 1030[35]; **4**, 884[103], 892[145]; **5**, 32[6,6c], 57[52], 1037[5], 1165[11,13,15], 1166[11,15], 1167[11,15], 1170[15], 1171[15], 1175[15], 1178[11,15], 1179[15], 1183[54]
Touré, S., **2**, 1102[125], 1103[125]
Toure, V., **3**, 380[8]
Tournayan, L., **8**, 436[73]
Tournilhac, F., **4**, 469[136]
Touru, T., **4**, 261[284]
Tourund, E., **4**, 48[140]
Tourwe, D., **6**, 707[46], 712[71]
Touzin, A. M., **4**, 111[152a], 173[33]; **5**, 676[5]; **7**, 229[118]
Touzot, P., **6**, 428[86]
Tovaglieri, M., **5**, 1138[65]
Tovrog, B. S., **7**, 452[46]
Towart, R., **2**, 385[321]
Towle, J. L., **8**, 568[484]
Town, C. M., **2**, 465[107]
Towney, P. O., **7**, 760[37], 761[37]
Towns, T. G., **5**, 901[24]
Townsend, C. A., **3**, 209[14]; **4**, 250[139], 497[97]; **5**, 15[109], 736[145], 737[145]; **7**, 355[44]
Townsend, J. M., **4**, 611[349]; **7**, 160[54]; **8**, 460[249], 988[30]
Townsend, L. B., **2**, 555[140]; **6**, 554[711,733,738]
Towson, J. C., **1**, 86[42], 834[128], 838[159,162]; **6**, 150[114]; **7**, 778[399,401,401a]
Toy, A., **5**, 1089[85], 1093[96], 1098[118,119], 1099[96c,118,119], 1101[133], 1112[96c,118,119,133], 1113[85]
Toy, M. S., **3**, 640[102]
Toyama, S., **1**, 851[47], 853[47]
Toyama, T., **3**, 923[43]
Toye, J., **2**, 60[18,18b]; **4**, 117[191]; **5**, 109[214]
Toyoda, H., **8**, 698[142], 709[45,45a]
Toyoda, J., **4**, 433[123], 1089[138], 1091[138]; **6**, 66[8]
Toyoda, T., **1**, 436[151,152]; **2**, 240[13], 244[29], 256[13], 257[13b], 478[15], 479[16], 748[122]; **8**, 331[31]
Toyofuku, M., **4**, 590[94,95], 592[95], 633[95]; **5**, 297[59], 1196[38], 1197[38]; **6**, 86[98]
Toyohiko, A., **4**, 507[150]
Toyooka, N., **5**, 832[39]
Toyo'oka, T., **4**, 507[151]
Toyoshima, K., **5**, 72[166]; **8**, 477[32]
Toyoshima, T., **3**, 528[47]
Toyota, M., **4**, 30[88,88k,l], 121[209,209a,b], 261[299]; **8**, 945[128]
Tozawa, Y., **4**, 115[180e]
Tozuka, Z., **6**, 604[34]
Tozune, S., **3**, 923[43]
Traas, P. C., **2**, 782[17]; **3**, 251[79], 254[79]; **8**, 535[161], 542[229], 946[139]
Traber, R. P., **4**, 456[46], 457[46], 459[89,92], 460[89,92], 472[46], 473[89], 474[92]
Trachtenberg, E. N., **7**, 84[4], 85[4,6]
Tradivel, R., **7**, 810[87]
Traencker, H. J., **3**, 194[13]
Trafford, D. J. H., **6**, 996[106]
Trahanovsky, W. S., **1**, 630[9], 631[9], 634[9], 641[9], 656[9], 658[9], 672[9]; **2**, 710[27], 734[4]; **3**, 551[4], 552[4], 660[19,20], 661[20], 679[19], 699[20]; **5**, 639[120,121], 1025[84]; **7**, 54[4], 56[4], 64[4], 66[4], 71[4], 72[4], 75[4], 77[4], 78[4], 80[4], 120[7], 167[94], 237[38], 429[152], 444[52], 476[58], 481[58], 493[193], 542[8], 543[8], 671[2], 672[2], 673[2], 674[2], 675[2], 705[17,18], 769[223], 851[20,21,25]
Trainor, G., **1**, 294[39-41]
Trammell, G. L., **3**, 351[43a]
Trammell, M. H., **1**, 131[103]; **3**, 261[148], 264[148]; **5**, 611[71]
Tramontano, A., **1**, 405[25]; **7**, 418[128], 603[118-120,122]; **8**, 17[110], 101[121,122], 102[124], 206[168], 537[189]
Tramontini, M., **1**, 57[31], 59[35]; **2**, 894[4], 897[4], 933[4], 948[184], 953[1], 954[1d]; **7**, 777[367,368]; **8**, 13[69], 123[85]
Tramp, D., **8**, 880[59]
Tramper, J., **8**, 185[14], 206[14]
Tran, H. W., **6**, 81[76], 82[76], 818[106]
Tranchepain, I., **3**, 258[127]
Tranchepain, L., **7**, 487[146], 495[146]
Tran Huu Dau, M.-E., **1**, 49[8]
Tranne, A., **5**, 90[57], 95[57]
Trapani, G., **2**, 187[43]; **8**, 657[97]
Trapentsier, P. T., **4**, 48[140]
Trass, P. C., **2**, 782[21]
Trave, R., **3**, 386[57], 395[98]; **8**, 349[145,146]
Travers, S., **1**, 218[49], 220[49], 223[49]
Traverso, J., **6**, 570[952]
Traverso, J. N., **6**, 570[942]
Travis, E. G., **7**, 767[192]
Traxler, J. T., **8**, 950[160]
Traxler, M. D., **2**, 153[107], 210[111], 250[37], 261[37], 675[53]
Trayhanovsky, W. S., **4**, 524[62]
Traylor, P. S., **7**, 12[95], 13[95]
Traylor, T. G., **4**, 294[247], 302[332], 314[483], 315[483], 1099[176]; **5**, 71[163]; **7**, 12[95], 13[95], 595[14], 597[14], 600[78], 601[78]; **8**, 99[107], 750[63]
Traynard, J. C., **7**, 500[242]
Traynelis, V. J., **2**, 765[78]; **6**, 960[54]; **7**, 223[44], 661[45], 764[123]
Traynham, J. G., **3**, 379[3], 390[81,83], 392[81,83], 649[207]; **4**, 279[110], 280[123], 297[277]; **7**, 15[149]
Traynor, L., **2**, 153[109]
Traynor, S. G., **3**, 770[174]; **5**, 707[30,31]
Treadgold, R., **2**, 651[119]
Treanor, R. L., **3**, 1048[11]
Treasurywala, A. M., **6**, 921[48]
Trebellas, J. C., **5**, 800[75]
Trecarten, M., **7**, 395[21]
Trecker, D. J., **3**, 334[220]; **5**, 66[79], 1025[81]; **7**, 230[135,136], 766[174]
Treco, B. G. R. T., **3**, 213[51]
Trecourt, F., **1**, 474[94,96]
Trede, A., **5**, 422[88], 423[88]
Treder, W., **6**, 49[94]
Trefonas, L. M., **1**, 60[36], 75[36], 468[55]; **6**, 962[74]
Trehan, A., **1**, 765[165]
Trehan, I. R., **5**, 515[18], 547[18]
Trehan, S., **2**, 89[34]; **3**, 497[106]; **5**, 774[173], 780[173]
Treiber, A. J. H., **4**, 1000[15]
Treibs, W., **2**, 902[43]; **7**, 92[42,46], 93[42], 99[113], 154[14]
Treier, K., **5**, 442[185]
Trejo, W. H., **5**, 86[33]
Trekoval, J., **1**, 10[48], 41[194]
Tremble, J., **3**, 257[120]
Tremelling, M. J., **4**, 458[66,67], 463[66,67]
Tremerie, B., **5**, 109[220]
Tremper, A., **5**, 948[292], 949[284], 950[284]
Tremper, H. S., **8**, 216[65], 486[63], 487[63-65], 813[14], 814[14]
Trenbeath, S., **2**, 746[108], 762[56], 824[120]
Trend, J. E., **5**, 439[166]; **6**, 1026[88], 1027[88]
Trenkle, B., **5**, 689[73]
Trepka, R. D., **7**, 483[125]
Treppendahl, S., **6**, 547[657], 570[943,945]
Treptow, W., **8**, 933[46]
Treshchova, E. G., **2**, 534[32], 535[38]

Tressl, R., **8**, 190[84]
Trethewey, A. N., **5**, 442[184]
Tretter, J. R., **2**, 149[90]
Tretyakov, V. P., **7**, 12[98]
Trevoy, L. W., **6**, 959[42]
Triaca, W. E., **3**, 636[58]
Tribble, M. T., **3**, 854[79]
Triebe, F. M., **8**, 135[48]
Trifan, D., **3**, 653[225]
Trifilieff, E., **7**, 247[101], 842[27,28]
Trifonov, L. S., **3**, 1038[93]; **6**, 74[37]
Trifunac, A. D., **1**, 370[67]; **2**, 1004[61]; **3**, 258[124]
Triggle, C. R., **8**, 92[68]
Trigo, G. G., **6**, 501[189]
Trill, H., **4**, 1073[21], 1076[21], 1090[21], 1092[21], 1098[21], 1100[188], 1102[21]
Trimarco, P., **6**, 555[814], 712[73]
Trimble, L. A., **6**, 118[104]
Trimitsis, G., **3**, 197[33]
Trimm, D. L., **7**, 759[7,8]
Trimmer, M. S., **8**, 672[21], 673[25], 696[25]
Trimmer, R. W., **4**, 445[210]
Trinajstic, N., **5**, 903[39]
Trindle, C., **8**, 584[25]
Trinkl, K.-H., **5**, 742[160]
Trinks, R., **3**, 594[184]
Trinquier, G., **6**, 120[119], 172[9]
Tripathi, J. B. P., **4**, 541[111], 689[69]
Tripathy, P. K., **2**, 405[43,44]
Tripathy, R., **5**, 348[74a]
Trippett, S., **1**, 464[39], 835[133]; **5**, 847[135]
Trischmann, H., **6**, 651[134]
Tristram, E. W., **6**, 685[359]; **8**, 50[117], 66[117]
Trius, A., **1**, 477[141]; **2**, 359[164]; **4**, 590[92]
Trivedi, B. C., **8**, 861[224]
Trivedi, K. N., **2**, 381[303], 401[29]
Trivedi, N. J., **7**, 674[39]
Trivedi, S. V., **3**, 416[16], 417[16]
Trivic, S., **8**, 373[127]
Trka, A., **8**, 882[88]
Trkovnik, M., **6**, 554[736]
Trocha-Grimshaw, J., **3**, 677[83]; **7**, 769[211]
Troeger, J., **4**, 282[133], 288[133]
Troepol'skaya, T. V., **1**, 378[104]
Troesch, J., **4**, 868[12], 887[12]
Trofimov, B. A., **3**, 259[128]; **4**, 50[142,142g], 55[157], 57[157o], 309[411]; **7**, 194[6]; **8**, 770[33]
Troger, W. C., **8**, 765[13]
Trogolo, C., **4**, 370[28]; **8**, 856[163]
Troise, C. A., **5**, 857[230]
Troisi, L., **7**, 167[186]
Troitskaya, L. L., **8**, 187[48]
Troitskaya, V. I., **6**, 510[295]
Troka, E., **6**, 789[110]
Trolliet, M., **7**, 124[41]
Trombini, C., **1**, 188[73], 189[73], 192[82]; **2**, 35[130], 36[130], 507[19], 566[23], 657[161b]; **3**, 168[488], 169[488]; **6**, 685[350], 976[4]; **7**, 549[42], 841[14]; **8**, 124[90], 252[111], 797[40], 842[46], 843[46]
Trömel, M., **7**, 236[30]
Tromelin, A., **2**, 332[54]
Trometer, J. D., **1**, 767[164], 768[167]; **3**, 226[199]; **5**, 931[185], 934[185]; **6**, 10[44], 11[44], 12[44]
Tromm, P., **5**, 438[165]
Trommer, W. E., **8**, 52[147], 66[147]
Trompenaars, W. P., **5**, 686[47]
Tronchet, J. M. J., **1**, 759[131]; **4**, 35[98d,e]
Tronich, W., **6**, 175[67], 179[125], 180[128]
Troostwijk, C. B., **8**, 95[91]
Trope, A. F., **1**, 872[89]; **3**, 785[36,36b]
Trost, B. A., **6**, 146[89]
Trost, B. M., **1**, 188[69], 358[9], 359[9], 362[9b], 405[27], 586[17], 587[17], 630[32], 675[32], 715[268], 716[268], 717[268], 722[32], 750[108], 757[122], 770[184,189], 793[272], 794[275], 805[312], 820[9,18], 82; **2**, 25[99], 70[51], 74[77], 109[62], 183[16], 184[16,22], 186[36], 240[5], 369[253], 370[253], 371[253], 372[253], 455[6], 547[92], 567[30], 581[102], 582[106], 587[140], 608[78], 614[120], 616[138], 707[1], 710[19], 742[70], 897[13], 902[13], 908[61], 981[25], 982[25], 1047[2], 1102[123]; **3**, 3[13], 7[34], 11[54], 12[59,60], 16[54], 17[54,86], 26[54], 34[192], 39[192], 55[281], 56[287], 58[287], 75[52], 86[9,17,39,41,61], 87[41],88[17,61,135], 89[61], 90[135], 91[61], 94[17], 95[17], 124[17,61], 139[378], 154[378], 155[378], 168[489], 169[489], 174[531,532,533], 176[531,532,540], 177[39,531,532,533], 178[41,532,543,544], 179[41,531,533,543,546], 181[543], 228[214], 274[25], 283[50], 380[10], 423[78], 436[13], 446[83,84], 524[33], 564[5], 583[126], 607[5], 761[144], 762[144],766[157], 785[32-34,37], 792[66], 832[67], 891[42], 903[122], 905[137], 909[149], 918[21,23], 921[36], 933[61], 953[102,103], 979[12], 994[40,43,44], 996[43], 1000[43], 1008[66], 1040[106]; **4**, 6[20,20a], 25[76], 46[76], 115[184a], 128[222], 159[85], 187[99], 197[99], 231[273], 247[103], 337[35,36], 507[152], 518[10,11], 519[11], 586[2,3,6,10,13,16], 587[26,31,32], 588[2,69-74], 589[10,86], 590[2,3,6,10,13,16],591[106,107], 593[128,129,130,131,132,133,134], 594[137,140], 596[167], 598[193,200,202,203,204,205], 599[216,223,225], 602[252], 607[225,309], 608[321,324], 610[333], 611[355], 614[13,32,380], 615[32,72,380,388,389,390], 616[106,167], 619[137], 620[74], 621[167], 623[200,398], 624[216], 625[223,225], 626[309], 628[380,400,401,402], 629[412,413], 631[72,390], 633[106,107], 634[137,140],638[193,200,202,203,204,205], 642[223,225], 643[252], 644[252], 647[309], 651[429,430,431], 652[167,432,433], 653[438], 695[1-3], 790[35], 792[66], 836[6], 870[28], 876[64], 987[133,147], 989[143]; **5**, 16[113], 46[39], 53[47], 56[39], 57[52], 109[215], 211[64], 239[1], 244[16,17,19], 245[17,20,21], 246[22], 270[1d], 287[35], 298[35,60-63], 299[35,62,65-67,70], 300[35,62,63,72,73,75,76],301[62], 302[35,73,79], 303[66,80,81], 304[66,80,82], 307[35,88-92], 308[35,70,94,97], 309[79], 310[35,98], 311[35,102-105], 333[45], 338[53], 347[72,72a], 353[72a], 373[106,106a,b], 374[106a], 408[31], 435[149], 461[95], 464[95], 468[95], 488[198], 524[53], 563[89], 596[37], 598[37,38], 645[1], 648[1j], 651[1], 683[36a], 685[40], 847[134], 856[210], 901[19,20], 903[20], 904[54], 905[19,20,59],910[20,83,85], 911[83], 912[83], 919[19,20,130], 920[20], 921[20,133,140], 922[130,133,134,135,136], 934[188], 935[189], 951[20], 953[295,296], 1006[33], 1007[40], 1012[52,53], 1020[70], 1027[70], 1035[33a], 1037[5], 1125[55], 1143[91], 1183[49-54], 1185[1]; **6**, 11[46,49], 20[77], 21[79], 24[98], 85[88,90], 86[95,99,101], 133[7], 143[68-73], 147[85], 154[150,151], 161[179], 165[200,201,202], 175[65], 239[76], 254[165],542[600], 662[215], 821[114], 829[1-3], 831[11], 832[16,17], 833[20,21,25,26], 834[25,26,29], 842[81,82], 848[11,111,112], 849[113,118], 850[29], 854[141], 888[65], 893[83], 905[146], 990[84], 991[87], 1016[32-34], 1018[32,39,40], 1020[32,39,47], 1021[50], 1022[33], 1035[138], 1044[16b], 1048[16], 1049[35]; **7**, 92[41,41a], 94[41,60], 119[4], 124[4,43,51], 125[4,43,52,54], 126[43,52], 127[4,51],128[4], 172[130], 173[130,133], 246[93], 320[64], 355[45], 491[180], 493[201], 518[17], 629[48], 668[81], 675[57], 769[212]; **8**, 6[34], 93[76], 318[59], 322[59], 385[49], 528[64], 545[279], 836[2], 840[39], 842[2b], 843[2b,51,60,62], 844[2b,51,64,64a,68], 847[2b], 932[41], 934[56], 935[62], 945[126], 992[54], 993[60], 994[60]
Troster, K., **2**, 960[33]
Trostmann, U., **5**, 929[171], 930[171]
Trotter, J., **1**, 300[71]; **5**, 211[62,63], 608[66]
Trotter, J. W., **6**, 835[45]
Trotter, P. J., **6**, 208[61]

Troughton, E., **1**, 41[203]
Trova, M. P., **3**, 872[58]; **6**, 960[51]
Trowbridge, B. D., **3**, 201[76]
Trozzolo, A. M., **4**, 1090[141]
Trska, P., **2**, 553[128]
Truc, V. C., **8**, 351[166]
Truce, W. E., **2**, 765[79,82]; **3**, 86[35], 88[137], 95[137], 123[249], 158[440,444], 161[444], 164[35,444,474], 165[137], 167[137,444], 173[35], 180[474,549], 181[474,554], 317[120], 794[79]; **4**, 47[135], 48[138,138d,140], 50[142,142a], 66[138d], 771[250]; **5**, 476[147]; **7**, 206[69]; **8**, 839[25,25a,28], 914[95], 968[90]
Truchet, F., **5**, 336[50]
Trudell, M. L., **4**, 1035[37], 1037[37]; **5**, 864[262,263]; **7**, 340[46]
Truedell, B. A., **8**, 47[124], 66[124]
Truesdale, L. K., **1**, 328[21-23], 846[17]; **3**, 781[13]; **5**, 560[76], 814[138]; **6**, 147[87], 682[338,340]; **7**, 489[166], 704[9]; **8**, 695[119]
Truesdell, D., **3**, 380[10]; **4**, 709[45], 710[45]
Truesdell, J. W., **7**, 458[115]
Trufanov, A. G., **4**, 701[29], 702[29]
Truice, W. E., **7**, 816[11]
Truitt, P., **8**, 651[70]
Trullinger, D. P., **6**, 560[868]
Trulson, M. O., **5**, 702[13]
Trumbull, E. R., **3**, 760[139], 774[139]; **5**, 552[24]; **6**, 961[65], 1012[4], 1013[4]
Trumbull, P. A., **3**, 760[139], 774[139]
Trumper, P. K., **4**, 212[98]
Trupiano, F., **7**, 346[12]
Trus, B., **2**, 547[101], 548[101]; **8**, 542[237]
Trust, R. I., **3**, 363[87], 365[63,87]; **7**, 711[59]; **8**, 932[44]
Trybulski, E. J., **1**, 408[35], 430[35]; **3**, 288[63]; **4**, 372[64a]; **6**, 660[202], 677[316]; **7**, 678[73]; **8**, 881[68]
Trzeciak, A. M., **4**, 923[30], 924[30], 925[30]
Trzupek, L. S., **8**, 986[10]
Tsai, C. Y., **4**, 1101[193]
Tsai, D. J. S., **1**, 405[25], 490[27], 623[83]; **2**, 14[54], 996[48]; **3**, 987[29], 990[29], 993[29]; **6**, 875[17], 876[17,35], 877[17], 882[17], 885[17], 887[35]
Tsai, F.-Y., **4**, 38[110]
Tsai, H., **7**, 25[40]
Tsai, L., **4**, 73[32], 111[158a]
Tsai, M., **4**, 681[54]
Tsai, M. M., **7**, 66[72]
Tsai, M.-Y., **6**, 8[35], 1045[23]; **8**, 334[62]
Tsai, T. Y. R., **5**, 130[39], 326[24]; **6**, 46[69]
Tsai, Y.-M., **2**, 85[14,16], 368[240], 575[61], 873[25], 1060[84], 1061[95]; **4**, 794[78]; **5**, 277[16], 279[16]; **7**, 204[59], 6 77[68]
Tsai Lee, C. S., **6**, 959[48]
Tsamo, E., **1**, 359[23], 364[23]
Tsang, R., **2**, 124[203], 232[182]; **4**, 813[185], 815[185,192,193,194], 817[185,193,194], 820[192]; **5**, 837[70]
Tsangaris, M. N., **4**, 347[97]
Tsan-Hsi Yang, **3**, 310[93], 311[93]
Tsankova, E., **3**, 390[73], 396[110], 397[110], 752[98]; **8**, 857[190]
Tsao, C.-H., **8**, 859[210,211,212]
Tsaroom, S., **6**, 93[132]
Tsay, Y.-H., **1**, 14[77], 310[106]; **5**, 232[135,136], 1109[174]
Tschaen, D. M., **1**, 238[34], 404[21]; **2**, 542[84]; **5**, 99[130], 485[181]; **6**, 759[139]
Tschesche, R., **7**, 573[117]
Tschinke, V., **8**, 670[10], 671[10]
Tschudi, G., **8**, 330[28]
Tschugaeff, L., **7**, 775[351]
Tse, A., **1**, 461[14], 464[14]
Tse, C. H., **5**, 221[58], 226[58]
Tse, C.-W., **5**, 226[112], 900[12], 901[12], 903[12], 905[12], 907[12], 913[12], 921[12], 926[12], 943[12], 1006[33]; **7**, 815[3], 824[3], 8 33[3]
Tse, H. L. A., **2**, 588[151], 589[151]; **3**, 248[54]
Tsegenidis, T., **8**, 245[73]
Tselinskii, I. V., **1**, 34[228,232]; **7**, 750[133]
Tseng, C. C., **3**, 220[121], 222[121c,142]; **6**, 848[108]; **8**, 275[142]
Tseng, C.-P., **5**, 417[66]; **6**, 524[369], 533[369]; **7**, 752[145]
Tseng, J., **5**, 589[212]
Tseng, L., **4**, 48[139]
Tseng, L. T., **8**, 5[31]
Tseou, H.-F., **5**, 828[5], 847[5]
Tsetlin, Ya. S., **6**, 509[282]
Tsetlina, E. O., **6**, 577[979]
Tsipis, C. A., **8**, 766[18], 770[35,36], 771[35,36]
Tsipouras, A., **2**, 648[94], 649[94]
Tsitrina, A. Yu., **6**, 490[106]
Tso, H.-H., **3**, 172[513], 173[513,517]; **7**, 261[68]
Tsoi, L. A., **4**, 50[142]
Tsolas, O., **2**, 456[43], 466[43], 467[43]
Tsou, C. P., **2**, 378[286]; **4**, 38[110]
Tsou, T. T., **1**, 193[87], 310[105]
Tsubaki, K., **1**, 212[9], 213[9]; **2**, 442[13], 449[13,49], 450[13]; **3**, 221[131], 420[48,49]
Tsubata, K., **2**, 492[51], 613[111], 635[47], 640[47], 784[38], 971[94], 1051[33], 1052[51], 1066[118,119], 1070[118]; **7**, 227[77], 802[47]; **8**, 817[32]
Tsuboi, H., **6**, 620[122]
Tsuboi, S., **2**, 838[178]; **3**, 848[53,54]; **4**, 229[215]; **6**, 867[209]; **8**, 126[95], 185[26], 190[26], 193[102], 194[104a], 195[104b]
Tsuboi, T., **7**, 299[48]
Tsubokawa, N., **7**, 747[104]
Tsubokawa, S., **7**, 77[123]
Tsubokura, Y., **4**, 1089[138], 1091[138]
Tsuboniwa, N., **2**, 19[76]; **4**, 607[310], 626[310], 647[310]
Tsubuki, M., **5**, 693[114], 694[114], 847[136], 1032[100]; **7**, 243[68], 423[142]; **8**, 534[158], 537[158]
Tsubuki, T., **1**, 369[64a]
Tsuchida, E., **2**, 387[333]; **3**, 677[81], 686[81]
Tsuchida, K., **2**, 1096[99]
Tsuchida, M., **2**, 348[53]
Tsuchida, T., **7**, 25[44], 26[54-56]
Tsuchida, Y., **8**, 410[93]
Tsuchihashi, G., **1**, 89[58], 90[57,58], 151[53,53b], 152[53], 158[53], 168[53b], 184[52], 185[54], 240[41], 248[64], 420[84], 513[47], 515[57], 524[92], 526[92,99], 527[104], 561[165], 566[208], 568[239], 865[87]; **2**, 10[40], 29[106], 363[194]; **3**, 135[356,357,360,361,362], 136[356,357,360,361,362,371], 137[357,360,361], 138[357], 139[356,360,361,362,371], 140[371,371b], 142[357,360,361], 143[357,360,361,371,371b], 147[394], 155[429], 156[360,361,362], 303[56], 730[42,42b]; **4**, 10[33], 20[63], 21[63], 37[104], 213[100,101]; **5**, 436[153]; **6**, 14[52-55], 893[82], 926[67], 927[75]; **7**, 298[35], 762[78], 778[416]; **8**, 10[57,58], 188[49], 193[49,101], 561[411]
Tsuchihashi, G. I., **6**, 150[110]
Tsuchiya, H., **4**, 653[437]
Tsuchiya, T., **5**, 914[115]; **6**, 535[537,538,539], 538[537,538,539]; **7**, 719[5], 720[14], 732[5,57]; **8**, 641[26], 817[30,31]
Tsuda, K., **3**, 816[79,79a]; **4**, 462[106], 475[106]
Tsuda, M., **2**, 810[63], 824[63]; **7**, 851[24]
Tsuda, N., **6**, 27[114]
Tsuda, T., **1**, 85[28,29]; **2**, 114[115-117]; **3**, 217[86], 259[137]; **4**, 254[179,180,182], 591[112], 592[126,127], 611[359], 617[126], 618[126], 633[112,126,127]; **5**, 620[16], 935[191], 936[191], 1157[170,171], 1183[56]; **6**, 88[103]; **7**, 745[76]; **8**, 548[324], 549[324,325]

Tsuda, Y., **5**, 76[233], 323[15,16], 1022[73,73c,74]; **6**, 525[379], 657[175]; **8**, 37[100], 42[100], 66[100], 203[151], 245[72], 334[64], 540[200]
Tsudaka, Y., **2**, 463[82,83], 464[82,83,98]
Tsuge, O., **1**, 770[190], 836[140]; **2**, 482[21]; **3**, 201[84]; **4**, 16[52b,c], 75[43a], 100[43], 111[152c], 120[197], 1086[114,115]; **5**, 104[184], 451[41], 470[41], 485[41], 758[81]; **6**, 542[603]
Tsuge, S., **6**, 494[138]; **8**, 263[32], 267[32]
Tsugoshi, T., **4**, 14[46], 55[157], 57[157h], 249[114], 257[114]; **5**, 473[153], 477[153]
Tsuhako, A., **7**, 299[43]
Tsui, D. S. K., **6**, 561[879]
Tsuji, J., **1**, 552[82], 553[88-98], 642[109], 643[109]; **2**, 166[154], 184[26], 270[74], 482[32], 484[32], 710[24], 728[143,144,145], 810[64], 1102[123]; **3**, 2[8], 11[8], 16[8], 17[8], 26[8], 28[169], 51[271], 95[153], 107[153], 114[153], 115[153], 198[47,48], 380[10], 390[79], 524[34], 639[75], 1008[72], 1009[72], 1010[72,75], 1033[76], 1040[105]; **4**, 8[27], 104[137], 227[202,203,204,205,207,208], 240[39], 254[39], 255[196,200], 261[297,298], 552[2], 553[2,3,5,7,9], 586[1,8,11,12,15], 588[55,65], 589[8,11,75], 590[1,8,11,12,15,96-98,100,101], 591[8,113,116-118], 592[119-125], 593[135], 594[96], 597[172], 600[227,230,231,232,233,236,237,238,240], 606[302], 609[331], 611[346,360,361,362,363,364,365,366,367], 612[362,368], 613[117,369], 629[405,410,411], 633[96], 636[366,367], 637[172], 638[410], 643[237,238,240], 646[302], 650[424,425], 653[435], 753[171], 837[17], 945[88]; **5**, 46[39], 55[48], 56[39], 281[18,19], 304[83], 305[85], 830[31], 833[49], 935[190], 1118[19], 1137[55], 1200[48], 1201[48]; **6**, 11[46], 20[77], 85[88], 137[42,43], 641[61], 831[11], 848[11], 849[117], 866[204], 875[18], 968[108], 1022[57], 1032[118]; **7**, 94[55], 95[63], 107[156], 141[134], 142[134,135,136,137], 406[73], 450[7,13,14], 451[7], 452[39,43,62], 453[69,83-92], 454[83,84,92,96,100-102], 455[7,86-90,106], 456[69,108], 457[108,109], 458[7,112], 459[108], 460[102,116], 461[117,118], 462[39,43], 463[125,126], 465[130], 739[33], 746[81], 809[86]; **8**, 7[43], 287[21], 450[163], 451[163c], 478[47], 479[47,47a], 481[47], 527[40], 528[59], 557[388], 773[68], 778[68], 951[176], 959[24,30], 960[31], 961[39], 978[146]
Tsuji, K., **6**, 457[163]
Tsuji, M., **4**, 589[88], 599[218], 640[218]
Tsuji, T., **3**, 49[263], 51[271], 380[10], 753[103], 901[108]; **4**, 192[117], 951[1], 953[1h], 954[1h], 961[1h], 968[1], 979[1], 1001[19], 1006[19,106], 1007[19]; **5**, 71[120], 77[268], 78[272], 585[198], 721[99], 904[44], 905[44], 927[163]; **6**, 3[10], 30[10], 83[84], 127[157], 835[47], 966[99], 976[10]; **7**, 229[111]; **8**, 338[95], 357[95], 836[10a], 837[10a]
Tsuji, Y., **3**, 1018[10], 1021[10], 1028[49]; **4**, 941[84]; **8**, 591[59], 614[83]
Tsujihara, K., **7**, 124[46], 764[121]
Tsujii, J., **5**, 532[87]
Tsujimoto, A., **1**, 515[56]
Tsujimoto, K., **1**, 803[307]; **7**, 862[79], 877[127], 882[169]
Tsujimoto, N., **8**, 389[73]
Tsujino, Y., **6**, 717[110]
Tsujita, J., **7**, 77[121]
Tsujita, Y., **7**, 77[122]
Tsujitani, R., **4**, 1033[21], 1037[21], 1040[21]
Tsukada, Y., **6**, 614[97]
Tsukagoshi, S., **2**, 610[95], 611[95], 1059[82]
Tsukamoto, A., **7**, 474[40]
Tsukamoto, G., **8**, 338[82], 339[82]
Tsukamoto, H., **8**, 205[156]
Tsukamoto, M., **2**, 585[128,129]; **3**, 355[54], 357[54]; **4**, 10[32], 21[66,66a,b], 107[146a,b], 108[146d]; **5**, 844[127]
Tsukanaka, M., **4**, 1018[219]; **5**, 770[141,142], 771[142], 780[141]
Tsukasa, K., **1**, 466[42], 473[42]
Tsukervanik, I. P., **3**, 309[92a], 315[109,111], 320[134], 321[138,140]
Tsukida, K., **2**, 821[109]; **3**, 168[490], 169[490]; **5**, 720[96]
Tsukihara, K., **7**, 299[48]
Tsukiyama, K., **4**, 596[163], 614[374], 621[163], 637[163]
Tsukomoto, A., **8**, 271[105,106,112]
Tsukui, N., **4**, 13[44]
Tsumaki, H., **6**, 114[76], 440[77]
Tsuneda, K., **8**, 450[163], 452[187], 535[162]
Tsunekawa, H., **5**, 841[95]
Tsuneoka, K., **4**, 331[14], 344[14]
Tsunetsugu, J., **5**, 626[39]
Tsuno, S., **2**, 128[240]
Tsuno, T., **5**, 736[145], 737[145]
Tsuno, Y., **5**, 596[21], 597[21], 601[21], 603[21], 606[21], 608[21]; **6**, 799[19,20]; **8**, 803[90], 807[90]
Tsunoda, S., **5**, 137[74]
Tsunoda, T., **2**, 576[73], 577[73]; **3**, 100[201], 103[201], 107[201]; **5**, 829[20], 940[224]; **6**, 900[116]; **8**, 216[59], 217[59]
Tsunokawa, Y., **7**, 385[118]
Tsurata, M., **5**, 473[154], 479[154]
Tsurata, T., **7**, 108[175]
Tsurugi, J., **8**, 248[86]
Tsurumaki, M., **8**, 195[107]
Tsuruoka, M., **4**, 427[69]
Tsuruta, H., **5**, 196[15], 714[70]; **8**, 881[80], 882[80], 971[106]
Tsuruta, K., **8**, 625[164]
Tsuruta, M., **4**, 239[27], 257[27], 261[27]
Tsuruta, T., **1**, 243[60]; **4**, 95[96,100]; **8**, 988[32]
Tsushima, T., **2**, 708[4]; **6**, 421[31]; **7**, 529[11]
Tsutsui, M., **5**, 1185[1], 1186[3,3a]; **6**, 214[96]; **7**, 92[40], 94[55]; **8**, 678[63], 685[63], 686[63], 698[146], 850[120]
Tsutsui, N., **7**, 368[59]
Tsutsui, T., **3**, 592[173]
Tsutsumi, H., **6**, 54[132], 93[133]
Tsutsumi, K., **4**, 8[28,28a]; **5**, 221[62]; **8**, 996[71]
Tsutsumi, O., **8**, 16[97]
Tsutsumi, S., **3**, 483[8], 554[20,21]; **5**, 1138[63]; **7**, 125[60], 446[69], 829[58]; **8**, 991[44]
Tsuyama, K., **3**, 136[370], 137[370], 138[370], 139[370], 140[370]
Tsuyuki, T., **3**, 741[50]; **4**, 405[250a]; **8**, 330[47], 340[100]
Tsuzuki, H., **5**, 77[254,256,258]
Tsuzuki, K., **3**, 49[264]; **4**, 159[80], 246[96], 258[96], 260[96]; **5**, 841[87]; **6**, 859[169]
Tsvetkov, Y. E., **6**, 49[100]
Tsvetkova, N. M., **5**, 1198[45]
Tsyban, A. V., **5**, 34[10]
Tsykhanskaya, I. I., **8**, 770[34], 771[42], 782[104]
Tsyryapkin, V. A., **4**, 218[148]
Tu, C., **5**, 530[71]
Tu, C.-Y., **3**, 709[17]; **7**, 239[44]; **8**, 540[195], 931[38]
Tubery, F., **1**, 367[55]; **5**, 829[22]; **6**, 919[40]
Tubiana, W., **3**, 729[41]
Tubul, A., **2**, 710[22], 718[73]; **4**, 1002[56]
Tuchinda, P., **1**, 558[133]; **6**, 159[175]
Tuck, D. G., **1**, 215[35]
Tucker, B., **5**, 117[275]; **6**, 491[118], 531[427], 823[119]
Tucker, J. R., **4**, 983[117,119]; **5**, 1086[74]; **8**, 451[180]
Tucker, L. C. N., **7**, 262[79]
Tückmantel, W., **1**, 218[52]; **4**, 96[103c], 886[117]
Tuddenham, D., **6**, 8[39], 927[76]; **7**, 198[26], 401[59], 403[59], 406[59]; **8**, 510[93]
Tuddenham, R. M., **2**, 720[86]
Tuerck, K. H. W., **4**, 292[227]
Tueting, D., **1**, 443[182], 446[197], 457[182]; **2**, 727[130], 749[133]; **3**, 233[272], 470[211], 473[211], 475[211], 476[211]; **4**, 594[143], 619[143], 633[143]

Tufariello, J. J., **2**, 111[87], 242[17]; **4**, 1076[35,36], 1078[50]; **5**, 250[39,39a], 251[39a], 253[47], 255[49], 256[54,57], 260[39a]; **6**, 734[1], 807[58]; **7**, 605[143]; **8**, 394[120], 395[127], 797[37]
Tuggle, R. M., **8**, 14[80]
Tughan, G., **7**, 481[110,111], 482[115,116]
Tugov, I. I., **2**, 387[334]
Tugusheva, N. Z., **6**, 554[741]
Tuite, M. R. J., **2**, 360[165a]
Tukada, H., **4**, 190[107], 984[121]; **5**, 210[58], 1086[71]
Tuladhar, S. M., **4**, 368[16,16b], 370[16b], 378[16b], 1033[24], 1055[24]
Tuleen, D. J., **7**, 208[80]
Tuleen, D. L., **6**, 147[82]; **7**, 207[75], 208[75], 210[96], 2 11[96], 212[96]
Tulich, L., **8**, 675[40], 676[40], 677[40], 694[40]
Tulip, T. H., **8**, 682[82]
Tull, R., **6**, 431[109]
Tullar, B. F., **8**, 143[60], 148[60]
Tuller, F. N., **1**, 3[20]; **3**, 14[71], 15[71]
Tullio, D. D., **7**, 80[139]
Tully, M. T., **4**, 604[280]
Tulshian, D. B., **6**, 889[67]
Tumas, W., **5**, 1120[21]
Tumer, A. B., **7**, 158[39]
Tumer, S. U., **5**, 1105[164]
Tun, M. M., **5**, 736[143]
Tuncay, A., **3**, 512[203]
Tüncher, W., **7**, 747[103]
Tundo, P., **5**, 487[188]
Tune, D. J., **8**, 766[20]
Tunemoto, D., **3**, 159[458], 161[458], 162[458], 167[458], 202[88]; **4**, 1040[86,87], 1041[87], 1045[86]; **5**, 924[144]; **6**, 159[174], 845[95], 865[95]; **8**, 844[74]
Tung, J. S., **3**, 507[172]
Tung, R. D., **6**, 6[29]; **7**, 400[48]
Tunga, A., **6**, 638[39]
Tunick, A. A., **4**, 5[19,19e]; **5**, 322[12]
Tuohey, P. J., **5**, 1133[30]
Tuong, T. D., **1**, 276[79], 277[79d,81,82]
Turba, V., **5**, 1025[81]
Turbak, A. F., **8**, 754[84]
Turchi, A., **8**, 754[103]
Turchi, I. J., **1**, 838[158]; **5**, 491[206]; **6**, 552[697]; **7**, 162[64], 336[34]
Turchi, N. J., **4**, 1096[159], 1097[159], 1098[159]
Turchin, K. F., **8**, 599[101]
Turecek, F., **5**, 557[58]
Turedek, F., **3**, 709[15]; **4**, 364[2], 372[59], 374[92]
Turkel, R. M., **4**, 968[92], 977[92]
Turkenburg, L. A. M., **4**, 1002[59], 1018[226]
Turnbull, J. H., **5**, 727[120]; **6**, 462[18]; **8**, 526[34]
Turnbull, J. K., **7**, 64[63]
Turnbull, K., **4**, 295[255]; **6**, 76[41], 98[41], 245[117], 246[117], 247[117], 248[117], 249[117], 252[117], 253[117], 254[117], 256[117]; **7**, 21[21], 476[60], 487[60], 488[60], 491[60], 504[60]; **8**, 384[23], 404[19], 405[21], 410[19,87], 4 11[19]
Turnbull, L. B., **3**, 158[437], 159[437], 160[437], 166[437]
Turnbull, M. D., **2**, 811[68]
Turnbull, M. M., **3**, 219[102]; **4**, 576[18]
Turner, A. B., **3**, 660[13]; **7**, 101[133], 136[110,114], 145[168]; **8**, 937[81,82]
Turner, D. L., **2**, 102[24]
Turner, D. W., **3**, 681[97]
Turner, E. E., **8**, 144[77]
Turner, H. S., **8**, 370[95], 916[106,107], 917[107], 918[106,107], 920[106]
Turner, J. O., **5**, 754[62]; **7**, 95[70]
Turner, J. V., **3**, 934[70]; **4**, 1040[81], 1043[81]; **6**, 893[85], 895[91], 896[91,94]; **8**, 500[51]
Turner, M. K., **8**, 198[130]
Turner, N. J., **2**, 456[67], 462[67]
Turner, R. B., **5**, 709[46], 945[248]; **6**, 707[41]; **7**, 543[20]; **8**, 494[24], 726[191,192]
Turner, R. W., **3**, 416[22]; **4**, 578[20,21], 579[22]; **5**, 461[101], 463[101], 683[39], 684[39]; **7**, 500[235]
Turner, S. G., **4**, 258[244]
Turner, W. B., **5**, 514[7]
Turner, W. R., **7**, 603[112,113]
Turner, W. W., **8**, 604[4]
Turney, T. W., **4**, 710[49], 712[49]; **8**, 453[194,195]
Turnowsky, F., **4**, 36[102]
Turocy, G., **4**, 988[137]
Turos, E., **1**, 92[69], 400[10]; **5**, 426[104], 485[181], 843[118]
Turrell, A. G., **3**, 499[124], 507[124], 673[69]; **7**, 872[97], 888[97]
Turro, N. J., **1**, 847[24]; **3**, 892[52]; **4**, 1081[79]; **5**, 125[12,15], 153[23-25], 154[31], 159[54], 165[23,78,80,82], 166[91], 176[78], 185[158], 194[4], 196[4], 197[4], 198[4], 200[4], 202[4], 207[49], 209[4], 436[155], 596[28], 597[28,29], 636[91], 639[91]; **7**, 41[25], 49[62], 851[15]; **8**, 91[59]
Tursch, B., **1**, 100[88]; **5**, 456[86]; **6**, 914[27]
Turuta, A. M., **7**, 479[89]
Tustin, G. C., **3**, 1026[39]
Tuttle, N., **2**, 140[35]
Tvorogov, A. N., **2**, 597[96], 610[96]
Twain, M., **2**, 232[175]
Tweddle, N. J., **5**, 105[198]
Tweedie, V. L., **8**, 966[73]
Twelves, R. R., **2**, 764[62]
Twigg, M. V., **3**, 1017[8]
Twine, C. E., Jr., **3**, 125[305], 126[305], 127[305]
Twiss, P., **1**, 17[219], 36[234]
Twitchin, B., **6**, 818[107], 896[94]
Twohig, M. F., **4**, 1033[24], 1053[130], 1055[24,136], 1056[136]
Tyagi, M. P., **4**, 36[102]
Tychopoulos, V., **2**, 956[12], 958[25]
Tyfield, S. P., **6**, 690[397], 692[397]
Tyle, Z., **7**, 802[47]
Tyler, J. K., **5**, 576[146]
Tyler, P. C., **6**, 48[85], 978[25]
Tyltin, A. K., **6**, 509[283]
Tyman, J. H. P., **2**, 956[12], 958[25]
Tyner, M., III, **7**, 800[30]
Tyrala, A., **2**, 362[183]; **3**, 158[446], 159[446]; **4**, 432[113]
Tyrlik, S., **3**, 565[24], 570[24], 583[24]; **8**, 447[100], 797[36]
Tyrrell, H., **5**, 647[12], 648[12]
Tyrrell, N. D., **8**, 831[92]
Tyson, R. L., **4**, 615[392], 629[392]
Tzeng, D., **2**, 584[117]
Tzodikov, N. R., **7**, 168[105]
Tzougraki, C., **6**, 635[14b], 636[14]; **8**, 959[21]
Tzschoppe, D., **2**, 969[86]

U

Uang, B. J., **2**, 105[44], 671[48], 698[82]; **4**, 799[112]
Uaprasert, V., **4**, 298[295]
Ubasawa, M., **6**, 604[26], 626[167]
Ubochi, C. I., **4**, 425[23]
Ubukata, M., **5**, 564[96]
Uccella, N., **8**, 965[62]
Ucciani, E., **2**, 138[20]; **4**, 926[38], 928[38]
Uchann, R., **8**, 212[15]
Uchibayashi, M., **3**, 602[218], 607[218]
Uchida, A., **3**, 483[9]; **4**, 41[118], 46[132], 53[132,132a]
Uchida, H., **6**, 938[134]; **8**, 371[103]
Uchida, I., **7**, 255[38]
Uchida, K., **2**, 589[155,156], 1061[92], 1066[119], 1069[92,132], 1071[92]; **3**, 243[11], 254[97], 259[135], 799[103]; **7**, 180[158], 805[65]; **8**, 734[6], 755[119], 757[6]
Uchida, M., **2**, 176[185]
Uchida, N., **3**, 168[501], 169[501], 170[501]; **8**, 170[78]
Uchida, S., **3**, 677[81], 686[81]; **4**, 1056[141]; **7**, 429[155]; **8**, 366[48,49], 837[15a]
Uchida, T., **2**, 611[101]; **3**, 469[203], 470[203], 473[203]; **4**, 17[53], 161[86c], 840[34]; **5**, 1175[38], 1178[38]; **6**, 47[77]; **7**, 443[51a], 834[79]; **8**, 787[119]
Uchida, Y., **2**, 463[82,83], 464[82,83]; **3**, 862[3]; **5**, 1158[173], 1175[38], 1177[43], 1178[38,43], 1180[47], 1181[47]; **8**, 446[94], 447[97], 452[94], 554[366]
Uchikawa, M., **2**, 119[162,163]; **3**, 1000[55], 1004[60]; **6**, 852[135], 877[38], 883[38], 885[56], 887[38], 890[56]
Uchimaru, T., **2**, 112[96], 242[20], 816[87]; **4**, 261[292]; **6**, 960[63]; **7**, 679[74]
Uchino, H., **8**, 388[63]
Uchino, N., **6**, 233[40]; **8**, 450[164]
Uchio, R., **7**, 56[16]
Uchio, Y., **4**, 124[212]
Uchiro, H., **2**, 657[168]
Uchiumi, S., **7**, 451[19], 452[19], 454[19]
Uchiumi, T., **7**, 79[129]
Uchiyama, H., **1**, 159[79]; **2**, 23[88]; **3**, 45[248]; **7**, 371[68], 379[100]
Uchiyama, M., **5**, 552[1]; **6**, 18[65], 603[11,17], 608[11], 614[94], 620[129], 624[129,148]; **7**, 750[131]
Uchiyama, Y., **7**, 407[83]
Uda, H., **1**, 833[119]; **2**, 282[38], 363[191], 641[72]; **3**, 946[87]; **4**, 18[59], 30[88], 106[140b], 121[205c], 258[240,241,250], 261[240], 262[240]; **5**, 124[10], 130[10]; **6**, 144[77], 150[118,119], 151[119], 155[155], 156[155], 902[129], 1043[12]; **7**, 205[64]; **8**, 12[66]
Uda, J., **3**, 1027[44]
Udaka, S., **4**, 27[79,79d]
Udall, W., **4**, 4[13]
Ude, H., **6**, 1022[58]
Udenfriend, S., **7**, 11[89]
Udodong, U. E., **1**, 406[29], 732[18]; **4**, 391[182,182a]; **5**, 494[217], 579[164]; **6**, 40[40], 118[101]; **7**, 567[102], 584[102]; **8**, 540[195]
Udupa, K. S., **8**, 137[55]
Uebelhart, P., **6**, 677[313]; **7**, 410[103]
Uebersax, B., **5**, 632[59]
Ueberwasser, H., **8**, 974[124]
Ueda, C., **7**, 752[153]; **8**, 375[155]
Ueda, E., **3**, 464[175]
Ueda, H., **5**, 522[45]; **7**, 407[84a]; **8**, 559[400]
Ueda, J., **3**, 644[134a]
Ueda, K., **2**, 743[82]; **3**, 648[185], 900[97]; **4**, 1089[138], 1091[138]; **7**, 538[64], 678[70]
Ueda, M., **3**, 135[341,342,343], 136[341,342,343], 137[341,342]; **4**, 439[165]; **5**, 830[31]; **6**, 531[429]
Ueda, N., **6**, 936[106]
Ueda, S., **4**, 30[89]
Ueda, T., **1**, 792[270]; **2**, 889[57]; **6**, 530[415], 531[426]; **8**, 422[35], 458[225], 658[99]
Ueda, Y., **1**, 526[95]; **2**, 648[89,90,92], 649[89,92], 809[54], 824[54], 1058[72]; **5**, 107[200], 407[28,28b], 439[167]; **6**, 156[162], 1036[145]; **7**, 425[149c]; **8**, 117[72]
Uehara, C., **6**, 931[90]
Uehara, H., **1**, 425[102]; **3**, 421[54]
Uehara, K., **2**, 357[145], 358[145]
Uehling, D. E., **1**, 800[298]; **4**, 158[79]
Uei, M., **4**, 386[150], 387[150]; **5**, 539[107]; **6**, 994[97]; **7**, 247[102], 257[51]
Ueki, M., **6**, 644[85]
Ueki, S., **5**, 850[146]
Uematsu, M., **2**, 564[8]; **7**, 208[78]; **8**, 99[107]
Uemura, D., **5**, 86[34]; **6**, 8[37]; **7**, 440[40]
Uemura, M., **1**, 554[108]; **4**, 520[33-35], 546[129]; **5**, 1098[114], 1112[114]
Uemura, S., **1**, 648[126], 664[201], 672[201], 712[201], 714[201], 828[75], 862[77]; **2**, 598[16]; **3**, 87[111], 106[111], 114[111], 381[26,27,29], 382[26,27,29]; **4**, 315[520], 340[47], 341[56,58], 347[87], 349[58], 370[41], 387[158], 398[216], 399[216e], 401[41,158c,216e,232,232a], 405[41,158c,216e,232a], 406[41,158c,216e,232a]; **6**, 289[194-197], 291[201], 293[194-197,226,227], 467[52], 1030[104], 1031[110,112], 1032[121]; **7**, 95[64,67,172], 108[171,172], 128[68], 129[70], 154[17], 451[29], 495[207,208,212,213], 496[212,213,214], 505[288], 520[27], 521[33], 523[43], 524[52], 530[23,25], 534[40,41], 760[32], 771[264], 773[308,309], 774[326], 775[352c,354,355], 776[308,309,355,356,358,363]; **8**, 413[130], 476[24], 848[104], 849[104,114], 851[124]
Uenishi, J., **1**, 193[86]; **2**, 642[76], 643[76]; **3**, 231[253], 558[51]; **6**, 8[39], 632[4]; **7**, 200[41], 208[86], 211[86], 396[25]
Ueno, A., **8**, 212[9], 222[9]
Ueno, H., **4**, 629[405]; **6**, 849[117]; **7**, 458[112]
Ueno, K., **4**, 1086[114,115]
Ueno, M., **8**, 989[36]
Ueno, T., **4**, 744[135]
Ueno, Y., **1**, 642[116], 645[116]; **2**, 323[28], 333[28], 578[88], 587[88,149]; **3**, 169[509], 985[25]; **4**, 738[98], 744[135], 792[65,67], 823[231]; **6**, 677[323], 876[34], 885[34]; **7**, 318[58], 319[58], 320[58], 322[68], 533[33], 616[19], 764[109], 765[137], 771[265], 772[265], 773[265]; **8**, 413[135], 846[80,84]
Uerdingen, W., **6**, 563[904]
Uesaka, M., **5**, 474[158]
Uesato, M., **2**, 736[26]
Uesato, S., **5**, 468[132]
Ueshima, T., **7**, 248[112], 809[84]
Ueta, K., **6**, 1022[64]
Ueyama, M., **5**, 621[18]
Ueyama, N., **4**, 446[213]
Uff, B. C., **1**, 544[34], 551[34], 553[34]; **7**, 671[7]; **8**, 392[103], 478[37]
Ufimtsev, S. V., **6**, 535[534]
Uggeri, F., **3**, 778[6], 788[6,54], 789[6,55,57]; **5**, 768[124], 779[124]; **7**, 829[55]

Ughetto, G., **4**, 403[239], 404[239]
Ugi, I., **2**, 753[3], 1083[2a], 1084[8-10,17], 1085[23], 1086[32], 1087[16,33,34,36,37], 1088[37-39,45,49,50], 1089[39,50,53-57], 1090[36,55,63-69,71-73,75,76], 1091[64,69], 1092[65,71], 1093[71,83-85], 1094[71,86,87,89,91], 1095[89,91,93], 1096[71,94], 1098[45,71,103-105], 1099[66,103,106,109b,110], 1100[55,71,118], 1101[118], 1102[73,124,125,127], 1103[73,118b,125,129,131], 1104[134], 1106[16], 1108[63-66,80]; **5**, 77[251]; **6**, 242[85-88], 243[85-88], 249[139], 254[166], 291[207], 293[207], 294[207,231,240,243], 295[207,251,256], 489[88,97], 638[45], 639[46], 667[243]; **7**, 778[397]; **8**, 384[35], 830[84,85]
Uglova, E. V., **8**, 99[107]
Ugo, R., **4**, 1031[4]; **5**, 1152[140], 1154[140]; **7**, 108[173], 452[61]; **8**, 443[1], 446[95], 449[157], 450[157], 452[95a,b], 457[95a-c,218], 458[218], 683[92], 689[92]
Ugolini, A., **1**, 419[79], 797[292], 802[292]; **6**, 995[104], 996[104]; **7**, 400[46]
Ugozzoli, F., **6**, 195[224]
Ugrak, B. I., **4**, 347[103]
Uguagliati, P., **4**, 710[48]
Uguen, D., **3**, 159[454], 161[454], 167[481], 168[481], 170[481], 172[481], 882[103]; **4**, 593[136], 599[217], 763[212]; **5**, 305[84,86]; **6**, 157[170], 161[186]
Uh, H.-S., **2**, 547[96], 551[96]; **7**, 258[44]; **8**, 819[44]
Uhde, G., **7**, 84[3]; **8**, 929[27]
Uhl, J., **8**, 652[71]
Uhl, K., **8**, 755[134]
Uhle, F. C., **8**, 956[6]
Uhlenbrauck, H., **2**, 725[111], 726[111]
Uhlig, F., **2**, 753[4]
Uhlig, G. F., **4**, 45[126,126a]
Uhlig, H., **6**, 1044[20]
Uhlig, H. F., **2**, 785[42]
Uhlmann, E., **6**, 625[154]
Uhm, S. J., **1**, 212[6], 213[6]; **8**, 907[73]
Uhm, S. T., **4**, 969[64]
Uhmann, R., **2**, 1099[110]
Uhrhammer, R., **2**, 127[235]
Uhrick, D. A., **5**, 17[114,115]
Uhrig, J., **6**, 502[215], 531[215]
Ujhazy, J., **7**, 13[110]
Ujiie, A., **4**, 505[134]
Ujjainwalla, M., **4**, 342[65]
Ukai, A., **5**, 323[15]
Ukai, J., **1**, 161[80,86,87,90]; **2**, 22[87], 65[31], 74[71], 269[72], 615[126]; **3**, 446[85,86]
Ukai, T., **1**, 441[173]
Ukai, Y., **2**, 631[18]
Ukaji, Y., **1**, 85[30], 104[30]; **2**, 225[156]; **8**, 64[218], 67[218]
Ukharov, O. V., **6**, 104[2]
Ukimoto, K., **6**, 237[61]
Ukita, T., **1**, 894[160]; **3**, 168[492], 169[492]; **4**, 817[207]; **6**, 1004[138], 1065[90b]; **7**, 615[9], 621[34], 623[35], 624[36]
Ulan, J. G., **8**, 431[62]
Ulatowski, T. G., **1**, 86[42,44]
Ulbricht, T. L. V., **8**, 642[29]
Ulery, H. E., **3**, 342[15]
Ullah, M. I., **3**, 807[26]
Ullah, Z., **6**, 836[55]
Ullenius, C., **2**, 120[170,171]; **3**, 212[35], 512[193,194]; **4**, 18[56], 171[26,27], 176[51], 178[60], 201[9], 227[209], 229[213,214,217,218,221], 240[44], 256[207], 262[207], 525[66], 532[86-88], 534[87,88], 535[87], 537[88], 538[87,88], 539[87,88]; **5**, 804[96], 973[10]
Ullman, E. F., **5**, 712[60]
Ullrich, F.-W., **2**, 779[4], 780[4]
Ullrich, J. W., **2**, 1038[101]
Ulm, E. H., **8**, 50[117], 66[117]
Ulman, A., **7**, 14[127], 40[5]
Ulmschneider, D., **8**, 724[171]
Ulrich, H., **5**, 86[24], 89[24], 90[24], 102[24], 117[275], 451[39], 470[39], 485[39]; **6**, 487[1], 488[1], 489[1], 495[1], 498[1], 515[1], 523[1], 524[1], 525[1], 526[1], 527[1], 531[427], 539[1], 796[16], 823[119]
Ulrich, P., **3**, 250[70]; **4**, 384[143]; **6**, 648[118]
Ulsaker, G. A., **6**, 509[262]
Ultee, W., **6**, 1013[8], 1017[8]
Ultée, W. J., **7**, 765[140]
Umani-Ronchi, A., **1**, 188[73], 189[73], 192[82]; **2**, 35[130], 36[130], 507[19], 566[23], 657[161b]; **3**, 168[488], 169[488]; **6**, 685[350], 976[4]; **7**, 549[42], 841[14]; **8**, 36[80], 54[80], 66[80], 124[90], 252[111], 289[27], 550[331,333], 551[336], 797[40], 840[36], 842[46], 843[46], 844[36], 913[94], 914[94]
Umbrasas, B. N., **2**, 712[39]
Umbreit, M. A., **3**, 583[119]; **6**, 686[369], 980[42], 985[65]; **7**, 87[21], 843[44]; **8**, 888[121,124]
Umebayashi, H., **8**, 643[38]
Umeda, I., **5**, 293[44], 1187[8], 1190[24], 1200[54]
Umeda, N., **8**, 170[95]
Umehara, J., **7**, 384[114a]
Umehara, T., **5**, 137[75], 143[75]
Umei, Y., **2**, 580[99]; **8**, 964[59]
Umemoto, T., **4**, 1040[86], 1045[86]; **5**, 924[144]
Umen, M. J., **1**, 116[46], 118[46], 433[225], 683[218]; **3**, 14[73], 15[73], 249[63]; **4**, 91[88b]
Umeno, M., **8**, 764[10], 773[10]
Umeyama, K., **1**, 159[78], 160[78], 161[78]; **2**, 23[88]
Umezawa, B., **3**, 672[65]; **7**, 339[43]; **8**, 29[38], 49[115], 50[119], 66[38,115,119]
Umezawa, H., **2**, 917[85]; **6**, 49[95]; **7**, 489[173]
Umezawa, J., **7**, 429[155]
Umezawa, S., **6**, 60[145]; **8**, 817[30,31]
Umezawa, T., **8**, 406[38]
Umezawa, Y., **2**, 917[85]
Umezono, A., **6**, 66[4]
Umezu, K., **3**, 220[125]; **6**, 20[74]
Umezu, T., **7**, 451[19], 452[19], 454[19]
Umi, Y., **7**, 208[87]
Umino, N., **8**, 64[216], 67[216], 170[94], 250[101], 253[117]
Umpleby, J. D., **4**, 876[63]; **5**, 38[23a,b]
Umrigar, P., **7**, 372[70]
Unangst, P. C., **2**, 828[130]; **6**, 501[191]; **8**, 50[118], 66[118]
Uncuta, C., **6**, 819[110]
Unde, N. R., **6**, 687[381]
Underhill, E. W., **3**, 224[174]
Underiner, T. L., **3**, 222[138]
Underwood, G. R., **3**, 383[42]; **5**, 736[140]
Underwood, H. W., Jr, **2**, 757[11]
Underwood, J. M., **4**, 390[168], 395[168e]
Underwood, W. G. E., **6**, 1017[38], 1024[38]
Undheim, K., **2**, 365[209]; **5**, 1012[52]; **6**, 509[262]
Unelius, C. R., **3**, 489[61], 495[61], 504[61], 511[61], 515[61]
Uneyama, K., **2**, 19[77,77a], 655[135]; **3**, 125[297,298], 126[298], 128[297], 129[297], 130[297,298], 133[297], 137[298], 167[484], 168[484,495,499,500], 169[495,499,500], 361[74], 648[185]; **4**, 371[47]; **7**, 98[104,105], 537[61], 765[164], 770[256c], 771[256], 819[22]
Ungaro, R., **2**, 137[17], 960[32]
Ungemach, F., **2**, 1016[27], 1017[31,34,35], 1018[35]; **6**, 737[31,37,41], 746[89]
Unger, F. M., **2**, 464[97]
Unger, L. R., **6**, 737[38], 939[141], 942[141]
Ungváry, F., **8**, 458[222]

Unni, M. K., **7**, 605[144]; **8**, 707[23], 950[161]
Unno, K., **2**, 819[100], 824[100], 1020[46]; **5**, 839[82], 864[257]; **6**, 896[93]
Uno, F., **1**, 377[98]
Uno, H., **1**, 350[154], 359[12], 360[12], 361[12], 362[12]; **2**, 58[9]; **3**, 729[39]; **4**, 27[83], 155[70]; **8**, 613[78]
Uno, M., **3**, 454[121]; **4**, 850[86]
Unrau, A. M., **8**, 646[49]
Untch, K. B., **4**, 10[33,33b]
Untch, K. G., **4**, 45[126], 159[85], 187[100], 261[295], 262[295], 1011[163]; **5**, 797[67]; **6**, 234[51], 235[51], 236[51], 647[112a], 838[67], 902[132], 970[126]; **8**, 477[36]
Unterweger, W.-D., **6**, 679[328]
Unverzagt, C., **2**, 1102[123]; **6**, 633[8], 634[8], 640[55], 641[8,55], 646[8], 652[8], 671[55]
Uosaki, Y., **5**, 71[139,140,141,142], 77[257]; **7**, 407[82]
Uozumi, Y., **3**, 1038[95]
Upeslacis, J., **8**, 30[42], 66[42]
Uphoff, J., **5**, 1107[165]
Ura, T., **7**, 708[31]
Urabaniak, W., **8**, 774[73]
Urabe, A., **4**, 607[316,317]; **5**, 275[11]
Urabe, H., **2**, 609[85-87], 624[86], 728[139]; **3**, 12[63], 454[119]; **4**, 116[186]; **7**, 307[15], 310[15], 323[15]
Urakawa, C., **8**, 383[21]
Uramoto, M., **8**, 418[10]
Uramoto, Y., **8**, 783[107]
Urano, S., **1**, 836[140]
Urano, Y., **1**, 57[32], 557[131]
Urata, Y., **4**, 823[231]; **6**, 577[978]
Uray, G., **4**, 439[157]; **6**, 553[704]
Urayama, T., **3**, 1032[67]
Urbach, H., **3**, 851[68]; **4**, 91[89]; **6**, 284[171]
Urban, C., **2**, 463[92], 464[92]
Urban, R., **2**, 1090[71], 1092[71], 1093[71], 1094[71,89], 1095[89,93], 1096[71], 1098[71,103], 1099[103,106,108], 1100[71]; **8**, 384[35]
Urban, R. S., **3**, 823[16], 825[16]
Urbanec, J., **7**, 451[25]
Urbanek, F., **3**, 851[62]
Urbanek, T., **5**, 571[115]
Urbani, R., **6**, 714[84]
Urbaniak, W., **8**, 765[13], 774[72]
Urbanski, T., **2**, 321[10], 329[10], 365[215]; **8**, 363[5]
Urbas, L., **2**, 321[9], 325[9], 326[9], 327[9], 328[9], 329[9], 354[109]
Urben, P. G., **8**, 643[36]
Urbi, G. B., **4**, 764[215]
Urbschat, E., **6**, 429[90]
Urch, C. J., **7**, 621[32]
Urea, M., **4**, 1076[39]
Urech, E., **4**, 41[119,119a]
Urech, R., **4**, 100[121]; **8**, 505[81]
Uriarte, E., **5**, 416[57]
Uriarte, P. J., **1**, 307[93], 310[93]
Uribe, G., **8**, 16[101], 237[14], 240[14], 244[14], 537[180], 708[41]
Urleb, U., **6**, 554[743,744]
Uroda, J. C., **3**, 757[124]
Urpí, F., **6**, 77[54], 122[137]; **8**, 385[41]
ur Rahman, A., **2**, 1090[72,73], 1102[73], 1103[73]
Urrios, P., **1**, 543[26]
Urrutia, G., **5**, 75[213]
Urry, G. W., **7**, 882[173]
Ursini, O., **5**, 771[151], 772[151]
Urso, F., **3**, 1036[84]; **4**, 227[210]; **7**, 482[117]
Ursprung, J. J., **3**, 709[16]
Uryu, T., **7**, 453[76]
Urz, R., **1**, 142[25], 143[34], 145[41], 150[41], 151[41], 153[41], 154[41]
Usami, Y., **7**, 256[39]
Usgaonkar, R. N., **2**, 792[66]
Ushakova, N. I., **8**, 770[34]
Ushakova, R. L., **6**, 530[422]
Ushakova, T., **4**, 841[46]
Ushanov, V. Zh., **4**, 50[142]
Ushida, S., **1**, 179[24]; **8**, 95[87,89], 591[58]
Ushio, H., **1**, 223[84], 225[84d]
Ushio, K., **8**, 185[18,25], 190[18,65,68], 191[86,87], 195[108], 589[48]
Ushio, Y., **1**, 766[162]; **8**, 215[53], 217[53], 227[121], 240[32], 244[70], 620[133], 624[133]
Ushioda, S., **8**, 964[52]
Ushirogochi, A., **8**, 837[13c]
Usieli, V., **5**, 1133[29]
Usifer, D., **3**, 513[208]; **8**, 935[68]
Usik, N. V., **4**, 85[77d]
Uskokovic, M., **8**, 269[82]
Uskokovic, M. R., **1**, 780[229], 822[34]; **2**, 530[23], 534[31], 547[100], 548[100]; **4**, 31[92,92j], 370[29], 384[143], 390[171], 413[277], 1076[40], 1080[71]; **5**, 4[39], 5[39,42,44], 8[44], 129[33], 256[58], 257[58b], 260[63c,d], 264[63c,d], 265[63d], 433[137b], 835[59]; **6**, 266[48], 531[452], 913[24], 960[56]; **7**, 268[122], 564[92], 567[92], 678[72], 701[66]; **8**, 608[46], 722[150]
Usmai, A. A., **5**, 595[17,20], 596[17,20]
Usmani, A. A., **5**, 608[64], 609[64]; **6**, 764[8]
Usov, A. V., **6**, 509[245]
Usov, V. A., **6**, 509[282]
Uspenskaya, K. S., **7**, 7[39]
Ustynyuk, T. K., **3**, 839[7], 848[7]
Ustyugov, A. N., **5**, 76[246], 552[16]
Usui, S., **4**, 477[165]
Usui, T., **8**, 817[30]
Usui, Y., **6**, 566[930]
Utaka, M., **2**, 199[87], 838[178]; **3**, 843[25], 844[32,33]; **8**, 185[26], 190[26], 191[89,90], 193[102], 194[104a], 195[104b], 205[160]
Utamapanya, S., **5**, 565[98]
Utawanit, T., **7**, 158[37]
Utera, J., **4**, 609[329]
Utille, J. P., **7**, 247[104]
Utimoto, K., **1**, 185[76], 190[76], 191[76], 193[85], 195[85], 198[85], 202[103], 203[103], 205[107-109], 206[109], 343[111], 347[132,133,134], 749[79], 807[318,322,324], 808[322,323,324], 809[327,329], 810[327,329a,b]; **2**, 20[82], 59[15], 584[126], 589[154,155,156,157], 597[10], 599[26], 603[42], 726[123], 1067[126]; **3**, 243[11], 244[29], 254[96,97], 259[135], 274[22], 279[37], 445[74], 484[24], 501[24], 759[133], 799[100,103]; **4**, 23[71], 120[203], 162[92], 300[305], 393[190,198], 394[190,192,198], 411[198,266d], 567[47,49-51], 588[68], 637[68], 721[31], 725[31], 756[184], 770[248,249], 771[253], 789[32], 791[42], 796[96], 824[241], 879[85], 885[111], 886[117], 900[182]; **5**, 927[164], 938[164], 1125[59,60]; **6**, 237[60,67], 243[60], 564[913,914], 979[28]; **7**, 259[62], 275[147], 276[147], 379[101], 595[21], 601[85]; **8**, 699[149], 734[6], 755[119], 757[6], 785[113], 798[46], 807[46], 818[41], 820[41], 823[58], 937[86]
Utka, J., **1**, 303[81]
Utley, J. H. P., **3**, 577[89], 634[8,20,24], 635[40], 636[8], 637[65,66], 638[8,20], 639[8,24], 642[111], 643[8], 647[195], 649[8,20,24], 650[66], 655[8,20]; **6**, 176[83]; **7**, 801[44], 806[73]; **8**, 132[9], 133[15,16,23-25], 135[16], 137[54], 568[471], 612[77]
Utne, T., **6**, 685[359]
Utsunomiya, I., **2**, 1068[129]; **3**, 512[198]
Utz, R., **6**, 430[101], 637[32,32c]
Uusvuori, R., **7**, 686[96]
Uwano, A., **5**, 442[185]
Uyama, H., **1**, 544[39]; **8**, 302[96]

Uyehara, T., **1**, 100[89]; **2**, 120[175], 948[181]; **3**, 730[45]; **4**, 238[14], 247[106], 257[106], 260[106]; **5**, 64[33], 225[103,104], 226[104]
Uyeo, S., **1**, 123[75], 373[82]; **6**, 820[111]; **8**, 527[47], 836[4], 842[4], 993[59]
Uzan, R., **8**, 552[357,358]
Uzarewicz, A., **8**, 875[41]
Uzlova, L. A., **7**, 294[18]
Uzzell, P. S., **6**, 734[9]

V

Vaalburg, W., **8**, 53[132], 66[132]
Vaben, R., **5**, 185[165]
Vacca, A., **8**, 458[219]
Vacca, J. P., **5**, 410[41]; **7**, 164[81], 313[38]
Vaccariello, T., **8**, 541[207]
Vaccaro, W., **1**, 122[70], 376[93]
Vaccher, C., **8**, 660[108]
Vacher, B., **4**, 765[226]; **7**, 731[55]
Vaerman, J. L., **5**, 422[81]
Vagberg, J., **7**, 94[57]
Vagberg, J. O., **4**, 371[49], 565[45], 591[115], 592[115], 617[115], 618[115], 633[115]; **6**, 849[116]
Vagt, H., **5**, 451[5]
Vagt, U., **3**, 619[25]
Vahrenhorst, A., **1**, 669[181], 670[181]
Vaid, B. K., **7**, 833[76]
Vaid, R. K., **7**, 222[37], 227[37,81], 833[76]
Vaidya, N. A., **5**, 835[59]
Vaidya, S. P., **2**, 142[47]
Vaidyanathaswamy, R., **4**, 311[450]
Vaidyanathaswamy, S., **8**, 726[193]
Vairamani, M., **7**, 400[45]
Vais, A. L., **4**, 408[259f]
Vaissermann, J., **5**, 1105[162,163]
Vajda, J., **7**, 777[389]
Vajna de Pava, O., **8**, 187[47]
Vakilwala, M. V., **7**, 738[30]
Vakul'skaya, T. I., **4**, 461[100], 475[100]
Val, J. A. F., **7**, 775[352a]
Valade, J., **2**, 607[75], 608[75], 609[81,83]; **3**, 741[50]; **8**, 547[316,316j,317], 548[319]
Valcavi, U., **7**, 65[68]
Valderrama, J. A., **7**, 355[43]
Valdes, C., **5**, 433[136b]
Valdés, J. A., **2**, 361[178]
Valdez, E., **4**, 932[63]
Valeev, F. A., **8**, 676[55], 682[55], 689[55], 691[55]
Valencia, N., **8**, 446[74], 452[74], 457[74]
Valenta, Z., **2**, 809[52], 823[52]; **5**, 125[21], 128[21], 347[72,72c]
Valente, L. F., **1**, 214[23]; **3**, 231[241], 266[196], 443[51]; **4**, 889[135], 893[150,151]; **8**, 756[148]
Valenti, E., **5**, 1062[59]
Valentin, E., **4**, 20[63], 21[63]; **5**, 331[41]; **6**, 709[55], 710[57-59], 711[62]
Valentin, F., **6**, 36[19]
Valentine, D., **5**, 165[80]
Valentine, D., Jr., **8**, 152[168], 159[2], 459[228], 460[249], 541[212], 606[18], 861[225]
Valentine, J. R., **8**, 584[20]
Valentine, J. S., **6**, 2[8], 22[8]
Valero, M. J., **4**, 35[98e]
Valette, G., **3**, 31[184]; **6**, 720[131]
Valiant, J., **7**, 778[414]
Valicenti, J. A., **6**, 959[48]
Valkó, K., **7**, 746[93]
Valkovich, P. B., **5**, 718[95]
Valle, G., **1**, 305[84,85], 323[84]; **4**, 330[5], 744[131]; **5**, 370[102], 371[102]; **6**, 150[114], 575[968]
Valle, L., **1**, 774[213]
Valle, S., **2**, 381[300]
Vallee, Y., **2**, 214[130,131]; **5**, 556[51], 575[132,135]
Vallén, S., **2**, 233[188]
Vallino, M., **1**, 13[68]
Valls, G., **8**, 201[144]
Valls, J., **3**, 12[65]; **8**, 533[154]
Valnot, J.-Y., **8**, 587[37]
Valoti, E., **2**, 735[15]; **3**, 312[102]
Valpey, R. S., **1**, 849[28]; **2**, 838[169]; **4**, 629[415]; **5**, 21[154], 22[154]; **6**, 1004[140]
Valpuesta Fernandez, M., **2**, 386[329]
Valt-Taphanel, M.-H., **5**, 417[65]
Valvassori, A., **5**, 1025[81]
Valverde, S., **1**, 759[132]; **8**, 333[55]
Valyocsik, E. W., **3**, 296[14]
Vaman Rao, M., **4**, 739[108]
Van, D. L., **5**, 442[183], 444[188]
van Aarssen, B. G. K., **8**, 589[46]
van Aelst, S. V., **6**, 38[37]
VanAllan, J. A., **2**, 380[298]; **5**, 752[50]
Van Alphen, **2**, 770[4]
van Arkel, B., **5**, 649[22], 650[22]
Van Asch, A., **5**, 117[278]
Vanasse, B., **6**, 651[136]; **7**, 261[66]
van Asselt, A., **8**, 673[25], 696[25]
van Asten, J. J. A., **7**, 765[140]
Van Audenhove, M., **5**, 127[25], 131[44]
van Balen, H. C. J. G., **2**, 662[22], 663[22], 664[22]; **5**, 431[121], 434[121b]
Van Beek, G., **7**, 373[73]
van Bekkum, H., **4**, 538[106]; **6**, 705[21]; **8**, 287[18,19], 288[19], 427[50], 444[7], 447[114,115], 453[114,115], 499[43], 500[43]
van Bergen, T. J., **1**, 299[62]; **3**, 512[199], 969[131,132]; **4**, 430[97]; **8**, 93[74,77], 94[77]
Van Binst, G., **6**, 707[46], 712[71]
van Boeckel, C. A. A., **2**, 599[20]; **6**, 38[37], 57[139], 619[116], 652[145], 658[186b], 662[214]
van Boom, H., **6**, 57[139]
van Boom, J. H., **2**, 599[20]; **6**, 17[62], 602[3], 619[116], 620[130], 652[145], 658[187], 661[211], 662[214]
Van Camp, A., **5**, 113[236]
van Campen, M. G., **7**, 595[8], 599[8a]
van Campen, M. G., Jr., **8**, 724[172]
Vance, D. E., **6**, 436[23]
Vance, R., **1**, 92[64]
Vance, R. L., **5**, 857[227]
van Cleve, J. W., **6**, 46[71]
van Daalen, J. J., **7**, 40[6]
Van De Heisteeg, B. J. J., **1**, 746[70]; **5**, 1125[57]
Van De Mark, M. R., **7**, 43[42], 774[310], 802[50]
van de Mieroop, W. F., **1**, 30[154]
Van den Born, H. W., **7**, 805[66]
Van Den Bosch, C. B., **8**, 499[43], 500[43]
Van Den Bril, M., **6**, 517[324,325], 544[324,325], 552[325]
van den Broek, L. A. G. M., **5**, 441[177]; **7**, 763[96]; **8**, 405[31]
Vandenbulcke-Coyette, B., **5**, 422[81]
Van Den Elzen, R., **5**, 567[105]; **7**, 766[181]
van den Engh, M., **7**, 236[18], 237[38], 564[109], 851[20]
van den Goorbergh, J. A. M., **3**, 58[293]
Vandenheste, T., **1**, 240[43]
van de Putte, T., **8**, 444[7]
van der Baan, J. L., **1**, 218[54]; **2**, 378[287], 801[36], 981[23]; **4**, 595[152], 877[69], 884[108]; **5**, 46[37]; **6**, 86[99], 856[150]; **7**, 373[73]
Vanderbilt, B. M., **2**, 321[5], 326[5]
van der Ent, A., **8**, 445[23]

Vanderesse, R., **3**, 509[179]; **8**, 14[86], 802[76], 840[38], 878[48]
Van der Eycken, E., **3**, 900[93]; **7**, 301[58,59]
Van der Eycken, J., **2**, 201[92]; **5**, 362[93], 363[93b]; **7**, 301[59]
van der Gen, A., **1**, 546[52], 563[172,174], 774[212]; **2**, 101[19], 482[33], 484[33]; **3**, 20[121], 25[121], 58[293], 242[10], 361[75]; **4**, 18[62], 20[62h]; **6**, 134[32], 705[32-34]; **7**, 125[57], 235[1]
van der Goot, H., **4**, 110[151]
Vanderhaeghe, H., **5**, 92[72]
van der Hart, J. A., **5**, 647[19], 650[19], 652[19], 653[19], 656[19]
van der Heide, F. R., **3**, 1056[35], 1062[35]
van der Helm, D., **3**, 216[71]
van der Holst, J. P. J., **4**, 51[145b]
Vander Jagt, D. L., **3**, 829[56]; **4**, 301[329]
van der Kerk, G. J. M., **1**, 30[153], 214[27]; **2**, 123[195,196], 124[204], 125[204], 280[27]; **8**, 264[40], 547[316,316e], 589[46]
van der Kerk, S. M., **8**, 96[93]
Van der Knaap, Th. A., **8**, 863[236]
van der Leij, M., **3**, 868[41]
van der Linde, L. M., **4**, 18[62], 20[62h]
van der Loop, E. A. R. M., **3**, 855[83]
van der Louw, J., **1**, 218[54]; **2**, 981[23]; **4**, 595[152], 877[69], 884[108]; **5**, 46[37]; **6**, 86[99]
van der Marel, G. A., **1**, 737[30]; **6**, 17[62], 619[116]
van der Plas, H. C., **4**, 423[7], 424[17,18], 426[48], 434[133], 465[117], 1004[75], 1021[75]; **5**, 584[192]
Van der Plas, H. E., **4**, 484[23], 485[23]
Vanderpool, S., **5**, 911[94]
Van Der Puy, M., **1**, 632[66]; **4**, 445[209]; **6**, 133[1,3]
van der Saal, W., **2**, 809[53]
Vanderslice, C. W., **8**, 709[43], 812[3], 908[78], 909[78]
van der Steen, F. H., **2**, 296[86], 922[99], 923[99], 936[99]; **5**, 101[161]
Van Der Steen, R., **6**, 494[131]
van der Toorn, J. M., **4**, 538[106]
Van der Veek, A. P. M., **6**, 556[828]
Van der Veen, J. M., **5**, 95[89], 96[105,116]
van der Veen, R. A., **8**, 93[74]
van der Veen, R. H., **1**, 299[62]; **5**, 196[14]
Van Derveer, D., **1**, 507[19]; **2**, 221[145], 226[158]; **3**, 17[84], 564[7]; **4**, 31[94], 355[132], 985[131]; **5**, 348[74b]; **7**, 682[81]
Vander Velde, D., **1**, 838[167]; **5**, 24[165]
Van der Ven, S., **4**, 1000[12]
Vanderwalle, M., **8**, 528[77]
van der Weerdt, A. J. A., **5**, 768[132]
VanderWerf, C. A., **2**, 943[170], 970[89], 971[89]; **3**, 564[8]; **6**, 253[158]; **8**, 860[222]
Van Der Werf, J. F., **8**, 53[132], 66[132]
van der Westhuizen, J., **2**, 1059[74]; **3**, 831[61]; **5**, 407[28]
van der Zeijden, A. A. H., **1**, 25[127]
Vandewalle, E. M., **4**, 1040[75]
Vandewalle, M., **2**, 201[92], 838[179]; **3**, 713[28], 900[93]; **4**, 239[15], 259[268], 262[268]; **5**, 127[25], 131[44], 362[93], 363[93b], 924[146]; **6**, 25[103], 690[395]; **7**, 105[147], 301[58,59], 363[33]; **8**, 41[94], 66[94], 122[80], 528[76], 544[252]
van Dijk, J., **7**, 40[6]
van Dijk-Knepper, J. J., **5**, 649[22], 650[22]
Van Dine, G. W., **4**, 1016[209]
van Dongen, J. M. A. M., **7**, 352[30], 356[30]
Vandormael, J., **6**, 111[61]
Van Draanen, N., **2**, 223[151]
Van Duuren, B. L., **2**, 284[57]
van Duyne, G., **1**, 28[143], 29[144], 34[169]; **2**, 507[26,27], 508[27]; **3**, 33[189], 34[198], 39[198], 592[175]; **6**, 727[195,197]
Van Duyne, G. D., **5**, 736[143,145], 737[145]
van Duzer, J., **5**, 162[67]
van Duzer, J. H., **6**, 734[10], 735[10], 736[26]
van Dyk, M. S., **5**, 501[270]; **6**, 108[36]
van Echten, E., **1**, 571[280]
Van Eenoo, M., **6**, 213[90]
Van Eerden, J., **7**, 333[25]
van Eikeren, P., **4**, 311[451]; **8**, 94[79]
van Elburg, P., **1**, 563[174]; **4**, 744[133], 824[238]
Van Emster, K., **3**, 705[2]
van Ende, D., **1**, 571[274], 631[53,55], 633[71], 634[71], 636[71,100], 637[71], 639[100], 641[71,100], 642[71], 647[55], 656[53,55,71,100,154], 657[53,55,71], 658[53,55,71,154], 659[53,55], 664[71,100,171], 666[171], 672[55,71,100,171,199,20], 825[56], 828[56]; **3**, 86[50], 87[67,76,77,84,117], 89[140], 90[140], 91[140], 92[140], 107[140], 109[84,140], 116[140], 136[76,77], 137[77], 141[77], 144[76,77], 145[76,77]; **4**, 318[560], 349[113], 350[113]; **7**, 473[30]; **8**, 847[97], 848[97e], 849[97e]
Van Engen, D., **1**, 807[316]; **6**, 1054[47]
Van Epp, J., **4**, 820[214]
van Es, T., **8**, 298[78], 299[78,79]
van Eyk, S. J., **2**, 19[77,77b], 31[77b], 573[52]
Van Fossen, R. Y., **3**, 891[42]
Van Gemmern, R., **4**, 1006[98]
Vangermain, E., **3**, 307[87]
van Gerresheim, W., **8**, 96[93]
van Gogh, J., **8**, 447[114], 453[114]
van Halbeek, H., **6**, 34[8], 1066[91]
Van Haverbeke, Y., **2**, 351[80], 364[80]
van Helden, R., **3**, 302[49]; **4**, 600[228], 601[246,248], 643[248]
van Heteren, A., **1**, 23[123]
Van Heyningen, E., **4**, 89[84e]
Van Hijfte, L., **4**, 153[61e], 194[121]; **5**, 243[11]
Van Hoof, E., **2**, 838[179]
van Hooff, H. J. G., **8**, 94[81], 95[82,83]
van Hooidonk, C., **4**, 51[145b]
van Hoozer, R., **6**, 960[50]
Van Horn, D. E., **2**, 112[98], 113[103], 242[20], 244[30], 245[20e], 246[20e,34], 247[20e,34]; **3**, 230[239], 232[239a], 238[239a], 443[55], 486[42-44], 495[43,44], 498[42], 529[53], 530[54]; **4**, 145[35], 249[133], 250[133], 866[1], 867[1], 883[102], 884[102], 889[135,136], 890[136], 892[136,140,141], 893[136]; **5**, 1166[25]; **8**, 680[72], 683[72], 693[72,112], 755[117,120], 758[117]
Van Horn, W. F., **8**, 408[67]
Van Horssen, L. W., **3**, 552[9], 557[9]
VanHulle, M., **4**, 603[272], 626[272], 645[272]
van Hulsen, E., **1**, 161[85]; **2**, 10[42], 21[42], 68[43]
van Hummel, G. J., **4**, 45[126]; **5**, 676[4], 686[46-48], 687[46,48]; **8**, 98[104,105]
Vanier, N. R., **1**, 531[129], 632[66]; **2**, 102[27]; **5**, 841[93], 859[93,234]; **6**, 133[1,3]
Van Johnson, B., **6**, 581[988,991]
Van Kamp, H., **8**, 528[85]
van Kampen, E. J., **2**, 770[11]
Vankar, P. S., **6**, 938[123], 939[123], 942[123]; **7**, 220[17]
Vankar, Y., **7**, 201[43]
Vankar, Y. D., **5**, 504[275]; **6**, 109[38], 216[108], 219[108], 289[198], 726[188], 938[127], 944[127]; **7**, 299[47], 760[47]; **8**, 391[89], 406[48], 874[21], 881[21,73], 882[73], 988[33]
van Klingeren, B., **8**, 56[169], 66[169]
van Koeveringe, J. A., **2**, 780[9]
van Koten, G., **1**, 25[126,127], 28[140], 30[154], 41[269], 428[121], 429[121], 457[121]; **2**, 114[121], 296[86], 922[99,100], 923[99], 936[99,100]; **3**, 208[11], 210[11,11a,25,26], 211[32], 219[11a], 234[11a]; **4**, 170[12]; **5**, 101[161]
van Kruchten, E. M. G. A., **3**, 223[147]
Van Laak, K., **5**, 736[145,145r], 737[145]
van Leersum, P. T., **8**, 937[81,82]
Van Leeuwen, P. W. N. M., **3**, 1022[20]; **4**, 587[38], 921[21]

Van Lente, M. A., **6**, 212[80]
van Leusen, A. M., **1**, 571[280,281,282]; **2**, 1084[11]; **3**, 158[449], 174[449,523,524], 175[449,523,524], 909[151]; **4**, 14[47,47k], 16[52d]; **5**, 416[56], 713[62], 728[62], 729[62]; **6**, 489[91,95], 538[549]; **7**, 232[156]
van Leusen, D., **1**, 571[280,282]; **2**, 1084[11]; **3**, 174[524], 175[524], 920[34], 923[34b], 934[34], 1008[69]; **4**, 1036[47,49]; **5**, 917[123]; **6**, 489[91,95], 873[9], 874[9c]
van Lier, P. M., **8**, 94[81], 95[82,83], 967[82]
van Loo, R., **2**, 782[26]
van Look, G., **6**, 652[147], 653[147], 654[147], 655[147], 681[147]
Vanmaele, J., **7**, 105[147]
Vanmaele, L., **6**, 690[395]
van Meerssche, M., **1**, 38[259], 838[166]; **2**, 424[35]; **3**, 857[90]; **4**, 1040[75]; **5**, 109[217]
Van Mele, B., **5**, 571[114]
van Meurs, F., **4**, 538[106]
VanMiddlesworth, F., **1**, 429[126], 798[289]; **6**, 995[101]; **8**, 190[85]
van Mier, G. P. M., **1**, 23[121]
van Minnen-Pathuis, G., **8**, 447[114,115], 453[114,115], 499[43], 500[43]
van Mourik, G. L., **3**, 249[62]
van Niel, M. B., **1**, 568[243]
Vannoorenberghe, Y., **1**, 223[78], 224[78]
van Noort, P. C. M., **5**, 212[69]
Vannucchi, A., **6**, 18[66]
van Oeveren, A., **1**, 223[84], 225[84e]
Van Ool, P. J. J. M., **1**, 661[165], 663[165]
Van Os, C. P. A., **6**, 533[508]
Vanotti, E., **8**, 358[197]
Van Parys, M., **8**, 41[94], 66[94]
van Rantwijk, F., **8**, 418[1], 419[1], 420[1], 423[1], 427[50], 431[1], 432[1], 433[1], 437[1], 438[1], 439[1], 444[7], 447[115], 453[115]
van Reijendam, J. W., **6**, 809[65]
van Rheenen, V., **6**, 840[73], 901[122]; **7**, 439[26,27]
van Rossum, A. J. R., **5**, 71[161]
Van Roy, M. J. H. M., **6**, 556[828]
Van Royen, L. A., **5**, 539[106]
van Rozendaal, H. L. M., **5**, 441[176,176d]
van Santen, R. A., **8**, 131[5]
van Schaik, T. A. M., **1**, 774[212]; **2**, 482[33], 484[33]
van Seters, A. J. C., **1**, 858[61]
van Steen, B. J., **2**, 821[110]
van Straten, J., **2**, 535[40], 542[40]; **5**, 6[48], 15[48], 20[48]
van Tamelen, E. E., **2**, 159[127], 970[90]; **3**, 99[185,191], 103[191a], 107[191], 126[310], 201[74], 341[4], 342[4a,6], 365[94,95], 373[131], 717[45], 752[93,94]; **5**, 646[3], 716[88,89], 929[166]; **6**, 624[143], 896[92]; **7**, 379[98]; **8**, 252[111], 323[117], 472[4], 475[20], 476[23,25], 478[23], 935[66], 946[137], 958[17], 977[139]
van Tilborg, W. J. M., **3**, 568[46,47], 569[50]; **4**, 459[87]; **8**, 131[4,5], 132[4], 134[4]
Vanucci, C., **5**, 347[72,72b]
van Veen, A., **8**, 447[115], 453[115]
van Velzen, J. C., **5**, 3[26]
VanVerst, M. E., **6**, 822[118]
Van Vliet, A., **8**, 427[50]
van Vliet, M. R. P., **2**, 114[121], 922[100], 936[100]
van Vliet, N. P., **3**, 367[100], 371[116]
van Vugt, B. H., **7**, 95[70,70a]
van Vuuren, G., **5**, 422[84]
Van Wallendael, S., **4**, 598[207,210]; **6**, 86[99]
Van Wijk, A. M., **8**, 499[43], 500[43]
van Wijnen, W. Th., **1**, 528[114]
Van Wijngaarden, B. H., **8**, 431[63]
van Wijngaarden, L. J., **8**, 56[169], 66[169]
Van Zorge, J. A., **3**, 649[208]
van Zütphen, L., **3**, 324[150]
Van Zyl, C. M., **2**, 727[134]; **5**, 762[96]; **8**, 917[115], 918[115]
Vaquero, J. J., **4**, 5[17]; **7**, 35[101]
Varadarajan, A., **5**, 203[39,39g], 204[39l], 209[39], 210[39]
Varadarajan, R., **7**, 267[119,120]
Váradi, G., **5**, 1138[65,66,69]
Varaggnat, J., **8**, 462[264]
Varagnat, J., **2**, 286[61]; **6**, 157[168]
Vara Prasad, J. V. N., **5**, 552[34]; **7**, 595[127], 604[127]; **8**, 713[77], 715[77], 721[140], 722[146,149]
Varaprath, S., **3**, 426[82], 429[82]; **4**, 841[41]
Vara Presad, J. V. N., **4**, 873[45]
Varava, T. I., **8**, 447[104], 450[104]
Vardi, S., **3**, 640[101]
Varenne, J., **6**, 618[107]
Varga, L., **6**, 54[132]
Varga, S. L., **6**, 664[223]
Vargaftik, M. N., **7**, 451[38]
Vargas, F., **8**, 398[144]
Varie, D. L., **5**, 436[158,158a,g], 442[158]; **7**, 364[41b]
Varinas, B., **2**, 232[175]
Varkey, T., **8**, 626[169]
Varkony, M., **7**, 842[25]
Varkony, T. H., **7**, 40[2], 842[24,35]
Varley, J. H., **7**, 772[288]
Varley, M. J., **6**, 996[106]
Varma, K. R., **7**, 606[154]
Varma, M., **6**, 938[135], 939[135]; **7**, 144[149]
Varma, R. K., **6**, 806[54]; **8**, 537[185]
Varma, R. S., **2**, 141[39], 321[16,18], 324[18], 325[16,18]; **6**, 107[24,27], 938[135], 939[135,142]; **7**, 144[149]; **8**, 363[3,4], 373[137,138], 374[139,140], 375[4], 376[140,165,166], 377[137,167,168]
Varma, R. V., **7**, 686[100]
Varma, V., **3**, 1031[64]; **4**, 941[85]
Varney, M. D., **1**, 747[64]; **5**, 841[97], 842[111], 859[235], 1123[37]; **6**, 471[61]
Varnier, N., **6**, 859[163]
Varonky, T. H., **7**, 14[128]
Vartanyan, A. G., **3**, 318[123]
Vartanyan, R. S., **6**, 270[83]
Vartanyan, S. A., **3**, 318[123]
Varughese, K. I., **6**, 436[10,11]
Varvoglis, A., **4**, 425[22], 1032[11]; **6**, 172[8]
Vasella, A., **2**, 324[40], 334[40]; **4**, 719[21], 1079[63]; **5**, 255[50], 264[50b], 418[71]; **6**, 54[128], 115[81], 128[162], 561[878], 939[140], 1000[127]; **7**, 86[13], 493[185]
Vasi, I. G., **6**, 441[80]
Vasilenko, N. P., **6**, 509[255]
Vasil'ev, A. F., **2**, 854[236]
Vasil'ev, L. S., **7**, 595[18]
Vasil'eva, L. L., **1**, 520[68]
Vasil'eva, S. P., **8**, 216[55]
Vasilvitskii, A. E., **4**, 1059[153]
Vasilvitskii, L. E., **4**, 1058[151]
Vasilvitskii, V. L., **4**, 1063[172]
Vaska, L., **8**, 446[82]
Vasquez, P. C., **7**, 374[76]
Vasquez, R. E., **7**, 761[64]
Vass, G., **3**, 126[315,316,317]
Vassil, T. C., **8**, 50[117], 66[117]
Vassilan, A., **6**, 671[280]
Vassilatos, S. N., **1**, 464[36], 477[140], 478[36]
Vassil'ev, Y. B., **3**, 635[31,35], 640[98], 647[177], 648[177,178,181], 649[177]
Vastag, S., **8**, 152[177]

Vasudevan, A., **6**, 805[53]
Vasvári-Debreczy, L., **2**, 789[56]
Vatakencherry, P. A., **3**, 17[85], 729[40]
Vatele, J.-M., **2**, 663[23], 664[23], 681[62]; **5**, 432[127], 839[85]
Vater, H.-J., **5**, 683[38], 684[38]
Vather, S., **1**, 568[228]
Vathke, H., **5**, 596[34], 598[34]
Vathke-Ernst, H., **5**, 341[58], 596[34], 598[34]
Vaughan, C. W., **4**, 483[1], 484[1], 487[1], 488[1], 489[1], 491[1], 492[1], 493[1], 495[1], 506[1], 508[1]
Vaughan, K., **3**, 49[264]; **4**, 159[80], 246[96], 258[96], 260[96]
Vaughan, R. J., **3**, 897[94]
Vaughan, W. R., **2**, 278[10,11], 280[11], 960[36]; **3**, 710[21], 848[51]; **4**, 274[57], 282[57,141,142,143], 283[143]; **5**, 552[14]; **6**, 208[59]
Vaughn, T. H., **3**, 273[9]; **4**, 315[522]
Vaughn, W. E., **4**, 317[549]
Vaultier, M., **3**, 88[135], 90[135]; **4**, 38[109b]; **5**, 336[50]; **6**, 76[45], 254[165], 542[600], 821[114]; **8**, 385[42], 386[54]
Vavon, G., **2**, 142[44]
Vawter, E. J., **1**, 218[52]; **2**, 124[210]
Vaya, J., **6**, 538[553]; **7**, 355[42]
Vazeux, M., **3**, 751[89]
Vazquez, M. A., **7**, 163[75]
Vazquez de Miguel, L. M., **1**, 463[28], 469[59]
Vázquez Tato, M. P., **7**, 746[85]
Vdovin, V. M., **8**, 769[30]
Veach, C. D., **3**, 579[98]
Veal, P. L., **3**, 688[115]
Veal, W. R., **3**, 1056[35], 1062[35]; **4**, 1033[22], 1057[22c]
Veale, C. A., **3**, 217[94]; **4**, 795[87]; **6**, 448[108]; **7**, 407[84b]; **8**, 846[81]
Veber, D. F., **1**, 823[44b]; **2**, 962[51]; **3**, 644[167]; **6**, 635[23], 636[23], 664[223]
Vebrel, J., **2**, 969[86]
Vecchi, E., **8**, 875[29]
Vecsei, I., **5**, 1138[66]
Vedananda, T. R., **7**, 674[44], 682[84]
Vedejs, E., **1**, 243[56], 266[49], 357[5], 378[5,105], 755[116], 756[116,116d-f], 758[116,124,127], 761[116], 875[97], 876[101]; **2**, 64[25], 282[36], 478[14], 556[150]; **3**, 86[12], 125[300], 128[300], 129[300], 133[300], 862[2], 865[25,26], 918[24,26], 934[66], 957[109-111], 958[66,113], 960[114-118]; **4**, 125[216,216e], 259[273], 261[273], 603[269], 645[269], 1086[110,111,119], 1087[110,119]; **5**, 250[40], 436[158,158a,b,g], 437[158f], 438[158d,163], 442[158], 532[85], 804[93], 829[16], 894[44,46-48]; **6**, 140[55], 175[82], 509[246], 855[147], 873[6], 893[81], 897[101], 898[6,102,105,106,107a,b], 901[123], 905[144], 1006[146]; **7**, 124[47], 160[50-52], 161[52], 176[52], 180[52], 183[52], 187[52], 228[93], 255[33], 258[56], 580[144], 586[144], 630[50]; **8**, 309[9], 312[9], 374[147], 651[68], 845[78], 846[82], 885[104], 948[146]
Vedeneyev, V. I., **7**, 852[42]
Vedenyapin, A. A., **8**, 150[127]
Vederas, J. C., **3**, 227[213]; **5**, 86[22,31]; **6**, 118[104]; **7**, 184[172]; **8**, 205[155]
Veefkind, A. H., **4**, 869[23,24,26], 972[79]; **5**, 986[38]
Veenstra, L., **2**, 418[22]; **4**, 115[184d]
Veenstra, S. J., **1**, 850[31,32]; **2**, 258[49]; **8**, 171[107], 720[138], 721[138], 722[138]
Veeramani, K., **5**, 728[122]
Veeraraghavan, S., **1**, 468[53]
Vega, J. C., **2**, 840[184]; **6**, 453[137]
Vega, S., **6**, 524[362,367]
Vegh, D., **1**, 86[41]
Vehre, R., **4**, 1031[8], 1043[8]
Veinberg, A. Ya., **8**, 956[7]
Veisman, E. A., **5**, 552[16]
Veith, M., **1**, 38[182], 40[193]
Vejdelek, Z. J., **2**, 765[78]
Vekemans, J. A. J. M., **6**, 984[58]
Velarde, E., **6**, 217[114]; **8**, 526[24]
Velasco, D., **8**, 125[94]
Velezheva, V. S., **6**, 543[608,611,615]
Velgová, H., **8**, 882[88]
Velichko, F. K., **4**, 288[187]
Velluz, L., **3**, 12[65]; **8**, 201[144], 533[154]
Vel'moga, I. S., **6**, 607[48]
Velusamy, T. P., **2**, 782[23]
Venanzi, L. M., **4**, 401[234b], 403[239], 404[239]
Vendenyapin, A. A., **8**, 150[132]
Venegas, M. G., **5**, 241[4]
Venepalli, B. R., **5**, 136[70], 164[76]
Venet, M., **1**, 563[175], 564[175], 568[175]
Veniard, L., **3**, 964[125]
Venier, C. G., **7**, 763[99], 766[99]
Venit, J., **6**, 646[103]
Venit, J. J., **2**, 106[47], 189[46], 209[46], 920[96]
Venkataraman, H., **5**, 707[35]
Venkataraman, K., **8**, 659[104], 950[161]
Venkataraman, S., **1**, 520[76], 521[76], 522[76,79]; **2**, 75[81]; **4**, 12[37,37e,f], 119[194], 226[196,198,199,200]; **6**, 154[146], 900[119]; **7**, 552[58], 554[58]; **8**, 813[9]
Venkataramani, P. S., **3**, 595[191]; **8**, 528[60], 530[108]
Venkataramu, S. D., **8**, 52[147], 66[147]
Venkatesan, K., **1**, 38[260,262]
Venkateswaran, N., **6**, 140[60]
Venkateswaran, R. V., **2**, 360[170]; **3**, 783[28]
Venkateswarlu, A., **6**, 655[156]; **8**, 269[81]
Venkatramanan, M. K., **4**, 795[86]; **5**, 250[37], 252[37], 255[37]
Venkov, A., **2**, 971[95]; **6**, 744[75], 746[75,88]
Vennesland, B., **8**, 79[1]
Veno, H., **7**, 406[73]
Ventataram, U. V., **7**, 763[91], 769[91]
Venton, D. L., **7**, 220[23]
Venturas, S., **4**, 768[241]
Venturello, C., **5**, 1133[23]; **7**, 381[107], 708[30]
Venturello, P., **2**, 345[20]
Venturini, I., **2**, 637[58], 639[58,62], 640[58,62], 930[132,133], 932[132,133]; **5**, 102[176,178]
Venugopalan, B., **2**, 363[198]; **8**, 625[165]
Venus-Danilova, E. D., **3**, 723[10]; **4**, 304[358]
Venuti, M. C., **1**, 469[57]
Venzo, A., **4**, 527[67]
Vera, M., **1**, 42[204]; **2**, 524[77]
Veracini, C. A., **6**, 176[103]
Veraprath, S., **4**, 841[42]
Verardo, G., **1**, 832[115]
Verbeek, J., **1**, 471[62]
Verbicky, J. W., Jr., **5**, 333[45]
Verbit, L., **7**, 606[158]
Verboom, W., **1**, 461[15], 464[15]; **2**, 379[295]; **3**, 219[104]; **5**, 686[46,47,49], 687[46,49]; **8**, 33[58], 66[58], 98[103-105]
Verbrel, J., **4**, 519[27]
Vercauteren, J., **2**, 1017[33]; **6**, 735[20], 739[20]
Vercek, B., **6**, 554[725]
Verdegaal, C. H. M., **6**, 662[214]
Verdek, B., **6**, 554[713,716,747]
Verdini, A. S., **6**, 804[51]
Verdol, J. A., **2**, 529[20]
Verdone, J. A., **2**, 6[30]
Verenikin, O. V., **7**, 493[196]
Vereš, K., **8**, 896[9]

Vereshchagin, A. L., **6**, 525[377]
Vereshchagin, L. I., **4**, 55[156]; **7**, 774[325]
Veretenov, S. O., **5**, 1057[51], 1062[51]
Verevkin, S. P., **3**, 304[65]
Verge, J. P., **5**, 412[44], 498[236]
Verhart, G. G. J., **6**, 638[41,41b]
Verhé, R., **2**, 343[15], 353[102], 357[102], 380[102], 423[34], 424[35]; **4**, 70[7]; **5**, 904[45], 905[45], 925[45], 926[45], 943[45]; **6**, 500[182], 547[663]; **8**, 36[73], 38[73], 66[73]
Verheyden, J. P. H., **2**, 139[29]; **6**, 603[16], 662[217]
Verhoeven, J. W., **5**, 72[169]; **8**, 96[93]
Verhoeven, T. R., **1**, 794[275]; **3**, 228[214], 436[13], 524[33]; **4**, 586[13], 588[74], 590[13], 596[167], 608[324], 614[13], 615[388,389,390], 616[167], 620[74], 621[167], 628[400,401,402], 631[390], 652[167]; **5**, 46[39], 56[39]; **6**, 175[65], 831[11], 848[11], 991[87]; **7**, 358[5], 364[5,42], 368[5,42], 375[5], 376[5], 378[5], 629[48]; **8**, 843[62], 844[68], 993[60], 994[60]
Verhulst, J., **2**, 740[55]
Verimer, T., **2**, 760[43]
Verkade, J. G., **4**, 252[162]
Verkholetova, G. P., **5**, 752[10,19], 767[10,19]
Verkoyen, C., **6**, 487[67]; **8**, 637[10]
Verkruijsse, H. D., **2**, 81[1], 82[1], 96[1], 587[145]; **3**, 87[120], 88[120], 105[120]; **4**, 869[22]; **6**, 962[76], 963[77], 965[77]
Verlaak, J. M. J., **5**, 561[80,84]
Verlaque, P., **3**, 892[47]
Verlhac, J.-B., **1**, 438[158], 457[158], 479[150], 480[150]
Verma, A., **1**, 123[76]
Verma, A. G., **5**, 1083[57], 1142[86]
Verma, R. D., **7**, 498[228]
Verma, V. K., **6**, 538[564,565]
Vermeer, P., **1**, 428[116]; **2**, 85[19-21], 584[125], 587[146,147,148], 589[153]; **3**, 217[81,82,85], 219[104], 254[96], 491[68,69], 531[84]; **4**, 895[163], 897[171], 898[171,178], 899[171], 900[180,181,182], 905[209]; **5**, 772[164], 949[282]; **8**, 743[48]
Vermeeren, H. P., **7**, 765[140]
Vernhet, C., **4**, 248[112]
Verniere, C., **1**, 191[79,80], 192[83]; **2**, 35[129]
Vernin, G., **3**, 505[162], 507[162], 512[162]
Vernon, J. M., **4**, 245[83], 509[162]; **5**, 379[112], 383[112], 384[112], 582[182]; **7**, 477[73]
Vernon, P., **4**, 412[268b,d], 562[34]
Vernon, R. H., **7**, 775[347], 776[347]
Verny, M., **4**, 55[156]
Veronese, A. C., **2**, 369[245]
Verpeaux, J.-N., **1**, 804[308], 805[308]; **3**, 447[93], 448[94], 493[81]; **4**, 459[80,86], 469[80,86,136]; **6**, 162[188]; **8**, 839[23], 842[42a], 847[88,88d]
Versluis, L., **8**, 670[10], 671[10]
Vertalier, S., **7**, 100[125]
Vertino, P. M., **6**, 736[30]
Vertommen, L., **6**, 496[156], 514[156]
Veschambre, H., **3**, 903[126]; **4**, 374[90]; **7**, 60[46b]; **8**, 187[32], 188[32,51], 203[148], 205[148,162,163], 558[399], 559[401], 560[402], 881[75]
Veselovsky, V. V., **5**, 345[70], 346[70], 453[66]
Vesely, I., **8**, 200[138]
Vesely, J. A., **3**, 304[63]
Vessal, B., **7**, 561[85], 738[29], 760[27]
Vessiere, R., **3**, 866[30]; **4**, 48[137,137c,140], 55[156,156g]; **5**, 938[207,211]
Vest, G., **8**, 390[79]
Vestermann, A., **2**, 136[14]
Vestrager, N. O., **6**, 545[636]
Vestweber, M., **4**, 491[68]
Vethaviyaser, N., **6**, 46[68], 846[104]
Vetter, H., **5**, 1133[25]
Vetter, W., **3**, 628[47]; **4**, 30[89]; **8**, 214[41]
Veveris, A., **2**, 345[23]
Vevert, J.-P., **1**, 131[101,102]; **4**, 125[216,216h]; **5**, 859[234]; **6**, 645[96], 859[167], 984[57]
Veyrat, C., **5**, 410[41]
Veyrieres, A., **2**, 464[96]; **6**, 51[107], 662[217]
Veysoglu, T., **1**, 514[52]; **2**, 824[121]; **7**, 548[68], 555[68], 557[68]
Veyssières-Rambaud, S., **8**, 227[119]
Viader, J., **1**, 892[149]
Vial, C., **6**, 1067[106]
Viala, J., **4**, 55[156], 245[81], 249[81]; **6**, 11[45]
Vialle, J., **2**, 214[130]; **5**, 556[52]
Viallefont, P., **3**, 46[254], 215[66], 251[75]; **8**, 662[115]
Viana, M. N., **8**, 660[108]
Viand, M. C., **6**, 253[155]
Vianello, E., **6**, 575[966]
Viani, F., **3**, 147[393]; **4**, 382[131a,b], 384[131b]; **8**, 856[170,171]
Viard, B., **1**, 303[79]
Viau, R., **1**, 512[37,38]; **3**, 147[397], 149[397,400,403], 150[397,403], 151[397,403]; **7**, 766[181]
Viavattene, R. L., **6**, 962[74]
Vibuljan, P., **2**, 146[71]; **5**, 768[127]
Vicens, J., **7**, 496[215]
Vicens, J. J., **1**, 508[20], 512[44], 712[263]; **3**, 96[161,164], 98[161,164], 99[164], 149[404,406], 151[406], 152[404,406], 153[406]
Vicente, M., **6**, 156[162]
Vicentini, C. B., **7**, 143[140]
Vicentini, G., **7**, 774[336]
Vichi, E. J. S., **4**, 665[9], 688[9]
Vick, B. A., **5**, 780[202]
Vick, S. C., **5**, 762[103]
Vickers, J. A., **3**, 688[114], 690[114]
Vickersveen, L., **8**, 369[84]
Vickovic, I., **7**, 698[51]
Vickovic, J., **6**, 766[23]
Victor, R., **5**, 1133[29]
Vidal, J., **2**, 616[137]
Vidal, M., **5**, 677[9,10]; **8**, 860[223]
Vidal, M. C., **4**, 505[139]
Vidari, G., **2**, 547[114], 551[114]; **5**, 351[81]; **8**, 932[40]
Vidulich, G. A., **1**, 293[32]
Vidyasagar, V., **7**, 683[87]
Viehe, H. G., **1**, 366[43]; **2**, 1007[1,3], 1008[4]; **3**, 271[2], 272[2], 551[3], 552[3], 556[33], 870[49], 963[124]; **4**, 3[10], 41[10], 45[130], 47[10], 65[10e], 66[10,10e], 299[302], 303[349], 758[189,190,191], 768[235]; **5**, 70[111-113], 109[213], 116[253,256,257], 422[81], 451[12], 676[3], 689[74], 694[3a]; **6**, 429[87,89], 495[142,143,144], 496[143,156], 497[143], 499[144,174,176], 506[226], 514[156], 515[317], 521[344], 543[624], 546[317], 962[75], 964[84], 965[88], 966[98]
Vieira, P. C., **7**, 586[162], 844[56]
Vierfond, J.-M., **2**, 961[38,41,42]
Vieta, R. S., **5**, 790[35]
Vietmeyer, N. D., **1**, 377[97]; **5**, 707[39], 708[43], 709[39], 739[43]
Vieweg, H., **6**, 509[259]
Vig, O. P., **3**, 416[18], 417[18]; **5**, 515[18], 547[18]
Vigevani, A., **4**, 1085[103]; **5**, 487[188]; **6**, 712[73]; **8**, 347[141], 350[141], 358[197]
Vigevani, E., **6**, 487[54], 489[54]
Vignali, M., **5**, 1154[158]
Vigne, B., **7**, 59[43], 60[43,47a], 78[126]
Vigneron, J.-P., **2**, 232[173]; **3**, 263[173], 557[39]; **4**, 956[17]; **5**, 186[169]; **8**, 161[18], 166[51-54], 178[52], 179[52], 545[294,295,296]
Vignes, R. P., **7**, 530[24], 531[24]
Vijaya Bhaskar, K., **4**, 796[94]

Vijayakumaran, K., **7**, 272[142,143], 276[143,148]
Vijh, A. K., **3**, 634[6], 635[38], 636[6], 637[6,61], 655[6]
Vijn, R. J., **2**, 652[123a], 1049[22], 1050[22], 1064[108], 1065[117]
Vikas, M., **7**, 674[34]
Viktorova, E. A., **6**, 860[180]; **8**, 628[177]
Viktorova, L. S., **6**, 450[121]
Viladoms, P., **4**, 478[168]
Vilamajó, L., **2**, 381[300]
Vilaplana, M. J., **4**, 440[170]
Vilarrasa, J., **6**, 77[54], 122[137], 570[954]; **8**, 385[41], 636[4]
Vil'chevskaya, V. D., **4**, 315[518]
Vilcsek, H., **3**, 818[93]
Vilhuber, H. G., **8**, 342[110]
Vill, J. J., **3**, 760[137]
Villa, A. C., **5**, 1079[51]
Villa, C. A., **1**, 894[160]; **7**, 625[38]; **8**, 514[105]
Villa, M., **2**, 535[37]
Villa, M.-J., **6**, 25[100]
Villa, P., **3**, 771[192]; **8**, 872[7]
Villa, R., **4**, 152[58], 207[58]
Villacorta, G. M., **1**, 432[137], 456[137]; **3**, 211[27]; **4**, 177[58], 180[58a], 230[253]
Villadsen, B., **6**, 462[16]
Villalobos, A., **1**, 529[124], 799[297]; **6**, 835[44]; **8**, 448[143]
Villamaña, J., **4**, 290[200]; **8**, 856[176]
Villani, F. J., **2**, 138[24]; **8**, 28[37], 66[37]
Villani, F. J., Jr., **3**, 524[36]
Villani, R., **4**, 802[127]
Villarica, R. M., **2**, 841[186]; **3**, 55[283]; **8**, 526[26], 527[48]
Villarreal, M. C., **8**, 584[17]
Villaverde, M. C., **3**, 586[156], 610[156]
Ville, G., **5**, 1145[104]; **7**, 272[143], 276[143]
Villemin, D., **1**, 821[26]; **2**, 344[16], 353[16], 359[16], 360[16], 363[16]; **3**, 219[103], 530[66], 533[66]; **6**, 175[79]; **7**, 841[16], 842[16]
Villenave, J.-J., **3**, 1046[3]; **7**, 7[43], 95[77]
Villhauer, E. B., **4**, 629[406,407]; **7**, 261[67]
Villieras, J., **1**, 214[21], 428[116], 830[91], 873[92,95b]; **2**, 414[16], 415[16], 427[42], 596[3], 980[22], 981[22]; **3**, 202[87,91-95], 243[14], 246[34,35], 247[46], 249[14], 263[14], 440[42], 441[42], 442[42], 464[172], 470[223,225,226], 473[217,225,226], 476[217], 485[31], 486[31], 522[18,19]; **4**, 34[97], 35[97], 895[166], 897[172], 900[180], 903[195]; **6**, 849[120], 965[87,89]; **8**, 267[71]
Villiéras, J., **3**, 759[129], 788[49]
Vilsmaier, E., **2**, 368[233]; **4**, 55[157], 1004[74]; **5**, 596[26], 597[26]; **6**, 535[535,536], 538[535,536]
Vilsmeier, A., **2**, 779[3]
Vince, D. G., **1**, 276[79]
Vince, R., **8**, 87[30]
Vincens, M., **5**, 677[9,10]; **8**, 860[223]
Vincent, B. F., Jr., **8**, 60[181], 66[181], 73[248], 74[248], 373[136]
Vincent, B. R., **1**, 528[122], 804[308], 805[308]
Vincent, E. J., **8**, 444[8]
Vincent, F., **7**, 810[87]
Vincent, J. E., **1**, 418[76], 892[151]; **5**, 95[89], 96[114]; **8**, 844[64]
Vincent, M., **5**, 829[25], 930[175], 931[175], 932[175]
Vincent, M. A., **4**, 484[16]
Vincent, P., **8**, 679[66], 680[66], 681[66], 683[66], 694[66]
Vincze, I., **2**, 838[171]
Vinet, V., **2**, 648[92], 649[92]; **5**, 107[200], 439[167]
Viney, M. M., **8**, 457[214], 458[214]
Vineyard, B. D., **8**, 459[232], 460[232], 535[166]
Vingiello, F. A., **8**, 323[114]
Vinje, M. G., **4**, 955[12]
Vink, A. B., **6**, 620[130]
Vinod, T. K., **2**, 715[55]; **4**, 493[78], 878[76]
Vinograd, L. Kh., **2**, 277[4], 285[60]
Vinogradoff, A. P., **4**, 443[185]
Vinokur, E., **3**, 564[14], 607[14]
Vinokurov, V. A., **6**, 439[70]
Viola, A., **5**, 735[139], 787[5], 798[4]; **6**, 866[205]
Viola, H., **6**, 422[36], 423[41]
Viout, P., **2**, 432[57]
Vioux, A., **8**, 766[20]
Virgili, A., **4**, 616[393], 629[393]
Virgilio, J. A., **6**, 533[488], 550[488]; **8**, 895[5], 898[5]
Virkar, S. D., **3**, 878[91]
Virmani, V., **6**, 538[552], 550[552]
Visani, N., **2**, 655[134,134b]; **5**, 282[24]
Viscomi, G. C., **6**, 804[51]
Viscontini, M., **6**, 546[655], 552[655]
Visentin, G., **3**, 312[102]
Vishnuvajjala, B., **3**, 679[90], 683[90]
Vishwakarma, L. C., **1**, 838[157]; **7**, 162[65,68], 181[65], 184[169], 202[47]; **8**, 395[134]
Vishwanath, V. M., **2**, 740[63a]
Viski, P., **7**, 558[80], 559[80], 560[80], 561[80]
Viskocil, J. F., Jr., **3**, 382[36]
Vismara, E., **4**, 739[112], 764[112], 768[235,240,243], 770[244]
Visnick, M., **1**, 789[262b], 790[262]
Viso, A., **4**, 368[17]
Visscher, J., **8**, 95[91]
Visser, C. M., **8**, 82[6], 84[6]
Visser, C. P., **5**, 755[74], 760[74]
Visser, G. M., **6**, 619[116]
Visser, G. W., **5**, 686[46,47], 687[46]
Visser, G. W. M., **7**, 535[47]
Visser, J. P., **4**, 587[39]
Viswanatha, T., **8**, 609[55]
Viswanathan, M., **2**, 583[111]
Viswanathan, N., **8**, 339[96]
Vit, J., **8**, 267[62], 541[207]
Vít, Z., **2**, 268[65]
Vita-Finzi, P., **2**, 547[114], 551[114]
Vitagliano, A., **4**, 631[420,421], 638[422]
Vitale, A. C., **3**, 756[115]
Vite, G. D., **3**, 20[117]; **4**, 261[283], 813[184,185], 815[185], 817[185]
Vite, J. P., **5**, 455[80]
Vité, J. P., **6**, 677[323]
Viteva, L., **2**, 281[29]
Viteva, L. Z., **4**, 21[68]
Viti, S. M., **3**, 225[185], 264[181]; **6**, 2[3], 25[3], 88[105], 89[105]; **7**, 401[60], 406[77], 409[77], 414[77], 415[77], 421[77], 422[141], 423[77,141,141b,c], 748[114]; **8**, 879[51,53], 880[51]
Vitkovskii, V. Yu., **4**, 461[100], 475[100]; **6**, 550[674]
Vitt, S. V., **4**, 218[147]
Vittorelli, P., **5**, 837[67], 856[67], 857[67,227]
Vittulli, G., **4**, 602[258]
Vitulli, G., **5**, 36[15], 1154[151,158], 1155[151]
Vitullo, V. P., **3**, 804[4]
Vivona, N., **8**, 663[117]
Vizi-Orosz, A., **5**, 1138[66]
Vizsolyi, J. P., **6**, 607[44]
Vladuchick, S. A., **1**, 880[112]; **4**, 968[57]
Vladuchick, W. C., **2**, 616[138]; **5**, 333[45]; **8**, 844[64,64a]
Vladuchik, S. A., **5**, 905[57]
Vlasova, N. N., **7**, 762[75]
Vlasova, T. F., **6**, 502[208], 554[763,764,786,789,791,792]
Vlattas, I., **3**, 105[219], 113[219]; **7**, 228[93]; **8**, 374[147,148]
Vliegenthart, J. F. G., **6**, 34[8], 660[203]
Vlietstra, E. J., **6**, 489[94]

Vloon, W. J., **2**, 812[73]
Vo, N. H., **5**, 850[157]
Voaden, D. J., **6**, 655[158]
Vocelle, D., **3**, 903[126]; **6**, 711[68]
Vodnansky, J., **1**, 10[48]
Voelter, W., **6**, 91[129], 637[35]
Vofsi, D., **8**, 532[130]
Vogel, A. I., **2**, 770[3]; **3**, 499[111]
Vogel, C., **4**, 222[177]; **5**, 410[41,41b,f], 411[41f], 681[26]
Vogel, D., **4**, 294[243]; **6**, 283[168]
Vogel, D. E., **5**, 829[18], 830[30], 847[135], 1004[29]
Vogel, E., **1**, 424[99]; **2**, 112[99,100], 197[79], 242[20], 245[20f,31], 246[20f,31], 247[31], 436[67]; **3**, 593[179,180], 726[26]; **4**, 145[35], 1002[47,57], 1006[102], 1007[129]; **5**, 63[1,2], 677[12], 687[65], 702[10], 714[68,69], 715[80], 716[10], 791[43], 794[47], 803[87], 804[96], 805[43], 806[47], 824[47], 929[165], 971[1,1a,c], 1025[80]; **7**, 602[96], 725[33]; **8**, 397[143]
Vogel, F., **6**, 37[33]
Vogel, F. G. M., **6**, 37[33]; **8**, 15[90]
Vogel, J., **6**, 288[186,187]
Vogel, M., **2**, 138[22]
Vogel, P., **5**, 384[126a,b], 1096[108], 1098[108d], 1099[108d], 1112[108d]; **7**, 257[49]
Vogel, T., **8**, 621[145]
Vogel, U., **5**, 758[82]
Vogeli, U., **2**, 345[31]
Vogiazoglou, D., **3**, 257[121]
Vogl, M., **6**, 509[264]
Vogl, O., **2**, 387[335]; **6**, 556[832]
Vogler, H. C., **4**, 587[24,38]
Vogt, B. R., **3**, 623[33], 855[85], 901[113], 903[113]
Vogt, C., **4**, 844[62]
Vogt, H. R., **3**, 345[20]
Vogt, K., **5**, 433[135,135b]
Vogt, P. F., **1**, 530[127]; **2**, 416[17]
Vogt, R. R., **3**, 273[9]; **4**, 276[72], 283[72], 285[159], 288[72], 292[223], 303[351]
Vogt, W., **5**, 451[5]
Vögtle, F., **2**, 402[30]; **3**, 414[2]; **5**, 812[130]; **6**, 70[18]
Vohra, N., **3**, 416[18], 417[18]
Vohra, R., **3**, 1057[39]
Voigt, B., **6**, 538[563], 550[563]; **8**, 537[184]
Voigt, E., **3**, 909[156]
Voitenko, Z. V., **6**, 509[283]
Voitkevich, S. A., **3**, 634[11], 635[11]
Vojtko, J., **7**, 154[21], 451[27,36]
Volante, R. B., **6**, 22[83], 759[139]
Volante, R. P., **1**, 402[17], 791[296a], 799[296]; **2**, 227[160], 482[37], 483[37], 485[37], 821[104], 852[104]; **3**, 45[247], 201[80]; **5**, 99[130], 410[41], 850[160]; **7**, 228[92], 416[122]
Volatron, F., **1**, 820[14]
Volger, H. C., **5**, 1187[10]; **7**, 40[4], 452[60]
Vol'kenau, N. A., **4**, 521[40,41], 529[73,76]
Volker, E. J., **7**, 69[86]
Völker, H., **8**, 858[208]
Volkmann, R. A., **1**, 350[150,151], 358[9], 359[9,24], 362[9b], 364[24], 385[119,121], 386[119,121]; **2**, 939[160], 946[176,177], 947[177,178], 948[178]; **3**, 358[67]; **4**, 54[152,152a,b]
Volkov, A. N., **4**, 50[142,142g], 55[157], 57[157o]
Volkova, K. A., **4**, 50[142,142g]
Volkova, L., **4**, 841[46]
Vollendorf, N. W., **4**, 982[114]
Vollhardt, J., **1**, 37[244,247,248,249,250,252], 528[120,121], 531[131]; **2**, 76[84]; **3**, 159[466], 166[466], 174[530]
Vollhardt, K. P. C., **1**, 733[14], 786[14]; **2**, 725[118], 962[50]; **3**, 255[106], 440[46], 537[90], 538[90-93], 539[100]; **4**, 239[16], 691[77], 898[176]; **5**, 385[130], 389[139], 435[151], 513[5], 514[5], 524[5e], 527[5], 691[83], 692[83,90], 693[83], 1134[36,37,39-41], 1143[91-94], 1144[93,95-100], 1145[104], 1149[126], 1150[127], 1151[128,129,130,132,133,134,135,136], 1154[154,156,159], 1156[165], 1183[59]; **6**, 757[129], 807[62]; **7**, 338[39,40]; **8**, 447[127], 463[127]
Vollmann, H., **3**, 887[12]
Vollmer, J. J., **7**, 710[50]
Volmer, M., **1**, 571[279]
Volodarsky, L. B., **6**, 114[75]
Volodkin, A. A., **3**, 814[69]
Voloshchuk, V. G., **7**, 773[304]
Volpe, A. A., **4**, 1001[33]
Volpe, T., **4**, 7[25], 221[162,163]
Volpi, E., **1**, 857[57]
Vol'pin, M. E., **2**, 727[136]; **3**, 297[25], 334[25]; **4**, 485[29], 840[33], 841[46], 1005[91]; **5**, 1174[34]; **6**, 836[53]; **7**, 7[51]; **8**, 457[215], 557[385]
Voltman, A., **3**, 810[48]
Volund, A., **6**, 637[31]
Volynets, N. F., **6**, 501[207]
Volz, H., **2**, 961[40]; **6**, 910[5], 1067[99]; **7**, 223[45]
von, W., **8**, 92[66]
von Angerer, E., **5**, 1070[27], 1073[27]
von Auwers, K., **3**, 324[148]; **8**, 141[34]
von Baeyer, A., **5**, 899[1]
von Bézard, D. A., **3**, 438[29]
Von Binst, G., **6**, 707[45]
von Boroldingen, L. A., **6**, 139[49]
von Brachel, H., **5**, 3[24], 6[24], 7[24]
von Braun, J., **3**, 324[151]; **8**, 301[90]
von Bruchhausen, F., **5**, 342[62a]
von Bülow, B. G., **6**, 961[72]
von Daacke, A., **4**, 102[128b,c]
von der Brüggen, U., **2**, 612[107], 629[1], 635[1]
von der Eltz, H.-U., **3**, 904[135]
von der Emden, W., **6**, 451[126]
von der Haar, F.-G., **5**, 721[101]
von Euler, H., **5**, 451[3]; **6**, 602[2]; **8**, 88[36]
VonGeldern, T. W., **1**, 771[196]; **5**, 645[1], 648[1i], 651[1], 663[37], 666[37]; **6**, 1045[30]
von Gustorf, E. K., **5**, 1130[9], 1131[15]
von Heeswijk, W. A. R., **6**, 660[203]
von Hessling, G., **4**, 663[1]
von Hinrichs, E., **2**, 1096[94]
von Hove, L., **8**, 348[130], 349[130]
von Ilsemann, G., **7**, 262[78], 362[25]
von Itzstein, M., **2**, 345[35], 351[35], 363[35]
von Jouanne, J., **5**, 77[255,259]
von Kiedrowski, G., **2**, 348[57], 357[57], 371[57,262]; **5**, 468[121-123,126], 531[81], 534[95]
von Kutepow, N., **4**, 939[74]
von Liebig, J., **3**, 821[1], 822[5], 834[5]
von Niessen, W., **4**, 484[16]
von Philipsborn, W., **2**, 345[31], 722[96]; **5**, 1154[149]; **6**, 690[400], 692[400], 708[47]
von Rein, F. W., **4**, 878[77], 879[77]
von Rudloff, E., **7**, 586[164], 710[55]
von Schickh, O., **2**, 342[6]; **6**, 104[9]; **7**, 752[142]
von Schnering, H.-G., **2**, 35[127]; **4**, 1022[253]; **5**, 155[36], 156[36], 157[36], 200[30], 206[45,46], 224[101]; **6**, 121[130], 863[191]; **8**, 354[170]
von Schriltz, D. M., **2**, 268[69]
von Strandtmann, M., **6**, 182[138], 1021[51]; **7**, 198[27]
von Tschammer, H., **7**, 700[61]
VonVoigtlander, P. F., **8**, 53[129], 66[129]

Vonwiller, S. C., **1**, 520[72], 779[223]; **2**, 66[33], 75[33]; **4**, 12[37], 119[192b,193], 159[85], 226[190,191,193,194,195], 259[265]; **6**, 154[145], 864[192]
von Zychlinski, H., **2**, 1090[72], 1093[81], 1094[87]
Voorbergen, P., **1**, 13[69]
Vöpel, K. H., **2**, 366[218], 782[30]
Vo-Quang, L., **4**, 48[137], 1003[64]; **8**, 798[51]
Vo-Quang, Y., **4**, 48[137], 1000[7]; **8**, 798[51]
Vora, M., **3**, 619[24]
Vora, M. M., **2**, 828[134]
Vora, V. C., **7**, 71[95]
Vorbrüggen, H., **1**, 773[203,203a]; **2**, 358[153], 381[308], 889[56,58]; **4**, 23[70], 433[120], 768[235], 964[50], 965[50]; **6**, 20[73], 22[73], 49[97], 193[215], 637[32,32a], 657[176], 667[242], 669[32a], 677[310]; **8**, 392[102]
vor der Bruck, D., **5**, 404[19], 405[19]
Vorländer, D., **2**, 147[76]; **5**, 752[49]; **6**, 970[121]
Vornberger, W., **6**, 180[128]
Vorobeva, L. I., **7**, 75[114]
Voronenkov, V. V., **3**, 306[81]
Voronkov, M. G., **4**, 291[208], 461[100], 475[100]; **6**, 509[245,282], 550[673,674], 577[979]; **8**, 763[1], 765[12], 769[1b,25], 770[33,34,38], 771[1b,25,42], 782[104], 785[1]
Voronov, V. K., **4**, 317[554]
Vorpagel, E. R., **8**, 332[38], 339[38]
Vorspohl, K., **4**, 760[196]
Vos, A., **1**, 299[62]
Vos, G. J. M., **4**, 45[126]; **5**, 676[4], 686[48], 687[48]
Vos, M., **1**, 10[55], 11[55b]
Voser, W., **7**, 128[171]
Voskresenskaya, T. P., **8**, 608[47]
Voss, E., **2**, 356[132]; **5**, 458[72,73], 459[72], 461[106]
Voss, G., **6**, 488[12], 508[12], 509[12], 512[12], 545[12], 6 73[289]
Voss, J., **5**, 436[157]; **6**, 419[5], 420[5], 425[5], 436[6,30], 437[6], 444[6,98], 445[6], 448[6,111], 449[6], 450[6], 453[6], 454[6], 455[6], 456[6], 475[91,92], 478[104], 482[91]; **8**, 303[100-102], 304[101]
Voss, S., **4**, 1092[144], 1093[144], 1102[199]
Voss, W., **4**, 47[133]
Vostrikova, O. S., **4**, 875[56]; **8**, 697[134], 698[140]
Vostrowsky, O., **1**, 755[115], 808[320], 812[115], 813[115]; **3**, 644[159]; **6**, 172[25], 174[58,61], 188[181]
Voticky, Z., **6**, 524[368]
Vötter, H.-D., **5**, 497[224]
Vottero, P., **1**, 212[4]
Vougioukas, A. E., **1**, 328[27], 343[27], 546[57]; **2**, 310[32], 311[32], 654[130]; **6**, 237[61]
Vowinkel, E., **6**, 244[111]; **8**, 815[23], 912[91,92]
Voyer, R., **5**, 88[48]
Voyle, M., **4**, 161[87,87b], 524[60]
Vrachnou-Astra, E., **3**, 565[20]
Vranesic, B., **7**, 380[103]; **8**, 11[60]
Vredenburgh, W. A., **8**, 564[442]
Vrencur, D. J., **3**, 180[549]
Vretblad, P., **2**, 1104[132]
Vrielink, J. J., **4**, 52[146]
Vries, T. R., **4**, 12[39]
Vriesema, B. K., **3**, 229[224,224a]; **6**, 70[18]; **8**, 84[12]
Vrieze, K., **2**, 114[121]
Vrijhof, P., **5**, 584[192]
Vroegop, P. J., **5**, 708[41]
Vu, B., **7**, 829[60]
Vuilhorgne, M., **2**, 1018[45]
Vuillerme, J.-P., **7**, 92[51]
Vukov, R., **7**, 722[21]
Vul'fson, N. S., **2**, 277[4], 285[60], 811[70], 813[70], 814[70]
Vullioud, C., **5**, 362[93], 363[93b]
Vullo, A. L., **6**, 178[121]
Vuorinen, E., **4**, 1064[174]
Vuper, M., **5**, 199[27]
Vuuren, G., **4**, 1063[170]
Vuuren, P. J., **4**, 1024[265]
Vyas, D. M., **5**, 439[168], 736[143]; **6**, 569[937]
Vyazankin, N. S., **2**, 365[214]; **8**, 546[310]
Vyazankina, O. A., **8**, 546[310]
Vyazgin, A. S., **4**, 50[142]
Vyaznikovtsev, L. V., **4**, 50[142]
Vysotskii, A. V., **3**, 328[172]
Vystrdil, A., **3**, 757[124]
V'yunov, K. A., **4**, 329[1], 344[1], 350[1], 351[1]

W

Waack, R., **1**, 22[114]; **2**, 124[206]; **4**, 96[103b], 868[14]
Waali, E. E., **2**, 598[14]; **4**, 1012[167]
Waba, M., **2**, 556[151]
Wachtel, H., **8**, 615[92]
Wachter, E., **2**, 1104[133]
Wachter, J., **1**, 310[107]
Wachtmeister, C. A., **3**, 660[9]
Wackerle, L., **2**, 1094[89], 1095[89]; **6**, 495[141]
Wackher, R. C., **4**, 270[8]
Wada, A., **4**, 261[286]
Wada, E., **4**, 111[152c]; **5**, 758[81]; **7**, 263[88]
Wada, F., **3**, 495[96], 497[103], 530[78], 535[78], 1026[41]; **4**, 856[99,101]
Wada, I., **2**, 651[113,114], 657[164]
Wada, M., **1**, 120[65], 236[30], 237[30], 350[152,153], 361[35,35a,b], 362[35a,b]; **2**, 24[93], 73[63], 113[112], 244[27], 245[27], 613[110], 615[127], 655[138], 656[138], 657[138], 905[56], 906[56], 920[95], 921[95]; **4**, 446[212]; **5**, 841[97]; **6**, 931[91]
Wada, R., **8**, 145[88]
Wada, T., **1**, 619[60]; **6**, 606[41]
Wada, Y., **2**, 555[138]; **5**, 622[23]
Waddell, T. G., **3**, 735[22], 770[173]
Wade, J. J., **2**, 477[10]; **5**, 514[8], 527[8]
Wade, K., **1**, 2[8], 6[33], 139[4]; **3**, 208[1], 763[151]; **8**, 99[107]
Wade, L. E., Jr., **5**, 67[94], 856[197]
Wade, L. G., **8**, 278
Wade, L. G., Jr., **2**, 757[13], 759[13]; **6**, 667[239]
Wade, M. J., **6**, 478[105]
Wade, P. A., **4**, 12[42], 429[83], 438[83], 441[83], 590[102]; **7**, 220[25], 665[69]; **8**, 70[224]
Wade, R. C., **8**, 374[141]
Wade, R. H., **1**, 846[16], 851[16]; **2**, 323[27]
Wade, R. S., **6**, 544[628]; **8**, 908[75]
Wade, T. N., **7**, 498[221]
Wade, W. S., **2**, 904[51]
Wadehn, J., **2**, 364[206]
Wadenstorfer, E., **2**, 657[160], 1052[50], 1053[50], 1067[50]
Wadgoonkar, P. P., **7**, 602[104,104b]
Wadia, M. S., **7**, 384[112]
Wadman, S., **3**, 218[100,100b], 229[234], 444[66,68]
Wadsworth, A., **5**, 1043[24], 1046[24], 1049[24], 1051[24,36b]
Wadsworth, A. H., **1**, 570[263]; **3**, 290[70]; **6**, 176[95]
Wadsworth, D. H., **4**, 50[142,142e]
Wadsworth, W. S., **7**, 396[22]
Wadsworth, W. S., Jr., **1**, 722[270], 761[139]
Waefler, J. P., **3**, 639[78]
Waegell, B., **3**, 380[8,13], 729[41], 839[12], 840[12], 853[70], 854[12]; **4**, 954[10]; **5**, 641[133]; **6**, 86[101], 776[54,56]; **7**, 59[42], 95[65], 452[47,48], 503[280,281]; **8**, 802[87]
Wagatsuma, M., **7**, 137[123], 139[123]
Wagatsuma, N., **7**, 453[76]
Wagenhofer, H., **3**, 893[53b]; **4**, 1081[75]
Wagenknecht, J. H., **8**, 285[3], 532[132], 974[126]
Wager, J. S., **6**, 546[646]
Wagle, D. R., **4**, 553[6]; **5**, 86[14,18], 95[95], 96[105,116,119,121], 98[121]; **7**, 454[99]
Waglund, T., **2**, 465[107]
Wagner, A., **3**, 582[111]; **5**, 431[123,123a,b], 432[133]; **6**, 540[588]; **7**, 768[200]; **8**, 273[125]
Wagner, A. F., **2**, 284[56]
Wagner, C. D., **3**, 332[207]
Wagner, D., **6**, 603[16], 662[217]; **7**, 495[206]
Wagner, D. P., **8**, 287[17]
Wagner, E. C., **2**, 954[7], 958[26]
Wagner, E. R., **3**, 492[73], 497[73]
Wagner, F., **6**, 515[315]
Wagner, G., **3**, 705[1]; **5**, 850[152]; **6**, 424[60], 507[233,238,239], 509[259], 515[238,239], 538[561,563], 550[563]
Wagner, H., **6**, 41[43]
Wagner, H.-U., **1**, 385[118]; **5**, 468[134]; **6**, 226[10], 256[10], 257[10], 575[970], 832[19]
Wagner, K., **6**, 637[32,32c]
Wagner, K.-G., **7**, 740[45]
Wagner, P., **4**, 121[207]
Wagner, P.-H., **2**, 739[43b]
Wagner, P. J., **3**, 1048[12]; **4**, 743[127], 744[127]; **5**, 127[23], 178[134], 219[38]; **7**, 41[26]
Wagner, R., **1**, 884[130]; **5**, 796[57], 815[57]; **7**, 397[31]; **8**, 652[71]
Wagner, R. B., **3**, 843[28], 844[30]
Wagner, R.-M., **5**, 710[56,56c], 719[56], 744[56]
Wagner, S. D., **1**, 189[74]
Wagner, U., **5**, 402[4], 403[4], 424[95]
Wagner, W. J., **8**, 942[115]
Wagner, W. M., **4**, 1000[12]
Wagner-Jauregg, T., **5**, 451[25,26], 491[26], 645[1], 651[1]
Wagnon, J., **1**, 107[4]; **3**, 419[45], 420[45]; **4**, 176[48]
Wagstaff, K., **8**, 83[7]
Wahl, A. R., **7**, 558[77]
Wahl, G. H., Jr., **3**, 736[23], 906[143]
Wahren, M., **4**, 485[29]
Wahren, R., **8**, 550[334]
Wai, J. S. M., **1**, 763[146]; **4**, 255[202]; **7**, 430[159], 442[46b,c], 489[165]
Waid, K., **6**, 190[197]
Waigh, R. D., **8**, 839[26a], 840[26]
Wailes, P. C., **1**, 139[4]; **3**, 824[18]; **4**, 153[62a]; **8**, 447[119a,b], 675[35-37], 679[37], 684[35], 691[36], 754[114]
Waiss, A. C., Jr., **3**, 666[45]
Waite, A. C., **8**, 860[221]
Waite, H., **4**, 310[436]
Waitkins, G. R., **7**, 84[1], 85[1], 108[1]
Waits, H. P., **4**, 725[46]
Wajirum, N., **5**, 578[149]; **6**, 690[388]
Wakabayashi, H., **4**, 107[140d]
Wakabayashi, N., **5**, 835[59], 862[250]
Wakabayashi, S., **1**, 520[71,75]; **2**, 374[275,277]; **4**, 258[248], 261[248]; **6**, 149[104], 603[15], 619[120], 624[15,120], 902[130,131], 939[137]; **7**, 454[96]; **8**, 353[152]
Wakabayashi, T., **2**, 885[50]; **8**, 493[22]
Wakabayashi, Y., **1**, 347[132,134]; **4**, 23[71], 162[92], 567[49]; **6**, 237[67], 564[913,914]; **8**, 168[70], 545[302]
Wakamatsu, H., **3**, 1027[44]; **5**, 1138[70]; **8**, 292[39]
Wakamatsu, K., **1**, 95[78], 450[213], 738[40]; **2**, 19[76], 59[15]; **4**, 391[179], 607[310], 626[310], 647[310], 770[248], 886[117]; **6**, 998[117]; **7**, 162[58], 243[65,66]
Wakamatsu, T., **5**, 808[109]; **6**, 8[38]; **8**, 244[57], 249[97], 253[97], 527[47], 620[132]
Wakamiya, T., **4**, 958[27]
Wakamoto, K., **4**, 878[83]
Wakasa, N., **3**, 229[230], 246[33], 257[33]

Wakasa, T., **7**, 414[108]
Wakasugi, T., **8**, 18[126]
Wakatsuka, H., **7**, 693[30], 694[30]
Wakatsuki, Y., **5**, 1139[74], 1140[74], 1141[81], 1142[74,87-89], 1145[103], 1146[74], 1152[137,138,142], 1153[103,148], 1154[150], 1158[137]
Wakefield, B. J., **1**, 2[6,7], 125[84], 412[50], 631[51], 672[51]; **3**, 86[4], 419[34,35]; **4**, 257[218], 426[53], 867[6], 968[57]; **5**, 829[25]; **6**, 546[654], 557[834]; **7**, 329[2]; **8**, 198[130]
Waki, M., **2**, 1094[88]
Wakimasu, M., **6**, 644[82]
Wakisaka, K., **5**, 724[111]
Wakisaka, M., **8**, 405[24]
Wakita, Y., **5**, 634[78]
Wakselman, C., **2**, 209[108], 749[129]; **3**, 257[121]; **4**, 1020[236]; **6**, 526[397,398], 527[408]
Wakselman, M., **6**, 639[48]
Walach, P., **8**, 863[234]
Walba, D. M., **1**, 429[125]; **2**, 547[101], 548[101]; **7**, 257[50], 268[124], 553[60]; **8**, 544[260]
Walborsky, H. M., **3**, 255[111], 439[38], 482[3], 587[146], 610[146], 1040[107]; **4**, 39[112], 200[7], 1007[107]; **5**, 30[3], 355[87a], 901[26]; **6**, 242[91], 243[91], 295[247,248], 727[194], 790[119], 985[66]; 7, 712[65]; **8**, 74[249], 830[87]
Wald, L., **5**, 423[89]
Walde, A., **7**, 99[107]
Walder, L., **4**, 130[226a,b], 764[222], 765[222], 808[155]
Walding, M., **6**, 46[67], 47[67]
Waldmann, H., **2**, 1102[123]; **5**, 366[97], 411[43,43c]; **6**, 46[63], 641[63], 643[80], 666[245,248], 667[245], 670[63,266,267], 671[266]
Waldmann, H. J., **2**, 456[22,33,52], 457[33], 458[33,52], 459[33,52], 460[33], 461[33], 462[33], 466[33]
Waldner, A., **6**, 937[117], 939[117], 940[117]
Waldron, J. J., **4**, 91[90], 92[90c], 261[294]
Waldvogel, G., **7**, 86[16a]
Walenta, R., **4**, 211[88]
Walinsky, S. W., **3**, 936[73]; **5**, 439[169], 910[88]
Walker, A., **5**, 116[269]; **6**, 783[87]
Walker, A. M., **8**, 83[7]
Walker, B., **6**, 455[155]
Walker, B. H., **7**, 252[10]
Walker, B. J., **1**, 755[115], 758[127], 761[137], 812[115], 813[115]; **7**, 396[22]
Walker, C., **5**, 531[75], 549[75]; **7**, 105[149]
Walker, C. B., Jr., **5**, 765[111]
Walker, D. G., **1**, 791[269]; **2**, 486[30], 488[30], 821[106]; **5**, 494[215,216], 579[162,163]
Walker, D. L., **7**, 318[53], 319[53]
Walker, D. M., **2**, 855[243]; **5**, 143[103]
Walker, E., **8**, 244[52]
Walker, E. F., **2**, 916[84]
Walker, E. R. H., **3**, 715[43]; **8**, 2[3], 37[102], 66[102], 237[9], 238[9], 240[9], 241[9], 244[9], 245[9], 247[9], 278, 541[209]
Walker, F. J., **3**, 225[185], 264[181]; **6**, 2[3], 8[39], 25[3], 88[105], 89[105,112], 927[71,76]; **7**, 198[26], 401[59], 402[63], 403[59,65], 406[59,77], 409[77], 414[77], 415[77], 421[77], 423[77]
Walker, G. N., **7**, 543[15]; **8**, 70[227], 71[227]
Walker, H. G., **7**, 84[1], 85[1], 108[1]
Walker, H. G., Jr., **2**, 735[16]
Walker, H. M., **6**, 280[142]
Walker, J., **3**, 633[4]; **8**, 916[98]
Walker, J. A., **3**, 528[46], 799[102], 939[77]; **5**, 829[12], 948[272], 1001[13]
Walker, J. C., **2**, 125[215,216], 127[232], 272[79], 564[9]; **3**, 47[257]; **4**, 82[62b,e], 217[129,130,131,132], 231[130,132], 243[74,75], 257[74,75], 260[75]; **5**, 367[100]
Walker, K. A. M., **8**, 152[173], 277[152], 445[13,14], 452[13,184,184a], 533[147], 535[147]
Walker, K. E., **8**, 376[161]
Walker, L. E., **5**, 70[109,110]
Walker, M. A., **1**, 835[133]
Walker, O. J., **3**, 636[45,57]
Walker, P., **7**, 135[105], 136[105], 137[105], 145[105]
Walker, P. J. C., **4**, 518[9], 542[9], 689[71]; **8**, 444[10]
Walker, R. A., **8**, 937[77]
Walker, R. T., **1**, 569[252]
Walker, S. M., **5**, 639[123]
Walker, T., **8**, 987[23]
Walker, W. E., **4**, 589[77], 590[77], 591[77], 597[181], 598[181], 638[181]
Walker, W. E., Jr., **3**, 407[150]
Walkowicz, C., **7**, 675[56]
Walkup, R. D., **2**, 214[132], 603[49]; **4**, 395[207a], 396[207a,b], 558[16]
Walkup, R. E., **2**, 780[10]
Wall, A., **3**, 878[93,94], 879[93,94], 880[94], 881[94]; **4**, 335[25], 359[159], 771[251]; **5**, 440[174], 441[174]; **6**, 161[182]
Wall, D. K., **3**, 888[18], 893[55]
Wall, M. E., **6**, 685[360]
Wall, R. G., **5**, 2[13], 3[13]
Wall, R. T., **2**, 897[14]
Wall, W. F., **8**, 198[130]
Wallace, B., **2**, 534[33-35], 535[34,35]
Wallace, D. J., **5**, 129[35]
Wallace, E., **8**, 103[130]
Wallace, G. M., **6**, 551[688]
Wallace, I. H., **5**, 862[251]
Wallace, I. H. M., **2**, 606[67], 619[67]
Wallace, M. A., **5**, 765[111]
Wallace, P., **1**, 782[232]
Wallace, P. M., **5**, 419[74]; **8**, 253[117], 395[129]
Wallace, R. G., **4**, 502[119]; **6**, 119[112], 765[16]
Wallace, R. H., **1**, 272[65]; **3**, 42[233], 572[64]; **8**, 109[7,13], 110[7], 112[7], 113[35]
Wallace, R. W., **6**, 1016[27]
Wallace, T. J., **7**, 759[5,13,16], 760[17,21,34-36], 761[36]
Wallace, T. W., **3**, 502[141]; **4**, 215[118], 497[96]; **5**, 683[39,115], 684[39], 692[94], 693[113,115], 694[115]; **6**, 152[140], 153[140]; **8**, 540[194]
Wallach, O., **2**, 141[38], 142[45], 152[104]; **3**, 822[4], 831[4], 832[4]; **4**, 239[33]; **8**, 526[21,22]
Wallbaum, F., **3**, 809[42]
Wallbillich, G., **4**, 1083[92], 1085[100,101]
Wallenberger, F. T., **3**, 574[80]
Wallenfels, K., **2**, 359[159]; **6**, 225[4], 226[4], 228[4], 230[4], 231[4], 232[4], 233[4], 234[4], 235[4], 236[4], 238[4,73], 239[4], 240[4], 241[4], 258[4], 489[87]
Waller, B., **6**, 646[99,99b]
Waller, C. W., **8**, 973[120]
Waller, F. J., **4**, 945[89]
Walley, R. J., **8**, 143[58]
Wallfahrer, U., **5**, 491[209]
Wallin, A. P., **1**, 480[159]
Wallin, S., **6**, 660[206]; **8**, 224[105], 969[96]
Walling, C., **3**, 154[417], 649[202]; **4**, 279[102,109], 282[109], 316[537], 716[1], 717[7,10], 725[10], 751[1], 752[162], 765[162]; **6**, 1022[62]; **7**, 11[86], 13[111], 17[169], 41[16], 95[81], 860[70]
Wallingford, V. H., **2**, 800[15]
Wallis, A. F. A., **3**, 690[126], 691[130], 693[130]

Wallis, C. J., **1**, 774[209], 779[225]; **2**, 596[4]
Wallis, E. S., **3**, 892[50]; **6**, 795[6], 796[6], 801[6]
Wallis, J. D., **1**, 823[40]; **4**, 55[157], 57[157j]
Wallis, J. M., **7**, 863[82], 864[86], 874[108]
Wallis, T. G., **3**, 872[56]
Walliser, F. M., **5**, 21[145], 572[123]
Wallner, A., **4**, 820[219]
Wallo, A., **8**, 242[44], 252[44], 455[205]
Walls, F., **1**, 552[80]; **8**, 368[67], 374[67]
Walpole, A. L., **3**, 351[38]
Walsgrove, T. C., **4**, 14[47,47h]; **7**, 362[27]
Walsh, A. D., **5**, 900[4]; **7**, 10[78]
Walsh, C., **8**, 87[31], 185[23]
Walsh, C. T., **2**, 904[49]
Walsh, E. B., **5**, 473[153], 477[153], 484[179]
Walsh, E. J., **1**, 885[134]; **2**, 759[29]; **4**, 1060[161], 1063[161a]
Walsh, E. J., Jr., **5**, 787[6,9], 794[49], 856[209]; **8**, 264[43-45], 825[65]
Walsh, E. N., **5**, 949[276]
Walsh, R., **3**, 382[36]; **5**, 160[55], 571[115], 585[199], 905[61], 906[61]
Walsh, T. F., **5**, 154[33]
Walsh, W. L., **8**, 736[23], 737[23]
Walshaw, K. B., **8**, 530[97]
Walshe, N. D. A., **1**, 797[293]; **6**, 996[107]
Walt, D. R., **5**, 1096[111], 1098[111]
Walter, D., **3**, 423[81], 975[4], 979[4]
Walter, D. S., **4**, 161[91]
Walter, G. J., **8**, 536[167]
Walter, H. A., **7**, 657[24]
Walter, L., **7**, 878[145]
Walter, P., **4**, 48[139]
Walter, R., **6**, 637[30]; **7**, 771[268], 772[268]
Walter, S., **1**, 38[183]
Walter, W., **2**, 867[16]; **3**, 105[218]; **6**, 419[4,5], 420[4,5,17-19], 421[18,19,30], 422[30], 423[38], 424[18,19,30,38], 425[5], 432[123], 482[124], 488[12], 508[12], 509[12,251], 512[12], 539[580], 540[589], 541[597], 545[12], 722[144]
Walter, W. F., **8**, 836[1,1b], 837[1], 847[1b], 964[53]
Walters, C. P., **6**, 509[253]; **8**, 37[99], 42[99], 66[99]
Walters, M. E., **3**, 616[13]; **8**, 531[126], 931[37]
Walters, R. L., **8**, 121[78]
Walters, T. R., **7**, 737[17]
Walther, D., **1**, 320[160]; **5**, 1157[168]
Walther, S., **4**, 1008[134]
Walther, W., **7**, 709[43], 710[43]
Walthew, J. M., **4**, 496[87]
Walton, D. R. M., **2**, 725[112], 743[78]; **3**, 219[104], 419[31], 500[126], 512[126], 522[17], 555[31]; **4**, 115[182]; **6**, 426[77]; **8**, 766[20]
Walton, E., **2**, 284[56]
Walton, E. S., **5**, 787[10]
Walton, J. C., **3**, 380[9]; **4**, 717[8], 729[63], 791[51]; **7**, 860[71]
Walton, R., **4**, 815[194], 817[194]
Walts, A. E., **2**, 6[25], 12[25], 13[25,57], 26[25], 27[25], 30[25], 31[25], 35[125], 41[25,125], 42[25]; **5**, 534[91]; **7**, 401[62]
Walz, P., **2**, 614[119]
Walzer, E., **2**, 547[113], 551[113]; **3**, 56[285]; **4**, 229[235,236], 1055[138]
Wambsgans, A., **8**, 74[245], 176[136], 393[110]
Wamhoff, H., **4**, 45[130,130c-e]; **5**, 484[179]; **6**, 428[86]; **7**, 739[38], 748[108]
Wamprecht, C., **6**, 526[396], 575[967]
Wamsley, E. J., **4**, 24[72,72a]
Wan, P., **4**, 298[293], 300[293]; **7**, 247[97]
Wanagat, U., **7**, 598[59]
Wanat, R. A., **1**, 28[143], 29[144], 34[169]; **2**, 507[26,27], 508[27,32]; **3**, 33[189], 34[198,199], 39[198,199]; **6**, 727[195,196,197]
Wand, M. D., **1**, 429[125]
Wander, J. D., **7**, 703[1], 709[1], 710[1]
Wandrey, C., **8**, 204[154]
Wang, A., **6**, 707[44]
Wang, B. S. L., **8**, 840[33]
Wang, C. H., **7**, 883[178]
Wang, C. J., **5**, 257[59,59a]
Wang, C.-L. J., **2**, 160[133], 369[249]; **3**, 212[39], 253[92]; **4**, 75[42b], 377[145a], 384[145a]; **6**, 76[45], 182[139]; **7**, 358[10], 371[10], 377[90], 674[45]
Wang, D., **2**, 553[132], 554[132], 567[26]; **7**, 423[144]; **8**, 777[82b]
Wang, D. G., **3**, 888[16]
Wang, E. J., **4**, 966[55]
Wang, H. H., **4**, 95[98]
Wang, I. C., **8**, 295[60]
Wang, J. L., **7**, 367[56]
Wang, J.-T., **1**, 165[112b]
Wang, J.-X., **5**, 1138[60]
Wang, K. K., **2**, 84[12], 88[28]; **3**, 215[60]; **4**, 395[204]; **8**, 2[14], 724[152,155-157,160-162,178], 725[178], 726[178,178b], 727[178]
Wang, L. C., **4**, 964[45]
Wang, M. O., **5**, 151[18]
Wang, N., **4**, 73[32], 111[158a]; **6**, 116[93], 938[125], 940[125]
Wang, N.-Y., **8**, 188[52]
Wang, P. C., **2**, 1090[70], 1100[70]; **6**, 247[131]
Wang, P. J., **5**, 799[73]
Wang, R. H. S., **2**, 387[337]
Wang, S., **1**, 339[89]; **4**, 815[191]
Wang, S.-S., **6**, 666[234], 667[234], 671[275]; **8**, 64[215], 394[118]
Wang, S. Y., **8**, 566[457], 568[468]
Wang, T., **3**, 163[472]; **5**, 636[99]
Wang, T.-F., **2**, 547[102,103], 548[102], 549[103]; **3**, 368[105]; **5**, 789[30]; **8**, 341[103], 928[24]
Wang, W., **4**, 871[34]; **7**, 883[178]
Wang, W.-L., **1**, 273[69]
Wang, X., **3**, 999[51], 1000[51b]; **7**, 374[77b]
Wang, Y., **1**, 188[95], 198[95], 838[167]; **2**, 579[94], 772[16]; **5**, 680[21], 1031[97]; **7**, 313[35]
Wang, Y. F., **2**, 909[63]; **4**, 262[310]; **8**, 191[94]
Wang, Z., **1**, 768[172]; **2**, 505[9], 786[48]; **5**, 841[87]; **7**, 583[156]; **8**, 224[101]
Wang-Griffin, L., **8**, 97[98]
Waninge, J. K., **8**, 95[91]
Wanless, G. G., **4**, 276[77], 284[77], 288[77], 289[77]
Wann, S. R., **8**, 237[17], 240[17], 249[17]
Wannagat, U., **2**, 183[13]
Wanner, K. T., **2**, 657[160], 1052[50], 1053[50], 1067[50,125]; **3**, 42[232]
Wanner, M. J., **2**, 718[81]; **6**, 181[132], 182[132]
Wanzke, W., **6**, 456[159]
Wanzlick, H.-W., **4**, 126[217d]
Warawa, E. J., **7**, 253[19], 254[19]
Warburg, O., **6**, 642[64]; **8**, 589[52]
Warburton, W. K., **8**, 663[116]
Ward, A. D., **2**, 809[55]; **4**, 340[50]; **7**, 534[42], 772[298]; **8**, 56[167], 66[167]
Ward, B., **5**, 736[145], 737[145]
Ward, D. D., **5**, 350[79]
Ward, D. E., **8**, 16[106], 17[106], 18[124]
Ward, D. G., **1**, 180[42], 181[42]
Ward, G., **8**, 140[26]
Ward, H. P., **6**, 120[125]
Ward, H. R., **4**, 1010[148], 1013[178]; **5**, 802[85]; **6**, 970[125]
Ward, J., **5**, 1116[8]

Ward, J. A., **4**, 904[204]
Ward, J. G., **7**, 27[66], 32[94]
Ward, J. S., **8**, 623[151]
Ward, M. D., **7**, 17[179]
Ward, N. D., **8**, 478[37]
Ward, P., **3**, 160[471], 161[471], 163[471], 803[1]
Ward, R. S., **1**, 566[217,218]; **3**, 692[137], 816[84]; **4**, 113[176], 239[20], 249[127], 258[20,127]
Ward, R. W., **8**, 626[175], 629[175]
Ward, S. E., **4**, 811[172]; **5**, 723[106]
Ward, S. J., **8**, 623[150]
Ward, T. J., **8**, 28[34], 66[34]
Wardell, J. L., **1**, 2[8]; **4**, 72[24], 867[7]; **8**, 412[109], 413[127], 851[132]
Wardle, R. B., **5**, 1123[39]; **6**, 814[96]
Wardleworth, J. M., **3**, 511[189]
Ware, A. C., **7**, 646[25]
Ware, D. W., **4**, 386[153,153b]
Ware, J. C., **7**, 595[14], 597[14], 600[78], 601[78]
Ware, R. S., **3**, 681[95]; **7**, 801[41]
Wariishi, K., **2**, 629[1], 635[1], 804[42]
Warin, R., **4**, 1033[16,16d], 1052[16d]
Waring, A. J., **3**, 803[1], 804[6]; **4**, 3[8], 6[8], 99[118d]; **5**, 223[74,80]; **7**, 108[180], 671[10], 673[10], 687[10]; **8**, 984[2]
Waring, C., **7**, 732[56]
Waring, P., **8**, 906[66], 909[66]
Warita, Y., **3**, 871[54]
Warkentin, J., **3**, 896[67]; **4**, 284[152], 815[191], 1089[132]; **5**, 901[27]; **6**, 1059[68]
Warman, D., **2**, 740[63a]
Warmus, J. S., **5**, 516[24], 517[24], 518[24], 524[50,51], 539[50], 548[50c]
Warne, T. M., Jr., **4**, 18[62], 20[62a,c]
Warner, A. M., **7**, 474[36]
Warner, C. R., **8**, 781[97]
Warner, D. T., **2**, 156[115]; **4**, 239[36,37], 243[37]
Warner, J. C., **5**, 492[249]
Warner, P., **1**, 343[114]; **2**, 125[216,217,221,222,224], 127[228], 271[77], 272[77,79,80], 315[42,44], 316[42,44], 317[44]; **4**, 217[131], 1012[175], 1021[241]; **6**, 1036[144]
Warner, R. W., **3**, 174[532], 176[532], 177[532], 178[532]; **4**, 629[412,413]; **6**, 849[118]
Warnet, R. J., **2**, 823[112]; **3**, 571[58], 596[193], 728[37], 1051[22], 1052[22]
Warnhoff, E. W., **1**, 853[48]; **3**, 380[7], 784[30], 854[73]; **4**, 1031[6], 1043[6], 1052[6], 1063[6]; **5**, 905[56]; **7**, 123[31], 673[25]; **8**, 90[48]
Warning, K., **2**, 1051[37]
Warnock, J., **7**, 281[174], 282[174]
Warnock, W. J., **5**, 257[59]
Warpehoski, M. A., **3**, 969[134]; **7**, 91[35]
Warr, J. C., **3**, 888[16]
Warrellow, G. J., **7**, 35[102]; **8**, 28[34], 66[34]
Warren, C. B., **5**, 571[116]
Warren, C. D., **3**, 273[7]; **6**, 652[143]
Warren, J. D., **6**, 554[710], 809[70]
Warren, K. S., **3**, 822[10], 829[10]
Warren, P., **3**, 670[60]
Warren, R. F. O., **8**, 212[24,25]
Warren, S., **1**, 570[269], 774[205,206,207,209,210,211], 776[206,214,216], 777[216b,217], 778[220,221], 779[225], 780[227], 781[230,230b,231], 782[232,233], 814[205b,216,217]; **2**, 73[65], 74[65], 202[95], 596[4]; **3**, 123[251], 124[261], 126[261], 201[76], 946[87]; **4**, 371[52], 378[52b]; **6**, 25[100], 830[5], 902[126], 932[94,95]; **7**, 369[61]; **8**, 13[67,68], 15[92], 864[244]
Warren, S. G., **4**, 731[70]; **6**, 614[87,88]
Warrener, R. N., **1**, 554[102]; **4**, 14[47,471]; **5**, 580[168], 632[61]; **7**, 380[102], 821[27]
Warrier, U., **1**, 243[54]
Warrington, B. H., **3**, 688[114], 690[114]
Warshawsky, A., **2**, 1074[143]; **3**, 302[51]; **5**, 407[25]
Wartski, L., **1**, 556[124,125], 561[162]; **2**, 428[46]; **4**, 10[34], 73[32], 112[158d,159], 113[164,168,168e], 240[48], 245[95], 259[270,271,272]; **5**, 410[41]; **8**, 850[118]
Warwel, S., **5**, 1119[20]; **8**, 755[130], 758[130]
Warwick, P. J., **3**, 530[77], 535[77]
Washausen, P., **7**, 62[50b]
Washburn, W., **3**, 1058[40]
Washburn, W. N., **7**, 40[8], 43[8,36]
Washburne, S. S., **6**, 809[68,69]
Washioka, Y., **3**, 1032[67]
Washiyama, H., **7**, 835[84]
Washiyama, M., **4**, 93[94]
Wasielweski, M. R., **7**, 854[49], 855[49]
Wasley, J. W. F., **8**, 32[54], 66[54], 376[163]
Wasmuth, D., **3**, 43[240], 44[240]
Wassenaer, S., **4**, 1093[150]
Wasserman, A., **4**, 1069[1]
Wasserman, E., **3**, 628[46]
Wasserman, H. H., **1**, 630[32], 675[32], 722[32]; **4**, 350[119], 510[171]; **5**, 86[16], 162[67], 513[4], 689[72], 692[92], 780[200], 945[253]; **6**, 93[133], 531[439,440,441,442], 612[73], 675[297], 734[10], 735[10], 736[26,27], 756[123], 899[111]; **7**, 96[88], 97[88], 98[88,103], 110[88], 111[88], 180[156], 183[156], 816[10], 818[10]; **8**, 32[55], 66[55], 374[144], 389[70]
Wassermann, A., **5**, 552[25,26], 594[1], 601[1], 604[1]
Wassink, B., **8**, 674[32]
Wassmundt, F. W., **3**, 503[149], 512[149]
Wassmuth, H., **5**, 412[45], 417[59]
Wasson, R., **4**, 35[99,99a]; **7**, 179[152]; **8**, 640[25]
Wasylishen, R., **3**, 382[36]
Wat, E. K. W., **3**, 430[92]
Watabe, T., **5**, 394[145b]; **7**, 172[124]
Watabe, Y., **6**, 442[88]
Watabu, H., **8**, 191[90]
Watanabe, A., **3**, 771[182]
Watanabe, E., **7**, 58[55], 62[55], 63[55]
Watanabe, F., **6**, 745[87]; **8**, 492[16], 508[16], 509[16]
Watanabe, H., **3**, 244[27,28], 464[177], 771[191]; **4**, 373[83]; **5**, 442[185]; **7**, 458[111]; **8**, 37[100], 42[100], 66[100], 188[50], 249[91], 294[54], 764[5]
Watanabe, I., **6**, 711[64]; **8**, 817[30]
Watanabe, J., **3**, 136[371], 139[371], 140[371,371b], 143[371,371b]
Watanabe, K., **1**, 546[56], 569[261]; **2**, 150[94,95], 225[156], 455[17], 885[50]; **3**, 426[83]; **4**, 1002[46]; **5**, 108[206]; **6**, 46[66], 124[145], 425[62], 533[503], 936[111], 1049[36]; **7**, 56[17,18], 57[18], 253[23], 765[149], 773[149,301]; **8**, 533[149], 803[90], 807[90], 917[117], 920[117]
Watanabe, K.-I., **8**, 253[113]
Watanabe, M., **1**, 223[83], 224[83], 328[29], 466[48,49], 474[104], 477[137,138]; **4**, 380[122], 445[204], 497[99], 738[98], 792[67]; **5**, 422[83]; **6**, 96[151]; **7**, 196[29], 355[47]; **8**, 244[57], 249[97], 253[97], 620[132], 817[34]
Watanabe, N., **2**, 967[76]; **4**, 341[58], 349[58], 1057[142]; **5**, 356[89]; **6**, 291[201]; **7**, 308[18], 496[214]
Watanabe, R., **3**, 543[117]; **6**, 640[56], 641[56]
Watanabe, S., **3**, 259[131]; **4**, 313[460], 1023[257]; **6**, 566[929,930]; **7**, 745[76]
Watanabe, T., **1**, 328[29], 391[151]; **2**, 388[344]; **3**, 726[25], 848[54]; **4**, 239[26], 251[26], 257[26]; **5**, 356[89], 583[183]; **6**, 523[351], 524[351]; **7**, 473[33], 501[33], 502[33]; **8**, 418[10], 709[45]
Watanabe, W. H., **5**, 830[29]

Watanabe, Y., **1**, 506[15]; **2**, 357[146], 358[146,151]; **3**, 507[175], 960[116], 1018[10], 1021[10,16], 1028[49]; **4**, 313[470], 602[264], 609[264], 644[264], 792[64,65], 930[52], 941[84]; **5**, 421[78], 894[48]; **6**, 83[84], 139[48], 206[37], 438[56], 602[7], 603[7], 605[37], 898[107b], 901[123], 966[99]; **7**, 245[77,78], 762[83]; **8**, 36[81], 54[81,161], 55[180], 66[81,161,180], 289[24], 291[35-37], 292[36], 293[46], 395[126], 533[150], 568[475], 591[59], 614[83], 652[80], 846[84]
Watanuki, M., **7**, 73[106]
Waterfield, A., **6**, 497[158]
Waterhouse, I., **1**, 793[273], 794[273c,276], 804[273]; **5**, 859[233], 888[25]; **6**, 993[92], 994[92], 997[112], 998[119]; **8**, 823[53]
Waterhouse, J., **4**, 231[269]
Waterman, E. L., **1**, 189[74]; **4**, 560[28]
Waters, D. L., **5**, 856[210], 1003[22]
Waters, J. A., **2**, 960[36]
Waters, R. M., **5**, 835[59], 862[250]; **6**, 655[158]
Waters, W. A., **3**, 660[16], 661[21]; **4**, 717[7]; **7**, 85[5], 98[101], 154[23], 157[33], 158[33b], 338[37], 530[12], 707[27], 709[46], 850[4], 851[19]
Waters, W. L., **3**, 568[42]; **4**, 302[332], 311[448], 315[515]
Waterson, D., **2**, 186[32], 584[119]; **3**, 17[92], 18[92,94], 42[94]; **4**, 243[64], 247[64], 255[64], 260[64], 382[134,134b]; **6**, 16[61]; **7**, 646[24]
Watkin, D., **1**, 243[55]
Watkins, B. F., **7**, 810[90]
Watkins, E. K., **4**, 874[52]
Watkins, J. C., **5**, 274[8,9], 277[8]
Watkins, J. J., **1**, 117[53,54], 333[59]; **3**, 213[41,42b]
Watkins, N. G., **6**, 790[116]
Watkinson, I. A., **8**, 561[412]
Watson, B. T., **4**, 485[30]; **5**, 1175[39,42], 1178[39,42]; **8**, 675[41], 679[41,67], 684[41]
Watson, D. G., **1**, 2[3], 37[3]; **6**, 436[9]
Watson, D. R., **8**, 843[48], 846[48]
Watson, E. R., **4**, 47[133]
Watson, F., **5**, 478[163]
Watson, G., **7**, 94[55]
Watson, J. M., **2**, 814[79]; **5**, 830[36], 834[53]
Watson, K. G., **6**, 428[81]
Watson, K. N., **5**, 485[184]
Watson, L. S., **2**, 971[92]; **5**, 780[201]
Watson, P., **1**, 278[85]
Watson, P. L., **7**, 3[6]
Watson, R., **8**, 568[472]
Watson, R. A., **1**, 218[52]; **2**, 124[209]; **4**, 96[105], 115[180c], 688[67]
Watson, S. P., **1**, 837[151,152]
Watson, T. R., **5**, 468[135]
Watson, W. H., **1**, 838[158]; **5**, 205[41], 207[41], 211[67]; **6**, 150[114], 744[72]; **7**, 162[64], 778[398,401,401a]; **8**, 36[88], 66[88]
Watson, W. P., **7**, 771[263]
Watt, D. R., **7**, 160[49]
Watt, D. S., **1**, 337[80], 542[9], 544[9], 551[9], 552[9], 553[9], 554[9], 555[9], 557[9], 560[9], 827[67], 828[69]; **2**, 73[62], 420[24], 690[70]; **3**, 39[218], 48[218], 247[47], 253[93]; **4**, 161[87,87b], 793[69], 1033[23]; **5**, 569[111]; **7**, 172[125], 174[140], 229[110]; **8**, 332[40], 333[57], 334[62], 345[127], 528[57,78], 986[14]
Watt, I., **5**, 63[8]; **8**, 90[49]
Watt, W., **5**, 1098[132], 1101[133], 1112[132,133]
Wattanabe, Y., **5**, 256[55,56]
Wattanasin, S., **2**, 73[69], 150[96], 151[96]; **3**, 131[328,329], 132[329], 135[328,329]; **5**, 514[8], 527[8,8b]; **6**, 94[142]; **8**, 114[55], 973[121]
Watthey, J. W. H., **5**, 702[10], 716[10]
Wattimena, F., **8**, 285[9], 292[9], 293[9]
Wattley, R. V., **2**, 227[160], 821[104], 852[104]
Watts, C. R., **4**, 1075[30], 1097[164]; **5**, 247[26], 248[26a], 249[26a], 625[30,32], 626[32,40], 630[40]
Watts, J. C., **6**, 270[77]
Watts, L., **3**, 855[82]; **4**, 701[26]
Watts, O., **4**, 82[62a]
Watts, P. H., **1**, 294[47]
Watts, W. E., **2**, 962[44]; **3**, 822[8], 836[8], 921[35]; **4**, 82[61], 393[187], 518[2], 519[22], 521[2,39,47], 522[47], 530[47], 541[109,110,112], 953[8], 954[8p], 961[8p]; **5**, 794[45], 984[33], 1037[6], 1039[6], 1040[6], 1049[6], 1138[68], 1165[8], 1183[8]; **6**, 786[95]
Watzel, R., **8**, 528[71], 971[108]
Waugh, F., **2**, 725[112]
Waugh, M. A., **2**, 524[78]; **3**, 34[196]; **6**, 1066[97]; **7**, 701[64]
Wauquier, J. P., **8**, 419[22], 420[22], 430[22], 436[22]
Wautier, H., **3**, 120[244], 142[244]
Wawer, I., **6**, 576[975]
Wawrzak, Z., **4**, 83[65c]
Wawrzyniewicz, W., **4**, 1001[20]
Wawzonek, S., **3**, 924[40]; **4**, 868[18]; **6**, 209[66]; **7**, 806[73]; **8**, 236[4], 242[4], 247[4], 248[4], 249[4]
Way, T. F., **5**, 73[204]
Wayaku, M., **3**, 1041[111]; **7**, 107[168]
Wayda, A., **6**, 9[40]; **8**, 458[224]
Waykole, L., **1**, 571[279]; **7**, 131[86]; **8**, 61[191], 66[191]
Wayland, B. B., **8**, 669[7], 670[7], 671[7]
Waymouth, R. M., **6**, 831[10], 832[10], 848[10]
Wayne, W., **2**, 141[40], 240[8]
Weakley, T. J. R., **6**, 836[55]
Weatherbee, C., **2**, 957[16], 969[83,83a]
Weaver, B. N., **8**, 70[227], 71[227]
Weaver, D. F., **1**, 528[117]
Weaver, J., **4**, 712[70]
Weaver, M. A., **2**, 387[337]
Weaver, T. D., **2**, 841[186]; **3**, 55[283]; **8**, 527[48]
Weaver, W. M., **6**, 203[1]; **7**, 655[18]
Weaver, W. W., **7**, 291[2]
Weavers, R. T., **4**, 803[134]; **8**, 333[54]
Webb, A. D., **2**, 173[180], 832[153]
Webb, C. F., **5**, 345[71a], 346[71a]
Webb, F. J., **1**, 506[1,2], 630[37,38], 631[38], 636[38]
Webb, I. D., **3**, 428[89]
Webb, J. L., **4**, 270[13]
Webb, K. S., **4**, 18[61], 249[130], 257[130], 262[130]; **6**, 1064[90a]; **7**, 625[42], 627[42,43]
Webb, M., **8**, 198[132]
Webb, M. B., **5**, 1166[17], 1167[17]
Webb, R. F., **6**, 620[131], 625[131]
Webb, R. R., **2**, 696[79];
Webb, R. R., Jr., **4**, 295[251], 398[216], 399[216b], 404[216b], 405[252], 741[125]; **5**, 434[141], 576[139,140,141]; **6**, 115[80], 960[57]; **8**, 540[195]
Webb, T. H., **1**, 759[134]
Webb, T. R., **4**, 428[75]; **6**, 938[126]
Webb, W. G., **8**, 330[29]
Webb, W. P., **3**, 818[95]
Webber, A., **6**, 176[83]; **8**, 595[74,75]
Webber, G. M., **3**, 816[79]
Webber, S. E., **1**, 808[319]; **3**, 217[94]
Weber, A., **3**, 174[526]; **4**, 1007[120], 1008[134]; **5**, 9[73], 595[15], 596[15,36], 598[36]; **6**, 233[47]
Weber, A. E., **1**, 400[11]; **2**, 113[107], 116[135], 254[43], 255[44], 256[45]; **8**, 386[53]
Weber, E., **1**, 158[74], 180[34], 294[38], 304[38], 615[51]
Weber, E. J., **2**, 3[11], 4[15], 6[35], 568[32], 573[49], 978[12]

Weber, G. F., **5**, 459[91]
Weber, H. P., **5**, 71[124]
Weber, J. C., **6**, 283[162]
Weber, J. V., **6**, 463[25]
Weber, K., **4**, 871[35], 876[35b]
Weber, K.-H., **3**, 890[34]
Weber, L., **1**, 36[237]; **4**, 119[195], 387[163c], 587[32], 588[73], 614[32,380], 615[32,380], 628[380]; **6**, 154[151]; **8**, 530[93], 843[51], 844[51]
Weber, M., **3**, 872[59]
Weber, R., **2**, 1017[34,35], 1018[35]; **6**, 462[13], 737[41]; **7**, 506[303]
Weber, T., **1**, 237[31], 239[31], 359[18], 380[18], 381[18]; **2**, 22[87]; **3**, 40[223], 41[223,227], 42[223]; **4**, 209[67]
Weber, W., **5**, 78[283]
Weber, W. P., **1**, 343[121], 345[121], 544[30], 580[1], 731[4], 815[4]; **2**, 564[5], 584[117]; **4**, 155[65], 868[11], 1001[26]; **5**, 718[95]; **6**, 237[59], 257[59], 1001[131]; **7**, 444[53], 616[14]; **8**, 318[59], 322[59], 406[35]
Weberndorfer, V., **4**, 1085[100]
Webers, W., **4**, 298[290]
Weber-Schilling, C. A., **4**, 126[217d]
Webster, G. R. B., **2**, 170[174]
Webster, J. A., **8**, 764[3], 776[3]
Webster, N. J. G., **5**, 137[78,79]; **7**, 549[43]
Webster, O. W., **3**, 855[84]; **5**, 71[147], 78[147], 430[116], 486[196]
Webster, P., **5**, 109[220,221]
Wechter, W. J., **7**, 595[10]; **8**, 316[56]
Weckerle, W., **1**, 55[25]; **8**, 568[469]
Wedegaertner, D. K., **4**, 719[25]
Wedemann, P., **5**, 293[46,49], 294[50], 296[54], 1190[25,26,28], 1191[25,26,28], 1194[33,34], 1195[34], 1197[42]
Wedemeyer, K.-F., **3**, 804[2], 810[2], 812[2]
Wedergaertner, D. K., **6**, 714[82]
Wee, A., **7**, 564[96], 565[96], 568[96], 569[96], 570[96]
Wee, A. G. H., **3**, 589[162], 610[162]; **4**, 373[82]; **6**, 494[134]
Weedon, A. C., **3**, 590[163]; **5**, 123[1], 126[1,22], 127[22], 128[22]; **8**, 17[114], 21[114], 115[66]
Weedon, B. C. L., **2**, 143[51], 821[108]; **3**, 553[12], 577[89], 633[5], 634[5,24], 635[5,33], 639[5,24], 640[107,107a], 642[112], 643[5,117,125,130], 644[112,147,148,148a,150,151,153,166], 647[33,107], 649[24]; **7**, 306[9], 801[44]; **8**, 974[128]
Weeks, P. D., **1**, 385[119], 386[119]; **2**, 939[160]; **3**, 20[118], 31[118]; **7**, 258[56], 630[50]
Weeks, R. P., **4**, 603[269], 645[269]
Weerasooriya, U., **2**, 597[5]; **3**, 1049[16], 1053[16]; **6**, 67[15], 69[15], 705[35,36]
Weerasuria, D. V., **4**, 683[60], 687[60]
Weeratunga, G., **4**, 73[35]
Weerawarna, K. S., **7**, 771[281]
Weese, K. J., **1**, 488[13]
Weetman, J., **2**, 681[59]; **5**, 355[87c], 356[87c], 365[87c,96c], 543[113]
Wege, D., **4**, 489[58], 1016[202]; **5**, 580[167], 584[193], 1130[6]
Wegener, G., **5**, 589[213]
Wegener, J., **8**, 58[176], 66[176]
Weglein, R. C., **3**, 709[14]
Wegler, R., **5**, 86[36], 87[37], 88[45]
Wegmann, B., **6**, 52[115]
Wegmann, H., **4**, 1076[39]
Wegner, E., **4**, 532[88], 534[88], 537[88], 538[88], 539[88]
Wegner, G., **4**, 152[54], 202[17,23]
Wegrzym, M., **6**, 550[677]
Wehinger, E., **2**, 377[281], 384[281]; **8**, 592[64]
Wehlacz, J. T., **4**, 878[75], 898[75]
Wehle, D., **1**, 672[190,191], 674[190,191], 714[190,191], 715[190], 718[190,191], 722[190,191], 867[80]; **4**, 784[15]; **8**, 335[67]
Wehman, A. T., **6**, 692[406]
Wehman, E., **3**, 210[25]
Wehmeyer, R. M., **1**, 426[111]; **2**, 121[191]; **3**, 209[19], 226[202], 263[171]; **4**, 175[41]; **5**, 386[132], 387[132c], 691[83], 692[83], 693[83]; **6**, 977[17]
Wehner, G., **1**, 548[67]; **3**, 197[40]; **4**, 113[169]; **6**, 229[24], 961[67]
Wehrli, H., **5**, 222[63], 229[122,123]
Wehrli, P., **2**, 866[9]
Wehrli, P. A., **3**, 168[491], 169[491], 171[491]; **6**, 490[110]
Wehrli, S., **3**, 380[7]; **7**, 544[39], 553[39], 556[39]
Wei, C.-P., **5**, 225[89], 231[133]
Wei, J., **1**, 390[142]; **2**, 603[48]; **7**, 330[10]; **8**, 690[103]
Wei, T.-Y., **7**, 103[141,142], 266[111], 267[111,116], 277[116]
Wei, Y., **4**, 884[103]
Wei, Z. W., **2**, 553[132], 554[132]
Wei, Z. Y., **2**, 567[26]
Weibel, F. R., **1**, 564[202]
Weiberth, F. J., **1**, 124[83]
Weichert, U., **6**, 48[82], 51[82]
Weichsel, Ch., **7**, 204[58]
Weidenbaum, K., **8**, 451[180]
Weidenhaupt, H.-J., **5**, 1037[4], 1132[18]
Weider, P., **4**, 561[29]; **8**, 395[131]
Weidhaup, K., **3**, 362[90]
Weidinger, H., **5**, 488[194]; **6**, 428[84], 488[8], 495[8], 499[8], 543[8], 566[8]
Weidlein, J., **8**, 756[145]
Weidler-Kubanek, A. M., **5**, 114[240]
Weidlich, H. A., **3**, 615[9]
Weidmann, B., **1**, 140[9], 141[9], 142[9,23], 144[9,39], 145[9,39,43], 146[39,43,44], 148[44], 149[9,39], 152[39], 155[43], 165[44,112a], 234[23], 330[45]; **2**, 5[17], 6[17], 22[17], 23[17d], 630[4]
Weidmann, H., **1**, 271[62,62b]; **2**, 280[21]; **3**, 570[55,129], 582[55], 583[55,129], 630[57], 631[57]; **6**, 978[22]; **7**, 32[92]; **8**, 812[5]
Weidmann, U., **2**, 25[97], 26[97,97a], 27[97], 31[97], 40[141]
Weidner, C. H., **1**, 765[166]; **5**, 362[93], 363[93h]
Weier, A., **5**, 1049[33], 1052[33]
Weigand, W., **2**, 1086[28], 1096[28]
Weigel, L. O., **1**, 523[83]; **2**, 228[163]; **6**, 8[36], 91[121]; **7**, 358[11]; **8**, 287[20], 288[20], 844[72,72c]
Weigel, T. M., **8**, 344[121,121b]
Weigele, M., **6**, 543[621]
Weigelt, L., **8**, 450[166]
Weigert, F. J., **3**, 1039[96]
Weigold, H., **1**, 139[4]; **4**, 153[62a]; **8**, 447[119a,b], 675[35-37], 679[37], 684[35], 691[36], 754[114]
Weigt, E., **5**, 225[113,114], 227[113,114], 228[114], 230[114], 232[135,136], 233[113,114]
Weihe, G. R., **7**, 73[103]
Weihrauch, T., **6**, 531[455]
Weijers, C. A. G. M., **7**, 429[150b]
Weijland, J., **8**, 148[107]
Weijlard, J., **7**, 710[49]
Weil, D. A., **6**, 708[48]
Weil, E., **6**, 501[203], 531[203]
Weil, E. D., **4**, 317[550]
Weil, R. A. N., **4**, 10[31]
Weiler, J., **4**, 953[9]
Weiler, L., **1**, 92[65], 569[250], 751[89,91]; **2**, 189[48,49], 832[151]; **3**, 58[290], 99[188], 107[188], 110[188], 396[103]; **4**, 799[116]
Weiller, B. H., **7**, 4[15]
Weill-Raynal, J., **1**, 367[56], 368[56], 370[56]; **3**, 521[3]; **4**, 144[22]; **7**, 804[63]; **8**, 541[212]
Weimann, G., **6**, 611[68]
Weimer, D. F., **1**, 767[180]
Weinberg, D. S., **2**, 528[14]; **6**, 209[71]
Weinberg, H. R., **3**, 634[10], 638[10]; **7**, 802[46]

Weinberg, J. S., **7**, 384[116]
Weinberg, N. L., **3**, 634[8,10], 636[8], 638[8,10], 639[8], 643[8], 649[8], 655[8]; **4**, 129[225]; **6**, 561[871]; **7**, 799[25,28], 800[28a], 801[36], 802[46], 806[73]
Weinberg, R. B., **3**, 380[10]
Weinberger, B., **3**, 1026[39]
Weinelt, A., **2**, 1095[93]
Weiner, B. Z., **4**, 426[65], 441[65]
Weiner, M. L., **3**, 843[23]
Weiner, S. A., **3**, 661[24]
Weingarten, H., **6**, 546[646], 582[992], 705[23]; **7**, 798[23]
Weingartner, T. F., **2**, 332[60]; **4**, 104[136b]
Weinges, K., **3**, 872[59,61-63], 874[68]; **4**, 799[113]
Weingold, D. H., **8**, 240[33]
Weinhardt, K., **1**, 88[53]; **8**, 587[40]
Weinig, P., **1**, 167[115]
Weininger, S. J., **4**, 915[14]
Weinkauff, D. J., **8**, 459[232], 460[232], 535[166]
Weinman, S., **5**, 740[151]
Weinreb, S., **6**, 756[123]
Weinreb, S. M., **1**, 92[68,69], 93[68], 238[34], 376[94], 377[95], 399[4], 400[10], 404[21], 405[4]; **2**, 542[84], 662[4], 1026[68], 1049[23], 1050[23], 1054[61], 1056[65,66], 1059[65], 1070[65,66], 1071[61], 1074[65,66], 1079[159]; **3**, 257[115], 499[121], 511[191]; **4**, 14[47,47o], 802[127]; **5**, 402[1], 403[1,10], 404[1], 406[23,23b], 410[1], 413[1,1b,50], 414[52,54], 415[55], 417[1], 422[1,87], 423[91], 424[87,92,97,98], 425[1,87,99,100], 426[1,87,104], 429[1], 430[1], 433[1], 434[1], 435[1], 436[1], 438[1], 440[1], 444[1], 451[15,19,20], 453[15], 454[15], 461[15], 464[15], 468[15], 469[15], 470[15], 473[15], 480[15], 485[15,181], 486[15], 491[15], 494[214], 499[15], 501[15], 508[15], 510[15], 511[15], 531[77,79,82], 539[108], 567[104]; **6**, 647[113], 705[25], 745[78], 756[78,128], 814[88], 894[90], 900[112], 906[148], 1035[137]; **7**, 183[166], 248[111], 486[143], 491[181], 548[56], 552[56], 748[112], 801[44]; **8**, 27[32], 66[32], 272[121], 394[121]
Weinschneider, S., **5**, 436[157]
Weinshenker, N. M., **3**, 380[13]; **4**, 370[26], 738[98]; **5**, 339[56], 347[56]; **6**, 21[78]; **7**, 294[13]; **8**, 163[40], 269[80], 800[68], 957[9]
Weinstein, B., **2**, 1090[71], 1092[71], 1093[71], 1094[71], 1096[71], 1098[71], 1100[71]; **3**, 305[75b]; **5**, 803[88]; **6**, 670[272]; **7**, 750[136]; **8**, 35[63], 66[63], 144[78], 496[33], 531[116]
Weinstein, R. M., **1**, 273[69], 411[49]; **2**, 567[29]; **4**, 115[177]
Weinstein, S. Y., **6**, 3[14]
Weinstock, I., **2**, 127[229]
Weinstock, J., **5**, 404[20], 405[20]; **6**, 811[74]; **7**, 236[16]
Weinstock, L. M., **1**, 425[103]; **2**, 648[97], 649[97b]; **3**, 380[11]; **4**, 230[250,251], 767[234], 1033[30]; **6**, 431[109], 452[132]; **7**, 493[188]; **8**, 272[122]
Weintraub, P. M., **3**, 890[31], 903[124]
Weintz, H.-J., **5**, 297[56,58], 1195[36], 1196[37]
Weiper, A., **3**, 642[113], 643[113], 644[113]
Weir, J. R., **3**, 539[98]
Weir, T. R., **8**, 959[28]
Weis, C. D., **2**, 534[36]; **3**, 381[32]; **4**, 270[10]; **7**, 747[102]
Weisbach, J. A., **8**, 568[471]
Weisbuch, F., **8**, 532[130]
Weise, G., **6**, 523[349]
Weisenfeld, R. B., **3**, 88[128], 105[217], 124[128]; **4**, 116[185c], 854[95]; **8**, 842[47]
Weiser, R., **8**, 273[124,125]
Weisgraber, K. H., **3**, 665[40]
Wei-shan, Z., **7**, 844[58]
Weisman, G. R., **6**, 581[988,991]; **8**, 70[235], 71[235], 166[65], 178[65], 179[65]
Weismiller, M. C., **1**, 838[162]; **7**, 778[399,401,401b], 779[401b]
Weiss, A., **6**, 502[210], 724[155]
Weiss, D. S., **3**, 815[73]
Weiss, E., **1**, 3[21], 9[40], 10[52], 12[60,62-64], 16[83-85], 18[92,93], 19[104], 20[106,108,109], 21[110], 22[112,116], 25[131], 34[224], 36[174], 38[253,255,256,257], 39[186,188,189,190], 40[192], 41[272]; **8**, 696[125]
Weiss, F., **5**, 244[18], 787[7]
Weiss, H. A., **7**, 760[35]
Weiss, J., **3**, 636[45]; **6**, 979[27]; **7**, 850[1]
Weiss, K., **4**, 976[100], 980[106], 981[106]; **5**, 1065[1], 1066[1,1a], 1074[1], 1083[1], 1084[1], 1093[1]; **6**, 671[279]
Weiss, K. T., **5**, 491[208]
Weiss, L. B., **4**, 369[19,23], 374[19]
Weiss, M. J., **4**, 91[88g], 141[15], 142[15]; **6**, 648[124]; **8**, 527[41,45], 528[45], 529[45], 530[45], 531[111], 564[443], 614[86]
Weiss, M. M., **5**, 418[70]
Weiss, R., **2**, 397[9]; **6**, 94[145], 96[148], 500[179]; **7**, 11[87], 107[162], 422[139], 452[45], 740[45]
Weiss, R. G., **3**, 1048[11]
Weiss, R. H., **4**, 473[147], 474[147]
Weiss, U., **2**, 381[306,308]; **3**, 380[7]; **7**, 544[39], 553[39], 556[39]
Weissbach, A., **2**, 466[120], 469[120]
Weissbart, D., **6**, 26[110]
Weissberger, A., **3**, 53[272], 892[50]; **4**, 51[144c]; **5**, 86[10], 118[10], 491[207]; **6**, 212[83]; **7**, 358[6], 372[6], 470[6], 472[6], 473[6], 474[6], 476[6], 516[2]; **8**, 366[40], 524[7], 530[7], 625[158], 626[158]
Weissberger, E., **5**, 1130[4,6]
Weissenfels, M., **2**, 785[42,46], 792[67]; **6**, 489[81,83]; **7**, 92[46]
Weissensteiner, W., **5**, 1133[30]
Weissermel, K., **3**, 1039[98]
Weissflog, E., **3**, 134[339], 135[339]
Weissman, P. M., **8**, 64[201], 66[201], 74[251], 260[4], 267[64,65], 541[206], 544[206,271]
Weissman, S. A., **8**, 722[145]
Weisz, I., **6**, 653[151]
Weith, W., **6**, 294[237]
Weitkamp, A. W., **8**, 439[80]
Weitkamp, H., **2**, 554[133]
Weitl, F. L., **8**, 861[224]
Weitz, E., **2**, 144[59]; **4**, 35[99]; **7**, 850[1]
Weitz, H. M., **2**, 321[13]; **6**, 105[16]
Weitzberg, M., **4**, 116[188c], 213[115,117], 215[117]
Weitzenböck, R., **2**, 156[116]
Weitzer, H., **8**, 174[124]
Welbourn, A. P., **8**, 615[94], 618[94]
Welch, A. J., **5**, 1134[45], 1136[54]
Welch, A. S., **4**, 629[409]
Welch, E., **2**, 504[5]
Welch, J., **2**, 523[88-90]; **3**, 373[129]; **7**, 231[150,151], 235[4]
Welch, J. T., **2**, 103[31,32], 209[108], 211[116], 631[13]; **5**, 828[8], 841[95]; **6**, 204[7], 216[7,108], 219[108], 858[161], 861[182]
Welch, M., **3**, 51[270]; **8**, 508[87]
Welch, M. C., **6**, 980[30]
Welch, M. J., **4**, 445[207]; **6**, 219[118]
Welch, R. W., **3**, 320[133]; **4**, 276[71], 283[71], 313[464]
Welch, S. C., **1**, 506[16], 826[59]; **2**, 102[23], 542[81,82]; **4**, 83[65b], 952[5]; **6**, 835[45]; **7**, 111[191]; **8**, 121[78], 531[126], 542[237], 931[37]
Welch, W. M., **2**, 939[160]
Welch, W. W., **1**, 385[119], 386[119]
Welcher, R. P., **4**, 348[107]
Welke, S., **1**, 749[78], 816[78]
Welker, M. E., **2**, 125[218], 127[226,227,230,231], 271[78], 315[40,41,43], 316[41,43], 933[141], 934[141,142]; **3**, 47[257]; **4**, 82[62c,d], 217[127,128], 231[127,128]; **8**, 447[127], 463[127]
Weller, A., **5**, 650[26]; **7**, 854[58], 855[58]
Weller, H. N., **6**, 1059[70], 1066[70]; **7**, 318[61], 376[83]

Weller, J. W., **7**, 12[99], 13[123]
Weller, T., **1**, 287[14]; **2**, 321[12], 324[12], 326[12], 329[12]; **4**, 78[52a], 110[150b]; **6**, 107[24], 911[16]; **8**, 363[1], 374[1]
Wellington, C. A., **5**, 910[80]
Wellmann, J., **8**, 797[30]
Wells, B. D., **8**, 36[51], 66[51]
Wells, D., **3**, 117[235,236], 155[235,236], 156[235,236], 946[87]; **5**, 272[4], 273[4], 275[4], 637[107]; **6**, 997[110]
Wells, E. E., Jr., **4**, 967[56]
Wells, G. J., **1**, 672[202], 700[202], 701[202], 705[202]; **2**, 572[41]; **3**, 586[154], 864[21]; **5**, 71[119], 856[210], 910[81], 912[81], 954[300]; **6**, 146[88]
Wells, J. N., **2**, 956[13], 958[13]
Wells, J. S., **5**, 86[33]
Wells, P. B., **8**, 431[61]
Wells, R. J., **4**, 6[23]; **8**, 625[162]
Wells, R. L., **2**, 834[155]
Wells, W. W., **6**, 653[150]
Welner, S., **3**, 628[50]
Welsh, C. E., **4**, 277[86]
Welter, T. R., **5**, 209[55]
Welvart, Z., **1**, 555[111,122], 557[111]; **8**, 37[91], 40[91], 44[91], 66[91]
Welzel, J., **5**, 442[183]
Welzel, P., **3**, 583[118]; **6**, 7[34]; **7**, 47[53]
Wember, M., **2**, 464[93]
Wemple, J., **2**, 214[128], 419[23]; **6**, 439[72]; **7**, 764[117]
Wendeborn, S. V., **5**, 736[145], 737[145]
Wendel, I., **2**, 162[140]
Wendelborn, D. F., **6**, 1026[88], 1027[88]
Wender, I., **5**, 1037[3], 1132[22], 1133[27], 1138[67], 1146[110]; **8**, 372[122], 452[189c,190], 455[206], 456[208], 608[44], 699[148], 763[1], 785[1]
Wender, P. A., **1**, 383[108,109], 464[35], 885[133a,b]; **2**, 479[19], 481[19], 553[125]; **3**, 31[182], 226[195], 242[7,8], 257[7,8], 264[183], 380[10]; **4**, 192[116], 611[356], 983[116], 1009[142]; **5**, 20[132], 123[1], 125[18], 126[1], 128[18,32], 130[32], 145[108], 249[36], 431[119], 639[125,126], 640[128,129], 641[130], 645[1], 647[15], 648[1i,j,r], 651[1], 653[1r,15], 656[15,30], 657[15], 660[30], 662[36], 663[37], 664[39], 665[41], 666[37,42], 667[43], 670[46], 736[142e,f,145], 737[145], 789[17], 803[89], 814[139], 825[89a], 864[256], 916[116,117], 924[148], 956[117,306], 976[19], 979[19], 982[30], 983[30], 984[32], 1005[31], 1026[85]; **6**, 9[43], 721[134,135], 1044[16a], 1045[30], 1048[16]; **8**, 123[82], 566[454]
Wenderoth, B., **1**, 83[26], 143[34,35], 145[41,42], 146[42], 148[42,46], 149[42], 150[41,46], 151[41,53,53a], 152[53,53a], 153[41,59], 154[41,59], 155[42], 156[35], 157[35,53a], 158[53,53a], 161[35], 165[46], 170[42], 215; **2**, 22[86], 307[14,15], 310[15], 640[68], 641[68]; **3**, 421[58]; **8**, 315[53], 802[80], 806[99]
Wenders, A., **6**, 551[684]
Wendisch, D., **4**, 951[1], 968[1], 979[1], 1000[9], 1016[9]; **5**, 714[68], 904[43], 905[43], 1188[15]
Wendlberger, G., **6**, 650[133c]
Wendler, J., **5**, 216[12], 219[12], 221[12]
Wendler, N. L., **2**, 160[135], 746[110]; **3**, 689[119]; **7**, 675[59]; **8**, 357[198], 358[198], 945[132]
Wendschuh, P. H., **5**, 707[39], 708[43], 709[39], 739[43]
Wendt, G. R., **6**, 425[63]
Wenger, E., **4**, 527[68], 532[68], 534[68], 536[68], 545[125], 546[125]
Wenger, R., **2**, 1015[22]
Wenhong, H., **1**, 543[15]
Wenis, E., **7**, 666[77]
Wenisch, F., **3**, 634[28]
Wenk, P., **7**, 694[31]
Wenkert, E., **1**, 748[73], 812[73], 846[18a], 847[18]; **2**, 384[317,318], 587[146], 823[112], 1015[20], 1018[44]; **3**, 229[233], 246[38], 380[7], 444[65], 446[81], 447[90-92], 448[97], 456[90,126], 457[133], 492[75], 493[83], 503[75,146], 509[178], 513[83,146,208], 572[68], 715[36], 818[95], 857[91,92], 908[146], 1051[22], 1052[22]; **4**, 876[63], 1033[26], 1035[26a], 1036[44], 1040[77], 1046[26a,b,110,116], 1048[120], 1051[26a], 1058[149,152], 1059[152], 1060[160]; **5**, 38[23a], 790[35], 903[37,38], 941[37,38], 942[229,231,232], 952[37,38], 964[324]; **6**, 675[300], 1042[4,6]; **7**, 221[31], 227[31]; **8**, 313[23,24], 528[86], 531[113], 838[20], 842[40,41], 880[63], 935[61,68]
Wenner, W., **8**, 140[11]
Wennerbeck, I., **2**, 365[210]
Wennerström, O., **3**, 499[119,120], 593[180,183], 594[184]
Wensing, M., **1**, 828[80], 831[100]
Wentland, M. P., **5**, 959[314,315,316]
Wentrup, C., **4**, 483[4], 484[4,4e,15], 495[4], 953[8], 954[81], 961[81], 1084[96]; **6**, 244[114], 440[74]; **7**, 21[17]
Wenzel, M., **8**, 49[113], 66[113]
Wepplo, P. J., **4**, 10[32,32f], 109[148]
Werbel, L. M., **6**, 533[505], 554[709]
Werbitzky, O., **5**, 418[71]
Werbitzy, D., **8**, 395[133]
Werhahn, R., **2**, 1104[133]
Wermeckes, B., **1**, 25[129]; **7**, 248[108]
Wermer, J. R., **1**, 13[67]
Wermuth, C. G., **6**, 509[274]; **8**, 60[192], 66[192]
Werner, G., **5**, 496[219,220], 497[223], 583[189]
Werner, H., **4**, 587[33,34]; **5**, 1085[64]
Werner, J. A., **3**, 942[81a], 1008[68]; **6**, 874[16]
Werner, L. H., **3**, 629[51]
Werner, N. D., **6**, 294[239]
Werner, W., **7**, 473[35]
Wernert, G. T., **3**, 71[28]
Wernick, D. L., **2**, 684[65]
Werntz, J. H., **4**, 315[531]
Werres, F., **5**, 198[23]
Wershofen, S., **7**, 400[47], 429[151]
Wersin, G., **6**, 637[28]
Werst, G., **8**, 299[80]
Werstiuk, N. H., **2**, 55[1], 441[2], 442[7], 443[2]; **3**, 86[24], 95[24], 159[24], 164[477]; **4**, 117[190], 163[95]; **6**, 833[23], 862[23], 961[73]
Werstiuk, W. H., **8**, 121[77]
Werth, R. G., **3**, 726[22]
Werthemann, L., **5**, 828[7], 839[7], 882[13], 888[13], 891[37], 892[13,37], 893[13]
Werumeus Buning, G. H., **8**, 95[91]
Wesberg, H. H., **7**, 722[21]
Wess, G., **4**, 1040[71], 1043[71]
Wessel, H. P., **6**, 533[500], 550[500], 652[142]
Wesseler, E. P., **3**, 319[131]; **7**, 738[31]
Wessels, F. L., **6**, 515[235]
Wessely, F., **2**, 355[122,123]; **3**, 807[24], 812[54,56,59], 813[65,66], 814[67,68], 817[86-88,90-92], 818[93]
Wesslén, B., **2**, 146[65]
Wessling, D., **3**, 753[104]
West, A. C., **6**, 47[78]
West, C. T., **8**, 88[41], 89[41], 105[41], 318[64], 319[70], 487[67], 546[312], 801[74]
West, D., **1**, 22[114]
West, D. E., **4**, 423[5], 426[5]
West, F. G., **3**, 918[26]; **4**, 1086[110,119], 1087[110,119]; **5**, 250[40]
West, H. D., **4**, 310[430]
West, J. P., **3**, 320[133]; **7**, 15[152]
West, P., **2**, 5[20], 6[20], 21[20]; **7**, 7[46]
West, R., **3**, 889[24]; **4**, 1015[200]; **5**, 199[28]; **8**, 764[2], 776[78], 777[2a]
West, W., **2**, 681[58], 683[58]; **6**, 502[217], 560[870]; **7**, 650[51]
Westberg, H. H., **3**, 651[218]

Westbrook, K., **7**, 49[62]
Westdorp, I., **6**, 985[64]
Westdrop, I., **3**, 584[130]
Wester, R. T., **1**, 744[57]; **4**, 650[426]; **5**, 1123[36]; **6**, 14[58], 16[58]
Westerberg, D. A., **8**, 51[121], 66[121]
Westerduin, P., **6**, 619[116]
Westerhof, P., **8**, 528[85,87]
Westerink, B. H. C., **2**, 902[46]
Westerlund, C., **3**, 495[95]; **8**, 384[37]
Westerman, I. J., **5**, 79[287]
Westerman, P. W., **2**, 735[10], 738[10]; **8**, 354[163], 724[173]
Westermann, J., **1**, 83[26], 141[21], 145[21,41,42], 146[21,42], 148[42,46], 149[21,42,49a], 150[41,46], 151[21,41,53,53a], 152[21,53,53a], 153[41], 154[41], 155[42], 157[53a], 158[53,53a], 162[103], 165[46], 169[117-120], 170; **2**, 22[86], 35[131]; **3**, 421[57,58]
Westheimer, F. H., **3**, 897[94]; **6**, 705[20]; **7**, 252[9]; **8**, 79[1], 82[1b], 561[414]
Westheimer, F. W., **7**, 236[19]
Westinger, B., **2**, 1090[73], 1099[109,109b,113], 1102[73], 1103[73]
Westling, M., **4**, 116[189]; **6**, 295[252,253,254]
Westman, T. L., **3**, 898[80]
Westmijze, H., **1**, 428[116]; **2**, 85[19-21], 587[146,147,148], 589[153]; **3**, 217[82,85], 219[104], 254[96]; **4**, 895[163], 897[171], 898[171], 899[171], 900[180,182], 905[209]; **8**, 743[48]
Westphalen, K.-O., **3**, 53[274]; **6**, 531[431,433]
Westrum, L. J., **2**, 1049[24]
Westwood, D., **1**, 568[243]
Westwood, K. T., **2**, 784[36], 792[62]; **6**, 489[84]
Westwood, R., **5**, 687[60]
Westwood, S., **7**, 580[144], 586[144]
Westwood, S. W., **1**, 894[155]; **4**, 822[224]
Wetli, M., **5**, 20[139]
Wetmore, S. I., **4**, 1081[77], 1082[87-90], 1083[87], 1103[87]
Wettach, R. M., **7**, 155[31a]
Wetter, H., **1**, 610[43], 635[86], 636[86], 672[86]; **2**, 587[138]; **5**, 836[64]; **6**, 656[172]; **7**, 673[26]
Wetter, H. F., **8**, 459[228], 776[81a,b]
Wetter, W. P., **4**, 960[35]
Wetterham, K. E., **7**, 839[2]
Wettlaufer, D. G., **1**, 359[22], 383[22], 384[22]; **3**, 217[88]; **4**, 206[43], 428[72]; **5**, 373[105]; **6**, 150[132], 151[133], 161[132,178]
Wettstein, A., **6**, 685[357,361]; **7**, 41[20], 128[171]; **8**, 268[74], 974[124]
Wetzel, D. M., **3**, 613[4], 619[4]; **8**, 505[79]
Wetzel, J. C., **8**, 407[58]
Wetzel, P., **5**, 645[1], 651[1]
Weuster, P., **2**, 510[44], 516[59], 830[139]; **6**, 722[141], 724[141]
Weuthen, M., **5**, 185[165]
Wewers, D., **1**, 36[237]
Wexler, A. J., **5**, 1032[98]
Wexler, B. A., **3**, 709[17]; **8**, 931[38]
Wexler, P. A., **5**, 1145[104]
Wexter, B. A., **7**, 239[44]
Wey, J. E., **5**, 386[132], 387[132b]
Weyenberg, D. R., **1**, 411[49]; **5**, 950[294]; **8**, 518[128,129], 564[445], 565[447]
Weyer, K., **8**, 738[29,32], 754[29]
Weyerstahl, P., **2**, 598[13]; **3**, 124[268,285], 125[268,285], 126[268,285], 127[268], 131[268], 584[132], 752[92]; **4**, 152[55], 1000[10], 1001[31,37,39], 1002[10,58], 1015[37], 1016[10], 1023[259]; **7**, 359[19]
Weygand, C., **8**, 533[143]
Weygand, F., **1**, 373[87], 374[87], 844[3c]; **3**, 887[4], 888[4], 893[4], 897[4,76], 898[76], 900[4], 903[4]; **6**, 437[37], 635[19], 636[19], 642[68], 668[252], 669[252]; **7**, 213[101,102]; **8**, 269[94], 270[96,97,100]
Weymuth, C., **4**, 764[222], 765[222], 808[155]
Weyna, P. L., **4**, 1005[87], 1020[87]
Whalen, D. L., **3**, 903[120]
Whalen, R., **7**, 16[160]
Whaley, A. M., **4**, 270[16], 271[16]; **6**, 204[10]
Whaley, W. M., **2**, 1016[26]; **6**, 736[25]
Whalley, W., **1**, 880[117]; **5**, 796[54]; **7**, 833[72]; **8**, 964[57]
Whan, D. A., **8**, 431[64]
Whang, J. J., **4**, 1077[53]
Whangbo, M.-H., **1**, 506[8]; **3**, 211[33]; **4**, 52[146], 170[17]; **6**, 133[4]
Wharry, D. L., **8**, 437[76]
Wharton, P. S., **3**, 391[88], 653[226], 892[52]; **5**, 794[48], 809[48,112]; **6**, 837[59], 1042[9], 1044[9], 1054[48], 1055[52a]; **8**, 341[102,106], 926[16], 927[19]
Wheatley, P. J., **4**, 519[13], 520[13]
Wheeler, D. M. S., **1**, 243[53]; **2**, 746[114]; **8**, 541[207]
Wheeler, H. L., **2**, 407[49]
Wheeler, M. M., **1**, 243[53]; **8**, 541[207]
Wheeler, N. G., **7**, 720[8]
Wheeler, O. H., **7**, 738[23]; **8**, 239[25], 240[25], 241[25]
Wheeler, R. A., **7**, 422[140]
Wheeler, T. N., **4**, 1050[124]
Whelan, J., **4**, 564[43], 599[221], 624[221], 641[221], 653[445]; **6**, 450[117]
Wheland, R. C., **5**, 65[61], 1188[15]
Whetstone, R. B., **8**, 428[52]
Whimp, P. O., **4**, 298[292]
Whipple, E. B., **5**, 168[103]
Whitby, R., **3**, 229[234], 444[66,67]; **4**, 878[83], 1089[128], 1092[128], 1093[128]
Whitcombe, G. P., **7**, 311[30], 312[30]
Whitcombe, M. J., **8**, 347[144]
White, A. A., **1**, 464[35]
White, A. D., **4**, 675[41]; **7**, 311[30], 312[30], 489[172]
White, A. H., **1**, 13[73], 16[89], 17[207,209,210,212,217,218,219], 36[233,234], 37[177,178], 520[73]; **2**, 606[69]; **5**, 144[104]
White, A. M., **3**, 333[212]; **4**, 305[364,366,367], 306[366]
White, A. M. S., **8**, 314[38], 315[44]
White, A. W., **4**, 192[116]; **5**, 814[139]
White, A. W. C., **7**, 231[137]
White, C., **8**, 445[34,54,54c], 454[200]
White, C. T., **2**, 221[145], 226[158]
White, D. A., **3**, 642[114]; **4**, 587[36], 665[8], 670[8], 674[8]
White, D. E., **7**, 254[27]
White, D. H., **3**, 592[175]; **6**, 835[45]; **7**, 723[27]
White, D. L., **4**, 695[4]
White, D. M., **6**, 653[149]
White, D. N. J., **8**, 724[170]
White, D. R., **1**, 827[66]; **2**, 420[25], 427[41]; **7**, 160[50]
White, E. H., **2**, 1102[120]; **6**, 291[199], 843[90]
White, E. N., **8**, 329[22], 338[22]
White, F. H., **1**, 514[54]; **2**, 1026[66]; **3**, 72[40], 81[40]; **4**, 380[119]; **6**, 740[64]; **7**, 224[51], 274[136]
White, F. L., **6**, 477[100]
White, G. L., **2**, 711[31]
White, J., **2**, 780[11]; **6**, 487[58-60], 489[59]
White, J. B., **1**, 248[69], 884[130]; **5**, 796[57], 815[57]
White, J. C., **5**, 63[11], 1023[78]
White, J. D., **1**, 131[103], 243[54], 259[30], 403[19]; **2**, 287[67], 421[26], 545[87], 631[15], 843[195]; **3**, 215[59], 261[148], 264[148], 351[43a], 683[103], 714[32]; **4**, 331[10], 345[10], 373[75],

1040[97,98], 1043[97,98]; **5**, 124[11], 129[11], 516[21], 531[21], 611[71,72], 712[57e], 839[81]; **6**, 7[32], 23[92], 655[164a]; **7**, 399[39], 400[43]; **8**, 113[34], 478[46], 481[46], 624[156], 798[58], 856[174]
White, J. F., **3**, 300[41]
White, J. L., **7**, 845[77]
White, K. B., **4**, 30[88]; **7**, 179[153]
White, K. S., **5**, 847[137]; **6**, 856[158,159], 857[159]
White, L. S., **5**, 196[12]
White, M. S. A., **8**, 545[281]
White, P. S., **2**, 809[52], 823[52]
White, R. E., **6**, 515[235]
White, R. F., **7**, 429[156]
White, R. H., **6**, 437[36]
White, R. L., Jr., **6**, 515[235]
White, R. W., **6**, 228[32]
White, S., **3**, 36[209], 37[212]; **6**, 723[147], 725[171], 728[171]
White, W. A., **6**, 705[23]
White, W. N., **5**, 854[174], 856[217]; **8**, 499[41]
Whitear, B., **4**, 44[125]
Whitehead, A., **6**, 709[53], 711[69]
Whitehead, C. W., **6**, 570[941,942,952]
Whitehead, J. F., **1**, 741[45]
Whitehead, M. A., **7**, 26[59]
Whitehouse, N. R., **2**, 1048[10]
Whitehouse, R. D., **3**, 114[234]; **6**, 1026[82], 1029[82]; **7**, 771[284], 772[284]
Whitehurst, J. S., **5**, 832[40]; **7**, 463[127]
Whiteley, C., **2**, 435[64]
Whiteley, R. N., **4**, 980[108]
Whitesell, J. K., **1**, 66[53-56], 72[74], 87[50]; **2**, 102[25], 475[2], 536[41-48], 537[42,45,48], 538[60], 539[60]; **3**, 28[172], 30[172], 31[172], 35[208], 36[210], 382[37], 384[51], 393[37], 781[14]; **4**, 100[120]; **5**, 424[96], 725[116]; **6**, 704[17], 705[22], 712[75], 717[112,114], 719[17], 725[17,112,172], 728[172], 838[65], 846[102], 900[113]; **7**, 543[16], 674[38]
Whitesell, M. A., **2**, 475[2]; **3**, 28[172], 30[172], 31[172], 36[210]; **4**, 100[120]; **5**, 477[160]; **6**, 704[17], 719[17], 725[17,172], 728[172]; **8**, 57[173], 66[173]
Whitesides, G. M., **1**, 116[46], 118[46], 143[32], 426[109], 431[134], 798[287]; **2**, 5[20], 6[20], 21[20], 455[20], 456[20,22,24,30-33,39-41,50-52,67,76], 457[33], 458[33,52], 459[33,50,52], 460[33,51], 461[33,50,76], 462[33,50,51,67], 463[89], 464[89], 465[106], 466[33], 684[65]; **3**, 248[55], 251[55], 269[55], 418[24,26], 419[47], 422[71], 423[71], 482[3,6], 494[87], 502[87], 557[46], 599[206]; **4**, 70[5], 148[47a], 169[2], 170[20], 176[46], 256[205]; **5**, 1131[14], 1145[104], 1173[32]; **6**, 49[94], 665[228], 1067[106]; **7**, 79[132], 80[132], 429[151], 632[58], 637[58]; **8**, 87[35], 113[32], 183[2], 185[2,15], 187[31], 189[58,59,61], 195[31], 196[31], 200[137], 204[31], 478[43], 480[43], 551[346], 851[133], 852[133b,138], 986[10]
Whitfield, F. B., **8**, 542[227]
Whitfield, G. H., **4**, 600[229]
Whitham, G. H., **1**, 776[215], 787[253]; **3**, 380[4], 616[12], 735[18]; **4**, 301[322], 302[322], 354[129], 799[118]; **5**, 255[50]; **6**, 687[374], 901[120]; **7**, 95[79], 470[2]; **8**, 505[74]
Whiting, A., **5**, 376[109]
Whiting, D. A., **2**, 183[20]; **3**, 688[116]; **6**, 2[2], 23[2]; **7**, 131[88], 329[4], 343[4]
Whiting, J., **6**, 923[56]
Whiting, M. C., **3**, 554[18,19]; **4**, 55[156], 519[15], 522[15,48], 526[48]; **6**, 961[73]
Whitlock, B. J., **2**, 842[190]; **3**, 557[40,43]
Whitlock, H. W., **4**, 373[66]
Whitlock, H. W., Jr., **2**, 842[190]; **3**, 557[40,41,43], 693[148], 694[148]; **8**, 530[103]
Whitman, B., **8**, 320[84]
Whitman, G. H., **5**, 475[143]; **6**, 899[108]; **7**, 92[45]
Whitman, P. J., **3**, 905[137]; **5**, 408[31], 488[198]
Whitmire, K. H., **2**, 1066[122]
Whitmore, F. C., **3**, 415[7]; **4**, 71[15], 272[36,38,39], 273[36,38,39], 287[38,39]; **7**, 100[114]; **8**, 328[14], 329[14]
Whitney, C. C., **3**, 483[12]; **4**, 889[137]; **7**, 597[51]; **8**, 716[89], 754[77], 755[132], 756[141]
Whitney, R. A., **3**, 135[363], 136[363], 139[363], 142[363], 156[363]; **4**, 301[316], 303[316], 310[316]; **8**, 852[142], 853[142b], 857[142b]
Whitney, S., **1**, 107[6], 110[6], 428[121], 429[121], 457[121]; **3**, 209[15]; **4**, 184[82]
Whitney, S. E., **5**, 382[122], 580[169]
Whitney, T. A., **1**, 528[115]
Whittaker, G., **3**, 639[84]
Whittaker, M., **2**, 315[45], 316[45]; **4**, 85[70]; **7**, 453[66]; **8**, 95[90]
Whittamore, P. R. O., **4**, 497[97]
Whitten, C. E., **1**, 108[8], 116[8], 359[21], 383[21], 384[21], 426[110], 429[122], 432[135,136]; **2**, 121[187], 489[50], 491[50]; **3**, 8[41], 9[41], 15[41], 20[41], 212[39], 226[39a,d,201], 250[70]; **4**, 77[50], 176[49], 178[49,61], 205[37-40], 245[87], 256[206,211]; **5**, 595[13]
Whittle, A. J., **2**, 166[153], 185[29]; **3**, 342[11]; **4**, 391[176]; **6**, 7[32]; **8**, 798[58], 849[113]
Whittle, J. R., **7**, 498[227]
Whittle, R. L., **7**, 486[143]
Whittle, R. R., **1**, 15[79], 255[18], 755[116], 756[116], 758[116], 761[116]; **5**, 255[52], 260[52], 264[52], 425[99,100], 426[104]; **6**, 900[112], 906[148]
Whittleton, S. N., **8**, 90[49]
Whitworth, S. M., **8**, 90[49]
Whritenow, D. C., **1**, 434[140]; **2**, 249[84], 264[58]
Whybrow, D., **3**, 431[98]
Wibaut, J. P., **4**, 280[122], 285[122]
Wibberley, D. G., **7**, 739[34]
Wiberg, K. B., **1**, 290[21], 321[21], 322[21]; **2**, 829[135], 977[5]; **3**, 381[19], 406[145], 890[34], 892[51], 901[105,110], 905[138]; **4**, 1009[144]; **5**, 66[75,76], 467[118], 675[2], 802[84], 901[25]; **6**, 717[116]; **7**, 12[97,102], 41[17], 85[8], 92[40,47], 99[112], 100[8,112], 235[6], 240[56], 252[5], 558[78], 562[78], 706[21], 851[19]; **8**, 224[111]
Wiberg, K. E., **7**, 703[2], 706[21], 709[2], 710[2], 712[2]
Wiberg, N., **5**, 850[152]; **8**, 472[8,9]
Wiberg, W. B., **5**, 1185[2]
Wicha, J., **1**, 329[39], 806[314]; **2**, 382[315]; **6**, 989[81]; **7**, 649[44]; **8**, 163[42]
Wicher, J., **1**, 162[92]
Wichmann, R., **8**, 204[154]
Wichterle, O., **5**, 418[68]; **8**, 652[75]
Wick, A. E., **5**, 828[6], 836[6], 888[26], 893[26]
Wickenkamp, R., **3**, 495[93b]
Wicker, R. J., **8**, 141[39], 142[39]
Wickham, G., **1**, 610[42]; **2**, 587[140]; **3**, 586[154], 864[21]; **5**, 71[119]; **7**, 616[17]; **8**, 852[141], 857[141]
Wickramaratne, M., **5**, 839[77]
Wickremesinghe, L. K. G., **8**, 341[105]
Wicks, G. E., **7**, 752[147]
Widdowson, D. A., **1**, 463[21,22]; **2**, 204[99]; **3**, 244[24], 464[170], 494[88], 503[148], 514[210], 681[97], 753[99]; **4**, 523[55], 524[55,63-65], 525[55,63-65], 526[55,64], 674[35], 688[35]; **6**, 690[394,399], 691[399], 692[399]; **7**, 123[35], 144[35]; **8**, 977[139]
Widener, R. K., **1**, 492[41], 493[41], 495[41], 822[30]; **4**, 74[40a], 239[22]; **6**, 190[192]
Widera, R., **4**, 387[163a-c]; **6**, 526[393]
Widiger, G. N., **1**, 359[25], 364[25]; **3**, 570[54]
Widlanski, T. S., **5**, 855[186]
Widler, L., **1**, 142[23], 146[44], 148[44], 158[70,71], 165[44]; **2**, 5[17], 6[17], 22[17,17c,87], 23[17b], 630[4]; **8**, 9[50]

Widmer, E., **8**, 205[161], 560[406]
Widmer, U., **2**, 652[127]; **3**, 809[39]; **5**, 709[45], 799[72]
Wiebecke, G. H., **2**, 204[99]
Wieber, G. M., **4**, 572[4]
Wiechert, R., **2**, 167[160], 360[171], 902[40]; **4**, 182[76]; **6**, 718[117]; **7**, 47[55], 74[111,112], 75[111,112], 86[16a], 383[111], 773[305]; **8**, 331[32], 881[72], 882[72]
Wiechman, B., **2**, 728[137]
Wieczorek, J. J., **3**, 816[79]
Wieczorek, J. S., **8**, 391[91]
Wiedeman, P. E., **7**, 255[37]
Wiedeman, W., **4**, 1016[206]
Wiedemann, D., **6**, 555[808]
Wiedemann, W., **5**, 714[69]
Wiedhaup, K., **3**, 361[75]
Wiegand, G. H., **8**, 303[100], 413[124]
Wiegers, K. E., **6**, 955[24]; **8**, 2[12]
Wieglepp, H., **7**, 65[69]
Wiegman, T., **8**, 53[132], 66[132]
Wiegrebe, W., **5**, 410[38]
Wiel, J.-B., **5**, 561[83]; **8**, 543[248,249]
Wieland, D. M., **3**, 223[158], 263[177]; **4**, 170[19]; **6**, 4[19], 9[41], 11[47]
Wieland, H., **4**, 1007[129]; **5**, 451[1]
Wieland, P., **3**, 816[79]; **6**, 685[361], 1059[64,65]; **7**, 41[20]; **8**, 974[124]
Wieland, T., **4**, 124[214b]; **6**, 438[40]
Wiemann, J., **3**, 578[92], 610[92]; **4**, 30[87]; **8**, 527[53], 532[130]
Wiemer, D. F., **2**, 103[33-36]; **3**, 395[102], 396[102]; **8**, 238[23], 261[6], 336[68], 927[21]
Wienand, A., **4**, 980[102]; **5**, 1086[75]
Wienreb, S. M., **2**, 720[85], 1054[59], 1070[59]
Wierenga, W., **2**, 801[35]; **3**, 99[185]; **5**, 603[51], 612[74]; **6**, 116[94]
Wiering, P. G., **5**, 77[263]
Wieringa, J. H., **7**, 98[97]
Wierschke, S. G., **1**, 297[58], 580[2], 581[2], 582[2]
Wiersdorff, W.-W., **5**, 391[143], 721[101]
Wiersema, A. K., **3**, 666[45]
Wiersum, U. E., **5**, 580[166], 584[192]
Wierzba, M., **4**, 83[65c]
Wierzchowski, R., **2**, 455[12]; **3**, 45[251]
Wiesboeck, R., **6**, 951[5]
Wieschollek, R., **5**, 1126[66]
Wiese, D., **8**, 56[168], 66[168]
Wiesemann, T. L., **4**, 240[46]
Wiesenfeld, A. W., **8**, 364[13]
Wieser, K., **7**, 506[293]
Wiesner, K., **5**, 130[38,39], 143[96], 326[24]; **6**, 46[69]
Wiesner, M., **6**, 46[57]
Wiessler, M., **5**, 680[23], 686[42]
Wiest, H., **5**, 491[205]
Wietfeld, B., **1**, 892[149]; **5**, 219[41], 225[41], 226[41,107], 228[41], 229[41], 234[41,140], 235[41]
Wife, R. L., **7**, 40[4], 48[59]
Wigand, P., **2**, 139[27,27a]
Wigfield, D. C., **8**, 2[6,13], 5[6], 26[29], 37[29], 66[29], 334[63]
Wigfield, Y. Y., **1**, 512[37]; **3**, 149[400]
Wiggins, D. W., **6**, 502[211]
Wiggins, J. M., **5**, 474[156]
Wightman, R., **6**, 624[136]
Wightman, R. H., **6**, 625[156]; **7**, 299[50]; **8**, 910[81]
Wightman, R. M., **3**, 577[88]; **7**, 852[40], 854[45]; **8**, 134[37]
Wigley, D. E., **5**, 1145[104]
Wijekoon, D., **7**, 822[34]
Wijers, H. E., **3**, 106[222], 113[222]
Wijesekera, T. P., **2**, 743[86]
Wijkens, P., **3**, 217[81]; **4**, 900[181]
Wijnberg, J. B. P. A., **2**, 1049[17,20], 1050[17,20], 1064[112]; **5**, 402[5]; **6**, 745[80,82], 746[94]; **8**, 273[128]
Wijnen, M. H. J., **7**, 16[165]
Wijsman, A., **3**, 242[10]; **6**, 134[32]
Wikel, J. H., **3**, 926[46], 928[46]
Wikholm, R. J., **3**, 828[48], 829[48]
Wikström, H., **7**, 331[15], 831[64]
Wilberg, K. B., **7**, 530[12]
Wilbey, M. D., **8**, 446[85]
Wilby, A. H., **6**, 119[109]; **8**, 81[4], 91[4], 104[4], 367[57], 440[83], 551[339], 958[19], 959[29]
Wilchek, M., **8**, 297[68]
Wilckens, M., **2**, 68[40]
Wilcock, J. D., **6**, 740[63]; **8**, 273[126]
Wilcott, R. M., **6**, 689[387]
Wilcox, C. F., Jr., **8**, 898[23], 899[23]
Wilcox, C. S., **1**, 744[58], 759[134], 832[106]; **2**, 102[27], 655[131]; **3**, 3[13]; **4**, 732[76], 791[44]; **5**, 841[86,93,104,106,107], 856[208], 857[222], 859[93,234,241], 872[222], 1123[38], 1133[28]; **6**, 858[162], 859[163,164], 978[21]
Wilcox, M., **8**, 593[67]
Wilcsek, R. J., **7**, 603[114]
Wilczynski, R., **8**, 455[205]
Wild, D., **5**, 687[60]; **8**, 398[144]
Wild, H., **1**, 242[46]
Wild, H.-J., **4**, 259[267]; **5**, 768[126], 779[126]
Wild, J., **2**, 157[120]; **8**, 544[275]
Wild, S. B., **4**, 520[30], 665[8], 670[8], 674[8], 688[65]
Wild, U. P., **1**, 286[11]
Wilde, H., **3**, 890[33]
Wilde, P. D., **3**, 840[15]; **5**, 595[16], 596[16]
Wilde, R. G., **2**, 556[150]; **5**, 436[158,158g], 437[158f], 438[158d], 442[158], 532[85]
Wildeman, J., **4**, 14[47,47k]; **6**, 538[549]
Wilder, L., **1**, 180[31]
Wildman, W. C., **8**, 968[86]
Wilds, A. L., **1**, 144[40]; **2**, 838[163,175]; **3**, 810[51], 888[19], 891[19], 900[19]; **8**, 88[37]
Wildsmith, E., **6**, 83[79]; **8**, 371[104], 393[112]
Wilen, S. H., **5**, 88[50]
Wiley, D. W., **3**, 888[21]; **5**, 76[232]
Wiley, G. A., **6**, 205[25,26], 210[25]
Wiley, J. R., **7**, 713[74]
Wiley, M. R., **1**, 296[55], 769[194]; **4**, 744[129]
Wiley, P., **4**, 51[144c]
Wiley, P. F., **4**, 1095[154]; **5**, 491[207]
Wiley, R. H., **2**, 345[38]; **4**, 51[144c]; **5**, 752[41]; **7**, 720[8]
Wilgus, H. S., III, **3**, 747[72]
Wilhelm, D., **1**, 29[147]; **4**, 597[169], 621[169], 637[169]; **8**, 724[168]
Wilhelm, E., **5**, 598[33]; **6**, 195[225], 558[841]
Wilhelm, H., **8**, 771[43]
Wilhelm, M., **8**, 659[103]
Wilhelm, R. S., **1**, 107[5], 110[5], 124[80], 131[5], 428[121], 429[121], 457[121]; **2**, 119[166], 120[183,184]; **3**, 213[46,46c,54], 214[56,57], 216[46], 223[46], 250[72,73], 264[72,186], 265[72], 491[70]; **4**, 148[49], 170[3], 176[3], 178[3,62], 180[62], 196[3], 197[3], 256[214,215]; **5**, 249[36], 931[186]; **6**, 4[22], 9[22], 10[22]
Wilk, M., **5**, 478[162]
Wilka, E.-M., **1**, 70[63], 141[22]; **2**, 120[179]; **3**, 125[290]; **4**, 229[238]
Wilke, G., **3**, 423[81], 587[150]; **4**, 601[245], 888[132], 889[132,137]; **5**, 35[11], 809[113], 1142[86], 1197[39]; **8**, 735[12,13], 738[30], 740[12,13,30], 741[13], 747[56], 752[56], 753[30], 756[13], 757[13]
Wilke, M., **7**, 358[14]

Wilke, S., **5**, 429[115]; **6**, 117[97]
Wilkening, D., **3**, 849[55]; **6**, 94[140]
Wilker, J. C., **4**, 439[168]
Wilkes, M. C., **5**, 1025[84]
Wilkie, C. A., **1**, 4[28]
Wilkins, C., **7**, 446[63]
Wilkins, C. K., Jr., **5**, 404[14]
Wilkins, C. W., Jr., **3**, 438[35]; **4**, 868[17], 869[17], 877[67]
Wilkins, J. M., **1**, 116[46,49], 118[46,49]; **2**, 120[169]; **3**, 8[43], 249[63]; **4**, 91[88b], 170[23], 178[23]
Wilkins, R. F., **2**, 901[39], 948[183], 959[31], 960[31], 962[45], 964[45,60,61], 965[63], 966[61,71], 967[61,63,71]
Wilkinson, D. L., **5**, 568[110]
Wilkinson, G., **1**, 2[7,8], 125[84], 139[3], 140[7], 193[88], 211[2], 212[2], 214[2,23], 215[31,31a], 222[2], 225[2], 231[1], 416[66], 422[94], 428[121], 429[121], 440[169], 451[216], 457[121]; **2**, 712[42]; **3**, 208[1,11], 210[11,11a], 219[11a], 228[214], 234[11a], 436[9,13], 524[33]; **4**, 518[2], 521[2], 586[13,14], 587[17], 590[13,14], 614[13], 663[2], 689[71], 735[84], 770[84], 887[123], 888[123], 914[1], 922[1], 925[1], 926[1], 932[1], 939[73], 941[1], 943[1]; **5**, 46[39], 56[39], 272[1], 641[131], 1115[1], 1116[1], 1126[1d], 1134[43]; **6**, 831[11], 832[12], 848[11], 865[12]; **7**, 108[171], 335[28], 358[8a], 594[5], 595[5], 598[5], 614[3], 629[48], 816[6a,b], 824[6], 825[6], 827[6a], 829[6a], 831[6a], 832[6a], 833[6a], 844[59]; **8**, 99[110], 100[114], 152[163,165,166,167], 278, 375[153], 443[1,2,5], 444[5b], 445[5,21,23,27,28,33,56-58], 446[68], 449[5a], 452[5b,57,58,184], 456[5a,21], 524[13], 568[462,478,479], 674[33], 708[42], 715[42], 717[42], 728[42], 734[3], 747[3], 753[3], 759[3], 851[132]
Wilkinson, G. W., **1**, 580[1]
Wilkinson, J., **4**, 588[53]
Wilkinson, J. B., **3**, 735[21]
Wilkinson, J. M., **5**, 125[19], 128[19,30], 134[30]
Wilkinson, P. A., **6**, 675[299]
Wilkinson, R. W., **8**, 973[120]
Wilkinson, S. G., **6**, 2[1], 28[1]; **7**, 41[18], 84[1], 85[1], 108[1]
Wilkinson, S. P., **5**, 1130[6]
Wilks, H. M., **8**, 206[172]
Wilks, T. J., **4**, 33[95,95a]
Will, B., **7**, 384[114c], 399[38], 400[38], 406[38], 409[38], 415[38]
Will, S. G., **7**, 356[51]
Willard, A. K., **1**, 126[88]; **2**, 101[20], 102[20], 182[9], 200[88], 604[55], 935[150]; **5**, 828[9], 840[9], 841[9,9c], 847[9], 856[9], 859[9], 886[20], 893[20], 1001[12]; **6**, 858[162], 860[178]
Willard, G. F., **6**, 1016[27]
Willard, J. E., **3**, 299[33]
Willard, K. E., **8**, 47[124], 66[124]
Willard, N., **2**, 718[81]
Willard, P. G., **4**, 313[473]
Willcott, M. R., III, **5**, 714[65], 900[11], 901[11], 906[11], 907[11], 910[11]
Wille, G., **6**, 619[116]
Willem, R., **3**, 587[148]
Willenz, J., **3**, 892[50]
Willert, I., **3**, 128[322], 130[322], 134[322]
Willey, F. G., **6**, 968[117]
Willey, P. R., **1**, 786[249]; **8**, 842[47]
Willfahrt, J., **6**, 67[9], 72[9]
Willhalm, B., **3**, 572[64]; **8**, 109[10], 110[10], 112[10], 116[10], 120[10]
Willi, A. V., **6**, 950[1]
Willi, M. R., **8**, 388[62]
Williams, A., **8**, 864[243]
Williams, A. C., **5**, 1183[58]
Williams, A. D., **3**, 600[216]; **4**, 17[54], 63[54]; **6**, 1053[46]; **8**, 843[59c]
Williams, A. L., **7**, 78[127]
Williams, A. R., **7**, 760[37], 761[37]
Williams, B. J., **6**, 671[276]
Williams, C. C., **3**, 145[386]; **6**, 135[24]
Williams, C. H., **6**, 420[24], 451[127]
Williams, C. N., **2**, 963[55]
Williams, D., **7**, 444[55]
Williams, D. H., **2**, 346[44]; **6**, 690[395]; **8**, 344[119], 345[119]
Williams, D. J., **1**, 832[114]; **2**, 204[99], 742[77]; **3**, 902[119], 934[64], 953[64]; **4**, 379[115], 380[115g], 381[126b], 382[126], 383[126], 390[175a], 391[115g]; **5**, 374[107], 376[107b], 528[67], 534[92]; **6**, 842[79], 995[103]; **7**, 112[198], 132[92], 133[92], 134[92], 352[33], 523[48], 820[26]; **8**, 847[100b]
Williams, D. L., **3**, 125[305], 126[305], 127[305]; **8**, 566[457], 568[468]
Williams, D. L. H., **4**, 366[5]; **7**, 493[194], 500[194], 746[84]
Williams, D. R., **1**, 109[15], 205[105], 514[53-55], 795[282], 820[13], 823[13]; **3**, 31[185], 32[185], 36[209], 37[212]; **4**, 380[119], 403[238], 404[238,244], 405[238], 406[238], 893[155]; **5**, 527[63], 534[90], 535[90], 769[136]; **6**, 25[100], 723[147], 725[171,173], 726[175], 728[171], 842[78], 997[113], 1030[108]; **7**, 131[79], 160[54], 274[136], 300[54], 410[97b], 503[276], 544[38], 551[38]; **8**, 846[86]
Williams, E., **6**, 439[71], 466[43], 656[171]; **7**, 674[45]
Williams, E. A., **5**, 474[158]
Williams, E. G., **2**, 553[130], 554[130]; **7**, 237[35]
Williams, E. H., **8**, 794[15]
Williams, F. J., **3**, 905[138]
Williams, G. D., **4**, 983[119]; **5**, 1086[74]
Williams, G. H., **3**, 505[162], 507[162], 512[162]
Williams, G. J., **6**, 716[94]; **7**, 174[139], 710[51]; **8**, 726[188]
Williams, G. M., **4**, 696[5]; **8**, 677[60], 844[66]
Williams, H. J., **4**, 189[103]; **8**, 543[243], 545[280]
Williams, H. W. R., **7**, 750[126]
Williams, I. D., **1**, 314[129]; **7**, 421[138], 424[138]
Williams, I. H., **1**, 314[136,137], 315[136,137]; **8**, 89[43]
Williams, J. C., Jr., **2**, 482[28], 483[28]
Williams, J. E., **6**, 133[4]
Williams, J. E., Jr., **1**, 506[7]
Williams, J. F., **2**, 456[58,63], 458[58,63], 465[103]
Williams, J. G., **5**, 731[130c]; **6**, 899[108]
Williams, J. H., **3**, 572[66]; **8**, 312[22], 321[22], 566[450]
Williams, J. K., **5**, 76[232]
Williams, J. L., **5**, 925[156]
Williams, J. L. R., **8**, 496[34]
Williams, J. M., **1**, 109[14], 127[92], 427[114]; **2**, 1[1], 225[155]; **3**, 18[97], 42[97], 43[97]; **5**, 96[122], 98[122]; **7**, 390[9], 401[61b], 407[61b]; **8**, 431[66], 459[226]
Williams, J. O., **5**, 418[70]
Williams, J. R., **2**, 553[126]; **3**, 390[85], 392[85], 693[142]; **5**, 131[45], 223[67-69,71], 596[28], 597[28]; **6**, 737[38], 939[141], 942[141], 1029[102]; **7**, 767[196]
Williams, J. W., **8**, 301[89]
Williams, K. J., **4**, 313[467]
Williams, L., **1**, 488[5], 492[5], 495[44], 498[49]; **3**, 199[66]; **7**, 595[27]
Williams, L. H., **6**, 104[6]
Williams, M. E., **3**, 505[168]
Williams, M. J., **4**, 405[248]; **8**, 856[184]
Williams, M. T., **7**, 743[65]
Williams, N. R., **3**, 734[12]; **8**, 514[112], 932[43]
Williams, P. A., **4**, 688[65]
Williams, P. D., **5**, 249[36], 531[76]
Williams, P. H., **2**, 139[31]; **5**, 432[130], 433[130b]; **7**, 446[70]
Williams, R. B., **1**, 373[92], 375[92], 376[92]
Williams, R. E., **4**, 571[1], 572[1]
Williams, R. G., **5**, 715[82], 739[82]

Williams, R. M., **1**, 123[74], 373[92], 375[92], 376[92], 404[23]; **2**, 582[107], 649[104], 1052[48], 1075[48], 1076[48,152]; **3**, 277[26], 790[61]; **6**, 960[52]; **7**, 183[167], 226[68], 230[125], 399[33], 551[53]; **8**, 655[86]
Williams, R. O., **5**, 791[27], 799[27], 829[25]; **8**, 338[88]
Williams, R. V., **3**, 746[68]; **6**, 1003[134]
Williams, R. W., **7**, 545[29]
Williams, T. H., **7**, 678[72]
Williams, T. M., **2**, 323[25,29], 333[25]
Williams, T. R., **8**, 410[91]
Williams, T. W., **3**, 690[124]
Williams, V. Z., Jr., **3**, 334[220]
Williams, W. G., **4**, 785[23]
Williamson, D., **7**, 821[31]
Williamson, D. H., **4**, 665[11], 666[11], 667[11], 669[11]; **5**, 916[119]; **8**, 139[2], 152[2], 154[2], 443[1], 447[1a]
Williamson, H., **3**, 511[189]
Williamson, K. L., **3**, 739[43]; **4**, 24[72,72b]; **5**, 347[72,72d,e]; **7**, 167[97,100]; **8**, 898[25], 899[25]
Williamson, M., **5**, 1202[56]
Williamson, R., **2**, 740[61], 756[6], 760[6]
Williamson, S. A., **2**, 1079[158]; **5**, 467[116], 528[68], 531[68,80]
Williams-Smith, D. L., **8**, 890[141]
Williard, P. G., **1**, 10[55], 19[105], 26[135,136], 27[138,139], 28[141], 29[149], 30[152], 32[157], 37[176,181], 38[184], 41[201], 43[136,141]; **2**, 100[5-8], 1096[98]; **3**, 58[291], 751[87]; **4**, 520[38], 542[38]; **5**, 137[82,83], 221[60], 1102[147]; **6**, 937[117], 939[117], 940[117]; **7**, 362[28,30]; **8**, 543[242], 940[106]
Willis, B. J., **2**, 784[39a]; **4**, 259[276], 1093[149]
Willis, C. J., **7**, 488[162]
Willis, C. L., **4**, 868[19]; **7**, 826[46]; **8**, 537[182]
Willis, H. B., **3**, 415[6]
Willis, J. P., **2**, 1018[40]; **6**, 751[106]; **7**, 264[89], 275[89]
Willis, P. A., **4**, 820[216]
Willis, W. W., Jr., **1**, 632[65,69], 633[65,69], 635[69,85], 636[65,85,103], 638[65], 640[85], 642[85], 643[85], 644[65], 645[65], 646[65], 647[65], 648[65,103], 669[65], 672[65,85,103], 682[85], 695[65], 700[65,85], 705; **2**, 76[87]; **3**, 87[100,107,109], 104[100], 105[107], 106[107], 110[100], 114[107], 117[100], 120[107], 157[107]; **6**, 966[97], 1028[95]; **7**, 770[249], 771[282]; **8**, 409[84]
Willison, D., **3**, 1025[33,33a]; **5**, 1043[25], 1048[25a], 1051[36a], 1055[45], 1056[25a], 1062[45]
Willman, K., **4**, 682[58]
Willmes, A., **6**, 189[187]
Willmott, W. E., **4**, 1032[10], 1063[10,10a]
Willner, I., **8**, 97[96]
Willnow, P., **2**, 456[45]
Wills, G. O., **3**, 721[5]
Wills, K. D., **8**, 626[172]
Wills, M., **5**, 362[93], 363[93g]
Wills, M. T., **5**, 787[10]
Willson, C. G., **4**, 310[432]
Willson, T., **6**, 1006[145]
Willson, T. M., **1**, 340[93], 615[52], 616[52], 694[238], 697[238], 698[238]
Willy, W. E., **6**, 212[81]; **8**, 542[228]
Wilms, H., **7**, 581[142]
Wilputte-Steinert, L., **8**, 451[182]
Wilshire, C., **1**, 861[73]
Wilshire, J. F. K., **3**, 329[186]
Wilson, A., **5**, 63[4]
Wilson, A. N., **2**, 284[56]; **7**, 306[7]
Wilson, B. D., **8**, 965[64]
Wilson, B. M., **6**, 675[299]; **8**, 987[23]
Wilson, C. A., II, **7**, 771[271]
Wilson, C. L., **3**, 706[5]
Wilson, C. V., **7**, 718[1], 731[1]; **8**, 251[105]
Wilson, D. A., **7**, 294[17]
Wilson, D. M., **2**, 735[13]
Wilson, D. R., **1**, 49[3], 50[3]
Wilson, E. A., **7**, 759[6]
Wilson, E. R., **3**, 422[70], 635[39]
Wilson, F. B., **3**, 382[36]
Wilson, F. G., **5**, 752[51]
Wilson, G. E., Jr., **2**, 805[44], 815[44]; **7**, 207[73]
Wilson, G. S., **7**, 765[161]
Wilson, J. D., **6**, 546[646]; **7**, 798[23]
Wilson, J. G., **7**, 666[73,76]
Wilson, J. M., **4**, 426[42], 427[42]; **6**, 799[24]
Wilson, J. S., **8**, 819[43], 820[43]
Wilson, J. W., **1**, 488[5-7,10,11], 490[25], 492[5,36,37,39], 494[37], 495[44,46], 498[51], 499[51], 501[36,37], 502[39]; **2**, 57[4]; **3**, 199[65,66]; **6**, 211[79]; **7**, 595[27]; **8**, 170[93]
Wilson, J. Z., **1**, 248[66], 742[49]; **2**, 911[69]; **5**, 435[150]
Wilson, K. D., **1**, 480[158]; **4**, 121[205d,e]
Wilson, K. E., **8**, 16[99], 542[230], 543[230]
Wilson, K. J., **6**, 450[118]
Wilson, L. A., **5**, 468[135]
Wilson, M., **8**, 198[132]
Wilson, M. A., **8**, 941[112]
Wilson, M. E., **5**, 1131[14], 1173[32]
Wilson, P., **5**, 181[155]
Wilson, R., **8**, 389[72]
Wilson, R. M., **5**, 154[33,34]
Wilson, S. E., **4**, 887[128,130,130b]; **8**, 483[54], 485[54]
Wilson, S. L., **7**, 169[109]
Wilson, S. R., **1**, 894[160]; **2**, 106[46]; **3**, 380[11], 616[13]; **4**, 290[206], 295[253], 398[218,218c], 399[218a-c]; **5**, 523[48], 524[54,54f], 526[57], 534[54], 605[59], 788[12], 829[18], 835[60], 847[135], 954[298], 1001[14], 1002[18], 1003[21], 1016[63]; **6**, 865[202]; **7**, 428[148g], 625[38]; **8**, 514[105], 545[282]
Wilson, S. T., **8**, 446[66]
Wilson, T., **1**, 340[93]; **2**, 564[9], 568[32], 630[8]
Wilson, T. M., **2**, 564[9], 568[32], 630[8], 655[145]
Wilson, W., **7**, 120[15]; **8**, 526[34]
Wilt, J. W., **3**, 714[33], 853[71]; **7**, 648[39]; **8**, 476[26], 941[114], 942[115]
Wilt, M. H., **7**, 544[37]
Wilton, D. C., **8**, 561[412]
Wilwerding, J. J., **5**, 341[58], 520[38]
Wilzbach, K. E., **5**, 646[3,8], 662[35]
Wimalasena, K., **7**, 99[108]
Wimmer, E., **1**, 92[64]; **5**, 857[227]
Winans, C. F., **8**, 143[56], 144[66]
Winberg, H. E., **8**, 652[77]
Winch, B. L., **3**, 790[58]
Winckler, H., **5**, 451[4]
Wincott, F. E., **2**, 578[84], 701[85]; **4**, 33[96], 34[96e]; **7**, 237[37]
Wincott, F. J., **1**, 103[95]
Windaus, A., **6**, 685[357]
Winders, J. A., **8**, 198[130]
Windholz, T. B., **6**, 659[197a]; **8**, 495[28]
Windhövel, U. F., **6**, 734[5]
Windle, J. J., **3**, 666[45]
Wineman, R. J., **8**, 991[49]
Wing, R. E., **7**, 235[1]
Wing, R. M., **4**, 608[320], 646[320]; **5**, 925[156]
Wingard, L. B., Jr., **2**, 1104[132]
Wingard, R. E., Jr., **3**, 867[33], 873[33]; **5**, 609[67]; **6**, 247[131]
Wingbermühle, D., **1**, 234[22], 253[8], 749[78], 816[78]
Wingfield, M., **2**, 852[233]
Wingler, F., **1**, 215[34]

Wing-Por Leung, **1**, 17[219], 36[234]
Wing-Wah Sy, **7**, 502[264]
Winiarski, J., **4**, 424[15], 426[15], 431[15], 432[15,114]
Winiski, A. P., **8**, 70[233]
Winitz, M., **8**, 145[88]
Wink, D. J., **8**, 447[130]
Winkelman, D. V., **7**, 761[62]
Winken, T., **5**, 161[62]
Winkhaus, G., **4**, 689[71]
Winkler, H. J. S., **4**, 519[13], 520[13]
Winkler, J., **5**, 418[71]; **7**, 505[284]
Winkler, J. D., **2**, 1010[8]; **3**, 24[144], 58[291], 218[97], 380[9]; **4**, 24[73,73a,b], 791[50], 801[125], 820[50]; **5**, 134[64,66], 137[66,82,83], 141[92]
Winkler, M., **5**, 1144[98]
Winkler, P., **6**, 531[428]
Winkler, R. R., **2**, 153[109]
Winkler, T., **1**, 830[96]; **2**, 534[36]; **3**, 381[32]; **4**, 754[175]; **5**, 439[166], 837[67], 856[67], 857[67]; **6**, 487[73], 489[73]
Winn, L. S., **3**, 539[100]
Winnick, F. M., **5**, 21[148]
Winnik, M., **7**, 42[32]
Winstein, S., **3**, 406[142], 653[225], 706[6]; **4**, 367[11], 386[153], 972[78]; **5**, 585[202], 702[10], 716[10], 906[63]; **6**, 284[173]; **7**, 92[38,50], 94[56]
Winter, B., **5**, 217[25], 222[25]
Winter, C. A., **6**, 219[123]
Winter, M., **6**, 612[70], 1059[64]
Winter, M. J., **1**, 310[109]; **4**, 976[100]; **5**, 1144[101], 1145[104]
Winter, R., **3**, 896[67]
Winter, R. A. E., **4**, 349[109]; **6**, 686[372], 982[49]
Winter, R. E. K., **3**, 620[29]; **5**, 678[17], 680[24], 683[24c]; **7**, 228[93]; **8**, 374[147]
Winter, S. R., **3**, 1024[27], 1028[51]
Winter, W., **7**, 777[364]
Winter, W. J., **5**, 74[208]
Winterfeldt, E., **1**, 753[101]; **2**, 381[308], 651[121], 1026[67], 1028[67]; **4**, 3[10], 41[10], 47[10], 48[137,137a], 52[147,147e], 55[156,156n], 65[10e], 66[10,10e], 200[4], 211[88,89,91], 222[177,178]; **5**, 945[247]; **6**, 501[204], 740[63], 745[81], 746[93], 834[36], 855[36]; **7**, 679[74,74b]; **8**, 58[176], 66[176], 100[115], 244[71], 247[71], 251[71], 253[71], 266[58,59], 269[59], 279, 543[239]
Winternitz, F., **3**, 854[77]; **7**, 71[100]
Wintgens, V., **5**, 639[122]
Wintner, C., **7**, 482[118]
Winton, P. M., **3**, 416[22]
Wintrop, S. O., **8**, 654[83]
Winwick, T., **7**, 356[51]
Winzenberg, K., **1**, 129[93], 779[224]; **2**, 651[120]; **5**, 342[63]; **6**, 807[58]
Wipf, D. O., **7**, 854[45]
Wipf, P., **3**, 994[43], 996[43], 1000[43]; **5**, 482[172], 842[108], 843[108], 844[130], 859[238], 863[238]; **6**, 540[582], 834[35], 855[35], 858[162]
Wipff, G., **1**, 49[8], 631[59], 632[59]; **4**, 538[104], 539[104]; **5**, 468[127]; **6**, 133[4]; **8**, 3[21], 89[43]
Wippel, H. G., **1**, 761[138], 773[204]
Wireko, F., **5**, 211[62,63]
Wiriyachitra, P., **7**, 330[11], 774[334]
Wirkus, M., **6**, 531[430]
Wirth, D. D., **5**, 520[39], 704[22], 1020[69], 1023[69]; **7**, 879[148], 880[148]
Wirth, E., **3**, 829[54]
Wirth, M. M., **7**, 10[79]
Wirth, R. K., **4**, 558[18]
Wirth, R. P., **6**, 667[239]
Wirth, U., **6**, 247[133]
Wirth, W., **1**, 466[45]
Wirth, W.-D., **3**, 414[1]
Wirth-Peitz, F., **2**, 464[93]
Wirthwein, R., **4**, 491[68], 492[72]
Wirtz, K. R., **5**, 689[75]
Wirz, B., **8**, 185[10]
Wirz, J., **4**, 300[307]
Wischhöfer, E., **2**, 1102[127]
Wise, D. S., **2**, 555[140]
Wise, S., **5**, 67[93], 70[116], 514[9,10], 527[9]
Wiseman, J. R., **5**, 394[147], 634[73]; **6**, 982[51]; **7**, 123[34], 355[46]
Wishka, D. G., **1**, 343[108]
Wisloff Nilssen, E., **1**, 488[8]
Wismontski-Knittel, T., **5**, 729[123]
Wisnieff, T. J., **4**, 1060[161], 1063[161a]
Wisotsky, M. J., **6**, 263[21]
Wisowaty, J. C., **2**, 759[31]
Wissinger, J. E., **1**, 881[119]; **5**, 806[103], 1026[86]; **7**, 673[20]
Wissocq, F., **4**, 95[102f]
Wistrand, L.-G., **2**, 1051[34], 1066[121]
Wiszniewski, V., **1**, 564[190]
Witham, G. H., **4**, 301[321]
Withers, G. P., **3**, 759[131,132], 760[135]
Witkop, B., **2**, 1095[93]; **3**, 804[8], 810[46]; **6**, 914[28]; **7**, 546[31]
Witt, J. R., **1**, 33[164]
Witt, K. E., **3**, 588[158]
Witte, H., **7**, 601[86,88]
Witte, J. F., **3**, 201[81]
Witte, K., **8**, 364[16]
Witte, L., **2**, 354[114], 357[114]
Wittek, P. J., **2**, 171[178,179], 173[180], 832[153]
Witten, C. H., **8**, 301[89]
Wittenberg, D., **3**, 381[23], 382[23], 565[17]
Wittenberger, S., **3**, 960[118]; **5**, 436[158,158g], 442[158]; **6**, 898[105], 905[144]
Wittenbrink, C., **3**, 644[164]
Wittenbrook, L. S., **7**, 212[100]
Wittereen, J. G., **4**, 18[62], 20[62h]
Witteveen, J. G., **5**, 768[132]; **6**, 714[87]
Witthake, P., **2**, 780[9]
Wittig, G., **1**, 213[17], 215[34]; **2**, 124[205], 182[5], 476[5], 477[6,7], 478[7], 482[23], 483[23]; **3**, 914[7], 915[13], 924[7,41], 965[13], 975[1]; **4**, 70[10], 96[103a], 98[103a], 140[6,7b], 483[2], 484[2], 488[55], 492[69], 500[103], 869[21], 872[43], 962[38]; **5**, 69[99,100], 379[113b], 380[113b,d,114,115], 381[115]; **6**, 685[362], 719[128], 720[128], 961[66], 967[103], 968[115,116]; **8**, 273[123], 298[72], 885[102]
Wittig, U., **5**, 291[43], 1190[29], 1192[29]
Wittle, E. L., **2**, 759[33]
Wittman, G., **8**, 418[5], 420[5], 423[5], 439[5], 441[5], 442[5]
Wittman, M. D., **1**, 131[99], 777[219]; **3**, 1007[64]; **5**, 827[2], 829[2], 867[2b]; **7**, 546[30], 580[30], 737[12]
Wittmann, M. D., **6**, 859[174], 878[40], 883[40]
Wittmann, R., **6**, 608[49], 610[61]
Wityak, J., **8**, 648[59]
Witz, M., **7**, 747[95]
Witzel, B., **6**, 487[57], 489[57]
Witzel, D., **6**, 233[43]
Witzel, T., **5**, 159[51], 189[51]
Witzeman, J. S., **5**, 805[100]
Witzgall, K., **4**, 115[182,182e]; **6**, 172[11], 179[127]
Wix, G., **7**, 70[93]
Wladislaw, B., **3**, 644[148,148a]
Wladkowski, B. D., **6**, 245[124]

Wlassics, I. D., **6**, 455[149]
Wlostowska, J., **6**, 575[971]
Wnuk, S., **2**, 348[62], 354[62]; **6**, 936[110]
Wocholski, C. K., **8**, 242[41]
Woderer, A., **8**, 563[436]
Woell, J. B., **1**, 308[96]; **3**, 1028[50], 1030[60]; **4**, 939[79], 941[79]
Woerde, H. W., **8**, 150[139]
Woessner, W. D., **3**, 124[272], 125[272], 126[272], 144[383], 145[383,386]; **4**, 11[36]; **6**, 135[24]
Woff, E., **5**, 595[20], 596[20]
Wogan, G. N., **3**, 499[121]
Woggon, W.-D., **6**, 538[571]; **7**, 86[15]
Wohl, A., **3**, 640[104]
Wohl, R. A., **2**, 1023[56], 1026[56]; **6**, 271[90]; **7**, 506[299]
Wöhler, F., **3**, 822[5], 834[5]; **6**, 233[42]
Wohllebe, J., **3**, 851[63]
Wojcicki, A., **5**, 272[2], 275[2], 277[13]
Wojciechowski, K., **4**, 433[117,118]; **6**, 84[85], 533[485], 645[97]
Wojciechowski, M., **6**, 558[859]
Wojcik, B., **8**, 814[18]
Wojnarowski, T., **8**, 756[146]
Wojtkowski, P., **4**, 145[34]
Wojtkowski, P. J., **2**, 111[87]
Wojtkowski, P. W., **2**, 242[15,17]
Wokaun, A., **1**, 286[11]
Wolanin, D. J., **5**, 669[45]
Wolber, G. J., **1**, 476[126]; **3**, 66[15], 74[15]
Wolcke, U., **2**, 740[62]; **7**, 678[69]
Wolcott, R. G., **6**, 277[126]
Wolczanski, P. T., **7**, 3[8]; **8**, 675[50], 676[51,53]
Wold, S., **2**, 1099[115]; **3**, 721[5]
Wolf, A., **3**, 382[38]; **6**, 602[2]
Wolf, A. P., **3**, 1040[109]; **4**, 445[203,204,206]; **5**, 1148[119,120]; **8**, 344[123]
Wolf, D. E., **2**, 284[56]; **7**, 778[414]
Wolf, F. J., **7**, 710[49]
Wolf, G., **2**, 64[24]; **8**, 568[471]
Wolf, G. C., **4**, 771[250]
Wolf, H., **3**, 348[28], 373[130]; **6**, 500[179]
Wolf, H. J., **7**, 67[78], 68[81]
Wolf, H. R., **5**, 221[62]
Wolf, J. F., **7**, 248[110], 801[44]
Wolf, K., **6**, 518[332]
Wolf, K. U., **8**, 57[172], 66[172]
Wolf, M., **4**, 89[84b]
Wolf, N., **8**, 812[5]
Wolf, U., **4**, 54[152]
Wolf, W., **3**, 505[164], 507[164,173], 512[164], 515[164]; **6**, 182[141]
Wolf, W. A., **4**, 1033[33]
Wolfarth, E. F., **5**, 854[174]
Wolfbeis, O. S., **2**, 362[184]; **6**, 553[704]
Wolfe, J. F., **2**, 166[152], 189[54]; **4**, 426[59], 452[16], 455[43], 456[47], 457[49], 458[65], 462[65,104], 463[43,110], 464[118], 465[65,104,116,118-120], 466[65,104,119], 467[119,130,131], 468[104,110], 469[104,110,116,134], 472[134], 473[134], 475[134], 477[49], 503[125]; **6**, 2[4], 110[56], 438[61]
Wolfe, J. R., **5**, 637[108]
Wolfe, J. W., **4**, 457[50], 477[50]
Wolfe, S., **1**, 506[8,11], 512[42,43], 528[116,117]; **2**, 414[14]; **3**, 147[399], 154[418], 155[418]; **4**, 314[481,482], 603[274]; **6**, 133[4], 207[47], 924[62], 926[66]; **7**, 92[47], 94[56], 196[11], 199[11], 236[28], 237[28], 571[115], 768[204], 844[57]
Wölfel, C., **1**, 368[63], 391[63]
Wölfel, G., **2**, 1004[58], 1005[58]; **6**, 83[81]
Wolff, C., **6**, 979[27]; **8**, 342[110], 912[92]
Wolff, H., **6**, 795[10], 798[10], 817[10], 820[10]; **7**, 666[71], 690[13]
Wolff, J. J., **2**, 1072[140]; **7**, 318[60]
Wolff, L., **3**, 887[1a], 890[1], 891[1], 893[1a], 894[1], 900[1a]; **8**, 328[2], 336[2]
Wolff, M. A., **8**, 704[8]
Wolff, M. E., **4**, 814[188]; **7**, 16[161], 236[14-16]
Wolff, O., **5**, 451[5]
Wolff, R. E., **3**, 30[175]
Wolff, S., **3**, 19[103], 901[115,116]; **4**, 611[343], 797[104]; **5**, 21[144], 133[56], 136[56,69-72], 141[56], 164[75,76], 176[75], 918[127]; **6**, 93[133], 836[58]; **7**, 140[131]; **8**, 248[83]
Wolff, W. A., **6**, 125[147]
Wolfram, J., **7**, 691[20]
Wolfrom, M. L., **1**, 55[26], 153[57]; **6**, 660[202]; **8**, 269[85,86], 293[47]
Wolgemuth, R. L., **1**, 554[101]
Wolin, R. L., **4**, 809[164]
Wolinsky, J., **2**, 23[89], 159[128], 529[20], 710[23]; **3**, 849[60], 850[61], 1050[17]; **4**, 30[89]; **5**, 3[25]
Wollenberg, R. H., **2**, 588[152]; **3**, 247[49], 250[49]; **4**, 143[18]; **7**, 26[59], 620[27]
Wollheim, R., **3**, 890[35], 903[127]
Wollmann, T. A., **2**, 112[101], 258[48,50], 261[48]; **8**, 159[108], 171[108], 178[108], 179[108]
Wollowitz, S., **5**, 948[290]; **6**, 154[148], 903[136], 904[141], 1026[88], 1027[88]; **7**, 770[249]; **8**, 409[84]
Wollthan, H., **4**, 868[13]
Wollweber, H., **5**, 316[2], 318[2], 426[105], 428[105], 429[105], 451[11], 513[2], 518[2], 594[1], 601[1], 604[1]
Wolman, Y., **2**, 1094[88,89], 1095[89]
Wolner, D., **7**, 47[57]
Wolochowicz, I., **3**, 565[24], 570[24], 583[24]; **8**, 797[36]
Wolowyk, M. W., **8**, 92[68]
Wolschann, P., **2**, 346[47]
Wolsieffer, L. A., **8**, 880[71], 881[71]
Wolter, A., **6**, 625[160]
Woltermann, A., **1**, 645[124], 669[124], 680[124], 755[113]; **3**, 105[214], 509[196], 512[196]; **4**, 120[200]
Woltersdorf, O. W., Jr., **5**, 780[201]
Wolterson, J. A., **5**, 603[51], 612[74]
Wong, A. C., **5**, 581[173]
Wong, C. F., **6**, 914[26]; **7**, 222[40]
Wong, C. H., **2**, 455[20], 456[20,24,27,34,49-51], 459[50], 460[49,51], 461[49,50], 462[50,51], 463[34,79,91], 464[91]; **6**, 49[94], 665[228]; **7**, 79[132], 80[132], 312[33], 316[45]; **8**, 183[2,3], 185[2,12], 187[34]
Wong, C. K., **1**, 699[247,254]; **3**, 107[225,226], 109[226], 386[60]; **4**, 372[55], 398[216], 399[216a], 401[216a], 405[216a], 410[216a]; **6**, 470[58]; **7**, 495[211]; **8**, 847[97,97d], 849[97d,107,110,115]
Wong, C. L., **7**, 852[37]
Wong, C. M., **1**, 563[186]; **2**, 762[55]; **3**, 854[73]; **4**, 12[42]; **6**, 134[17]; **7**, 760[48], 765[150]; **8**, 371[107], 987[20]
Wong, C. S., **8**, 447[116], 847[94,95]
Wong, F., **8**, 337[81]
Wong, G. S. K., **4**, 792[60], 795[86]
Wong, H., **3**, 168[491], 169[491], 171[491]; **4**, 398[216], 399[216d]; **6**, 538[573], 923[58]; **8**, 303[97]
Wong, H. M. C., **5**, 63[15]
Wong, H. N. C., **5**, 123[1], 126[1], 221[58], 226[58,112], 900[12], 901[12], 903[12], 905[12], 907[12], 913[12], 921[12], 926[12], 943[12], 1006[33]; **7**, 815[3,4], 824[3,4], 833[3,4]; **8**, 884[98], 885[98]
Wong, H. S., **4**, 629[416]
Wong, J., **2**, 353[100]; **4**, 262[311]; **7**, 31[87]
Wong, J.-W., **5**, 541[111]
Wong, J. Y., **4**, 738[98]; **8**, 800[68]
Wong, K. C. K., **4**, 955[12]
Wong, L. C. H., **2**, 780[10]
Wong, M. G., **4**, 538[103]

Wong, M. K. Y., **7**, 256[24]
Wong, M. S., **7**, 763[101]
Wong, M. Y. H., **8**, 90[48]
Wong, P. C., **4**, 37[107]; **5**, 645[1], 650[1q,25], 651[1]; **7**, 875[111]
Wong, P. K., **1**, 442[178]; **2**, 934[143]; **4**, 614[381]
Wong, R. Y., **3**, 672[66]
Wong, S. S., **1**, 37[239]
Wong, T., **8**, 884[98], 885[98]
Wong, W., **8**, 858[206]
Wong, W. C., **8**, 36[97], 42[97], 66[97]
Wong, W. S. D., **8**, 54[158], 66[158]
Wonnacott, A., **1**, 166[113], 314[128], 323[128], 341[98], 780[228]; **3**, 174[527,527a,b], 175[527a-c]
Woo, E. P., **2**, 387[336]
Woo, J. C., **8**, 445[51]
Woo, P. W. K., **8**, 843[48], 846[48]
Woo, S. H., **2**, 846[202]; **6**, 680[330]
Woo, S. L., **3**, 280[40]
Woo, S.-O., **3**, 689[122]
Wood, A., **1**, 425[105], 449[209]
Wood, A. E., **7**, 767[192]
Wood, A. F., **3**, 688[116]
Wood, A. M., **3**, 383[49]
Wood, C. Y., **2**, 1049[16]; **8**, 187[34]
Wood, D. C., **4**, 709[44,47], 710[44,47], 712[70]
Wood, G., **3**, 564[8], 727[28]
Wood, G. P., **6**, 463[23]
Wood, G. W., **3**, 380[5], 381[17]; **7**, 582[149]
Wood, H. B., **8**, 269[85]
Wood, H. C. S., **2**, 456[26]; **3**, 246[35]; **8**, 369[76]
Wood, J., **6**, 219[122]
Wood, J. L., **4**, 348[108,108a], 349[108a]; **7**, 105[151]
Wood, K. V., **8**, 629[186]
Wood, L. L., **3**, 814[71]
Wood, M. L., **2**, 963[54]
Wood, R. D., **4**, 398[213]; **6**, 5[27]; **7**, 406[72], 503[272]
Wood, S. E., **8**, 673[30]
Wood, S. G., **6**, 474[83]
Wood, T. R., **6**, 270[77]
Wood, W. A., **2**, 466[117], 468[117]; **8**, 36[68], 66[68]
Woodard, R. W., **7**, 574[126]
Woodard, S. S., **3**, 225[185], 264[181]; **6**, 2[3], 25[3], 88[105], 89[105]; **7**, 391[13], 406[77], 409[77], 411[13], 412[13], 413[13], 414[77], 415[77], 420[135,136], 421[77,136], 423[77]
Woodbridge, D. T., **7**, 765[157], 769[242,243], 771[242], 773[242]
Woodburn, H. M., **6**, 546[651]
Woodbury, R. P., **2**, 279[12], 605[60]; **7**, 144[152]
Woodcock, D., **2**, 1102[120]; **8**, 963[44], 972[114]
Woodgate, P. D., **1**, 753[102]; **3**, 325[157], 675[74]; **4**, 347[96], 350[121], 351[126], 354[126-128], 369[21,22], 370[21,22], 371[21], 377[21,22], 545[126], 1058[149]; **7**, 121[24], 502[261], 530[20], 531[20], 706[25]; **8**, 309[10], 311[10], 312[10], 313[10], 944[123]
Woodhouse, D. I., **2**, 722[92]; **4**, 698[14]
Woodin, R. L., **1**, 287[16]
Woodman, D. J., **6**, 547[666]
Woods, G. F., **3**, 846[45]; **4**, 5[17]; **6**, 967[101]; **7**, 582[149]
Woods, J. C., **6**, 802[39]
Woods, J. M., **7**, 737[17]
Woods, L. A., **1**, 107[7]; **3**, 208[9], 244[19]; **4**, 148[47b]
Woods, M. C., **3**, 380[4]; **7**, 254[27]
Woods, S. G., **4**, 443[185]
Woods, T. L., **5**, 639[120]
Woods, W. G., **8**, 720[125]
Woodward, B., **8**, 650[64]
Woodward, P., **3**, 380[10]; **4**, 712[70]; **5**, 1136[54], 1146[107]
Woodward, P. R., **6**, 443[93]
Woodward, R. A., **8**, 851[126], 858[126]
Woodward, R. B., **2**, 149[92], 156[117], 214[129], 358[153], 542[83], 773[24], 1022[53]; **3**, 781[14], 810[44], 914[9]; **4**, 44[124], 50[142], 120[202], 373[68], 1016[204], 1070[10], 1075[10], 1093[10]; **5**, 64[25], 66[25], 318[4], 341[61a], 451[50,51], 552[20], 618[3,4], 619[3], 620[15], 621[15], 622[15], 632[63], 635[3], 678[13], 699[1], 732[132,132a], 743[1], 754[61,69], 758[69], 760[61], 794[51], 819[155], 830[27], 857[223], 1002[19], 1009[19], 1186[6]; **6**, 46[74], 667[242], 672[284], 680[332], 681[332], 736[23], 1011[1]; **7**, 157[33], 438[14], 444[14]; **8**, 80[3], 844[76], 974[122]
Woodworth, R. C., **4**, 959[30]
Wool, I. G., **6**, 624[142]
Wooldridge, K. R. H., **8**, 656[87]
Woolford, R. G., **3**, 634[30], 639[30a,76,79], 644[30b], 649[209]
Woolhouse, A. D., **1**, 357[4]; **4**, 1104[213]; **6**, 570[955]; **7**, 470[4], 472[4], 473[4], 474[4], 476[4]
Woolias, M., **4**, 503[128]; **8**, 502[58], 505[58], 509[90]
Woolsey, N. F., **2**, 422[30], 423[30]; **4**, 872[37]; **7**, 723[24], 724[24]
Woosley, M. H., **5**, 797[58]
Woosley, R. W., **5**, 15[104]
Wooster, C. B., **8**, 308[4], 490[7], 509[89]
Worakun, T., **3**, 484[21]; **4**, 848[78], 854[93]
Wormser, H. C., **3**, 507[173]
Woroch, E. L., **4**, 31[92,92e]
Worrall, W. S., **2**, 420[24]
Worster, P. M., **6**, 726[181]
Worth, B. R., **4**, 674[35], 688[35]; **6**, 690[399], 691[399], 692[399]; **8**, 625[162], 821[48]
Worth, L., Jr., **5**, 736[142i,j]
Worthington, P. A., **7**, 53[1], 63[1]
Wosniak, L., **7**, 752[150]
Wotiz, J. H., **8**, 452[189c]
Wotring, L. L., **6**, 554[711]
Wouters, G., **2**, 742[75]
Wovkulich, P. M., **2**, 388[340], 530[23], 534[31], 547[100], 548[100], 651[122]; **3**, 24[149], 25[149], 26[165]; **4**, 390[171], 1076[40]; **5**, 4[39], 5[39,42], 256[58], 257[58b], 260[63c,d], 264[63c,d], 265[63d]; **7**, 268[122]
Woyrsch, O.-F., **6**, 120[122]
Wozniak, J., **7**, 132[100], 146[100]
Wrackmeyer, B., **4**, 886[119]
Wragg, R. T., **3**, 134[337]; **6**, 432[124]
Wray, S. K., **8**, 36[52], 66[52]
Wren, D., **6**, 941[152]
Wren, I. M., **4**, 307[395]
Wriede, P. A., **5**, 159[54], 165[82]
Wriede, U., **6**, 426[74]
Wright, B., **8**, 563[433]
Wright, B. T., **4**, 785[21], 790[40], 791[21,40]
Wright, C. D., **8**, 965[61]
Wright, D. B., **5**, 9[67,69]
Wright, D. R., **6**, 1013[10]
Wright, D. S., **1**, 6[33]; **3**, 763[151]
Wright, G. A., **5**, 1135[47]
Wright, G. C., **6**, 515[235]
Wright, G. F., **4**, 272[35], 273[35], 294[246], 310[433], 311[443]; **6**, 208[63], 283[165]; **7**, 92[42], 93[42], 746[86]
Wright, I. G., **2**, 170[174]
Wright, J., **4**, 85[75]; **6**, 883[60], 884[60]; **7**, 552[57]
Wright, J. J. K., **7**, 30[81], 31[88]
Wright, J. L., **4**, 379[115], 380[115g], 391[115g]; **7**, 523[48]
Wright, J. N., **6**, 74[28]
Wright, J. R., **3**, 402[128]
Wright, J. S., **5**, 72[170]

Wright, L. D., **7**, 778[414]
Wright, M. E., **2**, 727[132]; **5**, 757[78], 762[102], 791[37]
Wright, M. J., **8**, 887[115]
Wright, N. C. A., **8**, 545[290]
Wright, P. W., **1**, 780[229]; **5**, 859[233], 888[25]
Wright, S. C., **8**, 674[32]
Wright, S. H. B., **4**, 85[77c]
Wright, S. J., **7**, 58[56], 62[56], 63[56]
Wright, S. W., **6**, 452[136]; **8**, 616[104]
Wright, W. D., **5**, 1007[42]
Wrighton, M. S., **8**, 451[181,182], 567[459,460], 765[11,14], 767[21,22], 773[21,22], 778[11g]
Wrigley, T. I., **3**, 741[53]
Wristers, H. J., **5**, 797[63]
Wristers, J., **8**, 454[203], 611[65]
Wrobel, J., **2**, 538[64]
Wrobel, J. E., **1**, 329[36], 343[105]; **3**, 262[169]; **6**, 5[23]; **7**, 403[64]; **8**, 36[78], 37[78], 39[78], 44[78], 46[78], 54[78], 66[78]
Wróbel, J. T., **6**, 745[83]
Wroblowsky, H.-J., **6**, 111[64]
Wrubel, J., **4**, 436[141]
Wrzeciono, U., **2**, 787[52]
Wu, A., **3**, 50[268], 334[221,221b]; **4**, 391[183]
Wu, A. B., **1**, 3[21], 45[21e]
Wu, C., **4**, 82[62g], 218[133]; **5**, 71[143], 277[12]
Wu, C. C., **6**, 4[18]
Wu, C. N., **3**, 755[112]; **7**, 799[25], 801[36]
Wu, D., **4**, 243[76]
Wu, D. K., **2**, 420[25]
Wu, E. S. C., **1**, 188[67], 189[67]; **2**, 23[91]
Wu, G. S., **2**, 1017[31]; **3**, 629[53]; **6**, 737[31], 746[89]
Wu, H.-J., **5**, 170[112], 176[112], 461[99], 462[99]
Wu, J. C., **3**, 328[180]
Wu, J. P., **7**, 246[87]
Wu, L., **5**, 94[82]
Wu, S., **2**, 1049[22], 1050[22]; **4**, 795[80]; **6**, 65[2]
Wu, S. H., **5**, 736[142j]
Wu, T.-C., **1**, 188[95], 198[93,95], 199[93]; **3**, 209[19], 263[171], 353[50], 354[50], 565[17]; **5**, 516[26], 517[26], 518[26], 628[45]; **6**, 2[3], 25[3]
Wu, T.-T., **5**, 406[23,23b]; **6**, 814[88]
Wu, W.-S., **7**, 854[54], 855[54]
Wu, X., **5**, 24[169]
Wu, Y., **1**, 191[77]
Wu, Y.-D., **1**, 49[8], 80[23]; **2**, 24[96], 258[50]; **3**, 985[26b]; **4**, 379[117], 1079[65]; **5**, 79[292], 260[65,68], 261[65,68], 262[68]; **7**, 439[36]; **8**, 5[25], 6[34], 7[35], 89[43], 171[109], 723[151], 724[151]
Wu, Y.-J., **3**, 590[164]
Wu, Y.-L., **4**, 14[47,47h,i], 24[74]; **5**, 342[61c], 531[74], 532[74c]; **7**, 362[27]
Wu, Y. W., **6**, 832[14]
Wu, Y.-Y., **7**, 279[168,169], 280[168,169]
Wu, Z.-M., **7**, 105[150]
Wucherpfennig, W., **5**, 423[89]
Wudl, F., **3**, 528[46]
Wüest, H., **2**, 477[8,9]; **3**, 936[72], 953[101]; **4**, 115[183]; **5**, 832[40]; **6**, 834[33], 853[33], 875[24], 879[24]; **7**, 85[11]
Wuest, J. D., **4**, 426[46]; **5**, 850[152]; **6**, 581[989]; **8**, 85[19], 98[102], 720[133]
Wuhrmann, J.-J., **2**, 734[8,9], 738[8,9]
Wujciak, D. W., **5**, 876[2]
Wulff, C., **7**, 17[176]
Wulff, G., **6**, 34[9], 35[9], 49[9]
Wulff, K., **6**, 269[76]
Wulff, W., **4**, 115[179b], 257[224], 517[5], 518[5], 535[93], 538[93], 539[93]
Wulff, W. D., **2**, 588[151], 589[151]; **4**, 981[111]; **5**, 587[208], 588[208], 1065[3], 1066[3], 1067[10,11], 1070[15,17,20,22-25,29], 1071[20], 1072[20,22-25], 1073[10], 1074[17,20,25,29], 1075[11], 1076[41,43-45], 1077[48], 1079[48], 1080[52], 1085[63], 1086[22], 1089[82,84,86,87], 1090[86,87,90], 1091[90], 1092[82,94], 1093[3,96], 1094[3,82,87,94,98,99,102], 1096[3,98,99,106,108,108c], 1098[15,82,87,96a,98,99,106,108a-d,117,130], 1099[3,15,24,82,87,90,99,106,108c,d,117], 1100[87,116,117], 1101[3,43,87,90,117,143], 1102[3,43,102,146b,147], 1104[156], 1110[20], 1111[17,20,82], 1112[3,15,48,82,87,96a,98,99,106,108a-d,116,117,130], 1113[82,87,94,146,156], 1183[57]; **7**, 350[20]; **8**, 911[88], 933[52]
Wulfman, D. S., **1**, 832[110]; **4**, 953[8,8i], 954[8i,k], 1031[5], 1032[5], 1033[25], 1035[5], 1036[51], 1102[198]; **6**, 208[52], 211[52], 239[78]
Wulfmann, C. E., **2**, 149[84]
Wulvik, E. A., **8**, 864[240]
Wunderli, A., **3**, 809[40]; **4**, 1084[95]
Wunderlich, K., **6**, 565[920,921]; **8**, 918[120]
Wunderly, S., **6**, 921[48]
Wunderly, S. W., **5**, 154[33,34]
Wunner, J., **4**, 91[89]
Wünsch, E., **6**, 635[11], 636[11a], 637[28,31], 638[44], 644[87], 645[11], 650[133a,c], 664[218], 665[225], 669[265]
Wünsch, J. R., **2**, 373[274]
Wünschel, P., **8**, 114[57]
Wuonola, M. A., **6**, 76[45]; **8**, 844[76]
Wurster, C., **6**, 937[113]
Wursthorn, K. R., **1**, 411[49]; **2**, 584[127], 587[143]; **6**, 175[66], 182[66]
Würthwein, E. U., **1**, 372[77]; **2**, 64[24], 358[152], 1052[49], 1053[49,52,53], 1055[53]; **6**, 501[195,196,197,198], 535[527,528], 760[141]
Wurtz, A., **2**, 134[4]; **3**, 633[3]
Wurziger, H., **5**, 356[84], 451[36]
Wüst, H. H., **6**, 985[66]
Wüster, J., **4**, 1009[146]
Wustrack, R., **6**, 512[305]
Wustrow, D. J., **4**, 573[7], 614[377]
Wuts, P. G. M., **1**, 177[22]; **2**, 13[52,56], 14[56a], 15[63,64], 19[78], 26[100], 27[100b], 29[103], 35[128], 541[80], 547[96], 551[96], 996[49], 999[52]; **3**, 135[365,366], 136[365,366], 139[365,366], 142[365,366], 356[55], 1033[74]; **4**, 923[32]; **6**, 46[75], 750[104], 859[165]
Wuttke, F., **8**, 187[43]
Wutz, P. G. M., **5**, 20[137], 859[234]
Wyatt, J., **3**, 676[78], 686[78]
Wyckoff, C., **2**, 366[220]
Wyckoff, R. C., **7**, 16[160]
Wydra, R., **8**, 886[110]
Wykpiel, W., **1**, 476[124], 477[124], 478[147], 483[168]
Wykypiel, W., **3**, 66[11], 71[30], 74[11]; **7**, 225[57]
Wyler-Nelfer, S., **2**, 318[51]
Wylie, A. G., **8**, 369[76]
Wylie, R. D., **5**, 141[89]; **6**, 1050[40]
Wyman, P. A., **8**, 626[175], 629[175]
Wymann, W. E., **6**, 809[64]
Wynalda, D. J., **5**, 157[41]
Wynberg, H., **1**, 223[78], 224[78], 317[148]; **2**, 435[61], 769[1], 770[1], 771[1], 773[1,22,23,25], 774[23]; **3**, 551[1], 586[151], 649[208], 665[39], 689[120], 832[68a]; **4**, 12[39], 26[77], 27[77b], 230[241,242,243,249], 231[242,249,261,270,271,272]; **5**, 86[30], 88[30,46,47,50]; **6**, 134[16], 1013[14]; **7**, 98[97], 292[4]; **8**, 99[113], 349[134], 530[101], 837[11], 839[11]
Wynn, H., **8**, 92[68], 189[60], 928[22]
Wynne-Jones, W. F. K., **3**, 634[25a]

Wysocki, R. J., Jr., **5**, 78[271], 266[76], 267[76,76b], 268[76]
Wysong, E., **1**, 367[54]
Wyss, P. C., **6**, 48[86]
Wythes, M. J., **1**, 449[209]
Wyvratt, M. J., **3**, 623[32,33]; **4**, 373[67]; **5**, 347[73a]; **6**, 1063[81]

Xan, J., **7**, 759[6]
Xenakis, D., **6**, 790[111]
Xi, S.-K., **2**, 765[73]
Xia, C., **4**, 452[12]
Xia, Y., **1**, 772[201]; **3**, 509[179]
Xian, Y. T., **4**, 605[294,298]; **8**, 21[143]
Xiang, J.-N., **5**, 859[238], 863[238]
Xiang, Y. B., **1**, 413[55]; **2**, 259[83], 264[83]; **4**, 1040[70,71], 1043[71], 1044[70]; **5**, 377[110,110b], 378[110b]
Xiao, C., **1**, 212[5], 213[5], 214[5c]; **4**, 880[92,93], 881[93], 882[92,93]
Xiao, Y., **2**, 146[70]
Xie, G., **7**, 446[64]
Xie, J., **3**, 669[54,55]
Xie, Z.-F., **1**, 893[153]; **8**, 198[133,134]
Xing, W. K., **8**, 371[111]
Xing, Y., **5**, 211[66]
Xu, B., **3**, 638[93]
Xu, G., **2**, 146[70]
Xu, L., **4**, 1011[164]; **7**, 446[64]
Xu, M. R., **4**, 119[194], 226[199]; **8**, 813[9]
Xu, R. X., **2**, 743[84]
Xu, S. L., **5**, 690[80,80c], 733[136,136f], 734[136f]
Xu, X., **1**, 41[265]
Xu, X.-J., **5**, 1039[11], 1133[26], 1146[26]
Xu, Y., **4**, 854[94]; **7**, 579[133], 580[133]
Xu, Y. C., **5**, 1076[43], 1089[84], 1096[108], 1098[108d], 1099[108d], 1101[43], 1102[43,146b,147], 1112[108d], 1113[146]
Xuan, T., **5**, 429[112]
Xuong, N. D., **8**, 328[4], 340[4]
Xuong, N. T., **6**, 436[8]

Y

Yablokov, V. A., **7**, 641[2]
Yablokova, N. V., **7**, 641[2]
Yablonovskaya, S. D., **2**, 662[15], 664[15]
Yabuki, Y., **6**, 926[69]; **7**, 197[19]
Yabuta, K., **1**, 236[30], 237[30]
Yadagiri, P., **7**, 87[18,18a], 260[84], 713[72]
Yadav, A. K., **3**, 640[106], 644[156]
Yadav, J., **5**, 1096[110], 1098[110]
Yadav, J. S., **2**, 762[57]; **3**, 737[33]; **7**, 90[29], 246[95], 415[115d], 683[87]
Yadav, V. K., **4**, 24[73,73d], 790[37], 795[37]
Yaegashi, T., **6**, 507[240], 515[240]
Yaeger, D. B., **4**, 347[90]
Yaeger, E. B., **7**, 800[34]
Yaffe, A. D., **7**, 8[63]
Yagadiri, P., **7**, 415[115d]
Yagi, H., **3**, 665[40]
Yagi, M., **8**, 459[244]
Yagi, Y., **3**, 918[27], 968[128]; **6**, 531[460]
Yagihara, T., **3**, 923[43], 1008[71]
Yaginuma, F., **8**, 369[79]
Yaginuma, Y., **5**, 406[22]
Yagupol'skii, L. M., **6**, 496[154], 510[295], 556[824]; **7**, 773[304]
Yagupol'skii, Yu. L., **6**, 496[154]
Yagupsky, G., **8**, 445[27,28], 451[180]
Yagupsky, M., **8**, 445[28]
Yahata, N., **3**, 136[370,371], 137[370], 138[370,371a], 139[370,371], 140[370,371,371a], 143[371,371a], 144[371a]; **4**, 18[58], 259[257]; **8**, 277[155]
Yahia, F., **6**, 530[418]
Yahner, J. A., **4**, 111[155a], 441[172,176,178]
Yajima, H., **2**, 1099[112b]
Yajima, T., **6**, 490[111]
Yakhontov, L. N., **6**, 554[770,771]; **8**, 388[62], 599[101]
Yako, K., **4**, 1040[86], 1045[86]
Yako, T., **4**, 610[337]
Yakobson, G. G., **2**, 343[12], 359[12]; **5**, 422[86]; **6**, 525[386], 527[386,407]
Yakomoto, H., **6**, 542[601]
Yakovenko, V. S., **4**, 314[487]
Yakovlev, V. B., **8**, 606[27]
Yakubchik, A. I., **8**, 449[155]
Yakubov, A. P., **4**, 468[132], 469[132]
Yakura, T., **6**, 936[112]; **7**, 829[56,56c]
Yakushkina, N. I., **4**, 310[426]
Yale, H. L., **6**, 570[946], 795[14], 798[14], 821[14]; **7**, 202[49]
Yalpani, M., **6**, 473[75], 480[109], 969[118]
Yam, T. M., **8**, 840[33]
Yamabe, K., **6**, 509[267]
Yamachika, N. J., **7**, 528[6]
Yamada, F., **2**, 967[77]; **7**, 87[18], 335[29]
Yamada, H., **1**, 552[82], 568[234]; **3**, 100[195], 103[195], 107[195], 365[97], 390[70]; **4**, 230[252], 231[252], 511[178]; **5**, 839[76]; **6**, 5[26]; **7**, 145[165], 539[67], 708[31], 713[70]; **8**, 190[66,67,74], 204[152], 857[189]
Yamada, J., **1**, 329[30], 334[30]; **2**, 120[175], 579[89,90], 625[163,164]; **3**, 443[62], 730[45]; **4**, 591[118]
Yamada, K., **1**, 738[40], 739[38]; **2**, 819[99], 953[3b]; **3**, 469[206], 470[206], 473[206], 489[58,63], 495[58,63], 504[63], 511[58,63], 515[58,63], 530[57], 740[46]; **4**, 27[79,79a], 315[526,527], 391[179], 394[199], 434[132]; **5**, 74[206]; **6**, 531[425], 998[117]; **7**, 242[62], 243[65,66], 407[82], 618[22]; **8**, 176[132,133,134], 787[119], 826[70], 857[196], 885[105]
Yamada, M., **1**, 738[40], 795[282], 833[120]; **3**, 934[63]; **4**, 8[30a], 391[179], 1056[141,141b]; **5**, 464[108,109], 466[109]; **6**, 206[39], 442[88], 997[113], 998[117], 1022[56]; **7**, 162[58], 243[66]; **8**, 163[37,38], 545[297], 546[303]
Yamada, N., **3**, 464[175]; **4**, 589[78]; **8**, 652[73]
Yamada, S., **1**, 262[37], 359[19], 382[19,19a,b], 894[157]; **2**, 116[131], 1018[37]; **3**, 46[253], 99[184], 169[510], 172[510,511,512,515,516], 173[510,512,515,516], 741[50]; **4**, 85[75], 210[80,81,83,84], 221[149,150,151,152,153,154,155,156], 817[204,205]; **5**, 524[54], 534[54]; **6**, 4[18], 251[148], 423[46], 438[45], 548[669], 716[96-100], 717[105-107], 718[120], 721[136], 725[169], 726[184], 738[47,48], 797[17,18], 799[23], 811[17,18,78], 812[18,82,83], 813[83-85], 816[18], 1067[104]; **7**, 27[76], 29[78], 174[135], 362[31], 377[31], 410[92], 683[88], 835[84]; **8**, 55[179], 66[179], 146[99,100], 620[138]
Yamada, S.-I., **6**, 438[46], 645[95]; **8**, 241[37], 249[96], 541[207], 580[4,5,7], 581[5], 587[5]
Yamada, T., **1**, 314[124-126], 546[55]; **2**, 30[112a], 31[112], 116[142,143], 166[154], 270[74], 920[94], 921[94], 922[94,105], 924[105], 1089[57], 1102[123]; **3**, 639[75]; **4**, 16[52c], 591[113], 611[362], 612[362]; **5**, 377[110], 378[110a]; **6**, 542[603]; **7**, 155[29,29c], 166[89], 223[46], 442[49], 453[89,90], 455[89,90,106]
Yamada, X., **7**, 162[58]
Yamada, Y., **1**, 223[81], 224[81]; **2**, 370[256], 866[9]; **3**, 311[99], 342[12], 347[12], 381[30], 382[30], 421[59], 422[59], 644[146], 1023[23]; **4**, 30[88,88h,j], 33[96], 36[103,103a], 121[207], 253[175], 258[175], 290[205], 373[70], 404[245], 599[214,220], 642[220], 744[135], 812[177]; **5**, 79[286]; **6**, 76[46], 467[51], 614[96], 646[100b]; **7**, 73[105], 248[109], 314[39], 391[13], 411[13], 412[13], 413[13], 476[64], 803[55]; **8**, 384[36], 604[3], 643[38]
Yamaga, H., **1**, 876[98]
Yamagata, T., **6**, 866[208]
Yamagata, Y., **8**, 153[186]
Yamagatsu, T., **3**, 300[43]
Yamagisawa, A., **2**, 609[84]
Yamagishi, A., **7**, 778[412]; **8**, 444[6]
Yamagishi, T., **8**, 460[254]
Yamagiwa, S., **2**, 363[191]; **3**, 946[87]; **4**, 106[140b]; **7**, 205[64]
Yamago, S., **2**, 633[34b], 634[34b], 640[34]; **5**, 266[75], 268[75], 310[101]
Yamaguchi, A., **8**, 248[82]
Yamaguchi, H., **3**, 45[248], 135[353], 136[353], 137[353], 141[353], 142[353], 541[106], 558[50], 586[138]; **6**, 685[348], 959[45]; **7**, 91[34], 92[41,41b], 93[41b], 94[41], 297[30], 310[28], 340[45], 761[60], 765[60]; **8**, 161[16], 366[48,49], 413[131], 806[121]
Yamaguchi, K., **1**, 390[145], 391[145]; **2**, 1068[129]; **3**, 512[198], 867[36]; **4**, 1070[8]; **5**, 75[226,228], 248[29]; **6**, 914[28]; **7**, 743[64]
Yamaguchi, M., **1**, 19[98], 256[22], 257[23], 258[24,25], 259[24], 260[32,32b], 261[33], 266[47], 268[53,53b], 270[25], 275[25,76], 343[100-103,112,113], 419[78], 421[89], 489[21], 497[21], 751[110,111], 829[82-84], 831; **2**, 5[22], 10[22,22c,45,45a], 31[109], 70[52], 119[159,162,163], 174[183], 176[185], 244[28], 304[9], 305[9], 725[121], 780[7], 832[152], 846[204,205]; **3**, 45[248], 225[187], 277[33], 279[35,36], 286[58], 1000[55], 1001[56], 1004[60]; **4**, 10[32], 21[66,66a-c], 62[66c], 107[146a-c], 108[146d], 218[143], 606[305], 607[305,311,313], 626[311], 647[305], 648[311]; **5**, 151[17], 166[17], 308[96], 366[98]; **6**, 5[25,27], 7[31,33], 8[39], 175[76], 648[116], 668[260], 852[135], 877[38,39], 878[39], 880[46], 883[38,39], 885[56], 887[38,39], 890[56], 980[41]; **7**, 246[94], 379[99], 382[99], 400[44], 408[44], 419[133]; **8**, 11[61,62], 145[83], 365[32], 412[108b], 446[81],

460[254], 781[96], 797[33], 883[95], 884[95], 960[32], 987[22], 992[22b], 994[22]
Yamaguchi, R., **2**, 86[25], 1067[126]; **3**, 903[125]; **4**, 111[155b]; **6**, 624[139]; **7**, 218[9], 794[5], 878[145]
Yamaguchi, S., **7**, 806[71]; **8**, 16[98], 161[23], 165[48,49], 170[72], 178[49], 179[49]
Yamaguchi, T., **4**, 298[288]; **6**, 489[92]; **7**, 693[28]
Yamaguchi, Y., **1**, 544[39]; **4**, 130[226c]; **7**, 501[255], 662[52]; **8**, 302[96]
Yamaichi, A., **1**, 511[29], 566[213]; **4**, 10[34], 113[163]
Yamaji, T., **1**, 192[82]; **2**, 566[23]; **3**, 244[27], 464[177]; **7**, 458[111]
Yamakado, Y., **1**, 790[264]; **2**, 91[44]
Yamakami, N., **3**, 1027[44]; **8**, 292[39]
Yamakawa, K., **1**, 524[93], 526[93,95], 828[68]; **2**, 417[19], 429[48]; **4**, 6[20], 784[16], 1103[205]; **6**, 91[122], 93[134], 156[162]; **7**, 63[59], 132[96], 158[36a,b], 175[36b], 425[149c], 773[302]
Yamakawa, M., **5**, 293[48], 1186[3,3a], 1190[24]; **8**, 678[62], 683[62], 686[62]
Yamakawa, T., **4**, 590[101], 606[302], 609[331], 646[302]; **6**, 866[204]; **7**, 453[88], 455[88]; **8**, 557[388], 959[24], 961[39]
Yamakawa, Y., **4**, 313[470]
Yamamori, T., **1**, 569[248]
Yamamoto, A., **1**, 174[13], 176[17,18], 202[13], 440[190], 445[190], 714[267], 717[267]; **3**, 528[47], 530[80], 535[80]; **4**, 560[27], 589[84], 590[95], 592[95], 595[153], 598[84], 610[339], 615[153], 619[153], 626[339], 633[95], 649[339], 653[439,440]; **5**, 305[87], 850[147]; **6**, 86[100], 88[104], 842[83,84]; **7**, 582[138]; **8**, 391[90], 445[62], 838[20], 963[42]
Yamamoto, B. R., **4**, 18[59], 121[208], 993[161]
Yamamoto, G., **2**, 438[69b]
Yamamoto, H., **1**, 64[46], 77[3], 78[3,10,12,13,18], 79[21], 80[21], 81[21], 82[21], 83[27], 88[51,52,54,55], 92[60,61], 98[84], 99[84,85], 100[18], 161[80-83,86,87,90], 165[108-111], 266[48], 283[3], 316[3], 333[60,61]; **2**, 22[87], 65[31], 72[56,59], 74[71], 84[47], 88[33], 91[44,46,47], 93[46,47], 94[50], 96[57,58], 103[29], 114[122], 269[71,72], 282[40], 541[74], 556[155], 600[27], 615[126], 631[18], 655[148], 685[67], 995[45]; **3**, 96[162,166,167], 99[162,163,166,167], 103[166,167], 104[162,163], 105[166,216], 112[216], 120[162,163,166], 121[162,163], 131[162,163], 155[166], 202[97], 229[225], 246[35], 349[33], 354[60], 358[33], 446[85,86,88], 469[203], 470[203], 473[203], 483[13], 750[86], 787[47,48]; **4**, 107[140d], 140[8,11], 143[21], 209[64-66], 239[28], 254[177], 304[356], 745[141], 753[167], 814[186], 840[34], 883[95], 968[59], 969[59], 974[90], 976[96,97], 1007[119,124]; **5**, 355[87b], 377[111,111a,b], 434[147], 609[68], 611[72], 850[150,153,154,155]; **6**, 5[23], 14[52], 65[2], 91[126], 254[161], 291[217], 528[414], 767[26,28], 768[28], 769[28,29,32], 770[33-35], 771[37,38], 837[60], 849[123], 850[125], 856[152], 861[181], 865[200], 960[64]; **7**, 318[51], 537[59], 642[9], 696[38,43,44], 697[43,45-47,49]; **8**, 18[130], 43[108], 47[108], 64[213,220], 66[108,213], 67[220], 95[85], 100[117], 223[99,100], 224[99,100], 227[120], 269[81], 356[185], 388[60], 394[119], 419[17], 430[17], 545[284], 609[54], 659[106], 660[107], 787[119], 929[30], 968[85], 986[16]
Yamamoto, I., **1**, 779[222]; **6**, 609[51]; **8**, 899[28]
Yamamoto, J., **7**, 299[48]
Yamamoto, K., **2**, 632[25a-c], 638[25b,c], 640[25c], 654[25c], 728[143,144,145]; **3**, 228[222], 592[173], 593[181,182], 628[49]; **4**, 104[137], 227[206,207,208], 255[200], 600[237], 611[355], 643[237], 650[424], 653[435]; **5**, 55[48], 281[19], 953[296]; **6**, 531[438], 932[97]; **7**, 425[149a], 450[7], 451[7], 455[7], 458[7]; **8**, 55[180], 66[180], 153[188], 155[188], 164[46], 173[112,113], 178[46], 179[46], 289[24], 556[373], 609[53], 780[95], 782[105], 783[107], 784[110]
Yamamoto, M., **1**, 752[97]; **2**, 153[105]; **3**, 740[46]; **4**, 315[526,527,528], 394[189,199]; **5**, 369[101], 370[101b]; **6**, 20[75]; **7**, 257[47], 618[22]; **8**, 150[133,135], 151[133,135,145], 857[196], 968[85]
Yamamoto, N., **3**, 771[191]; **4**, 104[137], 227[208]; **8**, 241[37]
Yamamoto, S., **1**, 117[55], 124[82], 329[32]; **4**, 148[50], 149[50b], 179[64], 182[64c], 184[64c], 858[110]; **5**, 474[158], 487[193]; **6**, 3[10], 30[10], 835[47], 848[107]; **8**, 600[107], 899[28]
Yamamoto, T., **1**, 174[13], 202[13]; **2**, 584[118], 656[152]; **3**, 530[80], 535[80]; **4**, 163[97], 560[27], 589[84], 595[151], 598[84], 599[212], 604[288], 640[212], 641[212], 647[288]; **6**, 86[100]; **7**, 131[80]; **8**, 783[109], 838[20], 840[30], 935[60], 963[42]
Yamamoto, W., **2**, 580[99]; **7**, 208[87]
Yamamoto, Y., **1**, 78[7,9], 95[7], 100[89], 110[22], 113[28], 115[22], 117[55], 121[66], 124[82], 143[36], 158[36,73,74], 159[36,75], 180[33,38,40], 181[38,40], 185[60], 221[68], 223[74], 313[116], 314[130], 327[15]; **2**, 2[6], 3[6,6a], 4[12,12a,14], 5[18], 6[12,18,30], 10[12b,44,45b,46-48], 11[44,47,49,50], 13[56], 15[66], 18[30b,69], 22[45b], 24[18,18b], 30[18b,107], 31[107], 32[6a,120,120a,b], 57[5], 58[8,10], 61[8], 67[8,38], 68[39], 71[8], 72[8], 76[8], 95[55], 117[146], 119[146,157], 120[175], 128[241], 252[40], 257[40], 302[3], 303[3], 313[38], 314[38], 489[49], 490[49], 564[4,10], 566[18], 573[18,55,56], 574[18,58], 576[70], 579[89,90], 611[101], 625[163,164], 632[28a], 640[28], 743[79], 826[122], 907[59], 948[181], 977[8], 978[11], 979[11,15,16], 983[15,16], 984[15,16,29,30], 985[15,16,29,30], 986[15,16,31], 987[15,30,31], 988[30], 989[11], 990[11], 991[11], 992[11,15,37], 993[11,37], 995[46], 1102[121a], 1103[121]; **3**, 43[236], 87[97,98], 99[98,183], 100[98,183], 105[183], 157[98], 196[31], 212[36], 221[36], 222[136], 226[36,200], 232[267], 249[67], 262[168], 354[62], 443[62], 483[10], 510[185], 623[39], 751[88]; **4**, 89[84d], 145[35], 148[50], 149[50b], 150[53], 155[68c], 170[5], 179[5,64], 182[64a,c], 184[5,64a-c], 185[88], 186[5,88], 187[5], 188[101], 189[101], 201[11], 230[244], 231[244], 238[14], 244[79], 245[79], 247[106], 255[79], 257[106], 260[106], 388[162], 401[162a], 653[441], 739[111], 959[33], 969[68]; **5**, 277[13], 451[43], 485[43]; **6**, 4[19], 5[23], 533[492], 535[530], 569[938], 637[28], 832[13], 833[21], 843[91], 848[107,108], 864[13,194], 1016[26]; **7**, 226[70], 314[39], 417[130c], 453[70], 579[134]; **8**, 170[78,79], 353[159], 540[200], 676[79], 725[181], 778[85], 806[102], 807[116], 876[44], 877[44]
Yamamura, A., **8**, 159[108], 171[108], 178[108], 179[108]
Yamamura, K., **4**, 837[12]
Yamamura, M., **3**, 436[5], 437[5,22], 438[22], 485[36], 491[36], 494[36], 497[36]
Yamamura, S., **2**, 553[124]; **3**, 390[74,86], 392[74], 395[100], 396[113,115], 397[116], 398[113], 665[38], 676[79], 690[125], 691[128,131], 693[128,139], 694[151], 695[79], 696[151], 697[38,131,139,155,156], 698[128], 769[171]; **5**, 569[113]; **7**, 337[35,36]; **8**, 309[11-13], 310[11,13], 311[12], 803[95]
Yamamura, Y., **6**, 771[37]; **7**, 697[49]
Yamamuro, A., **2**, 258[50]; **7**, 172[124]
Yamana, M., **3**, 218[98]
Yamana, Y., **4**, 1024[264]
Yamanaka, A., **2**, 157[119]
Yamanaka, E., **2**, 1017[31], 1021[50]; **6**, 737[31], 746[89,90], 916[31]; **8**, 31[47], 66[47]
Yamanaka, H., **1**, 561[164]; **2**, 353[101], 364[205], 598[15]; **3**, 460[143], 461[143,147,148], 512[199], 530[64], 533[64], 541[108-111,113,114], 543[108-111,113,117,118]; **4**, 433[121], 1006[96], 1017[211], 1021[211]; **5**, 736[142h]; **6**, 498[167], 530[423], 543[604], 801[30]; **8**, 902[45]
Yamanaka, T., **6**, 1066[97]; **7**, 21[15], 537[61]
Yamanaka, Y., **1**, 162[104], 277[83]; **3**, 829[57]
Yamane, A., **2**, 889[57]
Yamane, H., **8**, 611[62]
Yamane, K., **3**, 818[97]
Yamane, S., **2**, 784[38], 1052[51]; **6**, 801[37]; **7**, 802[49]
Yamane, T., **4**, 841[49]
Yamane, Y., **8**, 323[113]
Yamanochi, T., **8**, 611[62]
Yamanoi, K., **6**, 124[145]
Yamanoi, T., **8**, 170[86-88], 178[86], 179[86]

Yamanouchi, A., **7**, 144[158]
Yamanouchi, T., **4**, 988[139]
Yamaoka, H., **5**, 79[286]
Yamaoka, S., **3**, 571[74], 574[74]; **8**, 836[3]
Yamasaki, H., **4**, 1007[109]
Yamasaki, K., **7**, 862[78], 877[127]
Yamasaki, N., **2**, 117[144,145], 922[102], 923[102]
Yamasaki, R. B., **8**, 36[67], 66[67]
Yamasaki, T., **6**, 119[116]
Yamasaki, Y., **4**, 302[337]
Yamashina, N., **3**, 231[248], 443[56]
Yamashina, S., **6**, 536[545], 538[545]
Yamashiro, R., **8**, 609[54]
Yamashita, A., **1**, 188[66], 189[66]; **2**, 23[91]; **3**, 423[79]; **4**, 517[5], 518[5], 543[123]; **5**, 712[57b], 1089[85], 1090[91], 1093[96], 1096[107], 1098[107,118,119,132], 1099[96c,107,118,119], 1101[91,133], 1104[155], 1112[91,96c,107,118,119,132,133], 1113[85,107,155]; **7**, 301[60]; **8**, 542[223], 544[223], 548[322]
Yamashita, D. S., **5**, 736[143]
Yamashita, H., **3**, 1016[3], 1039[101]; **4**, 231[263]; **6**, 839[69]; **7**, 77[123]
Yamashita, J., **1**, 750[88]; **3**, 466[187], 500[133]; **8**, 795[23], 906[65], 907[65], 908[65], 909[65], 910[65]
Yamashita, K., **2**, 198[84]; **7**, 415[115b]
Yamashita, M., **1**, 527[104], 561[165], 568[239], 569[249], 865[87]; **2**, 357[146], 358[146,151], 505[8]; **3**, 34[195], 35[202], 135[356,360,361,362], 136[356,360,361,362,371], 137[360,361], 139[356,360,361,362,371], 140[371,371b], 142[360,361], 143[360,361,371,371b], 156[360,361,362], 1029[52]; **4**, 10[33], 115[178], 1040[89,90], 1045[89,90]; **6**, 760[142]; **7**, 455[104], 550[51], 675[53]; **8**, 36[81], 54[81,161], 66[81,161], 291[35-37], 292[36], 293[46], 330[47], 411[105], 550[332], 563[429]
Yamashita, S., **2**, 976[2], 981[2], 982[2]; **4**, 972[80]
Yamashita, T., **2**, 558[160]; **4**, 124[212], 211[87], 252[163]; **8**, 87[34], 334[64]
Yamashita, Y., **2**, 765[72]; **3**, 244[22], 465[179], 494[89], 516[89]; **5**, 206[47], 819[154]
Yamataka, H., **1**, 116[50], 118[50]
Yamataka, K., **3**, 640[99], 672[65]; **5**, 422[82]
Yamato, H., **3**, 848[53], 923[44], 934[44], 954[44], 1008[70,71]; **7**, 350[27], 355[27]; **8**, 97[95]
Yamato, M., **2**, 167[157], 642[76], 643[76], 780[13], 836[160]
Yamato, T., **3**, 329[184,185]; **7**, 354[36], 355[36], 674[39]; **8**, 216[57], 251[104], 253[104]
Yamatomo, S., **4**, 298[288]
Yamauchi, M., **2**, 388[344]; **5**, 356[89]
Yamauchi, T., **3**, 789[70]; **6**, 17[63], 18[63]; **7**, 773[309], 776[309], 829[56]
Yamaura, M., **1**, 733[20]; **8**, 374[150]
Yamaura, Y., **3**, 1047[6]
Yamawaki, A., **3**, 843[25]
Yamawaki, J., **6**, 66[5]
Yamawaki, K., **7**, 708[31]
Yamaya, M., **3**, 243[11]; **8**, 734[6], 757[6]
Yamazaki, A., **6**, 626[166]
Yamazaki, H., **5**, 1133[31], 1135[52], 1137[57,58], 1139[74], 1140[74], 1141[81], 1142[74,87-89], 1145[103], 1146[74], 1152[137,138,142], 1153[103,148], 1154[150], 1155[160,161], 1156[160], 1158[137]; **8**, 807[117]
Yamazaki, M., **4**, 434[132]; **6**, 778[62], 780[62]; **7**, 455[105]
Yamazaki, N., **1**, 55[24], 390[145], 391[145]; **6**, 81[71,72], 441[78], 443[78]; **7**, 297[31]; **8**, 170[82-84,96,97]
Yamazaki, S., **6**, 531[438], 932[97]; **7**, 227[89]; **8**, 64[218], 67[218]
Yamazaki, T., **1**, 287[18]; **2**, 555[138], 656[152], 913[73]; **3**, 639[81], 853[72]; **4**, 27[82], 102[130], 216[123], 222[166], 398[217], 399[217b], 401[217b], 403[217b], 404[217b], 413[278a,b], 802[128]; **5**, 829[22], 841[87], 843[115]; **6**, 509[261,266,267,268], 529[464], 538[560]; **7**, 406[78b], 503[273]
Yamazaki, Y., **4**, 38[109a], 406[253], 408[253]; **6**, 89[118], 101[118]; **7**, 59[39,40], 503[271]; **8**, 189[62]
Yamoto, H., **3**, 919[32]
Yamoto, Y., **8**, 531[110]
Yamuguti, T., **4**, 30[89]
Yamura, A., **8**, 321[93]
Yan, T.-H., **1**, 672[202], 700[202], 701[202], 705[202]; **2**, 572[41]; **3**, 586[155], 610[155]; **5**, 856[210], 910[81], 912[81], 954[299,300]; **6**, 146[88]
Yanada, K., **7**, 73[104]; **8**, 366[48,49,52]
Yanagi, A., **2**, 743[79]
Yanagi, K., **4**, 379[114,114b], 382[114b], 383[114b], 413[114b]; **6**, 842[84]
Yanagi, T., **1**, 802[304]; **3**, 504[154], 511[154], 515[154]; **5**, 117[273], 767[120]
Yanagi, Z., **4**, 379[114,114b], 382[114b], 383[114b], 413[114b]
Yanagida, H., **2**, 73[66]
Yanagida, M., **1**, 766[161]
Yanagida, N., **2**, 617[144]; **4**, 239[19]; **6**, 22[86]
Yanagihara, N., **4**, 394[192], 567[51]; **6**, 237[60], 243[60]
Yanagino, H., **5**, 832[39]
Yanagisawa, A., **1**, 347[129]; **3**, 9[47], 10[47]; **4**, 211[90], 239[28], 255[195,197], 259[197], 260[197], 393[197,197b,c], 394[197b,c], 745[141], 753[164], 883[95]; **7**, 406[75]; **8**, 163[39], 836[7]
Yanagisawa, E., **1**, 561[160]
Yanagisawa, K., **3**, 437[22], 438[22], 485[36], 491[36], 494[36], 497[36]
Yanagita, M., **3**, 810[45]; **4**, 6[20], 23[70]
Yanagiya, M., **2**, 1050[28]; **3**, 100[199]; **5**, 564[97]; **7**, 168[101]
Yanai, H., **4**, 847[76]
Yanai, T., **2**, 360[168]
Yanami, T., **4**, 12[43], 158[78], 161[89b]; **7**, 220[21], 458[114]
Yanase, M., **3**, 224[181]; **7**, 299[45]
Yanaura, S., **1**, 359[14], 363[14], 384[14]
Yanbikov, Ya. M., **5**, 752[4], 767[4]
Yandovskii, V. N., **7**, 750[133]
Yang, B. V., **5**, 311[105]
Yang, C. C., **6**, 666[233], 667[233]
Yang, C.-P., **1**, 275[77]; **4**, 795[79]; **6**, 755[121]
Yang, C.-T., **2**, 959[28]
Yang, D., **8**, 20[140], 344[121,121b]
Yang, D. C., **4**, 981[111]; **5**, 1070[17,20,22-25], 1071[20], 1072[20,22-25], 1074[17,20,25], 1076[45], 1085[63], 1086[22], 1089[86], 1090[86], 1096[108,108c], 1098[108b,c], 1099[24,108c], 1101[143], 1110[20], 1111[17,20], 1112[108b,c], 1116[11], 1117[11], 1118[11], 1183[57]
Yang, D.-D. H., **2**, 530[21]
Yang, D. T. C., **8**, 356[187], 357[193,194,195], 358[195]
Yang, H., **5**, 682[33], 691[33], 1032[99]
Yang, J., **3**, 669[55]; **5**, 424[97]
Yang, J. J., **4**, 925[35]
Yang, L., **4**, 408[257a]; **7**, 246[82]
Yang, L.-W., **8**, 847[94]
Yang, N. C., **2**, 530[21]; **3**, 383[44]; **4**, 35[99]; **5**, 636[94-97,99], 637[103], 646[10], 647[10c], 654[27], 741[154], 817[146]; **6**, 687[375]; **7**, 95[74], 330[12]
Yang, R., **8**, 460[249]
Yang, S., **8**, 17[111]
Yang, S.-M., **4**, 744[137]
Yang, T.-K., **1**, 390[144]; **2**, 935[146,148], 936[146,148], 937[148], 999[39], 1049[21], 1050[21], 1063[104]; **5**, 100[149,150], 101[150]; **6**, 746[96]
Yang, T. X., **8**, 553[362]
Yang, W. J., **1**, 569[256]; **2**, 78[93]

Yang, Y., **3**, 159[464], 161[464], 166[464], 232[259]
Yang, Y.-L., **3**, 730[44]
Yang, Y.-T., **5**, 828[5], 847[5]
Yang, Z., **4**, 842[52], 858[111]
Yang, Z.-u., **3**, 323[146]
Yang, Z.-Y., **3**, 530[67], 531[67], 533[67], 538[93]; **5**, 1151[129]
Yang-Chung, G., **7**, 299[39]
Yaniv, R., **8**, 94[79]
Yankee, E. W., **2**, 1049[19]; **6**, 653[151]
Yannakopoulou, K., **3**, 282[47]
Yano, S., **1**, 243[56]; **4**, 1032[9], 1051[9]; **8**, 609[53]
Yano, T., **1**, 123[75], 373[82]; **3**, 249[66], 465[178], 470[178], 473[178]; **5**, 176[128], 621[18]
Yano, Y., **7**, 225[59], 761[64]
Yanotovski, M. Ts., **8**, 956[7]
Yanovskaya, L. A., **1**, 555[112]; **4**, 951[1], 968[1], 979[1]
Yanuck, M., **5**, 517[27], 538[27]
Yanuck, M. D., **2**, 770[10], 771[10]; **5**, 829[20], 1039[17], 1050[17], 1052[17]
Yao, A. N., **3**, 918[21], 922[39], 939[39]
Yaouanc, J.-J., **2**, 64[26]; **7**, 745[80]
Yarbro, C. L., **6**, 660[199]
Yardley, J. P., **8**, 972[112,113]
Yardley, Y. P., **7**, 153[11]
Yarnell, T. M., **1**, 608[36,37]; **3**, 370[110], 373[127]; **8**, 545[293]
Yarovenko, N. N., **6**, 217[113]
Yarrow, D. J., **4**, 691[76]
Yarrow, J. M., **7**, 583[153]
Yaser, H. K., **6**, 838[65], 846[102]
Yashima, T., **6**, 533[504]
Yashin, R., **8**, 108[1], 118[1]
Yashiro, M., **7**, 452[41]
Yashkina, L. V., **6**, 542[598]
Yashunskii, D. V., **1**, 95[75,76]; **6**, 466[40-42], 469[40,54]; **8**, 694[118]
Yasin, Y. M. G., **1**, 38[263]
Yasuda, A., **4**, 116[186]; **6**, 837[60], 960[64]; **7**, 180[158]; **8**, 459[244], 535[166], 929[30]
Yasuda, H., **1**, 19[98], 162[93-95,98,100-102], 163[94,106], 164[94], 180[32], 214[27]; **2**, 5[18], 6[18], 24[18,18a], 60[17], 611[102], 643[79], 644[79a]; **4**, 160[86a], 161[91]; **5**, 1148[114], 1172[28-30], 1182[28-30]; **6**, 935[104]; **7**, 453[92], 454[92]
Yasuda, K., **2**, 846[208]; **3**, 41[224]; **4**, 21[69], 222[173,174,175]
Yasuda, M., **4**, 180[67]; **5**, 492[240], 634[71,72,80,81], 650[25], 819[156]; **7**, 347[17], 355[17]; **8**, 517[126], 562[421]
Yasuda, N., **6**, 93[133]; **8**, 830[88]
Yasuda, S., **7**, 155[29,29c]
Yasuda, T., **7**, 537[61]
Yasuda, Y., **4**, 95[96]
Yasuhara, M., **7**, 773[300]
Yasui, S., **8**, 909[80], 917[118], 918[118], 919[118], 977[141]
Yasumoto, F., **6**, 619[115]
Yasumoto, S., **6**, 447[105], 450[105]
Yasuoka, H., **8**, 616[97]
Yasuoka, N., **1**, 19[98]; **3**, 672[65]; **5**, 422[82], 1200[49]
Yatabe, M., **2**, 232[177]; **8**, 460[254], 535[166], 765[11]
Yatagai, H., **1**, 117[55], 124[82], 158[74], 180[38], 181[36,38], 329[32], 502[55], 509[24], 635[82], 636[82], 637[82], 640[82], 672[82], 678[82], 679[82], 680[82], 681[82], 700[82], 705[82]; **2**, 4[12,12a], 6[12,30], 10[12b,44,45b], 11[44,50], 18[30b], 22[45b], 57[5], 58[8], 61[8], 67[8,38], 71[8], 72[8], 76[8], 117[146], 119[146], 566[18], 573[18,55], 574[18,58], 977[8]; **3**, 87[98], 99[98,183], 100[98,183], 105[183], 157[98], 196[31], 249[67], 470[195], 473[195], 483[10]; **4**, 145[35], 148[50], 149[50b], 179[64], 182[64c], 184[64c], 185[88], 186[88]; **6**, 848[107]; **7**, 226[70], 453[70]; **8**, 725[181]
Yatagai, M., **8**, 460[254]
Yates, G. B., **3**, 635[40], 637[65,66], 650[66]
Yates, J. B., **4**, 744[129,130], 745[139]; **5**, 576[138,141,142]; **6**, 115[80]
Yates, J. T., Jr., **8**, 421[31]
Yates, K., **4**, 298[293,294], 300[293,294]; **5**, 703[16b,18,18b]
Yates, M. J., **8**, 502[61]
Yates, P., **1**, 284[4]; **2**, 142[46]; **3**, 597[198], 888[21], 890[33,35], 891[38,41b], 903[121], 906[142]; **5**, 225[87,88,116], 227[116], 231[87,88,132], 233[116], 339[55], 572[123], 791[27], 799[27]; **6**, 557[837]
Yates, P. M., **5**, 21[145,148]
Yates, R. L., **1**, 506[11]; **5**, 75[220,223]; **8**, 437[76]
Yatsimirsky, A. K., **3**, 499[125], 669[53]
Yatsu, I., **7**, 761[64]
Yau, C.-C., **3**, 18[99], 19[100]
Yau, L., **6**, 624[138]
Yavari, I., **3**, 382[36]
Yax, E., **3**, 564[7]
Yazawa, N., **7**, 680[76]
Yeakey, E., **3**, 924[40]
Yeates, C., **1**, 801[302]; **3**, 229[234], 264[178], 444[66,67]; **4**, 390[168], 878[83]; **6**, 994[98]
Yeats, R. B., **8**, 269[76]
Yee, K. C., **3**, 8[42], 756[115]
Yee, Y. K., **2**, 844[197]; **4**, 10[34], 106[140a]
Yefsah, R., **5**, 1105[159,161,162]
Yeh, C.-L., **6**, 920[46]
Yeh, M. C. H., **1**, 212[5], 213[5], 214[5b]
Yeh, M. C. P., **1**, 115[41], 212[5,8,10,11,13], 213[5,8,10,11,13,14,17,17b,18], 214[5c,19], 215[10], 216[10], 217[10,13], 221[10], 432[138], 433[138]; **2**, 442[14], 449[14], 450[14,50]; **3**, 209[18]; **4**, 175[42], 880[92,93], 881[93], 882[92,93]
Yeh, M. H., **5**, 65[73]
Yeh, M. K., **1**, 476[122]; **3**, 35[203], 66[13], 197[35], 500[127], 501[127]
Yeh, M.-Y., **6**, 821[113]
Yeh, R.-H., **4**, 38[110]
Yelm, K. E., **2**, 1002[55]; **3**, 248[51]; **6**, 904[141]
Yen, C. C., **4**, 373[73]
Yen, H.-K., **7**, 220[25]
Yen, V. Q., **4**, 303[348]
Yen, Y., **8**, 531[121]
Yencha, A. J., **4**, 348[108], 349[108c]
Yeo, H. M., **3**, 693[142]
Yeo, Y. K., **8**, 384[28]
Yeoh, H. T.-L., **1**, 248[70]
Yeong, Y. C., **3**, 744[62]
Yerino, L., Jr., **5**, 154[33]
Yeske, P. E., **4**, 16[52a]
Yeung, B.-W. A., **1**, 766[156]; **4**, 813[185], 815[185], 817[185]
Yeung, L. L., **8**, 847[95]
Yevich, J. P., **8**, 253[121]
Yi, C. S., **2**, 828[134]; **3**, 629[54]
Yi, D., **2**, 146[70]
Yi, Q., **1**, 561[164]
Yi, Y., **2**, 772[19]
Yiannikouros, G., **6**, 448[107]
Yiannios, C. N., **7**, 760[33]
Yi-Fong, W., **4**, 382[133], 388[133]
Yijun, C., **5**, 1107[168,169]
Yim, N. C. F., **4**, 31[93]; **7**, 543[12], 551[12]
Yimenu, T., **2**, 524[78]; **3**, 34[196]
Yip, R. W., **5**, 638[116]
Yip, Y.-C., **5**, 221[58], 226[58,112], 900[12], 901[12], 903[12], 905[12], 907[12], 913[12], 921[12], 926[12], 943[12], 1006[33]; **7**, 815[3], 824[3], 833[3]; **8**, 847[95]

Yiswanathan, N., **7**, 221[32]
Ykman, P., **5**, 71[154]
Yo, C.-M., **6**, 865[200]
Yocklovich, S. G., **6**, 1024[78]
Yocklovidi, S. G., **7**, 162[60]
Yoda, H., **4**, 40[114]; **7**, 407[83]; **8**, 798[59]
Yoda, N., **2**, 142[46], 232[177]; **7**, 801[45]; **8**, 459[242], 460[254], 461[259], 535[166]
Yoder, J. E., **3**, 572[68]; **8**, 313[24]
Yodono, M., **4**, 837[9], 839[27]
Yogai, S., **6**, 1053[46]
Yogo, T., **3**, 446[76]
Yohannes, D., **4**, 433[119]; **5**, 492[238], 498[238]
Yokohama, S., **8**, 647[56]
Yokoi, M., **6**, 450[119], 454[119]
Yokokawa, C., **8**, 568[475]
Yokokawa, N., **8**, 881[83]
Yokomatsu, T., **2**, 363[193], 1060[91], 1073[141,142]; **4**, 401[233], 505[138]; **6**, 749[98], 751[108], 780[72], 879[43]; **8**, 934[57]
Yokomoto, M., **8**, 551[335]
Yokoo, K., **1**, 162[104], 254[14,14b], 277[80,82,83], 278[14a]; **8**, 881[76], 882[76]
Yokoo, S., **3**, 100[195], 103[195], 107[195], 390[69,70]
Yokota, K., **3**, 228[219], 246[37], 442[50], 639[85]; **4**, 596[159], 635[159]; **5**, 841[87], 843[115]
Yokota, K.-i., **3**, 470[219], 471[219], 472[219]
Yokota, N., **4**, 85[71], 203[30]
Yokota, Y., **7**, 751[138]
Yokoyama, A., **7**, 239[49]
Yokoyama, B., **5**, 812[133]
Yokoyama, H., **3**, 262[164], 466[191]; **6**, 1022[60]
Yokoyama, K., **2**, 615[128], 634[35], 640[35]; **5**, 282[25,27,31,32], 283[25,27], 285[25,27,31,32], 601[46,47], 603[47], 605[46,63]
Yokoyama, M., **1**, 232[13,14], 233[13,14], 234[13,14], 243[57], 253[9], 259[29], 276[9], 278[9], 332[52], 561[160], 829[87]; **2**, 311[33], 312[33,35], 353[101]; **3**, 567[34], 570[34]; **6**, 214[94], 428[82], 438[49], 457[163]; **7**, 843[48]; **8**, 113[47], 405[23]
Yokoyama, S., **1**, 86[35], 223[79], 224[79]; **4**, 230[254,255]; **6**, 76[44]; **8**, 241[36], 384[34], 412[110]
Yokoyama, T., **4**, 958[26], 960[26]; **8**, 285[7]
Yokoyama, Y., **1**, 562[167], 642[117], 645[117], 686[117]; **2**, 650[108], 736[26], 743[81], 780[10]; **4**, 112[158c]; **5**, 172[118]; **6**, 438[45,46], 439[71], 466[43], 1029[103]
Yokozawa, T., **2**, 635[40], 640[40]; **5**, 841[95]
Yokyama, T., **1**, 543[29]
Yolanda, G. Q., **1**, 552[81]
Yom-Tov, B., **2**, 765[70]
Yon, G. H., **8**, 563[435]
Yonashiro, M., **3**, 39[217]; **7**, 586[162], 844[56]
Yoneda, F., **2**, 357[143], 792[64], 877[38]; **3**, 216[72], 623[39]; **4**, 435[138], 1009[140]; **6**, 489[79], 533[486], 554[706,707,708]; **7**, 761[52,63,65]; **8**, 97[95], 562[422]
Yoneda, N., **1**, 367[53]; **2**, 780[10]; **3**, 326[163]; **6**, 489[90,92], 547[659,660], 559[863], 738[46]; **7**, 14[129]
Yoneda, R., **1**, 544[32,37], 548[66], 560[37,155], 561[37,156]; **2**, 801[25]; **6**, 227[20], 236[20]; **7**, 172[124]
Yoneda, T., **8**, 190[65]
Yonehara, H., **2**, 367[222]
Yonekawa, Y., **3**, 124[259], 125[259]; **8**, 190[77], 196[77]
Yonekura, M., **4**, 333[23], 398[215]; **7**, 493[199], 800[34]
Yonemitsu, O., **1**, 183[51], 763[145], 858[60], 894[156]; **2**, 10[40], 801[37]; **5**, 834[57]; **6**, 23[93], 652[140], 660[209], 777[61], 930[84]; **7**, 244[71], 245[73,80], 246[81,83-86], 370[66], 686[97]; **8**, 963[49]
Yonemura, K., **2**, 1052[45]
Yoneta, A., **8**, 406[38]
Yoneyama, K., **7**, 423[145], 424[145b]
Yoneyoshi, Y., **1**, 317[153]; **4**, 1038[56,57], 1039[56]
Yonezawa, K., **5**, 282[21], 595[12]
Yonezawa, T., **7**, 862[78], 877[127]
Yoo, B., **4**, 819[210]
Yoo, B. K., **7**, 309[21]
Yoo, S., **4**, 372[64a]; **6**, 660[202], 682[341]; **7**, 678[73]; **8**, 881[68]
Yoo, S.-e., **1**, 408[35], 430[35,132]; **3**, 288[63]; **4**, 24[75]
Yoon, D. C., **7**, 765[160]
Yoon, J., **1**, 535[144]; **4**, 670[23], 672[29], 687[23], 689[68], 696[5]
Yoon, M. S., **8**, 14[83]
Yoon, N. M., **4**, 254[178]; **8**, 16[99,107], 17[107], 18[121], 64[201], 66[201], 74[251], 214[31], 237[15,16,19], 238[22], 240[15,34], 241[22], 242[22], 244[15,16,22,51], 247[22,34], 249[15,51,95], 250[34], 251[22,104], 253[16,22,104], 260[3], 261[7,8,12,13], 263[8,27], 269[27], 272[114], 273[8,27], 275[27], 279, 412[112], 544[269], 709[50], 710[60], 721[60,143], 806[103], 875[38,39], 876[38,42], 877[42]
Yoon, U. C., **7**, 876[124]
Yoon, Y.-J., **6**, 524[371]
Yordy, J. D., **3**, 595[191]
Yorifuji, T., **1**, 417[69]
Yoritaka, K., **4**, 383[140]
Yorke, M., **8**, 264[45]
Yorke, S. C., **2**, 746[109]; **7**, 355[41]
Yorozu, K., **4**, 1086[114,115]
Yoshida, A., **1**, 233[20]; **2**, 213[125], 649[100], 1059[74]
Yoshida, E., **1**, 563[180]
Yoshida, H., **6**, 134[27], 186[172], 489[92]; **8**, 170[71]
Yoshida, J., **3**, 198[52], 469[203], 470[203], 473[203], 483[16], 500[129]; **4**, 599[225], 607[225], 625[225], 642[225], 759[192], 763[192], 840[34], 1032[9], 1051[9]; **6**, 16[60], 676[303]; **7**, 453[93], 455[93], 641[3], 642[8-11], 650[47-49], 765[163]; **8**, 770[32], 787[119]
Yoshida, K., **2**, 232[179], 635[45], 640[45], 646[86], 929[126], 930[126], 931[126], 1022[52]; **3**, 741[52]; **5**, 102[175], 181[154], 344[67b,c], 345[67e], 854[175]; **6**, 8[38]; **7**, 105[148], 225[59], 255[38], 353[35], 355[35], 794[7e], 801[36], 803[53,54], 850[9], 871[9], 878[143]
Yoshida, M., **2**, 917[86], 920[86]; **4**, 719[26]; **6**, 930[87]; **7**, 209[89], 779[424]; **8**, 817[30], 916[103], 917[103], 918[103], 919[103], 920[103]
Yoshida, S., **6**, 214[92]; **8**, 187[39]
Yoshida, T., **1**, 85[28,29], 563[178], 749[77]; **3**, 217[86], 259[137], 489[57], 495[57], 523[23], 795[81,85], 799[101]; **4**, 164[99,99b,c], 249[133], 250[133], 254[182], 883[102], 884[102], 889[136], 890[136,138], 892[136], 893[136], 964[49]; **5**, 1124[47], 1125[63]; **6**, 1021[52]; **7**, 73[104], 596[38], 708[31]; **8**, 146[93], 148[103,104], 153[184], 170[72], 252[110], 458[225], 675[42], 677[42], 678[42], 681[42], 685[42], 697[42], 801[72], 881[80], 882[80]
Yoshida, Z., **1**, 212[9], 213[9,15], 215[41], 216[41], 217[47], 327[12], 448[207], 449[210]; **2**, 185[30], 211[115], 215[115,133,134], 216[133,135], 217[30,136], 442[13,14], 449[13,14,49], 450[13,14,51], 867[15];; **3**, 221[131], 231[242], 420[48,49], 443[61], 445[61], 463[157,159], 1023[23], 1032[70], 1033[71], 1040[103]; **4**, 75[44a,b], 76[44c], 85[73], 100[122], 163[97], 249[117,118], 257[117,118], 377[104], 379[104j,115,115b], 380[104j,115b], 402[236], 403[237], 404[237], 408[259d], 511[178], 557[14], 558[19], 562[33], 564[41,42], 599[214,220], 606[308], 607[308], 642[220], 841[45], 858[109], 948[95]; **5**, 249[36], 438[164], 791[26], 889[32]; **6**, 425[65], 509[248,250], 538[567], 846[105]; **7**, 314[39], 315[43], 503[273]; **8**, 856[185]
Yoshidomi, M., **4**, 837[10]
Yoshifuji, M., **1**, 481[161]; **3**, 71[31-33], 483[14]; **4**, 532[83], 543[83], 545[83]; **7**, 225[65]; **8**, 675[44], 693[110]
Yoshifujii, S., **6**, 531[425]
Yoshihara, K., **1**, 882[121]; **5**, 809[122]
Yoshihara, M., **6**, 538[556]; **8**, 407[54]

Yoshii, E., **1**, 101[92], 243[56], 554[100,104]; **4**, 351[125]; **5**, 260[71], 534[93]; **6**, 150[126], 152[136], 624[144], 994[96]; **7**, 153[11]; **8**, 241[37], 546[309], 557[380,381], 782[102]
Yoshikawa, K., **3**, 244[25], 267[25], 494[84]; **7**, 759[11]; **8**, 698[138]
Yoshikawa, M., **6**, 602[5,10], 616[105]; **7**, 314[41], 315[41]
Yoshikawa, S., **6**, 534[516], 717[110]; **7**, 315[42]; **8**, 239[28]
Yoshikawa, Y., **7**, 764[123]
Yoshikoshi, A., **1**, 851[39], 852[39], 855[52]; **2**, 321[14], 325[14], 541[75]; **3**, 956[107]; **4**, 8[28,28a], 12[43], 18[59], 111[155b], 158[78], 161[89a,b], 243[69], 244[69], 245[69], 258[69], 307[398], 308[398], 370[44,45]; **6**, 107[26], 648[125], 1042[6], 1046[32a], 1047[32b]; **7**, 218[9], 220[21], 458[114], 564[87], 565[87]; **8**, 836[10c,d], 837[10d]
Yoshima, S., **7**, 657[35]
Yoshimoto, H., **8**, 36[85], 39[85], 66[85], 370[89]
Yoshimoto, K., **6**, 657[175]; **8**, 334[64]
Yoshimoto, M., **3**, 24[147], 919[33], 954[33]; **6**, 897[97]
Yoshimura, I., **1**, 853[49], 876[49]
Yoshimura, J., **1**, 371[74], 733[20]; **4**, 85[77e]; **6**, 560[869]; **8**, 154[192,193,194], 374[150]
Yoshimura, K., **5**, 209[54]
Yoshimura, N., **4**, 613[371]; **8**, 215[53], 217[53], 240[32], 244[70]
Yoshimura, R., **8**, 244[54]
Yoshimura, T., **4**, 587[43]; **6**, 677[318,318b], 753[117], 910[9], 924[9,61,63], 925[64], 926[65]; **7**, 470[10,11,13], 763[92]; **8**, 390[84], 391[84], 410[89]
Yoshimura, Y., **1**, 792[270]; **7**, 196[13]
Yoshinaga, K., **6**, 441[85]; **8**, 552[353,354]
Yoshinaga, Y., **1**, 192[82], 385[115]; **2**, 566[23], 939[159]; **5**, 100[157]
Yoshinaka, A., **1**, 336[73]
Yoshinari, T., **8**, 625[164]
Yoshino, A., **1**, 834[124]; **7**, 862[78], 877[127]
Yoshino, T., **6**, 49[95]; **7**, 356[49]
Yoshioka, H., **4**, 314[496], 599[220], 642[220]; **5**, 136[68], 137[74]; **6**, 221[132], 1050[37]; **7**, 606[157]
Yoshioka, K., **2**, 157[122]; **5**, 92[67]
Yoshioka, M., **1**, 223[84], 225[84b]; **4**, 16[52b], 23[71,71a], 162[92], 373[76]
Yoshioka, S., **2**, 167[157]
Yoshioka, T., **1**, 192[82]; **2**, 566[23]; **6**, 652[140], 660[209]; **7**, 244[71], 245[73], 246[81]; **8**, 320[81,82]
Yoshioka, Y., **3**, 867[36]
Yoshitake, J., **2**, 728[145]
Yoshitake, M., **4**, 315[527], 394[199]; **8**, 857[196]
Yoshiura, K., **3**, 1032[67]
Yoshizawa, A., **4**, 1020[234]; **5**, 847[136]
Yoshizawa, S., **3**, 647[194]
Yoshizumi, E., **6**, 559[861]
Yost, R. S., **4**, 73[33]
Yost, W. L., **3**, 566[33]
Yost, Y., **7**, 737[14]
You, M.-L., **2**, 105[40]
Youg, R. N., **1**, 632[64]
Youn, I. K., **8**, 563[435], 615[93]
Youn, J.-H., **2**, 1090[76]
Younathan, J., **4**, 452[22], 473[22]
Young, A. E., **5**, 30[3]
Young, B. Y., **6**, 643[76]
Young, C. A., **4**, 285[159]
Young, D., **2**, 587[140]; **7**, 616[17]
Young, D. E., **8**, 707[25]
Young, D. M., **3**, 613[4], 619[4]; **8**, 115[61], 505[79], 510[96]
Young, D. P., **7**, 446[65]
Young, D. W., **3**, 499[124], 507[124], 898[84]; **5**, 71[156]; **6**, 684[345], 685[345], 687[345]; **7**, 154[20], 279[166], 673[28], 845[81,82], 872[97], 888[97]
Young, E. I., **8**, 898[25], 899[25]
Young, F., **3**, 794[79]
Young, G. B., **3**, 323[146]
Young, G. T., **6**, 668[254], 669[254]; **8**, 974[129,130]
Young, G. W., **8**, 852[140], 856[162]
Young, H., **8**, 269[91]
Young, I. G., **5**, 855[184]
Young, I. M., **8**, 901[39]
Young, J. F., **8**, 152[163,165,166,167], 443[2,5], 445[5], 449[5a], 456[5a]
Young, J.-J., **2**, 378[284]; **4**, 1101[193]
Young, K., **2**, 541[79]
Young, L. B., **3**, 305[75b]; **4**, 313[469]
Young, L. H., **7**, 705[17]
Young, M. G., **6**, 644[93]
Young, M. W., **1**, 652[140], 712[264], 714[264]; **3**, 87[63], 114[63], 117[63]; **6**, 1026[81,83], 1027[81,83], 1028[83], 1029[81], 1030[81], 1031[81], 1032[120], 1033[81]; **7**, 87[22], 124[38], 128[38,69], 129[38], 146[170], 528[3], 769[228], 771[228,269], 775[353], 779[269]
Young, P. A., **7**, 198[27]
Young, P. R., **8**, 410[90]
Young, R. C., **4**, 33[95,95a]
Young, R. G., **8**, 368[68]
Young, R. J., **5**, 404[17]; **7**, 803[52]
Young, R. N., **1**, 821[28]; **3**, 289[67], 771[185]; **4**, 1059[155]; **6**, 26[109], 927[70]; **7**, 223[42], 360[21], 693[26]
Young, R. W., **8**, 88[42]
Young, S., **8**, 198[135]
Young, S. D., **2**, 94[54], 193[63], 194[68], 205[102,103], 206[102b,103], 219[68,144], 221[145], 223[151], 904[51]; **3**, 1055[32]
Young, W. C., **4**, 288[184]
Young, W. G., **5**, 43[33]; **6**, 284[173], 951[2]; **7**, 92[38,50], 94[56]
Young, W. J., **4**, 485[31]
Youngdahl, K., **8**, 22[146], 289[28]
Young Hwan Chang, **7**, 478[86]
Youngs, W. J., **8**, 458[225]
Youssef, A. S. A., **4**, 245[83]
Youssefyeh, R. D., **2**, 139[29]; **3**, 818[95]
Yovell, J., **5**, 71[149,150], 1006[35]
Yoxall, C. T., **6**, 714[84]
Ystenes, M., **7**, 240[58]
Yu, C.-A., **7**, 80[140]
Yu, C.-C., **2**, 1087[35]; **8**, 390[80]
Yu, C.-F., **3**, 173[517]
Yu, C.-M., **2**, 47[154]; **6**, 859[174], 864[198]
Yu, C.-N., **6**, 515[235]
Yu, L.-C., **2**, 107[57]; **3**, 124[253], 126[253], 937[75]; **4**, 113[171], 259[258]; **5**, 1023[77]; **6**, 141[62], 445[99], 893[80]
Yu, M., **4**, 1023[256]
Yu, P.-S., **2**, 419[23]
Yu, S.-G., **4**, 744[137]
Yu, T., **4**, 1011[164]
Yu, T.-Y., **2**, 748[128]
Yu, W. H. S., **7**, 16[165]
Yu, Y. S., **2**, 538[59], 539[59]
Yuan, H. S. H., **1**, 41[267]
Yuan, J.-J., **8**, 859[210]
Yuan, K., **2**, 357[137]
Yuan, S.-S., **3**, 730[44]
Yuan, W., **7**, 545[26]
Yuan Ke, Y., **2**, 643[79], 644[79a], 647[88b]
Yuasa, M., **2**, 387[333]; **8**, 50[119], 66[119]

Yuasa, Y., **2**, 374[275], 1060[88,91], 1073[141]; **4**, 410[261], 502[121,122], 847[77]; **6**, 749[98]
Yue, S., **1**, 268[56]; **3**, 599[211]
Yue, S. T., **1**, 851[44]
Yuen, P.-W., **5**, 404[18]
Yuh, Y., **3**, 382[36]
Yuhara, M., **4**, 591[118], 611[363], 613[369]; **6**, 641[61]
Yukawa, H., **2**, 1049[13]
Yukawa, Y., **6**, 799[19,20]
Yukhno, Yu. M., **6**, 535[534]
Yuki, H., **1**, 14[74-76]
Yuki, Y., **1**, 54[21], 339[85]
Yukimoto, Y., **2**, 904[51]
Yukizaki, H., **3**, 246[39], 257[39]
Yuldashev, Kh. Yu., **3**, 321[140]
Yulina, V. I., **3**, 306[81]
Yun, L. M., **7**, 7[47]
Yunker, M., **6**, 987[69]; **8**, 219[79]
Yura, T., **1**, 834[121b]; **2**, 116[132,139]; **4**, 230[256,257]; **6**, 26[108]
Yura, Y., **1**, 368[59], 369[59]; **4**, 113[164]; **5**, 717[94]
Yurchenko, A. G., **3**, 302[50], 383[47]
Yur'ev, V. P., **8**, 699[150]
Yur'eva, V. S., **7**, 774[335]
Yus, M., **1**, 830[94]; **3**, 253[90], 788[50]; **4**, 290[200], 291[211,212,213,214,215,217,218], 295[248,249,254], 302[338], 315[511], 349[110], 351[124], 354[110], 405[251], 735[82], 741[82], 799[111]; **5**, 755[71], 780[71]; **7**, 519[23], 533[35,36], 534[35], 630[53,54], 632[60]; **8**, 851[132], 856[176,179,183], 857[187]
Yusá, M., **4**, 291[216]
Yus-Astiz, M., **4**, 291[219], 303[341], 315[513]
Yuste, F., **1**, 544[35], 552[80]; **8**, 368[67], 374[67]
Yusufoglu, A., **7**, 429[151]
Yusupov, N. U., **3**, 303[58]
Yuzawa, T., **5**, 564[97]
Yvonne, G. R., **1**, 552[81]

Z

Zabicky, J., **2**, 348[52], 350[70], 363[52], 365[211]; **6**, 261[13], 419[5], 420[5,19], 421[19], 424[19], 425[5], 487[2], 488[2,12], 489[2], 508[12], 509[12], 512[12], 515[2], 523[2], 524[2], 545[12], 763[4]; **7**, 689[6]; **8**, 794[4]
Zabirov, N. G., **6**, 432[119], 538[570]
Zabkiewicz, J. A., **3**, 382[36]
Zablocka, M., **3**, 201[78,79]
Zablocki, J., **2**, 1028[79]; **6**, 734[14]
Zabza, A., **4**, 357[148]; **6**, 498[165]
Zacharewicz, W., **7**, 84[3]
Zacharias, G., **6**, 554[726]
Zacharie, B., **2**, 873[26]
Zaczek, N. M., **8**, 542[225]
Zadok, E., **7**, 14[130]
Zador, M., **7**, 447[73]
Zadrozny, R., **7**, 473[27]
Zaera, F., **8**, 421[30]
Zafiriadis, Z., **8**, 142[51]
Zagatti, P., **3**, 263[173]
Zagdoun, R., **8**, 270[95]
Zago, P., **3**, 595[189], 606[189]; **8**, 113[44]
Zagorevskii, V. A., **6**, 279[137]; **8**, 618[121], 619[121], 621[142], 627[176]
Zahalka, H. A., **4**, 553[4]; **7**, 451[31]
Zahler, R., **1**, 743[52], 744[52], 747[52], 749[52], 811[52]; **2**, 597[9]; **3**, 512[201]; **4**, 423[1], 425[1]; **5**, 1115[2], 1116[2], 1122[2a], 1123[2a]; **8**, 676[80]
Zahn, H., **6**, 635[13], 636[13], 668[253], 669[253]
Zahn, T., **5**, 635[86]
Zahnow, E. W., **7**, 801[36]
Záhorszky, U. I., **3**, 664[27]
Zahr, S., **2**, 1094[89], 1095[89]
Zahra, J.-P., **3**, 577[91]; **4**, 155[68f]
Zahradnik, R., **8**, 318[59], 322[59]
Zähringer, U., **6**, 33[7], 40[7], 57[7]
Zaichenko, N. L., **1**, 837[147]
Zaidlewicz, M., **6**, 865[201]; **8**, 708[42], 711[63], 715[42], 717[42], 719[120], 721[140], 722[145], 728[42], 875[41]
Zaiko, E. J., **4**, 31[94,94a], 519[26]; **6**, 1027[89]; **8**, 501[54], 502[54]
Zaikov, G. E., **7**, 542[7], 543[7]
Zaitseva, G. S., **2**, 726[122]
Zajac, W. W., Jr., **6**, 105[17], 106[18,19], 107[20,21], 675[299]; **7**, 218[7], 737[17]; **8**, 213[30], 214[33], 217[67], 218[69], 222[94,95], 925[10]
Zajacek, J. G., **8**, 898[23], 899[23]
Zajdel, W. J., **1**, 476[111,125], 477[111,125], 483[167]; **3**, 65[2], 66[12], 68[2,22], 69[2,22], 71[2], 74[12], 194[11]; **6**, 65[2]
Zakharkin, L. I., **1**, 212[4], 215[32]; **3**, 303[57], 898[81]; **5**, 768[122,135]; **8**, 214[39,50], 260[2], 266[54,56], 267[54,61,67], 271[108], 272[115], 274[135,139], 275[140], 698[147], 735[14], 741[36,38], 742[46], 746[55], 747[14], 748[14], 753[55], 754[78,80,87,99,106,113], 755[106,113]
Zakharov, E. P., **8**, 609[51]
Zakharova, I. A., **8**, 770[34]
Zakir, U., **4**, 297[272,273,274]
Zakrezewski, J., **7**, 6[29]
Zakrzewski, J., **6**, 271[89]; **7**, 741[48], 747[48]
Zaks, A., **8**, 185[13], 206[13]
Zakutansky, J., **4**, 980[110], 982[110]
Zalar, F. V., **3**, 649[206]
Zaleska, B., **2**, 378[290]
Zalesov, V. S., **3**, 887[10], 888[10], 889[10], 890[10], 893[10], 897[10], 900[10], 903[10]
Zalkow, L. H., **7**, 154[16], 174[136,137]; **8**, 340[89]
Zalkow, L. W., **7**, 478[85]
Zalukaev, L. P., **8**, 318[66], 546[308]
Zamarlik, H., **3**, 342[10]; **4**, 95[102f]
Zambon, R., **3**, 273[16], 274[16]
Zambonelli, L., **4**, 403[239], 404[239]
Zamboni, R., **4**, 316[540]; **5**, 94[88], 95[88]; **6**, 26[109], 93[132]; **8**, 540[195]
Zambri, P. M., **7**, 674[51]
Zamecnik, J., **3**, 20[105]; **8**, 331[30]
Zamojski, A., **2**, 662[14,18], 663[27,28], 664[14,18,27,28]; **5**, 168[104], 169[104,107,108], 170[110], 171[115]; **6**, 1013[17]
Zamureenko, V. A., **8**, 535[160]
Zanarotti, A., **3**, 691[127,132,134]; **4**, 1085[102]
Zanasi, R., **3**, 386[57]
Zander, K., **6**, 1044[19]
Zander, W., **5**, 703[19], 705[19]
Zandstra, H. R., **1**, 861[71]
Zanello, P., **8**, 458[219]
Zang, H. X., **4**, 605[296]
Zang, K., **1**, 749[317], 807[317]
Zanger, M., **8**, 860[222]
Zani, C. L., **2**, 745[97]
Zani, P., **5**, 440[173]
Zanirato, P., **7**, 477[69]; **8**, 385[46]
Zann, D., **4**, 180[66]
Zao, S. H., **7**, 777[378], 778[378]
Zapata, A., **3**, 983[21], 984[21,21a]
Zapevalov, A. Y., **3**, 644[137]
Zard, S. Z., **4**, 747[153], 748[156,157,160], 765[223], 768[239], 824[234]; **6**, 442[87], 938[130], 942[130]; **7**, 132[100], 146[100], 719[6], 720[6], 725[33], 726[6,35-37], 727[38], 728[41], 730[46,47,49], 731[52], 732[59]; **8**, 392[109], 393[113], 818[40]
Zarecki, A., **3**, 849[59]
Zaretskaya, I. I., **5**, 752[1,2,5,6,12,28,29,32,35-38], 754[32,35,36], 756[35,36], 767[6]
Zaretskii, V. I., **2**, 811[70], 813[70], 814[70]
Zaretzkii, Z., **7**, 40[7]
Zarges, W., **1**, 32[160], 33[162]
Zarin, P., **8**, 587[31]
Zaro, J., **6**, 1016[27]
Zask, A., **4**, 384[143], 539[108], 557[15]; **5**, 1094[98], 1096[98], 1098[98], 1112[98]; **7**, 238[43], 359[17]
Zaslona, A., **1**, 774[210]; **3**, 1057[37]; **4**, 1061[167]
Zass, E., **6**, 831[7]
Zassinovich, G., **8**, 91[56], 552[359]
Zatorski, A., **3**, 953[105]; **6**, 134[36], 150[117]; **7**, 197[22], 765[132]
Zaubitzer, T., **2**, 384[318]
Zaugg, H. E., **1**, 371[75,76]; **2**, 971[91], 1048[5], 1049[5], 1052[5], 1053[5], 1070[5]; **5**, 86[19], 485[180]; **7**, 804[63]; **8**, 292[41]
Zavada, J., **6**, 953[7], 1013[9]; **8**, 726[191]
Zavarin, E., **3**, 686[111]
Zavgorodnii, S. V., **3**, 305[70,74]
Zavitsas, A. A., **7**, 95[81]
Zavoranu, D., **8**, 124[89]
Zav'yalov, S. I., **8**, 610[57]
Zawacky, S., **3**, 717[45], 752[94]; **8**, 346[124], 946[137]
Zawadzki, S., **6**, 76[45], 79[64], 116[91], 267[55]; **7**, 500[245]; **8**, 385[44]
Zawisza, T., **2**, 360[172]

Zawoiski, S., **3**, 530[65], 533[65]
Zaworotko, M. J., **1**, 215[33]
Zayed, A., **2**, 760[38]; **8**, 478[38]
Zbaida, D., **5**, 78[274]
Zbaida, S., **1**, 836[141]; **5**, 959[319]
Zbiral, E., **1**, 623[76], 788[258]; **3**, 199[54], 812[56], 813[65], 814[68], 817[86,89], 939[76]; **4**, 252[158]; **6**, 206[40], 210[40], 1061[72]; **7**, 488[152], 491[182], 498[223], 506[152], 508[311], 588[172,173]; **8**, 860[223]
Zbozny, M., **8**, 334[63]
Zbur Wilson, J. A., **1**, 752[94]
Zdunneck, P., **1**, 215[32]
Zdanovich, V. I., **8**, 318[63], 486[61]
Zderic, S. A., **7**, 603[115,118,120]; **8**, 101[121,122], 102[126]
Zebovitz, T., **4**, 845[68], 847[68]
Zecchi, G., **4**, 1085[98,99,103]; **6**, 252[152]
Zecchini, G. P., **1**, 734[23]; **8**, 17[118]
Zech, K., **8**, 140[17]
Zechmeister, L., **5**, 451[9]; **8**, 365[31]
Zee, J., **3**, 846[44]
Zee, S.-H., **8**, 713[77], 715[77]
Zee-Cheng, K.-Y., **7**, 109[181]
Zeegers, P. J., **6**, 110[55]
Zeeh, B., **2**, 1086[26]; **6**, 111[64], 295[255]
Zeelen, F. J., **3**, 361[76], 367[76,100,103], 371[116]
Zefirov, N. S., **3**, 864[21]; **4**, 310[427], 330[5], 335[26], 342[67], 347[101,103,106], 356[144], 357[146], 969[66]; **6**, 2[9], 3[9]; **7**, 494[202]
Zeghdoudi, R., **8**, 881[70]
Zehani, S., **8**, 97[97]
Zehavi, U., **6**, 636[22], 651[138]
Zehnder, B., **1**, 564[198]
Zeidler, F., **4**, 272[31], 279[31], 280[31], 287[31]
Zeidler, U., **7**, 706[24]
Zeifman, Yu. V., **5**, 113[234]; **6**, 498[169,170], 500[170], 527[409], 547[667], 552[667]
Zeigler, F. E., **3**, 588[160], 610[160]
Zeigler, J. M., **6**, 504[223]
Zeilstra, J. J., **2**, 332[56]
Zein, N., **5**, 736[145], 737[145]
Zeinalov, F. K., **6**, 462[10]
Zeiss, H. H., **1**, 174[2,10,14], 175[16], 179[14]; **4**, 519[13], 520[13], 877[70]
Zeiss, H.-J., **1**, 180[37], 181[37]; **2**, 3[10], 6[10b], 12[10], 13[10], 14[10], 25[97], 26[97,101], 27[97,101], 31[97], 33[97b,121], 41[97b,101,121a], 42[101], 43[97b], 995[42]
Zelawski, Z. S., **2**, 808[50]
Zelenin, K. N., **6**, 487[41], 489[41], 515[310,311,312,313]
Zell, R., **8**, 205[159,161], 560[405,406]
Zelle, R. E., **1**, 134[113], 329[35]; **2**, 570[38], 652[126], 704[88]; **3**, 220[124]; **5**, 350[77]; **6**, 899[110], 900[110], 989[80], 995[80]; **7**, 246[89]
Zeller, E., **1**, 559[149]
Zeller, J. R., **1**, 534[140,141], 740[43], 741[43]
Zeller, K., **1**, 844[5b]
Zeller, K.-P., **3**, 887[11], 891[42], 892[11], 893[11], 897[11], 898[11], 900[11], 903[11], 905[11], 909[155]; **4**, 373[85], 1032[12]; **7**, 390[7], 632[57]; **8**, 851[132]
Zeller, P., **2**, 612[105]; **6**, 965[85]
Zeller, W. E., **5**, 519[102], 537[99], 538[102]
Zeltner, P., **2**, 965[70]
Zemach, D., **8**, 517[125]
Zembayashi, M., **3**, 437[21,25], 440[25], 448[25], 449[25,99,100], 450[25], 451[25], 452[25], 464[171], 484[26], 487[45], 492[26,74], 494[26], 495[26], 500[129], 503[26], 513[26]
Zemlidka, J., **2**, 785[39c]
Zemlyanova, T. G., **2**, 787[52]
Zen, S., **6**, 47[77]
Zeng, L.-M., **4**, 4[14]
Zeng, Y., **2**, 146[70]
Zenitani, Y., **8**, 320[81]
Zenk, M. H., **3**, 77[59]
Zenki, S., **1**, 391[151], 392[155]; **7**, 745[76]
Zenkovich, I. G., **7**, 483[127]
Zepp, R. G., **7**, 41[26]
Zerban, G., **4**, 1006[104]
Zerby, G. A., **3**, 136[373], 137[373]
Zercher, C. K., **5**, 736[145], 737[145]
Zerger, R. P., **1**, 9[42], 10[53]
Zervas, L., **6**, 632[3], 635[3], 644[86], 664[220], 666[232], 667[232]
Zervos, M., **1**, 556[124,125]; **4**, 113[168,168e], 245[95]
Zetterberg, K., **4**, 560[21,22,26], 598[184,185,188,190], 622[185], 623[190], 631[420], 638[185,190]; **7**, 474[44,45], 504[282]
Zetzsche, F., **8**, 286[11]
Zeugner, H., **6**, 525[376]
Zeuli, E., **8**, 875[29]
Zeuner, F., **4**, 434[126]
Zeuner, S., **4**, 985[128]; **5**, 1086[66]
Zhadanov, B. V., **6**, 490[106]
Zhai, D., **1**, 123[74], 373[92], 375[92], 376[92]; **2**, 582[107], 649[104], 1052[48], 1075[48], 1076[48]; **3**, 277[26]; **7**, 545[29]; **8**, 655[86]
Zhai, W., **1**, 404[23]; **2**, 582[107], 1076[152]; **3**, 277[26]; **7**, 545[29]; **8**, 655[86]
Zhang, C., **3**, 669[54,55]; **8**, 447[135]
Zhang, J., **1**, 255[19], 258[19], 271[19]; **2**, 355[127]
Zhang, K., **8**, 669[8]
Zhang, L., **2**, 772[18]; **3**, 638[93]
Zhang, L.-H., **5**, 864[262]; **6**, 737[42], 738[42,43]
Zhang, N., **4**, 380[124]; **7**, 283[188], 285[188]
Zhang, P., **2**, 772[16]; **6**, 184[150]
Zhang, W., **7**, 428[148g]
Zhang, W.-Y., **4**, 212[99]
Zhang, X.-a., **2**, 198[85], 632[27], 640[27]
Zhang, Y., **3**, 21[128], 251[78], 254[78]; **4**, 854[92]; **5**, 1166[17], 1167[17]; **8**, 382[6], 406[45], 879[55], 880[55]
Zhang, Y.-L., **5**, 75[229]
Zhang, Y.-Z., **3**, 503[148]
Zhao, C., **4**, 842[52]
Zhao, K., **3**, 125[302], 134[302]
Zhao, M., **3**, 638[93]
Zhao, S. H., **6**, 150[113]; **7**, 425[146], 777[381]
Zhao, Y.-F., **2**, 765[73]
Zhdan, P. A., **8**, 608[47]
Zhdankin, V. V., **4**, 347[101,103], 356[144]; **6**, 2[9], 3[9]
Zhdanov, Yu. A., **4**, 314[492,493]; **6**, 552[696]; **7**, 294[18]
Zhelkovskaya, V. P., **3**, 643[131]
Zhemaiduk, L. P., **7**, 543[18], 579[18], 581[18]
Zheng, S.-Q., **4**, 278[100], 286[100]
Zheng, Z.-I., **6**, 842[80], 997[114]
Zhesko, T. E., **6**, 276[120]
Zhestkov, V. P., **6**, 487[68], 489[68]
Zhi, L., **4**, 1089[137], 1090[137], 1091[137]
Zhidkova, A. M., **6**, 488[34], 502[208], 554[734,789,790,791]
Zhi-min, W., **7**, 844[58]
Zhou, B., **7**, 579[133], 580[133]
Zhou, B. N., **8**, 190[85]
Zhou, H., **8**, 678[63], 680[70], 685[63], 686[63], 691[70]
Zhou, J., **3**, 669[54,55]
Zhou, W. S., **7**, 166[86b]
Zhou, X., **8**, 413[122]
Zhou, X.-R., **7**, 105[150]
Zhou, Z., **4**, 1039[65]
Zhu, J., **2**, 357[137]; **4**, 753[168]; **6**, 845[99]

Zhu, Z., **4**, 629[408]
Zhuk, D. S., **5**, 938[208]
Zhukov, A. G., **3**, 302[50]
Zhuraleva, I. A., **6**, 538[555]
Zhuravleva, E. F., **4**, 1051[126]
Zhurkovich, I. K., **8**, 500[49]
Zia, A., **6**, 255[169]
Zibarev, A. V., **5**, 422[86]
Zibuck, R., **3**, 224[173]; **6**, 136[39]; **8**, 354[162]
Zicmanis, A., **1**, 543[23]; **2**, 345[23]
Ziegenbein, W., **3**, 271[2], 272[2]
Zieger, H. E., **8**, 314[26]
Ziegler, C., Jr., **4**, 754[178], 755[178], 844[64], 845[64,68], 846[73], 847[68,73], 848[73]
Ziegler, C. B., **4**, 375[97], 746[142]
Ziegler, C. B., Jr., **6**, 198[236]
Ziegler, D., **7**, 306[6]
Ziegler, E., **2**, 367[229]; **6**, 275[109,110], 524[356]
Ziegler, F., **4**, 854[95]
Ziegler, F. D., **5**, 864[256]
Ziegler, F. E., **1**, 567[220], 744[57]; **2**, 6[36], 72[79], 204[98], 353[96], 388[96], 547[102,103,107], 548[102], 549[103], 550[107]; **3**, 131[325], 215[64], 219[114], 226[198], 368[105], 499[140], 501[140], 502[140], 994[40]; **4**, 10[34], 40[113], 53[113], 113[164,176], 206[50], 249[124], 258[124], 259[256], 380[123], 650[426]; **5**, 11[85,86], 128[32], 130[32], 531[75], 549[75], 789[30], 790[23], 791[23], 827[2], 829[2], 832[42], 853[172], 867[2c], 877[8], 885[18,19], 886[24], 1000[11], 1123[36]; **6**, 10[44], 11[44], 12[44], 14[58], 16[58], 834[35], 842[80], 855[35], 997[114]; **7**, 257[48], 376[81]; **8**, 341[103], 928[24]
Ziegler, K., **1**, 139[6]; **3**, 194[5]; **4**, 866[4], 867[8], 868[13], 887[121,122], 888[133]; **5**, 66[77], 451[8]; **8**, 100[116], 260[1], 734[2,7,8], 735[10,11], 736[19], 737[2], 738[29,31], 739[11,19,33], 744[50], 753[2,31], 754[29,97,98,102,104], 755[104], 756[50], 758[165]
Ziegler, M. L., **5**, 635[86]
Ziegler, R., **4**, 207[61], 208[61]
Ziegler, T., **1**, 546[53]; **2**, 463[78]; **8**, 670[10,14], 671[10]
Zielinski, J., **6**, 471[63], 789[110]
Zielinski, M. B., **7**, 819[19]
Zielinski, W., **6**, 525[383,384], 771[42]
Ziemnicka-Merchant, B., **6**, 441[85]
Zienty, F. B., **4**, 35[98c]
Zierke, T., **2**, 33[122]
Zietz, J. R., Jr., **4**, 887[123], 888[123]; **8**, 100[114]
Ziffer, H., **3**, 734[8], 767[164]; **5**, 176[130], 218[28], 223[67-70]; **8**, 187[41], 203[147]
Zigman, A. R., **4**, 868[18]; **8**, 476[26]
Zigna, A. M., **4**, 605[294,297], 647[297]
Zikra, N., **3**, 249[65]
Zil'berman, E. N., **4**, 292[232]; **6**, 261[9]; **8**, 298[73]
Zilch, H., **2**, 874[27,28], 875[28]
Zilkha, A., **3**, 640[101]; **7**, 495[206]; **8**, 36[49], 66[49]
Zilniece, I., **1**, 543[23]
Zima, G., **1**, 669[183], 670[183], 671[183], 699[183]; **4**, 245[89], 340[49], 376[102], 377[102]; **6**, 1030[107]; **7**, 131[83,85], 520[24,26], 521[34]; **8**, 850[119]
Zimaity, M. T., **4**, 45[126]
Zimenkovskii, B. S., **8**, 657[95]
Zimero, C., **3**, 572[66]
Zimin, M. G., **6**, 432[119]
Zimmer, A., **2**, 163[150]
Zimmer, C., **4**, 280[130], 281[130], 282[130], 286[130]
Zimmer, H., **1**, 544[38], 562[38,169]; **2**, 505[7]; **3**, 459[138]; **6**, 705[31], 965[90]; **7**, 143[145], 346[6]
Zimmer-Gasser, B., **6**, 175[67]
Zimmerman, C. A., **2**, 110[74]
Zimmerman, D. C., **5**, 780[202]
Zimmerman, D. M., **2**, 1024[59]
Zimmerman, G. A., **5**, 125[17]
Zimmerman, H. E., **1**, 528[112]; **2**, 153[107], 210[111], 250[37], 261[37], 397[10], 412[9], 413[9], 675[53]; **3**, 158[442], 164[442], 922[37], 924[37]; **5**, 125[17], 194[1-4], 195[10], 196[1-4], 197[1-4,19], 198[1-4], 199[24,25], 200[4], 202[2-4], 204[40], 207[51], 209[2-4,55], 210[1,2,59,60], 219[42], 703[19], 705[19], 914[110]; **8**, 358[196], 491[11], 526[20]
Zimmerman, I., **5**, 636[101]
Zimmerman, J., **3**, 41[230]; **7**, 657[25]
Zimmerman, J. E., **6**, 664[222]
Zimmerman, R. L., **5**, 947[260], 960[260]
Zimmerman, W. T., **2**, 370[259]; **8**, 940[109], 947[109], 952[109]
Zimmermann, D. C., **2**, 1049[19]
Zimmermann, G., **2**, 60[18], 62[18d]
Zimmermann, H., **6**, 960[52]
Zimmermann, H.-J., **2**, 112[89], 241[14]; **4**, 145[23]
Zimmermann, J., **2**, 645[82], 1052[47]; **4**, 207[60,61], 208[61]
Zimmermann, P., **6**, 53[121,122], 73[25]
Zimmermann, R., **6**, 171[3,5], 172[5], 177[110,111], 181[110,111], 182[110,111], 184[110,111], 185[110,111], 198[3,5], 199[5], 200[5,110,111], 201[5,110], 202[5]; **8**, 863[238]
Zimmermann, R. G., **3**, 439[39]
Zimniak, A., **4**, 541[115]
Zinczuk, J., **6**, 650[128]
Zingales, F., **4**, 710[48]
Zingaro, R. A., **8**, 413[134]
Zingqing, Z., **1**, 543[15]
Zinke, H., **7**, 712[61]
Zinke, P., **6**, 508[286], 537[286]
Zinke, P. W., **1**, 263[42], 264[42], 265[42], 266[42], 274[74]
Zinn, M. F., **1**, 463[23]
Zinner, G., **1**, 386[124]; **2**, 1088[41-43]; **6**, 116[84], 547[661], 922[53]; **7**, 738[21]; **8**, 63[197], 64[197], 66[197], 70[225], 71[225]
Zinner, H., **6**, 677[322]
Zinner, I. G., **2**, 1088[41]
Zinner, J., **6**, 39[38]
Zinnius, A., **1**, 6[32]
Ziólkowski, J. J., **4**, 923[30], 924[30], 925[30]
Ziólkowsky, J. J., **7**, 95[73a]
Zioudrou, C., **6**, 614[83], 619[83]
Zipkin, R., **1**, 779[226]
Zipkin, R. E., **3**, 278[31], 289[31], 558[51,52]; **5**, 743[164], 744[164]
Zippel, M., **2**, 211[113]; **5**, 75[231]
Zirka, A. A., **8**, 608[48], 629[182,183]
Zirngibl, L., **4**, 493[80]
Zirotti, C., **1**, 185[55], 186[55], 221[68]; **2**, 29[105], 30[113], 31[113,113a], 998[51], 999[51]; **8**, 195[111], 203[111,149]
Zitrin, C. A., **8**, 269[90]
Zitsman, J., **2**, 279[18]
Zizuashvili, J., **5**, 223[84], 224[84], 740[150]
Zlatkina, V. L., **4**, 1063[172]
Zlotin, S. G., **7**, 493[195]
Zobácová, A., **6**, 225[6], 226[6], 258[6]
Zobova, N. N., **5**, 104[183,185], 107[183], 451[40], 470[40], 485[40]
Zocchi, M., **4**, 602[257,260,261]; **5**, 1131[12]
Zoeckler, M. T., **5**, 1000[10], 1002[10], 1009[10]
Zoeller, J. R., **5**, 712[57c]
Zoghaib, W. M., **8**, 18[124]
Zolch, L., **4**, 434[126]
Zoller, G., **6**, 76[40], 77[40]
Zoller, L. W., III, **2**, 757[9]
Zoller, P., **8**, 957[15]
Zoller, U., **1**, 819[1], 834[1], 835[1]; **2**, 1074[144]; **3**, 867[37], 872[37], 883[37], 884[37]; **7**, 516[2]

Zollinger, H., **2**, 748[127]; **4**, 443[191]; **6**, 204[20]
Zollo, P. H. A., **1**, 419[77]
Zoltewicz, J. A., **4**, 423[4], 426[57], 493[81]; **6**, 432[117]
Zoltewicz, J. T., **4**, 457[53]
Zombeck, A., **4**, 306[371]
Zon, G., **7**, 478[86]; **8**, 411[102]
Zönnchen, W., **6**, 642[65]
Zook, H. D., **1**, 3[18,20,21], 42[20c]; **3**, 13[67], 723[10]
Zoorob, H. H., **2**, 760[38]
Zorc, B., **8**, 815[22]
Zordan, M., **2**, 564[11]
Zoretic, P. A., **1**, 656[155], 658[155]; **6**, 667[238], 904[142]; **7**, 125[56], 129[56], 130[56]; **8**, 938[92]
Zorrilla Benitez, F., **2**, 348[60], 357[60]
Zosel, K., **8**, 738[31], 744[50], 753[31], 756[50]
Zosimo-Landolfo, G., **8**, 61[187], 66[187]
Zotova, T. D., **6**, 531[453]
Zoutendam, P., **3**, 595[191]
Zovko, M. J., **3**, 499[114]
Zrimšek, Z., **6**, 480[114], 554[715,782]
Zschage, O., **2**, 39[134b]; **6**, 863[189]
Zschiesche, R., **4**, 27[79], 1048[119]; **5**, 539[105]
Zschoch, F., **2**, 141[43]
Zschocke, A., **2**, 282[35]; **5**, 457[89]
Zsely, M., **5**, 438[164], 534[94]
Zsigmond, A. G., **8**, 418[5], 420[5], 423[5], 439[5], 441[5], 442[5], 883[90]
Zsindely, J., **3**, 809[39,40]; **5**, 638[117], 799[72]
Zsiska, M., **6**, 734[8], 736[8]
Zsolnai, L., **6**, 524[359]; **8**, 690[102]
Zuanic, M., **4**, 955[15]
Zubareva, N. D., **8**, 150[132]
Zubay, G., **6**, 436[23]
Zuberi, S. S., **7**, 747[95]
Zubiani, G., **1**, 206[111], 489[19], 498[19,50], 830[92], 832[92]
Zubova, T. E., **6**, 538[555]
Zuccarello, F., **3**, 386[57]
Zuccarello, G., **3**, 883[106,107]; **5**, 736[145], 737[145]
Zuccaro, L., **2**, 213[127]
Zucker, P. A., **1**, 894[160]; **7**, 625[38]; **8**, 514[105]
Zuckerman, J. J., **4**, 867[7]
Zuech, E. A., **7**, 449[5], 450[5], 452[5]
Zueger, M. F., **8**, 185[29], 190[78]
Zuffa, J. L., **8**, 446[70,71]
Züger, M., **4**, 110[150b]
Zugravescu, I., **3**, 921[36]
Zulaica, E., **2**, 765[77]
Zuman, P., **3**, 566[32]
Zumbulyadis, N., **4**, 50[142]
Zumbulyadis, Z., **7**, 774[315]
Zunker, D., **4**, 395[205]
Zunnebeld, W. A., **5**, 402[5]
Zupan, M., **4**, 356[139]; **5**, 829[19]
Zupancic, N., **5**, 829[19]
Zuraw, P., **8**, 855[160]
Zuraw, P. J., **4**, 390[173b,c]
Zürcher, A., **8**, 530[89]
Zurer, P. S. J., **6**, 779[65]; **8**, 335[51]
Zurflüh, R., **7**, 120[13], 123[13]
Zurqiyah, A., **8**, 531[120], 536[173], 538[173], 542[173]
Zurr, D., **6**, 659[195]; **7**, 244[69,70]
Zushi, K., **7**, 451[29], 534[40]
Zutter, U., **6**, 937[117], 939[117], 940[117]
Zutterman, E., **5**, 924[146]
Zutterman, F., **1**, 648[134], 653[134], 659[160], 672[160,208], 674[160], 676[160], 698[160], 704[160], 715[160], 717[160], 718[160], 862[76]; **3**, 786[44]; **5**, 139[84]; **6**, 1063[86]
Zviely, M., **7**, 707[28]
Zvonkova, E. N., **6**, 271[88]
Zwainz, J. G., **2**, 367[229]
Zwanenberg, D. J., **2**, 757[19]
Zwanenburg, B., **1**, 828[71], 858[61]; **2**, 435[63a,b]; **3**, 750[82,83], 868[41]; **5**, 440[172,175], 441[176,176d,177], 560[78], 561[80,84,85], 562[87], 568[109]; **6**, 538[575]; **7**, 473[26]; **8**, 405[22], 836[2], 843[2f]
Zwanenburg, B. L., **6**, 249[142], 276[121], 277[121], 538[574]
Zwanenbury, B., **4**, 317[556]
Zwart, L., **5**, 79[285]
Zwaschka, F., **6**, 196[229]
Zweifel, G., **1**, 78[6], 95[6], 220[66], 489[16,17,20]; **2**, 82[7], 83[11], 91[48], 586[134], 587[145]; **3**, 199[57,58], 259[133,134], 483[12,13], 486[40], 489[57], 495[40,57], 497[40], 498[40], 503[40], 553[16]; **4**, 141[14], 887[130], 889[137], 893[148], 901[186]; **6**, 244[109]; **7**, 595[11,17], 596[35], 597[51], 600[73], 601[73,80]; **8**, 214[46], 706[16], 707[19], 708[33-35], 716[33-35,88,89,91-93], 717[93-97], 719[119], 724[33,177], 726[33,34,95,177,195], 727[34,93], 735[17], 736[22], 737[17], 740[22], 742[44], 743[164], 746[17], 753[17], 754[22,77,79], 755[118,124,129,132], 756[141,142,150], 757[79,124,164], 758[124,164], 761[17]
Zweig, A., **7**, 854[47], 855[47]
Zweig, J. S., **4**, 229[212]
Zwenger, C., **3**, 660[3]
Zwick, A., **6**, 669[265]
Zwick, W., **2**, 448[37]; **4**, 741[124,126]
Zwierzak, A., **6**, 76[45], 79[64], 83[77,80], 116[91], 267[55]; **7**, 483[123], 500[244,245]; **8**, 385[44], 857[188]
Zwiesler, M. L., **7**, 185[174]
Zwikker, J. W., **6**, 489[94]
Zwolinski, B., **7**, 852[36]
Zybill, C. E., **6**, 177[119], 178[119], 190[200,202], 196[202,227]
Zychlinski, H. v., **2**, 1094[89], 1095[89]
Zydowsky, T. M., **5**, 796[53]; **7**, 230[124]; **8**, 50[120], 66[120]
Zygo, K., **5**, 433[139]
Zyk, N. V., **4**, 335[26], 347[106], 356[144]
Zymalkowski, F., **6**, 225[6], 226[6], 258[6], 488[32], 489[32], 566[32], 567[32], 568[32], 571[32], 795[5]; **7**, 741[46], 746[46], 747[99,100], 748[99,100]
Zyontz, L., **2**, 280[26]
Zysman, A., **8**, 532[130]

Cumulative Subject Index

JOHN NEWTON
David John Services Ltd, Maidenhead, Berks, UK

A-23187 — *see* Calcimycin
A26771B
synthesis
via nitrile oxide cyclization, **4**, 1127
Abietic acid
allylic oxidation, **7**, 93
Birch reduction
dissolving metals, **8**, 500
dioxo ester
rearrangement, **3**, 834
Ab initio calculations
carbonyl compounds
reduction, **8**, 4
Absolute stereochemistry
control
Diels–Alder reaction, **2**, 680
Diels–Alder reactions
chiral auxiliary based methods, **2**, 681
Abstraction
hydrogen atom
recombination, **3**, 1046
ABX blood antigen oligosaccharides
synthesis
Diels–Alder reaction, **2**, 681
Acenaphthalene
hydrobromination, **4**, 280
hydroformylation, **4**, 919
Pauson–Khand reaction, **5**, 1047
Acenaphthene, perisuccinoyl-
synthesis
Friedel–Crafts reaction, **2**, 763
Acenaphthenes
hydrochlorination, **4**, 273
synthesis
Friedel–Crafts cycloalkylation, **3**, 325
Acenaphthoquinone
reaction with hydroxides, **3**, 828
Acenaphthylene
reduction, **8**, 568
Acenaphthyne
synthesis
Ramberg–Bäcklund rearrangement, **3**, 883
Acerogenin
related ethers
synthesis, **3**, 688
Acetal, benzylidene
diol protection
removal, **6**, 660
Acetal, 4-methoxybenzylidene
diol protection
cleavage, **6**, 660
Acetaldehyde
oxidation
palladium(II) catalysis, **4**, 552
reaction with 2-hydroxy-1,4-naphthoquinone and amines
Mannich reaction, **2**, 960
Acetaldehyde, 2-aryl-2,2-dimethoxy-
aldol reaction
five-membered rings from, **2**, 620
Acetaldehyde, chloro-
by-product
Wacker process, **7**, 451
Acetaldehyde, cyclohexylidene-
oxidation, **7**, 306
Acetaldehyde, diphenyl-
Knoevenagel reaction
α-naphthol synthesis, **2**, 354
Acetaldehyde, *p*-hydroxyphenyl-
synthesis
via ketocarbenoids and furans, **4**, 1060
Acetaldehyde, α-methoxyphenyl-
synthesis
chiral, **1**, 527
Acetaldehyde, trichloro-
Oppenauer oxidation
secondary alcohols, **7**, 320, 323
Acetals
acyclic
alcohol protection, **6**, 647
asymmetric epoxidation
compatibility, **7**, 401
bicyclic
reduction, **8**, 227
carbonyl group protection, **6**, 675
chiral
aldol-type reactions, **2**, 650
asymmetric synthesis, **1**, 347
conjugate additions, **4**, 208–210
nucleophilic addition reactions, **1**, 63
cyanation, **1**, 551
cyclic
diol protection, **6**, 659
cyclization
Lewis acid induced, **3**, 362
vinylsilanes, **1**, 585
enol ethers from, **2**, 598
heterolysis
N-acyliminium ion reactions, **2**, 1084
hydride donors
to carbonium ions, **8**, 91
intermolecular additions
allylsilanes, **1**, 610
stereochemistry, **1**, 615

Mannich reaction, **2**, 1013
cyclization, **2**, 1015
reactions with allylsilanes, **2**, 567
reactions with enol silanes
Lewis acid mediated, **2**, 635
reactions with organocopper compounds, **3**, 226
reactions with organometallic compounds
Lewis acid promotion, **1**, 345
reduction
metal hydrides, **8**, 267
to ethers, **8**, 211–232
silyl ketene
preparation, **2**, 599, 604
synthesis
palladium(II) catalysis, **4**, 553
thiol ester silyl ketene
aldol condensation, stereoselectivity, **2**, 634
type III ene reaction, **2**, 553
α,β-unsaturated
addition reactions with alkylaluminum compounds, **1**, 88
N,O-Acetals
chiral
conjugate additions, **4**, 210
Acetals, allylic
reaction with organocopper compounds, **3**, 227
Acetals, α-amino-
synthesis
via azirines, **6**, 787
Acetals, bis(2,2,2-trichloroethyl)-
carbonyl group protection
removal, **6**, 677
Acetals, dithio-
synthesis
via oxidative cleavage of alkenes, **7**, 588
Acetals, halo-
radical cyclizations, **4**, 792
Acetals, 2-halo-
rearrangements, **3**, 788
Acetals, α-hydroxy
chiral
addition reactions with alkylaluminum compounds, **1**, 89
Acetals, α-keto
cyclic
nucleophilic addition reactions, **1**, 63
Peterson alkenation, **1**, 791
Acetals, propargylic
reaction with organocopper compounds, **3**, 227
Acetals, silyl ketene
amination, **6**, 118
Claisen rearrangement, **6**, 858
reaction with imines, **5**, 102
S,N-Acetals, *S*-trimethylsilyl
aldol condensation
stereoselectivity, **2**, 634
reaction with aldehydes
stereoselectivity, **2**, 632
Acetamidation
electrochemical
aromatic compounds, **7**, 800
Acetamide
catalyst
Knoevenagel reaction, **2**, 343
Acetamide, adamantyl-
synthesis, **6**, 401
Acetamide, (*N*-alkenyl)iodo-
cyclization
palladium catalysts, **4**, 843
Acetamide, α-allyloxy-
Wittig rearrangement, **3**, 1004
Acetamide, *N*-(2-bromocyclohexyl)-
synthesis
via Ritter reaction, **6**, 288
Acetamide, cyano-
Knoevenagel reaction, **2**, 361
Acetamide, dimethyl-
dimethyl acetal
Eschenmoser rearrangement, **5**, 891
Acetamide, *N,N*-dimethyl-
Vilsmeier–Haack reaction, **2**, 779
Acetamide, fluoro-
lithium enolates
stereoselectivity, **2**, 211
Acetamide, α-sulfinyl-
enolates
aldol reaction, stereoselectivity, **2**, 228
Acetamide, thiocyano-
Knoevenagel reaction, **2**, 361
Acetamide, trifluoro-
alkylation
alkyl halides, **6**, 83
Acetamides, fluorinated
synthesis, **7**, 498
Acetamides, phosphono-
Hofmann reaction
substituent effect, **6**, 801
Acetamidine, β-sulfonyl-
synthesis, **6**, 550
S-Acetamidomethyl group
thiol protection, **6**, 664
Acetate enolates
chiral
diastereofacial selectivity, **2**, 226
enantioselective aldol reaction, **2**, 315
Acetates
alcohol protection
carbohydrates, **6**, 657
deprotection, **6**, 657
nucleophilic addition to π-allylpalladium complexes
regioselectivity, **4**, 637
stereochemistry, **4**, 621
photochemical deoxygenation, **8**, 817
reduction
silanes, **8**, 824, 825
Acetates, alkylidenecyano-
addition reactions
with organomagnesium compounds, **4**, 89
with organozinc compounds, **4**, 95
Acetates, alkylideneisocyanato-
addition reactions
with organomagnesium compounds, **4**, 89
Acetates, alkylidenephosphono-
addition reactions
with organomagnesium compounds, **4**, 89
with organozinc compounds, **4**, 95
Acetates, 2-halocyano-
syn hydroxylation
alkenes, **7**, 445
Acetates, methoxy-
alcohol protection
nucleoside synthesis, **6**, 658
Acetates, β-nitro-
synthesis, **7**, 493

Acetates, phenoxy-
alcohol protection
nucleoside synthesis, **6**, 658
Acetates, 2,2,2-trialkoxy-
synthesis, **6**, 556
Acetic acid
t-butyl ester
enantioselective aldol reaction, **2**, 308
Ritter reaction, **6**, 269
Acetic acid, acylimino-
8-(–)-phenylmenthyl ester
synthesis, **2**, 996
Acetic acid, α-allyloxy-
esters, Wittig rearrangement, **3**, 1008
zirconium enolates, **3**, 1000
8–phenylmenthyl ester
Wittig rearrangement, **3**, 1001
Acetic acid, α-aminophenyl-
catalyst
Knoevenagel reaction, **2**, 343
Acetic acid, aryl-
esters
Knoevenagel reaction, **2**, 362
Knoevenagel reaction, **2**, 362
synthesis, **4**, 429
Perkin reaction, **2**, 406
Vilsmeier–Haack reaction, **2**, 786
Acetic acid, arylsulfinyl-
methyl ester
Knoevenagel reaction, stereochemistry, **2**, 350
Acetic acid, benzoyl-
ethyl ester, oxime
hydrogenation, **8**, 149
Acetic acid, bis(3′-thienyl)-
Friedel–Crafts reaction, **2**, 759
Acetic acid, bromo-
Vilsmeier–Haack reaction, **2**, 786
Acetic acid, α-bromo-
t-butyl ester
Reformatsky reagent, crystallographic study, **2**, 280
Acetic acid, 4-carboxy-β-phenyl-
Friedel–Crafts reaction, **2**, 756
Acetic acid, cyano-
esters
Knoevenagel reaction, **2**, 360
Knoevenagel reaction, **2**, 360
Vilsmeier–Haack reaction, **2**, 786
Acetic acid, 2,2-dialkyl-
synthesis, **3**, 53
Acetic acid, diazo-
esters
synthesis, **6**, 124
ethyl ester
C—H insertion reactions, **3**, 1051
Acetic acid, ethylnitro-
Knoevenagel reaction, **2**, 364
Acetic acid, fluoro-
toxicity, **6**, 216
Acetic acid, *p*-fluorophenyl-
hydrogenolysis, **8**, 903
Acetic acid, *p*-hydroxymethylphenyl-
carboxy-protecting groups
anchoring, **6**, 670
Acetic acid, iododifluoro-
silyl ketene acetal
preparation, **2**, 604
Acetic acid, isocyano-
esters
Knoevenagel reaction, **2**, 360
Acetic acid, methoxy-
ortho ester
diol protection, **6**, 660
Acetic acid, *N*-methoxyimino-
8-(–)-phenylmenthyl ester
reaction with allyl organometallic compounds, **2**, 995, 996
Acetic acid, methoxyphenyl-
methyl ester
synthesis, **5**, 1084
Acetic acid, 2-naphthylcyclopentyl-
Friedel–Crafts reaction, **2**, 761
Acetic acid, *o*-β-phenethylphenyl-
Friedel–Crafts reaction, **2**, 753
Acetic acid, phenyl-
acyl cyanide synthesis, **6**, 317
ethyl ester
acetylation, **2**, 734
acyloin coupling reaction, **3**, 619
solvent for reductive decarboxylation, **7**, 720
methyl ester
acyloin coupling reaction, **3**, 619
Schmidt reaction, **6**, 817
synthesis
via oxidative cleavage of 3-phenylpropene, **7**, 583
Acetic acid, phenylsulfinyl-
Knoevenagel reaction
activated methylenes, **2**, 363
Pummerer rearrangement, **7**, 194
Acetic acid, trialkyl-
esters
synthesis, **3**, 644
Acetic acid, tributylstannyl-
ethyl ester
reaction with benzaldehyde, **2**, 611
Acetic acid, trichloro-
reaction with thionyl chloride
N,*N*-dimethylformamide catalyst, **6**, 302
Acetic acid, trifluoro-
Beckmann rearrangement, **7**, 695
catalysis
epoxide ring opening, **3**, 738
Friedel–Crafts reaction, **2**, 736
Acetic acid, trimethylsilyl-
ethyl ester
acyloin coupling reaction, **3**, 619
Knoevenagel reaction, **2**, 369
Acetic anhydride
activator
DMSO oxidation of alcohols, **7**, 294
hydrogenation
ruthenium catalyst, **8**, 239
Perkin reaction, **2**, 400
synthesis
via ketene, **6**, 332
titanium tetrachloride complex
crystal structure, **1**, 303
Acetic anhydride, trifluoro-
activator
DMSO oxidation of alcohols, **7**, 295
Friedel–Crafts reaction, **2**, 754
reactions with boron-stabilized carbanions
synthesis of alkenes, **1**, 499
Acetic nitrate, trifluoro-
nitration with, **6**, 110

Acetidinone, acetoxy-
reaction with dienes, **2**, 1058
Acetimidate, arylsulfinyl-*N*-methoxy-
metallation, **2**, 488
Acetimidate, trichloro-
benzyl ester
reaction with alcohols, **6**, 23
glycoside synthesis, **6**, 34, 49, 50
4-methoxybenzyl ester
reaction with alcohols, **6**, 23
Acetimidates, trichloromethyl-
rearrangements, **6**, 843
Acetimide, *N*-hydroxymethylchloro-
amidomethylation with, **2**, 971
Acetimidic acid, trichloro-
allyl ester
alcohol protection, **6**, 652
benzyl ester
alcohol protection, **6**, 651
t-butyl ester
alcohol protection, **6**, 650
carboxy group protection, **6**, 668
Acetoacetates
Michael addition, **4**, 3
Acetoacetic acid
allyl esters
π-allylpalladium complexes from, **4**, 589
esters
synthesis, **6**, 332
ethyl ester
Reformatsky reaction, **2**, 284
ethyl ester, oxime
hydrogenation, **8**, 149
methyl ester
γ-alkylation, **3**, 58
enol silyl ethers, **2**, 606
hydrogenation, chirally modified catalyst, **8**, 150
Acetodiazoacetic acid
ethyl ester
synthesis, **3**, 889
Acetone
aldol reaction
aliphatic aldehydes, **2**, 143
aromatic aldehydes, **2**, 143
dimerization, **2**, 134
hydrogenation
catalytic, **8**, 141
lithium bromide complex
crystal structure, **1**, 299
phenylhydrazone
catalytic hydrogenation, **8**, 143
photolysis
with 1-methylthio-1-propyne, **5**, 163
reduction
dissolving metals, **8**, 114, 526
self-condensation, **2**, 141
sodium cation complexes
theoretical studies, **1**, 287
Acetone, acetyl-
enantioselective hydrogenation, **8**, 151
Acetone, 1-(*N*-acetyl-2-piperidyl)-
phenylhydrazone
catalytic hydrogenation, **8**, 143
Acetone, benzoyl-
aldol reactions
unsaturated β-diketones, synthesis, **2**, 189
Acetone, benzyl-
reduction
borohydrides, **8**, 537
Acetone, benzylidene-
hydrogenation, **8**, 551
kinetics, **8**, 535
iron complexes, **4**, 688
reduction
transfer hydrogenation, **8**, 552, 554
Acetone, dibenzylidene-
Nazarov cyclization, **5**, 752
thermal cyclization, **5**, 754
Acetone, α,α′-dibromo-
[4 + 3] cycloaddition reactions, **5**, 603
Acetone, dihydroxy-
arsenate monoester
aldolase substrate, **2**, 461
Acetone, dimethoxy-
dimerization, **2**, 140
Acetone, 1,3-diphenyl-
tosylhydrazone
organolithium indicator, **6**, 784
Acetone, geranyl-
allylic oxidation, **7**, 94
cyclization, **3**, 349
synthesis
via Carroll rearrangement, **5**, 835
via Claisen rearrangement, **5**, 828
Acetone, hexachloro-
hydride transfer
with 1,4-dihydropyridines, **8**, 93
Acetone, hexafluoro-
carbonylchlorobis(triphenylphosphine)iridium complex
crystal structure, **1**, 310
ene reaction, **2**, 538
Knoevenagel reaction, **2**, 366
Acetone, hydroxy-
Wittig reaction, **1**, 757
Acetone, phenyl-
enolate
reaction with propionaldehyde, **2**, 235
Acetone, tetrabromo-
[4 + 3] cycloaddition reactions, **5**, 603
Acetone, 1,1,1-trifluoroacetyl-
Knoevenagel reaction, **2**, 357
Acetone, triphenylphosphoanyldi-
chlorotrimethyltin complex
crystal structure, **1**, 305
Acetone cyanohydrin
catalyst
benzoin condensation, **1**, 543
Acetone cyanohydrin nitrate
nitration with, **6**, 110
Acetonides
diol protection, **6**, 660
Acetonitrile
decyanation, **8**, 252
Ritter reaction
to *N*-*t*-butyl acetamide, **6**, 261
Acetonitrile, 2-alkoxy-
synthesis
via sulfoxides, **6**, 239
Acetonitrile, alkoxydialkylamino-
amide acetal synthesis, **6**, 574
Acetonitrile, α-allyloxy-α-substituted
synthesis, **1**, 551
Acetonitrile, aryl-

Knoevenagel reaction, **2**, 362
synthesis
via hydroxybenzyl alcohols, **6**, 235
Acetonitrile, bromo-
coupling reactions
with arylzinc reagents, **3**, 260, 466
Acetonitrile, dialkylaminophenyl-
oxidative decyanation
phase transfer, **6**, 402
Acetonitrile, dichloro-
alkylation, **3**, 794
Acetonitrile, diethoxyphosphoryl-
oxide
reaction with alkenes, **3**, 201
Acetonitrile, 1,3-dioxolan-2-yl-
synthesis
via Wacker oxidation, **7**, 451
Acetonitrile, diphenyl-
aromatic nucleophilic substitution, **4**, 429
Acetonitrile, ethylthio-
synthesis, **6**, 231
Acetonitrile, 3-indolyl-
synthesis
Mannich reaction, **2**, 967
Acetonitrile, methoxy-
boron trifluoride complex
NMR, **1**, 292
Acetonitrile, phenyl-
aromatic nucleophilic substitution, **4**, 429
hydrogenation, **8**, 252
lithium enolate
crystal structure, **1**, 32
reduction, **8**, 253
synthesis
via $S_{RN}1$ reaction, **4**, 468
Acetonitrile, phenylselenyl-
conjugate addition reactions, **4**, 111
Acetonitrile, phenylsulfinyl-
Knoevenagel reaction
activated methylenes, **2**, 363
Acetonitrile, phenylsulfonyl-
conjugate addition reactions, **4**, 112
Acetonitrile, α-silyl-
Peterson alkenation, **1**, 790
Acetonitrile, trichloro-
O-alkyl trichloroacetimidate synthesis, **6**, 50
Knoevenagel reaction, **2**, 368
Acetonitrile, trihalo-
reactions with amines, **6**, 546
Acetonitrile, trimethoxy-
synthesis, **6**, 556
Acetonitrile, trimethylsilyl-
conjugate addition reactions, **4**, 111
Knoevenagel reaction, **2**, 369
Acetonitriles
Vilsmeier–Haack reaction, **2**, 789
Acetophenone
aldol reaction
benzaldehyde, **2**, 150
nucleophilic addition reactions
stereoselectivity, **1**, 69
oxidative rearrangement
solid support, **7**, 845
oxime
Beckmann rearrangement, **7**, 696
reaction with allylic organometallic compounds, **1**, 156
reduction
chloroborane, **7**, 603
synthesis
Friedel–Crafts reaction, **2**, 740
Acetophenone, *O*-alkyl-2-enoxycarbonyl-α-diazo-
reaction with rhodium acetate
carbonyl ylide intermediate, **4**, 1091
Acetophenone, benzylidene-
hydrogenation
catalytic, **8**, 142
oxide
benzylic acid rearrangement, **3**, 830
Acetophenone, bromo-
reactions with 2-bromocyclohexanone, **1**, 202
Acetophenone, *p*-bromo-
hydrogenation, **8**, 907
Acetophenone, 1-chloro-
reductions
dialkylzinc, **1**, 319
Acetophenone, 4-chlorotrifluoromethyl-
reduction
hydride transfer, **8**, 94
Acetophenone, diazo-
rearrangements, **3**, 887
Acetophenone, ω,ω-dichloro-
synthesis
Houben–Hoesch synthesis, **2**, 747
Acetophenone, 2′,6′-dihydroxy-
synthesis, **7**, 338
Acetophenone, 3,5-dihydroxy-
Mannich reaction, **2**, 956
Acetophenone, 2,4-diisopropyl-
Friedel–Crafts reaction, **2**, 738
Acetophenone, *o*-(dimethylaminomethyl)-
lithium enolate
crystal structure, **1**, 28
Acetophenone, enolate
reaction with π-allylpalladium complexes, **4**, 591
Acetophenone, *p*-ethyl-
ethylation
Friedel–Crafts reaction, **3**, 301
Acetophenone, *p*-fluoro-
reduction, **8**, 903
Acetophenone, 2-hydroxy-
Vilsmeier–Haack reaction, **2**, 790
Acetophenone, 4-hydroxy-
Mannich reaction, **2**, 956
Acetophenone, 2-hydroxy-2-phenyl-
reduction, **8**, 924
Acetophenone, methoxy-
tin(IV) chloride complexes
crystal structure, **1**, 306
Acetophenone, *p*-methoxy-
oxime
Beckmann rearrangement, **7**, 692
Acetophenone, 4-methyl-
synthesis
Friedel–Crafts reaction, **2**, 738
Acetophenone, nitro-
hydrogenation
catalytic, **8**, 141
Acetophenone, 2-phenyl-
reduction, **8**, 924
Acetophenone, 2,3,5,6-tetramethyl-
Acetophenone
Friedel–Crafts reaction, **2**, 745
Acetophenone, trifluoro-

electrochemical reduction, **8**, 987
reaction with 1,4-dihydropyridine, **8**, 93
Acetophenone, 2,4,6-trimethyl-
rearrangement, **2**, 745
Acetophenone imine, trichloro-
reduction, **6**, 500
Acetophenones
alkynes from, **8**, 950
Birch reduction
dissolving metals, **8**, 508
electropinacolization
induction of chirality, **8**, 134
electroreduction, **8**, 131
hydrogenation
asymmetric, **8**, 152
catalytic, **8**, 141
platinum oxide catalyst, **8**, 319
hydrosilylation
asymmetric, **8**, 174
O-methyloxime
reduction, **8**, 176
oxime
hydrogenation, **8**, 149
reduction
chirally modified lithium aluminum hydride, **8**, 168
dissolving metals, **8**, 115
ionic hydrogenation, **8**, 319
lithium aluminum hydride, **8**, 166
lithium amalgam, **8**, 115
modified lithium aluminum hydride, **8**, 164
stereospecific pinacolization
electroreduction, **8**, 133
1,4-Acetoxychlorination
palladium(II) catalysis, **4**, 565
Acetoxylation
electrochemical
aromatic compounds, **7**, 799
α-Acetoxylation
electrochemical
amides, **7**, 804
carbamates, **7**, 804
ketones, **7**, 798
Pummerer rearrangement, **7**, 196
Acetoxymercuration
vinylallenes
cyclopentenone synthesis, **5**, 774
trans-Acetoxypalladation
dienes, **4**, 565
Acetoxythallation
vinylallenes
cyclopentenone synthesis, **5**, 774
1,4-Acetoxytrifluoroacetoxylation
1,3-cyclohexadiene
palladium(II) catalysis, **4**, 565
Acetylation
base-catalyzed
ester synthesis, **6**, 327
Acetyl chloride, 1-phenanthryl-
Friedel–Crafts reaction, **2**, 757
Acetyl-CoA
structure, **6**, 436
Acetylene (*see also* Alkynes)
hydrosilylation, **8**, 769
monometallation, **3**, 271
trimerization, **5**, 1145
Acetylene, alkoxy-
reaction with ketenes
cyclobutenone synthesis, **5**, 689
Acetylene, alkylthio-
reaction with ketenes
cyclobutenone synthesis, **5**, 689
Acetylene, bis(trimethylsilyl)-
acylation
Friedel–Crafts reaction, **2**, 725
cycloaddition reactions, **5**, 1149
o-quinodimethane precursor
Diels–Alder reactions, **5**, 389
Acetylene, bis(trimethylstannyl)-
cycloaddition reactions, **5**, 1149
Acetylene, *t*-butyl-
trimerization
palladium catalysis, **5**, 1148
Acetylene, di-*t*-butyl-
hydrogenation
palladium-catalyzed, **8**, 431
Acetylene, di-*t*-butyl-
synthesis
Ramberg–Bäcklund rearrangement, **3**, 883
Acetylene, dichloro-
Michael addition, **4**, 42
Acetylene, dicyano-
ene reactions, **5**, 6
Acetylene, dilithio-
synthesis, **3**, 271
Acetylene, diphenyl-
acetoxymercuration, **8**, 858
carbolithiation, **4**, 872
hydrogenation, **8**, 440
hydrogenation to *trans*-stilbene
homogeneous catalysis, **8**, 458, 459
hydrozirconation, **8**, 688
photolysis
with methyl *p*-cyanobenzoate, **5**, 163
reaction with *t*-butyllithium, **4**, 872
reaction with carbene complexes, **5**, 1089
reaction with tetrahydropyridine carbene complexes, **5**, 1105
reduction, **8**, 485
transfer hydrogenation, **8**, 552
synthesis
Ramberg–Bäcklund rearrangement, **3**, 883
Acetylene, divinyl-
dimerization, **5**, 63
Acetylene, ethoxy-
carboboration, **4**, 886
carbocupration, **4**, 900
hydrozirconation, **3**, 498
reaction with alcohols, **6**, 559
reaction with dialkylallylboranes, **5**, 34
reaction with diphenylketene, **5**, 732
Acetylene, hexamethyldistannyl-
carboboration, **4**, 886
Acetylene, lithio-
synthesis, **3**, 271
Acetylene, phenyl-
carbocupration, **4**, 897
carbozincation, **4**, 883
cocycloaddition
3-hexyne, **5**, 1146
hydroalumination, **8**, 735
hydrochlorination, **4**, 277
hydrogenation to ethylbenzene

homogeneous catalysis, **8**, 456
hydrogenation to styrene
homogeneous catalysis, **8**, 457
hydrosilylation, **8**, 770
reduction, **8**, 485
Acetylene, silyl-
hydrosilylation, **8**, 771
Acetylene, sodio-
synthesis, **3**, 271
Acetylene, tolylsulfonyl-
Diels–Alder reactions, **5**, 324
Acetylene, trimethylsilyl-
carboboration, **4**, 886
in terminal alkyne synthesis, **3**, 531
Acetylene, trimethylsilylethoxy-
acid anhydride synthesis, **6**, 315
Acetylene, trimethylsilylmethoxy-
cycloaddition reactions, **5**, 1149
Acetylene, vinyl-
hydrochlorination, **4**, 278
hydrofluorination, **4**, 271
Acetylenedicarbonyl chloride
synthesis
via retro Diels–Alder reaction, **5**, 552
Acetylenedicarboxylic acid
dialkyl esters
ene reactions, **5**, 6
reaction with enamines, cyclobutene ring expansion, **5**, 687
dimethyl ester
Diels–Alder reactions, **5**, 347
hydrogenation to dimethyl fumarate, **8**, 458
hydrogenation to dimethyl maleate, **8**, 458, 459
Acetylenes — *see* Alkynes
1-Acetylethyl group
phosphoric acid protecting group, **6**, 625
N-Acetyl group
amine-protecting group, **6**, 642
Acetylides
organometallic
coupling reactions, 1-haloalkynes, **3**, 553
oxidative coupling reactions, **3**, 554
$S_{RN}1$ reactions, **4**, 472
Acetyl iodide, trifluoro-
deoxygenation
epoxides, **8**, 890
Acetylium tetrafluoroborate
formation
Friedel–Crafts reaction, **2**, 734
N-Acetylneuraminic acid aldolase
cloning, **2**, 464
organic synthesis
use in, **2**, 463
substrate specificity, **2**, 463
Acetyl nitrate
nitration with, **6**, 105, 106
Acetyl nitrate, trifluoro-
nitration with, **6**, 106
synthesis, **6**, 109
Acetylthiosulfenyl chloride
reactions with alkenes, **7**, 516
Acid anhydrides
acid halide synthesis, **6**, 307
acyloin coupling reaction, **3**, 617
amide synthesis, **6**, 383
α-amino-*N*-carboxylic
peptide synthesis, **6**, 383
α-amino-*N*-thiocarboxylic
peptide synthesis, **6**, 383
Curtius reaction, **6**, 810
synthesis, **6**, 301–318
via carboxylic acids, **6**, 309
via carboxylic acid salts, **6**, 314
Acid bromides
alkenic
divinyl ketones from, **5**, 777
reduction
metal hydrides, **8**, 264
synthesis
via acid chlorides, **6**, 306
Acid chlorides
acylation
alkylrhodium(I) complexes, **1**, 450
lithium dialkylcuprates, **1**, 428
organostannanes, **1**, 446
palladium complex catalysis, **1**, 436
synthesis of ketones, **1**, 414
acyloin coupling reaction, **3**, 617
acyl transfer
ester synthesis, **6**, 327
adducts
dimethylformamide, **6**, 493
alkenic
divinyl ketones from, **5**, 777
aromatic
thioamide adducts, **6**, 493
coupling reactions
with sp^3 organometallics, **3**, 463
Curtius reaction, **6**, 807
reaction with organoaluminum reagents
ketone synthesis, **1**, 95
reduction, **8**, 286
Reformatsky reaction, **2**, 296
synthesis
via carboxylic acids, **6**, 302
Tebbe reaction, **1**, 743
α,β-unsaturated
reaction with diazomethane, **3**, 889
vinyl substitutions
palladium complexes, **4**, 835
Acid cyanides
α-acylation, **4**, 261
Claisen condensation, **2**, 801
decarbonylation
palladium-catalyzed, **3**, 1041
Acid fluorides
amide synthesis, **6**, 383
reduction
metal hydrides, **8**, 264
synthesis, **6**, 306
via acid chlorides, **6**, 306
Acid halides
acid anhydride synthesis, **6**, 314
acid halide synthesis, **6**, 306
acylation
thiols, **6**, 440
aliphatic
divinyl ketones from, **5**, 775
amide synthesis, **6**, 383
decarbonylation, **3**, 1040
Friedel–Crafts reaction
bimolecular aromatic, **2**, 740

halogenation, **7**, 122
halogen transfer agents
acid halide synthesis, **6**, 304
α-ketonitrile synthesis, **6**, 317
methylenation
Tebbe reagent, **5**, 1124
nitrile synthesis, **6**, 233
Pummerer rearrangement, **7**, 203
reactions with organocopper reagents, **3**, 226
reduction, **8**, 239
hydrides, **8**, 262
stability
presence of Lewis acids, **2**, 709
synthesis, **6**, 301–318
via acid anhydrides, **6**, 307
via acid halides, **6**, 306
via acyl amides, **6**, 308
via aldehydes, **6**, 308
via carboxylic acid esters, **6**, 307
via carboxylic acids, **6**, 302
tandem vicinal dialkylations, **4**, 261
α,β-unsaturated
acylations by, **2**, 710
vinylic acylations
palladium complexes, **4**, 856
Acidic chalcogenides
catalysts
Friedel–Crafts reaction, **3**, 296
Acidic oxides
catalysts
Friedel–Crafts reaction, **3**, 296
Acidic sulfides
catalysts
Friedel–Crafts reaction, **3**, 296
Acid iodides
synthesis
via acid chlorides, **6**, 306
Acids
α-halogenation, **7**, 122
Aclacinomycin
synthesis
Friedel–Crafts reaction, **2**, 762
Aconitium alkaloids
synthesis, **8**, 945
β-Acoradiene
precursor
synthesis *via* intramolecular ene reaction, **5**, 11
synthesis
via photocycloaddition, **5**, 139
Acoragermacrone
synthesis
via cyclization, **1**, 553
via isoacoragermacrone, **7**, 619
β-Acorenol
precursor
synthesis *via* intramolecular ene reaction, **5**, 11
Acorenone
precursor
synthesis *via* intramolecular ene reaction, **5**, 11
synthesis
via arene–metal complexes, **4**, 543
via cyclopropane ring opening, **4**, 1043
via photochemical cycloaddition, **5**, 129
Acorenone B
precursor
synthesis *via* intramolecular ene reaction, **5**, 11
synthesis
via arene–metal complexes, **4**, 543
Acosamine
amino sugars, **2**, 323
Acridine
electroreduction, **8**, 594
hydrogenation
palladium catalysts, **8**, 598
regioselective reduction, **8**, 600
Acridine, dihydro-
hydride transfer
with 2,3,5,6-tetracyanobenzoquinone, **8**, 93
Acridine, 1,8-dioxodecahydro-
fluorimetric analysis
aldehydes, **2**, 354
Acridine, perhydro-
synthesis, **8**, 598
Acridinium ions
hydride acceptors
reduction with formic acid, **8**, 84
Acridinium salts, 10-methyl-
reduction
dihydropyridine, **8**, 589
Acridizinium cations
Diels–Alder reactions, **5**, 499
Acridonecarboxylic acids
synthesis
Friedel–Crafts reaction, **2**, 759
Acridones
photochemical ring opening, **5**, 712
synthesis, **7**, 333
via arynes, **4**, 497
Acrolein
conjugate additions
organocuprates, **4**, 183
cyclic acetal
hydroformylation, **4**, 923
[2 + 2] cycloaddition reactions, **5**, 72
hydroxyethylene, **5**, 73
dimer
nucleophilic addition reactions, **1**, 52
ene reactions
intermolecular, **5**, 3
Lewis acid catalysis, **5**, 5
Lewis acid complexes
conformation, **1**, 288
lithium cation complexes
structure, **1**, 289
Acrolein, β-chloro-
synthesis
Vilsmeier–Haack reaction, **2**, 785
Acrolein, β-dimethylamino-
synthesis
Vilsmeier–Haack reaction, **2**, 784
Acrolein, β-ethoxy-
synthesis
Vilsmeier–Haack reaction, **2**, 784
Acrolein, α-fluoro-
synthesis
via cyclopropane ring opening, **4**, 1020
Acrolein, α-halo-
synthesis
via dihalocarbene, **4**, 1005
Acrolein, α-lithio-
synthesis, **3**, 253
Acrolein, 2-siloxy-

generation of oxyallyl cations
[4 + 3] cycloaddition reactions, **5**, 597
Acrolein, 2-(trimethylsiloxy)-
[4 + 3] cycloaddition reactions, **5**, 606
Acrolein acetals
Diels–Alder reactions, **5**, 341
Acrylaldehyde, trimethyl-
synthesis
via hydroformylation, **4**, 924
Acrylamide, α-acyloxy-
synthesis, **2**, 1087
Acrylamide, α-cyano-
synthesis
Knoevenagel reaction, **2**, 361
Acrylamide, cyclohexenyl-
ene reactions
intramolecular, **5**, 15
Acrylamide, hydrophenyl-
magnesium salt
intramolecular ene reactions, **5**, 15
Acrylates
addition reactions
benzeneselenenyl chloride, **7**, 520
anionic polymerization, **4**, 246
Diels–Alder reactions, **5**, 355
ene reactions
Lewis acid catalysis, **5**, 4
optically active
cycloaddition reactions with nitrile oxides, **5**, 263
α-substituted
ene reactions, **5**, 4
synthesis
rearrangement of epoxides, **3**, 760
via retro Diels–Alder reaction, **5**, 553
vinyl substitutions
heterocyclic compounds, **4**, 837
Acrylates, α-halo-
ene reactions, **5**, 5
Acrylates, β-lithio-
synthesis, **3**, 253
Acrylates, 3-polyhydroxyalkyl-
synthesis
Knoevenagel reaction, **2**, 385
Acrylates, α-vinyl-
synthesis
copper catalysts, **3**, 217
Acrylic acid, α-acylamino-
asymmetric hydrogenation
homogeneous catalysis, **8**, 460
Acrylic acid, 3-aroyl-
synthesis, **2**, 744
Acrylic acid, 2-(diethylphosphono)-
ethyl ester
addition reaction with enolates, **4**, 103
Acrylic acid, α-formylamino-
reaction of isocyanoacetate
non-Knoevenagel product, **2**, 361
Acrylic acid, α-(methylthio)-
methyl ester
addition reaction with enolates, **4**, 109
Acrylic acid, β-nitro-
ethyl ester
Diels–Alder reactions, **5**, 320
methyl ester
Diels–Alder reactions, **5**, 320
Acrylic acid, perfluoro-
oxidative rearrangement, **7**, 816
Acrylic acid, α-phenylsulfinyl-
Pummerer rearrangement, **2**, 363
Acrylic acid, β-(2,6,6-trimethylcyclohexyl)-
synthesis
via oxidative cleavage, **7**, 587
Acrylic acids
acid chloride synthesis, **6**, 304
asymmetric hydrogenation
homogeneous catalysis, **8**, 461
borane complexes
structure, **1**, 289
configuration
Knoevenagel reaction product, **2**, 345
Diels–Alder reactions
chiral catalysis, **5**, 377
Lewis acid complexes
conformation, **1**, 288
tandem vicinal difunctionalization, **4**, 247
trisubstituted
asymmetric hydrogenation, **8**, 461
Acrylic esters
asymmetric hydrogenation
homogeneous catalysis, **8**, 461
Acrylonitrile
dimerization, **5**, 63
stereochemistry, **5**, 67
ene reactions
intermolecular, **5**, 3
hydroformylation, **4**, 926
oxidation
Wacker process, **7**, 451, 452
reactions with Yamamoto's reagent, **1**, 124
Ritter reaction, **6**, 265
mechanism, **6**, 263
synthesis
via aluminum compounds, **6**, 241
Acrylonitrile, 2-acetoxy-
cycloaddition reactions, **5**, 267
preparation
Darzens glycidic ester condensation, **2**, 419
Acrylonitrile, α-chloro-
Diels–Alder reactions
Lewis acid promoted, **5**, 339
Acrylonitrile, β-chloro-
synthesis
Vilsmeier–Haack reaction, **2**, 785
Acrylonitrile, 2-(*N*-methylanilino)-
addition reactions
with enolates, **4**, 100
with organolithium compounds, **4**, 79
Acrylonitrile, phenylseleno-
radical cyclization, **4**, 733
Acryloyl chloride
ene reactions
thermal, **5**, 3
synthesis
via acrylic acid, **6**, 302
Acryloyl chloride, β,β-dimethyl-
Nazarov cyclization, **5**, 778
Acryloylmethyl lactate
titanium tetrachloride complex
crystal structure, **1**, 303
Actinic activation
electron-transfer equilibria, **7**, 850
Actinide complexes

hydrogenation
alkenes, **8**, 447
hydrometallation, **8**, 696
Actinidine
synthesis, **3**, 599
via Diels–Alder reaction, **5**, 492
Actinobolin
synthesis, **1**, 404
ene reaction, **2**, 542
via cyclofunctionalization of cycloalkenes, **4**, 373
Active hydrogen compounds
aromatic nucleophilic substitution, **4**, 429–433
Active metals
reduction
acetals, **8**, 212
Active methylene compounds
diazo transfer, **6**, 125
Acuminatolide
synthesis
Knoevenagel reaction, **2**, 373
via Diels–Alder reaction, **5**, 468
Acyclase
substrate specificity
synthetic applicability, **2**, 456
Acyclic stereoselective synthesis
allyl metal reagents, **2**, 2
crotyl metal reagents, **2**, 2
Acyl amides
acid halide synthesis, **6**, 308
Acylamino radicals
cyclizations, **4**, 794
Acyl anions
addition reactions, **4**, 113–115
equivalents, **1**, 542
conjugate additions, **4**, 162
selenium containing, alkylation, **3**, 134, 136
sulfur containing, alkylation, **3**, 134
synthetic utility, **2**, 55
lithium
generation, **1**, 273
masked equivalents
benzoin condensation, **1**, 544
samarium
generation, **1**, 273
Acylating agents
Reformatsky reaction, **2**, 296
Acylation
acid catalyzed, **2**, 797
N-acylimidazoles, **6**, 333
amines, **6**, 382
arenes, **6**, 445
boron-stabilized carbanions, **1**, 497
carbanions, **6**, 445
Claisen condensation and, **2**, 817
enzymatic, **6**, 340
esters, **2**, 795–863
Friedel–Crafts reaction
bimolecular aromatic, **2**, 739
hydrogen sulfide
imidates and orthoesters, **6**, 450
imidothioates, **6**, 455
intermolecular
alkenes, **2**, 709
ketenes, **6**, 332
ketones, **2**, 795–863
mixed anhydrides
ester synthesis, **6**, 328
nitriles, **2**, 795–863
organocopper reagents, **1**, 426
organometallic compounds, **1**, 399
palladium catalysis
mechanism, **1**, 438
thiols
acyl halides, **6**, 440
anhydrides, ketenes and esters, **6**, 443
carboxylic acids, **6**, 437
O-Acylation
anomeric
glycoside synthesis, **6**, 59
glycoside synthesis, **6**, 49
Acyl hypofluorites
decarboxylative fluorination, **7**, 723
Acyl hypohalites
carboxyl radicals from, **7**, 718
Hunsdiecker reaction, **7**, 723
synthesis, **7**, 718
Acyl hypoiodites
synthesis, **7**, 723
N-Acyliminium ions
acyclic, **2**, 1070
intermolecular reactions, **2**, 1070
intramolecular reactions, **2**, 1071
addition reactions, **2**, 1047–1079
electrophilicity, **2**, 1056
generation, **2**, 1084
reactions
as carbocations, **2**, 1053
as Diels–Alder dienes, **2**, 1054, 1055
reviews, **2**, 1048
N-Acyliminium salts
stability, **2**, 1053
Acylimonium ions
initiators
polyene cyclization, **3**, 342
Acyl isocyanates
2-azetidinones from, **5**, 104
Acylmetallation
alkynes, **4**, 905
Acyl nitrates
decomposition
nitroalkanes, **7**, 729
Acylnitroso compounds
reactions with alkenes, **6**, 115
Acyloin rearrangement
2-hydroxy ketones, **3**, 791
Acyloins
coupling reactions, **3**, 613–631
heterocyclic systems, **3**, 629
cyclic
synthesis, **3**, 620
hydrogenation
catalytic, **8**, 142
synthesis
epoxide ring opening, **3**, 753
unsaturated
ene reactions, **5**, 23
unsymmetrical
synthesis, **1**, 551
Acyloxallyl cations
initiators
polyene cyclization, **3**, 343
α-Acyloxycarboxamides

synthesis, **2**, 1084
Acyloxymercuration
demercuration
alkenes, **4**, 314–316
Acyloxy radicals
cyclization, **4**, 812
Acyl phosphates
phosphorylation, **6**, 607
synthesis, **6**, 331
Acyl radicals
addition to alkenes, **4**, 740
cyclizations, **4**, 796, 798
samarium
generation, **1**, 273
Acyl tosylates
synthesis, **6**, 329
Acyl transfer
anhydrides
ester synthesis, **6**, 327
intramolecular
ketones, **2**, 845
to alchols
ester synthesis, **6**, 324
Acyl transfer agents
selenol esters, **6**, 461, 468
β-Acylvinyl carbocations
Diels–Alder reactions, **5**, 502
Acyl xanthates
photolysis
radical addition reactions, **4**, 749
Adaline
synthesis
Mannich reaction, **2**, 1014
Adamantane
alkylation
Friedel–Crafts reaction, **3**, 334
anodic oxidation, **7**, 794
arylation
Friedel–Crafts reaction, **3**, 322
functionalization
alkylthio, **7**, 14
oxidation
silver trifluoroacetate, **7**, 13
solid support, **7**, 842
the 'Gif' system, **7**, 13
oxidative rearrangement, **7**, 823
reactions with carbonium ions, **7**, 9
rearrangements, **3**, 854
synthesis
Friedel–Crafts reaction, **3**, 334
Adamantane, adamantylidene-
reaction with bromine, **4**, 330
kinetics, **4**, 344
Adamantane, amino-
quaternary
synthesis, **7**, 505
Adamantane, 1-amino-
synthesis
via 1-bromoadamantane, **6**, 270
Adamantane, 1-bromo-
reaction with naphthalene
Friedel–Crafts reaction, **3**, 302
Ritter reaction, **6**, 269
2,4,6,8-Adamantane, 1,3-dilithio-5,7-dimethyl-
methylation, **3**, 134
Adamantane, 1-hydroxymethyl-
Ritter reaction
effect of conditions, **6**, 264
Adamantane-1-carboxylic acid
ethyl ester
acyloin coupling reaction, **3**, 619
synthesis, **7**, 727
Adamantane-1,3-diol, 2-nitro-
synthesis
Henry reaction, **2**, 329
Adamantanethione *S*-methylide
cycloadditions, **4**, 1074
1-Adamantanol
synthesis
via solid support oxidation, **7**, 842
2-Adamantanol
oxidation
solid support, **7**, 842
Adamantanol, 3-proto-
synthesis
via intramolecular Barbier reaction, **1**, 262
2-Adamantanone
synthesis
via solid support oxidation, **7**, 842
Adamantanones
Peterson alkenation
enol ether preparation, **2**, 597
reduction
ionic hydrogenation, **8**, 319
stereoselectivity, **8**, 5
titanocene dichloride, **8**, 323
Adamantene
dimerization, **5**, 65
Adams' catalyst
hydrogenation, **8**, 418
hydrogenolysis
epoxides, **8**, 882
Addition–fragmentation
intramolecular
expansion, **1**, 892
Addition reactions
C—halogen bond formation, **7**, 527–539
C—N bond formation, **7**, 469–508
C—O bond formation
epoxidation, **7**, 357–385, 389–436
glycols, **7**, 437–447
Wacker oxidation, **7**, 449–466
C—S bond formation, **7**, 515–524
C—Se bond formation, **7**, 515–524
electrochemical, **4**, 129
radicals, **4**, 727–731
Adenosine, 8-bromo-
coupling reactions
with Grignard reagents, **3**, 462
Adenosine, 6-*N*-(3,3-dimethylallyl)-
allylic oxidation, **7**, 88
Adenosine 59-phosphate, *O*-(*N*-acetylthioleucyl)-
synthesis, **6**, 450
Adipic acid
synthesis
via oxidative cleavage of cyclohexene, **7**, 587
Adiponitrile
synthesis, **7**, 8
Adipoyl chloride
Friedel–Crafts reaction, **2**, 741
Adociane, diisocyano-
synthesis

via organostannane acylation, **1**, 446
Adociane, 7,20-diisocyano-
synthesis
via conjugate addition, **4**, 218
Adrene
synthesis
via Michael addition, **4**, 29
β-Adrenergic blocking agents
synthesis, **7**, 397
Adrenosterone
synthesis, **3**, 24
Adriamycin
synthesis, **7**, 341
Aerothionin
biosynthesis, **3**, 689
synthesis, **7**, 337
A-factor
synthesis
via conjugate addition, **4**, 215
Aflatoxin B_1
epoxidation, **7**, 374
Aflavinine, 3-demethyl-
synthesis
Mannich reaction, **2**, 911
Africane
biosynthesis, **3**, 404
Africanol
biosynthesis, **3**, 404
synthesis
via methyllithium addition to unsaturated acid, **1**, 413
α-Agarofuran, 19-keto-
reduction
dissolving metals, **8**, 118
Ajmalicine
microbial hydroxylation, **7**, 65
synthesis, **6**, 740
Knoevenagel reaction, **2**, 372
Aklavinone
synthesis
via cyclofunctionalization of cycloalkenes, **4**, 373
via Diels–Alder reactions, **5**, 327, 342, 393
Alamaridine
synthesis
Mannich reaction, **2**, 913
Alamethicine
synthesis, **2**, 1096
Alane, alkenyloxy-
preparation, **2**, 268
Alane, alkenyloxydialkyl-
homochiral
aldol reactions, **2**, 271
Alane, alkenyloxydiethyl-
aldol reactions
imines, **2**, 271
Alane, alkoxy-
synthesis, **8**, 214
Alane, chloro-
reduction
acetals, **8**, 214
Alane, chlorodiethyl-
aldol reactions, **2**, 272
Alane, crotyldiethyl-
reaction with aldehydes, **2**, 31
Alane, dialkylchloro-
aldol reactions
zinc coreagent, **2**, 269
Alane, dibromo-
selective ketone reduction, **8**, 18
Alane, dichloro-
reduction
acetals, **8**, 214
Alane, diethyl(phenylethynyl)-
reaction with epoxides
regioselectivity, **6**, 7
Alane, diethyl[(trimethylsilyl)ethynyl]-
reaction with epoxides
regioselectivity, **6**, 7
Alane, diisobutyl-
reaction with epoxides
regioselectivity, **6**, 7
Alane, diisobutylphenoxy-
aldol reaction, **2**, 271
Alane, dimethyl-4,4-dimethylpent-2-en-2-oxy-
aldol reactions, **2**, 268
Alane, methylalkenyl-
synthesis, **3**, 529
Alane, α-silyl-
allylation, **3**, 259
Alane, β-stannyl-
allylation, **3**, 259
Alane, triisobutyl-
reaction with epoxides
regioselectivity, **6**, 7
Alane, trimethyl-
reaction with epoxides
regioselectivity, **6**, 7
Alanes
chirally modified
asymmetric reduction, **8**, 169
reduction
acetals, **8**, 213
amides, **8**, 251
carboxylic acids, **8**, 238, 260
dimesityl ketone, **8**, 3
esters, **8**, 244
nitriles, **8**, 253
pyridines, **8**, 580
semipinacol rearrangement, **3**, 730
synthesis, **8**, 214
Alanes, alkenyl-
alkylation, **3**, 259
conjugate additions
α,β-enones, **4**, 141
reactions
nickel catalysis, **3**, 230
Alanes, allyl-
carboalumination, **4**, 891
Alanes, benzyl-
carboalumination, **4**, 891
Alangimaridine
synthesis
Mannich reaction, **2**, 913
Alaninal, phenyl-
nucleophilic addition reactions
stereoselectivity, **1**, 56
Alaninamide, phenyl-
reduction, **8**, 249
Alanine
asymmetric synthesis, **8**, 146
bislactim ether
lithium salt, crystal structure, **1**, 34

synthesis, **3**, 53
catalyst
Knoevenagel reaction, **2**, 343, 358
synthesis
via reductive amination, **8**, 144
Alanine, *N*-benzyloxy-
synthesis, **6**, 113
Alanine, *N*-carbamoyl-
Hofmann rearrangement, **6**, 802
Alaninol, *S*-phenyl-
methyl ether
lithiated imine, **3**, 37
Albene
synthesis, **8**, 932
via [3 + 2] cycloaddition reactions, **5**, 308
Alcohol dehydrogenase
hydride transfer, **8**, 82
Alcohols
acyl transfer to
ester synthesis, **6**, 324
addition to activated alkynes, **4**, 48
aliphatic saturated
anodic oxidation, **7**, 802
alkanenitrile synthesis, **6**, 234
alkynic
Ritter reaction, **6**, 268
π-allylpalladium complexes from, **4**, 588
anti-Markovnikov, **7**, 643
arene alkylation
Friedel–Crafts reaction, **3**, 309
axial
synthesis, **1**, 116
azide synthesis, **6**, 252
Birch reduction
proton source, **8**, 492
β-chiral
synthesis, **3**, 797
chiral synthesis
via aldehydes, **1**, 70
deoxygenation, **8**, 812, 818
deuterated
synthesis, *via* enzyme reduction, **8**, 203
dimerization
mercury-photosensitized, **7**, 5
dissolving metals, reductions
chemoselectivity, **8**, 113
ester synthesis
hydroxy group activation, **6**, 333
homoallylic tertiary
synthesis, ene reaction, **2**, 538
hydride donors, **8**, 88
catalysis, **8**, 91
photochemical reactions, **8**, 91
transfer hydrogenation, **8**, 551
hydrobromination, **4**, 282
inversion, **6**, 18, 21
oxidation, **7**, 299, 305–325
activated DMSO, **7**, 291–302
chromium reagents, **7**, 251–286
solid support, **7**, 841, 846
primary
oxidation, **7**, 305
synthesis, *via* oxidative cleavage of alkenes, **7**, 541
protecting groups, **6**, 646
reactions with alkenes, **4**, 297–316
palladium(II) catalysis, **4**, 553
reduction
ionic hydrogenation, **8**, 487
silanes, **8**, 216
to alkanes, **8**, 811–832
Ritter reaction, **6**, 267
secondary
synthesis, *via* oxidative cleavage of alkenes, **7**, 541
solvents for reduction
dissolving metals, **8**, 111
synthesis
via carboxylic acids, **8**, 235–254
via enantiomeric reduction of carbonyl compounds, **8**, 185
via epoxide reduction, **8**, 871
via hydrogen transfer, **8**, 110
via β-hydroxyalkyl selenides, **1**, 699, 718
via metal hydride reduction, **8**, 1–22
via organoboranes, **3**, 793
via organocerium compounds, **1**, 231
via oxidative cleavage of alkenes, **7**, 543
via reduction of hydroperoxides, **8**, 396
via substitution processes, **6**, 1–28
tertiary
from cyanoboronates, **3**, 798
from triorganylboranes, **3**, 780
synthesis, **1**, 66
thioacylation
anhydrides, thioketenes, thioesters and dithioesters, **6**, 449
thioacyl halides, **6**, 448
tritiated
synthesis, *via* enzyme reduction, **8**, 203
Vilsmeier–Haack reaction, **2**, 790
Alcohols, β-alkoxy
synthesis, **7**, 632
Alcohols, alkynic
asymmetric epoxidation
kinetic resolution, **7**, 423
oxidation, **7**, 300
Alcohols, amino
chiral aziridines from, **7**, 473
polymer-bound
selective ketone reduction, **8**, 18
synthesis
Knoevenagel reaction, reduction, **2**, 360
Alcohols, 1,3-amino
synthesis
via 1,3-dipolar cycloadditions, **4**, 1078
Alcohols, 2-amino
diastereoselective synthesis, **3**, 596
Lewis acid catalysts, **1**, 317
rearrangements, **3**, 778, 781
semipinacol rearrangements, **3**, 777
threo
synthesis, **1**, 380
synthesis
via O-silylated cyanohydrins, **1**, 548
Alcohols, γ-amino-
synthesis
via 1,3-dipolar cycloadditions, **4**, 1078
Alcohols, 2-amino-1,2-diaryl
rearrangement, **3**, 782
Alcohols, azido
cyclization, **7**, 473
Alcohols, 1,2-azido
synthesis

via epoxides, **6**, 93
Alcohols, γ-chloro
synthesis
ene reaction, **2**, 531
Alcohols, *erythro*-1,2-diamino
synthesis
Henry reaction, **2**, 335
Alcohols, epoxy
reduction
metal hydrides, **8**, 879
Alcohols, 2,3-epoxy
C(2)-amination
regioselective, **6**, 89
reactions with organocopper compounds, **3**, 225
reactions with organometallic compounds
regioselectivity, **1**, 343
rearrangement
to 1,2-epoxy-3-alkanols, **6**, 89
ring opening
stereochemistry, **6**, 5
Alcohols, α,β-epoxy-
alkene stereoselective synthesis, **7**, 369
synthesis, **7**, 378, 403
Alcohols, 2-nitro
diastereomeric mixtures
Henry reaction, **2**, 322
in synthesis, **2**, 323
in oxidation, **2**, 323
reductive denitration, **2**, 323
Alcohols, *threo*-nitro
synthesis
Henry reaction, **2**, 337
Alcohols, β-(phenylthio)
synthesis
organochromium-mediated, **1**, 203
Alcoholysis
acid chlorides
mechanism, **6**, 328
Aldehyde dehydrogenase
coimmobilized
diol oxidation, **7**, 316
Aldehydes
achiral
reactions with chiral allyl organometallics, **2**, 33–40
reactions with type I crotyl organometallics, **2**, 9–19
reactions with type III crotyl organometallics, **2**, 19–24
acid halide synthesis, **6**, 308
acyclic
synthesis *via* retro Diels–Alder reactions, **5**, 573
tandem vicinal difunctionalization, **4**, 243–245
addition reactions
cyanides, **1**, 460
1,2-addition reactions
acyl anions, **1**, 546
cyanohydrin ethers, **1**, 551
cyanohydrins, **1**, 548
α-(dialkylamino)nitriles, **1**, 554
hydrazones, **2**, 511
phosphonate carbanions, **1**, 562
aldol reactions
boron-mediated, **2**, 251
mixed, **2**, 139
syn/*anti* ratios, **2**, 266
with ketones, **2**, 142–156
aliphatic
ene and Prins reactions, **2**, 537
McFadyen–Stephens aldehyde synthesis, **8**, 297
Perkin reaction, **2**, 400
reactions with boron-stabilized carbanions, **1**, 499
alkenic
electroreduction, **8**, 134
α-alkylated
enantioselective synthesis, **3**, 35
synthesis, **3**, 26
alkylation, **3**, 20
α-alkylation, **4**, 260
alkyl enol ether derivatives
alkylation, **3**, 25
alkylidene transfer, **4**, 976
analysis
Knoevenagel reaction, **2**, 354
aromatic
ene and Prins reactions, **2**, 537
hydrogen donors, **8**, 557
hydrogenolysis, **8**, 319
aryl
methylenation, **1**, 738
β-aryl-α,β-unsaturated
synthesis, **2**, 139
asymmetric synthesis
hydroformylation, **4**, 931
bisulfite adducts
oxidation, **6**, 402
boron trifluoride complexes
NMR, **1**, 292
chiral
reactions with allyl organometallics, **2**, 24–32
α-chiral
Lewis acid complexes, **1**, 298
chiral β-alkoxy
aldol reaction, chelation control, **2**, 221
chiral α-methyl
reactions with allylboron compounds, **2**, 42
cycloaddition with diynes
bicyclic α-pyran synthesis, **5**, 1157
dehydrogenation
palladium catalysts, **7**, 140, 141
deuterated
synthesis, **8**, 271
dialkylzinc addition reactions, **1**, 317
Diels–Alder reactions, **2**, 662; **5**, 433
electron deficient
Diels–Alder reactions, **5**, 431
ene reaction, **2**, 534
enantioselective addition
alkyllithium, **1**, 72
organolithium, **1**, 70
enol acetates
halogenation, **7**, 121
enolates
addition reactions with alkenic π-systems, **4**, 99–105
arylation, **4**, 466
synthesis, **2**, 101
enol silyl ethers of, **2**, 599
geminal dialkylation
titanium(IV) reagents, **1**, 167
halogenation, **7**, 120
homologation, **3**, 897

diazo compounds, **6**, 129
hydride transfer, **8**, 86
hydrogenation
catalytic, **8**, 140
α-hydroxylation, **7**, 186
intermolecular additions
allylsilanes, **1**, 610
intermolecular pinacol coupling reactions, **3**, 570
intramolecular additions
allylsilanes, stereochemistry, **1**, 615
allyltrimethylsilane, **1**, 612
keto
aldol cyclization, **2**, 158
cyclization, regiochemistry, **2**, 159
Lewis acid complexes
rotational barriers, **1**, 290
metal enolates
alkylation, **3**, 3
α-methoxy
aldol reaction, stereoselective addition, **2**, 222
methylenation
Tebbe reagent, **5**, 1123
titanium isopropoxide, **5**, 1125
Meyers synthesis, **6**, 274
nonalkenic
electroreduction, **8**, 131
nucleophilic addition reactions
butyllithium, **1**, 70
optically active
synthesis, hydroformylation of prochiral alkenes, **3**, 1022
γ-oxo
synthesis, **3**, 103
photolysis
benzoin formation, **1**, 544
protection
via titanium reagents, **1**, 170
radical cyclizations, **4**, 817
reactions with activated dienes, **2**, 661–706
reactions with allenylsilanes, **1**, 599
reactions with allylic organocadmium compounds, **1**, 226
reactions with allyl metal compounds
synthesis of homoallylic alcohols, **6**, 864
reactions with arynes, **4**, 510
reactions with boron enolates, **2**, 250
reactions with boron stabilized carbanions, **1**, 498
reactions with α-bromo ketones, **1**, 202
reactions with chloromethyleniminium salts, **2**, 785
reactions with diazoalkanes, **1**, 845
reactions with dithioacetals, **1**, 564
reactions with nitriles, **6**, 270
reactions with organoaluminum reagents
discrimination between ketones and, **1**, 83
reactions with organocadmium compounds, **1**, 225
reactions with organocuprates, **1**, 108
reactions with organometallic compounds
chemoselectivity, **1**, 145
Cram *versus* anti-Cram selectivities, **1**, 80
Lewis acid promotion, **1**, 326
pinacolic coupling reactions, **1**, 270
reactions with type I crotylboron compounds, **2**, 10–15
reactions with zinc ester dienolates, **2**, 286
reduction
cathodic, **8**, 131
chiral boron reagents, **8**, 101
diimide, **8**, 478
dissolving metals, **8**, 307–323
electrochemical, **8**, 131
samarium diiodide, **8**, 115
selective, **8**, 16
Reformatsky reaction, **2**, 281
saturated metal enolates
alkylation, **3**, 20
Schiff bases
Mannich reaction, **2**, 954
selenenylation, **7**, 131
self-reactions, **2**, 136
sulfenylation, **7**, 125
synthesis
alkylboronic esters, **3**, 797
carbonylation, **3**, 1021
α-heterosubstituted sulfides and selenides, **3**, 141
organoboranes, **3**, 793
via alkenes, **7**, 602
via carboxylic acid reduction, **8**, 259–279, 283–304
via oxidative cleavage of alkenes, **7**, 541
via selective oxidation of primary alcohols, **7**, 305
tandem vicinal difunctionalization, **4**, 242–246
tri-*n*-butyltin enolates
alkylation, **3**, 20
unconjugated unsaturated
hydrogenation, **8**, 439
α,β-unsaturated
addition reactions with organozinc compounds, **4**, 95
aldol reactions, **2**, 137
alkylation, Cope rearrangement, **5**, 789
conjugate additions, **4**, 183, 208–212
Diels–Alder reactions, chiral catalysis, **5**, 377, 464
dienolates, alkylation, **3**, 25
electroreduction, **8**, 134
ene reactions, **5**, 5
enzymic reduction, **8**, 205
Henry reaction, regioselectivity, **2**, 330
Henry reaction, stereoselectivity, **2**, 330
hydrobromination, **4**, 282
hydroformylation, **4**, 924
hydrogenation, homogeneous catalysis, **8**, 453
imine protection, **4**, 252
preparation, directed aldol reaction, **2**, 477
preparation from epoxy sulfoxides, **2**, 417
reaction with organolithium compounds, **4**, 72
synthesis *via* bis(methylthio)allyllithium, **6**, 138
synthesis *via* retro Diels–Alder reactions, **5**, 553, 573
α,β,γ,δ-unsaturated
synthesis, **6**, 903
β,γ-unsaturated
isomerization, **6**, 896
optically active, synthesis, **6**, 855
stereoselective synthesis, **6**, 851
synthesis, **3**, 934
γ,δ-unsaturated
synthesis, **3**, 103
synthesis *via* Claisen rearrangement, **5**, 830
unsaturated aliphatic
hydrogenation, **8**, 140
Aldehydes, α-alkoxy
aldol reaction
stereoselective nonchelation, **2**, 307

chiral
reaction with enol silanes, **2**, 640
Diels–Alder reactions
$TiCl_4$-catalyzed, **2**, 667
N,N-dimethylhydrazones
reactions with organometallic compounds, **1**, 380
reactions with organochromium compounds, **1**, 198
reactions with organocuprates, **1**, 108
reactions with organozinc compounds
1,2-asymmetric induction, **1**, 336
stereoselectivity, **1**, 221
Aldehydes, β-alkoxy
aldol reaction
chelation control, **2**, 152
reactions with organocuprates, **1**, 108
reaction with allyl organometallic compounds, **2**, 985
reaction with enol silanes
chelation control with $TiCl_4$, **2**, 646
Aldehydes, α-alkoxy chiral
reactions with organochromium compounds
addition to crotyl halides, **1**, 185
Aldehydes, β-alkoxy-γ-hydroxy
nucleophilic addition reactions
stereoselectivity, **1**, 59
Aldehydes, β-alkoxy-α-methyl
reaction with allylchromium
stereoselectivity, **1**, 183
Aldehydes, alkynic
electroreduction, **8**, 134
Knoevenagel reaction, **2**, 365
Aldehydes, allenic
intramolecular ene reaction
type I, **2**, 547
reduction, **8**, 114
Aldehydes, β-allylsiloxy
intramolecular additions
Lewis acid catalyzed, **1**, 615
Aldehydes, amino
nucleophilic addition reactions
stereoselectivity, **1**, 56
Aldehydes, α-amino
dibenzyl protected
nucleophilic addition reactions, **1**, 56
reaction with enol silanes
chelation control with $TiCl_4$, **2**, 646
statine synthesis, **2**, 223
synthesis
use of protecting groups, **6**, 644
via ester reduction, **8**, 266
Aldehydes, β-amino
synthesis
Mannich reaction, **2**, 896
Aldehydes, α-aryl
synthesis
via rearrangement of arylalkenes, **7**, 828
Aldehydes, α-bromo
synthesis
via haloborane addition to alkynes, **4**, 358
Aldehydes, α,β-dialkoxy
reactions with organocuprates, **1**, 108
reaction with enol silanes
stereoselection, **2**, 642
Aldehydes, *N,N*-dibenzyl-α-amino
carbonyl compound complexes
nonchelation-controlled addition, **1**, 460
Aldehydes, α,β-dibenzyloxy
reactions with organometallic compounds
Lewis acids, **1**, 338
Aldehydes, α,β-dihydroxy
protected
synthesis, **7**, 442
reactions with organometallic compounds
Lewis acids, **1**, 337
synthesis, **7**, 441
Aldehydes, α,β-epoxy
imines
condensation to β-lactams, **5**, 96
reactions with organometallic compounds
Lewis acids, **1**, 339
Aldehydes, α-halo
reduction
stereoselectivity, **8**, 3
Aldehydes, 2-hydroxy
chiral
synthesis, **1**, 64, 69
oxidative cleavage, **7**, 709
synthesis
via 1,3-dioxathianes, **1**, 62
via formaldehyde dimethyl dithioacetal *S*-oxide, **1**, 526
via keto aminals, **1**, 64, 65
Aldehydes, keto
synthesis
via Kornblum oxidation, **7**, 654
via thio-Claisen rearrangement, **6**, 861
via Wacker oxidation, **7**, 455
Aldehydes, 1,4-keto
synthesis
via nickel-catalyzed acylation, **1**, 452
Aldehydes, β-keto
γ-alkylation, **3**, 58
Aldehydes, γ-keto
synthesis
via γ-oxo sulfone acetals, **6**, 159
Aldehydes, α-methyl
chiral
reaction with enol silanes, **2**, 640
Aldehydes, α-methyl-β-alkoxy
reaction with enol silanes
stereoselectivity, **2**, 643
Aldehydes, α-methyl-β,γ-unsaturated
reactions with crotylchromium
stereoselectivity, **1**, 184
Aldehydes, α-nitroso-
synthesis, **6**, 104
Aldehydes, α-(phenylthio)
synthesis, **1**, 570
Aldehydes, β-siloxy
NMR, **1**, 297
Aldehydes, α-triisopropylsilyl
synthesis
from vinylsilanes, **2**, 58
Alder's *endo* rule
Diels–Alder reaction, **5**, 318
Aldimines
aromatic
reactions with organometallic compounds, **1**, 383
chiral
reaction with allyl organometallics, **2**, 32
stereochemistry in nucleophilic addition reactions, **1**, 359
chiral α,β-unsaturated

reactions with organometallic compounds, **1**, 382
N-heterosubstituted
homoallylamines from, **2**, 994
imine anions from, **2**, 477
lithiated α,β-unsaturated
alkylation, **3**, 33
metallated
aldol reaction, **2**, 477
metallation, **2**, 476
pinacol coupling reactions, **3**, 580
reactions with allenic organometallic compounds
syn–anti selectivity, **2**, 993
reactions with crotyl-9-BBN, **2**, 15
reactions with crotyl organometallic compounds
regioselectivity, **2**, 989
reduction
metal hydrides, **8**, 272
α,β-unsaturated
addition reactions with organomagnesium compounds, **4**, 85
Aldimines, α-alkoxy-
reaction with allyl organometallic compounds, **2**, 987
Aldimines, *N*-isopropyl-
reaction with crotyl organometallic compounds
syn–anti selectivity, **2**, 992
Aldimines, *N*-phenylsulfonyl-
Diels–Alder reactions, **5**, 474
Aldimines, *N*-propyl-
reaction with allyl organometallic compounds, **2**, 983
Aldimines, *N*-*n*-propyl-
reaction with crotyl organometallic compounds
syn–anti selectivity, **2**, 992
Alditols, amino-
synthesis
via cyclization of allylic substrates, **4**, 404
Aldolase
asymmetric synthesis
summary of enzymes available, **2**, 467
organic synthesis
carbon–carbon bond formation, **2**, 456
substrate specificity
synthetic applicability, **2**, 456
Aldol reactions
acetyliron enolates
diastereofacial selectivity, **2**, 316
acid and general base catalysis, **2**, 133
acyl–transition metal complexes, **2**, 314
addition
driving force, **2**, 135
stereochemistry, **2**, 153
aldehydes
cross-addition, **2**, 139
self-addition, **2**, 136
alkenyloxydialkylalanes
homochiral, **2**, 271
alkenyloxydialkylboranes
homochiral, **2**, 248
aluminum-mediated, **2**, 268
2,3-*anti* products
from hindered aryl esters, **2**, 201
anti-selective, **2**, 256
anti/syn selectivity, **2**, 258
background, **2**, 134
boric acid
catalyst, **2**, 138
boron enolates
from homochiral acyl sultam, **2**, 253
boron-mediated, **2**, 240
kinetics, **2**, 246
α-bromo ketones
with aldehydes, **2**, 424
cascade
cyclic compound synthesis, **2**, 619
cations, **2**, 135
η^1-*C*-bound metal enolates, **2**, 312
cerium enolates, **1**, 243; **2**, 312
chiral auxiliary
recycling, **2**, 232
condensation
acylic stereocontrol, allyl metal reagents, **2**, 2
double asymmetric synthesis, **2**, 2
cross-coupling
aluminum-mediated, **2**, 268
crossed
from boryl enolates, **2**, 242
lithium dimethylhydrazone anions, **2**, 511
cyclizations
enantioselective, **2**, 167
intramolecular, aluminum-mediated, **2**, 269
ring-size selectivity, **2**, 165
stereochemistry, **2**, 166
diastereofacial selectivity, **2**, 217
diastereoselective
alkenyloxyboranes, **2**, 244
boron ligands, less polar solvents, **2**, 247
dicyclopentadienylchlorozirconium enolates
stereoselectivity, **2**, 305
syn:anti selectivity, **2**, 303
directed
alkenyloxyboranes, **2**, 242
electrochemistry, **2**, 138
enantiomerically pure
preparation, **2**, 232
enantioselective
use of hydrazones, **2**, 514
enol ethers, **2**, 611
enolsilanes
rhodium(I) catalyzed, **2**, 311
enones, **2**, 152
enzymatic, **2**, 455–470
equilibration
thermodynamic control, **2**, 234
Group I and II enolates, **2**, 181–235
Group III enolates, **2**, 239–275
imine anions
directed, **2**, 477
immolative process
chiral auxiliaries, **2**, 232
indirect
homoallylic alcohol synthesis, **6**, 864
intramolecular, **2**, 156–176
Reformatsky reaction product, **2**, 282
intramolecular diasteroselective
silyl enol ethers, **2**, 651
ketones
asymmetric, boron reagents, **2**, 264
cross-addition, **2**, 142
external chiral reagents, **2**, 262
self-addition, **2**, 140
with aldehydes, **2**, 142–156
kinetic control, **2**, 154
kinetic stereoselectivity

enolate stereochemistry and structure, **2**, 190
lanthanide metal enolates, **2**, 301
lithium-mediated, **2**, 239
mechanism
X-ray structure of intermediates, **1**, 4
mediated by alkenyloxydialkoxyboranes, **2**, 266
α-mercurio ketones
η^3-metal enolates, **2**, 312
syn:anti selectivity, **2**, 313
metallated alkimines
with carbonyl compounds, **2**, 477
methyl isocyanoacetate
diastereoselectivity, **2**, 318
enantioselectivity, **2**, 318
η^1-*O*-bound metal enolates, **2**, 302
open transition states, **2**, 155
propionyliron enolates
stereoselectivity, **2**, 317
radical cyclization, **4**, 791
reactions with aldehydes
boron-mediated, **2**, 251
reversibility, **2**, 134
limitations, **2**, 136
simple diastereoselection
use of preformed enolates, **2**, 190
solvent effects, **2**, 153
stereoselection
addition to chiral aldehydes, **2**, 217
cation, **2**, 191
enolate geometry, **2**, 190
stereoselective
allyl rearrangement, **6**, 833
stereoselectivity
chiral aldehydes, steric effects on facial preference, **2**, 221
chiral enolates, **2**, 223
restoring energy, **2**, 154
3,4-stereoselectivity, **2**, 248
substitution effect
enone formation, **2**, 146
syn/*anti* ratios, **2**, 266
syn-selective, 249
thermochemistry, **2**, 134
thermodynamic control, **2**, 154
thioates, **2**, 258
titanium enolates
enantioselectivity, **2**, 309
syn:anti selectivity, **2**, 306
syn stereoselectivity, **2**, 305
transannular cyclizations, **2**, 169
transition metal enolates, **2**, 301–318
unsymmetrical ketones
regioselectivity, **2**, 144
vinylaminodichloroboranes
with carbonyl compounds, **2**, 479
Wittig directed
use of lithium diisopropylamide, **2**, 182
Aldol-type reactions
α-bromo ketones
with aldehydes, **1**, 202
Aldonolactones
reduction
electrochemical, **8**, 292
formation of aldoses, **8**, 292
Aldonolactone sugars
synthesis
via Paterno–Büchi reaction, **5**, 158
Aldosterone
synthesis, **7**, 236
Aldoxan
synthesis, **2**, 138
Aldoxime ethers
reactions with organometallic compounds, **1**, 385
Aldoximes
Beckmann rearrangement, **6**, 763, 775; **7**, 695
dianions
alkylation, **3**, 35
oxidation
nitrile oxides from, **4**, 1078
reactions with allylboronates, **2**, 15
A. leucotreta
sex pheromone
synthesis, **2**, 78
Alicyclic compounds
synthesis
via reduction of aromatic compounds, **8**, 490
Aliquat-336
rhodium trichloride ion-pair
hydrogenation, **8**, 535
Alka-2,4-dienoic acid
ethyl ester
preparation, ene reaction, **2**, 535
1,*n*-Alkadiynes
hydroalumination
locoselectivity, **8**, 742
Alkali carbonates
phosphonium ylide synthesis, **6**, 175
Alkali hydroxides
phosphonium ylide synthesis, **6**, 175
Alkali metal cyanides
amide acetal synthesis, **6**, 573
2,2-bis(dialkylamino)carbonitrile synthesis, **6**, 578
Alkali metal enolates
carbonyl compounds
deprotonation, **2**, 100
α,β-unsaturated, **2**, 106
enol acetates
stable enolate equivalents, **2**, 108
silyl enol ethers
stable enolate equivalents, **2**, 108
synthesis, **2**, 100
from amide bases, **2**, 100
from ketenes, **2**, 107
miscellaneous methods, **2**, 109
Alkali metal fluorides
catalyst
Knoevenagel reaction, **2**, 343
Alkali metals
deselenations, **8**, 848
desulfurizations, **8**, 842
liquid ammonia
carbonyl compound reduction, **8**, 308
reduction
alkyl halides, **8**, 795
benzylic compounds, **8**, 971
P—C bonds, **8**, 858
Alkaloids
dehydrogenation
microbial, **7**, 65
hydroxylation
microbial, **7**, 65
synthesis

Dieckmann reaction, **2**, 829
via 1,3-dipolar cycloadditions, **4**, 1077
Alkanal, 3-phenyl-
chiral synthesis, **2**, 68
Alkane-1-boronates, 1-lithio-1-phenylthio-
reactions with carbonyl compounds, **1**, 501
Alkanecarbaldehydes
Baeyer–Villiger reaction, **7**, 684
Alkanephosphonates, 1,2-epoxy-
preparation, **2**, 427
Alkanes
acylation
Friedel–Crafts reaction, **2**, 727
alkylation
Friedel–Crafts reaction, **3**, 332
anodic oxidation, **7**, 793
Ritter reaction, **6**, 282
arene alkylation
Friedel–Crafts reaction, **3**, 322
carbonylation
transition metal catalysis, **7**, 6
cracking, **7**, 7
dehydrodimerization, **7**, 5
dehydrogenation
transition metal catalysis, **7**, 6
electrochemical oxidation, **7**, 8
functionalization, **7**, 2
electrophilic addition reactions, **7**, 7
silyl substituent, **7**, 8
hydroxylation
photolytic method, **7**, 12
isomerization, **7**, 5
microbial oxidation, **7**, 56
nitration, **7**, 8
reactions with alkylpotassium, **7**, 2
synthesis
via alcohols and amines, **8**, 811–832
via alkyl halide reduction, **8**, 793–807
via enzyme reduction of alkenes, **8**, 205
via trialkylboranes, **7**, 603
thermolysis, **7**, 7
Alkanes, azido-
synthesis, **7**, 607
Alkanes, bis(5-deazaflavin-10-yl)-
synthesis, **4**, 435
Alkanes, 1,1-bis(dialkoxyboryl)-
oxidation
formation of aldehydes, **7**, 600
Alkanes, 2,2-bis(dialkoxyboryl)-
oxidation
formation of ketones, **7**, 600
Alkanes, 1,1-bis(ethylthio)-
alkylation, **3**, 123
Alkanes, chloroalkoxy-
synthesis, **8**, 214
Alkanes, chlorophenyl-
cycloalkylations
Friedel–Crafts reaction, **3**, 324
Alkanes, diazo-
addition to ketones, **3**, 783
fluorination, **6**, 219
Alkanes, 1,1-diboryl-
synthesis, **1**, 489
Alkanes, 1,1-dibromo-
reagent from
enol ether synthesis, **2**, 597
Alkanes, 1,2-dibromo-
reductive elimination, **8**, 806
Alkanes, α,ω-dibromo-
monoarylation
with aryl Grignard reagents, **3**, 464
Alkanes, α,ω-dichloro-
benzene alkylation by
Friedel–Crafts reaction, **3**, 318
Alkanes, α,ω-diethynyl-
oxidative coupling, **3**, 557
Alkanes, difluoro-
synthesis, **4**, 271
Alkanes, *gem*-dihalo-
cyclopropanation, **4**, 961–976
dialkylation with
1,2-dicarbanionic species, **4**, 976
Alkanes, dimesitylboryl(trimethylsilyl)-
cleavage
synthesis of α-boryl carbanions, **1**, 490
Alkanes, 1,1-diseleno-
carbonyl compound synthesis from, **3**, 142
Alkanes, 1,1-disulfinyl-
reaction with allylic epoxides
synthesis of macrolides, **3**, 177
Alkanes, fluoro-
synthesis, **4**, 270
Alkanes, 1-fluoro-2-amino-
synthesis, **7**, 498
Alkanes, halo-
imidoyl halide synthesis, **6**, 527
Alkanes, (hydroxyalkyl)nitro-
anions
formation, **2**, 323
Alkanes, β-hydroxyaryl-
synthesis
Friedel–Crafts reaction, **3**, 313
Alkanes, 1-(indol-3-yl)-2-nitro-
reduction, **8**, 375
Alkanes, 1-lithio-1-(phenylseleno)cyano-
reaction with cyclohexenone, **1**, 686
Alkanes, 1-metallo-1,1-bis(alkylthio)-
in synthesis, **3**, 123
Alkanes, 1-metallo-1,1-bis(dithio)-
alkylation, **3**, 121
Alkanes, 1-metallo(phenylthio)-
in synthesis, **3**, 123
Alkanes, nitro-
acyl anion synthons, **2**, 324
aliphatic
reduction, **8**, 374
aryl radical traps, **4**, 472
α,α doubly deprotonated
Henry reaction, **2**, 335
functionalized
Henry reaction, **2**, 331
Michael addition, **4**, 12
synthesis, **2**, 321; **6**, 104
via decomposition of acyl nitrates, **7**, 729
tandem vicinal difunctionalization, **4**, 259
Alkanes, nitroso-
synthesis
via oxidation of amines, **7**, 737
Alkanes, α-phenylselenonitro-
metallation, **1**, 642
Alkanes, polyhalo-
reaction with alkenes

radical addition reactions, **4**, 753
Alkanes, 2-pyridyldi-
synthesis
via cocycloaddition, **5**, 1155
Alkanes, 1,1,1-trihalo-
aminal ester synthesis, **6**, 574
ortho acid synthesis, **6**, 556
tris(dialkylamino)alkane synthesis, **6**, 579
Alkanes, tris(dialkoxyboryl)-
synthesis
via production of boron-stabilized carbanions, **1**, 489
Alkanes, tris(dialkylamino)-
2,2-bis(dialkylamino)carbonitrile synthesis, **6**, 577
synthesis, **6**, 579
Alkanesulfonic acid, perfluoro-
catalyst
Friedel–Crafts reaction, **3**, 297
coupling reactions
with sp^3 organometallics, **3**, 455
Alkanesulfonic acids
synthesis, **7**, 14
Alkanesulfonyl bromide, α-halo-
reaction with alkenes, **3**, 879
Alkanethioates, *S*-2-methylbutyl esters
synthesis, **6**, 441
Alkanethioates, perfluoro-
O-alkyl esters
synthesis, **6**, 449
Alkanethiolates
reactions with aryl halides, **4**, 475
Alkanoates
enolates, **3**, 45
Alkanoic acid, ω-chloro-
benzene alkylation
Friedel–Crafts reaction, **3**, 303
Alkanoic acid, 2-oxo-
esters
synthesis, allylic anions, **2**, 60
Alkanoic acid, 5-oxo-
3,4-disubstituted
synthesis, **2**, 520
Alkanoic acids
α,β-disubstituted
synthesis *via* conjugate addition to sultams, **4**, 204
Alkanoic acids, aryl-
esters
synthesis, **3**, 778
methyl esters
anodic oxidation, **7**, 811
optically active esters
synthesis, Friedel–Crafts reaction, **3**, 312
oxidation, **7**, 336
synthesis, **3**, 788; **7**, 827
Friedel–Crafts reaction, **3**, 316
via oxidative rearrangement of aryl ketones, **7**, 829
Alkanoic acids, perfluoro-
decarboxylation, **7**, 930
Alkanols, aryl-
cycloalkylation
Friedel–Crafts reaction, **3**, 325
oxidation, **7**, 336
Alkanols, azido-
synthesis, **6**, 253
Alkanones, α-aryl-
synthesis, **7**, 827
Alkanoyl chloride, ω-trimethylsilylethynyl-
cyclization, **2**, 726
Alkatrienes
synthesis, **3**, 644
2-Alkenamides, 2-acylamino-
synthesis
Erlenmeyer azlactone synthesis, **2**, 405
Alkenation
allenic phosphonates
to cumulatrienes, **6**, 845
Alkenations
alkyl-*gem*-dichromium reagents, **1**, 205
carbonyl compounds
phosphorus stabilized, **1**, 755
(*E*)-selective, **1**, 758
sulfur stabilized
Julia coupling, **1**, 792
Alkenes
activated
conjugate additions catalyzed by Lewis acids, **4**, 139–164
acyclic
diastereoselective hydroxylation, **7**, 441
epoxidation, **7**, 359, 368, 378
Pauson–Khand reaction, **5**, 1043–1046
acylation, **2**, 709
acyloxymercuration-demercuration, **4**, 314
addition reactions, **7**, 493
carbon-centered radicals, **4**, 735–765
carbon nucleophiles, **4**, 571–583
cleavage, **7**, 506
dihalocarbenes, **4**, 1002–1004
ketocarbenoids, **4**, 1034–1050
nitrogen and halogen, **7**, 498
nitrogen and oxygen, **7**, 488
nitrogen and sulfur, **7**, 493
nitrogen nucleophiles, **4**, 559–563
oxygen nucleophiles, **4**, 552–559
reactive carbanions, **4**, 69–130
two nitrogen atoms, **7**, 484
alkoxymercuration–demercuration, **4**, 309
alkylation
Friedel–Crafts reaction, **3**, 331
palladium(II) catalysis, **4**, 571–580, 842
π-allylpalladium complexes from, **4**, 587
amination, **4**, 290–297; **7**, 470
aminomercuration–demercuration, **4**, 290
anodic oxidation, **7**, 794
arene alkylation
Friedel–Crafts reaction, **3**, 304
arylation by palladium complexes, **4**, 843–848
mechanism, **4**, 843
regiochemistry, **4**, 845
stereochemistry, **4**, 845
asymmetric dihydroxylation, **7**, 429
asymmetric hydrogenation
chiral catalysts, **8**, 459
homogeneous catalysis, **8**, 463
2-azetidinones from, **5**, 102–108
aziridines from, **7**, 470
benzylation
palladium complexes, **4**, 842
bicyclic
hydrochlorination, **4**, 273
bicyclic oxides
opening, **3**, 734

1,2-bifunctionalization, **7**, 533
bishydroxylation, **7**, 867
bridged bicyclic
  Pauson–Khand reaction, **5**, 1049–1051
bridgehead
  cycloadditions, **5**, 64
captodative
  radical addition reactions, **4**, 758
carboalumination, **4**, 887
carboboration, **4**, 885
carbocupration, **4**, 895
carbolithiation, **4**, 867–872
carbomagnesiation, **4**, 873, 874–877
carbometallation, **4**, 865–906
carbonylation
  palladium salt catalyst, **3**, 1030
carbozincation, **4**, 879, 880–883
conjugate additions
  catalyzed by Lewis acids, **4**, 140
conjugated
  hydrogenation, **8**, 449, 452
  Peterson alkenation, **1**, 789
  transfer hydrogenation, **8**, 453
coupling reactions, **3**, 482
  crossed, **3**, 484
  with aryl compounds, **3**, 492
  with carbene complexes, **5**, 1084
  with heteroaryl compounds, **3**, 497
cyclic
  epoxidation, **7**, 361, 364, 376
  hydroboration, stereofacial selectivity, **8**, 713
  ring contraction, **7**, 831
  ring expansion, **7**, 831
cyclization
  zirconium-promoted, **5**, 1164
[2 + 2] cycloadditions
  thermal, **5**, 63–79
cyclopropanation, **5**, 1084
  alkyl diazoacetate, **4**, 1035
deuterium-labeled
  synthesis, **3**, 867
dibromides
  protection, **6**, 685
dicarboxylation, **4**, 946–949
dichlorides
  protection, **6**, 685
difunctional
  coupling reactions with sp^3 organometallics, **3**, 448
dimerization, **3**, 482
  via 1,3-diradicals, **5**, 63–67
divinyl ketones from, **5**, 777
electrochemical oxidation, **7**, 98
electron deficient
  asymmetric nucleophilic addition, **4**, 199–232
  ene reactions, **5**, 2–6
  epoxidation, **7**, 372
  stabilized nucleophiles and, **4**, 1–58
electrophilic addition
  X–Y reagents, **4**, 329–359
endocyclic
  synthesis *via* retro Diels–Alder reactions, **5**, 560
ene reactions, **5**, 1–25
  intramolecular, **5**, 9–20
  Lewis acid catalysis, **5**, 4
epoxidation, **7**, 358, 390
  solid support, **7**, 841
esterification, **4**, 312
exocyclic
  regioselective synthesis, **5**, 1182
  synthesis *via* retro Diels–Alder reactions, **5**, 560
Friedel–Crafts reaction
  mechanism, **2**, 708
functionalized
  carbolithiation, **4**, 869
  carbomagnesiation, **4**, 877
  hydroformylation, **4**, 922–927
fused bicyclic
  Pauson–Khand reaction, **5**, 1046–1049
halogen derivatives
  Diels–Alder reactions, **5**, 327
halohydrins
  protection, **6**, 685
hydroalumination, **8**, 733–758
hydroboration, **7**, 595; **8**, 703–727
hydrobromination, **4**, 279–287
hydrocarboxylation, **4**, 932–946
hydrochlorination, **4**, 272–278
  stereochemistry, **4**, 272
hydrogenation, **8**, 421
  apparent *anti* addition, **8**, 427
  association constants, **8**, 425
  catalyst hindrance, **8**, 427
  conformational analysis, **8**, 429
  haptophilicity, **8**, 429
  heterogeneous catalysis, **8**, 417–442
  homogeneous catalysis, **8**, 443–463
  intramolecular nonbonding interactions, **8**, 428
  rate constants, **8**, 444
  stereochemistry, **8**, 426
  syn addition, **8**, 426
hydroiodination, **4**, 287
hydrosilylation, **8**, 763–789
  trichlorosilane, **7**, 642
hydroxylation
  anti, **7**, 438
  syn, enantioselective, **7**, 441
  Woodward's procedure, **7**, 444
hydroxymercuration–demercuration, **4**, 300
hydrozirconation, **4**, 153; **8**, 667–699
iminium ion cyclization, **2**, 1023
internal
  oxidation, **7**, 462
intramolecular carbomagnesiation, **4**, 876
isomerization
  hydroformylation, **4**, 918
metal-activated
  addition reactions, **4**, 551–565
  nucleophilic attack, **4**, 551–568
metathesis, **5**, 1115–1126
  catalysts, **5**, 1116
  functionalization, **5**, 1116
  polymerization, **5**, 1116
monocyclic
  Pauson–Khand reaction, **5**, 1046–1049
no directing groups
  epoxidations, **7**, 375
one-carbon homologation
  via Ramberg–Bäcklund rearrangement, **3**, 862
oxidation
  nitrogen addition, **7**, 469–508
  permanganate, **7**, 444, 844
  Wacker process, **7**, 449

oxidative rearrangement, **7**, 816, 828
solid support, **7**, 845
peroxymercuration–demercuration, **4**, 306
photoaddition reactions
with ynones, **5**, 164
photosensitized oxygenation, **7**, 96
pinacol coupling reactions
with carbonyl compounds, **3**, 598
polyfluorinated
cycloaddition reactions with ketenimines, **5**, 113
polymerization, **5**, 1115
protection, **6**, 684
radical addition reactions, **4**, 715–772
radical cyclizations, **4**, 779
carbon-centered radicals, **4**, 789
reactions with *N*-acyliminium ions
intramolecular, **2**, 1062
reactions with alcohols, **4**, 307
reactions with π-allylpalladium complexes
regioselectivity, **4**, 644
reactions with arynes, **4**, 510
reactions with carbon monoxide, **4**, 913–949
reactions with chloromethyleniminium salts
Vilsmeier–Haack reaction, **2**, 781
reactions with dialkyldithiophosphoric acids, **4**, 317
reactions with dienes
transition metal catalysis, **4**, 709–712
reactions with HX reagents, **4**, 269–319
reactions with hydrogen peroxide, **4**, 305, 306
reactions with ketocarbenes, **4**, 1031–1064
reactions with ketyls
organosamarium reagents, **1**, 268
reaction with Kolbe radicals, **3**, 646
reaction with nitrile oxides, **5**, 260
reduction
enzymes and microorganisms, **8**, 205
noncatalytic chemical methods, **8**, 471–487
reductive ozonolysis, **8**, 398
remote carboxyl groups
synthesis, **3**, 862
Ritter reaction, **6**, 267
silicon-mediated formation
Peterson alkenes, **1**, 782
stereochemistry
in coupling reactions, **3**, 436
steroidal
hydroxylation, **7**, 445
strained
reaction with π-allylpalladium complexes, **4**, 602
substituted
hydrosilylation, **8**, 776
synthesis *via* retro Diels–Alder reaction, **5**, 553–565
sulfur derivatives
Diels–Alder reactions, **5**, 324–327
synthesis, **8**, 959
alkenylalkyldimethoxyboronates, **3**, 799
alkenylboranes, **3**, 795
α-alkylation of γ-substituted allyl phosphonates, **3**, 202
alkylboranes, **3**, 795
π-allylnickel halides, **3**, 426
carboxylic acids, **3**, 652
1,1-dibromoalkanes, deprotonation, **3**, 202
sulfides or selenides, **3**, 114
via alkyne hydroboration/protonolysis, **8**, 726
via carbonyl compounds, **1**, 729–809
via deoxygenation of alcohols, **8**, 822
via dissolving metal reductions, **8**, 528
via elimination from diazo compounds, **6**, 128
via β-hydroxyalkyl selenides, **1**, 700, 721
via Julia coupling, **1**, 804
via ketones, **8**, 923–951
via metal carbene complexes, **1**, 807
via organoaluminum reagents, **1**, 92
via organoboranes, **7**, 603
via reaction of boron-stabilized carbanions with ketones, **1**, 498
via reductive β-elimination of vicinal dibromides, **8**, 797
via 2,3-sigmatropic rearrangement, **6**, 873, 877
via vinyl halides, **8**, 895–920
terminal
allylic oxidation, **7**, 95
oxidation to methyl ketones, **7**, 452
thioimidate synthesis, **6**, 540
trans-
synthesis, **8**, 478
unactivated
photocycloaddition reactions, **5**, 145–147
unfunctionalized
hydroformylation, **4**, 919–922
unsymmetrical
Friedel–Crafts acylations, **2**, 709
vinylation
stereospecific, **4**, 852
vinyl substitution with palladium complexes, **4**, 851–854
mechanism, **4**, 851
1-Alkenes
hydrogenation
homogeneous catalysis, **8**, 445
2-Alkenes
allylic oxidation, **7**, 93
(*E*)-Alkenes
synthesis
via Horner–Wadsworth–Emmons reaction, **1**, 762
via Julia coupling, **1**, 794
(*Z*)-Alkenes
synthesis
via Horner–Wadsworth–Emmons reaction, **1**, 763
Alkenes, 2-alkyl-4-hydroxy-
t-butyl carbonates
cyclization, **4**, 386
Alkenes, β-(alkylthio)-
addition reactions, **4**, 126
1-Alkenes, 1-alkynyl-2-halo-
synthesis
via haloborane addition to alkynes, **4**, 358
Alkenes, aryl-
oxidative rearrangment, **7**, 828
Alkenes, 1,2-bis(trimethylsilyloxy)-
synthesis
via acyloin condensation, **2**, 601
Alkenes, ω-bromo-
synthesis, **3**, 247
Alkenes, 1-bromo-1-(trimethylsilyl)-
cyclization, **1**, 589
2-Alkenes, 2-chloro-1,1,1-trifluoro-
Oshima–Takai reaction, **1**, 751
2-Alkenes, 1,4-diamino-
synthesis, **7**, 504

Alkenes, 1,1-diaryl-
  synthesis, **3**, 864
Alkenes, 1,2-dichloro-
  ozonolysis
    formation of methyl esters, **7**, 574
Alkenes, dideuterio-
  synthesis, **8**, 726
Alkenes, α,α-difluoro-
  addition reactions, **4**, 127
Alkenes, 1,1-dihalo-
  amidine synthesis, **6**, 550
Alkenes, 1,1-diiodo-
  synthesis
    via carboalumination, **4**, 890
Alkenes, 1,1-diseleno-
  reduction, **3**, 106
Alkenes, 1,1-disilyl-
  acylated vinylsilanes from, **2**, 718
Alkenes, 2,3-disilyl-
  acylation
    Friedel–Crafts reaction, **2**, 718
Alkenes, disubstituted
  synthesis
    via Horner reaction, **1**, 778
    via tandem vicinal difunctionalization, **4**, 250
Alkenes, fluoro-
  hydroformylation, **4**, 927
  synthesis, **3**, 420
Alkenes, halo-
  imidoyl halide synthesis, **6**, 527
  ortho acid synthesis, **6**, 556
Alkenes, 1-halo-2-bromo-
  synthesis
    via haloborane addition to alkynes, **4**, 358
Alkenes, γ-hydroxy-
  oxidative cleavage
    synthesis of lactones, **7**, 574
  selective oxidation, **7**, 454
Alkenes, ω-hydroxy-
  cyclization
    palladium(II) catalysis, **4**, 557
Alkenes, iodo-
  synthesis, **7**, 606
Alkenes, 3-methyl-5-hydroxy-
  cyclizations
    stereoselectivity, **4**, 380
Alkenes, nitro-
  conjugated
    Diels–Alder reactions, **2**, 325
    synthesis, **6**, 107
    transformations, **2**, 324
  Diels–Alder reactions, **5**, 320–322
  hydrogenation, **8**, 439
  Michael acceptors, **4**, 262
  Michael additions, **4**, 12, 18
    chiral enolates, **4**, 218
  reduction, **8**, 375
  synthesis, **7**, 493, 534
  tandem vicinal difunctionalization, **4**, 253
Alkenes, α-nitro-
  addition reactions
    with enolates, **4**, 100
    with organolithium compounds, **4**, 77
    with organomagnesium compounds, **4**, 85
  π-allylpalladium complexes from, **4**, 589
  conjugate additions
    Lewis acids, **4**, 142
  Henry reaction, **2**, 334
  synthesis
    via addition to 2-nitroallyl pivalate, **4**, 78
Alkenes, perfluoro-
  reactions with amines, **6**, 498
  reaction with nitric oxide, **7**, 488
Alkenes, perfluorochloro-
  reactions with amines, **6**, 498
Alkenes, phenylthio-
  synthesis
    via 1-lithio-1-phenylthioalkane-1-boronates, **1**, 501
Alkenes, β-sulfonylnitro-
  Diels–Alder reactions, **5**, 320; **6**, 161
Alkenes, tetrasubstituted
  synthesis, **3**, 864
    via tandem vicinal difunctionalization, **4**, 250
Alkenes, trialkylsilyl-
  divinyl ketones from, **5**, 777
Alkenes, trisubstituted
  Julia coupling, **1**, 797
  synthesis, **1**, 797
    from thiols and activated alkynes, **4**, 50
    via tandem vicinal difunctionalization, **4**, 250
2-Alkenoic acid
  deconjugated alkylation, **3**, 51
6-Alkenoic acid
  Kolbe electrolysis, **3**, 640
2-Alkenoic acids, 2-acylamino-
  synthesis
    Erlenmeyer azlactone synthesis, **2**, 405
2-Alkenoic acids, 2-alkyl-
  methyl esters
    synthesis *via* retro Diels–Alder reaction, **5**, 553
2-Alkenolides, 2-sulfinyl-
  conjugate additions, **4**, 213
2-Alkenones
  tandem vicinal difunctionalization, **4**, 242
β-Alkenylamines
  sulfenoamination, **4**, 401
Alkenyl bromides
  coupling reactions
    with Grignard reagents and alkyllithium reagents, **3**, 437
Alkenyl chlorides
  coupling reactions
    with Grignard reagents, **3**, 437
Alkenyl complexes
  benzannulation, **5**, 1100
Alkenyl groups
  addition reactions
    with alkenic π-systems, **4**, 72–99
  conjugate additions
    catalyzed by Lewis acids, **4**, 140–158
Alkenyl halides
  coupling reactions with sp^3 organometallics, **3**, 436
  reactions with ketones
    organosamarium compounds, **1**, 258
  reaction with 1-alkynes, **3**, 539
  reaction with organocopper compounds, **3**, 217
  synthesis
    via metal carbene complexes, **1**, 807
  tandem vicinal difunctionalizations, **4**, 260
Alkenyl iodides
  coupling reactions
    with Grignard reagents, **3**, 439

1-Alkenyllithiums, 1-seleno-
  synthesis, **1**, 666
1-Alkenyl metals, 1-seleno-
  synthesis, **1**, 644
Alkenyl pentafluorosilicates
  coupling
    butadiene synthesis, **3**, 483
Alkenynes
  hydroalumination
    locoselectivity, **8**, 742
1-Alken-3-ynes, 1-methoxy-
  synthesis, **2**, 89
Alkoxides
  alkali metal anions
    crystal structures, **1**, 37
  aromatic nucleophilic substitution, **4**, 437
  phosphonium ylide synthesis, **6**, 174
  phosphorylation, **6**, 603
  reaction with π-allylpalladium complexes
    stereochemistry, **4**, 622
  tandem vicinal difunctionalization, **4**, 257
Alkoxides, amino-
  o-lithiated
    hydroxylation, **7**, 333
Alkoxides, α-amino-
  lithiation
    addition reactions, **1**, 463
α-Alkoxyaldimines
  reaction with allyl organometallic compounds
    chelation control, **2**, 984, 988
β-Alkoxyaldimines
  reaction with allyl organometallic compounds
    1,3-asymmetric induction, **2**, 985, 988
α-Alkoxyalkyl esters
  carboxy-protecting groups, **6**, 666
Alkoxyamines, *N*-(homoallyl)-
  synthesis
    from aldoxime ethers, **2**, 995
Alkoxycarbonylation
  ketones, **2**, 839
α-Alkoxycarboxamides
  synthesis, **2**, 1086
Alkoxy groups
  cyanide exchange
    nitrile synthesis, **6**, 237
Alkoxymercuration
  demercuration
    alkenes, **4**, 309–312
  oxidative demercuration, **7**, 631, 632
Alkoxymethylation
  α-alkoxycarboxylic acid chlorides
    samarium diiodide, **1**, 259
Alkoxy radicals
  cyclization, **4**, 812
  fragmentation reactions, **4**, 816, 817
Alkyl alcohols
  bromination, **6**, 209
  chlorination
    displacement of hydroxy group, **6**, 204
  fluorination, **6**, 216
  iodination, **6**, 213
N-Alkylamides
  acyclic
    synthesis, **1**, 376
Alkylamides, *N*-α-chloro-
  acyliminium ions from, **2**, 971
Alkylamine, β-(2- or 3-pyrrolyl)-
  synthesis, **8**, 376
Alkylamines, α-ferrocenyl-
  stereoselective synthesis
    Ugi reaction, **2**, 1098
Alkyl anion synthons
  reagents, **2**, 324
Alkylarsino compounds
  halogenolysis, **3**, 203
Alkylation
  acyclic ketone enolates
    extraannular chirality transfer, **3**, 17
  acyl anion equivalents
    sulfur or selenium derivatives, **3**, 134
  aldehydes
    metal enolates, **3**, 3
  alkanes
    Friedel–Crafts reaction, **3**, 332
  alkenes
    Friedel–Crafts reaction, **3**, 331
  alkyl sulfonates, sultones and sulfonamides, **3**, 179
  alkynes
    Friedel–Crafts reaction, **3**, 332
  alkynides, **3**, 272
    alkyl halides, **3**, 272
    epoxides, **3**, 277
  alkynyl carbanions, **3**, 271–292
  allene carbanions, **3**, 256
  amides, **6**, 399
  amines
    alkyl halides, **6**, 65
    sulfonates, **6**, 72
  angular
    1-decalone lithium 1(9)-enolate, **3**, 16
  anomeric
    glycoside synthesis, **6**, 34
  arenes, **4**, 426
    Friedel–Crafts reaction, **3**, 298
    polyfunctional alkylating agents, **3**, 317
    with alcohols, **3**, 309
    with alkanes, **3**, 322
    with alkenes, **3**, 304
    with alkyl halides, **3**, 299
    with epoxides, **3**, 309
    with esters, **3**, 309
    with ethers, **3**, 309
    with lactones, **3**, 309
  aryl carbanions, **3**, 259
  axial
    4-*t*-butylcyclohexanone, **3**, 13
  azides, **6**, 76
  Beckmann rearrangement, **6**, 769
  carbanions
    boron stabilized, **1**, 495
    heteroatom-stabilized, **3**, 193–204
    nitrogen-stabilized, **3**, 65–82
    nonstabilized, **3**, 207–233
    sulfur- and selenium-containing, **3**, 85–181
  carbonyl compound nitrogen derivatives
    regiochemistry, **3**, 28
    stereochemistry, **3**, 28
  Claisen condensation and, **2**, 817
  cyanohydrin ethers, **1**, 552
  cyanohydrins, **1**, 550
  α-(dialkylamino)nitriles, **1**, 557
  diastereoselective

acyclic carboxylic acids, **3**, 44
acyclic enolates of carboxylic acid derivatives, **3**, 42
carboxylic acid enolates, **3**, 39
β-dicarbonyl compounds, **3**, 54, 58
dienolates
α,β-unsaturated carboxylic acids, **3**, 50
1,3-dithiane lithio derivatives, **1**, 568
1,1-(dithio)allyl metals, **3**, 131
1,1-(dithio)propargyl metals, **3**, 131
enantioselective synthesis, **3**, 35
enolates, **3**, 1–58
stereochemistry, **3**, 12
sterically hindered, **1**, 3
enols, **3**, 1–58
equatorial
4-*t*-butylcyclohexanone, **3**, 13
Friedel–Crafts, **3**, 293–335
heteroaromatic carbanions, **3**, 260
intramolecular
tandem carbanionic addition, **4**, 986
ketones
metal enolates, **3**, 3
masked carboxylic acid anions
asymmetric syntheses, **3**, 53
metal dienolates
α,β-unsaturated ketones, **3**, 21
metal enolates
carboxylic acid derivatives, **3**, 39
α-metalloalkyl selenoxides, **3**, 157
α-metalloalkyl sulfones, **3**, 158
α-metalloalkyl sulfoxides and selenoxides, **3**, 147
1-metallo-1,1-bis(dithio)alkanes
synthetic applications, **3**, 121
α-metalloorthoselenoformates, **3**, 144
α-metalloorthothioformates, **3**, 144
α-metallovinyl selenides, **3**, 104
α-metallovinyl selenoxides, **3**, 157
α-metallovinyl sulfides, **3**, 104
α-metallovinyl sulfone, **3**, 173
organomercury compounds
palladium complexes, **4**, 838
phosphonate carbanions, **1**, 563
phosphonium ylides, **6**, 182
S_N2' process, **3**, 257
α-selenoalkyllithium, **3**, 88
α-selenoallyllithium, **3**, 95
α-selenobenzyl metal, **3**, 94
α-selenopropargylic lithium derivatives, **3**, 104
silyl enol ethers, **3**, 25
sp^2 centers
epoxides, **3**, 262
stabilized metal enolates, **3**, 54
sulfur ylides, **3**, 178
synthesis
saturated aldehyde metal enolates, **3**, 20
α-thioalkyllithium, **3**, 88
α-thioallyllithium, **3**, 95
α-thiobenzyl metal, **3**, 94
α-thiopropargylic lithium derivatives, **3**, 104
vinyl- and aryl-lithium compounds, **3**, 247
vinyl carbanions, **3**, 241–266
alkyl halides, **3**, 242
heteroatom-substituted, **3**, 252
vinyl Grignard reagents, **3**, 242
α-Alkylation
enhancement, **4**, 260
Pummerer rearrangement
preparation of α-alkylated sulfides, **7**, 199
O-Alkylation
amides
deprotection, **6**, 672
anomeric
glycoside synthesis, **6**, 54
Alkylative amination
aldehydes
alkyltitanium(IV) complexes, **1**, 170
Alkylbenzyloxyamines
synthesis, **6**, 112
Alkyl carbenoids
insertion reactions, **3**, 1051
Alkyl 2-chloromethyl-4-nitrophenyl hydrogen phosphate
phosphorylation, **6**, 608
Alkyl fluorides
cleavage
metal–ammonia, **8**, 530
Friedel–Crafts reactions, **3**, 294
mixture with antimony fluoride
Friedel–Crafts reaction, intermediate, **3**, 299
primary
reduction with lithium aluminum hydride, **8**, 803
reduction
dissolving metals, **8**, 795
synthesis
via Ireland silyl ester enolate rearrangement, **5**, 841
Alkyl fluorosulfonates, β-nitroperfluoro-
synthesis, **7**, 493
Alkyl groups
addition reactions
with alkenic π-systems, **4**, 72–99
conjugate additions
catalyzed by Lewis acids, **4**, 140–158
Alkyl halides
alcohol synthesis, **6**, 2
alkylation
amine, **6**, 65
arenes, **3**, 299
sulfur- and selenium-stabilized carbanions, **3**, 86
vinyl carbanions, **3**, 242
carbonylation
formation of esters, **3**, 1028
catalytic hydrogenolysis, **8**, 794
coupling reactions
sodium metal, **3**, 414
with sp^3 carbon centers, **3**, 426
with sp^2 organometallics, **3**, 464
Friedel–Crafts reactions
alkylating agents, **3**, 294
α-functionalization, **4**, 260
haloalkylation, **3**, 118
nitrile synthesis, **6**, 226
oxidation
dimethyl sulfoxide, **7**, 291
reactions with π-allylnickel halides, **3**, 424
reactions with organocopper reagents, **3**, 215
reduction
to alkanes, **8**, 793–807
reduction potentials, **8**, 985
secondary
coupling reactions with sp^2 organometallics, **3**, 466
vinyl substitutions

palladium complexes, **4**, 842–856
Alkyl hydroperoxides
epoxidation, **7**, 375
Alkylidenation
alkyl-*gem*-dichromium reagents, **1**, 205
carbonyl compounds, **5**, 1122–1126
titanium metallacycles, **5**, 1124
α-Alkylidenation
sulfur oxidative removal, **3**, 26
Alkylidene carbenes
insertion reactions, **3**, 1049
Alkylidene carbenoids
insertion reactions, **3**, 1050
Alkylidene transfer
cyclopropane synthesis, **4**, 951–994
methylenation *versus* Tebbe reaction, **1**, 749
Alkyl iodides, perfluoro-
reaction with alkenes
palladium complexes, **4**, 842
Alkyl isocyanates
2-azetidinones from, **5**, 103
Alkyl metals, 1,1-bis(seleno)-
reactions with carbonyl compounds, **1**, 723
reactions with enals, **1**, 686, 687
Alkyl metals, α-seleno-
carbonyl compound homologation, **1**, 724
functionalized
reactions, **1**, 723
reactions with carbonyl compounds, **1**, 723
reactions with enals, **1**, 683
reactions with enones
regiochemistry, **1**, 682
synthesis, **1**, 658, 666, 669
via metallation of selenides, **1**, 635
Alkyl metals, α-selenoxy-
reactions with carbonyl compounds, **1**, 723
reactions with enals, **1**, 683
Alkyl metals, 1-silyl-1-seleno-
reactions with carbonyl compounds, **1**, 723
Alkyl nitrates, β-bromo-
synthesis, **7**, 533
Alkyl nitrite
reoxidant
Wacker process, **7**, 452
Alkyl radicals
heterocyclic formation
radical reactions, **4**, 792
substituted
carbocycle formation *via* cyclization, **4**, 791
Alkyl radicals, dichloro-
radical cyclizations, **4**, 792
Alkyl sulfides
reactions with π-allylpalladium complexes, **4**, 599
Alkyl sulfinates
reactions with organocopper reagents, **3**, 215
Alkyl sulfonates
reaction with superoxides
alcohol inversion, **6**, 22
Alkyl thiocyanates
trimerization, **5**, 1154
Alkyl tosylates
coupling reactions
with sp^2 organometallics, **3**, 466
Alkyl triflates
alkylation
carbonyl phosphine carbene complexes, **5**, 1076
N-Alkyl-*N*-vinylnitrosonium ions
imidate synthesis
amide protection, **6**, 672
Alkyne, ferrocenyl-
synthesis, **8**, 950
Alkyne, vinyl-
hydrosilylation, **8**, 772
Alkyneboron difluorides
reaction with oxiranes, **3**, 279
Alkyne insertion
metal carbene complexes
cyclopropanation, **4**, 980
Alkynes (*see also* specific compounds under Acetylene)
acetoxymercuration, **8**, 850
activated
conjugate additions catalyzed by Lewis acids, **4**, 139–164
acylation
Friedel–Crafts reaction, **2**, 723
simple, **2**, 723
acyloxymercuration, **4**, 315
addition reactions
benzeneselenenyl chloride, **7**, 521
carbon-centered radicals, **4**, 735–765
carbon nucleophiles, **4**, 571–583
dihalocarbenes, **4**, 1005
ketocarbenoids, **4**, 1050–1052
reactive carbanions, **4**, 69–130
alkoxymercuration, **4**, 312–316
alkenes from
hydroboration/protonolysis, **8**, 726
alkylated
synthesis, **3**, 799
alkylation
Friedel–Crafts reaction, **3**, 332
via cationic iron complexes, **4**, 582
aminomercuration-demercuration, **4**, 292
benzannulation
functionality, **5**, 1098
carboalumination, **4**, 888
regioselective, **4**, 890
carboboration, **4**, 884, 886
carbocupration, **4**, 896–901
carbolithiation, **4**, 872
carbomagnesiation, **4**, 877–879
carbometallation, **4**, 262, 865–906
carbonylation
nickel tetracarbonyl catalyst, **3**, 1027
carbozincation, **4**, 883
conjugate additions
Lewis acid catalyzed, **4**, 164
conjugated
hydrosilylation, **8**, 772
Cope rearrangement, **5**, 797
coupling with carbene complexes, **5**, 1089
cyclic
synthesis, **3**, 553, 556
cyclizations
formaldiminium ions, **2**, 1029
nitrogen nucleophiles, **4**, 411–413
zirconium-promoted, **5**, 1164
cyclotrimerization
regioselectivity, **5**, 1144–1151
disubstituted
hydrosilylation, **8**, 771
divinyl ketones from, **5**, 777

electrocyclization, **5**, 735–737
electron deficient
  ene reactions, **5**, 6–9
  stabilized nucleophiles and, **4**, 1–58
electrophilic addition
  X–Y reagents, **4**, 329–359
electrophilic heteroatom cyclizations, **4**, 393–397
four-membered heterocyclic compounds from, **5**, 116
functionalized
  carboalumination, **4**, 892
  carbomagnesiation, **4**, 878
  carbozincation, **4**, 884
  synthesis *via* retro Diels–Alder reaction, **5**, 565
heterocyclic
  coupling reactions with alkenyl bromides, **3**, 539
hydration, **4**, 299
hydroalumination, **8**, 733–758
  substituent control, regiochemistry, **8**, 750
  substituent effects, **8**, 749
hydroboration, **8**, 703–727
  organopalladium catalysis, **3**, 231
hydrobromination, **4**, 285
hydrocarboxylation, **4**, 932–946
hydrochlorination, **4**, 277
hydroesterification
  formation of α,β-unsaturated esters, **3**, 1030
hydrofluorination, **4**, 271
hydroformylation, **4**, 922
hydrogenation
  heterogeneous catalysis, **8**, 417–442
  homogeneous catalysis, **8**, 443–463
  mechanism, **8**, 431
  regioselectivity, **8**, 432
  stereoselectivity, **8**, 432
hydrogenation to *cis*-alkenes
  homogeneous catalysis, **8**, 457
hydrogenation to *trans*-alkenes
  homogeneous catalysis, **8**, 458
hydrogenation to saturated hydrocarbons
  homogeneous catalysis, **8**, 456
hydroiodination, **4**, 288
hydrosilylation, **8**, 763–789
  chlorodimethylsilane, **7**, 643
  (diethoxymethyl)silane, **7**, 643
hydroxylation, **7**, 439
hydrozirconation, **4**, 153; **8**, 667–699
intermolecular addition
  carbon nucleophiles, **4**, 41–46
  heteronucleophiles, **4**, 47–53
internal
  Pauson–Khand cycloadditions, **5**, 1041
intramolecular addition
  carbon nucleophiles, **4**, 46
  heteronucleophiles, **4**, 53
mercury-catalyzed hydration, **4**, 303
metal-activated
  heteroatom nucleophilic addition, **4**, 567
  nucleophilic attack, **4**, 551–568
metallation, **3**, 271
monosubstituted
  hydrosilylation, **8**, 770
nonfunctionalized
  carbozincation, **4**, 883
octacarbonyldicobalt complexes
  Pauson–Khand reaction, **5**, 1037
oxidation
  solid support, **7**, 844
oxidative rearrangement, **7**, 833
pinacol coupling reactions
  with carbonyl compounds, **3**, 602
protection, **6**, 684
radical addition reactions, **4**, 715–772
radical cyclizations
  carbon-centered radicals, **4**, 789
reactions with *N*-acyliminium ions
  intramolecular, **2**, 1062
reactions with alcohols, **4**, 309
reactions with carbon monoxide, **4**, 913–949
reactions with carboxylic acids, **4**, 313
reactions with Fischer carbene complexes
  alkyne concentration, **5**, 1099
  solvents, **5**, 1099
reactions with HX reagents, **4**, 269–319
reactions with iminium ions, **2**, 1028
reactions with ketocarbenes, **4**, 1031–1064
reactions with ketyls
  organosamarium compounds, **1**, 268
reaction with carbene complexes
  regiochemistry, **5**, 1093
reduction
  diimide, **8**, 477
  noncatalytic chemical methods, **8**, 471–487
semihydrogenation
  heterogeneous catalysis, **8**, 430
synthesis
  organoboranes, **3**, 780
  organocopper compounds, **3**, 217
  Ramberg–Bäcklund rearrangement, **3**, 883
  via aldehydes, **7**, 620
  via Julia coupling, **1**, 802, 805
  via oxidation of bishydrazones, **7**, 742
  via 2,3-sigmatropic rearrangement, **6**, 873
π-systems
  addition reactions, **4**, 128
  nucleophile addition, **4**, 41–53
tandem vicinal difunctionalization, **4**, 242, 249
terminal
  coupling reactions, **3**, 551
  ene reaction with formaldehyde, **2**, 531
  hydroboration, **8**, 708
  hydrogenation to alkenes, **8**, 457
  hydrozirconation, **8**, 684
  oxidative homocoupling, **3**, 552
  reaction with sp^2 carbon halides, **3**, 530
  stereospecific synthesis, **3**, 539
  synthesis, **3**, 531
trimerization
  Pauson–Khand reaction, **5**, 1038
thiylation, **4**, 317
ylidic rearrangements, **3**, 963
1-Alkynes
  hydroalumination
    asymmetrical diene synthesis, **3**, 486
  hydrozirconation
    asymmetrical diene synthesis, **3**, 486
  reaction with alkenyl halides, **3**, 539
  synthesis
    from dichloromethyllithium, **3**, 202
  vinylation, **3**, 521
4,5-Alkynes
  cyclization, **3**, 344
6,7-Alkynes

cyclization
selectivity, **3**, 344
Alkynes, alkoxy-
synthesis
organocopper compounds, **3**, 217
Alkynes, 1-alkoxy-
acid anhydride synthesis, **6**, 315
ortho acid synthesis, **6**, 556
Alkynes, amino-
cyclization
palladium(II) catalysis, **4**, 567
Alkynes, 1-amino-
reaction with nitriles, **6**, 401
Alkynes, α-amino-
synthesis, **3**, 282
Alkynes, aryl-
conjugated
one-pot synthesis, **3**, 539
hydrobromination, **4**, 285
one-pot synthesis, **3**, 541
Alkynes, 1-azido-
synthesis
failure, **6**, 247
Alkynes, bromo-
Chodkiewicz–Cadiot reaction, **3**, 553
reaction with trialkylaluminum, **3**, 285
Alkynes, chloro-
reaction with tertiary enolates, **3**, 284
Alkynes, ω-cyano-
cocycloaddition with alkynes, **5**, 1154
Alkynes, cyclopropyl-
rearrangement, **5**, 947
Alkynes, dialkyl-
cyclization
selectivity, **3**, 344
hydrogenation to *trans*-alkenes
homogeneous catalysis, **8**, 458
Alkynes, diaryl-
hydrobromination, **4**, 286
hydrogenation to *trans*-alkenes
homogeneous catalysis, **8**, 458
Alkynes, dihydroxy-
intramolecular oxypalladation, **4**, 394
Alkynes, β,γ-dihydroxy-
synthesis
via Payne rearrangement, Lewis acids, **1**, 343
Alkynes, halo-
coupling reactions
organometallic acetylides, **3**, 553
electrophilic substitution, **3**, 284
Alkynes, 1-halo-
hydroboration
protonolysis, **8**, 726
1-Alkynes, 1-halo-
hydrobromination, **4**, 286
Alkynes, ω-isocyanato-
cocycloaddition with silylated alkynes, **5**, 1156
Alkynes, 1-nitro-2-(trialkylsilyl)-
synthesis, **6**, 109
Alkynes, phenyl-
hydration, **4**, 300
Alkynes, silyl-
carbomagnesiation, **4**, 879
Alkynes, silylstannyl-
reaction with alkenyl iodide, **3**, 539
Alkynes, stannyl-
reactions with steroidal aldehydes
Cram selective, **1**, 335
Alkynes, trifluoromethyl-
ene reactions, **5**, 7
Alkynes, trimethoxymethyl-
synthesis, **6**, 556
Alkynes, trimethylsilyl-
ene reactions, **5**, 23
Alkynes, 1-trimethylsilyl-
carbozincation, **4**, 884
Alkynes, trimethylsilylmethyl-
synthesis, **3**, 281
β-Alkynic alcohols
synthesis
regioselectivity, **2**, 92
Alkynic chloride
hydrogenolysis, **8**, 898
Alkynides
alkylation, **3**, 272
alkyl halides, **3**, 272
sulfates, **3**, 272
synthesis, **3**, 272
Alkynoic acids
hydrobromination, **4**, 285
α,β-Alkynoic acids
hydroboration
protonolysis, **8**, 726
Alkynones
pyrolysis, **3**, 1049
silyl enol ethers
cyclization, **5**, 22
1-Alkyn-3-ones
hydrobromination
stereochemistry, **4**, 285
Alkynyl alcohols
cyclofunctionalization, **4**, 393
divinyl ketones from
cyclization, **5**, 767–769
synthesis
via alkynylcerium reagents, **1**, 243
Alkynylation
oxiranes and oxetanes
use of boron trifluoride, **1**, 343
vinyl organometallic reagents, **3**, 521
Alkynyl complexes
[3 + 2] cycloaddition reactions
diazoalkanes, **5**, 1070
Alkynyl groups
addition reactions
with alkenic π-systems, **4**, 72–99
conjugate additions
catalyzed by Lewis acids, **4**, 140–158
Alkynyl halides
cross-coupling reactions
organometallic reagents, **3**, 522
reactions with 1-alkenyl metals, **3**, 529
reaction with organocopper compounds, **3**, 219
tandem vicinal difunctionalizations, **4**, 260
Alkynyl organometallic compounds
[3 + 2] cycloaddition reactions, **5**, 277
Allene, 3,3-dialkyl-1-lithio-1-(phenylthio)-
reaction with ketones, **2**, 90
Allene, 1,3-dimethyl(*t*-butyldimethylsilyl)-
[3 + 2] cycloaddition reactions
with cyclohexanecarbaldehyde, **5**, 279
Allene, α-lithio-α-methoxy-

reaction with potassium *t*-butoxide, **2**, 88
Allene, 1-methyl-1-(trimethylsilyl)-
[3 + 2] cycloaddition reactions
with methyl vinyl ketone, **5**, 277
Allene, tetrachloro-
hydrochlorination, **4**, 277
Allene, tetrafluoro-
hydrobromination, **4**, 285
hydrochlorination, **4**, 277
Allene, tetramethyl-
laser photolysis
with benzophenone, **5**, 154
photocycloaddition reactions
with acetone, **5**, 167
Allene, trimethylsilyl-
[3 + 2] cycloaddition reactions
titanium tetrachloride catalyst, **5**, 277
reaction with carbonyl compounds, **2**, 84
Allene carbanions
alkylation, **3**, 256
Allene-1,3-dicarboxylic acids
reduction
zinc, **8**, 563
Allenes
acyloxymercuration, **4**, 315
addition reactions
carbon-centered radicals, **4**, 765
selenium electrophiles, **7**, 520
addition to 3,4-dimethylcyclohexenone
photochemical cycloaddition, **5**, 130
addition to octalone
photochemical cycloaddition, **5**, 130
alkoxymercuration, **4**, 311
π-allylpalladium complexes from, **4**, 587
aminomercuration, **4**, 292
carboboration, **4**, 885
carbomagnesiation, **4**, 875
conjugated
thiylation, **4**, 317
Cope rearrangement, **5**, 797
cyclic
synthesis *via* dihalocyclopropanes, **4**, 1010
cyclizations
nitrogen nucleophiles, **4**, 411–413
electrocyclization, **5**, 734
electrophilic heteroatom cyclizations, **4**, 393–397
ene reactions, **5**, 9
intramolecular, **5**, 19
epoxides
rearrangement, **3**, 741
exocyclic
synthesis, **2**, 89
hetero
reactions with vinylidenephosphoranes, **6**, 194
synthesis, **6**, 867
hydration, **4**, 299
hydroboration, **8**, 708, 714, 720
hydrobromination, **4**, 284
hydrochlorination, **4**, 276
hydrogenation, **8**, 434
hydrogenation to alkenes
homogeneous catalysis, **8**, 450
hydroxylation–carbonylation
palladium(II) catalysis, **4**, 558
mercury-catalyzed hydration, **4**, 303
photocycloaddition reactions, **5**, 133, 145
with carbonyl compounds, **5**, 167
pinacol coupling reactions
with carbonyl compounds, **3**, 605
radical cyclization
carbon-centered radicals, **4**, 789
reactions with alcohols, **4**, 308
reactions with π-allylpalladium complexes, **4**, 601
reactions with carboxylic acids, **4**, 313
reactions with Fischer carbene complexes, **5**, 1107
reduction
diimide, **8**, 477
synthesis, **4**, 868
sp^2–sp^2 coupling, **3**, 491
via α,β-alkynic ketones, **8**, 357
via Doering–Moore–Skattebøl reaction, **4**, 1009–1012
via 2,3-sigmatropic rearrangement, **6**, 873
π-systems
nucleophile addition, **4**, 53–58
tandem vicinal difunctionalization, **4**, 253
vinylic
hydrobromination, **4**, 285
ylidic rearrangements, **3**, 963
Allenes, ω-amino-
aminocarbonylation
palladium(II) catalysis, **4**, 562
Allenes, bromo-
coupling reactions
alkyl Grignard reagents, **3**, 439
dimerization, **3**, 491
reaction with alkynes, **3**, 531
reaction with cyanocuprates, **3**, 491
reaction with lithium dialkylcuprates, **3**, 217
Allenes, cyano-
synthesis
via substituted 2-propynols, **6**, 235
Allenes, dienyl-
electrocyclization, **5**, 734
Allenes, iodo-
reaction with arylchlorozinc, **3**, 491
Allenes, β-keto-
synthesis
via Claisen rearrangement, **5**, 828
Allenes, methoxy-
deprotonation, **3**, 256
Allenes, vinyl-
anthracene adduct
retro Diels–Alder reaction, **5**, 589
electrocyclization, **5**, 707
epoxidation
cyclopentenone synthesis, **5**, 772
solvolysis
cyclopentenone synthesis, **5**, 772–775
solvometallation
cyclopentenone synthesis, **5**, 774
synthesis
via electrocyclization, **5**, 708
Allenic acids
enzymic reduction
specificity, **8**, 205
hydrobromination, **4**, 285
Allenic alcohols
synthesis
via samarium diiodide, **1**, 257
Allenic alcohols, alkoxy-
solvolysis

cyclopentenone synthesis, **5**, 774
Allenic alcohols, vinyl-
epoxidation
cyclopentenone synthesis, **5**, 773
Allenic esters
Diels–Alder reactions, **5**, 358
Allenic organometallic compounds
reactions with aldimines
syn–anti selectivity, **2**, 993
reactions with imines, **2**, 975–1004
Allenic phosphonates
alkenation
to cumulatrienes, **6**, 845
Allenic sulfoxides
conjugate addition of nucleophiles, **6**, 840
Allenoxides
[4 + 3] cycloaddition reactions, **5**, 597
α-Allenyl alcohol
synthesis
regioselective, **2**, 92
Allenyl organometallics, **2**, 81–97
[3 + 2] cycloaddition reactions, **5**, 277–281
diasteroselective reactions, **2**, 91–96
enantioselective reactions, **2**, 96
heteroatom-substituted, **2**, 88
regioselective reactions, **2**, 82–91
nonheteroatom-substituted, **2**, 82–88
synthesis, **2**, 81
α-Allenyl phosphates
reaction with organocopper reagents, **3**, 223
Allenyl phosphoryl compounds
synthesis
via rearrangement, **6**, 844
Allenyl systems
Paterno–Büchi reaction, **5**, 165–168
Allethrolone
synthesis, **7**, 795
via Michael addition, **4**, 10
Allobetulone, 2-diazo-
photolysis, **3**, 903
Allodolicholactone
synthesis
via photoisomerizations, **5**, 231
Allodunnione
synthesis, **3**, 828
Allogeraniol
cyclization, **3**, 345
α-Allokainic acid
synthesis
via intramolecular ene reaction, **5**, 13
D-Allonselenoamide, 2,5-anhydro-3,4,6-tri-*O*-benzoyl-
synthesis, **6**, 477
Allopumiliotoxin A
synthesis
Mannich reaction, **2**, 1015
Allopumiliotoxin 323B′
synthesis
enantioselective, **2**, 1028
L-Allose
synthesis, **7**, 402
Alloxan
rearrangement, **3**, 822, 834
labeling studies, **3**, 823
Alloxanic acid
synthesis, **3**, 822
Allyl acetal
hydroformylation, **4**, 924
Allyl acetate, 2-(trimethylsilylmethyl)-
cycloaddition
palladium catalysis, **4**, 593
[4 + 3] cycloaddition reactions, **5**, 598
Allyl acetates
allylic transposition
palladium(II) catalysis, **4**, 576
cyclic ether synthesis, **6**, 24
dicarboxylation, **4**, 948
electrolysis, **8**, 976
hydrogenolysis
palladium-catalyzed, **6**, 866
oxidation
palladium(II) catalysis, **4**, 553
reactions with carbonyl compounds
samarium diiodide, **1**, 256
rearrangement
oxygen–oxygen transposition, **6**, 835
palladium catalysis, **4**, 596
reduction, **8**, 960
substituted, **8**, 960
synthesis
via alcohols, **6**, 835
transition metal catalyzed reactions, **6**, 847
Allyl alcohol, 1,1-dimethyl-
asymmetric epoxidation, **7**, 417
Allyl alcohol, 3,3-dimethyl-
asymmetric epoxidation, **7**, 409
Allyl alcohol, 2-ethoxy-
[4 + 3] cycloaddition reactions, **5**, 597
Allyl alcohol, 2-silylmethyl-
[4 + 3] cycloaddition reactions, **5**, 598
Allyl alcohol, stannyl-
asymmetric epoxidation, **7**, 413
Allyl alcohol, 3-trimethylsilyl-
asymmetric epoxidation, **7**, 413
Allyl alcohols
acyclic
synthesis *via* retro Diels–Alder reaction, **5**, 554
addition reactions
benzeneselenenyl chloride, **7**, 520
π-allylpalladium complexes from, **4**, 590
arene alkylation
Friedel–Crafts reaction, **3**, 322
arylation
via palladium catalysts, **4**, 848
asymmetric epoxidation, **7**, 397
molecular sieves, **7**, 396
asymmetric hydrogenation
homogeneous catalysis, **8**, 462
carbolithiation, **4**, 869
chlorination
displacement of hydroxy group, **6**, 206
cycloaddition reactions, **5**, 261
[4 + 3] cycloaddition reactions, **5**, 598
1,3-diene synthesis, **6**, 154
N,N-diisopropyl carbamates
oxaallylic anions, **3**, 196
(2,3*E*)-disubstituted
asymmetric epoxidation, **7**, 406
(2,3*Z*)-disubstituted
asymmetric epoxidation, **7**, 408
1,1-disubstituted
asymmetric epoxidation, **7**, 417
3,3-disubstituted

asymmetric epoxidation, **7**, 409
enzymic reduction
specificity, **8**, 205
epimerization, **6**, 839
epoxidation, **7**, 370, 378, 391
halomethylsilyl ethers
radical cyclization, **7**, 648
homogeneous hydrogenation
diastereoselectivity, **8**, 447
homologous β,γ-unsaturated amide synthesis, **6**, 853
hydrocarboxylation, **4**, 941
hydroformylation, **4**, 923
hydrogenolysis, **8**, 956
hydroxylation, **7**, 439
intramolecular hydrosilylation, **7**, 645
(3*Z*)-monosubstituted
asymmetric epoxidation, **7**, 405
nitrile synthesis, **6**, 234
optically active
synthesis, **6**, 839
oxidation, **7**, 306, 307, 318
Collins reagent, **7**, 258
4-(dimethylamino)pyridinium chlorochromate, **7**, 269
DMSO, **7**, 296
solid support, **7**, 841
oxidative rearrangement, **7**, 821
photocycloaddition reactions
copper-catalyzed, **5**, 147
rearrangement
oxidation, **6**, 836
reduction
dissolving metals, **8**, 971
(3*E*)-substituted
asymmetric peroxidation, **7**, 400
1,3-sigmatropic rearrangements
oxyanion-accelerated, **5**, 1002
1-substituted
asymmetric epoxidation, **7**, 409, 413
2-substituted
asymmetric epoxidation, **7**, 398
synthesis, **1**, 708; **7**, 84, 396
Knoevenagel reaction, **2**, 374
stereoselective, **6**, 838
via β-hydroxyalkyl selenides, **1**, 721
via organocerium compounds, **1**, 235
via organocopper reagents, **6**, 848
via oxidation of allylstannanes, **7**, 616
d^3-synthons, **6**, 838
tertiary
oxidative rearrangement with pyridinium chlorochromate, **7**, 263
transformation reactions, **6**, 850
to γ,δ-unsaturated carbonyls, **6**, 855
transition metal catalyzed reactions, **6**, 847
2,3,3-trisubstituted
asymmetric epoxidation, **7**, 409
vinylation
palladium complexes, **4**, 854
Allyl alcohols, β-fluoro-
synthesis
via cyclopropane ring opening, **4**, 1020
Allyl alcohols, nitro-
enantiomers
synthesis, **2**, 328
synthesis
via acetoxyselenation, **4**, 340
Allylamines
addition reactions
nitrogen nucleophiles, **4**, 562
carbolithiation, **4**, 871
carbonylation
formation of pyrrolidones, **3**, 1037
equilibration
enamines, **6**, 706, 707
hydrocarboxylation, **4**, 941
γ-lithiation, **1**, 477
metallated chiral
homoenolate equivalents, **2**, 62
oxidation
palladium(II) catalysis, **4**, 559
reduction
tributylstannanes, **8**, 961
synthesis, **1**, 559; **3**, 258; **6**, 843
via allyl selenides, **6**, 905
via Horner reaction, **1**, 774
transformation reactions
to γ,δ-unsaturated carbonyls, **6**, 855
vinylation
palladium complexes, **4**, 854
Allylamines, 2-aryl-
synthesis, **3**, 492
Allyl arsenites
π-allylpalladium complexes from, **4**, 590
Allylation
aldehydes
asymmetric, **6**, 865
carbonyl compounds
preparation of 1,4-dicarbonyl compounds, **7**, 455
enolates
palladium-catalyzed regioselective, **3**, 12
organometallic reagents
carbon–carbon bond forming reaction, **6**, 847
α-selenoalkyl metals, **3**, 91
sulfur- and selenium-stabilized carbanions, **3**, 88
Allyl borates
π-allylpalladium complexes from, **4**, 590
Allyl boronates
indirect aldol reaction, **6**, 864
Allyl bromide, 2-methoxy-
generation of 2-methoxyallyl cation
[4 + 3] cycloaddition reactions, **5**, 597
Allyl bromide, 2-siloxy-
2-siloxyallyl cation generation
[4 + 3] cycloaddition reactions, **5**, 603
Allyl bromides
alkylation
cyclic carbene complexes, **5**, 1076
hydrobromination, **4**, 280
reduction
lithium aluminum hydride, **8**, 965
Allyl carbamates
π-allylpalladium complexes from, **4**, 589, 592
Claisen-type rearrangement
palladium(II) catalysis, **4**, 564
metallated
homoaldol reaction, **6**, 863
Allyl carbonates
alcohol protection
cleavage, **6**, 659
π-allylpalladium complexes from, **4**, 589
palladium enolates

allylation, **4**, 592
transition metal catalyzed reactions, **6**, 847
Allyl carbonates, 2-(cyanomethyl)-
cycloaddition
palladium catalysis, **4**, 593
Allyl carbonates, ethyl-2-(sulfonylmethyl)-
cycloaddition
palladium catalysis, **4**, 593
Allyl carbonates, (methoxycarbonyl)methyl-
cycloaddition
palladium catalysis, **4**, 593
Allyl cations
[4 + 3] cycloaddition reactions, **5**, 601
β-heteroatom-substituted
[4 + 3] cycloaddition reactions, **5**, 594
initiators
polyene cyclization, **3**, 342
Allyl cations, 2-amino-
[4 + 3] cycloaddition reactions, **5**, 597
Allyl cations, 2-methoxy-
[4 + 3] cycloaddition reactions, **5**, 597
Allyl cations, 2-methyl-
[4 + 3] cycloaddition reactions, **5**, 603
Allyl cations, 2-silylmethyl-
[4 + 3] cycloaddition reactions, **5**, 598
Allyl cations, 2-(trimethylsiloxy)-
[4 + 3] cycloaddition reactions, **5**, 597, 606
Allyl cations, 2-trimethylsilylmethyl-
[4 + 3] cycloaddition reactions, **5**, 598
Allyl chloride
hydroboration, **8**, 713
hydrobromination, **4**, 280
hydroiodination, **4**, 288
Allyl chloride, 2-(trimethylsiloxy)-
reaction with silver perchlorate
generation of oxyallyl cations, **5**, 597
Allyl chlorides, siloxy-
siloxyallyl cation generation
[4 + 3] cycloaddition reactions, **5**, 606
π-Allyl complexes
reactions with nitrogen nucleophiles, **6**, 85
Allyl compounds
hydrogenation
heterogeneous catalysis, **8**, 439
metal complexes
nucleophilic addition, **4**, 585–654
microbial oxidation, **7**, 77
Allyl cyanide
synthesis
via allylic bromide, **6**, 230
Allyl ester enolates
Claisen rearrangement, **6**, 858
Allyl esters
amine-protecting group, **6**, 640
polymer-bound protecting group, **6**, 671
carboxy-protecting groups, **6**, 670
rearrangements
palladium(II) catalysis, **4**, 563
regioselective oxidation, **7**, 464
Allyl ethers
hydroformylation, **4**, 923
isomerization
enol ether preparation, **2**, 599
photocycloaddition reactions
copper-catalyzed, **5**, 147
regioselective oxidation, **7**, 464
Allyl hetero compounds
reduction
1,3-heteroatom–hydrogen transposition reaction, **6**, 865
transformation reactions, **6**, 853
Allylic alcohols, 1-(trimethylsilyl)-
rearrangement
formation of lithium homoenolates, **3**, 197
Allylic alkylation
palladium-catalyzed, **6**, 848, 849
nucleophiles, **4**, 590–600
Allylic amides
synthesis
via Horner reaction, **1**, 774
Allylic anions
1,4-addition reaction with conjugated enones, **6**, 863
boron-substituted, **2**, 56
halogen-substituted, **2**, 77
heteroatom-stabilized, **2**, 55–78
synthetic utility, **2**, 55
heteroatom-substituted
homoenolate equivalents, **6**, 833
homoenolate anion equivalent, **6**, 862
nitrogen-substituted, **2**, 60
N-nitroso-*N*-alkyl-, **2**, 61
1-oxy-
rearrangement, **2**, 69
oxygen-substituted, **2**, 66
phosphine-substituted, **2**, 64
selenium-substituted, **2**, 76
silicon-substituted, **2**, 57
sulfur-substituted, **2**, 71
reaction with electrophiles, **2**, 73
Allylic electrophiles
reaction with organocopper compounds, **3**, 220
Allylic halides
π-allylpalladium complexes from, **4**, 588
arene haloalkylation
Friedel–Crafts reaction, **3**, 321
Barbier-type reactions
organosamarium compounds, **1**, 256
carbonylation
formation of aldehydes, **3**, 1021
coupling reactions
with sp^3 carbon centers, **3**, 428
with sp^2 organometallics, **3**, 467
generation of allyl cations
[4 + 3] cycloaddition reactions, **5**, 597
haloalkylation, **3**, 118
β-heteroatom-substituted
[4 + 3] cycloaddition reactions, **5**, 597
hydrogenolysis, **8**, 955–981
hydrosilylation, **8**, 775
β-keto esters
cyclization, **1**, 265
reaction with ethyl diazoacetate, **3**, 925
reaction with vinyltin compounds
organopalladium catalysis, **3**, 232
1,5-Allylic hydrogen transfer
heptenyl radicals, **4**, 786
Allylic hydroxylation
Δ^4-steroids, **7**, 77
Allylic iodides
reaction with peracids
preparation of alcohols, **6**, 3
synthesis

via rearrangement of allylic alcohols, **6**, 835
Allylic oxidation, **7**, 83
allylic alcohols from, **7**, 84
metallation, **7**, 99
selenium dioxide
mechanism, **7**, 85
α,β-unsaturated carbonyl compounds, **7**, 99
with rearrangement, **7**, 817
Allylic phosphate esters
reactions with carbonyl compounds
samarium diiodide, **1**, 256
Allylic phosphine oxides
lithiated
γ-selective conjugate addition to cyclic enones, **6**, 863
Allylic phosphonates
lithiated
γ-selective conjugate addition to cyclic enones, **6**, 863
reduction
rearrangement, **6**, 865
Allylic silanes
protodesilylation
double bond shift, **6**, 865
Allylic substitution
carbon nucleophiles, **6**, 847
Allylic sulfides
chlorination, **7**, 209
Allylic sulfonyl carbanions
synthesis, **2**, 76
Allylic sulfoxides
α-lithiation, **2**, 74
monohapto, **2**, 5
trihapto, **2**, 5
Allylic transposition
palladium(II) catalysis, **4**, 563
Allylimidates
π-allylpalladium complexes from, **4**, 590
Claisen-type rearrangement
palladium(II) catalysis, **4**, 564
Allyl iodide
reaction with chlorosulfonyl isocyanate, **5**, 105
Allyl isocyanide
synthesis, **2**, 1083
Allyl metal compounds
protonation
1,3-heteroatom–hydrogen transposition reaction, **6**, 865
reactions with aldehydes
synthesis of homoallylic alcohols, **6**, 864
reactions with electrophiles, **6**, 832
Allyl nitro compounds
reduction, **8**, 962
Allyl organometallic compounds
chiral
C(1) or C(4) stereocenters, **2**, 38
conventional auxiliaries, **2**, 33
enantioselective, **2**, 33
reactions with achiral aldehydes, **2**, 33–40
reactions with chiral C═X electrophiles, **2**, 40–45
[3 + 2] cycloaddition reactions, **5**, 272–277
reactions with aldimines
1,3-asymmetric induction, **2**, 986
reactions with α-alkoxyaldimines
chelation control, **2**, 984, 988
reactions with β-alkoxyaldimines
1,3-asymmetric induction, **2**, 985, 988
reactions with *gem*-amino ethers, **2**, 1004
reactions with chiral C═N electrophiles
relative diastereoselectivity, **2**, 32
reactions with glyoxylate aldimines
1,3-asymmetric induction, **2**, 987
reactions with imines, **2**, 975–1004
reviews, **2**, 980
reactions with α-phenylaldimine
Cram selectivity, **2**, 984
reactions with 8-phenylmenthyl-*N*-methoxyiminoacetate
diastereoselectivity, **2**, 996
reactions with sulfenimines
Cram selectivity, **2**, 998
type I
stereochemical integrity, **2**, 5
uncatalyzed reactions
C═X electrophiles, **2**, 1–49
N-Allyloxycarbonyl group
protecting group
amines, **6**, 633, 640
Allyl phosphoryl compounds
synthesis
via rearrangement, **6**, 844
Allyl rearrangement
functional group transformation, **6**, 829–867
intermolecular, **6**, 830
intramolecular, **6**, 833
substitution reactions, **6**, 830
tertiary halides, **6**, 835
Allyl shifts
cyclohexadienones, **3**, 809
Allyl sulfenate
allyl sulfoxide
transposition reaction, **6**, 837
Allyl sulfones
reductive desulfurization, **8**, 840
Allyl sulfoxides
allyl sulfenate
transposition reaction, **6**, 837
lithiated
γ-selective conjugate addition to cyclic enones, **6**, 863
propargyl sulfenate
transposition reaction, **6**, 837
Allyl systems
C—C bond formation, **6**, 862
isomerization
1,3-hydrogen–hydrogen transpositions, **6**, 866
Allyl thiol
dianions
reactions with carbonyl compounds, **1**, 826
Allyl transfer
palladium(0)-catalyzed
amine protection, **6**, 640
Alnusone
synthesis, **3**, 126, 505; **6**, 134
Alpine borane
reaction with aldehydes, **7**, 603
Alstonine, tetrahydro-
microbial hydroxylation, **7**, 65
synthesis
Knoevenagel reaction, **2**, 373
Altholactones
synthesis, **7**, 712

Alumina
  Beckmann rearrangement, **6**, 765
  catalyst
    carbonyl epoxidation, **1**, 821
    Knoevenagel reaction, **2**, 344
  solid support
    chloral, **7**, 841
    oxidants, **7**, 840
Aluminates, tetraalkyl-
  coupling reactions
    with acyl chlorides and acid anhydrides, **3**, 463
  reactions with chiral keto esters
    stereoselectivity, **1**, 87
Aluminum
  reduction
    epoxides, **8**, 881
    thioimidates, **8**, 302
Aluminum, alkenyl-
  alkylation, **3**, 259
  coupling reactions
    with allylic chlorides, **3**, 475
    with aryl halides, **3**, 495
    with vinyl halides, **3**, 486
  in synthesis, **4**, 893
Aluminum, alkoxydichloro-
  catalysts
    Diels–Alder reactions, **5**, 376
Aluminum, alkyl-
  addition reactions
    masked carbonyl compounds, **1**, 88
  hydride donor
    reduction of carbonyls, **8**, 99
Aluminum, alkylthioallyl-
  reaction with allylic halides, **3**, 99
Aluminum, allyl-
  metalloene reactions, **5**, 31
Aluminum, chlorodihydrido-
  reduction
    enones, **8**, 545
Aluminum, crotyl-
  reaction with imines
    syn–anti selectivity, **2**, 989
  reaction with iminium salts, **2**, 1000
  synthesis, **2**, 9
Aluminum, crotyldiethyl-
  synthesis, **2**, 9
Aluminum, cyanodiethyl-
  reaction with conjugated ketones
    1,4-addition, **2**, 599
Aluminum, cyclohexylmethyl-
  synthesis, **8**, 758
Aluminum, dialkoxy-
  chiral catalysts
    Diels–Alder reactions, **5**, 376
Aluminum, (2,6-di-*t*-butyl-4-methyl)phenoxydiethyl-
  methyl toluate complex
    crystal structure, **1**, 301
Aluminum, (2,6-di-*t*-butyl-4-methyl)phenoxymethyl-
  ketone complexes, **1**, 283
Aluminum, dichloroethyl-
  catalyst
    Friedel–Crafts reaction, **2**, 709
Aluminum, dichloromenthyl-
  catalysts
    Diels–Alder reactions, **5**, 376
Aluminum, diethyl-
  enolates
    aldol reaction, stereoselective, **2**, 315
    regioselective synthesis, **2**, 114
  2,2,6,6-tetramethylpiperidide
    aldol reaction, **2**, 271
    aluminum enolates, **2**, 114
Aluminum, diethyl(1-hexynyl)-
  alkylation
    oxime mesylates, **6**, 769
Aluminum, dihydridoiodo-
  reduction
    enones, **8**, 545
Aluminum, dimethylchloro-
  aldol reaction
    catalysis, **2**, 269
Aluminum, dimethylphenylsilyldiethyl-
  deoxygenation
    epoxides, **8**, 886
Aluminum, hydridodiisobutyl-
  aluminum enolates
    synthesis, **2**, 114
Aluminum, propargyl-
  reactions with aldimines, **2**, 992
Aluminum, sulfatobis(diethyl)-
  catalyst
    allylstannane reaction with acetals, **2**, 578
Aluminum, trialkyl-
  conjugate additions
    α,β-unsaturated ketals, **4**, 209
  optically active
    reduction of ketones, **8**, 100
Aluminum, trialkynyl-
  conjugate additions
    α,β-enones, **4**, 143
Aluminum, triethyl-
  hydride donor
    reduction of carbonyls, **8**, 100
  reaction of allylic anions with carbonyl compounds
    regioselectivity, **2**, 67
  reaction with thioallyl anions
    α-selectivity, **2**, 71
Aluminum, triisobutyl-
  hydride donor
    reduction of carbonyls, **8**, 100
  reduction
    unsaturated ketones, **8**, 558, 564
  synthesis, **8**, 735
Aluminum, trimethyl-
  aldol reactions, **2**, 269
  Beckmann reaction, **7**, 697
  complex with benzophenone, **1**, 78
  conjugate additions
    α,β-enones, **4**, 140
  coupling reactions
    with difunctional alkenes, **3**, 449
  reaction with benzophenone
    role of Lewis acid, **1**, 325
  reaction with 2,6-di-*t*-butyl-4-alkylphenol, **1**, 78
Aluminum, tris(2-methylbutyl)-
  reduction
    unsaturated ketones, **8**, 564
Aluminum, tris(trimethylsilyl)-
  reactions with acyclic enones
    site selectivity, **1**, 83
  reactions with π-allylpalladium complexes
    regioselectivity, **4**, 642

stereochemistry, **4**, 625
Aluminum, vinyl-
reaction with vinyloxiranes, **5**, 936
Aluminum alkoxide
phosphorylation, **6**, 603
Aluminum alkynide, ethyl-
reaction with 3,4-epoxycyclopentene, **3**, 279
Aluminum alkynides
alkylation, **3**, 274
Aluminum amalgam
desulfurization, **8**, 844
reduction
aliphatic nitro compounds, **8**, 374
carbonyl compounds, **8**, 116
enones, **8**, 525
reductive cleavage
α-alkylthio ketone, **8**, 994
reductive dimerization
unsaturated carbonyl compounds, **8**, 532
Aluminum amide, diethyl-
oxirane ring-opening, **6**, 91
Aluminum amides
reactions with esters, **1**, 93
Aluminum bisphenoxides, methyl-
Claisen rearrangement
catalysis, **5**, 850
Aluminum bromide
catalyst
Friedel–Crafts reaction, **2**, 735, 741
Aluminum catalyst
Diels–Alder reaction
absolute stereochemistry, **2**, 685
Aluminum chloride
Beckmann rearrangement, **6**, 770
catalyst
Friedel–Crafts reaction, **2**, 709, 735
hydrosilylation, **8**, 765
Friedel–Crafts alkylations
catalyst, **3**, 294
propylene oxide, **3**, 769
lithium aluminum hydride
alkyl halide reduction, **8**, 803
epoxide reduction, **8**, 875
oxidative cleavage of alkenes
with ethanethiol, **7**, 588
Aluminum chloride, dialkyl-
conjugate additions
α,β-enones, **4**, 140
Aluminum chloride, diethyl-
Beckmann rearrangement, **6**, 768
Aluminum compounds
aldol reactions, **2**, 239, 268
Claisen rearrangement
catalysis, **5**, 850
Lewis acid complexes
structure, **1**, 287
nitrile synthesis, **6**, 241
Aluminum cyanide, diethyl-
conjugate additions
Lewis acid catalyzed, **4**, 162
Aluminum 2,6-di-*t*-butyl-4-methylphenoxide, diisobutyl-
reduction
enones, **8**, 545
Aluminum dichloride, alkoxy-
catalyst
Diels–Alder reaction, **2**, 663
Aluminum ene reactions, **5**, 31–33
Aluminum enolates
aldol reactions
from chiral acyliron complexes, **2**, 239
synthesis, **2**, 114
Aluminum hydrazide, dimethyl-
reactions with esters
carboxylic acid hydrazides, **1**, 93
Aluminum hydride, bis(diisopropylamino)-
reduction
enones, **8**, 543
Aluminum hydride, bis(4-methyl-1-piperazinyl)-
reduction
amides, **8**, 272
Aluminum hydride, bis(*N*-methylpiperidino)-
reduction
esters, **8**, 266
Aluminum hydride, di-*t*-butoxy-
reduction
enones, **8**, 543
Aluminum hydride, dichloro-
hydroalumination, **8**, 736
reduction
enones, **8**, 545
Aluminum hydride, diisobutyl- (DIBAL-H)
hydride donor, **8**, 100
hydroalumination, **8**, 736
reaction with 1-alkynylsilanes, **8**, 734
reduction
acetals, **8**, 214
amides, **8**, 272
carbonyl compounds, **8**, 20, 315
carboxylic acids, **8**, 238, 260
enones, **8**, 16, 544
epoxides, **8**, 880
esters, **8**, 244, 266
imines, **8**, 36
keto sulfides, **8**, 12
lactones, **8**, 269
oximes, **6**, 769
pyridines, **8**, 584
unsaturated carbonyl compounds, **8**, 543
Aluminum hydride, diisopropoxy-
reduction
enones, **8**, 543
Aluminum hydride, tri-*t*-alkoxy-
reduction
aldehydes, **8**, 17
Aluminum hydrides
reduction
pyridines, **8**, 583
pyridinium salts, **8**, 587
unsaturated carbonyl compounds, **8**, 541, 543
sources, **8**, 736
Aluminum hydrides, alkoxy-
reduction
carbonyl compounds, **8**, 2
quinones, **8**, 19
Aluminum iodide, diethyl-
Beckmann rearrangement, **6**, 767
Aluminum isopropoxide
crotonaldehyde reduction
in isopropyl alcohol, **8**, 88
epoxide ring opening, **3**, 770
Aluminum oxide

aldol reactions
self-condensation, **2**, 268
catalyst
Knoevenagel reaction, **2**, 359
Aluminum phenoxide
catalyst
Friedel–Crafts reaction, **3**, 296
Dowex resin bound
catalyst, Friedel–Crafts reaction, **3**, 297
Aluminum phenoxide, diisobutyl-
aldol reaction catalyst, **2**, 166
Aluminum phosphate
catalyst
Knoevenagel reaction, **2**, 345, 359
Aluminum reagents
organopalladium catalysis, **3**, 230
Aluminum selenide
reaction with nitriles, **6**, 477
Aluminum selenolate
reaction with esters, **6**, 466
Aluminum selenomethylate, dimethyl-
reaction with oxime sulfonates, **6**, 768
Aluminum thiolates, dialkyl-
Beckmann rearrangement, **6**, 767
Aluminum tribromide
catalyst
Friedel–Crafts reaction, **3**, 295
Aluminum tri-*t*-butoxide
oxidation
secondary alcohols, **7**, 323
Aluminum trichloride
catalyst
Friedel–Crafts reaction, **3**, 295
graphite-intercalated, catalyst
Friedel–Crafts reaction, **3**, 298
tetramethylurea complex
crystal structure, **1**, 301
Amadori rearrangement, **6**, 789
Amalgams
C—P bond cleavage, **8**, 863
Amaryllidaceae alkaloids
synthesis
Mannich reaction, **2**, 1032, 1042
use of imine anions, **2**, 480
via Diels–Alder reactions, **5**, 323
Amberlite IR-112
catalyst
Friedel–Crafts reaction, **3**, 296
Amidation
alkenes, **4**, 292
Amide acetals
azavinylogs
2-alkoxy-2-dialkylaminocarbonitrile synthesis, **6**, 573
ortho acid synthesis, **6**, 561
spirocyclic
synthesis, **6**, 568
synthesis, **6**, 566
Amide chlorides
chlorination, **6**, 499
self-condensation, **6**, 499
Amide fluorides
synthesis, **6**, 496
Amide group
O-alkylation
deprotection, **6**, 642
Amide halides
amide acetal synthesis, **6**, 566
synthesis, **6**, 495
tris(dialkylamino)alkane synthesis, **6**, 579, 580
Amides
acetalization, **6**, 569
activated
macrolactonization, **6**, 373
acylation, **6**, 504, 542
addition reactions
alkenes, **4**, 559
adducts
acylating reagents, **6**, 487
carbonic acids, **6**, 491
carboxylic acid derivatives, **6**, 493
sulfur compounds, **6**, 490
alkali metal anions
crystal structures, **1**, 37
alkoxymethyleniminium salt synthesis, **6**, 501
alkylation, **6**, 399
α-allenic
bridged azabicyclic systems, **2**, 89
amidine synthesis, **6**, 543
amidinium salt synthesis, **6**, 517
aminal ester synthesis, **6**, 575
anodic oxidation, **7**, 804
aromatic
Birch reduction, **8**, 507
arylation, **6**, 399
asymmetric hydroxylation, **7**, 183
α-bromo-
Reformatsky reaction, **2**, 292
chiral
asymmetric aldol reactions, **2**, 231
conjugate additions, **4**, 202
cyclic
deprotonation, **3**, 66
tandem vicinal difunctionalization, **4**, 249
dehydrogenation
copper(II) bromide, **7**, 144
deprotonation, **3**, 65
α-deprotonation, **1**, 476
enolates
addition reactions, **4**, 106–111
arylation, **4**, 466
stereoselectivity, **2**, 211
β-halo-α,β-unsaturated
addition reactions, **4**, 125
homologous β,γ-unsaturated
from allylic alcohols, **6**, 853
α-hydroxylation, **7**, 183
imidoyl halide synthesis, **6**, 523
lithiation
addition reactions, **1**, 464
lithium enolates
crystal structures, **1**, 30
methylenation
Tebbe reaction, **1**, 748
Tebbe reagent, **5**, 1124
microbial hydroxylation, **7**, 59
nucleophilic addition to π-allylpalladium complexes
regioselectivity, **4**, 639
oxidation
electrochemical, **2**, 1051
N-phenyl-β-bromo-α,β-unsaturated
synthesis *via* haloborane addition to alkynes, **4**, 358

protecting groups, **6**, 672
reactions with alkenes, **4**, 292–295
reactions with π-allylpalladium complexes, **4**, 598
stereochemistry, **4**, 623
reactions with organocopper complexes, **1**, 124
reaction with benzophenone dianion
organoytterbium compounds, **1**, 280
reduction, **8**, 248, 293
metal hydrides, **8**, 269
sulfenylation, **7**, 125
synthesis, **6**, 381–417
carbonylation, **3**, 1034
via hydration of alkynes, **4**, 300
via hydroformylation, **4**, 941
via ketones, **7**, 694
via Ritter reaction, **6**, 261
tandem vicinal difunctionalization, **4**, 246–249
tertiary
dehydrogenation, **7**, 122, 144
tin enolates
synthesis, **2**, 116
tris(dialkylamino)alkane synthesis, **6**, 579
unsaturated
lithiation, **1**, 480
α,β-unsaturated
chelated, Diels–Alder reactions, **5**, 365–367
Diels–Alder reactions, **5**, 360–365, 464
stereoselective conjugate reduction, **8**, 537
tandem vicinal difunctionalization, **4**, 257
γ,δ-unsaturated
stereoselective synthesis *via* Claisen rearrangement, **5**, 828
Vilsmeier–Haack reaction, **2**, 786
vinylogous
reduction, **8**, 55
synthesis, Eschenmoser coupling reaction, **2**, 865, 867
Amides, acyclic
tandem vicinal difunctionalization, **4**, 247–249
Amides, *N*-alkyl-
chlorination, **6**, 208
Amides, α-alkyl-β-keto
reduction, **8**, 11
Amides, alkynic
tandem vicinal difunctionalization, **4**, 249
Amides, *N*-allyl-
hydroformylation, **4**, 926
Amides, α-amino-
synthesis
Lewis acid catalysis, **1**, 349
Amides, bis(trimethylsilyl)-
crystal structures, **1**, 37
Amides, *t*-butyl-
reduction
metal hydrides, **8**, 271
Amides, dehydro-
synthesis
Erlenmeyer azlactone synthesis, **2**, 406
Amides, 2,2-dialkoxy-
dehydration, **6**, 566
Amides, *N*,*N*-dialkyl-
deprotonation
with lithium dialkylamides, **3**, 45
Amides, diethyl-
reduction
aluminates, **8**, 272
Amides, dimethyl-
reduction
metal hydrides, **8**, 271
Amides, *N*-halo-
radical reactions
alkenes, **7**, 503
Amides, hydroxy
γ-lactone synthesis, **6**, 353
Amides, β-hydroxy-
synthesis
via desulfurization, **1**, 523
Amides, *N*-(1-hydroxyalkyl)-
synthesis, **2**, 1049
Amides, α-keto-
asymmetric hydrogenation, **8**, 153
Amides, β-keto-
cyclization reactions
organosamarium compounds, **1**, 263
intermolecular pinacolic coupling reactions
organosamarium compounds, **1**, 271
Amides, β-keto-2-[2-(trimethylsilyl)methyl]-
cycloaddition reactions, **5**, 247
Amides, methoxy-
arylation
reactivity, **2**, 1053
Amides, *N*-methoxy-*N*-methyl-
acylation with, **1**, 399
synthesis
via acid chlorides, **1**, 399
Amides, *N*-methylenium
Diels–Alder reactions, **5**, 501
Amides, *N*-methyl-*N*-methylenium cations
Diels–Alder reactions, **5**, 501
Amides, trimethylsilyl-
oxirane ring opening, **6**, 91
Amides, vinyl-
hydroformylation, **4**, 926
Amides, vinylogous
reaction with oxonium salts, **6**, 502
synthesis
Mannich reaction, **2**, 903
via Beckmann reaction, **7**, 697
Amide thioacetals
synthesis, **6**, 568
Amidines
alkylation, **6**, 552
reduction, **8**, 302
synthesis, **6**, 542; **7**, 476
via alkenes, **7**, 494
via reduction of amidoximes, **8**, 394
thiolysis, **6**, 430
tris(dialkylamino)alkane synthesis, **6**, 579
Amidines, α-amino-
synthesis, **6**, 555
Amidines, α-keto-
synthesis, **6**, 556
Amidines, methylthio-
alkylation, **3**, 88
Amidinium salts
amide acetal synthesis, **6**, 568
amidine synthesis, **6**, 543
aminal ester synthesis, **6**, 575
2,2-bis(dialkylamino)carbonitrile synthesis, **6**, 578
synthesis, **6**, 512
tris(dialkylamino)alkane synthesis, **6**, 580
Amidinium salts, azavinylogous

synthesis, **6**, 522
Amidinium salts, vinylogous
synthesis, **6**, 522
Amidoalkylation
comparison with aminoalkylation, **2**, 971
electrochemical, **7**, 804
α-Amidoalkylation
amides, **1**, 371
Amidohydrolases
phthaloyl group removal
amine protection, **6**, 643
Amidomercuration, **8**, 854
alkenes, **4**, 741
demercuration
alkenes, **4**, 294
Amidorazones
acyl anion equivalents, **6**, 783
Amidoselenation
alkenes, **7**, 495, 523
Ritter reaction, **6**, 289
1-Amido-2-sulfenyl compounds
synthesis, **7**, 494
Amidoximes
reduction
synthesis of amidines, **8**, 394
Amidyl radicals
cyclizations, **4**, 812
Aminal esters
2,2-bis(dialkylamino)carbonitrile synthesis, **6**, 577
synthesis, **6**, 574
Aminals
deamination
imine synthesis, **6**, 719
reaction with enol silanes
Lewis acid mediated, **2**, 635
Amination
alkenes, **4**, 290–297
amines
primary, **7**, 741
secondary, **7**, 746
electrophilic, **6**, 119
hydrazine synthesis, **6**, 118
Amine nucleophiles
nucleophilic addition to π-allylpalladium complexes
regioselectivity, **4**, 638–640
stereochemistry, **4**, 622–624
Amine oxalates
Mannich reaction, **2**, 896
Amine oxides
asymmetric epoxidation
kinetic resolution, **7**, 423
deoxygenation, **8**, 390
Meisenheimer rearrangement, **6**, 843
oxidation with
halides, **7**, 663
polymers
alkyl iodide oxidation, **7**, 663
Amine *N*-oxides
deprotonation
azomethine ylide generation, **4**, 1089
Amines
acylation, **6**, 382
addition reactions
alkenes, **4**, 559
aliphatic
anodic oxidation, **7**, 803
alkylation, **6**, 65
alkyl halides, **6**, 65
sulfonates, **6**, 72
amidine synthesis, **6**, 543
amidinium salt synthesis, **6**, 513
aromatic
alkylation, **6**, 66
anodic oxidation, **7**, 804
Birch reduction, **8**, 498
hydrogenolysis, **8**, 916
aromatic nucleophilic substitution, **4**, 433
biological oxidation, **7**, 736
2,2-bis(dialkylamino)carbonitrile synthesis, **6**, 577
chiral auxiliary
aldol reaction, **2**, 233
cyclic
synthesis, **6**, 69
deamination
alcohol synthesis, **6**, 3
dehydrogenation, **7**, 738
dimerization
mercury-photosensitized, **7**, 5
enantioselective syntheses
via chiral *N,N*-dialkylhydrazones, **1**, 379
heteroaromatic
N-oxidation, **7**, 749
hydride transfer, **8**, 88
catalysis, **8**, 91
imidoyl halide synthesis, **6**, 527
lithium aluminum hydride modifiers, **8**, 168
macrocyclic
synthesis, **3**, 969
metallation
addition reactions, **1**, 463
nucleophilic addition to π-allylpalladium complexes, **4**, 598
regioselectivity, **4**, 638
stereochemistry, **4**, 622
optically active
Mannich reactions, **2**, 1025
oxidation
amide synthesis, **6**, 402
primary
catalyst, Knoevenagel reaction, **2**, 343
Mannich reaction, **2**, 968
oxidation, **7**, 736, 842
synthesis, **7**, 606
Vilsmeier–Haack reaction, **2**, 791
protecting groups for, **6**, 635
chelation, **6**, 645
silicon-based, **6**, 646
reactions with 1,1-dihaloalkenes, **6**, 498
reaction with alkenes, **4**, 290–297
rearrangements, **6**, 892
diastereoselectivity, **6**, 893
regioselectivity, **6**, 893
reduction
to alkanes, **8**, 811–832
reductive amination
methylation with formaldehyde, **8**, 47
reductive deamination, **8**, 826
secondary
catalyst, Knoevenagel reaction, **2**, 343
Mannich reaction, **2**, 956
oxidation, **7**, 745
synthesis, **7**, 607

secondary, chiral
aldol reaction, chiral auxiliary, **2**, 234
solvents for reduction
dissolving metals, **8**, 113
synthesis, **8**, 374
via carboxylic acid degradation, **6**, 795
via carboxylic acids, **8**, 235–254
via enzyme reduction of imines, **8**, 204
via imine reduction, **8**, 25–74
via oximes, oxime ethers and oxime esters, **8**, 64
via reduction of azides and triazines, **8**, 383
via reduction of hydroxylamines, **8**, 394
via reductive cleavage, **8**, 383
via substitution processes, **6**, 65–98
synthons
alkylation, **6**, 65
tertiary
catalyst, Knoevenagel reaction, **2**, 343
Reimer–Tiemann reaction, **2**, 772
synthesis, **7**, 607
thioacylation, **6**, 420
O-alkyl thiocarboxylates, **6**, 420
carbon disulfide, **6**, 428
dithiocarboxylates, **6**, 423
3*H*-1,2-dithiol-3-ones, **6**, 421
thioacyl chlorides, **6**, 422
thioamides, **6**, 424
thioketenes, **6**, 426
thiophosgene, **6**, 423
Amines, allylic tertiary
reduction
dissolving metals, **8**, 971
Amines, *N*-(arylthiomethyl)-
synthesis
via tertiary amine precursors, **1**, 370
Amines, bis(methoxymethyl)-
reduction
to dimethylamines, **8**, 27
Amines, *N,N*-bis(trimethylsilyl)-
desilylation
to primary amines, **1**, 369
Amines, *N*-(cyanomethyl)-
N-methylenamines from, **2**, 941
Amines, di-*t*-alkyl-
synthesis, **7**, 737
Amines, α,α′-dichloro-
synthesis, **6**, 495
Amines, α,α-difluoro-
synthesis, **6**, 495
Amines, halo-
reaction with alkenes, **7**, 471
Amines, *N*-halo-
radical cyclizations, **4**, 812
Amines, homoallylic
alkylation
palladium(II) catalysis, **4**, 573
Amines, β-hydroxy-
asymmetric epoxidation
kinetic resolution, **7**, 423
Amines, *N*-methylen-
reactions with organometallic compounds, **1**, 361
Amines, perchloryl-
synthesis
via chlorination of secondary amines, **7**, 747
Amines, perfluoro-*N*-bromo-
addition reactions
alkenes, **7**, 500
Amines, β-phenoxy-
synthesis, **7**, 490
Aminium cation radicals
Diels–Alder reactions, **5**, 520
Aminium ions
synthesis
via oxidation of secondary amines, **7**, 745
via oxidation of tertiary amines, **7**, 749
α-Amino acid chlorides, *N*-(trifluoroacetyl)-
Friedel–Crafts reaction
bimolecular aromatic, **2**, 740
Amino acids
acylating agents, **1**, 413
asymmetric synthesis, **3**, 53
reductive cleavage of hydrazines, **8**, 388
use of 2,5-diketopiperazines, **2**, 498
via azides, **6**, 77
deamination, **8**, 831
enantioselective synthesis
reaction of imines with allyl organometallic compounds, **2**, 986
fluorinated
synthesis *via* hydroformylation, **4**, 927
N-methylation
retrograde Diels–Alder reaction, **5**, 552
Strecker synthesis, **1**, 460
synthesis, **1**, 373
stereoselective, **8**, 647
via Ireland silyl ester enolate rearrangement, **5**, 841
via Lewis acid catalysis, **1**, 349
via reduction of keto acids, **8**, 386
thioacylation, **6**, 424
O-methyl thiocarboxylates, **6**, 420
unsaturated fluorinated
hydrogenolysis, **8**, 896
α-Amino acids
N-acylated
electrochemical oxidation, **2**, 1051
asymmetric aldol cyclizations, **2**, 167
asymmetric hydrogenation
modifying reagents, **8**, 150
asymmetric synthesis
from α-keto acids, **8**, 145
Ugi reaction, **2**, 1098
deamination–substitution
preparation of chiral alcohols, **6**, 3
esters
deamination–substitution, **6**, 4
imines, alkylation, **3**, 46
β-hydroxy-
optically pure, **2**, 254
synthesis, **8**, 144; **3**, 796
N-acyliminium ions, **2**, 1075
glycine cation equivalents, **2**, 1074
oxazolones, **2**, 396
Ugi reaction, **2**, 1095
β-Amino acids
synthesis
Mannich reaction, **2**, 916, 922
from 2-methyloxazoline, **2**, 492
α-Amino acids, acyl-
cyclodehydration, **2**, 403
Amino acids, *N*-acyl-
synthesis
cobalt-catalyzed carbonylation, **3**, 1027

Amino acids, *N*-*t*-butoxycarbonyl-
N-ethylation, **6**, 71
Amino acids, dehydro-
enantioselective catalytic hydrogenation
Monsanto procedure, **2**, 233
reduction, **2**, 406
synthesis, **7**, 122
Erlenmeyer azlactone synthesis, **2**, 406
α-Amino acids, γ-hydroxy-
synthesis, **7**, 490
α-Amino acids, *N*-hydroxy-
synthesis, **6**, 113
esters, **6**, 116
via oxime ethers, **1**, 386
α-Amino acids, *syn*-β-hydroxy-
enantioselective aldol reaction
chiral titanium enolates, **2**, 309
α-Amino acids, β,γ-unsaturated
synthesis, **3**, 117
Aminoacyl anions
reactions with electrophiles
carbonylation reaction, **3**, 1017
Amino alcohols
diazotization, **1**, 846
lithium aluminum hydride modifiers, **8**, 168
resolution, **7**, 493
synthesis
via cyclization of allylic substrates, **4**, 406
via reduction of cyclohexene oxide, **8**, 383
vicinal
synthesis, **6**, 715
Amino alcohols, alkynic
divinyl ketones from
cyclization, **5**, 769
Amino alcohols, dialkyl-
lithium aluminum hydride modifier, **8**, 164
Amino alcohols, monoalkyl-
lithium aluminum hydride modifiers, **8**, 168
Aminoalkylation
arylamines
Mannich reaction, **2**, 961
carbocyclic compounds
Mannich reaction, **2**, 961
phenols
Mannich reaction, **2**, 956
α-Aminoalkylation
synthesis, **2**, 953
ipso-Aminoalkylation–destannylation, **2**, 962
Aminocarbonylation
alkenes
palladium(II) catalysis, **4**, 561
Amino-Claisen rearrangement, **6**, 860
N-allyl ketene *N,O*-acetals, **6**, 861
Amino cyclitols
synthesis, **7**, 712
β-Amino esters
synthesis
from chiral silyl ketene acetals, **2**, 638
α-Amino esters, *O*-benzyl-*N*-hydroxy-
synthesis, **6**, 114
Amino ethers
chiral auxiliary
aldol reaction, **2**, 233
gem-Amino ethers
reactions with allyl organometallic compounds, **2**, 1003, 1004
reactions with crotyl organometallic compounds
dependence of product type on metal, **2**, 1005
reactions with propargyl organometallic compounds
dependence of product type on metal, **2**, 1005
gem-Amino ethers, *N*-bis(trimethylsilyl)-
reactions with allyl organometallic compounds, **2**, 1003
gem-Amino ethers, *N*-(trimethylsilyl)-
reactions with allyl organometallic compounds, **2**, 1003
α-Amino ketones
diazotization
synthesis of α-diazo ketones, **3**, 890
Aminomercuration
demercuration
alkenes, **4**, 290–292
Aminomercuration–oxidation, **7**, 638
Aminomethylation
Grignard reagents, **3**, 258
Aminonitrenes
synthesis
via oxidation of 1,1-disubstituted hydrazines, **7**, 742
Aminopalladation
aziridine synthesis, **7**, 474
Amino polyols
synthesis
stereoselective, **8**, 647
Amino radicals
cyclizations, **4**, 795
α-Amino radicals
ω-unsaturated
reductive cyclization, **1**, 275
Amino sugars
synthesis, **7**, 712; **8**, 388
stereoselective, **8**, 647
via intramolecular Diels–Alder reaction, **5**, 425
via palladium catalysis, **4**, 598
Aminosulfenylations
alkenes, **7**, 493
Aminotelluration
alkenes, **4**, 343
Aminyl radicals
cyclizations, **4**, 811
metal complexes
cyclizations, **4**, 812
synthesis
via oxidation of anilines, **7**, 739
via secondary amines, **7**, 745
Amminimium radicals
cations
cyclizations, **4**, 812
Ammonia
alkali metal reductions
benzylic compounds, **8**, 971
carbonyl compounds, **8**, 308
catalyst
Knoevenagel reaction, **2**, 343
dissolving metal reductions
added proton source, **8**, 112
chemoselectivity, **8**, 113
no added proton source, **8**, 112
Ammonium acetate
catalyst
Knoevenagel reaction, **2**, 343
Ammonium acetate, ethylenedi-

catalyst
Knoevenagel reaction, **2**, 343
Ammonium azides, tetra-*n*-butyl-
reaction with epoxides
ring opening, **6**, 91
Ammonium borohydride, tetraalkyl-
triazolyl ketone reduction, **8**, 13
Ammonium bromide, phenacylbenzyldimethyl-
Stevens rearrangement, **3**, 913
Ammonium carboxylates
dehydration, **6**, 382
Ammonium cation, tetraalkyl-
electroreduction
mediator, **8**, 132
Ammonium chloride
Claisen rearrangement
catalysis, **5**, 850
zinc
nitro compound reduction, **8**, 366
Ammonium chloride, dibenzyldimethyl-
photolysis
Ritter reaction, **6**, 280
Ammonium chlorochromate, benzyltriethyl-
oxidation
alcohols, **7**, 283
Ammonium chlorochromate, benzyltrimethyl-
oxidation
alcohols, **7**, 283
Ammonium chlorochromate, tetra-*n*-butyl-
oxidation
alcohols, **7**, 283
Ammonium chlorochromate, trimethyl-
oxidation
alcohols, **7**, 283
Ammonium chromate
resin support
alcohol oxidation, **7**, 280
Ammonium compounds, *p*-iodophenyltrimethyl-
$S_{RN}1$ reactions, **4**, 460
Ammonium cyanide, tetrabutyl-
catalyst
benzoin condensation, **1**, 543
Ammonium cyanide, triethyl-
nitrile synthesis, **6**, 234
Ammonium cyanoborohydride, tetrabutyl-
reduction
enones, **8**, 538
Ammonium cyanoborohydrides, tetraalkyl-
reductive amination
nonpolar solvents, **8**, 54
Ammonium dichromate
oxidation
solid support, **7**, 845
Ammonium dichromate, bis(tetrabutyl)-
oxidation
alcohols, **7**, 286
Ammonium enolates
enantioselective aldol reaction
gold catalysis, **2**, 317
Ammonium fluoride, benzyltrimethyl-
catalyst
allylsilane reactions with aldehydes, **2**, 571
Ammonium fluoride, *t*-butyl-
catalyst
allylsilane reactions with aldehydes, **2**, 571
Ammonium fluoride, tetrabutyl-
catalyst
enol silane reaction with aldehydes, **2**, 633
fluorination
alkyl alcohol derivatives, **6**, 219
Ammonium fluoride, tetra-*n*-butyl-
Henry reaction
silyl nitronates, **2**, 335
Henry reaction, high pressure
catalyst, **2**, 329
Ammonium formate
hydride donor
carbonyl compound reduction, **8**, 320
reductive alkylation of amines, **8**, 84
reduction
hydride transfer, **8**, 84
Ammonium formate, trialkyl-
hydrogen donor, **8**, 557
Ammonium halides, benzyldimethyl(trimethyl-silylmethyl)-
desilylation, **4**, 430
Ammonium hydroxide, benzyltriethyl-
aldol reaction
piperonal with *N*-crotonylpiperidine, **2**, 153
Ammonium hydroxide, tetraethyl-
hydroxylation
tetrasubstituted alkenes, **7**, 439
Ammonium iodide, (iodomethyl)trimethyl-
Eschenmoser's salt from, **2**, 899
Ammonium molybdate
oxidation
secondary alcohols, **7**, 320
Ammonium permanganate, benzyltriethyl-
oxidation
amines, **6**, 402
ethers, **7**, 236
Ammonium permanganate, benzyltrimethyl-
alkane oxidation, **7**, 12
Ammonium perruthenate, tetra-*n*-propyl-
oxidation
primary alcohols, **7**, 311
Ammonium persulfate
alkene hydroxylation, **7**, 447
Ammonium radical cations, alkyl-
alkane oxidation, **7**, 17
Ammonium salts
reduction
dissolving metals, **8**, 828
synthesis
via substitution processes, **6**, 65–98
Ammonium salts, allyltrialkyl-
reaction with Grignard reagents, **3**, 246
Ammonium salts, tetraalkyl-
intermolecular pinacol coupling reactions, **3**, 568
Ammonium salts, trialkyl-
reaction with activated alkynes, **4**, 49
Ammonium tetrabromooxomolybdate, benzyltrimethyl-
oxidation
secondary alcohols, **7**, 321
Ammonium tetrafluoroborate, *N*-sulfinyldimethyl-
alkylmercaptomethyleniminium salt synthesis, **6**, 512
Ammonium triacetoxyborohydride, tetra-*n*-butyl-
selective aldehyde reduction, **8**, 16
Ammonium triacetoxyborohydride, tetramethyl-
ketone reduction, **8**, 9
Ammonium trifluoroacetate, dibenzyl-
aldol reaction

regioselective, **2**, 156
Ammonium ylides, allylic
rearrangements, **6**, 854
Ammonium ylides, cyclic
2,3-sigmatropic rearrangements, **6**, 855
Ammonolysis
aryl halides, **4**, 434
Amoxycillin
synthesis
2-arylglycines, **3**, 303
Amphetamine
hydrogenation, **8**, 146
synthesis, **7**, 502
Amphimedine
synthesis
organopalladium catalysts, **3**, 232
Amphotericin B
synthesis, **1**, 564
use of aldol reaction, **2**, 195
via cuprate 1,2-addition, **1**, 126
via Horner–Wadsworth–Emmons reaction, **1**, 772
via Wittig reaction, **1**, 763
Ampicillin
synthesis
2-arylglycines, **3**, 303
α-Amyrin acetate
allylic oxidation, **7**, 112
Anacyclin
synthesis, **3**, 558
Anatoxin *a*
enantioselective synthesis
Mannich reaction, **2**, 1012
synthesis, **2**, 1069; **8**, 604
Eschenmoser coupling reaction, **2**, 879
Mannich reaction, **2**, 1012
via acylation of precursor, **1**, 403
via dibromocyclopropyl compounds, **4**, 1023
Anbadons
Knoevenagel reaction, **2**, 346
Ancistrocladine
synthesis, **3**, 506; **6**, 738
Ancistrocladisine
synthesis, **6**, 738
Androstadienedione
rearrangement, **3**, 804, 810
1,4-Androstadiene-3,17-dione
hydrogenation, **8**, 535
homogeneous catalysis, **8**, 452
Androstadienone, 4-methyl-
rearrangement, **3**, 805
5α-Androstane, 3β-acetoxy-
reduction
lithium aluminum hydride, **8**, 345
Androstane, 3-acetyl-3-bromo-
synthesis
via 17β-hydroxy-5α-androstan-3-one, **1**, 530
Androstane-2,17-dione
synthesis
ene reaction, **2**, 552
Androstane-3,17-dione
carbonyl group protection, **6**, 675
reactions with organometallic reagents
regioselectivity, **1**, 152
5α-Androstanes
microbial hydroxylation, **7**, 72
Androstanone
selenol esters
synthesis, **6**, 462
Androstan-11-one
reduction
dissolving metals, **8**, 118
Androstan-17-one
reduction
dissolving metals, **8**, 122
5α-Androstan-3-one
microbial hydroxylation, **7**, 69
5-Androstan-17-one
microbial hydroxylation, **7**, 71, 73
5-Androstan-17-one, 12,12-difluoro-
microbial hydroxylation, **7**, 73
5α-Androstan-3-one, 16β-hydroxy-
microbial hydroxylation, **7**, 71
Androstenedione
enzymatic reduction, **8**, 561
Androstene-3,17-dione
boron trifluoride complex
NMR, **1**, 293
Androst-4-ene-3,17-dione
microbial hydroxylation, **7**, 74
9α,10β-Androst-4-ene-3,17-dione
microbial hydroxylation, **7**, 71
9β,10α-Androst-4-ene-3,17-dione
microbial hydroxylation, **7**, 71
Androsten-3-ol-17-one
hydroxylation, **7**, 11
Androstenone
epoxide
rearrangement, **3**, 738
5-Androsten-16-one
reduction
dissolving metals, **8**, 122
5-Androsten-17-one, 3β-acetoxy-
reduction, **8**, 937
5β-Androst-9(11)-en-12-one, 3α,17β-diacetoxy-11-hydroxy-
rearrangement, **3**, 833
Androst-5-en-17-one, 1β,3β-dihydroxy-
synthesis, **7**, 73
4-Androsten-3-one, 17β-hydroxy-
reduction, **8**, 935
Anemonin
synthesis, **7**, 619
Angelicin
synthesis, **5**, 1096, 1099
regioselective, **5**, 1094
Anguidine
synthesis
ene reaction, **2**, 550
Angustidine
synthesis
Mannich reaction, **2**, 913
Angustine
synthesis
Mannich reaction, **2**, 913
Angustine, 13b,14-dihydro-
synthesis
Mannich reaction, **2**, 913
Anhydrides
acylation, **1**, 423
thiols, **6**, 443
acyl transfer
ester synthesis, **6**, 327

cyclic
reduction, **8**, 291
macrolactonization, **6**, 369
methylenation
Tebbe reagent, **5**, 1124
mixed
acylation, **6**, 328
reduction, **8**, 239
to aldehydes, **8**, 291
Tebbe reaction, **1**, 743
thioacylation
alcohols and phenols, **6**, 449
Vilsmeier–Haack reaction, **2**, 792
Anilides
metallation
addition reactions, **1**, 463
Anilides, *N*-alkyl-
α,β-unsaturated
photoinduced cyclization, **4**, 477
Anilides, *N*-methyl-
reduction
metal hydrides, **8**, 270, 272
Aniline, *N*-acyl-*o*-chloro-
photoinduced cyclization, **4**, 477
Aniline, *o*-alkyl-
metal complexes
addition reactions, **4**, 534
Aniline, *N*-allyl-
oxamination, **7**, 489
Aniline, benzylidene-
reaction with crotyl organometallic compounds
syn–anti selectivity, **2**, 991
reaction with silyl ketene acetals
stereoselectivity, **2**, 638
Aniline, *N*-benzylidine-
reactions with organometallic compounds, **1**, 361
reactions with sulfinyl-stabilized carbanions, **1**, 515
Aniline, bromo-
hydrogenation, **8**, 907
Aniline, 4-*n*-butylnitro-
synthesis, **4**, 433
Aniline, *N,N*-diethyl-
thexylborane complex
hydroboration, **8**, 709
Aniline, *N,N*-dimethyl-
Birch reduction
dissolving metals, **8**, 498
reaction with formaldehyde
Mannich reaction, **2**, 961
Rosenmund reduction, **8**, 287
Aniline, 2,3-dinitro-
reaction with piperidine, **4**, 423
Aniline, *N*-diphenylmethylene-
reaction with organometallic reagents, **2**, 975
Aniline, *N*-ethyl-
lithium aluminum hydride modifier, **8**, 166
Aniline, *N*-methyl-
thiobenzoylation, **6**, 424
Aniline, *p*-nitro-
N-alkylation, **6**, 66
synthesis, **6**, 110
Aniline, 2-nitroso-
synthesis
via oxidation of *o*-phenylenediamine, **7**, 737
Aniline, pentachloro-
oxidation
sodium hypochlorite, **7**, 738
Aniline, *N*-sulfinyl-
[3 + 2] cycloaddition reactions
with η^1-butynyliron complexes, **5**, 277
Aniline, 2,4,6-trimethyl-
synthesis
via $S_{RN}1$ reaction, **4**, 472
Aniline derivatives
formylation, **3**, 969
Aniline pivalamides
ortho lithiation
addition reactions, **1**, 464
Anilines
N-alkylation, **6**, 66
isopropylation
Friedel–Crafts reaction, **3**, 302
meta metallation
addition reactions, **1**, 463
ortho alkylation, **4**, 430
synthesis, **4**, 434; **8**, 367
tertiary
hydride donors, **8**, 98
Anilines, *o*-alkynyl-
synthesis, **3**, 543
Anilines, *o*-allyl-
N-substituted
carbonylation, **3**, 1038
Anilines, azido-
synthesis
via phthaloyl intermediate, **6**, 255
Anion exchange reactions
amide halide synthesis, **6**, 500
Anion exchange resins
aldol reaction, **2**, 138
chromic acid
alcohol oxidation, **7**, 280
Anisaldehyde
reduction
boranes, **8**, 316
Anisatin
synthesis, **7**, 242
Anisoic acids
Birch reduction
dissolving metals, **8**, 501
Anisole
meta-acylation, **4**, 532
Friedel–Crafts alkylation, **3**, 300
oxidative coupling, **3**, 669
photocycloaddition reactions
with vinylene carbonate, **5**, 653
Anisole, *p*-chloro-
hydrogenolysis, **8**, 906
Anisole, (*m*-cyanoalkyl)-
metal complexes
addition–protonation reactions, **4**, 543
Anisole, dihydro-
reactions with iron carbonyls, **4**, 665
Anisole, 2,4-dimethyl-
amidoalkylation, **2**, 971
Anisole, 2,6-dimethyl-
benzylation, **3**, 300
Anisole, *p*-fluoro-
catalytic hydrogenation, **8**, 903
Anisole, *p*-iodo-
reaction with phenylselenides, **4**, 454
Anisole, *o*-lithio-

acylation, **1**, 404
Anisole, 2-methoxythio-
metallated
alkylation, **3**, 135
Anisole, 2-phenyl-
synthesis
via benzyne, **4**, 510
Anisole, trimethylsilylmethylthio-
alkylation, **3**, 137
metallated
alkylation, **3**, 135
Anisoles
Birch reduction
dissolving metals, **8**, 493
electrochemical reduction, **8**, 517
reductive silylation, **8**, 518
Anisomycin
synthesis, **3**, 77; **8**, 605
Eschenmoser coupling reaction, **2**, 889
Annulation
intramolecular Barbier process
samarium diiodide, **1**, 262
Michael ring closure, **4**, 121, 260
[4 + 2] Annulation
oxyanion-accelerated
vinylcyclobutane rearrangement, **5**, 1020
α,α′-Annulation
bicyclic ketoester synthesis, **4**, 8
Annulations
two-alkyne, **5**, 1102
[3 + 2] Annulations
allenylsilanes, **1**, 596
[10]Annulene
disrotatory ring closure, **5**, 716
synthesis
via [6 + 4] cycloaddition, **5**, 623
[12]Annulene
electrocyclization, **5**, 717
[14]Annulene
synthesis
Knoevenagel reaction, **2**, 377
[16]Annulene
disrotatory ring closure, **5**, 716
Annulenes
synthesis, **3**, 594
Annulenes, dehydro-
synthesis, **3**, 556
[13]Annulenone, 6,8-bis(dihydro)-
synthesis
Knoevenagel reaction, **2**, 376
Annulenones
synthesis
aldol reaction, **2**, 152
Anodic α-acetoxylation
ketones, **7**, 798
Anodic hydroxylation
aromatic compounds, **7**, 800
Anodic α-methoxylation
ketones, **7**, 798
Anodic oxidation
alkanes, **7**, 793
benzylic position
aromatic compounds, **7**, 801
1,2-diols, **7**, 707
double mediatory systems, **7**, 809
electrochemical, **7**, 790
heteromediatory systems, **7**, 808
homomediatory systems, **7**, 808
mediators, **7**, 807
unsaturated compounds, **7**, 794
Ansamycin
occurrence, **2**, 1
Antamanide
lithium salt complexes
crystal structure, **1**, 300
Antamanide, perhydro-
lithium salt complexes
crystal structure, **1**, 300
Antheridic acid
synthesis, **7**, 90
Antheridiogen-An
synthesis
via vinylcyclopropane rearrangement, **5**, 1014
via vinylcyclopropane thermolysis, **4**, 1048
Antheridiogens
synthesis
via cyclofunctionalization of cycloalkene, **4**, 373
Anthracene, 9-bromo-
charge-transfer osmylation, **7**, 864
$S_{RN}1$ reaction, **4**, 461
Anthracene, 9-cyano-
photocycloaddition reactions
cycloheptatriene, **5**, 636
2,5-dimethyl-2,4-hexadiene, **5**, 636
Anthracene, 9,10-dibromo-
charge-transfer osmylation, **7**, 864
Anthracene, 9,10-dihydro-
synthesis
Friedel–Crafts reaction, **2**, 761
Anthracene, 9,10-dimethyl-
Diels–Alder reactions, **5**, 71
acyl nitroso compounds, **5**, 419
tetracyanoethylene, **5**, 76
Anthracene, 9-methyl-
hydrogenation
homogeneous catalysis, **8**, 455
Anthracene, 9-nitro-
charge-transfer osmylation, **7**, 864
Anthracene, 9-trifluoroacetyl-
hydrogenation
homogeneous catalysis, **8**, 455
Anthracene hydride
reaction with chalcone, **8**, 563
Anthracenes
anodic oxidation, **7**, 799
Ritter reaction, **6**, 282
Birch reduction
dissolving metals, **8**, 497
charge-transfer osmylation, **7**, 864
[4 + 3] cycloaddition reactions, **5**, 608
Diels–Alder reactions
benzynes, **5**, 383
Lewis acid promoted, **5**, 339
selenocarbonyl dienophiles, **5**, 442
hydrogenation, **8**, 438
homogeneous catalysis, **8**, 454, 455
osmium tetroxide complex
time-resolved spectra, **7**, 865
ozonization
[4 + 3] cycloadditions, **4**, 1075
photocycloaddition reactions
2,5-dimethyl-2,3-hexadiene, **5**, 636

photolyses, **5**, 637
radical cation
absorption spectrum, **7**, 865
reaction with tetracyanoethylene
thermochemistry, **5**, 76
retrograde Diels–Alder reactions, **5**, 552
thermal osmylation, **7**, 863
Vilsmeier–Haack reaction, **2**, 779
Anthracyclines
synthesis, **5**, 1098; **7**, 341
Dieckmann reaction, **2**, 824
Friedel–Crafts reaction, **2**, 761, 762
via benzocyclobutene ring opening, **5**, 693
via Diels–Alder reactions, **5**, 327, 375, 393
via Michael addition, **4**, 14, 27
Anthracyclines, demethoxy-
synthesis
via Diels–Alder reaction, **5**, 338
Anthracyclinone, 11-deoxy-
synthesis
via protected acetaldehyde cyanohydrin, **1**, 554
Anthracyclinones
synthesis, **7**, 345
via annulation of arynes, **1**, 554
via Diels–Alder reaction, **5**, 384
via oxyanion-accelerated rearrangement, **5**, 1022
Anthramycin
synthesis, **3**, 487
palladium-catalyzed carbonylation, **3**, 1038
via directed lithiation, **1**, 469
Anthranilic acid, *N*-(3-trifluoromethylphenyl)-
Friedel–Crafts reaction, **2**, 759
Anthranilic acids
benzyne from *via* diazotization
benzocyclobutene synthesis, **5**, 692
synthesis
via Hofmann reaction, **6**, 802
Anthranilohydroxamic acid
Lossen reaction, **6**, 824
Anthranol
retro Diels–Alder reaction, **5**, 564
Anthraquinone, bis(bromomethyl)-
Diels–Alder reactions, **5**, 394
9,10-Anthraquinone, 1,4-dihydroxy-5-methoxy-
Friedel–Crafts reaction, **2**, 762
Anthraquinones
biomimetic synthesis, **2**, 176
charge-transfer osmylation, **7**, 864
reaction with allylzinc bromide, **1**, 218
reduction
silanes, **8**, 318
synthesis, **7**, 341
Friedel–Crafts reaction, **2**, 754
via annulation of arynes, **1**, 554
via arene–metal complexes, **4**, 546
via arynes, **4**, 497
via benzocyclobutene ring opening, **5**, 693
via [2 + 2 + 2] cycloadditions, **5**, 1148
via Michael addition, **4**, 27
Anthrasteroids
synthesis, **7**, 833
Anthrone
synthesis, **2**, 173
9-Anthrone, 10-arylmethylene-
reduction, **8**, 950
9-Anthrylmethyl esters
carboxy-protecting groups
photolysis, **6**, 668
Antibiotic A 23187 — *see* Calcimycin
Antibiotic CC-1066
synthesis
via cyclopropanation, **4**, 1043
Antibiotics
synthesis
Eschenmoser coupling reaction, **2**, 887
Antibiotic X-206
synthesis
via higher order cuprate, **1**, 130
Antibiotic X-296
synthesis, **7**, 245
Anti Cram–Felkin stereochemical control
Diels–Alder reactions, **2**, 677
Antidepressants
synthesis, **7**, 397
Antimonic acid, fluoro-
catalyst
Friedel–Crafts reaction, **3**, 297
Antimony, alkylbis(phenylthio)-
synthesis, **7**, 728
Antimony compounds, crotyl-
type III
reactions with aldehydes, **2**, 24
Antimony pentachloride
activator
DMSO oxidation of alcohols, **7**, 299
catalyst
Friedel–Crafts reaction, **2**, 714
reaction with alkenes, **7**, 530
Antimony trifluoride
fluorination, **6**, 220
mixture with alkyl fluoride
Friedel–Crafts reaction, intermediate, **3**, 299
Antimycin A_3
synthesis
via activated amides, **6**, 373
via macrolactonization, **6**, 369
Antirhine
synthesis
via Baeyer–Villiger reaction, **7**, 682
Antofine
synthesis
via selective *ortho* lithiation, **1**, 466
Aobamine
synthesis, **1**, 564
Aphidicolin
synthesis, **3**, 717; **7**, 633; **8**, 946
epoxide ring opening, **3**, 752
rearrangement of allylic epoxides, **3**, 762
via conjugate addition, **4**, 215
β-1-Apiofuranoside
asymmetric synthesis
via photocycloaddition, **5**, 185
Apiose
synthesis
via Paterno–Büchi reaction, **5**, 158
Aplasmomycin
synthesis
via organocuprates, **4**, 176
via oxalate acylation, **1**, 425
Aplysiatoxin
synthesis, **3**, 126, 168; **7**, 246
Aplysiatoxin, debromo-

synthesis, **3**, 126; **7**, 246
Aplysin
synthesis, **3**, 783
Apocamphane-1-carboxylic acid
decarboxylation, **7**, 732
Apogossypol
hexamethyl ether
synthesis, **3**, 665
Apomitomycin
synthesis
Eschenmoser coupling reaction, **2**, 888
Apopinene, 2-ethyl-
hydroboration, **8**, 722
Aporphines
11-substituted
synthesis *via* arynes, **4**, 513
synthesis, **3**, 507
via arynes, **4**, 504
via Diels–Alder reaction, **5**, 384
via photocyclization, **5**, 724
Aquillochin
synthesis, **3**, 691
Arabinol
synthesis, **7**, 645
Arabinose
selective monoacetylation
enzymatic, **6**, 340
synthesis, **8**, 292
Arabinose, 2,5-anhydro-
Knoevenagel reaction, **2**, 385
β-L-Arabinoside, methyl 3,4-*O*-benzylidene-
reduction, **8**, 226
Arachidonic acid
eicosanoid metabolites
synthesis, **3**, 217
lipoxygenase metabolites
synthesis, **7**, 712
synthesis, **7**, 731
via (*Z*)-selective alkenation, **1**, 758
Arachidonic acid, 3-dehydro-
synthesis, **3**, 250
Arachidonic acid, 5,6-dehydro-
synthesis, **3**, 247
Arene–alkene photocycloaddition reactions
[3 + 2] and [5 + 2], **5**, 645–671
exciplex pathway, **5**, 649
mechanism, **5**, 648–654
Arenecarbaldehydes
Baeyer–Villiger reaction, **7**, 684
Arenecarbodithioates, 2-dialkylaminoethyl
synthesis
via ethyl arenecarbodithioates, **6**, 454
Arenecarbodithioates, vinyl
thioarylation
thioxoester synthesis, **6**, 450
transesterification, **6**, 454
Arenecarbonitriles
synthesis, **4**, 457
via $S_{RN}1$ reaction, **4**, 471
Arenecarbothioates, *O*-ethyl
synthesis
via imidates, **6**, 452
Arenecarbothioates, *S*-phenyl esters
synthesis, **6**, 441
Arenediazocyanide
Diels–Alder reactions, **5**, 428
Arenediazonium salts
carbonylation, **3**, 1026
generation
radical addition reactions, **4**, 757
hydrogenolysis, **8**, 916
radical cyclizations, **4**, 804
vinylation
palladium complexes, **4**, 835, 842, 856
Arene oxides
microbial hydroxylation, **7**, 78
Arenes
acylation, **6**, 445
alkylation
Friedel–Crafts reaction, **3**, 298
amination, **7**, 10
[4 + 3] cycloaddition reactions, **5**, 608
η^5-cyclohexadienyl complexes
addition reactions, **4**, 531–547
dithiocarboxylation, **6**, 456
electron deficient
nucleophilic addition, substitution by, **4**, 423–447
Friedel–Crafts acylation
via thiol esters, **6**, 445
Mannich reaction, **2**, 1016
metal complexes
cine–tele substitution, **4**, 527
cyclization, **4**, 524
nucleophilic addition reactions, **4**, 517–547
nucleophilic substitution, **4**, 521–531
synthesis, **4**, 519–521
nitrile synthesis, **6**, 240
osmylation
charge transfer, **7**, 865
electron transfer, **7**, 866
polyhalogenated
diaryne equivalents, **4**, 496
radical cations
electrophilic aromatic substitution, **7**, 870
time-resolved spectra, **7**, 864
regiospecific alkylation
Friedel–Crafts reaction, **3**, 303
synthesis
via aryl halides, **8**, 895–920
thioacylation, **6**, 453
vinyl substitutions
palladium complexes, **4**, 835–837
Arenes, alkoxy-
photocycloaddition reactions, **5**, 652
Arenes, alkyl-
photocycloaddition reactions, **5**, 652
Arenes, bromo-
carbonylation
palladium catalysts, **3**, 1018
Arenes, methoxy-
oxidative demethylation, **7**, 346, 350
Arenes, methyl-
intramolecular isomerization
Friedel–Crafts reaction, **3**, 328
Arenes, nitro-
addition reactions
with organomagnesium compounds, **4**, 85
Arenes, thiocyano-
synthesis, **4**, 443
Arenesulfenyl sulfamates
synthesis
via oxosulfenylation of alkenes, **4**, 335

Arenesulfonamides, *N,N*-dichloro-
reactions with alkenes, **7**, 498
Arenesulfonyl halides
addition reactions
alkenes, **7**, 518
Arenesulfonylhydrazones
Bamford–Stevens reaction, **6**, 776
synthesis, **8**, 940
Arenethioates, *S*-(2-oxoalkyl)
synthesis
via acylation of dipole-stabilized carbanions, **6**, 446
Arene thiols
dimerization
nicotinium dichromate, **7**, 277
Arenethiosulfenyl chlorides
reaction with alkenes, **7**, 516
Argentilactone
synthesis, **3**, 168
Arginine acid
hydrogenation
catalytic, **8**, 145
Aristeromycin
synthesis
via Diels–Alder reaction, **5**, 370
Aristolactone
synthesis
Wittig rearrangement, **3**, 1010
Aristoteline
synthesis
via Ritter reaction, mercuration, **6**, 284
Arndt–Eistert synthesis
diazoalkanes, **1**, 844
diazo compounds, **6**, 127
α-diazo ketones
synthesis, **3**, 888
Wolff rearrangement, **3**, 897
Arnottianamide, methyl-
synthesis
via Diels–Alder reaction, **5**, 500
Arnottinin
synthesis, **7**, 823
Aromadendrene
synthesis, **3**, 390
Aromatic compounds
activated
thioimidate synthesis, **6**, 540
anodic oxidation, **7**, 799
hydrogenation
homogeneous catalysis, **8**, 453
radical addition reactions, **4**, 766–770
reactions with chloromethyleniminium-based salts
Vilsmeier–Haack reaction, **2**, 779
reaction with ketocarbenes, **4**, 1031–1064
reduction
Benkeser reduction, **8**, 516
Birch reduction, **8**, 490
dissolving metals, **8**, 489–519
electrochemical methods, **8**, 517
photochemical methods, **8**, 517
reductive silylations, **8**, 517
synthesis
via 2,3-sigmatropic rearrangement, **6**, 873
Aromatic compounds, fluoro-
synthesis, **4**, 445
Aromatic compounds, nitro
irradiation, **7**, 43
Aromatic compounds, vinyl-
hydroformylation, **4**, 932
regioselectivity, **4**, 919
Aromatic halides
reactions with ketones
organosamarium compounds, **1**, 258
Aromatic hydrocarbons
hydrogenation
heterogeneous catalysis, **8**, 436
mechanism, **8**, 437
stereochemistry, **8**, 437
structure–reactivity, **8**, 436
nuclear hydroxylation
microbial, **7**, 78
Aromaticin
synthesis, **7**, 313
σ-Aromaticity
cyclopropanes, **5**, 900
Aromatic substitution
electrochemically induced, **4**, 453
electron-transfer, **7**, 872
Aromatin
synthesis, **1**, 566; **7**, 313
via Cope–Claisen rearrangement, **5**, 886
Aromatization
alkanes, **7**, 6
photochemical
cyclohexadienones, **3**, 813
quinones, **7**, 136
steroids
microbial, **7**, 67
Aroyl chlorides
acyloin coupling reaction, **3**, 617
Aroyl cyanides
synthesis
via phase transfer catalysis, **6**, 317
Arsabenzaldehyde
Knoevenagel reaction, **2**, 369
Arsenic ylides
cyclopropanation, **4**, 987
Arsenides
$S_{RN}1$ reactions, **4**, 474
Arsine, triphenyl-
platinum complex
in hydrocarboxylation, **4**, 939
Arsonium ylides
epoxidation, **1**, 825
reaction with carbonyl compounds
formation of alkenes and epoxides, **3**, 203
synthesis, **1**, 825
Arteannuin B
synthesis
Mannich reaction, **2**, 904
Artemesia ketone
synthesis, **3**, 869
Friedel–Crafts acylation of allylsilanes, **2**, 716
use of ylidic rearrangements, **3**, 964
via sequential dialkylation, **1**, 557
Artemisinin
synthesis
via Paterno–Büchi reaction, **5**, 155
Arthrobacter simplex
dehydrogenation, **7**, 145
Arylamide ions
arylation, **4**, 470
Arylamine, *N*-nitrosoacyl-

rearrangement
aryne generation, **4**, 487
Arylamines
aminoalkylation
Mannich reaction, **2**, 961
aromatic nucleophilic substitution, **4**, 433
synthesis
via $S_{RN}1$ reaction, **4**, 472
vinylation
palladium complexes, **4**, 856
Aryl anions
aryne generation, **4**, 486–488
Arylation
amides, **6**, 399
carbon nucleophilies, **4**, 429
intramolecular homolytic, **4**, 476
organomercury compounds
palladium complexes, **4**, 838
α-Arylation
metal enolates
regioselectivity, **3**, 12
Pummerer rearrangement
preparation of α-arylated sulfides, **7**, 199
Aryl bromides
hydrogenolysis, **8**, 906
Aryl carbanions
alkylation, **3**, 259
Aryl chlorides
hydrogenolysis, **8**, 904
vinyl substitutions
palladium complexes, **4**, 835
Aryl complexes
benzannulation, **5**, 1100
Aryl compounds
coupling reactions, **3**, 499
crossed, **3**, 501
intramolecular, **3**, 505
with alkenes, **3**, 492
dimerization, **3**, 499
Aryl compounds, allyl-
chiral
synthesis, **3**, 246
Aryl cyanates
synthesis
via phenols, **8**, 912
Aryl cyanides
synthesis, **6**, 241
Aryl fluorides
hydrogenolysis, **8**, 903
Aryl groups
addition reactions
with alkenic π-systems, **4**, 72–99
conjugate additions
catalyzed by Lewis acids, **4**, 140–158
Aryl halides
ammonolysis, **4**, 434
carbonylation
palladium catalysts, **3**, 1021
cross-coupling reactions
organometallic reagents, **3**, 522
deprotonation
aryne generation, **4**, 486
nitrile synthesis, **6**, 231
reaction with organocopper compounds, **3**, 219
reaction with phenoxides, **4**, 469
reduction, **8**, 895–920
synthesis, **7**, 340
vinyl substitutions
palladium complexes, **4**, 842–856
Arylic oxidation, **7**, 329
Aryl iodides
hydrogenolysis, **8**, 908
Aryl isocyanates
2-azetidinones from, **5**, 103
Aryloxy radicals
oxidative coupling, **3**, 660
Aryl phosphates
coupling reactions
with sp^3 organometallics, **3**, 455
reaction with alkenylaluminum, **3**, 492
Aryl radicals
coupling with cyanides, **4**, 471
cyclizations, **4**, 796–798
nucleophilic coupling, **4**, 451–480
Aryl scrambling
radical nucleophilic substitution, **4**, 454
Aryl sulfides
hydrogenolysis, **8**, 914
Aryl sulfinates
vinyl substitutions
palladium complexes, **4**, 856
Arylsulfonyl chlorides
vinyl substitutions
palladium complexes, **4**, 835, 856
2-Arylsulfonylethyl carbonate
alcohol protection
cleavage, **6**, 659
Aryl triflates
carbonylation
formation of esters, **3**, 1029
coupling reactions
with sp^3 organometallics, **3**, 455
with vinylstannane, **3**, 495
cross-coupling reactions
with terminal alkynes, **3**, 531
reaction with cyanocuprates, **3**, 219
reduction, **8**, 933
Aryne, oxazolinyl-
reaction with alkyllithiums, **4**, 494
Arynes
chemoselectivity, **4**, 492
Diels–Alder reaction, **4**, 512
1,3-dipolar additions, **4**, 512
generation, **4**, 485–490
metal complexes, **4**, 485
nucleophilic addition, **4**, 491–513
regioselectivity, **4**, 492–495
nucleophilic coupling, **4**, 483–513
reaction with ambident anions, **4**, 492
soft acids, **4**, 491
soft electrophiles, **4**, 484
zirconium complexes, **4**, 485
Arynic substitution
in synthesis, **4**, 495–513
Asatone
synthesis, **3**, 697
Asatone, bisdemethoxy-
synthesis, **3**, 697
Ascorbic acid
intermolecular redox reactions
via enediols, **8**, 88
Aspartame

esters, **6**, 324
synthesis, **6**, 384
enzymatic, **6**, 399
in UV light, **6**, 402
Aspartamine
synthesis, **3**, 543
Aspartate proteases
peptide synthesis, **6**, 395
Aspartic acid
lithium aluminum hydride modifiers, **8**, 168
synthesis
via reductive amination, **8**, 144
Asperdiol
synthesis, **7**, 647
via chromium(II) ion mediation, **1**, 187
Aspergillus awamori
hydrocarbon hydroxylation, **7**, 59
Aspergillus niger
hydrocarbon hydroxylation, **7**, 62
reduction
unsaturated carbonyl compounds, **8**, 558
(±)-Asperlin
synthesis
stereocontrolled, **2**, 94
Aspicilin
synthesis
via enone reduction, **8**, 545
via macrolactonization, **6**, 373
Aspidodispermine, deoxy-
synthesis, **7**, 175
Aspidosperma alkaloids
synthesis
iminium ion–arene cyclization, **2**, 1022
Mannich reactions, **2**, 1043
via annulation, **5**, 1100
via cyclohexadienyl complexes, **4**, 680
Aspidosperma alkaloids, deethyl-
synthesis
via retro Diels–Alder reactions, **5**, 581
Aspidospermidine, 6,7-dehydro-
synthesis
via Diels–Alder reaction, **5**, 372
Aspidospermidine, 16-methoxy-1,2,6,7-tetradehydro-
synthesis
Mannich reactions, **2**, 1043
Aspidospermine
synthesis
via cyclohexadienyl complexes, **4**, 679
via diene protection, **6**, 690
Aspidospermine alkaloids
synthesis, **6**, 754
Asteltoxin
synthesis
via photocycloaddition, **5**, 172
Asteriscanolide
synthesis
via [4 + 4] cycloaddition, **5**, 641
Asteromurin A
synthesis, **7**, 243
Asymmetric catalysts
Darzens glycidic ester condensation, **2**, 435
Asymmetric dihydroxylation
alkenes, **7**, 429
Asymmetric epoxidation
absolute configuration, **7**, 391
alcohol-free dichloromethane, **7**, 394
catalysis
titanium complexes, **7**, 422
catalyst preparation, **7**, 394
competing side reactions, **7**, 394
concentration, **7**, 394
diastereoselectivity, **7**, 397
enantiofacial selectivity, **7**, 397
enantioselectivity, **7**, 391
mechanism, **7**, 395
methods, **7**, 425
molecular sieves, **7**, 396
oxidant, **7**, 394
solvent, **7**, 394
stoichiometry
catalytic reaction, **7**, 393
ratio of titanium to tartrate, **7**, 393
1-substituted allyl alcohols
kinetics, **7**, 411
substrate structure, **7**, 397
titanium tartrate catalysis
mechanism, **7**, 420
Asymmetric hydrogenation
alkenes
homogeneous catalysis, **8**, 459
enamides
homogeneous catalysis, **8**, 460
Asymmetric hydroxylation
ketones, **7**, 162
Asymmetric synthesis
[3 + 2] cycloaddition reactions, **5**, 305
double
allyl organometallics, **2**, 40–45
enol ethers, **2**, 629–657
matched double
allyl organometallics, **2**, 41
mismatched double
allyl organometallics, **2**, 41
Atherosperminine
synthesis, **3**, 586
Atisine
synthesis
via Michael addition, **4**, 30
Atisiran-15-one
synthesis, **3**, 715
Atom transfer reactions
radical addition reactions, **4**, 751–758
radical cyclizations, **4**, 801–805, 824
radicals, **4**, 726
Atractyligenin
synthesis
via cyclopropane ring opening, **4**, 1044
Atrolactic acid
preparation of chiral reagent
asymmetric synthesis, **2**, 224
synthesis, **3**, 829
Atromentin
synthesis, **3**, 828
Attalpugite
solid support
oxidants, **7**, 845
Aucubigenone
synthesis
via Pauson–Khand reaction, **5**, 1062
Auraptene, 3,6-epoxy-
synthesis, **7**, 406
Auraptenol

oxidative rearrangement, **7**, 823
Aurentiacin
synthesis
Friedel–Crafts reaction, **2**, 735, 760
Autoxidation
alkanes, **7**, 10
dienes, **7**, 861
zirconium compounds
mechanism, **8**, 691
Avenaciolide
synthesis
via photocycloaddition, **5**, 171
Avermectin A_{1a}
synthesis, **2**, 577; **7**, 237
via heteronucleophile addition, **4**, 34
Avermectin A_{2a}
allylic oxidation, **7**, 93
Avermectin B_{1a}
selective hydrobromination, **4**, 356
Avermectin B_{1a}
synthesis
via acylation of alkynide, **1**, 419
Avermectins
synthesis, **1**, 569; **7**, 300
Diels–Alder reaction, **2**, 701
via Julia coupling, **1**, 797, 801, 802
via organoaluminum reagents, **1**, 103
via organostannane acylation, **1**, 447
Azaacetals
reduction, **8**, 228
to ethers, **8**, 211–232
Azaadamantane
synthesis
via Ritter reaction, **6**, 284
Azaalditol
synthesis, **7**, 638
Azaalkenes
synthesis
via benzoin condensation, **1**, 545
Azaallyl enolates
crystal structures, **1**, 28
Azaallyllithium reagents
silylation
preparation from hydrazones, **2**, 507
Azaallyl metal reagents
carboxylic acids
1,2-additions, **2**, 516
formation
from hydrazones, **2**, 506
from hydrazones
structure, **2**, 507
1-Azaallyl system
synthesis
via protonation–deprotonation, **6**, 722
Azaallyltitanium reagents
preparation
from hydrazones, **2**, 507
Azaanion-accelerated rearrangements
small rings, **5**, 1000–1004
Azaanthranols
synthesis
Friedel–Crafts reaction, **2**, 759
Azaaromatic compounds, amino-
synthesis, **4**, 434
Azabicyclic alkaloids
synthesis
chiral, **1**, 558
Azabicyclic systems
bridged α-allenic amides
synthesis, **2**, 89
2-Azabicyclo[2.2.*n*]alkenes
flash vapor pyrolysis, **5**, 576
Azabicyclohexene
synthesis
via nitrile ylide cycloaddition, **4**, 1083
2-Azabicyclo[3.1.0]hex-3-enes
synthesis
via ketocarbenoids and pyrroles, **4**, 1061
2-Azabicyclo[3.3.1]nonane
synthesis
Dieckmann reaction, **2**, 819
9-Azabicyclo[4.2.1]nonane
synthesis
Mannich reaction, **2**, 1012
Azabicyclo[3.3.1]nonanes
synthesis
Mannich reaction, **2**, 1024
via intramolecular Ritter reaction, **6**, 278
3-Azabicyclo[3.3.1]nonene
synthesis
via Ritter reaction, **6**, 284
Azabicyclo[3.2.1]octane
synthesis, **8**, 124
3-Azabicyclo[3.3.0]octane
synthesis
via intramolecular Ritter reaction, **6**, 273
1-Azabicyclo[2.2.2]octan-3-one
lithium enolate
aldol reaction with benzaldehyde, **2**, 198
8-Azabicyclo[3.2.1]oct-6-ene
Pauson–Khand reaction, **5**, 1051
2-Aza-1,3-butadiene
cycloaddition reactions, **6**, 757
1-Aza-1,3-butadiene, *N*-acyl-
Diels–Alder reactions, **5**, 473
1-Aza-1,3-butadiene, 2-*t*-butyldimethylsiloxy-
Diels–Alder reactions, **5**, 473
1-Aza-1,3-butadiene, *N*-phenylsulfonyl-
Diels–Alder reactions, **5**, 474
1-Aza-1,3-butadiene, 2-trimethylsiloxy-
Diels–Alder reactions, **5**, 473
2-Aza-1,3-butadiene, 3-trimethylsiloxy-
Diels–Alder reactions, **5**, 480
Azabutadienes
Diels–Alder reactions, **5**, 470–491
synthesis
via photocycloaddition, **5**, 161
1-Azabutadienes
cycloaddition reactions, **6**, 757
1-Aza-1,3-butadienes
Diels–Alder reactions, **5**, 473–480
2-Aza-1,3-butadienes
Diels–Alder reactions, **5**, 480–484
hetero
Diels–Alder reactions, **5**, 485
2-Aza-1,3-butadienes, 1,3-bis(*t*-butyldimethylsiloxy)-
Diels–Alder reactions, **5**, 480
Aza-Cope rearrangement, **5**, 877
cationic, **2**, 1072, 1077
palladium(II) catalysis, **4**, 563–565
3-Aza-Cope rearrangement, **6**, 860
Azacyclic systems

synthesis
via Ireland rearrangement, **5**, 843
2-Azacycloalkanones
reductive elimination, **8**, 926
Azacycloheptane, 2,2-disubstituted
synthesis
from allyl organometallic compounds, **2**, 995
Azacycloheptanes
synthesis
acyloin coupling reaction, **3**, 629
1-Azacyclohexan-3-one
Wolff–Kishner reduction, **8**, 926
1-Azacyclohexene
synthesis
via intramolecular Ritter reaction, **6**, 278
Azacyclopentane, tetramethyldisilyl-
in *N*-tetrazol-5-yl-β-lactam synthesis, **2**, 920
Azacyclopropanes
synthesis
via oxidation of β-stannyl phenylhydrazones, **7**, 628
Azadienes
cationic
Diels–Alder reactions, **5**, 492–501
five-membered ring heteroaromatic
Diels–Alder reactions, **5**, 491
heteroaromatic
Diels–Alder reactions, **5**, 411, 491–492
reactions with organometallic compounds, **1**, 382
six-membered ring heteroaromatic
Diels–Alder reactions, **5**, 491
1-Azadienes
isomerization, **6**, 721
reaction with Grignard reagents, **6**, 721
2-Azadienes
imine anions from
reactions, **2**, 479
reactions with organometallic compounds, **1**, 383
synthesis
via geminal disubstitution, **6**, 722
via isomerization of 1-azadienes, **6**, 721
2-Aza-1,3-dienes
synthesis
via retro Diels–Alder reactions, **5**, 559
Aza-di-π-methane rearrangements
photoisomerizations, **5**, 201, 220
Azadiradione
synthesis, **7**, 634
Azadispiro ketocyclic hydroxamic acids
oximes
synthesis, **2**, 329
Azaenolates
addition reactions
with alkenic π-systems, **4**, 99–113
chiral
conjugate additions, **4**, 221–226
Azafulvene
dimer
dilithiation, **1**, 473
Azafumarates
intramolecular cyclization, **5**, 414
4-Azaheptanedioic acid, 2,4-dimethyl-
dimethyl ester
Dieckmann reaction, **2**, 811
Azaheteroaromatic compounds, 2-methyl-
reactions
with aldehydes, **2**, 495
Azahexenyl radicals
cyclizations, **4**, 811
4-Azaindoles
synthesis
via $S_{RN}1$ reaction, **4**, 478
Azaketals
reduction, **8**, 228
Azalomycin B — *see* Elaiophylin
1-Aza-2-oxabicyclo[2.2.1]heptane
synthesis
via nitrone cyclization, **4**, 1115
7-Aza-8-oxabicyclo[4.2.1]nonanes
bridged
synthesis *via* nitrone cyclization, **4**, 1114
1-Aza-8-oxabicyclo[3.2.1]octane
synthesis
via nitrone cyclization, **4**, 1115
1-Aza-1-oxa-di-π-methane rearrangements
photoisomerizations, **5**, 202
3-Aza[10]paracyclophane
synthesis
acyloin coupling reaction, **3**, 629
2-Aza-1,3-pentadiene, 1-phenyl-
Diels–Alder reactions, **5**, 480
2-Azapentadienyl anion
reaction with carbonyl compounds
regioselectivity, **2**, 64
2-Azapropenium salts, 3-chloro-
synthesis, **6**, 517
Azaprostaglandins
synthesis, **8**, 944
9-Azaprostaglandins
skeleton of, synthesis
Dieckmann reaction, **2**, 823
Aza-*o*-quinodimethane
cycloaddition reactions, **6**, 757
1-Aza-4-silacyclohexane, *N*-aryl-
synthesis
via aminomercuration of allylic substrate, **4**, 405
Azaspirocycles
synthesis
via cyclohexadienyl complexes, **4**, 679
via palladium catalysis, **4**, 598
Azasulfenylation
alkenes, **4**, 332
2-Azasulfides
synthesis
from alkenes, **4**, 337
1-Azatrienes
electrocyclic ring closure, **5**, 741
Azelaic acid
Kolbe electrolysis, **3**, 640
Azepine, dihydro-
synthesis
via intermolecular addition, **4**, 48
Azepine, *N*-ethoxycarbonyl-
cycloaddition reactions
dienes, **5**, 634
Azepine, *N*-methoxycarbonyl-
synthesis, **7**, 507
Azepine, perhydro-
formamidines
alkylation, **3**, 72
Azepines
acylation

via tricarbonyliron complex, **4**, 707
synthesis
via cyclobutene ring expansion, **5**, 687
via heteronucleophile addition, **4**, 36
Azepines, 2,3-dihydro-
synthesis
via cyclobutene ring expansion, **5**, 687
Azepines, *N*-substituted
dimerization
via [6 + 4] cycloaddition, **5**, 634
Azepinone
synthesis
pinacol rearrangement, **3**, 729
Azetidine-2,4-diones
synthesis
rhodium-catalyzed carbonylation, **3**, 1037
Azetidine hydrazones
synthesis, **5**, 110
Azetidines, *N*-alkoxy-
synthesis, **8**, 60
Azetidines, 2-imino-
synthesis
via ketenimines, **5**, 113
Azetidines, vinyl-
cycloaddition reactions, **5**, 257
Azetidinethiones
synthesis
via azetidiminium salts, **5**, 110
2-Azetidinethiones
synthesis
via imines and thioketenes, **5**, 115
Azetidinimines
synthesis
via azetidiminium salts, **5**, 110
2-Azetidiniminium salts
synthesis
via keteniminium salts, **5**, 108–113
Azetidinone, 4-acetoxy-
acid-induced reaction, **2**, 1059
chiral
reaction with silyl ketene acetals, **2**, 647
reaction with allylsilane, **2**, 1060
reaction with enol silanes
Lewis acid mediated, **2**, 635
reaction with tin(II) enol ethers
chiral synthesis, **2**, 611
synthesis, **3**, 651
Azetidinone, 3-acyloxy-
synthesis, **2**, 1084
Azetidinone, 4-allenyl-
cyclization, **2**, 1061
Azetidinone, 3-amino-
synthesis, **2**, 941
Azetidinone, diaryl-
lithium enolates
aldol reactions, **2**, 212
2-Azetidinone, 3-(1-hydroxyethyl)-
synthesis, **7**, 647
Azetidinones
synthesis, **6**, 759
2-Azetidinones
α-amidoalkylation, **1**, 372
synthesis
via enolates and imines, **5**, 100–102
via isocyanates, **5**, 102–108
via ketenes and carbonyls, **5**, 90–100
via lithium phenylethynolate cycloaddition, **5**, 116
2-Azetidinones, 4-(phenylthio)-
synthesis
via Pummerer rearrangement, **7**, 201
Azetidinones, vinyl-
cycloaddition reactions, **5**, 257
Azetines
synthesis
via retro Diels–Alder reactions, **5**, 581
Azetinones
Diels–Alder reactions, **5**, 407
Azides
alkyl
amine synthesis, **6**, 76
alkylation, **6**, 76
azide transfer reactions, **6**, 256
Beckmann reaction, **7**, 696
cyclizations, **4**, 1157–1159
nitrogen exchange reactions, **6**, 254
nucleophilic addition to π-allylpalladium complexes, **4**, 598
regioselectivity, **4**, 640
oxidation, **7**, 752
protecting group
amines, **6**, 633
reactions with alkenes, **4**, 295–297
1,3-dipolar cycloadditions, **4**, 1099–1101
reduction
synthesis of secondary amines, **8**, 386
reductive alkylation
synthesis of secondary amines, **8**, 386
reductive cleavage
synthesis of amines, **8**, 383
synthesis, **6**, 245
via nitrosation of hydrazines and hydrazides, **7**, 744
via oxygen exchange reactions, **6**, 252
via sulfur exchange reactions, **6**, 252
Azides, acyl
Curtius reaction, **6**, 797
rearrangement, **3**, 908
synthesis, **6**, 249
Azides, alkoxycarbonyl
reactions, **7**, 477
Azides, alkyl
synthesis, **6**, 245
Azides, arenesulfonyl
reactions with alkenes, **7**, 483
Azides, aroyl
synthesis
via aroyl chlorides, **6**, 251
Azides, aryl
reactions with organoboranes, **7**, 607
synthesis, **6**, 248
Azides, carbamoyl
synthesis, **6**, 251
Azides, cycloalkenyl
cyclizations, **4**, 1158
Azides, 1,2-dichloro
synthesis, **7**, 507
Azides, diethylphosphoryl
reaction with norbornene, **7**, 483
Azides, α,β-epoxyacyl
synthesis
via acid halides, **6**, 249
Azides, ethoxycarbonyl
nitrenes from, **7**, 478

Azides, guanyl
  synthesis, **6**, 252
Azides, imidoyl
  synthesis
    via imidoyl halides, **6**, 252
Azides, 2-iodoalkyl
  aziridine synthesis, **7**, 474
  reactions with organoboranes, **7**, 607
Azides, phenyl
  reaction with octafluoroisobutene, **6**, 500
Azides, phenylselenenyl
  reactions with alkenes, **7**, 496, 522
Azides, propargylic
  synthesis, **6**, 247
Azides, thioacyl
  synthesis, **6**, 251
Azides, trimethylsilyl
  azide synthesis, **6**, 249
  oxidative cleavage of alkenes
    introduction of nitrogen, **7**, 588
Azides, vinyl
  synthesis
    via 2-azido halides, **6**, 247
Azidoalkenes
  alicyclic-bridged
    cyclizations, **4**, 1158
  aryl-bridged
    cyclizations, **4**, 1157
  cyclizations, **4**, 1157
  open-chain
    cyclizations, **4**, 1157
Azidoalkynes
  cyclizations, **4**, 1158
4-Azidobutyryl esters
  amine protection, **6**, 646
Azido groups
  amine protection, **6**, 646
Azidomercuration
  demercuration
    alkenes, **4**, 297
Azidoselenation
  alkenes, **7**, 496
  cyclohexadiene, **7**, 506
Azido sphingosine glycosylation method, **6**, 53
Azines
  Vilsmeier–Haack reaction, **2**, 792
Azines, α-cyano-
  synthesis, **6**, 241
Aziridine, 1-acyl-
  reaction with lithium aluminum hydride, **6**, 98
  reaction with organolithium compounds, **6**, 94
1-Aziridine, 2-amino-
  synthesis, **6**, 787
Aziridine, 1-arylsulfonyl-
  reaction with dimethyloxosulfonium methylide, **6**, 97
Aziridine, dienyl-
  radical opening, **5**, 938
  rearrangement
    transition metal catalyzed, **5**, 938
Aziridine, 1-(diphenylacetyl)-
  synthesis, **6**, 94
Aziridine, divinyl-
  rearrangement, **5**, 948
Aziridine, 1-ethoxycarbonyl-
  reaction with lithium amides, **6**, 94
  thermal rearrangement, **6**, 98
Aziridine, 1-methyl-
  reaction with lithium dimethylcuprate, **6**, 94
Aziridine, 2-methyl-
  arene alkylation by
    Friedel–Crafts reaction, **3**, 316
Aziridine, methylene-
  synthesis
    via dibromocyclopropyl compounds, **4**, 1022
Aziridine, 1-phenyl-2,3-bis(methoxycarbonyl)-
  ring opening
    azomethine ylide generation, **4**, 1085
Aziridine, 1-substituted-2,2-dimethyl-
  reaction with carbon nucleophiles
    regioselectivity, **6**, 94
Aziridine, 1-substituted-2-phenyl-
  reaction with carbon nucleophiles
    regioselectivity, **6**, 94
Aziridine, 1-tosyl-
  reaction with Grignard reagents, **6**, 94
Aziridine, 1,2,3-triphenyl-
  reactions with alkenes
    synthesis of heterocycles, **4**, 1085
2-Aziridineacetic acid
  synthesis, **2**, 943
2-Aziridinecarboxylic acid
  esters
    amino acid synthesis, **6**, 96
  methyl esters
    reaction with ethoxycarbonylmethylene-
      triphenylphosphorane, **6**, 96
2-Aziridinecarboxylic acid, 2-chloro-
  isopropyl ester
    preparation, **2**, 429
2-Aziridinecarboxylic acid, 3-methyl-
  β-amino acid synthesis, **6**, 96
Aziridines
  addition to activated alkynes, **4**, 48
  amides
    reduction, **8**, 271
  arene alkylation by
    Friedel–Crafts reaction, **3**, 316
  asymmetric synthesis, **1**, 837
  carbonylation
    formation of β-lactams, **3**, 1036
  chiral
    reduction, **6**, 98
    synthesis, **6**, 75
  cleavage
    use of Lewis acids, **1**, 343
  hazards, **7**, 470
  imino ester synthesis, **6**, 535
  nitrogen unsubstituted
    synthesis, **7**, 470
  phosphorylation, **7**, 483
  preparation, **2**, 428
  pyrolysis
    azomethine ylide generation, **4**, 1085
  quaternized, **7**, 484
  reaction with organocopper
    reagents, **3**, 224
  resolved
    synthesis, **7**, 482
  ring opening, **6**, 93; **7**, 470
  N-substituted with O or S, **7**, 483
  synthesis, **1**, 834; **6**, 755; **7**, 744
    via alkenes, **7**, 470, 472

via *N*-aminolactams, **7**, 744
via bromine azide addition to alkene, **4**, 349
via ketoximes, **1**, 387
via lithiohalo methyl phenyl sulfoxides, **1**, 526
via nitrones, **1**, 836
thallated
ring opening, **7**, 491
Aziridines, *N*-acyl-
synthesis, **7**, 477
Aziridines, *N*-acylamino-
synthesis, **7**, 482
Aziridines, *N*-alkenyl-
synthesis, **7**, 474
Aziridines, *N*-alkyl-
synthesis, **7**, 474
Aziridines, 2-amino-
synthesis, **7**, 476
Aziridines, *N*-amino-
decomposition, **7**, 482
synthesis, **7**, 480
Aziridines, *N*-aryl-
synthesis, **2**, 429; **7**, 476
Aziridines, aryloxysulfonyl-
synthesis, **7**, 484
Aziridines, *N*-arylsulfinyl-
synthesis, **7**, 483
Aziridines, 2-chloro-
synthesis, **7**, 479
Aziridines, *N*-chloro-
synthesis, **7**, 747
Aziridines, *N*-cyano-
synthesis, **7**, 477, 479
Aziridines, 2,2-dihalo-
synthesis, **6**, 498
Aziridines, *N*-heteroaryl-
synthesis, **7**, 476
Aziridines, imidoyl-
synthesis, **7**, 479
Aziridines, *S*-(–)-2-methyl-
synthesis, **7**, 473
Aziridines, 2-phenyl-
reaction with alkenes, **7**, 498
Aziridines, 2-phenylsulfonyl-
synthesis
via aromatic imines, **1**, 835
Aziridines, *N*-phosphonyl-
synthesis, **7**, 480
Aziridines, *N*-phthalimido-
cleavage, **7**, 482
ring opening, **7**, 487, 493
Aziridines, substituted
preparation
Darzens glycidic ester condensation, **2**, 428
Aziridines, *N*-sulfenyl-
synthesis, **7**, 483
Aziridines, sulfonyl-
synthesis, **7**, 477
Aziridines, 1,2,3-triphenyl-
ozonolysis, **7**, 474
Aziridines, vinyl-
photochemical rearrangements, **5**, 938
rearrangements, **5**, 909, 937
synthesis
reaction of allyllithium with aldimines, **2**, 982
reaction of chloro(methyl)allyllithium with imines, **2**, 982
Aziridinium salts
synthesis
via diazoalkanes, **1**, 836
2*H*-Azirine
Neber reaction, **6**, 786
Azirine, 2-aryl-
carbonylation
formation of isocyanates, **3**, 1039
2*H*-Azirine, 2-phenyl-
reaction with enolates, **2**, 942
Azirines
carbonylation
formation of bicyclic β-lactams, **3**, 1036
cycloaddition reactions
fulvenes, **5**, 630
Diels–Alder reactions, **5**, 413
reaction with arynes, **4**, 510
rearrangement
stereochemistry, **5**, 948
synthesis, **7**, 506
Azirines, aryl-
photolysis
nitrile ylides from, **4**, 1081
Azlactones — *see* Oxazolones
Azoalkanes
denitrogenation, **5**, 205
Azoalkenes
Diels–Alder reactions, **5**, 486
Azobenzene
reduction
synthesis of hydrobenzenes, **8**, 382
Azobenzene, 4,4′-dinitro-
synthesis, **8**, 370
Azobisisobutyronitrile
radical initiator, **4**, 725
Azocines, 1,2-dihydro-
synthesis
via cyclobutene ring expansion, **5**, 687
Azocinone
synthesis
Thorpe reactions, **2**, 851
Azo compounds
acyclic
Diels–Alder reactions, **5**, 428
cyclic
Diels–Alder reactions, **5**, 429
Diels–Alder reactions, **5**, 426
oxidation
synthesis of azoxy compounds, **7**, 750
radical initiators, **4**, 725
reduction
synthesis of hydrazo compounds, **8**, 382
reductive cleavage
synthesis of amines, **8**, 383
synthesis, **8**, 364
via primary arylamines, **7**, 738
Azo compounds, α-carbonyl-
synthesis
via oxidation of arylhydrazones of aldehydes, **7**, 747
Azodicarboxylates
Diels–Alder reactions, **5**, 486
Azodicarboxylic acids
diethyl ester
alcohol inversion, **6**, 22
amino alcohol cyclization, **6**, 74

Beckmann rearrangement reagent, **7**, 692
Diels–Alder reactions, **5**, 426
Mitsunobu reaction, ester synthesis, **6**, 333
diisopropyl ester
alcohol inversion, **6**, 22
dimethyl ester
alcohol inversion, **6**, 22
esters
electrophilic *N*-amino amination, **6**, 118
Azodicarboxylic esters
reduction, **8**, 388
Azolides
acid halide synthesis, **6**, 308
Azomethane
synthesis
via retro Diels–Alder reactions, **5**, 576
Azomethine imines
aryl-bridged
intramolecular cycloadditions, **4**, 1146
cyclizations, **4**, 1144–1150
sydnones, **4**, 1149
1,3-dipolar cycloadditions, **4**, 1095
open-chain
intramolecular cycloadditions, **4**, 1146
Azomethine imines, alkenyl
intramolecular cycloadditions, **4**, 1146
Azomethine imines, alkynyl
cycloadditions, **4**, 1147
Azomethine imines, cycloalkenyl
intramolecular cycloadditions, **4**, 1146
Azomethines
deprotonation
minimization, **1**, 357
enolizable
reactions with organometallic compounds, **1**, 361
metallated
reactions, **2**, 495
nonenolizable
reactions with organometallic compounds, **1**, 360
nucleophilic addition reactions
stereochemistry, **1**, 358, 362
reactions with dihalocarbenes, **6**, 498
reactivity
correlation with structure, **1**, 357
Azomethine ylides
cyclizations, **4**, 1134–1141
cycloaddition reactions
diasteroselective, **5**, 260
1,3-dipolar cycloadditions, **4**, 1085–1089
reactions with benzaldehyde, **5**, 265
tandem Michael–cyclization reactions, **4**, 1137
Azomethine ylides, alkenyl
cyclic
cycloadditions, **4**, 1136
cyclizations, **4**, 1134–1136
open-chain
cyclizations, **4**, 1134–1137
Azomethine ylides, alkynyl
cyclic
intramolecular cycloadditions, **4**, 1140
intramolecular cycloadditions, **4**, 1139–1141
open-chain
intramolecular cycloadditions, **4**, 1139
Azomethine ylides, allenyl
intramolecular cycloadditions, **4**, 1139–1141
Azomethinylide
synthesis, **6**, 572
Azoxybenzene
reaction with dihalocarbenes, **6**, 498
reduction
synthesis of hydrobenzenes, **8**, 382
Azoxybenzene, 2,2′-dicyano
synthesis, **8**, 365
Azoxybenzene, 3,3′-diiodo-
reduction, **8**, 365
Azoxybenzene, 3-trifluoromethyl-
synthesis, **8**, 364
Azoxy compounds
deoxygenation, **8**, 390
reduction
synthesis of hydrazo compounds, **8**, 382
reductive cleavage
synthesis of amines, **8**, 383
synthesis, **8**, 364
via oxidation of azo compounds, **7**, 750
via oxidation of primary amines, **7**, 736
Azulene, dichloro-
synthesis
via dihalocyclopropyl compounds, **4**, 1017
Azulene, 1,3-dimethyl-
synthesis
via [3 + 2] cycloaddition reactions, **5**, 285
Azulene, hexahydro-
synthesis
via palladium-ene reaction, **5**, 50
Azulene, *cis*-4-keto-
synthesis
via cycloaddition reactions, **5**, 274
Azulene, 2-methylene-6-oxo-2,6-dihydro-
synthesis
Knoevenagel reaction, **2**, 366
Azulene, perhydro-
synthesis
via carbonyl ylides, **4**, 1093
via [4 + 3] cycloaddition reactions, **5**, 598
Azulenes
[3 + 2] annulations, **1**, 603
synthesis, **2**, 85
via σ-alkyliron complexes, **4**, 579
via [3 + 2] cycloaddition reactions, **5**, 285
via [6 + 4] cycloaddition reactions, **5**, 626, 629
via electrocyclization, **5**, 744
via ketocarbenoid reaction with benzenes, **4**, 1052
via ketocarbenoids, **4**, 1055
Vilsmeier–Haack reaction, **2**, 780
Azulenes, hydro-
synthesis, **3**, 394
transannular ene reaction, **2**, 553
via Cope rearrangement, **5**, 803, 810
via cycloaddition reactions, **5**, 274
via [6 + 4] cycloaddition reactions, **5**, 626, 629
Azulenone, hydro-
aldol cyclization, **2**, 169
synthesis
ene reaction, **2**, 552
Azulenone, oxidoperhydro-
synthesis
via [4 + 3] cycloaddition, **5**, 609

B

Bachrachotoxin
  synthesis, **7**, 105
Bacillus putrificus
  reduction
    unsaturated carbonyl compounds, **8**, 558
Bacillus sphaericus
  dehydrogenation, **7**, 145
Back electron transfer
  electron-transfer oxidation, **7**, 852
Bactobolin
  synthesis, **1**, 404
    via dichloromethylcerium reagent, **1**, 238
Baeyer–Villiger reaction, **7**, 671–686
  buffers, **7**, 674
  catalysts, **7**, 674
    substituent effects, **7**, 673
  chemoselectivity, **7**, 675
  compared to Beckmann reaction, **7**, 690
  competitive, **7**, 675
  conformation, **7**, 673
  electronic factors, **7**, 673
  mechanism, **7**, 671
  peroxy acid
    substituent effects, **7**, 673
  radical scavengers, **7**, 674
  reaction methods, **7**, 674
  regioselectivity, **7**, 673, 676
  side reactions, **7**, 685
  β-silicon atom
    regiochemistry, **7**, 673
  stereochemistry, **7**, 672
  stereoelectronic requirements, **7**, 672
  steric factors, **7**, 673
Baiyunol
  synthesis, **1**, 568
Baker's yeast
  reduction
    carbonyl compounds, **8**, 184
    unsaturated carbonyl compounds, **8**, 560
Baker–Venkataraman synthesis
  intramolecular acyl transfer, **2**, 845
Bakkenolide
  synthesis, **3**, 939
Bakkenolide A
  synthesis
    via 2,3-sigmatropic rearrangement, **6**, 854
Baldulin
  synthesis, **1**, 564
Baldwin's rules
  intramolecular addition
    heteronucleophiles, **4**, 37–41
  Mannich reaction, **2**, 1024, 1034
  polyene cyclization, **3**, 344
Balz–Schiemann reaction
  fluorination, **6**, 220
Bamford–Stevens reaction, **6**, 776
  aprotic, **6**, 777; **8**, 941
  protic, **6**, 776; **8**, 943
  sulfonylhydrazone decomposition, **4**, 954
Barbaralanes
  synthesis
    via [4 + 3] cycloaddition, **5**, 612
Barbaralone
  synthesis
    via cyclopropanation, **4**, 1041
Barbier–Grignard type addition
  allylic halides
    carbonyl compounds, **1**, 177
Barbier-type reactions
  intermolecular
    organosamarium compounds, **1**, 256
  iron(III) salt catalysts
    organosamarium compounds, **1**, 257
  organosamarium compounds, **1**, 255
  ytterbium diiodide, **1**, 278
Barbiturates, alkylidene-
  addition reactions
    with organozinc compounds, **4**, 95
Barbituric acid
  Knoevenagel reaction, **2**, 352, 357
Barbituric acid, 5-arylidene-1,3-dimethyl-
  oxidation
    thiols, **7**, 761
Barbituric acid, *N*,*N*-dimethyl-
  allyl transfer
    amine protection, **6**, 641
  Knoevenagel reaction, **2**, 357
    Michael reaction, **2**, 352
Barium
  reduction
    ammonia, **8**, 113
Barium manganate
  oxidation
    diols, **7**, 318
    primary alcohols, **7**, 307
Barrelene
  flash vapor pyrolysis, **5**, 571
  photoisomerization, **5**, 204
    to semibullvalene, **5**, 194
Barrelenones
  photorearrangement, **5**, 229
Barton reaction, **7**, 9
  intramolecular functionalization, **7**, 41
  thiohydroxamate esters
    radical addition reactions, **4**, 747–750
    radical cyclizations, **4**, 799, 824
Bastadin
  synthesis, **7**, 337
9-BBN (*see* 9-Borabicyclo[3.3.1]nonane)
Büchi rearrangement
  2,3-sigmatropic rearrangement, **6**, 834, 853
Beauveria sulfurescens
  hydrocarbon hydroxylation, **7**, 58, 59
  reduction
    unsaturated carbonyl compounds, **8**, 558
Beckmann reaction, **7**, 689–701
  addition reactions, **7**, 695
  fragmentation, **6**, 1066; **7**, 698
  intramolecular, **7**, 697
  mechanism, **7**, 690
  stereochemistry, **7**, 690
Beckmann rearrangement, **6**, 763, 773; **7**, 690
  alkylation, **6**, 769
  amide synthesis, **6**, 404

cyclization, **6**, 771
nitrilium ions
trapping, **6**, 766
organoaluminum promotion, **1**, 98
Ritter reaction, **6**, 291
synthetic utility, **6**, 763
Benkeser reduction
aromatic rings, **8**, 516
Benzalacetone
reduction
electrochemical, **8**, 532
iron hydrides, **8**, 550
molybdenum complexes, **8**, 551
Benzalacetone, *o*-dimethylaminomethyl-
synthesis, **4**, 837
Benzalacetophenone
addition reactions
with organomanganese compounds, **4**, 98
conjugate addition
with aryl metallics, **4**, 70
Benzalaniline
reduction
dissolving metals, **8**, 124
Reformatsky reaction, **2**, 294
Benzaldehyde
aldol reaction
butanone, **2**, 146
boron trifluoride complex, **2**, 247
crystal structure, **1**, 300
Diels–Alder reactions, **5**, 433
diethyl acetal
reduction, **8**, 267
hydrogenation
catalytic, **8**, 140
oxime
catalytic hydrogenation, **8**, 143
oxime ether
reactions with butyllithium, **1**, 385
photolysis
with 1-hexyne, **5**, 163
reactions with allylic copper reagents, **1**, 113
reactions with allylic organometallic compounds, **1**, 156
reactions with azomethine ylides, **5**, 265
reactions with chromium chloride, **1**, 193
reactions with dimesitylboryl carbanions, **1**, 499
reactions with 2-naphthol and amines
Mannich reaction, **2**, 960
reduction
Clemmensen reduction, **8**, 310
electrolysis, **8**, 321
ionic hydrogenation, **8**, 318
Wolff–Kishner reduction, **8**, 338
Reformatsky reactions
kinetic stereoselection, **2**, 291
Benzaldehyde, 2-acetoxy-5-nitro-
synthesis, **7**, 657
Benzaldehyde, 4-acetyl-
acylation
palladium complex catalysis, **1**, 437
Benzaldehyde, allyl-
synthesis, **3**, 255
Benzaldehyde, *o*-amino-
Knoevenagel reaction, **2**, 357
synthesis
Knoevenagel reaction, **2**, 359
Benzaldehyde, 2-bromo-
dimerization, **3**, 501
Benzaldehyde, 4-bromo-
synthesis
carbonylation, **3**, 1021
Benzaldehyde, 4-*t*-butyl-
tin(IV) chloride complex
crystal structure, **1**, 303
NMR, **1**, 294
Benzaldehyde, 2-carboxy-
reduction
hydrogen transfer, **8**, 320
Benzaldehyde, 4-chloro-
reaction with diethylzinc, **1**, 216
Benzaldehyde, 2,4-dichloro-
synthesis
Vilsmeier–Haack reaction, **2**, 786
Benzaldehyde, 4-dimethylamino-
dichlorodiphenyltin complex
crystal structure, **1**, 305
reduction
boranes, **8**, 316
Benzaldehyde, 3,5-dinitro-
synthesis
via acyl halide reduction, **8**, 263
Benzaldehyde, *o*-fluoro-
Perkin reaction, **2**, 401
Benzaldehyde, hydroxy-
electropinacolization, **3**, 568
synthesis
Reimer–Tiemann reaction, **2**, 771
Benzaldehyde, nitro-
Reformatsky reaction, **2**, 285
Benzaldehyde, 4-nitro-
reactions with boron-stabilized carbanions
synthesis of alkenes, **1**, 499
synthesis, **8**, 291
via acyl halide reduction, **8**, 263
Benzaldehyde, 3-phenoxy-
cyanohydrin
benzoin condensation, **1**, 546
Benzaldehyde imine, *N*-trichlorovinyl-
reaction with chlorine, **6**, 528
Benzaldoximes
Beckmann rearrangement, **7**, 695
Benzalquinaldine
reduction, **8**, 568
N-Benzamide
adenine-protecting groups, **6**, 643
Benzamide, *N*,*N*-allylbenzyl-
Wittig rearrangement, **3**, 979
Benzamide, *N*,*N*-dimethyl-
reaction with phenylytterbium(II) iodide
synthesis of benzophenone, **1**, 278
reduction, **8**, 249
Vilsmeier–Haack reaction, **2**, 779
Benzamide, 2-hydroxy-
synthesis, **7**, 333
Benzamide, 2-methoxy-
Birch reduction
dissolving metals, **8**, 507
reduction, **3**, 51
Benzamide, 4-methoxy-
cytosine-protecting group, **6**, 643
Benzamides
Birch reduction

dissolving metals, **8**, 507
lithiation
addition reactions, **1**, 464
metallation
addition reactions, **1**, 466
Benzannulation
alkynes
functionality, **5**, 1098
aminohexatrienes, **5**, 720
aryl *versus* alkenyl complexes, **5**, 1100
carbene complexes, **5**, 1098
[3 + 2 + 1] cycloadditions, **5**, 1093
Diels–Alder reactions
tandem, **5**, 1099
Benz[*a*]anthracene, 7-acetoxy-
hydrogenolysis, **8**, 911
Benz[*a*]anthracene, 7,12-diacetoxy-
hydrogenolysis, **8**, 911
Benz[*a*]anthracene, 7-methoxy-
hydrogenolysis, **8**, 910
Benzazepines
synthesis, **4**, 446
via cyclobutene ring expansion, **5**, 687
via $S_{RN}1$ reaction, **4**, 479
Benzazocinone
synthesis
Friedel–Crafts reaction, **2**, 753
Benzene
alkylation *via* Pummerer rearrangement
dimethyl sulfoxide, **7**, 200
anodic oxidation, **7**, 800
charge-transfer osmylation, **7**, 864
charge transfer transition energy
EDA complexes, **7**, 870
formylation
dichloromethyl alkyl ethers, **2**, 750
Gattermann–Koch reaction, **2**, 749
hydrogenation
heterogeneous catalysis, **8**, 436
homogeneous catalysis, **8**, 453
one nucleofuge
$S_{RN}1$ reactions, **4**, 459
reaction with rhenium
metal vapor synthesis, **7**, 4
reductive silylation, **8**, 517
solvent
radical reactions, **4**, 721
thermal osmylation, **7**, 863
two nucleofuges
$S_{RN}1$ reactions, **4**, 459
Benzene, alkyl-
nitration, **6**, 110
oxidative degradation
microbial, **7**, 57
synthesis
via alkyl radical addition, **7**, 732
transalkylation
Friedel–Crafts reaction, **3**, 327
Benzene, allyl-
addition reactions
nitrogen and halogen, **7**, 498
synthesis
vinyl carbanion alkylation, **3**, 242
Benzene, *o*-bis(chloromethyl)-
tetrahydrofuran complex
crystal structure, **1**, 16
Benzene, bis(dialkylamino)-
aromatic nucleophilic substitution, **4**, 429
Benzene, *p*-bis(phenylthio)-
synthesis, **4**, 460
Benzene, bis(trifluoroacetoxy)iodo-
Hofmann reaction, **6**, 796
oxidative rearrangement
aliphatic amides, **6**, 803
Benzene, 1,4-bis(trimethylstannyl)-
reaction with *N*,*N*-dimethylmethyleneiminium chloride, **2**, 962
Benzene, bromo-
reaction with phenoxides, **4**, 469
reduction, **8**, 907
dissolving metals, **8**, 526
Benzene, 1-bromo-2-chloro-
hydrogenolysis, **8**, 901
reduction, **8**, 908
Benzene, 1-bromo-4-chloro-
hydrogenolysis, **8**, 901
Benzene, 4-*t*-butoxynitro-
synthesis, **4**, 437
Benzene, chloro-
hydrogenolysis, **8**, 906
synthesis
via dichlorocarbene, **4**, 1017
Benzene, 5-chloro-2,4-dimethoxynitro-
reduction, **8**, 367
Benzene, 1-chloro-2,4-dinitro-
sulfodechlorination, **4**, 443
Benzene, *m*-chloroiodo-
$S_{RN}1$ reactions, **4**, 460
Benzene, cyano-
photocycloaddition reactions, **5**, 652
Benzene, dibromo-
monoalkylation
with primary alkyl Grignard reagents or benzylzinc halides, **3**, 457
Benzene, 1,2-dibromo-
$S_{RN}1$ reactions, **4**, 460
Benzene, 1,4-dibromo-
carbonylation
selective, **3**, 1026
Benzene, 1,2-di-*t*-butyl-
hydrogenation
heterogeneous catalysis, **8**, 438
isomerization
Friedel–Crafts reaction, **3**, 327
Benzene, 1,4-di-*t*-butyl-
hydrogenation
high pressure, **8**, 438
Benzene, dichloro-
dialkylation
coupling reactions with primary alkyl Grignard reagents, **3**, 450
monoalkylation
with primary alkyl Grignard reagents, **3**, 457
Benzene, *m*-diethynyl-
polymerization, **3**, 557
Benzene, 2,6-difluoronitroso-
synthesis
via oxidation of 2,6-difluoroaniline, **7**, 737
Benzene, 1,3-dimethoxy-
Mannich reaction, **2**, 961
use in Houben–Hoesch synthesis, **2**, 748
Benzene, 1,3-dimethyl-

alkylation with 2,5-dichloro-2,5-dimethylhexane
Friedel–Crafts reaction, **3**, 318
Benzene, 1,2-dinitro-
reductive coupling, **8**, 370
Benzene, 1,3-dinitro-
coupling, **8**, 369
Benzene, 1,4-dinitro-
reduction, **8**, 366
reductive coupling, **8**, 370
Benzene, ethyl-
hydroperoxide
propylene oxide synthesis, **7**, 375
microbial hydroxylation, **7**, 76
synthesis
Friedel–Crafts reaction, **3**, 304
Benzene, ethylenedioxy-
Birch reduction
dissolving metals, **8**, 514
Benzene, halo-
nitration, **6**, 111
Benzene, hexacyano-
synthesis
via 2,4,6-trifluorotricyanobenzene, **6**, 232
Benzene, hexaethyl-
synthesis
Friedel–Crafts reaction, **3**, 301
Benzene, hexamethyl-
EDA complex
with maleic anhydride, **7**, 856
Benzene, 2-hydroxy-3-methoxy-1-(methylsulfinyl)acetyl-
Pummerer rearrangement
intramolecular participation by hydroxy groups, **7**, 202
Benzene, hydroxy(tosyloxy)iodo-
Hofmann rearrangement, **6**, 805
oxidative rearrangement, **7**, 833
α-tosyloxy ketone synthesis, **7**, 155
Benzene, iodo-
reaction with nonanal
chromium(II) chloride catalysis, **1**, 193
Benzene, 2-iodonitro-
Ullmann reaction, **3**, 499
Benzene, iodosyl-
alkane oxidation, **7**, 11
diacetate
α-hydroxylation, **7**, 179
oxidative decarboxylation, **7**, 722
reaction with carboxylic acids and iodine, **7**, 723
diazidation, **7**, 488
Hofmann rearrangement, **6**, 806
α-hydroxylation
enones, **7**, 179
ketones, **7**, 155
reaction with silyl enol ethers, **7**, 166
Benzene, isopropyl-
synthesis
Friedel–Crafts reaction, **3**, 304
Benzene, *o*-mercaptonitro-
in peptide synthesis, **3**, 302
Benzene, 1,2-methylenedioxy-
oxidative trimerization, **3**, 669
Benzene, nitro-
amination, **4**, 436
reaction with lithium phenolate, **7**, 334
reaction with organometallic reagents, **7**, 331
Benzene, nitroso-
Diels–Alder reactions
with 1,2-dihydropyridine, **5**, 418
Benzene, pentafluoro-
hydrogenolysis, **8**, 904
Benzene, pentafluorobromo-
reduction, **8**, 907
Benzene, pentamethyl-
radical cation
side chain substitution, **7**, 871
thallation, **7**, 872
Benzene, *n*-pentyl-
synthesis, **3**, 415
Benzene, polyalkyl-
transalkylation
Friedel–Crafts reaction, **3**, 327
Benzene, 2-propenyl-
rearrangement, **7**, 828
Benzene, *n*-propyl-
synthesis, **3**, 415
Benzene, 1,2,4,5-tetradehydro-
synthesis, **7**, 743
Benzene, *N,N,N′,N′*-tetramethyl-1,4-diamino-
synthesis
Mannich reaction, **2**, 962
Benzene, 1,2,3,4-tetraphenyl-5,6-diethyl-
synthesis
via [2 + 2 + 2] cycloaddition, **5**, 1146
Benzene, 1,3,5-trialkyl-
sterically crowded
electron-transfer oxidation, **7**, 869
Benzene, 1,3,5-tribromo
monoalkylation
with primary alkyl Grignard reagents or benzylzinc halides, **3**, 457
Benzene, 1,2,4-tri-*t*-butyl-
synthesis
via [2 + 2 + 2] cycloaddition, **5**, 1146
Benzene, 1,3,5-tri-*t*-butyl-
hydrogenation
heterogeneous catalysis, **8**, 438
synthesis
via [2 + 2 + 2] cycloaddition, **5**, 1148
Benzene, trichloro-
dialkylation
coupling reactions with primary alkyl Grignard reagents, **3**, 450
monoalkylation
with primary alkyl Grignard reagents, **3**, 457
Benzene, trifluoromethyl-
photocycloaddition reactions, **5**, 652
Benzene, 3-trifluoromethylnitro-
reduction, **8**, 364
Benzene, 1,3,5-triformyl-
synthesis
Vilsmeier–Haack reaction, **2**, 786
Benzene, triisopropyl-
formylation
Gattermann–Koch reaction, **2**, 749
Benzene, 1,2,3-trimethoxy-
hydrogenolysis, **8**, 910
Benzene, 1,3,5-trimethoxy-
Mannich reaction, **2**, 961
Benzene, 1,2,4-trimethyl-
reaction with isoprene
Friedel–Crafts reaction, **3**, 322

Benzene, 1-(trimethylsilyl)-2-methoxy-3-(2-hexenyl)-
metal complexes
reactions, **4**, 539
Benzene, 1,2,4-trinitro-
coupling, **8**, 370
Benzene, 1,3,5-trinitro-
coupling, **8**, 369
Benzene, 1,2,4-triphenyl-
synthesis
via [2 + 2 + 2] cycloaddition, **5**, 1148
Benzene, 1,3,5-triphenyl-
synthesis
via [2 + 2 + 2] cycloaddition, **5**, 1148
Benzene, tris(dialkylamino)-
aromatic nucleophilic substitution, **4**, 429
Benzeneacetic acid, α-methyl-4-(2-thienylcarbonyl)-
synthesis
hydroformylation, **4**, 932
Benzenediacrylic acids
synthesis
Perkin reaction, **2**, 399
Benzenediazonium-2-carboxylates
aryne precursors, **4**, 487
Benzenediazonium fluoroborate, 4-methoxy-
reduction, **8**, 917
Benzene dichloride, iodo-
acyl halide synthesis
via aldehydes, **6**, 308
Benzenediimide
synthesis
via reduction of benzenediazonium cation, **8**, 383
Benzene-1,2-diselenol
synthesis, **6**, 464
Benzenes
[4 + 3] cycloaddition reactions, **5**, 608
derivatives
synthesis *via* retro Diels–Alder reaction, **5**, 571–573
Diels–Alder reactions
benzynes, **5**, 383
irradiation
fulvene generation, **5**, 646
photocycloaddition reactions
dienes, **5**, 636
furan, **5**, 637
vinyl acetate, **5**, 667
reactions with ketocarbenoids, **4**, 1052–1058
synthesis
via cyclotrimerization of alkynes, **5**, 1144–1151
Benzeneselenamide
synthesis, **6**, 476
Benzeneselenenamide, *N,N*-diethyl-
use in selenenylation, **7**, 131
Benzeneselenenyl bromide
reaction with lithium enolates, **7**, 129
selenenylation, **7**, 131
Benzeneselenenyl chloride
addition reactions
alkenes, **7**, 520
allylic alcohols, **7**, 520
chlorocyclohexene, **7**, 520
reaction with alkanes, **7**, 534
reaction with lithium enolates, **7**, 129
selenenylation, **7**, 131
Benzeneselenenyl iodide
reaction with dienes, **7**, 505
Benzeneselenenyl trichloride
selenenylation, **7**, 135
Benzeneseleninic acid
oxidation, **7**, 674
reaction with *N*-acylhydrazines, **6**, 467
selenenylation, **7**, 132
Benzeneseleninic anhydride
α-hydroxylation
enones, **7**, 175
ketones, **7**, 158
oxidation, **7**, 132
quinone synthesis, **7**, 355
Benzeneseleninyl chloride
dehydrogenation, **7**, 135
Benzeneselenocarboxamide
deoxygenation
epoxides, **8**, 887
Benzenesulfenamide, 2,4-dinitro-
oxidation
synthesis of aziridines, **7**, 744
Benzenesulfenyl chloride
carbocyclization
1,4-dienes, **7**, 517
reactions with alkenes, **7**, 516
reactions with dienes, **7**, 516
reaction with 3,4,6-tri-*O*-benzyl-D-glucal, **6**, 60
Benzenesulfenyl chloride, 2,4-dinitro-
reactions with alkenes, **7**, 516
Benzenesulfinic acid, 2-amino-
aryne precursor, **4**, 488
Benzenesulfonic acid, trimethyl-2,4,6-trinitro-
alkylation by, **3**, 16
Benzenesulfonyl azide, *p*-carboxy-
diazo transfer reaction, **4**, 1033
Benzenesulfonyl azide, *n*-dodecyl-
diazo transfer reaction, **4**, 1033
Benzenesulfonyl chloride
acid chloride synthesis, **6**, 304
Beckmann rearrangement, **6**, 764; **7**, 699
Benzenesulfonyl hydrazide
decomposition
diimide from, **8**, 472
Benzenesulfonylhydrazone, 2,4,6-triisopropyl-
Bamford–Stevens reaction, **6**, 778
Benzenetellurinyl acetate
reactions with alkenes, **7**, 497
Benzenetellurol
synthesis, **7**, 774; **8**, 370
Benzenethiol
reactions with nitriles, **6**, 511
Benzenethiolate
Michael addition
4-*t*-butyl-1-cyanocyclohexene, **6**, 140
reaction with carvone, **6**, 141
Benzene-1,3,5-tricarbaldehyde
synthesis, **7**, 657
Benzene-2,4,6-tricarbaldehyde, chloro-
synthesis
Vilsmeier–Haack reaction, **2**, 785
Benzenium ion, *t*-butyl-
stability, **3**, 301
Benzenoid hydrocarbons
Birch reduction
dissolving metals, **8**, 493
Benzensulfonyl azide, *p*-acetamido-
diazo transfer reaction, **4**, 1033

Benzhydrol
  synthesis
    via benzophenone and ytterbium, **1**, 279
Benzhydrylamine, 4,4′-dimethoxy-
  reactions with π-allylpalladium complexes, **4**, 598
    stereochemistry, **4**, 623
Benzil
  aldol reaction with aliphatic ketones, **2**, 142
  hydrogenation
    cobalt catalysts, **8**, 154
  Knoevenagel reaction, **2**, 367
  monooxime
    hydrogenation, **8**, 148, 149
  photolysis
    with 1-*t*-butylthio-1-propyne, **5**, 163
  reaction with organometallic reagents, **1**, 153
  rearrangements, **3**, 821–836
    nonhydroxylic solvents, **3**, 824
    reaction conditions, **3**, 825
  reduction
    metal ions, **8**, 116
  synthesis
    via oxidative rearrangement, **7**, 829
Benzil, decafluoro-
  rearrangement, **3**, 825
Benzilic acid
  esters
    rearrangements, **3**, 823
  pharmacological activity, **3**, 826
  rearrangements, **3**, 821–836
    chemistry, **3**, 825
    labeling studies, **3**, 822
    mechanism, **3**, 822, 824
    migratory aptitudes, **3**, 822
    reaction conditions, **3**, 825
  Ritter reaction
    with benzonitrile, **6**, 276
  synthesis
    reaction conditions, **3**, 825
Benzimidazates, 2-allyloxy-
  metallated
    epoxidation, **1**, 829
Benzimidazole, 1-methyl-
  quaternary salts
    benzoin condensation, catalysis, **1**, 543
Benzimidazole, 2-methylthio-
  reaction with Grignard reagents, **3**, 461
Benzimidazole, 1-(phenylthiomethyl)-
  lithiation, **1**, 471
Benzimidazoles
  microbial hydroxylation, **7**, 79
  reaction with chloroform, **6**, 579
  reduction, **8**, 638
Benzimidazoline
  reduction
    hydride transfer, **8**, 291
Benzimidazolin-2-one
  reaction with crotonic acid, **3**, 306
Benzimidazolium salts
  synthesis
    via carboxylic acids, **8**, 277
Benzinden-1-ol, 2-nitro-
  synthesis
    Henry reaction, **2**, 329
Benz[*f*]indole, tetrahydro-
  synthesis
    via intramolecular Ritter reaction, **6**, 273
Benzisoxazoles
  reductive cleavage, **8**, 649
  reduction, **8**, 649
  synthesis, **4**, 439; **8**, 649
    via oxidation of primary aromatic amines, **7**, 739
2,1-Benzisoxazoles, 3-amino-
  synthesis, **4**, 436
1,2-Benzisoxazoles, 3-phenyl-
  synthesis, **4**, 439
5-Benzisoxazolylmethoxycarbonyl group
  amine-protecting group
    cleavage, **6**, 639
Benzoates
  alcohol protection
    carbohydrates, **6**, 657
  reduction
    stannanes, **8**, 824
    to 1,4-dihydrobenzoates, **3**, 613
Benzoates, α-keto-
  reduction
    stannanes, **8**, 824
Benzoazepinethione
  synthesis
    Friedel–Crafts reaction, **2**, 765
Benzoazepinone
  synthesis
    Friedel–Crafts reaction, **2**, 765
    palladium-catalyzed carbonylation, **3**, 1038
Benzobarrelene
  photoisomerization, **5**, 198
  substituted
    photoisomerizations, **5**, 210
Benzobarrelene, tetrachloro-
  synthesis
    via Diels–Alder reaction, **5**, 383
Benzobarrelenone
  photorearrangement, **5**, 229
Benzobicyclo[3.2.1]octanone, bromo-
  Favorskii rearrangement, **3**, 853
Benzocarbazoles
  lithiation
    addition reactions, **1**, 463
  synthesis
    via thermolysis, **5**, 725
Benzo[*c*]cinnoline dioxide
  synthesis, **8**, 364
Benzocyclobutanes
  synthesis, **3**, 161
    Parham-type cyclization, **3**, 251
Benzocyclobutanol
  synthesis, **3**, 265
Benzocyclobutanols, 1-vinyl-
  1,3-rearrangements, **5**, 1022
Benzocyclobutene
  electrocyclization, **5**, 721
  synthesis
    via arynes, **4**, 500
    via thermolysis of benzothiophene dioxides, **5**, 693
Benzocyclobutene, 1-acetoxy-1-methyl-
  thermolysis, **5**, 681
Benzocyclobutene, *trans*-1-acetoxy-2-phenyl-
  ring opening, **5**, 682
  synthesis
    via thermolysis of benzothiophene dioxides, **5**, 693
Benzocyclobutene, 1-acetyl-

isomerization, **5**, 681
Benzocyclobutene, *trans*-1-alkoxy-2-phenyl-
ring opening, **5**, 682
Benzocyclobutene, 1,2-dimethoxy-
ring opening, **5**, 683
Benzocyclobutene, diphenyl-
one-electron transfer, **5**, 77
Benzocyclobutene, 7,8-diphenyl-
Diels–Alder reactions, **5**, 391
Benzocyclobutene, [(methoxycarbonyl)amino]-
cycloaddition reactions
fulvenes, **5**, 627
Benzocyclobutene, 1-methoxy-1-phenyl-
rearrangement
anthracene synthesis, **5**, 694
Benzocyclobutene, *trans*-1-methoxy-2-phenyl-
ring opening, **5**, 682
Benzocyclobutene, 3-methyl
synthesis
via benzynes, **5**, 692
Benzocyclobutene, 6-methyl-
synthesis
via benzynes, **5**, 692
Benzocyclobutene, 7-methyl-
Diels–Alder reactions, **5**, 391
Benzocyclobutene, 1-phenyl-
ring opening, **5**, 682
Benzocyclobutene, 1-vinyl-
isomerization, **5**, 680
Benzocyclobutenecarboxylic acid
esters
synthesis *via* benzyne cyclization, **5**, 692
Benzocyclobutene-1,2-dione, 3-hydroxy-5-methyl-
synthesis
via cycloaddition, **5**, 693
Benzocyclobutenediones
cycloaddition reactions
metal catalyzed, **5**, 1202
Diels–Alder reactions, **5**, 395
Benzocyclobutene-1,2-diones
anthraquinones from, **5**, 690
Benzocyclobutenes
cycloaddition reactions
fulvenes, **5**, 627
electrocyclic ring opening, **5**, 1032, 1151
o-quinodimethane precursors, **5**, 691–694
Diels–Alder reactions, **5**, 386
synthesis, **5**, 692
via [2 + 2 + 2] cycloadditions, **5**, 1148, 1149
thermolysis, **5**, 1031
Benzocyclobuten-7-ol, 7-phenyl-
Diels–Alder reactions, **5**, 388
Benzocyclobutenols
synthesis
via benzyne, **5**, 692
via intramolecular cyclization, **5**, 692
Benzocyclobutenols, 1-alkyl-1-cyano-
ring opening
morphinan synthesis, **5**, 694
Benzocyclobutenols, *trans*-2-aryl-
synthesis
via benzyne cyclization, **5**, 692
Benzocyclobuten-1-one, 3,6-dimethoxy-
synthesis
via benzyne, **5**, 692
Benzocyclobutenones
o-quinodimethane precursors
Diels–Alder reactions, **5**, 388
synthesis
via benzyne, **5**, 692
Benzocyclobutenyl carbamate
Diels–Alder reactions, **5**, 390
Benzocycloheptenone
reduction
stereoselectivity, **8**, 6
Benzocyclohexenes
synthesis
via [2 + 2 + 2] cycloadditions, **5**, 1149
Benzocyclononadienes
synthesis
via intramolecular ene reactions, **5**, 20
Benzocyclooctatetraene
tricarbonyliron complexes
reaction with tetracyanoethylene, **4**, 710
Benzocyclooctenone
synthesis
Friedel–Crafts reaction, **2**, 753
Benzocyclopentenes
synthesis
via [2 + 2 + 2] cycloadditions, **5**, 1149
Benzocyclopropene
cycloaddition reactions
metal catalyzed, **5**, 1199
synthesis
via dihalocyclopropyl compounds, **4**, 1015
Benzodiazepine
synthesis
via nitrile imine cyclization, **4**, 1151
1*H*-1,5-Benzodiazepine, 4-formyl-2,2-dimethyl-
oxidative cleavage
potassium permanganate, **7**, 559
1,3,2-Benzodioxaborole
hydroboration, **8**, 719
1,3,2-Benzodioxaphosphole, 2,2,2-tribromo-
acid halide synthesis, **6**, 302
1,3,2-Benzodioxaphosphole, 2,2,2-trichloro-
acid chloride synthesis, **6**, 307
acid halide synthesis, **6**, 302
Benzodipyrrole
reduction
borohydrides, **8**, 618
Benzodithioles
reduction, **8**, 659
Benzo-1,3-dithiole-2-thiones
reduction
DIBAL, **8**, 661
Benzo[*j*]fluoranthene, 8-methyl-
synthesis
Friedel–Crafts cycloalkylation, **3**, 325
Benzofluorene
synthesis
via thermolysis, **5**, 721
Benzofluorenone
hydrogenation
palladium catalyst, **8**, 319
synthesis
Friedel–Crafts reaction, **2**, 757
Benzofuran, 2-alkylsulfonyl-
synthesis
Knoevenagel reaction, **2**, 363
Benzofuran, benzoyl-
synthesis

via chalcone, **7**, 829
Benzofuran, 2-bromo-
reaction with arylzinc bromide, **3**, 514
Benzofuran, dihydro-
synthesis, **3**, 265
via α,β-unsaturated ester, **4**, 73
Benzofuran, 2,3-dimethyl-
reduction
borohydrides, **8**, 628
Benzo[*c*]furan, 1,3-diphenyl-
Birch reduction, **8**, 627
hydrogenation, **8**, 626
Benzofuran, 2-ethyl-3-methyl-
reduction
borohydrides, **8**, 627
Benzofuran, hydroxy-
synthesis
via FVP, **5**, 732
Benzofuran, 5-methoxy-
electrolysis, **8**, 628
Benzofuran, 3-methoxycarbonyl-
reduction
dissolving metals, **8**, 626
Benzofuran, 3-methyl-
reduction
borohydrides, **8**, 628
Benzofuran, 3-methylenedihydro-
metal complexes, **4**, 526
synthesis
via arene–metal complexes, **4**, 526
Benzofuran, octahydro-
angular acetoxylation, **7**, 153
Benzofuran, phenyl-
synthesis
via oxidative rearrangement, **7**, 829
Benzofuran, 3-phenyl-
photoreduction, **8**, 628
Benzofuran, tetrahydro-
retro Diels–Alder reaction, **5**, 579
synthesis
via oxyanion-accelerated rearrangement, **5**, 1018
Benzofuran, 2-trimethylstannyl-
synthesis, **3**, 514
Benzofuran-3-ols, 3-methyl-2,3-dihydro-
synthesis
Friedel–Crafts reaction, **3**, 312
Benzofuranone
C-acylation, **2**, 836
O-acylation, **2**, 836
Benzofuran-2(3*H*)-one, 3-acetyl-
synthesis
via ketocarbenoids, **4**, 1057
2*H*-Benzo[*b*]furan-3-one, 2-benzyl-
ring scission, **3**, 830
3*H*-Benzo[*b*]furan-2-one, 3-benzyl-3-hydroxy-
synthesis, **3**, 831
3(2*H*)-Benzofuranones, 4-(1,3-dithian-2-yl)-
4,5,6,7-tetrahydro-
synthesis, **1**, 566
Benzofurans
coupling reactions
with alkyl Grignard reagents, **3**, 444
[3 + 2] cycloaddition reactions, **5**, 307
synthesis, **3**, 494; **7**, 628
via palladium(II) catalysis, **4**, 557
via sequential Michael ring closure, **4**, 262
Benzo[*b*]furans
electrochemical reduction, **8**, 628
reduction, **8**, 624
2-substituted derivatives
Vilsmeier–Haack reaction, **2**, 780
synthesis
via intramolecular organochromium reaction, **1**, 188
Benzo[*c*]furans
reduction, **8**, 626
Benzofuroxans
synthesis
via oxidation of primary aromatic amines, **7**, 739
Benzoic acid
methyl ester
acylation of boron-stabilized carbanions, **2**, 244
thallation, **4**, 841
Benzoic acid, alkyl-
synthesis
via benzyne, **4**, 510
Birch reduction
dissolving metals, **8**, 500
Benzoic acid, aryl-
Birch reduction
enolate generation, **3**, 51
Benzoic acid, 2-*t*-butyl-
hydrogenation
heterogeneous catalysis, **8**, 438
Benzoic acid, 4-(4′-chlorobutyl)-
intramolecular reductive alkylation
dissolving metals, **8**, 505
Benzoic acid, *p*-cyano-
methyl ester
photolysis with diphenylacetylene, **5**, 163
Benzoic acid, 2,5-diethylbenzoyl-
Friedel–Crafts reaction, **2**, 761
Benzoic acid, dihydro-
dianions
conjugate addition reactions, **4**, 111
synthesis
via reductive alkylation, **8**, 500
Benzoic acid, 2,5-dihydroxy-4-methoxy-
synthesis, **7**, 340
Benzoic acid, 2,4-dihydroxy-6-methyl-
methyl ester
synthesis, **2**, 821
Benzoic acid, 4,6-dimethoxy-2-(4′-methoxybenzyl)-
Friedel–Crafts reaction, **2**, 761
Benzoic acid, 2,4-dimethoxy-6-(2′-naphthyl)-
Friedel–Crafts reaction, **2**, 757
Benzoic acid, 4-fluoro-
hydrogenolysis, **8**, 903
Benzoic acid, hydroxy-
alkylation, **6**, 2
Benzoic acid, 4-isopropyl-
Birch reduction
dissolving metals, **8**, 500
Benzoic acid, mercapto-
synthesis, **4**, 444
Benzoic acid, 2-methoxy-
Birch reduction
dissolving metals, **8**, 502
Benzoic acid, 3-methoxy-
Birch reduction
dissolving metals, **8**, 501
Benzoic acid, *p*-nitro-

ethyl ester
Claisen condensation, **2**, 798
Benzoic acid, pentafluoro-
hydrogenolysis, **8**, 901
Benzoic acid, 2-phenyl-
Birch reduction
dissolving metals, **8**, 504
Benzoic acid, poly(alkylthio)-
synthesis, **4**, 441
Benzoic acid, poly(methylthio)-
synthesis, **4**, 441
Benzoic acid, 2,3,4,5-tetrafluoro-
synthesis
via carbonation of
bis(pentafluorophenyl)ytterbium, **1**, 277
Benzoic acid, tetrahydro-
synthesis
via Birch reduction, **8**, 500
Benzoic acid, 2,4,6-triisopropyl-
alkyl esters
metallation, **3**, 194
Benzoic acid anhydride
synthesis
via 4-benzylpyridine, **6**, 310
Benzoic esters, dihydro-
reactions with iron carbonyls, **4**, 666
Benzoin
oxidation
solid support, **7**, 846
oxime
hydrogenation, **8**, 148
Benzoin, deoxy-
reaction with α-selenoalkyllithium, **1**, 675
Benzoin, 2,4-dihydroxydeoxy-
Vilsmeier–Haack reaction, **2**, 790
Benzoin, *threo*-hydro-
synthesis, **7**, 441
Benzoin condensation, **1**, 541–579
catalysts, **1**, 543
electrophiles, **1**, 544
Benzomorphans
asymmetric synthesis
hydrogenation, **8**, 461
Benzomorpholines
synthesis, **8**, 654
Benzonitrile
acylation
synthesis of acetophenone, **1**, 498
photochemical cycloadditions
alkenes, **5**, 161
Benzonitrile, 4-alkoxy-
synthesis, **4**, 438
Benzonitrile, 4-chloro-
electrochemically induced $S_{RN}1$ reactions, **4**, 469
reaction with phenoxides, **4**, 469
Benzonitrile, 4-methyl-
hydrogenation, **8**, 252
Benzonitrile, 4-nitro-
oxide
1,3-dipolar cycloadditions, **4**, 1072
synthesis
via oxidation of 4-aminobenzonitrile, **7**, 737
Benzonitrile oxide
cycloaddition reactions
fulvenes, **5**, 630
tropones, **5**, 626
reaction with (α-oxyallyl)silane, **5**, 262
Benzonitrile oxide, 4-nitro-
reaction with 3,4,4-trimethyl-1-pentene, **5**, 262
Benzonitrilohexafluoro-2-propanide
reaction with methyl acrylate, **4**, 1081
Benzonitrilo-2-propanide
reaction with methyl acrylate, **4**, 1081
Benzonorbornadiene, 1,2-bis(trimethylsilyl)-
photoisomerization, **5**, 204
Benzonorbornadiene, 1,2-dimethyl-
photoisomerization, **5**, 204
Benzonorbornadiene, 6-methoxy-
bridgehead-substituted
photoisomerization, **5**, 204
Benzonorbornadiene, 1-methoxy-4-substituted
photoisomerization, **5**, 203
Benzonorbornadiene oxide
reduction
lithium triethylborohydride, **8**, 875
Benzonorbornadienes
deuteriated
photoisomerization, **5**, 204
photoisomerization, **5**, 197, 203, 205
stoichiometric complexes
with β-cyclodextrin, **5**, 210
substituted
photoisomerizations, **5**, 210
9-Benzonorbornenones
reduction
stereoselectivity, **8**, 5
Benzonorcaradiene
synthesis
via photoisomerization, **5**, 212
Benzo[*c*]phenanthrenes
synthesis
via electrocyclization, **5**, 720
Benzo[*c*]phenanthridine
synthesis
via arynes, **4**, 505
via $S_{RN}1$ reaction, **4**, 479
Benzo[*c*]phenanthridones
synthesis
via $S_{RN}1$ reaction, **4**, 479
Benzo[*k*]phenanthridones
synthesis
via photolysis, **5**, 728
Benzophenone
anil
reactions with Grignard reagents, **1**, 383
complex with trimethylaluminum, **1**, 78
electroreduction
chromium chloride, **8**, 133
hydrazone
reduction, Cram modification, **8**, 335
reduction, Henbest modification, **8**, 336
ketone dianion
reactions with esters and amides, **1**, 280
oxime
Beckmann rearrangement, **7**, 692
oxime, *O*-acyl
carboxyl radicals from, **7**, 719
photolysis, **7**, 720
photolysis
with 2-methylbut-1-en-3-yne, **5**, 164
with *cis*-1,4-polyisoprene, **5**, 161
reactions with boron stabilized carbanions, **1**, 498

reactions with dialkoxyboryl stabilized carbanions, **1**, 501
reactions with diethylzinc, **1**, 216
reactions with trimethylaluminum
role of Lewis acid, **1**, 325
reaction with 2-butene
oxetane formation, **5**, 152
reduction
boranes, **8**, 316
dissolving metals, **8**, 115, 308
ionic hydrogenation, **8**, 319
Wolff–Kishner reduction, **8**, 338
steroid esters
photolyses, **7**, 43
synthesis
carbonylation of phenyllithium, **3**, 1017
Benzophenone, 2-bromo-
reduction
hydrogen iodide, **8**, 323
Benzophenone, 4-bromo-
reaction with phenoxides, **4**, 469
Benzophenone, dilithio-
crystal structure, **1**, 25
Benzophenone, 4,4′-dimethoxy-
reduction
ionic hydrogenation, **8**, 319
Benzophenone, 4-phenyl-
photolysis, **5**, 154
Benzophenone, 2,3,4′-trihydroxy-
oxidative coupling
mechanism, **3**, 661
Benzophenone-4-carboxylic acid
dodecyl ester
photoinsertion, **7**, 42
Benzopinacol
oxidative cleavage, **7**, 707
2-Benzopyran-3-one
cycloaddition reactions
tropones, **5**, 618
Benzopyrans, dihydro-
synthesis
via benzocyclobutenes, **5**, 691
Benzopyrazine
electrochemical reduction, **8**, 643
Benzo[*a*]pyrene
dihydrodiols
synthesis, **7**, 333
Benzo[*e*]pyrene, 11,12-dihydro-
functionalization
with *N*-bromoacetamide, **4**, 356
Benzopyrrolizidine, dimethyl-
synthesis, **2**, 1039
Benzopyrylium salts
synthesis
Vilsmeier–Haack reaction, **2**, 790
1,4-Benzoquinone, 2-alkyl-
synthesis, **7**, 930
1,4-Benzoquinone, 2-alkyl-3-(2-pyridylthio)-
synthesis, **7**, 930
1,4-Benzoquinone, 5,6-dichloro-2,3-dicyano-
ether group removal
alcohol protection, **6**, 652
Benzoquinone, 2,3-dichloro-2,6-dicyano-
phenolic coupling, **3**, 661
Benzoquinone, 2,3-dichloro-5,6-dicyano-
debenzylation
benzyl ethers, **7**, 244
dehydrogenation, **7**, 135
1,4-Benzoquinone, 2,6-dihydroxy-
benzilic rearrangement, **3**, 829
1,4-Benzoquinone, hydroxy-
rearrangements, **3**, 828
Benzoquinone, 2,3,5,6-tetracyano-
hydride transfer
with dihydroacridine, **8**, 93
Benzoquinones
laser photolysis
with tetramethylallene, **5**, 154
reoxidant
Wacker process, **7**, 451
synthesis
via cyclobutenone ring opening, **5**, 690
via metal-catalyzed cycloaddition, **5**, 1202
1,4-Benzoquinones
Diels–Alder reactions, **5**, 342, 451
Lewis acid promoted, **5**, 339
hydrogenation, **8**, 152
catalytic, **8**, 142
synthesis
via metal-catalyzed cycloaddition, **5**, 1200
1*H*-2-Benzoselenin, 6,8-di-*t*-butyl-3,4-dihydro-4,4-dimethyl-
synthesis, **6**, 475
5*H*-[1]Benzoselenino[2,3-*b*]pyridine
synthesis, **6**, 472
Benzoselenophene
alkylation, **2**, 817
metallation, **1**, 644
Benzoselenophene, 2-lithio-
synthesis, **1**, 668
Benzosuberanone
synthesis
Friedel–Crafts reaction, **2**, 763
Benzosuberones
synthesis
Friedel–Crafts reaction, **2**, 755
Benzo systems
photoisomerization, **5**, 197
1,2,3,4-Benzotetrazine
synthesis, **7**, 743
Benzothiadiazoles
synthesis
via diazotization of aromatic amines, **7**, 740
4*H*-1,4-Benzothiazine, 2,3-dihydro-*N*-acyl-
aldol reaction
stereoselectivity, **2**, 211
1,4-Benzothiazines
reduction, **8**, 658
1,3-Benzothiazin-4-one
reduction
LAH, **8**, 658
Benzothiazole, 2-alkyl-
metallated
reactions, **2**, 495
reactions with carbonyl compounds, **2**, 496
synthesis, **6**, 490
Benzothiazole, 2-allyloxy-
reaction with Grignard reagents, **3**, 246
Benzothiazole, 2-chloro-
coupling reactions
with Grignard reagents, **3**, 461
Benzothiazole, 2-hydroxy-

reaction with copper alkynides, **3**, 283
1,3-Benzothiazole, 2-methyl-
synthesis
via $S_{RN}1$ reaction, **4**, 477
Benzothiazole, 2-methylthio-
coupling reactions
with Grignard reagents, **3**, 461
Benzothiazole, vinyl-
in synthesis
masked carbonyl derivative, **2**, 497
Benzothiazole, 2-vinyl-
addition reactions
with organolithium compounds, **4**, 76
Benzothiazoles
aromatic nucleophilic substitution, **4**, 432
metallated
reactions, **2**, 495
metallation, **6**, 541
reduction, **8**, 657
tandem vicinal difunctionalization, **4**, 252
Benzothiazoline, 2-phenyl-
reduction
unsaturated carbonyl compounds, **8**, 563
Benzothiazolium iodide
reaction with amines, **6**, 84
Benzothiazolium salts
catalysts
benzoin condensation, **1**, 543
enolization (attempted), **2**, 865
reduction, **8**, 657
Benzothiazolium salts, 2-alkoxy-
alcohol inversion, **6**, 22
Benzothiazolium salts, fluoro-
reaction with alcohols
iodination, **6**, 214
Benzothiazolium salts, 2-halo-
carbothioate synthesis, **6**, 438
Benzothiazolone, 2-lithio-
reaction with bis(trimethylsilyl) peroxide, **7**, 330
Benzothiazolones
synthesis, **4**, 444
Benzothiepinones
synthesis
Friedel–Crafts reaction, **2**, 765
Benzo[*c*]thiophene, 1,3-dihydro-
2,2-dioxide
synthesis *via* ketocarbenoids, **4**, 1057
Benzothiophene, hydroxy-
synthesis
via FVP, **5**, 732
Benzothiophene, 2-methyl-
ionic hydrogenation, **8**, 630
Benzothiophene, 3-methyl-
ionic hydrogenation, **8**, 630
Benzo[*b*]thiophene, 4-phenyl-
synthesis
via photocyclization–oxidation, **5**, 720
Benzo[*b*]thiophene, 7-phenyl-
synthesis
via photocyclization—oxidation, **5**, 720
Benzothiophene 2,2-dioxides, 2,5-dihydro-
thermolysis
benzocyclobutene synthesis, **5**, 693
Benzothiophenes
coupling reactions
with primary alkyl Grignard reagents, **3**, 447
methylation, **3**, 456
reduction, **8**, 629
synthesis, **7**, 628
via sequential Michael ring closure, **4**, 262
via $S_{RN}1$ reaction, **4**, 479
Benzotriazine
synthesis
via oxidation of amino-3-phenylindazoles, **7**, 743
Benzotriazinones
synthesis
via diazotization of aromatic amines, **7**, 740
Benzotriazole, 1-amino-
aryne precursor, **4**, 488
benzyne from, **7**, 482
nitration, **7**, 745
oxidation
to 1,2-didehydrobenzene, **7**, 743
Benzotriazole, 2-amino-
oxidation, **7**, 743
Benzotriazole, 1-benzoyloxy-
selective benzoylation, **6**, 337
Benzotriazole, 1-chloro-
oxidation
sulfoxides, **7**, 767
Benzotriazole, hydroxy-
esters
amidation, **6**, 394
Benzotriazoles
pyridinium chlorochromate
allylic alcohol oxidation, **7**, 264
reduction, **8**, 661
synthesis
via diazotization of aromatic amines, **7**, 740
via oxidation of primary aromatic amines, **7**, 739
Benzotriazolide, chlorophosphorylnitro-
phosphorylation, **6**, 620
Benzotriazolide, phosphorobis(1-hydroxy)-
phosphorylating agent, **6**, 619
Benzotriazolide, phosphorobis(nitro)-
phosphorylating agent, **6**, 619
Benzotriazoline
N-substituted
reaction with lithium alkynides, **3**, 282
Benzotricyclo[3.1.0.0^{2,6}]hex-3-ene
photoisomerization, **5**, 212
Benzotropolone
rearrangements, **3**, 818
Benzotropone
formation
aldol reaction, **2**, 144
1,2,5-Benzoxadiazoles
reduction, **8**, 664
1,3-Benzoxathian-4-one
synthesis
via intramolecular Pummerer rearrangement, **7**, 196
1,3-Benzoxathiolium salts, 2-substituted
Friedel–Crafts reaction, **2**, 737
1,3-Benzoxathiolium tetrafluoroborates
2-substituted
synthesis, **8**, 277
Benzoxazepinones
reduction
enolate generation, **3**, 51
synthesis
Mannich reaction, **2**, 956
Benzoxazines

Mannich reaction
with phenols, **2**, 969
synthesis
Mannich reaction, **2**, 968
1,4-Benzoxazines
reduction, **8**, 653
1,3-Benzoxazines, 2-aryl-
ring–chain tautomerism, **2**, 969
1,4-Benzoxazin-3-one, *N*-alkyldihydro-
reduction
LAH, **8**, 654
Benzoxazole
Friedel–Crafts reaction, **2**, 743
reduction, **8**, 650
synthesis
via Beckmann reaction, **7**, 698
Benzoxazole, 2-allenyl-
synthesis, **2**, 86
Benzoxazole, 2-(1,1-dimethylpropargyl)-
synthesis, **2**, 87
Benzoxazole, 2-(1-methylpropargyl)-
synthesis, **2**, 86
Benzoxazole, 2-phenacyl-
synthesis, **6**, 534
Benzoxazolium salts, chloro-
chlorination
alkyl alcohols, **6**, 206
Benzoxepines
synthesis
via $S_{RN}1$ reaction, **4**, 479
1(2*H*)-Benzoxocin, 2,6-epoxy-
synthesis
via Wharton reaction, **8**, 928
Benzoyl *t*-butyl nitroxide
quinones
synthesis, **7**, 349
Benzoyl chloride
synthesis
via benzaldehyde, **6**, 308
Benzoyl chloride, 2-fluoro-
Friedel–Crafts reaction
toluene, **2**, 736
Benzoyl chloride, 4-nitro-
reduction
metal hydrides, **8**, 290
Benzoyl chloride, 2,4,6-trichloro-
mixed anhydride synthesis, **6**, 329
Benzoyl cyanide
synthesis
via benzoyl chloride, **6**, 233
Benzoyl hypobromite, *m*-chloro-
synthesis, **7**, 535
Benzoylium ion
NMR data
study of stability, **2**, 734
Benzoyl peroxide
α-hydroxylation
esters, **7**, 182
ketones, **7**, 163
reaction with enamines
generation of α-benzoyloxy ketones, **7**, 171
triphenylphosphine compound
reaction with alcohols, **6**, 22
N-Benzoylphenylalanyl group
removal
chymotrypsin, **6**, 643
Benzoylpropionates
alcohol protection
cleavage, **6**, 658
Benzoyl xanthate
photolysis
radical addition reactions, **4**, 749
Benzpinacol
synthesis
via triphenylchromium complex, **1**, 176
Benzpinacolone
label redistribution
pinacol rearrangement, mechanism of, **3**, 723
Benzylacetone
reduction
transfer hydrogenation, **8**, 555
Benzyl alcohol, 4-methoxy-
Birch reduction
dissolving metals, **8**, 514
Benzyl alcohols
esters
electrohydrogenolysis, **8**, 974
protecting groups, **8**, 956
hydrogen donor
transfer hydrogenation, **8**, 551
hydrogenolysis, **8**, 956
oxidation, **7**, 306, 318
4-(dimethylamino)pyridinium chlorochromate, **7**, 269
solid support, **7**, 841, 844
reduction
dissolving metals, **8**, 971
Lewis acid activated, **8**, 966
sodium borohydride, **8**, 968
Benzyl alcohols, hydroxy-
nitrile synthesis, **6**, 235
Benzylamine
imines
isomerization, **6**, 721
Benzylamine, *N*-acyl-*o*-chloro-
photoinduced cyclization, **4**, 477
Benzylamine, α-alkyl-
stereoselective synthesis, **3**, 76
Benzylamine, *N*,*N*-dialkyl-
metallation
addition reactions, **1**, 463
Benzylamine, 4-*N′*,*N′*-dimethylamino-*N*,*N*-dimethyl-
synthesis, Mannich reaction, **2**, 961
Benzylamine, methyl-
hydrogenation, **8**, 146
Benzylamines
hydrogenolysis, **8**, 957
hydrogenolytic asymmetric transamination, **8**, 147
reduction
dissolving metals, **8**, 971
Benzylation
sulfur- and selenium-stabilized carbanions, **3**, 88
Benzyl bromide
vinyl substitutions
palladium complexes, **4**, 835
Benzyl bromide, 2,6-dichloro-
oxidation, **7**, 665
Benzyl carbamates
protecting groups
peptide synthesis, **6**, 635
Benzyl carbonate

alcohol protection
cleavage, **6**, 659
Benzyl chloride
reaction with methyl acrylate
palladium complexes, **4**, 842
vinyl substitutions
palladium complexes, **4**, 835
Benzyl chloride, 4-nitro-
Hass–Bender reaction, **7**, 660
Benzyl cyanide, α,α-bis(imidazolyl)-
synthesis, **6**, 579
Benzyl esters
carboxy-protecting groups, **6**, 667
cleavage
trimethylsilyl chlorochromate, **7**, 285
Benzyl group
alcohol protection, **6**, 23
amino acid protecting group
hydrogenolysis, **8**, 958
N-Benzyl group
amine protection, **6**, 644
S-Benzyl group
thiol protection, **6**, 664
Benzylic acetals
reduction
Lewis acid activated, **8**, 966
Benzylic anions
trimethylsilyl-stabilized
Michael donors, **4**, 259
Benzylic compounds
microbial oxidation, **7**, 75
Benzylic electrophiles
reaction with organocopper compounds, **3**, 220
Benzylic ethers
reduction
Lewis acid activated, **8**, 966
Benzylic halides
Barbier-type reactions
organosamarium compounds, **1**, 256
carbonylation
formation of esters, **3**, 1028
palladium catalysts, **3**, 1021
cleavage
zinc, **8**, 972
hydrogenolysis, **8**, 955–981
Raney nickel, **8**, 964
Kornblum oxidation, **7**, 653
reduction
sodium borohydride, **8**, 967
Benzylic ketals
reduction, **8**, 971
Benzylic thiols
reduction
Lewis acid activated, **8**, 966
Benzylidene
transition metal complexes
reaction with alkenes, **4**, 980
Benzylidene acetal, 4-methoxy-
reductive cleavage
sodium cyanoborohydride, **8**, 969
Benzylidene acetals
hydrogenation, **8**, 212
Benzylidene transfer
Simmons–Smith reaction, **4**, 968
N-Benzylidenimines
amine protection, **6**, 645
α-Benzylidines
aldol reaction, **2**, 147
Benzyl iodide
vinyl substitutions
palladium complexes, **4**, 835
Benzyloxycarbonyl group
amino acid protecting group
hydrogenolysis, **8**, 958
deprotection, **8**, 957
protecting group
peptide synthesis, **6**, 632, 635
Benzyloxymethyl group
alcohol protection, **6**, 647
Benzyl phenylpropiolate
synthesis
via of 2-acyloxypyridinium salts, **6**, 331
Benzyl tellurocyanate
photooxidation, **7**, 777
Benzyl xanthate
photolysis
radical addition reactions, **4**, 748
Benzyne, 4-chloro-
reaction with ammonia, **4**, 494
Benzyne, 3,6-dimethoxy-
reactions with acetonitrile, **4**, 492
Benzyne, 3-fluoro-
Diels–Alder reactions, **5**, 382
Benzyne, 3-isopropyl-
addition reactions
lithium piperidide, **4**, 493
Benzyne, 3-methyl-
Diels–Alder reactions, **5**, 381
Benzyne, tetrachloro-
Diels–Alder reactions, **5**, 383
Benzynes
ab initio calculations, **4**, 483
carbocupration, **4**, 872
carbolithiation, **4**, 872
cyclization, **4**, 499
Diels–Alder reactions, **5**, 379–385
double cyclization
in synthesis, **4**, 505
electrophilicity, **4**, 484
enthalpy of formation, **4**, 484
infrared spectrum, **4**, 483
intramolecular trapping by carbanions, **5**, 692
microwave spectrum, **4**, 484
relative reactivity
towards nucleophiles, **4**, 491
structure, **4**, 483
substituent effects
kinetic stability, **4**, 492
substituted
generation, **4**, 489
nucleophilic addition, **4**, 494
regioselective generation, **4**, 489
synthesis, **7**, 743
in thermal isomerization, **5**, 736
tandem vicinal difunctionalization, **4**, 250
Berberines
synthesis
via directed metallation, **1**, 463
Berbin-8-one
synthesis
carbonylation, **3**, 1038
Bergamotene

synthesis, **3**, 108, 249
Berninamycinic acid
synthesis
via regioselective metallation, **1**, 474
Berson–Salem subjacent orbital effect
1,3-sigmatropic rearrangements, **5**, 1009
Beryllium, dialkyl-
hydride donor
reduction of carbonyls, **8**, 100
Beryllium, dimethyl-
crystal structure, **1**, 13
Beryllium acetylide
crystal structure, **1**, 21
Beryllium compounds
Lewis acid complexes
structure, **1**, 287
Betaines
sulfur ylide reactions
carbonyl epoxidation, **1**, 820
synthesis, **6**, 190
Betulaprenols
synthesis, **3**, 170
Betulinic acid
ring A contraction, **3**, 834
[10.10]Betweenanene
epoxidation, **7**, 364
Betweenanenes
synthesis, **3**, 591, 946
intramolecular acyloin coupling reaction, **3**, 627
Beyerene
rearrangement, **3**, 715
Biacetyl
photochemistry, **5**, 154
reactions with alkanes, **7**, 7
Biaryls
formation in phenol ether couplings, **3**, 668
nucleophilic substitution
organometallic compounds, **4**, 427
synthesis, **3**, 663
aryl radical insertion, **3**, 677
via $S_{RN}1$ reaction, **4**, 471
Biaryls, amino-
synthesis
via $S_{RN}1$ reaction, **4**, 469–471
Biaryls, hydroxy-
synthesis
via $S_{RN}1$ reaction, **4**, 469–471
Bicyclic alcohols
synthesis
via organoytterbium compounds, **1**, 278
via samarium diiodide, **1**, 262
Bicyclic compounds
inside–inside
synthesis *via* ene reaction with methyl propiolate, **5**, 8
Wagner–Meerwein rearrangment, **3**, 706
Bicyclization
dienes, **5**, 1172
diynes, **5**, 1171
enynes, **5**, 1165–1170
promoted by Group IV metals, **5**, 1169
zirconium-promoted, **5**, 1163–1183
polyalkenes
mechanism, **3**, 374
Bicyclo[5.*n*.0]alkanes
functionalized
synthesis *via* Cope rearrangement, **5**, 979–982
Bicycloalkanones
inside–outside
via photocycloaddition reactions, **5**, 137
Bicyclo[4.1.0]alkenes
synthesis
via photoisomerization, **5**, 211
Bicycloalkenes, vinyl-
Cope rearrangement, **5**, 812–819
Bicycloaromatization
general strategy, **2**, 623
Bicyclobutane
deprotonation
n-butyllithium, **1**, 10
Bicyclo[1.1.0]butane, 1-cyano-
cycloaddition reactions, **5**, 1186
Bicyclo[1.1.0]butane, 1-methoxycarbonyl-
cycloaddition reactions, **5**, 1186
Bicyclo[1.1.0]butane, 1-methyl-
cycloaddition reactions, **5**, 1186
Bicyclo[1.1.0]butane, 1,2,2-trimethyl-
synthesis
via dihalocyclopropanes, **4**, 1013
Bicyclo[1.1.0]butanes
cycloaddition reactions
metal-catalyzed, **5**, 1185
synthesis
Wurtz reaction, **3**, 414, 422
Bicyclo[4.4.0]decadiene
synthesis, **3**, 390
Bicyclo[5.3.0]deca-2,10-diene
synthesis, **3**, 399
Bicyclo[4.4.0]decane
synthesis, **3**, 389, 391
cis-Bicyclo[4.4.0]decane-3,9-dione
intramolecular aldol
equilibrium, **2**, 169
Bicyclo[5.3.0]decanes
synthesis
via Cope rearrangement, **5**, 982
via photocycloaddition, **5**, 669
Bicyclo[4.3.1]decan-10-one
synthesis, **3**, 58
Bicyclo[4.2.2]deca-3,7,9-triene
dimerization, **5**, 66
Bicyclo[5.3.0]decatrienone
synthesis
via ketocarbenoids, **4**, 1055
Bicyclo[4.4.0]decene
synthesis, **3**, 393
polyene cyclization, **3**, 345
Bicyclo[6.2.0]dec-2-ene
thermolysis, **5**, 686
Bicyclo[4.4.0]decenol
synthesis, **3**, 392
Bicyclo[4.4.0]decen-3-one
synthesis
via Lewis acid allylation, **4**, 155
Bicyclo[6.4.0]dodecanes, alkyl-
synthesis
via [4 + 4] cycloaddition, **5**, 640
Bicyclo[6.4.0]dodecen-3-ones
synthesis
via organosilanes and α,β-enones, **4**, 99
Bicyclogermacrene
transannular cyclization, **3**, 390

Bicyclo[3.2.0]hepta-3,6-diene, 1-methoxy-
reaction with hexacarbonylpropynedicobalt complex
Pauson–Khand reaction mechanism, **5**, 1039
Bicyclo[2.2.1]hepta-2,5-diene-2,3-dicarboxylic acid
dimethyl ester
hydrogenation, **8**, 440
Bicyclo[3.2.1]heptadienes
synthesis
via Cope rearrangement, **5**, 804
Bicyclo[3.2.0]hepta-1,4-dien-3-ones
synthesis
via [2 + 2 + 2] cycloaddition, **5**, 1134
Bicyclo[4.1.0]heptane, 7-alkoxy-7-phenyl-
synthesis, **6**, 475
Bicyclo[2.2.1]heptane, 7-carboxy-
microbial hydroxylation, **7**, 59
Bicyclo[2.2.1]heptane, 2-methoxycarbonyl-
synthesis
via cycloaddition of bicyclo[2.1.0]pentanes, **5**, 1186
Bicyclo[2.2.1]heptane-7-carboxylic acid
synthesis, **3**, 903
Bicyclo[4.1.0]heptane-3,4-diones, 7-halo-
tautomerism, **5**, 714
Bicyclo[2.2.1]heptanes
synthesis
via cycloaddition of bicyclo[2.1.0]pentanes, **5**, 1186
via [3 + 2] cycloaddition reactions, **5**, 286
Bicyclo[3.1.1]heptanes
synthesis, **3**, 901
Bicyclo[3.2.0]heptanes
synthesis
via [4 + 4] cycloaddition, **5**, 639
Bicyclo[4.1.0]heptanes
synthesis, **1**, 664
via photocycloaddition, **5**, 669
Bicyclo[2.2.1]heptanol
carbocations
rearrangement, **3**, 707
Bicyclo[2.2.1]heptanone
lithium enolate
exoalkylation, **3**, 17
synthesis, **3**, 19; **5**, 1104
via intramolecular ene reactions, **5**, 21
via tandem Michael reactions, **4**, 121
Bicyclo[2.2.1]heptan-2-one
oximes
reduction, dissolving metals, **8**, 124
reduction
dissolving metals, **8**, 116, 120
Bicyclo[2.2.1]heptan-6-one
synthesis, **6**, 144
Bicyclo[4.1.0]heptanone
Knoevenagel reaction, **2**, 368
Bicyclo[2.2.1]heptanone, α-diazo-
Wolff rearrangement, **3**, 900
Bicycloheptan-2-one, 7,7-dimethyl-
reduction
dissolving metals, **8**, 121
Bicyclo[2.2.1]heptan-2-one, 1-methyl-
reduction
dissolving metals, **8**, 121
Bicyclo[2.2.1]heptan-2-one, 6-methyl-
synthesis
via intramolecular ene reactions, **5**, 21
Bicyclo[2.2.1]hept-2-enes
electrophilic attack, **4**, 330
oxidative cleavage
potassium permanganate, **7**, 558
Bicyclo[2.2.1]hept-5-enes
thermolysis, **5**, 558
Bicyclo[3.2.0]hept-6-enes
Pauson–Khand reaction, **5**, 1046, 1052
rearrangement, **5**, 1016
Bicyclo[4.1.0]hept-2-enes
synthesis
via photoisomerization, **5**, 196
Bicyclo[4.1.0]hept-3-enes
photoisomerization, **5**, 196
trans-Bicyclo[4.1.0]hept-3-enes
synthesis
Ramberg–Bäcklund rearrangement, **3**, 876
Bicyclo[4.2.0]hept-2-enes
thermal isomerizations
via retro Diels–Alder reactions, **5**, 586
Bicyclo[4.1.0]hept-2-enes, 7-(1-alkenyl)-
Cope rearrangement, **5**, 991
Bicyclo[4.1.0]heptenes, dibromo-
rearrangement
norbornene derivative, **4**, 1012
Bicyclo[2.2.1]hept-2-enes, 5-methylene-
hydrogenation
heterogeneous catalysis, **8**, 433
Bicycloheptenes, 7-vinyl-
Cope rearrangement, **5**, 815
Bicyclo[4.1.0]hept-2-enes, 7-*endo*-vinyl-
Cope rearrangement, **5**, 991
exo-Bicyclo[3.2.0]hept-2-en-7-ol
rearrangement, **5**, 1016
Bicyclo[2.2.1]heptenols
Cope rearrangement
product aromatization, **5**, 791
Bicycloheptenols, 2-vinyl-
oxy-Cope rearrangement, **5**, 815
Bicyclo[2.2.1]hept-5-en-2-one
reduction
dissolving metals, **8**, 121
Bicyclo[2.2.1]heptenones
photoisomerizations, **5**, 224, 228
Bicyclo[4.1.0]heptenones, 7-halo-
tautomerism, **5**, 714
Bicyclo[3.2.0]hept-7-one, 1-phenyl-
reaction with vinyllithium, **5**, 1022
Bicyclo[8.8.8]hexacosane
synthesis
intramolecular acyloin coupling reaction, **3**, 628
Bicyclo[2.2.0]hexadiene
reduction
diimide, **8**, 475
Bicyclo[2.2.0]hexa-2,5-diene, hexamethyl-
hydrogenation
heterogeneous catalysis, **8**, 428
Bicyclo[2.1.1]hexane
synthesis, **3**, 900
Bicyclo[2.2.0]hexane
synthesis, **3**, 901
Bicyclo[3.1.0]hexane
synthesis, **1**, 664
via reductive cyclization, **4**, 1007
via ring opening, **5**, 708
Bicyclo[3.1.0]hexane, dibromo-

cyclopropyl–allyl rearrangement, **4**, 1018
Bicyclo[2.2.1]hexane, 1-vinyl-
Ritter reaction, **6**, 273
Bicyclo[2.1.1]hexane-2-carboxylic acid
synthesis, **3**, 903
Bicyclo[2.1.1]hexane-6-carboxylic acid, *exo*-1,5,5-trimethyl-
synthesis, **3**, 900
Bicyclo[2.2.0]hexan-2-ol
oxidative rearrangement, **7**, 834
Bicyclo[2.1.1]hexan-2-one
synthesis, **7**, 834
Bicyclo[3.1.0]hexan-2-ones, 6-(1-alkenyl)-
Cope rearrangement, **5**, 987
Bicyclo[3.1.0]hexan-2-ones, 6-vinyl-
enol derivatives
Cope rearrangements, **5**, 804
Bicyclo[2.2.0]hex-1(4)-ene
dimerization, **5**, 66
Bicyclo[3.1.0]hex-2-ene, 6-*endo*-vinyl-
Cope rearrangement, **5**, 984
Bicyclo[2.1.1]hexenes
synthesis
Ramberg–Bäcklund rearrangement, **3**, 874
Bicyclo[2.2.0]hex-2-enes
synthesis
via photolysis, **5**, 737
thermolysis, **5**, 678
Bicyclo[3.1.0]hex-2-enes, 6-(1-alkenyl)-
Cope rearrangement, **5**, 984–991
synthesis, **5**, 990
via Cope rearrangement, **5**, 985
Bicyclo[3.1.0]hexenone
synthesis
via photolysis, **5**, 730
Bicyclo[2.2.0]hexenones, amino-
rearrangement, **5**, 732
Bicyclohumulenediol
synthesis, **3**, 404
Bicyclomycin
synthesis
Ugi reaction, **2**, 1096
via Peterson methylenation, **1**, 732
Bicyclo[3.2.2]nona-2,6-diene
synthesis
via Cope rearrangement, **5**, 991, 993
cis-Bicyclo[4.3.0]nona-2,4-diene
photolysis, **5**, 737
trans-Bicyclo[4.3.0]nona-2,4-diene
synthesis
via thermal rearrangement, **5**, 716
Bicyclo[5.2.0]nona-2,8-diene
synthesis
via photoisomerization, **5**, 709
Bicyclo[3.2.2]nona-6,8-dien-3-one
synthesis
via [4 + 3] cycloaddition, **5**, 608
Bicyclo[3.3.1]nonane
functionalization
alkylthio, **7**, 14
synthesis
via Michael addition, **4**, 27
Bicyclo[4.2.1]nonane
bridged
synthesis *via* nitrone cyclization, **4**, 1114
Bicyclo[4.3.0]nonane
synthesis
via [3 + 2] cycloaddition reactions, **5**, 304
Bicyclo[3.3.1]nonanone
reduction
dissolving metals, **8**, 118
Bicyclo[3.3.1]nonan-3-one, 2-bromo-
Favorskii rearrangement, **3**, 853
Bicyclo[6.1.0]nonan-5-one, 4-diazo-*trans*-
irradiation, **3**, 905
Bicyclo[4.3.0]nonan-2-one, 1-methyl-
oxime
Beckmann fragmentation, **7**, 698
Bicyclo[4.3.0]nonan-3-ones
synthesis
via organosilanes and α,β-enones, **4**, 98
Bicyclo[3.2.2]nona-2,6,8-triene
photoisomerization, **5**, 196
Bicyclo[5.2.0]nonatrienones
synthesis
via ketocarbenoids, **4**, 1056
Bicyclo[3.3.1]non-2-ene
epoxide
transannular hydride shifts, **3**, 735
Bicyclo[3.2.2]non-6-en-3-one
synthesis
via Cope rearrangement, **5**, 992
Bicyclo[3.3.1]nonenone
synthesis
aldol cyclization, **2**, 162
Bicyclo[3.3.1]non-2-en-4-one, 5-(2-ethylallyl)-1-methyl-
synthesis, **3**, 23
Bicyclo[3.2.2]nonenone, 1-methoxy-
bridged
photoisomerizations, **5**, 226
Bicyclo[4.3.0]nonen-3-one, vinyl-
synthesis
via organosilanes and α,β-enones, **4**, 99
Bicyclo[3.2.2]nonenones
photoisomerizations, **5**, 225, 228
Bicyclo[2.2.2]octa-1,4-diene
Pauson–Khand reaction, **5**, 1049
Bicyclo[2.2.2]octa-2,5-diene
flash vapor pyrolysis, **5**, 571
Bicyclo[3.2.1]octadiene
synthesis
via cyclopropanation/Cope rearrangement, **4**, 1049
Bicyclo[3.2.1]octa-2,6-diene
photoisomerization, **5**, 205
substituted
synthesis *via* Cope rearrangement, **5**, 985–988
synthesis
via Cope rearrangement, **5**, 794, 984, 987
via [4 + 3] cycloaddition reaction, **5**, 597
Bicyclo[3.3.0]octadiene
synthesis, **3**, 489
Bicyclo[4.2.0]octa-2,4-diene
tautomerism, **5**, 714
Bicyclo[2.2.2]octa-2,5-diene, 1,4-bis(methoxycarbonyl)-
thermolysis, **5**, 571
Bicyclo[4.2.0]octadiene, 1-cyano-
synthesis
via photocycloaddition, **5**, 161
Bicyclo[2.2.1]octadienone
elimination reactions, **5**, 558
Bicyclo[3.3.0]octa-2,6-diol

synthesis, **3**, 382
Bicyclo[3.3.0]octanecarbaldehyde
synthesis, **3**, 383
Bicyclo[3.3.0]octanedione
synthesis
via photoisomerization, **5**, 233
Bicyclo[2.2.2]octanes
synthesis
via cyclopropane ring opening, **4**, 1043
via photocycloaddition, **5**, 657
Bicyclo[3.2.1]octanes
ring formation, **3**, 380
synthesis
via Cope rearrangement, **5**, 993
via cyclopropane ring opening, **4**, 1043
via photocycloaddition, **5**, 657
via Pummerer rearrangement, **7**, 199
Bicyclo[3.3.0]octanes
ring formation, **3**, 380
synthesis
via Claisen rearrangement, **5**, 833
via [3 + 2] cycloaddition reactions, **5**, 290, 304
via photocycloaddition, **5**, 654, 657
Bicyclo[4.2.0]octanes
aromatization
benzocyclobutene synthesis, **5**, 692
rearrangement, **3**, 714
synthesis, **3**, 382
via photocycloaddition, **5**, 657
Bicyclo[2.2.2]octanes, 2-*exo*-methylene-6-vinyl-
Cope rearrangement, **5**, 815
Bicyclo[3.3.0]octanol
synthesis, **3**, 384
Bicyclo[2.2.2]octanone
synthesis, **3**, 19
via intramolecular ene reactions, **5**, 21
via Michael addition, **4**, 30
via tandem Michael reactions, **4**, 121
Bicyclo[3.2.1]octan-2-one
Beckmann rearrangement, **7**, 695
Bicyclo[3.3.0]octanone
synthesis, **3**, 139
C—H insertion reactions, **3**, 1060
via [2 + 2 + 2] cycloaddition, **5**, 1131
cis-Bicyclo[3.3.0]octan-2-one
synthesis
via metal-catalyzed cycloaddition, **5**, 1192
Bicyclo[5.1.0]octanone
Knoevenagel reaction, **2**, 368
Bicyclo[3.2.1]octan-3-ones, 2-bromo-
Favorskii rearrangement, **3**, 852
Bicyclo[4.2.0]octa-2,4,7-triene
tautomerism, **5**, 715
Bicyclo[4.2.0]octa-1,3,5-triene
o-quinodimethane precursors
Diels–Alder reactions, **5**, 386
Bicyclo[2.2.2]octene
dimerization, **5**, 65
Pauson–Khand reaction, **5**, 1051
synthesis
via Cope rearrangement, **5**, 812
via cyclization of alkynes, **1**, 605
via Diels–Alder reactions, **5**, 329
via organosilanes and α,β-enones, **4**, 99
Bicyclo[3.2.1]oct-2-ene
allylic oxidation, **7**, 95
Bicyclo[3.3.0]octene
synthesis, **3**, 380
Bicyclo[3.3.0]oct-1-ene
Pauson–Khand reaction, **5**, 1052
Bicyclo[3.3.0]oct-2-ene
Pauson–Khand reaction, **5**, 1047
Bicyclo[4.2.0]oct-7-ene
thermolysis, **5**, 678
Bicyclo[2.2.2]oct-2-ene, 5,5-dicyanomethylene-
photoisomerization, **5**, 196
Bicyclo[2.2.2]octene, vinyl-
Cope rearrangement, **5**, 794
Bicyclo[2.2.2]oct-2-ene-2,3-dicarboxylic acid
dimethyl ester
hydrogenation, **8**, 427
Bicyclo[3.3.0]oct-1(5)-ene-2,6-dione
synthesis, **1**, 567
Bicyclo[2.2.2]octenol
methanesulfonates
rearrangement, **3**, 717
Bicyclooctenone
synthesis
via Cope rearrangement, **5**, 804
Bicyclo[2.2.2]octenone
photoisomerizations, **5**, 218, 224, 228
Bicyclo[2.2.2]oct-5-en-2-one
Baeyer–Villiger reaction, **7**, 683
photoisomerizations, **5**, 200
Bicyclo[3.2.1]oct-6-en-3-one
synthesis
via [4 + 3] cycloaddition, **5**, 603
Bicyclo[3.3.0]octenone
addition reaction with 2-nitrobut-2-ene, **4**, 102
synthesis
aldol cyclization, **2**, 162
Bicyclo[3.3.0]oct-1-en-3-one
synthesis
via Pauson–Khand reaction, **5**, 1053, 1060
Bicyclo[3.2.1]oct-6-en-3-one, 8-alkylidene-
synthesis
via [4 + 3] cycloaddition, **5**, 604
Bicyclo[2.2.2]octenone, 1-methoxy-
photoisomerization, **5**, 226, 233
Bicyclooctenone, (siloxymethyl)-
reactions with allylic sulfinyl carbanions, **1**, 522
Bicyclo[10.3.0]-$\Delta^{1,15}$-pentadecen-14-one
synthesis
via Wacker oxidation, **7**, 455
Bicyclo[2.1.0]pentanes
cycloaddition reactions
metal-catalyzed, **5**, 1186
diradicals
via photolytic rearrangement, **5**, 914
synthesis, **3**, 901
Bicyclo[2.1.0]pentan-2-one
vinylogous Wolff rearrangement, **3**, 906
Bicyclo[2.1.0]pent-2-ene
thermolysis, **5**, 678
Bicyclo[4.3.0]proline
synthesis, **7**, 731
Bicyclopropylidene
cycloaddition reactions, **5**, 71
dimerization, **5**, 65
reaction with tetracyanoethylene, **5**, 78
Bicyclo[4.3.0] rings
polyene cyclization, **3**, 359

Bicyclo[4.4.0] rings
  polyene bicyclization, **3**, 360
cis-Bicyclo[5.4.0]undeca-8,10-diene
  synthesis
    via photolysis, **5**, 717
trans-Bicyclo [7.2.0]undeca-2,10-diene
  synthesis
    via electrocyclization, **5**, 717
Bicyclo[5.4.0]undecane
  synthesis
    via Cope rearrangement, **5**, 815, 982
Bicyclo[6.3.0]undecane
  synthesis, **3**, 406
Bicyclo[4.4.1]undecanone
  synthesis, **3**, 58
    via [6 + 4] cycloaddition, **5**, 624
Bicyclo[4.4.1]undecene
  dimerization, **5**, 65
  synthesis
    via [6 + 4] cycloaddition, **5**, 635
Bicyclo[5.3.1]undecene
  synthesis
    via anionic oxy-Cope rearrangement, **1**, 884
Bicyclo[4.4.1]undecenone
  synthesis
    via [6 + 4] cycloaddition, **5**, 620
Bicyclo[5.3.1]undecenone
  synthesis
    via Cope rearrangement, **5**, 1028
Bicyclo[5.4.0]undecen-3-one
  synthesis
    via Lewis acid allylation, **4**, 155
Bicyclo[6.3.0]undecen-3-one
  synthesis
    via organosilanes and α,β-enones, **4**, 99
Bifunctional conjunctive reagents
  [3 + 2] cycloaddition reactions, **5**, 287
  trimethylenemethane from, **5**, 298–308
Bifurandiones
  synthesis
    via [2 + 2 + 2] cycloaddition, **5**, 1138
Bile acids
  microbial hydroxylation, **7**, 73
Bile pigments
  synthesis
    Eschenmoser coupling reaction, **2**, 874
Bilobalide
  synthesis
    via Diels–Alder reaction, **5**, 356
Bilobolide acetate
  synthesis, **8**, 824
Binaphthols
  Diels–Alder reactions, **5**, 376
1,1′-Binaphthyl, 2,2′-dihydroxy-
  asymmetric reduction
    aluminum hydrides, **8**, 545
  lithium aluminum hydride modifier, **8**, 162
    chiral modification of reducing agents, **8**, 159
  reduction
    aluminum hydrides, **8**, 545
Binaphthyl, (hydroxymethyl)-
  synthesis, **4**, 427
2,2′-Binaphthyl-3,3′-dicarboxylic acid
  Friedel–Crafts reaction, **2**, 757
Binaphthyls
  chiral synthesis, **4**, 427
  synthesis, **3**, 499, 503
    nickel catalysts, **3**, 229
    use of vanadium oxytrichloride, **3**, 664
2,2′-Binaphthyls, hydroxy-
  synthesis
    via $S_{RN}1$ reaction, **4**, 477
Binaphthyls, tetrahydroxy-
  synthesis
    use of potassium ferricyanide, **3**, 664
Biochemical reduction
  unsaturated carbonyl compounds, **8**, 558
Biomimetic reduction
  allylic compounds, **8**, 977
  NAD(P)H models, **8**, 561
Biomimetic synthesis
  Wagner–Meerwein rearrangement, **3**, 714
Biotin
  synthesis, **3**, 151; **8**, 608
    from thiazolines and enolates, **2**, 946
    via INOC reaction, **4**, 1080, 1128
    via stereocontrolled reaction, **1**, 350
9,9′-Biphenanthryl, 10,10′-dihydroxy-
  lithium aluminum hydride modifier, **8**, 164
Biphenol
  synthesis, **3**, 664
Biphenyl
  alkylation
    Friedel–Crafts reaction, **3**, 304
  Birch reduction
    dissolving metals, **8**, 496
  chiral synthesis, **4**, 427
  2,2′-dianion
    crystal structure, **1**, 25
  fluorination
    synthesis, **3**, 499
  formylation
    dichloromethyl alkyl ethers, **2**, 750
  hydrogenation
    palladium-catalyzed, **8**, 438
  microbial hydroxylation, **7**, 78
  oxidative rearrangement, **7**, 833
  polyoxygenated
    synthesis, **3**, 503
  synthesis
    Negishi method, **3**, 503
    Vilsmeier–Haack reaction, **2**, 782
  unsymmetrical
    synthesis, **2**, 623; **4**, 429
Biphenyl, amino-
  synthesis
    via $S_{RN}1$ reaction, **4**, 471
Biphenyl, 2-amino-
  lithiation
    addition reactions, **1**, 463
Biphenyl, 2,2′-dihydroxy-
  oxidative coupling, **3**, 666
Biphenyl, 4,4′-dihydroxy-
  synthesis
    use of vanadium tetrachloride, **3**, 664
Biphenyl, 4,4′-dimethoxy-
  synthesis, **3**, 669
Biphenyl, 2,2′-dinitro-
  reduction, **8**, 364
Biphenyl, 2,6-dinitro-
  synthesis, **3**, 501
Biphenyl, 2,2′-divinyl-

photochemistry, **5**, 728
Biphenyl, 4-formyl-
synthesis
Gattermann–Koch reaction, **2**, 749
Biphenyl, 4-halo-
$S_{RN}1$ reaction, **4**, 461
Biphenyl, 4-methoxy-
Birch reduction
dissolving metals, **8**, 514
Biphenyl, 2-methyl-
synthesis, **7**, 833
Biphenyl, 4-methyl-
Birch reduction
dissolving metals, **8**, 496
Biphenyl, 2-(α-styryl)-
photochemistry, **5**, 726
Biphenyl, 3,3′,4,4′-tetramethoxy-
synthesis, **3**, 668
Biphenyl, 2-vinyl-
photochemistry, **5**, 726
Biphenyl-2-carboxylic acid
Friedel–Crafts reaction, **2**, 757
Biphenyl-2-carboxylic acid, 2′,4,4′,6,6′– pentanitro-
Schmidt reaction, **6**, 819
Biphenylcarboxylic acids
Birch reduction
dissolving metals, **8**, 504
Biphenylenes
synthesis
via [2 + 2 + 2] cycloaddition, **5**, 1150
2-(4-Biphenyl)-2-propoxycarbonyl group
carboxy-protecting group, **6**, 668
acid stability, **6**, 637
2,2′-Bipyridine
chromium(VI) oxide complex
alcohol oxidation, **7**, 260
reduction
metal hydrides, **8**, 580
Bipyridines
synthesis
via cycloaddition, **5**, 1153
Bipyridinium chlorochromate
oxidation
alcohols, **7**, 267
2,2′-Bipyridyl
reaction with phenyllithium, **3**, 512
synthesis, **3**, 509
Birch reduction
acetals, **8**, 212
aromatic compounds, **8**, 490
aryl ethers
carbocyclic enol ether preparation, **2**, 599
chemoselectivity, **8**, 530
experimental procedures, **8**, 492
hydrogenolysis, **8**, 514
intermediates
intramolecular protonation, **8**, 495
limitations, **8**, 493
mechanism, **8**, 490
pyridines, **8**, 591
pyrroles, **8**, 605
scope, **8**, 493
secondary reactions, **8**, 493
substituent effects, **8**, 493
survey, **8**, 493
Bisabolene
synthesis
via Horner reaction, **1**, 780
γ-Bisabolene
synthesis, **3**, 215
Bisallylic alcohols
allylic rearrangements, **7**, 822
tertiary
synthesis, **1**, 118
Bisamides
N-acyliminium ion precursors, **2**, 1049
Bis-annulation
aromatic nucleophilic substitution
competing reaction, **4**, 432
Bisaziridines
ring opening, **7**, 487
Bisbenzocycloheptatriene
synthesis
in steganone synthesis, **3**, 673
α-Bisbololone
synthesis
via benzoin alkylation, **1**, 552
1,1-Bisboronates
oxidation
aldehyde formation, **7**, 597
1,2-Bisboronates
oxidation
1,2-diol formation, **7**, 597
2,6-Bis(*t*-butylphenyl) cyanate
synthesis, **6**, 243
Biscarbamates
N-acyliminium ion precursors, **2**, 1049
Bischler–Napieralski reaction
Ritter reaction, **6**, 291
1,1-Bis(dialkylboryl) compounds
oxidation, **7**, 600
Bis(1,3-dialkylimidazolidin-2-ylidene)
catalyst
benzoin condensation, **1**, 543
Bis(dimethylamino)methylation
1,4-bis(trimethylstannyl)-2-butyne reaction with
Eschenmoser's salt, **2**, 1000
Bisepoxides
synthesis, **7**, 384
1,3-Bishomocubanone
Baeyer–Villiger reaction, **7**, 686
Bisimines
Diels–Alder reactions, **5**, 425
Bislactim ethers
alkylation, **3**, 53
Michael additions
unsaturated esters, **4**, 222
Bislactones
synthesis
via cyclization of cycloalkeneacetic acids, **4**, 370
2,2′-Bis(methylenecyclopentane)
Cope rearrangement, **5**, 820
Bismuth, μ-bis(triphenyl)-
oxidation
secondary alcohols, **7**, 322
Bismuth, μ-oxobis(chlorotriphenyl)-
glycol cleavage, **7**, 704
oxidation
allylic alcohols, **7**, 307
primary alcohols, **7**, 310
secondary alcohols, **7**, 322
Bismuth carbonate, triphenyl-

biaryl synthesis, **3**, 505
glycol cleavage, **7**, 704
oxidation
primary alcohols, **7**, 310
secondary alcohols, **7**, 322
Bismuth compounds, crotyl-
type III
reactions with aldehydes, **2**, 24
Bismuth diacetate, triphenyl-
reaction with diols, **6**, 23
Bismuth reagents
oxidation
secondary alcohols, **7**, 318
Bisnorcholenol, 3-keto-
microbial hydroxylation, **7**, 70
Bisnorisocomene
synthesis
via Pauson–Khand reaction, **5**, 1062
A-Bisnorsteroids
synthesis, **3**, 901
2,2-Bisoxazoles
reduction
LAH, **8**, 650
Bisphosphoranes
open chain
synthesis, **6**, 191
silicon-bridged
synthesis, **6**, 180
Bisquinonemethides
synthesis, **3**, 698
Bisthiazoles
synthesis, **3**, 511
Bis(thiazolin-2-ylidene)
catalyst
benzoin condensation, **1**, 543
Bisthioacetals
carbonyl group regeneration, **7**, 846
Bisthiophenes
coupling reactions, **3**, 512
synthesis, **3**, 515
Bistriazoles
benzyne precursors
Diels–Alder reactions, **5**, 382
Bisulfenylation
cyclobutanones, **6**, 143
Bisureas
N-acyliminium ion precursors, **2**, 1049
Blaise reaction
nitriles
acylation, Reformatsky reagents, **2**, 297
zinc enolates, **2**, 297
Bleomycins
synthesis
Mannich reaction, **2**, 917, 920
via Diels–Alder reaction, **5**, 492
Block copolymers
styrene–ethylene–butene–styrene
Friedel–Crafts alkylation, **3**, 303
Blood group antigenic determinants
synthesis, **6**, 43
Boat-like transition states
Diels–Alder reactions
decatrienones, **5**, 539–543
Boldine
synthesis, **3**, 686, 815
Bombykol
synthesis, **3**, 489, 799
σ-Bond metathesis, **7**, 3
Bonds
C—C
reductive cleavage, **8**, 995
C—halogen
hydrogenolysis, **8**, 895
C—Hg
reduction, **8**, 850
C—N
hydrogenolysis, **8**, 915
reductive cleavage, **8**, 995
C—O
hydrogenolysis, **8**, 910
reductive cleavage, **8**, 991
C—P
reduction, **8**, 858
C—S
hydrogenolysis, **8**, 913
reduction, **8**, 835-870
reductive cleavage, **8**, 993
C—Se
reduction, **8**, 847
reductive cleavage, **8**, 996
C—Si
oxidative cleavage, **6**, 16
C—Zn, **1**, 212
Boraadamantane
hydride donor, **8**, 102
9-Borabicyclo[3.3.1]nonane
hydroboration, **8**, 712, 713
kinetics, **8**, 724
K-glucoride from, **8**, 169
reaction with α-pinene, **8**, 101
reduction
acyl halides, **8**, 240
carboxylic acids, **8**, 237
unsaturated carbonyl compounds, **8**, 537, 543
synthesis, **2**, 57; **8**, 708
synthesis of 1,1-diboryl compounds, **1**, 489
9-Borabicyclo[3.3.1]nonane, *B*-(1-alkenyl)-
conjugate additions
α,β-enones, **4**, 147
9-Borabicyclo[3.3.1]nonane, *B*-alkyl-
oxidation
use of carbonyl compounds, **7**, 603
9-Borabicyclo[3.3.1]nonane, *B*-(1-alkynyl)-
conjugate additions
α,β-enones, **4**, 147
9-Borabicyclononane, allyl-
NMR, **2**, 976
reactions with aldimines, **2**, 983
diastereoselectivity, **2**, 985
9-Borabicyclo[3.3.1]nonane, allyl-
reactions with allyl organometallics, **2**, 32
9-Borabicyclononane, crotyl-
NMR, **2**, 976
reaction with imines
syn–anti selectivity, **2**, 991
9-Borabicyclo[3.3.1]nonane, crotyl-
reactions with achiral aldimines, **2**, 15
reactions with carbonyl compounds, **2**, 10
reactions with pyruvate esters, **2**, 11
9-Borabicyclo[3.3.1]nonane, *B*-1-(2-ethoxy-
2-iodovinyl)-
conjugate additions

α,β-enones, **4**, 147
9-Borabicyclo[3.3.1]nonane, *B*-iodo-
reactions with alkynes and allenes, **4**, 358
9-Borabicyclo[3.3.1]nonane, *B*-methyl-
deprotonation
alkylation of anion, **3**, 199
reaction with lithium amides
deprotonation, **1**, 491
9-Borabicyclononane, pent-3-en-2-yl-
reactions with imines
syn–anti selectivity, **2**, 990, 992
9-Borabicyclo[3.3.1]nonane, (3-pinanyl)-
asymmetric reduction, **8**, 160
reaction with aldehydes, **7**, 603
reduction
alkynic ketones, **8**, 537
9-Borabicyclo[3.3.1]nonane, *B*-siamyl-
oxidation
use of carbonyl compounds, **7**, 603
4a-Boranaphthalene, perhydro-
synthesis, **8**, 708
9-Borabicyclo[3.3.1]non-9-yl triflate
reaction with *S*-phenyl propanethioate, **2**, 245
Boracyclanes
oxidation, **7**, 596
Boracyclanes, *B*-alkyl-
conjugate additions
alkenes, **4**, 146
Borane
acrylic acid complexes
structure, **1**, 289
t-butylamine complex
selective ketone reduction, **8**, 18
carbonyl reduction, **8**, 20
chirally modified
asymmetric reduction, **8**, 169
complexes
hydroboration, **8**, 705
dimethyl sulfide complex
carbonyl compound reduction, **8**, 20
carboxylic acid reduction, **8**, 237
ester reduction, **8**, 244
hydroboration, **8**, 708
diphenylamine complex
carboxylic acid reduction, **8**, 237
hydroboration, **8**, 708
disubstituted
hydroboration, **8**, 712
formaldehyde complex
rotational barriers, **1**, 290
heterocyclic
oxidation, **7**, 601
propanal complex
rotational barriers, **1**, 290
reagent formed with 2-aminoethanol
selective aldehyde and ketone reduction, **8**, 18
reductions, **8**, 369
acetals, **8**, 214
carbonyl compounds, **8**, 315
carboxylic acids, **8**, 261
imines, **8**, 26, 36
nitroalkenes, **8**, 376
pyridines, **8**, 580
tetrahydrofuran complex
amide reductions, **8**, 249
carboxylic acid reduction, **8**, 237
hydroboration, **8**, 705
nitrile reduction, **8**, 253
reductive animation, **8**, 54
thioxane complex
hydroboration, **8**, 708
triethylamine complex
hydroboration, **8**, 708
α-trimethylsilyl-substituted
reactions with aldehydes, **1**, 501
Boranes, acyloxy-
chiral catalyst
Diels–Alder reactions, **5**, 377–379
diborane enediolates, **2**, 113
Boranes, alkenyl-
1,3-butadiene synthesis, **3**, 483
coupling reactions, **3**, 489
sp^2 organometallics, **3**, 473
with benzyl bromide, **3**, 465
cross-coupling reactions with 1-alkynyl halides, **3**, 530
oxidation
using alkaline hydrogen peroxide, **7**, 596
protonolysis, **8**, 726
synthesis
via α-trimethylsilyl-substituted boranes, **1**, 501
Boranes, *B*-(1-alkenyl)alkoxyfluoro-
conjugate additions
α,β-enones, **4**, 147
Boranes, alkenylamino-
aldol reactions, **2**, 244
Boranes, alkenyldialkoxy-
oxidation
formation of aldehydes, **7**, 602
Boranes, alkenyldialkyl-
brominolysis
stereochemistry, **7**, 605
protonolysis, **8**, 724, 726
reaction with iodine
rearrangements, **7**, 606
Boranes, alkenyldihydroxy-
brominolysis, **7**, 605
iodinolysis
stereochemistry, **7**, 606
Boranes, alkenyloxy-
directed aldol reactions, **2**, 240
homochiral
aldol reactions, **2**, 240
Hooz' reaction, **2**, 244
oxidation, **7**, 602
reactions with ketones, **1**, 499
synthesis
via acylation of boron-stabilized carbanions, **1**, 497
Boranes, alkenyloxyalkylalkoxy-
aldol reactions, **2**, 240
Boranes, alkenyloxydialkoxy-
aldol reactions, **2**, 240, 266
syn/*anti* ratios, **2**, 266
Boranes, alkenyloxydialkyl-
aldol reactions, **2**, 240
homochiral, **2**, 248
chiral
facial selectivity, **2**, 261
synthesis, **2**, 240
Boranes, alkenyloxydichloro-
preparation, **2**, 244

Boranes, alkoxy-
reaction with organometallic compounds, **7**, 595
Boranes, β-alkoxyalkyl-
stability, **8**, 705
Boranes, alkoxydialkyl-
ketone reduction, **8**, 9
Boranes, alkyl-
oxidation
formation of aldehydes, **7**, 601
protonolysis
carboxylic acids, **8**, 725
Boranes, alkylbromo-
dimethyl sulfide complex
synthesis, **8**, 719
Boranes, alkyldichloro-
synthesis, **8**, 718
Boranes, alkyldiethoxy-
synthesis, **7**, 603
Boranes, alkyldihydroxy-
oxidation, **7**, 597
Boranes, alkyldimesityl-
reactions with bases, **1**, 492
Boranes, *B*-alkyldiphenyl-
conjugate additions
alkenes, **4**, 146
Boranes, (alkylethenyl)dialkyl-
brominolysis
stereochemistry, **7**, 605
Boranes, alkylhalo-
hydroboration, **8**, 719
Boranes, alkynyl-
coupling reactions, **3**, 523
protonolysis, **8**, 725
reaction with epoxides, **6**, 7
Boranes, allyl-
oxidation, **7**, 596
reaction with imines, **2**, 976
Boranes, allyldialkyl-
protonolysis, **8**, 725
Boranes, allyldiisopinocampheyl-
reactions with aldehydes
asymmetric synthesis, **2**, 33
Boranes, allyldimesityl-
anion
reactions, **2**, 56
reactions with lithium amides, **1**, 492
Boranes, allyldimethoxy-
reactions with aldimines, **2**, 982
Boranes, aryldihydroxy-
nitration and oxidation of the ring, **7**, 602
oxidation, **7**, 596, 602
use of potassium permanganate, **7**, 602
Boranes, (arylethenyl)dialkyl-
brominolysis
stereochemistry, **7**, 605
Boranes, benzyl-
protonolysis, **8**, 725
Boranes, binaphthoxy-
Diels–Alder reactions, **5**, 376
Boranes, bis(benzoyloxy)-
reduction
hydrazones, **8**, 357
Boranes, bromo-
synthesis, **8**, 711
Boranes, *t*-butyl-
synthesis, **8**, 710
Boranes, butyldihydroxy-
oxidation
formation of butanol, **7**, 602
Boranes, catechol-
brominolysis, **7**, 605
coupling reactions
with aryl iodides, **3**, 496
hydroboration, **8**, 719
reduction
hyrazones, **8**, 356
Boranes, chloro-
reaction with acetophenone, **7**, 603
Boranes, β-chloroalkyl-
stability, **8**, 705
Boranes, chlorodivinyl-
synthesis, **8**, 711
Boranes, crotyl-
reactions with imines, **2**, 17
synthesis, **2**, 44
Boranes, crotyldiisopinocampheyl-
boratropic shift, **2**, 10
reactions with aldehydes, **2**, 61
Boranes, cycloalkyl-
oxidation
formation of cycloalkanones, **7**, 601
Boranes, cyclopropyl-
oxidation, **7**, 598
synthesis
via boron-ene reaction, **5**, 33
Boranes, dialkoxy-
chiral catalysts
Diels–Alder reactions, **5**, 376
Boranes, dialkoxy(α-phenylthio)-
oxidation
formation of monothioacetals, **7**, 602
Boranes, dialkyl-
hydroboration, **8**, 715
regioselectivity, **8**, 717
stability, **8**, 717
synthesis, **8**, 717
Boranes, dialkylallyl-
reaction with ethoxyacetylene, **5**, 34
Boranes, dialkylbromo-
synthesis, **8**, 711
Boranes, dialkylchloro-
alkenyloxyboraness from, **2**, 244
synthesis, **8**, 711
Boranes, dialkylcrotyl-
isomerization, **2**, 5
reactions with carbonyl compounds, **2**, 10
Boranes, dialkyl(dialkylamino)-
synthesis, **7**, 607
Boranes, dialkylhalo-
synthesis, **8**, 711
Boranes, dialkyl(methylthio)-
synthesis, **8**, 711
Boranes, di-*s*-alkylmonoalkyl-
protonolysis, **8**, 725
Boranes, diaryl-
hydroboration, **8**, 715
Boranes, dibromo-
dimethyl sulfide complex
synthesis, **8**, 718
Boranes, dicaranyl-
chiral hydroboration, **8**, 721
Boranes, dichlorophenyl-

ethyl ketone enolization, **2**, 244
syn diastereoselectivity, **2**, 245
Boranes, dichloro(vinylamino)-
aldol reaction
directed, **2**, 479
reaction with carbonyl compounds, **2**, 478
Boranes, dicyclohexyl-
hydroboration
regioselectivity, **8**, 716
synthesis of 1,1-diboryl compounds, **1**, 489
Boranes, diethoxysiamyl-
oxidation
using alkaline hydrogen peroxide, **7**, 595
Boranes, (diethylamino)dichloro-
dihydridoborate from, **8**, 171
Boranes, dihalo-
hydroboration, **8**, 718
Boranes, dihydroxy[lithio(trimethylsilyl)methyl]-
pinacol derivative
acylation, **1**, 498
Boranes, diisopinocampheyl-
chiral hydroboration, **8**, 720
hydroboration, **8**, 712
synthesis, **8**, 716
Boranes, dilongifolyl-
chiral hydroboration, **8**, 721
Boranes, dimesityl-
deprotonation, **3**, 199
hydroboration, **8**, 716
Boranes, dimesitylmethyl-
reactions with bases, **1**, 492
reactions with styrene oxide, **1**, 496
Boranes, dimethyl-
synthesis, **8**, 717
Boranes, 2,3-dimethyl-2-butyl-
reduction
carboxylic acids, **8**, 261
Boranes, diphenyl-
hydroboration, **8**, 716
Boranes, diphenylhydroxy-
oxidation, **7**, 603
reaction with ethoxyacetylene
mercury(II) acetate, **2**, 242
Boranes, disiamyl-
reduction
acyl halides, **8**, 263
amides, **8**, 273
lactones, **8**, 269
nitriles, **8**, 275
synthesis of 1,1-diboryl compounds, **1**, 489
Boranes, ethyldimesityl-
reactions with epoxides, **1**, 497
Boranes, ethylenedioxychloro-
enolates
generation from carbonyl compounds, **2**, 266
Boranes, haloalkyl-
stability, **8**, 705
Boranes, iodo-
synthesis, **8**, 711
Boranes, α-lithiodimesitylmethyl-
acylation, **1**, 498
Boranes, methyl-
synthesis, **8**, 710
Boranes, monoalkyl-
redistribution, **8**, 710
Boranes, monoalkylchloro-
synthesis, **8**, 711
Boranes, monoaryl-
regioselectivity
hydroboration, **8**, 710
Boranes, monochloro-
dimethyl sulfide complex
hydroboration, **8**, 711
hydroboration, **8**, 710
reduction
acetals, **8**, 214
Boranes, monohalo-
hydroboration, **8**, 710
regioselectivity, **8**, 711
Boranes, monoisopinocamphenyl-
alkene hydroboration, **3**, 797
chiral hydroboration, **8**, 721
synthesis, **8**, 710
Boranes, peroxybis(diacetoxy)-
1-hydroxy-2-acetoxyalkene synthesis, **7**, 446
Boranes, phenyl-
alkylation, **3**, 260
Boranes, phenyldihydroxy-
oxidation, **7**, 602
Boranes, pyridyl-
coupling reactions
with vinyl bromides, **3**, 498
Boranes, 1-pyrrolyl-
reduction
enones, **8**, 16
tetrahydrofuran complex
reduction, unsaturated carbonyl compounds, **8**, 537
Boranes, secondary alkyl
oxidation
formation of ketones, **7**, 600, 601
Boranes, thexyl-
hydroboration, **2**, 251; **8**, 709
reduction
acyl halides, **8**, 263
amides, **8**, 273
carboxylic acids, **8**, 237
lactones, **8**, 269
nitriles, **8**, 275
Boranes, thexylbromo-
dimethyl sulfide complex
carboxylic acid reduction, **8**, 261
Boranes, thexylchloro-
dimethyl sulfide complex
carboxylic acid reduction, **8**, 261
hydroboration, **8**, 719
Boranes, thio-
reduction
carboxylic acids, **8**, 261
Boranes, trialkenyl-
protonolysis, **8**, 724
Boranes, trialkyl-
brominolysis, **7**, 604
chlorination, **7**, 604
hydride donor
reduction of carbonyls, **8**, 99, 101
iodinolysis, **7**, 606
ketone reduction, **8**, 9
oxidation, **7**, 602
carbonyl compounds, **7**, 603
protonolysis, **8**, 724, 725
reaction with alkenes, **4**, 884
reaction with aryl Grignard reagents, **3**, 243

reaction with carbon monoxide, **3**, 793
reaction with α,β-unsaturated carbonyl compounds, **2**, 241
Boranes, triallyl-
ene reactions, **5**, 33
reaction with alkynes, **4**, 886
Boranes, tributyl-
hydride donor
reduction of carbonyls, **8**, 101
protonolysis, **8**, 724
Boranes, tri-*n*-butyl-
oxidation, **7**, 599
Boranes, trichloro-
aldol reactions
syn selectivity, **2**, 245
alkenyloxydichloroboranes from, **2**, 244
Boranes, tricrotyl-
hydrolysis, **8**, 725
Boranes, triethyl-
Lewis acid co-catalyst
[3 + 2] cycloaddition reactions, **5**, 296
oxidation, **7**, 593
reaction with alkenes, **4**, 885
reaction with thioallyl anions
α-selectivity, **2**, 72
Boranes, trifluoro-
diethyl ether complex
carbonyl compound reduction, **8**, 319
hydroboration, **8**, 708
reductive cleavage
benzylic compounds, **8**, 969
sodium borohydride reduction
carbonyl compounds, **8**, 315
water complex
carbonyl compound reduction, **8**, 319
Boranes, trimethoxy-
aldol reactions
catalysis, **2**, 266
Boranes, trimethyl-
oxidation, **7**, 593
protonolysis, **8**, 724
Boranes, 1,1,2-trimethylpropyl-
synthesis, **8**, 706
Boranes, tri-*n*-octyl-
oxidation, **7**, 603
Boranes, triorganyl-
carbonylation
route to tertiary alcohols, **3**, 779
rearrangements, **3**, 793
Boranes, triphenyl-
brominolysis, **7**, 604
Boranes, tris(3,3-dimethyl-1-butyl-1,2-d_2)-
bromination, **7**, 604
Boranes, tri(secondary alkyl)
iodinolysis, **7**, 606
Boranes, tris-2-norbornyl-
brominolysis, **7**, 604
Boranes, tristannylmethyl-
synthesis, **1**, 494
Boranes, vinyl-
boron–lithium exchange, **3**, 254
oxidation
aldehyde formation, **7**, 597
reactions with organometallic compounds, **1**, 492
synthesis, **8**, 716
via hydroboration of 1-alkynes, **1**, 492
Boranes, vinyloxy-
conjugate additions
alkenes, **4**, 145
synthesis
enolate geometry, **2**, 111
9b-Boraphenalene, perhydro-
synthesis, **8**, 708
9-Boratabicyclo[3.3.1]nonane
selective aldehyde reduction, **8**, 17
9-Boratabicyclo[3.3.1]nonane, *B*-alkoxy-
reduction
cyclic carbonyl compounds, **8**, 14
9-Boratabicyclo[3.3.1]nonane, *B*-alkyl-
reduction
cyclic carbonyl compounds, **8**, 14
9-Boratabicyclo[3.3.1]nonane, *B*-siamyl-
selective aldehyde reduction, **8**, 17
Borates, alkenyltrialkyl-
synthesis, **8**, 724
Borates, alkynyl-
coupling reactions, **3**, 554
synthesis, **3**, 799
Borates, tetraorganyl-
rearrangement, **3**, 780
Borepane, 3,6-dimethyl-
hydroborating agent, **3**, 199
synthesis, **8**, 707
Boric acid
aldol reactions
catalysis of aldol condensation, **2**, 240
catalyst
aldol condensations, **2**, 138
Boric acid, allenyl-
synthesis, **2**, 96
Boric acid, tetrafluoro-
reaction with 1,3-dienes, **7**, 536
Borinane
synthesis, **8**, 707, 717
Borinane, 1-chloro-
synthesis, **8**, 711
Borinane, 3,5-dimethyl-
hydroboration
regioselectivity, **8**, 717
synthesis, **8**, 707
Borinate, enol
synthesis
enolate geometry, **2**, 111
Bornane sultams
boron enolates, **2**, 252
conjugate additions, **4**, 204
Diels–Alder reactions, **5**, 362
Bornane-10,2-sultams, *N*-enoyl-
conjugate additions
hydrides, **4**, 231
Borneol
asymmetric hydrogenation, **8**, 144
oxidation
DMSO, **7**, 298
reaction with bromine and dihydropyran, **4**, 345
Bornyl acetate
microbial hydroxylation, **7**, 62
Bornylene
reaction with hydrofluoric acid, **4**, 270
Bornyl fumarate
photocycloaddition reactions
stilbenes, **5**, 132

Bornyl propenoates
reaction with benzenesulfenyl chloride, **4**, 331
Borodin–Hunsdiecker reaction
brominative decarboxylation, **4**, 1006
Borohydrides
asymmetric reduction, **8**, 169
exchange resin
selective aldehyde reduction, **8**, 16
reductions, **8**, 369
benzo [*b*]furans, **8**, 627
indoles, **8**, 616
pyridines, **8**, 580
pyridinium salts, **8**, 584
unsaturated carbonyl compounds, **8**, 536
Borohydrides, alkylcyano-
reduction
imines, **8**, 36
Borohydrides, cyano-
zinc-modified
selective ketone reduction, **8**, 18
Borohydrides, dialkylcyano-
reduction
imines, **8**, 36
Borohydrides, monoalkyl-
reduction
cyclic ketones, **8**, 14
Borohydrides, thexyl-di-*s*-butyl-
reduction
unsaturated carbonyl compounds, **8**, 537
Borohydrides, tri-*s*-butyl-
reduction
unsaturated carbonyl compounds, **8**, 537
Borohydrides, triphenyl-
selective ketone reduction, **8**, 18
Borolane, crotyl-*trans*-2,5-dimethyl-
allylboranes from
reactions with aldehydes, **2**, 33
Borolane, 2,5-dimethyl-
aldol reactions
enantioselectivities, **2**, 258
asymmetric reduction, **8**, 159
chiral hydroboration, **8**, 721
diastereoselectivity, **2**, 42
Borolane, *B*-methoxy-2,5-dimethyl-
synthesis, **2**, 33
Boromycin
synthesis, **1**, 568
Boron, dichloro-
enolates
synthesis, **2**, 114
Boron, dimethoxy-
enolates
synthesis, **2**, 114
Boron alkynides
alkylation, **3**, 274
Boron–ate complexes, crotyl-
reactions with aldehydes, **2**, 11
Boronates
cyclic
diol protection, **6**, 662
Boronates, alkenyl-
coupling reactions, **3**, 489
with alkenyl halides, **3**, 496
reactions
organopalladium catalysts, **3**, 231
Boronates, allyl-
reaction with 2,3-*O*-isopropylidene-D-glyceraldehyde oxime
Cram selectivity, **2**, 995
reaction with phenylmenthyl-*N*-methoxyiminoacetate, **2**, 995
reaction with sulfenimines, **2**, 999
Boronates, 1-bromo-1-alkenyl-
from 1-alkynes, **3**, 490
Boronates, crotyl-
reactions with oximes
syn–anti selectivity, **2**, 996, 997
synthesis, **2**, 977
Boronates, cyano-
rearrangements, **3**, 798
Boronates, 1,3-dienyl-
Diels–Alder reactions, **5**, 336
Boronates, β-ethoxy-
coupling reactions
with aryl iodides, **3**, 496
Boronates, tetraorganyl-
rearrangements, **3**, 798
Boronates, [γ-(trimethylsilyl)allyl]-
reactions with oximes, **2**, 996
Boron bromides
reactions with alkenes, **4**, 357
Boron compounds
aldol reactions, **2**, 240
carbanions
stabilization, **1**, 487–503
Lewis acid complexes
structure, **1**, 287
organopalladium catalysis, **3**, 231
Boron compounds, alkenyl-
cleavage, **8**, 725
Boron compounds, allyl-
configurational stability, **2**, 5
protonolysis, **8**, 725
reactions with chiral α-methyl aldehydes, **2**, 42
Boron compounds, aromatic
oxidation to phenols, **7**, 596
Boron compounds, aryl-
protonolysis, **8**, 725
Boron compounds, crotyl-
reactions with chiral α-methyl aldehydes, **2**, 42
type I, **2**, 10–17
reactions with C═N electrophiles, **2**, 15
Boron-ene reactions, **5**, 33
Boron enolates
aldol reactions
diastereofacial preferences, **2**, 224, 231
from homochiral acyl sultam
aldol reactions, **2**, 253
reactions with aldehydes, **2**, 250
reactions with *N*,*N*-dimethyl(methylene)iminium salts, **2**, 909
synthesis, **2**, 111
Boron enol ethers
synthesis
enolate geometry, **2**, 111
Boron fluoride
hydrofluorination
alkenes, **4**, 271
Boronic acid, alkenyl-
biaryl synthesis, **3**, 504
synthesis, **3**, 489
Boronic acid, γ-alkoxyallyl-

reactions with achiral aldehydes
diastereoselectivity, **2**, 14
reactions with carbonyl compounds, **2**, 35
Boronic acid, allenyl-
reaction with aldehydes, **6**, 865
Boronic acid, allyl-
esters
synthesis, **2**, 6
pinacol ester
reactions with aldehydes, **2**, 25
reactions with achiral aldehydes, **2**, 13
reactions with aldehydes, **2**, 31
reactions with aldoximes, **2**, 15
reactions with imines, **2**, 15
reactions with ketones, **2**, 15
reactions with α-methyl chiral aldehydes, **2**, 26
reactions with sulfenimides, **2**, 15
synthesis, **2**, 12, 13
tartrate
diastereoselective mismatched double asymmetric reactions with aldehydes, **2**, 41
enantioselectivity, **2**, 35
synthesis, **2**, 35
Boronic acid, α-chloroallyl-
mismatched diastereoselective reactions with aldehydes, **2**, 42
Boronic acid, α-chlorocrotyl-
diastereofacial preference, **2**, 45
reactions with aldehydes, **2**, 39
Boronic acid, 1-chloroethyl-
ester
synthesis, **3**, 796
Boronic acid, crotyl-
chiral
double asymmetric reactions, **2**, 41
dimethyl esters
reactions with achiral aldehydes, **2**, 13
pinacol ester
reactions with aldehydes, **2**, 26
reactions with chiral aldehydes, **2**, 25
reactions with aldehydes, **2**, 29
reactions with oxime silyl ethers, **2**, 8
synthesis, **2**, 12, 13
tartramide
stereoselective reactions with aldehydes, **2**, 44
tartrate
double diastereoselectivity, **2**, 42
reaction stereochemistry, **2**, 7
reactions with achiral aldehydes, **2**, 20
synthesis, **2**, 13
Boronic acid, α-methoxycrotyl-
mismatched double diastereoselectivity, **2**, 45
Boronic acid, α-methylcrotyl-
reaction with benzaldehyde, **2**, 39
Boronic esters
chiral
asymmetric synthesis, **3**, 780
synthesis, **3**, 796
rearrangements, **3**, 780, 796
Boron oxide
aldol reactions
catalysis of aldol condensation, **2**, 240
Boron trichloride
catalyst
Friedel–Crafts reaction, **2**, 735
Boron trifluoride
Beckmann rearrangement, **7**, 695
benzaldehyde complex, **2**, 247
crystal structure, **1**, 300
catalyst
allylsilane reactions, **2**, 567
allylsilane reactions with acetals, **2**, 576
allylstannane reactions with aldehydes, **2**, 573
Diels–Alder reactions, **2**, 664, 665
Friedel–Crafts reactions, **2**, 735; **3**, 295
reaction with allylsilanes, diastereoselectivity, **2**, 570
Diels–Alder reaction catalysts
diastereofacial selectivity, **2**, 679
dimethyl ether complexes
coordination energy, **1**, 290
epoxide ring opening, **3**, 741
etherate
ketone α-acetoxylation, **7**, 153
organocuprate reactions, **1**, 115; **3**, 212
ethyl acetate complex
NMR, **1**, 292
mercury(II) trifluoroacetate
ionic dissociation, **7**, 872
organolithium reactions
Lewis acid promotion, **1**, 329
reactions with organocopper compounds
rate enhancement, **1**, 343
reactions with organolithium compounds
alkynylation, **1**, 343
triethylsilane
2-octanol reduction, **8**, 813
Boron trifluoride etherate
catalyst
tandem vicinal difunctionalization, **4**, 255
Borrerine
synthesis, **6**, 746
Boryl compounds, dimesityl-
properties, **1**, 492
reactions with epoxides, **1**, 496
synthesis, **1**, 494
Boryl compounds, ethylenedioxy-
organometallic compounds
synthesis, **1**, 494
Boryl enolates
aldol reactions, **2**, 239
Boryl triflate
kinetic enolization of carbonyl compounds, **2**, 247
Boryl triflate, dialkyl-
boron enolates, **2**, 112
enolization of carbonyl compounds, **2**, 242
metal exchange reaction
alkenyloxysilane, **2**, 245
Boryl triflate, diisopinocampheyl-
aldol reactions, **2**, 257
Boschnialactone
synthesis
via photoisomerizations, **5**, 230
Boschnialic acid
synthesis
via magnesium-ene reaction, **5**, 42
Bostrycoidin
synthesis
via regioselective lithiation, **1**, 474
Botryodiplodin
synthesis

rearrangement of epoxides, **3**, 768
use of enol esters, **2**, 613
via conjugate addition, **4**, 211
Bourbonene
synthesis
via photochemical cycloaddition, **5**, 124, 129
Bouveault–Blanc reduction
esters, **8**, 243
conversion to primary alcohols, **3**, 613
Bovine serum albumin
asymmetric catalyst
Darzens glycidic ester condensation, **2**, 435
monoclonal antibodies
Claisen rearrangement, **5**, 855
Bovolide
synthesis, **5**, 1092
Brassinolide
synthesis
side chain introduction, **1**, 552
via Baeyer–Villiger reaction, **7**, 680
via carboalumination, **4**, 893
Wittig rearrangement, **3**, 1000
Braun reagent
aldol reaction
chiral synthesis, **2**, 227
Brefeldin A
synthesis
use of alcohol protection, **6**, 648
via activated esters, **6**, 373
via σ-alkyliron complexes, **4**, 579
via conjugate addition, **4**, 211
via [3 + 2] cycloaddition reactions, **5**, 308
via Julia coupling, **1**, 805
via macrolactonization, **6**, 370
Brefeldin A seco acid
synthesis, **7**, 625
Brefeldin C
synthesis
via diisopropyl phosphonate, Wittig reaction, **1**, 763
Brefeldins
synthesis, **3**, 287
via alkenylchromium reagents, **1**, 200
Brevetoxine B
synthesis
via 1,2-dithietane, **6**, 448
Brevianamide A
synthesis, **3**, 790
exo-Brevicomin
synthesis
aluminum ate complexes, **2**, 67
via cyclofunctionalization of cycloalkene, **4**, 373
via zinc chelation, **1**, 222
Brevicomins
synthesis, **3**, 644; **6**, 145; **7**, 643
via 1,2-addition of ethylcopper reagents, **1**, 134
via Lewis acid mediated Grignard addition, **1**, 336
via Wacker process, **7**, 451
Bridged azacycles
synthesis
Mannich reaction, **2**, 1014
Bridged carbocyclic systems
synthesis
via palladium(II) catalysis, **4**, 573
Bridged rings
synthesis
via radical cyclizations, **4**, 791
Bridged systems
synthesis
via [2 + 3] cycloaddition reactions, **5**, 951
Bridgehead halides
reduction
tributylstannane, **8**, 798
Brigl's anhydride
disaccharide synthesis, **6**, 48
Brönsted acids
catalysts
Friedel–Crafts reaction, **3**, 297
glycosylation, **6**, 51
Brönsted–Lewis superacid
catalysts
Friedel–Crafts reaction, **3**, 297
Bromides
vinyl substitutions
palladium complexes, **4**, 835
Bromination
amines, **7**, 741
boryl-substituted carbanions, **1**, 501
ketones
bromine, **7**, 120
nucleophilic displacement, **6**, 209
secondary amines, **7**, 747
Bromine
bromination
ketones, **7**, 120
conjugate enolate trap, **4**, 262
in the presence of nickel carboxylates
oxidation, diols, **7**, 314
reaction with alkenes, **4**, 344–346
Ritter reaction, **6**, 288
Bromine azide
addition reactions
alkenes, **7**, 500
aziridine synthesis, **7**, 473
Bromine fluoride
reaction with alkenes, **4**, 347
Bromine perchlorate, bis(*sym*-collidine)-
intramolecular bromoalkylamine addition to alkenes, **7**, 536
Brominolysis
C—B bonds, **7**, 604
α-Bromocarboxylates
aldol reactions
intramolecular, **2**, 269
Bromohydrin
coupling reactions
with aryl Grignard reagents, **3**, 464
epoxide synthesis, **6**, 25
reaction with magnesium halides, **3**, 757
Bromolactonization
cycloheptadienes
palladium catalysis, **4**, 687
Bromonitro compounds
synthesis, **7**, 501
Brook–Claisen rearrangements
tandem, **5**, 843
Brook rearrangement
desulfonylation, **5**, 1014
1-oxyallyl anions, **2**, 69
Brown–Walker electrolysis
of halfester dimerization, **3**, 640
Bruceantin

synthesis, **8**, 925
Brunke steroid synthesis
diene cyclization, **3**, 373
Bryostatin
synthesis, **2**, 264
via acylation with thiol esters, **1**, 434
Bufadienolide
synthesis
Knoevenagel reaction, **2**, 382
Buflomedil
synthesis
via alkylation of cyanohydrin anions, **1**, 552
Bunte salts
Diels–Alder reactions, **5**, 436
Burseran
synthesis, **1**, 566
via conjugate addition, **4**, 211
1,3-Butadiene
1,4-acetamidoiodination, **7**, 505
acylation
Friedel–Crafts reaction, **2**, 720
addition of D_2
Pd/Al_2O_3 catalysis, **8**, 433
carbocupration, **4**, 895
carbomagnesiation, **4**, 874
chlorination, **7**, 530
cyclization, **5**, 675
cycloaddition reactions
tropones, **5**, 618
[4 + 3] cycloaddition reactions, **5**, 603
diarylation
palladium catalysts, **4**, 849
dicarboxylation, **4**, 949
dimerization, **5**, 63
via nickel-ene reaction, **5**, 56
hydration, **4**, 299
hydroboration, **8**, 707
hydrobromination, **4**, 283
hydrocarboxylation, **4**, 945
hydrochlorination, **4**, 276
hydrogenation
homogeneous catalysis, **8**, 449
hydrosilylation, **8**, 776
photocycloaddition reactions
benzene, **5**, 636
reaction with *t*-butyllithium, **4**, 868
reaction with ethyl diazopyruvate, **4**, 1048
selective reduction, **8**, 565, 567, 568
substituted acyclic
synthesis *via* retro Diels–Alder reaction, **5**, 565
symmetrical
synthesis, **3**, 482
zirconocene complex
reactions with carbonyl compounds, **1**, 163
(Z,Z)-
synthesis, **3**, 485
1,3-Butadiene, 1-acetoxy-
cycloaddition reactions
tropones, **5**, 620
Diels–Alder reactions, **5**, 376
1,3-Butadiene, 1-*N*-acylamino-
synthesis
via Curtius reaction, **6**, 811
1,3-Butadiene, alkoxy-
Diels–Alder reaction, **2**, 662; **5**, 329
1,3-Butadiene, 1-alkoxysilyl-
synthesis
via cyclobutanones, **5**, 677
1,3-Butadiene, 4-alkyl-2-amino-4-(substituted amino)-1,1,3-tricyano-
synthesis
via retro Diels–Alder reaction, **5**, 566
1,3-Butadiene, 4-amino-1,1-dicyano-
synthesis
Knoevenagel reaction, **2**, 359
1,3-Butadiene, bis-2,3-chloromethyl-
synthesis
via palladium(II) catalysis, **4**, 566
1,3-Butadiene, 2,3-bis[(*N*,*N*-dimethylamino)methyl]-
synthesis
from 1,4-bis(trimethylstannyl)-2-butyne, **2**, 1002
1,3-Butadiene, 1,3-bis(trimethylsiloxy)-
Diels–Alder reactions, **5**, 323
synthesis from 1,3-diketones, **2**, 606
1,3-Butadiene, 2,3-bis(trimethylsiloxy)-
synthesis
via cyclobutenes, **5**, 684
via 2,3-butanedione, **2**, 605
1,3-Butadiene, 2,3-bis(trimethylsilylmethyl)-
Diels–Alder reactions, **5**, 338
1,3-Butadiene, 1-bromo-
hydrobromination, **4**, 283
synthesis
vinylic coupling, **3**, 490
1,3-Butadiene, 2-*t*-butyl-
reaction with π-allylpalladium complexes, **4**, 601
1,3-Butadiene, chloro-
synthesis
vinylic coupling, **3**, 487
1,3-Butadiene, 1,4-diacyl-
cyclic
synthesis *via* ketocarbenoids and furans, **4**, 1060
1,3-Butadiene, 1-diethylamino-
cycloaddition reactions
6,6-dimethylfulvene, **5**, 626
1,3-Butadiene, 1,1-dimethoxy-3-silyloxy-
Diels–Alder reaction, **2**, 662
1,3-Butadiene, 1,1-dimethoxy-3-trimethylsiloxy-
Diels–Alder reactions, **5**, 330
1,3-Butadiene, 2,3-dimethyl-
cycloaddition reactions, **5**, 199
[4 + 3] cycloaddition reactions, **5**, 603
Diels–Alder reactions, **5**, 372, 380
hydrobromination, **4**, 283
hydrosilylation, **8**, 780
zirconocene complex
reactions with carbonyl compounds, **1**, 163
1,3-Butadiene, 2-(*N*-dimethylaminomethyl)-3-(trimethylsilylmethyl)-
Diels–Alder reactions, **5**, 338
1,3-Butadiene, 1,1-dithio-
synthesis
via 2,3-sigmatropic rearrangement, **6**, 854
1,3-Butadiene, 2-ethyldiethylamino-
cycloaddition reactions
fulvenes, **5**, 629
1,3-Butadiene, 2-fluoro-
synthesis
via cyclopropane ring opening, **4**, 1020
1,3-Butadiene, 2-formyl-

iron tricarbonyl complex
reactions with organocuprates, **1**, 115
1,3-Butadiene, 2-hydroxy-
synthesis
via retro Diels–Alder reactions, **5**, 557
1,3-Butadiene, 2-(1′-hydroxyalkyl)-
synthesis
via 1-methylselenocyclobutyllithium, **1**, 709
1,3-Butadiene, 1-methoxy-
hetero Diels–Alder reaction
$Eu(fod)_3$-catalyzed, **2**, 671
high pressure, **2**, 663
1,3-Butadiene, 2-methoxy-3-methyl-
iterative rearrangements, **5**, 891
1,3-Butadiene, 2-methoxy-1-(phenylthio)-
Diels–Alder reactions, **5**, 333
1,3-Butadiene, 2-methoxy-3-(phenylthio)-
Diels–Alder reaction, **6**, 146
synthesis
via cyclobutenes, **5**, 683
1,3-Butadiene, 1-methoxy-3-(trimethylsiloxy)-
Diels–Alder reactions, **5**, 329
$ZnCl_2$-catalyzed, **2**, 663
[2 + 2] photocycloaddition, **5**, 1022
1,3-Butadiene, 2-methyl-
synthesis
via cycloaddition of 1-methylbicyclo[1.1.0]butane, **5**, 1186
1,3-Butadiene, 2-methyl-1-nitro-
synthesis, **6**, 109
1,3-Butadiene, 2-methyl-4-(trimethylsiloxy)-
Diels–Alder reactions, **5**, 376
1,3-Butadiene, 1-phenyl-
arylation
palladium catalysts, **4**, 849
hydrobromination, **4**, 283
hydrogenation
homogeneous catalysis, **8**, 449
selective reduction, **8**, 567
1,3-Butadiene, 1-phenyl-4-(2′-thienyl)-
photocyclization–oxidation, **5**, 720
1,3-Butadiene, 1-(phenylthio)-
Diels–Alder reactions, **5**, 333
1,3-Butadiene, 2-[(phenylthio)methyl]-
Diels–Alder reactions, **5**, 338
1,3-Butadiene, silyl-
Diels–Alder reactions, **5**, 335
synthesis, **3**, 487
vinylic coupling, **3**, 485
1,3-Butadiene, 1-trialkylsilyl-
acylation
Friedel–Crafts reaction, **2**, 721
1,3-Butadiene, 2-trialkylsilyl-
iron tricarbonyl complexes
acylation, **2**, 723
1,3-Butadiene, 2-tributylstannyl-
Diels–Alder reactions, **5**, 335
1,3-Butadiene, 2-triethylsilyl-
Diels–Alder reactions, **5**, 335
1,3-Butadiene, 1-trimethylsiloxy-
cycloaddition reactions
tropones, **5**, 620
cyclodimerization
[4 + 4] cycloaddition, **5**, 641
1,3-Butadiene, 2-trimethylsiloxy-
Diels–Alder reactions, **5**, 320, 329
1,3-Butadiene, 2-(trimethylsilylmethyl)-
Diels–Alder reactions, **5**, 337, 338
isoprenylation with, **2**, 721
1,3-Butadiene, 1-(trimethylsilyl)oxy-
reaction with singlet oxygen, **2**, 1068
1,3-Butadiene-2-carboxylate, 1-amino-
synthesis
via enamines and alkynic esters, **4**, 45
1,3-Butadiene-2,3-dicarbonitrile
synthesis
via retro Diels–Alder reaction, **5**, 566
1,3-Butadiene-2,3-dicarboxylic acid
synthesis
via retro Diels–Alder reaction, **5**, 565
1,2-Butadienoic acid
methyl ester
reaction with *C*-methyl-*N*-phenylnitrone, **5**, 255
2,3-Butadienoic acid
esters
ene reactions, **5**, 9
1,3-Butadiyne
alkylation, **3**, 284
1,3-Butadiyne, 1-alkyl-4-(*N*,*N*-dialkylamino)-
synthesis, **3**, 284
1,3-Butadiyne, bis(trimethylsilyl)-
alkylation, **3**, 284
hydrosilylation, **8**, 773
1,3-Butadiyne, 1-(*N*,*N*-dialkylamino)-
lithium derivative
synthesis, **3**, 284
Butadiynes
synthesis, **3**, 551
Butadiynes, 1,4-dialkynyl-
synthesis, **3**, 554
Butadiynes, 1,4-diaryl-
synthesis, **3**, 554
Buta-1,3-diynes, 1-trimethylsilyl-
acylation
Friedel–Crafts reaction, **2**, 725
Butanal
synthesis, **8**, 297
hydroformylation of propene, **3**, 1015
via hydrocarbonylation, **4**, 914
Butanal, 2-ethyl-
reaction with organometallic compounds
chemoselectivity, **1**, 148
Butanal, 3-hydroxy-
reaction with tetraallylzirconium, **1**, 157
Butanal, 2-phenyl-
reaction with organometallic reagents
diastereoselectivity, **1**, 151
Butanamide, diethyl-
alkylation, **3**, 68
Butane
autoxidation, **7**, 11
isomerization
Friedel–Crafts reaction, **3**, 334
Butane, 1-chloro-3-methyl-3-phenyl-
synthesis
Friedel–Crafts reaction, **3**, 320
Butane, 1,3-dichloro-3-methyl-
benzene alkylation by
Friedel–Crafts reaction, **3**, 320
Butane, 2,3-dimethyl-
oxidation
ozone, **7**, 14

Butane, 1,2-epoxy-
benzene alkylation with
Friedel–Crafts reaction, **3**, 313
Butane, 2,3-epoxy-
reaction with magnesium halides
epoxide ring opening, **3**, 755
rearrangement
boron trifluoride catalyzed, **3**, 742
Butane, 3-lithio-1-methoxy-
intramolecular solvated tetramer, **1**, 10
Butane, 1,1,3,3-tetramethyl-
bromination, **7**, 15
1,4-Butanedinitrile, 2-aryl-
reduction, **8**, 253
1,4-Butanediol
synthesis
via hydrogenation, **8**, 236
2,3-Butanediol
boronic esters, **3**, 797
chiral acetals
reduction, **8**, 222
oxidative cleavage, **7**, 707
pinacol rearrangement, **3**, 725
2,3-Butanedione
disilyl enol ethers, **2**, 605
Butanesulfonic acid, nonafluoro-
Friedel–Crafts reaction
bimolecular aromatic, **2**, 739
1,2,3,4-Butanetetracarboxylic acid
synthesis
via photolysis, **5**, 723
Butanoic acid
synthesis
via oxidation of carbon–tin bonds, **7**, 614
Butanoic acid, 4-aroyl-
Friedel–Crafts reaction, **2**, 759
synthesis, **2**, 744
Butanoic acid, 3-benzoylamino-
dilithium dianions
alkylation, **3**, 43
Butanoic acid, 3,3-dimethyl-
methyl ester
lithium enolate, crystal structure, **1**, 30
Butanoic acid, 4-dimethylamino-
reaction with *O*-methyl-*N,N'*-dicyclohexylisourea, **6**, 74
Butanoic acid, 3-hydroxy-
chiral synthesis
via microbial hydroxylation, **7**, 57
ethyl esters
alkylation, **3**, 44
methyl ester, dianion
aldol reaction, **2**, 225
Butanoic acid, 3-methyl-
ethyl ester
reduction, **3**, 617
Butanoic acid, 2-methyl-3-oxo-
ethyl ester
synthesis *via* samarium diiodide, **1**, 266
Butanoic acid, 4'-(2-naphthyl)-
Friedel–Crafts reaction
cyclization, **2**, 754
Butanoic acid, (4-phenylsulfonyl)-
dianion
reactions with imines, **1**, 350
Butanoic acid, sulfinyl-
Pummerer rearrangement
intramolecular, **7**, 196
Butanoic acid, 3'-(5,6,7,8-tetrahydro-2-phenanthryl)-
Friedel–Crafts reaction, **2**, 757
Butanoic acid, 4-(2-thionaphthoxy)-
Friedel–Crafts reaction, **2**, 765
2-Butanol
synthesis
via oxidation of organoboranes, **7**, 595
t-Butanol
Ritter reaction
to *N*-*t*-butyl acetamide, **6**, 261
1-Butanol, 4-anilino-3-methylamino-
asymmetric reduction
aluminum hydrides, **8**, 545
lithium aluminum hydride modifiers, **8**, 168
1-Butanol, 3-chloro-2-methyl-
synthesis from 2-butene
Prins reaction, **2**, 528
2-Butanol, 4-dimethylamino-1,2-diphenyl-3-methyl-
aluminum complex
reactions with keto esters, **1**, 86
2-Butanol, 3-methyl-
synthesis
via oxidation of organoboranes, **7**, 595
2-Butanol, 1-piperidyl-3,3-dimethyl-
diethylzinc reaction with benzaldehyde, **1**, 225
1-Butanol, 4-(2,6-xylidino)-3-methylamino-
asymmetric reduction
aluminum hydrides, **8**, 545
Butanone
aldol reaction
aliphatic aldehydes, **2**, 144
2-Butanone
aldol reaction
aliphatic aldehydes, **2**, 144
enamines
proton NMR, **6**, 712
enolates
arylation, **4**, 466
2-Butanone, 3,3-dimethyl-
lithium enolate
X-ray diffraction analysis, **1**, 1
reaction with zirconocene/isoprene complex, **1**, 163
2-Butanone, 4-hydroxy-
hydrogenation, **8**, 151
3-Butanone, 1-methoxy-
synthesis
via ring cleavage of methylenecyclopropane, **7**, 825
2-Butanone, 3-methyl-
acetylation, **2**, 834
reaction with crotyltitanium compounds, **1**, 158
1-Butanone, 3-methyl-1-(3-methyl-2-furyl)-
synthesis, **1**, 553
2-Butanone, phenyl-
hydrogenation
catalytic, **8**, 142
Reformatsky reaction
stereoselectivity, **2**, 291
2-Butanone, 1-(trimethylsilyl)-
synthesis
via acylation of copper reagents, **1**, 436
Butanoyl chloride, γ-furyl-
Friedel–Crafts reaction, **2**, 759
Butanoyl chloride, heptafluoro-
Friedel–Crafts reaction

bimolecular aromatic, **2**, 739
Butanoyl chloride, 4-(2-naphthyloxy)-
Friedel–Crafts reaction
regioselective, **2**, 765
Butanoyl chloride, 4-(2′-thionaphthoxy)-
Friedel–Crafts reaction, **2**, 765
Butatriene
synthesis
via retro Diels–Alder reaction, **5**, 589
1,2,3-Butatriene, 1,4-diphenyl-
hydrogenation
palladium-catalyzed, **8**, 436
2-Butenal, 2-methyl-
Diels–Alder reactions, **5**, 378
3-Butenal, methyl-2-phenyl-
synthesis, **1**, 560
1-Butene
asymmetric hydroformylation, **4**, 930
hydroformylation, **4**, 930
oxidation
Wacker process, **7**, 452
2-Butene
aminomercuration, **4**, 290
asymmetric hydroformylation, **4**, 930
dicarboxylation, **4**, 946
ene reactions, **5**, 2
hydroformylation, **4**, 930
oxidation
Wacker process, **7**, 451
synthesis
Ramberg–Bäcklund rearrangement, **3**, 861
cis-2-Butene
cyclobutanones from, **5**, 1087
oxidation, **7**, 462
2-Butene, 1-bromo-
alkylation, **3**, 253
reaction with organochromium compounds
anti selectivity, **1**, 179
synthesis
via 1,3-butadiene, **5**, 903
2-Butene, 1-cyano-
synthesis
via 1-bromo-2-butene, **6**, 230
2-Butene, 2,3-dideutero-
hydrochlorination, **4**, 272
1-Butene, 3,3-dimethoxy-2-methyl-
iterative rearrangements, **5**, 892
1-Butene, 3,3-dimethyl-
amidomercuration, **4**, 294
oxidation
Wacker process, **7**, 450
Pauson–Khand cycloaddition, **5**, 1041
2-Butene, 2,3-dimethyl-
ene reactions
Lewis acid catalysis, **5**, 4
hydroboration, **8**, 713
mechanism, **8**, 724
hydroformylation, **4**, 919
hydrosilylation, **8**, 776
photochemical cycloadditions
benzonitrile, **5**, 161
2-Butene, 2,3-diphenyl-
hydrogenation
stereochemistry, **8**, 426
2-Butene, 1,4-disilyl-
unsymmetrically substituted
acylation, **2**, 718
1-Butene, 3,4-epoxy-
reaction with Grignard reagents, **6**, 9
2-Butene, 1-iodo-3-trimethylsilyl-
alkylation by, **3**, 11
1-Butene, 3-methoxy-
reaction with nitrile oxide, **7**, 439
1-Butene, 4-methoxy-
reaction with magnesium hydride, **1**, 14
2-Butene, 2-methyl-
ene reactions
Lewis acid catalysis, **5**, 4
photolysis
with benzonitrile, **5**, 161
1-Butene, 3-nitro-
synthesis, **6**, 107
1-Butene, 4-nitro-
addition reaction with enolates, **4**, 104
2-Butene, 2-nitro-3-phenyl-
synthesis, **6**, 108
1-Butene, 2-phenyl-
hydrogenation
homogeneous catalysis, **8**, 463
1-Butene, 4-phenyl-
synthesis
via organochromium reagent, **1**, 175
1-Butene, 2,3,3-trimethyl-
deuterated
ene reactions, **5**, 2
Butenedioic acid, difluoro-
hydrogenation, **8**, 896
Butenedioic acid, fluoro-
hydrogenation, **8**, 896
3-Butene-1,2-diol, isopropylidene-
reaction with (ethoxycarbonyl)formonitrile oxide, **5**, 262
2-Butene-1,4-diones
epoxidations, **7**, 382
2-Butene oxide
deoxygenation, **8**, 889
hydrogenolysis, **8**, 882
3-Butenoic acid
hydrobromination, **4**, 282
synthesis
carbonylation of allylic chlorides, **3**, 1027
2-Butenoic acid, 3-bromo-
reaction with amines, **6**, 67
2-Butenoic acid, 2,3-dihydroxy-2-methyl-
hydroxylation
enantioselective, **7**, 441
2-Butenoic acid, 2-methyl-
hydroxylation
enantioselective, **7**, 441
2-Butenoic acid, 3-phenyl-
asymmetric hydrogenation
homogeneous catalysis, **8**, 461
methyl ester
hydrogenation, **8**, 452
Butenoic acid chloride
synthesis
via allyl chloride and carbon monoxide, **6**, 309
3-Butenol, 2-amino-
chiral synthesis, **6**, 88
2-Buten-1-ol, 2-*t*-butyl-
asymmetric epoxidation, **7**, 409
3-Buten-1-ol, 2,2-dimethyl-

hydrocarboxylation, **4**, 941
3-Buten-2-ol, 2-methyl-
oxidation
Wacker process, **7**, 453
2-Buten-1-ol, 2-methyl-4-phenyl-
asymmetric epoxidation, **7**, 409
3-Buten-2-ol, 3-phenyl-
hydrogenation
homogeneous catalysis, **8**, 447
3-Buten-1-ol, 1-phenyl-2-methyl-
synthesis
via trihaptotitanium compound, **1**, 159
α,β-Butenolide
synthesis
Knoevenagel reaction, **2**, 381
Δ^1-Butenolide
synthesis
Reformatsky reaction, **2**, 284
Butenolide anions
reactions with acetals
Lewis acid promoted, **1**, 347
Butenolides
chiral synthesis, **6**, 152
synthesis, **3**, 905; **7**, 596
use of disilyl enol ether, **2**, 619
via cyclofunctionalization of alkynoic acids, **4**, 393
via hydrocarboxylation, **4**, 937
via ortho lithiation, **1**, 472
via oxidation of a cyanohydrin, **1**, 551
via Peterson alkenation, **1**, 791
tandem vicinal difunctionalization, **4**, 249
Butenolides, hydroxy-
synthesis
multicomponent carbonylation, **3**, 1020
γ-Butenolides, 4-substituted
tandem vicinal difunctionalization, **4**, 249
Butenolides, 4-ylidene-
synthesis, **7**, 619
2-Buten-1-one, 1-phenyl-4,4,4-trifluoro-
reduction
dihydropyridines, **8**, 561
But-2-enoyl chloride, 3-methyl-
reaction with silyl ketene acetals, **2**, 804
Butenyl acetate
dicarboxylation, **4**, 948
2-Butenyl acetate, 3-methyl-
hydroformylation, **4**, 924
3-Butenyl bromide
coupling reactions
with phenyl Grignard reagents, **3**, 464
2-Butenylene dicarbamate
cyclization, **6**, 88
Butenyl radicals
cyclizations, **4**, 785
But-1-en-3-yne, 2-methyl-
photolysis
with benzophenone, **5**, 164
t-Butyl alcohol
solvent
radical reactions, **4**, 721
synthesis
via ethyl acetate, **1**, 398
t-Butylamide, *t*-octyl-
lithium derivative
enolate preparation, **2**, 600
t-Butylamine
imines
deprotonation, **6**, 720
Butyl benzoate
benzoyl chloride synthesis, **6**, 307
t-Butyl chromate
oxidation
ethers, **7**, 236
t-Butyl esters
carboxy-protecting groups
stability, **6**, 668
protecting groups
cleavage, **6**, 635
peptides, **6**, 633
t-Butyl hydroperoxide
asymmetric epoxidation, **7**, 394
chromium trioxide
alcohol oxidation, **7**, 278
oxidation
primary alcohols, **7**, 310
secondary alcohols, **7**, 323
propylene oxide synthesis, **7**, 375
reoxidant
Wacker process, **7**, 452, 462
safety, **7**, 394
secondary oxidant
osmium tetroxide oxidation, **7**, 439
storage, **7**, 394
t-Butyl hypochlorite
alkane chlorination, **7**, 17
t-Butyl hypoiodite
reaction with carboxylic acids, **7**, 723
t-Butyl isobutyrate
lithium enolate
crystal structure, **1**, 30
Butyl nitrate
nitration with, **6**, 110
t-Butyloxycarbonyl azide
protecting group
amines, **6**, 637
t-Butyloxycarbonyl group
carboxy-protecting group, **6**, 669, 670
cleavage, **6**, 635, 636
t-Butyloxymethyl group
alcohol protection, **6**, 647
t-Butyl peroxide
oxidative cleavage of alkenes
with molybdenum dioxide diacetylacetonate, **7**, 587
t-Butyl propionate
lithium enolate
crystal structure, **1**, 30
t-Butyl trimethylsilylacetate
lithium anion
Peterson alkenation, **1**, 789
2-Butyne
hydrogenation to *cis*-2-butene
homogeneous catalysis, **8**, 458
hydrozirconation, **8**, 690
reaction with iron carbene complexes, **5**, 1089
2-Butyne, 1,4-bis(trimethylstannyl)-
reaction with Eschenmoser's salt, **2**, 1000
2-Butyne, 1,4-dichloro-
reaction with bromine, **4**, 346
reaction with selenenyl halides, **4**, 342
reaction with sulfenyl halides, **4**, 336
1-Butyne, 3,3-dimethyl-
trimerization

rhodium catalysis, **5**, 1146
2-Butyne, hexafluoro-
hydrobromination, **4**, 286
1-Butyne, 3-methoxy-3-methyl-
organocopper compounds, **3**, 212
1-Butyne, 3-methyl-3-methoxy-
acylation
nontransferable ligand, **1**, 430
1-Butyne, 1-trimethylsilyl-
deprotonation
formation of organolithium reagent from, **2**, 993
2-Butyne-1,4-diol
carbomagnesiation, **4**, 878
1-Butyn-3-ol, 3-methyl-
trimerization
nickel catalysis, **5**, 1146
3-Butyn-2-ol, 2-methyl-
in terminal alkyne synthesis, **3**, 531
3-Butynone
conjugate additions
trialkylboranes, **4**, 163
[3 + 2] cycloaddition reactions
with 1,3,3-trimethyl-1-(trimethylsilyl)allene, **5**, 278
ene reactions
Lewis acid catalysis, **5**, 8
photolysis
with isobutene, **5**, 164
Butyraldehyde, 3-methoxy(methoxy)-
α-alkoxyaldimines derived from
reaction with allyl organometallic compounds, **2**, 987
Butyrate, glycidyl-
synthesis
enzymatic resolution, **6**, 340
Butyric acid, γ-amido-
synthesis
via aziridines, **6**, 96
Butyric acid, α-amino-
asymmetric synthesis, **8**, 146
Butyric acid, 2-amino-4-phosphono-
synthesis
via intramolecular ester enolate addition reactions, **4**, 111
Butyric acid, γ-bromo-
reactions with samarium diiodide
lactone synthesis, **1**, 259
Butyric acid, 3-(dimethylphenylsilyl)-
ethyl ester
reaction with *N*-silylimines, **2**, 936
ethyl ester, enolate
Mannich reaction, **2**, 926
Butyric acid, 2,3-dioxo-
t-butyl ester
rearrangement, **3**, 822, 831
Butyric acid, α-halo-
aryl esters
cycloalkylation, **3**, 324
Butyric acid, 4-hydroseleno-
ring closure, **6**, 462
Butyric acid, 3-hydroxy-
enolates
thienamycin synthesis, **2**, 925
esters
reaction with imines, **5**, 102
methyl ester
β-lactam synthesis, **2**, 937
Butyric acid, 4-phenyl-
Schmidt reaction, **6**, 817
Butyric acid, 3-trichloromethyl-
synthesis
via conjugate addition to α,β-unsaturated carboxylic acid, **4**, 202
Butyric acid, triisopropylsiloxy-
cycloaddition with imines, **5**, 99
γ-Butyrolactone, 2-amino-
synthesis, **5**, 1080
γ-Butyrolactone, α,α-bis(phenylthio)-
use as enolate precursors, **2**, 186
γ-Butyrolactone, 2,4-disubstituted
synthesis
via [2 + 2 + 2] cycloaddition, **5**, 1138
γ-Butyrolactone, 4,5-*trans*-disubstituted
synthesis
via 1,2-addition of organocuprates, **1**, 110
Butyrolactone, 5-ethenyl-
synthesis, **3**, 245
γ-Butyrolactone, (*E*)-α-heptylidine-
synthesis
use of enolates, **2**, 186
Butyrolactone, hydroxy-
alkylation, **3**, 41
dianion
diastereofacial selectivity, **2**, 204
synthesis
via alkylation of protected cyanohydrin, **1**, 552
γ-Butyrolactone, β-keto-
synthesis
via Reformatsky-type reaction, **1**, 551
δ-Butyrolactone, β-keto-
synthesis
Blaise reaction, **2**, 298
Butyrolactone, menthyloxy-
Diels–Alder reactions, **5**, 371
Butyrolactone, α-methylene-
synthesis, **5**, 1076; **7**, 102, 239, 502
carbonylation of homoallylic alcohols, **3**, 1031
via retro Diels–Alder reactions, **5**, 578
γ-Butyrolactone, 2,3,3-trimethyl-
synthesis
via hydrocarboxylation, **4**, 941
γ-Butyrolactone 2-acetic acid esters
synthesis
carbonylation, **3**, 1040
Butyrolactones
bicyclic
synthesis *via* Michael addition, **4**, 24
synthesis *via* samarium diiodide, **1**, 269
lithium enolate
aldol reaction, diastereoselection, **2**, 204
synthesis
via ketyl–alkene coupling reaction, **1**, 268
γ-Butyrolactones
alkynic ketone synthesis from, **1**, 419
hydrogenation, **8**, 246
polysubstituted
synthesis, **3**, 843
β-substituted
synthesis *via* conjugate addition to oxazepines, **4**, 206
synthesis
enantioselectivity, **3**, 956
Perkin reaction, **2**, 401

via [3 + 2] cycloaddition reactions, **5**, 297
via metal-catalyzed cycloaddition, **5**, 1200
via α-sulfonyl carbanions, **6**, 159
Butyronitrile
reduction
lithium triethoxyaluminum hydride, **8**, 274
Butyronitrile, 3,3-dimethyl-2-oxo-
synthesis
via acyl halides, **6**, 233
Butyronitrile, phenyl-
synthesis
via organochromium reagent, **1**, 175
Butyrophenone
oxidative rearrangement
solid support, **7**, 845

C

C_{40} archaebacterial diol
synthesis
use of aldol reaction, **2**, 195
C_{19} gibberellins
synthesis
Knoevenagel reaction, **2**, 370
Cadiot–Chodkiewicz coupling, **3**, 553
alkynes
organocopper compounds, **3**, 219
Cadmium, γ-alkoxyallyl-
reaction with glyceraldehyde acetonide, **2**, 31
Cadmium, aryl-
alkylation, **3**, 260
Cadmium, dicrotyl-
reactions with aldehydes
stereoselectivity, **1**, 220
Cadmium, methyl-
addition reactions
chiral aldehydes, **1**, 221
Cadmium chloride
sodium borohydride modifier
acyl halide reduction, **8**, 263
Cadmium reagents, alkyl-
addition reactions, **1**, 225
Cadmium reagents, aryl-
addition reactions, **1**, 225
Caesalpinine
synthesis
via photocycloaddition, **5**, 176
Cafestol
synthesis
via cyclopropane ring opening, **4**, 1043
Caged compounds
transannular reactions, **3**, 382
Cage-like structures
synthesis, **3**, 854
Caglioti reactions
carbonyl deoxygenations, **8**, 343
Calameon
synthesis
transannular ene reaction, **2**, 553
Calciferol
synthesis
via precalciferol, **5**, 700
Calciferol, 1,25-dihydroxy-
A ring
synthesis *via* intramolecular ene reaction, **5**, 18
Calcimycin
synthesis, **1**, 568; **3**, 126, 139
aldol reaction of magnesium enolate, **2**, 219
final step, **1**, 409
introduction of 2-keto pyrrole, **1**, 409
model system, **1**, 410
Calcium
Birch reduction, **8**, 492
dissolving metal reductions
unsaturated hydrocarbons, **8**, 480
reduction
ammonia, **8**, 113
enones, **8**, 524
epoxides, **8**, 881
Calcium hydride
reduction
acyl halides, **8**, 262
Calcium hypochlorite
glycol cleavage, **7**, 706
oxidation
secondary alcohols, **7**, 318
Calichemicins
synthesis
copper catalysts, **3**, 217
Ramberg–Bäcklund rearrangement, **3**, 883
via electrocyclization, **5**, 736
California red scale female sex pheromone
A1 component
synthesis *via* ene reaction with methyl propiolate, **5**, 8
California red scale pheromone
synthesis
via conjugate addition to α,β-unsaturated acetal, **4**, 209
via conjugate addition to α,β-unsaturated carboxylic acid, **4**, 202
Calonectrin
synthesis
via cyclohexadienyl complexes, **4**, 680
Camphene
hydrozirconation, **8**, 689
reaction with hydrofluoric acid, **4**, 270
rearrangement, **3**, 705
Vilsmeier–Haack reaction, **2**, 782
Camphenic acid
synthesis
via [4 + 3] cycloaddition, **5**, 603
Camphenilol
rearrangement, **3**, 706
Campherenone
synthesis, **3**, 427
via [3 + 2] cycloaddition reactions, **5**, 286
Camphor
chiral enoates
conjugate additions, **4**, 202
enol silane derivative
Mannich reaction, **2**, 908
enzymic hydroxylation
cytochrome *P*-450, **7**, 80
ketal
reduction, **8**, 222
reaction with lithium aluminum hydride
chiral modification of reducing agents, **8**, 159
rearrangement, **3**, 710
reduction
dissolving metals, **8**, 109, 110, 120
dissolving metals/ammonia, **8**, 112
ytterbium/ammonia, **8**, 113
Ritter reaction
with acetonitrile, **6**, 270
silyl ketene acetals, derivatives of
stereoselective reactions, **2**, 636
synthesis
via [3 + 2] cycloaddition reactions, **5**, 286
via intramolecular ene reactions, **5**, 21

Camphor, 3-*endo*-bromo-
rearrangement, **3**, 711
Camphor, diazo-
Wolff rearrangement, **3**, 900
Camphor, 3,3-dibromo-
Wagner–Meerwein rearrangement, **3**, 712
Camphor, iodo-
oxime
Beckmann fragmentation, **6**, 774
Camphoric acid, monoperoxy-
oxaziridine synthesis, **1**, 838
Camphor quinone, dihydro-
oxime
Beckmann fragmentation, **7**, 700
Camphor-9-sulfonic acid
synthesis, **3**, 710
Camphor-10-sulfonic acid
synthesis
from camphor, **3**, 710
Camphor-9-sulfonic acid, 3-*endo*-bromo-
synthesis, **3**, 711
Camphor-10-sulfonyl chloride
conjugate additions
enoates, **4**, 201
Camptothecin
synthesis
via activated allene, **4**, 54
Canadine
synthesis
via 6-*exo*-trig cyclization, **4**, 39
via tandem vicinal difunctionalization, **4**, 251
Candida cloacae
hydrocarbon oxidation, **7**, 56
Cannabinoids
microbial hydroxylation, **7**, 66
synthesis, **3**, 127
Knoevenagel reaction, **2**, 372
via Diels–Alder reactions, **5**, 468
Cannabinol
photochemical ring opening, **5**, 727
Cannabinol, hexahydro-
synthesis
via Diels–Alder reaction, **5**, 468
Cannabinol, 3-hydroxyhexahydro-
synthesis
via Diels–Alder reaction, **5**, 468
Cannabinol, 7-oxohexahydro-
tosylhydrazone acetate
Bamford–Stevens reaction, **6**, 776
Cannabinol, Δ^1-tetrahydro-
biomimetic synthesis, **2**, 621
Cannabisativine
synthesis
oxime reactions with allyl organometallic compounds, **2**, 996
Cannabisativine, anhydro-
synthesis
via Diels–Alder reaction, **5**, 414
Cannithrene II
synthesis, **3**, 591
Cannivonine
synthesis
via Cope rearrangement, **5**, 814
Cannivonine, dihydro-
synthesis
via dienyliron complexes, **4**, 673
Cannizzaro reaction
catalysts, **8**, 86
transition metals, **8**, 86
electron transfer mechanism, **3**, 824
enolizable aldehydes
transition metal catalysts, **8**, 86
mechanism
ketyl radical, **8**, 86
tetracoordinate intermediate, **8**, 86
reduction of nonenolizable aldehydes
hydride transfer, **8**, 86
Cantharidin
synthesis
via Diels–Alder reaction, **5**, 342
Capnellane
synthesis, **3**, 389
Capnellene
synthesis, **3**, 384, 404
via carbonyl–alkyne cyclization, **3**, 602
via Tebbe reagent, **1**, 748
$\Delta^{8(13)}$-Capnellene
synthesis, **6**, 780
$\Delta^{9(12)}$-Capnellene
synthesis, **3**, 20, 288
via magnesium-ene reaction, **5**, 40
via Nazarov cyclization, **5**, 763, 779
via ring-opening metathesis polymerization, **5**, 1121
via Tebbe reagent, **5**, 1124
$\Delta^{9(12)}$-Capnellene-8β,10α-diol
synthesis, **3**, 603
Caproic acid
reduction
hydrides, **8**, 260
Caproic acid, ε-amino-
catalyst
Knoevenagel reaction, **2**, 343
Caprolactam, 2-chloro-
rearrangements, **3**, 849
Capsaicinoids
synthesis
via Julia coupling, **1**, 797
1-Carbacephem
synthesis
Dieckmann reaction, **2**, 824
Dinagel reaction, **2**, 824
Carbacyclins
synthesis
stereoselectivity, **1**, 535
via cycloalkenyl sulfone, **4**, 79
via zirconium-promoted bicyclization of enynes, **5**, 1166
Carbalumination, **8**, 756
intramolecular, **8**, 758
Carbamates
anodic oxidation, **7**, 804
epoxidation directed by, **7**, 367
α'-lithioalkyl
alkylation, **3**, 88
α-methoxylation, **7**, 805
oxidation
electrochemical, **2**, 1051
reduction, **8**, 254
Carbamates, 2-alkenyl-

homoaldol reaction, **6**, 863
Carbamates, allyl-*N*-phenyl-
reaction with lithium cuprates, **3**, 222
Carbamates, α-chloro-
Carbamates
imines
Diels–Alder reactions, **5**, 405
Carbamates, *N*-(3-diphenylpropyl)-
synthesis, **6**, 94
Carbamates, *N*-halo-
reaction with conjugated alkenynes, **7**, 505
Carbamates, *N*-(1-hydroxyalkyl)-
synthesis, **2**, 1049
Carbamates, *N*-methoxymethyl
synthesis, **7**, 650
Carbamates, vinylogous
reaction with Grignard reagents, **2**, 388
Carbamazepin
epoxide
ring opening, **3**, 737
Carbamic acid, 3-alken-1-ynyl-
synthesis
stereospecificity, **2**, 94
Carbamic acid, allylthio-
alkylation, **3**, 103
sigmatropic rearrangement, **3**, 103
Carbamic acid, *N*,*N*-dialkyl-
allyl esters
reactions with carbonyl compounds, **2**, 67
Carbamic acid, *N*,*N*-diisopropyl-2-alkynyl-
titanium reagent
reaction with aldehydes, **2**, 94
Carbamic acid, *N*,*N*-dimethyldithio-
methylthiomethyl ester
alkylation, **3**, 136
Carbamic acid, dithio-
α-alkylated allylic
rearrangement, **3**, 117
allyl ester
reduction, **3**, 108
Carbamic acid, γ-methylthioallyl-
alkylation, **3**, 103
Carbamoyl chloride, *N*,*N*-dialkyl-
adducts
amides, **6**, 492
Carbamycin B
synthesis
via photoisomerization, **5**, 232
Carbanion-accelerated rearrangements
small rings, **5**, 1004–1006
Carbanions
acylation, **6**, 445
aliphatic
crystal structures, **1**, 9
alkali metal cations, **1**, 1–42
aggregation state, **1**, 5
carbonyl addition reactions, **1**, 49–74
coordination geometry, **1**, 7
coordination number, **1**, 7
alkaline earth metal cations, **1**, 1–42
aggregation state, **1**, 5
carbonyl addition reactions, **1**, 49–74
coordination geometry, **1**, 7
coordination number, **1**, 7
α-alkoxy
from protected cyanohydrins, **3**, 197
silicon-stabilized, alkylation, **3**, 198
alkynic
crystal structure, **1**, 20
alkynyl
alkylation, **3**, 271–292
allylic
boron-stabilized, **1**, 502
crystal structure, **1**, 18
allylic heteroatom-stabilized
alkylation, **3**, 196
allylic sulfinyl
addition reactions with carbonyl compounds, **1**, 517
reactions with enones, **1**, 520
allylic sulfonyl
reactions with C=X bonds, **1**, 529
anodic oxidation, **7**, 805
antimony-stabilized
alkylation, **3**, 203
arsenic-stabilized
alkylation, **3**, 203
aryl
crystal structure, **1**, 21
benzylic α-alkoxy
alkylation, **3**, 196
bis(dialkoxyboryl) stabilization
reactions with aldehydes, **1**, 501
bismuth-stabilized
alkylation, **3**, 203
boron-stabilized, **1**, 487–503
acylation, **1**, 497
alkylation, **1**, 495; **3**, 199
calculations, **1**, 487
carboxylation, **1**, 498
crystal structure, **1**, 488
geometry, **1**, 488
halogenation, **1**, 501
nonallylic, **1**, 494
reactions with aldehydes and ketones, **1**, 498
reactions with epoxides, **1**, 496
reactions with metal halides, **1**, 494
synthesis, **1**, 489
crystallization, **1**, 41
crystal structures, **1**, 8
2D-HOESY NMR, **1**, 41
dithiocarboxylation, **6**, 456
electron-transfer equilibria, **7**, 850
α-epoxy-
phosphorus-stabilized, alkylation, **3**, 199
germanium-stabilized
alkylation, **3**, 203
halogen-stabilized
alkylation, **3**, 202
heteroaromatic
alkylation, **3**, 260
heteroatom-stabilized
alkylation, **3**, 192
α-heteroatom stabilized
addition reactions, **4**, 115–117
heteroatom-substituted
crystal structure, **1**, 34
hydride donors
reduction of carbonyls, **8**, 98
lead-stabilized

alkylation, **3**, 203
mixed metal cations
crystal structure, **1**, 39
nitrogen-stabilized
addition reactions, **4**, 116
alkylations, **3**, 65–82
carbonyl compound addition reactions, **1**, 459–482
nitro-stabilized
reactions, **2**, 321
nonstabilized
alkylations, **3**, 207–233
nucleophilic addition/electrophilic coupling, **4**, 237–263
organochromium(III) equivalents, **1**, 174
oxygen-stabilized
addition reactions, **4**, 116
alkylation, **3**, 193
phosphorus-stabilized
addition reactions, **4**, 115
alkylation, **3**, 200
phosphoryl-stabilized
Wittig reaction, **1**, 761
$S_{RN}1$ reactions, **4**, 471
selenium-containing
alkylation, **3**, 85–181
selenium-stabilized, **1**, 629–724
reactions with carbonyl compounds, **1**, 672
synthesis, **1**, 630, 635
silicon-stabilized, **1**, 579–625
addition reactions, **4**, 116
alkylation, **3**, 200
σ–π-bonded, **1**, 583
α-sulfenylated allylic, **1**, 508
sulfenyl stabilization, **1**, 506
addition to carbonyl compounds, **1**, 506
configuration, **1**, 506
sulfinyl stabilization, **1**, 512
addition to C=N bonds, **1**, 515
addition to carbonyl compounds, **1**, 513
addition to nonactivated C=C bonds, **1**, 516
configuration, **1**, 512
sulfonimidoyl stabilization, **1**, 531
configuration, **1**, 531
reactions with carbonyl compounds, **1**, 532
sulfonyl stabilization, **1**, 528
configuration, **1**, 528
reactions with carbonyl compounds, **1**, 529
sulfur-containing
alkylation, **3**, 85–181
sulfur-stabilized, **1**, 505–536
addition reactions, **4**, 115
thioacylation, **6**, 453
thioimidate synthesis, **6**, 540
tin-stabilized
alkylation, **3**, 203
vinylic
crystal structure, **1**, 19
Carbanions, α-seleno
reaction with carboxylic acid derivatives, **1**, 694
synthesis, **1**, 655
Carbanions, siloxy-
preparation, **2**, 601
Carbanions, α-silyl
addition reactions
imines, **1**, 624
ambient, **1**, 623
crystal structure, **1**, 16
functionalized
addition reactions, **1**, 621
stabilization
hyperconjugation, **1**, 582
synthesis
general methods, **1**, 618
Carbapenams
synthesis
Eschenmoser coupling reaction, **2**, 887
Carbapenems
chiral
synthesis, **2**, 611
synthesis
Ugi reaction, **2**, 1102, 1103
via cycloaddition with CSI, **5**, 105
via Diels–Alder reactions, **5**, 407
via intramolecular ester enolate addition reactions, **4**, 110
via Wittig cyclization, **1**, 434
Carbapenems, 1β-methyl-
synthesis, **2**, 1059
1-Carbapen-2-ene
synthesis, **7**, 620
6a-Carbaprostaglandin I_2
synthesis
via Johnson sulfoxime reaction, **1**, 742
Carbazole
hydrogenation, **8**, 612
reduction
dissolving metals, **8**, 614
synthesis
via intramolecular vinyl substitution, **4**, 847
via thermolysis, **5**, 725
Carbazole, *N*-acyl-
reduction
metal hydrides, **8**, 270, 273
Carbazole, allyl-
anion
γ-alkylation, **2**, 61
Carbazole, hexahydro-
synthesis, **7**, 524
Carbazole, hydroxy-
synthesis
via FVP, **5**, 732
Carbazole, *N*-methyltetrahydro-
aminoalkylation
Mannich reaction, **2**, 967
Carbazole, *N*-9-phenylnona-2,4,6,8-tetraenoyl-
reduction
metal hydrides, **8**, 273
Carbazole, 1,2,3,4-tetrahydro-
reduction
dissolving metals, **8**, 615
Carbazole aminals
lithiation
addition reactions, **1**, 463
Carbene complexes
alkene metathesis, **5**, 1115
alkylation, **5**, 1075
cleavage, **5**, 1083
coupling reactions
with alkenes, **5**, 1084
with alkynes, **5**, 1089

cumulenes, **5**, 1107
cycloaddition
reactions, **5**, 1065–1113
nucleophilic substitutions, **5**, 1083
reactions, **5**, 1067
Carbene complexes, alkyl aminoalkylation, **5**, 1076
Carbene complexes, alkyl pentacarbonyl-
alkylation, **5**, 1076
anions
reaction with carbonyl compounds, **5**, 1076
Carbene complexes, amino-
aldol reactions, **5**, 1080
cycloaddition reactions, **5**, 1074
synthesis, **5**, 1066
Carbene complexes, tetracarbonyl phosphine
alkylation, **5**, 1076
Carbene complexes, α,β-unsaturated
Michael additions, **5**, 1081
Carbenes
chromium complexes
alkene synthesis, **1**, 807
deoxygenation
epoxides, **8**, 890
generation, **4**, 961
metal complexes
alkene synthesis, **1**, 807
reaction with alkanes, **7**, 8, 10
reaction with alkenes, **4**, 953
reaction with nitriles, **6**, 401
thioimidate synthesis, **6**, 540
titanium–zinc complexes
reactions with esters, **1**, 809
transition metal complexes
use in synthesis, **4**, 976–986
Carbenes, alkynyl-
transition metal complexes
[2 + 2] cycloaddition reactions, **5**, 1067, 1068
cycloaddition reactions with 1,3-dienes, **5**, 1072
ene reactions, **5**, 1075
Carbenes, cyclopropyl-
ring enlargement
cyclobutene synthesis, **5**, 677
Carbenes, dichloro-
generation, **4**, 1000
Carbenes, difluoro-
generation, **4**, 1000, 1001
Carbenes, dihalo-
addition to π-bonds, **4**, 1002–1005
electronic configuration, **4**, 1002
generation, **4**, 1000–1002
reaction with imines, **6**, 498
reaction with enol ethers, **1**, 878
structure, **4**, 1000
Carbenes, diiodo-
generation, **4**, 1001
Carbenes, diphenyl-
transition metal complexes
reaction with alkenes, **4**, 980
Carbenes, keto-
addition to alkenes, **4**, 1031–1064
Carbenes, α-siloxy-
intermediates
in enol ether preparation, **2**, 601
Carbenes, vinyl-
adducts
Cope rearrangement, **5**, 804
[4 + 3] cycloaddition reactions, **5**, 599, 604
α,β-unsaturated
addition reaction with enolates, **4**, 104
Carbenium ions
electron-transfer equilibria, **7**, 850
non-Kolbe electrolysis, **3**, 649
Carbenium ions, trialkoxy-
orthoester synthesis, **6**, 562
Carbenium salts, dialkoxy-
2,2-bis(dialkoxy)carbonitrile synthesis, **6**, 565
Carbenoids
alkylation, **3**, 202
deoxygenation
epoxides, **8**, 890
displacement reactions, **2**, 1049
halogen-stabilized
epoxidation, **1**, 830
insertion, **3**, 1047
reaction with alkenes, **4**, 953
Carbenoids, keto-
addition to alkenes, **4**, 1034–1050
regioselectivity, **4**, 1035
addition to alkynes, **4**, 1050–1052
generation, **4**, 1032
Carbenoids, β-oxido-
rearrangement, **1**, 873
Carbenoids, vinyl-
[4 + 3] cycloaddition reactions, **5**, 599
Carbinolamines
reduction, **8**, 974
Carbinols, allylvinyl-
rearrangements, **1**, 885
Carbinols, azido-
synthesis, **6**, 253
Carbinols, bis(2-pyridyl)-
synthesis, **3**, 826
Carbinols, diethylphenyl-
synthesis
via triphenylchromium complex, **1**, 176
Carbinols, divinyl-
asymmetric epoxidation, **7**, 416
regioselective rearrangements
via Claisen rearrangement, **5**, 851
Carbinols, ethynyl-
selective reduction, **8**, 530
Carbinols, α-silyl-
preparation, **2**, 601
Carbinols, triphenyl-
reduction
dissolving metals, **8**, 526
Carboalkoxylation
halides
aryl and vinyl, **3**, 1028
Carboalumination
alkenes, **4**, 887–893
regioselectivity, **4**, 887
catalysis
transition metal complexes, **4**, 889
internal alkynes, **4**, 890
intramolecular, **4**, 887
vinylalanes, **3**, 266
Carboboration
alkenes, **4**, 884–887

alkynes, **4**, 886
intramolecular, **4**, 884
Carbocations
o-hydroxybenzyl
Diels–Alder reactions, **5**, 501
Carbocations, α-fluoro-
in fluorination of alkenes, **4**, 344
Carbocupration
alkenes, **4**, 893–903
intramolecular, **4**, 898
Wittig alkenation
diene synthesis, **4**, 262
Carbocycles
synthesis
via Ireland silyl ester enolate rearrangement, **5**, 841, 843
Carbocyclic compounds
aminoalkylation
Mannich reaction, **2**, 961
Vilsmeier–Haack reaction, **2**, 779
Carbocyclines
synthesis
via Pauson–Khand reaction, **5**, 1060
Carbocyclization
hydroalumination, **8**, 758
Carbodealumination, **8**, 755
Carbodiimides
acid anhydride synthesis, **6**, 384
amidine synthesis, **6**, 546
cycloaddition reactions
isocyanates, **5**, 1156
ketenes, **5**, 99
peptide synthesis
coupling reagents, **6**, 385
Carbodiimides, dicyclohexyl-
acid anhydride synthesis, **6**, 313
activator
alcohol oxidation, DMSO, **7**, 293
acylation
amino acids, **6**, 387
esterification, **6**, 334
thiol ester synthesis, **6**, 437
Carbodiimides, diisopropyl-
peptide synthesis
solid phase, **6**, 387
Carbodiimides, 1-(3-dimethylaminopropyl)-3-ethyl-
Pfitzner–Moffatt oxidation, **7**, 294
Carbodiimidinium iodide
reaction with alcohols
iodination, **6**, 214
Carbodiphosphoranes
reactions with halogen compounds
formation of diphosphaallyl cations, **6**, 190
synthesis, **6**, 196
Carbodiphosphoranes, hexaphenyl-
reactions with heteroallenes, **6**, 190
synthesis, **6**, 196
Carbodithioates, carboxymethyl
synthesis
via alkylation of sodium carbodithioates, **6**, 454
Carbodithioates, phenyl
synthesis
via intramolecular thioacylation, **6**, 454
Carbohydrates
epoxides
reduction, **8**, 875, 878
fused
synthesis *via* radical cyclization, **4**, 792
α-ketol rearrangement, **3**, 831
nucleophilic addition reactions
stereoselectivity, **1**, 55
oxidation, **7**, 294
Collins reagent, **7**, 259
DMSO, **7**, 295, 296
pyridinium chlorochromate, **7**, 265
permethylation
functionalization for analysis, **6**, 647
protected
cleavage, **8**, 959
regiodifferentation
hydroxy group protection, **6**, 660
Sharpless–Masamune synthesis
Pummerer rearrangement in, **7**, 196
synthesis
Dieckmann reaction, **2**, 827
hetero Diels–Alder reaction, **2**, 663
via 1,3-dipolar cycloadditions, **4**, 1077
via osmium tetroxide, **7**, 440
via 5-*exo*-trig cyclization, **4**, 38
Ugi reaction
chiral templates, **6**, 405
Carbohydrates, azido
synthesis
via sulfonates, **6**, 245
Carbohydrates, 4-methoxybenzyl ethers
oxidation, **7**, 237
Carbolines
reduction
borohydrides, **8**, 618
β-Carbolines
lithiated formamidines
reaction with benzaldehyde, **1**, 482
synthesis, **3**, 72
β-Carbolines, 1-alkyl-3-methoxycarbonyl-1,2,3,4-tetrahydro-
synthesis
Mannich reaction, **2**, 1017
β-Carbolines, 1-alkyltetrahydro-
synthesis, **6**, 738
β-Carbolines, *trans*-N_b-benzyl-3-methoxycarbonyl-1-substituted-1,2,3,4-tetrahydro-,
synthesis
Mannich reaction, **2**, 1017
β-Carbolines, 3,4-dihydro-
silylation, **1**, 366
γ-Carbolines, dihydro-
synthesis, **5**, 1109
γ-Carbolines, hexahydro-
synthesis, **8**, 613
β-Carbolines, tetrahydro-
1,3-disubstituted
synthesis, Mannich reaction, **2**, 1017
synthesis, **6**, 737
Mannich reaction, **2**, 1017
Carbolithiation
alkenes, **4**, 867–873
intramolecular, **4**, 871
regioselectivity, **4**, 868
Carbomagnesiation
alkenes, **4**, 873–879

regioselectivity, **4**, 874
alkynes, **4**, 877–879
catalysis
transition metal complexes, **4**, 875
heterocycle synthesis, **4**, 877
intramolecular
alkenes, **4**, 876
Carbomercuration
alkynes, **4**, 904
Carbometallation
alkenes, **4**, 865–906
chemoselectivity, **4**, 866
definition, **4**, 866
heteroconjugate addition reactions, **4**, 120
organotransition metal compounds, **5**, 1163
reaction conditions, **4**, 867
regioselectivity, **4**, 866
stereoselectivity, **4**, 867
Carbomycin B
synthesis
via cycloheptadienyliron complexes, **4**, 686
Carbon
chromium(VI) oxide intercalation
alcohol oxidation, **7**, 282
Carbonates
alcohol protection, **6**, 657
cyclic
diol protection, **6**, 662
enol esters, **6**, 395
[3 + 2] cycloaddition reactions, **5**, 303
reduction
stannane, **8**, 824
Carbonates, α-methoxy-
reaction with enol silanes
Lewis acid mediated, **2**, 635
Carbonation
organoaluminum compounds, **8**, 737
organoytterbium compounds, **1**, 277
Carbon–boron bonds
oxidation, **7**, 593–608
Carbon–carbon bonds
electrochemical oxidation, **7**, 794
formation
C—H insertion, **3**, 1045–1062
oxidation, **7**, 793
Carbon dioxide
conjugate enolate trapping, **4**, 261
reactions with π-allylpalladium complexes, **4**, 601
regioselectivity, **4**, 643
Carbon disulfide
Knoevenagel reaction, **2**, 364
thioacylation
amines, **6**, 428
Carbon–halogen bonds
oxidation, **7**, 653–669
Carbon–hydrogen bonds
cleavage, anodic oxidation, **7**, 793
oxidation, **7**, 793
Carbonic acid
derivatives
adducts with amides, **6**, 491
Knoevenagel reaction, **2**, 368
Carbonitriles, 2-alkoxy-2-dialkylamino-
2,2-bis(dialkylamino)carbonitrile synthesis, **6**, 578
synthesis, **6**, 573
Carbonitriles, 2′-azido-2-phenyl-,
1,3-dipolar cycloaddition, **4**, 1101
Carbonitriles, 2,2-bis(dialkoxy)-
synthesis, **6**, 564
Carbonitriles, 2,2-bis(dialkylamino)-
synthesis, **6**, 577
tris(dialkylamino)alkane synthesis, **6**, 580
Carbonitriles, 2,2-dihalo-
2,2-bis(dialkylamino)carbonitrile synthesis, **6**, 577
Carbonium ions
hydride acceptors, **8**, 91
Carbon–mercury bonds
oxidation, **7**, 631
ozonolysis, **7**, 637
Carbon–metal bonds
oxidation, **7**, 613–638
Carbon monoxide
addition reactions
alkenes, **4**, 913–949
hydrosilylation in the presence of, **8**, 788
reaction with alcohols and amines
mechanism, **3**, 1016
reaction with π-allylpalladium complexes, **4**, 600
regioselectivity, **4**, 643
stereochemistry, **4**, 625
reaction with nitriles, **6**, 401
reaction with zirconium compounds, **8**, 691
reductions
aromatic nitro compounds, **8**, 372
Carbon–nitrogen bonds
radical additions
cyclizations, **4**, 815–818
Carbon–nitrogen compounds
1,2-addition reactions
organoaluminum compounds, **1**, 98
Carbon nucleophiles
aromatic nucleophilic substitution, **4**, 426–433
Carbonochloridothioates
ketone synthesis from
Grignard reagents, **3**, 463
Carbon–oxygen bonds
radical additions
cyclizations, **4**, 815–818
Carbon–palladium bonds
oxidation, **7**, 629
Carbon–selenium bonds
formation, **7**, 619
Carbon–silicon bonds
oxidation, **7**, 641–650
Carbon–sulfur bonds
formation, **7**, 515
Carbon tetrahalides
imidoyl halide synthesis, **6**, 524
Carbon–tin bonds
oxidation, **7**, 614
unactivated
oxidation, **7**, 614
Carbonylation, **3**, 1015–1041
additive
mechanism, **3**, 1019
alkanes
transition metal catalysis, **7**, 6
catalysts, **3**, 1016
direct
mechanism, **3**, 1018

double, **3**, 1039
homoenolates, **2**, 451
mechanisms, **3**, 1016
multicomponent
mechanism, **3**, 1020
substitutive
mechanism, **3**, 1018
Carbonyl chloride
acid chloride synthesis, **6**, 304
Carbonyl compounds
acyclic
reduction, **8**, 7
addition reactions
carbanions, **1**, 49–74
carbon-centred radicals, **4**, 765
nitrogen-stabilized carbanions, **1**, 460
organochromium compounds, **1**, 177
α-silyl phosphonates, **1**, 622
sulfur-stabilized carbanions, **1**, 506
aliphatic
intermolecular pinacol coupling reactions, **3**, 570
alkenation
enol ether preparation, **2**, 596
alkene synthesis from, **1**, 730
α-alkylated
asymmetric synthesis, enantioselectivity, **3**, 30
diastereoselective synthesis, **3**, 34
alkylidenation, **5**, 1122–1126
α,β-alkynic
conjugate additions, **4**, 185–187
allylation
preparation of 1,4-dicarbonyl compounds, **7**, 455
aromatic
intermolecular pinacol coupling reactions, **3**, 564
α-arylated
synthesis *via* $S_{RN}1$ reaction, **4**, 466–468
asymmetric hydrogenation, **8**, 144
asymmetric reduction, **8**, 159
α-bromo-
bromomagnesium enolates, **2**, 110
chiral
reaction with enol silanes, **2**, 640
condensation
amide chlorides, **6**, 499
cyclic
nucleophilic addition reactions, **1**, 67
reduction, **8**, 5, 14
cyclizations
vinylsilanes, **1**, 585
[3 + 2] cycloaddition reactions, **5**, 307
cyclopropyl
rearrangements, **5**, 941
deprotonation
alkali metal enolates, **2**, 100
enolization, **3**, 1
magnesium dialkylamides, **2**, 110
stoichiometric, **2**, 182
derivatives
dienes, **6**, 757
dienophiles, **6**, 756
heterocyclic synthesis, **6**, 733–760
derivatization, **6**, 703–728
α-diazo-
reaction with trialkylboranes, **2**, 242
synthesis, **3**, 889
Diels–Alder reactions, **5**, 430
gem-dimethylation
Tebbe reagent, **5**, 1124
electroreduction, **8**, 131
asymmetric, **8**, 134
indirect, **8**, 132
enantiomeric reductions
enzymes and microorganisms, **8**, 185
enolizable
reaction with organometallic compounds, **1**, 150
epoxidation, **1**, 819
functional group transformations, **6**, 763–793
α-halo-
Kornblum oxidation, **7**, 653
α-heteroatom-substituted
deprotonation, **2**, 101
reduction to enolates, **2**, 186
homogeneous catalytic hydrogenation, **8**, 152
homologation
α-selenoalkyl metals, **1**, 724
hydrazones and arylsulfonylhydrazones
Wolff–Kishner reduction, **8**, 327–359
α-hydroxylation, **7**, 144
3-iodo-
reduction, homoenolate generation, **2**, 442
Julia coupling
sulfones, **1**, 806
Knoevenagel reaction, **2**, 364
Lewis acid complexes, **1**, 283–321
σ- *versus* π-(η^2)-bonding, **1**, 284
conformation, **1**, 285
effects on rate and reactivity, **1**, 284
NMR, **1**, 292
theoretical studies, **1**, 286
X-ray crystallography, **1**, 299
Mannich reaction, **2**, 1010
masked
addition to alkylaluminum, **1**, 88
metal hydride reduction
diastereoselectivity, **8**, 7
α-methyl-β-hydroxy-
construction, **2**, 249
nitrogen derivatives
alkylation, regiochemistry, **3**, 28
alkylation, stereochemistry, **3**, 28
nonconjugated
addition reactions, **1**, 314
nucleophilic addition reactions
chiral auxiliaries, **1**, 61
stereocontrol, **1**, 150
oxidation
orthoacid synthesis, **6**, 561
oxidation by, **7**, 603
pinacol coupling reactions
with alkenes, **3**, 598
with alkynes, **3**, 602
with allenes, **3**, 605
polyalkenic α,β-unsaturated
reaction with organocuprates, **4**, 181
polycyclic α,β-unsaturated
reaction with organocuprates, **4**, 181
prochiral
nucleophilic addition reactions, **1**, 68
protecting groups, **6**, 675

reactions with alkyl pentacarbonyl carbene anions, **5**, 1076
reactions with allenylsilanes
titanium tetrachloride, **1**, 595
reactions with allylic sulfinyl carbanions, **1**, 517
reactions with crotyl organometallics
Lewis acid catalyzed, **2**, 4
reactions with *N,N*-dimethyl(methylene)iminium salts, **2**, 901
reactions with α-halo sulfones, **1**, 530
reactions with nitriles, **6**, 401
reactions with organocerium compounds, **1**, 234
reactions with organosamarium 'ate' complexes, **1**, 254
reactions with organosamarium(III) reagents, **1**, 253
reactions with organotitanium compounds, **1**, 145
reactions with organozinc reagents, **1**, 215
reactions with organozirconium compounds, **1**, 145
reactions with selenium-stabilized carbanions, **1**, 672
reactions with α-selenoalkylmetals, **1**, 723
reactions with sulfinyl-stabilized carbanions, **1**, 513
reactions with sulfonimidoyl carbanions, **1**, 532
reactions with sulfonyl-stabilized carbanions, **1**, 529
reduction
catalytic hydrogenation, **8**, 139–155
chemoselectivity, **8**, 15
chirally modified hydride reagents, **8**, 159–180
dissolving metals, **8**, 107–123
dissolving metals, absence of proton donors, **8**, 109
dissolving metals, mechanism, **8**, 108
dissolving metals, presence of proton donors, **8**, 110
enzymes and microorganisms, **8**, 185
metal hydrides, **8**, 1–22
silanes, **8**, 216
stereoselectivity, **8**, 3
reductive amination
sodium cyanoborohydride, **8**, 47
reductive coupling reactions, **3**, 563
with alkenes, **3**, 583
regeneration from hydrazones, **2**, 523
α-substituted
enolates, **2**, 99
reduction, **8**, 983–996
α-sulfinyl-α,β-unsaturated
enantiomers, Michael reaction, **2**, 363
synthesis, Knoevenagel reaction, **2**, 363
synthesis
from alkyl vinyl sulfides and selenides, **3**, 120
via alcohol oxidation, **7**, 305
via β-hydroxyalkyl selenides, **1**, 712, 714, 721
via oxidative cleavage of alkenes, **7**, 544
1,2-transposition
Shapiro reaction, **6**, 780
α,β-unsaturated
addition reactions, **1**, 311
1,4-addition reactions, **1**, 546, 566
1,4-addition reactions with cyanohydrin ethers, **1**, 552
1,4-addition reactions with cyanohydrins, **1**, 548
1,4-addition reactions with α-(dialkylamino)nitriles, **1**, 556
addition reactions with organometallic compounds, **1**, 155
alkali metal enolates, **2**, 106
allylic oxidation, **7**, 99
conjugate addition, **4**, 228
1,4-conjugate addition of hydrazones, **2**, 517
cycloaddition reactions, metal catalyzed, **5**, 1197
Diels Alder reactions, **5**, 453
enolates, **2**, 187
enolates from, **2**, 184
hydrogenation, **8**, 439
hydrogenation, homogeneous catalysis, **8**, 452
Lewis acid complexes, **1**, 287
Lewis acid complexes, NMR, **1**, 294
Michael additions, **4**, 217
oxidation, palladium(II) catalysis, **4**, 553
pinacol coupling reactions, **3**, 577
preparation, use of imine anions, **2**, 482
protection, **7**, 146
reaction with allenylsilanes, **1**, 596
reaction with boron reagents, **2**, 112
reaction with lithium diallylcuprate, **2**, 120
reaction with organocerium compounds, **1**, 235, 239
reaction with organocuprates, **4**, 179–187
reaction with organometallic compounds, site selectivity, **1**, 81
reaction with trialkylboranes, **2**, 241
regioselective oxidation, **7**, 462
Simmons–Smith reaction, **4**, 968
synthesis, **3**, 161; **7**, 119
synthesis, palladium catalysis, **4**, 611
tandem vicinal difunctionalization, **4**, 253
β,γ-unsaturated
acyclic, photoisomerizations, **5**, 220
regioselective oxidation, **7**, 462
semicyclic, photoisomerizations, **5**, 221
synthesis, **5**, 941; **6**, 850
γ,δ-unsaturated
synthesis *via* allyl alcohols and allylamines, **6**, 855
via Claisen rearrangement, **6**, 855
δ,ε-unsaturated
synthesis *via* 2,3-Wittig–oxy-Cope rearrangement, **6**, 852
unsaturated acetals
substitution reactions, **6**, 849
α-unsubstituted-β-hydroxy-
construction, **2**, 260
Carbonyl compounds, α-alkoxy
chiral
reaction with organometallic compounds, **1**, 153
reactions with organometallic compounds
Lewis acids, **1**, 335
Carbonyl compounds, 2-alkoxy-3-trimethylsilylalkenyl
nucleophilic addition reactions, **1**, 58
Carbonyl compounds, α-alkyl
nucleophilic addition reactions, **1**, 50
Carbonyl compounds, 2-alkyl-3-trimethylsilylalkenyl
nucleophilic addition reactions
stereoselectivity, **1**, 58
Carbonyl compounds, allenic
synthesis, **6**, 852
Carbonyl compounds, α-amino
nucleophilic addition reactions
stereoselectivity, **1**, 56
Carbonyl compounds, α-arylsulfinyl-α,β-unsaturated
homochiral
conjugate additions, **4**, 213

Carbonyl compounds, α-benzyloxy
nucleophilic addition reactions
selectivity, **1**, 52
Carbonyl compounds, α-bromo-
oxidation
triflamides, **7**, 668
Carbonyl compounds, α-chloro-
reduction, **8**, 20
Carbonyl compounds, cyclic azo-
synthesis
via oxidation of hydrazides, **7**, 748
Carbonyl compounds, α,β-dihydroxy
nucleophilic addition reactions
stereoselectivity, **1**, 55
Carbonyl compounds, α-halo
electrochemical reduction, **8**, 987
nucleophilic addition reactions
selectivity, **1**, 50
reduction, **8**, 19
reductive cleavage, **8**, 987
Carbonyl compounds, α-hydroxy
reactions with organometallic compounds
Lewis acids, **1**, 335
synthesis
via cleavage of 1,3-oxathianes, **1**, 61
via keto aminals, **1**, 64
Carbonyl compounds, α-iodo-
synthesis, **7**, 535
Carbonyl compounds, α-nitroaryl
synthesis, **4**, 429
Carbonyl compounds, α-oximino-
synthesis
via nitrosochlorination of alkenes, **4**, 357
Carbonyl compounds, α-oxygenated
Wittig reaction
selectivity, **1**, 757
Carbonyl compounds, α-phenylselenenyl-
synthesis, **7**, 522
Carbonyl compounds, α-seleno
enolates
reactivity, **1**, 691
reactions with enals, **1**, 686
Carbonyl compounds, α-sulfinyl
reactions with carbonyl compounds, **1**, 523
Carbonyl dibromide
reaction with amides, **6**, 495
Carbonyl dichloride
acid anhydride synthesis, **6**, 312
Carbonyl difluoride
imidoyl halide synthesis, **6**, 523
reaction with amides, **6**, 495
reaction with tertiary amides, **6**, 495
Carbonyldiimidazolide
acid anhydride synthesis, **6**, 313
imidazolide synthesis, **6**, 308
Carbonyl methylenation
iodomethylenation
samarium diiodide, **1**, 261
Carbonyl oxides
existence, **4**, 1098
Carbonyloxy radicals
cyclizations, **4**, 798
Carbonyl sulfide
conjugate enolate trapping, **4**, 261
Carbonyl ylides
alkyne cyclizations, **4**, 1163
cyclic
alkene cyclizations, **4**, 1162
cyclizations, **4**, 1159–1163
1,3-dipolar cycloadditions, **4**, 1089–1093
open-chain
alkene cyclizations, **4**, 1161
photogeneration, **4**, 1090
Carbonyl ylides, aryl
cyclizations, **4**, 1161
Carbopalladation
alkenes, **4**, 903
Carbosulfenylation
alkenes, **4**, 331
selectivity, **6**, 142
Carbothioates
α,β-unsaturated
synthesis, **6**, 453
Carbothioates, β-hydrazono-
O-alkyl esters
synthesis, **6**, 453
Carbothioates, β-oxo-
O-alkyl esters
synthesis, **6**, 453
Carbothioates, *O*-trimethylsilyl esters
synthesis, **6**, 448
Carboxamides
3-lithiated
reaction with electrophiles, **2**, 442
Carboxamides, α-allyloxy-
Wittig rearrangement, **3**, 1004
Carboxamides, α-bromo-
reaction with amines, **6**, 67
Carboxonium salts
alkoxymethyleniminium salt synthesis, **6**, 506
orthoacid synthesis, **6**, 561
synthesis
via amide alkylation, **6**, 502
Carboxy group activation
esterification
mechanism, **6**, 326
ester sythesis, **6**, 324
Carboxy groups
enzymic reduction
specificity, **8**, 201
protection, **6**, 665
Carboxyhydrazides, α-bromo-
reaction with amines, **6**, 67
Carboxy inversion reaction, **7**, 728
Carboxylates
alkylation
preparation of alcohols, **6**, 2
in the presence of bromine
oxidation, diols, **7**, 314
oxidation
thiols, **7**, 760
reaction with alkyl sulfonates
inversion of alcohols, **6**, 21
reaction with π-allylpalladium complexes
stereochemistry, **4**, 622
reaction with nitriles, **6**, 401
Carboxylates, γ-bromo
γ-lactone synthesis, **6**, 359
Carboxylation
alkenes, **4**, 932–946

catalysts, **4**, 939
mechanism, **4**, 936–939
ketones, **2**, 841
Carboxylic acid azides
amide synthesis, **6**, 389
Carboxylic acid chlorides
arylation
palladium complexes, **4**, 857
reactions with benzylsamarium reagents, **1**, 253
Carboxylic acid chlorides, α-alkoxy-
reactions with ketones
samarium diiodide, **1**, 259
Carboxylic acid derivatives
reduction, **8**, 235–254
Carboxylic acid esters
acid halide synthesis, **6**, 307
Carboxylic acid esters, α-allyloxy-
Wittig rearrangement, **3**, 1000
Carboxylic acid halides
vinyl substitutions
palladium complexes, **4**, 856
Carboxylic acids
acid anhydride synthesis, **6**, 309
acid halide synthesis, **6**, 302
acyclic
diastereoselective alkylation, **3**, 44
acyclic enolates
diastereoselective alkylation, **3**, 42
acylation
preparation of ketones, **1**, 411
addition to alkenes
palladium(II) catalysis, **4**, 553
alkenic
divinyl ketones from, **5**, 776
amide synthesis, **6**, 382
amidine synthesis, **6**, 543
amidinium salt synthesis, **6**, 513
anodic oxidation, **7**, 805
aromatic
Birch reduction, **8**, 499
azide synthesis, **6**, 253
coupling reactions
with alkyl Grignard reagents, **3**, 463
Darzens glycidic ester condensation, **2**, 425
degradation
amine synthesis, **6**, 795
dehydrogenation, **7**, 137
pyridine *N*-oxide, **7**, 144
derivatives
amide adducts, **6**, 493
nucleophilic addition, **1**, 397–453
reactions with organoaluminum reagents, **1**, 92
reactions with α-seleno carbanions, **1**, 694
α,α-dialkyl
asymmetric synthesis, **3**, 53
enantiomeric synthesis
Claisen rearrangement, **5**, 864
endocyclic enolates
diastereoselective alkylation, **3**, 39
enolates
cycloalkylation, **3**, 48
intramolecular cyclization, **3**, 49
esters
hydroxy group activation, **6**, 333
exocyclic enolates
diastereoselective alkylation, **3**, 39
homochiral β-branched
conjugate additions, **4**, 202
α-hydroxylation, **7**, 185
Ivanov reaction, **2**, 210
Kolbe electrolysis
cross-coupling, **3**, 642
symmetrical coupling, **3**, 637
3-lithiated
reaction with electrophiles, **2**, 442
masked anions
alkylation, **3**, 53
metal enolates
alkylation, **3**, 39
2-[6-(2-methoxyethyl)pyridyl] ester
acylating agent, **1**, 453
reactions with alkenes, **4**, 312–316
reduction
metal hydrides, **8**, 259–279
to aldehydes, **8**, 283–304
salts
acid anhydride synthesis, **6**, 314
sulfenylation, **7**, 125
synthesis
carbonylation, **3**, 1026
from ketones, **2**, 420
homologation of ketones, **2**, 419
via microbial oxidation, **7**, 56
via organoytterbium compounds, **1**, 277
via oxidative cleavage of alkenes, **7**, 541, 574
α-(trimethylsilyl)-α,β-unsaturated
reaction with organolithium compounds, **4**, 74
unsaturated
dehydrative cyclization *via* Friedel–Crafts reaction, **2**, 711
γ-lactonization, **6**, 360
synthesis, **3**, 862
synthesis *via* Perkin reaction, **2**, 401
α,β-unsaturated
diastereoselective additions, **4**, 200–208
enzymic reduction, **8**, 205
hydrobromination, **4**, 282
hydrogenation, homogeneous catalysis, **8**, 453
hydroiodination, **4**, 288
reaction with allylic halides, **3**, 50
β,γ-unsaturated
isomerization, **6**, 896
γ,δ-unsaturated
synthesis, *via* Claisen rearrangement, **5**, 828
cis-α,β-unsaturated
carbonylation of alkynes, **3**, 1027
α,β-unsaturated dienolates
alkylation, **3**, 50
α,β-unsaturated-α-nitro-, esters
synthesis, Knoevenagel reaction, **2**, 364
Vilsmeier–Haack reaction, **2**, 786
Carboxylic acids, *N*-acetylamino-
hydrogenation, **8**, 535
Carboxylic acids, β-alkyl-
synthesis
Knoevenagel reaction, **2**, 363
Carboxylic acids, allyl-
synthesis
ene reaction, **2**, 539
Carboxylic acids, α-allyloxy-

Wittig rearrangement, **3**, 999
Carboxylic acids, *syn*-α-amino-β-hydroxy-
enantioselective aldol reaction
gold catalysis, **2**, 317
Carboxylic acids, β-bromo-
cyclization, **6**, 345
Carboxylic acids, 2,3-epoxy-
methyl esters
reaction with organocuprates, **6**, 11
ring opening, **6**, 11
Carboxylic acids, α,β-epoxy-
synthesis
via sulfur ylide reagents, **1**, 822
Carboxylic acids, α-halo-
dianions
Darzens glycidic ester condensation, **2**, 425
resolution, **6**, 340
Carboxylic acids, α-hydrazino-
synthesis, **6**, 118
Carboxylic acids, α-hydrazino-β-hydroxy-
esters
synthesis, **6**, 118
Carboxylic acids, α-hydroxy-
asymmetric synthesis
from chiral α-keto esters, **1**, 49
chiral
synthesis, **1**, 86
'enantiomerically pure'
synthesis, **7**, 316
optically active
synthesis, **6**, 852
synthesis, **1**, 62
enantiomerically enriched, **1**, 66
via organoytterbium compounds, **1**, 280
Carboxylic acids, β-hydroxy-
elimination
alkene synthesis, **2**, 597
β-lactone synthesis, **6**, 347
Carboxylic acids, γ-hydroxy-
cyclization
γ-lactone synthesis, **6**, 354
Carboxylic acids, α-keto-
preparation
Darzens glycidic ester condensation, **2**, 420
Carboxylic acids, *syn*-α-methyl-β-hydroxy-
aldol reaction
titanium enolates, chiral auxiliary, **2**, 308
zirconium enolates, chiral auxiliary, **2**, 304
synthesis, **2**, 272
Carboxylic acids, perfluoro-
hydrogenation, **8**, 242
Carboxylic acids, α-seleno-
metallation, **1**, 642
Carboxylic acids, β-silyl-
oxidative decarboxylation
formation of alkenes, **7**, 628
Carboxylic acids, β-stannyl-
oxidation, **7**, 628
oxidative decarboxylation
formation of alkenes, **7**, 628
Carboxylic anhydrides
Pummerer rearrangement, **7**, 196
α,*O*-Carboxylic dianions
conjugate addition reactions, **4**, 111–113
Carboxylic esters
α,β-unsaturated α-sulfinyl
synthesis, **2**, 388
Carboxylic esters, α-keto-
hydrogenation, **8**, 152
Carboxylic esters, α-nitro-
synthesis, **6**, 104
Carboxylic esters, 4-oxo-
synthesis
via benzoin condensation, **1**, 542
Carboxylic groups
protection
organometallic transformation, **6**, 673
Carboxyl radicals
generation
functional group compatability, **7**, 718
Carboxymethyleniminium salts
acylation, **1**, 423
Carbozincation
alkenes, **4**, 879–884
alkynes, **4**, 883
stereoselectivity, **4**, 880
Cardenolides
side chain elaboration
Pummerer rearrangement, **7**, 196
synthesis
Knoevenagel reaction, **2**, 382
Carene
epoxides
ring opening, **3**, 736
3-Carene
allylboranes from
reactions with aldehydes, **2**, 33
allylic oxidation, **7**, 102
oxidation, **7**, 97
pyridinium fluorochromate, **7**, 267
ozonolysis
experimental details, **7**, 544
Carenones
synthesis
via Wharton reaction, **8**, 927
Carminomycinone, 11-deoxy-
synthesis
via ortho lithiation, **1**, 464
Carnegine
synthesis, **6**, 152, 739
Carotene
synthesis
use of enol ethers, **2**, 612
β-Carotene
synthesis, **3**, 169, 585
Ramberg–Bäcklund rearrangement, **3**, 883
Carotenoids
synthesis, **8**, 560
Carpanone
synthesis, **3**, 698
via Diels–Alder reactions, **5**, 468
Carroll rearrangement
ester enolates, **5**, 835
variant of Claisen rearrangement, **5**, 834
α-Cartopterone
synthesis
via retro Diels–Alder reaction, **5**, 571
2,3-Carvene
oxiranes
rearrangement, **3**, 771

Carvenolide
  synthesis, **3**, 849
Carveol
  oxidation
    solid support, **7**, 841
  reduction
    aluminum hydrides, **8**, 542
  synthesis, **7**, 99
cis-Carveol
  Claisen–Cope rearrangement, **5**, 881
Carvomenthene
  oxiranes
    rearrangement, **3**, 771
Carvomenthone
  rearrangement, **3**, 832
Carvones
  aldol reaction
    benzaldehyde, **2**, 152
  hydrogenation
    homogeneous catalysis, **8**, 446
  Michael addition
    benzenethiolate, **6**, 141
  photochemical cycloadditions, **5**, 123
  reaction with trimethylsilyl cyanide
    1,2-addition, **2**, 599
  reduction, **8**, 563
    biochemical, **8**, 559
    borohydride, **8**, 536
    dissolving metals, **8**, 526
    iron hydrides, **8**, 550
    metal hydrides, **8**, 315
    molybdenum complex catalyst, **8**, 554
  silyl ketene acetal derivatives
    Cope–Claisen rearrangement, **5**, 886
  synthesis, **7**, 99
Carvones, dihydro-
  ozonolysis
    Criegee rearrangement, **6**, 14
  synthesis, **6**, 141
Caryolan-1-ol
  synthesis, **3**, 386
Caryophyllane
  synthesis, **3**, 389
Caryophyllene
  synthesis, **3**, 386, 399
    via photochemical cycloaddition, **5**, 124
α-Caryophyllene alcohol
  synthesis, **3**, 400, 713
α-Caryopterone
  synthesis
    via retro Diels–Alder reaction, **5**, 564
Casbene
  synthesis, **3**, 431; **7**, 94, 647
Casegravol
  synthesis, **7**, 823
Castelanolide
  synthesis, **8**, 932
Castro reaction
  copper(I) alkynides
    reaction with aryl halides, **3**, 522
Catalytic transfer hydrogenation
  heterogeneous catalysis, **8**, 440
Catechols
  oxidation
    solid support, **7**, 843
    oxidative trimerization, **3**, 669
Catenane
  synthesis
    intramolecular acyloin coupling reaction, **3**, 628
Catharanthine
  synthesis
    via palladium catalysis, **4**, 598
Cation-exchange resins
  acidic
    catalyst, Friedel–Crafts reaction, **3**, 296
  Ritter reaction
    initiator, **6**, 283
Cation-forming agents
  metathetic, catalysts
    Friedel–Crafts reaction, **3**, 298
Cationic cyclizations, **5**, 751–781
CBT
  synthesis, **3**, 1012
CC-1065
  synthesis
    Sommelet–Hauser rearrangement, **3**, 969
    via Diels–Alder reaction, **5**, 492
Cectopia juvenile hormone
  synthesis, **3**, 99, 107
Cedrane oxide
  ozonation, **7**, 247
Cedranoids
  synthesis
    via photoisomerization, **5**, 233
Cedrene
  synthesis
    via Nazarov cyclization, **5**, 779
    via photocycloaddition, **5**, 647, 657
Cedrenone
  synthesis
    via photocycloaddition, **5**, 659
Cedrenone, bromo-
  synthesis, **5**, 659
Cedrol
  microbial hydroxylation, **7**, 64
  synthesis
    via photoisomerizations, **5**, 231
Celacinnine
  synthesis
    via cleavage of hydrazide, **8**, 389
Celite
  silver carbonate support, **7**, 841
Cell wall constituents
  bacteria
    synthesis, **6**, 52
Cembranolides
  synthesis, **3**, 99; **7**, 89
    via Horner–Wadsworth–Emmons reaction, **1**, 772
    via sulfones, **6**, 158
β-2,7,11-Cembratriene-4,6-diol
  synthesis
    via failed Wharton reaction, **8**, 929
Cembrene
  synthesis, **3**, 431
    Friedel–Crafts reaction, **2**, 711
Cephalosporin C
  synthesis
    Knoevenagel reaction, **2**, 358
Cephalosporins
  bicyclic

synthesis *via* cyclization of enol thioether, **4**, 410
reaction with dichlorine monoxide, **7**, 537
rearrangements, **3**, 954
synthesis
organopalladium catalysts, **3**, 232
via cycloadditions of acid chlorides and imines, **5**, 92
Yoshimoto's transformation, **6**, 897
Cephalosporins, 7α-methoxy-
synthesis, **7**, 741
Cephalotaxine
synthesis, **7**, 155
via arynes, **4**, 502
Cephalotaxus alkaloids
synthesis
electron transfer induced photocyclizations, **2**, 1038
Cephams
synthesis
Ugi reaction, **2**, 1103
Cephem dioxides
allylic oxidation, **7**, 112
oxidative rearrangement, **7**, 820
Ceramides
synthesis, **6**, 53
Henry reaction, **2**, 331
Cerebrosides
synthesis, **6**, 54
Cerium
use in cycloalkanone coupling reactions, **3**, 570
use in pinacol coupling reactions, **3**, 567
Cerium, alkenyl-
reactions with enones, **1**, 240
Cerium, alkyl-
in synthesis, **1**, 237
Cerium, alkynyl-
reactions, **1**, 242
Cerium, allyl-
synthesis, **1**, 239
Cerium, aryl-
reactions with enones, **1**, 240
Cerium, trimethylsilylethynyl-
reactions, **1**, 242
Cerium, trimethylsilylmethyl-
synthesis, **1**, 238
Cerium, α-trimethylsilylvinyl-
reactions with enones, **1**, 240
Cerium ammonium nitrate
cyclohexadienyliron complexes
decomplexation, **4**, 674
nitration with, **6**, 110
oxidation
benzylic alcohols, **7**, 308
quinones, **7**, 350
secondary alcohols, **7**, 322
tetrahydrofuran, **7**, 237
Cerium chloride
Grignard reagent system, **1**, 244
lithium aluminum hydride
alkyl halide reduction, **8**, 803
preparation, **1**, 232
reduction
enones, **8**, 540
Cerium complexes
aldol reaction, **2**, 311
Cerium enolates
aldol reaction, **2**, 312
synthesis, **1**, 243
Cerium reagents
glycol cleavage, **7**, 705
oxidants
silica support, **7**, 843
Cerium sulfate
oxidation
secondary alcohols, **7**, 322
Ceroplastol
synthesis
Dieckmann reaction, **2**, 824
Ceroplastol II
synthesis
via allyl chromium reagents, **1**, 187
Cerorubenic acid
synthesis
via methylenation and thermolysis, **1**, 740
Cerorubenic acid-III
synthesis
via Cope rearrangement, **5**, 816
Cesium
reduction
ammonia, **8**, 113
carbonyl compounds, **8**, 109
Cesium acetate
Perkin reaction, **2**, 402
Cesium fluoride
catalyst
allylsilane reactions with aldehydes, **2**, 571
Knoevenagel reaction, **2**, 343
Cesium fluoroxysulfate
reaction with alkenes, **4**, 347
Cesium iodide
reduction
carbonyl compounds, **8**, 113
Chain extension, **1**, 843–899
Chalcogenides
synthetic
catalysts, Friedel–Crafts reaction, **3**, 296
Chalcones
aldol reaction, **2**, 150
aziridination, **7**, 471
deoxygenation
silanes, **8**, 546
oxidative rearrangement, **7**, 829
reduction
aluminum hydrides, **8**, 543, 545
biochemical, **8**, 561
borohydrides, **8**, 538
Chamigrene
synthesis
via cyclopropane ring opening, **4**, 1043
Chanoclavine I
synthesis
via INOC reaction, **4**, 1080
via nitrone cyclization, **4**, 1120
Chapman rearrangement
Beckmann rearrangement, **7**, 690
Charge-accelerated rearrangements
cyclobutanes, **5**, 1016
cyclopropanes, **5**, 1006–1016
small rings, **5**, 999–1033
Chelidonine
synthesis

via arynes, **4**, 500
via Diels–Alder reaction, **5**, 391
Chenodeoxycholic acid
precursor
synthesis *via* ene reaction with methyl acrylate, **5**, 5
Chichibabin reaction
bipyridines, **8**, 596
Chinchona alkaloids
catalysts
conjugate additions, **4**, 230
Chinensin
synthesis
via ortho directed addition, **1**, 468
2(1*H*)-Chinolon
synthesis
Knoevenagel reaction, **2**, 357
Chirality
self-reproduction
alkylation of enolates, **3**, 40
Chitobiose, α-fucosyl-
synthesis
protecting groups, **6**, 633
Chitobiosyl azide, α-fucosyl-
synthesis
protecting groups, **6**, 633
Chloral
aldol reaction
unsymmetrical ketones, **2**, 144
ene reaction
addition to alkenes, **2**, 534
endo/*exo* selectivity, **2**, 534
regioselectivity, **2**, 534
oxidant
alumina support, **7**, 841
N-sulfonyl imine
Diels–Alder reactions, **5**, 402
Chloramine
amination
amines, **7**, 741
secondary amines, **7**, 746
irradiation, **7**, 40
reactions with alkenes, **7**, 498, 537
reactions with organoboranes, **7**, 606
reaction with trialkylboranes, **7**, 607
selenium elimination, **7**, 129
Chloranil
dehydrogenation, **7**, 135
Chlorides
catalysts
allylsilane reactions with aldehydes, **2**, 571
Chlorination
alkanes
remote functionalization, **7**, 43
amines, **7**, 741
ionic
sulfides, **7**, 193
nucleophilic displacement, **6**, 204
secondary amines, **7**, 747
template-directed
β-cyclodextrin, **7**, 49
trimethylborane, **7**, 604
Chlorine
activator
DMSO oxidation of alcohols, **7**, 298
ligand transfer
oxidation of cyclobutyl radicals, **7**, 860
reaction with alkenes, **4**, 344
reaction with thioamides, **6**, 496
Ritter reaction, **6**, 288
Chlorinolysis
C—B bonds, **7**, 604
Chloroacetyl esters
alcohol protection
cleavage, **6**, 658
Chloroacetyl group
amine-protecting group, **6**, 642
α-Chloro acids
synthesis
via α-amino acids, **6**, 207
m-Chloro-*p*-acyloxybenzyloxycarbonyl group
amine-protecting group
cleavage, **6**, 639
Chloroamination
alkenes, **7**, 498
Chloroamphenicol
synthesis, **2**, 325
Chlorofluorocarbons
synthesis, **4**, 270; **6**, 220
Chloroform
reactions with amines, **6**, 521
reaction with nitroarenes, **4**, 432
Chloroformate
synthesis
via DMSO, **7**, 299
Chlorohydrin acetate
synthesis, **7**, 527
Chlorohydrins
by-product
Wacker process, **7**, 451
synthesis, **3**, 224; **8**, 20
Chloromethoxylation
alkenes, **4**, 355
Chloromethylation
arenes
Friedel–Crafts reaction, **3**, 321
Chloromethyleniminium ions
synthesis, **6**, 487
Chloromethyleniminium salts
formation
Vilsmeier–Haack reaction, **2**, 779
reaction with alkenes
Vilsmeier–Haack reaction, **2**, 781
reaction with aromatic compounds, **2**, 779
2-Chloromethyl-4-nitrophenyl esters
phosphoric acid protecting group, **6**, 623
4-Chloro-2-nitrophenyl esters
phosphoric acid protecting group, **6**, 622
Chlorophosphoric acid
diamides
amide adducts, **6**, 490
ester amides
amide adducts, **6**, 490
Chlorosulfamation
alkenes, **4**, 347
Chlorosulfonyl isocyanate
acid anhydride synthesis, **6**, 313
activator
DMSO oxidation of alcohols, **7**, 299
amide synthesis, **6**, 386
reaction with imines, **5**, 105

Chlorothricin
  synthesis
    via tandem vicinal difunctionalization, **4**, 243
Chlorothricolide
  synthesis
    via Ireland rearrangement, **5**, 842
Chokol A
  enantioselective synthesis
    via Johnson rearrangement, **5**, 839
  synthesis
    via magnesium-ene reaction, **5**, 44
    via methylcerium reagent, **1**, 237
Cholanic acid, 3α-hydroxy-7-keto-
  reduction
    dissolving metals, **8**, 117
5β-Cholanic acid, 3α,11α,15β-trihydroxy-
  microbial hydroxylation, **7**, 73
5β-Cholanic acid, 3α,11β,15β-trihydroxy-
  microbial hydroxylation, **7**, 73
5β-Cholanic acid, 3α,15β,18α-trihydroxy-
  microbial hydroxylation, **7**, 73
Cholecalciferol
  allylic oxidation, **7**, 90
Cholecalciferol, 1α,25*S*,26-trihydroxy-
  synthesis, **5**, 257
1,3-Cholestadiene
  photooxidation, **7**, 111
5α-Cholesta-1,3-diene
  photoisomerization, **5**, 706
5β-Cholesta-1,3-diene
  synthesis
    via photoisomerization, **5**, 706
Cholestane
  selenol esters
    synthesis, **6**, 462
Cholestane, 3β-alkyl-
  synthesis
    via alkyllithium addition, **1**, 377
Cholestane, 2,3-epoxy-
  reaction with Grignard reagents, **3**, 757
Cholestane, 3-halo-
  Michael addition, **4**, 130
Cholestane, 24-hydroxy-
  synthesis, **3**, 168
3β,5α,6β-Cholestanediol
  esters
    reduction, **8**, 816
5α-Cholestane-2,3-dione
  rearrangement, **3**, 834
5α-Cholestane-3,4-dione
  rearrangement, **3**, 834
5α-Cholestane-2,3-dione, 4,4-dimethyl-
  rearrangement, **3**, 834
3β-Cholestanol
  nitrile synthesis, **6**, 235
3β,5α-Cholestanol, 4-methylene-
  asymmetric epoxidation, **7**, 414
  epoxidation, **7**, 365
3β-Cholestanol, methyl ether
  oxidation, **7**, 239
Cholestan-1-one
  reduction
    dissolving metals, **8**, 119
5α-Cholestan-3-one
  Mannich reaction
    with iminium salts, **2**, 901
  reactions with organometallic reagents
    equatorial or axial, **1**, 152
  tosylhydrazone
    reactions with alkyllithium compounds, **1**, 377
Cholestan-6-one, 3β-acetoxy-5α-chloro-
  synthesis, **7**, 529
Cholestan-3-one, 2α-halo-
  reductive elimination, **8**, 926
Cholestan-3-ones, 5-vinyl-
  pyrolysis
    intramolecular ene reaction, **5**, 21
Cholest-5-ene
  allylic oxidation, **7**, 101
  synthesis, **8**, 819
Δ^5-Cholestene, 3-methylene-
  synthesis
    via ketone methylenation, **1**, 506
Cholest-4-ene-3,6-dione
  reduction
    transition metals, **8**, 531
3β-Cholest-8(14)-enol
  hydrogenation
    heterogeneous catalysis, **8**, 428
Cholestenone
  hydrogenation
    catalytic, **8**, 533
  reduction
    borohydride, **8**, 536
Cholest-1-en-3-one
  Clemmensen reduction, **8**, 311
Cholest-4-en-3-one
  hydrogenation
    homogeneous catalysis, **8**, 452
  oxime
    Beckmann rearrangement, **7**, 692
  reduction
    dissolving metals, **8**, 526
    electrochemical, **8**, 532
  reductive elimination, **8**, 930
Cholest-5-en-3-one
  1,2-propylenedioxy ketal
    reduction, **8**, 222
Cholest-5-en-7-one
  synthesis, **7**, 101
5α-Cholest-1-en-3-one, 2-hydroxy-
  rearrangements, **3**, 832
5α-Cholest-3-en-2-one, 3-hydroxy-
  rearrangements, **3**, 832
Cholesterol
  acetate
    photochemical epoxidation, **7**, 384
  ethers
    synthesis, **6**, 23
  oxidation
    chromium(VI), **7**, 820
    DMSO, **7**, 294
    solid support, **7**, 841
  oxidative rearrangement, **7**, 835
Cholesterol, (20*S*)-hydroxy-
  synthesis, **3**, 127
Cholesterol, 24-hydroxy-
  synthesis, **3**, 161
Cholesterol, 25-hydroxy-
  precursor

synthesis *via* ene reaction with methyl acrylate, **5**, 5
synthesis, **8**, 694
Cholesterol, (25*R*)-26-hydroxy-
synthesis, **3**, 983
Cholesterol, thiocarbonyl-
reduction
tributylstannane, **8**, 820
Cholesterol 3-acetate, 1α-hydroxy-
synthesis
via intramolecular photocycloaddition, **5**, 180
Cholesterol acetate, 24-oxo-
synthesis
via ene reaction, **5**, 6
Cholesteryl benzoate
allylic oxidation, **7**, 104
Choline, thioxobenzoyl-
thiobenzoylating agent
synthesis, **6**, 450
Chorismate
Claisen rearrangement
enzymatic, **5**, 855
Chorismate mutase-prephenate dehydrogenase
Claisen rearrangement, **5**, 855
Chorismic acid
dimethyl ester
synthesis, Mannich reaction, **2**, 904
synthesis
via cyclopropanation, **4**, 1036
Chroman, hydroxylamino-
synthesis, **8**, 374
3-Chromanamine
synthesis, **8**, 376
Chromanones
dehydrogenation
use of thallium trinitrate, **7**, 144
use of trityl perchlorate, **7**, 144
Mannich reaction
with preformed iminium salts, **2**, 902
synthesis
Friedel–Crafts reaction, **2**, 758
via ketocarbenoids, **4**, 1057
thia analogs
Mannich reaction, with preformed iminium salts, **2**, 902
synthesis, Friedel–Crafts reaction, **2**, 759
Chroman-4-ones, 2-alkyl-
synthesis
via conjugate addition, **4**, 215
Chromanones, 4-thio-
dehydrogenation
use of trityl perchlorate, **7**, 144
Chromates
oxidation
halides, **7**, 663
sigmatropic rearrangement, **7**, 821
Chromates, alkylammonium
oxidation
alcohols, **7**, 283
Chromates, hydridopentacarbonyl-
reduction
acyl chlorides, **8**, 289
Chromates, metal alkyl
catalytic oxidants
alcohols, **7**, 285
Chromene
Vilsmeier–Haack reaction, **2**, 782
Chromenes, 3-nitro-
reduction, **8**, 374
Chromenones
synthesis
Vilsmeier–Haack reaction, **2**, 791
Chromic acid
inert inorganic support
alcohol oxidation, **7**, 279
oxidation
ethers, **7**, 235, 236
organoboranes, **7**, 600
silica support, **7**, 844
α,β-unsaturated carbonyl compounds, **7**, 99
resin supports
alcohol oxidation, **7**, 280
Chromic anhydride
oxidation
alumina support, **7**, 844
solid-supported, **7**, 840
quinone synthesis, **7**, 355
Chromium, α-acyl-
reactions, **1**, 202
Chromium, alkenyl-
intramolecular addition reactions, **1**, 200
reactions, **1**, 193
Chromium, γ-alkoxyallylic
reactions with aldehydes, **1**, 185, 190
Chromium, alkyl-
addition to carbonyl compounds, **1**, 202
Chromium, (alkylbenzene)tricarbonyl-
substitution reactions, **4**, 538
Chromium, alkyl-*gem*-di-
alkenation, **1**, 205
Chromium, alkynyl-
reactions, **1**, 201
Chromium, allylic
asymmetric induction, **1**, 187
enantioselective addition reactions, **1**, 192
intramolecular addition reactions, **1**, 187
reactions
1,2-asymmetric induction, **1**, 179
carbonyl addition, **1**, 177
with achiral aldehydes, **2**, 20
with achiral carbonyl compounds, **2**, 19
substituted substrates, **1**, 189
Chromium, (anisole)tricarbonyl-
addition–protonation reactions, **4**, 543
addition reactions, **4**, 538
Chromium, (η-arene)tricarbonyl-
addition–oxidation reactions, **4**, 531–541
tandem vicinal difunctionalization, **4**, 253
Chromium, (η^6-benzene)tricarbonyl-
addition–oxidation, **4**, 532
reaction with 2-lithio-1,3-dithiane, **4**, 545
Chromium, benzyl-
reaction with acrylonitrile, **1**, 175
Chromium, (η^6-benzyl alcohol)tricarbonyl-
Ritter reaction, **6**, 287
Chromium, chloroarene tricarbonyl-
coupling reactions
with tetrabutyltin, **3**, 454
Chromium, (chlorobenzene)tricarbonyl-
nucleophilic addition reactions, **4**, 519
reaction with lithioisobutyronitrile, **4**, 526

Chromium, crotyl-
  2,3-asymmetric induction, **1**, 181
  reactions
    carbonyl addition, **1**, 177
    with achiral carbonyl compounds, **2**, 19
    with glyceraldehyde acetonide, **2**, 29
    with α-methyl chiral aldehydes, **2**, 29
  synthesis, **1**, 179
Chromium, (η-1,3,5-cycloheptatriene)tricarbonyl-
  cycloaddition reactions
    dienes, **5**, 633
Chromium, (η^4-cyclohexadienyl)tricarbonyl-
  reaction with benzyl bromide, **4**, 712
Chromium, (μ-cyclopentadienyl)dinitrosobis-
  reduction
    vicinal dibromides, **8**, 797
Chromium, dichlorotris(tetrahydrofuran)alkyl-
  synthesis, **1**, 202
Chromium, dichlorotris(tetrahydrofuran)-*p*-tolyl-
  synthesis, **1**, 174
Chromium, (diphenyl ether)tricarbonyl-
  nucleophilic substitution, **4**, 527
Chromium, (fluorobenzene)tricarbonyl-
  reaction with diethyl sodiomalonate, **4**, 526
  synthesis, **4**, 523
Chromium, (haloarene)tricarbonyl-
  halide exchange, **4**, 527
  nucleophilic substitution, **4**, 522–524
Chromium, hexacarbonyl-
  allylic oxidation, **7**, 107
  Ritter reaction
    stereospecific, **6**, 287
Chromium, (indole)tricarbonyl-
  substitution reactions, **4**, 539
Chromium, isopropenyl-
  cycloaddition reactions
    cyclopentadiene in, **5**, 1070
Chromium, methallyl-
  reactions
    carbonyl addition, **1**, 177
Chromium, (*N*-methyltetrahydroquinoline)tricarbonyl-
  addition reactions, **4**, 534
Chromium, (naphthalene)tricarbonyl-
  addition reactions, **4**, 536
Chromium, naphthol tricarbonyl-
  synthesis, **5**, 1093
Chromium, pentacarbonyl(methoxyarylcarbene)-
  selenol ester, **6**, 473
Chromium, pentacarbonyl(methoxymethyl)-
  reaction with amines
    β-lactam synthesis, **3**, 1037
Chromium, propargyl-
  reactions
    carbonyl addition, **1**, 177
    with carbonyl compounds, **1**, 191
Chromium, propynyl-
  cycloaddition reactions with cyclopentadiene, **5**, 1072
Chromium, (styrene)tricarbonyl-
  addition reactions, **4**, 546
Chromium, α-thioalkyl-
  synthesis, **1**, 202
Chromium, (*o*-(trimethylsilyl)anisole)tricarbonyl-
  metallation, **4**, 539
Chromium, triphenyltris(tetrahydrofuran)-
  reaction with carbon monoxide
    benzpinacol synthesis, **1**, 175
  synthesis, **1**, 174
Chromium carbyne complexes
  anchorage
    amino acids, **6**, 671
Chromium chloride (*see also* Chromium di- and tri-chloride)
  catalyst
    Friedel–Crafts reaction, **2**, 737
Chromium complexes
  chiral
    imines, **1**, 364
  octahedral configuration, **1**, 179
Chromium complexes, hydrido-
  reduction
    unsaturated carbonyl compounds, **8**, 551
Chromium complexes, vinylcarbene-
  reaction with isoprene, **5**, 1070
Chromium diacetate
  reduction
    epoxides, **8**, 883
Chromium diacetate, bis(ethylenediamine)-
  reduction
    α,β-unsaturated ketone, **8**, 531
Chromium dichloride
  electroreduction
    carbonyl compounds, **8**, 133
  reductions
    nitro compounds, **8**, 371
Chromium(II) perchlorate
  radical cyclizations
    nonchain methods, **4**, 808
Chromium perchlorate, benzylpentaaquo-
  synthesis, **1**, 174
Chromium reagents
  acidic
    alcohol oxidation, **7**, 252
  alkane oxidation, **7**, 12
  allylic oxidation, **7**, 95
  aqueous acetic acid
    alcohol oxidation, **7**, 252
  dimethylformamide
    alcohol oxidation, **7**, 252
  dimethyl sulfoxide
    alcohol oxidation, **7**, 252
  glycol cleavage, **7**, 706
  heterocyclic bases
    alcohol oxidation, **7**, 256
  hexavalent
    oxidative cleavage of alkenes, **7**, 571
  Jones oxidation
    alcohols, **7**, 253
  organoborane oxidation, **7**, 600
  oxidants
    solid-supported, **7**, 839
  oxidation
    alcohols, **7**, 251–286
    silica support, **7**, 844
  oxidative rearrangements, **7**, 816
  sulfuric acid
    alcohol oxidation, **7**, 252
  two phase oxidation
    alcohols, **7**, 253
Chromium salts

deoxygenation
epoxides, **8**, 888
organochromium compound synthesis from, **1**, 174
reduction
alkenes, **8**, 531
alkyl halides, **8**, 796
mechanism, **8**, 482
unsaturated hydrocarbons, **8**, 481
reductive cleavage
α-halocarbonyl compounds, **8**, 987
ketols, **8**, 992
Chromium(II) salts
use in intermolecular pinacol coupling reactions, **3**, 565
Chromium trichloride
catalyst
Wurtz reaction, **3**, 421
lithium aluminum hydride
unsaturated hydrocarbon reduction, **8**, 485
reduction
vicinal dibromides, **8**, 797
Chromium trioxide
t-butyl hydroperoxide
alcohol oxidation, **7**, 278
carbon intercalation
alcohol oxidation, **7**, 282
catalytic oxidation
alcohols, **7**, 278
crown ethers
alcohol oxidation, **7**, 278
diethyl ether
alcohol oxidation, **7**, 278
2,4-dimethylpentane-2,4-diol complex
alcohol oxidation, **7**, 278
3,5-dimethylpyrazole complex
alcohol oxidation, **7**, 260
allylic oxidation, **7**, 104
inert inorganic support
alcohol oxidation, **7**, 279, 280
oxidation
ethers, **7**, 237, 239
sulfoxides, **7**, 768
tetraalkylstannanes, **7**, 614
oxidative cleavage of alkenes, **7**, 542
synthesis of carbonyl compounds, **7**, 571
synthesis of carboxylic acids, **7**, 587
pyridine complex
alcohol oxidation, **7**, 256
allylic oxidation, **7**, 100
Chromones
reduction
aluminum hydrides, **8**, 544
synthesis, **7**, 136
Chromyl azide
azido alcohols from, **7**, 491
Chromyl chloride
alkene complexes, **7**, 528
inert inorganic support
alcohol oxidation, **7**, 279
oxidation
solid support, **7**, 845
oxidative halogenation, **7**, 527
reaction with silyl enol ethers
ketone α-hydroxylation, **7**, 166
Chromyl fluoride
synthesis, **7**, 528
Chromyl trichloroacetate
organoborane oxidation, **7**, 601
Chrysanthemic acid
synthesis, **7**, 96
via carbomagnesiation, **4**, 874
Chrysanthemum acid
synthesis
via Claisen rearrangement, **6**, 859
Chrysenes
synthesis
via electrocyclization, **5**, 720
Chrysenes, 5,6,11,12-tetrahydro-
synthesis
via FVP, **5**, 725
Chrysomelidial
synthesis
via [3 + 2] cycloaddition reactions, **5**, 309
Chymotrypsin
phthaloyl group removal
amine protection, **6**, 643
substrate specificity
synthetic applicability, **2**, 456
Cicaprost
synthesis
Knoevenagel reaction, **2**, 381
Cinerolone
synthesis
via cinerone, **7**, 54
Cinerone
microbial oxidation, **7**, 54
Cine substitution
in synthesis, **4**, 496
Cinnamaldehydes
dicarbonyl(triphenylphosphine)iron complexes
crystal structure, **1**, 309
hydrogenation
catalytic, **8**, 140
oxidative rearrangement
solid support, **7**, 845
reaction with diethylzinc, **1**, 217
reaction with organocopper compounds, **1**, 113
reduction
aluminum hydrides, **8**, 541, 544
borohydrides, **8**, 537
molybdenum complexes, **8**, 551
synthesis, **8**, 301
Vilsmeier–Haack reaction, **2**, 782
Cinnamamides, *N*,*N*-dialkyl-
addition reactions
with organomagnesium compounds, **4**, 84
Cinnamates
reaction with lithium dimethylcuprate, **4**, 171
reduction
borohydrides, **8**, 536
Cinnamic acid
acyl cyanides
synthesis, **6**, 317
configuration
Knoevenagel reaction product, **2**, 345
methyl ester
reaction with lithium bis(phenyldimethylsilyl)cuprate, **2**, 200
reduction
transfer hydrogenation, **8**, 552

synthesis
Perkin reaction, **2**, 395, 399
cis-Cinnamic acid
Friedel–Crafts reaction, **2**, 757
Cinnamic acid, α-acetylamino-
asymmetric hydrogenation
homogeneous catalysis, **8**, 460
Cinnamic acid, α-aryl-
synthesis
Perkin reaction, **2**, 400
Cinnamic acid, 2-benzamido-
Erlenmeyer azlactone synthesis, **2**, 403
Cinnamic acid, *p*-chloro-
reduction, **8**, 905
Cinnamic acid, hydroxy-
oxidative dimerization, **3**, 692
Cinnamic acid, α-phenyl-
stereoisomers
Perkin reaction, **2**, 397
Cinnamic acid anhydride
synthesis
via 4-benzylpyridine, **6**, 310
Cinnamolide
synthesis, **7**, 307
Cinnamoyl azides
Curtius rearrangement, **6**, 799
Cinnamoyl chloride
reduction
metal hydrides, **8**, 290
Cinnamoyl group, 2-nitrodihydro-
reductive cyclization, **8**, 367
Cinnamyl acetate
hydrogenolysis, **8**, 977
Cinnamyl alcohol
asymmetric epoxidation, **7**, 393
kinetics, **7**, 421
oxidation
solid support, **7**, 841
Cinnamyl alcohol, α-phenyl-
epoxidation, **7**, 424
Cinnamyl alcohol epoxide
deoxygenation, **8**, 886
Cinnamyl cinnamate
reduction
transfer hydrogenation, **8**, 554
Cinnamyl compounds
oxidative rearrangement, **7**, 829
Cinnamyl esters
carboxy-protecting groups, **6**, 666
Cinnamyloxycarbonyl group
amine-protecting group, **6**, 641
Cinnolines
reduction, **8**, 640
ring opening
cathodic reduction, **8**, 641
Cinnolines, 4-amino-
synthesis
Friedel–Crafts reaction, **2**, 758
Cinnolone, 3-cyano-
synthesis
Friedel–Crafts reaction, **2**, 758
Citral
aldol reaction
2-butanone, **2**, 146
oxidative rearrangement, **7**, 828
reduction, **8**, 563, 564
borohydrides, **8**, 540
thermal ene reaction, **2**, 540
trans-Citral
Perkin reaction, **2**, 400
Citromycinone
synthesis
ene reaction, **2**, 549
Citronellal
hydrogenation
catalytic, **8**, 533
reduction
borohydrides, **8**, 540
synthesis
via conjugate addition to crotonaldehyde
N,O-acetal, **4**, 210
via Diels–Alder reaction, **5**, 468
thermal cyclization
ene reaction, **2**, 540
Citronellal, hydroxy-
synthesis
rhodium-catalyzed hydroformylation, **3**, 1022
via hydroformylation, **4**, 923
Citronellate, orthodihydro-
methyl ester
Claisen rearrangement, **5**, 888
Citronellene
hydroformylation, **4**, 922
Citronellic acid
synthesis
via organocopper-mediated additions, **4**, 152
Citronellol
microbial hydroxylation, **7**, 62
oxidation
solid support, **7**, 841
synthesis *via* conjugate addition to α,β-unsaturated
carboxylic amides, **4**, 203
synthesis *via* asymmetric hydrogenation of geraniol
or nerol
homogeneous catalysis, **8**, 462
biochemical reduction, **8**, 560
Claisen–Claisen rearrangement, **5**, 888
Claisen condensation
in synthesis, **2**, 820
mechanism, **2**, 797
reduction and, **2**, 818
retro, **2**, 855
stereochemistry, **2**, 846
tandem reaction, **2**, 852
thiocarboxylic esters, **6**, 446
Claisen–Cope rearrangement, **5**, 876–884
Claisen rearrangement, **5**, 827–866
abnormal, **5**, 834
acyclic substrates
remote stereocontrol, **5**, 864
stereochemistry, **5**, 862
alkyl substituents
kinetics, **5**, 856
allylic systems, **6**, 834
allyl vinyl ethers, **5**, 832–834
discovery, **5**, 827
amide acetal, **6**, 406
aromatic, **5**, 834
arylsulfonyl carbanion-accelerated, **5**, 1004
asymmetric induction, **6**, 858

carbanion-accelerated, **5**, 829, 1004
catalysis, **5**, 850
charge-accelerated, **5**, 847–850
competitive, **5**, 850
with Wittig rearrangement, **5**, 851
cyclic substrates
remote stereocontrol, **5**, 864
stereochemistry, **5**, 863
elimination *vs.* rearrangement, **5**, 853
enzymatic, **5**, 855
ester enolate, **6**, 859
ketenes, **5**, 829
kinetics, **5**, 856
mechanism, **5**, 856–865
ortho, **5**, 834
oxyanion-accelerated, **5**, 1000
para, **5**, 834
phosphorus-stabilized anions, **5**, 847
propargyl vinyl systems, **6**, 862
remote stereocontrol, **5**, 864
ring expansion
alkylaluminum-catalyzed, **5**, 850
self-immolative process
stereochemistry, intrinsic transfer, **5**, 860
solvent effects, **5**, 854
stereochemistry
intrinsic transfer, **5**, 860–863
stereocontrol, **5**, 859; **6**, 856
synthetic aspects, **5**, 830–855
tandem ene reactions, **5**, 11
transition state structures, **5**, 857
vinylogous anomeric effect, **5**, 856
ynamine, **5**, 836
Claisen–Schmidt reaction
acetone
with aromatic aldehydes, **2**, 143
mixed aldol reaction, **2**, 134
Clavicipitic acids
synthesis
Mannich reaction, **2**, 967
via diazoalkene cyclization, **4**, 1157
Clavulones
synthesis
via retro Diels–Alder reactions, **5**, 562
Claycop
solid support
oxidants, **7**, 846
Clayfen
Ritter reaction
initiator, **6**, 283
solid support
oxidants, **7**, 846
Clays
solid supports
oxidants, **7**, 840, 845
stabilized pillared catalysts
Friedel–Crafts reaction, **3**, 296
Cleavage reactions
alkenes, **7**, 541–589
synthesis of alcohols, **7**, 543
Clemmensen reduction
carbonyl compounds, **8**, 307, 309
mechanism, **8**, 309
Cleomiscosin A
synthesis
silver oxide oxidation, **3**, 691
Cleomiscosin B
synthesis
silver oxide oxidation, **3**, 691
Cloke rearrangement, **5**, 945
Clostridium paraputrificum
reduction
unsaturated carbonyl compounds, **8**, 558
Clovene
synthesis, **3**, 23, 386
C—H insertion reactions, **3**, 1058
Coal
hydrocracking, **3**, 328
Cobalamin
catalyst
nitrile reduction, **8**, 299
reduction
unsaturated carbonyl compounds, **8**, 562
Cobalt
alkene epoxidation catalysis, **7**, 383
Cobalt, acetylenehexacarbonyl-
alkyne protection, **6**, 692
Cobalt, allyl-
fragmentation
radical reactions, **4**, 746
Cobalt, acyl-
aldol reaction, stereoselective, **2**, 314
deprotonation
reaction, **2**, 127
Cobalt, bis(dimethylglyoximate)chloro(pyridine)-
catalyst
partial reduction of pyridinium salts, **8**, 600
Cobalt, carbonylhydrido-
reduction
unsaturated carbonyl compounds, **8**, 551
Cobalt, carbonylhydridotris(tributylphosphine)-
hydrogenation
alkenes, **8**, 446
Cobalt, carbonylhydridotris(triphenylphosphine)-
hydrogenation
alkenes, **8**, 446
Cobalt(I), chlorotris(triphenylphosphine)-
catalyst
Wurtz reaction, **3**, 421
Cobalt, dicarbonylhydridobis(tributylphosphine)-
hydrogenation
alkenes, **8**, 446
Cobalt, dienyl-
synthesis, **4**, 691
Cobalt, octacarbonyldi-
catalyst
acetal hydrogenation, **8**, 212
alkyne trimerization, **5**, 1146
carbonylation of aryl and vinyl halides, **3**, 1026
hydrosilylation, **8**, 764
silane reaction with carbonyl compounds, **2**, 603
dehalogenation
α-halocarbonyl compounds, **8**, 991
deoxygenation
epoxides, **8**, 890
Pauson–Khand reaction, **5**, 1037
Cobalt, tetracarbonylhydrido-
catalyst
hydroformylation, **4**, 915
Cobalt, tricarbonylhydrido(tributylphosphine)-

catalyst
hydroformylation, **4**, 915
Cobalt, trihydridotris(triphenylphosphine)-
hydrogenation
alkenes, **8**, 446
Cobalt acetate
chalcone formation, **2**, 150
Cobaltacycloheptanes
β-hydride elimination
1-heptene synthesis, **5**, 1141
Cobaltacycloheptenes
β-hydride elimination
1,6-heptadiene synthesis, **5**, 1141
Cobaltacyclopentadienes
pyridone synthesis
via [2 + 2 + 2] cycloaddition, **5**, 1155
reaction with carbon disulfide
thiopyran-2-thione synthesis, **5**, 1158
reaction with nitriles, **5**, 1152
synthesis
via [2 + 2 + 2] cycloaddition, **5**, 1142
Cobalt chloride
lithium aluminum hydride
unsaturated hydrocarbon reduction, **8**, 485
Cobalt complexes
allylic oxidation, **7**, 95
catalysts
carbonyl compound hydrogenation, **8**, 154
hydroboration, **8**, 709
glycol cleavage, **7**, 706
Cobalt complexes, carbene
furans from, **5**, 1092
Cobalt enolates
aldol reaction, **2**, 314
Cobalt halides
lithium aluminum hydride
unsaturated hydrocarbon reduction, **8**, 483
Cobalt hydride
elimination
alkene synthesis, **4**, 805
Cobaltocene
pyridone synthesis
via [2 + 2 + 2] cycloaddition, **5**, 1155
Cobalt perchlorate
alkane oxidation, **7**, 12
Cobalt phthalocyanines
vinyl substitutions
palladium complexes, **4**, 841
Cobalt triacetate
allylic oxidation, **7**, 92
Cocaine
Mannich base, **2**, 894
oxygen analog
synthesis, **2**, 623
synthesis
via nitrone cyclization, **4**, 1120
Cocycloaddition reactions
alkynes, alkenes and carbon monoxide, **5**, 1037
Codamine, *N*-trifluoroacetyl-
oxidative coupling, **3**, 670
Codeinone, 14-bromo-
catalytic hydrogenation, **8**, 899
Codling moth constituent
synthesis
via tandem vicinal difunctionalization, **4**, 250
Coenzyme A
dithioesters
synthesis, **6**, 455
Colchicine
synthesis, **3**, 807
electrooxidation, **3**, 683
via [4 + 3] cycloaddition, **5**, 604
via isomerization, **5**, 714
Colletodiol
synthesis
via Horner–Wadsworth–Emmons reaction, **1**, 769
via macrolactonization, **6**, 375
Collins reagent
oxidation
alcohols, **7**, 256
ethers, **7**, 240
Communic acids
biomimetic conversion
pimaranes, **7**, 634
Compactin
microbial oxidation, **7**, 77
synthesis, **3**, 589; **7**, 247; **8**, 925, 945
via cyclofunctionalization of cycloalkene, **4**, 373
via Diels–Alder reaction, **5**, 350
via nitrile oxide cyclization, **4**, 1128
Compactin, dihydro-
synthesis, **3**, 161
polyalkene cyclization, **2**, 714
Complex reducing agents
desulfurizations, **8**, 840
reduction
alkyl halides, **8**, 802
unsaturated carbonyl compounds, **8**, 551
Computer programs
nucleophilic reactivity
arenes, **4**, 425
Concerted reactions
heterocyclic synthesis, **6**, 756
Conduramine F1
synthesis
via Diels–Alder reaction, **5**, 418
Conduritol A
synthesis
via retro Diels–Alder reactions, **5**, 564
Confertifolin
synthesis, **1**, 570
Confertin
synthesis
ene reaction, **2**, 551
via conjugate addition, **4**, 229
via Cope rearrangement, **5**, 982
via ketocarbenoids, **4**, 1055
via Wharton reaction, **8**, 929
Conhydrine
synthesis, **1**, 555
Conia reaction
thermal intramolecular ene reactions, **5**, 20–23
Coniine
synthesis, **1**, 559; **6**, 769
Conjugate addition–enolate trapping
definition, **4**, 238
Conjugate addition reactions
chiral catalysts, **4**, 230
organocuprates, **4**, 169–195
stereoselectivity, **4**, 187

Conjugate enolates
  definition, **4**, 238
σ-Conjugation
  cyclopropanes, **5**, 900
Consecutive rearrangements, **5**, 875–896
Contact ion pairs
  electron-transfer oxidation, **7**, 851, 854
  intermolecular interactions
    electron-transfer oxidation, **7**, 852
Copacamphene
  synthesis, **3**, 20, 712
Copaene
  synthesis, **3**, 20
Cope–Claisen rearrangement, **5**, 884–887
Cope rearrangement, **5**, 785–822
  alkyl substitution, **5**, 789
  allylic systems, **6**, 834
  amino alcohol synthesis, **7**, 493
  carbanion-accelerated, **5**, 1005
  catalysis, **5**, 798–803
  chirality
    transfer, **5**, 821
  chiral vinyl substituents
    double diastereoselection, **5**, 817
  cis-1,2-divinylcyclopropane, **5**, 971–996
  conformation, **5**, 794
  conjugating substituents, **5**, 789
  double bond configuration, **5**, 821
  equilibria, **5**, 789–796
  erythro–threo ratios, **5**, 821
  heterodivinylcyclopropane, **5**, 939
  metal catalysis, **5**, 799
  palladium(II) catalysis, **4**, 576
  photochemical initiation, **5**, 802
  product aromatization, **5**, 790
  ring strain, **5**, 791–794
  stereospecificity, **5**, 819–822
  thioallyl dianions
    reaction with aldehydes, **2**, 72
  transition state conformation, **5**, 819, 857
Copper
  activated powder, catalyst
    alkene dimerization, **3**, 482
  polymer complexes
    oxidative coupling catalyst, **3**, 559
  Ullmann reaction, **3**, 499
Copper, alkenyl-
  synthesis, **8**, 696
  transmetallation, **8**, 693
  use in synthesis, **4**, 901–903
Copper(I), alkenyl-
  coupling reactions
    with alkenyliodinium tosylates, **3**, 522
  substitution at C=C, 3, 522
  synthesis, **3**, 522
Copper, alkyl-
  reactions with aldimines
    Lewis acid pretreatment, **1**, 350
Copper(I), alkyl-
  tandem vicinal difunctionalization, **4**, 254
Copper(I), alkylalkynyl-
  tandem vicinal difunctionalization, **4**, 256
Copper(I), alkyl(heteroalkyl)-
  tandem vicinal difunctionalization, **4**, 256
Copper, alkylthioallyl-
  allylation, **3**, 99
Copper, allyl-
  magnesium bromide reagent
    reactions with α,β-dialkoxy aldehydes, **1**, 109
Copper, 2,2′-bipyridyl-
  chalcone formation, **2**, 150
Copper, μ-bis(cyanotrihydroborato)-
  tetrakis(triphenylphosphine)di-
  reduction
    acyl halides, **8**, 264
Copper, crotyl-
  reaction with benzaldehyde, **1**, 113
Copper(I), dialkyl-
  tandem vicinal difunctionalization, **4**, 254–256
Copper(I), dicyclohexylamido-
  conjugate additions
    nontransferable ligand, **4**, 177
Copper(I), (diethoxyphosphoryl)methyl-
  alkylation, **3**, 201
Copper(I), diphenylphosphido-
  conjugate additions
    nontransferable ligand, **4**, 177
Copper, germyl-
  reaction with alkynes, **4**, 901
Copper(I), hexynyl-
  tandem vicinal difunctionalization, **4**, 256
Copper, iodofluoroacetates
  synthesis, **3**, 421
Copper, (isopropylthio)allyl-
  reactions with acetone, **1**, 508
Copper, lithiodimethyl-
  copper enolates
    mechanism of reaction, **2**, 120
Copper, methyl-
  aluminum enolates
    catalysis, **2**, 114
  structure, **3**, 210
  synthesis, **3**, 208
Copper(I), pentynyl-
  tandem vicinal difunctionalization, **4**, 256
Copper, phenyl-
  stability, **3**, 210
  structure, **3**, 210
  synthesis, **3**, 208
Copper, γ-silylated vinyl-
  acylation of, **1**, 428
Copper, tetrahydroboratobis(triphenylphosphine)-
  reduction
    hydrazones, **8**, 347
    unsaturated aldehydes, **8**, 540
  sodium borohydride modifier
    acyl halide reduction, **8**, 264
Copper(I), tetrakis[iodo(tri-*n*-butylphosphine)-
  coupling, **3**, 418
Copper, trimethylsilylmethyl-
  carbocupration, **4**, 898
Copper, α-trimethylsilylmethyl-
  acylation of, **1**, 436
Copper, vinyl-
  reaction with alkynyl halides, **3**, 219
  reaction with selenol esters, **6**, 469
  reaction with vinyl halides, **3**, 217
Copper acetate
  oxidative decarboxylation, **7**, 722
  reoxidant

Wacker process, **7**, 451
Copper(I) acetylides
Cadiot–Chodkiewicz coupling, **3**, 219
hydrolysis, **3**, 210
reaction with propargylic electrophiles, **3**, 223
synthesis, **3**, 208, 209
Copper aldimines
conjugate additions
α,β-enones, **4**, 162
Copper(I) alkynides
reaction with aryl halides, **3**, 522
Copper borates
reaction with allylic halides, **3**, 221
Copper bromide
halogenation
carbonyl compounds, **7**, 120
ketone dehydrogenation, **7**, 144
Copper(I) bromide
purification, **3**, 209
Copper chloride
halogenation
carbonyl compounds, **7**, 120
Kharasch–Sosnovsky reaction, **7**, 95
oxidation
primary alcohols, **7**, 308
reaction with organoboranes, **7**, 604
reoxidant
Wacker process, **7**, 451
Copper(II) chloride
catalyst
Knoevenagel reaction, **2**, 345
cyclohexadienyliron complexes
decomplexation, **4**, 674
Copper chromite
catalyst
carboxylic acid hydrogenation, **8**, 236
ester hydrogenation, **8**, 242
hydrogenation, **8**, 963
Copper compounds
zero-valent
organocopper compounds from, **3**, 209
Copper(I) compounds
catalysts
Grignard couplings, **3**, 419
halogen atom transfer addition reactions, **4**, 754
lithium enolate polyalkylation, **3**, 6
Copper(I) cyanide
copper alkynide synthesis, **4**, 176
purification, **3**, 209
Copperdilithium, cyanobis(dimethylphenylsilyl)-
tandem vicinal difunctionalization, **4**, 257
Copper enolates
synthesis, **2**, 119
Copper(I) enolates
enantioselective aldol reaction
acyliron complexes, **2**, 316
Copper hydride
elimination
radical cyclizations, **4**, 807
reduction
alkyl halides, **8**, 801
unsaturated carbonyl compounds, **8**, 548, 550
Copper iodide
magnesium hydride
unsaturated hydrocarbon reduction, **8**, 483
Copper(I) iodide
in alkylation
α-thioalkyllithium, **3**, 88
purification, **3**, 209
vinyl Grignard reagent alkylation
catalyst, **3**, 243
Coppermagnesium dihalides, alkyl-
tandem vicinal difunctionalization, **4**, 255
Copper(I) methyltrialkylboronates
conjugate additions
acrylonitrile, **4**, 148
Copper nitrate
benzylic halide oxidation, **7**, 666
nitration with
clay-supported, **6**, 111
reoxidant
Wacker process, **7**, 451
solid support
clay, **7**, 846
Copper salts
amine complexes
reduction, aromatic nitro compounds, **8**, 373
Copper sulfate
oxidation
diols, **7**, 313
Copper(I) triflate
catalyst
alkene dimerization, **3**, 482
Friedel–Crafts reaction, **2**, 737
Ullmann reaction, **3**, 499
Copyrine, tetrahydro-
synthesis
via Ritter reaction, **6**, 279
Corannulene
synthesis
intramolecular acyloin coupling reaction, **3**, 625
Corey–Chaykovsky reaction
addition of methylene group to carbonyl compounds
dimethyloxosulfonium methylide, **1**, 820
Corey–Winter reaction
2-thioxo-1,3-dioxolanes, **6**, 686
Coriamyrtin
synthesis, **7**, 162, 243
Coriolin
synthesis, **3**, 603, 785; **7**, 240; **8**, 123
organocopper compounds, **3**, 221
via [4 + 4] cycloaddition, **5**, 639
via organocuprate conjugate addition, **4**, 194
via Pauson–Khand reaction, **5**, 1060
via photocycloaddition, **5**, 146, 665, 666
via photoisomerization, **5**, 232, 233
via tandem radical cyclization, **1**, 270
via Wacker oxidation, **7**, 455
Cornforth model
aldehyde reactions
with allylboronates, **2**, 26
carbonyl compounds
reduction, **8**, 3
Coronafacic acid
synthesis
via Cope rearrangement, **5**, 791
via ene reaction with methyl propiolate, **5**, 8
via α-silyl carbanions, **1**, 783
Cortexolone
microbial hydroxylation, **7**, 74

Corticosteroids
  oxidative cleavage
    sodium bismuthate, **7**, 704
  synthesis, **3**, 126
    via acylation of organolithiums, **1**, 412
Cortisol
  microbial dehydrogenation, **7**, 67
Cortisol, fluoro-
  hormonal activity, **6**, 216
Cortisone
  microbial dehydrogenation, **7**, 67
(±)-Cortisone
  synthesis
    use of copper homoenolate, **2**, 452
Cortisone, 6α-methylhydro-
  microbial dehydrogenation, **7**, 68
Corydalic acid
  methyl ester
    synthesis, Mannich reaction, **2**, 929
Corydalisol
  synthesis, **1**, 564
Corydine
  synthesis, **3**, 807
Corynanthe alkaloids
  synthesis
    Mannich reaction, **2**, 1031
Corynantheoid indole alkaloids
  synthesis
    via Diels–Alder reactions, **5**, 467
Corynebacterium equi
  epoxidation, **7**, 429
Corynebatesium equi
  reduction
    unsaturated carbonyl compounds, **8**, 561
Corynespora cassicola
  epoxidation, **7**, 429
Costal
  synthesis
    via Wharton reaction, **8**, 928
Costunolide
  synthesis, **8**, 945
    via chromium(II) ion mediation, **1**, 188
    via cyclization, **1**, 553
    Wittig rearrangement, **3**, 1010
Costunolide, dihydro-
  synthesis
    via Cope–Claisen rearrangement, **5**, 886
Costus, dehydro-
  lactone
    synthesis, Mannich reaction, **2**, 911
Coulson–Moffit model
  cyclopropane
    bonding, **5**, 900
Coumaran-2,3-diones
  synthesis
    Friedel–Crafts reaction, **2**, 757
Coumarin-3-carboxylic acid
  esters
    Knoevenagel reaction, **2**, 354
  synthesis
    Knoevenagel reaction, **2**, 357
Coumarin-3-carboxylic acid, 3,4-dihydro-3-substituted-
  esters
    Knoevenagel reaction, **2**, 355
Coumarins
  synthesis
    Knoevenagel reaction, **2**, 362
    Perkin reaction, **2**, 395, 401
    via arylcerium reagents, **1**, 242
    Vilsmeier–Haack reaction, **2**, 790
Coumarins, 3-acyl-
  synthesis
    Knoevenagel reaction, **2**, 359
Coumarins, dihydro-
  synthesis, **7**, 336
    via aromatic Claisen rearrangement, **5**, 834
Coumarins, 4-hydroxy-
  synthesis
    sulfur-assisted carbonylation, **3**, 1034
Coumarins, 7-methoxy-3-pyridyl-
  synthesis
    Knoevenagel reaction, **2**, 362
Coumestones
  synthesis
    via isoflavones, **7**, 831
Counter electrodes
  electrosynthesis, **8**, 130
Coupling reactions
  acyloins, **3**, 613–631
  alkenes, **3**, 482
  sp carbon centers, **3**, 551–559
  sp^2 and *sp* carbon centers, **3**, 521–549
  sp^2 carbon centers, **3**, 481–515
  sp^3 and sp^2 carbon centers, **3**, 435–476
  sp^3 carbon centers, **3**, 413–432
  sp^3 organometallics and alkenyl halides, **3**, 436
Crabtree's catalyst
  hydrogenation
    alkenes, **8**, 452
Cracking
  alkanes, **7**, 7
Cram–anti-Cram ratio
  aldol reactions, **2**, 248
Cram–Felkin stereochemical control
  Diels–Alder reactions, **2**, 677
Cram's rule
  carbonyl compounds
    reduction, **8**, 3
  chiral aldehyde reactions
    with pinacol crotylboronates, **2**, 25
Crassin acetate
  basic nucleus
    synthesis, **2**, 194
Crenulidine, acetoxy-
  synthesis
    via Claisen rearrangement, **5**, 833
Cresols
  cycloalkylation
    Friedel–Crafts reaction, **3**, 304
  disproportionation
    Friedel–Crafts reaction, **3**, 328
o-Cresols
  Mannich reaction
    with preformed iminium salts, **2**, 960
p-Cresols
  arylation, **4**, 469, 470
  Mannich reaction
    with methylamine and formaldehyde, **2**, 969
  oxidative coupling, **3**, 665
p-Cresols, 2,6-di-*t*-butyl-

enolates
synthesis, **2**, 105
Criegee rearrangement
alcohol synthesis, **6**, 14
Crinine
synthesis, **6**, 741
Mannich reactions, **2**, 1042
Crispatic acid
synthesis, **8**, 647
Croomine
synthesis
via iodocyclization of allylic substrate, **4**, 404
Crotepoxide
synthesis
via retro Diels–Alder reactions, **5**, 563
Crotonaldehyde
ene reactions
Lewis acid catalysis, **5**, 6
hydrogenation, **8**, 140
Lewis acid complexes
NMR, **1**, 294
reduction
aluminum isopropoxide in isopropyl alcohol, **8**, 88
synthesis
via retro Diels–Alder reaction, **5**, 553
Crotonamides, *N,N*-dialkyl-
addition reactions
with organomagnesium compounds, **4**, 84
Crotonates
alcohol protection
cleavage, **6**, 658
conjugate additions
amines, **4**, 231
synthesis, **3**, 263
Crotonates, ethyl α-methyl-β-bromo-
catalytic hydrogenation, **8**, 899
Crotonates, methoxy-
alcohol protection
cleavage, **6**, 658
Crotonic acid
methyl ester
reaction with lithium
bis(phenyldimethylsilyl)cuprate, **2**, 200
reaction with 1-pyrroline 1-oxide, **5**, 256
Crotonic acid, 3-amino-2,4-dicyano-
reaction with benzil, **3**, 824
Crotonic acid, γ-amino-α-fluoro-
hydrogenolysis, **8**, 896
Crotonic acid, 4-bromo-
esters
Reformatsky reaction, regioselectivity, **2**, 286, 287
Crotonic acid, 2-methyl-
ethyl ester
alkylation of enolates, **2**, 187
Crotonic acid, 3-methyl-
ethyl ester
alkylation of enolates, **2**, 187
methyl ester
enolates, aldol reaction, **2**, 188
Crotonic acid, β-phenylseleno-
synthesis
via alkoxyselenation, **4**, 340
Crotonyl azides
Curtius rearrangement, **6**, 799
Crotonyl chloride
Nazarov cyclization, **5**, 778
Crotyl acetal
reduction, **8**, 213
Crotyl addition
aldehydes
2,5-dimethylborolane, **2**, 258
Crotyl bromide
reaction with benzaldehyde
chromium dichloride mediated, **1**, 179
Crotyl cyanide
Michael additions
chiral imines, **4**, 221
Crotyl halides
addition to α-alkoxy chiral aldehydes
chromium mediated, **1**, 185
Crotyl organometallic compounds
configurational stability, **2**, 5
mechanistic classification
simple diastereoselectivity, **2**, 3
reactions with aldimines
regioselectivity, **2**, 989
reactions with *gem*-amino ethers
dependence of product type on metal, **2**, 1005
reactions with imines, **2**, 988
regioselectivity, **2**, 988
reversibility, **2**, 980
syn–anti selectivity, **2**, 990
reactions with iminium salts
dependence of product type on metal, **2**, 1001
reactions with ketones
diastereoselectivity, **2**, 8
type I
chair-like transition states, **2**, 4
reactions with achiral aldehydes, **2**, 9–19
reactions with achiral imines, **2**, 9–19
reactions with achiral ketones, **2**, 9–19
reactions with aldehydes, **2**, 4
reactions with aldehydes, diastereofacial selectivity, **2**, 29
reactions with C=X electrophiles, **2**, 6
type II
reactions with carbonyl compounds, **2**, 4
type III
reactions with achiral aldehydes, **2**, 19–24
reactions with achiral ketones, **2**, 19–24
reactions with aldehydes, **2**, 5
reactions with aldehydes, diastereofacial selectivity, **2**, 29
reactions with C=X electrophiles, **2**, 6
18-Crown-6
alcohol inversion
suppression of elimination reactions, **6**, 21
18-Crown-6, dicyclohexyl-
S-t-butyl thiocarboxylic esters
synthesis, **6**, 440
Crown ethers
1,2-additions to carbonyl compounds
lower order cuprates, **1**, 115
chromium(VI) oxide
alcohol oxidation, **7**, 278
dissolving metals
reductions, **8**, 524
in sulfide metallation, **3**, 86
phenolic
synthesis, **7**, 333

reactions with organometallic compounds
Lewis acids, **1**, 335
reduction
aluminum hydrides, **8**, 541
synthesis, **3**, 591
Crustecdysone, 2-deoxy-
synthesis, **8**, 534
Cryptands
reduction
aluminum hydrides, **8**, 541
[2.2.2]-Cryptands
thioallyl anion reactions
regioselectivity, **2**, 71
Cryptausoline
synthesis
via arynes, **4**, 504
Cryptopleurine
synthesis, **3**, 507
use of thallium trifluoroacetate, **3**, 670
via selective *ortho* lithiation, **1**, 466
Cryptopleurospermine
synthesis, **1**, 564
Cryptowoline
synthesis
via arynes, **4**, 504
Crystal growth
carbanions, **1**, 40
Cubane amides
lithiation, **1**, 480
Cubanes
isomerizations
metal catalyzed, **5**, 1188
reactions with transition metal complexes, **7**, 4
synthesis, **3**, 848, 854
via photochemical cycloaddition, **5**, 123
Cubebin
synthesis, **1**, 566
Cularine alkaloids
synthesis
via arynes, **4**, 505
Cumene
solvent
reductive decarboxylation, **7**, 720
synthesis
Friedel–Crafts reaction, **3**, 294
Cumulatrienes
synthesis
via allenic phosphonates, **6**, 845
Cumulenes
addition reactions, **7**, 506
coupling reactions
carbene complexes, **5**, 1107
cyclic
synthesis *via* dihalocyclopropanes, **4**, 1010
hydrogenation, **8**, 434
synthesis
3,2-sigmatropic rearrangement, **3**, 963
via dihalocyclopropanes, **4**, 1010
via retro Diels–Alder reaction, **5**, 589
Cumyl hydroperoxide
asymmetric epoxidation, **7**, 394
Cunninghamella blakesleeana
hydrocarbon hydroxylation, **7**, 58
Cuparene
synthesis, **3**, 588
epoxide ring opening, **3**, 744
synthesis, unsuccessful
diastereoselectivity, **1**, 150
Cuparenone
synthesis, **3**, 785, 786
C—H insertion reactions, **3**, 1060
α-Cuparenone
synthesis
via conjugate addition, **4**, 215
via [3 + 2] cycloaddition reactions, **5**, 284
via [4 + 3] cycloaddition reactions, **5**, 603
β-Cuparenone
synthesis
via addition with organozinc compounds, **4**, 95
via conjugate addition of aryl cyanohydrin, **1**, 552
Cuprates (*see also* under Lithium and Dilithium compounds)
carbanions
crystal structure, **1**, 40
Claisen rearrangement, **5**, 844
enolates
synthesis, **2**, 119
higher order, **3**, 213
properties, **4**, 170
synthesis, **3**, 209
immobilization
solid supports, **3**, 211
reaction with tosylates, **3**, 248
synthesis
via transmetallation, **4**, 175
Cuprates, alkylphenylthio-
lithium salt
reaction with α,β-unsaturated carbonyl compounds, **2**, 121
Cuprates, benzyl-
synthesis, **3**, 209
Cuprates, bis(phenyldimethylsilyl)-
lithium salt
conjugate addition to α,β-unsaturated esters, **2**, 186
Cuprates, (α-carbalkoxyvinyl)-
reaction with activated halides, **3**, 217
Cuprates, cyano(2-thienyl)-
alkylation, **3**, 251
stability, **3**, 213
Cuprates, dialkenyl-
transmetallation to alkenylzinc reagents, **4**, 903
Cuprates, dialkyl-
lithium salt
conjugate addition to enones, **2**, 185
reactions with dienyliron complexes, **4**, 670
Cuprates, dialkylcyano-
carbonylation with α,β-unsaturated ketones
1,4-diketone synthesis, **3**, 1024
dilithium salt
reaction with α,β-unsaturated ketones, **2**, 120
Cuprates, di-*t*-butylphosphido-
stability, **3**, 211
Cuprates, dicyclohexylamido-
stability, **3**, 211
Cuprates, dicyclohexylphosphido-
stability, **3**, 211
Cuprates, dimethyl-
lithium salt
spirocyclic aldol formation, **2**, 166
Cuprates, diphenylphosphido-

stability, **3**, 211
Cuprates, divinyl-
alkylation, **3**, 259
Cuprates, hydrido-
reduction
unsaturated carbonyl compounds, **8**, 549
Cuprates, phenylthio-
reaction with acyl halides, **3**, 226
Cuprates, phosphino-
conjugate additions
enones, **4**, 177
Cuprates, α-selenoalkyl-
allylation, **3**, 91
Cuprates, silyl-
conjugate additions, **4**, 231
reaction with alkynes, **4**, 901
Cuprates, vinyl-
acylation of, **1**, 428
synthesis, **3**, 209
Curcumene
synthesis
nickel catalysts, **3**, 229
via conjugate addition, **4**, 211
α-Curcumene
synthesis
via reductive silylation of anisole, **8**, 518
ar-Curcumene
synthesis, **6**, 455
β-Curcumin
synthesis, **3**, 127
Current density
electrosynthesis, **8**, 130
Current efficiency
electrosynthesis, **8**, 130
Current yield
electrosynthesis, **8**, 130
Curtin–Hammett principle
radical cyclizations, **4**, 815
Curtius rearrangement, **6**, 806
acyl azides, **3**, 908; **6**, 797; **7**, 477
stereoselectivity, **6**, 798
Curvularin
synthesis
via Wacker oxidation, **7**, 455
Curvulin
synthesis
palladium-catalyzed carbonylation, **3**, 1033
Cyanallyl
discovery, **6**, 242
Cyanamides
amidine synthesis, **6**, 546
trimerization, **5**, 1154
Cyanates
alkyl
synthesis, **6**, 244
aromatic
synthesis, **6**, 244
synthesis, **6**, 243
Cyanation
amide acetals, **6**, 573
electrochemical
aromatic compounds, **7**, 801
hydrocarbons, **3**, 1046
Cyanides
addition reactions
carbonyl compounds, **1**, 460
aromatic nucleophilic substitution, **4**, 433
Michael donors, **4**, 259
polymer-supported catalyst
benzoin condensation, **1**, 543
Cyanides, α-alkoxyacyl
chiral
reaction with silyl enol ethers, **2**, 646
Cyanides, β-alkoxyacyl
chiral
reaction with silyl enol ethers, **2**, 646
Cyanides, arylsulfonyl
Diels–Alder reactions, **5**, 416
Cyanides, *t*-butyldimethylsilyl
reaction with cyclic vinyloxiranes, **5**, 936
Cyanides, trimethylsilyl-
acyl cyanide synthesis, **6**, 317
reaction with conjugated ketones
regiochemistry of addition, **2**, 599
Cyano-*t*-butyloxycarbonyl group
protecting group
amines, **6**, 638
Cyanocarbonylation
aryl iodides, **6**, 318
Cyanodienes
synthesis
use of ylidic rearrangements, **3**, 963
β-Cyano esters
metal enolates
alkylation, **3**, 54
Cyanogen
reactions with amines, **6**, 546
Cyanogen azide
decomposition
formation of cyanonitrene, **7**, 10
reactions with alkenes, **7**, 480
Cyanogen chloride
adducts
amides, **6**, 492
reactions with alkanes, **7**, 7
Cyanohydrin anions
addition reactions
alkenes, palladium(II) catalysis, **4**, 572
Cyanohydrin esters
reactions with carbonyl compounds, **1**, 551
Cyanohydrin ethers
anion
rearrangement, **2**, 69
reactions with carbonyl compounds, **1**, 551
Cyanohydrins
addition reaction
with α,β-enones, **4**, 113
carbonyl group protection, **6**, 681
chiral
benzoin condensation, **1**, 546
intramolecular alkylation, **3**, 48
Michael addition, **4**, 12
optically active
synthesis, **1**, 347
protected, **1**, 544
α-alkoxy carbanions from, **3**, 197
O-protected
benzoin condensation, **1**, 547
Ritter reaction, **6**, 265
O-silyl-protected

benzoin condensation, **1**, 548
unsaturated anions
intramolecular reactions, **3**, 51
Cyanohydrins, *O*-allyl-
Wittig rearrangement, **3**, 998
Cyanohydrins, ketone
β,γ-unsaturated
synthesis, **3**, 998
Cyanohydrins, *O*-trimethylsilyl-
alkylation, **3**, 197
p-benzoquinone protection, **6**, 682
carbonyl group protection, **6**, 682
Cyanophosphates
acyl anion equivalents, **1**, 544, 560
Cyanoselenation
ketene *O,O*-acetals, **6**, 565
Cyanoselenenation
alkenes, **7**, 522
Cyanoselenenylation
alkenes, **4**, 341
Cyanosulfenylation
alkenes, **4**, 337
synthesis of α-methylthionitriles, **6**, 239
Cyanuric acid chloride
adducts
amides, **6**, 492
amide synthesis, **6**, 383
Cyanuric acid esters
acid anhydride synthesis, **6**, 313
Cyclic compounds
synthesis
aldol reaction cascade, **2**, 619
allylic halides, **3**, 429
Wurtz reaction, **3**, 422
Cyclic voltammetry
electrosynthesis, **8**, 131
Cyclic voltammograms
oxidation potentials, **7**, 852
Cyclitols, amino-
synthesis
via cyclofunctionalization, **4**, 375, 400
Cyclization–carbonylation
carboxylate ions
palladium(II) catalysis, **4**, 558
Cyclization–demercuration
mercury(II)-induced, **8**, 857
Cyclization-induced rearrangement
palladium(II) catalysis, **4**, 563
Cyclization reactions
Beckmann rearrangement, **6**, 771
carbonyl derivatives
electrophilic or radical attack, **6**, 755
donor radical cations, **7**, 876
electrophilic heteroatom, **4**, 363–414
5-*endo*
alkenyl systems, **4**, 377
5-*endo*-trigonal
intramolecular addition, **4**, 40
6-*endo*-trigonal
intramolecular addition, **4**, 40
5-*exo*
alkenyl systems, **4**, 377
5-*exo*-trigonal
intramolecular addition, **4**, 38
6-*exo*-trigonal
intramolecular addition, **4**, 39
nitrogen heterocycles, 397–413
polyenes, **3**, 341–375
initiation, **3**, 342
mechanism, **3**, 374
propagation, **3**, 343
termination, **3**, 345
radical cations
unimolecular reaction, **7**, 858
$S_{RN}1$ reactions, **4**, 476–480
sulfur compounds, **4**, 413
tandem semipinacol rearrangements, **3**, 792
transannular electrophilic, **3**, 379–407
Cycloacylation
γ-hydroxy acids
γ-lactone synthesis, **6**, 350
lactone synthesis, **6**, 342
β-lactone synthesis, **6**, 346
macrolactonization, **6**, 369
Cycloaddition reactions
alkynes
alkenes, **5**, 676
carbene transition metal complexes, **5**, 1065–1113
1,3-dipolar
intermolecular, **4**, 1069–1104
intramolecular, **4**, 1111–1166
donor radical cations, **7**, 879
hole catalyzed
diene oxidation, **7**, 861
phenol ethers, **3**, 696
photochemical, **5**, 123–148
mechanisms, **5**, 124
radical cations
bimolecular reaction, **7**, 859
small ring compounds
metal-catalyzed, **5**, 1185–1204
thermal, **5**, 239–270
transition metal catalysts, **5**, 271–312
Wolff rearrangement, **3**, 905
[2 + 1 + 1] Cycloaddition reactions
cyclobutanones, **5**, 1087
[2 + 2] Cycloaddition reactions
alkenes
thermal, **5**, 63–79
carbene transition metal complexes, **5**, 1067
diastereofacial selectivity, **5**, 79
intramolecular, **5**, 67–72
thermal
stereochemistry, **5**, 79
[2 + 2 + 1] Cycloaddition reactions
cyclopentenone synthesis
Pauson–Khand reaction, **5**, 1037
[2 + 2 + 1 + 1] Cycloaddition reactions
two-alkyne annulations, **5**, 1102
[2 + 2 + 2] Cycloaddition reactions, **5**, 1129–1158
[3 + 2] Cycloaddition reactions
carbene complexes, **5**, 1070
intramolecular, **5**, 304
methylenecyclopropanes, **5**, 1188
radical anions, **7**, 862
synthons, **5**, 271
[4 + 2] Cycloaddition reactions
carbene complexes, **5**, 1070
radical anions, **7**, 862
[4 + 3] Cycloaddition reactions, **5**, 593–613

intramolecular, **5**, 609
nonconcerted, **4**, 1075
[4 + 4] Cycloaddition reactions, **5**, 635–641
[6 + 4] Cycloaddition reactions, **5**, 617–635
Cycloalkadienes
molybdenum complexes
reactions with *N*-substituted sulfoximine carbanions, **1**, 535
monoepoxides
reaction with lithium homocuprates, **3**, 226
Cycloalkanecarboxylic acid, 1,2-dialkyl-
synthesis
intramolecular alkylation, **3**, 49
1,2-Cycloalkanediols, 1,2-divinyl-
oxy-Cope rearrangement, **5**, 796
1-Cycloalkanepropionate, 2-alkoxycarbonyl-
synthesis
via ester enolate addition, **4**, 107
Cycloalkanes
bridged
synthesis *via* 1,3-dipolar cycloadditions, **4**, 1077
condensed
synthesis *via* 1,3-dipolar cycloadditions, **4**, 1077
synthesis
Wurtz reaction, **3**, 422
Cycloalkanes, alkylidene-
synthesis, **1**, 669
Cycloalkanes, 1-azido-1-thiomethyl-
rearrangement, **6**, 542
Cycloalkanes, *trans,trans*-1,2-bis(alkylidene)-
synthesis
via diyne bicyclization, **5**, 1171
Cycloalkanes, divinyl-
bridged
Cope rearrangements, **5**, 812–819
Cycloalkanes, 1,2-divinyl-
Cope rearrangement, **5**, 791, 803–812
Cycloalkanes, methylene-
epoxidation, **7**, 361, 364
ring expansion, **7**, 831
Cycloalkanol, 2-methoxy-
oxidative cleavage, **7**, 705
Cycloalkanones
boron trifluoride complex
NMR, **1**, 293
pinacol coupling reactions
cerium-induced, **3**, 570
ring contraction, **7**, 832
spiroannelation, **3**, 88
synthesis, **7**, 601
via Michael addition, **4**, 14
Cycloalkanones, alkylidene-
Grignard additions
copper catalyzed, **4**, 91
peroxy acid oxidation, **7**, 684
Cycloalkanones, 2-alkyl-2-phenyl-
synthesis, **3**, 36
Cycloalkanones, dibromomonochloro-
rearrangement, **3**, 851
Cycloalkanones, 2,3-dihydroxy-
synthesis
via 2-cycloalkenones, **1**, 534
Cycloalkanones, 2-formyl-
Michael addition, **4**, 5
Cycloalkanones, 2-nitro-
2-substituted
synthesis, **2**, 331
Cycloalkanones, polyhalo-
larger ring
rearrangements, **3**, 850
rearrangements, **3**, 849
Cycloalkenecarbaldehyde
aldimines
reactions with Grignard compounds, **1**, 382
Cycloalkene oxides
reduction
lithium aluminum hydride, **8**, 872
Cycloalkenes
allylic oxidation
selenium dioxide, **7**, 91
chirality transfer
sulfoxide–sulfenate rearrangements, **6**, 900
hydroalumination, **8**, 739
ring contraction, **7**, 831
synthesis
intramolecular McMurry reaction, **3**, 588
via cycloadditions, **5**, 64
toluene alkylation with
Friedel–Crafts reaction, **3**, 304
transannular reactions, **3**, 379
Cycloalkenes, 1-alkyl-
synthesis, **3**, 247
Cycloalkenes, 1-chloro-2-hydroperfluoro-
reduction, **8**, 897
Cycloalkenes, 1,2-dialkyl-
asymmetric epoxidation
kinetic resolution, **7**, 416
Cycloalkenes, dideuterio-
synthesis, **8**, 726
Cycloalkenes, 2,3-divinyl-
synthesis
via Cope rearrangement, **5**, 797
Cycloalkenes, epoxy-
nucleophilic reactions
Lewis acids, **1**, 343
3-Cycloalkenes, 1-trimethylsiloxy-1-vinyl-
thermal rearrangements, **5**, 1001
3-Cycloalkenols, 1-vinyl-
oxy-Cope rearrangements, **5**, 1001
Cycloalkenones
synthesis, **1**, 669
2-Cycloalkenones
dimerization
base initiated, **4**, 239
Grignard additions
copper catalyzed, **4**, 91
tandem vicinal difunctionalization, **4**, 242, 245
2-Cycloalkenones, 2-arylsulfinyl-
addition reaction with enolates, **4**, 108
addition reaction with Grignard reagents, **4**, 86
tandem vicinal difunctionalization, **4**, 245
2-Cycloalkenones, bromo-
hydrogenolysis, **8**, 900
Cycloalkenones, 3-nitro-
synthesis, **6**, 109
2-Cycloalkenones, β-silyl-
synthesis, **7**, 107
2-Cycloalkenones, 2-sulfinyl-
conjugate additions, **4**, 213
Cycloalkylation

arenes
Friedel–Crafts reaction, **3**, 309
carboxylic acid derivatives
enolates, **3**, 48
Friedel–Crafts reaction, **3**, 323
lactone synthesis, **6**, 342
β-lactone synthesis, **6**, 345
γ-lactone synthesis, **6**, 357
macrolactonization, **6**, 375
saturated ketones, **3**, 18
Cycloalkylcarboxylic acids
synthesis, **3**, 845
Cycloalkylidene epoxides, α-methylene-
macrocyclic
reaction with organocopper compounds, **3**, 226
3α-5-Cyclo-5-α-androstan-6-one
Mannich reaction
with iminium salts, **2**, 901
Cycloaraneosene
synthesis
via intramolecular ene reactions, **5**, 24
Cycloaraneosine
synthesis
via allyl chromium reagents, **1**, 187
via chromium-initiated cyclization, **1**, 188
Cycloartenol
biosynthesis, **3**, 1048
Cycloazasulfenylation
alkenes, **4**, 332
Cyclobutabenzannulation
biphenylene synthesis, **5**, 1151
Cyclobutadienes
isolation
transition metal complexes, **6**, 692
push–pull
synthesis, **6**, 191
synthesis
via retro Diels–Alder reaction, **5**, 568
Cyclobutanediols
pinacol rearrangement, **3**, 727
Cyclobutane-1,2-diones
ring contraction, **3**, 832
Cyclobutanediones, tetramethyl-
irradiation
[4 + 3] cycloaddition reaction, **5**, 597
Cyclobutanes
alkenes from, **5**, 64
charge-accelerated rearrangements, **5**, 1016
oxidative rearrangement, **7**, 824, 833
ring formation
thermal, **5**, 63–79
strain energy, **5**, 900
synthesis
intramolecular acyloin coupling reaction, **3**, 620
via photochemical cycloaddition, **5**, 123
Wurtz reaction, **3**, 422
Cyclobutanes, alkylidene-
isomerization
1-alkylcyclobutenes, **5**, 677
Cyclobutanes, aryl-
rearrangement
oxyanion-accelerated, **5**, 1018
Cyclobutanes, 1-cyano-1-(methylthio)-
synthesis, **1**, 561
Cyclobutanes, cyclopropyl-
synthesis, **5**, 927
Cyclobutanes, 1,2-dicyano-
synthesis
via acrylonitrile dimerization, **5**, 63
Cyclobutanes, 1,1-dicyano-2-methoxy-
cleavage, **5**, 73
Cyclobutanes, 1,2-dimethylene-
synthesis, **3**, 873
Cyclobutanes, divinyl-
rearrangements, **5**, 1024–1030
anion-accelerated, **5**, 1027–1030
thermolysis
[4 + 4] cycloaddition, **5**, 639
Cyclobutanes, 1,2-divinyl-
Cope rearrangement, **5**, 791, 805, 821
palladium catalysts, **5**, 799
3,3-sigmatropic rearrangement, **5**, 1024
synthesis
via [2 + 2] cycloaddition, **5**, 1025
Cyclobutanes, 1-lithio-1-selenophenyl-
synthesis
via cyclobutanones, **5**, 677
Cyclobutanes, methylene-
oxidation
Wacker process, **7**, 453
Cyclobutanes, octylidene-
synthesis, **1**, 653
Cyclobutanes, 1,2,3,4-tetravinyl-
Cope rearrangement, **5**, 810
Cyclobutanes, vinyl-
rearrangements
azaanion-accelerated, **5**, 1023
oxyanion-accelerated, stereochemistry, **5**, 1018
thermal, **5**, 1016–1024
ring expansion
oxyanion-accelerated, **5**, 1017–1023
synthesis
via photoisomerization, **5**, 199
Cyclobutane-3-thione, 2,2,4,4-tetramethyl-1-oxo-
S-methylide
cycloadditions, **4**, 1074
Cyclobutanols
oxidation
solid support, **7**, 841
oxidative cleavage, **7**, 825
ring expansion, **7**, 843
ring strain
relief, **1**, 887
synthesis, **7**, 41
photochemically mediated, **3**, 1048
Cyclobutanols, 1-(1′-alkenyl)-
synthesis, **1**, 709
Cyclobutanols, 2-(2-furyl)-
rearrangement
oxyanion-accelerated, **5**, 1018
Cyclobutanols, phenyl-
rearrangement
oxyanion-accelerated, **5**, 1018
Cyclobutanols, 1-vinyl-
rearrangement
oxyanion-accelerated, **5**, 1022
1-Cyclobutanols, 1-vinyl-
oxidation
Wacker process, **7**, 453
Cyclobutanols, 2-vinyl-

rearrangement
oxyanion-accelerated, **5**, 1016–1022
Cyclobutanones
annulation
in preparation of spirocyclic ketones, **5**, 921
chemoselective epoxidation, **7**, 385
dimethyl acetals
selective reduction, **8**, 217
enolates
aldol reactions, **2**, 186
epoxides
rearrangements, **1**, 862
oxidation
Baeyer–Villiger reaction, **7**, 674
ring expansion, **7**, 675
polyalkylation
side reaction to monoalkylation, **3**, 4
reactions with diazomethane, **1**, 848
reduction
aluminum amalgam, **8**, 116
synthesis, **3**, 785; **5**, 951, 1087, 1107
rearrangement of epoxides, **3**, 765
rearrangement of oxaspiropentanes, **3**, 761
via ring expansion, **5**, 919
via vinylcyclopropanes, **5**, 919
Cyclobutanones, 2-bromo-
Favorskii rearrangements
mechanism, **3**, 848
Cyclobutanones, 2-diazo-3,4-bis(diphenylmethylene)-
irradiation
Wolff rearrangement, **3**, 900
Cyclobutanones, 2,3-divinyl-
synthesis
via cyclobutenone ring opening, **5**, 690
Cyclobutanones, hexafluoro-
ene reaction, **2**, 538
Cyclobutanones, 2-hydroxy-
synthesis
intramolecular acyloin coupling reaction, **3**, 620
Cyclobutanones, 2-methyl-
synthesis, **3**, 175
Cyclobutanones, permethyl-
reaction with α-selenoalkyllithium, **1**, 674
Cyclobutanones, substituted
enantioselective synthesis
acylcobalt aldol reaction, **2**, 314
Cyclobutanones, 2-vinyl-
divinyl ketones from
cyclization, **5**, 770
synthesis
via Cope rearrangement, **5**, 805
via ring expansion of cyclopropylcarbinols, **5**, 1020
Cyclobutene
cis-3,4-disubstituted
cycloaddition reactions, **5**, 257
[2 + 2 + 2] cycloaddition reactions, **5**, 1130
electrocyclic ring opening, **5**, 1030
anion-accelerated, **5**, 1032
rearrangements, **5**, 1030–1033
ring opening reactions, **5**, 675–694
stereochemistry, **5**, 678
substituent effects, **5**, 678–683
two-carbon ring expansion, **5**, 686–688
synthesis, **3**, 163; **5**, 676
intramolecular McMurry reaction, **3**, 588
Ramberg–Bäcklund rearrangement, **3**, 862, 871
via ene reactions with methyl propiolate, **5**, 7
uses in synthesis, **5**, 683–688
Cyclobutene, alkylidene-
synthesis, **3**, 116
Cyclobutene, 3-*t*-butyl-3-methoxy-
isomerization, **5**, 679
Cyclobutene, 1-chloro-2-hydroperfluoro-
hydrogenation, **8**, 899
Cyclobutene, 1-chloro-2-iodotetrafluoro-
hydrogenolysis, **8**, 900
Cyclobutene, 1,2-dialkyl-
synthesis
via Ramberg–Bäcklund reaction, **3**, 873
Cyclobutene, *cis*-3,4-dichloro-
synthesis
via retro Diels–Alder reaction, **5**, 677
Cyclobutene, 3,3-diethyl-
isomerization, **5**, 679
Cyclobutene, diimino-
synthesis
via diarylalkynes, **5**, 1130
Cyclobutene, 1,1-dimethoxy-
synthesis
via benzynes, **5**, 692
Cyclobutene, 3,3-dimethoxy-
synthesis
via retro Diels–Alder reaction, **5**, 677
Cyclobutene, *cis*-3,4-dimethoxy-
synthesis and ring opening, **5**, 684
Cyclobutene, 3,3-dimethyl-
isomerization, **5**, 679
Cyclobutene, 3,4-dimethyl-
reduction
diimide, **8**, 475
thermolysis, **5**, 678
Cyclobutene, dimethylene-
synthesis
via Cope rearrangement, **5**, 797
via retro Diels–Alder reactions, **5**, 560
Cyclobutene, 3-ethoxy-
ring opening, **5**, 1031
Cyclobutene, 3-ethyl-3-methyl-
isomerization, **5**, 679
Cyclobutene, hexafluoro-
hydrogenation, **8**, 897
Cyclobutene, *cis*-3-methoxy-4-chloro-
synthesis and ring opening, **5**, 684
Cyclobutene, 1-methyl-
oxidation, **7**, 462
Cyclobutene, 3-methyl-3-isopropyl-
isomerization, **5**, 679
Cyclobutene, 3-methyl-3-propyl-
isomerization, **5**, 679
Cyclobutene, 1,3,3,4,4-pentafluoro-
reduction, **8**, 897
Cyclobutene, perfluoro-
ring opening, **5**, 680
Cyclobutene, 3-phenyl-
ring opening, **5**, 682
Cyclobutene, vinyl-
synthesis
via ring opening, **5**, 708
Cyclobutene-3-carbaldehyde
ring opening, **5**, 680

Cyclobutenediones
cycloaddition reactions
metal catalyzed, **5**, 1202–1204
synthesis
via [2 + 2 + 2] cycloaddition, **5**, 1130
Cyclobutenones
electrocyclic ring opening, **5**, 1025
ring opening, **5**, 688–691
synthesis, **5**, 676, 689, 1089
Cyclobutenones, 4-alkynyl-
photolysis, **5**, 733
thermolysis
benzoquinone synthesis, **5**, 690
Cyclobutenones, amino-
thermolysis, **5**, 732
Cyclobutenones, aryl-
photolysis, **5**, 733
thermal ring opening, **5**, 732
2-Cyclobutenones, 4-aryl-4-hydroxy-
rearrangement
hydroquinone synthesis, **5**, 690
Cyclobutenones, isopropylidene-
synthesis
via retro Diels–Alder reactions, **5**, 560
Cyclobutenones, methylene-
synthesis
via retro Diels–Alder reactions, **5**, 560
Cyclobutenones, vinyl-
cleavage
dienylketene synthesis, **5**, 689
photolysis, **5**, 733
Cyclobutylcarbinol
oxidative rearrangement, **7**, 834
Cyclobutyl isocyanide
rearrangements, **6**, 294
Cyclobutyl radicals
oxidation, **7**, 860
β-Cyclocitral
synthesis
Reformatsky reaction, **2**, 287
Cyclocopacamphene
synthesis
via diazoalkene cyclization, **4**, 1154
Cyclocuprate
Wurtz coupling, **3**, 423
Cyclodecadienedione
aldol cyclization, **2**, 169
Cyclodecadienes
Cope rearrangement
palladium catalysts, **5**, 799
monoepoxides
transannular cyclization, **3**, 396
transannular reactions, **3**, 389
1,2-Cyclodecadienes
hydrobromination, **4**, 284
hydrogenation
palladium-catalyzed, **8**, 435
1,3-Cyclodecadienes
synthesis
via cyclobutene ring opening, **5**, 686
1,5-Cyclodecadienes
Cope rearrangement, **5**, 794
equilibrium, **5**, 809
1,6-Cyclodecadienes
synthesis
via Cope–Claisen rearrangement, **5**, 884
2,6-Cyclodecadienones
synthesis
from protected cyanohydrins, **3**, 198
via cyclization, **1**, 553
1,2-Cyclodecanedione
reduction, **8**, 950
Cyclodecanes
functionalized
synthesis *via* Cope rearrangement, **5**, 796
Cyclodecanone
reduction, **8**, 935
Cyclodecapentaene
synthesis
via photolysis, **5**, 716
1,3,5-Cyclodecatriene
irradiation, **5**, 717
ring closure, **5**, 715
synthesis
via photoisomerization, **5**, 706
Cyclodecenes
stereoselective synthesis
via [6 + 4] cycloaddition, **5**, 624
synthesis
via cyclodecane, **7**, 15
transannular cyclization, **3**, 388
Cyclodecenols
synthesis
Wittig rearrangement, **3**, 1009
transannular cyclization, **3**, 393
Cyclodecenones
functionalized
synthesis, **7**, 625
Cyclodec-5-ynol
transannular cyclization, **3**, 396
Cyclodextrins
catalysts
benzoin condensation, **1**, 543
Reimer–Tiemann reaction
regioselectivity, **2**, 771
β-Cyclodextrins
electroreduction, **8**, 131
Reimer–Tiemann reaction, **2**, 771
abnormal, **2**, 773
stoichiometric complexes
with benzonorbornadiene, **5**, 210
template-directed chlorination
aromatic compounds, **7**, 49
Cyclododecadienone, 2-chloro-
rearrangements, **3**, 849
Cyclododecane, cyano-
reduction, **8**, 253
1,2-Cyclododecanediol
oxidative cleavage, **7**, 708
1,2-Cyclododecanedione
synthesis, **8**, 551
1,6-Cyclododecanedione
aldol cyclization, **2**, 169
1,7-Cyclododecanedione
aldol cyclization, **2**, 169
Cyclododecanol
tartrate
reaction with allenylboronic acid, **2**, 96
Cyclododecanone
lithiated imines

alkylation, **3**, 37
Cyclododecanone, 1-bromo-
reaction with methyl iodide, **1**, 202
Cyclododecanone, 2,3-epoxy-
rearrangement
epoxide ring opening, **3**, 753
Cyclododecanone, 2-methyl-
synthesis, **3**, 37
Cyclododecatetraene
synthesis
via Cope rearrangement, **5**, 812
1,5,9-Cyclododecatriene
hydroboration, **8**, 708
hydrogenation
homogeneous catalysis, **8**, 451
hydrosilylation, **8**, 780
Cyclododecene
oxidative halogenation, **7**, 527
Cyclododecene oxides
deoxygenation, **8**, 888
Cycloeudesmol
synthesis
via cyclopropanation, **4**, 1043
Cyclofenchene
synthesis, **3**, 709
Cyclofunctionalization
electrophile-initiated
mechanism, **4**, 365–367
heteroatom, **4**, 363–414
oxygen nucleophiles, **4**, 367–397
regioselectivity, **4**, 367
stereoselectivity, **4**, 366, 379–385
Cyclohepta[*cd*]benzofuran, 7-methoxy-
hydrogenation, **8**, 625
2,4-Cycloheptadieneacetic acid
lactonization, **4**, 371
Cycloheptadienes
alkylation
stereocontrolled, *via* iron carbonyl complexes, **4**, 581
bridged
synthesis *via* Cope rearrangement, **5**, 803
multiple functionalization
stereochemistry, **4**, 685
synthesis
via Cope rearrangement, **4**, 1048
1,4-Cycloheptadienes
photoisomerization, **5**, 196, 211
synthesis
via Cope rearrangement, **5**, 791, 803, 971
cis-1,4-Cycloheptadienes, 6,7-dimethyl-
synthesis
via Cope rearrangement, **5**, 973
Cycloheptadienol
oxidative rearrangement, **7**, 823
2,4-Cycloheptadienol
potassium salt
1,5-rearrangement, **5**, 1003
Cycloheptadienones
Nazarov cyclization, **5**, 760
Cycloheptanes
functionalized
synthesis *via* Cope rearrangement, **5**, 976–978
synthesis
intramolecular acyloin coupling reaction, **3**, 626
via intramolecular ene reactions, **5**, 24
via Michael addition, **4**, 6
via photocycloaddition, **5**, 657
Cycloheptanes, 1,2-divinyl-
Cope rearrangement, **5**, 810
Cycloheptanes, 1,3,5-trimethylene-
synthesis
via metal-catalyzed cooligomerization, **5**, 1195
Cycloheptanoids
synthesis
via photocycloaddition, **5**, 670
Cycloheptanols
formation
type II intramolecular ene reaction, **2**, 551
Cycloheptanols, 3-methylene-
synthesis
ene reaction, **2**, 547
Cycloheptanone
dimethyl acetals
selective reduction, **8**, 217
α-hydroxylation, **7**, 166
oxime
catalytic hydrogenation, **8**, 143
reduction
aluminum amalgam, **8**, 116
synthesis
Friedel–Crafts reaction, **2**, 711
homoenolates, **2**, 448
via ring expansion, **5**, 907
Cycloheptanone, 5-ethoxycarbonyl-2-methyl-
synthesis, **3**, 783
Cycloheptanone, 2-methyl-
synthesis
via ring expansion, **1**, 851
Cycloheptanone, 2-phenyl-
synthesis
via ring expansion, **1**, 851
Cyclohepta[*b*]pyrrolidines
synthesis
Mannich cyclization, **2**, 1041
Cycloheptatriene
anodic oxidation, **7**, 796
cycloaddition reactions
dienes, **5**, 632
hydride donor
to carbonium ions, **8**, 91
hydrogenation
homogeneous catalysis, **8**, 451
photocycloaddition reactions
9-cyanoanthracene, **5**, 636
synthesis
via ketocarbenoid reaction with benzenes, **4**, 1052, 1057
tautomerism, **5**, 713
Cycloheptatriene, 7,7-dimethoxy-
synthesis, **7**, 796
Cycloheptatriene, 1-methoxy-
anodic oxidation, **7**, 796
Cycloheptatriene, 3-methoxy-
anodic oxidation, **7**, 796
Cycloheptatriene, 7-methoxy-
synthesis, **7**, 796
Cycloheptatrienecarboxamide, *N,N*-dimethyl-
lithium enolate
crystal structure, **1**, 32

2,4,6-Cycloheptatrien-1-one
cycloaddition reactions, **5**, 618–626
Cycloheptene
oxidation
Wacker process, **7**, 450
oxide
rearrangement, lithium perchlorate catalyzed, **3**, 761
Pauson–Khand reaction, **5**, 1049
reduction
transfer hydrogenation, **8**, 552
Cycloheptene, methylene-
synthesis
via [4 + 3] cycloaddition reactions, **5**, 598
Cycloheptene, 1-nitro-
synthesis, **6**, 107
Cycloheptene, 1-nitromethyl-
synthesis
Knoevenagel reaction, **2**, 365
2-Cycloheptenol
synthesis, **7**, 413
2-Cycloheptenones
alkyl-substituted
synthesis, **3**, 202
4-Cyclohepten-1-ones
substituted
synthesis *via* Cope rearrangement, **5**, 976
synthesis
via [4 + 3] cycloaddition, **5**, 603
Cycloheptenones, 2-chloro-
synthesis, **1**, 878
via dihalocyclopropyl compounds, **4**, 1018
1,9-Cyclohexadecadiene
synthesis
via alkene metathesis, **5**, 1119
Cyclohexadecanone
synthesis
from protected cyanohydrins, **3**, 198
Cyclohexadiene
Pauson–Khand reaction, **5**, 1049
photochemical ring opening, **5**, 710
photocycloaddition reactions
anthracene, **5**, 636
synthesis
via [2 + 2 + 2] cycloaddition, **5**, 1142–1144
via retro Diels–Alder reaction, **5**, 569
1,2-Cyclohexadiene
synthesis
via dihalocyclopropanes, **4**, 1010
via electrocyclization, **5**, 735
1,3-Cyclohexadiene
addition–protonation reactions, **4**, 542
alkylation
stereocontrolled, *via* iron carbonyl complexes, **4**, 581
anodic oxidation, **7**, 795
[4 + 3] cycloaddition reactions, **5**, 603
Diels–Alder reactions
imines, **5**, 404, 408
disproportionation
hydrogenation, **8**, 440
hydration, **4**, 299
hydroboration, **8**, 716
hydrobromination, **4**, 284
hydrocarboxylation, **4**, 945
hydrogenation
homogeneous catalysis, **8**, 451
multiple functionalization
stereochemistry, **4**, 685
photoaddition reactions
with propionaldehyde, **5**, 165
selective reduction, **8**, 567
synthesis
via electrocyclization, **5**, 699–745
via hydroformylation, **4**, 922
via retro Diels–Alder reaction, **5**, 568
1,4-Cyclohexadiene
disproportionation
hydrogenation, **8**, 440
photoisomerization, **5**, 196
synthesis, **3**, 653
via hydroformylation, **4**, 922
via radical cyclization, **4**, 810
1,3-Cyclohexadiene, *cis*-5,6-dimethyl-
synthesis
via 2,4,6-octatriene electrocyclization, **5**, 702
1,3-Cyclohexadiene, *trans*-5,6-dimethyl-
synthesis
via 2,4,6-octatriene electrocyclization, **5**, 702
1,3-Cyclohexadiene, *cis*-5,6-dimethyl-1,4-diphenyl-
photochemical ring opening, **5**, 739
Cyclohexadiene, 1,4-disilyl-
diacylation
Friedel–Crafts reaction, **2**, 717
1,3-Cyclohexadiene, 2,3-divinyl-
synthesis
via Cope rearrangement, **5**, 797
1,3-Cyclohexadiene, 2-fluoro-
synthesis
via dihalocyclopropyl compounds, **4**, 1017
Cyclohexadiene, methoxy-
synthesis
Birch reduction, **2**, 599
1,3-Cyclohexadiene, 1-methoxy-
Diels–Alder reactions, **5**, 376
1,3-Cyclohexadiene, 5-methyl-
synthesis
via dienetricarbonylmanganese anions, **4**, 704
1,4-Cyclohexadiene, 1,2,4,5-tetraphenyl-
synthesis
via metal-catalyzed cycloaddition, **5**, 1197
1,3-Cyclohexadiene, 1-trimethylsiloxy-
t-butylation, **3**, 27
1,3-Cyclohexadiene, 2-trimethylsiloxy-
Diels–Alder reactions
imines, **5**, 403
2,4-Cyclohexadieneacetic acid
cyclofunctionalization, **4**, 371
Cyclohexadiene amino acids
synthesis
Diels–Alder reactions, **5**, 320
1,3-Cyclohexadienecarboxylic acids
synthesis
via Diels–Alder reactions, **5**, 322
Cyclohexadienimines, *N*-alkyl-
synthesis
via photolysis, **5**, 731
Cyclohexadienones
alkyl shifts, **3**, 804
annulation, **5**, 1093, 1099

aromatization
bond cleavage, **3**, 816
conjugation
rearrangements, **3**, 803
oxygen migration, **3**, 812
photochemical aromatization, **3**, 815
reactions with nucleophiles
rearrangements, **3**, 817
rearrangements, **3**, 803
synthesis, **7**, 105
phenol ether coupling, **3**, 683
via Diels–Alder reactions, **5**, 329
via Robinson annulation, **4**, 43
2,4-Cyclohexadienones
1,2-diaryl shifts, **3**, 806
photo rearrangements, **5**, 223
2,5-Cyclohexadienones
photo rearrangement, **5**, 730
2,4-Cyclohexadienones, 5-allyl-
Cope rearrangement, **5**, 790
Cyclohexadienones, 4,4-dimethyl-
rearrangements, **3**, 804
2,5-Cyclohexadienones, 4,4-disubstituted
Nazarov cyclization, **5**, 760
Cyclohexadienones, 4-ethyl-4-methyl-
rearrangements, **3**, 804
Cyclohexadienones, 2-hydroxy-
synthesis, **7**, 835
2,5-Cyclohexadienones, 4-hydroxy-4-methyl-
methyl group shift, **3**, 804
rearrangement, **3**, 806
2,5-Cyclohexadienones, 2,4,4,6-tetrabromo-
6-*endo*-cyclization, **4**, 377
oxidation
thiols, **7**, 760
Cyclohexadienyl radicals
radical addition reactions
aromatic compounds, **4**, 766
Cyclohexane
acetoxylation
transition metal catalysis, **7**, 12
aminooxidation, **7**, 8
aromatization, **7**, 6
autoxidation, **7**, 11
benzene alkylation with
Friedel–Crafts reaction, **3**, 322
electrochemical oxidation, **7**, 793
functionalization, **7**, 7
isomerization, **7**, 5
oxidation
chloro(tetraphenylporphyrin)manganese catalyst, **7**, 12
rearrangement, **7**, 8
synthesis
intramolecular acyloin coupling reaction, **3**, 625
via [2 + 2 + 2] cycloaddition, **5**, 1141
via ene reactions, **5**, 9
via Michael addition, **4**, 27
Cyclohexane, alkylidene-
ene reactions
Lewis acid catalysis, **5**, 6
reduction
diimide, **8**, 476
Cyclohexane, 1-alkyl-3-tosyl-
reduction
steric control, **8**, 961
Cyclohexane, arylthio-
synthesis, **7**, 14
Cyclohexane, 1-azido-2-trifluoroacetoxy-
synthesis, **7**, 491
Cyclohexane, 1-bromo-4-*t*-butyl-
cyanohydrins, **1**, 550
Cyclohexane, *t*-butyl-
rearrangements
cycloalkanes, **6**, 895
Wittig rearrangement, **6**, 883
Cyclohexane, 4-*t*-butylmethylene-
hydrogenation
heterogeneous catalysis, **8**, 429
Cyclohexane, chloro-
synthesis, **7**, 14
Cyclohexane, α-chloronitroso-
Diels–Alder reactions, **5**, 418
Cyclohexane, cyano-
intramolecular cyclization, **3**, 48
Cyclohexane, cyclohexyl-
microbial hydroxylation, **7**, 58
Cyclohexane, cyclopropylidene-
cycloaddition reactions
carbon dioxide, metal catalyzed, **5**, 1196
Cyclohexane, 1,2-dimethyl-
oxidation
peracids, **7**, 13
Cyclohexane, 1,2-dimethylene-
[4 + 3] cycloaddition reactions, **5**, 603
Diels–Alder reactions, **5**, 338
synthesis
Ramberg–Bäcklund rearrangement, **3**, 879
Cyclohexane, 2,4-dioxo-carboxylic acid
dianion
aldol cyclization, **2**, 171
Cyclohexane, 1,2-divinyl-
Cope rearrangement, **5**, 794, 809, 821
Cyclohexane, methyl-
electrophilic reactions, **7**, 10
oxidation, **7**, 12
Cyclohexane, methylene-
ene reactions
Lewis acid catalysis, **5**, 4
epoxidation, **7**, 363
hydroboration
stereochemistry, **8**, 707
hydrogenation
heterogeneous catalysis, **8**, 429
rearrangement, **6**, 901
stereospecific rearrangement, **3**, 919
3-substituted
epoxidation, **7**, 363
synthesis
via boron-stabilized carbanions, **1**, 498
Cyclohexane, 5-methylene-3-vinylallylidene-
synthesis
via metal-catalyzed cyclodimerization, **5**, 1190
Cyclohexane, nitro-
reduction, **8**, 375
Cyclohexane, 1-vinyl-2-alkyl-
synthesis
via intramolecular ene reactions, **5**, 17
Cyclohexane, 1-vinyl-2-alkylidine-
synthesis

via intramolecular ene reaction, **5**, 18
Cyclohexanecarbaldehyde
[3 + 2] cycloaddition reactions
with 1,3-dimethyl(*t*-butyldimethylsilyl)allene, **5**, 279
synthesis, **8**, 291
Cyclohexanecarbaldehyde, 2-methyl-
synthesis
via hydroformylation, **4**, 919
Cyclohexanecarbonitrile, 1-piperidino-
reactions with Grignard reagents, **1**, 370
Cyclohexanecarboxylate, 4-*t*-butyl-
methyl ester
reaction with dimethylaluminum methylselenolate, **6**, 466
Cyclohexanecarboxylic acid
methyl ester
acyloin coupling reaction, **3**, 619
piperidide
reduction, **8**, 270
Cyclohexanecarboxylic acid, 2-oxo-
enzymic reduction
specificity, **8**, 197
Cyclohexanecarboxylic acid chloride
synthesis
via cyclohexane, **6**, 308
1,2-Cyclohexanediamine, *N,N,N′,N′*- tetramethyl-
hydroxylation
osmium tetroxide, **7**, 442
1,2-Cyclohexanedicarboxylic acid
dimethyl ester
acyloin coupling reaction, **3**, 623
intramolecular acyloin coupling reaction, **3**, 621
synthesis
intramolecular acyloin coupling reaction, **3**, 622
1,2-Cyclohexanediol
cis
synthesis, **7**, 444
oxidative cleavage, **7**, 704–708
trans
synthesis, **7**, 447
intramolecular pinacol coupling reactions, **3**, 575
1,3-Cyclohexanediol
catalytic hydrogenation, **8**, 814
1,2-Cyclohexanediol, 1,2-dimethyl-
pinacol rearrangement, **3**, 724, 761
1,2-Cyclohexanediol, 1-methyl-
oxidative cleavage, **7**, 708
1,3-Cyclohexanediol, *trans,trans*-2-nitro-
synthesis
Henry reaction, **2**, 327
1,2-Cyclohexanediol, 4-vinyl-
oxidative cleavage, **7**, 708
1,2-Cyclohexanedione
dianions
aldol reaction, **2**, 199
reaction with arylbiguanides, **3**, 832
rearrangement, **3**, 822
1,3-Cyclohexanedione
enol ethers
acylation, **2**, 835
enzymic reduction
specificity, **8**, 201
synthesis
via Knoevenagel and Claisen condensations, **4**, 2
1,4-Cyclohexanedione
Clemmensen reduction, **8**, 313
hydrogenation
catalytic, **8**, 142
1,3-Cyclohexanedione, 2-chloro-
ring contraction, **3**, 871, 875
1,3-Cyclohexanedione, 5,5-dimethyl-
Clemmensen reduction, **8**, 312
hydrogenation, **8**, 551
1,3-Cyclohexanedione, 2-methyl-
alkylation, **3**, 55
Michael addition, **4**, 20
1,3-Cyclohexanedione, 4-pentyl-
synthesis
via Michael addition, **4**, 6
Cyclohexanimines, 2-alkyl-
reduction, **8**, 43
Cyclohexanimines, 3,3,5-trimethyl-
reduction, **8**, 43
Cyclohexaniminium compounds, 2-methyl-
reduction, **8**, 43
Cyclohexanols
catalytic hydrogenation, **8**, 814
formation
ene reaction, **2**, 540
type II intramolecular ene reactions, **2**, 547
functionalized
synthesis, **7**, 625
hydrozirconation
diastereoselectivity, **8**, 689
isomerization
catalytic hydrogenation, **8**, 142
oxidation
solid support, **7**, 845
Cyclohexanols, 1-acetyl-
reductive cleavage
metal ions, **8**, 992
Cyclohexanols, 2-alkyl-3-stannyl-
synthesis, **7**, 623
Cyclohexanols, *trans*-2-azido-
synthesis
via cyclohexene oxide, **6**, 253
Cyclohexanols, 2-diethylamino-
synthesis, **6**, 89
Cyclohexanols, 1,2-divinyl-
Cope rearrangement, **5**, 796
Cyclohexanols, 2-methyl-
oxidation
solid support, **7**, 841
synthesis
epoxide ring opening, **3**, 753
Cyclohexanols, 3-methylene-
synthesis
ene reaction, **2**, 547
Cyclohexanols, 2-nitroalkyl-
synthesis
Henry reaction, **2**, 329
Cyclohexanols, 1-nitromethyl-
synthesis
Henry reaction, **2**, 329
Cyclohexanols, 3,3,5-trimethyl-
isomerization
catalytic hydrogenation, **8**, 141
Cyclohexanone
α-acetoxylation, **7**, 154

aldol reaction, **2**, 147
alkylation
asymmetric induction, **6**, 725, 726
allylation
Wacker oxidation, **7**, 455
t-butyldimethylsilyl enol ether
ene reactions, **5**, 1075
Darzens glycidic ester condensation, **2**, 428
dimethyl acetals
selective reduction, **8**, 217
dimethylhydrazone
lithiated, osmometry, **2**, 507
lithiated, X-ray structures, **2**, 507
4,4-disubstituted
synthesis *via* Michael addition, **4**, 26
2,2-disubstituted, chiral
synthesis *via* Claisen rearrangement, **5**, 832
enamines
axial alkylation, **3**, 30
enolates
aldol reaction, stereoselectivity, **2**, 197
enol ethers
reduction, **8**, 937
hydrogenation
catalytic, **8**, 141
α-hydroxylation
electrocatalytic method, **7**, 158
isotopically substituted
Baeyer–Villiger reaction, **7**, 672
keto aldehydes from, **1**, 461
lithiated dimethylhydrazone
crystal structure, **1**, 28
lithium enolate
reaction with benzaldehyde, **2**, 234
moderately hindered
reduction, dissolving metals, **8**, 119
nucleophilic addition reactions
lithium salts, **1**, 315
stereoselectivity, **1**, 67
one or no α-substituents, reduction
dissolving metals, stereoselectivity, **8**, 116
oxidation
Baeyer–Villiger reaction, **7**, 675
oxime
catalytic hydrogenation, **8**, 143
reduction, **8**, 393
pyrrolidine enamine
dialkylation, **3**, 29
methylation, **3**, 30
reactions with alkyllithium and alkyl Grignard reagents
stereoselectivity, **1**, 79
reactions with boron-stabilized carbanions, **1**, 498
reactions with 2-bromooctane
samarium diiodide, **1**, 259
reactions with dialkoxyboryl-stabilized carbanions, **1**, 501
reactions with diazomethane, **1**, 850
reactions with organometallic compounds
stereoselectivity, **1**, 333
reduction, **8**, 924
aluminum amalgam, **8**, 116
dissolving metals, **8**, 112
dissolving metals, stereoselectivity, **8**, 116
ionic hydrogenation, **8**, 318, 319
stereoselectivity, **8**, 5
Ritter reaction, **6**, 270
sterically hindered
reduction, dissolving metals, **8**, 118
substituted
expansion with ethyl diazoacetate, **1**, 853
nucleophilic addition reactions, **1**, 67
thiolate substitution
selectivity, **7**, 125
tri-*n*-butyltin enolates
alkylation, **3**, 7
Cyclohexanone, 2-alkyl-
oxime
catalytic hydrogenation, **8**, 143
Cyclohexanone, 3-alkyl-
1-enolates
alkylation, **3**, 15
Cyclohexanone, 2-allyl-
Baeyer–Villiger reaction, **7**, 675
expansion with diazomethane, **1**, 851
synthesis
via Wacker oxidation, **7**, 455
Cyclohexanone, 2-allyl-2-methyl-
synthesis
regioselective alkylation, **3**, 28
Cyclohexanone, benzylidene-
oxime
Beckmann rearrangement, **7**, 694
reduction
metal hydrides, **8**, 315
Cyclohexanone, 1,3-bisdiazo-
irradiation, **3**, 905
Cyclohexanone, 4,4-bis(ethoxycarbonyl)-
enamine
Michael addition, **4**, 8
Cyclohexanone, 2-bromo-
reaction with bromoacetophenone, **1**, 202
Cyclohexanone, 2-bromo-2-methyl-
synthesis, **6**, 710
Cyclohexanone, 2-bromo-6-methyl-
synthesis, **6**, 710
Cyclohexanone, 3-*t*-butyl-
lithium enolate
methylation, **3**, 15
Cyclohexanone, 4-*t*-butyl-
chiral lithium enolate
alkylation, **3**, 13
dimethyl acetal
selective reduction, **8**, 217
nucleophilic addition reactions
equatorial or axial, **1**, 152
methyllithium, **1**, 316
organometallic compounds, **1**, 156
use of Lewis acid, **1**, 283
reactions with *n*-butyllithium–ytterbium trichloride, **1**, 276
reactions with methyllithium
stereoselectivity, **1**, 79
reactions with methylzinc, **1**, 223
reactions with organocadmium compounds, **1**, 226
reactions with organometallic compounds
Lewis acids, **1**, 333
stereoselectivity, **1**, 333
reactions with α-selenoalkyllithium
stereochemistry, **1**, 677

reduction
dissolving metals, stereoselectivity, **8**, 117
stabilized metal enolates
metallation, **3**, 55
synthesis
via 4-*t*-butyl-1-ethylidenecyclohexane, **1**, 535
Cyclohexanone, 5-*t*-butyl-
α-methyl substituents
axial alkylation, **3**, 14
Cyclohexanone, 2-*n*-butyl-2-methyl-
synthesis
alkylation of unsymmetrical enolate, **3**, 8
Cyclohexanone, 1,2-[$^{14}C_2$]-2-chloro-
reaction with sodium pentylate, **3**, 840
Cyclohexanone, 2-chloromethylene-
synthesis
via dichlorocarbene, **4**, 1004
Cyclohexanone, 2,3-dialkyl-
1-enolates
alkylation, **3**, 15
synthesis, **3**, 8
Cyclohexanone, 3,5-dialkyl-
lithium 1-enolate
alkylation, **3**, 8
Cyclohexanone, 2-diazo-
photolysis, **3**, 903
Cyclohexanone, 2,6-dibromo-
[4 + 3] cycloaddition reactions, **5**, 608
Cyclohexanone, 2,4-di-*t*-butyl-
synthesis, **3**, 26
Cyclohexanone, 2,3-dihydroxy-3,5,5-trimethyl-
synthesis
via thermolysis of triols, **1**, 534
Cyclohexanone, 2,2-dimethyl-
lithium enolate
reaction with benzaldehyde, **2**, 198
palladation, **7**, 630
synthesis
alkylation of enolate, **3**, 2
Cyclohexanone, 2,4-dimethyl-
cis–trans isomerism
via pyrrolidine enamine, **6**, 709
lithium enolate
synthesis of cycloheximide, **2**, 198
Cyclohexanone, *cis*-2,5-dimethyl-
Knoevenagel reaction
stereochemistry, **2**, 352
Cyclohexanone, *trans*-2,5-dimethyl-
Knoevenagel reaction
stereochemistry, **2**, 352
Cyclohexanone, 2,6-dimethyl-
synthesis, **3**, 34
Cyclohexanone, 3,3-dimethyl-
α-acetoxylation, **7**, 154
Cyclohexanone, 2,2-diphenyl-
(*E*)-enone, **2**, 148
Cyclohexanone, 4,4-diphenyl-
Clemmensen reduction
mechanism, **8**, 310
Cyclohexanone, 4-ethoxycarbonyl-
rearrangement, **3**, 783
Cyclohexanone, 2-ethyl-
synthesis, **3**, 35
Cyclohexanone, 2-ethyl-4-methoxycarbonyl-
synthesis
Claisen condensation, **2**, 817
Cyclohexanone, 2-halo-
eliminations
Wolff–Kishner reductions, **8**, 341
Favorskii rearrangements, **3**, 848
Cyclohexanone, 2-isopropyl-
expansion with diazomethane, **1**, 851
Cyclohexanone, isopropylidene-
[3 + 2] cycloaddition reactions
with 1-methyl-1-(trimethylsilyl)allene, **5**, 278
Cyclohexanone, 2-methoxy-
titanium chloride complex
NMR, **1**, 295
Cyclohexanone, 2-methoxycarbonyl-
dimethylhydrazone
lithiated, X-ray structures, **2**, 508
lithiated anion, crystal structure, **1**, 34
Cyclohexanone, methyl-
reactions with organolithium compounds
Lewis acids, **1**, 333
reduction
aluminum amalgam, **8**, 116
Cyclohexanone, 2-methyl-
cyclohexylimine
deprotonation, **6**, 720
enamine
Michael addition, **4**, 6
enolate anion
preparation of kinetic enol ether, **2**, 599
enolates, **3**, 2
enol ethers
alkylation, **3**, 8
formylation, **2**, 837
lithiated dimethylhydrazones
crystal structure, **3**, 34
lithium enolate
alkylation, **3**, 8
Mannich bases
regiochemistry, **2**, 907
Michael addition, **4**, 6, 20
reductive amination
selectivity, **8**, 54
with ammonia, **8**, 54
regioselective alkylation, **3**, 8
ring expansion, **1**, 873
silyl enol ether
[3 + 2] cycloaddition reactions, **5**, 282
sulfenylation, **7**, 125
synthesis
stereochemistry, **6**, 725
TMS enol derivative
alkylation, **3**, 28
TMS enol ethers
t-butylation, **3**, 25
with diazomethane, **1**, 851
Cyclohexanone, 3-methyl-
enamine
regioisomeric, **6**, 710
reduction
dissolving metals, stereoselectivity, **8**, 116
Cyclohexanone, 4-methyl-
reduction, **8**, 934
selective reduction, **8**, 17
Cyclohexanone, 2-methyl-6-allyl-
synthesis

regioselective alkylation, **3**, 28
Cyclohexanone, 2-methyl-4-*t*-butyl-
synthesis, **3**, 32
Cyclohexanone, 2-methyl-6-butyl-
synthesis
alkylation of unsymmetrical enolate, **3**, 8
Cyclohexanone, 3-methyl-5-*t*-butyl-
lithium 1-enolate
stereoselectivity of alkylation, **3**, 15
Cyclohexanone, 2-methyl-2-nitro-
synthesis, **6**, 106, 107
Cyclohexanone, 3-methyl-2-nitro-
synthesis, **6**, 106
Cyclohexanone, 4-methyl-2-nitro-
synthesis, **6**, 106
Cyclohexanone, 2-methylsulfonyl-
reduction, **8**, 15
Cyclohexanone, 2-methylthio-
reduction, **8**, 15
Cyclohexanone, 2-methyl-3-(4-tosyloxybutyl)-
exocycloalkylation, **3**, 20
Cyclohexanone, 2-phenyl-
Reformatsky reaction
stereoselectivity, **2**, 291
synthesis, **3**, 257
Cyclohexanone, 3-phenyl-
synthesis
via thiocarbonyl ylides, **4**, 1095
Cyclohexanone, 4-phenylthio-
synthesis
via oxyanion-accelerated rearrangement, **5**, 1023
Cyclohexanone, β-silyl-
synthesis
via α,β-unsaturated acylsilanes, **1**, 598
Cyclohexanone, 4-substituted
reductive amination
selectivity, **8**, 54
Cyclohexanone, 2,2,6,6-tetramethyl-
palladation, **7**, 630
reaction with α-selenoalkyllithium, **1**, 674
Cyclohexanone, 3-(2-tosyloxyethyl)-
endocycloalkylation, **3**, 19
Cyclohexanone, 2,2,6-trimethyl-
reaction with α-selenoalkyllithium, **1**, 674
stereochemistry, **1**, 677
Cyclohexanone, 3,3,5-trimethyl-
reduction
dissolving metals, stereoselectivity, **8**, 117
Cyclohexanone, 2-vinyl-
cyclodecenones from, **1**, 880
Cyclohexanone, 3-vinyl-
synthesis, **7**, 457
Cyclohexanone enamine
reaction with dichlorocarbene, **4**, 1004
Cyclohexanonephenylimine, lithio-
crystal structure, **1**, 28
3,3-Cyclohexano-4-oxopentanal
synthesis
via Claisen rearrangement, oxidation, **7**, 456
1,2,4-Cyclohexatriene
synthesis
via electrocyclization, **5**, 735
Cyclohexene
allylic oxidation, **7**, 99
anodic oxidation, **7**, 794
aziridination, **7**, 470
bromination, **7**, 539
[2 + 2 + 2] cycloaddition reactions, **5**, 1130
diamination, **7**, 484
disproportionation
hydrogenation, **8**, 440
epoxidation, **7**, 374
functionalized
synthesis, **7**, 625
hydride donor
carbonyl compound reduction, **8**, 320
hydrogenolysis, **8**, 958
hydroalumination, **8**, 739
hydroboration, **8**, 716
stereochemistry, **8**, 707
hydrobromination, **4**, 279
hydroformylation, **4**, 914
phosphite-modified rhodium catalysts, **3**, 1022
hydrogenation
homogeneous catalysis, **8**, 446
hydroxylation, **7**, 444
oxidation
Wacker process, **7**, 451, 452
with heteropolyacids, **7**, 462
oxidative cleavage
ruthenium tetroxide, **7**, 587
oxidative rearrangement
solid support, **7**, 845
Pauson–Khand reaction, **5**, 1049
retrograde Diels–Alder reaction, **5**, 552
synthesis
Ramberg–Bäcklund rearrangement, **3**, 876
via [2 + 2 + 2] cycloaddition, **5**, 1141
via vinylcyclobutane rearrangement, **5**, 1016
Cyclohex-2-ene, *cis*-1-acetoxy-4-chloro-
synthesis
via palladium(II) catalysis, **4**, 565
Cyclohexene, 1-acetoxy-2-methyl-
reaction with tributylmethoxytin
preparation of organotin(IV) enol ethers, **2**, 608
Cyclohex-2-ene, *trans*-1-acetoxy-4-trifluoroacetoxy-
synthesis
via palladium(II) catalysis, **4**, 565
Cyclohexene, acetyl-
reduction
molybdenum complex catalyst, **8**, 554
Cyclohexene, 1-acetyl-
synthesis, **1**, 430
Cyclohexene, 1-alkoxy-
hydrogenation
palladium-catalyzed, **8**, 429
Cyclohexene, 1-alkyl-
allylic oxidation, **7**, 818
Cyclohexene, 6-azido-1-phenyl-
synthesis, **7**, 502
Cyclohexene, 1-benzyl-
oxide
syn-opening, **3**, 741
Cyclohexene, 3-*t*-butyl-
hydroxylation, **7**, 447
Cyclohexene, 4-*t*-butyl-
hydroxylation, **7**, 447
Cyclohexene, 4-*t*-butyl-1-cyano-
Michael addition
benzenethiolate, **6**, 140

Cyclohexene, 4-*t*-butyl-1-phenyl-
  hydrochlorination, **4**, 273
Cyclohexene, chloro-
  addition reactions
    benzeneselenenyl chloride, **7**, 520
Cyclohexene, 3-chloro-2-fluoro-
  synthesis
    via dihalocyclopropyl compounds, **4**, 1017
Cyclohex-2-ene, *cis*-1,4-diacetoxy-
  synthesis
    via palladium(II) catalysis, **4**, 565
Cyclohex-2-ene, *trans*-1,4-diacetoxy-
  synthesis
    via palladium(II) catalysis, **4**, 565
Cyclohexene, 1,2-dimethyl-
  hydrogenation, **8**, 426
  hydroxylation, **7**, 445
Cyclohexene, 1,6-dimethyl-
  epoxidation, **5**, 130
  reduction
    diimide, **8**, 476
Cyclohexene, 4,4-dimethyl-
  oxidative rearrangement, **7**, 817
Cyclohexenes, 3,5-dimethylene-
  synthesis
    via [2 + 2 + 2] cycloaddition, **5**, 1141
Cyclohexene, 3,6-dimethylene-
  synthesis
    via [2 + 2 + 2] cycloaddition, **5**, 1141
Cyclohexene, 1,2-divinyl-
  thermal cyclization, **5**, 713
Cyclohexene, 3,4-epoxy-
  reaction with lithium dimethylcuprate, **6**, 9
Cyclohexene, 1,2-epoxy-3-hydroxy-
  reaction with lithium dimethylcuprate
    regioselectivity, **6**, 8
Cyclohexene, methoxy-
  cycloaddition reactions
    with benzonitrile, **5**, 161
Cyclohexene, 1-methyl-
  acetoxylation
    electrochemical oxidation, **7**, 790
  allylic oxidation, **7**, 100
  ene reactions
    reaction with formaldehyde, **2**, 533
  hydroformylation, **4**, 919
2-Cyclohexene, 1-methyl-
  allylic oxidation, **7**, 101
Cyclohexene, 2-methyl-
  carbosulfenylation
    selectivity, **6**, 142
Cyclohexene, 1-nitro-
  synthesis, **6**, 107
Cyclohexene, 1-phenyl-
  hydroboration, **8**, 722
  nitro addition reactions, **7**, 488
Cyclohexene, 1-phenyl-4-*t*-butyl-
  hydrobromination, **4**, 280
Cyclohexene, 1-trimethylsiloxy-4-*t*-butyl-
  t-butylation
    diastereoselectivity, **3**, 26
Cyclohexene, vinyl-
  dicarboxylation, **4**, 948
  hydrocarboxylation, **4**, 939
Cyclohexene, 1-vinyl-
  diamination, **7**, 486
Cyclohexene, 4-vinyl-
  anodic oxidation, **7**, 796
1-Cyclohexeneacetic acid
  γ-lactones from, **4**, 371
1-Cyclohexeneacetic acid, 2-methyl-
  butenolides from, **4**, 371
3-Cyclohexenecarbaldehyde
  hydroformylation, **4**, 922
1-Cyclohexenecarbaldehyde, 3-hydroxy-
  synthesis, **1**, 564
1-Cyclohexenecarbonitrile
  reactions with Yamamoto's reagent, **1**, 124
1-Cyclohexenecarboxamide
  oxidation, **6**, 804
3-Cyclohexene-1-carboxamide
  rearrangement, **6**, 804
1,2-Cyclohexenedicarboxylic acid
  dimethyl ester
    hydrogenation, **8**, 426
Cyclohexene oxide
  anodic oxidation, **7**, 707
  anti opening, stereoelectronic aspects, **3**, 733
  initiators, polyene cyclization, **3**, 356
  reaction with Grignard reagents, ring opening, **3**, 754
  rearrangement, **3**, 760
  rearrangement, lithium halide catalyzed, **3**, 764
  reduction
    metal hydrides, **8**, 873
Cyclohexene oxide, 1,4-dialkyl-
  reduction
    lithium aluminum hydride, **8**, 875
Cyclohexene oxide, 1,2-dimethyl-
  rearrangement, lithium halide catalyzed, **3**, 764
  rearrangement, lithium percholate catalyzed, **3**, 761
Cyclohexene oxide, β-hydroxy-
  reduction
    metal hydrides, **8**, 873
Cyclohexene oxide, 1-methyl-
  cyclization, **3**, 342
  syn-opening, **3**, 741
  reaction with magnesium bromide, **3**, 757
  rearrangement, **3**, 753
  rearrangement, lithium halide catalyzed, **3**, 763
  rearrangement, lithium perchlorate catalyzed, **3**, 761
Cyclohexene oxide, 1-phenyl
  opening, **3**, 734
Cyclohexene oxide, 2-(trimethylgermyl)-
  reduction
    metal hydrides, **8**, 873
Cyclohexene oxide, 2-(trimethylsilyl)-
  reduction
    metal hydrides, **8**, 873
Cyclohexenocycloalkanones
  synthesis
    via copper catalyzed Grignard addition, **4**, 91
2-Cyclohexenol
  aziridination, **7**, 481
  synthesis, **7**, 413; **8**, 166
    via chiral reduction of cyclohexenone, **8**, 169
3-Cyclohexenol
  synthesis
    via [3 + 3] annulation, **5**, 1020
Cyclohexenol, allylic

epoxidation, **7**, 364
1-Cyclohexenol, 2-bromo-
synthesis
via dihalocyclopropyl compounds, **4**, 1018
2-Cyclohexenol, 6-(*N*-substituted amino)-3-aryl-
synthesis, **6**, 787
Cyclohexenol, vinyl-
allylic rearrangements, **7**, 822
2-Cyclohexenone
addition reaction
Lewis acid catalysis, **1**, 313
with organomagnesium compounds, **4**, 89
γ-alkylation, **4**, 674
asymmetric reduction, **8**, 166
boron trifluoride complex
NMR, **1**, 293, 294
conjugated
reduction, **8**, 6
enantioselective alkylation
organocuprates, **4**, 172
Grignard additions
copper catalyzed, **4**, 92
hydrogenation
homogeneous catalysis, **8**, 446
lithium enolates
methylation, **1**, 688
reaction with α-cyanobenzyllithium, **1**, 235
reaction with Grignard reagents, **4**, 254
reaction with lithiotributylstannane, **7**, 623
reaction with lithium dialkylcuprates, **4**, 173
reaction with organoaluminum reagents
site selectivity, **1**, 82, 85, 95
reaction with organometallic reagents, **1**, 155
reduction
aluminum hydrides, **8**, 542, 545
9-borabicyclo[3.3.1]nonane, **8**, 537
synthesis, **1**, 383
via Robinson annulation, **4**, 6
2-Cyclohexenone, 2-acetyl-
Diels–Alder reactions, **5**, 461
Cyclohexenone, 4-alkenyl-
photocycloaddition reactions, **5**, 144
2-Cyclohexenone, 3-alkoxy-
lithium dienolates
α′-alkylation, **3**, 21
Cyclohexenone, 4-alkoxycarbonyl-
synthesis
Dieckmann reaction, **2**, 807
Cyclohexenone, 4-alkyl-
photocycloaddition reactions, **5**, 142
2-Cyclohexenone, 4-alkyl-
synthesis, **3**, 21
2-Cyclohexenone, 3-amino-
extended dienolates
γ-alkylation, **3**, 24
Cyclohexenone, 2-benzyl-3-methyl-
alkylation
Cope rearrangement, **5**, 789
Cyclohexenone, 4-*t*-butyl-
photocycloaddition reactions, **5**, 130
2-Cyclohexenone, 5-*t*-butyl-
reduction
K-selectride, **8**, 536
2-Cyclohexenone, 4-(3-chloropropyl)-4-methyl-6-(2-ethylallyl)-
cycloalkylation, **3**, 23
Cyclohexenone, 3,4-dimethyl-
addition to allene
photochemical cycloaddition, **5**, 130
photocycloaddition reactions, **5**, 131
2-Cyclohexenone, 5,5-dimethyl-
reduction
borohydride, **8**, 536
2-Cyclohexenone, 4,4-diphenyl-
Clemmensen reduction, **8**, 312
2-Cyclohexenone, 4,4-disubstituted
synthesis
via cyclohexadienyliron complexes, **4**, 675
Cyclohexenone, α-epoxy-
synthesis
via retro Diels–Alder reactions, **5**, 563
2-Cyclohexenone, 2-hydroxy-
reduction
aluminum hydrides, **8**, 545
Cyclohexenone, 4-isopropyl-
photocycloaddition reactions, **5**, 130
Cyclohexenone, methyl-
reaction with 1,1-bis(methylseleno)-1-propyllithium, **1**, 689
Cyclohexenone, 2-methyl-
[3 + 2] cycloaddition reactions, **5**, 301
2-Cyclohexenone, 2-methyl-
hydrogenation
homogeneous catalysis, **8**, 462
Cyclohexenone, 3-methyl-
aldol reaction, **2**, 152
photocycloaddition reactions, **5**, 125
reaction with methylmagnesium iodide
enolates, **2**, 185
2-Cyclohexenone, 3-methyl-
Grignard additions
copper catalyzed, **4**, 92
Michael addition, **4**, 17
reduction
borohydride, **8**, 536
transfer hydrogenation, **8**, 552
2-Cyclohexenone, 5-methyl-
conjugate addition
organocuprates, **4**, 187
Cyclohexenone, 3-nitro-
Diels–Alder reactions, **5**, 320
Cyclohexenone, 5-substituted
synthesis
via Diels–Alder reactions, **5**, 324
2-Cyclohexenone, 5-substituted
synthesis
via arene–metal complexes, **4**, 543
2-Cyclohexenone, 3,3,5-trimethyl-
reduction
aluminum hydrides, **8**, 545
1,4-dihydronicotinamide, **8**, 562
2-Cyclohexenone, 3,5,5-trimethyl-
cleavage
ozonolysis with phase transfer agents, **7**, 548
2-Cyclohexenone, 5-trimethylsilyl-
reaction with Grignard reagents
copper catalyzed, **4**, 211
Cyclohexenones
α′-alkylation, **3**, 21
aromatization, **7**, 131

biochemical reduction, **8**, 558
Clemmensen reduction, **8**, 311
[3 + 2] cycloaddition reactions, **5**, 301
with η^1-butynyliron complexes, **5**, 277
fused
synthesis *via* ketone enolates, **4**, 99
photocycloaddition reactions, **5**, 125
pinacol coupling reactions, **3**, 577
reaction with lithium dimethylcuprate
enol ether preparation, **2**, 599
ring contraction, **7**, 832
spiroannulation, **3**, 22
synthesis
aldol cyclization, **2**, 162
carbonylation, **3**, 1025
via Diels–Alder reactions, **5**, 329
via Michael reaction, **4**, 2
tandem vicinal difunctionalization, **4**, 245
β-unsubstituted
reduction, **8**, 536
Vilsmeier–Haack reaction, **2**, 786
zirconium dienolates
aldol reaction, **2**, 303
3-Cyclohexenylamines
synthesis
via azaanion-accelerated rearrangement, **5**, 1023
Cycloheximide
synthesis
aldol reaction, **2**, 198
Cyclohexylamine
imines
deprotonation, **6**, 720
oxidation
m-chloroperbenzoic acid, **7**, 737
Cyclohexylamine, *N*-methyl-4-*t*-butyl-
reaction with allyl organometallic compounds, **2**, 983
Cyclohexylidene, *t*-butyl-
reduction, **8**, 231
Cyclohexylidene epoxides, α-alkenyl-
reaction with lithium homocuprates, **3**, 226
Cyclohexylimine, *N*-methyl-4-*t*-butyl-
reaction with allyl organometallic compounds
stereochemistry, **2**, 983
Cyclohexyl isocyanide
reduction
dissolving metals, **8**, 830
Cyclohexyl radicals
addition to methyl acrylate, **4**, 736
Cyclohexyne
anthracene adduct
retro Diels–Alder reaction, **5**, 589
Cyclomethylenomycin A
synthesis
via Pauson–Khand reaction, **5**, 1051
Cycloneosamandione
synthesis, **7**, 169
1,2-Cyclononadiene
hydrobromination, **4**, 284
hydrochlorination, **4**, 276
hydrogenation
palladium-catalyzed, **8**, 435
reaction with iodine azide, **7**, 506
synthesis
via dihalocyclopropanes, **4**, 1010
1,5-Cyclononadiene
Cope rearrangement, **5**, 794
equilibrium, **5**, 806
transannular cyclization, **3**, 386
Cyclononatetraene
thermal rearrangement, **5**, 716
1,2,3-Cyclononatriene
synthesis
via dihalocyclopropanes, **4**, 1010
1,2,6-Cyclononatriene
hydrobromination, **4**, 284
1,3,5-Cyclononatriene
electrocyclization, **5**, 702
photoisomerization, **5**, 709
photolysis, **5**, 737
1,3,6-Cyclononatriene
synthesis
via photoisomerization, **5**, 709
Cyclononatrienols
Cope rearrangement, **5**, 806
6-Cyclononenol, 2,3-epoxy-
synthesis, **7**, 413
Cyclononenones
synthesis
via Cope rearrangement, **5**, 796
Cyclooctadecane, 1,9-bis(3-methoxycarbonylpropyl)-
acyloin coupling reaction
intramolecular, **3**, 628
Cyclooctadienes
bridged
synthesis *via* Cope rearrangement, **5**, 806
monoepoxides
transannular hydride shifts, **3**, 735
synthesis
via [4 + 4] cycloaddition, **5**, 639, 640
transannular reactions, **3**, 383
1,3-Cyclooctadiene
anodic oxidation, **7**, 795
dimerization, **5**, 66
hydrocarboxylation, **4**, 945
hydrogenation
homogeneous catalysis, **8**, 450
oxidation
palladium(II) catalysis, **4**, 559
photoaddition reactions
with acetone, **5**, 166
reaction with *N*-acyliminium ions, **2**, 1070
synthesis
via Cope rearrangement, **5**, 805
1,4-Cyclooctadiene
hydrogenation
homogeneous catalysis, **8**, 450
1,5-Cyclooctadiene
anodic oxidation, **7**, 796
bridged
Cope rearrangement, **5**, 816
cycloreversion reactions, **5**, 64
dimerization, **5**, 66
hydroboration, **8**, 708, 714
hydrogenation
heterogeneous catalysis, **8**, 433
homogeneous catalysis, **8**, 449, 450
ozonolysis, **8**, 399
synthesis
via Cope rearrangement, **5**, 791
via divinylcyclobutane rearrangements, **5**, 1025

transannular cyclization, **3**, 381
1,5-Cyclooctadiene, 1,5-dimethyl-
transannular cyclization, **3**, 382
2,4-Cyclooctadienol
oxidative rearrangement, **7**, 823
Cyclooctadienones
synthesis
via cyclobutenone ring opening, **5**, 690
2,4-Cyclooctadienones
dimerization, **5**, 66
4*H*,5*H*,9*H*,10*H*-Cycloocta[1,2-*b*:6,5-*b'*]difuran
synthesis
via [4 + 4] cycloaddition, **5**, 639
Cyclooctanecarbaldehyde, 5-methylene-
transannular cyclization, **3**, 383
1,2-Cyclooctanediol
cis
synthesis, **7**, 444
Cyclooctanes
fused
synthesis *via* Cope rearrangement, **5**, 806
synthesis
via cyclization of 1,8-dialdehydes, **3**, 575
via [4 + 4] cycloaddition, **5**, 639
via divinylcyclobutane rearrangements, **5**, 1024
via intramolecular ene reactions, **5**, 24
via Michael addition, **4**, 6
Cyclooctanes, 1,2-divinyl-
Cope rearrangement, **5**, 810–812
Cyclooctanone
enol ester from
O-acylation, **2**, 598
reduction, **8**, 950
synthesis, **3**, 781
Cyclooctanone, 2,8-dibromo-
rearrangement, **3**, 850
Cyclooctanone, 2,2,8-tribromo-
rearrangement, **3**, 850
Cyclooctatetraene
cycloaddition reactions
dienes, **5**, 634
monoepoxide
rearrangement, **3**, 757
synthesis
via photoisomerization, **5**, 205
tautomerism, **5**, 715
1,3,5-Cyclooctatriene
tautomerism, **5**, 714
Cyclooctene
epoxidation
oxygen, **7**, 383
oxide
solvolysis, **3**, 735
Pauson–Khand reaction, **5**, 1049
photoaddition reactions
with acetone, **5**, 166
photocycloaddition reactions
with toluene, **5**, 655
ring-opening metathesis polymerization, **5**, 1120
transannular electrophilic cyclization, **3**, 380
4-Cyclooctene, hydroperoxy-
synthesis, **7**, 728
Cyclooctene, 6-iodo-
reactions with lithium cuprates
mechanism, **3**, 213
Cyclooctene, 1-methyl-
hydroboration, **8**, 714, 718
Cyclooctene, methylene-
Cope rearrangement, **5**, 794
5-Cyclooctene, 3-methyleneallylidene-
synthesis
via metal-catalyzed cyclodimerization, **5**, 1190
Cyclooctene, 1-nitro-
synthesis, **6**, 107
Cyclooctene, 1-phenyl-
oxidation, **7**, 384
Cyclooctenol
acetate
synthesis, **2**, 598
Cyclooctenones
synthesis
via Cope rearrangement, **5**, 1028
4-Cyclooctenones
electrochemical transannulation, **3**, 600
Cyclooctyne
synthesis
via oxidation of bishydrazones, **7**, 742
Cyclopalladated complexes
N,N-dialkylbenzylamine
vinyl substitutions, **4**, 837
vinyl substitutions, **4**, 835, 837
Cyclopalladation–oxidation, **7**, 630
2-Cyclopentadecanone
synthesis
via cyclization, **1**, 553
2-Cyclopentadecenone
synthesis, **3**, 51
Cyclopentadiene
anodic oxidation, **7**, 795
cycloaddition reactions
isopropenyl chromium complexes, **5**, 1070
propynyl chromium complexes, **5**, 1072
tropones, **5**, 618, 621
[4 + 3] cycloaddition reactions, **5**, 603–605
Diels–Alder reactions, **5**, 380–383, 451
comparison of promoters, **5**, 345
imines, **5**, 403
Lewis acid promoted, **5**, 340
water promoted, **5**, 344
Pauson–Khand reaction, **5**, 1046
retrograde Diels–Alder reactions, **5**, 552
selective reduction, **8**, 567
synthesis
Ramberg–Bäcklund rearrangement, **3**, 874, 875
via [3 + 2] cycloaddition reactions, **5**, 278, 1090
via retro Diels–Alder reaction, **5**, 568
1,3-Cyclopentadiene
hydrochlorination, **4**, 276
1,3-Cyclopentadiene, 1-amino-
synthesis
via [2 + 2 + 2] cycloaddition, **5**, 1131
Cyclopentadiene, hexachloro-
hydrogenolysis, **8**, 898
Cyclopentadiene, hexamethyl-
cycloaddition with *C,N*-diphenylnitrone, **4**, 1075
Cyclopentadiene, 5-(methoxymethyl)-
Diels–Alder reactions, **5**, 353
Cyclopentadiene, phenyl-
chromium tricarbonyl complex, **4**, 527
1,3-Cyclopentadiene, C-5 substituted

Diels–Alder reactions, **5**, 347
Cyclopentadienone
[4 + 3] cycloaddition reactions, **5**, 603
substituted
synthesis *via* retro Diels–Alder reaction, **5**, 568
synthesis
via [2 + 2 + 2] cycloaddition, **5**, 1133–1135
Cyclopentadienone, bis(trimethylsilyl)-
synthesis
via [2 + 2 + 2] cycloaddition, **5**, 1134
Cyclopentadienone, 2,5-di-*t*-butyl-3,4-dimethyl-
synthesis
via [2 + 2 + 2] cycloaddition, **5**, 1135
Cyclopentadienone, 2,5-dimethoxycarbonyl-3,4-diphenyl-
cycloaddition reactions
cyclooctatetraene, **5**, 634
N-ethoxycarbonylazepine, **5**, 634
Cyclopentadienone, 2,5-dimethyl-3,4-diphenyl-
cycloaddition reactions
cycloheptatriene, **5**, 632
fulvenes, **5**, 626
tropones, **5**, 620, 622
Cyclopentadienone, tetra-*t*-butoxy-
synthesis
via [2 + 2 + 2] cycloaddition, **5**, 1133
Cyclopentadienone, tetrakis(dimethylamino)-
iron complex
synthesis, **5**, 1133
Cyclopentadienone, tetrakis(trifluoromethyl)-
synthesis
via [2 + 2 + 2] cycloaddition, **5**, 1134
Cyclopentadienone, tetraphenyl-
[4 + 3] cycloaddition reactions, **5**, 604
reduction, **8**, 557
synthesis, **2**, 142
Cyclopentadienone epoxides
synthesis
via retro Diels–Alder reactions, **5**, 561
Cyclopentadienyl anion
Vilsmeier–Haack reaction, **2**, 780
Cyclopentadienylmethyl metal complexes
synthesis
via [2 + 2 + 2] cycloaddition, **5**, 1147
Cyclopentane
annulation, **1**, 892
via free radical reaction, **5**, 926
functionalization, **7**, 7
reaction with transition metal complexes, **7**, 3
reaction with tungsten
metal vapor synthesis, **7**, 4
stereoselective annulations
intramolecular diastereoselective additions, **2**, 651
stereospecific synthesis, **3**, 653
synthesis, **3**, 647; **6**, 127
intramolecular acyloin coupling reaction, **3**, 623
via ene reactions, **5**, 9
via intramolecular ene reactions, **5**, 21
via metal-catalyzed cycloaddition, **5**, 1200
via Michael addition, **4**, 24
via photocycloaddition, **5**, 657
Cyclopentane, acetyl-
synthesis
polyene cyclization, **3**, 347
Cyclopentane, alkylidene-
synthesis
via [3 + 2] cycloaddition reactions, **5**, 290
via metal-catalyzed codimerizations, **5**, 1189
Cyclopentane, benzylidene-
synthesis, **1**, 663
Cyclopentane, dimethylene-
synthesis, **5**, 1107
Cyclopentane, 1,3-dimethylene-
synthesis
via metal-catalyzed cooligomerization, **5**, 1195
Cyclopentane, 1,2-dimethylene-3,3,4,4,5,5-hexamethyl-
[4 + 3] cycloaddition reactions, **5**, 600
Cyclopentane, 2,2-dimethylmethylene-
synthesis
via metal-catalyzed cycloaddition, **5**, 1190
Cyclopentane, (diphenylmethylene)-
synthesis
via metal-catalyzed cycloadditions, **5**, 1189
Cyclopentane, divinyl-
synthesis
via palladium-ene reaction, **5**, 48
Cyclopentane, 1,2-divinyl-
Cope rearrangement, **5**, 794, 796, 806, 821
Cyclopentane, ethylidene-
synthesis
via [3 + 2] cycloaddition reactions, **5**, 290
Cyclopentane, (iodomethylene)-
synthesis
via radical cyclization, **4**, 803
Cyclopentane, 3-methoxycarbonylmethylene-
synthesis
via metal-catalyzed cycloaddition, **5**, 1190
Cyclopentane, methylene-
synthesis
via [3 + 2] cycloaddition reactions, **5**, 287
via metal-catalyzed cycloadditions, **5**, 1188
thio-Wittig rearrangement, **6**, 895
Cyclopentane, 2-methylmethylene-
synthesis
via metal-catalyzed cycloaddition, **5**, 1190
Cyclopentane, 2-methylvinyl-
synthesis
via magnesium-ene reaction, **5**, 38
via nickel-ene reaction, **5**, 56
Cyclopentane, silylmethylene-
synthesis
via metal-catalyzed cycloaddition, **5**, 1190, 1192
Cyclopentane, sulfonyl(methylene)-
synthesis
via [3 + 2] cycloaddition reactions, **5**, 305
Cyclopentane, vinyl-
synthesis
via [3 + 2] cycloaddition reactions, **5**, 281
Cyclopentane, 1-vinyl-2-alkyl-
synthesis
via intramolecular ene reactions, **5**, 10–15
Cyclopentane, 1-vinyl-2-alkylidine-
synthesis
via intramolecular ene reactions, **5**, 15–17
Cyclopentane, 2-vinyl-1-methylene-
synthesis
via intramolecular ene reaction, **5**, 15
Cyclopentane, ylidene-
synthesis, **3**, 251

Cyclopentaneacetic acid, vinyl-
  synthesis
    via palladium-ene reactions, **5**, 55
Cyclopentanecarbaldehyde
  synthesis, **3**, 769
Cyclopentanecarboxylates, 2-hydroxy-
  synthesis
    via intramolecular Barbier cyclization, **1**, 264
Cyclopentanecarboxylic acid
  esters
    synthesis *via* [3 + 2] cycloaddition reactions, **5**, 282
  methyl ester
    synthesis, **3**, 903
  polyfunctionalized
    synthesis, **3**, 848
Cyclopentanecarboxylic acid, 1-hydroxy-2-isopropyl-5-methyl-
  synthesis, **3**, 831
Cyclopentanecarboxylic acid, 3-methylene-
  esters
    synthesis *via* metal-catalyzed codimerization, **5**, 1191
1,3-Cyclopentanedialdehyde
  synthesis
    via oxidative cleavage of alkenes, **7**, 558
1,3-Cyclopentanedicarboxylic acid
  absolute configuration
    Schmidt reaction, **6**, 818
1,2-Cyclopentanediol
  oxidation
    sodium bismuthate, **7**, 704
  oxidative cleavage, **7**, 705, 708
  synthesis
    intramolecular pinacol coupling reactions, **3**, 574
1,2-Cyclopentanediol, *cis*-4-methylene-
  synthesis, **5**, 246
1,3-Cyclopentanedione
  Clemmensen reduction, **8**, 313
  2,2-disubstituted
    synthesis, pinacol rearrangement, **3**, 728
  enzymic reduction
    specificity, **8**, 201
1,3-Cyclopentanedione, 4-hydroxy-
  synthesis, **3**, 829
1,3-Cyclopentanedione, 2-methyl-
  enolates
    alkylation, **3**, 55
Cyclopentane-1,2,4-trione
  synthesis
    ketone oxallylation, **2**, 838
Cyclopentannulation
  methyl-3-phenylsulfonyl orthopropionate, **6**, 164
Cyclopentanoid monoterpenes
  synthesis, **3**, 850
Cyclopentanoids
  fused
    synthesis *via* cyclopropane ring opening, **4**, 1048
  polycondensed
    synthesis *via* photoisomerizations, **5**, 229
  synthesis
    via [3 + 2] cycloaddition reactions, **5**, 287, 561
    via retro Diels–Alder reactions, **5**, 561
Cyclopentanol, *cis*-2-alkenyl-
  synthesis
    ene reaction, **2**, 547
Cyclopentanol, 3-allyl-
  synthesis
    via carbomagnesiation, **4**, 877
Cyclopentanol, dimethyl-
  preparative electrolysis
    from 6-hepten-2-one, **8**, 134
Cyclopentanol, divinyl-
  rearrangements, **1**, 881
Cyclopentanol, *cis*-2-propargyl-
  cyclofunctionalization, **4**, 393
Cyclopentanols
  formation
    type II intramolecular ene reaction, **2**, 551
  intramolecular ene reaction, **2**, 542
  synthesis
    via samarium diiodide, **1**, 261
Cyclopentanone, 3-alkenyl-
  1-enolate
    alkylation, **3**, 17
Cyclopentanone, 3-alkyl-
  1-enolate
    alkylation, **3**, 17
Cyclopentanone, 2-allyl-
  synthesis
    alkylation of enolate, **3**, 6
Cyclopentanone, 2-benzyl-
  synthesis
    alkylation of enolate, **3**, 6
Cyclopentanone, 3-(2-bromoethyl)-
  endocycloalkylation, **3**, 19
Cyclopentanone, 2,3-dialkyl-
  synthesis
    conjugate addition–enolate alkylation, **3**, 9
Cyclopentanone, 2-diazo-
  synthesis, **3**, 900
Cyclopentanone, 2,5-dibromo-
  [4 + 3] cycloaddition reactions, **5**, 603
Cyclopentanone, 2,2-dimethyl-
  reduction
    chloroborane, **7**, 603
Cyclopentanone, 2-ethoxycarbonyl-
  tosylhydrazone
    synthesis, **2**, 513
Cyclopentanone, 3-formyl-
  synthesis, **1**, 527
Cyclopentanone, 2-halo-
  rearrangements, **3**, 848
Cyclopentanone, 2-methyl-
  synthesis
    alkylation of enolate, **3**, 4
Cyclopentanone, 3-methyl-3-phenyl-
  1-enolate
    alkylation, **3**, 17
Cyclopentanone, 2-nitro-
  synthesis, **6**, 105
Cyclopentanone, 2-nitro-3,4-dimethyl-
  synthesis, **6**, 106
Cyclopentanone, permethyl-
  reaction with α-selenoalkyllithium, **1**, 674
Cyclopentanone, 2-phenyl-
  annulation
    via [3 + 2] cycloaddition reactions, **5**, 304
  reaction with organometallic compounds, **1**, 150
Cyclopentanone, 3-phenyl-
  deprotonation, **3**, 17

1-enolate
alkylation, **3**, 17
synthesis
via [4 + 3] cycloaddition, **5**, 601
Cyclopentanone, β-substituted
synthesis
Knoevenagel reaction, **2**, 363
Cyclopentanone, 3-(2-tosyloxyethyl)-
endocycloalkylation, **3**, 19
Cyclopentanone, 2,2,5-trimethyl-
aldol reaction
isovaleraldehyde, **2**, 154
Cyclopentanone, 2-undecyl-
oxime mesylate
Beckmann rearrangement, **6**, 770
Cyclopentanone, 3-vinyl-
synthesis
via conjugate addition, **4**, 215
Cyclopentanone enamine
reaction with dihalocarbenes, **4**, 1004
Cyclopentanones
aldol reaction, **2**, 141, 147
annulation
intramolecular Barbier process, **1**, 262
boron trifluoride complex
NMR, **1**, 293
dehydrogenation
use of phenylselenium trichloride, **7**, 135
dimethyl acetals
selective reduction, **8**, 217
enolate
Michael additions, **5**, 1082
formation
type II intramolecular ene reaction, **2**, 551
lithium enolates
crystal structure, **1**, 26
X-ray diffraction analysis, **1**, 1, 3
magnesium enolates
aldol reaction, **2**, 199
polyalkylation
side reaction to monoalkylation, **3**, 4
reactions with ethyl diazoacetate, **1**, 849
reactions with organoaluminum reagents
stereoselectivity, **1**, 79
reduction
aluminum amalgam, **8**, 116
dissolving metals, **8**, 122
Reformatsky reaction
addition of carbon nucleophiles, **2**, 282
Ritter reaction, **6**, 270
substituted
nucleophilic addition reactions, **1**, 67
synthesis, **1**, 862
synthesis
carbonylation, **3**, 1024
Dieckmann cyclization, **2**, 796
ene reaction, **2**, 544
Friedel–Crafts reaction, **2**, 756
via [2 + 2 + 2] cycloaddition, **5**, 1130
via [3 + 2] cycloaddition, **5**, 283–286
via Michael addition, **4**, 18
Cyclopentapyrazoles
synthesis, **3**, 831
1*H*-Cyclopenta[*c*]pyrroles
synthesis
via metal-catalyzed cycloaddition, **5**, 1194
Cyclopenta[*b*]pyrrolidines
synthesis
Mannich cyclization, **2**, 1041
1*H*-Cyclopenta[*c*]pyrrolo-1,3-diones, 5-alkylidenehexahydro-
synthesis
via metal-catalyzed cycloaddition, **5**, 1194
Cyclopentene, 1-acetoxy-
Pauson–Khand reaction, **5**, 1048
Cyclopentene, 1-acetyl-2-methyl-
synthesis, **7**, 8
cyclohexane acetylation, **2**, 728
Cyclopentene, 1-aryl-
thermal ene reaction
mechanistic studies, **2**, 539
Cyclopentene, 3-(3-butynyl)-
Pauson–Khand reaction, **5**, 1057, 1058
Cyclopentene, 3-chloro-
hydroboration, **8**, 705
Cyclopentene, 1-chloro-2-hydroperfluoro-
hydrogenation, **8**, 899
Cyclopentene, 1-chloro-2-iodohexafluoro-
hydrogenolysis, **8**, 900
Cyclopentene, 1-chloroperfluoro-
reduction, **8**, 897
Cyclopentene, dichloro-
synthesis
via dichlorocyclopropyl compounds, **4**, 1023
Cyclopentene, 1,2-dimethoxy-
synthesis, **5**, 1083
Cyclopentene, 1,2-dimethyl-
hydrochlorination, **4**, 272
Cyclopentene, 1,5-dimethyl-
reduction
diimide, **8**, 476
1-Cyclopentene, 1,2-disubstituted
ozonolysis, **4**, 1099
Cyclopentene, 3,4-epoxy-
reaction with ethylaluminum alkynide, **3**, 279
reaction with Grignard reagents, **3**, 265
Cyclopentene, 1,3,3,4,4,5,5-heptafluoro-
reduction, **8**, 897
Cyclopentene, 4-hydroxy-4-(1-hexynyl)-
synthesis, **3**, 279
Cyclopentene, methoxy-
cycloaddition reactions
with benzonitrile, **5**, 161
Cyclopentene, 1-methyl-
cyclopropanation, **5**, 1085
hydrochlorination, **4**, 272
Pauson–Khand reaction, **5**, 1046
Cyclopentene, 3-methyl-
synthesis
Ramberg–Bäcklund rearrangement, **3**, 874
Cyclopentene, 3-methylene-
annulation, **5**, 774
Cyclopentene, 4-methylene-
synthesis
via metal-catalyzed cycloaddition, **5**, 1194
via retro Diels–Alder reactions, **5**, 563
2-Cyclopentene, 1-methylimino-3-methyl-5,5-diphenyl-
synthesis
via metal-catalyzed cycloaddition, **5**, 1195
Cyclopentene, 1-(4-pentynyl)-

Pauson–Khand reaction, **5**, 1057, 1062
Cyclopentene, 1-phenyl-
hydroboration, **8**, 722
3-Cyclopentene, 2-phenylsulfonyl-
methyl ester, acetate
synthesis, **3**, 654
Cyclopentene, 3-substituted 1-vinyl-
Diels–Alder reactions, **5**, 349
Cyclopentene, (trimethylsilyl)-
annulations, **1**, 596
Cyclopentene, vinyl-
synthesis
nickel-catalyzed rearrangement, **5**, 917
2-Cyclopenteneacetic acid
cyclofunctionalization, **4**, 370
Cyclopentene-1-carboxylic acid
synthesis, **3**, 905
Cyclopentene-3-carboxylic acid
esters
synthesis, **7**, 832
2-Cyclopentenedicarboxylates
synthesis
via addition reactions with organozinc compounds, **4**, 95
3,5-Cyclopentenediols
synthesis
prostaglandin precursor, **3**, 155
2-Cyclopentene-1,4-dione
reduction, **8**, 163
aluminum hydrides, **8**, 544
Cyclopentene-1,2-diones
synthesis
via metal-catalyzed cycloaddition, **5**, 1200
4-Cyclopentene-1,3-diones, 2-alkylidene-
synthesis
via cyclobutenone ring opening, **5**, 690
Cyclopentenes
annulation, **5**, 951
opening of cyclopropyl ketones, **5**, 925
use of vinylcyclopropane, **5**, 919
carbonylation
cobalt carbonyl catalyst, **3**, 1024
[2 + 2 + 2] cycloaddition reactions, **5**, 1130
diamination, **7**, 484
hydrocarboxylation
dicarboxylation, **4**, 947
irradiation
with *m*-xylene, **5**, 651
oxidation
Wacker process, **7**, 451, 452
Wacker process with heteropolyacids, **7**, 462
oxidative cleavage
ozone, **7**, 558
oxide
rearrangement, lithium halide catalyzed, **3**, 764
Pauson–Khand reaction, **5**, 1046
rearrangement
vinylcyclopropane, **5**, 907
synthesis
Ramberg–Bäcklund rearrangement, **3**, 874
selectivity, **5**, 907
via [4 + 1] annulation, **5**, 1008
via [3 + 2] cycloaddition reactions, **5**, 277
via Michael addition, **4**, 16
via reaction of allenylsilanes with α,β-unsaturated carbonyl compounds, **1**, 596
via vinylcyclopropane rearrangement, **5**, 1012
via vinylcyclopropane thermolysis, **4**, 1048
Cyclopentenocycloalkanones
synthesis
via copper catalyzed Grignard addition, **4**, 91
1-Cyclopentenol, 2,3-epoxy-
synthesis, **7**, 413
2-Cyclopentenol, 4-oxo-
acetate
conjugate additions, **4**, 211
Cyclopentenol, vinyl-
allylic rearrangements, **7**, 822
Cyclopentenols
synthesis
via retro Diels–Alder reactions, **5**, 562
Cyclopentenone, 3-alkyl-
lithium dienolates
methylation, **3**, 22
Cyclopentenone, 3-alkyl-4-(hydroxyalkyl)-
synthesis
via Pauson–Khand reaction, **5**, 1057
2-Cyclopentenone, 3-alkylidene-
synthesis
via Nazarov cyclization, **5**, 777
Cyclopentenone, 3-amino-
extended dienolates
γ-alkylation, **3**, 24
Cyclopentenone, 5-aryl-
synthesis
via Pauson–Khand reaction, **5**, 1045
Cyclopentenone, 2-bromo-
reduction
aluminum hydrides, **8**, 545
Cyclopentenone, 3-*n*-butyl-
photocycloaddition reactions, **5**, 125, 127
2-Cyclopentenone, 5-chloro-
synthesis
alkyne acylation, **2**, 725
via Nazarov cyclization, **5**, 777
Cyclopentenone, dialkyl-
reduction
dissolving metals, **8**, 122
Cyclopentenone, 2,5-dialkyl-
synthesis
via [3 + 2] cycloaddition reactions, **5**, 285
Cyclopentenone, 4,4-dialkyl-
synthesis, **3**, 42
Cyclopentenone, 4,5-dialkyl-
synthesis, **2**, 726
Cyclopentenone, 4,5-dihydroxy-
enolates
alkylation, **3**, 11
2-Cyclopentenone, 2,5-dimethyl-
synthesis
via vinylallene epoxidation, **5**, 772
2-Cyclopentenone, 4,4-dimethyl-
dimerization
base catalyzed, **4**, 239
synthesis
via Wacker oxidation, **7**, 456
Cyclopentenone, 4,4-disubstituted 3-methyl-
synthesis
via Nazarov cyclization, **5**, 767
Cyclopentenone, 2-ethoxycarbonyl-

[3 + 2] cycloaddition reactions, **5**, 273
2-Cyclopentenone, 4-hydroxy-
conjugate additions
Lewis acids, **4**, 143
synthesis, **2**, 142; **3**, 10
via Nazarov cyclization, **5**, 771
2-Cyclopentenone, 5-methoxy-
reaction with Gilman reagents, **4**, 211
2-Cyclopentenone, 2-[6-(methoxycarbonyl)-1-hexyl]-
reactions with dithioacetal oxides, **1**, 528
2-Cyclopentenone, 2-methyl-
conjugate additions
chiral organocopper compounds, **4**, 227
tandem vicinal difunctionalization, **4**, 245
2-Cyclopentenone, 4-methyl-
synthesis
via Nazarov cyclization, **5**, 767
Cyclopentenone, 2-methylene-
synthesis
carbonylation of 1-iodo-1,4-dienes, **3**, 1025
Cyclopentenone, 4-methylene-
synthesis
via [2 + 2 + 2] cycloaddition, **5**, 1131
Cyclopentenone, 5-methylene-
synthesis
via Nazarov cyclization, **5**, 780
via retro Diels–Alder reactions, **5**, 560
2-Cyclopentenone, 2-pentyl-
synthesis
via double bond migration, **7**, 457
2-Cyclopentenone, 5-pentyl-
synthesis
via Claisen rearrangement, oxidation, **7**, 457
2-Cyclopentenone, 3-phenylthio-
synthesis
via Nazarov cyclization, **5**, 778
2-Cyclopentenone, 5-phenylthio-
synthesis
via Nazarov cyclization, **5**, 778
Cyclopentenones
addition reactions
with α-silyl ester enolates, **4**, 107
α′-alkylation, **3**, 21
annulations
regiospecific, **1**, 584
Wacker oxidation, **7**, 455
[3 + 2] cycloaddition reactions, **5**, 301
palladium catalyzed, **5**, 281
dialkylation, **4**, 255
dicarboxylation, **4**, 948
α-enolate
reaction with aldehydes, **2**, 198
functionalized
synthesis *via* retro Diels–Alder reactions, **5**, 560
isomerization, **5**, 762
photocycloaddition reactions
stereochemical scrambling, **5**, 128
reaction with 1-phenylselenoallyllithium
regiochemical control, **1**, 691
reduction
aluminum hydrides, **8**, 543
biochemical, **8**, 558
dissolving metals, **8**, 123
molecular orbital calculations, **8**, 16
selective reduction
borohydrides, **8**, 539
4-substituted
reactions with allylic sulfinyl carbanions, **1**, 521
synthesis, **1**, 555; **3**, 936; **5**, 1105; **7**, 797, 802, 819
allenyl organoaluminum, **2**, 89
from 2-chloro-1,3-cyclohexanediones, **3**, 871
Ramberg–Bäcklund rearrangement, **3**, 868, 874, 875
via conjugate addition to α-nitroalkenes, **4**, 143
via [2 + 2 + 2] cycloaddition, **5**, 1131–1133
via [3 + 2] cycloaddition, **5**, 283–286
via cyclopropane ring opening, **4**, 1046
via dihalocyclopropyl compounds, **4**, 1018
via divinyl ketones, **1**, 430
via hydration of dienynes/ring closure, **5**, 752
via Nazarov cyclization, **5**, 757
via Pauson–Khand reaction, **5**, 1037
via three-carbon annulation, **1**, 548
via vinylallene epoxidation, **5**, 772
tandem vicinal difunctionalization, **4**, 245
zirconium dienolates
aldol reaction, **2**, 303
Cyclopentylamines, 2-methyl-
reduction
stereoselectivity, **8**, 55
Cyclopentylmethyl radicals
synthesis, **7**, 731
Cyclopeptides
synthesis, **6**, 389
Ugi reaction, **2**, 1095
Cyclophanedienes
synthesis
via arynes, **4**, 507
Cyclophanediones
synthesis, **6**, 134
[*m.m*]-*meta*-Cyclophanediones
synthesis
via $S_{RN}1$ reaction, **4**, 477
Cyclophanes
synthesis, **3**, 557, 591, 594
1,2-rearrangement, **3**, 927
unsaturated
synthesis, **3**, 877
m-Cyclophanes
synthesis
coupling reactions, **3**, 452
via cycloaromatization reaction, **2**, 622
Cyclopropa[*c*]cinnolines
synthesis
via nitrilimine 1,1-cycloaddition, **4**, 1084
Cyclopropanation
acrylaldehyde
via enolate alkylation, **4**, 239
alkenes, **5**, 1084
alkyl diazoacetate, **4**, 1035
intramolecular, **4**, 1040–1043
asymmetric, **4**, 961, 1038
cobalt catalysts, **4**, 1040
diastereoselectivity, **4**, 1037
enantioselective, **4**, 980, 987
sequential Michael ring closure, **4**, 262
Simmons–Smith methylenating agent
hydroalumination adducts, **8**, 756
via conjugate addition, **4**, 258
Cyclopropane, 1-acetoxy-3-alkyldifluoro-

ring opening, **4**, 1020
Cyclopropane, 1-alkyl-1-halo-
synthesis
via lithium carbenoids, **4**, 1008
Cyclopropane, alkylidene-
reactions with alkenes
metal catalyzed, **5**, 1191
synthesis, **1**, 652; **3**, 116
via lithium–halogen exchange, **4**, 1008
Cyclopropane, allyl-
synthesis.
via boron-ene reaction, **5**, 33
via cycloaddition of bicyclo[1.1.0]butanes, **5**, 1185
Cyclopropane, allylidene-
synthesis, **1**, 652
via Peterson alkenation, **1**, 786
Cyclopropane, 1-*p*-anisyl-2-vinyl-
rearrangement
cyclopentene synthesis, **5**, 1014
Cyclopropane, aryl-
synthesis, **3**, 120
Cyclopropane, 2,3-bis(alkoxycarbonyl)-1-(2-methyl-1-propenyl)-
synthesis
via metal-catalyzed cycloaddition, **5**, 1197
Cyclopropane, 1,1-bis(benzenesulfonyl)-
use in synthesis, **6**, 161
Cyclopropane, 1,1-bis(methylthio)-
ketones from, **3**, 124
Cyclopropane, 1,1-bis(phenylseleno)-
synthesis, **1**, 638; **3**, 136
Cyclopropane, 1,1-bis(seleno)-
synthesis, **1**, 657
Cyclopropane, bromo-
reaction with lithium in diethyl ether
crystal structure, **1**, 10
synthesis
via bromocarbene, **5**, 1012
Cyclopropane, 1-bromo-1-tributylstannyl-
synthesis
by transmetallation, **3**, 196
Cyclopropane, butylidene-
cycloaddition reactions
carbon dioxide, metal catalyzed, **5**, 1196
Cyclopropane, chlorofluoro-2-(trimethylsilyl)methyl-
rearrangement
2-fluoro-1,3-butadiene, **4**, 1020
Cyclopropane, 1-cyano-2,2-dihalo-
synthesis
via dihalocarbene, **4**, 1002
Cyclopropane, deutero-
synthesis
via cyclopropanation, **4**, 1039
Cyclopropane, *gem*-dialkyl-
synthesis
via organocuprates, **4**, 1009
Cyclopropane, dibromo-
lithium–halogen exchange, **4**, 1007–1009
Cyclopropane, dibromotetramethyl-
rearrangement
1,2,2-trimethylbicyclo[1.1.0]butane, **4**, 1013
Cyclopropane, dibromovinyl-
rearrangement, **5**, 950
Cyclopropane, dichloro-
synthesis, **4**, 1000
Cyclopropane, dienyl-
rearrangement
palladium catalysis, **5**, 917
Cyclopropane, difluoro-
ring opening, **4**, 1020
Cyclopropanes, dihalo-
electrocyclic ring opening, **4**, 1016–1020
elimination/addition reactions, **4**, 1014–1016
elimination reactions, **4**, 1014–1016
monoreduction
selective, **8**, 806
ring expansion, **4**, 1017–1020
solvolysis, **4**, 1021
synthesis, **4**, 999–1025
transformations, **4**, 1006
Cyclopropane, dimethyl-
synthesis, **3**, 216
Cyclopropane, 1-dimethylamino-2-vinyl-
rearrangement
activation energy, **5**, 1007
Cyclopropane, 2,2-dimethyl-1-methylene-
codimerization
metal catalyzed, **5**, 1191
cycloaddition reactions, **5**, 1190
[3 + 2] cycloaddition reactions
nickel catalyzed, **5**, 294
Cyclopropane, diphenylidene-
cycloaddition reactions, **5**, 1189, 1190
with unsaturated ketones, **5**, 1192
[3 + 2] cycloaddition reactions
nickel catalyzed, **5**, 294
Cyclopropane, 1,1-diphenyl-2-isocyano-2-methyl-
reduction, **8**, 830
Cyclopropane, 1,1-dithio-
synthesis, **3**, 124
Cyclopropane, divinyl-
Cope rearrangement, **4**, 1048
synthesis
via cyclopropanation, **4**, 1049
Cyclopropane, 1,2-divinyl-
Cope rearrangement, **5**, 791, 803–805, 820
enantiospecificity, **5**, 973–976
mechanism, **5**, 972
stereospecificity, **5**, 973
substituent effects, **5**, 973
rearrangements, **5**, 971–996
ring cleavage
selectivity, **5**, 912
tricyclic
Cope rearrangement, **5**, 993–996
Cyclopropane, 1-ethoxy-1-lithio-
synthesis
metallation, **3**, 194
Cyclopropane, 1-ethoxy-1-trimethylsiloxy-
cycloaddition reactions
aldehydes, metal catalyzed, **5**, 1200
Cyclopropane, hexylidene-
[3 + 2] cycloaddition reactions, **5**, 290
Cyclopropane, 1-hydroxyalkyl-
ring opening, **4**, 1043
Cyclopropane, 1-(1′-hydroxyalkyl)-1-(methylseleno)-
rearrangement, **1**, 717
Cyclopropane, isopropylidene-
cycloaddition reactions, **5**, 1189, 1190
carbon dioxide, metal catalyzed, **5**, 1196

[3 + 2] cycloaddition reactions, **5**, 290
Cyclopropane, keto vinyl-
free radical 1,6-addition reactions
alkyl boranes, **5**, 926
rearrangement, **5**, 909
Cyclopropane, lithiobromo-
reaction with catechol borane, **4**, 1008
Cyclopropane, (1-methoxy-1-phenylthio)-
synthesis
via Pummerer rearrangement, **6**, 146
Cyclopropane, 1-methoxy-2-vinyl-
rearrangement
activation energy, **5**, 1007
Cyclopropane, methylene-
addition to dichlorocarbene, **4**, 1002
π-allylpalladium complexes from, **4**, 587
codimerization
metal catalyzed, **5**, 1191
cooligomerization
allene, metal catalyzed, **5**, 1195
cycloaddition reactions
alkynes, metal catalyzed, **5**, 1194, 1195
carbon dioxide, metal catalyzed, **5**, 1196
2-cyclopentenones, metal catalyzed, **5**, 1193
metal catalyzed, **5**, 1188, 1193
[3 + 2] cycloaddition reactions, **5**, 288
diastereoselectivity, **5**, 290
distal ring-opening, **5**, 288
ketenimines, metal catalyzed, **5**, 1195
metal catalyzed, **5**, 1194
nickel catalyzed, **5**, 293
palladium catalyzed, **5**, 289
proximal ring-opening, **5**, 288
oxidative cleavage, **7**, 825
oxidative rearrangement, **7**, 833
reaction with alkenes
metal catalyzed, **5**, 1191
Cyclopropane, 1-methylene-2-vinyl-
codimerization
with norbornene, **5**, 1190
Cope rearrangement, **5**, 794
cyclodimerization, **5**, 1190
rearrangement, **5**, 947
Cyclopropane, 2-methylmethylene-
cycloaddition reactions, **5**, 1190
Cyclopropane, *cis*-methylvinyl-
flash vacuum pyrolysis, **5**, 906
Cyclopropane, monohalo-
synthesis
via reductive dehalogenation, **4**, 1006
Cyclopropane, 1-oxido-1-(1′-phenylselenoxyalkyl)-
rearrangement, **1**, 715
Cyclopropane, oxyvinyl-
ring expansion, **5**, 919
Cyclopropane, 1-phenyl-1-methylseleno-
synthesis, **1**, 669
Cyclopropane, 1-phenyl-2,3-phenacyl-
synthesis, **1**, 655
Cyclopropane, phenylseleno-
metallation, **1**, 641
Cyclopropane, phenylthio-
allylation, **3**, 88
Cyclopropane, 1-phenylthio-1-(trimethylsiloxy)-
synthesis
via silyl-Pummerer rearrangement, **6**, 146
Cyclopropane, propenylidene-
[2 + 2] cycloaddition reactions
tetracyanoethylene, **5**, 76
Cyclopropane, silavinyl-
rearrangement, **5**, 950
Cyclopropane, siloxy-
coupling reactions
with Grignard reagents, **3**, 460
with sp^3 organometallics, **3**, 455
Cyclopropane, silyloxy-
cleavage
iron(III) chloride, **2**, 444
1,6-diketones from, **2**, 445
homoenolate precursor, **2**, 442
synthesis, **2**, 443
Cyclopropane, 1-silyloxy-2-carboalkoxy-
ring cleavage
via homoenolates, **4**, 120
Cyclopropane, tetramethyl-
anodic oxidation, **7**, 794
Cyclopropane, trialkylsilyloxy-
rearrangement, **1**, 879
Cyclopropane, 2-(trimethylsilyl)methylene-
cycloaddition reactions, **5**, 1190
with unsaturated ketones, **5**, 1192
Cyclopropane, vinyl-
π-allylpalladium complexes from, **4**, 590
bonding, **5**, 901
cycloaddition reactions, **5**, 926
metal catalyzed, **5**, 1200
palladium catalysis, **4**, 593
[2 + 2] cycloaddition reactions, **5**, 71
electron deficient
[3 + 2] cycloaddition reactions, **5**, 281
free-radical polymerization, **5**, 926
one-electron transfer, **5**, 77
Paterno–Büchi reaction, **5**, 157
photochemical rearrangement, **5**, 913
radical annulation, **4**, 824
radical reactions, **5**, 926
reaction with nucleophiles, **5**, 921
rearrangements, **5**, 211, 267, 899–965, 1006
carbanion-accelerated, **5**, 1012–1014
carbocation-accelerated, **5**, 1014–1016
copper catalysis, **5**, 917
cyclopentene, **5**, 907
oxyanion-accelerated, **5**, 1007–1011
platinum catalysis, **5**, 917
rhodium catalysis, **5**, 916, 917
stereoselectivity, **5**, 1007
stereospecificity, **5**, 907
substituent effects, **5**, 904
1,5-sigmatropic shift
hydrogen, **5**, 906
strain energy, **5**, 901
α-sulfonyl carbanions
rearrangements, **5**, 1012
synthesis, **5**, 905
via metal-catalyzed cycloaddition, **5**, 1197–1199
via photoisomerization, **5**, 194
via ylide addition to carbonyl, **5**, 951
thermolysis, **4**, 1048
Cyclopropane, vinyldihalo-
rearrangement
cyclopentadienes, **4**, 1012

Cyclopropane, vinylmethylene-
synthesis
via dihalocyclopropyl compounds, **4**, 1015
Cyclopropane-1-acetaldehyde, 2,2-dimethyl-3-(2′-oxo)-propyl-
dimethyl acetal
synthesis, *via* ozonolysis of 3-carene, **7**, 548
Cyclopropanecarbaldehyde, 1-(arylthio)-
reduction
aluminum hydrides, **8**, 544
Cyclopropanecarbodithioate, methyl
synthesis
via methyl cyclodithioformate, **6**, 456
Cyclopropanecarboxylic acid, 2-allyl-
methyl ester
synthesis *via* cycloaddition of bicyclo[1.1.0]butanes, **5**, 1186
synthesis
via magnesium-ene reaction, **5**, 30
Cyclopropanecarboxylic acid, 1-amino-
synthesis
via ketocarbenoid addition to alkynes, **4**, 1050
Cyclopropanecarboxylic acid, 2-*t*-butyl-2-(trimethylsilyloxy)-
esters
reactions with *N,N*-dimethyl(methylene)iminium salts, **2**, 911
1-Cyclopropanecarboxylic acid, 2-hydroxymethyl-1-amino-
synthesis, **1**, 559
Cyclopropanecarboxylic acid, 2-siloxy-
methyl ester cycloaddition reactions
carbonyl compounds, metal catalyzed, **5**, 1200
Cyclopropanecarboxylic acid, 2-silyloxy-
homoenolate equivalents
reactions with *N,N*-dimethyl(methylene)iminium salts, **2**, 911
reactions with carbonyl compounds, **2**, 448
Cyclopropanecarboxylic acid anhydride
synthesis, **6**, 311
Cyclopropanecarboxylic acids
ethyl ester
Friedel–Crafts reaction, **2**, 756
synthesis, **3**, 848
zinc carbenoid, **2**, 444
Cyclopropanedicarboxylates, 1-alkyl-2-halo-
synthesis
via addition with organozinc compounds, **4**, 95
1,1-Cyclopropanedicarboxylic acid, 2-alkenyl-
intermolecular alkylation, **3**, 56
Cyclopropanediols
synthesis
intramolecular pinacol coupling, **3**, 572
Cyclopropane-2,3-dioxopropionic acid
ethyl ester
rearrangement, **3**, 831
Cyclopropane ketal
synthesis
via dihalocyclopropyl compounds, **4**, 1015
Cyclopropanes
bonding, **5**, 900
charge-accelerated rearrangements, **5**, 1006–1016
cleavage, **2**, 444
catalytic hydrogenation, **4**, 1043
[2 + 1] cycloadditions, **5**, 1084
diradical opening, **5**, 900
enantioselectivity, **4**, 952
energetics, **5**, 900
energy content
effect on synthesis, **5**, 904
formation
arene–alkene photocycloadditions, **5**, 649
metal homoenolate reaction, **2**, 443
from Δ^1-pyrazolines, **4**, 1102
functionalized
synthesis, **4**, 1031
lithiation, **1**, 480
neighboring group
epoxide ring opening, **3**, 736, 752, 753
optically active
synthesis *via* conjugate addition to oxazepines, **4**, 206
oxidative rearrangement, **7**, 823, 833
polarized
1,3-dipolar synthetic equivalents, **5**, 266
reactions with transition metal complexes, **7**, 4
reactivity, **5**, 901
ring expansion, **3**, 785
stereochemistry, **4**, 952
stereoselective synthesis
Knoevenagel reaction, **2**, 360
strain energy, **5**, 900
substituents
stereospecificity, **4**, 952
synthesis, **3**, 163; **4**, 951; **6**, 556
Darzens glycidic ester condensation, **2**, 432
from enones, **2**, 431
reduction of malonate, **3**, 620
via alkylidene transfer, **4**, 951–994
via diazo ketones, **6**, 126
via 1,3-eliminative cyclization of γ-stannyl alcohols, **7**, 621
via enamines and diazomethane, **6**, 716
via Michael reaction, **4**, 2
via reductive dehalogenation, **4**, 1006
Wurtz reaction, **3**, 422
Cyclopropanesulfomorpholine
synthesis, **3**, 181
Cyclopropanesulfonic acid
t-butyl esters
synthesis, **3**, 180
neopentyl ester
synthesis, **3**, 181
Cyclopropane-1,1,2-tricarboxylic acid
triethyl ester
synthesis *via* Michael reaction, **4**, 2
Cyclopropanols
oxidation
lead tetraacetate, **7**, 824
oxidative cleavage, **7**, 824
rearrangement, **1**, 874
synthesis
via dissolving metal reductions, **8**, 528
via organosamarium compounds, **1**, 261
Cyclopropanols, vinyl-
lithium salts
rearrangements, **5**, 1007
pyrolysis, **5**, 920
salts
synthesis, **5**, 1007

Cyclopropanone, 2,2-dimethyl-
[4 + 3] cycloaddition reactions, **5**, 597
Cyclopropanone, 2,3-dimethyl-
Lewis acid complexes
structure, **1**, 287
Cyclopropanone, diphenyl-
cycloaddition reactions
metal catalyzed, **5**, 1200
Cyclopropanones
cycloaddition reactions
metal catalyzed, **5**, 1200
[4 + 3] cycloaddition reactions, **5**, 597
N,N-dialkylhydrazones
preparation, **2**, 505
Favorskii rearrangement, **3**, 840
reactions with diazomethane, **1**, 847
Cycloproparenes
synthesis
via Peterson alkenation, **1**, 786
Cyclopropene, 1,2-dibromo-
synthesis
via dihalocyclopropyl compounds, **4**, 1015
Cyclopropene, 3,3-dicyclopropyl-
cycloaddition reactions
metal catalyzed, **5**, 1198
dimerization, **5**, 65
Cyclopropene, 3,3-difluoro-
synthesis
via retro Diels–Alder reactions, **5**, 560
Cyclopropene, 3,3-dimethoxy-
cycloaddition reactions
metal catalyzed, **5**, 1199
dimerization, **5**, 65
Cyclopropene, 3,3-dimethyl-
cyclodimerization
metal catalyzed, **5**, 1197
Cyclopropene, 1,2-diphenyl-
Cope rearrangement, **5**, 794
cycloaddition reactions
metal catalyzed, **5**, 1197
Cyclopropene, 1,2-diphenyl-3-methyl-3-*o*-vinylphenyl-
intramolecular [2 + 2] cycloadditions, **5**, 67
Cyclopropene, halo-
synthesis
via dihalocyclopropyl compounds, **4**, 1015
Cyclopropene, 3-methoxycarbonyl-1-propyl-
cycloaddition reactions
metal catalyzed, **5**, 1198
Cyclopropene, 1-methyl-
carboboration, **4**, 885
dimerization, **5**, 1197
reactions with triallylboranes, **5**, 33
Cyclopropene, 3-methyl-3-cyclopropyl-
cycloaddition reactions
metal catalyzed, **5**, 1198
Cyclopropene, methylene-
cycloadditions, **5**, 64
dimerization, **5**, 65
Cyclopropene, tetrachloro-
synthesis
via dihalocyclopropyl compounds, **4**, 1015
Cyclopropene, trihalo-
aminolysis, **6**, 521
Cyclopropene, vinyl-
π-allylpalladium complexes from, **4**, 587
rearrangement, **5**, 947
Cyclopropenes
π-allylpalladium complexes from, **4**, 587
carbocupration, **4**, 895
carbomagnesiation, **4**, 874
carbozincation, **4**, 880
cycloaddition reactions, **5**, 64
alkanes, metal catalyzed, **5**, 1197–1199
oxidative cleavage, **7**, 825
synthesis
intramolecular McMurry reaction, **3**, 588
via [2 + 1] cycloadditions, **5**, 1089
Cyclopropenone, 3-acyl-
ring enlargement
cyclobutene synthesis, **5**, 677
Cyclopropenone, diphenyl-
cycloaddition reactions
ketenes, metal catalyzed, **5**, 1200
Cyclopropenone, 3-hydroxymethyl-
ring enlargement
cyclobutene synthesis, **5**, 677
Cyclopropenone ketals
1,3-dipolar synthetic equivalents, **5**, 266
Cyclopropenones
cycloaddition reactions
metal catalyzed, **5**, 1200
synthesis
via dihalocarbene, **4**, 1005
Cyclopropyl aldehydes
rearrangement, **5**, 909
Cyclopropylamine, 2-phenyl-
synthesis
via Curtius reaction, **6**, 811
Cyclopropyl bromides
reduction
lithium aluminum hydride, **8**, 802
Cyclopropylcarbinols
oxidative rearrangement, **7**, 825
spiro-fused
oxidative rearrangement, **7**, 834
Cyclopropylcarbinols, dichloro-
solvolysis
divinyl ketones from, **5**, 770
Cyclopropylcarbinyl anion
reactivity, **5**, 901
Cyclopropylcarbinyl cation
reactivity, **5**, 901
Cyclopropylcarbinyl radical
reactivity, **5**, 901
Cyclopropyl compounds
[3 + 2] cycloaddition reactions
palladium catalyzed, **5**, 281
Cyclopropyl compounds, 1-bromo-
synthesis
via lithium–halogen exchange, **4**, 1007
Cyclopropyl compounds, fluorobromo-
synthesis
via brominative decarboxylation, **4**, 1006
Cyclopropyl esters
rearrangement, **5**, 909
Cyclopropylimines
rearrangements, **5**, 909, 945
use in alkaloid synthesis, **5**, 952
synthesis, **5**, 946

Cyclopropyl ketones
  rearrangement, **5**, 909
Cyclopropyl-π-methane rearrangements
  photoisomerizations, **5**, 198
Cyclopropylselenonyl anions
  synthesis, **1**, 828
Cycloreversion reactions
  cyclobutanes, **5**, 64
Cyclosarkomycin
  synthesis
    via Pauson–Khand reaction, **5**, 1051
Cyclosativene
  synthesis, **7**, 517
Cycloseychellene
  synthesis
    Prins reaction, **2**, 542
Cyclosporin A
  aldol reaction
    synthesis of MeBMT, **2**, 219
  synthesis, **6**, 385
1,8-Cyclotetradecadiene
  synthesis
    alkene metathesis, **5**, 1119
Cyclotetradecene, 1-triethylsilyloxy-
  synthesis, **8**, 557
1,2-Cyclotridecadiene
  hydrobromination, **4**, 284
  hydrogenation
    homogeneous catalysis, **8**, 450
Cycloundecadiene
  Cope rearrangement
    equilibrium, **5**, 810
1,2-Cycloundecadiene
  reaction with iodine azide, **7**, 506
Cycloundecanone, 2-bromo-
  rearrangements, **3**, 849
1,3,5-Cycloundecatriene
  irradiation, **5**, 717
Cycloundecene
  hydroalumination, **8**, 739
  transannular cyclization, **3**, 398
p-Cymene
  solvent
    reductive decarboxylation, **7**, 720
Cysteine
  protecting groups
    use in peptide synthesis, **6**, 664
Cysteine, *N*-benzoyl-
  lithium borohydride modifier, **8**, 169
Cysteine, 4-picolyl-
  cleavage, **8**, 974
Cysteine proteases
  peptide synthesis, **6**, 395
Cystine, *N*,*N*′-dibenzoyl-
  lithium borohydride modifier, **8**, 169
Cytidine 5′-monophosphoneuraminic acid
  synthesis
    enzymatic methods, **2**, 464
Cytochalasin B
  synthesis
    via Diels–Alder reaction, **5**, 351
Cytochalasins
  3,2-sigmatropic rearrangement
    synthesis, stereocontrol, **3**, 960
  synthesis, **7**, 183
    via iterative rearrangements, **5**, 894
    via SmI_2-promoted macrocyclization, **1**, 266
Cytochrome *P*-450
  alkane hydroxylation, **7**, 11
  alkene epoxidation catalysis, **7**, 382
  camphor hydroxylation
    catalyst, **7**, 80
Cytosine
  fluorination, **7**, 535
Cytovaricin
  synthesis, **1**, 401

D

Dactylol
synthesis, **3**, 404
DAHP synthetase
organic synthesis
use in, **2**, 466
Dakin oxidation
aryl aldehydes
synthesis of phenols, **7**, 674
Dakin–West reaction
N-acyl-α-amino ketones, **3**, 889
α-Damascone
synthesis
via Grignard reagent and base, **1**, 417
β-Damascone
microbial oxidation, **7**, 77
δ-Damascone
synthesis
Knoevenagel reaction, **2**, 370
Damsin
synthesis, **3**, 20; **7**, 313
Damsinic acid
synthesis
via Cope rearrangement, **5**, 982
Dane salt
α-amido β-lactams from, **5**, 95
Danishefsky's diene
cycloaddition reactions, **5**, 1072
Diels–Alder reactions, **5**, 329
Daphnane diterpenoids
synthesis
via Cope rearrangement, **5**, 984
Daphneticin
synthesis, **3**, 691
Darvon alcohol
lithium aluminum hydride modifier, **8**, 164
Darzens aldehyde synthesis
epoxide ring opening, **3**, 738
Darzens glycidic ester condensation, **2**, 409–439
aromatic nucleophilic substitution
competing reaction, **4**, 432
asymmetric catalysts, **2**, 435
asymmetric variants, **2**, 435
cis:trans isomer ratios, **2**, 414
intramolecular, **2**, 427, 434
mechanism, **2**, 411
modifications, **2**, 427
phase-transfer conditions, **2**, 429
solid–liquid systems, **2**, 434
stereochemistry, **2**, 412
Darzens–Nenitzescu reaction
alkene acylation, **5**, 777
Daucenone
synthesis, **3**, 586
Daucon
synthesis
via [4 + 3] cycloaddition, **5**, 609
Daunomycin
synthesis, **7**, 341
Daunomycin, 11-deoxy-
synthesis
carbonyl group protection, **6**, 680
Daunomycin, 4-dimethoxy-
synthesis
via arynes, **4**, 497
Daunomycinone
synthesis, **5**, 1096, 1098, 1099
via alkynylcerium reagents, **1**, 242
via annulation, **1**, 554
via Diels–Alder reaction, **5**, 393
Daunomycinone, demethoxy-
synthesis, **7**, 351, 352
Daunomycinone, 4-demethoxy-
synthesis, **5**, 1096
via Diels–Alder reaction, **5**, 375, 384
via oxyanion-accelerated rearrangement, **5**, 1023
Daunomycinone, 7-deoxy-
synthesis
via chiral acetals, **1**, 64
Daunomycinone, 11-deoxy-
synthesis, **1**, 567; **5**, 1096
Daunomycinone, dideoxy-
synthesis
via aromatic Claisen rearrangement, **5**, 834
Daunomycinone, 1-methoxy-
synthesis
via Diels–Alder reactions, **5**, 396
Daunosamine
amino sugars, **2**, 323
synthesis, **1**, 349
Diels–Alder reaction, **2**, 689
Mannich reaction, **2**, 924
via Baeyer–Villiger reaction, **7**, 678
Dauricine
synthesis, **3**, 687
Davanone
synthesis
aldol reaction, **2**, 202
Davy's reagent
thiocarboxylic ester synthesis, **6**, 437
DB 2073
synthesis
via cyclobutenone ring opening, **5**, 689
via electrocyclization, **5**, 732
DCC — *see* Carbodiimide, dicyclohexyl-
DDATHF
synthesis
via Diels–Alder reaction, **5**, 492
Deacylation
enzymatic, **6**, 340
Dealkylation
Friedel–Crafts reaction, **3**, 327
Deamination
amines
alcohol synthesis, **6**, 3
diazenes, **8**, 828
hydrazines
potassium superoxide, **7**, 744
4-Deazafervenulin, 3-chloro-
synthesis, **7**, 342
4-Deazafervenulin 2-oxide
reaction with Vilsmeier–Haack reagent, **7**, 342
5-Deazaflavin, 1,5-dihydro-
reduction
unsaturated carbonyl compounds, **8**, 562

5-Deazaflavins
reduction
unsaturated carbonyl compounds, **8**, 562
synthesis, **4**, 435
Debenzhydrylation
Friedel–Crafts reaction, **3**, 328
De-*t*-butylation
Friedel–Crafts reaction, **3**, 329
Decaborane
reduction
acetals, **8**, 214
trans-Decaladienone
synthesis, **5**, 1100
Decaladienones
aromatization, **3**, 810
9,10-Decalindiol
oxidative cleavage, **7**, 704, 708
Decalindione
intramolecular aldolization, **2**, 169
cis-Decalindione
synthesis
via Michael addition, **4**, 27
Decalindiones
enzymic reductions
synthesis of hydroxy ketones, **8**, 188
Decalins
aromatization, **7**, 7
oxidation
benzyltrimethylammonium permanganate, **7**, 12
synthesis
intramolecular cyclization of cyanocyclohexanes, **3**, 48
polyene bicyclization, **3**, 360
polyene cyclization, **3**, 350
transannular ene reaction, **2**, 553
cis-Decalins
synthesis, **3**, 360
via Cope rearrangement, **5**, 812
trans-Decalins
synthesis, **3**, 360
trans-Decalone
synthesis
via Michael addition, **4**, 27
Decalone, α-acyl-
ring contraction, **7**, 686
Decalone, 9-methyl-
synthesis
exocycloalkylation, **3**, 20
Reimer–Tiemann reaction, **2**, 773
Decalone, 10-methyl-
lithium enolate
alkylation, **3**, 2, 15
angular alkylation, **3**, 16
stabilized metal enolates
metallation, **3**, 55
synthesis
Reimer–Tiemann reaction, **2**, 773
1-Decalone, 9-nitro-
synthesis, **6**, 107
Decalones
reduction
enol ether preparation, **2**, 599
synthesis
regiospecific alkylation, **3**, 11
1-Decalones
alkylation, **3**, 11
enzymic reduction
specificity, **8**, 197
lithium 1(9)-enolate
angular alkylation, **3**, 16
reduction
dissolving metals, **8**, 120
dissolving metals/ammonia, **8**, 112
synthesis
exocycloalkylation, **3**, 20
TMS enol ether
phenylthiomethylation, **3**, 26
2-Decalones
enzymic reduction
specificity, **8**, 197
lithium 2-enolate
alkylation, **3**, 16, 21
lithium 1(2)-enolates
methylation, **3**, 15
3-substituted enolates
alkylation, diastereoselectivity, **3**, 55
cis-β-Decalones
synthesis
via Cope rearrangement, **5**, 814
Decanamide
hydrogenation, **8**, 248
Decane
autoxidation, **7**, 10
1,10-Decanedioic acid
dimethyl ester
intramolecular acyloin coupling reaction, **3**, 626
1,2-Decanediol
oxidative cleavage, **7**, 706
Decanesulfonic acid, perfluoro-
catalyst, solid superacid
Friedel–Crafts reaction, **3**, 298, 305
1-Decanol
synthesis
via hydrogenation, **8**, 236
5-Decanone, 6-benzylidene-
synthesis
via photocycloaddition, **5**, 163
Decarbonylation, **3**, 1015–1041
acyl radicals, **7**, 718
mechanism, **3**, 1020, 1040
reductive decarboxylation, **7**, 721
Decarboxylation
carboxyl radicals, **7**, 717
Decarboxylative amination, **7**, 729
Decarboxylative chalcogenation, **7**, 725
Decarboxylative fluorination
acyl hypofluorites, **7**, 723
Decarboxylative halogenation, **7**, 723
Decarboxylative iodination, **7**, 724
Decarboxylative oxygenation, **7**, 727
Decarboxylative phosphorylation, **7**, 725
Decarboxylative selenation, **7**, 726
Decarboxylative telluration, **7**, 726
Decatetraenes
electrocyclization, **5**, 743
1,6,8-Decatriene
Diels–Alder reactions
intramolecular, **5**, 522
1,7,9-Decatriene
Diels–Alder reactions
intramolecular, **5**, 519
geminal substituents

Diels–Alder reactions, **5**, 524
monosubstituted
Diels–Alder reactions, **5**, 533
1,7,9-Decatriene, sulfonyl-
Diels–Alder reactions
intramolecular, **5**, 522
Decatrienes
Diels–Alder reactions
diastereoselection, **5**, 515–527
twist asynchronicity, **5**, 516
heteroatom substituted
Diels–Alder reactions, **5**, 527–532
Decatrienones
Diels–Alder reactions
boat-like transition states, **5**, 539–543
Decatrien-3-ones
Diels–Alder reactions
intramolecular, **5**, 519
stereoselectivity, **5**, 518
1-Decene
benzene alkylation with
Friedel–Crafts reaction, **3**, 304
epoxidation, **7**, 375
oxidation
Wacker process, **7**, 451, 452
synthesis, **3**, 248
1-Decene, 10-nitro-
cyclization
via nitrile oxide, **4**, 1127
2-Decenoic acid, 10-hydroxy-
synthesis
Knoevenagel reaction, **2**, 381
Decyanation
isocyanides
tributylstannane, **8**, 830
5-Decyne
photolysis
with benzaldehyde, **5**, 163
reduction
dissolving metals, **8**, 479
Defucogilvocarcin V
synthesis, **7**, 347
Degradation reactions, **6**, 795–825
Dehydrating agents
enamine synthesis, **6**, 705
Dehydration
dienyliron complexes, **4**, 668
reduction
ketones, **8**, 924
Dehydrodimerization
alkanes, **7**, 5
Dehydrogenases
hydrogenation
unsaturated ketones, **8**, 561
Dehydrogenation
activated C—H bonds
oxidation, **7**, 119–146
alkanes
transition metal catalysis, **7**, 6
nitrogen compounds, **7**, 742
steroids
microbial, **7**, 66, 67
Dehydrohalogenation
mechanism, **7**, 122
Delphinine
synthesis, **6**, 402
Deltamethrin
synthesis
via chiral cyanohydrins, **1**, 546
Demercuration
acyloxymercuration
alkenes, **4**, 314–316
alkoxymercuration
alkenes, **4**, 309–312
amidomercuration
alkenes, **4**, 294
aminomercuration
alkenes, **4**, 290–292
azidomercuration
alkenes, **4**, 297
hydroxymercuration
alkenes, **4**, 300–305
peroxymercuration
alkenes, **4**, 306
reduction, **8**, 850
Demethylation
nucleoside 5′-phosphoric acid methyl ester, **6**, 624
Dendrobatid alkaloids
synthesis
enantioselective, **2**, 1028
Eschenmoser coupling reaction, **2**, 876
Dendrolasin
synthesis, **3**, 99; **6**, 145
from furan-3-carbaldehyde, **3**, 195
reduction of sulfides, **3**, 107
via carboalumination, **4**, 893
via tandem Claisen–Cope rearrangement, **5**, 879
Denudatin A
synthesis, **3**, 694
Deoxygenation
alcohols, **8**, 812
to alkanes, **8**, 811
benzoates
photosensitization, **8**, 817
carbonyl compounds
via hydrazones, **8**, 328
epoxides, **8**, 884
free-radical
alcohols, **8**, 818
Deoxymercuration, **8**, 853
Deplancheine
asymmetric synthesis, **3**, 81
synthesis
via iminium ion–vinylsilane cyclization, **1**, 592
Depresosterol
synthesis
use of homoenolates, **2**, 452
Deprotonation
donor radical cations, **7**, 877
radical cations
bimolecular reaction, **7**, 859
Depsipeptides
strained cyclic
synthesis, **6**, 638
Deselenation
nucleophilic attack, **8**, 847
Deselenative coupling
selenol esters, **6**, 475
Desilylation
π-allyl complexes
[3 + 2] cycloaddition reactions, **5**, 300
phosphonium salts

fluoride ion induced, **6**, 175
Desmosterol
synthesis, **3**, 427
Desmotroposantonin
synthesis, **3**, 804
Desulfurization
o-aminobenzyl sulfide, **8**, 976
benzylic compounds
rhodium complexes, **8**, 963
definition, **8**, 835
p-phenylsulfonylphenyl *p*-tolyl sulfide, **8**, 914
Deuterium
labeling
hydrozirconation, **8**, 691
Deuterolysis
demercuration, **8**, 850
Dewar benzene
1,4-bridged
synthesis, **3**, 872
rearrangement, **7**, 854
Dexamethasone
synthesis
industrial scale, **6**, 219
Dextrorphan
synthesis, **3**, 77
Diacetone alcohol
synthesis, **2**, 140
1,4-Diacetoxylation
cycloheptadienes
palladium catalysis, **4**, 686
palladium(II) catalysis, **4**, 565
Diacylation
Friedel–Crafts reaction, **2**, 712
C,N-Diacyliminium ions
addition reactions, **2**, 1074
intermolecular reactions, **2**, 1074
intramolecular reactions, **2**, 1076
Diacyl peroxides
allylic oxidation, **7**, 96
Dialdehydes
Henry reaction, **2**, 326
intramolecular aldol reaction, **2**, 156
Knoevenagel reaction, **2**, 365
Dialkyl arylphosphonates
synthesis
via $S_{RN}1$ reaction, **4**, 473
Dialkylation
1,2-dicarbanionic species, **4**, 976
enolates
equilibration, **3**, 4
2,3-Dialkylation
alkadienoates, **4**, 253
2,6-Dialkylation
alkadienoates, **4**, 253
cis-Dialkylation
Michael acceptors, **4**, 243
Dialkylative enone transposition, **7**, 615
Dialkyl phosphates
esterification, **6**, 615
Dialkyl phosphonates
alkylation, **3**, 201
Dialkyl sulfates
amide alkylation, **6**, 503
Diallenes
synthesis
via dihalocyclopropanes, **4**, 1010
vic-Dials
synthesis, **7**, 307
Diamantane
synthesis
Friedel–Crafts reaction, **3**, 334
Diamination
alkenes
palladium(II) catalysis, **4**, 560
Diamines
chiral auxiliary
aldol reaction, **2**, 233
chiral catalysts
enantioselective addition of alkyllithium to aldehydes, **1**, 72
synthesis, **7**, 479
via alkenes, **7**, 484
via aziridine ring opening, **7**, 487
via reductive cleavage of cyclic hydrazines, **8**, 388
1,2-Diamines
coupling reactions
with imines, **3**, 564
reactions with iminium salts, **6**, 515
synthesis, **6**, 94
Diamines, vicinal
synthesis
via Diels–Alder reactions, **5**, 426
2,2′-Diaminobiphenyl
phosphoric acid protecting group, **6**, 625
Diamondoid hydrocarbons
synthesis
Friedel–Crafts reaction, **3**, 334
Dianions
γ-acylation, **2**, 832
Dianthranol
synthesis
via photolysis, **5**, 729
Dianthrones
photolysis, **5**, 729
Diarylamines
synthesis, **4**, 434
Diarylboryl hexachloroantimonate
catalyst
Friedel–Crafts reaction, **2**, 744
Diaryne
polyhalogenated arenes as equivalents, **4**, 496
Diastereofacial differentiation
asymmetric synthesis, **3**, 72
Diastereofacial selectivity
aldol reaction, **2**, 217, 218
allyl organometallic compounds, **2**, 2
reaction with aldimines, **2**, 978
chiral auxiliaries
aldol reaction, **2**, 232
Cram's rule
chiral electrophiles, **2**, 639
cyclopropanes, **4**, 952
Diels–Alder reactions, **2**, 677
Lewis acids, **2**, 678
in enolate–imine condensations
Mannich reaction, **2**, 922
Diastereoselective addition
achiral carbon nucleophiles
chiral alkenes, **4**, 200–218
Diastereoselective reactions
allenyl organometallics, **2**, 91–96
propargyl organometallics, **2**, 91–96

Diastereoselectivity
relative
allyl organometallics, **2**, 3, 24–33
simple
allyl organometallics, **2**, 2–24
Diazaalkenes
insertion reactions, **3**, 1049
1,8-Diazaanthraquinone
synthesis, **7**, 355
Diazaazulene
synthesis
via [6 + 4] cycloaddition, **5**, 627
1,2-Diazabicyclo[3.1.0]hex-2-ene
synthesis
via diazoalkane cycloaddition, **4**, 1103
Diazabicyclooctane
in sulfide metallation, **3**, 86
1,4-Diazabicyclo[2.2.2]octane
reduction
aluminum hydrides, **8**, 543
thioallyl anion preparation, **2**, 71
1,2-Diaza-1,3-butadienes
Diels–Alder reactions, **5**, 486
hetero
Diels–Alder reactions, **5**, 486
1,3-Diaza-1,3-butadienes
Diels–Alder reactions, **5**, 486
1,4-Diaza-1,3-butadienes
Diels–Alder reactions, **5**, 486
2,3-Diaza-1,3-butadienes
Diels–Alder reactions, **5**, 491
Diazanaphthalenes
oxidation
hydrogen peroxide and sodium tungstate, **7**, 750
2,3-Diaza-5-norbornene
Pauson–Khand reaction, **5**, 1050
Diaza[3.2.1]octane
synthesis
via azomethine imine, **4**, 1096
1,3,2-Diazaphosphorinane, 2-benzyl-2-oxo-
lithium carbanion
crystal structure, **1**, 36
Diazaquinomycin A
synthesis, **7**, 355
Diazaspiroalkanes
reduction, **8**, 229
1,1-Diazene
synthesis
via oxidation of 1,1-disubstituted hydrazines, **7**, 742
Diazenes
deamination, **8**, 828
Diazides
synthesis, **7**, 487
via acetals, **6**, 254
via dihalides, **6**, 247
Diazine
oxidation, **7**, 750
1,2-Diazines
Diels–Alder reactions, **5**, 491
synthesis
via retro Diels–Alder reactions, **5**, 583
1,3-Diazines
Diels–Alder reactions, **5**, 491
1,4-Diazines
Diels–Alder reactions, **5**, 491
Diaziridine, 1-benzyl-
reaction with lithium dimethylcuprate, **6**, 95
Diaziridines
synthesis
via imines and oximes, **1**, 838
Diazirine, chloro-
synthesis
via oxidation of amidines, **7**, 739
Diazirines
carbene precursors, **4**, 961
synthesis
via imines and oximes, **1**, 838
Diazoacetates
ketocarbene precursors, **4**, 1033
Diazoacetoacetates
ketocarbene precursors, **4**, 1033
α-Diazo aldehydes
synthesis, **3**, 890
Diazoalkanes
chain extension, **1**, 844
[3 + 2] cycloaddition reactions
alkynyl complexes, **5**, 1070
cyclopropane synthesis, **4**, 953–961
1,3-dipolar cycloadditions, **4**, 1101–1104
epoxidations, **1**, 832
higher
synthesis, **6**, 121
photochemical reactions with alkenes, **4**, 954
properties, **6**, 120
reactions with alkenes
metal-catalyzed, **4**, 954–961
reactions with carboxylic acids, **6**, 337
synthesis, **6**, 120, 778
Diazoalkanes, aryl-
synthesis, **6**, 121
Diazoalkenes
aryl-bridged
cyclizations, **4**, 1153
cyclization, **4**, 1151–1155
open-chain
cyclizations, **4**, 1152
3-Diazoalkenes
cyclization, **4**, 1156
Diazoalkynes
cyclization, **4**, 1156
Diazo compounds
C—H insertion reactions, **6**, 127
cyclization, **4**, 1151–1157
decomposition
catalysts, **4**, 1032
heteroatom–hydrogen insertion reactions, **6**, 127
ketocarbenes from, **4**, 1032
reaction with sulfenyl halides
formation of α-chlorosulfides, **7**, 213
reduction
synthesis of hydrazines, **8**, 382
reductive cleavage
synthesis of amines, **8**, 383
synthesis
via oxidation of hydrazones, **7**, 742
via oximes, **7**, 751
synthetic applications, **6**, 126
Diazo compounds, alkyl-
Diels–Alder reactions, **5**, 430
Diazo coupling
amide halide synthesis, **6**, 499

Diazocycloalkenes
cyclizations, **4**, 1154
2,3-Diazo-Dewar benzene
nitrogen extrusion, **5**, 568
Diazo esters
photolysis, **3**, 894
Diazo groups
nitrile synthesis, **6**, 239
Diazo insertion reactions
rhodium-catalyzed, **3**, 1051
Diazo ketones
Darzens glycidic ester condensation, **2**, 422
ketocarbene precursors, **4**, 1033
secondary
synthesis, **3**, 889
synthesis
via oxidation of 1,2-diketone monohydrazones, **7**, 742
α,β-unsaturated
synthesis, **3**, 890
β,γ-unsaturated
vinylogous Wolff rearrangement, **3**, 906
UV spectra, **3**, 891
α-Diazo ketones
rearrangements, **3**, 887
synthesis, **3**, 888
Diazo ketones, epoxy
preparation
Darzens glycidic ester condensation, **2**, 422
Diazomalonates
ketocarbene precursors, **4**, 1033
Diazomalonic acid
dimethyl ester
deoxygenation, epoxides, **8**, 890
Diazomethane
cycloaddition reactions
fulvenes, **5**, 630
reaction with enamines
synthesis of substituted cyclopropanes, **6**, 716
reaction with hydroalumination adducts, **8**, 756
synthesis, **6**, 120
Diazomethane, α-acyl-
reactions with aliphatic ketones
hydroxide-catalyzed, **1**, 846
Diazomethane, phenyl-
cycloaddition with styrene, **4**, 1103
Diazomethane, phenylsulfonyl-
reactions with cyclohexanones, **1**, 851
Diazomethane, trimethylsilyl-
[3 + 2] cycloaddition reaction
alkynyl carbene complexes, **5**, 1070
reaction with methyl tetrolate, **5**, 1070
synthesis, **6**, 121
trifluoroborane complex
2-methylcyclohexanone homologation, **1**, 851
Diazomethane, vinyl-
ketocarbene precursors, **4**, 1033
Diazonium salts
amine deamination
alcohol synthesis, **6**, 3
bromination, **6**, 211
chlorination, **6**, 208
Diels–Alder reactions, **5**, 430
fluorination, **6**, 220
iodination, **6**, 215
reduction, **8**, 916
synthesis of hydrazines, **8**, 382
reductive cleavage, **8**, 383
Diazonium salts — *see also* Arenediazonium salts
Diazonium salts, aryl-
coupling reactions
with alkenes, **3**, 497
hydride acceptors, **8**, 91
synthesis, **7**, 340
Diazonium tetrafluoroborate
synthesis
via diazotization, **7**, 740
o-Diazo oxides
photochemical decomposition, **3**, 904
Diazopyruvates
ketocarbene precursors, **4**, 1033
Diazotization
amines
primary, **7**, 740
Diazo transfer
active methylene compounds, **6**, 125
sulfonyl azides, **4**, 1033
Diazotype offset photocopying
o-diazo oxides, **3**, 904
9-Diazoxanthene
cycloaddition with methyl acrylate, **4**, 1103
DIBAL-H — *see* Aluminum hydride, diisobutyl-
Dibenzobarrelene
photoisomerization, **5**, 198
substituted
photoisomerizations, **5**, 210
6,7,8,9-Dibenzobicyclo[3.2.2]nonan-3-one
synthesis
via [4 + 3] cycloaddition, **5**, 608
Dibenzochroman-4-one
synthesis
Friedel–Crafts reaction, **2**, 759
Dibenzo-18-crown-6
polymer supported
catalyst, Knoevenagel reaction, **2**, 345
Dibenzocyclobutane
side product in coupling reaction, **3**, 505
Dibenzo[*a,e*]cyclooctadiene, 5,6,11,12-tetrahydro-
synthesis
via [4 + 4] cycloaddition, **5**, 639
Dibenzocyclooctanone
hydrazone
reduction, Henbest modification, **8**, 336
2,3,6,7-Dibenzocyclooctanone
synthesis
Friedel–Crafts reaction, **2**, 753
Dibenzofurans
photochemical ring opening, **5**, 712
reduction
dissolving metals, **8**, 626
Dibenzofurans, hydroxy-
synthesis
via FVP, **5**, 732
Dibenzophosphepin
C—P bond cleavage, **8**, 864
Dibenzophosphole oxide
Horner reaction, **1**, 776
Dibenzophosphole ylides
alkene synthesis
(*E*)-selective, **1**, 758
Dibenzo systems
photoisomerization, **5**, 197

Dibenzothiepinones
synthesis
Friedel–Crafts reaction, **2**, 765
Dibenzothiophene
Birch reduction, **8**, 629
electrochemical reduction, **8**, 611
methylation, **3**, 456
Dibenzoyl peroxydicarbonate
α-hydroxylation
oxazolidinones, **7**, 184
Dibenzylamine, *N*-nitroso-
synthesis
via oxidative deacylation, **7**, 749
1,6-Diboracyclododecane
synthesis, **8**, 707
Diborane
hydroboration, **8**, 705
imine reduction, **6**, 724
reaction with organometallic compounds, **7**, 595
reduction
acyl halides, **8**, 240, 263
carbonyl compounds, **8**, 1, 315
carboxylic acids, **8**, 237, 261
epoxides, **8**, 875
lactones, **8**, 269
1,1-Diboryl compounds
oxidation
alcohol formation, **7**, 596
synthesis and cleavage, **1**, 489
1,2-Diboryl compounds
oxidation
formation of alkenes, **7**, 601
Diboryl enediolates
aldol reaction, **2**, 245
Dibromides
geminal
double alkyl substitution, **3**, 216
vicinal
reaction with dialkyl cuprates, **3**, 216
reduction, **8**, 797
reduction with tributylstannanes, **8**, 798
α,α-Dibromo compounds
ketocarbenes from, **4**, 1032
Dibromohydrins
rearrangement, **1**, 874
Di-*t*-butylamine
synthesis, **7**, 737
Dibutylamine, *N*-chloro-
reaction with butadiene, **7**, 505
Di-*t*-butylsilylene group
diol protection, **6**, 662
Dicarbacondensation
definition, **4**, 238
1,2-Dicarbanionic compounds
dialkylation, **4**, 976
Dicarbonyl compounds
methylenation
Tebbe reagent, **1**, 743
monoprotection, **6**, 684
1,2-Dicarbonyl compounds
Baeyer–Villiger reaction, **7**, 684
diazo-coupling reactions, **3**, 893
oxidation, **7**, 153
oxidative cleavage, **7**, 709
synthesis, **7**, 439, 664
1,3-Dicarbonyl compounds
α-alkenylation, **7**, 620
alkylation, **3**, 54
α-alk-1-ynylation, **7**, 620
aromatic $S_{RN}1$ reactions, **4**, 467
α-arylated
synthesis *via* $S_{RN}1$ reaction, **4**, 467
cyclic 2-diazo-
Wolff rearrangement, **3**, 903
dehydrogenation, **2**, 388
dianions
γ-alkylation, **3**, 58
dienolates
γ-alkylation, **3**, 1
selenenylation, **7**, 131
synthesis
Eschenmoser coupling reaction, **2**, 865
1,4-Dicarbonyl compounds
dehydrogenation
use of selenium dioxide, **7**, 132
synthesis
conjugate addition, **2**, 330
via Claisen rearrangement, **6**, 860
via Wacker oxidation, **7**, 455
synthesis, **5**, 941
use of cyclopropanes, **5**, 903
1,5-Dicarbonyl compounds
synthesis
conjugate addition, **2**, 331
Eschenmoser coupling reaction, **2**, 875, 876
from hydrazones, **2**, 517
via Claisen rearrangement, **6**, 860
via Wacker oxidation, **7**, 458
1,2-Dicarbonyl compounds, 1-alkyl-2-aryl-
benzylic rearrangement, **3**, 829
1,3-Dicarbonyl compounds, alkylidene-
Diels–Alder reactions, **5**, 467
1,3-Dicarbonyl compounds, arylidene-
Diels–Alder reactions, **5**, 467
1,3-Dicarbonyl compounds, 2-ethynyl-
synthesis, **3**, 286
1,3-Dicarbonyl compounds, 5-nitro-
Henry reaction
cyclization, **2**, 334
1,3-Dicarbonyl compounds, 6-nitro-
Henry reaction
intramolecular, **2**, 334
Dicarbonyl compounds, α-seleno-
oxidative *syn* elimination, **2**, 388
synthesis, **2**, 388
Dicarboxylation
alkenes, **4**, 946–949
mechanism, **4**, 946
Dicarboxylic acids
monodecarboxylation, **7**, 727
1,4-Dicarboxylic acids
di-*t*-butyl peroxy esters
pyrolysis, **7**, 722
oxidative decarboxylation, **7**, 722
Dichloramine-T
reaction with trialkylborane, **7**, 604
Dichlorides
geminal
solvolysis, divinyl ketones from, **5**, 770
vicinal
reduction with tributylstannane, **8**, 798
Dichlorine oxide

ene-type chlorination, **7**, 537
Dichlorohydrins
rearrangement, **1**, 873
Dichromates
oxidation
halides, **7**, 663
Diconiferyl alcohol, dehydro-
synthesis, **3**, 693
Dicranenone A
synthesis
via cyclopropane ring opening, **4**, 1046
Dictyopterene
synthesis
via Cope rearrangement, **5**, 803
Dictyopterene C′
synthesis
via Cope rearrangement, **5**, 973
Dicyanogen triselenide
decarboxylative selenation, **7**, 726
Dicyanomethylation
aryl iodides
with sodium salt of malononitrile, **3**, 454
Dicyclododecyl tartrate
asymmetric epoxidation
kinetic resolution, **7**, 395
kinetics, **7**, 413
Dicyclohexylamine
Mannich reaction with phenols
steric hindrance, **2**, 956
Dicyclohexyl tartrate
asymmetric epoxidation, **7**, 395
kinetics, **7**, 411
Dicyclopenta[*a,d*]cyclooctanes
synthesis
via divinylcyclobutane rearrangement, **5**, 1029
Dicyclopentadiene
hydroformylation, **4**, 922
hydrosilylation, **8**, 781
oxidation
palladium(II) catalysis, **4**, 559
oxidative cleavage
potassium permanganate, **7**, 559
reaction injection molding, **5**, 1120
reactions with nitrogen oxides, **7**, 488
ring opening metathesis polymerization, **5**, 1120
Dicyclopentadiene, tetrahydro-
isomerization, **7**, 5
Dicyclopentylamine
Mannich reaction with phenols
steric hindrance, **2**, 956
Dicyclopropylimine
rearrangement, **5**, 945
Didemnin
antibiotics
synthesis, **6**, 446
Dideoxygenation
vicinal, **8**, 818
Dideuteriomethylenation
modified Tebbe reagent, **5**, 1125
Dieckmann reaction, **2**, 806
ester groups
regioselectivity, **2**, 815
β-heteroatoms
regioselectivity, **2**, 812
in synthesis, **2**, 822
mechanism, **2**, 797, 806
Baldwin's rules, **2**, 807
regioselectivity, **2**, 808
retro, **2**, 855
ring size, **2**, 808
α-substitution
regioselectivity, **2**, 811
β-substitution
regioselectivity, **2**, 813
tandem reaction, **2**, 852
thiocarboxylic esters, **6**, 446
Diels–Alder reactions
activated dienes, **2**, 662
activation enthalpy, **5**, 317
activation entropy, **5**, 317
anthracene with maleic anhydride
aluminum chloride catalysis, **1**, 284
asymmetric, **4**, 1079
mechanisms, **1**, 311
benzannulation
tandem, **5**, 1099
benzynes
regiochemistry, **5**, 390–393
stereochemistry, **5**, 390–393
cationic heterodienes, **5**, 492–507
chiral catalysts, **5**, 376–379
absolute stereochemistry, **2**, 681
clays as promoters, **5**, 345
diastereoface selective, **5**, 347–352
diastereoselectivity
conformational effects, **5**, 526
relative, **5**, 514, 532–543
simple, **5**, 514, 515–532
enantioselective
chiral Lewis acids, **2**, 654
endocyclic dienophiles, **5**, 371
π-facial control, **5**, 319
heterodienes, **5**, 451–507
heterodienophiles, **5**, 320–328, 402–444
high pressure reactions, **5**, 341–343
mechanism, **2**, 664
imines
intramolecular, **5**, 413–416
intermolecular, **5**, 315–396
reactivity, **5**, 316
intramolecular, **5**, 513–546
asymmetric, **5**, 543–546
carbonyl compounds, **5**, 435
diastereoselectivity, **5**, 522–526
Lewis acid catalyzed, **5**, 519–522
inverse electron demand, **5**, 452
Lewis acid catalysts, **2**, 663
Lewis acid catalyzed reactions, **5**, 339–346
conditions, **2**, 673
mechanism, **2**, 665
mechanism, **2**, 664; **5**, 451
medium promoted, **5**, 339–346
neutral, **5**, 452
normal, **5**, 452
radical cyclizations, **4**, 791
regiochemistry, **5**, 317
retrograde, **5**, 551–589
intramolecular, **5**, 584–587
silica gel as promoter, **5**, 345
stereochemistry, **5**, 318
stereoface selective, **5**, 352–379
tandem ene reactions

alkynes, **5**, 7
thermal reactions
mechanism, **2**, 664
transannular cycloadditions, **5**, 532
ultrasound promoted, **5**, 341–343
volume of activation, **5**, 458
water promoted, **5**, 344
zeolites as promoters, **5**, 345
Dienamides
Diels–Alder reactions, **5**, 331
Dienamines
acrylic
Diels–Alder reactions, **5**, 331
Vilsmeier–Haack reaction, **2**, 783
2,4-Diene-1,6-diones
synthesis
via dienetricarbonyliron complexes, **4**, 701
Dienes
activated
Diels–Alder reaction, **2**, 662
reactions with aldehydes, **2**, 661–706
acyloxymercuration, **4**, 315
addition reactions
carbon-centered radicals, **4**, 765
alkoxymercuration, **4**, 311
alkylation
via iron carbonyl complexes, **4**, 580–582
allylic hydroxy
stereospecific synthesis, **8**, 727
amidomercuration, **4**, 295
arene alkylation
Friedel–Crafts reaction, **3**, 322
arylation
palladium complexes, **4**, 849
autoxidation, **7**, 861
bicyclization, **5**, 1172
boron-substituted
Diels–Alder reactions, **5**, 335–337
carboalumination, **4**, 887
carbolithiation, **4**, 867–872
carbonyl derivatives
cycloaddition reactions, **6**, 757
chiral
Diels–Alder reactions, **5**, 348–350, 373–376
conjugated
addition reactions with selenium electrophiles, **7**, 520
alkoxymercuration, **4**, 311
aminomercuration–demercuration, **4**, 291
anodic oxidation, **7**, 795
[3 + 2] cycloaddition reactions, **5**, 297
hydrobromination, **4**, 283
hydrocarboxylation, **4**, 945
hydrochlorination, **4**, 276
hydroformylation, **4**, 922
hydrogenation, **8**, 433
hydrogenation mechanism, **8**, 433
partial reduction, **8**, 564
reaction with chlorosulfonyl isocyanate, **5**, 105
Ritter reaction, **4**, 293
stereospecific synthesis, **4**, 1020
synthesis, **3**, 878
thiylation, **4**, 317
conjugated acyclic
Pauson–Khand reaction, **5**, 1044
cyclic
photoisomerizations, **5**, 196
cyclic hydroboration, **8**, 709, 711
cycloaddition reactions
fulvenes, **5**, 626–630
tropones, **5**, 618–625
[3 + 2] cycloaddition reactions, **5**, 307
[4 + 3] cycloaddition reactions, **4**, 1075; **5**, 601
cyclopropanation
regioselectivity, **4**, 1035
dialkoxylation
palladium(II) catalysis, **4**, 565
1,4-diamination, **7**, 504
from π-allylpalladium complexes, **4**, 608–610
functionalized
carbomagnesiation, **4**, 877
heteroatom-substituted
Diels–Alder reactions, **5**, 328–339
heterodienophile additions, **5**, 401–444
hydroalumination
locoselectivity, **8**, 742
hydroboration, **8**, 705, 707, 716
hydrofluorination, **4**, 271
hydrogenation
regioselectivity, **8**, 433
stereoselectivity, **8**, 433
to saturated hydrocarbons
homogeneous catalysis, **8**, 449
hydroiodination, **4**, 288
hydroxymercuration–demercuration, **4**, 303
hydrozirconation, **8**, 676, 684
iron tricarbonyl complexes
acylation, **2**, 721
metal-activated
heteroatom nucleophilic addition, **4**, 565
monoepoxides
rearrangement, **3**, 770
'skipped'
synthesis, **3**, 244, 265
nitrogen-substituted
Diels–Alder reactions, **5**, 331–333
nonconjugated
alkoxymercuration, **4**, 311
aminomercuration–demercuration, **4**, 291
hydroboration, **8**, 714
hydrobromination, **4**, 283
hydrochlorination, **4**, 276
hydroformylation, **4**, 922
reactions with hydrogen sulfide, **4**, 317
Ritter reaction, **4**, 293
oxidation
singlet oxygen, **7**, 97
oxidative rearrangement, **7**, 832
oxygen-substituted
Diels–Alder reactions, **5**, 329–331, 434
peroxymercuration–demercuration, **4**, 307
polycyclic
photoisomerization, **5**, 196
radical cyclization
carbon-centered radicals, **4**, 789
reactions with π-allylpalladium complexes
regioselectivity, **4**, 643
reactions with carbon electrophiles
transition metal catalysis, **4**, 695–712
reactions with carboxylic acids, **4**, 313
1,2-reduction to alkenes
homogeneous catalysis, **8**, 449

1,4-reduction to alkenes
homogeneous catalysis, **8**, 451
regioselective hydroxylation, **7**, 438
silicon-substituted
Diels–Alder reactions, **5**, 335–337
sulfur-substituted
Diels–Alder reactions, **5**, 333
synthesis, **8**, 727
3,2-sigmatropic rearrangement, **3**, 964
via hydroalumination, **8**, 757
via β-hydroxyalkyl selenides, **1**, 705
via Julia coupling, **1**, 800
tin-substituted
Diels–Alder reactions, **5**, 335–337
1,4-transfer of chirality
palladium(II) catalysis, **4**, 576
unactivated
photocycloaddition reactions, **5**, 145
vinylation
palladium complexes, **4**, 839, 855
1,2-Dienes
chloropalladation, **4**, 565
1,3-Dienes
acylation
Friedel–Crafts reaction, **2**, 720, 721
π-allylpalladium complexes from, **4**, 587
carbocupration, **4**, 895
conjugated
heteroatom nucleophilic addition, **4**, 565
coupling with carbene complexes, **5**, 1084
cyclic
[4 + 3] cycloaddition reactions, **5**, 603–605
cycloaddition reactions with alkynyl carbene complexes, **5**, 1072
1,4-diazides from, **7**, 504
hydroboration, **8**, 720
hydrosilylation, **8**, 778
hydrozirconation
regioselectivity, **8**, 685
insertion reactions
allylpalladium compounds, **5**, 35
open-chain
[4 + 3] cycloaddition reactions, **5**, 603
photocycloaddition reactions, **5**, 635–638
protection, **6**, 690
reaction with π-allylpalladium complexes, **4**, 601
reaction with dihalocarbenes, **4**, 1002
reaction with 5-ethoxy-2-pyrrolidinone, **2**, 1057
reaction with iron carbene complexes, **5**, 1088
reaction with Kolbe radicals, **3**, 647
reaction with trifluoroacetyl nitrate, **7**, 505
synthesis
copper catalysts, **3**, 217
from alkenes, **3**, 879
organopalladium catalysis, **3**, 232
via allylic alcohols, **6**, 154
via carboalumination, **4**, 889
via cyclobutenes, **5**, 683–686, 1030
via Pauson–Khand reaction, **5**, 1039, 1043
Vilsmeier–Haack reaction, **2**, 782
(*E*)-1,3-Dienes
synthesis
allylic anions, **2**, 65
(*Z*)-1,3-Dienes
synthesis
allylic anions, **2**, 65
1,4-Dienes
acyclic
photoisomerizations, **5**, 195
oxidation
pyridinium dichromate, **7**, 276
photoisomerization, **5**, 194–213
cis–trans, **5**, 207
retro-ene reaction, **5**, 907
synthesis, **3**, 249
coupling reactions of allylic halides, **3**, 473
via π-allylpalladium complexes in, **4**, 595
via boron-ene reaction, **5**, 34
via palladium-ene reactions, **5**, 56
1,5-Dienes
addition reactions
nitrogen nucleophiles, **4**, 562
carboalumination, **4**, 887
cyclic
synthesis, **3**, 429
medium rings
conformation, **3**, 386
synthesis, **3**, 428
organopalladium catalysis, **3**, 231
phosphonium ylide alkylation, **3**, 201
via palladium-ene reaction, **5**, 48
thermal rearrangement, **5**, 786
1,6-Dienes
ene reactions, **5**, 10–15
1,7-Dienes
ene reactions
intramolecular, **5**, 17
synthesis, **1**, 663
α,ω-Dienes
cycloaddition with nickel(0) complexes, **5**, 1131
dihydroboration, **8**, 714
hydroboration, **8**, 711
synthesis
alkene metathesis, **5**, 1117
(*E,Z*)-Dienes
synthesis
via Horner reaction, **1**, 779
Dienes, acetoxy-
palladiumene reactions, **5**, 46
Dienes, acyl-
synthesis
carbonylation, **3**, 1024
1,5-Dienes, 1-acyl-
Cope rearrangement
catalysis, **5**, 799
1,5-Dienes, 2-acyl-
Cope rearrangement
catalysis, **5**, 798
1,5-Dienes, 3-acyl-
Cope rearrangement
catalysis, **5**, 799
Dienes, *N*-acylamino-
Diels–Alder reactions, **5**, 331
1,3-Dienes, *N*-acylamino-
synthesis
via thermal rearrangement, **6**, 843
Dienes, 2-alkoxy-
Diels–Alder reactions, **5**, 348
1,3-Dienes, 1-amino-
Diels–Alder reactions, **5**, 331
Dienes, conjugated
acylation

Friedel–Crafts reaction, **2**, 720
1,3-Dienes, 2-dialkylamino-
Diels–Alder reactions, **5**, 331
1,4-Dienes, 2-ethoxy-
synthesis
via boron-ene reaction, **5**, 34
1,3-Dienes, 3-hydroxy-
synthesis
via α,β-unsaturated aldehydes, **7**, 458
1,3-Dienes, 1-iodo-
carbonylation
palladium catalysts, **3**, 1030
1,4-Dienes, 1-iodo-
carbonylation
palladium catalyst, **3**, 1025
Dienes, 1-(*O*-methylmandeloxy)-
Diels–Alder reactions, **5**, 373
1,3-Dienes, 1-nitro-
synthesis, **6**, 109
via electrophilic nitration, **4**, 356
Dienes, phenylsulfonyl-
synthesis, **7**, 519
1,3-Dienes, 2-(phenylsulfonyl)-
Diels–Alder reactions, **5**, 333
Dienes, siloxy-
Diels–Alder reactions, **5**, 329, 407
1,3-Dienes, 1-silyl-
formylation, **2**, 728
synthesis
organopalladium catalysis, **3**, 232, 233
1,3-Dienes, 2-(silylmethyl)-
Diels–Alder reactions, **5**, 337–339
Dienes, silyloxy-
synthesis
via γ-silylated vinylcopper, **1**, 428
1,3-Dienes, 2-(stannylmethyl)-
Diels–Alder reactions, **5**, 337–339
1,3-Dienes, 2-(thiomethyl)-
Diels–Alder reactions, **5**, 337–339
Dienes, trimethylsilyloxy-
Diels–Alder reaction, **2**, 663
Dienoates
synthesis
via nickel-ene reaction, **5**, 36
Dienoic acids
synthesis, **3**, 882
2,5-Dienoic acids
synthesis
nickel-catalyzed carbonylation, **3**, 1027
Dienolates
addition reactions, **4**, 106–111
copper(I)
alkylation, **3**, 50
diastereoselective alkylation
solvent effects, **3**, 24
extended
alkylation, stereochemistry, **3**, 23
monoalkylation, **3**, 23
heteroannular extended
equatorial alkylation, **3**, 24
Michael additions, **4**, 30
α,β-unsaturated carboxylic acids
alkylation, **3**, 50
Dienols
hydrozirconation
regioselectivity, **8**, 686
synthesis
via vinyl epoxides, **6**, 11
1,5-Dien-3-ols
synthesis
Wittig rearrangement, **3**, 994
β,β′-Dienols
synthesis
via oxidation of β-hydroxy-γ-alkenyl selenides, **1**, 709
β,δ-Dienols
synthesis
via 1-lithio-3-alkenyl phenyl selenoxides, **1**, 709
Dienone–phenol rearrangements, **3**, 803–820
intracyclic migrations, **3**, 804
Dienones
conjugated
synthesis, **6**, 841
epoxidation, **7**, 372
Robinson annulation, **4**, 8
2,4-Dienones
rearrangements, **3**, 803
α,β,δ,γ-Dienones
synthesis
via conjugate additions, **4**, 147
o-Dienones
synthesis
via Claisen rearrangement, **5**, 834
Dienophiles
chiral
Diels–Alder reactions, **5**, 350–352, 354–373
Diels–Alder reactions
intramolecular, **5**, 531
Dienyl systems
Paterno–Büchi reaction, **5**, 165–168
Dienynes
conjugated
one-pot synthesis, **3**, 539
divinyl ketones from
cyclization, **5**, 767
hydration, **5**, 752
1,11-Dien-6-ynes
intramolecular [2 + 2 + 2] cycloaddition, **5**, 1141
1,13-Dien-7-ynes
intramolecular [2 + 2 + 2] cycloaddition, **5**, 1141
Diesters
synthesis
oxidative carbonylation of alkynes, **3**, 1030
unsymmetrical
acylation, **2**, 799
vicinal
reduction, **3**, 614
β-Diesters
enolates
reaction with allylic acetate, **3**, 56
metal enolates
alkylation, **3**, 54
Diesters, 2-aryl-4-oxo-
synthesis
use of SAMP/RAMP, **2**, 520
Diethylamine
Mannich reaction with phenols
steric hindrance, **2**, 956
reaction with 2-naphthol and benzaldehyde
Mannich reaction, **2**, 960
Diethylamine, trimethylsilyl-
alcohol protection, **6**, 653

Diethyl benzenedicarboxylate
reduction
electrochemical, **8**, 243
Diethyl carbonate
alkoxycarbonylation
ketones, **2**, 839
Diethyl dicarbonate
alkoxycarbonylation
ketones, **2**, 839
Diethyl malonate
proton donor
electroreduction of retinal, **8**, 134
Diethyl phenylphosphonate
synthesis
via $S_{RN}1$ reaction, **4**, 473
Diethyl phthalate
reduction
electrochemical, **8**, 243
titanium tetrachloride complex
crystal structure, **1**, 303
Diethyl tartrate
asymmetric epoxidation, **7**, 395
hydrogenation, **8**, 242
Diethyl tetradecanedioate
hydrogenation, **8**, 242
α,ω-Diethynyl monomers
oxidative polymerization, **3**, 557
Difluoramine
deamination, **8**, 829
Dihalides
vicinal
reduction, **8**, 803
Dihalocyclopropanation
alkenes, **4**, 1002
Dihalohydrins
rearrangement, **1**, 873
Dihydrofolate reductase
inhibitors
synthesis, Knoevenagel reaction, **2**, 385
p-(Dihydroxyboryl)benzyloxycarbonyl group
amine-protecting group
cleavage, **6**, 639
Diimidazole, carbonyl-
amide synthesis, **6**, 389
Diimidazole, *N,N'*-oxalyl-
amide synthesis, **6**, 389
Diimide, dideuterio-
synthesis, **8**, 473
Diimides
disproportionation, **8**, 473
isomers
relative energies, **8**, 473
reduction
mechanism, **8**, 472
stereoselectivity, **8**, 475
unsaturated hydrocarbons, **8**, 472
relative reactivities
reduction, **8**, 474
synthesis, **8**, 472
Diimides, alkoxycarbonylaroyl-
Diels–Alder reactions, **5**, 486
Diimides, arylaroyl-
Diels–Alder reactions, **5**, 486
Diimides, diaroyl-
Diels–Alder reactions, **5**, 486
1,3-Diimines
reduction
dissolving metals, **8**, 124
Diisoeugenol, dehydro-
synthesis, **3**, 693
Diisophorane
Ritter reaction, **6**, 268
Diisopropylamine
Mannich reaction with phenols
steric hindrance, **2**, 956
Diisopropylsilylene group
diol protection, **6**, 662
10-Diisopropylsulfonamide isobornyl esters
β-lactams from, **2**, 924
Diisopropyl tartrate
asymmetric epoxidation, **7**, 395
Diketene
coupling reactions
with sp^3 organometallics, **3**, 446
synthesis, **6**, 332
Diketone, α-(4-isobutylphenyl)-β-methyl
benzylic acid rearrangement, **3**, 829
Diketones
aldol cyclization, **2**, 161–166
bicyclic
enzymic reduction, specificity, **8**, 201
decarbonylation, **3**, 1041
dianions
aldol reactions, **2**, 189
enzymic reduction
enantiotropic specificity, **8**, 188
specificity, **8**, 193
macrocyclic
transannular aldol cyclization reactions, **2**, 169
monocyclic
enzymic reduction, specificity, **8**, 201
polycyclic
enzymic reduction, specificity, **8**, 201
Reformatsky reaction, **2**, 283
synthesis
bis-dithiane dialkylation, **3**, 128
1,2-Diketones
alicyclic
benzilic acid rearrangement, **3**, 831
ring contraction rearrangement, **3**, 831
aliphatic
benzilic acid rearrangement, **3**, 831
alkynes from, **8**, 950
aromatic
DMSO oxidation, **7**, 295
Knoevenagel reaction, **2**, 367
monoketals
reactions with arynes, **4**, 496
reactions with π-allylnickel halides, **3**, 424
synthesis
carbonylation of lithium amides, **3**, 1017
via acylstannanes, **1**, 438
via benzoin condensation, **1**, 546
via Kornblum oxidation, **7**, 654
via organosamarium compounds, **1**, 273
via Swern oxidation, **7**, 300
1,3-Diketones
γ-alkylation, **3**, 58
cleavage, **2**, 855
cyclic enolates
alkylation, **3**, 55
dianions

aldol reactions, **2**, 189
disilyl enol ethers, **2**, 605
enantioselective hydrogenation, **8**, 151
Knoevenagel reaction
α,β-unsaturated products, **2**, 357
Mannich reaction
with preformed iminium salts, **2**, 904
metal enolates
alkylation, **3**, 54
monoreduction, **8**, 938
reduction, **8**, 13
Clemmensen reduction, **8**, 312
electrolysis, **8**, 321
sulfenylation, **7**, 125
synthesis
use of hydrazone anions, **2**, 516
Vilsmeier–Haack reaction, **2**, 786
1,4-Diketones
Clemmensen reduction, **8**, 313
disilyl enol ethers
regioselectivity, **2**, 606
synthesis
via acylation of organozincs, **1**, 448
via benzoin condensation, **1**, 542
via nickel-catalyzed acylation, **1**, 452
via γ-oxo sulfone acetals, **6**, 159
1,5-Diketones
acyclic
regiochemical cyclization, **2**, 163
synthesis, **1**, 558
via acylation of organozincs, **1**, 448
1,6-Diketones
β,γ-unsaturated
photoisomerizations, **5**, 227
synthesis
from silyloxycyclopropane, **2**, 445
1,2-Diketones, bisaryl
rearrangements, **3**, 825
Diketones, diphenyl
alkynes from, **8**, 951
1,3-Diketones, α-methylene-
synthesis
Mannich reaction, **2**, 905
Dilactones
cyclization
via Friedel–Crafts reaction, **2**, 711
synthesis
palladium-catalyzed carbonylation, **3**, 1032
Dilantin
anticonvulsant
synthesis, **3**, 826
Dilithium alkylcyano(2-thienyl)cuprate
alkylation, **3**, 261
Dilithium cyanocuprates
1,2-additions, **1**, 107
Dilithium tetrachlorocuprate
alkene dimerization, **3**, 482
alkylation
Grignard reagents, **3**, 244
vinyl Grignard reagent alkylation
catalyst, **3**, 243
Wurtz coupling, **3**, 415
Dilithium trialkylcuprates
reactions with tosylhydrazones, **1**, 378
Dimanganese heptoxide
oxidation
ethers, **7**, 236
Dimedone
allyl transfer
amine protection, **6**, 641
Knoevenagel reaction
Michael reaction, **2**, 352
synthesis
Claisen condensation, **2**, 796
ketone acylation, **2**, 843
via Michael addition, **4**, 6
Dimenthyl succinate
asymmetric cyclopropanation, **4**, 976
Dimerization
α-alkenes
hydroalumination, **8**, 744
disproportionation
radical anions, **7**, 861, 884
dissolving metal
reductions, **8**, 527
donor radical cations, **7**, 879
radical cations
bimolecular reaction, **7**, 859
Dimesoperiodate
oxidant
solid support, **7**, 843
Di-π-methane rearrangements
chemoselectivity, **5**, 206
hetero-substituted, **5**, 199–202
lamps and filters, **5**, 212
mechanism, **5**, 202–206
benzo–vinyl bridging, **5**, 203
divinyl bridging, **5**, 203
nomenclature, **5**, 194
photochemical apparatus, **5**, 212
photochemical hazards, **5**, 213
photochemical reaction conditions, **5**, 212
photoisomerizations, **5**, 193–213
practical aspects, **5**, 212
regioselectivity, **5**, 209
scope, **5**, 195–202
selectivity, **5**, 206–211
stereoselectivity, **5**, 210
synthetic utility, **5**, 211
2-(3,5-Dimethoxyphenyl)-2-propoxycarbonyl group
protecting group
acid stability, **6**, 637
cleavage, **6**, 636
Dimethylamine
reaction with π-allylpalladium complexes
stereochemistry, **4**, 623
Dimethylamine, *N*-chloro-
reactions with organoboranes, **7**, 607
reaction with trialkylborane, **7**, 604
Dimethylamine, trimethylsilyl-
enamine synthesis
via cyclohexanone, **6**, 705
2-Dimethylamino-4-nitrophenyl dihydrogen phosphate
phosphorylation, **6**, 609
Dimethyl diazomalonate
reaction with benzaldehyde
carbonyl ylide intermediate, **4**, 1090
Dimethyl fumarate
Diels–Alder reactions
Lewis acid promoted, **5**, 339
Dimethyl itaconate
ene reactions

Lewis acid catalysis, **5**, 5
hydroformylation, **4**, 925, 932
Dimethyl maleate
[3 + 2] cycloaddition reactions, **5**, 300
Diels–Alder reactions, **5**, 392
reduction, **8**, 563
Dimethyl muconate
[3 + 2] cycloaddition reactions, **5**, 297, 307
Dimethyl phthalate
hydrogenation
homogeneous catalysis, **8**, 454
Dimethyl succinate
synthesis
via dicarboxylation, **4**, 947
Dimethyl sulfoxide
oxidation, **7**, 769
Dimethyl tartrate
asymmetric epoxidation, **7**, 395
Dimethyl terephthalate
hydrogenation
homogeneous catalysis, **8**, 454
Dimsyl anions
$S_{RN}1$ reactions, **4**, 472
Dinitriles
Ritter reaction, **6**, 265
synthesis
via displacement reaction, **6**, 229
α,ω-Dinitriles
Ritter reaction
cyclization, **6**, 280
Dinitriles, 2-aryl-4-oxo-
synthesis
use of SAMP/RAMP, **2**, 520
2,4-Dinitrobenzenesulfenyl esters
carboxy-protecting groups
photolytic cleavage, **6**, 668
Dinitrogen tetroxide
nitration with, **6**, 107, 110
oxidation
hydrazines, **7**, 744
thiols, **7**, 761
reaction with cumulenes, **7**, 506
Dinitrogen trioxide
reactions with alkenes, **7**, 488
Diolides
synthesis
via Mitsunobu conditions, **6**, 368
Diols
monodeoxygenation, **8**, 818, 820
ortho acid synthesis, **6**, 560
oxidation
lactone synthesis, **7**, 312
prochiral
oxidation by enzymes, **7**, 316
synthesis
chiral, **1**, 66
vicinal
epoxide synthesis, **6**, 26
1,2-Diols
alkene protection, **6**, 686
cleavage
chromium oxides, **7**, 282
erythro
synthesis, **1**, 191
protecting groups, **6**, 659
reduction, **8**, 814
regioselective benzylation, **6**, 651
synthesis, **7**, 645, 647
coupling reactions, **3**, 597
from β,γ-epoxy alcohols, **3**, 264
reductive coupling of carbonyl compounds, **3**, 563
via dimesitylboryl carbanions, **1**, 499
1,3-Diols
monoethers
synthesis, **3**, 979
protecting groups, **6**, 659
synthesis, **7**, 645, 649
via reduction of β-hydroxy ketones, **8**, 8
via reaction of epoxides with boron-stabilized carbanions, **1**, 497
1,4-Diols
synthesis
alkenyltrialkylboronates, **3**, 799
Diols, 3-azido-
synthesis
via epoxy alcohols, **6**, 254
Diols, bis(dialkylamino)-
synthesis, **8**, 166
Diols, chloro-
synthesis
via asymmetric epoxidation, **7**, 424
1,3-Diols, 4-phenylthio-
cyclization, **6**, 25
1,4-Diols, 2-phenylthio-
reaction with dimethyl sulfate
cyclization, **6**, 25
Diols, vicinal
oxidation
α-diketones, **7**, 300
2,4-Dione, 3-substituted
synthesis
Knoevenagel reaction, **2**, 358
1,3-Diones
synthesis
from 2,3-epoxyketones, **3**, 771
1,6-Diones
synthesis
coupling of α,β-unsaturated carbonyl compounds, **3**, 577
Diorganocuprates
properties, **4**, 170
structure, **4**, 170
1,1-Diorganometallics
oxidation, **4**, 882
synthesis
via carbozincation, **4**, 879
Diorganozinc reagents
enantioselective addition reactions, **1**, 223
Diosphenols
rearrangements, **3**, 832
synthesis
via Claisen rearrangement, **5**, 848
Dioxabicyclo[2.2.1]heptane
reduction, **8**, 227
2,7-Dioxabicyclo[4.1.0]heptane
preparation
Darzens glycidic ester condensation, **2**, 427
2,7-Dioxabicyclo[3.2.0]hept-3-ene
photocycloaddition reactions, **5**, 170
synthesis
via photocycloaddition, **5**, 168
2,7-Dioxabicyclo[3.2.0]hept-3-ene-6-carboxylic acid

menthyl ester
synthesis *via* photocycloaddition, **5**, 169
Dioxabicyclo[3.3.1]nonane
reduction, **8**, 227
2,9-Dioxabicyclo[3.3.1]nonane
synthesis
via Wacker oxidation, **7**, 451
3,7-Dioxabicyclo[3.3.0]octane
synthesis
use of disilyl enol ether, **2**, 617
6,8-Dioxabicyclo[3.2.1]octane
reduction, **8**, 227
synthesis, **7**, 828
3,7-Dioxabicyclo[3.3.0]oct-1(5)-ene
synthesis
via retro Diels–Alder reactions, **5**, 579
Dioxaborinane, trimethyl-
hydroboration, **8**, 719
1,3,2-Dioxaborinanes
reactions with allyl organometallics, **2**, 32
Dioxalones
enolates
diastereoselective alkylation, **3**, 40
1,4-Dioxane, 2,2,3-trimethoxy-
synthesis, **6**, 560
1,3-Dioxanes
carbonyl group protection, **6**, 677
chiral carbonyl equivalent
Lewis acid promoted reactions, **1**, 347
reduction, **8**, 221, 659
synthesis
Prins reaction, **2**, 528
1,3-Dioxanes, 5,5-dibromo-
carbonyl group protection
removal, **6**, 677
1,2-Dioxanes, *trans*-3,6-disubstituted
synthesis
via mercuricyclization of hydroperoxides, **4**, 390
1,3-Dioxanes, 5-methylene-
carbonyl group protection
removal, **6**, 677
1,6-Dioxaspiro[4.5]decane
reduction, **8**, 220
1,4-Dioxaspiro[4,5]decane, 6-acetyl-6-allyl-
oxidative cleavage
sodium periodate and osmium tetroxide, **7**, 564
Dioxaspiro[3,3]heptanes
synthesis
via photocycloaddition, **5**, 167
1,7-Dioxaspiro[5.5]undecane, 2-ethyl-8-methyl-
synthesis, **7**, 625
1,7-Dioxaspiro[5.5]undecane, 4-hydroxy-
synthesis, **7**, 237
1,7-Dioxaspiro[5.5]undecane, 2-hydroxymethyl-8-methyl-
synthesis, **7**, 635
1,3-Dioxathiane
nucleophilic addition reactions
chiral auxiliary, **1**, 62
1,4-Dioxenes
synthesis, **3**, 651
Dioxetane
alkene oxygenation, **7**, 96
1,2-Dioxetanes
excited states
thermal generation, **5**, 198
reduction
with glutathione, **8**, 398
p-Dioxin
detoxification, **7**, 845
Dioxin, 2-chloro-
acid chloride synthesis, **6**, 305
Dioxindole, 3-phenyl-
synthesis, **3**, 835
Dioxinone
conjugate additions
dialkylcuprates, **4**, 207
Dioxirane, dialkyl-
epoxidizing agent, **7**, 374
Dioxirane, dimethyl-
epoxidization
alkenes, **7**, 167
oxidant
reaction with quadricyclane, **3**, 736
oxidation
primary amines, **7**, 737
pyridine, **7**, 750
secondary amines, **7**, 745
oxygen atom transfer, **8**, 398
Dioxirane, methyl-
oxygen atom transfer, **8**, 398
Dioxiranes
alkane oxidation, **7**, 13
synthesis
via potassium peroxymonosulfate, **1**, 834
Dioxolane, amino-
synthesis, **6**, 572
1,3-Dioxolane, 2-aryl-
reduction
sodium borohydride, **8**, 215
1,3-Dioxolane, 2-(2-bromoethyl)-
Grignard reagents
acylation, **1**, 452
1,3-Dioxolane, 4-bromomethyl-
carbonyl group protection
removal, **6**, 677
1,2-Dioxolane, *cis*-3,5-dialkyl-
synthesis
via mercuricyclization of hydroperoxides, **4**, 390
1,3-Dioxolane, 2-dialkylamino-
synthesis, **6**, 569
1,3-Dioxolane, 2,4-dimethyl-
reduction, **8**, 221
1,3-Dioxolane, 2-dimethylamino-
transacetalization
carbonyl group protection, **6**, 677
1,3-Dioxolane, 2,2-dimethyl-4-methylene-
lithium allenolates
synthesis, **2**, 109
1,3-Dioxolane, 2-ethoxy-
decomposition , **6**, 687
1,3-Dioxolane, 2,2,4,4,5,5-hexamethyl-
reduction, **8**, 221
1,3-Dioxolane, 2-(2-methoxycarbonylethyl)-2-methyl-
acyloin coupling reaction, **3**, 619
1,3-Dioxolane, 2-phenyl-
fragmentation
alkene protection, **6**, 687
rearrangement, **6**, 687
1,3-Dioxolane, 2,2,4,4-tetramethyl-
reduction, **8**, 221
1,3-Dioxolane, 2-thioxo-

desulfurization
alkene protection, **6**, 686
1,3-Dioxolane, 2-vinyl-
reduction
lithium aluminum hydride, **8**, 213
1,3-Dioxolanes
carbonyl group protection, **6**, 677
reduction, **8**, 221, 659
Dioxolanones
chiral
aldol reaction, **2**, 208
enolates
aldol reaction, **2**, 206
1,3-Dioxolan-4-ones
addition reactions with nitroalkenes, **4**, 109
Michael additions
nitroalkenes, **4**, 218
thermolysis
carbonyl ylide generation, **4**, 1089
1,3-Dioxolan-2-ylium cations
hydroxylation
alkenes, **7**, 445
anti hydroxylation
alkenes, **7**, 447
reaction with silyl ketene acetals, **2**, 804
1,3-Dioxole, 2,2-diisopropyl-
cycloaddition reactions
ethyl pyruvate, **5**, 160
1,3-Dioxole, 2,2-dimethyl-
photocycloaddition reaction
methyl phenyl glyoxylate, **5**, 160
Dioxolenes
iron complexes
reaction with organocopper compounds, **3**, 218
Dioxolenium ions
Diels–Alder reactions
intramolecular, **5**, 519
Dioxolenones
photocycloaddition reactions, **5**, 134, 137
1,3-Dioxoles
photolysis, **5**, 154
Dioxygen trapping
Paterno–Büchi reaction, **5**, 155
Dipentene
allylic oxidation, **7**, 99
Dipeptides
hydroxyethylene isosteres
synthesis *via* Carroll rearrangement, **5**, 836
N-methylation
retrograde Diels–Alder reaction, **5**, 552
synthesis
via asymmetric hydrogenation of dehydropeptides, 460
Diphenic acid
Schmidt reaction, **6**, 820
Diphenoquinone
synthesis, **3**, 664
Diphenylamine
synthesis
via $S_{RN}1$ reaction, **4**, 471
Diphenylamine-2,2′-dicarboxylic acids
Friedel–Crafts reaction, **2**, 759
Diphenyliodonium-2-carboxylates
aryne precursors, **4**, 488
Diphenylphosphinyl group
amine-protecting group, **6**, 644
Diphenyl phosphorazidate
acyl azide synthesis, **6**, 251
Diphenyl sulfoxide
oxidation, **7**, 769
Diphenylthiophosphinyl group
amine-protecting group, **6**, 644
Diphosgene
imidoyl halide synthesis, **6**, 523
reaction with amides, **6**, 495
Diphosphaallyl cations
synthesis, **6**, 190
1,2-Diphosphines
chiral catalysts
asymmetric hydrogenation of alkenes, **8**, 459
1,4-Diphosphines
chiral catalysts
asymmetric hydrogenation of alkenes, **8**, 459
Diphosphonates, methylidene-
synthesis
Knoevenagel reaction, **2**, 363
Diphosphorus tetraiodide
deoxygenation
epoxides, **8**, 886
iodination
alkyl alcohols, **6**, 213
reaction with dimethylformamide, **6**, 495
Diploda gossypina
epoxidation, **7**, 429
Diplodialide A
synthesis
Eschenmoser coupling reaction, **2**, 890
Diplodialide B
synthesis
via Wacker oxidation, **7**, 454
Diplodialides
synthesis, **3**, 286
Diploicin
synthesis
manganese dioxide oxidation, **3**, 688
1,3-Dipolar additions
regiospecificity, **4**, 1070
Dipolar cycloaddition
diazoalkanes, **6**, 126
1,3-Dipolar cycloaddition reactions, **5**, 247
absolute stereoselection, **5**, 260
frontier molecular orbital theory, **4**, 1073
intermolecular, **4**, 1069–1104
intramolecular, **4**, 1111–1166
mechanism, **4**, 1070–1072
nitrones, **4**, 1076–1078
nonconcerted, **4**, 1073–1075
regioselectivity, **5**, 247
relative stereoselection, **5**, 254
stepwise mechanism, **4**, 1072
stereoselectivity, **5**, 254
stereospecificity, **4**, 1072
synthetic equivalents, **5**, 266
[3 + 2] Dipolar cycloadditions
regiochemical control, **4**, 1073
1,3-Dipoles
classification, **4**, 1071
cycloaddition reactions, **4**, 730
fulvenes, **5**, 630
tropones, **5**, 625
structure, **4**, 1070
Dipropargylamines

intramolecular cycloaddition reactions, **5**, 1154
Dipyridones
photodimerization, **5**, 638
1,2-Dipyrrolidinylethane
alkene hydroxylation
osmium tetroxide, **7**, 442
Diquinane enone
photocycloaddition reactions, **5**, 144
Diquinanes
synthesis
via intramolecular ene reaction, **5**, 11
via photoisomerizations, **5**, 226
via rhodium-catalyzed rearrangement, **5**, 916
o-Diquinomethanes
Diels–Alder reactions, **5**, 524
Disaccharides, furanosyl
synthesis
stereoselectivity, **4**, 384
Diselenide, diaryl
reaction with carbon monoxide, **6**, 467
Diselenide, dimesityl
oxidation
allylic alcohols, **7**, 307
Diselenide, diphenyl
reaction with lithium enolates, **7**, 129
reduction, **6**, 463
Ritter reaction, **6**, 289
use in selenenylation, **7**, 131
Diselenide, 2,2′-dipyridyl
addition reactions with alkenes, **7**, 495
Diselenides
oxidation, **7**, 769
primary alcohols, **7**, 310
Diselenoacetals
acyl anion equivalents, **1**, 571
Diselenoketals
deselenation
nickel boride, **8**, 848
tin hydrides, **8**, 846
7,8-Disilabicyclo[2.2.2]octa-2,5-dienes
thermolysis
via retro Diels–Alder reaction, **5**, 587
Disilane, 1,1,1-trichloro-2,2,2-trimethyl-
reaction with π-allylpalladium complexes
stereochemistry, **4**, 626
Disilazane, hexamethyl-
alcohol protection, **6**, 653
amine synthesis, **6**, 83
nomenclature, **2**, 182
Disilazine, hexamethyl-
nomenclature, **2**, 183
Disilenes
generation
via retro Diels–Alder reaction, **5**, 587
Disiloxane, 1,3-bis(dimethylethynyl)-
oxidative coupling, **3**, 557
Disiloxane, 1,1,3,3-tetramethyl-
hydrosilylation, **8**, 19
1,4-Disilylcyclohexa-2,5-dienes
diacylation
Friedel–Crafts reaction, **2**, 717
Disparlure
synthesis, **3**, 224, 286, 644
Dispermol
synthesis, **7**, 331
Disproportionation
Friedel–Crafts reaction, **3**, 327
Dissolving metal conjugate reduction
synthesis
α-alkylated ketones, **4**, 254
Dissolving metals
reduction
acyl halides, **8**, 240
amides, **8**, 248
aromatic rings, **8**, 489–519
benzo[*b*]furans, **8**, 626
benzo[*b*]thiophenes, **8**, 629
benzylic compounds, **8**, 971
C═X to CHXH, **8**, 107–126
carbonyl compounds, **8**, 307–323
carboxylic acids, **8**, 236
chemoselectivity, **8**, 113, 530
conjugated dienes, **8**, 564
enones, **8**, 524
epoxides, **8**, 880
esters, **8**, 242
furans, **8**, 607
imines, **8**, 123
indoles, **8**, 614
isocyanides, **8**, 830
lactones, **8**, 247
mechanism, **8**, 525
nitriles, **8**, 252
oximes, **8**, 124
pyridines, **8**, 595
pyrroles, **8**, 605
stereochemistry, **8**, 525
stereoselectivity, **8**, 116
stereoselectivity, unsaturated hydrocarbons, **8**, 478
thioketones, **8**, 126
thiophenes, **8**, 609
unsaturated hydrocarbons, **8**, 478
Distannoxane, hexabutyl-
oxidation
sulfides, **7**, 764
Disuccinoyl peroxide
anti hydroxylation
alkenes, **7**, 446
Disulfides
hydrogenolysis, **8**, 914
reduction
sodium borohydride, **8**, 369
synthesis
via thiols, **7**, 758
tandem vicinal difunctionalization, **4**, 262
thiol protection, **6**, 665
Disulfides, dialkyl
reactions with trialkylboranes, **7**, 607
Disulfides, diaryl
reactions with trialkylboranes, **7**, 607
Disulfides, diphenyl
thiol carboxylic esters
synthesis, **6**, 439
Disulfones
desulfurization
eliminative, **8**, 839
1,1-Disulfones
alkylation, **3**, 176
N,N-Disulfonimides
reduction, **8**, 827
Ditellurides
oxidation, **7**, 774

Diterpene alkaloids
synthesis
via arynes, **4**, 500
via benzocyclobutene ring opening, **5**, 693
Diterpenes
microbial hydroxylation, **7**, 64
Diterpenoids
Horner–Wadsworth–Emmons reaction, **1**, 763
tetracyclic
synthesis, **3**, 715
1,3,2-Dithiaborinane
dimethyl sulfide complex
carboxylic acid reduction, **8**, 261
1,3,2-Dithiaborolane
hydroboration, **8**, 719, 720
1,4-Dithiadiene monosulfoxide
oxidation, **7**, 766
1,3,2,4-Dithiadiphosphetane 2,4-disulfide
thiocarboxylic ester synthesis, **6**, 437
1,3,2,4-Dithiadiphosphetane 2,4-disulfide, 2,4-bis[4-methoxyphenyl]-
see Lawesson's reagent
1,3-Dithiane, 2-aryl-
metal complexes
substitution reactions, **4**, 539
reaction with 2-cyclohexenone, **1**, 511
1,3-Dithiane, 2-chloro-
reaction with aryl Grignard reagents, **3**, 242
synthesis
via sulfide chlorination with NCS, **7**, 207
1,3-Dithiane, 2-dichloromethyl-
dihydrohalogenation
synthesis of ketene dithioacetals, **6**, 134
1,3-Dithiane, 2-ethylidene-
allyllithium derivative
reaction with aldehydes, **1**, 511
1,3-Dithiane, 2-ethynyl-
alkylation, **3**, 132
silyl and lithio derivatives, **2**, 90
1,3-Dithiane, 2-hydroxymethyl-
carbanions
reactions with epoxycyclohexanone, **1**, 511
1,3-Dithiane, 1-lithio-
reaction with epoxides, **3**, 127
1,3-Dithiane, 2-lithio-
in synthesis, **3**, 126
reaction with nitroarenes, **4**, 428
reaction with oxiranes, **3**, 128
1,3-Dithiane, 2-metallo-
in synthesis, **3**, 124
1,3-Dithiane, 2-methyl-
metal complexes
addition reactions, **4**, 535
1,3-Dithiane, 2-phenyl-
carbanions, **1**, 511
Stevens rearrangement, **3**, 925
1,3-Dithiane, 2-(1-propen-1-yl)-
crotyllithium derivative
reactions with aldehydes, **1**, 512
1,3-Dithiane, 2-(*p*-substituted)aryl-2-lithio-
reactions with *t*-butylbenzene, **4**, 537
1,3-Dithiane *S,S'*-dioxides
reaction with butyllithium, **1**, 526
1,3-Dithianes
acyl anion equivalents, **1**, 563; **3**, 144
addition reactions, **4**, 113
carbanions
crystal structure, **1**, 36
carbonyl group protection, **6**, 679
desulfurization
organolithium compounds, **8**, 847
formyl anion equivalents, **1**, 510
metallated
tandem vicinal difunctionalization, **4**, 258
Michael addition
kinetic *vs*. thermodynamic results, **4**, 10
1-oxide
alkylation, **3**, 137
sulfides from
use in synthesis, **6**, 134
1,3-Dithianyl anion, 2-methyl-
carbonyl group protection, **6**, 679
Dithianylidene anions, vinylogous
tandem vicinal difunctionalization, **4**, 258
1,3-Dithian-2-ylmethoxycarbonyl group
amine-protecting group, **6**, 639
1,3-Dithianylmethyl esters
carboxy-protecting groups, **6**, 666
Dithiatropazine
synthesis
acyloin coupling reaction, **3**, 618
1,2-Dithietane
synthesis
via thioxocarboxylic ester as precursor, **6**, 448
1,3-Dithietane
1,1-dioxide
synthesis, **7**, 768
4*H*-1,3-Dithiins
synthesis
via dimerization of thioketones, **5**, 556
Dithioacetal *S,S'*-dioxides
additions to C=X bonds, **1**, 526
Dithioacetal *S*-oxides
additions to C=X bonds, **1**, 526
Dithioacetals
acyl anion equivalents, **1**, 544, 563
carbonyl group protection, **6**, 677
reduction, **8**, 989
vinylsilane terminated cyclizations, **1**, 586
Dithioacetals
reaction with allylsilanes, **2**, 580
reaction with allylstannanes
carbon–sulfur bond cleavage, **2**, 581
reaction with vinylsilanes, **2**, 582
Dithioacetals, α-oxoketene
conjugate additions
organocuprates, **4**, 191
Dithioacrylates, methyl
synthesis
via flash vacuum thermolysis, **6**, 455
Dithiocarbamates
synthesis
via carbon disulfide, **6**, 428
Dithiocarbamates, *S*-(dialkylaminomethyl)-
iminium salts
generation *in situ*, **1**, 370
Dithiocarbamic acid
allylic esters
3,3-sigmatropic rearrangement, **6**, 846

Dithiocarbonates
deoxygenation, **8**, 818
Dithiocarbonates, *S*-allyl
nucleophilic addition to π-allylpalladium complexes
regioselectivity, **4**, 641
Dithiocarbonates, *O*-allyl *S*-alkyl-
nucleophilic addition to π-allylpalladium complexes
regioselectivity, **4**, 641
stereochemistry, **4**, 624
Dithiocarbonates, *O*-allyl *S*-methyl
Claisen-type rearrangement
palladium(II) catalysis, **4**, 564
Dithiocarbonates, *S,S*-dimethyl, **6**, 846
reactions with carbanions, **6**, 446
Dithiocarboxylates
thioacylation
amines, **6**, 423
Dithiocarboxylation
arenes, **6**, 456
carbanions, **6**, 456
Dithiocarboxylic acids
synthesis
via acylation of hydrogen sulfide, **6**, 455
via dithiocarboxylation of arenes and carbanions, **6**, 456
thioacylation
amines, **6**, 421
Dithiocarboxylic esters
synthesis, **6**, 453
via thioacylation of thiols, **6**, 453
Dithiocinnamates, methyl
synthesis, **6**, 455
Dithioenolates
addition reactions, **4**, 106–111
Dithioesters
Diels–Alder reactions, **5**, 438
reduction, **8**, 303
thioacylation
alcohols and phenols, **6**, 449
Dithioformic acid, cyano-
methyl ester
Diels–Alder reactions, **5**, 439
Dithioketals
carbonyl group protection, **6**, 677
desulfurization, **8**, 836
alkali metals, **8**, 842
LAH–$CuCl_2$, **8**, 840
Dithiolactones
synthesis, **6**, 453
via acylation of hydrogen sulfide, **6**, 455
via dithiocarboxylation of arenes and carbanions, **6**, 456
via thioacylation of thiols, **6**, 453
γ-Dithiolactones, β,γ-unsaturated
synthesis
allenylsilver compounds, **2**, 85
1,3-Dithiolane, 2-alkylidene-
cycloreversion
synthesis of thioketenes, **6**, 426
1,3-Dithiolane, 2,2-diaryl-
desulfurization
organolithium compounds, **8**, 847
1,2-Dithiolane, 4-methylthio-
synthesis
via bromine addition to alkene, **4**, 345
Dithiolanes
carbonyl group protection, **6**, 679
1,3-Dithiolanes
reduction, **8**, 231
Dithioles
reduction, **8**, 659
1,3-Dithiolones
intramolecular cycloadditions, **4**, 1163
3*H*-1,2-Dithiol-3-ones
synthesis of thioamides, **6**, 421
Dithiomalonates, *S,S'*-dialkyl
synthesis, **6**, 437
Dithiomalonates, *O,O*-diethyl
transesterification, **6**, 454
Dithiooxalate, *O,O'*-diethyl
reaction with glycols
synthesis of thioxoesters, **6**, 449
Dithiooxamides
synthesis
via thiolysis of imidoyl chlorides, **6**, 428
via thiolysis of 1,1,1-trihalides, **6**, 432
thioacylation
primary amines, **6**, 425
Dithioparabanic acid
O,N-acetals
synthesis, **6**, 576
Dithiophosphinic acids, diphenyl-
thioacylation
alcohols, **6**, 449
Dithiophosphoric acid
thiolysis, **6**, 432
Dithiopropionic acid
deprotonation
aldol reaction, **2**, 214
Dithiosuccinyl group
amine-protecting group, **6**, 643
[6]Ditriaxane, dimethyl-
synthesis, **6**, 777
Divided cells
electrosynthesis, **8**, 130
Dixanthates
vicinal
radical decomposition, **6**, 687
1,3-Diyl trapping reaction
intermolecular
trimethylenemethane, **5**, 240
intramolecular
trimethylenemethane, **5**, 241
trimethylenemethane, **5**, 239
Diynes
bicyclization, **5**, 1171
zirconium-promoted, **5**, 1164
conjugated
hydration, **4**, 300
hydroboration, **8**, 716
synthesis, **3**, 525, 554
hydroalumination
locoselectivity, **8**, 742
hydroboration
protonolysis, **8**, 727
intramolecular cycloaddition reactions
bicyclic 2-pyrone synthesis, **5**, 1157
semihydrogenation, **8**, 433
silylated
Friedel–Crafts acylation, **2**, 725

synthesis
phase-transfer catalysts, **3**, 559
1,4-Diynes
synthesis
organocopper compounds, **3**, 223
1,5-Diynes
Cope rearrangement, **5**, 797
α,ω-Diynes
cycloaddition reactions
isocyanates, **5**, 1156
intramolecular cycloaddition reactions
nitriles, **5**, 1154
1,3-Diynes, 1-trimethylsilyl-
synthesis, **3**, 553
9,11-Dodecadien-1-ol
acetate
synthesis, **2**, 76
synthesis *via* retro Diels–Alder reaction, **5**, 567
1,11-Dodecadien-3-one, 7-acetoxy-
trisannelation reagent
synthesis, **7**, 461
9,11-Dodecadien-1-yl acetate
synthesis, **1**, 680
Dodeca-3,5-diyn-1-ol
synthesis, **3**, 273
Dodecahedrane, 1,16-dimethyl-
synthesis
Friedel–Crafts reaction, **3**, 334
Dodecahedranes
Ritter reaction, **6**, 283
synthesis
via cyclofunctionalization of cycloalkenes, **4**, 373
via Diels–Alder reaction, **5**, 347
via Nazarov cyclization, **5**, 768
via Pauson–Khand reaction, **5**, 1062
Dodecanal
reduction
titanocene dichloride, **8**, 323
Dodecanamide
reduction, **8**, 249
Dodecane, 1-bromo-
Kornblum oxidation
solvent, **7**, 654
Dodecanedioic acid, 4,9-dioxo-
synthesis
via dialdehydes, **1**, 547
Dodecanenitrile
hydrogenation, **8**, 252
Dodecanol, 3,7,11-trimethyl-
synthesis
via asymmetric hydrogenation, **8**, 463
6-Dodecanone
reduction
titanocene dichloride, **8**, 323
Dodecanone, dibromo-
rearrangement, **3**, 851
Dodeca-2,6,10-triene-1,12-diol
asymmetric epoxidation, **7**, 404
1-Dodecene
oxidation
Wacker process, **7**, 450, 451
1-Dodecyl acetate
reduction
silanes, **8**, 246
Dodecylamine, dimethyl-
α-deprotonation, **1**, 476; **3**, 65
Doering–Moore–Skattebøl reaction
allene synthesis, **4**, 1009–1012
Doering reaction
dichlorocarbene addition, **4**, 1000
β-Dolabrin
synthesis
via tricarbonyl(tropone)iron complex, **4**, 707
Dolastane
synthesis, **3**, 488
Domesticine
synthesis
photochemical oxidation, **3**, 677
D-Dopa
synthesis
via L-serine, **1**, 413
L-Dopa
manufacture, **2**, 406
synthesis
via enzymic hydroxylation, **7**, 79
via microbial methods, **7**, 78
via L-tyrosine, **7**, 678
Dopamine β-monooxygenase
oxidation, **7**, 99
Dopamine receptor stimulating compounds
synthesis, **7**, 831
Double asymmetric synthesis
aldol reactions, **2**, 2, 248
Double diastereofacial selectivity
Diels–Alder reaction, **2**, 686
Double diastereoselection
aldol reaction, **2**, 232
Double stereodifferentiation
consonant, **2**, 232
dissonant, **2**, 232
matched pairs, **2**, 232
mismatched pairs, **2**, 232
Douglas fir tussock moth
sex pheromone
synthesis, **3**, 161
Dowex **3**, 50
catalyst
Friedel–Crafts reaction, **3**, 296
Drimanes
synthesis
via Diels–Alder reaction, **5**, 331
Drimatriene sulfoxide
synthesis
via electrocyclization, **5**, 735
Drimenyl acetate
allylic oxidation, **7**, 90
Drimnanes
synthesis
farnesol bicyclization, **3**, 342
Drynap
reductions
nitro compounds, **8**, 365
Dubamine
synthesis, **3**, 514
Dunnione
rearrangement, **3**, 828
Durene
thallation, **7**, 872
Durene, acetyl-

Friedel–Crafts reaction
reversibility, **2**, 745
Dyes
synthesis
Knoevenagel reaction, **2**, 387
Dynemicin A
synthesis
via electrocyclization, **5**, 736
Dysidenin, dimethyl-
synthesis
Ugi reaction, **2**, 1096

E

Ebelactone A
synthesis
via Ireland rearrangement, **5**, 842
Ecdysone
synthesis
Wittig rearrangement, **3**, 1000
Ecdysone, 20-hydroxy-
side chain
synthesis, **8**, 537
Echinocandin D
synthesis, **2**, 256
via reductive alkylation of azides, **8**, 386
Edman degradation
immobilization of enzymes
Ugi reaction, **2**, 1104
Egomaketone
synthesis, **6**, 455
palladium-catalyzed carbonylation, **3**, 1023
Eicosatetraenoic acid, 12-hydroxy-
synthesis
via ketocarbenoids and furans, **4**, 1059
1-Eicosene
oxidative cleavage
phase transfer assisted, **7**, 578
Eight-membered rings
synthesis
aldol condensation, **2**, 651
aldol reaction cascade, **2**, 623
via [4 + 4] cycloaddition, **5**, 635
Elaeocarpine
Mannich base, **2**, 894
Elaeokanines
synthesis, **6**, 756
via acyliminium ion terminated cyclization, **1**, 592
via pyrolytic dehydrosulfinylation, **1**, 515
via retro Diels–Alder reaction, **5**, 567
Elaidic acid
Kolbe electrolysis, **3**, 642
Elaiophylin
synthesis, **1**, 569; **2**, 263
Elbs persulfate oxidation
hydroquinones, **7**, 340
Eldanolide
synthesis, **1**, 565; **3**, 796
via cerium reagents, **1**, 240
Electrochemical oxidation, **7**, 789–811
alkenes
palladium(II) catalysis, **4**, 553
amount of electricity, **7**, 793
constant current method, **7**, 792
controlled potential method, **7**, 792
diaphragm, **7**, 792
ethers, **7**, 247
organoboranes, **7**, 602
supporting electrolytes, **7**, 793
techniques, **7**, 792
Electrochemical pinacolization
aromatic compounds, **3**, 567
Electrochemical reduction
acyl halides, **8**, 240
alkyl halides
chromium(II) salt catalyst, **8**, 797
allylic compounds, **8**, 974
amides, **8**, 248, 294
aromatic rings, **8**, 517
asymmetric
carbonyl compounds, **8**, 134
imines, **8**, 137
benzo[*b*]thiophene, **8**, 630
carbonyl compounds, **8**, 307
carboxylic acids, **8**, 236
cleavage
α-halo ketones, **8**, 987
C—N bonds, **8**, 995
demercuration, **8**, 857
dimerization, **8**, 527
epoxides, **8**, 884
esters, **8**, 242
indirect
carbonyl compounds, **8**, 132
indoles, **8**, 624
ketones
stereocontrol, **8**, 133
lactones, **8**, 247
mesylates, **8**, 817
nitriles, **8**, 252
nitro compounds, **8**, 366
pyridines, **8**, 591
pyridinium salts, **8**, 594
thioamides, **8**, 303
transition metal ions
carbonyl compounds, **8**, 133
α,β-unsaturated ketones, **8**, 532
Electrochemical reductive cleavage
C—S bonds, **8**, 994
α-oxygenated carbonyl compounds, **8**, 992
Electrochemistry
aldol reaction, **2**, 138
Ritter reaction, **6**, 281
Electrocyclic processes
higher order, **5**, 743
Electrocyclization
1,3,5-hexatrienes, **5**, 706–730
orbital correlation diagram, **5**, 703
six-electron
1,3,5-hexatriene, **5**, 699
stereochemistry, **5**, 703–706
Electrode reaction
indirect
electrosynthesis, **8**, 131
Electrodes
electrochemical oxidation, **7**, 792
Electrohydrodimerization
enones, **8**, 532
Electrolysis
carbonyl compounds, **8**, 321
oxidation, **7**, 791
Electron acceptors
reduction potentials, **7**, 854
Electron transfer
acceptor radical anions, **7**, 884
donor radical cations, **7**, 882
radical anions
bimolecular reaction, **7**, 861

radical cations
bimolecular reaction, **7**, 860
radicals, **4**, 726
Electron-transfer oxidation, **7**, 849–889
chain process, **7**, 860
formulation, **7**, 852
photochemical activation, **7**, 862
radical ions, **7**, 854
synthetic transformations, **7**, 873
thermal activation, **7**, 862
Electron transfer reduction
alcohols, **8**, 815
C—halogen bonds, **8**, 985
C—O bonds, **8**, 991
C—S bonds, **8**, 993
enones, **8**, 524
Electrooxidation
halide salts, **7**, 537
Electrophilic addition
acceptor radical anions, **7**, 884
radical anions
bimolecular reaction, **7**, 861
Electrophilic aromatic substitution
arene radical cations, **7**, 870
Electrophilic coupling
nucleophilic addition
carbanions, **4**, 237–263
Electrophilic oxidation
electron-transfer oxidation *versus*, **7**, 868
Electroreductive cyclization
Schiff bases, **8**, 136
Electrosynthesis
principles, **8**, 129
β-Elemene
synthesis, **7**, 94
Elemol
synthesis, **3**, 431; **6**, 145; **8**, 945
α-Eleostearate
hydrogenation
homogeneous catalysis, **8**, 451
Eleutherinol
synthesis, **2**, 171
Ellipticine
synthesis
via Diels–Alder reaction, **5**, 384
via electrocyclization, **5**, 721
Ellipticine, 9-hydroxy-
synthesis
via Baeyer–Villiger reaction, **7**, 684
Eloidisine
synthesis
Ugi reaction, **2**, 1097
Elsholtzia ketone
synthesis, **3**, 999
Elsholzione, dehydro-
synthesis
alkenylsilane acylation, **2**, 713
Emde degradation
amines, **6**, 70
pyridines, **8**, 597
Emodin
synthesis
dianion γ-acylation, **2**, 832
Enalapril
synthesis, **6**, 384
Enals
Michael acceptors, **4**, 261
Enamides
asymmetric hydrogenation
homogeneous catalysis, **8**, 460
cycloaddition with dihalocarbenes, **4**, 1004
Diels–Alder reactions, **5**, 322–324
α-hydroxylation, **7**, 170
ozonolysis, **7**, 171
protonation, **2**, 1052
tandem vicinal difunctionalization, **4**, 249
Vilsmeier–Haack reaction, **2**, 783
Enamides, *N,N*-dialkyl-
reactions with Grignards reagents, **4**, 257
Enamidines
Vilsmeier–Haack reaction, **2**, 792
Enamines
addition of carbene complexes, **4**, 980
alkylation, **3**, 28; **6**, 714
reversibility, **3**, 29
allylation
palladium catalysis, **4**, 654
anodic oxidation, **7**, 798
carbonyl group derivatization, **6**, 705
chiral
conjugate additions, **4**, 221–226
via allylamines, **6**, 866
[2 + 2] cycloaddition reactions, **5**, 71
[3 + 2] cycloaddition reactions
iron catalyzed, **5**, 285
cyclohexanone
axial alkylation, **3**, 30
dicarbonyl compound monoprotection, **6**, 684
Diels–Alder reactions, **5**, 322–324
hindered aldehyde
C-alkylation, **3**, 30
hydrogenation
heterogeneous catalysis, **8**, 439
hydrogenolysis, **8**, 915
α-hydroxylation, **7**, 170
infrared spectra, **6**, 711
Knoevenagel reaction, **2**, 367
Michael addition, **4**, 5
NMR
carbon **6**, 13, 712
nitrogen **6**, 15, 708
proton, **6**, 712
photoelectron spectra, **6**, 711
properties
chemical, **6**, 707
protonation, **6**, 717
reactions, **6**, 713
with alkynic esters, **4**, 45
with arynes, **4**, 510
with dihalocarbenes, **4**, 1004
with isocyanates, **5**, 103
with molecular bromine, **6**, 710
reduction, **8**, 938
hydrides, **8**, 55
stereochemistry, **8**, 55
regiochemistry, **6**, 709
proton NMR, **6**, 712
Simmons–Smith reaction, **4**, 968
stereochemistry, **6**, 716
structures, **6**, 707
synthesis, **6**, 705

via amide methylenation using Tebbe reagent, **5**, 1124
via palladium(II) catalysis, **4**, 560
via retro Diels–Alder reactions, **5**, 558
via Wittig–Horner type reaction, **6**, 69
ultraviolet spectra, **6**, 711
Vilsmeier–Haack reaction, **2**, 783
X-ray structure
single-crystal, **6**, 708
Enamines, *N*,*N*-bis(trimethylsilyl)-
anion formation
methyllithium, **6**, 722
Enamines, chloro-
acyloxyiminium salts, **6**, 493
cyclic
generation of 2-aminoallyl cations, **5**, 597
[4 + 3] cycloaddition reactions, **5**, 608
Favorskii rearrangement, **3**, 857
reactions with amines, **6**, 520
reactions with carboxylic acids, **1**, 424
reactions with hydrogen halides, **6**, 497
synthesis, **5**, 109
Enamines, -cyano-
addition reactions
with cycloalkenones, **4**, 117
cleavage, **2**, 857
Enamines, α-halo
reaction with carboxylic acids, **6**, 493
Enamines, morpholino
α-acetoxylation, **7**, 170
Enamines, nitro-
addition reactions, **4**, 124
Enamines, tetramethyl-α-chloro-
β-lactams from, **5**, 112
Enamines, *N*-tosyl-
synthesis
via palladium(II) catalysis, **4**, 561
Enamines, *N*-trimethylsilyl-
anion formation
methyllithium, **6**, 722
Enamino ketones
synthesis, **6**, 770
Enaminones
addition reactions, **4**, 123
reduction
borohydrides, **8**, 540
Enantioselective reactions
achiral carbon nucleophiles
achiral substrates, **4**, 228–231
alkylation, **3**, 35
allenyl organometallics, **2**, 96
propargyl organometallics, **2**, 96
Endiandric acids
synthesis, **3**, 558
via electrocyclization, **5**, 743
Endomyces reessii
β-hydroxylation, **7**, 56
oxidative rearrangement, **7**, 829
Endoperoxides
synthesis
via mercuricyclization of hydroperoxides, **4**, 390
Endothiopeptides
synthesis
via thioacylation, **6**, 420, 423
Ene carbamates
protonation, **2**, 1052
Vilsmeier–Haack reaction, **2**, 783
Enediols
synthesis
via retro Diels–Alder reactions, **5**, 557
Ene diones
synthesis
via palladium catalysis, **4**, 611
Enediynes
synthesis
cobalt-catalyzed cyclizations, **3**, 255
1-Ene-6,12-diynes
intramolecular cycloadditions, **5**, 1144
7-Ene-1,13-diynes
intramolecular cycloadditions, **5**, 1144
Enephosphinilation
ketones, **6**, 782
Ene reactions, **2**, 527–558
aliphatic Friedel–Crafts reaction
mechanism, **2**, 708
alkenes
enophiles, **5**, 1–25
Lewis acid catalysis, **5**, 4–6
alkynes
Lewis acid catalysis, **5**, 7–9
carbene complexes, **5**, 1075
enantioselective, **5**, 13
intermolecular, **2**, 528; **5**, 2–9
intramolecular, **2**, 540
alkenes, **5**, 9–20
enols, **5**, 20–23
type I, **2**, 540
type II, **2**, 547
Lewis acid catalysis, **5**, 1
formaldehyde, addition to alkenes, **2**, 530
regioselectivity, **2**, 534
tandem Claisen rearrangement, **5**, 11
tandem Diels–Alder reaction
alkynes, **5**, 7
thermal
alkenes, **5**, 2–4
alkynes, **5**, 6
cis/trans selectivity, **5**, 3
endo/exo selectivity, **5**, 3
formaldehyde, **2**, 529
regioselectivity, **5**, 3
transannular, **2**, 553; **5**, 20
with singlet oxygen, **7**, 818
type I
asymmetric induction, **2**, 541
1,2-disubstituted double bonds, **2**, 541
type III
acetals, **2**, 553
Enkephalins
synthesis
kinetically controlled, **6**, 399
Enmein
synthesis
via Birch reduction, **8**, 496
Enoates
conjugate additions
isopropylmagnesium bromide, **4**, 172
Diels–Alder reactions, **5**, 354–359
reactions with α-selenoalkyl metals
regiochemistry, **1**, 682
Enol acetates
alkali metal enolates

reaction, **2**, 108
anodic oxidation, **7**, 797
dihalocyclopropanation, **4**, 1005
electrochemical acetoxylation, **7**, 170
α-hydroxylation
ketones, **7**, 167
iodination, **7**, 121
nitration, **6**, 106
Vilsmeier–Haack reaction, **2**, 783
Enolates
2-acetidinone synthesis, **5**, 100–102
acyclic, carboxylic acid derivatives
diastereoselective alkylation, **3**, 42
addition reactions
carbon-centered radicals, **4**, 765
with alkenic π-systems, **4**, 99–113
addition to π-allylpalladium complexes, **4**, 591–594
regioselectivity, **4**, 632
stereochemistry, **4**, 616–618
aggregation
geometry, **3**, 4
alkylation, **3**, 1–58
stereochemistry, **3**, 12
C-alkylation
tandem vicinal difunctionalization, **4**, 240
allylation
palladium-catalyzed regioselective, **3**, 12
α-allyloxy
Wittig rearrangement, **3**, 996
aluminum
masked, facially selective sigmatropic protocol, **1**, 91
amination, **6**, 118
carbonyl compounds
halogenation, **7**, 120
carboxylic acids
cycloalkylation, **3**, 48
chiral
conjugate additions, **4**, 217–221
endocyclic
stereochemical alkylation, **3**, 41
equilibration
alkylation, **3**, 4
equivalents
uses, **4**, 238
fluorinated
formation, **2**, 115
α-fluoro-
synthesis, **2**, 103
generation *in situ*
acylation, **2**, 830
geometry
prediction, **2**, 101
Group I
aldol reactions, **2**, 181–235
Group II
aldol reactions, **2**, 181–235
Group III
aldol reactions, **2**, 239–275
halogen-substituted
reaction with trialkylboranes, **2**, 242
α-hydroxylation, **7**, 159
lithium
α-alkylation, **3**, 3
Michael addition, **4**, 258
preformed
acylation, **2**, 830
Claisen condensation, **2**, 799
reactions, **2**, 797
reactions with arynes, **4**, 496
reactions with aziridines
synthesis of γ-amido ketones, **6**, 96
reactions with *N,N*-dimethyl(methylene)iminium salts, **2**, 909
reaction transition states
stereochemistry, **1**, 2
regio-defined
aldol reactions, **2**, 182
selenenylation
low temperature reaction, **7**, 129
α-silyl-
synthesis, **2**, 103
stabilized
intramolecular alkylation, **3**, 55
stereoisomers
nomenclature, **2**, 100
sterically congested
stereoselectivity of alkylation, **3**, 15
structures
experimental studies, **2**, 100
thermodynamic/kinetic control, **2**, 101
2-substituted
distortion, **3**, 14
sulfenylation, **7**, 124
α-sulfinyl acetate
aldol reactions, asymmetric, **2**, 227
synthesis, **2**, 99–128
metallic potassium, **2**, 105
synthesis from carbonyl compounds
α-heteroatom-substituted, **2**, 186
tetrasubstituted
from ketenes, **2**, 196
transmetallation
tri-*n*-butyltin chloride, **3**, 10
trapped
acylation, **2**, 832
vinylogous
aldol reactions, **2**, 152
Enol carboxylates
coupling reactions
with sp^3 organometallics, **3**, 444
Enol esters
acid halide synthesis, **6**, 307
alkyl
formation, **2**, 596
conversion to enolates, **2**, 184
halogenation, **7**, 530
reaction with arylsulfonyl peroxides, **7**, 169
reaction with carbonyl compounds
use of Lewis acid catalysts, **2**, 612
α-sulfonyloxylation, **7**, 171
Enol ethers
2-acetidinone synthesis, **5**, 100–102
addition reactions, **2**, 595–625
aldol reactions, **2**, 611
anodic oxidation, **7**, 797, 803
asymmetric synthesis, **2**, 629–657
coupling reactions
with sp^3 organometallics, **3**, 444
cyclic
photoreactions with benzonitrile, **5**, 161
cycloaddition reactions
alkynic carbene complexes, **5**, 1067

dicarbonyl compound monoprotection, **6**, 684
disilyl
synthesis, **2**, 605
ene reactions, **5**, 1075
germyl
formation, **2**, 610
reactions, **2**, 624
halogenation, **7**, 121, 530
α-hydroxy
intramolecular hydrosilylation, **7**, 645
Mannich reaction, **2**, 1013
organotin(IV)
formation, **2**, 608
oxidation
pyridinium chlorochromate, **7**, 267
oxidative rearrangement, **7**, 816
polysilyl
synthesis, **2**, 605
preformed
acylation, **2**, 830
reaction with acetals, **2**, 612
reaction with *N*-acyliminium ions, **2**, 1064
reaction with benzeneselenenyl chloride, **7**, 520
reaction with carbonyl compounds
catalyzed, **2**, 612
reaction with dihalocarbenes, **4**, 1005
rearrangements, **3**, 789
reduction, **8**, 937
Simmons–Smith reaction, **4**, 968
stannyl
from enol silyl ethers, **2**, 609
reactions, **2**, 624
synthesis, **2**, 607
steroids
dehydrogenation, **7**, 136
synthesis, **2**, 595–625
allylic anions, **2**, 66
via esters using Tebbe methylenation reagent, **5**, 1123
tin(II)
formation, **2**, 609
Vilsmeier–Haack reaction, **2**, 783
Enol ethers, alkyl
ene reactions, **5**, 1075
formation, **2**, 596
α-hydroxylation
ketones, **7**, 167
Enol ethers, trimethylsilyl
rhodium enolates
aldol reaction, **2**, 310
synthesis from α-trimethylsilyl epoxides
reaction with Grignard reagents, **3**, 759
trichlorotitanium enolates
syn selective aldol reaction, **2**, 310
Enolization
kinetic
carbonyl compounds, **2**, 247
Enol lactones
dihalocyclopropanation, **4**, 1005
Friedel–Crafts reaction, **2**, 744
synthesis
via retro Diels–Alder reactions, **5**, 561
Enol phosphates
1,2-addition reactions
organoaluminum compounds, **1**, 92
coupling reactions
with sp^3 organometallics, **3**, 444
enol equivalents, **2**, 610
reaction with dialkylcuprates, **3**, 218
reduction
titanium salts, **8**, 531
Enol pyruvates
Mannich reaction, **2**, 904
Enols
alkylation, **3**, 1–58
ene reactions
intramolecular, **5**, 20–23
hydrogenolysis, **8**, 910
oxidative rearrangement, **7**, 816, 828
reactions with α-selenoalkyl metals
regiochemistry, **1**, 682
silylated
oxidative rearrangements, **7**, 816
synthesis
via retro Diels–Alder reactions, **5**, 557
Enol silanes
aldol reaction
rhodium(I) catalyzed, **2**, 311
cyclic
reaction with aldehydes, stereoselectivity, **2**, 632
heteroatom substituted
reaction with aldehydes, **2**, 642
reaction with aldehydes, diastereoselectivity, **2**, 643
reaction with acetals
Lewis acid mediated, **2**, 635
reaction with *N*-acyliminium ions, **2**, 1066, 1067, 1070
reaction with aldehydes
diastereoselectivity, **2**, 630, 646
fluoride catalyst, **2**, 633
fluoride ion catalyzed, **2**, 634
reaction with *N,N*-dimethyl(methylene)iminium salts
Mannich reaction, **2**, 905
reaction with chiral acetals
diastereoselectivity, **2**, 651
reaction with chiral α-alkoxy aldehydes
diastereoselectivity, **2**, 643
reaction with chiral azetinones
Lewis acid mediated, **2**, 649
reaction with chiral α-methyl aldehydes
diastereoselectivity, **2**, 640
reaction with dimethyl acetals
diastereoselectivity, **2**, 635
reaction with glycine cation equivalents, **2**, 1075
reaction with imines
Lewis acid mediated, **2**, 635
Enol silanes, nonstereogenic
reaction with aldehydes
diastereoselectivity, **2**, 640, 644
reaction with chiral α,β-dialkoxy aldehydes
reaction with aldehydes, diastereoselectivity, **2**, 644
Enol silanes, stereogenic
reaction with aldehydes
diastereoselectivity, **2**, 641, 645
reaction with chiral azetinones
Lewis acid mediated, **2**, 649
reaction with chiral α,β-dialkoxy aldehydes
diastereoselectivity, **2**, 644
Enol stannanes
reaction with π-allylpalladium complexes, **4**, 591
Enol sulfonates
coupling reactions

with sp^3 organometallics, **3**, 444
Enol triflates
coupling reactions
with lithium diarylcuprates, **3**, 492
with lithium divinylcuprates, **3**, 487
with sp^3 organometallics, **3**, 445
vinylation
palladium complexes, **4**, 859
vinyl carbanion equivalents, **1**, 195
Enones
acyclic
reaction with tris(trimethylsilyl)aluminum, **1**, 83
tandem vicinal difunctionalization, **4**, 243
β-alkylthio-α,β-unsaturated
formylation, **2**, 838
anti-Bredt's bridgehead
evidence for, **4**, 31
asymmetric reduction
Lewis acid coordination, **1**, 319
conjugate additions
organocuprates, **4**, 179
conjugated
Barbier reaction, **1**, 263
cyclopropanes from, **2**, 431
hydrosilylation, **8**, 781
reaction with hydroalumination adducts, **8**, 758
reaction with zinc ester dieneolates, **2**, 287
reaction with zinc ester enolates, **2**, 285
synthesis, **3**, 844
cuprate complex
spectroscopy, **4**, 171
cyclic
conjugate additions with chiral sulfinyl anions, **4**, 226
synthesis, **7**, 711
deoxygenation, **8**, 545
β-dialkylamino conjugated
reduction, **8**, 540
α,β-dialkylation
conjugate addition–enolate alkylation, **3**, 8
electrochemical reduction
yohimbine hydrochloride, **8**, 532
epoxidation, **4**, 35
from β-mercurio ketones, **2**, 443
hydrogenation
catalytic, **8**, 533
hydrosilylation
asymmetric, **8**, 784
Michael additions
protection, **6**, 687
partial reduction, **8**, 523–568
photochemical addition to alcohols
radical reactions, **4**, 753
reactions with allylic sulfinyl carbanions, **1**, 520
reactions with α-selenoalkyl metals
regiochemistry, **1**, 682
reduction
chemoselectivity, **8**, 15
synthesis
allylic oxidation, **7**, 113
via vinyl epoxides, **6**, 11
vicinal dialkylation, **3**, 8
α,β-Enones
addition reactions
with organomagnesium compounds, **4**, 83
with organozinc compounds, **4**, 95
alicyclic
addition reactions with organomagnesium compounds, **4**, 89
conjugate additions
trimethylaluminum, **4**, 140
vinyl groups, **4**, 141
Grignard additions
copper catalyzed, **4**, 91
protection device
β-stannylenol silylenol ether, **7**, 619
β,γ-Enones
bridged
photoisomerizations, **5**, 224–228
bridged bicyclic
photoisomerizations, **5**, 228
photoisomerizations, **5**, 215
reactions with organocerium reagents, **1**, 240
γ,δ-Enones
synthesis
via conjugate additions, **4**, 147
Enones, γ-acetoxy-
reaction with lithium dimethylcuprate, **4**, 171
Enones, α-alkoxy-
reduction
lithium aluminum hydride, **8**, 8
Enones, β-alkoxy-
cyclic
synthesis *via* Michael addition, **4**, 44
α,β-Enones, β-alkylthio-
addition reactions, **4**, 126
conjugate additions
organocuprates, **4**, 190
Enones, β'-amino-
divinyl ketones from
cyclization, **5**, 766
Enones, chloro-
synthesis
via dihalocarbene, **4**, 1005
via dihalocyclopropyl compounds, **4**, 1018
Enones, β'-chloro-
divinyl ketones from
cyclization, **5**, 766
Enones, β,β-disubstituted
Michael addition, **4**, 17
tandem vicinal difunctionalization, **4**, 244
Enones, α-fluoro-
synthesis
via electrophilic fluorination, **4**, 344
Enones, β-halo
addition reactions, **4**, 125
Enones, α'-hydroxy-
divinyl ketones from
cyclization, **5**, 766
Enones, β-hydroxy-
preparation, **2**, 674
Enones, β-iodo-
conjugate additions
organocuprates, **4**, 173
α,β-Enones, δ-(iodoacetoxy)-
intramolecular cyclization
via silyl ketene acetals, **4**, 161
Enones, β'-substituted
divinyl ketones from
cyclization, **5**, 766
Enones, β-(2-vinylcyclopropyl)-
synthesis, **5**, 979

Enynes
acyclic
Pauson–Khand reaction, **5**, 1053–1055
acyclic heteroatom-containing
Pauson–Khand reaction, **5**, 1055
bicyclization, **5**, 1165–1170
zirconium-promoted, **5**, 1163–1183
bicyclization–carbonylation, **5**, 1165
carbomagnesiation, **4**, 875
conjugated
synthesis, **3**, 878
cyclic
Pauson–Khand reaction, **5**, 1057–1060
fluorinated
synthesis, **3**, 525
functionalized
carbomagnesiation, **4**, 877
hydroboration, **8**, 717
intramolecular cycloaddition with isocyanides, **5**, 1132
in vitamin D synthesis, **3**, 545
reaction with Fischer carbene complexes, **5**, 1104
reaction with lithium organometallics, **4**, 868
semihydrogenation, **8**, 432
skipped
synthesis, **3**, 274
stereospecific synthesis, **3**, 539
synthesis
palladium catalysis, **3**, 217
via hydroalumination, **8**, 757
via Sakurai–Hosomi allylsilane conjugate addition, **5**, 1166
1,3-Enynes
synthesis
from alkynes, **3**, 880
lithium propargyls, **2**, 91
stereoselective, **3**, 522
1,4-Enynes
synthesis
via boron-ene reaction, **5**, 34
1,5-Enynes
Cope rearrangement, **5**, 797
synthesis, **3**, 104, 107
organopalladium catalysis, **3**, 231
1,6-Enynes
cyclization *via* intramolecular ene reaction
palladium catalysis, **5**, 16
ene reactions
intramolecular, **5**, 15–17
1,7-Enynes
ene reactions
intramolecular, **5**, 18
α,ω-Enynes
intramolecular cycloaddition reactions, **5**, 1143
Enynes, 1-chloro-
stereospecific synthesis, **3**, 539
Enynes, dithienyl-
synthesis, **3**, 527
Enynes, halo-
hydroiodination, **4**, 289
Enynols
divinyl ketones from
cyclization, **5**, 768
Enzymes
aldol reaction
use in, **2**, 456
cofactors
regenerating systems, **2**, 456
deactivation
oxidation, **2**, 456
dehydrogenation
carbonyl compounds, **7**, 145
experimental methodology
reduction, **8**, 185
immobilization
Ugi reaction, **2**, 1104
oxidation
diols, **7**, 316
sulfides, **7**, 194
unactivated C—H bonds, **7**, 79
peptide synthesis, **6**, 395
reduction
carbonyl compounds, **8**, 185
diastereotopic face distinctions, **8**, 192
epoxides, **8**, 884
specificity, **8**, 193
unsaturated carbonyl compounds, **8**, 558
sources
reduction, **8**, 184
substrate specificity, **2**, 456
use in organic chemistry
cofactor regeneration, **2**, 456
hollow fiber reactors, **2**, 456
membranes, **2**, 456
stability, **2**, 456
use in synthesis
immobilization, **2**, 456
Ephedrine
Diels–Alder reactions
intramolecular, **5**, 545
lithium aluminum hydride modifier, **8**, 166
Mannich reaction, **2**, 962
reaction with 2,2′-bis(bromomethyl)-1,1′-binaphthyl
N-alkylation, **6**, 71
synthesis
via benzoin condensation, **1**, 543
via conjugate addition, **4**, 227
Ephedrine, *N*-methyl-
asymmetric reduction
aluminum hydrides, **8**, 546
chiral silyl ketene acetals from
aldol condensation, **2**, 930
N-ethylaniline complex
reduction, unsaturated carbonyl compounds, **8**, 545
lithium aluminum hydride modifier, **8**, 166
silyl ketene acetals, derivatives of
reaction with imines, **2**, 638
stereoselective reactions, **2**, 636
Ephedrine amides
enolates
diastereoselective alkylation, **3**, 45
Ephedrine amides, *N*-methyl-
β-substituted α,β-unsaturated
addition reactions with organomagnesium compounds, **4**, 85
6a-Epipretazettine
synthesis, **3**, 683
4-Epibrefeldin C
synthesis
via alkenylchromium reagents, **1**, 200
Epicampherenone
synthesis

via [3 + 2] cycloaddition reactions, **5**, 286
14-Epicorynoline
synthesis
via Diels–Alder reaction, **5**, 500
Epielwesine
synthesis
Mannich reaction, **2**, 1032
via iminium ion–vinylsilane cyclization, **1**, 592
Epiepoformine
synthesis
via retro Diels–Alder reactions, **5**, 564
Epiepoxydon, **1**, 819–839
synthesis
via retro Diels–Alder reactions, **5**, 564
5-Epi-α-eudesmol
synthesis
via nitrone cyclization, **4**, 1115
Epimodhephene
synthesis
via intramolecular ene reactions, **5**, 22
Epinephrine (adrenalin)
vicinal amino alcohols
biological importance, **2**, 323
Epiophinocarpine
synthesis
from *trans* tetracyclic lactams, **2**, 946
Epipentenomycin
synthesis
via retro Diels–Alder reactions, **5**, 561
Epiprecapnelladiene
synthesis
via photocycloaddition, **5**, 139
15-Epi-$\Delta^{8(12)}$-prostaglandin E_1
synthesis, **8**, 561
Epi-β-santalene
synthesis, **3**, 427
4-Epishikimate, methyltriacetyl-
asymmetric synthesis, **6**, 161
4-Epishikimic acid
synthesis
via Diels–Alder reaction, **5**, 373
Episulfides
formation
Eschenmoser coupling reaction, **2**, 867
Ritter reaction, **6**, 277
Episulfonium ions
synthesis, **7**, 493
Epoformine
synthesis
via retro Diels–Alder reactions, **5**, 564
Epoxidation, **1**, 819–839
addition reactions, **7**, 357–385
alkenes, **7**, 390
solid support, **7**, 841
asymmetric methods, **7**, 389–436
titanium-catalyzed, **7**, 390
chemoselective, **7**, 384
intramolecular, **1**, 822
peracids, **7**, 375
steroids
microbial, **7**, 66
template-directed, **7**, 43
Epoxides (*see also* Oxiranes)
α-acetoxy steroidal
rearrangement, **3**, 739
alkenes
protection, **6**, 685
alkylation, **3**, 262
alkynides, **3**, 277
with sulfur- and selenium-stabilized carbanions, **3**, 86
allylic
rearrangement, **3**, 762
α-amino
rearrangement, **3**, 740
amino alcohol synthesis, **7**, 493
arene alkylation
Friedel–Crafts reaction, **3**, 309
asymmetric
diols, **7**, 390
preparation, **2**, 435
azide synthesis, **6**, 253
bromination, **6**, 211
carbene precursors, **4**, 961
α-carbonyl
rearrangement, **3**, 738
chlorination, **6**, 207
α-chloro
acid-catalyzed rearrangement, **3**, 739
thermal rearrangement, **3**, 739
cleavage
samarium triiodide, **1**, 260
deoxygenation, **8**, 884
α-electron withdrawing group
rearrangement, **3**, 746
fluorination, **6**, 219
formation
semipinacol rearrangement, **3**, 778
Friedel–Crafts reactions, **3**, 769
homochiral
synthesis, **7**, 429
hydrogenation
heterogeneous catalysis, **8**, 439
iodination, **6**, 214
nucleophilic opening
titanium-assisted, **7**, 405
opening
anti, **3**, 734
hydroxy neighboring group, **3**, 735
stereochemistry, **3**, 733
syn, **3**, 734
ortho acid synthesis, **6**, 560
oxidative rearrangement, **7**, 826
reactions with dialkoxyboryl carbanions, **1**, 496
reactions with hydroalumination adducts, **8**, 758
reactions with lithiodithiane, **1**, 569
reactions with nitriles, **6**, 271
reactions with organocerium compounds, **1**, 233
reactions with organocopper compounds, **3**, 223
reactions with organometallic compounds
alcohol synthesis, **6**, 4
Lewis acid promotion, **1**, 342
reactions with α-selenoalkyllithium, **3**, 91
rearrangements
acid-catalyzed, **3**, 733–771
protic acid catalyzed, **3**, 734
reduction, **8**, 871–891
ring opening
boron trifluoride catalyzed, **3**, 741
magnesium halide catalysis, **3**, 754
mechanism in aqueous acid, **3**, 736
nitrogen nucleophiles, **6**, 88

regioselectivity, **7**, 390
with Grignard reagents, **3**, 466
stereospecific deoxygenation
selenoamides, **6**, 481
α-substituted
rearrangement, **3**, 738
synthesis
via 1-chloroalkyl *p*-tolyl sulfoxide, **1**, 526
via cyclofunctionalization of allylic alcohols, **4**, 367
via Darzens glycidic ester condensation, **2**, 409
via β-hydroxyalkyl selenides, **1**, 712, 718, 721
via β-substituted alcohols, **6**, 25
α-trimethylsilyl
reaction with Grignard reagents, **3**, 759
β-trimethylsilyl
synthesis, **3**, 759
α,β-unsaturated
preparation, **2**, 421
Epoxides, acyclic vinyl
reaction with organocopper compounds, **3**, 226
Epoxides, α-lithio
from transmetallation, **3**, 198
Epoxides, nitro-
reduction
sodium borohydride, **8**, 874
Epoxides, vinyl
π-allylpalladium complexes from, **4**, 589
cyclic
ring opening, **6**, 9
reaction with nitrogen nucleophiles, **6**, 86
reaction with organocuprates, **3**, 225
ring opening, **7**, 491
organometallic reagents, **6**, 9
Epoxydon
synthesis
via retro Diels–Alder reactions, **5**, 564
2,3-Epoxysqualene
synthesis, **3**, 178
Equilenin
synthesis
ketone oxalylation, **2**, 838
Equilenin, 11-oxo-
methyl ether
synthesis *via* conjugate addition, **4**, 215
Equilenin ketal
Birch reduction
dissolving metals, **8**, 497
Equilibrium constants
aldol additions, **2**, 134
Equisetin
synthesis
via Ireland rearrangement, **5**, 843
Eremophilane
biosynthesis, **3**, 388
synthesis, **8**, 528
Eremophilone
hydrogenation
Wilkinson catalyst, **8**, 445
synthesis
rearrangement of allylic epoxides, **3**, 762
Ergoline, 2-bromo-
reduction
borohydrides, **8**, 618
Ergosterol
acetate
oxidative halogenation, **7**, 529
diene protection, **6**, 691
25,28-dihydroxylated
synthesis, **3**, 983
selective reduction, **8**, 565
Ergot alkaloids
synthesis
Mannich reaction, **2**, 967
via INOC reaction, **4**, 1080
Erigerol
synthesis
via Diels–Alder reaction, **5**, 329
Eriolanin
synthesis, **8**, 925
Erlenmeyer azlactone synthesis, **2**, 395, 402–407
lead acetate, **2**, 402
Erucic acid
Kolbe electrolysis, **3**, 642
Erybidine, *O*-methyl-
synthesis, **3**, 816
Erysodienone
synthesis, **3**, 816
use of alkaline ferricyanide, **3**, 681
Erythramine
related structure
synthesis *via* azomethine ylide cyclization, **4**, 1140
Erythrina alkaloids
synthesis, **3**, 505; **6**, 746
electron transfer induced photocyclizations, **2**, 1038
via Diels–Alder reactions, **5**, 323
Erythrinan
skeleton
synthesis *via* azomethine ylide cyclization, **4**, 1136
synthesis
N-acyliminium ions, **2**, 1056
cis-Erythrinan, 15,16-dimethoxy-
synthesis, **2**, 1038
Erythro compounds
aldol diastereomers
thermodynamics, **2**, 153
Erythromycin
oxime
Beckmann reaction, **7**, 698
partial synthesis
stereocontrol, **3**, 960
synthesis
Woodward, **2**, 214, 221
zirconium enolates, **2**, 303
Erythromycin A
oxime
Beckmann rearrangement, **6**, 766
D-Erythronolactone
reduction
disiamylborane, **8**, 269
Erythronolide A
synthesis, **1**, 430
aldol reaction, **2**, 205
aldol reaction of lithium enolate, **2**, 219
use of lithium enolate, **2**, 194
via sulfur ylide reagents, **1**, 824
Erythronolide A, 9,9-dihydro-
synthesis, **7**, 246
via macrolactonization, **6**, 370
Erythronolide B
synthesis, **1**, 430; **3**, 288
via Baeyer–Villiger reaction, **7**, 678

via cyclofunctionalization of cyclohexadienone, **4**, 372
Erythronolide B, 6-deoxy-
synthesis, **2**, 253
Diels–Alder reaction, **2**, 700
via cuprate acylation, **1**, 436
via macrolactonization, **6**, 372
Erythronolides
synthesis, **1**, 564
via Grignard addition, **1**, 408
via Horner–Wadsworth–Emmons reaction, **1**, 772
via macrolactonization, **6**, 371
via reactions of organocuprates and homochiral aldehydes, **1**, 125
via Wittig reaction, **1**, 757
Erythro–threo diastereoselectivity
Michael addition, **4**, 21
Eschenmoser amide acetal rearrangement
variant of Claisen rearrangement, **5**, 836–838
Eschenmoser coupling reaction, **2**, 865–890
carbon–carbon bond formation, **2**, 869
Knoevenagel modification, **2**, 873
mechanism, **2**, 867
Robinson annelation, **2**, 885
sulfide contraction, **2**, 869
synthesis, **2**, 876
thio-Wittig modification, **2**, 874
Eschenmoser fragmentation, **8**, 948
definition, **6**, 1043
Eschenmoser's salt
Mannich reaction, **2**, 899
Eserethole
synthesis
via azomethine ylide cyclization, **4**, 1088, 1136
Esperamicin
synthesis, **3**, 545
copper catalysts, **3**, 217
via electrocyclization, **5**, 736
Esperamicin A
synthesis, **3**, 27
Estafiatin
synthesis, **7**, 363
Ester enolates
acyclic
alkylation, **3**, 42
addition reactions, **4**, 106–111
alkenes, palladium(II) catalysis, **4**, 572
arylation, **4**, 466
stereoselectivity, **2**, 200
synthesis, **2**, 101
Ester enolates, bromo
Darzens glycidic ester condensation, **2**, 427
Esterification
alkylative, **6**, 335
Esters
activated
macrolactonization, **6**, 373
synthesis, **6**, 323–376
acyclic
synthesis *via* retro Diels–Alder reactions, **5**, 573
tandem vicinal difunctionalization, **4**, 247–249
acylation, **2**, 795–863; **6**, 328
thiols, **6**, 443
acylation of organometallic reagents, **1**, 416
acyloin coupling reaction
heterogeneous conditions, **3**, 614
homogeneous conditions, **3**, 615
necessary reaction conditions, **3**, 614
preferred reaction conditions, **3**, 617
with ketones, **3**, 630
alcohol protection, **6**, 657
alkenic
divinyl ketones from, **5**, 776
α-alkoxy-
(*Z*)- and (*E*)-enolates, **2**, 102
β-alkoxy-α,β-unsaturated
addition reactions, **4**, 125
alkylidenation
dihaloalkane reagents, **5**, 1125
alkyl-substutited bromo
Reformatsky reaction, **2**, 289
β-(alkylthio)-α,β-unsaturated
addition reactions, **4**, 126
α-amino
hydrogenation, **8**, 242
zinc ester enolates, preparation, **2**, 296
β-amino
synthesis *via* palladium(II) catalysis, **4**, 560
arene alkylation
Friedel–Crafts reaction, **3**, 309
aromatic carboxylic
Birch reduction, **8**, 505
asymmetric epoxidation
compatibility, **7**, 401
asymmetric hydroxylation, **7**, 181
boron trifluoride complex
NMR, **1**, 292
α-bromo
Reformatsky reaction, cerium metal, **2**, 312
carboxy-protecting groups, **6**, 665
chiral
diastereoselective additions, **4**, 200–202
chiral α-alkoxy
lithium enolates, **2**, 227
chiral β-amino thiol enolates
diasterofacial preference, **2**, 225
cleavage
deprotection, **6**, 665
lithium chloride, **6**, 206
cyclic
tandem vicinal difunctionalization, **4**, 249
Darzens glycidic ester condensation
phase-transfer catalysis, **2**, 429
dehydrogenation, **7**, 144
use of benzeneseleninyl chloride, **7**, 135
β,δ-diketo
reduction, **8**, 9
β,β-disubstituted
Michael addition, **4**, 17
electrochemical amidation, **6**, 392
epoxide synthesis
diazomethane, **1**, 832
β-halo-α,β-unsaturated
addition reactions, **4**, 125
hindered aryl
anti aldols, **2**, 201
deprotonation, **3**, 194
homochiral β-hydroxy
synthesis *via* conjugate addition to sultams, **4**, 204
hydrogenation, **8**, 242
α-hydroxyalkyl-α,β-unsaturated
synthesis *via* heteronucleophile addition, **4**, 34

α-hydroxylation, **7**, 179
syn-3-hydroxy-2-methyl
synthesis, **2**, 252
γ-hydroxy-α,β-unsaturated
hydroxylation, **7**, 439
iodination, **6**, 214; **7**, 121
α-iodo
Reformatsky reaction, cerium metal, **2**, 312
Julia coupling, **1**, 803
lithium enolates
crystal structures, **1**, 30
methylenation
Tebbe reaction, **1**, 747, **5**, 1123
α-methylthio
deprotonation, **2**, 103
mixed
acylation, **2**, 799
ortho ester synthesis, **6**, 560
polyunsaturated
tandem vicinal difunctionalization, **4**, 253
reactions with benzophenone dianion
organoytterbium compounds, **1**, 280
reactions with organoaluminum reagents, **1**, 92
reduction
alkali metals, **3**, 613
metal hydrides, **8**, 266
silanes, **8**, 824
stannane, **8**, 824
to aldehydes, **8**, 292
Reformatsky reaction, **2**, 296
selenenylation, **7**, 129, 131
sulfenylation, **7**, 125
selective, **7**, 125
sulfinylation, **7**, 127
α-sulfinyl-β-hydroxy
aldol reaction, **2**, 227
α-sulfonyl-α,β-unsaturated
synthesis, Knoevenagel reaction, **2**, 362
synthesis, **6**, 323–376
carbonylation, **3**, 1028
via ethers, **7**, 236
via hydration of alkynes, **4**, 300
via oxidative cleavage of alkenes, **7**, 574
tandem vicinal dialkylations, **4**, 261
tandem vicinal difunctionalization, **4**, 246–249
α-trimethylsilyl
Reformatsky reaction, **2**, 294
β-trimethylsilyl(amino)
cyclization, **2**, 935
unactivated
aminolysis, **6**, 389
Esters, *p*-alkoxybenzyl
anchoring groups, **6**, 671
Esters, alkynic
hydrostannation, **8**, 548
hydrozirconation, **8**, 683
reaction with allylic alcohols, **6**, 856
tandem vicinal difunctionalization, **4**, 247
Esters, allenic
hydrochlorination, **4**, 277
tandem vicinal difunctionalization, **4**, 249
thermal rearrangement
to dienoic ester, **6**, 867
Esters, β-amino-α,β-unsaturated
functionalized, synthesis, **6**, 67
Esters, bis(trimethylsilyl)
Peterson alkenation, **1**, 791
Esters, *t*-butyl
synthesis, **6**, 337
Esters, dialkoxybenzyl
anchoring groups, **6**, 671
Esters, α-diazo
C—H insertion reactions, **3**, 1054
higher
synthesis, **6**, 125
synthesis, **6**, 122, 124
Esters, dienoic
thermal rearrangement
via β-allenic ester, **6**, 867
Esters, β-enamino
synthesis
Knoevenagel reaction, Meldrum's acid, **2**, 356
Esters, α-fluoro-α,β-unsaturated
Oshima–Takai reaction, **1**, 751
Esters, α-halo
Darzens glycidic ester condensation, **2**, 432
reduction
Alpine borane, **7**, 603
Esters, hydroxy
Ritter reaction, **6**, 268
synthesis, **6**, 877
Esters, 1-hydroxy
chiral
synthesis, **1**, 66, 86
oxidation
synthesis of α-keto esters, **7**, 324
Esters, 2-hydroxy
alkylation, **3**, 43
chiral titanium enolates
enantioselective synthesis, **2**, 309
dianions
alkylation, **2**, 225
enantioselective
aldol reaction, acetyliron complex, **2**, 315
synthesis
via organoaluminum reagents, **1**, 84
Esters, 4-hydroxy
dianions
aldol reaction, **2**, 225
synthesis
homoaldol reaction, **2**, 445
Esters, 1,2-keto
synthesis
Knoevenagel reaction, oxidation, **2**, 360
Esters, 1,3-keto
aldol reaction, **2**, 209
γ-alkylation, **3**, 58
diazo transfer, **6**, 125
intermolecular pinacolic coupling reactions
organosamarium compounds, **1**, 271
intramolecular Barbier cyclization
samarium diiodide, **1**, 264
Knoevenagel reaction, **2**, 359
synthesis, **3**, 783, 784
Esters, 1,4-keto
synthesis
homoenolates, **2**, 449
Esters, 1,6-keto
synthesis
zinc homoenolate, **2**, 448
synthesis, Reformatsky reaction, **2**, 296
Esters, β-keto-2-[2-(trimethylsilyl)methyl]-

cycloaddition reactions, **5**, 247
Esters, α-keto-β,γ-unsaturated
Diels–Alder reactions, **5**, 461
Esters, α-nitroso-
synthesis, **6**, 104
Esters, γ-oxocarboxylic acid
alcohol protection
cleavage, **6**, 658
Esters, α-seleno
metallation, **1**, 642
Esters, γ-stannyl-α,β-unsaturated
coupling reactions
with alkenyl halides, **3**, 443
Esters unsaturated
Reformatsky reaction, **2**, 294
Esters, α,β-unsaturated
addition of carbene complexes, **4**, 980
chelated
Diels–Alder reactions, **5**, 365–367
conjugate additions, **4**, 184
conjugate addition to lithium bis(phenyldimethylsilyl)cuprate, **2**, 186
dehydrogenation, **7**, 142
Diels–Alder reactions, **5**, 461
enzymic reduction, **8**, 205
Grignard additions, copper catalyzed, **4**, 91
hydrobromination, **4**, 282
hydroformylation, **4**, 925
Michael acceptors, **4**, 261
reactions with 1,1-bis(seleno)alkyllithium, **1**, 694
reactions with organolithium compounds, **4**, 72
stereochemistry, **7**, 396
synthesis, **3**, 865
Ramberg–Bäcklund rearrangement, **3**, 870
synthesis from β-hydroxyalkyl selenides, **1**, 705
synthesis *via* retro Diels–Alder reaction, **5**, 553
Esters, β,γ-unsaturated
synthesis
coupling reactions, **3**, 443
synthesis *via* tandem vicinal difunctionalization, **4**, 249
Esters, vinyl
cycloaddition reactions, **5**, 255
Darzens glycidic ester condensation, **2**, 421
Estradiol
bistrimethylsilyl ether
reductive silylation, **8**, 518
synthesis
via benzocyclobutene ring opening, **5**, 693
Estradiol, 2-hydroxy-
synthesis, **7**, 331
Estra-1,3,5(10)-trien-17β-ol, 3-methoxy-
acetate
reaction with mercury(II) acetate, **7**, 331
Estratrienone
synthesis, **7**, 338
Estrogenic steroids
synthesis
via arynes, **4**, 501
Estrogens
synthesis, **7**, 331
Estrone
Birch reduction
dissolving metals, **8**, 493
cyclization, **3**, 371
methyl ether
synthesis, **3**, 1061
synthesis, **7**, 338
polyene bicyclization, **3**, 360
polyene cyclization, **3**, 366
via Baeyer–Villiger reaction, **7**, 682
via benzocyclobutene ring opening, **5**, 693
via Cope rearrangement, **5**, 790
8α-Estrone
Mannich reaction
with iminium salts, **2**, 902
Estrone, C,18-bisnor-13α,17α-dehydro-
synthesis
via photoisomerization, **5**, 232
Estrone methyl ether
synthesis
via conjugate addition, **4**, 215
Ethane
ethylation
Friedel–Crafts reaction, **3**, 333
Ethane, 2-arylnitro-
double deprotonation
Henry reaction, **2**, 337
Ethane, azido-
synthesis
via ethyl iodide, **6**, 245
Ethane, 1,2-bis(oxazolinyl)-
dilithation, **4**, 976
Ethane, 1,2-diaryl-
dimerization, **3**, 673
Ethane, 1,2-dibromo-
lactone bromination, **7**, 121
reduction
dissolving metals, **8**, 526
Ethane, 1,2-dibromotetrachloro-
alkane bromination, **7**, 15
Ethane, 1,2-dihalo-
arene alkylation
Friedel–Crafts reaction, **3**, 318
Ethane, 1,2-dihalo-2-phenyl-
arene alkylation
Friedel–Crafts reaction, **3**, 318
Ethane, 1,2-diisocyano-1,2-diphenyl-
reduction, **8**, 831
Ethane, dimethoxy-
alkali metal stabilized carbanions
crystal structure, **1**, 5
Ethane, 1,2-diphenyl-
synthesis, **3**, 638
Ethane, hexafluoro-
synthesis, **3**, 640
Ethane, hexamethyl-
synthesis, **3**, 415
Ethane, nitro-
addition reaction with enolates, **4**, 104
Ethane, pentaalkoxy-
synthesis, **6**, 556
Ethane, 2-substituted-1,1-dimethyl-1-nitro-
reduction, **8**, 375
Ethane, 1,1,1-trifluoro-2,2-diaryl-
synthesis
Friedel–Crafts reaction, **3**, 311
Ethanediol, 1,2-dicyclohexyl-
boronic esters, **3**, 796
Ethanediol, diisopropyl-
boronic esters, **3**, 797
Ethanethiol

oxidative cleavage of alkenes
synthesis of dithioacetals, **7**, 588
9,10-Ethanoanthracene, 9,10-dihydro-
retro Diels–Alder reaction, **5**, 589
Ethanol, 2-amino-2-phenyl-
hydrogenation, **8**, 146
Ethanol, 2-aryl-
synthesis
via microbial methods, **7**, 76
Ethanol, 2-bromo-
acetate
reaction with aryl Grignard reagents, **3**, 243
reaction with aryl Grignard reagents, **3**, 243
Ethanol, 1-cyclohexyl-
hydrogenation
catalytic, **8**, 141
Ethanol, α-(2-hydroxyphenyl)-
lactate
Friedel–Crafts reaction, **3**, 311
Ethanol, 1-(4-methylphenyl)-
Birch reduction
dissolving metals, **8**, 515
Ethanol, 1-phenyl-
absolute configuration, **8**, 160
hydrogen donor
styryl ketones, **8**, 552
transfer hydrogenation, **8**, 552
Ethanol, 2-phenylthio-1,2-diphenyl-
synthesis
via benzylphenyl sulphide, **1**, 506
Ethers
acyclic
synthesis, **6**, 22
alcohol protection, **6**, 647
π-allylpalladium complexes from, **4**, 588
arene alkylation
Friedel–Crafts reaction, **3**, 309
asymmetric epoxidation
compatibility, **7**, 401
bridged-ring
synthesis *via* cyclofunctionalization, **4**, 373
cleavage
lithium bromide, **6**, 210
lithium chloride, **6**, 206
cyclic
synthesis, **6**, 22
cyclic allylic
Wittig rearrangement, **3**, 1008
cyclic propargylic
Wittig rearrangement, **3**, 1008
epoxidation directed by, **7**, 367
hydride donors
to carbonium ions, **8**, 91
hydrobromination, **4**, 282
iodination, **6**, 214
oxidation
activated C—H bonds, **7**, 235–248
mechanism, **7**, 236
selectivity, **7**, 238
reactions with arynes, **4**, 507
rearrangements, **6**, 874
diastereocontrol, **6**, 880
(E)/(Z) selectivity, **6**, 875
saturated aliphatic
anodic oxidation, **7**, 803
synthesis
via carboxylic acids, **8**, 235–254
via reduction, **8**, 211–232
via electrophile cyclization, **7**, 523
via substitution processes, **6**, 1–28
Wittig rearrangement
absolute configuration, **6**, 884
Ethers, alkyl haloalkyl
arene haloalkylation by
Friedel–Crafts reaction, **3**, 321
Ethers, alkyl methyl
synthesis
via trialkylboranes, **7**, 603
Ethers, alkyl vinyl
reaction with tetracyanoethylene
solvent effects, **5**, 76
Ethers, alkynic
carbometallation
enol ether preparation, **2**, 596
Ethers, allenyl methyl
metallation, **2**, 596
Ethers, allyl
cycloaddition reactions, **5**, 260
inside alkoxy effect
cycloaddition reactions, **5**, 260
isomerization
vinyl ether synthesis, **6**, 866
isomerization to propargyl ether
alcohol protection, **6**, 652
oxidation
palladium(II) catalysis, **4**, 553
Pauson–Khand reaction
regiocontrol, **5**, 1044
thermolysis
retro-ene reaction, **6**, 866
Ethers, allyl benzyl
Wittig rearrangement, **3**, 989
Ethers, allylic
Wittig rearrangement
mechanism, **3**, 977
Ethers, allyl lithiomethyl
Wittig rearrangement, **3**, 982
Ethers, allyl methyl
reduction
LAH/$TiCl_4$, **8**, 967
Ethers, allyl propargyl
carbonylation
use of cobalt complexes catalysts, **3**, 1025
Pauson–Khand reaction, **5**, 1055
Wittig rearrangement, **3**, 984
Ethers, allyl silyl
reaction with aryl Grignard reagents, **3**, 246
Ethers, allyl thiophenyl
desulfurization, **8**, 840
Ethers, allyl vinyl
Claisen rearrangement, **5**, 832–834, 1001
discovery, **5**, 827
synthesis
via allyl formate alkenation, **6**, 856
via Claisen rearrangement, **5**, 830–832; **6**, 856
via Wittig-type alkenation, **5**, 830
Ethers, aryl
oxidation
radical cation reactions, **3**, 662
synthesis, **3**, 686
Ethers, aryl alkyl
synthesis

C—O coupling, **3**, 690
Ethers, α-arylamino
reaction with lithium alkynides, **3**, 282
Ethers, aryl 4-cyanophenyl
synthesis, **4**, 439
Ethers, aryl fluoroalkyl
synthesis, **4**, 438
Ethers, aryl silyl
Birch reduction
dissolving metals, **8**, 494
Ethers, aryl tetrazolyl
substitution reactions
nickel catalysts, **3**, 229
Ethers, α-azido
synthesis
via acetals, **6**, 254
Ethers, benzaldoxime trimethylsilyl
reaction with crotyl boronates
syn–anti selectivity, **2**, 996
Ethers, benzenoid
Birch reduction
dissolving metals, **8**, 493
Ethers, benzothiazolyl
reaction with organocopper compounds, **3**, 222
Ethers, benzyl
alcohol protection, **6**, 650
α-alkoxy carbanions from
Wittig rearrangement, **3**, 197
oxidation
Jones reagent, **7**, 240
Ethers, benzyl chloromethyl
reaction with carbonyl compounds
samarium diiodide, **1**, 259
Ethers, benzyl *trans*-crotyl
Wittig rearrangement, **3**, 976
Ethers, benzyl ethyl
oxidation, **7**, 240
Ethers, benzyl methyl
deprotonation
by *n*-butyllithium, **3**, 197
oxidation, **7**, 240
reductive cleavage, **8**, 974
Ethers, biaryl
reductive fission
dissolving metals, **8**, 514
Ethers, bis-γ,γ-(dimethyl)allyl
Wittig rearrangement
mechanism, **3**, 977
Ethers, bisallyl vinyl
Claisen rearrangement
catalysis, **5**, 850
Ethers, bornyl bromotetrahydropyranyl
synthesis
via bromine addition to alkene, **4**, 345
Ethers, *t*-butyl
alcohol protection
amino acids, **6**, 650
Ethers, *n*-butyl dimethylaminomethyl
N,N-dimethyl(methylene)iminium salt
preparation from, **2**, 901
Ethers, *t*-butyldimethylsilyl
alcohol protection, **6**, 655
cleavage, **6**, 655
stability, **6**, 655
Ethers, *t*-butyldiphenylsilyl
alcohol protection
removal, **6**, 656
Ethers, *t*-butyl methyl
potassium salts
synthesis, **3**, 194
Ethers, *n*-butyl vinyl
reaction with tetracyanoethylene
thermochemistry, **5**, 76
Ethers, α-chlorodialkyl
Grignard reagents
preparation, **3**, 194
Ethers, chlorofluorocyclopropyl
rearrangement
to α-fluoroacrolein, **4**, 1020
Ethers, chloromethyl (–)-menthyl
allyl organometallics synthesis, **2**, 39
Ethers, chloromethyl methyl
α-halometallation, **3**, 194
Ethers, crotyl
Wittig rearrangement, **3**, 1004
Ethers, crotyl propargyl
ene reactions
intramolecular, **5**, 16
Ethers, crotyl propenyl
Claisen rearrangement
transition state structures, **5**, 857
Ethers, α-cyano
synthesis
via acetals and ketals, **6**, 238
Ethers, cyclohexenyl
substituted
hydrogenation, **8**, 439
Ethers, 2-cyclohexen-1-yl methyl
synthesis, **3**, 651
Ethers, *n*-decyl methyl
oxidation, **7**, 239
Ethers, dialkyl
cleavage
bromotrimethylsilane, **6**, 210
chlorination, **6**, 207
Ethers, diallyl
Wittig rearrangement, **3**, 976, 991
Ethers, α,α-dibromomethyl methyl
acid bromide synthesis, **6**, 305
Ethers, di-*n*-butyl
oxidation, **7**, 236
Ethers, dichloromethyl methyl
acid chloride synthesis, **6**, 305
anion
trialkylcarbinol synthesis, **3**, 794
Ethers, diethyl
oxidation, **7**, 235
Ethers, α,α-dihalo
ortho acid synthesis, **6**, 556
Ethers, dimethoxybenzyl
alcohol protection
oxidative deprotection, **6**, 651
Ethers, dimethoxytrityl
alcohol protection, **6**, 650
Ethers, dimethyl
boron trifluoride complexes
coordination energy, **1**, 290
deprotonation
with *n*-butyllithium, **3**, 194
potassium salts
synthesis, **3**, 194
Ethers, dipropargyl

intramolecular cycloaddition reactions
pyridoxine synthesis, **5**, 1154
Wittig rearrangement, **3**, 991
Ethers, *N,N*-(disubstituted)aminomethyl
reactions with Grignard reagents, **1**, 368
Ethers, divinyl
photoisomerization, **5**, 200
Ethers, epoxy
reaction with organocopper compounds, **3**, 225
Ritter reaction
to oxazolines, **6**, 276
Ethers, 1-ethoxyethyl
alcohol protection, **6**, 649
Ethers, ethyl propenyl
[2 + 2] cycloaddition reactions, **5**, 1067
Ethers, ethyl vinyl
cycloaddition reactions
propynyl tungsten complexes, **5**, 1073
Ethers, farnesyl
synthesis, **3**, 429
Ethers, geranyl
synthesis, **3**, 429
Ethers, α-halo
reaction with aryl Grignard reagents, **3**, 242
Ethers, halomethylsilyl
allylic alcohols
radical cyclization, **7**, 648
Ethers, β-halovinyl
coupling reactions
with aryl Grignard reagents, **3**, 492
Ethers, imidium
alcohol synthesis, **6**, 20
Ethers, imino-
alcohol inversion, **6**, 22
alcohol synthesis, **6**, 20
Ethers, iodomethyl ethyl
reaction with enol silyl ether
regioselectivity, **2**, 616
Ethers, η^1-iron allyl vinyl
Claisen rearrangement, **5**, 1075
Ethers, 1-isopropyl-2-butenyl benzyl
Wittig rearrangement, **3**, 990
Ethers, ketoxime methyl
deprotonation, **3**, 35
Ethers, (4-methoxybenzyloxy)methyl
alcohol protecting group, **7**, 246
Ethers, 2-methoxyethoxymethyl
alcohol protection
removal, **6**, 648
Ethers, methoxymethyl
alcohol protection, **6**, 647
Ethers, 2-methoxyphenoxymethyl
alcohol protection, **6**, 648
nucleophilic addition reactions, **1**, 51
Ethers, 4-methoxytetrahydropyranyl
alcohol protection
oligonucleotide synthesis, **6**, 650
Ethers, methyl
cleavage
iodotrimethylsilane, **6**, 647
Ethers, methyl cyclohexenyl
ene reactions, **5**, 1075
Ethers, methyl propenyl
metallation, **2**, 596
Ethers, (methylthio)methyl
alcohol protection, **6**, 647
synthesis
via Pummerer rearrangement, **7**, 292
Ethers, methyl tropyl
synthesis
via 1,3-sigmatropic shift, **5**, 1003
Ethers, methyl vinyl
reaction with *t*-butyllithium, **3**, 252
Ethers, monomethoxybenzyl
alcohol protection
oxidative deprotection, **6**, 651
Ethers, monomethoxytrityl
alcohol protection, **6**, 650
Ethers, neopentyl
Diels–Alder reactions, **5**, 356
Ethers, *o*-nitrobenzyl
alcohol protection
photolytic deprotection, **6**, 651
Ethers, 4-nitrophenyl
synthesis, **4**, 438
Ethers, 2-nitrovinyl ethyl
synthesis, **6**, 109
Ethers, 2-octenyl vinyl
3,3-sigmatropic rearrangement, **7**, 457
Ethers, α-(phenylthio)
α-lithio ether synthesis
reductive lithiation, **6**, 145
Ethers, phenyl vinyl
hydroformylation, **4**, 923
Ethers, propargylic
rearrangement, **6**, 852
thermolysis
retro-ene reaction, **6**, 866
Wittig rearrangement, **3**, 986
Ethers, *trans*-propenyl ethyl
reaction with tetracyanoethylene, **5**, 78
Ethers, propenyl methyl
[2 + 2] cycloaddition reactions, **5**, 75
Ethers, α-silyl
electrochemical oxidation
acetal formation, **6**, 676
Ethers, *erythro*-silylnitro-
synthesis
Henry reaction, **2**, 335
Ethers, tetrahydropyranyl
alcohol protection, **6**, 648
nucleophilic addition reactions, **1**, 51
Ethers, thexyldimethylsilyl
alcohol protection
removal, **6**, 656
Ethers, trialkylsilyl
stability
alcohol-protecting groups, **6**, 653
Ethers, (trialkylstannyl)methyl allylic
lithiation, **3**, 982
Ethers, tribenzylsilyl
alcohol protection
prostaglandin epoxidation, **6**, 657
Ethers, 2,2,2-trichloroethoxymethyl
alcohol protection, **6**, 648
Ethers, trichloroethyl
alcohol protection, **6**, 648
Ethers, triethylsilyl
alcohol protection
removal, **6**, 656
Ethers, triisopropylsilyl
alcohol protection

removal, **6**, 656
epoxidations, **7**, 382
Ethers, trimethylsilyl
alcohol protection, **6**, 653
Ethers, 2-(trimethylsilyl)ethoxymethyl
alcohol protection, **6**, 648
Ethers, trimethylsilyl vinyl
reaction with boryl triflate
boron enolates from, **2**, 113
titanium enolates
synthesis, **2**, 117
Ethers, trityl
alcohol protection, **6**, 650
Ethers, tri-*p*-xylylsilyl
alcohol protection
prostaglandin epoxidation, **6**, 657
Ethers, vinyl
alkoxymercuration, **8**, 853
cycloaddition reactions, **5**, 255
[2 + 2] cycloaddition reactions, **5**, 73
cyclopropanation, **4**, 1035, 1046
Diels–Alder reactions, **5**, 372
α-hydroxylation, **7**, 169
hydrozirconation, **8**, 683
Pauson–Khand reaction, **5**, 1045
reactions with arynes, **4**, 510
reactions with ketene acetals, **5**, 684
synthesis
via acetal hydrogenation, **8**, 212
via allyl ethers, **6**, 866
via β-hydroxyalkyl selenides, **1**, 705
Ethoxycarbonylation
dimesitylboron stabilized carbanion, **1**, 498
Ethyl acetate
reaction with bromomethylmagnesium, **1**, 398
titanium tetrachloride complex
crystal structure, **1**, 302
Ethyl acetate, 2-methoxy-
boron trofluoride complex
NMR, **1**, 293
Ethyl acetoacetate
synthesis
Claisen condensation, **2**, 796
Ethyl alaninate
hydrogenation, **8**, 242
Ethylamine, β-aryl-
synthesis
Friedel–Crafts reaction, **3**, 316
Ethylamine, *N*-aryltrichloro-
cyclization, **6**, 500
Ethylamine, 2-(1-cyclohexenyl)-
enzymatic hydroxylation, **7**, 99
Ethylamine, cyclohexyl-
synthesis
via reductive alkylation of azidocyclohexane, **8**, 386
Ethylamine, diisopropyl-
Rosenmund reduction, **8**, 287
Ethylamine, 2-methoxy-1,2-diphenyl-
imine anion alkylation, **6**, 726
Ethylamine, *N*-methyl-*N*-phenyl-
lithium aluminum hydride modifier, **8**, 171
Ethylamine, phenyl-
aldimines derived from
reaction with allyl organometallic compounds, **2**, 985, 986
α-alkoxyaldimines
reaction with allyl organometallic compounds, **2**, 987
β-alkoxyaldimines
reaction with allyl organometallic compounds, **2**, 987
Ethylamine, 1-phenyl-
conjugate additions
methyl vinyl ketone, **4**, 221
imine anion
reactions, **6**, 725
Ethylamine, 2-phenyl-
synthesis
hydroformylation, **4**, 919
Ethylamine, thienyl-
synthesis, **8**, 376
Ethyl anisate
titanium tetrachloride complex
crystal structure, **1**, 303
Ethyl benzoate
hydrogenation, **8**, 242
reduction
electrochemical, **8**, 243
metal hydrides, **8**, 244
Ethyl bromoacetate
coupling reactions
with arylzinc reagents, **3**, 466
Ethyl *n*-butyrate
reduction
metal hydrides, **8**, 266
Ethyl chloroformate
acid anhydride synthesis, **6**, 312
Ethyl cinnamate
reduction
transfer hydrogenation, **8**, 552
tin(IV) chloride complex
crystal structure, **1**, 305
Ethyl diazoacetate
ketone homologation, **3**, 783
reactions with ketones
Lewis acid catalyzed, **1**, 846
Ethylenamine
synthesis
via retro Diels–Alder reactions, **5**, 558
Ethylene
carboalumination, **4**, 887
carboboration, **4**, 885
carbolithiation, **4**, 867
carbomagnesiation, **4**, 874
carbozincation, **4**, 880
dialkylation
via σ-alkyliron complexes, **4**, 576
dicarboxylation, **4**, 947
hydrosilylation, **8**, 773
monosubstituted
hydrosilylation, **8**, 774
oligomerization
lithium hydride, **8**, 734
oxidation
Wacker process, **7**, 449
Pauson–Khand reaction, **5**, 1043
Ethylene, alkoxy-
reaction with tetracyanoethylene, **5**, 71
Ethylene, 1,1-bis(benzenesulfonyl)-
reaction with ketones
addition, **4**, 102

Ethylene, 1,2-bis(tri-*n*-butylstannyl)-
acylation
Friedel–Crafts reaction, **2**, 726
alkylation, **3**, 247
Ethylene, 1,2-bis(trifluoromethyl)-1,2-dicyano-
[2 + 2] cycloaddition reactions, **5**, 75
reaction with tetramethoxyethylene, **5**, 75
Ethylene, 1-bromo-2-phenylthio-
coupling reaction
with alkyl Grignard reagents, **3**, 449
with secondary alkyl Grignard reagents, **3**, 441
reaction with Grignard reagents
palladium catalysts, **3**, 230
tandem couplings, **3**, 492
Ethylene, bromotrifluoro-
hydrogenolysis, **8**, 900
Ethylene, chlorotrifluoro-
hydrogenation, **8**, 898
hydrogenolysis, **8**, 900
Ethylene, 1-cyano-1-alkoxycarbonyl-
[2 + 2] cycloaddition reactions, **5**, 73
Ethylene, cyclopropyl-
hydroiodination, **4**, 287
photoaddition reactions
aromatic carbonyl compounds, **5**, 165
Ethylene, 1,1-dichloro-
coupling reactions
with alkyl Grignard reagents, **3**, 448
Ethylene, 1,2-dichloro-
coupling reactions
with alkyl Grignard reagents, **3**, 449
with vinylic Grignard reagents, **3**, 487
Ethylene, 1,1-dichloro-2,2-difluoro-
addition reactions with conjugated dienes, **5**, 69
Ethylene, 1,1-dicyano-
carbonyl group protection, **6**, 680
[2 + 2] cycloaddition reactions
hydroxyethylene, **5**, 72
Ethylene, 1,1-dicyano-2,2-bis(trifluoromethyl)-
reaction with tricarbonyl(cycloheptatriene)iron complexes, **4**, 710
reaction with tricarbonyl(cyclooctatetraene)iron complexes, **4**, 709
Ethylene, 1,2-dicyclopropyl-
hydration, **4**, 298
Ethylene, 1,1-difluoro-
addition reactions
benzeneselenenyl chloride, **7**, 520
reaction with butadiene, **5**, 70
Ethylene, 1,1-difluoro-2,2-dichloro-
intramolecular [2 + 2] cycloadditions, **5**, 69
reaction with butadiene, **5**, 71
Ethylene, 1-dimesitylboryl-1-trimethylsilyl-
reactions with organometallic compounds, **1**, 492
Ethylene, 1,1-dimethoxy-
reactions with arynes, **4**, 510
Ethylene, diphenylarseno-
reaction with organolithium compounds
formation of α-arseno anions, **3**, 203
Ethylene, 1-halo-2-trimethylsilyl-
acylation
Friedel–Crafts reaction, **2**, 715
Ethylene, hydroxy-
[2 + 2] cycloaddition reactions, **5**, 72
Ethylene, iodotrifluoro-
hydrogenolysis, **8**, 900
Ethylene, nitro-
Diels–Alder reactions, **5**, 320
ene reactions
thermal, **5**, 3
Ethylene, 1-nitro-2-(3,4-methylenedioxyphenyl)-
reaction with azomethine ylides, **5**, 265
Ethylene, 2-nitro-1-(trimethylsilyl)-
synthesis
via nitryl iodide, **4**, 357
Ethylene, polychloro-
coupling reactions, **3**, 487
Ethylene, siloxy-
preparation, **2**, 600
Ethylene, β-sulfinylnitro-
Diels–Alder reactions, **5**, 320
Ethylene, tetraamino-
oxidation, **6**, 519
Ethylene, tetracyano-
adduct with 7-methylenenorbornadiene, **5**, 65
cycloaddition reactions, **5**, 273
alkenes, **5**, 71
[2 + 2] cycloaddition reactions
hydroxyethylene, **5**, 72
propenylidenecyclopropane, **5**, 76
Diels–Alder reaction
9,10-dimethylanthracene, **5**, 76
ene reactions
intermolecular, **5**, 3
polymerization initiation, **5**, 74
reaction with anthracene
thermochemistry, **5**, 76
reaction with *p*-methoxystyrene
solvent effects, **5**, 75
Ethylene, tetrafluoro-
cycloaddition reactions, **5**, 70
intramolecular [2 + 2] cycloadditions, **5**, 69
reaction with nitric oxide, **7**, 488
Ethylene, tetrahalo-
hydrobromination, **4**, 280
Ethylene, tetramethoxy-
reaction with 1,2-bis(trifluoromethyl)-1,2-dicyanoethylene, **5**, 75
Ethylene, tetramethyl-
photolysis
with 3-pentyn-2-one, **5**, 164
Ethylene, tetraphenyl-
Wurtz reaction
catalyst, **3**, 414
Ethylene, tetravinyl-
synthesis
via photolysis, **5**, 738
Ethylene, triamino-
oxidation, **6**, 519
Ethylene, tribenzoyl-
hydrobromination, **4**, 282
Ethylene, trichloro-
synthesis, **4**, 270
Ethylene, trifluoro-
reaction with butadiene, **5**, 70
Ethylene, 1-(trimethylsilyl)cyclopropyl-
synthesis
via reductive lithiation, **6**, 146
Ethylenediamine
solvent for reduction
dissolving metals, **8**, 113
Ethylenediamine, *N*,*N'*-benzylidene-

[4 + 3] cycloaddition reactions, **5**, 598
Ethylenediamine, *N,N,N′,N′*-tetramethyl-
alkali metal stabilized carbanions
crystal structure, **1**, 5
deprotonation, **1**, 476
in sulfide metallation, **3**, 86
lithium aluminum hydride modifiers, **8**, 168
Ethylene-1,1′-dicarbonitrile, 2-benzoyl-2-phenyl-
synthesis, **3**, 826
Ethylene-1,2-diols
synthesis
via retro Diels–Alder reactions, **5**, 557
Ethylene oxide
phosphonium ylide synthesis, **6**, 175
synthesis
via oxidation of ethylene, **7**, 384
Ethylene oxide, tetracyano-
reactions with alkenes
via carbonyl ylides, **4**, 1090
Ethylenetricarboxylates, 1-allylic 2,2-dimethyl
cyclization
intramolecular ene reaction, **5**, 12
Diels–Alder reactions, **5**, 461
Ethyl fluoroacetate
aldol reaction
diastereoselection, **2**, 209
Ethyl 3-furoate
[4 + 3] cycloaddition with 1-phenyl-2-oxyallyl, **5**, 601
Ethyl halides
arene alkylation
Friedel–Crafts reaction, **3**, 300
Ethyl hexanoate
reduction
metal hydrides, **8**, 244
Ethylidene transfer
Simmons–Smith reaction, **4**, 968
Ethyl iodide
ethylation with
stereochemistry, **3**, 14
Ethyl levulinate
reaction with ate complexes, **1**, 156
reaction with methyltitanium triisopropoxide, **1**, 141
Ethyl mandelate
synthesis
via hydride transfer to ethyl phenylglyoxylate, **8**, 85
Ethyl oleate
metathesis
tungsten catalysts, **5**, 1118
Ethyl pentanoate
hydrogenation, **8**, 242
Ethyl phenylglyoxylate
reduction
hydride transfer, **8**, 85, 93
Ethyl propiolate
Diels–Alder reactions, **5**, 320
Ethyl (trimethylsilyl)acetate
Peterson alkenation, **1**, 789
Eucannabinolide
synthesis
via Cope rearrangement, **5**, 809
Eudesmane
rearrangement, **3**, 388
synthesis, **3**, 396
Eudesmol
synthesis, **3**, 20; **6**, 777; **8**, 943
Eugenol, methyl-
reactions with nitriles, **6**, 272
Europium salts
use in intermolecular pinacol coupling reactions, **3**, 565
Euryfuran
synthesis, **1**, 570
Eusiderin
synthesis
use of silver oxide, **3**, 691
Evans' chiral auxiliary
use in amine synthesis, **6**, 77
Exaltone
synthesis
via cyclization, **1**, 553
via intramolecular Barbier reaction, **1**, 262

F

Fabianine
  synthesis
    via Diels–Alder reaction, **5**, 492
Faranal
  synthesis, **8**, 556
Faranal, dehydro-
  reduction, **8**, 556
α-Farnesene
  synthesis
    via carboalumination, **4**, 893
Farnesol
  bicyclization, **3**, 342
  cyclization, **3**, 360
  derivatives
    reduction, **8**, 961
  peroxy ester
    intramolecular epoxidation, **7**, 381
  synthesis, **3**, 170
    via carboalumination, **4**, 893
Farnesol, 10,11-epoxy-
  synthesis, **3**, 99
Fastigilin-C
  synthesis
    via Claisen rearrangement, **5**, 851
Fatty acid alcohols
  synthesis
    alkene metathesis, **5**, 1117
Fatty acids
  synthesis, **3**, 643
  unsaturated
    hydrofluorination, **4**, 271
Favorskii rearrangement, **3**, 839–857
  in synthesis, **3**, 842
  Lewis acids, **3**, 856
  mechanism, **3**, 840
  reaction conditions, **3**, 840
  side-products, **3**, 840
  stereospecificity, **3**, 848
Fawcettimine
  synthesis, **2**, 157
Felkin–Anh addition
  single stereocenter imines
    reaction with allyl organometallic reagents, **2**, 983
Felkin–Anh paradigm
  chiral aldehyde reactions
    with pinacol crotylboronates, **2**, 25
Felkin model
  aldol reaction
    asymmetric induction, **2**, 219
Fenchenes
  synthesis
    from fenchyl alcohol, **3**, 709
Fenchone
  reduction
    dissolving metals, **8**, 121
  Tebbe reaction, **1**, 743
Fenchyl alcohol
  rearrangement, **3**, 709
[4.4.4.5]Fenestrane
  synthesis, **3**, 901
Fenestranes
  synthesis
    Dieckmann reaction, **2**, 829
    via photocycloaddition reactions, **5**, 136
Fenton's reagent
  alkane hydroxylation, **7**, 11
Ferensimycin
  synthesis
    via *N*-methoxy-*N*-methylamide chemistry, **1**, 402
Ferrates, acyltetracarbonyl-
  reduction
    acyl chlorides, **8**, 289
Ferrates, hydrido-
  reduction
    imines, **8**, 36
Ferrates, tetracarbonyl-
  reduction
    acyl chlorides, **8**, 289
    nitroarenes, **8**, 371
Ferrates, tetracarbonylhydrido-
  dehalogenation
    α-halocarbonyl compounds, **8**, 991
  reduction
    acyl chlorides, **8**, 289
    imidoyl chlorides, **8**, 301
    unsaturated carbonyl compounds, **8**, 550
Ferrier-type rearrangements
  Claisen rearrangements
    competition, **5**, 850
Ferrocene
  Mannich reaction, **2**, 961
Ferrocenecarbothioates, *O*-alkyl
  synthesis
    via *S*-methyl ferrocenecarbodithioate, **6**, 450
Ferrocenophanes
  synthesis, **3**, 594
Ferrocenylcarbaldehyde
  Knoevenagel reaction, **2**, 365
Ferruginol
  synthesis, **3**, 169
Ferryl radicals
  Fenton's reagent
    hydroxylation of alkanes, **7**, 11
Ferulic acid
  oxidation, **3**, 693
Fervenulin
  analogs
    synthesis, **7**, 342
Fervenulone, 2-methyl-
  synthesis, **7**, 342
Finkelstein reaction
  chlorine/bromine exchange, **6**, 212
  iodination, **6**, 216
Finkelstein-type reaction
  alkyl tosylates
    organosamarium compounds, **1**, 257
Fischer carbene complexes
  reactions with alkynes
    alkyne concentration, **5**, 1099
    mechanisms, **5**, 1094
    solvents, **5**, 1099
Fischer–Helferich method
  glycosides
    synthesis, **6**, 34, 35

Fischer–Spei esterification
  acid catalysis, **6**, 325
Fittig synthesis
  Perkin transformation, **2**, 401
Five-membered rings
  formation
    polyene cyclization, **3**, 347
  synthesis
    aldol reaction cascade, **2**, 620
    Friedel–Crafts reaction, **2**, 756
FK-506
  synthesis, **1**, 799
    via acylation of dithiane, **1**, 425
    via Ireland rearrangement, **5**, 843
    via *N*-methoxy-*N*-methylamide chemistry, **1**, 402
    via organoaluminum reagents, **1**, 101
    via Schlessinger method, **1**, 791
Flash vacuum pyrolysis
  alkene protection, **6**, 689
Flash vapor pyrolysis
  retrograde Diels–Alder reactions, **5**, 552
Flattening rule
  reduction
    cyclic ketones, **8**, 7
Flavanone, 3-hydroxy-
  ring scission, **3**, 831
Flavanones
  bromination, **7**, 120
  dehydrogenation
    use of thallium trinitrate, **7**, 144
  reduction
    aluminum hydrides, **8**, 545
    metal hydrides, **8**, 314
Flavenes
  synthesis
    via aromatic Claisen rearrangement, **5**, 834
Flavinantine, methyl-
  synthesis, **3**, 81
    anodic oxidation, **3**, 685
    electrooxidation, **3**, 685
Flavins
  oxidation
    sulfides, **7**, 763
    thiols, **7**, 761
Flavobacterium spp.
  reduction
    unsaturated carbonyl compounds, **8**, 560
Flavones
  intramolecular acyl transfer, **2**, 845
  synthesis, **7**, 120, 136
4-Flavones
  synthesis
    Knoevenagel reaction, **2**, 379
Flavopereirine
  synthesis
    via 3-lithiation of an indole, **1**, 474
Flexibilene
  synthesis, **3**, 591
Fluorene
  Birch reduction
    dissolving metals, **8**, 496
  synthesis, **3**, 543
Fluorene, diazo-
  synthesis
    via fluorenone hydrazone, **7**, 742
Fluorene, 9-diazo-
  deoxygenation
    epoxides, **8**, 890
Fluorene, 9-(difluoromethylene)-
  cycloaddition reactions, **5**, 70
Fluorene, 9,9-disubstituted
  synthesis
    via alkyllithium addition, **1**, 377
Fluorene, 1-methyl-
  synthesis, **8**, 140
Fluorene, tetrahydro-
  hydrogenation
    heterogeneous catalysis, **8**, 430
Fluorene-1-carbaldehyde
  hydrogenation
    catalytic, **8**, 140
Fluorenecarboxylic acid, 9-hydroxy-
  synthesis, **3**, 828
Fluoren-2-ol, 7-methoxy-
  Birch reduction
    dissolving metals, **8**, 497
9-Fluorenone
  synthesis, **3**, 828
Fluorenonecarboxylic acid
  reduction
    hydrogen iodide, **8**, 323
Fluorenone-4-carboxylic acid
  synthesis
    Friedel–Crafts reaction, **2**, 757
Fluorenones
  reduction
    dissolving metals, **8**, 115
    Wolff–Kishner reduction, **8**, 338
  synthesis
    Friedel–Crafts reaction, **2**, 757
  tosylhydrazone
    reactions with alkyllithium, **1**, 377
9-Fluorenyl anions
  aromatic nucleophilic substitution, **4**, 429
9-Fluorenylmethoxycarbonyl group
  protecting group
    hydrogenolysis, **6**, 638
    peptide synthesis, **6**, 638
9-Fluorenylmethyl carbonate
  alcohol protection
    cleavage, **6**, 659
Fluorides
  catalyst
    enol silane reaction with aldehydes, **2**, 633
Fluorides, 1,2-iodo-
  synthesis, **7**, 536
Fluorides, 1,2-nitro-
  synthesis
    via electrophilic nitration, **4**, 356
Fluorination
  alkanes, **7**, 15
  nucleophilic displacement, **6**, 216
  secondary amines, **7**, 747
Fluorine
  reactions with alkenes, **4**, 344
Fluorodesulfonylation
  arylfluorosulfonyl fluorides, **4**, 445
Fluorohydrin
  synthesis
    epoxide ring opening, **3**, 749
Fluorohydrocarbons
  synthesis, **3**, 640

Fluoromethylenation
  carbonyl compounds
    sulfoximines, **1**, 741
Fluoronitration
  alkenes, **7**, 498
Fluorosulfonic acid esters
  amide alkylation, **6**, 502
Fluorosulfuric acid
  catalyst
    Friedel–Crafts reaction, **3**, 297
Formaldehyde
  aldol reaction, **2**, 139
  borane complexes
    rotational barriers, **1**, 290
  conjugate enolate trapping, **4**, 261
  Diels–Alder reactions, **2**, 662; **5**, 433
  ene reaction
    chlorodimethylaluminum catalyzed, **2**, 531
    Lewis acid catalyzed alkene addition, **2**, 530
  lithium salt complexes
    theoretical studies, **1**, 286
  Prins reaction
    addition to alkenes, **2**, 528
  proton complexes
    theoretical studies, **1**, 286
  reaction with phenols
    Mannich reaction, **2**, 956
  reaction with water
    Lewis acids, **1**, 315
  reduction
    metal hydrides, **8**, 2
  thermal ene reaction
    addition to alkenes, **2**, 529
Formaldehyde dimethyl dithioacetal *S*-oxide
  reaction with carbonyl compounds, **1**, 526
Formaldehyde dithioacetals
  formyl anion equivalents, **1**, 510
Formaldehyde di-*p*-tolyl dithioacetal *S*-oxide
  reaction with enones, **1**, 527
  synthesis
    via menthyl *p*-toluenesulfinate, **1**, 526
Formaldehyde imines
  synthesis
    Mannich reaction, **2**, 915
  trimerization, **1**, 361
Formaldiminium ions
  alkyne cyclization, **2**, 1029
Formaldine, methylthio-
  alkylation, **3**, 137
  metallated
    alkylation, **3**, 135
Formaldines
  α-heterosubstituted
    carbonyl compound synthesis from, **3**, 141
Formaldoxime ethers
  reactions with organometallic compounds, **1**, 385
Formamides
  dehydration, **6**, 243
  reduction
    metal hydrides, **8**, 249
  synthesis, **3**, 420
Formamides, *N*-*t*-alkyl-
  Ritter reaction, **6**, 266
Formamides, dimethyl-
  acid chloride synthesis, **6**, 302
Formamides, *N*,*N*-dimethyl-
  adducts
    phosphorus oxychloride, **6**, 487
  dialkyl acetals
    reaction with carbene complexes, **5**, 1079
Formamides, *N*-trimethylsilyl-*N*-alkyl-
  reactions with organocopper complexes, **1**, 124
Formamidines
  alkylation, **3**, 68
  lithiation, **1**, 482
  synthesis, **6**, 490
Formamidinium chloride, chloro-
  synthesis, **6**, 331
Formamidinium chloride, *N*,*N*,*N*′,*N*′- tetramethylchloro-
  acid anhydride synthesis, **6**, 313
Formamidinium-*N*,*N*-dialkyl dithiocarbaminates
  synthesis, **6**, 518
Formamidinium perchlorate, *N*,*N*,*N*′,*N*′- tetramethyl-
  synthesis, **6**, 518
Formamidinium salts, *N*,*N*′-dialkyl-*N*,*N*′-diaryl-
  synthesis, **6**, 518
Formamidinium salts, *N*,*N*′-diaryl-
  synthesis, **6**, 512
Formamidinium salts, *N*,*N*,*N*′,*N*′-tetrasubstituted
  synthesis, **6**, 512
Formanilide, *N*-methyl-
  adducts
    phosphorus oxychloride, **6**, 487
  Vilsmeier–Haack reaction, **2**, 779
Formates
  hydride donor
    hydrogenolysis, **8**, 958
  reduction
    hydride transfer, **8**, 84
Formates, alkyl chloro-
  anhydride synthesis, **6**, 329
  dimethylformamide adducts, **6**, 491
  nitrile synthesis, **6**, 234
  reactions with amides, **6**, 504
Formates, alkylthio chloro-
  reaction with thioamides, **6**, 508
Formates, azido-
  synthesis, **6**, 251
Formates, chloro-
  reduction
    silanes, **8**, 825
Formates, chlorothio-
  ketone synthesis from
    Grignard reagents, **3**, 463
Formates, cyano-
  Diels–Alder reactions, **5**, 416
  synthesis
    via chloroformates, **6**, 233
Formates, α-metalloorthoseleno-
  ester precursors, **3**, 144
Formates, α-metalloorthothio-
  ester precursors, **3**, 144
Formates, trichloromethyl chloro-
  amide synthesis
    elevated temperature, **6**, 383
Formates, triphenylmethyl
  Ritter reaction, **6**, 269
Formic acid
  amides
    catalytic hydrogenation, **8**, 144
  hydride transfer, **8**, 84, 557
    to carbonium ions, **8**, 91

hydrogenation
nitriles, **8**, 299
reduction
carboxylic acids, **8**, 285
pyridinium salts, **8**, 590
Formic acid, azodi-
hydrolysis of dipotassium or disodium salt
diimide from, **8**, 472
Formic acid, benzoyl-
amides
catalytic hydrogenation, **8**, 145
bornyl ester
asymmetric hydrogenation, **8**, 144
menthyl ester
asymmetric hydrogenation, **8**, 144
methyl ester
hydrogenation, **8**, 151
phenethyl ester
asymmetric hydrogenation, **8**, 144
Formic acid, ethylcyano-
Knoevenagel reaction, **2**, 368
Formic acid anhydride
synthesis
via chlorosulfonyl isocyanate, **6**, 313
Formiminium salts
reduction
carboxylic acids, **8**, 285
Formonitrile, (ethoxycarbonyl)-
oxide
reaction with isopropylidene-3-butene-1,2-diol, **5**, 262
N-Formylamines
adducts
phosphorus oxychloride, **6**, 487
Formylation
aliphatic, **2**, 728
amines, **6**, 384
aromatic nucleophilic substitution and hydrolysis, **4**, 432
carbonyl compounds
samarium diiodide, **1**, 274
Gattermann and related reactions, **2**, 749
ketones, **2**, 837
Formyl chloride
generation, **2**, 749
synthesis
via 1-dimethylamino-1-chloro-2- methylpropene, **6**, 306
N-Formylenamines
synthesis
via reductive cleavage, **8**, 393
Formyl fluoride
formylation
modified Gattermann–Koch reaction, **2**, 749
N-Formyl group
amine-protecting group, **6**, 642
Formyl iodide
synthesis, **6**, 306
Forskolin
microbial hydroxylation, **7**, 64
synthesis, **7**, 105; **8**, 171
via alkynide addition, **1**, 421
via Cope rearrangement, **5**, 814
via 6-*exo*-trig cyclization, **4**, 40
Forster reaction
diazo compounds
synthesis from oximes, **7**, 751
Forsythide
aglucone dimethyl ester
synthesis *via* photoisomerization, **5**, 231
Fosfomycin
microbial epoxidation, **7**, 429
Four component condensation — *see* Ugi reaction
Four-membered rings
synthesis
aldol reaction cascade, **2**, 619
Fragmentation reactions, **6**, 1041–1069
acceptor radical anions, **7**, 882
enolate assisted, **6**, 1056
mechanism, **6**, 1043
metal asisted, **6**, 1061
radical anions
unimolecular decomposition, **7**, 861
radical cyclization, **4**, 824
seven-center, **6**, 1042
silicon assisted, **6**, 1061
stereochemistry, **6**, 1043
α-Fragmentation reactions
donor radical cations, **7**, 873
radical cations
unimolecular reaction, **7**, 857
β-Fragmentation reactions
alkoxy radicals, **4**, 815–818
donor radical cations, **7**, 874
radical cations
unimolecular reaction, **7**, 857
Fragranol
synthesis, **3**, 103
Fredericamycin A
synthesis, **7**, 340
organocopper compounds, **3**, 219
Free radicals
electron-transfer equilibria, **7**, 850
oxidation
Ritter reaction, **6**, 280
Fremy's salt
oxidation
primary amines, **7**, 737
secondary amines, **7**, 746
quinone synthesis, **7**, 143, 346
Frenolicin, deoxy-
synthesis, **5**, 1096
via arene–metal complexes, **4**, 539
via cyclobutenone ring opening, **5**, 690
regioselective, **5**, 1094
Friedelan-3-one
reduction
dissolving metals, **8**, 117
with lactone
reduction, dissolving metals, **8**, 117
Friedelan-7-one
reduction
dissolving metals, **8**, 119
Friedel–Crafts reaction
acylation
arenes with thiol esters, **6**, 445
aliphatic, **2**, 707–731; **3**, 294
catalysts, **2**, 709
mechanism, **2**, 708
reaction temperatures, **2**, 709
alkylating agents, **3**, 294
alkylation, **3**, 293–335

arene alkylation
kinetics, **3**, 300
asymmetric alkylation, **3**, 302
bimolecular aromatic, **2**, 733–750
catalysts, **2**, 735
mechanism, **2**, 734
reagent systems, **2**, 735
solvents, **2**, 738
stoichiometry, **2**, 739
use of protic acid, **2**, 711
Bouveault procedure, **2**, 738
catalysts, **3**, 294, 295
cocatalysts, **3**, 295
dithiocarboxylation, **6**, 456
Elbs procedure, **2**, 738
epoxides, **3**, 769
intramolecular aromatic, **2**, 753–766
electron density, **2**, 754
ring size, **2**, 755
Perrier procedure, **2**, 738
rearrangement, **2**, 745
Ritter reaction
initiators, **6**, 283
thioacylation
arenes and carbanions, **6**, 453
transacylation, **2**, 745
Friedelin
backbone rearrangement, **3**, 709
Fries reaction, **2**, 745
Frontalin
synthesis
via chiral auxiliary, **1**, 65
via Wacker oxidation, **7**, 451
Frontier Molecular Orbital Theory
1,3-dipolar cycloadditions, **4**, 1073
radical reactions, **4**, 727
Fructose
separation from glucose
Knoevenagel reaction, **2**, 354
(+)-Fructose, 1-alkylamino-1-deoxy-
synthesis
via (+)-glucosylamine, **6**, 789
Fructose, deoxy-
synthesis
FDP aldolase, **2**, 462
D-Fructose-1,6-diphosphate aldolase
catalytic action, **2**, 456
characteristics, **2**, 461
substrate preparation
dihydroxyacetone phosphate, **2**, 461
substrate specificity, **2**, 456
use in organic syntheses, **2**, 457, 462
Frullanolide
synthesis, **3**, 1031
α-L-Fucopyranosides
synthesis, **6**, 42
Fucose
synthesis
Diels–Alder reaction, **2**, 689
Fucoside, allyl
selective cleavage, **6**, 652
Fulgides
photochemical ring closure, **5**, 722
Fulmonitrile oxide
reaction with acetylene
ab initio calculations, **4**, 1070
Fulvene, 6-amino-
cycloaddition reactions
α-pyrones, **5**, 626
Fulvene, bis(methylthio)-
reaction with alcoholates, **6**, 557
Fulvene, 6,6-dialkyl-
[4 + 3] cycloaddition reactions, **5**, 604
Fulvene, 2,3-diformyl-6-dimethylamino-
Knoevenagel reaction, **2**, 366
Fulvene, 9,10-dihydro-
Diels–Alder reactions, **5**, 347
Fulvene, 6,6-dimethyl-
cycloaddition reactions
benzocyclobutenes, **5**, 627
dienes, **5**, 626
nitropyridyl betaines, **5**, 630
tropones, **5**, 631
retro Diels–Alder reaction, **5**, 563
Fulvene, 6-(dimethylamino)-
cycloaddition reactions
dienes, **5**, 627
thiophenes, **5**, 629
[4 + 3] cycloaddition reactions, **5**, 604
Fulvene, 6,6-diphenyl-
cycloaddition reactions
dienes, **5**, 627
Fulvene, 6-methyl-
cycloaddition reactions
dienes, **5**, 629
Fulvenes
anthracene adduct
retro Diels–Alder reaction, **5**, 589
cycloaddition reactions, **5**, 626
[4 + 3] cycloaddition reactions, **5**, 604
Pauson–Khand reaction, **5**, 1046
retrograde Diels–Alder reactions, **5**, 552
synthesis
via benzene irradiation, **5**, 646
via lithium–halogen exchange, **4**, 1008
tandem vicinal difunctionalization, **4**, 242, 253
Vilsmeier–Haack reaction, **2**, 782
Fumarates, dimenthyl-
Diels–Alder reactions, **5**, 355
Fumarates, iodo-
dimerization, **3**, 482
Fumaric acid, cyano-
dimethyl ester
[2 + 2] cycloaddition reactions, **5**, 73
Fumaric acid, 2,3-dicyano-
dimethyl ester
cycloadditions, **4**, 1074
Fumaronitrile
Ritter reaction, **6**, 265
synthesis
via 1,2-diiodoethylene, **6**, 231
Fumaryl chloride
synthesis
via maleic anhydride, **6**, 304
Functional group transformations
allyl rearrangement, **6**, 829–867
Furan, 2-alkenyldihydro-
synthesis
via cyclization of γ-allenic ketones, **4**, 397
Furan, 2-alkenyltetrahydro-
synthesis
via cyclization of γ-allenic alcohols, **4**, 395

Furan, alkylidenetetrahydro-
  synthesis
    via [3 + 2] cycloaddition reactions, **5**, 283
  tetrasubstituted
    synthesis, **1**, 591
Furan, aminomethyl-
  synthesis, **3**, 258
Furan, 2,5-bis(trimethylsiloxy)-
  reaction with carbonyl compounds
    titanium tetrachloride catalyst, **2**, 617
  synthesis
    from succinic anhydrides, **2**, 607
Furan, 3-bromomethyl-
  alkylation by
    cuprates, **3**, 250
Furan, 2-(bromomethyl)tetrahydro-
  reaction with ketones
    samarium diiodide, **1**, 259
Furan, 2,2-dialkoxydihydro-
  synthesis, **6**, 559
Furan, dihydro-
  [2 + 3] annulation, **5**, 930
  coupling reactions
    with alkyl Grignard reagents, **3**, 444
  Pauson–Khand reaction, **5**, 1046
  reaction with Grignard reagents
    nickel catalysts, **3**, 229
  reaction with organocopper compounds, **3**, 218
  synthesis
    from allylic anions and carbonyls, **2**, 60
    lithium allenes, **2**, 88
    ring formation, **6**, 24
    selectivity, **5**, 907
    via allenylsilanes, **1**, 599
    via [3 + 2] cycloaddition reactions, **5**, 279
    via cyclopropanation, **4**, 1035, 1046, 1049
    via metal-catalyzed cycloaddition, **5**, 1200
    via [2 + 3] reaction, **5**, 951
    via rearrangements, **5**, 952
    via retro Diels–Alder reactions, **5**, 579
    via vinyloxiranes, **5**, 929
Furan, 2,5-dihydro-
  synthesis
    allenyllithium compounds, **2**, 89
Furan, 4,5-dihydro-
  synthesis
    Knoevenagel reaction, **2**, 380
Furan, 2,5-dihydro-3,4-dimethyl-
  synthesis
    via retro Diels–Alder reactions, **5**, 579
Furan, 2,3-dihydro-2,3-dimethylene-
  synthesis
    via retro Diels–Alder reactions, **5**, 579
Furan, 2,2-dimethoxy-2,3-dihydro-
  synthesis
    via ring opening of dichlorocyclopropyl
      compounds, **4**, 1022
Furan, 2,5-dimethoxy-2,5-dihydro-
  synthesis, **7**, 802
Furan, dimethyl-
  hydrogen donor, **8**, 557
Furan, 2,3-dimethylene-
  dimerization, **5**, 638
    via [4 + 4] cycloaddition, **5**, 639
Furan, 2,5-dimethyltetrahydro-
  synthesis
    via 2,5-hexanediol, **6**, 25
Furan, hydroxydihydro-
  synthesis
    from benzoin and DMAD, **4**, 52
Furan, 2-lithio-
  alkylation, **3**, 261
  reaction with propylene oxide, **3**, 264
Furan, 3-lithio-
  reaction with epoxides, **3**, 264
Furan, 2-methoxycarbonyl-
  Diels–Alder reactions, **5**, 382
Furan, 2-methoxy-2,5-dihydro-
  synthesis, **2**, 89
Furan, 2-methyl-
  [4 + 3] cycloaddition reactions, **5**, 606
  hydrogenation, **8**, 606
  Mannich reaction
    with formaldehyde and dimethylamine, **2**, 964
Furan, 3-methyl-
  [4 + 3] cycloaddition with 1-phenyl-2-oxyallyl, **5**, 601
Furan, 4-methyl-
  synthesis
    via activated allene, **4**, 54
Furan, 2-methyl-4,5-dihydrotetrahydro-
  carboboration, **4**, 885
Furan, 3-methylenetetrahydro-
  synthesis
    via [3 + 2] cycloaddition reactions, **5**, 307
    via metal-catalyzed cycloaddition, **5**, 1196
Furan, 2-methyl-3-phenyl-
  synthesis
    via 3-phenyl-4-oxopentanal, **7**, 456
Furan, 2-methyltetrahydro-
  alkylation
    Friedel–Crafts reaction, **3**, 317
  benzene alkylation
    Friedel–Crafts reaction, **3**, 315
  nucleophilic addition reactions
    Grignard reagents, **1**, 72
Furan, 3-silyldihydro-
  synthesis, **2**, 575
Furan, tetrahydro-
  annulation, **1**, 891
  arene alkylation by
    Friedel–Crafts reaction, **3**, 315
  conjugate additions
    organocuprates, solvent effects, **4**, 178
  deprotonation, **3**, 194
  cis-2,5-disubstituted
    synthesis, stereoselectivity, **4**, 383
  trans-2,5-disubstituted
    synthesis *via* cyclization of γ-alkenyl alcohols, **4**,
      378
  cis fused
    synthesis *via* cyclization, **4**, 371
  oxidation, **7**, 236
    electrochemical, **7**, 248
  polycyclic
    oxidation, **7**, 239
  potassium salts
    synthesis, **3**, 194
  solvent for reduction
    dissolving metals, **8**, 112
  spirocyclic
    synthesis, stereochemistry, **4**, 390
  3-substituted

synthesis, **3**, 647
synthesis, **3**, 792
palladium(II) catalysis, **4**, 558
ring formation, **6**, 24
via cyclopropane ring opening, **4**, 1046
via electrophile cyclization, **7**, 523
via metal-catalyzed cycloaddition, **5**, 1200
via palladium-ene reactions, **5**, 51
via vinyloxiranes, **5**, 927
Furan, tetramethylenetetrahydro-
synthesis
via retro Diels–Alder reactions, **5**, 579
Furan, 1-trimethylsiloxy-
aldol reaction
regiochemistry, **2**, 625
Furan, 2-trimethylsiloxy-
aldol condensation
stereoselectivity, **2**, 634
cyclic
reaction with aldehydes, stereoselectivity, **2**, 632
reaction with carbonyl compounds
tin(IV) chloride catalyst, **2**, 617
Furan, vinyl-
cyclopropanation, **4**, 1059
Furanacetic acid lactones, *cis*-3-hydroxytetrahydro-
synthesis
palladium-catalyzed oxycarbonylation, **3**, 1033
Furanal, tetrahydro-
nucleophilic addition reactions
selectivity, **1**, 53
Furan-3-carbodithioate, ethyl
synthesis
via nitrile, **6**, 455
Furancembraolides
synthesis
(*Z*)-selectivity, **1**, 767
Furandiones
synthesis
via [2 + 2 + 2] cycloaddition, **5**, 1138
Furaneol
synthesis
FDP aldolase, **2**, 462
Furanether B
synthesis
via Pauson–Khand reaction, **5**, 1052
3-Furanmethanol
synthesis
via photocycloaddition, **5**, 169
Furanocyclopropane
synthesis
via ketocarbenoids and furans, **4**, 1058, 1059
Furanols
asymmetric epoxidation
kinetic resolution, **7**, 423
Furanomycin
synthesis
Ugi reaction, **2**, 1100
Furanonapthoquinones
synthesis
Friedel–Crafts reaction, **2**, 744
3(2*H*)-Furanone
dienolate
reaction at γ-position, **2**, 189
Furanone, acyl-
synthesis
Knoevenagel reaction, **2**, 359
Furan-2-one, 3,5-dimethylenetetrahydro-
synthesis, **6**, 784
3(2*H*)-Furanone, 2,2-disubstituted-5-alkyl-
extended dienolates
γ-alkylation, **3**, 24
Furanones
synthesis
via C—H insertion reactions, **3**, 1056
via [2 + 2 + 2] cycloaddition, **5**, 1136–1138
via [3 + 2] cycloaddition reactions, **5**, 286
via cyclopropane ring opening, **4**, 1046
via dibromocyclopropyl compounds, **4**, 1023
via palladium(II)-catalyzed acylation, **1**, 450
Furan-2(5*H*)-ones
5-substituted
synthesis, **1**, 514
Furanones, tetrahydro-
synthesis, **5**, 943
Furanose
synthesis, **6**, 35
Furanosides
reductive ring cleavage, **8**, 218
L-*ido*-Furanosides
synthesis
via vinylmagnesium bromide addition to α-nitroalkenes, **4**, 85
Furanoterpenes
synthesis
via retro Diels–Alder reactions, **5**, 579
Furans
acylation
Friedel–Crafts reaction, **2**, 744
anodic oxidation, **7**, 802
coupling reactions
with sp^3 organometallics, **3**, 459
[4 + 3] cycloaddition reactions, **5**, 605–607
Diels–Alder reactions, **5**, 342, 380–383
comparison of promoters, **5**, 345
intermolecular dimerization, **3**, 509
γ-lactone synthesis, **6**, 365
lithiation, **1**, 472
Mannich reaction
with formaldehyde and secondary amines, **2**, 964
oxidation
pyridinium chlorochromate, **7**, 267
photocycloaddition reactions
benzene, **5**, 637
carbonyl compounds, **5**, 168–178
reactions with ketocarbenoids, **4**, 1058–1061
reduction, **8**, 603–630
retrograde Diels–Alder reactions, **5**, 552
synthesis
via activated alkynes, **4**, 52
via alkynes, palladium(II) catalysis, **4**, 557, 567
via σ-alkyliron complexes, **4**, 576
via allenyl organoaluminum, **2**, 88
via [2 + 2 + 2] cycloaddition, **5**, 1092, 1136
via cyclopropane ring opening, **4**, 1046
via Diels–Alder reactions, **5**, 491
via ketocarbenoid addition to alkynes, **4**, 1051
via Knoevenagel reaction, **2**, 380
Vilsmeier–Haack reaction, **2**, 780
Furanyl sulfides, tetrahydro-
reduction, **8**, 230
Furfural
aldol reaction, **2**, 134

2-Furfuryl alcohol
  Claisen–Cope rearrangement, **5**, 879
Furfuryl alcohols, tetrahydro-
  synthesis, **7**, 632
Furfurylidene carbinols
  electrocyclization, **5**, 771
2,2′-Furil
  rearrangement, **3**, 826
Furofuran lignans
  synthesis
    Knoevenagel reaction, **2**, 372
2-Furoic acid, 5-alkyl-
  reduction
    dissolving metals, **8**, 607
2-Furoic acid, 5-phenyl-
  reduction
    dissolving metals, **8**, 607
Furoic acids
  Birch reduction, **8**, 607
Furopyran
  hydrogenation, **8**, 625
Furo[3,4-*c*]pyridine
  synthesis
    via retro Diels–Alder reactions, **5**, 584
Furopyridines
  synthesis, **3**, 543
Furoxans
  ring opening, **8**, 664
  synthesis
    via 1,3-dipolar cycloadditions, **4**, 1079
Furst–Plattner rule
  epoxides
    opening, **3**, 734
Furylamine, tetrahydro-
  chiral catalysts
    nucleophilic addition reactions, **1**, 72
2-Furylcarbinols
  solvolysis
    divinyl ketones from, **5**, 771
3-Furylmethyl benzoate, 2-methyl-
  flash vacuum pyrolysis
    [4 + 4] cycloaddition, **5**, 639
Fuscinic acid
  dimethyl ether
    oxidation, **3**, 831
Fused rings
  radical cyclizations, **4**, 791
Fusicocca-2,8,10-triene
  synthesis
    via allyl chromium reagents, **1**, 187
Fusicoccins
  synthesis, **3**, 575; **7**, 710
Futoene
  synthesis, **3**, 696

G

Gabaculine
synthesis
via cyclohexadienyl complexes, **4**, 682
Gabriel synthesis
amines, **6**, 79
modified, **6**, 81
aziridines, **7**, 472
β-D-Galactopyranose, 1,6-anhydro-
benzylidene acetal
reduction, **8**, 227
Galactopyranose, cyclohexylidene-
Paterno–Büchi photocycloaddition reaction
with furan, **5**, 187
α-D-Galactopyranosides
synthesis, **6**, 42
β-D-Galactopyranosides
synthesis, **6**, 41
α-D-Galactopyranosides, 2-acetamino-2-deoxy-
synthesis, **6**, 42
Galactopyranosides, benzylidene-
reduction, **8**, 230
Galactopyranosides, methyl 3,4-*O*-benzylidene-
reduction, **8**, 227
β-D-Galactopyranosylamine, 2,3,4,6-tetra-*O*-pivaloyl-
Ugi reaction
highly stereoselective reaction, **2**, 1099
β-Galactosamine, 2-deoxy-2-phthalimido-
reactivity, **6**, 42
Galactose
reduction, **8**, 224
D-Galactose, 2,3,4-tri-*O*-benzyl-
glycoside synthesis, **6**, 57
D-Galactose oxidase
oxidation
diols, **7**, 312
Galbulin
synthesis, **3**, 696
Gallium, trimethyl-
lithium alkynides
reaction with oxiranes, **3**, 279
reactions with epoxides
Lewis acid, catalytic, **1**, 343
Gallium trichloride
polystyrene-divinylbenzene copolymer beads, catalyst
Friedel–Crafts reaction, **3**, 298
Gascardic acid
synthesis
Dieckmann reaction, **2**, 815, 824
via Johnson methylenation, **1**, 738
Gattermann reactions, **2**, 749
bromination, **6**, 211
chlorination, **6**, 208
thiocarbonyl compounds, **3**, 582
Geiparvarin
synthesis
from 3(2*H*)-furanone, **2**, 189
Geissoschizine
synthesis, **1**, 593; **6**, 739, 743
Mannich reaction, **2**, 1031
Gelsemine
synthesis, **2**, 1069, 1072; **7**, 318
N-acyliminium ions, **2**, 1065
Geneserine
synthesis
via benzocyclobutene ring opening, **5**, 681
Geodiamolide A
synthesis
via Johnson rearrangement, **5**, 839
Geodoxin
synthesis
use of lead dioxide, **3**, 690
Geotrichum candidum
reduction
unsaturated carbonyl compounds, **8**, 560
Gephyrotoxin
synthesis, **8**, 652
Eschenmoser coupling reaction, **2**, 876, 877
via reductive cleavage of tetrahydrooxazines, **8**, 395
Gephyrotoxin, perhydro-
synthesis
Eschenmoser coupling reaction, **2**, 877
Mannich cyclization, **2**, 1041
Gephyrotoxin-223AB
synthesis, **1**, 559
via Diels–Alder reaction, **5**, 421
via organoaluminum-promoted Beckmann rearrangement, **1**, 104
Geranial
asymmetric reduction
aluminum hydrides, **8**, 545
biochemical reduction, **8**, 559
hydrogenation
homogeneous catalysis, **8**, 462
Geraniol
asymmetric epoxidation, **7**, 395, 409
asymmetric hydrogenation
synthesis of citronellol, **8**, 462
aziridination, **7**, 481
biochemical reduction, **8**, 559, 560
chlorination
displacement of hydroxy group, **6**, 205
cyclization, **3**, 345, 347, 352
epoxidation, **7**, 368
microbial hydroxylation, **7**, 62
oxidation, **7**, 306
synthesis
via carboalumination, **4**, 893
Geraniol, geranyl-
cyclization, **3**, 362
Geraniol, tetrahydro-
oxidation
solid support, **7**, 841
Geranyl
synthesis, **3**, 428
Geranyl acetate
allylic oxidation, **7**, 89
allylic oxidative rearrangement, **7**, 109
reduction, **8**, 960
Geranylacetone
cyclization, **3**, 346
Geranyl chloride
aziridination, **7**, 481
Germacranes

allylic oxidation, **7**, 88
synthesis, **7**, 625
via Cope rearrangement, **1**, 882; **5**, 796
Germacranolides
synthesis
via Cope rearrangement, **5**, 809
transannular cyclization, **3**, 396
Germacrene, dihydro-
synthesis, **1**, 561
Germacrenes
transannular reactions, **3**, 389
synthesis
Germacrone lactones
synthesis
via cyclization, **1**, 553
Germacrones
Cope rearrangement, **5**, 809
intramolecular cyclization
epoxide ring opening, **3**, 769
synthesis
from protected cyanohydrins, **3**, 198
via cyclization, **1**, 553
Germane, allyltrimethyl-
ene reactions, **5**, 2
Germane, chlorotrimethyl-
reaction with ketone enolates
preparation of enol germyl ethers, **2**, 610
Germanium hydride, tributyl-
hydrogen donor
radical reactions, **4**, 738
Germanium hydrides
quinone reduction, **8**, 19
Germylcupration
alkynes, **4**, 901
Gibbane
synthesis, **2**, 167
Gibberellic acid
synthesis, **2**, 156; **3**, 572, 602
use of alcohol protection, **6**, 648
via Baeyer–Villiger reaction, **7**, 677
Gibberellin, 11-hydroxy-
synthesis
via intramolecular photocycloaddition, **5**, 180
Gibberellin A_1
reduction
borohydrides, **8**, 537
Gibberellin A_3
allylic oxidation, **7**, 90
ketone
reduction, **8**, 537
Gibberellin A_7
allylic oxidation, **7**, 90
Gibberellin A_1, 2-deuterio-
methyl ester
synthesis, **8**, 537
Gibberellin A_5, hydroxy-
synthesis, **8**, 537
Gibberellin A_{20}, hydroxy-
synthesis, **8**, 537
Gibberellins
epoxides
oxidative rearrangement, **7**, 826
methylenation
modified Tebbe reagent, **5**, 1124
rearrangement, **3**, 715
reduction
borohydrides, **8**, 537
synthesis, **7**, 301
carbonyl methylenation step, **1**, 749
rearrangement of epoxides, **3**, 766
via Birch reduction, **8**, 500, 503
via cyclofunctionalization of cycloalkene, **4**, 373
Wagner–Meerwein rearrangement, **3**, 715
Giese method
radical addition reactions
alkenes, **4**, 735–742
Gif system
alkane oxidation, **7**, 13
Gilman cuprates
1,2-additions, **1**, 107
reactions with ketones
comparison with aldehydes, **1**, 116
Gilman reagents
conjugate additions
N-enoylsultams, **4**, 204
reaction with epoxides, **3**, 223
tandem vicinal difunctionalization, **4**, 253
Gingerol
enantioselective synthesis
use of SAMP/RAMP, **2**, 514
use of α-sulfinylhydrazones, **2**, 515
synthesis
via α-sulfinyl hydrazones, **1**, 524
(±)-[6]-Gingerol
synthesis
regioselective deprotonation, **2**, 183
Ginkgolide
synthesis, **7**, 182
via [3 + 2] cycloaddition reactions, **5**, 311
Ginkgolide B
synthesis, **3**, 546; **8**, 171
via Baeyer–Villiger reaction, **7**, 680
via tandem vicinal difunctionalization, **4**, 256
organocopper compounds, **3**, 220
Ginkgolide B–kadsurenone hybrid
synthesis
via photocycloaddition, **5**, 176
Glacosporone
synthesis, **6**, 136
Glaucine
synthesis
use of vanadium oxytrifluoride, **3**, 670
Gloeosporone
synthesis, **1**, 568; **3**, 281
use of hydrazones, **2**, 505
via lactone acylation, **1**, 420
D-Glucal, 3,4,6-tri-*O*-benzyl-
reaction with phenylsulfenyl chloride, **6**, 60
Glucals
Paterno–Büchi reaction, **5**, 158
Glucofuranose
Paterno–Büchi photocycloaddition reaction with furan, **5**, 187
1,2-α-D-Glucofuranose
asymmetric hydrogen transfer, **8**, 552
D-Glucofuranose, 1,2,5,6-di-*O*-isopropylidene-
reduction
tributylstannane, **8**, 820
titanium enolates
chiral reagent, **2**, 308

α-Glucofuranose, 5-*O*,6-*O*-disilyl-3-*O*-acryloyl-1,2-*O*-isopropylidene-
Diels–Alder reaction, **5**, 366
D-Gluco-D-guloheptano-γ-lactone
reduction
sodium borohydride, **8**, 269
Gluconobacter roseus
enzymes
diol oxidation, **7**, 316
D-Glucono-1,5-lactones
reduction, **8**, 292
Glucopyranolactones
alkynic ketone synthesis from, **1**, 419
Glucopyranose
thallium alkoxide
phosphorylation, **6**, 603
D-Glucopyranose, 2,3,4,6-tetra-*O*-benzyl-
Wittig reaction, **7**, 635
α-Glucopyranosides
synthesis, **6**, 39
β-Glucopyranosides
synthesis, **6**, 38, 41
β-D-Glucopyranosides, 2-deoxy-
synthesis, **6**, 61
β-Glucopyranosides, 1,3-dienyltetraacetyl-
Diels–Alder reactions, **5**, 374–376
α-D-Glucopyranosides, methyl-4,6-*O*-benzylidene-
reduction, **8**, 224
β-D-Glucopyranosylamine, tetra-*O*-methyl-
synthesis
Ugi reaction, **2**, 1099
Glucopyranosyl bromide, tetraacetyl-
alkylation, **5**, 374
stability, **6**, 38
α-D-Glucopyranosyl bromide, 2,3,4,6-tetra-*O*-benzyl-
stability, **6**, 38
α-D-Glucopyranosyl fluoride
synthesis, **6**, 46
Glucopyranosyl halides, 2-bromo-2-deoxy-
synthesis, **6**, 61
Glucopyranosyl halides, 2-deoxy-2-phenylthio-
synthesis, **6**, 61
Glucopyranosyl radical
synthesis
via Paterno–Büchi reaction, **5**, 159
α-D-Glucopyranuronic acid
synthesis, **6**, 43
β-Glucosamine, 2-deoxy-2-phthalimido-
reactivity, **6**, 42
Glucose
hydrogenation
catalytic, **8**, 140
reduction
nitro compounds, **8**, 366
separation from fructose
Knoevenagel reaction, **2**, 354
D-Glucose
diethyl dithioacetal
oxidative cleavage, **7**, 710
L-Glucose
synthesis
Diels–Alder reaction, **2**, 690
Glucose, 3-deoxy-
synthesis, **8**, 819
Glucose, 5-deoxy-
monodeoxygenation, **8**, 820
Glucose, 1,2,4,6-di-*O*-benzylidene
reduction, **8**, 226
D-Glucose, 2,3,4,6-tetra-*O*-benzyl-
glycoside synthesis, **6**, 57
reaction with trichloroacetonitrile, **6**, 50
synthesis, **6**, 57
D-Glucose, 2,3,4-tri-*O*-benzyl-
glycoside synthesis, **6**, 57
Glucuronic acid, 4-deoxy-
synthesis
Diels–Alder reaction, **2**, 692
Glutamic acid
asymmetric synthesis, **8**, 146
enantiomers
synthesis *via* conjugate addition, **4**, 222
synthesis, **8**, 149
via reductive amination, **8**, 144
Glutamic acid, 3-hydroxy-
synthesis, **1**, 119
Glutamic acid, 4-methylene-
synthesis, **6**, 96
Glutarates
disilyl ketene acetals, **2**, 606
2,3-disubstituted
synthesis *via* ester enolate addition, **4**, 107
erythro
synthesis *via* Michael addition, **4**, 21
Glutaric acid
diethyl ester
acyloin coupling reaction, **3**, 623
Glutaric acid, α-keto-
diethyl ester, oxime acetate
hydrogenation, **8**, 149
Glutaric acid, 3-methyl-
racemization, **2**, 742
Glutaric acid, perfluoro-
Kolbe electrolysis, **3**, 640
Glutaric esters
synthesis
dicarboxylation, **4**, 947
Glutathione
catalyst
methylglyoxal reduction, **8**, 87
reduction of 1,2-dioxetanes, **8**, 398
Glycal, 2-nitro-
synthesis, **6**, 108
Glycals
pyranoid
Ireland–Claisen rearrangement, **5**, 859
synthesis
via isocyanate cycloaddition, **5**, 108
Glycamines
synthesis
via electroreduction of oximes, **8**, 137
Glyceraldehyde
reaction with hemoglobin
in presence of $NaBH_3CN$, **6**, 790
rearrangement, **3**, 831
L-Glyceraldehyde
synthesis, **1**, 568
Glyceraldehyde, cyclohexylidene-
nucleophilic addition reactions
stereoselectivity, **1**, 55
Glyceraldehyde, 2,3-*O*,*O*-dibenzyl-
nucleophilic addition reactions
stereoselectivity, **1**, 55

Glyceraldehyde, 2,3-*O*-isopropylidene-
  Knoevenagel reaction, **2**, 385
  nucleophilic addition reactions, **1**, 53
  oxime
    reaction with allyl boronates, **2**, 995
  reactions with crotyl bromide/chromium(II) chloride, **1**, 185
  reactions with organocuprates, **1**, 110
  reactions with organometallic compounds, **1**, 153
    Lewis acids, **1**, 339
  synthesis, **7**, 713
Glyceraldehyde acetonide *N,N*-dimethylhydrazone
  reactions with organocopper complexes, **1**, 121
Glyceraldehyde acetonides
  imines
    condensation to β-lactams, **5**, 96
  reactions with allylboronates, **2**, 26
  reactions with allyl organometallics, **2**, 41
Glyceric acid
  reaction with pivaldehyde, **3**, 40
D-Glycerose, 2,2′-*O*-methylenebis-
  intramolecular aldolization, **2**, 167
Glycidic acids
  decarboxylation, **2**, 426
Glycidic esters
  preparation, **2**, 409
  reaction with organocuprates, **3**, 225
Glycidic esters, 3-phenyl-
  rearrangement
    migratory preferences, **3**, 747
Glycidic esters, 3-substituted
  reaction with organometallic compounds, **6**, 11
Glycidic thiol esters
  preparation
    Darzens glycidic ester condensation, **2**, 418
Glycidol
  synthesis, **7**, 397
Glycidonitriles
  preparation, **2**, 419
Glycidyl tosylate
  reaction with lithium cyanodiphenylcuprates, **3**, 224
Glycinamide, *N*-phthaloyl-
  iminium salts from, **5**, 112
Glycinate esters
  reactions with enolizable imines
    Mannich reaction, **2**, 922
Glycinates, aroyl phenyl
  tandem rearrangements, **5**, 877
Glycine
  bislactim ethers from, **3**, 53
  t-butyl ester, camphor imine
    alkylation, **3**, 46
  cation equivalents
    addition reactions, **2**, 1074
    halogenation, **2**, 1052
Glycine, *N*-acyl-
  Erlenmeyer azlactone synthesis, **2**, 402
Glycine, α-alkenyl-
  methyl ester
    preparation, **2**, 499
Glycine, 2-aryl-
  esters
    synthesis, Friedel–Crafts reaction, **3**, 303
Glycine, *N,N*-bissilyl-
  zinc enolates
    reaction with *N*-silylimines, **2**, 936
Glycine, cyclopentenyl-
  synthesis
    N-acyliminium ions, **2**, 1076
Glycine, cyclopentyl-
  synthesis
    N-acyliminium ions, **2**, 1076
Glycine, *N,N*-dimethyl-
  t-butyl ester
    lithium enolates, **2**, 221
  methyl ester
    lithium enolates, **2**, 221
Glycine, α,α-di-*n*-propyl-
  synthesis
    Ugi reaction, **2**, 1096
Glycine, α-halo-*N*-(*t*-butoxycarbonyl)-
  electrophilic glycinates, **1**, 373
Glycine, neopentyl-
  synthesis
    via Hofmann reaction, **6**, 801
Glycine, phenyl-
  asymmetric synthesis, **8**, 146
  ethyl ester
    hydrogenation, **8**, 146
  synthesis, **8**, 148
Glycine, *N*-4-quinolylmethyl-
  Friedel–Crafts reaction, **2**, 759
Glycine, vinyl-
  synthesis, **7**, 722
Glycinoeclepin A
  synthesis, **2**, 159
    via Baeyer–Villiger reaction, **7**, 680
    via cyclofunctionalization of cycloalkene, **4**, 373
    via tandem vicinal difunctionalization, **4**, 245
Glycolic acid
  synthesis, **3**, 822
    via intramolecular disproportionation of glyoxal, **8**, 87
Glycolipids
  synthesis
    Diels–Alder reaction, **2**, 692
  synthesis, **6**, 33
Glycols
  acetals
    stereoselectivity, **2**, 578
  catalytic hydrogenation, **8**, 814
  cleavage reactions, **7**, 703–714
  oxidation, **7**, 803
    solid support, **7**, 843
  oxidative cleavage
    solid support, **7**, 841
  synthesis, **7**, 437–447
Glycopeptides
  synthesis
    carboxy-protecting groups, **6**, 666
    protecting groups, **6**, 633
O-Glycopeptides
  synthesis
    tumor associated antigen structure, **6**, 639
Glycophospholipids
  synthesis, **6**, 33, 51
Glycoproteins
  synthesis, **6**, 33
α-D-Glycopyranosides
  synthesis, **6**, 42
Glycosides

synthesis, **6**, 33
α/β-selectivity, **6**, 38
via alkenylchromium reagents, **1**, 198
via trichloroacetimidates, **6**, 51
α-Glycosides
stereoselective construction
benzyl-type protecting groups, **6**, 652
C-Glycosides
synthesis, **6**, 46
copper catalysts, **3**, 216
Prins reaction, **2**, 555
via Ireland silyl ester enolate rearrangement, **5**, 841
N-Glycosides
Amadori rearrangement, **6**, 789
1,2-*trans*-Glycosides
stereoselective construction
neighboring-group assistance, **6**, 657
Glycosides, amino-
deamination, **8**, 831
Glycosides, *C*-aryl
synthesis
Friedel–Crafts reaction, **3**, 303
Glycosides, 2-deoxy-
synthesis
via alkoxyselenation, **4**, 339
via heteroatom cyclization, **4**, 391
Glycosides, *C*-methyl-
synthesis
via alkenylchromium reagents, **1**, 198
Glycosphingolipids
synthesis, **6**, 33, 53
C-Glycosyl compounds
synthesis
via Paterno–Büchi reaction, **5**, 158
Glycosyl fluorides
Friedel–Crafts reaction, **3**, 303
glycoside synthesis, **6**, 46
Glycosyl halides
reactivity, **6**, 38
stability, **6**, 38
synthesis, **6**, 37
α-Glycosyl halides
reaction with dialkyl homocuprates, **3**, 216
β-Glycosyl halides
synthesis, **6**, 42
Glycosyl hydrolases
glycoside synthesis, **6**, 49
β-Glycosyl imidates
synthesis, **6**, 54
Glycosyl phosphates
glycoside synthesis, **6**, 49
synthesis, **6**, 51
Glycosyl pyrophosphates
glycoside synthesis, **6**, 49
Glycosyl sulfonates
glycoside synthesis, **6**, 49
Glycosyl transferases
glycoside synthesis, **6**, 49
Glycoxylic acid, *p*-bromophenyl-
menthyl ester
crystal structure, **5**, 186
Glycyrrhetinic acid
allylic oxidation, **7**, 87
Glyoxal
benzilic acid rearrangement, **3**, 831
rearrangement, **3**, 822
reduction
intramolecular disproportionation, **8**, 87
Glyoxal, methyl-
reduction
synthesis of lactic acid, **8**, 87
Glyoxal, phenyl-
benzilic acid rearrangement, **3**, 829
reaction with enol silyl ether, **2**, 616
Glyoxalase inhibitor I
synthesis
via Diels–Alder reaction, **5**, 370
Glyoxaldehyde
Diels–Alder reaction, **2**, 662
Glyoxalic acid
hydrogenation, **8**, 236
2,4,6-triisopropylbenzenesulfonylhydrazone
diazoacetate synthesis, **6**, 124
Glyoxylates
N-acylimines
Diels–Alder reactions, **5**, 405
aldimines
reaction with allyl organometallic compounds, **2**, 987
ene reaction
endo/*exo* selectivity, **2**, 535
homoallylic
intramolecular ene reaction, **2**, 542
menthyl
N-substituted imines, organometallic addition reactions, **1**, 363
trans-2-phenylcyclohexyl
ene reaction, **2**, 536
8-phenylmenthyl
ene reaction, **2**, 536
synthesis
via Kornblum oxidation, **7**, 654
Glyoxylic acid
methyl ester
Diels–Alder reactions, **5**, 431
methyl phenyl ester
photocycloaddition reactions, **5**, 160
Glyoxylic acid, α-naphthyl-
menthyl ester
asymmetric hydrogenation, **8**, 144
Glyoxylic acid, phenyl-
asymmetric electroreduction, **8**, 134
esters
photocycloaddition reactions, **5**, 185
ethyl ester
reduction, hydride transfer, **8**, 85, 93
2-(1-methyl-1-phenylethyl)-5-methylcyclohexyl ester
crystal structure, **5**, 185
Glyoxylic thioamide, *N*,*N*-dimethyl-
synthesis, **6**, 489
Gnididone
synthesis
Dieckmann reaction, **2**, 824
via retro Diels–Alder reactions, **5**, 579
Gold complexes
enantioselective aldol reaction
catalysis, **2**, 317
ferrocenylphosphine
aldol reaction, **2**, 318
Gomberg–Bachmann–Hey process, **3**, 505
Gomberg–Bachmann process, **3**, 505
Gomberg process, **3**, 505

β-Gorgonene
synthesis
via Peterson alkenation, **1**, 731
Gorgosterol, demethyl-
synthesis
use of homoenolates, **2**, 452
Grahamimycin A
synthesis, **3**, 575
Granaticin
synthesis
via annulation, **1**, 554
via organoaluminum reagents, **1**, 101
Grandisol
synthesis, **3**, 48, 103, 785; **7**, 239
alkenylsilane acylation, **2**, 713
Graphite bisulfate
esterification
catalyst, **6**, 325
Grayanotoxins
synthesis
via photocycloaddition, **5**, 670
Griffin fragmentation
photocycloreversion, **5**, 199
Grifolin
synthesis
via cyclobutenone ring opening, **5**, 689
via electrocyclization, **5**, 732
Grignard reactions
abnormal, **1**, 244
Grignard reagents
acylation, **1**, 399
alkenyl
configuration in coupling reactions, **3**, 464
alkynic
coupling reaction with 1-haloalkynes, **3**, 553
alkynide
alkylation, **3**, 272
allylic
carbomagnesiation, mechanism, **4**, 874
coupling reactions with heteroaromatic halides, **3**, 461
coupling reaction with bromobenzene, **3**, 451
intramolecular carbomagnesiation, **4**, 876
amination, **6**, 118
anodic dimerization, **7**, 805
aromatic nucleophilic substitution, **4**, 427
aryl
alkylation, **3**, 242
dimerization, **3**, 499
asymmetric
nucleophilic addition reactions, **1**, 69
Beckmann rearrangement, **6**, 770
2-butadienyl
coupling reactions with alkyl halides, **3**, 465
carbomagnesiation, **4**, 874
cerium chloride system, **1**, 244
coupling, **3**, 415
cross-coupling reactions
with organic halides, **3**, 436
crystal structure, **1**, 13
cyclopropyl
coupling reactions with bromobenzene, **3**, 452
desulfurization, **8**, 840
hydride transfer
carbonyl reduction, **8**, 99
nitrile reduction, **8**, 300
imine anion synthesis, **6**, 719
ketone synthesis, **6**, 446
nitrile synthesis, **6**, 241
nucleophilic addition reactions
α-alkoxy acyclic ketones, **1**, 50
carbonyl compounds, **1**, 49
chiral ketones, **1**, 58
nucleophilic addition to π-allylpalladium complexes, **4**, 596
regioselectivity, **4**, 635–637
stereochemistry, **4**, 620
phosphonium ylide synthesis, **6**, 194
primary alkyl
coupling reactions with alkenyl halides, **3**, 436
coupling reactions with aromatic halides, **3**, 450
propargyl
physical properties, **2**, 81
structure, **2**, 81
reactions with alkenyl halides
organonickel catlysis, **3**, 228
reactions with α-alkoxy acyclic ketones
cyclic chelate model, **1**, 51
reactions with epoxides
alcohol synthesis, **6**, 4
ring opening, **3**, 754
secondary alkyl
coupling reactions with alkenyl halides, **3**, 440
coupling reactions with aromatic halides, **3**, 452
α-sulfonyl
alkylation, **3**, 159
tandem vicinal difunctionalization, **4**, 257
tertiary alkyl
coupling reactions with alkenyl halides, **3**, 441
coupling reactions with aromatic halides, **3**, 452
vinyl
alkylation, **3**, 242
coupling reactions, **3**, 484
Grignard reagents, alkyl
reaction with cyclohexanone
stereoselectivity, **1**, 79
Grignard reagents, allyldimethylsilylmethyl-
hydroxymethylation, **7**, 647
Griseofulvin
synthesis
via Michael addition, **4**, 27, 44
Griseofulvin, dehydro-
reduction
synthesis of griseofulvin, **8**, 452
synthesis
oxidation of griseophenone A, **3**, 689
Griseofulvoxin, dehydro-
synthesis
use of manganese dioxide, **3**, 690
Grob fragmentation, **2**, 1047
definition, **6**, 1042
intramolecular [2 + 2] photocycloaddition, **6**, 1062
Group transfer reactions
radicals, **4**, 726
Grundmann method
reduction
aroyl chlorides, **8**, 291
Guaiacol
hydrogenolysis, **8**, 912
Guaiacols, 4-alkyl-
Reimer–Tiemann reaction, **2**, 773
Guaiane

rearrangement, **3**, 388
synthesis, **3**, 396
via photocycloaddition, **5**, 669
via vinylcyclopropane thermolysis, **4**, 1048
Guaianolides
synthesis
via Pauson–Khand reaction, **5**, 1052
Guaianolide sesquiterpenes
synthesis
via cycloaddition reactions, **5**, 275
Guaipyridine
synthesis
via Diels–Alder reaction, **5**, 492
Guanidates, acylphosphoro-
phosphorylation, **6**, 614
Guanidines
N-substituted
reduction, **8**, 639
Guanidinium salts
tris(dialkylamino)alkane synthesis, **6**, 582
Guanine
amine protection, **6**, 642
Gyrinidal
synthesis, **7**, 109

H

Haagenolide
synthesis
via cyclization, **1**, 553
Wittig rearrangement, **3**, 1010
Hafnabicycles
synthesis, **5**, 1170
Hafnacycles
three-membered
synthesis, **5**, 1175
Hafnium
bicyclization catalyst
enynes, **5**, 1169
hydrometallation, **8**, 676
Halides
aromatic
coupling reactions with primary alkyl Grignard reagents, **3**, 450
coupling reactions with secondary and tertiary alkyl Grignard reagents, **3**, 452
coupling reaction with sp^3 organometallics, **3**, 450
double carbonylation, palladium-catalyzed, **3**, 1039
aromatic nucleophilic substitution, **4**, 445
carbanions
crystal structure, **1**, 38
carbonylation, **3**, 1021
dehalogenation
metal hydrides, **8**, 684
heteroaromatic
coupling reactions with sp^3 organometallics, **3**, 459
oxidation, **7**, 653
primary
homologation, phenylthiomethyllithium, **6**, 139
reactions with organocerium compounds, **1**, 233
synthesis, **6**, 203–221
vinyl substitutions
palladium complexes, **4**, 842
Halide salts
reductive cleavage
α-halo ketones, **8**, 988
Haloalkylation
alkyl and allyl halides, **3**, 118
arenes
Friedel–Crafts reaction, **3**, 320
Haloamides
reactions with alkenes, **4**, 355
Haloamination
alkenes, **4**, 355
Halocarbonyl group
acid halide synthesis, **6**, 308
Halodealumination, **8**, 754
2-Haloethyloxycarbonyl groups
amine-protecting group
cleavage, **6**, 639
Halofunctionalization
alkenes, **7**, 533
Halogenation
alkanes, **7**, 15
amines, **7**, 741
anodic oxidation, **7**, 810
boryl-substituted carbanions, **1**, 501
electrochemical
aromatic compounds, **7**, 800
enzyme-catalyzed, **7**, 539
ionic
sulfides, **7**, 193
nucleophilic substitution, **6**, 203
phosphonium ylides, **6**, 177
secondary amines, **7**, 747
sulfides, **7**, 206
regioselectivity, **7**, 210
α,β-unsaturated carbonyl compound synthesis, **7**, 120
Halogenation–dehydrohalogenation, **7**, 120
Halogenative cleavage
zirconium compounds, **8**, 691
Halogen atom transfer addition reactions
radical reactions, **4**, 753–755
Halogen atom transfer reactions
radical cyclizations, **4**, 802–804
Halogen azides
reactions with alkenes, **4**, 349
Halogen exchange
amide halides, **6**, 500
hydrogen fluoride, **4**, 270
Halogen isocyanates
reactions with alkenes, **4**, 351
Halogen nitrates
reactions with alkenes, **4**, 350
Halogenoetherification
alkenes, **7**, 535
Halogens
activator
DMSO oxidation of alcohols, **7**, 298
nucleofuge
in aromatic $S_{RN}1$ reactions, **4**, 457
oxidation
sulfides, **7**, 763
sulfoxides, **7**, 767
thiols, **7**, 760
reactions with alkenes, **4**, 344–348
reactions with α-chloroenamines, **6**, 497
Halogen thiocyanates
reactions with alkenes, **4**, 351
Halohydrin esters
alkene hydroxylation, **7**, 444
Halohydrins
rearrangements, **3**, 787
semipinacol rearrangements, **3**, 777
synthesis
epoxide ring opening, **3**, 754
Halolactonization
γ,δ-enoic acids, **6**, 361
δ-lactone synthesis, **6**, 366
synthesis
β-lactones, **4**, 368
Halometallic reagents
oxidative halogenation, **7**, 527
Halomethylation
carbonyl compounds
samarium diiodide, **1**, 260
Halomethyl compounds
oxidation, **7**, 666
Halonium ions
amino alcohol synthesis, **7**, 492
cyclic

aziridine synthesis, **7**, 473
Halonium ions, dialkyl-
preparation of
Friedel–Crafts reaction, intermediate, **3**, 299
Haloperoxidases
cytosine halogenation, **7**, 539
Halopropenylation
alkyl and allyl halides, **3**, 118
Hantzsch esters
hydride donors, **8**, 92
Harman, tetrahydro-
synthesis, **3**, 81
Hass–Bender reaction
benzylic halides, **7**, 659
Hastanicine
synthesis
via cyclopropane ring opening, **4**, 1045
Hasubanan alkaloids
synthesis
via Diels–Alder reactions, **5**, 323
Heathcock's reagent
stereoselective reaction
enol silanes and aldehydes, **2**, 642
Heck reaction, **4**, 903
Hederagenin, methyl-
deoxygenation, **8**, 821
Hedycaryol
Cope rearrangement, **5**, 809
transannular cyclization, **3**, 390
(±)-Helenalin
synthesis, **2**, 160
Helenanolides
synthesis, **7**, 164
Helianthrone
photolysis, **5**, 729
Helical molecules
synthesis
lithium allenes, **2**, 88
Heliotridane, trihydroxy-
synthesis, **5**, 940
Heliotridine
synthesis
via Diels–Alder reaction, **5**, 421
Hell–Vollard–Zelinski reaction
conditions
halogenation of acids, **7**, 122
Helminthogermacrene
synthesis, **7**, 94
Helminthosphoral
synthesis
keto aldehydes, **2**, 158
Hemiacetals, amino-
reaction with enol ethers
use in alkaloid synthesis, **2**, 613
Hemoglobin
reaction with glyceraldehyde
in presence of $NaBH_3CN$, **6**, 790
11-Heneicosene
synthesis, **3**, 644
6-Henicosen-11-one
synthesis, **1**, 563
Henry reaction
basicity, **2**, 325
carbonyl component
concentration, **2**, 325
dialdehydes, **2**, 326
diastereoselectivity
tetrabutylammonium fluoride catalyst, **2**, 335
heterogeneous phase method, **2**, 330
intramolecular
6-nitro-1,3-dicarbonyl compounds, **2**, 334
ketones, **2**, 329
nitroalkanes
functionalized, **2**, 331
oxaallylic anions, **2**, 321–340
procedures, **2**, 325
reaction conditions, **2**, 323
regioselectivity
erythro-sphingosine, **2**, 331
reviews, **2**, 321
silyl nitronates, **2**, 335
solvent-free method
heterogeneous, **2**, 330
stereoselective
bicyclic trimethylsilyl nitronates, **2**, 336
α,β-unsaturated carbonyl compounds
regioselectivity, **2**, 330
stereoselectivity, **2**, 330
utility, **2**, 322
Hentriacontane-14,16-dione
synthesis, **8**, 645
1,6-Heptadiene
chlorination, **7**, 532
hydrocarboxylation, **4**, 941
2,5-Heptadiene
synthesis
via retro Diels–Alder reaction, **5**, 567
1,5-Heptadiene, 2,6-dimethyl-
hydroformylation, **4**, 922
Heptadiene, diphenyl-
intramolecular [2 + 2] cycloadditions, **5**, 67
2,6-Heptadienoic acid
synthesis
via nickel-ene reaction, **5**, 36
4,6-Heptadienoic acid
sodium salt
Diels–Alder reactions, **5**, 344
1,4-Heptadienol, 4-methyl-
synthesis
via carboboration, **4**, 885
3,5-Heptadien-2-one
hydrogenation
nickel catalyst, **8**, 535
1,6-Heptadiyne
thermal isomerization, **5**, 736
Heptafulvene, 8,8-dimethyl-
cycloaddition reactions
dienes, **5**, 634
Heptahendecafulvadiene
pericyclic reactions, **5**, 744
Heptanal
reaction with allylic organometallic compounds, **1**, 156
reductive allylation, **3**, 109
Heptanal, 2-ethyl-
synthesis
via hydroformylation, **4**, 918
Heptanal, 3-methyl-
synthesis
hydroformylation of 2-methyl-1-hexene, **3**, 1022
Heptane, 3-methyl-

oxidation
transition metal catalysis, **7**, 12
Heptane, 1-methylseleno-
synthesis, **1**, 663
Heptane, tricyclic
synthesis, **7**, 517
Heptanedioic acid, 3-ethoxycarbonyl-
diethyl ester
Dieckmann reaction, **2**, 808
Heptanedioic acid, 2-methyl-
diethyl ester
Dieckmann reaction, **2**, 811
Heptanoic acid, 4-amino-3-hydroxy-6-methyl-
synthesis, **1**, 119
n-Heptanol
oxidation
4-(dimethylamino)pyridinium chlorochromate, **7**, 269
Heptanol, 5,6-epoxy-
ring opening
stereospecificity, **3**, 751
2-Heptanone
lithium 2-enolates
benzylation, **3**, 7
reaction with allylic organometallic compounds, **1**, 156
4-Heptanone
aldol reaction, **2**, 144
4-Heptanone, 3,5-dibromo-2,6-dimethyl-
[4 + 3] cycloaddition reactions, **5**, 603
1,3,5-Heptatrienes
thermal reactions, **5**, 707
Heptatrienones, amino-
electrocyclization, **5**, 710
1-Heptene
hydroxylation
osmium tetroxide, **7**, 442
1-Heptene, 1-acetoxy-
photocycloaddition reactions, **5**, 127
Heptene, 2-chloro-
hydrogenation, **8**, 898
6-Heptenoic acid
radical decarboxylation, **7**, 731
6-Hepten-2-ol
synthesis
via reduction of 6-hepten-2-one, **8**, 134
6-Hepten-2-ol, 2,6-dimethyl-
hydroformylation, **4**, 923
2-Hepten-1-ol, 2-methyl-
asymmetric epoxidation, **7**, 409
6-Hepten-2-one
preparative electrolysis, **8**, 134
4-Heptenone, 2-hydroxy-
synthesis
via [4 + 3] cycloaddition, **5**, 603
Heptenyl radicals
cyclizations, **4**, 785
stereoselectivity, **4**, 789
1-Hepten-6-ynes
Pauson–Khand reaction, **5**, 1053
1-Hepten-6-ynes, 7-(trimethylsilyl)-
bicyclization
mechanism, **5**, 1178
reaction with cyclopentadienylzirconium complexes, **5**, 1165
Heptulosonic acid, 3-deoxy-D-*arabino*-
7-phosphate
shikimate pathway, **2**, 462
7-phosphonate
synthesis, enzymes, **2**, 466
Heratomin
synthesis, **5**, 1096
5-HETE
synthesis, **3**, 289
12-HETE
synthesis, **3**, 289
copper-catalyzed, **3**, 216
Heteroaromatic compounds
coupling reactions, **3**, 509
with aryl compounds, **3**, 512
hydrogenation
homogeneous catalysis, **8**, 453
$S_{RN}1$ reaction, **4**, 462
Heteroarynes
in synthesis, **4**, 503
intermediates
nucleophilic substitution, **4**, 485
Hetero-Cope rearrangement
allylic systems, **6**, 834
carbanion-accelerated, **5**, 1004
Heterocuprates, **3**, 211
acylation, **1**, 431
synthesis, **3**, 209
Heterocyclic compounds
benzilic acid rearrangement, **3**, 834
four-membered
synthesis, **5**, 85–118
hydride transfer, **8**, 92
organomercury compounds
palladium complexes, **4**, 839
synthesis
Dieckmann reaction, **2**, 829
via carbonyl compound derivatives, **6**, 733–760
via dihalocyclopropanes, **4**, 1021–1023
unsaturated
synthesis *via* retro Diels–Alder reactions, **5**, 577–584
Vilsmeier–Haack reaction, **2**, 780
vinyl substitutions
palladium complexes, **4**, 835–837
Hetero Diels–Alder reaction
aldehydes, **2**, 662
enantioselective
chiral Lewis acids, **2**, 654
heterocyclic synthesis, **6**, 756
high pressure, **2**, 663
Heterodienes
cationic
Diels–Alder reactions, **5**, 492–507
Diels–Alder reactions, **5**, 451–507
intramolecular, **5**, 531
Heteroelectrocyclization
applications, **5**, 740–743
Heteronucleophiles
addition reactions
allenes, **4**, 55
conjugate addition
intermolecular, **4**, 30–37
Heteropolyacids
reoxidants
Wacker process, **7**, 452

Heterotropanone
synthesis
via retro Diels–Alder reaction, **5**, 569
Heterotropantrione
synthesis, **3**, 697
Heteroyohimboid indole alkaloids
synthesis
via Diels–Alder reactions, **5**, 467
Heusler–Kalvoda reaction, **7**, 41
Hexadecanedioic acid
dimethyl ester
synthesis, **3**, 642
2,15-Hexadecanedione
macrocyclization, **2**, 166
Hexadecan-5-olide, 6-acetoxy-
synthesis
via Payne rearrangement, Lewis acids, **1**, 343
synthesis, **7**, 623
Hexadecanoyl chloride, 16-phenyl-
Friedel–Crafts reaction, **2**, 753
Hexadecatrienal
synthesis
via vinyl iodides, **1**, 808
1-Hexadecene
epoxidation, **7**, 429
11-Hexadecynoic acid
synthesis, **3**, 646
1,5-Hexadiene
dicarboxylation, **4**, 948
hydroalumination, **8**, 758
hydrocarboxylation, **4**, 941
photocycloaddition reactions, **5**, 136
reaction with chlorosulfonyl isocyanate, **5**, 105
synthesis
via Claisen–Cope rearrangement, **5**, 883
2,4-Hexadiene
cycloaddition products, **5**, 69
Diels–Alder reactions
imines, **5**, 408
hydrobromination, **4**, 283
isomerization, **5**, 74
selective reduction, **8**, 568
zirconocene complex
reactions with carbonyl compounds, **1**, 163
1,5-Hexadiene, 2,5-dimethyl-
hydroboration, **8**, 707
1,5-Hexadiene, 3,4-dimethyl-
Claisen rearrangement
transition state structures, **5**, 857
Cope rearrangement, **5**, 820
2,3-Hexadiene, 2,5-dimethyl-
photocycloaddition reactions
anthracene, **5**, 636
2,4-Hexadiene, 2,5-dimethyl-
cycloaddition reactions, **5**, 71
[2 + 2] cycloaddition reactions
tetracyanoethylene, **5**, 76
epoxidation, **7**, 359
photocycloaddition reactions
9-cyanoanthracene, **5**, 636
selective reduction, **8**, 567
1,5-Hexadiene, 3,4-diphenyl-
Cope rearrangement, **5**, 799
1,5-Hexadiene, 3-hydroxy-
thermal rearrangements, **5**, 1000
1,5-Hexadiene, 2-methyl-
hydroboration, **8**, 714
1,5-Hexadiene, 2-methyl-3-phenyl-
Cope rearrangement
palladium catalysts, **5**, 799
2,4-Hexadienoic acid
ethyl ester
cycloaddition reactions, tropones, **5**, 620
3,5-Hexadienoic acid
sodium salt
Diels–Alder reactions, **5**, 344
3,5-Hexadienoic acid, 6-methoxy-
sodium salt
Diels–Alder reactions, **5**, 344
1,5-Hexadien-3-ol
synthesis
via 2,3-Wittig rearrangement, **5**, 888
2,4-Hexadien-3-ol
asymmetric epoxidation
kinetic resolution, **7**, 414
substituent effect, **7**, 421
Hexadiynene
synthesis, **3**, 528
Hexa-1,5-diyn-3-ene, 3-alkyl-4-(1-alkynyl)-
synthesis, **3**, 554
1,5-Hexadiynes
cooligomerization with alkynes
benzocyclobutene synthesis, **5**, 692
Hexafluorophosphonium nitrite
reactions with alkanes, **7**, 10
Hexafuranos-5-ulose
photocycloaddition reactions, **5**, 185
Hexalins
photochemical ring opening, **5**, 739
ring opening, **5**, 708
trans-$\Delta^{1,3}$-Hexalins
photoisomerization, **5**, 706
Hexamethylenetetramine
N-alkylation, **6**, 85
Hexamethylphosphoramide
in sulfide metallation, **3**, 86
Hexanal, 2-ethyl-
potassium enolates
alkylation, **3**, 20
Hexanal, 5-oxo-
synthesis
via Wacker oxidation, **7**, 458
Hexanal, 2-propyl-
synthesis
via hydroformylation, **4**, 918
Hexanamide, *N,N*-dimethyl-
reduction, **8**, 249
Hexane, 1-chloro-
reaction with 2-methyl-2-propylpentanoate
effect of solvent on rate, **6**, 2
Hexane, 3-chloro-1,1-bis(methylseleno)-
metallation, **1**, 638
Hexane, 3-chloro-1,1-bis(phenylseleno)-
metallation, **1**, 638
Hexane, 2,5-dichloro-2,5-dimethyl-
alkylation of 1,3-dimethylbenzene
Friedel–Crafts reaction, **3**, 318
Hexane, 2,4-dihalo-
benzene alkylation by
Friedel–Crafts reaction, **3**, 318
Hexanedioic acid
dimethyl ester

acyloin coupling reaction, **3**, 625
Hexanedioic acid, 3,4-diphenyl-
diethyl ester
acyloin coupling reaction, **3**, 615
Hexanedioic acid, 3-methyl-
diethyl ester
Dieckmann reaction, **2**, 813
Hexanedioic acid, 2,2,5,5-tetramethyl-
dimethyl ester
acyloin coupling reaction, **3**, 625
1,2-Hexanediol
oxidative cleavage, **7**, 708
2,5-Hexanediol
cyclodehydration, **6**, 25
Hexanediol, 2,5-dimethyl-
nickel acetate
cyclic ketone reduction, **8**, 14
1,2,6-Hexanetriol
stannylation, **6**, 18
Hexanoic acid, 3,5-dioxo-
methyl ester
dienol silyl ether, **2**, 607
Hexanoic acid, 2-ethyl-
allyl trapping reagent, **6**, 641
ethyl ester
acyloin coupling reaction, **3**, 619
Hexanoic acid, 5-hydroxy-
t-butyl ester
deprotonation, **2**, 225
Hexanoic acid, 5-oxo-3-phenyl-
methyl ester
stereochemistry, **2**, 520
Hexanoic acid, 2-phenyl-
Schmidt reaction, **6**, 818
Hexanoic acid, 6-phenyl-
Friedel–Crafts reaction, **2**, 753
Hexanoic acid, 2-phenyl-2-methyl-
Schmidt reaction, **6**, 818
Hexanoic anhydride
reduction
borane, **8**, 240
1-Hexanol
synthesis
via hydrogenation, **8**, 236
Hexanol, 4,5-epoxy-
ring opening
stereospecificity, **3**, 751
Hexanol, 2-ethyl-
oxidation
solid support, **7**, 841
synthesis
via hydrocarbonylation, **4**, 914
Hexano-γ-lactones, tetraacyl-
reduction
disiamylborane, **8**, 269
2-Hexanone, 6-bromo-3,3-dimethyl-
terminal lithium enolates
cycloalkylation, **3**, 18
3-Hexanone, 4-diazo-2,2,5,5-tetramethyl-
synthesis, **3**, 894
3-Hexanone, 4-hydroxy-2,2,5,5-tetramethyl-
synthesis
acyloin coupling reaction, **3**, 619
4-Hexanone, 3-methyl-
synthesis, **3**, 37
Hexanoyl chloride
reduction
metal hydrides, **8**, 240
Hexaquinacene
synthesis
via Nazarov cyclization, **5**, 768
Hexatrienes
annulated
electrocyclization, **5**, 711–730
cyclic
electrocyclization, **5**, 711–730
1,3,5-Hexatrienes
arylation
palladium catalysts, **4**, 850
electrocyclizations, **5**, 699, 706–730
monoannulated
electrocyclization, **5**, 711–721
vinylation
palladium complexes, **4**, 856
Hexatrienes, amino-
benzannulation, **5**, 720
cyclization, **5**, 718
1,3,5-Hexatrienes, 1-dialkylamino-
electrocyclization, **5**, 710
2-Hexenal
reaction with organoaluminum reagents
site selectivity, **1**, 85
reaction with organometallic compounds
chemoselectivity, **1**, 148
1-Hexene
ene reactions
Lewis acid catalysis, **5**, 4
hydrogenation
homogeneous catalysis, **8**, 445, 447
3-Hexene
cis
epoxidation, **7**, 374
diamination, **7**, 484
3-Hexene, 2-acetoxy-5-chloro-
reaction with diethylamine, **6**, 85
1-Hexene, 6-bromo-3-methyl-1-trimethylsilyl-
5-*exo*-trig closure
via Grignard reagents, **4**, 120
2-Hexene, 2,5-dimethyl-4,5-epoxy-
synthesis
via photoisomerization, **5**, 201
1-Hexene, 2-methyl-
hydroformation
phosphite-modified rhodium catalysts, **3**, 1022
3-Hexene, 1-nitro-
synthesis, **6**, 104
2-Hexene, 6-phenyl-
intramolecular cycloaddition, **5**, 649
photocycloaddition reactions, **5**, 654, 658
3-Hexene-1,6-dioic acid
Schmidt reaction, **6**, 818
2-Hexenedioic acid, 5-amino-2-fluoro-
hydrogenolysis, **8**, 896
3-Hexene-1,5-diyne, 1,6-dideuterio-
thermal isomerization, **5**, 736
4-Hexenoic acid, 2-acetyl-2-methyl-6-bromo-
ethyl ester
cyclization, **1**, 266
5-Hexenoic acid, 3-(*N*-acylamino)-
iodolactonization
stereoselectivity, **4**, 382
5-Hexenoic acid, 2-alkyl-

cyclization
stereoselectivity, **4**, 383
5-Hexenoic acid, 2-amino-4-methyl-
synthesis
via ene reaction of acrylate esters, **5**, 4
2-Hexenoic acid, 4,5-epoxy-
reaction with nitrogen nucleophiles, **6**, 87
5-Hexenoic acid, 3-hydroxy-
selenolactonization
stereoselectivity, **4**, 382
4-Hexenoic acid, 5-methyl-
hydrobromination, **4**, 282
5-Hexenoic acid, 3-methyl-
iodolactonization
stereoselectivity, **4**, 382
1-Hexen-3-ol
hydrogen donor
transfer hydrogenation, **8**, 552
2-Hexen-1-ol
epoxidation, **7**, 395
1-Hexenol, 3-chloro-
aziridination, **7**, 481
5-Hexen-2-ol, 3-methyl-
synthesis
carbomagnesiation, **4**, 877
2-Hexenylamine
allylic hydroxylation, **7**, 99
Hexenyl radicals
cyclizations, **4**, 781–785
accelerating substituents, **4**, 783
decelerating substituents, **4**, 783
stereoselectivity, **4**, 787–789
substituent effects, **4**, 783
5-Hexenyl radicals
cyclization, **4**, 781–783;**7**, 731
Hexenyl radicals, 3-methyl-
cyclization
stereoselectivity, **4**, 787
1-Hexen-5-ynes
Pauson–Khand reaction, **5**, 1053
Hexofuranose
photocycloaddition
with furan, **5**, 170
Hexopyranosides
synthesis, **6**, 51
L-Hexoses
synthesis, **7**, 402
1-Hexyne
hydrogenation to hexane
homogeneous catalysis, **8**, 456
hydrogenation to 1-hexene
homogeneous catalysis, **8**, 457
hydrosilylation, **8**, 770
photolysis
with benzaldehyde, **5**, 163
3-Hexyne
benzylation
Friedel–Crafts reaction, **3**, 332
cocycloaddition
phenylacetylene, **5**, 1146
hydrogenation to *cis*-hex-3-ene
homogeneous catalysis, **8**, 458
reduction
dissolving metals, **8**, 479
1-Hexyne, 1-bromo-
boronate
reaction with 2-thienyllithium, **3**, 498
1-Hexyne, 1-chloro-
hydrogenolysis, **8**, 898
3-Hexyne-1,6-dioic acid
Schmidt reaction, **6**, 818
Hibiscone C
synthesis
via photocycloaddition, **5**, 145
Hikosamine
synthesis
Diels–Alder reaction, **2**, 694
Himachalene
synthesis, **1**, 558
β-Himachalene
synthesis
via Cope rearrangement, **5**, 803, 983
Hinesol
synthesis
via cyclopropane ring opening, **4**, 1043
Hinokinin
synthesis, **1**, 566
via retro Diels–Alder reactions, **5**, 578
Hinokitiol
synthesis
via [4 + 3] cycloaddition, **5**, 609
Hippuric acid
Erlenmeyer azlactone synthesis, **2**, 402
Hirsutane
biosynthesis, **3**, 404
synthesis, **3**, 389
Hirsutene
synthesis, **3**, 402, 590; **7**, 524
via carbonyl–alkyne cyclization, **3**, 602
via conjugate addition, **4**, 226
via [3 + 2] cycloaddition reactions, **5**, 310
via Nazarov cyclization, **5**, 763, 779
via nitrone cyclization, **4**, 1120
via photocycloaddition, **5**, 665, 666
Hirsutic acid
synthesis, **3**, 783; **6**, 778
via intramolecular addition, **4**, 46
via Michael addition, **4**, 25
via Pauson–Khand reaction, **5**, 1060
Histrionicotoxin
synthesis
N-acyliminium ion reactions, **2**, 1049
Eschenmoser coupling reaction, **2**, 876, 878
Histrionicotoxin, deamylperhydro-
synthesis
via diazoalkene cyclization, **4**, 1158
Histrionicotoxin, perhydro-
structure, **1**, 364
synthesis, **6**, 764
Dieckmann reaction, **2**, 824
spirocyclization, **2**, 1064
via cyclohexadienyl complexes, **4**, 679
via palladium catalysis, **4**, 598
HLCE
acylation
enzymatic, **6**, 340
Hobartine
synthesis
via nitrene cyclization, **4**, 1119
via Ritter reaction, mercuration, **6**, 284
13-HODE
synthesis, **3**, 488

Hofmann elimination
  ylide preparation, **3**, 918
Hofmann–Löffler–Freytag reaction
  cyclization
    nitrogen-centered radicals, **4**, 814
  intramolecular functionalization, **7**, 40
Hofmann rearrangement, **3**, 908; **6**, 800
  amides
    amine synthesis, **6**, 796
  methoxide method, **6**, 801
  oxidative
    lead tetraacetate, **6**, 802
  stereoselectivity, **6**, 798
Hog pancreatic lipase
  epoxide hydrolysis, **7**, 429
Hole-catalyzed cycloadditions
  oxidation
    dienes, **7**, 861
Homoadamantane
  rearrangements, **3**, 854
Homoadamantane, 3-acetamido-
  hydrolysis, **6**, 266
3-Homoadamantanol
  Ritter reaction
    effect of conditions, **6**, 264
3-Homoadamantene
  dimerization, **5**, 65
Homoaldol reaction
  asymmetric, **6**, 863
  hetero-substituted allylic anions, **6**, 863
  homoenolate and carbonyl compound, **2**, 445
Homoallyl acetates
  oxidation, **7**, 464
Homoallyl alcohols
  aldol equivalents, **2**, 2
  anti
    synthesis, **3**, 984; **6**, 883
  asymmetric epoxidation, **7**, 419
  asymmetric hydrogenation
    homogeneous catalysis, **8**, 462
  asymmetric synthesis, **2**, 33
  carbonylation
    γ-lactone synthesis, **6**, 363
  cyclization
    1,3-asymmetric induction, **4**, 386
  epoxidation, **7**, 366, 371
  homogeneous hydrogenation
    diastereoselectivity, **8**, 447
  intramolecular hydrosilylation, **7**, 645
  reduction
    borohydride, **8**, 536
  1,3-sigmatropic rearrangements
    anion-accelerated, **5**, 1003
  syn
    synthesis, **6**, 879, 883
  synthesis, **3**, 263; **8**, 758
    allylsilanes, **2**, 567
    Prins reaction, **2**, 564
    use of tosylhydrazones, **2**, 513
    via allyl alcohols, **6**, 850
    via allyl metal compounds and aldehydes, **6**, 864
    via ether rearrangement, **6**, 876
    via 2,3-sigmatropic rearrangement, **6**, 877
  trans configuration
    synthesis, **6**, 877
Homoallyl alcohols, dichloro-
  geminal
    solvolysis, divinyl ketones from, **5**, 771
Homoallylamines
  synthesis
    allyl organometallic reagent reactions with imines, **2**, 981
    from *N*-heterosubstituted aldimines, **2**, 994
Homoallyl chloroformates
  cyclization
    palladium complexes, **4**, 857
Homoallyl esters
  regioselective oxidation, **7**, 464
Homoallyl ethers
  regioselective oxidation, **7**, 464
Homoallyl hetero systems
  synthesis, **6**, 853
Homo-5α-androstanes
  microbial hydroxylation, **7**, 72
D-Homo-4-androstene-3,17a-dione
  synthesis, **7**, 461
Homoaporphine
  synthesis, **3**, 807
Homoaporphine alkaloids
  synthesis
    via cyclopropane ring opening, **4**, 1020
    via dichlorocyclopropyl compounds, **4**, 1023
Homoazulene, 1-trifluoroacetyl-
  synthesis
    Friedel–Crafts reaction, **2**, 744
Homobarrelene
  photoisomerization, **5**, 196
Homocarboxylic anhydrides
  Friedel–Crafts reaction, **2**, 744
Homocubane
  rearrangements, **3**, 854
Homocubanecarboxylic acid
  synthesis, **3**, 903
1(9)-Homocubene, 9-phenyl-
  Bamford–Stevens reaction, **6**, 779
9-Homocubylidene, 1-phenyl-
  Bamford–Stevens reaction, **6**, 779
Homocuprates, **3**, 211
  mixed, **3**, 212
  synthesis, **3**, 209
Homocuprates, trialkylsilylmetal-
  tandem vicinal difunctionalization, **4**, 255
Homoenolate anions
  allylic anions as, **6**, 862
Homoenolates
  addition reactions, **4**, 117–120
  chiral
    anion equivalents, **6**, 863
  conjugate addition, **2**, 448; **4**, 163
  elimination reaction, **2**, 443
  equivalents, **2**, 442
  esters
    chiral, **2**, 452
  radicals
    reaction, **2**, 448
  reaction with carbonyl compounds
    'homoaldol' reaction, **2**, 445
  substitution reactions, **2**, 449
  synthetic utility, **2**, 55
  tautomerism, **2**, 441
Homoestrone
  synthesis

via benzocyclobutene ring opening, **5**, 693
Homo-Favorskii rearrangement, **3**, 857
Homofernascene
synthesis, **1**, 568
Homogeraniol
asymmetric hydrogenation, **8**, 463
Homoglaucine
synthesis
anodic oxidation, **3**, 673
Homolaudanosine
synthesis, **3**, 79; **7**, 712
Homologation
Wolff rearrangement, **3**, 897
Homolytic addition
donor radical cations, **7**, 881
radical cations
bimolecular reaction, **7**, 860
D-Homo-19-norandrost-4-en-3-one
synthesis
via trisannulation, **7**, 461
Homophthalimide, *N*-chloro-4,4-dialkyl-
Hofmann rearrangement, **6**, 802
D-Homoprogesterone
microbial hydroxylation, **7**, 70
Homopropargylic alcohols
carbomagnesiation, **4**, 879
synthesis, **2**, 84
via allenylsilanes and carbonyl compounds, **1**, 595
via samarium diiodide, **1**, 257
threo-Homopropargylic alcohols
synthesis
diastereoselective, **2**, 91
Homoprotoberberine, 2,3,9,10,11-pentamethoxy-
synthesis, **7**, 712
Homosecodaphniphyllic acid
methyl ester
synthesis, Mannich reaction, **2**, 1024
D-Homosteroids
synthesis
polyene cyclizations, **3**, 369
4,5-Homotropones
synthesis
[4 + 3] cycloaddition reactions, **5**, 609
4-Homotwistane
Ritter reaction, **6**, 270
Homo-tyrosine
synthesis
via oxalate esters, **1**, 425
Hooker oxidation
2-hydroxy-3-alkyl-1,4-naphthoquinones, **3**, 828
Hooz reaction
α-diazocarbonyl compounds
trialkylborane, **2**, 244
Hopane
epoxide
rearrangement, **3**, 745
Horner–Emmons reaction
α,β-unsaturated aldehydes
advantage of Peterson alkenation, **2**, 486
α,β-unsaturated esters
stereochemistry, **7**, 396
Horner reaction
phosphine oxides, **1**, 761, 773
Horner–Wadsworth–Emmons reaction
asymmetric, **1**, 773
mechanism, **1**, 761
phosphonate carbanion
reaction with carbonyl derivative, **1**, 761
(*E*)-selectivity
phosphonate size, **1**, 762
Horner–Wittig reaction
enol ether preparation, **2**, 596
Horse liver alcohol dehydrogenase
coimmobilized
diol oxidation, **7**, 316
Horseradish peroxidase
aromatic hydroxylation, **7**, 79
Hostapon process, **7**, 14
Houben–Hoesch synthesis
intramolecular, **2**, 758
nitriles, **2**, 747
Hüchel molecular orbital calculations
Claisen rearrangement, **5**, 856
Human leukocyte elastase
ynenol lactone inhibitors
synthesis, **3**, 217
Human trisaccharide blood group antigens
synthesis, **2**, 663
Humulene
rearrangement, **3**, 389
synthesis, **3**, 431, 591
coupling reaction of alkenylboranes, **3**, 473
via cyclization, **1**, 553
transannular cyclization, **3**, 399
Wagner–Meerwein rearrangement, **3**, 714
Humulene 1,2-epoxide
transannular cyclization, **3**, 402
Humulene 4,5-epoxide
transannular cyclization, **3**, 404
Humulene 8,9-epoxide
transannular reactions, **3**, 405
Humulene epoxides
transannular cyclization, **3**, 402
Humulol
synthesis, **3**, 399
Hunsdiecker reaction, **7**, 717–732
Hybridalactone
synthesis, **3**, 290
Hycanthone
synthesis
Friedel–Crafts reaction, **2**, 758
Hydantoin
peptide synthesis
via ester fragments, **6**, 399
Perkin reaction, **2**, 406
reduction, **8**, 639
Hydantoin, dehydro-
Diels–Alder reactions, **5**, 406
synthesis
via *N*-chlorination, **5**, 406
Hydantoin, 1,3-dibromo-5,5-dimethyl-
bromination
alkyl alcohols, **6**, 209
Hydantoin, 5,5′-diphenyl-
synthesis, **3**, 826
Hydantoin, 2,4-dithio-
desulfurization, **8**, 639
Hydantoin, 5-ethoxy-
synthesis, **5**, 1109
Hydantoin, methoxy-
Diels–Alder reactions, **5**, 406
Hydantoin, 5-methoxy-

reactions with alkenes, **2**, 1074
Hydantoin, 3-methyl-5,5-diphenyl-
synthesis, **3**, 826
Hydrangenol
synthesis
via directed lithiation, **1**, 477
Hydrastine
synthesis, **2**, 1085
Hydration
alkenes, **4**, 297–316
Hydrazarenes
oxidation
solid support, **7**, 843
Hydrazides
acid halide synthesis, **6**, 308
Curtius reaction, **6**, 806
hydrogenation
Raney nickel, **6**, 403
reductive cleavage, **8**, 388
Hydrazides, arenesulfonyl-
decomposition
aldehydes, **8**, 297
McFadyen–Stephens aldehyde synthesis, **8**, 297
Hydrazides, azido-
synthesis, **6**, 252
Hydrazides, 2,4,6-triisopropylbenzenesulfonyl-
McFadyen–Stephens aldehyde synthesis, **8**, 297
Hydrazine hydrate
reductions
aliphatic nitro compounds, **8**, 375
Hydrazines
chiral
synthesis, **2**, 514
diimide synthesis from, **8**, 472
oxidation, **7**, 742, 747
diimide from, **8**, 472
solid support, **7**, 846
photolysis, **7**, 9
reduction
silbenes, **8**, 568
ultrasonic irradiation, **8**, 368
reductive cleavage, **8**, 388
synthesis
via hydrazones, **8**, 70
via oxidation of secondary amines, **7**, 745
via reduction of diazo compounds and diazonium salts, **8**, 382
Vilsmeier–Haack reaction, **2**, 792
Hydrazines, acyl-
imidoyl halide synthesis, **6**, 489
Hydrazines, alkyl-
synthesis, **6**, 116
Hydrazines, *N*-alkyl-*N*-aryl-
synthesis, **6**, 116
Hydrazines, *N*-aryl-
synthesis, **6**, 119
Hydrazines, *N*,*N'*-disubstituted
azomethine imines from, **4**, 1095
Hydrazines, *N*,*N*-disubstituted
synthesis, **6**, 119
Hydrazines, 1-methyl-1-phenyl-
oxidation
potassium superoxide, **7**, 744
Hydrazines, monoalkyl-
synthesis
via amination of primary alkylamines, **7**, 741
Hydrazines, polysilyl-
reaction with carbonyl compounds, **6**, 116
Hydrazines, tetrafluoro-
reactions with alkenes, **7**, 485
Hydrazino compounds
synthesis, **6**, 116
Hydrazobenzene
synthesis
via reduction of azobenzenes and azoxybenzenes, **8**, 382
Hydrazo compounds
reduction, **8**, 364
synthesis
via reduction of azo and azoxy compounds, **8**, 382
Hydrazoic acid
Schmidt reaction, **6**, 798
synthesis, **6**, 245
Hydrazones
acyl anion equivalents
reactions, **2**, 523
anions, **2**, 503–524
thermal stability, **2**, 507
asymmetric hydrogenation, **8**, 145
asymmetric hydroxylation, **7**, 187
azaallylcopper derivatives
use in synthesis, **2**, 507
azaallylmagnesium bromide derivatives
use in synthesis, **2**, 507
azaallyl metal reagents from, **2**, 506
aziridine synthesis, **1**, 835
azomethine imine precursors, **4**, 1096
carbonyl compounds from, **2**, 523
carbonyl group derivatization, **6**, 726
carbonyl group protection, **6**, 682
chiral
X-ray structure, **2**, 508
cleavage
regeneration of carbonyl groups, **2**, 523
sodium perborate, **2**, 524
cyclizations, **4**, 1148
cyclopropanation, **4**, 954
dehydrogenation, **7**, 144
deprotonation, **3**, 34
regiochemistry, **2**, 509
stereochemistry, **2**, 509
hydrogenation
catalytic, **8**, 143
α-hydroxylation, **7**, 187
infrared spectra, **6**, 727
lithiated
structure, **6**, 727
metallated
metal enolate equivalents, **3**, 30
Michael additions
unsaturated esters, **4**, 222
NMR
carbon **6**, 13, 727
proton, **6**, 727
oxidation, **7**, 742
potassium salts
preparation, **2**, 507
preparation
from ketones and aldehydes, **2**, 504
properties
chemical, **6**, 727
reactions, **6**, 727

reactions with π-allylpalladium complexes, **4**, 603
regioselectivity, **4**, 644
reactions with carbonyl compounds
heterocycle synthesis, **2**, 520
reactions with organocerium reagents
diastereoselectivity, **1**, 239
reactions with organometallic compounds, **1**, 377
reduction
to hydrazines, **8**, 70
reductive cleavage, **8**, 387
reductive elimination, **8**, 939
SAMP
optically pure amine synthesis, **1**, 380
sodium salts
preparation, **2**, 507
spectra, **6**, 727
stereochemistry, **6**, 728
structure, **6**, 727
sulfinylation, **7**, 128
synthesis, **6**, 726
via halides, **7**, 668
titanated
syn selective aldol additions, **2**, 512
ultraviolet spectra, **6**, 727
unsymmetrical
deprotonation, **3**, 34
Vilsmeier–Haack reaction, **2**, 791
Wolff–Kishner reduction, **8**, 328
Hydrazones, α-alkoxy-
acyclic
reactions with organocopper reagents, **1**, 121
Hydrazones, *N*-arylsulfonyl-
reduction
hydrides, **8**, 343
to arylsulfonylhydrazines, **8**, 70
Hydrazones, bis(tosyl)-
reactions with organocopper complexes, **1**, 122
Hydrazones, *N*,*N*-dialkyl-
carbonyl group protection, **6**, 684
chiral
reactions with organometallic compounds, **1**, 379
hydrolysis
copper-catalyzed, **2**, 524
metallated
alkylation, **3**, 34
structure, **6**, 727
Hydrazones, *N*,*N*-dimethyl-
anions
aggregation, **2**, 508
cyclic α,β-unsaturated
alkylation, **3**, 34
deprotonation, **3**, 34
formation, **2**, 504
β-hydroxy
synthesis, **2**, 512
lithiated
axial alkylation, **3**, 34
oxidation
Clayfen, **7**, 846
quaternary salts
Neber rearrangement, **6**, 787
sulfenylation, **7**, 127
α,β-unsaturated δ-hydroxy
synthesis, **2**, 512
Hydrazones, β-hydroxytosyl-
synthesis, **2**, 513
Hydrazones, β-stannyl
oxidation, **7**, 628
Hydrazones, β-stannyl phenyl-
oxidation, **7**, 628
Hydrazones, α-sulfinyl-
chiral
stereospecific aldol synthesis, **2**, 514
Hydrazones, α-sulfinyl dimethyl-
chiral
enantioselective aldol reactions, **2**, 515
Hydrazones, sulfonyl-
decomposition
cyclopropanation, **4**, 954
Hydrazones, tosyl-
acid-catalyzed cyclization, **4**, 1156
diazoalkanes from, **4**, 1101
dilithio dianions
aldol reaction, **2**, 513
reactions with organometallic compounds, **1**, 377
unsaturated
synthesis, **8**, 929
Hydrazones, triisopropylphenyl-
reactions with alcohols
diazoacetate synthesis, **4**, 1033
Hydrazones, α,β-unsaturated *N*,*N*-dimethyl-
Diels–Alder reactions, **5**, 473
Hydrazonyl halides
base treatment
nitrilimines from, **4**, 1083
Hydrazulene
synthesis, **3**, 406
Hydride abstraction
dienyliron complexes
directing effects, **4**, 667
steric effects, **4**, 669
frontier molecular orbitals
dienyliron complexes, **4**, 667
Hydride acceptors
carbonium ions, **8**, 91
Hydride donors
reactivity, **8**, 80
reduction
catalysts, **8**, 82
mechanism, **8**, 81
structural types, **8**, 80
structure, **8**, 80
tertiary anilines, **8**, 98
β-Hydride elimination
hydroformylation, **4**, 918
Hydride reagents
chirally modified
carbonyl compound reduction, **8**, 159–180
Hydrides
aromatic nucleophilic substitution, **4**, 444
delivery from carbon
reduction, **8**, 79–103
desulfurizations, **8**, 839
reduction
alcohols, **8**, 812
cyclic imines, stereoselectivity, **8**, 37
imines, chemoselectivity, **8**, 37
reductive deamination
amines, **8**, 826
tandem vicinal difunctionalization, **4**, 254
Hydride shifts
in alkyne acylation, **2**, 725

Hydride sources
hydrogenolysis
palladium, **8**, 958
Hydride transfer
activation, **8**, 82
alcohols, **8**, 88
aldehydes, **8**, 86
amines, **8**, 88
ammonium formate
transition metal catalyst, **8**, 84
carbonyls, **8**, 323
cation effects, **8**, 90
formic acid, **8**, 84
from transition metal alkyls, **8**, 103
heterocycles, **8**, 92
catalysis, **8**, 97
hydrocarbons, **8**, 91
intramolecular, **8**, 90
organometallics
reduction of carbonyls, **8**, 98
reagents, **7**, 244
Hydrindanediols
pinacol rearrangement, **3**, 727
Hydrindanes
synthesis, **3**, 359, 386, 602, 1052
intramolecular cyclization of cyanocyclohexanes, **3**, 48
via retro Diels–Alder reaction, **5**, 572
Hydrindanones
angular alkylation
stereochemistry, **3**, 17
synthesis
regiospecific alkylation, **3**, 11
via cycloaddition reactions, **5**, 273
via Michael addition, **4**, 24
1-Hydrindanones
alkylation, **3**, 11
trans-Hydrindene
synthesis
Knoevenagel reaction, **2**, 370
Hydrindene acid
Birch reduction
dissolving metals, **8**, 500
Hydrindenediones
synthesis
via intramolecular addition, **4**, 46
Hydrindenones
hydrogenation
stereoselectivity, **8**, 534
synthesis
via Michael addition, **4**, 24
Hydrindinone
synthesis, **1**, 585
Hydroalumination
adducts
chemical derivatives, **8**, 753
alkenes, **8**, 692, 698
alkynes
reactivity, **8**, 738
substituent control, regiochemistry, **8**, 750
1-alkynes
asymmetrical diene synthesis, **3**, 486
symmetrical diene synthesis, **3**, 483
chemoselectivity, **8**, 734
history, **8**, 734
in organic synthesis, **8**, 757
interfering functional groups, **8**, 742
kinetic rate expressions, **8**, 747
locoselectivity, **8**, 734, 742, 744
mechanism, **8**, 747
metal promoters
alkenes, **8**, 751
reaction rates, **8**, 747
rearrangement, **8**, 676
regioselectivity, **8**, 734, 745
scope, **8**, 739
side reactions, **8**, 744
solvent effects, **8**, 747
stereoselectivity, **8**, 734, 746
substituent effects
alkynes, **8**, 749
thermodynamics, **8**, 670
transition metal catalysts, **8**, 747
unsaturated hydrocarbons, **8**, 733–758
vinylalanes, **3**, 266
Hydrobenzamide
Mannich reaction, **2**, 916
synthesis
Mannich reaction, **2**, 916
Hydroborates
synthesis
via alkyldimesitylboranes, **1**, 492
Hydroboration
acyclic alkenes, **8**, 704
alkenes, **4**, 357
alkynes
organopalladium catalysis, **3**, 231
chiral, **8**, 720
dimethylborolane
enantioselectivity, **2**, 258
fundamentals, **8**, 704
mechanism, **8**, 724
unsaturated hydrocarbons, **8**, 703–727
with thexylborane, **2**, 251
Hydroboration–oxidation
enamines, **6**, 715
Hydrobromination
alkenes, **4**, 279–287
stereochemistry, **4**, 279
Hydrocarbons
acid halide synthesis, **6**, 308
acyclic
enantioselective hydroxylation, **7**, 57
microbial oxidation, **7**, 56
cyclic
microbial oxidation, **7**, 58
dimerization
mercury-catalyzed, **3**, 1047
hydride transfer, **8**, 91
oxidation
metalloporphyrin-catalyzed, **7**, 50
polyunsaturated substituted
synthesis *via* retro Diels–Alder reaction, **5**, 565–573
Ritter reaction, **6**, 270
Hydrocarboxylation
alkenes, **4**, 939–941
asymmetric, **4**, 945
catalysts, **3**, 1027
conjugated dienes, **4**, 945
mechanism, **3**, 1019
Hydrochloric acid

reaction with tertiary alkyl alcohols
displacement of hydroxy group, **6**, 204
Hydrochlorination
alkenes, **4**, 272–278
Hydrocortisone
oxidation
solid supports, **7**, 845
Hydrocyanation
alkenes
hydrozirconation, **8**, 694
alkynes
hydrozirconation, **8**, 688
zirconium compounds, **8**, 692
Hydrocyanic acid
alkanenitrile synthesis, **6**, 234
reaction with orthoesters, **6**, 564
Hydrodimerization
enones, **8**, 532
Hydroesterification
alkenes
asymmetric, **4**, 945
catalysts, **3**, 1029
mechanism, **3**, 1019
styrene
palladium catalyst, **3**, 1030
Hydrofluorination
alkenes, **4**, 270–272
Hydroformylation
alkenes, **4**, 913–949
asymmetric, **4**, 927–932
catalysts, **4**, 915
cobalt, **3**, 1021
rhodium, **3**, 1021
formation of 1-propanal, **3**, 1015
functionalized alkenes, **4**, 922–927
mechanism, **3**, 1019; **4**, 915
regioselectivity, **4**, 916–919
stereoselectivity, **4**, 916–919
unfunctionalized alkenes, **4**, 919–922
Hydrogallation
alkenes, **8**, 698
Hydrogenation
acetals, **8**, 212
acyl halides
Rosenmund reaction, **8**, 239
aldonolactones, **8**, 292
alkenes, **8**, 421
comparison with Wacker oxidation, **7**, 450
double bond migration, **8**, 422
heterogeneous catalysis, **8**, 417–442
homogeneous catalysis, **8**, 443–463
mechanism, **8**, 422
structure–reactivity, **8**, 424
alkynes
heterogeneous catalysis, **8**, 417–442
homogeneous catalysis, **8**, 443–463
amides, **8**, 248
anhydrides, **8**, 292
aromatic compounds
homogeneous catalysis, **8**, 453
carboxylic acids, **8**, 236
catalysts
chirally modified, **8**, 149
heterogeneous, **8**, 417–442
kinetics, **8**, 419
mechanism, **8**, 420
transport phenomena, **8**, 419
catalytic
aromatic carbonyl compounds, **8**, 319
benzo[*b*]furans, **8**, 624
benzo[*b*]thiophenes, **8**, 629
carbonyl compound reduction, **8**, 139–155
conjugated dienes, **8**, 565
enones, **8**, 533
furans, **8**, 606
indoles, **8**, 612
pyridines, **8**, 597
pyrroles, **8**, 604
thiophenes, **8**, 608
unsaturated carbonyl compounds, **8**, 533
diastereoselective asymmetric, **8**, 144
electrocatalytic
ketones and aldehydes, **8**, 135
esters, **8**, 242
heteroaromatic compounds
homogeneous catalysis, **8**, 453
heterogeneous catalysis
alkenes and alkynes, **8**, 417–442
chiral catalyst, **8**, 149
homogeneous catalysis
alkenes and alkynes, **8**, 443–463
conjugated alkenes, **8**, 449
imidoyl chlorides, **8**, 301
ionic
alkenes, **8**, 486
carbonyl compound reduction, **8**, 317
mechanism, **8**, 486
isomerization modifiers
alkenes, **8**, 423
lactones, **8**, 246
nitriles, **8**, 251, 298
nitroso compounds
aromatic, **8**, 372
zirconium compounds, **8**, 690
Hydrogen atom transfer reactions
intramolecular cyclization, **4**, 820
radical addition reactions, **4**, 752
radical cyclizations, **4**, 801
Hydrogen bromide
reaction with alkyl alcohols, **6**, 209
Hydrogen cyanide
Ritter reaction, **6**, 266
Hydrogen fluoride
fluorination
alkyl alcohols, **6**, 216
pyridine complex
hydrofluorination, **4**, 271
trialkylamine complex
hydrofluorination, **4**, 271
Hydrogen halides
addition reactions
nitriles, **6**, 497
addition to propiolic acid, **4**, 51
reactions with alkenes, **4**, 270–290
reactions with α-chloroenamines, **6**, 497
Hydrogen iodide
iodination
alkyl alcohols, **6**, 213
reduction
allylic compounds, **8**, 978
Hydrogenolysis
allyl halides, **8**, 955–981

amines, **8**, 826
aromatic carbonyl compounds, **8**, 319
Birch reduction
dissolving metals, **8**, 514
catalytic
alcohols, **8**, 814
alkyl halides, **8**, 794
C—N bonds, **8**, 915
C—O bonds, **8**, 910
C—S bonds, **8**, 913
epoxides, **8**, 881
vinyl halides, **8**, 895
Hydrogen peroxide
acidic
organoborane oxidation, **7**, 597
alkaline
organoborane oxidation, **7**, 595
Baeyer–Villiger reaction, **7**, 674
epoxidations with, **7**, 381
glycol cleavage, **7**, 708
hydroxylation
alkenes, **7**, 438, 446
α-hydroxylation
ketones, **7**, 163
oxidation
primary amines, **7**, 737
selenides, **7**, 771
sulfides, **7**, 194, 762
sulfoxides, **7**, 766
thiols, **7**, 760
oxidative hydrolysis
ozonides, **7**, 574
reoxidant
Wacker process, **7**, 452, 462
silylated
oxidation, **7**, 674
Hydrogen selenide
reaction with nitriles, **6**, 476
Hydrogen sulfide
acylation
imidates and orthoesters, **6**, 450
imidothioates, **6**, 455
carbon monoxide
reduction, aromatic nitro compounds, **8**, 372
demercurations, **8**, 857
reduction
carbonyl compounds, **8**, 323
Hydrogen telluride
reductions
aromatic compounds, **8**, 370
Hydrogen transfer
intramolecular
stereoselectivity, **6**, 865
reduction
carbonyl compounds, **8**, 320
Hydrogermylation
alkenes, **8**, 699
radical addition reactions, **4**, 770
radical reactions
rate, **4**, 738
Hydroindoles
synthesis
Mannich cyclization, **2**, 1041
Hydroindolones
synthesis
Mannich reaction, **2**, 1011
Hydroiodination
alkenes, **4**, 287–290
Hydrolithiation
catalytic, **8**, 697
Hydrolysis
esters
enantiotopically selective, **6**, 342
hydrazones
regeneration of carbonyl groups, **2**, 524
Hydromagnesiation
catalytic, **8**, 697
silylalkynes, **4**, 879
unsaturated hydrocarbons, **8**, 751
Hydrometallation, **8**, 695
catalytic, **8**, 697
mechanism, **8**, 671
unsaturated hydrocarbons, **8**, 667–699
Hydroperoxides
mercuricyclization, **4**, 390
reduction
synthesis of alcohols, **8**, 396
Hydroperoxides, alkyl
oxidation
organoboranes, **7**, 602
trialkylborane, **7**, 599
Hydroperoxides, *t*-butyl
allylic oxidation, **7**, 96
oxidation
primary amines, **7**, 737
selenium reoxidant
allylic oxidation, **7**, 88
Hydroperoxides, trityl
epoxidation, **7**, 376
Hydroquinolones
synthesis
Mannich reaction, **2**, 1011
Hydroquinones
electrochemical reoxidation
Wacker process, **7**, 452
oxidation
chromium(VI) oxide, **7**, 278
solid support, **7**, 843
synthesis, **7**, 339, 340
Hydroquinones, cyano-
synthesis
via haloquinones, **6**, 231
Hydroquinones, 2-methyl-
Mannich reaction, **2**, 969
Hydroquinones, silyl-protected
oxidation
pyridinium chlorochromate, **7**, 264
Hydrosilanes
hydrosilylation
unsaturated hydrocarbons, **8**, 765
Hydrosilylation
acetylene, **8**, 769
alkenes, **8**, 699
alcohol synthesis, **6**, 17
trichlorosilane, **7**, 642
alkynes
chlorodimethylsilane, **7**, 643
(diethoxymethyl)silane, **7**, 643
asymmetric, **8**, 173
alkenes, **8**, 782
chiral catalyst, **7**, 642
carbon monoxide, **8**, 788

catalysts
unsaturated hydrocarbons, **8**, 764
conjugated alkynes, **8**, 772
conjugated enones, **8**, 781
cyclic polyenes, **8**, 780
disubstituted alkynes, **8**, 771
ethylene, **8**, 773
intramolecular, **8**, 788
allyl alcohols, **7**, 645
carbonyl compounds, **8**, 9
isoprene, **8**, 779
mechanism
unsaturated hydrocarbons, **8**, 765
monosubstituted alkynes, **8**, 770
monosubstituted ethylenes, **8**, 774
organofluorosilicates
synthesis, **7**, 642
α-oxy ketones, **8**, 8
polyenes, **8**, 778
solvents
unsaturated hydrocarbons, **8**, 765
substituted ethylenes, **8**, 776
unsaturated hydrocarbons, **8**, 763–789
1,4-Hydrosilylation
α,β-unsaturated carbonyl compounds
enol ether preparation, **2**, 603
Hydrostannylation
alkenes, **8**, 699
radical addition reactions, **4**, 770
carbonyl compounds, **8**, 21
radical cyclization, **4**, 796
unsaturated esters, **8**, 548
1,4-Hydrostannylation
O-stannyl ketene acetal, **2**, 609
Hydrotitanation
alkenes, **8**, 696
Hydroxamates, *O*-acyl seleno-
decomposition
synthesis of alkyl 2-pyridyl selenides, **7**, 726
photolysis, **7**, 722
Hydroxamates, *O*-acyl thio-
carboxyl radicals from, **7**, 719
decomposition
noralkyl hydroperoxides, **7**, 727
fragmentation
thiophilic radicals, **7**, 719
photolysis, **7**, 731
alkyl 2-pyridyl sulfides, **7**, 726
decarboxylative iodination, **7**, 725
reaction with tris(phenylthio)phosphorus, **7**, 727
reductive decarboxylation, **7**, 720, 721
Hydroxamic acid chlorides
base treatment
nitrile oxides from, **4**, 1078
Hydroxamic acids
N-acylimines
reactions with organometallic compounds, **1**, 376
Lossen reaction, **6**, 821
nitroso derivatives
Diels–Alder reactions, **5**, 420
oxidation
periodate, **6**, 402
reduction
titanium trichloride, **6**, 402; **8**, 395
Hydroxamic acids, *O*-acyl-
Lossen reaction, **6**, 798
α-Hydroxy acids
absolute configuration
synthesis, **6**, 882
asymmetric hydrogenation
modifying reagents, **8**, 150
oxidative cleavage, **7**, 709
synthesis
double carbonylation, **3**, 1039
β-Hydroxy acids
6-membered ring *O,O*-acetals, endocyclic enolates
alkylation, **3**, 41
γ-Hydroxy acids
cycloacylation
γ-lactone synthesis, **6**, 350
HGA lactonization, **6**, 358
γ-Hydroxyalkyl bromides
synthesis, **3**, 120
γ-Hydroxyalkyl iodides
synthesis, **3**, 120
α-Hydroxycarboxamides
synthesis, **2**, 1086
Hydroxy esters
macrolactonization, **6**, 369
α-Hydroxy esters
synthesis
via reduction of α-keto esters, **8**, 169
Hydroxy group activation
ester synthesis, **6**, 333
Hydroxylactonizations
γ,δ-enoic acids, **6**, 361
Hydroxylamine, *N*-alkyl-
synthesis, **6**, 112, 115
Hydroxylamine, *N*-allyl-
synthesis, **6**, 113
Hydroxylamine, *N*-allyl-*N*-aryl-
synthesis, **6**, 115
Hydroxylamine, *O*-aryl-
synthesis, **6**, 114
Hydroxylamine, *N*-aryl-*O*-acetyl-
hydrazine synthesis, **6**, 119
Hydroxylamine, *O*-(arylsulfonyl)-
reaction with alkenes, **7**, 471
synthesis, **6**, 116
Hydroxylamine, *O*-benzyl-
N-alkylation, **6**, 83
synthesis, **6**, 112
Hydroxylamine, *N,N*-dialkyl-
phenacyl bromide oxidation, **7**, 663
reductive cleavage
synthesis of secondary amines, **8**, 395
Hydroxylamine, *N,O*-dimethyl-
reaction with acyl chlorides, **8**, 272
Hydroxylamine, *O*-(2,4-dinitrophenyl)-
amination
secondary amines, **7**, 746
electrophilic *N*-aminations, **6**, 119
Hydroxylamine, *N*-(diphenylphosphinyl)-
synthesis, **6**, 114
Hydroxylamine, *O*-(diphenylphosphinyl)-, **6**, 119
amination
secondary amines, **7**, 746
electrophilic *N*-aminations, **6**, 119
synthesis, **6**, 114
Hydroxylamine, *N*-(homoallyl)-
reaction with allyl organometallic compounds, **2**, 994
Hydroxylamine, *O*-mesityl-

amination
secondary amines, **7**, 746
Hydroxylamine, *O*-mesitylenesulfonyl-
amination
pyridines, **7**, 750
secondary amines, **7**, 746
electrophilic *N*-aminations, **6**, 119
reactions with organoboranes, **7**, 606
Hydroxylamine, *O*-(mesitylsulfonyl)-
Beckmann rearrangement, **7**, 694
Hydroxylamine, *N*-phenyl-
synthesis, **8**, 366
Hydroxylamine, *O*-phosphinyl-
synthesis, **6**, 116
Hydroxylamine, tri-*t*-butyl-
reduction
synthesis of di-*t*-butylamine, **8**, 395
Hydroxylamine, *O*-trimethylsilyl-
synthesis, **6**, 114
Hydroxylamine, tris(trimethylsilyl)-
reaction with acid chlorides, **6**, 114
Hydroxylamine ethers
synthesis
via oxime ethers, **8**, 60
Hydroxylamines
allylic
synthesis, **6**, 115
amine oxidation
intermediate, **7**, 738
N,N-disubstituted
reactions with organometallic compounds, **1**, 391
enzymic reduction, **8**, 395
oxidation, **7**, 742, 747
with halides, **7**, 663
reduction
metal hydrides, **8**, 27
synthesis of amines, **8**, 394
synthesis, **6**, 111; **8**, 366, 373
via oxidation of primary amines, **7**, 736
via oxidation of secondary amines, **7**, 745
via oximes, **8**, 60
Vilsmeier–Haack reaction, **2**, 792
Hydroxylamine-*O*-sulfonic acid
amination
amines, **7**, 741
secondary amines, **7**, 746
Beckmann rearrangement, **6**, 764
deamination
amino acids, **8**, 828
electrophilic *N*-aminations, **6**, 119
Lossen reaction, **6**, 825
reactions with organoboranes, **7**, 606
Hydroxylation
α to carbonyl, **7**, 152
α to cyanide, **7**, 186
anti
alkenes, **7**, 438, 446
activated C—H bonds
oxidation, **7**, 151–187
alkanes, **7**, 11
alkenes, **7**, 437
anodic
aromatic compounds, **7**, 800
regioselective
dienes, **7**, 438
steroids, **7**, 132
microbial, **7**, 66, 68
microbial, chemoselectivity, **7**, 69
microbial, regioselectivity, **7**, 70
microbial, stereoselectivity, **7**, 72
syn
alkenes, **7**, 438, 439
β-Hydroxylation
aliphatic carboxylic acids
microorganisms, **7**, 56
Hydroxylation–carbonylation
allenes
palladium(II) catalysis, **4**, 558
Hydroxymercuration, **8**, 854
demercuration
alkenes, **4**, 300–305
Hydroxymethylation
nucleophilic, **7**, 647
samarium diiodide
Barbier-type reaction, **1**, 259
Hydroxyselenenation
alkenes, **7**, 522
Hydroxysulfenylation
alkenes, **7**, 518
Hydrozirconation
1-alkynes
asymmetrical diene synthesis, **3**, 486
symmetrical diene synthesis, **3**, 483
chemoselectivity, **8**, 683
chlorohydridobis(cyclopentadienyl)zirconium, **8**, 675
conditions, **8**, 676
diastereoselectivity, **8**, 688
enantioselectivity, **8**, 690
mechanism, **8**, 668
regioselectivity, **8**, 684
synthetic utilization, **8**, 690
thermodynamics, **8**, 669
unsaturated hydrocarbons, **8**, 667
Hygrine
synthesis, **8**, 273
Hyperacyloin condensation
synthesis of phenanthraquinone, **3**, 619
Hypercornine
synthesis, **1**, 564
Hypericin
synthesis, **3**, 699
Hypnophilin
synthesis, **3**, 603
organocopper compounds, **3**, 221
via tandem radical cyclization, **1**, 270
Hypochlorite
irradiation, **7**, 41
Hypochlorite, *t*-butyl
oxidation
sulfides to sulfoxides, **7**, 194
Hypofluorous acid
reaction with alkenes, **4**, 347
Hypohalites
alkoxy radicals from, **4**, 812
reaction with alkenes, **4**, 347
Hypophosphorus acid
reduction
pyrroles, **8**, 606

I

Ibogamine
synthesis, **7**, 476
via Diels–Alder reaction, **5**, 373
via palladium catalysis, **4**, 598
Ibuprofen
methyl ester
synthesis, **7**, 829
Icosanoic acid, 10,16-dimethyl-
synthesis, **3**, 644
L-Idose, 2-deoxy-
synthesis
FDP aldolase, **2**, 462
Ikarugamycin
synthesis
(*Z*)-selectivity *via* Wittig reaction, **1**, 765
via Cope rearrangement, **5**, 817
via electrocyclization, **5**, 725
Illudane
biosynthesis, **3**, 404
Iloprost
synthesis
Knoevenagel reaction, **2**, 381
Imidates
acyclic
addition reactions, **2**, 488
acylation
hydrogen sulfide, **6**, 450
cyclic
reduction, **8**, 302
imidate synthesis, **6**, 534
metallated
addition reactions, **2**, 488
reduction, **2**, 1050; **8**, 302
sulfhydrolysis, **6**, 450
synthesis, **6**, 529
thiolysis, **6**, 429
transesterification, **6**, 534
tris(dialkylamino)alkane synthesis, **6**, 579
Imidates, allylic *N*-phenyl-
rearrangements
oxygen–nitrogen transposition, **6**, 843
Imidazole, 1-acyl-
Claisen condensation, **2**, 801
Imidazole, 1,1′-carbonyldi-
Beckmann rearrangement, **7**, 692
Imidazole, dihydro-
synthesis
via cyclization of methylisourea, **4**, 388
Imidazole, mercapto-
oxidation, **7**, 760
Imidazole, 1-methoyl-
acylation, **6**, 516
Imidazole, *N*-methyl-
hydroxyalkylation
protection, **6**, 682
Imidazole, 1-methyl-5-chloro-
phosphorylation, **6**, 601
Imidazole, 2-(5-norbornen-2-yl)-
synthesis
via retro Diels–Alder reaction, **5**, 557
Imidazole, *N*-phosphoryl-
phosphorylation, **6**, 614
Imidazole, *N*,*N*′-thionyldi-
amidine synthesis, **6**, 546
Imidazolecarboxylic acids
electrolytic reduction, **8**, 285
Imidazoles
N-alkyl
lithiation, **1**, 477
[2 + 2 + 2] cycloaddition reactions, **5**, 1143
Diels–Alder reactions, **5**, 491
metallation
addition reactions, **1**, 471
reaction with chloroform, **6**, 579
reduction, **8**, 638
synthesis, **6**, 517
1,2,4-trisubstituted
N-acylimines from, **1**, 376
Imidazoles, acyl-
acylation, **1**, 423
reduction
metal hydrides, **8**, 271
Imidazoles, 2-acyl-
alkylation, **6**, 516
Imidazoles, *N*-acyl-
acylation, **6**, 333
Imidazoles, 1-benzyl-2-alkyl-4,5-dihydro-
methiodide salt
reactions with organometallic compounds, **1**, 366
Imidazoles, *N*-hydroxy-
reduction
titanium(III) chloride, **8**, 395
Imidazoles, thioacyl-
thioarylation, **6**, 450
Imidazoles, 2-vinyl-
synthesis
via retro Diels–Alder reactions, **5**, 557
Imidazole-1-thiocarbonyl compounds
deoxygenation, **8**, 818
Imidazolides
acid anhydride synthesis, **6**, 313
acid halide synthesis, **6**, 308
Curtius reaction, **6**, 810
Imidazolides, imidoyl-
amidine synthesis, **6**, 551
Imidazolidine, 1,3-dimethyl-2-phenyl-
lithiation
addition reactions, **1**, 463
2-Imidazolidinone, 1-acyl-
synthesis
via Curtius reaction, **6**, 814
Imidazolidin-2-ones
bicyclic
synthesis, **2**, 1062
reduction
LAH, **8**, 639
1,3-Imidazolidin-4-ones
addition reactions with nitroalkenes, **4**, 109
Imidazolidiones
enolates
diastereoselective alkylation, **3**, 45
Imidazoline, 2-alkenyl-
preparation, **2**, 494
2-Imidazoline, 2-alkenyl-

addition reactions
with organolithium compounds, **4**, 76
Imidazoline, 2-alkyl-
preparation, **2**, 494
Imidazoline, 2-methyl-
metallated
reactions, **2**, 494
Imidazolines
conjugate additions, **4**, 207
reaction with isocyanates, **6**, 579
reduction, **8**, 638
synthesis, **7**, 479
via intramolecular Ritter reaction, **6**, 277
2-Imidazolines
reductive decyclization, **8**, 638
Imidazolinones
synthesis, **7**, 486
Imidazolium dichromate
oxidation
alcohols, **7**, 278
Imidazolium salts
reduction, **8**, 638
Imidazol-2-one, *N*-phenyl-
metallation, **1**, 464
Imidazolones
lithium compounds
oxidation, **7**, 330
Imidazo[4,5-*b*]pyridines, 3-aryl-2-methyl-
synthesis, **4**, 436
1*H*-Imidazo[1,5-*a*]pyrrole, 2,3-dihydro-
synthesis
Mannich reaction, **2**, 968
Imides
addition reactions
Grignard reagents, **2**, 1049
alkylation
O-alkylisourea, **6**, 74
chiral
asymmetric aldol reactions, **2**, 231
conjugate additions, **4**, 202
chiral 2-oxazolidones
diastereoselective alkylation, **3**, 45
cyclic
reduction, **2**, 1049
reduction, metal hydrides, **8**, 273
homologated
synthesis, Eschenmoser coupling reaction, **2**, 874
reduction, **8**, 254
diastereoselective, **2**, 1049
regioselectivity, **2**, 1049
synthesis, **6**, 409
Vilsmeier–Haack reaction, **2**, 792
Imides, α-allenyl-
reaction with organocopper reagents, **3**, 223
Imidocarboxylic acids
thiolysis, **6**, 428
Imido ester hydrochlorides
synthesis, **6**, 507
Imido esters
reaction with hydrogen selenide, **6**, 472, 473
Imidothioates
acylation
hydrogen sulfide, **6**, 455
hydrolysis
synthesis of thiol esters, **6**, 444
sulfhydrolysis, **6**, 455
synthesis of dithiocarboxylic esters, **6**, 453
Imidoyl chlorides
coupling reactions
with primary alkyl Grignard reagents, **3**, 463
reduction, **8**, 300
metal hydrides, **8**, 272
selenol ester synthesis, **6**, 473
synthesis, **6**, 489, 767
thiolysis, **6**, 428
Imidoyl compounds
amidine synthesis, **6**, 550
amidinium salt synthesis, **6**, 515
synthesis, **6**, 523
Imidoyl cyanide
synthesis, **6**, 768
Imidoyl halides
base treatment
nitrile ylides from, **4**, 1081
imidate synthesis, **6**, 532
imidoyl halide synthesis, **6**, 527
reactions with hydrogen halides, **6**, 497
synthesis, **6**, 523
via amides, **6**, 489
thioimidate synthesis, **6**, 539
Imidoyl halides, keto-
synthesis
via Ritter reaction, **6**, 295
Imidoyl iodide
synthesis, **7**, 696
Imine anions
isomerization, **6**, 723
protonation, **6**, 721
X-ray structure
single-crystal, **6**, 723
Imines
2-acetidinone synthesis, **5**, 100–102
achiral
reactions with allyl organometallic compounds, **2**, 980
reactions with type I crotyl metallics, **2**, 9–19
activated
reactions with allenylsilanes, **1**, 602
synthesis of substituted amines, **1**, 357
amine protection, **6**, 645
anions
aldol reaction, **2**, 477
heteroatom-stabilized, **2**, 482
phosphorus-stabilized, **2**, 482
silicon-stabilized, **2**, 482
aryl-substituted
photoisomerization, **5**, 202
asymmetric hydrogenation, **8**, 145
asymmetric reduction, **8**, 176
2-azetidiniminium salts from, **5**, 108–113
carbonyl group derivatization, **6**, 719
chiral
conjugate additions, **4**, 210
reaction with silyl ketene acetals, **2**, 647
coupling reactions
with 1,2-diamines, **3**, 564
cyclic
Diels–Alder reactions, **5**, 406–408
homogeneous hydrogenation, **8**, 155
reactions with enolates, **2**, 942
reactions with organometallic compounds, **1**, 364
cyclization

tin(IV) chloride promotion, **2**, 1024
cycloaddition reactions
vinylidene complexes, **5**, 1068
[3 + 2] cycloaddition reactions, **5**, 307
Darzens glycidic ester condensation, **2**, 422
deprotonation
regiochemistry, **6**, 720
regioselectivity, **1**, 357
diastereoselective addition reactions
chiral silyl ketene acetals, **2**, 638
Diels–Alder reactions, **5**, 402–416
electrophilicity
methods for increase, **1**, 357
electroreduction, **8**, 135
asymmetric, **8**, 137
mechanism and products, **8**, 135
endocyclic anions
from unsaturated heterocycles, **2**, 481
Erlenmeyer azlactone synthesis, **2**, 404
esters
N-alkylation, **6**, 83
halomagnesium derivatives
alkylation, **3**, 31
heterocyclic synthesis, **6**, 734
N-heterosubstituted
reaction with allyl organometallic compounds, **2**, 994
homogeneous catalytic hydrogenation, **8**, 152
hydrosilylation, **8**, 180
infrared spectra, **6**, 724
Knoevenagel reaction, **2**, 367
lithiated
alkylation, **3**, 31
axial alkylation, **3**, 32
Mannich reaction, **2**, 915, 970
metallated
metal enolate equivalents, **3**, 30
metallated chiral
asymmetric alkylation, **3**, 35
metallation
sulfenylation of aldehydes, **7**, 125
NMR
carbon, **6**, 13, 724
proton, **6**, 724
nucleophilic addition reactions, **1**, 355–393
one stereocenter
reaction with allyl organometallic compounds, **2**, 983
oxidation, **6**, 527
mechanism, **1**, 837
N-oxidation, **7**, 750
α-phosphorus stabilized imines
aldol reaction, **2**, 483
N-phosphorus substituted
reduction, **8**, 74
pinacol coupling reactions, **3**, 579
intermolecular, **3**, 579
intramolecular, **3**, 581
with ketones, **3**, 596
properties
chemical, **6**, 723
proton abstraction, **1**, 356
reactions, **6**, 724
reactions with acid chlorides, **2**, 1050
reactions with allenic titanium reagents, **2**, 95
reactions with allylboronates, **2**, 15
reactions with allyl organometallic reagents, **2**, 975–1004
reactions with carboxylic acid derivatives
Mannich reaction, **2**, 917
reactions with crotyl organometallic compounds
syn–anti selectivity, **2**, 989, 990
reactions with enolates
Mannich reaction, **2**, 919
reactions with enol silanes
Lewis acid mediated, **2**, 635
reactions with Fischer carbene complexes, **5**, 1107
reactions with highly acidic active methylene compounds
Mannich reaction, **2**, 916
reactions with ketene bis(trimethylsilyl)acetals, **2**, 930
reactions with ketones, **2**, 933
reactions with organocerium compounds, **1**, 236
reactions with organocopper complexes, **1**, 119
reactions with organometallic compounds, **1**, 360
Lewis acid promotion, **1**, 349
reactions with pent-3-ene-2-yl-9-borabicyclononane
syn–anti selectivity, **2**, 992
reactions with propargyl organometallic compounds
variation of yield with metal, **2**, 993
reactions with α-silylbenzylic anions, **1**, 624
reactions with silyl ketene acetals, **2**, 929
reactions with sulfinyl-stabilized carbanions, **1**, 515
reactions with vinyl silyl ketene acetals, **2**, 930
reactions with ylides, **1**, 835
reduction
diimide, **8**, 478
dissolving metals, **8**, 123
enzymes and microorganisms, **8**, 204
mechanism, **8**, 26
metal hydrides, **8**, 25–74
synthesis of amines, **6**, 724
Reformatsky reaction, **2**, 294
spectra, **6**, 724
stereochemistry, **6**, 725
structure, **6**, 723
N-sulfur substituted
reduction, **8**, 74
synthesis, **6**, 719
via aziridine thermolysis, **5**, 938
via carboxylic acids, **8**, 284
via reactions of amides and organocuprates, **1**, 124
via reduction of oximes, **8**, 392
tandem vicinal difunctionalization, **4**, 252
α-trialkylsilyl-stabilized anions
aldol reaction, **2**, 484
two stereocenters
reaction with allyl organometallic compounds, **2**, 987
Vilsmeier–Haack reaction, **2**, 792
Imines, acyclic *N*-alkyl-
Mannich reaction, **2**, 916
Imines, acyclic *N*-aryl-
Mannich reaction, **2**, 916
Imines, *C*-acyl-
Diels–Alder reactions, **5**, 408
Imines, *N*-acyl-
acyclic
Diels–Alder reactions, **5**, 404
Diels–Alder reactions, **5**, 404–408, 485
protonation, **2**, 1052
reactions with organocopper complexes, **1**, 122

reactions with organometallic compounds, **1**, 371, 373
reactivity, **1**, 371
Imines, alkyl-
Diels–Alder reactions, **5**, 409–411
Imines, aryl-
Diels–Alder reactions, **5**, 409–411
Imines, *N*-(*t*-butyldimethylsilyl)-
reactions with silyl ketene acetals, **2**, 938
Imines, α-chloro-
preparation, **2**, 422
Imines, *N*-chloro-
hydrazones
Neber reaction, **6**, 786
Imines, cyclopropyl-
rearrangements, **5**, 941
Imines, diphenylphosphinyl-
prochiral
asymmetric reduction, **8**, 176
Imines, epoxy-
synthesis
via Sharpless epoxidation, **5**, 98
Imines, α-halo-
masked α-halocarbonyl compounds
Darzens condensation, **2**, 422
Imines, 2-hydroxy-
rearrangement, **3**, 790
semipinacol rearrangement, **3**, 778
Imines, *N*-silyl-
preparation, **2**, 935
reactions with enolates, **2**, 934
Imines, sulfinyl-
reduction
lithium aluminum hydride, **8**, 74
Imines, *N*-sulfonyl-
Diels–Alder reactions, **5**, 402–404
reduction
sodium cyanoborohydride, **8**, 74
Imines, thione
Diels–Alder reactions, **5**, 441
Imines, *N*-trialkylsilyl-
enolizable carbonyl compounds
reactions with organometallic compounds, **1**, 391
Imines, *N*-trimethylsilyl-
in situ synthesis, **1**, 390
reactions with organometallic compounds, **1**, 390
Imines, α,β-unsaturated
Diels–Alder reactions, **5**, 473
Imines, α,β-unsaturated *N*-phenylsulfonyl-
Diels–Alder reactions, **5**, 473, 474
Iminimium ions, acyl-
cyclization
heterocyclic synthesis, **6**, 746
synthesis, **6**, 744
Iminium bromide, bromomethylene-
synthesis, **6**, 495
Iminium chloride, α-chloro-
synthesis, **5**, 108
Iminium chlorides, *aci*-nitro-
reactions with organocopper complexes
synthesis of ketoximes, **1**, 121
Iminium ions
chiral
reaction with enol silanes, **2**, 649
cyclization, **6**, 736
enantioselective, **2**, 1027
endocyclic, *N*-acyl group in the ring, **2**, 1057
intermolecular reactions, **2**, 1057
intramolecular reactions, **2**, 1062
endocyclic, *N*-acyl group outside the ring, **2**, 1066
intermolecular reactions, **2**, 1066
intramolecular reactions, **2**, 1069
generation
Mannich reactions, **2**, 1008
Pictet–Spengler cyclization, **2**, 1021
heterocyclic synthesis, **6**, 734
intramolecular cyclization, **2**, 1007
intramolecular Mannich reactions, **2**, 1007
Mannich reactions, **2**, 954
intermediate, **2**, 895
silyl enol ethers, **2**, 1015
nucleophilic additions
stereochemistry, **2**, 1008
synperiplanar, **2**, 1013
photochemistry, **2**, 1037
reaction with allyl organometallic reagents, **2**, 975–1004
synthesis, **6**, 734
with alkenes, **2**, 1023
Iminium salts
amidine synthesis, **6**, 542
amidinium salt synthesis, **6**, 514
cyclic
Mannich reaction, **2**, 912
synthesis, **6**, 503
in situ generation, **1**, 367
reactions with allyl organometallic compounds, **2**, 1002
oxidation, **7**, 664
preformed
Mannich reaction, **2**, 898, 956, 960
reactions, **2**, 899
reactions with allyl organometallic compounds, **2**, 1000
synthesis, Mannich reaction, **2**, 898
reactions with crotyl organometallic compounds
dependence of product type on metal, **2**, 1001
reactions with halogen-substituted allylic anions
regioselectivity, **2**, 77
reactions with organometallic compounds, **1**, 365
reactions with propargyl organometallic reagents
dependence of product type on metal, **2**, 1001
reactions with unsymmetrical methyl ketones
regiochemistry, **2**, 902
silicon stabilization
cyclizations, **1**, 592
synthesis, **6**, 485–583
trimethylsilyl
nucleophilic addition, **1**, 391
ω-unsaturated
reduction by samarium diiodide, **1**, 275
Iminium salts, *N*-acyl-
[3 + 2] cycloaddition reactions
with 1,3-dimethyl(*t*-butyldimethylsilyl)allene, **5**, 279
generation
Mannich reactions, **2**, 1008
reactions with allenylsilanes, **1**, 598
reactions with organometallic compounds, **1**, 371, 373
reactivity, **1**, 371
silicon stabilization
cyclizations, **1**, 592
Iminium salts, acyloxy-

synthesis, **6**, 493
Iminium salts, alkoxymethylene-
amide acetal synthesis, **6**, 567, 573
amidine synthesis, **6**, 543
imidate synthesis, **6**, 529
ortho acid synthesis, **6**, 561
synthesis, **6**, 501
Iminium salts, alkylmercaptomethylene-
amidine synthesis, **6**, 543
synthesis, **6**, 508
thioimidate synthesis, **6**, 536
Iminium salts, aryloxymethylene-
synthesis, **6**, 505
Iminium salts, bromomethylene-
synthesis, **6**, 495
Iminium salts, *N*,*N*-dialkyl-
acyclic
Mannich reaction, **2**, 898
Iminium salts, dihalomethylene-
amide halide synthesis, **6**, 498
Iminium salts, *N*,*N*-dimethyl(methylene)-
chloride
synthesis, **2**, 900
generation *in situ*, **2**, 901
Mannich reaction, **2**, 899
preparation, **2**, 899
reactions with enol silanes
Mannich reaction, **2**, 905
triflate
synthesis, **2**, 901
trifluoroacetate
synthesis, **2**, 900
Iminium salts, *N*,*N*-disilyl-
Mannich reaction, **2**, 913
Iminium salts, halomethylene-
alkoxymethyleniminium salt synthesis, **6**, 505
amide halide synthesis, **6**, 499
amidine synthesis, **6**, 543
synthesis, **6**, 495
Iminium salts, (methylthio)alkylidene-
Knoevenagel reaction, **2**, 368
Iminium salts, *N*-silyl-
Mannich reaction, **2**, 913
Iminium salts, α-thio-
formation
Eschenmoser coupling reaction, **2**, 867
Iminium salts, trimethylsiloxymethylene-
synthesis, **6**, 502
α-Imino acids
reduction
enzymes, **8**, 204
Iminodicarboxylic acid
di-*t*-butyl ester
Gabriel synthesis, **6**, 81
methyl *t*-butyl ester
Gabriel synthesis, **6**, 81
Iminodicarboxylic acids
synthesis, **8**, 146
Imino esters
acylation, **6**, 504
alkylation, **6**, 504
cyclic
aminal ester synthesis, **6**, 575
reaction with silyl ketene acetals
stereoselectivity, **2**, 638
Imino esters, *N*-acyl-
reactions with amides, **6**, 569
Immonium cations
Diels–Alder reactions, **5**, 409–411, 492, 500
initiators
polyene cyclization, **3**, 343
Immonium ions, *N*-alkylaryl-
Diels–Alder reactions, **5**, 500
Incensole
synthesis
via cyclofunctionalization of cycloalkene, **4**, 373
Indacrinone
synthesis
via Nazarov cyclization, **5**, 780
Indane
intermolecular *meta* cycloaddition
to vinyl acetate, **5**, 667
trans-Indane, 2-benzylidene-1-diphenylmethylene-
photochemical reactions, **5**, 721
Indane, 1-phenyl-
synthesis
via photoisomerization, **5**, 208
Indane-2-carboxylic acid
Birch reduction
dissolving metals, **8**, 500
Indane-6-carboxylic acid, 1-oxo-
synthesis
Friedel–Crafts reaction, **2**, 756
Indanedione
dehydrodimers
C—C cleavage, **8**, 995
Knoevenagel reaction, **2**, 358
1,3-Indanedione, 2-diazo-
synthesis, **3**, 893
Indanedione, perhydro-
synthesis
via dissolving metal reductions, **8**, 528
1,2,3-Indanetrione
thermal ene reaction, **2**, 539
Indanomycin
synthesis
via cuprate 1,2-addition, **1**, 126
Indan-1-one, 2-alkyl-
alkylations
via Michael addition, **4**, 230
Indan-1-one, 2,6-dimethyl-
synthesis
Friedel–Crafts reaction, **2**, 756
Indan-1-one, 6-methoxy-
synthesis
Friedel–Crafts reaction, **2**, 756
Indan-1-one, 2-methyl-
synthesis
Friedel–Crafts reaction, **2**, 756
Indanone, perhydro-
synthesis, **3**, 832
Indan-2-one, perhydro-
synthesis, **5**, 1173
Indanones
aldol reaction, **2**, 141
angularly substituted
synthesis *via* Nazarov cyclization, **5**, 760
Birch reduction
dissolving metals, **8**, 509
oxime
Beckmann rearrangement, **7**, 691
reduction

dissolving metals, **8**, 123
synthesis
Friedel–Crafts reaction, **2**, 754, 755
via [2 + 2 + 2] cycloaddition, **5**, 1133
Indazolediones
benzilic acid rearrangement, **3**, 831
Indazoles
reduction, **8**, 636
Indazolinone
synthesis
via reduction of methyl 2-azidobenzoate, **8**, 386
Indazolium salts
reduction
borohydride, **8**, 637
Indene, 3-chloro-1-dimethylamino-
synthesis
Vilsmeier–Haack reaction, **2**, 786
cis-Indene, 8,9-dihydro-
synthesis
via thermal rearrangement, **5**, 716
trans-Indene, 8,9-dihydro-
synthesis
via photoisomerization, **5**, 716
Indene, 1,1-dimethyl-
hydroalumination, **8**, 744
Indene, 1-dimethylamino-
synthesis
Vilsmeier–Haack reaction, **2**, 782
Indene, hexahydro-
cis-annulated
synthesis *via* palladium-ene reaction, **5**, 50
Indene, 2-methyl-
hydrozirconation
diastereoselectivity, **8**, 688
Indene, 2-nitrohydroxy-
synthesis
Henry reaction, **2**, 329
Indene, 2-vinyl-
synthesis
via photoisomerization, **5**, 212
via retro Diels–Alder reactions, **5**, 584
Indenecarboxylic acid
synthesis, **3**, 904
Indenes
anions
phenylation, **4**, 472
hydrobromination, **4**, 280
ozonolysis
in ammonia, **7**, 507
Pauson–Khand reaction, **5**, 1047
photooxidation, **7**, 98
synthesis
via [3 + 2] cycloaddition reactions, **5**, 1090
via dihalocyclopropanes, **4**, 1012
via photoisomerization, **5**, 197
Vilsmeier–Haack reaction, **2**, 782
Indenes, hydro-
synthesis
via Cope rearrangement, **5**, 812
Indenoisoquinoline, tetrahydro-
synthesis
via Neber rearrangement, **6**, 787
1-Indenol, 2-nitro-
synthesis
Henry reaction, **2**, 329
Inden-1-one, 2-acetylamino-
synthesis
Friedel–Crafts reaction, **2**, 757
Indenone, 2,3-diethyl-
synthesis
Friedel–Crafts reaction, **3**, 332
Indenone-3-carboxylic acid, 2-alkyl-
synthesis, **3**, 828
Indenones
synthesis
Friedel–Crafts reaction, **2**, 757
via [2 + 2 + 2] cycloaddition, **5**, 1135
Indenopyran-1,9-dione
synthesis
Knoevenagel reaction, **2**, 378
Indenopyridazine-3,9-dione
synthesis
Knoevenagel reaction, **2**, 378
Indenopyridine-1,3-dione
synthesis
Knoevenagel reaction, **2**, 378
Indeno[2,1-*b*]thiophen-8-one
synthesis
Friedel–Crafts reaction, **2**, 758
Indium compounds, crotyl-
type III
reactions with aldehydes, **2**, 24
Indole, *N*-acetyl-
hydrogenation, **8**, 613
Indole, 3-acetyl-1-benzenesulfonyl-
synthesis
Friedel–Crafts reaction, **2**, 744
Indole, 1-acetyl-2,3-dihydro-7-hydroxy-
synthesis, **7**, 335
Indole, 1-acetyl-4-trimethylsilyl-
Friedel–Crafts reaction, **2**, 742
Indole, acyl-
reduction
metal hydrides, **8**, 270
Indole, 3-alkyl-
synthesis
via $S_{RN}1$ reaction, **4**, 478
Indole, *N*-alkyl-
reduction
sodium borohydride, **8**, 616
Indole, 4-(benzyloxy)-
synthesis, **8**, 368
Indole, 1,4-bis(trimethylsilyl)-
Mannich reaction
intermediate, **2**, 968
Indole, 4-bromo-3-iodo-
synthesis, **3**, 498
Indole, dihydro-
lithiated formamidines
reaction with benzaldehyde, **1**, 482
Indole, 5,6-dimethoxy-
reduction
borohydrides, **8**, 618
Indole, 2,3-dimethyl-
reduction
dissolving metals, **8**, 615
stereochemistry, **8**, 624
stereoselective reduction, **8**, 624
Indole, 3-*N*,*N*-dimethylaminomethyl-
synthesis
Mannich reaction, **2**, 967
Indole, 2,3-diphenyl-

reduction
borohydrides, **8**, 618
synthesis
via benzyne, **4**, 510
Indole, 2-ethoxycarbonyl-5-hydroxy-
Mannich reaction, **2**, 967
Indole, hexahydro-
synthesis, **6**, 742
Indole, 7-methoxy-
synthesis, **7**, 335
Indole, 5-methoxydihydro-
synthesis
via arene–metal complexes, **4**, 523
Indole, 5-methoxy-1-methyl-
reduction
dissolving metals, **8**, 614
Indole, 2-methyl-
hydrogenation, **8**, 612
Indole, 3-methyl-
synthesis
via hydroformylation, **4**, 926
via intramolecular vinyl substitution, **4**, 846
Indole, 5-nitro-
reduction
borohydrides, **8**, 618
Indole, 2-oxy-
Vilsmeier–Haack reaction, **2**, 787
Indole, 2-phenyl-
synthesis, **3**, 513
Indole, 3-phenyl-
synthesis, **3**, 512
Indole, *N*-phenyl-
reduction
dissolving metals, **8**, 614
Indole, 1,2,3-trialkyl-
aminoalkylation
Mannich reaction, **2**, 967
β-Indoleacetic acid
synthesis
via intramolecular vinyl substitution, **4**, 846
Indoleacetic acid, dihydro-
ester, synthesis
carbonylation, **3**, 1038
Indole alkaloids
pentacyclic
synthesis *via* Michael addition, **4**, 25
synthesis, **3**, 81
iminium ion–arene cyclization, **2**, 1021
Knoevenagel reaction, **2**, 372, 384
via oxaziridines, **1**, 838
Indole-3-carbaldehyde
thallation, **7**, 335
Indole-2-carboxylates, *N*-alkyl-
reduction
borohydrides, **8**, 618
Indole-2-carboxylic acid
reduction
dissolving metals, **8**, 614
Indole-3-carboxylic acid
ethyl ester
reduction, dissolving metals, **8**, 615
Indole-3-carboxylic acid, 1-methyl-
Baeyer–Villiger reaction, **7**, 678
Indole-2,3-quinodimethane
synthesis
Knoevenagel reaction, **2**, 377
Indoles
coupling reactions, **3**, 511
cyclization
palladium catalysts, **4**, 836
[2 + 2 + 2] cycloaddition reactions, **5**, 1143
Friedel–Crafts acylation, **2**, 742, 743
Mannich reactions, **2**, 966
with imines, **2**, 970
with 1-piperidine, **2**, 970
metal complexes
addition reactions, **4**, 535
meta metallation
addition reactions, **1**, 463
reaction with copper(II) chloride, **7**, 532
reaction with dihalocarbenes, **4**, 1004
reduction
hydrides, **8**, 55
selective reduction, **8**, 530
2-substituted
lithiation, **1**, 474
N-substituted
lithiation, **1**, 473
synthesis, **4**, 429; **7**, 335
Houben–Hoesch synthesis, **2**, 748
via alkynes, palladium(II) catalysis, **4**, 560, 567
via cyclization of β-aminoalkynes, **4**, 411
via nitrogen-stabilized carbanions, **1**, 464
via $S_{RN}1$ reaction, **4**, 478
Vilsmeier–Haack reaction, **2**, 780
1*H*-Indoles, 3-(1-dialkylamino)alkyl-
synthesis
via vinylogous iminium salts, **1**, 367
Indoles, dihydro-
synthesis
via electrocyclic ring closure, **5**, 713
Indoles, hydroxy-
synthesis
via FVP, **5**, 732
Indoles, 2-substituted
synthesis
via hetero-Cope rearrangement, **5**, 1004
Indoles, vinyl-
thermolysis, **5**, 725
Indoline
Cope rearrangement, **5**, 790
Indoline, *N*-methyl-
metal complexes
addition reactions, **4**, 535
Indoline, 3-vinyl-
oxidative cleavage
ozone, **7**, 544
2-Indolinones
synthesis, **4**, 429
via ketocarbenoids, **4**, 1057
3*H*-Indolium salts, 3-(1-pyrrolidinylmethylene)-
nucleophilic addition reactions, **1**, 367
Indolizidine, 1,2-dihydroxy-
synthesis
via isoascorbic acid, **1**, 594
Indolizidines
synthesis, **6**, 746
chiral, **1**, 558
Mannich cyclization, **2**, 1041
Mannich reaction, **2**, 1010
via cyclization of β-allenylamine, **4**, 411
Indolizidinone

synthesis
via ketocarbenoids and pyrroles, 4, 1061
Indolizine, 8-acetoxy-3-acetyl-
Mannich reaction
with iminium salts, 2, 962
Indolizine, amino-
synthesis, 3, 541
Indolizine, 1,2-diphenyl-
Mannich reaction
with formaldehyde and dicyclohexylamine, 2, 962
Indolizines
Mannich reaction, 2, 962
synthesis
Perkin reaction, 2, 399
Vilsmeier–Haack reaction, 2, 780
5(1*H*)-Indolizinone, 2,3-dihydro-
Mannich reaction
with *N*,*N*-dimethylmethyleniminium chloride, 2, 962
Indolizinosuberenones
synthesis
Friedel–Crafts reaction, 2, 765
Indolmycin
synthesis
via conjugate addition to oxazepines, 4, 206
Indoloazepinone
synthesis
Friedel–Crafts reaction, 2, 765
Indoloisoquinoline
synthesis, 6, 771
Indolomorphan
synthesis, 8, 621
Indolone, 2-phenyl-
rearrangement, 3, 835
Indolones
Diels–Alder reactions, 5, 408
Indoloquinolizidine
ketones
synthesis, Mannich reaction, 2, 1028
Indoloquinolizidine, ethylidene-
synthesis
Mannich reaction, 2, 1031
trans-Indoloquinolizine
synthesis
Polonovsky–Potier method, 2, 1021
Indolo[2,3-*a*]quinolizine, octahydro-
synthesis
Mannich reaction, 2, 1018
Indoloquinolizine alkaloids
synthesis
Mannich reaction, 2, 1018
via 3-lithiation of an indole, 1, 474
Indolosuberenones
synthesis
Friedel–Crafts reaction, 2, 765
Infrared laser beams
alkene protection, 6, 689
Ingenane diterpenoids
synthesis
via Cope rearrangement, 5, 984
Ingenol
synthesis
via [6 + 4] cycloaddition, 5, 624
via Ireland rearrangement, 5, 843
via photocycloaddition reactions, 5, 137
Ingramycin
synthesis
via macrolactonization, 6, 373
Initiators
low temperature
radical reactions, 4, 721
INOC reactions (*see also* 'intramolecular cycloaddition' under Nitrile oxides and derivatives)
intramolecular nitrile oxide cycloaddition, 4, 1080, 1124
tandem Diels–Alder, 4, 1132
tandem Michael reactions, 4, 1132
Inomycin
synthesis, 1, 569
Inosamines
synthesis
via Diels–Alder reaction, 5, 418
neo-Inositol, 1,4-dideoxy-1,4-dinitro-
synthesis
Henry reaction, 2, 326
Inositol phosphates
synthesis, 7, 245
Insect antifeedants
intermediate
synthesis, 2, 185
Insect pheromones
γ-lactone
synthesis, 8, 166
synthesis
via carbocupration, 4, 903
via ene reaction, 5, 8
via photocycloaddition, 5, 165
Insertion reactions
C—H
carbon–carbon bond formation, 3, 1045–1062
intermolecular, 3, 1046
intramolecular, carbacycles, 3, 1048
intramolecular, heterocycles, 3, 1056
photochemical, 3, 1048, 1057
Insulin
transpeptidation
kinetically controlled, 6, 399
Integerrinecic acid
Baeyer–Villiger reaction, 7, 679
Interface reactions
electrochemical oxidation, 7, 790
Interhalogens
reaction with alkenes, 4, 347
Intermolecular coupling
electrochemical
aromatic compounds, 7, 801
Intersaccharidic bonds
stability, 6, 634
Intramolecular addition
Baldwin's rules
heteronucleophiles, 4, 37–41
Intramolecular coupling
electrochemical
aromatic compounds, 7, 801
Intramolecular functionalization
C—H bonds, 7, 40
Intramolecular reactions
dissolving metals
reductions, 8, 528
Invictolide
synthesis

using zirconium-promoted bicyclization of enynes, **5**, 1166
Iodides
Kornblum oxidation, **7**, 654
reactions with carbonyl compounds
organosamarium compounds, **1**, 257
vinyl substitutions
palladium complexes, **4**, 835
Iodinanes
aziridination, **7**, 477
Iodination
electrochemical, **7**, 810
nucleophilic displacement, **6**, 213
secondary amines, **7**, 747
Iodine
catalyst
Friedel–Crafts reaction, **2**, 737
conjugate enolate trap, **4**, 262
hypervalent
enone α-hydroxylation, **7**, 179
ketone α-hydroxylation, **7**, 155
reaction with carboxylic acids, **7**, 723
reaction with alkenes, **4**, 346
silver benzoate
alkene hydroxylation, **7**, 447
Iodine acetate
glycol cleavage, **7**, 706
Iodine atom transfer reactions
radical cyclizations, **4**, 803
Iodine azide
addition reactions
alkenes, **7**, 502
aziridine synthesis, **7**, 473
azirine synthesis, **7**, 502
reactions with allenes, **7**, 506
Ritter reaction, **6**, 289
Iodine fluoride
reaction with alkenes, **4**, 347
Iodine isocyanate
addition reactions
alkenes, **7**, 501
aziridination, **7**, 473
Iodine monochloride
alkane chlorination, **7**, 16
Iodine pentafluoride
Hofmann rearrangement, **6**, 803
Iodine reagents
glycol cleavage, **7**, 706
oxidative rearrangment, **7**, 828
Iodine tetrafluoroborate, bis(*sym*-collidine)-
α-iodocarbonyl compound synthesis
from alkenes, **7**, 535
Iodine tetrafluoroborate, bis(pyridine)-
reaction with 1,3-dienes, **7**, 536
Iodine triacetate
glycol cleavage, **7**, 706
Iodinium tosylates, alkenyl-
coupling reactions
with alkenylcopper(I) compounds, **3**, 522
Iodinolysis
C—B bonds, **7**, 606
Iodocarbonylation
epoxide synthesis, **6**, 26
Iodocarbonyl compounds
radical cyclizations, **4**, 802
reactions with alkenes
radical addition reactions, **4**, 754
Iodohydrin
deoxygenation
epoxides, **8**, 891
synthesis
via iodomethylation with samarium diiodide, **1**, 260
Iodolactamization
alkenes, **7**, 503
Iodolactonization
epoxide synthesis, **6**, 26
lactone synthesis, **7**, 523
Iodomethylation
carbonyl compounds
samarium diiodide, **1**, 260
Iodomethylenation
carbonyl methylenation, **1**, 261
Iodonium salts, aryl-
arene substitution reactions, **4**, 425
Iodonium tosylates, alkynylphenyl-
reaction with vinylcopper compounds, **3**, 219
tandem vicinal difunctionalization, **4**, 260
Iodonium ylides
ketocarbenes from, **4**, 1032
Ion exchange resin
catalyst
Knoevenagel reaction, **2**, 345
Ionic hydrogenation
benzothiophenes, **8**, 629
furans, **8**, 608
indoles, **8**, 623
thiophenes, **8**, 610
tosylates, **8**, 813
β-Ionine
silyl ether
oxidative cleavage, **7**, 587
Ionization potentials
electron donors, **7**, 853
measurement
gas-phase, **7**, 852
Ionomycin
synthesis
stereoselectivity, **4**, 384
use of hydrazones, **2**, 505
β-Ionone
irradiation, **5**, 741
pinacols
synthesis, *via* electroreduction, **8**, 135
synthesis
via Carroll rearrangement, **5**, 835
ψ-Ionone
cyclization, **3**, 349
Ionones
pinacol coupling reactions, **3**, 577
Ionophore antibiotics
noncyclic
synthesis *via* [4 + 3] cycloaddition, **5**, 612
synthesis, **2**, 248
via sulfones, **6**, 158
Ipalbidine
synthesis
Eschenmoser coupling reaction, **2**, 881
via Diels–Alder reaction, **5**, 411
via ketocarbenoids and pyrroles, **4**, 1061
Ipsdienol
synthesis

acylation in, **2**, 721
Ipsenol
synthesis
acylation in, **2**, 721
ene reaction, **2**, 538
via retro Diels–Alder reaction, **5**, 555
Iptycenes
synthesis
via Diels–Alder reaction, **5**, 383
Ireland–Claisen rearrangement
ring formation, **4**, 791
stereochemistry
control, **6**, 859
Ireland silyl ester enolate rearrangement
kinetics, **5**, 856
variant of Claisen rearrangement, **5**, 840–847
Iridium
allylic oxidation
catalyst, **7**, 108
catalyst
carbonyl compound hydrogenolysis, **8**, 320
reduction
transfer hydrogenation, **8**, 366
Iridium, bis[chlorobis(cyclooctene)]-
catalyst
hydrosilylation, **8**, 764
Iridium, cyclooctadienebis(trialkylphosphine)-
hydrogenation
alkenes, **8**, 446
Iridium, cyclooctadiene(trialkylphosphine)pyridyl-
hydrogenation
alkenes, **8**, 446
Iridium, tetrakis(diethylphenylphosphine)-
catalyst
hydrogenation, **8**, 534
Iridium chloride
allylic oxidation, **7**, 95
Iridium chloride, (3,4,7,8-tetramethyl-1,10-phenanthroline)(cyclo-1,5-octadiene)-
transfer hydrogenation, **8**, 552
Iridium complexes
hydride transfer
catalyst, **8**, 91
Iridium tetrafluoroborate, diacetonatodihydrido-(triphenylphosphine)-
crystal structure, **1**, 307
Iridodiol, dehydro-
synthesis
via conjugate addition, **4**, 218
Iridoids
synthesis
Knoevenagel reaction, **2**, 358, 372
use of α,α-bissulfenylated lactones, **2**, 186
via Ireland silyl ester enolate rearrangement, **5**, 841
Iridolactones
synthesis, **3**, 850
Iridomyrmecin
synthesis, **3**, 384
via magnesium-ene reaction, **5**, 41
via photoisomerizations, **5**, 231
Iron
reduction
enones, **8**, 524
Iron, acyl complexes
aldol reactions, **2**, 272
enantioselective, **2**, 315
enolates
synthesis and use, **2**, 125
reactions with π-allylpalladium complexes
regioselectivity, **4**, 642
α,β-unsaturated
conjugate additions, organolithium compounds, **4**, 217
Iron, alkoxycyclohexadienyl-
nucleophilic addition
regiocontrol, **4**, 674
Iron, (arene)cyclopentadienyl-
addition–oxidation reactions, **4**, 541
synthesis, **4**, 521
Iron, (η^6-benzene)cyclopentadienyl-
addition–oxidation reactions, **4**, 541
Iron, (benzocyclooctatetraene)tricarbonyl-
reaction with tetracyanoethylene, **4**, 710
Iron, (benzylideneacetone)tricarbonyl-
reactions with dienes, **4**, 665
Iron, butadienetricarbonyl-
acetylation, **4**, 697
synthesis, **4**, 663
Iron, carbonylcyclopentadienylethoxy-carbonyl(triphenylphosphine)-
transmetallation
stereoselective addition to symmetrical ketones, **1**, 119
Iron, carbonylcyclopentadienylmethoxy-carbonyl(triphenylphosphine)-
ketone–imine reactions, **2**, 933
Iron, cyclobutadienetricarbonyl
synthesis and reactions, **4**, 701
Iron, cyclohexadienyl-
nucleophilic addition
regiocontrol, **4**, 674
synthesis, **4**, 663
via nucleophilic addition, **4**, 664
Iron, cyclohexadienylmethoxycarbonyl-
nucleophilic addition
regiocontrol, **4**, 674
Iron, cyclopentadienyl(fluoroarene)-
nucleophilic substitution, **4**, 530
Iron, cyclopentadienyl(halobenzene)-
nucleophilic substitution, **4**, 529–531
Iron, cyclopentadienyl(η^6-nitrobenzene)-
nucleophilic substitution, **4**, 530
Iron, dicarbonylcycloheptadienyl(triphenylphosphine)-
nucleophilic addition, **4**, 673
Iron, dicarbonylcycloheptadienyl(triphenyl phosphite)-hexafluorophosphate
synthesis and reactions, **4**, 674
Iron, dicarbonylcyclohexadiene(triphenylphosphine)-
electrophilic substitution, **4**, 698
Iron, dicarbonylcyclopentadienyl-
alkene complexes
reactions with nucleophiles, **4**, 562
Lewis acid, **1**, 307
nucleophilic addition
alkenes, **4**, 576
Iron, dicarbonylcyclopentadienyl(cinnamaldehyde)-
crystal structure, **1**, 308
Iron, dicarbonylcyclopentadienyl(cyclohexenone)-
crystal structure, **1**, 308, 314
Iron, dicarbonylcyclopentadienyl(4-methoxy-3-butenone)-
crystal structure, **1**, 308

Iron, dicarbonylcyclopentadienyl(tropone)-
crystal structure, **1**, 308
Iron, (η^6-*o*-dichlorobenzene)cyclopentadienyl-
nucleophilic substitution, **4**, 529
Iron, dodecacarbonyltri-
reactions with dienes, **4**, 665
Iron, nonacarbonyldi-
[3 + 2] cycloaddition reactions
with α,α'-dibromo ketone, **5**, 282
dehalogenation
α-halocarbonyl compounds, **8**, 991
reactions with dienes, **4**, 665
Iron, pentacarbonyl-
catalyst
carbonylation of alkyl and aralkyl halides, **3**, 1026
deoxygenation
epoxides, **8**, 890
pinacol coupling reactions
aromatic aldehydes, **3**, 565
reactions with dienes, **4**, 665
Iron, tricarbonyl(cycloheptatriene)-
formylation, **4**, 706
reactions with acyl tetrafluoroborates, **4**, 707
reactions with dienophiles, **4**, 710
reaction with phosphoryl trichloride, **4**, 706
Iron, tricarbonyl(η^4-cyclohexadiene)-
reactions with carbanions, **4**, 580
reactions with nucleophiles, **4**, 670
synthesis, **4**, 665–670
Iron, tricarbonyl(cyclooctatetraene)-
formylation, **4**, 706
Friedel–Crafts acetylation, **4**, 706
reaction with tetracyanoethylene, **4**, 709
Iron, tricarbonyl(2,4,6-cyclooctatrienone)-
reaction with tetracyanoethylene, **4**, 710
Iron, tricarbonyl(η^4-diene)-
oxidative cyclization, **4**, 670
reactions with carbon electrophiles, **4**, 697–702
Iron, tricarbonyl(heptafulvalene)-
reaction with tetracyanoethylene, **4**, 710
Iron, tricarbonyl(heptafulvene)-
reaction with phosphoryl trichloride, **4**, 707
Iron, tricarbonyl(1,3,5-heptatriene)-
derivatives
reaction with tetracyanoethylene, **4**, 710
Iron, tricarbonyl(1-hydroxymethylcyclohexadiene)-
synthesis, **4**, 669
Iron, tricarbonyl(*N*-methoxycarbonylazepine)-
reaction with tetracyanoethylene, **4**, 711
Iron, tricarbonyl(1-methylcyclohexadienyl)-
synthesis, **4**, 669
Iron, tricarbonyl(*trans*-pentadiene)-
hydride abstraction, **4**, 663
Iron, tricarbonylpentadienol-
reaction with acid, **4**, 664
Iron, tricarbonyl(η^4-tetraene)-
acylation, **4**, 706–709
alkylation, **4**, 706–709
electrophilic reactions, **4**, 706
Iron, tricarbonyl(η^4-triene)-
acylation, **4**, 706–709
alkylation, **4**, 706–709
electrophilic reactions, **4**, 706
Iron, tricarbonyl(tropone)-
reaction with tetracyanoethylene, **4**, 710
synthesis, **4**, 707
Iron, tricarbonyl(vinylcyclobutadiene)-
derivatives
reaction with tetracyanoethylene, **4**, 710
Iron carbonyls
dehalogenation
α-halocarbonyl compounds, **8**, 991
reductive cleavage
ketol acetates, **8**, 993
Iron chlorides
cyclohexadienyliron complexes
decomplexation, **4**, 674
epoxide ring opening, **3**, 770
lithium aluminum hydride
unsaturated hydrocarbon reduction, **8**, 485
reaction with organoboranes, **7**, 604
silica support
dehydration, **7**, 843
Iron clusters
reductions
nitroarenes, **8**, 371
Iron complexes
allylic oxidation, **7**, 95
carbonylation
formation of asymmetric iron acyls, **3**, 1029
catalysts
aryl Grignard reagent reaction with alkenyl halides, **3**, 494
α,β-unsaturated acyl
Diels–Alder reaction, **5**, 367–369
Iron complexes, acryloyl-
[3 + 2] cycloaddition reactions
with allyltributyltin, **5**, 277
Iron complexes, allenyl-
[3 + 2] cycloaddition reactions, **5**, 279
Iron complexes, η^1-allyl-
[3 + 2] cycloaddition reactions, **5**, 272
with toluenesulfonyl isocyanate, **5**, 275
Iron complexes, η^1-2-butenyl-
[3 + 2] cycloaddition reactions, **5**, 273
Iron complexes, η^1-butynyl-
[3 + 2] cycloaddition reactions
with cyclohexenone, **5**, 277
Iron complexes, carbene
reactions with alkenes, **5**, 1088
reactions with 1,3-dienes, **5**, 1088
Iron complexes, cycloheptadienyl-
hydride abstraction, **4**, 686
Iron complexes, cyclohexadienyl-
nucleophilic addition
steric hindrance, **4**, 675
trimethylsilyl-substituted
enolate nucleophilic addition, **4**, 677
Iron complexes, cyclopentadienylcarbene
cyclopropanation, **5**, 1086
Iron complexes, dicarbonyl-η^5-cyclopentadienyl-
[3 + 2] cycloaddition reactions, **5**, 272
Iron complexes, dienetricarbonyl-
acylated
cleavage, **4**, 702
formylation, **4**, 701
Iron complexes, dienyl-
addition of chiral nucleophiles, **4**, 688
enantiomerically enriched
synthesis, **4**, 687
nucleophilic additions, **4**, 670–674
resolution, **4**, 687

stability, **4**, 664
symmetrical
reaction with chiral nucleophiles, **4**, 689
synthesis, **4**, 665–689
trimethylsilyl-substituted
hydride abstraction, **4**, 667, 669
X-ray crystallography, **4**, 664
Iron complexes, η^1-2-methallyl-
[3 + 2] cycloaddition reactions, **5**, 273
Iron complexes, η^1-1-propynyl-
[3 + 2] cycloaddition reactions
with cyclohexenone, **5**, 280
Iron complexes, 2-propynyl-
[3 + 2] cycloaddition reactions, **5**, 277
Iron complexes, α-thioalkyl-
alkylation, **5**, 1086
Iron complexes, tricarbonyl(4-methyltropone)-
[3 + 2] cycloaddition reactions, **5**, 274
Iron complexes, tricarbonyltropylium-
[3 + 2] cycloaddition reactions, **5**, 274
substituted
[3 + 2] cycloaddition reactions, **5**, 274
Iron complexes, α,β-unsaturated acyl-
[3 + 2] cycloaddition reactions
with allylstannanes, **5**, 277
Michael acceptors
tandem vicinal difunctionalization, **4**, 243
cis-γ-Irone
precursor
synthesis *via* intramolecular ene reaction, **5**, 18
ψ-Irone
cyclization
Lewis acid induced, **3**, 349
Iron enolates
acetyl
aldol reaction, diastereofacial selectivity, **2**, 316
aldol reaction, **2**, 315
propionyl
aldol reaction, **2**, 317
Irones
synthesis
via hydroformylation, **4**, 924
Iron–graphite
reduction
vicinal dibromides, **8**, 797
Iron hydrides
reduction
unsaturated carbonyl compounds, **8**, 550
Iron nitrate
nitration with
clay-supported, **6**, 111
reduction
dissolving metals, **8**, 526
solid support
clay, **7**, 846
Iron oxide
catalysts
reduction, **8**, 366
Iron perchlorate, 2,6-dichlorophenylporphyrin-
aziridination, **7**, 484
Iron polyphthalocyanine
reduction
α-halo ketones, **8**, 994
Iron porphyrins
alkene epoxidation catalysis, **7**, 382
γ-Irradiation
hydrosilylation
unsaturated hydrocarbons, **8**, 764
Isatin, 5-bromo-1-piperidyl-
reaction with naphthols
Mannich reaction, **2**, 958
Ishwarane
synthesis, **3**, 20
Ishwarone
synthesis, **3**, 20
Isoacoragermacrone
isomerization, **7**, 619
Isoalloxazines
synthesis, **4**, 436
Isoamides, *O*-acyl-
synthesis
via Ritter reaction, **6**, 293
Isoamijiol
synthesis, **3**, 586, 603
Isoasatone, demethoxy-
synthesis, **3**, 697
Isoatisirene
synthesis, **6**, 780
Isobenzofuran, 1,3-diphenyl-
Diels–Alder reactions
selenoaldehydes, **5**, 442
Isobenzofuranone
synthesis, **7**, 340
Isobenzofurans
cycloaddition reactions, **1**, 464
Diels–Alder reactions, **5**, 413
synthesis, **7**, 340
via Diels–Alder reactions, **5**, 382
via retro Diels–Alder reactions, **5**, 580
Isoboldine
synthesis, **3**, 679
Isoborneol, 3-*trans*-benzylidene-
epoxidation, **7**, 365
Isoborneol, 10-mercapto-
Michael addition, **5**, 370
Isoborneol-10-diisopropylsulfonamide
esters
reaction with imines, **5**, 102
Isobornylamine
imine anion from cyclohexanone
alkylation, **6**, 725
Isobutene
arene alkylation
Friedel–Crafts reaction, **3**, 306
ene reactions
Lewis acid catalysis, **5**, 4
photolysis
with 3-butyn-2-one, **5**, 164
Ritter reaction
mechanism, **6**, 263
Isobutene, octafluoro-
reaction with phenyl azide, **6**, 500
Isobutylamides, ω-alkynic
synthesis, **8**, 694
Isobutylamine, *n*-butyl-
methylation, **3**, 30
Isobutyraldehyde
aldol reactions
diastereofacial selectivity, **2**, 264
potassium enolates
alkylation, **3**, 20
Isobutyramides

alkylation, **6**, 501
Isobutyric acid,
α-lithiated esters
crystallization, **1**, 41
Isobutyric acid, α-amino-
peptides
synthesis, **2**, 1096
Isobutyric acid, α-bromo-
ethyl ester
acylation, Reformatsky reaction, **2**, 296
Reformatsky reaction, **2**, 278
Isobutyric acid, isobutyryl-
ethyl ester
Reformatsky reaction, **2**, 278
Isobutyrophenone
enolate
Michael additions, **5**, 1082
Isobutyryl group
guanine-protecting group, **6**, 642
Isocalamendiol
synthesis
transannular ene reaction, **2**, 553
Isocarbacyclin
synthesis, **1**, 568; **3**, 139
via 1,2-addition of silylcuprate, **1**, 133
via Claisen rearrangement, **5**, 833
Isocaryophyllene
transannular cyclization, **3**, 387
Isochroman-3-one
synthesis
via benzocyclobutene ring opening, **5**, 681
Isochromanones
synthesis
via directed metallation, **1**, 463
via hetero electrocyclization, **5**, 741
Isocomene
synthesis, **3**, 385, 713
ene reaction, **2**, 546
via Carroll rearrangement, **5**, 835
via intramolecular ene reaction, **5**, 11
via Pauson–Khand reaction, **5**, 1062
via photocycloaddition, **5**, 143, 660, 662
Isocorytuberine
synthesis, **3**, 807
Isocoumarins
synthesis, **3**, 543
via bromocyclization of phenylethynylbenzoate ester, **4**, 395
via orthothallation/palladium catalysts, **4**, 841
via palladium(II) catalysis, **4**, 558
via $S_{RN}1$ reaction, **4**, 479
Isocyanate, chlorosulfonyl
acid anhydride synthesis, **6**, 313
amide synthesis, **6**, 386
Isocyanates
amide synthesis, **6**, 399
2-azetidinones from, **5**, 102–108
cycloaddition reactions
heterocycle synthesis, **5**, 1158
with alkynes, **5**, 1155
reactions with organoytterbium reagents, **1**, 278
reactions with ytterbium ketone dianions, **1**, 280
reduction, **8**, 254
triphenylstannane, **8**, 74
solvolysis
to give amines, **6**, 801
synthesis
carbonylation, **3**, 1039
solvolytic conversion, **6**, 796
Isocyanates, phenyl
reduction, **8**, 254
Isocyanates, trichloroacetyl
reaction with dihydropyrans
glycal synthesis, **5**, 108
Isocyanides
acidic hydrolysis, **6**, 294
addition reactions
carbon-centered radicals, **4**, 765
amide synthesis, **6**, 387
amidine synthesis, **6**, 546
amidinium salt synthesis, **6**, 517
chemistry, **2**, 1083
imidate synthesis, **6**, 533
imidoyl halide synthesis, **6**, 526
isomerization, **6**, 294
ortho-lithiated aryl
synthesis, **3**, 255
metal–ammonia reduction, **8**, 830
properties, **6**, 293
reactions with π-allylpalladium complexes, **4**, 600
regioselectivity, **4**, 643
reactions with Fischer carbene complexes, **5**, 1109
reactions with Grignard reagents, **1**, 544
reduction, **8**, 830
tributylstannane, **8**, 831
Ritter reaction, **6**, 293
substitution reactions, **6**, 261–296
synthesis, **6**, 242
via amides, **6**, 489
thioimidate synthesis, **6**, 540
Isocyanides, 4-nitrophenyl
O-acyl thiohydroxamate photolysis, **7**, 731
Isocyanides, tosylmethyl
acyl anion equivalents, **1**, 571
Isocyanoacetates
Aldol reactions
Lewis acid asymmetric induction, **1**, 320
(±)-2-Isocyanopupukeanane
synthesis, **2**, 161
Isocyanuric acid, trichloro-
sulfide chlorination, **7**, 207
Isocycloseychellene
synthesis
Prins reaction, **2**, 542
Isoderminin
synthesis, **1**, 570
Isodicyclopentadiene
cycloaddition reactions
tropones, **5**, 618
Isodysidenin, dimethyl-
synthesis
Ugi reaction, **2**, 1096
Isoegomaketone
synthesis
alkenylsilane acylation, **2**, 713
Isoellipticine
synthesis
via Diels–Alder reaction, **5**, 385
Isoeugenol
oxidation, **3**, 690
Isoflavanones
synthesis

via isoflavones, **7**, 831
Isoflavans
synthesis
via isoflavones, **7**, 831
Isoflavones
reduction
DIBAL-H, **8**, 544
synthesis, **7**, 827
via chalcone, **7**, 829
Vilsmeier–Haack reaction, **2**, 790
Isogermacrone
epoxide
rearrangement, **3**, 752
Isogibberellin
from gibberellin
Wagner–Meerwein rearrangement, **3**, 715
Isognididione
synthesis
via retro Diels–Alder reactions, **5**, 579
Isoharringtonine
synthesis, **3**, 596
Isoheterotropanone
synthesis
via retro Diels–Alder reaction, **5**, 569
Isoheterotropantrione
synthesis, **3**, 697
Isoindoles
reduction, **8**, 624
substituted
synthesis *via* retro Diels–Alder reactions, **5**, 582
synthesis
via arynes, **4**, 503
via retro Diels–Alder reactions, **5**, 582
via 5-*exo*-trig cyclization, **4**, 38
Isoindoline
alkylation
stereoselective and regioselective, **3**, 77
Isoindolinones
synthesis
carbonylation of *o*-bromoaminoalkyl benzenes, **3**, 1037
Isoingenol
synthesis
via [6 + 4] cycloaddition, **5**, 624
Isoiridomyrmecin
synthesis
via photocycloaddition, **5**, 667
via photoisomerizations, **5**, 231
Isokaurene epoxide
rearrangement, **3**, 715
Isokhusimone
synthesis, **3**, 590
Isolinderalactone
Cope rearrangement, **5**, 809
Isolineatin
synthesis
via cyclofunctionalization of cycloalkene, **4**, 373
Isolobophytolide
synthesis
via photocycloaddition, **5**, 173
Isolongifolene
rearrangement, **3**, 737
synthesis, **3**, 713
Isolongifolene, 9-oxo-
reduction
aluminum hydrides, **8**, 542
Isomaltose
C-analog
synthesis, **1**, 198
Isomerization
configurational
hydroalumination, **8**, 744
Friedel–Crafts reaction, **3**, 327
Isomunchnones
cycloadditions, **4**, 1163
Isonicotinamide, 1-benzyl-1,2-dihydro-
reduction
dihydropyridine, **8**, 589
Isonicotinium dichromate
oxidation
alcohols, **7**, 277
(Isonicotinyl)oxycarbonyl group
protecting group
cleavage, **6**, 635
Isonitrin B
synthesis
via photocycloaddition, **5**, 167
Isonocardicin A
synthesis
Ugi reaction, **2**, 1101
Isooctopine
synthesis, **8**, 145
Isopelletierine
synthesis
Schopf reaction, **2**, 943
Isophorone
acetone self-condensation, **2**, 141
conjugate addition
with alkylmagnesium halides, **4**, 70
hydrogenation
homogeneous catalysis, **8**, 462
reaction with organocuprates, **4**, 180
reduction, **8**, 563
synthesis
via Michael reaction, **4**, 2
Isophotosantonic lactone, isodihydro-*O*-acetyl-
α-acetylation, **7**, 153
Isophthalic acid
acid dichloride synthesis, **6**, 302
Isopilocereine
synthesis
use of ferricyanide, **3**, 686
Isopimarene
oxidative rearrangement, **7**, 820
Isopiperitenol
synthesis
Wittig rearrangement, **3**, 1010
Isopiperitenone
synthesis
via Cope rearrangement, **5**, 817
Isopodophyllotoxone
synthesis, **3**, 695
Isoprene
anodic oxidation, **7**, 795
[4 + 3] cycloaddition reactions, **5**, 603
Diels–Alder reactions
Lewis acid promoted, **5**, 339
hydrobromination, **4**, 283
hydrocarboxylation, **4**, 945
hydrochlorination, **4**, 276
hydrogenation
homogeneous catalysis, **8**, 449, 451

hydrosilylation, **8**, 779
reaction with 1,2,4-trimethylbenzene
Friedel–Crafts reaction, **3**, 322
reaction with vinylchromium carbene complexes, **5**, 1070
selective reduction, **8**, 567, 568
zirconocene complex
reactions with carbonyl compounds, **1**, 163
Isoprene, stannyl-
Diels–Alder reactions, **5**, 337
Isoprene monoepoxide
coupling reactions
with alkenyl Grignard reagents, **3**, 476
Isoprenoids
conjugated
synthesis, **3**, 882
microbial hydroxylation, **7**, 62
synthesis
via Ramberg–Bäcklund reaction, **6**, 161
Isopropenyl acetate
reaction with acetals
in synthesis of botryodiplodin, **2**, 612
reaction with triethylmethoxytin
preparation of organotin(IV) enol ethers, **2**, 608
Isopropenyl acetoacetate
synthesis
via retro Diels–Alder reactions, **5**, 558
Isopropoxy group
cyclohexadienyliron complexes
directing effect, **4**, 675
Isopropylamine, β-phenyl-
synthesis, **8**, 376
Isopropylidene ketals
protecting group
carbohydrates, **6**, 631
Isoproterenol
synthesis, **2**, 1086
Isopulegol
oxidation
solid support, **7**, 841
Isopulegone
oxidation, **7**, 154
Isopyrocalciferol
synthesis
via electrocyclization, **5**, 700
Isoquinoline
electroreduction, **8**, 594
hydrogenation
nickel catalysts, **8**, 597, 598
2-oxide
deoxygenation, **8**, 391
reactions with allenic tin, **2**, 86
reduction
borohydrides, **8**, 581
dihydropyridine, **8**, 589
dissolving metals, **8**, 596
homogeneous catalysis, **8**, 600
metal hydrides, **8**, 580
reductive alkylation
borohydrides, **8**, 581
Reissert compounds, **8**, 295
synthesis
aldol cyclization, **2**, 173
Isoquinoline, *N*-acyl-1-alkylidene-1,2,3,4-tetrahydro-
hydrogenation
synthesis of isoquinoline alkaloids, **8**, 461
Isoquinoline, 1-alkyl-
regioselective synthesis, **4**, 446
Isoquinoline, 4-alkyl-
synthesis
use of imine anions, **2**, 482
Isoquinoline, benzyl-
asymmetric synthesis, **3**, 81
Isoquinoline, 3-bromo-
$S_{RN}1$ reaction, **4**, 462
Isoquinoline, 2-chloro-
coupling reactions
with Grignard reagents, **3**, 461
Isoquinoline, 4-chloro-
coupling reactions
with Grignard reagents, **3**, 461
Isoquinoline, 6-chlorotetrahydro-
synthesis
via arene–metal complexes, **4**, 523
Isoquinoline, cyano-
reduction
borohydrides, **8**, 581
synthesis, **4**, 433
Isoquinoline, dihydro-
synthesis, **6**, 272, 771
via diazoalkene cyclization, **4**, 1157
via Ritter reaction, **4**, 293; **6**, 295
Isoquinoline, 3,4-dihydro-
reaction with phthalide enolates
synthesis of protoberberine alkaloids, **2**, 946
silylation, **1**, 366
N-silyliminium salts
Mannich reaction, **2**, 913
Isoquinoline, 3,4-dihydro-6,7-dialkoxy-
reactions with organometallic compounds, **1**, 366
Isoquinoline, 3,4-dihydro-6,7-dimethoxy-
reactions with sulfinyl-stabilized carbanions, **1**, 516
Isoquinoline, hydro-
synthesis
Mannich reactions, **2**, 1023
Isoquinoline, *p*-hydroxybenzyltetrahydro-
anodic oxidation, **3**, 666
Isoquinoline, 10-hydroxydecahydro-
synthesis
stereochemistry, **2**, 1023
Isoquinoline, hydroxytetrahydro-
oxidation, **7**, 339
Isoquinoline, 5-nitro-
reduction
borohydrides, **8**, 582
Isoquinoline, 1-nitroso-
synthesis
via oxidation of sulfimides, **7**, 752
Isoquinoline, octahydro-
synthesis, **6**, 757
Isoquinoline, perhydro-
synthesis
N-acyliminium ions, **2**, 1073
Isoquinoline, pivaloyl-
lithiated
reaction with cyclohexanone, **1**, 481
Isoquinoline, *N*-pivaloyltetrahydro-
bromomagnesium derivative
crystal structure, **1**, 35
Isoquinoline, tetrahydro-
alkylation, **3**, 71
asymmetric synthesis

N-acyliminium ions, **2**, 1067
lithiated
reactions with aldehydes, **1**, 341
metallation, **1**, 481
oxidation
formation of nitrone, **7**, 745
oxidative coupling, **3**, 665
synthesis, **6**, 736, 738
via Diels–Alder reactions, **5**, 322
Isoquinoline, tetrasubstituted
synthesis
via Beckmann rearrangement, **7**, 695
Isoquinoline, 1-trichloromethyltetrahydro-
synthesis, **6**, 736
Isoquinoline, 1-trimethylstannyl-
Friedel–Crafts reaction, **2**, 743
1,3(2,4)-Isoquinolinediones
ring contraction, **3**, 835
Isoquinoline formamidine, tetrahydro-
alkylation
selectivity, **3**, 75
Isoquinolinephosphonates, dimethyl-
synthesis, **4**, 446
Isoquinolines
(*g*)-fused
synthesis, **1**, 475
synthesis, **1**, 482; **3**, 77; **6**, 401, 751, 757, 771
via dihalocarbene, **4**, 1004
via $S_{RN}1$ reaction, **4**, 478
Isoquinolines, 5,6-dihydro-
synthesis
via FVP, **5**, 718
Isoquinolinium cations
Diels–Alder reactions, **5**, 499, 500
Isoquinolinium cations, 2,4-dinitrophenyl-
Diels–Alder reactions, **5**, 500
Isoquinolinium salts
reduction
aluminum hydrides, **8**, 587
borohydrides, **8**, 587
Isoquinolinium salts, *N*-alkyl-3,4-dihydro-
Mannich reaction, **2**, 912
Isoquinolinodioxopyrroline
Diels–Alder reactions, **5**, 323
Isoquinolinoid alkaloids
tetracyclic
synthesis *via* [2 + 2 + 2] cycloaddition, **5**, 1150
3(2*H*)-Isoquinolinones, 1,4-dihydro-
synthesis
via $S_{RN}1$ reaction, **4**, 477
Isoquinolin-1-ones, 1,2,3,4-tetrahydro-
N-substituted, synthesis
carbonylation, **3**, 1038
1(2*H*)-Isoquinolone, 3-aryl-4-hydroxy-3,4-dihydro-
synthesis
Mannich reaction, **2**, 927
Isoquinolones
synthesis
via arynes, **4**, 503
Isoquinolones, dihydro-
synthesis
Mannich reaction, **2**, 928, 956
via $S_{RN}1$ reaction, **4**, 479
Isoquinuclidines
synthesis, **6**, 86
via Wittig reaction, **1**, 757
Isorenieratene
synthesis, **3**, 585
Isoretronecanol
synthesis
Eschenmoser coupling reaction, **2**, 881
via Baeyer–Villiger reaction, **7**, 677
Isorotenone
synthesis, **7**, 157
Isosalutaridine
synthesis, **3**, 679
Isoselenazoles
reduction, **8**, 658
Isoselenocyanates
reduction
tributylstannanes, **8**, 830
Isoseychellene
synthesis
Prins reaction, **2**, 542
Isosilybin
synthesis, **3**, 691
Isositsirikine
synthesis, **1**, 593
Mannich reaction, **2**, 1031
Isosparteine
ethylmagnesium bromide complex
crystal structure, **1**, 13
Isostegane
synthesis, **1**, 566
use of vanadium oxytrifluoride, **3**, 675
Isosteviol
from steviol
Wagner–Meerwein rearrangement, **3**, 715
Isotetralin
reaction with dihalocarbenes, **4**, 1002
reduction
Wilkinson catalyst, **8**, 445
Isothiazole, 4-nitro-
catalytic hydrogenation, **8**, 656
Isothiazoles
reduction, **8**, 656
Isothiocyanates
amide synthesis, **6**, 399
reduction, **8**, 830
tributylstannane, **8**, 831
Isothiocyanates, allyl-
synthesis
via allylthiocyanates, **6**, 846
Isothiocyanates, β-*trans*-phenylselenoalkyl
synthesis, **7**, 496
Isourea, *O*-alkyl-
imide alkylation, **6**, 74
phthalimide alkylation, **6**, 80
Isourea, alkylaryl-
hydrogenolysis, **8**, 912
Isourea, *O*-alkyl-*N*,*N*-dicyclohexyl-
hydrogenation, **8**, 815
Isourea, allyl-
π-allylpalladium complexes from, **4**, 590
Isourea, *O*-aryl-*N*,*N*′-dialkyl-
hydrogenolysis, **8**, 912
Isourea, *O*-aryl-*N*,*N*-dialkyl-
hydrogenolysis, **8**, 912
Isourea, *O*-aryl-*N*,*N*′-dicyclohexyl-
synthesis, **8**, 913
Isourea, *O*-aryl-*N*,*N*-diethyl-
hydrogenolysis, **8**, 912

Isourea, *O*-geranyl-
  reaction with phthalimide
    N-allylation, **6**, 86
Isourea, *O*-linalyl-
  reaction with phthalimide
    N-allylation, **6**, 86
Isourea, *O*-methyl-*N,N'*-dicyclohexyl-
  alkylation, **6**, 74
Isovaleraldehyde
  aldol reaction
    2,2,5-trimethylcyclopentanone, **2**, 154
Isoxazole, amino-
  synthesis
    via activated allene, **4**, 56
Isoxazole, 5-chloro-
  dechlorination
    sodium borohydride, **8**, 646
Isoxazole, 5-cyano-
  reduction, **8**, 646
Isoxazole, sulfinyl-4,5-dihydro-
  metallated
    reaction with aldehydes, **2**, 487
Isoxazole, 3-*p*-tolylsulfinylmethyl-4,5-dihydro-
  metallated
    reaction with aldehydes, **2**, 486
Isoxazoles
  Beckmann fragmentation, **6**, 775
  Diels–Alder reactions, **5**, 491
  hydrogenation
    over Pd or Pt, **6**, 403
  rearrangement, **6**, 543
  reduction, **8**, 644
  reductive cleavage, **8**, 392
  synthesis
    via Horner reaction, **1**, 779
Isoxazoles, 3-aryl-
  synthesis
    via retro Diels–Alder reactions, **5**, 584
Isoxazoles, 4,5-dihydro-
  reductive cleavage, **8**, 392
Isoxazoles, 4-(oxoalkyl)-
  pyridines from, **8**, 645
Isoxazoles, 4-silyl-
  synthesis
    via [3 + 2] annulations, **1**, 602
Isoxazoles, tetrahydro-
  reduction, **8**, 395
Isoxazolidine, *N*-methyl-
  synthesis
    via nitrile oxide cyclization, **4**, 1131
Isoxazolidine, 2-phenyl-3,5-dioxo-
  Knoevenagel reaction, **2**, 357
Isoxazolidines
  bicyclic
    synthesis *via* 1,3-dipolar cycloadditions, **4**, 1077
  bridged
    synthesis, **1**, 393
    synthesis *via* nitrone cyclization, **4**, 1114
  fused
    synthesis *via* nitrone cyclization, **4**, 1113, 1114
  reduction, **8**, 648
    ring opening, **8**, 648
  synthesis
    via 1,3-dipolar cycloadditions, **4**, 1076
2-Isoxazoline, 3,5-diphenyl-
  reduction
    LAH, **8**, 647
Isoxazoline, methoxycarbonyl-
  synthesis
    gold(I) enolate, **2**, 233
2-Isoxazoline, 2-methyl-
  reduction, **8**, 647
Isoxazoline-4-carboxylic acids
  esters of
    reduction, **8**, 647
Isoxazolines
  5,9-fused bicyclic
    synthesis *via* nitrile oxide cyclization, **4**, 1127
  in tetrahydropyran cyclization
    stereoselectivity, **4**, 383
  reduction, **8**, 70, 647
  5-substituted
    synthesis *via* 1,3-dipolar cycloadditions, **4**, 1079
  synthesis
    via 1,3-dipolar cycloadditions, **4**, 1078
Δ^2-Isoxazolines
  synthesis, **7**, 628
Isoxazolin-5-one
  hydrogenation
    palladium catalyst, **8**, 649
Isoxazolium salts
  reduction, **8**, 644, 646
Isoxazolones
  Knoevenagel reaction, **2**, 364
    stereoselectivity, **2**, 351
Isoxazolylsulfonamides
  hydrogenation, **8**, 645
Itaconic acid
  asymmetric hydrogenation
    homogeneous catalysis, **8**, 461
  esters
    hydrogenation, **8**, 449
  transfer hydrogenation
    triethylammonium formate, **8**, 84
Iterative rearrangements, **5**, 891–896
Iturinic acid
  synthesis
    Eschenmoser coupling reaction, **2**, 875
Ivalin
  synthesis
    via conjugate addition to α,β-unsaturated imine, **4**, 211
Ivanov reaction
  carboxylic acid dianions
    reaction with aldehyde or ketone, **2**, 210
  stereoselectivity
    effect of counterion, **2**, 211
  Zimmerman–Traxler transition states, **2**, 153

J

Japanese hop ether
  synthesis
    via Pauson–Khand reaction, **5**, 1051
Jasmonate, methyl-
  synthesis
    via conjugate addition, **4**, 215
Jasmone
  precursor synthesis, **1**, 558
  synthesis
    alkene protection, **6**, 689
    via Grignard addition, **1**, 407
    via Nazarov cyclization, **5**, 780
    via retro Diels–Alder reactions, **5**, 561
    via thioesters, **6**, 439
    via Wacker oxidation, **7**, 454
cis-Jasmone
  synthesis
    aldol cyclization, **2**, 161
    via [3 + 2] cycloaddition reactions, **5**, 308
Jasmone, dihydro-
  synthesis, **1**, 563; **3**, 869; **7**, 457
    via [3 + 2] cycloaddition reactions, **5**, 308
    via dialkylative enone transposition, **7**, 615
    via Wacker oxidation, **7**, 454
Jasmonic acid
  biosynthesis
    via Nazarov cyclization, **5**, 780
  methyl ester
    asymmetric synthesis, **6**, 150
    synthesis, **3**, 653
Jasmonic acid, dihydro-
  methyl ester
    synthesis, **2**, 710
Jasmonoids
  synthesis, **1**, 566
    conjugate addition, **2**, 331
Jatropholones
  synthesis
    via Diels–Alder reaction, **5**, 342
Jatrophone
  synthesis, **3**, 26
Jatrophone, 2β-hydroxy-
  synthesis
    via [3 + 2] cycloaddition reactions, **5**, 311
Johnson–Faulkner rearrangement
  aldols, **5**, 839
Johnson ortho ester rearrangement
  allyl alcohols
    remote stereocontrol, **5**, 864
  variant of Claisen rearrangement, **5**, 839
Johnson reaction
  use of *N*-methylphenylsulfonimidoylmethyllithium, **1**, 737
Jones oxidation
  chromium(VI) reagents
    alcohols, **7**, 253
    ethers, **7**, 240
Jones reagent
  cyclohexadienyliron complexes
    decomplexation, **4**, 674
Joubertiamine, 3-*O*′-methoxy-4′-*O*-methyl-
  synthesis
    stereocontrolled, *via* Eschenmoser rearrangement, **5**, 838
Joubertiamine, *O*-methyl-
  synthesis
    via cyclohexadienyliron complexes, **4**, 674
Juglone
  Diels–Alder reactions, **5**, 373, 376
Julia coupling
  allylsilanes, **2**, 586
  reductive cleavage, **1**, 794
  E/*Z*-selectivity, **1**, 793
  sulfur-stabilized alkenations, **1**, 792
Julolidine
  synthesis
    Eschenmoser coupling reaction, **2**, 881
Juncusol
  synthesis
    via retro Diels–Alder reaction, **5**, 572
Junenol
  synthesis
    ene reaction, **2**, 541
Justicidin
  synthesis
    via ortho-directed addition, **1**, 468
Juvabione
  synthesis, **2**, 91; **8**, 948
    from protected cyanohydrins, **3**, 198
    via Cope rearrangement, **5**, 821
Juvenile hormone
  synthesis, **3**, 99, 107
    via iterative rearrangements, **5**, 891

K

K-glucoride
  synthesis, **8**, 169
K-selectride — *see* Potassium tri-*s*-butylborohydride
Kahweol
  synthesis
    via cyclopropane ring opening, **4**, 1044
Kainic acid
  synthesis
    via Diels–Alder reaction, **5**, 468
α-Kainic acid
  synthesis
    via intramolecular ene reaction, **5**, 14
    via Ireland rearrangement, **5**, 843
Karachine
  synthesis
    Mannich reaction, **2**, 1013
Karahanaenone
  synthesis
    pinacol rearrangement, **3**, 728
    via Cope rearrangement, **5**, 803, 976
    via [4 + 3] cycloaddition, **5**, 603
ent-Kaurane
  microbial hydroxylation, **7**, 64
Kaurene
  rearrangement, **3**, 715
  synthesis
    via Birch reduction, **8**, 500
KDO synthetase
  organic synthesis
    use in, **2**, 465
Kessanol
  synthesis
    ene reaction, **2**, 551
    Knoevenagel reaction, **2**, 381
    via cyclofunctionalization of cycloalkene, **4**, 373
Ketal, cycloheptylidene
  diol protection
    removal, **6**, 660
Ketal, cyclohexylidene
  diol protection
    removal, **6**, 660
Ketal, cyclopentylidene
  diol protection
    removal, **6**, 660
Ketal, isopropylidene
  diol protection
    removal, **6**, 660
Ketals
  acyclic
    selective reduction, **8**, 217
  asymmetric epoxidation
    compatibility, **7**, 401
  carbonyl group protection, **6**, 675
  chiral
    conjugate additions, **4**, 208–210
  cyclic
    diol protection, **6**, 659
  reduction
    metal hydrides, **8**, 267
  α,β-unsaturated
    addition reactions with alkylaluminum compounds, **1**, 88
Ketals, cyclopropenone
  vinylcarbene generation
    [4 + 3] cycloaddition reactions, **5**, 599
Ketals, α-hydroxy
  chiral
    addition reactions with alkylaluminum compounds, **1**, 89
Ketene, dichloro-
  generation, **5**, 86
Ketene, diphenyl-
  reaction with π-allylpalladium complexes, **4**, 602
  reaction with benzoquinone, **5**, 86
  reaction with ethoxyacetylene, **5**, 732
  reaction with tricarbonyl(cycloheptatriene)iron complexes, **4**, 710
Ketene, vinyl-
  synthesis
    via cyclobutenone, **5**, 675
Ketene acetals
  [2 + 2] cycloaddition reactions, **5**, 71
  Diels–Alder reactions, **5**, 461
  ortho acid synthesis, **6**, 556
  preparation, **2**, 605
    Eschenmoser coupling reaction, **2**, 869
  reactions with isocyanates, **5**, 103
  reaction with vinyl ethers, **5**, 684
  synthesis
    via Horner reaction, **1**, 774
Ketene *N,O*-acetals
  protonation, **6**, 505
Ketene *O,O*-acetals
  2,2-bis(dialkoxy)carbonitrile synthesis, **6**, 564
Ketene *S,N*-acetals
  alkylmercaptomethyleniminium salt synthesis, **6**, 511
Ketene *S,S*-acetals
  hydrolysis
    synthesis of thiol esters, **6**, 444
Ketene *N,O*-acetals, *N*-allyl-
  amino-Claisen rearrangement, **6**, 861
Ketene acetals, bis(trimethylsilyl)-
  reaction with imines, **5**, 102
Ketene *O*-alkyl *O′*-silyl acetals
  Vilsmeier–Haack reaction, **2**, 792
Ketene aminals
  amidinium salt synthesis, **6**, 518
  2,2-bis(dialkylamino)carbonitrile synthesis, **6**, 577
  reactions with isocyanates, **5**, 103
  tris(dialkylamino)alkane synthesis, **6**, 582
Ketene dithioacetals
  alkynylsilane cyclization reactions, **1**, 608
  coupling reactions
    with alkyl Grignard reagents, **3**, 448
  deprotonation
    γ-selectivity, **2**, 72
  synthesis, **6**, 134
Ketene-*N*-methylimine, diphenyl-
  cycloaddition reactions
    metal catalyzed, **5**, 1195
Ketenes
  acetals
    silyl enol derivatives, **3**, 50
  acylation, **6**, 332

thiols, **6**, 443
alkali metal enolates, **2**, 107
bis(trimethylsilyl) acetals
aldol condensation, stereoselectivity, **2**, 634
reaction with aldehydes, **2**, 632
boron enolates
stereoselectivity, **2**, 112
carbene precursors, **4**, 961
dithioacetal monoxide
addition reaction with enolates, **4**, 100, 109
electrocyclization, **5**, 730–734
formation
lithium ester enolates, **2**, 278
hydration, **4**, 299
intramolecular [2 + 2] cycloaddition, **5**, 1021
Perkin reaction, **2**, 399
reaction with boron reagents
production of alkenyloxyboranes, **2**, 242
reaction with carbonyl compounds
chemoselectivity, **5**, 86
regioselectivity, **5**, 86
stereoselectivity, **5**, 87–89
reaction with dienes
transition metal catalysis, **4**, 709–712
reaction with imines
chemoselectivity, **5**, 92–99
regioselectivity, **5**, 92–99
stereoselectivity, **5**, 95–99
reaction with nitriles, **6**, 401
reaction with silylamines, **2**, 605
silyl acetals
reaction with nitroarenes, **4**, 429
rhodium enolates, aldol reaction, **2**, 310
synthesis
via retro Diels–Alder reactions, **5**, 558
thioacetal monoxides
Michael addition, **4**, 10
tin enolates
synthesis, **2**, 117
Ketenes, cyano-
generation, **5**, 90
Ketenes, diacyl *S,S*-acetals
synthesis
Knoevenagel rection, **2**, 364
Ketenes, dimethyl-
synthesis
via retro Diels–Alder reactions, **5**, 558
Ketenes, diphenyl-
[3 + 2] cycloaddition reactions
with η^1-butynyliron complexes, **5**, 277
Ketenes, divinyl-
Diels–Alder reactions, **5**, 395
Ketenes, β-keto
aminals
alkylation, **6**, 518
Ketenes, methylene-
synthesis
Knoevenagel reaction, Meldrum's acid, **2**, 356
Ketenes, β-(methylmercaptothiocarbonyl)-
aminals
alkylation, **6**, 519
Ketenes, vinyl-
[2 + 2] cycloaddition
1,3-dienes, **5**, 1020
intramolecular cycloadditions, **5**, 1029
synthesis
via cyclobutenone ring opening, **5**, 688, 689
Ketene selenoacetals
synthesis
via β-hydroxyalkyl selenides, **1**, 705
Ketene thioacetals
preparation
from aldehyde dimethylhydrazones, **2**, 517
reaction with *N*-acyliminium ions, **2**, 1064
Ketene thioketals
synthesis
via Horner reaction, **1**, 774
Ketenimines
alkenylaminoboranes from, **2**, 244
amidine synthesis, **6**, 546
cycloaddition reactions, **5**, 113
[3 + 2] cycloaddition reactions, **5**, 297
imidoyl halide synthesis, **6**, 526
reactions with enolates
Mannich reaction, **2**, 927
reactions with Fischer carbene complexes, **5**, 1109
synthesis
via amides, **6**, 489
unsymmetrical
deprotonation, **2**, 476
Keteniminium salts
2-azetidiniminium salts from, **5**, 108–113
Keteniminium tetrafluoroborate, tetramethyl-
β-lactams from, **5**, 112
Ketimines
metallated
alkylation, **3**, 31
unsymmetrical
deprotonation, **3**, 32
Ketoacetates
synthesis
via solid support oxidation of acetates, **7**, 842
Keto acids
enzymic reductions
lactate dehydrogenases, **8**, 189
synthesis
via oxidation of alkylidene cycloalkanones, **7**, 684
α-Keto acids
synthesis, **7**, 661
double carbonylation, **3**, 1039
oxazolones, **2**, 396
β-Keto acids
π-allylpalladium complexes from, **4**, 592
enzymic reduction
synthesis of β-hydroxy esters, **8**, 190
Knoevenagel reaction, **2**, 359
reaction with allyl carboxylates
palladium catalysis, **4**, 618
γ-Keto acids
(2*H*)-pyridazinones
reduction, **8**, 343
synthesis
via acylation of boron-stabilized carbanions, **1**, 497
β-Keto aldehydes
metal enolates
alkylation, **3**, 54
Ketoamides
Reformatsky reaction
regioselectivity, **2**, 284
α-Ketoamides, γ-amino-
synthesis
via palladium(II) catalysis, **4**, 560

Ketocarbenes
rearrangement
inhibition by copper, **3**, 896
Wolff rearrangement, **3**, 893
Keto esters
dianions
aldol reactions, **2**, 189
enzymic reductions
lactate dehydrogenases, **8**, 189
reactions with organoaluminum–ate complexes
facial selectivity, **1**, 86
Reformatsky reaction
regioselectivity, **2**, 284
synthesis
double carbonylation, **3**, 1039
α-Keto esters
synthesis, **7**, 661
via oxalic acid derivatives, **1**, 425
β-Keto esters
cleavage, **2**, 855
decarboxylation, **2**, 817
metal enolates
alkylation, **3**, 54
sulfenylation, **7**, 125
synthesis
Claisen condensation, **2**, 817
γ-Keto esters
synthesis
via ester enolate addition reactions, **4**, 109
α-Ketohydrazones
oxidation
synthesis of α-diazo ketones, **3**, 890
α-Ketol acetates
reductive cleavage
iron carbonyls, **8**, 993
metals, **8**, 991
α-Ketol rearrangement
pinacol rearrangement
comparison with, **3**, 722
Ketols
cleavage
mechanism, **8**, 984
deoxygenation
metal ions, **8**, 992
α-Ketols
reductive cleavage
metal ions, **8**, 992
metals, **8**, 991
synthesis
via samarium acyl anions, **1**, 273
Ketone enolates
addition reactions
alkenes, palladium(II) catalysis, **4**, 572
Michael additions, **5**, 1082
Ketones
γ-acetoxy-α,β-unsaturated
reaction with cuprates, **4**, 179
achiral
reactions with type I crotyl organometallics, **2**, 9–19
reactions with type III crotyl organometallics, **2**, 19–24
acyclic
aldol reaction, **2**, 143
α-alkylated, synthesis, **3**, 26
lithiated imines, **3**, 37
reduction, **8**, 2
regiospecific alkylation, **3**, 3
synthesis *via* retro Diels–Alder reactions, **5**, 573
tandem vicinal difunctionalization, **4**, 243–245
acyclic aliphatic
Baeyer–Villiger reaction, **7**, 676
acyclic enolates
alkylation, **3**, 17
acylation, **2**, 795–863
acid catalysis, **2**, 832
by esters, **2**, 829
regiochemistry, **2**, 835
acyloin coupling reactions
with esters, **3**, 630
1,2-addition reactions
acyl anions, **1**, 546
cyanohydrin ethers, **1**, 551
cyanohydrins, **1**, 548
α-(dialkylamino)nitriles, **1**, 555
hydrazones, **2**, 511
phosphonate carbanions, **1**, 562
addition to diazoalkanes, **3**, 783
aldol reactions
external chiral reagents, **2**, 262
self-addition, **2**, 140
with aldehydes, **2**, 142–156
aliphatic
Perkin reaction, **2**, 400
alkenic
electroreduction, **8**, 134
synthesis *via* Claisen rearrangement, **5**, 827
γ-alkoxy-α,β-unsaturated
conjugate additions, organocuprates, **4**, 179
alkyl–aryl
aldol reaction, **2**, 150
α-alkylated
enantioselective synthesis, **3**, 35
synthesis *via* dissolving metal conjugate reduction, **4**, 254
α′-alkylated
synthesis, **3**, 28
α-alkylation, **4**, 260
alkyl enol ether derivatives
alkylation, **3**, 25
alkylidenation
dihaloalkane reagents, **5**, 1125
α-alkyl-β,γ-unsaturated
synthesis, **3**, 23
alkynes from, **8**, 950
alkynic
cyclization, catalysts, **5**, 22
electroreduction, **8**, 134
synthesis, **1**, 405
synthesis from lactones, **1**, 418
α,β-alkynic
reduction, **8**, 357, 545
allenic
reduction, **8**, 114
amides from
Beckmann rearrangement, **7**, 694
aromatic
Birch reduction, **8**, 508
hydrogenolysis, **8**, 319
reactions with boron-stabilized carbanions, **1**, 498
reduction, **8**, 114, 115

aromatic methyl
  aldol reaction, **2**, 150
arylation
  regiochemistry, **4**, 465
asymmetric aldol reaction
  boron reagents, **2**, 264
Barbier-type reactions
  organosamarium compounds, **1**, 256
bicyclic
  alkylation, **3**, 11
  enzymic reduction, **8**, 197
  synthesis *via* palladium catalysts, **4**, 841
bicyclic β,γ-unsaturated
  photoisomerizations, **5**, 222
boron trifluoride complex
  NMR, **1**, 292
bridged bicyclic
  enzymic reduction, specificity, **8**, 200
bromination
  bromine, **7**, 120
chiral enolates
  aldol stereoselection, **2**, 224
chiral β-hydroxy
  aldol reaction, stereoselectivity, **2**, 224
cross-coupling reactions
  organoytterbium compounds, **1**, 279
cyclic
  aldol reaction, **2**, 147
  α-alkylated, synthesis, **3**, 26
  axial selectivity of alkyl addition, **1**, 78
  dehydrogenation, **7**, 132
  dehydrogenation using palladium(II) chloride, **7**, 140
  homologation, **3**, 781
  nucleophilic addition reactions, **1**, 67
  reactions with diazoalkanes, **1**, 847
  reduction, **8**, 5, 14
  regiospecific alkylation, **3**, 3
  ring contraction, **7**, 831
  ring expansion, **7**, 831
  stereocontrol, cathodic reduction, **8**, 133
  tandem vicinal difunctionalization, **4**, 245
cyclic 2-alkoxycarbonyl
  synthesis, **2**, 806
β-cyclopropyl-α,β-unsaturated
  reaction with cuprates, **4**, 180
Darzens glycidic ester condensation, **2**, 424
dehydrogenation, **7**, 144
  benzeneseleninyl chloride, **7**, 135
  copper(II) bromide, **7**, 144
  palladium catalysts, **7**, 141
deprotonation
  regioselectivity, **2**, 183
dienolates
  intramolecular γ-alkylation, **3**, 25
dimethylthioacetal *S,S*-dioxides
  ketone synthesis from, **3**, 143
electron deficient
  Diels–Alder reactions, **5**, 432
  ene reaction, **2**, 538
electroreduction, **8**, 132
  stereocontrol, **8**, 133
enantioselective reduction
  Lewis acid coordination, **1**, 317
enolate geometry
  effect of base, **2**, 192
enolates
  addition reactions with alkenic π-systems, **4**, 99–105
  bromination, **7**, 120
  crystal structures, **1**, 26
  deprotonation regioselectivity, **2**, 101
enolizable
  methylenation using Tebbe reagent, **5**, 1123
  reactions with organocerium compounds, **1**, 234
  reactions with organosamarium(III) reagents, **1**, 253
enol silyl ethers of, **2**, 599
fluoro
  synthesis, epoxide ring opening, **3**, 748
α-formyl α,β-unsaturated
  synthesis, **2**, 838
geminal dialkylation
  titanium(IV) reagents, **1**, 167
halogenation, **7**, 120
Henry reaction, **2**, 329
homologation, **3**, 783
  diazo compounds, **6**, 129
  to enones, **7**, 821
hydrogenation
  catalytic, **8**, 141
α-hydroxylation, **7**, 152
hydrozirconation, **8**, 683
intermolecular acylation, **2**, 837
intermolecular additions
  allylsilanes, **1**, 610
intermolecular pinacol coupling reactions, **3**, 570
intramolecular acylation, **2**, 843
intramolecular additions
  allyltrimethylsilane, **1**, 612
Lewis acid complexes
  rotational barriers, **1**, 290
macrocyclic
  lithiated imines, **3**, 37
5–7-membered cyclic
  syn selective aldol reaction, titanium enolates, **2**, 306
metal enolates
  alkylation, **3**, 3
α-metallated
  formation, **3**, 3
O-metallated tautomers
  formation, **3**, 3
1-methoxy-substituted cyclic β,γ-unsaturated
  photoisomerizations, **5**, 226
methylenation, **1**, 532
  phenylthiomethyllithium, **6**, 139
  Tebbe reaction, **1**, 746; **5**, 1123
α-methyl α,β-unsaturated
  synthesis, **3**, 33
mixed aldol reaction, **2**, 142
monocyclic β,γ-unsaturated
  photoisomerizations, **5**, 222
nonalkenic
  electroreduction, **8**, 131
nonconjugated alkenic
  electroreduction, **8**, 134
optical resolution, **1**, 534
photolysis, **7**, 41
polyunsaturated
  tandem vicinal difunctionalization, **4**, 253
radical cyclizations, **4**, 817

reactions with allylic organocadmium compounds, **1**, 226
reactions with arynes, **4**, 510
reactions with boron-stabilized carbanions, **1**, 498
reactions with chloromethyleneiminium salts, **2**, 785
reactions with diazoalkanes, **1**, 845
homologation, **3**, 778
reactions with dienes
transition metal catalysis, **4**, 709–712
reactions with dithioacetals, **1**, 564
reactions with organoaluminum reagents
discrimination between aldehydes and, **1**, 83
reactions with organocadmium compounds, **1**, 225
reactions with organocerium reagents, **1**, 233
reactions with organocopper compounds, **1**, 116
reactions with organometallic compounds
chemoselectivity, **1**, 145
Lewis acid promotion, **1**, 326
reactions with samarium diiodide
pinacolic coupling reactions, **1**, 271
reactions with trialkylaluminum
synthesis of aluminum enolates, **2**, 114
reactions with type I crotylboron compounds, **2**, 10–15
reactions with zinc ester dieneolates, **2**, 286
reduction, **8**, 923–951
Alpine borane, **7**, 603
cathodic, **8**, 131
chiral boron reagents, **8**, 101
diimide, **8**, 478
2,5-dimethylborolane, **2**, 258
dissolving metals, **8**, 307–323
dissolving metals, stereoselectivity, **8**, 116
ionic hydrogenation, **8**, 487
samarium diiodide, **8**, 115
selective, **8**, 18
reductive coupling
nitriles, **1**, 273
Reformatsky reaction, **2**, 281
saturated
cycloalkylation, **3**, 18
saturated heterocyclic
aldol reaction, **2**, 149
selenenylation, **7**, 129, 131
kinetic product, **7**, 130
self-condensation, **2**, 141
spirocyclic β,γ-unsaturated
photoisomerizations, **5**, 222
steroidal
synthesis, regiospecific alkylation, **3**, 11
steroids
dehydrogenation, **7**, 132, 136
α-substituted
reductive elimination, **8**, 925
β-substituted α,β-unsaturated
reaction with cuprates, **4**, 180
sulfenylation, **7**, 125
sulfinylation, **7**, 127
synthesis
α-alkoxy carbanions, **3**, 197
alkylboronic esters, **3**, 797
alkynylborates, **3**, 799
carbonylation, **3**, 1023
coupling reactions with organometallics, **3**, 463
cyanoboronates, **3**, 798
α-heterosubstituted sulfides and selenides, **3**, 141
intramolecular dehydrative acylations, **2**, 711
intramolecular Friedel–Crafts reaction, **2**, 710
organoboranes, **3**, 780, 793
palladium mediated, **2**, 749
syn selective aldol reaction, zirconium enolates, **2**, 302
via acylation of boron-stabilized carbanions, **1**, 497
via alkenes, **7**, 600
via carboxylic acid derivatives, **1**, 398
via β-hydroxyalkyl selenides, mechanism, **1**, 718
via oxidation of secondary alcohols, **7**, 318
via oxidative cleavage of alkenes, **7**, 541
via Wacker oxidation of alkenes, **7**, 450
Vilsmeier synthesis, **2**, 748
tandem vicinal difunctionalization, **4**, 242–246
tin enolates
synthesis, **2**, 116
β-tosyloxy-α,β-unsaturated
cyclopropanation, **4**, 976
tricyclic
synthesis, regiospecific alkylation, **3**, 11
unconjugated unsaturated
hydrogenation, **8**, 439
unsaturated
reduction, diimide, **8**, 476
α,β-unsaturated
addition reaction with Grignard reagents, **4**, 83
Baeyer–Villiger reaction, **7**, 684
1,3-carbonyl group transposition, **6**, 836
conjugate additions, **4**, 208–212
conjugate reduction, **4**, 239
dehydrogenation, **7**, 142
deprotonation, **2**, 105
dimethylhydrazones, deprotonation, **2**, 506
dissolving metal reduction, **8**, 481
electroreduction, **8**, 134
ene reactions, **5**, 5
enzymic reduction, **8**, 191, 205
Henry reaction, regioselectivity, **2**, 330
Henry reaction, stereoselectivity, **2**, 330
hydrazones, γ-deprotonation, **2**, 509
hydride additions, **2**, 106
hydrobromination, **4**, 282
hydroformylation, **4**, 924
hydrogenation, **8**, 533
hydrogenation, homogeneous catalysis, **8**, 452
α-hydroxylation, **7**, 174
metal dienolates, alkylation, **3**, 21
methylation, **3**, 23
Michael acceptors, **4**, 261
partial reduction, **8**, 526
reaction with organocuprates, **1**, 116
reaction with organolithium compounds, **4**, 72
reaction with vinyl zirconium reagents, **1**, 155
reduction, metal hydrides, **8**, 15
sp^2 center, hydroxylation, **7**, 179
synthesis, **3**, 880, 894
synthesis *via* allylic anions, **2**, 61
synthesis *via* cyclopropane ring opening, **4**, 1020
synthesis *via* Knoevenagel reaction, **2**, 359
synthesis *via* Mannich bases, **2**, 897
synthesis *via* Ramberg–Bäcklund rearrangement, **3**, 870
synthesis *via* retro Diels–Alder reactions, **5**, 553, 573
α,β;β′,γ′;δ,ε-unsaturated

photoisomerizations, **5**, 229
β,γ-unsaturated
allylic oxidation, **7**, 819
formation of, Friedel–Crafts reaction, **2**, 708
rearrangement, **5**, 216
stereoselective synthesis, **6**, 851
synthesis *via* acylation of π-allylnickel complexes, **1**, 453
synthesis *via* π-allylnickel halides, **3**, 424
$\beta,\gamma;\beta',\gamma'$-unsaturated
photoisomerizations, **5**, 229
γ,δ-unsaturated
synthesis *via* Claisen rearrangement, **5**, 1004
unsymmetrical
Michael addition, **4**, 6
regioselective alkylation, **3**, 2
synthesis, **3**, 199
unsymmetrical enamines
alkylation, **3**, 30
unsymmetrical enolates
regioselective alkylation, **3**, 7
1,2-Ketones
transposition
Baeyer–Villiger reaction, **7**, 684
Ketones, α-acetoxy
reduction, **8**, 935
Ketones, acyclic diaryl
Baeyer–Villiger reaction, **7**, 678
Ketones, β-acylamide
synthesis, **6**, 271
Ketones, 1-adamantyl ethyl
aldol reaction
stereoselectivity, **2**, 193
Ketones, α-alkoxy
cyclic
nucleophilic addition reactions, **1**, 52
reactions with organocuprates, **1**, 108
Ketones, β-alkoxy
reduction, **8**, 9
synthesis
via palladium(II) catalysis, **4**, 553
Ketones, α-alkoxy acyclic
nucleophilic addition reactions
Grignard reagents, **1**, 51
Ketones, alkoxyethynyl vinyl
Michael addition, **4**, 44
Ketones, alkyl phenyl
Baeyer–Villiger reaction
regiochemistry, **7**, 673
Ketones, 2-(alkylsulfinyl)-1-alkenyl
reaction with amines, **6**, 69
Ketones, α-alkylthio
reductive cleavage, **8**, 993
Ketones, 2-(alkylthio)-1-alkenyl
reaction with amines, **6**, 69
Ketones, alkynic
tandem vicinal difunctionalization, **4**, 245
Ketones, α,β-alkynyl
allenolates
1,4-addition, **2**, 116
reduction
Alpine borane, **7**, 603
Ketones, alkynyl trifluoromethyl
conjugate additions
organocuprates, **4**, 194
Ketones, allenyl
synthesis, **3**, 991
tandem vicinal difunctionalization, **4**, 245
Ketones, α-allyloxy
Claisen rearrangement, **5**, 847
enolates
Claisen rearrangements, **5**, 1001
Wittig rearrangement, **3**, 996
Ketones, allyl vinyl
cyclization, **5**, 755–761
Rupe rearrangement, **5**, 768
synthesis
via hydration of dienynes, **5**, 752
Ketones, β-amido
preparation
Friedel–Crafts acylations, **2**, 709
Ketones, γ-amido
synthesis
via aziridines, **6**, 96
Ketones, amino
reduction, **8**, 13
synthesis, **7**, 506
Ketones, α-amino-
expansion, **1**, 889
hydrosilylation, **8**, 13
rearrangement, **3**, 790
reduction
electrolysis, **8**, 321
reductive cleavage, **8**, 995
synthesis, **6**, 786, 787
Friedel–Crafts reaction, **2**, 756
Wolff–Kishner reduction, **8**, 927
Ketones, β-amino
synthesis
Mannich reaction, **2**, 902
Ketones, 2-amino-1-alkenyl
synthesis
via substitution processes, **6**, 69
Ketones, anthracene-9,10-diyl-bis(styryl
reduction
aluminum hydrides, **8**, 543
Ketones, 9-anthryl styryl
reduction
aluminum hydrides, **8**, 543
Ketones, aryl
oxidative rearrangement, **7**, 829
reduction
hydride transfer, **8**, 91
synthesis
via rearrangement of arylalkenes, **7**, 828
Ketones, α-aryl
preparation
via $S_{RN}1$ reaction, **4**, 463–466
synthesis
Friedel–Crafts reaction, **3**, 306
Ketones, aryl alkyl
Baeyer–Villiger reaction, **7**, 678
electroreduction, **8**, 131
Ketones, aryl methyl
synthesis
carbonylation, **3**, 1024
Ketones, γ-aryl-α-trifluoromethyl
cycloalkylation
Friedel–Crafts reaction, **3**, 324
Ketones, β-asymmetric amino
nucleophilic addition reactions
stereoselectivity, **1**, 60

Ketones, aziridinyl
  synthesis
    via 1,3-dipolar cycloaddition, **4**, 1101
Ketones, benzyl phenyl
  synthesis
    via oxidative rearrangement, **7**, 829
Ketones, bicyclic
  preparation
    Friedel–Crafts reaction, **2**, 711
Ketones, bicyclic halo
  Favorskii rearrangement, **3**, 851
Ketones, β,β-bis(alkylthio)-α,β-unsaturated
  reduction, **8**, 542
  selective reduction, **8**, 540
Ketones, bis(phenylethynyl)
  Michael/anti-Michael addition
    with ethyl acetoacetate, **4**, 46
Ketones, bridged bicyclic
  Baeyer–Villiger reaction, **7**, 682
Ketones, bridged polycyclic
  Baeyer–Villiger reaction, **7**, 682
Ketones, α-bromo
  enolates
    synthesis, **2**, 109
  reactions with aldehydes, **1**, 202
  synthesis, **7**, 533
Ketones, α-bromoalkyl
  [4 + 3] cycloaddition reactions, **5**, 595
Ketones, α-bromobenzyl
  [4 + 3] cycloaddition reactions, **5**, 595
Ketones, α-bromo-β-hydroxy
  synthesis, **6**, 26
Ketones, 4-*t*-butylcyclohexylmethyl
  exocyclic enolate
    methylation, **3**, 16
Ketones, *t*-butyl ethyl
  aldol reaction
    stereoselectivity, **2**, 193, 195
  bromomagnesium enolate
    reaction with benzaldehyde, **2**, 234
  magnesium bromide enolate
    crystal structure, **1**, 29
Ketones, chloro
  Reimer–Tiemann reaction
    abnormal, **2**, 773
Ketones, α-chloro
  homologation, **3**, 787
  Reformatsky reaction, **2**, 285
  synthesis, **7**, 527, 538
    via diazo ketones, **6**, 207
Ketones, α-chlorodibenzyl
  reaction with furan
    [4 + 3] cycloaddition, **5**, 594
Ketones, β-chlorovinyl
  reaction with thioamides, **6**, 508
Ketones, 2-cyano
  cyclic
    synthesis, **6**, 240
Ketones, cyclobutyl phenyl
  reduction
    silanes, **8**, 318
Ketones, cyclopentenyl methyl
  regioselective aldol cyclization, **2**, 159
Ketones, cyclopropyl
  reduction, **8**, 21
    dissolving metals, **8**, 114
  ring opening
    cyclopentene annulation, **5**, 925
Ketones, cyclopropyl methyl
  preparation
    acylation of homoallylic silanes, **2**, 719
  reduction
    dissolving metals, **8**, 309
Ketones, cyclopropyl phenyl
  reduction
    dissolving metals, **8**, 309
    hydrogen transfer, **8**, 320
    silanes, **8**, 318
Ketones, dialkenyl
  selenocyclization, **4**, 390
Ketones, dialkyl
  electroreduction, **8**, 132
Ketones, diallyl
  synthesis
    carbonylation, **3**, 1023
Ketones, diaryl
  electroreduction, **8**, 131
Ketones, diazo
  alkylation, **3**, 794
  bromination, **6**, 211
  C—H insertion reactions, **3**, 1054
    copper-catalyzed, **3**, 1051
  α-chlorination, **6**, 207
  fluorination, **6**, 219
Ketones, α-diazo
  acyclic
    synthesis, **6**, 122
  boron enolates, **2**, 111
  cyclic
    synthesis, **6**, 123
  Mannich reactions
    with preformed iminium salts, **2**, 903
  synthesis, **6**, 122
Ketones, dibenzyl
  α,α′-keto dianion
    crystal structure, **1**, 29
  Vilsmeier-Haack reaction, **2**, 785
Ketones, α,α′-dibromo
  [4 + 3] cycloaddition reactions, **5**, 603
Ketones, α,α′-dibromomethyl alkyl
  [4 + 3] cycloaddition reactions, **5**, 605
Ketones, di-*t*-butyl
  reaction with α-selenoalkyllithium, **1**, 674
Ketones, β,β-dichlorovinyl
  aminolysis, **6**, 521
Ketones, dicyclopropyl
  reactions with organocuprates in presence of crown ethers
    anti-Cram selectivities, **1**, 115
Ketones, α,α-dihalo
  reduction, **8**, 20
    metal salts, **8**, 987
Ketones, α,α′-dihalo
  [4 + 3] cycloaddition reactions, **5**, 594, 603
Ketones, diisobutyl
  aldol reaction, **2**, 144
  reaction with 2-pentenylzinc bromide, **1**, 219
Ketones, diisopropyl
  aldol reaction, **2**, 144
Ketones, dimesityl
  reduction
    alane, **8**, 3

Ketones, divinyl
synthesis, **1**, 430; **3**, 844
Ketones, α,α′-divinyl
cyclization, **5**, 755–761
in situ generation, **5**, 766–770, 775–778
photocyclization, **5**, 760
ring closure
2-cyclopentenone synthesis, **5**, 752
silicon-directed cyclization, **5**, 761–766
solvolysis
cyclization, **5**, 770–775
Ketones, α,β-epoxy
reduction, **8**, 11, 21, 992
ring opening
to 1,3-diones, **3**, 771
synthesis
tin enolates, **2**, 424
via α-bromo-β-hydroxy ketones, **6**, 26
Wharton reaction, **8**, 927
Ketones, γ,δ-epoxy
reduction, **8**, 11
Ketones, *cis*-epoxy
preparation, **2**, 424
Ketones, 2-ethenyl
synthesis
vinylmagnesium halide alkylation, **3**, 242
Ketones, α-ethoxy
synthesis
via 1,2-diketones, **1**, 217
Ketones, ethyl
aldol reactions
external chiral reagents, **2**, 262
enolization, **2**, 244
from carbohydrates
aldol reaction, stereoselection, **2**, 226
stereoselective aldol reaction
titanium enolate, chiral auxiliary, **2**, 307
Ketones, ethyl cyclohexyl
lithium enolates
aldol reaction, facial selectivity, **2**, 221
Ketones, ethyl mesityl
aldol reaction
stereoselectivity, **2**, 193
Ketones, ethyl trityl
aldol reactions
aluminum-mediated, **2**, 269
Ketones, ethynyl methyl
Robinson annulation, **4**, 43
Ketones, ethynyl phenyl
Michael addition, **4**, 41
Ketones, α-fluoro
synthesis, **7**, 538
Ketones, α-formyl
dehydrogenation, **7**, 136
Ketones, fused ring bicyclic
Baeyer–Villiger reaction, **7**, 680
Ketones, fused ring polycyclic
Baeyer–Villiger reaction, **7**, 680
Ketones, α-halo
acyclic
Favorskii rearrangement, **3**, 842
alkene synthesis from, **3**, 871
[4 + 3] cycloaddition reactions, **5**, 595
homologation, **3**, 783
reaction with diiodomethane
organosamarium compounds, **1**, 261
reaction with vinyl and aryl Grignard reagents, **3**, 242
rearrangements, **3**, 788, 828, 839
reduction, **8**, 19
Alpine borane, **7**, 603
stereoselectivity, **8**, 3
reductive cleavage
metals, **8**, 986
reductive elimination, **8**, 925
reductive silylation, **2**, 600
Ketones, 2-haloalkyl aryl
ketals
rearrangement, **3**, 789
Ketones, 2-halocycloalkyl
rearrangements, **3**, 845
Ketones, 2,4,6-heptatrienyl 4-methoxyphenyl
synthesis
via monoacylation, iron(III) catalyzed, **1**, 416
Ketones, heteroaryl
synthesis
via $S_{RN}1$ reaction, **4**, 468
Ketones, α-hydroperoxy
synthesis, **7**, 156, 159
Ketones, hydroxy
α,β-unsaturated
Diels–Alder reactions, **5**, 359
Ketones, α-hydroxy
Diels–Alder reactions
intramolecular asymmetric, **5**, 543
(*Z*)-enolates, **2**, 102
hydrazones
asymmetric synthesis, **2**, 506
rearrangements, **3**, 791
reduction
diastereoselectivity, **8**, 7
synthesis
carbonylation of lithium amides, **3**, 1017
via benzoin condensation, **1**, 542
via diaryl ketone dianions, **1**, 280
via α-keto acetals, **1**, 63
Ketones, β-hydroxy
aldol reaction
cerium enolates, **2**, 311
reaction with allenylboronic acid, **2**, 97
reduction, **8**, 8
synthesis
via cerium reagents, **1**, 244
via 1,3-dipolar cycloadditions, **4**, 1078
via α,β-epoxy ketones, **3**, 264
Ketones, γ-hydroxy
reduction, **8**, 10
synthesis
via acylation of boron-stabilized carbanions, **1**, 497
Ketones, hydroxymethyl
synthesis, **2**, 838
Ketones, α-hydroxy(trifluoromethyl)-
synthesis, **1**, 543
Ketones, α-iodo
synthesis, **7**, 530
Ketones, ω-iodoalkyl
reaction with samarium diiodide
synthesis of cyclopentanols, **1**, 261
zinc compounds from
coupling reactions with alkenyl halides, **3**, 443
Ketones, α-mercurio
aldol reaction
syn:anti selectivity, **2**, 313

synthesis, **2**, 128
Ketones, mesityl
acetone self-condensation, **2**, 141
Ketones, mesityl phenyl
reduction kinetics, **8**, 2
Ketones, methyl
aldol reactions, **2**, 264
photostimulated ring closure, **4**, 476
self-condensation, **2**, 244
synthesis
via palladium(II) catalysis, **4**, 552
Ketones, methyl cyclohexyl
enolates, **2**, 264
Ketones, α-methylene
synthesis
formylation, **2**, 838
from Mannich bases, **2**, 897
via β-keto sulfides, **6**, 141
Ketones, methyl neopentyl
enolates, **2**, 264
Ketones, α-methyl-β,γ-unsaturated
reduction, **8**, 10
Ketones, methyl vinyl
aldol reaction, **2**, 152
[3 + 2] cycloaddition reactions
palladium catalyzed, **5**, 281
with 1-methyl-1-(trimethylsilyl)allene, **5**, 277
Diels–Alder reactions
comparison of promoters, **5**, 345
water promoted, **5**, 344
ene reactions
intermolecular, **5**, 3
Lewis acid catalysis, **5**, 5
reduction
transfer hydrogenation, **8**, 552
Robinson annulation
phenol synthesis, **4**, 8
Ketones, monocyclic
Baeyer–Villiger reaction, **7**, 678
Ketones, monocyclic halo
Favorskii rearrangements, **3**, 848
Ketones, naphthyl
Birch reduction
dissolving metals, **8**, 508
Ketones, α-nitrato
reduction, **7**, 154
Ketones, α-nitro
cyclic
synthesis, **6**, 105
preparation, **2**, 323
Ketones, α-oxy
hydrosilylation, **8**, 8
Ketones, phenyl
synthesis
via acylation of boron-stabilized carbanions, **1**, 497
Ketones, phenylthienyl
synthesis, **3**, 515
Ketones, α-phenylthio
reduction, **8**, 15
synthesis
via acylation of boron-stabilized carbanions, **1**, 497
Ketones, polycyclic
preparation
Friedel–Crafts reaction, **2**, 711
Ketones, polycyclic halo
Favorskii rearrangement, **3**, 854
Ketones, polyhalo
Favorskii rearrangement, **3**, 843
Ketones, α-seleno
metallation, **1**, 642
Ketones, α-silyl
rearrangement
enol ether preparation, **2**, 601
Ketones, β-silyl
synthesis, **1**, 436
Ketones, β-silyl divinyl
cyclization, **5**, 762
Ketones, β-silyloxy
intramolecular hydrosilylation, **7**, 645
Ketones, silyl vinyl
synthesis, **2**, 76
alkoxyallene, **2**, 88
Ketones, spirocyclic
Baeyer–Villiger reaction, **7**, 678
Ketones, 2-substituted
N-(9-phenylfluorene-9-yl)amino
deprotonation, alkylation, **3**, 44
Ketones, α-substituted β-hydroxy
reduction, **8**, 9
Ketones, α-sulfinyl
enolates
aldol reaction, stereoselectivity, **2**, 229
Ketones, tetrabromo
[4 + 3] cycloaddition reactions, **5**, 603
Ketones, tetrahydrofurfuryl
reactions with Grignard reagents, **3**, 996
Ketones, tetramethyldibromo
[3 + 2] cycloaddition reactions
with α-methylstyrene, **5**, 283
Ketones, tetrasubstituted dibromo
[4 + 3] cycloaddition reactions, **5**, 603
Ketones, α-tosyl
reduction, **8**, 926
Ketones, α-(trialkylsilyl)vinyl
reaction with enolates, **4**, 100
Ketones, α-triazolyl
reduction, **8**, 13
Ketones, tribromo
[4 + 3] cycloaddition reactions, **5**, 603
Ketones, trichlorostannyl
enone synthesis, **2**, 443
Ketones, 1,1,1-trifluoromethyl
ene reaction, **2**, 538, 539
Ketones, α-trimethylsilyloxy
aldol reaction
stereoselectivity, maximization, **2**, 193
Ketones, β-trimethylsilyloxy
preparation
from silicon compounds, **2**, 269
Ketones, vinyl
arylation
orgnothallium compounds, **4**, 841
Ketones, β-(2-vinylcyclopropyl)
α,β-unsaturated
Cope rearrangement, **5**, 979–984
Ketones, α-vinyl β-hydroxy
reduction, **8**, 10
Ketonitriles
Reformatsky reaction
regioselectivity, **2**, 284
Ketosteroids
ecdysone side

synthesis, **7**, 243
Ketoximes
Beckmann rearrangement, **6**, 763; **7**, 691
α-hydroxylation, **7**, 187
reactions with organometallic compounds, **1**, 387
synthesis
via nitroiminium chlorides, **1**, 121
unsymmetrical
deprotonation, **3**, 35
α-Ketoximes
oxidation
synthesis of α-diazo ketones, **3**, 890
Ketoximes, α-hydroxy-
photoreaction, **6**, 765
Ketoximes, *O*-substituted
Beckmann rearrangement, **7**, 693
Ketoximes, *O*-tosyl-
Beckmann rearrangement, **7**, 693
Ketoximes, *O*-unsubstituted
Beckmann rearrangement, **7**, 691
Ketyls
Barbier-type coupling reactions
samarium diiodide, **1**, 263
radical cyclizations, **4**, 809
reactions with alkenes
organosamarium reagents, **1**, 268
reactions with alkynes
organosamarium compounds, **1**, 268
Keyhole limpet hemocyanin
monoclonal antibodies
Claisen rearrangement, **5**, 855
Kharasch method
radical addition reactions, **4**, 751–758, 770
Kharasch–Sosnovsky reaction
allylic oxidation, **7**, 84, 95
Khellin
oxidation, **7**, 462
synthesis, **5**, 1096, 1098
Sommelet–Hauser rearrangement, **3**, 969
Khellinone
synthesis
via cyclobutenone ring opening, **5**, 689
Khusimone
synthesis
via magnesium-ene reaction, **5**, 45
Kiliani–Fischer synthesis
sugars, **1**, 460
Kishner–Leonard eliminations
reduction of hydrazones, **8**, 341
Kjellmanianone
synthesis, **7**, 175, 176
Knoevenagel reaction, **2**, 341–388, 401
active methylene compound
variation, **2**, 354
analytical applications, **2**, 354
competitive reactions, **2**, 352
conditions, **2**, 343
Doebner modification, **2**, 356
kinetics, **2**, 347
limitation, **2**, 354
mechanism, **2**, 347
products
acidity, **2**, 346
Lewis acidity, **2**, 346
physical properties, **2**, 345
spectroscopy, **2**, 345
pyridine derivatives
synthesis, **2**, 378
scope, **2**, 354
sequential reactions, **2**, 369
solvents, **2**, 345
stereochemistry, **2**, 350
steric effects, **2**, 369
synthetic alternatives, **2**, 388
synthetic applications, **2**, 375
Knorr pyrrole synthesis
Knoevenagel reaction, **2**, 376
Köbrich reagents
carbenoids
epoxidation, **1**, 830
Koch–Haaf reaction, **7**, 8
Ritter-type reaction
carboxylic acid synthesis, **6**, 291
Koch reaction
carboxylic acid synthesis, **3**, 1017
Koenigs–Knorr method
glycoside synthesis, **6**, 34, 37
catalysts, **6**, 39
solvents, **6**, 40
Kolbe electrolysis
anode material, **3**, 635
cathode material, **3**, 636
critical potential, **3**, 634
current densities, **3**, 634
distribution of active species
electrochemical oxidation, **7**, 791
ionic additives, **3**, 634
mechanism, **3**, 636
pH, **3**, 634
reaction conditions, **3**, 634
solvents, **3**, 635
Kolbe radicals
addition to double bonds, **3**, 646
Kolbe reactions, **3**, 633–658
radical addition reactions, **4**, 759
radical cyclizations, **4**, 805
reaction conditions, dimerization, **7**, 806
Kornblum oxidation
activated halides, **7**, 653
limitations, **7**, 654
Kostanecki reaction
intramolecular acyl transfer, **2**, 845
Kreysiginone
synthesis
use of vanadium oxytrifluoride, **3**, 681
Krief–Reich synthesis
allylsilanes, **2**, 586
Kröhnke oxidation
activated halides
carbonyl compounds, **7**, 657
Kushimone
synthesis
via conjugate addition, **4**, 218

L

L-selectride — *see* Lithium tri-*s*-butylborohydride
Labda-7,14-dien-13-ol
  synthesis, **3**, 168
Lactaldehyde, 3-nitro-
  cyclization of glyoxal with nitromethane, **2**, 327
Lactam acetals
  synthesis, **6**, 566
β-Lactam antibiotics
  synthesis, **6**, 388
    Mannich reaction, **2**, 915
    Reformatsky reaction, **2**, 296
    use of ester protecting groups, **6**, 670
Lactamidines
  aminal ester synthesis, **6**, 575
  reaction with isocyanates, **6**, 579
Lactams
  ω-allenyl
    synthesis, **2**, 89
  bicyclic
    alkylation, **3**, 42
  enolates
    stereoselectivity, **2**, 211
  α-hydroxylation, **7**, 183
  microbial hydroxylation, **7**, 60
  reduction, **8**, 248
    metal hydrides, **8**, 273
  selenenylation, **7**, 129
  steroids
    dehydrogenation, **7**, 132
  sulfenylation, **7**, 125
  synthesis, **6**, 407
    carbonylation, **3**, 1035
    via Ritter reaction, **4**, 293
    via unsaturated amides, **7**, 524
  Vilsmeier–Haack reaction, **2**, 786
  vinylogous
    synthesis, Knoevenagel reaction, **2**, 368
β-Lactams
  aldol reactions
    syn stereoselectivity, **2**, 304
    synthesis of thienamycin, **2**, 212
  β-allylated
    synthesis *via* radical addition reactions, **4**, 745
  bicyclic
    synthesis *via* cyclization of allenic amines, **4**, 410
  enantiopure
    synthesis *via* chiral ketenes or imines, **5**, 96
  homochiral
    synthesis, **2**, 256
    synthesis *via* conjugate addition, **4**, 231
  in enolate–imine condensations
    mechanism, **2**, 917
  stereoselective synthesis, **7**, 517
  sulfenylated
    synthesis, *via* Pummerer rearrangement, **7**, 202
  synthesis, **3**, 902; **5**, 1107; **6**, 389, 405, 759, 783; **7**, 729
    C—H insertion reactions, **3**, 1056
    Mannich reaction, **2**, 917
    Ugi reaction, **2**, 1101
    use of silyl enol ethers, **2**, 636
    via cyclization of β,γ-unsaturated amides, **4**, 398
    via 1,3-dipolar cycloadditions, **4**, 1076
    via intramolecular photocycloaddition, **5**, 181
    via ketenes and carbonyls, **5**, 90
    via ketocarbenoids, **4**, 1057
    via palladium(II) catalysis, **4**, 563
    via radical cyclization, **4**, 795
    via tandem vicinal difunctionalization, **4**, 243
  4-unsubstituted
    synthesis, **2**, 941
γ-Lactams
  synthesis
    C—H insertion reactions, **3**, 1057
    Knoevenagel reaction, **2**, 367
ε-Lactams
  Beckmann rearrangement, **7**, 691
γ-Lactams, *N,O*-acetal-
  bicyclic
    Diels–Alder reactions, **5**, 372
Lactams, α-acetoxy-
  reaction with tin(II) enol ethers, **2**, 611
Lactams, *N*-alkyl-
  heterocyclic
    metallation, **1**, 478
β-Lactams, 3-alkylidene-
  synthesis
    via cycloaddition with CSI, **5**, 107
β-Lactams, α-amido-
  synthesis
    via Dane salts, **5**, 95
β-Lactams, 3-amino-
  synthesis
    via homochiral ketenes, **5**, 98
β-Lactams, aza-
  synthesis, **3**, 902
β-Lactams, *N*-benzyl-
  Wittig rearrangement, **3**, 979
Lactams, halo-
  rearrangements, **3**, 849
Lactams, hydroxy-
  synthesis
    via cyclic imide reduction, **8**, 273
Lactams, iodo-
  synthesis
    via iodocyclization of allylic imidates, **4**, 403
δ-Lactams, α-methylene-
  synthesis
    allyl organometallic compounds, **2**, 980
β-Lactams, α-phenylseleno-
  metallation, **1**, 642
Lactams, α-silyl-
  Peterson alkenation, **1**, 790
β-Lactams, *N*-tetrazol-5-yl-
  synthesis
    Mannich reaction, **2**, 920
Lactate dehydrogenase
  substrate specificity
    synthetic applicability, **2**, 456
Lactate dehydrogenases
  reduction
    keto acids, **8**, 189
Lactic acid, ethyl ester, acrylate
  Diels–Alder reactions, **5**, 365

Lactic acid, ethyl ester, fumarate
  Diels–Alder reactions, **5**, 365
Lactic aldehyde, *N*-silylimine
  synthesis, **2**, 937
Lactim ethers
  alkylation, **6**, 505
  reactions with amides, **6**, 569
γ-Lactols
  synthesis
    via benzaldehyde, **1**, 502
Lactones
  alkynic ketone synthesis from, **1**, 418
  γ-amino-
    synthesis *via* [2 + 2 + 2] cycloaddition, **5**, 1138
  aminolysis, **6**, 389
  β-amino-α,β-unsaturated
    functionalized, synthesis, **6**, 67
  arene alkylation
    Friedel–Crafts reaction, **3**, 309
  α,α-bissulfenylated
    enolate precursors, **2**, 186
  bridged-ring
    synthesis *via* cyclofunctionalization, **4**, 373
  bromination, **7**, 121
  Darzens glycidic ester condensation, **2**, 420
  dehydrative cyclization
    via Friedel–Crafts reaction, **2**, 711
  enolates
    stereoselectivity, **2**, 200
  five-membered ring
    synthesis, carbonylation of allylic alcohols, **3**, 1032
  α-hydroxylation, **7**, 179
  α-iodo-α,β-unsaturated
    synthesis, **7**, 536
  macrocyclic
    synthesis, **3**, 431; **6**, 438
  methylenation
    Tebbe reaction, **1**, 744; **5**, 1123
  of silyl ketene acetals, **2**, 605
  reactions with organocerium reagents, **1**, 239
  reduction, **8**, 246
    metal hydrides, **8**, 268
  selenenylation, **7**, 129
  steroids
    dehydrogenation, **7**, 132, 136
  sulfenylation, **7**, 125
  synthesis, **6**, 323–376; **7**, 517
    carbonylation, **3**, 1031
    epoxide ring opening, **3**, 752
    Eschenmoser coupling reaction, **2**, 890
    via amides, **7**, 524
    via [3 + 2] cycloaddition reactions, **5**, 297
    via diols, **7**, 312
    via ethers, **7**, 236
    via monodecarboxylation of dicarboxylic acids, **7**, 727
    via organosamarium compounds, **1**, 259, 266
    via oxidative cleavage of alkenes, **7**, 574
    via selenolactonization, **7**, 523
  tandem vicinal difunctionalization, **4**, 249
  unsaturated
    enolate Claisen rearrangement, **6**, 859
    synthesis, from alkynes, palladium(II) catalysis, **4**, 567
  β,γ-unsaturated
    synthesis *via* [2 + 2 + 2] cycloaddition, **5**, 1138
  Vilsmeier–Haack reaction, **2**, 786
α-Lactones
  synthesis, **6**, 342
β-Lactones
  azide synthesis, **6**, 253
  Grignard reagent alkylation, **3**, 245
  quinone synthesis
    Perkin reaction, **2**, 399
  reaction with organocopper compounds, **3**, 227
  β-substituted enolates
    alkylation, **3**, 41
  synthesis, **6**, 342
    via C—C connections, **6**, 350
    via cyclization of β,γ-unsaturated acids, **4**, 368
    via cycloacylation, **6**, 346
    via cycloalkylation, **6**, 345
    via ketenes and carbonyls, **5**, 86
γ-Lactones
  bicyclic
    synthesis, **6**, 356
  Jones reagent, **2**, 57
  nitrile synthesis, **6**, 236
  opening
    chlorination, **6**, 206
  β- and γ-substituted
    alkylation, **3**, 41
  synthesis, **6**, 350
    carbonylation, **3**, 1031
    stereoselectivity, **4**, 381, 382
    via σ-alkyliron complexes, **4**, 576
    via cyclization of cycloalkeneacetic acids, **4**, 369
    via cyclization of β,γ-unsaturated acids, **4**, 376
    via iodolactonization of benzoic acids, **4**, 374
    via metal-catalyzed cycloaddition, **5**, 1196, 1200
    via rearrangements, **6**, 362
  unsaturated
    synthesis, **6**, 352
δ-Lactones
  chiral
    preparation, **2**, 520
  enolates
    alkylation, **3**, 41
  synthesis, **6**, 365
    carbonylation of homoallylic alcohols, **3**, 1033
    stereoselectivity, **4**, 381, 382
    via cyclization of 4,5-hexadienoic acid, **4**, 396
    via cyclofunctionalization, **4**, 372
    via lactonization of 5,7-dienoic acids, **4**, 378
  unsaturated
    synthesis, Knoevenagel reaction, **2**, 381
γ-Lactones, γ-alkylidene-
  synthesis, **7**, 524
    via cyclofunctionalization of alkynoic acids, **4**, 393
δ-Lactones, δ-alkylidene-
  synthesis
    via cyclofunctionalization of alkynoic acids, **4**, 393
Lactones, allylic
  reaction with sodium malonate
    stereospecific reaction, **6**, 848
Lactones, 2-(arylsulfinyl)-α,β-unsaturated
  addition reaction with enolates, **4**, 108
δ-Lactones, α-carboxy-
  synthesis
    Knoevenagel reaction, Meldrum's acid, **2**, 356
Lactones, epoxy-
  preparation

Darzens glycidic ester condensation, **2**, 421
Lactones, γ-hydroxy-
synthesis
via [2 + 2 + 2] cycloaddition, **5**, 1138
γ-Lactones, 2-hydroxy-
synthesis
from protected cyanohydrins, **3**, 198
γ-Lactones, 4-hydroxy-
synthesis
homoaldol reaction, **2**, 445
Lactones, imino-
synthesis, **7**, 524
Lactones, α-methylene-
synthesis, **5**, 942; **7**, 129
Knoevenagel reaction, Meldrum's acid, **2**, 356
Mannich reaction, **2**, 904
via σ-alkyliron complexes, **4**, 576
via allyl chromium reagent, **1**, 189
via dehydrogenation reactions, **7**, 125
via hydrocarboxylation, **4**, 937, 941
Lactones, β-methylene-
synthesis
via allenic sulfoxide, **6**, 841
γ-Lactones, α-methylene-
synthesis, **6**, 784
δ-Lactones, α-methylene-
synthesis
Mannich reaction, **2**, 911
Lactones, α-silyl-
Peterson alkenation, **1**, 790
γ-Lactones, β-(*p*-tolylsulfinyl)-
synthesis
via 3-(*p*-tolylsulfinyl)propionic acid, **1**, 513
Lactones, unsaturated macrocyclic
epoxidation, **7**, 361
Lactones, vinyl
ring-opening and coupling reactions
with Grignard reagents, **3**, 476
Lactonization
enantioselective, **6**, 337
Ladenburg reduction
pyridines, **8**, 595
5α-Lanosta-2,8-diene
synthesis, **8**, 935
Lanostanol, 7,11-dioxo-
acetate
Wolff–Kishner reduction, **8**, 330
Lanostanol, 11-oxo-
acetate
oxidative rearrangement, **7**, 832
Lanost-8-en-3-one
cyclopalladation–oxidation, **7**, 630
reduction, **8**, 935
Lanthanide, dichlorocyclopentadienyltris-
(tetrahydrofuran)-
enone reduction, **8**, 16
Lanthanide catalysts
Diels–Alder reactions, **2**, 667
Lanthanide complexes
Diels–Alder reaction catalysts
diastereofacial selectivity, **2**, 679
β-diketonate, chiral
Diels–Alder reactions, absolute stereochemistry, **2**, 682
Lanthanide compounds
toxicity, **1**, 252
Lanthanide homoenolates
reactions with carbonyl compounds, **2**, 446
Lanthanide metal enolates
aldol reaction, **2**, 301
structure, **2**, 301
Lanthanide oxides
dissociation energies, **1**, 252
Lanthanides
hard acids, **1**, 252
hydrometallation, **8**, 696
ionic radii, **1**, 252
Lanthanide shift reagents
carbonyl compound complexes
NMR, **1**, 294
Lanthanide triflates
amidine synthesis, **6**, 546
Lanthanide trihalides
catalysts
Friedel–Crafts reaction, **3**, 295
Lanthanoid chlorides
selective aldehyde reduction, **8**, 17
Lanthanoid complexes
hydrogenation
alkenes, **8**, 447
Lanthanoid compounds
reaction with epoxides
regioselectivity, **6**, 9
Lanthanoid ions
reduction
enones, **8**, 538
Lanthanoids
use in pinacol coupling reactions, **3**, 567
Lanthanoid trichlorides
Friedel–Crafts catalysts, **3**, 302
benzylation of benzene, **3**, 302
Lanthanum nickel hydrides
reduction
unsaturated carbonyl compounds, **8**, 551
Lantin
synthesis
Diels–Alder reaction, **2**, 699
Lasiodiplodin
synthesis
via Diels–Alder reactions, **5**, 330
Lasiodiplodin methyl ether
synthesis
via Wacker oxidation, **7**, 454
Lasubine II
synthesis
via nitrone cyclization, **4**, 1116
Laudanosine
oxidation, **3**, 685
oxidative coupling, **3**, 670
synthesis, **3**, 80
Laudanosoline
methiodide
oxidative coupling, **3**, 666
Laureacetal-C
synthesis
via intramolecular photocycloaddition, **5**, 180
Laurencin
synthesis, **3**, 126
Laurenene
synthesis, **3**, 385
via photocycloaddition, **5**, 144, 662, 663
via Wacker oxidation, **7**, 455

Laurenyne
synthesis, **6**, 752
type III ene reaction, **2**, 555
via cyclization, **1**, 591
Lavandulal
synthesis, **6**, 455
Lavandulol
cyclization, **3**, 345
synthesis, **2**, 578
ene reaction, **2**, 530
Lavendamycin
pharmacophores
synthesis, **7**, 347
synthesis
via Curtius reaction, **6**, 814
via Diels–Alder reaction, **5**, 492
Lawesson's reagent, **2**, 867
Eschenmoser coupling reaction, **2**, 867
thiocarboxylic ester synthesis, **6**, 437
Lead, allyl-
reaction with aldimines, **2**, 982
Lead, aryl-
vinyl substitutions
palladium complexes, **4**, 841
Lead acetate
Erlenmeyer azlactone synthesis, **2**, 402
Lead azide
azidation, **7**, 488
Lead carboxylates
synthesis, **7**, 719
Lead nitrate
benzylic halide oxidation, **7**, 666
Lead phenyliododiacetate
oxidative cleavage of alkenes
with trimethylsilyl azide, **7**, 588
Lead salts
decarboxylative halogenation, **7**, 724
oxidative rearrangements, **7**, 816
Lead tetraacetate
adamantane functionalization, **7**, 14
alkane oxidation, **7**, 13
allylic oxidation, **7**, 92
decarboxylative halogenation, **7**, 724
glycol cleavage, **7**, 708
mechanism, **7**, 709
Hofmann rearrangement, **6**, 796
oxidative, **6**, 802
α-hydroxylation
ketones, **7**, 152
ketone α-acetoxylation, **7**, 145
oxidation
aromatic compounds, **7**, 338
organoboranes, **7**, 602
oxidative cleavage of alkenes
with trimethylsilyl azide, **7**, 588
oxidative decarboxylation, **7**, 722
oxidative rearrangement, **7**, 827
quinones
synthesis, **7**, 352
reductive decarboxylation, **7**, 720
silyloxycyclopropane
oxidation, **2**, 445
Lead tetrabenzoate
α-hydroxylation
ketones, **7**, 167
reaction with silyl dienol ethers, **7**, 178
Lead tetrakisfluoroacetate
oxidation
aromatic compounds, **7**, 338
Lead triacetate, alk-1-enyl-
synthesis, **7**, 620
Lead triacetate, aryl-
aromatic arylation reactions, **3**, 505
Lead trifluoroacetate
alkane oxidation, **7**, 13
Lemieux–Johnson oxidation, **7**, 711
Lemieux–von Rudloff oxidation, **7**, 710
oxidative cleavage of alkenes
with permanganate and periodate, **7**, 586
Lepidine
radical addition reactions, **4**, 768
Lepidine, 2-chloro-
hydrogenation, **8**, 905
Leucarins
synthesis
Reimer–Tiemann reaction, **2**, 774
Leucine
t-butyl ester
imine anion alkylation, **6**, 726
t-butyl ester, enamines
alkylation, **3**, 36
synthesis
via reductive amination, **8**, 144
Leucine, *N*-*t*-butyloxycarbonyl-
synthesis, **6**, 816
L-Leucine, (*S*)-4-hydroxy-5-methyl-3-oxohexanoyl-
esters
synthesis, **6**, 446
Leuckart reaction
reductive alkylation of amines
ammonium formate, **8**, 84
Leukotriene A_4
synthesis, **3**, 289
synthesis of intermediate
via organocopper reagents, **1**, 131
Leukotriene B_4
synthesis, **3**, 489
synthesis of analogs
via carbocupration/1,2-addition, **1**, 131
Leukotriene B_4, 14,15-dehydro-
synthesis, **3**, 289
Leukotriene D, 5-deoxy-
precursor synthesis
via sulfoxide–sulfenate rearrangement, **5**, 890
Leukotrienes
synthesis, **3**, 289
organocopper compounds, **3**, 217
via D-arabinose, **7**, 242
via sulfur ylide reagents, **1**, 821
Levoglucosenone
Diels–Alder reaction, **5**, 350
Levulinic acid
ethyl ester
reaction with enol silyl ether, **2**, 616
Levulinic acid, 3-arylidene-
Friedel–Crafts reaction, **2**, 760
Levulinic acid esters
alcohol protection
cleavage, **6**, 658
Lewis acids
carbonyl compound complexes, **1**, 283–321
σ- *versus* π-(η^2)-bonding, **1**, 284

conformation, **1**, 285
NMR, **1**, 292
theoretical studies, **1**, 286
X-ray crystallography, **1**, 299
carbonyl compound reduction
metal hydrides, **8**, 314
Friedel–Crafts reaction
catalysts, **3**, 295
hydroalumination
alkynes, **8**, 750
promoters
Diels–Alder reactions, **5**, 339–341
reactions
structural models, **1**, 311
with organometallic compounds, **1**, 325–353
Ritter reaction
initiators, **6**, 283
transition metals, **1**, 307
Lewis superacids
catalysts
Friedel–Crafts reaction, **3**, 297
Libocedrol
synthesis
use of alkaline ferricyanide, **3**, 686
Lieben's rule
aldol reaction
aldehydes, **2**, 139
Lignan, aryltetralin-
one-pot synthesis
ene reaction, **2**, 533
Lignans
synthesis
via conjugate addition, **4**, 258
via tandem vicinal difunctionalization, **4**, 249
Ligularone
synthesis
via retro Diels–Alder reactions, **5**, 579
Lilac alcohol
synthesis
via [4 + 3] cycloaddition, **5**, 611
Limaspermine
synthesis
via cyclohexadienyl complexes, **4**, 680
Limonene
anodic oxidation, **7**, 796
hydroboration
protonolysis, **8**, 726
hydroformation
phosphite-modified rhodium catalysts, **3**, 1022
hydroformylation, **4**, 922
hydrogen transfer
carbonyl compound reduction, **8**, 320
oxiranes
rearrangement, **3**, 771
synthesis, **7**, 429
from α- and β-pinene, **3**, 708
via stereospecific Ritter reaction, **6**, 278
Vilsmeier–Haack reaction, **2**, 782
Limonene, tetramethyl-
epoxidation, **7**, 362
Linalool
cyclization, **3**, 352
microbial hydroxylation, **7**, 62
synthesis, **3**, 170
via retro Diels–Alder reaction, **5**, 555
Linalool, dehydro-
ene reaction, **5**, 15
Linaloyl oxide
synthesis, **3**, 126
Lincomycins
chlorination
displacement of hydroxy group, **6**, 205
Lincosamine
synthesis
Diels–Alder reaction, **2**, 694
Linderalactone
Cope rearrangement, **5**, 809
Lipases
acylation, **6**, 340
substrate specificity
synthetic applicability, **2**, 456
Lipoamide
immobilization, **8**, 369
reduction
isoxazoles, **8**, 645
Lipoamide A_2
iron complex
reduced form, **8**, 649
Lipoic acid
synthesis, **7**, 90
(+)-α-Lipoic acid
synthesis
via chiral acetals, **2**, 651
Lipopolysaccharides
synthesis, **6**, 57
Lipotoxins
synthesis
via metal carbene complexes, **1**, 808
Lipstatin
β-lactone, **6**, 342
Lipstatin, tetrahydro-
synthesis
chiral reaction, **2**, 652
Liquid crystal properties
S-aryl arenecarbothioates
4,4′-disubstituted, **6**, 441
Lithiation
nitrogen compounds
addition reactions, **1**, 461
Lithiodithianes
acylation, **1**, 568
Lithium
amalgam
reduction, **8**, 115
Birch reduction, **8**, 492
alkyl halides, **8**, 795
deoxygenation
epoxides, **8**, 889
in alcohol
alkyl halide reduction, **8**, 795
liquid ammonia
carbonyl compound reduction, **8**, 308
α,β-unsaturated ketone reduction, **8**, 478
methylamine
amide reduction, **8**, 294
carboxylic acid reduction, **8**, 284
reduction
ammonia, **8**, 113
carbonyl compounds, **8**, 109
enones, **8**, 524
epoxides, **8**, 880
reductive cleavage

α-alkylthio ketone, **8**, 993
α-ketals, **8**, 991
reductive dimerization
unsaturated carbonyl compounds, **8**, 532
Lithium, alkenyl-
coupling reactions
with alkenyl halides, **3**, 485
Lithium, α-alkoxy-
carbanions
epoxidation, **1**, 829
Lithium, alkoxyallyl-
alkoxyallylaluminum compounds from, **2**, 10
Lithium, alkyl-
C—P bond cleavage, **8**, 859
enantioselective addition
aldehydes, **1**, 72
enone additions, **4**, 243
primary
coupling reactions with alkenyl halides, **3**, 436
reactions with dienylcobalt complexes, **4**, 691
reaction with cyclohexanone
stereoselectivity, **1**, 79
tandem vicinal difunctionalization, **4**, 257
Lithium, allenyl-
reaction with epoxides, **3**, 264
Lithium, allyl-
configurational stability, **2**, 21
crystal structure, **1**, 18
reactions with glyceraldehyde acetonide, **2**, 29
reaction with dienes, **4**, 868
tetramethylethylenediamine complex
crystal structure, **1**, 18
Wurtz coupling, **3**, 419
Lithium, allylsulfonyl-
reaction with epoxides, **6**, 7
Lithium, allylthiophenyl-
reaction with epoxides, **6**, 7
Lithium, aryl-
alkylation, **3**, 247
coupling reactions
with alkenyl halides, **3**, 494
vinyl substitutions
palladium complexes, **4**, 841
Lithium, benzyl-
crystal structure, **1**, 11
Lithium, benzyldithiocarbamato-
alkylation, **3**, 95
Lithium, benzylphenylthio-
alkylation, **3**, 95
Lithium, benzylthiothiazolino-
alkylation, **3**, 95
Lithium, bis(methylthio)allyl-
α,β-unsaturated aldehyde synthesis, **6**, 138
Lithium, bis(phenylthio)benzyl-
synthesis, **3**, 123
Lithium, bis(phenylthio)methyl-
synthesis, **3**, 123
Lithium, 1,1-bis(seleno)alkyl-
reactions, **1**, 694
reactivity
reactions with carbonyl compounds, **1**, 672
Lithium, *t*-butoxymethyl-
synthesis, **3**, 194
Lithium, *n*-butyl-
mixed aggregate complex with *t*-butoxide
crystal structure, **1**, 10
nucleophilic addition reactions
stereoselectivity, **1**, 70
Lithium, *t*-butyl-
coupling
dihalides, **3**, 419
Lithium, 3-(*t*-butyldimethylsiloxy)allenyl-
reactions, **2**, 89
Lithium, *t*-butylethynyl-
crystal structure, **1**, 20
Lithium, *o*-(*t*-butylthio)phenyl-
crystal structure, **1**, 23
Lithium, (2-carbamoylallyl)-
Michael reactions, **4**, 121
Lithium, chloro(methyl)allyl-
reaction with aldimines, **2**, 982
Lithium, crotyl-
configurational stability, **2**, 21
crotyl organometallics from, **2**, 5
reaction with imines
regioselectivity, **2**, 988
syn–anti selectivity, **2**, 989
reaction with iminium salts, **2**, 1000
structure, **2**, 977
Lithium, 1-cyano-2,2-dimethylcyclopropyl-
crystal structure, **1**, 32
Lithium, cyclohexadienyl-
alkylation, **3**, 255
Lithium, cyclohexyl-
crystal structure, **1**, 9
Lithium, cyclopentenyl-
synthesis, **3**, 247
Lithium, (dialkoxyphosphoryl)trimethylsilylalkyl-
alkylation, **3**, 201
Lithium, dibromomethyl-
addition to esters, **1**, 874
Lithium, 1,1-dichloroallyl-
synthesis
alkylation, **3**, 202
Lithium, diethoxymethyl-
synthesis
by transmetallation, **3**, 196
Lithium, (diethoxyphosphoryl)dichloromethyl-
alkylation, **3**, 202
Lithium, 1,1-difluoroallyl-
synthesis
alkylation, **3**, 202
Lithium, 2,6-dimethoxyphenyl-
crystal structure, **1**, 23
Lithium, *o*-(dimethylaminomethyl)phenyl-
crystal structure, **1**, 25
Lithium, 2,6-dimethylaminophenyl-
crystal structure, **1**, 23, 24
Lithium, 5,5-dimethyl-2-hexenyl-
synthesis
via carbolithiation, **4**, 868
Lithium, dimethylphenylsilyl-
deoxygenation
epoxides, **8**, 886
Lithium, (dimethylphosphoryl)methyl-
alkylation, **3**, 201
Lithium, diphenylarsinomethyl-
alkylation
with epoxides, **3**, 203
synthesis
by transmetallation, **3**, 203
Lithium, diphenylcyclopropylcarbinyl-

ring opening, **4**, 872
Lithium, (diphenylphosphinoyl)alkyl-
alkylation
with epoxides, **3**, 201
Lithium, 1,1-(diseleno)alkyl-
synthesis, **3**, 87
Lithium, 1,1-(diseleno)benzyl-
synthesis, **3**, 87
Lithium, 1,1-(dithio)allyl-
alkylation, **3**, 131
synthesis, **3**, 131
Lithium, 1,1-(dithio)propargyl-
alkylation, **3**, 131
Lithium, ethyl-
crystal structure, **1**, 9
Lithium, furyl-
nucleophilic addition reactions
factors affecting stereoselectivity, **1**, 54
Lithium, 2-furyl-
coupling reactions
with alkenyl bromides, **3**, 497
Lithium, glycosyl-
synthesis
by transmetallation, **3**, 196
Lithium, 5-hexen-1-yl-
cyclization, **4**, 871
Lithium, indenyl-
tetramethylethylenediamine complex
crystal structure, **1**, 19
Lithium, mesityl-
crystal structure, **1**, 23
Lithium, [methoxyl(phenylthio)(trimethylsilyl)methyl]-
tandem vicinal difunctionalization, **4**, 259
Lithium, *o*-methoxyphenyl-
crystal structure, **1**, 23
Lithium, 1-methoxy-1-phenylselenomethyl-
reactivity
reactions with carbonyl compounds, **1**, 672
Lithium, methoxy(phenylthio)(trimethylsilyl)methyl-
Peterson alkenation, **1**, 787
Lithium, α-methoxyvinyl-
acyl anion equivalent, **1**, 544
alkylation
alkyl enol ethers preparation, **2**, 596
Lithium, methyl-
crystal structure, **1**, 9
tetramethylethylenediamine complex
crystal structure, **1**, 10
Lithium, 3-methyl-3-methoxy-1-butynyl-
conjugate additions
nontransferable ligand, **4**, 176
Lithium, 1-methylseleno-2,2-dimethylpropyl-
reaction with heptanal
stereochemistry, **1**, 677
Lithium, methylselenomethyl-
alkylation, **3**, 90
Lithium, methylthiomethyl-
epoxidation
2-cyclohexenone, **1**, 826
Lithium, 1-octen-2-yl-
synthesis, **6**, 781
Lithium, 1-pentynyl-
conjugate additions
nontransferable ligand, **4**, 176
Lithium, perfluoroalkyl-
reactions with imines
Lewis acid pretreatment, **1**, 350
Lithium, phenyl-
addition reactions
alkenes, palladium(II) catalysis, **4**, 572
crystal structure, **1**, 22
Lithium, phenylethynyl-
crystal structure, **1**, 20
Lithium, 1-phenyl-2-methylseleno-2-oct-5-enyl-
cyclization, **1**, 663
Lithium, 1-phenylseleno-1-hexyl-
alkylation, **3**, 90
Lithium, phenylselenomethyl-
alkylation, **3**, 90
synthesis, **1**, 666
Lithium, 2-phenylseleno-2-propyl-
stability, **1**, 632
synthesis, **1**, 634
Lithium, 1-phenylseleno-1-thioalkyl-
reactivity
reactions with carbonyl compounds, **1**, 672
Lithium, 1-phenylselenovinyl-
reactivity
reactions with carbonyl compounds, **1**, 672
Lithium, α-(phenylsulfonyl)allyl-
X-ray structure, **1**, 528
Lithium, α-(phenylsulfonyl)benzyl-
X-ray structure, **1**, 528
Lithium, (phenylthio)methyl-
homologation
primary halides, **6**, 139
ketones
methylenation, **6**, 139
synthesis
via thioanisole, **1**, 506
Lithium, 1-phosphonato-1-phenylselenoalkyl-
reactivity
reactions with carbonyl compounds, **1**, 672
Lithium, α-selenoalkyl-
acyl anion equivalents
synthesis, **3**, 121
alkylation, **3**, 88
allylation, **3**, 91
nucleophilicity
reactions with carbonyl compounds, **1**, 672
reactions, **1**, 694; **3**, 88
reactions with carbonyl compounds
reactivity, **1**, 672
stereochemistry, **1**, 677
synthesis, **1**, 655; **3**, 87
via selenium–lithium exchange, **1**, 631
Lithium, α-selenoallenyl-
synthesis, **3**, 87
Lithium, α-selenoallyl-
alkylation, **3**, 95
ambident reactivity, **1**, 678
reactivity
reactions with carbonyl compounds, **1**, 672
synthesis, **3**, 87
Lithium, α-selenobenzyl-
alkylation, **3**, 94, 95
reactions with alkenes, **1**, 664
reactivity
reactions with carbonyl compounds, **1**, 672
synthesis, **3**, 87
Lithium, 1-selenocyclobutyl-
alkylation, **3**, 90

Lithium, 1-selenocyclopropyl-
alkylation, **3**, 90
Lithium, α-selenocyclopropyl-
reactivity
reactions with carbonyl compounds, **1**, 672
Lithium, selenomethyl-
synthesis, **1**, 631
Lithium, α-selenopropargyl-
alkylation, **3**, 104
synthesis, **3**, 87
Lithium, 1-seleno-1-silylalkyl-
reactivity
reactions with carbonyl compounds, **1**, 672
Lithium, α-selenovinyl-
synthesis, **3**, 87
Lithium, α-selenoxyalkyl-
reactions, **1**, 694
Lithium, 2,3,5,6-tetrakis(dimethylaminomethyl)phenyl-
crystal structure, **1**, 25
Lithium, (tetramethylcyclopropyl)methyl-
crystal structure, **1**, 9
Lithium, 2-thienyl-
coupling reactions
with alkenyl bromides, **3**, 497
Lithium, α-thioalkyl-
acyl anion equivalents
synthesis, **3**, 121
alkylation, **3**, 88
reactions, **3**, 88
synthesis, **3**, 87
Lithium, α-thioallyl-
alkylation, **3**, 95
reaction with allyl halides, **3**, 99
reaction with epoxides, **3**, 100
Lithium, α-thiobenzyl-
alkylation, **3**, 94
Lithium, α-thiopropargyl-
alkylation, **3**, 104
Lithium, trialkylsilyldichloromethyl-
alkylation, **3**, 200
Lithium, trialkylstannyl-
tandem vicinal difunctionalization, **4**, 257
Lithium, trialkylstannylmethyl-
reactions with carbonyl compounds
methylenation, **1**, 754
Lithium, triarylstannylmethyl-
reactions with carbonyl compounds
methylenation, **1**, 755
Lithium, 3-triethylsilyloxypentadienyl-
alkylation, **3**, 196
Lithium, trimethylsilyl-
tandem vicinal difunctionalization, **4**, 257
Lithium, trimethylsilylallyl-
alkylation
regioselectivity, **3**, 200
reaction with dichloroethylaluminum, **2**, 10
Lithium, trimethylsilyl(phenylthio)methyl-
alkylation, **3**, 137
Lithium, triphenylmethyl-
ketone deprotonation, **2**, 183
Lithium, tris(phenylthio)methyl-
reactions with enones, **6**, 140
reaction with carvone, **6**, 141
tandem vicinal difunctionalization, **4**, 259
Lithium, tri-*n*-stannyl-
sequential Michael ring closure, **4**, 262
Lithium, tris(trimethylsilyl)methyl-
alkylation, **3**, 200
tetrahydrofuran complex
crystal structure, **1**, 16
Lithium, vinyl-
alkylation, **3**, 247
intramolecular carbolithiation, **4**, 872
oxidation
with silyl peroxides, **2**, 603
reaction with alkyl halides, **3**, 247
Lithium acetylide
ethylenediamine complex
reaction with epoxides, **6**, 7
Lithium alkenyltrialkylalanate
conjugate additions
α,β-enones, **4**, 142
Lithium alkoxytriaryloxyaluminates
reactions with diaryl phosphorochloridates
hydroxy group activation, **6**, 18
Lithium alkylcuprates
reduction
tosylates, **8**, 813
Lithium alkylcyano(2-thienyl)cuprate
preparation, **3**, 213
Lithium alkynides
alkylation
alkyl halides, **3**, 272
reaction with methyl triflate, **3**, 281
Lithium alkynylboronates
conjugate additions
alkylideneacetoacetates, **4**, 148
Lithium alkynylcuprates
reaction with haloallenes, **3**, 274
Lithium aluminum deuteride
reduction
epoxides, **8**, 872
Lithium aluminum hydride
alcohol modifiers
reduction, **8**, 161
alkyl halide reduction
mechanism, **8**, 802
chiral alkoxy derivatives
synthesis, **8**, 159
chirally modified
reduction, **8**, 160
copper chloride
desulfurization, **8**, 840
C—P bond cleavage, **8**, 863
cyclic ketone reduction
stereochemistry, **8**, 5
demercurations, **8**, 851
derivatives
benzylic halide reduction, **8**, 967
hydroalumination, **8**, 736
metal salt systems
alkyl halide reduction, **8**, 803
reaction with ethylene, **8**, 735
reduction
acetals, **8**, 213
acyl halides, **8**, 241, 263
aliphatic nitro compounds, **8**, 374
alkyl halides, **8**, 802
amides, **8**, 269
benzylic halides, **8**, 965
carbonyl compounds, **8**, 2, 5, 18, 313
carboxylic acids, **8**, 237

epoxides, **8**, 872
esters, **8**, 245, 267
hydrazones, **8**, 345
imines, **8**, 26, 36
lactones, **8**, 247, 268
nitriles, **8**, 274
phosphonium salts, **8**, 860
pyridines, **8**, 579, 583
tosylates, **8**, 812
unsaturated carbonyl compounds, **8**, 536, 542, 545
unsaturated hydrocarbons, **8**, 483
vicinal dibromides, **8**, 797
reduction kinetics, **8**, 2
selective ketone reduction, **8**, 18
transition metal halides
unsaturated hydrocarbon reduction, **8**, 485
Lithium aluminum hydride, bipyridylnickel-
desulfurizations, **8**, 840
Lithium aluminum hydride, dicyclopentadienylnickel-
desulfurizations, **8**, 840
Lithium amides
chiral catalysts
nucleophilic addition reactions, **1**, 72
reaction with carbon monoxide
mechanism, **3**, 1017
Lithium benzyl oxide
acyloxazolidinones
cleavage, **2**, 438
Lithium bis(benzyldimethyl)silylamide
aldol reaction
stereoselectivity, **2**, 59
Lithium bis(phenyldimethylsilyl)cuprate
introduction of hydroxy groups, **7**, 646
Lithium bis(trimethylsilyl)amide
Darzens glycidic ester condensation, **2**, 427
Lithium borohydride
ethyl acetate system
hydroboration, **8**, 709
reduction, **8**, 880
1,3-diketones, **8**, 13
epoxides, **8**, 875
esters, **8**, 244
Lewis acids, esters, **8**, 244
unsaturated carbonyl compounds, **8**, 536
Lithium bromide
acetone complex
crystal structure, **1**, 299
reaction with ethers, **6**, 210
Lithium bronze
reduction
enones, **8**, 526
Lithium *t*-butoxyaluminum hydride
reduction kinetics, **8**, 2
Lithium *n*-butylborohydride
synthesis, **8**, 538
Lithium *n*-butyldiisobutylaluminum hydride
reduction
amides, **8**, 272
unsaturated carbonyl compounds, **8**, 544
Lithium *t*-butyl-*t*-octylamide
enolate formation
hindered base, **2**, 182
Lithium cations
acrolein complexes
structure, **1**, 289
Lithium chlorate
biomimetic reduction
allylic compounds, **8**, 977
Lithium chloride
reaction with esters, **6**, 206
Lithium cobalt phthalocyanine
TcBoc removal, **6**, 638
Lithium compounds
aldol reactions
comparison with boron compounds, **2**, 239
amination
alkyl or aryl, **6**, 118
use in intermolecular pinacol coupling reactions
aliphatic carbonyl compounds, **3**, 570
Lithium dialkenylcuprates
acylation of, **1**, 428
Lithium dialkylamides
ester enolization
Claisen rearrangement, **5**, 828
imine anion synthesis, **6**, 719
Lithium dialkylcuprates
acylation of, **1**, 428
conjugate additions
enolate synthesis, **3**, 8
Lithium dialkylcyanocuprates
structure, **3**, 213
Lithium diallenylcuprates
alkylation, **3**, 256
Lithium diallylcuprates
reactions with carbonyl compounds
formation of 1,2-adducts, **1**, 113
Lithium di-*t*-butoxyaluminum hydride
reduction
pyridines, **8**, 580
Lithium 9,9-di-*n*-butyl-9-borabicyclo[3.3.1]nonanate
reduction
epoxides, **8**, 876
Lithium dibutylcuprate
reactions with ketones, **1**, 116
Lithium diethoxyaluminum hydride
reduction
amides, **8**, 271
Lithium diisopropylamide
aldehyde reduction, **8**, 88
Claisen condensation, **2**, 182
deprotonation of *N*-allylamide
γ-selectivity, **2**, 61
Lithium dimesitylborohydride
reduction
cyclohexanones, **8**, 14
Lithium dimethylcuprate
mixture with water
structure, **3**, 212
reactions with aldehydes, **1**, 108
reactions with α,β-dialkoxy aldehydes, **1**, 109
reactions with dienyliron complexes, **4**, 673
reactions with epoxides
regioselectivity, **6**, 5
reactions with ketones, **1**, 116
stability, **3**, 211
synthesis, **3**, 208
tetrahydrofuran solution
structure, **3**, 211
Lithium dimethylcyanocuprate
reaction with epoxides
regioselectivity, **6**, 5
Lithium diorganocopper compounds

coupling, **3**, 419
Lithium diphenylcuprate
reaction with alkyl bromide, **3**, 248
reaction with allylic acetate, **3**, 257
Lithium diphenylmethane
lithium (12-crown-4) complex
crystal structure, **1**, 11
Lithium diphenylphosphide
reduction
epoxides, **8**, 885
Lithium diphenylthiophosphides
reactions with π-allylpalladium complexes
regioselectivity, **4**, 642
stereochemistry, **4**, 625
Lithium divinylcuprates
coupling reactions
with enol triflates, **3**, 487
vinyl halide coupling, **3**, 482
Lithium ene reactions
intramolecular, **5**, 37–46
Lithium enolates
aldol reactions
diastereofacial selectivity, **2**, 217
alkylation, **3**, 2
Claisen rearrangement, **5**, 847
α-methyl substituents
axial alkylation, **3**, 14
synthesis, **2**, 100
thiol carboxylic esters
acylation, **6**, 446
Lithium halides
catalysts
epoxide rearrangement, **3**, 760, 763
Lithium halocarbenoids
alkylation, **3**, 202
Lithium halocarbenoids, cyclopropylidene-
synthesis by halogen–lithium exchange
alkylation, **3**, 202
Lithium–halogen exchange
dihalocyclopropanes, **4**, 1007–1009
Lithium hexamethyldisilazane
aldol reaction
stereoselectivity, **2**, 192
enolate formation, **2**, 182
Lithium hexamethyldisilazide
crystal structure, **1**, 6
Lithium hexamethyldisilylamide
ketone enolates
synthesis, **3**, 3
Lithium hydride
reaction with ethylene, **8**, 734
reduction
acyl halides, **8**, 262
carbonyl compounds, **8**, 22
Lithium iodide
β-alkoxy ketone reduction, **8**, 9
catalyst
aldol reaction, **2**, 146
iodination
esters, **6**, 215
Lithium iodophenylcuprate
coupling with allylic alcohols, **3**, 259
Lithium isohexylcyanocuprate
reaction with epoxides, **6**, 9
Lithium isopropoxide
hydride donor
reduction of steroidal ketones, **8**, 89
Lithium *N*-isopropylcyclohexylamide
ester enolates
generation, **2**, 182
Lithium 3-lithiopropoxide
acylation, **1**, 404
Lithium/magnesium acetylide
crystal structure, **1**, 39
Lithium methoxide
reaction with formaldehyde
transition state, **8**, 88
Lithium methoxyaluminum hydride
reduction kinetics, **8**, 2
Lithium naphthalide
reduction, **3**, 263
Lithium *t*-octyl-*t*-butylamide
ketone enolate synthesis, **3**, 3
Lithium organo(fluorosilyl)amides
crystal structure, **1**, 38
Lithium pentamethyltricuprate
structure, **3**, 213
Lithium perchlorate
catalyst
epoxide ring opening, **3**, 760, 761
Lithium phenylethynolate
cycloaddition with carbonyls
2-oxetanone synthesis, **5**, 116
Lithium phenylthio(2-vinylcyclopropyl)cuprate
conjugate additions
β-iodoenone, **4**, 173
Lithium salts
catalysts
epoxide ring opening, **3**, 760
formaldehyde complexes
theoretical studies, **1**, 286
Lithium tetraethylaluminate
synthesis, **8**, 735
Lithium tetrakis(dihydro-*N*-pyridyl)aluminate
reduction
unsaturated carbonyl compounds, **8**, 536
Lithium tetramethyldiphenyldisilylamide
ketone enolates
synthesis, **3**, 3
Lithium 2,2,6,6-tetramethylpiperidide
enolate formation
regioselectivity, **2**, 182
reaction with allyl selenide, **2**, 76
Lithium thiocyanate
catalysts
epoxide ring opening, **3**, 767
Lithium trialkylaluminum hydride
hydroalumination, **8**, 736
Lithium tri-*n*-alkylborohydride
hydroboration, **8**, 718
Lithium tri-*t*-butoxyaluminum hydride
reduction
acyl halides, **8**, 263
carboxylic acids, **8**, 260, 261
esters, **8**, 267
lactones, **8**, 268
nitriles, **8**, 274
pyridines, **8**, 580
unsaturated carbonyl compounds, **8**, 542–544
Lithium tri-*t*-butylberyllate
crystal structure, **1**, 13
Lithium tri-*s*-butylborohydride

reduction
cyclic ketones, **8**, 15
cyclohexanones, **8**, 14
imines, **8**, 36
keto sulfides, **8**, 12
unsaturated carbonyl compounds, **8**, 536
Lithium triethoxyaluminum hydride
reduction
amides, **8**, 271
nitriles, **8**, 274
Lithium triethylborohydride (Super Hydride)
reduction
alkyl halides, **8**, 804, 805
allylic leaving group, **8**, 960
epoxides, **8**, 875
imines, **8**, 36
ketones, **8**, 10
nitroalkenes, **8**, 377
tosylates, **8**, 813
unsaturated carbonyl compounds, **8**, 536
selective ketone reduction, **8**, 18
Lithium trimethoxyaluminum hydride
cyclic ketone reduction
stereochemistry, **8**, 5
reduction
acyl halides, **8**, 263
carboxylic acids, **8**, 260
nitriles, **8**, 274
unsaturated carbonyl compounds, **8**, 542
Lithium trimethyldicuprate
structure, **3**, 213
Lithium tri-*t*-pentyloxyaluminum hydride
reduction
acyl halides, **8**, 263
Lithium triphenylcuprate
structure, **3**, 213
Lithium triphenylmethane
crystal structure, **1**, 11
Lithium tris(*t*-butylthio)aluminum hydride
reduction
unsaturated carbonyl compounds, **8**, 543
Lithium tris[(3-ethyl-3-pentyl)oxy]aluminum hydride
reduction
aldehydes, **8**, 17
Lithium trisiamylborohydride
reduction
cyclohexanones, **8**, 14
Lithium tris(*trans*-methylcyclopentyl)borohydride
reduction
cyclohexanones, **8**, 14
Lithocholic acid
microbial hydroxylation, **7**, 73
Liver alcohol dehydrogenase
metal complex
models, **8**, 82, 97
Living polymers
ring opening metathesis polymerization, **5**, 1120
Locopodine
synthesis
via conjugate addition, **4**, 240
Locorenine
related structure
synthesis *via* azomethine ylide cyclization, **4**, 1140
Loganin, deoxy-
synthesis
Knoevenagel reaction, **2**, 372
via Diels–Alder reaction, **5**, 468
Loganin, 1-*O*-methyl-
synthesis
via Diels–Alder reactions, **5**, 363
Loganin aglycone
synthesis, **7**, 301
Loganins
aglucone 6-acetate
synthesis *via* photoisomerization, **5**, 232
biosynthesis, **5**, 468
synthesis, **3**, 599
via [3 + 2] cycloaddition reactions, **5**, 310
via photochemical cycloaddition, **5**, 129
Loline
synthesis
via transannular cyclization, **4**, 398
Longicamphor
reduction
dissolving metals, **8**, 121
Longifolene
hydroboration, **8**, 721
synthesis, **3**, 599
pinacol rearrangement, **3**, 729
via diazoalkene cyclization, **4**, 1154
Wagner–Meerwein rearrangement, **3**, 713
Looplure
synthesis, **3**, 644
Lophotoxin
synthesis, **3**, 497
Loroxanthin
synthesis, **6**, 782
Lossen reaction, **3**, 908
Lossen rearrangement, **6**, 821
hydroxamic acids, **6**, 798
stereoselectivity, **6**, 798
Lubimin, oxy-
synthesis, **7**, 178
Luciduline
synthesis, **8**, 945
via Cope rearrangement, **5**, 814
via Diels–Alder reaction, **5**, 351
Luciferin aldehyde
synthesis
epoxide ring opening, **3**, 743
Lukes reduction
formic acid
pyridinium salts, **8**, 590
Lumiflavin, 4a-hydroperoxy-
oxidation
sulfides, **7**, 763
Lumisantonin
photochemistry, **5**, 730
Lumisterol
synthesis
via photochemical ring closure, **5**, 700
Lupanone oxide
cyclopalladation–oxidation, **7**, 630
Lupeol
reduction–alkylation, **8**, 527
Lupinine
synthesis
Eschenmoser coupling reaction, **2**, 881
Lycodine
synthesis
Mannich reaction, **2**, 1013
Lycodoline

synthesis
Mannich reaction, **2**, 1012
Lycopodine
Mannich base, **2**, 894
synthesis, **2**, 159
Mannich reaction, **2**, 1013
use of cuprates derived from hydrazones, **2**, 518
via heteronucleophile addition, **4**, 31
Lycoranes
synthesis
via arynes, **4**, 502, 503
Lycorenine alkaloids
synthesis, **1**, 568
Lycoricidine
synthesis
via aryllithium addition to α-nitroalkane, **4**, 78
Lycorine alkaloids
synthesis, **7**, 336
Lysergic acid
synthesis, **6**, 757
via arynes, **4**, 501
via Diels–Alder reaction, **5**, 414
via spirolactonization, **6**, 357
Lysergic acid, 2-methyl-
synthesis, **3**, 126
Lythrancepine alkaloids
synthesis
Eschenmoser coupling reaction, **2**, 881

M

Macbecin 1
synthesis
(*E*)- and (*Z*)-selectivity, **1**, 764
Macbecins
synthesis of segment
via Wittig or $CrCl_2$ reaction, **1**, 808
McMurry reaction, **3**, 583
intermolecular, **3**, 585
intramolecular, **3**, 588
Macrocyclic compounds
synthesis
alkene metathesis, **5**, 1118–1120
Eschenmoser coupling reaction, **2**, 890
Macrocyclic ethers
Wittig ring contractions, **3**, 1010
Macrocyclization
radical reactions, **4**, 791
Macrolactams
synthesis
C—H insertion reactions, **3**, 1057
Macrolactonization
enzymatic, **6**, 376
δ-lactone synthesis, **6**, 368
Macrolide antibiotics
synthesis, **2**, 248; **7**, 57
via cycloheptadienyliron complexes, **4**, 686
via (*Z*)-selective Wittig reaction, **1**, 763
via thiol esters, **6**, 438
Macrolides
synthesis
via thallium(I) thiolates, **6**, 440
synthesis, **3**, 286
C—H insertion reactions, **3**, 1058
Macrolides, oximino-
synthesis, **7**, 507
Macromolecules
synthesis
Ugi reaction, **2**, 1104
Macrophyllate, ethyl-
synthesis
use of alkaline potassium ferricyanide, **3**, 665
Maesopsin
ring scission, **3**, 831
Maesopsin, 4,4′,6-tri-*O*-methyl-
irradiation, **3**, 831
Magic acid
catalyst
Friedel–Crafts reaction, **3**, 297
Magnesium
deoxygenation
epoxides, **8**, 889
desulfurization
sulfones, **8**, 843
graphite-suspended
use in intermolecular pinacol coupling reactions, **3**, 570
reduction
enones, **8**, 524
nitro compounds, **8**, 365
reductive dimerization
unsaturated carbonyl compounds, **8**, 532
Magnesium, 1-alkenyl-
allylation, **5**, 32
Magnesium, 2-alkenyl-
precursors
synthesis for magnesium-ene reaction, **5**, 37
Magnesium, alkylbromo-
boron trifluoride complex
reactions with acetals, **1**, 346
Magnesium, alkynylhalo-
cross-coupling reactions
with vinyl iodides, **3**, 527
Magnesium, allenylbromo-
synthesis, **2**, 81
Magnesium, allylbromo-
reaction with aldoxime ethers, **2**, 995
reaction with *N*-diphenylmethyleneaniline, **2**, 976
reaction with *N*-methyl-4-*t*-butylcyclohexylamine, **2**, 983
reaction with sulfenimine, **2**, 998
Magnesium, allylchloro-
crystal structure, **1**, 18
Magnesium, aryl-
vinyl substitutions
palladium complexes, **4**, 841
Magnesium, bis(2,4-dimethyl-2,4-pentadienyl)-
crystal structure, **1**, 18
Magnesium, bis(indenyl)-
tetramethylethylenediamine complex
crystal structure, **1**, 19
Magnesium, bis(phenylethynyl)-
tetramethylethylenediamine complex
crystal structure, **1**, 21
Magnesium, bromocrotyl-
crotyl organometallics from, **2**, 5
Magnesium, bromodecyl-
nucleophilic addition reactions
acrolein dimer, **1**, 52
Magnesium, bromodiisopropylamino-
Claisen condensation, **2**, 182
Magnesium, bromoethynyl-
synthesis, **3**, 271
Magnesium, bromomethyl-
reaction with ethyl acetate, **1**, 398
tetrahydrofuran solvate
crystal structure, **1**, 13
Magnesium, bromophenyl-
diethyl etherate
crystal structure, **1**, 25
Magnesium, bromo(α-silylvinyl)-
alkylation, **3**, 244
Magnesium, bromo(2-thienyl)-
reaction with vinyloxirane, **3**, 265
Magnesium, bromo-2,4,6-trimethylphenoxy-
catalyst
aldol reaction, **2**, 137
Magnesium, bromovinyl-
alkylation, **3**, 243
Magnesium, 1,2-butadienylhalo-
reaction with aldehydes, **2**, 91
Magnesium, chloro(diisopropoxymethyl)silylmethyl-
hydroxymethylation with, **3**, 200; **7**, 647
Magnesium, chloroethyl-
crystal structure, **1**, 13

Magnesium, chloro(phenyldimethylsilyl)methyl-
  Peterson reaction, **1**, 737
Magnesium, chloroprenyl-
  alkylation
    copper catalysis, **3**, 215
Magnesium, crotyl-
  reaction with imines
    regioselectivity, **2**, 988
    syn–anti selectivity, **2**, 989
  reaction with iminium salts, **2**, 1000
  structure, **2**, 977
Magnesium, 15-crown-4-xylylchloro-
  crystal structure, **1**, 26
Magnesium, dialkyl-
  crystal structure, **1**, 13
  nucleophilic addition reactions
    stereoselectivity, **1**, 72
Magnesium, diallyl-
  carbomagnesiation
    allylic alcohols, **4**, 877
Magnesium, diethyl-
  18-crown-6 complex
    crystal structure, **1**, 15
  2,1,1-cryptand complex
    crystal structure, **1**, 15
Magnesium, dimethyl-
  crystal structure, **1**, 16
Magnesium, dineopentyl
  2,1,1-cryptand complex
    crystal structure, **1**, 15
Magnesium, diphenyl-
  tetramethylethylenediamine complex
    crystal structure, **1**, 25
Magnesium, ethyl-
  diethyl ether solvate
    crystal structure, **1**, 13
Magnesium, ethyl-3-(*N*-cyclohexyl-*N*-methylamino)propyl-
  crystal structure, **1**, 14
Magnesium, ethyl-3-(*N,N*-dimethylamino)propyl-
  crystal structure, **1**, 14
Magnesium, ethynylidenebis(bromo-
  synthesis, **3**, 271
Magnesium, pentamethylene-
  crystal structure, **1**, 16
Magnesium, propargyl-
  reactions with aldimines, **2**, 992
Magnesium alkoxide
  phosphorylation, **6**, 603
Magnesium amides
  reactions with π-allylpalladium complexes, **4**, 598
Magnesium amides, halo-
  oxirane ring-opening, **6**, 91
Magnesium bromide
  catalyst
    allylstannane reaction with carbonyl compounds, **2**, 573
    Diels–Alder reaction, **2**, 667
    Friedel–Crafts reaction, **2**, 737
  Diels–Alder reaction catalysts
    diastereofacial selectivity, **2**, 679
  Tebbe reaction, **1**, 746
Magnesium bromide, alkenyl-
  allylzincation, **4**, 880
  reaction with epoxides, **6**, 5
Magnesium bromide, allyl-
  reaction with homopropargylic alcohols, **4**, 879
Magnesium bromide, isobornyloxy-
  hydride donor
    use in chiral syntheses, **8**, 89
Magnesium bromide, isobutyl-
  reduction
    carboxylic acids, **8**, 284
Magnesium bromide, 2,7-octadienyl-
  cyclization
    magnesium-ene reaction, **5**, 38
Magnesium carbonate, methyl-
  ketone carboxylation
    Stile's reagent, **2**, 841
Magnesium chloride
  sodium cyanoborohydride
    reductive amination, **8**, 54
Magnesium chloride, (2-alkenyl)allyl-
  ene reactions, **5**, 43
Magnesium cyclopropanolate
  cycloheptanone synthesis, **2**, 448
Magnesium ene reactions, **5**, 30
  intramolecular, **5**, 37–46, 59
  ring size, **5**, 60
Magnesium enolates
  aldol reactions
    diastereofacial selectivity, **2**, 217
  Claisen rearrangement, **5**, 847
  synthesis, **2**, 110
    Reformatsky reaction, **2**, 186
  thiol carboxylic esters
    acylation, **6**, 446
Magnesium ester enolates
  reactions with nitriles
    Blaise reaction, **2**, 298
Magnesium halides
  epoxide ring opening, **3**, 754
Magnesium halides, allyl-
  carbomagnesiation, **4**, 874
Magnesium hydride
  reduction
    cyclic carbonyl compounds, **8**, 14
Magnesium monoperoxyphthalate
  Baeyer–Villiger reaction, **7**, 674
  epoxidizing agent, **7**, 374
Magnesium oxide
  catalyst
    Knoevenagel reaction, **2**, 345
Magnesium perchlorate
  catalyst
    Friedel–Crafts reaction, **2**, 737
Magnocurarine methiodide
  model reaction
    dimerization, **3**, 687
Mahubenolides
  synthesis, **6**, 784
Makomakine
  synthesis
    use of ammonium ylides, **3**, 955
    via Ritter reaction, mercuration, **6**, 284
Maleates, iodo-
  dimerization, **3**, 482
Maleic acid
  hydride transfer
    with 1,4-dihydropyridines, **8**, 93
Maleic acid bis(dimethylamide)
  dications, **6**, 501

Maleic acid dinitrile, 1,2-diamino-
reactions with amines, **6**, 517
Maleic anhydride
alkylated
synthesis, **7**, 930
benzene irradiation
fulvene trap, **5**, 646
[3 + 2] cycloaddition reactions
with η^1-allylpalladium complexes, **5**, 275
Diels–Alder reactions
Lewis acid promoted, **5**, 339
EDA complex
with hexamethylbenzene, **7**, 856
ene reactions
intermolecular, **5**, 2
Maleonitrile
synthesis
via 1,2-diiodoethylene, **6**, 231
Malic acid
diethyl ester
alkylation, **3**, 44
Malonamides
Knoevenagel reaction, **2**, 357
Malonate, diethyl
reaction with π-allylpalladium complexes, **4**, 590
Malonate, 5-methyl (5*R*)-methoxycarbonyl-(3*E*)-decenyl-
cyclization
palladium catalysis, **4**, 650
Malonates
Michael addition, **4**, 3
sulfenylation, **7**, 125
Malonates, acyl-
reduction
sodium borohydride, **8**, 277
Malonates, acylamino-
synthesis, **1**, 373
Malonates, alkylidene-
addition reactions
with organomagnesium compounds, **4**, 89
with organozinc compounds, **4**, 95
Malonates, isopropylidenemethylene-
addition reaction
with organomagnesium compounds, **4**, 89
Malonic acid
diethyl ester
Claisen condensation, **2**, 801
esters
Knoevenagel reaction, **2**, 354
Knoevenagel reaction, **2**, 356
Malonic acid, alkyl-
synthesis
via disubstituted organopotassium compounds, **7**, 3
Malonic acid, benzylidene-
dimethyl ester
[3 + 2] cycloaddition reactions, **5**, 302
Malonic acid, (ω-bromoalkyl)-
diethyl ester
intramolecular alkylation, **3**, 55
Malonic acid, methylene-
diesters
synthesis *via* retro Diels–Alder reaction, **5**, 553
dimenthyl ester
Diels–Alder reactions, **5**, 356
dimethyl ester
cycloaddition reactions, **5**, 272
Malonic acid, oxo-
dialkyl esters
ene reaction, **2**, 538
diethyl ester
ene reaction, **2**, 538
Malonic acid, thioxo-
diethyl ester
Diels–Alder reactions, **5**, 436
Malonic acid dibromide
synthesis
via oxalyl bromide, **6**, 308
Malonic esters, acylimino-
reactions with organometallic compounds, **1**, 373
Malonodialdehyde
Knoevenagel reaction
active methylene compound, **2**, 358
Malonodiamides
Knoevenagel reaction, **2**, 357
Malonodinitrile
Knoevenagel reaction, **2**, 358
ylidene
Knoevenagel reaction, **2**, 359
Malonodinitrile, 2-chlorobenzylidene-
synthesis
Knoevenagel reaction, **2**, 385
Malononitrile
Vilsmeier–Haack reaction, **2**, 789
Malononitrile, alkylidene-
tandem vicinal difunctionalization, **4**, 251
Malononitrile, benzylidene-
cycloaddition reactions, **5**, 273
Malyngolide
synthesis
via chiral auxiliary, **1**, 65
via conjugate addition to α,β-unsaturated carboxylic amides, **4**, 202
Mandelic acid
boron enolate
diastereofacial preference, **2**, 232
homochiral
from alkenyloxyboranes, **2**, 249
menthyl ester
synthesis, **1**, 223
Mandelic acid, hexahydro-
synthesis
ketone oxalylation, **2**, 838
Manganacycles
synthesis
via carbomanganation, **4**, 906
Manganese, alkyl-
deoxygenation
epoxides, **8**, 889
reactions with carbonyl compounds
Lewis acid promotion, **1**, 331
Manganese, alkylpentacarbonyl-
reaction with alkynes, **4**, 905
Manganese, arenetricarbonyl-
addition–oxidation reactions, **4**, 542
nucleophilic reactions, **4**, 689
synthesis, **4**, 520
Manganese, butenetetracarbonyl-
anion
synthesis, **4**, 703
Manganese, chloro(tetraphenylporphyrin)-
alkane oxidation, **7**, 11
Manganese, dicarbonyldienylnitrosyl-

synthesis, **4**, 689
Manganese, tricarbonylcycloheptadienyl-
synthesis, **4**, 689
Manganese, tricarbonylcyclohexadiene-
anion
reaction with methyl iodide, **4**, 704
synthesis, **4**, 702
Manganese, tricarbonyl(η^4-diene)-
anions
reactions with carbon electrophiles, **4**, 702–705
Manganese, tricarbonyl(halobenzene)-
nucleophilic substitution, **4**, 531
Manganese, tricarbonyl(1-methylbutadiene)-
anion
synthesis, **4**, 704
Manganese, tricarbonyl(η^4-polyene)-
anion
synthesis, **4**, 703
Manganese, tricarbonyl(η^6-(6-substituted)cyclohexadienyl)-
addition–oxidation reactions, **4**, 542
Manganese acetate
radical addition reactions, **4**, 763
radical cyclizations
nonchain methods, **4**, 806
reaction with alkenes, **7**, 532
Manganese azide
1,2-diazides from alkenes and, **7**, 487
Manganese chloride
lithium aluminum hydride
unsaturated hydrocarbon reduction, **8**, 485
Manganese complexes
allylic oxidation, **7**, 95
Manganese complexes, dienyl-
synthesis, **4**, 689–691
Manganese compounds, crotyl-
type III
reactions with aldehydes, **2**, 24
Manganese dioxide
glycol cleavage, **7**, 708
oxidation
p-aminophenol, **7**, 349
diols, **7**, 318
primary alcohols, **7**, 306
primary arylamines, **7**, 738
secondary alcohols, **7**, 324
quinone synthesis, **7**, 142, 350, 355
Manganese enolates
synthesis and reaction, **2**, 127
Manganese triacetate
allylic oxidation, **7**, 92
α′-hydroxylation
enones, **7**, 174
α-oxidation
enones, **7**, 154
Manicone
enantioselective synthesis
use of α-sulfinylhydrazones, **2**, 516
Mannich bases
addition reactions
acyl anions, **1**, 547
deamination, **2**, 897, 933
description, **2**, 894
Mannich cyclization
molecular rearrangements, **2**, 1040
Mannich reaction
Baldwin's rules, **2**, 1008
bimolecular aliphatic, **2**, 893–951
bimolecular aromatic, **2**, 953–973
classical, **2**, 893
intramolecular, **2**, 1007–1044
limitations, **2**, 896
mechanism, **2**, 895, 954
regiochemistry, **2**, 896
reviews, **2**, 894
scope, **2**, 896
steric factors, **2**, 896
titanium tetrachloride mediated, **2**, 897
Ugi reaction and, **2**, 1090
with preformed iminium salts, **2**, 898
Mannitol
chiral sulfur methylide, **1**, 825
D-Mannofuranose, 2,3:5,6-di-*O*-isopropylidene-
transfer hydrogenation, **8**, 552
β-D-Mannofuranosides
synthesis, **6**, 56
Mannonojimycin, deoxy-
synthesis
FDP aldolase, **2**, 463
via aminomercuration–oxidation, **7**, 638
D-Manno-2-octulosonate, 3-deoxy-
lipopolysaccharides, **6**, 57
α-Mannopyranosides
synthesis, **6**, 39
β-Mannopyranosides
synthesis, **6**, 39, 43
D-Mannose, 2,3,4,6-tetra-*O*-benzyl-
glycoside synthesis, **6**, 57
α-D-Mannoside
reduction, **8**, 226
Manoalide
synthesis
carbonyl group protection, **6**, 677
Marasmane
biosynthesis, **3**, 404
Maritimol
synthesis, **3**, 717
Marmine
synthesis, **7**, 406
Matrine
synthesis, **6**, 746
Matsutake alcohol
synthesis
via retro Diels–Alder reaction, **5**, 554
Maysine
synthesis, **7**, 57
Maytansine
precursor
synthesis *via* nitrile oxide cyclization, **4**, 1132
synthesis, **3**, 126; **7**, 380
Mazur oxidation, **7**, 842
McFadyen–Stephens aldehyde synthesis, **8**, 297
Mecambrine
synthesis
alkaline photolysis, **3**, 686
Meerwein arylation, **3**, 505
atom transfer reactions
radical addition reactions, **4**, 757
intramolecular, **4**, 804
Meerwein–Ponndorf reaction
organosamarium compounds, **1**, 258
Meerwein–Ponndorf–Verley reaction

electron transfer mechanism, **3**, 824
reduction of crotonaldehyde
aluminum isopropoxide in isopropyl alcohol, **8**, 88
transition state, **8**, 88
Meisenheimer rearrangement
amine oxides, **6**, 834, 843
Meldrum's acid
flash vapour pyrolysis, **5**, 732
imidoylation, **2**, 356
Knoevenagel reaction
active methylene compound, **2**, 355
Michael reaction, **2**, 352
Melodinus alkaloids
synthesis
Mannich reactions, **2**, 1042
Meloscine
biomimetic synthesis, **6**, 755
synthesis
Mannich reactions, **2**, 1042
Menthadiene
cyclization, **3**, 349
Menthol
asymmetric hydrogenation, **8**, 144
esterification
enzymatic, **6**, 341
lithium aluminum hydride modifier, **8**, 161
oxidation
solid support, **7**, 841, 845
(–)-Menthol, β-4-deoxy-L-glycoside
synthesis
Diels–Alder reaction, **2**, 692
Menthol, phenyl-
crotonate ester
addition reactions with organocopper reagents, **1**, 313
glyoxalate esters
nucleophilic addition reactions, **1**, 65
Menthol, 8-phenyl-
chiral malonic esters
intermolecular alkylation, **3**, 56
conjugate additions
organocuprates, **4**, 201
Menthone
oxime
Beckmann rearrangement, **7**, 691
photocycloaddition
with furan, **5**, 170
rearrangement, **3**, 831
reduction
dissolving metals, **8**, 111
dissolving metals, stereoselectivity, **8**, 116
electrolysis, **8**, 321
lithium/ammonia/ethanol mixture, **8**, 112
Menthyl acetate
chiral enolates
asymmetric induction, **2**, 225
Menthyl crotonate
addition reaction
phenylmagnesium bromide, **4**, 200
Menthyl esters
Mannich reaction, **2**, 919
Reformatsky reaction, **2**, 922
synthesis
Mannich reaction, **2**, 924
Mercuration
activation barriers, **7**, 869
charge transfer excitation energies
EDA complexes, **7**, 870
EDA complexes
intermediates, **7**, 868
Mercurilactonization
δ-lactone synthesis, **6**, 366
reductive demercuration, **8**, 853
unsaturated carboxylic acids, **6**, 361
Mercury
reduction
α-bromo ketones, **8**, 986
Mercury, allenyl-
synthesis, **2**, 85
Mercury, arylchloro-
reaction with vinyl cuprates, **3**, 497
Mercury, bis(bromomethyl)-
addition to alkenes, **4**, 968
Mercury, bromo-5-hexenyl-
reductions, **8**, 852
Mercury, bromo-4-methylcyclohexyl-
reductions, **8**, 852
Mercury, chlorovinyl-
coupling reactions
with vinyl cuprates, **3**, 489
Mercury, cycloperoxy-
demercuration, **8**, 855
Mercury, diaryl-
extrusion of mercury, **3**, 501
Mercury, diphenyl-
acid anhydride synthesis, **6**, 312
cleavage
acidic, **8**, 850
Mercury, iodo(iodomethyl)-
addition to alkenes, **4**, 968
Mercury, propargyl-
synthesis, **2**, 85
Mercury, vinyl-
dimerization
diene synthesis, **3**, 484
Mercury acetate
α-acetoxylation
ketones, **7**, 154
allylic oxidation, **7**, 92, 108
dehydrogenation
steroids, **7**, 93
ketone α-acetoxylation, **7**, 145
Mercury acetate, cinnamyl-
solvolysis, **7**, 92
Mercury acetate, crotyl-
solvolysis, **7**, 92
Mercury bis(trifluoro)acetate
polyene cyclization, **3**, 342
Mercury carboxylates
acid anhydride synthesis, **6**, 315
Mercury enolates
aldol reaction
syn stereoselective, **2**, 314
Mercury hydride
radical addition reactions
alkenes, **4**, 741
radical cyclizations, **4**, 799
Mercury nitrate
oxidation
halides, **7**, 665
reaction with alkenes, **7**, 533
Mercury nitrite

nitration with, **6**, 108
Mercury oxide
allylic oxidation, **7**, 93
decarboxylative halogenation, **7**, 724
Mercury salts
catalysts
Cope rearrangement, **5**, 802
decarboxylative halogenation, **7**, 724
halofunctionalization
alkenes, **7**, 533
reactions with alkanes, **7**, 3
Ritter reaction, **6**, 283
Mercury-sensitized photoreactions
di-π-methane rearrangement, **5**, 195
Mercury trifluoroacetate
electrophilic oxidation, **7**, 868
Mercury trifluoroacetate, pentamethylphenyl-
synthesis, **7**, 870
Mercury(II) acetate
oxidation of amines, **2**, 1021
Mercury(II) salts, aryl-
dimerization, **3**, 501
Merrifield synthesis
peptides, **6**, 670
Mesaconitine
synthesis, **6**, 402
Δ^7-Mesembrenone
synthesis
Eschenmoser coupling reaction, **2**, 885
Mesembrine
synthesis
N-acyliminium ions, **2**, 1065
Mannich reaction, **2**, 1010
via enamines, **6**, 717
Mesitonitrile oxide
cycloaddition reactions
tropones, **5**, 626
Mesitylene
acylation
Friedel–Crafts reaction, **2**, 735
amidoalkylation, **2**, 971
formation, **2**, 141
Mesitylene, diacetyl-
synthesis
Friedel–Crafts reaction, **2**, 734
Mesitylene-2-sulfonylhydrazone
fragmentation, **6**, 779
O-Mesitylenesulfonylhydroxylamine
Beckmann rearrangement, **6**, 764
Mesityl nitrile oxide
use in 1,3-dipolar cycloadditions, **4**, 1079
Mesityl oxide
synthesis, **8**, 533
Mesoionic compounds
1,3-dipolar cycloadditions, **4**, 1096–1098
Mesonaphthodianthrone
synthesis
via photolysis, **5**, 729
Mesoxylates
N-acylimines
Diels–Alder reactions, **5**, 405
Diels–Alder reactions, **5**, 432
Mesylates
alcohols
hydroxy group activation, **6**, 19
bromination, **6**, 210
chlorination, **6**, 206
α-Mesyloxylation
ketones, **7**, 155
[3.3]Metacyclophane
synthesis
via arene–metal complexes, **4**, 540
[2.2]Metacyclophane, octamethyl-
synthesis, **6**, 778
[2,2]Metacyclophane-4,9-diene, 15,16-dimethyl-
synthesis
via electrocyclization, **5**, 705
Metacyclophanes
synthesis, **3**, 126; **7**, 354
5-Metacyclophanes
synthesis
via dihalocyclopropyl compounds, **4**, 1017
Metal acetates
allylic oxidation, **7**, 92
Metal alkoxides
catalysts
Friedel–Crafts reaction, **3**, 296
Metal alkyls
catalysts
Friedel–Crafts reaction, **3**, 296
Metal aluminides
synthesis, **8**, 839
Metal amides
amidine synthesis, **6**, 546
tandem vicinal difunctionalization, **4**, 257
Metal borides
deselenations, **8**, 848
Metal carbonyls
deoxygenation
epoxides, **8**, 890
Metal complexes
cationic pentadienyl
nucleophilic addition, **4**, 663–692
dienyl
nucleophilic addition, stereocontrol, **4**, 685
Metal dienolates
α,β-unsaturated ketones
alkylation, **3**, 21
Metal enolates
C-alkylation, **3**, 4
O-alkylation
competition with *C*-alkylation, **3**, 4
carboxylic acid derivative
alkylation, **3**, 39
chirality transfer, **3**, 13
molecular aggregates
dependence on solvent, **3**, 3
saturated aldehydes
alkylation, **3**, 20
stabilized
alkylations, **3**, 54
Metal homoenolates, **2**, 441–453
Metal hydrides
demercurations, **8**, 851
radical addition reactions
alkenes, **4**, 735–742
reduction
acetals, **8**, 213
acyl halides, **8**, 240
alkyl halides, **8**, 798
amides, **8**, 249
arylsulfonylhydrazones, **8**, 343

carbonyl compounds, **8**, 1–22, 313
carboxylic acids, **8**, 237, 259–279
epoxides, **8**, 872
esters, **8**, 244
imines, **8**, 25–74
lactones, **8**, 247
nitriles, **8**, 253
pyridines, **8**, 579
unsaturated carbonyl compounds, **8**, 536
transition metal halides
reduction, mechanism, **8**, 483
unsaturated hydrocarbon reductions, **8**, 483
Metal ions
oxidation
thiols, **7**, 759
Metallacycles
alkene metathesis, **5**, 1115
Metallacyclobutane, 2-methylene-
[3 + 2] cycloaddition reactions, **5**, 293
Metallacyclobutane complexes
Tebbe reaction, **1**, 748
Metallacyclopentadienes
reactions with alkynes
benzene synthesis, **5**, 1144
Metallacyclopentenes
reactions with alkenes, **5**, 1142
Metallaenolates
structure, **2**, 125
Metallation
acyclic systems
addition reactions, **1**, 477
nitrogen stabilization, **1**, 461
alkynes, **3**, 271
carbocyclic systems
addition reactions, **1**, 461, 480
heterocyclic systems
addition reactions, **1**, 470, 480
Metallic oxidants
ethers, **7**, 236
Metallocarbene
insertion, **3**, 1047
Metallocene dichlorides
deoxygenation
epoxides, **8**, 889
Metallocenes
bent
hydrometallation, **8**, 669
Metallo-Claisen reaction, **4**, 880
Metallodealumination, **8**, 754
Metalloenamines, **2**, 475–501
Metallo-ene reactions, **5**, 29–60
intermolecular, **5**, 30–37
intramolecular, **5**, 37–59
regioselectivity, **5**, 30, 60
ring size, **5**, 59
stereoselectivity, **5**, 30, 60
Metalloproteases
peptide synthesis, **6**, 395
Metals
reductive cleavage
α-halo ketones, **8**, 986
Metazocine
synthesis, **3**, 77
Methacrylates, thienyl-
synthesis, **7**, 596
Methacrylic acid, α-phenylthio-
methyl ester
Michael addition, **6**, 144
α-Methacrylothioamide, *N,N*-dimethyl-
addition reactions
with organomagnesium compounds, **4**, 85
Methallyl cation
[4 + 3] cycloaddition reactions, **5**, 597
β-Methallyl iodide
generation of methallyl cation
[4 + 3] cycloaddition reactions, **5**, 597
Methane
ethylation
Friedel–Crafts reaction, **3**, 333
oxidation
ozone, **7**, 14
reaction with elemental sulfur, **7**, 14
Methane, alkoxybis(sulfonyl)-
alkylation, **3**, 177
Methane, alkoxychloryl-
reaction with allylsilanes, **2**, 580
Methane, alkoxydialkylamino-
preformed
Mannich reaction, **2**, 956
reaction with phenols
Mannich reaction, **2**, 958, 959
Methane, arylbis(methylseleno)-
metallation, **1**, 641
Methane, arylbis(phenylseleno)-
metallation, **1**, 641
Methane, bis(dialkylamino)-
preformed
Mannich reaction, **2**, 956
reaction with phenols
Mannich reaction, **2**, 956, 958
Methane, bis(*N,N*-dimethyldithiocarbamato)-
methylthiomethyl ester
alkylation, **3**, 136
Methane, bis(1,3,2-dioxaborin-2-yl)-
deprotonation
alkylation of anion, **3**, 199
Methane, bis(2,6-dioxo-4,4-dimethylcyclohexyl)-
analysis of aldehydes
Knoevenagel reaction, **2**, 354
Methane, bis(methylsulfonyl)-3-(2,6-dimethoxypyridyl)sulfonyl-
potassium salt
structure, **1**, 528
Methane, bis(phenylseleno)-
metallation, **1**, 641
Methane, borylstannyl-
cleavage
synthesis of boron-stabilized carbanions, **1**, 490
Methane, bromochloro-
lithium–bromine exchange
sonication, **1**, 830
reaction with 1,2-bis(oxazolinyl)ethane, **4**, 976
Methane, chlorodifluoro-
reaction with amides, **6**, 579
Methane, chloroiodo-
epoxidation, **1**, 830
Methane, chloromethoxy-
reaction with vinylsilanes
carbon–oxygen bond cleavage, **2**, 581
Methane, chloro(phenylseleno)-
metallation, **1**, 641
Methane, chloro(phenylthio)-

reaction with allylsilanes, **2**, 580
Methane, dialkoxyhalo-
2,2-bis(dialkoxy)carbonitrile synthesis, **6**, 565
Methane, diazo-
C-acylation, **3**, 888
Methane, dibenzoyl-
synthesis
Claisen condensation, **2**, 796
Methane, dibromo-
Simmons–Smith reaction, **4**, 968
Methane, dichlorodiphenyl-
synthesis
Friedel–Crafts reaction, **3**, 320
Methane, diiodo-
reaction with α-halo ketones
organosamarium compounds, **1**, 261
Simmons–Smith reaction, **4**, 968
Methane, dimethoxy-
solvent
Reformatsky reagent, **2**, 279
Methane, dimethylene-
[3 + 2] cycloaddition reactions, **5**, 272–282
Methane, di-*N*-morpholinyl-
in Mannich reaction of phenols with morpholine, **2**, 958
Methane, dipiperidyl-
reaction with naphthols
Mannich reaction, **2**, 958
Methane, ethoxy-*N*-morpholinyl-
Mannich reaction with 2-naphthol
nonprotic solvent, **2**, 959
Methane, iodo-
carbonylation
formation of acetyl iodide, **3**, 1018
Methane, methoxybis(trimethylsilyl)-
deprotonation
with *s*-butyllithium, **3**, 198
Methane, methoxy(trimethylsilyl)-
deprotonation
with *s*-butyllithium, **3**, 198
Methane, phenylnitro-
nitronate carbanion
crystal structure, **1**, 34
synthesis, **6**, 105
Methane, phenyl(trimethylsilyl)phenylseleno-
metallation, **1**, 642
Methane, polyhalo-
reaction with ketones, **3**, 787
Methane, tetrachloro-
acid chloride synthesis, **6**, 303
alkane chlorination, **7**, 15
Methane, tetranitro-
fragmentation
unstable radical anions, **7**, 855
nitration with, **6**, 107
Methane, triamino-
synthesis, **6**, 579, 580
Methane, trichlorobromo-
alkane bromination, **7**, 15
Methane, triformyl-
synthesis
Vilsmeier–Haack reaction, **2**, 786
Methane, trimethylene-
complexes
synthesis, **5**, 1107
[4 + 3] cycloaddition reactions, **5**, 598
1,3-diyl trapping reaction, **5**, 239
stereoselectivity, **5**, 242
reactions, **5**, 240
regioselectivity, **5**, 240
synthetic equivalents, **5**, 244
synthons
[3 + 2] cycloaddition reactions, **5**, 287–312
Methane, triphenyl-
dyes
synthesis, Reimer–Tiemann reaction, **2**, 774
Methane, tris(dimethoxyboryl)-
cleavage
alkylation of anion from, **3**, 199
Methane, tris(dimethylamino)-
synthesis, **6**, 579
Methane, tris(formylamino)-
synthesis, **6**, 503
Methane, tris(methylseleno)-
metallation, **1**, 641
Methane, tris(methylthio)-
ketone homologation, **1**, 878
Methane, tris(phenylseleno)-
metallation, **1**, 641
Methane, tris(phenylthio)-
reaction with nitroarenes, **4**, 432
Methane, vinyl(trimethylsilyl)phenylseleno-
metallation, **1**, 642
Methane monooxygenase
hydrocarbon hydroxylation
catalyst, **7**, 80
Methane phosphonate, 1-(trimethylsiloxy)phenyl-
diethyl ester
acyl anion equivalents, **1**, 544
Methanesulfenyl chloride
reactions with alkenes, **7**, 516
reactions with dienes, **7**, 516
Methanesulfinic acid, aminoimino-
diaryl disulfide reduction, **4**, 443
Methanesulfinic acid, trifluoro-
Ramberg–Bäcklund rearrangement, **3**, 868
Methanesulfonamide, trifluoro-
amine synthesis, **6**, 83
Methanesulfonate, trifluoro-
vinyl ester
reaction with homoenolates, **2**, 449
Methanesulfonate, trimethylsilyltrifluoro-
reaction with amides, **6**, 502
Methanesulfonates
octyl esters
nitrile synthesis, **6**, 236
Methanesulfonic acid
Beckmann rearrangement, **7**, 691
with phosphorus pentoxide, **6**, 764
diisobutylaluminum salt
reactions with carbonyl compounds, **2**, 68
Methanesulfonic acid, trifluoro-
Beckmann rearrangement, **7**, 695
catalyst
Friedel–Crafts reaction, **3**, 297
esters
conversion to amides by carbonylation, **3**, 1035
Friedel–Crafts reaction, **2**, 754
bimolecular aromatic, **2**, 739
ionic hydrogenation
carbonyl compounds, **8**, 319
trifluoroacetyl ester

Friedel–Crafts reaction, **2**, 740
Methanesulfonic acid esters, trifluoro-
amide alkylation, **6**, 502
Methanesulfonic anhydride, acetyl-
Friedel–Crafts reaction
bimolecular aromatic, **2**, 739
Methanesulfonic anhydride, trifluoro-
activator
DMSO oxidation of alcohols, **7**, 299
reactions with amides, **6**, 504
Methanesulfonyl azide
diazo transfer reaction, **4**, 1033
Methanesulfonyl bromide, bromo-
reaction with alkenes, **4**, 359
Methanesulfonyl chloride
2-hydroxyselenide elimination reactions, **3**, 787
Methanesulfonyl chloride, trichloro-
alkane chlorination, **7**, 16
oxidation
thiols, **7**, 761
Methanesulfuryl chloride
synthesis, **7**, 14
Methanethiol, phenyl-
dianions
reactions with carbonyl compounds, **1**, 826
Methanimine
synthesis
via retro Diels–Alder reactions, **5**, 576
Methaniminium chloride, *N*,*N*-dimethylchlorosulfite
Curtius reaction, **6**, 810
1,6-Methano[10]annulene amide
lithiation
addition reactions, **1**, 466
1,6-Methano[10]annulen-11-one
synthesis
via [6 + 4] cycloaddition, **5**, 623
4,9-Methano[11]annulenone
oxime
Beckmann rearrangement, **6**, 764
Methanols, trialkyl-
synthesis, **3**, 793, 794
2,4-Methanoproline
synthesis
via intramolecular photocycloaddition, **5**, 179
9(0)-Methanoprostacyclin
synthesis, **6**, 780
Methionine
N-(benzyloxycarbonyl) groups
cleavage, **8**, 959
Methoxatin
synthesis, **7**, 349
4-Methoxybenzyl esters
carboxy-protecting groups
cleavage, **6**, 668
4-Methoxybenzyl group
alcohol protection, **6**, 23
ether protection, **6**, 634
4-Methoxybenzyloxycarbonyl group
amino acid protecting group
hydrogenolysis, **8**, 958
protecting group
cleavage, **6**, 635
Methoxylamine
oxidation
synthesis of aziridines, **7**, 744
α-Methoxylation
electrochemical
amides, **7**, 804
aromatic compounds, **7**, 799
carbamates, **7**, 804
ketones, **7**, 798
Methoxymercuration
carboxy-protecting groups
deprotection, **6**, 666
4-Methoxy-2,3,6-trimethylphenylsulfonyl group
arginine guanidino protection, **6**, 644
Methyl α-acetamidoacrylate
ene reactions
Lewis acid catalysis, **5**, 5
Methyl acetate, methoxy-
boron trifluoride complex
NMR, **1**, 292
Methyl acrylate
borane complexes
structure, **1**, 289
[3 + 2] cycloaddition reactions
with electron deficient vinylcyclopropanes, **5**, 281
Diels–Alder reactions, **5**, 461
Lewis acid promoted, **5**, 339
α-silapyran, **5**, 1074
ene reactions
intermolecular, **5**, 3
Lewis acid complexes
conformation, **1**, 288
oxidation
Wacker process, **7**, 451
reaction with iron carbonyl, **5**, 1131
reaction with vinyl chromium carbene complexes, **5**, 1070
synthon
tandem vicinal difunctionalization, **4**, 247, 256
Methylalumination
zirconium catalysis, **4**, 890
Methylamine, alkoxy-
Mannich reaction
intermediate, **2**, 895
Methylamine, bis(*p*-methoxyphenyl)-
reaction with π-allyl complexes, **6**, 86
Methylamine, *N*,*N*-bis(trimethylsilyl)methoxy-
cleavage
generation of *N*-silyliminium salts, **2**, 913
formaldehyde imine equivalent, **1**, 368
Methylamine, cyano-
iminium ion precursors, **4**, 1088
Methylamine, hydroxy-
Mannich reaction
intermediate, **2**, 895
Methyl benzoate
reduction
electrochemical, **8**, 242
Methyl α-bromomethacrylate
ene reactions
Lewis acid catalysis, **5**, 5
Methyl ceriferate
synthesis, **3**, 99
Methyl crotonate
Diels–Alder reactions
Lewis acid promoted, **5**, 340
Lewis acid complexes
NMR, **1**, 294
reaction with Danishefsky's diene, **5**, 1072
Methylcupration

alkynes, **4**, 898
Methyl α-cyanoacrylate
ene reactions
Lewis acid catalysis, **5**, 5
Methyl cyanoformate
alkoxycarbonylation
ketones, **2**, 839
Methyl dihydrojasmonate
synthesis, **7**, 457
S-Methyldithiocarbonyl compounds
deoxygenation, **8**, 818
Methyl β-eleostearate
hydrogenation
homogeneous catalysis, **8**, 451
Methylenation
carbonyl compounds, **1**, 731
samarium induced, **1**, 751
silicon stabilized, **1**, 731
sulfur stabilized, **1**, 737
Tebbe reagent, **1**, 743
titanium stabilized, **1**, 743
titanium–zinc, **1**, 749
Methyleneamine, *N*-cyclohexyl-
reaction with allylmagnesium bromide, **2**, 980
N-Methyleneamines
synthesis
from *N*-(cyanomethyl)amines, **2**, 941
Methylenediamines
Mannich reaction, acid-catalyzed
intermediate, **2**, 895
Mannich reaction, base-catalyzed
intermediate, **2**, 895
Methylene groups
activated
oxidation, **7**, 267
Methyleneiminium salts
Mannich reaction
reviews, **2**, 894
Methylene transfer
cyclopropane synthesis, **4**, 951–994
Methylenium cations, thioamido-
Diels–Alder reactions, **5**, 504
Methylenomycin A, deepoxy-2,3-didehydro-
synthesis
via Nazarov cyclization, **5**, 780
Methylenomycin A, deepoxy-4,5-didehydro-
synthesis, **7**, 243
Methylenomycin B
synthesis
2-butyne acylation, **2**, 725
via Nazarov cyclization, **5**, 780
via nickel-ene reaction, **5**, 36
via Pauson–Khand reaction, **5**, 1051
Methyl esters
carboxy-protecting groups, **6**, 665
Methyl 10-fluorofarnesoate
regioselective epoxidation, **7**, 359
Methyl iodide
alkylation with, **3**, 14
α-functionalization, **4**, 260
Methyl isocyanoacetate
aldol reaction
ferrocenylphosphine–gold complexes, **2**, 318
enantioselective aldol reaction
gold catalysis, **2**, 317
Methyl jasmonate
synthesis, **7**, 59
via palladium catalysis, **4**, 612
via Pummerer rearrangement, **7**, 206
Methyl linoleate
peroxymercurials
reduction, **8**, 855
Methyl lithioacetate
reaction with cycloheptadienyliron complexes, **4**, 674
Methyl methacrylate
Diels–Alder reactions
Lewis acid promoted, **5**, 340
ene reactions
Lewis acid catalysis, **5**, 5
thermal, **5**, 3
hydroformylation, **4**, 932
reaction with cyclopentadiene, **5**, 1071
tandem vicinal difunctionalization, **4**, 247
Methyl 3-nitroacrylate
nitration with, **6**, 108
Methylococcus spp.
hydrocarbon hydroxylation, **7**, 56
Methyl octanoate
reduction
electrochemical, **8**, 243
Methyl oleate
peroxymercurials
reduction, **8**, 855
Methylols
reduction
to *N*-methylamides, **8**, 27
Methylosinus spp.
hydrocarbon hydroxylation, **7**, 56
Methylotropic bacteria
hydrocarbon hydroxylation, **7**, 56
Methyl oxalates
reduction
stannanes, **8**, 824
α-Methylphenacyl esters
carboxy-protecting groups, **6**, 666
Methyl propiolate
ene reactions, **5**, 6
Lewis acid catalysis, **5**, 7
Methyl retinoate
synthesis
via Julia coupling, **1**, 803
Methyl shikimate
synthesis
via cyclohexadienyl complexes, **4**, 684
Methyl sorbate
1,4-hydrogenation
homogeneous catalysis, **8**, 451
reduction
copper hydrides, **8**, 549
Methylsulfonylethoxycarbonyl group
protecting group
removal, **6**, 638
2-[4-(Methylsulfonyl)phenylsulfonyl]ethoxycarbonyl group
protecting group
amines, **6**, 638
Methyl α-D-tetranitroside
synthesis
via iodine isocyanate addition to alkene, **4**, 351
2-Methylthioethoxycarbonyl group
amine-protecting group, **6**, 639, 666
2-Methylthioethyl esters

carboxy-protecting group, **6**, 639, 666
Methyl toluate
(2,6-di-*t*-butyl-4-methyl)phenoxydiethylaluminum complex
crystal structure, **1**, 301
Methyl trifluoroacetate
hydrogenation, **8**, 242
Methyl undecylenate
amidomercurial
reduction, **8**, 858
Methyl vernolate
rearrangement, **3**, 752
Methymycin
synthesis
via macrolactonization, **6**, 372
Methynolide
synthesis, **7**, 246
stereocontrol, **3**, 960
via iterative rearrangements, **5**, 894
via macrolactonization, **6**, 369
Mevalonolactone
synthesis, **7**, 312, 316
Mevalonolactone, anhydro-
synthesis, **7**, 240
Mevinic acids
synthesis
via Horner–Wadsworth–Emmons reaction, **1**, 772
Mevinolin
analogs
synthesis, *via* chiral acetals, **2**, 651
synthesis, **3**, 589
ene reaction, **2**, 548
via an alkynic ketone, **1**, 405
via cyclofunctionalization of cycloalkene, **4**, 373
via organocuprate conjugate addition, **4**, 194
Meyer–Schuster reaction
propargylic alcohols, **6**, 836
Meytansine
synthesis
via t-butyl (*p*-tolylsulfinyl)acetate, **1**, 523
Michael acceptors
conjugate enolate anion addition, **4**, 261
Michael addition
abnormal, **4**, 4
antiparallel addition, **4**, 23
closed transition state model
stereochemistry, **4**, 21
definition, **4**, 258
dienolate double, **4**, 30
enantioselective, **6**, 849
intermolecular, **4**, 3–23
diastereoselectivity, **4**, 18
intramolecular, **4**, 24–30
Knoevenagel products
side reaction, **2**, 352
mechanism, **4**, 1
open transition state model
stereochemistry, **4**, 21
radical cyclization, **4**, 791
sequential, **4**, 261
stereochemistry
solvent effect, **4**, 20
tandem, **4**, 121
under aprotic conditions, **4**, 10
α,β-unsaturated carbene complexes, **5**, 1081
Michael–Michael ring closure reactions, **7**, 625
Michael ring closure
annulation, **4**, 260
sequential, **4**, 262
Microbial dehydrogenation
carbonyl compounds, **7**, 145
Microbial epoxidation, **7**, 429
Microbial hydroxylation
ketones, **7**, 158
Microbial oxidation
alternatives, **7**, 79
enantiotopic discrimination, **7**, 57
mechanism, **7**, 56
nonsteroidal substrates, **7**, 56
steroids, **7**, 56
unactivated C—H bonds, **7**, 53–80
Micrococcus flavus
β-hydroxylation, **7**, 56
Microorganisms
cultures
collections, **7**, 55
immobilized
steroid dehydrogenation, **7**, 68
mutation, **7**, 56
oxidation
unactivated C—H bonds, **7**, 53
uses, **7**, 55
reduction
carbonyl compounds, **8**, 185
sources, **7**, 55
reduction, **8**, 184
taxonomy, **7**, 55
Milbemycin β_1
synthesis
via Julia coupling, **1**, 801
Milbemycin β_3
synthesis
via activated esters, **6**, 373
via Julia coupling, **1**, 801
via lithium cuprate, **1**, 128
via macrolactonization, **6**, 375
via macrolide ring closure, **6**, 369
Milbemycin E
spiroacetal fragment
synthesis, **1**, 568
Milbemycins
synthesis, **7**, 300
spiroketal portion, **1**, 419
via carboalumination, **4**, 893
via Horner reaction, **1**, 779
via Julia coupling, **1**, 797, 801
Minelsin
anticholinergenic and spasmolytic agent, **3**, 826
Minisci reaction
alkenes, **7**, 498
nucleophilic radical addition reactions, **4**, 768
Mislow allyl sulfoxide–allyl sulfenate rearrangement
2,3-sigmatropic rearrangement, **6**, 834
Mitomycin C
synthesis, **7**, 353
Mitomycins
synthesis
via Baeyer–Villiger reaction, **7**, 684
via Peterson methylenation, **1**, 732
Mitosane
synthesis
via selenoamination of allylic arylamines, **4**, 403

Mitosene, 7-methoxy-
synthesis
Mannich reaction, **2**, 1015
synthesis, **3**, 261
Mitsunobu reaction
1-*O*-activation
glycoside synthesis, **6**, 49
activation of alcohols, **7**, 752
bromides
alkyl alcohols, **6**, 210
ester synthesis, **6**, 333
fluorination
alkyl alcohols, **6**, 218
MK-801
synthesis, **3**, 71
Modhephene
synthesis, **5**, 924
retrosynthetic analysis, **4**, 732
via [3 + 2] cycloaddition reactions, **5**, 310
via intramolecular ene reactions, **5**, 11, 2
via Nazarov cyclization, **5**, 779
via photocycloaddition, **5**, 666
via photoisomerization, **5**, 233
Mokupalide
synthesis, **3**, 99
via carboalumination, **4**, 893
Molecular sieves
asymmetric epoxidation, **7**, 396
enamine synthesis
water removal, **6**, 705
Molybdenates, decacarbonylhydridobis-
reduction
aldehydes, **8**, 17
Molybdenum
oxidation
secondary alcohols, **7**, 320
Molybdenum, arenetricarbonyl-
catalyst
Friedel–Crafts reaction, **3**, 300
Molybdenum, η^3-crotyl-
reaction with benzaldehyde
diastereoselectivity, **2**, 35
Molybdenum, η^3-cyclopentadienylcrotyl-
configurational stability, **2**, 6
Molybdenum, dicyclopentadienyltetracarbonyl-
(acetaldehyde)-
crystal structure, **1**, 310
Molybdenum, hexacarbonyl-
dehalogenation
α-halocarbonyl compounds, **8**, 991
α-hydroxylation
ketones, **7**, 167
transfer hydrogenation
unsaturated ketones, **8**, 554
Molybdenum acetylacetonate complexes
deoxygenation
epoxides, **8**, 889
Molybdenum catalysts
alkene metathesis, **5**, 1118
Molybdenum complexes
cycloalkadiene complexes
reactions with *N*-substituted sulfoximine
carbanions, **1**, 535
Molybdenum complexes, alkylidene-
carbonyl alkylidenation, **5**, 1126
Molybdenum complexes, carbene
chemistry, **5**, 1091
Molybdenum complexes, hydrido-
reduction
unsaturated carbonyl compounds, **8**, 551
Molybdenum complexes, oxo-
deoxygenation
epoxides, **8**, 889
Molybdenum complexes, peroxy-
epoxidations with, **7**, 382
α-hydroxylation
amides, **7**, 183
enones, **7**, 175
esters, **7**, 180
ketones, **7**, 160
ketoximes, **7**, 187
Molybdenum dioxide diacetylacetonate
oxidative cleavage of alkenes
with *t*-butyl peroxide, **7**, 587
Molybdenum enolates
aldol reaction, **2**, 312
synthesis and reaction, **2**, 127
Molybdenum oxide
activator
DMSO oxidation of alcohols, **7**, 299
Molybdenum pentachloride
catalyst
Friedel–Crafts reaction, **2**, 737
reaction with alkenes, **7**, 530
Molybdenum pentoxide
oxidation
alkenyloxyboranes, **7**, 602
Molybdenum salts
reduction
alkenes, **8**, 531
Molybdenum trioxide
catalyst
carbonyl compound hydrogenolysis, **8**, 320
Monacolin-K
microbial oxidation, **7**, 77
Monensin
synthesis
stereoselectivity, **4**, 384
via alkynide addition, **1**, 420
via Claisen rearrangement, **5**, 853
via Lewis acid chelation-controlled addition, **1**, 336
synthesis by Still
use of magnesium enolate, **2**, 194
Monensin B
synthesis, **7**, 361
Monensin lactone
synthesis
Diels–Alder reaction, **2**, 701
Monic acid C
synthesis
via carbosulfenylation of alkenes, **4**, 331
Monoacylation
polyols
selective, **6**, 337
Monobactam antibiotics
synthesis
from *N*-methyleneamines, **2**, 941
Mannich reaction, **2**, 913
Monoclonal antibodies
synthetic protein catalysts
Claisen rearrangement, **5**, 855
Monocyclofarnesol

synthesis
via carboalumination, **4**, 893
Monomerine I
synthesis
conjugate addition, **2**, 330
Monoperoxysuccinic acid
anti hydroxylation
alkenes, **7**, 446
Monosaccharides
asymmetric hydrogen transfer, **8**, 552
reduction
unsaturated carbonyl compounds, **8**, 545
synthesis
Diels–Alder reaction, **2**, 688
Monoterpenes
synthesis
via DMSO, **7**, 301
Monothiofumarate, *O,O*-dimethyl
synthesis
via sulfhydrolysis of orthoesters, **6**, 452
Monothiomaleate, *O,O*-dimethyl
synthesis
via sulfhydrolysis of orthoesters, **6**, 452
Monothiomalonates
S-alkyl ester
synthesis, **6**, 438
Monsanto process
acetic acid production, **3**, 1018
Montmorillonite clays
catalyst
allylsilane, reaction with acetals, **2**, 576
enol ether, reaction with acetals, **2**, 612
enol silanes, reaction with acetals, **2**, 635
Friedel–Crafts reaction, **3**, 300
Morphan, phenyl-
synthesis
Mannich reaction, **2**, 1024
Morphans
synthesis
via radical cyclization, **4**, 812
Morphinans
asymmetric synthesis
hydrogenation, **8**, 461
synthesis
via benzocyclobutene ring opening, **5**, 693
synthesis, **6**, 163
Morphine alkaloids
synthesis, **7**, 801
Morphines
asymmetric synthesis
hydrogenation, **8**, 461
synthesis, **6**, 163
via cycloalkenyl sulfone, **4**, 79
via vinylic sulfones, **4**, 251
Morphinoid analgesics
synthesis
via diazonium ions, **1**, 836
Morpholidite, chlorophosphoro-
phosphorylation, **6**, 620
Morpholine
N-alkylation, **6**, 66
allyl transfer
amine protection, **6**, 640
Morpholine, *N*-formyl-
Vilsmeier–Haack reaction, **2**, 779
Morpholine *N*-oxide, *N*-methyl-
asymmetric dihydroxylation, **7**, 429
oxidation
primary alcohols, **7**, 309, 311
2-Morpholinoethyl isocyanide
amide synthesis, **6**, 387
Morpholinones
reduction, **8**, 653
Mosher–Yamaguchi reagent
reduction
unsaturated carbonyl compounds, **8**, 545
Moth pheromones
synthesis
via dienetricarbonyliron complexes, **4**, 701
MSD-92, 4-deaza-
synthesis, **7**, 342
Mukaiyama reaction
asymmetric synthesis
use of silyl enol ethers, **2**, 629
mechanism, **2**, 630
Mukapolide
synthesis
reduction of sulfides, **3**, 107
Mukulol
synthesis
via cyclization, **1**, 553
Multifidene
synthesis
alkene protection, **6**, 689
via Cope rearrangement, **5**, 806
via retro Diels–Alder reactions, **5**, 563
Multifloramine
synthesis, **3**, 807
use of ferricyanide, **3**, 681
Munchnones
cycloaddition reactions, **4**, 1137–1139
1,3-dipolar cycloadditions, **4**, 1096
Munchnones, *C*-alkenyl
azomethine ylides
cycloadditions, **4**, 1139
Munchnones, *N*-alkenyl
azomethine ylides
cycloadditions, **4**, 1139
Munchnones, trifluoroacetyl-
cyclization, **4**, 1139
Murein
synthesis, **6**, 52
Muscalure
synthesis, **3**, 644
Muscarine
synthesis, **6**, 764
via [3 + 2] cycloaddition reactions, **5**, 286
via [4 + 3] cycloaddition reactions, **5**, 605
Muscone
synthesis, **2**, 270; **3**, 168, 787; **7**, 57; **8**, 557
alkynylsilane acylation, **2**, 726
Dieckmann reaction, **2**, 824
via cyclization, **1**, 553
via dihalocyclopropyl compounds, **4**, 1018
via intramolecular Barbier reaction, **1**, 262
via Julia coupling, **1**, 803
via Raphael–Nazarov cyclization, **5**, 779
via Wacker oxidation, **7**, 455
via Wagner–Meerwein rearrangement, **7**, 806
(±)-Muscone
synthesis, **2**, 166
Muscopyridine

synthesis
coupling reactions, **3**, 460
via Raphael–Nazarov cyclization, **5**, 779
Mus musculus pheromone
synthesis
via cyclofunctionalization of cycloalkene, **4**, 373
Mustard gas
synthesis
via electrophilic addition, **4**, 330
Muxone
synthesis
via cyclobutene ring expansion, **5**, 687
Mycinolide V
synthesis
via macrolactonization, **6**, 370
Mycinomycin
synthesis, **3**, 797
Mycophenolic acid
synthesis
via cyclobutenone ring opening, **5**, 689
via electrocyclization, **5**, 732
Myoporone, 7-hydroxy-
synthesis, **7**, 827
Myrcene
hydrosilylation, **8**, 780
synthesis, **3**, 429
Myrtenal
optically active ligand from
synthesis of homoallyl alcohols, **1**, 612
Myrtenol
synthesis, **7**, 92, 99
Mytloxanthin
synthesis
Claisen condensation, **2**, 821

N

Nafion **7**, 511
chromium(III) oxidants
alcohol oxidation, **7**, 282
Nafion-H
catalyst
Friedel–Crafts reaction, **2**, 736
Nafion resin
catalyst, solid superacid
Friedel–Crafts reaction, **3**, 298
Nagata's reagent
Michael addition, **4**, 23
Nagilactones
synthesis, **7**, 331
Nametkin rearrangement, **3**, 706
Nanaomycin
synthesis, **5**, 1096
regioselective, **5**, 1094
Nanaomycin A
synthesis
via cyclobutenone ring opening, **5**, 690
via metal-catalyzed cycloaddition, **5**, 1203
Naphthacene
hydrogenation
homogeneous catalysis, **8**, 455
Naphthaldehyde
tandem vicinal difunctionalization, **4**, 243
1-Naphthaldehyde
formylation
modified Gattermann–Koch reaction, **2**, 749
imines
tandem vicinal difunctionalization, **4**, 252
2-Naphthaldehyde, 1-chloro-3,4-dihydro-
hydrogenation, **8**, 898
Naphthalene
alkylation
1-bromoadamantane, **3**, 302
Friedel–Crafts reaction, **3**, 304
anodic oxidation, **7**, 799
Benkeser reduction
dissolving metals, **8**, 516
Birch reduction
dissolving metals, **8**, 496
carbolithiation, **4**, 871
charge-transfer osmylation, **7**, 864
competitive alkylation
Friedel–Crafts reaction, **3**, 300
[4 + 3] cycloaddition reactions, **5**, 608
formylation
dichloromethyl alkyl ethers, **2**, 750
Gattermann–Koch reaction, **2**, 749
hydrogenation
heterogeneous catalysis, **8**, 439
homogeneous catalysis, **8**, 454
palladium-catalyzed, **8**, 438
isopropylation
Friedel–Crafts reaction, **3**, 304
reductive silylation, **8**, 518
regioselective isopropylation
Friedel–Crafts reaction, **3**, 305
synthesis, **7**, 628
via benzyne Diels–Alder reactions, **5**, 381
via electrocyclization, **5**, 720
via ketocarbenoids, **4**, 1056
via sequential Michael ring closure, **4**, 262
thermal osmylation, **7**, 863
Naphthalene, acetyl-
Birch reduction
dissolving metals, **8**, 503, 510
reduction
ionic hydrogenation, **8**, 319
Naphthalene, 2-(1-adamantyl)-
synthesis
Friedel–Crafts reaction, **3**, 302
Naphthalene, 1-alkanoyl-6-methoxy-
synthesis
via Birch reduction, **8**, 510
Naphthalene, alkoxy-
Birch reduction
dissolving metals, **8**, 496
Naphthalene, alkyl-
Birch reduction
dissolving metals, **8**, 496
Naphthalene, 1-alkyl-2-nitro-
synthesis, **4**, 429
Naphthalene, *N*-arenesulfenylimino-1,4-dihydro-
thermolysis
sulfenylnitrenes from, **7**, 483
Naphthalene, benzoyl-
Wolff–Kishner reduction, **8**, 338
Naphthalene, 1-bromo-
reduction, **8**, 908
Naphthalene, *t*-butyl-
synthesis
Friedel–Crafts reaction, **3**, 311
Naphthalene, 3-butyl-2-methyl-1-nitro-
synthesis, **4**, 428
Naphthalene, chloro-
hydrogenolysis, **8**, 906
Naphthalene, 2-chloro-
synthesis
via dichlorocarbene, **4**, 1016
Naphthalene, dihydro-
metal complexes
addition reactions, **4**, 546
Pauson–Khand reaction, **5**, 1049
synthesis
via thermolysis, **5**, 713
Naphthalene, 1,2-dihydro-
synthesis
via FVP, **5**, 718
Vilsmeier–Haack reaction, **2**, 782
Naphthalene, 1,4-dihydro-
photoisomerization, **5**, 197
Naphthalene, 9,10-dihydro-
hydride transfer, **8**, 92
cis-Naphthalene, 9,10-dihydro-
synthesis
via thermal isomerization, **5**, 716
trans-Naphthalene, 9,10-dihydro-
photolysis, **5**, 716
Naphthalene, dihydrothienyl-
synthesis, **3**, 497
Naphthalene, 1,4-dimethoxy-
metal complexes

addition reactions, **4**, 536
Naphthalene, 1,6-dimethoxy-
Birch reduction
dissolving metals, **8**, 503
Naphthalene, 1,8-dimethylamino-
proton sponge
cyclization reactions, **4**, 843
Naphthalene, 1,8-divinyl-
isomerization, **5**, 68
Naphthalene, 1-fluoro-
hydrogenolysis, **8**, 904
Naphthalene, halo-
$S_{RN}1$ reaction, **4**, 461
Naphthalene, hexahydro-
synthesis
via Diels–Alder reaction, **5**, 331
Naphthalene, iodo-
coupling with naphthoxides, **4**, 470
Naphthalene, 1-methoxy-4-nitro-
synthesis, **6**, 111
Naphthalene, methyl-
isomerization
Friedel–Crafts reaction, **3**, 327
Naphthalene, 2-methyl-
Friedel–Crafts reaction
isobutyryl fluoride, **2**, 735
Naphthalene, octahydro-
cis-fused
synthesis *via* palladium-ene reaction, **5**, 50
Naphthalene, tetrafluoro-
hydrogenolysis, **8**, 904
Naphthalene, tetrahydro-
chiral derivatives
synthesis, **3**, 327
synthesis
Friedel–Crafts reaction, **3**, 311
Naphthalene, 2-trimethylsilyl-
Birch reduction
dissolving metals, **8**, 513
1,2-Naphthalenedicarboxylic anhydride
reduction
borane, **8**, 240
1,5-Naphthalenedisulfonate
reduction, **8**, 918
stability, **8**, 916
Naphthalenephosphonate, dimethyl-
synthesis, **4**, 446
Naphthalenesulfonyl azide
diazo transfer reaction, **4**, 1033
2-Naphthalenetellurenyl iodide
synthesis, **7**, 774
Naphthalen-1,4-imines, 1,4-dihydro-
synthesis
via Diels–Alder reactions, **5**, 382
Naphthalen-2-ol, 4a-decahydro-
synthesis, **7**, 413
Naphthalen-2-ol, 4a-methyl-2,3,4,4a,5,6,7,8-octahydro-
synthesis, **7**, 413
Naphthalen-1-ol, 2-(*N*-substituted amino)-1,2,3,4-tetrahydro-
synthesis, **6**, 787
Naphthalen-1(2*H*)-one, 7-acetyl-3,4-dihydro-
synthesis
Friedel–Crafts reaction, **2**, 760
Naphthalen-1(2*H*)-one, 3,4-dihydro-5,8-dimethoxy-
synthesis, **2**, 763
Naphthalin, 1,2-dihydro-
synthesis
via conjugate addition to oxazolines, **4**, 206
1,2-Naphthalyne
addition reactions, **4**, 493
coupling reactions
selectivity, **4**, 492
generation, **4**, 489
Naphthene, 2-diazo-1-oxo-
ring contraction, **3**, 902
Naphtho[*b*]cyclopropene
cycloaddition reactions
metal catalyzed, **5**, 1199
Naphtho[1,8-*cd*]-1,2-diselenole
oxidation, **7**, 770
Naphthoic acids
oxazolines from
tandem vicinal difunctionalization, **4**, 252
1-Naphthoic acids
Birch reduction
dissolving metals, **8**, 502
2-Naphthoic acids
Birch reduction
dissolving metals, **8**, 502, 503
2-Naphthoic acids, 3-mercapto-
flash pyrolysis
synthesis of β-thiolactones, **6**, 440
1-Naphthoic acid, 2-methoxy-
Birch reduction
dissolving metals, **8**, 502
1-Naphthoic acids, 4-methoxy-
Birch reduction
dissolving metals, **8**, 503
2-Naphthoic acids, methoxy-
Birch reduction
dissolving metals, **8**, 503
Naphthoic acids, tetrahydro-
Birch reduction
dissolving metals, **8**, 503
1,8-Naphthoic anhydride
reduction
borane, **8**, 240
Naphthol
hydrogenation, **8**, 912
Reimer–Tiemann reaction
normal, **2**, 769
synthesis, **7**, 144
via FVP, **5**, 732
1-Naphthol
oxidation
solid support, **7**, 843
reaction with dipiperidylmethane
Mannich reaction, **2**, 958
synthesis
Knoevenagel reaction, **2**, 354
2-Naphthol
Birch reduction
dissolving metals, **8**, 493, 497
Mannich reaction with ethoxy-*N*-morpholinylmethane
nonprotic solvent, **2**, 959
oxidative dimerization, **3**, 665
reaction with benzaldehyde
Mannich reaction, **2**, 960
reaction with benzoxazines
Mannich reaction, **2**, 970
reaction with dipiperidylmethane

Mannich reaction, **2**, 958
2-Naphthol, 6-methoxy-
Birch reduction
dissolving metals, **8**, 497
1-Naphthol, 2-methyl-
Mannich reactions
with preformed salts, **2**, 960
Naphthoquinones
synthesis
via 'one-pot' *ortho* lithiation, **1**, 466
via metal-catalyzed cycloaddition, **5**, 1202
1,4-Naphthoquinones
in microbial dehydrogenation
steroids, **7**, 67
synthesis, **7**, 345
Naphthoquinones, 2-alkyl-
asymmetric epoxidation, **7**, 425
1,4-Naphthoquinones, 2,3-dichloro-
monoalkylation
with tetraalkyltins or alkylzirconium complexes, **3**, 458
1,4-Naphthoquinones, 2-hydroxy-
reaction with acetaldehyde and amines
Mannich reaction, **2**, 960
Naphthoquinones, tetrahydro-
Diels–Alder reactions, **5**, 394
Naphtho[2,1-*d*]thiazolium salts
catalysts
benzoin condensation, **1**, 543
Naphthoxazine
synthesis, **6**, 787
Naphthoxides
arylation, **4**, 470
coupling with iodonaphthalenes, **4**, 470
Naphthylamines
amine–amine exchange reactions, **4**, 435
1-Naphthylimine
reactions with organometallic compounds, **1**, 383
Naphthylimine, *N*-cyclohexyl-
addition reactions
with organolithium compounds, **4**, 76
1,8-Naphthyridinium chlorochromate
oxidation
alcohols, **7**, 270
1,8-Naphthyridinium dichromate
oxidation
alcohols, **7**, 278
Naproxen
asymmetric synthesis, **3**, 789
synthesis, **3**, 1022; **7**, 506
via hydroformylation, **4**, 932
Napthaldehyde-9-carboxylic acid
synthesis, **3**, 828
Narbomycin
synthesis
via cuprate acylation, **1**, 436
Nargenicin A_1, 18-deoxy-
synthesis
via macrolactonization, **6**, 370
Narwedine
synthesis, **3**, 683
Nauclefine
synthesis
Mannich reaction, **2**, 913
Nazarov cyclizations, **5**, 751–781
abnormal, **5**, 760
cyclopentenones by, **2**, 710
mechanism, **5**, 754
stereochemistry, **5**, 754
tin-directed, **5**, 765
Nazarov-type cyclization reactions
vinylsilanes, **1**, 585
Neber rearrangement, **6**, 786
β-Necrodol
synthesis
via conjugate addition to sultam, **4**, 204
via magnesium-ene reaction, **5**, 45
Nef reaction
nitroalkanes, **2**, 324
solid support, **7**, 842, 844
Neocarzinostatin
synthesis
via electrocyclization, **5**, 736
Neoclovene
synthesis, **3**, 386
Neohexene
synthesis
via Phillips Triolefin Process, **5**, 1117
Neolignan
synthesis
use of silver oxide, **3**, 691
Neomethynolide
synthesis
via alkyne acylation by lactones, **1**, 421
Neopentane
synthesis, **3**, 415
Neopentyl alcohol
reaction with dichlorotriphenylphosphorus, **6**, 205
Neopentyl bromide
nitrile synthesis, **6**, 229
Neopentyl compounds
deoxygenation, **8**, 820
Neopentyl iodide
synthesis, **6**, 213
Neopentyl tosylate
reaction with lithium bromide, **6**, 210
Neosporol
synthesis
via Claisen rearrangement, **5**, 832
Nepetalactone, dihydro-
synthesis
via Cope rearrangement, **5**, 812
Neral
asymmetric reduction
aluminum hydrides, **8**, 545
hydrogenation
homogeneous catalysis, **8**, 462
Nerol
asymmetric hydrogenation
synthesis of citronellol, **8**, 462
oxidation, **7**, 306
synthesis
stereoselectivity, **3**, 180
Nerol, neryl-
synthesis, **3**, 170
Nerolidol
synthesis, **3**, 170
via retro Diels–Alder reaction, **5**, 555
Neryl acetate
allylic oxidation, **7**, 89
Neuraminic acid, *N*-acetyl-
2α-glycoside

synthesis *via* carbosulfenylation of alkenes, **4**, 331
synthesis
Diels–Alder reaction, **2**, 694
(*Z*)-selectivity, **1**, 765
Nezukone
synthesis
via [4 + 3] cycloaddition, **5**, 609
via cyclopropanation/Cope rearrangement, **4**, 1049
via oxidation of carbon–tin bonds, **7**, 615
Nickel
alumina
hydrogenation catalyst, **8**, 319
catalyst
cross-coupling reactions, **3**, 523
hydrosilation, **8**, 556
hydrogenation catalyst
pyridines, **8**, 597
Nickel, acyl-
reactions with π-allylpalladium complexes
regioselectivity, **4**, 642
Nickel, π-allylhalo-
chemoselectivity, **3**, 424
preparation, **3**, 423
reactions, **3**, 423
Nickel, bipyridyl(cyclooctadiene)-
desulfurization, **8**, 838
Nickel, bis(acrylonitrile)-
catalyst
bicyclo[1.1.0]butane cycloaddition reactions, **5**, 1186
[3 + 2] cycloaddition reactions, **5**, 293
Nickel, bis(1,5-cyclooctadiene)-
alkenyl halide dimerization
diene synthesis, **3**, 483
catalyst
Ullmann reaction, **3**, 500
Nickel, bis(*N*-methylsalicylaldimine)-
catalyst
reduction, unsaturated ketones, **8**, 558
Nickel, dichloro(1,2-bis(diphenylphosphino)ethane)-
catalyst
Grignard reagents, **3**, 228
Nickel, dichloro(1,3-bis(diphenylphosphino)propane)-
catalyst
crossed alkene coupling, **3**, 484
Grignard reagents, **3**, 228
Nickel, dichlorobis(trialkylphosphine)
catalyst
Ullmann reaction, **3**, 500
Nickel, dichlorobis(triphenylphosphine)-
catalyst
crossed alkene coupling, **3**, 484
Grignard reagents, **3**, 228
Nickel, *trans*-dichlorobis(triphenylphosphine)-
nitrile synthesis, **6**, 232
Nickel, phosphine
catalyst
epoxide hydrogenation, **8**, 882
Nickel, phosphinecarbonyls
catalysts
alkyne trimerization, **5**, 1145
Nickel, tetracarbonyl-
reduction
alkyl halides, **8**, 797
Nickel, tetrakis(triphenylphosphine)-
catalyst
crossed alkene coupling, **3**, 484
Ullmann reaction, **3**, 500
Nickel, tris(triphenylphosphine)-
nitrile synthesis, **6**, 232
Nickel acetate
2,5-dimethylhexanediol
cyclic ketone reduction, **8**, 14
sodium hydride
unsaturated hydrocarbon reduction, **8**, 483
Nickelacyclopentenediones
synthesis
via phenylacetylenes, **5**, 1130
5-Nickelafuranones
2-pyrones from
via [2 + 2 + 2] cycloaddition, **5**, 1157
synthesis
via [2 + 2 + 2] cycloaddition, **5**, 1138
Nickel/aluminum alloy
reduction
aromatic nitro compounds, **8**, 373
Nickel benzoate
oxidation
diols, **7**, 316
Nickel borate
catalyst
epoxide hydrogenation, **8**, 882
Nickel boride
catalysts
aliphatic nitro compound reduction, **8**, 375
C—Se bond cleavage, **8**, 996
deselenations, **8**, 848
desulfurizations, **8**, 839
reduction
benzylic dithioacetals, **8**, 968
Nickel catalysis
acylation, **1**, 450
carbanion alkylations, **3**, 227
cycloaddition reactions
methylenecyclopropanes, **5**, 1188
Nickel chloride
catalysts
aliphatic nitro compound reduction, **8**, 375
lithium aluminum hydride
unsaturated hydrocarbon reduction, **8**, 485
Nickel complexes
catalysts
desulfurizations, **8**, 836
Grignard reagent alkylation, **3**, 244
hydrosilylation, **8**, 764
Wurtz reaction, **3**, 421
Nickel complexes, π-allyl-
regioselectivity, **3**, 426
stereoselectivity, **3**, 426
Nickel-ene reactions, **5**, 35–37, 56–59
Nickel 2-ethylhexanoate
oxidation
diols, **7**, 316
Nickel peroxide
aromatization, **7**, 143
oxidation
primary arylamines, **7**, 738
Nickel salts
catalysts
hydroalumination, **8**, 752
Nickel sulfide
catalyst

silane reaction with carbonyl compounds, **2**, 603
Nicotelline
synthesis, **3**, 510
Nicotinaldehyde acetal
synthesis, **6**, 557
Nicotinamide
electroreduction, **8**, 592
reduction
borohydrides, **8**, 580
Nicotinamide, 1-benzyl-1,4-dihydro-
demercurations, **8**, 858
reduction
aryl bromides, **8**, 908
unsaturated carbonyl compounds, **8**, 562
Nicotinamide, 1-(2,6-dichlorobenzyl)-1,4-dihydro-
reductions
aryl nitroso compounds, **8**, 373
Nicotinamide, 1,4-dihydro-
biomimetic reducing agents, **8**, 977
heterocycle reduction
catalysis, **8**, 97
hydride donors, **8**, 92
reaction with water, **8**, 94
Nicotinamide, 1-phenyl-1,4-dihydro-
biomimetic reduction
allylic compounds, **8**, 977
Nicotinamide, 1-propyl-1,4-dihydro-
biomimetic reducing agents, **8**, 977
Nicotinates, 5-aryl-
synthesis, **3**, 515
Nicotine adenine dinucleotide
biomimetic reducing agents, **8**, 977
models
biomimetic reductions, **8**, 561
reduction
aryl nitroso compounds, **8**, 373
Nicotinic acid
hydrogenation, **8**, 599
microbial hydroxylation, **7**, 79
Nicotinic acid, 6-hydroxy-
synthesis
via microbial hydroxylation, **7**, 79
Nicotinium dichromate
oxidation
alcohols, **7**, 277
Nifedepin
synthesis
Knoevenagel reaction, **2**, 377
NIH shift
microbial hydroxylation
aromatic compounds, **7**, 78
Nikkomycin
synthesis
Ugi reaction, **2**, 1096
Niobates, carbonyldicyclopentadienylhydrido-
reduction
acyl chlorides, **8**, 290
Niobium
catalysts
alkylidenation, carbonyl compounds, **5**, 1125
hydrometallation
mechanism, **8**, 672
Nitrate esters
alkoxy radicals from, **4**, 813
Nitrates
alcohol inversion, **6**, 21
oxidation
halides, **7**, 664
Nitration
electrochemical
aromatic compounds, **7**, 800
secondary amines, **7**, 746
Nitrenes
alkenic
intramolecular cyclization, **7**, 476
reactions with enamines
stereochemical control, **6**, 717
synthesis
via alkenes, **7**, 470
Nitrenes, amino-
synthesis
via oxidation of 1,1-disubstituted hydrazines, **7**, 742
Nitrenes, aryl-
aziridines from, **7**, 476
Nitrenes, benzamido-
synthesis, **7**, 482
Nitrenes, cyano-
synthesis, **7**, 479
via decomposition of cyanogen azide, **7**, 10
Nitrenes, ethoxycarbonyl-
reactions with alkanes, **7**, 10
synthesis, **7**, 478
Nitric acid
quinone synthesis, **7**, 355
Nitric oxide
reactions with alkenes, **7**, 488
α-Nitrile anions
addition reactions
with alkenic π-systems, **4**, 99–113
conjugate addition reactions, **4**, 111–113
Nitrile esters
alkoxy radicals from, **4**, 812
Nitrile imines
aryl-bridged
cyclizations, **4**, 1150
cyclizations, **4**, 1150
open-chain
cyclizations, **4**, 1150
Nitrile imines, alkenyl
cyclizations, **4**, 1150
Nitrile imines, alkynyl
cyclizations, **4**, 1151
Nitrile imines, cycloalkenyl
cyclizations, **4**, 1151
Nitrile oxides
alicyclic-bridged
cycloadditions, **4**, 1129–1131
aryl-bridged
cycloadditions, **4**, 1131
cyclizations, **4**, 1124–1134
cycloaddition reactions, **5**, 257
diastereoselective, **5**, 260
tropones, **5**, 626
with acrylates, **5**, 263
deoxygenation, **8**, 390
1,3-dipolar cycloadditions, **4**, 1070, 1078–1081
intramolecular cycloaddition, **4**, 1124
reaction with alkenes, **5**, 260
tandem reaction sequences
cyclizations, **4**, 1132
Nitrile oxides, alkenyl

cyclic
intramolecular cycloaddition, **4**, 1127–1132
cyclization, **4**, 1125, 1126
long-chain
cyclization, **4**, 1127
open-chain
cyclization, **4**, 1125–1127
Nitrile oxides, alkynyl
INOC reactions, **4**, 1133
Nitrile oxides, *t*-butyl
use in 1,3-dipolar cycloadditions, **4**, 1079
Nitrile oxides, cycloalkenyl
intramolecular cycloadditions, **4**, 1128
Nitrile oxides, furanyl-
cyclization, **4**, 1129
Nitriles
acylation, **2**, 795–863
alkoxymethyleniminium salt synthesis, **6**, 506
allenic
hydrochlorination, **4**, 277
amide synthesis, **6**, 400
hydration, **6**, 400
amidine synthesis, **6**, 546
amidinium salt synthesis, **6**, 516
β-amino-α,β-unsaturated
functionalized, synthesis, **6**, 67
α-aryl
synthesis *via* $S_{RN}1$ reaction, **4**, 468
bisdithioester synthesis, **6**, 455
Blaise reaction
acylation, Reformatsky reagents, **2**, 297
boron trifluoride complex
NMR, **1**, 292
carbanions
intramolecular alkylation, **3**, 49
cocycloaddition reactions
alkynes, **5**, 1152
Darzens glycidic ester condensation, **2**, 419
phase-transfer catalysis, **2**, 429
Diels–Alder reactions, **5**, 416
Houben–Hoesch synthesis, **2**, 747
hydrozirconation, **8**, 683
imidate synthesis, **6**, 533
imidoyl halide synthesis, **6**, 526
intramolecular alkylation, **3**, 48
lithium enolate
crystal structure, **1**, 32
metallation
addition reactions, **1**, 468
radical additions
alkoxy radicals, **4**, 815
reactions with amides, **6**, 569
reactions with arynes, **4**, 497
reactions with diaryl ketone dianions
organoytterbium compounds, **1**, 280
reactions with hydrogen halides, **6**, 497
reactions with organocerium compounds, **1**, 236
reactions with organocopper complexes, **1**, 123
reactions with thiols, **6**, 511
reduction, **8**, 251, 298
metal hydrides, **8**, 274
reductive coupling
ketones, **1**, 273
substitution reactions, **6**, 261–296
synthesis, **6**, 225–255
via amides, **6**, 489
via amines, **7**, 739
via oxidative cleavage of alkenes, **7**, 542, 588
tandem vicinal difunctionalization, **4**, 251
thioimidate synthesis, **6**, 540
thiolysis, **6**, 430
a,β-unsaturated
hydrobromination, **4**, 282
hydrogenation, homogeneous catalysis, **8**, 452
synthesis, **1**, 560, 774
synthesis *via* Ramberg–Bäcklund rearrangement, **3**, 870
tandem vicinal difunctionalization, **4**, 251
Vilsmeier–Haack reaction, **2**, 789
Nitriles, alkane
synthesis
via alcohols, **6**, 234
Nitriles, α-amino-
acyl anion equivalents, **1**, 559
synthesis
via Lewis acid catalysis, **1**, 349
Nitriles, (aminoaryl)alkyl-
synthesis, **8**, 368
Nitriles, α-(arylseleno)-
acyl anion equivalents, **1**, 562
Nitriles, γ-bromo-β-oxo-
dehydrohalogenation
generation of oxyallyl cations, **5**, 595
Nitriles, α-(dialkylamino)-
acyl anion equivalents, **1**, 544, 554
Nitriles, *N*,*N*-(disubstituted)aminomethyl-
reactions with Grignard reagents, **1**, 370
Nitriles, epoxy-
aromatic
α-cleavage, **3**, 748
Nitriles, α-keto-
O,*N*-acetals
O-ethyl arenecarbothioate synthesis, **6**, 452
reduction
Alpine borane, **7**, 603
synthesis, **6**, 316
via acid halides, **6**, 317
Nitriles, β-keto
Knoevenagel reaction, **2**, 361
Nitriles, α-methylthio-
synthesis
via cyanosulfenylation, **6**, 239
Nitriles, (nitroaryl)alkyl-
reduction, **8**, 368
Nitriles, 4-oxo-
synthesis
via benzoin condensation, **1**, 542
Nitriles, α-seleno-
metallation, **1**, 642
Nitriles, β-trimethylsiloxy-
synthesis
via oxiranes, **6**, 237
Nitrile-stabilized anions
addition reactions
alkenes, palladium(II) catalysis, **4**, 572
Nitrile sulfides
cyclizations, **4**, 1165
Nitrile ylides
aryl-bridged
intramolecular cycloadditions, **4**, 1144
cyclizations, **4**, 1141–1144
cycloaddition reactions

fulvenes, **5**, 630
1,3-dipolar cycloadditions, **4**, 1081–1083
open-chain
intramolecular cycloadditions, **4**, 1143
structure, **4**, 1082
Nitrile ylides, alkenyl
intramolecular cycloadditions, **4**, 1142–1144
Nitrile ylides, alkynyl
intramolecular cycloadditions, **4**, 1144
Nitrilimines
1,3-dipolar cycloadditions, **4**, 1083–1085
Nitrilimines, diphenyl-
cycloaddition reactions
tropones, **5**, 625
Nitrilium ions
intramolecular Ritter reaction, **6**, 278
cyclization, **6**, 272
Nitrilium salts
alkoxymethyleniminium salt synthesis, **6**, 507
amidine synthesis, **6**, 543
amidinium salt synthesis, **6**, 516
imidate synthesis, **6**, 529
synthesis
via nitriles, **8**, 275
Nitrilium salts, *N*-alkyl-
Houben–Hoesch synthesis, **2**, 748
Nitrimines
reduction
sodium cyanoborohydride, **8**, 74
Nitrites
oxidation
halides, **7**, 664
reaction with alkyl sulfonates, **6**, 22
trapping
aryl radicals, **4**, 453
Nitroacetamidation
alkenes, **4**, 356
Nitro alcohols
reduction, **8**, 374
2-Nitro alcohols
O-trialkylsilyl ethers
synthesis, **2**, 335
Nitroaldol reaction — *see* Henry reaction
o-Nitrobenzhydryl esters
carboxy-protecting groups
photolytic cleavage, **6**, 668
o-Nitrobenzyl group
phosphoric acid protecting group, **6**, 624
p-Nitrobenzyloxycarbonyl group
protecting group
cleavage, **6**, 635
o-Nitrocinnamoyl group
amine-protecting group, **6**, 642
p-Nitrocinnamyloxycarbonyl group
amine-protecting group, **6**, 641
Nitro compounds
aliphatic
reduction, **8**, 373
synthesis, **6**, 104
aromatic
alkylation, **4**, 428
reduction, **8**, 364, 366, 367, 371
synthesis, **6**, 110
reactions with alkenes, **7**, 488
reactions with organocerium compounds, **1**, 233
reduction, **8**, 363–379
ammonium formate, **8**, 84
synthesis, **6**, 103–132; **7**, 493
via nitroso compounds, **7**, 752
via *N*-oxidation of oximes, **7**, 751
via oxidation of primary amines, **7**, 736
via solid support oxidation of amines, **7**, 842
α,β-unsaturated
hydrogenation, homogeneous catalysis, **8**, 452
Nitrogen
extrusion
diene synthesis *via* retro Diels–Alder reaction, **5**, 567
Nitrogen-centered radicals
cyclizations, **4**, 811–814
Nitrogen compounds
oxidation, **7**, 735–753
Nitrogen dioxide
reactions with alkenes, **7**, 488
Nitrogen groups
functionalization
oxidative cleavage, **7**, 588
Nitrogen nucleophiles
addition reactions
alkenes, **4**, 559–563
aromatic nucleophilic substitution, **4**, 433–437
reactions with π-allylpalladium complexes, **4**, 598
Nitrogen trichloride
reaction with organoboranes, **7**, 604
Nitrogen ylides
preparation, **3**, 918
rearrangement, **6**, 855
Nitro groups
arenes
nucleophilic addition, substitution by, **4**, 425
Nitromercuration
alkenes, **7**, 501, 534
regioselectivity, **6**, 108
Nitronates
addition reactions
carbon-centered radicals, **4**, 765
reduction
borane, **8**, 74
Nitronates, bicyclic trimethylsilyl
Henry reaction
stereoselective, **2**, 336
Nitronates, silyl
Henry reaction, **2**, 335
Nitrones
acyclic chiral
reactions with organometallic compounds, **1**, 391
chiral
reaction with silyl ketene acetals, **2**, 647
cyclic
exo transition state, **5**, 255
intramolecular cycloaddition, **4**, 1120
reactions with organometallic compounds, **1**, 393
synthesis *via* cyclization of δ-allenylamine, **4**, 412
thiolactam synthesis, **6**, 428
cyclizations, **4**, 1113–1124
cycloaddition reactions, **5**, 254
diasteroselective, **5**, 260
deoxygenation, **8**, 390
α-hydroxylation, **7**, 186
E/*Z*-isomerization, **5**, 255
optically active
cycloaddition reactions, **5**, 264

reactions with enol silanes
Lewis acid mediated, **2**, 635
reactions with organometallic compounds, **1**, 391
reduction
lithium aluminum hydride, **8**, 64
reversible cycloaddition reactions, **5**, 256
synthesis
via oxidation of imines, **7**, 750
tandem Michael–cyclization reactions, **4**, 1121
Nitrones, alkenyl-
alicyclic-bridged
cyclization, **4**, 1120
aryl-bridged
cyclization, **4**, 1119
cyclic
cycloaddition, **4**, 1117–1120
open-chain
cyclizations, **4**, 1113–1117
Nitrones, *C*-(5-alkenyl)-
cyclization, **4**, 1113
Nitrones, *C*-(6-alkenyl)-
cyclization, **4**, 1114
Nitrones, *N*-(alkenyl)-
cyclization, **4**, 1115–1117
Nitrones, alkynyl-
cycloadditions, **4**, 1124
Nitrones, allenyl-
cycloadditions, **4**, 1124
Nitrones, *C*-(cycloalkenyl)-
cyclization, **4**, 1117–1119
Nitrones, *N*-(cycloalkenyl)-
intramolecular cycloaddition, **4**, 1119
Nitrones, *C,N*-diphenyl-
[4 + 3] cycloaddition reactions, **5**, 600
reactions with diethyl methylenemalonate, **4**, 1077
Nitrones, *N*-methyl-
reactions with diethyl methylenemalonate, **4**, 1077
Nitronic acids
Henry reaction
acid strength, **2**, 322
Nitronic acids, α-hydroxy-
preparation, **2**, 323
Nitronic esters
tandem Diels–Alder–cyclization reactions, **4**, 1122–1124
Nitronium hexafluorophosphate
nitration with, **6**, 109
Nitronium tetrafluoroborate
nitration with, **6**, 105, 107–109
hydrazines, **7**, 745
reactions with alkenes, **4**, 356; **7**, 488
o-Nitrophenoxyacetyl group
amine-protecting group, **6**, 642
o-Nitrophenylacetyl group
amine-protecting group, **6**, 642
3-(*o*-Nitrophenyl)propionyl group
amine-protecting group, **6**, 642
o-Nitrophenylsulfenyl group
amine-protecting group
peptides, **6**, 644
Nitrosamine anions
deprotonation, **1**, 476
Nitrosamines
anions
alkylation, **3**, 66
deprotonation, **3**, 65
photoaddition to alkenes, **7**, 488
reductive cleavage, **8**, 388, 389
synthesis
via secondary amines, **7**, 746
N-Nitrosamines
reduction, **6**, 119
Nitrosamines, diphenyl-
synthesis
via oxidation of 1,1-diphenylhydrazine, **7**, 744
Nitrosation
secondary amines, **7**, 746
β-Nitroselenation
alkenes, **7**, 496
Nitroselenenylation
alkenes, **6**, 109
Nitroso compounds
aromatic
reduction, **8**, 364, 366, 367
Diels–Alder reactions, **5**, 417–422
oxidation, **7**, 751
reactions with alkenes, **7**, 488
reduction, **8**, 363–379
synthesis, **6**, 103–132
via nitro compound reduction, **8**, 364
via oxidation of *N*-alkylhydroxylamines, **7**, 748
via oxidation of primary amines, **7**, 736
Nitroso compounds, acyl-
Diels–Alder reactions, **5**, 419–421, 485
synthesis
via oxidation of hydroxamic acids and *N*-acylhydroxylamines, **7**, 748
Nitroso compounds, aryl-
Diels–Alder reactions, **5**, 417
Nitroso compounds, α-chloro-
Diels–Alder reactions, **5**, 418
Nitroso compounds, cyano-
Diels–Alder reactions, **5**, 421
Nitroso compounds, sulfonyl-
Diels–Alder reactions, **5**, 421
Nitroso compounds, vinyl-
Diels–Alder reactions, **5**, 422, 485
Nitrosonium fluoroborate
Ritter reaction, **6**, 287
Nitrosonium hexafluorophosphate
Ritter reaction, **6**, 270
Nitrosonium ions, *N*-alkyl-*N*-vinyl
imidate synthesis
amide protection, **6**, 672
Nitrosonium ions, vinyl-
Diels–Alder reactions, **5**, 501
intramolecular, **5**, 539
Nitrosonium ions, *N*-vinyl-*N*-cyclohexyl-
Diels–Alder reactions, **5**, 501
Nitrosyl chloride
alkane chlorination, **7**, 15
aziridine synthesis, **7**, 474
imidoyl halide synthesis, **6**, 526
Nitrosyl cyanide
Diels–Alder reactions, **5**, 421
Nitrosyl fluoride
allylic oxidation, **7**, 113
Nitrosyl halides
reactions with alkenes, **4**, 357; **7**, 500
Nitrosyl hydrogen sulfate
addition to alkenes, **7**, 493
Nitrosylsulfuric acid

synthesis
via nitrosating agent, 7, 740
Nitrous oxide
methane oxidation, 7, 14
oxidative rearrangement, 7, 833
Nitroxides
synthesis
via oxidation of secondary amines, 7, 745
Nitryl chloride
addition reactions
alkenes, 7, 500
nitration with, 6, 108
Nitryl fluoride
nitration with, 6, 109
Nitryl fluorosulfonate
addition to perfluoroalkenes, 7, 493
Nitryl halides
reactions with alkenes, 4, 357
Nitryl iodide
addition reactions
alkenes, 7, 502
nitration with, 6, 108
reaction with isoprene, 7, 505
synthesis, 7, 534
Nitryl tetrafluoroborate
addition to alkenes, 7, 493
Nocardia corallina
epoxidation, 7, 429
Nocardicins
synthesis
Ugi reaction, 2, 1101
Nodusmycin
synthesis
via macrolactonization, 6, 373
Nojirimycin, 1α-cyano-1-deoxy-
1α-amino derivative
synthesis, 1, 364
Nojirimycin, deoxy-
synthesis
FDP aldolase, 2, 463
Nojirimycin, 1-deoxy-
synthesis
via aminomercuration–oxidation, 7, 638
Nonactic acid
methyl ester
synthesis, 1, 131
synthesis
via chiral acetals, 2, 651
via [4 + 3] cycloaddition, 5, 611
1,2-Nonadiene
hydrogenation
homogeneous catalysis, 8, 450
1,8-Nonadiyne
oxidative polymerization, 3, 552
Nonanal
reaction with iodobenzene
chromium(II) chloride, 1, 193
synthesis
via hydroformylation, 4, 918
1,9-Nonanedioic acid, 5-methylene-
dimethyl ester
intramolecular acyloin coupling reaction, 3, 625
2,5-Nonanedione
aldol cyclization, 2, 161
Nonanoic acid, 2-methyl-
dimethyl ester
intramolecular acyloin coupling reaction, 3, 626
2,4,6,8-Nonatetraenaldehyde, 9-phenyl-
synthesis, 8, 273
Nonatrienes
Diels–Alder reactions
diastereoselection, 5, 515–527
twist asynchronicity, 5, 516
heteroatom substituted
Diels–Alder reactions, 5, 527–532
1,4,8-Nonatrienes
hydroboration, 8, 708
1,6,8-Nonatrienes
cis-fused
Diels–Alder reactions, 5, 524
Nonatrienes, amido-
Diels–Alder reactions
intramolecular, 5, 529
1,6,8-Nonatrienes, sulfonyl-
Diels–Alder reactions
intramolecular, 5, 522
trans-Non-6-enal
synthesis
via photocycloaddition, 5, 165
1-Nonene, 3-acetoxy-
oxidation
Wacker process, 7, 453
2-Nonene, 1-acetoxy-
Wacker oxidation, 7, 453
1-Nonene, 6,7-dihydroxy-
Wacker oxidation
synthesis of brevicomin, 7, 451
8-Nonenoate, (*R*)-3-oxo-7-(methoxycarbonyloxy)-
palladium complex
chirality transfer, 4, 649
trans-Non-6-en-1-ol
synthesis
via photocycloaddition/reduction, 5, 165
Non-Kolbe electrolysis, 3, 634
carbenium ions, 3, 649
experimental procedure, 3, 654
Nonmetallodealumination, 8, 754
Nonylamine, 2-hydroxy-
synthesis
chiral, 1, 559
Nookatone
synthesis
via Raphael–Nazarov cyclization, 5, 779
via Wacker oxidation, 7, 458
Nopol
synthesis
ene reaction, 2, 529
Nopol benzyl ether
reduction
9-borobicyclo[3.3.1]nonane, 8, 102
Noradamantane
synthesis, 3, 854
epoxide ring opening, 3, 746
Norbornadiene
anodic oxidation, 7, 796
carbolithiation, 4, 869
carbonylation
cobalt carbonyl catalyst, 3, 1024
[2 + 2 + 2] cycloaddition reactions, 5, 1130
dissolving metal reductions, 8, 481
homo-Diels–Alder cycloaddition, 5, 1141
hydrobromination, 4, 283

hydrochlorination, **4**, 276
hydrogenation
homogeneous catalysis, **8**, 449
hydrosilylation, **8**, 781
oxidation
palladium(II) catalysis, **4**, 559
oxidative halogenation, **7**, 528
oxide
rearrangement, **3**, 736
Pauson–Khand reaction, **5**, 1049
photocyclization
chemoselectivity, **5**, 206
synthesis
via photoisomerization, **5**, 205
Norbornadiene, 7-alkoxy-
reduction
diimide, **8**, 475
Norbornadiene, 2,3-dimethoxycarbonyl-
[3 + 2] cycloaddition reactions
with methylenecyclopropane, **5**, 289
Norbornadiene, 7-methylene-
adduct with tetracyanoethylene, **5**, 65
Norbornadienol
oxidative rearrangement, **7**, 824
7-Norbornadienol
sodium salt
1,3-sigmatropic shift, **5**, 1003
Norbornane
carbocations
rearrangement, **3**, 707
Norbornane, 2-bromo-
synthesis, **7**, 604
Norbornane, 1-iodo-
bromide substitution, **6**, 3
Norbornane-2-carboxylic acid
enolates
diastereoselective alkylation, **3**, 39
Norbornanethiocarboxamides, 3-oxo-
synthesis
via dithiocarboxylic acids, **6**, 421
Norbornanone
synthesis
via sequential Michael ring closure, **4**, 262
Norbornene
aziridination, **7**, 479
[2 + 2 + 2] cycloaddition reactions, **5**, 1130
deuterium addition, **8**, 427
hydrocarboxylation, **4**, 939
hydroformylation, **4**, 932
hydrozirconation, **8**, 689
metallo-allylation/methoxycarbonylation
nickel-ene reaction, **5**, 36
oxidative halogenation, **7**, 528
oxide
rearrangement, lithium halide catalyzed, **3**, 764
rearrangement, lithium perchlorate catalyzed, **3**, 761
reduction, dissolving metals, **8**, 880
2,3-*exo*-oxides
rearrangement, **3**, 740
Pauson–Khand reaction, **5**, 1049
reaction with lithium organometallics, **4**, 869
ring opening metathesis polymerization, **5**, 1120
synthesis
via vinylcyclopropane rearrangement, **5**, 1013
1-Norbornene
dimerization, **5**, 65
Norbornene, *endo,endo*-5,6-bis(methoxycarbonyl)
living polymer synthesis, **5**, 1121
Norbornene, 2-chloro-
exo-oxide
rearrangement, **3**, 739
5-Norbornene-2-carboxylic acid
synthesis
via Diels–Alder reaction, **5**, 365, 366
Norbornene, *exo*-methylene-
synthesis
via Diels–Alder reactions, **5**, 324, 358
Norbornen-2-ols
Pauson–Khand cycloaddition
regioselectivity, **5**, 1042
Norbornenone
Baeyer–Villiger reaction, **7**, 682
homologation of ketones, **3**, 783
2-Norbornenones
Pauson–Khand cycloaddition
regioselectivity, **5**, 1042
Norcamphor
Baeyer–Villiger reaction, **7**, 682
ethylene ketal
reduction, **8**, 222
ketals
selective reduction, **8**, 218
Norcaradiene
tautomerism, **5**, 713
Norcarane
oxidation, **7**, 12
Norcarane, dibromo-
rearrangement
bicyclobutane derivative, **4**, 1013
Norcarane, dichloro-
synthesis
via dichlorocarbene, **4**, 1000
Norcardicin
synthesis, **6**, 760
A-Nor-5α-cholestan-3β-ol, 3α-carboxy-
synthesis, **3**, 834
A-Nor-5α-cholestan-2β-ol-2-carboxylic acid
rearrangements, **3**, 832
synthesis, **3**, 833
Norcoralydine
synthesis, **3**, 81
Norephedrine
N-acyl-2-oxazolidone from, **2**, 251
18-Nor-D-homo steroids
angular alkylation, **3**, 17
Norjasmone, dihydro-
synthesis
via double bond migration, **7**, 457
Norlongifolane, 3-keto-
semicarbazone
reduction, **8**, 338
Norpectinatone
synthesis
via conjugate addition to α,β-unsaturated carboxylic acid, **4**, 202
Norpinane, 2-ethylidene-
π-allylpalladium complexes from, **4**, 587
A-Nor-5α-pregnan-2-ol-20-one-2-carboxylic acid
synthesis, **3**, 833
Norpyrenophorin
synthesis, **3**, 126

Norrish type II reaction
  cyclobutanol, **3**, 1048
Norsecurinine
  synthesis
    via Horner–Wadsworth–Emmons reaction, **1**, 769
Norsterepolide
  synthesis
    via Raphael–Nazarov cyclization, **5**, 779
Norsteroids
  synthesis
    via benzocyclobutene ring opening, **5**, 693
19-Norsteroids
  synthesis
    polyene cyclization, **3**, 371
A-Norsteroids
  synthesis, **3**, 903
D-Norsteroids
  synthesis, **3**, 901
19-Nortestosterone
  synthesis, **7**, 460
Nortricyclene, 3-methoxy-
  synthesis, **3**, 653
A-Nortriterpenes
  synthesis, **3**, 903
Novobiocin
  microbial oxidation, **7**, 77
Nozaki protocol
  application
    1,4- and 1,2-addition, **1**, 101
Nuciferal
  synthesis, **3**, 161
Nuclear magnetic resonance
  carbanions, **1**, 41
  carbonyl compounds
    Lewis acid complexes, **1**, 292
  Knoevenagel reaction products
    structure determination, coupling constants, **2**, 345
Nucleophilic addition
  arene–metal complexes, **4**, 517–547
  donor radical cations, **7**, 878
  electrophilic coupling
    carbanions, **4**, 237–263
  radical cations
    bimolecular reaction, **7**, 859
Nucleophilic aromatic substitution
  diasteroselectivity, **4**, 426
  enantioselectivity, **4**, 426
  regioselectivity, **4**, 426
  solid-state, **4**, 445
Nucleophilic coupling
  aryl radicals, **4**, 451–480
  arynes, **4**, 483–513
Nucleophilic/electrophilic carbacondensation
  definition, **4**, 238
Nucleosides
  amino sugars
    synthesis, **7**, 712
  analogs
    synthesis, Eschenmoser coupling reaction, **2**, 889
  5′-hydroxyl group
    selective masking, **6**, 657
  phosphorylation, **6**, 603
  synthesis
    via Peterson alkenation, **1**, 792
C-Nucleosides
  synthesis
    via Baeyer–Villiger reaction, **7**, 682
    via [4 + 3] cycloaddition, **5**, 605, 611
    via organomercury compounds, **4**, 839
Nucleosides, 6-alkylpurine
  synthesis
    coupling reactions, **3**, 462
Nucleosides, 4-amino-5-aminocarbonylimidazolyl-
  alkylation, **6**, 501
C-Nucleosides, 3-deoxy-
  synthesis, **1**, 113
Nucleotides
  amide-type protecting groups, **6**, 642
Nystatin
  synthesis
    use of aldol reaction, **2**, 195
Nystatin, *N*-(deoxyfructosyl)-
  synthesis, **6**, 789

O

Obaflorin
  β-lactone, **6**, 342
Obtusilactone
  synthesis, **3**, 844; **6**, 784
Occidentalol
  synthesis, **8**, 924
    via retro Diels–Alder reaction, **5**, 569
Ochratoxins
  synthesis
    via ortho lithiation, **1**, 470
Ochromycinone
  synthesis, **1**, 567
Ocimene
  synthesis
    via carboalumination, **4**, 893
Ocimenones
  synthesis
    diene acylation, **2**, 720
Ocoteine
  intracoupling reaction
    with benzyltetrahydroisoquinoline, **3**, 670
  synthesis, **3**, 81
Octadecane, 9,10-epoxy
  synthesis, **1**, 718
Octadecene, 7,8-epoxy-2-methyl-
  synthesis
    via t-butyl 5-methylhexyl sulfoxide, **1**, 514
9-Octadecen-18-olide
  synthesis
    alkene metathesis, **5**, 1118
1,7-Octadiene
  microbial epoxidation, **7**, 429
2,6-Octadiene
  cyclization, **3**, 342
1,7-Octadiene, 3-acetoxy-
  cyclization
    palladium-ene reaction, **5**, 50
  synthesis
    via palladium-catalyzed oxidation, **7**, 460
Octadiene, 4,5-dimethyl-
  Cope rearrangement, **5**, 821
1,7-Octadien-3-one
  synthesis
    via hydrolysis and oxidation, **7**, 460
2,7-Octadienyl acetates, 4-alkyl-4-hydroxy-
  cyclization
    palladium-ene reaction, **5**, 47
1,7-Octadiyne, 1,8-diethoxy-
  bicyclization, **5**, 1171
$\Delta^{9(10)}$-Octalin
  reduction
    trialkylsilane, **8**, 486
Δ^{4}-Octalin, 4-(3-butenyl)-3-oxo-
  synthesis
    via Michael addition and aldol condensation, **7**, 460
Δ^{5}-Octalin, 4,4,10-trimethyl-
  allylic oxidation, **7**, 100
Octalinediones
  synthesis
    via intramolecular addition, **4**, 46
Octalin-1-one
  synthesis
    via homoenolate addition reaction, **4**, 120
1(9)-Octalin-2-one
  α′-alkylation, **3**, 21
  cross conjugated lithium dienolate
    metallation, **3**, 21
1(9)-Octalin-2-one, 10-methyl-
  cyclohexylamine
    methylation, **3**, 33
Octalins
  conformation, **3**, 354
  2,3-sigmatropic rearrangement
    chirality transfer, **6**, 893
β-Octalone
  hydrogenation
    catalytic, **8**, 533
$\Delta^{3,4}$-2-Octalone
  synthesis
    via cyclohexanone, **7**, 460
Octalone, methyl-
  synthesis, **7**, 464
Octalones
  addition to allene
    photochemical cycloaddition, **5**, 130
  aldol cyclization, **2**, 162
  Clemmensen reduction, **8**, 312
  Nazarov cyclization, **5**, 757
  reduction
    dissolving metals, **8**, 525
  synthesis
    metal–ammonia reduction, **2**, 184
    via Robinson annulation, **4**, 7
Octanal, 2-methyl-
  synthesis
    via hydroformylation, **4**, 918
Octane, 2-bromo-
  reaction with cyclohexanone
    samarium diiodide, **1**, 259
Octane, 1-cyano-
  synthesis
    via 2-octyl sulfonate, **6**, 236
Octane, 1,2-epoxy-
  hydride migration
    epoxide ring opening, **3**, 742
Octane, 2-iodo-
  Kornblum oxidation
    solvent, **7**, 655
Octane, methoxy-
  synthesis, **7**, 603
1,8-Octanedioic acid, 2,7-dimethyl-
  dimethyl ester
    synthesis, **3**, 623
1,2-Octanediol
  oxidative cleavage, **7**, 708
3,4-Octanedione, 2-acetoxy-
  cycloaddition reactions, **5**, 247
2-Octanol
  catalytic hydrogenation, **8**, 814
  oxidation
    solid support, **7**, 845
2-Octanone
  reduction
    samarium diiodide, **8**, 115

4-Octanone, 5-hydroxy-
synthesis
acyloin coupling reaction, **3**, 619
3-Octanone, 2-hydroxy-2,6-dimethyl-
Wolff–Kishner reduction, **8**, 926
Octatetraynediamines
synthesis, **3**, 555
1,3,5-Octatriene
intermediate
2,4,6-octatriene electrocyclization, **5**, 702
2,4,6-Octatriene
electrocyclization, **5**, 702
selective reduction, **8**, 568
1,4,7-Octatriene, 2,7-dimethyl-
synthesis
via cycloaddition of 1-methylbicyclo[1.1.0]butane, **5**, 1186
2-*trans*-4-*trans*-6-*trans*-Octatrienoic acid, 3,7-dimethyl-
synthesis
via sulfones, **6**, 157
1,4,7-Octatriyne-3,6-diol, 3,6-di-*t*-butyl-
synthesis, **3**, 557
Octavalene, 3-phenyl-5-bromo-
synthesis
via dihalocyclopropyl compounds, **4**, 1017
trans-2-Octenal
synthesis, **6**, 139
6-Octenal, 7-methyl-
ene reaction, **2**, 541
Octene
hydrozirconation, **8**, 673
1-Octene
carbolithiation, **4**, 868
ene reactions
Lewis acid catalysis, **5**, 4
hydrosilylation, **8**, 763, 774
oxidation
Wacker process, **7**, 451, 452
4-Octene
hydroformylation, **4**, 918
1-Octene, 1,3-bis(methylthio)-
synthesis, **6**, 139
4-Octene oxide
deoxygenation, **8**, 888
1-Octen-3-ol
synthesis
via retro Diels–Alder reaction, **5**, 554
2-Octenol
acetate
oxidation, **7**, 464
6-Octen-3-one, 8-bromo-4-methyl-
cyclization
samarium diiodide, **1**, 266
1-Octen-3-one, 1-halo-
reduction, **8**, 163
Octenyl radicals
cyclization, **4**, 786
6-Octen-1-yne
cyclization
intramolecular ene reaction, **5**, 15
1-Octen-7-yne, 8-(trimethylsilyl)-
reaction with cyclopentadienylzirconium complexes, **5**, 1165
Octosyl acid A
synthesis, **7**, 245
Diels–Alder reaction, **2**, 696
D-*manno*-2-Octulopyranosate, 3-deoxy-
synthesis
Diels–Alder reaction, **2**, 692
2-Octylamine
synthesis, **6**, 80
Octyl nitrite
photolysis, **7**, 9
Octyne
reduction
dissolving metals, **8**, 479
4-Octyne
hydroalumination, **8**, 748
2-Octyne, 1-bromo-
reaction with Grignard reagents, **3**, 273
2-Octyne, 1-iodo-
reaction with 3-butyn-1-ol, **3**, 273
3-Octyn-2-one
photocycloaddition reactions, **5**, 164
Olah's reagent
fluorination
alkyl alcohols, **6**, 216
Olamide, *N*-methyl-
Ritter reaction, **6**, 268
Oleandomycin
seco acid
synthesis, **2**, 264
13(18)-Oleanene
synthesis, **3**, 709
Oleic acid
Kolbe electrolysis, **3**, 642
Oligomerization
aliphatic alkenes
Friedel–Crafts reaction, **3**, 331
template-controlled, **4**, 825
Oligonucleotides
synthesis
need for alcohol protection, **6**, 650
Oligopeptides
synthesis
via asymmetric hydrogenation of dehydropeptides, **8**, 460
Oligosaccharides
side chain cleavage, **8**, 219
synthesis, **6**, 33; **7**, 245
via heteroatom cyclization, **4**, 391
Olive fly pheromone
synthesis, **7**, 237
Oliventolic acid
methyl ester
synthesis, **2**, 621
Onium compounds
Ritter reaction, **6**, 287
α-Onocerin
synthesis, **3**, 638
Oogoniol steroids
synthesis
via microbial methods, **7**, 73
Ophiobolane
synthesis
via divinylcyclobutane rearrangement, **5**, 1029
Ophiobolin C
synthesis
via Brook–Claisen rearrangement, **5**, 843
Ophiobolins
Dieckmann reaction, **2**, 824
synthesis, **3**, 575; **7**, 710

via alkenylchromium reagents, **1**, 200
via Cope rearrangement, **5**, 806
via Nazarov cyclization, **5**, 759
via oxy-Cope rearrangement, **1**, 883
Oppenauer oxidation
primary alcohols, **7**, 309
trichloroacetaldehyde
secondary alcohols, **7**, 320
Oppolzer's chiral auxiliary
use in amine synthesis, **6**, 77
Oppolzer's chiral sultam
reactions with nitrile oxides, **4**, 1079
Orantine, *O*-methyl-
synthesis
via iodine azide addition to alkene, **4**, 350
Orcinol
synthesis, **2**, 170
Orellanine
synthesis, **3**, 509
Organic conductors
S-aryl arenecarbothioates, **6**, 441
Organic oxides
oxidation
thiols, **7**, 760
Organoaluminum reagents, **1**, 77–105
acylation
palladium catalysis, **1**, 450
1,2-addition reactions
carbon–nitrogen compounds, **1**, 98
alkene protection, **6**, 690
Claisen rearrangement
catalysis, **5**, 850
ate complexes
reactions with keto esters, **1**, 86
ate complexes, silyl
acyl silane synthesis, **1**, 97
chiral
site selective addition reactions, **1**, 78
stereoselective addition reactions, **1**, 78
conjugate additions
alkenes, **4**, 140–144
nucleophilic addition to π-allylpalladium complexes, **4**, 595
regioselectivity, **4**, 635
stereochemistry, **4**, 620
reactions with acid derivatives, **1**, 92
reactions with epoxides
alcohol synthesis, **6**, 4
reactions with α,β-unsaturated carbonyl compounds
site selectivity, **1**, 81
tandem vicinal difunctionalization, **4**, 257
Organobismuth reagents
pentavalent
glycol cleavage, **7**, 704
Organoboranes
autoxidation, **7**, 598
deprotonation, **1**, 490
electrocyclic reactions, **7**, 594
group transfer
radical addition reactions, **4**, 756
ionic reactions
stereochemistry, **7**, 594
oxidations, **7**, 594
carbonyl compounds, **7**, 603
pyridinium chlorochromate, **7**, 264
protonolysis, **8**, 724
radical reactions, **7**, 594
reactivity, **7**, 593
rearrangements, **3**, 779
α-substituted
cleavage, **1**, 490
synthesis, **8**, 703
Wurtz coupling, **3**, 418
Organoboranes, dialkoxy(α-phenylthio)-
oxidation, **7**, 604
Organoboron compounds
conjugate additions
alkenes, **4**, 144–148
oxidation, **7**, 330
rearrangements, **3**, 793
Organoboronic acids
vinyl substitutions
palladium complexes, **4**, 841
Organocadmium reagents, **1**, 211–227
addition reactions, **1**, 225
with alkenic π-systems, **4**, 98
diastereoselective addition reactions, **1**, 220
reactions with carbonyl compounds
Lewis acid promotion, **1**, 326
reactions with imines
Lewis acid promotion, **1**, 349
Organocadmium reagents, allylic
addition reactions, **1**, 226
Organocadmium reagents, benzylic
addition reactions, **1**, 226
Organocerium reagents, **1**, 231–248
reactions, **1**, 233
synthesis, **1**, 232, 233
thermal stability, **1**, 233
Organochromium reagents, **1**, 173–207
carbanion equivalents, **1**, 174
C—C bond forming reactions, **1**, 175
reactions with carbonyl compounds
Lewis acid promotion, **1**, 331
structure, **1**, 174
synthesis, **1**, 174
Organocobalt complexes
radical cyclizations
nonchain methods, **4**, 805
Organocopper, allylic reagents
reaction with benzaldehyde, **1**, 113
Organocopper reagents, **1**, 107–136
acylation, **1**, 426
palladium catalysis, **1**, 450
stoichiometric, **1**, 426
1,2-additions
aldehydes and ketones, **1**, 108
imines, nitriles and amides, **1**, 119
alkylations
nonstabilized carbanions, **3**, 208
alkynyl
reactions with enones, **1**, 118
association with boron trifluoride
increased reactivity, **1**, 347
catalysts, **3**, 210
preparation, **3**, 208
conjugate additions, **4**, 228, 240
alkenes, **4**, 148–153
coupling, **3**, 415
cross-coupling reactions
unsaturated halides, **3**, 522
enolates

acylation, **2**, 832
from chiral carbanions
conjugate additions, **4**, 227
natural product synthesis, **1**, 125
reactions with aldehydes, **1**, 108
reactions with amides, **1**, 124
reactions with electrophiles
mechanism, **3**, 213
reactions with epoxides
rates, **1**, 343
reactions with imines, **1**, 119
reactions with ketones, **1**, 116
reactions with nitriles, **1**, 123
synthesis, **3**, 208, 419
tandem vicinal dialkylation, **4**, 254–257
Organocuprates (*see also* Cuprates)
conjugate addition reactions, **4**, 169–195
alkenes, **4**, 148–153
Lewis acid effects, **4**, 179
mechanism, **4**, 170
reagent variations, **4**, 173
solvent effects, **4**, 178
nontransferable ligands, **4**, 175–177
organozinc compounds in synthesis, **4**, 175
reactions with epoxides
alcohol synthesis, **6**, 4
reactions with α-seleno-α,β-unsaturated ketones, **1**, 669
synthesis, **4**, 170
triorganotin groups
transfer, **4**, 174
Organofluorosilicates
synthesis, **7**, 642
Organoiron phthalocyanines
vinyl substitutions
palladium complexes, **4**, 841
Organolithium reagents
acylation, **1**, 399
addition reactions
with alkenic π-systems, **4**, 72–83
aggregation, **4**, 257
aromatic nucleophilic substitution, **4**, 427
asymmetric
nucleophilic addition reactions, **1**, 69
chiral dipole-stabilized
stereoselective alkylation, **3**, 75
conjugate additions, **4**, 229
cyclization, **4**, 871
deselenations, **8**, 849
enantioselective addition
aldehydes, **1**, 70
indicator
1,3-diphenylacetone tosylhydrazone, **6**, 784
ketone synthesis
from carboxylic acids, **1**, 411
nucleophilic addition reactions
carbonyl compounds, **1**, 49
chiral ketones, **1**, 58
nucleophilic addition to π-allylpalladium complexes, **4**, 596
regioselectivity, **4**, 635–637
stereochemistry, **4**, 620
oxidation, **7**, 330
phosphonium ylide synthesis, **6**, 174
reactions with acetals, **1**, 347
reactions with carbonyl compounds
Lewis acid promotion, **1**, 329
reactions with epoxides
use of Lewis acids, **1**, 343
Wurtz coupling, **3**, 419
Organomagnesium compounds
addition reactions
copper catalyzed, **4**, 89–93
with alkenic π-systems, **4**, 83–89
oxidation, **7**, 330
primary
coupling reactions with alkenyl halides, **3**, 436
Wurtz coupling, **3**, 415
Organomanganese compounds
addition reactions
with alkenic π-systems, **4**, 98
Organomercury compounds, **1**, 211–227
acylation
palladium catalysis, **1**, 450
addition reactions, **1**, 225
addition to alkenes, **4**, 968
reaction with π-allylpalladium complexes
stereochemistry, **4**, 620
vinyl substitutions
palladium complexes, **4**, 838
Organometallic compounds
acylation
palladium catalysis, **1**, 450
alkyl
reactions with selenides, **1**, 630
alkynyl
[3 + 2] cycloaddition reactions, **5**, 277
allenyl
[3 + 2] cycloaddition reactions, **5**, 277–281
allyl
cycloaddition reactions, **5**, 272–277
isotopic perturbation techniques, **2**, 977
allyl and propargyl/allenic
reactions with imines, **2**, 975–1004
aromatic nucleophilic substitution, **4**, 427–429
cross-coupling reactions
with unsaturated halides, **3**, 522
crotyl
isotopic perturbation techniques, **2**, 977
reactions with aldimines, regiochemistry, **2**, 978
reactions with aldimines, stereochemistry, **2**, 978
hydride transfer
reduction of carbonyls, **8**, 98
nitrile synthesis, **6**, 241
oxidation, **7**, 613
reactions
Lewis acids, **1**, 325–353
reactions with aldehydes
Cram *versus* anti-Cram selectivities, **1**, 80
reactions with cyclic ketones
stereoselectivity, **1**, 333
reactions with epoxides
alcohol synthesis, **6**, 4
Organometallic compounds, alkenyl-
carbozincation, **4**, 880
reaction with 1-alkynyl halides, **3**, 529
Organometallic compounds, aryl-
reaction with oxygen, **7**, 329
Organometallic compounds, α-silyl-
addition reactions, **1**, 618
Organometallic compounds, vinyl-
acylation, **1**, 401

alkynylation, **3**, 521
Organometallic polymers
synthesis, **3**, 557
Organonickel compounds
acylation, **1**, 451
catalysts
carbanion alkylations, **3**, 228
Organopalladium compounds
catalysts
Grignard reactions, **3**, 230
synthesis, **4**, 834
vinyl substitutions, **4**, 833–861
Organophosphoric acids
derivatives, **6**, 601–627
Organophosphorus reagents
amide synthesis, **6**, 389
Organosamarium 'ate' complexes, **1**, 253
Organosamarium halides
reactions with aldehydes, **1**, 254
Organosamarium reagents
Barbier-type reactions, **1**, 255
carbonyl addition reactions, **1**, 253
reactions with enolizable reagents, **1**, 253
synthesis
via transmetallation, **1**, 253, 254
Organoselenium reagents
carbanions, **1**, 629–724
oxidation
allylic alcohols, **7**, 307
Organosilanes
addition reactions
with alkenic π-systems, **4**, 98
conjugate additions
alkenes, **4**, 155–158
Mannich reactions, **2**, 1030
reactions with carbonyl compounds
Lewis acid promotion, **1**, 327
reductive cleavage
benzylic compounds, **8**, 969
Organosilicon compounds
bond energies, **1**, 582
carbanions
field effects, **1**, 580
hyperconjugation, **1**, 581
inductive effects, **1**, 580
p–d π-bonding, **1**, 581
reactions with carbonyl compounds, **1**, 579–625
selectivity, **1**, 580
nucleophilic substitution reactions, **1**, 582
reactivity
carbanions, **1**, 580
Organosilver compounds
carbometallation
enynes, **4**, 905
Organosodium compounds
coupling, **3**, 414
vinyl substitutions
palladium complexes, **4**, 841
Organostannanes
acylation
acid chlorides, **1**, 446
palladium complex catalysis, **1**, 436
α-alkoxy organolithiums from, **3**, 195
conjugate additions
alkenes, **4**, 155–158
hydride donors
reduction of carbonyls, **8**, 98
reactions with carbonyl compounds
Lewis acid promotion, **1**, 327
toxicity, **7**, 614
Organotellurides
dehalogenation
α-halocarbonyl compounds, **8**, 990
Organothallium compounds
arylation
vinyl ketones, **4**, 841
reactions with π-allylpalladium complexes, **4**, 595
Organotin compounds
coupling reactions
with alkenyl halides, **3**, 442
with aromatic halides, **3**, 452
3-iodo-2-[(trimethylsilyl)methyl]propene
trimethylenemethane synthetic equivalent, **5**, 246
nucleophilic addition to π-allylpalladium complexes, **4**, 594
regioselectivity, **4**, 633
stereochemistry, **4**, 619
primary alkyl
coupling reactions with aromatic halides, **3**, 453
Organotitanium reagents, **1**, 139–170
properties, **1**, 140
reactions with carbonyl compounds
Lewis acid promotion, **1**, 330
reactivity, **1**, 144
synthesis, **1**, 142
Organoytterbium reagents
reactions with carbonyl compounds
synthesis of alcohols, **1**, 277
use, **1**, 276
Organozinc reagents, **1**, 211–227
acylation
palladium catalysis, **1**, 448
addition reactions, **1**, 215
with alkenic π-systems, **4**, 93–97
conjugate additions, **4**, 229
copper-catalyzed reactions, **3**, 221
coupling reactions
with alkenyl halides, **3**, 442
with aromatic halides, **3**, 452
diastereoselective addition reactions, **1**, 220
hydride donors
reduction of carbonyls, **8**, 99
1-methoxy-2-butyne, **2**, 91
nickel catalysts, **3**, 228
nucleophilic addition to π-allylpalladium complexes, **4**, 595
regioselectivity, **4**, 634
stereochemistry, **4**, 619
perfluoroalkyl
with alkenyl halides, **3**, 444
primary alkyl
coupling reactions with alkenyl halides, **3**, 442
coupling reactions with aromatic halides, **3**, 453
reactions with carbonyl compounds
Lewis acid promotion, **1**, 326
reactions with imines
Lewis acid promotion, **1**, 349
secondary alkyl
coupling reactions with alkenyl halides, **3**, 442
Wurtz coupling, **3**, 420
synthesis, **1**, 211, **8**, 698
transmetallation, **1**, 214

Organozinc reagents, allylic
  synthesis, **1**, 212
Organozinc reagents, benzylic
  synthesis, **1**, 212
Organozirconium reagents, **1**, 139–170
  conjugate additions
    alkenes, **4**, 153–155
  nucleophilic addition to π-allylpalladium complexes, **4**, 595
    regioselectivity, **4**, 635
    stereochemistry, **4**, 620
  properties, **1**, 140
  reactivity, **1**, 144
  synthesis, **1**, 142
  tandem vicinal difunctionalization, **4**, 257
Orientalone
  synthesis
    use of potassium ferricyanide, **3**, 680
Ornithine, *N*-benzyl-*threo*-β-hydroxy-
  synthesis, **8**, 648
Ornithine, *threo*-β-hydroxy-
  synthesis, **8**, 648
Orsellinic acid
  synthesis, **2**, 170
Ortho acids
  synthesis, **6**, 556
Ortho amides
  alkoxymethyleniminium salt synthesis, **6**, 505
  alkylmercaptomethyleniminium salt synthesis, **6**, 511
  amide acetal synthesis, **6**, 571
  amidine synthesis, **6**, 553
  amidinium salt synthesis, **6**, 518
  aminal ester synthesis, **6**, 575
  hydride donating ability, **8**, 85
  imidate synthesis, **6**, 533
  tris(dialkylamino)alkane synthesis, **6**, 581
Orthocarbonates
  nitrile synthesis, **6**, 238
Orthocarbonic acids
  alkylmercaptomethyleniminium salt synthesis, **6**, 512
  derivatives
    tris(dialkylamino)alkane synthesis, **6**, 582
Orthocarbonic esters
  ortho ester synthesis, **6**, 562
Ortho esters
  acylation
    hydrogen sulfide, **6**, 450
  2-*O*-acylglycosyl halides, **6**, 49
  amide acetal synthesis, **6**, 570
  amidine synthesis, **6**, 553
  amidinium salt synthesis, **6**, 518
  aminal ester synthesis, **6**, 574
  bicyclic
    synthesis, **6**, 561
  2,2-bis(dialkoxy)carbonitrile synthesis, **6**, 564
  carboxy group protection
    organometallic transformation, **6**, 673
  cyclic
    diol protection, **6**, 659
    synthesis, **6**, 557
  diol protection, **6**, 660
  hydride donating ability, **8**, 85
  imidate synthesis, **6**, 533
  Knoevenagel reaction, **2**, 368
  ortho ester synthesis, **6**, 563
  reduction
    metal hydrides, **8**, 266
  spirocyclic
    synthesis, **6**, 560
  sugar
    synthesis, **6**, 561
  sulfhydrolysis, **6**, 450
  synthesis, **6**, 485–583
  tandem vicinal dialkylations, **4**, 261
  tris(dialkylamino)alkane synthesis, **6**, 581
Ortho esters, β-keto
  synthesis, **6**, 556
Ortho esters, triaryl
  synthesis, **6**, 556
Ortho esters, vinyl
  Diels–Alder reactions, **5**, 341
Orthoformates, tri-*t*-butyl
  synthesis, **6**, 556
Orthoformic acid
  synthesis
    Reimer–Tiemann reaction, **2**, 774
Orthoformylation
  aromatic compounds, **3**, 969
Orthopropionic acid, methyl-3-phenylsulfonyl
  cyclopentannulation, **6**, 164
Orthothioformates
  Michael donors, **4**, 259
Orthotrithiobenzoate, 3-iodotriethyl-
  reaction with methanol, **6**, 564
Ortyn
  anticholinergenic and spasmolytic agent, **3**, 826
Osmium
  catalyst
    carbonyl compound hydrogenolysis, **8**, 320
    oxyamination, **7**, 489
Osmium, bromohydridocarbonyltris(triphenylphosphine)-
  hydrogenation
    alkenes, **8**, 449
Osmium, pentaamine-
  acetone complex
    crystal structure, **1**, 310
Osmium *t*-alkylimides
  reactions with alkenes, **7**, 485
Osmium tetroxide
  alkene oxidation
    diasteroselectivity, **7**, 439
    stoichiometry, **7**, 439
  asymmetric dihydroxylation, **7**, 429
  α-hydroxylation
    ketones, **7**, 166
  syn hydroxylation
    alkenes, **7**, 439
  osmylation
    arenes, **7**, 863
  oxidation
    alkenes, mechanism, **7**, 438
    primary alcohols, **7**, 310
    sulfoxides, **7**, 768
  oxidative cleavage of alkenes
    catalysts, **7**, 542
    synthesis of carbonyl compounds, **7**, 564
  reaction with alkyl enol ethers, **7**, 170
  reaction with vinyl cyanide, **7**, 172
Osmylation
  arenes, **7**, 862
    electron transfer, **7**, 866

charge-transfer
arenes, **7**, 865
features, **7**, 865
thermal
features, **7**, 865
Oudemansins
synthesis
via Horner reaction, **1**, 777
Ovalicin
synthesis, **6**, 784
Ovatodiolide
transannular cyclization, **3**, 407
Overlap control
Darzens glycidic ester condensation, **2**, 413
Perkin reaction, **2**, 398
Oxaallylic anions
aldol reaction
Group III enolates, **2**, 1
1-Oxa-2-aza-di-π-methane rearrangements
photoisomerizations, **5**, 202
Oxabetweenallenes
synthesis
organocopper compounds, **3**, 223
7-Oxabicyclo[2.2.1]heptane, tetramethylene-
Diels–Alder reactions, **5**, 384
7-Oxabicyclo[2.2.1]hept-5-en-2-one
reactions with organocuprates, **1**, 117
9-Oxabicyclo[3.3.1]nonane
synthesis, **2**, 623
8-Oxabicyclo[3.2.1]oct-6-ene
Pauson–Khand reaction, **5**, 1050, 1051
3-Oxabicyclo[3.3.0]oct-6-en-7-ones
synthesis
use of cobalt complexes, **3**, 1025
8-Oxabicyclo[3.2.1]oct-6-en-3-ones
[4 + 3] cycloaddition reactions
tropone synthesis, **5**, 609
synthesis
via [4 + 3] cycloaddition, **5**, 594, 597, 605
Oxabicyclo[2.2.2]octyl ortho esters
carboxy group protection
organometallic transformation, **6**, 673
1-Oxa-1,3-butadienes
cationic
Diels–Alder reactions, **5**, 501
Diels–Alder reactions, **5**, 453–458
intramolecular, **5**, 464–468
electron-deficient
Diels–Alder reactions, **5**, 458–464
electron-donating substituted
Diels–Alder reactions, **5**, 464
hetero
Diels–Alder reactions, **5**, 468
Oxacepham
synthesis, **5**, 1107
Ugi reaction, **2**, 1103
4-Oxa-5α-cholestan-3-one
hydrogenation, **8**, 247
Oxacines, dihydro-
synthesis
via cyclobutene ring expansion, **5**, 687
2-Oxacyclopentylidene
transition metal complexes
synthesis, **5**, 1076
4*H*-1,3,4-Oxadiazinium salts
synthesis
via iodocyclization of allylbenzohydrazides, **4**, 391
Oxadiazin-5-ones
reduction, **8**, 663
1,2,4-Oxadiazoles
reduction, **8**, 663
1,2,5-Oxadiazoles
reduction, **8**, 664
1,3,4-Oxadiazoles
Diels–Alder reactions, **5**, 491
reduction, **8**, 664
1,2,4-Oxadiazoles, 3-azido-
synthesis, **6**, 245
1,3,4-Oxadiazoles, 2,5-diaryl-
synthesis, **6**, 490
1,2,4-Oxadiazoline, 3-phenyl-
reduction, **8**, 663
Δ^3-1,3,4-Oxadiazolines
thermolysis
carbonyl ylide generation, **4**, 1089
Oxadiazolin-5-ones
carbon dioxide elimination
nitrilimines from, **4**, 1084
1,3,4-Oxadiazolium salts
reduction
sodium sulfide, **8**, 664
Oxa-di-π-methane rearrangements
applications, **5**, 229–235
mechanism, **5**, 216–219
photoisomerizations, **5**, 200, 215–235
limitations, **5**, 228
substrates, **5**, 219–228
Oxahydrindene
synthesis, **7**, 300
via heteronucleophile addition, **4**, 34
Oxalacetic acid
hydrogenolytic asymmetric transamination, **8**, 147
Oxalates, bisthioxo-
dialkyl esters
synthesis, **6**, 450
Oxalates, methyl
reduction
stannanes, **8**, 824
Oxalic acid
reactions with Grignard reagents
synthesis of α-keto esters, **1**, 425
2-Oxalin-5-one, 4-arylmethylene-
synthesis
Perkin reaction, **2**, 404
Oxalylation
ketones, **2**, 838
Oxalyl chloride
acid chloride synthesis, **6**, 304
acid halide synthesis, **6**, 308
activator
DMSO oxidation of alcohols, **7**, 296
alcohol oxidation
dimethyl sulfoxide, **7**, 291
chloromethyleneiminium salt preparation, **2**, 779
reactions with alkanes, **7**, 7
Oxalyl chloride, ethyl-
Friedel–Crafts reaction, **2**, 741
Oxalyl dichloride
imidoyl halide synthesis, **6**, 523
Oxalyl dihalides
reaction with amides, **6**, 495
Oxamates

synthesis
carbonylation of amines and alcohols, **3**, 1040
Oxametallacyclic compounds
ytterbium, **1**, 279
2-Oxa-3-metalla-1,5-diene
3,3-sigmatropic rearrangement, **2**, 6
Oxamination
alkenes, **7**, 488
vicinal
palladium(II) catalysis, **4**, 560
7-Oxanorbornadiene
Pauson–Khand reaction, **5**, 1050
Oxapenam
synthesis, **5**, 1107
1,2-Oxaphosphetanes
intermediates in Wittig reaction, **1**, 755
Oxaphospholene, methylene-
synthesis
from activated allene, **4**, 57
Oxaporphine
synthesis
oxidative coupling, **3**, 670
Oxasecoalkylation
chain extension
via Grob fragmentation, **6**, 1048
1-Oxa-2-silacyclohexa-3,5-diene, 2,2-dimethyl-
Diels–Alder reaction, **5**, 587
Oxasilatane
Peterson alkenation, **1**, 785
Oxaspirocyclopentane
use in synthesis, **5**, 919
Oxaspiro[2.0.*n*]heptanes
synthesis, **1**, 712
Oxaspiro[2.0.*n*]hexanes
synthesis, **1**, 712
Oxaspirolactone
synthesis
via cyclofunctionalization of hydroxyoctynoic acid, **4**, 394
1-Oxaspiro[2.6]nonane
solvolysis
transannular hydride shifts, **3**, 735
Oxaspiro[2.0.*n*]octanes
synthesis, **1**, 712
Oxaspiropentanes
rearrangement
lithium perchlorate catalyzed, **3**, 761
synthesis
via diphenylcyclopropylsulfonium halides, **1**, 820
1,3-Oxathiane
carbonyl group protection, **6**, 680
metallated
alkylation, **3**, 135
nucleophilic addition reactions
stereoselectivity, **1**, 61
1,3-Oxathiane, 2-lithio-
alkylation, **3**, 137
Oxathianes
chiral
nucleophilic addition reactions, **1**, 63
reduction, **8**, 231
1,4-Oxathiocine
synthesis
via ketocarbenoids and thiophenes, **4**, 1063
Oxathiolanes
reduction, **8**, 231
1,3-Oxathiolanes
carbonyl group protection, **6**, 680
Oxathiolanes, 4,4-dimethyl-
3,3-dioxide
alkylation, **3**, 136
1,2,3,4-Oxatriazoles
synthesis
via acyl azides, **6**, 251
7-Oxatricyclo[4.2.0.0]octane
synthesis
via Paterno–Büchi reaction, **5**, 157
Oxazaborolidine
synthesis, **8**, 171
Oxazaborolidine, *B*-methylated
synthesis, **8**, 171
1,3,5-Oxazaphospholes, 4,5-dihydro-
nitrile ylides from, **4**, 1081
1,3,2-Oxazaphosphorinane
Claisen rearrangement, **5**, 847
Oxazepanedione
Knoevenagel reaction, **2**, 357
Oxazepane-5,7-dione
Knoevenagel reaction
stereoselectivity, **2**, 351
Oxazepines
addition reaction
with organomagnesium compounds, **4**, 89
Michael additions, **4**, 206
reductive alkylation
Birch reduction, **8**, 508
1,2-Oxazepines, dihydro-
synthesis
via cyclization of β-allenic oximes, **4**, 397
1,2-Oxazine-3,6-dione, tetrahydro-
photochemical decarboxylation, **7**, 729
1,2-Oxazines
chiral
deprotonation, **2**, 486
reduction, **8**, 652
1,3-Oxazines
reduction, **8**, 653
synthesis, **6**, 534
1,4-Oxazines
reduction, **8**, 653
1,3-Oxazines, 2-alkyldihydro-
alkylation, **3**, 53
1,3-Oxazines, allyloxymethyl-
Wittig rearrangement, **3**, 1005
Oxazines, dihydro-
synthesis
via cyclization of methylisourea, **4**, 388
via Diels–Alder reaction, **5**, 418
via iodocyclization of pentenol imidates, **4**, 408
1,2-Oxazines, 3,6-dihydro-
synthesis
via Diels–Alder reactions, **5**, 417
1,3-Oxazines, dihydro-
reaction with carbonyl compounds
two-carbon homologation, **2**, 493
reduction
sodium borohydride, **8**, 275
synthesis
via carboxylic acids, **8**, 275
via Ritter reaction, **6**, 273, 295
α,β-unsaturated
preparation, **2**, 493

1,3-Oxazines, 5,6-dihydro-
bicyclic
synthesis, **2**, 1071
synthesis
via photocycloaddition, **5**, 161
1,3-Oxazines, 2-methyldihydro-
metallated
reactions, **2**, 492
1,3-Oxazines, 2-styryldihydro-
addition reactions
with organolithium compounds, **4**, 76
Oxazines, tetrahydro-
synthesis
via cyclization of pentenol derivatives, **4**, 408
via allylic systems, 6-*endo* cyclization, **4**, 386
via iodocyclization of unsaturated amine oxides, **4**, 391
1,4-Oxazines, tetrahydro-
reduction, **8**, 653
1,3-Oxazines, 4,4,6-trimethyl-5,6-dihydro-
methiodide salt
reactions with organometallic compounds, **1**, 366
1,3-Oxazin-2-one
synthesis
N-acyliminium ions, **2**, 1054
6*H*-1,3-Oxazin-6-one
synthesis
via retro Diels–Alder reactions, **5**, 584
Oxazin-2-one, 3-bromotetrahydro-
synthesis, **1**, 376
Oxazinone, tetrahydro-
synthesis
via bromocyclization of allylamine carbamates, **4**, 387
Oxazinones
synthesis, **2**, 1071
Oxazirconacycloheptenes
synthesis
via dienylzirconium reagents, **1**, 162
Oxaziridines
reduction, **8**, 395
spirocyclic
synthesis, **1**, 838
synthesis, **1**, 834
via imines, **1**, 837
via oxidation of imines, **7**, 750
Oxaziridines, 2-aryl-3-sulfamyl-
oxidation
sulfides, **7**, 778
Oxaziridines, 2-arylsulfonyl-3-phenyl-
α-hydroxylation
ketones, **7**, 162
Oxaziridines, camphorsulfonyl-
α-hydroxylation
ketones, **7**, 162
Oxaziridines, pentamethylene-
reaction with alkenes, **7**, 470
Oxaziridines, 2-sulfamyl-
oxidation of sulfides
synthesis of sulfoxides, **6**, 150
Oxaziridines, 2-sulfonyl-
α-hydroxylation
amides, **7**, 183
enones, **7**, 176
esters, **7**, 181
ketones, **7**, 162
oxidation
selenides, **7**, 772
Oxazole, 3-acetyl-2,3-dihydro-2,2-dimethyl-
photochemistry, **5**, 160
Oxazole, dihydro-
enolates
diastereoselective alkylation, **3**, 40
Oxazole, 4,5-diphenyl-
synthesis
protected carboxy groups, **6**, 675
Oxazole, 5-methyl-
metallation, **1**, 477
Oxazole, 2-stannyl-
coupling reactions, **3**, 511
with bromobenzenes, **3**, 514
1,3-Oxazole-2,4-diones
synthesis
via [2 + 2 + 2] cycloaddition, **5**, 1141
Oxazoles
Diels–Alder reactions, **5**, 382, 491
metallation
addition reactions, **1**, 471
reduction, **8**, 650
synthesis
from acyclic imidate salts, **2**, 488
via Ritter-type reactions, **6**, 275
Oxazolide, phosphorobis-
phosphorylating agent, **6**, 619
Oxazolidine, 4,5-dialkyl-
synthesis
via heterocyclization of acylaminomethyl ethers, **4**, 407
Oxazolidine, α,β-ethylenic
conjugate additions
organocopper compounds, **4**, 210
1,3-Oxazolidine, 3-methyl-
Mannich reaction, **2**, 965
1,2,4-triazole-catalyzed, **2**, 965
Oxazolidine, 5-phenyl-
synthesis, **7**, 492
1,3-Oxazolidine-4-carboxylic acid
methyl ester
addition reactions with nitroalkenes, **4**, 109
Oxazolidine-*N*-oxyl, 4,4-dimethyl
nitroxide
synthesis, **1**, 393
Oxazolidines
enolates
diastereoselective alkylation, **3**, 40
Mannich reaction, **2**, 965
synthesis
via cyclization of amidals, **4**, 408
synthesis, **7**, 492
Oxazolidinimides
acylation, **2**, 846
Oxazolidinium salts
synthesis
via bromocyclization of dialkylaminomethyl ethers, **4**, 407
Oxazolidinone enolates, *N*-(α-bromoacetyl)-
Darzens glycidic ester condensation, **2**, 437
Oxazolidinones
asymmetric hydroxylation, **7**, 184
enolates
α-hydroxylation, **7**, 184
synthesis, **1**, 273

via bromocyclization of allylamine carbamates, **4**, 387
via bromocyclization of thiocarbamidate, **4**, 406
via cyclization of allylamines, **4**, 389
via Ritter reaction, **6**, 279
Oxazolidinones, *N*-acyl-
chiral
Diels–Alder reactions, **5**, 361
Diels–Alder reactions
intramolecular asymmetric, **5**, 543
α,β-unsaturated
Diels–Alder reactions, **1**, 312
Oxazolidinones, *N*-acryloyl-
[2 + 2] cycloaddition reaction
1,1-bis(methylthio)ethylene, **5**, 24
Diels–Alder reactions, **5**, 376
Oxazolidinones, *N*-crotonoyl-
Diels–Alder reactions, **5**, 376
Oxazolidinones, 2,7,9-decatrienoyl-
Diels–Alder reactions, **5**, 362
Oxazolidinones, *N*-(α-haloacetyl)-
asymmetric
Darzens glycidic ester condensation, **2**, 436
preparation, **2**, 436
1,3-Oxazolidin-2-ones, 4-isopropyl-3-phenacyl-
reduction
sodium borohydride, **8**, 652
Oxazolidinones, 2,8,10-undecatrienoyl-
Diels–Alder reactions, **5**, 362
2-Oxazolidones
synthesis
from organic isocyanates and terminal alkene epoxides, **3**, 765
2-Oxazolidones, *N*-acyl-
chiral
conversion to boron enolate, **2**, 250
2-Oxazolidones, 3-(2-bromopropionyl)-
Reformatsky reaction
stereoselectivity, **2**, 292
Oxazolidones, *N*-(α,β-unsaturated acyl)-
conjugate additions
Lewis acids, **4**, 140
2-Oxazolidones, 4-vinyl-
hydrolysis, **6**, 88
Oxazolidylacetyl chloride, phenyl-
synthesis
via phenylglycine, **5**, 98
Oxazoline anions
addition reactions
alkenes, palladium(II) catalysis, **4**, 572
1,3-Oxazoline-4-carboxylic acid
methyl ester
addition reactions with nitroalkenes, **4**, 109
Oxazolines
acylation, **2**, 805
alkenic
iodolactamization, **7**, 503
arene substitution reactions, **4**, 425
aromatic
nucleophilic addition reactions, **1**, 69
carboxy group protection
organometallic transformation, **6**, 674
chiral
conjugate additions, **4**, 204–206
dehydrogenation
use of benzeneseleninic anhydride, **7**, 132
enantioselective aldol reaction
gold catalysis, **2**, 317
glycosides
synthesis, **6**, 42
homochiral
boron azaenolates, **2**, 252
metallation, **3**, 261
addition reactions, **1**, 468
optically active
N-allylketene *N,O*-acetal synthesis, **6**, 862
reduction, **8**, 650
ring opening, **7**, 487
synthesis, **7**, 477, 490, 493
via Ritter reaction, **4**, 293
tandem vicinal difunctionalization, **4**, 252
α,β-unsaturated
preparation, **2**, 491
1,2-Oxazolines
synthesis
via cyclization of allylic amides, **4**, 386
1,3-Oxazolines
alkylation, **6**, 505
synthesis, **6**, 534
via iodocyclization of trichloroacetimidates, **4**, 407
2-Oxazolines
reaction with amines, **6**, 74
synthesis
via Ritter reaction, **6**, 295
Oxazolines, 2-alkyl-
alkylation, **3**, 53
metallated achiral
reactions with carbonyl compounds, **2**, 489
Oxazolines, allyloxymethyl-
Wittig rearrangement, **3**, 1005
Oxazolines, allyloxymethyl anions
Wittig rearrangement
lithium cation chelation, **3**, 1006
Oxazolines, aryl-
vicinal dialkylations, **4**, 257
Oxazolines, chiral
nucleophilic addition reactions
remote asymmetric induction, **1**, 60
Oxazolines, 4,5-dialkyl-
synthesis
via stereoselective cyclization, **4**, 386
2-Oxazolines, 4,4-dimethyl-
methiodide salt
reactions with organometallic compounds, **1**, 366
Oxazolines, 2-ethyl-
metallated
reactions, **2**, 490
Oxazolines, 2-methyl-
metallated
reactions, **2**, 489
1,2-Oxazolines, 5-methylene-
synthesis
via cyclization of propargylamines amides, **4**, 387
Oxazolines, 1-naphthyl-
vicinal dialkylations, **4**, 257
Oxazolines, α-(phenylthio)vinyl-
addition reactions
with organolithium compounds, **4**, 76
Oxazolines, 3-pyridyl-
reaction with organometallic compounds, **4**, 427
Oxazolines, 4-pyridyl-
ortho-metallation, **4**, 428

Oxazolines, 2,4,4-trialkyl-
N-oxide
reactions with Grignard reagents, **1**, 393
Oxazolines, vinyl-
addition reactions
with organolithium compounds, **4**, 76
Oxazolinium salts, N-methyl-
electrochemical addition reactions, **4**, 130
Oxazolinones
reduction, **8**, 650
Oxazolin-4-ones
cycloadditions, **4**, 1163
2-Oxazolin-5-ones
reduction
sodium borohydride, **8**, 651
3-Oxazolin-2-ones
reduction
LAH, **8**, 652
4-Oxazolin-2-ones
photocycloaddition reaction, **5**, 160
Oxazolin-5-ones, 4-benzylidene-2-methyl-
hydrolysis, **2**, 406
Oxazolin-5-ones, 4-benzylidene-2-phenyl-
hydrolysis, **2**, 406
Δ^2-Oxazolium 5-oxides
azomethine ylides from, **4**, 1097
Oxazolium salts
electroreduction
acyl carbanion equivalents, **1**, 544
reduction, **8**, 650
2-Oxazolone, 3-ketopinyl-
cleavage
lower order cuprate, **1**, 119
Oxazol-5-one, 2-phenyl-
Perkin reaction, **2**, 403
Oxazolone, triphenyl-
synthesis
via Ritter-type reactions, **6**, 276
Oxazolones
Friedel–Crafts reaction, **2**, 744
geometric isomers, **2**, 403
Michael addition, **4**, 12
Perkin reaction, **2**, 405
synthesis
Erlenmeyer, **2**, 396
via Ritter-type reaction, **6**, 276
Oxazol-2-ylacetic acid, 4-cyano-5-methyl-
ethyl ester
synthesis, **6**, 775
Oxepane
synthesis
via photochemical reaction, **6**, 448
Oxepine, dihydro-
synthesis
via Cope rearrangement, **4**, 1049
Oxepinobenzofuran
synthesis
ferricyanide oxidation, **3**, 666
Oxepins
Prins reaction, **2**, 564
synthesis
via oxirane rearrangement, **5**, 929
via photocycloaddition, **5**, 165
3-Oxetanecarboxylic acid
ester
Wolff rearrangement, **3**, 902
Oxetanes
alkylation with
sulfur- and selenium-stabilized carbanions, **3**, 86
alkynylation
use of boron trifluoride, **1**, 343
arene alkylation by
Friedel–Crafts reaction, **3**, 314
coupling reactions
with phenyllithium, **3**, 466
reaction with lithiodithiane, **1**, 569
reaction with α-selenoalkyllithium, **3**, 91
Ritter reaction, **6**, 276
strained tricyclic
decomposition, **5**, 178
synthesis
via epoxides, **1**, 820
via cyclofunctionalization of allylic alcohols, **4**, 368
via Paterno–Büchi reaction, **5**, 151
unsaturated
stereoselective synthesis *via* photocycloaddition, **5**, 176
Oxetanes, alkoxy-
synthesis
via photocycloaddition, **5**, 161
Oxetanes, 2-alkoxy-
synthesis
via thermal cycloaddition, **5**, 151
Oxetanes, 3-alkoxy-
synthesis
via Paterno–Büchi reaction, **5**, 151, 159
Oxetanes, 2-alkylidene-
synthesis
via photocycloaddition, **5**, 167
Oxetanes, *trans*-4-alkyl-3-methylthio-
synthesis
via Paterno–Büchi reaction, **5**, 160
Oxetanes, alkynyl-
synthesis
via photocycloaddition, **5**, 176
Oxetanes, α-chloro
synthesis, **7**, 725
Oxetanes, 2-imino-
synthesis
via ketenimines, **5**, 114
via lithium (N-phenyl)phenylethynamide cycloaddition, **5**, 117
via photocycloaddition, **5**, 167
Oxetanes, 3-methoxy-
synthesis, **1**, 670
Oxetanes, 2-methyl-
benzene alkylation by
Friedel–Crafts reaction, **3**, 314
Oxetanes, 2-methylene-
synthesis
via retro Diels–Alder reactions, **5**, 577
Oxetanes, 3-methylene-
synthesis
via retro Diels–Alder reactions, **5**, 577
Oxetanes, 3-siloxy-
synthesis
via Paterno–Büchi reaction, **5**, 158
Oxetanes, 2-thioxo-
synthesis
via lithium thioalkynolate cycloaddition, **5**, 117
Oxetanocin

synthesis
via Paterno–Büchi reaction, **5**, 151
2-Oxetanones
synthesis
via ketenes and carbonyls, **5**, 86–89
via lithium phenylethynolate cycloaddition, **5**, 116
Oxetenes
synthesis
via photocycloaddition, **5**, 162
via ynamines and carbonyls, **5**, 116
Oxetin
synthesis
via Paterno–Büchi reaction, **5**, 151
Oxidation
activated C—H bonds, **7**, 83–113
alcohols, **7**, 251–286, 291–302, 305–325
dehydrogenation, **7**, 119–146
ethers, **7**, 235–248
hydroxylation, **7**, 151–187
quinone synthesis, **7**, 345–356
sulfur compounds, **7**, 193–214
vinylic, **7**, 329–344
allylic stannanes, **7**, 616
π-allylpalladium complexes, **7**, 629
arenes
nucleophilic displacement of hydrogen by, **4**, 423
azo compounds
synthesis of azoxy compounds, **7**, 750
biomimetic, **7**, 40
by pyridinium salts
of primary and secondary alcohols, **8**, 96
carbon–boron bonds, **7**, 593–608
carbon–carbon bonds
microbial, **7**, 66
carbon–halogen bonds, **7**, 653–669
carbon–hydrogen bonds
remote functionalization, **7**, 39–51
carbon–mercury bonds, **7**, 631
carbon–metal bonds, **7**, 613–638
carbon–palladium bonds, **7**, 629
carbon–silicon bonds, **7**, 641–650
carbon–tin bonds, **7**, 614
unactivated, **7**, 614
definition, **7**, 39
electrochemical, **7**, 707, 789–811
alkenes, **7**, 98
enzymatic, **7**, 99
hydroalumination adducts, **8**, 753
nitrogen compounds, **7**, 735–753
nitroso compounds, **7**, 751
oximes, **7**, 751
phosphorus compounds, **7**, 735–753
primary alcohols, **7**, 305
primary amines, **7**, 736
secondary amines, **7**, 745
selenides
to selenones, **7**, 773
to selenoxides, **7**, 770
selenium compounds, **7**, 757–779
selenols, **7**, 769
solid-supported reagents, **7**, 839–847
alumina, **7**, 841
clay, **7**, 845
silica, **7**, 842
spores, **7**, 80
sulfoxides
to sulfones, **7**, 766
sulfur compounds, **7**, 757–779
tellurium compounds, **7**, 757–779
tertiary nitrogen compounds, **7**, 748
γ-trialkylstannyl alcohols, **7**, 621
trigonal nitrogen compounds, **7**, 749
unactivated C—H bonds, **7**, 1–17
microbial methods, **7**, 53–80
vinylstannanes, **7**, 620
zirconium compounds
to alcohols, **8**, 691
Oxidation potentials
definition
electrosynthesis, **8**, 129
electron donors, **7**, 853
electron-transfer oxidation
driving force, **7**, 852
organic compounds, **7**, 852
Oxidative cleavage
alkenes, **7**, 541
nitrogen and sulfur functionalization, **7**, 588
phase transfer catalysis, **7**, 559
Oxidative coupling
copper–polymer complex catalysts, **3**, 559
organometallic acetylides, **3**, 554
phenols, **3**, 659–700
terminal alkynes, **3**, 552
Oxidative cyclization
dienyliron complexes
stereocontrol, **4**, 686
Oxidative decarboxylation
aliphatic carboxylic acids, **7**, 722
1,4-dihydrobenzoic acids, **8**, 500
Oxidative demercuration
alkoxymercuration, **7**, 631
Oxidative demethylation
methoxyarenes, **7**, 346
Oxidative desilylation
C—Si to C—O, **8**, 788
Oxidative halogenation
halometallic reagents, **7**, 527
Oxidative homocoupling reactions
terminal alkynes, **3**, 552
Oxidative phosphorylation, **6**, 614
Oxidative rearrangements, **7**, 815–836
skeletal, **7**, 827
Oxides
oxidation
thiols, **7**, 761
N-Oxides
oxidation with, **7**, 661
reactions with arynes, **4**, 508
synthesis
via oxidation of tertiary amines, **7**, 748
2-Oxidoallyl, 1-hydroxy-
[4 + 3] cycloaddition reactions, **5**, 597
Oxidobenzopyrylium ylides
cycloadditions, **4**, 1163
1-Oxido-2-pyridylmethyl group
phosphoric acid protecting group, **6**, 624
Oxidopyrylium ylides
cycloadditions, **4**, 1163
Oxidoreductases
dehydrogenation
carbonyl compounds, **7**, 145
sources, **8**, 184

2,3-Oxidosqualene
  synthesis, **3**, 99
Oxime acetates
  α-hydroxylation, **7**, 186
  photoisomerization, **5**, 202
Oxime esters
  reduction
    hydrides, **8**, 60
    to amines, **8**, 64
    to hydroxylamine esters, **8**, 60
Oxime ether anions
  structure, **6**, 727
Oxime ethers
  asymmetric reduction
    Lewis acid coordination, **1**, 317
  O-benzyl
    reactions with silyl ketene acetals, **2**, 940
  boron trifluoride activated
    reactions with organometallic compounds, **1**, 385
  nucleophilic radical addition
    tin pinacolate, **4**, 765
  reactions with allyl organometallic compounds, **2**, 994
  reactions with enolates, **2**, 939
  reactions with organometallic compounds, **1**, 385
  reduction
    hydrides, **8**, 60
    to amines, **8**, 64
    to hydroxylamine ethers, **8**, 60
Oximes
  addition reactions
    carbon-centred radicals, **4**, 765
  alkenic mesylates
    intramolecular cyclization, **6**, 771
  anions
    acyclic, **2**, 386
    cyclic, **2**, 386
    stabilized, **2**, 486
  aromatic
    oxidation, **7**, 276
  aziridine synthesis, **1**, 835
  Beckmann rearrangement, **6**, 404, 763
  carbamates
    Neber reaction, **6**, 787
  carbonyl group derivatization, **6**, 726
  carbonyl group protection, **6**, 682
  cleavage
    trimethylsilyl chlorochromate, **7**, 285
  cyclic
    stereoselective reduction, **8**, 64
  cyclization, **4**, 1120
  deoxygenation
    titanium(III) chloride, **8**, 371
  dianions
    structure, **6**, 727
  Diels–Alder reactions, **5**, 412
  electroreduction, **8**, 135, 137
  hydrogenation
    asymmetric, **8**, 145
    catalytic, **8**, 143
    homogeneous, **8**, 155
  infrared spectra, **6**, 727
  isomerization
    Beckmann rearrangement, **7**, 691
  metallated
    addition reactions, **2**, 486
    metal enolate equivalents, **3**, 30
  methanesulfonates
    Beckmann rearrangement, **7**, 693
  N-nitrosation
    synthesis of *N*-nitrimines, **7**, 751
  NMR
    carbon **6**, 13, 727
    proton, **6**, 727
  oxidation, **7**, 751
  pinacol coupling reactions
    with ketones, **3**, 596
  properties
    chemical, **6**, 727
  radical additions
    alkoxy radicals, **4**, 815
  reactions, **6**, 727
  reactions with allyl organometallic compounds, **2**, 994
    diastereoselective, **2**, 32
  reactions with crotyl boronates
    syn–anti selectivity, **2**, 997
  reactions with organometallic compounds, **1**, 385
  reduction, **8**, 176
    dissolving metals, **8**, 124
    hydrides, **8**, 60
    stereoselectivity, **8**, 64
    synthesis of imines, **8**, 392
    to amines, **8**, 64
    to hydroxylamines, **8**, 60
  spectra, **6**, 727
  stereochemistry, **6**, 728
  structure, **6**, 727
  synthesis, **6**, 726
    via trimethylsilylamines, **7**, 737
  tosylates
    Beckmann rearrangement, **1**, 387
  ultraviolet spectra, **6**, 727
  Vilsmeier–Haack reaction, **2**, 792
Oximes, *O*-acyl-
  carboxyl radicals from, **7**, 719
Oximes, azido
  synthesis, **6**, 252
Oximes, α-hydroxyamino
  synthesis, **6**, 114
Oximes, *O*-methyl-
  reduction, **8**, 176
Oximes, β-stannyl
  oxidation, **7**, 628
Oximes, 2-sulfato-
  synthesis, **7**, 493
Oxime sulfonates
  rearrangement, **6**, 542
Oxindole, 3-alkylidene-
  synthesis
    via $S_{RN}1$ reaction, **4**, 477
Oxindoles
  synthesis
    via $S_{RN}1$ reaction, **4**, 477
    Sommelet–Hauser rearrangement, **3**, 969
Oxiranemethanol, 2-methyl-
  synthesis
    via asymmetric epoxidation, **7**, 398
    via 4-nitrobenzoate derivative, **7**, 398
Oxiranes (*see also* Epoxides)
  alkynic
    isomerization, **5**, 929
  alkynylation
    use of boron trifluoride, **1**, 343

dienyl
isomerization, **5**, 929
in alkene oxidation
hydrogen peroxide, 7, 446
neighboring group
epoxide opening, **3**, 735
oxidative cleavage, **7**, 709
ring opening
carbon nucleophiles, **6**, 4
carbonyl ylide generation, **4**, 1089
Lewis acids, **1**, 345
Oxiranes, 2-acyl-
preparation, **2**, 423
Oxiranes, 1,2-diaryl-
photofragmentation
carbonyl ylide generation, **4**, 1090
Oxiranes, 1,2-di-*n*-propyl-
reactions with boryl compounds, **1**, 497
Oxiranes, divinyl-
flash vacuum pyrolysis
product control, **5**, 930
synthesis
via [2 + 3] annulation, **5**, 930
Oxiranes, *trans*-divinyl-
rearrangement, **5**, 929
Oxiranes, 2-imidoyl-
preparation, **2**, 423
Oxiranes, 1-methyl-2-pentyl-
reaction with boryl compounds, **1**, 496
Oxiranes, phenyl-
cleavage
pyridinium chlorochromate, **7**, 267
Oxiranes, (*p*-tolylsulfinyl)-
synthesis
via chloromethyl *p*-tolyl sulfoxide, **1**, 524
Oxiranes, vinyl-
cyclic
nucleophilic opening, **5**, 931
nucleophilic opening, **5**, 931
optically pure
racemization, **5**, 929
radical addition reactions
alkenes, **5**, 931
radical polymerization, **5**, 931
reaction with aryl Grignard reagents, **3**, 265
reaction with organocopper reagents, **6**, 848
reaction with sodium phenoxide
ring opening reaction, **5**, 936
rearrangements, **5**, 909, 928
synthesis, **1**, 510, 712
from allylic ethers, **2**, 70
via photoisomerization, **5**, 200
thermal isomerization, **5**, 929
Oxiranes, vinylalkynyl-
rearrangement
to vinylcyclopropyl aldehydes, **5**, 931
Oxocane
2,8-disubstituted
synthesis, **7**, 679
Oxocarbenium ions
cyclization, **6**, 750
heterocyclic synthesis, **6**, 749
synthesis, **6**, 749
Δ^4-Oxocene, 2,8-disubstituted
synthesis
via cyclization of oxonium ions, **1**, 591
Oxocenes
synthesis, **6**, 752
type III ene reaction, **2**, 555
via cyclization of acetals, **1**, 589
Oxocenone
synthesis
via activated alkynes, **4**, 53
Oxocineole
enzymic reduction
diastereotopic face distinction, **8**, 192
2*H*-Oxocins, 3,6,7,8-tetrahydro-
synthesis
via cyclization of acetals, **1**, 589
Oxocrinine
synthesis, **3**, 683
Oxone — *see* Potassium hydrogen persulfate
Oxonium hexachloroantimonate, *O*-acetyldiethyl-
Friedel–Crafts reaction, **2**, 737
Oxonium ions
chiral
reaction with enol silanes, **2**, 650
initiators
polyene cyclization, **3**, 343, 354
Oxonium ylides
rearrangements, **3**, 942; **6**, 874, 881
Wittig rearrangement, **3**, 1008
Oxo process
hydroformylation of alkenes, **4**, 914
Oxosulfenylation
alkenes, **4**, 335, 337
Oxosulfonium ylides
addition reactions, **4**, 115
2-Oxyallyl, 1-phenyl-
[4 + 3] cycloaddition with 3-methylfuran, **5**, 601
Oxyallyl cations
[4 + 3] cycloaddition reactions, **5**, 594
polyene cyclization, **3**, 354
2-Oxyallyl synthons
[3 + 2] cycloaddition reactions, **5**, 282–287
[4 + 3] cycloaddition reactions, **5**, 603
Oxyamidation
alkenes, **7**, 488
Oxyanion-accelerated rearrangements
small rings, **5**, 1000–1004
Oxy-Cope reactions
palladium(II) catalysis, **4**, 563–565
Oxy-Cope rearrangements, **1**, 880; **5**, 785–822
allylic systems, **6**, 834, 863
anionic, **5**, 785–822
following Wittig rearrangement, **3**, 994
3-hydroxy-1,5-hexadienes, **5**, 1000
irreversibility, **5**, 795
oxyanion-accelerated, **5**, 1000
product aromatization, **5**, 791
trienes, **5**, 889
Oxygen
epoxidations using, **7**, 384
molecular
amide α-hydroxylation, **7**, 183
enone α-hydroxylation, **7**, 175
ester α-hydroxylation, **7**, 180
ketone α-hydroxylation, **7**, 156, 159
oxidation
ethers, **7**, 247
singlet

allylic oxidation, **7**, 96, 110
ester α-hydroxylation, **7**, 182
ketone α-hydroxylation, **7**, 165, 169
oxidative rearrangements, **7**, 816
reaction with bis-silyl ketene acetals, **7**, 185
reaction with silyl dienol ethers, **7**, 177
triplet
radical reactions, **4**, 720
Oxygenase
aromatic hydroxylation
catalyst, **7**, 80
Oxygen-centered radicals
cyclizations, **4**, 811–814
Oxygen heterocycles
ring opening, **8**, 957
Oxygen nucleophiles
addition reactions
alkenes, **4**, 552–559
aromatic nucleophilic substitution, **4**, 437–440
nucleophilic addition to π-allylpalladium complexes, **4**, 596–598
regioselectivity, **4**, 637
stereochemistry, **4**, 621
Oxymercuration
alkenes, **4**, 741
synthesis of ketones, **7**, 451
humulene, **3**, 400
oxidative demercuration, **7**, 632
13-Oxyprostanoids
synthesis
via conjugate addition of aryl cyanohydrin, **1**, 552
Oxy radicals
cyclizations, **4**, 811
α-Oxy radicals
addition reactions
tin hydride catalysis, **4**, 739
Oxytelluration
alkenes, **4**, 343
terminal alkenes
enol ether preparation, **2**, 598
Oxythallation
alkenes
synthesis of ketones, **7**, 451
Ozone
alkane oxidation, **7**, 14
silica support, **7**, 842
oxidation
ethers, **7**, 247
primary amines, **7**, 737
selenides, **7**, 771
oxidative cleavage of alkenes
catalysts, **7**, 542
synthesis of alcohols, **7**, 543
synthesis of carbonyl compounds, **7**, 544
synthesis of carboxylic acids, **7**, 574
reactions with alkenes
1,3-dipolar cycloadditions, **4**, 1098
Ozonization
ethers, **7**, 247
methylene groups
solid support, **7**, 842
Ozonolysis
cyclic alkenes
in ammonia, **7**, 507
hydrazones
regeneration of carbonyl groups, **2**, 524
silyl enol ethers, **7**, 166
vinyl silane
generation of α-hydroxy ketones, **7**, 172

P

Paliclavine
  synthesis
    via nitrile oxide cyclization, **4**, 1131
Palladation
  alkenes, **7**, 490
  vinyl substitutions, **4**, 835
Palladium
  allylic oxidation, **7**, 94
    catalyst, **7**, 107
  barium sulfate
    epoxide hydrogenolysis, **8**, 882
  carbon
    catalyst, alkyl halide hydrogenolysis, **8**, 794
  catalyst
    acyl chloride reduction, **8**, 286
    carbanion alkylations, **3**, 227
    Cope rearrangement, **5**, 799
    cross-coupling reactions, **3**, 523
    cycloaddition reactions, methylenecyclopropanes, **5**, 1188
    synthesis of enynes, **3**, 217
  charcoal
    epoxide hydrogenation, **8**, 882
  dehydrogenation, **7**, 139
    mechanism, **7**, 141
  hydrogenation catalyst
    pyridines, **8**, 597
  hydrogenolysis
    allyl halides, **8**, 956
  oxidative rearrangment, **7**, 828
  polymer-bound
    catalyst, hydrogenation, **8**, 418
  reduction
    transfer hydrogenation, **8**, 366
Palladium, η^3-allyl-
  [3 + 2] cycloaddition reactions, **5**, 300
Palladium, π-allyl-
  reactions with nucleophiles, **6**, 20
Palladium, allylchloro-
  catalyst
    TASF reaction with organic halides, **3**, 233
Palladium, η^3-allylcyclopentadienyl-
  [3 + 2] cycloaddition reactions
    methylenecyclopropane, **5**, 289
Palladium, benzyl(chloro)bis(triphenylphosphine)-
  catalyst
    acylation, **1**, 440
Palladium, bis(acetonitrile)dichloro-
  catalyst
    vinyl iodide reaction with organotin compounds, **3**, 232
Palladium, bis(dibenzylideneacetone)-
  catalyst
    vinyl substitutions, **4**, 835
  [3 + 2] cycloaddition reactions
    methylenecyclopropane, **5**, 289
Palladium, bis(phenylphosphine)pentakis-
  hydrogenation
    alkenes, **8**, 447
Palladium, dichlorobis(triphenylphospine)-
  catalysis
    halide carbonylation, **3**, 1021
    vinyl iodide reaction with organotin compounds, **3**, 232
Palladium, dichloro(DPPP)-
  desulfurizations
    allyl sulfones, **8**, 840
Palladium, phenylbis(triphenylphosphine)-
  catalysis
    arylmagnesium halide reaction with alkyl halides, **3**, 244
Palladium, tetrakis(triphenylphosphine)-
  catalyst
    acyl halide reduction, **8**, 265
    aryl halide reaction with organotin compounds, **3**, 232
    coupling reactions between organolithium and vinyl halides, **3**, 485
    [3 + 2] cycloaddition reactions, **5**, 299
    halide carbonylation, **3**, 1021
    vinylic Grignard coupling, **3**, 485
    vinyl substitutions, **4**, 835
  desulfurizations
    allyl sulfones, **8**, 840
  nitrile synthesis, **6**, 232
Palladium, η^3-trimethylenemethane-, **5**, 244
  [3 + 2] cycloaddition reactions, **5**, 300
Palladium acetate
  allylic oxidation, **7**, 94
  catalyst
    [3 + 2] cycloaddition reactions, **5**, 299
  diazo compound decomposition catalyst, **4**, 1033
  oxidation
    diols, **7**, 314
Palladium bis(trifluoroacetate)
  allylic oxidation, **7**, 94
Palladium chloride
  allylic oxidation, **7**, 95
  catalysts
    alkene dimerization, **3**, 482
    alkenyl halide dimerization, **3**, 484
    alkyne trimerization, **5**, 1147
  metal hydride reduction
    carbonyl compounds, **8**, 315
Palladium chloride, bis(benzonitrile)-
  diazo compound decomposition catalyst, **4**, 1033
Palladium complexes
  acylation
    catalysis, **1**, 436
  catalysts
    hydrosilylation, **8**, 764
  ferrocene
    catalyst, Grignard reagent alkylation, **3**, 244
Palladium complexes, π-allyl-
  addition of carbon nucleophiles
    functional group effects, **4**, 629
    ligand effects, **4**, 631
    regioselectivity, **4**, 627–632
    stereochemistry, **4**, 615–621
    substituent effects, **4**, 627–629
  addition of enolates
    regioselectivity, **4**, 632
  chemoselectivity, **4**, 587–614
  diastereoselectivity, **4**, 614–627

mechanism, **4**, 614
in synthesis, **4**, 585–654
nucleophilic addition reactions, **4**, 610
chirality transfer, **4**, 649–651
enantioselectivity, **4**, 649–654
regioselectivity, **4**, 627–649
oxidation, **4**, 603; **7**, 629
regioselectivity, **4**, 645
stereochemistry, **4**, 625
photochemistry, **4**, 610
precursors, **4**, 587–590
reactions, **3**, 423; **4**, 600–614
regioselectivity, **4**, 643–649
reduction, **4**, 604–606
regioselectivity, **4**, 646
stereochemistry, **4**, 626
transformation to dienes, **4**, 608–610
umpolung, **4**, 606–608
regioselectivity, **4**, 647–649
stereochemistry, **4**, 626
Palladium complexes, montmorillonitesilyl-
reduction
nitroaromatics, **8**, 372
Palladium complexes, nitro-
alkene oxidation, **7**, 452
Palladium complexes, oxa-π-allyl-
reactions, **4**, 611–614
Palladium-ene reactions, **5**, 35–37
intramolecular, **5**, 46–56
stereoselectivity, **5**, 60
Palladium enolates
allylation, **4**, 592
Palladium homoenolates
β-elimination
α,β-unsaturated carbonyl compounds, **2**, 443
substitution reactions, **2**, 450
Palladium salts
catalysts
alkene addition reactions, **4**, 551
oxidative addition to allyl acetate, **4**, 614
Ritter reaction, **6**, 284
Palustric ester
photochemical ring opening, **5**, 712
Palygorskite
solid support
oxidants, **7**, 845
Palytoxin
synthesis
use of protecting groups, **6**, 632
via alkenylchromium reagents, **1**, 197
β-Panasinsene
synthesis
via Johnson methylenation, **1**, 739
via ketone methylenation, optical resolution, **1**, 533
Paniculide A
synthesis
via Diels–Alder reaction of alkynic ketone, **1**, 406
via retro Diels–Alder reactions, **5**, 579
β-Panisene
synthesis
via copper-catalyzed photocycloaddition, **5**, 147
Pantolactone
Diels–Alder reactions, **5**, 365
Pantolactone, keto-
asymmetric hydrogenation, **8**, 152
Pantoyllactone
synthesis
via enzymic reduction, **8**, 190
Parabanic acid
O,N-acetals
synthesis, **6**, 576
aminals
synthesis, **6**, 581
Paraconic ester
synthesis
via oxidation of lactol, **6**, 357
via Stobbe reaction, **6**, 356
[8]Paracyclophane, 4-carboxy-
synthesis, **3**, 905
[3.3]Paracyclophanediene
synthesis, **3**, 877
Paracyclophanes
hydrogenation, **8**, 437
synthesis
via intramolecular acyloin coupling reaction, **3**, 628
[10]Paracyclophanes
synthesis
via ene reaction with methyl propiolate, **5**, 8
Paraffins
dehydrogenation
nitrile oxides from, **4**, 1078
Paraldol
synthesis, **2**, 138
Parham cyclization
bromoaromatic carboxylic acids, **1**, 412
Parham-type cyclization
benzocyclobutanes, **3**, 251
Parinaric acid
synthesis, **3**, 116
(±)-Parthenin
synthesis, **2**, 161
Passerini reaction, **2**, 1083–1106
amide synthesis, **6**, 405
isocyanides, **6**, 295
mechanism, **2**, 1085
Patchoulol
microbial hydroxylation, **7**, 64
Paterno–Büchi photocycloaddition reaction, **5**, 151–188
dienyl systems, **5**, 165–168
excited-state asymmetric synthesis, **5**, 183
heterocyclic synthesis, **6**, 759
intramolecular, **5**, 178–183
imides with alkenes, **5**, 181
mechanism, **5**, 152–157
spectroscopy, **5**, 153
transannular, **5**, 179
Pauson–Khand reaction, **5**, 1037–1062
alkene regioselectivity
electronic effects, **5**, 1042
bicyclization–carbonylation of enynes, **5**, 1165
intermolecular, **5**, 1043–1053
intramolecular, **5**, 1053–1062
mechanism, **5**, 1039–1043
scope, **5**, 1038
Payne rearrangement
epoxides
hydroxy neighboring group, **3**, 735
epoxy alcohols, **7**, 402
2,3-epoxy alcohols, **6**, 89
PDE-I
synthesis
via Eschenmoser coupling reaction, **2**, 885

PDE-II
  synthesis
    via Eschenmoser coupling reaction, **2**, 885
Pederine
  asymmetric synthesis, **2**, 846
  synthesis
    via [4 + 3] cycloaddition, **5**, 612
    via reduction of imidates, **2**, 1050
(–)-Peduncularine
  synthesis
    via (*S*)-malic acid, **2**, 1065
Pelargonic acid
  acid chloride synthesis, **6**, 303
Pelletierine
  synthesis, **1**, 393; **8**, 273
    via Schopf reaction, **2**, 943
Penam, 6-bromo-
  enolates
    aldol reaction, **2**, 212
Penams
  synthesis
    via Ugi reaction, **2**, 1102
Penems
  synthesis, **6**, 899
    via cycloaddition with CSI, **5**, 105
Penicillanic acid, 6-acylaminothio-
  thiol ester synthesis, **6**, 438
Penicillin, 6-aminomethyl-
  synthesis
    via enolate–oxime ether condensation, **2**, 939
Penicillin acylase
  phenylacetyl group removal, **6**, 643
Penicillinate, diazo-
  decomposition
    rhodium(II) catalyzed, **4**, 1053
Penicillin G
  Curtius reaction, **6**, 812
  synthesis
    via Ugi reaction, **2**, 1103
Penicillins
  reaction with dichlorine monooxide, **7**, 537
  synthesis
    via cycloadditions of acid chlorides and imines, **5**, 92
    via Dieckmann reaction, **2**, 824
    via oxazolones, **2**, 396
Penicillins, diazo-
  rearrangement, **3**, 934
Penicillins, semisynthetic
  synthesis
    via 2-arylglycines, **3**, 303
Penicillin sulfoxide
  methyl ester
    Pummerer rearrangement, **7**, 205
Penicillium concavo-rugulosum
  hydrocarbon hydroxylation, **7**, 59
Penicillium decumbens
  reduction
    unsaturated carbonyl compounds, **8**, 558
Penicillium spinulosum
  epoxidation, **7**, 429
2,4,6,8,10,12,14-Pentadecaneheptaone
  aldol cyclization, **2**, 171
Pentadienal, 2-methyl-5-phenyl-
  biochemical reduction, **8**, 560
1,3-Pentadiene
  di-π-methane rearrangement, **5**, 195
  1,4-hydrogenation
    homogeneous catalysis, **8**, 451
  phenylation, **4**, 472
  photoaddition reactions
    with acetaldehyde, **5**, 165
  selective reduction, **8**, 568
  zirconocene complex
    reactions with carbonyl compounds, **1**, 163
1,4-Pentadiene
  hydroboration, **8**, 707
  hydrocarboxylation, **4**, 941
2,4-Pentadiene
  Diels–Alder reactions
    Lewis acid promoted, **5**, 339
1,3-Pentadiene, 5-amino-
  synthesis
    via palladium catalysis, **4**, 598
1,3-Pentadiene, 2,4-dimethyl-
  photocycloaddition reactions
    benzene, **5**, 636
1,4-Pentadiene, 2,4-dimethyl-
  hydroboration, **8**, 707
1,4-Pentadiene, 3,3-dimethyl-
  photoisomerization, **5**, 195
1,3-Pentadiene, 1-ethoxy-4-methyl-
  Diels–Alder reactions, **5**, 329
1,2-Pentadiene, 3-ethyl-
  hydrogenation
    homogeneous catalysis, **8**, 450
1,3-Pentadiene, 2-methyl-
  hydrobromination, **4**, 283
1,3-Pentadiene, 3-methyl-
  zirconocene complex
    reactions with carbonyl compounds, **1**, 163
1,3-Pentadiene, 4-methyl-
  cycloaddition reactions, **5**, 71
  selective reduction, **8**, 568
2,4-Pentadienoic acid
  lactones
    synthesis *via* [2 + 2 + 2] cycloaddition, **5**, 1138
  methyl ester
    [3 + 2] cycloaddition reactions, **5**, 297
    Diels–Alder reactions with imines, **5**, 409
2,4-Pentadienoic acid, trichloro-
  esters
    reduction, **8**, 267
1,4-Pentadien-3-ol
  asymmetric epoxidation, **7**, 416
1,4-Pentadien-3-ol, 1,1,3,5,5-pentaphenyl-
  phenyldimethylsilyl ethers
    photoisomerization, **5**, 195
  photoisomerization, **5**, 195
2,4-Pentadieno-4-lactone
  synthesis
    via carbonylation of alkynes, **3**, 1032
Pentadienylation
  carbonyl compounds
    regioselective reaction, **2**, 59
Pentaene
  synthesis
    via Horner reaction, **1**, 779
Pentalenane
  synthesis, **3**, 389
Pentalene
  Pauson–Khand reaction, **5**, 1047

synthesis
via conjugate addition, **4**, 226
via pinacol rearrangement, **3**, 726
via vinylcyclopropane thermolysis, **4**, 1048
Pentalene, hexahydro-
synthesis
via palladium-ene reaction, **5**, 50
Pentalenene
synthesis, **3**, 20, 384, 400
via ene reaction, **2**, 546
via Pauson–Khand reaction, **5**, 1061, 1062
via stereoselective cuprate reaction, **1**, 133
via Wacker oxidation, **7**, 455
Pentalenene, hydroxy-
synthesis, **3**, 400
Pentalenic acid
synthesis, **3**, 400; **7**, 109
Pentalenolactone E
synthesis, **3**, 400
via C—H insertion reactions, **3**, 1059, 1060
Pentalenolactone E methyl ester
synthesis
via Pauson–Khand reaction, **5**, 1061
Pentalenolactone F
synthesis, **3**, 400
Pentalenolactone G
synthesis, **3**, 766
via photoisomerization, **5**, 234
Pentalenolactones
synthesis, **3**, 400
via Prins reaction, **2**, 553
1-Pentalol, 4-phenyl-
synthesis
via Friedel–Crafts reaction, **3**, 315
2,2,5,7,8-Pentamethylchroman-6-sulfonyl group
amine protection, **6**, 644
Pentanal, 3,4-dimethyl-
synthesis
via hydroformylation, **4**, 919
Pentanal, 4-methyl-
photocycloaddition reactions
with furan, **5**, 169
Pentanal, 3-phenyl-4-oxo-
synthesis
via Claisen rearrangement, oxidation, **7**, 456
Pentanals, 4-oxo-
synthesis
via Claisen rearrangement, oxidation, **7**, 456
Pentane, 1-bromo-
reaction with 2-methyl-2-propylpentanoate, **6**, 2
Pentane, 3-chloro-2,2-dimethoxy-3-methyl-
iterative rearrangements, **5**, 891
Pentane, 2,3-epoxy-
resolution, **7**, 429
Pentane, 3-hydroxy-2,2-dimethoxy-3-methyl-
iterative rearrangements, **5**, 891
Pentane, 1-iodo-
reaction with 2-methyl-2-propylpentanoate
effect of counterion on rate, **6**, 2
Pentane, 2-methyl-
hydroxylation, **7**, 12
Pentane, 2,2,4-trimethyl-
benzene alkylation with
Friedel–Crafts reaction, **3**, 322
1,5-Pentanedioic acid
dimethyl ester
acyloin coupling reaction, **3**, 615
2,4-Pentanediol
chiral acetals
reduction, **8**, 222
2,4-Pentanediol, 2,4-dimethyl-
chromium trioxide complex
alcohol oxidation, **7**, 278
2,4-Pentanedione, 3,3-dimethyl-
titanium tetrachloride complex
crystal structure, **1**, 303
2,4-Pentanedione, 1-phenyl-
dianion
aldol reaction, **2**, 190
Pentanenitrile
hydrogenation, **8**, 252
Pentanoic acid, 3-diazo-2,4-dioxo-
methyl ester
Wolff rearrangement, **3**, 897
Pentanoic acid, 2-methyl-2-propyl-
alkylation, **6**, 2
Pentanoic acid, 5-oxo-
synthesis
via oxidative cleavage of cyclopentene, **7**, 558
2-Pentanol, 3-bromo-
reaction with magnesium halides, **3**, 755
3-Pentanol, 2-bromo-
reaction with magnesium halides, **3**, 755
3-Pentanol, 2,4-dimethyl-
tartrate
reaction with allenylboronic acid, **2**, 96
1-Pentanol, 3,4-epoxy-
ring opening
stereospecificity, **3**, 751
Pentanol, 1-phenyl-
borane modifier
asymmetric reduction, **8**, 170
2-Pentanone
aldol reaction
aliphatic aldehydes, **2**, 144
reduction, **8**, 924
3-Pentanone
aldol reaction, **2**, 144
benzylamine imine
deprotonation, **6**, 722
enolate
preparation, **2**, 263
lithium enolates
aldol reaction, facial selectivity, **2**, 221
3-Pentanone, 2-benzyloxy-
reaction with trichloromethyltitanium
NMR, **1**, 295
tin(IV) chloride complexes
crystal structure, **1**, 306
2-Pentanone, 5-bromo-3,3-dimethyl-
terminal lithium enolates
cycloalkylation, **3**, 18
3-Pentanone, dibenzylidene-
thermal cyclization, **5**, 754
3-Pentanone, 2,4-dibromo-
[3 + 2] cycloaddition reactions
with styrene, **5**, 283
[4 + 3] cycloaddition reactions, **5**, 603
oxyallyl cation generation
[4 + 3] cycloaddition reactions, **5**, 603
3-Pentanone, 2,2-dimethyl-
bromozinc enolates

spectra, **2**, 280
ethylzinc enolates
spectra, **2**, 280
2-Pentanone, 3-methyl-
lithium enolate
aldol reaction, **2**, 223
Pentaprismane
synthesis
via Baeyer–Villiger reaction, **7**, 683
Pentatetraene
synthesis
via retro Diels–Alder reaction, **5**, 589
Pentazocine
synthesis, **8**, 314
4-Pentenal, 2,2-dimethyl-
hydrogenation
catalytic, **8**, 140
4-Pentenal, 3-phenyl-
synthesis
via Claisen rearrangement, oxidation, **7**, 456
4-Pentenal, 2-*p*-tolyl-2-methyl-
synthesis
via Wacker oxidation, **7**, 455
3-Pentene, 1-bromo-
synthesis
via vinylcyclopropane, **5**, 903
1-Pentene, 3,3,4,4,5,5,5-heptafluoro-
hydrobromination, **4**, 280
1-Pentene, 2-methyl-
hydroformylation, **4**, 922
1-Pentene, 3-methyl-
hydroformylation, **4**, 922
1-Pentene, 4-methyl-
oxidation
Wacker process, **7**, 451
2-Pentene, 3-methyl-
hydroesterification, **4**, 936
1-Pentene, 1,1,5-trichloro-
hydrogenation, **8**, 898
1-Pentene, 3,4,4-trimethyl-
reaction with *p*-nitrobenzonitrile oxide, **5**, 262
4-Pentene-1,3-diol
synthesis
via formation of lactones on oxycarbonylation, **3**, 1033
3-Pentenoic acid
hydrobromination, **4**, 282
4-Pentenoic acid
hydrobromination, **4**, 282
Pentenoic acid, 2-alkyl-3-methyl-
iodolactonization, **4**, 380
2-Pentenoic acid, 5-amino-
synthesis
via Mannich reaction, **2**, 930
4-Pentenoic acid, 2-amino-
iodolactonization
stereoselectivity, **4**, 382
4-Pentenoic acid, 2,4-dimethyl-3-hydroxy-
cyclofunctionalization
stereoselectivity, **4**, 379
4-Pentenoic acid, 3-hydroxy-
palladium-catalyzed carbonylation
formation of dilactones, **3**, 1032
3-Pentenoic acid, 4-methyl-
hydrobromination, **4**, 282
4-Penten-2-ol, 2,3-dimethyl-
synthesis
via trihaptotitanium compound, **1**, 159
4-Penten-2-ol, 2-methyl-
dicarboxylation, **4**, 948
4-Penten-2-ol, 4-phenyl-
hydrogenation
homogeneous catalysis, **8**, 447
Pentenomycin
synthesis
via retro Diels–Alder reactions, **5**, 561
2-Pentenone
Lewis acid complexes
NMR, **1**, 294
3-Pentenone
addition reaction
with organomagnesium compounds, **4**, 89
Lewis acid complexes
NMR, **1**, 294
3-Penten-2-one
conjugate addition reactions, **4**, 169
ene reactions
Lewis acid catalysis, **5**, 6
Robinson annulation, **4**, 18
3-Penten-2-one, 3-methyl-
[3 + 2] cycloaddition reactions, **5**, 278
3-Penten-2-one, 4-methyl-
addition reactions
with organomanganese compounds, **4**, 98
5,5-Pentenones
fused
synthesis, **1**, 585
3-Pentenonitrile, 4-methyl-
Ritter reaction, **6**, 279
Pentenyl radicals
cyclizations, **4**, 785
Pentyl nitrate
nitration with, **6**, 105
Pentyl nitrite
diazotization, **7**, 740
nitration with, **6**, 106
1-Pentyne
hydrogenation to 1-pentene
homogeneous catalysis, **8**, 457
hydrosilylation, **8**, 771
reaction with borane, **1**, 489
2-Pentyne
hydrogenation to *cis*-2-pentene
homogeneous catalysis, **8**, 457
reaction with carbene complexes
regiochemistry, **5**, 1093
4-Pentynoic acid
hydroiodination, **4**, 289
3-Pentyn-2-one
photolysis
with tetramethylethylene, **5**, 164
Peptides
asymmetric synthesis
Ugi reaction, **2**, 1098
carboxylic acids
Lossen reaction, **6**, 822
conformationally constrained mimics
synthesis *via* Claisen rearrangement, **5**, 832
coupling
Ugi reaction, **2**, 1094
sterically hindered
synthesis, **6**, 276

synthesis
benzyloxycarbonyl protecting group, **6**, 632
enzymatic, **6**, 395
Friedel–Crafts benzylation, **3**, 302
racemization, **2**, 403
solid phase, **6**, 382
thioacylation
dithiocarboxylates, **6**, 423
Peptides, dehydro-
asymmetric hydrogenation
synthesis of dipeptides and oligopeptides, **8**, 460
Peptidoglycan
synthesis, **6**, 52
Peracetic acid
anti hydroxylation
alkenes, **7**, 446
Baeyer–Villiger reaction, **7**, 674
chromium oxide cooxidant
alcohol oxidation, **7**, 279
epoxidizing agent, **7**, 372
oxidation
selenides, **7**, 771
sulfoxides, **7**, 766
Peracetic acid, trifluoro-
anti hydroxylation
alkenes, **7**, 446
Baeyer–Villiger reaction, **7**, 674
boron trifluoride mixture
oxidant, **3**, 753
epoxidizing agent, **7**, 373
oxidation
organoboranes, **7**, 599
sulfoxides, **7**, 766
Perbenzoic acid
oxidation
organoboranes, **7**, 599
sulfoxides, **7**, 766
Perbenzoic acid, *m*-chloro-
Baeyer–Villiger reaction, **7**, 674
epoxidations, **7**, 359
oxidation
allylstannanes, **7**, 616
primary amines, **7**, 737
selenides, **7**, 771
sulfides to sulfoxides, **7**, 194
oxidative halogenation
alkenes, **7**, 535
Perbenzoic acid, 3,5-dinitro-
epoxidizing agent, **7**, 373
Perbenzoic acid, 4-nitro-
epoxidizing agent, **7**, 373
oxidation
ethers, **7**, 247
Perbenzoic acid, 2-sulfo-
anti hydroxylation
alkenes, **7**, 446
Perchlorates, 1,2-nitro-
synthesis
via electrophilic nitration, **4**, 356
Perchloric acid
catalyst
Friedel–Crafts reaction, **2**, 736
Perchlorocarbonyl compounds
radical cyclizations, **4**, 802
Perepoxide
alkene oxygenation, **7**, 96
Perezone, 6,15-hydroxy-
synthesis, **8**, 537
Performic acid
anti hydroxylation
alkenes, **7**, 446
epoxidizing agent, **7**, 372
Perhalo-substituted radicals
radical cyclizations, **4**, 802
Pericyclic reactions
[4 + 4] and [6 + 4] cycloadditions, **5**, 617
Claisen rearrangement, **5**, 856
Perillaketone
synthesis
via photocycloaddition, **5**, 168
Perillene
synthesis
via tandem Claisen–Cope rearrangement, **5**, 879
Perinaphthenones
synthesis
via Friedel–Crafts cycloalkylation, **3**, 325
1,3-Perinaphthindanedione
Knoevenagel reaction, **2**, 358
Periodates
glycol cleavage, **7**, 708
oxidants
silica support, **7**, 843
oxidative cleavage of alkenes
with permanganate, **7**, 586
Periodic acid
glycol cleavage, **7**, 708
mechanism, **7**, 709
oxidant
solid-supported, **7**, 841
Periodinane
oxidation
primary alcohols, **7**, 311
secondary alcohols, **7**, 324
Periplanone B
synthesis, **1**, 553; **7**, 619
via Cope rearrangement, **5**, 809
Periselectivity
cycloaddition reactions
control, **5**, 617
tropones, **5**, 620
cyclopentadiene
reaction with fulvenes, **5**, 626
[4]Peristylane
synthesis
via Diels–Alder reactions, **5**, 324
Perkin reaction, **2**, 395–407
Fittig extension, **2**, 401
mechanism, **2**, 396
scope, **2**, 399
Perlolidine
synthesis
via arynes, **4**, 505
Permutit Q
catalyst
Friedel–Crafts reaction, **3**, 297
Peroxide, allyl *t*-butyl
reaction with methyl propionate
radical addition, **4**, 753
Peroxides
alkoxy radicals from, **4**, 812
allylic oxidation, **7**, 95
catalysts

hydrosilylation, **8**, 764
disubstituted
reduction, **8**, 396
hydroalumination adducts, **8**, 753
oxidation
selenides, **7**, 771
sulfides, **7**, 762
sulfoxides, **7**, 766
photolysis, **3**, 642
radical initiators, **4**, 725
reactions with alkenes, **4**, 305–307
reductive cleavage, **8**, 396
Peroxides, arylsulfonyl
reaction with enol esters, **7**, 169
Peroxides, bis(trimethylsilyl)
hydroxylation
aryllithium, **7**, 330
oxidation
allylic alcohols, **7**, 308
reaction with lithium phenolate, **7**, 334
Peroxides, *t*-butyl *exo*-2-norbornyl
synthesis, **8**, 855
Peroxides, hexamethyldisilyl
reaction with enol acetates, **7**, 169
Peroxy acids
alkane oxidation, **7**, 13
allylic oxidation, **7**, 96
anti hydroxylation
alkenes, **7**, 446
decomposition
alcohols, **7**, 727
epoxidations, **7**, 358
intramolecular, **7**, 375
α-hydroxylation
esters, **7**, 182
ketones, **7**, 158
silyl ketene acetals, **7**, 185
oxidation
ethers, **7**, 247
organoboranes, **7**, 599
selenides, **7**, 771
sulfides, **7**, 762
sulfoxides, **7**, 766
thiols, **7**, 760
reaction with enol acetate, **7**, 167
reaction with silyl dienol ethers, **7**, 177
reaction with silyl enol ethers
ketone α-hydroxylation, **7**, 163
Peroxyarsenic acid
polymer bound
oxidation, **7**, 674
Peroxycamphoric acid
asymmetric epoxidation, **7**, 390
Peroxycarbonic acid, *o*-trichloroethyl-
cyclobutanones
chemoselective epoxidation, **7**, 385
Peroxycarboximidic acids
epoxidizing agents, **7**, 373
Peroxycarboxylic acids
anti hydroxylation
alkenes, **7**, 438
Peroxydodecanoic acid
oxidation
sulfoxides, **7**, 766
Peroxy esters
allylic oxidation, **7**, 95
t-butyl
pyrolysis, **7**, 720
α-hydroxylation
ketones, **7**, 158
reductive decarboxylation, **7**, 720
silyl-protected
epoxidations utilizing, **7**, 381
Peroxymercuration
demercuration
alkenes, **4**, 306
Peroxymercury compounds
demercuration, **8**, 854
Peroxyphosphates
allylic oxidation, **7**, 96
Peroxyphosphonates
allylic oxidation, **7**, 96
Peroxyphosphoric acid
oxidation
aryl ketones, **7**, 674
ethers, **7**, 247
Peroxyphthalic acid, monomagnesium salt
oxidation
sulfides to sulfoxides, **7**, 194
Peroxysulfuric acid
Baeyer–Villiger reaction, **7**, 674
silylated
oxidation, **7**, 674
Perrhenyl chloride
reaction with alkenes, **7**, 530
Persulfate
decarboxylation
chloroform solvent, **7**, 720
Perylene
hydrogenation
homogeneous catalysis, **8**, 455
Peshawarine
synthesis, **1**, 564
Pestalotin
synthesis, **3**, 278
chelation control, **2**, 641
via Diels–Alder reaction, **2**, 699
Petasalbine
synthesis
via retro Diels–Alder reactions, **5**, 579
Peterson alkenation, **1**, 731, 786
carbonyl compounds
enol ether preparation, **2**, 597
catalysis
cerium, **1**, 734
titanium, **1**, 734
chemoselectivity, **1**, 734
definition, **1**, 785
elimination conditions, **1**, 732
heterosubstituted alkene synthesis, **1**, 786
Hudrlik version, **3**, 224
Lewis acid catalysis, **1**, 792
mechanism, **1**, 620, 785
phosphorus substituted alkenes, **1**, 788
reactivity of metal anions, **1**, 732
silicon-stabilized imine anions, **2**, 482
silicon substitution, **1**, 737
α-silyl organometallic compounds, **1**, 620
stereochemistry, **1**, 621
sulfur substituted alkene synthesis, **1**, 786
terminal dienes
stereocontrolled synthesis, **2**, 58

Peterson reagent
addition to aldehydes and ketones, **1**, 238
aromatic nucleophilic substitution, **4**, 429
Pethidine
synthesis, **3**, 845
[2.2]Phanes
synthesis, **3**, 414
Phase transfer catalysis
alkene oxidation
palladium(II) catalysis, **4**, 553
diyne synthesis, **3**, 559
nitrile synthesis, **6**, 233
nucleophilic substitution, **4**, 426
oxidative cleavage of alkenes, **7**, 542
synthesis of carbonyl compounds, **7**, 559
synthesis of carboxylic acids, **7**, 578
sulfur ylide reactions, **1**, 821
α-Phellandrene
photochemical isomerization, **5**, 738
Phellandric acid
synthesis
via Birch reduction, **8**, 500
Phenacyl azide
synthesis, **7**, 506
Phenacyl bromide
oxidation
N,N-dialkylhydroxy amines, **7**, 663
Phenacyl esters
carboxy-protecting groups, **6**, 666
Phenacyl sulfides
photolysis
thioaldehyde generation, **5**, 436
Phenanthraquinone
photolysis
with benzophenone, **5**, 156
synthesis
via acyloin coupling reaction, **3**, 619
Phenanthrene, 9-acetoxy-
hydrolysis, **8**, 911
Phenanthrene, 4,5-bis(dimethylamino)-
synthesis
via Ramberg–Bäcklund rearrangement, **3**, 876
Phenanthrene, 9-bromo-
$S_{RN}1$ reaction, **4**, 461
Phenanthrene, 9-diazo-10-oxo-
Wolff rearrangement, **3**, 903
Phenanthrene, 9,10-dihydro-
Birch reduction
dissolving metals, **8**, 497
Phenanthrene, 9-ethoxy-
hydrogenolysis, **8**, 911
Phenanthrene, hydro-
synthesis
via epoxide ring opening, **3**, 753
Phenanthrene, 9-methoxy-
hydrogenolysis, **8**, 911
Phenanthrene, octahydro-
synthesis
via Friedel–Crafts cycloalkylation, **3**, 325
Phenanthrene, perhydro-
synthesis, **3**, 578, 640
Phenanthrene, 4-styryl-
photochemical irradiation, **5**, 729
Phenanthrene, 4-vinyl-
photochemistry, **5**, 726
Phenanthrene amide, trimethoxy-
metallation, **1**, 466
Phenanthrene-4,5-dicarboxylic acid
Schmidt reaction, **6**, 819
Phenanthrenes
automerization
Friedel–Crafts reaction, **3**, 331
Birch reduction
dissolving metals, **8**, 497
carbolithiation, **4**, 871
epoxidation, **7**, 374
hydrogenation, **8**, 438
heterogeneous catalysis, **8**, 439
oxidative rearrangement, **7**, 833
synthesis, **3**, 507
via benzyne Diels–Alder reactions, **5**, 381
via electrocyclization, **5**, 720
via Ramberg–Bäcklund rearrangement, **3**, 876
via regiospecific alkylation, **3**, 11
thermal osmylation, **7**, 863
Phenanthrenes, dihydro-
synthesis
via electrocyclization, **5**, 718
via retro Diels–Alder reaction, **5**, 572
Phenanthrenes, 9,10-dihydro-
synthesis
via photolysis, **5**, 723
1(2*H*)-Phenanthrenone, 3,4-dihydro-
rearrangement, **2**, 766
4(1*H*)-Phenanthrenone, dihydro-
Birch reduction
dissolving metals, **8**, 511
2(3*H*)-Phenanthrenone, 4a-methyl-4,4a,9,10-tetrahydro-
photolysis, **8**, 563
Phenanthrenones, hydro-
Birch reduction
dissolving metals, **8**, 510
3(2*H*)-Phenanthrenones, 1,9,10,10a-tetrahydro-
electroreduction
pinacolization, **8**, 135
Phenanthride
synthesis
via directed metallation, **1**, 463
Phenanthridines
hydrogenation
homogeneous catalysis, **8**, 456
reduction
dihydropyridine, **8**, 589
Reissert compounds, **8**, 295
synthesis
via arynes, **4**, 505
via organopalladium catalysts, **3**, 231
5-Phenanthroic acid, 4-formyl-
lactol
Schmidt reaction, **6**, 819
Phenanthrol
synthesis
via ketocarbenoids, **4**, 1056
1,10-Phenanthroline
reduction
metal hydrides, **8**, 580
Phenanthrols
synthesis
via organopalladium catalysts, **3**, 231
Phenanthroquinones
benzilic acid rearrangement, **3**, 828
synthesis, **3**, 828

9,10-Phenanthryne
  relative reactivity
    towards nucleophiles, **4**, 491
Phenazine-1-carboxylic acids
  synthesis, **4**, 435
Phenazine-1,6-dicarboxylic acid
  synthesis, **3**, 699
Phencyclone
  cycloaddition reactions
    cycloheptatriene, **5**, 632
Phenethylamine, *N*-acetyl-
  aldol reaction
    diastereofacial preference, **2**, 231
Phenethylamines
  α-substituted
    synthesis, **1**, 369
(–)-Phenmenthol
  Diels–Alder reaction
    diastereoselectivity, **2**, 688
Phenol, 2-alkyl-
  oxidative rearrangement, **7**, 835
Phenol, 2-allyl-
  cyclization
    palladium(II) catalysis, **4**, 557
  synthesis
    via Claisen rearrangement, **5**, 834
Phenol, 2-amino-
  o-diazo oxides from, **3**, 904
Phenol, 4-amino-
  catalyst
    Knoevenagel reaction, **2**, 343
Phenol, 3-bromo-2-chloro-
  hydrogenolysis, **8**, 902
  reduction, **8**, 908
Phenol, 2-*t*-butyl-
  Mannich reaction
    with preformed iminium salts, **2**, 960
Phenol, 4-chloro-
  hydrogenolysis, **8**, 912
Phenol, 2,6-di-*t*-butyl-
  aromatic nucleophilic substitution, **4**, 429
Phenol, 2,6-di-*t*-butyl-4-alkyl-
  reaction with trimethylaluminum, **1**, 78
Phenol, 2,6-di-*t*-butyl-4-methyl-
  enolates
    decomposition, **2**, 196
  reaction with diisobutylaluminum hydride, **8**, 100
Phenol, 2,4-dichloro-
  Mannich reaction
    with methylamine and formaldehyde, **2**, 969
Phenol, 2,4-dimethoxy-
  arylation, **4**, 470
Phenol, 2,5-dimethyl-
  Mannich reaction
    nonprotic solvent, **2**, 959
    with formaldehyde, **2**, 957
    with preformed iminium salts, **2**, 960
Phenol, 3,5-dimethyl-
  lithium aluminum hydride modifier, **8**, 166
Phenol, 4-fluoro-
  metallation, **7**, 333
Phenol, 3-methoxy-
  benzoylation
    Friedel–Crafts reaction, **2**, 735
Phenol, 4-methoxy-
  arylation, **4**, 470
Phenol, 4-nitro-
  esters
    amidation, **6**, 394
  hydrogenolysis, **8**, 912
  reaction with formaldehyde
    Mannich reaction, **2**, 956
Phenol, 4-*t*-octyl-
  synthesis
    via Friedel–Crafts reaction, **3**, 307
Phenol, 3-pentadecyl-
  reaction with formaldehyde
    Mannich reaction, **2**, 956
Phenol, pentafluoro-
  esters
    amidation, **6**, 394
Phenol, phenyl-
  biphenyls from
    hydrogenation, **8**, 912
Phenol, 2,4,6-tribromo-
  reduction, **8**, 908
Phenolates
  reaction with π-allylpalladium complexes
    stereochemistry, **4**, 622
Phenolates, cyano-
  irradiation, **8**, 300
Phenol ethers
  oxidative coupling, **3**, 659–700
    biaryls, **3**, 668
    mechanism, **3**, 660
  trimerization, **3**, 669
Phenolphthalein
  formylation
    Reimer–Tiemann reaction, **2**, 770
Phenols
  alkylation
    branched alkenes, **3**, 304
  o-alkylation, **4**, 430
  o-alkylation by 1-hexene
    Friedel–Crafts reaction, **3**, 306
  alkylation with isobutene
    Friedel–Crafts reaction, **3**, 306
  aminoalkylation
    Mannich reaction, **2**, 956
  binding to titanium(IV) compounds
    asymmetric epoxidation, **7**, 409
  Birch reduction
    dissolving metals, **8**, 497
  cycloalkylation
    Friedel–Crafts reaction, **3**, 304
  deoxygenation
    Birch reduction, **8**, 514
  electrochemical reduction, **4**, 439
  hydrogenation
    homogeneous catalysis, **8**, 454
  hydrogenolysis, **8**, 910
  Mannich reaction
    with preformed iminium salts, **2**, 960
    with primary amines, **2**, 968
  o-methylation
    reduction of Mannich bases, **2**, 953
  nitration, **6**, 110, 111
  oxidative coupling, **3**, 659–700
    electron transfer, **3**, 661
    mechanism, **3**, 660
    radical substitution, **3**, 661
  phenol coupling, **3**, 663

postoxidative coupling, **3**, 662
reaction with formaldehyde
Mannich reaction, **2**, 956
reduction
dissolving metals, **8**, 493
Reimer–Tiemann reaction
normal, **2**, 769
synthesis, **7**, 131, 800
via Diels–Alder reactions, **5**, 329
via nitroarenes, **4**, 438
via radical cyclizations, **4**, 807
thioacylation
anhydrides, thioketenes, thioesters and dithioesters, **6**, 449
thioacyl halides, **6**, 448
Vilsmeier–Haack reaction, **2**, 790
Phenol triflate
reduction by hydride transfer
selectivity, **8**, 84
Phenothiazine, *N*-acyl-
aldol reactions
stereoselectivity, **2**, 211
Phenothiazines
lithiation
addition reactions, **1**, 469
reduction, **8**, 659
Phenothiazine sulfoxide
Pummerer rearrangement, **7**, 202
Phenoxazines
reduction, **8**, 653
Phenoxide, bis(2,6-di-*t*-butyl-4-methyl-
methylaluminum complex
reactions of organolithium compounds, **1**, 333
Phenoxide, bis(2,4,6-tri-*t*-butyl-
methylaluminum complex
reactions of organolithium compounds, **1**, 333
Phenoxides
arylation, **4**, 495
reactions with arynes, **4**, 492
Phenoxythiocarbonyl compounds
deoxygenation, **8**, 819
Phenylacetyl group
removal
peptides, **6**, 643
Phenylalanine
asymmetric synthesis, **8**, 146
deamination
stereochemistry, **6**, 3
ethyl ester
deamination–substitution, **6**, 4
synthesis, **8**, 149
via Mannich reaction, **2**, 916
via reductive amination, **8**, 144
L-Phenylalanine
enantioselective aldol cyclizations, **2**, 167
α-Phenylaldimine
reaction with allyl organometallic compounds
Cram selectivity, **2**, 984
2-Phenylcarbonitrile, 2′-azido-
1,3-dipolar cycloaddition, **4**, 1101
Phenyl chloroformate
nitrile synthesis, **6**, 234
Phenyl dichlorophosphate
activator
DMSO oxidation of alcohols, **7**, 299
adducts
dimethylformamide, **6**, 490
S-Phenyldithiocarbonyl compounds
deoxygenation, **8**, 820
2-Phenylethoxycarbonyl group
protecting group
hydrogenolysis, **6**, 638
Phenyliodonium chloride
alkane chlorination, **7**, 16
Phenyl isocyanate
[3 + 2] cycloaddition reactions
palladium-catalyzed, **5**, 281
(–)-Phenylmenthyl esters, α-bromo-*N*-Boc-glycine
reactions with Grignard reagents, **1**, 376
3-Phenylpropionyl group
removal
chymotrypsin, **6**, 643
Phenyl radicals
addition reactions
tin hydride catalysis, **4**, 739
Phenylseleno etherification
intramolecular
lactones, **5**, 833
α-Phenylsulfonyl esters
enolates
reaction with allylic acetate, **3**, 56
Phenylsulfonyl groups
radical addition reactions
alkenes, **4**, 771
Phenylthioalkylation
silyl enol ethers, **3**, 25
Phenylthiomethylstannylations
silyl enol ethers, **3**, 26
β-Phenylthio radicals
radical cyclization, **4**, 825
Phenyl thiovinyl compounds
desulfurization, **8**, 840
Pheromones
synthesis, **3**, 643, 644
carbonyl protection, **6**, 677
via alkene metathesis, **5**, 1117
via ene reaction, **5**, 8
via photocycloaddition, **5**, 165
Phillips Triolefin Process
alkene metathesis, **5**, 1117
Phorocantholide
synthesis, **7**, 627
via Cope rearrangement, **5**, 808
Phorone
acetone self-condensation, **2**, 141
Phosgene
activator
DMSO oxidation of alcohols, **7**, 299
adducts
amides, **6**, 491
chloromethyleniminium salt preparation, **2**, 779
imidoyl halide synthesis, **6**, 523
reaction with amides, **6**, 495
Phosgene iminium salts
amide halide synthesis, **6**, 498
Phosinimides
synthesis
via reaction of phosphines with azides, **7**, 752
Phosphaalkenes
synthesis
via retro Diels–Alder reactions, **5**, 577
Phosphacumulene ylides

reactions with acidic compounds, **6**, 192, 193
Phosphanamide, triphenyl-
amide adducts
in carbon tetrachloride, **6**, 489
Phosphanions
arylation, **4**, 473
Phosphate extension
3,4-epoxy alcohol synthesis, **6**, 26
Phosphates, bromo-
phosphorylation, **6**, 601
Phosphates, chloro-
phosphorylation, **6**, 601
Phosphates, cyclic
alkene protection, **6**, 687
Phosphates, *O,O*-diethyl
reduction
catalytic hydrogenation, **8**, 817
Phosphates, halo-
phosphorylation, **6**, 601
Phosphates, α-keto-
synthesis, **7**, 155
Phosphazines
diazo-coupling reactions, **3**, 893
Phosphimides
oxidation
ozone, **7**, 752
Phosphinates
O-glycosylation, **6**, 51
Phosphine, *o*-anisylcyclohexylmethyl-
asymmetric hydrogenation
alkenes, **8**, 460
Phosphine, *p*-anisyldiphenyl-
synthesis
via $S_{RN}1$ reaction, **4**, 473
Phosphine, bis(*N,N*-dimethyl-3-aminopropyl)phenyl-
Eschenmoser coupling reaction, **2**, 870
Phosphine, dichlorophenyl-
adducts
dimethylformamide, **6**, 490
Phosphine, diphenyl-*p*-styryl-
acid chloride synthesis, **6**, 304
Phosphine, ferrocenyl-
enantioselective aldol reaction
catalysis, **2**, 317
Phosphine, isopropenyl-
synthesis
via retro Diels–Alder reactions, **5**, 560
Phosphine, phenyl-
reduction
ultrasonics, **8**, 859
Phosphine, prop-1-enyl-
synthesis
via retro Diels–Alder reactions, **5**, 560
Phosphine, trialkyl-
reaction with alkynes, **4**, 51
Phosphine, triaryl-
synthesis
via $S_{RN}1$ reaction, **4**, 473
vinyl substitutions
palladium complexes, **4**, 841
Phosphine, triphenyl-
Beckmann rearrangement reagent, **7**, 692
catalyst
acid chloride synthesis, **6**, 302, 303
deoxygenation
epoxides, **8**, 885
hydrogenolysis
transition metals, **8**, 859
imidoyl halide synthesis, **6**, 524
Mitsunobu reaction
ester synthesis, **6**, 333
palladium complexes
vinyl substitution reactions, **4**, 835
reactions with dienyliron complexes, **4**, 672
synthesis
via $S_{RN}1$ reaction, **4**, 473
Phosphine, tri-*o*-tolyl-
palladium complexes
vinyl substitution reactions, **4**, 835
Phosphine dichloride, triphenyl-
reaction with lithium carboxylates, **1**, 424
Phosphine dihalide, triphenyl-
amide adducts, **6**, 489
Phosphine hydriodide, triphenyl-
deoxygenation
epoxides, **8**, 886
Phosphine oxide, alkyl-
alkylation, **3**, 201
Phosphine oxide, α-diazo-
Wolff rearrangement, **3**, 909
Phosphine oxide, ethyl[(menthoxycarbonyl)-
methyl]phenyl-
deprotonation
alkylation, decarboxylation, **3**, 201
Phosphine oxide, tri-*n*-butyl-
catalyst
Pauson–Khand reaction, **5**, 1048
Phosphine oxides
C—P bond cleavage, **8**, 864
Horner reaction, **1**, 773
perfluorinated
hydrolysis, **8**, 864
ylide synthesis, **6**, 173
Phosphine oxides, alkyldiphenyl-
synthesis, **8**, 860
Phosphine oxides, β-keto
reduction, **8**, 12
Phosphines
π-allylpalladium complexes, **4**, 588
amination
reaction with *O*-diphenylphosphinyl-
hydroxylamine, **7**, 752
dehalogenation
α-halocarbonyl compounds, **8**, 990
halogenation, **7**, 752
α-lithiated tertiary
phosphonium ylide synthesis, **6**, 172
oxidation
phosphine oxides, **7**, 752
reactions with π-allylpalladium complexes
enantioselectivity, **4**, 651–654
reactions with arynes, **4**, 508
reduction
nitro compounds, **8**, 366
tertiary
alkylidenephosphorane synthesis, **6**, 171
Phosphines, alkylbis(phenylthio)-
synthesis, **7**, 727
Phosphines, β-(dimethylamino)alkyl-
nickel compounds
Grignard reagent catalysts, **3**, 228
Phosphines, ferrocenyl-

chiral catalysts
asymmetric hydrogenation of alkenes, **8**, 459
nickel compounds
Grignard reagent catalysts, **3**, 228
Phosphines, tris(dialkylamino)
adducts
amides, **6**, 489
Phosphines, vinyl-
synthesis
via retro Diels–Alder reactions, **5**, 560
Phosphine selenides
synthesis
via oxidation of phosphines, **7**, 752
Phosphine sulfides
synthesis
via oxidation of phosphines, **7**, 752
Phosphinic anhydride, diphenyl-
synthesis
via oxidation with perbenzoic acid, **7**, 753
Phosphinimides
synthesis
via phosphines and azides, **8**, 385
Phosphinite, chlorodiphenyl-
mixed anhydride with carboxylic acids
acylation, **1**, 424
Phosphinodithioic acid
thiolysis, **6**, 432
Phosphinothioic amide, phenyl-
reductive elimination, **1**, 742
Phosphinothricin
synthesis
via intramolecular ester enolate addition reactions, **4**, 111
Phosphite diiodide
iodination
alkyl alcohols, **6**, 213
Phosphites
oxidation
synthesis of phosphates, **7**, 753
reactions with π-allylpalladium complexes, **4**, 599
Phosphites, trialkyl
reaction with acyl halides, **8**, 278
Phosphites, trimethylsilyl-
diethyl ester
reaction with aldehydes, **3**, 199
Phosphites, triphenyl
ozonide
oxidative rearrangements, **7**, 819
reaction with dienyliron complexes, **4**, 673
Phosphites, tris(*o*-phenylphenyl)
co-catalyst
alkyne co-dimerization, **5**, 296
Phosphodiesterases
analogs
synthesis, **2**, 885
Phosphole, 1-phenyl-3,4-dimethyl-
thermolysis, **8**, 865
Phospholen
reaction with aroyl chloride, **8**, 290
Phospholiposterol
synthesis, **6**, 620
Phosphonate carbanions
α-heterosubstituted
acyl anion equivalents, **1**, 562
Phosphonates
O-glycosylation, **6**, 51
Horner–Wadsworth–Emmons reaction, **1**, 761
Knoevenagel reaction, **2**, 363
synthesis, **2**, 103
α,β-unsaturated
addition reaction with enolates, **4**, 102
Phosphonates, α-acetoxy allyl
transposition reactions, **6**, 845
Phosphonates, acylamino-
synthesis, **1**, 373
Phosphonates, acylimino-
reactions with organometallic compounds, **1**, 373
Phosphonates, alkoxycarbonyl-
anion
Knoevenagel reaction, **2**, 363
Phosphonates, alkylbis(phenylthio)-
synthesis, **7**, 727
Phosphonates, allenic
reaction with allylic alcohols, **6**, 856
Phosphonates, cyano-
anion
Knoevenagel reaction, **2**, 363
Phosphonates, diazomethyl-
Horner–Wittig reaction, **2**, 597
Phosphonates, α,α-difluoroalkyl
alkylation, **3**, 202
Phosphonates, α-fluoro
alkylation, **3**, 202
Phosphonates, β-keto-
3-lithiated
reaction with electrophiles, **2**, 442
synthesis
via propargylic alcohols, **6**, 845
synthesis, **2**, 103
Phosphonates, phenylselenomethyl
metallation, **1**, 641
Phosphonates, α-silyl-
addition reactions
carbonyl compounds, **1**, 622
Phosphonates, (trimethylsilyl)alkyl
lithio anion
Peterson alkenation, **1**, 788
Phosphonates, vinyl-
dialkyl esters
tandem vicinal difunctionalization, **4**, 252
Michael addition, **4**, 15
Phosphonic acid
hydrogenation
nitriles, **8**, 298
Phosphonic acid, bis(dimethylamido)-
allyl ester
deprotonation, **3**, 199
Phosphonic acid, ethoxymethyl(2-trimethylsilyl)-
diethyl ester
alkylation, **3**, 199
Phosphonic acid, 2-methyl-1-vinyl-
microbial epoxidation, **7**, 429
Phosphonic acid, *P*-nitrophenylmethyl-
P—C bond cleavage, **8**, 865
Phosphonic acids
synthesis
via phosphines, **7**, 753
Phosphonic diamides, *N,N,N',N'*-tetramethyl-
ortho metallation, **1**, 464
2-Phosphonioethyl carbonate
alcohol protection
cleavage, **6**, 659

Phosphonium 1,2-bisylides
synthesis
via alkynes, **6**, 172
Phosphonium 1,3-bisylides
boron-bridged
synthesis, **6**, 181
Phosphonium bromide, vinyltriphenyl-
Diels–Alder reactions, **5**, 328
Michael addition, **4**, 18
phosphonium ylide synthesis, **6**, 176
Phosphonium fluorides
phosphonium ylide synthesis, **6**, 175
Phosphonium hydriodide
reduction
indoles, **8**, 624
Phosphonium iodide, methyltriphenoxy-
deoxygenation
epoxides, **8**, 886
Phosphonium perchlorate, chlorotris(dimethyl-amino)-
acid anhydride synthesis, **6**, 311
Phosphonium permanganate, triphenylmethyl-
reaction with vinyl cyanide, **7**, 172
Phosphonium salts
base hydrolysis, **8**, 863
phosphonium ylide synthesis, **6**, 173
electrochemistry, **6**, 176
reduction
lithium aluminum hydride, **8**, 860
Phosphonium salts, alkoxy-
N-alkylation
phthalimides, **6**, 80
alkylation by, **6**, 20
amine alkylation, **6**, 74
Phosphonium salts, alkylthio-
N-alkylation
phthalimides, **6**, 81
Phosphonium salts, allylic
bond cleavage, **8**, 863
Phosphonium salts, 1,3-butadienyl-
phosphonium ylide synthesis, **6**, 176
Phosphonium salts, 1,3,5-butatrienyl-
phosphonium ylide synthesis, **6**, 176
Phosphonium salts, cycloalkyltriphenyl-
alkylation, **6**, 184
Phosphonium salts, cyclopropyl-
cycloaddition reactions, **5**, 268
phosphonium ylide synthesis, **6**, 176
Phosphonium salts, cyclopropyltriphenyl-
phosphonium ylide synthesis, **6**, 176
Phosphonium salts, phenylselenomethyl
metallation, **1**, 641
Phosphonium salts, tetraalkyl-
phosphonium ylide synthesis, **6**, 173
Phosphonium salts, ureidomethyl-
hydrolysis, **8**, 862
Phosphonium salts, vinyl-
Diels–Alder reactions, **5**, 328
hydrolysis, **8**, 862, 863
Michael addition, **4**, 18
phosphonium ylide synthesis, **6**, 173
tandem vicinal difunctionalization, **4**, 252
Phosphonium salts, vinyltriphenyl-
phosphonium ylide synthesis, **6**, 176
Phosphonium semi-ylides
formation, **6**, 172
Phosphonium tetrafluoroborate, ethoxycarbonyl-cyclopropyl-
cycloaddition reactions, **5**, 268
Phosphonium ylides
addition reactions, **4**, 115
alkene synthesis, **1**, 755
conversions, **6**, 177
cyclopropanation, **4**, 987
ester substituted
synthesis, **6**, 186
four membered
synthesis, **6**, 194
synthesis, **6**, 171–198
tandem vicinal difunctionalization, **4**, 259
Phosphonium ylides, acyl-
synthesis, **6**, 185
Phosphonium ylides, (alkylthio)thiocarbonyl-
synthesis, **6**, 187
Phosphonium ylides, allylic tributyl-
synthesis
via palladium(0) catalysis, **1**, 759
Phosphonium ylides, allyloxytrimethylphenyl-
Claisen rearrangement, **5**, 830
Phosphonium ylides, dialkylboryl-
synthesis, **6**, 181
Phosphonium ylides, germanyl-
synthesis, **6**, 180
Phosphonium ylides, *P*-halo-
synthesis, **6**, 172
Phosphonium ylides, mercury-substituted
synthesis, **6**, 181
Phosphonium ylides, stannyl-
synthesis, **6**, 180
Phosphonium ylides, trimethylstibino-
synthesis, **6**, 179
Phosphonoacetic acid, triethyl-
Knoevenagel reaction
(*E*) product, **2**, 363
titanated
Knoevenagel reaction, (*Z*) product, **2**, 363
Phosphonoamidate, *N*-*t*-butyl α-chloro-
rearrangement, **8**, 864
Phosphonothionates, *O*-allyl
rearrangements, **4**, 642
Phosphonyl anions
chiral
conjugate additions, **4**, 226
Phosphonyl chloride, alkyl-
synthesis, **7**, 10
Phosphoramidate, *N*,*N*-dibromo-
addition reactions
alkenes, **7**, 500
Phosphoramidate, *N*-(trimethylsilyl)-
diethyl ester
reaction with alkyl bromides, **6**, 83
Phosphoramidates
cyclic
alkene protection, **6**, 687
phosphorylation, **6**, 614
Phosphoramide, hexamethyl-
acid anhydride synthesis, **6**, 311
dissolving metals
reduction, **8**, 524
reaction with thionyl chloride, **6**, 302
reductions
aromatic rings, **8**, 490

Phosphoramidic acid, *N*-(*t*-butoxycarbonyl)-
diethyl ester
reaction with alkyl halides, **6**, 82
Phosphoramidites
phosphorylation, **6**, 618
Phosphoramidoazidic acid, *N*-phenyl-
phenyl ester
Curtius reaction, **6**, 816
Phosphorane, acetylmethylenetriphenyl-
reactions with organolithium compounds, **6**, 189
Phosphorane, (acylalkoxycarbonylmethylene)-
thermolysis, **8**, 863
Phosphorane, 1-acylmethylene-
alkylation, **6**, 182
Phosphorane, acylmethylenetriphenyl-
alkylation, **6**, 182
Phosphorane, 1,2-alkadienylidene-
synthesis, **6**, 197
Phosphorane, 2,4-alkadienylidene-
synthesis, **6**, 184
Phosphorane, alkoxycarbonylhalomethylenetriphenyl-
synthesis, **6**, 172
Phosphorane, alkylidenetrialkyl-
lithium salt complexes, **6**, 175
Phosphorane, alkylidenetriphenyl-
alkylation
intramolecular, **6**, 183
alkynyl-substituted
synthesis, **6**, 185
formylation, **6**, 186
Phosphorane, alkylthiocarbonylalkylidene-
synthesis, **6**, 187
Phosphorane, alkylthiomethylene-
synthesis, **6**, 177
Phosphorane, allylidenetriphenyl-
reactions with chloro compounds, **6**, 189
synthesis, **6**, 184
Phosphorane, arylazomethylene-
synthesis, **6**, 178
Phosphorane, *N*-aryliminovinylidenetriphenyl-
dimerization, **6**, 195
Phosphorane, bisalkylidene-
exocyclic
synthesis, **6**, 191
Phosphorane, bis(ethylthio)vinylidenetriphenyl-
reactions with heteroallenes, **6**, 195
Phosphorane, bis(phenylseleno)methylenetriphenyl-
synthesis
via carbenoid method, **6**, 171
Phosphorane, bis(phenylthio)methylenetriphenyl-
synthesis
via carbenoid method, **6**, 171
Phosphorane, carbamoylmethylene-
synthesis, **6**, 187
Phosphorane, cyanomethylenetriphenyl-
alkylation, **6**, 182
Phosphorane, cycloalkylidenetriphenyl-
synthesis, **6**, 184
Phosphorane, cyclopentadienylidenetriphenyl-
reactions, **6**, 189
Phosphorane, diacetylmethylenetriphenyl-
deprotonation
selectivity, **6**, 189
Phosphorane, dialkylborylalkylidenetriphenyl-
reactions with polar compounds, **6**, 188
Phosphorane, dibromomethylenetriphenyl-
synthesis, **6**, 172
Phosphorane, dibromotriphenyl-
acid halide synthesis, **6**, 302
bromination
alkyl alcohols, **6**, 209
Phosphorane, dichloromethylene-
synthesis, **6**, 172
Phosphorane, dichlorotriphenyl-
acid halide synthesis, **6**, 302
reaction with neopentyl alcohol, **6**, 205
Phosphorane, diethoxyphosphinomethylenetriphenyl-
synthesis, **6**, 179
Phosphorane, diethoxythiovinylidenetriphenyl-
reactions with heteroallenes, **6**, 195
Phosphorane, diethoxytriphenyl-
reactions with 2-amino alcohols, **6**, 74
Phosphorane, diethoxyvinylidenetriphenyl-
reactions with alcohols, **6**, 193
reactions with carbonyl compounds, **6**, 193
Phosphorane, difluoromethylene-
synthesis, **6**, 172
Phosphorane, difluoromethylenetriphenyl-
synthesis, **6**, 172
Phosphorane, dihalomethylene-
synthesis, **6**, 172
Phosphorane, dihalotriorgano-
reaction with activated methylene compounds, **6**, 173
Phosphorane, diiodomethylenetriphenyl-
synthesis, **6**, 172
Phosphorane, 1-dimethylarsinomethylenetrimethyl-
synthesis, **6**, 179
Phosphorane, diphenoxyphosphinomethylenetriphenyl-
synthesis, **6**, 179
Phosphorane, 1-diphenylphosphinylalkylidene-
synthesis, **6**, 179
Phosphorane, ethoxyiminocarbonylmethylene-
synthesis, **6**, 193
Phosphorane, ethoxyvinylidenetriphenyl-
reactions with alcohols, **6**, 193
Phosphorane, ethylidenetriphenyl-
Wittig reaction, **1**, 757
Phosphorane, formyl-
Michael addition, **4**, 16
Phosphorane, formylalkylidene-
synthesis, **6**, 186
Phosphorane, 1-formylmethylene-
alkylation, **6**, 182
Phosphorane, ω-haloalkylidenetriphenyl-
alkylation
intramolecular, **6**, 183
Phosphorane, 2-iminoalkylidenetriphenyl-
synthesis, **6**, 186
Phosphorane, iminovinylidene-
reactions with acidic compounds, **6**, 193
synthesis, **6**, 197
Phosphorane, iminovinylidenetriphenyl-
phosphonium ylide synthesis, **6**, 191
reactions with alkyl halides, **6**, 191
reactions with heteroallenes
cycloaddition, **6**, 194
synthesis, **6**, 196
Phosphorane, 4-oxo-2-alkenylidenetriphenyl-
synthesis, **6**, 184
Phosphorane, oxovinylidene-
phosphonium ylide synthesis, **6**, 191
reactions with acidic compounds, **6**, 193

Phosphorane, oxovinylidenetriphenyl-
  reactions with alkyl halides, **6**, 191
  reactions with halogen compounds, **6**, 194
  reactions with heteroallenes
    cycloaddition, **6**, 194
  synthesis, **6**, 196, 197
Phosphorane, phenylfluoro-
  fluorination
    alkyl alcohols, **6**, 217
Phosphorane, *N*-phenyliminovinylidenetriphenyl-
  cycloaddition, **6**, 194
  reactions with carboxylic acids, **6**, 193
  reactions with heteroallenes, **6**, 195
Phosphorane, 1-phenylselenoalkylidenetriphenyl-
  synthesis, **6**, 178
Phosphorane, 1-phosphinylmethylene-
  synthesis, **6**, 190
Phosphorane, propadienylidene-
  synthesis, **6**, 197, 198
Phosphorane, α-silylalkylidene-
  synthesis, **6**, 179
Phosphorane, 1-sulfinylalkylidene-
  synthesis, **6**, 178
Phosphorane, 1-sulfonylalkylidene-
  synthesis, **6**, 178
Phosphorane, thioxovinylidene-
  reactions with acidic compounds, **6**, 193
  synthesis, **6**, 197
Phosphorane, thioxovinylidenetriphenyl-
  reaction with heteroallenes
    cycloaddition, **6**, 194
  synthesis, **6**, 197
Phosphorane, α-trimethylsilylalkylidene-
  α-trimethylsilylphosphonium salt synthesis, **6**, 188
Phosphorane, vinylidene-
  cycloaddition, **6**, 194
  phosphonium ylide synthesis, **6**, 191
  synthesis, **6**, 196, 197
Phosphorane dihalides, triaryl-
  imidoyl halide synthesis, **6**, 524
Phosphorane halides, tetraaryl-
  imidoyl halide synthesis, **6**, 524
Phosphoranes
  cumulated
    synthesis, **6**, 196
  orthoester
    synthesis, **6**, 193
Phosphoranes, acyl-
  charge-directed conjugate addition, **4**, 243
Phosphoranes, (acylalkylidene)-
  reduction
    zinc/acetic acid, **8**, 863
Phosphoranes, alkylidene-
  alkenylation, **6**, 184
  synthesis, **6**, 171
Phosphoranes, imino-
  amidine synthesis, **6**, 546
Phosphorazidate, diphenyl-
  amide synthesis, **6**, 389
Phosphorazidic acid
  diethyl ester
    Curtius reaction, **6**, 816
  di-*p*-nitrophenyl ester
    Curtius reaction, **6**, 816
  diphenyl ester
    Curtius reaction, **6**, 797, 811
Phosphoric acid
  arenesulfonic anhydrides
    phosphorylation, **6**, 603
  catalyst
    Friedel–Crafts reaction, **2**, 736
  protecting groups, **6**, 621
Phosphoric acid, dichloro-
  Friedel–Crafts reaction, **2**, 754
Phosphoric acid esters
  triphenylphosphonium salts
    phosphorylation, **6**, 615
Phosphoric acids, dialkyldithio-
  reactions with alkenes, **4**, 317
Phosphorin, 2-(2′-pyridyl)-
  synthesis, **8**, 865
Phosphorochloridates
  phosphorylation, **6**, 601
Phosphorochloridites
  phosphorylation, **6**, 616
Phosphorodiamidates, *N,N,N′N′*-tetramethyl-
  deoxygenation, **8**, 817
Phosphorodiamidates, vinyl *N,N,N′,N′*- tetramethyl-
  ketone reduction, **8**, 932
Phosphorodiamidic acid, tetramethyl-
  allyl ester
    deprotonation, **2**, 64
Phosphorodichloridates
  phosphorylation, **6**, 601
Phosphorofluoridates
  phosphorylation, **6**, 601
Phosphoroguanidate
  stability, **6**, 614
Phosphorohydrazidates
  phosphorylation, **6**, 614
Phosphoroselenoic acid, *O,O*-dialkyl
  deoxygenation
    epoxides, **8**, 887
Phosphorothioates
  phosphorylation, **6**, 614
Phosphorothioites
  phosphorylation, **6**, 618
Phosphorous acid, bis(dimethylamino)-
  butyllithium
    epoxide reduction, **8**, 885
Phosphorus
  yellow
    reduction, nitro compounds, **8**, 366
Phosphorus, tris(phenylthio)-
  reaction with *O*-acyl thiohydroxamates, **7**, 727
Phosphorus acid halides
  acid anhydride synthesis, **6**, 310
Phosphorus chlorides
  acid chloride synthesis, **6**, 302
Phosphorus compounds
  Diels–Alder reactions, **5**, 444
  oxidation, **7**, 735–753
  pentavalent
    phosphorylation, **6**, 601
  reactions with amides, **6**, 495
  trivalent
    phosphorylation, **6**, 616
Phosphorus halides
  polymer-bound
    acid halide synthesis, **6**, 303
Phosphorus iodide
  metal hydride reduction

carbonyl compounds, **8**, 315
Phosphorus nitrile chloride
adducts
dimethylformamide, **6**, 490
Phosphorus nucleophiles
aromatic nucleophilic substitution, **4**, 446
Phosphorus oxychloride
adducts
amides, **6**, 487
phosphorylation, **6**, 601
reaction with amides, **8**, 301
Phosphorus pentabromide
reaction with amides, **6**, 495
Phosphorus pentachloride
anion-exchange resin-bound
acid chloride synthesis, **6**, 303
chlorination
alkyl alcohols, **6**, 204
Phosphorus pentahalides
imidoyl halide synthesis, **6**, 524
Phosphorus pentasulfide
Eschenmoser coupling reaction, **2**, 867
thiocarboxylic ester synthesis, **6**, 437
Phosphorus pentoxide
activator
DMSO oxidation of alcohols, **7**, 299
Phosphorus sulfur trichloride
adducts
amides, **6**, 487
Phosphorus tribromide
bromination
alkyl alcohols, **6**, 209
Phosphorus trichloride
chlorination
alkyl alcohols, **6**, 204
Phosphorus trihalides
adducts
amides, **6**, 490
imidoyl halide synthesis, **6**, 524
Phosphorus triiodide
iodination
alkyl alcohols, **6**, 213
Phosphorus ylides
alkylation
formation of phosphonium salts, **3**, 200
solubilizer
lithium halides, **3**, 760
Phosphorylating agents
bifunctional
unsymmetrical phosphotriesters, **6**, 618
Phosphorylation
decarboxylative chalcogenation, **7**, 727
in synthesis, **6**, 601
Phosphoryl 4-nitrophenoxide
phosphorylation, **6**, 608
Phosphoryl phenoxide
phosphorylation, **6**, 608
Phosphoryl trichloride
chloromethyleniminium salt preparation, **2**, 779
Phosphoryl 2,4,6-trinitrophenoxide
phosphorylation, **6**, 608
Phosphotriesters
unsymmetrical
synthesis, **6**, 618
Photochemical electron transfer
charge transfer, **7**, 850
Photochemical pinacolization
aromatic compounds, **3**, 567
Photochemical reactions
Ritter reaction, **6**, 280
Photochemical reduction
allylic compounds, **8**, 978
aromatic rings, **8**, 517
Photochlorination
alkanes, **7**, 15
Photocyclizations
electron transfer induced
Mannich reactions, **2**, 1037
Photocycloaddition reactions
1,3-dienes, **5**, 635–638
enantioselectivity, **5**, 132
intermolecular, **5**, 125–132
regiochemistry, **5**, 125–127
intramolecular, **5**, 133–145
regioselectivity, **5**, 133–137
stereoselectivity, **5**, 137–145
stereochemistry, **5**, 128–132
[2 + 2] Photocycloaddition reactions
copper catalysis, **5**, 147
[3 + 2] Photocycloaddition reactions
arene–alkene, **5**, 645–671
[5 + 2] Photocycloaddition reactions
arene–alkene, **5**, 645–671
Photoelectrochemical oxidation
halide salts, **7**, 539
Photoelectron spectra
ionization potentials, **7**, 852
Photoisomerizations
di-π-methane, **5**, 193–213
fragmentations, **5**, 209
radical-type rearrangement, **5**, 208
1,3-sigmatropic shift, **5**, 207
Photolithography
diazo ketones, **3**, 887
Photo Reimer–Tiemann reaction, **2**, 772
o-Phthalaldehyde
reaction with organometallic reagents, **1**, 154
Phthalazines
reduction, **8**, 640
ring opening
cathodic reduction, **8**, 641
Phthalic acid
dimethyl ester
synthesis *via* retro Diels–Alder reaction, **5**, 571
Schmidt reaction, **6**, 819
Phthalic anhydride
hydrogenation, **8**, 239
phthalic acid dichloride synthesis, **6**, 307
reduction
electrochemical, **8**, 240
Schmidt reaction, **6**, 819
Phthalide, 1,2-dihydro-
photolysis, **5**, 739
Phthalide enolates
reaction with 3,4-dihydroisoquinolines
synthesis of protoberberine alkaloids, **2**, 946
reaction with Schiff bases
Mannich reaction, **2**, 927
Phthalide isoquinoline alkaloids
synthesis
via Mannich reaction, **2**, 912
Phthalide isoquinolines

synthesis
via Mannich reaction, **2**, 894
Phthalides
optically active
synthesis, **1**, 60
reactions with arynes, **4**, 497
synthesis
via carbonylation of halides, **3**, 1033
Phthalimide, *N*-amino-
oxidation, **7**, 742
reaction with alkenes, **7**, 481
Phthalimide, *N*-bromo-
addition reactions
alkenes, **7**, 500
Phthalimide, *N*-hydroxy-
catalyst
thiol ester synthesis, **6**, 437
Phthalimide, *N*-hydroxymethyl-
amidoalkylation with, **2**, 971
Phthalimide, *N*-phenylseleno-
ether synthesis, **7**, 523
selenol ester synthesis, **6**, 466
Phthalimide, *N*-vinyl-
hydrocarboxylation, **4**, 941
Phthalimides
alkylation, **6**, 80
Beckmann rearrangement, **6**, 770
Hofmann reaction, **6**, 802
photochemistry, **7**, 42
reaction with allylic esters, **6**, 86
reduction
sodium borohydride, **8**, 274
Phthalimides, *N*-alkyl-
reduction, **8**, 254
Phthalimidine-3-carboxylic acid, 3-hydroxy-
synthesis, **3**, 835
Phthalimidines, 3-hydroxy-
synthesis, **8**, 274
Phthalimidoketo aldehyde
synthesis, **7**, 657
Phthalonimides
rearrangements, **3**, 835
Phthaloyl chloride
acid chloride synthesis, **6**, 304
reaction with aluminum chloride, **2**, 754
Phthaloyl group
protecting group
amines, **6**, 643
trisaccharides, **6**, 634
Phyllanthocin
synthesis
diastereoselectivity, **1**, 822
via [3 + 2] cycloaddition reactions, **5**, 311
Phyllanthoside
synthesis
via ene reaction, **2**, 541
Phyllocladane
synthesis
via [3 + 2] cycloaddition reactions, **5**, 311
Phyllocladene
synthesis
via Birch reduction, **8**, 500
Phyllostine
synthesis
via retro Diels–Alder reactions, **5**, 564
Phyltetralin
synthesis
via conjugate addition to oxazolines, **4**, 206
Phytol
synthesis, **7**, 109
Piceatannol, dihydro-
oxidative coupling, **3**, 666
2-Picoline
nitration
nitronium tetrafluoroborate, **7**, 750
Picoline *N*-oxide
oxidation with, **7**, 661
Picolyl anions
phenylation, **4**, 472
4-Picolyl esters
carboxy-protecting groups
cleavage, **6**, 668
Picric acid
chlorination, **6**, 208
Picrolichenic acid
synthesis
via manganese dioxide oxidation, **3**, 679
Picropodophyllone
synthesis, **3**, 696
Picrotoxinin
synthesis, **7**, 162, 243
via cyclofunctionalization of cycloalkenes, **4**, 373
via hydrazones, **2**, 509
via Johnson methylenation, **1**, 738
via regiospecific alkylation of hydrazone anions, **2**, 518
Pictet–Gams isoquinoline synthesis
Ritter reaction, **6**, 291
Pictet–Hubert (Morgan–Walls) phenanthridine synthesis
Ritter reaction, **6**, 291
Pictet–Spengler condensation
iminium ions
heterocyclic synthesis, **6**, 736
mechanism, **2**, 1020
synthesis of aromatic alkaloids, **2**, 1016
Pikronolide
synthesis, **7**, 246
Pilocerine
synthesis, **3**, 687
Pimaranes
synthesis
via biomimetic conversion of communic acids, **7**, 634
Pimelate, 4-hydroxy-
enantioselective lactonization, **6**, 337
Pimelate dehydrogenase
reduction
catalyst, **8**, 205
Pimelic acid, 4-keto-
from siloxycyclopropanes, **2**, 451
Pinacol coupling reactions, **3**, 563–605
intermolecular, **3**, 564
intramolecular, **3**, 572
mixed, **3**, 595
organosamarium compounds, **1**, 270
Pinacolones
enolates, **2**, 264
Kishner–Leonard elimination, **8**, 341
label redistribution
pinacol rearrangement, mechanism of, **3**, 723
lithium enolates
α-chiral aldehydes, **2**, 218

crystal structure, **1**, 26
X-ray diffraction analysis, **1**, 3
potassium enolate
crystal structure, **1**, 26
reduction
chloroborane, **7**, 603
sodium enolate
crystal structure, **1**, 26
synthesis
via pinacol rearrangement, **3**, 721
Pinacol rearrangement, **3**, 721–730
applications, **3**, 726
definition, **3**, 721
mechanism, **3**, 722
migratory aptitudes, **3**, 726
Pinacols
oxidative cleavage, **7**, 707
synthesis
via dissolving metals, **8**, 109
via organoytterbium compounds, **1**, 279
unsymmetrical
synthesis, using ytterbium, **1**, 279
Pinacol-type reactions
β-hydroxy sulfides, **1**, 861
3-Pinanecarbaldehyde
synthesis
via hydroformylation, **4**, 919
Pinanediol
boronic esters, **3**, 796
3-Pinanone, 2-hydroxy-
glycinate esters, enolates
alkylation, **3**, 46
α-Pinene
allylboranes from
reactions with aldehydes, **2**, 33
allylic oxidation, **7**, 99
hydroboration, **8**, 704, 709
hydroformylation, **4**, 919
hydrosilylation, **8**, 777
metallation
oxidation, **7**, 99
oxide
rearrangement, **3**, 771
photooxidation, **7**, 111
rearrangement, **3**, 705
reduction
diimide, **8**, 475
Ritter reaction, **6**, 289
β-Pinene
ene reactions, **5**, 2
hydride donor
reduction of carbonyls, **8**, 100
hydroalumination, **8**, 739
hydrozirconation, **8**, 689
photooxidation, **7**, 111
reduction
9-borobicyclo[3.3.1]nonane, **8**, 102
synthesis
via stereospecific Ritter reaction, **6**, 278
δ-Pinene
synthesis
via methyllithium addition, **1**, 377
α-Pinene, 7-trimethylsilyl-
acylation
Friedel–Crafts reaction, **2**, 717
Pinenes
fragmentation, **3**, 708
Pinidine, dihydro-
synthesis, **1**, 559
Pinocarveol
synthesis, **7**, 92, 99
Pinosylvin
synthesis, **2**, 170
Pipecolic acid
synthesis, **2**, 1074
via *N*-acyliminium ions, **2**, 1078
via Ireland rearrangement, **5**, 843
Piperazine, 2,5-diketo-
bislactam ethers
regiochemistry of deprotonation, **2**, 499
bislactim ethers
metallated, reactions, **2**, 498
1,4-Piperazine-2,5-dione, *N*,*N*-dibenzyl-
reduction, **8**, 249
Piperazines
metallated
diastereoselective reactions, **2**, 499
Piperazines, diketo-
synthesis, **6**, 392
Pipercide, dehydro-
synthesis
via palladium-catalyzed coupling reactions, **3**, 545
Δ^1-Piperideine
reactions with organometallic compounds, **1**, 364
synthesis
via allylboranes, **2**, 982
Piperidides
reduction
aluminates, **8**, 272
Piperidine, 2-alkenyl-
synthesis
via cyclization of δ-allenylamines, **4**, 412
Piperidine, 3-alkylidene-
synthesis
via Mannich reaction, **2**, 1030
Piperidine, *N*-allyl-
heat of hydrogenation, **6**, 707
Piperidine, 3-amino-
synthesis, **8**, 598
Piperidine, *N*-benzoyl-
hydrogenation, **8**, 248
Piperidine, *N*-benzyl-2-cyano-6-methyl-
alkylation, **1**, 557
Piperidine, 1-benzyl-2,6-dicyano-
alkylation, **1**, 557
Piperidine, *N*-chloro-
addition reactions, **7**, 499
Piperidine, dehydro-
asymmetric alkylation, **3**, 77
Piperidine, *N*-formyl-
Vilsmeier–Haack reaction, **2**, 779
Piperidine, 4-hydroxy-
synthesis, **2**, 1002
via Mannich reaction, **2**, 1027, 1029
Piperidine, *N*-methyl-
deprotonation, **1**, 476; **3**, 65
Piperidine, *N*-nitroso-
photoaddition to alkenes, **7**, 490
synthesis
via nitrosation of 1-methylpiperidine, **7**, 749
Piperidine, pentylidene-
synthesis

via Mannich reaction, **2**, 1027
Piperidine, *N*-propenyl-
heat of hydrogenation, **6**, 707
Piperidine, α-propyl-
synthesis, **1**, 558
Piperidine, *N*-thiocarbamoyl-
amine-protecting group, **6**, 642
Piperidine alkaloids
synthesis
via enol ethers, **2**, 613
Piperidine amides
alkylation, **3**, 69
lithiation
carbonyl addition reactions, **1**, 483
Piperidinediones
Tebbe reaction, **1**, 745
Piperidine formamidine, α-allylic
metalation, **3**, 70
Piperidine formamidines
alkylation, **3**, 69
Piperidines
N-alkylation, **6**, 66
anodic oxidation, **7**, 804
lithiated formamidines
reaction with benzaldehyde, **1**, 482
reaction with 2-naphthol and benzaldehyde
Mannich reaction, **2**, 960
synthesis
via N-acyliminium ions, **2**, 1066
via 6-*exo*-cyclization, **4**, 404
via Mannich reactions, **2**, 1023
via palladium-ene reactions, **5**, 51
via solvomercuration of amines, **4**, 290
trimer
synthesis, **2**, 970
Piperidines, *N*-(2-nitroalkyl)-
synthesis, **7**, 490
Piperidinium acetate
catalyst
Knoevenagel reaction, **2**, 343
regioselective aldol cyclization, **2**, 159
Piperidinium selenocarboxylates
synthesis, **6**, 465
Piperidinol
synthesis
via nitrone cyclization, **4**, 1116
Piperidinyl-1-oxyl, 2,2,6,6-tetramethyl-
oxidation
primary alcohols, **7**, 308
2-Piperidone
bridged
microbial hydroxylation, **7**, 60
2-Piperidone, 1-methyl-
reduction
lithium aluminum hydride, **8**, 273
2-Piperidone, 5-phenyl-
reduction, **8**, 249
Piperidones
synthesis
via Knoevenagel reaction, **2**, 361
Piperine
synthesis, **2**, 153
Piperitone
oxiranes
rearrangement, **3**, 771
Piperonal
aldol reaction
N-crotonylpiperidine, **2**, 153
Piperylene
anodic oxidation, **7**, 795
Pipitzol
synthesis
via organocuprate conjugate addition, **4**, 191
via photocycloaddition, **5**, 660
Pirprofen
chemoselective epoxidation, **7**, 384
Pivalaldehyde
synthesis
via epoxide ring opening, **3**, 742
Pivalate, 2-nitropropenyl-
Michael addition, **4**, 14
Pivalates
photochemical deoxygenation, **8**, 817
Pivalic acid
diethylboryl ester
aldol reactions, **2**, 244
2-nitroallyl ester
addition reactions with organolithium compounds, **4**, 78
synthesis
via Friedel–Crafts dealkylation, **3**, 330
Pivalic acid dimethylamide
alkylation, **6**, 501
Pivaloyl azide
nitrenes from, **7**, 477
Pivaloyl chloride
Friedel–Crafts reaction
bimolecular aromatic, **2**, 740
Pivaloyl group
alcohol protection
glycosylation, **6**, 657
Pivaloyl thioamide
alkylation, **2**, 868
Platelet activating factor
antagonist ligands
synthesis *via* Paterno–Büchi reaction, **5**, 152
Platinic acid, hexachloro-
catalyst
hydrosilylation, **8**, 556, 764
Platinum
carbon
catalysts, hydrosilylation, **8**, 764
catalyst
hydrogenation, **8**, 418
hydrogenation of pyridines, **8**, 597
hydrogenolysis
benzylic alcohols, **8**, 963
Platinum, carbonylhydrido(trichlorostannate)-bis(triphenylphosphine)-
catalyst
hydroformylation, **4**, 915
Platinum, dichlorobis(triphenylphosphine)-
catalyst
acyl halide reduction, **8**, 265
Platinum, trichloromethylbis(triphenylphosphine)-
synthesis, **7**, 4
Platinum complexes, η^1-allyl-
[3 + 2] cycloaddition reactions
tetracyanoethylene, **5**, 275
Platinum complexes, halomethyl(arylphosphine)-
metallocyclization
Friedel–Crafts reaction, **3**, 323

Platinum dimers
  catalysts
    hydrosilylation, **8**, 557
Platinum-ene cyclizations, **5**, 56–59
Platinum oxide
  catalyst
    carbonyl compound hydrogenolysis, **8**, 319
Pleraplysillin **3**, 1
  synthesis, **3**, 487
Pleurotin
  synthesis, **7**, 350
Plumbagin
  synthesis
    via retro Diels–Alder reaction, **5**, 564
Plumbanes, alkyl-
  reactions with aldehydes
    Lewis acid promotion, **1**, 329
Plumbemycin
  synthesis
    via Ugi reaction, **2**, 1097
Plysiatoxin, debromo-
  synthesis, **3**, 168
Podocarpic acid, dimethoxy-
  synthesis
    via Baeyer–Villiger reaction, **7**, 678
Podophyllotoxin
  synthesis
    via arynes, **4**, 501
    via benzyne cyclization, **5**, 692
    via Knoevenagel reaction, **2**, 381
Podophyllum lignans
  synthesis
    via aldol reaction, **2**, 201
Podorhizol
  synthesis, **1**, 566
    via aldol reaction, **2**, 204
Podorhizon
  synthesis
    via conjugate addition, **4**, 215
Podototarin
  synthesis
    via alkaline potassium ferricyanide, **3**, 665
Poison-dart frog alkaloids
  synthesis
    via Eschenmoser coupling reaction, **2**, 876
Poitediol
  synthesis
    via alkynylvinylcyclobutanol rearrangement, **5**, 1026
    via Cope rearrangement, **5**, 806
Polarity
  inversion
    electrochemical oxidation, **7**, 790
Polarography
  electrosynthesis, **8**, 131
Polonovski reaction
  fragmentation, **6**, 1067
Polonovski–Potier cyclization
  diastereoselection, **2**, 1021
Polyacylations
  Friedel–Crafts reaction, **2**, 712
Polyalkylation
  enolates
    equilibration, **3**, 4
Polyamides
  carboxy-protecting groups, **6**, 670
Polyamines
  reduction
    aluminum hydrides, **8**, 541
Polyarylenealkenylenes
  synthesis
    via Knoevenagel reaction, **2**, 388
Polybenzimidazole
  palladium chloride complex
    reduction, **8**, 372
Polybutadiene
  hydrogenation
    homogeneous catalysis, **8**, 449
Polycyclic aromatic hydrocarbons
  hydrogenation
    chemoselectivity, **8**, 439
    heterogeneous catalysis, **8**, 438
    regioselectivity, **8**, 438
    stereoselectivity, **8**, 439
  $S_{RN}1$ reaction, **4**, 461
Polycyclic hydrocarbons
  fused
    Birch reduction, dissolving metals, **8**, 496
Polycyclopentanoids
  synthesis
    via magnesium-ene reaction, **5**, 40
Polydithioesters
  synthesis
    via polyacrylonitrile, **6**, 456
Polyenepolyynes
  cyclic
    synthesis, **3**, 556
Polyenes
  addition reactions, **7**, 504
  alkoxymercuration, **4**, 311
  bicyclizations, **3**, 359
  conjugated
    synthesis, **3**, 878
  cyclic
    hydrosilylation, **8**, 780
  cyclization, **2**, 714; **3**, 341–375
    mechanism, **3**, 374
    with iminium ion initiators, **2**, 1026
  hydrogenation
    regioselectivity, **8**, 433
    stereoselectivity, **8**, 433
  hydrogenation to saturated hydrocarbons
    homogeneous catalysis, **8**, 449
  hydrosilylation, **8**, 778
  monocyclization, **3**, 347
  reactions with carbon electrophiles
    transition metal catalysis, **4**, 695–712
  synthesis
    via hydroalumination, **8**, 757
    via Julia coupling, **1**, 802
  tetracyclization, **3**, 362
  tricyclization, **3**, 362
  Vilsmeier–Haack reaction, **2**, 782
Polyenes, ω,ω′-biazulenyl-
  synthesis, **3**, 586
Polyenes, bis-2-thienyl-
  synthesis, **3**, 586
Polyether antibiotics
  occurrence, **2**, 1
  synthesis
    via Ireland silyl ester enolate rearrangement, **5**, 840
Polyethers

dissolving metals
reductions, **8**, 524
Poly(ethylene glycol)
carboxy-protecting group
polymer support, **6**, 670
reductions in
acyl halides, **8**, 240
Reimer–Tiemann reaction, **2**, 772
solvent
Wacker oxidation, **7**, 451
Polyglymes
aromatic nucleophilic substitution
sulfur nucleophiles, **4**, 443
Polygodial
synthesis, **7**, 91, 307
Polyheteroarylenealkenylenes
synthesis
via Knoevenagel reaction, **2**, 388
Polyhexamethylene thioterephthalates
synthesis
via phenyl carboxylates, **6**, 443
cis-1,4-Polyisoprene
photolysis
with benzophenone, **5**, 161
Polyketide aromatic compounds
synthesis
via Michael addition, **4**, 14
Polyketides
aldol condensation
biomimetic synthesis, **2**, 619
cyclization to aromatic rings, **2**, 170–176
synthesis, **2**, 248
Polymer esters
anchoring groups
carboxylic acids, **6**, 670
Polymers
chromium(VI) oxidants support
alcohol oxidation, **7**, 280
monodispersed
synthesis, **5**, 1121
synthesis
via alkene metathesis, **5**, 1120–1122
via α,ω-diethynyl monomers, **3**, 557
via Knoevenagel reaction, **2**, 387
Polymethinium salts
Knoevenagel reaction, **2**, 358
Polymethylhydrosiloxane
hydrosilylation
unsaturated hydrocarbons, **8**, 765
reduction
allylic amines, **8**, 961
Polynorbornene
synthesis
via ring opening metathesis polymerization, **5**, 1121
Polynuclear aromatic halides
vinyl substitution
palladium complexes, **4**, 845
Polyols
monodeoxygenation, **8**, 820
regioselective substitution
hydroxy group, **6**, 79
selective monoacylation, **6**, 337
synthesis
via epoxides, **6**, 8
1,3-Polyols
synthesis, **1**, 569
Polyoxins
synthesis
via Ugi reaction, **2**, 1097
Polyoxochromium dichloride
oxidative halogenation, **7**, 530
Polypentenamer
synthesis
via ring opening metathesis polymerization, **5**, 1120
Poly(1,4-phenylene sulfide)
synthesis, **4**, 461
Polyphosphoric acid
Beckmann rearrangement, **6**, 763
catalyst
Friedel–Crafts reaction, **2**, 736
Erlenmeyer azlactone synthesis, **2**, 403
Friedel–Crafts reaction
acylation of alkenes, **2**, 711
Polyphosphorus acid trimethylsilyl ester
amidine synthesis, **6**, 546
Polyporic acid
rearrangement, **3**, 828
Polyquinanes
synthesis
via Nazarov cyclization, **5**, 779
via photoisomerizations, **5**, 226
Polyquinenes
synthesis
via bisannulation, **1**, 262
Polystachins
synthesis
via cinnamyl compounds, **7**, 831
Polystyrene, chloromethyl-
anchoring groups
amino acids, **6**, 670
Polystyrene, hydroxycrotonylaminomethyl-
carboxy-protecting groups
anchorage, **6**, 671
Polystyrenes
hydroxylation
thallium, **7**, 333
Polythiolactones
macrocyclic
synthesis, **6**, 441
Poly(vinyl alcohol)
hydrogen donor
transfer hydrogenation, **8**, 551
Poly(vinylbenzophenone)
photocycloaddition reactions
alkenes, **5**, 161
Poly(vinylpyridinium chlorochromate)
oxidation
alcohols, **7**, 282
Poly(vinylpyridinium dichromate)
oxidation
alcohols, **7**, 282
Polyynes
linear
synthesis, **3**, 555
platinum polymer, **3**, 558
synthesis, **3**, 551
Pomeranz–Fritsch synthesis
isoquinolines, **6**, 751
Ponzio reaction
oxidation of oximes

dinitrogen tetroxide, **7**, 751
Porphycenes
synthesis, **3**, 594
Porphyrin, 5-formyloctaethyl-
Knoevenagel reactions, **2**, 354
Porphyrin, octamethyl-
synthesis
via Diels–Alder reaction, **5**, 492
Porphyrins
aziridination catalysts, **7**, 477
Knoevenagel reaction, **2**, 357
manganese complexes
aziridination catalysts, **7**, 484
catalyst for radical-based processes, **7**, 8
pinacol rearrangement, **3**, 729
Vilsmeier–Haack reaction, **2**, 780
Porphyrins, 2-alkoxy-5,10,15,20-tetraphenyl-
synthesis, **4**, 437
Porphyrins, tetraphenyl-
synthesis
via Knoevenagel reaction, **2**, 387
Potassium
alcohols as solvents
reduction, **8**, 111
Birch reduction, **8**, 492
crown ethers
alkyl fluoride reduction, **8**, 795
reduction
ammonia, **8**, 113
carbonyl compounds, **8**, 109
enones, **8**, 524
reductive dimerization
unsaturated carbonyl compounds, **8**, 532
Potassium, crotyl-
crotylboronates from, **2**, 13
crotyl organometallics from, **2**, 5
structure, **2**, 977
Potassium, methyl-
synthesis
crystal structure, **1**, 12
Potassium, phenylthioallyl-
methylation
selectivity, **3**, 99
Potassium, phenylthioprenyl-
methylation
selectivity, **3**, 99
Potassium, trimethylsilyl-
deoxygenation
epoxides, **8**, 886
Potassium, triphenylmethyl-
ketone deprotonation, **2**, 183
Potassium borohydride
reduction
epoxides, **8**, 875
imines, **8**, 36
nitro compounds, **8**, 366
Potassium *t*-butoxide
xonotlite
catalyst, Knoevenagel reaction, **2**, 345, 359
Potassium carbonylferrate
halide carbonylation
formation of aldehydes, **3**, 1021
Potassium dichromate
oxidant
solid support, **7**, 841, 845
Potassium diisopropylamide
reaction with *N,N*-dimethylhydrazones, **2**, 506
Potassium enolates
α,α-disubstituted aldehydes
alkylation, **3**, 20
nitration, **6**, 105
synthesis, **2**, 100
Potassium fluoride
catalyst
allylsilane reactions with aldehydes, **2**, 571
Knoevenagel reaction, **2**, 343, 359
Potassium hexamethyldisilazane
enolate formation, **2**, 182
Potassium hydrogen persulfate (oxone)
oxidation
sulfides, **7**, 765
sulfoxides, **7**, 769
Potassium iodate
hydroxylation
alkenes, **7**, 445
Potassium nitrodisulfonate — *see* Fremy's salt
Potassium nitrosodisulfonate
quinone synthesis, **7**, 143
Potassium pentacyanocobaltate
hydrogenation
alkenes, **8**, 449
Potassium pentacyanohydridocobaltate
catalyst
hydrogenation, **8**, 535
Potassium permanganate
aqueous
oxidative cleavage of alkenes, **7**, 558
basic
alkane oxidation, **7**, 12
catalytic oxidative cleavage
alkenes, **7**, 542
heterogeneous oxidation
alkenes, **7**, 586
hydroxylation
alkenes, **7**, 444
mixed solvent systems
oxidative cleavage of alkenes, **7**, 558
oxidation
diols, **7**, 313
sulfoxides, **7**, 768
oxidative cleavage of alkenes, **7**, 542
phase transfer assisted, **7**, 559
synthesis of carbonyl compounds, **7**, 558
synthesis of carboxylic acids, **7**, 578
with periodate, **7**, 586
reaction with vinyl cyanide, **7**, 172
solid support
clay, **7**, 845
silica, **7**, 844
Potassium selenocyanate
deoxygenation
epoxides, **8**, 887
Potassium superoxide
ketone α-hydroxylation, **7**, 157
oxidation
hydrazines, **7**, 744
primary amines, **7**, 738
Potassium tetracarbonylhydridoferrate
reductive amination
carbonyl compounds, **8**, 54
Potassium tri-*s*-butylborohydride
reduction

acyl halides, **8**, 242
benzyloxy ynones, **8**, 7
nitroalkenes, **8**, 377
unsaturated carbonyl compounds, **8**, 536
Potassium triethylborohydride
amide reduction, **8**, 11
Potassium triisopropoxyborohydride
alkene reduction, **3**, 797
selective ketone reduction, **8**, 18
Potassium triphenylborohydride
reduction
alkyl halides, **8**, 805
Potassium trisiamylborohydride
reduction
cyclohexanones, **8**, 14
Precalciferol
synthesis
via photochemical ring opening, **5**, 700
Precapnelladiene, **8**, 942
synthesis
via Claisen rearrangement, **5**, 831
Preclavulone A
biosynthesis
via Nazarov cyclization, **5**, 780
Precocene, 7-ethoxy-
synthesis, **5**, 1096, 1098
Prefulvene
formation
via benzene irradiation, **5**, 649
Pregna-14,16-dien-20-one
reduction
hydrosilylation, **8**, 557
Pregna-5,16-dien-20-one, 3β-acetyloxy-
dienyltricarbonyliron complexes
asymmetric synthesis, **4**, 688
5α-Pregnane
allylic oxidation, **7**, 100
5β-Pregnan-12-one
reduction
dissolving metals, **8**, 119
5α-Pregnan-6-one, 3β,20α-diacetoxy-
Mannich reaction
with iminium salts, **2**, 901
5α-Pregn-1-en-2-ol-3,20-dione
rearrangement, **3**, 833
Pregnenolene
oxidation
Bornstein's reagent, **7**, 533
Pregnenolone, 17α-bromo-
rearrangements, **3**, 846
Preisocalamendiol
cyclization
transannular ene reaction, **2**, 553
Prelog–Djerassi lactone
synthesis, **8**, 857
anti-Cram selectivity, **2**, 573
via [4 + 3] cycloaddition, **5**, 611
via cycloheptadienyliron complexes, **4**, 686
via dichlorocarbene, **4**, 1005
via Diels–Alder reaction, **2**, 700
via ene reaction, **2**, 534
via hydroformylation, **4**, 923
Prelog–Djerassi lactonic acid
synthesis, **2**, 251, 259; **7**, 300
via aldol reaction, γ-position, **2**, 189
via chiral reagent, **2**, 224
via dihalocyclopropyl compounds, **4**, 1018
Premonensin
asymmetric synthesis, **2**, 846
synthesis, **1**, 429
Prephenic acid
synthesis
via Diels–Alder reactions, **5**, 324
Pretetramid
synthesis, **2**, 173
Previtamin D
synthesis
via photolysis, **5**, 737
Prévost reaction
hydroxylation
alkenes, **7**, 438, 447
Prezizaene
synthesis
via Cope rearrangement, **5**, 989
Prezizanol
synthesis
via Cope rearrangement, **5**, 989
Primetin
synthesis, **7**, 341
Prins reaction, **2**, 527–558
control, **2**, 563
formaldehyde
addition to alkenes, **2**, 528
intermolecular, **2**, 528
intramolecular, **2**, 540
type I, **2**, 540
type II, **2**, 547
mechanism, **2**, 564
Pristane
microbial hydroxylation, **7**, 62
Prodigiosin
synthesis
via Diels–Alder reaction, **5**, 492
Progesterone
allylic oxidation, **7**, 96
enone
reduction, **8**, 549
hydrogenation
homogeneous catalysis, **8**, 452
microbial hydroxylation, **7**, 68, 70, 73
reduction–alkylation, **8**, 527
Progesterone, 9α-bromo-11β-hydroxy-
reaction with chromium(II) acetate, **1**, 175
Progesterone, 11α-hydroxy-
enantiospecific synthesis, **3**, 371
oxidation
DMSO, **7**, 295
synthesis, **3**, 126
Progesterone, 11-keto-
synthesis
via Cope rearrangement, **5**, 790
Progesterone, 21-methyl-
synthesis
via acylation of organocadmiums, **1**, 447
Progestin
synthesis, **3**, 846
Proline
borane modifier
asymmetric reduction, **8**, 170
chiral catalysts
nucleophilic addition reactions, **1**, 72
enantioselective aldol cyclization, **2**, 167

lithium aluminum hydride modifiers, **8**, 168
peptides
synthesis, **2**, 1097
Proline, *N*-acryloyl-
benzyl ester
Diels–Alder reaction, **5**, 365
Diels–Alder reactions, **5**, 366
Proline, *N*-benzyloxycarbonyl-
Curtius reaction, **6**, 813
Proline, 3,4-dehydro-
synthesis, **8**, 606
Proline, *N*-hydroxy-
synthesis
via oxidation of pyrrolidine, **7**, 745
Proline, *N*-pyruvoyl-
catalytic hydrogenation, **8**, 145
Prolinol
reaction with 2,2′-bis(bromomethyl)-1,1′-binaphthyl
N-alkylation, **6**, 71
Prolinolamides
addition reactions
with organomagnesium compounds, **4**, 85
alkylation, **3**, 45
reductive alkylation
Birch reduction, **8**, 508
Prolyl chloride, *N*-(trifluoroacetyl)-
Friedel–Crafts reaction
bimolecular aromatic, **2**, 740
Propacin
synthesis, **3**, 691
1,2-Propadiene
hydrochlorination, **4**, 276
1,2-Propadiene, 1-phenyl-
hydrochlorination, **4**, 276
Propadienethione
synthesis
via retro Diels–Alder reactions, **5**, 575
Propanal
borane complexes
rotational barriers, **1**, 290
Propanal, 2-acetoxy-
synthesis
via hydroformylation, **4**, 932
Propanal, 2-cyclohexyl-
aldol reaction
simple diastereoselection, **2**, 214
Propanal, 2-hydroxy-
synthesis
via hydroformylation, **4**, 932
Propanal, 2′-(2-methoxy-6-naphthyl)-
synthesis
via rhodium-catalyzed hydroformylation, **3**, 1022
Propanal, 2-methyl-
synthesis, **8**, 297
Propanal, 2-methyl-3-phenyl-
lithiation
with tributylstannyllithium, **3**, 195
Propanal, 2-phenyl-
acetal
synthesis *via* hydroformylation of styrene, **3**, 1022
addition reactions with bromomethylmagnesium, **1**, 317
aldol reaction
simple diastereoselection, **2**, 214
reaction with lithium enolates
stereoselection, **2**, 217
reaction with methyl pyrrolidine complex anions, **5**, 1080
reaction with methyltitanium triisopropoxide, **1**, 141
reaction with organometallic reagents
diastereoselectivity, **1**, 151
Lewis acids, **1**, 334
Propanal, 3-phenyl-
acetal
synthesis *via* hydroformylation of styrene, **3**, 1022
synthesis, **8**, 297
Propane
propylation
Friedel–Crafts reaction, **3**, 333
reaction with rhenium
metal vapor synthesis, **7**, 4
Propane, 1,3-bis(methylthio)-2-methoxy-
synthesis, **6**, 139
Propane, 2,2-bis(phenylseleno)-
stability, **1**, 632
Propane, 3-chloro-1,1-bis(methylseleno)-
metallation, **1**, 638
Propane, 3-chloro-1,1-bis(phenylseleno)-
metallation, **1**, 638
Propane, 1-chloro-2,3-diamino-
synthesis, **6**, 94
Propane, 3-chloro-2-methyl-1-phenylthio-
metallation, **3**, 89
Propane, 1-chloro-2-phenyl-
benzene alkylation
Friedel–Crafts reaction, **3**, 300, 302
Propane, 2-chloro-1-phenyl-
benzene alkylation
Friedel–Crafts reaction, **3**, 300, 302
Propane, 1,3-diiodo-
[3 + 2] cycloaddition reactions
copper-catalyzed, **5**, 282
Propane, 1-(2,5-dimethoxy-4-methylphenyl)-2-amino-
synthesis, **8**, 375
Propane, 1-dimethylamino-3-lithio-
intramolecular solvated tetramer, **1**, 10
Propane, 1,1-diphenyl-
synthesis
via Friedel–Crafts reaction, **3**, 311
Propane, 1,2-diphenyl-
synthesis
via Friedel–Crafts reaction, **3**, 300
Propane, 1,3-disubstituted 2-methylene-
bifunctional conjunctive reagent, **5**, 298
Propane, 2-lithio-2-phenylseleno-
synthesis, **1**, 634
Propane, 2-methyl-2-nitro-
synthesis
via oxidation of *t*-butylamine, **7**, 737
Propane, 2-nitro-
aromatic nucleophilic substitution, **4**, 429
Propane, 1-phenyl-2,2-dialkoxy-
synthesis
via Wacker oxidation, **7**, 452
Propane, 2-phenylseleno-2-phenylthio-
stability, **1**, 632
Propane, 1,1,1,2-tetrachloro-2-methyl-
nitrile synthesis, **6**, 229
1,2-Propanediol
pinacol rearrangement, **3**, 725
1,3-Propanediol, 2-*aci*-nitro-
synthesis, **2**, 323

1,2-Propanediol, 1-phenyl-
synthesis, **7**, 442
1,3-Propanediol, 1-phenyl-
oxidation
solid support, **7**, 841
1,3-Propanediol, DL-*threo*-1-phenyl-2-nitro-
synthesis
via Henry reaction, **2**, 325
1,3-Propanedione, 1,3-diphenyl-
Knoevenagel reaction, **2**, 357
Propanedithioates, 2-aryl-
methyl ester
synthesis, **6**, 455
Propanedithioates, 3,3,3-trialkyl-
alkyl esters
synthesis, **6**, 455
1,3-Propanedithiol
demercurations, **8**, 857
Propanethioic acid
3-(3-ethyl)pentyl ester
reaction with borolanyl triflate, **2**, 259
S-phenyl ester
reaction with 9-borabicyclo[3.3.1]non-9-yl triflate, **2**, 245
Propanoic acid, β-(3-acenaphthoyl)-
Friedel–Crafts reaction, **2**, 763
Propanoic acid, 3-aroyl-
synthesis, **2**, 744
Propanoic acid, 2,2-dimethyl-
protonolysis
organoboranes, **8**, 724
Propanoic acid, 3′-(2-naphthyl)-
Friedel–Crafts reaction
cyclization, **2**, 754
Propanoic acid, β-phenyl-
Friedel–Crafts reaction, **2**, 756
Propanoic acids, 2-aryl-
chiral synthesis
microbial oxidation, **7**, 57
1-Propanol, 2,3-diamino-
synthesis, **6**, 94
2-Propanol, 1,3-diamino-
vicinal diamine synthesis, **6**, 94
1-Propanol, 2,3-epoxy-
opening
Payne rearrangement, **3**, 735
2-Propanol, 2-hydroperoxyhexafluoro-
oxidation
sulfides, **7**, 763
1-Propanol, 2-methyl-
synthesis
via hydrogenation, **8**, 236
1-Propanol, 2-methyl-3-ethoxy-
synthesis
via hydroformylation, **4**, 923
1-Propanol, 1-phenyl-
perdeuterated
synthesis, **1**, 223
1-Propanol, 2-phenyl-
benzene alkylation with
Friedel–Crafts reaction, **3**, 311
synthesis
via Friedel–Crafts reaction, **3**, 313
2-Propanol, 1-phenyl-
benzene alkylation with
Friedel–Crafts reaction, **3**, 311
synthesis
via Friedel–Crafts reaction, **3**, 313
1-Propanol, 1-(3-phenyl-1-indolizidinyl)-
synthesis
via γ-diketones, **1**, 547
2-Propanone, 1-aryl-
synthesis, **7**, 828
2-Propanone, 1,3-bisdiazo-1,3-diphenyl-
Wolff rearrangement
intermediates, **3**, 905
2-Propanone, 1,3-dibromo-1-phenyl-
[4 + 3] cycloaddition reactions, **5**, 605
2-Propanone, 1-phenyl-
tris(dimethylamino)sulfonium enolate
formation, **2**, 135
1-Propanones, 2-alkoxy-1-(1,3-dithian-2-yl)-
synthesis, **1**, 568
Propanoyl chloride, 2-methoxy-β-phenyl-
Friedel–Crafts reaction, **2**, 756
Propanoyl chloride, 4-methoxy-β-phenyl-
Friedel–Crafts reaction, **2**, 756
Propanoyl chloride, β-thienyl-
Friedel–Crafts reaction, **2**, 758
Propanoyl–iron complexes
chiral
asymmetric aldol reactions, **2**, 272
Propargyl alcohol, aryl-
synthesis, **3**, 537
Propargyl aldehyde
cycloaddition reactions
carbene complexes, **5**, 1073
Propargyl cation
[4 + 3] cycloaddition reactions, **5**, 598
Propargylic acetals
carbocupration, **4**, 899
cyclization, **5**, 769
reduction, **8**, 213
reductive cleavage, **8**, 214
Propargylic acetates
reactions with carbonyl compounds
organosamarium compounds, **1**, 257
Propargylic acid
dilithium salt
reaction with oxiranes, **3**, 278
Propargylic alcohols
allylic alcohols from, **7**, 396
dilithiated
reaction with α,ω-dibromoalkanes, **3**, 281
imidate esters
thermal rearrangement, **6**, 843
isomerization
to α,β-unsaturated carbonyl compounds, **6**, 836
nitrile synthesis, **6**, 234
optically active
synthesis, **1**, 347
trimerization, **5**, 1145
Propargylic amines
carbomagnesiation, **4**, 878
divinyl ketone from, **5**, 753
Propargylic anion equivalents
allenylsilanes
synthesis of substituted alkynes, **1**, 595
Propargylic electrophiles
reaction with organocopper compounds, **3**, 220
Propargylic esters
rearrangement

allenic esters, **6**, 836
Propargylic rearrangements
functional group transformation, **6**, 830
Propargyl organometallic compounds
diasteroselective reactions, **2**, 91–96
enantioselective reactions, **2**, 96
heteroatom substituted, **2**, 88
nonheteroatom substituted
regioselective reactions, **2**, 82–88
reactions with *gem*-amino ethers
dependence of product type on metal, **2**, 1005
reactions with imines, **2**, 975–1004
variation of yield with metal, **2**, 993
reactions with iminium salts, **2**, 1000
dependence of product type on metal, **2**, 1001
regioselective reactions, **2**, 82–91
synthesis, **2**, 81
Propargyl sulfenate
allene sulfoxide
transposition reaction, **6**, 837
Propargyl systems
isomerization
1,3-hydrogen–hydrogen transpositions, **6**, 866
[1.3.4]Propellane
synthesis
via Diels–Alder reaction, **5**, 372
[3.3.1]Propellane
solvolysis, **4**, 1021
[4.4.1]Propellane
solvolysis, **4**, 1021
[4.4.4]Propellane
synthesis
via cyclopropanation, **4**, 1041
Propellanes
synthesis, **3**, 573
via Cope rearrangement, **5**, 814
via [3 + 2] cycloaddition reactions, **5**, 310
via cyclopropane ring opening, **5**, 924
via dihalocyclopropanes, **4**, 1009
via photocycloaddition, **5**, 666
[3.3.3]Propellanes
synthesis
via intramolecular ene reactions, **5**, 11, 21
via photoisomerization, **5**, 233
[4.3.2]Propellanes
rearrangement, **8**, 931
[4,3,2]Propellanols
rearrangement, **3**, 709
[4.2.2]Propella-2,4,7,9-tetraene
isomerization
via retro Diels–Alder reactions, **5**, 585
Propene
disproportionation, **5**, 1116
Propene, 3-acetoxy-3-phenyl-
synthesis, **7**, 95
1-Propene, 1,1-bis(methylseleno)-
synthesis, **1**, 638
Propene, 1-bromo-
hydroiodination, **4**, 288
Propene, 2-*t*-butyl-
photooxygenation, **7**, 399
Propene, 3-chloro-
arene alkylation by
Friedel–Crafts reaction, **3**, 321
Propene, 2-chloromethyl-3-chloro-
bifunctional conjunctive reagent, **5**, 299
Propene, 2-chloromethyl-3-trimethylsilyl-
bifunctional conjunctive reagent, **5**, 299
Propene, 3-diazo-
[3 + 2] cycloaddition reactions
alkynyl carbene complexes, **5**, 1070
Propene, 1,3-dichloro-
hydrogenation, **8**, 898
Propene, 1-dimethylamino-1-chloro-2-methyl-
acid halide synthesis, **6**, 305
Propene, 1,3-diphenyl-
photoisomerization, **5**, 208
Propene, 1,3-disilyl-
acylation
Friedel–Crafts reaction, **2**, 718
Propene, 1-ethoxy-
Diels–Alder reactions, **5**, 461, 473
Propene, 2-iodo-
reaction with chromium chloride, **1**, 193
Propene, 3-iodo-2-(trimethylsilyl)methyl-
[4 + 3] cycloaddition reactions, **5**, 599
trimethylenemethane synthetic equivalent, **5**, 246
Propene, 1-lithio-1-seleno-3-methyl-
reactions with benzaldehyde, **1**, 679
Propene, 2-methoxy-
Eschenmoser rearrangement, **5**, 891
Propene, methylaryl-
photoisomerization, **5**, 197
Propene, 3-nitro-
α,α-double-deprotonation
synthesis of nitro alcohols, **2**, 63
Propene, 1-phenyl-
allylic oxidation, **7**, 95
diamination, **7**, 484
hydrobromination, **4**, 280
oxidative rearrangement
solid support, **7**, 845
synthesis
via palladium catalysts, **4**, 840
via vinyl carbanion alkylation, **3**, 242
Propene, 2-phenyl-
Vilsmeier–Haack reaction, **2**, 782
Propene, 3-phenyl-
oxidation
Wacker process, **7**, 452
oxidative cleavage
phase transfer assisted, **7**, 583
photoisomerization, **5**, 197
Propene, 1-phenylpentafluoro-
hydrogenation, **8**, 896
Propene, 1,1,2-trichloro-
synthesis, **4**, 270
Propene, 3,3,3-trifluoro-
hydrobromination, **4**, 280
hydroformylation, **4**, 927
Propene, 3,3,3-trifluoro-2-trifluoromethyl-
hydrobromination, **4**, 280
Propene oxide
Friedel–Crafts reaction, **3**, 321
hydrogenolysis, **8**, 882
Propene oxide, 2-phenyl-
reduction
borohydrides, **8**, 875
Propenes, 3-aryl-
Pauson–Khand reaction, **5**, 1045
2-Propenethiol
lithiated

reactions with carbonyl compounds, **1**, 510
Propenoic acid, 2-(6-methoxy-2-naphthyl)-
hydrogenation
homogeneous catalysis, **8**, 461
2-Propenoic acid, 3-(3-methylenecyclopentyl)-
methyl ester
synthesis *via* metal-catalyzed cycloaddition, **5**, 1192
2-Propen-1-ol, 2-methyl-
asymmetric epoxidation, **7**, 398
2-Propen-1-ol, 2-nitro-
pivalate
multiple coupling reagent, **2**, 325
2-Propen-1-ol, 2-(trimethylsilyl)methyl-
bifunctional conjunctive reagent, **5**, 299
2-Propen-1-one, 1-(4′-methoxyphenyl)-3-phenyl-
reactions with organocerium compounds, **1**, 235
2-Propen-1-yl, 2-(trimethylsilyl)methyl-
acetate
[3 + 2] cycloaddition reactions, **5**, 298
β-Propiolactone
enolates
diastereofacial selectivity, **2**, 205
reaction with phenyl Grignard reagents, **3**, 466
synthesis
via carbonylation of ethylene, **3**, 1031
β-Propiolactone, β-ethynyl-
reaction with organocopper compounds, **3**, 227
Propiolamidines, 3-amino-
synthesis, **6**, 550
Propiolic acid
addition of hydrogen halides, **4**, 51
hydrobromination, **4**, 285
Propiolic acid, phenyl-
hydrobromination, **4**, 286
Propionaldehyde, 2-(methoxy)methoxy-
aldimine derivatives
reaction with allyl organometallic compounds, **2**, 984
α-alkoxyaldimines derivatives
reaction with allyl organometallic compounds, **2**, 987
Propionaldehyde, 2-phenyl-
reactions with allylsilanes
diastereofacial selectivity, **2**, 570
Propionaldehyde, 3-phenyl-
acetal
reaction with isopropenyl acetate, **2**, 612
Propionaldehyde diethyl acetal
carbocupration, **4**, 900
Propionaldehydes
anion equivalent
addition reactions, **4**, 117
3-substituted
synthesis, **6**, 849
Propionamide, β-aroyl-*N*-alkyl-
addition reactions
with organomagnesium compounds, **4**, 84
Propionamide, *N*,*N*-dimethyl-
dimethyl acetal
rearrangement, stereochemistry, **5**, 837
lithium enolate
crystal structure, **1**, 31
Propionamide, 2-phenyl-
Hofmann rearrangement, **6**, 804
Propionamide, *N*-phenyl-3-chloro-
synthesis, **7**, 696
Propionamide, 3-stannyl-
lithiation
dianionic homoenolate, **2**, 447
Propionamides, 3-phenylsulfinyl-
Pummerer rearrangement, **7**, 201
formation of sulfenylated β-lactam, **7**, 202
Propionamidine, α-arylamino-
synthesis, **6**, 555
Propionate enolate
enantioselective aldol reaction
acyliron complexes, **2**, 316
Propionates
esters, from carbohydrates
aldol reaction, stereoselection, **2**, 226
Propionic acid, 2-aryl-
synthesis
via hydroformylation, **4**, 932
Propionic acid, 2-alkoxy-
esters
aldol reaction, **2**, 205
Propionic acid, 3-amino-2,2-dimethyl-3-phenyl-
synthesis
via Mannich reaction, **2**, 922
Propionic acid, 2-aryl-
synthesis, **3**, 244
Propionic acid, 2-azido-3-(benzyloxy)-
benzyl ester
serine synthesis, **6**, 77
Propionic acid, 2-bromo-
t-butyl ester
catalyst, Grignard reagent alkylation, **3**, 244
ethyl ester
Reformatsky reaction, stereoselectivity, **2**, 291
hydrolysis, **6**, 342
methyl ester
reaction with zinc, **2**, 279
Reformatsky reaction, **2**, 293
reaction with 1-phenylethylamine, **6**, 67
Propionic acid, 3-bromo-
methyl ester
reaction with samarium, **1**, 254
Propionic acid, 2-chloro-2-methyl-
methyl ester
nitrile synthesis, **6**, 229
Propionic acid, 2-cyano-2-methyl-3-phenyl-
rearrangements, **6**, 799
Propionic acid, 3-(cyclopent-2-enyl)-
methyl ester
synthesis *via* cycloaddition of bicyclo[2.1.0]pentane, **5**, 1186
Propionic acid, 3,3-dialkyl-
optically active
synthesis *via* conjugate addition to oxazolines, **4**, 204
Propionic acid, 3-(3,4-dimethoxyphenyl)-
oxidation, **7**, 336
Propionic acid, 2-halo-
aryl esters
cycloalkylation, **3**, 324
Propionic acid, 3-lithio-
synthesis and reaction, **2**, 447
Propionic acid, 2-(mesyloxy)-
benzene alkylation with
Friedel–Crafts reaction, **3**, 312
Propionic acid, 3-methoxy-1,2-diaryl-

synthesis, **7**, 829
Propionic acid, methyl-2-(chlorosulfonyloxy)-
benzene alkylation with
Friedel–Crafts reaction, **3**, 312
Propionic acid, methyl-2-phenyl-
synthesis
via Friedel–Crafts reaction, **3**, 312
Propionic acid, 2-phenyl-
rearrangements, **6**, 799
Propionic acid, 3-phenyl-
ethyl ester
acyloin coupling reaction, **3**, 619
Schmidt reaction, **6**, 817
Propionic acid, 2-phenyl-2-(*t*-butylthio)-
synthesis
via arene–metal complex, **4**, 527
Propionic acid, 3-thienyl-
ethyl ester
acyloin coupling reaction, **3**, 619
Propionic acid, 3-(*p*-tolylsulfinyl)-
dianions
reactions with carbonyl compounds, **1**, 513
Propionic acid, 3-(2,3,4-trimethoxyphenyl)-
oxidation, **7**, 337
Propionic acid, 3-trimethylsilyl-
ethyl ester
acyloin coupling reaction, **3**, 619
Propionic acids
synthesis, **4**, 429
Propionitrile, 2,2-bis(dimethylamino)-
synthesis, **6**, 577
Propionitrile, 2,2-dimethoxy-
synthesis
via Wacker oxidation, **7**, 451, 452
Propionitrile, 3-hydroxy-
synthesis
via ethylene oxide, **6**, 236
Propionitrile, 3-oxo-3-phenyl-
synthesis
via phenacyl bromide, **6**, 231
Propionyl chloride
Friedel–Crafts reaction
bimolecular aromatic, **2**, 740
Propiophenone
aldol reactions
diasteroselective, **2**, 244
oxidative rearrangement
solid support, **7**, 845
reduction
lithium aluminum hydride, **8**, 166
tin enolates, **2**, 610
Propranolol
synthesis, **6**, 341
Propylure
synthesis, **3**, 799
2-Propynal, trimethylsilyl-
Knoevenagel reaction, **2**, 365
Propyne
hydroiodination, **4**, 288
trimerization
potassium chromate catalysis, **5**, 1148
Propyne, 1,3-bis(triisopropylsilyl)-
anion
enynes from, **2**, 91
Propyne, 1,3-bis(trimethylsilyl)-
dilithium anion
reaction with aliphatic carbonyl compounds, **2**, 91
Peterson alkenation, **1**, 790
reaction with chloral
Lewis acid promotion, **1**, 328
Propyne, 3-(*t*-butyldimethylsilyl)-1-(trimethylsilyl)-
anion
1,3-enynes from, **2**, 91
Propyne, 1-*t*-butylthio-
photolysis
with benzil, **5**, 163
Propyne, dilithio-
alkylation, **3**, 281
Propyne, 1-methylthio-
photolysis
with acetone, **5**, 163
Propyne, 1-phenyl-
hydroalumination, **8**, 737
Propyne, 3,3,3-trifluoro-
hydrobromination, **4**, 285
hydroiodination, **4**, 288
synthesis, **4**, 271
Propynoate esters
conjugate additions
Lewis acid catalyzed, **4**, 164
Prostacyclin
analogs
synthesis *via* Knoevenagel reaction, **2**, 381
synthesis
via cyclofunctionalization of
propargylcyclopentanol, **4**, 393
via rearrangement of epoxides, **3**, 767
Prostacyclins
synthesis
via Pauson–Khand reaction, **5**, 1051
Prostaglandin, 11-deoxy-
synthesis
via enolate alkylation, **3**, 9
Prostaglandin, 9-fluoromethylene-
synthesis
via Johnson methylenation, halogen incorporation,
1, 741
Prostaglandin, 5-oxo-
synthesis
via hydration of alkynes, **4**, 300
Prostaglandin A_2
synthesis
via Johnson rearrangement, **5**, 839
Prostaglandin D_1
methyl ester
synthesis, **1**, 570
Prostaglandin E_1
synthesis
via Diels–Alder reaction, **5**, 492
via Michael addition, **4**, 13
Prostaglandin E_2
synthesis
via Diels–Alder reaction, **5**, 492
Prostaglandin endoperoxide
synthesis
via palladium-ene reaction, **5**, 35
Prostaglandin $F_{2\alpha}$
synthesis, **3**, 290, 781; **6**, 139; **8**, 163
via cyclopropane ring opening, **4**, 1045
Δ^5-Prostaglandin $F_{1\alpha}$, 11-deoxy-6,11-α-epoxy-
synthesis, **7**, 633
Prostaglandin I_2, 5-hydroxy-

synthesis, **7**, 633
Δ^6-Prostaglandin I_1, 9(*O*)-thia-
synthesis, **7**, 621
Prostaglandins
hydroxy group
protection, **6**, 653
microbial hydroxylation, **7**, 66
precursors
synthesis, **3**, 139
reduction
hydride transfer, **8**, 100
stereoselective synthesis
via cyclopropane ring opening, **4**, 1046
synthesis, **1**, 569; **3**, 103, 126, 279, 289, 649; **7**, 59, 180, 824; **8**, 163, 171, 269, 560, 561, 695
via addition reactions with organozincates, **4**, 97
via asymmetric reduction, **8**, 546
via Baeyer–Villiger reaction, **7**, 682, 686
via borohyride reduction, **8**, 537
via carbomercuration, **4**, 904
via catalytic hydrogenation, **8**, 567
via 1,4-chirality transfer, **6**, 9
via conjugate addition, **2**, 330
via conjugate addition–enolate alkylation, **3**, 9
via conjugate addition to α,β-enones, **4**, 141, 142
via copper catalyzed Grignard additions, **4**, 91
via cyclopropane ring opening, **4**, 1045
via Dieckmann reaction, **2**, 823
via Diels–Alder reaction, **5**, 353
via dihydropyrans, **7**, 831
via DMSO, **7**, 302
via enol stannyl ether, **2**, 609
via enone reduction, **8**, 545
via intramolecular ene reactions, **5**, 16
via Michael addition, **4**, 10
via microbial oxidation, **7**, 54
via Nazarov cyclization, **5**, 780
via organoborane Michael addition, **4**, 145
via organocuprate conjugate addition, **4**, 187
via Paterno–Büchi reaction, **5**, 157
via Prins reaction, **2**, 529
via protected cyanohydrins, **1**, 553; **3**, 198
via tandem vicinal difunctionalization, **4**, 245
via vinylic sulfones, **4**, 251
via vinylzirconium(IV) complexes, **1**, 155
via Wacker oxidation, **7**, 454
Prostaglandins, 5,6-didehydro-
synthesis
via enolate alkylation, **3**, 10
Prostanoic acid
synthesis
via 1,4-addition of allylic sulfoxides to enones, **1**, 520
Prostanoids
synthesis, **1**, 566
via cyclopropane ring opening, **5**, 924
via 1,3-dipolar cycloadditions, **4**, 1077
via electrocyclization, **5**, 771
via organoaluminum reagents, **1**, 103
via Pauson–Khand reaction, **5**, 1051
Proteases
peptide synthesis, **6**, 395
phthaloyl group removal
amine protection, **6**, 643
Protecting groups, **6**, 631–693
N-acyl, **6**, 642
alcohols, **6**, 646
N-alkyl, **6**, 644
N-alkylidene, **6**, 644
amines, **6**, 635
carbonyl compounds, **6**, 675
carboxy, **6**, 665
interdependence, **6**, 633
orthogonal stability, **6**, 633
photosensitive, **6**, 668
polymer esters
carboxylic acids, **6**, 670
principal demands, **6**, 631
silyl
alcohol protection, **6**, 652
thiols, **6**, 664
two-step
amines, **6**, 639
Protoberberine
synthesis
via Mannich reaction, **2**, 894, 912
via photoinduced iminium ion–benzylsilane cyclization, **2**, 1040
Protoberberine alkaloids
synthesis
via phthalide enolates, **2**, 946
Protodealumination, **8**, 737
Protodesilylation
Prins reaction, **2**, 564
Protodezirconation
zirconium compounds, **8**, 690
Protoemetinol
synthesis
via Mannich reaction, **2**, 913
Protoilludane
biosynthesis, **3**, 404
synthesis, **3**, 389
6-Protoilludene
synthesis
via magnesium-ene reaction, **5**, 40
Protolichesterinic acid ester
synthesis, **6**, 354
Protonation
acceptor radical anions, **7**, 884
radical anions
bimolecular reaction, **7**, 861
Protonolysis
demercuration, **8**, 850
hydroalumination adducts, **8**, 753
zirconium compounds, **8**, 690
Protons
formaldehyde complexes
theoretical studies, **1**, 286
Protoporphorins
reaction with tetracyanoethylene, **5**, 71
Protostephanone
enantioselective synthesis, **3**, 685
synthesis, **3**, 685
Provitamin D
photochemical ring opening, **5**, 700
photolysis, **5**, 737
Proxicromil
synthesis, **7**, 338
PS-5
synthesis, **6**, 759
via Mannich reaction, **2**, 922, 924
via reactions of enol silanes, **2**, 648

via silyl enol ethers, **2**, 637
PS-5, 1-β-methyl-
synthesis
via diastereoselective reaction, **2**, 652
Pschorr reaction
radical cyclizations, **4**, 811
ring closure, **3**, 507
Pseudocumene
radical cations
oxidation, **7**, 870
Pseudocumene, iodo-
reaction with amides, **4**, 452
Pseudocytidine
synthesis
via Baeyer–Villiger reaction, **7**, 682
via [4 + 3] cycloaddition, **5**, 611
Pseudoguaiane
biosynthesis, **3**, 388
synthesis
via photocycloaddition, **5**, 669
via silyl enol ethers, **2**, 614
Pseudoguaianolides
synthesis
via Pauson–Khand reaction, **5**, 1052
Pseudohalides
synthesis, **6**, 225–255
Pseudohalogens
reactions with alkenes, **4**, 348–356
Pseudoisocytidine
synthesis
via [4 + 3] cycloaddition, **5**, 611
Pseudomonas oleovorans
epoxidation, **7**, 429
Pseudomonas ovalis
reduction
unsaturated carbonyl compounds, **8**, 559
Pseudomonate B, methyl deoxy-
synthesis
via Julia coupling, **1**, 795
Pseudomonic acid
synthesis
via Claisen–Claisen rearrangement, **5**, 888
via ene reaction, **2**, 531
Pseudomonic acid A
synthesis
via Diels–Alder reaction, **5**, 435
Pseudomonic acid C
synthesis
via alkoxyselenation, **4**, 339
via Julia coupling, **1**, 794, 795
Pseudomonic acid esters
synthesis
via Peterson alkenation, **1**, 791
Pseudopericyclic reactions
hetero electrocyclization, **5**, 741
(–)-Pseudopterosin A
synthesis
via cycloaromatization reaction, **2**, 622
Pseudopyranoses
synthesis
via Knoevenagel reaction, **2**, 386
Pseudo sugars
synthesis
via Knoevenagel reaction, **2**, 386
Pseudouridine
synthesis
via [4 + 3] cycloaddition, **5**, 611
Pseudouridine, 2-thio-
synthesis
via [4 + 3] cycloaddition, **5**, 611
Psoralen
tritiation, **8**, 626
Pteridines, substituted
synthesis
via organocopper compounds, **3**, 219
Pterins, substituted
synthesis
via organocopper compounds, **3**, 219
Pterocarpans
synthesis
via isoflavones, **7**, 831
Pterodactyladiene
synthesis, **3**, 872
Puerarin, 7,4′-di-*O*-methyl-
synthesis, **7**, 830
Pulchellon
synthesis, **8**, 935
Pulegone
dienyltricarbonyliron complexes
asymmetric synthesis, **4**, 688
oxidation
peroxy acid, **7**, 684
oxiranes
rearrangement, **3**, 771
reduction, **8**, 563
molybdenum complex catalyst, **8**, 554
synthesis
via ene reaction, **2**, 540
Pulo'upone
synthesis
via Diels–Alder reactions, **5**, 364
via intramolecular Diels–Alder reaction, **5**, 545
Pumiliotoxin A
synthesis, **6**, 742
enantioselective, **2**, 1028
via ene reaction, **2**, 550
via Mannich reaction, **2**, 1030
Pumiliotoxin C
synthesis, **6**, 756, 769
via Diels–Alder reactions, **5**, 333, 360
via Eschenmoser coupling reaction, **2**, 876
via nitrone cyclization, **4**, 1117
Pumiliotoxin 251D
synthesis
via Mannich reaction, **2**, 1031
Pumiliotoxins
synthesis
via iminium ion–vinylsilane cyclization, **1**, 593
Pummerer rearrangement, **7**, 194
abnormal reactions, **7**, 203
α-alkylation
preparation of α-alkylated sulfides, **7**, 199
α-arylation
preparation of α-arylated sulfides, **7**, 199
asymmetric reaction
α-acetoxylation, **7**, 199
β-elimination, **7**, 204
examples, **7**, 196
hydroxylic solvents, **7**, 202
intramolecular
α-acetoxylation, **7**, 196
participation by hydroxy groups, **7**, 202

preparation of α-alkylated and α-arylated sulfides, **7**, 199
mechanism, **7**, 195
(methylthio)methyl ethers, **7**, 292
nitrogen participation, **7**, 201
oxidation
halides, **7**, 667
oxidative rearrangement, **7**, 826
α-phenylsulfinylacrylates, **2**, 363
sulfoxides
formation of α-functionalized sulfides, **7**, 193
transannular reactions, **7**, 205
trimethylsilyl triflate, **7**, 202
vinylic sulfoxides, **6**, 151
vinylogous, **7**, 204
Pummerer's ketone
synthesis
use of silver carbonate, **3**, 664
Punaglandins
synthesis
via retro Diels–Alder reactions, **5**, 562
Punctatin A
synthesis, **3**, 984
via photochemical C—H insertion reactions, **3**, 1058
Pupukeanane, isocyano-
synthesis, **7**, 318
Purine, 9-alkyl-6-iodo-
$S_{RN}1$ reaction, **4**, 462
Purine, 6-chloro-
coupling reactions
with primary alkyl Grignard reagents, **3**, 462
Purine, 6-methylthio-
coupling reactions
with primary alkyl Grignard reagents, **3**, 462
Purines
analogs
synthesis *via* Eschenmoser coupling reaction, **2**, 889
synthesis
via Eschenmoser coupling reaction, **2**, 889
Push–pull alkenes
addition reactions, **4**, 122–128
Pyllodulcin
synthesis
via directed lithiation, **1**, 477
Pyran
synthesis
via palladium(II) catalysis, **4**, 557
Pyran, 2-alkenyldihydro-
synthesis
via cyclization of δ-allenic ketones, **4**, 397
Pyran, 2-alkenyltetrahydro-
synthesis
via cyclization of δ-allenic alcohols, **4**, 396
Pyran, 3-alkyl-4-chlorotetrahydro-
synthesis from 1-alkenes
Prins reaction, **2**, 528
Pyran, 2-alkyltetrahydro-
synthesis
via Lewis acid promoted reaction, **1**, 346, 347
Pyran, 4-chlorotetrahydro-
formation
type III ene reaction, **2**, 553
Pyran, *cis*-2,6-dialkyl-4-chlorotetrahydro-
synthesis
via ene reaction, **2**, 554
Pyran, dihydro-
allylic oxidation, **7**, 103
oxidation
pyridinium chlorochromate, **7**, 267
reaction with dichlorocarbene, **4**, 1005
ring contraction, **7**, 831
synthesis
via allylic anions and epoxides, **2**, 60
Pyran, tetrahydro-
arene alkylation by
Friedel–Crafts reaction, **3**, 315
synthesis
stereoselectivity, **4**, 381, 384
via electrophile cyclization, **7**, 523
2*H*-Pyran-3,5-diol, 3,4,5,6-tetrahydro-4-nitro-
synthesis
via Henry reaction, **2**, 327
2*H*-Pyran-3-ol, tetrahydro-2,2,6-trimethyl-
synthesis
via sulcatol, **7**, 634
Pyranonaphthyridine
synthesis
via Knoevenagel reaction, **2**, 380
Pyranones *see* Pyrones
Pyranooxepin
synthesis
via Friedel–Crafts reaction, **2**, 765
Pyrano[2,3-*b*]pyridine
synthesis
via Knoevenagel reaction, **2**, 380
Pyranopyrone
aldol cyclization, **2**, 170
Pyranoquinoline
synthesis
via Perkin reaction, **2**, 401
Pyranose
synthesis, **6**, 35
Diels–Alder reaction, **2**, 690
Pyranoside, 2,3-anhydro-4-*O*-tosyl-
reaction with sodium azide, **6**, 91
Pyranosides
methylenation, **1**, 737
reductive ring cleavage, **8**, 218
C-Pyranosides
synthesis
via cuprate 1,2-addition, **1**, 126
Pyranosides, methyl-
2,3-unsaturated
reduction, **8**, 219
Pyrans
synthesis
via Knoevenagel reaction, **2**, 379, 380
via photolysis, **5**, 741
Pyrans, dihydro-
coupling reactions
with alkyl Grignard reagents, **3**, 444
[2 + 2] cycloaddition reactions
methyl tetrolate, **5**, 1067
metallation, **3**, 252
Pauson–Khand reaction, **5**, 45, 1048
reaction with Grignard reagents
nickel catalysts, **3**, 229
reaction with organocopper compounds, **3**, 218
synthesis, **1**, 589
via Diels–Alder reaction, **5**, 435

Pyrans, 3,4-dihydro-
reaction with dimethyl acetylenedicarboxylate
dihydrooxacine synthesis, **5**, 687
reaction with isocyanates
glycal synthesis, **5**, 108
synthesis
via Diels–Alder reaction, **5**, 453
Pyrans, 5,6-dihydro-
synthesis
via Diels–Alder reactions, **5**, 430
via vinylsilane acetals, **1**, 589
Pyranulose acetate
synthesis, **4**, 1092
Pyranylamines, tetrahydro-
reduction, **8**, 228
Pyranyl sulfides, tetrahydro-
reduction, **8**, 230
Pyrazine, chloro-
$S_{RN}1$ reaction, **4**, 462
synthesis
via dichlorocarbene insertion, **4**, 1021
Pyrazine, 2,5-dibenzyl-
hydrogenation, **8**, 643
Pyrazine, 2,5-diboradihydro-
oxidation
use of chromyl trichloroacetate, **7**, 601
Pyrazine, tetrachloro-
oxidation
hydrogen peroxide, **7**, 750
Pyrazines
amination, **4**, 436
Diels–Alder reactions, **5**, 491
reduction, **8**, 643
Vilsmeier–Haack reaction, **2**, 789
Pyrazines, 2-acyloxy-
acylating agent, **1**, 422
Pyrazinethiol
synthesis, **7**, 667
Pyrazinium chlorochromate
oxidation
alcohols, **7**, 271
Pyrazino[1,2-*a*]indole, 1,2,3,4-tetrahydro-
synthesis
via Ritter reaction, palladium, **6**, 284
Pyrazino[1,2-*a*]quinoline, hexahydro-
synthesis
via Knoevenagel reaction, **2**, 379
Pyrazinyl sulfoxide
Pummerer rearrangement, **7**, 667
Pyrazole, 1-acetyl-
Friedel–Crafts reaction, **2**, 744
5*H*-Pyrazole, 5-acetyl-3,4-diethoxycarbonyl-5-methyl-
synthesis
via cycloaddition, **3**, 893
Pyrazole, 4-acyl-
synthesis
via hydrazone anions, **2**, 523
Pyrazole, amino-
synthesis
via activated allene, **4**, 56
1,2-Pyrazole, 4-(3-butenyl)-
synthesis
via retro Diels–Alder reactions, **5**, 582
Pyrazole, 3,5-dimethyl-
chromium trioxide complex
alcohol oxidation, **7**, 260
allylic oxidation, **7**, 104
pyridinium chlorochromate
allylic alcohol oxidation, **7**, 264
Pyrazole, 3,5-dimethyl-*N*-acyl-
reduction
metal hydrides, **8**, 271
Pyrazole, nitro-
reduction, **8**, 636
Pyrazole, 4-(2′-styryl)-
synthesis
via retro Diels–Alder reactions, **5**, 584
Pyrazole, tetrahydro-
synthesis
via amination, **7**, 741
Pyrazole, triaryl-
synthesis
via hydrazone anions, **2**, 522
Pyrazole-4-carbaldehyde
synthesis
via Vilsmeier–Haack reaction, **2**, 791
Pyrazoles
reduction, **8**, 636
synthesis, **3**, 905; **5**, 1070; **6**, 117
via dihalocyclopropyl compounds, **4**, 1023
via hydrazone anions, **2**, 522
via Knoevenagel reaction, **2**, 362, 379
via retro Diels–Alder reactions, **5**, 582
Vilsmeier–Haack reaction, **2**, 780
Pyrazoles, *N*-acyl-
reduction, **8**, 636, 965
metal hydrides, **8**, 271
Pyrazoles, *N*-alkyl
lithiation, **1**, 477
Pyrazoles, 3,5-dialkyl-
synthesis, **1**, 557
3,5-Pyrazolidinedione, 4-diazo-
decomposition, **3**, 902
3,5-Pyrazolidinedione, 1,2-dimethyl-
Knoevenagel reaction, **2**, 357
Pyrazolidines
reduction, **8**, 636
Pyrazoline, tosyl-
reaction with trimethylaluminum
pyrazolol synthesis, **1**, 98
Pyrazolines
nitrogen extrusion
cyclopropane formation, **4**, 954
reduction, **8**, 636
synthesis
via azacyclopropanes, **7**, 628
1-Pyrazolines
synthesis
via diazo compounds, **4**, 953
2-Pyrazolines
5-substituted
synthesis *via* nitrilimine cycloaddition, **4**, 1084
1-Pyrazolines, 3-substituted
synthesis
via diazoalkane cycloaddition, **4**, 1102
Pyrazolinone
stereochemistry
epoxidation, **7**, 372
2-Pyrazolin-5-one
synthesis
via hydrazone anions, **2**, 523
via intermolecular addition, **4**, 51

4-Pyrazolinone, *N*-sulfonyl-
reduction
L-selectride, **8**, 637
5-Pyrazolinones
catalytic hydrogenation, **8**, 637
Pyrazoloimidazolidinone
reduction, **8**, 636
Pyrazolols
synthesis
via tosylpyrazolines, **1**, 98
Pyrazolone, benzylidene-
Knoevenagel reaction
stereoselectivity, **2**, 351
Pyrazolones
Knoevenagel reaction, **2**, 364
stereoselectivity, **2**, 351
5-Pyrazolones, 4-arylidene-
Diels–Alder reactions, **5**, 454
Pyrazolopyridines
synthesis
via Vilsmeier–Haack reaction, **2**, 787
Pyrene, 15,16-dimethyldihydro-
electrocyclization, **5**, 705
Pyrene, tetrahydro-
synthesis
via photolysis, **5**, 728
Pyrene-3-aldehyde
Wolff–Kishner reduction, **8**, 338
Pyrenes
synthesis
via electrocyclization, **5**, 720
Pyrenochaetic acid A
synthesis
via retro Diels–Alder reaction, **5**, 571
Pyrenophorin
synthesis, **1**, 569; **3**, 126
via macrolactonization, **6**, 375
via organostannane acylation, **1**, 446
Pyrethrin
esters, **6**, 324
Pyrethrin-I
pyrolysis, **5**, 717
Pyrethroids
enantioselective synthesis
via cyclopropanation, **4**, 1039
synthesis, **3**, 848
via chiral cyanohydrins, **1**, 546
Pyrethrolone
synthesis
via Wacker oxidation, **7**, 455
Pyrethrolone, tetrahydro-
methyl ether
synthesis *via* cyclopropane ring opening, **4**, 1046
Pyridazine, chloro-
alkylation
with primary alkyl Grignard reagents, **3**, 461
Pyridazine, *N,N*-diacylhexahydro-
ring scission, **8**, 640
Pyridazine, 4,5-dibenzoyl-
synthesis, **7**, 777
Pyridazine, 4-nitro-, 1-oxide
ammonia adducts
oxidation, **4**, 434
Pyridazine *N*-oxides
deoxygenation, **8**, 390, 640
reaction with aryl Grignard regents, **3**, 494
Pyridazines
ammonia adducts
oxidation, **4**, 434
Diels–Alder reactions, **5**, 491
reduction, **8**, 640
synthesis
via Knoevenagel reaction, **2**, 362, 379
Pyridazines, dihydro-
reduction, **8**, 640
Pyridazines, tetrahydro-
reduction, **8**, 640
Pyridazinium salts
reduction, **8**, 640
3-Pyridazinone
reduction
zinc, **8**, 563
6-Pyridazinone, 1-alkyl-
reduction
LAH, **8**, 641
6-Pyridazinone, 4,5-dihydro-
reduction
excess LAH, **8**, 641
Pyridazones
synthesis
via hydrazone anions, **2**, 522
2,2′-Pyridil
rearrangement, **3**, 826
2,2′-Pyridilic acid
synthesis, **3**, 826
Pyridine, 2-acetyl-
reduction
hydride transfer, **8**, 94
Pyridine, 3-acetyl-
hydrogenation
catalytic, **8**, 141
Pyridine, 4-acetyl-
hydrogenation
catalytic, **8**, 141
Pyridine, 1-acyl-4-benzylidene-1,4-dihydro-
synthesis
use in mixed anhydride synthesis, **6**, 310
Pyridine, 6-alkoxy-2-hydroxy-
synthesis
via Knoevenagel reaction, **2**, 378
Pyridine, 2-alkylthio-
synthesis
via [2 + 2 + 2] cycloaddition, **5**, 1154
Pyridine, amino-
hydrogenation, **8**, 598
Pyridine, 2-amino-
synthesis
via [2 + 2 + 2] cycloaddition, **5**, 1154
Pyridine, 2-benzoyl-
reduction
hydride transfer, **8**, 94
Pyridine, 4-benzoyl-
reduction
dissolving metals, **8**, 115
Pyridine, 4-benzyl-
acid anhydride synthesis, **6**, 310
Pyridine, bis(alkylsulfonyl)-
synthesis, **4**, 443
Pyridine, bis(alkylthio)-
synthesis, **4**, 443
Pyridine, bis(5-hexynyl)methyl-
synthesis

via cycloaddition, **5**, 1154
Pyridine, 2-bromo-
coupling reactions
with primary alkyl Grignard reagents, **3**, 460
Pyridine, 3-bromo-
dehydrohalogenation
pyridyne generation, **4**, 489
$S_{RN}1$ reaction, **4**, 462
Pyridine, 4-bromo-
$S_{RN}1$ reaction, **4**, 462
Pyridine, 2-chloro-
coupling reactions
with primary alkyl Grignard reagents, **3**, 460
Pyridine, 3-chloro-
synthesis
via dichlorocarbene insertion, **4**, 1021
via dihalocarbene, **4**, 1004
Pyridine, 2-chloro-3,5-dinitro-
glycoside synthesis, **6**, 54
Pyridine, chlorofluoro-
hydrogenolysis, **8**, 901
Pyridine, 2-cinnamoyl-
reduction
dihydropyridines, **8**, 561
Pyridine, 3-cyano-
reduction
borohydrides, **8**, 580
Pyridine, cyanodihydro-
dimerization, **5**, 64
Pyridine, 3-cyano-1,4,5,6-tetrahydro-
synthesis, **8**, 580
Pyridine, 2,3-dehydro-
generation, **4**, 489
Pyridine, 2,6-dibromo-
oxidation
hydrogen peroxide in trifluoroacetic acid, **7**, 750
reaction with carbanions
via $S_{RN}1$ reaction, **4**, 468
Pyridine, 2,6-dichloro-
coupling reactions
with primary alkyl Grignard reagents, **3**, 460
Pyridine, 3,5-dichlorotrifluoro-
hydrogenation, **8**, 905
Pyridine, 3,5-dicyano-
reduction
aluminum hydrides, **8**, 583
borohydrides, **8**, 580
Pyridine, 3,5-diethoxycarbonyl-
reduction
borohydrides, **8**, 580
Pyridine, 3,5-diethoxycarbonyl-2,6-dimethyl-1,4-dihydro-
reduction
unsaturated carbonyl compounds, **8**, 561
Pyridine, 2,6-dihydroxy-
synthesis
via Knoevenagel reaction, **2**, 378
Pyridine, 2,6-dimethyl-
hydrogenation
nickel catalysts, **8**, 597
α-lithiated
crystal structure, **1**, 12
Rosenmund reduction, **8**, 287
Pyridine, 4-dimethylamino-
catalyst
thiol ester synthesis, **6**, 437
esterification
carbohydrates, **6**, 657
1-*N*-oxide
oxidation with, **7**, 662
Pyridine, *N*,*N*-dimethylamino-
catalyst
acylation, **6**, 327
Pyridine, 2,6-disubstituted-1,2,5,6-tetrahydro-
synthesis
via Mannich reaction, **2**, 1035
Pyridine, hydroxy-
Reimer–Tiemann reaction
normal, **2**, 770
Pyridine, 3-hydroxy-
electroreduction, **8**, 593
hydrogenation, **8**, 598
Pyridine, 3-hydroxy-*N*-phenyl-
betaine
reaction with alkynes, **4**, 48
Pyridine, iodo-
coupling reactions
with alkylzinc reagents, **3**, 460
Pyridine, 3-iodo-
$S_{RN}1$ reaction, **4**, 462
Pyridine, mercapto-
synthesis, **4**, 441
Pyridine, 2-mercapto-
N-oxide
O-acyl thiohydroxamates from, **7**, 719
polymer-bonded
synthesis *via* Friedel–Crafts reaction, **3**, 302
Pyridine, 2-methoxy-
ortho metallation, **1**, 474
Pyridine, 3-methoxy-
reduction
borohydrides, **8**, 584
Pyridine, β-methoxyvinyl-
synthesis, **2**, 598
Pyridine, methyl-
microbial oxidation, **7**, 75
Pyridine, 2-methyl-
hydrogenation
nickel catalysts, **8**, 597
Pyridine, 4-methyl-
Vilsmeier–Haack reaction, **2**, 789
Pyridine, 2-methyl-3-acyl-
synthesis, **1**, 560
Pyridine, 2-methyl-3-formyl-
synthesis, **1**, 560
Pyridine, 6-methyl-2,3,4,5-tetrahydro-
N-oxide
reaction with allylmagnesium bromide, **1**, 393
Pyridine, *N*-methyltetrahydro-
hydroformylation, **4**, 927
Pyridine, 3-methylthio-
reduction
borohydrides, **8**, 584
Pyridine, 6-methyl-2-vinyl-
Michael addition, **4**, 10
Pyridine, 3-nitro-
electroreduction, **8**, 593
reduction
borohydrides, **8**, 580
Pyridine, 4-nitro-
aromatic nucleophilic substitution, **4**, 432
N-oxide

aromatic nucleophilic substitution, **4**, 432
Pyridine, 2-nitroso-
synthesis
via oxidation of sulfimides, **7**, 752
Pyridine, pentachloro-
dehalogenation
aluminum hydrides, **8**, 584
Pyridine, pentafluoro-
alcohol protecting group, **4**, 439
catalytic hydrogenation, **8**, 903
oxidation
hydrogen peroxide, **7**, 750
Pyridine, 2-phenyl-
hydrogenation
nickel catalysts, **8**, 598
Pyridine, 4-phenyl-
synthesis, **3**, 513
Pyridine, 4-pyrrolidino-
catalyst
acylation, **6**, 327
Pyridine, tetrachloro-4-substituted
dehalogenation
aluminum hydrides, **8**, 584
Pyridine, 3-trialkylstannyl-
reduction
borohydrides, **8**, 585
Pyridine, 3-trimethylsilyl-
reduction
borohydrides, **8**, 585
Pyridine, 2-trimethylstannyl-
Friedel–Crafts reaction, **2**, 743
Pyridinecarbaldehyde
synthesis, **7**, 656
Pyridine-2-carbaldehyde
aldol reaction
4-(*N,N*-dimethylamino)-2-butanone, **2**, 147
Pyridine-3-carbaldehyde
synthesis
via Vilsmeier–Haack reaction, **2**, 783
Pyridine-3-carbaldehyde, 4-phenyl-
synthesis
via Vilsmeier–Haack reaction, **2**, 782
Pyridinecarbohydroxamic acid
Lossen reaction, **6**, 824
2-Pyridinecarboxylic acid
Curtius reaction, **6**, 812
4-Pyridinecarboxylic acid, 2,6-diphenyl-
synthesis
via oxidative cleavage of alkenes, **7**, 578
Pyridinecarboxylic acids
electrolytic reduction, **8**, 285
hydrogenation, **8**, 599
Pyridinedicarbaldehyde
synthesis, **7**, 656
2,6-Pyridinedicarbothioate, *O,O′*-dimethyl
reaction with glycols
synthesis of thioxoesters, **6**, 449
Pyridine-3,5-dicarboxylic acid
hydride transfer, **8**, 82
hydrogenation, **8**, 599
Pyridine-3,5-dicarboxylic acid, 1,4-dihydro-
hydride donors, **8**, 92
Pyridine *N*-oxides
dehydrogenation, **7**, 144
deoxygenation, **8**, 391, 600
oxidation with, **7**, 661
reduction, **8**, 392, 587
Pyridine *N*-oxides, 2-acylthio-
synthesis, **6**, 441
Pyridines
acylating agents, **1**, 422
borane
reductive amination, **8**, 54
catalyst
acid chloride synthesis, **6**, 302
coupling reactions
with sp^3 organometallics, **3**, 460
derivatives
synthesis, **2**, 521
hydrogenation
homogeneous catalysis, **8**, 455, 600
metallation
addition reactions, **1**, 471
N-oxidation
m-chloroperbenzoic acid, **7**, 749
N-oxides
Knoevenagel reaction, **2**, 364
reactions with alkyl radicals, **4**, 768
reduction, **8**, 579–600
aluminum hydrides, **8**, 583
borohydrides, **8**, 580
dihydropyridine, **8**, 589
regioselective cyanation, **4**, 433
2-substituted
lithiation, **1**, 474
synthesis
via [2 + 2 + 2] cycloaddition, **5**, 1152–1155
via Diels–Alder reactions, **5**, 412, 416, 491
via metal catalysts, **5**, 1153
via nitrogen-stabilized carbanions, **1**, 461
via Reimer–Tiemann reaction, **2**, 773
via retro Diels–Alder reactions, **5**, 583
via tandem vicinal difunctionalization, **4**, 251
Pyridines, 4-acyloxy-
reaction with thiols
synthesis of carbothioates, **6**, 443
Pyridines, 3-alkoxy-
metallation, **1**, 474
Pyridines, alkyl-
oxide
reduction, **8**, 392
synthesis, **4**, 428
via alkyl radical addition, **7**, 732
Pyridines, 2-alkyl-
synthesis
via [2 + 2 + 2] cycloaddition, **5**, 1152
Pyridines, 3-alkyl-
synthesis
via imine anions, **2**, 482
Pyridines, 4-alkyl-
regioselective synthesis, **4**, 446
Pyridines, 2-aryl-
synthesis, **3**, 512; **4**, 430
via [2 + 2 + 2] cycloaddition, **5**, 1152
Pyridines, 4-aryl-
synthesis, **3**, 512; **4**, 428
Pyridines, 2-(ω-cyanoalkyl)-
synthesis
via cocycloaddition, **5**, 1155
Pyridine, dihydro-
analysis of aldehydes
Knoevenagel reaction, **2**, 354

aromatization
solid support, **7**, 846
Hantzsch synthesis
Knoevenagel reaction, **2**, 376
optically active
synthesis, **2**, 521
oxamination, **7**, 489
oxidation, **4**, 428
reaction with singlet oxygen, **2**, 1068
synthesis
via conjugate addition, **4**, 215
via conjugate addition to oxazolines, **4**, 206
via ketocarbenoids and pyrroles, **4**, 1061
via Knoevenagel reaction, **2**, 384
Pyridines, 1,2-dihydro-
Diels–Alder reactions
acyl nitroso compounds, **5**, 420
nitrosobenzene, **5**, 418
reduction
borohydrides, **8**, 585
synthesis
via hetero electrocyclization, **5**, 741
Pyridines, 1,4-dihydro-
acid stability, **8**, 95
chiral
intramolecular reductions by, **8**, 95
chiral macrocyclic
reduction, model, **8**, 95
enantioselective reductions, **8**, 93
hydride donor, **8**, 92
macrocyclic
enantioselective reductions, **8**, 95
redox potentials, **8**, 93
reduction
pyridinium salts, **8**, 589
reduction potential
thermodynamic activation, **8**, 82
synthesis
via retro Diels–Alder reactions, **5**, 583
Pyridines, 5,6-dihydro-
synthesis
via intramolecular Ritter reaction, **6**, 273
via retro Diels–Alder reactions, **5**, 583
Pyridines, halo-
coupling reactions, **3**, 509
oxide
reduction, **8**, 392
reaction with magnesium dialkylcuprates, **3**, 219
Pyridines, 2-halo-
$S_{RN}1$ reaction, **4**, 462
Pyridines, 2-imino-
synthesis
via [2 + 2 + 2] cycloaddition, **5**, 1156
Pyridines, substituted
synthesis
via organocopper compounds, **3**, 219
Pyridines, tetrahydro-
carbene complexes
reactions with diphenylacetylene, **5**, 1105
Schopf reaction, **2**, 943
synthesis
via N-acyliminium ions, **2**, 1072
via Diels–Alder reactions, **5**, 404, 406
via Mannich reaction, **2**, 1034
Pyridines, 1,2,3,6-tetrahydro-
oxamination, **7**, 489
Pyridines, vinyl-
Michael addition, **4**, 10
synthesis, **3**, 498
2(1*H*)-Pyridinethione, 3-formyl-
aldolization, **2**, 150
2-Pyridinethione, *N*-hydroxy-
carbamates
radical cyclization, **4**, 812
2-Pyridinethiones
synthesis
via [2 + 2 + 2] cycloaddition, **5**, 1158
Pyridinethiones, *N*-alkoxy-
alkoxy radicals from, **4**, 812
Pyridinium bromide, 1-methyl-4-carbamoyl-
reduction
dithionite, **8**, 589
Pyridinium chloride, 2-benzoylthio-1-methyl-
acid anhydride synthesis, **6**, 310
Pyridinium chlorochromate
allylic oxidation, **7**, 103
organoborane oxidation, **7**, 601
oxidation
alcohols, **7**, 260
2-nitro alcohols, **2**, 323
solid-supported, **7**, 841
oxidative halogenation reagent, **7**, 530
Pyridinium chlorochromate, 4-(dimethylamino)-
oxidation
alcohols, **7**, 269
Pyridinium chromate
inert inorganic support
alcohol oxidation, **7**, 279
oxidation
solid support, **7**, 845
Pyridinium dichromate
allylic oxidation, **7**, 103
oxidation
alcohols, **7**, 272
Pyridinium dichromate, 3-carboxy-
oxidation
alcohols, **7**, 277
Pyridinium dichromate, 4-carboxy-
oxidation
alcohols, **7**, 277
Pyridinium fluorochromate
oxidation
alcohols, **7**, 267
Pyridinium iodide, 1-methyl-4-cyano-
reduction
borohydrides, **8**, 587
Pyridinium salts
arene substitution reactions, **4**, 425
oxidation
of primary and secondary alcohols, **8**, 96
reactions with alkyl radicals, **4**, 768
reduction
aluminum hydrides, **8**, 587
borohydrides, **8**, 584
dihydropyridine, **8**, 589
regioselectivity, **8**, 92
Pyridinium salts, 2-acyloxy-
acylation
alcohols, **6**, 331
Pyridinium salts, 2-acylthio-*N*-alkyl-
acylating agents, **6**, 442
Pyridinium salts, alkoxy-

reduction
aluminum hydrides, **8**, 587
Pyridinium salts, *N*-alkyl-
reductive deamination, **8**, 827
Pyridinium salts, 1-alkyl-3-carbamoyl-
partial reduction, **8**, 600
Pyridinium salts, 1-amino-
reduction
borohydrides, **8**, 587
Pyridinium salts, *N*-(aryloxy)-
rearrangement, **4**, 430
Pyridinium salts, azo-
arene substitution reactions, **4**, 425
Pyridinium salts, 4-cyano-1-methyl-
polarographic reduction, **8**, 595
Pyridinium salts, 5,6-dihydro-
reactions with organometallic compounds, **1**, 367
Pyridinium salts, 1,3-dimethyl-
reduction
borohydrides, **8**, 587
Pyridinium salts, *N*-(2,6-dimethyl-4-oxopyridin-1-yl)-
aromatic nucleophilic substitution, **4**, 430
Pyridinium salts, fluoro-
reaction with alcohols
iodination, **6**, 214
Pyridinium salts, 2-halo-
carbothioate synthesis, **6**, 438
Pyridinium salts, 4-nitroaryl-
electroreduction, **8**, 595
Pyridinium sulfonate, 2-fluoro-1-methyl-
activator
DMSO oxidation of alcohols, **7**, 299
Pyridinium *p*-toluenesulfonate
catalyst
Diels–Alder reaction, **2**, 683
Pyridinium tosylate, 2-fluoro-1-methyl-
glycoside synthesis, **6**, 54
Pyridinotropolone
rearrangements, **3**, 818
Pyridin-1-yl, 2,6-dimethyl-4-oxo-
arene substitution reactions, **4**, 425
Pyrido[2,3-*d*]benzopyran, 10-hydroxy-10-methyl-6-nitro-
synthesis, **4**, 430
Pyrido[4,3-*b*]carbazole
synthesis
via Knoevenagel reaction, **2**, 379
Pyrido[3,2-*d*]coumarins
synthesis, **4**, 430
Pyridodipyrimidine
alcohol oxidation, **8**, 97
2-Pyridone, 1-alkyl-
reduction
aluminum hydrides, **8**, 583
2-Pyridone, 5,6-dihydro-
synthesis
via Mannich reaction, **2**, 930
Pyridone, *N*-hydroxy-
oxidation with, **7**, 662
2-Pyridone, *N*-methyl-
photodimerization, **5**, 637
4-Pyridone, 1-methyl-
reduction
dissolving metals, **8**, 597
Pyridones
electroreduction, **8**, 593
hydrogenation, **8**, 598
synthesis
via Knoevenagel reaction, **2**, 361
2-Pyridones
photodimerization, **5**, 637, 638
synthesis
via [2 + 2 + 2] cycloaddition, **5**, 1155–1157
via δ-ketonitriles, **6**, 280
via Mannich reaction, **2**, 916
4-Pyridones
reduction
dissolving metals, **8**, 597
1*H*-Pyrido[3,2,1-*kl*]phenothiazine
Mannich reaction
with preformed iminium salts, **2**, 902
synthesis
via Friedel–Crafts reaction, **2**, 759
Pyridopyrimidines
synthesis, **3**, 543
Pyrido[2,1-*b*]quinazoline
synthesis
via carbonylation, **3**, 1038
Pyridoxine
synthesis
via [2 + 2 + 2] cycloaddition, **5**, 1154
Pyridylamides, *N*-methylamino-
acylation
Grignard reagents, **1**, 422
Pyridyl betaine, nitro-
cycloaddition reactions
fulvenes, **5**, 630
2-Pyridylcarboxylates
acylation, **1**, 434
Pyridylethoxycarbonyl group
amine-protecting group, **6**, 639
Pyridylethyl esters
carboxy-protecting groups, **6**, 666
3-Pyridyl isocyanide
O-acyl thiohydroxamate photolysis, **7**, 731
2-Pyridyl ketone-*O*-acyloximes
acylation
Grignard reagents, **1**, 422
Pyridylnitriles
cycloaddition reactions, **5**, 1152
Pyridylsulfonyloxy group
alcohol inversion, **6**, 22
Pyridyl thioesters
use in synthesis
lactones, **6**, 438
Pyridyl triflate
coupling reactions
with Grignard reagents, **3**, 460
2,3-Pyridynes
nucleophilic addition, **4**, 494
3,4-Pyridynes
Diels–Alder reactions, **5**, 384
nucleophilic addition, **4**, 494
Pyrimidine, 5-bromo-6-chloro-2,4-dimethyl-
hydrogenolysis, **8**, 902
Pyrimidine, 2-chloro-
$S_{RN}1$ reaction, **4**, 462
Pyrimidine, 5-chloro-
synthesis
via dichlorocarbene insertion, **4**, 1021
Pyrimidine, 2-chloro-4,6-dimethyl-
hydrogenolysis, **8**, 906

Pyrimidine, 4-dialkylboryloxy-2-isopropyl-6-methyl-
  boryl enolates from, **2**, 244
Pyrimidine, 1,3-dialkyl-5,5-dimethylhexahydro-
  alkylation, **6**, 523
Pyrimidine, 2,4-diamino-5-(3,4,5-triethylbenzyl)-
  synthesis, **3**, 301
Pyrimidine, hydroxy-
  Reimer–Tiemann reaction
    normal, **2**, 770
Pyrimidine, 4-methoxy-
  1-oxide
    cyanation, **4**, 433
Pyrimidine, 6-methyl-
  Vilsmeier–Haack reaction, **2**, 789
Pyrimidine, 2-methylthio-
  alkylation
    with primary alkyl Grignard reagents, **3**, 461
Pyrimidine, perhydro-
  synthesis
    via Mannich reaction, **2**, 916
Pyrimidine, thio-
  synthesis
    via Eschenmoser coupling reaction, **2**, 889
Pyrimidine, trichloro-
  alkylation
    with primary alkyl Grignard reagents, **3**, 461
5-Pyrimidinecarbonitriles, 2-aryloxy-4-amino-
  synthesis, **4**, 440
Pyrimidinediones
  isoalloxazines from, **4**, 436
  synthesis
    via [2 + 2 + 2] cycloaddition, **5**, 1158
Pyrimidines
  analogs
    synthesis *via* Eschenmoser coupling reaction, **2**, 889
  Diels–Alder reactions, **5**, 491
  heterocyclic fused
    synthesis, **3**, 543
  N-oxides
    cyanation, **4**, 433
    Knoevenagel reaction, **2**, 364
  reduction, **8**, 642
  synthesis
    via Vilsmeier–Haack reaction, **2**, 787
Pyrimidines, *N*-benzyldihydro-
  reduction
    sodium borohydride, **8**, 642
Pyrimidine-2-thione, 1-aryl-
  reduction
    LAH, **8**, 642
Pyrimidine-2(1*H*)-thiones
  reduction
    sodium borohydride, **8**, 642
Pyrimidinium salts, dihydro-
  reduction, **8**, 642
Pyrimidinone
  synthesis
    via retro Diels–Alder reactions, **5**, 583
Pyrimidin-2(1*H*)-ones
  reduction
    sodium borohydride, **8**, 642
Pyrimidoblamic acid
  synthesis
    via Diels–Alder reaction, **5**, 492
Pyrimidopyridine
  synthesis
    via Peterson alkenation, **1**, 792
Pyroangolensolide
  synthesis, **7**, 174
Pyrocalciferol
  synthesis
    via electrocyclization, **5**, 700
Pyrogermacrone
  synthesis
    via Cope rearrangement, **5**, 809
Pyrone, α-cuprio-
  reactivity, **3**, 217
Pyrone, dihydro-
  optically pure
    Diels–Alder reaction, **2**, 688
  reduction
    DIBAL-H, **8**, 544
  synthesis
    via Diels–Alder reaction, **2**, 665
    via Knoevenagel reaction, **2**, 359
2-Pyrone, 3,4-dihydro-4,4-dimethyl-
  photolysis
    with benzoquinone, **5**, 156
Pyrone, 3,5-diphenyl-
  synthesis
    via Vilsmeier–Haack reaction, **2**, 785
Pyrone, 4-hydroxy-6-methyl-
  Knoevenagel reaction product
    Michael reaction, **2**, 359
2-Pyrone, 4-isopropyl-6-methyl-
  alkylation, **3**, 24
4-Pyrone, 2-methoxy-
  reduction
    stereoselectivity, **8**, 5
Pyrone-3-carboxylate
  synthesis
    via Michael addition, **4**, 41
Pyrones
  [3 + 2] cycloaddition reactions, **5**, 307
  Diels–Alder reactions, **5**, 330
  enzymic reduction
    specificity, **8**, 196
  synthesis
    via activated alkynes, **4**, 53
    via [2 + 2 + 1 + 1] cycloadditions, **5**, 1102
    via [2 + 2 + 2] cycloaddition, **5**, 1157
    via dibromocyclopropyl compounds, **4**, 1023
    via Michael addition, **4**, 41
    via tandem vicinal difunctionalization, **4**, 251
2-Pyrones
  6-conjugated
    synthesis, **7**, 109
  [4 + 3] cycloaddition reactions, **5**, 604
  photodimerization, **5**, 638
  synthesis
    via organocuprate conjugate addition, **4**, 192
4-Pyrones
  aldol cyclization, **2**, 170
  reduction
    borohydrides, **8**, 540
  synthesis
    via palladium(II) catalysis, **4**, 557, 558
4-Pyrones, 2,3-dihydro-
  α′-acetoxylation, **7**, 175
4-Pyrones, tetrahydro-
  cyclization, **5**, 766

synthesis
via intramolecular diastereoselective additions, **2**, 651
α-Pyrone sulfone
Diels–Alder reaction, **6**, 161
Pyrophosphates
phosphorylation, **6**, 605
Pyrophosphates, tetraalkyl
synthesis, **6**, 607
Pyrrole, 2-acetyl-
hydrogenation, **8**, 604
Pyrrole, 3-acetyl-
oxidative rearrangement
solid support, **7**, 846
Pyrrole, 2-acetyl-1-methyl-
Friedel–Crafts reaction
rearrangement, **2**, 745
Pyrrole, acyl-
reduction
metal hydrides, **8**, 270
Pyrrole, 1-alkyl-
reactions with carbenoids, **4**, 1061
Pyrrole, 1-benzenesulfonyl-
Friedel–Crafts reaction, **2**, 743
Pyrrole, 2-benzyl-
reduction
dissolving metals, **8**, 605
Pyrrole, 2,5-bis(piperidylmethyl)-
synthesis
via Mannich reaction, **2**, 965
Pyrrole, 2,5-bis(trimethylsiloxy)-
reaction with carbonyl compounds, **2**, 620
Pyrrole, 2-(4′-chlorobenzoyl)-1,3,5-trimethyl-
rearrangement, **2**, 745
Pyrrole, 3-(4′-chlorobenzoyl)-1,2,4-trimethyl-
rearrangement, **2**, 745
Pyrrole, 4-cyano-3,3-diaryl-5-methyl-2-oxo-2,3-dihydro-
synthesis, **3**, 826
Pyrrole, 2,5-dimethyl-
Birch reduction, **8**, 605
Pyrrole, 1-dimethylamino-
Friedel–Crafts acylation, **2**, 737
Pyrrole, 3,4-dimethyl-2,5-diphenyl-
synthesis, **6**, 789
Pyrrole, 3-hydroxy-
synthesis
via activated alkynes, **4**, 52
Pyrrole, 2-lithio-
alkylation, **3**, 261
Pyrrole, 2-lithio-*N*-(*N*′,*N*′-dimethylamino)
acylation, **1**, 410
Pyrrole, 1-methyl-
Friedel–Crafts reaction, **2**, 743
Mannich reaction
nonprotic solvent, **2**, 965
with formaldehyde and dimethylamine hydrochloride, **2**, 965
photocycloaddition reactions
with carbonyl compounds, **5**, 176
Pyrrole, 2-methyl-
Mannich reaction
with formaldehyde and secondary amines, **2**, 965
Pyrrole, 1-methyl-2,3,5-tris(3-pyridyl)-
synthesis
via γ-diketones, **1**, 547
Pyrrole, nitro-1-alkyl-
aromatic nucleophilic substitution, **4**, 432
Pyrrole, 1-phenyl-2-lithio-
crystal structure, **1**, 35
Pyrrole, 2-(2-pyrrolidyl)-
synthesis
via Mannich reaction, **2**, 971
Pyrrole, 1-trimethylsilyl-
Diels–Alder reactions, **5**, 382
Pyrrole 1-aspartates
Friedel–Crafts reaction, **2**, 757
Pyrrole-2-carbaldehyde
synthesis
via Reimer–Tiemann reaction, **2**, 770
via Vilsmeier–Haack reaction, **2**, 787
Pyrrole-2-carbaldehydes, 5-substituted
synthesis
via dithiation of azafulvene dimer, **1**, 473
metallation, **1**, 473
Pyrrole-2-carboxylic acid
reduction, **8**, 606
Reimer–Tiemann reaction, **2**, 771
Pyrrole-2-carboxylic acid, 4,5-dimethyl-
ethyl ester
Mannich reaction, **2**, 965
Pyrrole-3-carboxylic acid, 2,4-dimethyl-
ethyl ester
Mannich reaction, **2**, 968
Pyrrole-3-carboxylic acid, 2,5-dimethyl-
ethyl ester
Mannich reaction, **2**, 965
Pyrroles
acylation
bimolecular aromatic, **2**, 739
[2 + 2 + 2] cycloaddition reactions, **5**, 1143
[4 + 3] cycloaddition reactions, **5**, 608
Diels–Alder reactions, **5**, 382, 491
Friedel–Crafts acylation, **2**, 742
Mannich reaction
with formaldehyde and secondary amines, **2**, 962, 965
with imines, **2**, 970
with primary amine hydrochlorides, **2**, 968
metallation, **1**, 473
reaction with dihalocarbenes, **4**, 1004
reactions with ketocarbenoids, **4**, 1061–1063
reduction, **8**, 603–630
N-substituted
lithiation, **1**, 473
synthesis, **2**, 943
regiocontrolled, **1**, 552
via alkynes, palladium(II) catalysis, **4**, 567
via anilino ketones and activated alkynes, **4**, 52
via C—H insertion reactions, **3**, 1057
via cyclization of β-aminoalkynes, **4**, 411
via [2 + 2 + 2] cycloaddition, **5**, 1140
via [3 + 2] cycloaddition reactions, **5**, 297
via Diels–Alder reaction, **5**, 428
via dipolar cycloadditions with munchnones, **4**, 1097
via metal-catalyzed cycloadditions, **5**, 1195
via Michael addition, **4**, 16
via nitrogen-stabilized carbanions, **1**, 461
via palladium catalysis, **4**, 598
via retro Diels–Alder reactions, **5**, 581
Vilsmeier–Haack reaction, **2**, 780

from succinimides, **2**, 607
Pyrroles, 1-amino-
retrograde Diels–Alder reactions, **5**, 571
Pyrroles, 1,2-dienyl-
electrocyclic ring closure, **5**, 713
Pyrroles, dihydro-
annulated
synthesis *via* Knoevenagel reaction, **2**, 378
carbene complexes
reaction with alkynes, **5**, 1105
Schopf reaction, **2**, 943
synthesis
via aminomercuration of dienes, **4**, 291
Pyrroles, 2-keto-
introduction into natural products, **1**, 409
Pyrroles, 1-phenyl-
synthesis
via [2 + 2 + 2] cycloaddition, **5**, 1140
Pyrroles, 1,2,5-trisubstituted
synthesis, **1**, 559
Pyrrolidides
reduction
aluminates, **8**, 272
Pyrrolidine, acyl-
synthesis
via Mannich cyclization, **2**, 1041
Pyrrolidine, allyl-
anion
γ-alkylation, **2**, 61
Pyrrolidine, 1-amino-2-(methoxymethyl)-
N,N-dimethylhydrazine replacement
chiral auxiliary, **2**, 514
lithiated hydrazone
asymmetric alkylation, **3**, 37
lithiated hydrazone enolate
crystal structure, **1**, 29
synthesis, **6**, 119
Pyrrolidine, 5-butyl-2-heptyl-
synthesis
via Eschenmoser coupling reaction, **2**, 881
Pyrrolidine, 2,5-dimethyl-
cyclohexanone enamine from
alkylation, **3**, 35
synthesis, **6**, 717
Pyrrolidine, *cis*-dimethyl-
synthesis
via cyclization of 3-methyl-4-pentenylamine, **4**, 403
Pyrrolidine, 2,3-dioxo-
dimerization, **2**, 141
Pyrrolidine, 2-(diphenylhydroxymethyl)-
reduction, **8**, 171
Pyrrolidine, divinyl-
synthesis
via palladium-ene reactions, **5**, 53
Pyrrolidine, hydroxy-
HGA lactonization, **6**, 358
Pyrrolidine, 2-methoxy-
synthesis, **3**, 651
Pyrrolidine, 2-methoxymethyl-
chiral
copper ligand, **2**, 120
enamine
reaction with nitrostyrenes, **6**, 716
α-hydroxylation, **7**, 184
iminium salts from, **5**, 111
Pyrrolidine, methyl-
carbene complexes
reaction with 2-phenylpropanal, **5**, 1080
Pyrrolidine, 1-methyl-
deprotonation, **1**, 476; **3**, 65
N-oxide
azomethine ylides from, **4**, 1089
reduction
lithium aluminum hydride, **8**, 273
Pyrrolidine, 3-methylene-
synthesis
via allyl organometallic compounds, **2**, 981
via crotyl organometallic compounds, **2**, 982
Pyrrolidine, 1-phenylsulfenyl-
reaction with 1-octene, **7**, 493
Pyrrolidine, 1-propionyl-
enolates
stereoselectivity, **2**, 211
Pyrrolidine alkaloids
synthesis
via cyclofunctionalization, **4**, 401
via enol ethers, **2**, 613
via Eschenmoser coupling reaction, **2**, 881
Pyrrolidine amides, 2-(1-hydroxy-1-methylethyl)-
addition reactions
with organomagnesium compounds, **4**, 85
Pyrrolidine-2,4-dione, 3-diazo-
ring contraction
route to β-lactams, **3**, 902
Pyrrolidines
alkylation, **3**, 69
N-alkylation, **6**, 66
annulation, **1**, 889
chiral auxiliaries
nucleophilic addition reactions, **1**, 64, 65
Diels–Alder reactions, **5**, 366
enamines
alkylation, **6**, 714
lithiated formamidines
reaction with benzaldehyde, **1**, 482
synthesis, **1**, 669; **3**, 647; **6**, 740
chiral, **1**, 558
via N-acyliminium ions, **2**, 1066
via alkenes, **7**, 476
via cyclization of γ-allenylamines, **4**, 412
via cyclization of vinylic substrates, **4**, 398
via [3 + 2] cycloaddition reactions, **5**, 307
via ene reactions, **5**, 10
via α-methoxy carbamates, **1**, 377
via Michael addition, **4**, 24
via palladium-ene reactions, **5**, 51
via solvomercuration of amines, **4**, 290
Pyrrolidines, 2,5-disubstituted
synthesis
via cyclization of allylic substrates, **4**, 403
Pyrrolidinium ions, dimethyl-
electroreduction of nonalkenic carbonyl compounds
mediator, **8**, 133
electroreduction of nonconjugated alkenic ketones
mediator, **8**, 134
Pyrrolidinium tetrafluoroborate, dimethyl-
electropinacolization
aliphatic carbonyl compounds, **3**, 570
Pyrrolidinium tetrafluoroborate, 1-vinyl-2-ethoxy-
Diels–Alder reactions, **5**, 500
Pyrrolidinometacyclophanes

synthesis
via azomethine ylide cyclization, **4**, 1136
3-Pyrrolid-2-ones
Vilsmeier–Haack reaction, **2**, 787
3-Pyrrolidone
phenylacetyl amide
microbial hydroxylation, **7**, 60
Pyrrolidone, 5-acetoxy-
N-acyliminium ion intermediate, **2**, 1059
Pyrrolidone, 5-acetoxy-1-methyl-
N-acyliminium ion intermediate, **2**, 1059
3-Pyrrolidone, 1-aryl-
synthesis
via Diels–Alder reactions, **5**, 417
2-Pyrrolidone, 5-ethoxy-
reaction with 1,3-dienes, **2**, 1057
Pyrrolidone, 1-methyl-
oxy-Cope rearrangement
effect on, **5**, 787
reaction with arynes, **4**, 495
Vilsmeier–Haack reaction, **2**, 779
2-Pyrrolidone hydrotribromide
bromination
flavanones, **7**, 120
Pyrrolidones
enolates
synthesis, **2**, 105
synthesis
via carbonylation of allylamines, **3**, 1037
via cyclization of vinylic substrates, **4**, 398
1-Pyrroline, 2-acetyl-
synthesis, **8**, 604
1-Pyrroline, 2-alkyl-
deprotonation
thermodynamic considerations, **1**, 358
Pyrroline, 1-amino-2-methoxymethyl-
hydrazones
conversion to anions, **6**, 728
3-Pyrroline, 1-methyl-
synthesis
via retro Diels–Alder reactions, **5**, 581
Pyrroline alkaloids
synthesis
via cyclopropane ring opening, **5**, 921
1-Pyrroline 1-oxide
reaction with methyl crotonate, **5**, 256
Pyrrolines
annulation
via cyclopropane ring opening, **5**, 921
synthesis
via Beckmann reaction, **7**, 697
via [3 + 2] cycloaddition reactions, **5**, 297
via Michael addition, **4**, 16
via rearrangements, **5**, 952
via vinylaziridine ring opening, **5**, 937
trimer
Mannich reaction, **2**, 970
1-Pyrrolines
reactions with organometallic compounds, **1**, 364
synthesis
via intramolecular Ritter reaction, **6**, 273
via vinylaziridine ring opening, **5**, 937
2-Pyrrolines
asymmetric alkylation, **3**, 77
2-functionalization
metallation, **1**, 473
synthesis
via vinylaziridine ring opening, **5**, 937
3-Pyrrolines
synthesis
via cyclization of α-aminoallenes, **4**, 411
via vinylaziridine ring opening, **5**, 937
Pyrrolines, 2,5-dialkyl-
reduction, **8**, 47
synthesis
via *N*-substituted allylic anions, **2**, 62
Pyrrolines, dioxo-
Diels–Alder reactions, **5**, 323
3-Pyrrolines, 2-methylene-
synthesis
via metal-catalyzed cycloadditions, **5**, 1195
Pyrrolin-2-one-5-spiro-5′-thiolen-4′-one
synthesis
via metal-catalyzed cycloaddition, **5**, 1200
Pyrrolizidine alkaloids
synthesis
via cyclization of δ-allenylamine, **4**, 412
via [4 + 3] cycloaddition, **5**, 605
via cyclopropane ring opening, **5**, 921
via Eschenmoser coupling reaction, **2**, 881
Pyrrolizidines
functionalized
synthesis, **5**, 951
synthesis, **6**, 746
chiral, **1**, 558
via *N*-acyliminium ions, **2**, 1063
via aziridine thermolysis, **5**, 940
via epoxides, **3**, 736
via Mannich cyclization, **2**, 1041
via nitrone cyclization, **4**, 1120
Pyrrolizidine sulfide
homologation, **3**, 493
Pyrrolizidone
synthesis, **5**, 945
via Baeyer–Villiger reaction, **7**, 677
Pyrrolo[3,4-*b*]indole, 2,4-dihydro-
synthesis
via diazoalkene cyclization, **4**, 1157
via Knoevenagel reaction, **2**, 377
Pyrrolopyrazoline
synthesis
via diazoalkene cyclization, **4**, 1153
Pyrrolopyridines
lithiation, **1**, 471
Pyrrolo[2,3-*b*]pyridines
synthesis
via $S_{RN}1$ reaction, **4**, 478
1*H*-Pyrrolo[3,2-*b*]pyridines
synthesis
via $S_{RN}1$ reaction, **4**, 478
Pyrrolo[2,3-*b*]pyridines, 2-alkyl-
synthesis
via $S_{RN}1$ reaction, **4**, 478
Pyrrolo[2,3-*c*]pyridines, 2-alkyl-
synthesis
via $S_{RN}1$ reaction, **4**, 478
Pyrrolo[3,2-*c*]pyridines, 2-alkyl-
synthesis
via $S_{RN}1$ reaction, **4**, 478
Pyrrolo[1,2-*a*]quinoline
synthesis
via intramolecular hydride transfer, **8**, 98

Pyrrolothiophene
synthesis
via 5-*exo*-trig cyclization, **4**, 38
2-Pyrrolylacetate
synthesis
via ketocarbenoids and pyrroles, **4**, 1061
ϵ-Pyrromycinone
synthesis
via cyclofunctionalization of cycloalkene, **4**, 373
Pyruvates
ene reaction, **2**, 538
synthesis
via β-cleavage of epoxides, **3**, 759
Pyruvic acid
amides
catalytic hydrogenation, **8**, 145
hydrogenation
catalytic, **8**, 145
menthyl ester
asymmetric hydrogenation, **8**, 144
methyl ester
hydrogenation, modified metal catalyst, **8**, 151
thermal ene reaction, **2**, 539
trans-2-phenylcyclohexyl ester
ene reaction, **2**, 539
Pyruvic acid, phenyl-
ethyl ester, oxime acetate
hydrogenation, **8**, 149
synthesis
via cobalt-catalyzed double carbonylation, **3**, 1039
Pyruvyl chloride
synthesis
via dichloromethyl methyl ether, **6**, 305
Pyrylium salts
hydride acceptors, **8**, 91
synthesis
via Diels–Alder reactions, **5**, 502
via Friedel–Crafts acylation, **2**, 712
Pyrylium salts, 3-oxido-
cycloaddition
carbonyl ylide intermediate, **4**, 1092
Pyrylium ylides, oxido-
unsaturated side chain
dipolar cycloaddition, **4**, 1093

Q

Qinghaosu
  synthesis
    via Paterno–Büchi reaction, **5**, 155
Quadricyclane
  cycloaddition reactions, **5**, 1187
  reaction with dimethyldioxirane, **3**, 736
  synthesis
    via photocyclization, **5**, 206
Quadrone
  synthesis, **3**, 573, 709; **7**, 105, 817
    via Cope rearrangement, **5**, 804, 994
    via cyclopropanation/Cope rearrangement, **4**, 1049
    via cyclopropane ring opening, **4**, 1045
    via organostannane acylation, **1**, 447
    via palladium(II) catalysis, **4**, 573
    via Pauson–Khand reaction, **5**, 1060
    via photocycloaddition, **5**, 669
    via Wharton reaction, **8**, 927
Quadrone, decarboxy-
  synthesis
    via photocycloaddition, **5**, 667
    via Wacker oxidation, **7**, 455
Quadrone, dedimethyl-
  synthesis
    via photocycloaddition, **5**, 667
Quasi-Favorskii rearrangement
  2-arylalkanoic acids, **3**, 788
Quassin
  synthesis
    via Diels–Alder reaction, **5**, 351
Quassinoids
  oxidation, **7**, 239
  synthesis, **7**, 174; **8**, 929
    via cyclofunctionalization of cycloalkene, **4**, 373
    via Diels–Alder reactions, **5**, 344
    via Wharton reaction, **8**, 928
Quaternary centers
  contiguous
    synthesis *via* Ireland rearrangement, **5**, 841
Queen bee substance
  synthesis
    via Wacker oxidation, **7**, 454
Quercus lactone
  synthesis, **1**, 565
    via conjugate addition, **2**, 330
2,2′-Quinaldil
  rearrangement, **3**, 826
Quinaldine
  reaction with alkyl radicals, **4**, 770
Quinanes
  synthesis
    via retro Diels–Alder reactions, **5**, 560
Quinazoline, 4-chloro-
  $S_{RN}1$ reaction, **4**, 462
Quinazoline, 3,4-dihydro-
  synthesis
    via intramolecular Ritter reaction, **6**, 277
Quinazolines
  oxidation, **7**, 480
  reaction with allenylmagnesium bromide, **2**, 86
  reduction, **8**, 642
  synthesis, **6**, 273
Quinidine, dihydro-
  asymmetric dihydroxylation, **7**, 429
Quinine
  lithium aluminum hydride modifier, **8**, 164
Quinine, dihydro-
  asymmetric dihydroxylation, **7**, 429
Quinisatine
  rearrangement, **3**, 835
Quinocarcin
  synthesis, **1**, 404; **2**, 1069
p-Quinodimethane
  synthesis
    via ketocarbenoids, **4**, 1054
o-Quinodimethane, 7-butyl-
  Diels–Alder reactions, **5**, 391
o-Quinodimethane, diacetoxy-
  Diels–Alder reactions, **5**, 395
o-Quinodimethane, 7,8-dibromo-
  Diels–Alder reactions, **5**, 394
Quinodimethanes
  precursors
    synthesis, **3**, 255
  synthesis, **3**, 161, 173
o-Quinodimethanes
  Diels–Alder reactions, **5**, 385–396
    imines, **5**, 410
  synthesis, **5**, 386–390
    via benzocyclobutenes, **5**, 675, 691
    via benzocyclobutene thermolysis, **5**, 1031
    via electrocyclic ring opening, **5**, 1151
    via thermolysis, **5**, 741
5,8-Quinoflavone
  synthesis, **7**, 341
o-Quinol
  acetates
    extracyclic migrations, **3**, 813
    synthesis, **7**, 338
  diacetates
    rearrangements, **3**, 812
Quinoline, 1-alkoxycarbonyl-2-(2-alkynyl)-1,2-dihydro-
  synthesis
    via lithium allenes, **2**, 86
Quinoline, 2-bromo-
  $S_{RN}1$ reaction, **4**, 462
Quinoline, 3-bromo-
  coupling reactions
    with Grignard reagents, **3**, 461
  $S_{RN}1$ reaction, **4**, 462
Quinoline, 2-chloro-
  coupling reactions
    with Grignard reagents, **3**, 461
  oxidation
    peroxymaleic acid, **7**, 750
  $S_{RN}1$ reaction, **4**, 462
  reactions with benzyl sulfides, **4**, 475
Quinoline, 3-chloro-
  synthesis
    via dihalocarbene, **4**, 1004, 1021
Quinoline, 5-chloro-7-iodo-
  $S_{RN}1$ reactions, **4**, 460
Quinoline, cyano-
  reduction

borohydrides, **8**, 581
Quinoline, 2-cyano-
synthesis, **4**, 433
Quinoline, 3,4-dihydro-
reaction with trimethylsilyl triflate, **1**, 391
Quinoline, 3-dimethylamino-
reduction
borohydrides, **8**, 581
Quinoline, 2-halo-
$S_{RN}1$ reactions, **4**, 458
Quinoline, 8-hydroxy-
esters
reaction with Grignard reagents, **1**, 422
Quinoline, 2-iodo-
$S_{RN}1$ reaction, **4**, 462
Quinoline, mercapto-
synthesis, **4**, 441
Quinoline, 6-methoxytetrahydro-
synthesis
via arene–metal complexes, **4**, 523
Quinoline, 2-methyl-
reduction
ruthenium phosphine/formic acid complex, **8**, 591
Quinoline, 4-methyl-
reduction
homogeneous catalysis, **8**, 600
Quinoline, *N*-methyltetrahydro-
synthesis
via Diels–Alder reaction, **5**, 500
Quinoline, 2-methylthio-
synthesis
lithium allenes, **2**, 86
Quinoline, 4-naphthyl-
synthesis, **4**, 428
Quinoline, 8-nitro-
aromatic nucleophilic substitution, **4**, 432
Quinoline, 8-oxy-
dihydroboronite
selective aldehyde reduction, **8**, 17
Quinoline, 2-trimethylstannyl-
Friedel–Crafts reaction, **2**, 743
Quinoline-*N*-borane
reduction
aluminum hydrides, **8**, 584
4-Quinolinecarbohydroxamic acid
Lossen reaction, **6**, 824
Quinoline-3-carbonyl chloride, 2-phenyl-
Friedel–Crafts reaction, **2**, 757
Quinolinecarboxylic acid
reductive decarboxylation, **7**, 720
Quinolinediones, tetrahydro-
synthesis
via conjugate addition, **4**, 222
via hydrazone anions, **2**, 520
Quinoline-S
catalyst
Rosenmund reduction, **8**, 286
Quinolines
acylating agents, **1**, 422
coupling reactions
with sp^3 organometallics, **3**, 460
electroreduction, **8**, 594
hydrogenation
homogeneous catalysis, **8**, 456
nickel catalysts, **8**, 597, 598
N-oxides
deoxygenation, **8**, 391
Knoevenagel reaction, **2**, 364
reduction
aluminum hydrides, **8**, 584
borohydrides, **8**, 580
dihydropyridine, **8**, 589
dissolving metals, **8**, 596
formates, **8**, 591
regioselective, **8**, 600
sodium hydride, **8**, 588
regioselective cyanation, **4**, 433
Reissert compounds, **8**, 295
synthesis, **7**, 628
via lithium allenes, **2**, 86
via Reimer–Tiemann reaction, **2**, 773
via sequential Michael ring closure, **4**, 262
via tandem vicinal difunctionalization, **4**, 251
via Vilsmeier–Haack reaction, **2**, 787
Quinolines, chloro-
synthesis, **3**, 513
Quinolines, dihydro-
synthesis
via benzocyclobutenes, **5**, 691
via FVP, **5**, 718
Quinolines, 2,3-disubstituted
synthesis, **7**, 627
Quinolines, halo-
coupling reactions, **3**, 509
Quinolines, hydroxy-
Reimer–Tiemann reaction
normal, **2**, 770
Quinolines, octahydro-
cis-fused
synthesis *via* palladium-ene reactions, **5**, 51
Quinolines, tetrahydro-
lithiated formamidines
reaction with benzaldehyde, **1**, 482
microbial hydroxylation, **7**, 75
oxidation, **7**, 745
2-Quinolinethiol
synthesis
via $S_{RN}1$ reaction, **4**, 475
Quinolinium chlorochromate
oxidation
alcohols, **7**, 271
Quinolinium dichromate
oxidation
alcohols, **7**, 277
Quinolinium salts
reduction
aluminum hydrides, **8**, 587
borohydrides, **8**, 587
Quinolinium salts, 1-methyl-
reduction
formates, **8**, 591
Quinolizidine alkaloids
synthesis
chiral, **1**, 559
via Eschenmoser coupling reaction, **2**, 881
Quinolizidines
synthesis, **1**, 559; **6**, 746
via γ-diketones, **1**, 547
via Mannich reaction, **2**, 1009, 1010
via nitrone cyclization, **4**, 1120
2-Quinolone, 4-methoxycarbonyl-
synthesis

via intermolecular vinyl substitution, **4**, 846
Quinolones
synthesis
via activated alkynes, **4**, 52
2-Quinolones
synthesis
via Vilsmeier–Haack reaction, **2**, 787
2-Quinolones, 3,4-dihydro-1-hydroxy-
synthesis
via oxidation of tetrahydroquinolines, **7**, 745
Quinolones, 4-phenyl-3-vinyl-
photolysis, **5**, 728
8-Quinolyl phosphate
hydrolysis, **6**, 624
o-Quinomethide imines
Diels–Alder reactions, **5**, 473
Quinone, 1,2-dicyano-4,5-dichloro-
cycloaddition reactions, **5**, 273
Quinone diacetals
synthesis, **7**, 799
Quinone diazides
synthesis, **6**, 122
Quinone epoxides
synthesis
via retro Diels–Alder reactions, **5**, 563
o-Quinone methides
Diels–Alder reactions, **5**, 468
Quinones
addition reactions
carbon-centred radicals, **4**, 765
arenes from, **8**, 949
aromatization, **7**, 136
benzilic rearrangement, **3**, 828
hydride transfer
with 1,4-dihydropyridines, **8**, 93
hydrogenation, **8**, 152
intramolecular cycloaddition
nitrones, **4**, 1119
Perkin reaction, **2**, 399
reactions with π-allylnickel halides, **3**, 424
reduction
hydroquinones, **8**, 19
silanes, **8**, 318
synthesis, **7**, 143, 345–356, 800
via electrocyclization, **5**, 733
via metal-catalyzed cycloaddition, **5**, 1202–1204
via solid support oxidation, **7**, 841
use in dehydrogenation
imines, **7**, 138
vinyl substitutions
heterocyclic compounds, **4**, 837
o-Quinones
Diels–Alder reactions, **5**, 468
p-Quinones
Diels–Alder reactions, **5**, 330, 341
radical alkylation, **7**, 930
synthesis, **7**, 346
Quinones, azido-
synthesis
via haloquinones, **6**, 247
Quinones, cyano-
synthesis, **6**, 238
Quinones, hydroxy-
benzilic rearrangement, **3**, 828
Quinonoid α-diazo ketones
dipolar cycloaddition, **4**, 1104
Quinoxaline, 2-chloro-
$S_{RN}1$ reaction, **4**, 462
Quinoxalines
1,4-dioxides
deoxygenation, **8**, 391
reduction, **8**, 643
synthesis, stereocontrol, **3**, 960

R

Racemization
amino acids
oxazolinones, **6**, 635
Radical addition reactions
alkenes, **4**, 715–772
alkynes, **4**, 715–772
Radical anions
chemistry, **7**, 861
Radical cations
bimolecular reactions, **7**, 858
chemistry, **7**, 857
electron-transfer oxidation, **7**, 850
unimolecular reactions, **7**, 857
Radical cyclizations
acyl radicals, **4**, 798
chain methods, **4**, 790–799
fragmentation method, **4**, 799
nonchain methods, **4**, 805–809
manganese(III) acetate, **4**, 806
stereoselectivity, **4**, 787–789
via alkene addition, **4**, 779–827
Radical ions
electron-transfer oxidation
reactive intermediates, **7**, 854
Radical–radical coupling
nonchain methods, **4**, 758–762
cyclizations, **4**, 805
Radical reactions
chain, **4**, 724
initiation, **4**, 724
elementary, **4**, 726–731
fragmentation method, **4**, 742–747
heterocyclic synthesis
carbonyl derivatives, **6**, 755
methods, **4**, 724
nonchain methods, **4**, 725, 758–765
propagation, **4**, 725
protecting groups, **4**, 721
rate constants, **4**, 722
reaction concentrations, **4**, 722
sequential, **4**, 818–827
intramolecular transformations, **4**, 820
tandem, **4**, 819–827
Radical relay chlorination, **7**, 46
catalytic turnover, **7**, 50
selectivity, **7**, 47
template-directed, **7**, 47
Radicals
ambiphilic
reactions, **4**, 730
bond dissociation energies, **4**, 717
carbon-centered
addition reactions, **4**, 735–765
addition to multiple bonds, **4**, 765–770
cyclizations to aromatic rings, **4**, 809–811
cyclizations to carbon–carbon multiple bonds, **4**, 789–809
carbonyl-substituted
addition reactions, **4**, 740
cyclizations, **4**, 785
electrophilic
reactions, **4**, 729
elimination, **4**, 721
ether-substituted
cyclizations, **4**, 795
fragmentation, **4**, 721
heteroatom-centered
addition reactions, **4**, 770–772
reactions, **4**, 731
initiators, **4**, 721
nitrogen-centered
cyclizations, **4**, 811–814
nucleophilic
addition to alkenes, **4**, 755
reactions, **4**, 728
oxygen-centered
cyclizations, **4**, 811–814
persistent, **4**, 717
reactions
in synthesis, **4**, 720–722
with solvents, **4**, 719
selective coupling, **4**, 718
stereochemistry, **4**, 719
structure, **4**, 719
transiency, **4**, 717–719
Radio frequency plasma reactions
di-π-methane rearrangements, **5**, 195
Radiolysis
Ritter reaction, **6**, 280
Radiopharmaceuticals
synthesis, **4**, 445
Ramberg–Bäcklund rearrangement, **3**, 861–883
conjugated dienoic acids
synthesis, **6**, 841
functional group compatibility, **3**, 865
mechanism, **3**, 866
Michael-induced, **3**, 880
phase-transfer conditions, **3**, 863
reaction conditions, **3**, 862
scope, **3**, 862
stereoselectivity, **3**, 862
synthesis of alkenes, **3**, 163; **6**, 161
uses, **3**, 862
variations, **3**, 868
Raney nickel
deselenations, **8**, 847
desulfurizations, **8**, 836
α-alkylthio carbonyl compounds, **8**, 995
mechanism, **8**, 837
hydrogenation, **8**, 418
alcohols, **8**, 815
pyridines, **8**, 597
hydrogenolysis
alkyl halides, **8**, 794
benzylic alcohols, **8**, 963
carbonyl compounds, **8**, 320
reduction, **8**, 366
epoxides, **8**, 881
Rearrangements
alcohol synthesis, **6**, 14
charge-accelerated
small rings, **5**, 999–1033
donor radical cations, **7**, 875
radical cations

unimolecular reaction, **7**, 858
vinylcyclopropanes, **5**, 899–965
1,3-Rearrangements
homologations, **1**, 885
3,3-Rearrangements
cationic variations, **1**, 889
Recifeiolide
synthesis, **3**, 286
via ene reaction, **2**, 538
via Wacker oxidation, **7**, 455
Recombination
hydrogen atom abstraction, **3**, 1046
Red Al — *see* Sodium bis(2-methoxyethoxy)aluminum hydride
Red bollworm moth
sex pheromone
synthesis, **3**, 169
Redox reactions
radical addition, **4**, 726, 762–765
Reduction
acetals, azaacetals and thioacetals
to ethers, **8**, 211–232
alcohols
to alkanes, **8**, 811–832
alkenes
enzymes and microorganisms, **8**, 205
alkenes and alkynes
noncatalytic chemical methods, **8**, 471–487
alkyl halides, **8**, 793–807
aromatic rings
dissolving metals, **8**, 489–519
benzo[*b*]furans, **8**, 624
benzo[*b*]thiophenes, **8**, 629
carbonyl compound arylsulfonylhydrazones
hydrides, **8**, 343
carbonyl compounds
enantiomeric distinctions, **8**, 185
metal hydrides, **8**, 1–22
carboxylic acid derivatives, **8**, 235–254
carboxylic acids
metal hydrides, **8**, 237, 259–279
to aldehydes, **8**, 283–304
C—halogen bonds, **8**, 985
C=N
dissolving metals, **8**, 123
C=N to CHNH
metal hydrides, **8**, 25–74
C=S
dissolving metals, **8**, 126
C=X to CH_2
dissolving metals, **8**, 307–323
Wolff–Kishner reduction, **8**, 327–359
C=X to CHXH
catalytic hydrogenation, **8**, 139–155
chirally modified hydride reagents, **8**, 159–180
dissolving metals, **8**, 107–126
electrolytically, **8**, 129–137
enzymes and microorganisms, **8**, 183–207
hydride delivery from carbon, **8**, 79–103
enones, **8**, 523–568
epoxides, **8**, 871–891
furans, **8**, 606
heterocycles, **8**, 603–630
Hg—C bonds, **8**, 850
imines
enzymes and microorganisms, **8**, 204
metal hydrides, **8**, 25–74
indoles, **8**, 612
isocyanides, **8**, 830
ketones, **8**, 923–951
dissolving metals, stereoselectivity, **8**, 116
metal hydrides
unsaturated carbonyl compounds, **8**, 536
nitro compounds, **8**, 363–379
nitroso compounds, **8**, 363–379
N—N bonds, **8**, 381–399
N=N bonds, **8**, 381–399
N—O bonds, **8**, 381
one-electron
pyridines, **8**, 591
O—O bonds, **8**, 381
partial and complete
heterocycles, **8**, 635–666
P—C bonds, **8**, 858
pyridines, **8**, 579–600
pyrroles, **8**, 604
S—C bonds, **8**, 835–870
Se—C bonds, **8**, 847
selective
acetals, **8**, 216
selenides
use in synthesis, **3**, 106
styrenes, **8**, 523–568
α-substituted carbonyl compounds, **8**, 983–996
sulfides
use in synthesis, **3**, 106
thiophenes, **8**, 608
transition metal hydrides, **8**, 548
vinyl halides, **8**, 895–920
Reduction potentials
electron acceptors, **7**, 855
electron-transfer oxidation
driving force, **7**, 852
electrosynthesis, **8**, 129
metal oxidants, **7**, 854
oxidants
electron acceptors, **7**, 854
Reductive alkylations
benzoic acids
Birch reduction, **8**, 499
metal–ammonia reduction, **8**, 527
Reductive cleavage
α-halo ketones
halide salts, **8**, 988
metals, **8**, 986
nitrogen compounds, **8**, 383
sulfur compounds
α-halo ketones, **8**, 989
Reductive deamination
amines, **8**, 826
to alkanes, **8**, 811
Reductive decarboxylation, **7**, 720
Reductive decyanation
nitriles
electrolysis, **8**, 252
Reductive dehalogenation
alkyl halides, **8**, 794
dihalocyclopropanes, **4**, 1006
Reductive desulfurization
thiocarbonyl group
Raney nickel, **6**, 447
Reductive dimerization

alkynes
hydroalumination, **8**, 744
electrochemical
unsaturated carbonyl compounds, **8**, 532
Reductive elimination
acylation
organostannanes, **1**, 444
hydrazones, **8**, 939
ketones, **8**, 925
Reductive ozonolysis
alkenes, **8**, 398
Reductive silylations
aromatic rings, **8**, 517
Reductones
intermolecular redox reactions
via enediols, **8**, 88
Reed reaction, **7**, 14
Reference electrodes
electrosynthesis, **8**, 130
Reformatsky reaction
cerium enolates
generation and reaction, **2**, 312
chemoselectivity, **2**, 283
kinetic stereoselection, **2**, 291
magnesium enolates
preparation, **2**, 186
regioselectivity, **2**, 285, 288
stereoselectivity, **2**, 289
thermodynamic stereoselection, **2**, 289
zinc enolates, **2**, 122, 277–298
Reformatsky reagents
t-butyl bromoacetate
crystal structure, **1**, 30
coupling reactions
with alkenyl halides, **3**, 443
with aromatic halides, **3**, 454
in enolate–imine condensations
stereoselectivity, **2**, 918
NMR spectral data
enolates, **2**, 281
zinc enolates
isolation and stability, **2**, 278
Reformatsky-type reaction
organosamarium compounds, **1**, 266
Reforming
alkanes, **7**, 7
Reframoline
synthesis, **3**, 81
Regioselectivity
aldol cyclization, **2**, 156
homoenolate anion equivalents
allylic anions, **2**, 55
Reike powders
reactive zinc
Reformatsky reaction, **2**, 282
Reimer–Tiemann reaction, **2**, 769–775
abnormal, **2**, 773
high pressure, **2**, 772
industrial applications, **2**, 772
limitations, **2**, 770
mechanism, **2**, 774
normal, **2**, 769
regioselectivity, **2**, 771
scope, **2**, 770
Reissert compounds
reduction
amides, **8**, 295
synthesis
via heterocyclic amines, **8**, 295
Remote functionalization
chlorination, **7**, 43
oxidation
C—H bonds, **7**, 39–51
Remote oxidations
alkanes, **7**, 42
photochemical, **7**, 42
prospects, **7**, 50
Reorganization energy
electron-transfer oxidation, **7**, 852
Reserpine
precursor
synthesis, **7**, 677
synthesis, **7**, 647
diastereoselection, **2**, 1022
via Cope rearrangement, **5**, 814
via cyclofunctionalization of cycloalkenes, **4**, 373
via Diels–Alder reaction, **5**, 341
Resins
chromium(VI) oxidants support
alcohol oxidation, **7**, 280
Resorcinol, 4,6-dinitro-
synthesis, **6**, 110
Resorcinols
synthesis
via aldol cyclization, **2**, 170
Resorcylic acid
synthesis, **2**, 171
Resorcylide
synthesis
via Wacker oxidation, **7**, 455
Resorcylide, dihydroxy-
synthesis
via cyclization, **1**, 553
Rethrolones
synthesis, **3**, 126
Rethronoids
synthesis, **1**, 566
Reticuline
intracoupling to morphines, **3**, 679
synthesis, **3**, 79
Retigeranic acid
synthesis
via photocycloaddition, **5**, 664
Retinal
electroreduction
pinacolization, **8**, 134
pinacol coupling reactions, **3**, 577
cis-Retinal
synthesis
(*Z*)-selectivity, **1**, 765
13-*cis*-Retinal, 13-*t*-butyl-
Schiff base
heterocyclization, **5**, 742
Retinoic acid
synthesis, **7**, 109
Retinol
oxidation, **7**, 311
Retroaddition reactions
cyclobutanes, **5**, 64
Retro-aldol reaction
anti aldols
thermodynamic stereoselectivity, **2**, 195

equilibration
effect of counterion, **2**, 235
thermodynamic control, **2**, 235
solvent effect, **2**, 196
stereochemical homogeneity
loss of, **2**, 192
Retro-Dieckmann reaction, **2**, 806
Retro-Diels–Alder reaction
alkene protection, **6**, 689
enamine synthesis, **6**, 706
Retroelectrocyclization
triene synthesis, **5**, 737–740
Retro-ene reactions, **6**, 832
1,3-heteroatom–hydrogen transposition reaction, **6**, 865
Retro-Knoevenagel reaction, **2**, 349
Retronecic acid
synthesis, **3**, 420
Retronecine
synthesis
via azomethine ylide cycloaddition, **4**, 1087
via cycloazasulfenylation of alkenes, **4**, 333
via Diels–Alder reaction, **5**, 421
Retro-Ritter reaction, **6**, 263
Retrosynthetic analysis
radical reactions, **4**, 731
Reversed micelles
nitrile synthesis, **6**, 229
α-Rhamnopyranosides
synthesis, **6**, 41
β-Rhamnopyranosides
synthesis, **6**, 45
α-L-Rhamnoside, methyl 2,3-benzylidene-
reduction, **8**, 226
Rhenium
catalysts
alkene metathesis, **5**, 1118
hydrogenolysis, benzylic alcohols, **8**, 963
metal vapor synthesis
reactions with alkanes, **7**, 4
Rhenium, cyclopentadienylnitroso(triphenylphosphine)-
crystal structure, **1**, 309
Rhenium, cyclopentadienylnitroso(triphenylphosphine)-(acetophenone)-
crystal structure, **1**, 309
Rhenium, cyclopentadienylnitroso(triphenylphosphine)-(phenylacetaldehyde)-
crystal structure, **1**, 309
Rhenium acyl complexes
deprotonation
reaction, **2**, 127
Rhenium enolates
aldol reactions, **2**, 312
photochemical aldol reaction, **2**, 312
Rhenium oxide
catalyst
carboxylic acid hydrogenation, **8**, 236
Rhizopus nigricans
reduction
unsaturated carbonyl compounds, **8**, 558
Rhodacyclopentadienes
synthesis
via [2 + 2 + 2] cycloaddition, **5**, 1135
Rhodanine
Perkin reaction, **2**, 406
Rhodinose
synthesis
via Lewis acid mediated Grignard addition, **1**, 336
Rhodium
acyclation catalyst, **1**, 450
allylic oxidation catalyst, **7**, 107
Cope rearrangement catalyst, **5**, 802
hydrogenation catalyst
pyridines, **8**, 597
hydrogenolysis catalyst
benzylic alcohols, **8**, 963
carbonyl compounds, **8**, 320
hydrometallation
mechanism, **8**, 672
pentamethylcyclopentadienyl derivatives
hydrogenation of alkenes, **8**, 445
reduction
transfer hydrogenation, **8**, 366
Rhodium, acetatotris(triphenylphosphine)-
hydrogenation
alkenes, **8**, 445
Rhodium, bis(acetonitrile)(1,5-cyclooctadiene)-
hydrogenation
alkenes, **8**, 445
Rhodium, bis(acetylacetone)-
catalyst
hydrosilation, **8**, 556
Rhodium, bromotris(triphenylphosphine)-
hydrogenation
alkenes, **8**, 445
Rhodium, carbonylhydridotris(triphenylphosphine)-
hydrogenation
alkenes, **8**, 445
Rhodium, chloro(carbonyl)bis(triphenylphosphine)-
catalyst
acylation, **1**, 451
Rhodium, chlorodicarbonylbis[bis(diphenylphosphino)-methane]di-
hydrogenation
alkenes, **8**, 445
Rhodium, chlorotris(triphenylphosphine)-
catalyst
decarbonylation, **3**, 1040
silane reaction with carbonyl compounds, **2**, 603
hydrogenation catalyst, **8**, 152, 535
alkenes, **8**, 443, 445
homogeneous catalysis, **8**, 443
hydrosilylation
α,β-unsaturated carbonyl compounds, **8**, 20, 555
reduction
unsaturated esters, **8**, 555
Rhodium, chlorotris(triphenylphosphine *m*-trisulfonate)-
hydrogenation
alkenes, **8**, 445
Rhodium, dicarbonylchlorobis-
lithium chloride salt
catalyst, alkenyl halide dimerization, **3**, 484
Rhodium, dodecacarbonyltetrakis-
catalyst
hydroformylation, **4**, 915
Rhodium, hexadecacarbonylhexa-
catalyst
hydrogenation, **8**, 600
Rhodium, iodotris(triphenylphosphine)-
hydrogenation
alkenes, **8**, 445
Rhodium, nitrosotris(triphenylphosphine)-

hydrogenation
alkenes, **8**, 445
Rhodium acetate
allylic oxidation, **7**, 95
catalyst
C—H insertion reactions, **3**, 1051
deoxygenation
epoxides, **8**, 890
Rhodium carboxylates
dimeric
diazo compound decomposition catalysts, **4**, 1033
Rhodium chloride
allylic oxidation, **7**, 95
Rhodium cluster, tetrakis(μ-acetato)di-
catalyst
hydrosilation, **8**, 556
Rhodium clusters
hydrogenation
alkenes, **8**, 445
Rhodium complexes
carbonyl
reduction, aromatic nitro compounds, **8**, 372
catalysts
hydroboration, **8**, 709
hydrosilylation, **8**, 764
enantioselective aldol reaction
catalysis, **2**, 311
homogeneous hydrogenation, **8**, 152
hydride transfer
catalyst, **8**, 91
polymer bound
catalyst, hydrogenation, **8**, 419
Rhodium complexes, alkyl-
acylation
acid chlorides, **1**, 450
Rhodium enolates
aldol reaction, **2**, 310
Rhodium hydride, tetrakis(triphenylphosphine)-
oxidation
diols, **7**, 314
Rhodium trichloride
Aliquat-336 ion-pair
hydrogenation, **8**, 535
Riboflavin
synthesis, **8**, 292
D-Ribofuranose, 2,3-*O*-isopropylidene-
Knoevenagel reaction, **2**, 386
Ribofuranoside
synthesis
via Baeyer–Villiger reaction, **7**, 684
β-D-Ribofuranosyl-1-carbonitrile, 2,3,5-tri-*O*-benzoyl-
reaction with hydrogen selenide, **6**, 477
Ribofuranosyl cyanide
synthesis
via Lewis acid promoted reaction, **1**, 347
Ribonuclease A
synthesis, **6**, 384
Ribonucleosides
phosphorylation, **6**, 601
Ribose
synthesis, **8**, 292
D-Ribose
selective monoacetylation
enzymatic, **6**, 340
D-Ribose, 2-deoxy-
selective monoacetylation
enzymatic, **6**, 340
synthesis, **5**, 263
Ribulose
synthesis
via Lewis acids, nonchelation selectivity, **1**, 339
Ricinelaidic acid
synthesis
via ene reaction, **2**, 538
Ricinoleic acid
synthesis
via ene reaction, **2**, 538
Rieke copper
acylation, **1**, 426
Rifamycin
ansa bridge
synthesis, **1**, 182
Rifamycin S
syn selective aldol reaction
zirconium enolates, **2**, 303
synthesis, **2**, 264
via Baeyer–Villiger reaction, **7**, 683
via Diels–Alder reaction, **2**, 703
via Wittig reaction, **1**, 762
Rimuene
synthesis, **3**, 21
Ring contractions
Wolff rearrangement, **3**, 900
ylides
3,2-sigmatropic rearrangements, **3**, 954
Ring expansion, **1**, 843–899
via Claisen rearrangement, **5**, 831
via iterative sigmatropic processes, **5**, 894–896
via silyloxycyclopropanes, **2**, 445
via Wagner–Meerwein reactions, **3**, 713
via ylides, **3**, 957
Ring-growing reactions
3,2-rearrangement, **3**, 957
Ring opening metathesis polymerization
alkene metathesis, **5**, 1120
Ring opening reactions
cyclobutenes, **5**, 675–694
two-carbon ring expansion, **5**, 686–688
cyclobutenones, **5**, 688–691
epoxides
with nitrogen nucleophiles, **6**, 88
Ristosamine
amino sugars, **2**, 323
synthesis
via aldol reaction, **2**, 195
Ritter reaction
acetonitrile
reaction with methyl phenyl sulfoxide, **7**, 201
acids
concentration, **6**, 264
alkenes, **4**, 292–294
amination
alkenes, **4**, 290
carbenium ion source, **6**, 267
extensions, **6**, 280
initial description, **6**, 261
intramolecular, **6**, 272
Lewis acids
catalyst, **6**, 264
mechanism, **6**, 261
metallic reagents, **6**, 283
modified, **7**, 488, 490

nitriles, **6**, 265
physical techniques, **6**, 280
reaction conditions, **6**, 263
solvents
polarity, **6**, 264
vinylogous, **7**, 505
Ritter-type reactions, **6**, 261–296
amide synthesis, **6**, 401
isocyanides, **6**, 293
Robinson annulation
aldol reaction, **2**, 156
cyclohexenone synthesis, **4**, 2, 6
1,5-diketone cyclization, **2**, 162
Rocaglamide
synthesis, **1**, 564
via [3 + 2] cycloaddition reactions, **5**, 311
Roflamycin
synthesis, **1**, 568
Rosaramicin
synthesis
via Wacker oxidation, **7**, 454
Rosenmund reduction
acyl chlorides, **8**, 259, 286
mechanism, **8**, 287
Rosettane
synthesis
via photocycloaddition, **5**, 662, 663
Rotaxanes
diethylmagnesium/18-crown-6 complex
crystal structure, **1**, 15
synthesis
via intramolecular acyloin coupling reaction, **3**, 628
Rothins
synthesis, **6**, 780
Royleanone
synthesis
via metal-catalyzed cycloaddition, **5**, 1203
Rubidium
reduction
ammonia, **8**, 113
carbonyl compounds, **8**, 109
Rubidium fluoride
catalyst
Knoevenagel reaction, **2**, 343
Rubradirins
synthesis, **7**, 346
via Horner–Wadsworth–Emmons reaction, **1**, 772
Rudmollin
synthesis
via photocycloaddition, **5**, 669
Rule of diaxial opening
epoxides, **3**, 734
Rule of five
intramolecular photocycloaddition reactions, **5**, 133
Rupe rearrangement
alkynic alcohols, **5**, 768
Ruthenium
alkene metethesis catalyst, **5**, 1118
hydrogenation catalyst, **8**, 418
pyridines, **8**, 597
Ruthenium, carbonyldihydridotris(triphenylphosphine)-
hydrogenation
benzylideneacetone, **8**, 551
Ruthenium, chlorohydridotris(triphenylphosphine)-
hydrogenation
alkenes, **8**, 445
transfer hydrogenation
α,β-unsaturated ketones, **8**, 551
Ruthenium, cyclopentadienyltris(dimethyl sulfoxide)
nucleophilic substitution, **4**, 531
Ruthenium, decacarbonyl(isocyanide)tri-
hydrogenation
alkenes, **8**, 446
Ruthenium, dichlorobis(triphenylphosphine)-
formic acid complex
2-methylquinoline reduction, **8**, 591
oxidation
allylic alcohols, **7**, 308
Ruthenium, dichlorotris(triphenylphosphine)-
hydrogenation catalyst, **8**, 369, 535
alkenes, **8**, 445
anhydrides, **8**, 239
oxidation
primary alcohols, **7**, 309, 310
transfer hydrogenation, **8**, 557
α,β-unsaturated ketones, **8**, 551
Ruthenium, dihydridotetrakis(triphenylphosphine)-
hydrogenation
alkenes, **8**, 445
oxidation
diols, **7**, 314
Ruthenium, octacarbonyltetrahydridobis(2,3-*O*-isopropylidene-2,3-dihydroxy-1,4-bis(diphenylphosphino)butane)tetrakis-
transfer hydrogenation, **8**, 552
Ruthenium, tetrachlorotris(2,3-*O*-isopropylidene-2,3-dihydroxy-1,4-bis(diphenylphosphino)butane)bis-
transfer hydrogenation, **8**, 552
Ruthenium, tris(acetonitrile)chloro-[bis(diphenylphosphino)methane]
hydrogenation
alkenes, **8**, 446
Ruthenium, tris(triphenylphosphine)-
reductions
aliphatic nitro compounds, **8**, 374
Ruthenium complexes
catalysts
hydrosilylation, **8**, 764
hydrogenation
alkenes, **8**, 154
oxidation
primary alcohols, **7**, 309
secondary alcohols, **7**, 324
transfer hydrogenation
silanes, **8**, 554
Ruthenium dichlorate, dioxygen(6,6′-dichlorobipyridyl)-
oxidation
ethers, **7**, 236
Ruthenium dioxide
catalyst
carboxylic acid hydrogenation, **8**, 236
hydrated
oxidation, allylic alcohols, **7**, 308
oxidation
ethers, **7**, 235, 238
oxidative cleavage of alkenes
catalysts, **7**, 542
periodate cleavage of alkenes
catalyst, **7**, 587
Ruthenium tetroxide
asymmetric dihydroxylation, **7**, 431
oxidation

benzyl ethers, **7**, 240
benzyl methyl ether, **7**, 240
ethers, **7**, 236, 237
organoboranes, **7**, 602
oxidative cleavage of alkenes, **7**, 542
synthesis of carbonyl compounds, **7**, 564
synthesis of carboxylic acids, **7**, 587
Ruthenium trichloride
catalyst
ether oxidation, **7**, 238
periodate cleavage of alkenes
catalyst, **7**, 587

S

Saccharides
coupling
via heteroatom cyclization, **4**, 391
synthesis
via trichloroacetimidates, **6**, 51
Saccharides, acyl
reduction
metal hydrides, **8**, 271
Saccharin, chloro-
reduction
metal hydrides, **8**, 271
Saccharomyces cerevisiae
reduction
unsaturated carbonyl compounds, **8**, 559
Saframycin B
synthesis, **7**, 350
Sakurai cyclization
γ-lactone formation, **6**, 357
Salaün reagent
solid support, **7**, 843
Salcomine
cobalt(II) complex
oxidation, quinones, **7**, 354
oxygen
quinone synthesis, **7**, 355
Salicylaldehyde
dichlorodimethyltin complex
crystal structure, **1**, 305
Knoevenagel reaction, **2**, 357
reaction with malonic esters, **2**, 354
synthesis, **8**, 285
via Reimer–Tiemann reaction, **2**, 772
Vilsmeier–Haack reaction, **2**, 790
Salicylaldehyde, fluoro-
synthesis, **8**, 285
Salicylamides
synthesis, **4**, 434
Salinomycin
synthesis, **7**, 245
Salsolidine
synthesis, **3**, 78
Samarium
acyl anions and radicals
generation, **1**, 273
oxidation state
stability, **1**, 252
reaction with methyl β-bromopropionate, **1**, 254
Samarium, allyl-
reactions with carbonyl compounds, **1**, 256
Samarium, benzyl-
reactions with carbonyl compounds, **1**, 253
Samarium, dicyclopentadienyl-
intermolecular Barbier-type reactions, **1**, 258
intermolecular pinacolic coupling reactions
organosamarium compounds, **1**, 271
Meerwein–Ponndorf oxidation
aldehydes, **1**, 258
reactions promoted by, **1**, 255
reactions with benzylic halides, **1**, 253
synthesis
via samarium diiodide, **1**, 255
Samarium, ethyliodo-
reaction with benzaldehyde, **1**, 254
Samarium chlorides
reduction
enones, **8**, 540
toxicity, **1**, 252
Samarium diiodide
Barbier-type reaction
mechanism, **1**, 258
characterization, **1**, 255
deoxygenation
epoxides, **8**, 889
intermolecular pinacol coupling reactions
aliphatic carbonyl compounds, **3**, 570
intramolecular pinacol coupling reactions, **3**, 574
iodohydrin synthesis, **1**, 831
pinacol coupling reactions, **3**, 567
radical cyclizations
nonchain methods, **4**, 809
reactions promoted by, **1**, 255
reactions with acyl halides
preparation of samarium acyl anions, **1**, 273
reduction
alkyl halides, **8**, 797
epoxides, **8**, 883
reductive cleavage
α-alkylthio ketones, **8**, 994
synthesis, **8**, 115
via oxidation of samarium, **1**, 255
Samarium reagents, **1**, 251–280
acyl anion chemistry, **1**, 273
acyl radical chemistry, **1**, 273
ketyl–alkene coupling reactions, **1**, 268
pinacolic coupling reactions, **1**, 270
Reformatsky-type reactions, **1**, 266
Samarium salts
reduction
carbonyl compounds, **8**, 115
reductive cleavage
ketols, **8**, 992
Sanadaol
synthesis
via ene reaction, **2**, 553
Sandmeyer reaction
bromination, **6**, 211
chlorination, **6**, 208
Sanitoxins
synthesis
via Blaise reaction, **2**, 297
α-Santalene
synthesis, **3**, 161, 427, 712
β-Santalene
synthesis, **3**, 712
via Diels–Alder reactions, **5**, 358
α-Santalol
synthesis
via Wittig reaction, **1**, 757
Santene
synthesis, **3**, 706
Santonin
rearrangement, **3**, 804
synthesis, **8**, 530
Sapogenins

steroidal
reduction, **8**, 220
Sarcophine
transannular cyclization, **3**, 407
Sarcosine
reaction with 2,4-dimethylphenol
Mannich reaction, **2**, 956
Sarett oxidation
alcohols
chromium(VI) oxide/pyridine complex, **7**, 256
Sarkomycin
synthesis, **3**, 937
via σ-alkyliron complexes, **4**, 579
via [3 + 2] cycloaddition reactions, **5**, 308
via cyclopropanes, **5**, 907
via nitrile oxide cyclization, **4**, 1126
via Pauson–Khand reaction, **5**, 1051
via retro Diels–Alder reaction, **5**, 560, 568
via tandem vicinal difunctionalization, **4**, 259
via vinylcyclopropane thermolysis, **4**, 1048
Sarracenin
synthesis
via photocycloaddition, **5**, 129, 166
Sativene
synthesis, **3**, 20, 712
Saxitoxin
synthesis
via azomethine imine cyclization, **4**, 1147
via Eschenmoser coupling reaction, **2**, 879
Scandine
biomimetic synthesis, **6**, 755
Scandium, bis(cyclopentadienyl)hydrido-
hydrometallation
alkenes, **8**, 696
Schiff bases
catalytic hydrogenation, **8**, 143
electroreduction, **8**, 136
electroreductive cyclization, **8**, 136
homogeneous hydrogenation, **8**, 155
reactions with organocopper complexes, **1**, 119
Schizandrin, deoxy-
synthesis
via vanadium oxytrifluoride, **3**, 676
Schizandrin C
synthesis
via thallium trifluoroacetate, **3**, 669
Schmidt reaction, **3**, 908; **6**, 817
amide synthesis, **6**, 404
carboxylic acids, **6**, 817
hydrazoic acids, **6**, 798
ketones, **6**, 820
stereoselectivity, **6**, 798
Schopf reaction, **2**, 943
Schweizer's reagent
phosphonium ylide synthesis, **6**, 176
reaction with divinylcuprates, **3**, 259
Sclerin
synthesis
via cycloaromatization reaction, **2**, 621
Scopine
synthesis
via [4 + 3] cycloaddition, **5**, 609
Sebacic acid
synthesis, **3**, 640
7,12-Sechoishwaran-12-ol
synthesis
via nitrone cyclization, **4**, 1120
6,7-Secoagroclavine
synthesis
via Mannich reaction, **2**, 967
Secoalkylation
chain extension
via Grob fragmentation, **6**, 1048
Secoiridoids
synthesis
via Knoevenagel reaction, **2**, 358
Secoisoquinoline alkaloids
synthesis, **1**, 552
Secologanin
aglycone
synthesis *via* Claisen condensation, **2**, 822
synthesis, **3**, 599
via Knoevenagel reaction, **2**, 371
16,17-Secopregnanes
reduction
dissolving metals, **8**, 114
Secosulfenylation
cyclobutanones, **6**, 143
7,16-Secotrinervita-7,11-diene, 3α-acetoxy-15β-hydroxy-
synthesis
via organoaluminum reagents, **1**, 100
1,2,3-Selenadiazoles
aryne reactions, **4**, 509
reactions with amines, **6**, 480
Wolff rearrangement, **3**, 909
1,2,3-Selenadiazoles, 4-aryl-
arylethynylselenolate synthesis, **6**, 473
Selenaldehydes
Diels–Alder reactions, **5**, 442
synthesis, **5**, 443; **6**, 475
Selenamides
aliphatic
synthesis, **6**, 477
reactions, **6**, 481
reactivity, **6**, 461
synthesis, **6**, 476
via sulfenylation of primary amines, **7**, 741
Selenates
rearrangement, **6**, 904
Selenation
decarboxylative chalcogenation, **7**, 726
electrochemical, **7**, 819
Selenazofurin
synthesis, **6**, 474
Selenazole-4-carboxamide, 2β-D-ribofuranosyl-
synthesis, **6**, 478
Selenazole-4-carboxylate, 2-(2,3,5-tri-*O*-benzoyl-D-ribofuranosyl)-
ethyl ester
synthesis, **6**, 477
1,3-Selenazoles
synthesis, **6**, 474, 481
Selenazolium hydroxide, anhydro-2,3,5-triphenyl-4-hydroxy-
synthesis, **6**, 481
Selenenamides, *N*-acetyl-
selenol ester synthesis, **6**, 466
Selenenic acid, aryl-
allylic oxidation
alkenes, **7**, 91
Selenenic acids

synthesis, **7**, 770
Selenenyl bromide, phenyl-
reaction with alkenyldihydroxyboranes, **7**, 608
Selenenyl bromide, 2-pyridyl-
dehydrogenation
carbonyl compounds, **7**, 128
Selenenylenones, 2-phenyl-
synthesis, **7**, 521
Selenenyl halides
reactions with alkenes, **4**, 339–342
Selenenyl pseudohalides
reactions with alkenes, **4**, 339–342
Selenide, methyl phenyl
metallation, **1**, 641
Selenide, phenyl trimethylsilyl
dehalogenation
benzoin acetates, **8**, 993
Selenide, phenyl trimethylsilylmethyl
metallation, **1**, 641
Selenide, triphenylphosphine
deoxygenation
epoxides, **8**, 887
Selenides
addition to alkynes, **4**, 50
alkenes from, **3**, 114
alkyl and allyl halides from, **3**, 118
alkylated
use in synthesis, **3**, 106
allylic
oxidative rearrangement, **3**, 117
deselenation
nickel boride, **8**, 848
halogenation, **7**, 772
α-heterosubstituted
carbonyl compound synthesis from, **3**, 141
metallation
synthesis of selenoalkyl metals, **1**, 635
oxidation, **7**, 129, 770
to selenones, **7**, 773
photooxidation, **7**, 774
reactions with alkenes, **4**, 317–319
reactions with alkyl metals, **1**, 630
reduction
use in synthesis, **3**, 106
Selenides, acetamido
synthesis, **7**, 495
Selenides, acyl phenyl
reaction with tri-*n*-butyltin hydride
reductive decarboxylation, **7**, 721
Selenides, 2-adamantyl phenyl
synthesis
via adamantane, **7**, 14
Selenides, alkenyl
coupling reactions
with sp^3 organometallics, **3**, 446
Selenides, alkenyl phenyl
synthesis, **7**, 608
Selenides, alkenyl pyridyl
metallation, **1**, 648
Selenides, alkyl phenyl
oxidation, **7**, 773
Selenides, alkyl 2-pyridyl
synthesis, **7**, 726
Selenides, alkyl vinyl
carbonyl compounds from, **3**, 120
Selenides, allenyl phenyl
synthesis, **3**, 106
Selenides, allyl
metallation, **1**, 640
oxidation, **3**, 117
rearrangement, **6**, 904
synthesis
via β-hydroxyalkyl selenides, **1**, 705
Selenides, β-amido phenyl
synthesis
via Ritter reaction, **6**, 289
Selenides, aryl
coupling reactions
with Grignard reagents, **3**, 456
tandem vicinal difunctionalization, **4**, 257
Selenides, aryl alkyl
synthesis, **4**, 447
Selenides, aryl 1-(2-methyl-1-propenyl)
metallation, **1**, 647
Selenides, aryl 1-propenyl
metallation, **1**, 647
Selenides, aryl vinyl
alkylation, **3**, 106
Selenides, benzyl
metallation, **1**, 640
Selenides, diaryl
synthesis, **4**, 443, 447
via $S_{RN}1$ reaction, **4**, 476
Selenides, diphenyl
synthesis
via $S_{RN}1$ reaction, **4**, 476
Selenides, homoallyl
synthesis, **3**, 91
Selenides, β-hydroxy
deoxygenation, **8**, 887
elimination reactions, **3**, 787
epoxide synthesis, **6**, 26
oxidation
solid support, **7**, 841
pinacol-type reactions, **1**, 861
rearrangement, **3**, 786
semipinacol rearrangements, **3**, 777
synthesis
via selenium-stabilized anions, **1**, 828
Selenides, β-hydroxy-γ-alkenyl
rearrangement, **1**, 717
Selenides, β-hydroxyalkyl
epoxide synthesis from
mechanism, **1**, 718
in synthesis, **1**, 696, 721
reactions with carbonyl compounds, **1**, 673
rearrangement, **1**, 714
reduction, **1**, 699
reductive elimination, **1**, 700
synthesis, **1**, 650
Selenides, γ-hydroxyalkyl
oxidation, **3**, 120
Selenides, β-hydroxyphenyl
oxidative rearrangement, **7**, 819
Ritter reaction, **6**, 289
Selenides, β-hydroxy-α-trimethylsilylalkyl
reductive elimination, **1**, 705
Selenides, α-lithio
epoxidation, **1**, 828
Selenides, α-metalloalkyl
synthesis
via metallation, **1**, 630

Selenides, α-metalloallenyl phenyl
  synthesis, **1**, 646
Selenides, α-metallovinyl
  alkylation, **3**, 104
  synthesis
    via metallation, **1**, 630
Selenides, 1-metallovinyl aryl
  synthesis, **1**, 646
Selenides, 4-nitrophenyl methyl
  synthesis, **4**, 447
Selenides, nor-alkyl-2-pyridyl
  synthesis, **7**, 722
Selenides, phenyl
  reduction, **6**, 470
Selenides, propargylic
  metallation, **1**, 640
Selenides, propargyl phenyl
  oxidative rearrangement, **7**, 826
Selenides, α-silylalkyl
  carbonyl compound synthesis from, **3**, 141
Selenides, trimethylsilyl
  selenol ester synthesis, **6**, 463
Selenides, vinyl
  metallation, **1**, 644
  reactions with organometallic compounds, **1**, 669
  reaction with Grignard reagents, **3**, 493
  synthesis, **3**, 253
    via β-hydroxyalkyl selenides, **1**, 705
Seleninic acid, allyl-
  in allylic oxidation
    selenium dioxide, **7**, 85
Seleninic acid, phenyl-
  hydroxylation
    alkenes, **7**, 446
Seleninic acids
  oxidation, **7**, 770
  synthesis, **7**, 770
Seleninic anhydride, 2-pyridine
  allylic oxidation, **7**, 110
Selenium
  carbanions
    synthesis, **1**, 630
    synthesis *via* metallation, **1**, 635
  carbanions stabilized by
    alkylation, **3**, 85–181
  dehydrogenation
    carbonyl compounds, **7**, 128
  halogen displacement, **7**, 124
  reductions, **8**, 370
    carbonyl compounds, **8**, 323
    nitro compounds, **8**, 366
Selenium compounds
  oxidation, **7**, 757–779
    secondary alcohols, **7**, 323
  reactions with arynes, **4**, 508
  tetravalent
    reaction with alkenes, **4**, 342
Selenium dioxide
  allylic oxidation, **7**, 84
    α,β-unsaturated carbonyl compounds, **7**, 108
  anti hydroxylation
    alkenes, **7**, 446
  oxidant
    silica support, **7**, 843
  oxidative rearrangement, **7**, 829, 832
Selenium imides
  Diels–Alder reactions
    diamines from, **7**, 486
Selenium insertion reaction
  hydroalumination adducts, **8**, 754
Selenium nucleophiles
  aromatic nucleophilic substitution, **4**, 447
Selenium tetrabromide
  reactions with alkenes, **4**, 342
Selenium tetrachloride
  reactions with alkenes, **4**, 342
Selenium ylides
  epoxidation, **1**, 825
  reactions with enals, **1**, 683
Selenoacetals
  carbonyl group regeneration, **7**, 846
  synthesis, **1**, 656
Selenobenzaldehyde, 2,4,6-tri-*t*-butyl-
  synthesis, **6**, 475
Selenobenzoate, *p*-nitrobenzyl
  synthesis, **6**, 465
Selenobenzoate, trimethylsilyl
  synthesis, **6**, 473
Selenobenzoic acid
  synthesis, **6**, 465
γ-Selenobutyrolactone
  synthesis, **6**, 462
Selenocarbamates, β-phenyl-
  synthesis, **7**, 495
Selenocarbonates, *Se*-phenyl-
  reduction
    stannanes, **8**, 825
Selenocarbonyl compounds
  Diels–Alder reactions, **5**, 442
Selenocarboxylates
  alkylation, **6**, 464
Selenocarboxylates, *Se*-acylmethyl
  selenium extrusion, **6**, 469
Selenocyanates
  alkyl
    synthesis, **7**, 608
  oxidation, **7**, 770
  reaction with carboxylic acids, **6**, 466
Selenocyclizations, **7**, 495
Seleno-1,3-dienes,1-phenyl-
  synthesis
    via methoxyselenation, **4**, 339
Seleno esters
  synthesis, **1**, 95
Selenoformamide
  synthesis, **6**, 480
Selenoketones
  synthesis, **5**, 442
Selenolactams
  synthesis, **6**, 478
Selenolactones
  metallation, **1**, 642
Selenolactonization, **7**, 523
Selenolates, arylethynyl-
  reaction with alcohols, **6**, 473
Selenol esters
  aromatic
    synthesis, **6**, 462
  Friedel–Crafts reaction, **2**, 737
  reactions, **6**, 468, 474
    with cuprates, **6**, 469

with isocyanides, **6**, 470
reactivity, **6**, 461
synthesis, **6**, 461–481
Selenols
acylation, **6**, 462
oxidation, **7**, 769
radical additions
alkenes, **4**, 770
reductions
aromatic compounds, **8**, 370
synthesis, **6**, 462
Selenones
epoxidation, **1**, 828
metallation, **1**, 650
oxidation, **7**, 773
Selenones, α-metalloalkyl
reactivity
reactions with carbonyl compounds, **1**, 672
synthesis, **1**, 648; **3**, 87
via metallation, **1**, 630
Selenones, vinyl
reactions with organometallic compounds, **1**, 669
Selenonic acids
oxidation
to selenoxides, **7**, 770
Selenonium bromide, phenacylmethyl(dimethyl)-
metallation, **1**, 655
Selenonium salts
metallation, **1**, 651
Selenonium salts, allyldimethyl-
metallation, **1**, 653
Selenonium salts, allylmethylphenyl-
metallation, **1**, 653
Selenonium salts, α-metalloalkyl-
synthesis, **1**, 648; **3**, 87
via metallation, **1**, 630
Selenonium ylides
reactivity
reactions with carbonyl compounds, **1**, 672
Selenophene, 2-lithio-
synthesis, **1**, 668
Selenophenes
coupling reactions
with primary alkyl Grignard reagents, **3**, 447
metallation, **1**, 644
synthesis, **6**, 481
via [2 + 2 + 2] cycloaddition, **5**, 1139
Vilsmeier–Haack reaction, **2**, 780
Selenophenols
conjugate additions
enones, **4**, 231
reduction
imines, **8**, 36
Selenophthalimide, *N*-phenyl-
addition reactions
alkenes, **7**, 522
Selenosuccinimide, *N*-phenyl-
addition reactions
alkenes, **7**, 522
Selenosulfides
synthesis, **7**, 519
Selenosulfonates
addition reactions
alkenes, **7**, 523
Selenosulfonation
alkenes, **4**, 341
Selenosulfones
synthesis, **7**, 519
Selenothiolactonization
alkenes, **7**, 520
Selenoxide, 2-azidocyclohexyl phenyl
synthesis, **7**, 772
Selenoxide, benzyl phenyl
synthesis, **7**, 772
Selenoxide, di-4-anisyl
Kornblum oxidation, **7**, 657
Selenoxide, dimethyl
Kornblum oxidation, **7**, 657
Selenoxide, methyl phenyl
synthesis, **7**, 772
Selenoxides
chiral
synthesis, **7**, 777, 779
elimination
carbonyl compounds, **7**, 128, 146
metallation, **1**, 649
oxidation, **7**, 657, 770
to selenones, **7**, 773
rearrangement, **6**, 904
alcohol synthesis, **6**, 14
Selenoxides, alkyl
alkylation, **3**, 147, 157
Selenoxides, allyl
rearrangement, **3**, 117
Selenoxides, α-lithioalkyl
synthesis
via alkylation, **3**, 157
Selenoxides, α-metalloalkyl
alkylation, **3**, 147, 157
reactivity
reactions with carbonyl compounds, **1**, 672
synthesis, **1**, 648; **3**, 87
via metallation, **1**, 630
Selenoxides, α-metallovinyl
alkylation, **3**, 157
reactivity
reactions with carbonyl compounds, **1**, 672
synthesis, **1**, 630; **3**, 87
Selenoxides, β-oxidoalkyl
synthesis, **1**, 650
Selenoxides, vinyl
reactions with organometallic compounds, **1**, 669
2-Selenoxobenzothiazole, 3-methyl-
deoxygenation
epoxides, **8**, 887
Selenuranes
reactions with aldehydes, **1**, 651
Selinene
synthesis, **6**, 777; **8**, 943
via oxyanion-accelerated rearrangement, **5**, 1020
via reductive lithiation, **6**, 146
Semibenzilic pathway
Favorskii rearrangement, **3**, 840
mechanism, **3**, 828, 836
Semibullvalene
synthesis, **3**, 640
via photoisomerization, **5**, 194, 204
Semibullvalenes, dihydro-
synthesis
via retro electrocyclization, **5**, 737
Semicarbazones
reduction, **8**, 336

Vilsmeier–Haack reaction, **2**, 791
Semipinacol rearrangements, **3**, 777–799
definition, **3**, 777
pinacol rearrangement
comparison with, **3**, 722
tandem cyclization reactions, **3**, 792
Semi-ylides
phosphonium
formation, **6**, 172
Senecioic acid
dicopper(I) dianion
alkylation, **3**, 50
Senecioic acid, 4-bromo-
trimethylsilyl ester
Reformatsky reaction, **2**, 286
Senecioic acid amide, *N*-isopropyl-
dianions
alkylation, **3**, 50
Senepoxyde
synthesis
alkene protection, **6**, 689
via retro Diels–Alder reactions, **5**, 563
Senoxepin
synthesis
via Peterson alkenation, **1**, 733
Senoxydene
precursor
synthesis *via* intramolecular ene reactions, **5**, 22
Senoxyn-4-en-3-one, 8-oxy-
synthesis, **3**, 404
Sepiolite
solid support
oxidants, **7**, 845
Septamycin
synthesis
A-ring fragment, **1**, 429
Sequential rearrangements, **5**, 876–891
Serine
enantioselective aldol cyclizations, **2**, 169
hydroxy groups
protection, **6**, 650
β-lactone
reaction with organocopper compounds, **3**, 227
reaction with pivaldehyde, **3**, 40
Serine, phenyl-
synthesis, **8**, 148
Serine proteases
peptide synthesis, **6**, 395
Sesamin
synthesis
via Diels–Alder reaction, **5**, 468
via Knoevenagel reaction, **2**, 373
Sesamol, benzyl-
oxidative coupling, **3**, 669
Sesamolin
synthesis
via Diels–Alder reaction, **5**, 468
via Knoevenagel reaction, **2**, 373
Sesbanine
synthesis
via regioselective lithiation, **1**, 474
Sesquicarene
synthesis, **3**, 288
Sesquifenchene
synthesis, **3**, 161
Sesquinorbornene
reaction with methanol, **5**, 74
Sesquiterpenes
hydrazulene-based
synthesis, **7**, 301
marine
synthesis, **2**, 710
microbial hydroxylation, **7**, 63
polycyclic
biosynthesis, **3**, 388
synthesis, **3**, 288
via photoisomerizations, **5**, 230
Seven-membered rings
synthesis
via aldol reaction cascade, **2**, 623
via [4 + 3] cycloadditions, **5**, 593
via Friedel–Crafts reaction, **2**, 763
via intramolecular aldolization of keto aldehydes, **2**, 160
via polyene cyclization, **3**, 357
Sex pheromones
bark beetle
synthesis, **1**, 218
Seychellene
synthesis, **3**, 20
via Diels–Alder reactions, **5**, 329
via Prins reaction, **2**, 542
via radical cyclization, **4**, 796
Shapiro reaction, **6**, 779; **8**, 944
limitations, **8**, 948
regioselectivity, **8**, 944
stereoselectivity, **8**, 948
vinyllithium generation, **3**, 251
Shell Higher Olefin Process
alkene metathesis, **5**, 1117
Shikimic acid
synthesis
via cyclofunctionalization of cycloalkene, **4**, 373
via cyclohexadienyl complexes, **4**, 683
via Diels–Alder reactions, **5**, 335, 360, 363
Shikimic acid pathway
Claisen rearrangement, **5**, 855
Showdomycin
synthesis
via [4 + 3] cycloaddition, **5**, 611
via retro-Dieckmann reaction, **2**, 855
Sialic acids
synthesis
via enzymatic method, **2**, 463, 464
Sibirine
synthesis
via nitrile oxide cyclization, **4**, 1129
Sibirosamine
synthesis
stereospecific, **1**, 413
Sigmatropic rearrangements
alcohol synthesis, **6**, 14
Baldwin's rules, **3**, 915
carbene complexes, **5**, 1075
1,2-Sigmatropic rearrangements, **3**, 921
chirality transfer, **3**, 927
1,3-Sigmatropic rearrangements
allylic alcohols, **5**, 1001
aza-anion accelerated, **5**, 1003
oxyanion-accelerated
stereochemistry, **5**, 1002
1,5-Sigmatropic rearrangements

carbanion-accelerated, **5**, 1005
oxyanion-accelerated, **5**, 1003
2,3-Sigmatropic rearrangements, **6**, 873–908
allylic systems, **6**, 834
3,2-Sigmatropic rearrangements, **3**, 932
ylides
stereocontrol, **3**, 943
3,3-Sigmatropic rearrangements
allylic systems, **6**, 834
aluminum enolates, **1**, 91
erythro–threo ratio, **3**, 949
homologations, **1**, 880
Silabarrelene
photorearrangement, **5**, 199
7-Silabicyclo[2.2.2]octadiene
thermolysis
retro Diels–Alder reaction, **5**, 587
Silacycles
intramolecular hydrosilylation, **8**, 774
Silacyclopentene
acetylation
Friedel–Crafts reaction, **2**, 717
Silane, 2-acetoxymethyl-3-allyltrimethyl-
trimethylenemethane synthetic equivalent, **5**, 244
Silane, allyl(diethylamino)dimethyl-
reaction with BuLi/TMEDA, **2**, 59
Silane, allyloxy-
siloxy carbanions from, **2**, 601
Silane, allyltriisopropyl-
deprotonation
γ-selectivity, **2**, 58
Silane, allyltrimethyl-
alcohol protection, **6**, 654
ene reactions, **5**, 2
intramolecular additions
carbonyl compounds, **1**, 612
dissolving metals, **8**, 513
Silane, aryltrimethyl-
metal/metal exchange, **7**, 649
Silane, benzyl-
Birch reduction
dissolving metals, **8**, 513
Mannich reaction, **2**, 1035
Silane, benzyltrimethyl-
C—Si bond cleavage, **7**, 649
Silane, *t*-butyldimethylchloro-
O-silylation with, **2**, 604
Silane, chloro-
hydrosilylation
unsaturated hydrocarbons, **8**, 765
Silane, chlorodimethyl-
hydrosilylation
alkynes, **7**, 643
Silane, (chloromethyl)trimethyl-
Darzens glycidic ester condensation, **2**, 426
Silane, chlorotrimethyl-
acyloin coupling reaction
trapping agent, **3**, 615
alcohol protection, **6**, 653
reaction with conjugated ketones
1,4-addition, **2**, 599
Silane, crotyl-
reaction with achiral carbonyl compounds, **2**, 17
reaction with iminium salts, **2**, 1002
synthesis, **2**, 977
Silane, crotyltrimethyl-
configurational stability, **2**, 6
Silane, cyanotrimethyl-
Beckmann rearrangement, **6**, 768
Silane, cyclopropyl-
acylation
Friedel–Crafts reaction, **2**, 728
Silane, (diethoxymethyl)-
hydrosilylation
alkynes, **7**, 643
Silane, diiododimethyl-
reduction
benzylic alcohols, **8**, 979
Silane, dimethylphenyl-
hydrosilylation
carbonyl compounds, **8**, 21
ketone reduction, **8**, 8
oxidation, **7**, 646
Silane, diphenyl-
reduction
carbonyl compounds, **8**, 322
Silane, diphenyl(4-pentenyl)-
ring closure, **8**, 774
Silane, ethynyl-
hydroalumination, **8**, 748
reaction with acetals, **2**, 579
reaction with aldehydes, **2**, 575
reaction with ketals, **2**, 579
reaction with ketones, **2**, 575
Silane, fluorotrimethyl-
aldol reactions
catalytic cycle, **2**, 633
Silane, hydrido-
ionic hydrogenation
unsaturated carbonyl compounds, **8**, 546
Silane, (iodomethyl)trimethyl-
reaction with sulfonyl carbanions, **5**, 1014
Silane, iodotrimethyl-
Beckmann rearrangement, **6**, 767
dehalogenation
α-halo ketones, **8**, 988
ester cleavage, **6**, 665
iodination
alkyl alcohols, **6**, 214
methyl ether cleavage, **6**, 647
Silane, methoxybis(trimethylsilyl)methyl-
methoxycarbonyl anion synthon, **7**, 650
Silane, methoxy(trimethylsilyl)methyl-
formyl anion synthon, **7**, 650
Silane, methyldiphenylchloro-
reaction with lithium ester enolates
regiochemistry of silylation, **2**, 604
Silane, methyltrichloro-
dehalogenation
α-halo ketones, **8**, 988
Silane, (2-nitroethenyl)trimethyl-
synthesis, **6**, 107, 109
Silane, nitrovinyl-
synthesis, **6**, 108
Silane, (α-oxyallyl)-
reaction with benzonitrile oxide, **5**, 262
Silane, (pentadienyl)trimethyl-
acylation
Friedel–Crafts reaction, **2**, 721
Silane, phenyl-
transfer hydrogenation
molybdenum complex catalyst, **8**, 554

Silane, α-phenylthiomethyltrimethyl-
reaction with alkyl halides
synthesis of aldehydes, **6**, 139
Silane, 1-phenylthiovinyl-
Nazarov cyclization, **5**, 778
Silane, 2-phenylthiovinyl-
Nazarov cyclization, **5**, 778
Silane, propargyltrimethyl-
condensation with acyl cyanide, **2**, 85
reaction with *N*-acyliminium ions, **2**, 1061, 1071
reaction with ω-ethoxy lactams, **2**, 89
reaction with glycine cation equivalents, **2**, 1075
Silane, 2-propynyl-
reaction with cyclic *N*-acyliminium ions, **2**, 89
Silane, trichloro-
addition to alkenes, **7**, 642
reduction
carbonyl compounds, **8**, 322
carboxylic acids, **8**, 238
Silane, triethyl-
ionic hydrogenation
carbonyl compounds, **8**, 319
oligosaccharide side chain cleavage, **8**, 219
reaction with *N*-acyliminium ions
reduction, **2**, 1077
reaction with unsaturated esters
rhodium catalysts, **8**, 555
reduction
acyl halides, **8**, 265
alcohols, **8**, 813
carbocations, **8**, 275
carbonyl compounds, **8**, 318
1,4-dihydropyridine, **8**, 589
Silane, triisopropyl-
reaction with acyl chloride
reductive decarboxylation, **7**, 721
Silane, trimethylvinyl-
Friedel–Crafts acylation, **2**, 712
Silane, tris(trimethylsilyl)-
halide chain reductions
propagation, **4**, 739
reduction
alkyl halides, **8**, 801
Silane, vinyl(alkoxy)-
synthesis, **7**, 644
Silane, vinyltrichloro-
ene reactions
thermal, **5**, 3
Silane, vinyltrimethyl-
ethylene equivalent
alkene acylation, **5**, 777
Silanes
acylation
Friedel–Crafts reaction, **2**, 712, 728
hypervalent
acyl halide reduction, **8**, 265
reduction
acetals, **8**, 216
acyl halides, **8**, 265
alkyl halides, **8**, 801
carbonyl compounds, **8**, 318
carboxylic acids, **8**, 261
esters, **8**, 824
imines, **8**, 36
synthesis, **2**, 582
via Ireland silyl ester enolate rearrangement, **5**, 841
vinyl substitution
palladium complexes, **4**, 840
Silanes, acyl-
reaction with sulfonyl carbanions, **5**, 1014
rearrangement
enol ether preparation, **2**, 601
synthesis, **7**, 598
via Claisen rearrangement, **5**, 850
via organoaluminium reagents, **1**, 97
α,β-unsaturated
reactions with allenylsilanes, **1**, 598
Silanes, alkenyl-
acylation
Friedel–Crafts reaction, **2**, 712
carbomagnesiation
intramolecular, **4**, 879
formylation, **2**, 728
intramolecular acylation
Friedel–Crafts reaction, **2**, 714
synthesis
via carbocupration, **4**, 900
via metal carbene complexes, **1**, 808
Silanes, alkenyloxy-
metal exchange reaction
dialkylboryl triflate, **2**, 245
Silanes, alkoxy-
hydrosilylation
unsaturated hydrocarbons, **8**, 765
Silanes, alkynyl-
acylation
Friedel–Crafts reaction, **2**, 725
carboalumination, **4**, 892
coupling reactions
with aryl halides, **3**, 538
hydroalumination, **8**, 741
hydroboration, **8**, 708
reaction with diisobutylaluminum hydride, **8**, 734
reaction with electrophilic π-systems, **1**, 604
Silanes, allenyl-
[3 + 2] annulations, **1**, 596
reactions with acetals, **2**, 579
reactions with activated imines, **1**, 602
reactions with acyliminium salts, **1**, 598; **2**, 1061
reactions with aldehydes, **1**, 599; **2**, 575
reactions with carbonyl compounds
synthesis of substituted alkynes, **1**, 595
titanium tetrachloride, **1**, 595
reactions with imines
syn–anti selectivity, **2**, 992
reactions with ketals, **2**, 579
reactions with ketones, **2**, 575
reactions with tropylium ions, **1**, 603
reactions with α,β-unsaturated acylsilanes, **1**, 598
reactions with α,β-unsaturated carbonyl compounds, **1**, 596
synthesis, **2**, 587
Silanes, allyl-
acylation
Friedel–Crafts reaction, **2**, 716
addition reactions
stereochemistry, **1**, 610
allylations
Lewis acid promoted, **1**, 346
allylic rearrangements, **7**, 822
π-allylpalladium complexes from, **4**, 588
chiral

reaction with aldehydes, stereospecifically, **2**, 568
conjugate additions to α,β-enones
Lewis acid catalyzed, **4**, 155
[4 + 3] cycloaddition reactions, **5**, 598
electrophilic substitutions
allylic rearrangement, **6**, 832
epoxidation, **7**, 360
intermolecular additions
aldehydes, ketones and acetals, **1**, 610
internal additions
stereochemistry, **1**, 615
Mannich reaction, **2**, 1032
metallated
additions to aldehydes, **1**, 113
reactions
fluoride ion catalysis, **2**, 565
reactions with acetals, **2**, 576
reactions with *N*-acyliminium ions, **2**, 1060, 1064, 1066, 1070, 1071
reactions with aldehydes, **2**, 567
reactions with alkoxymethyl chlorides, **2**, 580
reactions with carbonyl compounds, **1**, 610; **2**, 563–590
reactions with glycine cation equivalents, **2**, 1075
reactions with imines, **2**, 976
reactions with iminium salts, **2**, 1002
reactions with ketals, **2**, 576
reactions with ketones, **2**, 567
reactions with phenylthiomethyl chlorides, **2**, 580
reactions with vinyloxiranes
regioselectivity, **5**, 936
synthesis, **2**, 582
via coupling reactions, **3**, 437, 445
via [3 + 2] cycloaddition reactions, **5**, 304
via esters, **1**, 244
via β-hydroxyalkyl selenides, **1**, 705
via nickel catalysts, **3**, 229
via Peterson methylation, **1**, 238, 735
Silanes, allylamino-
metallated
addition reactions, **1**, 624
Silanes, aryl-
Birch reduction
dissolving metals, **8**, 513
defluorosilylation
aryne generation, **4**, 487
Silanes, bisallyl-
synthesis, **3**, 482
Silanes, chiral acyl-
nucleophilic addition reactions
stereoselectivity, **1**, 57
Silanes, dienyl-
synthesis
via zinc-ene reactions, **5**, 32
Silanes, dimethylfluoro-
reaction with alkenyl iodides
organopalladium catalysis, **3**, 233
Silanes, α,β-epoxy-
Peterson reaction, **1**, 737
reaction with organocopper compounds, **3**, 224
synthesis, **7**, 643
via Darzens glycidic ester condensation, **2**, 426
via vinylsilanes, **2**, 57
synthesis and rearrangement
enol ether preparation, **2**, 601
Silanes, α,β-epoxyalkyl-
deprotonation, **3**, 198
Silanes, homoallylic
intermolecular acylation
Friedel–Crafts reaction, **2**, 719
Silanes, 2-hydroxy-
synthesis, **7**, 643
via trimethylsilylmethylcerium reagent, **1**, 238
Silanes, β-hydroxyalkyl(1-naphthyl)phenylmethyl-
synthesis, **1**, 785
Silanes, β-keto(aldehydo)-
synthesis from α-trimethylsilyl epoxides
reaction with Grignard reagents, **3**, 759
Silanes, organo-
hydride donor
ionic hydrogenation, **8**, 486
Silanes, propargyl-
acylation
Friedel–Crafts reaction, **2**, 726
electrophilic additions
formation of β-silyl carbocations, **1**, 616
reaction with acetals, **2**, 579
reaction with *N*-acyliminium ions, **2**, 1071
reaction with aldehydes, **2**, 575
reaction with ketals, **2**, 579
reaction with ketones, **2**, 575
synthesis, **2**, 587
Silanes, trialkyl-
nucleophilic addition reactions
stereoselectivity, **1**, 57
reduction
acetals, **8**, 216
Silanes, vinyl-
conjugate addition
organocuprates, **4**, 191
coupling reactions
with aryl bromides, **3**, 495
with organic halides, **8**, 786
cyclization reactions
acetal- and carbonyl-initiated, **1**, 585
Nazarov type, **1**, 585
divinyl ketones from, **5**, 777
epoxidation, **2**, 58, 601
hydrosilylation, **8**, 774
hydroxylation
generation of α-hydroxy ketones, **7**, 172
intramolecular addition, **1**, 584
Mannich reaction, **2**, 1030
oxidative rearrangement, **7**, 816
reaction with acetals, **2**, 579
reaction with *N*-acyliminium ions, **2**, 1064
reaction with aldehydes, **2**, 575
reaction with carbonyl compounds, **2**, 563–590
reaction with electrophiles, **8**, 785
reaction with glycine cation equivalents, **2**, 1074
reaction with ketals, **2**, 579
reaction with ketones, **2**, 575
reaction with methoxymethyl chloride
carbon–oxygen bond cleavage, **2**, 581
synthesis, **2**, 588; **8**, 769
via β-hydroxyalkyl selenides, **1**, 705
via Peterson reaction, **1**, 786
vinyl anion equivalents, **1**, 583
Silanol, (3*E*)-phenylethenyldimethyl-
asymmetric epoxidation, **7**, 423
Silanones
generation

via retro Diels–Alder reaction, **5**, 587
2-Silapropene, 2-methyl-
synthesis
via retro Diels–Alder reaction, **5**, 587
Sila-Pummerer rearrangement
β-elimination, **7**, 204
2-Silapyrans
Diels–Alder reaction, **5**, 587
methyl acrylate, **5**, 1074
Silasemibullvalene
synthesis
via photoisomerization, **5**, 199
Silenes
generation
via retro Diels–Alder reaction, **5**, 587
Silica
solid support
oxidants, **7**, 840
oxidation, **7**, 842
Silica gel
catalyst
Pauson–Khand reaction, **5**, 1056
Silicates, crotyl
pentacoordinate
configurational stability, **2**, 6
reactions with achiral aldehydes, **2**, 17
Silicates, hydrido-
carbonyl compound reduction, **8**, 20
Silicates, organopentafluoro-
synthesis, **7**, 642
Silicates, pentafluoro-
synthetic reactions, **8**, 787
Silicon hydrides
reduction
carbonyl compounds, **8**, 20
unsaturated carbonyl compounds, **8**, 546
Silicon reagents
Darzens glycidic ester condensation, **2**, 426
organopalladium catalysis, **3**, 233
reactions with achiral carbonyl compounds, **2**, 17
Silphinene
synthesis
via Nazarov cyclization, **5**, 779
via photocycloaddition, **5**, 662
Silphiperfol-5-ene
synthesis
via photocycloaddition, **5**, 664
Silphiperfol-6-en-5-one
synthesis
via photoisomerization, **5**, 232
Silver
catalysts
Grignard reagent coupling, **3**, 418
Silver, allenyl-
synthesis, **2**, 85
Silver acetate
allylic oxidation, **7**, 92
Silver benzoate
iodine
alkene hydroxylation, **7**, 447
Wolff rearrangement
initiator, **3**, 891
Silver carbonate
on celite
oxidant, **7**, 841
oxidation
diols, **7**, 318
α,ω-diols, **7**, 312
secondary alcohols, **7**, 320
Silver carboxylates
reaction with halogens, **7**, 723
synthesis, **7**, 718
Silver cyanide
isocyanide synthesis, **6**, 243
Silver dichromate, tetrakis(pyridine)-
oxidation
alcohols, **7**, 286
Silver homoenolates
substitution reactions, **2**, 450
Silver nitrate
in halohydrin rearrangements
formation of aldehydes, **3**, 758
oxidation
halides, **7**, 664
Silver oxide
quinone synthesis, **7**, 355
reaction with acyl chloride
preparation of silver carboxylates, **7**, 723
Wolff rearrangement
initiator, **3**, 891
Silver permanganate, bispyridine-
oxidation
primary arylamines, **7**, 738
Silver salts
catalysts
Cope rearrangement, **5**, 802
Kornblum oxidation, **7**, 656
Ritter reaction
initiators, **6**, 283
Wurtz coupling, **3**, 422
Silver tetrafluoroborate
activator
DMSO oxidation of alcohols, **7**, 299
Silver trifluoroacetate
alkane oxidation, **7**, 13
Silybin
synthesis, **3**, 691
Silylamine, α-cyano-
azomethine ylides from, **4**, 1088
Silylamines
reaction with ketenes, **2**, 605
Silylation
alcohol protection, **6**, 654
1-*O*-Silylation
glycoside synthesis, **6**, 49
C-Silylation
Claisen rearrangements
competition, **5**, 850
Silyl carbonates
synthons
[3 + 2] cycloaddition reactions, **5**, 304
Silyl chromate, bis(triphenyl-
oxidative cleavage
alkenes, **7**, 571
Silyl compounds, titanated
reactions with carbonyl compounds, **1**, 161
Silyl cyanides, trialkyl-
reactions with carbonyl compounds
Lewis acid promotion, **1**, 328
Silyl dienol ethers
cross-conjugated
alkylation, **3**, 28

extended
γ-alkylation, **3**, 27
homoannular
alkylation, **3**, 27
α'-hydroxylation, **7**, 177
Silylenes
dicarbonyl compound monoprotection, **6**, 684
Silyl enol ethers
aldehyde
allylation, **3**, 28
aldol reactions, **8**, 786
Lewis acid promoted, **1**, 346
alkali metal enolates
reaction, **2**, 108
alkylation, **3**, 25
amination, **6**, 118
asymmetric synthesis, **2**, 629
Beckmann rearrangement, **6**, 770; **7**, 697
chiral
diastereoselective aldol additions, **2**, 636
chlorination, **7**, 530
cleavage
methylmagnesium bromide, **2**, 110
conjugate additions
alkenes, **4**, 158–162
conversion to enolates, **2**, 184
coupling reactions
with aryl Grignard reagents, **3**, 492
with primary bromides, **3**, 454
with primary alkyl Grignard reagents, **3**, 445
cyclic
synthesis, **2**, 601
cycloalkylation, **3**, 27
dehydrogenation
palladium catalysts, **7**, 141
quinones, **7**, 137
dihalocyclopropanation, **4**, 1005
ene reactions, **5**, 1075
halogenation, **7**, 121
α-hydroxylation
ketones, **7**, 163
intramolecular alkylation, **3**, 26
Mannich reactions
iminium ions, **2**, 1015
ozonolysis, **7**, 166
reactions, **2**, 613
reactions with aldehydes
Lewis acid mediated, **2**, 630
reactions with carbonyl compounds
catalysts, **2**, 614
chemoselectiviy, **2**, 615
regioselectivity, **2**, 616
reactions with α-chloromethyl phenyl sulfides, **6**, 141
reactions with nitroarenes, **4**, 429
reduction, **8**, 935
regiospecific synthesis, **2**, 599
sulfenylation, **7**, 125
α-sulfonyloxygenation, **7**, 145
synthesis, **2**, 599
via lithium homoenolates, **3**, 197
via oxidative cleavage, **7**, 587
vinyl substitution
palladium complexes, **4**, 840
Silylepoxy ethers
rearrangements
alcohol synthesis, **6**, 14
Silyl groups, 2-furyldimethyl-
desilylation, **7**, 647
Silyl halides, trialkyl-
reaction between aldehydes and organocuprates, **1**, 112
Silyl-hydroformylation
cycloalkenes
enol ether preparation, **2**, 603
Silylimines, *N*-trimethyl-
reaction with allyl organometallic compounds, **2**, 999
reaction with silylketene acetals, **5**, 102
Silyl ketene acetals
chiral
aldol reaction, **2**, 636
diastereoselective addition to imines, **2**, 638, 639
diastereoselective aldol additions, **2**, 636
reaction with aldehydes, diastereoselectivity, **2**, 637
Claisen condensation, **2**, 803
conjugate additions
alkenes, **4**, 158–162
α,β-enones, **4**, 162
dehydrogenation, **7**, 142
diastereoselective addition reactions
chiral aldehydes, **2**, 652
from butyrolactone
reaction with aldehydes, **2**, 632
α-hydroxylation, **7**, 182
reactions with aldehydes
Lewis acid mediated, **2**, 630
reactions with imines, **2**, 929
diastereoselectivity, **2**, 636
reactions with oxime ethers, **2**, 940
reactions with *N*-silylimines, **2**, 937
synthesis
via Ireland silyl ester enolate rearrangement, **5**, 841
thiol esters
reaction with aldehydes, **2**, 644
Silyl ketene acetals, bis-
α-hydroxylation, **7**, 185
Silylmetallation
alkynes, **8**, 771
Silylmethyl radicals
cyclizations, **4**, 794; **7**, 648
Silylonium salts, α-trimethyl-
desilylation
azomethine ylide generation, **4**, 1086
Silyl perbenzoates, triorgano-
rearrangement, **7**, 641
Silyl peroxide
rearrangement, **7**, 641
Silyl polyphosphate, trimethyl-
cyclization
alkenic oximes, **6**, 771
Silyl triflate, trimethyl-
reduction
acetals, **8**, 216
Simmons–Smith cyclopropanation, **4**, 968
asymmetric, **4**, 968
hydroalumination adducts, **8**, 756
unsaturated ketones, **1**, 533
Simonini complex
alkene hydroxylation, **7**, 447
Sinapic acid
oxidation, **3**, 692
Sinefugin
synthesis

via Ugi reaction, **2**, 1096
α-Sinensal
synthesis, **1**, 560; **3**, 936
β-Sinensal
synthesis, **3**, 429
via tandem Claisen–Cope rearrangement, **5**, 878
Single electron transfer
desulfurization
electropositive metals, **8**, 842
Sinularene
synthesis
via Cope rearrangement, **5**, 989
via magnesium-ene reaction, **5**, 41
Sinularene, 12-acetoxy-
synthesis
via magnesium-ene reaction, **5**, 41
Sirenin
synthesis, **3**, 288, 788; **7**, 86
Six-membered rings
synthesis
via aldol reaction cascade, **2**, 620
via Friedel–Crafts reaction, **2**, 758
via polyene cyclization, **3**, 349
Skattebøl rearrangement
heterocyclic version, **4**, 1021
vinylcyclopropylidene–cyclopentylidene compounds, **4**, 1012
Skeletal reorganizations, **1**, 843–899
α-Skytanthine
synthesis
via conjugate addition to α,β-unsaturated carboxylic acid, **4**, 202
via magnesium-ene reaction, **5**, 41
δ-Skytanthine
synthesis
via magnesium-ene reaction, **5**, 41
Slaframine
synthesis
via Diels–Alder reaction, **5**, 414
Small ring compounds
cycloaddition reactions
metal-catalyzed, **5**, 1185–1204
Sodium
Birch reduction, **8**, 492
alkyl halides, **8**, 795
ethanol as solvent
reduction, **8**, 111
in alcohol
alkyl halide reduction, **8**, 795
liquid ammonia
amide reduction, **8**, 293
carbonyl compound reduction, **8**, 308
reduction
amidines, **8**, 302
ammonia, **8**, 113
carbonyl compounds, **8**, 109
enones, **8**, 524
epoxides, **8**, 880
reductive dimerization
unsaturated carbonyl compounds, **8**, 532
Sodium, benzyl-
tetramethylethylenediamine solvate
crystal structure, **1**, 13
Sodium, indenyl-
tetramethylethylenediamine complex
crystal structure, **1**, 19
Sodium, methyl-
synthesis
crystal structure, **1**, 12
Sodium acetate
Rosenmund reduction, **8**, 287
Sodium acetoxyborohydride
hydroboration, **8**, 709
Sodium aluminum hydride
reduction
amides, **8**, 271
esters, **8**, 267
imines, **8**, 36
nitriles, **8**, 274
reduction kinetics, **8**, 2
Sodium amalgam
C—P bond cleavage, **8**, 863
demercuration, **8**, 857
stereoselectivity, **8**, 857
desulfurization, **8**, 843
reduction
aldonolactones, **8**, 292
enones, **8**, 525
reductive cleavage
α-alkylthio ketone, **8**, 993
reductive dimerization
unsaturated carbonyl compounds, **8**, 532
Sodium amide
phosphonium ylide synthesis, **6**, 174
Sodium arsenate
reduction
nitro compounds, **8**, 366
Sodium azide
reaction with π-allyl complexes, **6**, 86
reaction with trialkylboranes, **7**, 607
Sodium benzoate
reduction
dissolving metals, **8**, 526
Sodium bis(2-methoxyethoxy)aluminum hydride
allylic alcohol synthesis
reduction, **7**, 397
reduction
amides, **8**, 271
aromatic nitriles, **8**, 274
benzylic halides, **8**, 967
carbonyl compounds, **8**, 314
carboxylic acids, **8**, 238
epoxides, **8**, 879
esters, **8**, 267
imines, **8**, 36
lactones, **8**, 268
pyridines, **8**, 584
α-siloxy ketones, **8**, 7
unsaturated carbonyl compounds, **8**, 542–544
Sodium bis(2-methoxyethoxy)ethoxyaluminum hydride
reduction
lactones, **8**, 268
Sodium bis(2-methoxyethoxy)-*N*-methylpiperidinoaluminum hydride
reduction
esters, **8**, 267
Sodium bismuthate
glycol cleavage, **7**, 703
Sodium bis(trimethylsilyl)amide
phosphonium ylide synthesis, **6**, 174
Sodium borodeuteride
labeling

demercuration, **8**, 852
reduction
gibberellins, **8**, 537
Sodium borohydride
cerium chloride complex
cyclic ketone reduction, **8**, 15
enone reduction, **8**, 539
chirally modified
reduction, **8**, 160
demercurations, **8**, 851
hydroboration, **8**, 708
liquid ammonia
reductive amination, **8**, 54
reduction
acetals, **8**, 215
acyl halides, **8**, 240, 263
alkyl halides, **8**, 803
N-alkylphthalimides, **8**, 254
amides, **8**, 249
benzylic alcohols, **8**, 962
benzylic halides, **8**, 967
carbonyl compounds, **8**, 2, 313
carboxylic acids, **8**, 237
enones, **8**, 15
epoxides, **8**, 874
esters, **8**, 244, 267
hydrazones, **8**, 345
imines, **8**, 26
imines, chemoselectivity, **8**, 37
indoles, **8**, 616
ketones, **8**, 9
keto sulfides, **8**, 12
lactones, **8**, 269
nitriles, **8**, 253
nitro compounds, **8**, 366
pyridines, **8**, 579
tosylates, **8**, 812
unsaturated carbonyl compounds, **8**, 536
unsaturated hydrocarbons, **8**, 485
reductive demercuration, **7**, 632
selective aldehyde reduction, **8**, 16
selective ketone reduction, **8**, 18
trifluoroacetic acid
carbonyl compound reduction, **8**, 315
Sodium bromate
reduction
dissolving metals, **8**, 526
Sodium bromite
oxidation
secondary alcohols, **7**, 322
Sodium cation complexes
acetone
theoretical studies, **1**, 287
crystal structure, **1**, 299
Sodium chromoglycate
synthesis, **7**, 338
Sodium cyanoborohydride
boron trifluoride mixture
epoxide reduction, **3**, 753
reduction
acetals, **8**, 216
alkyl halides, **8**, 806
allylic leaving group, **8**, 960
benzylic compounds, **8**, 969
carbonyl compounds, **8**, 314
enones, **8**, 538
epoxides, **8**, 876
hydrazones, **8**, 350
imines, **8**, 26, 36
imines, chemoselectivity, **8**, 37
pyridines, **8**, 580
tosylates, **8**, 812
unsaturated carbonyl compounds, **8**, 536
reductive amination, **8**, 26, 47
biochemical applications, **8**, 47
Sodium (cyclopentadienyl)dicarbonylferrate
deoxygenation
epoxides, **8**, 890
Sodium dichromate
oxidation
alcohols, **7**, 252
Sodium *O*,*O*-diethyl phosphorotelluroate
deoxygenation
epoxides, **8**, 887
Sodium diisobutylaluminum hydride
reduction
aromatic nitriles, **8**, 274
Sodium dithionite
demercurations, **8**, 857
reduction
dienoic carboxylic acids, **8**, 563
imines, **8**, 36
1-methyl-4-carbamoylpyridinium bromide, **8**, 589
pyridines, **8**, 589
Sodium enolates
Claisen rearrangement, **5**, 847
synthesis, **2**, 100
Sodium hexamethyldisilazane
enolate formation, **2**, 182
Sodium hexamethyldisilazide
crystal structure, **1**, 37
Sodium hydride
phosphonium ylide synthesis, **6**, 175
reduction
acyl halides, **8**, 262
cyclic carbonyl compounds, **8**, 14
enones, **8**, 16
epoxides, **8**, 879
quinoline, **8**, 588
unsaturated hydrocarbons, **8**, 485
Sodium hydrogen telluride
reduction, **8**, 880
aromatic compounds, **8**, 370
Sodium hypochlorite
oxidation
organoboranes, **7**, 602
primary arylamines, **7**, 738
secondary alcohols, **7**, 318
Sodium metaperiodate
oxidant
solid support, **7**, 842
Sodium methoxide
oxidant
solid support, **7**, 842
Sodium methylsulfinate
phosphonium ylide synthesis, **6**, 175
Sodium naphthalenide
reductive cleavage
aryl–phosphorus bonds, **8**, 859
Sodium nitrite
oxidation
halides, **7**, 665

reduction
dissolving metals, **8**, 526
Sodium octacarbonylhydridodiferrate
reduction
unsaturated carbonyl compounds, **8**, 550
Sodium perborate
1-hydroxy-1-acetoxyalkene synthesis, **7**, 446
oxidation, **7**, 674
organoboranes, **7**, 602
primary amines, **7**, 737
primary arylamines, **7**, 738
Sodium percarbonate
oxidation
primary amines, **7**, 737
Sodium periodate
oxidation
ethers, **7**, 238
selenides, **7**, 772
sulfides to sulfoxides, **7**, 194
sulfoxides, **7**, 769
oxidative cleavage of alkenes
synthesis of carbonyl compounds, **7**, 564
with catalysts, **7**, 542
oxidative rearrangement
phenols, **7**, 835
Sodium permanganate
oxidation
primary amines, **7**, 737
Sodium persulfate
oxidative decarboxylation, **7**, 722
Sodium–potassium alloy
ester reduction
heterogeneous conditions, **3**, 615
Sodium selenoisocyanate
reaction with trialkylboranes, **7**, 608
Sodium sulfide
reduction
dibromoalkanes, **8**, 806
nitro compounds, **8**, 370
Sodium telluride
synthesis, **8**, 370
Sodium tetracarbonylcobaltate
catalyst
carbonylation of alkyl halides, **3**, 1029
Sodium tetracarbonylferrate
catalyst
carbonylation of alkyl and aralkyl halides, **3**, 1026
halide carbonylation
formation of aldehydes, **3**, 1021
ketone synthesis
carbonylation, **3**, 1024
reduction
anhydrides, **8**, 291, 293
thiol esters, **8**, 293
Sodium tetrachloroaluminate
catalyst
Friedel–Crafts reaction, **2**, 756
Sodium tetraphenylborate
oxidation
organoboranes, **7**, 603
Sodium *p*-toluenesulfonamide
reaction with π-allyl complexes, **6**, 86
Sodium triacetoxyborohydride
reductive amination, **8**, 54
Sodium tri-*t*-butoxyaluminum hydride
reduction
amides, **8**, 271
Sodium triethoxyaluminum hydride
reduction
nitriles, **8**, 274
Sodium triethylborohydride
reduction
isoquinoline, **8**, 583
Sodium trimethoxyborohydride
demercuration, **8**, 853
reduction
acyl halides, **8**, 263
unsaturated carbonyl compounds, **8**, 536
Solanesol
synthesis, **3**, 170
Solasodine
reduction, **8**, 228
Solavetivone
synthesis
via ketocarbenoids, **4**, 1056
Solenopsin
synthesis, **1**, 558
Solenopsin A
synthesis, **6**, 770
Solenopsin B
synthesis, **6**, 771
Solid-supported reagents
oxidation, **7**, 839–847
alumina, **7**, 841
clay, **7**, 845
silica, **7**, 842
Solinopis invicta
trail pheromone component
synthesis, **1**, 568
Solvent cage
electron-transfer oxidation, **7**, 852
Solvent effects
radical reactions, **4**, 720
Solvents
electrochemical oxidation, **7**, 792
radical reactions, **4**, 720
Solvent-separated ion pairs
electron-transfer oxidation, **7**, 851
Sommelet–Hauser rearrangement, **3**, 965; **6**, 854
aminomethylation
aromatic compounds, **6**, 893
asymmetry, **3**, 969
ylidic, **3**, 914
Sommelet–Hauser-type rearrangement
ylides, **4**, 430
Sommelet oxidation
benzaldehydes
synthesis, **7**, 666
Sommelet rearrangement
ammonium ylides, **6**, 834
Sonication
hydroboration, **8**, 716
Sonn–Müller reduction
imidoyl chlorides, **8**, 300
Sonochemistry
nitrile synthesis, **6**, 234
Sorbamide, *N,N*-dialkyl-
addition reactions
with organomagnesium compounds, **4**, 84
Sorbamide, *N,N*-diethyl-
conjugate additions
organomagnesium reagents, **4**, 183

Sorbic acid
sodium salt
hydrogenation, **8**, 450
Southern corn rootworm pheromone
synthesis
via conjugate addition to α,β-unsaturated carboxylic acid, **4**, 202
Soybean lipoxygenase
irreversible inhibitors
synthesis, **3**, 217
Sparteine
ethylmagnesium bromide complex
crystal structure, **1**, 13
Sparteine, 6-benzyl-
ethylmagnesium bromide complex
crystal structure, **1**, 14
Specionin
synthesis, **7**, 301
via ene reaction, **2**, 537
Spectinomycin
Mannich reaction, **2**, 903
synthesis
via Diels–Alder reaction, **2**, 696
Sphingolipids
amino alcohols and, **2**, 323
erythro-Sphingosine
synthesis
via Henry reaction, **2**, 331
threo-Sphingosine
synthesis
via intramolecular Diels–Alder reaction, **5**, 425
Sphingosines
synthesis, **6**, 53
Sphondin
synthesis, **5**, 1096, 1099
regioselective, **5**, 1094
Spiroacetal pheromones
synthesis, **2**, 331
via dihalocarbene insertion, **4**, 1022
Spiroacetals
hydroxymercuration, **8**, 854
synthesis, **3**, 252
via cyclization of enol ethers, **4**, 390
Spiro[*n*,4]alkenones
synthesis
via [3 + 2] cycloaddition reactions, **5**, 285
Spiroannulation
conjugate additions
bisorganocuprates, **4**, 192
Wurtz coupling, **3**, 423
Spiroazepinedione
synthesis
via intramolecular photocycloaddition, **5**, 181
Spirobenzylisoquinoline alkaloids
synthesis
via photoinduced iminium ion–benzylsilane cyclization, **2**, 1040
Spiro compounds
synthesis
via cyclopropane ring opening, **4**, 1043
via radical cyclizations, **4**, 791
Spirocyclizations
N-acyliminium ions, **2**, 1064
polyenes, **3**, 354
Spirocyclobutanone
annulation
via ring expansion, **5**, 919
synthesis
via rearrangement of vinylcyclopropane, **5**, 919
Spirocyclohexa-1,4-diene
oxidative rearrangement, **7**, 833
Spiro[5.5]cyclohexadiene
synthesis
via arene–metal complexes, **4**, 541
Spirocyclohexanone
oxime
Ritter reaction, **6**, 279
Spirocyclopentanes
π-allylpalladium complexes from, **4**, 587
Spirocyclopropanes
synthesis
via dihalocyclopropanes, **4**, 1014
Spiro[4,5]decadienones
synthesis
via vinylsilanes, **1**, 584
Spirodienones
oxygen migration, **3**, 813
synthesis, **3**, 679; **7**, 136
via aryl radical insertion, **3**, 686
via C—C phenol–phenol coupling, **3**, 679
via ketocarbenoids, **4**, 1056
Spirodihydrofuranone
synthesis
via lithium allenes, **2**, 88
Spiroethers
synthesis, **3**, 688
Spiro[2.4]hepta-4,6-diene
cycloaddition reactions
tropones, **5**, 621
Spiro[2.4]heptane, 2-methylene-
synthesis
via metal-catalyzed cooligomerization, **5**, 1195
Spiro[4.11]hexadecenone
synthesis
via [3 + 2] cycloaddition reactions, **5**, 285
Spiroindolenine
synthesis, **6**, 737
Spiroketals
chiral
reaction with silyl enol ethers, **2**, 651
reduction, **8**, 220
synthesis, **8**, 837
via organocerium reagents, **1**, 239
Spirolactones
synthesis
via oxidation of hydroxyalkenes, **7**, 267
Spirolactonization
Reformatsky reaction, **6**, 357
Spiro[3.5]nonanone, 5-methylene-
divinylcyclobutanols from, **5**, 1028
Spiro[4,4]nonatetraene
synthesis, **2**, 710
Spiropyrrolidinones
synthesis
via intramolecular vinyl substitution, **4**, 847
Spirorenone
synthesis
via microbial methods, **7**, 74
Spirothiazines
synthesis
via thiol addition to alkenes, **4**, 317
Spirovetevanes

synthesis
via tandem vicinal difunctionalization, **4**, 242
Sporamine
synthesis, **7**, 536
Spores
oxidation, **7**, 80
Squalene
photocrosslinking reactions
poly(vinylbenzophenone), **5**, 161
synthesis, **3**, 99, 170; **6**, 145; **7**, 87
via arynes, **4**, 507
via iterative rearrangements, **5**, 892
via phosphonium ylides, **3**, 201
via reduction of sulfides, **3**, 107
via 3,2-sigmatropic rearrangement, **3**, 943
via sulfones, **6**, 157
via sulfur ylides, 3,2-rearrangement, **3**, 933
via ylides, **3**, 919
Squalene, 2,3-epoxy-
synthesis, **3**, 126
Squalene, 1-hydroxy-
asymmetric epoxidation, **7**, 409
Squalene, perhydro-
synthesis, **3**, 586
Squalenoids
oxacyclic
synthesis *via exo* alkene cyclization, **4**, 378
synthesis
(*Z*)-selectivity, **1**, 767
Squaric acid
dialkyl esters
cyclobutenones from, **5**, 689
2-Stanna-1,3-dioxolane
synthesis, **3**, 571
Stannane, [2-(acetoxymethyl)-3-allyl]-tri-*n*-butyl-
reactions with crotyl organometallic compounds, **2**, 982
Stannane, acyl-
asymmetric reduction
to α-alkoxy organostannanes, **3**, 196
prochiral
enantioselective reduction, **8**, 164
Stannane, 1-adamantyltrimethyl-
oxidation
formation of tertiary alcohol, **7**, 614
Stannane, (1-alkynyl)tributyl-
acylation
platinum catalyzed, **1**, 447
Stannane, allenyl-
reaction with *N*-acyliminium ions, **2**, 1061
reaction with aldehydes, **2**, 575
reaction with ketones, **2**, 575
synthesis, **2**, 587
Stannane, allylchloro-
acylation
Friedel–Crafts reaction, **2**, 726
Stannane, allyltri-*n*-butyl-
allylation, **4**, 743
reaction with aldimines, **2**, 986
reaction with α-alkylimines, **2**, 981
Stannane, allyltrimethyl-
radical reactions
fragmentation methods, **4**, 744
Stannane, aryltri-*n*-butyl-
dimerization, **3**, 500
Stannane, crotyl-
reaction with aldehydes, **2**, 4
reaction with iminium salts, **2**, 1002
synthesis, **2**, 977
Stannane, crotyltrialkyl-
isomerization, **2**, 6
reactions with aldehydes, **2**, 18
Stannane, crotyltri-*n*-butyl-
reaction with α-alkylamines
syn–anti selectivity, **2**, 989
stability
boron trifluoride etherate, **2**, 977
Stannane, dienylmethyl-
reaction with aldehydes, **2**, 575
Stannane, diphenyl-
reduction
unsaturated carbonyl compounds, **8**, 548
Stannane, ethoxy-α-chloromethyltributyl-
reaction with Grignard reagents
preparation of *O*-ethyl organostannanes, **3**, 196
Stannane, ethynyl-
reaction with aldehydes, **2**, 575
reaction with ketones, **2**, 575
Stannane, α-hydroxy-
synthesis
via enantioselective reduction of acylstannone, **8**, 164
Stannane, γ-hydroxy-
fragmentation, **1**, 894
synthesis
via sequential Michael ring closure, **4**, 262
Stannane, propargyl-
reaction with aldehydes, **2**, 575
reaction with ketones, **2**, 575
synthesis, **2**, 587
Stannane, pyridyl-
coupling reactions
with bromopyridine, **3**, 510
Stannane, β-silylvinyl-
coupling reactions
with alkenyl halides, **3**, 495
Stannane, α-sulfonylalkyl-
coupling reactions
with alkenyl halides, **3**, 443
Stannane, 4-tetrahydropyranyloxyphenyltrimethyl-
reaction with *N,N*-dimethylmethyleniminium chloride, **2**, 962
Stannane, 3-thienyltrimethyl-
Mannich reaction, **2**, 963
Stannane, trialkyl-
reduction
unsaturated nitriles, **8**, 548
Stannane, triaryl-
reduction
unsaturated nitriles, **8**, 548
Stannane, tri-*n*-butyl-
decyanation
isocyanides, **8**, 830
deoxygenation
thioesters, **8**, 818
radical reduction
allylic groups, **8**, 969
reduction, **8**, 961
acyl halides, **8**, 264
alkyl halides, **8**, 798
unsaturated carbonyl compounds, **8**, 548
Stannane, tributyldeutero-

reduction
alkyl halides, **8**, 798
Stannane, tributyltritio-
reduction
alkyl halides, **8**, 798
Stannane, triphenyl-
reduction
acyl halides, **8**, 264
carbonyl compounds, **8**, 322
isocyanates, **8**, 74
unsaturated carbonyl compounds, **8**, 548
Stannane, vinyl-
coupling reactions
butadiene synthesis, **3**, 483
with alkenyl halides, **3**, 495
with vinyl iodides, **3**, 488
oxidation, **7**, 620
radical cyclizations, **4**, 799
radical reactions
fragmentation methods, **4**, 743–746
reaction with aldehydes, **2**, 575
reaction with carbonyl compounds, **2**, 563–590
reaction with ketones, **2**, 575
synthesis, **2**, 588
Stannane, [(2,6-xylylimino)(trialkylsilyl)methyl]-
transmetallation, **1**, 546
Stannanes
acylation
Lewis acid catalyzed, **2**, 726
non-Lewis acid catalyzed, **2**, 727
dehalogenation
α-halocarbonyl compounds, **8**, 991
reduction
acyl halides, **8**, 264
esters, **8**, 824
O-thiocarbonyl compounds, **8**, 818
synthesis, **2**, 582
via Ireland silyl ester enolate rearrangement, **5**, 841
toxicity, **8**, 800
Stannanes, alk-1-ynyltrialkyl-
oxidation, **7**, 620
Stannanes, allyl-
π-allylpalladium complexes from, **4**, 588
conjugate additions to α,β-enones
Lewis acid catalyzed, **4**, 155
[3 + 2] cycloaddition reactions
with acyliron complexes, **5**, 277
electrophilic substitutions
allylic rearrangement, **6**, 832
Lewis acid catalyzed reactions
regiospecificity, **2**, 565
oxidation, **7**, 616
Prins reaction
mechanism, **2**, 564
radical cyclizations, **4**, 799
radical reactions
fragmentation methods, **4**, 743–746
reaction with acetals, **2**, 578
reaction with *N*-acyliminium ions, **2**, 1060, 1064, 1067
reaction with aldehydes, **2**, 572
reaction with aldimines
syn–anti selectivity, **2**, 983, 991
reaction with amines, **2**, 1002
reaction with carbonyl compounds, **2**, 563–590
reaction with imines, **2**, 976
reaction with iminium salts, **2**, 1002
reaction with ketals, **2**, 578
reaction with ketones, **2**, 572
reaction with vinyloxiranes
regioselectivity, **5**, 936
synthesis, **2**, 587
Stannanes, α-amino-
transmetallation, **1**, 476, 479
Stannanes, cyclohexenyl-
hydroxylation, **7**, 616
synthesis
via Diels–Alder reaction, **5**, 335
Stannanes, dialkoxy-
diol protection, **6**, 662
Stannanes, 1,2-epoxy-
synthesis
via oxidation of vinylstannanes, **7**, 620
Stannanes, tetraalkyl-
oxidation
chromium trioxide, **7**, 614
Stannanes, tetrasubstituted
synthesis, **1**, 445
γ-Stannyl alcohols
cyclic
1,4-fragmentation, **7**, 621
1,3-eliminative cyclization
formation of cyclopropanes, **7**, 621
γ-Stannyl alcohols, trialkyl-
oxidation, **7**, 621
Stannylcupration
alkynes, **4**, 901
Stannylene, dialkyl-
reactions with polyols, **6**, 18
O-Stannyl ketene acetal
formation
by 1,4-hydrostannation, **2**, 609
Stannyl thiolates
polythiolactone synthesis, **6**, 441
Statine
analogs
synthesis *via* aldol reaction, **2**, 223
synthesis
via N-acyliminium ions, **2**, 1060
Staudinger reaction
heterocyclic synthesis, **6**, 759
5*H*-thieno[2,3-*c*]pyrrole synthesis, **2**, 378
Stearic acid, *trans*-2-epoxy-
methyl ester
Ritter reaction, **6**, 271
Steganacin
synthesis, **1**, 566
via vanadium oxytrifluoride, **3**, 674
Stegane
synthesis
via vanadium oxytrifluoride, **3**, 675
Steganone
synthesis, **3**, 150, 501
via cyclobutene ring expansion, **5**, 687
via ring expansion, **3**, 674
via thallium trifluoroacetate, **3**, 673
Stemodin
synthesis, **3**, 717
Stemodinone, deoxy-
synthesis
via samarium diiodide, **1**, 259
Stemodione, 2-deoxy-

synthesis
via ene reaction, **2**, 545
Stephen reduction
nitriles, **8**, 298
Stephens–Castro coupling
alkynic ketones, **3**, 226
copper acetylide intermediates, **3**, 217
Stereodifferentiation
double, **2**, 32
Stereoelectronics
reactions of chiral carbonyl compounds with nucleophiles, **2**, 24
Stereoselective synthesis
allyl organometallics
uncatalyzed, **2**, 1
Stereoselectivity
aldol reaction
kinetic and thermodynamic control, **2**, 154
syn–anti
in enolate–imine condensations, **2**, 918
Stereospecificity
anti
epoxide ring opening, **3**, 733
Sterepolide
synthesis, **7**, 246
Sterigmatocystin, dihydro-*O*-methyl-
synthesis
via Friedel–Crafts reaction, **2**, 760
Steroid-5-enes
addition reactions
nitrosyl chloride, **7**, 500
Steroids
A-ring aromatic
synthesis *via* [2 + 2 + 2] cycloaddition, **5**, 1151
aromatic
synthesis, **3**, 366
B-ring aromatic
synthesis *via* [2 + 2 + 2] cycloaddition, **5**, 1151
carbonyl compounds
NMR, **1**, 293
hydroxylation
metalloporphyrin, **7**, 50
iodoaryl esters
radical relay chlorination, **7**, 46
ketones
dehydrogenation, **7**, 132
dehydrogenation, selenium dioxide, **7**, 128
oxidation, **7**, 675
exo-methylene
epoxides, opening, **3**, 743
microbial dehydrogenation, **7**, 145
microbial oxidation, **7**, 66
nonaromatic, synthesis
polyene cyclization, **3**, 369
synthesis
Sarett, **2**, 158
via Dieckmann reaction, **2**, 823
via 1,5-diketone cyclization, **2**, 163
via Ireland silyl ester enolate rearrangement, **5**, 841
via palladium catalyzed oxidation, **7**, 460
via polyene cyclization, **3**, 362
Woodward's, **2**, 156
total synthesis
1,5-diketone cyclization, **2**, 162
unsaturated
hydrofluorination, **4**, 271
Steroids, 17α-bromo
rearrangements, **3**, 846
Steroids, 21-bromo
rearrangements, **3**, 846
Steroids, *trans*-β-cyanohydroxy
synthesis
via epoxides, **6**, 237
Steroids, halo
ring A contractions, **3**, 854
Steroids, 19-hydroxy
synthesis
via microbial methods, **7**, 74
Steroids, Δ^4-3-keto
microbial hydroxylation, **7**, 72
Steroids, 11-keto
homochiral
synthesis, **4**, 218
reduction
dissolving metals, **8**, 118
Steroids, 12-keto
reduction
dissolving metals, **8**, 119
dissolving metals/ammonia, **8**, 112
Steroids, nitro
reduction, **8**, 374
Sterpurene
synthesis, **3**, 402, 714
Sterpuric acid
synthesis, **7**, 164
Stevens rearrangement, **3**, 913–971; **6**, 854
ammonium ylides, **6**, 834
benzyldimethyl(trimethylsilylmethyl)ammonium halides, **4**, 430
Steviol
rearrangement, **3**, 715
Stibides
$S_{RN}1$ reactions, **4**, 474
Stibine, diphenyl-
selective ketone reduction, **8**, 18
Stibonium triflate, tetraphenyl-
oxirane ring-opening, **6**, 89
4-Stilbazole, 3-cyano-
Ritter reaction, **6**, 279
Stilbene, *o*-bromo-
photocyclization, **5**, 724
Stilbene, cyano-
synthesis, **1**, 561
Stilbene, difluoro-
hydrogenation, **8**, 896
Stilbene, α-fluoro-
hydrogenation, **8**, 896
Stilbene oxide
deoxygenation, **8**, 886
reaction with Grignard reagents
epoxide ring opening, **3**, 755
Stilbene oxide, α-cyano-
ring opening
carbonyl ylide generation, **4**, 1090
Stilbenes
cleavage by sodium hydrazide, **7**, 506
nitro addition reactions, **7**, 488
oxidation
osmium tetroxide, **7**, 441
solid support, **7**, 841
photocyclization, **5**, 723
bornyl fumarate, **5**, 132

reduction
hydrazine, **8**, 568
synthesis, **3**, 497
via Horner reaction, **1**, 776
via Knoevenagel reaction, **2**, 362
via palladium catalysts, **4**, 840
via Ramberg–Bäcklund rearrangement, **3**, 864, 865
Stilbenes, chloro-
hydrogenation, **8**, 899
peroxy acid reaction
epoxides as reactive intermediates, **3**, 739
Stilbestrol, diethyl-
photolysis, **5**, 723
Stiles' reagent
ketone carboxylation
methylmagnesium carbonate, **2**, 841
Stille acylation
rate
factors affecting, **1**, 442
Still–Wittig rearrangement, **3**, 983; **6**, 879
ethers, **6**, 875
Stobbe reaction
succinic esters
deprotonation, **6**, 355
Stork enamine reaction, **3**, 28
Stork–Eschenmoser hypothesis
polyalkene cyclization, **3**, 341
Strecker synthesis
amino acids, **1**, 460
Streptazoline
synthesis
via *N*-acyliminium ions, **2**, 1064
Streptogramin
synthesis
via Peterson alkenation, **1**, 791
Streptonigrin
synthesis, **1**, 560; **7**, 347
via Curtius reaction, **6**, 814
via Diels–Alder reaction, **5**, 406, 492
Streptovaricin
synthesis
(*Z*)-selectivity, **1**, 764
Strictosidine
derivatives
synthesis *via* Knoevenagel reaction, **2**, 373
Strigol
synthesis
via Raphael–Nazarov cyclization, **5**, 779
Strontium
reduction
ammonia, **8**, 113
Styrene, bromo-
catalytic hydrogenation, **8**, 900
reaction with aldehydes
chromium(II) chloride, **1**, 193
Styrene, *t*-butyl peroxy-
synthesis, **8**, 855
Styrene, α-cyclopropyl-
[3 + 2] cycloaddition reactions
with 2,4-dibromopentan-3-one, **5**, 283
Styrene, *cis*-β-deuterio-
[3 + 2] cycloaddition reactions
iron catalyzed, **5**, 285
Styrene, dicyano-
oxidative cleavage
synthesis of dithioacetal, **7**, 588
Styrene, α-ethoxy-
reduction, **8**, 937
Styrene, 4-methoxy-
[2 + 2] cycloaddition reactions, **5**, 73
hydroboration, **8**, 713
reaction with tetracyanoethylene, **5**, 71
solvent effects, **5**, 75
Styrene, α-methyl-
[3 + 2] cycloaddition reactions
with tetramethyldibromo ketones, **5**, 283
hydroesterification, **4**, 945
Styrene, β-methyl-
epoxidation, **7**, 383
oxidation, **7**, 464
Styrene, *trans*-β-methyl-*p*-methoxy-
reaction with tetracyanoethylene
solvent effects, **5**, 76
thermochemistry, **5**, 76
Styrene, β-methyl-β-nitro-
reduction, **8**, 376
Styrene, β-nitro-
conjugate additions, **4**, 224
synthesis
via acid catalysis, **2**, 326
via nitryl iodide to alkene, **4**, 357
Styrene, 2-nitro-
hydroformylation, **4**, 926
Styrene, 4-nitro-
reduction, **8**, 364
Styrene, pentafluoro-
hydroformylation, **4**, 927
Styrene, β-tetrahydropyranyl-
oxidation
regioselectivity, **7**, 464
Styrene, 4-(2-thienylcarbonyl)-
hydroformylation, **4**, 932
Styrene, trifluoro-
dimerization, **5**, 64
Styrene, 4-(trifluoromethyl)-
hydroboration, **8**, 713
Styrene oxide
optically pure
synthesis, **1**, 833
reaction with organocopper compounds, **3**, 224
rearrangement, lithium halide catalyzed, **3**, 764
rearrangement, lithium perchlorate catalyzed, **3**, 761
reduction
lithium aluminum hydride, **8**, 875
synthesis, **7**, 423
Styrene oxide, β-methyl-
reduction
lithium aluminum hydride, **8**, 872
Styrenes
anodic oxidation, **7**, 796
carboalumination, **4**, 887
cleavage by sodium hydrazide, **7**, 506
conjugated
partial reduction, **8**, 564
cyclobutanones from, **5**, 1087
cyclopropanation, **4**, 1035
dimerization, **5**, 63
hydration, **4**, 298
hydroboration, **8**, 704, 718
hydroesterification
palladium catalyst, **3**, 1030
hydroformylation, **4**, 919, 930–932

platinum catalysts, **3**, 1022
hydrogenation
homogeneous catalysis, **8**, 453
hydrometallation, **8**, 672
hydrosilylation
asymmetric, **8**, 783
hydrozirconation
regioselectivity, **8**, 685
oxidation
Wacker process, 7, 451, 452
oxidative rearrangement
solid support, **7**, 845
partial reduction, **8**, 523–568
Pauson–Khand reaction, **5**, 1045
synthesis, **3**, 495
via Friedel–Crafts reaction, **3**, 294
via vinylic coupling, **3**, 485
Vilsmeier–Haack reaction, **2**, 782
Substitution, radical nucleophilic, unimolecular
reactions, **4**, 463–476
aromatic substrates, **4**, 458
association, **4**, 453
cyclizations, **4**, 476–480
definition, **4**, 452
fragmentation, **4**, 454
intramolecular
ring closure, **4**, 476
mechanism, **4**, 452–462
nucleofuges, **4**, 457
photostimulated, **4**, 452
propagation, **4**, 453
reviews, **4**, 452
solvents, **4**, 456
termination, **4**, 455
Succinaldehyde
3-substituted esters
synthesis *via* conjugate addition to imidazoline, **4**, 207
Succinaldehyde, 3-alkyl-
methyl esters
synthesis *via* copper catalyzed Grignard additions, **4**, 93
Succinic acid
diesters, dianion enolates
stereochemistry, **2**, 103
diethyl ester
disilyl ketene acetals, **2**, 606
2,3-disubstituted
synthesis, **3**, 638
Succinic acid, α-benzyl-β-phenyl-
synthesis, **3**, 828
Succinic acid, 2-methyl-
dimethyl ester
intramolecular acyloin coupling reaction, **3**, 621
Succinic anhydride
disilyl enol ethers
synthesis, **2**, 607
hydrogenation, **8**, 239
Succinimide, *N*-benzenesulfonyloxy-
Lossen reaction, **6**, 822
Succinimide, *N*-bromo-
activator
DMSO oxidation of alcohols, **7**, 299
addition reactions
alkenes, **7**, 500
alkane bromination, **7**, 16
allylic oxidation, **7**, 112
oxidation
aldehydes, **6**, 308
secondary alcohols, **7**, 318
Succinimide, *N*-chloro-
activator
DMSO oxidation of alcohols, **7**, 299
decarboxylative halogenation, **7**, 724
diisopropyl sulfide
oxidation of secondary diols, **7**, 318
oxidation
primary alcohols, **7**, 309
sulfide chlorination
formation of α-chlorosulfides, **7**, 207
Succinimide, *N*-iodo-
oxidative cleavage, **7**, 706
Succinimide, *N*-methyl-
reduction, **8**, 254
Succinimide, *N*-methyl-2-hydroxy-
Diels–Alder reactions, **5**, 365
Succinimides
Tebbe reaction, **1**, 745
Succinimides, *exo*-methylene-
synthesis
via [2 + 2 + 2] cycloaddition, **5**, 1141
Succinoin, di-*t*-butyl-
synthesis
via intramolecular acyloin coupling reaction, **3**, 621
Succinonitrile, diimino-
Diels–Alder reactions, **5**, 486
Sugar aldehydes
Knoevenagel reaction, **2**, 385
Wittig reaction, **1**, 759
Sugar dialdehydes
Henry reactions
cyclization, **2**, 328
Sugars
acetals and acetates of
reaction with allylsilanes, **2**, 577
anhydro
glycosyl donors, **6**, 48
branched
synthesis, **2**, 139
bromides
reaction with ethynylstannanes, **2**, 582
chlorination
displacement of hydroxy group, **6**, 205
2-deoxy
glycosides, **6**, 59
saccharides, **6**, 59
Kiliani–Fischer synthesis, **1**, 460
synthetic application
Knoevenagel reaction, **2**, 385
thioacetals
reaction with allylstannanes, **2**, 581
Sugars, amino
synthesis
via cyclofunctionalization, **4**, 375, 400
Sugars, aminodeoxy
synthesis
via Henry reaction, **2**, 330
Sugars, 2-azido-2-deoxy-
glycoside synthesis, **6**, 42
Sugars, branched-chain
synthesis
via Paterno–Büchi reaction, **5**, 158

Sugars, deoxyamino-
  synthesis
    via Peterson methylenation, **1**, 732
Sugars, *C*-methyldeoxy-
  synthesis, **8**, 694
Sugars, 3-*C*-methylene-
  synthesis
    via Peterson methylenation, **1**, 732
Sulcatol
  formation of
    tetrahydro-2,2,6-trimethyl-2*H*-pyran-3-ol, **7**, 634
  synthesis
    enzymatic resolution, **6**, 340
Sulfamates, *N,N*-dimethyl-
  catalytic hydrogenation, **8**, 817
Sulfamides
  synthesis
    via amines, **7**, 739
Sulfamides, diaryl
  reactions with organometallic compounds, **1**, 390
Sulfate esters
  cyclic
    synthesis, **7**, 431
Sulfates
  chlorination
    nucleophilic displacement, **6**, 206
Sulfenamides
  ketone sulfenylation, **7**, 125
  synthesis
    via sulfenylation of primary amines, **7**, 741
Sulfenamides, nitroaryl-
  synthesis, **7**, 483
Sulfenates
  rearrangements
    chirality transfer, **6**, 899
    diastereoselectivity, **6**, 900
    from sulfoxides, **6**, 899
    stereochemistry, **6**, 899
Sulfenates, propargylic
  rearrangement, **6**, 903
Sulfenes
  Diels–Alder reactions, **5**, 440–442
Sulfenimide, trityl-
  reaction with aldehydes, **2**, 940
Sulfenimides
  amine synthesis, **6**, 83
  reactions with allylboronates, **2**, 15
Sulfenimine, phenyl-
  reactions with allyl organometallics
    diasteroselective, **2**, 32
Sulfenimines
  reactions with allyl organometallic compounds, **2**, 998
    Cram selectivity, **2**, 998, 999
  reactions with enolates, **2**, 940
Sulfenimines, *S*-aryl-
  reactions with organometallic compounds, **1**, 389
Sulfenimines, *S*-trityl-
  reduction
    sodium cyanoborohydride, **8**, 74
  synthesis
    via condensation of aldehydes with
      tritylsulfenamide, **2**, 940
Sulfenylation
  amines, **7**, 741
  esters, **7**, 125
Sulfenyl chlorides
  reactions with phosphonium ylides, **6**, 177
  tandem vicinal difunctionalization, **4**, 262
2-Sulfenyl compounds, 1-amido-
  synthesis, **7**, 494
Sulfenyl groups
  carbonyl compounds, **7**, 124
Sulfenyl halides
  reactions with alkenes, **4**, 330–337
Sulfhydrolysis
  imidates, **6**, 450
  imidothioates, **6**, 455
Sulfide, benzyl *t*-butyl
  chlorination
    regioselectivity, **7**, 212
Sulfide, benzyl ethyl
  chlorination
    regioselectivity, **7**, 210
Sulfide, benzyl isopropyl
  chlorination
    regioselectivity, **7**, 210
Sulfide, benzyl *p*-methoxybenzyl
  chlorination
    regioselectivity, **7**, 212
Sulfide, benzyl methyl
  chlorination, **7**, 210
Sulfide, benzyl *p*-methylbenzyl
  chlorination
    selectivity, **7**, 212
Sulfide, bis(α-bromobenzyl)
  dehydrogenation
    ylide generation for [4 + 3] cycloaddition, **5**, 600
Sulfide, chloro cyclopropyl
  synthesis
    via sulfide chlorination, **7**, 209
Sulfide, chloromethyl phenyl
  reaction with silyl enol ethers, **6**, 141
  reaction with silyl ketene acetal
    regioselectivity, **2**, 617
  synthesis, **7**, 212
Sulfide, crotyl phenyl
  chlorination, **7**, 210
Sulfide, diisopropyl
  oxidation
    primary alcohols, **7**, 309
Sulfide, dimethyl
  chlorine activator
    DMSO oxidation of alcohols, **7**, 297
  diborane complex
    carboxylic acid reduction, **8**, 261
  oxidative cleavage
    alkenes, ozone, **7**, 544
  solvent
    alkylcopper compound reactions, **3**, 210
Sulfide, di-*n*-propyl
  oxidation
    4-(dimethylamino)pyridinium chlorochromate, **7**, 269
Sulfide, ethyl methyl
  chlorination
    regioselectivity, **7**, 212
Sulfide, 1-methoxycyclopropyl phenyl
  reductive lithiation, **6**, 146
Sulfide, methyl phenyl
  Friedel–Crafts acylation, **2**, 741
Sulfide, 1-naphthyl ethyl
  desulfurization, **8**, 914

Sulfide, 1-naphthyl isopropyl
desulfurization, **8**, 914
Sulfide, 1-naphthyl phenyl
desulfurization, **8**, 914
Sulfide, 1-(trimethylsilyl)cyclopropyl phenyl
reductive lithiation, **6**, 145
Sulfide contraction — *see* Eschenmoser coupling reaction
Sulfides
alkenes from, **3**, 114
alkyl and allyl halides from, **3**, 118
alkylated
use in synthesis, **3**, 106
anions
reaction with boranes, **3**, 795
annulation
stereospecific, **6**, 144
benzylic
reduction, **8**, 964
carbanions
crystal structure, **1**, 36
chemoselective epoxidation, **7**, 384
cleavage
metal–ammonia, **8**, 531
Darzens glycidic ester condensation, **2**, 417
desulfurization, **8**, 836, 842
LAH–$CuCl_2$, **8**, 840
tin hydrides, **8**, 846
α-halogenation, **7**, 206
regioselectivity, **7**, 210
heteroaromatic
coupling reactions with sp^3 organometallics, **3**, 459
α-heterosubstituted
carbonyl compound synthesis from, **3**, 141
γ-lithiated
synthesis, **4**, 869
metallation
use of additives, **3**, 86
Michael addition
stereospecific, **6**, 144
oxidation, **7**, 124
asymmetric, **6**, 150
bipyridinium chlorochromate, **7**, 267
pyridinium chlorochromate, **7**, 267
solid support, **7**, 842, 843
to sulfoxides, **7**, 193, 762
reactions with alkenes, **4**, 316
reactions with π-allylpalladium complexes
stereochemistry, **4**, 624
rearrangements, **6**, 892
diastereoselectivity, **6**, 893
regioselectivity, **6**, 893
reduction
use in synthesis, **3**, 106
synthesis, **6**, 133–167
via oxidative cleavage of alkenes, **7**, 542
tandem vicinal difunctionalization, **4**, 257
α-thiometallation, **3**, 196
Wittig rearrangement, **3**, 978
Sulfides, β-acetamidinovinyl
synthesis
via alkynes, **4**, 336
Sulfides, acetamido
synthesis, **7**, 494
Sulfides, α-acetoxy
synthesis
via Pummerer rearrangement to carbohydrates, **7**, 196
Sulfides, alkenyl
coupling reactions
with sp^3 organometallics, **3**, 446
synthesis
via metal carbene complexes, **1**, 808
Sulfides, alkoxyaryl alkyl
synthesis, **4**, 441
Sulfides, alkyl
ionic halogenation
mechanism, **7**, 195
oxidation, **7**, 193
synthesis
via Pummerer rearrangement, **7**, 199
Sulfides, alkyl aryl
desulfurization, **8**, 847
synthesis, **4**, 444; **7**, 726
via $S_{RN}1$ reaction, **4**, 474
Sulfides, alkyl 2-pyridyl
synthesis, **7**, 726
Sulfides, alkyl vinyl
carbonyl compounds from, **3**, 120
Sulfides, 1-alkynyl
metallation, **3**, 106
Sulfides, alkynyl allyl
sigmatropic rearrangement
synthesis of thioketenes, **6**, 426
Sulfides, alkynyl silyl
thioacylation, **6**, 426
Sulfides, allyl
oxidation, **3**, 116
radical addition reactions
irradiation, **4**, 745
reaction with allylic bromides, **6**, 145
reduction
selectivity, **3**, 107
2,3-sigmatropic rearrangement, **6**, 846
synthesis, **7**, 517
via β-hydroxyalkyl selenides, **1**, 705
use in synthesis, **6**, 138
Sulfides, allyl benzyl
metallation
selectivity, **3**, 99
Sulfides, allyl phenyl
chlorination, **7**, 209
Sulfides, allyl 2-pyridyl
reduction, **3**, 108
Sulfides, amino
synthesis, **7**, 495
Sulfides, *o*-aminobenzyl
desulfurization, **8**, 976
Sulfides, aryl
coupling reactions
with Grignard reagents, **3**, 456
synthesis
via Pummerer rearrangement, **7**, 199
Sulfides, aryldiazo phenyl
$S_{RN}1$ reactions, **4**, 471
Sulfides, α-azido
synthesis
via thioketals, **6**, 254
Sulfides, benzothiazolyl alkyl
desulfurization
tin hydrides, **8**, 846
Sulfides, benzyl α-chlorobenzyl

Ramberg–Bäcklund rearrangement, **3**, 870
Sulfides, benzyl α,α-dichlorobenzyl
Ramberg–Bäcklund rearrangement, **3**, 870
Sulfides, benzylic
in zearalenone synthesis, **6**, 137
use in synthesis, **6**, 138
Sulfides, bis-
reaction with vinylmagnesium halides
regioselectivity, **3**, 493
Sulfides, bis(β-chloroethyl)
synthesis
via electrophilic addition, **4**, 330
Sulfides, bis(trimethylsilyl)
reaction with bromine, **4**, 331
Sulfides, α-bromosilyl silyl
bromo-desilylation
thiocarbonyl ylide generation, **4**, 1095
Sulfides, *t*-butyl
thiol protection, **6**, 664
Sulfides, β-carbonyl aryl
Knoevenagel reaction, **2**, 363
Sulfides, α-chloro
cyclic
synthesis, **6**, 142
in synthesis, **7**, 214
solvolysis, **7**, 214
stereoselective synthesis, **6**, 142
synthesis, **7**, 212
via sulfide chlorination, **7**, 206
vicinal functionalization
alkenes, **6**, 141
Sulfides, α-chlorophenacyl phenacyl
Ramberg–Bäcklund rearrangement, **3**, 870
Sulfides, α-cyano
synthesis
via thioacetals and thioketals, **6**, 238
Sulfides, β-cyano aryl
Knoevenagel reaction, **2**, 363
Sulfides, cycloalkyl phenyl
synthesis, **3**, 88
Sulfides, cyclopropyl phenyl
reaction with butyllithium, **6**, 143
Sulfides, dialkyl
reactions with arynes, **4**, 507
synthesis, **7**, 607
Sulfides, diaryl
synthesis, **4**, 457
via $S_{RN}1$ reaction, **4**, 474
unsymmetrical
synthesis, **4**, 443
Sulfides, diazo
nitrile synthesis, **6**, 240
Sulfides, 1,3-dienyl
alkylation, **3**, 105
Sulfides, α,α-dihalo
hydrolysis
synthesis of thiol esters, **6**, 444
Sulfides, divinyl
electrocyclic ring closure
thiocarbonyl ylide generation, **4**, 1093
Sulfides, α,β-epoxy-
synthesis
via Darzens glycidic ester condensation, **2**, 417
Sulfides, β-fluoro
synthesis
via alkenes, **4**, 331
Sulfides, haloalkyl phenyl
rearrangement, **3**, 88
Sulfides, homoallylic
alkylation
palladium(II) catalysis, **4**, 573
Sulfides, hydroxy-
elimination reactions, **3**, 786
Sulfides, β-hydroxy
oxidation
solid support, **7**, 841
pinacol-type reactions, **1**, 861
rearrangement, **3**, 784
semipinacol rearrangements, **3**, 777, 778
synthesis
via reduction of β-keto sulfides, **8**, 12
Sulfides, β-keto
Knoevenagel reaction
stereochemistry, **2**, 363
reduction, **8**, 12
synthesis
via silyl enol ethers, **6**, 141
Sulfides, β-ketophenyl
synthesis
via alkynes, **4**, 336
Sulfides, α-lithio
anions
epoxidation, **1**, 827
Sulfides, α-metallovinyl
alkylation, **3**, 104
Sulfides, α-methoxyalkenyl phenyl
carbonyl compound synthesis from, **3**, 141
Sulfides, α-methoxy allyl
α-methylenated acyl anion equivalent, **3**, 144
Sulfides, methyl
desulfurization, **8**, 958
Sulfides, 3-methyl-2-butenyl phenyl
allylic carbanions, **1**, 508
Sulfides, methyl (trimethylsilyl)methyl
Peterson alkenation, **1**, 787
Sulfides, β-nitro
synthesis
via alkenes, **7**, 493
Sulfides, phenyl (trimethylsilyl)methyl
Peterson alkenation, **1**, 787
Sulfides, 2-pyridyl
coupling reactions
with Grignard reagents, **3**, 460
Sulfides, α-silylalkyl phenyl
carbonyl compound synthesis from, **3**, 141
Sulfides, thioacyl diphenylthiophosphinyl
thioacylation
thiols, **6**, 454
Sulfides, α-thiomethylcyanomethyl
Wittig rearrangement, **3**, 978
Sulfides, trimethylsilyl alkyl
reactions with π-allylpalladium complexes
regioselectivity, **4**, 642
Sulfides, vinyl
Diels–Alder reactions, **5**, 326
Paterno–Büchi reaction
with benzophenone, **5**, 160
reaction with alkenylaluminum, **3**, 492
reaction with Grignard reagents
nickel catalysts, **3**, 229
synthesis, **7**, 517
Sulfilimine, diphenyl-

reaction with alkenes, **7**, 470
Sulfimides
oxidation
synthesis of nitroso compounds, **7**, 752
Sulfinamides
unsaturated
synthesis, **6**, 841
Sulfinate, menthyl
sulfoxide synthesis
optically active, **6**, 148
Sulfinates
arylation
palladium complexes, **4**, 858
vinyl substitutions
palladium complexes, **4**, 842
Sulfinates, phenyl-
reaction with π-allylpalladium complexes
stereochemistry, **4**, 624
Sulfine, α-chloro-
Diels–Alder reactions, **5**, 441
Sulfines
Diels–Alder reactions, **5**, 440–442
Sulfines, α-oxo-
Diels–Alder reactions, **5**, 441
Sulfinic acids
synthesis
via thiols, **7**, 759
Sulfinic acids, allylic
fragmentation, **6**, 866
to terminal alkenes, **6**, 842
retro-ene reactions, **5**, 424
Sulfinyl anions
chiral
conjugate additions, **4**, 226
Sulfinyl chlorides
tandem vicinal difunctionalization, **4**, 262
Sulfinyl compounds, 1,3-dicarbonyl-2-phenyl-
pyrolysis, **2**, 388
Sulfinyl compounds, α,β-unsaturated
synthesis
via Knoevenagel reaction, **2**, 363
N-Sulfinyl dienophiles
Diels–Alder reactions, **5**, 422
intramolecular, **5**, 425
Sulfite esters
cyclic
asymmetric dihydroxylation, **7**, 431
Sulfites
aromatic nucleophilic substitution, **4**, 443
4-Sulfobenzyl esters
carboxy-protecting groups
cleavage, **6**, 668
Sulfolane, 3-methyl-
solvent
Wacker oxidation, **7**, 450
3-Sulfolene
reaction with alkyl iodides
selectivity, **3**, 172
Sulfolenes
1,3-dienes from, **3**, 173
Sulfonamide, phenacyl-
reduction
dissolving metals, **8**, 994
Sulfonamides
amidomercuration, **4**, 295
Darzens glycidic ester condensation
phase-transfer catalysis, **2**, 429
deamination, **8**, 828
desulfurization, **8**, 836
reactions with π-allylpalladium complexes, **4**, 598
Sulfonamides, alkyl
alkylation, **3**, 179
Sulfonamides, chloro-
adducts
dimethylformamide, **6**, 490
Sulfonamides, *N*,*N*-dibromo-
reactions with alkenes, **7**, 483
Sulfonamides, *N*,*N*-dihalo-
addition reactions
alkenes, **7**, 499
Sulfonamides, homoallylic
synthesis
via retro-ene reactions, **5**, 425
Sulfonamides, *N*-sulfinyl-
Diels–Alder reactions
dienes, **5**, 403
Sulfonamidomercuration
alkenes, **8**, 856
Sulfonates
alkylation
vinyl carbanions, **3**, 242
alkyl esters
alkylation, **6**, 23
amine alkylation, **6**, 72
cyclic
alcohol synthesis, **6**, 19
iodination, **6**, 214
nitrile synthesis, **6**, 235
reduction
lithium aluminum hydride, **8**, 812
Sulfonates, alkyl
alkylation, **3**, 179
reactions with carboxylates
inversion of alcohols, **6**, 21
Sulfonates, allylic
reduction, **8**, 974
Sulfonates, *S*-(dialkylaminomethyl)dithio-
iminium salts
generation *in situ*, **1**, 370
Sulfonates, dialkylaminotrifluoro-
halogen transfer agents
acid fluoride synthesis, **6**, 307
Sulfonation
alcohols
hydroxy group activation, **6**, 18
Sulfone, allyl phenyl
1,1-dilithiated
reaction with benzaldehyde, **2**, 76
Sulfone, benzyl phenyl
lithium salt
crystal structure, **1**, 528
Sulfone, α-bromoethyl ethyl
Ramberg–Bäcklund rearrangement, **3**, 861
Sulfone, 1,3-butadienyl tosyl
reaction with dialkylcuprates, **6**, 161
Sulfone, α-chloroethyl ethyl
Ramberg–Bäcklund rearrangement, **3**, 861
Sulfone, chloromethyl phenyl
aromatic nucleophilic substitution, **4**, 432
Darzens-type reactions, **1**, 530
epoxidation, **1**, 827

reaction with quinoxaline, **4**, 432
Sulfone, dibenzyl
Ramberg–Bäcklund rearrangement, **3**, 864
Sulfone, ethynyl *p*-tolyl
ene reactions
Lewis acid catalysis, **5**, 8
Sulfone, methoxymethyl phenyl
anions
reaction with cyclic ketones, **3**, 785
lithium anion
addition to ketones, **1**, 865
Sulfone, methyl α-bromovinyl
Diels–Alder reactions, **5**, 324
Sulfone, methyl methylthiomethyl
alkylation, **3**, 139
Sulfone, methyl phenyl
alkylation, **3**, 159
Sulfone, methylthiomethyl *p*-tolyl
methylthiomethyl ester
alkylation, **3**, 136
Sulfone, methylthiomethyl *p*-tosyl
alkylation, **3**, 139
Sulfone, phenylthiomethyl phenyl
anions
reaction with cyclic ketones, **3**, 785
Sulfone, phenyl (trimethylsilyl)methyl
Peterson alkenation, **1**, 787
Sulfone, α-triflyldimethyl
alkylation, **3**, 177
Sulfone acetals, γ-oxo
acylation, **6**, 159
Sulfones
alkylation, **3**, 158
carbanions
crystal structure, **1**, 36
chlorination
mechanism, **3**, 864
cyclic
diene protection, **6**, 690
Darzens glycidic ester condensation, **2**, 415
phase-transfer catalysis, **2**, 429
desulfurization, **8**, 837
chemoselective, **8**, 836
metal–ammonia, **8**, 842
Friedel–Crafts cyclization, **6**, 165
hydrobromination, **4**, 282
hydrogenolysis, **8**, 914
hydroiodination, **4**, 288
(*E*)-isomers
synthesis *via* Knoevenagel reaction, **2**, 363
Julia coupling
carbonyl compounds, **1**, 806
Knoevenagel reaction
activated methylenes, **2**, 362
synthesis, **6**, 133–167
via sulfoxides, **7**, 766
tandem vicinal difunctionalization, **4**, 251
use in synthesis, **6**, 157
Sulfones, acetoxyphenyl-
o-quinodimethane precursor
Diels–Alder reactions, **5**, 392
Sulfones, alkenyl
coupling reactions
with sp^3 organometallics, **3**, 446
hydroxylation, **7**, 441
Sulfones, alkyl
in synthesis, **3**, 160
α-metallo, α-heterosubstituted
alkylation, **3**, 174
Sulfones, alkynyl
synthesis, **7**, 519
Sulfones, allenyl
hetero-Cope rearrangement, **5**, 1004
reaction with allylic alcohols, **6**, 856, 857
synthesis, **7**, 519
Sulfones, allyl
π-allylpalladium complexes from, **4**, 589
in synthesis, **3**, 169
radical cyclizations, **4**, 799
reduction, **8**, 975
Sulfones, allyl dienyl
synthesis, **6**, 161
Sulfones, aryl
coupling reactions
with Grignard reagents, **3**, 456
Sulfones, β-azidovinyl
synthesis
via iodine azide addition to alkene, **4**, 350
Sulfones, bis(aryloxynitrophenyl)
synthesis, **4**, 439
Sulfones, bis(4-chloro-3-nitrophenyl)
polycondensation, **4**, 439
Sulfones, γ-bromo-α,β-unsaturated phenyl
addition reaction
with organomagnesium compounds, **4**, 89
Sulfones, α-chloro
synthesis, **3**, 864
Sulfones, cycloalkenyl
addition reactions
with organolithium compounds, **4**, 78
Sulfones, cyclopentyl
conjugate additions
organocuprates, **4**, 192
Sulfones, di-*s*-alkyl
Ramberg–Bäcklund rearrangement, **3**, 864
Sulfones, α-diazo
Wolff rearrangement, **3**, 909
Sulfones, γ,γ-dimethylallenyl phenyl
reaction with butyllithium, **2**, 91
Sulfones, epoxy
Darzens glycidic ester condensation, **2**, 416
synthesis, **2**, 415
via Darzens glycidic ester condensation, **2**, 431
Sulfones, α-halo
reactions with carbonyl compounds, **1**, 530
reactions with trialkylboranes, **3**, 794
synthesis, **3**, 862
Sulfones, α-haloalkyl
Ramberg–Bäcklund reaction, **6**, 161
Sulfones, β-halovinyl
addition reactions, **4**, 127
Sulfones, α-hydroxy
o-quinodimethane precursors
Diels–Alder reactions, **5**, 389
Sulfones, *syn*-hydroxy
synthesis
via aldehydes, **6**, 164
Sulfones, γ-hydroxy-α,β-unsaturated
addition reaction
with organomagnesium compounds, **4**, 89
Sulfones, α-iodovinyl
synthesis

via iodine azide addition to alkene, **4**, 350
Sulfones, α-isocyanoalkyl
alkylation, **3**, 175
Sulfones, α-keto
desulfurization, **8**, 843
Sulfones, β-keto
metal enolates
alkylation, **3**, 54
Sulfones, α-metalloalkyl
reactions, **3**, 158
Sulfones, α-metalloallyl
reactions, **3**, 168
Sulfones, α-metallovinyl
reactions, **3**, 173
Sulfones, α-sulfinyl
reactions, **3**, 176
Sulfones, thiomethyl
Michael addition, **4**, 18
Sulfones, α-tosyloxy
Ramberg–Bäcklund rearrangement, **3**, 868
Sulfones, α-(trimethylsilyl)vinyl phenyl
addition reactions
with organolithium compounds, **4**, 79
Sulfones, (*E*)-α,β-unsaturated
synthesis
via Knoevenagel reaction, **2**, 362
Sulfones, vinyl
addition reaction with enolates, **4**, 102
deprotonation, **3**, 253
desulfurization, **8**, 842
Diels–Alder reactions, **5**, 324
functionalization
Michael addition, **4**, 13
tandem difunctionalization, **4**, 251
Peterson alkenation, **1**, 786
reaction with Grignard reagents, **3**, 493
selectivity, **6**, 162
heteroconjugate addition, **6**, 164
stereoselective reduction
sodium dithionate, **8**, 847
synthesis, **7**, 517, 523
via Julia coupling, **1**, 805
tandem vicinal difunctionalization, **4**, 257
Sulfones, vinyl amino
synthesis, **6**, 163
Sulfones, vinyl phenyl
desulfurization, **8**, 840
Sulfonic acids
Knoevenagel reaction
activated methylenes, **2**, 362
synthesis
via thiols, **7**, 759
Sulfonic acids, 2-amino
synthesis, **7**, 495
Sulfonimide, *N*,*N*-bis(trifluoromethane)-
reduction, **8**, 827
Sulfonimines
reactions with organometallic compounds, **1**, 390
Sulfonium, tris(dimethylamino)-
difluorotrimethylsiliconate
catalyst, stereoselectivity, **2**, 634
Sulfonium benzylide, diphenyl-
reactions with aldehydes
synthesis of *trans*-stilbene oxides, **1**, 824
Sulfonium fluoride, tris(diethylamino)-
catalyst
allylsilane reactions with aldehydes, **2**, 572
Sulfonium fluoroborate, dimethyl(methylthio)-
catalyst
allylstannane reaction with thioacetals, **2**, 581
reactions with alkenes, **7**, 493
Sulfonium methylide, dimethyl-
cyclopropanation, **4**, 987
epoxidation
carbonyl compounds, **1**, 820
Sulfonium methylide, dimethyloxy-
cyclopropanation, **4**, 987
epoxidation
carbonyl compounds, **1**, 820
Sulfonium methylides
synthesis
via sulfides, **6**, 893
Sulfonium salts
polymeric resins
phase transfer catalysts, **1**, 821
reactions with alkenes, **4**, 337
sulfur ylides from, **1**, 820
Sulfonium salts, alkyl diphenyl
O-alkylation
amide protection, **6**, 672
Sulfonium salts, chloro-
reactions with alkenes, **4**, 337
Sulfonium salts, β-hydroxy-
epoxide synthesis, **6**, 26
Sulfonium salts, α-metalloalkyl
synthesis, **3**, 87
Sulfonium salts, oxy-
reactions with alkenes, **4**, 337
Sulfonium tetrafluoroborates, alkyldiphenyl-
amide alkylation, **6**, 502
Sulfonium ylides
addition reactions, **4**, 115
2,3-rearrangements, **6**, 873
ring expansions, **6**, 898
synthesis
via sulfides, **6**, 893
Sulfonium ylides, acyl-
Wolff rearrangement, **3**, 909
Sulfonium ylides, allylic
rearrangements, **6**, 854
Sulfonium ylides, cyclic
2,3-sigmatropic rearrangements, **6**, 855
Sulfonyl chloride
arylation
palladium complexes, **4**, 858
Sulfonyl groups
substitutions
organoaluminum reagents, **6**, 165
Sulfonyl halogenides
adducts
amides, **6**, 490
Sulfonylimines
reaction with allyl organometallic compounds, **2**, 999
Sulfoquinovosyl phosphate
synthesis, **6**, 51
Sulfoxide, benzyl *t*-butyl
carbanion, **1**, 512
reactions with carbonyl compounds, **1**, 513
Sulfoxide, benzyl methyl
carbanion, **1**, 512
Sulfoxide, bis(trimethylsilylmethyl)
disiloxane release from

thiocarbonyl ylide generation, **4**, 1095
Sulfoxide, *t*-butyl thiomethyl
alkylation, **3**, 139
Sulfoxide, chiral vinyl
[3 + 2] cycloaddition reactions
asymmetric induction, **5**, 301
Sulfoxide, chloromethyl phenyl
Darzens-type reactions, **1**, 530
lithiation
butyllithium, **1**, 524
Sulfoxide, cyclopropyl phenyl
methylation
Pummerer rearrangement, **7**, 202
Sulfoxide, dibenzyl
Pummerer rearrangement, **7**, 194
Sulfoxide, dimethyl
activated
reagents, **7**, 293
anion
conjugate additions, **4**, 177
oxidation, **7**, 653
alcohols, **7**, 291–302
mechanism, **7**, 292
Sulfoxide, ethyl ethylthio
alkylation, **3**, 139
Sulfoxide, methyl 2-chlorophenyl
lithium anion
ring expansion with cyclobutanones, **1**, 862
Sulfoxide, methyl methylthio
alkylation, **3**, 139
Sulfoxide, methyl methylthiomethyl
alkylation, **3**, 139
metallated
alkylation, **3**, 135
Sulfoxide, methyl thiomethyl
alkylation, **3**, 137
Sulfoxide, methyl *p*-tolyl
carbanions
reactions with carbonyl compounds, **1**, 513
epoxide synthesis, **1**, 833
α-lithiated
reactions with aldehydes, **1**, 341
Sulfoxide, phenyl thiomethyl
alkylation, **3**, 139
Sulfoxide elimination
carbonyl compound dehydrogenation
choice of reagent, **7**, 146
dehydrogenation, **7**, 124
Sulfoxides
alkenes from
sulfenic acid elimination, **3**, 154
alkylated
in synthesis, **3**, 154
carbanions
crystal structure, **1**, 36
chiral
nucleophilic addition reactions, **1**, 69
synthesis, **7**, 777, 778
Darzens glycidic ester condensation, **2**, 416
desulfurization, **8**, 837
homochiral
synthesis, **6**, 900
hydrogenolysis, **8**, 914
Knoevenagel reaction
activated methylenes, **2**, 362
nitrile synthesis, **6**, 239
optically active
synthesis, **6**, 149
oxidation
to sulfones, **7**, 766
Pummerer rearrangement
α-acetoxylation of alkyl sulfides, **7**, 196
rearrangements
alcohol synthesis, **6**, 14
chirality transfer, **6**, 899
diastereoselectivity, **6**, 900
stereochemistry, **6**, 899
to sulfenates, **6**, 899
2,3-rearrangements, **6**, 873
reduction
as part of Pummerer rearrangement, **7**, 193
synthesis, **6**, 133–167
via sulfides, **7**, 762
tandem vicinal difunctionalization, **4**, 251
α,β-unsaturated
addition reactions with organomagnesium compounds, **4**, 86
use in synthesis
chirality, **6**, 148
Sulfoxides, 1-alkenyl aryl
alkylation, **3**, 155
Sulfoxides, alk-1-enyl phenyl
Pummerer rearrangement
with thionyl chloride, **7**, 205
Sulfoxides, alkyl
alkylation, **3**, 147
reduction, **3**, 155
Sulfoxides, alkyl aryl
carbanions
reactions with carbonyl compounds, **1**, 513
Sulfoxides, alkynyl
synthesis, **7**, 763
Sulfoxides, allenyl
desulfurization, **8**, 847
electrocyclic ring-closure, **6**, 903
intramolecular cycloaddition, **6**, 903
rearrangement
to conjugated dienones, **6**, 841
synthesis
via propargylic sulfenates, **6**, 155
Sulfoxides, allyl
alkylation, **3**, 155
metallation, **3**, 155
Michael addition, **4**, 12
rearrangements, **6**, 152, 899
stability, **6**, 902
Sulfoxides, allyl aryl
reactions with aromatic aldehydes, **1**, 517
Sulfoxides, allyl *p*-tolyl
reactions with carbonyl compounds, **1**, 519
Sulfoxides, aryl
coupling reactions
with Grignard reagents, **3**, 456
Sulfoxides, aryl vinyl
isomerization, **6**, 839
Sulfoxides, α-chloro
cyclobutene synthesis from, **3**, 872
optically active
synthesis, **6**, 156
Sulfoxides, cyclohexyl phenyl
reaction with trifluoroacetic anhydride

β-elimination of α-thiocarbocation intermediate, **7**, 204
Sulfoxides, cyclopentenone
Pummerer rearrangement
with dichloro ketene, **7**, 206
Sulfoxides, epoxy
reaction with amines, **6**, 91
synthesis
via Darzens glycidic ester condensation, **2**, 416
via α-halo sulfoxides, **1**, 524
Sulfoxides, α-halo
reactions with carbonyl compounds, **1**, 524
Sulfoxides, β-hydroxy
chiral
in synthesis, **6**, 156
epoxide synthesis, **6**, 26
homoallylic
synthesis, **6**, 156
synthesis
via organoaluminum reagents, **1**, 84
via α-sulfinyl carbanions, **1**, 514
Sulfoxides, indolizidinyl
reaction with butanal
via α-sulfinyl carbanion, **1**, 514
Sulfoxides, α-keto
desulfurization, **8**, 847
Sulfoxides, β-keto
allylic
reduction, **6**, 156
metal enolates
alkylation, **3**, 54
optically active
synthesis, **6**, 155
propargylic
reduction, **6**, 156
Pummerer rearrangement, **7**, 194
reduction, **8**, 12
synthesis
via allenic sulfoxides, **6**, 840
Sulfoxides, α-lithio
anions
epoxidation, **1**, 827
Sulfoxides, α-metalloalkyl
alkylation, **3**, 147
Sulfoxides, silyl
thermolysis
aryne generation, **4**, 488
Sulfoxides, thioacetal
carbonyl compound synthesis from, **3**, 142
Sulfoxides, vinyl
addition reaction with enolates, **4**, 100, 102
asymmetric synthesis
Michael-type, **6**, 150
chiral
conjugate additions, **4**, 213
conjugate additions
silyl ketene acetals, **4**, 161
Diels–Alder reactions, **5**, 324, 369–371
asymmetric, **6**, 150
dipolar cycloaddition with nitrones
chiral induction, **6**, 152
Michael addition, **4**, 13
Michael-type addition
tertiary allylic alcohols, **5**, 830
optically active
synthesis, **6**, 150
Pummerer rearrangement, **6**, 151
synthesis
via reactions of allyl phenyl sulfoxide with cyclic ketones, **1**, 520
use
asymmetric synthesis, **4**, 251
Sulfoximides
alkylation, **3**, 173
carbanions
crystal structure, **1**, 36
Sulfoximine, *N*,*S*-dimethyl-*S*-phenyl-
acidity
pK_a value, **1**, 531
lithium derivative
reaction with benzaldehyde, **1**, 532
reactions with ketones, **1**, 535
Sulfoximine, *S*-methyl-*S*-phenyl-
reactions with cycloalkadiene–molybdenum complexes, **1**, 535
Sulfoximine, *S*-methyl-*S*-phenyl-*N*-phenylsulfonyl-
acidity, **1**, 531
Sulfoximine, *S*-methyl-*S*-phenyl-*N*-silyl-
reactions with aldehydes, **1**, 532
Sulfoximine, *S*-methyl-*S*-phenyl-*N*-tosyl-
optically pure
synthesis, **1**, 788
Sulfoximine, *S*-phenyl-*N*-trimethylsilyl-*S*-trimethylsilylmethyl-
X-ray structure, **1**, 531
Sulfoximines
Darzens glycidic ester condensation, **2**, 417
N-substituted
α-carbanions, **1**, 535
ylides
carbonyl epoxidation, **1**, 820
Sulfoximines, alkenyl-
reaction with organozinc reagents
nickel catalysis, **3**, 230
synthesis
via silylation of β-hydroxysulfoximines, **1**, 536
Sulfoximines, alkyl α-chloroalkyl
Ramberg–Bäcklund rearrangement, **3**, 871
Sulfoximines, cycloalkenyl-
syn hydroxylation
diastereoselectivity, **7**, 440
Sulfoximines, epoxy
Darzens glycidic ester condensation, **2**, 418
synthesis
via Darzens glycidic ester condensation, **2**, 417
Sulfoximines, α-halo *N*-tosyl
Ramberg–Bäcklund rearrangement, **3**, 870
Sulfoximines, β-hydroxy-
reductive elimination, **1**, 738
synthesis
stereoselectivity, **1**, 536
via prochiral carbonyl compounds, **1**, 532
Sulfoximines, (β-hydroxyalkyl)-
synthesis
via *N*-silyl-*S*-methyl-*S*-phenylsulfoximine, **1**, 532
Sulfoximine ylides
addition reactions, **4**, 115
Sulfoxonium salts
ylides from
in Pummerer rearrangement, **7**, 195
Sulfoxonium ylides
3,2-sigmatropic rearrangement, **3**, 939

Sulfoxonium ylides, β-keto-
Wolff rearrangement, **3**, 909
Sulfur
carbanions stabilized by
alkylation, **3**, 85–181
dehydrogenation with, **7**, 124
electrophilic
reactions with alkenes, **7**, 516
halogen displacement, **7**, 124
radical
reactions with alkenes, **7**, 518
reductions, **8**, 370
Sulfuration
decarboxylative chalcogenation, **7**, 726
Sulfur-based rearrangements, **5**, 889–891
Sulfur compounds
adducts
amides, **6**, 490
oxidation, **7**, 757–779
activated C—H bonds, **7**, 193–214
reactions with amides, **6**, 496
reductive cleavage
α-halo ketones, **8**, 989
Sulfur compounds, vinylic
chiral
conjugate additions, **4**, 213–217
Sulfur dichloride
reactions with dienes, **7**, 516
Sulfur dioxide
bisimides
Diels–Alder additions to dienes, **7**, 486
extrusion
diene synthesis *via* retro Diels–Alder reaction, **5**, 567
in Mannich reaction
nonprotic solvent, **2**, 959
reaction with π-allylpalladium complexes, **4**, 601
Sulfur dioxide insertion reaction
hydroalumination adducts, **8**, 754
Sulfur-extrusion reaction — *see* Eschenmoser coupling reaction
Sulfur groups
functionalization
oxidative cleavage, **7**, 588
Sulfur heterocycles
synthesis, **7**, 524
Sulfuric acid
catalyst
carboxylic acid acylations, **2**, 711
Sulfur insertion reaction
hydroalumination adducts, **8**, 754
Sulfur monochloride
reactions with dienes, **7**, 516
Sulfur monosulfide
reaction with alkenes, **7**, 516
Sulfur nucleophiles
aromatic nucleophilic substitution, **4**, 441–444
nucleophilic addition to π-allylpalladium complexes, **4**, 599
regioselectivity, **4**, 640–642
stereochemistry, **4**, 624
Sulfur tetrafluoride
fluorination
alkyl alcohols, **6**, 216
reaction with amides, **6**, 496
Sulfur trifluoride, diethylamino-
fluorination
alkyl alcohols, **6**, 217
Sulfur trioxide
alkane functionalization, **7**, 14
pyridine
activator, DMSO oxidation of alcohols, **7**, 296
Sulfur trioxide insertion reaction
hydroalumination adducts, **8**, 754
Sulfuryl bromide
bromination
alkyl alcohols, **6**, 209
Sulfuryl chloride
adducts
dimethylformamide, **6**, 491
alkane chlorination, **7**, 16
alkane chlorosulfonation, **7**, 14
chlorination
alkyl alcohols, **6**, 204
chloromethyleniminium salt preparation, **2**, 779
oxidative rearrangement
gibberellin epoxides, **7**, 826
reaction with hydroalumination adducts, **8**, 754
sulfide halogenation, **7**, 206
Sulfuryl chloride fluoride
amide synthesis, **6**, 388
Sulfur ylides
alkylation, **3**, 178
cyclopropanation, **4**, 987
epoxidation
carbonyl compounds, **1**, 820
ketocarbenes from, **4**, 1032
reaction with trialkylboranes, **2**, 242
sigmatropic rearrangement, **5**, 894
synthesis, **3**, 918
tandem vicinal difunctionalization, **4**, 258
Sultam, 1-cyclohexenoyl-
conjugate additions
organocuprates, **4**, 204
Sultams
tandem vicinal difunctionalization, **4**, 249
Sultams, *N*-acyl-
aldol reaction
diastereofacial preference, **2**, 231
Diels–Alder reactions
intramolecular asymmetric, **5**, 543
homochiral
aldol reactions, **2**, 253
Sultams, *N*-crotonyl-
Diels–Alder reactions, **5**, 365
Sultams, *N*-enoyl-
addition reactions
with organomagnesium compounds, **4**, 85
chiral
conjugate additions, **4**, 204
conjugate additions
Grignard reagents, **4**, 204
Diels–Alder reactions, **5**, 365
Sultones
arene alkylation
Friedel–Crafts reaction, **3**, 317
Sultones, alkyl
alkylation, **3**, 179
Superacids
catalysts
Friedel–Crafts reaction, **3**, 297
solid, catalysts

Friedel–Crafts reaction, **3**, 297
Super Deuteride
deuteration
alkyl halides, **8**, 805
Super enamines
Henry reaction, **2**, 337
Super Hydride — *see* Lithium triethylborohydride
Superoxides
reaction with alkyl sulfonates
alcohol inversion, **6**, 22
Supporting electrolytes
electrosynthesis, **8**, 130
Suprofen
synthesis
via hydroformylation, **4**, 932
Surfactants
nonionic
synthesis, **6**, 37
Surugatoxin
synthesis
via Knoevenagel reaction, **2**, 384
Suzuki couplings
alkenylboron species, **3**, 489
Swern oxidation
alcohols, **7**, 291
DMSO, **7**, 296
primary alcohols, **7**, 396
Sydnone, *C*-methyl-*N*-phenyl-
1,3-dipolar cycloadditions, **4**, 1096
Sydnone, *N*-phenyl-
dipolar cycloaddition reaction with styrene, **4**, 1097
Sydnones
azomethine imine cyclizations, **4**, 1149
1,3-dipolar cycloadditions, **4**, 1096
photolysis
nitrilimines from, **4**, 1084
tandem intermolecular–intramolecular
cycloadditions, **4**, 1149
Sydnones, 4-acetyl-3-aryl-
Schmidt reaction, **6**, 821
Sydowic acid
synthesis
via 1,2-addition of trimethylaluminum, **1**, 104
Synthetase
organic synthesis
carbon–carbon bond formation, **2**, 456
Syringaresinol
synthesis, **3**, 693

T

Tabersonine, 16-methoxy-
  synthesis
    via Mannich reactions, **2**, 1043
Tagatose
  synthesis
    via Lewis acids, nonchelation selectivity, **1**, 339
Tagetones
  synthesis
    via aliphatic acylation, **2**, 718
    via diene acylation, **2**, 720
    via 1,6-addition, **1**, 554
Talaromycin A
  synthesis
    via functionalized alkyne addition, **1**, 419
    via radical cyclization, **4**, 794
Talaromycin B
  synthesis, **1**, 568; **7**, 237
Talaromycins
  synthesis, **7**, 239
Talose
  synthesis
    via Diels–Alder reaction, **2**, 689
Tamoxifen
  synthesis, **3**, 585
Tamura reagent
  Beckmann rearrangement, **6**, 764
Tandem rearrangements, **5**, 876–891
Tantalates, carbonyldicyclopentadienylhydrido-
  reduction
    acyl chlorides, **8**, 290
Tantalum
  hydrometallation
    mechanism, **8**, 672
Tantalum, *t*-butylalkylidene-
  t-butylalkene synthesis, **1**, 743
Tantalum catalysts
  alkene metathesis, **5**, 1118
  alkylidenation
    carbonyl compounds, **5**, 1122, 1125
Tartaric acid
  acetal
    stereospecific bromination, **3**, 789
Tartaric acid, monoacyl-
  catalyst
    Diels–Alder reactions, **5**, 377
Tartaric acid diamide, *N,N,N′,N′*-tetramethyl-
  α,β-unsaturated ketal derivatives
    conjugate additions, **4**, 209
Tartramide, dibenzyl-
  catalyst
    asymmetric epoxidation, **7**, 424
Tartramide, dicyclohexyl-
  asymmetric epoxidation
    homoallylic alcohols, **7**, 419
Tartrates
  chiral
    asymmetric epoxidation, **7**, 390
  esters
    asymmetric epoxidation, **7**, 395
  polymer-linked
    asymmetric epoxidation, **7**, 395
Taurolithocholic acid
  microbial hydroxylation, **7**, 73
Taxanes
  synthesis, **3**, 832; **7**, 242
    via Cope rearrangement, **5**, 796
    via [4 + 4] cycloaddition, **5**, 640
    via epoxide ring opening, **3**, 744
Tebbe reaction
  titanium-stabilized methylenation, **1**, 743
Tebbe reagent
  alkene synthesis, **1**, 807
  allyl vinyl ethers, **5**, 830
  enol ether synthesis, **2**, 597
  hydrozirconation, **8**, 676
  methylenation, **5**, 1122
  reaction with norbornene, **5**, 1121
  synthesis, **5**, 1124
Tellurapyrylium dyes
  photooxidation, **7**, 777
Telluration
  decarboxylative chalcogenation, **7**, 726
Telluride, dialkyl
  reductions
    nitro compounds, **8**, 371
Tellurides
  addition to alkynes, **4**, 50
  aromatic
    synthesis, **4**, 447
  oxidation, **7**, 776
    to telluroxides, **7**, 775
  reductions
    nitro compounds, **8**, 366
Tellurides, alkenyl
  coupling reactions
    with sp^3 organometallics, **3**, 446
Tellurides, aryl phenyl
  synthesis
    via $S_{RN}1$ reaction, **4**, 476
Tellurides, diaryl
  symmetrical
    synthesis, **4**, 447
  synthesis
    via $S_{RN}1$ reaction, **4**, 476
Tellurides, diphenyl
  synthesis
    via $S_{RN}1$ reaction, **4**, 476
Tellurinic acid
  synthesis, **7**, 775
Tellurinyl acetates
  reactions with alkenes, **4**, 343
Tellurium
  reductions, **8**, 370
    nitro compounds, **8**, 366
    unsaturated carbonyl compounds, **8**, 563
Tellurium compounds
  catalyst
    Wurtz reaction, **3**, 421
  oxazoline synthesis, **7**, 492
  oxidation, **7**, 757–779
    to ditellurides, **7**, 774
  photooxidation, **7**, 777
  reactions with alkenes, **4**, 343
  reactions with arynes, **4**, 508

Tellurium nucleophiles
  aromatic nucleophilic substitution, **4**, 447
Tellurium tetrachloride
  reaction with alkenes, **7**, 534
Tellurium triacetate, phenyl-
  synthesis, **7**, 774
Tellurium trichloride, 2-naphthyl-
  reaction with alkenes, **7**, 534
Tellurium ylides
  epoxidation, **1**, 825
1-Tellurochromene
  oxidation, **7**, 774
Tellurolactamization
  alkenes, **7**, 497
Tellurols
  oxidation
    to ditellurides, **7**, 774
Tellurone, bis(4-methoxyphenyl)
  synthesis, **7**, 776
Tellurone, dodecyl 4-methoxyphenyl
  synthesis, **7**, 776
Tellurones
  synthesis, **7**, 776
Tellurophene
  coupling reactions
    with primary alkyl Grignard reagents, **3**, 447
Tellurophenopyridazine
  photooxidation, **7**, 777
Telluroxides
  synthesis, **7**, 775
Teloidine
  synthesis
    via [4 + 3] cycloaddition, **5**, 609
Terephthalic acid
  acid dichloride synthesis, **6**, 302
  dimethyl ester
    synthesis *via* retro Diels–Alder reaction, **5**, 571
Terephthalic acid bis(dimethylamide)
  dications, **6**, 501
Terephthaloyl chloride
  acyloin coupling reaction, **3**, 617
Ternaphthyls
  synthesis
    via nickel catalysts, **3**, 229
Terpene oxides
  epoxide ring opening
    zinc bromide catalysis, **3**, 771
Terpenes
  cyclic
    biogenetic origins, **3**, 380
  epoxidation
    microbial, **7**, 429
  hydroxylation
    microbial, **7**, 62
  ketones
    dehydrogenation, **7**, 132
  synthesis, **3**, 428
    via Ireland silyl ester enolate rearrangement, **5**, 841
    via photoisomerizations, **5**, 230
Terpenes, polybromomono-
  Favorskii rearrangements, **3**, 849
Terpenoids
  synthesis
    via Dieckmann condensation, **2**, 824
    via 1,3-dipolar cycloadditions, **4**, 1077
Terphenyl
  synthesis, **3**, 503
Terpineol
  synthesis
    via linalool cyclization, **3**, 352
    via α- and β-pinene, **3**, 708
Terpyridine
  synthesis, **3**, 512
2,2′:6′,2″-Terpyridinium
  hydrochloride chlorochromate
    synthesis, **7**, 269
Terramycin
  synthesis, **7**, 160
Terrein
  synthesis
    via retro Diels–Alder reactions, **5**, 561
Testosterone
  hydrogenation
    catalytic, **8**, 533
    homogeneous catalysis, **8**, 452
  oxidation
    DMSO, **7**, 295, 296
  synthesis
    via polyene cyclization, **3**, 371
Testosterone, 17-methyl-
  hydrogenation
    homogeneous catalysis, **8**, 452
Tetracycles
  polyene cyclization, **3**, 369
Tetracyclic triterpenes
  synthesis
    via arynes, **4**, 501
Tetracyclines
  oxygenation, **7**, 157
  synthesis, **3**, 809
    via conjugate addition of aryl cyanohydrin, **1**, 552
    via oxyanion-accelerated rearrangement, **5**, 1022
Tetracyclone
  benzyne assay
    Diels–Alder reaction, **5**, 380
  synthesis, **2**, 142
Tetracyclo[4.3.0.0^{2,4}.0^{3,7}]non-8-ene
  cycloaddition reactions, **5**, 1187
Tetracyclo[3.3.0.0^{2,8}.0^{4,6}]octane
  synthesis
    via metal-catalyzed cycloaddition, **5**, 1187
Tetracycloundecanone
  cis,syn,cis
    synthesis *via* photoisomerization, **5**, 233
Tetraenes
  conjugated
    synthesis, **3**, 880
η^4-Tetraenes
  transition metal complexes
    reactions with electrophiles, **4**, 706
  tricarbonylmanganese complexes, **4**, 712
Tetrahydrofolate
  one-carbon transfer agent, **2**, 955
Tetrahydrofolate, *N*(5),*N*(10)-methylene-
  synthesis from tetrahydrofolate, **2**, 955
1,3-(1,1,3,3-Tetraisopropyldisiloxanylidene) group
  diol protection, **6**, 662
Tetralin, 1,4-dihydro-
  reduction
    Wilkinson catalyst, **8**, 445
Tetralin, 1,4-dimethyl-
  synthesis

via Friedel–Crafts reaction, **3**, 318
Tetralin, 5-hydroxy-2-(di-*n*-propylamino)-
synthesis, **7**, 331
Tetralin, 8-hydroxy-2-(di-*n*-propylamino)-
synthesis, **7**, 331
Tetralin, 5-nitro-
synthesis
via electrocyclization, **5**, 719
Tetralins
synthesis, **7**, 331
via silicon-stabilized cyclizations, **1**, 585
Tetralone, 6-methoxy-
reduction
hydrogen transfer, **8**, 320
Tetralones
dehydrogenation, **7**, 144
synthesis
via electrocyclization, **5**, 719
via enolate addition/cyclization, **4**, 258
via Friedel–Crafts cycloalkylation, **3**, 326
1-Tetralones
Birch reduction
dissolving metals, **8**, 509
reduction
dissolving metals, **8**, 114
ionic hydrogenation, **8**, 319
synthesis
via oxyanion-accelerated rearrangement, **5**, 1022
via thermal ring opening, **5**, 711
2-Tetralones
methylation
organometallic compounds, **1**, 150
pyrrolidine enamine
monomethylation, **3**, 29
synthesis
via ketocarbenoids, **4**, 1055
1-Tetralones, 2-oximo-
reduction
chemoselective, **8**, 125
1,4-Tetramethylene diradical
alkene dimerizations, **5**, 72
N,N,N',N'-Tetramethyl(methylene)diamine
Mannich base, **2**, 909
N,N,N',N'-Tetramethyl(methylene)diamine-
N,N-dimethyl(methylene)iminium salt
preparation from, **2**, 901
*Lactoneo*tetraosyl ceramide
synthesis, **6**, 53
Tetrapeptides
phospho analogs
synthesis, **2**, 1097
Tetraphenyl
synthesis, **3**, 501
1,3,5,7-Tetrathiacyclooctane
tetraanion
methylation, **3**, 134
Tetrathiomalonate, diethyl
synthesis
via O,O-diethyl dithiomalonate, **6**, 454
Tetrazene, tetramethyl-
zinc chloride complex
reaction with α-methylstyrene, **7**, 485
Tetrazenes
synthesis
via oxidation of 1,1-disubstituted hydrazines, **7**, 742
via oxidation of secondary amines with Fremy's salt, **7**, 746
Tetrazine, 3,6-diphenyl-
cycloaddition reactions
fulvenes, **5**, 627
Tetrazines
Diels–Alder reactions, **5**, 411, 413
1,2,4,5-Tetrazines
Diels–Alder reactions, **5**, 491
Tetrazole, 1-(2-bromocyclohexyl)-2-methyl-
synthesis, **7**, 501
5-Tetrazolediazonium chloride
thermal decomposition, **8**, 890
Tetrazoles
amination, **4**, 436
photochemical decomposition
nitrilimines from, **4**, 1084
Tetrazoles, α-hydroxyalkyl-
synthesis, **2**, 1086
Tetrazolophenanthridine
synthesis
via 1,3-dipolar cycloaddition, **4**, 1101
Tetrodotoxin
synthesis, **7**, 169
Tetrolic acid
hydrobromination, **4**, 286
Tetrolic acid, methyl ester
[2 + 2] cycloaddition reactions, **5**, 1067
cycloaddition reactions with chromium propynyl complexes, **5**, 1072
reaction with trimethylsilyldiazomethane, **5**, 1070
Tetronates
synthesis
via Ireland silyl ester enolate rearrangement, **5**, 841
Tetronic acids
α-allylation
via Claisen rearrangement, **5**, 832
synthesis
via Blaise reaction, **2**, 298
via Reformatsky-type reaction, **1**, 551
Thalictuberine
synthesis, **3**, 586
Thaliphorphine acetate
synthesis, **3**, 672
Thallation
activation barriers, **7**, 869
charge transfer excitation energies
EDA complexes, **7**, 870
durene, **7**, 872
EDA complexes
intermediates, **7**, 868
electrophilic aromatic, **7**, 335
Thallium, aryl-
carbonylation
palladium catalysts, **3**, 1033
Thallium acetate
anti hydroxylation
alkenes, **7**, 447
syn hydroxylation
alkenes, **7**, 445
Thallium carboxylates
acid anhydride synthesis, **6**, 315
decarboxylative iodination, **7**, 724
Thallium cyanide
acyl cyanide synthesis, **6**, 317
Thallium reagents

decarboxylative halogenation, **7**, 724
oxidants
solid-supported, **7**, 839
oxidative rearrangment, **7**, 828
reactions with aromatic compounds, **7**, 335
Thallium salts
catalysts
alkyl halide coupling, **3**, 418
Thallium sulfate
α-hydroxylation
ketones, **7**, 154
anti hydroxylation
alkenes, **7**, 447
Thallium thiolates
acylation
thiol ester synthesis, **6**, 440
Thallium triacetate
α-acetoxylation
ketones, **7**, 154
morpholino enamines, **7**, 170
allylic oxidation, **7**, 92
α-hydroxylation
carboxylic acids, **7**, 185
syn hydroxylation
alkenes, **7**, 445
reaction with alkenes, **7**, 534
Thallium trifluoroacetate
electrophilic oxidation, **7**, 868
quinones
synthesis, **7**, 354
Thallium trifluoroacetate, aryl-
biaryl synthesis, **3**, 505
Thallium trinitrate
α-acetoxylation
ketones, **7**, 154
chromanone dehydrogenation, **7**, 144
oxidative rearrangement, **7**, 827
solid support
clay, **7**, 845
Thallium tris(trifluoroacetate)
dimerization, **3**, 499
Thebaine
synthesis
via oxidation by thallium tris(trifluoroacetate), **3**, 680
Thermochemical measurements
carbanions, **1**, 41
Thermochemistry
aldol reaction, **2**, 134
Thermolysin
peptide synthesis, **6**, 399
2-Thiaadamantane
synthesis
via Baeyer–Villiger reaction, **7**, 683
5-Thiabicyclo[2.1.1]hexane, 2-bromo-
synthesis
via bromine addition to alkene, **4**, 346
Thiabutadienes
cationic
Diels–Alder reactions, **5**, 504–507
Diels–Alder reactions, **5**, 469
Thiacyclodec-4-ene *S*-oxide
synthesis
via 1,6-dibromo-3,4-hexanediol, **1**, 517
Thiacyclohexane, 4-*t*-butyl-
metallation, **3**, 151
Thiacyclooctene
sigmatropic rearrangement, **5**, 896
1-Thiadecalin
synthesis *via* transannular addition
α-sulfinyl carbanions to nonactivated double bonds, **1**, 517
Thiadecalin, β-hydroxy-
synthesis
via ketone enolate addition to sulfones, **4**, 102
1,2,6-Thiadiazine 1,1-diones
synthesis, **8**, 645
1,2,3-Thiadiazoles
flash-vacuum pyrolysis
synthesis of thioketenes, **6**, 426
flash vacuum thermolysis
synthesis of thioketenes, **6**, 449
Wolff rearrangement, **3**, 909
Thiadiazolidines
synthesis
via Diels–Alder reactions, **5**, 426
Δ^3-1,3,4-Thiadiazoline
thiocarbonyl ylides from, **4**, 1093
Thiadiazolines
synthesis, **7**, 486
Thiamine
acyloin formation
catalysis, **1**, 542
Thiane
chlorination
formation of 3,4-dihydro-2*H*-thiin, **7**, 206
Thiane, α-hydroxy-
synthesis, **8**, 934
Thiane-1,3-diol, 4-nitro-
synthesis
via Henry reaction, **2**, 327
Thiane *S*-oxides
carbanions
NMR, **1**, 513
Thianthrene, 2,7-dinitro-
synthesis, **4**, 443
Thiaprostacyclins
synthesis
via sulfur heterocyclization, **4**, 413
1,2,3,4-Thiatriazoles
synthesis
via thioacyl azides, **6**, 251
1,2,3,4-Thiatriazoles, 5-alkoxy-
degradation, **6**, 244
Thiazanes
reduction, **8**, 231
1,3-Thiazine, 5-acyl-
reduction
sodium cyanoborohydride, **8**, 658
6*H*-1,3-Thiazine, 2-phenyl-
electroreduction
regioselectivity, **8**, 136
Thiazine 1-imine, 3,6-dihydro-
synthesis
via Diels–Alder reactions, **5**, 422
Thiazine imines, dihydro-
Diels–Alder reactions, **5**, 426
Thiazine 1-oxide, 3,6-dihydro-
synthesis
via Diels–Alder reactions, **5**, 422, 424
Thiazine oxides, dihydro-
Diels–Alder reactions, **5**, 424

Thiazines
cycloaddition reactions
acid chlorides, **5**, 92
synthesis
via Ritter reaction, **6**, 276
1,3-Thiazines
reduction, **8**, 658
1,3-Thiazines, dihydro-
reduction, **8**, 231
synthesis
via carboxylic acids, **8**, 275
1,3-Thiazinium salts
synthesis, **6**, 508
4*H*-1,3-Thiazinium salts, 5,6-dihydro-
synthesis
via Diels–Alder reactions, **5**, 504
Thiazole, 2-acetylamino-4-methyl-
Mannich reaction
with formaldehyde and dimethylamine, **2**, 962
Thiazole, 2-bromo-
coupling reactions, **3**, 511
$S_{RN}1$ reaction, **4**, 462
Thiazole, 2-chloro-
$S_{RN}1$ reaction, **4**, 462
reaction with pinacolone enolates, **4**, 464
1,3-Thiazole, 2-hydroxyethyl-4-methyl-
alkylation, **1**, 543
Thiazole, 2-oxopropyl-
Knoevenagel reaction, **2**, 364
Thiazole-2-carboxylic acid
ethyl ester
reduction, **8**, 293
Thiazolecarboxylic acids
electrolytic reduction, **8**, 285
Thiazoles
amination, **4**, 436
Diels–Alder reactions, **5**, 491
metallation, **1**, 477
reduction, **8**, 656
1,3-Thiazolidine, *N,N'*-carbonyldi-2-thione-
amide synthesis, **6**, 389
Thiazolidine, β-hydroxy-
synthesis
via metallated 2-methylthiazoline, **2**, 494
Thiazolidines
reduction, **8**, 231
1,3-Thiazolidine-2-thione, 3-acyl-
tin(II) ester enolates from, **2**, 610
Thiazolidin-2-one, 3-acyl-
reduction
metal hydrides, **8**, 272
Thiazolidinones
reduction, **8**, 231
Thiazoline, 2-alkyl-
alkylation, **3**, 53
Thiazoline, α-lithioalkylthio-
alkylation, **3**, 88
Thiazoline, 2-methyl-
metallated
reactions, **2**, 494
Thiazolines
cycloaddition reactions
acid chlorides, **5**, 92
reduction, **8**, 656
synthesis
via Ritter reaction, **6**, 276
1,3-Thiazolines
reactions with organometallic compounds, **1**, 364
stereoselective, **1**, 350, 359
1,3-Thiazolines, 2-methyl-
metallation, **6**, 541
Thiazoline-2-thiones
ring opening, **8**, 657
1,3-Thiazolin-5-one, 4-alkylidene-
reduction
sodium borohydride, **8**, 656
Thiazolium carboxylates
catalysts
benzoin condensation, **1**, 543
Thiazolium salts
catalysts
benzoin condensation, **1**, 543
carbonyl condensation reactions, **1**, 542
1,3-Thiazolones
intramolecular cycloadditions, **4**, 1163
Thiazyne, fluoro-
reaction with perfluoropropene, **7**, 483
Thienamycin
1β-methyl analog
synthesis, **2**, 1058
synthesis, **3**, 1036; **7**, 647; **8**, 647
via aldol reaction, **2**, 212
via Baeyer–Villiger reaction, **7**, 680
via enolate–imine condensations, **2**, 925
via Mannich reaction, **2**, 926, 933, 936, 937
via reactions of enol silanes, **2**, 648
via simple diasteroselection, **2**, 201
synthesis of precursor
via higher order cuprate, **1**, 133
Thienamycin, 1-β-methyl-
synthesis
via diastereoselective reaction, **2**, 653
2-Thieniumbutadienes
Diels–Alder reactions, **5**, 504
Thienium salts
Diels–Alder reactions, **5**, 439
Thienium salts, aryl-
Diels–Alder reactions, **5**, 504
Thieno[*b*]azepinones
synthesis
via Friedel–Crafts reaction, **2**, 765
Thieno[3,4-*b*]furan
synthesis
via retro Diels–Alder reactions, **5**, 584
Thienopyridines
synthesis, **3**, 543
via Vilsmeier–Haack reaction, **2**, 787
5*H*-Thieno[2,3-*c*]pyrrole
synthesis
via Knoevenagel reaction, **2**, 378
Thieno[*b*]suberanone
synthesis
via Friedel–Crafts reaction, **2**, 765
Thietanes, 2-imino-
synthesis
via ketenimines, **5**, 114
2-Thietanones
synthesis
via ketenes and carbonyls, **5**, 89
2*H*-Thiin, 3,4-dihydro-
synthesis
via chlorination of thiane, **7**, 206

Thiirane, 2,3-bis(trimethylsilyl)-
synthesis
via thiocyanogen addition to alkene, **4**, 349
Thiirane, vinyl-
rearrangement, **5**, 909, 931
Thiirane 1,1-dioxide
Ramberg–Bäcklund rearrangement, **3**, 866, 867
Thiiranes
nitrile synthesis, **6**, 238
synthesis, **1**, 819, 834; **7**, 515
via alkenes, **4**, 331
Thioacetal monoxides
Michael addition, **4**, 10
Thioacetals
reaction with allylsilanes, **2**, 580
reduction, **8**, 229, 935
to ethers, **8**, 211–232
O-silyl ketene
synthesis, **2**, 605
use in synthesis, **6**, 134
Thioacetamides, trichloro-
synthesis
via thiolysis of trichlorovinylamines, **6**, 429
Thioacetanilide, *o*-iodo-
ring closure
via $S_{RN}1$ reaction, **4**, 477
Thioacetates
conjugate additions
enones, **4**, 231
Thioacetonitrile, phenyl-
conjugate addition reactions, **4**, 111
Thioacetophenone
pinacol coupling reactions, **3**, 582
Thioacrolein
synthesis
via retro Diels–Alder reaction, **5**, 575
Thioacrylamide, α-chloro-
synthesis
via α,α-dichlorination of propionimidium salts, **6**, 429
Thioacrylamide, α-cyano-
synthesis
via Knoevenagel reaction, **2**, 361
Thioacrylamide, *N*,*N*-dimethyl-
synthesis
via retro Diels–Alder reaction, **5**, 556
Thioacrylamides
synthesis
via retro Diels–Alder reaction, **5**, 556
Thioacrylates, *S*-methyl
synthesis
via DCC, **6**, 437
Thioacyl anhydrides
thioacylation
thiols, **6**, 453
Thioacylation
alcohols and phenols
anhydrides, thioketenes, thioesters and dithioesters, **6**, 449
thioacyl halides, **6**, 448
amines, **6**, 420
arenes, **6**, 453
carbanions, **6**, 453
thiols
thioacyl halides, thioacyl anhydrides and thioketenes, **6**, 453
Thioacyl chlorides
thioacylation
amines, **6**, 422
Thioacyl halides
thioacylation
alcohols and phenols, **6**, 448
thiols, **6**, 453
Thioaldehydes
Diels–Alder reactions, **5**, 436
S-oxides
synthesis *via* retro Diels–Alder reactions, **5**, 575
Thioalkoxides
aromatic nucleophilic substitution, **4**, 443
Thioallyl anions
synthesis and reactions
regioselectivity, **2**, 71
Thioamides
acylation, **6**, 508
adducts
acylating reagents, **6**, 487
acyl chlorides, **6**, 493
alkylation
reversible, **2**, 868
with Meerwein's reagent, **6**, 403
alkylmercaptomethyleniminium salt synthesis, **6**, 508
amidinium salt synthesis, **6**, 517
desulfurization, **6**, 403
enolates
aldol reactions, stereoselectivity, **2**, 214
imidoyl halide synthesis, **6**, 523
primary amines
aldol reaction, stereoselectivity, **2**, 215
reactions with chlorine, **6**, 496
rearrangement, **6**, 511
reduction, **8**, 303
secondary amines
aldol reaction, stereoselectivity, **2**, 215
S-silylation, **6**, 425
stability, **6**, 419
synthesis, **2**, 216; **6**, 419–432
via 3*H*-1,2-dithiol-3-ones, **6**, 421
via Eschenmoser coupling reaction, **2**, 867
tandem vicinal difunctionalization, **4**, 257
thioacylation
amines, **6**, 424
α,β-unsaturated
reactivity
synthesis, **6**, 425
Thioamides, β-amino-
synthesis
via thioamides, **6**, 425
Thioamides, α-chloro-
synthesis
via thiolysis of imidoyl chlorides, **6**, 429
Thioamides, *N*,*N*-dialkyl-
α,β-unsaturated
addition reactions with enolates, **4**, 100
addition reactions with organomagnesium compounds, **4**, 85
Thioamides, *N*,*N*-dimethyl-
synthesis
via dithiocarbamate and nitriles, **6**, 431
Thioamides, α-hydroxy-
synthesis
via phosphinodithioic acid, **6**, 432
Thioanhydrides, bisthioacyl-

thioacylation
alcohols, **6**, 449
Thio anions
aromatic nucleophilic substitution, **4**, 441
Thioanisole
acylation
Friedel–Crafts reaction, **2**, 744
deprotonation
reaction with carbonyl compounds, **1**, 826
oxidation
solid support, **7**, 844
Thioasparagine
synthesis
via nitrile thiolysis, **6**, 430
Thioates
aldol reactions, **2**, 258
Thioates, *S*-(2-pyridyl)-
acylation, **1**, 407
Thiobenzaldehyde
ene reaction, **2**, 555
Thiobenzaldehyde dianions
alkylation and allylation
selectivity, **3**, 95
Thiobenzamides
synthesis
via thiolysis of imidoyl chlorides, **6**, 428
Thiobenzamides, *o*-(2-oxoalkyl)-
synthesis
via ring opening of thiolactones, **6**, 420
Thiobenzamides, 2-(thioacylamino)-
synthesis
via dithiocarboxylates, **6**, 424
Thiobenzanilide
reduction, **8**, 303
Thiobenzanilide, 4-hydroxy-
reduction, **8**, 303
Thiobenzanilide, *o*-iodo-
ring closure
via $S_{RN}1$ reaction, **4**, 477
Thiobenzanilide, 4-methoxy-
reduction, **8**, 303
Thiobenzoates, *O*-deoxyribosyl
synthesis, **6**, 450
Thiobenzophenone
hydride transfer
with 1,4-dihydropyridines, **8**, 93
reactions with ketenimines, **5**, 114
reduction
dissolving metals, **8**, 126
Thiobenzoyl azolides
synthesis
via dithiocarboxylic acids, **6**, 421
via thioacyl chlorides, **6**, 422
Thiobenzoyl compounds
deoxygenation, **8**, 818
Thiocarbamates, *N*,*N*-dimethyl-
catalytic hydrogenation, **8**, 817
Thiocarbamoyl chlorides
nitrile synthesis, **6**, 234
synthesis
via thiophosgene, **6**, 423
Thiocarbenium ions
cyclization, **6**, 754
heterocyclic synthesis, **6**, 753
synthesis, **6**, 753
Thiocarbonates
synthesis
via allylic xanthates, **6**, 842
Thiocarbonyl compounds
Diels–Alder reactions, **5**, 435–442
ene reaction, **2**, 555
pinacol coupling reactions, **3**, 582
stability, **6**, 419
O-Thiocarbonyl compounds
reduction
stannanes, **8**, 818
Thiocarbonyl groups
carbonyl group regeneration, **7**, 846
Thiocarbonyl ylides
cyclizations, **4**, 1163–1165
1,3-dipolar cycloadditions, **4**, 1074, 1093–1095
Thiocarboxylates, 2-acetoxy-
O-methyl ester
synthesis, **6**, 451
Thiocarboxylates, *O*-alkyl
thioacylation
amines, **6**, 420
Thiocarboxylates, 2-hydroxy-
O-methyl ester
synthesis, **6**, 451
Thiocarboxylates, 2-mesyloxy-
O-methyl ester
synthesis, **6**, 451
Thiocarboxylates, *O*-methyl
thioacylation
amino acids, **6**, 420
Thiocarboxylic acids
Eschenmoser coupling reaction, **2**, 875
esters
NMR, **6**, 436
spectroscopy, **6**, 436
S-esters
synthesis, **6**, 436
thiolysis, **6**, 432
4-Thiochromanone
dehydrogenation
use of trityl perchlorate, **7**, 144
Thiocinnamates, *S*-phenyl esters
synthesis, **6**, 441
Thio-Claisen rearrangement, **5**, 889; **6**, 847, 860
Thio compounds
α,β-unsaturated
synthesis *via* retro Diels–Alder reaction, **5**, 556
Thiocoumarins
synthesis
ease of formation, **6**, 443
via Lewis acid cyclization, **6**, 443
Thiocoumarins, dihydro-
synthesis
ease of formation, **6**, 443
Thiocresol
reduction, **8**, 568
Thiocyanates
alkyl
synthesis, **7**, 608
aromatic nucleophilic substitution, **4**, 443
Ritter reaction, **6**, 291
thiol carboxylic esters
synthesis, **6**, 439
Thiocyanates, alkyl
synthesis of imidothioates, **6**, 444
Thiocyanates, allyl

rearrangement
to allyl isothiocyanates, **6**, 846
Thiocyanogen
reactions with alkenes, **4**, 348
Thiocyanogen chloride
reactions with alkenes, **7**, 516
Thioenamides
tandem vicinal difunctionalization, **4**, 249
Thioenamides, *N*,*N*-dialkyl-
reactions with Grignards reagents, **4**, 257
Thioesters
acylation, **1**, 433
lithium dialkylcuprates, **1**, 428
amide synthesis, **6**, 395
Claisen condensation, **6**, 446
deoxygenation, **8**, 818
desulfurization, **8**, 838
Dieckmann condensation, **6**, 446
Diels–Alder reactions, **5**, 438
enolates
aldol reactions, stereoselectivity, **2**, 214
Michael donors, **4**, 259
3-hydroxy
synthesis, **2**, 260, 261
α′-lithioalkyl
alkylation, **3**, 88
NMR, **6**, 436
2-pyridyl
synthesis, **1**, 407
reactions with organocuprates, **1**, 116
reduction, **8**, 303
reductive coupling, **3**, 618
spectroscopy, **6**, 436
synthesis, **6**, 435–457
via hydration of alkynes, **4**, 300
Thioesters, β-amino
synthesis
via Mannich reaction, **2**, 920, 922
Thioformaldehyde
synthesis
via retro Diels–Alder reactions, **5**, 575
Thioformamide, *N*,*N*-dimethyl-
adducts
phosphorus oxychloride, **6**, 489
Thioformamides
synthesis
via carbon disulfide, **6**, 428
via thioacylation, **6**, 420
Thioformate, *O*-ethyl
synthesis
via sulfhydrolysis of orthoesters, **6**, 452
Thioformates, *O*-*t*-alkyl
synthesis
via sulfhydrolysis of imidates, **6**, 451
Thioformates, chloro-
S-phenyl ester
coupling reactions, **6**, 446
thiol carboxylic esters
synthesis, **6**, 439
Thioformohydrazides
synthesis
via thioformylation of hydrazines, **6**, 420
Thioformohydroxamic acids
synthesis
via thioformylation of hydroxylamines, **6**, 420
Thioglycolic acid, *S*-thioacyl-
thioacylation
amines, **6**, 423
Thioglycosides
heterocyclic
synthesis, **6**, 46
synthesis, **6**, 46, 48
Thiohydrazides
synthesis
via dithiocarboxylic acids, **6**, 421
via thioacylation of hydrazines, **6**, 424
via thioamides, **6**, 424
Thiohydroxamate esters
radical addition reactions, **4**, 747–750
synthesis, **8**, 825
Thiohydroxamic acids
synthesis
via dithiocarboxylic acids, **6**, 421
via thioacylation of hydroxylamines, **6**, 424
Thioic acid, isobutyl-
S-*t*-butyl ester
Mannich reaction, **2**, 922
Thioimidates
reduction, **8**, 302
synthesis, **6**, 536, 767
via organoaluminum reagents, **1**, 98
thioimidate synthesis, **6**, 541
Thioimidates, *S*-allyl-
Claisen-type rearrangement
palladium(II) catalysis, **4**, 564
rearrangement
sulfur–nitrogen transposition, **6**, 846
Thioimidates, *N*-phenyl-
addition reactions
with organomagnesium compounds, **4**, 85
Thioimidates, *N*-trimethylsilylmethyl-
synthesis, **6**, 542
Thioimides
homologated
synthesis *via* Eschenmoser coupling reaction, **2**, 875
Thioimido esters
acylation, **6**, 510
alkylation, **6**, 510
Thioiminium salts
aromatic nucleophilic substitution, **4**, 443
reactions with active methylene compounds, **2**, 873
Thioketals
cleavage
metal–ammonia, **8**, 531
reduction, **8**, 229
Schmidt reaction, **6**, 821
Thioketenes
cycloaddition reactions, **5**, 115
synthesis
via retro Diels–Alder reactions, **5**, 575
thioacylation
alcohols and phenols, **6**, 449
amines, **6**, 426
thiols, **6**, 453
Thioketenes, bis(trimethylsilyl)-
thioacylation, **6**, 426
Thioketones
Diels–Alder reactions, **5**, 436
reduction
dissolving metals, **8**, 126
α,β-unsaturated acyclic

synthesis *via* retro Diels–Alder reaction, **5**, 556
Thioketones, 4-alkyl
synthesis
via conjugate addition, **4**, 258
Thioketones, α-chloro-α-phenyl
synthesis, **7**, 213
Thiolactams
dehydrogenation
p-toluenesulfinyl chloride, **7**, 128
desulfurization
with MCPBA, **6**, 403
synthesis, **6**, 419–432
via lactams, **6**, 430
via thiolysis of amidines, **6**, 430
Thiolactim ethers
synthesis, **6**, 541
Thiolactones
synthesis, **6**, 435–457
five- and six-membered, **6**, 443
via flash pyrolysis, **6**, 440
via iodocyclization of γ,δ-unsaturated thioamides, **4**, 413
via ketenes and carbonyls, **5**, 89
Thiolane
chlorination
formation of 2-chlorothiolane, **7**, 207
formation of 2,3-dichlorothiolane, **7**, 206
Thiolane, 2-chloro-
synthesis
via chlorination of thiolane, **7**, 207
Thiolane, 2,3-dichloro-
synthesis
via chlorination of thiolane, **7**, 207
Thiolates
reduction
α-alkylthiocarbonyl compounds, **8**, 995
α-Thiolation
difunctionalization, **4**, 262
Thiolenones
synthesis, **7**, 596
Thiolen-2-one-5-spiro-5′-thiolen-4′-one
synthesis
via metal-catalyzed cycloaddition, **5**, 1200
Thiol esters
acylation, **2**, 805
boryl enolates from, **2**, 242
Darzens glycidic ester condensation, **2**, 418
Dieckmann reaction, **2**, 815
enolates
acylation, **6**, 446
macrolactonization, **6**, 370
reaction with aldehydes
Lewis acid mediated, **2**, 631
reduction, **8**, 293
synthesis, **6**, 437
via acylation of arenes and carbanions, **6**, 445
via hydrolysis of imidothioates, thioorthoesters and ketene *S,S*-acetals, **6**, 444
synthesis from thiols
via acylation with acyl halides, **6**, 440
via acylation with anhydrides, ketenes and esters, **6**, 443
via acylation with carboxylic acids, **6**, 437
Thiol esters, β-hydroxy
β-lactone synthesis, **6**, 346
Thiol lactones
synthesis, **6**, 437
via acylation of arenes and carbanions, **6**, 445
via hydrolysis of imidothioates, thioorthoesters and ketene *S,S*-acetals, **6**, 444
synthesis from thiols
via acylation with acyl halides, **6**, 440
via acylation with anhydrides, ketenes and esters, **6**, 443
via acylation with carboxylic acids, **6**, 437
γ-Thiol lactones
synthesis
via phosphorus pentoxide, **6**, 440
δ-Thiol lactones
synthesis
via phosphorus pentoxide, **6**, 440
Thiols
acylation
acyl halides, **6**, 440
anhydrides, ketenes and esters, **6**, 443
carboxylic acids, **6**, 437
addition to alkynes, **4**, 50
conjugate additions
enantioselectivity, **4**, 231
dehalogenation
α-halo ketones, **8**, 989
desulfurization
tin hydrides, **8**, 846
dimerization
pyridinium chlorochromate, **7**, 267
oxidation, **7**, 758
chromium(VI) oxide, **7**, 278
oxygen, **7**, 759
oxidative coupling
solid support, **7**, 846
protecting groups, **6**, 664
radical additions
alkenes, **4**, 770
reactions with 4-acyloxypyridines
synthesis of carbothioates, **6**, 443
reactions with alkenes, **4**, 316
reactions with nitriles, **6**, 511
reduction
Raney nickel, **8**, 964
thioacylation
thioacyl halides, thioacyl anhydrides and thioketenes, **6**, 453
transesterification, **6**, 454
Thiols, aryl
coupling reactions
with Grignard reagents, **3**, 456
Thiols, benzene-
hydrogenolysis, **8**, 914
Thiols, benzyl
desulfurization, **8**, 978
Thiols, tertiary
reductive decarboxylation, **7**, 720
O-acyl thiohydroxamates, **7**, 721
Thiolsubtilisin
peptide synthesis, **6**, 399
Thiolysis
amidines, **6**, 430
imidates, **6**, 429
imidocarboxylic acids, **6**, 428
imidoyl chlorides, **6**, 428
nitriles, **6**, 430
1,1,1-trihalides, **6**, 432

Thiomalonamides
synthesis
via thiolysis of phosgene immonium chloride, **6**, 429
Thiomalonates
S-alkyl esters
synthesis, **6**, 437
α-Thionitriles
acyl anion equivalents, **1**, 561
Thionitroso compounds
Diels–Alder reactions, **5**, 422
Thionium ions
chiral
reaction with enol silanes, **2**, 649
Thionyl bromide
acid bromide synthesis, **6**, 305
Thionyl chloride
acid anhydride synthesis, **6**, 309
activator
DMSO oxidation of alcohols, **7**, 298
adducts
dimethylformamide, **6**, 491
amide synthesis, **6**, 383
reaction with amides, **6**, 496
reaction with carboxylic acids, **6**, 302
reaction with hydroalumination adducts, **8**, 754
reaction with organoboranes, **7**, 604
reaction with sulfoxides
Pummerer rearrangement, **7**, 203
Thionyl fluoride
synthesis
via thionyl chloride, **6**, 307
Thionyl halides
imidoyl halide synthesis, **6**, 526
Thioorthoesters
hydrolysis
synthesis of thiol esters, **6**, 444
3-Thiophenamine
synthesis
via $S_{RN}1$ reaction, **4**, 473
Thiophene, 2-acetamido-
synthesis
via iodocyclization of γ,δ-unsaturated thioamides, **4**, 413
Thiophene, 2-acetyl-3-hydroxy-
synthesis
via activated alkynes, **4**, 52
Thiophene, 2-acyl-
reduction
dissolving metals, **8**, 609
Thiophene, 2-alkenyl-
cyclopropanation, **4**, 1063
Thiophene, 2-alkyl-
reduction
ionic hydrogenation, **8**, 610
Thiophene, 2-benzoyl-
reduction
ionic hydrogenation, **8**, 610
Thiophene, 2-bromo-
halogen–metal exchange
with Grignard reagents, **3**, 459
$S_{RN}1$ reaction, **4**, 462
Thiophene, 3-bromo-
coupling reactions
with Grignard reagents, **3**, 459, 513
$S_{RN}1$ reaction, **4**, 462
reaction with acetone enolates, **4**, 464
Thiophene, 2-chloro-
$S_{RN}1$ reaction, **4**, 462
Thiophene, 2-chlorocarbonyl-
synthesis
via thiophene and phosgene, **6**, 308
Thiophene, 3-cyanomethyl-
Vilsmeier–Haack reaction, **2**, 780
Thiophene, diaryl-
synthesis, **3**, 513
Thiophene, dihydro-
synthesis
via Knoevenagel reaction, **2**, 360
Thiophene, 3,4-dimethoxy-
Mannich reaction, **2**, 963
Thiophene, 2,5-dimethyl-
reduction
ionic hydrogenation, **8**, 610
Thiophene, 2,3-dimethylene-
dimerization
via [4 + 4] cycloaddition, **5**, 639
Thiophene, 2-iodo-
$S_{RN}1$ reaction, **4**, 462
Thiophene, 2-lithio-
alkylation, **3**, 261
conjugate additions
nontransferable ligand, **4**, 176
Thiophene, 2-methoxy-
Mannich reaction
with formaldehyde and secondary amines, **2**, 963
Thiophene, 3-methoxy-
Mannich reaction
with formaldehyde and secondary amines, **2**, 963
Thiophene, nitro-
aromatic nucleophilic substitution, **4**, 432
Thiophene, tetrahydro-
reductive desulfurization, **8**, 836
synthesis
via tandem vicinal difunctionalization, **4**, 251
Thiophene, 2,3,5-tribromo-
reduction, **8**, 908
Thiophene, vinyl-
synthesis, **3**, 497
2-Thiopheneacetonitrile
reduction
ionic hydrogenation, **8**, 611
Thiophene alcohols
asymmetric epoxidation
kinetic resolution, **7**, 423
2-Thiophenecarboxylic acid
reduction
dissolving metals, **8**, 609
3-Thiophenecarboxylic acid
Birch reduction, **8**, 609
2-Thiophenecarboxylic acid, 5-methyl-
electrochemical reduction, **8**, 611
Thiophene dioxide, 2-ethyl-5-methyl-
cycloaddition reactions
fulvenes, **5**, 628
Thiophene dioxide, 2-substituted 2,5-dihydro-
thermolysis, **5**, 567
Thiophenes
acylation
Friedel–Crafts reaction, **2**, 737
acyloin coupling reaction
esters, **3**, 615

coupling reactions
with sp^3 organometallics, **3**, 459
with primary alkyl Grignard reagents, **3**, 447
desulfurization, **8**, 837
hydrogenation
homogeneous catalysis, **8**, 456
intermolecular dimerization, **3**, 509
intramolecular cycloaddition
nitrones, **4**, 1119
lithiation, **1**, 472
Mannich reaction
with *N,N*-dimethylmethyleniminium chloride, **2**, 963
with formaldehyde and secondary amines, **2**, 963
with iminium salts, **2**, 962
metallation
addition reactions, **1**, 471
photocycloaddition reactions
with benzaldehyde, **5**, 176
reactions with ketocarbenoids, **4**, 1063
reduction, **8**, 603–630
electrochemical, **8**, 611
synthesis
via activated alkynes, **4**, 52
via [2 + 2 + 2] cycloaddition, **5**, 1139
via hydrogen sulfide and polyynes, **4**, 317
2,3,5-trisubstituted
synthesis, **2**, 847
Vilsmeier–Haack reaction, **2**, 780
Thiophenes, 2-oxazolinyl-
lithiation, **1**, 472
Thiophenocyclopropane
synthesis
via ketocarbenoids and thiophenes, **4**, 1063
Thiophenol
electrochemical reduction, **4**, 439
synthesis, **4**, 441
Thiophenophane
2,5-bridged
synthesis *via* acyloin coupling reaction, **3**, 630
Thiophenoxides
arylation, **4**, 495
reactions with arynes, **4**, 492
α-Thiophenylenones
Michael addition
phenol synthesis, **4**, 8
Thiophiles
Eschenmoser coupling reaction
effect on rate and yield, **2**, 870
Thiophosgene
Diels–Alder reactions, **5**, 439
thioacylation
amines, **6**, 423
Thiophosphates
phosphoric acid protecting group, **6**, 625
Thiopinacols
synthesis, **3**, 582
Thiopropionic acid
t-butyl ester, lithium enolate
Woodward erythromycin synthesis, **2**, 214
Thiopyran
synthesis
via Knoevenagel reaction, **2**, 378
Thiopyran, dihydro-
synthesis
via ketocarbenoids and thiophenes, **4**, 1063
Thiopyran, 2,3-dihydro-
ring-opening coupling reaction
with primary alkyl Grignard reagents, **3**, 447
2*H*-Thiopyran, 3,4-dihydro-
synthesis
via Diels–Alder reaction, **5**, 469
via Knoevenagel reaction, **2**, 361
Thiopyran-2-thiones
synthesis
via [2 + 2 + 2] cycloaddition, **5**, 1158
Thiopyridone
synthesis
via Knoevenagel reaction, **2**, 378
2-Thiopyridones, *N*-acyloxy-
synthesis
via alkyl radical sources, **6**, 442
Thiosorbamide, *N,N*-dimethyl-
addition reactions
with enolates, **4**, 100
with organolithium compounds, **4**, 76
Thiosphondin
synthesis, **5**, 1096
Thiosulfonates
synthesis, **7**, 726
4-Thiouracil
cleavage
thioamide synthesis, **6**, 425
Thiourea, tetramethyl-
catalyst
Rosenmund reduction, **8**, 286
Thioureas
amidinium salt synthesis, **6**, 517
S-dioxide
synthesis *via* ozonolysis of 3-carene, **7**, 548
oxidative cleavage
alkenes, ozone, **7**, 544
synthesis
via carbon disulfide, **6**, 428
via dithiocarboxylates, **6**, 424
via thiophosgene, **6**, 423
Thiourethanes, *O*-alkyl
synthesis
via dithiocarboxylates, **6**, 424
via thiophosgene, **6**, 423
Thiovaleramide
aldol reaction
stereoselectivity, **2**, 215
Thio-Wittig rearrangement, **6**, 853
methylenecyclopentane, **6**, 895
1,3-Thioxanes
reduction, **8**, 230
Thioxanthone
reduction
boranes, **8**, 316
dissolving metals, **8**, 115
Thioxoarachidonate, methyl
synthesis, **6**, 452
Thioxoesters
amidinium salt synthesis, **6**, 517
synthesis, **6**, 446
via acylation of hydrogen sulfide, **6**, 450
via thioacylation of arenes and carbanions, **6**, 453
synthesis from alcohols and phenols
via thioacylation with anhydrides, thioketenes, thioesters and dithioesters, **6**, 449
via thioacylation with thioacyl halides, **6**, 448

thioacylation
alcohols and phenols, **6**, 449
Thioxolactones
synthesis, **6**, 446
via acylation of hydrogen sulfide, **6**, 450
via thioacylation of arenes and carbanions, **6**, 453
synthesis from alcohols and phenols
via thioacylation with anhydrides, thioketenes, thioesters and dithioesters, **6**, 449
via thioacylation with thioacyl halides, **6**, 448
Thiyl radicals
addition reactions
alkenes, **7**, 519
Thorpe reactions, **2**, 848
conditions, **2**, 849
in synthesis, **2**, 851
mechanism, **2**, 848
regioselectivity, **2**, 851
scope, **2**, 849
Thorpe–Ziegler reactions, **2**, 848
Threo compounds
aldol diastereomers
thermodynamics, **2**, 153
Threonine
hydroxy groups
protection, **6**, 650
Mannich reaction
with formaldehyde and 2,4-dimethylphenol, **2**, 968
synthesis, **8**, 148
via hydroformylation, **4**, 932
L-Threose, 2,3-*O*-cyclohexylidine-4-deoxy-
chiral imine
β-aminoamides from, **2**, 924
Thromboxane A_2
analog
synthesis *via* intramolecular photocycloaddition, **5**, 180
synthesis
via [4 + 3] cycloaddition, **5**, 605
via Paterno–Büchi reaction, **5**, 151
Thromboxane B_2
synthesis
stereocontrolled, *via* Eschenmoser rearrangement, **5**, 837
Thromboxanes
synthesis
via [4 + 3] cycloaddition, **5**, 612
Thrysiferol
synthesis, **7**, 633
α-Thujaplicin
synthesis
via [4 + 3] cycloaddition, **5**, 609
β-Thujaplicin
synthesis
via [4 + 3] cycloaddition, **5**, 609
via tricarbonyl(tropone)iron complex, **4**, 707
Thujaplicins
synthesis
via dihalocyclopropyl compounds, **4**, 1018
Thujopsene
synthesis, **7**, 100
Thymidine, 5′-*O*-acetyl-
oxidation
Collins reagent, **7**, 259
Thymidine, 5′-*O*-trityl-
oxidation
Collins reagent, **7**, 259
Thymol
hydrogenation, **8**, 912
Thymol, methoxy-
oxidation
via alkaline ferricyanide, **3**, 686
Thymoquinone, libocedroxy-
synthesis
via alkaline ferricyanide, **3**, 686
Tiffeneau–Demjanov rearrangement
2-amino alcohols, **3**, 781
diazonium ion rearrangement, **1**, 846
pinacol rearrangement
comparison with, **3**, 722
Tiglaldehyde
aldimine, anion
regiochemistry, **2**, 478
Tigliane diterpenoids
synthesis
via Cope rearrangement, **5**, 984
Tiglic acid
allylation, **3**, 50
allylic oxidation, **7**, 818
hydrogenation, **8**, 552
homogeneous catalysis, **8**, 461
mercurated
demercuration, **8**, 857
Tiglic acid, γ-iodo-
t-butyl ester
alkylation by, **3**, 11
Tiglic aldehyde
Lewis acid complexes
NMR, **1**, 294
Tin
reduction
enones, **8**, 524
Tin, acetonyltributyl-
reaction with aldehydes
aldol reaction, **2**, 611
Tin, α-alkoxyallyl-
anions
synthesis, **2**, 71
Tin, alkynyl-
coupling reactions, **3**, 529
Tin, allenyl-
reactions with isoquinoline, **2**, 86
Tin, allyl-
carbonylation
palladium catalysts, **3**, 1023
coupling reactions
with acyl chlorides, **3**, 463
with aromatic halides, **3**, 453
Tin, allyltributyl-
[3 + 2] cycloaddition reactions
with acyliron complexes, **5**, 277
Tin, aryl-
vinyl substitutions
palladium complexes, **4**, 841
Tin, benzyl-
coupling reactions
with acyl chlorides, **3**, 463
Tin, bis(trimethylsilylpropargyl)diiodo-
reaction with carbonyl compounds, **2**, 82
Tin, chlorotrimethyl-
triphenylphosphoanyldiacetone complex
crystal structure, **1**, 305

Tin, chlorotriphenyl-
tetramethylurea complex
crystal structure, **1**, 305
transmetallation
conjugate enolates, **4**, 260
Tin, crotyl-
reactions with achiral carbonyl compounds, **2**, 18
Tin, cyanomethyltributyl-
cyanomethylation
aryl bromides, **3**, 454
Tin, 1-cyclohexenyloxytributyl-
reaction with benzaldehyde
aldol reaction, **2**, 611
Tin, (α-deuteriobenzyl)tributyl-
acylation, **1**, 444
Tin, diallenyldibromo-
synthesis, **2**, 82
Tin, dichlorodimethyl-
salicylaldehyde complex
crystal structure, **1**, 305
tetramethylurea complex
crystal structure, **1**, 305
Tin, dichlorodiphenyl-
p-dimethylaminobenzaldehyde complex
crystal structure, **1**, 305
Tin, dimethylhalo-
oxidation
retention of configuration, **7**, 615
Tin, hexabutyldi-
photolysis
radical addition reactions, **4**, 754
radical addition reactions
irradiation, **4**, 745
Tin, hydridophenyl-
reaction with α,β-unsaturated carbonyl compounds, **2**, 609
Tin, hydroxymethyl-
coupling reactions
with aromatic halides, **3**, 453
Tin, methoxymethyl-
coupling reactions
with aromatic halides, **3**, 453
Tin, methyl-
coupling reactions
with aromatic halides, **3**, 453
Tin, sulfidobis(trimethyl)-
reaction with α-mercurated ketones
synthesis of enol stannyl ethers, **2**, 609
Tin, tetraphenyl-
reaction with aryl halides, **3**, 504
Tin, trialkylamino-
reaction with carbonyl compounds
synthesis of enol stannyl ethers, **2**, 609
Tin, tri-*n*-butylchloro-
organotin(IV) enol ethers from, **2**, 608
radical reactions, **4**, 738
Tin, tributylmethoxy-
reaction with 2-methyl-1-acetoxycyclohexene
preparation of organotin(IV) enol ethers, **2**, 608
Tin, triethylmethoxy-
reaction with isopropenyl acetate
preparation of organotin(IV) enol ethers, **2**, 608
Tin alkoxides, trialkyl-
reactions with polyols, **6**, 18
Tin compounds
acylation
Friedel–Crafts reaction, **2**, 726
lithium exchange
formation of α-alkoxylithiums, **3**, 195
organopalladium catalysts, **3**, 231
Tin dichloride
catalyst
hydrocarboxylation, **4**, 939
reduction
allylic compounds, **8**, 979
imidoyl chlorides, **8**, 301
nitro compounds, **8**, 365, 371
Tin enolates
aldol reactions, **2**, 255
chiral auxiliary, **2**, 233
stereoselective, acetyliron complex, **2**, 315
α,β-epoxy ketones
synthesis, **2**, 424
synthesis, **2**, 116, 610
Tin enol ethers
formation, **2**, 608
Tin ester enolates
formation, **2**, 610
Tin hydride, tri-*n*-butyl-
allyl trapping reagent, **6**, 641
hydrostannation
carbonyl compounds, **8**, 21
radical reactions, **4**, 738
reaction with acyl phenyl selenides
reductive decarboxylation, **7**, 721
reaction with α,β-unsaturated carbonyl compounds, **2**, 609
reduction
acyl halides, **8**, 265
aldehydes, **8**, 17
carbonyl compounds, **8**, 20
thione thiolates, **8**, 268
transfer hydrogenation, **8**, 553
Tin hydrides
1,4-addition
to α,β-unsaturated carbonyl compounds, **2**, 609
deselenations, **8**, 849
desulfurization, **8**, 844
radical addition reactions
alkenes, **4**, 735–740
syringe pump addition, **4**, 738
radical cyclizations, **4**, 790
catalytic, **4**, 790–796
syringe pump addition, **4**, 790
reduction
quinones, **8**, 19
unsaturated carbonyl compounds, **8**, 547
Tin oxide, bis(tri-*n*-butyl-
oxidation
secondary alcohols, **7**, 320
oxygen transfer agent
alkyl halides, **6**, 3
Tin oxide, dibutyl-
diol protection, **6**, 662
Tin oxyperoxide, dibutyl-
epoxidizing agent, **7**, 379
Tin pinacolate
nucleophilic radical addition
oxime ethers, **4**, 765
radical addition reactions, **4**, 760
Tin tetrachloride
4-*t*-butylbenzaldehyde complex

crystal structure, **1**, 303
catalyst
allylsilane reaction with acetals, **2**, 576
allylstannane reaction with acetals, **2**, 578
epoxide ring opening, **3**, 770
ethyl cinnamate complex
crystal structure, **1**, 305
ketone complexes
crystal structure, **1**, 306
methoxyacetophenone complexes
crystal structure, **1**, 306
Tin triflate
aldol reaction
α-bromo-β-hydroxy ketone synthesis, **6**, 26
Tin triflate, tributyl-
hydrostannation
carbonyl compounds, **8**, 21
transfer hydrogenation, **8**, 553
Tipson–Cohen reaction
alkene protection, **6**, 687
Tirandamycic acid
synthesis, **6**, 750
Tirandamycin
dioxabicyclononane unit
synthesis, **1**, 564
synthesis
via Diels–Alder reaction, **2**, 702
Tirandamycin A
synthesis, **7**, 246
Tischtschenko reaction
aldol reaction, **2**, 137
hydride transfer
aluminum tri-*t*-butoxide, **2**, 138
Titanabicycles
generation, **5**, 1171
Titanacyclic compounds
bicyclization, **5**, 1169
Titanium, 2-alkenyltriphenoxy-
reactions with ketones
diastereoselectivity, **2**, 23
Titanium, alkyl-
reactions with carbonyl compounds, **1**, 145
Titanium, (alkylthio)allyl-
reactions with carbonyl compounds, **1**, 508
Titanium, alkyltris(dimethylamino)-
reaction with carbonyl compounds
chemoselectivity, **1**, 149
Titanium, allenyl-
β-alkynic alcohols from, **2**, 92
configurational stability, **2**, 94
reactions with aldehydes, **2**, 91
diastereoselectivity, **2**, 35
reactions with imines, **2**, 95
Titanium, allyl-
heterosubstituted
reactions with carbonyl compounds, **1**, 161
phosphorus-containing
reactions with carbonyl compounds, **1**, 161
reaction with carbonyl compounds, **1**, 143, 156
Titanium, allyltriisopropoxy-
reaction with allyl-9-borabicyclo[3.3.1]nonane, **2**, 32
reaction with carbonyl compounds, **1**, 156
Titanium, aryl-
reactions with carbonyl compounds, **1**, 145
Titanium, bis(cyclopentadienyl)chloro-
enolates
aldol reaction, *anti* stereoselectivity, **2**, 309
Titanium, η^3-bis(cyclopentadienyl)crotyl-
configurational stability, **2**, 6
Titanium, bis(dibenzyltartramide)tetraalkoxybis-
X-ray crystallography, **7**, 421
Titanium, chloromethyl-
alkyl halide methylation, **3**, 421
Titanium, chlorotris(dimethylamino)-
reaction with aldehydes
diastereoselectivity, **2**, 68
reaction with crotyl carbamate anions
stereoselectivity, **2**, 68
Titanium, crotyl-
reactions with carbonyl compounds, **1**, 158, 340
synthesis, **2**, 5
Titanium, η^1-crotyl-
reactions with achiral carbonyl compounds, **2**, 22
Titanium, η^3-crotyl-
reactions with aldehydes
diastereoselectivity, **2**, 23
Titanium, cyclopentadienyldialkoxy-
enolates
enantioselective aldol reaction, **2**, 308
Titanium, dialkoxy-
chiral catalysts
Diels–Alder reactions, **5**, 376
Titanium, dialkyl-
synthesis, **1**, 143
Titanium, diaryl-
synthesis, **1**, 143
Titanium, dichlorodiisopropoxymethyl-
synthesis, **1**, 142
Titanium, dichlorodimethyl-
reaction with carbonyl compounds
chemoselectivity, **1**, 149
Titanium, dichlorodiphenyl-
reaction with carbonyl compounds
chemoselectivity, **1**, 149
Titanium, dienyl-
reactions with carbonyl compounds, **1**, 162
Titanium, methyl-
chiral ligands
reactions with aromatic aldehydes, **1**, 165
Titanium, methyl(acylpyrrolidinylmethoxy)-
diisopropoxy-
reactions with carbonyl compounds, **1**, 166
Titanium, monoalkyl-
synthesis, **1**, 142
Titanium, monoaryl-
synthesis, **1**, 142
Titanium, phenyl-
chiral ligands
reactions with aromatic aldehydes, **1**, 165
Titanium, propargyl-
reactions with carbonyl compounds, **1**, 165
reaction with aldehydes, **2**, 92
Titanium, tetraisopropoxy-
additive to lithium borohydride
reduction, epoxides, **8**, 880
catalyst
glycolacetal reactions with allylsilanes, **2**, 578
sodium cyanoborohydride
reductive amination, **8**, 54
thioallyl anions
reaction with aldehydes, **2**, 72
transesterification catalyst, **6**, 339

Titanium, tetrakis(dimethylamino)-
  amidine synthesis, **6**, 546
Titanium, trialkoxy-
  enolates
    aldol reaction, *syn* stereoselectivity, **2**, 305
Titanium, trichloro-
  enolates
    stereochemistry of reaction, **2**, 630
Titanium, trichloromethyl-
  properties, **1**, 141
  reaction with 2-benzyloxy-3-pentanone
    NMR, **1**, 295
  reaction with carbonyl compounds
    chemoselectivity, **1**, 149
  synthesis, **1**, 142
Titanium, triisopropoxy-
  enolates
    aldol reaction, *syn:anti* selectivity, **2**, 306
Titanium, triisopropoxymethyl-
  properties, **1**, 141
  reaction with alkoxy ketones, **1**, 153
  reaction with carbonyl compounds
    chemoselectivity, **1**, 145
Titanium, tris(dialkylamino)-
  enolates
    aldol reaction, *syn* stereoselectivity, **2**, 305
Titanium, tris(diethylamino)-
  enolates
    aldol reaction, *syn:anti* selectivity, **2**, 306
Titanium alkoxides
  asymmetric epoxidation, **7**, 390, 395
Titanium alkynides
  reactions with epoxides, **5**, 936
Titanium aluminum methylene
  Tebbe reagent
    alkylidenation, **5**, 1122
Titanium ate complexes
  synthesis, **1**, 143
Titanium *t*-butoxide
  asymmetric epoxidation, **7**, 395
Titanium catalysts
  alkene metathesis, **5**, 1118
  alkylidenation
    carbonyl compounds, **5**, 1122
  bicyclization
    enynes, **5**, 1169
Titanium complexes, alkenyl-
  hydroalumination, **5**, 1124
Titanium compounds
  catalysts
    aliphatic nitro compound reduction, **8**, 375
    asymmetric epoxidation, **7**, 422
    hydroboration, **8**, 709
  chiral Ti atom
    reactions with carbonyl compounds, **1**, 167
  use in intermolecular pinacol coupling reactions
    aliphatic carbonyl compounds, **3**, 570
  use in intramolecular pinacol coupling reactions, **3**, 572
  use in pinacol coupling reactions, **3**, 565
  use in reductive coupling reactions
    carbonyls with alkenes, **3**, 583
Titanium ditriflate
  Claisen condensation, **2**, 802
Titanium enolates
  aldol reaction
    diastereofacial preference, **2**, 224
    enantioselectivity, **2**, 309
    syn stereoselectivity, **2**, 305
  synthesis, **2**, 117
Titanium homoenolates
  reactions, **2**, 445
Titanium isopropoxide
  asymmetric epoxidation, **7**, 395
  epoxide ring opening, **3**, 770
  nucleophilic attack
    epoxides, **7**, 405
Titanium isopropoxide, phenyl-
  synthesis, **1**, 139
Titanium oxametallacycles
  carbonyl methylenation, **5**, 1122
Titanium propionate, 3,3,3-trichloro-
  synthesis, **5**, 1200
Titanium reagents, chirally modified
  enantioselective addition
    carbonyl compounds, **1**, 165
Titanium salts
  reduction
    alkenes, **8**, 531
    carbonyl compounds, **8**, 113
  reductive cleavage
    α-halocarbonyl compounds, **8**, 987
    ketols, **8**, 992
Titanium tartramide complexes
  catalyst
    asymmetric epoxidation, **7**, 424
Titanium tartrate
  catalyst
    asymmetric epoxidation, **7**, 390, 422, 423, 425
    asymmetric epoxidation, mechanism, **7**, 420
    asymmetric epoxidation, reaction variables, **7**, 393
Titanium tartrate, dichlorodiisopropoxy-
  catalyst
    asymmetric epoxidation, **7**, 424
Titanium tetrachloride
  acetic anhydride complex
    crystal structure, **1**, 303
  acryloylmethyl lactate complex
    crystal structure, **1**, 303
  allenylsilanes
    reactions with carbonyl compounds, **1**, 595
  carbonyl compound complexes
    NMR, **1**, 294
  catalyst
    allylsilane reactions, **2**, 567
    allylsilane reactions with acetals, **2**, 576
    allylsilane reactions, diastereoselectivity, **2**, 570
    allylstannane reactions with carbonyl compounds, **2**, 573
    Diels–Alder reaction, **2**, 667
    glycolacetal reactions with allylsilanes, **2**, 578
    Knoevenagel reaction, **2**, 343
  Diels–Alder reaction catalysts
    diastereofacial selectivity, **2**, 679
  diethyl phthalate complex
    crystal structure, **1**, 303
  3,3-dimethyl-2,4-pentanedione complex
    crystal structure, **1**, 303
  enamine synthesis
    dehydrating agent, **6**, 705
  ethyl acetate complex
    crystal structure, **1**, 302

ethyl anisate complex
crystal structure, **1**, 303
lithium aluminum hydride
unsaturated hydrocarbon reduction, **8**, 483
methylenation
carbonyl compounds, **1**, 749
reduction
triazolyl ketones, **8**, 13
vicinal dibromides, **8**, 797
Titanium trichloride
catalyst
Wurtz reaction, **3**, 421
deoxygenation
epoxides, **8**, 889
lithium aluminum hydride
unsaturated hydrocarbon reduction, **8**, 485
reduction
alkyl halides, **8**, 797
carbonyl compounds, **8**, 116
nitro compounds, **8**, 371
vicinal dibromides, **8**, 797
Titanocene
benzyne complex
synthesis, **5**, 1174
diphenylacetylene complex
synthesis, **5**, 1174
synthesis, **1**, 139
Titanocene, crotyl-
reaction with carbonyl compounds, **1**, 158
synthesis, **1**, 143
Titanocene, dimethyl-
synthesis, **8**, 755
Titanocene dichloride
deoxygenation
epoxides, **8**, 889
hydroalumination, **8**, 751
reduction
carbonyl compounds, **8**, 323
Titanocyclobutane
Tebbe reaction, **1**, 743
α-Tocopherol
oxidative coupling
cycloaddition, **3**, 698
synthesis, **3**, 644; **7**, 347; **8**, 560
via cuprate 1,2-addition, **1**, 130
via iterative Claisen rearrangement, **5**, 892
Tollens' reaction
formaldehyde, **2**, 139
o-Tolualdehyde
synthesis, **8**, 301
o-Toluamide, *N*,*N*-dimethyl-
enolates
Mannich reaction, **2**, 928
Toluene
alkylation
Friedel–Crafts reaction, **3**, 300, 304
with *n*-butene, **3**, 304
electrochemical reduction, **8**, 517
hydrogenation
heterogeneous catalysis, **8**, 436
isopropylation
Friedel–Crafts reaction, **3**, 311
photocycloaddition reactions
with cyclooctene, **5**, 655
Toluene, *n*-alkyl-
isomerization
Friedel–Crafts reaction, **3**, 327
Toluene, dihydro-
reaction with pentacarbonyliron, **4**, 668
Toluene, 4-dodecenoyl-
synthesis
via Friedel–Crafts reaction, **2**, 736
Toluene, *p*-fluoro-
catalytic hydrogenation, **8**, 903
Toluene, perfluoro-
alcohol protecting group, **4**, 439
hydrogenolysis, **8**, 901
Toluene, *p*-trimethylsilyl-
Birch reduction
dissolving metals, **8**, 513
Toluene, 2,4,6-trinitro-
Vilsmeier–Haack reaction, **2**, 789
p-Toluenesulfinate, menthyl
synthesis, **6**, 149
p-Toluenesulfinyl chloride
dehydrogenation
thiolactams, **7**, 128
p-Toluenesulfonamide, *N*-sulfinyl-
Diels–Alder reactions, **5**, 424
p-Toluenesulfonates
nucleophilic addition to π-allylpalladium complexes
regioselectivity, **4**, 640
p-Toluenesulfonic anhydride, acetyl-
Friedel–Crafts reaction
bimolecular aromatic, **2**, 739
p-Toluenesulfonylacetonitrile
Michael donor, **4**, 259–262
1-*O*-Toluenesulfonylation
glycoside synthesis, **6**, 49
Toluenesulfonyl azide
diazo transfer reaction, **4**, 1033
p-Toluenesulfonyl chloride
activator
DMSO oxidation of alcohols, **7**, 299
Beckmann rearrangement, **6**, 764
Toluenesulfonyl isocyanates
[3 + 2] cycloaddition reactions
with η^1-allyliron complexes, **5**, 275
reaction with dihydropyrans
glycal synthesis, **5**, 108
o-Toluidine, *N*,*N*-dimethyl-
metal complexes
addition reactions, **4**, 535
p-Toluidine, *N*-ethyl-
Mannich reaction, **2**, 961
Tolylethoxycarbonyl group
protecting group
removal, **6**, 638
p-Tolyl isocyanide
isomerization
kinetics, **6**, 294
Tolylsulfonyl group
amine protecting group
removal, **6**, 644
Tomatidine
reduction, **8**, 228
Topological rule
Michael addition
stereochemistry, **4**, 21
Torreyal
synthesis
via tandem Claisen–Cope rearrangement, **5**, 879

Torulopsis apicola
hydrocarbon oxidation, **7**, 56
Torulopsis candida
β-hydroxylation, **7**, 56
Torulopsis gropengiesseri
hydrocarbon oxidation, **7**, 56
Tosamides
addition reactions
alkenes, **4**, 559
Tosylates
alcohols
hydroxy group activation, **6**, 19
bromination, **6**, 210
chlorination, **6**, 206
Kornblum oxidation
carbonyl compounds, **7**, 654
reactions with carbonyl compounds
organosamarium compounds, **1**, 257
reduction
lithium aluminum hydride, **8**, 812
Tosylhydrazones
hydroxide ion assisted decomposition
synthesis of α-diazo ketones, **3**, 890
Totara-8,11,13-triene, 13-methoxy-
bromination, **7**, 331
Totarol
metabolites, **7**, 331
oxidative coupling
alkaline potassium ferricyanide, **3**, 665
Totarol, 12-hydroxy-
synthesis, **7**, 331
Trachelanthamidine
synthesis
via Eschenmoser coupling reaction, **2**, 881
Trachelanthic acid
synthesis, **1**, 568
Trail pheromones
synthesis
via Eschenmoser coupling reaction, **2**, 881
Trajectory analysis
carbonyl compounds
reduction, **8**, 3
Transacetalization
carbonyl group protection, **6**, 676
Transalkylation
Friedel–Crafts reaction, **3**, 327
Transamination
hydrogenolytic asymmetric, **8**, 146
Transannular alkylation
3,3-like rearrangement, **1**, 890
oxy-Cope rearrangement, **1**, 883
Transannular cyclizations
electrophilic, **3**, 379–407
imines, **3**, 581
Transbenzhydrylation
Friedel–Crafts reaction, **3**, 328
Transbenzylation
Friedel–Crafts reaction, **3**, 329
Trans-*t*-butylation
Friedel–Crafts reaction, **3**, 329
Transesterification
S-alkyl thiocarboxylates, **6**, 443
thiols, **6**, 454
synthesis of esters, **6**, 339
Transfer hydrogenation
alcohols
hydrogen donors, **8**, 551
conjugated alkene bonds
homogeneous catalysis, **8**, 453
nitroarenes, **8**, 367
Transition metal alkyls
hydride transfer, **8**, 103
Transition metal carbonyls
desulfurizations, **8**, 847
Transition metal complexes
acyl
aldol reaction, **2**, 314
alkylidene
alkene metathesis, **5**, 1118
carbenes
cycloaddition reactions, **5**, 1065–1113
catalysts
hydrosilylation, **8**, 764
Claisen rearrangement
catalysis, **5**, 850
η^4-diene
reaction with electrophiles, **4**, 697–705
epoxidation catalysis, **7**, 382
α-hydroxylation
esters, **7**, 182
ketones, **7**, 152
methyl pyrrolidine
alkylation, **5**, 1076
oxidation
sulfoxides, **7**, 768
Transition metal enolates
acyl
aldol reaction, **2**, 301
aldol reaction, **2**, 301–318
structure, **2**, 301
Transition metal halides
metal hydrides
reduction, mechanism, **8**, 483
unsaturated hydrocarbon reductions, **8**, 483
reactions with organolithium compounds
complex Lewis acid reagent, **1**, 330
Transition metal hydrides
reduction
carbonyl compounds, **8**, 22
unsaturated carbonyl compounds, **8**, 548
Transition metal ions
electroreduction
carbonyl compounds, **8**, 133
Transition metal nucleophiles
reactions with π-allylpalladium complexes, **4**, 600
Transition metals
catalysts
cycloaddition reactions, **5**, 271–312
Lewis acids, **1**, 307
reductions
nitro compounds, **8**, 371
Transition states, boat-like
Diels–Alder reactions
decatrienones, **5**, 539–543
Transketolase
organic synthesis
carbon–carbon bond formation, **2**, 456
use in, **2**, 464, 465
Transmetallation
acylation
organostannanes, **1**, 444
zirconium compounds, **8**, 692

Transposition reactions
  1,3-heteroatom, **6**, 834
  1,3-heteroatom–carbon, **6**, 847
  1,3-heteroatom–hydrogen, **6**, 865
  1,3-hydrogen–hydrogen, **6**, 866
  oxygen–halogen, **6**, 834
  oxygen–nitrogen, **6**, 842
  oxygen–oxygen, **6**, 835
  oxygen–phosphorus, **6**, 844
  oxygen–sulfur, **6**, 837
  sulfur–nitrogen, **6**, 846
  sulfur–sulfur, **6**, 846
Transthioacetalization
  carbonyl group protection, **6**, 679
Transylidation
  phosphonium salt deprotonation, **6**, 175
  phosphonium ylides, **6**, 177
Trehalose
  synthesis, **6**, 54
Treibe's reaction
  allylic oxidation, **7**, 637
Trialkylamines
  addition to activated alkynes, **4**, 48
Trialkylsilyl groups
  nucleophilic addition reactions
    stereoselectivity, **1**, 58
Triamantane
  synthesis
    via Friedel–Crafts reaction, **3**, 334
Triannulenes
  synthesis
    via intramolecular acyloin coupling reaction, **3**, 627
Triarylamines
  synthesis, **4**, 434
Triarylmethyl groups
  alcohol protection, **6**, 23
Triazenes
  reductive cleavage
    synthesis of amines, **8**, 383
1,2,4-Triazine
  aromatic nucleophilic substitution, **4**, 432
  reaction with nitronate anions, **4**, 424
1,3,5-Triazine
  amidine synthesis, **6**, 554
1,3,5-Triazine, 6-methyl-
  Knoevenagel reaction, **2**, 364
1,3,5-Triazine, 1,3,5-trialkylhexahydro-
  cleavage
    generation of *N*-silyliminium salts, **2**, 914
  synthesis
    via trimerization of formaldehyde imines, **1**, 361
Triazines
  oxidation, **7**, 750
  synthesis
    via Diels–Alder reactions, **5**, 412
1,2,3-Triazines
  Diels–Alder reactions, **5**, 491
1,2,4-Triazines
  Diels–Alder reactions, **5**, 491
1,3,5-Triazines
  Diels–Alder reactions, **5**, 491
Triazinetriones
  synthesis
    via [2 + 2 + 2] cycloaddition, **5**, 1158
1,2,3-Triazole, 1-acyloxy-
  synthesis, **6**, 329
1,2,4-Triazole, 4-amino-
  nitrobenzene amination, **4**, 436
4*H*-1,2,4-Triazole, 4-(4-chlorophenyl)-
  quaternary salts of
    benzoin condensation, catalysis, **1**, 543
Triazole, phosphoryl-
  phosphorylation, **6**, 614
Triazoles
  N-alkyl
    lithiation, **1**, 477
  synthesis
    via azide cyclization, **4**, 1157
    via hydrazoic acid and alkynes, **4**, 296
1,2,3-Triazoles
  reduction, **8**, 661
  synthesis
    via deamination of 1-aminotriazoles, **7**, 744
    via 1,3-dipolar cycloadditions, **4**, 1099, 1100
  Wolff rearrangement, **3**, 909
1,2,4-Triazoles
  metallation
    addition reactions, **1**, 471
1,2,3-Triazoles, 1-aryl-5-amino-
  oxidation, permanganate
    amide synthesis, **6**, 402
Triazolide, phosphorobis-
  phosphorylating agent, **6**, 619
Triazoline, *N*-vinyl-
  decomposition
    aziridine synthesis, **7**, 475
Triazolinedione
  cycloaddition reactions, **5**, 206
1,2,4-Triazoline-3,5-dione, 4-phenyl-
  Diels–Alder reactions, **5**, 428
  diene protection, **6**, 690
  oxidative rearrangement, **7**, 833
1,2,4-Triazoline-3,5-diones
  Diels–Alder reactions, **5**, 429
Triazolines
  aziridine synthesis, **7**, 475
  synthesis
    via azide cyclization, **4**, 1157
    via 1,3-dipolar cycloadditions, **4**, 1099
  thermolysis
    photolysis, **7**, 476
Δ^2-Triazolines, 5-substituted
  synthesis
    via 1,3-dipolar cycloadditions, **4**, 1099
1,2,4-Triazolin-5-one, 1-aryl-
  synthesis
    via Curtius reaction, **6**, 815
1,2,4-Triazolium chloride, 3-methylthio-1,4-diphenyl-
  masked carboxylate equivalent, **8**, 662
1,2,4-Triazolium salts
  reduction, **8**, 662
    metal hydrides, **8**, 276
4*H*-[1,2,4]Triazolo[4,3-*a*][1,4]benzodiazepine
  Mannich reaction
    with *N,N*-dimethylmethyleniminium chloride, **2**, 962
Triazolocephem
  synthesis
    via diazoalkene cyclization, **4**, 1159
1,2,4-Triazol-5-ones
  reduction
    LAH, **8**, 662

Triazolopenam
synthesis
via diazoalkene cyclization, **4**, 1159
2,2,2-Tribromoethyl group
phosphoric acid protecting group, **6**, 625
1,2,3-Tricarbonyl compounds
diazo-coupling reactions, **3**, 893
β-Trichloro-*t*-butyloxycarbonyl group
protecting group
amines, **6**, 638
2,2,2-Trichloro-1,1-dimethyl group
phosphoric acid protecting group, **6**, 625
2,2,2-Trichloroethyl carbonate
alcohol protection
cleavage, **6**, 659
2,2,2-Trichloroethyl group
phosphoric acid protecting group, **6**, 625
Trichloromethyl chloroformate
activator
DMSO oxidation of alcohols, **7**, 299
Trichodermol
synthesis
via cyclohexadienyl complexes, **4**, 680, 682
Trichodiene
synthesis
via cyclohexadienyl complexes, **4**, 682
via Diels–Alder reaction, **5**, 338
via Nazarov cyclization, **5**, 758
Trichosporum fermentans
β-hydroxylation, **7**, 56
Trichostatin A
synthesis
via protected cyanohydrins, **3**, 198
Trichothec-9-ene, 12,13-epoxy-
synthesis
via 4-methyl-2-cyclohexenone, **1**, 522
Trichothecenes
epoxide opening, **3**, 736
synthesis, **3**, 714
via cyclohexadienyl complexes, **4**, 680, 681
via sulfur ylide reagents, **1**, 822
Trichoviridine
metabolite
synthesis *via* photocycloaddition, **5**, 167
Tricosene
synthesis
via alkene metathesis, **5**, 1118
Tricyclic compounds
Cope rearrangements, **5**, 818
Tricyclic diterpenoids
synthesis, **3**, 325
Tricyclo[4.2.2.$0^{2,5}$]deca-3,7-diene-9,10-dicarboxylic acid
oxidation, **7**, 462
Tricyclo[5.2.1.$0^{4,10}$]decane
synthesis
via Pauson–Khand reaction, **5**, 1058
Tricyclo[6.2.0.$0^{2,7}$]decane
synthesis
via thermal ene reaction, **2**, 553
Tricyclo[5.2.1.$0^{2,6}$]decane, 4-allylidene-
synthesis
via metal-catalyzed codimerization, **5**, 1190
Tricyclo[5.2.1.$0^{2,6}$]decane, 4-methylene-3-vinyl-
synthesis
via metal-catalyzed codimerization, **5**, 1190
Tricyclodecanes
dihetero
synthesis *via* cyclofunctionalization, **4**, 373
Tricyclo[5.2.1.$0^{4,10}$]decane-2,5,8-trione
synthesis
via Pauson–Khand reaction, **5**, 1062
Tricyclodecenols, *exo*-methylene-
synthesis
via retro Diels–Alder reactions, **5**, 562
Tricyclo[5.3.0.$0^{2,6}$]dec-4-en-3-ones
synthesis
via Pauson–Khand reaction, **5**, 1046
Tricyclo[2.2.1.$0^{2,6}$]heptan-3-one
oximes
reduction, dissolving metals, **8**, 124
Tricyclo[3.1.0.$0^{2,4}$]hexane, 3,3,6,6-tetramethyl-
synthesis
via metal catalyzed cyclodimerization, **5**, 1197
Tricyclohumuladiol
synthesis, **3**, 399, 402
Tricyclo[4.3.0.$0^{2,5}$]non-3-ene
synthesis
via photolysis, **5**, 737
Tricyclo[3.2.2.$0^{2,4}$]non-6-ene-8,9-dione
irradiation, **5**, 713
Tricyclo[4.2.1.$0^{2,5}$]non-7-enes
synthesis
via cycloaddition of quadricyclane, **5**, 1187
syn-Tricyclo[4.2.0.$0^{2,5}$]octa-3,7-dienes
synthesis
via isomerization of cubenes, **5**, 1188
Tricyclo[4.2.0.$0^{1,4}$]octane
synthesis, **3**, 901
Tricyclo[3.3.0.$0^{2,8}$]octane-4,7-diones
photoisomerizations, **5**, 227
Tricyclo[3.2.1.$0^{2,7}$]octan-6-one
synthesis
via tandem Michael reactions, **4**, 121
Tricyclo[3.3.0.$0^{2,8}$]octan-3-one
synthesis
via photoisomerization, **5**, 200
Tricyclo[3.2.1.$0^{3,6}$]octan-4-one, 6,7,7-trimethyl-
synthesis, **3**, 906
Tricyclooctanones
synthesis
via Michael addition, **4**, 18
exo-Tricyclo[3.2.1.$0^{2,4}$]oct-6-ene
cycloaddition reactions, **5**, 1187
Tricyclo[2.1.1.$0^{4,5}$]pentane
synthesis, **3**, 894
Tricyclopentanoids
angularly-fused
synthesis *via* photocycloaddition, **5**, 662
Tricyclo[6.4.2.0.$2^{3,6}$]tetradeca-1(8),4,13-triene
isomerization
via retro Diels–Alder reactions, **5**, 585
Tricyclo[5.4.0.$0^{2,8}$]undecane
synthesis
via bromomethoxylation of alkene, **4**, 355
Tricyclo[6.3.0.$0^{1,5}$]undecane
synthesis
via Pauson–Khand reaction, **5**, 1057
Tricyclo[5.2.2.$0^{2,6}$]undecanediones
synthesis
via photoisomerization, **5**, 233
Tricyclo[6.3.0.$0^{2,6}$]undecanes

synthesis
via photocycloaddition, **5**, 664
trans-Tricyclo[5.4.0.0^{4,6}]undec-2-ene
synthesis
via thermal rearrangement, **5**, 716
Tridecanenitrile
reduction, **8**, 253
Trienes
arylation
palladium complexes, **4**, 849
bis-annulated
electrocyclic reactions, **5**, 721–728
conjugated
synthesis, **3**, 879, 882
cyclic
synthesis *via* retro Diels–Alder reactions, **5**, 573
cycloaddition reactions
dienes, **5**, 632–635
Diels–Alder reactions
boat-like transition states, **5**, 539–543
intramolecular asymmetric, **5**, 543
hydroboration, **8**, 705
hydrogenation to saturated hydrocarbons
homogeneous catalysis, **8**, 449
1,2-reduction to alkenes
homogeneous catalysis, **8**, 449
saturated connecting chains
Diels–Alder reactions, **5**, 533–539
vinylation
palladium complexes, **4**, 855
η^4-Trienes
transition metal complexes
reactions with electrophiles, **4**, 706
tricarbonylmanganese complexes, **4**, 712
1,2,7-Trienes
cyclization
via nickel-ene reaction, **5**, 57
1,3,5-Trienes
Vilsmeier–Haack reaction, **2**, 782
Trienoic acids
synthesis, **3**, 882
Trienones
Robinson annulation, **4**, 8
Triethylamine
alcohol oxidation
DMSO, 7, 292
α-deprotonation, **1**, 476; **3**, 65
reaction with arynes, **4**, 505
Triethylamine, 2-chloro-1,1,2-trifluoro-
fluorination
alkyl alcohols, **6**, 217
Triethyloxonium tetrafluoroborate
O-alkylation of enolates
regioselectivity, **2**, 597
Triflamide, *N*-4-acetoxyphenyl-
oxidation
halides, **7**, 668
Triflamides
oxidation
alkyl halides, **7**, 668
Triflate, trimethylsilyl
promoter
Diels–Alder reactions, **5**, 341
synthesis, **7**, 650
Triflates
alcohols
hydroxy group activation, **6**, 19
bromination, **6**, 210
catalysts
Friedel–Crafts reaction, **3**, 295
catalytic hydrogenation, **8**, 817
chlorination, **6**, 206
fluorination, **6**, 218
glycoside synthesis, **6**, 56
vinyl substitutions
palladium complexes, **4**, 835, 842, 858
Triflates, alkyl
alkynation
carbonyl phosphine carbene complexes, **5**, 1077
Triflic acid
promoter
Diels–Alder reactions, **5**, 341
Triflic hydrazides
oxidation
alkyl halides, **7**, 668
Triflone, mesyl-
alkylation, **3**, 868
Trifluoroacetylation
heterocycles
comparison of rates, **2**, 735
Trifluoroacetyl group
amine-protecting group, **6**, 642
Triglycerides
esters, **6**, 324
1,1,1-Trihalides
thiolysis, **6**, 432
Trihaptotitanium compounds
synthesis, **1**, 159
Triisopropylsilyl protecting groups
nucleophilic addition reactions
effect on stereoselectivity, **1**, 62
Trikentriorhodin
synthesis
via Claisen condensation, **2**, 821
Triketones
aldol cyclization, **2**, 165
Trimethylamine
deprotonation, **1**, 476
Trimethylamine *N*-oxide
cyclohexadienyliron complexes, decomplexation, **4**, 674
oxidant
C—Si bonds, **7**, 641
oxidation of organoboron derivatives, **7**, 597
secondary oxidant
osmium tetroxide oxidation, **7**, 439
Trimethyl α-phosphonoacrylate
ene reactions
Lewis acid catalysis, **5**, 5
Trimethylsilyl bromide
halogen transfer agent
acid halide synthesis, **6**, 306
4-(Trimethylsilyl)-2-buten-1-yl esters
carboxy-protecting groups, **6**, 670
Trimethylsilyl chloride
ketone synthesis
via carboxylic acids, **1**, 411
Trimethylsilyl chlorochromate
oxidation
alcohols, **7**, 283
Trimethylsilyl cyanide
acyl cyanide synthesis, **6**, 234

isocyanide synthesis, **6**, 243
nitrile synthesis, **6**, 229
Trimethylsilyl enolates
exchange reaction
dialkylboryl triflates, **2**, 244
Trimethylsilyl esters
acid halide synthesis, **6**, 307
2-(Trimethylsilyl)ethyl carbonate
alcohol protection
cleavage, **6**, 659
2-Trimethylsilylethylethoxycarbonyl group
amine-protecting group
cleavage, **6**, 639
Trimethylsilyl group
Beckmann fragmentation
stereochemistry, **6**, 774
Trimethylsilyl iodide
catalyst
enol silanes, reaction with acetals, **2**, 635
halogen transfer agent
acid halide synthesis, **6**, 306
Trimethylsilylmethoxymethyl carbanion
Peterson alkenation, **2**, 597
Trimethylsilyl perchlorate
carboxy-protecting group
chemoselective differentiation, **6**, 669
Trimethylsilylthioaldehyde *S*-methylide
thiocarbonyl ylide generation from, **4**, 1095
Trimethylsilyl triflate
catalyst
allylsilane reaction with acetals, **2**, 576
enol silanes, reaction with acetals, **2**, 635
glycoside synthesis, **6**, 49
Trinoranastreptene
synthesis
via Julia coupling, **1**, 804
Triols
synthesis, **7**, 645
1,2,6-Triols
glycol cleavage
δ-lactone synthesis, **6**, 366
Triones
enzymic reduction
specificity, **8**, 193
1,2,3-Triones
synthesis, **7**, 656
Triorganocuprates
properties, **4**, 170
1,1,1-Triorganometallics
synthesis
via carbozincation, **4**, 879
3,4,10-Trioxa-3-adamantyl group
carboxy group protection
organometallic transformation, **6**, 673
3,6,8-Trioxabicyclo[3.2.1]octane
reduction, **8**, 227
Trioxane
aldol reaction
ketones, **2**, 146
synthesis
via Paterno–Büchi reaction, **5**, 154
1,2,4-Trioxane
synthesis
via Paterno–Büchi reaction, **5**, 157
Tripeptides
phospho analogs
synthesis, **2**, 1097
Triphenylene, dihydro-
ring opening, **5**, 729
Triphenylenes
synthesis
via electrocyclization, **5**, 720
via photolysis, **5**, 729
Triphenylmethane carbanion
lithium (12-crown-4) complex
crystal structure, **1**, 11
N-Triphenylmethyl group
amine protection, **6**, 644
Triphenylmethyl tetrafluoroborate
oxidation
diols, **7**, 316
Triphenylphosphonioethoxycarbonyl group
protecting group
amines, **6**, 638
Triphosgene
reaction with amides, **6**, 495
Triphyophylline, *O*-methyltetradehydro-
synthesis
via photochemical oxidation, **3**, 677
Triple asymmetric synthesis
aldol reaction, **2**, 265
Triprolidine
microbial hydroxylation, **7**, 76
Triptycenes
photofragmentation, **5**, 209
synthesis
via Diels–Alder reaction, **5**, 383
Triquinacenes
substituted derivatives
synthesis *via* tricarbonyl(cyclooctatetraene)iron complexes, **4**, 710
synthesis
via Pauson–Khand reaction, **5**, 1058, 1062
Triquinanes
angularly fused
synthesis *via* Pauson–Khand reaction, **5**, 1047, 1052, 1057, 1061
synthesis *via* photocycloaddition, **5**, 662
cyclization, **5**, 759
linearly fused
synthesis *via* photocycloaddition reactions, **5**, 145
synthesis *via* Pauson–Khand reaction, **5**, 1052, 1060
ring opening, **5**, 926
synthesis, **3**, 384; **5**, 951
via cyclopropane ring opening, **4**, 1048
via Nazarov cyclization, **5**, 763
via photochemical rearrangement, **5**, 916
via vinylcyclopropane thermolysis, **4**, 1048
Triquinane sesquiterpenes
synthesis
via organocopper compounds, **3**, 221
Tris(cetylpyridinium) 12-tungstophosphate
glycol cleavage, **7**, 708
Tris(diethylamino)sulfonium difluorotrimethylsilicate
reaction with organic halides
palladium catalysis, **3**, 233
Tris(3,6-dioxaheptyl)amine
catalyst
chloropyridine dechlorination, **4**, 439
Trishomoallylic alcohol
asymmetric epoxidation, **7**, 419

Trishomocubane, fluoro-
  Ritter reaction, **6**, 270
Trisporone, deoxy-
  synthesis, **3**, 169
Triterpenes
  acyclic
    microbial hydroxylation, **7**, 62
  synthesis
    via benzocyclobutene ring opening, **5**, 693
    via polyalkene cyclization, **3**, 364
2,6,7-Trithiabicyclo[2.2.2]octane, 1-lithio-4-methyl-
  alkylation, **3**, 145
1,3,5-Trithiane
  metallation, **3**, 134
Trithiodicarbonic acid
  O,O-diethyl ester
    alkoxycarbonylation, **2**, 840
Trithiomalonate, *O,S*-diethyl
  synthesis
    via O,O-diethyl dithiomalonate, **6**, 454
S-Trityl group
  thiol protection, **6**, 664
Trityl hydroperoxide
  asymmetric epoxidation, **7**, 394
    trishomoallylic alcohol, **7**, 419
Tritylone ethers
  Wolff–Kishner reduction, **8**, 343
Tritylon group
  alcohol protection, **6**, 650
Trityl perchlorate
  catalyst
    aldol reaction, **2**, 632
    allylsilane reaction with acetals, **2**, 576
    enol silanes, reaction with acetals, **2**, 635
Trityl tetrafluoroborate
  hydride transfer reagent, **7**, 244
Trityl triflate
  catalyst
    aldol reaction, **2**, 632
Trogodermal
  synthesis, **3**, 243
Tropaldehyde, hydro-
  synthesis
    via hydroformylation, **4**, 930
Tropane alkaloids
  synthesis
    via [4 + 3] cycloaddition, **5**, 609
Tropanediol
  synthesis
    via [4 + 3] cycloaddition, **5**, 609
Tropidine
  hydroformylation, **4**, 927
  photochemistry
    retro Diels–Alder reaction, **5**, 586
Tropine
  synthesis
    via [4 + 3] cycloaddition, **5**, 609
α-Tropolone
  synthesis
    via dihalocyclopropyl compounds, **4**, 1018
Tropolone, amino-
  rearrangements, **3**, 818
Tropolone, 4-methyl-
  reaction with dipiperidylmethane
    Mannich reaction, **2**, 958
Tropolonecarboxylic acid
  rearrangements, **3**, 818
Tropolone–phenol rearrangements, **3**, 817
Tropolones
  synthesis
    via ketocarbenoid reaction with benzenes, **4**, 1052
γ-Tropolones
  synthesis
    [4 + 3] cycloaddition reactions, **5**, 609
Tropone, 2-chloro-
  cycloaddition reactions, **5**, 620
    fulvenes, **5**, 631
Tropone, 2,7-dialkyl-
  synthesis
    via [4 + 3] cycloaddition reactions, **5**, 609
Tropone, 2-methyl-
  iron tricarbonyl complex
    [3 + 2] cycloaddition reactions, **5**, 274
Tropones
  cycloaddition reactions, **5**, 618–626
    fulvenes, **5**, 631
  substituted
    synthesis *via* retro Diels–Alder reactions, **5**, 573
  synthesis, **7**, 796
    via [4 + 3] cycloaddition, **5**, 604
    via dihalocyclopropyl compounds, **4**, 1018
    via ketocarbenoid reaction with benzenes, **4**, 1052
Tropones, ethoxycarbonyl-
  cycloaddition reactions
    dienes, **5**, 621
Tropones, methoxy-
  cycloaddition reactions
    dienes, **5**, 621
Troponoids
  synthesis
    via [4 + 3] cycloaddition, **5**, 605, 609
[4](2,7)Troponophane
  synthesis
    via [6 + 4] cycloaddition, **5**, 623
Tropylium ions
  reactions with allenylsilanes, **1**, 603
  reactions with isocyanides, **6**, 294
Tryptamine
  thioacylation
    dithiocarboxylates, **6**, 423
Tryptamine, α-alkyl-
  synthesis, **8**, 375
Tryptophans
  enantioselective aldol cyclizations, **2**, 169
  synthesis, **7**, 335
    via Mannich reaction, **2**, 967
  thioacylation
    dithiocarboxylates, **6**, 423
Tuberculostearic acid
  synthesis, **3**, 644
Tulipalin A
  synthesis
    via retro Diels–Alder reactions, **5**, 577
Tungstates, hydridopentacarbonyl-
  reduction
    acyl chlorides, **8**, 289
Tungsten
  metal vapor synthesis
    reactions with alkanes, **7**, 4
Tungsten acid
  anti hydroxylation
    alkenes, **7**, 446

Tungsten catalysts
  alkene metathesis, **5**, 1118
  alkylidenation
    carbonyl compounds, **5**, 1122
Tungsten complexes, alkyl carbene
  coupling reactions
    acyclic products, **5**, 1103
Tungsten complexes, alkylidene-
  carbonyl alkylidenation, **5**, 1125
Tungsten complexes, peroxo
  epoxidations with, **7**, 382
Tungsten complexes, propenyl-
  reaction with Danishefsky's diene, **5**, 1072
Tungsten complexes, propynyl-
  [2 + 2] cycloaddition reactions, **5**, 1067
  cycloaddition reactions
    ethyl vinyl ether, **5**, 1073
Tungsten complexes, vinyl-
  cycloaddition reactions, **5**, 1072
Tungsten enolates
  aldol reaction, **2**, 312
  synthesis and reaction, **2**, 127
Tungsten halides
  deoxygenation
    epoxides, **8**, 888
Tungsten hexachloride
  catalyst
    alkene metathesis, **5**, 1116
  deoxygenation
    epoxides, **8**, 889
Tungsten hexafluoroantimate, tricarbonyl-
  nitroso(trimethylphosphine)-
  crystal structure, **1**, 309
Tungsten oxide
  anti hydroxylation
    alkenes, **7**, 446
Tungsten salts
  reduction
    alkenes, **8**, 531
Tunicaminyluracil
  synthesis
    via Diels–Alder reaction, **2**, 697
    via retro Diels–Alder reaction, **5**, 553
Turmerone
  synthesis, **2**, 804; **3**, 28, 126; **6**, 455
    via conjugate addition to oxazolines, **4**, 206
Turneforcidine
  synthesis
    via cycloazasulfenylation of alkenes, **4**, 333
Tutin
  synthesis, **7**, 243
Twistane
  synthesis
    via cyclofunctionalization of cycloalkenes, **4**, 373
Twist asynchronicity
  Diels–Alder reactions, **5**, 516
Twistbrendane
  synthesis, **3**, 854
exo-Twistbrendan-2-ol
  brosylate
    acetolysis, **3**, 709
Tyllophorine
  synthesis
    via conjugate addition, **4**, 231
Tylonolide
  synthesis, **2**, 257; **7**, 246
    via macrolactonization, **6**, 370
Tylophorine
  synthesis
    via Friedel–Crafts reaction, **2**, 740
    via thallium trifluoroacetate, **3**, 670
    via vanadium oxytrifluoride, **3**, 670
Tylosin
  aglycones
    synthesis, **1**, 436
  synthesis
    via cycloheptadienyliron complexes, **4**, 686
L-Tyrosine
  microbial hydroxylation, **7**, 78
  synthesis, **3**, 816
Tyrosine, 4-picolyl-
  cleavage, **8**, 974
Tyrosine *O*-methyl ether
  enantioselective aldol cyclizations, **2**, 167

U

Ugi reaction, **2**, 1083–1106
  amide synthesis, **6**, 405
  conditions, **2**, 1089
  general features, **2**, 1087
  limitations, **2**, 1087
  mechanism, **2**, 1090
  preparative advantages, **2**, 1089
  scope, **2**, 1087
  side reactions, **2**, 1092
  stereochemistry, **2**, 1090
  syntheses, **2**, 1094
Ullmann reaction
  biaryl synthesis, **3**, 482, 499
  organocopper compounds, **3**, 209, 219
Ultrasonic irradiation
  C—P bond cleavage, **8**, 858
  hydrosilylation
    unsaturated hydrocarbons, **8**, 764
  nitrene generation, **7**, 477
  reduction
    dissolving metals, **8**, 109
  Reformatsky reaction, **2**, 279, 296
  Reimer–Tiemann reaction, **2**, 772
Ultraviolet irradiation
  hydrosilylation
    unsaturated hydrocarbons, **8**, 764
Umbelliferone
  synthesis
    via Vilsmeier–Haack reaction, **2**, 790
Umbellone
  photochemistry, **5**, 730
Umpolung
  β-acyl anions
    homoenolates, **2**, 442
Undeca-2,5-diyl-1-ol
  synthesis, **3**, 273
2-Undecanone
  reduction
    ionic hydrogenation, **8**, 318
Undecanone, dibromo-
  rearrangement, **3**, 851
2,7,9-Undecatriene, 2-methyl-
  Diels–Alder reactions
    intramolecular, **5**, 522
10-Undecenal
  synthesis, **8**, 297
Undecenoic acid
  oxidation
    Wacker process, **7**, 450
1-Undecen-3-ol
  oxidation
    Wacker process, **7**, 453
Undivided cells
  electrosynthesis, **8**, 130
Unsaturated compounds
  anodic oxidation, **7**, 794
α,β-Unsaturated esters
  Dieckmann reaction, **2**, 817
Untriacontane, 3-methyl-
  synthesis, **3**, 414
Upial
  synthesis, **7**, 817
Uracil
  fluorination, **7**, 535
Uracil, (azidofuranosyl)-
  cyclization, **4**, 1158
Uracil, *N*-benzyl-
  reduction
    L-selectride, **8**, 642
Uracil, 5-bromo-
  reduction, **8**, 908
Uracil, dihydro-
  dehydrogenation
    copper(II) bromide, **7**, 144
    use of enzymes, **7**, 146
Uranium complexes
  carbonyl methylenation, **5**, 1126
Urazole
  synthesis
    via cycloaddition, **5**, 206
Urea
  Vilsmeier–Haack reaction, **2**, 791
Urea, *N*-nitroso-
  carbene precursors, **4**, 961
Urea, tetramethyl-
  aluminum trichloride complex
    crystal structure, **1**, 301
  chlorotriphenyltin complex
    crystal structure, **1**, 305
  dichlorodimethyltin complex
    crystal structure, **1**, 305
  lithium halide sensitizer
    epoxide ring opening, **3**, 763
Urea nitrate
  nitration with, **6**, 110
Urethane, *N,N*-dichloro-
  reactions with alkenes, **7**, 498
Urethanes
  Diels–Alder reactions
    intramolecular, **5**, 527
  lithiation
    addition reactions, **1**, 469
  protecting groups
    peptide synthesis, **6**, 635
  vinylogous
    synthesis *via* Eschenmoser coupling reaction, **2**, 865, 867
    synthesis *via* Knoevenagel reaction, **2**, 368
Urethanes, α-cyano-
  synthesis, **1**, 559
Uridine, alkenyldeoxy-
  synthesis, **8**, 694
Uridine, allyl-
  synthesis
    via allylation of mercury intermediates, **3**, 476
β-Uridine, 3′-*O*-benzyl-2′-deoxy-5-trifluoromethyl-
  synthesis
    via reductive desulfurization, **6**, 447
Uridine, 2-deoxy-
  quinone derivatives
    synthesis, **7**, 350
Uronic acid, amino-

synthesis
via bromolactamization of silyl imidate, **4**, 399
Uroporphyrins
synthesis
via Knoevenagel reaction, **2**, 376
Ursolinic acid
ring A contraction, **3**, 834

V

Valeraldehyde
  synthesis
    via hydroformylation, **4**, 922
Valerane
  synthesis, **3**, 20
Valeric acid, δ-bromo-
  reactions with samarium diiodide
    lactone synthesis, **1**, 259
Valeric acid, 5-(2,3-dimethoxyphenyl)-
  Friedel–Crafts reaction, **2**, 764
Valeric acid, 5-(4-isopropylphenyl)-
  Friedel–Crafts reaction, **2**, 764
γ-Valerolactone
  acylation, **1**, 418
  benzene alkylation by
    Friedel–Crafts reaction, **3**, 317
  hydrogenation, **8**, 246
Valerolactone, 5-ethenyl-
  synthesis, **3**, 245
Valerolactone, β-hydroxy-
  synthesis
    via SmI_2-promoted reductive cyclizations, **1**, 267
Valeryl chloride, 5-(2-acetoxy-3-methoxyphenyl)-
  Friedel–Crafts reaction, **2**, 764
Valine
  N-acyl-2-oxazolidone from, **2**, 251
  bislactim ethers from, **3**, 53
  borane modifier
    asymmetric reduction, **8**, 170
  t-butyl ester
    imine anion alkylation, **6**, 726
  t-butyl ester, enamines
    alkylation, **3**, 36
  enantioselective aldol cyclizations, **2**, 169
Valinol
  imines
    reactions with organometallic compounds, **1**, 363
Valinol, *O*-*t*-butyl-
  imines
    reactions with organolithium reagents, **1**, 383
Vallerenal
  synthesis
    via Nazarov cyclization, **5**, 780
Vallesiachotamine
  synthesis
    via Knoevenagel reaction, **2**, 384
Vanadates, tricarbonylcyclopentadienylhydrido-
  reduction
    acyl chlorides, **8**, 289
Vanadium compounds
  glycol cleavage, **7**, 707
  use in intermolecular pinacol coupling reactions, **3**, 565
Vanadium dichloride
  reduction
    carbonyl compounds, **8**, 116
Vanadium salts
  reduction
    alkenes, **8**, 531
Vanadium sulfate
  reductions
    nitro compounds, **8**, 371
Vanadium trichloride
  catalyst
    Wurtz reaction, **3**, 421
  lithium aluminum hydride
    unsaturated hydrocarbon reduction, **8**, 485
  reduction
    vicinal dibromides, **8**, 797
Vanadyl acetylacetonate
  allylic oxidation, **7**, 95
Vanadyl bisacetylacetonate
  glycol cleavage, **7**, 707
  oxidation
    secondary alcohols, **7**, 321
Vancosamine
  amino sugars, **2**, 323
Vapor-phase irradiation
  di-π-methane rearrangement, **5**, 195
Vaska's complex
  hydrogenation
    alkenes, **8**, 446
    methyl acrylate, **8**, 453
Velleral
  synthesis
    via cyclobutene ring expansion, **5**, 687
Venustatriol
  synthesis, **7**, 633
    via cerium reagent, **1**, 237
Veraguensin
  synthesis, **3**, 693
Veratrole
  electrolytic oxidation, **3**, 668
  Friedel–Crafts acylation, **2**, 737
Veratrole, 4-methyl-
  oxidative coupling, **3**, 669
Veratronitrile
  intramolecular Ritter reaction, **6**, 272
Verbenene
  synthesis
    via methyllithium reaction, **1**, 377
Verbenol
  allylic oxidation, **7**, 99
  asymmetric epoxidation, **7**, 414
  synthesis, **3**, 126
Verbenone
  allylic oxidation, **7**, 99
  tosylhydrazone
    reaction with methyllithium, **1**, 377
Vermiculin antibiotics
  synthesis, **3**, 126
Vermiculine
  synthesis, **1**, 568; **8**, 647
    via macrolactonization, **6**, 371
Vernolepin
  synthesis, **3**, 280; **7**, 105
    via cyclopropane ring opening, **5**, 924
    via Diels–Alder reactions, **5**, 330, 345
    via Knoevenagel reaction, **2**, 381
    via Mannich reaction, **2**, 911
Vernomenin
  synthesis
    via cyclopropane ring opening, **5**, 924
Verrucarin

synthesis
via carboalumination, **4**, 893
Verrucarine E
synthesis
via retro Diels–Alder reactions, **5**, 581
Verrucarinic acid
synthesis, **7**, 240
Verrucarol
synthesis, **6**, 143
via cyclohexadienyl complexes, **4**, 680
via ene reaction, **2**, 547
Verticillene
synthesis, **3**, 591
Veticadinol
synthesis
via intramolecular ene reaction, **5**, 17
via Knoevenagel reaction, **2**, 373
α-Vetispirene
synthesis, **3**, 586
via Wacker oxidation, **7**, 455
β-Vetivone
synthesis, **3**, 20, 22
via conjugate addition, **4**, 211
Vicarious nucleophilic substitution
arenes, **4**, 424
Vicinal dialkylation
tandem
definition, **4**, 238
Vicinal difunctionalization
tandem, **4**, 238
definition, **4**, 238
electrophiles, **4**, 259
nucleophiles, **4**, 253–259
stereochemistry, **4**, 240–242
α,β-unsaturated substrates, **4**, 242–253
Vigneron–Jacquet complex
reduction
unsaturated carbonyl compounds, **8**, 545
Vilsmeier–Haack reaction, **2**, 777–792
solvents, **2**, 779
Vilsmeier synthesis, **2**, 748
Vincamine
synthesis, **6**, 746
via Diels–Alder reaction, **5**, 409
via Mannich reaction, **2**, 1015
Vindoline
synthesis
via Michael addition, **4**, 25
Vindorosine
synthesis
via Michael addition, **4**, 25
Vineomycinone B_2
synthesis
via Diels–Alder reaction, **2**, 698
Vinyl acetal
hydroformylation, **4**, 924
Vinyl acetate
hydroformylation, **4**, 924, 932
intermolecular *meta* cycloaddition
to indane, **5**, 667
photocycloaddition reactions
to benzene, **5**, 667
reaction with chlorosulfonyl isocyanate, **5**, 105
reduction, **8**, 934
synthesis
via palladium(II) catalysis, **4**, 553
Vinyl acetoacetate
synthesis
via retro Diels–Alder reactions, **5**, 558
Vinyl alcohols
oxidation
solid support, **7**, 841
synthesis
via retro Diels–Alder reactions, **5**, 557
Vinyl aluminate
alkylation
copper-catalyzed, **3**, 215
Vinylamines
synthesis, **6**, 67
Vinylation
1-alkynes, **3**, 521
alkynic iodides
palladium-catalyzed, **3**, 544
metal enolates
regioselectivity, **3**, 12
organomercury compounds
palladium complexes, **4**, 839
Vinyl bromides
hydrobromination, **4**, 280
hydrogenolysis, **8**, 899
reaction with aldehydes, **1**, 193
Vinyl carbanions
alkylation, **3**, 241–266
heteroatom substituted
alkylation, **3**, 252
Vinyl cations
nitrile-trapped
isoquinoline synthesis, **6**, 401
Vinyl chlorides
cleavage
metal–ammonia, **8**, 530
hydrobromination, **4**, 280
hydrogenolysis, **8**, 897
reaction with *N*-acyliminium ions, **2**, 1064
synthesis, **4**, 277
Vinyl compounds
hydrogenation
heterogeneous catalysis, **8**, 439
$S_{RN}1$ reaction, **4**, 462
Vinyl cyanide
hydroxylation, **7**, 172
Vinylene carbonate
photocycloaddition reactions
with anisole, **5**, 653
Vinyl epoxides
radical cyclization
carbon-centered radicals, **4**, 789
reaction with arylstannanes
organopalladium catalysis, **3**, 232
transition metal catalyzed reactions, **6**, 847
Vinyl esters
reduction, **8**, 930
synthesis
via retro Diels–Alder reactions, **5**, 557
Vinyl ethers
hydrosilylation, **8**, 775
reduction, **8**, 934
diimide, **8**, 476
synthesis
via Horner reaction, **1**, 774
via palladium(II) catalysis, **4**, 553
via retro Diels–Alder reactions, **5**, 557

vinylic acylations
palladium complexes, **4**, 857
Vinyl fluorides
hydrogenolysis, **8**, 896
synthesis, **4**, 271
Vinyl halides
carbonylation
formation of esters, **3**, 1028
cross-coupling reactions
organometallic reagents, **3**, 522
cyclocarbonylation
formation of α-methylene lactones, **3**, 1032
hydrogenolysis, **8**, 895
hydrosilylation, **8**, 775
nitrile synthesis, **6**, 231
oxidative rearrangement, **7**, 816
reaction with lithium dialkylcuprates, **3**, 217
reduction, **8**, 895–920, 937
chromium(II) salts, **1**, 193
diimide, **8**, 476
α-substituted
arene alkylation, **3**, 322
synthesis, **3**, 788
vinyl substitutions
palladium complexes, **4**, 842–856
Vinylic oxidation, **7**, 329
Vinylidenamine
synthesis
via retro Diels–Alder reactions, **5**, 576
Vinylidene complexes
cycloaddition reactions
imines, **5**, 1068
Vinyl iodides
carbonylation
formation of ketones, **3**, 1023
hydrogenolysis, **8**, 900
reactions with benzaldehyde
chromium(II) chloride, **1**, 193
reactions with organotin compounds
organopalladium catalysis, **3**, 232
Vinyl lactones
transition metal catalyzed reactions, **6**, 847
Vinyl magnesiocuprates
synthesis
via alkylmagnesium halide reactions with alkynes, **3**, 243
Vinyl metals, α-seleno-
synthesis, **1**, 644, 665
Vinyl phosphates
phosphorylation, **6**, 611
reduction, **8**, 930
Vinyl pivalate
hydroformylation, **4**, 924
Vinyl radicals
addition reactions
tin hydride catalysis, **4**, 739
cyclizations, **4**, 796–798
structure, **4**, 719
Vinyl selenides
reduction, **8**, 934
Vinyl substitutions
intermolecular
palladium complexes, **4**, 845
intramolecular
palladium complexes, **4**, 846
organopalladium compounds, **4**, 833–861
Vinyl sulfides
desulfurization, **8**, 837
hydrogenolysis, **8**, 913
hydroxylation, **7**, 173
oxidative rearrangement, **7**, 816
reduction, **8**, 934
synthesis
via Horner reaction, **1**, 774
Vinyl sulfones
hydrogenolysis, **8**, 913
Vinyl sulfoxides
hydrogenolysis, **8**, 913
reduction, **8**, 934
synthesis
diastereoselectivity, **2**, 75
Vinyl triflates
carbonylation
formation of aldehydes, **3**, 1021
formation of ketones, **3**, 1023
cross-coupling reactions, **3**, 529
intermediate in dolastane synthesis, **3**, 488
reaction with tin compounds
organopalladium catalysts, **3**, 232
reduction, **8**, 933
synthesis
via organocopper compounds, **3**, 218
Virantmycin
synthesis, **7**, 406
Virescenol A
homo-Favorskii rearrangement, **3**, 857
Virginiamycin M_2
synthesis
via aldol reaction, **2**, 189
(+)-Viridifloric acid
synthesis
via aldol reaction, **2**, 206
Virolin
synthesis
via silver oxide, **3**, 691
Vitamin A
acetate
synthesis *via* enol ethers, **2**, 616
epoxide ring opening, **3**, 757
synthesis, **2**, 410; **3**, 169, 170
via carboalumination, **4**, 893
via hydroformlyation, **4**, 924
via Julia coupling, **1**, 803
via organocopper compounds, **3**, 223
via Reformatsky reaction, **2**, 287
via sulfones, **6**, 157
Vitamin A aldehyde
synthesis
via hydroformylation, **4**, 924
Vitamin B_1
catalyst
benzoin condensation, **1**, 543
Vitamin B_6
synthesis, **7**, 338
Vitamin B_{12}
catalyst
radical cyclizations, nonchain methods, **4**, 807
reductive radical addition, **4**, 765
synthesis
via Eschenmoser coupling reaction, **2**, 866

Vitamin D
  interconversions, **5**, 700
  synthesis, **3**, 109, 545
    via Claisen–Claisen rearrangement, **5**, 888
    via Claisen rearrangement, **5**, 859
    via Horner reaction, **1**, 780
    via Horner–Wittig process, **1**, 779
    via organocopper compounds, **3**, 223
    via organopalladium catalysts, **3**, 232
    via photolysis, **5**, 737
Vitamin D_3
  epoxidation, **7**, 362, 376
  hydrozirconation, **8**, 689
  precursor
    synthesis *via* Diels–Alder reaction, **5**, 349
  synthesis, **3**, 168, 173
Vitamin D_3, 1α,25-dihydroxy-
  precursor synthesis
    via Johnson rearrangement, **5**, 839
  synthesis, **3**, 984
Vitamin D_2, 22,23-epoxy-
  synthesis
    via retro Diels–Alder reaction, **5**, 569
Vitamin E
  asymmetric synthesis, **6**, 152
  sidechain
    synthesis *via* aldol reaction, **2**, 195
  synthesis, **5**, 1095, 1098
    via dihalocyclopropanes, **4**, 1011
    via iterative Claisen rearrangement, **5**, 892
Vitamin K
  synthesis, **5**, 1095
Volume of activation
  [2 + 2] cycloaddition reactions, **5**, 77
von Braun amide degradation
  alkyl bromide synthesis
    from tertiary amines, **6**, 212
  Ritter reaction, **6**, 291
von Richter rearrangement
  aromatic nitro halides, **6**, 240
Vorbrüggen–Eschenmoser reaction
  ester synthesis, **6**, 334

W

Wacker oxidation
addition reactions
C—O bond formation, **7**, 449–466
palladium(II) catalysis, **4**, 552
reaction conditions, **7**, 450
reoxidants, **7**, 451
scope, **7**, 450
solvents, **7**, 450
Wagner–Meerwein rearrangements, **3**, 705–717
bicyclic systems, **3**, 706
definition, **3**, 706
Ritter reaction, **6**, 291
stereoelectronic features, **3**, 709
use in synthesis, **3**, 710
Walburganal
synthesis
via transketalization, **6**, 677
Wallemia C
synthesis
via Claisen condensation, **2**, 821
Walsh model
cyclopropane
bonding, **5**, 900
Warburganal
synthesis, **7**, 87
Water
cocatalyst
Friedel–Crafts reaction, **2**, 735
reaction with formaldehyde
Lewis acids, **1**, 314
solvent for reduction
dissolving metals, **8**, 111
Waxes
esters, **6**, 324
Wenker synthesis
aziridines, **7**, 472
Wharton rearrangement
allylic alcohols
oxygen–oxygen transposition, **6**, 837
definition, **6**, 1042
α,β-epoxy ketones
fragmentation, **8**, 341
reduction of ketones, **8**, 927
Wheland intermediate
Friedel–Crafts reaction
arene alkylation, **3**, 298
Widdrol
synthesis, **7**, 100
Wieland–Miescher diketones
synthesis, **2**, 167
Wilkinson catalyst — *see* Rhodium, chlorotris(triphenylphosphine)
Willardiin
synthesis
via Ugi reaction, **2**, 1096
Willgerodt reaction
amide synthesis, **6**, 404
Kindler modification
alternative, **7**, 829
thioamide synthesis, **6**, 405
Wilsonirine, *N*-trifluoroacetyl-
synthesis
via diphenyl selenoxide, **3**, 666
via vanadium oxytrifluoride, **3**, 670
Withafenin A
synthesis, **7**, 366
Wittig–Horner reactions
selectivity
Knoevenagel reaction, **2**, 353
2,3-Wittig–oxy-Cope rearrangement
tandem
δ,ε-unsaturated carbonyl compounds, **6**, 852
Peterson methylenation compared with, **1**, 731
Wittig rearrangement, **3**, 975–1012; **6**, 873
absolute configuration, **6**, 884
alkene synthesis, **1**, 755
asymmetric induction
simple diastereoselectivity, **6**, 889
aziridine synthesis, **7**, 474
ethers
chelation, **6**, 887
chirality transfer, **6**, 884
chirality transfer out of cycle, **6**, 887
diastereoselectivity, **6**, 880
1,2-rearrangement
electron transfer mechanism, **3**, 824
mechanism, **3**, 979
2,3-sigmatropic rearrangement, **3**, 981; **6**, 834
α-(allyloxy)carbanions, **6**, 850
anionic, asymmetric induction, **6**, 852
aza version, **6**, 853
3,3-Claisen rearrangement, competition, **5**, 851
diallyl ethers, **5**, 888
thio version, **6**, 853, 895
transfer of chirality, **6**, 852
stereochemistry, **3**, 943
stereocontrol
allylic C—O bond, **6**, 889
ethers, **6**, 889
sulfones
chain elongation, **6**, 890
tandem and sequential rearrangements, **3**, 994
Wittig-type alkenation
allyl vinyl ethers, **5**, 830
Wolff–Kishner reduction
Barton modification, **8**, 330
carbonyl compounds, **8**, 307
hydrazones and arylsulfonylhydrazones, **8**, 327–359
chemoselectivity, **8**, 338
Cram modification, **8**, 335
Henbest modification, **8**, 336
Huang–Minlon modification, **8**, 329
isomerization of double bonds, **8**, 340
limitations, **8**, 338
mechanism, **8**, 328
modified procedures, **8**, 329
Nagata and Itazaki modification, **8**, 332
scope, **8**, 338
side reactions, **8**, 342
steric effects, **8**, 340
Wolff rearrangement, **3**, 887–909
chemistry, **3**, 897
competing reactions, **3**, 893

diazo compounds, **4**, 1032; **6**, 127
α-diazo ketones, **1**, 844
initiation, **3**, 891
photolysis, **3**, 891
thermolysis, **3**, 891
transition metal catalysts, **3**, 891
mechanism, **3**, 891
stereochemistry, **3**, 891
vinylogous, **3**, 906
Woodward–Hoffmann rules
alkene dimerization, **5**, 64
Claisen rearrangement, **5**, 857
1,3-sigmatropic rearrangements
stereochemistry, **5**, 1009
Working electrodes
electrosynthesis, **8**, 130
Wurtz reaction
classical, **3**, 414
coupling reactions, **3**, 413
intramolecular, **3**, 422
variants, **3**, 414
Wuweizisu-C
synthesis
via vanadium oxytrifluoride, **3**, 676

X

X-206
  synthesis, **1**, 409; **2**, 263
X-14547A
  synthesis
    final step, **1**, 409
    introduction of 2-ketopyrrole, **1**, 409
    via Julia coupling, **1**, 800
X-14881
  synthesis, **1**, 567
Xanthates, allylic
  synthesis
    via rearrangement, **6**, 842
9*H*-Xanthene, 4,6-dioxo-2,2,8,8-tetramethyl-1,2,3,4,5,6,7,8-octahydro-
  analysis of aldehydes
    Knoevenagel reaction, **2**, 354
Xanthen-9-one, 3,6-diethoxy-
  reduction
    boranes, **8**, 316
Xanthenones
  photochemical ring opening, **5**, 712
Xanthobacter Py2
  epoxides
    resolution, **7**, 429
Xanthocillin
  synthesis, **2**, 1084
Xanthone, 2,6-dihydroxy-
  synthesis
    via ferricyanide, **3**, 688
Xanthones
  reduction
    boranes, **8**, 316
    dissolving metals, **8**, 115
  synthesis
    via electrocyclization, **5**, 719
    via Friedel–Crafts reaction, **2**, 758
Xenon difluoride
  decarboxylative fluorination, **7**, 723
Xonotlite
  catalyst
    Knoevenagel reaction, **2**, 345, 359
X-ray crystallography
  carbonyl compounds
    Lewis acid complexes, **1**, 299
m-Xylene
  irradiation
    with cyclopentene, **5**, 651
  radical cations
    oxidation, **7**, 870
p-Xylene
  formylation
    Gattermann–Koch reaction, **2**, 749
  Friedel–Crafts acetylation, **2**, 738
  hydrogenation
    homogeneous catalysis, **8**, 454
  radical cations
    oxidation, **7**, 870
Xylenes
  Beckmann rearrangement
    solvent, **6**, 763
  isomerization
    Friedel–Crafts reaction, **3**, 327
  reduction
    photochemical method, **8**, 517
Xylenols
  transalkylation
    Friedel–Crafts reaction, **3**, 329
Xylitol
  synthesis, **7**, 645
(–)-Xylomollin
  synthesis
    via ene reaction, **2**, 537
Xylopinine
  synthesis
    via arynes, **4**, 501
    via electron transfer induced photocyclization, **2**, 1040
    via Mannich reaction, **2**, 1035
D-Xylose
  selective monoacetylation
    enzymatic, **6**, 340
Xylose, amino-
  synthesis
    via Diels–Alder reaction, **5**, 428
D-Xylose, 2,3,4-tri-*O*-benzyl-
  glycoside synthesis, **6**, 57
m-Xylylene
  synthesis, **6**, 778
o-Xylylenes
  cycloaddition reactions
    tropones, **5**, 622
  Diels–Alder reactions, **5**, 385
  dimerization, **5**, 638
  synthesis
    via electrocyclic ring opening, **5**, 1151
    via retro Diels–Alder reaction, **5**, 588

Y

Yamamoto's reagent
reactions with ketones, **1**, 117
reactions with nitriles, **1**, 124
Yangonin
synthesis, **7**, 109
Yeast
benzaldehyde reaction with acetaldehyde, **1**, 543
Yeast lipase
acylation
enzymatic, **6**, 340
Ylangene
synthesis, **3**, 20
Ylangocamphor
reduction
dissolving metals, **8**, 121
Ylides
alkaline hydrolysis, **8**, 863
ammonium
rearrangements, **6**, 854, 855
ring expansion, **6**, 897
cyclic
ring contraction, **6**, 897
cyclopropane synthesis, **4**, 986
direct formation, **3**, 919
electrocyclic closures
in oxirane rearrangement, **5**, 929
nonstabilized
Wittig reaction, **1**, 757
phosphonium
synthesis, **6**, 171–198
semistabilized
Wittig reaction, **1**, 758
stabilized
Wittig reaction, **1**, 759
sulfonium
rearrangements, **6**, 854, 855, 873
ring expansion, **6**, 897
synthesis, **6**, 893
synthesis
via diazo compounds, **6**, 128
Ylidic rearrangements
definition, **3**, 916
Ynamines
acid anhydride synthesis, **6**, 315
amidine synthesis, **6**, 550
cycloaddition reactions
ketenimines, **5**, 113
reactions with carbonyl compounds, **5**, 116
reactions with hydrogen halides, **6**, 497
reactions with ketenes
cyclobutenone synthesis, **5**, 689
retrograde Diels–Alder reactions, **5**, 557
Ynamines, silyl-
reactions with arynes, **4**, 510
Ynediols
divinyl ketones from
cyclization, **5**, 768
Ynolates
synthesis, **2**, 109
Ynones
photoaddition reactions
with alkenes, **5**, 164
α,β-Ynones, α′-amino-
synthesis
via acylisoxazolidides as leaving groups, **1**, 405
Yohimbane
derivatives
synthesis *via* Knoevenagel reaction, **2**, 382
Yohimbine
oxidation
DMSO, **7**, 295
synthesis
via arynes, **4**, 501
via Thorpe reaction, **2**, 851
Yohimbine hydrochloride
electrochemical reduction
enones, **8**, 532
Yohimbone
asymmetric synthesis, **3**, 81
synthesis, **3**, 72
Mannich reaction, **2**, 1034
Yomogi alcohol
synthesis
via 3,2-rearrangement, **3**, 933
Ytterbium
dissolving metal reductions
unsaturated hydrocarbons, **8**, 481
oxidation state
stability, **1**, 252
reduction
ammonia, **8**, 113
use in pinacol coupling reactions, **3**, 567
Ytterbium, dialkynyl-
synthesis, **1**, 276
Ytterbium, diaryl-
polyfluorinated
synthesis, **1**, 276
Ytterbium, phenyliodo-
reaction with *N*,*N*-dimethylbenzamide
synthesis of benzophenone, **1**, 278
Ytterbium chloride
toxicity, **1**, 252
Ytterbium dibromide
solubility, **1**, 278
Ytterbium diiodide
Barbier-type reactions, **1**, 278
reduction
carbonyl compounds, **8**, 115
solubility, **1**, 278
Ytterbium reagents, **1**, 251–280
Barbier-type reactions, **1**, 278
Ytterbium salts
redox potentials, **1**, 278
Yttrium compounds
reaction with epoxides
regioselectivity, **6**, 9

Z

Zearalenone
derivative
synthesis, **6**, 440
microbial hydroxylation, **7**, 59
synthesis, **1**, 568; **3**, 49; **6**, 136
via cyclization, **1**, 553
via macrolactonization, **6**, 369, 370
via palladium-catalyzed carbonylation, **3**, 1033
via Wacker oxidation, **7**, 454
Zeatine, β-D-ribofuranoside
synthesis, **7**, 88
Zeolites
asymmetric epoxidation, **7**, 396
catalysis
Friedel–Crafts reaction, **2**, 736
modified, catalysts
Friedel–Crafts reaction, **3**, 296
polyfunctional catalysts
acidity, Friedel–Crafts reaction, **3**, 305
shape selective catalysts
Friedel–Crafts reaction, **3**, 296
solid supports
oxidants, **7**, 840
Zerumbone
8,9-epoxide
transannular cyclization, **3**, 406
Ziegler catalysts
hydrogenation
alkenes, **8**, 447
Ziegler–Natta catalysts
metal alkyls, **3**, 296
Zimmerman–Traxler model
aldol reaction, **2**, 6, 261
stereoselectivity, **2**, 155, 197
steric interactions, **2**, 200
Ivanov reaction, **2**, 210
Zinc
activation
Reformatsky reaction, **2**, 282
ammonium chloride
nitro compound reduction, **8**, 366
Clemmensen reduction, **8**, 309
desulfurization
ammonium chloride, **8**, 843
dissolving metal reductions
unsaturated hydrocarbons, **8**, 480
reduction
alkyl halides, **8**, 795
benzylic compounds, **8**, 972
α-bromo ketones, **8**, 986
enones, **8**, 524
epoxides, **8**, 881
nitriles, **8**, 299
nitro compounds, **8**, 364
potassium hydroxide/dimethyl sulfoxide, **8**, 113
vicinal dibromides, **8**, 797
reductive cleavage
α-alkylthio ketone, **8**, 993
ketol acetates, **8**, 991
reductive dimerization
unsaturated carbonyl compounds, **8**, 532
Zinc, alkenyl-
coupling reactions
with aryl iodides, **3**, 495
Zinc, alkyl-
addition reactions, **1**, 216
Zinc, alkyliodo-
synthesis, **1**, 212
Zinc alkynylbromo-
carbozincation, **4**, 883
Zinc, alkynylchloro-
reaction with alkenyl halides
palladium-catalyzed, **3**, 524
Zinc, allenylbromo-
addition reactions, **1**, 220
synthesis, **2**, 81
Zinc, allyl-
addition reactions, **1**, 218
reaction with aldehydes, **2**, 23, 29, 91
Zinc, allylbromo-
carbozincation, **4**, 880
reaction with aldoxime ethers, **2**, 995
reaction with *N*-methyl-4-*t*-butylcyclohexylamine
dependence of product ratio on solvent, **2**, 983
reaction with phenylmenthyl-
N-methoxyiminoacetate, **2**, 995
Zinc, allylchloro-
reaction with aldehydes, **2**, 31
Zinc, aryl-
addition reactions, **1**, 216
alkylation, **3**, 260
Zinc, arylchloro-
coupling reactions
with alkenyl bromides, **3**, 495
Zinc, benzyl-
coupling reactions
with aromatic halides, **3**, 453
Zinc, bis(2-methylbutyl)-
hydride donor
reaction with phenyl isopropyl ketone, **8**, 99
Zinc, bis(perdeuteroethyl)-
reaction with benzaldehyde, **1**, 223
Zinc, bromo(carboxyethyl)-
t-butyl ester
synthesis, **2**, 279
Zinc, bromo(carboxyisopropyl)-
t-butyl ester
synthesis, **2**, 279
Zinc, bromo(carboxymethyl)-
t-butyl ester
synthesis, **2**, 279
spectra, **2**, 281
Zinc, bromo(diethoxyphosphoryl)difluoromethyl-
alkylation
with allylic halides, **3**, 202
Zinc, bromo(propargyl)-
reaction with alkynes, **4**, 883
Zinc, chloro-2-furyl-
coupling reactions
with alkenyl iodides, **3**, 497
Zinc, chloroheteroaryl-
alkylation, **3**, 261
Zinc, chloropyridyl-
coupling reactions, **3**, 510

Zinc, cinnamyl-
reactions with aldehydes, **2**, 23
Zinc, cinnamylbromo-
synthesis, **1**, 214
Zinc, crotyl-
reaction with aldehydes, **2**, 23
reaction with imines
regioselectivity, **2**, 988
syn–anti selectivity, **2**, 989
reaction with iminium salts, **2**, 1000
Zinc, dialkyl-
hydride donor
reduction of carbonyls, **8**, 99
reaction with alkynes, **4**, 883
reduction
aroyl chlorides, **8**, 291
synthesis, **1**, 212, 215
Zinc, diallyl-
reactions with α-alkoxyaldehydes
stereoselectivity, **1**, 221
reactions with oximes
diastereoselective, **2**, 32
reactions with sulfenimine, **2**, 998
Zinc, dibenzyl-
synthesis, **1**, 215
Zinc, dibutyl-
reaction with benzaldehyde, **1**, 216
Zinc, di-*t*-butyl-
synthesis
via transmetallation, **1**, 214
Zinc, dicrotyl-
metallo-ene reactions, **5**, 31
reactions with aldehydes
stereoselectivity, **1**, 220
Zinc, diethyl-
carbozincation, **4**, 884
enantioselective addition reactions, **1**, 223
reaction with benzaldehyde, **1**, 223
reaction with 1,2-diketones, **1**, 217
Zinc, dimethyl-
Tebbe reaction, **1**, 746
Zinc, divinyl-
enantioselective addition reactions, **1**, 223
synthesis
via transmetallation, **1**, 214
Zinc, ethyl-
enolate
synthesis, **2**, 123
Zinc, homoallyl-
coupling reactions
with aromatic halides, **3**, 453
Zinc, homopropargyl-
coupling reactions
with aromatic halides, **3**, 453
Zinc, methyl-
addition reactions
chiral aldehydes, **1**, 221
Zinc, methylenedi-
Tebbe reaction, **1**, 746
Zinc, 2-pentenylbromo-
reaction with diisobutyl ketone, **1**, 219
Zinc, phenylethyl-
coupling reactions
with aromatic halides, **3**, 453
Zinc, propargyl-
addition reactions, **1**, 218
reactions with aldimines, **2**, 992
Zinc, silylmethyl-
coupling reactions
with aromatic halides, **3**, 453
Zinc acetate
catalyst
Knoevenagel reaction, **2**, 345
Zinc amalgam
reduction
enones, **8**, 525
Zincate, triorgano-
lithium salt
reaction with α,β-unsaturated carbonyl compounds, **2**, 124
Zinc borohydride
ketone reduction, **8**, 11
diastereoselectivity, **8**, 7
reduction
acetals, **8**, 215
Zinc chloride
catalyst
Diels–Alder reaction, **2**, 664, 665, 679
Friedel–Crafts reaction, **2**, 709
vinylic Grignard coupling, **3**, 485
enolates
stereoselection, **2**, 204
transfer hydrogenation, **8**, 553
Zinc compounds
3-iodo-2-[(trimethylsilyl)methyl]propene
trimethylenemethane synthetic equivalent, **5**, 246
Zinc–copper couple
deoxygenation
epoxides, **8**, 888
Zinc cyanoborohydride
reductive amination
imines, **8**, 53
Zinc dialkylamide
ketone deprotonation
synthesis of zinc ester enolates, **2**, 280
Zinc dichromate
oxidation
ethers, **7**, 236
Zinc-ene reactions, **5**, 31–33
intramolecular, **5**, 37–46
Zinc enolates
aldol reaction
thermodynamic control, **2**, 289
Blaise reaction, **2**, 297
isolation
Reformatsky reaction, **2**, 278
Reformatsky reaction, **2**, 277–298
stability
Reformatsky reaction, **2**, 278
structure, **2**, 280
synthesis, **2**, 122
Zinc ester dieneolates
reaction with carbonyl compounds, **2**, 286
reaction with conjugated enones, **2**, 287
Zinc ester enolates
reaction with conjugated enones, **2**, 285
Zinc halides
epoxide ring opening, **3**, 771
Zinc halides, allyl-
reactions with silylated alkynes, **5**, 32
Zinc homoenolates
acylation, **2**, 449

cyclopropane synthesis, **2**, 443
reactions, **2**, 447, 448
substitution reactions
allylation, **2**, 449
Zinc iodide
reduction
benzylic compounds, **8**, 969
sodium cyanoborohydride reduction
carbonyl compounds, **8**, 315
Zinc ketone enolates
crystallography
Reformatsky reagent, **2**, 280
structure, **2**, 125
synthesis, **2**, 280
Zincophorin
synthesis, **7**, 246
via chiral reaction, **2**, 652
via Diels–Alder reaction, **2**, 704
Zinc oxide
catalyst
Knoevenagel reaction, **2**, 345
Zinc permanganate
oxidation
ethers, **7**, 236, 237
solid support, **7**, 844
Zinc reagents
organopalladium catalysis, **3**, 230
Zingiberenol
synthesis
via Diels–Alder reactions, **5**, 324
Zirconabicycles
reactions, **5**, 1165–1170
synthesis, **5**, 1171, 1173
Zirconacycles
five-membered
synthesis, **5**, 1173–1182
three-membered
synthesis, **5**, 1173–1182
Zirconacyclopentadienes
synthesis, **5**, 1165, 1178–1182
Zirconacyclopentanes
synthesis, **5**, 1178–1182
Zirconacyclopentenes
synthesis, **5**, 1178–1182
1-Zircona-3-cyclopentenes
synthesis, **5**, 1172
Zirconacyclopropanes
reactions with alkenes, **5**, 1180
synthesis, **5**, 1173–1177
Zirconacyclopropenes
synthesis, **5**, 1173–1177
Zirconium, alkenyl-
coupling reactions
with aryl iodides, **3**, 495
reactions, **8**, 690
nickel catalysis, **3**, 230
Zirconium, alkyl-
reactions, **8**, 690
reactions with carbonyl compounds, **1**, 145
Zirconium, alkyltributoxy-
reaction with carbonyl compounds
chemoselectivity, **1**, 149
Zirconium, alkynyl-
hydrozirconation, **8**, 682
Zirconium, aryl-
reactions with carbonyl compounds, **1**, 145
Zirconium, aryltributoxy-
reaction with carbonyl compounds
chemoselectivity, **1**, 149
Zirconium, *t*-butyltributoxy-
reaction with carbonyl compounds
chemoselectivity, **1**, 149
Zirconium, chlorocyclopentadienylhydrido-
catalyst
hydroesterification, **3**, 1030
Zirconium, chlorodicyclopentadienyl-
enolates
aldol reaction, *syn:anti* selectivity, **2**, 303
aldol reaction, *syn* stereoselectivity, **2**, 302
aldol reaction, stereoselectivity, **2**, 305
Zirconium, chlorodicyclopentadienylcrotyl-
reactions with α-methyl chiral aldehydes, **2**, 29
Zirconium, chlorohydridobis(cyclopentadienyl)-
hydrometallation, **8**, 673
hydrozirconation, **8**, 675
alkenes, **4**, 153
purity, **8**, 675
synthesis, **8**, 675
Zirconium, crotyl-
reactions with aldehydes, **2**, 24
synthesis, **2**, 5
Zirconium, crotyltrialkoxy-
reactions with aldehydes
diastereoselectivity, **2**, 24
Zirconium, dialkyl-
synthesis, **1**, 143
Zirconium, diaryl-
synthesis, **1**, 143
Zirconium, dienyl-
reactions with carbonyl compounds, **1**, 162
Zirconium, monoalkyl-
synthesis, **1**, 142
Zirconium, tetraallyl-
reaction with aldol, **2**, 31
reaction with carbonyl compounds, **1**, 157
Zirconium, tetramethyl-
methylation
carbonyl compounds, **1**, 150
Zirconium, vinyl-
hydrozirconation, **8**, 682
reaction with α,β-unsaturated ketones, **1**, 155
synthesis, **1**, 143
Zirconium catalysts
alkylidenation
carbonyl compounds, **5**, 1122
bicyclization, **5**, 1165–1170
enynes, **5**, 1163–1183
Friedel–Crafts reactions, **2**, 737
Zirconium complexes, alkyl-
monoalkylation
2,3-dichloro-1,4-naphthoquinone, **3**, 458
Zirconium compounds
aluminum complexes
carbonyl alkylidenation, **5**, 1125
Zirconium enolates
aldol reactions
diastereofacial preferences, **2**, 231
stereoselectivity, **2**, 302
diastereoselectivity
reversed, **2**, 198, 208
synthesis, **2**, 119
Zirconium hydride

catalysts
carbonylation, **3**, 1027
Zirconium hydride, bis(cyclopentadienyl)-
oxidation
primary alcohols, **7**, 309
Zirconium reagents
organopalladium catalysis, **3**, 230
Zirconium reagents, allylic
reaction with carbonyl compounds, **1**, 156
Zirconium tetrahalides
lithium aluminum hydride
unsaturated hydrocarbon reduction, **8**, 483
Zirconocene
alkyne complex
synthesis, **5**, 1175
benzyne complex
reaction with stilbene, **5**, 1178
synthesis, **5**, 1174
1-butene complex
reaction with stilbene, **5**, 1180
synthesis, **5**, 1175, 1178
cycloalkyne complex
synthesis, **5**, 1175
stilbene complex
synthesis, **5**, 1174, 1177, 1180
Zirconocene, crotyl-
reaction with carbonyl compounds, **1**, 158
synthesis, **1**, 143
Zirconocene, diene-
reactions with carbonyl compounds, **1**, 162
Zirconocene, isoprene-
reactions with carbonyl compounds, **1**, 163
Zirconocene dichloride
synthesis, **1**, 143
ZSM-5 zeolite
catalyst
Friedel–Crafts reaction, **3**, 305
Zygosporin
3,2-sigmatropic rearrangement
synthesis, stereocontrol, **3**, 960